Environment

The Science Behind the Stories

5TH EDITION

Jay Withgott

Matthew Laposata

PEARSON

Boston Columbus Indianapolis New York San Francisco Upper Saddle River
Amsterdam Cape Town Dubai London Madrid Milan Munich Paris Montréal Toronto
Delhi Mexico City São Paulo Sydney Hong Kong Seoul Singapore Taipei Tokyo

Editor-in-Chief: Beth Wilbur
Executive Director of Development: Deborah Gale
Acquisitions Editor: Alison Rodal
Project/Development Editor: Anna Amato
Editorial Assistant: Libby Reiser
Associate Media Producer: Daniel Ross
Marketing Manager: Amee Mosley
Managing Editor: Michael Early
Project Manager: Shannon Tozier
Production Management: Kelly Keeler, Cenveo Publishers, Inc.
Compositor: Cenveo Publishers, Inc.
Illustrators: Imagineeringart.com Inc.
Design Manager: Derek Bacchus
Interior and Cover Designer: Tandem Creative, Inc.
Text Permissions Project Manager: Joseph Croscup and Michael Farmer
Text Permissions Specialist: Electronic Publishing Services
Photo Editor: Travis Amos
Photo Permissions Management: Q2A/Bill Smith
Photo Researcher: Zoe Milgram, Q2A/Bill Smith
Manufacturing Buyer: Jeffery Sargent
Text Printer: Courier Kendallville
Cover Printer: Lehigh-Phoenix Color/Hagerstown

Cover Photo Credit: mark ferguson/Alamy

Credits and acknowledgments for materials borrowed from other sources and reproduced, with permission, in this textbook appear on the appropriate page within the text or on page CR-1.

Library of Congress Cataloging-in-Publication Data
Withgott, Jay.
Environment : the science behind the stories / Jay Withgott, Matt Laposata. -- Fifth edition.
pages cm
Previous editions cataloged
under Brennan, Scott
Includes bibliographical references and index.
ISBN 978-0-321-89742-8
1. Environmental sciences. I. Title.
GE105.B74 2013 363.7-- dc23

2013004851

ISBN 10: 0-321-89742-0; ISBN 13: 978-0-321-89742-8 (Student Edition)
ISBN 10: 0-321-92757-5; ISBN 13: 978-0-321-92757-6 (Books a la Carte)
ISBN 10: 0-13-354014-6; ISBN 13: 978-0-13-354014-7 (School Edition)

PEARSON www.pearsonhighered.com

3 4 5 6 7 8 9 10—CKV—17 16 15 14

About the Authors

Jay Withgott has authored *Environment: The Science Behind the Stories* as well as its brief version, *Essential Environment*, since their inception. In dedicating himself to these books, he works to keep abreast of a diverse and rapidly changing field and continually seeks to develop new and better ways to help today's students learn environmental science.

As a researcher, Jay has published scientific papers in ecology, evolution, animal behavior, and conservation biology in journals ranging from *Evolution* to *Proceedings of the National Academy of Sciences*. As an instructor, he has taught university lab courses in ecology and other disciplines. As a science writer, he has authored articles for numerous journals and magazines including *Science, New Scientist, BioScience, Smithsonian,* and *Natural History*. By combining his scientific training with prior experience as a newspaper reporter and editor, he strives to make science accessible and engaging for general audiences. Jay holds degrees from Yale University, the University of Arkansas, and the University of Arizona.

Jay lives with his wife, biologist Susan Masta, in Portland, Oregon.

Matthew Laposata is a professor of environmental science at Kennesaw State University (KSU). He holds a bachelor's degree in biology education from Indiana University of Pennsylvania, a master's degree in biology from Bowling Green State University, and a doctorate in ecology from The Pennsylvania State University.

Matt is the coordinator of KSU's two-semester general education science sequence titled Science, Society, and the Environment, which enrolls roughly 6000 students per year. He focuses exclusively on introductory environmental science courses and has enjoyed teaching and interacting with thousands of nonscience majors during his career. He is an active scholar in environmental science education and has received grants from state, federal, and private sources to develop and evaluate innovative curricular materials. His scholarly work has received numerous awards, including the Georgia Board of Regents' highest award for the Scholarship of Teaching and Learning.

Matt resides in suburban Atlanta with his wife, Lisa, and children, Lauren, Cameron, and Saffron.

ABOUT OUR SUSTAINABILITY INITIATIVES

This book is carefully crafted to minimize environmental impact. The materials used to manufacture this book originated from sources committed to responsible forestry practices. The paper is Forest Stewardship Council® (FSC®) certified. The printing, binding, cover, and paper come from facilities that minimize waste, energy consumption, and the use of harmful chemicals.

Pearson closes the loop by recycling every out-of-date text returned to our warehouse. We pulp the books, and the pulp is used to produce items such as paper coffee cups and shopping bags. In addition, Pearson has become the first climate-neutral educational publishing company.

The future holds great promise for reducing our impact on Earth's environment, and Pearson is proud to be leading the way. We strive to publish the best books with the most up-to-date and accurate content, and to do so in ways that minimize our environmental impact.

PEARSON

Brief Contents

Contents

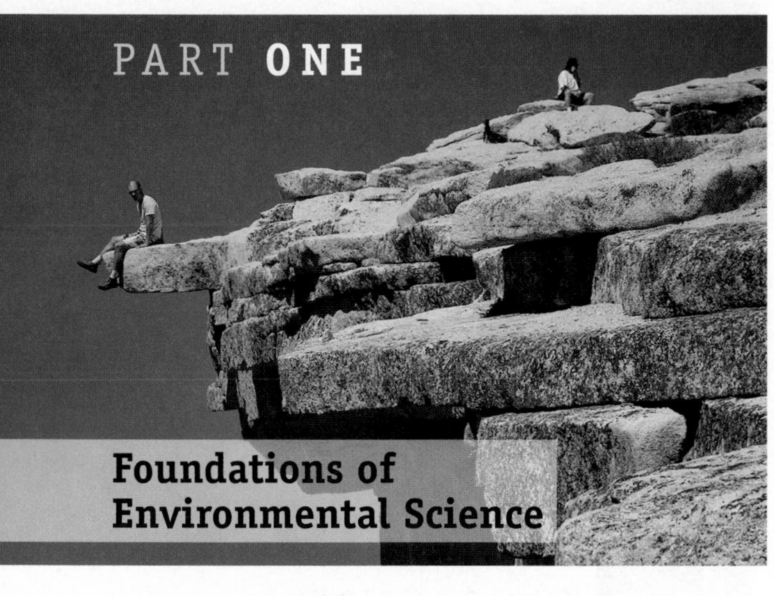

PART ONE

Foundations of Environmental Science

Preface

Dear Student,

You are coming of age at a unique and momentous time in history. Within your lifetime, our global society must chart a promising course for a sustainable future. The stakes could not be higher.

Today we live long lives enriched with astonishing technologies, in societies more free, just, and equal than ever before. We enjoy wealth on a scale our ancestors could hardly have dreamed of. Yet we have purchased these wonderful things at a price. By exploiting Earth's resources and ecological services, we are depleting our planet's bank account and running up its credit card. We are altering our planet's land, air, water, nutrient cycles, biodiversity, and climate at dizzying speeds. More than ever before, the future of our society rests with how we treat the world around us.

Your future is being shaped by the phenomena you will learn about in your environmental science course. Environmental science gives us a big-picture understanding of the world and our place within it. Environmental science also offers hope and solutions, revealing ways to address the problems we create. Environmental science is not simply some subject you learn in college. Rather, it provides you basic literacy in the foremost issues of the 21st century, and it relates to everything around you over your entire lifetime.

We have written this book because today's students will shape tomorrow's world. At this unique moment in history, students of your generation are key to achieving a sustainable future for our civilization. The many environmental challenges that face us can seem overwhelming, but you should feel encouraged and motivated. Remember that each dilemma is also an opportunity. For every problem that human carelessness has created, human ingenuity can devise a solution. Now is the time for innovation, creativity, and the fresh perspectives that a new generation can offer. Your own ideas and energy *will* make a difference.

—Jay Withgott and Matthew Laposata

Dear Instructor,

You perform one of our society's most vital jobs by educating today's students—the citizens and leaders of tomorrow—on the fundamentals of the world around them, the nature of science, and the most central issues of our time. We have written this book to assist you in this endeavor because we feel that the crucial role of environmental science in today's world makes it imperative to engage, educate, and inspire a broad audience of students.

In *Environment: The Science Behind the Stories*, we strive to implement a diversity of modern teaching approaches and to show how science can inform efforts to bring about a sustainable society. We aim to encourage critical thinking and to maintain a balanced approach as we flesh out the vibrant social debate that accompanies environmental issues. As we assess the challenges facing our civilization and our planet, we focus on providing forward-looking solutions, for we truly feel there are many reasons for optimism.

In crafting the fifth edition of this text, we have incorporated the most current information from this fast-moving field and have streamlined our presentation to promote learning. We have examined every line with care to make sure all content is accurate, clear, and up-to-date. Moreover, we have introduced a number of major changes that are new to this edition.

New to This Edition

With the fifth edition we welcome Dr. Matthew Laposata as an author. Professor of environmental science at Kennesaw State University in Georgia, Matt teaches and coordinates his university's environmental science courses while actively engaging in outside projects to promote environmental science education. Matt's ideas, energy, and commitment to outstanding teaching have already enlivened and strengthened this book as well as its brief version, *Essential Environment*. Please welcome him to our author team!

This fifth edition includes an array of revisions that together enhance our content and presentation while strengthening our commitment to teach science in an engaging and accessible way.

- **CENTRAL CASE STUDY** Ten of our 23 *Central Case Studies* are new to this edition, providing a wealth of fresh stories and new ways to frame issues in environmental science. Students will travel from Pennsylvania to Hawai'i and from Africa to Japan as they learn how debates over hydraulic fracturing, oil sands extraction, air pollution, and wildlife conservation are affecting people's lives.

- **Chapter 2:** The Tohoku Earthquake: Has it Shaken the World's Trust in Nuclear Power?
- **Chapter 3:** Saving Hawaii's Native Forest Birds
- **Chapter 5:** The Vanishing Oysters of the Chesapeake Bay
- **Chapter 6:** Costa Rica Values its Ecosystem Services
- **Chapter 7:** Hydrofracking the Marcellus Shale
- **Chapter 9:** Iowa's Farmers Practice No-Till Agriculture
- **Chapter 11:** Will We Slice through the Serengeti?
- **Chapter 15:** Starving the Louisiana Coast of Sediment
- **Chapter 17:** Clearing the Air in L.A. and Mexico City
- **Chapter 19:** Alberta's Oil Sands and the Keystone XL Pipeline

THE SCIENCE BEHIND THE STORY Fully 18 of our 42 *Science Behind the Story* features are new to this edition, providing a current and exciting selection of scientific studies to highlight. Students will follow researchers as they help to restore an oyster fishery; monitor animal populations; evaluate energy sources; and assess impacts of smog, aquifer contamination, fallout from Fukushima, and oil from the *Deepwater Horizon* spill. Selected features are supported by new "Process of Science" exercises online in *MasteringEnvironmentalScience* that use these examples to help students explore how scientists conduct their work.

- **Chapter 2:** Tracking Fukushima's Nuclear Legacy
- **Chapter 3:** Hawaii: Species Factory and Lab of Evolution
- **Chapter 3:** Monitoring Bird Populations at Hakalau Forest
- **Chapter 4:** Chronicling Ecological Recovery at Mount St. Helens
- **Chapter 5:** "Turning the Tide" for Native Oysters in Chesapeake Bay
- **Chapter 6:** Do Payments Help Preserve Forest?
- **Chapter 7:** Does Fracking Contaminate Drinking Water?
- **Chapter 8:** Did Soap Operas Reduce Fertility in Brazil?
- **Chapter 9:** Can No-Till Farming Help Us Fight Climate Change?
- **Chapter 11:** Wildlife Declines in African Reserves
- **Chapter 16:** Predicting the Oceans' "Garbage Patches"
- **Chapter 17:** Measuring the Health Impacts of Mexico City's Air Pollution
- **Chapter 18:** How Do Climate Models Work?
- **Chapter 19:** Discovering Impacts of the Gulf Oil Spill
- **Chapter 20:** Health Impacts of Chernobyl and Fukushima
- **Chapter 20:** Assessing EROI Values of Energy Sources
- **Chapter 21:** Comparing Energy Sources
- **Chapter 21:** What are the Impacts of Solar and Wind Development?

FAQ This new feature highlights questions frequently posed by students in introductory environmental science courses. Some *FAQs* address widely held misconceptions, whereas others fill in common conceptual gaps in student knowledge. This feature addresses not only the questions students ask, but also the questions they sometimes hesitate to ask. In so doing, it shows students they are not alone in having these questions, and it helps to foster an environment of open inquiry in the classroom.

DATA Q Each chapter now contains questions that help students to actively engage with graphs and other data-driven figures. The questions accompany several figures in each chapter, challenging students to practice quantitative skills of interpretation and analysis. To encourage students to test their understanding as they read, answers are provided in Appendix A.

Currency and coverage of topical issues To live up to our book's hard-won reputation for currency, we've incorporated the most recent data possible throughout, and we've enhanced coverage of issues now gaining prominence. As climate change and energy concerns play ever-larger roles in today's world, our coverage has evolved. This edition highlights how renewable energy is growing, yet also how we continue reaching further for fossil fuels with deep offshore drilling, Arctic drilling, hydraulic fracturing for oil and shale gas, and extraction of oil sands. These choices make energy returned on investment (EROI) ratios crucially important, especially as climate change gathers force. Climate change connections continue to proliferate among topics throughout our text, and our climate change chapter includes new coverage of climate modeling, geoengineering, research into jet stream effects on extreme weather, impacts of Hurricane Sandy and other events, the latest climate predictions for the United States and the world, efforts toward carbon neutrality, and political responses at all levels.

This edition also expands its coverage of a diversity of topics including the valuation of ecosystem services, introduced species and their ecological impacts on islands, prospects for nuclear power and safety after Fukushima, advanced biofuels, hormone-disrupting substances, impacts on coastal wetlands, plastic pollution in the oceans, environmental policy, ocean acidification, sustainable agriculture, green-collar jobs, and the rebound effect in energy conservation. We continue to use sustainability as an organizing theme throughout the book, and we aid these efforts by moving primary coverage of sustainable development to Chapter 6 and previewing Chapter 24's campus sustainability coverage in Chapter 1.

- **Enhanced style elements** We have updated and improved the look and clarity of our visual presentation throughout the text. A more open layout, more engaging photo treatments, improved maps in the case studies, and redesigned table styles all make the book more inviting and accessible for learning. This edition includes over 30% new photos, graphs, and illustrations, while existing figures have been revised to reflect current data or for better clarity or pedagogy.

Existing Features

We have also retained the major features that made the first four editions of our book unique and that are proving so successful in classrooms across North America:

- **An emphasis on science and data analysis** We have maintained and strengthened our commitment to a rigorous presentation of modern scientific research while at the same time making science clear, accessible, and engaging to students. Explaining and illustrating the *process* of science remains a foundational goal of this endeavor. We also continue to provide an abundance of clearly cited data-rich graphs, with accompanying tools for data analysis. In our text, our figures, and numerous print and online features, we aim to challenge students and to assist them with the vital skills of data analysis and interpretation.

- **An emphasis on solutions** For many students, today's deluge of environmental dilemmas can lead them to believe that there is no hope or that they cannot personally make a difference in tackling these challenges. We have aimed to counter this impression by highlighting innovative solutions being developed around the world. While being careful not to paint too rosy a picture of the challenges that lie ahead, we demonstrate that there is ample reason for optimism, and we encourage action. Our campus sustainability coverage (Chapters 1 and 24) shows students how their peers are applying principles and lessons from environmental science to forge sustainable solutions on their own campuses. To recognize the efforts of faculty and students in encouraging sustainable practices on campus and in their communities, Pearson Education will continue to grant Sustainable Solutions Awards to exceptional campus programs. See www.masteringenvironmentalscience.com for entry details and for profiles of previous winners.

- **Central Case Studies integrated throughout the text.** We integrate each chapter's *Central Case Study* into the main text, weaving information and elaboration throughout the chapter. In this way, compelling stories about real people and real places help to teach foundational concepts by giving students a tangible framework with which to incorporate novel ideas. We are gratified that students and instructors using our book have so consistently applauded this approach, and we hope it continues to bring further success in environmental science education.

- **The Science Behind the Story** Because we strive to engage students in the scientific process of testing and discovery, we feature *The Science Behind the Story* boxes in each chapter. By guiding students through key research efforts, this feature shows not merely *what* scientists discovered, but *how* they discovered it.

- **Weighing the Issues** These questions aim to help develop the critical-thinking skills students need to navigate multifaceted issues at the juncture of science, policy, and ethics. They serve as stopping points for students to reflect on what they have read, wrestle with complex dilemmas, and engage in spirited classroom discussion.

- **Diverse end-of-chapter features** *Reviewing Objectives* summarizes each chapter's main points and relates them to the chapter's learning objectives, enabling students to confirm that they have understood the most crucial ideas and to review concepts by turning to specified page numbers. *Testing Your Comprehension* provides concise study questions on main topics, while *Seeking Solutions* encourages broader creative thinking aimed at finding solutions. "Think It Through" questions place students in a scenario and empower them to make decisions to resolve problems. *Calculating Ecological Footprints* enables students to quantify the impacts of their own choices and measure how individual impacts scale up to the societal level.

MasteringEnvironmentalScience

With this edition we are thrilled to offer expanded opportunities through *MasteringEnvironmentalScience*, our powerful yet easy-to-use online learning and assessment platform. We have developed new content and activities specifically to support features in the textbook, thus strengthening the connection between these online and print resources. This approach encourages students to practice their science literacy skills in an interactive environment with a diverse set of automatically graded exercises. Students benefit from self-paced activities that feature immediate wrong-answer feedback, while instructors can gauge student performance with informative diagnostics. By enabling assessment of student learning outside the classroom, *MasteringEnvironmentalScience* helps the instructor to maximize the impact of in-classroom time. As a result, both educators and learners benefit from an integrated text and online solution.

- **New to this edition** Informed by instructor feedback and instructors' desires for students to leave their environmental science course with a mastery of science literacy skills, the following are additions to *MasteringEnvironmentalScience*. The first three were created specifically for the fifth edition by our textbook's co-author Matthew Laposata:

- *Process of Science* activities help students navigate the scientific method, guiding them through in-depth explorations of experimental design using *Science Behind the Story* features from the fifth edition. These activities encourage students to think like a scientist and to practice basic skills in experimental design.
- *Interpreting Graphs and Data: Data Q* activities pair with the new in-text *Data Analysis Questions* and coach students to further develop skills related to presenting, interpreting, and thinking critically about environmental science data.
- *"First Impressions" Pre-Quizzes* help instructors determine their students' existing knowledge of environmental issues and core content areas at the outset of the academic term, providing class-specific data that can then be employed for powerful teachable moments throughout the term. Assessment items in the Test Bank connect to each quiz item, so instructors can formally assess student understanding.
- *More Video Field Trips* have been added to the existing library in *MasteringEnvironmentalScience*. With three new videos you can now kick off your class period with a short visit to a wind farm, a site tackling invasive species, or a sustainable college campus.

Existing features *MasteringEnvironmentalScience* also retains its popular existing features. These include existing *Interpreting Graphs and Data* exercises and the interactive *GraphIt!* program, each of which guides students in exploring how to present and interpret data and how to create graphs; interactive *Causes and Consequences* exercises, which let students probe the causes behind major issues, their consequences, and possible solutions; and *Viewpoints*, paired essays authored by invited experts who present divergent points of view on topical questions.

Environment: The Science Behind the Stories has grown from our experiences in teaching, research, and writing. We have been guided in our efforts by input from the hundreds of instructors across North America who have served as reviewers and advisors. The participation of so many learned, thoughtful, and committed experts and educators has improved this volume in countless ways.

We sincerely hope that our efforts are worthy of the immense importance of our subject matter. We invite you to let us know how well we have achieved our goals and where you feel we have fallen short. Please write to us in care of our editor Alison Rodal (alison.rodal@pearson.com) at Pearson Education. We value your feedback and are eager to know how we can serve you better.

—Jay Withgott and Matthew Laposata

Instructor Supplements

Instructor Resource Center on DVD with TestGen (0-321-93954-9)

This powerful media package is organized chapter-by-chapter and includes all teaching resources in one convenient location. You'll find Video Field Trips, PowerPoint presentations, Active Lecture questions to facilitate class discussions (for use with or without clickers), and an image library that includes all art and tables from the text.

Included on the IRDVD, the test bank includes hundreds of multiple-choice questions plus unique graphing, and scenario-based questions to test students' critical-thinking abilities.

Instructor Guide (0-321-92782-6)

This comprehensive resource provides chapter outlines, key terms, and teaching tips for lecture and classroom activities.

Blackboard Open Access (0-321-92779-6)

MasteringEnvironmentalScience™ for Environment: The Science Behind the Stories (0-321-92752-4)

The *MasteringEnvironmentalScience* platform is the most effective and widely used online tutorial, homework, and assessment system for the sciences.

ENGAGE WITH REAL PEOPLE, REAL PLACES, AND REAL DATA

Integrated Central Case Studies highlight the real people, real places, and real data behind environmental issues. The Integrated Central Case Studies provide contextual framework to make science memorable and engaging.

NEW! 30% of the Central Case Studies in the book are entirely new. New case studies focus on hydraulic fracturing (Ch07), sustainable agriculture (Ch09), wildlife conservation (Ch12), air pollution (Ch17), oil sands extraction (Ch19), and more!

Central Case Studies draw students into the chapter with engaging topics that begin and are woven throughout each chapter.

CENTRAL **CASE STUDY**

The Tohoku Earthquake: Has It Shaken the World's Trust in Nuclear Power?

"This used to be one of the best places for a business. I'm amazed at how little is left."

—Takahiro Chiba, surveying the devastated downtown area of Ishinomaki, Japan, where his family's sushi restaurant was located

"Fukushima should not just contain lessons for Japan, but for all 31 countries with nuclear power."

—Tatsujiro Suzuki, Vice-chairman, Japan Atomic Energy Commission

At 2:46 p.m. on March 11, 2011, the land along the northeastern coast of the Japanese island of Honshu began to shake violently—and continued to shake for six minutes. These tremors were caused when a large section of the seafloor along a fault line 125 km (77 mi) offshore suddenly lurched, releasing huge amounts of energy through the crust and generating an earthquake of magnitude 9.0 on the Richter scale (a scale used to measure the strength of earthquakes). Little did any- as strong ocean surges followed the 1923 Tokyo–Yokohama earthquake, pushing walls of debris in front of them and drowning victims still trapped in the wreckage from the earthquake.

The Japanese had built seawalls to protect against tsunamis, but the Tohoku quake caused the island of Honshu to sink, lowering the height of the seawalls by up to 2 m (6.5 ft) in some locations. Waves reaching up to 15 m (49 ft) in height then overwhelmed these defenses (FIGURE 2.1). The raging water swept up to 9.6

New topographical maps help students see the political and environmental context of stories.

NEW! Video Field trips include:

- Invasive Species: Lionfish
- Solutions: Sustainability on Campus
- Wind Power

Video Field Trips offer fascinating tours of real environmental issues and solutions employed to cope with them. These videos are assignable in MasteringEnvironmentalScience and are on the Instructor Resource DVD.

Current Events from the *New York Times* are regularly updated and invite you to connect course topics with current environmental issues.

ABC News video clips invite you to engage in the current conversation about the environment and sustainability.

INTERPRET AND ANALYZE STORIES USING SCIENTIFIC LITERACY SKILLS

Science Behind the Story features highlight how scientists develop hypotheses, test predictions, analyze and interpret data, and share findings.

THE SCIENCE BEHIND THE STORY

Did Soap Operas Reduce Fertility in Brazil?

Over the past 50 years, the South American nation of Brazil experienced the second-largest drop in fertility among developing nations with large populations—second only to China. In the 1960s, the average woman in Brazil had six children. Today, Brazil's total fertility rate is 1.9 children per woman, which is lower than that of the United States. Brazil's drastic decrease in fertility is interesting because, unlike China, it occurred without governmental policies that advocated controls on its citizens' reproduction.

So how did Brazil accomplish this? A major factor was change in society's view of women. It began with a civil rights movement in the 1960s, which gave females equal access to education and the opportunity to pursue careers outside the home. These efforts have been highly successful. Women now comprise 40% of the workforce in Brazil and graduate from college in greater numbers than men. And in 2010, Brazilians elected a woman, Dilma Rousseff, as their nation's president.

Although the Brazilian government

Brazilian soap operas, called *telenovelas*, are a surprising cultural force for promoting lower fertility. Here, residents gather outside a cafe in Rio de Janeiro to watch the popular program *Avenida Brasil*.

China; the procedure is illegal except in rare circumstances.

As Brazil's economy grew with industrialization, people's nutrition and access to health care improved, greatly reducing infant mortality rates. Families no longer needed to have more children than they desired for fear one or more would die at a young age.

characters, settings, and plot lines with which everyday Brazilians can identify.

Telenovelas do not overtly address fertility issues, but they do promote a vision of the "ideal" Brazilian family. This family is typically middle or upper class, materialistic, individualistic, and full of empow-

NEW! 30% of the Science Behind the Stories are entirely new in the text.

Topics include:

- Tracking Fukushima's nuclear legacy (Ch02)
- Tracking Populations of Hakalau's Forest Birds (Ch03)
- Does Fracking Contaminate Drinking Water? (Ch07)

And More!

NEW! Data Analysis Q's are paired with select figures in each chapter to help you develop your scientific literacy skills. These questions allow you to check your own understanding of environmental data as you read through each chapter.

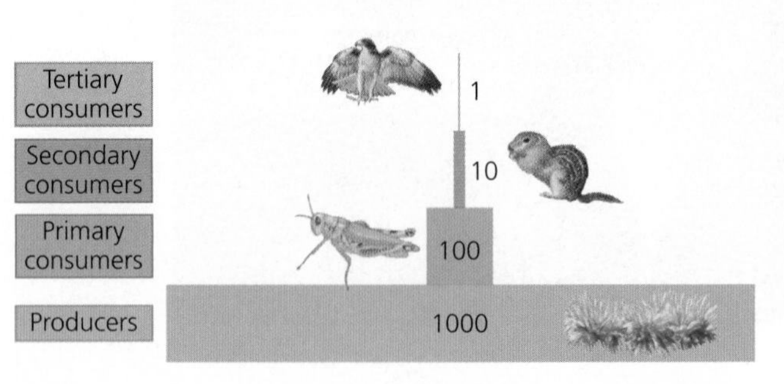

FIGURE 4.10 Lower trophic levels generally contain more organisms, energy content, and biomass than higher trophic levels. The tenfold ratio shown here is typical, but the shape of the pyramid may vary greatly.

DATA Q Using the ratios shown in this example, let's suppose that a system has 3000 grasshoppers. How many rodents would be expected?

USE MasteringEnvironmentalScience®
TO PRACTICE SCIENTIFIC LITERACY SKILLS

NEW! Process of Science Coaching Activities, created by coauthor Matt Laposata, help you practice the process of science demonstrated in the Science Behind the Story feature. These activities allow you to think like a scientist and put the scientific method into practice.

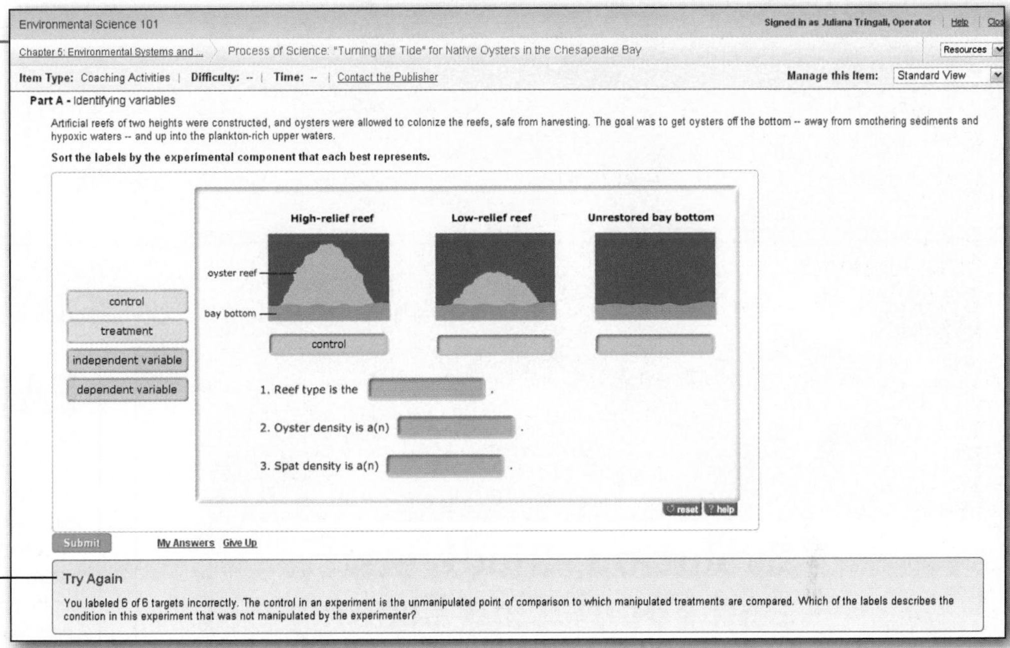

Wrong Answer Feedback Gain a better understanding of the process of science with specific wrong-feedback.

Withgott/Laposata 5e — Signed in as Lee Ann Doctor, Instructor | Help | Close

Chapter 8: Climate Change › Interpreting Graphs and Data: Projections of Global Warming — Resources

Item Type: Coaching Activities | Difficulty: 1 | Time: 3m | Learning Outcomes | Contact the Publisher — Manage this Item: Standard View

Interpreting Graphs and Data: Projections of Global Warming

Scientists used computer models of global circulation to forecast the amount of global warming likely to result from several different scenarios.

Can you interpret the graph to answer these questions? Note that the shading around the graph lines indicates uncertainty in the predictions.

Business as Usual (no actions taken to reduce CO_2 emissions)
Sustainable World (significant actions taken to reduce CO_2 emissions)
Today's World (immediate cessation of CO_2 emissions)
20th century data

GLOBAL SURFACE WARMING (°C): 6.0, 5.0, 4.0, 3.0, 2.0, 1.0, 0, −1.0
DATE: 1900, 2000, 2100

© Pearson Education, Inc.

Source: IPCC. Climate change 2007: Synthesis Report. Geneva, Switzerland

Part A

What information is presented on the y-axis of the graph?

- global surface warming, in °Celsius
- global surface warming, in °Fahrenheit
- time, in 100-year intervals
- global surface temperature, in °Celsius

Submit — My Answers — Give Up

Part B

What does the yellow line represent?

- **Sustainable World:** The amount of global warming that is likely to occur if governments and individuals take significant actions to slow the increase in CO2 emissions.
- **Today's World:** The amount of global warming that is likely to occur if CO2 emissions cease immediately and CO2 concentrations continue at their current level.
- **Business as Usual:** The amount of global warming that is likely to occur if governments and individuals take no action to slow the increase in CO2 emissions.

Submit — My Answers — Give Up

Part C

Which line is *not* a computer-generated forecast?

- the black line representing 20th century data
- the yellow line representing Today's World
- the red line representing Business as Usual
- the blue line representing a Sustainable World

Submit — My Answers — Give Up

Expanded! Interpreting Graphs and Data Activities help you develop basic data analysis skills and practice interpreting environmental data.

Keep Practicing

GraphIt Activities help you analyze an environmental issue and understand the research data.

IDENTIFY LEARNING GOALS AND RECOGNIZE MISCONCEPTIONS

Evolution, Biodiversity, and Population Ecology

Upon completing this chapter, you will be able to:

- Explain natural selection and cite evidence for this process
- Describe how evolution influences biodiversity
- Discuss reasons for species extinction and mass extinction events
- List the levels of ecological organization
- Outline the characteristics of populations that help predict population growth
- Assess logistic growth, carrying capacity, limiting factors, and other fundamental concepts in population ecology
- Identify efforts and challenges involved in the conservation of biodiversity

Learning Objectives at the beginning of each chapter define what you should be able to do after completing the chapter. MasteringEnvironmentalScience also links assessments to learning objectives so professors can track students' progress.

NEW! FAQs highlight and correct common misconceptions about environmental isssues.

FAQ **Isn't evolution based on just one man's beliefs?**

Because Charles Darwin contributed so much to our early understanding of evolution, many people assume the concept itself hinges on his ideas. But scientists and laypeople had been observing nature and puzzling over fossils for a long time, and the notion of evolution was being discussed long before Darwin. Once he and Alfred Russel Wallace independently proposed the concept of natural selection, scientists finally gained a precise and feasible mechanism to explain how and why organisms change across generations. Later, geneticists discovered Gregor Mendel's research and worked out how traits are inherited—and modern evolutionary biology was born. Twentieth-century scientists Fisher, Wright, Dobzhansky, Simpson, Mayr, and others ran experiments and developed sophisticated mathematical models, documenting phenomena with extensive evidence and making evolutionary biology into one of science's strongest fields. Since then, evolutionary research by thousands of scientists has driven our understanding of biology and has facilitated spectacular advances in agriculture, medicine, and biotechnology.

Reviewing Objectives at the end of each chapter use a learning objective framework to help you review concepts and prepare for exams.

Reviewing Objectives

You should now be able to:

- **Explain natural selection and cite evidence for this process**
  - Because organisms produce excess young, individuals vary in their traits, and many traits are inherited, some individuals will prove better at surviving and reproducing. Their genes will be passed on and become more prominent in future generations. (p. 50)
  - Mutations and recombination provide the genetic variation for natural selection. (p. 50)

USE MasteringEnvironmentalScience® TO ENHANCE YOUR UNDERSTANDING AND IMPROVE YOUR GRADE

Concept Review Activities guide you through an understanding of complex content.

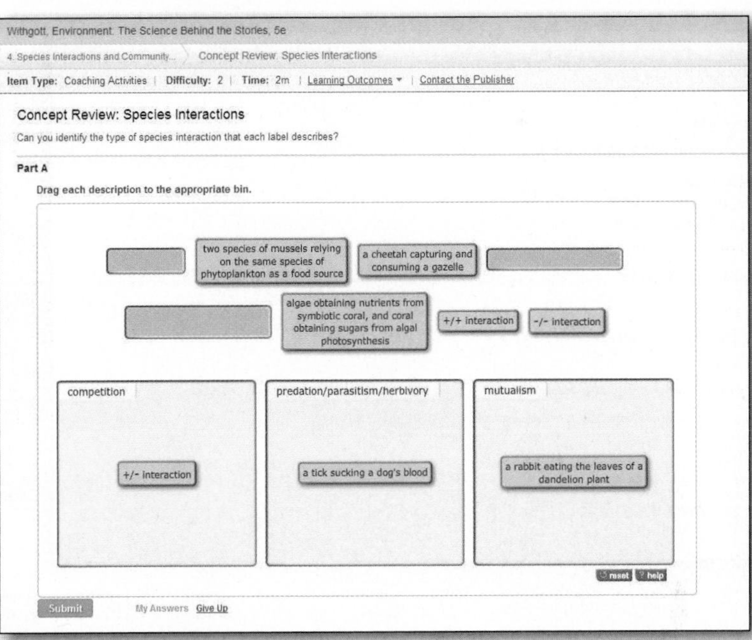

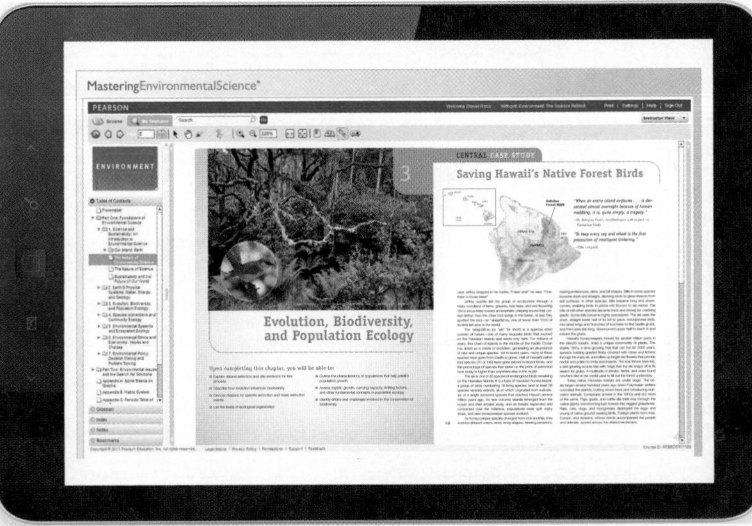

The **Pearson eText** gives you access to your text whenever and wherever you can access the Internet so you can view it on your computer and/or your mobile device (including IOS and Android)

Additional MasteringEnvironmentalScience Resources Include:

- **Learning Catalytics™** allows you to communicate with your instructor in real time via your mobile device during class to enhance your understanding and comprehension of the course material.
- **First impressions quiz questions** help you assess your understanding of environmental issues and key concepts from the very first day of class.
- **BioFlix coaching activities** use dynamic 3-D animations to teach tough topics in Environmental Science. Activities are highly interactive, automatically graded, and include a wide variety of question types.
- **Study Area** offers a 24/7 self-study resource for you presented in an easy-to-understand chapter guide design.

FOR INSTRUCTORS: ENGAGE AND ASSESS STUDENTS WITH MasteringEnvironmentalScience®

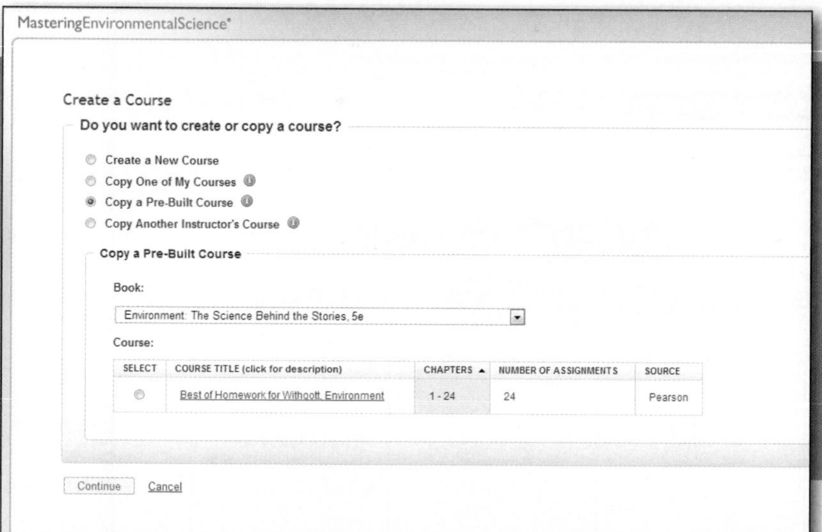

Quickly select pre-built assignments created by Matthew Laposata
that offer a wide range of interactive, engaging, assignable activities—including animations, videos, and news articles.

With **MasteringEnvironmental Science™** students benefit from activities that feature immediate wrong-answer feedback to help keep them on track.

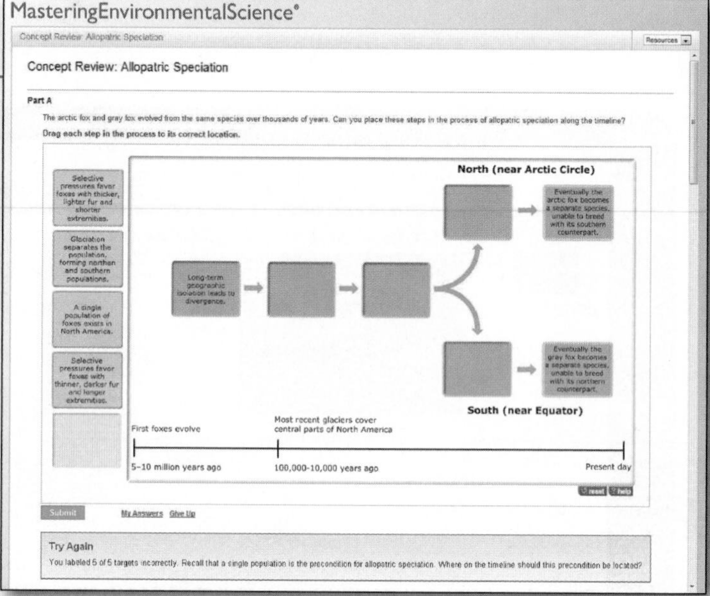

NEW! Learning Catalytics

Learning Catalytics™ is a "bring your own device" student engagement, assessment, and classroom intelligence system. With Learning Catalytics you can infuse your lectures with opportunities for active learning, assess student understanding in real time, and adjust your lecture accordingly.

TRACK STUDENT PERFORMANCE AND HELP STUDENTS SUCCEED

Gradebook

Get easy-to-interpret insights into student performance using the gradebook.

- Every assignment is automatically graded.
- Shades of red highlight vulnerable students and challenging assignments.

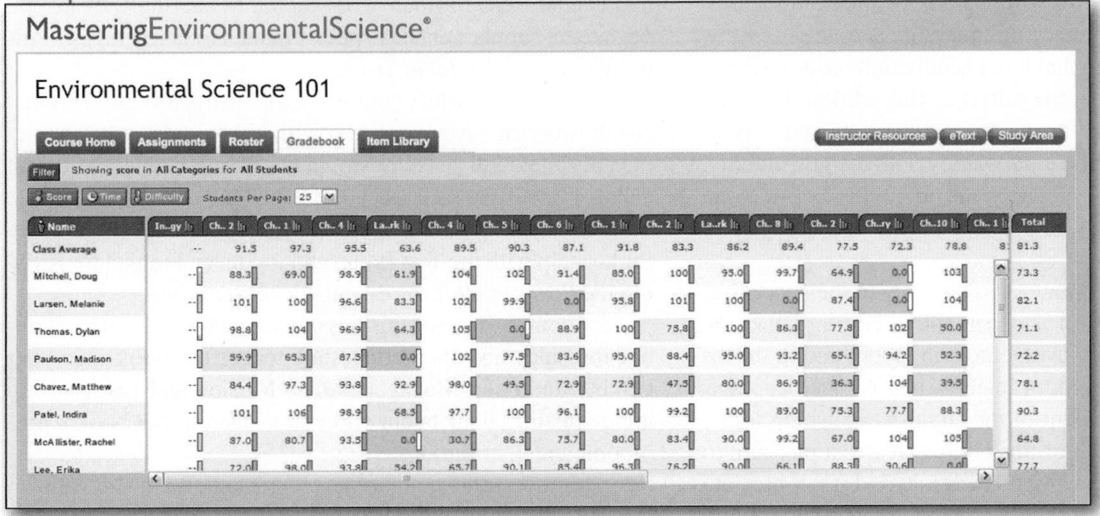

One-click Diagnostics

show where students are struggling or progressing.

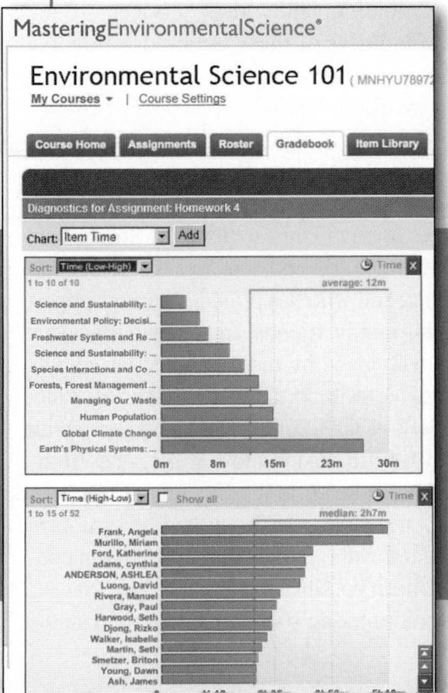

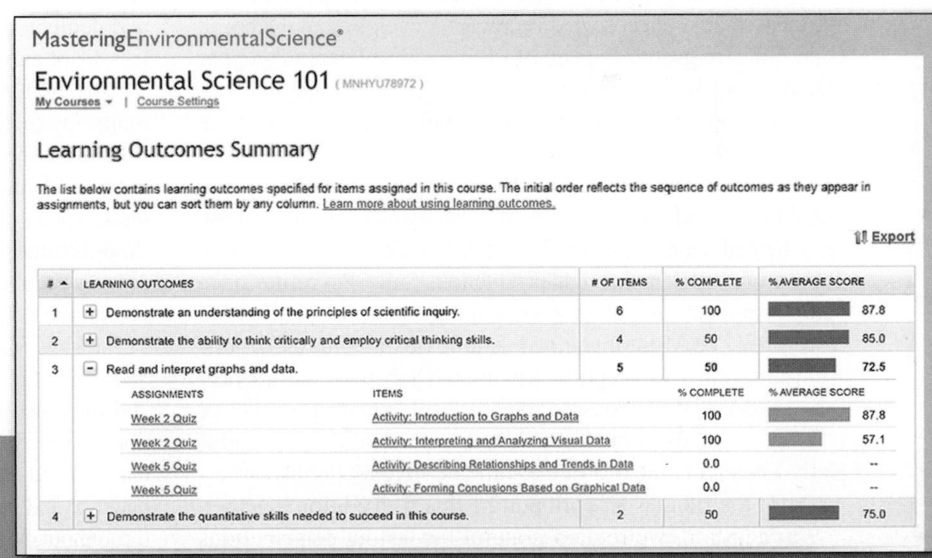

Track student performance and concept mastery against publisher-provided learning outcomes, or create your own.

Acknowledgments

A textbook is the product of *many* more minds and hearts than one might guess from the names on the cover. The two of us are exceedingly fortunate to be supported and guided by a tremendous publishing team and by a small army of experts in environmental science who have generously shared their time and expertise. The strengths of this book result from the collective labor and dedication of innumerable people.

We would first like to thank our acquisitions editor, Alison Rodal. Alison joined us at the outset of this edition, bringing a fresh perspective to the book along with skills and experience from multiple aspects of publishing. Her insight, alacrity, and efficacy have greatly enhanced the outcome. We—and the instructors and students who use this book—are fortunate to have her at the helm.

Project editor Anna Amato was also key to the success of this edition. Anna's careful and perceptive editing benefited all of us, and her creative involvement in layout and in the art program improved the book in many ways. We also appreciated her skillful management of the endless publishing logistics during the preparation of this edition. It was a pleasure to work with Alison and Anna on this edition, and we appreciate their patience with us and their dedication to top-quality work.

We wish to thank our editor-in-chief Beth Wilbur for her strong and steady support of this book through its five editions. We also thank executive director of development Deborah Gale. Sincere gratitude is due to Beth and to Pearson's upper management for continuing to invest the resources and top-notch personnel that our books are enjoying now and have enjoyed over the past decade.

Editorial assistants Rachel Brickner and Libby Reiser provided timely and effective help. We also thank Camille Herrera, who helped launch production of this edition, and Shannon Tozier, who saw it through production. Sally Peyrefitte once again provided meticulous copy editing of our text, and photo researcher Zoe Milgram helped acquire quality photos. Wynne Au Yeung did an exceptionally smooth job with the art program, and Yvo Riezebos designed the brilliant new text interior and the cover. We send a huge thank-you to production editor Kelly Keeler and the rest of the staff at Cenveo® Publisher Services for their fantastic work putting this fifth edition together.

In addition, we remain grateful for lasting contributions to the book's earlier editions by Nora Lally-Graves, Mary Ann Murray, Susan Teahan, Tim Flem, and Dan Kaveney, as well as by Etienne Benson, Russell Chun, Jonathan Frye, April Lynch, Kristy Manning, and many others. Needless to say, Scott Brennan was instrumental. His ideas, words, and voice have reverberated through the editions even as the particulars have evolved many times over. And as much as anyone, our former editor Chalon Bridges deserves credit for making this book what it is. Chalon's heartfelt commitment to quality educational publishing has long inspired us all, and our efforts continue to owe a great deal to her astute guidance and vision.

As we move deeper into the electronic age, *Mastering-EnvironmentalScience* plays an ever-larger role in what we do. As we worked to expand our online offerings with this edition, we thank Kayla Rihani, Julie Stoughton, Steven Frankel, Karen Sheh, Tania Mlawer, Juliana Golden, Lee Ann Doctor, and Daniel Ross for their work on the Mastering website and our media supplements. A special thanks to Eric Flagg for his tremendous *Video Field Trips.*

As always, a select number of top instructors from around North America have teamed up with us to produce the supplementary materials used by so many educators, and we remain deeply grateful for their valuable help. Our thanks go to Danielle DuCharme for updating our Instructor's Guide, to Todd Tracy for his help with the Test Bank, and to Steven Frankel for revising the PowerPoint lectures and clicker questions.

Of course, none of this has any impact on education without the sales and marketing staff to get the book into your hands. Marketing Managers Amee Mosley and Lauren Harp are dedicating their talent and enthusiasm to the book's promotion and distribution.

Moreover, the many sales representatives who help to communicate our vision, deliver our product to instructors, and work with instructors to assure their satisfaction, are absolutely vital. We have been blessed with an amazingly sharp and dedicated sales force, and we deeply appreciate their tireless work and commitment.

In the lists of reviewers that follow, we acknowledge the many instructors and outside experts who have helped us to maximize the quality and accuracy of our content and presentation through their chapter reviews, feature reviews, class tests, focus group participation, and other services. If the thoughtfulness and thoroughness of these hundreds of people are any indication, we feel confident that the teaching of environmental science is in excellent hands!

Lastly, we each owe personal debts to the people nearest and dearest to us. Jay thanks his parents and his many teachers and mentors over the years for making his own life and education so enriching. He gives loving thanks to his wife, Susan, who has endured this book's writing and revision over the years with patience and understanding, and who has provided caring support throughout. Matt thanks his family, friends, and colleagues, and is grateful for his children, who give him three reasons to care passionately about the future. Most importantly, he thanks his wife, Lisa, for blessing every day of his life for the past 25 years with her keen insight, passion for life, unconscious grace, and effortless beauty—and for understanding him in ways no one else ever could. The talents, input, and advice of Susan and of Lisa have been vital to this project, and without their support our own contributions would not have been possible.

We dedicate this book to today's students, who will shape tomorrow's world.

—Jay Withgott and Matthew Laposata

Reviewers

We wish to express special thanks to the dedicated reviewers who shared their time and expertise to help make this fifth edition the best it could be. Their efforts built on those of the roughly 600 instructors and outside experts who have reviewed material for the previous four editions of this book through chapter reviews, pre-revision reviews, feature consultation, student reviews, class testing, and focus groups. Our sincere gratitude goes out to all of them.

Reviewers for the Fifth Edition

Charles Acosta, *Northern Kentucky University*
Eric Atkinson, *Northwest College*
Terrence Bensel, *Allegheny College*
Jill Bessetti, *Columbia College*
Donna Bivans, *Pitt Community College*
Philip A. Clifford, *Volunteer State Community College*
Dolores M. Eggers, *University of North Carolina at Asheville*
James English, *Gardner-Webb University*
Jonathan Fingerut, *St. Joseph's University*
Robyn Fischer, *Aurora University*
Eric J. Fitch, *Marietta College*
Michael Freake, *Lee University*
Karen Gaines, *Eastern Illinois University*
Katharine A. Gehl, *Asheville Buncombe Technical Community College*
Jeffrey J. Gordon, *Bowling Green State University*
Jennifer A. Hanselman, *Westfield State University*
William Hopper, *Florida Memorial University*
Jack Jeffrey, *formerly USFWS, Pepeekeo, Hawaii*
Richard R. Jurin, *University of Northern Colorado*
Karen Klein, *Northampton Community College*
Ned Knight, *Linfield College*
George Kraemer, *Purchase College*
Diana Kropf Gomez, *Dallas County Community College*
Max Kummerow, *Curtin University*
Hugh Lefcort, *Gonzaga University*
Jeffrey Mahr, *Georgia Perimeter College, Newton*
Allan Matthias, *University of Arizona*
Chuck McClaugherty, *University of Mount Union*
Annabelle McKie-Voerste, *Dalton State College*
Alberto Mestas-Nuñez, *Texas A&M University–Corpus Christi*
William C. Miller, *Temple University*
Bruce Olszewski, *San Jose State University*
Anthony Overton, *East Carolina University*
Gulni Ozbay, *Delaware State University*
Clayton Penniman, *Central Connecticut State University*
Thomas Pliske, *Florida International University*
Daniel Ratcliff, *Rose State College*
Irene Rossell, *University of North Carolina at Asheville*
Dana Royer, *Wesleyan University*
Steve Rudnick, *University of Massachusetts–Boston*
Dork Sahagian, *Lehigh University*
Andrew Shella, *Terra State Community College*
Jeff Slepski, *Mt. San Jacinto College*
Rik Smith, *Columbia Basin College*
Michelle Stevens, *California State University–Sacramento*
Keith S. Summerville, *Drake University*
Todd Tracy, *Northwestern College*
Ray Williams, *Rio Hondo College*
Sharon Walsh, *New Mexico State University*

Reviewers for Previous Editions

Matthew Abbott, *Des Moines Area Community College*; David Aborne, *University of Tennessee–Chattanooga;* Jeffrey Albert, *Watson Institute of International Studies;* Shamim Ahsan, *Metropolitan State College of Denver;* Isoken T. Aighewi, *University of Maryland–Eastern Shore;* John V. Aliff, *Georgia Perimeter College;* Mary E. Allen, *Hartwick College;* Deniz Z. Altin, *Georgia Perimeter College*; Dula Amarasiriwardena, *Hampshire College;* Gary I. Anderson, *Santa Rosa Junior College;* Mark W. Anderson, *The University of Maine*; Corey Andries, *Albuquerque Technical Vocational Institute;* David M. Armstrong, *University of Colorado–Boulder;* David L. Arnold, *Ball State University;* Joseph Arruda, *Pittsburg State University;* Thomas W. H. Backman, *Linfield College;* Timothy J. Bailey, *Pittsburg State University;* Stokes Baker, *University of Detroit;* Kenneth Banks, *University of North Texas;* Narinder Bansal, *Ohlone College;* Jon Barbour, *University of Colorado–Denver;* Reuben Barret, *Prairie State College;* Morgan Barrows, *Saddleback College;* Henry Bart, *LaSalle University;* James Bartalome, *University of California–Berkeley;* Marilynn Bartels, *Black Hawk College;* David Bass, *University of Central Oklahoma;* Christy Bazan, *Illinois State University;* Christopher Beals, *Volunteer State Community College;* Laura Beaton, *York College, City University of New York*; Hans T. Beck, *Northern Illinois University;* Richard Beckwitt, *Framingham State College;* Barbara Bekken, *Virginia Polytechnic Institute and State University;* Elizabeth Bell, *Santa Clara University;* Timothy Bell, *Chicago State University;* David Belt, *Johnson County Community College;* Gary Beluzo, *Holyoke Community College;* Terrence Bensel, *Allegheny College;* Bob Bennett, *University of Arkansas;* William B. N. Berry, *University of California, Berkeley;* Kristina Beuning, *University of Wisconsin–Eau Claire;* Peter Biesmeyer, *North Country Community College;* Donna Bivans, *Pitt Community College;* Grady Price Blount, *Texas A&M University–Corpus Christi;* Marsha Bollinger, *Winthrop University;* Lisa K. Bonneau, *Metropolitan Community College–Blue River*; Bruno Borsari, *Winona State University;* Richard D. Bowden, *Allegheny College;* Frederick J. Brenner, *Grove City College;* Nancy Broshot, *Linfield College;* Bonnie L. Brown, *Virginia Commonwealth University;* David Brown, *California State University–Chico;* Evert Brown, *Casper College;* Hugh Brown, *Ball State University;* J. Christopher Brown, *University of Kansas;* Dan Buresh, *Sitting Bull College;* Dale Burnside, *Lenoir-Rhyne University*; Lee Burras, *Iowa State University;* Hauke Busch, *Augusta State University;* Christina Buttington, *University of Wisconsin–Milwaukee;* Charles E. Button, *University of Cincinnati, Clermont College;* John S. Campbell, *Northwest College;* Myra Carmen Hall, *Georgia Perimeter College*; Mike Carney, *Jenks High School;* Kelly S. Cartwright, *College of Lake County;* Jon Cawley, *Roanoke College;* Michelle Cawthorn, *Georgia Southern University;* Linda Chalker-Scott, *University of Washington;* Brad S. Chandler, *Palo Alto College;* Paul Chandler, *Ball State University;* David A. Charlet, *Community College of Southern Nevada;* Sudip Chattopadhyay, *San Francisco State University;* Tait Chirenje, *Richard Stockton College;* Luanne Clark, *Lansing Community College*; Richard Clements, *Chattanooga State Technical Community College;* Kenneth E. Clifton, *Lewis and Clark College;* Reggie Cobb, *Nash Community College;* John E. Cochran, *Columbia Basin College;* Donna Cohen, *MassBay Community College*; Luke W. Cole, *Center on Race, Poverty, and the Environment;* Mandy L. Comes, *Rockingham Community College;* Thomas L. Crisman, *University of Florida;* Jessica Crowe, *South*

Georgia College; Ann Cutter, *Randolph Community College;* Gregory A. Dahlem, *Northern Kentucky University;* Randi Darling, *Westfield State College;* Mary E. Davis, *University of Massachusetts, Boston;* Thomas A. Davis, *Loras College;* Lola M. Deets, *Pennsylvania State University–Erie;* Ed DeGrauw, *Portland Community College;* Roger del Moral, *University of Washington;* Bob Dennison, *Lead teacher at HISD and Robert E. Lee High School*; Michael L. Denniston, *Georgia Perimeter College*; Doreen Dewell, *Whatcom Community College*; Craig Diamond, *Florida State University;* Darren Divine, *Community College of Southern Nevada;* Stephanie Dockstader, *Monroe Community College;* Toby Dogwiler, *Winona State University;* Jeffrey Dorale, *University of Iowa;* Tracey Dosch, *Waubonsee Community College;* Michael L. Draney, *University of Wisconsin–Green Bay;* Iver W. Duedall, *Florida Institute of Technology;* Dee Eggers, *University of North Carolina–Asheville;* Jane Ellis, *Presbyterian College;* Amy Ellwein, *University of New Mexico*; JodyLee Estrada Duek, *Pima Community College;* Jeffrey R. Dunk, *Humboldt State University;* Jean W. Dupon, *Menlo College;* Robert M. East, Jr., *Washington & Jefferson College;* Margaret L. Edwards-Wilson, *Ferris State University;* Anne H. Ehrlich, *Stanford University;* Thomas R. Embich, *Harrisburg Area Community College;* Kenneth Engelbrecht, *Metropolitan State College of Denver;* Bill Epperly, *Robert Morris College;* Corey Etchberger, *Johnson County Community College;* W. F. J. Evans, *Trent University;* Paul Fader, *Freed Hardeman University;* Joseph Fail, *Johnson C. Smith University;* Bonnie Fancher, *Switzerland County High School;* Jiasong Fang, *Iowa State University;* Marsha Fanning, *Lenoir Rhyne College*; Leslie Fay, *Rock Valley College;* Debra A. Feikert, *Antelope Valley College;* M. Siobhan Fennessy, *Kenyon College;* Francette Fey, *Macomb Community College;* Steven Fields, *Winthrop University;* Brad Fiero, *Pima Community College;* Dane Fisher, *Pfeiffer University;* David G. Fisher, *Maharishi University of Management;* Linda M. Fitzhugh, *Gulf Coast Community College;* Doug Flournoy, *Indian Hills Community College–Ottumwa;* Johanna Foster, *Johnson County Community College;* Chris Fox, *Catonsville Community College;* Nancy Frank, *University of Wisconsin–Milwaukee;* Steven Frankel, *Northeastern Illinois University;* Arthur Fredeen, *University of Northern British Columbia;* Chad Freed, *Widener University;* Robert Frye, *University of Arizona;* Laura Furlong, *Northwestern College;* Navida Gangully, *Oak Ridge High School;* Sandi B. Gardner, *Triton College;* Kristen S. Genet, *Anoka Ramsey Community College;* Stephen Getchell, *Mohawk Valley Community College;* Marcia Gillette, *Indiana University–Kokomo;* Scott Gleeson, *University of Kentucky*; Sue Glenn, *Gloucester County College;* Thad Godish, *Ball State University;* Nisse Goldberg, *Jacksonville University;* Michele Goldsmith, *Emerson College;* Jeffrey J. Gordon, *Bowling Green State University;* John G. Graveel, *Purdue University;* Jack Greene, *Millikan High School;* Cheryl Greengrove, *University of Washington;* Amy R. Gregory, *University of Cincinnati, Clermont College;* Carol Griffin, *Grand Valley State University;* Carl W. Grobe, *Westfield State College;* Sherri Gross, *Ithaca College;* David E. Grunklee, *Hawkeye Community College;* Judy Guinan, *Radford University;* Gian Gupta, *University of Maryland, Eastern Shore;* Mark Gustafson, *Texas Lutheran University;* Daniel Guthrie, *Claremont College;* Sue Habeck, *Tacoma Community College;* David Hacker, *New Mexico Highlands University;* Greg Haenel, *Elon University;* Mark Hammer, *Wayne State University;* Grace Hanners, *Huntingtown High School;* Michael Hanson, *Bellevue Community College;* Alton Harestad, *Simon Fraser University;* Barbara Harvey, *Kirkwood Community College;* David Hassenzahl, *University of Nevada Las Vegas;* Jill Haukos, *South Plains College;* Keith Hench, *Kirkwood Community College;* George Hinman, *Washington State University;* Jason Hlebakos, *Mount San Jacinto College;* Joseph Hobbs, *University of Missouri–Columbia;* Jason Hoeksema, *Cabrillo College;* Curtis Hollabaugh, *University of West Georgia;* Robert D. Hollister, *Grand Valley State University;* David Hong, *Diamond Bar High School;* Catherine Hooey, *Pittsburgh State University;* Kathleen Hornberger, *Widener University;* Debra Howell, *Chabot College;* April Huff, *North Seattle Community College;* Pamela Davey Huggins, *Fairmont State University;* Barbara Hunnicutt, *Seminole Community College;* Jonathan E. Hutchins, *Buena Vista University;* James M. Hutcheon, *Asian University for Women;* Don Hyder, *San Juan College*; Daniel Hyke, *Alhambra High School;* Juana Ibanez, *University of New Orleans;* Walter Illman, *University of Iowa;* Neil Ingraham, *California State University, Fresno*; Daniel Ippolito, *Anderson University;* Bonnie Jacobs, *Southern Methodist University;* Jason Janke, *Metropolitan State College of Denver*; Nan Jenks-Jay, *Middlebury College;* Linda Jensen-Carey, *Southwestern Michigan College*; Stephen R. Johnson, *William Penn University;* Gail F. Johnston, *Lindenwood University;* Gina Johnston, *California State University, Chico;* Paul Jurena, *University of Texas–San Antonio;* Richard R. Jurin, *University of Northern Colorado;* Thomas M. Justice, *McLennan Community College;* Stanley S. Kabala, *Duquesne University;* Brian Kaestner, *Saint Mary's Hall;* Steve Kahl, *Plymouth State University;* David M. Kargbo, *Temple University*; Susan Karr, *Carson–Newman College*; Jerry H. Kavouras, *Lewis University;* Carol Kearns, *University of Colorado–Boulder;* Richard R. Keenan, *Providence Senior High School;* Dawn G. Keller, *Hawkeye Community College;* Myung-Hoon Kim, *Georgia Perimeter College*; Kevin King, *Clinton Community College;* John C. Kinworthy, *Concordia University;* Cindy Klevickis, *James Madison University;* Ned J. Knight, *Linfield College;* David Knowles, *East Carolina University;* Penelope M. Koines, *University of Maryland;* Alexander Kolovos, *University of North Carolina–Chapel Hill;* Erica Kosal, *North Carolina Wesleyan College;* Steven Kosztya, *Baldwin Wallace College;* Robert J. Koester, *Ball State University;* Tom Kozel, *Anderson College;* Robert G. Kremer, *Metropolitan State College of Denver;* Jim Krest, *University of South Florida–South Florida;* Sushma Krishnamurthy, *Texas A&M International University;* Diana Kropf-Gomez, *Richland College (DCCCD);* Jerome Kruegar, *South Dakota State University;* James Kubicki, *The Pennsylvania State University;* Frank T. Kuserk, *Moravian College;* Diane M. LaCole, *Georgia Perimeter College;* Troy A. Ladine, *East Texas Baptist University;* William R. Lammela, *Nazareth College;* Vic Landrum, *Washburn University;* Tom Langen, *Clarkson University;* Andrew Lapinski, *Reading Area Community College;* Matthew Laposata, *Kennesaw State University*; Michael T. Lares, *University of Mary;* Kim D. B. Largen, *George Mason University;* John Latto, *University of California–Berkeley;* Lissa Leege, *Georgia Southern University;* James Lehner, *Taft School;* Kurt Leuschner, *College of the Desert*; Stephen D. Lewis, *California State University, Fresno;* Chun Liang, *Miami University;* John Logue, *University of South Carolina–Sumter;* John F. Looney, Jr., *University of Massachusetts–Boston;* Joseph Luczkovich, *East Carolina University;* Linda Lusby, *Acadia University;* Richard A. Lutz, *Rutgers University;* Jennifer Lyman, *Rocky Mountain College;* Les M. Lynn, *Bergen Community College;* Timothy F. Lyon, *Ball State University;* Sue Ellen Lyons, *Holy Cross School;* Ian R. MacDonald, *Texas A&M University;* James G. March, *Washington and Jefferson College;* Blase Maffia, *University of Miami;* Robert L. Mahler, *University of Idaho;* Keith Malmos, *Valencia Community College;* Kenneth Mantai, *State University of New York–Fredonia;* Anthony J.M. Marcattilio, *St. Cloud State University;* Heidi Marcum, *Baylor University;* Nancy Markee, *University of Nevada–Reno;* Patrick S. Market, *University of Missouri–Columbia;* Michael D. Marlen, *Southwestern Illinois College;* Kimberly Marsella, *Skidmore College*; Steven R. Martin, *Humboldt State University;* John Mathwig, *College of Lake County;* Allan Matthias, *University of Arizona;* Robert Mauck, *Kenyon College;* Brian Maurer, *Michigan State University*; Bill Mautz, *University of New Hampshire;* Kathy McCann Evans, *Reading Area Community College*; Debbie McClinton, *Brevard Community College;* Paul McDaniel, *University of Idaho;* Jake McDonald, *University of New Mexico;* Gregory McIsaac, *Cornell University;* Robert M.L. McKay, *Bowling Green State University*; Dan McNally, *Bryant University;* Richard McNeil, *Cornell University;* Julie Meents, *Columbia College;* Alberto Mestas-Nunez, *Texas A&M University–Corpus Christi;* Mike L. Meyer, *New Mexico Highlands University;* Steven J. Meyer, *University of Wisconsin–Green Bay;* Patrick Michaels, *Cato Institute;* Christopher Migliaccio, *Miami Dade Community College;* Matthew R. Milnes, *University of California, Irvine;* Kiran Misra, *Edinboro University of Pennsylvania;* Mark Mitch, *New England College;* Lori Moore, *Northwest Iowa Community College;* Paul Montagna, *University of Texas–Austin;* Brian W. Moores, *Randolph-Macon College;* James T. Morris, *University of South Carolina;* Sherri Morris, *Bradley University;* Mary Murphy, *Penn State Abington;* William M. Murphy, *California State University–Chico;* Carla S. Murray, *Carl Sandburg College;* Rao Mylavarapu, *University of Florida;* Jane Nadel-Klein, *Trinity College;* Muthena Naseri, *Moorpark College;* Michael J. Neilson, *University of Alabama–Birmingham;* Benjamin Neimark, *Temple University*; Michael Nicodemus, *Abilene Christian University;* Richard A. Niesenbaum, *Muhlenberg College;* Moti Nissani, *Wayne State University;* Richard B. Norgaard, *University of California–Berkeley;* John Novak, *Colgate University;* Mark P. Oemke, *Alma College;* Niamh O'Leary, *Wells College;* Bruce Olszewski, *San Jose State University;* Brian O'Neill, *Brown University;* Nancy Ostiguy, *Penn State University;* David R. Ownby, *Stephen F. Austin State University;* Eric Pallant, *Allegheny College;* Philip Parker, *University of Wisconsin–Platteville;* Tommy Parker, *University of Louisville;* Brian Peck, *Simpson College;* Brian D. Peer, *Simpson College;* Clayton Penniman, *Central Connecticut State;* Christopher Pennuto, *Buffalo*

State College; Donald J. Perkey, *University of Alabama–Huntsville;* Barry Perlmutter, *College of Southern Nevada;* Shana Petermann, *Minnesota State Community and Technical College–Moorhead;* Craig D. Phelps, *Rutgers University;* Neal Phillip, *Bronx Community College*; Frank X. Phillips, *McNeese State University;* Raymond Pierotti, *University of Kansas;* Elizabeth Pixley, *Monroe Community College;* John Pleasants, *Iowa State University*; Thomas E. Pliske, *Florida International University;* Gerald Pollack, *Georgia Perimeter College*; Mike Priano, *Westchester Community College*; Daryl Prigmore, *University of Colorado;* Avram G. Primack, *Miami University of Ohio;* Alison Purcell, *Humboldt State University*; Sarah Quast, *Middlesex Community College;* Daniel Ratcliff, *Rose State College*; Patricia Bolin Ratliff, *Eastern Oklahoma State College;* Loren A. Raymond, *Appalachian State University;* Barbara Reynolds, *University of North Carolina–Asheville;* Thomas J. Rice, *California Polytechnic State University;* Samuel K. Riffell, *Mississippi State University;* Kayla Rihani, *Northeast Illinois University*; James Riley, *University of Arizona*; Gary Ritchison, *Eastern Kentucky University;* Virginia Rivers, *Truckee Meadows Community College;* Roger Robbins, *East Carolina University;* Tom Robertson, *Portland Community College, Rock Creek Campus;* Mark Robson, *University of Medicine and Dentistry of New Jersey;* Carlton Lee Rockett, *Bowling Green State University;* Angel M. Rodriguez, *Broward Community College;* Deanne Roquet, *Lake Superior College;* Armin Rosencranz, *Stanford University;* Irene Rossell, *University of North Carolina*; Robert E. Roth, *The Ohio State University;* George E. Rough, *South Puget Sound Community College;* Steven Rudnick, *University of Massachusetts–Boston;* John Rueter, *Portland State University;* Christopher T. Ruhland, *Minnesota State University;* Shamili A. Sandiford; *College of DuPage;* Robert Sanford, *University of Southern Maine;* Ronald Sass, *Rice University;* Carl Schafer, *University of Connecticut;* Jeffery A. Schneider, *State University of New York–Oswego;* Kimberly Schulte, *Georgia Perimeter College*; Edward G. Schultz, III, *Valencia Community College;* Mark Schwartz, *University of California–Davis;* Jennifer Scrafford, *Loyola College;* Richard Seigel, *Towson University;* Julie Seiter, *University of Nevada–Las Vegas;* Wendy E. Sera, *NDAA's National Ocean Service;* Maureen Sevigny, *Oregon Institute of Technology;* Rebecca Sheesley, *University of Wisconsin–Madison;* Pamela Shlachtman, *Miami Palmetto Senior High School;* Brian Shmaefsky, *Kingwood College;* William Shockner, *Community College of Baltimore County;* Christian V. Shorey, *University of Iowa;* Elizabeth Shrader, *Community College of Baltimore County*; Robert Sidorsky, *Northfield Mt. Hermon High School;* Linda Sigismondi, *University of Rio Grande;* Gary Silverman, *Bowling Green State University*; Jeffrey Simmons, *West Virginia Wesleyan College;* Cynthia Simon, *University of New England;* Jan Simpkin, *College of Southern Idaho;* Michael Singer, *Wesleyan University;* Diane Sklensky, *Le Moyne College;* Ben Smith, *Palos Verdes Peninsula High School;* Mark Smith, *Chaffey College;* Patricia L. Smith, *Valencia Community College;* Sherilyn Smith, *Le Moyne College;* Debra Socci, *Seminole Community College;* Roy Sofield, *Chattanooga State Technical Community College;* Annelle Soponis, *Reading Area Community College*; Douglas J. Spieles, *Denison University;* Ravi Srinivas, *University of St. Thomas;* Bruce Stallsmith, *University of Alabama–Huntsville;* Jon G. Stanley, *Metropolitan State College of Denver;* Ninian Stein, *Wheaton College*; Jeff Steinmetz, *Queens University of Charlotte;* Richard Stevens, *Louisiana State University*; Bill Stewart, *Middle Tennessee State University;* Dion C. Stewart, *Georgia Perimeter College*; Julie Stoughton, *University of Nevada–Reno*; Richard J. Strange, *University of Tennessee;* Robert Strikwerda, *Indiana University–Kokomo;* Richard Stringer, *Harrisburg Area Community College;* Norm Strobel, *Bluegrass Community Technical College*; Andrew Suarez, *University of Illinois;* Keith S. Summerville, *Drake University;* Ronald Sundell, *Northern Michigan University;* Bruce Sundrud, *Harrisburg Area Community College;* Jim Swan, *Albuquerque Technical Vocational Institute;* Mark L. Taper, *Montana State University;* Todd Tarrant, *Michigan State University*; Max R. Terman, *Tabor College;* Julienne Thomas, *Robert Morris College;* Patricia Terry, *University of Wisconsin–Green Bay;* Jamey Thompson, *Hudson Valley Community College;* Rudi Thompson, *University of North Texas;* Todd Tracy, *Northwestern College;* Amy Treonis, *Creighton University;* Adrian Treves, *Wildlife Conservation Society;* Frederick R. Troeh, *Iowa State University;* Virginia Turner, *Robert Morris College;* Michael Tveten, *Pima Community College;* Thomas Tyning, *Berkshire Community College;* Charles Umbanhowar, *St. Olaf College;* G. Peter van Walsum, *Baylor University;* Callie A. Vanderbilt, *San Juan College;* Elichia A. Venso, *Salisbury University;* Rob Viens, *Bellevue Community College;* Michael Vorwerk, *Westfield State College;* Caryl Waggett, *Allegheny College;* Maud M. Walsh, *Louisiana State University;* Daniel W. Ward, *Waubonsee Community College;* Darrell Watson, *The University of Mary Hardin Baylor;* Phillip L. Watson, *Ferris State University;* Lisa Weasel, *Portland State University;* Kathryn Weatherhead, *Hilton Head High School;* John F. Weishampel, *University of Central Florida;* Peter Weishampel, *Northland College;* Barry Welch, *San Antonio College;* Kelly Wessell, *Tompkins Cortland Community College*; James W.C. White, *University of Colorado;* Susan Whitehead, *Becker College;* Jeffrey Wilcox, *University of North Carolina at Asheville;* Richard D. Wilk, *Union College;* Donald L. Williams, *Park University;* Justin Williams, *Sam Houston University;* Ray E. Williams, *Rio Hondo College;* Roberta Williams, *University of Nevada–Las Vegas;* Dwina Willis, *Freed-Hardeman University;* Shaun Wilson, *East Carolina University;* Tom Wilson, *University of Arizona;* James Winebrake, *Rochester Institute of Technology;* Danielle Wirth, *Des Moines Area Community College;* Lorne Wolfe, *Georgia Southern University;* Brian G. Wolff, *Minnesota State Colleges and Universities*; Marjorie Wonham, *University of Alberta;* Wes Wood, *Auburn University;* Jessica Wooten, *Franklin University;* Jeffrey S. Wooters, *Pensacola Junior College;* Joan G. Wright, *Truckee Meadows Community College;* Michael Wright, *Truckee Meadows Community College;* S. Rebecca Yeomans, *South Georgia College;* Karen Zagula, *Waketech Community College*; Lynne Zeman, *Kirkwood Community College;* Zhihong Zhang, *Chatham College.*

Suppliers of Student Reviews

Christine Brady, *California State Polytechnic University, Pomona*
Steven Rudnick, *University of Massachusetts–Boston*
Ninian R. Stein, *Wheaton College*
Todd Tracy, *Northwestern College*
Lorne Wolfe, *Georgia Southern University*

Class Testers

David Aborne, *University of Tennessee–Chattanooga;* Reuben Barret, *Prairie State College;* Morgan Barrows, *Saddleback College;* Henry Bart, *LaSalle University;* James Bartalome, *University of California–Berkeley;* Christy Bazan, *Illinois State University;* Richard Beckwitt, *Framingham State College;* Elizabeth Bell, *Santa Clara University;* Peter Biesmeyer, *North Country Community College;* Donna Bivans, *Pitt Community College;* Evert Brown, *Casper College;* Christina Buttington, *University of Wisconsin–Milwaukee;* Tait Chirenje, *Richard Stockton College;* Reggie Cobb, *Nash Community College;* Ann Cutter, *Randolph Community College;* Lola Deets, *Pennsylvania State University–Erie;* Ed DeGrauw, *Portland Community College;* Stephanie Dockstader, *Monroe Community College;* Dee Eggers, *University of North Carolina–Asheville;* Jane Ellis, *Presbyterian College;* Paul Fader, *Freed Hardeman University;* Joseph Fail, *Johnson C. Smith University;* Brad Fiero, *Pima Community College, West Campus;* Dane Fisher, *Pfeiffer University;* Chad Freed, *Widener University;* Stephen Getchell, *Mohawk Valley Community College;* Sue Glenn, *Gloucester County College;* Sue Habeck, *Tacoma Community College;* Mark Hammer, *Wayne State University;* Michael Hanson, *Bellevue Community College;* David Hassenzahl, *Oakland Community College;* Kathleen Hornberger, *Widener University;* Paul Jurena, *University of Texas–San Antonio;* Dawn Keller, *Hawkeye Community College;* David Knowles, *East Carolina University;* Erica Kosal, *Wesleyan College;* John Logue, *University of Southern Carolina Sumter;* Keith Malmos, *Valencia Community College;* Nancy Markee, *University of Nevada–Reno;* Bill Mautz, *University of New Hampshire;* Julie Meents, *Columbia College;* Lori Moore, *Northwest Iowa Community College;* Elizabeth Pixley, *Monroe Community College;* John Novak, *Colgate University;* Brian Peck, *Simpson College;* Sarah Quast, *Middlesex Community College;* Roger Robbins, *East Carolina University;* Mark Schwartz, *University of California–Davis;* Julie Seiter, *University of Nevada–Las Vegas;* Brian Shmaefsky, *Kingwood College;* Diane Sklensky, *Le Moyne College;* Mark Smith, *Fullerton College;* Patricia Smith, *Valencia Community College East;* Sherilyn Smith, *Le Moyne College;* Jim Swan, *Albuquerque Technical Vocational Institute;* Amy Treonis, *Creighton University;* Darrell Watson, *The University of Mary Hardin Baylor;* Barry Welch, *San Antonio College;* Susan Whitehead, *Becker College;* Roberta Williams, *University of Nevada–Las Vegas;* Justin Williams, *Sam Houston University;* Tom Wilson, *University of Arizona.*

Environment

The Science Behind the Stories

5TH EDITION

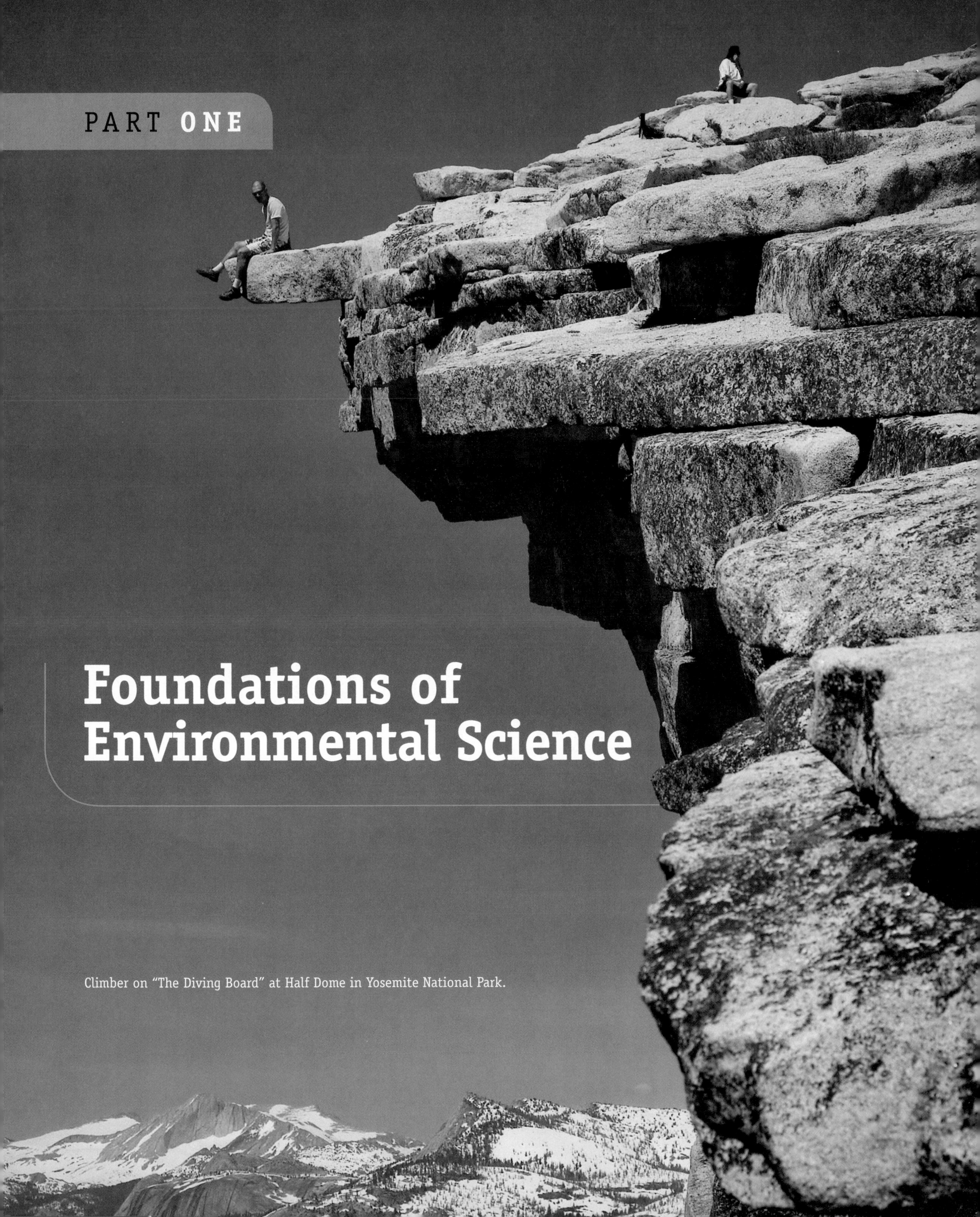

PART ONE

Foundations of Environmental Science

Climber on "The Diving Board" at Half Dome in Yosemite National Park.

1

Our Island, Earth

Science and Sustainability: An Introduction to Environmental Science

Upon completing this chapter, you will be able to:

- Define the term *environment* and describe the field of environmental science
- Explain the importance of natural resources and ecosystem services to our lives
- Discuss the consequences of population growth and resource consumption
- Characterize the nature of environmental science
- Understand the scientific method and the process of science
- Diagnose and illustrate some of the pressures on the global environment
- Articulate the concept of sustainability and describe campus sustainability efforts

Our Island, Earth

Viewed from space, our home planet resembles a small blue marble suspended in a vast inky-black void. Earth may seem enormous to us as we go about our lives on its surface, but the astronaut's view reveals that our planet is finite and limited. With this perspective, it becomes clear that as our population, technological power, and resource consumption all increase, so does our capacity to alter our surroundings and damage the very systems that keep us alive. Finding ways to live peacefully, healthfully, and sustainably on our diverse and complex planet is our society's prime challenge today. Meeting this challenge will require a solid scientific understanding of natural and social systems alike. The field of environmental science is crucial in this endeavor.

Our environment surrounds us

A photograph of Earth offers a revealing perspective, but it cannot convey the complexity of our environment. Our **environment** consists of all the living and nonliving things around us. It includes the continents, oceans, clouds, and ice caps you can see in the photo of Earth from space, as well as the animals, plants, forests, and farms that comprise the landscapes surrounding us. In a more inclusive sense, it also encompasses the structures, urban centers, and living spaces that people have created. In its broadest sense, our environment includes the complex webs of social relationships and institutions that shape our daily lives.

People commonly use the term *environment* in the first, most narrow sense—to mean a nonhuman or "natural" world apart from human society. This usage is unfortunate, because it masks the vital fact that people exist within the environment and are part of nature. As one of many species on Earth, we share with others the same dependence on a healthy, functioning planet. The limitations of language make it all too easy to speak of "people and nature," or "humans and the environment," as though they were separate and did not interact. However, the fundamental insight of environmental science is that we are part of the "natural" world and that our interactions with the rest of it matter a great deal.

Environmental science explores our interactions with the world

Understanding our relationship with the world around us is vital because we depend utterly on our environment for air, water, food, shelter, and everything else essential for living. Moreover, we modify our environment. Many of our actions have enriched our lives, bringing us better health, longer life spans, and greater material wealth, mobility, and leisure time—yet they have also often degraded the natural systems that sustain us. Impacts such as air and water pollution, soil erosion, and species extinction compromise our well-being and jeopardize our ability to build a society that will survive and thrive in the long term.

Environmental science is the study of how the natural world works, how our environment affects us, and how we affect our environment. We need to understand how we interact with our environment in order to devise solutions to our most pressing challenges. It can be daunting to reflect on the sheer magnitude of environmental dilemmas that confront us today, but these problems also bring countless opportunities for creative solutions.

Environmental scientists study the issues most centrally important to our world and its future. Right now, global conditions are changing more quickly than ever. Right now, through science, we are gaining knowledge more rapidly than ever. And right now, the window of opportunity for acting to solve problems is still open. With such bountiful opportunities, this particular moment in history is indeed an exciting time to be alive—and to be studying environmental science.

We rely on natural resources

An island by definition is finite and bounded, and its inhabitants must cope with limitations in the materials they need. On our island—planet Earth—human beings, like all living things, ultimately face environmental constraints. Specifically, there are limits to many of our **natural resources,** the substances and energy sources that we take from our environment and that we need in order to survive.

Natural resources that are replenished over short periods are known as **renewable natural resources.** Some renewable natural resources, such as sunlight, wind, and wave energy, are perpetually renewed and essentially inexhaustible. In contrast, **nonrenewable natural resources,** such as minerals and crude oil, are in finite supply and are formed much more slowly than we use them. Once we deplete a nonrenewable resource, it is no longer available.

We can view the renewability of natural resources as a continuum (**FIGURE 1.1**). Renewable resources such as timber, water, and soil renew themselves over months, years, or decades, and can be depleted if we use them faster than they are replenished. For example, pumping groundwater faster than it is restored can deplete underground aquifers and turn lush landscapes into deserts. Populations of animals and plants we harvest from the wild may vanish if we overharvest them.

We rely on ecosystem services

If we think of natural resources as "goods" produced by nature, then it is also true that Earth's natural systems provide "services" on which we depend. Our planet's ecological systems purify air and water, cycle nutrients, regulate climate, pollinate plants, and receive and recycle our waste. Such essential services are commonly called **ecosystem services.** Ecosystem services arise from the normal functioning of natural systems and are not meant for our benefit, yet we could not survive without them. Later we will examine the countless and profound ways that ecosystem services support our lives and civilization (pp. 116–117, 152, 290).

Just as we may deplete natural resources, we may degrade ecosystem services. This can occur when we deplete resources, destroy habitat, or generate pollution. In recent years, our depletion of nature's goods and our disruption of nature's services have intensified, driven by rising affluence and a human population that grows larger every day.

Renewable natural resources

- Sunlight
- Wind energy
- Wave energy
- Geothermal energy

- Fresh water
- Forest products
- Agricultural crops
- Soils

Nonrenewable natural resources

- Crude oil
- Natural gas
- Coal
- Copper, aluminum, and other metals

FIGURE 1.1 Natural resources lie along a continuum from perpetually renewable to nonrenewable. Perpetually renewable, or inexhaustible, resources such as sunlight and wind energy **(left)**, will always be there for us. Renewable resources such as timber, soils, and fresh water **(center)** may be replenished on intermediate time scales, if we are careful not to deplete them. Nonrenewable resources such as oil and coal **(right)** exist in limited amounts that could one day be gone.

Population growth amplifies our impact

For nearly all of human history, fewer than a million people populated Earth at any one time. Today our population has grown beyond 7 *billion* people. This means that for every one person who used to exist, several thousand people exist today! **FIGURE 1.2** shows just how recently and suddenly this dramatic change has come about.

Two phenomena triggered our remarkable increase in population size. The first was our transition from a hunter-gatherer lifestyle to an agricultural way of life. This change began around 10,000 years ago and is known as the **agricultural revolution.** As people began to grow crops, domesticate animals, and live sedentary lives on farms and in villages, they produced more food to meet their nutritional needs and began having more children.

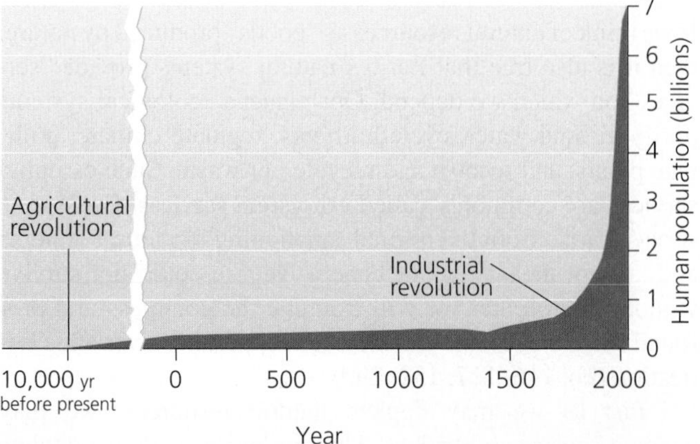

FIGURE 1.2 The global human population increased after the agricultural revolution and then skyrocketed as a result of the industrial revolution. *Data compiled from U.S. Census Bureau, U.N. Population Division, and other sources.*

The second notable phenomenon, known as the **industrial revolution,** began in the mid-1700s. It entailed a shift from rural life, animal-powered agriculture, and handcrafted goods toward an urban society provisioned by the mass production of factory-made goods and powered by **fossil fuels** (nonrenewable energy sources including oil, coal, and natural gas; pp. 524–526). Industrialization brought technological advances and improvements in sanitation and medicine, and it enhanced agricultural production through the use of fossil-fuel-powered equipment and synthetic pesticides and fertilizers (pp. 218, 247).

The factors driving population growth have brought us better lives in many ways. Yet as our world fills up with people, population growth has begun to threaten our well-being. We must ask how well the planet can accommodate 7 billion of us—or the 9 billion forecast by 2050. Already our sheer numbers, unparalleled in history, are putting unprecedented stress on natural systems and the availability of resources.

Resource consumption exerts social and environmental pressures

Besides stimulating population growth, industrialization increased the amount of resources each one of us consumes. As we mined energy sources and manufactured ever-greater numbers of goods, we enhanced the material affluence of many of the world's people. In the process, however, human society has consumed more and more of the planet's limited resources.

One way to quantify resource consumption is to use the concept of the "ecological footprint," developed in the 1990s by environmental scientists Mathis Wackernagel and William Rees. An **ecological footprint** expresses environmental impact in terms

FIGURE 1.3 **An "ecological footprint" represents the total area of biologically productive land and water needed to produce the resources and dispose of the waste for a given person or population.** *Adapted from an illustration by Philip Testemale in Wackernagel, M., and W. Rees, 1996.* Our ecological footprint: Reducing human impact on the Earth. *Gabriola Island, British Columbia: New Society Publishers.*

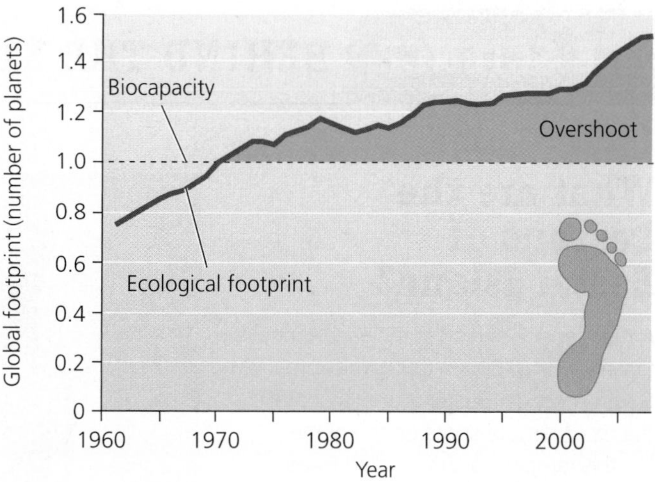

FIGURE 1.4 **Data indicate that we have overshot Earth's biocapacity—its capacity to support us—by 50%.** We are using renewable natural resources 50% faster than they are being replenished. *Data from WWF, 2012.* Living planet report 2012. *WWF International, Gland, Switzerland.*

DATA Q How much larger is the global ecological footprint today than it was half a century ago?

of the cumulative area of biologically productive land and water required to provide the resources a person or population consumes and to dispose of or recycle the waste the person or population produces (**FIGURE 1.3**). It measures the total area of Earth's biologically productive surface that a given person or population "uses" once all direct and indirect impacts are totaled up.

For humanity as a whole, Wackernagel and his colleagues calculate that our species is now using 50% more of the planet's resources than are available on a sustainable basis. That is, we are depleting renewable resources by using them 50% faster than they are being replenished. This is essentially like drawing the money out of a bank account rather than living off the interest the money makes. This excess use has been termed **overshoot** because we are overshooting, or surpassing, Earth's capacity to sustainably support us (**FIGURE 1.4**).

Moreover, people from wealthy nations such as the United States have much larger ecological footprints than do people from poorer nations. If all the world's people consumed resources at the rate of U.S. citizens, we would need the equivalent of four planet Earths.

Environmental science can help us avoid past mistakes

It remains to be seen what consequences resource consumption and population growth will have for today's global society, but historical evidence shows that civilizations can crumble when pressures from population and consumption overwhelm resource availability. Historians have inferred that environmental degradation contributed to the fall of the Greek and Roman empires; the Angkor civilization of Southeast Asia; and the Maya, Anasazi, and other civilizations of the New World. In Iraq and other regions of the Middle East, areas that are barren desert today were lush enough to support the origin of agriculture when great ancient societies thrived there. Easter Island has long been held up as the most striking case of a society's self-destruction after depleting its resources, although new research disputes this interpretation (see **THE SCIENCE BEHIND THE STORY**, pp. 6–7).

In his 2005 book *Collapse*, scientist and author Jared Diamond synthesized existing research and formulated general reasons why civilizations succeed and persist, or fail and collapse. Success and persistence, he argued, depend largely on how societies interact with their environments and on how they respond to problems.

In today's globalized society, the stakes are higher than ever because our environmental impacts are global. If we cannot forge sustainable solutions to our problems, then the resulting societal collapse will be global. Fortunately, environmental science holds keys to building a better world. By studying environmental science, you will learn to evaluate the many changes happening around us and to think critically and creatively about ways to respond.

The Nature of Environmental Science

Environmental scientists aim to comprehend how Earth's natural systems function, how these systems affect people, and how we are influencing those systems. Many environmental

What are the Lessons of Easter Island?

A mere speck of land in the vast Pacific Ocean, fully 3750 km (2325 mi) from South America, Easter Island is one of the most remote spots on the globe. Yet this far-flung island—called Rapa Nui by its inhabitants—is the focus of an intense debate among scientists seeking to clarify its enigmatic history and decipher the lessons it has to offer us.

Ever since European explorers stumbled upon Rapa Nui on Easter Sunday, 1722, outsiders have been struck by the island's barren landscape. Early European accounts suggested that the 2000–3000 people living on the island seemed impoverished, subsisting on a few meager crops and possessing only stone tools. Yet the forlorn island also featured hundreds of gigantic statues of carved rock. How could people without wheels or ropes, on an island without trees, have moved 90-ton statues 10 m (33 ft) high as far as 10 km (6.2 mi) from the quarry where they were chiseled to the sites where they were erected? Apparently some calamity must have befallen a once-mighty civilization.

Many researchers have set out to solve Easter Island's mysteries. A key discovery was that the island was once lushly forested. Scientist John Flenley and his colleagues drilled cores deep into lake sediments and examined ancient pollen grains preserved there, seeking to reconstruct, layer by layer, the history of vegetation in the region. Finding a great deal of palm pollen, they inferred that when Polynesian people colonized the island (A.D. 300–900, they estimated), it was covered with palm trees similar to the Chilean wine palm—a tall, slow-growing tree that can live for centuries.

Archaeologists found ancient palm nut casings buried in soil near carbon-lined channels made by palm roots. Researchers deciphering script

Easter Island's immense statues

on stone tablets discerned characters etched in the form of palm trees.

By studying pollen and the remains of wood from charcoal, archaeologist Catherine Orliac found that at least 21 other plant species—now gone—had also been common. Clearly the island had supported a diverse forest. Forest plants would have provided fuelwood, building material for houses and canoes, fruit to eat, fiber for clothing—and, researchers guessed, logs and fibrous rope to help move statues.

Pollen analysis showed that trees declined, replaced by ferns and grasses. Then between 1400 and 1600, pollen levels plummeted. Charcoal in the soil proved the forest had been burned, likely in slash-and-burn farming. Researchers concluded that the islanders, desperate for forest resources and cropland, had deforested their own island.

With the forest gone, soil eroded away (data from lake bottoms showed a great deal of sediment accumulating). Erosion would have lowered yields of bananas, sugarcane, and sweet potatoes, perhaps leading to starvation and population decline.

Further evidence indicated that wild animals disappeared. Archaeologist David Steadman analyzed 6500 bones and found that at least 31 bird species provided food for the islanders. Today, only one native bird species is left. Remains from charcoal fires show that early islanders feasted on fish, sharks, porpoises, turtles, octopus, and shellfish—but in later years they consumed little seafood.

As resources declined, researchers concluded, people fell into clan warfare, revealed by unearthed weapons and skulls with head wounds. Rapa Nui appeared to be a tragic case of ecological suicide: A once-flourishing civilization depleted its resources and destroyed itself. In this interpretation—advanced by Flenley and writer Paul Bahn, and by scientist Jared Diamond in his best-selling 2005 book *Collapse*—Rapa Nui seemed to offer a clear lesson: We on our global island, planet Earth, had better learn to use our limited resources sustainably.

When Terry Hunt and Carl Lipo began research on Rapa Nui in 2001, they expected simply to help fill gaps in a well-understood history. But science is a process of discovery, and sometimes evidence leads researchers far from where they anticipated. For Hunt, an anthropologist at University of Hawaii at Manoa, and Lipo, an archaeologist at California State University, Long Beach, their work ended up convincing them that nearly everything about the traditional "ecocide" interpretation was wrong.

First, their radiocarbon dating (p. 24) indicated that people had not colonized the island until about A.D. 1200. This finding suggested that deforestation occurred suddenly, soon after arrival. How could so few people have destroyed so much forest so fast?

Hunt and Lipo's answer: rats. When Polynesians settled new islands, they brought crop plants, domestic animals such as chickens, and rats. Whether rats were stowaways or were brought intentionally as food is not known. In either case, rats can multiply quickly, and they soon overran Rapa Nui.

Rats ate palm nuts (researchers see their tooth marks on old nut casings). Hunt and Lipo suggest they ate so many nuts and shoots that the trees could not regenerate. With no young trees growing, the palm went extinct once mature trees died.

Diamond and others counter that over 20 additional plant species went extinct on Rapa Nui, that plenty of palm nuts escaped rat damage, and that most plants survived rats on other islands. Moreover, people brought the rats, so the forest was still destroyed as a result of human colonization.

Despite the forest loss, Hunt and Lipo argue that islanders were able to persist and thrive. Archaeology shows how islanders adapted to Rapa Nui's poor soil and windy weather by developing rock gardens to protect crop plants and nourish the soil. Tools that previous researchers viewed as weapons were actually farm implements, Hunt and Lipo concluded; lethal injuries were rare; and no evidence of battle or defensive fortresses was uncovered.

Hunt, Lipo, and others also unearthed old roads and inferred how the statues were transported. It had been thought that a powerful central authority forced armies of laborers to move them, but Hunt and Lipo concluded that small numbers of people could move them by tilting and rocking them upright like refrigerators. Indeed, the distribution of statues on the island suggested the work of family groups. Islanders had adapted to their resource-poor environment by becoming a peaceful and cooperative society, with the statues providing a harmless outlet for competition over status and prestige.

Altogether, the evidence led Hunt and Lipo to propose that far from destroying their environment, the islanders had acted as responsible stewards. The collapse of this sustainable civilization, they argue, came with the arrival of Europeans, who unwittingly brought contagious diseases to which the islanders had never been exposed. Indeed, historical journals of sequential European voyages depict a society falling into disarray as if reeling from epidemics, its statues tumbling around it.

Peruvian ships then began raiding Rapa Nui and taking islanders away into slavery. Foreigners acquired the land, forced the remaining people into labor, and introduced thousands of sheep, which destroyed the few native plants left on the island. Thus, the collapse of Rapa Nui civilization resulted from a barrage of disease, violence, and slave-raids following foreign contact. Before that, Hunt and Lipo say, Rapa Nui's people boasted 500 years of a peaceful and resilient society.

Hunt and Lipo's interpretation, put forth in a 2011 book, *The Statues That Walked*, represents a paradigm shift (p. 13) in how we view Easter Island. Debate between the two camps remains heated. Meanwhile, research continues as scientists look for new ways to test the differing hypotheses. In the long-term, data from additional studies should lead us closer and closer to the truth.

Like the people of Rapa Nui, we are all stranded together on an island with limited resources. What is the lesson of Easter Island for our global island, Earth? Perhaps there are two: That any island population must learn to live within its means—but that with care and ingenuity, there is hope that we can. ■

Were the haunting statues of Easter Island (Rapa Nui) erected by a civilization that collapsed after devastating its environment, or by a sustainable civilization that fell because of outside influence?

scientists are motivated by a desire to develop solutions to environmental problems. These solutions (such as new technologies, policy decisions, or resource management strategies) are *applications* of environmental science. The study of such applications and their consequences is, in turn, also part of environmental science.

Environmental science is interdisciplinary

Studying our interactions with our environment is a complex endeavor that requires expertise from many disciplines, including ecology, earth science, chemistry, biology, geography, economics, political science, demography, ethics, and others. Environmental science is thus an **interdisciplinary field**—one that borrows techniques from multiple disciplines and brings their research results together into a broad synthesis (**FIGURE 1.5**).

Traditional established disciplines are valuable because their scholars delve deeply into topics, developing expertise in particular areas and uncovering new knowledge. In contrast, interdisciplinary fields are valuable because their practitioners consolidate and synthesize the specialized knowledge from many disciplines and make sense of it in a broad context to better serve the multifaceted interests of society.

Environmental science is especially broad because it encompasses not only the **natural sciences** (disciplines that examine the natural world), but also the **social sciences** (disciplines that address human interactions and institutions). Most environmental science programs focus more on the natural sciences, whereas programs that emphasize the social sciences often use the term **environmental studies.** Whichever approach one takes, these fields bring together many diverse perspectives and sources of knowledge.

Just as an interdisciplinary approach to studying issues can help us better understand them, an integrated approach to addressing environmental problems can produce effective solutions for society. For example, we used to add lead to gasoline to make cars run more smoothly, even though researchers knew that lead emissions from tailpipes caused health problems, including brain damage and premature death. In 1970 air pollution was severe, and motor vehicles accounted for 78% of U.S. lead emissions. In response, environmental scientists, engineers, medical researchers, and policymakers all merged their knowledge and skills into a process that eventually brought about a ban on leaded gasoline. By 1996 all gasoline sold in the United States was unleaded, and the nation's largest source of atmospheric lead pollution had been completely eliminated.

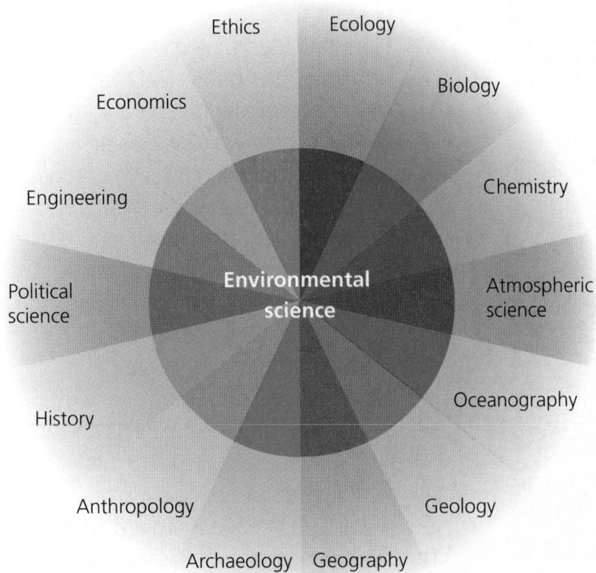

FIGURE 1.5 Environmental science is an interdisciplinary pursuit. It involves input from many different established fields of study across the natural sciences and social sciences.

People vary in how they perceive environmental problems

Environmental science arose in the latter half of the 20th century as people sought to better understand environmental problems, their origins, and their solutions. However, the perception of what constitutes a problem may vary from one person to another, or from one situation to another. A person's age, gender, class, race, nationality, employment, income, and educational background can all affect whether he or she considers a given environmental condition or change to be a "problem."

For instance, Americans today are more likely to view the application of the pesticide DDT as a problem than in the 1940s and 1950s, because today more is known about the health risks of pesticides (**FIGURE 1.6**). However, a person living

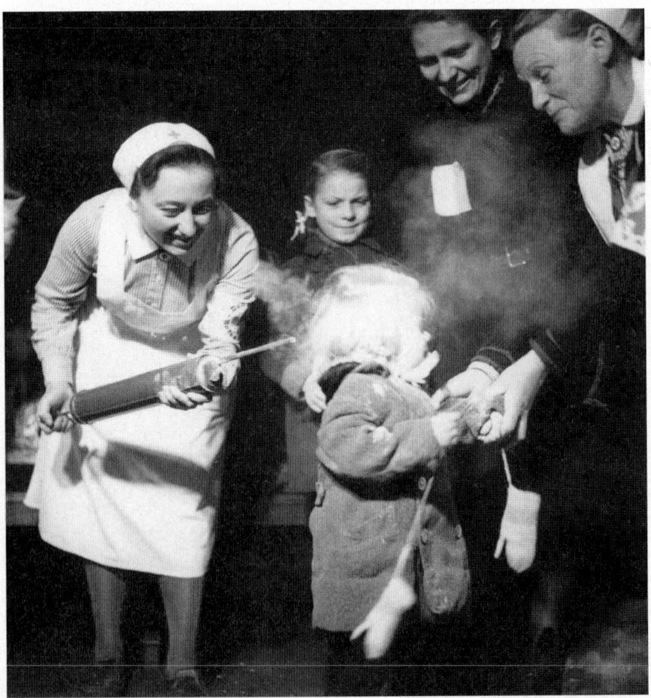

FIGURE 1.6 How a person or a society defines an environmental problem can vary with time and circumstance. In 1945, health hazards of the pesticide DDT were not yet known, so children were doused with the chemical to treat head lice. Today, knowing of its toxicity to people, many developed nations have banned DDT. However, in developing countries where malaria is a public health threat, DDT is welcomed as a means of eradicating mosquitoes that transmit the disease.

today in a malaria-infested village in Africa may welcome the use of DDT if it kills mosquitoes that transmit malaria, because he or she may view malaria as a more immediate health threat. Thus an African and an American who have each knowledgeably assessed the pros and cons may, because of differences in their circumstances, differ in their attitude toward DDT.

People also vary in their awareness of problems. For example, in many cultures women are responsible for collecting water and fuelwood, and as a result they perceive environmental degradation that affects these resources more readily than men do. Moreover, in most societies information about environmental health risks tends to reach wealthy people more readily than poor people. Thus, who you are, where you live, and what you do influences how you perceive your environment, how change affects you, and how you react to change.

FIGURE 1.7 Environmental scientists play roles very different from those of environmental activists, like those shown here. Although many environmental scientists search for solutions to environmental problems, they aim to keep their research objective and to avoid advocacy.

Environmental science is not the same as environmentalism

Although many environmental scientists are interested in solving problems, it would be incorrect to confuse environmental science with environmentalism or environmental activism. They are very different. Environmental science involves the scientific study of the environment and our interactions with it. In contrast, **environmentalism** is a social movement dedicated to protecting the natural world—and, by extension, people—from undesirable changes brought about by human actions (**FIGURE 1.7**).

> **FAQ** **Aren't environmental scientists also environmentalists?**
>
> Not necessarily. Although environmental scientists search for solutions to environmental problems, they strive to keep their research rigorously objective and free from advocacy. Of course, like all human beings, scientists are motivated by personal values and interests—and like any human endeavor, science can never be entirely free of social influence. Yet while personal values and social concerns may help shape the questions scientists ask, scientists do their utmost to carry out their work impartially and to interpret their results with wide-open minds. Remaining open to whatever conclusions the data demand is a hallmark of the effective scientist.

The Nature of Science

Modern scientists describe **science** as a systematic process for learning about the world and testing our understanding of it. The term *science* is also commonly used to refer to the accumulated body of knowledge that arises from this dynamic process of questioning, observation, testing, and discovery.

Knowledge gained from science can be applied to address societal needs—for instance, to develop technology or to inform policy and management decisions (**FIGURE 1.8**). Many scientists are motivated by the potential for developing useful applications, whereas others are motivated simply by a desire to understand how the world works.

Why does science matter? As astronomer and author Carl Sagan wrote in his 1995 treatise, *The Demon Haunted World: Science as a Candle in the Dark*, "We've arranged a global civilization in which the most crucial elements—transportation, communications, and all other industries; agriculture, medicine, education, entertainment, protecting the environment; and even the key democratic institution of voting—profoundly depend on science and technology." Indeed, from the food we eat to the clothing we wear to the healthcare we depend on, virtually everything in our lives has been improved by the application of science. Sagan and countless other thinkers have argued that science is essential if we hope to develop solutions to the challenges we face.

Scientists test ideas by critically examining evidence

Science is all about asking and answering questions. Scientists examine how the world works by making observations, taking measurements, and testing whether their ideas are supported by evidence. The effective scientist thinks critically and does not simply accept conventional wisdom from others. The scientist becomes excited by novel ideas but is skeptical and judges ideas by the strength of evidence that supports them. In these ways, scientists are good role models for all of us, because we can all benefit from learning to think critically in our everyday lives.

A great deal of scientific work is **observational science** or **descriptive science,** research in which scientists gather basic information about organisms, materials, systems, or processes that are not yet well known. In this approach, researchers explore new frontiers of knowledge by observing and measuring phenomena to gain a better understanding of them. Such research is common in traditional fields such as astronomy, paleontology, and taxonomy, as well as in newer, fast-growing fields such as molecular biology and genomics.

(a) Chevy Volt, an electric hybrid car

(b) Prescribed burning

FIGURE 1.8 **Scientific knowledge can be applied in engineering and technology and in policy and management decisions.** Energy-efficient electric automobiles such as the Chevy Volt **(a)** are technological advances made possible by materials and energy research. Prescribed burning **(b)**, shown here in the Ouachita National Forest, Arkansas, is a management practice to restore healthy forests that is informed by scientific research into forest ecology.

Once enough basic information is known about a subject, scientists can begin posing questions that seek deeper explanations about how and why things are the way they are. At this point they may pursue **hypothesis-driven science,** research that proceeds in a more targeted and structured manner, using experiments to test hypotheses within a framework traditionally known as the scientific method.

The scientific method is a traditional approach to research

The **scientific method** is a technique for testing ideas with observations. There is nothing mysterious or intimidating about the scientific method; it is merely a formalized version of the way any of us might naturally use logic to resolve a question.

Because science is an active, creative process, innovative researchers regularly depart from the traditional scientific method when particular situations demand it. Moreover, scientists in different fields approach their work differently because they deal with dissimilar types of information. A natural scientist, such as a chemist, conducts research quite differently than a social scientist, such as a sociologist. Nonetheless, scientists of all persuasions broadly agree on fundamental elements of the process of scientific inquiry.

As practiced by individual researchers or research teams, the scientific method (FIGURE 1.9) typically consists of the steps outlined below.

Make observations Advances in science typically begin with the observation of some phenomenon that the scientist wishes to explain. Observations set the scientific method in motion and also function throughout the process.

Ask questions Curiosity is a fundamental human characteristic. Just observe the explorations of young children in a new environment—they want to touch, taste, watch, and listen to everything that catches their attention, and as soon as they can speak, they begin asking questions. Scientists, in this respect, are kids at heart. Why are certain plants or animals less common today than they once were? Why are storms becoming more severe and flooding more frequent? What is causing excessive growth of algae in local ponds? When pesticides poison fish or frogs, are people also affected? All of these are questions environmental scientists ask.

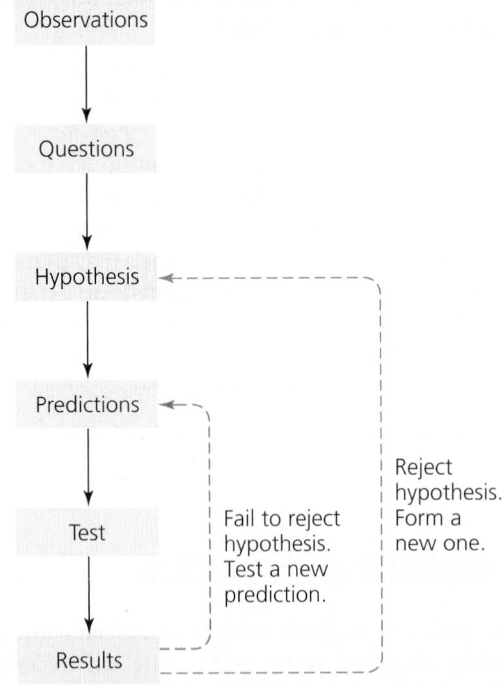

FIGURE 1.9 **The scientific method is the traditional experimental approach that scientists use to learn how the world works.** This simple diagram, although useful and instructive, cannot convey the true dynamic and creative nature of science. Researchers often pursue their work in ways that vary legitimately from this model.

Develop a hypothesis Scientists address their questions by devising explanations that they can test. A **hypothesis** is a statement that attempts to explain a phenomenon or answer a scientific question. For example, a scientist investigating the question of why algae are growing excessively in local ponds might observe that chemical fertilizers are being applied on farm fields nearby. The scientist might then propose a hypothesis as follows: "Agricultural fertilizers running into ponds cause the amount of algae in the ponds to increase."

Make predictions The scientist next uses the hypothesis to generate **predictions,** specific statements that can be directly and unequivocally tested. In our algae example, a researcher might predict: "If agricultural fertilizers are added to a pond, the quantity of algae in the pond will increase."

Test the predictions Scientists test predictions by gathering evidence that could potentially refute them and thus disprove the hypothesis. The strongest form of evidence comes from experimentation. An **experiment** is an activity designed to test the validity of a prediction or a hypothesis. It involves manipulating **variables,** or conditions that can change.

For example, a scientist could test the prediction linking algal growth to fertilizer by selecting two identical ponds and adding fertilizer to one of them. In this example, fertilizer input is an **independent variable,** a variable the scientist manipulates, whereas the quantity of algae that results is the **dependent variable,** one that depends on the fertilizer input. If the two ponds are identical except for a single independent variable (fertilizer input), then any differences that arise between the ponds can be attributed to that variable. Such an experiment is known as a **controlled experiment** because the scientist controls for the effects of all variables except the one he or she is testing. In our example, the pond left unfertilized serves as a **control,** an unmanipulated point of comparison for the manipulated **treatment** pond.

Whenever possible, it is best to replicate one's experiment; that is, to stage multiple tests of the same comparison. Our scientist could perform a replicated experiment on, say, 10 pairs of ponds, adding fertilizer to one of each pair.

FIGURE 1.10 Researchers gather data in order to test predictions in experiments. Here, Dr. Jennifer Smith of the Scripps Institution of Oceanography in San Diego photographs coral at a remote reef in the South Pacific. Data from analysis of the photos will help her test hypotheses about how human impacts affect the condition and community structure of coral reefs.

Analyze and interpret results Scientists record **data,** or information, from their studies (FIGURE 1.10). They particularly value quantitative data (information expressed using numbers), because numbers provide precision and are easy to compare. The scientist running the fertilization experiment, for instance, might quantify the area of water surface covered by algae in each pond or might measure the dry weight of algae in a certain volume of water taken from each. It is vital, however, to collect data that is representative. Because it is impractical to measure a pond's total algal growth, our researcher would instead sample from multiple areas of the pond. These areas must be selected in a random manner, since choosing areas with the most growth or the least growth, or areas most convenient to sample, would not provide a representative sample.

Even with the precision that numbers provide, a scientist's results may not be clear-cut. Data from treatments and controls may vary only slightly, or replicates may yield different results. The researcher must therefore analyze the data using statistical tests. With these mathematical methods, scientists can determine objectively and precisely the strength and reliability of patterns they find.

Some research, especially in the social sciences, involves data that is qualitative, or not expressible in terms of numbers. Research involving historical texts, personal interviews, surveys, case studies, or descriptive observation of behavior can include qualitative data on which quantitative statistical analysis may not be possible.

If experiments disprove a hypothesis, the scientist will reject it and may formulate a new hypothesis to replace it. If experiments fail to disprove a hypothesis, this lends support to the hypothesis but does not *prove* it is correct. The scientist may choose to generate new predictions to test the hypothesis in different ways and further assess its likelihood of being true. Thus, the scientific method loops back on itself, often giving rise to repeated rounds of hypothesis revision and new experimentation (see Figure 1.9).

If repeated tests fail to reject a hypothesis, evidence in favor of it accumulates, and the researcher may eventually conclude that the hypothesis is well supported. Ideally, the scientist would want to test all possible explanations. For instance, our researcher might formulate an additional hypothesis, proposing that algae increase in fertilized ponds because chemical fertilizers diminish the numbers of fish or invertebrate animals that eat algae. It is possible, of course, that both hypotheses could be correct and that each may explain some portion of the initial observation that local ponds were experiencing algal blooms.

We test hypotheses in different ways

An experiment in which the researcher actively chooses and manipulates the independent variable is known as a **manipulative experiment.** A manipulative experiment provides the strongest type of evidence a scientist can obtain, because it can reveal causal relationships, showing that changes in an independent variable cause changes in a dependent variable. In practice, however, we cannot run manipulative experiments for all questions, especially for processes involving large spatial scales or long time scales. For example, in studying the effects of global climate change (Chapter 18), we cannot run a manipulative experiment adding carbon dioxide to 10 treatment planets and 10 control planets and then compare the results!

Thus, in environmental science, it is common for researchers to run **natural experiments,** which compare how dependent variables are expressed in naturally different contexts. In such experiments, the independent variable varies naturally, and researchers test their hypotheses by searching for **correlation,** or statistical association among variables.

For instance, let's suppose our scientist studying algae surveys 50 ponds, 25 of which happen to be fed by fertilizer runoff from nearby farm fields and 25 of which are not. Let's say he or she finds seven times more algal growth in the fertilized ponds than in the unfertilized ponds. The scientist would conclude that algal growth is correlated with fertilizer input; that is, that one tends to increase along with the other.

This type of evidence is not as strong as the causal demonstration that manipulative experiments can provide, but often a natural experiment is the only feasible approach for studying a subject of immense scale, such as an ecosystem or a planet. Because many questions in environmental science are complex and exist on large scales, they must be addressed with correlative data. As such, environmental scientists cannot always provide clear-cut, black-and-white answers to questions from policymakers and the public. Nonetheless, good correlative studies can make for very strong science, and they preserve the real-world complexity that manipulative experiments often sacrifice. Whenever possible, scientists try to integrate natural and manipulative experiments to gain the advantages of each.

The scientific process continues beyond the scientific method

Scientific work takes place within the context of a community of peers. To have impact, a researcher's work must be published and made accessible to this community. Thus, the scientific method is embedded within a larger process involving the scientific community as a whole (**FIGURE 1.11**).

Peer review When a researcher's work is done and the results analyzed, he or she writes up the findings and submits them to a journal (a scholarly publication in which scientists share their work). The journal's editor asks several other scientists who specialize in the subject area to examine the manuscript, provide comments and criticism (generally anonymously), and judge whether the work merits publication in the journal. This procedure, known as **peer review,** is an essential part of the scientific process.

Peer review is a valuable guard against faulty research contaminating the literature on which all scientists rely. However, because scientists are human, personal biases and politics can sometimes creep into the review process. Fortunately, just as individual scientists strive to remain objective in conducting their research, the scientific community does its best to ensure fair review of all work. Winston Churchill once called democracy the worst form of government, except

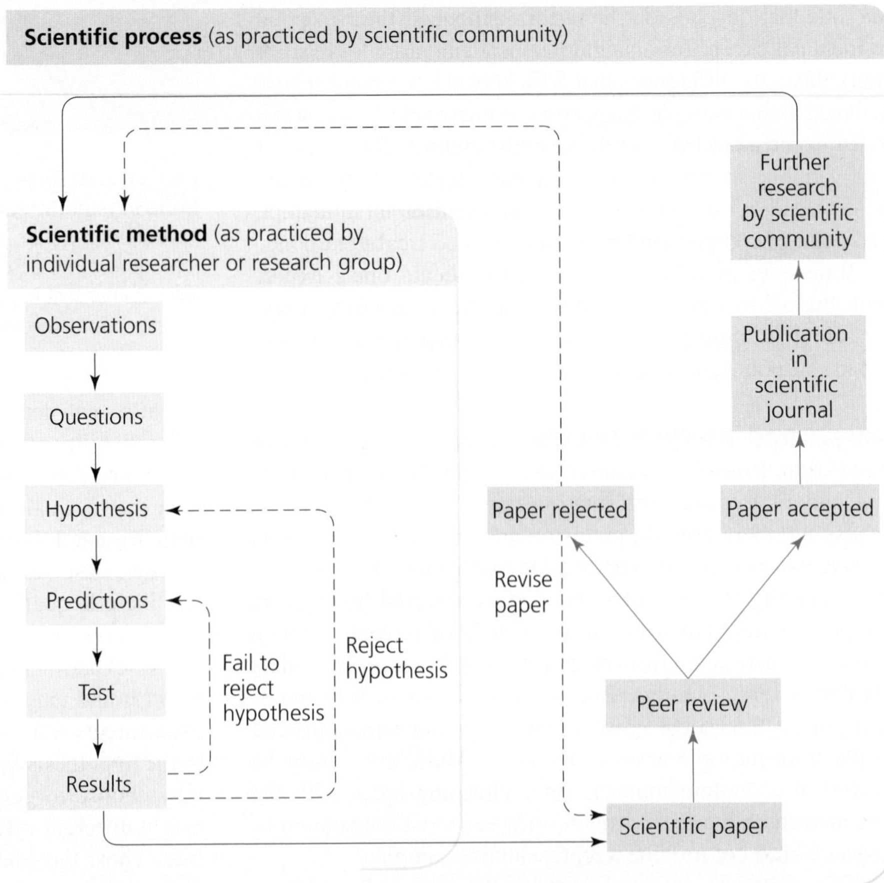

FIGURE 1.11 The scientific method followed by individual research teams exists within the overall process of science at the level of the scientific community. This process includes peer review and publication of research, acquisition of funding, and the elaboration of theory through the cumulative work of many researchers.

for all the others that had been tried. The same might be said about peer review; it is an imperfect system, yet it is the best we have.

Conference presentations Scientists frequently present their work at professional conferences, where they interact with colleagues and receive comments on their research. Such feedback can help improve a researcher's work before it is submitted for publication.

Grants and funding To fund their research, most scientists need to spend enormous amounts of time requesting money from private foundations or from government agencies such as the National Science Foundation. Grant applications undergo peer review just as scientific papers do, and competition for funding is generally intense.

Scientists' reliance on funding sources can occasionally lead to conflicts of interest. A researcher who obtains data showing his or her funding source in an unfavorable light may be reluctant to publish the results for fear of losing funding—or worse yet, could be tempted to doctor the results. This situation can arise, for instance, when an industry funds research to test its products for safety or environmental impact. Most scientists resist these pressures, but when you are critically assessing a scientific study, it is always a good idea to note where the researchers obtained their funding.

WEIGHING THE ISSUES

FOLLOW THE MONEY Let us say you are a research scientist, and you want to study the impacts of chemicals released into lakes by pulp-and-paper mills. Obtaining research funding has been difficult. Then a representative from a large pulp-and-paper company contacts you. The company also is interested in how its chemical effluents affect water bodies, and it would like to fund your research. What are the benefits and drawbacks of this offer? Would you accept the offer?

Repeatability The careful scientist may test a hypothesis repeatedly in various ways. Following publication, other scientists may attempt to reproduce the results in their own experiments and analyses. Scientists are inherently cautious about accepting a novel hypothesis, so the more a result can be reproduced by different research teams, the more confidence scientists will have that it provides the correct explanation for an observed phenomenon.

Theories If a hypothesis survives repeated testing by numerous research teams and continues to predict experimental outcomes and observations accurately, it may potentially be incorporated into a theory. A **theory** is a widely accepted, well-tested explanation of one or more cause-and-effect relationships that has been extensively validated by a great amount of research. Whereas a hypothesis is a simple explanatory statement that may be disproven by a single experiment, a theory consolidates many related hypotheses that have been supported by a large body of experimental and observational data.

Note that scientific use of the word *theory* differs from popular usage of the word. In everyday language when we say something is "just a theory," we are suggesting it is a speculative idea without much substance. However, scientists mean just the opposite when they use the term. To them, a theory is a conceptual framework that effectively explains a phenomenon and has undergone extensive and rigorous testing, such that confidence in it is extremely strong.

For example, Darwin's theory of evolution by natural selection (pp. 50–53) has been supported and elaborated by many thousands of studies over 150 years of intensive research. Large bodies of research have shown repeatedly and in great detail how plants and animals change over generations, or evolve, expressing characteristics that best promote survival and reproduction. Because of its strong support and explanatory power, evolutionary theory is the central unifying principle of modern biology. Other prominent scientific theories include atomic theory, cell theory, big bang theory, plate tectonics, and general relativity.

Applications Knowledge gained from scientific research may be applied to help meet society's needs and address society's problems. As discussed earlier (see Figure 1.8), scientific research informs and facilitates new technologies, engineering approaches, policy decisions, and management strategies. Still, even when research is able to provide clear information, deciding on the optimal social response to a problem can be difficult. Moreover, many predicaments addressed by environmental science are so-called **wicked problems:** problems complex enough to have no simple solution and whose very nature changes over time. For this reason, society will benefit if it trains and funds scientists to continue studying such problems as they evolve.

Science goes through "paradigm shifts"

As the scientific community accumulates data in an area of research, interpretations sometimes may change. Thomas Kuhn's influential 1962 book *The Structure of Scientific Revolutions* argued that science goes through periodic upheavals in thought, in which one scientific **paradigm,** or dominant view, is abandoned for another. For example, before the 16th century, European scientists believed that Earth was at the center of the universe. Their data on the movements of planets fit that concept somewhat well—yet the idea eventually was disproved after Nicolaus Copernicus showed that placing the sun at the center of the solar system explained the planetary data much better.

Another paradigm shift occurred in the 1960s, when geologists accepted plate tectonics (pp. 34–36). By this time, evidence for the movement of continents and the action of tectonic plates had accumulated and become overwhelmingly convincing. Paradigm shifts demonstrate the strength and vitality of science, showing science to be a process that refines and improves itself through time.

Understanding how science works is vital to assessing how scientific interpretations progress through time as information accrues. This is especially relevant in environmental science—a young field that is changing rapidly as we attain vast amounts of new information, as human impacts on our planet multiply, and as we gather lessons from the consequences of our actions.

Sustainability and Our Future

Throughout this book you will encounter environmental scientists asking questions, testing hypotheses, conducting experiments, analyzing data, and drawing conclusions about the causes and consequences of environmental change. Environmental scientists who study the condition of our environment and the consequences of our impacts are addressing the most centrally important issues of our time.

Achieving sustainable solutions is vital

The primary challenge in our increasingly populated world is how to live within our planet's means, such that Earth and its resources can sustain us—and all life—for the future. This is the challenge of **sustainability,** a guiding principle of modern environmental science. Sustainability means leaving our children and grandchildren a world as rich and full as the world we live in now. It means conserving Earth's resources so that our descendants may enjoy them as we have. It means developing solutions that work in the long term. Sustainability requires maintaining ecological systems, because we cannot sustain human civilization without sustaining the natural systems that nourish it.

We can think of our planet's resources as a bank account. If we deplete resources, we draw down the bank account. However, we can choose instead to use the interest and leave the principal intact so that we can continue using the interest far into the future. Currently we are drawing down Earth's **natural capital,** its accumulated wealth of resources. Recall (p. 5) that one research group estimates that we are withdrawing our planet's natural capital 50% faster than it is being replenished. To live off nature's interest—its replenishable resources—is sustainable. To draw down resources faster than they are replaced is to eat into nature's capital—the bank account for our planet and our civilization—and we cannot get away with this for long.

Population and consumption drive environmental impact

We modify our environment in many ways, but the steep and sudden rise in human population (Chapter 8) has amplified nearly all of our impacts. We add about 80 million people to the planet each year—that's over 200,000 per day. The rate of population growth is now slowing, but our absolute numbers continue to increase.

Our consumption of resources has risen even faster than our population. The modern rise in affluence has been a positive development for humanity, and our conversion of the planet's natural capital has made life better for most of us so far. However, like rising population, rising per capita consumption magnifies the demands we make on our environment.

The world's citizens have not benefited equally from our overall rise in affluence. Today the 20 wealthiest nations boast over 55 times the per capita income of the 20 poorest nations—three times the gap that existed just two generations ago. The ecological footprint of the average citizen of a developed nation such as the United States is considerably larger than that of the average resident of a developing country (**FIGURE 1.12**). Within the United States, the richest 10% of people claim fully half the income, and the richest 1% claim nearly a quarter of all income.

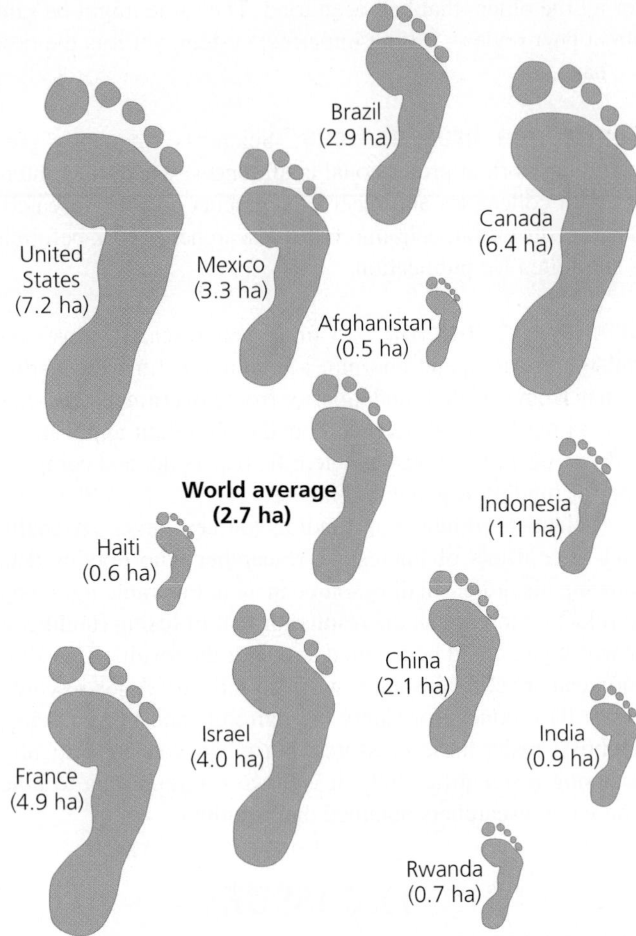

FIGURE 1.12 The citizens of some nations have much larger ecological footprints than the citizens of others. Ecological footprints for average citizens of several nations are shown, along with the world's average per capita footprint of 2.7 hectares. One hectare (ha) = 2.47 acres. *Data from Global Footprint Network, in: WWF, 2012.* Living planet report 2012. *WWF International, Gland, Switzerland.*

DATA **Q** Which nation has the largest footprint, and how many times larger is it than that of the nation with the smallest footprint?

WEIGHING THE ISSUES

ECOLOGICAL FOOTPRINTS What do you think accounts for the variation in sizes of per capita ecological footprints among societies? Do you feel that nations with larger footprints have a moral obligation to reduce their environmental impact, so as to leave more resources available for nations with smaller footprints? Why or why not?

Our dramatic growth in population and consumption is intensifying the many environmental impacts we examine in this book, including erosion and other impacts from agriculture (Chapters 9 and 10), deforestation (Chapter 12), toxic substances (Chapter 14), fresh water depletion (Chapter 15), fisheries declines (Chapter 16), air and water pollution (Chapters 15–17), waste generation (Chapter 22), mineral extraction and mining impacts (Chapter 23), and of course, global climate change (Chapter 18). These impacts degrade our health and quality of life, and they alter the ecosystems and

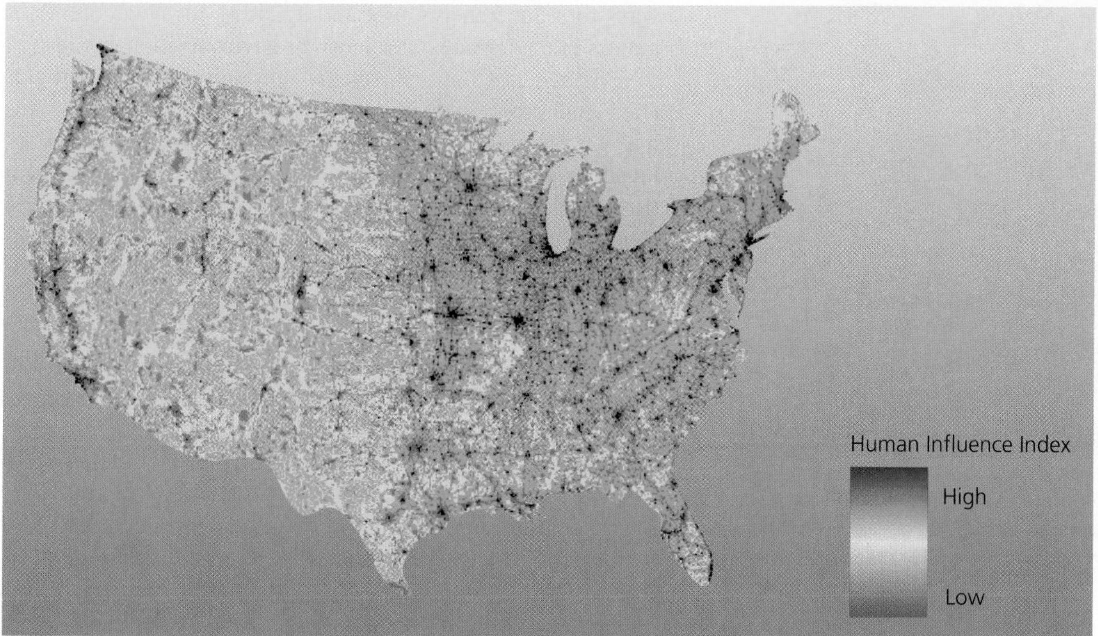

FIGURE 1.13 Human settlement, roads and transportation networks, nighttime light pollution, and agriculture and other land uses all influence terrestrial ecosystems. A U.S. map summarizing these influences shows that we live in a highly modified environment and suggests we would be wise to carefully nurture natural systems and manage remaining resources. *Used by permission of the Center for International Earth Science Information Network (CIESIN), The Earth Institute, Columbia University. © 2012.*

DATA Q Can you find where you live on the map? Do the ecosystems in your area experience a more-than-average, less-than-average, or average amount of human influence? What human impacts do you think most affect natural systems in your region?

landscapes in which we live (FIGURE 1.13). They are also driving the loss of Earth's biodiversity (Chapter 11)—perhaps our greatest problem, because extinction is irreversible. Once a species becomes extinct, it is lost forever.

The most comprehensive scientific assessment of the condition of the world's ecological systems and their capacity to continue supporting our civilization was completed in 2005, when over 2000 leading environmental scientists from nearly 100 nations completed the Millennium Ecosystem Assessment (TABLE 1.1). The Millennium Ecosystem Assessment makes clear that our degradation of environmental systems is having negative impacts on all of us, but that with care and diligence we can still turn many of these trends around.

Our energy choices will influence our future enormously

Our reliance on fossil fuels to power our civilization has intensified virtually every impact we exert on our environment, from habitat alteration to air pollution to climate change. Fossil fuels have also helped to bring us the material affluence we enjoy. By exploiting the richly concentrated energy in coal, oil, and natural gas, we have been able to power the machinery of the industrial revolution, produce the chemicals that boost agricultural yields, run the vehicles and transportation networks of our mobile society, and manufacture and distribute our countless consumer products. The lifestyles we lead today are a direct result of the availability of fossil fuels (Chapter 19).

However, in extracting coal, oil, and natural gas, we are splurging on a one-time bonanza, for these fuels are nonrenewable and in finite supply. Scientists calculate that we have depleted roughly half the world's conventional oil supplies and that a crisis could hit once supply begins to decline while demand continues to rise (pp. 532–533). How we handle future fossil fuel shortages will greatly influence the nature of our lives in the 21st century.

TABLE 1.1 Main Findings of the Millennium Ecosystem Assessment
• Over the past 50 years, people have altered ecosystems more rapidly and extensively than ever, largely to meet growing demands for food, fresh water, timber, fiber, and fuel. This has caused a substantial and largely irreversible loss in the diversity of life on Earth.
• Changes to ecosystems have contributed to substantial net gains in human well-being and economic development. However, these gains have been achieved at growing costs, including the degradation of ecosystems and the services they provide and, for some people, the worsening of poverty.
• This degradation could grow significantly worse during the first half of this century.
• We can reverse the degradation of ecosystems while meeting increasing demands for their services, but doing so will require that we significantly modify many policies, institutions, and practices.

Adapted from Millennium Ecosystem Assessment, 2005. Ecosystems and human well-being: biodiversity synthesis. *World Resources Institute, Washington, DC.*

FIGURE 1.14 We can develop clean and renewable energy sources for our sustainable use. Just as a flowering plant gathers energy from the sun, rooftop panels like these harness solar energy.

Sustainable solutions abound

Humanity's challenge is to develop solutions that enhance our quality of life while protecting and restoring the environment that supports us. Fortunately, many workable solutions are at hand. For instance:

- Renewable energy sources (Chapters 20 and 21) are being developed to replace fossil fuels (FIGURE 1.14), and energy-efficiency efforts are gaining ground.
- In response to agricultural impacts, scientists and others have developed and promoted soil conservation, high-efficiency irrigation, and organic agriculture (Chapters 9 and 10).
- Laws and new technologies have reduced the pollution emitted by industry and automobiles in wealthier countries (Chapters 15–17).
- Conservation biologists are helping to protect habitat and safeguard endangered species (Chapter 11).
- Recycling is helping us conserve resources and alleviate waste disposal problems (Chapter 22).
- Governments, businesses, and individuals are taking steps to reduce emissions of the greenhouse gases that drive climate change (Chapter 18).

These are a few of the many efforts we will examine while exploring sustainable solutions in the course of this book.

Students are promoting solutions on campus

As a college student, you can help to design and implement sustainable solutions on your own campus. Proponents of **campus sustainability** seek ways to help colleges and universities reduce their ecological footprints. Student-run organizations often play a key role in initiating recycling programs, finding ways to reduce energy use, and agitating for new courses or majors in environmental science or environmental studies.

We tend to think of colleges and universities as enlightened and progressive institutions that benefit society. This may be true, but colleges and universities are also centers of lavish resource consumption. Institutions of higher education feature extensive infrastructure including classrooms, offices, research labs, residential housing, dining establishments, sports arenas, vehicle fleets, and road networks. The 4500 campuses in the United States interact with thousands of businesses and spend $400 billion each year on products and services. The ecological footprint of a typical college or university is substantial, and together these institutions generate perhaps 2% of U.S. carbon emissions.

Reducing the size of this footprint is challenging. Colleges and universities tend to be bastions of tradition, where institutional habits are deeply ingrained and where bureaucratic inertia can block the best intentions for positive change. Nonetheless, faculty, staff, administrators, and students are progressing on a variety of fronts to make the operations of educational institutions more sustainable (FIGURE 1.15).

Students are often the ones who initiate change, although support from faculty, staff, and administrators is crucial for success. Students often feel freest to express themselves. Students also arrive on campus with new ideas and perspectives, and they generally are less attached to traditional ways of doing things.

Campus sustainability efforts are diverse

Students are advancing sustainability efforts on their campuses by promoting efficient transportation options, running recycling programs, planting trees and restoring native plants, growing organic gardens, and fostering sustainable dining halls. They are working with faculty and administrators to improve energy efficiency and water conservation in campus buildings and to ensure that new buildings meet certification guidelines for sustainable construction. To help address global climate change, students are urging their institutions to reduce greenhouse gas emissions and to use and invest in renewable energy.

In response, nearly 700 university presidents have signed onto the American College and University Presidents' Climate Commitment: a public pledge to inventory emissions, set target dates and milestones for becoming carbon-neutral, take immediate steps to lower emissions, and integrate sustainability into the curriculum.

Students who take the initiative to promote sustainable practices on their campuses accomplish several things at once:

- Students can truly make a difference by reducing the ecological footprint of a campus. The consumptive impact of educating, feeding, and housing hundreds or thousands of students is immense, so considerable opportunity exists for reducing the waste of resources.
- Students who act to advance campus sustainability can serve as models for their peers, helping to make them aware that they, too, can address problems.

(a) Urging divestment from fossil fuels

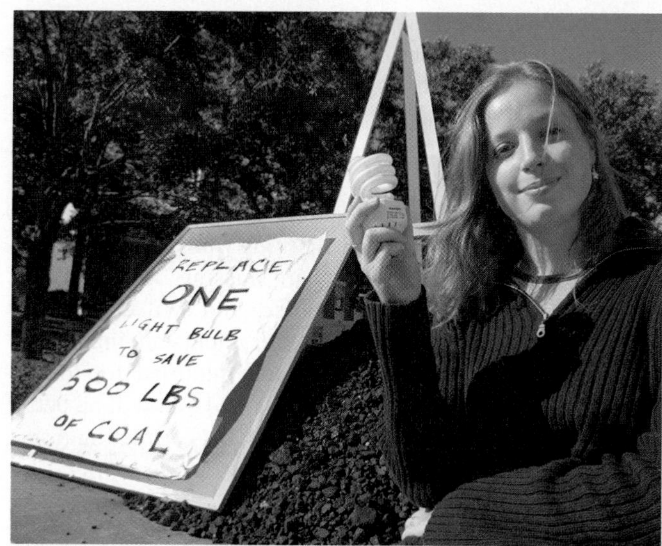

(b) Installing energy-efficient CFLs at Dickinson

(c) Recycling at Davidson

(d) Collecting electronic waste at UT Austin

FIGURE 1.15 Students are helping to make their campuses more sustainable in all kinds of ways.

- Students can learn and grow as a result. Colleges and universities are microcosms of society at large, and the challenges, successes, and failures encountered while working with others as part of a team can serve as valuable preparation for similar efforts in our broader society.

In our final chapter and throughout this book you will encounter examples of campus sustainability efforts. Should you wish to pursue such efforts on your own campus, information and links available in the **Selected Sources and References** point you toward organizations and resources that can help.

You will also see that the general concept of sustainability is infused throughout this book with one issue after another, and that the vital concept of sustainable development is tackled in Chapter 6. Our final chapter (Chapter 24) rounds out our discussion by presenting a summary of approaches to sustainability—on college and university campuses and in the world at large. All along the way, we will explore a wide array of issues in our far-reaching search for sustainable solutions.

Environmental science prepares you for the future

By taking a course in environmental science, you are helping to prepare yourself for a lifetime in a world increasingly dominated by concerns over sustainability. As society's concerns have evolved, colleges and universities have adapted their academic curricula. The course for which you are using this book right now likely did not exist a generation ago. As our society comes to appreciate the looming challenges of creating a sustainable future, colleges and universities are attempting to better train students to confront these challenges.

Still, the latest nationwide survey of campus sustainability efforts (in 2008) found that the percentage of American colleges and universities that require all students to take at least one course related to environmental science or sustainability had actually *decreased*, from 8% in 2001 to just 4% in 2008. At most schools, fewer than half of students take even a single course on the basic functions of Earth's natural systems, and still fewer take courses on the links between human activity and sustainability. As a result, the report stated, "students are slightly less likely to be environmentally literate when they graduate in 2008 than in 2001."

This surprising finding suggests that students like you are in a privileged minority, benefiting from a valuable education that most of your peers are missing. As a result of your environmental science course, you will come away from your college years with a better understanding of how the world works. You will be better qualified for tomorrow's green-collar job opportunities. And you will be better prepared to navigate the many challenges of creating a sustainable future.

Conclusion

Finding effective ways of living peacefully, healthfully, and sustainably on our diverse and complex planet will require a thorough scientific understanding of both natural and social systems. Environmental science helps us understand our intricate relationship with our environment and informs our attempts to solve and prevent environmental problems. Many of the trends detailed in this book may cause us worry, but others give us reason for optimism. Solving environmental problems can move us toward health, longevity, peace, and prosperity. Science in general, and environmental science in particular, can aid us in our efforts to develop balanced and workable solutions to the many challenges we face today and to create a better world for ourselves and our children.

Reviewing Objectives

You should now be able to:

Define the term *environment* and describe the field of environmental science

- Our environment consists of everything around us, including living and nonliving things. (p. 3)
- People are part of the environment and are not separate from nature. (p. 3)
- Environmental science is the study of how the natural world works, how our environment affects us, and how we affect our environment. (p. 3)

Explain the importance of natural resources and ecosystem services to our lives

- Some resources are inexhaustible or perpetually renewable, others are nonrenewable, and still others are renewable if we do not overexploit them. (pp. 3–4)
- Ecosystem services are benefits we receive from the processes and normal functioning of natural systems. (p. 3)
- Resources and ecosystem services are essential to human life and civilization, yet we are depleting and degrading many of them. (p. 3)

Discuss the consequences of population growth and resource consumption

- Rapid growth of the human population magnifies our environmental impacts. (p. 4)
- An ecological footprint quantifies a person's or nation's resource consumption in terms of area of biologically productive land and water. (pp. 4–5)

Characterize the nature of environmental science

- Environmental science employs approaches and insights from numerous disciplines in the natural sciences and social sciences. (p. 8)
- Environmental scientists are not advocates for environmental causes; instead, they study scientific issues objectively. (p. 9)

Understand the scientific method and the process of science

- Science is a process of using observations to test ideas. (pp. 9–10)
- The scientific method consists of making observations, formulating questions, stating a hypothesis, generating predictions, testing predictions, and analyzing results. (pp. 10–12)
- There are different ways to test questions scientifically (for example, with manipulative experiments to determine causation or with natural experiments and correlation). (p. 12)
- Scientific research occurs within a larger process that includes peer review, journal publication, and interaction with colleagues. (pp. 12–13)
- Science goes through paradigm shifts. This openness to change is what gives science its strength. (p. 13)

Diagnose and illustrate some of the pressures on the global environment

- Rising human population and intensifying per capita consumption magnify human impacts on the environment. (p. 14)
- Human activities are having diverse impacts, including resource depletion, air and water pollution, climate change, habitat destruction, and biodiversity loss. (pp. 14–15)

- Fossil fuel use drives many of our impacts. Replacing fossil fuels with clean renewable energy can reduce these impacts. (pp. 15–16)
- We are developing sustainable solutions that promote our quality of life while protecting and restoring our environment. (p. 16)

Articulate the concept of sustainability and describe campus sustainability efforts

- Sustainability means living within our planet's means, such that Earth's resources can sustain us—and all life—for the future. (p. 14)
- Many college students are taking action to promote sustainable solutions on their campuses, ranging from recycling to energy efficiency to water conservation to transportation, and more. (pp. 16–17)
- Your environmental science course will help prepare you for the challenges and opportunities in a world striving for a sustainable future. (pp. 17–18)

Testing Your Comprehension

1. What do renewable resources and nonrenewable resources have in common? How are they different? Identify two renewable and two nonrenewable resources.
2. How and why did the agricultural revolution affect human population size? How and why did the industrial revolution affect human population size? Explain what benefits and what environmental impacts have resulted.
3. What is an *ecological footprint*? Explain what is meant by the term *overshoot*.
4. What is *environmental science*? Name several disciplines that environmental science draws upon.
5. What are the two meanings of *science*? Name three applications of science.
6. Describe the scientific method. What is its typical sequence of steps?
7. Explain the difference between correlation and causation, and state how these concepts relate to manipulative and natural experiments.
8. What needs to occur before a researcher's results are published? Why is this process important?
9. Give examples of three major environmental problems in the world today, along with their causes. How are these problems interrelated? Can you name a potential solution for each?
10. Describe in your own words what you think is meant by the term *sustainability*. Name three ways that students, faculty, or administrators are seeking to make their campuses more sustainable.

Seeking Solutions

1. Many resources are renewable if we use them in moderation but can become nonrenewable if we overexploit them. Order the following resources on a continuum of renewability (see Figure 1.1), from most renewable to least renewable: soils, timber, fresh water, food crops, and biodiversity. What factors influenced your choices? For each resource, what might constitute overexploitation, and what might constitute sustainable use?
2. What do you think is the lesson of Easter Island? What more would you like to learn or understand about this island and its people? What similarities do you perceive between the history of Easter Island and the modern history of our society? What differences do you see between their predicament and ours?
3. What environmental problem do *you* feel most acutely yourself? Do you think there are people in the world who do not view your issue as a problem? Who might they be, and why might they take a different view?
4. If the human population were to stabilize tomorrow and never reach 8 billion people, would that solve our environmental problems? Which types of problems might get better, and which might become worse?
5. Find out what sustainability efforts are being made on your campus. What results have these efforts produced so far? What further efforts would you like to see pursued on your campus? Do you foresee any obstacles to these efforts? How could these obstacles be overcome? How could you become involved?
6. **THINK IT THROUGH** You have become head of a major funding agency that disburses funding to scientists pursuing research in environmental science. You must give your staff several priorities to determine what types of scientific research to fund. What environmental problems would you most like to see addressed with research? Describe the research you think would need to be completed so that workable solutions to these problems could be developed. What else, beyond scientific research, might be needed to develop sustainable solutions?

Calculating Ecological Footprints

Mathis Wackernagel and his colleagues at the Global Footprint Network have continually been refining the method of calculating ecological footprints—the amount of biologically productive land and water required to produce the energy and natural resources we consume and to absorb the wastes we generate. According to their most recent data, there are 1.8 hectares (4.4 acres) available for each person in the world, yet we use on average 2.7 ha (6.7 acres) per person, creating a global ecological deficit, or overshoot (p. 5), of 50%.

Compare the ecological footprints of each nation listed in the table. Calculate their proportional relationships to the world population's average ecological footprint and to the area available globally to meet our ecological demands.

Nation	Ecological footprint (hectares per person)	Proportion relative to world average footprint	Proportion relative to world area available
Bangladesh	0.7	0.3 (0.7 ÷ 2.7)	0.4 (0.7 ÷ 1.8)
Tanzania	1.2		
Colombia	1.8		
Thailand	2.4		
Mexico	3.3		
Sweden	5.7		
United States	7.2		
World average	2.7	1.0 (2.7 ÷ 2.7)	1.5 (2.7 ÷ 1.8)
Your personal footprint (see Question 4)			

Data from Global Footprint Network, in: WWF, 2012. Living planet report 2012. *WWF International, Gland, Switzerland.*

1. Why do you think the ecological footprint for people in Bangladesh is so small?
2. Why is it so large for people in the United States?
3. Based on the data in the table, how do you think average per capita income affects ecological footprints?
4. Go to an online footprint calculator such as the one at http://www.myfootprint.org or http://www.footprintnetwork.org/en/index.php/GFN/page/personal_footprint and take the test to determine your own personal ecological footprint. Enter the value you obtain in the table and calculate the other values as you did for each nation. How does your footprint compare to those of the average person in the United States? How does it compare to that of people from other nations? Name three actions you could take to reduce your footprint. (*Note*: Save this number—you will calculate your footprint again in Chapter 24 at the end of your course!)

MasteringENVIRONMENTALSCIENCE™

STUDENTS

Go to **MasteringEnvironmentalScience** for assignments, the etext, and the Study Area with practice tests, videos, current events, and activities.

INSTRUCTORS

Go to **MasteringEnvironmentalScience** for automatically graded activities, current events, videos, and reading questions that you can assign to your students, plus Instructor Resources.

Rescue workers in northeastern Japan

2

Earth's Physical Systems: Matter, Energy, and Geology

Upon completing this chapter, you will be able to:

- Explain the fundamentals of matter and chemistry and apply them to real-world situations
- Differentiate among forms of energy and explain the basics of energy flow
- Distinguish photosynthesis, cellular respiration, and chemosynthesis and summarize their importance to living things
- Explain how plate tectonics and the rock cycle shape the landscape around us and the earth beneath our feet
- List major types of geologic hazards and describe ways to mitigate their impacts

CENTRAL CASE STUDY

The Tohoku Earthquake: Has It Shaken the World's Trust in Nuclear Power?

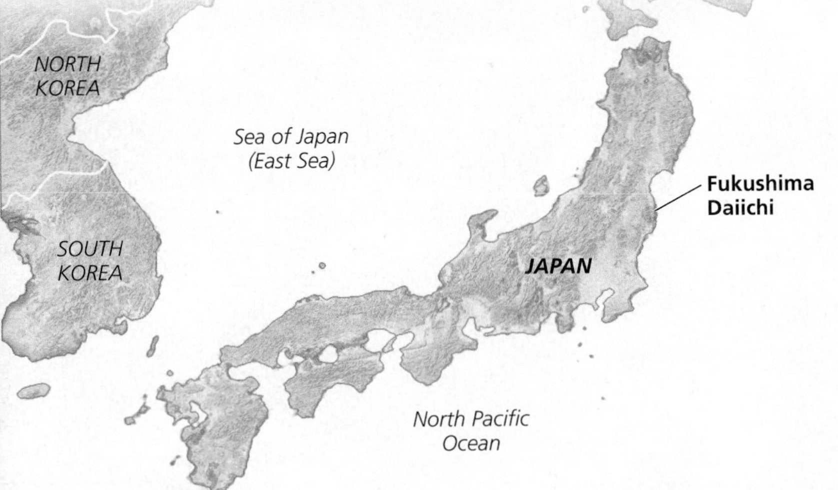

"This used to be one of the best places for a business. I'm amazed at how little is left."

—Takahiro Chiba, surveying the devastated downtown area of Ishinomaki, Japan, where his family's sushi restaurant was located

"Fukushima should not just contain lessons for Japan, but for all 31 countries with nuclear power."

—Tatsujiro Suzuki, Vice-chairman, Japan Atomic Energy Commission

At 2:46 p.m. on March 11, 2011, the land along the northeastern coast of the Japanese island of Honshu began to shake violently—and continued to shake for six minutes. These tremors were caused when a large section of the seafloor along a fault line 125 km (77 mi) offshore suddenly lurched, releasing huge amounts of energy through the crust and generating an earthquake of magnitude 9.0 on the Richter scale (a scale used to measure the strength of earthquakes). Little did anyone know at the time that this quake would initiate a series of events that would affect not only Japan, but also the future of nuclear power around the world.

The Tohoku earthquake, as it was later named, was not the first major earthquake to strike Japan. The city of Kobe experienced substantial damage from a quake in 1995 that claimed over 5500 lives. And in 1923, an earthquake devastated the cities of Tokyo and Yokohama, resulting in over 142,000 deaths. Losses of life and property from the Tohoku quake were far less extensive than the losses from these earlier events, thanks to new stringent building codes that enable buildings to resist crumbling and toppling over during earthquakes. But even when the earth stopped shaking, the residents of northeastern Japan knew that further danger might still await them—from a tsunami.

A **tsunami** ("harbor wave" in English) is a powerful surge of seawater generated when an offshore earthquake displaces large volumes of rocks and sediment on the ocean bottom, suddenly pushing the overlying ocean water upward. This upward movement of water creates waves that speed outward from the earthquake site in all directions. These waves are hardly noticeable at sea, but can rear up to staggering heights when they enter the shallow waters near shore and can sweep inland with great force. The fear of a tsunami was well founded, as strong ocean surges followed the 1923 Tokyo–Yokohama earthquake, pushing walls of debris in front of them and drowning victims still trapped in the wreckage from the earthquake.

The Japanese had built seawalls to protect against tsunamis, but the Tohoku quake caused the island of Honshu to sink, lowering the height of the seawalls by up to 2 m (6.5 ft) in some locations. Waves reaching up to 15 m (49 ft) in height then overwhelmed these defenses (**FIGURE 2.1**). The raging water swept up to 9.6 km (6 mi) inland, scoured buildings from their foundations, and

FIGURE 2.1 Tsunami waves overtop a seawall following the Tohoku earthquake in 2011. The tsunami caused a greater loss of life and property than the earthquake that generated it and led to a meltdown at the Fukushima Daiichi nuclear power plant.

inundated towns, villages, and productive agricultural land. As the water's energy faded, it receded, carrying structural debris, vehicles, livestock, and human bodies out to sea.

When the tsunami overtopped the 5.7-m (19-ft) seawall protecting the Fukushima Daiichi nuclear power plant, it flooded the diesel-powered emergency generators responsible for circulating water to cool the plant's nuclear reactors. With the local electrical grid knocked out by the earthquake and the backup generators off-line, the nuclear fuel in the cores of the three active reactors at the plant began to overheat. The water that normally kept the nuclear fuel submerged within the reactor cores boiled off, exposing the nuclear material to the air and further elevating temperatures inside the cores. As the overheated nuclear fuel melted (called a nuclear meltdown), chemical reactions within the reactors generated hydrogen gas, which set off explosions in each of the three reactor buildings, releasing radioactive material into the air.

As events worsened over several tense days, Japanese authorities became desperate to cool the reactors, contain the radioactive emissions, and prevent a full-blown catastrophe that could render large portions of their nation uninhabitable. They sent in teams of engineers and soldiers who risked their lives amid the radiation and flooded the reactor cores with seawater pumped in from the ocean.

The 1–2–3 punch of the earthquake–tsunami–nuclear accident left over 19,000 people dead or missing and caused $300 billion in material damage. Around 340,000 people were displaced from their homes, including 100,000 people from towns near the Fukushima Daiichi plant where radioactive fallout contaminated the soil to unsafe levels. A 20-km (12-mi) area around the Fukushima Daiichi plant has been permanently evacuated, and the full extent of nuclear contamination is still being determined (see **THE SCIENCE BEHIND THE STORY**, pp. 26–27). Some of the greatest concerns center on contaminated food and water, so domestically produced crops and seafood will require testing for radiation for years to come. Full recovery from these events is expected to take decades.

One of the longest-lasting legacies of these events may be the impact on the future of nuclear power in Japan and around the world. The Japanese government had championed a view that nuclear power was perfectly safe, but the events at Fukushima have shaken public support for nuclear power in Japan. In the summer of 2012, over 100,000 people marched in Tokyo to protest the restarting of nuclear reactors that had been shut down after the Tohoku quake, and public opinion surveys found 70% of Japanese wished for their nation to rely less on nuclear power.

In North America and Europe, the events at Fukushima Daiichi caused a public already skeptical of the safety of nuclear power to become further wary of its use. But with the challenges facing us in climate change (Chapter 18), many energy analysts, scientists, and policymakers caution that we should not abandon a carbon-free source of energy like nuclear power, but rather refocus efforts on maximizing its safety. Whatever the eventual outcome of these analyses, the events of March 11, 2011, will not soon be forgotten—in Japan or elsewhere.

Matter, Chemistry, and the Environment

The tragic events in northeastern Japan were the result of large-scale forces generated by the powerful geological processes that shape the surface of our planet. Environmental scientists regularly study these types of processes to understand how our planet works. Because all large-scale processes are made up of small-scale components, however, environmental science—the broadest of scientific fields—must also study small-scale phenomena. At the smallest scale, an understanding of matter itself helps us to fully appreciate all the processes of our world.

All material in the universe that has mass and occupies space—solid, liquid, and gas alike—is called **matter.** In our quick tour of matter in the pages that follow, we examine types of matter and some of the important ways they interact—phenomena that together we term **chemistry.** Once you examine any environmental issue, you will likely discover chemistry playing a central role. Knowledge of chemistry is crucial for understanding how gases such as carbon dioxide and methane contribute to global climate change, how pollutants such as sulfur dioxide and nitric oxide cause acid rain, and how pesticides and other compounds we release into the environment affect the health of wildlife and people. Such knowledge is central, too, in understanding water pollution and wastewater treatment, hazardous waste and its cleanup and disposal, atmospheric ozone depletion, and most energy issues. Moreover, countless applications of chemistry can help us address these environmental problems.

Matter is conserved

To appreciate the chemistry involved in environmental science, we must begin with a grasp of the fundamentals. Matter may be transformed from one type of substance into others, but it cannot be created or destroyed. This principle is referred to as the **law of conservation of matter.** In environmental science, this principle helps us understand that the amount of matter stays constant as it is recycled in ecosystems and nutrient cycles (p. 117). It also makes it clear to us that we cannot simply wish away "undesirable" matter, such as nuclear waste and toxic pollutants. Since harmful substances like these can't be destroyed, we must take prudent steps to mitigate their impacts on the environment.

Atoms and elements are chemical building blocks

The nuclear reactor at Fukushima used **uranium** to power its reactors, and uranium is an example of an element. An **element** is a fundamental type of matter, a chemical substance with a given set of properties that cannot be broken down into substances with other properties. Chemists currently recognize 92 elements occurring in nature, as well as more than 20 others that they have created in the lab. Elements especially abundant on our planet include **hydrogen** (in water), **oxygen** (in the air), **silicon** (in Earth's crust), **nitrogen** (in the air), and **carbon** (in living organisms) (TABLE 2.1). Each element is assigned an abbreviation, or chemical symbol (for instance,

TABLE 2.1 Earth's Most Abundant Chemical Elements, by Mass

Earth's crust		Oceans		Air		Organisms	
Oxygen (O)	49.5%	Oxygen (O)	88.3%	Nitrogen (N)	78.1%	Oxygen (O)	65.0%
Silicon (Si)	25.7%	Hydrogen (H)	11.0%	Oxygen (O)	21.0%	Carbon (C)	18.5%
Aluminum (Al)	7.4%	Chlorine (Cl)	1.9%	Argon (Ar)	0.9%	Hydrogen (H)	9.5%
Iron (Fe)	4.7%	Sodium (Na)	1.1%	Other	<0.1%	Nitrogen (N)	3.3%
Calcium (Ca)	3.6%	Magnesium (Mg)	0.1%			Calcium (Ca)	1.5%
Sodium (Na)	2.8%	Sulfur (S)	0.1%			Phosphorus (P)	1.0%
Potassium (K)	2.6%	Calcium (Ca)	<0.1%			Potassium (K)	0.4%
Magnesium (Mg)	2.1%	Potassium (K)	<0.1%			Sulfur (S)	0.3%
Other	1.6%	Bromine (Br)	<0.1%			Other	0.5%

H for hydrogen and O for oxygen). The *periodic table of the elements* (see **APPENDIX D**) organizes the elements according to their chemical properties and behavior.

Atoms are the smallest units that maintain the chemical properties of the element. Atoms of each element hold a defined number of **protons** (positively charged particles) in the atom's nucleus (its dense center), and this number is called the element's *atomic number*. (Elemental carbon, for instance, has six protons in its nucleus; thus, its atomic number is 6.) Most atoms also contain **neutrons** (particles lacking electric charge) in their nuclei, and an element's *mass number* denotes the combined number of protons and neutrons in the atom. An atom's nucleus is surrounded by negatively charged particles known as **electrons,** which balance the positive charge of the protons (**FIGURE 2.2**).

Isotopes Although all atoms of a given element contain the same number of protons, they do not necessarily contain the same number of neutrons. Atoms of the same element with differing numbers of neutrons are referred to as **isotopes** (**FIGURE 2.3a**). Thus, isotopes of a given element have the same atomic number, but different mass numbers. Isotopes are denoted by their elemental symbol preceded by the mass number. For example, ^{14}C (carbon-14) is an isotope of carbon with eight neutrons (and six protons) in the nucleus rather than the six neutrons (and six protons) of ^{12}C (carbon-12), the most abundant carbon isotope.

Some isotopes, called *radioisotopes*, are **radioactive** and "decay" by changing their chemical identity as they shed subatomic particles and emit high-energy radiation. The radiation released by radioisotopes harms organisms because it focuses a great deal of energy in a very small area, which can be damaging to vital macromolecules within cells. Radiation can break apart large, biochemically important molecules (such as enzymes or components of cell membranes) or change the structure of DNA. Changes in DNA sequence can cause immediate cell death or increase the probability of the organism developing cancerous tumors at a later time.

The greatest danger from radioisotopes occurs when they enter the bodies of organisms through the lungs, skin, or digestive system. When the radioactive material is within the organism's tissues, nearby cells are exposed to high levels of harmful radiation for long periods, greatly increasing adverse impacts. It is therefore important after nuclear accidents, like that at Fukushima, to regularly test food and water supplies for radioisotopes and to determine the eventual fate of radioactive particles released into the environment (see **The Science behind the Story**, pp. 26–27).

Radioisotopes decay into lighter and lighter radioisotopes until they become stable isotopes (isotopes that are not radioactive). Each radioisotope decays at a rate determined by that isotope's **half-life,** the amount of time it takes for one-half the atoms to give off radiation and decay. Different radioisotopes have very different half-lives, ranging from fractions of a second to billions of years. The radioisotope uranium-235 (^{235}U), used in commercial nuclear power plants like Fukushima, decays into a series of daughter isotopes, eventually forming lead-207 (^{207}Pb), and has a half-life

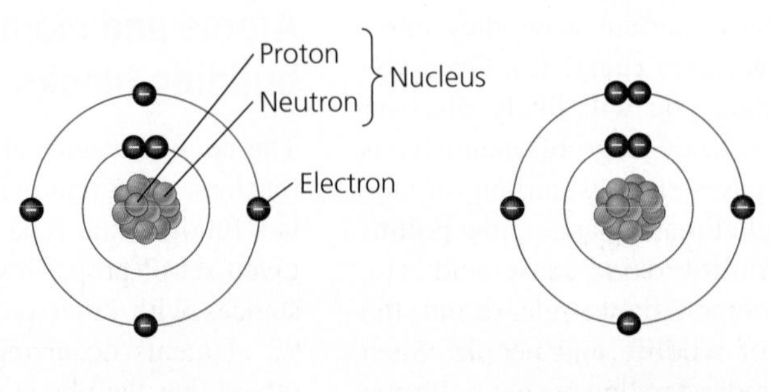

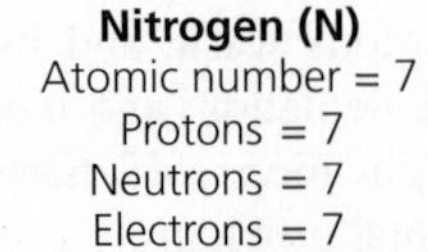

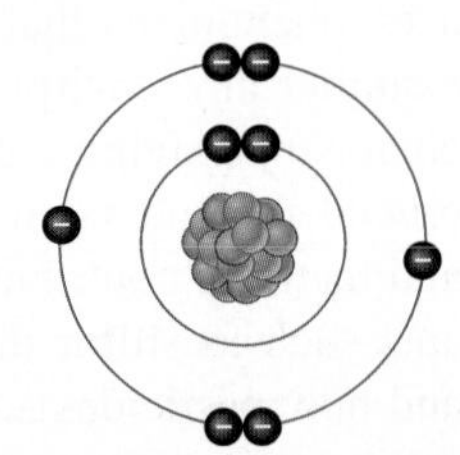

FIGURE 2.2 In an atom, protons and neutrons stay in the nucleus, and electrons move around the nucleus. Each chemical element has its own particular number of protons. Carbon possesses six protons, nitrogen seven, and oxygen eight. These schematic diagrams are meant to clearly show and compare numbers of electrons for these three elements. In reality, however, electrons do *not* orbit the nucleus in rings as shown; they move through space in more complex ways.

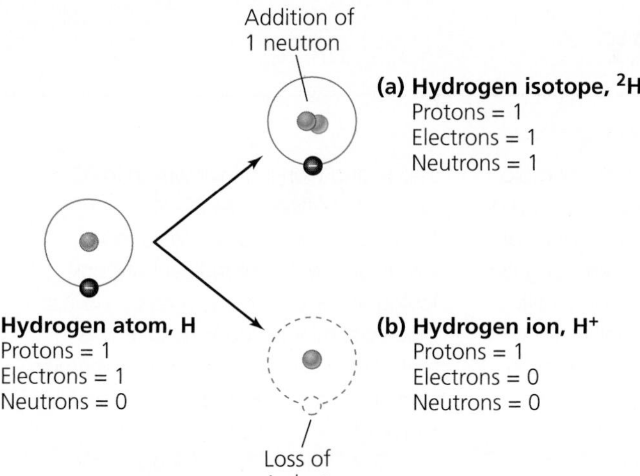

FIGURE 2.3 Hydrogen has a mass number of 1 because a typical atom of this element contains one proton and no neutrons. Deuterium (hydrogen-2 or ^2H), an isotope of hydrogen **(a)**, contains a neutron as well as a proton and thus has greater mass than a typical hydrogen atom; its mass number is 2. The hydrogen ion, H^+ **(b)**, occurs when an electron is lost; it therefore has a positive charge.

FAQ When something is irradiated, does it become radioactive?

Thanks to comic books and movies, many people believe that when an organism is exposed to high-energy radiation from a source outside the body, such as the sun or nuclear waste, the organism becomes a *source* of ionizing radiation—that is, it becomes "radioactive." In reality, this does not happen.

An irradiated organism suffers damage from radiation, but it does not absorb the ionizing radiation, store it, and then re-emit it to the environment. The radiation simply enters the organism's cells, causes damage, and passes through the organism. So even after experiencing substantial impacts from radiation poisoning, the organism is no more radioactive than it was before exposure because it was only exposed to radiation (a form of energy) and was not contaminated with radioactive particles (a form of matter) that emit harmful radiation.

The damaging effects of ionizing radiation can be put to positive use, though. Radiation has been used to sterilize medical supplies since the 1950s to prevent patients from receiving secondary infections while in the hospital. Raw meat can also be irradiated to kill harmful microbes that may be lurking within the meat, such as *Salmonella* or disease-causing strains of *Escherichia coli (E. coli)*. In both cases, the irradiated material does not become radioactive, just free of living microorganisms or viruses that could cause sickness or death.

of about 700 million years. Radioisotopes released into the environment from the Fukushima nuclear power plant accident included iodine-131 (half-life of 8 days), cesium-134 (half-life of 2 years), and cesium-137 (half-life of 30 years), showing the long-term implications of nuclear accidents.

Ions Atoms may also gain or lose electrons, thereby becoming **ions**, electrically charged atoms or combinations of atoms (FIGURE 2.3b). Ions are denoted by their elemental symbol followed by their ionic charge. For instance, a common ion used by mussels and clams to form shells is Ca^{2+}, a calcium atom that has lost two electrons and thus has a charge of positive 2. The damaging radiation emitted by radioisotopes is called **ionizing radiation** because it generates ions when it strikes molecules, and these ions affect the stability and functionality of biological molecules.

Atoms bond to form molecules and compounds

Atoms can bond together and form **molecules,** combinations of two or more atoms. Common molecules containing only a single element include those of hydrogen (H_2) and oxygen (O_2), each of which exists as a gas at room temperature. A molecule composed of atoms of two or more different elements is called a **compound.** One compound is **water;** it is composed of two hydrogen atoms bonded to one oxygen atom, and it is denoted by the chemical formula H_2O. Another compound is **carbon dioxide,** consisting of one carbon atom bonded to two oxygen atoms; its chemical formula is CO_2.

Atoms bond together because of an attraction for one another's electrons. Because the strength of this attraction varies among elements, atoms may be held together in different ways. When electrons are shared between atoms, a **covalent bond** forms. For example, two atoms of hydrogen share electrons equally as they bind together to form hydrogen gas, H_2. However, in a water molecule, oxygen attracts shared electrons more strongly than does hydrogen. The result is that water has a partial negative charge at its oxygen end and partial positive charges at its hydrogen ends. This arrangement allows water molecules to adhere to one another in a type of weakly attractive interaction called a *hydrogen bond* (FIGURE 2.4).

In compounds in which the strength of attraction is sufficiently unequal, an electron may be transferred from one atom to another. This creates oppositely charged ions that form **ionic**

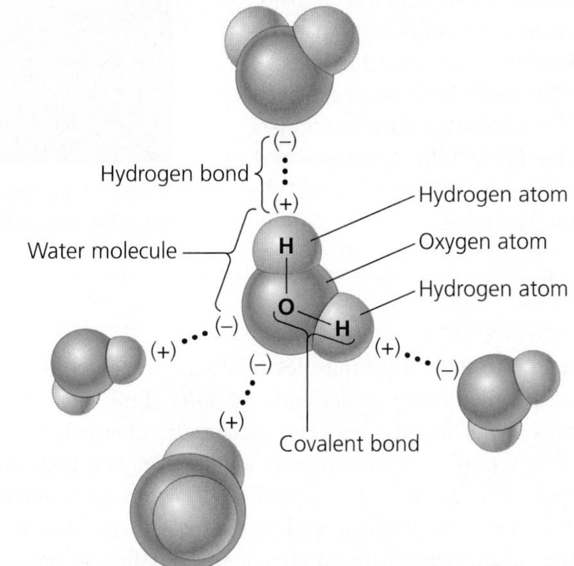

FIGURE 2.4 By enabling water molecules to adhere loosely to one another, hydrogen bonds give water several unique properties crucial for life.

Tracking Fukushima's Nuclear Legacy

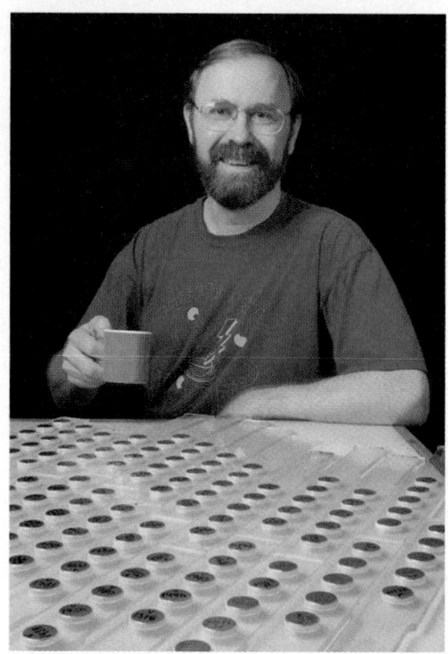

Dr. Ken Buesseler, Woods Hole Oceanographic Institution

When Dr. Ken Buesseler, a senior scientist at the Woods Hole Oceanographic Institution in Massachusetts, observed the events unfolding in northeastern Japan in 2011, he saw an opportunity to use his expertise to determine the movements and fate of the radioisotopes released into the Pacific Ocean from the crippled Fukushima Daiichi nuclear power plant. Dr. Buesseler had done this before when he used radioisotopes released from the Chernobyl nuclear accident (Chapter 20) in 1986 as tracers to study the poorly-understood currents in the nearby Black Sea.

Scientists equate the input of radioisotopes into waters to "pouring dye into the ocean." By knowing the half-lives of relevant radioisotopes and their daughter isotopes and by tracking their quantities across bodies of water, scientists can study large-scale circulation patterns in inland seas and the ocean. Quick action was needed, so Buesseler scrambled to assemble a team of scientists and secure resources to mount an expedition to sample the waters offshore from Fukushima before the radioisotopes dispersed throughout the Pacific.

Thanks to a $3.7-million award from the Gordon and Betty Moore Foundation, Buesseler and his colleagues were able to charter the research vessel *Ka'imikai-o-Kanaloa* from the University of Hawaii and zigzag their way through the waters of the north Pacific in June 2011—starting 640 km (400 mi) off the coast of Japan and working their way in to 32 km (20 mi) offshore. At 30 sites along the cruise route, the scientists tested the air and water for radioactivity, collected water samples to measure levels of several radioisotopes (**FIGURE 1**), and tested plankton and other free-swimming organisms to determine the extent of radioisotope uptake by organisms.

Buesseler's group found water radiation levels of more than 100,000 becquerels per cubic meter (Bq/m^3; a becquerel is a unit of measurement for radioactivity) in early April, up from a pre-accident level of about 1.5 becquerels per cubic meter. These post-accident radiation levels were about 100 times greater than those found in the Black Sea after the Chernobyl incident (**FIGURE 2**). So while the release of radioisotopes from Chernobyl was roughly five times larger than that from Fukushima, the

FIGURE 1 An international team of scientists tracked the fate of radioactive material released into the ocean from the Fukushima nuclear power plant.

bonds due to their differing electrical charges. Such associations are called *ionic compounds*, or *salts*. Table salt (NaCl) contains ionic bonds between positively charged sodium ions (Na^+), each of which donates an electron, and negatively charged chloride ions (Cl^-), each of which receives an electron.

Elements, molecules, and compounds can also come together in mixtures without chemically bonding or reacting. Homogeneous mixtures of substances are called *solutions*, a term most often applied to liquids, but also applicable to some gases and solids. Air in the atmosphere is a solution formed of constituents such as nitrogen, oxygen, water vapor, carbon dioxide, **methane** (CH_4), and **ozone** (O_3). Ocean water, plant sap, petroleum, and metal alloys such as brass are all solutions.

Water's chemistry facilitates life

Water has unique properties that give it an amazing capacity to support life. Water's ability to form loose connections of hydrogen bonds gives it several properties that help to support life and stabilize Earth's climate:

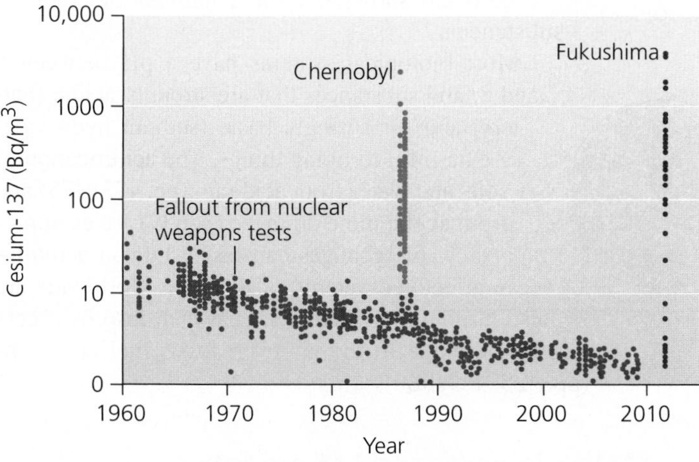

FIGURE 2 Ocean concentrations of radioactive cesium-137 show evidence of the nuclear accidents at Chernobyl and Fukushima. While Chernobyl released five times the radioactive material as Fukushima, the impacts on the ocean were greater from the Fukushima accident due to its coastal location. The EPA standard for safe drinking water is 10,000 Bq/m^3. *Note the Y axis is logarithmic, such that each unit is 10 times greater than the previous unit. Source: Ken Buesseler, Woods Hole Oceanographic Institution.*

DATA Q What was the trend in ocean radioactivity levels from cesium-137 from 1960 to 2010? Did releases from the Chernobyl accident significantly alter this trend? What long-term trends would you therefore expect to see following the releases from Fukushima in 2011?

impact on the ocean was far greater from Fukushima. Despite this, the team concluded that radioactivity levels in waters in their sampling areas did not pose an immediate threat to humans or other free-swimming organisms.

The study, published in the *Proceedings of the National Academy of Sciences* in 2012, found that prevailing currents in the area had a major effect on the distribution of radioactive material originating from Fukushima. The Kuroshio current, which flows from south to north along the eastern coast of Japan, helped carry radioactive material quickly away from the coastline and out to sea, but the current also concentrated radioactivity in some areas near shore. Buoys deployed by the expedition revealed that the current created an eddy—a mass of swirling water—near the coast of Fukushima that mixed minimally with waters in the Kuroshio current. This served to concentrate the water near shore, likely creating a "hotspot" of radioactivity in coastal waters.

Another surprise was that radioactivity levels in the water did not decline to pre-accident levels over time, but remained high many months after the accident. This observation suggests radioactive material may still be leaking from the plant into the ocean and/or radioactively-contaminated groundwater may be feeding into the ocean. Radioactivity levels in locally captured fish had not declined one and a half years after the accident, further suggesting continued releases of radioactive material into the marine food web.

Due to the relatively long half-life of radioisotopes (half-life of 30 years), radioactive material could accumulate in offshore sediments and be ingested by species that live on the ocean bottom or that filter water for food. Shellfish, bottom-dwelling fish, and other creatures are therefore likely to experience greater exposure to radioactive compounds, and for a longer duration, than species living in the open water. Sediments in areas where radiation was concentrated are likely to remain contaminated for several decades, posing a long-term threat to humans consuming seafood from these areas. Accordingly, the selling of 36 types of fish and other types of seafood harvested off the coast of Fukushima was banned in June 2012 by the Japanese government once fishing resumed elsewhere in the region.

Although the work of Buesseler and his team has shed light on the movements of radioisotopes released into the ocean by the Fukushima accident, there are still many questions that remain, such as how radioactive contaminants will move through aquatic food webs. Continuous monitoring of radioisotope concentrations in aquatic organisms, likely for many decades, will be necessary to determine threats to human health and to ecosystems. ■

- Water's cohesion (think of how a water droplet holds together) facilitates the transport of nutrients and waste in organisms.
- Water molecules bond well with ions and other partially charged molecules, so water can hold in solution, or dissolve, many other molecules, including chemicals vital for life.
- Water molecules in ice are farther apart than in liquid water, so ice is less dense. This enables ice to float on water, insulating lakes and ponds and preventing them from freezing solid in winter (**FIGURE 2.5**).
- Heating weakens hydrogen bonds before it speeds molecular motion, so water can absorb a great deal of heat with only small changes in its temperature. This capacity to resist change helps stabilize water bodies, organisms, and climate systems.

Hydrogen ions determine acidity

In any aqueous solution, a small number of water molecules split apart, each forming a hydrogen ion (H^+) and a hydroxide

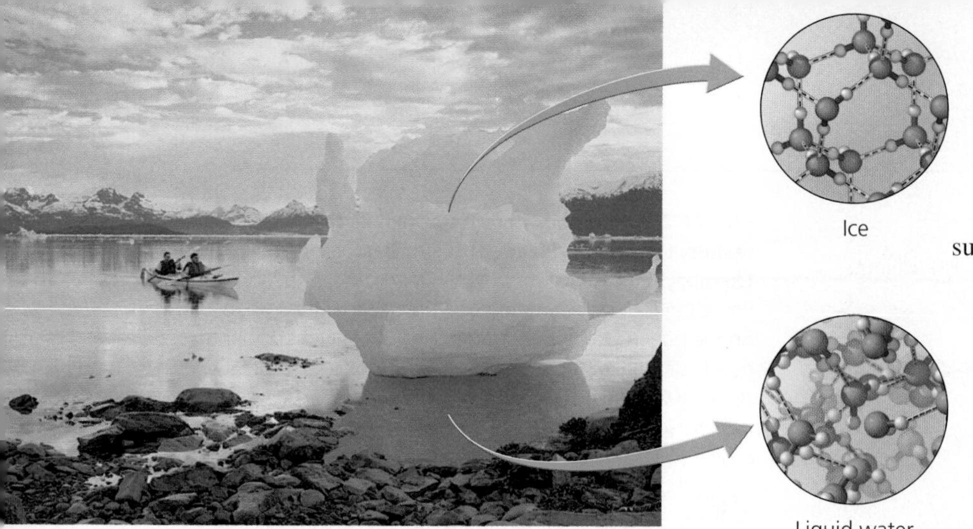

FIGURE 2.5 Ice floats in liquid water because ice is less dense. In ice, molecules are connected by stable hydrogen bonds, forming a spacious crystal lattice. In liquid water, hydrogen bonds frequently break and reform, and the molecules are closer together and less organized.

ion (OH^-). The product of hydrogen and hydroxide ion concentrations is always the same; as one increases, the other decreases. Pure water contains equal numbers of these ions. Solutions in which the H^+ concentration is greater than the OH^- concentration are **acidic,** whereas solutions in which the OH^- concentration exceeds the H^+ concentration are **basic,** or alkaline.

The **pH** scale (**FIGURE 2.6**) quantifies the acidity or alkalinity of solutions. It runs from 0 to 14; pure water has a hydrogen ion concentration of 10^{-7} and a pH of 7. Solutions with a pH less than 7 are acidic, and those with a pH greater than 7 are basic. Because the pH scale is logarithmic, each step on the scale represents a 10-fold difference in hydrogen ion concentration. Thus, a substance with a pH of 6 contains 10 times as many hydrogen ions as a substance with a pH of 7 and 100 times as many hydrogen ions as a substance with a pH of 8. Figure 2.6 shows pH for a number of common substances.

Most biological systems have a pH between 6 and 8, and substances that are strongly acidic (battery acid) or strongly basic (sodium hydroxide) are harmful to living things. The acidification of soils and water from acid rain (pp. 473–475) and from acidic mine drainage (p. 639) are examples of how pH changes caused by human activities can affect ecosystems. The oceans worldwide are also turning more acidic as seawater absorbs excess carbon dioxide in the air from fossil fuel emissions (pp. 428–429, 488).

Matter is composed of organic and inorganic compounds

Beyond their need for water, living things also depend on organic compounds, which they create and of which they are created. **Organic compounds** consist of carbon atoms (and generally hydrogen atoms) joined by covalent bonds, and they may also include other elements, such as nitrogen, oxygen, sulfur, and phosphorus. Inorganic compounds, in contrast, lack carbon–carbon bonds.

Carbon's unusual ability to bond together in chains, rings, and other structures to build elaborate molecules has resulted in millions of different organic compounds. One class of such compounds that is important in environmental science is **hydrocarbons**, which consist solely of bonded atoms of carbon and hydrogen (although other elements may enter these compounds as impurities). Fossil fuels and the many petroleum products we make from them (Chapter 19) consist largely of hydrocarbons.

The simplest hydrocarbon is methane (CH_4), the key component of natural gas; it has one carbon atom bonded to four hydrogen atoms (**FIGURE 2.7a**). Adding another carbon atom and two more hydrogen atoms gives us ethane (C_2H_6), the next-simplest hydrocarbon (**FIGURE 2.7b**). The smallest (and

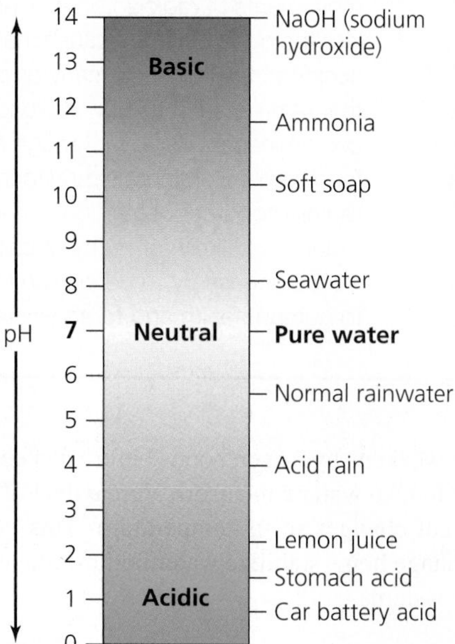

FIGURE 2.6 The pH scale measures how acidic or basic (alkaline) a solution is. The pH of pure water is 7, the midpoint of the scale. Acidic solutions have a pH less than 7, whereas basic solutions have a pH greater than 7.

(a) Methane, CH_4 **(b) Ethane, C_2H_6** **(c) Naphthalene, $C_{10}H_8$**

FIGURE 2.7 Hydrocarbons have a diversity of chemical structures. The simplest hydrocarbon is methane **(a)**. Many hydrocarbons consist of linear chains of carbon atoms with hydrogen atoms attached; the shortest of these is ethane **(b)**. Volatile hydrocarbons with multiple rings, such as naphthalene **(c)**, are called polycyclic aromatic hydrocarbons (PAHs).

therefore lightest-weight) hydrocarbons (those consisting of four or fewer carbon atoms) exist in a gaseous state at moderate temperatures and pressures. Larger (and therefore heavier) hydrocarbons are liquids, and those containing over 20 carbon atoms are normally solids.

Some hydrocarbons from petroleum are known to pose health hazards to wildlife and people. For example, some polycyclic aromatic hydrocarbons, or PAHs (**FIGURE 2.7c**), can evaporate from spilled oil and gasoline and can mix with water, putting the eggs and young of fish and other aquatic creatures at risk. PAHs also occur in particulate form in various combustion products, including cigarette smoke, wood smoke, and charred meat.

At the same time, some hydrocarbons from petroleum have become ubiquitous in our modern lifestyle because they are moldable into nearly any shape and resist chemical breakdown. While these **plastics** have many benefits for manufactured goods, they can be a persistent source of pollution due to their longevity in the environment (pp. 432–436).

Macromolecules are building blocks of life

Just as carbon atoms in hydrocarbons may be strung together in chains, organic compounds sometimes combine to form long chains of repeated molecules. These chains are called **polymers,** and there are three types of polymers that are essential to life: proteins, nucleic acids, and carbohydrates. Along with lipids (which are not polymers), these types of molecules are referred to as **macromolecules** because of their large sizes.

Proteins consist of long chains of organic molecules called *amino acids*. The many types of proteins serve various functions. Some help produce tissues and provide structural support; for example, animals use proteins to generate skin, hair, muscles, and tendons. Some proteins help store energy, whereas others transport substances. Some act in the immune system to defend the organism against foreign attackers. Still others are hormones, molecules that act as chemical messengers within an organism. Proteins can also serve as *enzymes*, molecules that catalyze, or promote, certain chemical reactions.

Nucleic acids direct the production of proteins. The two nucleic acids—**deoxyribonucleic acid (DNA)** and **ribonucleic acid (RNA)**—carry the hereditary information for organisms and are responsible for passing traits from parents to offspring. Nucleic acids are composed of a series of nucleotides, each of which contains a sugar molecule, a phosphate group, and a nitrogenous base. DNA contains four types of nucleotides and can be pictured as a ladder twisted into a spiral, giving the molecule a shape called a double helix (**FIGURE 2.8**). Regions of DNA coding for particular proteins that perform particular functions are called **genes.**

Carbohydrates include simple sugars that are three to seven carbon atoms long. Glucose ($C_6H_{12}O_6$) fuels living cells and serves as a building block for complex carbohydrates, such as starch. Plants use starch to store energy, and animals eat plants to acquire starch. Plants and animals also use complex carbohydrates to build structure. Insects and crustaceans form hard shells from the carbohydrate chitin. Cellulose, the most abundant organic compound on Earth, is a complex carbohydrate found in the cell walls of leaves, bark, stems, and roots.

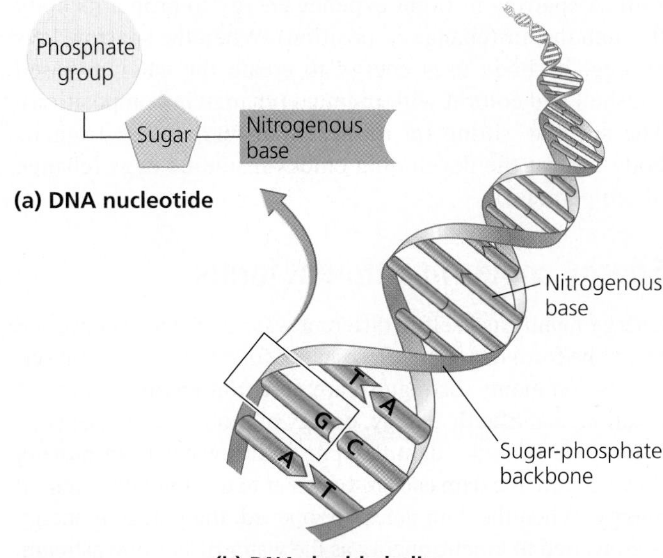

FIGURE 2.8 Nucleic acids encode genetic information in the sequence of nucleotides, small molecules that pair together like rungs of a ladder. DNA includes four types of nucleotides **(a)**, each with a different nitrogenous base: adenine (A), guanine (G), cytosine (C), and thymine (T). Adenine (A) pairs with thymine (T), and cytosine (C) pairs with guanine (G). In RNA, thymine is replaced by uracil (U). DNA **(b)** twists into the shape of a double helix.

Lipids include fats and oils (for energy storage), phospholipids (for cell membranes), waxes (for structure), and steroids (for hormone production). Although chemically diverse, these compounds are grouped together because they do not dissolve in water.

All of these macromolecules are found in cells, the most basic unit of organismal organization. Organisms range in complexity from single-celled bacteria to plants and animals that contain millions of cells. Cells vary greatly in size, shape, and function. Biologists classify organisms into two groups based on the structure of their cells. The cells of *eukaryotes* (plants, animals, fungi, and protists) contain a membrane-enclosed nucleus and various membrane-enclosed *organelles* that perform specific functions. *Prokaryotes* (bacteria and archaea) are generally single-celled, and their cells lack membrane-enclosed organelles and a nucleus.

Energy: An Introduction

Creating and maintaining organized complexity—of a cell, an organism, or an ecological system—requires energy. Energy is needed to organize matter into complex forms, to build and maintain cellular structure, to govern species' interactions, and to drive the geologic forces that shape our planet. Energy is somehow involved in nearly every chemical, biological, and physical phenomenon.

But what, exactly, is energy? **Energy** is the capacity to change the position, physical composition, or temperature of matter—in other words, a force that can accomplish work. As we saw with Japan's 2011 earthquake and tsunami, geologic events involve some of the most dramatic releases of energy in nature. Energy is omnipresent in living things as well. A sparrow in flight expends energy to propel its body through the air (change of position). When the sparrow lays an egg, its body uses energy to create the calcium-based eggshell and color it with pigment (change in composition). The sparrow sitting on its nest transfers energy from its body to heat the developing chicks inside its eggs (change of temperature).

Energy comes in different forms

Energy manifests itself in different ways and can be converted from one form to another. Two major forms of energy that scientists commonly distinguish are **potential energy,** energy of position; and **kinetic energy,** energy of motion. Consider river water held behind a dam. By preventing water from moving downstream, the dam causes the water to accumulate potential energy. When the dam gates are opened, the potential energy is converted to kinetic energy as the water rushes downstream.

Energy conversions take place at the atomic level every time a chemical bond is broken or formed. Chemical energy is essentially potential energy stored in the bonds between atoms. Bonds differ in their amounts of chemical energy, depending on the atoms they hold together. Converting molecules with high-energy bonds (such as the carbon–carbon bonds of fossil fuels) into molecules with lower-energy bonds (such as the bonds in water or carbon dioxide) releases energy by changing potential energy into kinetic energy and produces motion, action, or heat. Just as automobile engines split the hydrocarbons of gasoline to release chemical energy and generate movement, our bodies split glucose molecules in our food for the same purpose (FIGURE 2.9).

Besides occurring as chemical energy, potential energy can occur as nuclear energy, the energy that holds atomic nuclei together. Nuclear power plants utilize this energy when they break apart the nuclei of large atoms within their reactors. Mechanical energy, such as the energy stored in a compressed spring, is yet another type of potential energy. Kinetic energy can also express itself in different forms, including thermal energy, light energy, sound energy, and electrical energy—all of which involve the movement of atoms, subatomic particles, molecules, or objects.

Energy is always conserved, but it changes in quality

Although energy can change from one form to another, it cannot be created or destroyed. Just as matter is conserved (p. 23), the total energy in the universe remains constant and thus is said to be conserved. Scientists refer to this principle as the **first law of thermodynamics.** The potential energy of the water behind a dam will equal the kinetic energy of its eventual movement downstream. Likewise, we obtain energy from the food we eat and then expend it in exercise, apply it toward maintaining our body and all its functions, or store it in fat. We do not somehow create additional energy or end up with less energy than the food gives us. Any particular system in nature can temporarily increase or decrease in energy, but the total amount in the universe always remains constant.

While the overall amount of energy is conserved in any conversion of energy, the **second law of thermodynamics** states that the nature of energy will change from a more-ordered state to a less-ordered state as long as no force counteracts this tendency. That is, systems tend to move toward increasing disorder, or *entropy*. For instance, a log of firewood—the highly organized and structurally complex product of many years of tree growth—transforms in the campfire to a residue of carbon ash, smoke, and gases such as carbon dioxide and water vapor, as well as the light and the heat of

FIGURE 2.9 Energy is released when potential energy is converted to kinetic energy. Potential energy stored in sugars (such as glucose) in the food we eat, combined with oxygen, becomes kinetic energy when we exercise, releasing carbon dioxide, water, and heat as by-products.

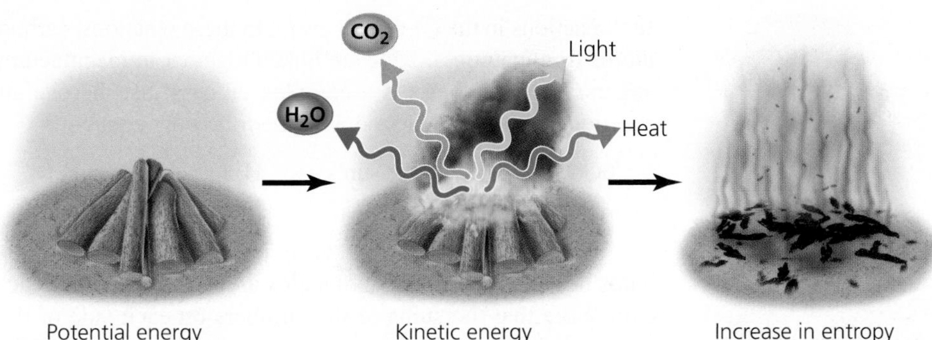

FIGURE 2.10 The burning of firewood demonstrates energy conversion from a more-ordered to a less-ordered state. This increase in entropy reflects the second law of thermodynamics.

the flame (FIGURE 2.10). With the help of oxygen, the complex biological polymers that make up the wood are converted into a disorganized assortment of rudimentary molecules and heat and light energy. When energy transforms from a more-ordered state to less-ordered state, it cannot accomplish tasks as efficiently. For example, the level of energy available in ash (a less-ordered state of wood) is far lower than that available in a log of firewood (the more-ordered state of wood).

If the second law of thermodynamics specifies that systems tend to move toward disorder, then how does any system ever hold together? The order of an object or system can be increased by the input of energy from outside the system. Living organisms, for example, maintain their highly ordered structure and function by consuming energy. When they die and these inputs of energy cease, the organisms undergo decomposition and attain a less-ordered state.

Some energy sources are easier to harness than others

The nature of an energy source helps determine how easily people can harness it. Sources such as fossil fuels and the electricity we produce in power plants contain concentrated energy that we can readily release. It is relatively easy for us to gain large amounts of energy efficiently from such sources. In contrast, sunlight and the heat stored in ocean water are more diffuse. Each day the world's oceans absorb heat energy from the sun equivalent to that of 250 billion barrels of oil—more than 3000 times as much as our global society uses in a year. However, because this energy is spread across such vast areas, it is difficult to harness efficiently.

In each attempt we make to harness energy, some portion escapes. We can express our degree of success in capturing energy in terms of the **energy conversion efficiency,** the ratio of the useful output of energy to the amount we need to input. When we burn gasoline in an automobile engine, only about 16% of the energy released is used to power the automobile, and the rest of the energy is converted to heat and escapes without being used (p. 508). Incandescent light bulbs are even less efficient; only 5% of their energy is converted to light, while the remainder escapes as heat.

Light energy from the sun powers most living systems

The energy that powers Earth's biological systems comes primarily from the sun. The sun releases radiation from large portions of the electromagnetic spectrum, although our atmosphere filters much of this out and we see only some of this radiation as visible light (FIGURE 2.11). Most of the sun's energy is reflected, or else absorbed and re-emitted, by the atmosphere, land, or water (p. 485). Solar energy drives winds, ocean currents, weather, and climate patterns. A small amount (less than 1% of the total) powers plant growth, and a still smaller amount flows from plants into the organisms that eat them and the organisms that decompose dead organic matter. A minuscule percentage of this energy is eventually deposited below ground in the chemical bonds in fossil fuels (because fossil fuels are derived from ancient plants; pp. 524–526).

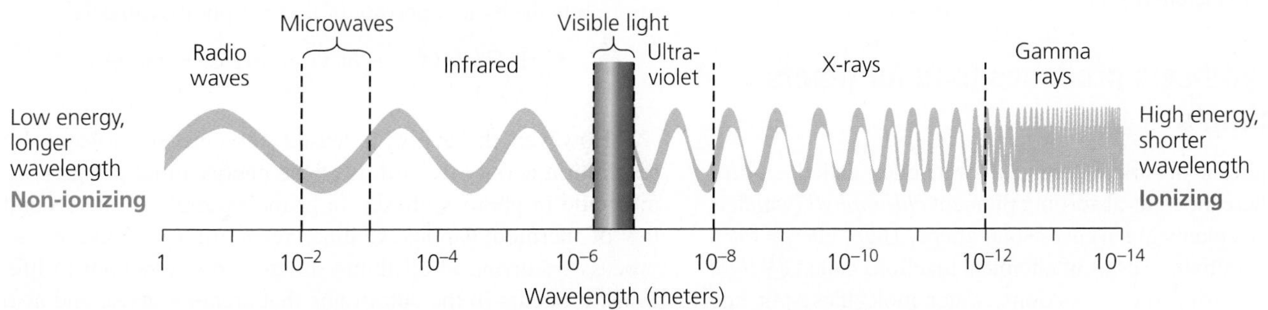

FIGURE 2.11 The sun emits radiation from many portions of the electromagnetic spectrum. Visible light makes up only a small proportion of this energy. Some radiation that reaches our planet is reflected back; some is absorbed by air, land, and water; and a small amount powers photosynthesis.

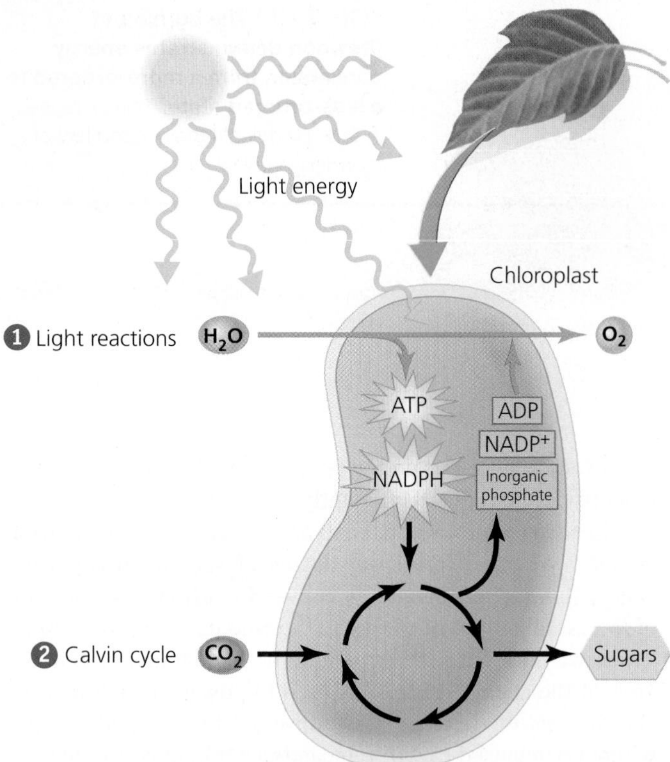

FIGURE 2.12 In photosynthesis, autotrophs such as plants, algae, and cyanobacteria use sunlight to convert water and carbon dioxide into oxygen and sugar. In the light reactions, water is converted to oxygen in the presence of sunlight, creating high-energy molecules (ATP and NADPH). These molecules help drive reactions in the Calvin cycle, in which carbon dioxide is used to produce sugars. Molecules of ADP, NADP+, and inorganic phosphate created in the Calvin cycle, in turn, help power the light reactions, creating an endless loop.

Some organisms use the sun's radiation directly to produce their own food. Such organisms, called **autotrophs** or **primary producers,** include green plants, algae, and cyanobacteria (a type of bacteria named for their cyan, or blue-green, color). Autotrophs turn light energy from the sun into chemical energy in a process called **photosynthesis** (**FIGURE 2.12**). In photosynthesis, sunlight powers a series of chemical reactions that convert carbon dioxide and water into sugars, transforming diffuse energy from the sun into concentrated energy the organism can use. Photosynthesis is an example of moving toward a state of lower entropy, and as such, it requires a substantial input of outside energy.

Photosynthesis produces food for plants and animals

Photosynthesis occurs within cell organelles called *chloroplasts*, where the light-absorbing pigment *chlorophyll* (which is what makes plants green) uses solar energy (light energy from the sun) to initiate a series of chemical reactions called ❶ *light reactions*. During these reactions, water molecules split and react to form hydrogen ions (H^+) and molecular oxygen (O_2), thus creating the oxygen that we breathe. The light reactions also produce small, high-energy molecules that are used to fuel reactions in the ❷ *Calvin cycle*. In these reactions, carbon atoms from carbon dioxide are linked together to manufacture sugars. Photosynthesis is a complex process, but the overall reaction can be summarized with the following equation:

$$6CO_2 + 12H_2O + \underset{\text{energy}}{\text{the sun's}} \rightarrow \underset{\text{(sugar)}}{C_6H_{12}O_6} + 6O_2 + 6H_2O$$

The number preceding each molecular formula indicates how many of those molecules are involved in the reaction. Note that the sums of the numbers on each side of the equation for each element are equal; that is, there are 6 C, 24 H, and 24 O atoms on each side. This illustrates how chemical equations are balanced, with each atom recycled and matter conserved. No atoms are lost; they are simply rearranged among molecules. Note also that water appears on both sides of the equation. The reason is that for every 12 water molecules that are input and split in the process, 6 water molecules are newly created. We can streamline the photosynthesis equation by showing only the net loss of 6 water molecules:

$$6CO_2 + 6H_2O + \text{the sun's energy} \rightarrow \underset{\text{(sugar)}}{C_6H_{12}O_6} + 6O_2$$

Thus, in photosynthesis, water, carbon dioxide, and light energy from the sun are transformed to produce sugar (glucose) and oxygen. To accomplish this, green plants draw up water from the ground through their roots, absorb carbon dioxide from the air through their leaves, and harness sunlight. With these ingredients, they create sugars for their growth and maintenance, and they release oxygen as a by-product.

Animals, in turn, depend on the sugars and oxygen from photosynthesis. Animals survive by eating plants, or by eating animals that have eaten plants, and by taking in oxygen. In fact, it is thought that animals appeared on Earth's surface only after the planet's atmosphere had been supplied with oxygen by cyanobacteria, the earliest autotrophs.

Cellular respiration releases chemical energy

Organisms make use of the chemical energy created by photosynthesis in a process called **cellular respiration,** which is vital to life. To release the chemical energy of glucose, cells employ oxygen to convert glucose back into its original starting materials, water and carbon dioxide. The energy released during this process is used to form chemical bonds or to perform other tasks within cells. The net equation for cellular respiration is the exact opposite of that for photosynthesis:

$$\underset{\text{(sugar)}}{C_6H_{12}O_6} + 6O_2 \rightarrow 6CO_2 + 6H_2O + \text{energy}$$

However, the energy released per glucose molecule in respiration is only two-thirds of the energy input per glucose molecule in photosynthesis—a prime example of the second law of thermodynamics. Cellular respiration is a continuous process occurring in all living things and is essential to life. Thus, it occurs in the autotrophs that create glucose and also in **heterotrophs,** organisms that gain their energy by feeding on other organisms. Heterotrophs include most animals, as well as the fungi and microbes that decompose organic matter.

FIGURE 2.13 **Geyser in the Black Rock Desert of Nevada.** Geysers propel scalding water into the air, powered by geothermal energy from deep below ground. The bright colors of the rocks are from colonies of bacteria that thrive in the hot, mineral-laden water.

FIGURE 2.14 **Life thrives in darkness around hydrothermal vents on the ocean floor.** Vents send spouts of hot, mineral-rich water into the cold blackness of the deep sea. Specialized biological communities thrive in these unusual conditions, where organisms, such as the giant tubeworms is this image, survive thanks to bacteria that produce food from hydrogen sulfide through the process of chemosynthesis.

Geothermal energy also powers Earth's systems

Although the sun is life's primary energy source, it is not the only one for our planet. A minor additional energy source is the gravitational pull of the moon, which in conjunction with the sun's pull causes ocean tides (p. 427). Another significant energy source is geothermal heating emanating from inside Earth, powered primarily by radioactivity (p. 24). Radiation from radioisotopes deep inside our planet heats the inner Earth, and this heat gradually makes its way to the surface. There it heats magma that erupts from volcanoes, drives plate tectonics (p. 34), and warms water, which in some locations shoots out of the ground in the form of geysers (FIGURE 2.13). In places such as "The Geysers" region of northern California, this geothermal energy can be harnessed to produce electricity.

Although we harness geothermal energy for our own use today, geothermal energy was powering biological communities long before people appeared on Earth. On the ocean floor, jets of geothermally heated water—essentially underwater geysers—gush into the icy-cold depths. In one of the more amazing scientific discoveries of recent decades, scientists realized that these **hydrothermal vents** can host entire communities of specialized organisms that thrive in the extreme high-temperature, high-pressure conditions. Gigantic clams, immense tubeworms, and odd mussels, shrimps, crabs, and fish all flourish in the seemingly hostile environment of near-scalding water that shoots out of tall chimneys of encrusted minerals (FIGURE 2.14).

These locations are so deep underwater that they completely lack sunlight, so the energy flow of these communities cannot be fueled through photosynthesis. Instead, bacteria in deep-sea vents use the chemical-bond energy of hydrogen sulfide (H_2S) to transform inorganic carbon into organic carbon compounds in a process called **chemosynthesis:**

$$6CO_2 + 6H_2O + 3H_2S \rightarrow \underset{\text{(sugar)}}{C_6H_{12}O_6} + 3H_2SO_4$$

Chemosynthesis occurs in various ways, but note how this particular reaction for chemosynthesis closely resembles the photosynthesis reaction. These two processes use different energy sources, but each uses water and carbon dioxide to produce sugar and a by-product, and each produces potential energy that is later released during respiration. Energy from chemosynthesis passes through the deep-sea-vent animal community as heterotrophs such as clams, mussels, and shrimp gain nutrition from chemoautotrophic bacteria. Hydrothermal vent communities excited scientists because they were novel and unexpected, and they showed just how much we still have to learn about the workings of our planet.

Geology: The Physical Basis for Environmental Science

If we want to understand how our planet functions, a good way to start is to examine the rocks, soil, and sediments beneath our feet. The physical processes that take place at and below Earth's surface shape the landscape around us and lay the foundation for most environmental systems and for life.

Understanding the physical nature of our planet also benefits our society, for without the study of Earth's rocks and the processes that shape them, we would have no copper, iron, or steel for our industries, products, and technologies; no energy from fossil fuels; no uranium for nuclear power plants; and no geothermal power generation. These are just a few examples of how we draw on resources and processes from beneath the surface of our planet and put them to use in our everyday lives.

Our planet is dynamic, and this dynamism is what motivates **geology,** the study of Earth's physical features, processes, and history. A human lifetime is just a blink of an eye in the long course of geologic time, and the Earth we experience is merely a snapshot in our changing planet's long history. We can begin to grasp this long-term dynamism as we consider two processes of fundamental importance—plate tectonics and the rock cycle. After examining the geologic processes that shape our planet, we will close our chapter by exploring the catastrophic hazards we sometimes face from these processes.

Earth consists of layers

Most geologic processes take place near Earth's surface, but our planet consists of multiple layers (**FIGURE 2.15**). At Earth's center is a dense **core** consisting mostly of iron, solid in the inner core and molten in the outer core. Surrounding the core is a thick layer of less dense, elastic rock called the **mantle.** A portion of the upper mantle called the **asthenosphere** contains especially soft rock, melted in some areas. The harder rock above the asthenosphere is what we know as the **lithosphere.** The lithosphere includes both the uppermost mantle and the entirety of Earth's third major layer, the **crust,** the thin, brittle, low-density layer of rock that covers Earth's surface. The intense heat in the inner Earth rises from core to mantle to crust, and it eventually dissipates at the surface. In regions where the asthenosphere's molten rock approaches to within a few miles of the surface, we can harness geothermal energy by drilling boreholes into the crust (see Figure 21.17, p. 598).

In fact, we don't even need to drill deep or find geysers to take advantage of geothermal energy. Anywhere in the world, the soil and rock just below the surface are fairly constant in temperature, cooler than the air above the surface in summer and warmer than the air in winter. Because of this, you can use a geothermal heat pump (also called a ground-source heat pump) to heat and cool a home (Figure 21.19, p. 600). Over 600,000 U.S. homes now use these highly efficient systems.

The heat from the inner layers of Earth also drives convection currents that flow in loops in the mantle, pushing the mantle's soft rock cyclically upward (as it warms) and downward (as it cools), like a gigantic conveyor belt system. As the mantle material moves, it drags large plates of lithosphere along its surface. This movement of lithospheric plates is known as **plate tectonics,** a process of extraordinary importance to our planet.

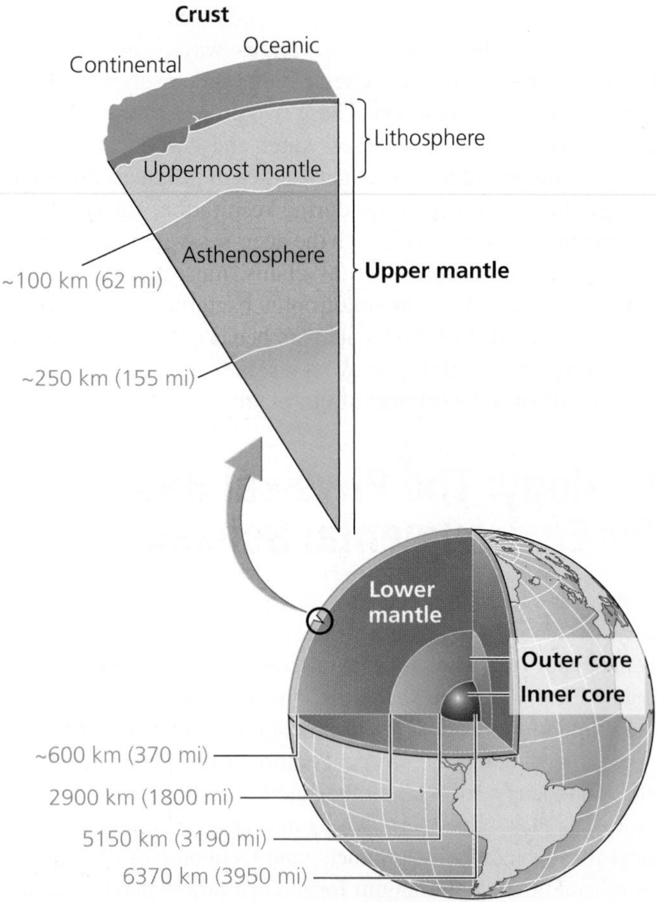

FIGURE 2.15 Earth's three primary layers—core, mantle, and crust—are themselves layered. The inner core of solid iron is surrounded by an outer core of molten iron, and the rocky mantle includes the molten asthenosphere near its upper edge. At Earth's surface, dense and thin oceanic crust abuts lighter, thicker continental crust. The lithosphere consists of the crust and the uppermost mantle above the asthenosphere.

Plate tectonics shapes Earth's geography

Our planet's surface consists of about 15 major tectonic plates, which fit together like pieces of a jigsaw puzzle (**FIGURE 2.16**). Imagine peeling an orange and then placing the pieces of peel back onto the fruit; the ragged pieces of peel are like the lithospheric plates riding atop Earth's surface. However, the plates are thinner relative to the planet's size, more like the skin of an apple. These plates move at rates of roughly 2–15 cm (1–6 in.) per year. This slow movement has influenced Earth's climate and life's evolution throughout our planet's history as the continents combined, separated, and recombined in various configurations. By studying ancient rock formations throughout the world, geologists have determined that at least twice, all landmasses were joined together in a "supercontinent." Scientists have dubbed the one that occurred about 225 million years ago *Pangaea* (see Figure 2.16).

There are three types of plate boundaries

The processes that occur at the boundaries between plates have major consequences. We can categorize these boundaries into three types: divergent, transform, and convergent plate boundaries.

At **divergent plate boundaries,** tectonic plates push apart from one another as **magma** (rock heated to a molten, liquid state) rises upward to the surface, creating new lithosphere as it cools (**FIGURE 2.17a**). An example is the Mid-Atlantic Ridge, part of a 74,000-km (46,000-mi) system of divergent plate boundaries slicing across the floors of the world's oceans.

Where two plates meet, they may slip and grind alongside one another, forming a **transform plate boundary** (**FIGURE 2.17b**). This movement creates friction that generates earthquakes (p. 40) along strike-slip faults. *Faults* are fractures in Earth's crust, and at strike-slip faults, each landmass moves horizontally in opposite directions. The Pacific Plate and the North American Plate, for example, slide past one another along California's San Andreas Fault. Southern California is slowly inching its way northward along this fault, and

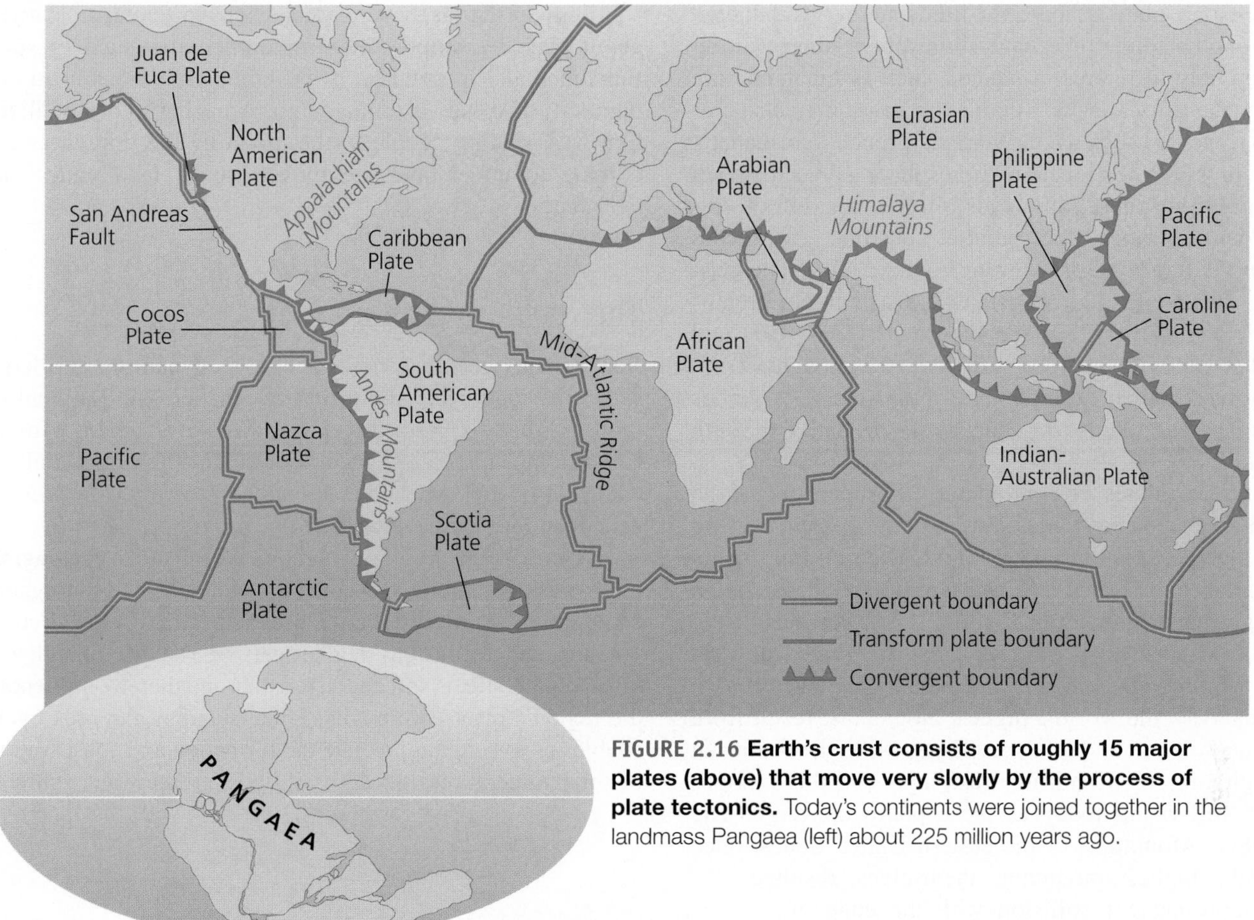

FIGURE 2.16 Earth's crust consists of roughly 15 major plates (above) that move very slowly by the process of plate tectonics. Today's continents were joined together in the landmass Pangaea (left) about 225 million years ago.

the site of Los Angeles will eventually reach that of modern-day San Francisco.

Convergent plate boundaries, where two plates converge or come together, can give rise to different outcomes (FIGURE 2.17c). As plates of newly formed lithosphere push outward from divergent plate boundaries, this oceanic lithosphere gradually cools, becoming denser. Eventually (after millions of years), it becomes denser than the asthenosphere beneath it and dives downward into the asthenosphere in a process called **subduction.** As the lithospheric plate descends, it slides beneath a neighboring plate that is less dense, forming a convergent plate boundary. The subducted plate is heated and pressurized as it sinks, and water vapor escapes, helping to melt rock (by lowering its melting temperature). The molten rock rises, and this magma may erupt through the surface at volcanoes (pp. 40–42).

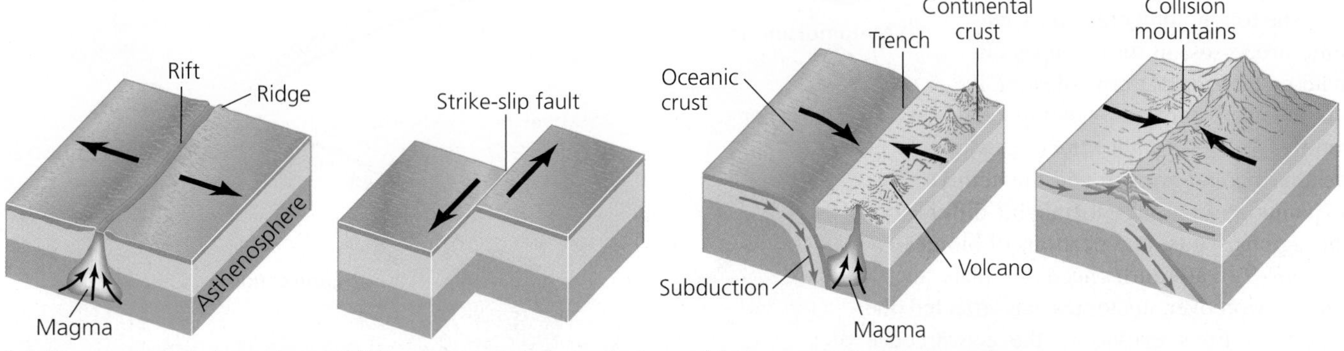

FIGURE 2.17 There are three types of boundaries between tectonic plates, generating different geologic processes. At a divergent plate boundary, such as a mid-ocean ridge on the seafloor **(a)**, the two plates move gradually away from the boundary in the manner of conveyor belts, and magma from beneath the crust may extrude as lava. At a transform plate boundary **(b)**, two plates slide alongside one another, creating friction that leads to earthquakes. Where plates collide at a convergent plate boundary **(c)**, one plate is subducted beneath another, leading to either volcanism or the formation of mountain ranges.

When one plate of oceanic lithosphere is subducted beneath another plate of oceanic lithosphere, the resulting volcanism may form arcs of islands, such as Japan and the Aleutian Islands of Alaska. Subduction zones may also create deep trenches, such as the Mariana Trench, our planet's deepest abyss. When oceanic lithosphere slides beneath continental lithosphere, this leads to the formation of volcanic mountain ranges that parallel coastlines (a process shown in the left-hand drawing of Figure 2.17c). The Cascades in the Pacific Northwest, where Mount St. Helens erupted violently in 1980 (see Figure 17.11, p. 457) and renewed its activity in 2004, are fueled by magma from subduction. Another example is South America's Andes Mountains, where the Nazca Plate slides beneath the South American Plate.

When two plates of continental lithosphere meet, the continental crust on both sides resists subduction and instead crushes together, bending, buckling, and deforming layers of rock from both plates in a **continental collision** (shown in the right-hand drawing of Figure 2.17c). Portions of the accumulating masses of buckled crust are forced upward as they are pressed together, and mountain ranges result. The Himalayas, the world's highest mountains, result from the Indian-Australian Plate's collision with the Eurasian Plate beginning 40–50 million years ago, and these mountains are still rising today as these plates converge. The Appalachian Mountains of the eastern United States, once the world's highest mountains themselves, resulted from a more-ancient collision with the edge of what is today Africa.

Tectonics produces Earth's landforms

In these ways, the processes of plate tectonics create the landforms around us. Tectonic movements build mountains; shape the geography of oceans, islands, and continents; and give rise to earthquakes and volcanoes.

The topography created by tectonic processes, in turn, shapes climate by altering patterns of rainfall, wind, ocean currents, heating, and cooling—all of which affect rates of weathering and erosion and the ability of plants and animals to inhabit different regions. Thus, the locations of biomes (pp. 93–99) are influenced by plate tectonics. Moreover, tectonics has affected the history of life's evolution; the convergence of landmasses into supercontinents such as Pangaea is thought to have contributed to widespread extinctions by reducing the area of species-rich coastal regions and by creating an arid continental interior with extreme temperature swings.

Only in the last several decades have scientists learned about plate tectonics—this environmental system of such fundamental importance was completely unknown to humanity just half a century ago. Amazingly, our civilization was sending people to the moon by the time we were coming to understand the movement of land under our very feet.

The rock cycle alters rock

Just as plate tectonics shows geology's dynamism at a large scale, the rock cycle shows it at a smaller one. We tend to think of rock as pretty solid stuff. Yet over geologic time, rocks and the minerals that comprise them are heated, melted, cooled, broken down, and reassembled in a very slow process called the **rock cycle** (**FIGURE 2.18**).

A **rock** is any solid aggregation of minerals. A **mineral,** in turn, is any naturally occurring solid element or inorganic compound with a crystal structure, a specific chemical composition, and distinct physical properties. The type of rock in a given region affects soil characteristics and thereby influences the region's plant community. Understanding the rock cycle enables us to better appreciate the formation and conservation of soils, mineral resources, fossil fuels, groundwater sources,

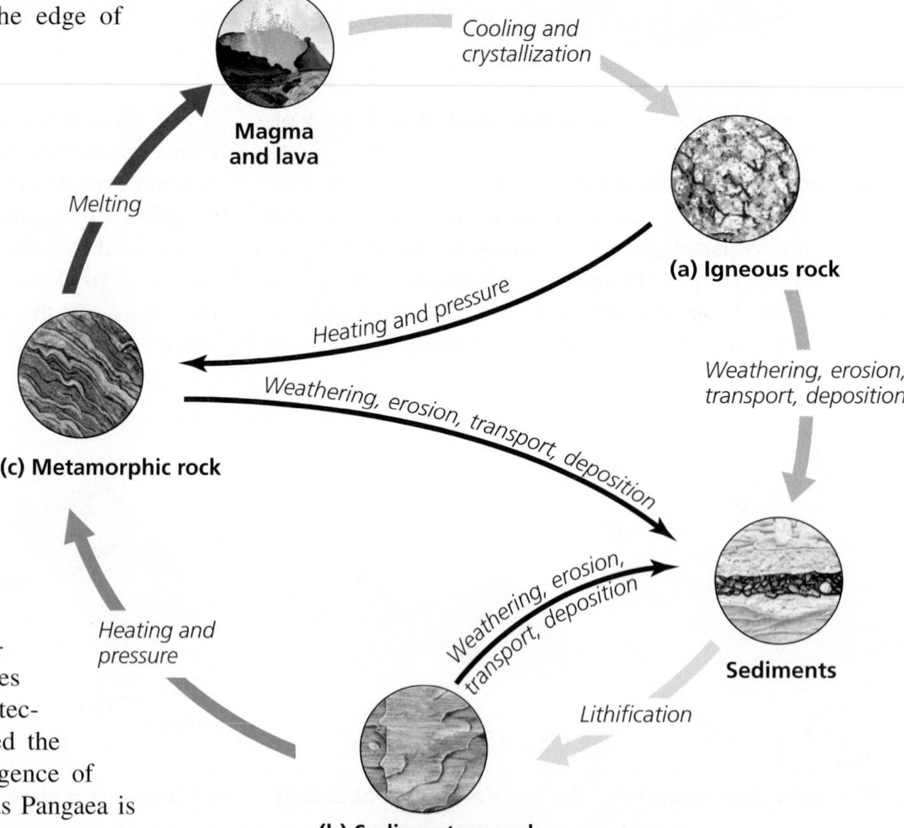

FIGURE 2.18 The rock cycle. Igneous rock **(a)** is formed when rock melts and the resulting magma or lava then cools. Sedimentary rock **(b)** is formed when rock is weathered and eroded and the resulting sediments are compressed to form new rock. Metamorphic rock **(c)** is formed when rock is subjected to intense heat and pressure underground. Through these processes, each type of rock can be converted into either of the other two types.

(a) Intrusive igneous rock: Granite at Yosemite National Park

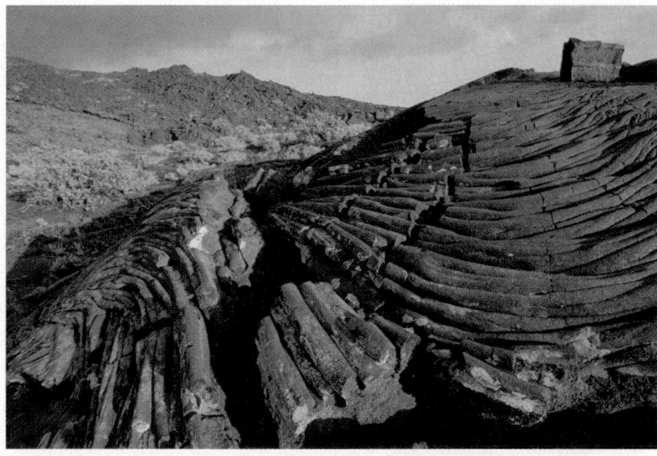

(b) Extrusive igneous rock: Basalt in the Canary Islands

(c) Sedimentary rock: Sandstone in Arizona

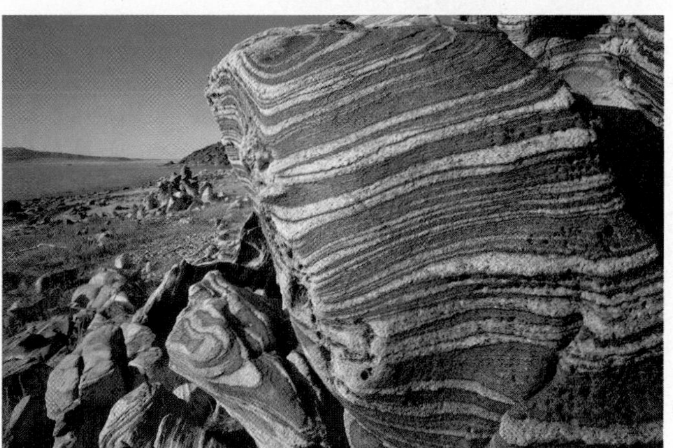

(d) Metamorphic rock: Gneiss in Utah

FIGURE 2.19 Examples of rock types. The towering rock formations of Yosemite National Park are made of granite **(a)**, a type of intrusive igneous rock. The lava flows on the Canary Islands form basalt **(b)**, a type of extrusive igneous rock. The layered formation in Paria Canyon, Arizona, is an example of sandstone **(c)**, a type of sedimentary rock. Gneiss (pronounced "nice") **(d)**, at Antelope Island, Utah, is a type of metamorphic rock.

geothermal energy sources, and other natural resources—all of which we discuss in later chapters.

Igneous rock All rocks can melt. At high enough temperatures, rock will enter the molten, liquid state called magma. If magma is released through the lithosphere (as in a volcanic eruption), it may flow or spatter across Earth's surface as **lava.** Rock that forms when magma or lava cools is called **igneous rock** (from the Latin *ignis*, meaning "fire") (FIGURE 2.18a).

Igneous rock comes in two main classes because magma can solidify in different ways. When magma cools slowly and solidifies while it is below Earth's surface, it forms intrusive igneous rock. This process created the famous rock formations at Yosemite National Park (FIGURE 2.19a). Granite is the best-known type of intrusive rock. A slow cooling process allows minerals of different types to aggregate into large crystals, giving granite its multicolored, coarse-grained appearance. In contrast, when molten rock is ejected from a volcano, it cools quickly, so minerals have little time to grow into coarse crystals. This class of igneous rock is called extrusive igneous rock, and its most common representative is basalt, the principal rock type of the Japanese islands (FIGURE 2.19b).

Sedimentary rock All exposed rock weathers away with time. The relentless forces of wind, water, freezing, and thawing eat away at rocks, stripping off one tiny grain (or large chunk) after another. Through weathering (p. 219) and erosion (pp. 222–223), particles of rock blown by wind or washed away by water come to rest downhill, downstream, or downwind from their sources, eventually forming **sediments.** Alternatively, some sediments form chemically from the precipitation of substances out of solution.

Sediment layers accumulate over time, causing the weight and pressure of overlying layers to increase. **Sedimentary rock** (FIGURE 2.18b) is formed as sediments are physically pressed together (*compaction*) and as dissolved

Have We Brought on a New Geologic Epoch?

Geologist Jan Zalasiewicz, University of Leicester

Have the impacts of human beings on Earth been so profound as to warrant creating a new time period in the geologic record?

The geologic timescale (**APPENDIX E**) shows the full span of Earth's history—all 4.5 *billion* years of it—and focuses in on the most recent 543 million years. Geologists have subdivided Earth's history into 3 eras and 11 periods. The most recent period, the Quaternary period, occupies a thin slice of time at the top of the scale because this period began "only" 1.8 million years ago.

Geologists divide this immensely long timescale using evidence from *stratigraphy*, the study of *strata*, or layers, of sedimentary rock. (As we have seen, sedimentary rock is laid down in chronological sequence, so by studying it researchers can infer how conditions changed through time.) Where scientists find fossil evidence for major and sudden changes in the physical, chemical, or biological conditions present on Earth between one set of layers and the next, they create a boundary between geologic time periods. For instance, fossil evidence for mass extinctions (pp. 59–61, 281–283) determines several boundaries, such as that between the Permian and Triassic periods.

Traditionally, time periods are defined and named based on changes actually seen in the geologic record. So when a group of geologists suggested naming the current portion of our planet's history the "Anthropocene" (from the Greek word *anthropos*, meaning "human"), scientists sat up and took notice.

We live in the Holocene epoch, the most recent slice of the Quaternary period. The Holocene epoch began about 11,500 years ago with a warming trend that melted glaciers and brought Earth out of its most recent ice age. Since then, Earth's climate has been remarkably constant, and this constancy provided our species with the long-term stability we needed to develop agriculture and civilization. However, since the industrial revolution (p. 4), human activity has had major impacts on Earth's basic processes. The question for geologists is: Have those effects been strong enough to warrant naming a new geologic epoch after ourselves?

The idea began to catch on after being proposed by Nobel Prize–winning chemist Paul Crutzen (p. 470) in 2000. In 2008, a team of 21 scientists from the Stratigraphy Commission of the Geological Society of London, led by geologist Jan Zalasiewicz of the University of Leicester (U.K.), formally advocated the proposal in a paper in *GSA Today*, a journal of the Geological Society of America (GSA).

In their GSA paper, the British geologists reviewed a broad sweep of published scientific evidence and advanced several reasons to rename the past 200 or so years the *Anthropocene* (**FIGURE 1**).

First, humans have caused a sharp increase in erosion worldwide. By clearing forests and raising crops, we have sent immense amounts of soil downwind and downstream from continents into oceans. This rapid deposition of sediment in the oceans will be noticeable in the stratigraphic record far into the future as today's sediments become compacted into tomorrow's sedimentary rock layers.

Second, our species has rapidly altered the composition of the atmosphere by emitting greenhouse gases as a result of deforestation, agriculture, and especially our combustion of coal, oil, and natural gas. These releases have already brought carbon dioxide and methane to their highest levels in at least 800,000 years (pp. 486–487).

minerals seep through sediments and act as a kind of glue, binding sediment particles together (*cementation*). The formation of rock through these processes of compaction and cementation is termed *lithification*. Examples of sedimentary rock include sandstone, made of cemented sand particles; shale, comprised of still smaller mud particles; and limestone, formed as dissolved calcite precipitates from water or as calcite from marine organisms settles to the bottom.

These processes also create the fossils of organisms (p. 55) and the fossil fuels we use for energy. Because sedimentary layers pile up in chronological order (**FIGURE 2.19c**), geologists and paleontologists can assign relative dates to fossils they find in sedimentary rock and thereby make inferences about Earth's history (see **THE SCIENCE BEHIND THE STORY** above).

Metamorphic rock Geologic forces may bend, uplift, compress, or stretch rock. When any type of rock is subjected to great heat or pressure, it may alter its form, becoming **metamorphic rock** (from the Greek for "changed form") (**FIGURE 2.18c**). The forces that metamorphose rock generally occur deep underground, at temperatures lower than the rock's melting point, but high enough to change its appearance and physical properties. Metamorphic rock (**FIGURE 2.19d**) includes rock such as slate, formed when shale is subjected to heat and pressure, and marble, formed when limestone is heated and pressurized.

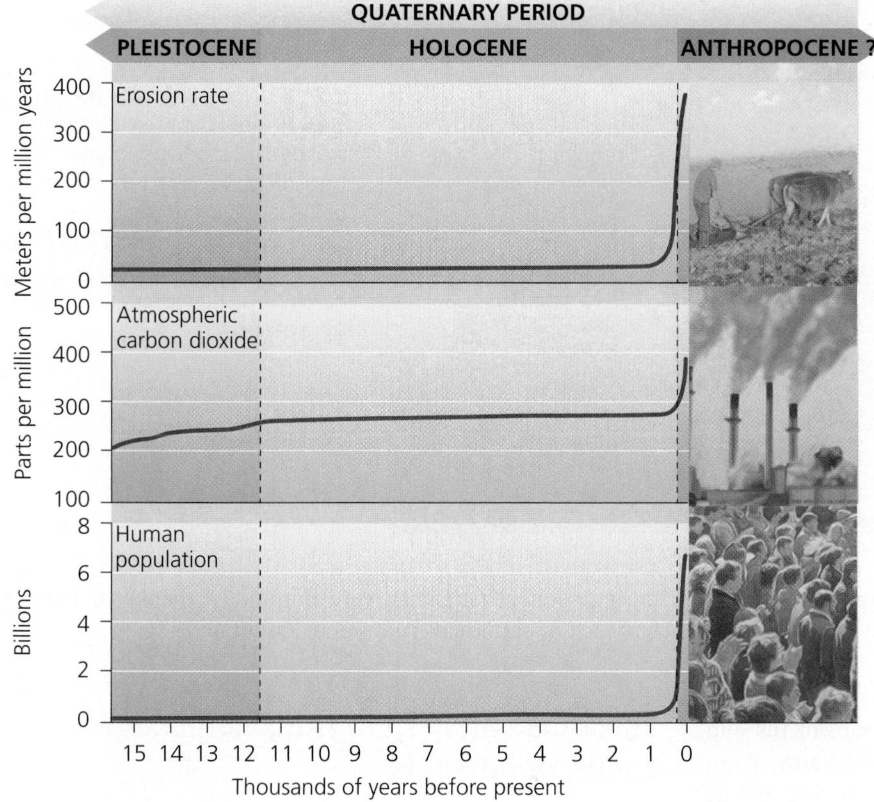

FIGURE 1 Global soil erosion rates (top) and atmospheric carbon dioxide concentrations (middle) have increased sharply in just the last few hundred years, along with human population (bottom). These patterns have persuaded some geologists that we should recognize a new epoch in Earth history and call it the Anthropocene. *Adapted from Zalasiewicz, J., et al., 2008. Are we now living in the Anthropocene?* GSA Today 18*(2), 4–8, Figure 1.*

As greenhouse gas concentrations rise, so does temperature. Earth's temperature has risen 0.7 °C (1.3 °F) in the past century and is predicted to rise by 1.8–4.0 °C (3.2–7.2 °F) in the current century (p. 496). Rising temperatures are melting polar ice. The influx of meltwater into the oceans, combined with the fact that warmed water expands in volume, means that sea level is rising. Global sea level rose 17 cm (6.7 in.) in the last century and will likely rise by much more in this century (pp. 498–501). Moreover, increasing atmospheric carbon dioxide acidifies ocean water (pp. 428–429, 488), which kills off coral reefs (and will even likely dissolve some of the geologic strata being formed).

All these changes—along with the pollution and habitat disturbance we are inflicting on Earth's biotic communities—are causing extinctions of animals and plants (pp. 281–283). If we step back to view our impacts in deep geologic time, it becomes clear that we are setting in motion a rapid new mass extinction event.

Finally, our recent explosion in population has intensified our impact—changes are happening in the blink of an eye, geologically speaking. The entire period of our society's existence so far could end up represented in as little as a millimeter of rock in the future, so the changes we are bringing about now would appear sudden to a geologist of the far future.

Many scientists resist the idea of renaming our current epoch because they question whether the proposal holds true to the tradition of defining time periods based on changes actually seen in the geologic record. But supporters of the idea maintain that today's changes will be readily visible in future stratigraphy and that we should recognize this unprecedented time of rapid change in Earth's history by designating it the Anthropocene. ■

FAQ Technically speaking, is coal a rock?

Coal is a geologic resource of great importance in environmental science, but many people are unsure of what coal is or where it comes from. Coal is hard and is found underground, like many types of rocks, so some people would classify it as a rock. Others, however, argue coal is not a rock because it is the modified remains of organic material, mostly woody plants, from hundreds of millions of years ago.

According to geologists, coal is a rock because it is a solid aggregation of minerals, particularly carbon. Carbon is classified as a mineral because it is a naturally occurring solid element with a crystalline structure. Other rocks, such as some types of limestone, also contain the remains of ancient organisms (such as the shells of marine organisms) and are categorized with coal as organic sedimentary rocks.

Geologic and Natural Hazards

Plate tectonics gives rise to creative forces that shape our planet—yet some of the consequences of tectonic movement can also pose hazards to us. Earthquakes and volcanoes are examples of geologic hazards. We can see how such hazards relate to tectonic processes by examining a map of the circum-Pacific belt, or "ring of fire" (**FIGURE 2.20**). Nine out of 10 earthquakes and over half the world's volcanoes occur along this 40,000-km (25,000-mi) arc of subduction zones and fault systems. Like many locations along the circum-Pacific belt, Japan has experienced earthquakes and volcanism frequently in its past. As a result, it had put precautions in place for such dangers.

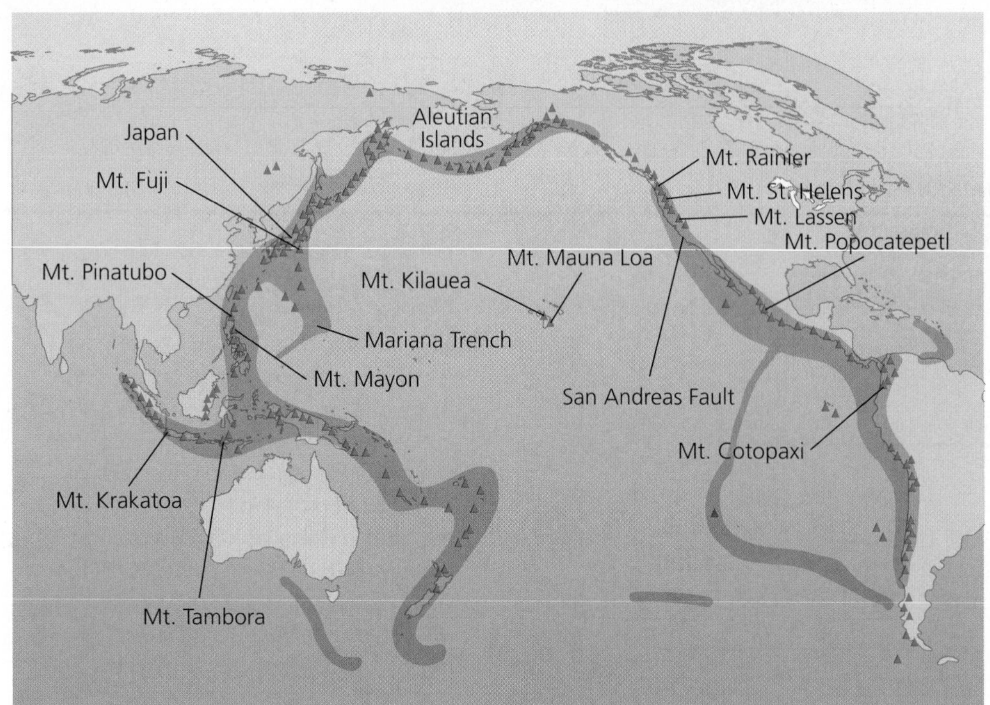

FIGURE 2.20 Most of our planet's volcanoes and earthquakes occur along the circum-Pacific belt, or "ring of fire." In this map, red symbols indicate major volcanoes, and gray-shaded areas indicate areas of greatest earthquake risk.

DATA Q What similarities do you note between the "ring of fire" around the edges of the Pacific Ocean and the boundaries of the tectonic plates shown in Figure 2.16? Which of type of plate boundary (see Figure 2.17) is most common along the length of the "ring of fire"?

Earthquakes result from movement at plate boundaries and faults

Along tectonic plate boundaries, and in other places where faults occur, Earth may relieve built-up pressure in fits and starts. Each release of energy causes what we know as an **earthquake.** Most earthquakes are barely perceptible, but as shown by the Tohoku quake of 2011, they are occasionally powerful enough to cause significant losses of human life and property (TABLE 2.2). Damage is generally greatest where soils are loose or saturated with water—areas of cities built atop landfill are particularly susceptible. For instance, during the 1989 Loma Prieta earthquake, which shook San Francisco and Oakland while their baseball teams were playing in the World Series, one of the areas hardest hit was San Francisco's Marina District, a neighborhood built atop soil and debris dumped into the bay, including rubble from the city's 1906 earthquake.

To minimize damage from earthquakes, engineers have developed ways to protect buildings from shaking. They do this by strengthening structural components while also designing points at which a structure can move and sway harmlessly with ground motion. Just as a flexible tree trunk bends in a storm while a brittle one breaks, buildings with built-in flexibility are more likely to withstand an earthquake's violent shaking. Such designs are an important part of new building codes in California, Japan, and other quake-prone regions, and many older structures are being retrofitted to meet these codes.

Such designs are more expensive to build than conventional designs, so many buildings in poorer nations do not have such protections. One example is Haiti, where a 7.0 magnitude earthquake in 2010 devastated huge portions of the capital city of Port-au-Prince and claimed an estimated 230,000 lives (FIGURE 2.21). While the Tohoku earthquake released over 950 times more energy than the earthquake that struck Haiti, mortality and property damage from the Tohoku quake (not including the damage and loss of life caused by the subsequent tsunami) were minimized thanks to Japan's embrace of earthquake-conscious building designs.

Volcanoes arise from rifts, subduction zones, or hotspots

Where molten rock, hot gas, or ash erupts through Earth's surface, a **volcano** is formed, often creating a mountain over time as cooled lava accumulates. As we have seen, lava can extrude

TABLE 2.2 Examples of Large or Recent Earthquakes

Year	Location	Fatalities	Magnitude[1]
1556	Shaanxi Province, China	830,000	~8
1755	Lisbon, Portugal	70,000[2]	8.7
1906	San Francisco, California	3,000	7.8
1923	Kwanto, Japan	143,000	7.9
1964	Anchorage, Alaska	128[2]	9.2
1976	Tangshan, China	255,000+	7.5
1985	Michoacan, Mexico	9,500	8.0
1989	Loma Prieta, California	63	6.9
1994	Northridge, California	60	6.7
1995	Kobe, Japan	5,502	6.9
2004	Northern Sumatra	228,000[2]	9.1
2005	Kashmir, Pakistan	86,000	7.6
2008	Sichuan Province, China	50,000+	7.9
2010	Port-au-Prince, Haiti	236,000	7.0
2010	Maule, Chile	500	8.8
2011	Northern Japan	19,000[2]	9.0

[1]*Measured by moment magnitude; each full unit is roughly 32 times as powerful as the preceding full unit.*

[2]*Includes deaths from the resulting tsunami.*

FIGURE 2.21 **The 2010 earthquake in Haiti devastated the capital city of Port-au-Prince and killed an estimated 230,000 people.** One reason for the extensive loss of life was that many of Haiti's buildings were not constructed to withstand earthquakes.

along mid-ocean ridges or over subduction zones as one tectonic plate dives beneath another. Due to its position along subduction zones, Japan has over 100 active volcanoes, which is 10% of the world total and more than any other nation. Mount Fuji, one of Japan's most prominent and recognizable natural features, is one such active volcano.

Lava may also be emitted at *hotspots*, localized areas where plugs of molten rock from the mantle erupt through the crust. As a tectonic plate moves across a hotspot, repeated eruptions from this source may create a linear series of volcanoes. The Hawaiian Islands provide an example of this process (FIGURE 2.22a).

At some volcanoes, lava flows slowly downhill, such as at Mount Kilauea in Hawaii (FIGURE 2.22b), which has been erupting continuously since 1983! At other times, a volcano may let loose large amounts of ash and cinder in a sudden explosion, such as during the 1980 eruption of Mount St. Helens (see Figure 17.11, p. 457). Sometimes a volcano can unleash a *pyroclastic flow*—a fast-moving cloud of toxic gas, ash, and rock fragments that races down the slopes at speeds up to 725 km/hr (450 mph), enveloping everything in its path. Such a flow buried the inhabitants of the ancient Roman cities of Pompeii and Herculaneum in A.D. 79, when Mount Vesuvius erupted.

Besides affecting people, volcanic eruptions exert environmental impacts (TABLE 2.3). Ash blocks sunlight, while sulfur emissions lead to a sulfuric acid haze that blocks radiation and cools the atmosphere. Large eruptions—such as that of Mount Pinatubo in the Philippines in 1991 (p. 457)—can depress temperatures throughout the world. When Indonesia's Mount Tambora erupted in 1815, it cooled average global temperatures by 0.4–0.7 °C (0.7–1.3 °F), enough to cause crop failures worldwide and make 1816 "the year without a summer."

One of the world's largest volcanoes—so large it is called a "supervolcano"—lies in the United States. The entire basin of Yellowstone National Park is an ancient supervolcano that has at times erupted so massively as to cover large parts of

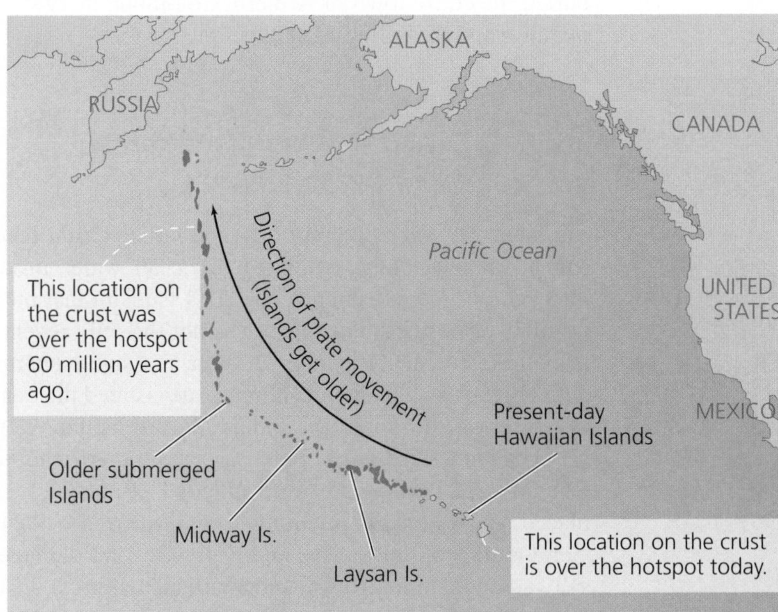

(a) Current and former Hawaiian Islands, formed as crust moves over a volcanic hotspot

(b) Mt. Kilauea erupting

FIGURE 2.22 **The Hawaiian Islands are the product of a hotspot on Earth's mantle.** The Hawaiian Islands **(a)** have been formed by repeated eruptions from a hotspot of magma in the mantle as the Pacific Plate passes over the hotspot. The Big Island of Hawaii is most recently formed, and it is still volcanically active. The other islands are older and have already begun eroding away. To their northwest stretches a long series of former islands, now submerged. The active volcano Kilauea **(b)**, on the Big Island's southeast coast, is currently located above the edge of the hotspot.

TABLE 2.3 Examples of Notable Volcanic Eruptions

Year	Location	Impacts	Magnitude[2]
640,000 B.P.[1]	Yellowstone Caldera, Wyoming, U.S.	Most recent "mega-eruption" at site of Yellowstone National Park	8
6870 B.P.	Mount Mazama, Oregon, U.S.	Created Crater Lake	7
A.D. 79	Mount Vesuvius, Italy	Buried Pompeii and Herculaneum	5
1815	Mount Tambora, Indonesia	Created "year without a summer"; killed at least 70,000 people	7
1883	Krakatau, Indonesia	Killed over 36,000 people; heard 5000 km (3000 mi) away; affected weather for 5 years	6
1980	Mount St. Helens, Washington, U.S.	Blew top off mountain; sent ash 19 km (12 mi) into sky and into 11 U.S. states; 57 people killed	5
1983–present	Kilauea, Hawaii, U.S.	Continuous lava flow	1
1991	Mount Pinatubo, Philippines	Sulfuric aerosols lowered world temperature 0.5 °C (0.9 °F)	6
2010	Eyjafjallajokull, Iceland	Ash cloud disrupted air travel throughout Europe	1

[1]B.P. = years before the present.

[2]Measured by the Volcanic Explosivity Index, which ranges from 0 (least powerful) to 8 (most powerful).

the continent deeply in ash. Although another eruption is not expected imminently, the region is still geothermally active; the park hosts fully two-thirds of the world's geysers.

Landslides are a form of mass wasting

On a smaller scale than volcanoes or earthquakes, another type of geologic hazard, the **landslide,** occurs when large amounts of rock or soil collapse and flow downhill. Landslides are a severe and often sudden manifestation of the more general phenomenon of **mass wasting**, the downslope movement of soil and rock due to gravity. Mass wasting occurs naturally, and heavy rains may saturate soils and trigger mudslides of soil, rock, and water. However, mass wasting can also be brought about by human land use practices that expose or loosen soil, making slopes more prone to collapse.

Most often, mass wasting erodes unstable hillsides, damaging property one structure at a time (FIGURE 2.23). Occasionally, mass wasting events can be colossal and deadly; mudslides that followed the torrential rainfall of Hurricane Mitch in Nicaragua and Honduras in 1998 killed over 11,000 people. Mudslides caused when volcanic eruptions melt snow and send huge volumes of destabilized mud racing downhill are called *lahars*, and these are particularly dangerous. A lahar buried the entire town of Armero, Colombia, in 1985 following an eruption, killing 21,000 people.

FIGURE 2.23 **Landslides are frequent in sloping areas along the California coast, particularly after heavy winter rains saturate soils.** Homes built on unstable slopes, such as these in Laguna Beach, Orange County, can be damaged or destroyed when slopes give way.

Tsunamis can follow earthquakes, volcanoes, or landslides

Earthquakes, volcanic eruptions, and large coastal landslides can all displace huge volumes of ocean water instantaneously and trigger a tsunami. The 2011 tsunami that inundated portions of northeastern Japan was not the only recent major tsunami event. In December 2004, a massive tsunami, triggered by an earthquake off Sumatra, devastated the coastlines of countries all around the Indian Ocean, including Indonesia, Thailand, Sri Lanka, India, and several African nations. Roughly 228,000 people were killed, 1–2 million were displaced, and whole communities were destroyed (FIGURE 2.24). Coral reefs, coastal forests, and wetlands were damaged, and saltwater contaminated soil and aquifers, making it difficult to restore the affected areas.

Those of us who live in the United States and Canada should not consider tsunamis to be something that occurs only in faraway places. A large tsunami struck North America's Pacific coast in 1700 following a huge earthquake in the Pacific Northwest, and one following the Alaskan earthquake of 1964 drowned over 100 people from Alaska to California. Residents of Boston, New York, Washington, D.C., and other cities along the Atlantic coast could be at risk if an unstable portion of a Canary Island volcano that researchers are monitoring were to one day slump into the sea.

One of the best protections against tsunamis is advance warning. Since the 2004 tsunami, nations and international agencies have stepped up efforts to develop systems to give coastal residents notice of approaching tsunamis so they can move inland or to higher ground. We can also lessen the impacts of tsunamis when they occur if we preserve coastal ecosystems, such as coral reefs and mangrove forests (pp. 430–432), which help protect coastlines by absorbing wave energy.

FIGURE 2.24 The December 2004 tsunami completely destroyed Banda Aceh, at the northern tip of Sumatra in Indonesia. This site was near the epicenter of the earthquake, and the influx of water—up to 30 m (100 ft) high in places—arrived within minutes.

We can worsen or mitigate the impacts of natural hazards

Aside from geologic hazards, people face other types of natural hazards. Heavy rains can lead to flooding that ravages low-lying areas near rivers and streams (p. 400). Coastal erosion can eat away at beaches (p. 390). Wildfire can threaten life and property in fire-prone areas. Tornadoes and hurricanes (p. 456) can cause extensive damage and loss of life.

Although we refer to such phenomena as "natural hazards," the magnitude of their impacts upon us often depends on choices we make. We tend to worsen the impacts of so-called natural hazards in various ways:

- As our population grows, more people live in areas susceptible to natural disasters.
- Many of us choose to live in areas that we deem attractive, but that are also prone to hazards. For instance, coastlines are vulnerable to tsunamis and erosion by storms, and mountainous areas are prone to volcanoes and mass wasting.
- We use and engineer landscapes around us in ways that can increase the frequency or severity of natural hazards. Damming and diking rivers to control floods can sometimes lead to catastrophic flooding (p. 401), and suppressing natural wildfire puts forests at risk of larger, truly damaging fires (pp. 320–321). Mining practices (pp. 639–643), the clearing of forests for agriculture, and clear-cutting on slopes (p. 317) can each induce mass wasting, increase runoff, compact soil, and change drainage patterns.
- As we change Earth's climate by emitting greenhouse gases (Chapter 18), we alter patterns of precipitation, thereby increasing risks of drought, fire, flooding, and mudslides locally and regionally. Rising sea levels induced by global warming increase coastal erosion. Some research suggests that warming ocean temperatures may increase the power and duration of hurricanes.

We can often reduce or mitigate the impacts of hazards through the thoughtful use of technology, engineering, and policy, informed by a solid understanding of geology and ecology. Examples of this, as we've noted, include building earthquake-resistant structures; designing early warning systems for earthquakes, tsunamis, and volcanoes based on analysis of previous events; and conserving reefs and shoreline vegetation to protect against tsunamis and coastal erosion. In addition, better forestry, agriculture, and mining practices can help prevent mass wasting. Zoning regulations, building codes, and insurance incentives that discourage development in areas prone to landslides, floods, fires, and storm surges can help keep us out of harm's way and can decrease taxpayer expense in cleaning up after natural disasters. Finally, mitigating global climate change may help reduce the frequency of natural hazards in many regions.

WEIGHING THE ISSUES

YOUR RISK FROM NATURAL HAZARDS What types of natural hazards (earthquakes? tsunamis? flooding? landslides? fires? hurricanes? tornadoes?) occur in the area where you live? Name three things you personally can do to minimize your risk from these hazards. If a natural disaster strikes, should people be allowed to rebuild in the same areas if those areas are prone to experience the hazard again? Discuss one example you are familiar with from news accounts or personal experience.

Conclusion

The Tohoku earthquake of 2011 demonstrated that a firm understanding of the chemical basis of matter, the nature of energy, and Earth's geologic processes is needed to comprehend such events and their impacts in the modern world. After all, these physical phenomena are in some way tied to nearly every significant process in environmental science.

An understanding of matter is essential for all science. Knowledge of chemistry guides our predictions of how radioisotopes released from the Fukushima Daiichi plant will behave in the environment, but it also provides tools for analyzing agricultural practices, managing water resources, reforming energy policy, conducting toxicological studies, and finding ways to mitigate global climate change.

Likewise, an understanding of energy is both of fundamental scientific importance and of considerable practical relevance. For example, can Japan (whose nuclear reactors supplied one-third of its electricity prior to the Tohoku quake) or any other nation transition from fossil fuels to energy sources that do not emit greenhouse gases if nuclear power is not a part of that transition? Can we develop new technologies and approaches that maximize energy conversion efficiency? As population increase and economic growth lead to ever-increasing demand for energy, understanding the nature of energy is vital to a sustainable future.

Physical processes of geology such as plate tectonics and the rock cycle are centrally important because they shape Earth's terrain and form the foundation for living systems that overlie the landscape. But geologic processes also generate phenomena that can threaten our lives and property, including earthquakes, volcanoes, landslides, and tsunamis. The Tohoku quake also showed how geological processes in one location (an earthquake and tsunami in Japan) can initiate events whose impacts (global perceptions of the safety of nuclear power) go far beyond that location.

Reviewing Objectives

You should now be able to:

Explain the fundamentals of matter and chemistry and apply them to real-world situations

- Understanding matter and chemistry is important for developing solutions to environmental problems. (p. 23)
- Matter in the universe is conserved; it cannot be created or destroyed. (p. 23)
- Atoms can form molecules and compounds, and changes at the atomic level can result in alternate forms of elements, such as ions and isotopes. (pp. 23–26)
- Water's chemistry facilitates life. (pp. 26–27)
- The pH scale quantifies acidity and alkalinity. (pp. 27–28)
- Living things depend on organic compounds, which are carbon-based. (pp. 28–29)
- Macromolecules, including proteins, nucleic acids, carbohydrates, and lipids, are key building blocks of life. (p. 29)

Differentiate among forms of energy and explain the basics of energy flow

- Energy can convert from one form to another—for instance, from potential to kinetic energy, and vice versa. Chemical energy is potential energy in the bonds between atoms. (pp. 29–31)
- The total amount of energy in the universe is conserved; it cannot be created or destroyed. (p. 30)
- Systems tend to increase in entropy, or disorder, unless energy is added to build or maintain order and complexity. (pp. 30–31)
- Earth's systems are powered mainly by radiation from the sun, geothermal heating from the planet's core, and gravitational interactions among Earth, the sun, and the moon. (pp. 31–33)

Distinguish photosynthesis, cellular respiration, and chemosynthesis and summarize their importance to living things

- In photosynthesis, autotrophs use carbon dioxide, water, and solar energy to produce oxygen and the sugars they need. (p. 32)
- In cellular respiration, organisms extract energy from sugars by converting them in the presence of oxygen into carbon dioxide and water. (p. 32)
- In chemosynthesis, specialized autotrophs use carbon dioxide, water, and chemical energy from minerals to produce sugars. (p. 33)

Explain how plate tectonics and the rock cycle shape the landscape around us and the earth beneath our feet

- Earth consists of distinct layers that differ in composition, temperature, density, and other characteristics. (p. 34)
- Plate tectonics is a fundamental system that shapes Earth's physical geography and produces earthquakes and volcanoes. (p. 34)
- Tectonic plates meet at three types of boundaries: divergent, transform, and convergent. (pp. 34–36)
- Matter is cycled within the lithosphere, and rocks transform from one type to another. (pp. 36–39)

List major types of geologic hazards and describe ways to mitigate their impacts

- The circum-Pacific belt, or "ring of fire," spawns most of the world's volcanoes and earthquakes. (pp. 39–40)
- Earthquakes result from movement at faults and plate boundaries. We cannot prevent them, but we can build structures and cities in safer ways. (p. 40)

- Volcanoes arise from heating by magma at rifts, subduction zones, and hotspots. (pp. 40–42)
- Landslides and other forms of mass wasting can occur on small or large scales; damage can be minimized by understanding their risks. (p. 42)
- Tsunamis can flood coastlines and cause immense damage. Early warning systems will be key in minimizing future losses. (pp. 42–43)
- We often worsen impacts from natural hazards, but we can reduce them through better land use practices. (p. 43)

Testing Your Comprehension

1. What are the basic building blocks of matter? Provide several examples using chemicals common in Earth's physical or biological systems.
2. How does an ion differ from an isotope? Now differentiate among an atom, a molecule, and a compound.
3. Describe two major forms of energy, and give examples of each.
4. Compare and contrast the first and second laws of thermodynamics.
5. Describe the three major sources of energy that power Earth's environmental systems.
6. What substances are produced by photosynthesis? By cellular respiration? By chemosynthesis?
7. Name the primary layers that make up our planet. Which portions does the lithosphere include?
8. Describe what occurs at a divergent plate boundary. What happens at a transform plate boundary? Compare and contrast the types of processes that can occur at a convergent plate boundary.
9. Name the three main types of rocks, and describe how each type may be converted to the others via the rock cycle.
10. What causes earthquakes? What are tsunamis, and what causes them? How does a Hawaiian volcano such as Kilauea differ from a volcano in the Cascades of North America such as Mount St. Helens?

Seeking Solutions

1. Think of an example of an environmental problem not mentioned in this chapter that a good knowledge of chemistry might help us solve. How could chemistry help us address the problem?
2. Think about the ways we harness and use energy sources in our society—both renewable sources such as geothermal energy and nonrenewable sources such as coal, oil, and natural gas. What implications does the first law of thermodynamics have for our energy usage? How is the second law of thermodynamics relevant to our use of energy?
3. Referring to the chemical reactions for photosynthesis and respiration, provide an argument for why increasing amounts of carbon dioxide in the atmosphere due to global climate change (Chapter 18) might potentially increase amounts of oxygen in the atmosphere. Now give an argument for why increasing amounts of carbon dioxide might potentially decrease amounts of atmospheric oxygen. What would you need to know to determine which of these two outcomes might occur?
4. Describe how plate tectonics accounts for the formation of (a) mountains, (b) volcanoes, and (c) earthquakes. Why do you think it took so long for scientists to discover an environmental system of such fundamental importance as plate tectonics?
5. For each of the following "natural hazards," describe one thing that can be done to minimize or mitigate its impact on our lives and property: (a) earthquakes, (b) landslides, (c) flooding
6. **THINK IT THROUGH** You live in a coastal community in a geologically active region prone to earthquakes, and you sit on the board of your regional electric utility. The utility is considering constructing a nuclear power plant 10 miles up the coast from your town. Some of your fellow citizens are supportive of the project, as it will bring employment to the region and provide a carbon-free source of electricity. However, some local residents fear a repeat of the events of 2011 in northeastern Japan if there should be a significant earthquake along one of the onshore or offshore faults. As a member of the utility board, what specific information would you insist on obtaining from geologists and other scientists before you cast your vote on the project? What assurances from scientific research would you feel you need before voting in favor of the project?

Calculating Ecological Footprints

The second law of thermodynamics has profound implications for human impacts on the environment, as it affects the efficiency with which we produce our food. In ecological systems, a rough rule of thumb is that when energy is transferred from plants to plant-eaters or from prey to predator, the efficiency is only about 10% (p. 82). Much of this inefficiency is a consequence of the second law of thermodynamics. Another way to think of this is that eating 1 Calorie of meat from an animal is the ecological equivalent of eating 10 Calories of plant material. So when we raise animals for meat using grain, it is less energetically efficient than if we ate the grain ourselves.

Humans are considered omnivores because we can eat both plants and animals. The choices we make about what to eat have significant ecological consequences. With this in mind, calculate the ecological energy requirements for four different diets, each of which provides a total of 2000 dietary Calories per day.

Diet	Source of Calories	Number of Calories consumed	Ecologically equivalent Calories	Total ecologically equivalent Calories
100% plant	Plant			
0% animal	Animal			
90% plant	Plant	1800	1800	3800
10% animal	Animal	200	2000	
50% plant	Plant			
50% animal	Animal			
0% plant	Plant			
100% animal	Animal			

1. How many ecologically equivalent Calories would it take to support you for a year for each of the four diets listed?
2. How does the ecological impact from a diet consisting strictly of animal products (e.g., milk, other dairy products, eggs, and meat) compare with that of a strictly vegetarian diet? How many additional ecologically equivalent Calories do you consume each day by including as little as 10% of your Calories from animal sources?
3. What percentages of the Calories in your own diet do you think come from plant versus animal sources? Estimate the ecological impact of your diet, relative to a strictly vegetarian one.
4. Describe some challenges of providing food for the growing human population, especially as people in many poorer nations develop a taste for an American-style diet rich in animal protein and fat.

STUDENTS

Go to **MasteringEnvironmentalScience** for assignments, the etext, and the Study Area with practice tests, videos, current events, and activities.

INSTRUCTORS

Go to **MasteringEnvironmentalScience** for automatically graded activities, current events, videos, and reading questions that you can assign to your students, plus Instructor Resources.

Native Hawaiian forest at Hakalau Forest NWR, and the endangered 'akiapōlā'au

Evolution, Biodiversity, and Population Ecology

Upon completing this chapter, you will be able to:

- Explain natural selection and cite evidence for this process
- Describe how evolution influences biodiversity
- Discuss reasons for species extinction and mass extinction events
- List the levels of ecological organization
- Outline the characteristics of populations that help predict population growth
- Assess logistic growth, carrying capacity, limiting factors, and other fundamental concepts in population ecology
- Identify efforts and challenges involved in the conservation of biodiversity

CENTRAL CASE STUDY

Saving Hawaii's Native Forest Birds

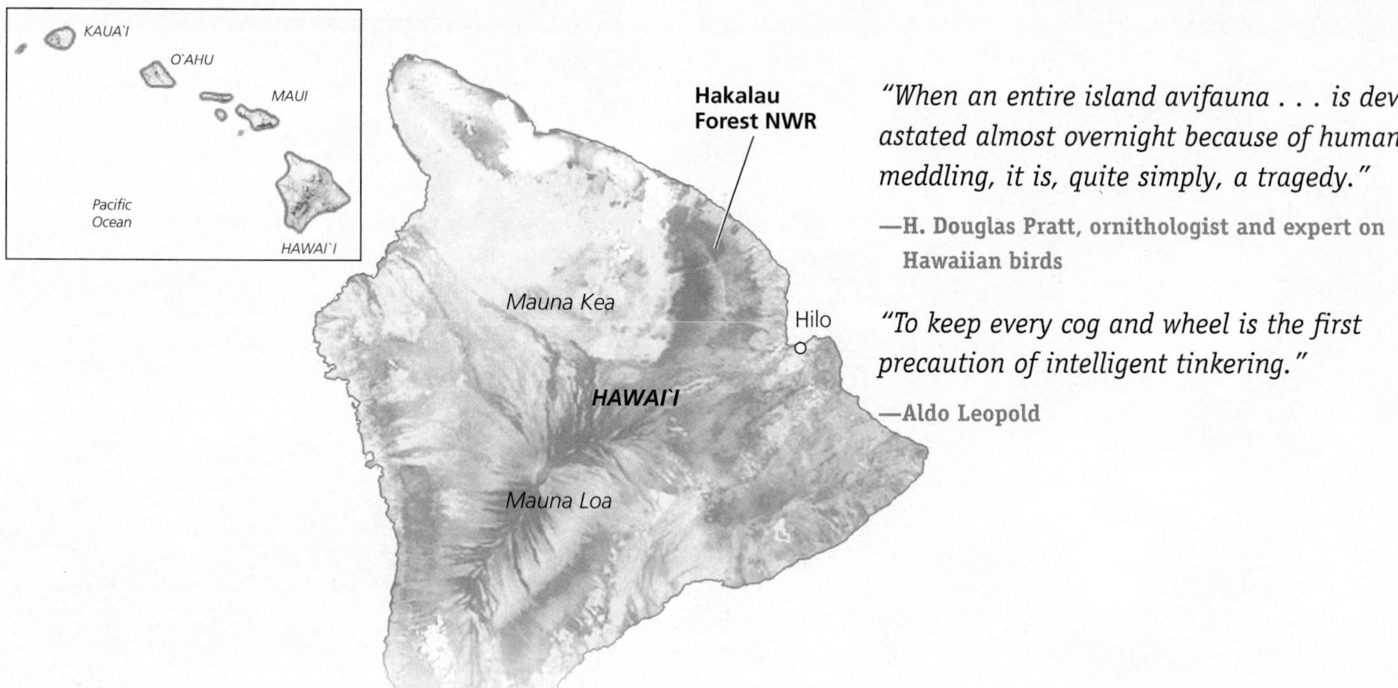

"When an entire island avifauna . . . is devastated almost overnight because of human meddling, it is, quite simply, a tragedy."

—H. Douglas Pratt, ornithologist and expert on Hawaiian birds

"To keep every cog and wheel is the first precaution of intelligent tinkering."

—Aldo Leopold

Jack Jeffrey stopped in his tracks. "I hear one!" he said. "Over there in those trees!"

Jeffrey quickly led his group of ecotourists through a misty woodland of ferns, grasses, koa trees, and red-flowering 'ōhi'a-lehua trees toward an emphatic chirping sound that carried farther than the other bird songs in the forest. At last they spotted the bird—an 'akiapōlā'au, one of fewer than 1500 of its kind left alive in the world.

The 'akiapōlā'au (or "aki" for short) is a sparrow-sized wonder of nature—one of many exquisite birds that evolved on the Hawaiian Islands and exists only here. For millions of years, this chain of islands in the middle of the Pacific Ocean has acted as a cradle of evolution, generating an abundance of new and unique species. Yet in recent years, many of these species have gone from cradle to grave. Half of Hawaii's native bird species (70 of 140) have gone extinct in recent times, and the percentage of species that teeter on the brink of extinction here today is higher than anywhere else in the world.

The aki is one of 30 species of endangered birds remaining on the Hawaiian Islands. It is a type of Hawaiian honeycreeper, a group of birds numbering 18 living species (and at least 38 species recently extinct), all of which originated from individuals of a single ancestral species that reached Hawai'i several million years ago. As new volcanic islands emerged from the ocean and then eroded away, and as forests expanded and contracted over the millennia, populations were split many times, and new honeycreeper species evolved.

As honeycreeper species diverged from one another, they evolved different colors, sizes, body shapes, feeding behaviors, mating preferences, diets, and bill shapes. Bills in some species became short and straight, allowing birds to glean insects from leaf surfaces. In other species, bills became long and downcurved, enabling birds to probe into flowers to sip nectar. The bills of still other species became thick and strong for cracking seeds. Some bills became highly specialized: The aki uses the short, straight lower half of its bill to peck, woodpecker-style, into dead twigs and branches of koa trees to find beetle grubs, and then uses the long, downcurved upper half to reach in and extract the grubs.

Hawaii's honeycreepers thrived for several million years in the island's forests, amid a unique community of plants. The stately 'ōhi'a, a slow-growing tree that can live for 2000 years, spreads twisting gnarled limbs covered with moss and lichens through the misty air, and offers up bright red flowers that provide nectar and pollen to birds and insects. The koa thrives here too, a fast-growing acacia tree with twigs that the aki snaps off in its search for grubs. A multitude of shrubs, herbs, and vines found nowhere else in the world used to fill out the forest understory.

Today native Hawaiian forests are under siege. The crisis began 750 or more years ago when Polynesian settlers colonized the islands, cutting down trees and introducing nonnative animals. Europeans arrived in the 1800s and did more of the same. Pigs, goats, and cattle ate their way through the native plants, transforming lush forests into ragged grasslands. Rats, cats, dogs, and mongooses destroyed the eggs and young of native ground-nesting birds. Foreign plants from Asia, Europe, and America, whose seeds accompanied the people and animals, spread across the altered landscape.

The newly introduced organisms wreaked havoc because native Hawaiian organisms were unprepared to resist them. Hawai'i is so isolated that only one mammal—a bat—had ever arrived naturally. As a result, plants had faced no pressure to invest in defenses (such as thick bark, spines, or chemical toxins) against plant-eating livestock. Likewise, nothing had led birds to evolve defenses against voracious nest predators such as rats and mongooses. The vulnerable native species were easy pickings, and many were soon wiped off the face of the planet.

With the arrival of people, domestic animals, and invasive plants also came diseases. The native fauna were not adapted to resist diseases they had never encountered. Avian pox and avian malaria spread through Hawaii's birds. Malaria and the mosquitoes that carry the disease killed off the natives everywhere except on the high slopes of the mountains, where it becomes too cold for the malaria parasite to survive. Today few native forest birds exist anywhere on the Hawaiian Islands below 1500 m (4500 ft) in elevation.

The aki being watched by Jack Jeffrey's group inhabits the Hakalau Forest National Wildlife Refuge, which sits high on the slopes of Mauna Kea, a volcano on the island of Hawai'i, the largest island in the chain. At Hakalau, native birds find one of the few remaining patches of malaria-free native forest on the island.

Conservation biologists and managers have worked hard to keep the Hakalau Forest protected. Jack Jeffrey was the refuge biologist here for 20 years, and led a number of innovative projects to save native plants and birds from extinction. Managers at Hakalau have fenced out pigs to safeguard forested areas, and refuge staff and volunteers have planted thousands of native plants in areas deforested by cattle grazing. Young restored native forest is now regrowing on thousands of acres. More birds are using this restored forest year by year.

Today global climate change is throwing up a new challenge. As temperatures warm, mosquitoes move upslope, and malaria and pox spread deeper into the remaining forests, so that even protected areas such as Hakalau are not immune. The next generation of managers will need to innovate novel strategies to fend off extinction for the island's native species.

Plenty of challenges remain, but the restoration successes at Hakalau Forest so far provide hope that through responsible management we can save Hawaii's native flora and fauna and preserve the priceless bounty of millions of years of evolution on this extraordinary chain of islands. ●

Evolution: The Source of Earth's Biodiversity

The honeycreepers and the other native animals and plants of Hawai'i help reveal how our world became populated with the remarkable diversity of life we see today. Scientific study shows us that our planet has progressed from a stark world inhabited solely by microbes to a lush cornucopia of millions of species (FIGURE 3.1).

A **species** is a particular type of organism or, more precisely, a population or group of populations whose members share characteristics and can freely breed with one another and produce fertile offspring. A **population** is a group of individuals of a particular species that live in a particular area. Over vast spans of time, the process of biological evolution

(a) 'I'iwi (*Vestiaria coccinea*)

(b) Nēnē (*Branta sandvicensis*)

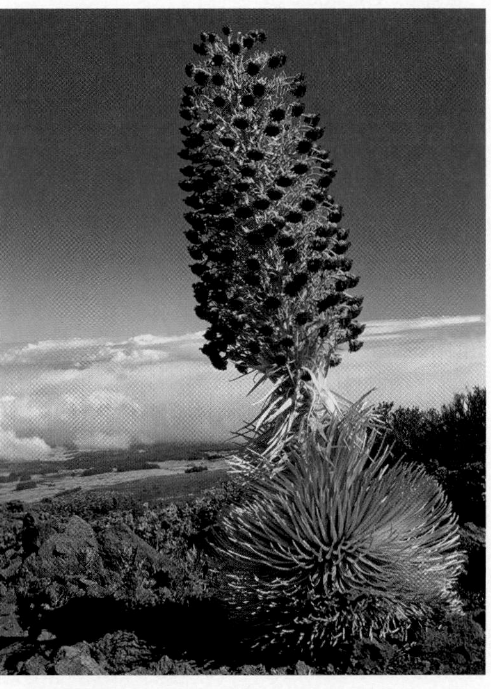

(c) Haleakala silversword (*Argyroxiphium sandwichense*)

(d) Happyface spider (*Theridion grallator*)

FIGURE 3.1 **Hawai'i hosts a treasure trove of biological diversity.**

has shaped populations and species, giving us the vibrant abundance of life that enriches Earth today.

Evolution in the broad sense means change over time, and biological evolution consists of change in populations of organisms across generations. Changes in genes (p. 29) often lead to modifications in the appearance or behavior of organisms from generation to generation. Biological evolution results from random genetic changes and may be directed by natural selection. **Natural selection** is the process by which inherited characteristics that enhance survival and reproduction are passed on more frequently to future generations than those that do not, thus altering the genetic makeup of populations through time.

Evolution is one of the best-supported and most illuminating concepts in all of science, and it is the very foundation of modern biology. Perceiving how species adapt to their environments and change over time is crucial for comprehending ecology and learning the history of life. Evolutionary processes influence many aspects of environmental science, including pesticide resistance, agriculture, medicine, and environmental health.

Natural selection shapes organisms and diversity

In 1858, **Charles Darwin** and **Alfred Russel Wallace** each independently proposed the concept of natural selection as a mechanism for evolution and as a way to explain the great variety of living things. Both Darwin and Wallace were exceptionally keen naturalists from England who had studied plants and animals in such exotic locales as the Galápagos Islands (Darwin) and the Malay Archipelago (Wallace). In the century and a half since then, many thousands of scientists have refined our understanding of natural selection and evolution.

Natural selection is a simple concept that offers a powerful explanation for patterns evident in nature. The idea of natural selection follows logically from a few straightforward premises that are readily apparent to anyone who observes the life around us:

- Organisms face a constant struggle to survive and reproduce.
- Organisms tend to produce more offspring than can survive.
- Individuals of a species vary in their characteristics.

Variation is due to differences in genes, the environments in which genes are expressed, and the interactions between genes and environment. As a result of this variation, some individuals of a species will be better suited to their environment than others and will be better able to reproduce.

Many characteristics are passed from parent to offspring through the genes, and a parent that produces many offspring will pass on more genes to the next generation than a parent that produces few or no offspring. In the next generation, therefore, the genes of better-adapted individuals will outnumber those of individuals that are less well adapted. From one generation to another through time, characteristics, or traits, that lead to better and better reproductive success in a given environment will evolve in the population. This process is termed **adaptation,** and a trait that promotes success is also called an **adaptation** or an **adaptive trait.**

Selection acts on genetic variation

For an organism to pass a trait along to future generations, genes in the organism's DNA (p. 29) must code for the trait. In an organism's lifetime, its DNA will be copied millions of times by millions of cells. In all this copying and recopying, sometimes a mistake is made. Accidental changes in DNA, called **mutations,** give rise to genetic variation among individuals. If a mutation occurs in a sperm or egg cell, it may be passed on to the next generation. Most mutations have little effect, but some can be deadly, whereas others can be beneficial. Those that are not lethal provide the genetic variation on which natural selection acts.

Genetic variation is also generated as organisms mix their genetic material through *recombination* during sexual reproduction. When organisms reproduce sexually, a portion of each parent's genes contributes to the genes of the offspring. This process produces novel combinations of genes, generating variation among individuals.

When natural selection acts on genetic variation by favoring certain variants, it can drive a feature in a particular direction (FIGURE 3.2). Because such evolutionary change

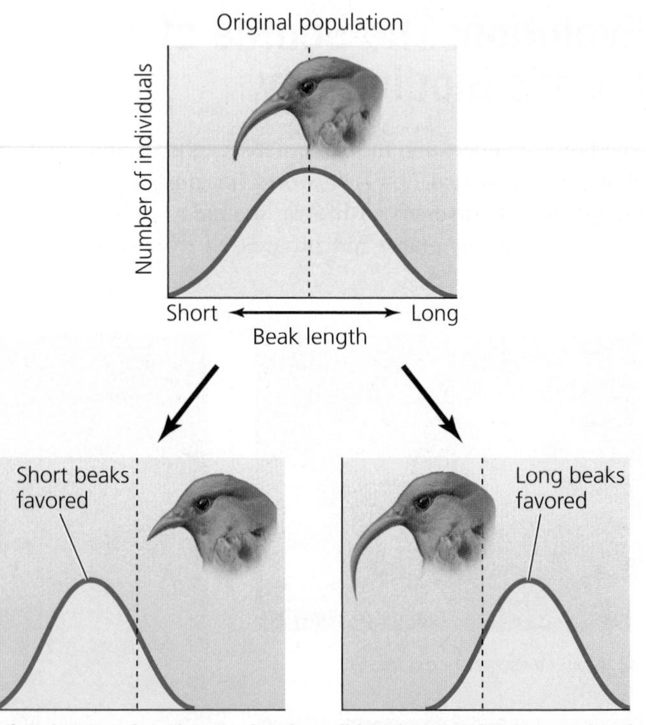

FIGURE 3.2 Natural selection can drive a feature in various directions. Let's consider the 'i'iwi, a Hawaiian honeycreeper, and assume its population possesses genetic variation for the length of its curved bill. In an environment where flowers grow shorter nectar tubes **(a)**, birds with shorter bills could feed perfectly well and could avoid investing in a long bill, so natural selection would favor a decrease in bill length across the population. But in an environment where flowers have longer tubes **(b)**, birds with longer bills could feed more effectively and pass on more genes, causing the population to shift toward a longer average bill length.

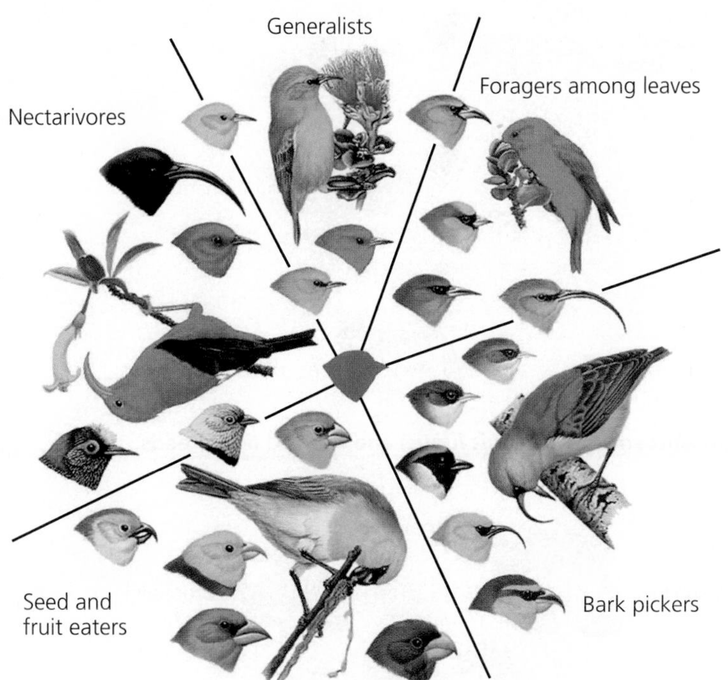

(a) Divergent evolution of Hawaiian honeycreepers

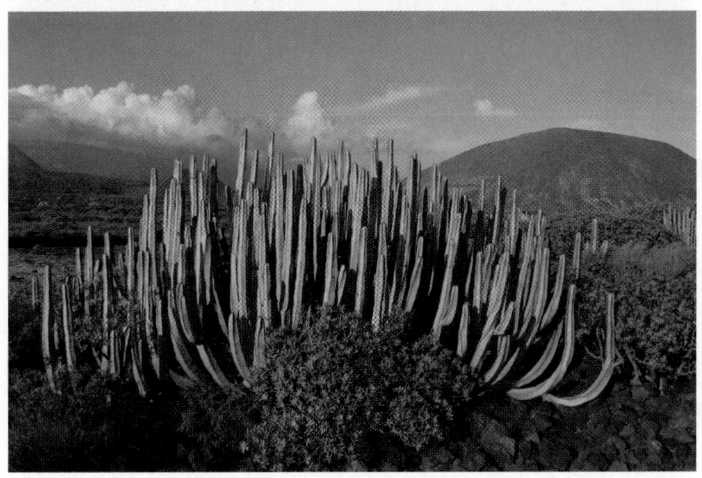

(b) Convergent evolution of a cactus in Arizona (top) and a euphorb (spurge) in the Canary Islands (bottom)

FIGURE 3.3 Natural selection can cause closely related species to diverge in appearance or distantly related species to converge in appearance. Hawaiian honeycreepers **(a)** diversified as they adapted to different food resources and habitats, as indicated by the diversity of their plumage colors and bill shapes. In contrast, cacti of the Americas and euphorbs of Africa **(b)** became more similar to one another as they independently adapted to arid environments through the evolution of tough succulent tissues to hold water, thorns to keep thirsty animals away, and photosynthetic stems without leaves to reduce surface area and water loss.

generally requires a great deal of time, a species cannot always adapt to environmental conditions that change quickly. For instance, the warming of our global climate today (Chapter 18) is occurring too rapidly for most species to adapt, and we may lose many species to extinction as a result.

However, genetic variation can sometimes help protect a population against novel challenges. One of the honeycreeper species of the Hakalau Forest, the ʻamakihi, has in recent years been discovered in ʻōhiʻa trees at very low elevations, well within the zone where avian malaria has killed off all other honeycreepers. Researchers studying this population have determined that some of the ʻamakihis living here when malaria arrived had genes that by chance gave them a natural resistance to the disease. These resistant birds survived malaria's onslaught, and their descendants reestablished a population that continues to grow today. Similarly, the ʻapapane (another Hawaiian honeycreeper) and the ʻōmaʻo (a native Hawaiian thrush) also are showing some degree of resistance to malaria. Scientists hope that perhaps some individuals of the rarer native birds of Hakalau might also harbor resistance genes that may help them persist in the face of malaria.

Selective pressures from the environment influence adaptation

Environmental conditions determine what pressures natural selection will exert, and these selective pressures affect which members of a population will survive and reproduce. Over many generations, this results in the evolution of traits that enable success within the environment in question. Closely related species that live in very different environments and thus experience very different selective pressures tend to diverge in their traits as the differing pressures drive the evolution of different adaptations (**FIGURE 3.3a**). Conversely, sometimes very unrelated species may acquire similar traits as they adapt to selective pressures from similar environments; this is called **convergent evolution** (**FIGURE 3.3b**).

However, environments change over time, and organisms may move to new locations and encounter new conditions. In either case, a trait that promotes success at one time or place may not do so at another. Hawaiian honeycreepers such as the ʻapapane and the ʻiʻiwi fly long distances in search of flowering trees. This behavior had long helped them to find the best resources across a diverse landscape. However, once malaria arrived, the strategy backfired, as birds from malaria-free areas would sometimes fly into death zones. Today, thousands of honeycreepers from mountain forests die each year when they fly downslope and are bitten by malarial mosquitoes. As environmental conditions vary in time and space, adaptation becomes a moving target.

FAQ Isn't evolution based on just one man's beliefs?

Because Charles Darwin contributed so much to our early understanding of evolution, many people assume the concept itself hinges on his ideas. But scientists and laypeople had been observing nature and puzzling over fossils for a long time, and the notion of evolution was being discussed long before Darwin. Once he and Alfred Russel Wallace independently proposed the concept of natural selection, scientists finally gained a precise and feasible mechanism to explain how and why organisms change across generations. Later, geneticists discovered Gregor Mendel's research and worked out how traits are inherited—and modern evolutionary biology was born. Twentieth-century scientists Fisher, Wright, Dobzhansky, Simpson, Mayr, and others ran experiments and developed sophisticated mathematical models, documenting phenomena with extensive evidence and making evolutionary biology into one of science's strongest fields. Since then, evolutionary research by thousands of scientists has driven our understanding of biology and has facilitated spectacular advances in agriculture, medicine, and biotechnology.

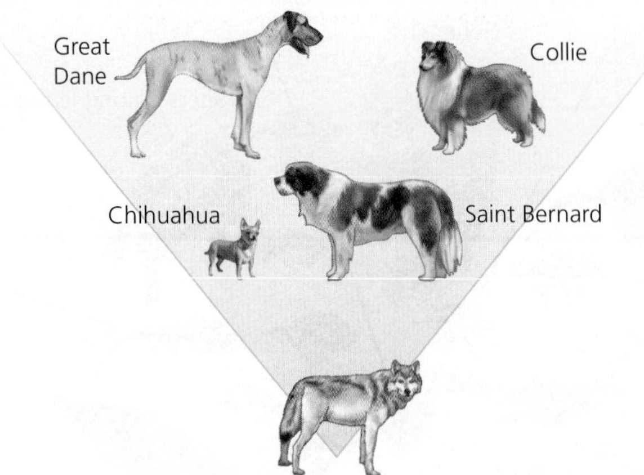

(a) Ancestral wolf (*Canis lupus*) and derived dog breeds

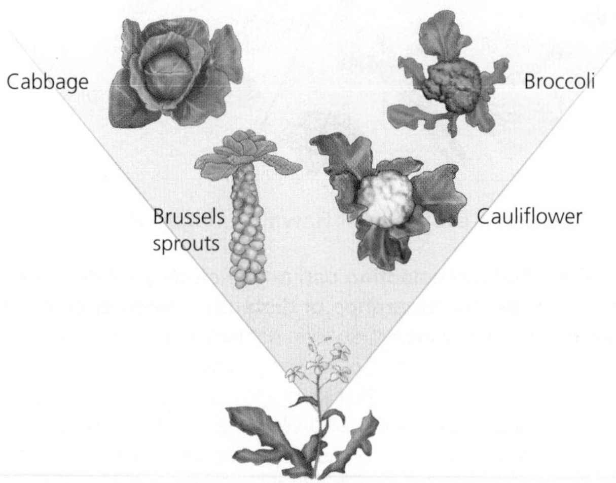

(b) Ancestral *Brassica oleracea* and derived crops

FIGURE 3.4 Selective breeding, or artificial selection, has resulted in our many breeds of dogs and varieties of crops. With dogs **(a)**, we began with the gray wolf *(Canis lupus)* as the ancestral wild species, and by breeding like with like and selecting for the traits we prefer, we have produced breeds as different as Great Danes and Chihuahuas**.** By this same process we have created our immense variety of crop plants **(b)**. Cabbage, brussels sprouts, broccoli, and cauliflower were all generated from a single ancestral species, *Brassica oleracea*.

Evidence of selection is all around us

The results of natural selection are all around us, visible in every adaptation of every organism. In addition, scientists have demonstrated the rapid evolution of traits by selection in countless lab experiments, mostly with fast-reproducing organisms such as bacteria, yeast, and fruit flies.

The evidence for selection that may be most familiar to us is that which Darwin himself cited prominently in his work 150 years ago: our breeding of domesticated animals. In domesticated dogs, cats, and livestock, we have conducted selection under our own direction. We have chosen animals with traits we like and bred them together, while not breeding those with variants we do not like. Through such *selective breeding,* we have been able to augment particular traits we prefer.

Consider the great diversity of dog breeds (FIGURE 3.4a). People generated every type of dog alive today by starting with a single ancestral species and selecting for particular desired traits as individuals were bred together. From Great Dane to Chihuahua, all dogs are able to interbreed and produce viable offspring, yet breeders maintain striking differences among them by allowing only like individuals to breed with like. This process of selection conducted under human direction is termed **artificial selection.**

Artificial selection has given us the many crop plants we depend on for food, all of which people domesticated from wild ancestors and carefully bred over years, centuries, or millennia (FIGURE 3.4b). Through selective breeding, we have created corn with bigger, sweeter kernels; wheat and rice with larger and more numerous grains; and apples, pears, and oranges with better taste. We have diversified single types into many—for instance, breeding variants of the plant *Brassica oleracea* to create broccoli, cauliflower, cabbage, and brussels sprouts. Our entire agricultural system is based on artificial selection. We depend on a working understanding of evolution for the very food we eat.

Evolution generates biodiversity

Just as artificial selection helps us create new types of pets, farm animals, and crop plants, natural selection serves to elaborate and diversify traits in wild organisms. Over the long term, natural selection helps lead to the formation of new species and whole new types of organisms. Life's complexity can be expressed as **biological diversity,** or **biodiversity** for short, which refers to the variety of life across all levels of biological organization, including the diversity of species, genes, populations, and communities (we will introduce communities shortly: p. 60 and Chapter 4).

Scientists have described about 1.8 million species, but many more remain undiscovered or unnamed. Estimates for the total number of species in the world vary, but they range from 3 million up to 100 million. Hawaii's insect fauna provides one example of how much we have yet to learn. Scientists studying fruit flies in the Hawaiian Islands have described over 500 species of them, but they have also identified about 500 others that have not yet been formally named and described. Still more fruit fly species probably exist but have not yet been found.

Subtropical islands such as Hawai'i are by no means the only places rich in biodiversity, however. Step outside anywhere, and you will find many species within close reach. They may not always be large and conspicuous like Yellowstone's bears or the Serengeti's elephants, but they will be there. Plants poke up from cracks in asphalt in every city in the world, and even Antarctic ice harbors microbes. A handful of backyard soil may contain an entire miniature world of life, including insects, mites, millipedes, nematode worms, plant seeds, fungi, and millions upon millions of bacteria. (We will examine Earth's biodiversity in detail in Chapter 11.)

Speciation produces new types of organisms

How did Earth come to have so many species? The process by which new species are generated is termed **speciation.** Speciation can occur in a number of ways, but the main mode is generally thought to be *allopatric speciation*, whereby species form from populations that become physically separated over some geographic distance. To understand allopatric speciation, begin by picturing a population of organisms. Individuals within the population possess many similarities that unify them as a species because they are able to breed with one another and share genetic information. However, if the population is broken up into two or more isolated areas, individuals from one area cannot reproduce with individuals from the others.

When a mutation arises in the DNA of an organism in one of these newly isolated populations, it cannot spread to the other populations. Over time, each population will independently accumulate its own set of mutations. Eventually, the populations may diverge, growing so different that their members can no longer mate with one another. Once this has happened, there is no turning back; the two populations can no longer share genetic information, and they will embark on their own independent evolutionary trajectories as separate species (**FIGURE 3.5**). The populations will continue diverging in their characteristics as chance mutations accumulate that cause them to differ more and more. If environmental conditions happen to differ for the two populations, then natural selection may accelerate the divergence.

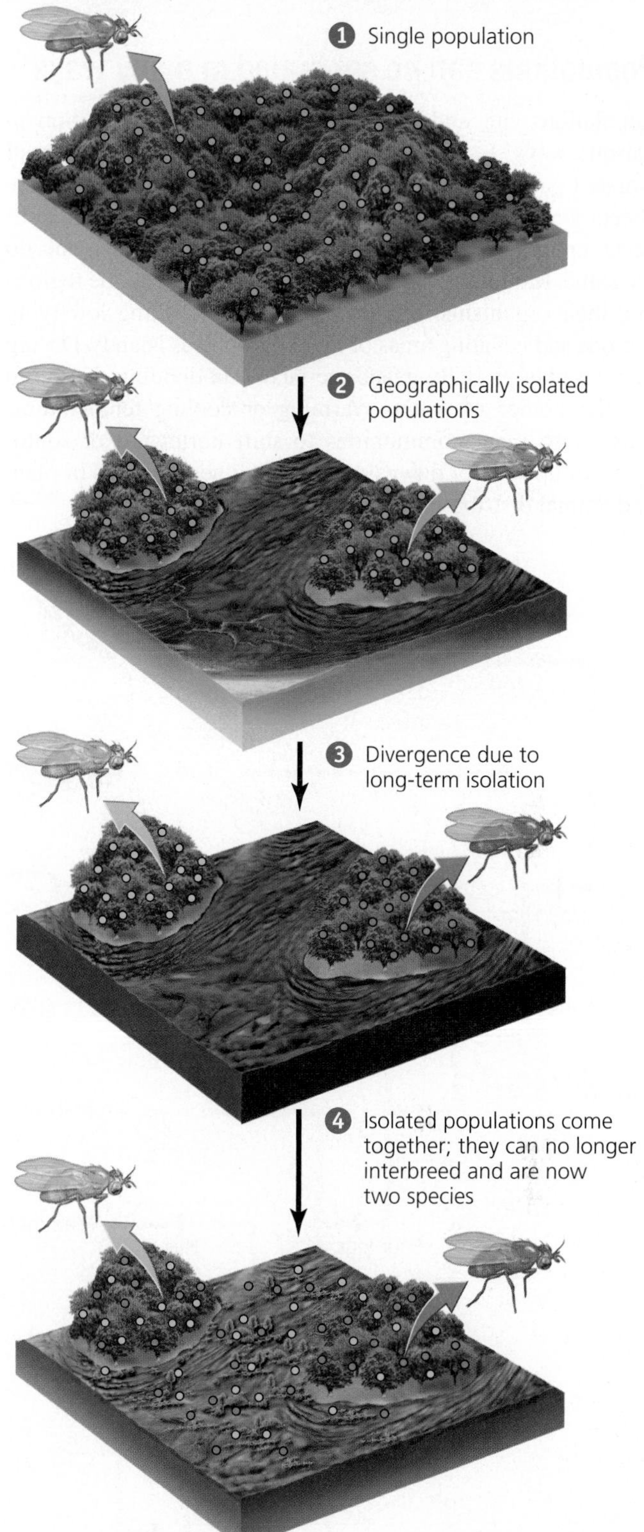

FIGURE 3.5 In allopatric speciation, species form from populations that become physically separated over some geographic distance. This long, slow process begins when a geographic barrier splits a population—as when forest 1 is destroyed by lava flowing from a volcano but isolated patches of forest 2 are left. In Hawai'i such forested patches are called *kipukas*. Hawaiian fruit flies are weak fliers and become isolated in kipukas. Over the centuries, each population accumulates its own independent set of genetic changes 3, until individuals become genetically distinct from and unable to breed with individuals from the other population. The two populations now represent separate species and will remain so even when the geographic barrier disappears 4, new forest grows over the eroding lava rock, and the new species intermix.

Populations can be separated in many ways

Populations can undergo long-term geographic isolation in various ways. Lava flows can destroy forest, leaving small isolated patches intact (as shown in Figure 3.5). Glacial ice sheets may move across continents during ice ages and split populations in two. Major rivers may change course and do the same. Mountain ranges may be uplifted and divide regions and their organisms. Sea level may rise, flooding low-lying regions and isolating areas of higher ground as islands. Drying climate may partially evaporate lakes, subdividing them into smaller bodies of water. Warming or cooling temperatures may cause plant communities to shift northward or southward, or upslope or downslope, creating new patterns of plant and animal distribution.

Alternatively, sometimes new areas are created and organisms colonize them, establishing isolated populations. Hawai'i provides an example. As shown in Figure 2.22 (Chapter 2, p. 41), the Pacific tectonic plate moves over a volcanic "hotspot" that extrudes magma into the ocean, building volcanoes that form islands once they break the water's surface. The plate inches northwest, dragging each island with it, while new islands are formed at the hotspot. The result, over millions of years, is a long string of islands, an *archipelago*. As each new island is formed, plants and animals that colonize it may undergo allopatric speciation if they are isolated enough from their source population (see **THE SCIENCE BEHIND THE STORY**, pp. 56–57).

For speciation to occur, populations must remain isolated for a very long time, generally thousands of generations. Then, if the geologic or climatic process that has isolated populations reverses itself—if the glacier recedes, or the river returns to its old course, or warm temperatures turn cool again—then the populations may come back together. This is the moment of truth for speciation. If the populations have not diverged enough, their members will begin interbreeding and reestablish gene flow, mixing the mutations that each population accrued while isolated. However, if the populations have diverged sufficiently, they will not interbreed, and two species will have been formed, each destined to continue on its own evolutionary path.

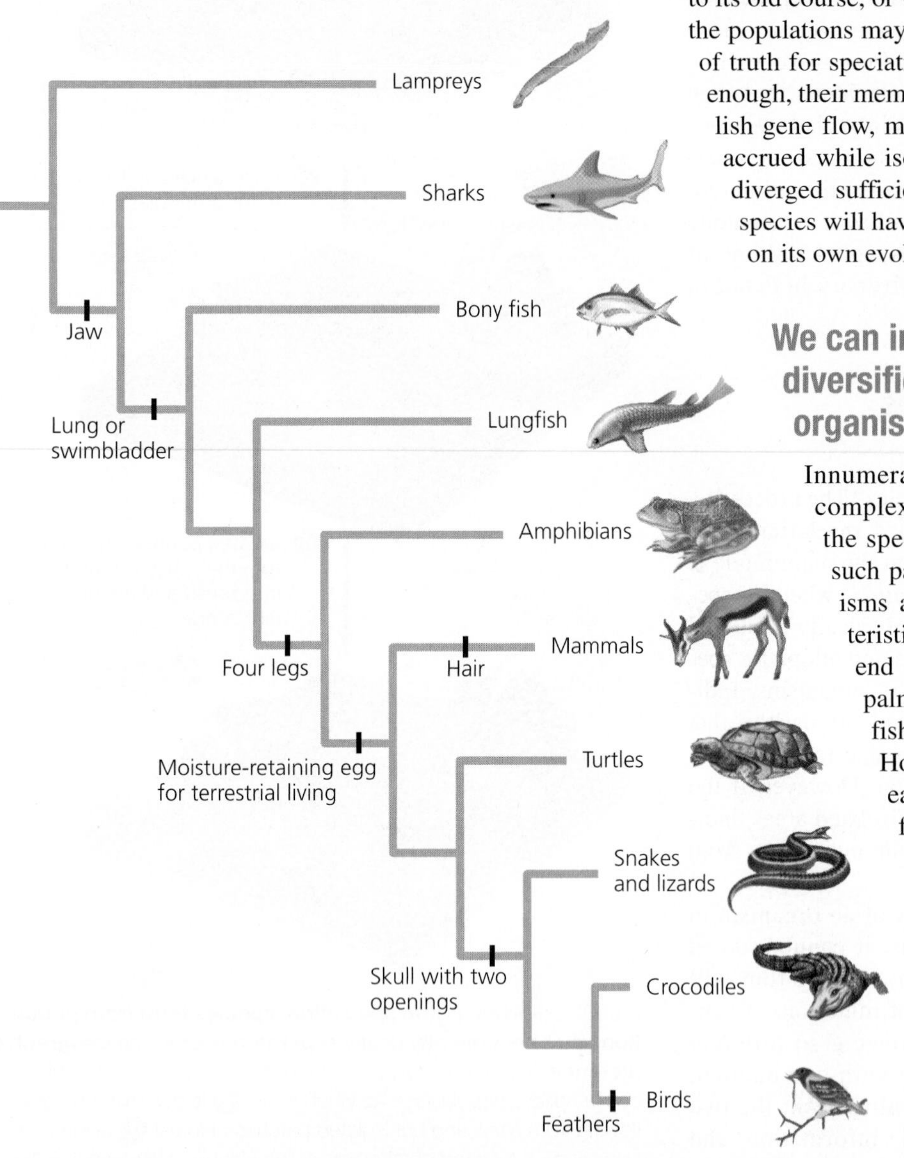

FIGURE 3.6 Phylogenetic trees show the history of life's divergence. The tree here illustrates relationships among groups of vertebrates—just one small portion of the huge and complex "tree of life." Each branch results from a speciation event; as you follow the tree left to right from its trunk to the tips of its branches, you proceed forward in time, tracing life's history. In this tree, major traits are "mapped" on to indicate when they originated. For instance, all vertebrates to the right of the hash mark indicating the origin of jaws possess jaws, whereas lampreys diverged before jaws originated and thus lack them.

We can infer the history of life's diversification by comparing organisms

Innumerable speciation events have generated complex patterns of diversity at levels above the species level. Evolutionary biologists study such patterns, examining how groups of organisms arose and how they evolved the characteristics they show. For instance, how did we end up with plants as different as mosses, palm trees, daisies, and redwoods? Why do fish swim, snakes slither, and sparrows sing? How and why did birds, bats, and insects each independently evolve the ability to fly? To address such questions, we need to know how the major groups diverged from one another over time.

Scientists represent this history of divergence by using branching, tree-like diagrams called **phylogenetic trees.** Similar to family genealogies, these diagrams illustrate scientists' hypotheses as to how divergence took place (**FIGURE 3.6**). Phylogenetic trees can show relationships among species, groups of species, populations, or genes. Scientists construct these trees by analyzing patterns of similarity among the genes or external traits of present-day organisms and by inferring which groups share similarities because they are related.

Once we have a phylogenetic tree, we can map traits onto the tree according to which organisms possess them, and we can thereby trace how the traits have evolved. For instance, phylogenetic research shows that birds, bats, and insects are distantly related, with many flightless groups between them. (Note in Figure 3.6 that birds and mammals are separated on the tree and that insects are outside the tree.) Therefore, it is far simpler to conclude that these three very different groups evolved flight independently than it would be to conclude that the many flightless groups between them each lost an ancestral ability to fly. Because phylogenetic trees help biologists make such inferences about so many traits, they have become one of the modern biologist's most powerful tools.

Knowing how organisms are related to one another also helps scientists to classify them and name them, so that we can make sense of the life around us and communicate effectively. *Taxonomists* use an organism's physical appearance and genetic makeup to determine its species. These scientists then group species by their similarity into a hierarchy of categories meant to reflect evolutionary relationships. Related species are grouped together into *genera* (singular, *genus*), related genera are grouped into families, and so on (**FIGURE 3.7**). Each species is given a two-part Latin or Latinized scientific name denoting its genus and species.

For instance, the ʻakiapōlāʻau, *Hemignathus munroi*, is similar to other Hawaiian honeycreepers in the genus *Hemignathus*. These species are closely related in evolutionary terms, as indicated by the genus name they share. They are more distantly related to honeycreepers in other genera, but all honeycreepers are classified together in the family Fringillidae. This system of naming and classification was devised by Swedish botanist Carl Linnaeus (1707–1778) long before Darwin's work on evolution. Today biologists use evolutionary information from phylogenetic trees to help classify organisms under the Linnaean system's rules.

The fossil record teaches us about life's long history

Scientists also decipher life's history by studying fossils. As organisms die, some are buried by sediment. Under certain conditions, the hard parts of their bodies—such as bones, shells, and teeth—may be preserved, as sediments are compressed into rock (pp. 37–38). Minerals replace the organic material, leaving behind a **fossil,** an imprint in stone of the dead organism (**FIGURE 3.8**, p. 58). In countless locations throughout the world, geologic processes across millions of years have buried sediments and later brought sedimentary rock layers to the surface, revealing assemblages of fossilized plants and animals from different time periods. By dating the rock layers that contain fossils, paleontologists (scientists

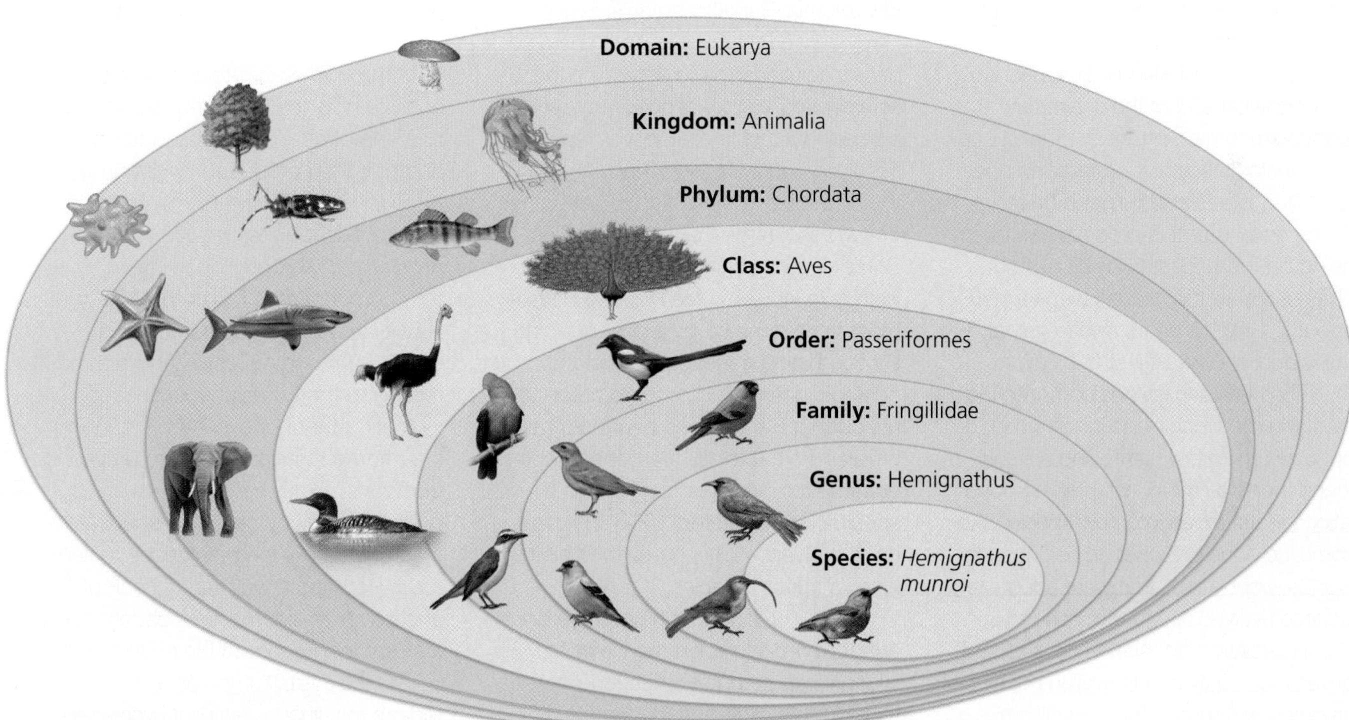

FIGURE 3.7 Taxonomists classify organisms using a hierarchical system meant to reflect evolutionary relationships. Species similar in appearance, behavior, and genetics (because they share recent common ancestry) are placed in the same genus. Organisms of similar genera are placed in the same family. Families are placed within orders, orders within classes, classes within phyla, phyla within kingdoms, and kingdoms within domains. For example, honeycreepers belong to the class Aves, along with peacocks, loons, and ostriches. However, the differences between these species, which have diverged across millions of years of evolution, are great enough that they are placed in different orders, families, and genera.

Hawaii: Species Factory and Lab of Evolution

For scientists who study how species form, no place on Earth is more fascinating and informative than the Hawaiian Islands, often called a "natural laboratory of evolution."

The key to this laboratory lies in the process that drives Hawaii's geologic history. Turn back to Figure 2.22 (Chapter 2, p. 41), and examine it closely. Deep beneath the Pacific Ocean, a volcanic "hotspot" spurts magma as the Pacific Plate slides across it in tectonic motion like a conveyor belt. Mountains of lava accumulate underwater until eventually a volcano rises above the waves, building an island. As the tectonic plate moves northwest, it carries each newly formed island with it, creating a long chain, or *archipelago*. Over several million years, each island gradually subsides, erodes, and disappears beneath the waves. As old islands disappear on the northwest end of the chain, new islands are formed on the southeast end.

Geologists analyzing radioisotopes (p. 24) in the islands' rocks have determined that this process has been going on for at least 85 million years. They estimate that Kaua'i was formed about 5.1 million years ago (mya), and the island of Hawai'i just 0.43 mya.

The Hawaiian Islands comprise the most remote archipelago in the world, but over time a few plants and animals found their way there, establishing populations that evolved into new species. As some individuals hopped to neighboring islands, populations that were adequately isolated evolved into further species. Such speciation by "island-hopping" has driven the radiation of Hawaiian honeycreepers and many other organisms.

For instance, the barren and windswept high volcanic slopes of Hawai'i are graced by some of the most striking flowering plants in the world, the silverswords (see Figure 3.1b). These spectacular plants have spiky, silvery leaves and tall stalks that explode into bloom with flowers once in the plant's long life before it dies. Researchers have discovered that Hawaii's 28 species of silverswords all evolved from a modest tarweed plant from California that reached Hawai'i and diversified by island-hopping. University of California–Berkeley botanist Bruce Baldwin and other researchers analyzed genetic relationships to determine the silverswords' history of speciation, and learned that their radiation was rapid, taking place in just 5 million years.

Dr. Heather Lerner, of Earlham College

The best-understood radiation has occurred with the Hawaiian fruit flies. Some of these insects speciate within islands in kipukas (see Figure 3.5), but most have done so by island-hopping. By combining genetic analysis and geologic dating, researchers determined that the process began 25 mya on islands that today are beneath the ocean. From a single original fruit fly species, an estimated 1000 species have evolved—fully one-sixth of all the world's fruit fly species.

Other groups that have undergone adaptive radiations on the Hawaiian Islands include damselflies, crickets, mirid bugs, spiders, and multiple families of plants. Scientists propose that once a species colonizes an island, it can often spread and evolve rapidly because competitors are few and there tend to be unoccupied niches (p. 61).

The Hawaiian honeycreepers are so diverse that researchers have long puzzled over what type of bird gave rise to their radiation—and whether there was just one colonizing ancestor or many. In 2011, to clarify how the honeycreeper radiation took place, one research team combined genetic sequencing technology with resources from museum collections and our knowledge of Hawaiian geology.

Heather Lerner and five colleagues first took tissue samples from bird specimens in museum collections. Working with Robert Fleischer and Helen James at the Smithsonian Institution, Lerner, now at Earlham College in Indiana, sampled 19 species of honeycreepers plus 28 diverse types of finches from around the Pacific Rim that experts had identified as possible ancestors.

Lerner's team obtained data from 13 genes and from mitochondrial genomes by sequencing DNA (p. 29) from each tissue sample. They ran the data through computer programs to analyze how the DNA sequences—and thus the birds—were related to one another, then produced phylogenetic trees (p. 54) showing the relationships. They published their results in the journal *Current Biology.*

Lerner's team found that the Hawaiian honeycreepers apparently derive from one ancestor, and are most related to the Eurasian rosefinches, indicating that honeycreepers evolved after some rosefinch-like bird arrived from Asia. Today's rosefinches are partly nomadic; when food supplies crash, flocks fly long distances to find food. Perhaps a wandering flock of ancestral rosefinches was caught up in a storm long ago and blown to Hawai'i.

Once this common ancestor of today's rosefinches and honeycreepers arrived on an ancient Hawaiian island, its progeny adapted to conditions there by natural selection, resulting in modified bill shape, diet, and coloration. Every once in a great while, wandering birds colonized

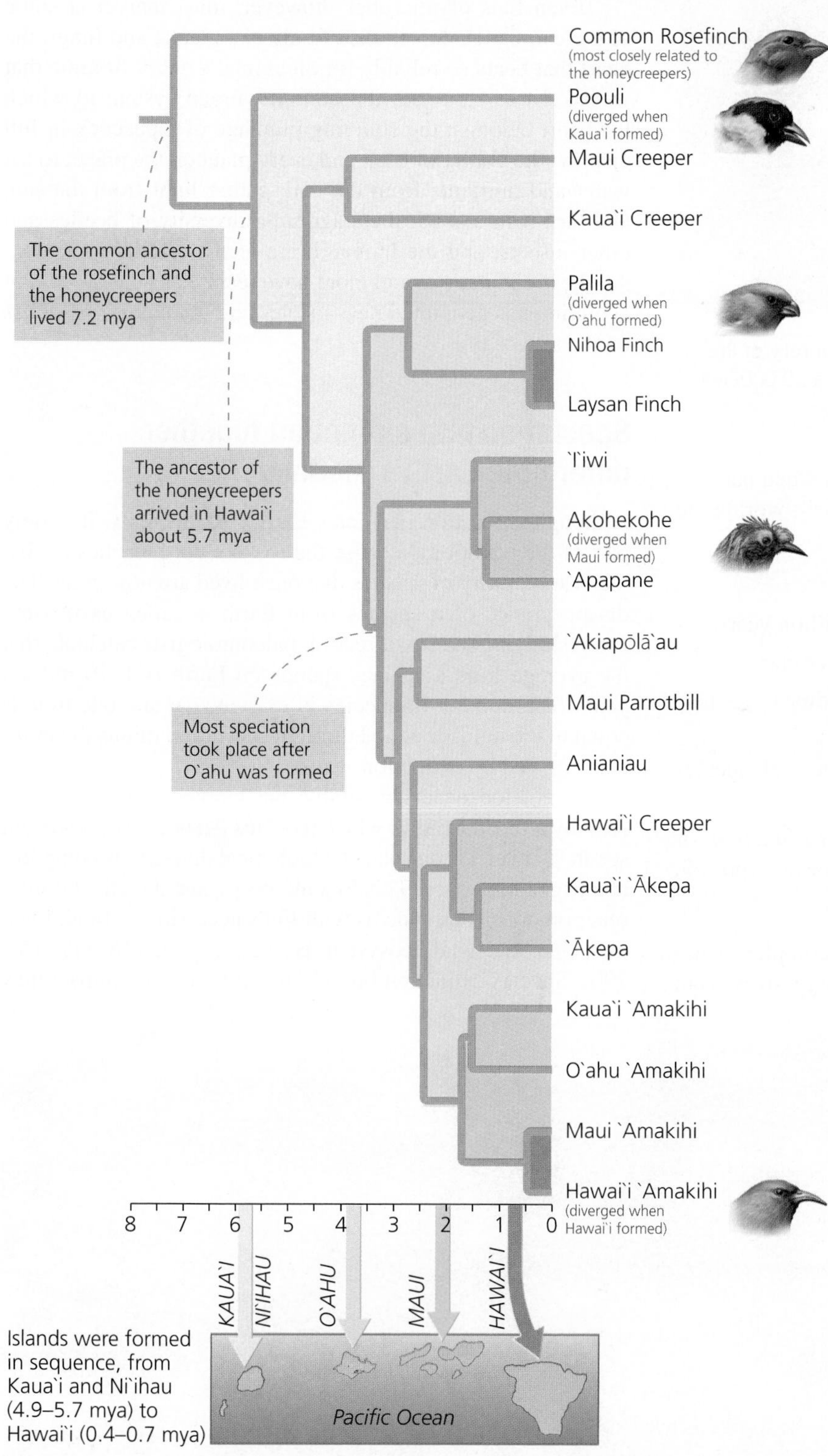

FIGURE 1 Using gene sequences, researchers generated this phylogenetic tree showing relationships among the Hawaiian honeycreepers. They then matched the history of the birds' diversification with the known geologic history of the islands' formation. *Adapted from: Lerner, H.R.L., et al. 2011. Multilocus resolution of phylogeny and timescale in the extant adaptive radiation of Hawaiian honeycreepers.* Curr Biol. *21: 1838–1844.*

other islands, founding populations that each adapted to local conditions and might eventually evolve into separate species.

Because the age of each island is known, Lerner's team could calibrate rates of evolutionary change in the DNA sequences of the birds, and thus measure the age of each divergence. That is, they could tell how "old" each bird species is. They found that the rosefinch-like ancestor arrived by 5.7 mya, about the time that the oldest of today's main islands (Kaua'i and Ni'ihau) were forming. After O'ahu emerged 4.0–3.7 mya, the speciation process went into overdrive, giving rise to many new species with distinctively different colors, bill shapes, and habits. By the time Maui arose 2.4–1.9 mya, most of the major differences in body form and appearance had evolved (**FIGURE 1**).

Thus, most major innovations arose midway through the process, when O'ahu and Kaua'i were the main islands in the chain. After this burst of innovation, major changes were fewer, perhaps because most possibilities had been explored, or perhaps because the newer islands of Maui and Hawai'i were too close together to isolate populations adequately.

The team's data show that the age of each honeycreeper species does not neatly match the age of the island(s) it inhabits today. Instead, the island-hopping process was complex, with some birds hopping "backwards" from newer islands to older ones. Moreover, within each island there is great variation in climate, topography, and vegetation, as windward slopes catch moisture from trade winds over the ocean and become lush and green, whereas leeward slopes in the rainshadow (p. 100) are arid. The varied habitats and rugged topography create barriers that can lead to speciation within islands.

For all these reasons, the "natural laboratory" of Hawai'i still has much to teach us about how the honeycreepers and other groups have evolved, and how new species are formed. ■

FIGURE 3.8 **The fossil record helps reveal the history of life on Earth.** Here, a paleontologist in India excavates a 50,000-year-old fossilized elephant skull.

who study the history of Earth's life) can learn when particular organisms lived. The cumulative body of fossils worldwide is known as the **fossil record.**

The fossil record shows that:

- Life has existed on Earth for at least 3.5 billion years.
- Earlier types of organisms evolved into later ones.
- The number of species existing at any one time has generally increased through time.
- The species living today are a tiny fraction of all species that ever lived; the vast majority are extinct.
- There have been several episodes of *mass extinction*, or simultaneous loss of great numbers of species (pp. 59, 281–283).

Across life's 3.5 billion years on Earth, complex structures have evolved from simple ones, and large sizes from small ones. However, simplicity and small size have also evolved when favored by natural selection; it is easy to argue that Earth still belongs to the bacteria and other microbes, some of them little changed over eons.

Even fans of microbes, however, must marvel at some of the exquisite adaptations of animals, plants, and fungi: the heart that beats so reliably for an animal's entire lifetime that we take it for granted; the complex organ system to which the heart belongs; the stunning plumage of a peacock in full display; the ability of each and every plant on the planet to lift water and nutrients from the soil, gather light from the sun, and turn it into food; the staggering diversity of beetles and other insects; and the human brain and its ability to reason. All these adaptations and more have resulted as the process of evolution has generated new species and whole new branches on the tree of life.

Speciation and extinction together determine Earth's biodiversity

Although speciation generates Earth's biodiversity, it is only part of the equation, because the fossil record teaches us that the vast majority of species that once lived are now gone. The disappearance of a species from Earth is called **extinction.** From studying the fossil record, paleontologists calculate that the average time a species spends on Earth is 1–10 million years. The number of species in existence at any one time is equal to the number added through speciation minus the number removed by extinction.

Extinction occurs naturally, but human impact can profoundly affect the rate at which it occurs (FIGURE 3.9). As we will see in Chapter 11, our planet's biological diversity is being lost at a frightening pace. This loss affects people directly, because other organisms provide us with life's necessities—food, fiber, medicine, and vital ecosystem services (pp. 3, 116–117, 152, 290). Species extinction brought about by human impact may

FIGURE 3.9 **Until 10,000 years ago, the North American continent teemed with "megafauna"—** mammoths, camels, giant ground sloths, lions, saber-toothed cats, and various types of horses, antelope, and bears. Nearly all these large mammals went extinct suddenly once people arrived on the continent. Similar extinctions occurred in other areas simultaneously with human arrival, suggesting to many scientists that overhunting or other human impacts were responsible. (See also Figure 11.8, p. 282.)

well be the single biggest problem we face, because the loss of a species is irreversible.

Some species are especially vulnerable to extinction

In general, extinction occurs when environmental conditions change rapidly or severely enough that a species cannot adapt genetically to the change; the slow process of natural selection simply does not have enough time to work. All manner of events can cause extinction—climate change, the arrival of new species, severe weather events, and more. In general, small populations are vulnerable to extinction because fluctuations in their size could, by chance, bring the population size to zero. Species narrowly specialized to some particular resource or way of life are also vulnerable, because environmental changes that make that resource or way of life unavailable can doom them. Species that are **endemic** to a region, meaning that they occur nowhere else on the planet, also face elevated risks of extinction because all their members belong to a single, sometimes small, population.

Island-dwelling species are frequently vulnerable. Because islands are smaller than mainland areas and are isolated by water, only some species reach islands, whereas many others never do. As a result, some of the pressures and challenges faced daily by organisms on the mainland simply don't exist on islands. For instance, only one land mammal—a bat—ever reached Hawai'i naturally, so Hawaii's birds evolved for millions of years without having to defend against the threat of predation by mammals. Likewise, Hawaii's plants did not need to protect themselves against plant-eating mammals. Because defenses are costly to invest in, most island birds and plants lost any defenses their ancestors may have had.

Eventually, people—first Polynesians and then Europeans—arrived in Hawai'i, bringing cattle, goats, pigs, rats, dogs, cats, and mongooses. Hawaii's native organisms were completely unprepared. Rats, cats, and mongooses preyed on ground-nesting seabirds, ducks, geese, an ibis, and flightless rails, driving a number of these birds extinct. Livestock ate through the vegetation, turning lush forests into desolate grasslands. All in all, half of Hawaii's native birds were driven extinct within just decades after human arrival.

On a mainland, "islands" of habitat (such as forested mountaintops) can host endemic species that are vulnerable to extinction (**FIGURE 3.10**). In the United States, many amphibians are limited to very small ranges. The Yosemite toad is restricted to a small region of the Sierra Nevada in California, the Houston toad occupies just a few areas of Texas woodland, and the Florida bog frog lives in a tiny region of Florida wetland. Fully 40 salamander species in the United States are restricted to areas the size of a typical county, and some of these live atop single mountains.

Earth has seen several episodes of mass extinction

Most extinction occurs gradually, one species at a time. The rate at which this type of extinction occurs is referred to as the *background extinction rate*. However, Earth has seen five events of staggering proportions that killed off massive numbers of species at once. These episodes, called **mass extinction events,** have occurred at widely spaced intervals in Earth's history and have wiped out 50–95% of our planet's species each time.

The best-known mass extinction occurred 66 million years ago and brought an end to the dinosaurs (although birds are modern representatives of dinosaurs). Evidence suggests that the impact of a gigantic asteroid caused this event, called the Cretaceous–Paleogene, or K–Pg, event. Still more catastrophic was the mass extinction at the end of the Permian period 250 million years ago (see **APPENDIX E** for Earth's geologic periods). Paleontologists estimate that 75–95% of all species may have perished during this event, described by one researcher as the "mother of all mass extinctions." Hypotheses as to what caused the end-Permian extinction event include an asteroid impact, massive volcanism, methane releases and global warming, or some combination of factors.

The sixth mass extinction is upon us

Many biologists have concluded that Earth is currently entering its sixth mass extinction event—and that we are the cause. Indeed, the Millennium Ecosystem Assessment (p. 15) estimated that today's extinction rate is 100–1000 times higher than the background rate, and rising (p. 283). Changes to Earth's natural systems set in motion by human population growth, development, and resource depletion have driven many species extinct and are threatening countless more. The alteration and outright destruction of natural habitats, the hunting and harvesting of species, and the introduction of species from one place to another where they can harm native

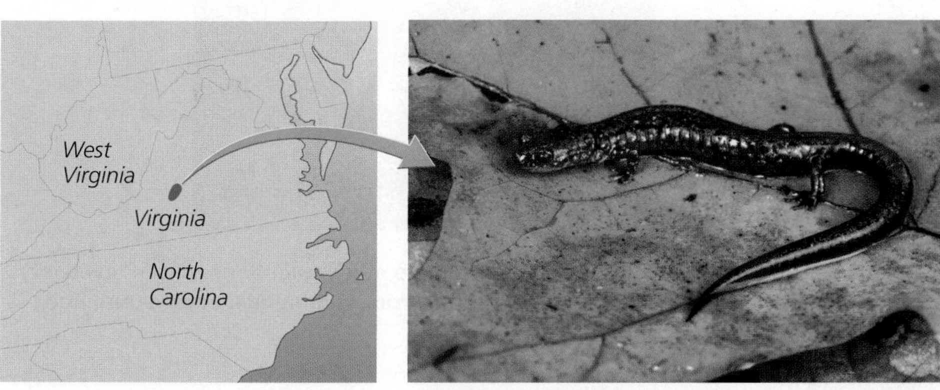

FIGURE 3.10 Small range sizes can leave species vulnerable to extinction if severe changes occur in their local environment. The Peaks of Otter salamander (*Plethodon hubrichti*) lives on only a few peaks in Virginia's Blue Ridge Mountains.

species—these processes and more have combined to threaten Earth's biodiversity (pp. 283–290).

When we look around us, it may not appear as though a human version of an asteroid impact is taking place, but we cannot judge such things on the timescale of a human lifetime. On the geologic timescale, extinction over 100 years or even 10,000 years appears instantaneous. In contrast, speciation is a slow enough process that it will take life millions of years to recover—by which time our own species will most likely not be around.

Levels of Ecological Organization

Extinction, speciation, and other evolutionary mechanisms and patterns play key roles in ecology. **Ecology** is the scientific study of the distribution and abundance of organisms, the interactions among organisms, and the relationships between organisms and their environments. It is often said that ecology provides the stage on which the play of evolution unfolds. The two are intertwined in many ways.

We study ecology at several levels

Life exists in a hierarchy of levels, from atoms, molecules, and cells (pp. 23–29) up through the **biosphere,** which is the cumulative total of living things on Earth and the areas they inhabit. Ecologists are scientists who study relationships at the higher levels of this hierarchy (FIGURE 3.11), namely at the levels of the organism, population, community, ecosystem, and biosphere.

At the level of the organism, the science of ecology describes relationships between an organism and its physical environment. Organismal ecology helps us understand, for example, what aspects of a Hawaiian honeycreeper's environment are important to it, and why. In contrast, **population ecology** examines the dynamics of population change and the factors that affect the distribution and abundance of members of a population. It helps us understand why populations of some species (such as endangered honeycreepers) decline while populations of others (such as ourselves) increase.

In ecology, a **community** consists of an assemblage of populations of interacting species that live in the same area. A population of ʻakiapōlāʻau, a population of koa trees, a population of wood-boring grubs, and a population of ferns, together with all the other interacting plant, animal, fungal, and microbial populations in the Hakalau Forest, would be considered a community. **Community ecology** focuses on patterns of species diversity and on interactions among species, ranging from one-to-one interactions to complex interrelationships involving the entire community.

Ecosystems encompass communities and the abiotic (nonliving) material and forces with which community members interact. Hakalau's cloud-forest ecosystem consists of its community plus the air, water, soil, nutrients, and energy used by the community's organisms. **Ecosystem ecology** reveals patterns, such as the flow of energy and nutrients, by studying living and nonliving components of systems in conjunction. Today's warming climate (Chapter 18) is having ecosystem-level consequences as it exerts impacts on the organisms of

Biosphere

The sum total of living things on Earth and the areas they inhabit

Ecosystem

A functional system consisting of a community, its nonliving environment, and the interactions between them

Community

A set of populations of different species living together in a particular area

Population

A group of individuals of a species that live in a particular area

Organism

An individual living thing

FIGURE 3.11 Ecologists study questions on the levels of the organism, population, community, ecosystem, and biosphere.

Hakalau and many other ecosystems throughout the world. As new technologies allow scientists to learn more about the complex dynamics of natural systems at a global scale, ecologists are increasingly expanding their horizons beyond ecosystems to the biosphere as a whole.

In the remainder of this chapter we explore ecology up through the population level. In Chapter 4 we examine community ecology, and in Chapter 5 we consider ecology at the levels of the ecosystem and biosphere.

Each organism has habitat needs

At the level of the organism, each individual relates to its environment in ways that tend to maximize its survival and reproduction. One key relationship involves the specific environment in which an organism lives, its **habitat.** A species' habitat consists of the living and nonliving elements around it, including rock, soil, leaf litter, humidity, plant life, and more. The ʻakiapōlāʻau (**FIGURE 3.12**) lives in a habitat of cool, moist, montane forest of native koa and ʻōhiʻa trees, where it is high enough in elevation to be safe from avian malaria.

Each organism thrives in certain habitats and not in others, leading to nonrandom patterns of **habitat use.** Mobile organisms actively select habitats in which to live from among the range of options they encounter, a process called **habitat selection.** In the case of plants and of rooted animals (such as sea anemones in the ocean), whose young disperse and settle passively, patterns of habitat use result from success in some habitats and failure in others.

Habitats are scale dependent. A tiny soil mite may use less than a square meter of soil in its lifetime. A vulture, elephant, or whale, in contrast, may traverse miles upon miles of air, land, or water in just a day. Species also may have different habitat needs in different seasons; many migratory birds use distinct breeding, wintering, and migratory habitats.

The criteria by which organisms favor some habitats over others can vary greatly. The soil mite may assess available habitats in terms of the chemistry, moisture, and texture of the soil and the percentage and type of organic matter. The vulture may ignore not only soil but also topography and vegetation, focusing solely on the abundance of dead animals in the area that it scavenges for food. For a whale, water temperature, salinity, and the density of marine microorganisms might be critical characteristics. Each species assesses habitats differently because each species has different needs.

FIGURE 3.12 The ʻakiapōlāʻau lives in a habitat of cool, moist, native forest on the slopes of Hawaiian volcanoes. It fills a unique niche by virtue of its odd bill, whose short, straight bottom half, and long, curved top half allows it to specialize on digging grubs out from native trees.

Habitat use is important in environmental science because the availability and quality of habitat are crucial to an organism's well-being. Indeed, because habitats provide everything an organism needs, including nutrition, shelter, breeding sites, and mates, the organism's very survival depends on the availability of suitable habitats. Often this need results in conflict with people who want to alter or develop a habitat for their own purposes.

Niche and specialization are key concepts in ecology

Another way in which an organism relates to its environment is through its niche. A species' **niche** reflects its use of resources and its functional role in a community. This includes its consumption of certain foods, its role in the flow of energy and matter, and its interactions with other organisms. The niche is a multidimensional concept, a kind of summary of everything an organism does. The pioneering ecologist Eugene Odum once wrote that "habitat is the organism's address, and the niche is its profession."

Organisms vary in the breadth of their niches. Species with narrow breadth, and thus very specific requirements, are said to be **specialists.** Those with broad tolerances, able to use a wide array of resources, are **generalists.** A native Hawaiian honeycreeper like the ʻakiapōlāʻau (see Figure 3.12) is a specialist, because its unique bill is exquisitely adapted for feeding on grubs that tunnel through the wood of native trees. In contrast, the common myna (a bird introduced to Hawaiʻi from Asia) is a generalist; its unremarkable bill allows it to eat many types of foods in many types of habitats. As a result, the common myna has spread through virtually all areas of the Hawaiian Islands where native birds have disappeared and where human development has altered the landscape. Today it is one of Hawaii's most numerous birds.

Specialists succeed over evolutionary time by being extremely good at the things they do, but they are vulnerable when conditions change and threaten the habitat or resource on which they have specialized. Generalists succeed by being able to live in many different places and to withstand variable conditions, but they may not thrive in any one situation as much as a specialist would. An organism's habitat preferences, niche, and degree of specialization each reflect adaptations of the species and are products of natural selection.

Population Ecology

Individuals of a species that inhabit a particular area make up a population. Species may consist of multiple populations that are geographically isolated from one another. This is the case with Hawaii's state bird, the nēnē, a goose that grazes in open grassy areas (see Figure 3.1d). Originally common throughout the Hawaiian Islands, the nēnē (pronounced "nay-nay") was nearly driven to extinction by human hunting; livestock and alien plants that destroyed and displaced the vegetation it fed on; and rats, cats, dogs, pigs, and mongooses that preyed on its eggs and young. Nēnēs disappeared from all islands except the island of Hawai'i, and in recent decades biologists and wildlife managers have labored to breed it in captivity and reintroduce it to protected areas on other islands. These efforts have met with success, and today nēnēs live in at least seven populations on four islands.

In contrast, the human species faces few threats from other animals, is a consummate generalist, and has spread into nearly every corner of the planet. As a result, it is difficult to define a distinct human population on anything less than the global scale. In the ecological sense of the word, all 7 billion of us comprise one population.

Populations show characteristics that help predict their dynamics

All populations—from humans to nēnēs—exhibit characteristics that help population ecologists predict the future dynamics of the population. Attributes such as density, distribution, sex ratio, age structure, and birth and death rates all help the ecologist understand how a population may grow or decline. The ability to predict growth or decline is useful in monitoring and managing threatened and endangered species (see THE SCIENCE BEHIND THE STORY, pp. 64–65). It is also useful in studying human populations (Chapter 8). Understanding human population dynamics is a central element of environmental science and is one of the prime challenges for our society today.

Population size Expressed as the number of individual organisms present at a given time, **population size** may increase, decrease, undergo cyclical change, or remain stable over time. Populations generally grow when resources are abundant and natural enemies are few. Populations can decline in response to loss of resources, negative impacts from other species, or natural disasters that kill large numbers of individuals. Researchers estimate that the nēnē population surpassed 25,000 birds before Europeans reached the Hawaiian Islands. By the 1950s, after two centuries of impacts from hunting, agriculture, non-native mammals, and invasive plants, the population was down to just 30 individuals. Since then, intensive conservation efforts have turned this decline around, and now over 2000 nēnēs live on the Hawaiian Islands.

The passenger pigeon, now extinct, illustrates the extremes of population size (FIGURE 3.13). Not long ago it was the most abundant bird in North America; flocks of passenger pigeons literally darkened the skies. In the early 1800s, ornithologist Alexander Wilson watched a flock of 2 billion birds 390 km (240 mi) long that took 5 hours to fly over and sounded like a tornado. Passenger pigeons nested in gigantic colonies in the forests of the upper Midwest and southern Canada. Once settlers arrived and began cutting the forests, however, the birds made easy targets for market hunters, who gunned down thousands at a time and shipped them to market by the wagonload. By the end of the 19th century, the passenger pigeon population had declined to such a low number that the birds could not form the large colonies they apparently needed in order to breed. In 1914, the last passenger pigeon on Earth died in the Cincinnati Zoo, bringing the continent's most numerous bird species to extinction within just a few decades.

Population density The flocks and breeding colonies of passenger pigeons showed high population density, another attribute that ecologists assess to understand populations. **Population density** describes the number of individuals in a population per unit area. High population density makes it easier for organisms to group together and find mates, but it can

(a) Passenger pigeon

(b) 19th-century lithograph of pigeon hunting in Iowa

FIGURE 3.13 **The passenger pigeon was once North America's most numerous bird.** Its flocks literally darkened the skies when millions of birds passed overhead. However, hunting and deforestation drove the species to extinction within just a few decades.

also lead to competition and conflict if space, food, or mates are in limited supply. Overcrowded organisms may become vulnerable to the predators that feed on them, and close contact among individuals can increase the transmission of infectious disease. For these reasons, organisms sometimes leave an area when densities become too high. In contrast, at low population densities, organisms benefit from more space and resources but may find it harder to locate mates and companions.

Population distribution **Population distribution** describes the spatial arrangement of organisms in an area. Ecologists define three distribution types: random, uniform, and clumped (FIGURE 3.14). In a *random distribution*, individuals are located haphazardly in no particular pattern. This type of distribution can occur when the resources an organism needs are plentiful throughout an area and other organisms do not strongly influence where members of a population settle.

A *uniform distribution* is one in which individuals are evenly spaced. This can occur when individuals hold territories or compete for space. In a desert where water is scarce, each plant needs space for its roots to gather moisture. Plants may even poison one another's roots as a means of competing for space. As a result, plants may end up growing at equal distances from one another.

In a *clumped distribution*, the pattern most common in nature, organisms arrange themselves according to the availability of the resources they need to survive. Many Hawaiian honeycreepers tend to cluster near actively flowering trees that offer nectar. Many desert plants grow in patches around isolated springs or along streambeds that flow with water after rains. Human beings also exhibit clumped distribution: People frequently aggregate in villages, towns, or cities. Clumped distributions often indicate that species are seeking certain habitats or resources that are themselves clumped.

Distributions can depend on the scale at which one measures them. At small scales a population may be distributed uniformly, yet this may occur within one patch of a larger, clumped distribution. At very large scales, all organisms show clumped or patchy distributions, because some parts of the total area they inhabit are bound to be more hospitable than others.

(a) Random: Distribution of organisms displays no pattern.

(b) Uniform: Individuals are spaced evenly.

(c) Clumped: Individuals concentrate in certain areas.

FIGURE 3.14 **Individuals in a population can spatially distribute themselves in three fundamental ways.**

Sex ratio A population's **sex ratio** is its proportion of males to females, and this can influence whether the population will increase or decrease in size over time. In monogamous species (in which each sex takes a single mate), a 1:1 sex ratio maximizes population growth, whereas an unbalanced ratio leaves many individuals of one sex without mates. Most species are not monogamous, however, so sex ratios may vary from one species to another.

Age structure Populations generally consist of individuals of different ages. **Age distribution,** or **age structure,** describes the relative numbers of organisms of each age within a population. By combining this information with data on the reproductive potential of individuals in each age class, a population ecologist can predict how the population may grow or shrink.

For many plants and animals that continue growing in size as they age, older individuals reproduce more: A tree that is large because it is old can produce more seeds, and a fish that is large because it is old may produce more eggs. In some animals, such as birds, the experience they gain with age often makes older individuals better breeders.

Human beings are unusual because we often survive past our reproductive years. A human population made up largely of older (post-reproductive) individuals will tend to decline over time, whereas one with many young people (of reproductive or pre-reproductive age) will tend to increase. We will use diagrams to explore these ideas further in Chapter 8 (pp. 197–199) as we study human population growth.

Monitoring Bird Populations at Hakalau Forest

Are populations of honeycreepers increasing or decreasing? It's a high-stakes question. The answer could tell us whether management efforts to save them are on the right track, or whether the birds may be headed for extinction.

At Hakalau Forest National Wildlife Refuge, on the rainy slope of Mauna Kea on the island of Hawai'i, biologists have been working hard for years to understand the dynamics of bird populations. But monitoring population trends is not easy, and there is still debate and uncertainty.

Established in 1985, the Hakalau Refuge is home to nine species of native forest birds, including four federally endangered ones: the Hawai'i 'ākepa, the Hawai'i creeper, the 'akiapōlā'au, and the 'io, or Hawaiian hawk. A number of non-native birds now occur here as well. Much of the region's native forest had been cleared for cattle ranching years ago, while free-roaming pigs and invasive plants had degraded the rest.

The biologists employed after 1985 to manage Hakalau fenced pigs and feral cattle out of the forest and labored to restore the area by removing invasive weeds and by raising and planting half a million native plants. Gradually the damaged forest recovered, and stands of young trees took root at higher elevations. But has the restored forest brought higher populations of native forest birds?

Beginning in 1987, federal biologists and trained observers conducted regular surveys of birds on the refuge. Following protocols commonly used by field ornithologists, they performed "point counts" by walking transects and stopping for 8 minutes at a time in predetermined spots and counting numbers of each

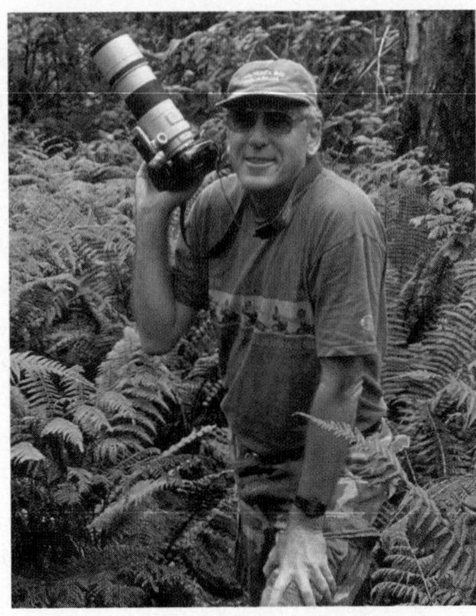

Jack Jeffrey in the Hakalau Forest

species seen or heard. With 343 points along 15 transects across the refuge, this sampling allowed them to estimate population densities for each bird species. Researchers then analyzed changes in these samples through time to make inferences about changes in population densities and population sizes.

After 21 years of point counts, refuge biologists summarized their analyses. At a 2008 workshop and in a 2009 technical report, they concluded that populations of most native birds were either stable or slowly increasing across most of the refuge.

Their report included a series of graphs similar to that shown in **FIGURE 1**. Densities of birds varied from year to year, which is normal and expected, as this may result from varying conditions (such as weather). Because of this variation, long-term studies are necessary; scientists expect that over time, random year-to-year variation will be overshadowed by true long-term trends that reveal the actual growth or decline of the population.

To interpret trends, researchers use a statistical method called *regression*. Linear regression mathematically analyzes how values change through time and determines a straight line that can be drawn through them to most accurately represent their trend. Linear regression thus places a "line of best fit" through the data points on a graph (see Figure 1).

With the data from Hakalau, linear regression led managers to conclude that over the 21-year period:

- In high-elevation pasture that managers were restoring to forest, populations of birds that used the young trees were rising sharply.
- In middle-elevation open forest being managed, populations of all native birds were either stable or increasing.

The long-term picture thus appeared bright.

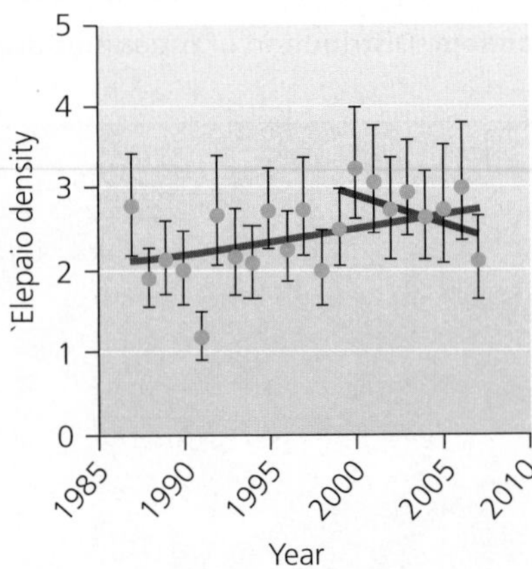

FIGURE 1 Population density data for the Hawai'i 'elepaio, a native forest bird, typify data gathered at Hakalau Forest NWR. Over a 21-year period, regression shows an increase (red line), yet over the most recent 9 years, regression shows a decrease (blue line). Thin black bars indicate 95% confidence intervals. *Data from Camp, Richard J., et al. 2009.* Passerine bird trends at Hakalau Forest National Wildlife Refuge, Hawai'i. *Hawai'i Cooperative Studies Unit Technical Report HCSU-011.*

DATA Q What is the trend in the data during the first 12 years of the period shown? Can you draw a line through the first 12 data points that you feel best represents this trend?

However, during the most recent 9 years of the 21-year period, many of the populations had decreased. Regressions run on the last 9 years alone showed apparent declines for many species (see Figure 1).

Biologist Leonard Freed of the University of Hawai'i at Manoa argued that federal biologists were overemphasizing the positive long-term trends and ignoring the negative near-term trends. For years, Freed and his colleagues had conducted research within Hakalau, focusing on the breeding biology of the Hawai'i 'ākepa (**FIGURE 2a**). Their research suggested that the 'ākepa began suffering competition for food once the non-native Japanese white-eye (**FIGURE 2b**) became abundant in the forest. This competition, along with attacks from parasitic lice in the nest, stunted the growth of young birds, Freed maintained, and threatened the 'ākepa population. Freed urged that white-eyes be trapped and killed in order to save the 'ākepa. Refuge biologists called Freed's results controversial and said they needed validation.

The debate came to a head in a pair of papers published back-to-back in the ornithological journal *Condor* in 2010. The team of federal biologists, led by Richard Camp and including Jack Jeffrey, presented their data and acknowledged that many populations showed downward trends in the most recent 9 years. Freed and his colleague Rebecca Cann reanalyzed the federal data using alternative methods and questioned whether the earlier apparent increases were reliable.

At stake is how to manage the forest and its birds. If Camp and his colleagues are right, then management actions taken so far seem to have been effective, boosting populations or holding them stable in the face of dire threats. If Freed and Cann are right, then management strategies may need to be rethought. Researchers on all sides are debating the issues diligently because they care deeply about the forest and its birds and are trying their utmost to save them during a time of crisis.

Many factors could account for the apparent recent declines in populations of 'ākepas and other native honeycreepers, including the simple fact that thicker forest vegetation has made it harder for counters to detect birds. Of concern, though, is the possibility that challenges from outside the refuge—such as malaria and pox being driven upslope by climate warming—might eventually overwhelm even the best management efforts.

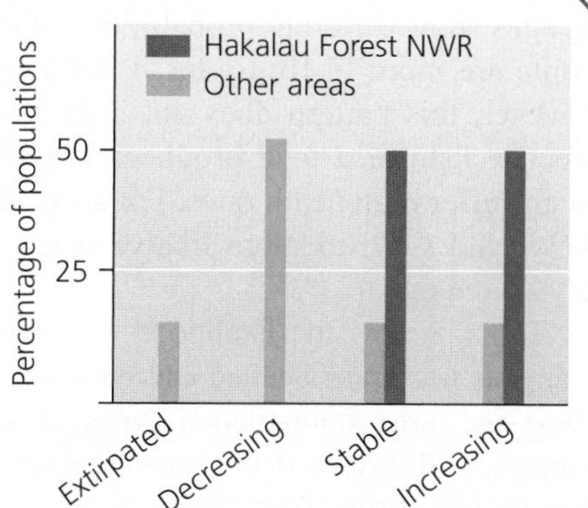

FIGURE 3 At Hakalau Forest NWR, native forest bird populations were judged to be stable or increasing, whereas at four other protected areas on the island of Hawai'i, most populations were judged to be decreasing, and some have recently vanished. *Data from Camp, Richard J., et al. 2009.* Passerine bird trends at Hakalau Forest National Wildlife Refuge, Hawai'i. *Hawai'i Cooperative Studies Unit Technical Report HCSU-011.*

Today researchers continue to survey Hakalau's birds, adding to their valuable long-term database of population trends. But government budget cuts are threatening their ability to analyze the data, as well as their capacity to safeguard the refuge. Amid cuts in funding and staffing, pigs broke through fences and began degrading the newly restored forest. Most funding was recently reinstated, but each time budget cuts are made, refuge staff must work hard at greater expense just to regain the progress previously made.

Despite the apparent recent declines in bird populations at Hakalau, populations there seem to be faring better than elsewhere on the island of Hawai'i (**FIGURE 3**). Moreover, the reforestation of Hakalau's upper zone is creating new habitat into which birds are moving. This success is a hopeful sign that research and careful management can help undo past damage and preserve endangered island species. ■

(a) Hawai`i `ākepa

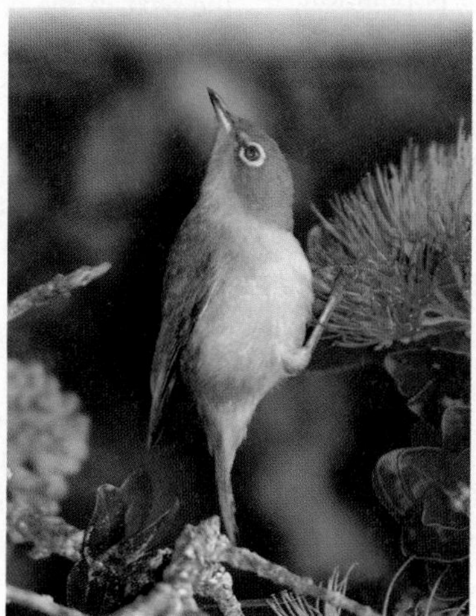

(b) Japanese white-eye

FIGURE 2 Work by biologist Leonard Freed suggests that the `ākepa is suffering competition from the non-native Japanese white-eye.

Birth and death rates All the preceding factors can influence the rates at which individuals within a population are born and die. Just as individuals of differing ages have different reproductive capacities, individuals of differing ages show different probabilities of dying. For instance, people are more likely to die at old ages than young ages. However, this pattern does not hold for all organisms. An insect, a fish, or a toad produces large numbers of young, which suffer high death rates. For such animals, death is less likely (and survival more likely) at an older age than at a very young age.

To show how the likelihood of survival varies with age, ecologists use graphs called **survivorship curves** (FIGURE 3.15). There are three fundamental types of survivorship curves. Humans, with higher death rates at older ages, show a type I survivorship curve. Toads, with highest death rates at young ages, show a type III survivorship curve. A type II survivorship curve is intermediate and indicates equal rates of death at all ages. Many birds are thought to show type II curves.

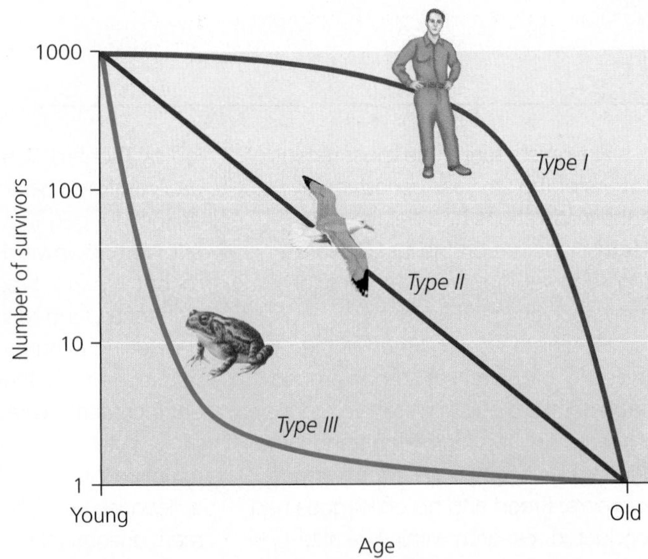

FIGURE 3.15 Survivorship curves show how an individual's likelihood of survival varies with age. In a type I survivorship curve, survival rates are high when organisms are young and decrease sharply when organisms are old. In a type II survivorship curve, survival rates are equivalent regardless of an organism's age. In a type III survivorship curve, most mortality takes place at young ages, and survival rates are greater at older ages.

DATA Q Which organism has the highest rate of survival at a young age: a toad, a bird, or a human being?

Populations may grow, shrink, or remain stable

Now that we have outlined some key attributes of populations, we are ready to take a quantitative view of population change by examining some simple mathematical concepts used by population ecologists and by **demographers** (scientists who study human populations). Population growth, or decline, is determined by four factors:

- Births within the population (*natality*)
- Deaths within the population (*mortality*)
- **Immigration** (arrival of individuals from outside the population)
- **Emigration** (departure of individuals from the population)

Births and immigration add individuals to a population, whereas deaths and emigration remove individuals. A convenient way to express rates of birth and death is to measure the number of births and deaths per 1000 individuals per year. These rates are termed the *crude birth rate* and the *crude death rate*.

If we are not interested in the effects of migration, we can measure the **rate of natural increase** by subtracting the crude death rate from the crude birth rate:

$$\text{(crude birth rate)} - \text{(crude death rate)} = \text{rate of natural increase}$$

The rate of natural increase reflects the degree to which a population is growing or shrinking as a result of its own internal factors.

To obtain an overall **population growth rate,** the total rate of change in a population's size per unit time, we must also take into account the effects of migration. Thus, we include terms for immigration and emigration (each expressed per 1000 individuals per year) in the formula, as follows:

$$\text{(crude birth rate} - \text{crude death rate)} + \text{(immigration rate} - \text{emigration rate)} = \text{population growth rate}$$

The resulting number tells us the net change in a population's size per 1000 individuals per year. For example, a population with a crude birth rate of 18 per 1000/yr, a crude death rate of 10 per 1000/yr, an immigration rate of 5 per 1000/yr, and an emigration rate of 7 per 1000/yr would have a population growth rate of 6 per 1000/yr:

$$(18/1000 - 10/1000) + (5/1000 - 7/1000) = 6/1000$$

Thus, a population of 1000 in one year will reach 1006 in the next. If the population is 1,000,000, it will reach 1,006,000 the next year. Such population increases are often expressed as percentages, which we can calculate using the following formula:

$$\text{population growth rate} \times 100\%$$

Thus, a growth rate of 6/1000 would be expressed as:

$$6/1000 \times 100\% = 0.6\%$$

By measuring population growth in terms of percentages, scientists can compare increases and decreases in species that have far different population sizes. They can also project changes that will occur in the population over longer periods, much like you might calculate the amount of interest your savings account will earn over time.

Unregulated populations increase by exponential growth

When a population increases by a fixed percentage each year, it is said to undergo **exponential growth.** Imagine you put money in a savings account at a fixed interest rate and leave it untouched for years. As the principal accrues interest and grows larger, you earn still more interest, and the sum grows by escalating amounts each year. The reason is that a fixed percentage of a small number makes for a small increase, but the same percentage of a large number produces a large increase. Thus, as savings accounts (or populations) become larger, each incremental increase likewise gets larger. Such acceleration is a characteristic of exponential growth.

We can visualize changes in population size by using population growth curves. The J-shaped curve in FIGURE 3.16 shows exponential growth. Populations of organisms increase exponentially unless they meet constraints. Each organism reproduces by a certain amount, and as populations get larger, there are more individuals reproducing by that amount. If there are adequate resources and no external limits, ecologists theoretically expect exponential growth.

Normally, exponential growth occurs in nature only when a population is small, competition is minimal, and environmental conditions are ideal for the organism in question. Most often, these conditions occur when the organism is introduced to a new environment that contains abundant resources to exploit. Mold growing on a piece of fruit, or bacteria colonizing a recently dead animal, are cases in point. Plants colonizing regions during primary succession (p. 85) after glaciers recede or volcanoes erupt may also grow exponentially. In Hawai'i, many of the species that colonized the islands from other locations underwent exponential growth for a time after their arrival. One current example of exponential growth in the mainland United States is the Eurasian collared dove (see Figure 3.16). Unlike its extinct relative the passenger pigeon, this species arrived here from Europe, thrives in human-disturbed areas, and has spread across North America in a matter of years.

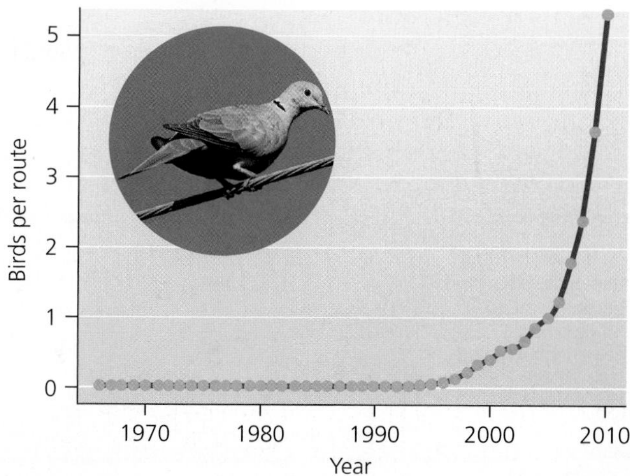

FIGURE 3.16 A population may grow exponentially for a time when colonizing an unoccupied environment or exploiting an unused resource. The Eurasian collared dove is currently spreading across the North American continent, propelled by exponential growth. *Data from Sauer, J.R., et al., 2011.* The North American Breeding Bird Survey, results and analysis 1966–2009. *v. 3.23.2011. USGS Patuxent Wildlife Research Center, Laurel, MD.*

Limiting factors restrain population growth

Exponential growth rarely lasts long. If even a single species were to increase exponentially for very many generations, it would blanket the planet's surface! Instead, every population eventually is constrained by **limiting factors**—physical, chemical, and biological attributes of the environment that restrain population growth. These limiting factors determine the **carrying capacity,** the maximum population size of a species that a given environment can sustain.

Ecologists use the S-shaped curve in FIGURE 3.17 to show how an initial exponential increase is slowed and eventually brought to a standstill by limiting factors. Called the **logistic growth curve,** it rises sharply at first but then begins to level off as the effects of limiting factors become stronger. Eventually the collective force of these factors stabilizes the population size at its carrying capacity.

We can witness this process by taking a closer look at the data for the Eurasian collared dove population, as gathered by thousands of volunteer birders and analyzed by government biologists through the Breeding Bird Survey, a long-running citizen science project. The dove first reached North America in Florida a few decades ago and subsequently spread north and west. Today its numbers are growing fastest in western areas it has recently reached, and slower in eastern areas where it has been present for longer. In Florida, it has apparently reached carrying capacity (FIGURE 3.18). Populations of other European birds that spread across North America in the past, such as the house sparrow and European starling, have peaked and are today beginning to decline.

Many factors influence a population's growth rate and carrying capacity. For animals in terrestrial environments, limiting factors include temperature extremes; prevalence of disease; abundance of predators; and the availability of food, water, mates, shelter, and suitable breeding sites. Plants are often limited by amounts of sunlight and moisture and the type of soil chemistry, in addition to disease and attack from plant-eating animals. In aquatic systems, limiting factors include salinity, sunlight, temperature, dissolved oxygen, fertilizers, and pollutants. To determine limiting factors, ecologists may conduct

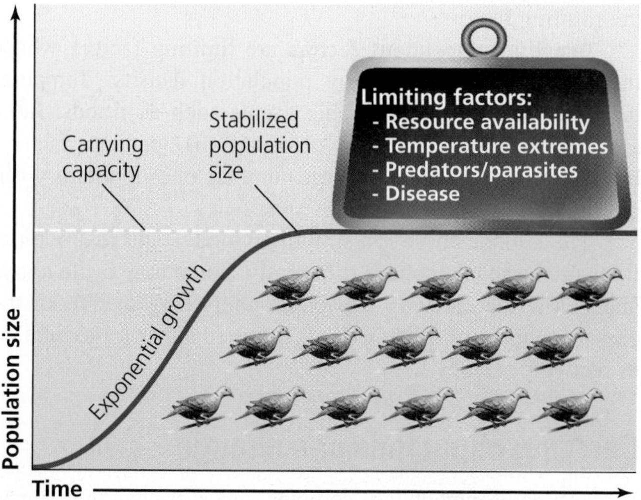

FIGURE 3.17 The logistic growth curve shows how population size may increase rapidly at first, then grow more slowly, and finally stabilize at a carrying capacity. Carrying capacity is determined both by the *biotic potential* of the organism and by various external limiting factors.

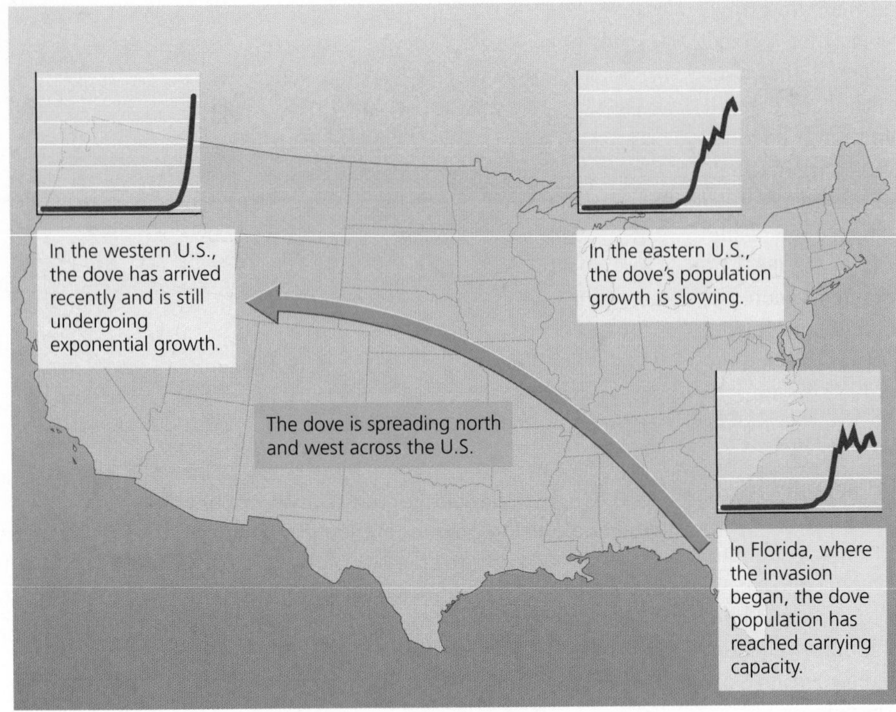

FIGURE 3.18 Exponential growth slows down over time and gives way to logistic growth. By breaking down the continent-wide data for the Eurasian collared dove from Figure 3.16, we can track its spread to the north and west of Florida, where it first arrived. Today its population growth is fastest in the west, slower in the east (where the species has been present for longer), and stable in Florida (where it has apparently reached carrying capacity). *Data from Sauer, J.R., et al., 2011.* The North American Breeding Bird Survey, results and analysis 1966–2009. *v. 3.23.2011. USGS Patuxent Wildlife Research Center, Laurel, MD.*

DATA Q Looking ahead several decades into the future, what do you predict the population growth graph for the western United States will look like?

experiments in which they increase or decrease a hypothesized limiting factor and observe its effects on population size.

The influence of some factors depends on population density

A population's density can enhance or diminish the impact of certain limiting factors. Recall that high population density can help organisms find mates but can also increase competition and the risk of predation and disease. Such factors are said to be **density-dependent factors,** because their influence rises and falls with population density. The logistic growth curve in Figure 3.17 represents the effects of density dependence. The larger the population size, the stronger the effects of the limiting factors.

Density-independent factors are limiting factors whose influence is not affected by population density. Temperature extremes and catastrophic events such as floods, fires, and landslides are examples of density-independent factors, because they can eliminate large numbers of individuals without regard to their density.

The logistic curve is a simplified model, and real populations in nature can behave differently. Some may cycle above and below the carrying capacity. Others may overshoot the carrying capacity and then crash, destined either for extinction or recovery (**FIGURE 3.19**).

Carrying capacities can change

Because environments are complex and ever-changing, carrying capacity can vary. If a fire destroys a forest, for example, the carrying capacities for most forest animals will decline, whereas carrying capacities for species that benefit from fire will increase. Our own species has proved capable of intentionally altering our environment so as to raise our carrying capacity. When our ancestors began to build shelters and use fire for heating and cooking, they eased the limiting factors of cold climates and were able to expand into new territory. As human civilization has developed, we have overcome limiting factors time and again through the development of new technologies and cultural institutions. People have managed so far to increase the planet's carrying capacity for our species, but we have done so by appropriating immense proportions of the planet's resources. In the process, we have reduced carrying capacities for countless other organisms that rely on those same resources.

WEIGHING THE ISSUES

CARRYING CAPACITY AND HUMAN POPULATION GROWTH

The global human population has passed 7 billion, and we have far exceeded our planet's historic carrying capacity for people. Name some specific means by which you think we have accomplished this. What limiting factors exist for the human population today? Do you think we can continue to raise our carrying capacity? Might Earth's future carrying capacity for us decrease? Why or why not?

Reproductive strategies vary among species

Limiting factors from an organism's environment help to regulate its population size, but the attributes of the organism also matter. For example, organisms differ in their *biotic potential*, or capacity to produce offspring. A fish with a short gestation period that lays thousands of eggs at a time has high biotic potential, whereas a whale with a long gestation period that gives birth to a single calf at a time has low biotic potential.

Giraffes, elephants, humans, and other large animals with low biotic potential produce relatively few offspring during their lifetimes. Species that take this approach to reproduction

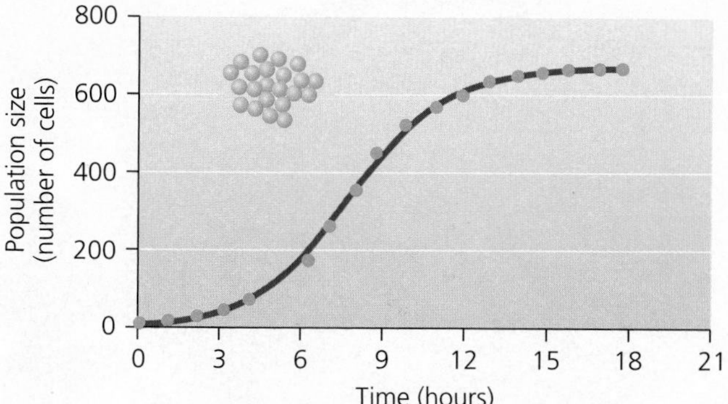

(a) Yeast cells, *Saccharomyces cerevisiae*

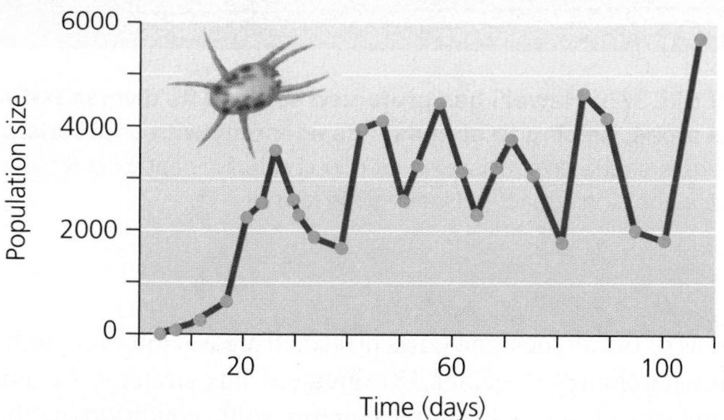

(b) Mite, *Eotetranychus sexmaculatus*

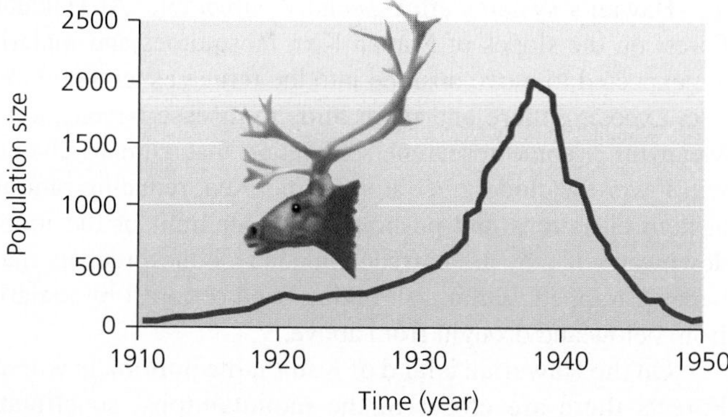

(c) St. Paul reindeer, *Rangifer tarandus*

FIGURE 3.19 Population growth in nature may depart from the logistic growth curve in various ways. Yeast cells from an early lab experiment show logistic growth **(a)** that closely matches the theoretical model. Some organisms, such as the mite shown here, show cycles **(b)** in which population fluctuates above and below the carrying capacity. Populations that rise too fast and deplete resources may crash just as suddenly **(c)**, such as the population of reindeer introduced to the Bering Sea island of St. Paul.
Data from: (a) Pearl, R., 1927. The growth of populations. Quarterly Review of Biology *2: 532–548; (b) Huffaker, C.B., 1958. Experimental studies on predation: Dispersion factors and predator-prey oscillations.* Hilgardia *27: 343–383, Figure 7, © 1958 by the Regents of the University of California; (c) Adapted from Scheffer, Victor B., 1951. The rise and fall of a reindeer herd.* The Scientific Monthly *73: 356–362, Fig. 1. Reprinted with permission from AAAS.*

require a long time to gestate and raise their young, but the considerable energy and resources they devote to caring for and protecting them helps give these few offspring a high likelihood of survival. Such species are said to be **K-selected** (because their populations tend to stabilize over time near carrying capacity, commonly abbreviated *K*). Because their populations stay close to carrying capacity, these organisms must compete to hold their own in a crowded world. Thus, natural selection favors investing in high-quality offspring that can be good competitors.

In contrast, species that are **r-selected** have high biotic potential and devote their energy and resources to producing many offspring in a short time. Their offspring do not require parental care after birth, so r-strategists simply leave their survival to chance. The abbreviation *r* denotes the per capita rate at which a population increases in the absence of limiting factors. Population sizes of r-selected species fluctuate greatly, such that they are often well below carrying capacity. This is why natural selection in these species favors traits that lead to rapid population growth. Many fish, plants, frogs, insects, and others are r-selected.

It is important to note, however, that *these are two extremes on a continuum* and that most species fall somewhere between the extremes of r-selected and K-selected species. Moreover, many organisms show combinations of traits that do not correspond to a place on the continuum. A redwood tree, for instance, is large and long-lived, yet it produces many small seeds and offers no parental care.

Conserving Biodiversity

Environmental changes that affect populations have been taking place as long as life has existed, but today human development, resource extraction, and population pressure are speeding the rate of change and altering the types of change. Science is crucial in helping us understand how we modify our environment. However, the threats to biodiversity have complex social, economic, and political roots, so environmental scientists recognize that we must also understand these aspects if we are to develop sustainable solutions.

Fortunately, millions of people around the world are taking action to safeguard biodiversity and to preserve and restore Earth's ecological and evolutionary processes. We will explore these efforts more fully in our discussion of biodiversity and conservation biology in Chapter 11. For now, let us see how Hawaiians have been confronting the challenges to their biodiversity.

Introduced species pose challenges for native populations and communities

On top of our direct effects on populations, communities, and ecosystems, human beings exert many indirect effects—for instance, by introducing species into areas where they do not occur naturally. Some such introduced species thrive in their new surroundings, killing or displacing native species (pp. 88–89, 286–289). Island species are particularly vulnerable to introduced species: They have evolved in isolation in small areas with a limited community of other species, and so they lack defenses against mainland species that are well adapted to deal with a broad array of enemies.

The Hawaiian Islands have been utterly transformed by impacts from introduced species. Cattle, goats, sheep, and pigs eat native vegetation, endangering plant populations and altering entire landscapes. Alien grasses, shrubs, and trees spread across the landscapes that livestock have altered. Rat, cats, dogs, and mongooses eat the eggs and young of ground-nesting birds with impunity, and have driven a number of them extinct. Forest birds suffering already from predation and habitat loss now also struggle against diseases like pox and malaria. Pigs have made the malaria problem worse, because they dig holes in the forest floor, where rainwater forms shallow pools in which mosquitoes breed.

As a result, biologists and land managers have found that trying to help a species in trouble often means trying to eradicate or control populations of another that is doing too well. For instance, in many areas pigs are being hunted and pig-free areas are being fenced off. However, pigs are clever animals, and some usually escape. Moreover, native Hawaiian people who hunt pigs have no incentive to get rid of every last pig; otherwise, they could no longer hunt them.

FIGURE 3.20 Hawai'i has protected some of its diverse natural areas, helping to stimulate its economy with ecotourism. Here, a scuba diver observes raccoon butterflyfish at a coral reef along the Kona coast of Hawaii's Big Island.

Innovative solutions are working

Amid all the challenges of Hawaii's extinction crisis, hard work is resulting in some inspirational success stories, and several species have been saved from imminent extinction already. At Hakalau Forest, ranchland is being restored to forest, invasive plants are being removed and native ones are being planted, and nēnē are being protected while new populations of them are being established.

Elsewhere across Hawai'i, tracts of public land are being managed with similar goals and techniques, and some private landholders have joined in conservation efforts. Early work at Hawai'i Volcanoes National Park inspired the work at Hakalau, as well as efforts by managers and volunteers from the Hawai'i Division of Forestry and Wildlife, The Nature Conservancy of Hawai'i, Kamehameha Schools, and local watershed protection groups. People are protecting land, removing alien mammals and weeds, and restoring native habitats. Offshore, conservation efforts are gaining steam as well, as Hawaiians strive to protect their fabulous coral reefs, seagrass beds, and beaches from pollution and overfishing. The northwesternmost Hawaiian Islands are now part of the largest federally declared marine reserve (p. 443) in the world.

Hawai'i and its citizens are reaping benefits from their conservation efforts—economic benefits as well as ecological ones. The islands' wildlife and natural areas draw tourists from around the world, a phenomenon called **ecotourism** (FIGURE 3.20). A large percentage of Hawaii's tourism is ecotourism, and tourism as a whole draws more than 7 million visitors to Hawai'i each year, provides thousands of jobs to Hawaiians, and pumps $12 billion annually into the state's economy.

Climate change now poses an extra challenge

Traditionally, people sought to conserve populations of threatened species by preserving and managing tracts of land (or areas of ocean) designated as protected areas. However, global climate change (Chapter 18) threatens this strategy. As temperatures climb and rainfall patterns shift, conditions within protected areas may turn unsuitable for the species they were meant to protect.

Hawaii's systems are especially vulnerable. At Hakalau Forest on the slopes of Mauna Kea, mosquitoes and malaria are expected to move upslope into the refuge as temperatures rise, exposing more and more birds to disease (FIGURE 3.21). Meanwhile, some researchers maintain that climate change will lower the cloud layer atop Mauna Kea, reducing rainfall at high elevations and pushing the upper limit of the forest downward. If they are correct, Hakalau's honeycreepers may become trapped within a shrinking band of forest by malaria from below and drought from above.

On the Hawaiian island of Kaua'i, the outlook is worse: Forests there are closer to the mountaintops, so climate warming is expected to shift the forests upward until they vanish, leaving their inhabitants nowhere to go. Already mosquitoes have moved upslope and bird populations are diminishing. Two honeycreeper species, the 'akeke'e and the 'akikiki, were recently added to the Endangered Species List (p. 296).

WEIGHING THE ISSUES

HOW BEST TO CONSERVE BIODIVERSITY? Most people view national parks and ecotourism as excellent ways to help keep ecological systems intact. Yet plenty of native Hawaiian creatures face declining populations and the threat of extinction despite living within a reserve, and climate change and disease pay no heed to park boundaries. What lessons can we learn from this about the conservation of biodiversity? Are parks and preserves sufficient? What other approaches might we pursue to save declining species?

(a) Today

(b) With 2°C of climate warming

FIGURE 3.21 **Researchers have modeled how a warming climate will affect the native birds of Hakalau Forest NWR.** Avian malaria cannot survive where temperatures dip below 13°C, and it peaks where summer temperatures average 17°C. Today **(a)**, 24% of Hakalau lies above (cooler than) the 13°C isotherm and is free of malaria. If climate warms by 2°C, however **(b)**, then the isotherms move upslope, and only 1% of Hakalau will remain cooler than 13°C and malaria-free. *Data from: Benning, T.L., et al. 2002. Interactions of climate change with biological invasions and land use in the Hawaiian Islands: Modeling the fate of endemic birds using a geographic information systems.* Proc Natl. Acad. Sci. *99: 14246–14249.*

The challenges of climate change mean that scientists and managers need to come up with new ways to help save declining populations. We will learn about the many efforts being made across the world in our exploration of biodiversity and conservation biology in Chapter 11. In Hawai'i, it remains to be seen how effectively management and ecotourism can stem the tide of challenges and help preserve natural systems in the long term. Resources and efforts to preserve habitat and protect endangered species will likely need to be stepped up. Programs to restore altered communities to their former condition—as is being done at Hakalau Forest—will also be necessary. The restoration of ecological communities is one phenomenon we will examine in our next chapter, as we shift from populations to communities.

Conclusion

The honeycreepers of Hakalau Forest National Wildlife Refuge, along with many other Hawaiian species, have helped to illuminate the fundamentals of evolution and population ecology that are integral to environmental science. The evolutionary processes of natural selection, speciation, and extinction help determine Earth's biodiversity. Understanding how ecological processes function at the population level is crucial to protecting biodiversity threatened by the mass extinction event that many biologists maintain is now underway. Population ecology also informs the study of human populations (Chapter 8), another key endeavor in environmental science.

Reviewing Objectives

You should now be able to:

Explain natural selection and cite evidence for this process

- Because organisms produce excess young, individuals vary in their traits, and many traits are inherited, some individuals will prove better at surviving and reproducing. Their genes will be passed on and become more prominent in future generations. (p. 50)
- Mutations and recombination provide the genetic variation for natural selection. (p. 50)
- We have produced our pets, farm animals, and crop plants by artificial selection. (p. 52)

Describe how evolution influences biodiversity

- Natural selection can act as a diversifying force as species adapt to their environments in myriad ways. (pp. 52–53)
- Speciation by geographic isolation (or other means) produces new species. (pp. 53–54)
- The branching patterns of phylogenetic trees reflect the historical pattern in which lineages of organisms have diverged. (p. 54)

- The fossil record informs us about life's history (pp. 55, 58)

Discuss reasons for species extinction and mass extinction events

- Extinction may occur when species that are highly specialized or that have small populations encounter rapid environmental change. (p. 59)
- Earth's life has experienced five known episodes of mass extinction, due to asteroid impact and possibly volcanism and other factors. (p. 59)
- Today, human impact may be initiating a sixth mass extinction. (pp. 59–60)

List the levels of ecological organization

- Ecologists study phenomena on the organismal, population, community, and ecosystem levels—and, increasingly, at the level of the biosphere. (pp. 60–61)
- Habitat, niche, and specialization are important ecological concepts. (p. 61)

Outline the characteristics of populations that help predict population growth

- Populations are characterized by population size, population density, population distribution, sex ratio, and age structure. (pp. 62–63)
- Birth and death rates, as well as immigration and emigration, determine how a population will grow or decline. (p. 66)

Assess logistic growth, carrying capacity, limiting factors, and other fundamental concepts in population ecology

- Populations unrestrained by limiting factors will undergo exponential growth. (pp. 66–67)
- Logistic growth describes the effects of density-dependent limiting factors; growth slows as population size increases, and population size levels off at a carrying capacity. (p. 67)
- Carrying capacity is the maximum size a population can attain over the long term in a given environment. (pp. 67–68)
- K-selection and r-selection describe theoretical endpoints in how organisms can allocate growth and reproduction. (pp. 68–69)

Identify efforts and challenges involved in the conservation of biodiversity

- Social and economic factors influence our impacts on natural systems in complex ways. (p. 69)
- Introduced species are one of many impacts that affect native species and systems, particularly on islands. (pp. 69–70)
- Extensive efforts to protect and restore species and habitats are needed to prevent further erosion of biodiversity. (p. 70)
- Climate change is challenging the effectiveness of protected areas. (pp. 70–71)

Testing Your Comprehension

1. Explain the premises and logic that support the concept of natural selection.
2. Describe two examples of evidence for natural selection.
3. Describe the steps involved in allopatric speciation.
4. Name three organisms that have become extinct or are threatened with extinction. For each, give a probable reason for its decline.
5. What is the difference between a species and a population? Between a population and a community?
6. Define and contrast the concepts of habitat and niche.
7. List and describe each of the five major population characteristics discussed. Briefly explain how each shapes population dynamics.
8. Can a species undergo exponential growth forever? Explain your answer.
9. Describe how limiting factors relate to carrying capacity.
10. Explain the difference between K-selected species and r-selected species. For each, give an example that was not mentioned in the chapter.

Seeking Solutions

1. In what ways has artificial selection changed people's quality of life? Give examples. How might artificial selection be used to improve our quality of life further? Can you envision a way it could be used to reduce our environmental impact?
2. In your region, what species are threatened with extinction? What reasons lie behind their endangerment? Suggest steps that could be taken to bolster their populations.
3. Do you think the human species can continue raising its global carrying capacity? How so, or why not? Do you

think we *should* try to keep raising our carrying capacity? Why or why not?

4. Describe some of the challenges facing native species of plants and animals in Hawai'i. What steps have been taken at Hakalau Forest NWR and elsewhere to address these challenges? What steps do you think biologists and managers will need to take in the future to safeguard native Hawaiian species, populations, and communities?
5. What are some advantages of ecotourism for a state like Hawai'i? Can you think of any potential disadvantages?
6. **THINK IT THROUGH** You are a population ecologist studying animals in a national park, and park managers are asking for advice on how to focus their limited conservation funds. How would you rate the following three species, from most vulnerable (and thus most in need of conservation attention) to least vulnerable? Give reasons for your choices.
   - A bird with an even (1:1) sex ratio that is a habitat generalist
   - A salamander endemic to the park that lives in high-elevation forest
   - A fish that specializes on a few types of invertebrate prey and has a large population size

Calculating Ecological Footprints

Americans love their coffee, whether it comes from Hawaii's Kona region or elsewhere in the world. In 2012, coffee consumption in the United States neared 3.8 billion pounds (out of 19.2 billion pounds produced globally). Next to petroleum, coffee is the most valuable (legal) commodity on the world market, and the United States is its leading importer. Given this data, estimate coffee consumption rates in the table below.

Most coffee in Kona and elsewhere is produced in large plantations, where coffee is the only tree species and is grown in full sun where natural forests have been cut. However, approximately 2% of coffee is produced in smaller groves where coffee trees are intermingled with other species under a partial canopy. These shade-grown coffee plantations maintain greater habitat diversity for birds and other tropical rainforest wildlife.

	Population	Pounds of coffee per day	Pounds of coffee per year
You (or the average American)	1	0.34	12.6
Your class			
Your hometown			
Your state			
United States			

Data from International Coffee Organization.

1. What percentage of global coffee production is consumed in the United States? If U.S. coffee drinkers consumed only shade-grown coffee, how much would shade-grown production need to increase to meet demand?
2. How much extra would you be willing to pay per pound for shade-grown coffee as opposed to standard coffee, if you knew that your money would help to prevent habitat loss for forest-dwelling birds, such as the many songbirds that migrate between Latin America and North America?
3. If everyone in the United States were willing to pay as much extra per pound for shade-grown coffee as you are, how much additional money would that generate for biodiversity conservation in the tropics each year?

MasteringENVIRONMENTALSCIENCE™

STUDENTS

Go to **MasteringEnvironmentalScience** for assignments, the etext, and the Study Area with practice tests, videos, current events, and activities.

INSTRUCTORS

Go to **MasteringEnvironmentalScience** for automatically graded activities, current events, videos, and reading questions that you can assign to your students, plus Instructor Resources.

4

Workers use a pressure hose to remove zebra mussels from the pump room of a power plant near Detroit.

Species Interactions and Community Ecology

Upon completing this chapter, you will be able to:

- Compare and contrast the major types of species interactions
- Characterize feeding relationships and energy flow, using them to construct trophic levels and food webs
- Distinguish characteristics of a keystone species
- Characterize disturbance, succession, and notions of community change
- Perceive and predict the potential impacts of invasive species in communities
- Explain the goals and methods of restoration ecology
- Describe biomes and identify the terrestrial biomes of the world

CENTRAL CASE STUDY

Black and White, and Spread All Over: Zebra Mussels Invade the Great Lakes

"We are seeing changes in the Great Lakes that are more rapid and more destructive than any time in [their] history."

—Andy Buchsbaum, National Wildlife Federation

"When you tear away the bottom of the food chain, everything that is above it is going to be disrupted."

—Tom Nalepa, National Oceanic and Atmospheric Administration

Things had been looking up for the Great Lakes. The pollution-fouled waters of Lake Erie and the other Great Lakes shared by Canada and the United States had become cleaner in the years following the Clean Water Act of 1970 and a binational agreement in 1972. As government regulation brought industrial discharges under control, people once again began to use the lakes for recreation, and populations of fish rebounded.

Then the zebra mussel arrived. Black-and-white-striped shellfish the size of a dime, zebra mussels attach to hard surfaces and feed on algae by filtering water through their gills. This mollusk, given the scientific name *Dreissena polymorpha*, is native to the Caspian Sea, Black Sea, and Azov Sea in western Asia and eastern Europe. In 1988, it was discovered in North American waters at Lake St. Clair, which connects Lake Erie with Lake Huron. People had brought it to this continent by accident when ships arriving from Europe discharged ballast water containing the mussels or their larvae into the Great Lakes.

Within just two years of their discovery, zebra mussels had multiplied and reached all five of the Great Lakes. The next year, these invaders entered New York's Hudson River to the east, and the Illinois River at Chicago to the west. From the Illinois River and its canals, they soon reached the Mississippi River, giving them access to a vast watershed covering 40% of the United States. In just three more years, they spread to 19 U.S. states and two Canadian provinces. By 2010, they had colonized waters in 30 U.S. states.

How could a mussel spread so quickly? The zebra mussel's larval stage is well adapted for long-distance dispersal. Its tiny larvae drift freely for several weeks, traveling as far as the currents take them. Adults that attach themselves to boats and ships may be transported from one place to another, even to isolated lakes and ponds well away from major rivers. Moreover, in North America the mussels encountered none of the particular species of predators, competitors, and parasites that had evolved with them in the Old World and limited their population growth there.

Zebra mussels clog water intake pipes at factories, power plants, municipal water supplies, and wastewater treatment facilities (**FIGURE 4.1a**). At one Michigan power plant, workers counted 700,000 mussels per square meter of pipe. Great densities of these organisms can damage boat engines, degrade docks, foul fishing gear, and sink buoys that ships use for navigation. Through such impacts, zebra mussels cost Great Lakes economies an estimated $5 billion in the first decade of the invasion, and they continue to impose costs of hundreds of millions of dollars each year.

Zebra mussels also have severe impacts on ecological systems. They eat primarily **phytoplankton:** microscopic photosynthetic algae, protists, and cyanobacteria that drift in open water. Because each mussel filters a liter or more of water every day, zebra mussels consume so much phytoplankton that they can deplete populations. Phytoplankton is the foundation of the Great Lakes food web, so its depletion is bad news for **zooplankton,** the tiny aquatic animals that eat phytoplankton—and for the fish that eat both. Researchers are finding that water bodies with zebra mussels contain fewer zooplankton and open-water fish than water bodies without them. Zebra mussels also suffocate native mollusks by attaching to their shells (**FIGURE 4.1b**).

However, zebra mussels benefit many bottom-feeding invertebrates and fish. By filtering algae and organic matter

(a) Zebra mussels clog water intake pipes of power plants and industrial facilities.

(b) Zebra mussels foul, starve, and suffocate native clams by adhering to their shells and sealing them shut.

FIGURE 4.1 **Invasive species can have severe economic and ecological impacts.**

from open water and depositing nutrients in feces that sink, they shift the community's nutrient balance to the bottom and benefit species that feed there. Once they clear the water, sunlight penetrates more deeply, spurring the growth of large-leafed underwater plants and algae. In the long term, however, eutrophication (pp. 108–109, 412) may ensue, bringing harm to the system.

In recent years, scientists have documented a surprising new twist: One invader is being displaced by another. The quagga mussel (*Dreissena bugensis*), a close relative of the zebra mussel from Ukraine, is spreading through the Great Lakes and beyond. This species, named after an extinct zebra-like animal, is replacing the zebra mussel in many locations.

In a second twist, today there are signs that the zebra mussel invasion may be losing steam. In a number of areas, zebra mussel populations have begun declining, including in part of Lake Huron, portions of the Mississippi River, the St. Croix River in Minnesota, Lake of the Ozarks in Missouri, and other locations. Some declines are due to displacement by quagga mussels, but for others the reasons are unknown. In some cases, predation by fish or ducks is being credited for driving down zebra mussel numbers, as native predators develop a taste for the invader. A similar process has already occurred in Europe, where waterfowl began preying heavily on zebra mussels and contributed to population crashes.

In areas where zebra mussels are declining, some of the native fish and invertebrates that suffered from their arrival are now recovering. Ecologists are monitoring populations closely to see how the situation develops. No one expects zebra mussels to disappear or be eradicated, but there is now hope that some of their impacts might be reversed. ■

Species Interactions

Interactions among species are the threads in the fabric of ecological communities. By interacting with many species in a variety of ways, zebra mussels have set in motion an array of changes in the communities they have invaded. To understand why invasive species introduced by people can cause so much disruption, we must first look at how species naturally interact. Ecologists organize species interactions into several fundamental categories (TABLE 4.1).

TABLE 4.1 Species Interactions: Effects on Their Participants

Type of interaction	Effect on Species 1	Effect on Species 2
Mutualism	+	+
Predation, parasitism, herbivory	+	–
Competition	–	–

"+" denotes a positive effect; "–" denotes a negative effect.

Competition can occur when resources are limited

When multiple organisms seek the same limited resource, their relationship is said to be one of **competition.** Competing organisms do not usually fight with one another directly and physically. Competition is generally more subtle and indirect, taking place as organisms vie with one another to procure resources. Such resources include food, water, space, shelter, mates, sunlight, and more. Competitive interactions can take place between members of the same species, called **intraspecific competition,** or between members of different species, called **interspecific competition.**

If individuals of the same species are competing for limited resources, then competition becomes more intense when there are more individuals per unit area (denser populations). This is density dependence (p. 68), and it can limit the growth of a population.

Whereas intraspecific competition is a population-level phenomenon, interspecific competition affects communities. Interspecific competition can give rise to different types of outcomes. If one species is a very effective competitor, it may

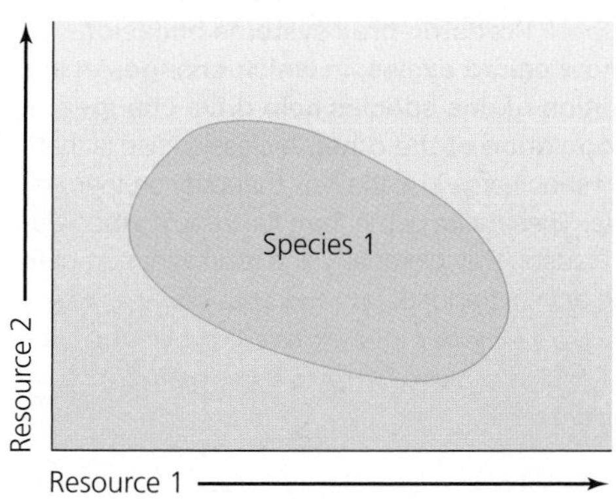

(a) Fundamental niche

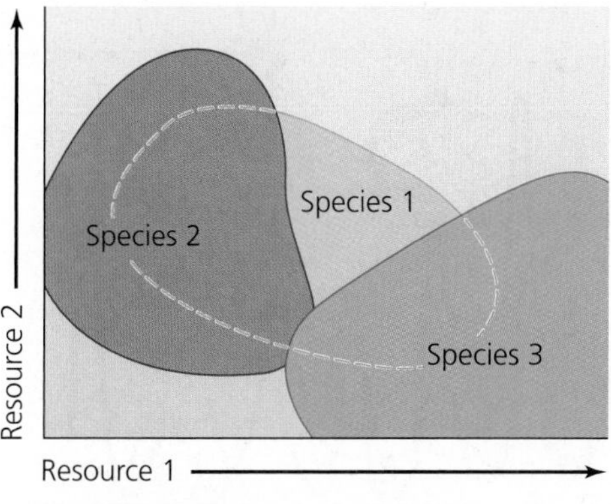

(b) Realized niche

FIGURE 4.2 Competition may force an organism to play a more limited ecological role or use fewer resources than it would in the absence of its competitor. With no competitors **(a),** an organism can exploit its full fundamental niche. When competitors restrict what an organism can do or what resources it can use **(b),** the organism is limited to a realized niche, which covers only part of its fundamental niche.

exclude another species from resource use entirely. This outcome, called **competitive exclusion,** occurred in parts of the Great Lakes as zebra mussels displaced native mussels—and it is happening now as quagga mussels displace zebra mussels.

Alternatively, if neither competitor fully excludes the other, the species may continue to live side by side. This result, called **species coexistence,** may produce a stable point of equilibrium, at which the relative population sizes of each remain fairly constant through time.

Coexisting species that use the same resources tend to adjust to their competitors to minimize competition with them. Individuals may do this by changing their behavior so as to use only a portion of the total array of resources they are capable of using. In such cases, individuals do not fulfill their entire niche. A species' niche reflects its use of resources and its functional role in a community, including its habitat use, food consumption, and other attributes (p. 61)—it is a kind of multidimensional summary of everything an organism does.

The full niche of a species is called its **fundamental niche** (FIGURE 4.2a). An individual that plays only part of its role or uses only some of its resources because of competition or other types of species interactions is said to display a **realized niche** (FIGURE 4.2b), the portion of its fundamental niche that is actually "realized," or fulfilled. The quagga mussel can occupy a wider range of water conditions and substrates than the zebra mussel, so it is thought to have a larger fundamental niche. The quagga mussel also appears to be reducing the zebra mussel's realized niche, as it displaces the zebra mussel in many areas.

Species experience similar adjustments over evolutionary time. Over many generations, the process of natural selection (pp. 50–53) may respond to competition by favoring individuals that use slightly different resources or that use shared resources in different ways. If two bird species eat the same type of seeds, individuals that prefer eating larger or smaller seeds might minimize competition and thereby survive and reproduce more effectively. If the seed-eating tendencies are inherited, then these preferences may be passed on to offspring, and over time natural selection may drive one species to specialize on larger seeds and the other to specialize on smaller seeds. Natural selection might also favor one bird species becoming more active in the morning and the other more active in the evening, thus minimizing interference. This process is called **resource partitioning** because the species partition, or divide, the resources they use in common by specializing in different ways (FIGURE 4.3).

Resource partitioning can lead to **character displacement,** in which competing species come to diverge in their physical characteristics because of the evolution of traits best suited to the range of resources they use. For birds that specialize on eating larger seeds, natural selection may favor the evolution of larger bills that enable them to make best use of this resource, whereas for birds specializing on smaller seeds, smaller bills may be favored. This is precisely what extensive recent research has revealed about the finches from the Galápagos Islands that were first described by Charles Darwin (p. 50).

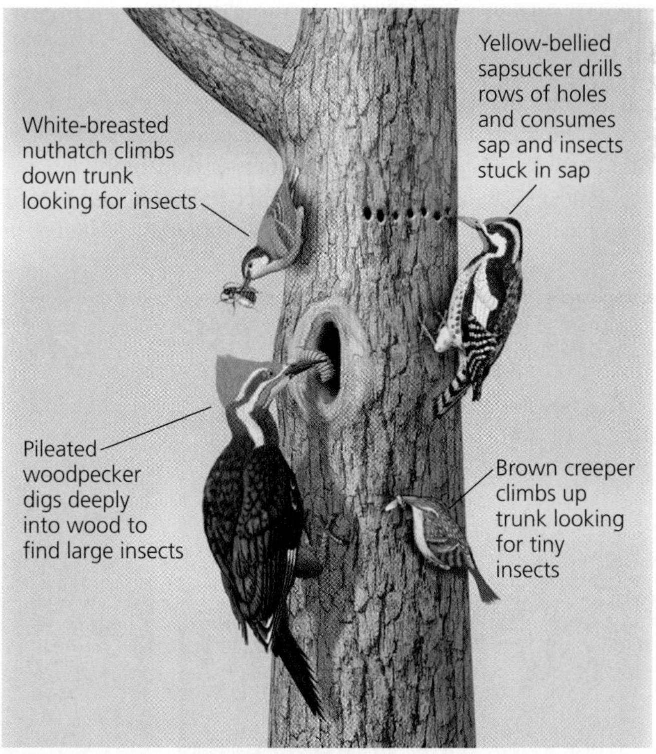

FIGURE 4.3 When species compete, they may partition resources, each specializing on a slightly different resource or way of attaining a shared resource. Many types of birds forage for insects on tree trunks, but they use different portions of the trunk and seek different foods in different ways.

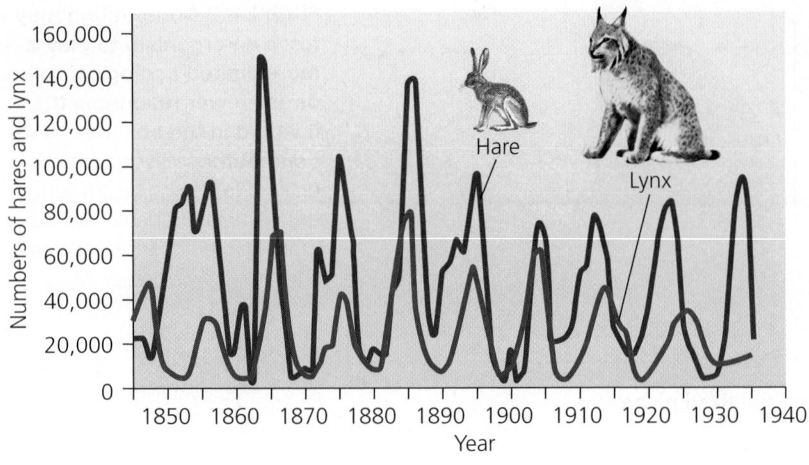

FIGURE 4.4 Predator–prey systems occasionally show paired cycles, in which changes in a population of one species help drive changes in a population of the other. A classic case is that of snowshoe hares and the lynx that prey on them in Canada. These data come from fur-trapping records of the Hudson Bay Company and represent numbers of each animal trapped. *Data from MacLulich, D.A., 1937. Fluctuation in the numbers of varying hare (*Lepus americanus*).* University of Toronto Studies in Biology Series 43, *Toronto: University of Toronto Press.*

Several types of interactions are exploitative

In competitive interactions, each participant exerts a negative effect on other participants, because each takes resources the others could have used. This is reflected in the two minus signs shown for competition in Table 4.1. In other types of interactions, some participants benefit while others are harmed; that is, one species exploits the other (note the +/– interactions in Table 4.1). Such exploitative interactions include predation, parasitism, and herbivory.

Predators kill and consume prey

Every living thing needs to procure food, and for most animals, that means eating other living organisms. **Predation** is the process by which individuals of one species—the **predator**—hunt, capture, kill, and consume individuals of another species, the **prey.** Interactions among predators and prey structure the food webs we will examine shortly, and they help shape community composition by influencing the relative numbers of predators and prey.

Predation by zebra mussels on phytoplankton has reduced phytoplankton populations by up to 90%, according to studies in the Great Lakes and Hudson River. Zebra mussels also consume the smaller types of zooplankton, and this predation has diminished zooplankton populations by up to 70% in Lake Erie and the Hudson River. Most predators are also prey, however. Zebra mussels have become a food source for muskrats; crayfish; a number of North American diving ducks; and a variety of fish, including flounder, sturgeon, eels, carp, and freshwater drum.

Predation can sometimes drive cyclical population dynamics. An increase in the population size of prey creates more food for predators, which may survive and reproduce more effectively as a result. As the predator population rises, intensified predation drives down the population of prey. Diminished numbers of prey in turn cause some predators to starve, and the predator population declines. This allows the prey population to rise again, starting the cycle anew (**FIGURE 4.4**).

Predation also has evolutionary ramifications. Individual predators that are more adept at capturing prey will likely live longer, healthier lives and be better able to provide for their offspring. Natural selection (pp. 50–53) will thereby lead to the evolution of adaptations (p. 50) that make predators better hunters. Prey face an even stronger selective pressure—the risk of immediate death. As a result, predation pressure has driven the evolution of an elaborate array of defenses against being eaten (**FIGURE 4.5**).

(a) Crypsis

(b) Warning coloration

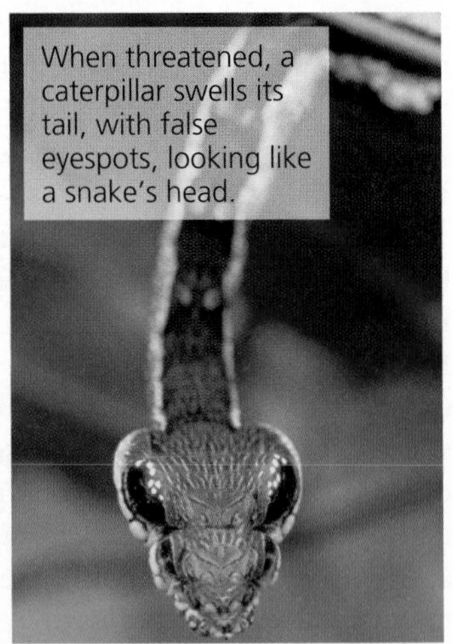

(c) Mimicry

FIGURE 4.5 Natural selection to avoid predation has resulted in fabulous adaptations. Some prey species use cryptic coloration **(a)** to blend into their background. Others are brightly colored **(b)** to warn predators they are toxic, distasteful, or dangerous. Others use mimicry **(c)** to fool predators.

FIGURE 4.6 A parasite benefits at the expense of its host. Parasitic sea lampreys gain nourishment by sucking blood from fish in the Great Lakes.

Parasites exploit living hosts

Organisms can exploit other organisms without killing them. **Parasitism** is a relationship in which one organism, the **parasite,** depends on another, the **host,** for nourishment or some other benefit while doing the host harm. Unlike predation, parasitism usually does not result in an organism's immediate death.

Many types of parasites live inside their hosts. For example, tapeworms live in their hosts' digestive tracts, robbing them of nutrition. Other types of parasites are free-living and come into contact with their hosts infrequently. Cuckoos of Eurasia and cowbirds of the Americas lay their eggs in other birds' nests and let the host species raise the parasite's young. Still other parasites live on the exterior of their hosts, such as fleas that bite their hosts, or ticks that attach themselves to the skin. The sea lamprey is a tube-shaped vertebrate that grasps the bodies of fish with a suction-cup mouth and a rasping tongue, sucking their blood for days or weeks (FIGURE 4.6). Sea lampreys invaded the Great Lakes from the Atlantic Ocean after people dug canals to connect the lakes for shipping, and the lampreys soon devastated economically important fisheries of chubs, lake herring, whitefish, and lake trout. Fisheries managers have since reduced lamprey populations by applying chemicals that selectively kill lamprey larvae.

Many insects parasitize other insects, killing them in the process, and are called **parasitoids.** Parasitoid wasps lay eggs on caterpillars, and when the eggs hatch, the wasp larvae burrow into the caterpillar's tissues and slowly consume them. The wasp larvae metamorphose into adults and fly from the body of the dying caterpillar.

Parasites that cause disease in their hosts are called **pathogens.** Common human pathogens include the protists that cause malaria and amoebic dysentery, the bacteria that cause pneumonia and tuberculosis, and the viruses that cause hepatitis and AIDS.

Just as predators and prey evolve in response to one another, so do parasites and hosts. **Coevolution** describes a long-term reciprocal process in which two (or more) types of organisms repeatedly respond by natural selection to the other's adaptations. This seesawing process can occur with any type of species interaction. Hosts and parasites often become locked in a duel of escalating adaptations, known as an *evolutionary arms race*. Like rival nations racing to stay ahead of one another in military technology, host and parasite may repeatedly evolve new responses to the other's latest advance. In the long run, though, it may not be in a parasite's best interest to do its host too much harm. A parasite might leave more offspring in the next generation—and thus be favored by natural selection—if it allows its host to live longer.

Herbivores exploit plants

In **herbivory,** animals feed on the tissues of plants. Insects that feed on plants are the most common type of herbivore; nearly every plant in the world is attacked by insects (FIGURE 4.7). Herbivory generally does not kill a plant outright but may affect its growth and reproduction.

Like animal prey, plants have evolved an impressive arsenal of defenses against the animals that feed on them. Many plants produce chemicals that are toxic or distasteful to herbivores. Others arm themselves with thorns, spines, or irritating hairs. In response, herbivores evolve ways to overcome

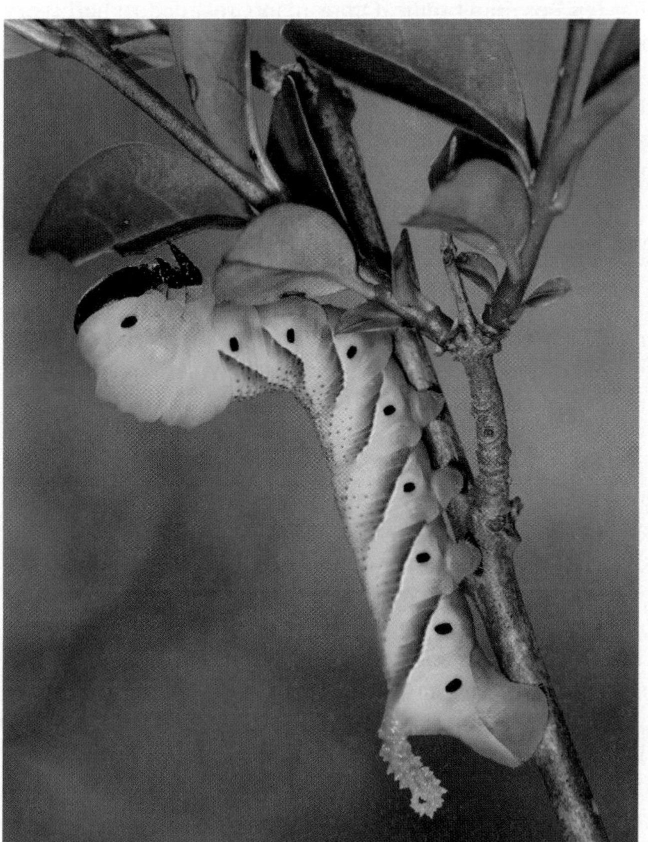

FIGURE 4.7 In herbivory, animals feed on plants. This larva (caterpillar) of the death's head hawk moth is feeding on leaves in Europe.

FIGURE 4.8 **In mutualism, organisms of different species benefit one another.** Hummingbirds visit flowers to gather nectar, and in the process they transfer pollen between flowers, helping the plant to reproduce.

these defenses, and the plant and the animal may embark on an evolutionary arms race.

Some plants recruit certain animals as allies to assist in their defense. Such plants may encourage ants to take up residence by providing swelled stems for the ants to nest in, or nectar-bearing structures for them to feed from. In return, the ants protect the plant by attacking insects that land or crawl on it. Other plants respond to herbivory by releasing volatile chemicals when they are bitten or pierced. The airborne chemicals attract predatory insects that may attack the herbivore. Such cooperative strategies—essentially trading food for protection—are examples of mutualism.

Mutualists help one another

Unlike exploitative interactions, **mutualism** is a relationship in which two or more species benefit from interacting with one another. Generally each partner provides some resource or service that the other needs.

Many mutualistic relationships—like many parasitic relationships—occur between organisms that live in close physical contact. Physically close association is called **symbiosis,** and symbiosis can be either mutualistic or parasitic. (Indeed, biologists hypothesize that many mutualistic associations evolved from parasitic ones.) Thousands of terrestrial plant species depend on mutualisms with fungi; plant roots and some fungi together form symbiotic associations called mycorrhizae. In these relationships, the plant provides energy and protection to the fungus while the fungus helps the plant absorb nutrients from the soil. In the ocean, coral polyps, the tiny animals that build coral reefs (p. 431), share beneficial arrangements with algae known as zooxanthellae (p. 431). The coral provide housing and nutrients for the algae in exchange for a steady supply of food that the algae produce through photosynthesis (p. 32).

You, too, are engaged in a symbiotic mutualism. Your digestive tract is filled with microbes that help you digest food and carry out other bodily functions—microbes that you are providing a place to live. Without these mutualistic microbes, none of us would survive for long.

Not all mutualists live in close proximity. **Pollination** (FIGURE 4.8), an interaction vital to agriculture and our food supply (p. 254), involves free-living organisms that may encounter each other only once. Bees, birds, bats, and other creatures transfer pollen (containing male sex cells) from flower to flower, fertilizing ovaries (containing female sex cells) that grow into fruits with seeds. Most pollinating animals visit flowers for their nectar, a reward the plant uses to entice them. The pollinators receive food, and the plants are pollinated and reproduce. Various types of bees pollinate 73% of our crops, one expert has estimated—from soybeans to potatoes to tomatoes to beans to cabbage to oranges.

Ecological Communities

A **community** is an assemblage of populations of organisms living in the same area at the same time (as we saw in Figure 3.11, p. 60). Members of a community interact with one another in the ways discussed above, and these species interactions have indirect effects that ripple outward to affect other community members. The strength of interactions also varies, and together species' interactions help determine the structure, function, and species composition of communities. Community ecology (p. 60) is the scientific study of species interactions and the dynamics of communities. Community ecologists study which species coexist, how they interact, how communities change through time, and why these patterns occur.

Energy passes among trophic levels

Some of the most important interactions among community members involve who eats whom. As organisms feed on one another, matter and energy move through the community from one **trophic level,** or rank in the feeding hierarchy, to another (FIGURE 4.9).

Producers **Producers,** or **autotrophs** ("self-feeders"), comprise the first trophic level. Terrestrial green plants, cyanobacteria, and algae capture solar energy and use photosynthesis to produce sugars (p. 32). The chemosynthetic bacteria of hot springs and deep-sea hydrothermal vents use geothermal energy in a similar way to produce food (p. 33).

Consumers Organisms that consume producers are known as **primary consumers** and comprise the second trophic level. Herbivorous grazing animals, such as deer and grasshoppers, are primary consumers. The third trophic level consists of **secondary consumers,** which prey on primary consumers. Wolves that prey on deer are considered

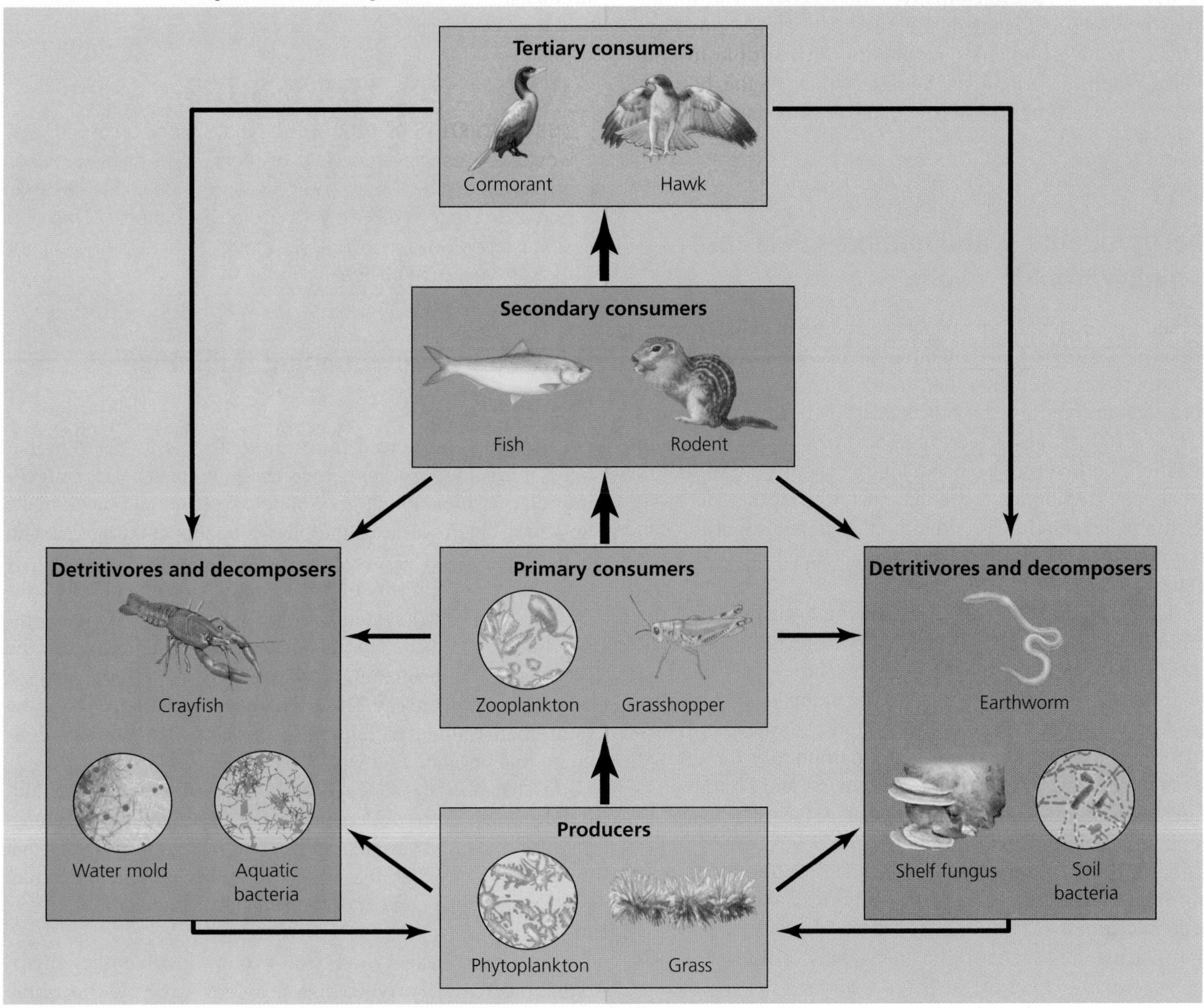

FIGURE 4.9 Ecologists organize species hierarchically by trophic level. The diagram shows aquatic (**left**) and terrestrial (**right**) examples at each level. Arrows indicate the direction of energy flow. Producers generate food by photosynthesis, primary consumers (herbivores) feed on producers, secondary consumers eat primary consumers, and tertiary consumers eat secondary consumers. Detritivores and decomposers feed on nonliving organic matter and "close the loop" by returning nutrients to the soil or the water column for use by producers.

secondary consumers, as are rodents and birds that prey on grasshoppers. Predators that feed at still higher trophic levels are known as **tertiary consumers.** Examples of tertiary consumers include hawks and owls that eat rodents that have eaten grasshoppers. Note that most primary consumers are **herbivores** because they consume plants, whereas secondary and tertiary consumers are **carnivores** because they eat animals. Animals that eat both plant and animal food are referred to as **omnivores.**

Detritivores and decomposers Detritivores and decomposers consume nonliving organic matter. **Detritivores,** such as millipedes and soil insects, scavenge the waste products or dead bodies of other community members. **Decomposers,** such as fungi and bacteria, break down leaf litter and other nonliving matter into simpler constituents that can be taken up and used by plants. These organisms enhance the topmost soil layers (p. 218) and play essential roles as the community's recyclers, making nutrients from organic matter available for reuse by living members of the community.

In Great Lakes communities, phytoplankton are the main producers, floating freely and conducting photosynthesis using sunlight that penetrates the upper layer of the water. Zooplankton are primary consumers, feeding on the phytoplankton. Phytoplankton-eating fish are primary consumers, and zooplankton-eating fish are secondary consumers. Tertiary consumers include larger fish and birds that feed on plankton-eating fish. (The left side of Figure 4.9 shows these relationships in a very generalized form.) Zebra mussels and quagga mussels, by eating both phytoplankton and zooplankton, function on multiple trophic levels. When an organism dies and sinks to the bottom, detritivores scavenge its tissues and decomposers recycle its nutrients.

Energy, biomass, and numbers decrease at higher trophic levels

At each trophic level, organisms use energy in cellular respiration (p. 32) to grow and maintain themselves. More energy goes toward maintenance than to building new tissues, and most ends up being given off as heat. Only a small amount of the energy is transferred to the next trophic level through predation, herbivory, or parasitism. A general rule of thumb is that each trophic level contains just 10% of the energy of the trophic level below it (although the actual proportion can vary greatly). This pattern can be visualized as a pyramid (FIGURE 4.10).

This pyramid-like pattern also tends to hold for the numbers of organisms at each trophic level; in general, fewer organisms exist at higher trophic levels than at lower ones. A grasshopper eats many plants in its lifetime, a rodent eats many grasshoppers, and a hawk eats many rodents. Thus, for every hawk in a community there must be many rodents, still more grasshoppers, and an immense number of plants. Moreover, because the difference in numbers of organisms among trophic levels tends to be large, the same pyramid-like relationship often holds true for **biomass,** the collective mass of living matter in a given place and time.

The pyramid pattern illustrates why eating at lower trophic levels—being vegan or vegetarian, for instance—decreases a person's ecological footprint. Each amount of meat or other animal product we eat requires the input of a considerably greater amount of plant material (see Figure 10.9, p. 249). Thus, when we eat animal products, we use up far more energy per calorie that we gain than when we eat plant products.

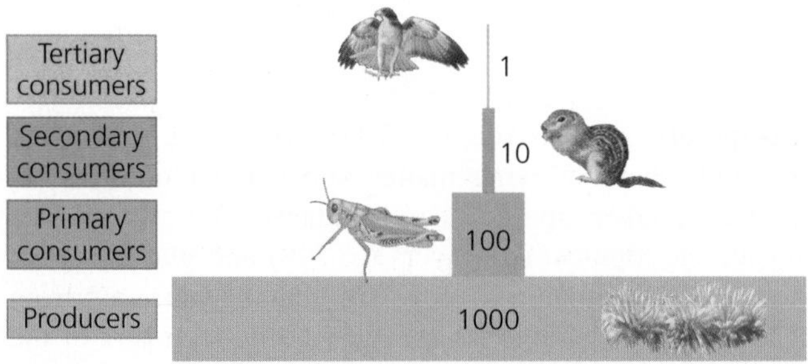

FIGURE 4.10 Lower trophic levels generally contain more organisms, energy content, and biomass than higher trophic levels. The tenfold ratio shown here is typical, but the shape of the pyramid may vary greatly.

DATA **Q** Using the ratios shown in this example, let's suppose that a system has 3000 grasshoppers. How many rodents would be expected?

WEIGHING THE ISSUES

THE FOOTPRINTS OF OUR DIETS What proportion of your diet would you estimate consists of meat, milk, eggs, or other animal products? Would you be willing to decrease this proportion in order to reduce your ecological footprint? Describe some other ways in which you could reduce your footprint through your food choices.

Food webs show feeding relationships and energy flow

As energy is transferred from lower trophic levels to higher ones, it is said to pass up a **food chain,** a linear series of feeding relationships. Plant, grasshopper, rodent, and hawk make up a food chain—as do phytoplankton, zooplankton, fish, and fish-eating birds.

Thinking in terms of food chains is conceptually useful, but ecological systems are far more complex than simple linear chains. A more accurate representation of the feeding relationships in a community is a **food web**—a visual map of energy flow that uses arrows to show the many paths along which energy passes as organisms consume one another.

FIGURE 4.11 shows a food web from a temperate deciduous forest of eastern North America. It is greatly simplified and leaves out the vast majority of species and interactions that occur. Note, however, that even within this simplified diagram we can pick out a number of food chains involving different sets of species.

A Great Lakes food web would involve the phytoplankton that photosynthesize near the water's surface, the zooplankton that eat them, fish that eat phytoplankton and zooplankton, larger fish that eat the smaller fish, and lampreys that parasitize the fish. It would include a number of native mussels and clams and, since 1988, the zebra mussels and quagga mussels that are displacing them. It would include diving ducks that formerly fed on native bivalves and now prey on the mussels.

This food web would also show that crayfish and other *benthic* (bottom-dwelling) invertebrate animals feed from the refuse of the non-native mussels. Although the mussels' waste promotes bacterial growth and pathogens that introduce disease to native bivalves, it also provides nutrients that nourish many benthic invertebrates. Finally, the food web would include underwater plants and macroscopic algae, whose growth is enhanced as the non-native mussels filter out phytoplankton, allowing sunlight to penetrate deeply into the water column. (Jump ahead to Figure 4.15a, p. 89, for an illustration of some of these effects.)

Overall, zebra and quagga mussels alter the Great Lakes food web by shifting productivity from open-water regions to benthic and *littoral* (nearshore) regions. In so doing, the mussels help benthic and littoral fishes and make life harder for open-water fishes (see THE SCIENCE BEHIND THE STORY, pp. 86–87).

FIGURE 4.11 Food webs represent feeding relationships in a community. This food web shows organisms on several trophic levels in eastern North America's temperate deciduous forest. Arrows lead from one organism to another to indicate the direction of energy flow as a result of predation, parasitism, or herbivory. Like most food web diagrams, this one is a simplification, because the actual community contains many more species and interactions than can be shown.

Some organisms play outsized roles in communities

"Some animals are more equal than others," George Orwell wrote in his classic novel *Animal Farm*. Although Orwell was making wry sociopolitical commentary, his remark hints at a truth in ecology. In communities, ecologists have found, some species exert greater influence than do others. A species that has strong or wide-reaching impact far out of proportion to its abundance is often called a **keystone species.** A keystone is the wedge-shaped stone at the top of an arch that is vital for holding the structure together; remove the keystone, and the arch will collapse (FIGURE 4.12a). In an ecological community, removal of a keystone species will likewise have major consequences.

Often, secondary or tertiary consumers near the tops of food chains are considered keystone species. Top predators control populations of herbivores, which otherwise would multiply and could greatly modify the plant community (FIGURE 4.12b). Thus, predators at high trophic levels can indirectly promote populations of organisms at low trophic levels by keeping species at intermediate trophic levels in check, a phenomenon ecologists refer to as a **trophic cascade.** In the United States, for example, government bounties promoted the hunting of wolves and mountain lions, which were largely exterminated by the middle of the 20th century. In the absence of these predators, deer populations grew unnaturally dense and have overgrazed forest-floor vegetation and eliminated tree seedlings, causing major changes in forest structure.

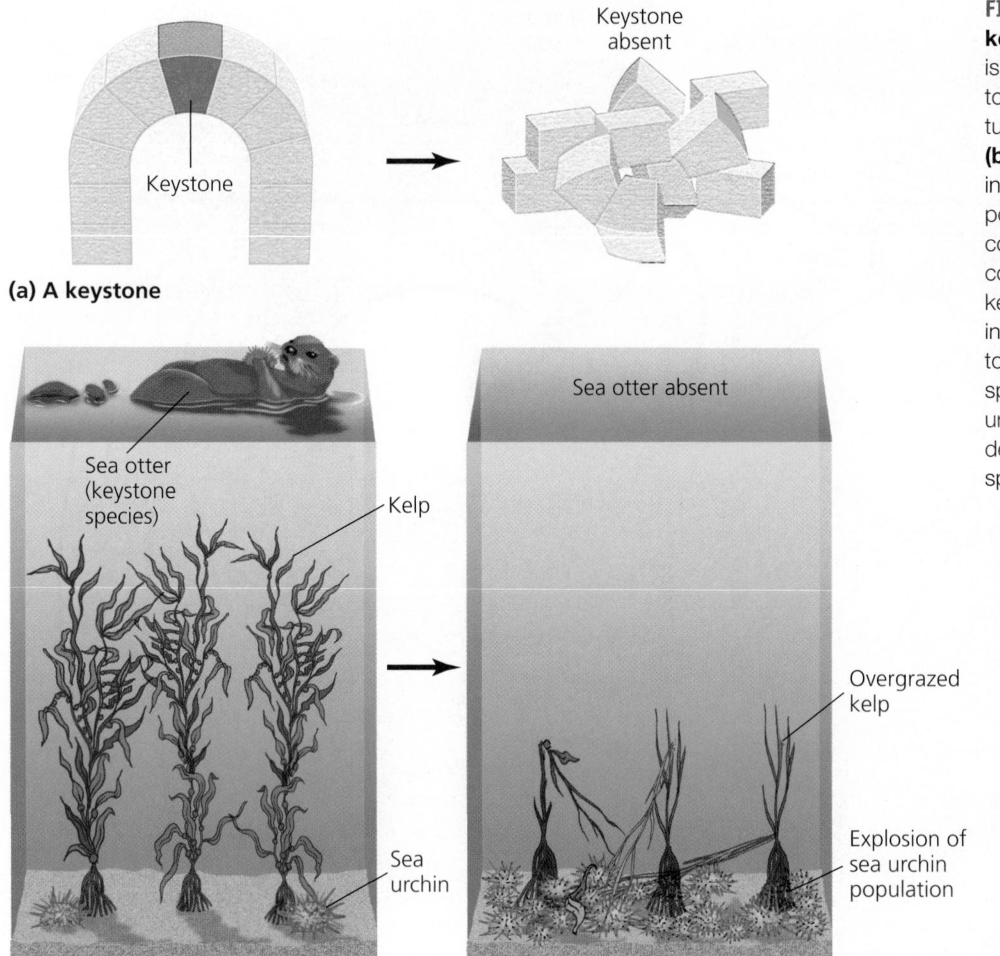

FIGURE 4.12 Sea otters are a keystone species. A keystone **(a)** is the wedge-shaped stone at the top of an arch that holds its structure together. A keystone species **(b)** is a species that exerts great influence on a community's composition and structure. Sea otters consume sea urchins that eat kelp in coastal waters of the Pacific. Otters keep urchin numbers down, allowing lush underwater forests of kelp to grow, providing habitat for many species. When otters are absent, urchins increase and devour the kelp, destroying habitat and depressing species diversity.

The removal of top predators in the United States was an uncontrolled large-scale experiment with unintended consequences, but ecologists have verified the keystone species concept in controlled scientific experiments. Classic research by marine biologist Robert Paine established that the predatory sea star *Pisaster ochraceus* shapes the community composition of intertidal organisms (p. 427) on North America's Pacific coast. When *Pisaster* is present in this community, species diversity is high, with various types of barnacles, mussels, and algae. When *Pisaster* is removed, the mussels it preys on become numerous and displace other species, suppressing species diversity.

Research off the U.S. Atlantic coast published in 2007 suggests that the reduction of shark populations by commercial fishing has allowed populations of certain skates and rays to increase, which in turn has depressed numbers of bay scallops and other bivalves they eat.

Animals at high trophic levels—such as wolves, sea stars, sharks, and sea otters (see Figure 4.12)—are often viewed as keystone species that can trigger trophic cascades. However, other types of organisms also exert strong community-wide effects. "Ecosystem engineers" physically modify the environment shared by community members. Beavers build dams across streams, creating ponds and swamps by flooding large areas of land. Prairie dogs dig burrows that aerate the soil and serve as homes for other animals. Ants disperse seeds, redistribute nutrients, and selectively protect or destroy different insects and plants within the radius of their colonies. And zebra and quagga mussels alter the communities they invade by filtering plankton out of the water.

Less conspicuous organisms toward the bottoms of food chains may exert still more impact. Remove the fungi that decompose dead matter, or the insects that control plant growth, or the phytoplankton that are the base of the marine food chain, and a community may change very rapidly indeed. However, because there are usually more species at lower trophic levels, it is less likely that any single one of them alone has wide influence. Often if one species is removed, other species that remain may be able to perform many of its functions.

Communities respond to disturbance in various ways

The removal of a keystone species is just one type of disturbance that can modify the composition, structure, or function of an ecological community. In ecological terms, a **disturbance** is an event that affects environmental conditions rapidly and drastically, resulting in changes to the community and ecosystem. A disturbance can be as localized as a tree falling in a forest, creating a gap in the canopy that lets in light and alters conditions for plants and animals in the gap. Or it can be as large and severe as a hurricane, tornado, or volcanic eruption. Some disturbances are sudden, such as landslides or floods, while others are more gradual, such as climate change. Some disturbances regularly reoccur and are considered normal aspects of a system (such as periodic fire, seasonal storms, or cyclic insect outbreaks). Organisms may, in fact, adapt to regular and predictable types of disturbance. For instance, many plants that grow in fire-prone regions have evolved ways of surviving fire and have seeds that depend on fire to germinate in the nutrient-rich soil that fire leaves behind. Today, human impacts are major sources of disturbance for ecological communities worldwide—from habitat alteration to pollution to the introduction of non-native species such as the zebra mussel.

Communities are dynamic systems and may respond to disturbance in several ways. A community that resists change and remains stable despite disturbance is said to show **resistance** to the disturbance. Alternatively, a community may show **resilience,** meaning that it changes in response to disturbance but later returns to its original state. Or, a community may be modified by disturbance permanently and never return to its original state.

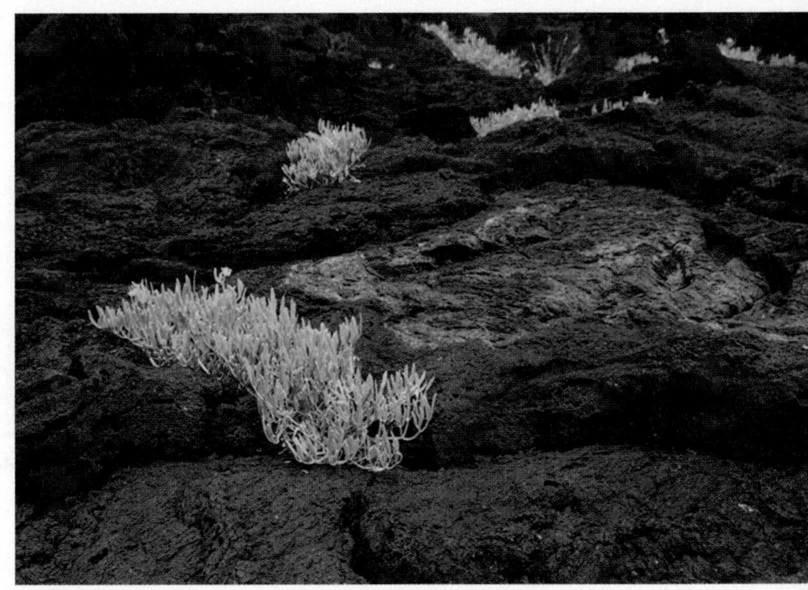

FIGURE 4.13 Primary succession begins as organisms colonize a lifeless new surface. Here, plants establish on a volcanic lava flow.

Succession follows severe disturbance

If a disturbance is severe enough to eliminate all or most of the species in a community, the affected site may then undergo a somewhat predictable series of changes that ecologists have traditionally called **succession.** In the conventional view of this process, there are two types of succession. **Primary succession** follows a disturbance so severe that no vegetation or soil life remains from the community that had occupied the site. In primary succession, a biotic community is built essentially from scratch. In contrast, **secondary succession** begins when a disturbance dramatically alters an existing community but does not destroy all living things or all organic matter in the soil. In secondary succession, vestiges of the previous community remain, and these building blocks help shape the process.

At terrestrial sites, primary succession takes place after a bare expanse of rock, sand, or sediment becomes newly exposed to the atmosphere. This can occur when glaciers retreat, lakes dry up, or volcanic lava or ash spreads across the landscape (FIGURE 4.13). Species that arrive first and colonize the new substrate are referred to as **pioneer species.** Pioneer species are well adapted for colonization, having traits such as spores or seeds that can travel long distances.

The pioneers best suited to colonizing bare rock are the mutualistic aggregates of fungi and algae known as lichens. In lichens, the algal component provides food and energy via photosynthesis while the fungal component grips the rock and captures the moisture that both organisms need. As lichens grow, they secrete acids that break down the rock surface, beginning the formation of soil. Small plants and insects arrive, providing more nutrients and habitat. As time passes, larger plants and animals establish themselves, vegetation increases, and species diversity rises.

Secondary succession begins when a fire, a hurricane, logging, or farming removes much of the biotic community. Consider a farmed field in eastern North America that has been abandoned (FIGURE 4.14). After farming ends, the site will be colonized by pioneer species of grasses, herbs, and forbs that disperse well or were already in the vicinity. Soon, shrubs and fast-growing trees

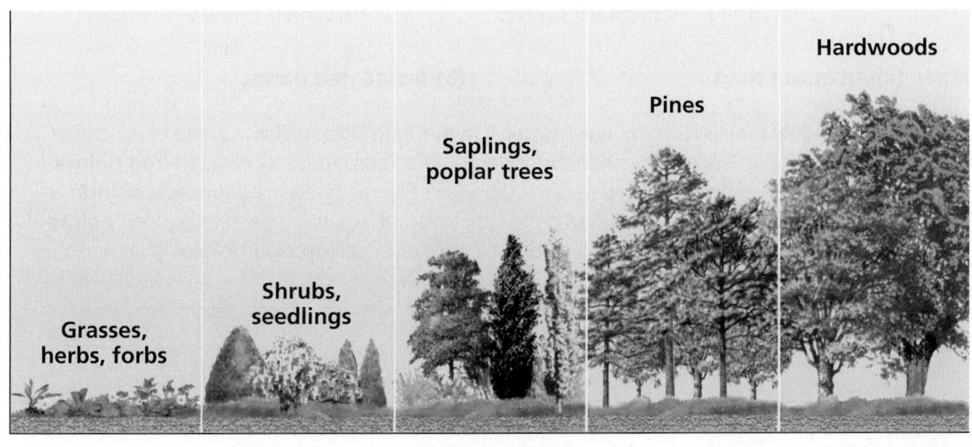

FIGURE 4.14 Secondary succession occurs after a disturbance (such as a fire, landslide, or farming) removes most vegetation from an area. Shown is a typical series of changes in a plant community of eastern North America following abandonment of a farmed field.

Determining Zebra Mussels' Impacts on Fish Communities

When zebra mussels appeared in the Great Lakes, people feared for sport fisheries and estimated that fish population declines could cost billions of dollars. The mussels would deplete the phytoplankton and zooplankton that fish depended on for food, people reasoned.

However, food webs are complicated systems, and disentangling them to infer the impacts of any one species is difficult. Thus, even 15 years after the arrival of zebra mussels, there was no solid evidence of widespread harm to fish populations.

So, aquatic ecologist David Strayer of the Institute of Ecosystem Studies in Millbrook, New York, joined Kathryn Hattala and Andrew Kahnle of New York State's Department of Environmental Conservation (DEC). They mined data sets on fish populations in the Hudson River, which zebra mussels had invaded in 1991.

Strayer and other scientists had been studying this community for years. Their data showed that after zebra mussels invaded the Hudson:

- Biomass of phytoplankton fell by 80%.
- Biomass of small zooplankton fell by 76%.
- Biomass of large zooplankton fell by 52%.

Zebra mussels increased filter-feeding in the community 30-fold, depleting phytoplankton and small zooplankton and leaving larger zooplankton with less phytoplankton to eat. Overall, zooplankton and invertebrate animals of the open water (which are eaten by open-water fish) declined by 70%.

However, Strayer had also found that *benthic,* or bottom-dwelling,

Dr. David Strayer samples aquatic invertebrates

invertebrates in shallow water (especially in the nearshore, or *littoral,* zone) had increased, because the mussels' shells provide habitat structure and their feces provide nutrients.

These contrasting trends in the benthic shallows and the open deep water led Strayer's team to hypothesize that zebra mussels would harm open-water fish that ate plankton but would help littoral-feeding fish. They predicted that following the zebra mussel invasion, larvae and juveniles of six common open-water fish species would decline in number, decline in growth rate, and shift downriver toward saltier water, where mussels are absent. Conversely, they predicted that larvae and juveniles of 10 littoral fish species would increase in number, increase in growth rate, and shift upriver to regions of greatest zebra mussel density.

To test their predictions, the researchers analyzed data from fish surveys carried out by DEC scientists over 26 years, spanning periods before and after the zebra mussel's arrival. Strayer's team compared data on abundance, growth, and distribution of young fish before and after 1991.

The results supported their predictions. Larvae and juveniles of open-water fish, such as American shad, blueback herring, and alewife, tended to decline in abundance in the years after zebra mussels were introduced (**FIGURE 1a**). Those of littoral fish, such as tessellated darter, bluegill, and largemouth bass, tended to increase (**FIGURE 1b**).

Growth rates showed the same trend: Open-water fish grew more slowly after zebra mussels invaded, whereas littoral fish grew more quickly.

In terms of distribution in the 248-km (154-mi) stretch of river studied, open-water fish shifted downstream

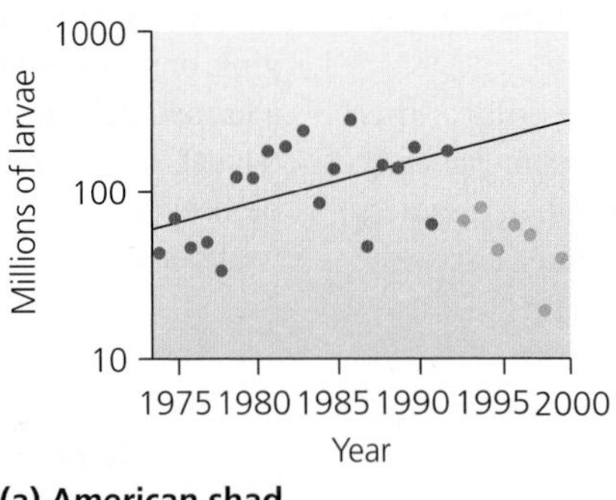

(a) American shad

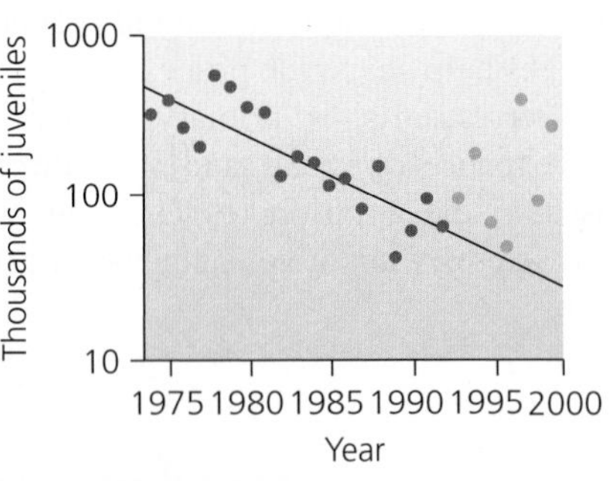

(b) Tessellated darter

FIGURE 1 Zebra mussels harm open-water fish but help littoral fish. Larvae of American shad **(a),** an open-water fish, were increasing before zebra mussels invaded (**red points and trend line**). After zebra mussels invaded, shad larvae decreased (**orange points**). In contrast, juveniles of the tessellated darter **(b),** a littoral fish, were decreasing (**red points and trend line**) but increased after zebra mussels invaded (**orange points**). *Source: Strayer, D., et al., 2004. Effects of an invasive bivalve (*Dreissena polymorpha*) on fish in the Hudson River estuary.* Canadian Journal of Fisheries and Aquatic Sciences *61: 924–941. © 2004. Reprinted by permission of NRC Research Press.*

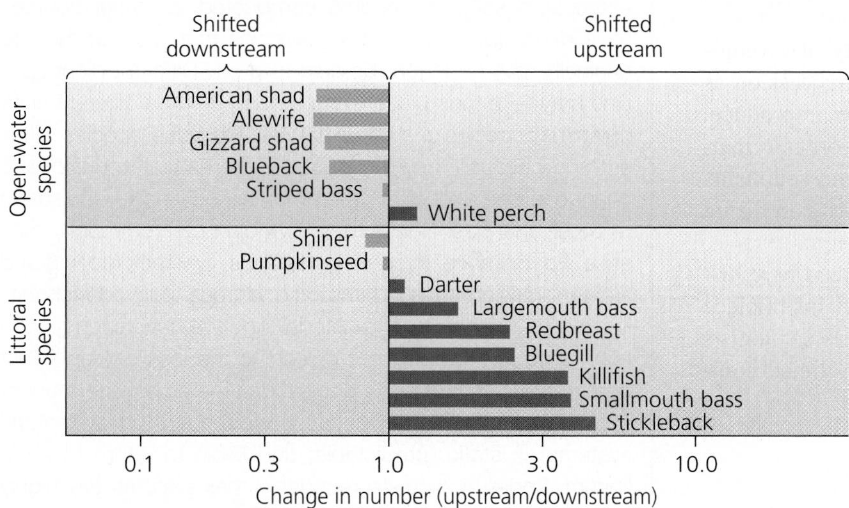

FIGURE 2 Different fish respond in different ways to zebra mussels. Following the arrival of zebra mussels, the young of open-water fish tended to shift downstream toward areas with fewer zebra mussels. Young of littoral fish tended to shift upstream toward areas with more zebra mussels. *Source: Strayer, D., et al., 2004. Effects of an invasive bivalve (*Dreissena polymorpha*) on fish in the Hudson River estuary.* Canadian Journal of Fisheries and Aquatic Sciences *61: 924–941. © 2004. Reprinted by permission of NRC Research Press.*

toward areas with fewer zebra mussels, whereas littoral fish shifted upstream toward areas with more zebra mussels (**FIGURE 2**). Overall, the data supported the hypothesis that the fish community would respond to changes in food resources caused by zebra mussels.

But then, surprisingly, as Strayer and his colleagues continued their research, some of the zebra mussel's impacts began to reverse. Populations of native mussels and clams in the Hudson that had crashed after the zebra mussel invaded (likely because of competition for food), began to stabilize and persist at about 4–22% of their pre-invasion population sizes. Populations of crustaceans, flatworms, and other invertebrates also rebounded. Several types of zooplankton began to increase (**FIGURE 3**).

To determine why these changes were occurring, Strayer's research teams stepped up monitoring efforts and ran experiments placing cages (to keep out predators) around some areas of mussels. They found that mussels within the cages grew larger than mussels outside, indicating that predators were feeding on mussels and preventing most of them from growing to their full size. Indeed, Strayer's teams determined that the zebra mussel's survival rate had fallen to less than 1% of what it was during the early years of the invasion. Predators such as the blue crab were eating more and more of the mussels, and large mussels were becoming rare. As the average size of the mussels decreased, their filtering capacity fell by more than 80%, and more zooplankton began to survive.

Do these trends suggest that the zebra mussel might just fade away and prove harmless in the long run? Strayer feels it is too early to answer that question; much remains unknown, and he is continuing his research. He cautions that zebra mussels remain abundant, that phytoplankton levels have not yet bounced back, and that there is no guarantee that zebra mussel impacts will continue to diminish. Nonetheless, the apparent turnaround in the Hudson River is intriguing for ecologists and provides hope that the Hudson's native systems may recover. ■

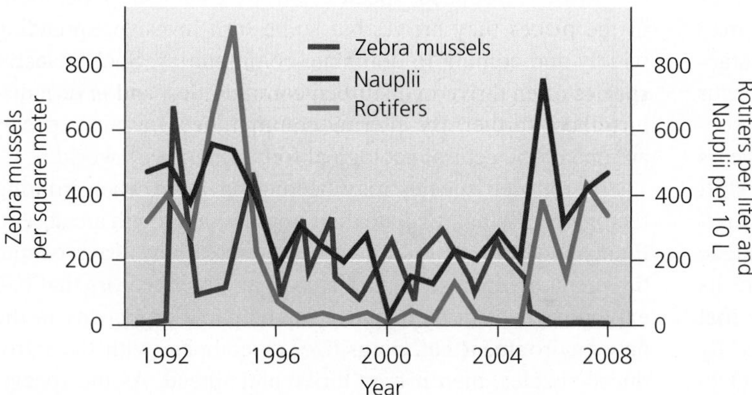

FIGURE 3 Zooplankton can bounce back once zebra mussels decline. Large, mature zebra mussels have decreased in density since their initial sudden increase upon colonizing the Hudson River (**red line**). Two types of zooplankton (**blue and green lines**) declined following the zebra mussel's introduction, but they recovered after 2005, once large mussels disappeared. *Source: Pace, M.L, et al., 2010. Recovery of native zooplankton associated with increased mortality of an invasive mussel.* Ecosphere *1(1): Article 3.*

such as aspens and poplars begin to grow. As time passes, pine trees rise above the pioneer trees and shrubs, forming a pine-dominated forest. This pine forest develops an understory of hardwood trees, because pine seedlings do not grow well under a canopy but some hardwood seedlings do. Eventually the hardwoods outgrow the pines, creating a hardwood forest.

Processes of succession occur in a diversity of ecological systems. For instance, ponds may undergo succession as algae, microbes, plants, and zooplankton grow, reproduce, and die, gradually filling the water body with organic matter. The pond acquires further organic matter and sediments from streams and surface runoff, and eventually it can fill in, becoming a bog (p. 395) or even a terrestrial system.

In the traditional view of succession described here, the process leads to a **climax community,** which remains in place until some disturbance restarts succession. Early ecologists felt that each region had its own characteristic climax community, determined by climate.

Communities may undergo shifts

Today, ecologists recognize that community change is far more variable and less predictable than early models of succession suggested. Conditions at one stage may promote progression to another stage, or organisms may, through competition, inhibit a community's progression to another stage. The trajectory of change can vary greatly according to chance factors, such as which particular species happen to gain an early foothold. And climax communities are not determined solely by climate, but vary with soil conditions and other factors from one time or place to another. Ecologists came to modify their views about how communities respond to disturbance after observing changes in long-term field studies at locations such as Mount St. Helens following its eruption (see THE SCIENCE BEHIND THE STORY, pp. 90–91).

Once a community is disturbed and changes are set in motion, there is no guarantee that the community will ever return to its original state. Sometimes communities may undergo a **phase shift,** or **regime shift,** in which the character of the community fundamentally changes. This can occur if some crucial climatic threshold is passed, a keystone species is lost, or a non-native species invades. In recent years many coral reef communities have undergone a phase shift and become dominated by algae, after people had overharvested populations of fish or turtles that eat algae. Once algae overgrow a reef, the community may never shift back to its original coral-dominated state. Phase shifts make clear that we cannot count on being able to reverse damage caused by human disturbance. Instead, some of the changes we set in motion may become permanent.

Moreover, many ecologists now think that human disturbance is creating communities that are wholly new and have not previously occurred on Earth. These **novel communities,** or **no-analog communities,** are composed of novel mixtures of plants and animals and have no analog or precedent. As we enter more deeply into an age of fast-changing climate, habitat alteration, species extinctions, and species invasions, many scientists predict that we can expect to see more and more novel communities.

FAQ **If we disturb a community, won't it return to its original state if we just leave the area alone?**

Probably not, if the disturbance has been substantial. For example, if soil has become compacted, or water sources have dried up, then the plant species that grew at the site originally may no longer be able to grow. Different plant species may take their place—and among them, a different suite of animal species may find habitat. Because species interact and because they rely on particular habitat conditions, a change in one aspect of a community can lead to a cascade of other changes. Sometimes a whole new community may arise. For instance, in some grasslands, livestock grazing and fire suppression have led shrubs and trees to invade, changing the grasslands to shrublands. And we have seen (p. 84) how removing sea otters can lead to the loss of kelp forest communities. In the past, people didn't realize how permanent such changes could be, because we tended to view natural systems as static, predictable, and liable to return to equilibrium. Today ecologists recognize that systems are highly dynamic and can sometimes undergo rapid, extreme, and long-lasting change.

Invasive species pose new threats to community stability

Traditional concepts of communities involve species understood to be native to an area. But what if a species not native to the area (a non-native, alien, or exotic species) arrives from elsewhere? In our age of global mobility and trade, people have moved countless organisms from place to place, intentionally or by accident, such that today most non-native arrivals in a community are **introduced species,** species introduced by people.

Most introduced species fail to establish populations in the places they arrive, but some turn invasive, spreading widely and coming to dominate communities. Such **invasive species** often thrive in disturbed communities, and in turn disturb them further. By altering communities, invasive species are one of the central ecological forces in today's world.

Introduced species may become invasive when limiting factors (p. 67) that regulate their population growth are absent. Plants and animals brought to a new area may leave behind the predators, parasites, herbivores, and competitors that had exploited them in their native land. If few organisms in the new environment eat, parasitize, or compete with the introduced species, then it may thrive and spread. As the species proliferates, it may exert diverse influences on other community members (FIGURE 4.15).

Zebra and quagga mussels spread with global trade, inadvertently transported in the ballast water of cargo ships. (To maintain stability at sea, ships take water into their hulls as they begin their voyage and then discharge that water at their destination.) Decades of unregulated exchange of ballast water have ferried countless species across the oceans. Other types of freshwater invaders include predatory fishes released from aquaria, bait buckets, stocking, or aquaculture;

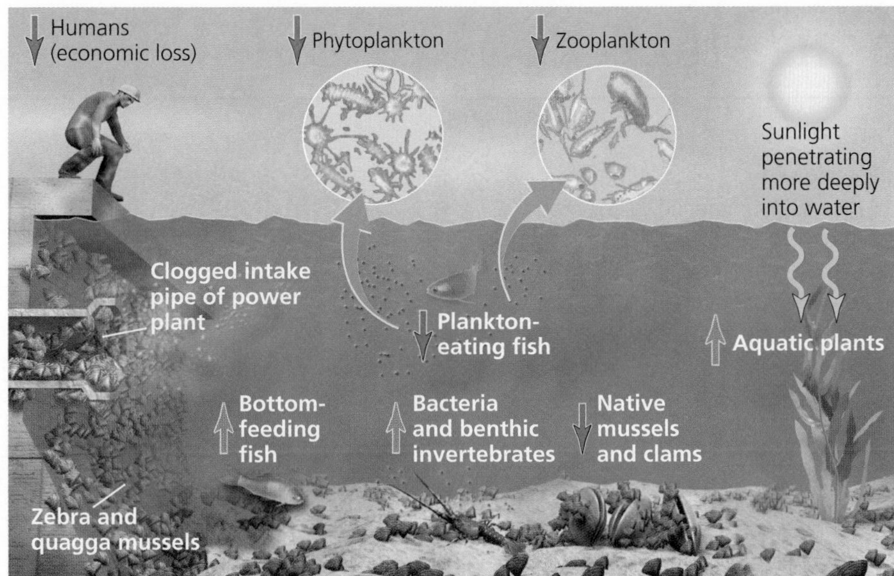

(a) Impacts of zebra and quagga mussels on a Great Lakes nearshore community

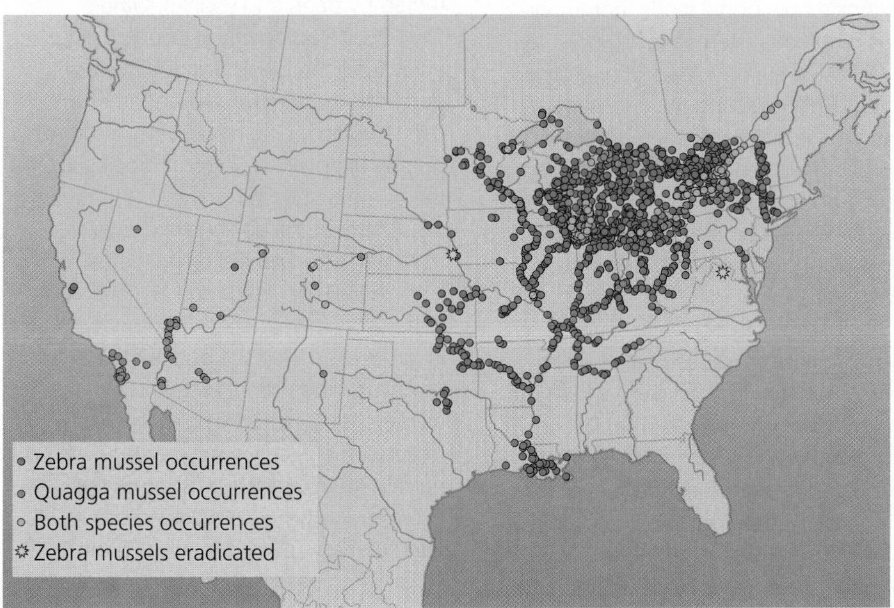

(b) Occurrence of zebra and quagga mussels in North America, as of December 2012

FIGURE 4.15 The zebra mussel and the quagga mussel are modifying ecological communities. By filtering phytoplankton and small zooplankton from open water, they exert impacts **(a)** on other species, both negative (red downward arrows) and positive (green upward arrows). The map **(b)** shows known occurrences of zebra and quagga mussels as of December 2012. In just two decades they spread across North America, assisted by accidental transportation on the hulls of boats. *Source (b): U.S. Geological Survey.*

aquatic plants that become ecosystem engineers; crayfish that disrupt the benthic portions of food webs; and disease pathogens.

Examples abound of invasive species that have had major ecological impacts (pp. 286–289). The chestnut blight, an Asian fungus, killed nearly every mature American chestnut, the dominant tree species of many eastern North America forests, between 1900 and 1930. Asian trees had evolved defenses against the fungus over millennia of coevolution, but the American chestnut had not. A different fungus caused Dutch elm disease, destroying most of the American elms that once gracefully lined the streets of many U.S. cities. Fish introduced into streams for sport compete with and exclude native fish. Grasses introduced in the American West for ranching have overrun entire regions, pushing out native vegetation. Hundreds of island-dwelling animals and plants worldwide have been driven extinct by goats, pigs, and rats introduced by human colonists (Chapter 3).

Ecologists generally view the impacts of invasive species—and introduced species in general—as overwhelmingly negative. Yet many people enjoy the beauty of introduced ornamental plants in their gardens. Some organisms are introduced intentionally to control pest populations through biocontrol (p. 256). And some introduced species that have turned invasive provide benefits to our economy, such as the European honeybee, which pollinates many of our crops (p. 254). Whatever view one takes, the impacts of invasive species on native species and ecological communities are significant, and they grow year by year with the increasing mobility of people and the globalization of our society.

Chronicling Ecological Recovery at Mount St. Helens

Step outside, look around, and you'll likely find secondary succession occurring somewhere nearby. The nearest weedy lot or overgrown field will show how plants colonize a disturbed area and begin building a new community from the foundations of the old. But finding primary succession is not as easy. It's unusual to come across a place where all life has been extinguished and a brand new community is being built from scratch as new organisms arrive from far away.

The eruption of Mount St. Helens offered ecologists a rare opportunity to study how communities recover from catastrophic disturbance. On May 18, 1980, this volcano in the state of Washington erupted in sudden and spectacular violence, with 500 times the force of the atomic blast at Hiroshima.

The massive explosion obliterated an entire landscape of forest as a scalding mix of gas, steam, ash, and rock was hurled outward for miles. A pyroclastic flow (p. 41) sped downslope, along with the largest landslide in recorded history. Rock and ash rained down for miles around, and mudslides and lahars (p. 42) raced down river valleys, devastating everything in their paths. Altogether, 4.1 km^3 (1.0 mi^3) of material was ejected from the mountain, severely altering 1650 km^2 (637 mi^2), an area larger than the entire city of Houston, Texas.

In the aftermath of the blast, ecologists moved in to take advantage of the natural experiment of a lifetime. For them, the eruption provided an extraordinary chance to study how primary succession unfolds on a fresh volcanic surface. Which organisms would arrive first? What kind of community would emerge? How long it would take? These researchers set up study plots to examine how populations, communities, and ecosystems would respond.

Today, over 30 years later, the barren gray moonscape that resulted from the blast is a vibrant green (**FIGURE 1**), carpeted with colorful flowers each summer. And what ecologists have learned has modified our view of primary succession and informed the entire study of disturbance ecology.

Given the ferocity and scale of the eruption, most scientists initially presumed that life had been wiped out completely over a large area. Based on traditional views of succession, they expected that pioneer species would colonize the area gradually, spreading slowly from the outside margins inward, and that over many years a community would be rebuilt in a systematic and predictable way.

Dr. Virginia Dale at Mount St. Helens.

Instead, researchers discovered that some plants and animals had survived the blast. Some were protected by deep snowbanks. Others were sheltered on steep slopes facing away from the blast. Still others were dormant underground when the eruption occurred. These survivors, it turned out, would play key roles in rebuilding the community.

Many of the ecologists drawn to Mount St. Helens studied plants. Virginia Dale of Oak Ridge National Laboratory in Tennessee and her colleagues examined the debris avalanche, a landslide of rock and ash as deep as a 15-story building. This region appeared barren, yet small numbers of plants of 20 species had survived, growing from bits of root or stem carried down in the avalanche. However, most plant regrowth occurred from seeds blown in from afar. Dale's team used sticky traps to sample these

(a) 1980

(b) 2012

FIGURE 1 Mount St. Helens (a) after the eruption in 1980, and (b) in 2012.

seeds as they chronicled the area's recovery.

Plant regrowth was very slow at Dale's study plots for several years and then accelerated. After 20 years, 150 species of plants covered 65% of the ground. One important pioneer species was the red alder. This tree germinates on debris, grows quickly, deals well with browsing by animals, and produces many seeds at a young age. As a result, it has become the dominant tree species on the debris avalanche. Because it fixes nitrogen (pp. 123, 126), the red alder enhances soil fertility and thereby helps other plants grow. Researchers predict that red alder will remain dominant for years or decades and that conifers such as Douglas fir (which today are moving in and beginning to seed) will eventually outgrow them and establish a conifer-dominated forest.

Patterns of plant growth have varied in different areas. Roger del Moral of the University of Washington and his colleagues compared ecological responses on a variety of surfaces, including barren pumice, mixed ash and rock, mudflows, and the "blowdown zone" where trees were toppled like matchsticks. Numbers of species and percent of plant cover increased in different ways on each surface (**FIGURE 2**), affected by a diversity of factors. Windblown seeds accounted for most regrowth, but plants that happened to survive in sheltered "refugia" within the impact zone helped to repopulate areas nearby.

Chance played a large role in determining which organisms survived and how vegetation recovered, del Moral and others found. Had the eruption occurred in late summer instead of spring, there would have been no snow, and many of the plants that survived would have died. Had the eruption occurred at night instead of in the morning, nocturnal animals would have been hit harder.

Animals played major roles in the recovery right from the beginning. In fact, researchers such as Patrick Sugg and John Edwards of the University of Washington showed that insects and spiders arrived in the impact zone in great numbers before plants did. Insects fly, while spiders disperse by "ballooning" on silken threads, so in summer the atmosphere is filled with an "aerial plankton" of windblown arthropods. Trapping and monitoring at Mount St. Helens in the months following the eruption showed that insects and spiders landed in the impact zone by the billions. Researchers estimated that over 1500 species arrived in the first few years, surviving by scavenging or by preying on other arthropods. Most individuals soon died, but the nutrients from their bodies enriched the soil, helping the community to develop.

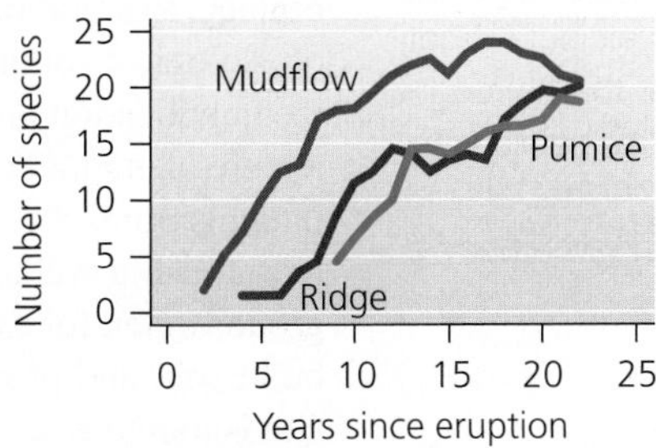

(a) Species richness

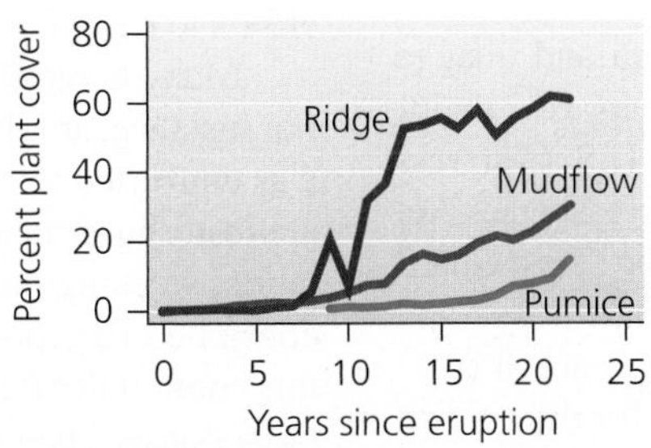

(b) Percent plant cover

FIGURE 2 Plants recovered differently at mudflow, ridge, and pumice sites at Mount St. Helens. In the 25 years after the eruption, **(a)** species richness of plants and **(b)** percentage of ground covered by plants both increased. *Data from del Moral, R., et al., 2005. Proximity, microsites, and biotic interactions during early succession. Pp. 93–109 in V. Dale et al., eds.,* Ecological responses to the 1980 eruption of Mount St. Helens. *Springer, New York.*

DATA Q Which substrate gained the greatest species richness, and what was its highest number of species? On which substrate was the increase in percent plant cover the slowest?

Once plants took hold, animals began exerting influence through herbivory. Caterpillars fed on plants, occasionally extinguishing small populations as they began to establish. As elk from surrounding forests moved into the region, Dale and her team fenced off "exclosure" plots to compare how plants grew within the ungrazed exclosures versus in plots grazed by elk outside the exclosures. Both types of plots saw increases in plant cover and similar amounts of species diversity, because elk herbivory can spur plant growth that compensates for what they eat. Non-native species did best in the grazed plots, whereas species important for forest recovery did best in the ungrazed exclosures. Overall, elk herbivory did not stall plant regrowth.

A surprising number of amphibians and small mammals survived in refugia in the impact zone, researchers found. Many of these species were underground or dormant at the time of the blast, and enough were able to survive once they made their way to the surface that populations slowly rebuilt and spread from these few chance survivors.

All told, research at Mount St. Helens has shown that succession is not a simple and predictable process. Instead, communities recover from disturbance in ways that are dynamic, complex, and highly dependent on chance factors affecting which species survive to repopulate the new landscape.

The results from Mount St. Helens also show life's resilience. Even when the vast majority of organisms perish in a natural disaster, a few may survive, and their descendants may eventually build a new community.

Ecological change at Mount St. Helens is still in its early stages and will continue for many decades more. All along, ecologists will continue to study and learn from this tremendous natural experiment. ■

WEIGHING THE ISSUES

ARE INVASIVE SPECIES ALL BAD? Some ethicists question the notion that all invasive species should automatically be considered bad. If we introduce a non-native species to a community and it greatly modifies the community, do you think that is a bad thing? What if it drives another species extinct? What if the invasive species arrived on its own, rather than through human intervention? What if it provides economic services to our society, as the European honeybee does? What ethical standard(s) (p. 136) would you apply to determine whether we should reject or welcome an invasive species?

We can respond to invasive species through control, eradication, or prevention

Scientific research and media attention to zebra and quagga mussels helped put invasive species on the map as a major environmental and economic issue. In 1990 the U.S. Congress passed legislation that led to the National Invasive Species Act of 1996. Among other things, this law directed the Coast Guard to ensure that ships dump their freshwater ballast at sea and exchange it with salt water before entering the Great Lakes.

Since then, funding has become available for the control and eradication of invasive species. Eradication (total elimination of the invasive population) is difficult, so managers instead usually aim to control these populations; that is, to limit their growth, spread, and impact. Managers have tried to control zebra mussels by removing them manually; applying toxic chemicals; drying them out; depriving them of oxygen; introducing predators and diseases; and stressing them with heat, sound, electricity, carbon dioxide, and ultraviolet light. However, most of these are localized and short-term fixes unable to make a dent in the immense populations at large in the environment. With one invasive species after another, managers are finding that control and eradication measures are so difficult and expensive that trying to prevent invasions in the first place (through strategies such as ballast water regulations) represents a better investment.

To prevent invasions, it helps to be able to predict where a given species might spread. By analyzing the biology of the organism, scientists can try to model the environmental conditions in which it will thrive. In 2007, researchers applied knowledge of how zebra and quagga mussels use calcium in water to create their shells to predict where the mussels might do best. The researchers mapped low-risk and high-risk regions across North America, and these mostly conformed to the areas of actual spread. A remaining question is how quagga mussels may differ from zebra mussels. As Figure 4.15b shows, quagga mussels have leapfrogged zebra mussels by spreading into some western states, and no one knows why they have so far been more successful in the West.

Altered communities can be restored

Invasive species are adding to the tremendous transformations that humans have already forced on ecological systems through habitat alteration, deforestation, pollution, climate change, the hunting of keystone species, and other activities. With so much of Earth's landscape altered by human impact, many communities and ecosystems are severely degraded. Because ecological systems support our civilization and all of life, when degraded systems cease to function, our health and well-being are threatened.

This realization has given rise to the science of **restoration ecology.** Restoration ecologists research the historical conditions of ecological communities as they existed before our industrialized civilization altered them. They then try to devise ways to restore altered areas to an earlier condition. In some cases, the intent is primarily to restore the functionality of the system—to reestablish a wetland's ability to filter pollutants and recharge groundwater, for example, or a forest's ability to cleanse the air, build soil, and provide habitat for wildlife. In other cases, the aim is to return a community to its natural "presettlement" condition. Either way, the science of restoration ecology informs the practice of **ecological restoration,** the actual on-the-ground efforts to carry out these visions and restore communities.

Many ecological restoration efforts are underway today. For instance, nearly all the tallgrass prairie in the United States was converted to agriculture in the 19th century. Now, people are restoring patches of prairie by planting native prairie vegetation, weeding out invaders and competitors, and introducing controlled fire to mimic the fires that historically maintained this community (**FIGURE 4.16**). The region outside Chicago, Illinois, boasts several of the largest prairie restoration projects so far, including a 184-ha (455-acre) area inside the massive ring of the Fermilab nuclear accelerator in Batavia.

The world's largest restoration project is the ongoing effort to restore parts of the Florida Everglades, a 7500-km^2 (4700-mi^2) ecosystem of marshes and seasonally flooded grasslands. This formerly vast wetland system has been drying out for decades because the water that feeds it has been managed for flood control and overdrawn for irrigation and development. Economically important fisheries have suffered greatly as a result, and the region's famed populations of wading birds have dropped by 90–95%. The 30-year, $7.8-billion restoration project intends to restore natural water flow by undoing the damming and diversions of 1600 km (1000 mi) of canals, 1150 km (720 mi) of

FIGURE 4.16 Ecological restoration is being used to restore prairies. Here ecologists from the Midewin National Tallgrass Prairie inspect native grasses in a prairie restoration area on the site of the former Joliet Arsenal in Illinois.

WEIGHING THE ISSUES

RESTORING "NATURAL" COMMUNITIES

Practitioners of ecological restoration in North America often aim to restore communities to their natural state. But what does "natural" mean? Does it mean the state of the community before industrialization? Before Europeans came to the New World? Before any people laid eyes on the community? Let's say Native Americans altered a forest community 8000 years ago by burning the underbrush regularly to improve hunting and continued doing so until Europeans arrived 400 years ago and cut down the forest for farming. Today the area's inhabitants want to restore the land to its "natural" forested state. Should they try to recreate the forest of the Native Americans or the forest that existed before Native Americans arrived? What values do you think underlie the desire for restoration?

levees, and 200 water control structures. Because the Everglades provides drinking water for millions of Florida citizens, as well as considerable tourism revenue, restoring its ecosystem services (pp. 3, 116–117, 152, 290) should prove economically beneficial as well as ecologically valuable. We will explore ecological restoration projects further in Chapter 11 (pp. 299, 302).

As our population grows and development spreads, ecological restoration is becoming an increasingly vital conservation strategy. However, restoration is difficult, time-consuming, and expensive, and it is not always successful. It is therefore best, whenever possible, to protect natural systems from degradation in the first place.

Earth's Biomes

Across the world, each location is home to different sets of species, leading to endless variety in community composition. However, communities in far-flung places often share strong similarities in their structure and function. This allows us to classify communities into broad types. A **biome** is a major regional complex of similar communities—a large-scale ecological unit recognized primarily by its dominant plant type and vegetation structure. The world contains a number of biomes, each covering large geographic areas (**FIGURE 4.17**).

Each biome encompasses a variety of communities that share similarities. For example, the eastern United States supports the temperate deciduous forest biome. From New Hampshire to the Great Lakes to eastern Texas, precipitation and temperature are similar enough that most of the region's natural plant cover consists of broad-leafed trees that lose their leaves in winter. Within this region, however, there exist many different types of temperate deciduous forest, such as oak–hickory, beech–maple, and aspen–birch forests, each sufficiently different to be designated a separate community.

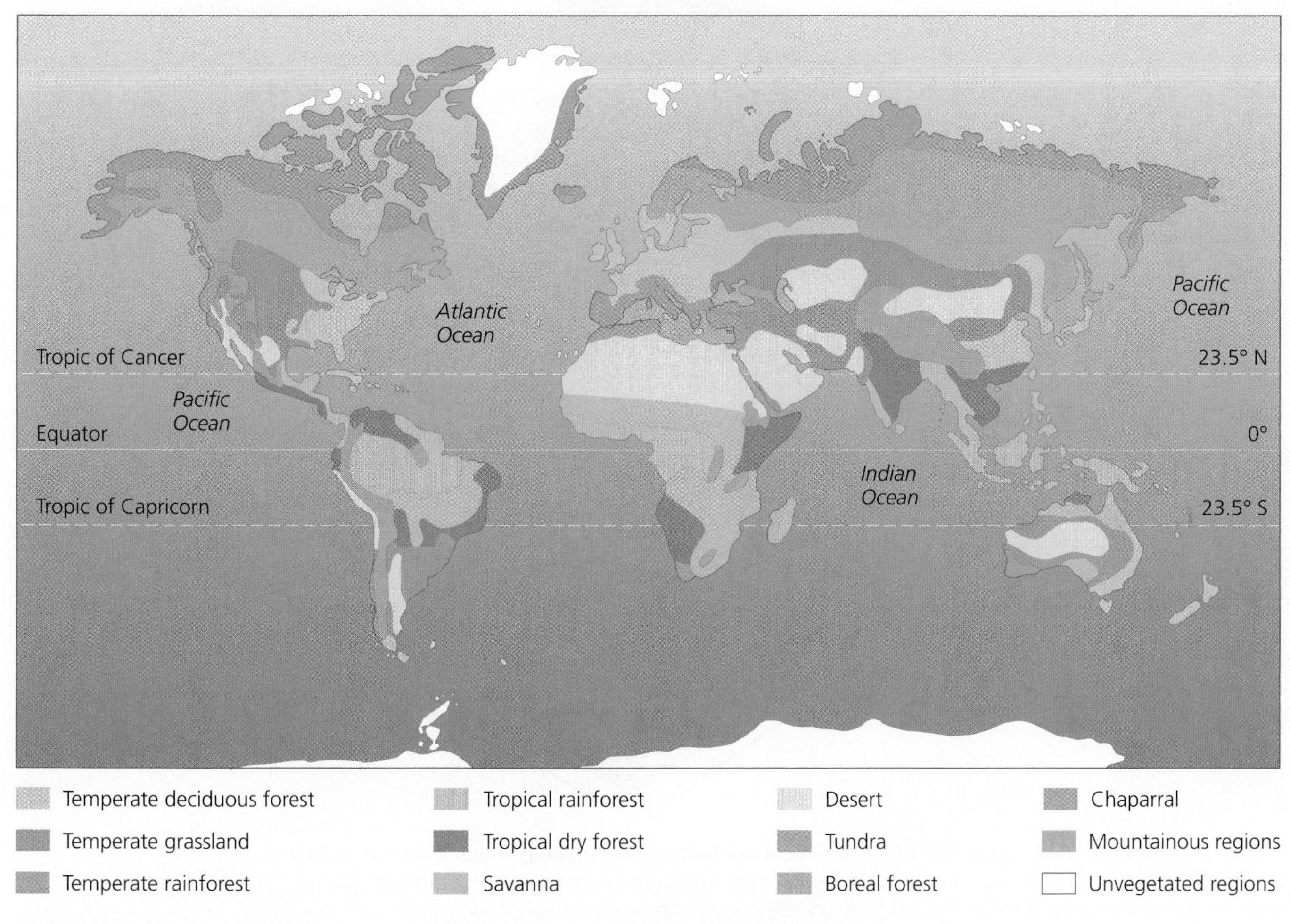

FIGURE 4.17 Biomes are distributed around the world, roughly correlated with latitude.

Climate influences the locations of biomes

Which biome covers each portion of the planet depends on a variety of abiotic factors, including temperature, precipitation, atmospheric and oceanic circulation patterns, and soil characteristics. Among these factors, temperature and precipitation exert the greatest influence (**FIGURE 4.18**). Because biomes are largely a function of climate, and because temperature and precipitation are the best indicators of an area's climate, scientists use **climate diagrams,** or **climatographs,** to depict such information.

Global climate patterns cause biomes to occur in large patches in different parts of the world. For instance, temperate deciduous forest occurs in eastern North America, Europe, and eastern China. Note in Figure 4.18 how patches representing the same biome tend to occur at similar latitudes. This is due to Earth's north–south gradients in temperature and to atmospheric circulation patterns (p. 455).

Aquatic and coastal systems resemble biomes

In our discussion of biomes, we will focus exclusively on terrestrial systems because the biome concept, as traditionally developed and applied, has been limited to terrestrial systems. However, areas equivalent to biomes also exist in the oceans, along coasts, and in freshwater systems. One might consider the shallows along the world's coastlines to represent one aquatic system, the continental shelves another, and the open ocean, the deep sea, coral reefs, and kelp forests as still others. Many coastal systems—such as salt marshes, rocky intertidal communities, mangrove forests, and estuaries—share both terrestrial and aquatic components. And freshwater systems such as those of the Great Lakes are widely distributed throughout the world.

Unlike terrestrial biomes, aquatic systems are shaped not by air temperature and precipitation, but by water temperature, salinity, dissolved nutrients, wave action, currents, depth, light levels, and type of substrate (e.g., sandy, muddy, or rocky

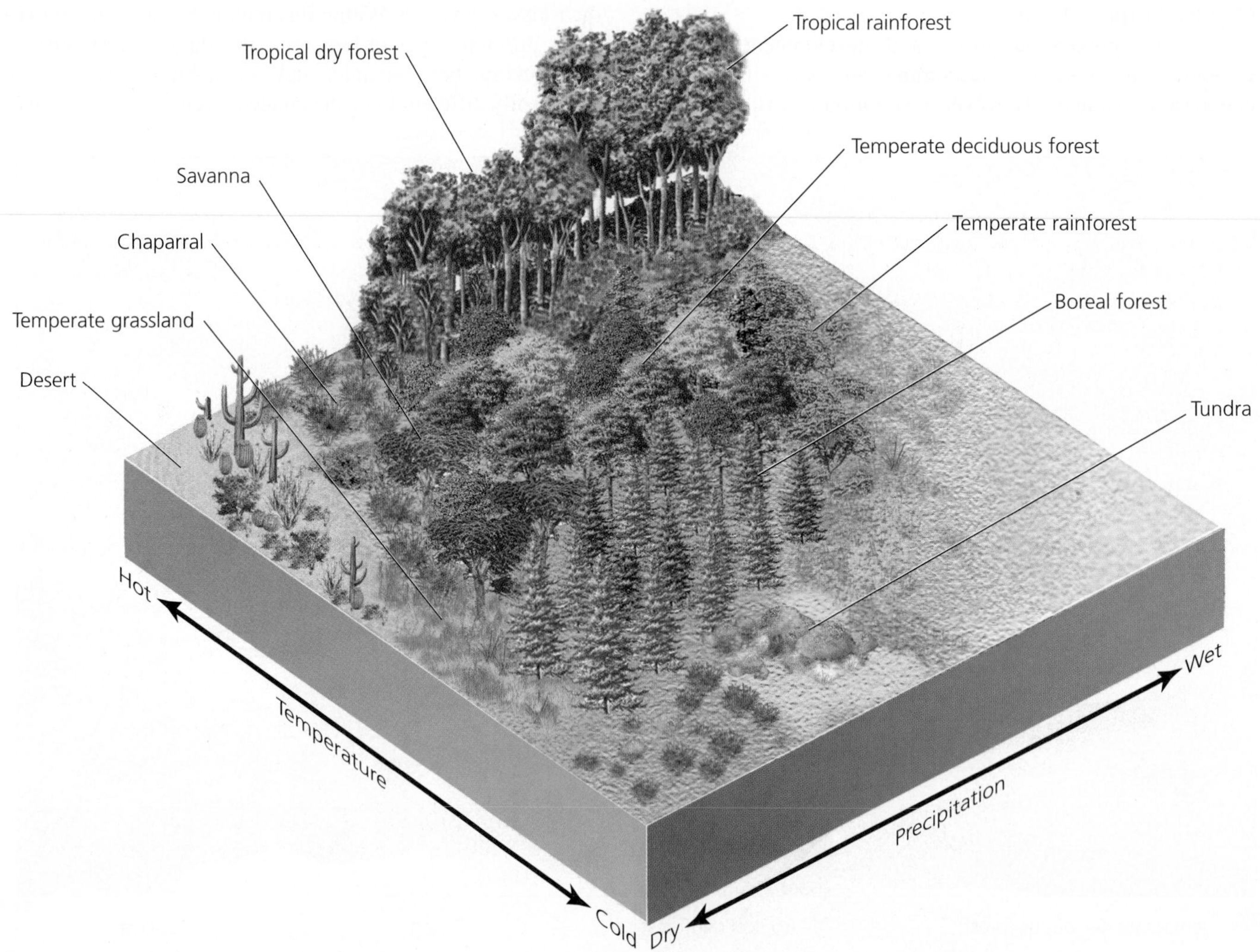

FIGURE 4.18 Temperature and precipitation are the main factors determining which biome occurs in an area. As precipitation increases, vegetation becomes taller and more luxuriant. As temperature increases, types of plant communities change. For instance, deserts occur in dry regions; tropical rainforests occur in warm, wet regions; and tundra occurs in the coldest regions.

(a) Temperate deciduous forest

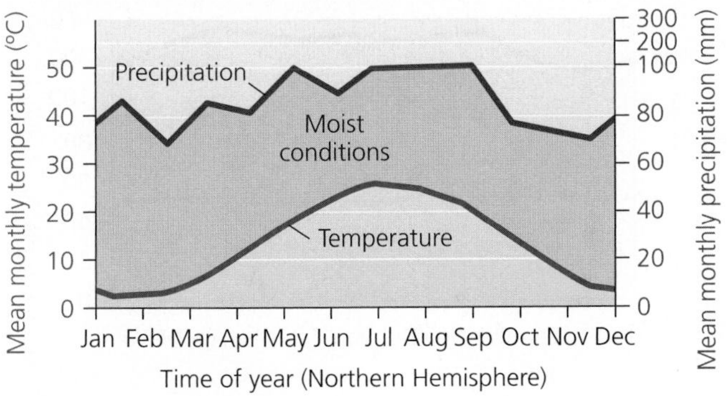

(b) Washington, D.C., USA

FIGURE 4.19 **Temperate deciduous forests (a) experience relatively stable seasonal precipitation but varied seasonal temperatures.** Scientists use climate diagrams **(b)** to illustrate average monthly precipitation and temperature. In these diagrams, the curves indicate precipitation (**blue**) and temperature (**red**) from month to month. When the precipitation curve lies above the temperature curve (as is the case year-round in the temperate deciduous forest around Washington, D.C., shown here), the region experiences relatively "moist" conditions, indicated with green coloration. *Climatograph here and in the following figures adapted from Breckle, S.W., and H. Walter, trans. by G. Lawlor. 2002.* Walter's vegetation of the Earth: The ecological systems of the geo-biosphere, *4th ed. Originally published by Eugen Ulmer KG, 1999, used by permission.*

bottom). Marine communities are also more clearly delineated by their animal life than by their plant life. We will examine freshwater, marine, and coastal systems in the greater detail they deserve in Chapters 15 and 16.

We can divide the world into ten terrestrial biomes

Temperate deciduous forest The **temperate deciduous forest** (FIGURE 4.19) that dominates the landscape around the central and southern Great Lakes is characterized by broad-leafed trees that are *deciduous*, meaning that they lose their leaves each fall and remain dormant during winter, when hard freezes would endanger leaves. These mid-latitude forests occur in much of Europe and eastern China as well as in eastern North America—all areas where precipitation is spread relatively evenly throughout the year.

(a) Temperate grassland

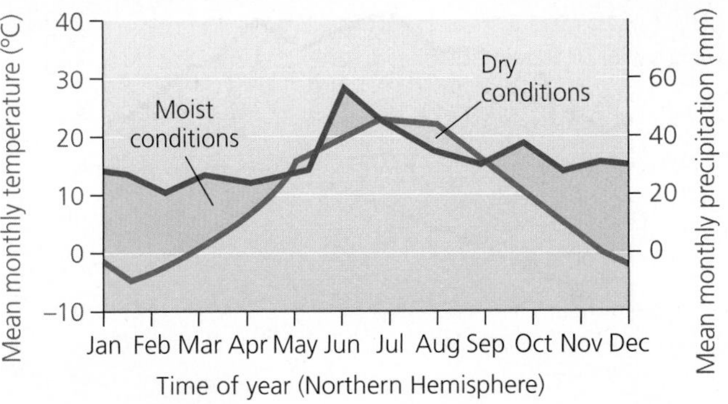

(b) Odessa, Ukraine

FIGURE 4.20 **Temperate grasslands experience seasonal temperature variation and too little precipitation for many trees to grow.** The climatograph indicates "moist" climate conditions (**green**) as well as "dry" climate conditions (**yellow,** when the temperature curve is above the precipitation curve). *Climatograph adapted from Breckle, S.W., 2002.*

DATA Q How do you expect evaporation rates may relate to the temperature and precipitation patterns from July to September to explain this period's dry conditions?

Soils of the temperate deciduous forest are relatively fertile, but this biome consists of far fewer tree species than are found in tropical rainforests. Oaks, beeches, and maples are a few of the most abundant types of trees in these forests. Some typical animals of the temperate deciduous forest of eastern North America are shown in Figure 4.11 (p. 83).

Temperate grassland Traveling westward from the Great Lakes, temperature differences between winter and summer become more extreme, rainfall diminishes, and we find **temperate grasslands** (FIGURE 4.20). This is because the limited amount of precipitation in the Great Plains region west of the Mississippi River can support grasses more easily than trees. Also known as *steppe* or *prairie*, temperate grasslands were once widespread throughout parts of North and South America and much of central Asia.

(a) Temperate rainforest

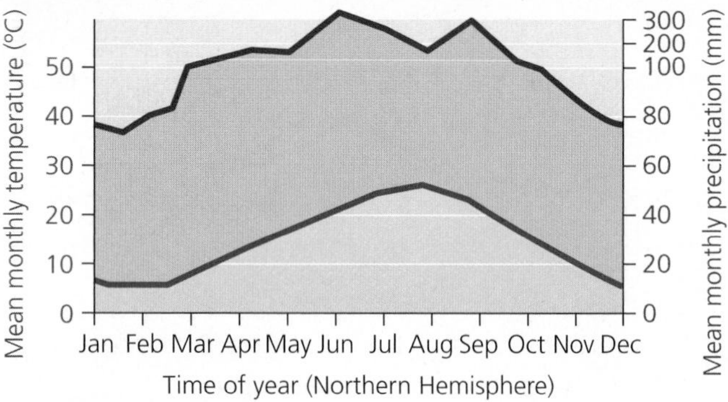

(b) Nagasaki, Japan

FIGURE 4.21 Temperate rainforests receive a great deal of precipitation and have moist, mossy interiors. *Climatograph adapted from Breckle, S.W., 2002.*

(a) Tropical rainforest

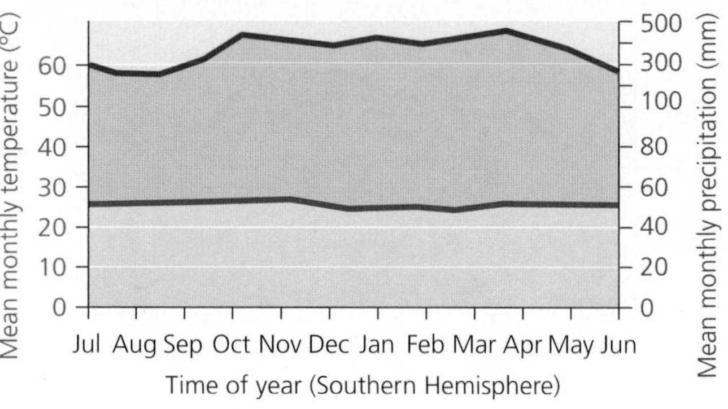

(b) Bogor, Java, Indonesia

FIGURE 4.22 Tropical rainforests, famed for their biodiversity, grow under constant, warm temperatures and a great deal of rain. *Climatograph adapted from Breckle, S.W., 2002.*

Today people have converted most of the world's grasslands for agriculture, and almost no undisturbed native grasslands exist in North America. Vertebrate animals of the native North American grasslands include American bison, prairie dogs, pronghorn antelope, and ground-nesting birds such as meadowlarks and prairie chickens. Because we have converted so many grasslands for farming and ranching, most of these animals exist today at only a small fraction of their historic population sizes.

Temperate rainforest Further west in North America, the topography becomes varied, and biome types intermix. The coastal Pacific Northwest region, with its heavy rainfall, features **temperate rainforest** (FIGURE 4.21). Coniferous trees, such as cedars, spruces, hemlocks, and Douglas fir, grow very tall in the temperate rainforest, and the forest interior is shaded and damp. Moisture-loving animals, such as the bright yellow banana slug, are common. The soils of temperate rainforests are usually quite fertile but are susceptible to landslides and erosion if forests are cleared.

We extract large amounts of lumber and other commercially valuable forest products from temperate rainforests. Logging has eliminated most old-growth trees in these forests, driving species such as the spotted owl and marbled murrelet toward extinction. Local people generally support timber extraction, but they also suffer the consequences of overharvesting.

Tropical rainforest In tropical regions we see the same pattern found in temperate regions: Areas of high rainfall grow rainforests, areas of intermediate rainfall host dry or deciduous forests, and areas of lower rainfall are dominated by grasses. However, tropical biomes differ from their temperate counterparts in other ways because they are closer to the equator and therefore warmer on average year-round. For one thing, they hold far greater biodiversity.

Tropical rainforest (FIGURE 4.22)—found in Central America, South America, Southeast Asia, west Africa, and other tropical regions—is characterized by year-round rain and uniformly warm temperatures. Tropical rainforests have dark, damp interiors, lush vegetation, and highly diverse biotic communities, with more species of insects, birds, amphibians, and other animals than any other biome.

These forests are not dominated by single species of trees, as are forests closer to the poles, but instead consist of very high numbers of tree species intermixed, each at a low density. Any given tree may be draped with vines, enveloped by strangler figs, and loaded with epiphytes (orchids and other plants that grow in trees), such that trees occasionally collapse under the weight of all the life they support.

(a) Tropical dry forest

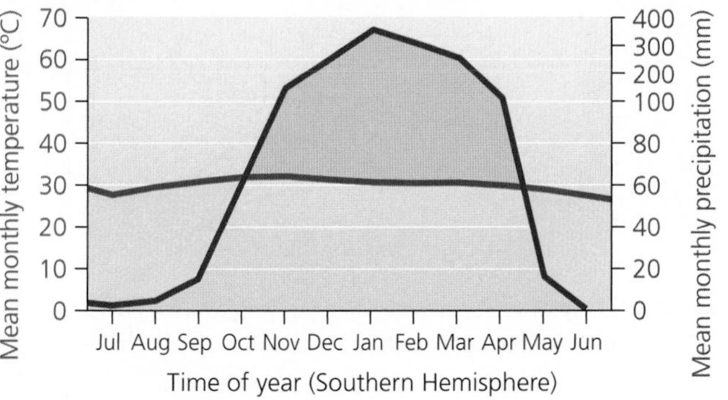

(b) Darwin, Australia

FIGURE 4.23 **Tropical dry forests experience significant seasonal variation in precipitation and relatively stable, warm temperatures.** *Climatograph adapted from Breckle, S.W., 2002.*

(a) Savanna

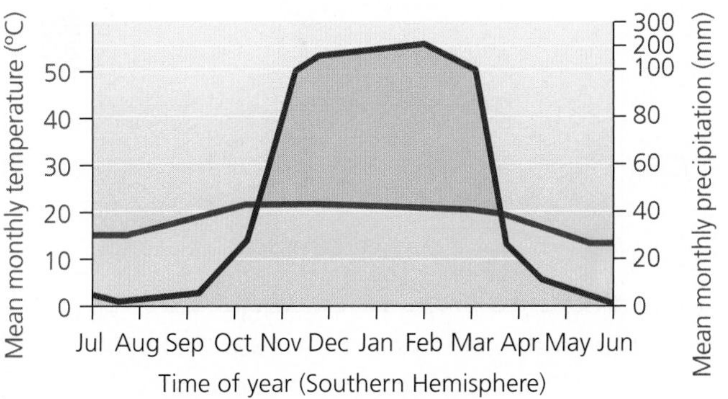

(b) Harare, Zimbabwe

FIGURE 4.24 **Savannas are grasslands with clusters of trees.** They experience slight seasonal variation in temperature but significant variation in rainfall. *Climatograph adapted from Breckle, S.W., 2002.*

Despite this profusion of life, tropical rainforests have poor, acidic soils that are low in organic matter. Nearly all nutrients present in this biome are contained in the plants, not in the soil. An unfortunate consequence is that once tropical rainforests are cleared, the nutrient-poor soil can support agriculture for only a short time (p. 221). As a result, farmed areas are abandoned quickly, and farmers move on and clear more forest.

Tropical dry forest Tropical areas that are warm year-round but where rainfall is lower overall and highly seasonal give rise to **tropical dry forest**, or **tropical deciduous forest** (FIGURE 4.23), a biome widespread in India, Africa, South America, and northern Australia. Wet and dry seasons each span about half a year in tropical dry forest. Organisms that inhabit tropical dry forest have adapted to seasonal fluctuations in precipitation and temperature. For instance, many plants are deciduous and leaf out and grow profusely with the rains, then drop their leaves during dry times of year.

Rains during the wet season can be heavy and, coupled with erosion-prone soils, can lead to severe soil loss where people have cleared forest. Across the globe, we have converted a great deal of tropical dry forest to agriculture. Clearing for farming or ranching is straightforward because vegetation is lower and canopies less dense than in tropical rainforest.

Savanna Drier tropical regions give rise to **savanna** (FIGURE 4.24), tropical grassland interspersed with clusters of acacias or other trees. The savanna biome is found across stretches of Africa, South America, Australia, India, and other dry tropical regions. Precipitation in savannas usually arrives during distinct rainy seasons, whereas in the dry season grazing animals concentrate near widely spaced water holes. Common herbivores on the African savanna include zebras, gazelles, and giraffes. Predators of these grazers include lions, hyenas, and other highly mobile carnivores.

Desert Where rainfall is very sparse, **desert** (FIGURE 4.25) forms. The driest biome on Earth, most deserts receive well under 25 cm (9.8 in.) of precipitation per year, much of it during isolated storms months or years apart. Some deserts, such as Africa's Sahara and Namib deserts, are mostly bare sand dunes; others, such as the Sonoran Desert of Arizona and northwest Mexico, receive more rain and are more heavily vegetated.

Deserts are not always hot; the high desert of the western United States is positively cold in winter. Because deserts have low humidity and little vegetation to insulate them from temperature extremes, sunlight readily heats them in the

(a) Desert

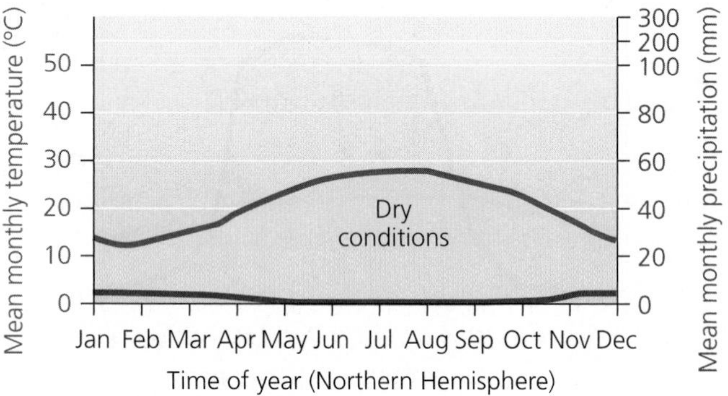

(b) Cairo, Egypt

FIGURE 4.25 **Deserts are dry year-round, but they are not always hot.** The temperature curve is consistently above the precipitation curve in this climatograph for Cairo, Egypt, indicating that the region experiences "dry" conditions all year. *Climatograph adapted from Breckle, S.W., 2002.*

(a) Tundra

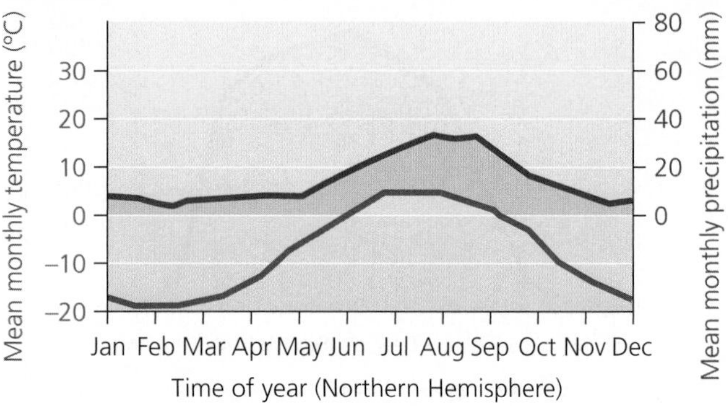

(b) Vaigach, Russia

FIGURE 4.26 **Tundra is a cold, dry biome found near the poles.** Alpine tundra occurs atop high mountains at lower latitudes. *Climatograph adapted from Breckle, S.W., 2002.*

daytime, but heat is quickly lost at night. As a result, temperatures vary greatly from day to night and across seasons of the year. Desert soils can be quite saline and are sometimes known as *lithosols*, or stone soils, for their high mineral and low organic-matter content.

Desert animals and plants show many adaptations to deal with a harsh climate. Most reptiles and mammals, such as rattlesnakes and kangaroo mice, are active in the cool of night. Many Australian desert birds are nomadic, wandering long distances to find areas of recent rainfall and plant growth. Desert plants tend to have thick, leathery leaves to reduce water loss, or green trunks so that the plant can photosynthesize without leaves, minimizing the surface area prone to water loss. The spines of cacti and other desert plants guard them from being eaten by herbivores desperate for the precious water these plants hold. Such traits have evolved by convergent evolution in deserts across the world (see Figure 3.3b, p. 51).

Tundra Nearly as dry as desert, **tundra** (FIGURE 4.26) occurs at very high latitudes in northern Russia, Canada, and Scandinavia. Extremely cold winters with little daylight and summers with lengthy days characterize this landscape of lichens and low, scrubby vegetation without trees. The great seasonal variation in temperature and day length results from this biome's high-latitude location, angled toward the sun in summer and away from the sun in winter.

Because of the cold climate, underground soil remains more or less permanently frozen and is called **permafrost.** During the long, cold winters, surface soil freezes as well. When the weather warms, the soil melts and produces pools of surface water, forming an ideal habitat for mosquitoes and other insects. The swarms of insects benefit bird species that migrate long distances to breed during the brief but productive summer. Caribou also migrate to the tundra to breed, then leave for the winter. Only a few animals, such as polar bears and musk oxen, can survive year-round here.

Most tundra remains intact and relatively unaltered by human occupation and development. However, atmospheric circulation patterns (p. 455) bring our airborne pollutants to this biome, and global climate change is heating high-latitude regions more intensely than other areas (p. 503). Climate change is melting sea ice, altering seasonal cycles to which animals have adapted, and melting permafrost, releasing methane gas that further worsens climate change.

Tundra also occurs as alpine tundra at the tops of tall mountains in temperate and tropical regions, where high elevation creates conditions similar to those of high latitude.

(a) Boreal forest

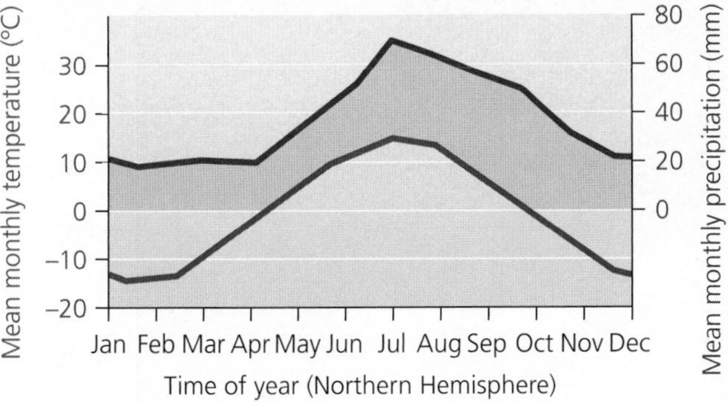

(b) Archangelsk, Russia

FIGURE 4.27 Boreal forest experiences long, cold winters, cool summers, and moderate precipitation. *Climatograph adapted from Breckle, S.W., 2002.*

(a) Chaparral

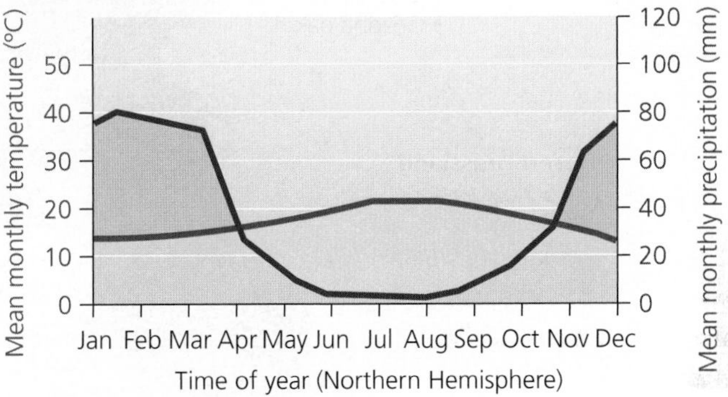

(b) Los Angeles, California, USA

FIGURE 4.28 Chaparral is a highly seasonal biome dominated by shrubs, influenced by marine weather, and dependent on fire. *Climatograph adapted from Breckle, S.W., 2002.*

Boreal forest The northern coniferous forest, or **boreal forest,** often called **taiga** (FIGURE 4.27), extends across much of Canada, Alaska, Russia, and Scandinavia. A few species of evergreen trees, such as black spruce, dominate large stretches of forest, interspersed with occasional bogs and lakes. These forests develop in cooler, drier regions than do temperate forests, and they experience long, cold winters and short, cool summers.

Soils are typically nutrient-poor and somewhat acidic. As a result of strong seasonal variation in day length, temperature, and precipitation, many organisms compress a year's worth of feeding, breeding, and rearing of young into a few warm, wet months. Year-round residents of boreal forest include mammals such as moose, wolves, bears, lynx, and rodents. This biome also hosts many insect-eating birds that migrate from the tropics to breed during the brief, intensely productive, summer season.

Chaparral In contrast to the boreal forest's broad, continuous distribution, **chaparral** (FIGURE 4.28) is limited to small patches widely flung around the globe. Chaparral consists mostly of evergreen shrubs and is densely thicketed. This biome is highly seasonal, with mild, wet winters and warm, dry summers—a climate induced by oceanic influences and often termed "Mediterranean." Besides ringing the Mediterranean Sea, chaparral occurs along the coasts of California, Chile, and southern Australia. Chaparral communities experience frequent fire, and their plants are adapted to resist fire or even to depend on it for germination of their seeds.

Altitude creates patterns analogous to latitude

As any hiker or skier knows, climbing in elevation causes a much more rapid change in climate than moving the same distance toward the poles. Vegetative communities change along mountain slopes in correspondence with this altitude-induced climate variation (FIGURE 4.29). It is often said that hiking up a mountain in the southwestern United States is like walking from Mexico to Canada. A hiker ascending one of southern Arizona's higher mountains would begin in Sonoran Desert or desert grassland and proceed through oak woodland, pine forest, and finally spruce–fir forest—the equivalent of passing through several biomes. A hiker scaling one of the great peaks of the Andes in Ecuador, near the equator, could begin in tropical rainforest and end amid glaciers in alpine tundra.

At higher altitudes, temperature, atmospheric pressure, and oxygen all decline, whereas ultraviolet radiation increases.

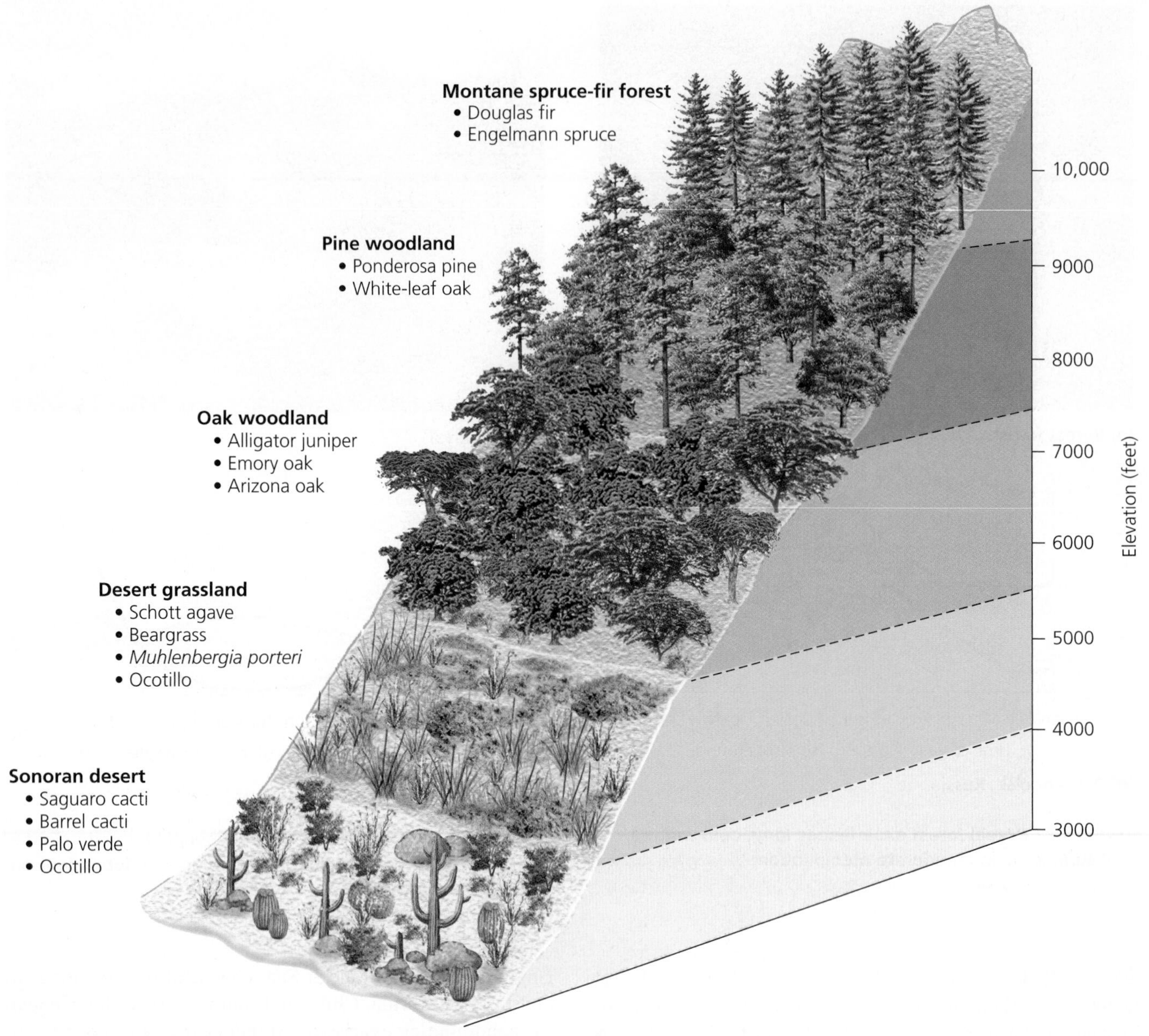

FIGURE 4.29 **As altitude increases, communities change in ways similar to how they change with latitude as one moves toward the poles.** Climbing a mountain in southern Arizona, as pictured here, takes the hiker through the equivalent of several biomes.

Moreover, mountains can alter local climate in other ways, such as through the **rainshadow** effect. When moisture-laden air ascends a steep slope, it releases precipitation as it cools. By the time it flows over the top of the mountain and down the other side, it can be very dry, creating a relatively arid region in the rainshadow. All these phenomena affect the nature and distribution of ecological communities in mountainous areas.

Conclusion

The natural world is so complex that we can visualize it in many ways and at various scales. Dividing the world's communities into major types, or biomes, is informative at the broadest geographic scales. Understanding how communities function at more local scales requires understanding how species interact with one another. Species interactions such as predation, parasitism, competition, and mutualism give rise to effects both weak and strong, direct and indirect. Feeding relationships can be represented by the concepts of trophic levels and food webs, and particularly influential species are sometimes called keystone species. People alter communities, in part by introducing non-native species that sometimes turn invasive. But more and more, through ecological restoration, we are also attempting to undo some of the changes we have caused.

Reviewing Objectives

You should now be able to:

Compare and contrast the major types of species interactions

- Competition results when individuals or species vie for limited resources. It can occur within or among species and can result in coexistence or exclusion. It also can lead to realized niches, resource partitioning, and character displacement. (pp. 76–77)
- In predation, an individual of one species kills and consumes an individual of another. Predation is the basis of food webs and can influence population dynamics and community composition. (p. 78)
- In parasitism, an individual of one species derives benefit by harming (but usually not killing) an individual of another. (p. 79)
- Herbivory is an exploitative interaction in which an animal feeds on a plant. (pp. 79–80)
- In mutualism, species benefit from one another. (p. 80)
- In some mutualistic and parasitic interactions, the participants are symbiotic, whereas in others they are free-living. (p. 80)

Characterize feeding relationships and energy flow, using them to construct trophic levels and food webs

- Energy is transferred among trophic levels in food chains. (pp. 80–81)
- Lower trophic levels generally contain more energy, biomass, and individuals. (p. 82)
- Food webs illustrate feeding relationships and energy flow among species in a community. (pp. 82–83)

Distinguish characteristics of a keystone species

- Keystone species exert impacts on communities that are far out of proportion to their abundance. (pp. 83–84)
- Top predators are frequently considered keystone species, but other types of organisms also exert strong effects on communities. (pp. 83–84)

Characterize disturbance, succession, and notions of community change

- Disturbances are varied, and communities respond to disturbance in different ways. (pp. 84–85)
- Succession describes a typical pattern of community change through time. (p. 85)
- Primary succession begins with an area devoid of life. Secondary succession begins with an area that has been severely disturbed but where remnants of the original community remain. (pp. 85, 88)
- Ecologists today view succession as being less predictable and deterministic than they did in the past. (p. 88)
- If disturbance is severe enough, communities may undergo phase shifts involving irreversible change—or novel communities may form. (p. 88)

Perceive and predict the potential impacts of invasive species in communities

- People have introduced countless species to new areas. Some of these non-native species may become invasive if they do not encounter limiting factors on their population growth. (pp. 88–89)
- Invasive species such as the zebra mussel have altered the composition, structure, and function of communities. (pp. 88–89)
- We can respond to invasive species with prevention, control, and eradication measures. (p. 92)

Explain the goals and methods of restoration ecology

- Restoration ecology is the science of restoring communities to a previous, more functional or more "natural" condition, variously defined as before human impact or before recent industrial impact. (p. 92)
- The growing practice of ecological restoration, informed by the science of restoration ecology, helps us restore ecological systems. (pp. 92–93)

Describe biomes and identify the terrestrial biomes of the world

- Biomes represent major classes of communities spanning large geographic areas. (p. 93)
- The distribution of biomes is determined by temperature, precipitation, and other factors. (p. 94)
- Aquatic and coastal systems can be classified in ways similar to terrestrial biomes, but these systems are determined by different factors. (pp. 94–95)
- Biomes include temperate deciduous forest, temperate grassland, temperate rainforest, tropical rainforest, tropical dry forest, savanna, desert, tundra, boreal forest, and chaparral. (pp. 95–99)
- Mountains, with their diversity of elevations and climate conditions, host mixtures of ecological communities. (pp. 99–100)

Testing Your Comprehension

1. How does competition lead to a realized niche? How does it promote resource partitioning?
2. Contrast the three main types of exploitative species interactions. How do predation, parasitism, and herbivory differ?
3. Give examples of symbiotic and nonsymbiotic mutualisms. Describe at least one way in which a mutualism affects your daily life.
4. Compare and contrast trophic levels, food chains, and food webs. How are these concepts related, and how do they differ?
5. What is meant by the term *keystone species*, and what types of organisms are most often considered keystone species?
6. Explain primary succession. How does it differ from secondary succession? Give an example of each.
7. Name five changes to Great Lakes communities that have occurred since the invasion of the zebra mussel.
8. What is restoration ecology? Why is it an important scientific pursuit in today's world?
9. What factors most strongly influence the type of biome that forms in a particular place on land? What factors determine the type of aquatic system that may form in a given location?
10. Draw a typical climate diagram for a tropical rainforest. Label all parts of the diagram and describe all of the types of information an ecologist could glean from such a diagram. Now draw a climate diagram for a desert. How does it differ from your rainforest climatograph, and what does this tell you about how the two biomes differ?

Seeking Solutions

1. Suppose you spot two species of birds feeding side by side, eating seeds from the same plant. You begin to wonder whether competition is at work. Describe how you might design scientific research to address this question. What observations would you try to make at the outset? Would you try to manipulate the system to test your hypothesis that the two birds are competing? If so, how?
2. Spend some time outside on your campus, in your yard, or in the nearest park or natural area. Find at least 10 species of organisms (plants, animals, or others), and observe each one long enough to watch it feed or to make an educated guess about how it derives its nutrition. Now, using Figure 4.11 as a model, draw a simple food web involving all the organisms you observed.
3. Can you think of one organism not mentioned in this chapter as a keystone species that you believe may be a keystone species? For what reasons do you suspect this? How could you experimentally test whether an organism is a keystone species?
4. Why do scientists consider invasive species to be a problem? What makes a species "invasive," and what ecological effects can invasive species have? Give examples.
5. Our map of biomes in Figure 4.17 appears to be a permanent record of stable patterns across the planet. But are the locations and identities of biomes truly permanent, or might they change over time? Provide reasons for your answers.
6. **THINK IT THROUGH** A federal agency has put you in charge of devising responses to the zebra mussel invasion. Based on what you know from this chapter, how would you seek to control this species' spread and reduce its impacts? What strategies would you consider pursuing immediately, and for which strategies would you commission further scientific research? For each of your ideas, name one benefit or advantage, and identify one obstacle it might face in being implemented. What additional steps might you suggest to deal with the unfolding quagga mussel invasion?

Calculating Ecological Footprints

In 2005, environmental scientists David Pimentel, Rodolfo Zuniga, and Doug Morrison of Cornell University reviewed scientific estimates for the economic and ecological costs inflicted by introduced and invasive species in the United States. They found that approximately 50,000 species have been introduced in the United States and that these account for over $120 billion in economic costs each year. These costs include direct losses and damage, as well as costs required for control of the invasive species. (Pimentel's group did not try to quantify monetary estimates for losses of biodiversity, ecosystem services, and aesthetics, which they say would drive total costs several times higher.) Calculate values missing from the table to determine the number of introduced species of each type of organism and the annual cost that each inflicts on our economy.

Group of organism	Percentage of total introduced	Number of species introduced	Percentage of total annual costs	Annual economic costs
Plants	50.0	25,000	27.2	
Microbes	40.0		20.2	
Arthropods	9.0		15.7	
Fish	0.28		4.2	
Birds	0.19		1.5	$1.9 billion
Mollusks	0.18		1.7	
Reptiles and amphibians	0.11		0.009	
Mammals	0.04	20	29.4	$37.5 billion
TOTAL	100	50,000	100	$127.4 billion

Source: Pimentel, D., R. Zuniga, and D. Morrison, 2005. Update on the environmental and economic costs associated with alien-invasive species in the United States. Ecological Economics *52: 273–288.*

1. Of the 50,000 species introduced into the United States, half are plants. Describe two ways in which non-native plants might be brought to a new environment. How might we help prevent non-native plants from establishing in new areas and posing threats to native communities?
2. Organisms that damage crop plants are the most costly of introduced species. Weeds, pathogenic microbes, and arthropods that attack crops together account for half of the costs documented by Pimentel and his colleagues. What steps can we—farmers, governments, and all of us as a society—take to minimize the impacts of invasive species on crops?
3. How might your own behavior influence the influx and ecological impacts of non-native species like those listed above? Name three things you could personally do to help reduce the impacts of invasive species.

5

An oysterman unloads his catch on the shores of the Chesapeake Bay.

Environmental Systems and Ecosystem Ecology

Upon completing this chapter, you will be able to:

- Describe the nature of environmental systems
- Define ecosystems and evaluate how living and nonliving entities interact in ecosystem-level ecology
- Outline the fundamentals of landscape ecology, GIS, and ecological modeling
- Assess ecosystem services and how they benefit our lives
- Compare and contrast how water, carbon, nitrogen, and phosphorus cycle through the environment
- Explain how human impact is affecting biogeochemical cycles

CENTRAL CASE STUDY

The Vanishing Oysters of the Chesapeake Bay

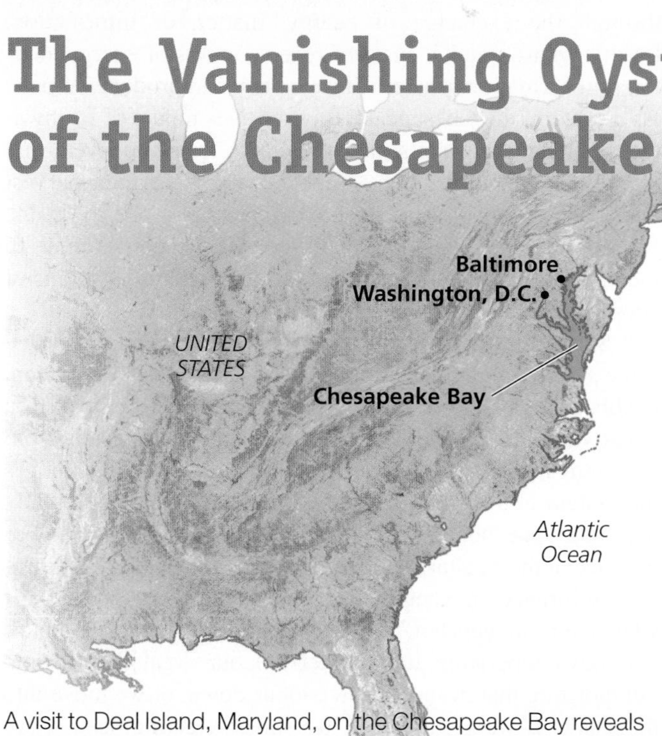

Baltimore
Washington, D.C.
UNITED STATES
Chesapeake Bay
Atlantic Ocean

"I'm 60. Danny's 58. We're the young ones."

—Grant Corbin, Oysterman in Deal Island, Maryland

"The Bay continues to be in serious trouble. And it's really no question why this is occurring. We simply haven't managed the Chesapeake Bay as a system the way science tells us we must."

—Will Baker, President, Chesapeake Bay Foundation

A visit to Deal Island, Maryland, on the Chesapeake Bay reveals a situation that is, unfortunately, all too common in modern America. The island, which was once bustling with productive industries and growing populations, is suffering. Economic opportunities in the community are few, and its populace is increasingly "graying" as more and more young people leave to find work elsewhere. In 1930, Deal Island had a population of 1237 residents. In 2010 it was a mere 471 people—and only 75 of them were under age 18.

Unlike other parts of the country with similar stories of economic decline, the demise of Deal Island and other bayside towns was not caused by the closing of a local factory, steel mill, or corporate headquarters. It was caused by the collapse of the Chesapeake Bay oyster fishery.

The Chesapeake Bay was once a thriving system of interacting plants, animals, and microbes. Blue crabs, scallops, and fish such as giant sturgeon, striped bass, and shad thrived in the bay. Nutrients carried to the bay by streams in its roughly 168,000 km^2 (64,000 mi^2) **watershed**—the land area that funnels water to the bay through rivers—nourished fields of underwater grasses that provided food and refuge to juvenile fish, shellfish, and crabs. Hundreds of millions of oysters kept the bay's water clear by filtering nutrients and phytoplankton (microscopic photosynthetic algae, protists, and cyanobacteria that drift near the surface) from the water column.

Although oysters had been eaten locally for some time, the intensive harvest of bay oysters for export began in the 1830s, and by the 1880s the bay boasted the world's largest oyster fishery. People flocked to the Chesapeake to work on oystering ships or in canneries, dockyards, and shipyards. Bayside towns prospered along with the oyster industry and developed a unique maritime culture that defined the region.

But by 2010 the bay's oyster populations had been reduced to a mere 1% of their historical abundance, and the oyster industry was all but ruined. Perpetual overharvesting, habitat destruction, virulent oyster diseases, and water pollution had nearly eradicated this economically and ecologically important species from bay waters. The monetary losses associated with the fishery collapse have been staggering, costing the economies of Maryland and Virginia an estimated $4 billion in lost economic activity from 1980 to 2010 alone.

One of the biggest impacts in recent decades on oysters is the pollution of the bay with high levels of the nutrients nitrogen and phosphorus from agricultural fertilizers, animal manure, stormwater runoff, and atmospheric compounds produced by fossil fuel combustion. Oysters naturally filter nutrients from water, but with so few oysters today, elevated nutrient levels have caused phytoplankton populations in the bay to increase. When phytoplankton die, settle to the bay bottom, and are decomposed by bacteria, oxygen in the water is depleted (a condition called **hypoxia**), which creates "dead zones" in the bay. Grasses, oysters, and other immobile organisms perish in dead zones when deprived of oxygen. Crabs, fish, and other mobile organisms are forced to flee to habitats where oxygen levels are higher, but they face smaller food supplies and increased predation pressure. Hypoxia, along with other human impacts on the Chesapeake Bay, cause it to be included on the Environmental Protection Agency's list of highly polluted waters.

Recent events in the Chesapeake have, at long last, given reason for hope for the recovery of the Chesapeake Bay system. The EPA agreed in 2010, for the first time in the region, to hold bay states to strict pollutant "budgets" that aim to substantially reduce inputs of nitrogen and phosphorus into the bay by 2025. Further, oyster restoration efforts are finally showing promise (see **THE SCIENCE BEHIND THE STORY**, pp. 118–119) in the Chesapeake. If these initiatives can begin to restore the bay to health, Deal Island and other communities may again enjoy the prosperity they once did on the scenic shores of the Chesapeake. ■

Earth's Environmental Systems

Understanding the rise and fall of the oyster industry in the Chesapeake Bay, as with many other human impacts on the environment, involves comprehending the complex networks of interlinked systems that comprise Earth's environment. These include physical systems ranging from matter and molecules up to magma and mountains (Chapter 2). They include biological systems ranging from organisms and populations (Chapter 3) to communities of interacting species (Chapter 4). In ecosystems they involve the interaction of living creatures with the nonliving entities around them. Earth's systems encompass cycles involving rock, air, and water that shape our landscapes and guide the flow of chemical elements and compounds that support life and regulate climate. We depend on these systems for our very survival.

Assessing questions holistically by taking a "systems approach" is helpful in environmental science, in which so many issues are multifaceted and complex. Such a broad and integrative approach poses challenges, because systems often show behavior that is difficult to predict. The scientific method (pp. 10–12) is easiest when researchers can isolate and manipulate small parts of complex systems, focusing on manageable components one at a time. However, environmental scientists are rising to the challenge of studying systems holistically, helping us to develop comprehensive solutions to complicated problems such as those faced in the Chesapeake Bay.

Systems involve feedback loops

A **system** is a network of relationships among parts, elements, or components that interact with and influence one another through the exchange of energy, matter, or information. Earth's environmental systems receive inputs of energy, matter, or information, process these inputs, and produce outputs. As a system, the Chesapeake Bay receives inputs of freshwater, sediments, nutrients, and pollutants from the rivers that empty into it. Oystermen, crabbers, and fishermen harvest some of the bay system's output: matter and energy in the form of seafood. This output subsequently becomes input to the nation's economic system and to the digestive systems of people who consume the seafood.

Sometimes a system's output can serve as input to that same system, a circular process described as a **feedback loop.** Feedback loops are of two types, negative and positive. In a **negative feedback loop** (FIGURE 5.1a), output that results from a system moving in one direction acts as input that moves the system in the other direction. Input and output essentially neutralize one another's effects, stabilizing the system. For instance, a thermostat stabilizes a room's temperature by turning the furnace on when the room gets cold and shutting it off when the room gets hot. Similarly, negative feedback regulates our body temperature. If we get too hot, our sweat glands pump out moisture that evaporates to cool us down, or we move into the shade. If we get too cold, we shiver, creating heat, or we move into the sun. Most systems in nature involve negative

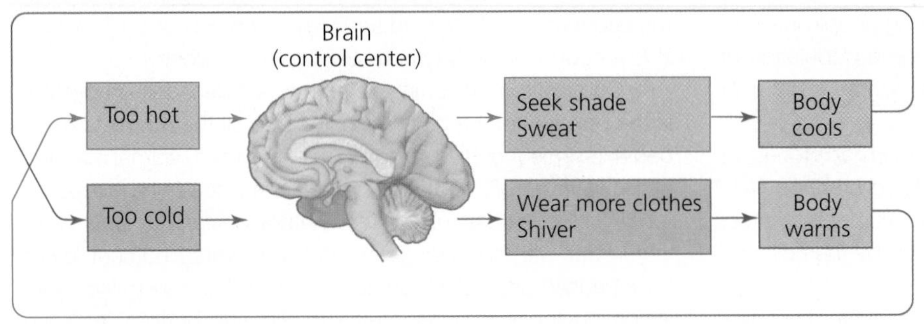

(a) Negative feedback

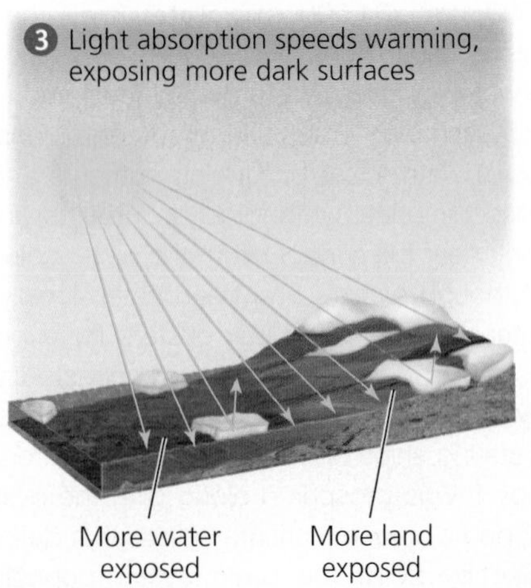

(b) Positive feedback

FIGURE 5.1 **Negative feedback loops exert a stabilizing influence on systems, whereas positive feedback loops have a destabilizing effect.** The human body's response to heat and cold **(a)** involves a negative feedback loop that keeps core body temperatures relatively stable. Positive feedback loops, in contrast, push systems away from equilibrium; for example, when Arctic glaciers and sea ice melt because of global warming **(b)**, darker surfaces are exposed, which absorb more sunlight, causing further warming and further melting.

feedback loops. Negative feedback enhances stability, and over time only those systems that are stable will persist.

Positive feedback loops have the opposite effect. Rather than stabilizing a system, they drive it further toward an extreme. In positive feedback, increased output leads to increased input, leading to further increased output. Exponential growth in a population (pp. 66–67) is one such example. The more individuals there are, the more offspring can be produced. Another example is the spread of cancer; as cells multiply out of control, the process is self-accelerating.

One positive feedback cycle of great concern to environmental scientists today involves the melting of glaciers and sea ice in the Arctic as a result of global warming (p. 498). Ice and snow, being white, reflect sunlight and keep surfaces cool. But if the climate warms enough to melt the ice and snow, darker surfaces of land and water are exposed, and these darker surfaces absorb sunlight. This absorption warms the surface, causing further melting, which in turn exposes more dark surface area, leading to further warming (FIGURE 5.1b). Runaway cycles of positive feedback are rare in nature, but they are common in natural systems altered by human impact, and they can destabilize those systems.

FAQ But isn't positive feedback "good" and negative feedback "bad"?

Understanding negative and positive feedback in systems can be difficult, because it goes against the way we use those terms in everyday language. In daily life, positive feedback (such as a complimentary comment on a writing assignment) is something that makes us feel good, whereas negative feedback (such as criticism on schoolwork) may make us feel bad. In essence, we have been trained to view positive feedback as a stabilizing force ("Keep up the good work, and you'll succeed") and negative feedback as a destabilizing force ("You need to change your approach if you're going to succeed").

In environmental systems, it's the opposite! Negative feedback resists change in systems, and in doing so it enhances stability, typically keeping conditions within ranges beneficial to life. Positive feedback exerts destabilizing effects that push conditions in systems to extremes, threatening organisms adapted to the system's normal conditions. Thus, negative feedback in environmental systems typically aids living things, whereas positive feedback often harms them.

Systems show several defining properties

In a system stabilized by negative feedback, when processes move in opposing directions at equivalent rates so that their effects balance out, they are said to be in **dynamic equilibrium.** Processes in dynamic equilibrium can contribute to **homeostasis,** the tendency of a system to maintain constant or stable internal conditions. A system (such as an organism) in homeostasis keeps its internal conditions within a range that allows it to function. However, the steady state of a homeostatic system may itself change slowly over time. For instance, Earth has experienced gradual changes in atmospheric composition and ocean chemistry over its long history, yet life persists and our planet remains, by most definitions, a homeostatic system.

It is difficult to understand systems fully just by focusing on their individual components because systems can show **emergent properties,** characteristics not evident in the components alone. Stating that systems possess emergent properties is a lot like saying, "The whole is more than the sum of its parts." For example, if you were to reduce a tree to its component parts (leaves, branches, trunk, bark, roots, fruit, and so on) you would not be able to predict the whole tree's emergent properties, which include the role the tree plays as habitat for birds, insects, fungi, and other organisms (FIGURE 5.2). You could analyze the tree's chloroplasts (photosynthetic cell organelles), diagram its branch structure, and evaluate the nutritional content of its fruit, but you would still be unable to understand the tree as habitat, as part of a forest landscape, or as a reservoir for carbon storage.

Systems seldom have well-defined boundaries, so deciding where one system ends and another begins can be difficult. Consider a smartphone. It is certainly a system—a network

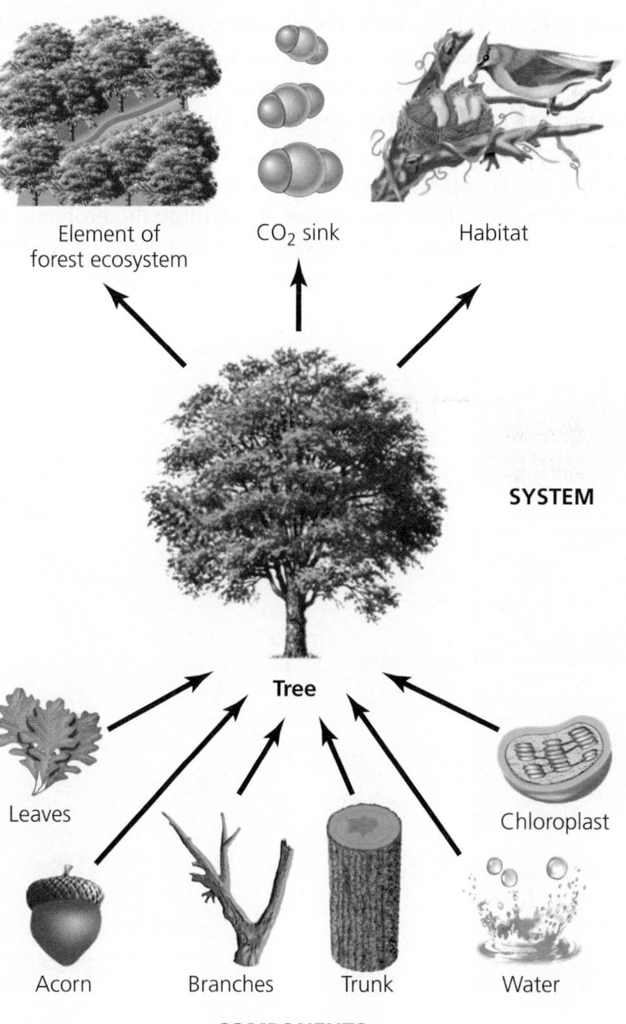

FIGURE 5.2 A system's emergent properties are not evident when we break the system down into its component parts. For example, a tree serves as wildlife habitat and plays roles in forest ecology and global climate regulation, but you would not know that from considering the tree only as a collection of leaves, branches, and chloroplasts.

of circuits and parts that interact and exchange energy and information—but where are its boundaries? Is the system merely the phone itself, or does it include the other phones you call, the websites you access on it, and the cellular and Wi-Fi networks that keep it connected? What about the energy grid that recharges the phone's battery, with its transmission lines and distant power plants?

No matter how we attempt to isolate or define a system, we soon see that it has connections to systems larger and smaller than itself. Systems may exchange energy, matter, and information with other systems, and they may contain or be contained within other systems. Thus, where we draw boundaries may depend on the spatial (space) or temporal (time) scale at which we choose to focus.

Environmental systems interact

The Chesapeake Bay and the rivers that empty into it are an example of interacting systems. On a map, these rivers are a branched and braided network of water channels surrounded by farms, cities, and forests (**FIGURE 5.3**). But where are the boundaries of this system? For a scientist interested in **runoff** (precipitation that flows over land and enters waterways) and the flow of water, sediment, or pollutants, it may make the most sense to view the bay's watershed as a system. However, for a scientist interested in the bay's dead zones, it may be best to view the watershed together with the bay as the system of interest, because their interaction is central to the problem. In environmental science, identifying the boundaries of systems depends on the questions being addressed.

The dead zones in the Chesapeake Bay are due to the extremely high levels of nitrogen and phosphorus delivered to its waters from the 6 states in its watershed and the 15 states in its **airshed**—the geographic area that produces air pollutants that are likely to end up in a waterway. In 2007, the bay received an estimated 127 million kg (281 million lb) of nitrogen and 8.3 million kg (18.2 million lb) of phosphorus, with roughly one-third of nitrogen inputs from atmospheric sources. Agriculture was a major source of these nutrients, contributing 40% of the nitrogen (**FIGURE 5.4a**) and 45% of the phosphorus (**FIGURE 5.4b**) entering the bay.

Elevated nitrogen and phosphorus inputs cause phytoplankton in the bay's waters to flourish. High densities lead to elevated mortality in phytoplankton populations, and dead phytoplankton settle to the bottom of the bay. The remains of dead phytoplankton are joined on the bottom by the waste products of zooplankton, tiny creatures that feed on phytoplankton. The abundance of organic material causes an explosion in populations of bacterial decomposers, which deplete the oxygen in bottom waters while consuming this material. Deprived of oxygen, organisms will flee if they can or will suffocate if they cannot. Oxygen replenishes slowly at the bottom because fresh water entering the bay from rivers remains naturally stratified in a layer at the surface and is slow to mix with the denser, saltier bay water. This limits the amount of oxygenated surface water that reaches the bottom-dwelling life that needs it. The process of nutrient overenrichment, blooms of algae, increased production of organic matter, and subsequent ecosystem degradation is known as **eutrophication** (**FIGURE 5.5**).

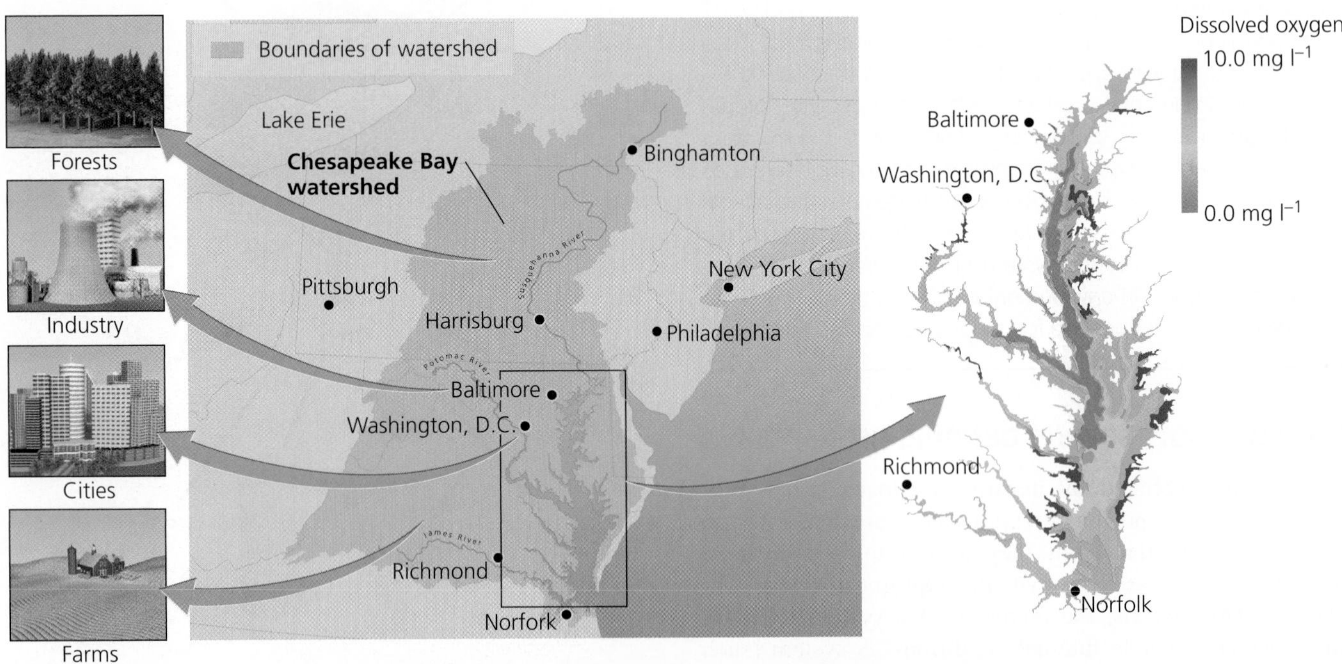

FIGURE 5.3 The Chesapeake Bay watershed encompasses 168,000 km2 (64,000 mi2) of land area in six states and the District of Columbia. Tens of thousands of streams carry water, sediment, and pollutants from a variety of sources downriver to the Chesapeake, where nutrient pollution has given rise to large areas of hypoxic waters. The zoomed-in map (**at right**) shows dissolved oxygen concentrations in the Chesapeake Bay in 2011. Oysters, crabs, and fish typically require a minimum of 3 mg/L of oxygen and are therefore excluded from large portions of the bay where oxygen levels are too low. *Source: Figure at right adapted from Chesapeake Bay Record Dead Zone Map, Chesapeake Bay Foundation.*

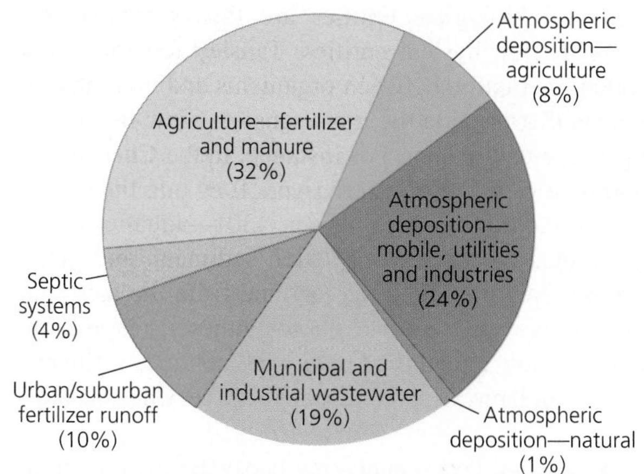

(a) Sources of nitrogen entering the Chesapeake Bay

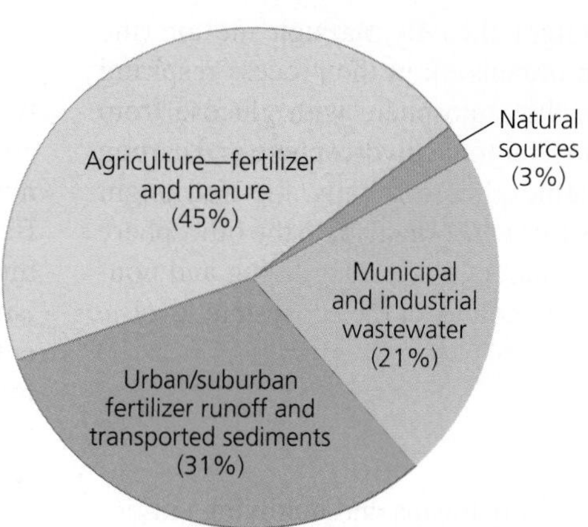

(b) Sources of phosphorus entering the Chesapeake Bay

FIGURE 5.4 The Chesapeake Bay receives inputs of nitrogen (a) and phosphorus (b) from many sources in its watershed. *Data from Chesapeake Bay Program Watershed Model Phase 4.3 (Chesapeake Bay Program Office, 2009).*

We may perceive Earth's systems in various ways

There are many ways to delineate natural systems. Categorizing environmental systems can help make Earth's dazzling complexity comprehensible to the human brain and accessible to problem solving. For instance, scientists sometimes divide Earth's components into structural spheres. The **lithosphere** (p. 34) is the rock and sediment beneath our feet, the planet's uppermost mantle and crust. The **atmosphere** (p. 450) is composed of the air surrounding our planet. The **hydrosphere** (p. 391) encompasses all water—salt or fresh, liquid, ice, or vapor—in surface bodies, underground, and in the atmosphere. The **biosphere** (p. 60) consists of all the planet's organisms and the abiotic (nonliving) portions of the environment with which they interact.

Picture a robin plucking an earthworm from the ground after a rain. You are witnessing an organism (the robin) consuming another organism (the earthworm) by removing it from part of the lithosphere (the soil) that the earthworm had been modifying, after rain (from the hydrosphere) moistened

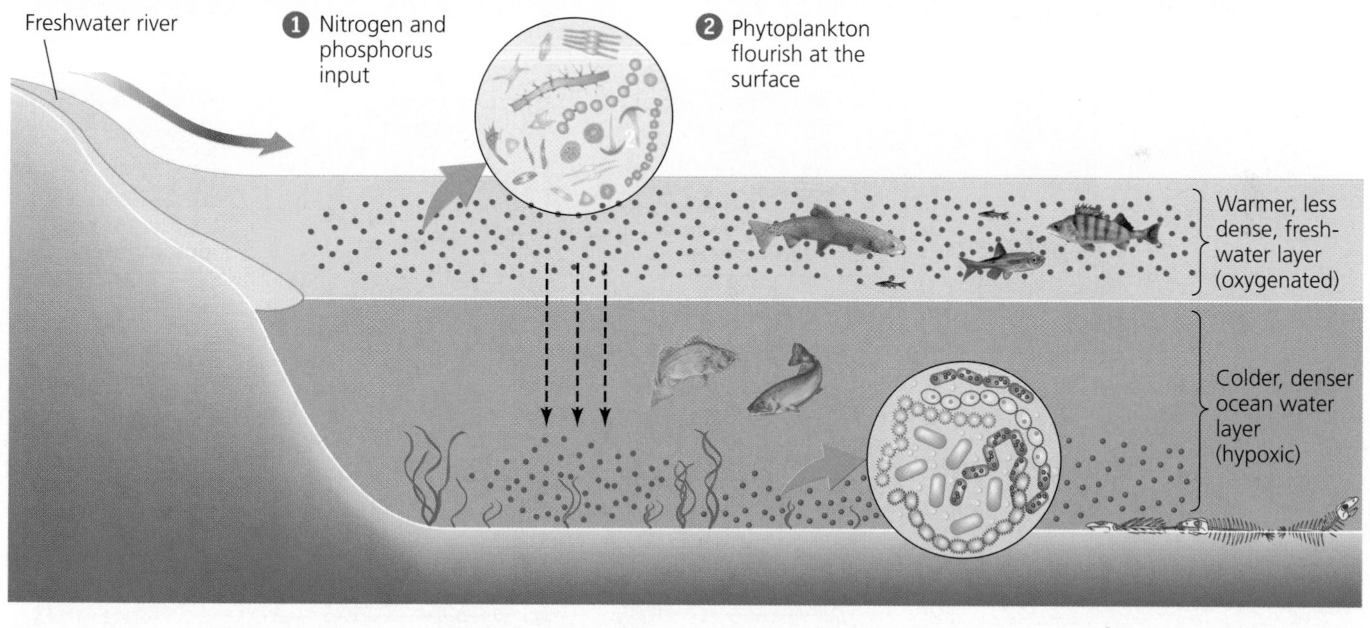

FIGURE 5.5 Excess nitrogen and phosphorus causes eutrophication in aquatic systems such as the Chesapeake Bay. Coupled with stratification (layering) of water, eutrophication can severely deplete dissolved oxygen. Nutrients from river water 1 boost growth of phytoplankton 2, which die and are decomposed at the bottom by bacteria 3. Stability of the surface layer prevents deeper water from absorbing oxygen to replace oxygen consumed by decomposers 4, and the oxygen depletion suffocates or drives away bottom-dwelling marine life 5. This process gives rise to hypoxic zones like those in the bay.

the ground. The robin might then fly through the air (the atmosphere) to a tree (an organism), in the process respiring (combining oxygen from the atmosphere with glucose from the organism, and adding water to the hydrosphere and carbon dioxide and heat to the atmosphere). Finally, the bird might defecate, adding nutrients from the organism to the lithosphere below. The study of such interactions among living and nonliving things is a key part of ecology at the ecosystem level.

Ecosystems

An **ecosystem** consists of all organisms and nonliving entities that occur and interact in a particular area at the same time. The ecosystem concept builds on the idea of the biological community (Chapter 4), but ecosystems include abiotic components as well as biotic ones. In ecosystems, energy flows and matter cycles among these components.

Ecosystems are systems of interacting living and nonliving entities

The ecosystem concept originated early last century with scientists such as British ecologist Arthur Tansley, who recognized that biological entities are tightly intertwined with chemical and physical entities. Tansley felt that there was so much interaction between organisms and their abiotic environments that it made the most sense to view living and nonliving elements together. For instance, in the Chesapeake Bay **estuary**—a water body where rivers flow into the ocean, mixing fresh water with saltwater (p. 430)—aquatic organisms are affected by the flow of water, sediment, and nutrients from the rivers that feed the bay and from the land that feeds those rivers. In turn, the photosynthesis, respiration, and decomposition that these organisms conduct influence the chemical and physical conditions of the Chesapeake's waters.

Ecologists soon began analyzing ecosystems as an engineer might analyze the operation of a machine. In this view, ecosystems are systems that receive inputs of energy, process and transform that energy while cycling matter internally, and produce outputs (such as heat, water flow, and animal waste products) that enter other ecosystems.

Energy flows in one direction through ecosystems. Most arrives as radiation from the sun, powers the system, and exits in the form of heat (**FIGURE 5.6a**). Matter, in contrast, is generally recycled within ecosystems (**FIGURE 5.6b**). Energy and matter pass among producers, consumers, and decomposers

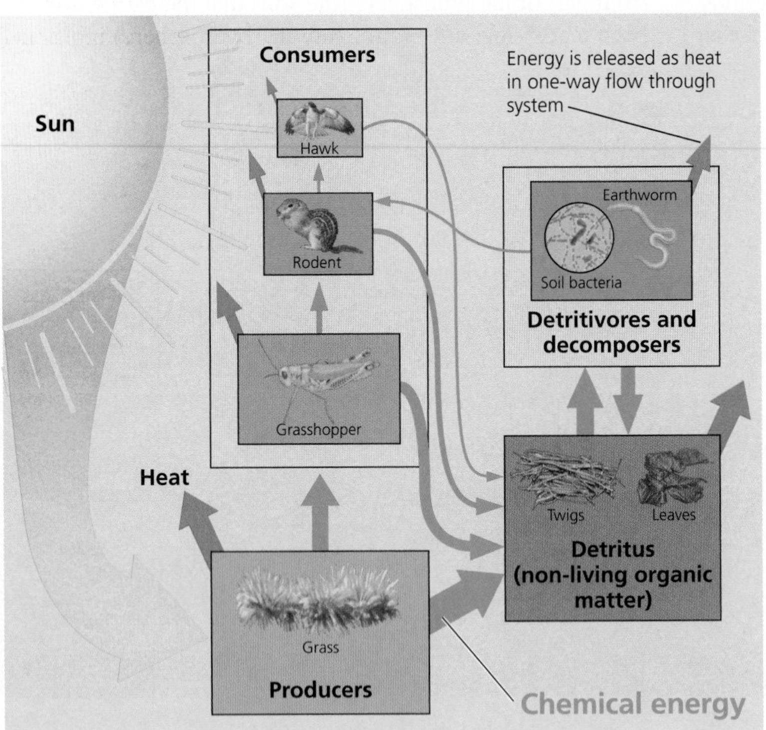

(a) Energy flowing through an ecosystem

(b) Matter cycling within an ecosystem

FIGURE 5.6 In systems, energy flows in one direction, whereas matter is recycled. In **(a)**, light energy from the sun **(yellow arrow)** drives photosynthesis in producers, which begins the transfer of chemical energy **(green arrows)** among trophic levels (pp. 80–82) and detritus. Energy exits the system through respiration in the form of heat **(red arrows).** In **(b)**, nutrients **(blue arrows)** move among trophic levels and detritus. In both diagrams, box sizes conceptually represent quantities of energy or matter content, and arrow widths represent relative magnitudes of energy or matter transfer. Such values may vary greatly among ecosystems. For simplicity, various abiotic components (such as water, air, and inorganic soil content) of ecosystems have been omitted.

DATA Q Based on the figure, which transfer of chemical energy is the largest in ecosystems? Which transfer is the largest for nutrient cycling in ecosystems?

through food-web relationships (pp. 82–83). Matter is recycled because when organisms die and decay, their nutrients remain in the system. In contrast, most energy that organisms take in drives cellular respiration and is released as heat.

Energy is converted to biomass

As autotrophs, such as green plants and phytoplankton, convert solar energy to the energy of chemical bonds in sugars through photosynthesis (p. 32), they perform **primary production.** Specifically, the total amount of chemical energy produced by autotrophs is termed **gross primary production.** Autotrophs use most of this production to power their own metabolism by cellular respiration (p. 32). The energy that remains after respiration and that is used to generate biomass (such as leaves, stems, and roots; p. 82) ecologists call **net primary production.** Thus, net primary production equals gross primary production minus the energy used in respiration.

Another way to think of net primary production is that it represents the energy or biomass available for consumption by heterotrophs. Some of this plant biomass is eaten by herbivores. Plant matter not eaten by herbivores becomes fodder for detritivores and decomposers once the plant dies or drops its leaves. Heterotrophs use the energy they gain from plant biomass for their own metabolism, growth, and reproduction. Some of this energy is used by heterotrophs to generate biomass in their bodies (such as skin, muscle, or bone), which is termed **secondary production.**

Ecosystems vary in the rate at which autotrophs convert energy to biomass. The rate at which this conversion occurs is termed **productivity,** and ecosystems whose plants convert solar energy to biomass rapidly are said to have high **net primary productivity.** Freshwater wetlands, tropical forests, coral reefs, and algal beds tend to have the highest net primary productivities, whereas deserts, tundra, and open ocean tend to have the lowest (**FIGURE 5.7**). Variation among ecosystems and among biomes (Chapter 4) in net primary productivity results in geographic patterns across the globe (**FIGURE 5.8**). In terrestrial ecosystems, net primary productivity tends to increase with temperature and precipitation. In aquatic ecosystems, net primary productivity tends to rise with light and the availability of nutrients.

Nutrients influence productivity

Nutrients are elements and compounds (pp. 23–26) that organisms consume and require for survival. Organisms need several dozen naturally occurring nutrients to survive. Elements and compounds required in relatively large amounts (such as nitrogen, carbon, and phosphorus) are called **macronutrients.** Nutrients needed in small amounts are called **micronutrients** (examples include zinc, copper, and iron).

Nutrients stimulate production by plants, and lack of nutrients can limit production. As mentioned earlier, the availability of nitrogen or phosphorus frequently is a limiting factor (p. 67) for plant or algal growth. When these nutrients are added to a system, producers show the greatest response to whichever nutrient has been in shortest supply. Nitrogen tends to be limiting in marine systems, and phosphorus in freshwater systems, though both contribute to eutrophication in all waters. Thus eutrophication in the Chesapeake Bay is driven by excess nitrogen, whereas eutrophication in the freshwater ponds and lakes in the bay's watershed are spurred by increases in phosphorus.

Canadian ecologist David Schindler and others demonstrated the effects of phosphorus on freshwater systems in the 1970s by experimentally manipulating entire lakes. In one experiment, his team bisected a 16-ha (40-acre) lake in Ontario with a plastic barrier. To one half the researchers added carbon, nitrate, and phosphate; to the other they added only carbon and nitrate. Soon after the experiment began, they witnessed a dramatic increase in algae in the half of the lake that received phosphate, whereas the other half (the control for the experiment, p. 11) continued to host algal

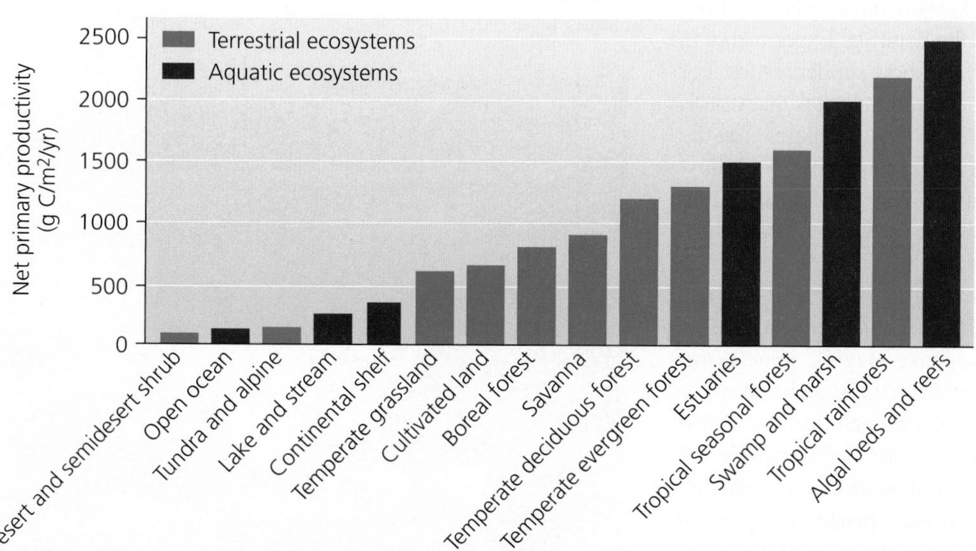

FIGURE 5.7 Net primary productivity varies greatly between ecosystem types. Freshwater wetlands, tropical forests, coral reefs, and algal beds show high values on average, whereas deserts, tundra, and the open ocean show low values. *Data from Whittaker, R.H., 1975.* Communities and ecosystems, *2nd ed. NewYork: MacMillan.*

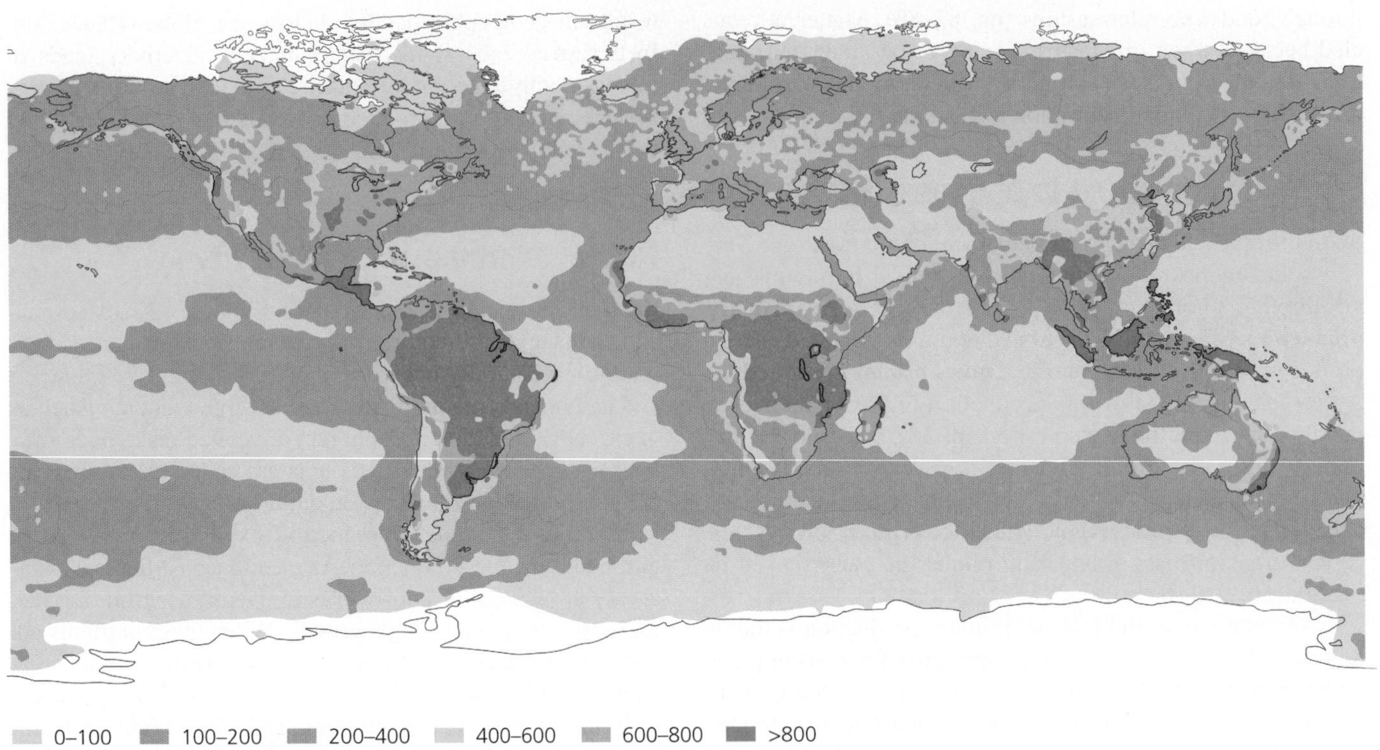

FIGURE 5.8 A world map of net primary production based on satellite data shows that on land, net primary production varies geographically with temperature and precipitation. In the world's oceans, net primary production (shown here as grams of carbon fixed per square meter per year) is highest around the margins of continents, where nutrients (of both natural and human origin) run off from land. *Data from Field, C.B., et al., 1998. Primary production of the biosphere: Integrating terrestrial and oceanic components.* Science *281: 237–240. Reprinted with permission from AAAS.*

levels typical for lakes in the region (**FIGURE 5.9**). This difference held until shortly after they stopped fertilizing seven years later. At that point, algae decreased to normal levels in the half that had previously received phosphate. Such experiments showed clearly that phosphorus addition can markedly increase primary productivity in freshwater lakes.

Similar experiments in coastal ocean waters show nitrogen to be the more important limiting factor for primary productivity. In experiments in the 1980s and 1990s, Swedish ecologist Edna Granéli took samples of ocean water from the Baltic Sea and added phosphate, nitrate, or nothing. Chlorophyll and phytoplankton increased greatly in the flasks with nitrate, whereas those with phosphate did not differ from the controls. Experiments in Long Island Sound by other researchers show similar results. For open ocean waters far from shore, research indicates that iron is a highly effective nutrient for stimulating phytoplankton growth.

Increased nutrient pollution from farms, cities, and industries has led to the development of over 500 documented hypoxic dead zones globally as of 2010 (**FIGURE 5.10**), including that of the Chesapeake Bay as well as a large dead zone that forms each year in the Gulf of Mexico off the Louisiana coast near the mouth of the Mississippi River (pp. 412–413). Some are seasonal (like the Chesapeake Bay's), some occur irregularly, and others are permanent. The increase in the number of dead zones—there were 162 documented in the 1980s and 49 in the 1960s—reflects how the activities of people are changing nutrient concentrations in waters around the world.

If one were to add up all the world's marine and coastal dead zones, they would cover an area the size of the state of Michigan. Scientists calculate that the amount of marine life missing from the oceans as a result of dead zones likely exceeds the total amount of shellfish harvested each year from the entire United States—a harvest worth over $2 billion.

The good news is that in locations where people have reduced nutrient runoff, dead zones have begun to disappear. In New York City, hypoxic zones at the mouths of the Hudson

FIGURE 5.9 The upper portion of this lake in Ontario was experimentally treated with the addition of phosphate. This treated portion experienced an immediate, dramatic, and prolonged algal bloom, identifiable by its opaque waters.

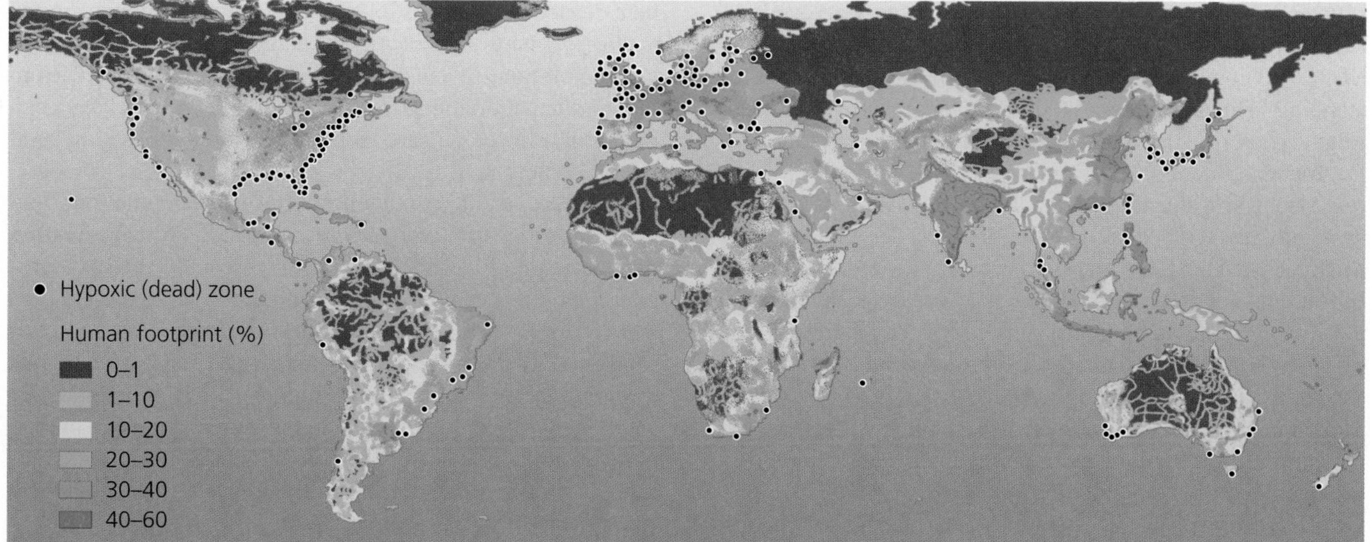

FIGURE 5.10 Over 500 marine dead zones have been recorded across the world. Dead zones as of 2010 (shown by dots on the map) occur mostly offshore from areas of land with the greatest human ecological footprints (here, expressed on a scale of 0 to 100, with higher numbers indicating bigger human footprints). *Data from World Resources Institute, 2010, http://www.wri.org/project/eutrophication/map and Diaz, R., and R. Rosenberg, 2008. Spreading dead zones and consequences for marine ecosystems.* Science *321: 926–929. Reprinted with permission from AAAS.*

and East rivers were nearly eliminated once sewage treatment was improved. The Black Sea, which borders Ukraine, Russia, Turkey, and eastern Europe, had long suffered one of the world's worst hypoxic zones. Then in the 1990s, after the Soviet Union collapsed, industrial agriculture in the region declined drastically. With fewer fertilizers draining into it, the Black Sea began to recover, and today fisheries are reviving. However, agricultural collapse is not a strategy anyone would choose to alleviate hypoxia. Rather, scientists are proposing a variety of innovative and economically acceptable ways to reduce nutrient runoff (p. 128).

Ecosystems interact with one another

Whether we stand at a river's mouth or peruse a satellite image, we can conceptualize ecosystems at different scales. An ecosystem can be as small as a puddle of water or as large as a bay, lake, or forest. For some purposes, scientists even view the entire biosphere as a single all-encompassing ecosystem. The term is most often used, however, to refer to systems of moderate geographic extent that are somewhat self-contained. For example, the tidal marshes in the Chesapeake where river water empties into the bay are an ecosystem, as are the sections of the bay dominated by oyster reefs.

Adjacent ecosystems may share components and interact extensively. For instance, a pond ecosystem is very different from a forest ecosystem that surrounds it, but salamanders that develop in the pond live their adult lives under logs on the forest floor until returning to the pond to breed. Rainwater that nourishes forest plants may eventually make its way to the pond, carrying with it nutrients from the forest's leaf litter. Likewise, rivers, tidal marshes, and open waters in estuaries all may interact, as do forests and prairie where they converge. Areas where ecosystems meet may consist of transitional zones called **ecotones,** in which elements of each ecosystem mix.

WEIGHING THE ISSUES

ECOSYSTEMS WHERE YOU LIVE Think about the area where you live, and briefly describe this region's ecosystems. How do these systems interact? For instance, does any water pass from one to another? Describe the boundaries of watersheds in your region. If one ecosystem were greatly modified (say, if a shopping mall were built atop a wetland or amid a forest), what impacts on nearby systems might result? (Note: If you live in a city, realize that urban areas can be thought of as ecosystems, too!)

Landscape ecologists study geographic patterns

Because components of different ecosystems may intermix, ecologists often find it useful to view these systems on a larger geographic scale that encompasses multiple ecosystems. For instance, if you are studying large mammals such as black bears, which move seasonally from mountains to valleys or between mountain ranges, you had better consider the overall landscape that includes all these areas. If you study fish such as salmon, which migrate between marine and freshwater ecosystems, you need to know how these systems interact.

In such a broad-scale approach, called **landscape ecology,** scientists study how landscape structure affects the abundance, distribution, and interaction of organisms. Landscape-level approaches are also helping scientists, citizens, planners, and policymakers to plan for sustainable regional development (pp. 342–350).

For a landscape ecologist, a landscape is made up of a spatial array of **patches.** Depending on the researcher's perspective, patches may consist of ecosystems or may simply be areas of habitat for a particular organism. Patches are spread spatially over a landscape in a **mosaic.** This metaphor reflects how natural systems often are arrayed across landscapes in complex patterns, like an intricate work of art. Thus, a forest ecologist may refer to a mosaic of forested patches left standing in an agricultural landscape. An amphibian biologist might speak of a mosaic of patches of pond habitat that frogs use for reproduction.

FIGURE 5.11 illustrates a landscape consisting of five ecosystem types, with ecotones along their borders (indicated by thick red lines). At this scale, we perceive a mosaic consisting of four patches and a river. However, we can view a landscape at different scales. The figure's inset shows a magnified view of an ecotone. At this finer resolution, we see that the ecotone consists of patches of forest and grassland in a complex arrangement. The scale at which an ecologist focuses will depend on the questions he or she is interested in, or on the organisms he or she is studying.

Every organism has specific habitat needs, so when its habitat is distributed in patches across a landscape, individuals may need to expend energy and risk predation traveling from one to another. If the patches are far apart, the organism's population may become divided into subpopulations, each occupying a different patch in the mosaic. Such a network of subpopulations, most of whose members stay within their respective patches but some of whom move among patches or mate with members of other patches, is called a **metapopulation.** When patches are still more isolated from one another, individuals may not be able to travel between them at all. In such a case, smaller subpopulations may be at risk of extinction.

Because of this extinction risk, metapopulations and landscape ecology are of great interest to **conservation biologists** (pp. 294–295), scientists who study the loss, protection, and restoration of biodiversity. Of particular concern is the fragmentation of habitat into small and isolated patches (pp. 284, 330–331)—something that often results from human development pressures. Establishing corridors of habitat (see **FIGURE 5.11**) to link patches and allow animals to move among them is one approach that conservation biologists pursue as they attempt to maintain biodiversity in the face of human impact (pp. 327–328).

Remote sensing helps us apply landscape ecology

As more scientists take a landscape perspective, they are benefiting from better and better remote-sensing technologies.

FIGURE 5.11 Landscape ecology deals with spatial patterns above the ecosystem level. This generalized diagram of a landscape shows a mosaic of patches of five ecosystem types (three terrestrial types, a marsh, and a river). Thick red lines indicate ecotones. A stretch of lowland broadleaf forest running along the river serves as a corridor connecting the large region of forest on the left to the smaller patch of forest alongside the marsh. The inset shows a magnified view of the forest-grassland ecotone and how it consists of patches on a smaller scale.

Satellites orbiting Earth are sending us more and better data than ever before on how the surface of our planet looks. By helping us monitor our planet from above, satellite imagery is making vital contributions to modern environmental science.

A common tool for research in landscape ecology is the **geographic information system (GIS).** A GIS consists of computer software that takes multiple types of data (for instance, on geology, hydrology, topography, vegetation, animal populations, and human infrastructure) and combines them on a common set of geographic coordinates. The idea is to create a complete picture of a landscape and to analyze how elements of the different data sets are arrayed spatially and how they may be correlated.

FIGURE 5.12 illustrates in a simplified way how different datasets of a GIS are combined, layer upon layer, to form a composite map. GIS has become a valuable tool used by geographers, landscape ecologists, resource managers, and conservation biologists. Principles of landscape ecology, and tools such as GIS, are increasingly used in regional planning processes (pp. 342–350).

GIS is being used to guide restoration efforts in the Chesapeake Bay. The *ChesapeakeStat* website, which was launched in 2010, enables scientists, educators, policymakers, and citizens to create customized composite maps that overlay parameters important to the bay's health. This tool is being used to assess the bay's current status, the effects of restoration efforts, and progress toward long-term goals.

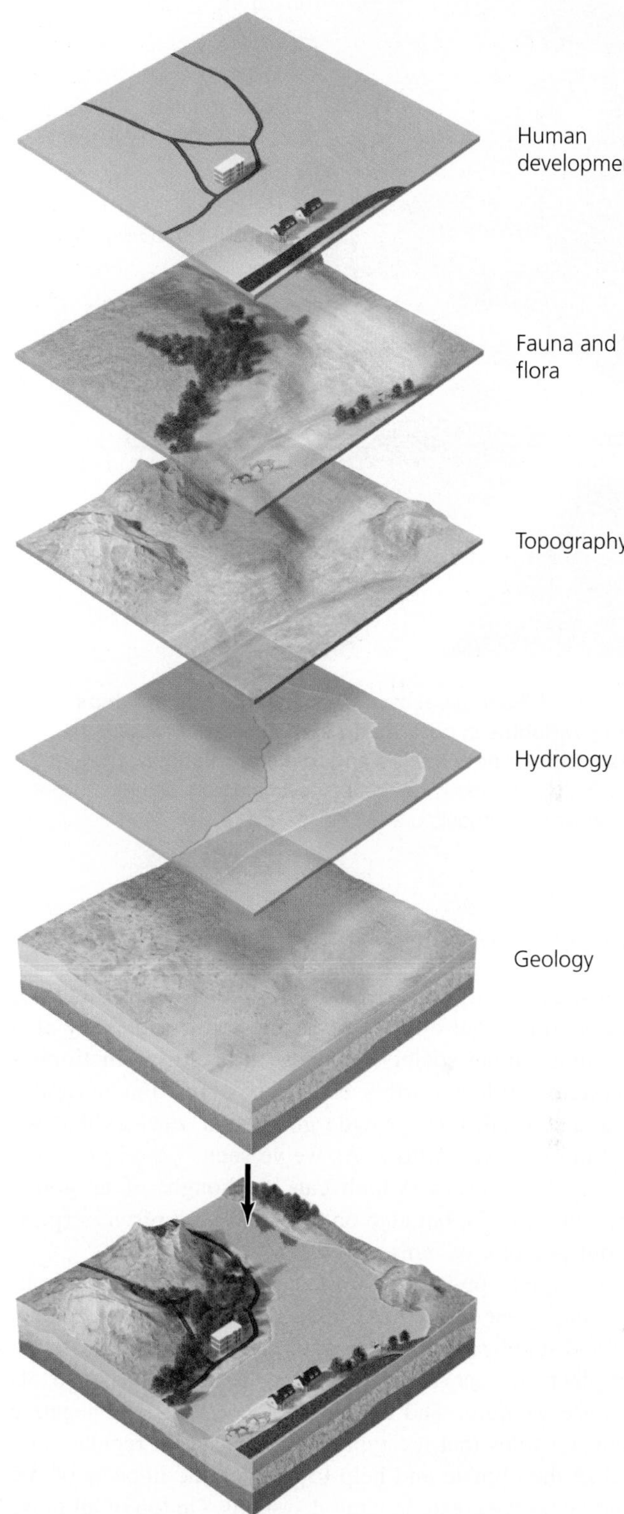

FIGURE 5.12 Geographic information systems (GIS) allow us to layer different types of data on natural landscape features and human land uses so as to produce maps integrating this information. GIS can be used to explore correlations among these data sets and to help in regional planning.

Modeling helps ecologists understand systems

Another way in which ecologists seek to make sense of the complex systems they study is by working with models. In science, a **model** is a simplified representation of a complex natural process, designed to help us understand how the process occurs and to make predictions. **Ecological modeling** is the practice of constructing and testing models that aim to explain and predict how ecological systems function.

Because ecological processes (for ecosystems, communities, or populations) involve so many factors, ecological models can be mathematically complicated. However, the general approach of ecological modeling is easy to understand (**FIGURE 5.13**). Researchers gather data from nature on relationships that interest them and then form a hypothesis about what those relationships are. They construct a model that attempts to explain the relationships in a generalized way so that people can use the model to make predictions about how the system will behave. Modelers test their predictions by gathering new data from natural systems, and they use this new data to refine the model, making it increasingly accurate.

Note that the process illustrated in Figure 5.13 resembles the scientific method (pp. 10–12); models are essentially hypotheses about how systems function. Accordingly, the use of models is a key part of ecological research and environmental regulation today. As just one example, the National Oceanic and Atmospheric Administration (NOAA) uses the Chesapeake Bay Fisheries Ecosystem model to examine trophic interactions among fish species in the bay, the effects of hypoxia on fish populations, and how the distribution of underwater grasses influences blue crab populations. Data from scientific journal articles (p. 12) and direct measurements are used to establish the model's parameters, which are then used to predict the effects of differing fish harvest levels on species and ecosystems in the Chesapeake Bay.

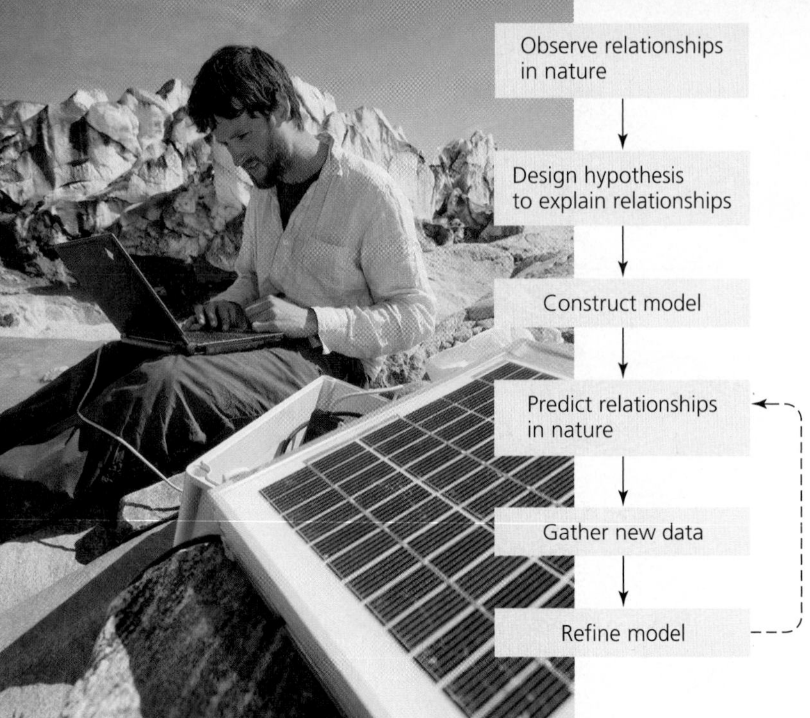

FIGURE 5.13 **Ecological modelers observe relationships among variables in nature and then construct models to explain those relationships and make predictions.** They test and refine the models by gathering new data from nature and seeing how well the models predict those data.

TABLE 5.1 Ecosystem Services
Ecological processes do many things that benefit us:
• Regulate oxygen, carbon dioxide, stratospheric ozone, and other atmospheric gases
• Regulate temperature and precipitation by means of ocean currents, cloud formation, and so on
• Protect against storms, floods, and droughts, mainly by means of vegetation
• Store and regulate water supplies in watersheds and aquifers
• Prevent soil erosion
• Form soil by weathering rock and accumulating organic material
• Cycle carbon, nitrogen, phosphorus, sulfur, and other nutrients
• Filter waste, remove toxins, recover nutrients, and control pollution
• Pollinate plant crops and wild plants so they reproduce
• Control crop pests with predators and parasites
• Provide habitat for organisms to breed, feed, rest, migrate, and winter
• Produce fish, game, crops, nuts, and fruits that people eat
• Supply lumber, fuel, metals, fodder, and fiber
• Furnish medicines, pets, ornamental plants, and genes for resistance to pathogens and crop pests
• Provide recreation such as ecotourism, fishing, hiking, birding, hunting, and kayaking
• Provide aesthetic, artistic, educational, spiritual, and scientific amenities

Ecosystem services sustain our world

When scientists try to understand how ecosystems function, it is not simply out of curiosity about the world. They also know that human society depends on healthy, functioning ecosystems. When Earth's ecosystems function normally and undisturbed, they provide goods and services that we could not survive without. As we've seen, we rely not just on natural resources (which can be thought of as goods from nature; p. 3), but also on the *ecosystem services* (p. 3) that our planet's systems provide. (TABLE 5.1).

Ecological processes form the soil that nourishes our crops, purify the water we drink and the air that we breathe, store and stabilize supplies of water that we use, pollinate the food plants we eat, and receive and break down (some of) the waste we dump and the pollution we emit. The negative feedback cycles that are typical of ecosystems regulate and stabilize the climate and help to dampen the impacts of the disturbances we create in natural systems. On top of all these services that are vital for our very existence, ecosystems also provide services that enhance the quality of our lives, ranging from recreational opportunities to pleasing scenery to inspiration and spiritual renewal. Ecosystem goods and ecosystem services (FIGURE 5.14) support our lives and society in profound and innumerable ways.

One of the most important ecosystem services is the cycling of nutrients. Through the processes that take place within and among ecosystems, the chemical elements and compounds that we need—carbon, nitrogen, phosphorus, water, and many more—cycle through our environment in complex ways.

FAQ If we had to pay for the services provided by nature, what would it cost?

Estimating the economic value of all of Earth's ecosystem services is no easy task, but a study published in 1997 by an international team of scientists and economists put a rough dollar figure on the many benefits nature provides us (p. 152). The team did not examine all of the biosphere's services, but rather focused on 17 key ecosystem services from 16 biomes. The authors defined *ecosystem services* as natural processes (such as nutrient cycling or regulation of the global climate) and renewable natural resources (such as timber or food crops, but not including nonrenewable minerals and fossil fuels) and calculated the economic value of these services.

The study determined that the value of these 17 ecosystem services ranged from $16 to $54 trillion per year, with an average of $33 trillion ($48 billion in 2013 dollars when adjusted for inflation). This is a stunning quantity, especially when one considers that the sum total of the gross national products of every nation in 1997 was only $18 trillion. That is, the ecosystem services provided by nature had a value 1.8 times higher than the entire economic activity of the planet. Such studies are useful because they allow us to more accurately conduct cost-benefit analyses of proposed projects by considering both the expected economic gains from development and the economic costs from the degradation of ecosystem services due to development.

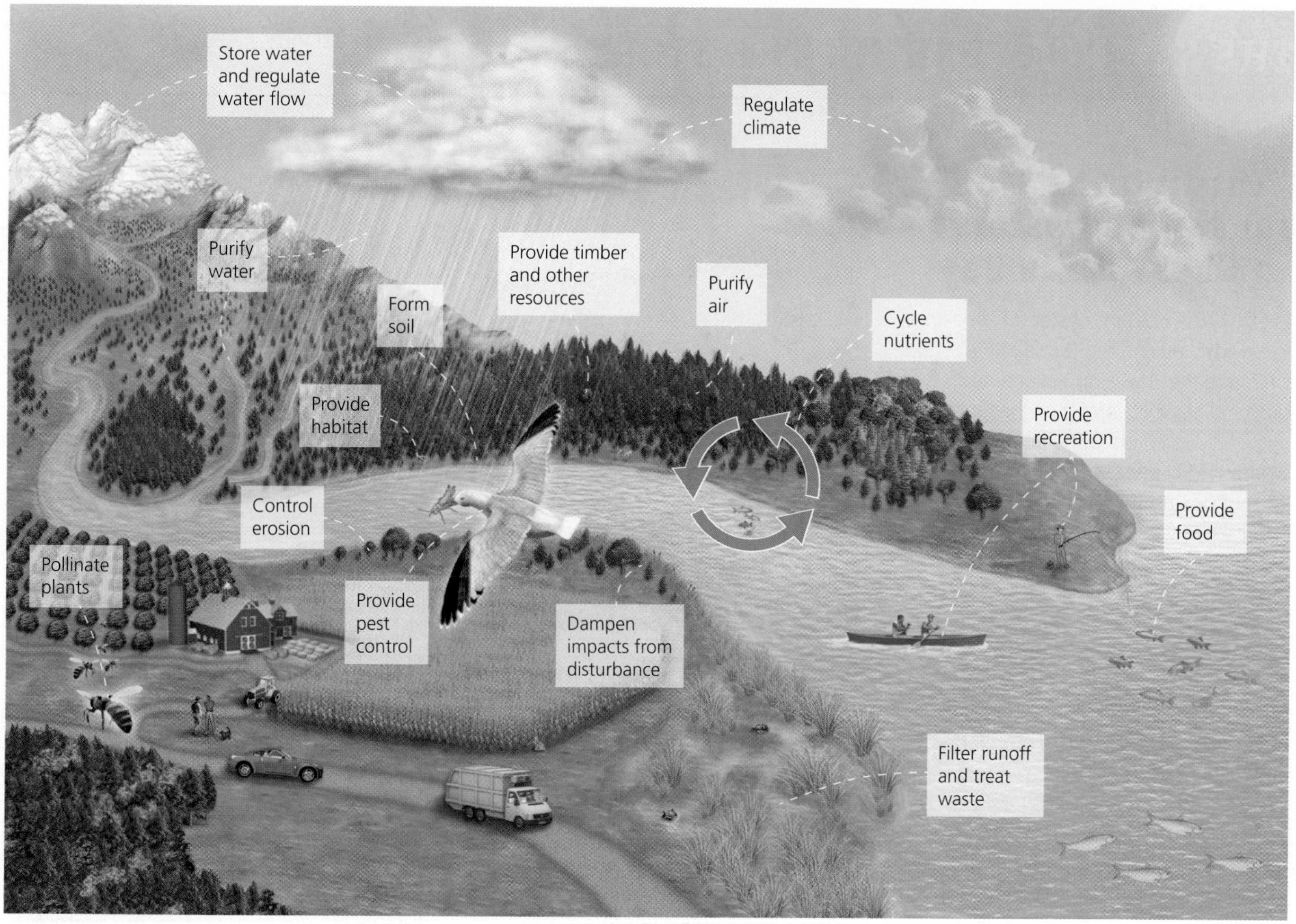

FIGURE 5.14 Ecological processes naturally provide countless services that we call *ecosystem services*. Our society, indeed our very survival, depends on these services.

Biogeochemical Cycles

Just as nitrogen and phosphorus from fertilizer on Pennsylvania corn fields end up in Chesapeake Bay oysters on our dinner plates, all nutrients move through the environment in intricate ways. Whereas energy enters an ecosystem from the sun, flows from organism to organism, and dissipates to the atmosphere as heat, the physical matter of an ecosystem is circulated over and over again.

Nutrients circulate through ecosystems in biogeochemical cycles

Nutrients move through ecosystems in **nutrient cycles,** also known as **biogeochemical cycles.** In these pathways, chemical elements or molecules travel through the atmosphere, hydrosphere, and lithosphere, and from one organism to another, in dynamic equilibrium. A carbon atom in your fingernail today might have helped compose the muscle of a cow a year ago, may have resided in a blade of grass a month before that, and may have been part of a dinosaur's tooth 100 million years ago. After we die, the nutrients in our bodies will spread widely through the environment, eventually being incorporated by an untold number of organisms far into the future.

Nutrients and other materials move from one **pool,** or **reservoir,** to another, remaining for varying amounts of time (the **residence time**) in each. The dinosaur, the grass, the cow, and you are each reservoirs for carbon atoms, as are sedimentary rocks and the atmosphere. The rate at which materials move between reservoirs is termed a **flux,** and the flux between any given reservoirs can change over time. When a reservoir releases more materials than it accepts, it is called a **source,** and when a reservoir accepts more materials than it releases, it is called a **sink.** FIGURE 5.15 illustrates these concepts in a simple manner.

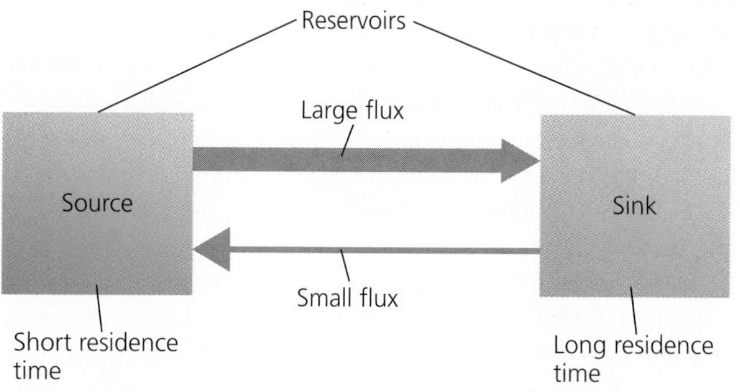

FIGURE 5.15 The main components of a biogeochemical cycle are reservoirs and fluxes. A source releases more materials than it accepts, and a sink accepts more materials than it releases.

"Turning the Tide" for Native Oysters in Chesapeake Bay

In 2001, the Eastern oyster (*Crassostrea virginica*) was in dire trouble in the Chesapeake Bay. Populations had dropped by 99%, and the Chesapeake's oyster industry, once the largest in the world, had collapsed. Poor water quality, reef destruction, virulent diseases spread by transplanted oysters, and 200 years of overharvesting all contributed to the collapse.

Restoration efforts had largely failed. Moreover, when scientists or resource managers proposed to rebuild oyster populations by significantly restricting oyster harvests or establishing oyster reef "sanctuaries," these initiatives were typically defeated by the politically powerful oyster industry. All this had occurred in a place whose very name (derived from the Algonquin word *Chesepiook*) means "great shellfish bay".

With the collapse of the native oyster fishery and with political obstacles blocking restoration projects for native oysters, support grew among the oyster industry, state resource managers, and some scientists for the introduction of Suminoe oysters (*Crassostrea ariakensis*) from Asia. This species seemed well suited for conditions in the bay and showed resistance to the parasitic diseases that were ravaging native oysters. Proponents argued that introducing Suminoe oysters would reestablish thriving oyster populations in the bay and revitalize the oyster fishery.

Introducing oysters would also improve the bay's water quality, proponents said, because as oysters feed, they filter phytoplankton and sediments from the water column.

Filter-feeding by oysters is an important ecological service in the bay because it reduces phytoplankton densities, clarifies waters, and supports the growth of underwater grasses that provide food and refuge for waterfowl and young crabs. Because introductions of invasive species can have profound ecological impacts (pp. 88–89), the Army Corps of Engineers was directed to coordinate an environmental impact statement (EIS, p. 174) on oyster restoration approaches in the Chesapeake.

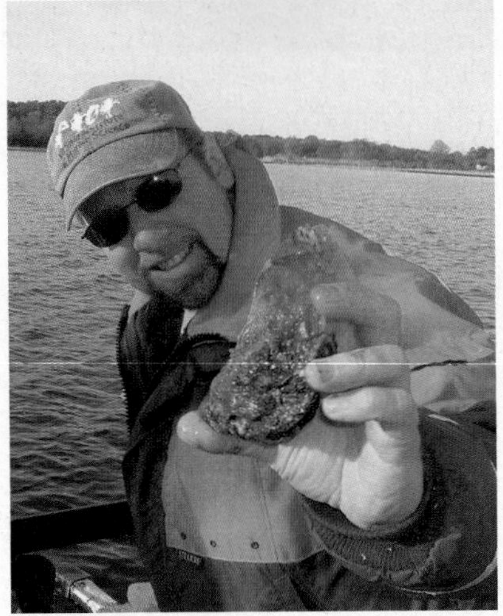

David Schulte, U.S. Army Corps of Engineers

It was in this politically charged, high-stakes environment that Dave Schulte, a scientist with the Corps and doctoral student at the College of William and Mary, set out to determine whether there was a viable approach to restoring native oyster populations. The work he and his team began would help turn the tide in favor of native oysters in the bay's restoration efforts.

One of the biggest impacts on native oysters was the destruction of oyster reefs by a century of intensive oyster harvesting. Oysters settle and grow best on the shells of other oysters, and over long periods this process forms reefs (underwater outcrops of living oysters and oyster shells) that solidify and become as hard as stone. Throughout the bay, massive reefs that at one time had jutted out of the water at low tide had been reduced to rubble on the bottom from a century of repeated scouring by metal dredgers used by oyster harvesting ships. The key, Schulte realized, was to construct artificial reefs like those that once existed, to get oysters off the bottom—away from smothering sediments and hypoxic waters—and up into the plankton-rich upper waters.

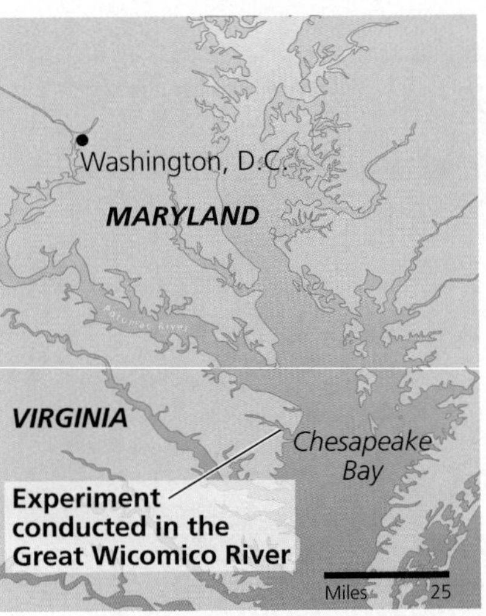

FIGURE 1 **Schulte's study was conducted in the Great Wicomico River in Virginia in the lower Chesapeake Bay.**

Armed with the resources available to the Corps, he opted to take a landscape ecology approach and restore patches of reef habitat on nine complexes of reefs covering a total of 35.3 hectares (87 acres) in an oyster sanctuary near the mouth of the Great Wicomico River (**FIGURE 1**) in the lower Chesapeake Bay. This approach was very different from the smaller-scale restoration efforts of the past.

Artificial reefs of two heights were constructed in 2004 (**FIGURE 2**), and oysters were allowed to colonize the reefs, safe from harvesting. Oyster populations on the constructed reefs were sampled in 2007, and the results were stunning. The reef complex supported an estimated 185 million oysters, a number nearly as large as the wild population of 200 million oysters estimated to live on the remaining degraded habitat in all of Maryland's waters.

Higher reefs supported an average of over 1000 oysters per square meter—four times more than the lower reefs and 170 times more than unrestored bottom (**FIGURE 3**). Like natural reefs, the constructed reefs began to solidify, providing a firm foundation for the settlement of new oysters. In 2009, Schulte's research made a splash when his team published its findings in the journal *Science*, bringing international attention to their study.

After reviewing eight alternative approaches to oyster restoration that involved one or more oyster species, the Corps advocated an approach that avoided the introduction of non-native oysters. Instead it proposed a combination of native oyster restoration, a temporary moratorium on oyster harvests (accompanied by a compensation program for the oyster industry), and enhanced support for oyster aquaculture in the bay region.

Schulte's restoration project cost roughly $3 million and will require substantial investments if it is to be repeated elsewhere in the bay. This is particularly true in upper portions of the bay, where oyster reproduction levels are lower (requiring restored reefs to be "seeded" with oysters), water conditions are poorer, and oysters are less resistant to disease. Many scientists contend that expanded reef restoration efforts are worth the cost because they enhance oyster populations and provide a vital service to the bay through water filtering.

These efforts are encouraged by the success of the project to date. By summer 2012, the majority of high-relief reef acreage was thriving, despite pressures from poachers and severe anoxic conditions in several years. Moreover, many of the low-relief reefs had accumulated enough shell to now be classified as high-relief reef. The oyster recruitment in 2012 was some of the best Schulte had seen during the project, boding well for the continuance of the reef complexes.

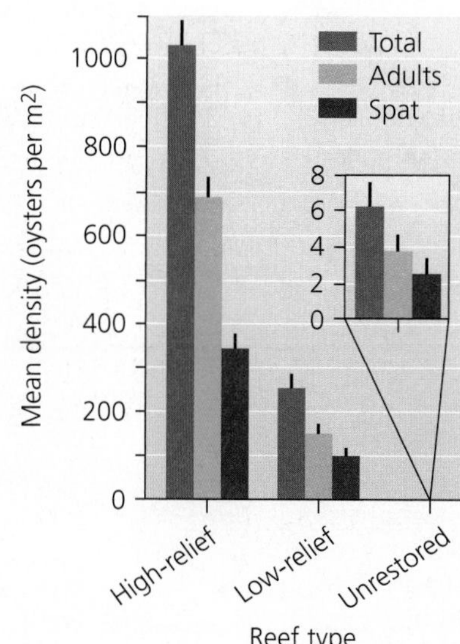

FIGURE 3 Reef height had a profound effect on the density of adult oysters and spat (newly settled oysters). Schulte's work suggested that native oyster populations could rebound in portions of Chesapeake Bay if they were provided elevated reefs and were protected from harvest. *Data from Schulte, D.M., R.P. Burke, and R.N. Lipicus, 2009. Unprecedented restoration of a native oyster meta population.* Science *325: 1124–1128.*

FIGURE 2 A water cannon blows oyster shells off a barge and onto the river bottom to create an artificial oyster reef for the experiment.

Some scientists also see value in promoting oyster farming, in which restoration efforts would be supported by businesses instead of taxpayers.

Regardless of how they will be funded, protected sites for oyster restoration efforts are being established. Maryland recently designated 3640 hectares (9000 acres) of new oyster sanctuaries—25% of existing oyster reefs in state waters—where restoration projects like Schulte's could be replicated. This movement toward increased protection for oyster populations, coupled with findings of growing resistance to disease in bay oysters, has given new hope that native oysters may once again thrive in the bay that bears their name. ■

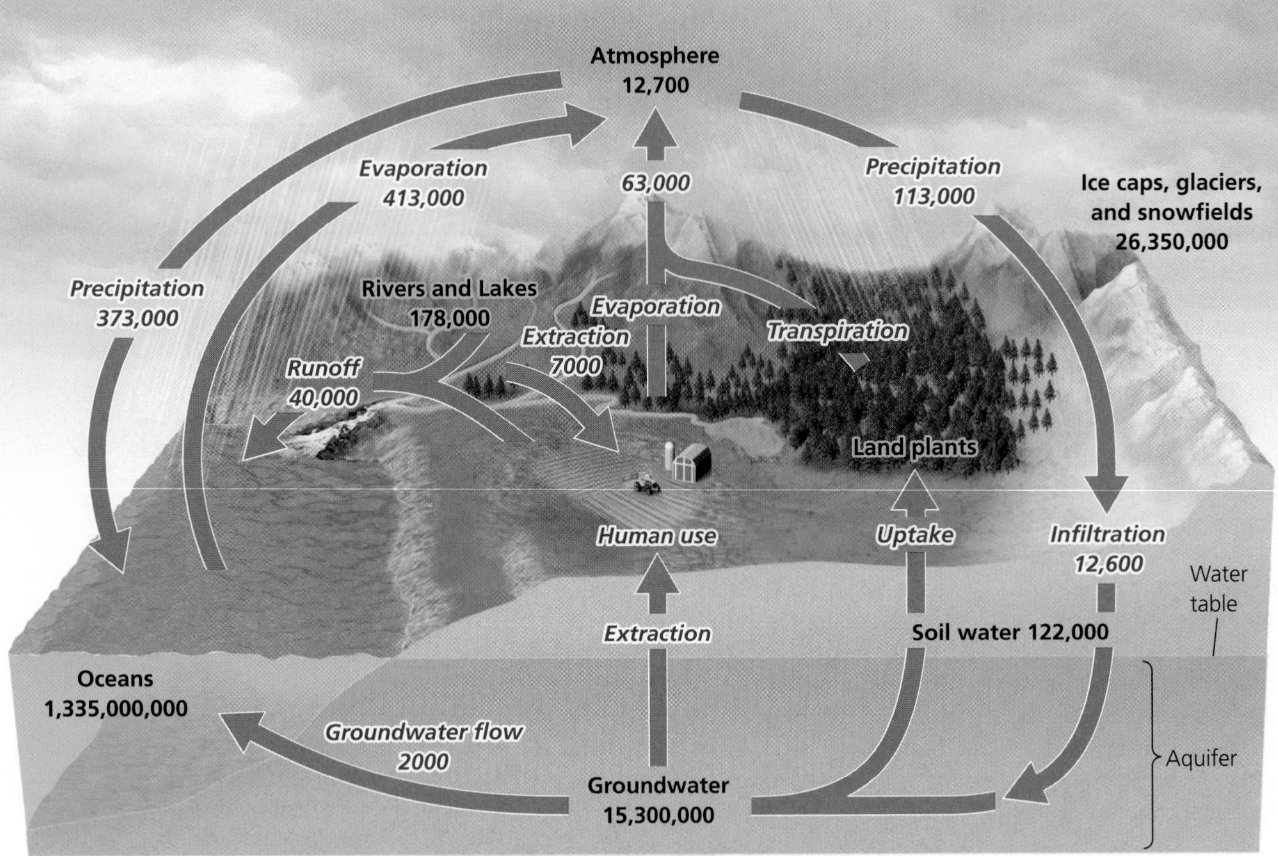

FIGURE 5.16 The water cycle, or hydrologic cycle, summarizes the many routes that water molecules take as they move through the environment. Gray arrows represent fluxes among reservoirs, or pools, for water. Oceans hold 97% of our planet's water, whereas most fresh water resides in groundwater and ice caps. Water vapor in the atmosphere condenses and falls to the surface as precipitation, then evaporates from land and transpires from plants to return to the atmosphere. Water flows downhill into rivers, eventually reaching the oceans. In the figure, pool names are printed in black type, and numbers in black type represent pool sizes expressed in units of cubic kilometers (km^3). Processes, printed in italic red type, give rise to fluxes, printed in italic red type and expressed in km^3 per year. *Data from* Schlesinger, W.H., 2013. Biogeochemistry: An analysis of global change, *3rd ed. Academic Press, London.*

Human activity has influenced certain fluxes. We have increased the flux of nitrogen from the atmosphere to reservoirs on Earth's surface, and we have shifted the flux of carbon in the opposite direction. As we discuss biogeochemical cycles, think about how they involve negative feedback loops that promote dynamic equilibrium, and also consider how some human actions can generate destabilizing positive feedback loops.

The water cycle affects all other cycles

Water is so integral to life and to Earth's fundamental processes that we frequently take it for granted. The essential medium for all manner of biochemical reactions (pp. 26–27), water plays key roles in nearly every environmental system, including each of the nutrient cycles we are about to discuss. Water carries nutrients, sediments, and pollutants from the continents to the oceans via surface runoff, streams, and rivers. These materials can then be carried thousands of miles on ocean currents. Water also carries atmospheric pollutants to the surface when they dissolve in falling rain or snow. The **water cycle,** or **hydrologic cycle** (FIGURE 5.16), summarizes how water—in liquid, gaseous, and solid forms—flows through our environment.

The oceans are the main reservoir in the water cycle, holding 97% of all water on Earth. The fresh water we depend on for our survival accounts for less than 3%, and two-thirds of this small amount is tied up in glaciers, snowfields, and ice caps (p. 391). Thus, considerably less than 1% of the planet's water is in forms that we can readily use—groundwater, surface fresh water, and rain from atmospheric water vapor.

Evaporation and transpiration Water moves from oceans, lakes, ponds, rivers, and moist soil into the atmosphere by **evaporation,** the conversion of a liquid to gaseous form. Warm temperatures and strong winds speed rates of evaporation. A greater degree of exposure has the same effect; an area logged of its forest or converted to agriculture or residential use will lose water more readily than a comparable area that remains vegetated. Water also enters the atmosphere by **transpiration,** the release of water vapor by plants through their leaves, or by evaporation from the surfaces of organisms (such as sweating in humans). Transpiration and evaporation act as natural processes of distillation, because water escaping into the air as a gas leaves behind its dissolved substances.

Precipitation, runoff, and surface water Water returns from the atmosphere to Earth's surface as **precipitation** when water vapor condenses and falls as rain or snow. This moisture may be taken up by plants and used by animals, but much of it flows as runoff into streams, rivers, lakes, ponds, and oceans. Amounts of precipitation vary greatly from region to region, helping give rise to our planet's variety of biomes (pp. 93–99).

Groundwater Some precipitation and surface water soaks down through soil and rock and becomes **groundwater,** water found beneath layers of soil. Groundwater recharges **aquifers,** spongelike regions of rock and soil that are underground reservoirs of water. The upper limit of groundwater held in an aquifer is referred to as the **water table.** (See Figure 15.4, p. 392, for an illustration of these features.) Aquifers can hold groundwater for long periods of time, so the water may be quite ancient. In some cases groundwater can take hundreds or even thousands of years to recharge fully after being depleted. Groundwater becomes surface water when it emerges from springs or flows into streams, rivers, lakes, or the ocean from the soil (p. 393).

Our impacts on the water cycle are extensive

Human activity affects every aspect of the water cycle. By damming rivers, we slow the movement of water from the land to the sea, and we increase evaporation by holding water in reservoirs. We remove natural vegetation by clear-cutting and developing land, which increases surface runoff, decreases infiltration and transpiration, and promotes soil erosion. Our withdrawals of surface water and groundwater for agriculture, industry, and domestic uses deplete rivers, lakes, and streams and lower water tables. And by emitting into the atmosphere pollutants that dissolve in water droplets, we change the chemical nature of precipitation, in effect sabotaging the natural distillation process that evaporation and transpiration provide. Water shortages have already given rise to conflicts worldwide, from the Middle East to the American West (pp. 407–408). (We will revisit the water cycle, water resources, and human impacts in more detail in Chapter 15).

WEIGHING THE ISSUES

YOUR WATER Has your region faced any water shortages or conflicts over water use? If not, can you describe how such problems affect some other region? What is the quality of your region's water, and what pollution threats does it face? Given your knowledge of the water cycle, what solutions would you propose for water shortages and/or water pollution in your region?

The carbon cycle circulates a vital organic nutrient

As the definitive component of organic molecules (p. 28), carbon is an ingredient in carbohydrates, fats, and proteins and occurs in the bones, cartilage, and shells of all living things. From DNA to fossil fuels, from plastics to pharmaceuticals, carbon (C) atoms are everywhere. The **carbon cycle** describes the routes that carbon atoms take through the environment (**FIGURE 5.17**).

Photosynthesis, respiration, and food webs Producers—including plants, algae, and cyanobacteria—pull carbon dioxide out of the atmosphere and out of surface water to use in photosynthesis. Photosynthesis (p. 32) breaks the bonds in carbon dioxide (CO_2) and water (H_2O) to produce oxygen (O_2) and carbohydrates (e.g., glucose, $C_6H_{12}O_6$). Autotrophs use some of the carbohydrates to fuel cellular respiration, thereby releasing some of the carbon back into the atmosphere and oceans as CO_2. When producers are eaten by primary consumers, which in turn are eaten by secondary and tertiary consumers, more carbohydrates are broken down in cellular respiration, producing carbon dioxide and water. The same process occurs as decomposers consume waste and dead organic matter. Cellular respiration from all these organisms releases carbon back into the atmosphere and oceans.

Organisms use carbon for structural growth, so a portion of the carbon an organism takes in becomes incorporated into its tissues (such as net primary production in plants; p. 111). The abundance of plants and the fact that they take in so much carbon dioxide for photosynthesis makes plants a major reservoir for carbon. Because CO_2 is a greenhouse gas of primary concern (p. 484), much research on global climate change is directed toward measuring the amount of CO_2 that plants store. Scientists are working hard to better understand exactly how this portion of the carbon cycle influences Earth's climate (see **THE SCIENCE BEHIND THE STORY**, pp. 124–125).

Sediment storage of carbon As aquatic organisms die, their remains may settle in sediments in ocean basins or freshwater wetlands. As sediment accumulates, older layers are buried more deeply, experiencing high pressure over long periods of time. These conditions can convert soft tissues into fossil fuels—coal, oil, and natural gas (p. 521)—and can turn shells and skeletons into sedimentary rock, such as limestone. Sedimentary rock (p. 37) comprises the largest reservoir in the carbon cycle. Although any given carbon atom spends a relatively short time in the atmosphere, carbon trapped in sedimentary rock may reside there for hundreds of millions of years.

Carbon trapped in sediments and fossil fuel deposits may eventually be released into the oceans or atmosphere by geologic processes such as uplift, erosion, and volcanic eruptions (pp. 40–42). It also reenters the atmosphere when we extract and burn fossil fuels.

The oceans The world's oceans are the second-largest reservoir in the carbon cycle. They absorb carbon-containing compounds from the atmosphere, from terrestrial runoff, from undersea volcanoes, and from the waste products and detritus of marine organisms. Some carbon atoms absorbed by the oceans—in the form of carbon dioxide, carbonate ions (CO_3^{2-}), and bicarbonate ions (HCO_3^-)—combine with calcium ions (Ca^{2+}) to form calcium carbonate ($CaCO_3$), an essential ingredient in the skeletons and shells of microscopic marine organisms. As these organisms die, their calcium carbonate shells sink to the ocean floor and begin to form sedimentary rock. The rates at which the oceans absorb and release carbon depend on

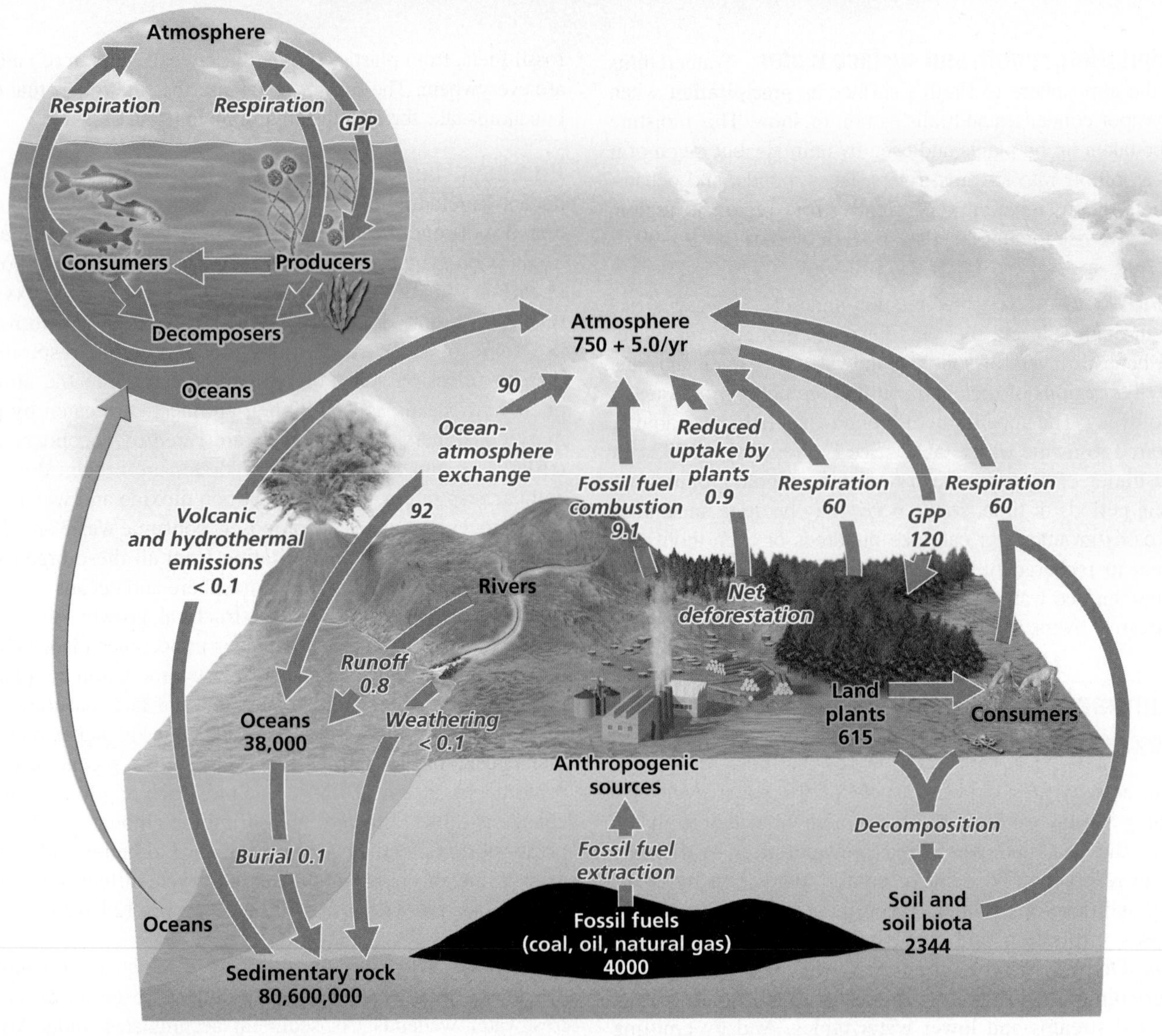

FIGURE 5.17 The carbon cycle summarizes the many routes that carbon atoms take as they move through the environment. Gray arrows represent fluxes among reservoirs, or pools, for carbon. In the carbon cycle, plants use carbon dioxide from the atmosphere for photosynthesis (gross primary production, or "GPP" in the figure). Carbon dioxide is returned to the atmosphere through cellular respiration by plants, their consumers, and decomposers. The oceans sequester carbon in their water and in deep sediments. The vast majority of the planet's carbon is stored in sedimentary rock. In the figure, pool names are printed in black type, and numbers in black type represent pool sizes expressed in petagrams (units of 10^{15} g) of carbon. Processes, printed in italic red type, give rise to fluxes, printed in italic red type and expressed in petagrams of carbon per year. *Data from Schlesinger, W.H., 2013.* Biogeochemistry: An analysis of global change, *3rd ed. Academic Press, London.*

many factors, including temperature and the numbers of marine organisms converting CO_2 into carbohydrates and carbonates.

We are shifting carbon from the lithosphere to the atmosphere

By mining fossil fuel deposits, we are essentially removing carbon from an underground reservoir with a residence time of millions of years. By combusting fossil fuels in our automobiles, homes, and industries, we release carbon dioxide and greatly increase the flux of carbon from the ground to the air. Since the mid-18th century, our fossil fuel combustion has added over 250 billion metric tons (276 billion tons) of carbon to the atmosphere. The movement of CO_2 from the atmosphere back to the hydrosphere, lithosphere, and biosphere has not kept pace.

In addition, cutting down forests removes carbon from the pool of vegetation and releases it to the air. And if less vegetation is left on the surface, there are fewer plants to draw CO_2 back out of the atmosphere.

As a result, scientists estimate that today's atmospheric carbon dioxide reservoir is the largest that Earth has experienced in the past 800,000 years, and likely in the past 20 million years (p. 486). The ongoing flux of carbon into the atmosphere is the driving force behind today's anthropogenic global climate change (Chapter 18).

Some of the excess CO_2 in the atmosphere is now being absorbed by ocean water. This is causing ocean water to become more acidic, leading to problems that threaten many marine organisms (pp. 428–429).

Our understanding of the carbon cycle is not yet complete. Scientists remain baffled by the so-called missing carbon sink.

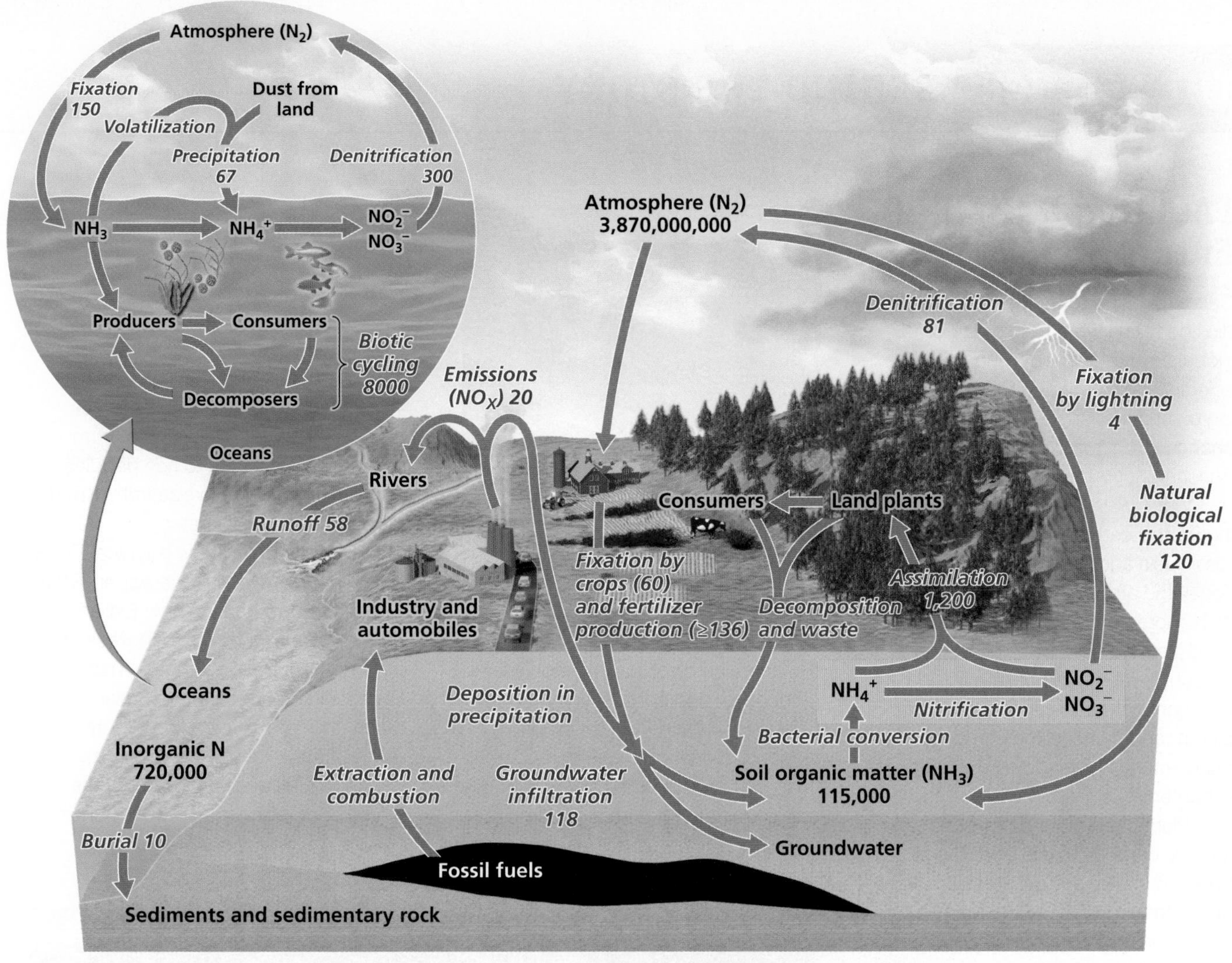

FIGURE 5.18 **The nitrogen cycle summarizes the many routes that nitrogen atoms take as they move through the environment.** Gray arrows represent fluxes among reservoirs, or pools, for nitrogen. In the nitrogen cycle, specialized bacteria play key roles in "fixing" atmospheric nitrogen and converting it to chemical forms that plants can use. Other types of bacteria convert nitrogen compounds back to the atmospheric gas, N_2. In the oceans, inorganic nitrogen is buried in sediments, whereas nitrogen compounds are cycled through food webs as they are on land. In the figure, pool names are printed in black type, and numbers in black type represent pool sizes expressed in teragrams (units of 10^{12} g) of nitrogen. Processes, printed in italic red type, give rise to fluxes, printed in italic red type and expressed in teragrams of nitrogen per year. *Data from Schlesinger, W.H., 2013.* Biogeochemistry: An analysis of global change, *3rd ed. Academic Press, London.*

Of the carbon dioxide we emit by fossil fuel combustion and deforestation, researchers have measured how much goes into the atmosphere and oceans, but there remain roughly 2.3–2.6 billion metric tons unaccounted for. Many scientists think this CO_2 must be taken up by plants or soils of the temperate and boreal forests (pp. 95–96, 99). They'd like to know for sure, though, because if certain forests are acting as a major sink for carbon (and thus restraining global climate change), we'd like to be able to keep it that way. For if forests that today are sinks were to turn into sources and begin releasing the "missing" carbon, climate change could accelerate drastically.

The nitrogen cycle involves specialized bacteria

Nitrogen (N) makes up 78% of our atmosphere by mass and is the sixth most abundant element on Earth. It is an essential ingredient in the proteins, DNA, and RNA that build our bodies. Despite its abundance in the air, nitrogen gas (N_2) is chemically inert and cannot cycle out of the atmosphere and into living organisms without assistance from lightning, highly specialized bacteria, or human intervention. For this reason, the element is relatively scarce in the lithosphere and hydrosphere and in organisms. However, once nitrogen undergoes the right kind of chemical change, it becomes biologically active and available to organisms, and it can act as a potent fertilizer. Its scarcity makes biologically active nitrogen a limiting factor for plant growth. For all these reasons the **nitrogen cycle** (FIGURE 5.18) is of vital importance to us and to all other organisms.

Nitrogen fixation To become biologically available, inert nitrogen gas (N_2) must be "fixed," or combined with hydrogen in nature to form ammonia (NH_3), whose water-soluble

FACE-ing a High-CO_2 Future

Can fumigating trees with carbon dioxide tell us what to expect from global climate change? Hundreds of scientists think so, and they are testing plants' responses to atmospheric change at unique outdoor Free-Air CO_2 Enrichment (FACE) facilities.

Our civilization is radically altering Earth's carbon cycle by burning fossil fuels and deforesting landscapes. Today the atmosphere contains over 40% more carbon dioxide (CO_2) than it did just two centuries ago, and the amount is rapidly increasing. Rising CO_2 concentrations are warming our planet, and global climate change brings many unwelcome consequences (Chapter 18).

Plants and other autotrophs remove carbon dioxide from the atmosphere to use in photosynthesis, and all organisms add CO_2 to the atmosphere by cellular respiration (p. 32). Will more CO_2 mean more plant growth, and will more plants be able to absorb and store much of the extra CO_2? Perhaps, but before we rely on forests and phytoplankton to save us from our own emissions, we'd better be sure they can do so.

Historically, if a researcher wanted to measure how plants respond to increased carbon dioxide, he or she would alter gas levels in a small enclosure such as a lab or a greenhouse. But can we really scale up results from such small indoor experiments and trust that they will show how entire forests will behave? Many scientists thought not, so they pioneered Free-Air CO_2 Enrichment. In FACE experiments, ambient levels of CO_2 encompassing areas of forest (or other vegetation) outdoors are precisely controlled. With their large scale and open-air conditions, FACE experiments include most factors that influence a plant community in the wild, such as variation in temperature, sunlight, precipitation, herbivorous insects, disease pathogens, and competition among plants. By measuring how plants respond to changing gas compositions in such real-world conditions, we can better learn how ecosystems may change in the carbon dioxide–soaked world that awaits us.

Aspen FACE site researcher Dr. Mark Kubiske of the U.S. Forest Service

Dozens of organizations have sponsored FACE facilities—36 sites in 17 nations so far, including U.S. sites in Arizona, California, Illinois, Minnesota, Nevada, North Carolina, Tennessee, Wisconsin, and Wyoming. The sites cover a variety of ecosystems, from forests to grasslands to rice paddies, and the plots range in size from 1 m to 30 m (3–98 ft) in diameter.

To understand how a typical FACE study works, let's visit the Aspen FACE Experiment at the Harshaw Experimental Forest (where aspen trees are common) near Rhinelander, Wisconsin. Here, tall steel and plastic towers and pipes ring 12 circular plots of forest 30 m (98 ft) in diameter (**FIGURE 1**). The pipes release CO_2, bathing the plants in an atmosphere 50% richer

FIGURE 1 At the Aspen FACE facility in Wisconsin, tall towers and pipes control the atmospheric composition around selected patches of trees.

in CO_2 than today's (equal to what is expected for the year 2050). Sensors monitor wind conditions, and computers control for the influence of wind by adjusting CO_2 releases, keeping ambient concentrations stable within each plot.

The pipes at the Aspen plots also release tropospheric ozone (O_3, a major pollutant in urban smog; p. 465), and researchers study how this gas and CO_2 affect plant growth, leaf and root conditions, soil carbon content, and much more. Pipes at some plots release normal air, serving as controls for the treatment plots.

Researchers using the Aspen FACE facility have been asking whether forest trees will sequester more carbon as CO_2 levels rise, whether this will change as trees grow, how CO_2 interacts with ozone, and how these gases affect trees' interactions with insects and diseases. They have learned a number of things so far, including:

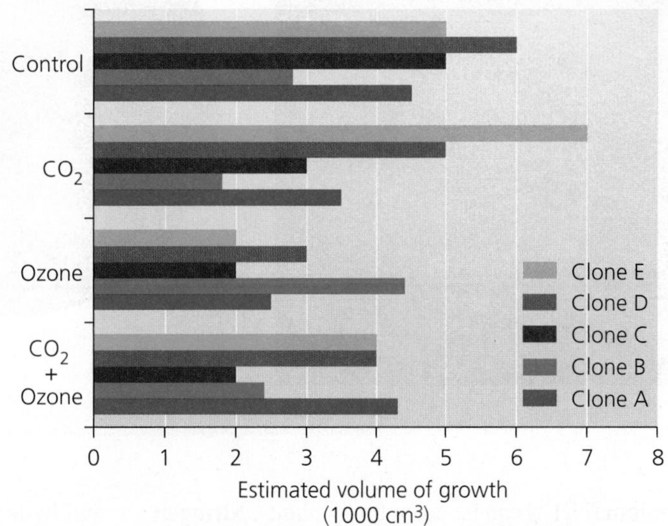

FIGURE 2 Data from five clones of aspens show that during the study period, trees supplied with carbon dioxide grew more than did the control trees, whereas those supplied with ozone grew less. Trees supplied with both gases did not grow differently from the controls. *Source: Isebrands, J.G., et al., 2001. Growth responses of* Populus tremuloides *clones to interacting elevated carbon dioxide and tropospheric ozone.* Environmental Pollution *115 (3): 359–371, Fig. 2.*

- Insects and diseases that attack aspen and birch trees increase as atmospheric levels of ozone and CO_2 rise.
- High CO_2 concentrations delay aspen leaf aging, which makes some aspens vulnerable to frost damage in winter.
- Elevated CO_2 levels increase photosynthesis and tree growth—but moderate levels of ozone offset this increased growth (**FIGURE 2**). Because many modelers have not taken ozone into account when estimating how much carbon trees can sequester, the Aspen FACE data suggest that their models may overestimate the amount of CO_2 that trees will pull out of the air.

Together, such results indicate that rising carbon dioxide levels could have a variety of negative impacts on trees and forests. Thus, the old expectation that more CO_2 makes for happier plants appears to be an oversimplification. Indeed, research from other FACE sites is showing that increased growth from enhanced CO_2 is often temporary and that growth rates later flatten out or decline. Recent work on crop plants has even shown that high CO_2 makes some crops less nutritious.

Obtaining solid answers to questions like these often takes years or decades, and FACE experiments are designed to monitor plots for the long term as the plants mature. Some FACE sites have been operating for 20 years and are beginning to produce data that could not be gathered in any other way.

Thus, researchers were shocked in 2008 when the U.S. Department of Energy (DOE), which funds Aspen and other major sites, announced it would cease funding. The DOE advised scientists to cut the trees down and dig up the soil to analyze carbon content. This analysis would provide urgently needed data on carbon sequestration, the DOE said, and then millions of dollars could be shifted toward a new and improved generation of FACE experiments.

The trees at the Aspen site were harvested for analysis in 2009, and researchers are now regrowing forest on the site with funding from the United States Forest Service. The knowledge gains from this project, and other FACE sites in which growth is ongoing, have been substantial. Data from the Aspen FACE project alone has generated more than 180 peer-reviewed scientific articles and is helping scientists today better predict how forest ecosystems will respond to the atmosphere of tomorrow. ■

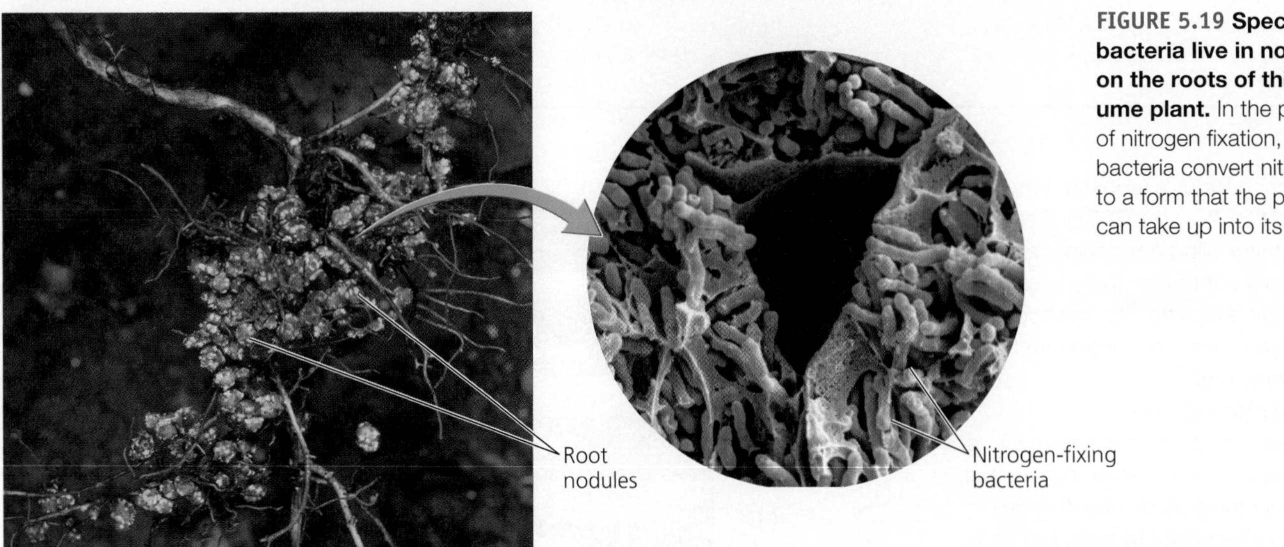

FIGURE 5.19 Specialized bacteria live in nodules on the roots of this legume plant. In the process of nitrogen fixation, the bacteria convert nitrogen to a form that the plant can take up into its roots.

ions of ammonium (NH_4^+) can be taken up by plants. **Nitrogen fixation** can be accomplished in two ways: by the intense energy of lightning strikes, or when air in the top layer of soil comes in contact with particular types of **nitrogen-fixing bacteria.** These bacteria live in a mutualistic relationship (p. 80) with many types of plants, including soybeans and other legumes, providing them nutrients by converting nitrogen to a usable form. Some farmers nourish soils by planting crops that host nitrogen-fixing bacteria among their roots (**FIGURE 5.19**).

Nitrification and denitrification Other types of specialized bacteria then perform a process known as **nitrification.** In this process, ammonium ions are first converted into nitrite ions (NO_2^-), then into nitrate ions (NO_3^-). Plants can take up these ions, which also become available after atmospheric deposition on soils or in water or after application of nitrate-based fertilizer.

Animals obtain the nitrogen they need by consuming plants or other animals. Decomposers obtain nitrogen from dead and decaying plant and animal matter and from animal urine and feces. Once decomposers process nitrogen-rich compounds, they release ammonium ions, making these available to nitrifying bacteria to convert again to nitrates and nitrites.

The next step in the nitrogen cycle occurs when **denitrifying bacteria** convert nitrates in soil or water to gaseous nitrogen via a multistep process. Denitrification thereby completes the cycle by releasing nitrogen back into the atmosphere as a gas.

We have greatly influenced the nitrogen cycle

Historically, nitrogen fixation was a **bottleneck,** a step that limited the flux of nitrogen out of the atmosphere. This changed with the research of two German chemists early in the 20th century. Fritz Haber found a way to combine nitrogen and hydrogen gases to synthesize ammonia, a key ingredient in modern explosives and agricultural fertilizers, and Carl Bosch devised methods to produce ammonia on an industrial scale. The **Haber-Bosch process** enabled people to overcome the limits on productivity long imposed by nitrogen scarcity in nature. By enhancing agriculture, the new fertilizers contributed to the past century's enormous increase in human population. Farmers, homeowners, and golf course managers alike all took advantage of fertilizers, dramatically altering the nitrogen cycle. Today, using the Haber-Bosch process, our species is fixing at least as much nitrogen as is being fixed naturally. We have effectively doubled the rate of nitrogen fixation on Earth, overwhelming nature's denitrification abilities.

By fixing atmospheric nitrogen with fertilizers, we increase nitrogen's flux from the atmosphere to Earth's surface. We also enhance this flux by cultivating legume crops whose roots host nitrogen-fixing bacteria. Moreover, we reduce nitrogen's return to the air when we destroy wetlands that filter nutrients; wetland plants host denitrifying bacteria that convert nitrates to nitrogen gas, so wetlands can mop up a great deal of nitrogen pollution.

When our farming practices speed runoff and allow soil erosion, nitrogen flows from farms into terrestrial and aquatic ecosystems, leading to nutrient pollution, eutrophication, and hypoxia. These impacts have become painfully evident to oystermen and scientists in the Chesapeake Bay, but hypoxia in waters is by no means the only human impact on the nitrogen cycle. When we burn forests and fields, we force nitrogen out of soils and vegetation and into the atmosphere. When we burn fossil fuels, we release nitric oxide (NO) into the atmosphere, where it reacts to form nitrogen dioxide (NO_2). This compound is a precursor to nitric acid (HNO_3), a key component of acid precipitation (pp. 473–475). We introduce another nitrogen-containing gas, nitrous oxide (N_2O), when anaerobic bacteria break down the tremendous volume of animal waste produced in agricultural feedlots (pp. 250–251). Oddly enough, the overapplication of nitrogen-based fertilizers can strip the soil of other essential nutrients, such as calcium and potassium, because fertilizer flushes them out. As these examples show, human activities have affected the nitrogen cycle in diverse and often far-reaching ways.

The phosphorus cycle circulates a limited nutrient

The element phosphorus (P) is a key component of cell membranes and of several molecules vital for life, including DNA, RNA, ATP, and ADP (pp. 29, 32). Although phosphorus is indispensable for life, the amount of phosphorus in organisms is dwarfed by the vast amounts in rocks, soil, sediments, and the oceans. Unlike the carbon and nitrogen cycles, the **phosphorus cycle** (FIGURE 5.20) has no appreciable atmospheric component besides the transport of tiny amounts in wind-blown dust and sea spray.

Geology and phosphorus availability The vast majority of Earth's phosphorus is contained within rocks and is released only by weathering (p. 219), which releases phosphate ions (PO_4^{3-}) into water. Phosphates dissolved in lakes or in the oceans precipitate into solid form, settle to the bottom, and reenter the lithosphere's phosphorus reservoir in sediments. Because most phosphorus is bound up in rock and only slowly released, environmental concentrations of phosphorus available to organisms tend to be very low. This scarcity explains why phosphorus is frequently a limiting factor for plant growth and why an influx of phosphorus can produce immediate and dramatic effects.

Food webs Aquatic producers take up phosphates from surrounding waters, whereas terrestrial producers take up phosphorus from soil water through their roots. Primary consumers acquire phosphorus from plant tissues and pass it on to secondary and tertiary consumers (Chapter 4). Consumers also pass phosphorus to the soil through the excretion of waste. Decomposers break down phosphorus-rich organisms and their wastes and, in so doing, return phosphorus to the soil.

We affect the phosphorus cycle

People increase phosphorus concentrations in surface waters through runoff of the phosphorus-rich fertilizers we apply to lawns and farmlands. A 2008 study determined that an average hectare of land in the Chesapeake Bay region received a

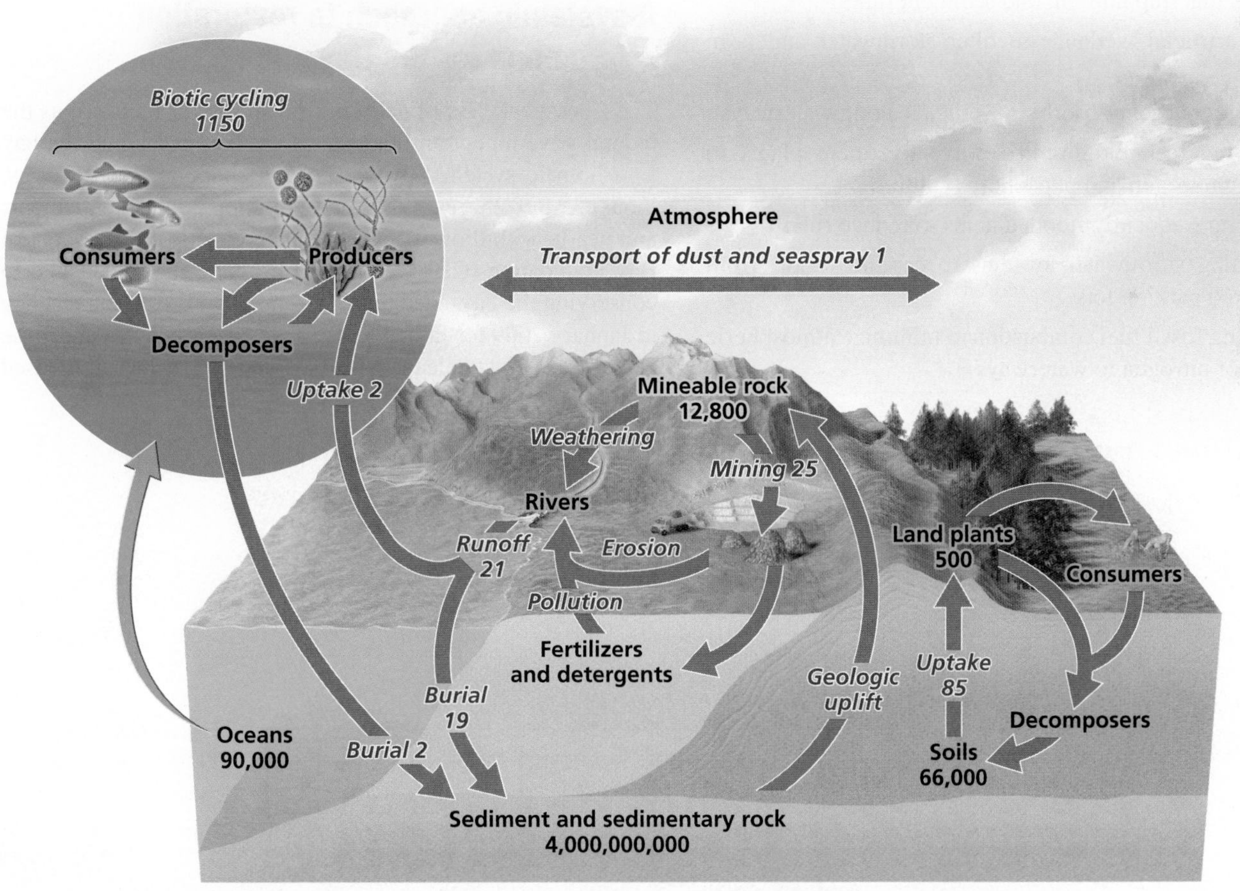

FIGURE 5.20 The phosphorus cycle summarizes the many routes that phosphorus atoms take as they move through the environment. Gray arrows represent fluxes among reservoirs, or pools, for phosphorus. Most phosphorus resides underground in rock and sediment. Rocks containing phosphorus are uplifted geologically and slowly weathered away. Small amounts of phosphorus cycle through food webs, where this nutrient is often a limiting factor for plant growth. In the figure, pool names are printed in black type, and numbers in black type represent pool sizes expressed in teragrams (units of 10^{12} g) of phosphorus. Processes, printed in italic red type, give rise to fluxes, printed in italic red type and expressed in teragrams of phosphorus per year. *Data from Schlesinger, W.H., 2013.* Biogeochemistry: An analysis of global change, *3rd ed. Academic Press, London.*

net input of 4.52 kg (10 lb) of phosphorus per year, promoting phosphorus accumulation in soils, runoff into waterways, and phytoplankton blooms and hypoxia in the bay. People also add phosphorus to waterways through releases of treated wastewater rich in phosphates from domestic use of phosphate detergents.

Tackling nutrient enrichment requires diverse approaches

With our reliance on synthetic fertilizers for food production and fossil fuels for energy, nutrient enrichment of ecosystems will be a challenge for many years to come. But there are a number of approaches available to control nutrient pollution in the Chesapeake Bay watershed, Mississippi River watershed, and other waterways affected by eutrophication:

- Reducing fertilizer use on farms and lawns
- Changing the timing of fertilizer application to minimize rainy-season runoff
- More effectively managing manure applications to farmland to reduce nutrient runoff
- Planting and maintaining vegetation "buffers" around streams that trap nutrient and sediment runoff
- Using artificial wetlands to filter stormwater and farm runoff
- Restoring nutrient-absorbing wetlands along waterways
- Improving technologies in sewage treatment plants to enhance nitrogen and phosphorus capture
- Restoring frequently flooded lands to reduce runoff
- Upgrading stormwater systems to capture runoff from roads and parking lots
- Reducing fossil fuel combustion to minimize atmospheric inputs of nitrogen to waterways

These approaches have widely varying costs for the same level of nutrient reduction. For example, embracing approaches such as planting vegetation buffers around streams, restoring wetlands, and practicing sustainable agriculture can reduce nutrient inputs into waterways at a fraction of the cost of other approaches, all while creating habitat for wildlife (FIGURE 5.21). Ultimately, the approaches embraced depend on the major sources of nutrients for a given waterway along with economic considerations.

WEIGHING THE ISSUES

NUTRIENT POLLUTION AND ITS FINANCIAL IMPACTS A sizable amount of the nitrogen and phosphorus that enters the Chesapeake Bay originates from farms and other sources far from the bay, yet it is people living near the bay, such as oystermen and crabbers, who bear many of the negative impacts. Who do you believe should be responsible for addressing this problem? Should environmental policies on this issue be developed and enforced by state governments, the federal government, both, or neither? Explain the reasons for your answer.

A systemic approach to restoration offers hope for the Bay

The Chesapeake Bay finally has prospects for recovery as the federal government and bay states are now managing the bay as a holistic system. Arriving at this endpoint was not easy, though. After 25 years of failed pollution control agreements and nearly $6 billion spent on cleanup efforts, the Chesapeake Bay Foundation (CBF), a nonprofit organization dedicated to conserving the bay, sued the Environmental Protection Agency in January 2009 for failing to use its available powers under the Clean Water Act to clean up the bay. The CBF's lawsuit focused

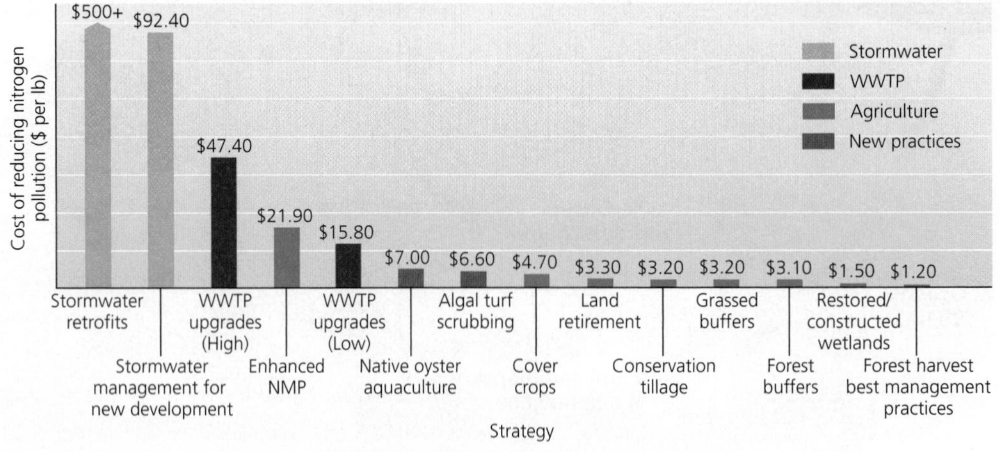

FIGURE 5.21 The cost per pound of reducing nitrogen inputs into the Chesapeake Bay varies widely. Approaches slowing runoff to waterways avoid nitrogen inputs for a few dollars per pound, whereas upgrades to wastewater treatment plants (WWTP), enhanced nutrient management plans (NMP—careful regulation of nutrient applications), and stormwater upgrades can be considerably more expensive.

DATA Q For what it costs to remove 1 pound of nitrogen by using enhanced nutrient management programs (NMP), how many pounds of nitrogen could be kept out of waterways by planting forested buffers around streams instead?

media attention on the plight of the bay, its ongoing water quality issues, and its depleted fisheries—and spurred action.

In May 2009, President Obama directed the EPA and other federal agencies through an executive order to establish a comprehensive plan for the restoration of the Chesapeake Bay. One year later, the EPA and the CBF announced a settlement in which the EPA agreed to work with surrounding states to provide aggressive pollution regulation in the bay. In December 2010, a comprehensive "pollution budget" was developed and implemented with the assistance of the District of Columbia, Delaware, Maryland, New York, Pennsylvania, Virginia, and West Virginia.

Existing efforts to reduce nutrient and sediment inputs into the bay and to limit harvests of oysters, crabs, and fish are already leading to modest improvement in some aspects of the bay's health. For example, the CBF's "2012 State of the Bay" report shows the bay's overall health has improved 10% since 2008 and cites recent studies and trends in nutrient concentrations that suggest pollution reduction strategies are working to reduce inputs of nitrogen and phosphorus to bay waters. The Chesapeake Bay remains highly degraded and much work is still needed. Still, the 17 million people living in the Chesapeake Bay watershed have reason to hope that the Chesapeake Bay of tomorrow may be healthier than it is today, thanks to the collaborative efforts of concerned citizens, advocacy organizations, and the federal and bay state governments.

Conclusion

Thinking in terms of systems is important in understanding Earth's dynamics, so that we may learn how to avoid disrupting its processes and how to mitigate any disruptions we cause. By studying the environment from a systems perspective and by integrating scientific findings with the policy process, people who care about the Chesapeake Bay and other waterways are working today to address dead zones around the world.

Earth hosts many interacting systems, and the way one perceives them depends on the questions in which one is interested. Life interacts with its nonliving environment in ecosystems, systems through which energy flows and matter is recycled. Understanding the biogeochemical cycles that describe the movement of nutrients within and among ecosystems is crucial because human activities are causing significant changes in the ways those cycles function.

Unperturbed ecosystems use renewable solar energy, recycle nutrients, and are stabilized by negative feedback loops. The environmental systems we see on Earth today are those that have survived the test of time. Our industrialized civilization is young in comparison. These natural systems therefore provide us a blueprint to mimic as we move towards greater sustainability in modern society.

Reviewing Objectives

You should now be able to:

Describe the nature of environmental systems

- Earth's natural systems are complex, so environmental scientists often take a holistic approach to studying environmental systems. (p. 106)
- Systems are networks of interacting components that generally involve feedback loops, show dynamic equilibrium, and result in emergent properties. (pp. 106–107)
- Negative feedback stabilizes systems, whereas positive feedback destabilizes systems. Positive feedback often results from human disturbance of natural systems. (pp. 106–107)
- Because environmental systems interact and overlap, one's delineation of a system depends on the questions in which one is interested. (p. 108)
- Hypoxia in the Chesapeake Bay, which results from nutrient pollution in the rivers that feed it, illustrates how systems are interrelated. (p. 108)

Define ecosystems and evaluate how living and nonliving entities interact in ecosystem-level ecology

- Ecosystems consist of all organisms and nonliving entities that occur and interact in a particular area at the same time. (pp. 110–111)
- Energy flows in one direction through ecosystems, whereas matter is recycled. (pp. 110–111)
- Energy is converted to biomass, and ecosystems vary in their productivity. (pp. 111–112)
- Input of nutrients can boost productivity, but an excess of nutrients can alter ecosystems and cause severe ecological and economic consequences. (pp. 111–113)

Outline the fundamentals of landscape ecology, GIS, and ecological modeling

- Landscape ecology studies how landscape structure influences organisms. (pp. 113–114)
- Landscapes consist of patches spatially arrayed in a mosaic. Organisms dependent on certain types of patches may occur in metapopulations. (pp. 113–114)
- With the help of remote sensing technology and GIS, landscape ecology is being increasingly used in conservation and regional planning. (pp. 114–115)
- Ecological modeling helps ecologists make sense of the complex systems they study. (pp. 115–116)

Assess ecosystem services and how they benefit our lives

- Ecosystems provide the "goods" we know of as natural resources. (p. 116)

- Ecological processes naturally provide services that we depend on for everyday living. (pp. 116–117)

Compare and contrast how water, carbon, nitrogen, and phosphorous cycle through the environment

- A source is a reservoir that contributes more of a material than it receives, and a sink is one that receives more than it provides. (p. 117)
- Water moves widely through the environment in the water cycle. (pp. 120–121)
- Most carbon is contained in sedimentary rock. Substantial amounts also occur in the oceans and in soil. Carbon flux between organisms and the atmosphere occurs via photosynthesis and respiration. (pp. 121–123)
- Nitrogen is a vital nutrient for plant growth. Most nitrogen is in the atmosphere, so it must be "fixed" by specialized bacteria or lightning before plants can use it. (pp. 123, 126)
- Phosphorus is most abundant in sedimentary rock, with substantial amounts in soil and the oceans. Phosphorus has no appreciable atmospheric reservoir. It is a key nutrient for plant growth. (p. 127)

Explain how human impact is affecting biogeochemical cycles

- People are affecting Earth's biogeochemical cycles by shifting carbon from fossil fuel reservoirs into the atmosphere, shifting nitrogen from the atmosphere to the planet's surface, and depleting groundwater supplies, among other impacts. (pp. 121–128)
- Policy can help us address problems with nutrient pollution. (p. 128)

Testing Your Comprehension

1. Which type of feedback loop is more common in nature, and which more commonly results from human action? How might the emergence of a positive feedback loop affect a system in homeostasis?
2. Describe how hypoxic conditions can develop in ecosystems such as the Chesapeake Bay.
3. What is the difference between an ecosystem and a community?
4. Describe the typical movement of energy through an ecosystem. Now describe the typical movement of matter through an ecosystem.
5. Explain net primary productivity. Name one ecosystem with high net primary productivity and one with low net primary productivity.
6. Why are patches in a landscape mosaic often important to people who are interested in conserving populations of rare animals?
7. What is the difference between evaporation and transpiration? Give examples of how the water cycle interacts with the carbon, phosphorus, and nitrogen cycles.
8. What role does each of the following play in the carbon cycle?
   - Cars
   - Photosynthesis
   - The oceans
   - Earth's crust
9. Distinguish the function performed by nitrogen-fixing bacteria from that performed by denitrifying bacteria.
10. How has human activity altered the carbon cycle? The phosphorus cycle? The nitrogen cycle? What environmental problems have arisen from these changes?

Seeking Solutions

1. Once vegetation is cleared from a riverbank, water begins to erode the bank away. This erosion may dislodge more vegetation. Would you expect this to result in a feedback process? If so, which type—negative or positive? Explain your answer. How might we halt or reverse this process?
2. Consider the ecosystem(s) that surround(s) your campus. Describe one way in which energy flows through and matter is recycled. Now pick one type of nutrient, and briefly describe how it moves through your ecosystem(s). Does the landscape contain patches? Can you describe any ecotones?
3. For a conservation biologist interested in sustaining populations of the organisms below, why would it be helpful to take a landscape ecology perspective? Explain your answer in each case.
   - A forest-breeding warbler that suffers poor nesting success in small, fragmented forest patches
   - A bighorn sheep that must move seasonally between mountains and lowlands
   - A toad that lives in upland areas but travels cross-country to breed in localized pools each spring

4. A simple change in the flux between just two reservoirs in a single nutrient cycle can potentially have major consequences for ecosystems and, indeed, for the entire Earth. Explain how this can be, using one example from the carbon cycle and one example from the nitrogen cycle.
5. How do you think we might solve the problem of eutrophication in the Chesapeake Bay? Assess several possible solutions, your reasons for believing they might work, and the likely hurdles we might face. Explain who should be responsible for implementing these solutions, and why.
6. **THINK IT THROUGH** You are an oysterman in the Chesapeake Bay, and your income is decreasing because the dead zone is making it harder to harvest oysters. One day your senator comes to town, and you have a one-minute audience with her. What steps would you urge her to take in Washington, D.C., to try to help alleviate the dead zone and bring back the oyster fishery?

   Now suppose you are a Pennsylvania farmer who has learned that the government is offering incentives to farmers to help reduce fertilizer runoff into the Chesapeake Bay. What types of approaches described in the text might you be willing to try, and why?

Calculating Ecological Footprints

In the United States, a common dream is to own a suburban home with a weed-free green lawn. Nationwide, Americans tend about 40.5 million acres of lawn grass. But conventional lawn care involves inputs of fertilizers, pesticides, and irrigation water, and using gasoline or electricity for mowing and other care—all of which raise environmental and health concerns. Using the figures for a typical lawn in the table, calculate the total amount of fertilizer, water, and gasoline used in lawn care across the nation each year.

	Acreage of lawn	Fertilizer used (lbs)	Water used (gal)	Gasoline used (gal)
For the typical 1/4-acre lawn	0.25	37	16,000	4.9
For all lawns in your hometown				
For all lawns in the United States	40,500,000			

Data from Chameides, B., 2008. http://www.nicholas.duke.edu/thegreengrok/lawns.

1. How much fertilizer is applied each year on lawns throughout the United States? Where does the nitrogen for this fertilizer come from? What becomes of the nitrogen and phosphorus applied to a suburban lawn that is not taken up by grass?
2. Leaving grass clippings on a lawn decreases the need for fertilizer by 50%. What else might a homeowner do to decrease fertilizer use in a yard and the environmental impacts of nutrient pollution?
3. How much gasoline could Americans save each year if they did not take care of lawns? At today's gas prices, how much money would this save?

6

A Costa Rican farmer tends his crops

Ethics, Economics, and Sustainable Development

Upon completing this chapter, you will be able to:

- Characterize the influences of culture and worldview on the choices people make
- Outline the nature and historical expansion of ethics in Western culture
- Compare major approaches in environmental ethics
- Explain how our economies exist within the environment and rely on ecosystem services
- Describe principles of classical and neoclassical economics and summarize their implications for the environment
- Illustrate aspects of environmental economics and ecological economics
- Describe how individuals and businesses can help move our economic system in a sustainable direction
- Explain the pursuit of sustainable development

Costa Rica Values Its Ecosystem Services

"Costa Rica's PSA program has been one of the conservation success stories of the last decade."

—Stefano Pagiola, The World Bank

"In the last 25 years, my home country has tripled its GDP while doubling the size of its forests."

—Carlos Manuel Rodriguez, former Minister of Energy and the Environment, Costa Rica

Few nations have utterly transformed their path of development in just a few decades. Costa Rica has. In the 1980s, this small Central American country of 4.7 million people was losing its forests faster than almost any other nation on Earth. Yet today Costa Rica has regained much of its forest cover, boasts a world-class park system, and stands as a global model for sustainable resource management.

How did Costa Rica do it? The story is complex, but one driving force came in 1996, when the nation began paying landholders to conserve forest on private land. The government program, called *Pago por Servicios Ambientales* (PSA)—translated as Payment for Environmental Services—was the first of its kind in the world.

Nature provides ecosystem services (pp. 3, 116–117), such as air and water purification, climate regulation, soil fertility, and pollination of crops. Historically we have taken these gifts for granted and have not paid for them in the marketplace. As a result, ecosystem services have diminished as we have degraded the natural systems that provide them. For this reason, many economists are urging us to create financial incentives for conserving ecosystem services.

In Costa Rica, which had lost over three-quarters of its forest, leaders were ready to try this approach in their national policy. As part of Forest Law 7575, passed in 1996, the government began paying farmers and ranchers to preserve forest on their land instead of cutting it down, to replant cleared areas with new forest, or to manage forests sustainably with limited harvesting. Since then, the harvesting option has been dropped, but payments are now also made for allowing forest to regenerate naturally and for establishing agroforestry systems.

Landholders granted PSA contracts received payments spread over five years if they pursued these practices. Payments were designed to be competitive with potential profits from farming or cattle ranching, and today these payments average \$78/hectare/year (\$32/acre/year).

The PSA program recognized four ecosystem services that forests provide:

1. Watershed protection: Forests cleanse water by filtering pollutants, and they conserve water and reduce soil erosion by slowing runoff.
2. Biodiversity (which is especially rich in tropical forests such as Costa Rica's).
3. Scenic beauty, for recreation and ecotourism.
4. Absorption and storage of carbon: By pulling greenhouse gases out of the atmosphere, forests slow global warming.

To fund the program, the government sought money from people and companies who benefited from these services. For watershed protection, irrigators, bottlers, municipal water suppliers, and utilities that generate hydropower all made voluntary payments into the program, and a tariff on water users was added in 2005. For biodiversity and scenery, the country tried to

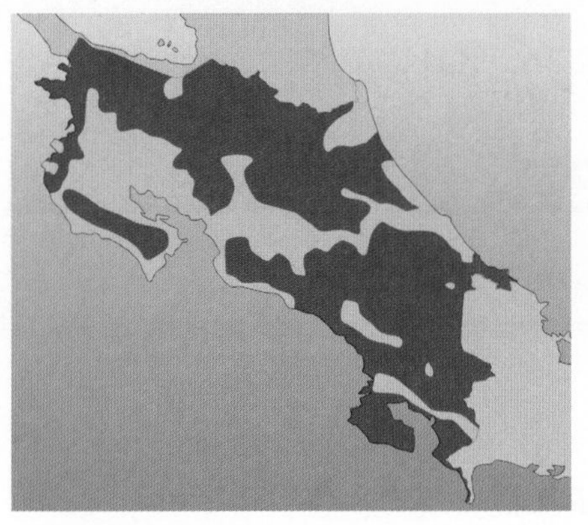

(a) 1940

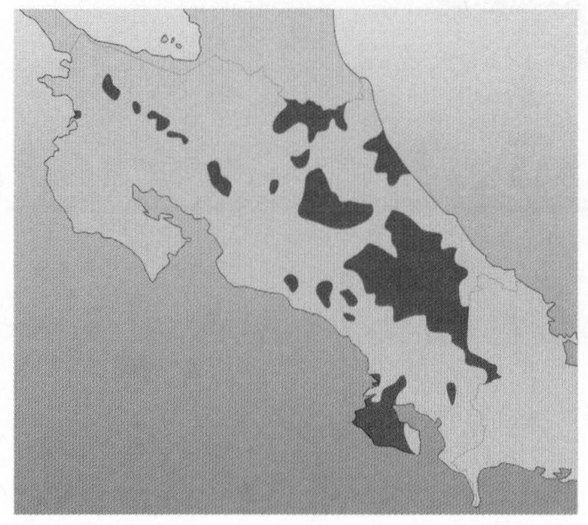

(b) 1987

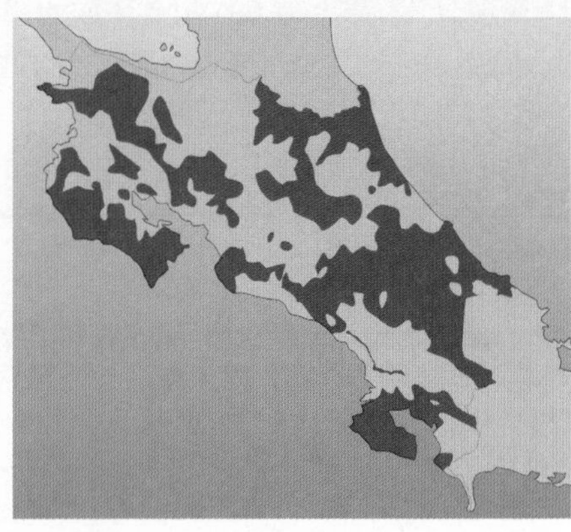

(c) 2005

FIGURE 6.1 Forest cover in Costa Rica decreased between 1940 and 1987, but it increased by 2005.
Data from FONAFIFO.

target ecotourism, while international lending agencies provided loans and donations. Because greenhouse gases are emitted when fossil fuels are burned, the nation used money from a 3.5% tax on fossil fuels to help fund the program. It also sought to sell carbon offsets on global carbon trading markets (p. 513).

Costa Rican landholders lined up for the payments. The agency created to administer the PSA program, the *Fondo Nacional de Financiamiento Forestal* (FONAFIFO), signed up landowners, sent forestry agents (called *regentes*) out to advise them on forest conservation and to monitor compliance, and processed millions of dollars in payments.

As the program proceeded, Costa Rica's rate of deforestation fell while new forest regrew, outpacing the loss of existing forest. As a result, the country's forest cover increased by 10% in the decade after 1996. Many policymakers, economists, and environmental advocates cheered the program's apparent success in safeguarding forests and the ecosystem services they provide.

However, some observers argued that forest loss had been slowing for other reasons and that the program itself had little effect. They contended that payments were being wasted on people who were not planning to cut down their trees. Critics also lamented that larger, wealthier landowners utilized the program most, while low-income small farmers were underrepresented. These concerns were born out by a number of researchers who have conducted studies of the program's effectiveness (see **THE SCIENCE BEHIND THE STORY**, pp. 144–145).

In response, the government modified its policies, seeking to make the program more accessible to small farmers and to target the payments to locations where forest is most at risk and environmental assets are greatest. By 2009, FONAFIFO had paid 57 billion colónes ($110 million) to over 8300 landholders and had registered over 670,000 ha (1.66 million acres)—13% of the nation's land area.

Today the program continues, and deforestation has been virtually eliminated. The nation's area of land covered by forest has risen from a low of 17% in 1983 to 52% today (**FIGURE 6.1**). In the years since the PSA program was initiated, Costa Ricans have enjoyed an increase in per capita income of over 50%—an enhancement in wealth surpassing the vast majority of other nations.

Many factors contributed toward Costa Rica's success in building a wealthier society while protecting its ecological assets. Back in 1948, Costa Rica became the first nation in the world to abolish its army. The country's leaders reinvested funds from the military budget into health and education instead. With a stable democratic government and a healthy and educated citizenry, the stage was set for well-managed development, including innovative advances in conservation. In the 1970s the nation began establishing national parks, eventually creating one of the world's finest systems of protected areas, covering one-quarter of Costa Rica's territory. Ecotourism at the parks today brings wealth into the country: Each year 2 million tourists from around the world inject $2 billion into Costa Rica's economy.

As a result, Costa Ricans understand the economic value of protecting their natural capital. With the economic value of nature so clear, an ethic of conservation has grown and flourished. In today's Costa Rica, an ethical appreciation for nature and an economic appreciation for nature go hand in hand, pointing the way toward truly sustainable development.

Culture, Worldview, and the Environment

Costa Rica's program of payment for environmental services has inspired similar approaches throughout the world. From Mexico to Australia to China, these programs are succeeding as people everywhere begin to better appreciate the contributions of ecological systems to human economies. The United States issues payments for ecological services through its Conservation Reserve Program (p. 239), which pays farmers to retain vegetation to prevent erosion, conserve soil, enhance wildlife habitat, and reduce water pollution. All such programs seek to defuse a dilemma facing many rural landholders, who often feel short-term economic pressure to clear forest for agriculture even though they may feel an ethical concern for the integrity of the forest and its flora and fauna. Such trade-offs between ethical concerns and short-term economic benefits arise frequently in today's world.

Ethics and economics involve values

Environmental science entails a firm understanding of the natural sciences. To address environmental problems, however, we also need to understand how people perceive, value, and relate to their environment philosophically and pragmatically. Ethics and economics are two very different disciplines, but each deals with questions of what we value and how those values influence our decisions and actions. To find sustainable solutions to environmental problems, we must aim to comprehend not only how natural systems work, but also how values shape human behavior.

Culture and worldview influence our decisions

Every action we take affects our environment. Whether we are growing food, building homes, manufacturing products, or fueling vehicles, we meet our needs by extracting resources and altering our surroundings. In deciding how to manipulate our environment to meet our needs, we rely on rational assessments of costs and benefits, but our decisions are also influenced by our culture and our worldview (**FIGURE 6.2**).

Culture can be defined as the ensemble of knowledge, beliefs, values, and learned ways of life shared by a group of people. Culture, together with personal experience and personal circumstance, influences each person's perception of the world and his or her place within it—the person's **worldview.** A worldview reflects beliefs about the meaning, operation, and essence of the world.

People with different worldviews can study the same situation yet draw dramatically different conclusions. For example, two Costa Rican ranchers owning identical landholdings may make different decisions about how to manage them. One might opt to receive payments and conserve forest on his land, whereas the other might opt to clear every hectare for cattle grazing.

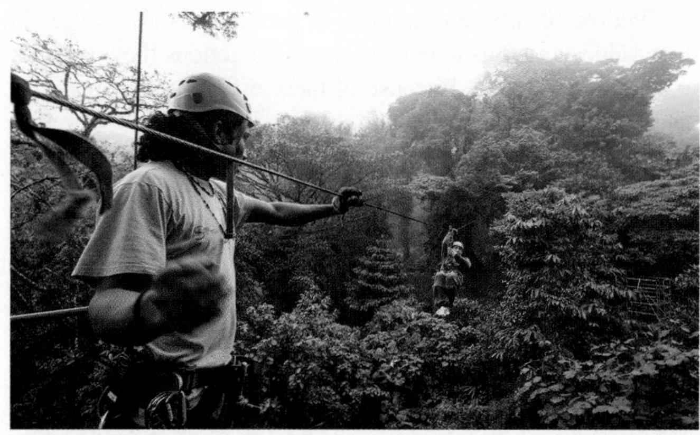

(a) Zip-lining at a Costa Rican national park

(b) A pepper farmer makes a living off the land

FIGURE 6.2 Differences in culture, wealth, and personal circumstance may lead people to interact differently with the landscape. Zip-lining in Costa Rica's rainforests **(a)** provides recreation to visiting ecotourists and employment to young Costa Ricans. Ecotourists may feel inspired by the majesty of the forest, and tourism workers may see economic advantage in preserving forest. A Costa Rican farm worker **(b)** needs agriculture that produces enough to live on, and may favor clearing forest if the short-term economic benefits outweigh the costs.

Many factors shape our worldviews

Many factors shape people's worldviews and perception of their environment. Religion and spiritual beliefs are among the most influential. A person's political ideology also may shape his or her attitudes. For instance, one's views on the proper role of government may guide whether one wants government to intervene in a market economy to protect environmental quality. Shared cultural experience is another factor. A community may share a particular outlook if its members have lived through similar experiences. Early European settlers in the Americas, facing the struggles of frontier life, viewed their environment as a hostile force because inclement weather and wild animals frequently destroyed crops and killed livestock. Many people still view nature as a hostile adversary to overcome.

As you progress through your course and through your life, you will encounter scientific data on the environmental impacts of our choices (where to live, how to make a living, what to eat, what to wear, how to spend our leisure time, and so on). You will also see that culture and worldviews play critical roles in such choices. Acquiring scientific understanding is vital in our search for sustainable solutions, but we also need to consider ethics and economics, because these disciplines help us understand how and why we value the things we value.

Environmental Ethics

Ethics is a branch of philosophy that involves the study of good and bad, of right and wrong. The term *ethics* can also refer to the set of moral principles or values held by a person or a society. Ethicists help clarify how people judge right from wrong by elucidating the criteria, standards, or rules that people use in making these judgments. Such criteria are grounded in values—for instance, promoting human welfare, maximizing individual freedom, or minimizing pain and suffering.

People of different cultures or worldviews may differ in their values and thus may disagree about actions they consider to be right or wrong. Because of these differences, some ethicists are **relativists,** believing that ethics do and should vary with social context. However, different societies show a remarkable extent of agreement on what moral standards are appropriate. For this reason, many ethicists are **universalists,** maintaining that there exist objective notions of right and wrong that hold across cultures and contexts. For both relativists and universalists, ethics is a prescriptive pursuit; rather than simply describing behavior, it prescribes how we ought to behave.

Ethical standards help us judge right from wrong

Ethical standards are the criteria that help differentiate right from wrong. We all employ ethical standards as tools for making countless decisions in our everyday lives. One classic ethical standard is the *categorical imperative* proposed by German philosopher Immanuel Kant, which advises us to treat others as we would prefer to be treated ourselves. In Christianity this standard is called the "Golden Rule," and most of the world's religions teach this same lesson. Another ethical standard is the *principle of utility,* elaborated by British philosophers Jeremy Bentham and John Stuart Mill. The utilitarian principle holds that something is right when it produces the greatest practical benefits for the most people.

We value things in two ways

People ascribe value to things in two main ways. One way is to value something for the pragmatic benefits it brings us if we put it to use. This is termed **instrumental value** (or **utilitarian value**). The other way is to value something for its intrinsic worth, to feel that something has a right to exist and is valuable for its own sake. This notion is termed **intrinsic value** or **inherent value.**

A person may ascribe instrumental value to a forest because we can harvest timber from it, hunt game in it, and drink clean water it has captured and filtered. A person may perceive intrinsic value in a forest because it provides homes for countless organisms that the person feels have an inherent right to live alongside us on our shared planet. A forest, an animal, a lake, or a mountain can have both intrinsic and instrumental value. However, different people may emphasize different types of value.

Paying money for ecosystem services is a utilitarian approach that attempts to quantify instrumental values by assigning market prices to them. For people who tend to view nature through the lens of intrinsic values, this approach makes them uneasy. Indeed, scientists who share the goal of conserving natural amenities sometimes disagree on the means of doing so. For instance, Stanford University biologist Gretchen Daily is a key proponent of assigning market values to ecosystem services, with the utilitarian goal of engaging market forces to assist in their conservation. Yet in 2006 a student in her department, Douglas McCauley, authored a commentary in the scientific journal *Nature* that argued eloquently against such an approach. McCauley warned that the "commodification of nature" distracted from the environment's intrinsic value, which he maintained was infinite and priceless. "Nature conservation must be framed as a moral issue," McCauley wrote. "We will make more progress in the long run by appealing to people's hearts rather than to their wallets."

Environmental ethics pertains to people and the environment

The application of ethical standards to relationships between people and nonhuman entities is known as **environmental ethics.** This branch of ethics arose once people began to perceive environmental change brought by industrialization. Our interactions with our environment can give rise to ethical questions that are difficult to resolve. Consider some examples:

1. Is the present generation obligated to conserve resources for future generations? If so, how much should we conserve?
2. Can we justify exposing some communities to a disproportionate share of pollution? If not, what actions are warranted to prevent this?
3. Are humans justified in driving species to extinction? If destroying a forest would drive extinct an insect species few people have heard of but would create jobs for 10,000 people, would that action be ethically admissible? What if it were an owl species? What if only 100 jobs would be created? What if it were a species harmful to people, such as a mosquito, bacterium, or virus?

The first question is central to the notion of sustainability (p. 14) and to the pursuit of sustainable development (pp. 156–157), which lie at the heart of environmental science. Sustainability means leaving our descendants a world in which they can meet their needs at least as well as we have met ours. From an ethical perspective, sustainability means treating future generations as we would prefer to be treated ourselves. The second question goes to the heart of environmental justice, which we will tackle shortly (pp. 140–141). The third set of questions involves intrinsic values (but also instrumental values) and typifies questions that arise in debates over endangered species management, habitat protection, and conservation biology (Chapter 11).

We have expanded our ethical consideration

Answers to questions like those above depend partly on what ethical standard(s) a person adopts. They also depend on the breadth and inclusiveness of the person's domain of ethical concern. A person who ascribes intrinsic value to insects and feels responsibility for their welfare would answer the third set of questions very differently from a person whose domain of ethical concern ends with human beings.

Many traditional non-Western cultures have long granted nonhuman entities intrinsic value and ethical standing. Australian Aborigines view their landscape as sacred and alive. Many native cultures across the Americas feature ethical systems that encompass both people and aspects of their environment.

As the history of Western cultures (European and European-derived societies) has progressed, people have granted intrinsic value and extended ethical consideration to more and more people and things. Today, concern for the welfare of domesticated animals is evident in the great care many people provide for their pets. Animal rights advocates voice concern for animals that are hunted, raised in pens, or used in laboratory testing. Most people now accept that wild animals merit ethical consideration. Increasing numbers of people today see intrinsic value in whole natural communities. Some go further and suggest that all of nature—living and nonliving things alike—should be ethically recognized.

What has helped broaden our ethical domain in these ways? Rising economic prosperity has played a role, by making us less anxious about our day-to-day survival. Science has also played a role, by demonstrating that people do not stand apart from nature, but rather are part of it. Ecology makes clear that organisms are interconnected and that what affects plants, animals, and ecosystems also affects people. Evolutionary biology shows that human beings, as one species out of millions, have evolved subject to the same pressures as other organisms.

We can simplify our continuum of attitudes toward the natural world by dividing it into three ethical perspectives: anthropocentrism, biocentrism, and ecocentrism (**FIGURE 6.3**).

Anthropocentrism People who have a human-centered view of our relationship with the environment display **anthropocentrism.** An anthropocentrist denies, overlooks, or devalues the notion that nonhuman things have intrinsic value. An anthropocentrist also evaluates the costs and benefits of actions solely according to their impact on people. For example, if cutting down a Costa Rican forest for farming or ranching would provide significant economic benefits while doing little harm to aesthetics or human health, the anthropocentrist would conclude this was worthwhile, even if it would destroy habitat for many plants and animals. Conversely, if protecting the forest would provide greater economic, spiritual, or other benefits to people, an anthropocentrist would favor its protection. In the anthropocentric perspective, anything not providing a readily apparent benefit to people is considered to be of negligible value.

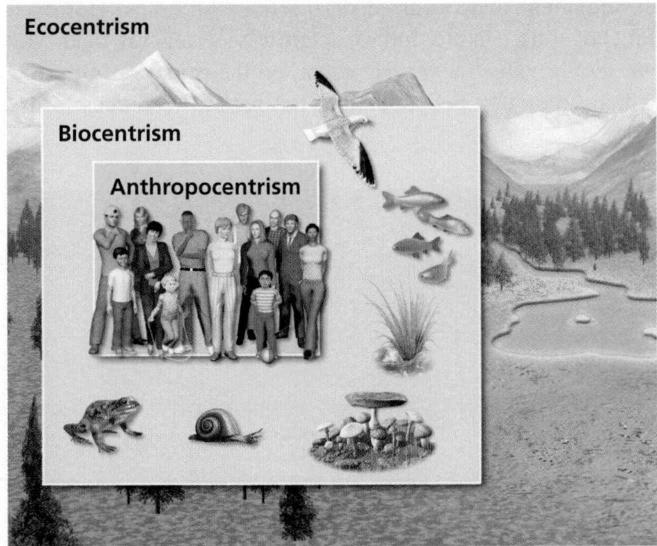

FIGURE 6.3 We can categorize people's ethical perspectives as anthropocentric, biocentric, or ecocentric. Anthropocentrists extend ethical standing only to human beings and judge actions in terms of their effects on people. Biocentrists value and consider all living things, human and otherwise. Ecocentrists extend ethical consideration to living and nonliving components of the environment holistically, valuing the larger functional systems of which they are a part.

Biocentrism **Biocentrism** ascribes intrinsic value to certain living things or to the biotic realm in general. In this perspective, human life and nonhuman life both have ethical standing, so a biocentrist evaluates actions in terms of their overall impact on living things. A biocentrist might oppose clearing a forest if this would destroy countless plants and animals, even if it would increase food production and generate economic growth for people. Some biocentrists advocate equal consideration for all living things, whereas others grant some types of organisms more consideration than others.

Ecocentrism **Ecocentrism** judges actions in terms of their effects on whole ecological systems, which consist of living and nonliving elements and their interrelationships. An ecocentrist values the well-being of entire species, communities, or ecosystems over the welfare of a given individual. Implicit in this view is that preserving systems generally protects their components, whereas protecting components may not safeguard the entire system. An ecocentrist would respond to a proposal to clear forest by broadly assessing the potential impacts on water quality, air quality, wildlife populations, soil structure, nutrient cycling, and ecosystem services. Ecocentrism is a more holistic perspective than biocentrism or anthropocentrism. It encompasses a wider variety of entities at a larger scale and seeks to preserve the connections that tie them together into functional systems.

Environmental ethics has ancient roots

Environmental ethics arose as an academic discipline in the 1970s, but people have contemplated our ethical relations with nature for thousands of years. The ancient Greek philosopher Plato argued that humanity had a moral obligation to our environment, writing, "The land is our ancestral home and we must cherish it even more than children cherish their mother."

Some ethicists and theologians have pointed to the religious traditions of Christianity, Judaism, and Islam as sources of anthropocentric hostility toward the environment. They point out biblical passages such as, "Be fruitful and multiply, and fill the earth and subdue it; and have dominion over the fish of the sea and over the birds of the air and over every living thing that moves upon the earth." Such wording, according to many scholars, has encouraged animosity and disregard toward nature.

Others interpret sacred texts of these religions to encourage benevolent human stewardship over nature. Consider the directive, "You shall not defile the land in which you live." If one views the natural world as God's creation, then surely it must be considered sinful to degrade that creation. Indeed, it must be seen as one's duty to act as a responsible steward of the world in which we live.

The industrial revolution inspired reaction

As industrialization spread in the 19th century, it amplified human impacts on the environment. In this period of social and economic transformation, agricultural economies became industrial ones, machines enhanced or replaced human and animal labor, and many people moved from farms to cities. Population rose dramatically, consumption of natural resources accelerated, and pollution intensified as we burned coal to fuel railroads, steamships, ironworks, and factories.

Some writers of the time drew attention to the drawbacks of industrialization. British critic John Ruskin called cities "little more than laboratories for the distillation into heaven of venomous smokes and smells." Ruskin worried that while people prized the material benefits that nature provided, they no longer appreciated its spiritual and aesthetic benefits. Motivated by such concerns, a number of citizens' groups sprang up in 19th-century England, forerunners of today's environmental organizations.

In the United States during the 1840s, a philosophical movement called **transcendentalism** flourished, espoused in New England by philosophers **Ralph Waldo Emerson** and **Henry David Thoreau** and by poet **Walt Whitman.** The transcendentalists viewed nature as a direct manifestation of the divine, emphasizing the soul's oneness with nature and God. Like Ruskin, the transcendentalists objected to an attraction to material things, and they promoted a holistic view of nature in which natural entities were symbols or messengers of deeper truths. Although Thoreau viewed nature as divine, he also observed the natural world closely and came to understand it in the manner of a scientist; indeed, he can be considered one of the first ecologists. His book *Walden,* in which he recorded his observations and thoughts while living at Walden Pond away from the bustle of urban Massachusetts, remains a classic of American literature.

Conservation and preservation arose with the 20th century

One admirer of Emerson and Thoreau was **John Muir** (1838–1914), a Scottish immigrant to the United States who made California's Yosemite Valley his wilderness home. Although Muir chose to live in isolation in his beloved Sierra Nevada for long stretches of time, he also became politically active and won fame as a tireless advocate for the preservation of wilderness (FIGURE 6.4).

Muir was motivated by the rapid deforestation and environmental degradation he witnessed throughout North America and by his belief that the natural world should be treated with the same respect that we give to cathedrals. Today he is associated with the **preservation ethic,** which holds that we should protect the natural environment in a pristine, unaltered state. Muir argued that nature deserved protection for its own sake (an ecocentrist argument resting on the notion of intrinsic value), but he also maintained that nature promoted human happiness (an anthropocentrist argument based on instrumental value). "Everybody needs beauty as well as bread," he wrote, "Places to play in and pray in, where nature may heal and give strength to body and soul alike."

FIGURE 6.4 **A pioneering advocate of the preservation ethic, John Muir helped establish the Sierra Club, a leading environmental organization.** Here Muir **(right)** is shown with President Theodore Roosevelt in Yosemite National Park in 1903. After this wilderness camping trip with Muir, the president expanded protection of areas in the Sierra Nevada.

Some of the factors that motivated Muir also inspired **Gifford Pinchot** (1865–1946), the first professionally trained American forester (FIGURE 6.5). Pinchot founded what would become the U.S. Forest Service and served as its chief in President Theodore Roosevelt's administration. Like Muir, Pinchot opposed the deforestation and unregulated economic development that occurred during their lifetimes. However, Pinchot took a more anthropocentric view of how and why we should value nature. He espoused the **conservation ethic,** which holds that people should put natural resources to use

FIGURE 6.5 **Gifford Pinchot, the first chief of what would become the U.S. Forest Service, was a leading proponent of the conservation ethic.** This ethic holds that people should use natural resources in ways that ensure the greatest good for the greatest number for the longest time.

but that we have a responsibility to manage them wisely. The conservation ethic employs a utilitarian standard, stating that we should allocate resources so as to provide the greatest good to the greatest number of people for the longest time. Whereas preservation aims to preserve nature for its own sake and for our aesthetic and spiritual benefit, conservation promotes the prudent, efficient, and sustainable extraction and use of natural resources for the good of present and future generations.

Pinchot and Muir came to represent different branches of the American environmental movement, and their contrasting ethical approaches often pitted them against one another on policy issues of the day. For instance, Pinchot supported damming Hetch Hetchy Valley in Yosemite National Park to provide drinking water for San Francisco, considering this "the highest possible use which could be made of" the valley. Muir was aghast, and proclaimed, "Dam Hetch Hetchy! As well dam for water tanks the people's cathedrals and churches, for no holier temple has ever been consecrated by the heart of man."

Despite their differences, both Muir and Pinchot represented reactions against a prevailing "development ethic," which holds that people should be masters of nature and which promotes economic development without regard to its negative consequences. Both Pinchot and Muir left legacies that reverberate today in the ethical approaches to environmentalism.

Today these schools of thought have spread globally, and people worldwide are wrestling with how to balance preservation, conservation, and economic development. In Costa Rica, leaders and citizens felt that development had gone too far and that precious ecosystem services were being degraded and lost, threatening the country's future. The nation responded by establishing an extensive system of national parks, eventually protecting 24% of its land area—one of the highest percentages of any nation in the world. Beyond this preservationist policy, many conservationist policies were implemented to encourage the sustainable use of soil, water, and forests. Costa Rica's program to pay for ecological services mixes these approaches, encouraging the preservation of forests within lands actively used for agricultural production to meet overall goals of conservation and sustainable development.

WEIGHING THE ISSUES

PRESERVATION AND CONSERVATION Do you identify more with the preservation ethic, the conservation ethic, both, or neither? Think of a forest, wetland or other important natural resource in your region. Give an example of a situation in which you might adopt a preservation ethic and an example of one in which you might adopt a conservation ethic. Are there conditions under which you'd follow neither, but instead adopt a "development ethic"?

Aldo Leopold's land ethic inspires many people

As a young forester and wildlife manager, **Aldo Leopold** (1887–1949; FIGURE 6.6) began his career in the conservationist camp after graduating from Yale Forestry School, which Pinchot had helped found just as Roosevelt and Pinchot were advancing conservation on the national stage. As a forest manager in Arizona and New Mexico, Leopold embraced the government policy of shooting predators, such as wolves, to increase populations of deer and other game animals.

At the same time, Leopold followed the development of ecological science. He eventually ceased to view certain species as "good" or "bad" and instead came to see that healthy ecological systems depend on the protection of all their interacting parts. Drawing an analogy to mechanical maintenance, he wrote, "to keep every cog and wheel is the first precaution of intelligent tinkering."

It was not just science that pulled Leopold from an anthropocentric perspective toward a more holistic one. One day he shot a wolf, and when he reached the animal, Leopold was transfixed by "a fierce green fire dying in her eyes." He perceived intrinsic value in the wolf, and the experience remained with him for the rest of his life, helping to lead him to an ecocentric ethical outlook. Years later, as a University of Wisconsin professor, Leopold argued that people should view themselves and "the land" as members of the same community and that we are obligated to treat the land in an ethical manner. In his 1949 essay "The Land Ethic," he wrote:

> All ethics so far evolved rest upon a single premise: that the individual is a member of a community of interdependent parts. . . . The land ethic simply enlarges the boundaries of the community to include soils, waters, plants, and animals, or collectively: the land. . . . A land ethic changes the role of *Homo sapiens* from conqueror of the land-community to plain member and citizen of it. . . . It implies respect for his fellow-members, and also respect for the community as such.

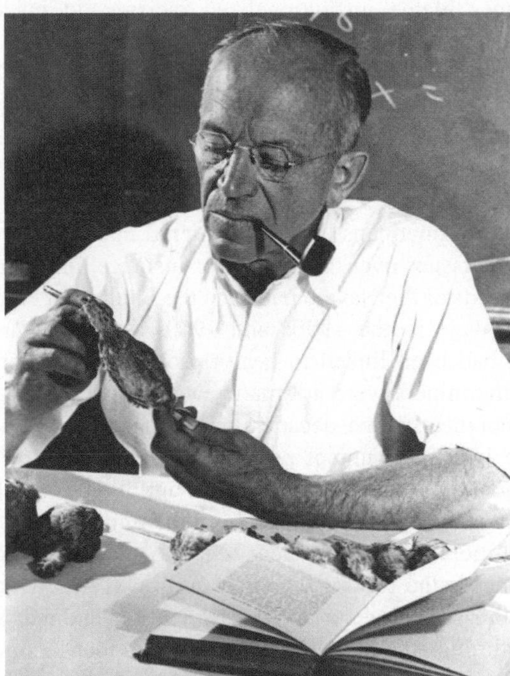

FIGURE 6.6 **Aldo Leopold, a wildlife manager and environmental philosopher, articulated a new relationship between people and the environment.** In his essay "The Land Ethic" he called on people to include the environment in their ethical outlook.

Leopold intended that the land ethic would help guide decision making. "A thing is right," he wrote, "when it tends to preserve the integrity, stability, and beauty of the biotic community. It is wrong when it tends otherwise." Leopold died before seeing "The Land Ethic" and his best-known book, *A Sand County Almanac,* in print, but today many view him as the most eloquent philosopher of environmental ethics.

Environmental justice seeks equal treatment for all races and classes

Our society's domain of ethical concern has been expanding from rich to poor, and from majority races and ethnic groups to minority ones. This ethical expansion involves applying a standard of equal treatment, and it has given rise to the environmental justice movement. **Environmental justice** involves the fair and equitable treatment of all people with respect to environmental policy and practice, regardless of their income, race, or ethnicity.

The struggle for environmental justice has been fueled by the recognition that poor people tend to be exposed to a greater share of pollution, hazards, and environmental degradation than are richer people. Environmental justice advocates also note that racial and ethnic minorities tend to suffer more exposure to most hazards than whites. Indeed, studies repeatedly document that poor and nonwhite communities each tend to bear heavier burdens of air pollution, lead poisoning, pesticide exposure, toxic waste exposure, and workplace hazards. This is thought to occur because lower-income and minority communities often have less access to information on environmental health risks, less political power with which to protect their interests, and less money to spend on avoiding or alleviating risks. Environmental justice proponents also sometimes blame institutionalized racism and inadequate government policies.

A protest in the 1980s by residents of Warren County, North Carolina, against a toxic waste dump in their community helped to ignite the environmental justice movement (**FIGURE 6.7**). The state had chosen to site the dump in the county with the highest percentage of African Americans.

Native Americans have encountered many environmental justice issues over the years. For instance, uranium mining on lands of the Navajo nation in the Southwest employed many Navajo in the 1950s and 1960s. Although uranium mining had been linked to health problems and premature death, the miners were not made aware of radiation and its risks. For nearly two decades neither the mining industry nor the U.S. government provided the miners information or safeguards. Many Navajo families built homes and bread-baking ovens out of waste rock from the mines, not realizing it was radioactive. Lung cancer began to appear among Navajo miners in the 1960s. A later generation of Americans perceived this as negligence and discrimination, and they sought justice through the Radiation Exposure Compensation Act of 1990, a federal law compensating Navajo miners who suffered health effects from unprotected work in the mines.

Similarly, white residents of the Appalachian region have long been the focus of environmental justice concerns. Mountaintop coal mining practices (pp. 641–645) in this economically neglected region provide jobs to local residents but also pollute water, bury streams, degrade forests, and cause flooding. The mostly low-income residents of affected Appalachian communities have historically had little political power to voice complaints over the impacts of these mining practices.

FIGURE 6.7 Environmental justice first gained prominence with this protest against a toxic waste dump in North Carolina.

Today, although our economies have grown, the gaps between rich and poor have widened. And despite much progress toward racial equality, significant inequities remain. Environmental laws have proliferated, but minorities and the poor still suffer substandard environmental conditions (**FIGURE 6.8**). Still, today more people are fighting environmental hazards in their communities and winning.

One ongoing success story is in California's San Joaquin Valley. The poor, mostly Latino, farm workers in this region who harvest much of the U.S. food supply of fruits and vegetables also suffer some of the nation's worst air pollution. Industrial agriculture generates pesticide emissions, dairy feedlot emissions, and windblown dust from eroding farmland, yet this pollution was not being regulated. Valley residents enlisted the help of organizations including the Center on Race, Poverty, and the Environment, a San Francisco–based environmental justice law firm. Together they persuaded California regulators to enforce Clean Air Act provisions and convinced California legislators to pass new legislation regulating agricultural emissions. Today the farm workers and their advocates continue working to strengthen and enforce clean air regulations.

Environmental justice is a key component in pursuing the environmental, economic, and social goals of sustainability and sustainable development (pp.14, 156–157). As we explore environmental issues from a scientific standpoint throughout this book, we will also encounter the social, political, ethical, and economic aspects of these issues, and the concept of environmental justice will arise again and again.

FIGURE 6.8 Hurricane Katrina revealed our ongoing need for environmental justice. Many of the people most affected by the storm were poor and nonwhite. These children are playing in the Lower Ninth Ward of New Orleans, where many homes were destroyed and water remained unsafe to drink long afterwards.

WEIGHING THE ISSUES

ENVIRONMENTAL JUSTICE Consider the place where you grew up. Where were the factories, waste dumps, and polluting facilities located, and who lived closest to them? Who lives nearest them in the town or city that hosts your campus? Do you think the concerns of environmental justice advocates are justified? If so, what could be done to ensure that poor communities do not suffer more hazards than wealthy ones?

Economics and the Environment

Questions of environmental justice, like questions of how to value ecosystem services, intertwine ethical issues with economic ones—and friction often develops between people's ethical concerns and their economic desires. Addressing ethical, economic, and environmental concerns together in a mutually productive way is a primary goal of the modern drive for sustainable development (pp. 156–157).

Is there a trade-off between the economy and the environment?

Measures to safeguard environmental quality frequently mesh well with ethical considerations, but we often hear it said that environmental protection runs counter to economic interests. People argue that environmental protection costs too much money, interferes with progress, or leads to job loss. However, growing numbers of economists assert that there doesn't need to be a trade-off—and that in fact, environmental protection generally *enhances* our economy.

The view one takes depends in part on whether one thinks in the short term or the long term. The economic judgments we make often pertain to short time scales, and in the short term many activities that cause environmental harm (such as natural resource extraction) may be economically profitable. In the longer term, however, environmental degradation generally feeds back and harms economies. The view one takes also depends on whether one stands to benefit directly. Often when resource extraction or development causes environmental degradation, a few private parties benefit economically, but the broader public is harmed.

Traditional economic schools of thought have underestimated or overlooked the contributions of the environment to our economies, and as a result, many people have equated environmental protection with economic sacrifice. Newer schools of thought recognize that economies are coupled to the environment and reliant on its goods and services. For people holding this worldview, our economic health depends on environmental protection. This is why Costa Rica and other nations have taken steps to better protect their natural assets.

Today, concern over climate change, pollution, fluctuating fossil fuel supplies, and dependence on foreign oil have led many economists and policymakers to recognize immense opportunities in revamping our economies with clean and renewable energy technologies. Building a green-energy economy presents a clear case of how economic advancement and environmental protection can go hand-in-hand.

Economics studies resource allocation

Like ethics, economics examines factors that guide human behavior. **Economics** is the study of how people decide to use potentially scarce resources to provide goods and services that are in demand. By this definition, environmental problems are also economic problems that can intensify as population and per capita resource consumption increase. For example, pollution may be viewed as a depletion of the resources of clean air, water, or soil. Indeed, the word *economics* and the word *ecology* come from the same Greek root, *oikos,* meaning "household." Economists traditionally have studied the household of human society, and ecologists the broader household of all life.

Several types of economies exist

An **economy** is a social system that converts resources into **goods,** material commodities manufactured for and bought by individuals and businesses; and **services,** work done for others as a form of business. The oldest type of economy is the **subsistence economy.** People in subsistence economies meet their daily needs by subsisting on what they can gather from nature or produce on their own (by hunting, fishing, or farming), rather than by working for wages and purchasing life's necessities.

A second type of economy is the **capitalist market economy.** In this system, interactions among buyers and sellers determine which goods and services are produced, how many are produced, and how these are distributed. Capitalist economies contrast with **centrally planned economies,** or *state socialist economies*, in which government determines how to allocate resources. In reality, however, virtually all national economies today are hybrid systems, often termed **mixed economies.** The United States—the world's strongest proponent of capitalism—has in fact borrowed a great deal from socialism, and China—the world's largest socialist state—hosts a robust market system. By combining aspects of both approaches, mixed economies encourage stability and prosperity. When things get out of balance—such as when unregulated financial practices contributed to the U.S. recession in 2008–2009—then prosperity may decline, and balance may need to be restored.

Economies rely on goods and services from the environment

Our societies and our economies exist within the natural environment and depend on it in vital ways. Economies receive inputs (such as natural resources and ecosystem services) from the environment, process them in various ways, and discharge outputs (such as waste) into the environment (**FIGURE 6.9**). The material inputs and the waste-absorbing capacity that Earth can provide are ultimately finite.

Although the interactions between economies and the environment are readily apparent, traditional economic schools of thought have long overlooked their importance. Many mainstream economists still adhere to a worldview that largely ignores the environment (considering only the yellow box in the middle of Figure 6.9). This conventional view, which continues to drive most policy decisions, implies that inputs such as natural resources are free and limitless and that wastes can be endlessly absorbed at no cost. In contrast, modern economists belonging to the fast-growing fields of environmental economics and ecological economics (p. 150) explicitly recognize that economies exist within the environment and depend crucially upon it for natural resources and ecosystem services.

Natural resources (pp. 3–4) are the various substances and forces that sustain our society and our everyday lives: the fresh water we drink, the trees that provide our lumber, the rocks that provide our metals, and the energy from the sun, wind, water, and fossil fuels. We can think of natural resources as "goods" produced by nature. Environmental systems also naturally function in a manner that supports our economies. Earth's ecological systems purify air and water, form soil, cycle nutrients, regulate climate, pollinate plants, and recycle waste. Such essential ecosystem services (pp. 3, 116–117, 152, 290) support the very life that makes our economic activity possible. Together, nature's resources and services make up the natural capital (p. 14) on which our economies and societies depend.

While our environment enables economic activity by providing ecosystem goods and services, economic activity

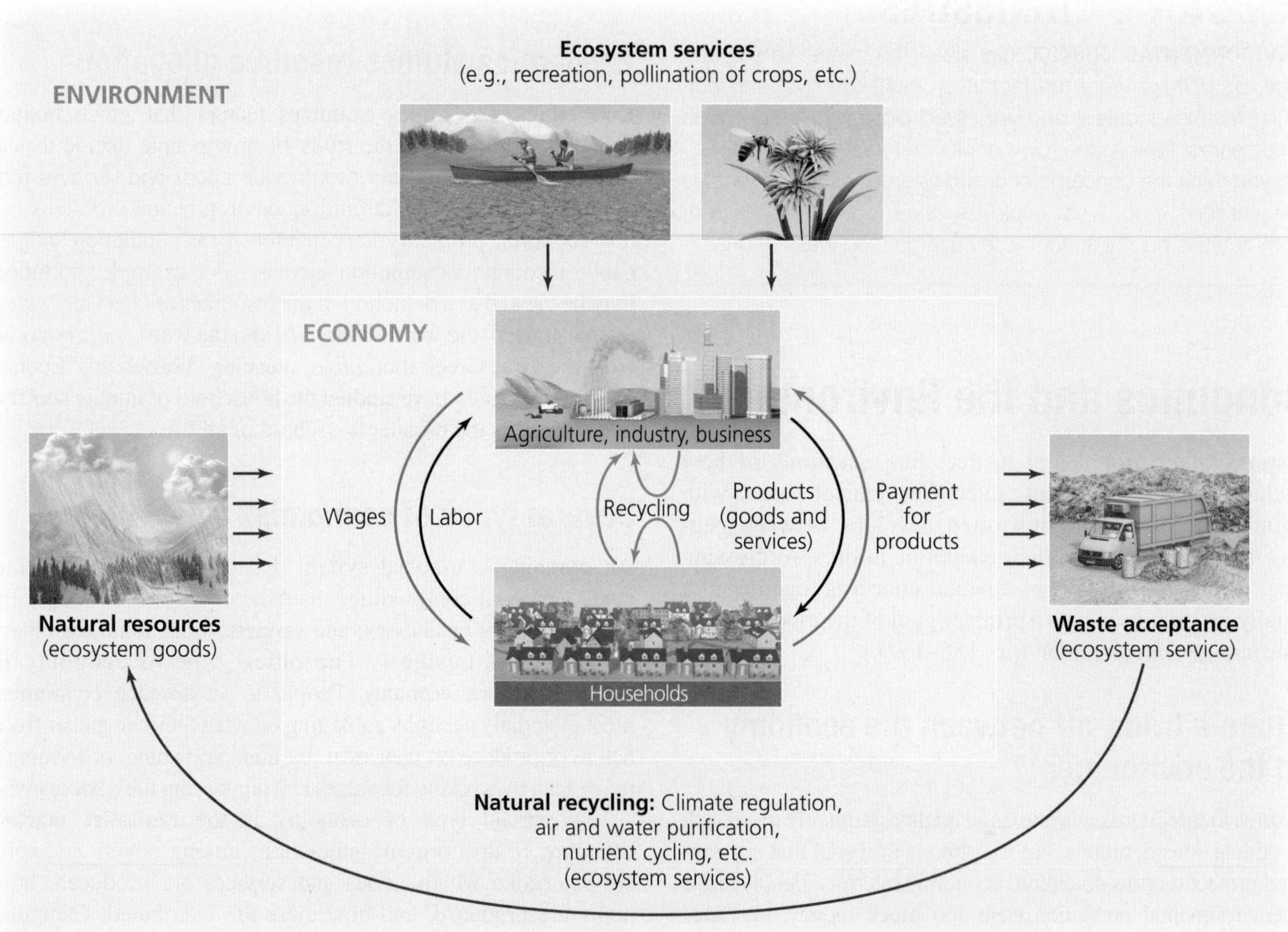

FIGURE 6.9 Economies exist within the natural environment, receiving resources from it, discharging waste into it, and benefiting from ecosystem services. Conventional neoclassical economics has focused only on processes of production and consumption between households and businesses (**yellow box in middle**), viewing the environment merely as an external factor that helps enable the production of goods. In contrast, environmental and ecological economists explicitly recognize that economies exist within the natural environment and depend on all that it offers.

can in turn affect the environment. When we deplete natural resources and generate pollution, we degrade the capacity of ecological systems to function. Scientists with the Millennium Ecosystem Assessment (p. 15) concluded in 2005 that 15 of 24 ecosystem services they surveyed globally were being degraded or used unsustainably. The degradation of ecosystem services can disrupt economies. In Costa Rica, rapid forest loss up through the 1980s was causing soil erosion, water pollution, and biodiversity loss that increasingly threatened the country's economic potential. Low-income small farmers were the first to feel these impacts. Indeed, across the world, ecological degradation is harming poor and marginalized people before wealthy ones, the Millennium Ecosystem Assessment found. As a result, restoring ecosystem services stands as a prime avenue for alleviating poverty.

The relationships among economic and environmental conditions are only now becoming widely recognized. Let's briefly examine how economic thought has evolved, tracing the path that is finally beginning to lead economies to become more compatible with the natural systems on which they depend.

Adam Smith proposed an "invisible hand"

Economics shares a common intellectual heritage with ethics, and practitioners of both study the relationship between individual action and societal well-being. Early philosophers had long believed that individuals acting in their own self-interest harm society. However, Scottish philosopher **Adam Smith** (1723–1790) argued that self-interested economic behavior could benefit society, as long as the behavior was constrained by the rule of law and private property rights and operated within a competitive marketplace. Known today as a founder of **classical economics,** Smith felt that when people pursue their own economic self-interest under these conditions, the marketplace will behave as if guided by "an invisible hand" to benefit society as a whole. In his 1776 book *Inquiry into the Nature and Causes of the Wealth of Nations,* Smith wrote:

> It is not from the benevolence of the butcher, the brewer, or the baker that we expect our dinner, but from their regard to their own self-interest. [Each individual] intends only his own security, only his own gain. And he is led in this by an invisible hand to promote an end which was no part of his intention. By pursuing his own interests he frequently promotes that of society more effectually than when he really intends to.

Smith's philosophy remains a pillar of free-market thought today, and many credit it for the tremendous gains in material wealth that industrialized nations have achieved. Others assert that free-market policies can worsen inequalities between rich and poor and intensify environmental degradation.

Neoclassical economics considers supply, demand, costs, and benefits

Economists subsequently adopted more quantitative approaches as they aimed to explain human behavior. **Neoclassical economics** examines the psychological factors underlying consumer choices, explaining market prices in terms of consumer preferences for units of particular commodities. Standard neoclassical models assume that people behave rationally and have access to full information.

In neoclassical economics, buyers desire a low price, whereas sellers desire a high price. This conflict results in a compromise price being reached and the "right" quantity of commodities being bought and sold (**FIGURE 6.10a**). This is phrased in terms of *supply*, the amount of a product offered for sale at a given price, and *demand*, the amount of a product people will buy at a given price if free to do so. Theoretically, the market moves toward an equilibrium point, a price at which supply

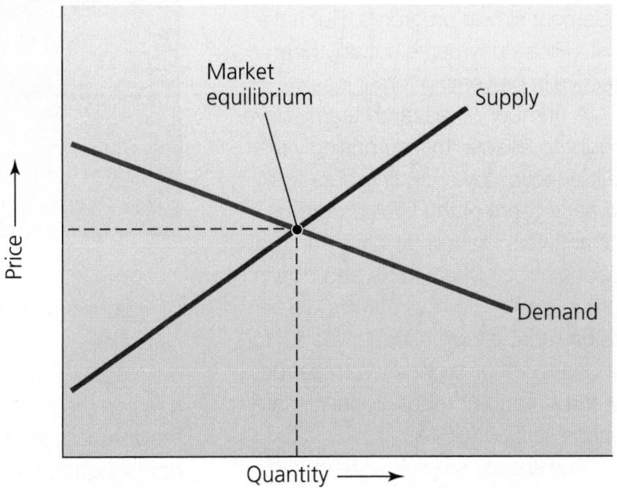

(a) Classic supply–demand curve

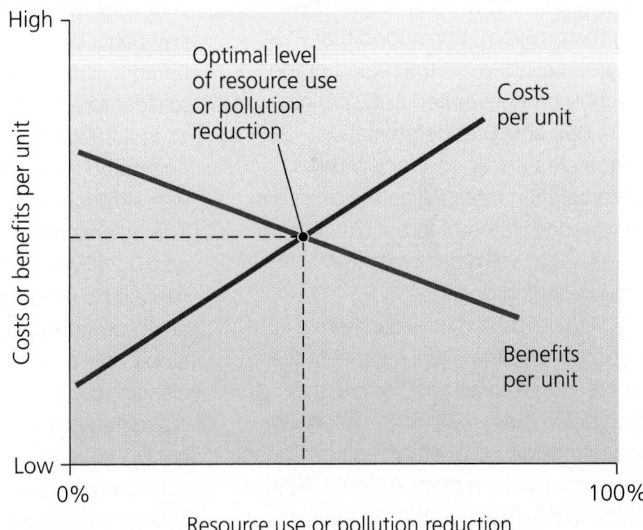

(b) Marginal benefit and cost curves

FIGURE 6.10 We can graph the fundamentals of supply and demand and costs and benefits. In a supply-and-demand graph **(a)**, the demand curve indicates the quantity of a given good (or service) that consumers desire at each price, and the supply curve indicates the quantity produced at each price. The market automatically moves toward an equilibrium point, at which supply equals demand. We can use a similar graph **(b)** to determine an "optimal" level of resource use or pollution control. In this graph, the cost per unit of resource use or pollution cleanup rises as the process proceeds and it becomes expensive to extract or clean up the remaining amounts. Meanwhile, the benefits per unit of resource use or pollution cleanup decrease. The point where the lines intersect gives the optimal level.

Do Payments Help Preserve Forest?

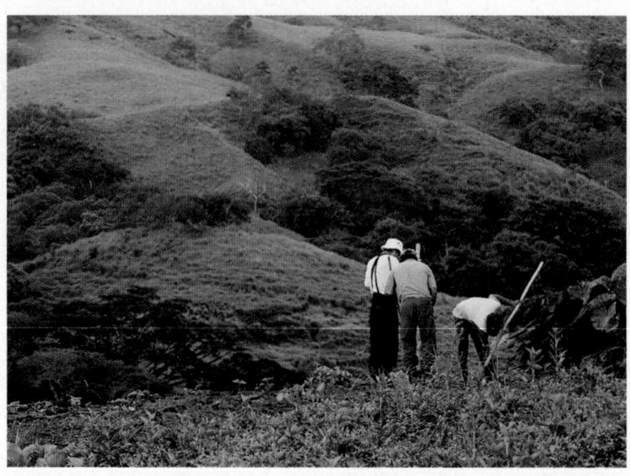

Costa Rican farmers take PSA payments into consideration when judging whether to clear forest.

Costa Rica's program to pay for ecosystem services has garnered international praise and has inspired other nations to implement similar programs. But have Costa Rica's payments actually been effective in preventing forest loss?

A number of research teams have sought to answer this surprisingly difficult question by analyzing data from the early years of the PSA program. The answers matter; for if we can draw lessons about what works and doesn't work in Costa Rica's program, we can design more effective incentives for forest conservation worldwide to sequester carbon and combat global climate change (pp. 314, 511).

Some early studies were quick to credit the PSA program for saving forests. A 2006 study conducted for FONAFIFO, the agency administering the program, concluded that PSA payments in the central region of the country had prevented 108,000 ha (267,000 acres) of deforestation—38% of the area under contract. Indeed, deforestation rates fell as the program proceeded; rates of forest clearance in 1997–2000 were half what they were in the preceding decade.

However, some researchers hypothesized that PSA payments were not responsible for this decline and that deforestation would have waned anyway because of other factors. To test this hypothesis, a team led by G. Arturo Sanchez-Azofeifa of the University of Alberta and Alexander Pfaff of Duke University worked with FONAFIFO's data on PSA payments, as well as data on land use and forest cover from satellite surveys of Costa Rica. They layered these data onto maps using a geographic information system (GIS) (p. 115), then explored the patterns revealed.

In 2007 in the journal *Conservation Biology*, they reported that only 7.7% of PSA contracts were located within 1 km of regions where forest was at greatest risk of clearance. PSA contracts were only slightly more likely to be near such a region than far from it. This meant, they argued, that PSA contracts were not being targeted to regions where they could have the most impact.

Moreover, since enrollment was voluntary, most landowners applying for payments likely had land unprofitable for agriculture and were not planning to clear forest in the first place (**FIGURE 1**). In a 2008 paper, these researchers compared lands under PSA contracts with similar lands not under contracts. The deforestation rate on non-PSA lands was 0.21%/yr, whereas PSA lands experienced no forest loss. However, their analyses indicated that the PSA lands stood only a 0.08%/yr likelihood of being cleared, suggesting that the program prevented only 0.08%/yr of forest loss, not 0.21%/yr.

Other research was bearing this out; at least two studies had found that many PSA participants, when interviewed, said they would have retained their forest even without the PSA program.

These researchers argued that the enhanced protection of forest during these years was most likely due to other factors. In particular, Forest Law 7575, besides establishing the PSA system, had banned forest clearing nationwide. This top-down government mandate, assuming it was enforceable, in theory made the PSA payments unnecessary. However, the PSA program made the mandate far more palatable to legislators, and most researchers feel that Forest Law 7575 might never have passed had it not included the PSA program.

Despite the PSA program's questionable impact in preserving existing forest, scientific studies show that it has been effective in regenerating new forest. In Costa Rica's Osa Peninsula,

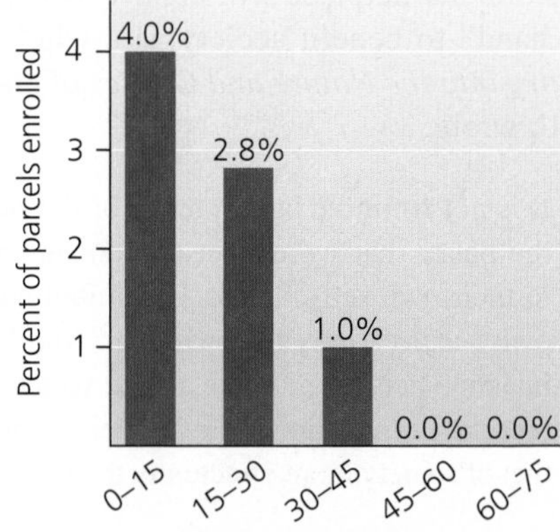

FIGURE 1 In areas at greater risk of deforestation, lower percentages of land parcels were enrolled in the PSA program. This is because land more profitable for agriculture was less often enrolled. *Data: Pfaff, A., et al. 2008.* Payments for environmental services: Empirical analysis for Costa Rica. *Working Papers Series SAN08-05, Terry Sanford Institute of Public Policy, Duke University.*

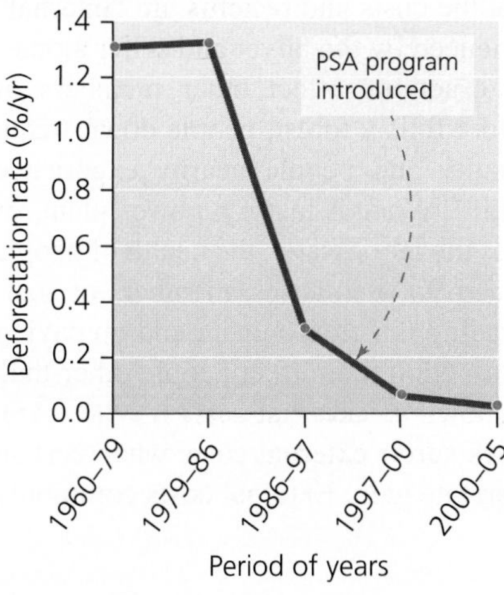

(a) Deforestation decreased

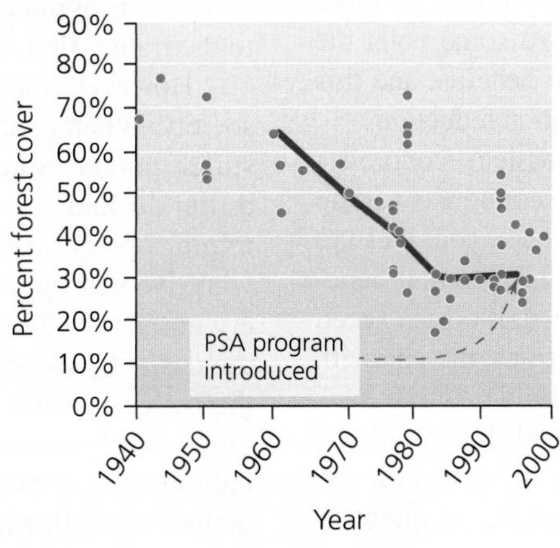

(b) Forest cover began increasing

FIGURE 2 Forest recovery was underway in Costa Rica before the PSA program began. Deforestation rates **(a)** had already dropped steeply, indicating that other factors were responsible. Forest cover **(b)** also began increasing shortly before the program's initiation (the graph gathers together estimates from many studies). *Data: From Daniels, Amy E., et al. 2010. Understanding the impacts of Costa Rica's PES: Are we asking the right questions?* Ecological Economics *69: 2116–2126.*

Rodrigo Sierra and Eric Russman of the University of Texas at Austin found in 2006 that PSA farms had five times more regrowing forest than did non-PSA farms. Interviews with farmers indicated that the program was encouraging them to let land grow back into forest if they did not need it for immediate production.

In the nation's northern Caribbean plain, a team led by Wayde Morse of the University of Idaho used a diversity of methods and analyses, combining on-the-ground interviews with satellite data, to produce the most thorough study of a region yet. Their published summary in 2009 found that PSA payments plus the clearance ban had helped to reduce deforestation rates from 1.43%/year to 0.10%/year and that the program encouraged forest regrowth still more. Meanwhile, dissertation work by Rodrigo Arriagada indicated that the regeneration of new forest seemed to be the PSA program's major effect at the national level as well.

Most researchers today hold that Costa Rica's success in forest protection results from a long history of conservation policies and economic developments. Indeed, deforestation rates had been dropping steeply before the PSA program was initiated (**FIGURE 2**). Major reasons include the following:

- The PSA program was preceded by other policies that encouraged forest cover (tax rebates and tax credits for timber production, not ecosystem services).
- The creation of national parks fed a boom in ecotourism, so Costa Ricans were seeing how conserving natural areas could bring economic benefits.
- Changing market prices for commodities discouraged ranching as the price of meat fell.
- After an economic crisis roiled Latin America in the 1980s, Costa Rica's government ended subsidies for ranchers and farmers that had encouraged expansion into forested areas.

To help the PSA program make better use of its money, most researchers today feel that PSA payments should be targeted. Instead of paying equal amounts to anyone who applies, FONAFIFO should prioritize applicants, or pay more money, in regions that are ecologically most valuable or that are at greatest risk of deforestation.

In a 2008 paper in the journal *Ecological Economics*, Tobias Wünscher of Bonn, Germany, and colleagues modeled and tested seven possible ways to target the payments, using data from Costa Rica's Nicoya Peninsula. They found that all seven approaches would improve on the existing program. The best approach paid some landowners more than others while selecting sites based on the perceived value of their ecological services and the apparent risk of deforestation. This approach more than doubled the economic efficiency of the program.

To determine how much money to offer each landowner, Wünscher's team suggested using auctions, having each applicant for PSA funds put in a bid stating how much they were requesting. Because applicants have outnumbered available contracts 3 to 1, FONAFIFO could favor the lower bids to keep costs low, while the auction system could make differential payments politically acceptable.

The Costa Rican government is responding to suggestions from researchers. It has begun aiming PSA payments toward regions of greater environmental value. To help fight poverty, it has tried to make the program more accessible to small low-income farmers in undeveloped regions. In 2006, the government also raised the payment amounts considerably. Researchers—and other nations—will be watching closely to see how the program develops. ■

equals demand. By applying similar reasoning to environmental issues, economists can determine "optimal" levels of resource use or pollution control (**FIGURE 6.10b**). For instance, reducing pollution emitted by a car or factory is often cost-effective at first, but as pollution is reduced, it becomes more and more costly to eliminate each remaining amount. At some point the cost per unit of reduction rises to match the benefits, and this break-even point is the optimal level of pollution reduction.

To evaluate an action or decision, neoclassical economists use **cost-benefit analysis.** In this approach, economists add up the estimated costs of a proposed action and compare these to the sum of benefits estimated to result from the action. If benefits exceed costs, the action should be pursued; if costs exceed benefits, it should not. Given a choice of actions, the one with the greatest excess of benefits over costs should be chosen.

This reasoning seems eminently logical, but problems arise when not all costs and benefits can be easily identified, defined, or quantified. It may be simple to quantify the value of bananas grown or cattle raised on a tract of land cleared for agriculture, yet difficult to assign monetary value to the complex ecological costs of clearing the forest. Because monetary benefits are usually more easily quantified than environmental costs, benefits tend to be overrepresented in traditional cost-benefit analyses. As a result, environmental advocates often feel such analyses are predisposed toward economic development and against environmental protection.

Neoclassical economics has environmental consequences

Today's market systems operate largely in accord with the principles of neoclassical economics. These systems have generated unprecedented material wealth for our societies. Alas, four fundamental assumptions of neoclassical economics often contribute to environmental degradation.

Are resources infinite or substitutable? One assumption is that natural resources and human resources (such as workers) are either infinite or largely substitutable and interchangeable. This implies that once we have depleted a resource, we will always be able to find a replacement for it. As a result, ecosystem goods and services are treated as free gifts of nature, endlessly abundant and resilient. The market imposes no penalties for depleting them, as they are assumed to be ultimately substitutable by human technological ingenuity.

Certainly it is true that many resources can be replaced. Our societies have transitioned from manual labor to animal labor to steam-driven power to fossil fuel power, and they may yet transition to renewable power sources such as wind and solar energy. However, Earth's material resources are ultimately limited. Nonrenewable resources (such as fossil fuels) can be depleted. Many renewable resources (such as soils, fish stocks, and forest products) can be used up if we exploit them faster than they are replenished. And even seemingly inexhaustible resources can become so polluted that we can no longer use them (such as contaminated water supplies).

Are costs and benefits internal? A second assumption of neoclassical economics is that all costs and benefits associated with an exchange of goods or services are borne by individuals engaging directly in the transaction. In other words, it is assumed that the costs and benefits are "internal" to the transaction, experienced by the buyer and seller alone.

However, many transactions affect other members of society. When a landholder fells a forest, people downstream suffer poorer water quality and people nearby experience dirtier air and less wildlife. When a factory, power plant, or mining operation pollutes the air or water, the health of people who live nearby is harmed. In such cases, members of society not involved in degrading the environment end up paying the costs. Costs of a transaction that affect people other than the buyer or seller are known as **external costs** (**FIGURE 6.11**). Often whole communities suffer external costs when certain individuals experience private gain. External costs commonly include the following:

- Health problems, stress, or anxiety experienced by people downstream or downwind from a pollution source
- Property damage
- Declines in desirable features of the environment, such as fewer fish in a stream
- Aesthetic damage, such as from air pollution, clear-cutting, or strip-mining
- Declining real estate values, lost tourism revenue, higher healthcare expenses, etc., resulting from these problems

If market prices do not take the social, environmental, or economic costs of environmental degradation into account, then taxpayers bear the burden of paying them. When economies ignore external costs, this creates a false impression of the consequences of our choices, and people continue to be unjustly subjected to the impacts of activities in which they did not participate. External costs are one reason that governments develop environmental policy (pp. 165, 168).

FIGURE 6.11 People commonly suffer external costs from air pollution. Here, residents of Sampit, Indonesia, cycle through smoke from fires set to clear forest for oil palm plantations.

Should we discount the future? Third, neoclassical economics grants an event in the future less value than one in the present. In economic terminology, future effects are *discounted*. **Discounting** is meant to reflect how human beings tend to grant more importance to present conditions than to future conditions. Just as you might rather have an ice cream cone today than be promised one next month, market demand is greater for goods and services that are received sooner. Economists quantify this importance by assigning discount rates when calculating costs and benefits. For example, applying a 5% annual discount rate to forestry decisions means that a stand of trees whose timber is worth $500,000 on the market today would drop in perceived value by 5% each year. From the perspective of today's market, having the timber in 10 years would be worth only $299,368. By this logic, the more quickly the trees are cut, the more they are worth.

Discounting encourages policymakers to play down the long-term consequences of decisions. Many environmental problems unfold gradually, yet discounting discourages us from addressing resource depletion, pollution buildup, and other cumulative impacts. Instead, discounting shunts the costs of dealing with such problems onto future generations.

The choice of what discount rate to use is subjective and involves ethical judgment. This has become apparent as economists debate how discount rates influence our estimates of the costs of global climate change to society (see THE SCIENCE BEHIND THE STORY, pp. 148–149).

Is all growth good? **Economic growth** can be defined as an increase in an economy's production and consumption of goods and services. Neoclassical economics assumes that economic growth is essential for maintaining social order, because a growing economy can alleviate the discontent of the poor by creating opportunities for poor people to become wealthier. A rising tide raises all boats, as the saying goes; if we make the overall economic pie larger, then each person's slice can become larger, even if some people still have much smaller slices than others.

To the extent that economic growth is a means to an end—a path to greater human well-being—it is a good thing. However, when growth becomes an end in itself it may no longer be the best route toward well-being. Sociologists have coined a word for the way that consumption and material affluence often fail to bring people contentment: **affluenza.** Moreover, people in poverty may feel poorer and less happy as the gap between rich and poor widens, even if they gain more wealth themselves. Critics of the growth paradigm fear that an endless pursuit of growth cannot be sustained and will eventually destroy our economic system, because resources to support growth are ultimately limited.

How sustainable is economic growth?

Our global economy is seven times the size it was just half a century ago. All measures of economic activity—trade, rates of production, amount and value of goods manufactured—are higher than ever before. This has brought many people much greater material wealth (although not equitably, and gaps between rich and poor are wide and growing).

Economic growth stems from two sources: (1) an increase in inputs to the economy (such as greater inputs of labor and natural resources) and (2) improvements in the efficiency of production due to better technologies and approaches (that is, ideas and equipment that enable us to produce more goods with fewer inputs). This second approach—whereby we produce more with less—is often termed **economic development.**

As our population and consumption rise, it is becoming clearer that we cannot sustain growth forever using the first approach. Nonrenewable resources on Earth are finite in quantity, and there are limits on the rates that we can harvest renewable resources. As for the second approach, we have used technological innovation to push back the limits on growth time and again. More-efficient technologies for extracting minerals, fossil fuels, and groundwater allow us to exploit these resources more fully with less waste. Better machinery in our factories speeds our manufacturing. We continue to make computer chips more powerful while also making them smaller (FIGURE 6.12). In these ways, we are producing more goods and services while using relatively fewer resources.

Can we conclude, then, that human ingenuity and improvements in technology will allow us to overcome all our environmental limitations and continue economic growth indefinitely? Answering "yes" are people sometimes referred to as **Cornucopians.** (In Greek mythology, a *cornucopia*—literally "horn of plenty"—is a magical goat's horn that overflowed with grain, fruit, and flowers.) Responding "no" are people called **Cassandras,** named after the mythical princess of Troy with the gift of prophecy, whose dire predictions were not believed.

The Cassandra view has been articulated most famously by a group of researchers who published a series of books including *The Limits to Growth* (1972), *Beyond the Limits* (1992), and *Limits to Growth: The Thirty-year Update* (2004). Starting from the premise that Earth's natural capital is finite, and using calculations of resource availability and consumption, they

FIGURE 6.12 **The miniaturization of computer chips is a striking advance in efficiency.** It allows us to store and use far more information with far less input of raw materials.

Ethics in Economics: Discounting and Global Climate Change

How much will it cost our society if we do nothing about global climate change? To address this question and to help decide how to respond to climate change, the British government commissioned economist Nicholas Stern to lead a team to assess the economic costs that a changing climate may impose on society. But once Stern's report was published in 2006, a debate ensued that had as much to do with ethics as with economics. The dispute centered on discounting (p. 147), an economic practice heavy with ethical implications.

To produce the *Stern Review on the Economics of Climate Change*, Stern's research team surveyed the burgeoning scientific literature on the impacts of rising temperatures, changing rainfall patterns, rising sea level, and increasing storminess (see Chapter 18). It then estimated the economic consequences of these climatic changes and tried to put a price tag on the global cost. The report concluded that without action to forestall it, climate change would cause losses in annual global gross domestic product (GDP) of 5–20% by the year 2200 (**FIGURE 1**).

Sir Nicholas Stern, head of the Government Economic Service, United Kingdom

Stern's team also calculated that by paying just 1% of GDP annually starting now, our society could stabilize atmospheric greenhouse gas concentrations and prevent most of these future economic losses. The bottom line: Spending a relatively small amount of money now will save us much larger expenses in the future. This conclusion caught the attention of governments worldwide; for the first time, leading economists were advancing a strong economic argument for tackling climate change immediately.

The *Stern Review*'s conclusions depend partly on how one chooses to weigh future impacts versus current impacts. Stern used two discount factors. One accounted for the likelihood that people in the future will be richer than we are today and thus better able to handle economic costs. The other considered whether the future should be discounted simply because it is the future. This latter discount factor (called a *pure time discount factor*) is essentially an ethical issue because it assigns explicit values to the welfare of future versus current generations.

The *Stern Review* used a pure time discount rate of 0.1%. This means that an impact occurring next year is judged 99.9% as important as one occurring this year. It means that the welfare of a person born 100 years from now is valued at 90% of one born today. This discount rate treats current and future generations *nearly* equally. Future generations are down-weighted only because of the (very small) possibility that our species could go extinct (in which case, there would be no future generations to be concerned about).

Some economists viewed this discount rate as too low. Yale University economist William Nordhaus proposed a discount rate starting at 3% and falling to 1% in 300 years. He maintained that such numbers are more objective because they reflect how people value things, as revealed by prices we pay in the marketplace. Indeed, when economists assess capital investments or construction projects (say, development of a railroad, dam, or highway) they

---

ran simulation models to predict how our economies would fare in the future. Model runs that used data matching our society's current consumption patterns consistently predicted economic collapse as resources become scarce (**FIGURE 6.13a**). The team experimented to see what parameters would produce model runs predicting a sustainable civilization (**FIGURE 6.13b**). They used these results to make recommendations for achieving sustainability.

Cornucopians, such as the economist Julian Simon and the statistician Bjorn Lomborg, counter that Cassandras underestimate the human capacity to innovate and the degree to which technologies can increase our access to resources. They also argue that market forces help us avoid resource depletion. As resources become scarce, prices rise, and individuals and firms gain incentive to turn to different resources, shift to different products, or reuse and recycle. All of these actions will alleviate pressure on the resource. As Simon put it:

> The natural world allows, and the developed world promotes through the marketplace, responses to human needs and shortages. . . . The main fuel to speed our progress is our stock of knowledge, and the brake is our lack of imagination. The ultimate resource is people—skilled, spirited, and hopeful people who will exert their wills and imaginations for their own benefit, and so, inevitably, for the benefit of us all.

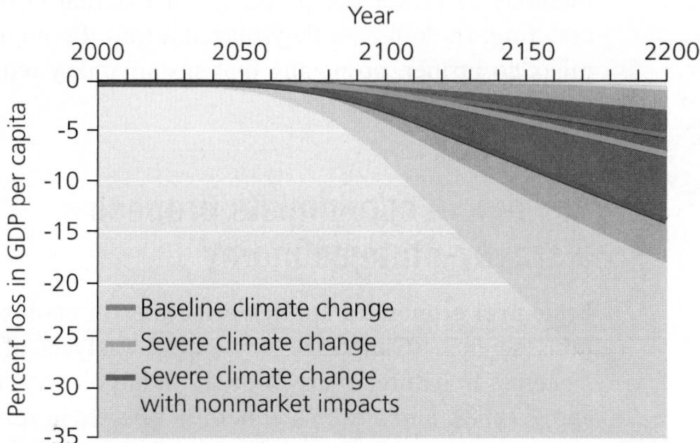

FIGURE 1 Baseline climate change forecast by the IPCC could decrease global per capita GDP by 5.3% annually by the year 2200. Severe climate change could bring annual losses of 7.3%—and adding nonmarket values (pp. 150–151) raises this figure to 13.8%. Gray-shaded areas show ranges of future values judged statistically to be 95% likely; darkest gray indicates where ranges overlap for all three lines, and lightest gray for just one line. *Data: HM Treasury, 2007.* Stern review on the economics of climate change. *London, U.K.*

typically choose discount rates close to those Nordhaus suggests. Nordhaus argued that the *Stern Review*'s near-zero discount rate overweights the future, forcing people today to pay too much to address hypothetical future impacts.

In their response to Nordhaus and other critics, Stern and his team argued that discount rates of 3% or 1% may be useful for assessing development projects but are too high for long-term environmental problems that directly affect human well-being. A 3% rate means that a person born in 1995 is valued only half as much as a person born in 1970. It means a grandchild is judged to be worth far less than a grandparent simply because of the dates they were born. For various reasons, Stern argued, the market should not be used to guide ethical decisions.

Stern's group also published sensitivity analyses that examined how their conclusions would vary under different discount rates. These showed that the report's main message—that it's cheaper to deal with climate change now than later—was robust across all discount rates up to at least 1.5%.

At the end of the day, the choice of a discount rate is an ethical decision on which well-intentioned people may differ. As governments, businesses, and individuals begin to invest in addressing climate change, the debate over the *Stern Review* reveals how ethics and economics remain intertwined.

Addressing global climate change is becoming urgent, as the climate is changing faster than scientists had predicted just a few years ago. Stern announced that newer scientific forecasts of accelerated change mean that it is becoming even more expensive to mitigate global warming than the *Stern Review* had calculated. Instead of 1% of GDP annually, the world would need to spend 2% of GDP each year in order to control climate change, Stern said—and that was in 2008.

Since then, Hurricane Sandy, the drought of 2012, and other climate-related disasters have made clear that the costs of inaction will be immense. These developments are all helping drive home the same message: The more quickly we get going, the better off we'll be. ●

What accounts for this divergence in views between Cornucopians and Cassandras? It may be because we are living in a unique time of transition. Throughout our long prehistory and most of our history, we have lived in a world with low numbers of people. In such an "empty world," we could always rely on being able to exploit more resources. Today, however, the human population is so large that it is beginning to strain Earth's systems and deplete its resources. In this new "full world," we are encountering limits. Because we are still learning what those limits are, people have a wide diversity of viewpoints.

As we will see with many issues, both Cornucopians and Cassandras make valid points. Cassandras have often underestimated our ability to innovate and adapt—yet ultimately, all nonrenewable resources are finite, and renewable resources can be exploited only at limited rates. If our population and consumption continue to grow and we do not shift to full reuse and recycling, we will continue depleting our natural capital, putting greater and greater demands on our capacity to innovate.

WEIGHING THE ISSUES

CORNUCOPIAN OR CASSANDRA? Would you consider yourself more of a Cornucopian or a Cassandra? Which aspect(s) of each point of view do you share? Do you think our society would be better off if everyone were a Cornucopian or if everyone were a Cassandra—or do we need both?

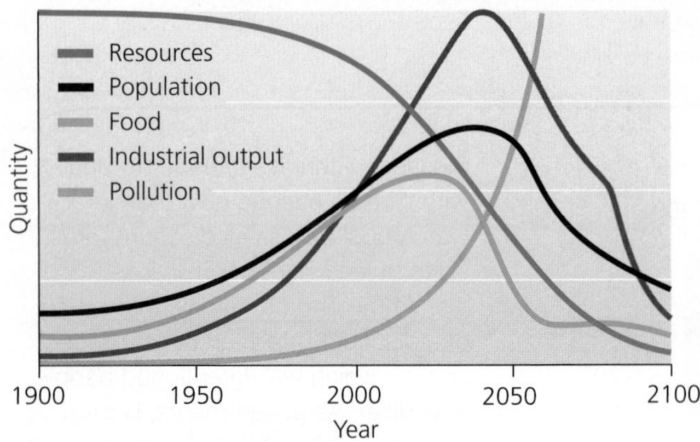

(a) Projection based on status quo policies

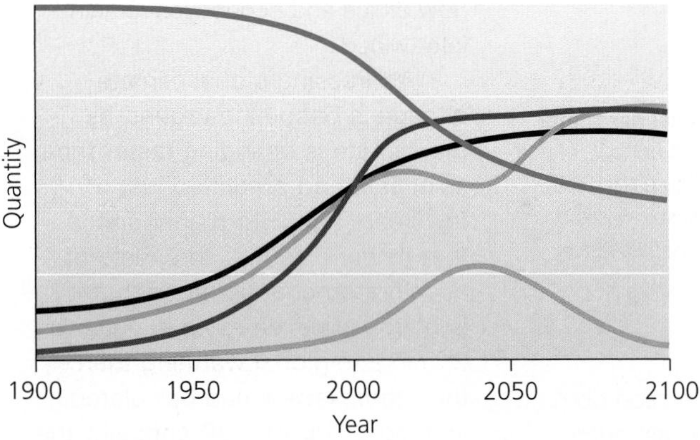

(b) Projection based on policies for sustainability

FIGURE 6.13 **Environmental scientists have used data to project future trends in human population, resource availability, food production, industrial output, and pollution.** Shown in **(a)** is a projection for a world in which society pursues policies typical of the 20th century, but allowing for twice as many resources to be discovered and extracted as are now accessible. In this projection, population and production rise until declining resources cause them to fall suddenly while pollution continues to rise. In contrast, **(b)** shows a scenario with policies aimed at sustainability. In this projection, population levels off below 8 billion, production and resource availability stabilize at medium-high levels, and pollution declines to low levels. *Data from Meadows, D., et al., 2004.* Limits to growth: The 30-year update. *White River Junction, VT: Chelsea Green Publishing. Used by permission.*

DATA Q Suppose the researchers showed you a third projection, one that simulated a world in which we used resources even faster and more intensively than we do today. Draw a graph showing how you predict each of the five trend lines in part (a) would change.

Environmental economists devise strategies for sustainability

The Cornucopian view has held considerable sway over mainstream economists, but economists in the newer field of **environmental economics** maintain that economic growth may be unsustainable if we do not reduce our demand for resources and make resource use far more efficient. Most environmental economists feel that, with effort, we can modify the principles of neoclassical economics to address environmental challenges, and thereby attain sustainability within our current economic systems.

Environmental economists were the first to develop methods to tackle the problems of external costs and discounting. In doing so, they blazed a trail. Ecological economists go further, proposing that sustainability requires more far-reaching changes.

Ecological economists propose a steady-state economy

Ecological economics applies principles of ecology and systems science (Chapters 3–5) to the analysis of economic systems. In nature, every population has a carrying capacity (pp. 67–68), and systems generally operate in self-renewing cycles, not in a linear or progressive manner. Ecological economists maintain that human societies, like natural populations, cannot permanently surpass their environmental limitations and that we should not expect endless economic growth. Many of these economists advocate economies that neither grow nor shrink, but rather are stable. Such **steady-state economies** are intended to mirror natural ecological systems.

To achieve a steady-state economy, proponents such as economist **Herman Daly** believe we will need to rethink our assumptions and fundamentally change the way we conduct economic transactions. Critics of a steady-state economy assert that halting growth will dampen our quality of life. Ecological economists respond that technological advances will continue, behavioral changes (such as greater use of recycling) will enhance sustainability, and wealth and happiness can continue to rise after economic growth has leveled off.

Attaining sustainability will certainly require the reforms pioneered by environmental economists and may require the fundamental shifts in thinking, values, and behavior advocated by ecological economists. These economists are actively developing strategies for sustainability, and we will now survey a few that they offer.

We can assign monetary value to ecosystem goods and services

As Costa Ricans know, ecosystems provide us essential resources and life-support services, including arable soil, waste treatment, clean water, and clean air. Yet we often abuse the very ecological systems that sustain us. Why is this? From the economist's perspective, people overexploit natural resources and systems largely because the market assigns these entities no quantitative monetary value, or assigns values that underestimate their true worth.

Ecosystem services are said to have **nonmarket values,** values not usually included in the price of a good or service (FIGURE 6.14). For example, the aesthetic and recreational pleasure we obtain from natural landscapes is something of real value. Yet because we do not generally pay money for this, its value is hard to quantify and appears in no traditional measures of economic worth. Or consider Earth's water cycle (pp. 120–121), by which rain fills our reservoirs with drinking water, rivers give us hydropower and flush away our waste, and water evaporates, purifying itself of contaminants and later falling as rain. This natural cycle is vital to our very existence, yet because we do not pay for it, markets impose no financial penalties when we disturb it.

In Costa Rica and elsewhere, environmental and ecological economists have sought ways to assign market values to ecosystem services. One technique, **contingent valuation,** uses surveys to determine how much people are willing to pay to protect or restore a resource. Such an approach was used to assess a proposal to develop a mine near world-famous Kakadu National Park in Australia in the 1990s. To determine how much the land was "worth" economically if preserved undeveloped, researchers interviewed 2000 citizens and asked how much they would be willing to pay to prevent mine development. On average, respondents said their households would pay \$80 to \$143 per year to prevent the predicted impacts. Multiplying these figures by the number of households in Australia, the researchers found that preservation was "worth" \$435 to \$777 million annually to Australia's population. These numbers exceeded the \$102 million in annual economic benefits expected from mine development, so the researchers concluded that it was best to preserve the land undeveloped.

Because contingent valuation relies on survey questions and not actual expenditures, critics point out that people may volunteer idealistic (inflated) values, knowing that they will not actually have to pay the price they name. As a result, many researchers prefer to use methods that measure people's preferences as revealed by data on actual behavior. For example, to gauge how much people value parks, researchers may calculate the amount of money, time, or effort people expend to travel to parks. Economists may compare housing prices for similar homes in different settings to infer the dollar value of landscapes, views, or peace and quiet. They may also assign environmental amenities value by calculating the cost required to restore natural systems that have been damaged, to replace their functions with technology, or to mitigate harm from pollution.

(a) Use value: The worth of something we use directly

(b) Existence value: The worth of knowing that something exists, even if we never experience it ourselves

FIGURE 6.14 Accounting for nonmarket values such as those shown here may help us make better environmental and economic decisions.

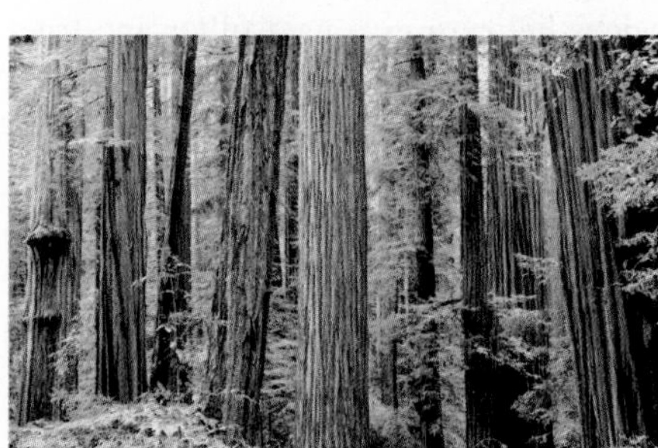

(c) Option value: The worth of something we might use later

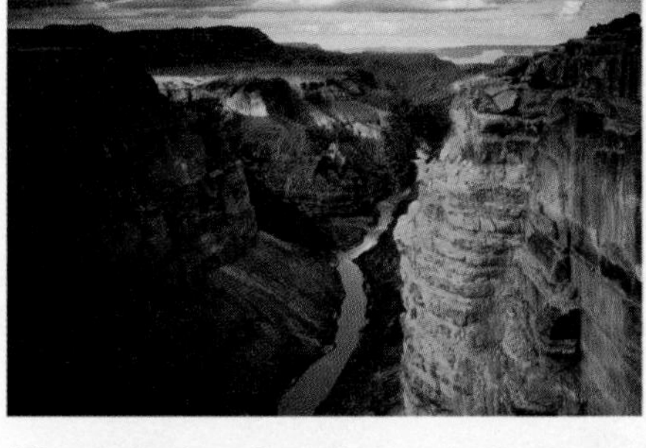

(d) Aesthetic value: The worth of something's beauty or emotional appeal

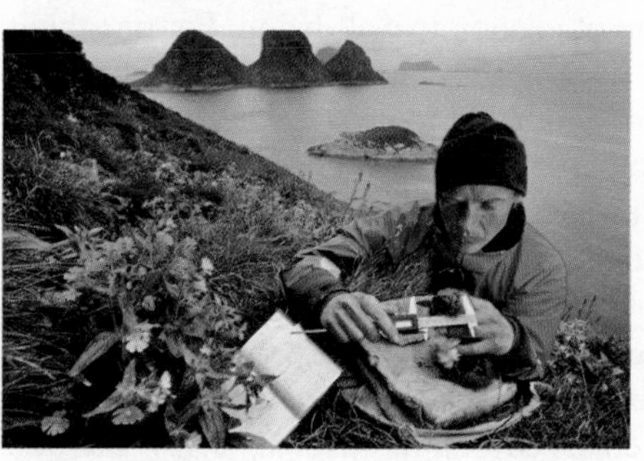

(e) Scientific value: The worth of something for research

(f) Educational value: The worth of something for teaching and learning

(g) Cultural value: The worth of something that sustains or helps define a culture

One 2004 study in Costa Rica showed how the value of particular services could be precisely measured. Taylor Ricketts of Stanford University, working with Gretchen Daily and others, studied native bees at a large coffee plantation. By carefully measuring bee pollination (p. 80, 254) and resulting coffee production in areas near forest and in areas far from forest, Ricketts calculated that forests were providing the farm with pollination services worth $60,000 per year. The study was persuasive and gained international attention, yet short-term economic incentives led the plantation owners to disregard its lessons. When the price of coffee dipped, the owners converted the land to pineapple, which does not need pollinators.

In 1997 a research team headed by environmental economist Robert Costanza set out to calculate the total economic value of all the services that oceans, forests, wetlands, and other systems provide across the world. Costanza's team combed the scientific literature and evaluated over 100 studies that used various valuation methods to estimate dollar values for 17 major ecosystem services, such as water purification, climate regulation, plant pollination, and pollution cleanup (**FIGURE 6.15**). To improve the accuracy of estimates, the researchers reevaluated the data using multiple valuation techniques. They then multiplied average estimates for each ecosystem by the global area it occupied. Their analysis, reported in the journal *Nature*, calculated that Earth's biosphere in total provides at least $33 trillion ($48 trillion in 2013 due to inflation) worth of ecosystem services each year—more than the GDP of all nations combined!

This research sparked excitement, but also debate. Some ethicists argued that we should not put dollar figures on services such as clean air and water, because they are priceless and we would perish without them. Others said that arguing for conservation purely on economic grounds risked not being able to justify it whenever it failed to deliver clear economic benefits.

In 2002, Costanza joined Andrew Balmford and 17 other colleagues to compare the benefits and costs of preserving natural systems intact versus converting wild lands for agriculture, logging, or fish farming. After reviewing many studies, they reported in the journal *Science* that a global network of nature reserves covering 15% of Earth's land surface and 30% of the ocean would be worth $4.4 to $5.2 trillion. This amount is 100 times greater than the amount those areas would be worth if they were converted for direct exploitative human use. This demonstrates, in their words, that "conservation in reserves represents a strikingly good bargain."

In 2010, researchers wrapped up a large U.N.-sponsored international effort to summarize and assess attempts to quantify the economic value of natural systems. *The Economics of Ecosystems and Biodiversity* study has published a number of fascinating reports that you can download online. Regarding measuring, or valuation, of nature's economic worth, this effort concludes:

> Valuation is seen not as a panacea, but rather as a tool to help recalibrate the faulty economic compass that has led us to decisions that are prejudicial to both current well-being and that of future generations. The invisibility of biodiversity values has often encouraged inefficient use or even destruction of the natural capital that is the foundation of our economies.

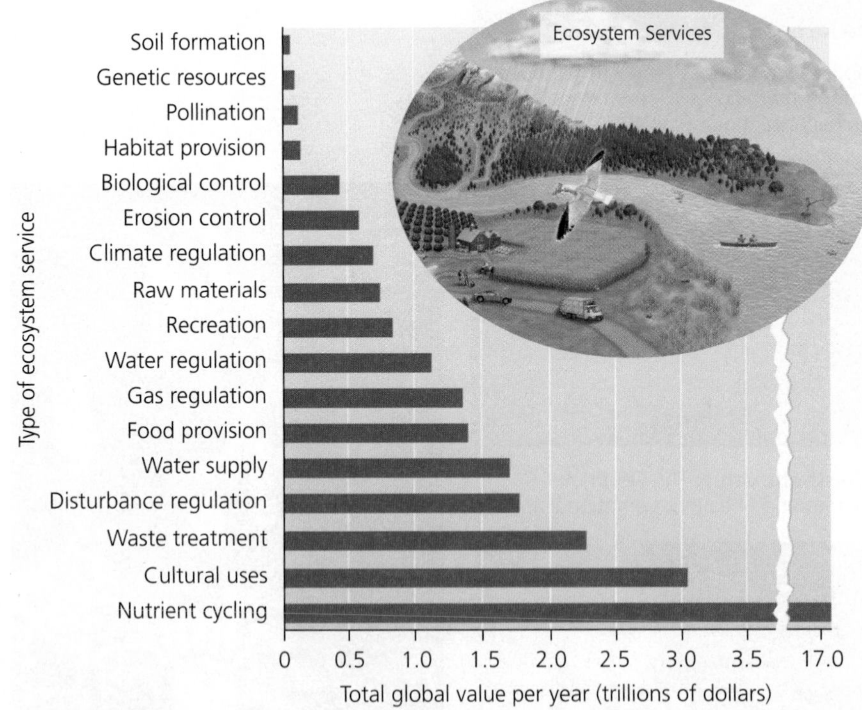

FIGURE 6.15 Environmental economists in 1997 estimated the value of the world's ecosystem services at more than $33 trillion ($48 trillion in 2013 dollars). Shown are subtotals for each ecosystem service. This amount is an underestimate because it does not include ecosystems for which adequate data were unavailable. *Data from Costanza, R., et al., 1997. The value of the world's ecosystem services and natural capital.* Nature *387: 253–260.*

We can measure progress with full cost accounting

If assigning market values to ecosystem services helps to give us a fuller and truer picture of costs and benefits, then we can take a similar approach to measuring the economic progress we make as a society. For decades, policymakers and the public have assessed each nation's economy by calculating its **Gross Domestic Product (GDP),** the total monetary value of final goods and services the nation produces each year. Governments regularly use GDP to make policy decisions that affect billions of people. However, GDP is a poor measure of economic well-being. For one thing, it does not account for nonmarket values. For another, it lumps together all economic activity, both desirable and undesirable. Thus, GDP can rise in response to economic activities that harm society.

For example, crime boosts GDP because crime forces people to invest in security measures and to replace stolen items. Oil spills (**FIGURE 6.16**) increase GDP because they require cleanups, which cost money and increase the production

FIGURE 6.16 **The *Exxon Valdez* oil spill, which devastated Alaskan beaches and fisheries in 1989, actually increased the nation's GDP.** The many monetary transactions resulting from cleanup efforts mean that pollution increases GDP—one of many reasons GDP is not a reliable indicator of well-being.

of goods and services. Natural disasters such as Hurricane Sandy, Hurricane Katrina, or Japan's 2011 earthquake and tsunami boost GDP becau... emergency measures and clean... War can augment GDP because ... involved in manufacturing weapon... Pollution causes GDP to increase wh... stance is manufactured and again when so... up the pollution.

Some economists have tried to develop i... differentiate between desirable and undesirable ... activity and that can function as better guides to ... well-being. One such alternative to the GDP is the ...ine **Progress Indicator (GPI).** To calculate GPI, economists begin with conventional economic activity and add to it positive contributions that are not paid for with money, such as volunteer work and parenting. They then subtract negative impacts, such as crime and pollution (FIGURE 6.17a). The resulting GPI thereby considers more activities than does the GDP, and it includes only those that enhance societal well-being.

The two indices can differ strikingly: FIGURE 6.17b compares per capita GPI and per capita GDP for the United States from 1950 to 2004. On a per-person basis, the nation's GDP rose greatly, but its GPI remained flat for 30 years. This difference suggests that U.S. citizens are spending more and more money but that their quality of life is not getting better.

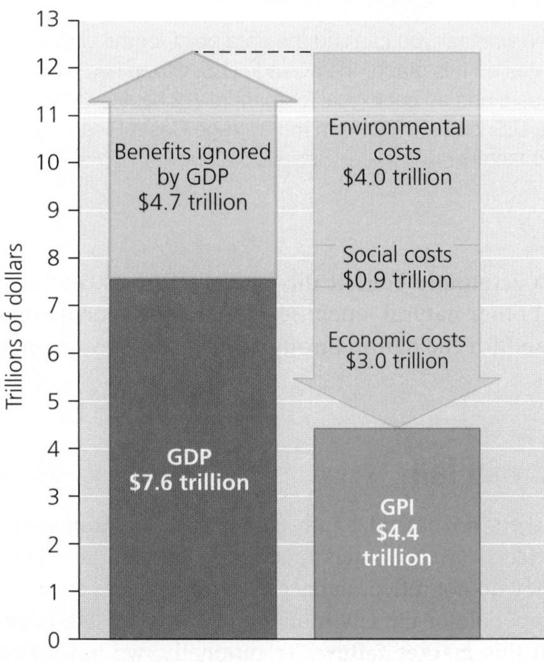

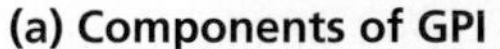

(a) Components of GPI

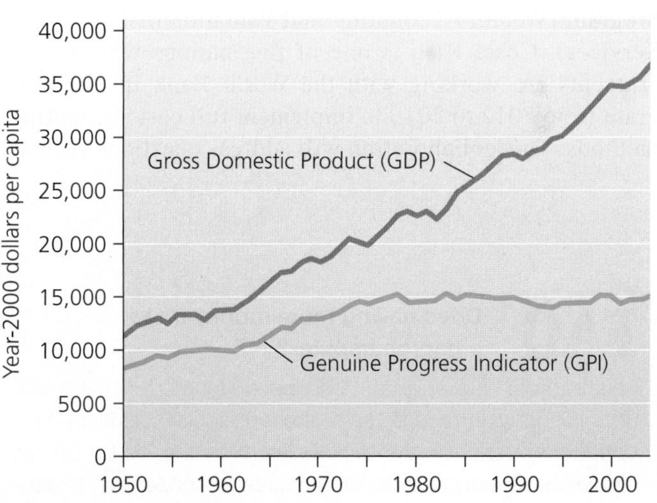

(b) Change in U.S. GDP vs. GPI

FIGURE 6.17 **Full cost accounting indicators such as the GPI attempt to measure progress and well-being more effectively than pure economic indicators such as the GDP.** We see in **(a)** how the Genuine Progress Indicator (**orange bar at right**) adds to the Gross Domestic Product (**red bar at left**) benefits such as volunteering and parenting (**upward-pointing gold arrow**). The GPI then subtracts external environmental costs such as pollution, social costs such as divorce and crime, and economic costs such as borrowing and the gap between rich and poor (**downward-pointing gold arrow**). Shown are values for the United States in 2004. We see in **(b)** that per capita U.S. GDP has increased dramatically since 1950, yet per capita U.S. GPI leveled off after 1975. GPI advocates say this indicates that we are spending more money but that our lives are not much better. *Data from Talberth, J., C. Cobb, and N. Slattery, 2007.* The Genuine Progress Indicator 2006: A tool for sustainable development. *Redefining Progress, Oakland, CA. By permission of John Talberth, Ph.D. All data are adjusted for inflation by using year-2000 dollars.*

...ch as the GPI that attempts to monetize ...or all costs and benefits is an example of **full ...ccounting,** also called **true cost accounting.** A variety ...f full cost accounting indicators have been devised. The Index of Sustainable Economic Welfare (ISEW) gave rise to the GPI. Net Economic Welfare (NEW) adjusts GDP by adding the value of leisure time and personal transactions while deducting costs of environmental degradation. The Human Development Index assesses standard of living, life expectancy, and education.

Critics of full cost accounting complain that the approach is subjective and too easily driven by ideology. Proponents respond that making a subjective attempt to measure progress is better than misapplying a more objective indicator such as the GDP to quantify well-being—something it was never meant to do.

In recent years, attempts to measure and pursue happiness (rather than economic output) as the prime goal of national policy are gaining ground. In 2012 a global conference on this topic was hosted by Bhutan, a small Asian nation that has pioneered this approach with its measure of Gross National Happiness (GNH).

Another recently devised indicator is the Happy Planet Index (HPI), which divides the number of happy years of life for the average citizen of a nation by that nation's ecological footprint. The HPI thus measures how much output (years of happiness) we gain per amount of input (resources used) we invest. By this measure, Costa Rica was calculated to be the top-performing nation in the world (**FIGURE 6.18**).

Today the World Bank (p. 178) is driving an international effort to promote full cost accounting and the valuation of ecosystem services. In 2010 it launched the WAVES program (Wealth Accounting and Valuation of Ecosystem Services). Costa Rica is one of five nations whose governments are working with the World Bank in this program from 2012 to 2015 to implement full cost accounting methods. This collaboration will address questions such as how much economic benefit the nation's forests, national parks, and other natural amenities generate for local communities and for the nation through tourism and watershed protection.

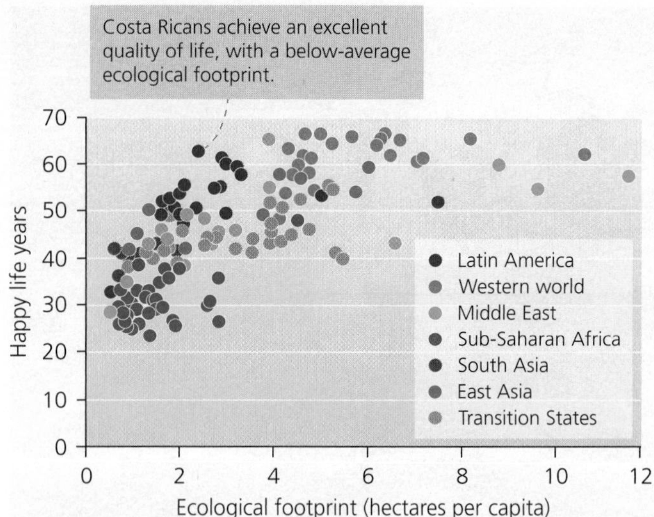

FIGURE 6.18 The Happy Planet Index assesses nations for "sustainable well-being" by measuring the happiness of their citizens together with their prospects for future happiness. The HPI takes the average number of years of happy life a nation's citizens experience and divides that by the nation's per capita ecological footprint. The overall pattern among 151 nations—a curve that rises steeply and then levels off—indicates that increases in resource use at low levels enhance people's well-being greatly, but that once ecological footprints are high, further resource consumption does not buy more happiness. *Data from new economics foundation, 2012.* The Happy Planet Index: 2012 report. *nef, London.*

DATA Q See whether you can find the data point for the United States on this graph; the average U.S. citizen has 61.5 happy life years and an ecological footprint of 7.2 ha. How does the average U.S. citizen differ from the average Costa Rican citizen in these two factors?

FAQ

Does having more money make a person happier?

This age-old question has long been debated in the realm of philosophy. In recent years, though, social scientists have conducted serious scientific research on the issue. So far, studies have found a surprising degree of consensus: In general, we become happier as we get wealthier, but once we gain a certain level of wealth, attaining further wealth no longer increases our happiness. Apparently financial security alleviates normal day-to-day economic worries, but once those worries are taken care of, our degree of happiness revolves around other aspects of our lives. Such research may help us guide our own decisions, but it also suggests that trying to enhance a society's happiness might best be achieved by raising many people's income up by a little, rather than by raising some people's income up by a lot.

Markets can fail

When markets do not take into account the environment's positive effects on economies (such as ecosystem services) or when they do not reflect the negative impacts of economic activity on people or the environment (external costs), economists call this **market failure.** Traditionally, we have tried to counteract market failure with government intervention. Governments can restrain corporate behavior through laws and regulations. They can tax environmentally harmful activities. Or, they can design economic incentives that use market mechanisms to promote fairness, resource conservation, and economic sustainability. Paying for ecosystem services represents one way of deploying economic incentives toward policy goals. We will examine legislation, regulation, green taxation, subsidies, and market incentives such as cap-and-trade permit trading in Chapter 7 in our discussion of environmental policy.

FIGURE 6.19 Ecolabeling gives consumers information on the environmental impact of products, enabling us to promote sustainable business practices through our purchasing decisions. Fair-trade and shade-grown coffee are examples of the many ecolabeled products now widely available.

Ecolabeling helps address market failure

Another way to mitigate market failure is to help consumers gain full and accurate information with which to make purchasing decisions. Increasingly, manufacturers are designating on their labels how their products were grown, harvested, or manufactured. This approach, called **ecolabeling,** serves to inform consumers which brands use environmentally benign processes (**FIGURE 6.19**). By preferentially buying ecolabeled products, each of us as consumers can provide businesses a powerful incentive to switch to more sustainable processes.

One early example of ecolabeling was to label cans of tuna as "dolphin-safe," indicating that the methods used to catch the tuna avoid the accidental capture of dolphins. Other examples include labeling recycled paper (pp. 617–619), organic foods (pp. 257–260), fair trade and shade-grown coffee (pp. 73, 260, 270), and sustainable forestry products (pp. 322–323).

In a similar vein, individuals who invest money in the stock market can pursue **socially responsible investing,** which entails investing in companies that have met criteria for environmental or social sustainability. A fast-growing trend, in 2012 U.S. investors sank $3.7 trillion into socially responsible investing, representing 11% of all investment.

Businesses are responding to sustainability concerns

As more consumers and investors express preferences for sustainable products and services, many industries, businesses, and corporations are "greening" their operations. By finding ways to enhance energy efficiency, reduce toxic substances, increase the use of recycled materials, and minimize greenhouse gas emissions, businesses often discover that they reduce costs and increase profit.

Some companies have cultivated an eco-conscious image from the start, such as Ben & Jerry's (ice cream), Patagonia (outdoor apparel), and Seventh Generation (household products). The phone company CREDO donates a portion of its proceeds to environmental and progressive nonprofit groups according to how its customers vote to distribute the funds. At the local level, entrepreneurs are starting thousands of sustainably oriented businesses all across the world.

Today corporate sustainability has gone mainstream, and some of the world's largest corporations have joined in, including McDonald's, Starbucks, Intel, Ford Motor Company, Toyota, IKEA, Dow, DuPont, and BASF. Carpeting company Interface, Inc., provides one of the best-known cases of a company changing its practices to promote sustainability (p. 623), but there are many more. IBM was judged the greenest large company in America in a survey run by *Newsweek* magazine and two research agencies in 2012. IBM's Smarter Planet initiative provides technology to help businesses and governments develop sustainable approaches, such as smart energy grids; green buildings; and efficient systems to manage traffic congestion, pollution, and water supplies.

Hewlett-Packard runs programs to reuse and recycle used toner cartridges, electronics, and plastics, and Dell reuses or recycles 98% of its nonhazardous waste. Sprint Nextel aims to recycle 90% of the phones it sells by crediting customers who turn in old phones. Nike, Inc., collects millions of used sneakers and recycles their materials to create synthetic surfaces for basketball courts, tennis courts, and running tracks. The Gap, Inc., cut energy use in its stores and distribution centers, runs alternative transportation programs for its employees, and has a sustainably designed headquarters building. Microsoft became carbon-neutral in 2012 by charging each of its divisions with monitoring and offsetting its greenhouse gas emissions.

One prime example of corporate sustainability is the publisher of your textbook, Pearson Education. In 2009, Pearson became the first global media company to become climate-neutral (p. 514), and it has remained that way since. Pearson has eliminated its net carbon emissions by installing solar and wind energy to power its facilities, by purchasing further clean renewable energy through utilities, and by funding forest conservation efforts. Pearson has also reduced its energy use by investing in LEED-certified green buildings (pp. 348–350), fuel-efficient fleet vehicles, videoconferencing in lieu of airplane travel, and other energy-saving strategies. Pearson has reduced its use of plastic and styrofoam by 85% and 50%, respectively, in recent years, and is using low-VOC inks and FSC-certified paper (pp. 307–308, 322). For its efforts, Pearson has been recognized by the Dow Jones Sustainability World Index for over a decade and has scored highest of any media company in its environmental dimension.

Wal-Mart provides the highest-profile example of corporate greening efforts (**FIGURE 6.20**). Advocates of sustainability had long criticized the world's largest retailer for its environmental and social impacts. The company responded by launching a quest to sell organic and sustainable products, reduce packaging and use recycled materials, enhance fuel efficiency in its truck fleet, reduce energy use in its stores, power itself with renewable energy, cut carbon dioxide emissions, and preserve one acre of natural land for every acre developed. It is also developing a "sustainability index" that

FIGURE 6.20 Wal-Mart has embarked on an extensive program to make its operations more sustainable. Here, a Wal-Mart cashier bags compact fluorescent bulbs in reusable canvas bags for a customer.

will rate the products it carries to help inform eco-conscious consumers. Many observers remain skeptical of Wal-Mart's commitment to sustainability and point to unfulfilled promises. Yet, if the corporation delivers on its stated goals, the social and environmental benefits could be substantial. Wal-Mart's vast global reach and its ability to persuade suppliers to alter their ways in order to retain its business mean that any change it enacts could have far-reaching impacts.

Of course, corporations exist to make money for their shareholders, so they cannot be expected to pursue goals that do not turn a profit. Moreover, many corporate greening efforts are more rhetoric than reality, pursued mostly for public relations purposes. Such corporate **greenwashing** can mislead consumers into thinking a company is acting more sustainably than it actually is. The bottled water industry presents a stark example of greenwashing. Advertising with words such as "pure" and "natural" and images of forests and alpine springs leads us to believe that bottled water is cleaner and healthier for us to drink. In reality, bottled water is often less safe than tap water, the plastic bottles are a major source of waste, considerable amounts of oil are burned to transport the bottles, and the industry depletes aquifers in local communities (pp. 399–401).

In the end, it is up to all of us in our roles as consumers to encourage trends in sustainability by rewarding those businesses that truly promote sustainable solutions. This is one way that each of us can express the ethical values we cherish through the economic system in which we live.

Sustainable Development

Today's search for sustainable solutions centers on **sustainable development,** economic advancement that uses resources in a manner that satisfies our current needs but does not compromise the future availability of resources. Sustainable development is an economic pursuit, but it is also an ethical pursuit because it asks today's generations to manage our resource use so that future generations can enjoy similar access to resources.

Sustainable development involves environmental protection, economic well-being, and social equity

Economists employ the term **development** to describe the use of natural resources for economic advancement (as opposed to simple subsistence, or survival). Development involves making purposeful changes intended to improve the quality of human life. Construction of homes, schools, hospitals, power plants, factories, and transportation networks are all examples of development. Sustainable development is defined by the United Nations (p. 178) as development that "meets the needs of the present without sacrificing the ability of future generations to meet their own needs." This definition is taken from the United Nations–sponsored Brundtland Commission (named after its chair, Norwegian prime minister Gro Harlem Brundtland), which published an influential 1987 report titled *Our Common Future.*

Prior to the Brundtland Report, "sustainable development" would have been widely viewed as an oxymoron—a phrase that contradicts itself. Advocates of development felt that protecting the environment threatened people's economic needs, whereas advocates of environmental protection held that development degrades the environment, thereby jeopardizing the very improvements for human life that were intended. In recent years, however, people of all persuasions have increasingly perceived how environmental quality supports our quality of life and have concluded that our civilization cannot exist without a healthy and functional natural environment.

We also now recognize that society's poorer people suffer the most from environmental degradation. This realization led advocates of environmental protection, economic development, and social justice to begin working together toward common goals. This cooperative approach gave rise to the modern drive for sustainable development. Today, it is widely recognized that sustainability does not mean simply protecting the environment against the ravages of human development. Instead, it means finding ways to promote social justice, economic well-being, and environmental quality at the same time (**FIGURE 6.21**). As a result, governments, businesses, industries, and organizations pursuing sustainable development try to satisfy a **triple bottom line,** a trio of goals including economic advancement, environmental protection, and social equity.

Programs that pay for ecosystem services are one example of a sustainable development approach that seeks to satisfy a triple bottom line. Costa Rica's PSA program aims to enhance its citizens' well-being by conserving the country's natural assets, while compensating affected landholders for any economic losses. The intention is to achieve a win-win-win result that pays off in economic, social, and environmental dimensions.

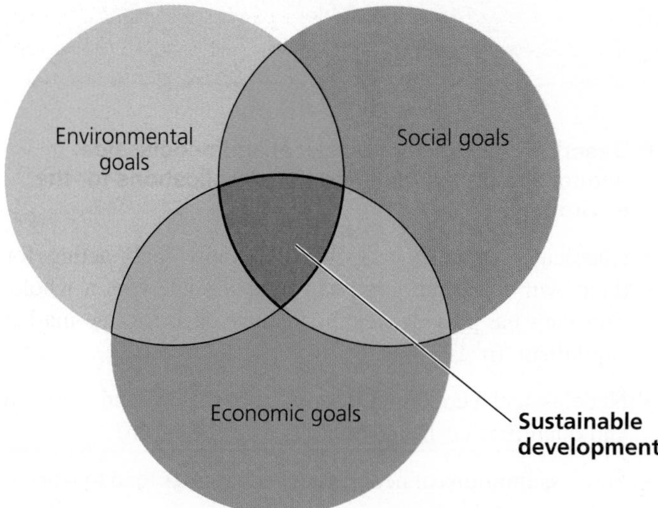

FIGURE 6.21 Sustainable development occurs where three sets of goals overlap: social, economic, and environmental goals.

TABLE 6.1 U.N. Millennium Development Goals for 2015
• Eradicate extreme poverty and hunger
• Achieve universal primary education
• Promote gender equality and empower women
• Reduce child mortality
• Improve maternal health
• Combat HIV/AIDS, malaria, and other diseases
• Ensure environmental sustainability
• Develop a global partnership for development

Source: United Nations. End poverty 2015: Millennium Development Goals. *© United Nations. Reproduced with permission.*

Of course, designing sustainable solutions to complex problems is never simple, and people differ in what they mean by "sustainable development." Proponents of a school of thought called *weak sustainability* feel that we can allow natural capital to decline as long as human-made capital increases to compensate for it. In contrast, proponents of *strong sustainability* insist that human-made capital cannot substitute for natural capital and that we must not allow natural capital to diminish.

Sustainable development is a global movement

Sustainable development has blossomed on the world stage as a distinctly international movement. For years the United Nations, the World Bank, and other international organizations (p. 178) have sponsored conferences, funded projects, published research, and facilitated collaboration across national borders among governments, businesses, and nonprofit organizations, all in the name of sustainable development.

In 1992, the Earth Summit at Río de Janeiro, Brazil, was the world's first major gathering focused on sustainable development, bringing together representatives from over 200 nations. This conference gave rise to several notable achievements, including the Convention on Biological Diversity (p. 297) and the Framework Convention on Climate Change (p. 510). Ten years later, nations reconvened in Johannesburg, South Africa, at the 2002 World Summit on Sustainable Development. Then in 2012, the world returned to Río de Janeiro for the Rio-Plus-20 conference (p. 177).

Meanwhile, in 2000 world leaders came together to adopt the United Nations Millennium Declaration, which set forth eight **Millennium Development Goals** for humanity (TABLE 6.1). Each broad goal for sustainable development has several specific underlying targets that may be met by implementing concrete strategies. For instance, strategies to "develop a global partnership for development" include working with governments and corporations of wealthy nations to provide poor nations with more financial aid, more debt relief, freer access to global markets, and better access to inexpensive drugs and to technologies such as cell phones and Internet access.

As we approach the 2015 target date for many of the Millennium Development Goals, we are making better progress on some than on others. We still have a long way to go to resolve the many challenges facing our world. Pursuing solutions that meet environmental, economic, and social goals simultaneously, addressing a triple bottom line, can help pave the way for a truly sustainable society that promotes economic and social well-being while limiting environmental impact.

Conclusion

The valuation of ecosystem services, ecolabeling and ecotourism, corporate sustainability, and alternative means of measuring growth have all helped to bring economic approaches to bear on environmental protection and resource conservation. As economics becomes more oriented toward sustainability, it renews some of its historic ties to ethics.

Environmental ethics has expanded people's domain of ethical concern outward to encompass more societies, cultures, creatures, and even nonliving entities. This ethical expansion involves the concept of distributional equity, or equal treatment, which is the aim of environmental justice. One type of distributional equity is equity among generations. Concern among today's generations for the well-being of tomorrow's generations is the basis of sustainability and sustainable development.

In pursuing sustainable development, we recognize that economic well-being and environmental well-being work in tandem and depend on one another. Equating economic well-being solely with economic growth, as many economists traditionally have, suggests that economic health entails a trade-off with environmental quality. However, if we can enhance economic well-being without intensifying resource depletion, then economic health and environmental quality can be mutually reinforcing, and sustainable development will be within reach.

Reviewing Objectives

You should now be able to:

Characterize the influences of culture and worldview on the choices people make

- Culture and personal experience influence a person's worldview. (p. 135)
- Factors such as religion, political ideology, and shared experiences shape our perspectives on the environment and the choices we make. (p. 135)

Outline the nature and historical expansion of ethics in Western culture

- Environmental ethics applies ethical standards to relationships between people and aspects of their environments. (p. 136)
- We can value things for utilitarian reasons (instrumental value) or for their own sake (intrinsic value). (p. 136)
- Our society's domain of ethical concern has been expanding, and we have granted more and more entities ethical standing. (pp. 136–137)
- Anthropocentrism values people above all else, whereas biocentrism values all life, and ecocentrism values ecological systems. (p. 137)
- The industrial revolution inspired philosophical reactions that fed into environmental ethics. (p. 138)

Compare major approaches in environmental ethics

- The preservation ethic, espoused by John Muir, values preserving natural systems intact. (p. 138)
- The conservation ethic, espoused by Gifford Pinchot, promotes responsible long-term use of resources. (pp. 138–139)
- The philosophy of Aldo Leopold, including his land ethic, has deeply influenced modern environmental ethics. (pp. 139–140)
- Environmental justice seeks equal treatment for people of all income levels, races, and ethnicities. (pp. 140–141)

Explain how our economies exist within the environment and rely on ecosystem services

- Economies depend on the ecological systems around them for natural resources and ecosystem services. (p. 142)
- Ecosystem services, provided for free by nature, enable our economies to prosper. The degradation of ecosystem services threatens our economic well-being. (pp. 142–143)

Describe principles of classical and neoclassical economics and summarize their implications for the environment

- Classical economics proposes that individuals acting for their own economic good can benefit society as a whole. This view has provided a philosophical basis for free-market capitalism. (p. 143)
- Neoclassical economics focuses on supply and demand and quantifies costs and benefits. (pp. 143, 146)
- Four assumptions of neoclassical economics tend to worsen environmental impacts. (pp. 146–147)
- Conventional economic theory promotes infinite economic growth, with little regard to potential environmental impact, despite the fact that resource consumption cannot rise indefinitely. (pp. 147–150)

Illustrate aspects of environmental economics and ecological economics

- Environmental economists advocate reforming economic practices to promote sustainability. Ecological economists agree but advocate going further by pursuing a steady-state economy. (p. 150)
- Assigning monetary value to ecosystem goods and services can help reduce external costs and make market prices reflect full costs and benefits. (pp. 150–152)
- Full cost accounting indicators aim to measure economic progress and human well-being more effectively than GDP. (pp. 152–154)

Describe how individuals and businesses can help move our economic system in a sustainable direction

- Consumer choice in the marketplace, facilitated by ecolabeling, can encourage businesses to pursue sustainable practices. (p. 155)
- Many corporations are modifying their operations to become more sustainable, and this often is financially profitable. (pp. 155–156)

Explain the pursuit of sustainable development

- Sustainable development promotes people's economic advancement while using resources in a way that satisfies today's needs without compromising the needs of future generations. (p. 156)
- Advocates of sustainable development pursue environmental, economic, and social goals in a coordinated way. (pp. 156–157)
- Sustainable development has thrived as a global movement on the international stage. (p. 157)

Testing Your Comprehension

1. What does the study of ethics encompass? Describe and differentiate instrumental value and intrinsic value. What is environmental ethics?
2. Compare and contrast anthropocentrism, biocentrism, and ecocentrism. Explain how individuals with each perspective might evaluate the development of a shopping mall atop a wetland in your town or city.
3. Differentiate the preservation ethic from the conservation ethic. Explain the contributions of John Muir and Gifford Pinchot in the history of environmental ethics.
4. Describe Aldo Leopold's land ethic. How did Leopold define the "community" to which ethical standards should be applied?
5. Define the concept of environmental justice. Give an example of an inequity relevant to environmental justice that you believe exists in your city, state, or country.
6. Name and describe two key contributions that the environment makes to the economy.
7. Describe four ways in which critics hold that neoclassical economic approaches can worsen environmental problems.
8. Compare and contrast the views of neoclassical economists, environmental economists, and ecological economists, particularly regarding the issue of economic growth.
9. What is contingent valuation, and what is one of its weaknesses? Describe an alternative method that avoids this weakness.
10. How can sustainable development be defined? What is meant by the triple bottom line? Why is it important to pursue sustainable development?

Seeking Solutions

1. Describe your worldview as it pertains to your relationship with your environment. How do you think your culture has influenced your worldview? How do you think your personal experience has affected it? How do you think your worldview shapes your decisions? Do you feel that you fit into any particular category discussed in this chapter? Why or why not?
2. Visit the U.S. EPA's online tool for assessing environmental justice concerns, at http://epamap14.epa.gov/ejmap/entry.html, and enter your city, county, or zip code. On the map page, check the six types of "Sites reporting to EPA" in the list of features on the right, and let it plot these on the map. Now click on "Demographics." Apply various demographic layers one by one, including at least "Per Capita Income," "Below Poverty (%)," and "Minority (%)." The site will map these parameters and show how the EPA-regulated facilities are located relative to them.

   What patterns do you see? Where are most facilities located, relative to wealth, race, and other factors? What do you think accounts for the patterns you find? Do you perceive any problems revealed by these data? What could be done to alleviate such problems?
3. Do you think that a steady-state economy is a practical alternative to our current approach that prioritizes economic growth? Why or why not?
4. Do you think we should attempt to quantify and assign market values to ecosystem services and other entities that have only nonmarket values? Why or why not?
5. Suppose you are a Costa Rican farmer who needs to decide whether to clear a stand of forest or apply to receive payments to preserve it through the PSA program. Describe all the types of information you would want to consider before making your decision. Now, describe what you think each of the following people would recommend to you if you were to go to them for advice:
   - a preservationist
   - a conservationist
   - a neoclassical economist
   - an ecological economist
6. **THINK IT THROUGH** A manufacturing facility on a river near your home provides jobs for 200 people in your community of 10,000 people, and it pays $2 million in taxes to the local government each year. Sales taxes from purchases made by plant employees and their families contribute an additional $1 million to local government coffers. However, newspaper reports and recent peer-reviewed studies in well-respected scientific journals reveal that the plant has been discharging large amounts of waste into the river, causing a 25% increase in cancer rates, a 30% reduction in riverfront property values, and a 75% decrease in native fish populations.

   The plant owner says the facility can stay in business only because there are no regulations mandating expensive treatment of waste from the plant. If such regulations were imposed, he says he would close the plant, lay off its employees, and relocate to a more business-friendly community.

   How would you recommend resolving this situation? What further information would you want to know before making a recommendation? In arriving at your recommendation, how did you weigh the costs and benefits associated with each of the plant's impacts?

Calculating Ecological Footprints

The population of the United States grew from 152,271,000 to 292,892,000 between 1950 and 2004, but the nation's Gross Domestic Product (GDP) grew still faster. Taking these two trends together, Figure 6.17b (p. 153) shows how per capita GDP rose over this half-century. During this period, the per capita Genuine Progress Indicator (GPI) grew as well, but more slowly than GDP.

Given the values of the major components of the GPI for the United States in 1950 and 2004, and the population figures above, calculate and enter the per capita rates for these components, as well as the overall GPI values. Refer to Figure 6.17a (p. 153) to check how to calculate the GPI.

Components of GPI	U.S. total in 1950 (trillions of dollars)	Per capita in 1950 (thousands of dollars)	U.S. total in 2004 (trillions of dollars)	Per capita in 2004 (thousands of dollars)
GDP	1.153		7.589	
Benefits	1.041		4.746	
Environmental costs	0.407		3.990	
Social and economic costs	0.476		3.926	
GPI				

Data from Talberth, J., C. Cobb, and N. Slattery, 2007. The Genuine Progress Indicator 2006: A tool for sustainable development. *Redefining Progress, Oakland, CA.*

1. How many times greater was the GDP in 2004 than in 1950? By how many times did the GPI increase between 1950 and 2004? What does this comparison tell you?
2. By how many times, respectively, did benefits, environmental costs, and social and economic costs increase between 1950 and 2004? How are trends in each of these components driving the overall trend between the GPI and the GDP? Which component has gotten worst over the years? How would you account for these trends?
3. There are many ways to define and measure the benefits and costs that go into the GPI. How do you think a person with a biocentric worldview would measure these differently from a person with an anthropocentric worldview? Whose GPI for the year 2005 would likely be higher?
4. Now consider your own life. Very roughly, what would you estimate are the values of the benefits, environmental costs, and social and economic costs you experience? What could you do to help improve these trends in your own personal accounting?

Rachel Farnelli rides a swing in her backyard overlooking a natural gas well in Dimock, Pennsylvania

7

Environmental Policy: Making Decisions and Solving Problems

Upon completing this chapter, you will be able to:

- Describe environmental policy and assess its societal context
- Discuss the role of science in policymaking
- Identify the institutions important to U.S. environmental policy and recognize major U.S. environmental laws
- List the institutions involved with international environmental policy and describe how nations handle transboundary issues
- Categorize the different approaches to environmental policy

CENTRAL CASE STUDY

Hydrofracking the Marcellus Shale

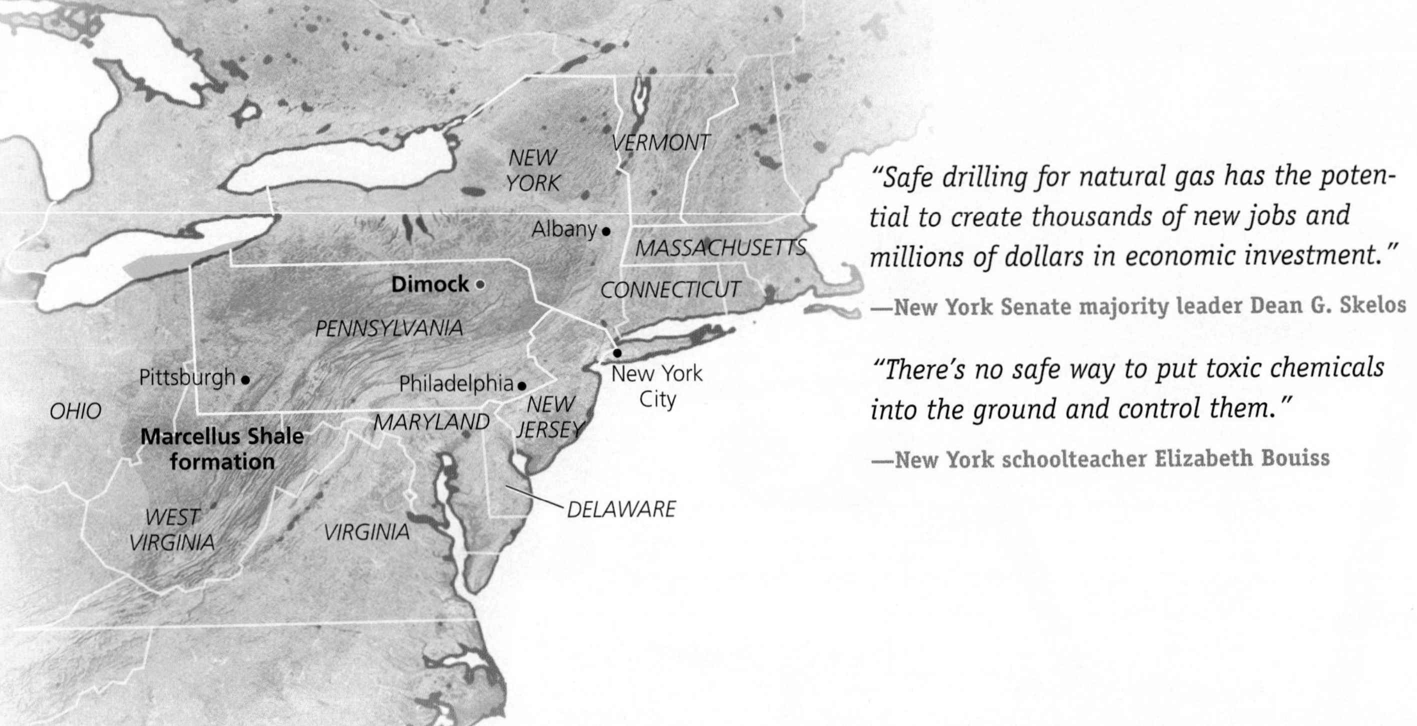

"Safe drilling for natural gas has the potential to create thousands of new jobs and millions of dollars in economic investment."

—New York Senate majority leader Dean G. Skelos

"There's no safe way to put toxic chemicals into the ground and control them."

—New York schoolteacher Elizabeth Bouiss

When the men from Cabot Oil and Gas Corporation came to the small town of Dimock in rural Pennsylvania, many of Dimock's 1500 residents were happy to sign on to the contracts the company offered. In exchange for the right to drill for natural gas on their land, Cabot would pay them royalties on sales of the gas extracted from drilling pads placed on their property. For some in the small, rural community, the gas money seemed like a ticket to economic security.

Soon the new drilling sites around Dimock were producing the most natural gas from anywhere in the Marcellus Shale, the vast gas-bearing rock formation that underlies portions of Pennsylvania, New York, West Virginia, and Ohio. Money and jobs from the gas boom kept Dimock economically afloat, even as other towns reeled from recession and cut funding for schools and basic services.

Yet despite the economic gains, some Dimock residents were having second thoughts about drilling. Their once-peaceful community was now experiencing constant noise and lights, heavy truck traffic, toxic wastewater spills, and air pollution from the drilling sites. Then people's drinking water began to turn brown, gray, or cloudy with sediment, while strange chemical smells began wafting from their water wells. On New Years Day, 2009, Norma Fiorentino's well exploded. Methane had built up in her well water, and a spark from a motorized pump set off a potentially lethal blast.

Residents blamed the drilling technique that Cabot Oil and Gas was using: hydraulic fracturing. Citizens who could no longer drink their own well water appealed for help. Retired schoolteacher Victoria Switzer said she approached Cabot, her local political leaders, and the Pennsylvania Department of Environmental Protection (DEP) but was turned away by them all. Then she went to the news media, and the story began to get out. Documentary filmmaker Josh Fox came to town and filmed residents setting their methane-contaminated tap water on fire. His 2010 film *Gasland* won numerous awards, and Dimock became Ground Zero in the burgeoning national debate over hydraulic fracturing.

In the United States, virtually all the easily accessible oil and natural gas has already been discovered and extracted. To extract more, we've needed to develop ever-more powerful technology to reach petroleum deposits that are deeper under the earth, deeper under the ocean, and at lower concentrations. In areas such as the Marcellus Shale, a great deal of natural gas is locked up deep underground in countless tiny bubbles dispersed throughout formations of shale. The technique of hydraulic fracturing is now making this **shale gas** accessible.

Hydraulic fracturing (also called **hydrofracking,** or simply **fracking**) involves drilling deep into the earth and then angling the drill horizontally when a shale formation is reached. An electric charge sets off targeted explosions that perforate the drilling pipe and create fractures in the shale. Drillers then pump a slurry of water, sand, and chemicals down through the pipe under great pressure. The sand lodges in the fractures and holds them open, while a portion of the liquids return to the surface. Bubbles of natural gas trapped in the shale migrate into the system of fractures and gradually rise to the surface through the drilling pipe (**FIGURE 7.1**).

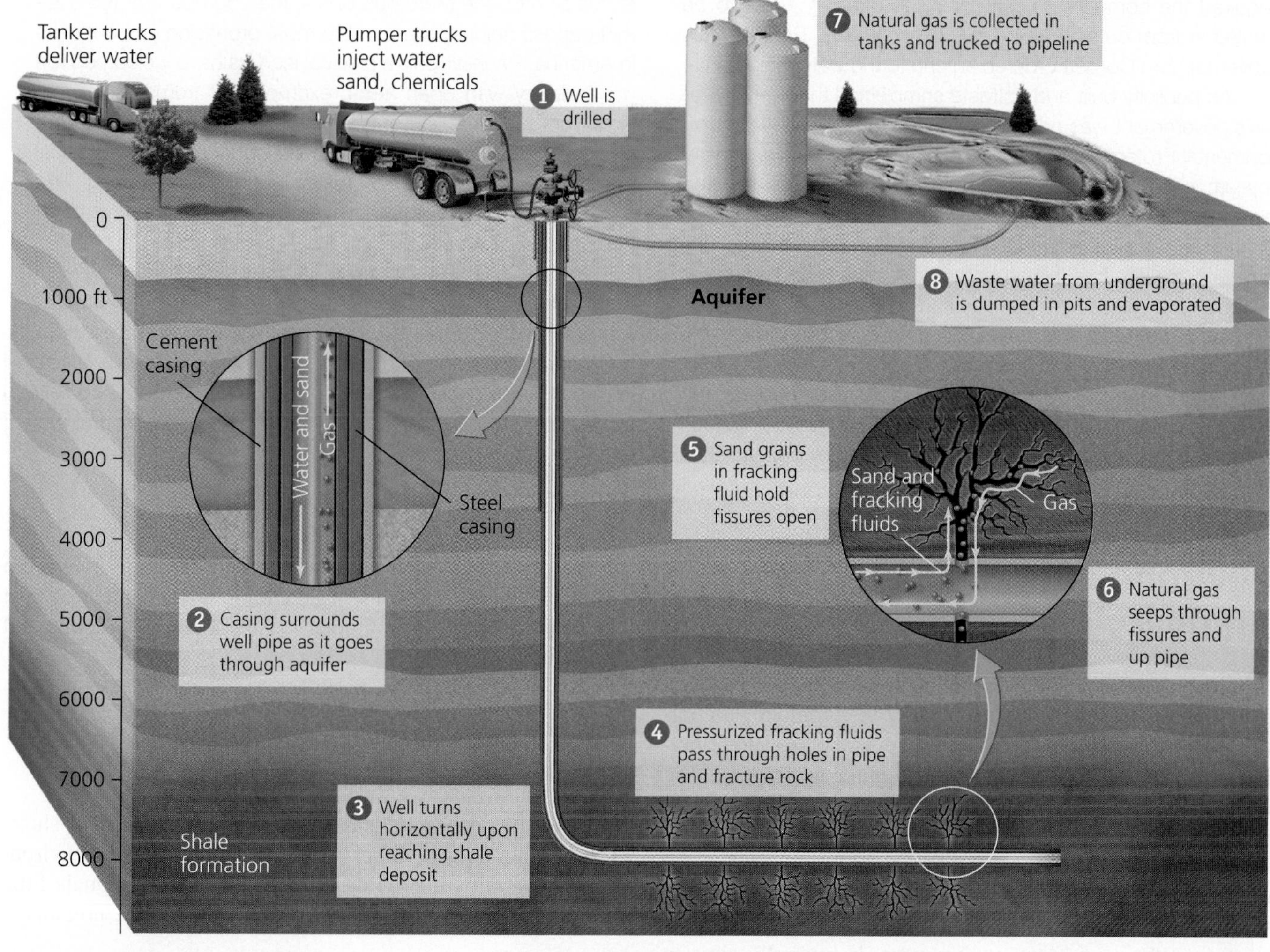

FIGURE 7.1 Hydraulic fracturing is used to extract natural gas trapped in shale deposits deep underground. A well is drilled ❶ with protective casing ❷ and is turned horizontally ❸ upon reaching a shale deposit. Pressurized fluids fracture the rock ❹, and sand lodges in the cracks ❺, holding them open for natural gas ❻ to seep into them and rise through the pipe to the surface ❼. Polluted fluids also rise to the surface and are piped to wastewater pits ❽.

By boosting production of natural gas, hydrofracking employs people and keeps the price of gas low. Expanded use of natural gas reduces the United States' dependence on burning coal for electricity. Because natural gas is cleaner-burning than oil or coal, burning it in place of these other fossil fuels reduces the greenhouse gas emissions that lead to climate change.

For all these reasons, policymakers have encouraged hydrofracking. They have freed it from many regulatory constraints that would normally apply to such a process. Drilling companies have been exempted from seven major federal environmental laws that protect public health, including the National Environmental Policy Act and the Safe Drinking Water Act.

As a result, gas companies do not need to report the chemical additives they use during the fracking process, nor do they need to test for chemical compounds that are drawn up from the ground in fracking wastewater. Much of this wastewater is radioactive because drillers add radioisotopes (p. 24) as tracers to the fracking fluids they inject, and because naturally occurring radioisotopes are brought up from deep underground. Consequently, no one—neither regulators nor policymakers nor scientists nor homeowners—has access to the data necessary to make fully informed judgments about any potential health or environmental effects of fracking.

In such a climate of uncertainty, people in places like Dimock are left to wonder, worry, and argue. Victoria Switzer and several dozen other families in town eventually brought lawsuits against Cabot, but many of their neighbors are angry with them. Residents whose water has not been affected—and those who decided that receiving gas payments was worth the health risks from pollution—blame their fellow townspeople who have spoken out about water quality for bringing global media attention to Dimock and driving down property values. Victoria Switzer's neighbor Anne Teel experienced water problems from the eight gas wells on her property but says Cabot helped fix them and that those bringing the lawsuit are doing their town a disservice.

When outside influences disrupt a community and cause neighbor to turn against neighbor, some wondered, can't government help by enforcing rules to protect the public welfare and to ensure that people are treated equally and justly?

After recognizing that for some Dimock families the water was undrinkable, Pennsylvania's DEP eventually fined Cabot and required the company to pay for clean drinking water to be hauled in from outside town. In 2011, however, Pennsylvania Governor Tom Corbett ordered an end to the water shipments.

As publicity built and activists complained that Pennsylvania's government was not protecting its citizens, the U.S. Environmental Protection Agency (EPA) stepped in, sending federal researchers to run tests of Dimock's water. Results showed elevated levels of several chemicals that could threaten health in 5 out of 64 wells tested, but the EPA stated that Cabot was mitigating these impacts. However, some residents and scientists suspected influence from industry and politicians and took issue with the EPA's conclusions. The EPA denied such influence, and then embarked on a nationwide study of the health and environmental impacts of hydraulic fracturing, due out in late 2014.

At the state level, governments have responded in various ways. Pennsylvania and New York offer a study in contrast. Both states sit atop the Marcellus Shale, the 250,000-km^2 (95,000-mi^2) deposit holding some of the most promising gas deposits in America. Pennsylvania's political leaders have welcomed the gas industry with open arms, exempting it from regulations. New York's leaders, in contrast, have placed a moratorium on fracking until further studies can be conducted and political agreements reached. Ohio, Texas, Louisiana, Wyoming, and other states have encouraged hydrofracking, whereas Vermont has banned it completely.

People differ in their views on the proper role of government, but most would agree that government policy should protect people's equality of opportunity, promote their economic advancement, and safeguard them from undue harm. As we discuss environmental policy, we will see how society is struggling to balance these aims in the case of natural gas extraction. ■

Environmental Policy: An Overview

When a society reaches broad agreement that a problem exists, it may urge its leaders to resolve the problem with policy. **Policy** consists of a formal set of general plans and principles intended to address problems and guide decision making in specific instances. **Public policy** is policy made by governments, including those at the local, state, federal, and international levels. Public policy consists of laws, regulations, orders, incentives, and practices intended to advance societal well-being. **Environmental policy** is policy that pertains to our interactions with our environment. It generally aims to regulate resource use or reduce pollution in order to promote human welfare and/or protect natural systems.

Forging effective policy requires input from science, ethics, and economics. Science provides information and analyses needed to identify and understand problems and devise potential solutions to them (Chapter 1). Ethics and economics offer criteria by which to assess problems and help clarify how society might address them (Chapter 6). Government interacts with individual citizens, organizations, and the private sector in various ways to formulate policy (**FIGURE 7.2**).

The ongoing debate over hydraulic fracturing of the Marcellus Shale illustrates how science, ethics, and economics each inform and motivate the making of policy. Scientific research enabled the technological advances that allow us to find and extract fossil fuels, and it provides data on emissions resulting from coal, oil, and natural gas. Science also documents that leaks and spills from gas drilling can pollute the air and contaminate groundwater, and that hydrofracking can sometimes cause minor earthquakes.

In economic terms, hydraulic fracturing helps us exploit a valuable fuel resource that powers our economy and can substitute for dirtier fuel sources. For communities near drilling sites, shale gas extraction supplies a short-term boost in jobs and income while introducing health risks and long-term costs of pollution cleanup. Ethically, hydrofracking can pose problems if drilling by private companies pollutes water and air on which the public relies. Such conflicts commonly result from development, and environmental policy tries to address them by balancing the benefits of economic advancement against the costs of impacts on human health, social well-being, and ecological systems.

Environmental policy addresses issues of fairness and resource use

Because market capitalism is driven by incentives for short-term economic gain rather than long-term social and environmental stability, it provides businesses and individuals little enticement to minimize environmental impact. Unregulated

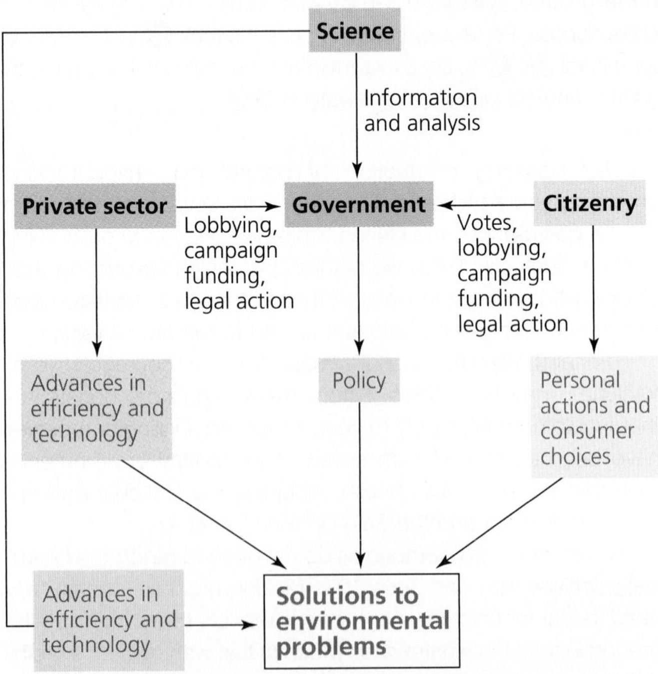

FIGURE 7.2 Policy plays a central role in how we address environmental problems. Citizens, businesses, and scientific research inform and influence policymakers. The public policy they create can help to produce lasting solutions.

market capitalism also provides businesses or individuals little incentive to equalize costs and benefits among parties. Market prices often do not reflect the value of environmental contributions to economies or the full costs imposed on the public by private parties when their actions degrade the environment (Chapter 6). Such market failure (p. 154) has traditionally been viewed as justification for government intervention.

In modern mixed economies (p. 141), governments typically intervene in the marketplace for several reasons:

- To provide social services, such as national defense, medical care, and education
- To provide "safety nets" (for the elderly, the poor, victims of natural disasters, and so on)
- To eliminate unfair advantages held by single buyers or sellers
- To manage publicly held resources
- To mitigate pollution and other threats to health and quality of life

Environmental policy aims to protect environmental quality and the natural resources people use, and also to promote equity or fairness in people's use of resources.

The tragedy of the commons When publicly accessible resources are open to unregulated exploitation, they tend to become overused and, as a result, are damaged or depleted. So argued environmental scientist Garrett Hardin in his 1968 essay "The Tragedy of the Commons." Basing his argument on an age-old scenario, Hardin explained that in a public pasture (or "common") open to unregulated grazing, each person who grazes animals will be motivated by self-interest to increase the number of his or her animals in the pasture. Because no single person owns the pasture, no one has incentive to expend effort taking care of it. Instead, each person takes what he or she can until the resource is depleted and overgrazing causes the pasture's food production to collapse. This is known as the **tragedy of the commons.**

The tragedy of the commons pertains to many types of resources held and used in common by the public: forests, fisheries, clean air, clean water—even the global climate. When such resources are being depleted or degraded, it is in society's interest to develop guidelines for their use. In Hardin's example of a common pasture, guidelines might limit the number of animals each person can graze or might require pasture users to help restore and manage the shared resource. These two concepts—restriction of use, and management—are central to environmental policy today.

Public oversight through government is a standard way to alleviate the tragedy of the commons, but this dilemma can also be addressed in other ways. One is a bottom-up cooperative approach, in which users of the resource band together and cooperate to prevent overexploitation. Indeed, many traditional societies over the centuries have developed ways to manage resources cooperatively and sustainably at the community level. Another approach is privatization, in which the resource is subdivided and allotments are sold into private ownership, so that each owner has incentive to conserve his or her portion of the resource in ways that maximize its productivity.

The cooperative approach may work if the resource is localized and its use is easily enforced, but often these conditions do not hold. Privatization may work if property rights can be clearly assigned (as with land), but it tends not to work with resources such as air or water. Privatization also opens the door to short-term profit-taking at the long-term expense of the resource. Thus, in many cases public oversight and regulation by democratic government are likely the best ways to avoid the tragedy of the commons.

Free riders A second reason we develop policy for publicly held resources is the **free rider** predicament. Let's say a community on a river suffers from water pollution that emanates from ten different factories. The problem could in theory be solved if every factory voluntarily agreed to reduce its own pollution. However, once they all begin reducing their pollution, it becomes tempting for any one of them to stop doing so. A factory that avoids the efforts others are making would in essence get a "free ride." If enough factories take a free ride, the whole enterprise will collapse. Because of the free rider problem, private voluntary efforts are often less effective than efforts mandated by public policy.

External costs Environmental policy also aims to promote fairness by dealing with external costs (p. 146), harmful impacts suffered by people not involved in the actions that created them. For example, a factory might reap greater profits by discharging waste into a river instead of paying for proper waste disposal or recycling. Its actions, however, impose external costs (water pollution, reduced fish populations, aesthetic degradation, or other problems) on downstream users of the river. Likewise, natural gas drilling operations that pollute groundwater or cause earthquakes may impose external costs on people living nearby (FIGURE 7.3). Contamination of drinking water by methane from natural gas, or by the many chemicals used in fracking fluids in the drilling process, can disrupt people's lives and affect their health (see THE SCIENCE BEHIND THE STORY, pp. 166–167).

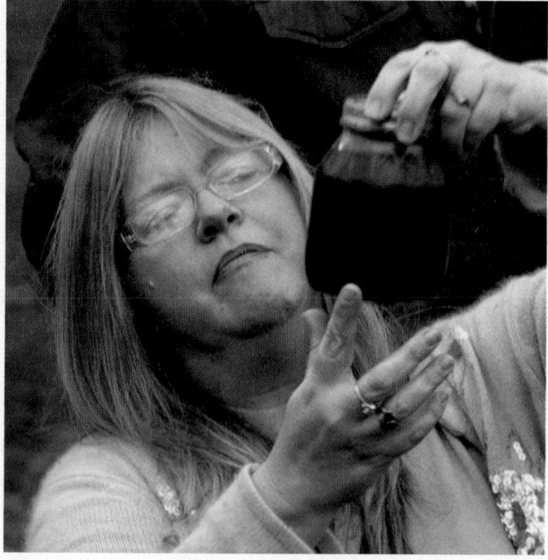

FIGURE 7.3 **Pollution from shale gas drilling operations may create external costs.** Dimock resident Patricia Farnelli holds polluted road runoff water from the fracking activity on her land.

Does Fracking Contaminate Drinking Water?

A Pennsylvania homeowner sets fire to her tap water

In 2010, gas industry executives testified to a Congressional panel that no one had ever proven a single case of groundwater contamination from hydraulic fracturing. Yet critics of the practice maintain that such pollution is commonplace, and many citizens living near drilling sites voice concerns about pollution and health. Moreover, numerous people are shown setting their methane-rich tap water on fire in the documentary film *Gasland*.

So, what scientific evidence is there on the issue? It's not a straightforward question. Determining whether clean groundwater has become dirty, what caused any contamination, and what health impacts might result are each complex pursuits.

Debate over a 2011 scientific paper that documented methane in drinking water near hydrofracking sites illustrates the degree to which researchers can disagree over data and interpretations—especially when the topic is politically charged. The paper, published in the prestigious scientific journal *Proceedings of the National Academy of Sciences* (*PNAS*) by Stephen Osborn and three colleagues from Duke University, elicited so many queries and critiques that the authors set up an online "frequently asked questions" webpage to address the barrage of correspondence.

In its research, Osborn's team visited private wells that draw drinking water from aquifers that lie above the Marcellus and Utica Shale formations in Pennsylvania and New York (**FIGURE 1**). With homeowners' permission, they collected samples of water from 68 wells. Half the wells were located within 1 km of active gas drilling sites, while half were more than 1 km away from the nearest drilling site. The researchers took the water samples to the lab and analyzed them for their chemical constituents.

Osborn and his colleagues found no evidence that well water was being contaminated by chemicals from the hydrofracking process or by salty fluids rising up from deep underground as a result of fracking. The chemistry of samples near and far from drilling sites was similar, and these data were also similar to 124 water samples collected from nearby regions by other researchers in past years.

However, when Osborn's team tested for methane, the main component of natural gas, they found that concentrations averaged 17 times higher at wells near natural gas drilling sites than at wells far from drilling sites (**FIGURE 2**). Although methane dissolved in drinking water is not considered toxic to drink, it is an asphyxiating gas and a hazard for fire and explosions. The average methane concentration these researchers documented in well water near drilling sites fell within

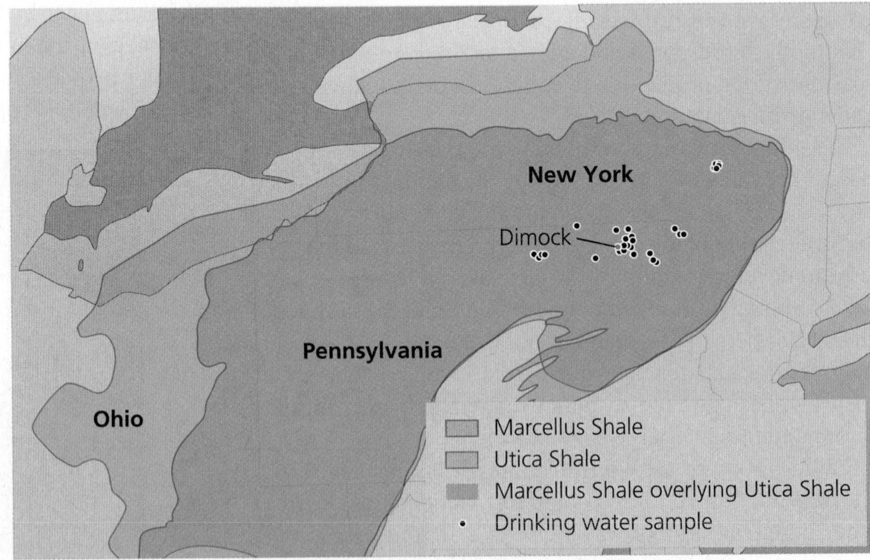

FIGURE 1 Methane migration was studied in water wells in northeastern Pennsylvania and adjacent New York, near the northeastern extent of the vast Marcellus Shale formation. *Data from Osborn, S.G. et al., 2011. Methane contamination of drinking water accompanying gas-well drilling and hydraulic fracturing.* PNAS *108: 8172–8176.*

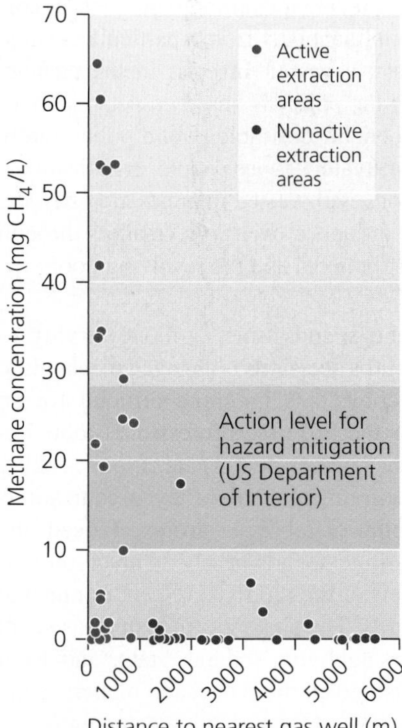

FIGURE 2 Methane concentrations were far higher at water wells near gas drilling sites (left end of graph) than at wells far from them (right end of graph). *Data from Osborn, S.G. et al., 2011. Methane contamination of drinking water accompanying gas-well drilling and hydraulic fracturing.* PNAS *108: 8172–8176.*

DATA Q Let's say you are going to sample two new water wells for methane, one just 250 m from a gas drilling site, and one 5000 m from a drilling site. Based on the data shown here, predict, for each of these wells, the likelihood that methane concentrations will be above the "Action level for hazard mitigation."

the range of the federal government's defined action level for hazard mitigation. Nine samples were above the action level.

The team then sought to determine where this methane had come from: Was it *thermogenic* methane produced deep underground along with fossil fuel formation (which would suggest contamination from drilling), or was it *biogenic* methane produced at shallow depths by microorganisms (p. 525)? To determine this, they analyzed ratios of isotopes (p. 24) of carbon and hydrogen in the methane as well as ratios of methane to other hydrocarbon gases.

Their analyses showed that the methane from well water near drilling sites was thermogenic gas that had originated deep underground. In contrast, methane present in small amounts in wells far from drilling sites was mostly of shallow biogenic origin. Moreover, the chemistry of the natural gas in wells near drilling sites was consistent with the chemistry of the Marcellus Shale formation underlying the region. This suggested that the methane near drilling sites likely resulted from drilling.

Still, the Marcellus Shale lies 1–2 km below the surface, whereas the drinking water wells extend only 36–190 m underground. How could methane from so far down have reached these shallow wells? Osborn's team proposed three possible ways:

1. Gas-rich solutions from deep underground were displaced by fracking and traveled upward through existing fractures to the shallow aquifer. (This explanation seemed unlikely to the researchers, given that fracking fluids and salty fluids from deep underground had not gotten into the wells.)
2. Fracking unintentionally created or enlarged fractures in shallow areas far above the shale formations, allowing gas to travel to the surface. (The researchers judged this explanation unlikely but possible.)
3. Some gas well casings were leaking, allowing gas pulled up from deep below to escape from the well casing near the surface and infiltrate shallow aquifers. (The team judged this to be the most likely explanation.)

In light of these findings, Osborn's team recommended more research into methane's underground movements, as well as "long-term coordinated sampling and monitoring" near drilling sites. They also urged that more data should be required—and made public—before drilling proceeds.

Several researchers wrote to *PNAS*, critiquing the paper. Some thought the results did not rule out natural seepage or the effects of past drilling. Some felt the fact that they did not find fracking fluids in the aquifers meant the methane could not have come from hydrofracking. Others complained that the sampling of homeowners was nonrandom. All pointed out that to truly determine whether fracking was responsible for methane in groundwater, one would need to make careful comparisons of wells before and after drilling occurred.

While countering most of these complaints, Osborn's team agreed that the sampling of homeowners was nonrandom, but said this is unavoidable when researchers need to seek their permission for sampling.

Osborn's team agreed with its critics that before-and-after comparisons of water quality in areas where drilling occurred would offer the best way to determine whether hydrofracking has impacts on water quality. They pointed out that such a comparison could easily be done, because the gas industry possesses archived data from many hundreds of water wells in Pennsylvania before drilling began. Osborn's team proposed that they would gladly collaborate with industry, landowners, independent analysis labs, and the Pennsylvania Department of Environmental Protection to conduct a joint study using such data. To date, the researchers have yet to receive an agreement from the gas industry to conduct such studies. ■

When governments take action to force industries to protect water quality or to reimburse residents for damage, they internalize costs. The costs are paid by the companies, which add them to the price of their products and pass them on to consumers. Higher market prices, in turn, may reduce demand for the products, and consumers may instead favor less-expensive products whose production imposes fewer costs on society.

These goals of environmental policy—to protect resources against the tragedy of the commons and to promote fairness by eliminating free riders and addressing external costs—are reflected in today's diversity of approaches to environmental policy. As an example, the **polluter-pays principle** specifies that a party responsible for pollution should be held responsible for covering the costs of its impacts. This principle helps protect resources such as clean water and air, promotes just treatment of all parties, and helps shift external costs into the market prices of goods and services.

Many factors hinder environmental policy

If the goals of environmental policy are so noble, why is it that environmental laws and regulations are often challenged and that policymakers frequently ignore or reject the ideas of environmental advocates?

Most environmental policy has come in the form of regulations handed down from government. Businesses and individuals sometimes view regulations as overly restrictive, bureaucratic, or costly. For instance, many landowners fear that zoning regulations (p. 343) or protections for endangered species (p. 296) will restrict how they can use their land. Developers complain of time and money lost to bureaucracy in obtaining permits; reviews by government agencies; surveys for endangered species; and required environmental controls, monitoring, and mitigation. In the eyes of such property owners and businesspeople, environmental regulation often means inconvenience and economic cost.

Another hurdle for environmental policy stems from the nature of environmental problems, which often develop gradually. The degradation of ecosystems and public health caused by human impact on the environment are long-term processes. In contrast, human behavior is geared toward addressing short-term needs, and this is reflected in our social institutions. Businesses usually opt for short-term economic gain over long-term concerns. The news media have a short attention span based on the daily news cycle, so new and sudden events are given more coverage than slowly developing long-term trends. Politicians often act out of short-term interest because they depend on re-election every few years. For all these reasons, environmental policy goals that seem admirable and that attract wide public support may end up being obstructed.

More broadly, policy in general can be held up for a variety of reasons—even if it is rational policy favored by a majority of people. Translating any given policy idea into reality is long, hard work. The checks and balances in a constitutional democracy seek to ensure that new policy is implemented only after it has gone through extensive review and discussion. In general this is a very good thing, but less desirable factors can also hinder the implementation of good policy. In democracies such as the United States, each person has a political voice and can make a difference. However, money often wields influence over policymakers, and some people and organizations wield more influence than others.

Vested interests People, organizations, industries, or corporations that stand to gain financially from a particular change in policy are said to have a **vested interest** in the change. Because their support for the policy is based on personal benefit rather than societal benefit, desirable public policy rarely emerges as a result. Unfortunately, those people, organizations, industries, and corporations with vested interests are often the ones that exert the most influence over policymakers through lobbying, campaign contributions, and the revolving door.

Lobbying Anyone who spends time or money trying to influence an elected official's decisions is engaged in **lobbying.** Although anyone can lobby, it is far more difficult for an ordinary citizen than for the full-time professional lobbyists employed by the many businesses and organizations seeking a voice in politics. Environmental advocacy organizations are not the most influential of lobbying groups. Indeed, the American Petroleum Institute spends nearly as much on lobbying as the entire budgets of the top five U.S. environmental advocacy groups combined. The shale gas industry has spent many millions of dollars lobbying federal, state, and local policymakers, and has also given these decision makers many millions in contributions toward their election campaigns.

Campaign contributions Supporting a candidate's election efforts with money is another way to make one's voice heard. Environmental policy often regulates the activities of corporations and industries, so they have a strong vested interest in shaping it. Corporations and industries may not legally make direct campaign contributions, but they are allowed to establish political action committees (PACs), which raise money to help candidates win elections, in hope of gaining access to those individuals once they are elected. In the wake of the controversial U.S. Supreme Court decision in the case *Citizens United v. Federal Election Commission* in 2010, corporations and unions are now also allowed to purchase political ads supporting or opposing candidates. As a result, corporations and unions can now exert more political influence than in the past.

The revolving door Some individuals employed in industry gain political influence when they take jobs with the government agencies responsible for regulating their industries. Conversely, businesses often hire former government officials who had regulated their industries. This movement of individuals between government and the private sector is known as the **revolving door.**

As an example, after Pennsylvania regulators tried to strengthen oversight of wastewater from gas drilling, three of the regulators left their government posts and went to work for the gas industry. As another example, before becoming U.S. vice president, Dick Cheney was chairman and CEO of Halliburton, a leading energy services company that helped to pioneer hydraulic fracturing. After taking office in 2001, Cheney convened an energy task force that met 40 times with industry officials, yet only once with environmental advocates, and never publicly revealed its proceedings. Many policies friendly to fossil fuel industries followed, including language inserted into the Energy Policy Act of 2005 that exempted hydrofracking from the Safe Drinking Water Act.

Defenders of the revolving door assert that corporate executives who take government jobs regulating their own industry bring with them an intimate knowledge of the industry that makes them highly qualified and likely to benefit society with well-informed policy. Critics contend that taking a job regulating your former employer is a clear conflict of interest that undermines the effectiveness of the regulatory process.

Science informs policy but is sometimes disregarded

Economic interests, ethical values, and political ideology all influence the policy process, but environmental policy that is effective is generally informed by scientific research. For instance, when deciding whether and how to regulate a substance that may pose a public health risk, regulatory agencies such as the Environmental Protection Agency (EPA) comb the scientific literature for information and may commission new studies to research unresolved questions. When trying to pass a bill to reduce pollution, a legislator may use data from scientific studies to quantify the cost of the pollution or the predicted benefits of its reduction. The more information a policymaker can glean from science, the better policy he or she will be able to craft. In today's world, a nation's strength depends on its commitment to science. This is why governments devote a portion of our taxes to fund scientific research.

Unfortunately, sometimes policymakers allow ideology to determine policy on scientific matters. In 2004, the nonpartisan Union of Concerned Scientists released a statement that faulted the U.S. presidential administration of George W. Bush for ignoring scientific advice; manipulating scientific information for political ends; censoring, suppressing, and editing reports from government scientists; placing people who were unqualified or had clear conflicts of interest in positions of power; and misleading the public by misrepresenting scientific knowledge. More than 12,000 American scientists signed on to this statement. Many government scientists working on politically sensitive issues such as climate change or endangered species protection said they had found their work suppressed or discredited and their jobs threatened. Many chose self-censorship.

Of course, either political party can politicize science, and the Union of Concerned Scientists has faulted the administration of Barack Obama and the administrations of other U.S. presidents when they have allowed politics to trump science.

Whenever taxpayer-funded science is suppressed or distorted for political ends we all lose. Abuses of power generally come to light only when conscientious government scientists risk their careers to alert the public and when journalists work hard to uncover and publicize these issues. We cannot take for granted that science will play a role in policy. As scientifically literate citizens of a democracy, we all need to stay vigilant and help ensure that our government representatives are making proper use of the tremendous scientific assets we have at our disposal.

U.S. Environmental Law and Policy

The United States provides a good focus for understanding environmental policy in constitutional democracies worldwide, for several reasons. First, the United States has pioneered innovative environmental policy. Second, U.S. policies have served as models—of both success and failure—for many other nations and international government bodies. Third, the United States exerts a great deal of influence on the affairs of other nations. Finally, understanding U.S. environmental policy at the federal level helps us to understand it at local, state, and international levels.

Federal policy arises from the three branches of government

Federal policy in the United States results from actions of the three branches of government—legislative, executive, and judicial—established under the U.S. Constitution. Statutory law, or **legislation,** is created by Congress, which consists of the Senate and the House of Representatives. For instance, to deal with water pollution across the United States, Congress passed the Federal Water Pollution Control Acts of 1965 and 1972, and then in 1977 passed the Clean Water Act. These laws regulated the discharge of wastes, especially from industry, into rivers and streams, and thereby improved water quality markedly across the nation.

Bills introduced by legislators may be shepherded from subcommittee through committee and on to passage by the Congress (FIGURE 7.4). If a bill passes through all of these steps, it may become law with the president's signature. Legislation is enacted (approved) or vetoed (rejected) by the president, whose veto may in turn be overridden by a two-thirds vote of Congress.

A bill can die in countless fashions along the way, however, and only a small proportion of bills ever become law. For instance, legislators from New York, Pennsylvania, and Colorado in 2009 and each year since have introduced a bill to restore Safe Drinking Water Act regulations on hydraulic fracturing and to require the gas industry to disclose the chemicals it uses in hydrofracking (chemicals said by leading scientists and the EPA to number over 1000). Amid opposition from the industry, this bill for the so-called FRAC Act (Fracturing Responsibility and Awareness to Chemicals Act) has so far been unable to secure enough votes to make it out of committee.

Once a law is enacted, its implementation and enforcement is assigned to an administrative agency within the executive branch. Administrative agencies (TABLE 7.1) may be established by Congress or by presidential order, and they are sometimes nicknamed the "fourth branch" of government. They create a great deal of policy in the form of **regulations,** specific rules or requirements intended to help achieve the objectives of the more broadly written statutory law. Besides issuing regulations,

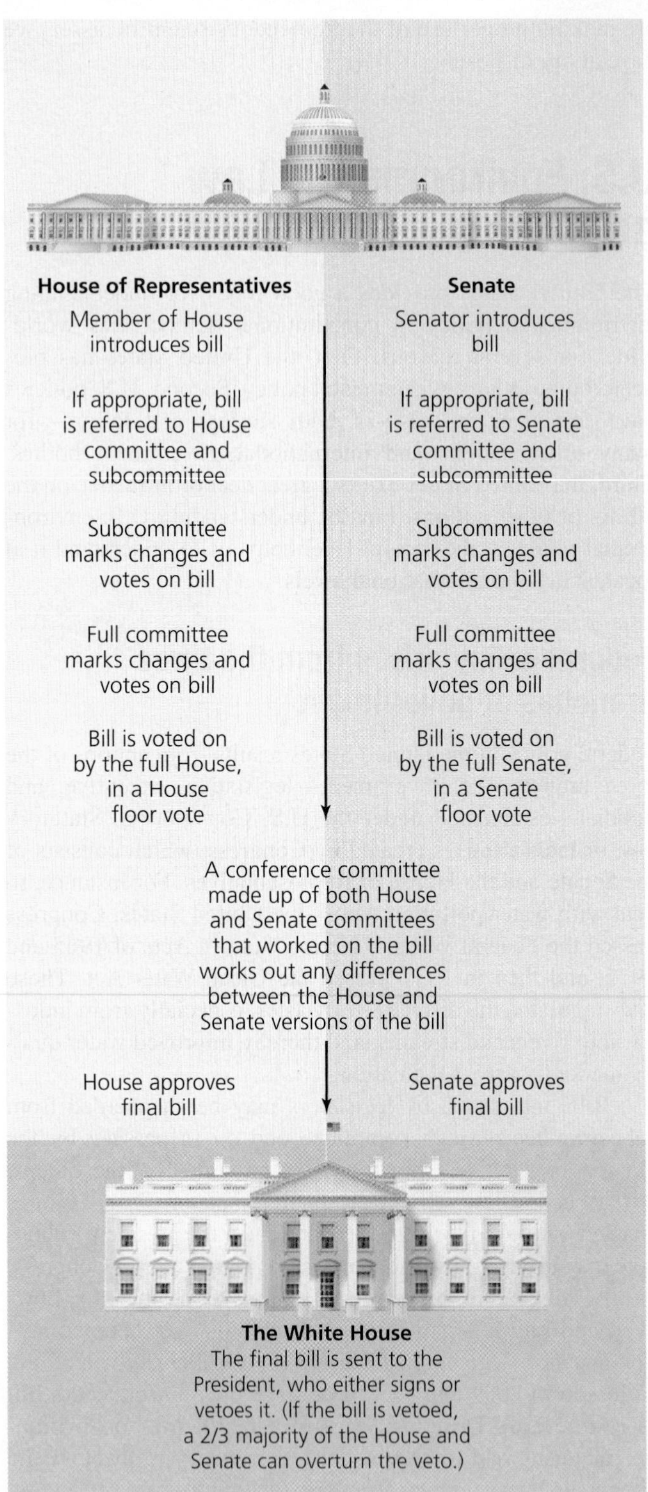

FIGURE 7.4 Before a bill becomes U.S. law, it must clear hurdles in both legislative bodies. If the bill passes the House and Senate, a conference committee works out differences between versions before the bill is sent to the president. The president may then sign or veto the bill.

administrative agencies monitor compliance with laws and regulations and enforce them when they are violated. For instance, the Environmental Protection Agency is an administrative agency that regulates some aspects of fossil fuel extraction and waste handling under various laws including the Clean Water Act. As some citizens call for stronger regulation of hydraulic fracturing, it is the EPA that would in most cases become involved.

The president may also issue *executive orders,* specific legal instructions for government agencies. For example, in 2012 President Obama issued an executive order that for the first time required oil and gas companies to publicly disclose the chemicals they use in hydraulic fracturing. However, the order required this disclosure only after drilling is complete. The White House had originally intended to require disclosure before a well was drilled, but heavy lobbying from industry in an election year caused it to bow to industry demands.

The judiciary, consisting of the Supreme Court and various lower courts, is charged with interpreting legislation. This is necessary because social norms, societal conditions, and technologies change over time, and because Congress must write laws broadly to ensure that they apply to varied circumstances throughout the nation. Decisions rendered by the courts make up a body of law known as *case law.* Previous rulings serve as *precedents,* or legal guides, for later cases, steering judicial decisions through time. The judiciary has been an important arena for environmental policy. Grassroots environmental advocates and nongovernmental organizations use lawsuits to help level the playing field with large corporations and government agencies. Conversely, the courts hear complaints from businesses and individuals challenging the constitutional validity of environmental laws they feel to be infringing on their rights. Individuals and organizations also lodge suits against government agencies when they feel the agencies are failing to enforce their own regulations.

Courts interpret the constitutionality of policy

The U.S. Constitution lays out several principles that have come to be especially relevant to environmental policy. One of these is from the Fifth Amendment, which ensures, in part, that private property shall not "be taken for public use without just compensation." Courts have interpreted this clause, known as the *takings clause*, to ban not only the literal taking of private property but also what is known as regulatory taking. A **regulatory taking** occurs when the government, by means of a law or regulation, deprives a property owner of all or some economic uses of his or her property. Many people cite the takings clause in opposing environmental regulations that restrict development on privately owned land. For example, some would contend that zoning regulations (p. 343) that prohibit a landowner from allowing gas drilling and hydrofracking in a residential neighborhood deprive the landowner of an economically valuable use of the land and, therefore, may violate the Fifth Amendment.

In a landmark case in 1992, the U.S. Supreme Court ruled that a state land use law intended to "prevent serious public harm" violated the takings clause. The case, *Lucas v. South Carolina Coastal Council,* involved a developer named Lucas who in 1986 purchased beachfront property in South Carolina for $975,000. In 1988, before Lucas began to build, South Carolina's legislature passed a law banning construction on eroding beaches (**FIGURE 7.5**). A state agency classified the Lucas property as an eroding beach and prohibited residential construction there. Lucas contested this decision, and the Supreme Court ruled in his favor, declaring that the state law deprived Lucas of

TABLE 7.1 Federal Administrative Agencies that Influence Environmental Policy

EXECUTIVE OFFICE OF THE PRESIDENT	**DEPARTMENT OF THE INTERIOR**
Council on Environmental Quality	Bureau of Indian Affairs Bureau of Land Management (BLM) Bureau of Reclamation Minerals Management Service National Park Service (NPS) Office of Surface Mining U.S. Fish and Wildlife Service (USFWS) U.S. Geological Survey (USGS)
DEPARTMENT OF AGRICULTURE (USDA)	
Natural Resources Conservation Service U.S. Forest Service (USFS)	
DEPARTMENT OF COMMERCE	**DEPARTMENT OF JUSTICE**
Bureau of the Census National Marine Fisheries Service National Oceanic and Atmospheric Administration (NOAA)	Environmental and Natural Resources Division
DEPARTMENT OF DEFENSE	**DEPARTMENT OF LABOR**
Army Corps of Engineers	Occupational Safety and Health Administration (OSHA)
DEPARTMENT OF ENERGY (DOE)	**DEPARTMENT OF STATE**
Energy Efficiency and Renewable Energy Energy Information Administration (EIA) Federal Energy Regulatory Commission (FERC) National Laboratories and Technology Centers Office of Environmental Management Office of Fossil Energy	Bureau of Oceans and International Environmental and Scientific Affairs Bureau of Population, Refugees, and Migration
	DEPARTMENT OF TRANSPORTATION
	Federal Transit Administration
DEPARTMENT OF HEALTH AND HUMAN SERVICES	**INDEPENDENT AGENCIES**
Agency for Toxic Substances and Disease Registry Centers for Disease Control and Prevention (CDC) Food and Drug Administration (FDA)	Consumer Product Safety Commission Environmental Protection Agency (EPA) National Aeronautics and Space Administration (NASA) National Transportation Research Center Nuclear Regulatory Commission (NRC) Tennessee Valley Authority (TVA) U.S. Agency for International Development (USAID)

Source: U.S. General Services Administration, Washington, D.C.

FIGURE 7.5 Should government be able to restrict development in areas where erosion, storms, and flooding pose risks to life and property, such as on this South Carolina beach? If so, does this constitute a taking of private property rights, and should the property owner be compensated? These questions were addressed in *Lucas v. South Carolina Coastal Council*.

all economically beneficial uses of his land. As a result, Lucas was allowed to build homes on the land. Regulatory taking remains a contentious area of law—a key issue in the sensitive balance between private rights and the public good.

State and local governments also make environmental policy

The structure of the federal government is mirrored at the state level with governors, legislatures, judiciaries, and agencies. States, counties, and municipalities all generate environmental policy of their own, and they frequently interact to address issues (FIGURE 7.6). For instance, in 2012 Pennsylvania's legislature passed a bill updating regulations on the natural gas industry, tightening safety standards, and imposing an "impact fee" on drillers. Pennsylvania had been the largest gas-producing state not to tax the industry, but as criticism of hydraulic fracturing grew and as state financial needs intensified, the fee became more desirable. Pennsylvania's law allows county and municipal governments to decide whether to impose the fee. Proceeds help to fund various state and

(a) Monitoring water quality

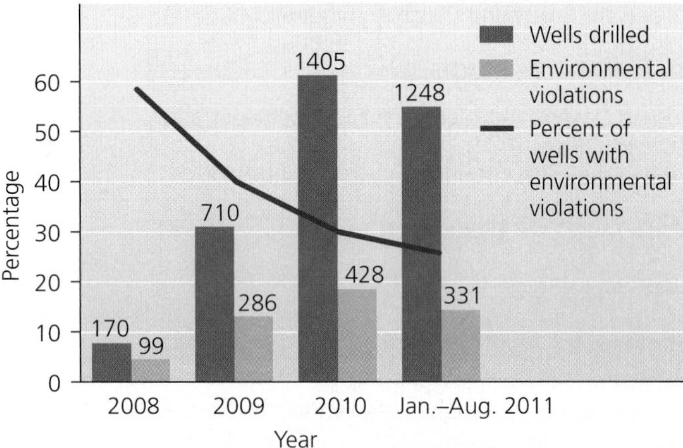

(b) Is fracking pollution decreasing or increasing?

FIGURE 7.6 **State and local governments administer a great deal of environmental policy.** Technicians take water samples from a home **(a)** to give regulators from Pennsylvania's Department of Environmental Protection (DEP) data on impacts of fracking. DEP staff also cite gas drillers for environmental violations. These data **(b)** showing a decline in the percent of wells cited for environmental violations were used by researchers to argue that fracking problems in Pennsylvania were decreasing as a result of effective regulation by the DEP. However, the researchers did not disclose that they received funding from the gas industry and that the report did not receive peer review (p. 12). In response, The State University of New York at Buffalo shut down the researchers' institute there. *Data from Considine, T., et al. 2012.* Environmental impacts during Marcellus Shale gas drilling: Causes, impacts, and remedies. *Shale Resources and Society Institute, State University of New York at Buffalo.*

local programs. Governor Tom Corbett signed the bill into law, but he had also asked legislators to prevent municipalities from regulating drilling. The legislators refused, and in the resulting compromise, the law forced towns to permit drilling in all areas, including residential zones, but allowed them to use their zoning standards to regulate noise, lighting, and structures.

Ideally, states and localities can act as laboratories for experimenting with novel policy concepts, so that ideas that succeed may be adopted elsewhere. Because Pennsylvania and New York have pursued different approaches to hydraulic fracturing so far, we will be able to compare results from the two approaches.

In New York, the state maintained a de facto moratorium on hydrofracking until 2012, when Governor Andrew Cuomo and the Department of Environmental Conservation (DEC) proposed banning fracking in the watersheds of New York City and Syracuse while allowing it in the rest of the state. Over 140 upstate New York towns responded by using their zoning powers to enact local bans on fracking, potentially setting up struggles among the governor, the legislature, and the towns to determine policy. Cuomo retreated from his stance in the face of widespread anti-fracking protests, and held off on a decision through much of 2013, as his administration assessed new scientific reports and as polls showed his state's citizens split evenly on whether to allow fracking.

For political leaders in New York, Pennsylvania, and 30 other states across the country, the benefits of shale gas development have been alluring. Development means jobs for rural communities, and proponents told New York policymakers that fracking could generate 50,000 jobs and $11 billion in economic output for their state. Moreover, because natural gas burns more cleanly than coal, it has gained wide appeal. Many advocates of sustainable energy view natural gas as a "bridge" between fossil fuels and cleaner renewable sources, although others think pursuing shale gas will simply commit us more deeply to fossil fuels. Policymakers must weigh all these considerations along with the potential risks to health and water supplies.

State laws cannot violate principles of the U.S. Constitution, and if state and federal laws conflict, federal laws take precedence. Federal policymakers may influence environmental policy at the level of the states by:

- Supplanting state law to force change. (This is uncommon.)
- Providing financial incentives to encourage change. (This can be effective if federal funding is adequate and states need the money.)
- Following an approach of "cooperative federalism," whereby a federal agency sets national standards and then works with state agencies to achieve them in each state. (This is most common.)

Sometimes relations between federal and state governments grow testy. When the U.S. EPA came in to test Dimock's water, the head of Pennsylvania's DEP complained that this was undue interference in state affairs. In recent years, pressure to weaken federal oversight and hand over power to the states has grown, but political scientists argue that retaining strong federal control over environmental policy is a good idea for several reasons:

- Citizens of all states should have equitable protection against environmental and health impacts.
- Dealing with environmental problems often requires an "economy of scale" in resources, such that one strong national effort is far more efficient than 50 state efforts.
- Many issues involve "transboundary" disputes that cross state lines, and nationwide efforts minimize disputes among states.

As we proceed through our discussion of federal policy, keep in mind that environmental policy is also created at the state and local levels.

Early U.S. environmental policy promoted development

U.S. environmental policy was created in three periods. Laws enacted during the first period, from the 1780s to the late 1800s, accompanied the westward expansion of the nation and were intended mainly to promote settlement and the extraction and use of the continent's abundant natural resources. Among these early laws were the *General Land Ordinances of 1785 and 1787,* which gave the federal government the right to manage unsettled lands and created a grid system for surveying them and readying them for private ownership. From 1785 to the 1870s, the government promoted settlement in the Midwest and West and doled out to its citizens the lands it had expropriated from Native Americans.

Western settlement provided U.S. citizens with means to achieve prosperity and also served to relieve crowding in Eastern cities. It expanded the geographical reach of the United States at a time when the young nation was still jostling with European powers for control of the continent. It also wholly displaced the millions of Native Americans who had long inhabited these lands. U.S. environmental policy of this era reflected the perception that Western lands were vast and inexhaustible in natural resources. The following are a few laws typical of this era:

- The *Homestead Act of 1862* allowed any citizen to claim, for a $16 fee, 65 ha (160 acres) of public land by living there for 5 years and cultivating the land or building a home (**FIGURE 7.7a**). Those who could pay $176 were granted a waiting period of only 14 months.
- The *General Mining Act of 1872* legalized and promoted mining by private individuals on public lands for just $5 per acre, subject to local customs, with no government oversight (**FIGURE 7.7b**). This law is still on the books today (pp. 182, 646).
- The *Timber Culture Act of 1873* granted 65 ha (160 acres) to any citizen promising to cultivate trees on one-quarter of that area (**FIGURE 7.7c**).

FIGURE 7.7 Early U.S. environmental policy promoted settlement and natural resource extraction. Settlers **(a)** moved west with the help of land policies such as the Homestead Act of 1862. Mining activities **(b)** on public lands were largely unregulated under laws such as the General Mining Act of 1872. The nation's ancient forests were cut **(c)**, even as the Timber Culture Act of 1873 promoted tree planting on settled agricultural lands.

(a) Settlers in Nebraska, circa 1860

(c) Loggers felling an old-growth tree, Washington

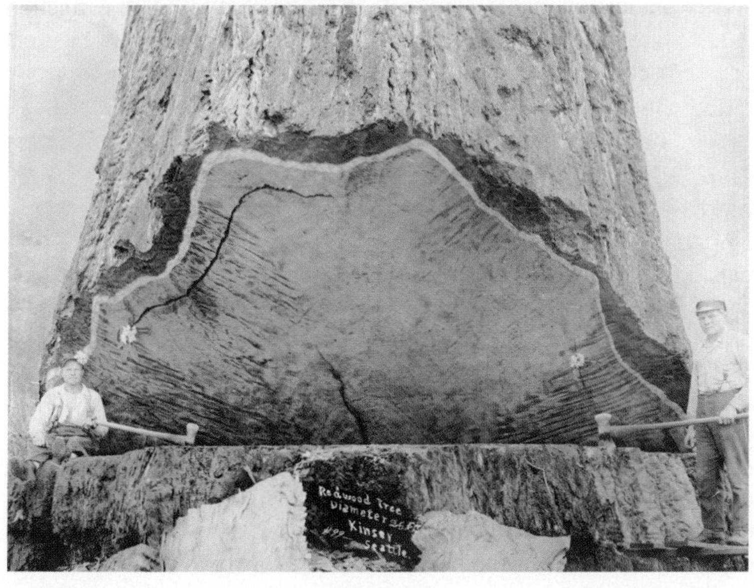

(b) Nineteenth-century mining operation, Alaska

In addition, many millions of acres of land were doled out to railroad companies, which built rail lines to transport people, resources, and goods across the continent. All these policies encouraged settlers, entrepreneurs, and land speculators to move west.

The second wave of U.S. environmental policy encouraged conservation

In the late 1800s, as the continent became more populated and its resources were increasingly exploited, public perception and government policy toward natural resources began to shift. Reflecting the emerging conservation and preservation ethics (pp. 138–139) in American society, laws of this period aimed to alleviate some of the environmental impacts of westward expansion.

In 1872, Congress designated Yellowstone as the world's first national park. In 1891, Congress authorized the president to create "forest reserves" in order to prevent overharvesting and protect forested watersheds. In 1903, President Theodore Roosevelt created the first national wildlife refuge. These acts enabled the creation of a national park system, national forest system, and national wildlife refuge system that still stand as global models (pp. 316, 323). These developments reflected a new understanding that the West's resources were exhaustible and required legal protection.

Land management policies continued through the 20th century, targeting soil conservation after the Dust Bowl years (pp. 224–225) and wilderness preservation with the Wilderness Act of 1964 (pp. 324–325), which sought to preserve pristine lands "untrammeled by man, where man himself is a visitor who does not remain."

The third wave of U.S environmental policy responded to pollution

Further social changes in the 20th century gave rise to the third major period of U.S. environmental policy. In a more densely populated nation driven by technology, heavy industry, and intensive resource consumption, Americans found themselves better off economically but living amid dirtier air, dirtier water, and more waste and toxic chemicals. During the 1960s and 1970s, several events triggered increased awareness of environmental problems, bringing about a shift in public priorities and important changes in public policy.

A landmark event was the 1962 publication of *Silent Spring,* a book by American scientist and writer Rachel Carson (**FIGURE 7.8**). *Silent Spring* awakened the public to the ecological and health impacts of pesticides and industrial chemicals (pp. 368–369). The book's title refers to Carson's warning that pesticides might kill so many birds that few would be left to sing in springtime.

Ohio's Cuyahoga River (**FIGURE 7.9**) also drew attention to the hazards of pollution. The Cuyahoga was so polluted with oil and industrial waste that the river actually caught fire near Cleveland a number of times in the 1950s and 1960s. This spectacle, coupled with an enormous oil spill off the Pacific coast near Santa Barbara, California, in 1969, moved the public to prompt

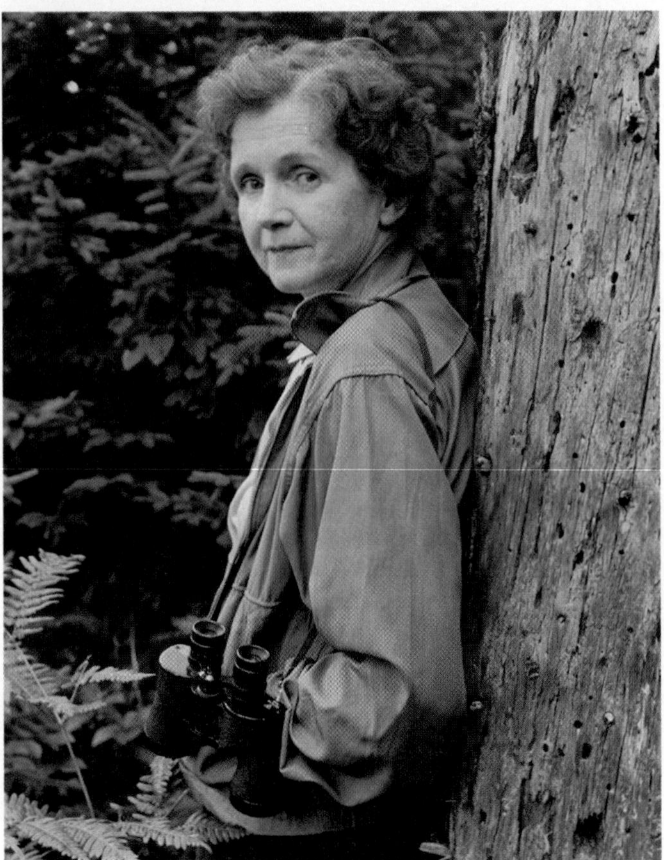

FIGURE 7.8 Scientist and writer Rachel Carson illuminated the problem of pollution from DDT and other pesticides in her 1962 book, *Silent Spring*.

Congress and the president to better safeguard the environment and public health. The first Earth Day event in 1970 helped to galvanize public support for action to address pollution problems.

Today, largely because of policies enacted since the 1960s, our health is better protected and the nation's air and water are considerably cleaner. Thanks to the many citizens who worked tirelessly in grassroots efforts, and to the policymakers who listened and chose to make a difference in people's lives, today we enjoy a cleaner environment where industrial chemicals, waste disposal, and resource extraction are more carefully regulated. All of us alive today owe a great deal to the dedicated people who designed policy to tackle pollution problems during this period.

NEPA and the EIS process grant citizens input

One of the foremost U.S. environmental laws is the **National Environmental Policy Act (NEPA)**, signed into law by Republican President Richard Nixon on January 1, 1970. NEPA created an agency called the Council on Environmental Quality and required that an **environmental impact statement (EIS)** be prepared for any major federal action that might significantly affect environmental quality. An EIS is a report of results from detailed studies that assess the potential impacts on the environment that would likely result from a development project undertaken or funded by the federal government.

FIGURE 7.9 **Ohio's Cuyahoga River was so polluted with oil and waste that the river caught fire multiple times in the 1950s and 1960s and would burn for days at a time.**

The EIS process forces government agencies and businesses that contract with them to evaluate environmental impacts before proceeding with a new dam, highway, or building project. The EIS process uses a cost-benefit approach (p. 146) and generally does not halt development projects. However, it does provide incentives to lessen environmental damage. NEPA also grants ordinary citizens input in the policy process by requiring that environmental impact statements be made publicly available and that public comment on them be solicited and considered.

State governments have adopted EIS processes as well, and examining New York's experience with regard to hydraulic fracturing provides an example of how the process works. In 1992 New York developed a Generic EIS that set parameters for permitting oil and gas wells in general. In 2009 a Draft Supplemental Generic EIS (SGEIS) for hydrofracking in particular was prepared and released for public review. The public submitted over 13,000 comments. In 2010 Governor David Paterson ordered the Department of Environmental Conservation (DEC) to conduct further review, and in 2011 the DEC released a Revised Draft SGEIS. This time, a record-setting 66,000 public comments were received. After reviewing these comments, the Final SGEIS will be prepared, which will lay out parameters for how the Department should review individual hydrofracking actions. The DEC would then assess proposed hydrofracking actions one-by-one with EIS processes, and would issue permits for those that meet the terms of the GEIS and the SGEIS.

Creation of the EPA marked a shift in environmental policy

Six months after signing NEPA into law, Nixon issued an executive order calling for a new integrated approach to environmental policy. "The Government's environmentally related activities have grown up piecemeal over the years," the order stated. "The time has come to organize them rationally and systematically." Nixon's order moved elements of agencies regulating water quality, air pollution, solid waste, and other issues into the newly created **Environmental Protection Agency (EPA)**. The order charged the EPA with conducting and evaluating research, monitoring environmental quality, setting and enforcing standards for pollution levels, assisting the states in meeting the standards, and educating the public.

FAQ **Isn't the EPA an advocate for the environment?**

Like all administrative agencies, the EPA is part of the executive branch and operates in line with the policies of the presidential administration in power at the time. As such, the EPA under one president may function very differently from the EPA under another one. Indeed, sometimes the agency may impede environmental regulations! The EPA employs many dedicated scientists who carry out careful research and make scientifically informed policy recommendations. They advise administrators appointed by the president, however, and policy decisions are ultimately made by these politically appointed administrators.

Other prominent laws followed

Ongoing public demand for a cleaner environment during this period resulted in a number of major laws that remain fundamental to U.S. environmental policy (**FIGURE 7.10**). These laws helped to clean up air and water, protect rare and endangered species, and control hazardous waste and toxic substances.

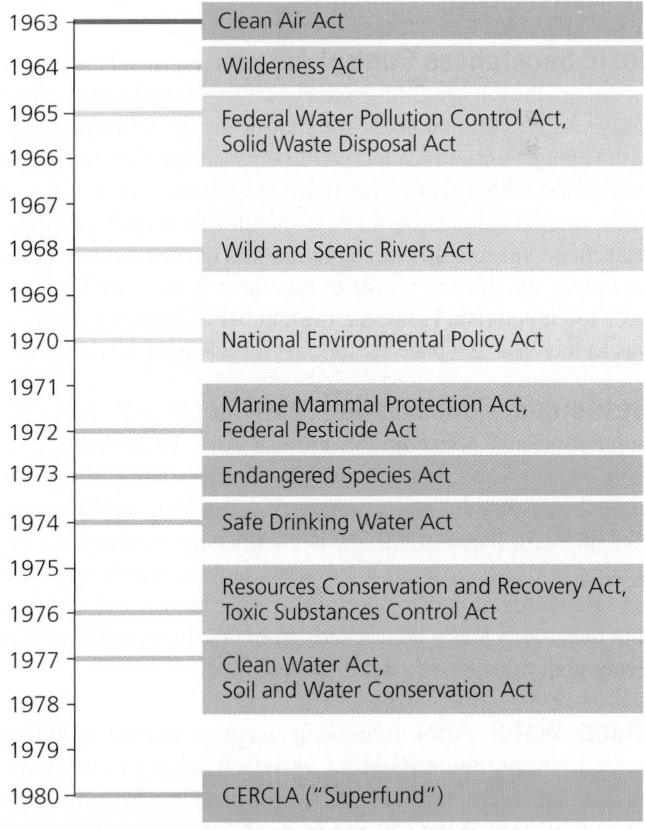

FIGURE 7.10 **Most major U.S. environmental laws were enacted in the 1960s and 1970s.**

You will encounter most of them in later chapters of this book, and they have already helped to shape the quality of your life.

Clean Air Act By the 1960s and 1970s, air pollution in the United States from automobile traffic and industry had become severe, and was contributing to tens of thousands of deaths every year. The Clean Air Act of 1963 and its major amendments of 1970 and 1990 turned this situation around. This legislation sets standards for air quality, imposes limits on emissions from new sources, enables citizens to sue violators, funds research on pollution control, and established an emissions trading program for sulfur dioxide. These measures have improved air quality markedly; the air we breathe today is far cleaner thanks to the Clean Air Act (pp. 458–462).

Endangered Species Act Habitat loss and other pressures had driven a number of species extinct and were threatening many more. The Endangered Species Act was passed in 1973 to protect species threatened with extinction. It forbids the destruction of individuals of listed species or their habitat on public and private land, and it provides funding for recovery efforts (p. 296). Subsequent agreements allowed for negotiation with private landholders.

Safe Drinking Water Act While surface waterways like the Cuyahoga River were being fouled with pollution, groundwater in aquifers across the nation was likewise becoming contaminated from industrial waste and other sources. The Safe Drinking Water Act of 1974 authorized the EPA to set quality standards for tap water provided by public water systems, and to work with states to protect drinking water sources from contamination.

Toxic Substances Control Act As industrial chemistry advanced during the 20th century, we began producing more and more novel chemicals for industrial and consumer use, even though virtually none were being adequately tested for potential health effects. The Toxic Substances Control Act of 1976 (p. 383) directs the EPA to monitor thousands of industrial chemicals manufactured or imported into the United States and gives the agency power to ban them if they are found to pose too much risk. However, the number of chemicals continues to increase at a pace far too fast for adequate testing.

Resources Conservation and Recovery Act As population and consumption grew, so did the generation of solid waste—and with industrial development, hazardous waste proliferated. The Resources Conservation and Recovery Act (p. 613), passed in 1976, is the primary federal law pertaining to the disposal of solid waste and hazardous waste. It sets standards, mandates permitting procedures, and requires that hazardous waste be tracked "from cradle to grave" as it is generated, transported, and disposed of.

Clean Water Act Prior to passage of federal pollution laws, water pollution problems were left largely to local and state governments or were addressed through lawsuits. The flaming waters of the Cuyahoga, however, indicated to many people that tough federal legislation was needed. Thanks to restrictions on pollutants by the Federal Water Pollution Control Acts of 1965 and 1972, and then the Clean Water Act of 1977, U.S. waterways finally began to recover. The Clean Water Act (pp. 413–414) regulates the discharge of wastes, especially from industry, into rivers and streams. It aims to protect wildlife as well as human health, and it established a system for granting permits for the discharge of pollutants.

Soil and Water Conservation Act Although U.S. farmers and policymakers learned lessons about the importance of conserving topsoil during the Dust Bowl (pp. 224–225), soil erosion and water pollution in agricultural areas worsened again later as production intensified on farms and rangeland. The Soil and Water Conservation Act, passed in 1977, directs the U.S. Department of Agriculture to survey and assess soil and water conditions across the nation periodically and prepare national plans for conservation. Results have been mixed, but the country has seen many success stories in soil and water conservation as a result.

CERCLA ("Superfund") By 1980, the United States was spotted with thousands of sites where years of unregulated pollution had severely contaminated land and water. Highly publicized instances of buried hazardous waste threatening people's health in locations such as Love Canal, New York (pp. 626, 627), led Congress that year to pass the Comprehensive Environmental Response Compensation and Liability Act (CERCLA), commonly called the Superfund Act (p. 627). This law provides a funded program to clean up hazardous waste from the nation's worst polluted sites. Costs have been staggering, but the EPA continues to progress through the many sites that remain.

Throughout the 1980s Congress strengthened, broadened, and elaborated upon the laws of the 1970s. For example, major amendments were made to the Clean Water Act in 1987 and to the Clean Air Act in 1990. Today thousands of federal, state, and local laws and regulations help protect health and environmental quality in the United States and abroad. Public enthusiasm for environmental protection remains strong, with polls repeatedly showing that an overwhelming majority of Americans favor environmental protection.

The social context for policy evolves over time

Historians suggest that major advances in environmental policy occurred in the 1960s and 1970s because several factors converged. First, environmental problems became widely and readily apparent and were directly affecting people's lives. Second, people could visualize policies to deal with the problems. Third, the political climate was ripe, with a supportive public and leaders who were willing to act. In addition, photographs from the space program allowed humanity to see, for the first time ever, images of Earth from space (see photos on pp. 2 and 671). It is hard for us today to comprehend the power of those images at the time, but they revolutionized many people's worldviews by making us aware of the finite nature of our planet.

By the 1990s, the political climate in the United States had changed. Although public support for the goals of environmental protection remained high, many citizens and policy experts began to feel that the legislative and regulatory means used to achieve these goals too often imposed economic burdens on businesses or individuals. Increasingly, attempts were made at the federal level to roll back or weaken environmental laws. These began with the administration of Ronald Reagan and culminated in an array of efforts by the George W. Bush administration and by the Congresses in power from 1994 through 2006.

As advocates of environmental protection watched their hard-won gains eroding, many began to feel that new perspectives and strategies were needed. In a provocative 2004 essay titled "The Death of Environmentalism," political consultants Michael Shellenberger and Ted Nordhaus argued that environmental advocates needed to appeal to people's core values and not simply offer technical policy fixes. They needed to stop labeling problems as "environmental" and start showing people why these problems are actually human issues that lie at the heart of our quality of life. They needed to be more responsive to people's needs and to articulate a positive, inspiring vision for the future. Many environmental advocates reacted defensively to Shellenberger and Nordhaus's suggestions, but their views opened a productive discussion, and in 2008 Barack Obama embraced a similar approach in his presidential campaign.

Today in the United States, legal protections for public health and environmental quality remain strong in some areas but have eroded in others. As energy issues move to the fore, people continue to experience impacts from fossil fuel use and extraction while striving to find a path toward cleaner energy. As policymakers sought to encourage natural gas extraction, hydraulic fracturing won exemptions from the Safe Drinking Water Act and at least six more of the nation's most fundamental environmental laws (p. 163). Fracking is exempted from key aspects of the National Environmental Policy Act, the Clean Air Act, the Clean Water Act, the Resource Conservation and Recovery Act, the Superfund Act, and the Emergency Planning and Community Right To Know Act.

Environmental policy advances today on the international stage

Amid the heightened partisanship of today's American politics, environmental policy has gotten caught in the political crosshairs. Despite the fact that some of the greatest early conservationists were Republicans, and despite the fact that the words *conservative* and *conservation* share the same root and original meaning, environmental issues have today become identified as a predominantly Democratic concern. The result is that significant bipartisan advances now rarely occur. Consequently, the United States now wields far less clout internationally on environmental policy.

Although U.S. leadership has waned internationally, many other nations are forging ahead with innovative environmental policy. Germany has used policy to make impressive strides with solar energy (pp. 581–582). Sweden maintains a thriving society while promoting progressive environmental policies. China, despite becoming the world's biggest polluter, is also taking significant steps toward renewable energy, reforestation, and pollution control.

Worldwide, we have embarked on a fourth wave of environmental policy, one focused on sustainability and sustainable development (pp. 14, 156–157). This approach aims to safeguard natural systems while raising living standards for the world's people. In 2012, the world's nations met in Río de Janeiro, Brazil, at the U.N.-sponsored Rio-Plus-20 conference, to explore the latest strategies for promoting economic vitality and social equity while safeguarding environmental quality. This conference built on the 1992 Earth Summit at Río de Janeiro and the 2002 World Summit on Sustainable Development in Johannesburg, South Africa, each of which unified leaders from 200 nations.

Moreover, the pressing issue of global climate change (Chapter 18) has come to dominate global discussion of environmental policy (**FIGURE 7.11**). A series of international conferences (pp. 510–511) in recent years has brought together representatives of all the world's nations to grapple with the

FAQ **If something is harmful, wouldn't the government have made it illegal?**

This assumption vastly overestimates the power of government. In fact, we live surrounded by risky or hazardous things that remain perfectly legal. Some (such as junk food, alcohol, or cigarettes) persist because they are popular. Others (such as fossil fuel pollution) persist because mitigating the problem is costly, complicated, or grand in scale. Still others (such as many toxic chemicals) go unregulated because they are released to consumers more quickly than scientists can determine their health effects. In many cases, financially valuable products or practices that harm health or the environment have politically powerful constituencies. Corporations and industries lobby policymakers to shield their products or practices from regulation, fearing that regulation could adversely affect sales and profits. Often these powerful societal, economic, and political pressures cause health and environmental hazards to go unaddressed by government.

FIGURE 7.11 Concerns over climate change are driving environmental policy in all nations today. Here, college students and activists in Washington, D.C., urge U.S. leaders to enact policies to help bring the atmosphere's carbon dioxide concentration back down to 350 parts per million.

vexing issues of how to reduce the greenhouse gas emissions that drive climate change. As we continue to feel the social, economic, and ecological effects of climate change and other environmental impacts, environmental policy and the search for sustainable solutions will become central parts of governance and everyday life for all of us in the years ahead.

International Environmental Policy

Environmental systems pay no heed to political boundaries, so environmental problems often are not restricted to the confines of particular nations. Because one nation's laws have no authority in other nations, international policy is vital to solving transboundary problems. However, international law is often more nebulous in its origins and weaker in its authority than national law. As environmental scientist Hilary French has noted, "The world is still composed of nation-states that view themselves as sovereign, meaning that international law has the force of moral suasion, but few if any real teeth."

International law includes customary law and conventional law

International law known as **customary law** arises from long-standing practices, or customs, held in common by most cultures. In contrast, international law known as **conventional law** arises from **conventions,** or **treaties** (written contracts among nations), into which nations enter. One example is the United Nations Framework Convention on Climate Change, a 1994 treaty that established a framework for agreements to reduce greenhouse gas emissions that contribute to global climate change. The Kyoto Protocol (a *protocol* is an amendment or addition to a convention) later specified the agreed-upon details of the emissions limits (p. 510). Another example is the Vienna Convention for the Protection of the Ozone Layer and its subsequent Montreal Protocol, an accord to reduce the emission of airborne chemicals that deplete the ozone layer (p. 472). TABLE 7.2 shows a selection of major environmental treaties.

Treaties such as the Montreal Protocol and the Kyoto Protocol are truly global in scope and have been ratified by most of the world's nations. However, treaties are also signed among pairs or groups of nations. The United States, Mexico, and Canada entered into the **North American Free Trade Agreement (NAFTA)** in 1994 (FIGURE 7.12). NAFTA eliminated trade barriers such as tariffs on imports and exports to make goods cheaper for everyone. Yet NAFTA also undermined protections for workers and the environment, although side agreements were negotiated to try to address these concerns. NAFTA's impacts on jobs and on environmental quality in the three nations have been complex and far-reaching.

Several organizations shape international environmental policy

A number of international organizations act to influence the policy and behavior of nations by providing funding, applying political or economic pressure, and directing media attention.

FIGURE 7.12 **The North American Free Trade Agreement (NAFTA) eliminated trade barriers to make goods cheaper.** However, some U.S. manufacturing jobs moved to Mexico (such as to this garment factory in Tehuacan), where wages are lower and health and environmental regulations are more lax. Overall, NAFTA's impacts on jobs and on environmental quality in the three nations have been mixed.

The United Nations Founded in 1945 and including representatives from all nations of the world, the **United Nations (U.N.)** seeks "to maintain international peace and security; to develop friendly relations among nations; to cooperate in solving international economic, social, cultural, and humanitarian problems and in promoting respect for human rights and fundamental freedoms; and to be a centre for harmonizing the actions of nations in attaining these ends."

Headquartered in New York City, the United Nations plays an active role in shaping international environmental policy by sponsoring conferences, coordinating treaties, and publishing research. An agency within it, the *United Nations Environment Programme* (*UNEP*), promotes sustainability with research and outreach activities that provide information to policymakers and scientists throughout the world.

The World Bank Established in 1944 and based in Washington, D.C., the **World Bank** is one of the largest sources of funding for economic development. This institution shapes environmental policy through its funding of dams, irrigation infrastructure, and other major development projects. In fiscal year 2012, the World Bank provided $35 billion in loans and support for projects designed to benefit the poorest people in the poorest countries.

Despite its admirable mission, the World Bank has frequently been criticized for funding unsustainable projects that cause more environmental problems than they solve, such as dams that flood valuable forests and farmland in order to provide electricity. Providing for the needs of growing human populations in poor nations while minimizing damage to the environmental systems on which people depend can be a tough balancing act. Environmental scientists today agree that the concept of sustainable development must be the guiding principle for such efforts.

TABLE 7.2 Major International Environmental Treaties

CONVENTION OR PROTOCOL	YEAR IT CAME INTO FORCE	NATIONS THAT HAVE RATIFIED IT	U.S. STATUS
Convention on International Trade in Endangered Species of Wild Fauna and Flora (CITES) (p. 297)	1975	175	Ratified
Ramsar Convention on Wetlands of International Importance (p. 405)	1975	159	Ratified
Protocol on Substances that Deplete the Ozone Layer (Montreal Protocol), of the Vienna Convention for the Protection of the Ozone Layer (p. 472)	1989	196	Ratified
Basel Convention on the Control of Transboundary Movements of Hazardous Wastes and Their Disposal (p. 626)	1992	172	Signed but has not ratified
Convention on Biological Diversity (p. 297)	1993	168	Signed but has not ratified
Stockholm Convention on Persistent Organic Pollutants (p. 384)	2004	152	Signed but has not ratified
Kyoto Protocol, of the UN Framework Convention on Climate Change (p. 510)	2005	184	Signed but has not ratified

The European Union The **European Union (EU)** seeks to promote Europe's unity and its economic and social progress (including environmental protection) and to "assert Europe's role in the world." The EU can sign binding treaties on behalf of its 27 member nations and can enact regulations that have the same authority as national laws. It can also issue *directives*, which are more advisory in nature. The EU's European Environment Agency addresses waste management, noise pollution, water pollution, air pollution, habitat degradation, and natural hazards. The EU also seeks to remove trade barriers among member nations. It has classified some nations' environmental regulations as barriers to trade, arguing that the stricter environmental laws of some northern European nations limit the import and sale of environmentally harmful products from other member nations.

The World Trade Organization Based in Geneva, Switzerland, the **World Trade Organization (WTO)** represents multinational corporations and promotes free trade by reducing obstacles to international commerce and enforcing fairness among nations in trading practices. The WTO has authority to impose financial penalties on nations that do not comply with its directives. These penalties can sometimes affect environmental policy.

Like the EU, the WTO has interpreted some national environmental laws as unfair barriers to trade. For instance, in 1995 the U.S. EPA issued regulations requiring cleaner-burning gasoline in U.S. cities, following Congress's amendments of the Clean Air Act. Brazil and Venezuela filed a complaint with the WTO, saying the new rules discriminated against the petroleum they exported to the United States, which did not burn as cleanly. The WTO agreed, ruling that even though the South American gasoline posed a threat to human health in the United States, the EPA rules were an illegal trade barrier. The ruling forced the United States to weaken its regulations of gasoline. Not surprisingly, critics have frequently charged that the WTO aggravates environmental problems.

WEIGHING THE ISSUES

TRADE BARRIERS AND ENVIRONMENTAL PROTECTION If Nation A has stricter laws for environmental protection than Nation B, and if these laws restrict the ability of Nation B to export its goods to Nation A, then by the policy of the WTO and the EU, Nation A's environmental protection laws can be overruled in the name of free trade. Do you think this is right? What if Nation A is a wealthy industrialized country and Nation B is a poor developing country that needs every economic boost it can get?

Nongovernmental Organizations A number of **nongovernmental organizations (NGOs)** have become international in scope and exert influence over international policy. These groups are diverse in their size and mission; those that advocate for aspects of environmental protection are known as environmental NGOs ("ENGOs"). Groups such as the Nature Conservancy focus on accomplishing conservation objectives on the ground (for example, purchasing and managing land and habitat for rare species) without becoming politically involved. Other groups, such as Conservation International, the World Wide Fund for Nature, Greenpeace, and Population Connection, attempt to shape policy through research, education, lobbying, or protest. NGOs apply more funding and expertise to environmental problems—and conduct more research intended to solve them—than do many national governments.

International institutions wield influence in a globalizing world

As globalization proceeds, our world becomes ever more interconnected. As a result, human societies and Earth's ecological systems are being altered at unprecedented rates and scales. Trade and technology have expanded the global reach of all societies. Those, such as the United States, that import and consume goods and resources from far and wide exert extensive impacts on the planet's environmental systems. Multinational corporations operate outside the reach of national laws and rarely have incentive to conserve resources or conduct their business sustainably in the nations where they operate. For all these reasons, in today's globalizing world the organizations and institutions that shape international policy are becoming increasingly influential.

Approaches to Environmental Policy

When most of us think of environmental policy, what comes to mind are major laws, such as the Clean Water Act, or government regulations, such as those limiting what an industry can dump into a water supply. However, environmental policy is diverse and can follow three major approaches (**FIGURE 7.13**). These approaches include lawsuits in the courts, command-and-control policy, and economic policy tools.

Conflicts can be addressed in court

Prior to the legislative push of recent decades, most environmental policy questions in the United States were addressed with lawsuits in the courts through *tort law*, which is law that deals with harm caused to one entity by another. (The word *tort* is French for a wrong or an injustice.) Most pollution issues were subject primarily to nuisance law, one form of tort law. Individuals suffering external costs from pollution would seek redress through lawsuits against polluters, one case at a time, much as some residents of Dimock, Pennsylvania, are doing today against Cabot Oil and Gas. The courts sometimes punished polluters by ordering them to stop their operations or pay damages to the affected individuals. However, as industrialization proceeded and population grew denser, pollution became harder to avoid and judges were reluctant to hinder industry, which was viewed to be promoting society's economic development.

In 1970, the U.S. Supreme Court heard the case *Boomer v. Atlantic Cement Company*. The court ruled that residents of Albany, New York, who were suffering pollution from a nearby cement plant were entitled to financial compensation. However, the court refused to shut down the plant. Instead, it allowed the plant to continue operating once it paid the residents. The court had calculated that the economic costs to the company of controlling its pollution were greater than the economic costs of the pollution to the residents, and the court based its decision on an attempt to minimize overall costs. In handing down this ruling, the justices essentially let the market decide between right and wrong. For people concerned about the pervasive spread of pollution through society, rulings like these showed that tort law was no longer a viable avenue for preventing pollution. People began to view legislation and regulation as more effective means of protecting public health and safety.

PROBLEM
Pollution from factory harms people's health

SOLUTIONS
Three policy approaches

1 People can sue factory in court.

2 Government can regulate emissions.

3 Economic policy tools can create incentives: A factory that pollutes less (right) will outcompete one that pollutes more (left) through permit-trading, avoiding green taxes, or selling ecolabeled products.

FIGURE 7.13 Three major policy approaches exist to resolve environmental problems. To address pollution from a factory, we might 1 seek damages through lawsuits, 2 limit pollution through legislation and regulation, or 3 reduce pollution using market-based strategies.

Command-and-control policy has improved our lives

Most environmental laws and regulations of recent decades use a **command-and-control** approach, in which a regulating agency prohibits certain actions—or sets rules, standards, or limits—and threatens punishment for those who violate these terms. This simple and direct approach to policymaking has brought citizens of the United States and many other nations cleaner air, cleaner water, safer workplaces, healthier neighborhoods, and many other advances. The relatively safe, healthy, comfortable lives most of us enjoy today owe much to the command-and-control environmental policy of the past few decades.

Even in plain financial terms, putting health and quality of life aside, command-and-control policy has been effective. For over a decade, the White House Office of Management and Budget has undertaken yearly analyses of U.S. regulatory policy to determine whether regulations result in more economic costs or more economic benefits. These analyses have consistently revealed that benefits outweigh costs, and that environmental regulations have been most beneficial of all. You can explore some of these data in *Calculating Ecological Footprints* (p. 186).

Despite these successes, many people have grown disenchanted with the top-down nature of the command-and-control approach and have come to view government mandates as restrictions on their freedom. Sometimes government actions are well intentioned but not well enough informed, so they can lead to unforeseen consequences. Regulatory policy can also fail if a government does not live up to its responsibilities to protect its citizens or treat them equitably. This may occur when leaders allow themselves to be unduly influenced by **interest groups,** small groups of people seeking private gain that work against the larger public interest.

Economic tools can help achieve policy goals

The most common critique of command-and-control policy is that it achieves its goals in a more costly and less efficient manner than the marketplace can. Whereas command-and-control policy mandates particular solutions to problems, private entities competing in a free market can innovate and may produce new or better solutions at lower cost. As a result, political scientists, economists, and policymakers today are exploring alternative policy approaches that aim to channel the innovation and economic efficiency of the market in directions that benefit the public. Such economic policy tools use financial incentives to encourage desired outcomes, discourage undesired outcomes, and set market dynamics in motion to achieve goals in an economically efficient manner.

Like regulation and like the tort law approach, economic policy tools aim to "internalize" external costs suffered by the public by building these costs into market prices. Each of these three approaches has strengths and weaknesses, and each is best suited to different conditions. The approaches may also be used together. For instance, government regulation is often needed to frame market-based efforts, and citizens can use the courts to ensure that regulations are enforced. Let's now explore several main types of economic policy tools: taxes, subsidies, and permit trading.

Green taxes discourage undesirable activities

The most straightforward economic policy tool—taxation—can be used to discourage undesirable activities. In taxation, money passes from private parties to the government, which uses it to pay for services to benefit the public. Taxing undesirable activities helps to internalize external costs by making them part of the cost of doing business. Taxes on environmentally harmful activities and products are called **green taxes.** When a business pays a green tax, it is essentially reimbursing the public for environmental damage it causes.

Under green taxation, a firm owning a factory that pollutes a waterway would pay taxes on the amount of pollution it discharges—the more pollution, the higher the tax payment. This gives firms a financial incentive to reduce pollution while allowing them the freedom to decide how best to do so. One polluter might choose to invest in technologies to reduce pollution if doing so is more affordable than paying the taxes. Another polluter might find abating its pollution more costly and could choose to pay the taxes instead—funds the government might then apply toward mitigating pollution in some other way.

Green taxes have yet to gain wide support in the United States, although similar "sin taxes" on cigarettes and alcohol are long-accepted tools of U.S. social policy. Taxes on pollution have been widely instituted in Europe, where many nations have adopted the polluter-pays principle (p. 168). Today there is debate worldwide about whether we should implement carbon taxes—taxes on gasoline, coal-based electricity, and fossil-fuel-intensive products according to the carbon emissions they produce—in order to fight global climate change (p. 513).

Green taxation provides incentive for industry to lower emissions not merely to a level specified in a regulation, but to still-lower levels. However, green taxes do have drawbacks: Businesses will most likely pass on their tax expenses to consumers, and these increased costs may affect low-income consumers disproportionately more than high-income ones.

Subsidies promote certain activities

Another type of economic policy tool is the **subsidy,** a government giveaway of money or resources that is intended to support and promote an industry or activity. Subsidies take many forms. A *tax break* is a common form of subsidy. Relieving the tax burden on an industry, firm, or individual assists it by reducing its expenses. Because a tax break deprives the government's treasury of funds it would otherwise collect, it has the same financial effect as a direct giveaway of money.

Subsidies can be used to promote environmentally sustainable activities, but all too often they are used to prop up unsustainable ones. From 2002 to 2008, the U.S. government

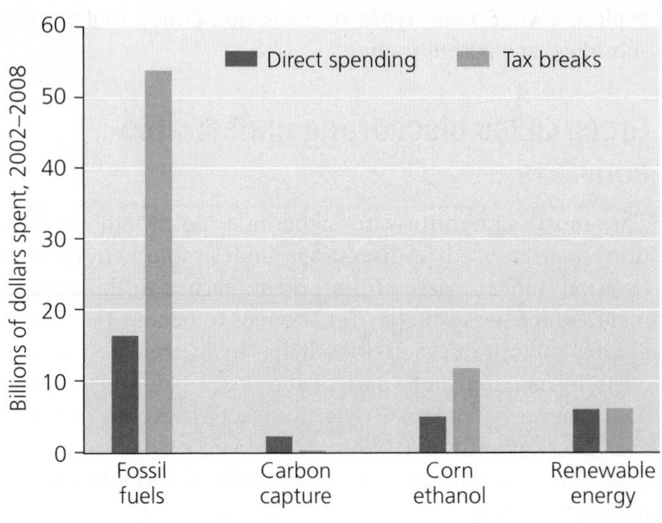

(a) U.S. energy subsidies

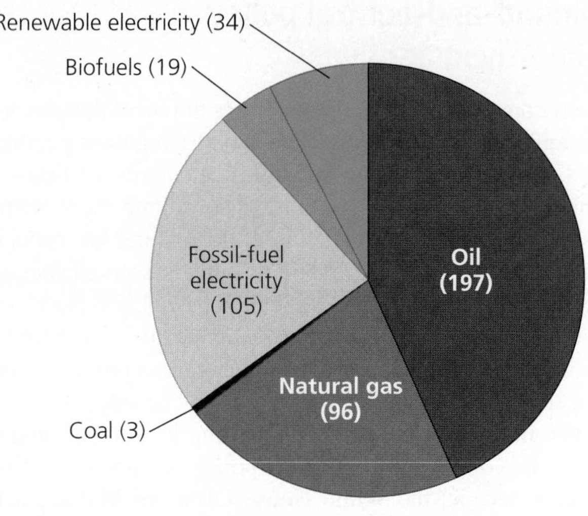

(b) Global energy subsidies

FIGURE 7.14 **The well-established fossil fuel industries receive more subsidies than the young and struggling renewable energy industries,** both **(a)** in the United States (from 2002 to 2008) and **(b)** globally (from 2007 to 2010). *Data from: (a) Environmental Law Institute, 2009.* Estimating U.S. government subsidies to energy sources: 2002–2008. *ELI, Washington, D.C.; and (b) International Energy Agency.*

DATA Q In the United States, how many dollars in total subsidies go to fossil fuels for every dollar that goes to renewable energy (excluding corn ethanol)?

gave $72 billion of its citizens' money—$240 from each man, woman, and child—to fossil fuel corporations (three-quarters of this in the form of tax breaks), while providing just $29 billion to renewable energy (FIGURE 7.14a). Moreover, most of the subsidies for renewable energy went toward corn ethanol, which is not widely viewed as a sustainable fuel (pp. 569–570).

One recent study calculated that from 1918 to the present, oil and gas have received 75 times more subsidies than new renewable energy sources in the United States, and 13 times more on a per-year basis (p. 585). Internationally in the period 2007–2010, fossil fuel energy subsidies outpaced renewable energy subsidies by a ratio of nearly 8 to 1, according to calculations by the International Energy Agency (FIGURE 7.14b).

In 2009, President Obama and other leaders of the Group of 20 (G-20) nations resolved to gradually phase out their collective $300 billion of annual fossil fuel subsidies. Doing so would hasten a shift to cleaner renewable energy sources and accomplish half the greenhouse gas emissions cuts needed to hold global warming to 2°C. However, since that time, fossil fuel subsidies have *grown*, not shrunk, rising by one estimate to $775 billion in 2012. A prime reason is that consumers have grown used to artificially low subsidized prices for gasoline and electricity and would punish policymakers who try to lift the subsidies. Economists are now trying to design feasible and politically palatable means of shifting subsidies from fossil fuels to renewable energy sources or toward rebates to taxpayers.

Many other environmentally harmful subsidies exist. Under the General Mining Act of 1872 (pp. 173, 646), mining companies extract $500 million to $1 billion in minerals from U.S. public lands each year without paying a penny in royalties to the taxpayers who own these lands. Since this law was enacted, the U.S. government has given away nearly $250 billion of mineral resources, and mining activities have polluted more than 40% of watersheds in the West. The 140-year-old law still allows mining companies to buy public lands for $5 or less per acre.

On the U.S. national forests, the U.S. Forest Service spends $35 million of taxpayer money each year building roads to allow private timber corporations access to cut trees, which the companies then sell at a profit (FIGURE 7.15).

Advocates of sustainable resource use have long urged governments to subsidize only environmentally sustainable

FIGURE 7.15 **When companies extract timber from U.S. national forests, taxpayers pay the costs of building and maintaining access roads.** "By absorbing these costs, the federal government shields the timber industry from the true cost of doing business," the nonprofit group Taxpayers for Common Sense concludes.

activities. This is beginning to happen, but lobbying from entrenched interests is keeping progress slow.

We can harness market dynamics to promote sustainability

With subsidies and green taxes, policymakers employ financial incentives in direct and selective ways. However, we may also pursue policy goals by establishing financial incentives and then letting marketplace dynamics run their course. One example is ecolabeling (p. 155), the practice whereby sellers who use sustainable practices in growing, harvesting, or manufacturing their products advertise this fact on their labels, hoping to win approval from buyers. In many cases, ecolabeling has grown from initial steps taken by government to require the disclosure of information to consumers. Once established, ecolabeling can spread in a free market as more and more businesses seek to win consumer confidence and to outcompete less sustainably produced brands. Another example of employing market dynamics for policy goals is permit trading.

Permit trading can save money and produce results

In the innovative market-based approach known as **permit trading,** the government creates a market in permits for an environmentally harmful activity, and companies, utilities, or industries then buy, sell, or trade rights to conduct the activity. To decrease emissions of air pollutants, a government might grant emissions permits and set up an **emissions trading system.** In a **cap-and-trade** emissions trading system, the government first determines the overall amount of pollution it will accept (i.e., it caps the total amount allowed) and then issues permits to polluters that allow them each to emit a certain fraction of that amount. Polluters may buy, sell, and trade these permits with other polluters.

Suppose, for example, you own an industrial plant with permits to release 10 units of pollution, but you find that you can make your plant more efficient and release only 5 units instead. You then have a surplus of permits, which might be very valuable to some other plant owner who is having trouble reducing pollution or who wants to expand production. In such a case, you can sell your extra permits. Doing so generates income for you and meets the needs of the other plant, while the total amount of pollution does not rise. By providing companies an economic incentive to find ways to reduce emissions, permit trading can reduce expenses for both industry and the public relative to a conventional regulatory system.

To lower emissions further, the government may reduce the amount of overall emissions allowed year by year. Environmental organizations may buy up permits and "retire" them, reducing the overall amount of pollution still more. To see an illustration of how a cap-and-trade system works, jump ahead to Figure 18.31 (p. 513).

A cap-and-trade system has been in place in the United States, established by the 1990 amendments to the Clean Air Act (p. 458) that mandated lower emissions of sulfur dioxide, a major contributor to acid deposition (pp. 473–475). Since

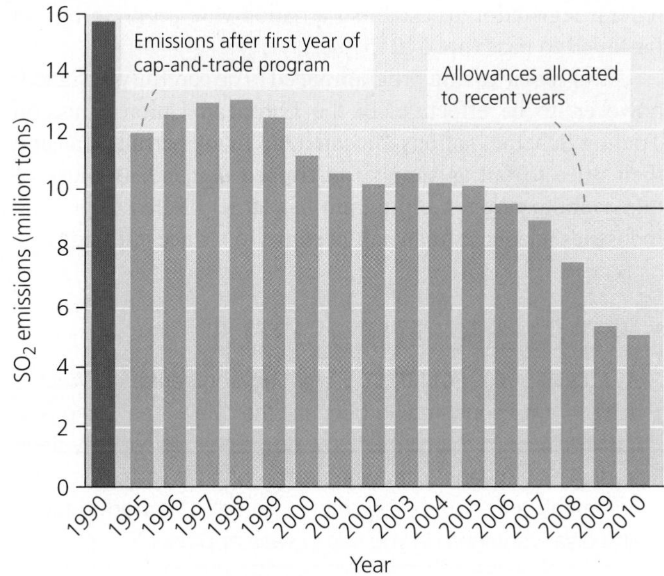

FIGURE 7.16 Permit trading has helped to reduce emissions of sulfur dioxide by 67% since 1990. As of 2010, emissions from U.S. sources participating in the program mandated by the 1990 Clean Air Act amendments had dropped well below the amount allocated in permits (black line). *Data from U.S. Environmental Protection Agency.*

then, sulfur dioxide emissions from sources in the program have declined by 67% (**FIGURE 7.16**), sulfate deposition has been reduced, and air quality and visibility have improved. The 67% reduction in pollution was greater than the amount actually required by the legislation, offering evidence that cap-and-trade systems can sometimes cut pollution more effectively than command-and-control regulation. Moreover, the cuts were attained at much less cost than was predicted and with no apparent effect on electricity supply or economic growth. Savings from the permit trading system have been estimated at billions of dollars per year, and the EPA calculates that the program's benefits outweigh its costs by about 40 to 1. Other similar programs have also shown success, including one in the Los Angeles basin to reduce smog (p. 458) and one among northeastern states aimed at nitrogen oxides.

Although cap-and-trade programs can reduce pollution, they do allow hotspots of pollution to occur around plants that buy permits to pollute more. Moreover, large firms can hoard permits, deterring smaller new firms from entering the market, thereby suppressing competition. Nonetheless, permit trading shows promise for safeguarding environmental quality while granting industries the flexibility to lessen their impacts in ways that are economically palatable.

In recent years, emissions trading programs have begun to address the greenhouse gas emissions that drive global climate change. In the European Union Emission Trading Scheme (p. 512), each participating nation allocates emissions permits to its industries according to their emissions at the start of the program. The industries then can trade permits freely, establishing a market whereby the price of a permit fluctuates according to supply and demand. In the United States, the Chicago Climate Exchange pioneered this approach, and carbon-trading programs are running in California and among the northeastern states, although federal legislation to establish a nationwide program has so far failed to pass (pp. 512–513).

Emissions trading programs need to be carefully designed, however, to be effective. In the European Union Emission Trading Scheme, nations allocated too many permits, causing their price to fall as supply outstripped demand. Because of this overallocation, the program has largely failed to provide industries adequate financial incentive to reduce emissions.

WEIGHING THE ISSUES

A LICENSE TO POLLUTE? Some environmental advocates oppose emissions trading because they view it as giving polluters "a license to pollute." How do you feel about emissions trading as a means of reducing air pollution? Would you favor command-and-control regulation instead? What advantages and disadvantages do you see in each approach?

Market incentives also operate at the local level

You may have already taken part in transactions involving financial incentives as policy tools. Many municipalities charge residents for waste disposal according to the amount of waste they generate. Other cities place taxes or disposal fees on items whose safe disposal is costly, such as tires and motor oil. Still others give rebates to residents who buy water-efficient toilets and appliances, because the rebates can cost the city less than upgrading its wastewater treatment system. Likewise, power utilities sometimes offer discounts to customers who buy high-efficiency lightbulbs and appliances, because doing so is cheaper for the utilities than expanding the generating capacity of their plants.

The creative use of economic policy tools is growing at all levels, while command-and-control regulation and legal action in the courts continue to play vital roles in environmental policymaking. As a result, we have a variety of effective policy strategies available to us as we seek sustainable solutions to our society's challenges.

Conclusion

Environmental policy is a problem-solving tool that makes use of science, ethics, and economics and that requires an astute understanding of the political process. Conventional command-and-control approaches of legislation and regulation remain the most common approaches to policymaking, but tort law retains influence, and innovative market-based policy tools are increasingly being developed. Historically, the United States has often led the way with environmental policy, but environmental issues often span political boundaries and require international cooperation. By integrating the fundaments of environmental policy introduced in this chapter with your knowledge of the natural sciences, you will be well equipped to develop your own creative solutions to many of our society's most challenging problems.

Reviewing Objectives

You should now be able to:

Describe environmental policy and assess its societal context

- Policy is a tool for decision making and problem solving that makes use of information from science and values from ethics and economics. (p. 164)
- Environmental policy aims to protect natural resources and environmental amenities from degradation or depletion and to promote equitable treatment of people. It addresses the tragedy of the commons, free riders, and external costs. (pp. 164–165, 168)
- In a democracy, anyone can have a voice in policy, but corporations and well-funded organizations tend to exert the most influence. (pp. 168–169)

Discuss the role of science in policymaking

- Data from scientific research are vital for informing policy. (p. 169)
- Policymakers may sometimes ignore or distort science for political ends, so we in the public need to remain vigilant. (p. 169)

Identify the institutions important to U.S. environmental policy and recognize major U.S. environmental laws

- The legislative, executive, and judicial branches, together with administrative agencies, all play roles in U.S. environmental policy. (pp. 169–171)
- Courts have constrained the power of government regulations by seeking to prevent regulatory takings. (pp. 170–171)
- State and local governments also implement environmental policy. (pp. 171–172)
- U.S. environmental policy came in three waves. The first promoted frontier expansion and resource extraction. The second aimed to mitigate impacts of the first through conservation. The third targeted pollution and gave us many of today's major environmental laws. (pp. 173–174)
- Far-reaching developments from 1970 were enactment of the National Environmental Policy Act (NEPA), establishment of the EIS process, and creation of the Environmental Protection Agency (EPA). (pp. 174–175)
- Major U.S. environmental laws include the Clean Air Act, the Clean Water Act, the Endangered Species Act, and others. (pp. 175–176)

- A fourth wave of environmental policy is now building internationally around sustainable development and global climate change. (pp. 177–178)

List the institutions involved with international environmental policy and describe how nations handle transboundary issues

- Many environmental problems cross political boundaries and must be addressed internationally. (p. 178)
- International policy includes customary law (law by shared traditional custom) and conventional law (law by treaty). (p. 178)
- The United Nations, World Bank, European Union, World Trade Organization, and nongovernmental organizations all shape international policy. (pp. 178–179)

Categorize the different approaches to environmental policy

- Tort law has been a traditional approach to resolving environmental disputes. (p. 180)
- Legislation from Congress and regulations from administrative agencies make up most federal policy. These top-down approaches are referred to as command-and-control. (p. 181)
- Economic policy tools include green taxes to discourage undesirable activities, subsidies to encourage targeted activities, and market-based approaches such as permit trading that harness market dynamics to achieve policy goals. (pp. 181–184)

Testing Your Comprehension

1. Describe two major justifications for environmental policy, and discuss three problems that environmental policy commonly seeks to address.
2. What is *the tragedy of the commons*? Explain how the concept might apply to an unregulated industry that is a source of water pollution.
3. Outline the primary responsibilities of the legislative, executive, and judicial branches of the U.S. government. What is the "fourth branch" of the U.S. government?
4. What is meant by a *regulatory taking*?
5. Summarize how the first, second, and third waves of environmental policy in U.S. history differed from one another. Describe what now appears to be the fourth wave.
6. What did the National Environmental Policy Act accomplish? Briefly describe the origin and mission of the U.S. Environmental Protection Agency.
7. What is the difference between customary law and conventional law? What challenges do transboundary environmental problems present?
8. Why are environmental regulations sometimes considered to be unfair barriers to trade?
9. Compare and contrast the three major approaches to environmental policy: tort law, command-and-control, and economic policy tools.
10. Explain how each of the following work: a green tax, a subsidy, and marketable emissions permits.

Seeking Solutions

1. Classical economist Adam Smith argued that individuals can benefit society by pursuing their own self-interest (Chapter 6; p. 143). Do you agree? Can you describe a situation in which a person acting in self-interest either (a) benefits society by addressing an environmental problem, or (b) harms society by causing an environmental problem? How might policy help in either situation? What are some advantages and disadvantages of environmental laws and regulations?
2. Reflect on the causes for the transitions in U.S. history from one type of environmental policy to another. Now peer into the future, and think about how life and society might be different in 25, 50, or 100 years. What would you predict about the environmental policy of the future, and why? What issues might it address? Do you predict we will have more or less environmental policy?
3. Compare the roles of the United Nations, the World Bank, the European Union, the World Trade Organization, and nongovernmental organizations. If you could gain the support of just one of these institutions for a policy you favored, which would you choose? Why?
4. Consider the main approaches to environmental policy—tort law, command-and-control, and economic policy tools. Describe an advantage and a disadvantage of each. Do you think any one approach is most effective? Could we do with just one approach, or does it help to have more?
5. Think of one environmental problem that you would like to see solved. From what you've learned about policymaking in this chapter, describe how you think you could best create policy to address this problem.
6. **THINK IT THROUGH** You have just been elected to Congress as the representative from the Pennsylvania district that includes the town of Dimock. What policy approaches would you choose to pursue to search for solutions to the debate over hydraulic fracturing that has been dividing your constituents? Give reasons for your choices.

Calculating Ecological Footprints

Critics of command-and-control regulation typically maintain that regulations are costly to business and industry, yet cost-benefit analyses (p. 146) have repeatedly shown that regulations bring citizens more benefits than costs, overall. Each year the U.S. Office of Management and Budget assesses costs and benefits of major federal regulations of administrative agencies. Results from the most recent report, covering the decade from 2000 to 2010, are shown below. This decade includes periods of both Republican and Democratic control of the presidency and of Congress. Subtract costs from benefits, and enter these values for each agency in the third column. Divide benefits by costs, and enter these values in the fourth column.

Costs and benefits of major U.S. federal regulations, 2000–2010 (in billions of dollars)

AGENCY	BENEFITS	COSTS	BENEFITS MINUS COSTS	BENEFIT: COST RATIO
Department of Energy	8.0 to 10.9	4.5 to 5.1	3.5 to 5.8	1.8 to 2.1
Department of Health and Human Services	18.0 to 40.5	3.7 to 5.2		
Department of Transportation	14.6 to 25.5	7.5 to 14.3		
Environmental Protection Agency (EPA)	81.8 to 550.7	23.3 to 28.5		
Other departments	5.4 to 9.1	3.1 to 3.8		
Total	**127.8 to 636.7**	**42.1 to 56.9**		

Data from U.S. Office of Management and Budget, 2011. 2011 report to Congress on the benefits and costs of federal regulations and unfunded mandates on state, local, and tribal entities. *OMB, Washington, D.C.*

1. How many of the agencies shown have regulations that exert more costs than benefits? How many have regulations that provide more benefits than costs?
2. Which agency's regulations have the greatest excess of benefits over costs? Which agency's regulations have the greatest maximum ratio of benefits to costs?
3. What percentage of total benefits from regulations comes from EPA regulations? Most of the benefits and costs from EPA regulations are from air pollution rules resulting from the Clean Air Act and its amendments. Judging solely by these data, would you say that Clean Air Act legislation has been a success or a failure for U.S. citizens? Why?

STUDENTS

Go to **MasteringEnvironmentalScience** for assignments, the etext, and the Study Area with practice tests, videos, current events, and activities.

INSTRUCTORS

Go to **MasteringEnvironmentalScience** for automatically graded activities, current events, videos, and reading questions that you can assign to your students, plus Instructor Resources.

PART TWO

Environmental Issues and the Search for Solutions

8

Crowded street in Shanghai, one of China's largest cities

Human Population

Upon completing this chapter, you will be able to:

- Perceive the scope of human population growth
- Assess divergent views on population growth
- Evaluate how human population, affluence, and technology affect the environment
- Explain and apply the fundamentals of demography
- Outline and assess the concept of demographic transition
- Describe how family planning, the status of women, and wealth and poverty affect population growth
- Characterize the dimensions of the HIV/AIDS epidemic

CENTRAL CASE STUDY

China's One-Child Policy

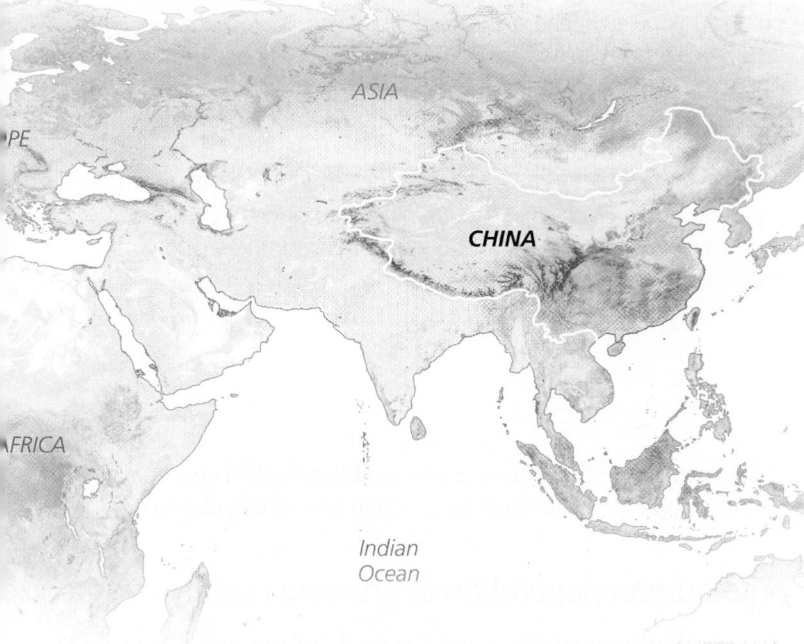

"The population problem concerns us, but it will concern our children and grandchildren even more. How we respond to the population threat may do more to shape the world in which they will live than anything else we do."

—Lester Brown, President, Earth Policy Institute

"As you improve health in a society, population growth goes down. . . . Before I learned about it, I thought it was paradoxical."

—Bill Gates, Chair, Microsoft Corporation

The People's Republic of China is the world's most populous nation, home to one-fifth of the 7 billion people living on Earth today.

When Mao Zedong founded the country's current regime six decades ago roughly 540 million people lived in a mostly rural, war-torn, impoverished nation. Mao believed population growth was desirable, and under his rule China grew and changed. By 1970, improvements in food production, food distribution, and public health allowed China's population to swell to 790 million people. At that time, the average Chinese woman gave birth to 5.8 children in her lifetime.

However, the country's burgeoning population and its industrial and agricultural development were eroding the nation's soils, depleting its water, leveling its forests, and polluting its air. Chinese leaders realized that the nation might not be able to feed its people if their numbers grew much larger. They saw that continued population growth could exhaust resources and threaten the stability and progress of Chinese society. The government decided to institute a population control program that prohibited most Chinese couples from having more than one child.

The program began with education and outreach efforts encouraging people to marry later and have fewer children (**FIGURE 8.1**). Along with these efforts, the Chinese government increased the availability of contraceptives and abortion. By 1975, China's annual population growth rate had dropped from 2.8% to 1.8%.

To further decrease the birth rate, in 1979 the government took the more drastic step of instituting a system of rewards and punishments to enforce a one-child limit. One-child families received better access to schools, medical care, housing, and government jobs, and mothers with only one child were given longer maternity leaves. In contrast, families with more than one child were subjected to monetary fines, employment discrimination, and social scorn and ridicule. In some cases, the fines exceeded half of a couple's annual income.

Population growth rates dropped still further, but public resistance to the policy was simmering. Beginning in 1984, the one-child policy was loosened, strengthened, and then loosened again as government leaders explored ways to maximize population control while minimizing public opposition. Today the one-child program is less strict than in past years and applies mostly to families in urban areas. Many farmers and ethnic minorities in rural areas are exempted, because success on the farm often depends on having multiple children.

In enforcing its policies, China has been conducting one of the largest and most controversial social experiments in history. In purely quantitative terms, the experiment has been a major success: The nation's growth rate is now down to 0.5%, making it easier for the country to deal with its many social, economic, and environmental challenges.

However, the one-child policy has also produced unintended consequences. Traditionally, Chinese culture has valued sons because they carry on the family name, assist with

FIGURE 8.1 Billboards and murals like this one in the Chinese city of Chengdu promote the national "one-child" policy.

farm labor in rural areas, and care for aging parents. Daughters, in contrast, will most likely marry and leave their parents, as the culture dictates. As a result, they cannot provide the same benefits to their parents as will sons. Thus, faced with being limited to just one child, many Chinese couples prefer a son to a daughter. Tragically, this has led to selective abortion, killing of female infants, an unbalanced sex ratio, and a black-market trade in teenaged girls for young men who cannot find wives.

Further problems are expected in the near future, including an aging population and a shrinking workforce. Moreover, China's policies have elicited criticism worldwide from people who oppose government intrusion into personal reproductive choices.

As other nations become more and more crowded, might their governments also feel forced to turn to drastic policies that restrict individual freedoms? In this chapter, we examine human population dynamics worldwide, consider their causes, and assess their consequences for the environment and our society.

Our World at Seven Billion

China receives a great deal of attention on population issues because of its unique reproductive policies and its status as the world's most populous nation. But China is not alone in dealing with population issues. India, China's neighbor, is also a population powerhouse, and is only slightly less populous than China. India lacks China's stringent reproductive policies, though, and soon will overtake China and possess the world's largest population (**FIGURE 8.2**).

Like India, many of the world's poorer nations continue to experience substantial population growth. Many of these nations are ill-equipped to handle such growth, and this leads to stresses on society, the environment, and people's well-being. In our world of now 7 *billion* people, one of our greatest challenges in this century is finding ways to slow the growth of the human population without the requirement of measures such as those used in China, but by establishing the conditions for all people that lead them to desire to have fewer children.

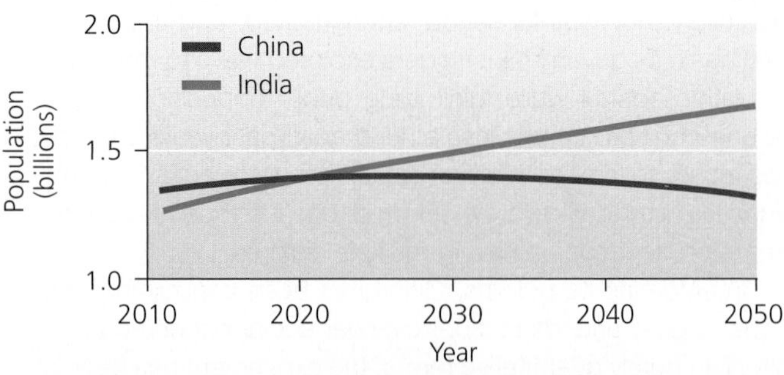

FIGURE 8.2 China and India will likely soon switch places as the most populous and second-most populous nations. China's rate of growth is now lower than India's as a result of China's aggressive population policies. *Data from Population Division of the Department of Economic and Social Affairs of the United Nations Secretariat, 2011.* World population prospects: The 2010 revision, *http://esa.un.org/wpp. © United Nations, 2011.*

The human population is growing rapidly

Our global population is now over 7 billion and grows by more than 70 million people each year. This is the equivalent of adding all the people of California, Texas, and New Jersey to the world annually—and it means that we add more than 2 people to the planet *every second.* Take a look at **FIGURE 8.3** and note just how recent and sudden our

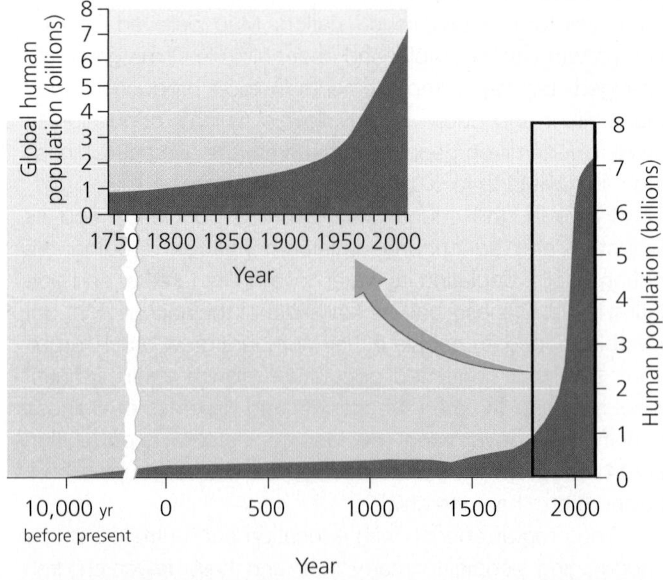

FIGURE 8.3 We have risen from fewer than 1 billion in 1800 to 7 billion today. Viewing global human population size over a long time scale **(bottom graph)** and growth since the industrial revolution **(inset top graph)** shows that nearly all growth has occurred in just the past 200 years. *Data from U.S. Bureau of Census.*

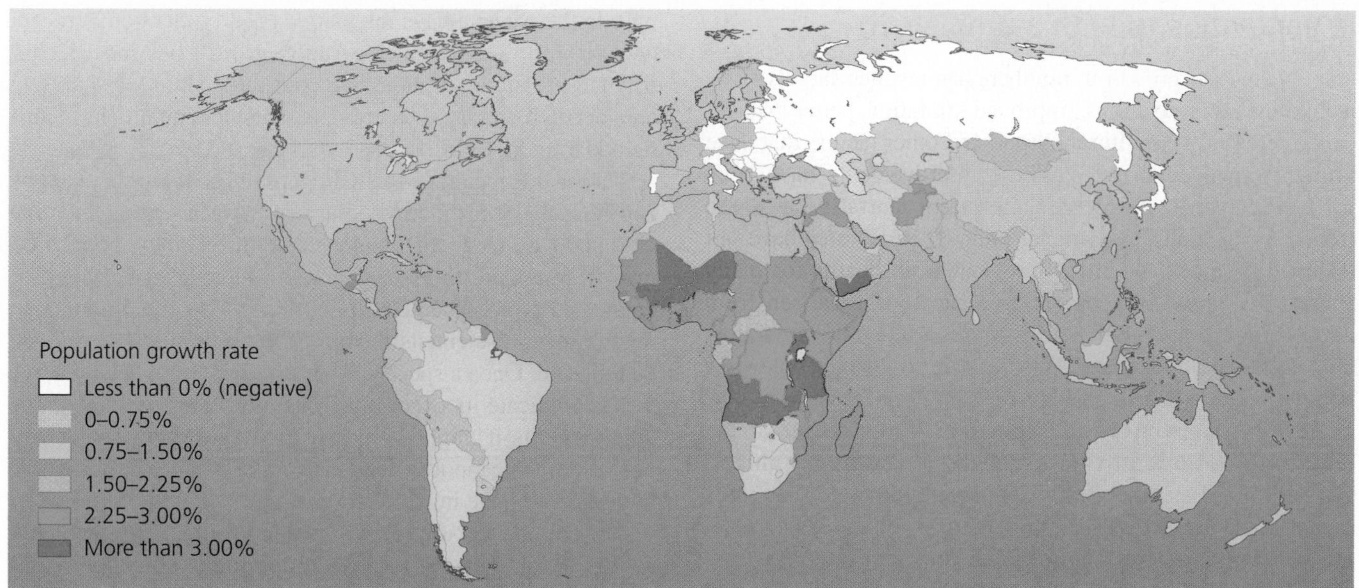

FIGURE 8.4 Population growth rates vary greatly from place to place. Population is growing fastest in poorer nations of the tropics and subtropics but is now beginning to decrease in some northern industrialized nations. Shown are rates of natural increase (p. 201) as of 2012. *Data from Population Reference Bureau, 2012. 2012 World population data sheet.*

DATA Q Which world region has the highest population growth rate? What are the geographic patterns in growth rate within this region?

rapid increase has been. It took until after 1800, virtually all of human history, for our population to reach 1 billion. Yet we reached 2 billion by 1930, and 3 billion in just 30 more years. Our population added its next billion in just 15 years, and it has taken only 12 years to add each of the next three installments of a billion people. Consider when you were born, and estimate the number of people added to the planet just since that time. No previous generations have *ever* lived amid so many other people.

You may not feel particularly crowded, but consider that we in the United States are not experiencing as much population growth and density as people in most other nations. In today's world, rates of annual growth vary greatly from region to region (**FIGURE 8.4**) and are highest in developing nations.

What accounts for our unprecedented growth? As you may recall, exponential growth—the increase in a quantity by a fixed percentage per unit time—accelerates increase in population size, just as compound interest accrues in a savings account (Chapter 3; pp. 66–67). The reason for this pattern is that a fixed percentage of a small number makes for a small increase, but that the same percentage of a large number produces a large increase. Thus, even if the growth *rate* remains steady, population *size* will increase by greater increments with each successive generation.

For much of the 20th century, the growth rate of the human population rose from year to year. This rate peaked at 2.1% during the 1960s and has declined to 1.2% since then. Although 1.2% may sound small, exponential growth endows small numbers with large consequences. A hypothetical population starting with one man and one woman that grows at 1.2% gives rise to a population of 2939 after 40 generations and 112,695 after 60 generations.

At a 1.2% annual growth rate, a population doubles in size in just 58 years. We can roughly estimate doubling times with a handy rule of thumb. Just take the number 70 (which is 100 times 0.7, the natural logarithm of 2) and divide it by the annual percentage growth rate: 70/1.2 = 58.3. Had China not instituted its one-child policy, and had its growth rate remained at 2.8%, it would have taken only 25 years to double in size.

FAQ How big is a billion?

Human beings have trouble conceptualizing huge numbers. As a result, we often fail to recognize the true magnitude of a number such as 7 billion. Although we know that a billion is bigger than a million, we tend to view both numbers as impossibly large and therefore similar in size. For example, guess (without calculating) how long it would take a banker to count out $1 million if she did it at a rate of a dollar a second for 8 hours a day, 7 days a week. Now guess how long it would take to count $1 billion at the same rate. The difference between your estimate and the answer may surprise you. Counting $1 million would take a mere 35 days, whereas counting $1 billion would take 95 years! Living 1 million seconds takes only 12 days, while living for 1 billion seconds requires more than 31 years. You couldn't live for 7 billion seconds if you tried, because that would take 221 years. Examples like these can help us appreciate the *b* in *billion*.

Is population growth a problem?

Our spectacular growth in numbers has resulted largely from technological innovations, improved sanitation, better medical care, increased agricultural output, and other factors that have brought down death rates. These improvements have been particularly successful in reducing **infant mortality rates,** the frequency of children dying in infancy. Birth rates have not declined as much, so births have outpaced deaths for many years now. Thus, our population explosion has arisen from a very good thing—our ability to keep more of our fellow human beings alive longer! Why then do so many people view population growth as a problem?

Let's start with a bit of history. At the outset of the industrial revolution (p. 4), population growth was universally regarded as a good thing. For parents, high birth rates meant more children to support them in old age. For society, it meant a greater pool of labor for factory work. However, British economist **Thomas Malthus** (1766–1834) had a different view. Malthus (FIGURE 8.5a) argued that unless population growth were controlled by laws or other social strictures, the number of people would eventually outgrow the available food supply. Malthus's most influential work, *An Essay on the Principle of Population,* published in 1798, argued that if society did not limit births (through abstinence and contraception, for instance), then rising death rates would reduce the population through war, disease, and starvation.

In our day, biologists Paul and Anne Ehrlich of Stanford University have been called "neo-Malthusians" because they too have warned that our population may grow faster than our ability to produce and distribute food. In his best-selling 1968 book, *The Population Bomb,* Paul Ehrlich (FIGURE 8.5b) predicted that population growth would unleash famine and conflict that would consume civilization by the end of the 20th century.

Although human population quadrupled in the past 100 years—the fastest it has ever grown (see Figure 8.3)—Ehrlich's forecasts have not fully materialized. This is due, in part, to the way we have intensified food production in recent decades (pp. 245–247). Population growth has indeed contributed to famine, disease, and conflict—but as we shall see, enhanced prosperity, education, and gender equality have also helped to reduce birth rates.

(a) Thomas Malthus

(b) Paul Ehrlich

FIGURE 8.5 **Thomas Malthus and Paul Ehrlich each argued that runaway population growth would surpass food supply and lead to disaster.**

Does this mean we can disregard the concerns of Malthus and Ehrlich? Some Cornucopians (p. 147) say yes. Under the Cornucopian view that many economists hold, population growth poses no problem if new resources can be found or created to replace depleted ones (pp. 147–149). In contrast, environmental scientists recognize that not all resources can be replaced. Once a species has gone extinct, for example, we cannot replicate its exact functions in an ecosystem or know what benefits it might have provided us. Land, too, is irreplaceable; we cannot expand Earth like a balloon to increase the space we have in which to live.

Even if resource substitution could hypothetically enable our population to grow indefinitely, could we maintain the quality of life that we desire for ourselves and our descendants? Unless the availability and quality of all resources keeps pace forever with population growth, the average person in the future will have less space in which to live, less food to eat, and less material wealth than the average person does today. Thus, population growth is indeed a problem if it depletes resources, stresses social systems, and degrades the natural environment, such that our quality of life declines (FIGURE 8.6).

Some national governments now fear falling populations

Despite these considerations, many policymakers find it difficult to let go of the notion that population growth increases a nation's economic, political, and military strength. This notion has held sway despite the clear fact that in today's world, population growth is correlated with poverty, not wealth: Strong and wealthy nations tend to have slow population growth, whereas nations with fast population growth tend to be weak and poor. While China and India struggle to get their population growth under control, many other national governments are offering financial and social incentives that encourage their own citizens to produce more children.

Much of the concern is that when birth rates decline, a population grows older. In aging populations, larger numbers of elderly people will need social services, but fewer workers will be available to pay taxes to fund these services. Such concerns about the sustainability of the Social Security program in the United States, for example, led some in the media to call for an emphasis on increased fertility when it was reported that the U.S. birthrate in 2011 had reached its lowest level in recorded history.

Moreover, in nations with declining birth rates that are also accepting large numbers of foreign immigrants, some native-born citizens may worry that their country's culture is at risk of being diluted or lost. For both these reasons, governments of nations now experiencing population declines (such as many in Europe) feel uneasy. According to the Population Reference Bureau, two of every three European governments

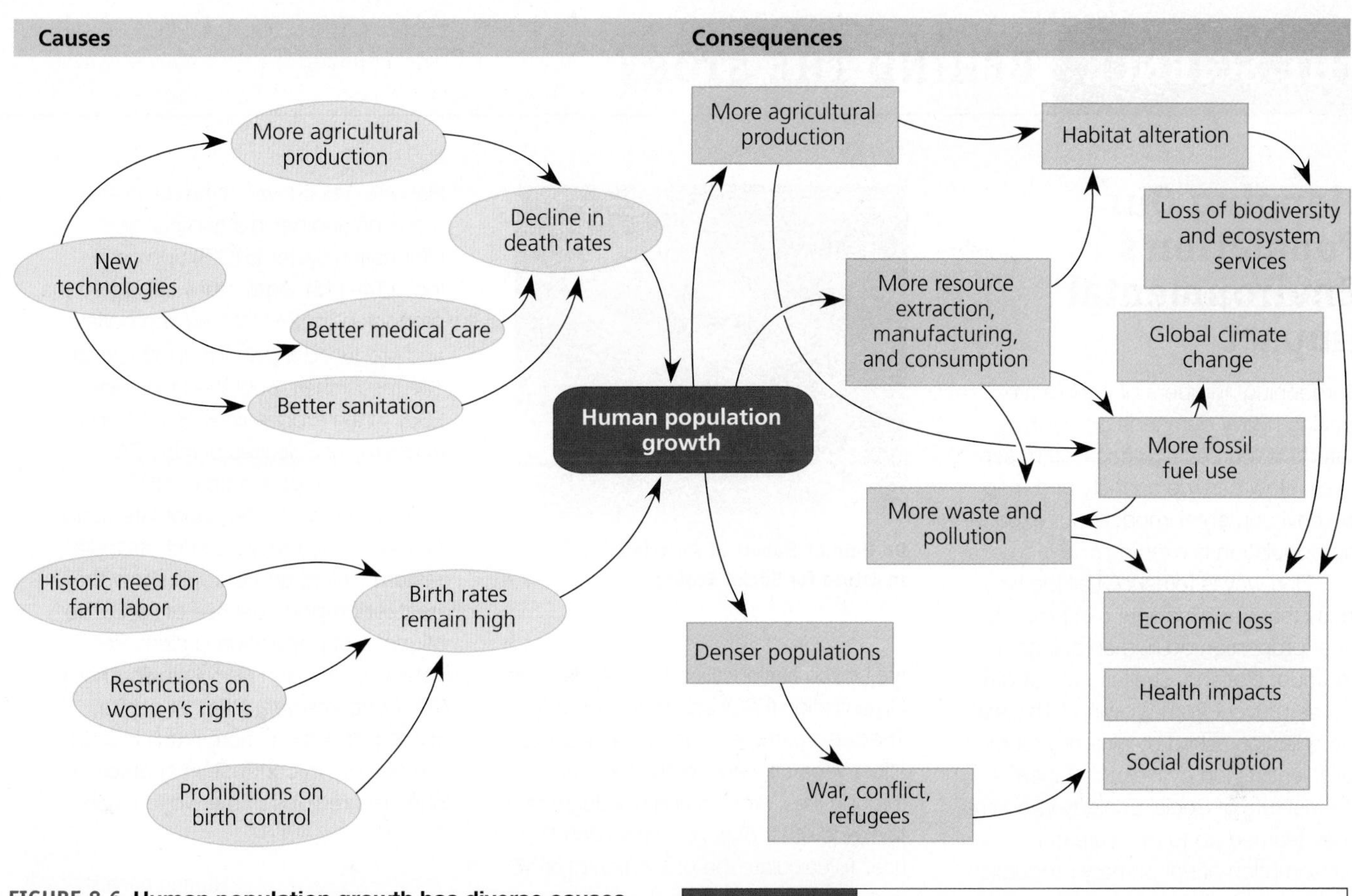

FIGURE 8.6 Human population growth has diverse causes and consequences. The consequences of rapid population growth are generally negative for society and the environment. Arrows in this concept map lead from causes to consequences. Note that items grouped within outlined boxes do not necessarily share any special relationship; the outlined boxes are merely to streamline the figure.

Solutions

As you progress through this chapter, try to identify as many solutions to human population growth as you can. What could you personally do to help address this issue? Consider how each action or solution might affect items in the concept map above.

now feel their birth rates are too low, and none states that its country's birth rate is too high. However, outside Europe, 49% of national governments still feel their birth rates are too high, and only 12% feel they are too low.

Population is one of several factors that affect the environment

One widely used formula gives us a handy way to think about population and other factors that affect environmental quality. Nicknamed the **IPAT model**, it is a variation of a formula proposed in 1974 by Paul Ehrlich and John Holdren, a Harvard University environmental scientist who today is President Obama's science advisor. The IPAT model represents how our total impact (I) on the environment results from the interaction among population (P), affluence (A), and technology (T):

$$I = P \times A \times T$$

Increased population intensifies impact on the environment as more individuals take up space, use resources, and generate waste. Increased affluence magnifies environmental impact through the greater per capita resource consumption that generally has accompanied enhanced wealth. Technology that enhances our abilities to exploit minerals, fossil fuels, old-growth forests, or fisheries generally increases impact, but technology to reduce smokestack emissions, harness renewable energy, or improve manufacturing efficiency can decrease impact.

We might also add a sensitivity factor (S) to the equation to denote how sensitive a given environment is to human pressures:

$$I = P \times A \times T \times S$$

For instance, the arid lands of western China are more sensitive to human disturbance than the moist regions of southeastern China. Plants grow more slowly in the arid west, making the land more vulnerable to deforestation and soil degradation. Thus, adding an additional person to western China has more environmental impact than adding one to southeastern China.

We could refine the IPAT equation further by adding terms for the effects of social institutions such as education, laws and their enforcement, stable and cohesive societies,

Mapping Our Population's Environmental Impact

Burgeoning numbers of people are making heavy demands on Earth's natural resources and ecosystem services. How can we quantify and map the environmental impacts our expanding population is exerting?

One way is to ask. Of all the biomass that Earth's plants can produce, what proportion do human beings use (for food, clothing, shelter, etc.) or otherwise prevent from growing? This was the question asked by nine environmental scientists led by Helmut Haberl of the Institute of Social Ecology in Austria. They teamed up to measure our consumption of net primary production (NPP; p. 111), the net amount of energy stored in plant matter as a result of photosynthesis. Human overuse of NPP diminishes resources for other species; alters habitats, communities, and ecosystems; and threatens our future ability to derive ecosystem services.

Haberl's team began with a well-established model that maps how vegetation varies with climate across the globe and used it to produce a detailed world map of "potential NPP"—vegetation that would exist if there were no human influence. The team then gathered data for the year 2000 on crop harvests, timber harvests, grazing pressure, and other human uses of vegetation from various global databases from the United Nations Food and Agriculture Organization (FAO) and other sources. They also gathered data on how people affect vegetation indirectly, such as through fires, erosion and soil degradation, and other changes due to land use. To calculate the proportion of NPP that people appropriate, the researchers divided the amounts used up in these impacts by the total "potential" amount.

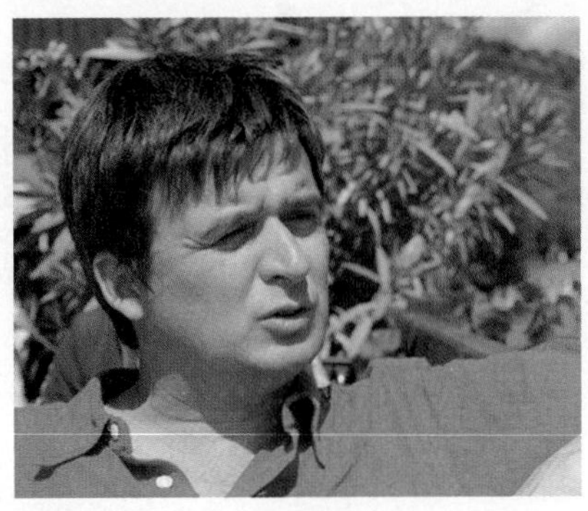

Dr. Helmut Haberl of Austria's Institute for Social Ecology

When all the data crunching was done, Haberl's group concluded that people harvest 12.5% of global NPP and that land use reduces it 9.6% further and fires 1.7% further (**FIGURE 1**). This makes us responsible for using up fully 23.8% of the planet's NPP—a staggeringly large amount for just a single species! Half of this use occurred on cropland, where 83.5% of NPP was used. In urban areas, 73.0% of NPP was consumed; on grazing land, 19.4%; and in forests, 6.6%.

To determine how human use of NPP varies across regions of the world, Haberl's group layered the data sets atop one another in a geographic information systems (GIS) approach (pp. 114–115). Again, they calculated the proportion of NPP that we appropriate and produced a global map (**FIGURE 2**). The researchers published their results in 2007 in the *Proceedings of the National Academy of Sciences of the USA*.

In their global map of NPP consumption, densely populated and heavily farmed regions such as India, eastern China, and Europe show the greatest proportional use of NPP. The influence of population is clear. For instance, although people in southern Asia consume very little per capita, dense populations here result in a 63% use of NPP. In contrast, in sparsely inhabited regions of the world (such

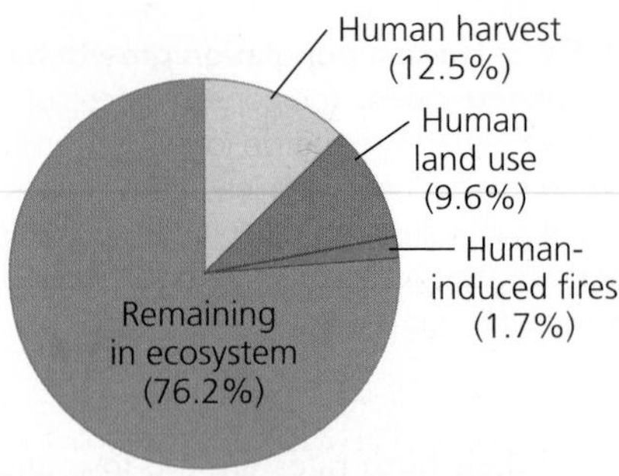

FIGURE 1 Humanity uses or causes Earth to lose 23.8% of the planet's net primary production. Direct harvesting (of crops, timber, etc.) accounts for most of this, and land use impacts and fire also contribute. *Data from Haberl, H., et al., 2007. Quantifying and mapping the human appropriation of net primary production in Earth's terrestrial ecosystems.* Proc. Natl. Acad. Sci. *104:12942–12947.*

---

and ethical standards that promote environmental well-being. Such factors all affect how population, affluence, and technology translate into environmental impact.

Impact can be thought of in various ways, but we can generally boil it down either to pollution or resource consumption. The depletion of resources by larger and hungrier populations has been a focus of scientists and philosophers since Malthus's time. Today, researchers calculate that humanity is appropriating for its own use nearly one-quarter of Earth's terrestrial net primary production (see **THE SCIENCE BEHIND THE STORY**, above).

One reason our population has kept growing, despite limited resources, is that we have developed technology—the *T* in the IPAT equation—time and again to increase efficiency, alleviate our strain on resources, and allow us to expand further. For instance, we have employed technological advances to increase global agricultural production faster than our population has risen (p. 245).

Modern-day China shows how all elements of the IPAT formula can combine to cause tremendous environmental impact in little time. The world's fastest-growing economy over the past two decades, China is "demonstrating what

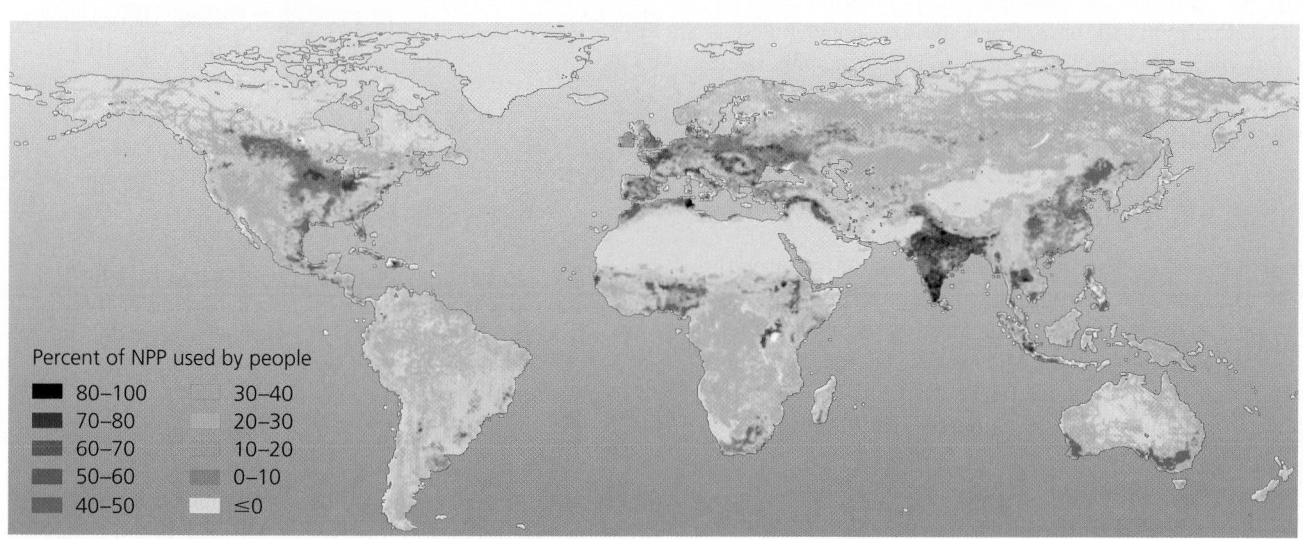

FIGURE 2 The proportion of Earth's net primary production that people appropriate varies from region to region. Regions that are densely populated or intensively farmed exert the heaviest impact. *Source: Haberl, H., et al., 2007. Quantifying and mapping the human appropriation of net primary production in Earth's terrestrial ecosystems.* Proc. Natl. Acad. Sci. *104:12942–12947, Fig 1b. © 2007 National Academy of Sciences, U.S.A. By permission.*

as the boreal forest, Arctic tundra, Himalayas, and Sahara Desert), humans consume almost no NPP. In North America, NPP use is heaviest in the East, Midwest, and Great Plains. In general, the map shows heavy appropriation of NPP in areas where population is dense relative to the area's vegetative production.

The map does not fully show the effects of resource consumption due to affluence. Wealthy societies tend to import food, fiber, energy, and products from other places, and this consumption can drive environmental degradation in poorer regions. For instance, North Americans and Europeans import timber logged from the Amazon basin, as well as soybeans and beef grown in areas where Amazonian forest was cleared. Through global trade, we redistribute the products we gain from the planet's NPP. As a result, the environmental impacts of our consumption are often felt far from where we consume products.

By showing areas of high and low impact, maps like the one produced in this project can help us to make better decisions and minimize our impacts on ecosystems and ecosystem services. Haberl's team also says its data show that we are using a great deal of Earth's plant matter already, so that schemes to expand biomass energy production (pp. 566–572) may not be wise.

Environmental scientists who have commented on the paper agree and say the team's data should give us pause. As Jonathan Foley of the University of Wisconsin and his colleagues put it, "Ultimately, we need to question how much of the biosphere's productivity we can appropriate before planetary systems begin to break down. 30%? 40%? 50%? More? . . . Or have we already crossed that threshold?" ●

happens when large numbers of poor people rapidly become more affluent," in the words of Earth Policy Institute president Lester Brown. While millions of Chinese are increasing their material wealth and their resource consumption, the country is battling unprecedented environmental challenges brought about by its rapid economic development. Intensive agriculture has expanded westward out of the country's moist rice-growing areas, causing farmland to erode and literally blow away, much like the Dust Bowl tragedy that befell the U.S. heartland in the 1930s (pp. 224–225). China has overpumped aquifers and has drawn so much water for irrigation from the Yellow River that the once-mighty waterway now dries up in many stretches. Although China is reducing its air pollution from industry and charcoal-burning homes, the country faces new urban pollution and congestion threats from rapidly rising numbers of automobiles. In August 2010, for example, a 100-km (60-mi) traffic jam formed on the outskirts of the capital city of Beijing and persisted for more than 10 days! Such issues are not unique to China. As the world's other industrializing countries strive to attain the material prosperity that industrialized nations enjoy, they too may soon face many of the same challenges as China.

Demography

It is a fallacy to think of people as being somehow outside nature. We exist within our environment as one species out of many. As such, all the principles of population ecology (Chapter 3) that apply to birds, frogs, and passenger pigeons apply to humans as well. The application of principles from population ecology to the study of statistical change in human populations is the focus of **demography.**

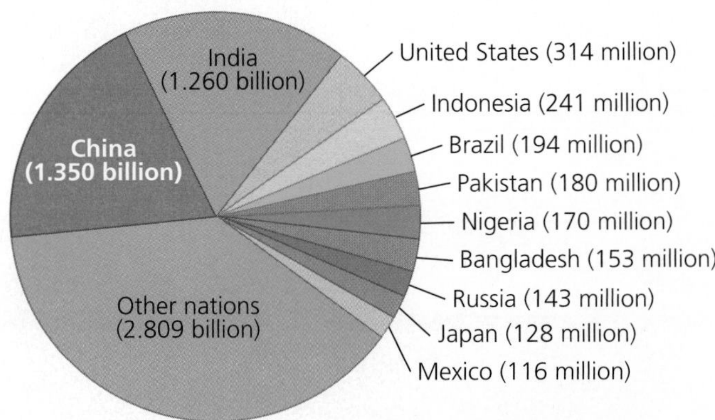

FIGURE 8.7 Almost one in five people in the world lives in China, and more than one of every six live in India. Three of every five people live in one of the 11 nations that have populations above 100 million. *Data from Population Reference Bureau,* 2012. 2012 World population data sheet.

Earth has a carrying capacity for us

Environmental factors set limits on our population growth (Chapter 3), and the environment has a carrying capacity (pp. 67–68) for our species, just as it does for every other. We happen to be a particularly successful species, however, and we have repeatedly raised this carrying capacity by developing technology to overcome the natural limits on our population growth.

Environmental scientists who have tried to pin a number to the human carrying capacity have come up with wildly differing estimates. The most rigorous estimates range from 1–2 billion people living prosperously in a healthy environment to 33 billion living in extreme poverty in a degraded world of intensive cultivation without natural areas. As our population climbs beyond 7 billion, we may yet continue to find ways to raise our carrying capacity. Given our knowledge of population ecology and logistic growth (p. 67), however, we have no reason to presume that human numbers can go on growing indefinitely.

Demography is the study of human population

Demographers study population size, density, distribution, age structure, sex ratio, and rates of birth, death, immigration, and emigration of people, just as population ecologists study these characteristics in other organisms. Each of these characteristics is useful for predicting population dynamics and environmental impacts.

Population size Our global human population of over 7 billion is spread among 200 nations with populations ranging up to China's 1.35 billion, India's 1.26 billion, and the 314 million of the United States (FIGURE 8.7). The United Nations Population Division estimates that by the year 2050, the global population will surpass 9 billion (FIGURE 8.8). However, population size alone—the absolute number of individuals—doesn't tell the whole story. Rather, a population's environmental impact depends on its density, distribution, and composition (as well as on affluence, technology, and other factors outlined earlier).

Population density and distribution People are distributed unevenly over our planet. In ecological terms, our distribution is clumped (p. 63) at all spatial scales. At the global scale (FIGURE 8.9), population density is highest in regions with temperate, subtropical, and tropical climates, such as China, Europe, Mexico, southern Africa, and India. Population density is lowest in regions with extreme-climate biomes, such as desert, rainforest, and tundra. Dense along seacoasts and rivers, human population is less dense away from water. At more local scales, we cluster together in cities and towns.

This uneven distribution means that certain areas bear more environmental impact than others. Just as the Yellow River experiences pressure from Chinese cities and farms,

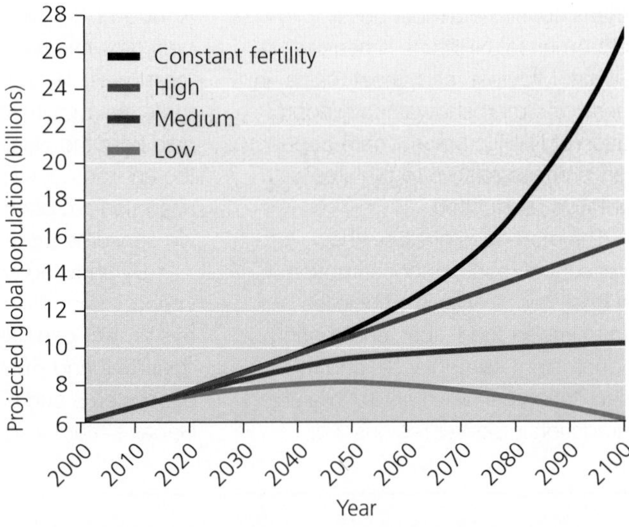

FIGURE 8.8 The United Nations predicts world population growth. In the latest projection (made in 2010), population is estimated to reach 11.0 billion in the year 2050 if fertility rates remain constant at 2005–2010 levels (top line in graph). However, U.N. demographers expect fertility rates to continue falling, so they arrived at a best guess (medium scenario) of 9.3 billion for 2050. In the high scenario, if women on average have 0.5 child more than in the medium scenario, population will reach 10.6 billion in 2050. In the low scenario, if women have 0.5 child fewer than in the medium scenario, the world will contain 8.1 billion people in 2050. *Adapted by permission from Population Division of the Department of Economic and Social Affairs of the United Nations Secretariat, 2011.* World population prospects: The 2010 revision. *http://esa.un.org/wpp, Fig 1. © United Nations, 2011.*

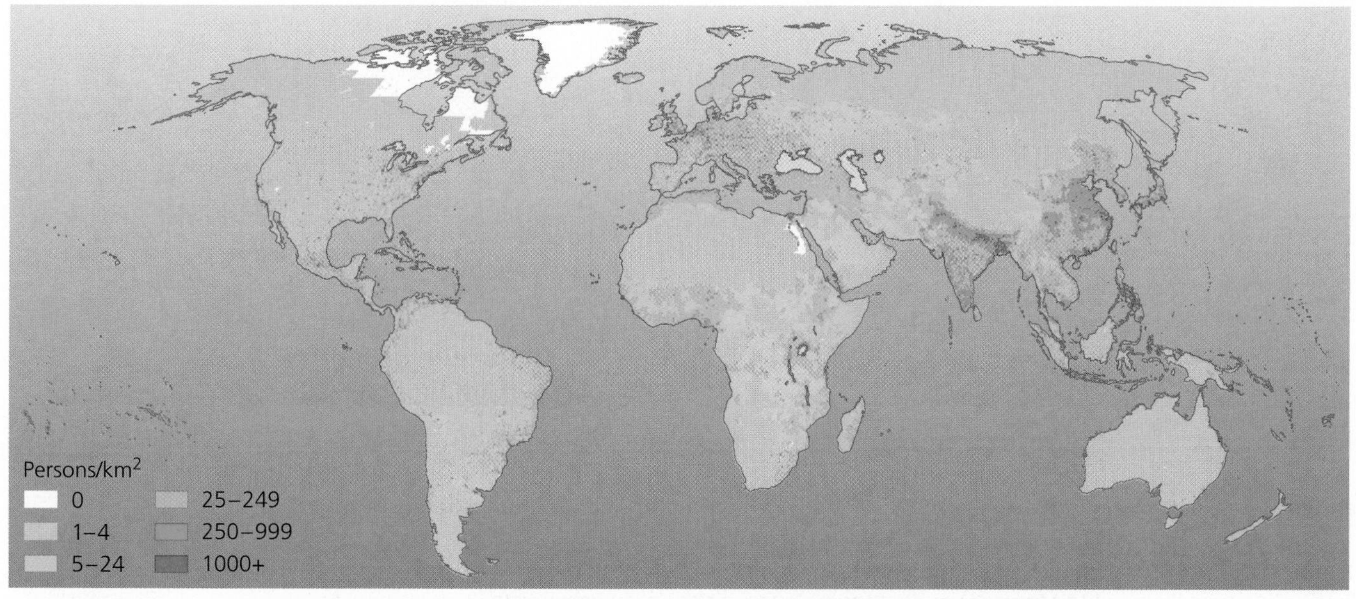

FIGURE 8.9 Human population density varies significantly from one region to another. Arctic and desert regions have the lowest population densities, whereas areas of India, Bangladesh, and eastern China have the densest populations. *Source:* The world: Population density, 2000. *Center for International Earth Science Information Network (CIESIN), Columbia University; and Centro Internacional de Agricultura Tropical (CIAT). 2005. Gridded Population of the World Version 3 (GPWv3). Palisades, NY: Socioeconomic Data and Applications Center (SEDAC), Columbia University.*

the world's other major rivers, from the Nile to the Danube to the Ganges to the Mississippi, all receive more than their share of human impact. At the same time, some areas with low population density are sensitive (a high *S* value in our revised IPAT model) and thus vulnerable to impact. Deserts and arid grasslands, for instance, are easily degraded by development that commandeers too much water.

Age structure Age structure (p. 63), describes the relative numbers of individuals of each age class within a population. Data on age structure are especially valuable to demographers trying to predict future dynamics of human populations. A population made up mostly of individuals past reproductive age will tend to decline over time. In contrast, a population with many individuals of reproductive age or pre-reproductive age is likely to increase. A population with an even age distribution will likely remain stable as births keep pace with deaths.

Age structure diagrams, often called population pyramids, are visual tools scientists use to illustrate age structure (FIGURE 8.10). The width of each horizontal bar represents the number of people in each age class. A pyramid with a wide base denotes a large proportion of people who have not yet reached reproductive age—and this indicates a population soon capable of rapid growth. In this respect, a wide base of a population pyramid is like an oversized engine on a rocket—the bigger the booster, the faster the increase.

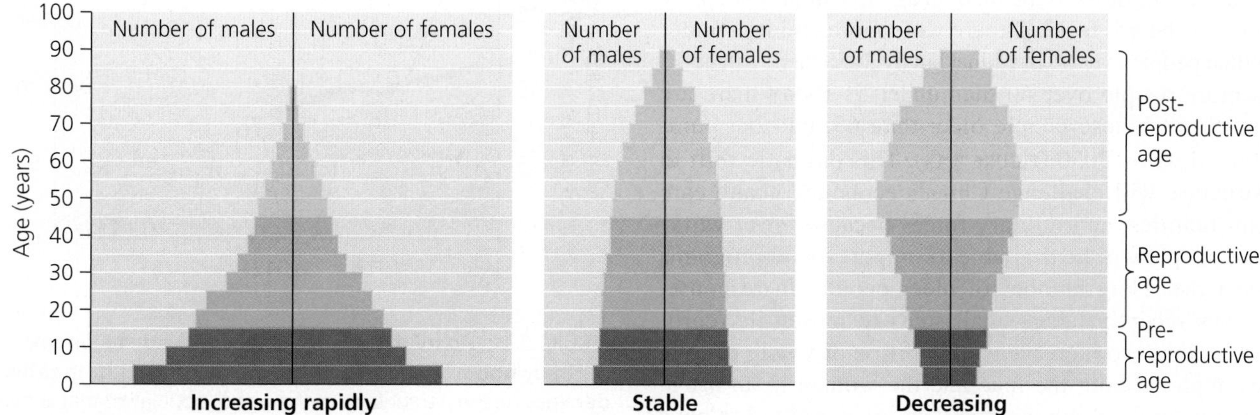

FIGURE 8.10 Age structure diagrams show numbers of individuals of different age classes in a population. A diagram like that on the left is weighted toward young age classes, indicating a population that will grow quickly. A diagram like that on the right is weighted toward old age classes, indicating a population that will decline. Populations with balanced age structures, like the one shown in the middle diagram, will remain relatively stable in size.

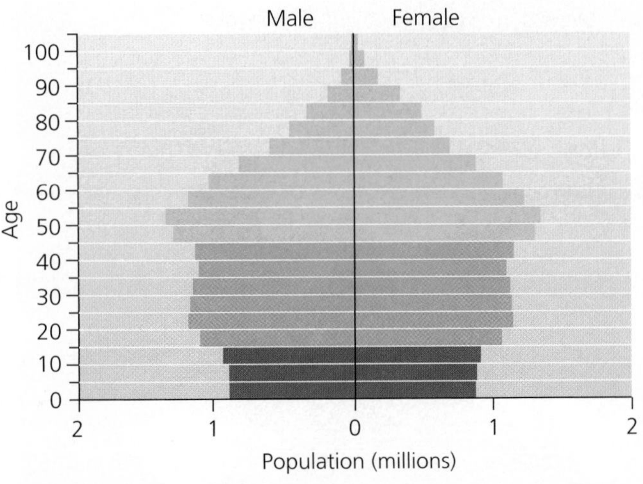

(a) Age structure diagram of Canada in 2012

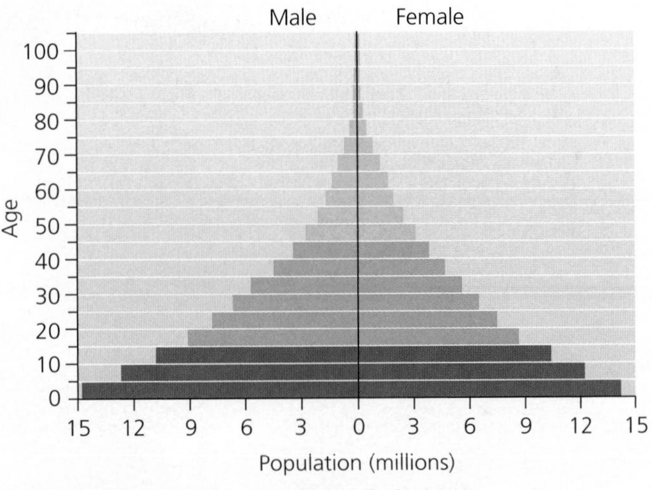

(b) Age structure diagram of Nigeria in 2012

FIGURE 8.11 Canada (a) shows a fairly balanced age structure, whereas Nigeria (b) shows an age distribution heavily weighted toward young people. Nigeria's population growth rate (2.6%) is over six times greater than Canada's (0.4%). *Data from United States Census Bureau International Database. http://www.census.gov/population/international/data/idb/.*

As an example, compare age structures for the nations of Canada and Nigeria (FIGURE 8.11). Nigeria's large concentration of individuals in young age groups portends a great deal of reproduction. Not surprisingly, Nigeria's population growth rate is much greater than Canada's.

Today, populations are aging in many nations (FIGURE 8.12). The global median age today is 28, but it will be 38 in the year 2050. Population aging is pronounced in the United States, where the "baby boom" generation is now approaching retirement age. In coming years you will likely be required to pay more taxes to support Social Security and Medicare benefits for your elders.

By causing dramatic reductions in the number of children born since 1970, China's one-child policy virtually guaranteed that the nation's population age structure would change. Indeed, in 1970 the median age in China was 20; by 2050 it will be 45. In 1970 there were more children under age 5 than people over 60 in China, but by 2050 there will be 12 times more people over 60 than under 5! Today there are 121 million Chinese people older than 65, but that number will triple by 2050 (FIGURE 8.13). This dramatic shift in age structure will challenge China's economy, health care system, families, and military forces because fewer working-age people will be available to support social programs to assist the rising number of older people. In response, China has recently taken small steps to loosen its reproductive policies to increase its proportion of young people. For example, if both the man and the woman in an urban couple are single children, then they are permitted to have a second child.

On the one hand, older populations will present new challenges for many nations as increasing numbers of elderly will require the care and assistance of relatively fewer working-age citizens.

On the other hand, a shift in age structure toward an older population reduces the proportion of dependent children. Fewer young adults may mean a decrease in crime rates. Moreover, older people are often productive members of society, contributing volunteer activities and services to their children and grandchildren. In many ways both good and bad, "graying" populations will affect societies in China, the United States, Europe, and other nations throughout our lifetimes.

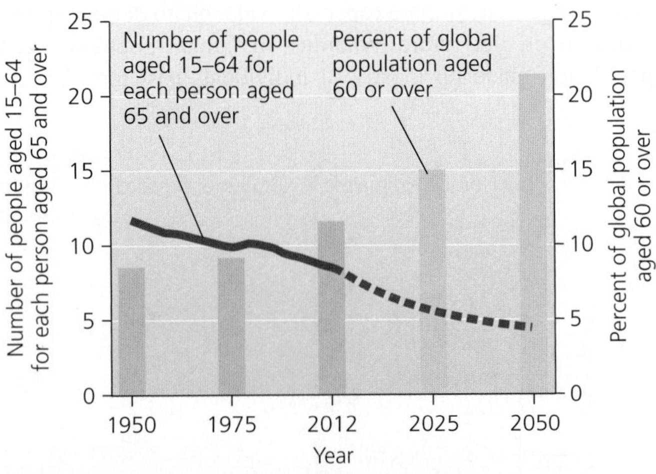

FIGURE 8.12 Populations are aging worldwide as people of "baby boom" generations grow older and as birth rates decline. As the percentage of the global population that is over 60 increases, fewer working-age people are available to support the elderly. *Data from Population Division of the Department of Economic and Social Affairs of the United Nations Secretariat, 2009.* World population prospects: The 2008 revision. *http://esa.un.org/wpp. © United Nations, 2009. By permission. 2012 update from United States Census Bureau International Database. http://www.census.gov/population/international/data/idb/.*

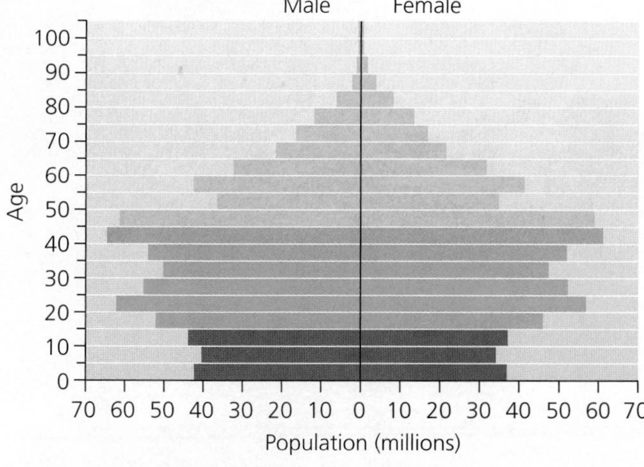

(a) Age structure diagram of China in 2012

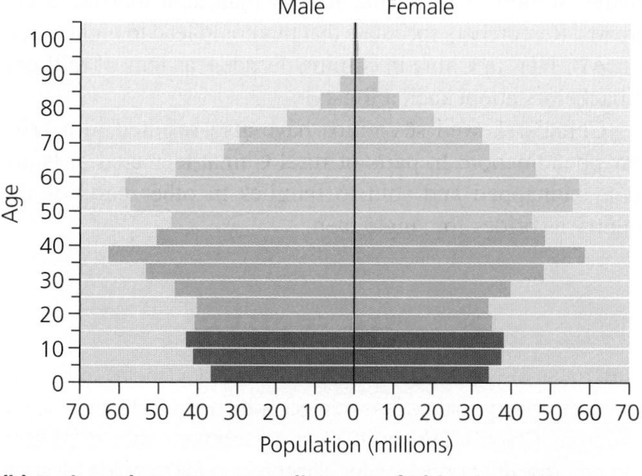

(b) Projected age structure diagram of China in 2025

(c) Young female factory workers in Hong Kong

(d) Elderly Chinese

FIGURE 8.13 **As China's population ages, older people will outnumber the young.** Population pyramids show the predicted graying of the Chinese population from 2012 **(a)** to 2025 **(b)**. Today's children may, as working-age adults **(c)**, face pressures to support greater numbers of elderly citizens **(d)** than has any previous generation. *Data in (a) and (b) from United States Census Bureau International Database. http://www.census.gov/population/international/data/idb.*

Sex ratios The ratio of males to females also can affect population dynamics. Note that population pyramids give data on sex ratios by representing numbers of males and females on opposite sides of each diagram. To understand the consequences of sex ratio variation, imagine two islands, one populated by 99 men and 1 woman and the other by 50 men and 50 women. Where would we be likely to see the greatest population increase over time? Of course, the island with an equal number of men and women would have a greater number of potential mothers and thus a greater potential for population growth.

The naturally occurring sex ratio at birth in human populations features a slight preponderance of males; for every 100 female infants born, about 106 male infants are born. This phenomenon is an evolutionary adaptation (p. 50) to the fact that males are slightly more prone to death during any given year of life. It tends to ensure that the ratio of men to women will be approximately equal when people reach reproductive age. Thus, a slightly uneven sex ratio at birth may be beneficial. However, a greatly distorted ratio can lead to problems.

In recent years, demographers have witnessed an unsettling trend in China: The ratio of newborn boys to girls has become strongly skewed. Today, roughly 120 boys are born for every 100 girls. Some provinces have reported sex ratios as high as 138 boys for every 100 girls. The leading hypothesis for these unusual sex ratios is that many parents, having learned the sex of their fetuses by ultrasound, are selectively aborting female fetuses.

Recall that Chinese culture has traditionally valued sons over daughters. Sociologists maintain that this cultural gender preference, combined with the government's one-child policy, has led some couples to selectively abort female fetuses or to abandon or kill female infants. The Chinese government reinforced this gender discrimination when in 1984 it exempted rural peasants from the one-child policy if their first child was a girl, but not if the first child was a boy.

China's skewed sex ratio may further lower population growth rates. However, it has the undesirable social consequence of leaving large numbers of Chinese men single. Many of these men find employment as migrant workers and tend to engage in more risky sexual activity than their married counterparts. Researchers speculate that this could lead to higher incidence of HIV in China in coming decades, as tens of millions of bachelors adopt such a lifestyle.

China's skewed sex ratios have also resulted in a grim new phenomenon. In parts of rural China, teenaged girls are being kidnapped and sold to families in other parts of the country as brides for single men.

WEIGHING THE ISSUES

CHINA'S REPRODUCTIVE POLICY Consider the benefits as well as the problems associated with a reproductive policy such as China's. Do you think a government should be able to enforce strict penalties for citizens who fail to abide by such a policy? If you disagree with China's policy, what alternatives can you suggest for dealing with the resource demands of a rapidly growing population?

Population change results from birth, death, immigration, and emigration

Rates of birth, death, immigration, and emigration determine whether a population grows, shrinks, or remains stable. The formula for measuring population growth (p. 66) also pertains to people: Birth and immigration add individuals to a population, whereas death and emigration remove individuals. Technological advances have led to a dramatic decline in human death rates, widening the gap between birth rates and death rates and resulting in the global human population expansion.

In today's ever more crowded world, immigration and emigration play increasingly important roles. Refugees, people forced to flee their home country or region, have become more numerous as a result of war, civil strife, and environmental degradation. The United Nations puts the number of refugees who flee to escape poor environmental conditions at 25 million per year and possibly many more. Often the movement of refugees causes environmental problems in the receiving region as these desperate victims try to eke out an existence with no livelihood and no cultural or economic attachment to the land or incentive to conserve its resources. The millions who fled Rwanda following the genocide there in the 1990s, for example, inadvertently destroyed large areas of forest while trying to obtain fuelwood, food, and shelter to stay alive in the Democratic Republic of Congo (**FIGURE 8.14**).

For most of the past 2000 years, China's population was relatively stable. The first significant increases resulted from enhanced agricultural production and a powerful government during the Qing Dynasty in the 1800s. Population growth began to outstrip food supplies by the 1850s, and quality of life for the average Chinese peasant began to decline. Over the next 100 years, China's population grew slowly, at about 0.3% per year, amid food shortages and political instability. Population growth rates rose again following Mao's establishment of the People's Republic in 1949, and they have declined since the advent of the one-child policy (**TABLE 8.1**).

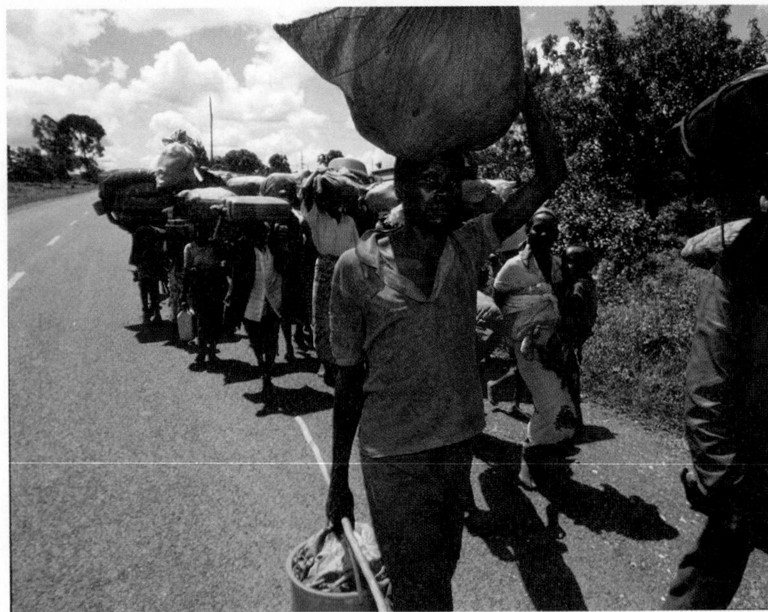

FIGURE 8.14 Immigration, including movements of refugees, can affect a nation's demographics. The flight of refugees from Rwanda into the Democratic Republic of Congo in 1994 following the Rwandan genocide caused unimaginable hardship for the refugees and tremendous stress on the environment into which they moved.

In recent decades, falling growth rates in many countries have led to an overall decline in the global growth rate (**FIGURE 8.15**). This decline has come about, in part, from a steep drop in birth rates. Note, however, that although the rate of growth is slowing, the absolute size of the population continues to increase. Even though our percentage increases are getting smaller year by year, these are percentages of ever-larger numbers, so we continue to add over 70 million people to the planet each year.

Total fertility rate influences population growth

One key statistic demographers calculate to examine a population's potential for growth is the **total fertility rate (TFR)**, the average number of children born per woman during her lifetime. **Replacement fertility** is the TFR that keeps the size of a population stable. For humans, replacement fertility roughly

TABLE 8.1 Trends in China's Population Growth

MEASURE	1950	1970	1990	2012
Total fertility rate	5.8	5.8	2.2	1.5
Rate of natural increase (% per year)	1.9	2.6	1.4	0.5
Doubling time (years)	37	27	49	140
Population (billions)	0.56	0.83	1.15	1.35

Data from China Population Information and Research Center; and Population Reference Bureau, 2012. 2012 World population data sheet.

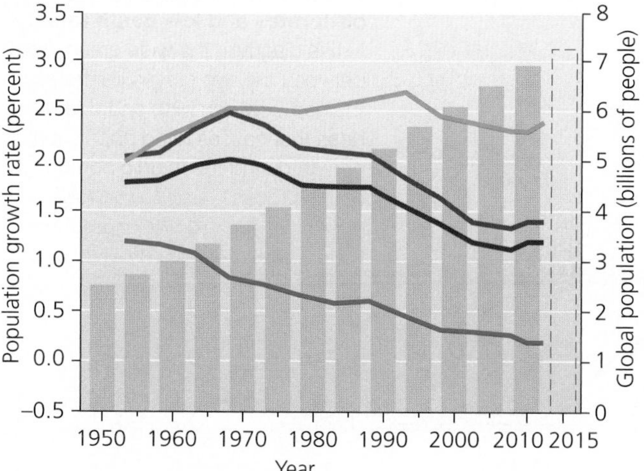

FIGURE 8.15 The annual growth rate of the global human population peaked in the late 1960s and has declined since then. Growth rates of developed nations have fallen since 1950, whereas those of developing nations have fallen since the global peak in the late 1960s. For the world's least developed nations, growth rates began to fall in the 1990s. Although growth rates are declining, global population size is still growing about the same amount each year, because smaller percentage increases of ever-larger numbers produce roughly equivalent additional amounts.
Data from Population Division of the Department of Economic and Social Affairs of the United Nations Secretariat, 2011. World population prospects: The 2010 revision. *http://esa.un.org/wpp. © United Nations, 2011. Data updates for 2011–2012 from Population Reference Bureau, 2011 and 2012* World population data sheets. *Global population in 2015 is projected.*

equals a TFR of 2.1. (Two children replace the mother and father, and the extra 0.1 accounts for the risk of a child dying before reaching reproductive age.) If the TFR drops below 2.1, population size (in the absence of immigration) will shrink.

Factors such as industrialization, improved women's rights, and quality health care have driven TFR downward in many nations in recent years. All these factors have come together in Europe, where TFR has dropped from 2.6 to 1.6 in the past half-century. Nearly every European nation now has a fertility rate below the replacement level, and populations are declining in 16 of 45 European nations. In 2012, Europe's overall annual **rate of natural increase** (also called the *natural rate of population change*)—change due to birth and death rates alone, excluding migration—was between 0.0% and 0.1%. Worldwide by 2012, 92 countries had fallen below the replacement fertility of 2.1. These countries make up roughly half of the world's population and include China (with a TFR of 1.5). TABLE 8.2 shows TFRs of major continental regions.

Many nations are experiencing the demographic transition

Many nations with lowered birth rates and TFRs are experiencing a common set of interrelated changes. In countries with good sanitation, effective health care, and reliable food supplies, more people than ever before are living long lives. As a result, over the past 50 years the life expectancy for the average person has increased from 46 to 70 years as the global death rate has dropped from 20 deaths per 1000 people to 8 deaths per 1000 people. Strictly speaking, **life expectancy** is the average number of years that an individual in a particular age group is likely to continue to live, but often people use this term to refer to the average number of years a person can expect to live from birth. Much of the increase in life expectancy is due to reduced rates of infant mortality. Societies going through these changes are generally those that have undergone urbanization and industrialization and have generated personal wealth for their citizens.

TABLE 8.2 Total Fertility Rates for Major Continental Regions

REGION	TOTAL FERTILITY RATE (TFR)
Africa	4.7
Australia and South Pacific	2.5
Latin America and Caribbean	2.2
Asia	2.2
North America	1.9
Europe	1.6

Data from Population Reference Bureau, 2012. 2012 World population data sheet.

To make sense of these trends, demographers developed a concept called the **demographic transition.** This is a model of economic and cultural change first proposed in the 1940s and 1950s by demographer Frank Notestein to explain the declining death rates and birth rates that have occurred in Western nations as they industrialized. Notestein believed nations move from a stable pre-industrial state of high birth and death rates to a stable post-industrial state of low birth and death rates (FIGURE 8.16). Industrialization, he proposed, causes these rates to fall by first decreasing mortality and then lessening the need for large families. Parents thereafter choose to invest in quality of life rather than quantity of children. Because death rates fall before birth rates fall, a period of net population growth results. Thus, under the demographic transition model, population growth is seen as a temporary phenomenon that occurs as societies move from one stage of development to another.

WEIGHING THE ISSUES

CONSEQUENCES OF LOW FERTILITY? In the United States, Canada, and almost every European nation, the total fertility rate is now at or below the replacement fertility rate (although some of these nations are still growing because of immigration). What economic or social consequences do you think might result from below-replacement fertility rates? Would you rather live in a society with a growing population, a shrinking population, or a stable population? Why?

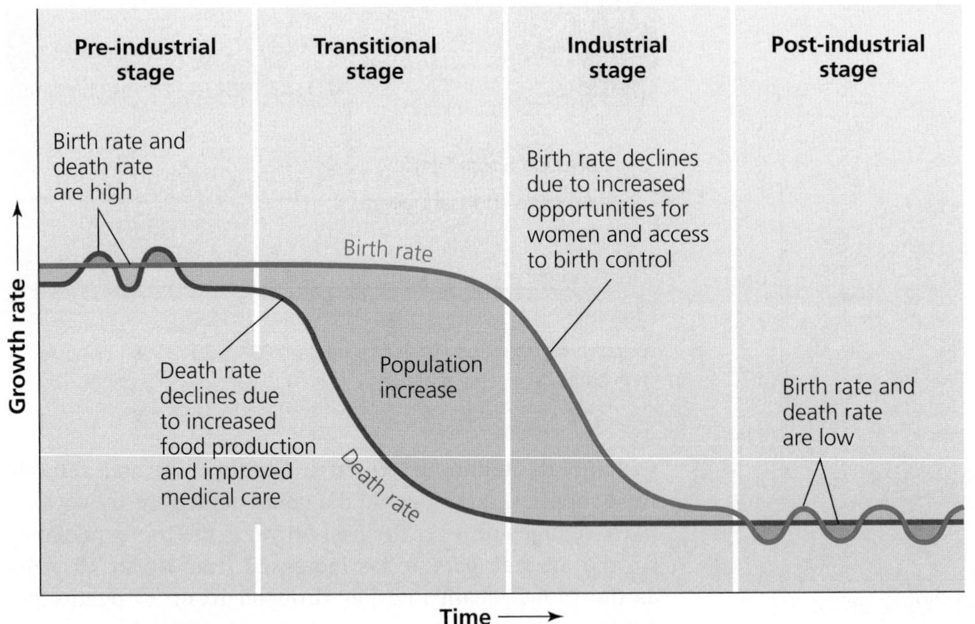

FIGURE 8.16 The demographic transition models a process that has taken some populations from a pre-industrial stage of high birth rates and high death rates to a post-industrial stage of low birth rates and low death rates. In this diagram, the wide green area between the two curves illustrates the gap between birth and death rates that causes rapid population growth during the middle portion of this process. *Adapted from Kent, M. and K. Crews, 1990.* World population: Fundamentals of growth. *By permission of the Population Reference Bureau.*

The pre-industrial stage The first stage of the demographic transition model is the **pre-industrial stage,** characterized by conditions that have defined most of human history. In pre-industrial societies, both death rates and birth rates are high. Death rates are high because disease is widespread, medical care rudimentary, and food supplies unreliable and difficult to obtain. Birth rates are high because people must compensate for infant mortality by having several children. In this stage, children are valuable as workers who can help meet a family's basic needs. Populations within the pre-industrial stage are not likely to experience much growth, which is why the human population grew relatively slowly until the industrial revolution.

Industrialization and falling death rates Industrialization initiates the second stage of the demographic transition, known as the **transitional stage.** This transition from the pre-industrial stage to the industrial stage is generally characterized by declining death rates due to increased food production and improved medical care. Birth rates in the transitional stage remain high, however, because people have not yet grown used to the new economic and social conditions. As a result, population growth surges.

The industrial stage and falling birth rates The third stage in the demographic transition is the **industrial stage.** Industrialization increases opportunities for employment outside the home, particularly for women. Children become less valuable, in economic terms, because they do not help meet family food needs as they did in the pre-industrial stage. If couples are aware of this, and if they have access to birth control, they may choose to have fewer children. Birth rates fall, closing the gap with death rates and reducing population growth.

The post-industrial stage In the final stage, the **post-industrial stage,** both birth and death rates have fallen to low and stable levels. Population sizes stabilize or decline slightly. The society enjoys the fruits of industrialization without the threat of runaway population growth. The United States is an example of a nation in this stage, although the U.S. population is growing faster than most other post-industrial nations because of a relatively high immigration rate.

Is the demographic transition a universal process?

The demographic transition has occurred in many European countries, the United States, Canada, Japan, and several other developed nations over the past 200 to 300 years. It is a model that may or may not apply to all developing nations as they industrialize now and in the future. On the one hand, note in Figure 8.15 (p. 201) how growth rates fell first for industrialized nations, then for less developed nations, and finally for least developed nations, suggesting that it may merely be a matter of time before all nations experience the transition. On the other hand, some developing nations may already be suffering too much from the impacts of large populations to replicate the developed world's transition. And some demographers assert that the transition will fail in cultures that place greater value on childbirth or grant women fewer freedoms.

Moreover, natural scientists estimate that for people of all nations to attain the material standard of living that North Americans now enjoy, we would need the natural resources of four-and-a-half more planet Earths. Whether developing nations (which include the vast majority of the planet's people) pass through the demographic transition is one of the most important and far-reaching questions for the future of our civilization and Earth's environment.

Population and Society

Demographic transition theory links the quantitative study of how populations change with the societal factors that influence (and are influenced by) population dynamics. There are many factors that affect fertility in a given society. They include public health issues, such as people's access to contraceptives and the rate of infant mortality. They also include cultural factors—such as the level of women's rights, the relative acceptance of contraceptive use, and even television programs (see **THE SCIENCE BEHIND THE STORY**, pp. 204–205). There are also effects from economic factors, such as the society's level of affluence, the importance of child labor, and the availability of governmental support for retirees. Let's now examine a few of these societal influences on fertility more closely.

Family planning is a key approach for controlling population growth

Perhaps the greatest single factor enabling a society to slow its population growth is the ability of women and couples to engage in **family planning,** the effort to plan the number and spacing of one's children. Family-planning programs and clinics offer information and counseling to potential mothers and fathers on reproductive issues. An important component of family planning is **birth control,** the effort to control the number of children a woman bears by reducing the frequency of pregnancy. Birth control relies on **contraception,** the deliberate attempt to prevent pregnancy despite engaging in sexual intercourse. Common methods of modern contraception in use today include condoms, spermicide, hormonal treatments (birth control pill/hormone injection), intrauterine devices (IUDs), and permanent sterilization through tubal ligation or vasectomy. Many family-planning organizations aid clients by offering free or discounted contraceptives.

Worldwide in 2012, 56% of women (aged 15–49) reported using contraceptives, with rates of use varying widely among nations. China, at 84%, had the highest rate of contraceptive use of any nation. Eight European nations showed rates of contraceptive use of 70% or more, as did Australia, Brazil, Canada, Colombia, Costa Rica, Cuba, Dominican Republic, Micronesia, New Zealand, Paraguay, Puerto Rico, South Korea, Thailand, the United States, and Uruguay. At the other end of the spectrum, 14 African nations had rates below 10%.

Low usage rates for contraceptives in some societies are caused by limited availability, especially in rural areas. In others, low usage may be due to religious doctrine or cultural influences that hinder family planning, denying counseling and contraceptives to people who might otherwise use them. This can result in family sizes that are larger than the parents desire and to elevated rates of population growth.

In a physiological sense, access to family planning (and the civil rights to demand its use) gives women control over their **reproductive window,** the period of their life, beginning with sexual maturity and ending with menopause, in which they may become pregnant. A woman can bear up to 25 children within this window (**FIGURE 8.17a**), but she may choose to delay the birth of her first child to pursue education and employment. She may also use contraception to delay her first child, space births within the window, and "close" her reproductive window after achieving her desired family size (**FIGURE 8.17b**).

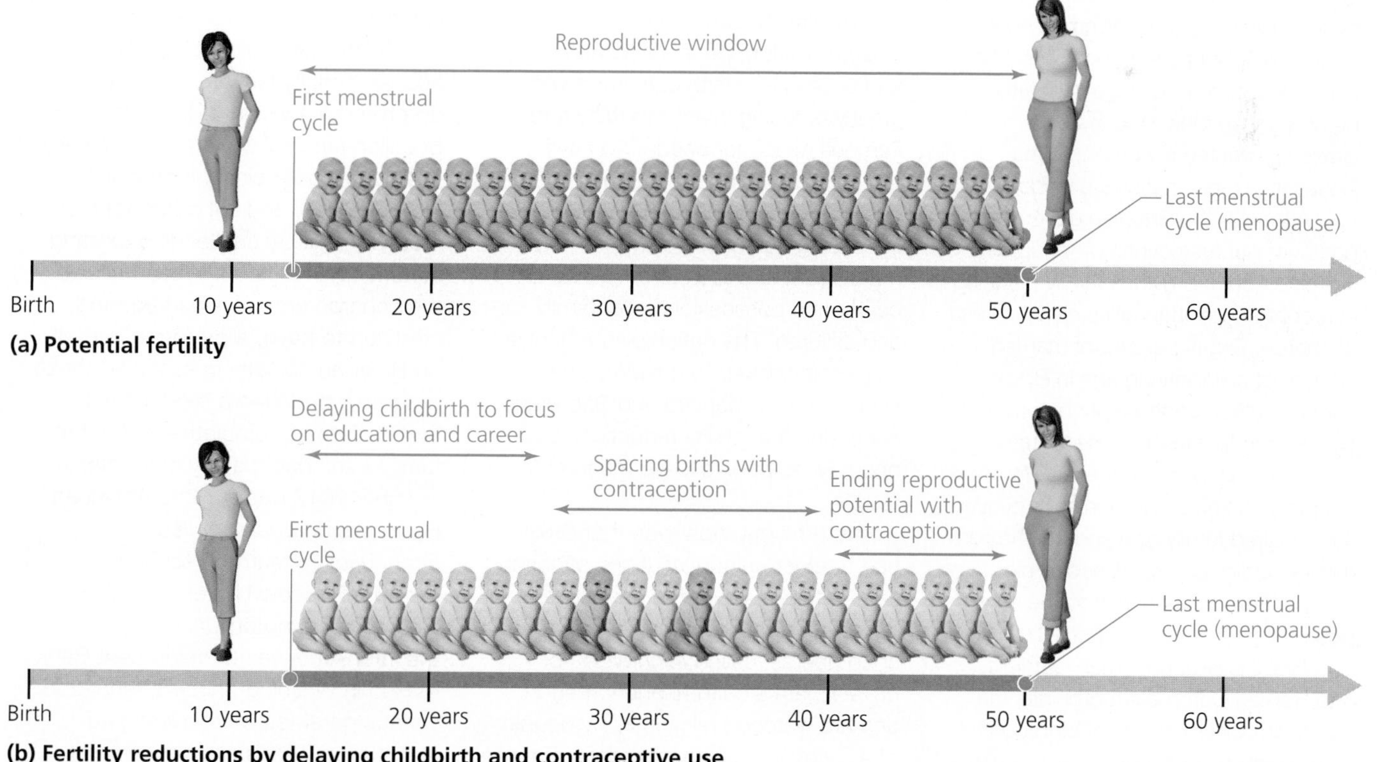

FIGURE 8.17 Women can potentially have very high fertility within their "reproductive window" but can choose to reduce the number of children they bear. They may do this by delaying the birth of their first child, or by using contraception to space pregnancies or to end their reproductive window at the time of their choosing.

Did Soap Operas Reduce Fertility in Brazil?

Brazilian soap operas, called *telenovelas*, are a surprising cultural force for promoting lower fertility. Here, residents gather outside a cafe in Rio de Janeiro to watch the popular program *Avenida Brasil*.

Over the past 50 years, the South American nation of Brazil experienced the second-largest drop in fertility among developing nations with large populations—second only to China. In the 1960s, the average woman in Brazil had six children. Today, Brazil's total fertility rate is 1.9 children per woman, which is lower than that of the United States. Brazil's drastic decrease in fertility is interesting because, unlike China, it occurred without governmental policies that advocated controls on its citizens' reproduction.

So how did Brazil accomplish this? A major factor was change in society's view of women. It began with a civil rights movement in the 1960s, which gave females equal access to education and the opportunity to pursue careers outside the home. These efforts have been highly successful. Women now comprise 40% of the workforce in Brazil and graduate from college in greater numbers than men. And in 2010, Brazilians elected a woman, Dilma Rousseff, as their nation's president.

Although the Brazilian government does not put restrictions on people's reproduction, it provides family planning and contraception to all its citizens free of charge. Eighty percent of married women of childbearing age in Brazil currently utilize contraception, a rate higher than that in the United States or Canada. Universal access to family planning has given women control over their desired family size and has helped reduce fertility across all economic groups, from the very rich to the very poor.

Brazil is largely Roman Catholic, and Roman Catholicism prohibits the use of artificial methods of birth control, so the high rates of contraceptive use in modern Brazil represent a significant shift from traditional values. Induced abortion is not utilized in Brazil as it is in China; the procedure is illegal except in rare circumstances.

As Brazil's economy grew with industrialization, people's nutrition and access to health care improved, greatly reducing infant mortality rates. Families no longer needed to have more children than they desired for fear one or more would die at a young age. Increasing personal wealth promoted materialism and greater emphasis on career and possessions over family and children. The nation also urbanized as people flocked to growing cities such as Río de Janeiro and São Paolo, conveying the fertility reductions that occur when people leave the farm for the city.

It turns out, however, that Brazil had a rather unique influence affecting its fertility rates over the past several decades—"soap operas." Brazilian soap operas, called *telenovelas* or *novelas*, are a cultural phenomenon and are watched religiously by people of all ages, races, and incomes. Each *novela* follows the activities of several fictional families, and these TV shows are wildly popular because they have characters, settings, and plot lines with which everyday Brazilians can identify.

Telenovelas do not overtly address fertility issues, but they do promote a vision of the "ideal" Brazilian family. This family is typically middle or upper class, materialistic, individualistic, and full of empowered women. By challenging existing cultural and religious values through their characters, *novelas* had, and continue to have, a profound impact on Brazilian society. In essence, these programs provided a model family for Brazilians to emulate—with small family sizes being a key characteristic.

In a 2012 paper in the *American Economic Journal: Applied Economics*, a team of researchers from Bocconi University in Italy, George Washington University, and the Inter-American Development Bank (based in Washington, D.C.) analyzed various parameters to investigate statistical relationships between *telenovelas* and fertility patterns in Brazil from 1965 to 2000. Rede Globo, the network that has a virtual monopoly

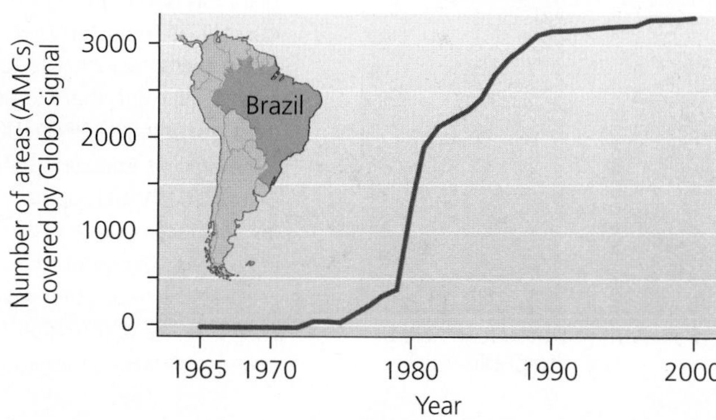

FIGURE 1 The Globo television network expanded over time and now reaches nearly all households in Brazil. Fertility declines across minimally comparable areas (AMCs) were correlated with the availability of Globo, and its *novelas*, over the time periods in the study. *Source: La Ferrara, E., et al., 2012.* Soap operas and fertility: Evidence from Brazil. *Am. Econ. J. Appl. Econ. 4: 1-31.*

on the most popular *novelas*, increased the number of areas that received its signal in Brazil over those 35 years (**FIGURE 1**), and currently it reaches 98% of Brazilian households. By combining data on Rede Globo broadcast range with demographic data, the researchers were able to compare changes in fertility patterns over time in areas of Brazil that received access to *novelas* with areas in Brazil that did not.

The team found that women in areas that received the Globo signal had significantly lower fertility than those in areas not served by Rede Globo, and they also found that fertility declines were related to the timing of receiving access to *novelas*. Diving deeper into census data, the researchers concluded that the greatest reductions in fertility occurred in women aged 25–34 and 35–44, but not in younger women (**FIGURE 2**). The authors hypothesized that this effect was likely due to the fact that women between 25 and 44 were closer in age to the main female characters in *novelas*, who typically had no children or only a single child. The depressive effect on fertility among women in areas served by Globo was therefore attributed to wider spacing of births and earlier ending of reproduction, not to delaying the birth of their first child.

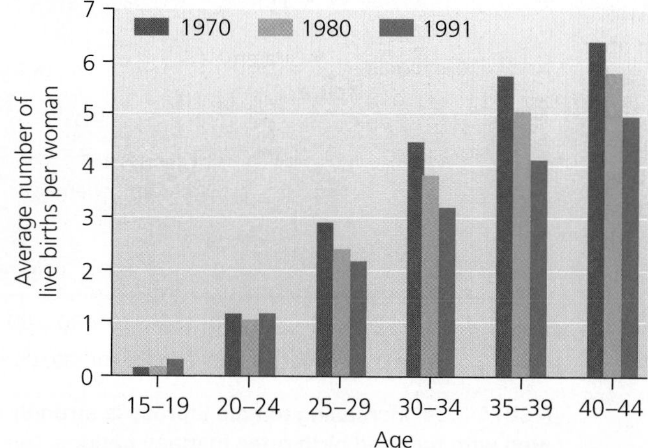

FIGURE 2 Fertility declines among Brazilian women between 1970 and 1991 were most pronounced in later age classes. The authors attribute some of this decline to women in those age classes emulating the low fertility of lead female characters in *novelas*. *Source: La Ferrara, E., et al., 2012.* Soap operas and fertility: Evidence from Brazil. *Am. Econ. J. Appl. Econ. 4: 1-31.*

Further evidence for the influence of *novelas* on fertility was found in school records. The researchers found that fifth graders living in areas reached by the Globo network were four times more likely to be named after the lead characters in *novelas* in the year they were born than were children in areas not served by Globo. These results were particularly compelling because most *novela* characters have relatively unusual names.

The researchers determined that access to television alone did not depress fertility. For example, comparisons of fertility rate in areas with access to a different television network, Sistema Brasileiro de Televisão, found no relationship. The study authors concluded that this was likely due to the network's reliance on programming imported from other nations, with which everyday Brazilians did not connect as they did with *novelas*.

Television's ability to change cultural attitudes is not limited to Brazil. A 2007 study found that access to cable television in rural villages in India correlated with lower fertility, reduced acceptance of violence against women, decreased preference for sons over daughters, and increases in attitudes about female autonomy.

The factors that affect human fertility can be complex and vary greatly from one society to another. The influence of *novelas* on fertility in Brazil also shows that the factors influencing fertility can sometimes come from unexpected sources. ■

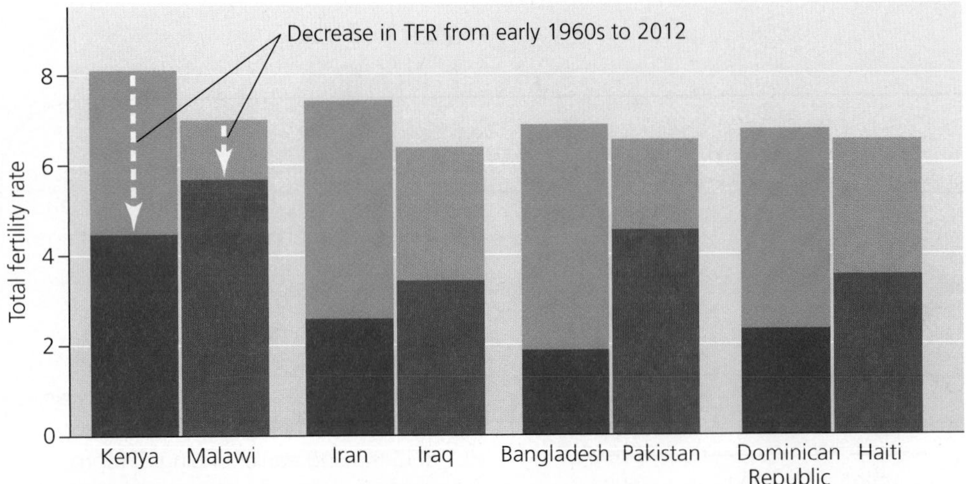

FIGURE 8.18 Four pairs of neighboring countries demonstrate the results of family planning initiatives. In each case, the nation that invested in family planning and established societal conditions that promoted low fertility (**blue bars**) reduced its total fertility rate (TFR) more than its neighbor (**red bars**) from 1960–1965 to 2012. *Data from from Population Reference Bureau, 2012* World population data sheet *and Population Division of the Department of Economic and Social Affairs of the United Nations Secretariat, 2011.* World population prospects: The 2010 revision, *http://esa.un.org/wpp. © United Nations, 2011.*

Family-planning programs are working around the world

Data show that funding and policies that encourage family planning can lower population growth rates in all types of nations, even those that are least industrialized. No nation has pursued a sustained population control program as intrusive as China's, but other rapidly growing nations have implemented programs that are less restrictive.

India was the first nation to implement population control policies. However, when some policymakers introduced forced sterilization in the 1970s, the resulting outcry brought down the government. Since then, India's efforts have been more modest and far less coercive, focusing on family planning and reproductive health care. This has greatly reduced rates of growth in India, but India will nonetheless likely overtake China and become the world's most populous nation in several decades because of China's more aggressive population initiatives.

The government of Thailand has reduced birth rates and slowed population growth. In the 1960s, Thailand's growth rate was 2.3%, but in 2012 it was 0.5%. This decline was achieved without a one-child policy, resulting instead from an education-based approach to family planning and the increased availability of contraceptives. Brazil, Cuba, Iran, Mexico, and many other developing countries have instituted active programs to reduce their population growth; these entail setting targets and providing incentives, education, contraception, and reproductive health care. Studies show that nations with such programs have lower fertility rates than similar nations without them (FIGURE 8.18).

Empowering women reduces fertility rates

Today, many social scientists and policymakers recognize that for population growth to slow and stabilize, women should be granted equal power to men in societies worldwide. This would have many benefits: Studies show that where women are freer to decide whether and when to have children, fertility rates fall, and children are better cared for, healthier, and better educated.

One benefit of equal rights for women is the ability to make reproductive decisions. In many societies, men restrict women's decision-making abilities, including decisions about how many children a woman will bear. Fertility rates have dropped most noticeably in nations where women have gained improved access to contraceptives and to family planning.

Equality for women also involves expanding educational opportunities for them. In many nations girls are discouraged from pursuing an education or are kept out of school altogether. Worldwide, over two-thirds of people who cannot read are women. And data clearly show that as women receive educational opportunities, fertility rates decline (FIGURE 8.19). Education helps more women delay childbirth as they pursue careers and gives them a greater say in reproductive decisions.

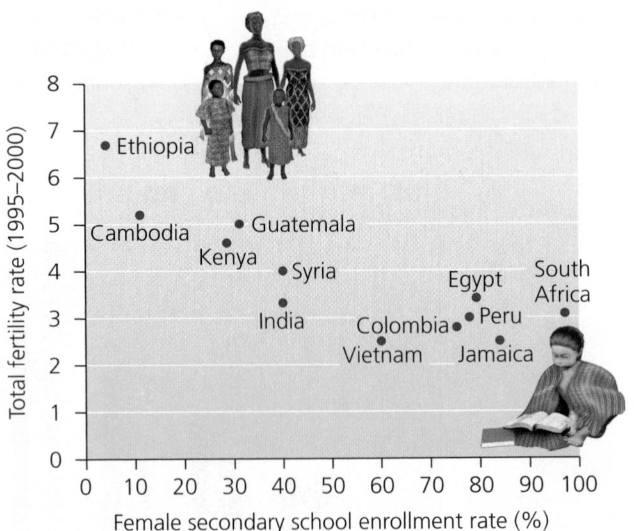

FIGURE 8.19 Increasing female literacy is strongly associated with reduced birth rates in many nations. *Data from McDonald, M., and D. Nierenberg, 2003.* Linking population, women, and biodiversity. *State of the world 2003. Washington, D.C.: Worldwatch Institute.*

DATA Q Is the relationship between total fertility rate and the rate of enrollment of girls in secondary school positive (as variable 1 increases, so does variable 2), negative (as variable 1 increases, variable 2 decreases), or is there no obvious relationship (increases in variable 1 are not correlated with changes in variable 2)?

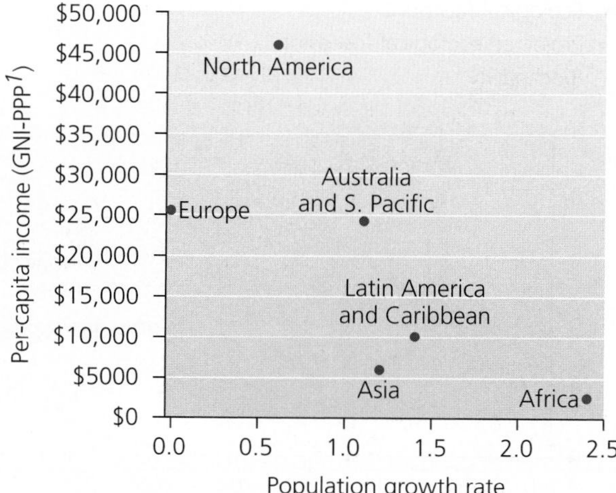

FIGURE 8.20 Poverty and population growth show a fairly strong correlation, despite the influence of many other factors. Per capita income is here measured in GNI PPP, or "gross national income in purchasing power parity." GNI PPP is a measure that standardizes income among nations by converting it to "international" dollars, which indicate the amount of goods and services one could buy in the United States with a given amount of money. *Data from Population Reference Bureau, 2010.* 2010 World population data sheet.

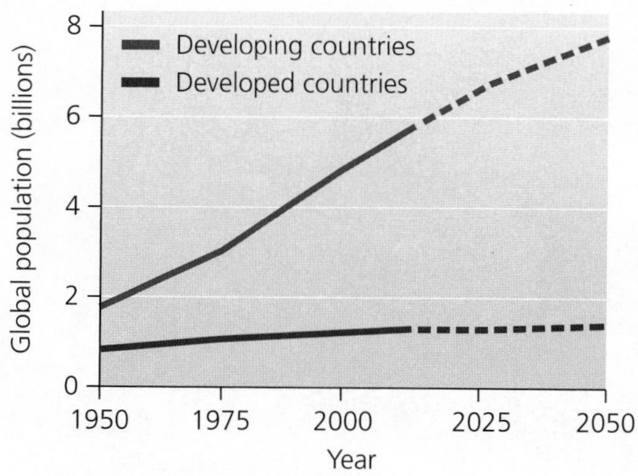

FIGURE 8.21 Over 99% of the next 1 billion people added to Earth's population will be born into developing countries. Dashed portions of the lines indicate projected future trends. *Data from Population Division of the Department of Economic and Social Affairs of the United Nations Secretariat, 2011.* World population prospects: The 2010 revision. *http://esa.un.org/wpp. © United Nations, 2011. Data updates for 2011–2012 from Population Reference Bureau, 2011 and 2012* World population data sheets.

Increasing affluence lowers fertility

Poorer societies tend to show higher population growth rates than do wealthier societies (**FIGURE 8.20**), as one would expect given the demographic transition model. There are many ways that growing affluence and reducing poverty lead to lower rates of population growth.

Historically, people tended to conceive many children, which helped ensure that at least some would survive. Today's improved medical care in wealthy nations has reduced infant mortality rates, making it less necessary to bear multiple children. Increasing urbanization has also driven TFR down; whereas rural families need children to contribute to farm labor, in urban areas children are usually excluded from the labor market, are required to go to school, and impose economic costs on their families. Moreover, if a government provides some form of social security, as most do these days, parents need fewer children to support them in their old age. Finally, with greater educational opportunities and changing roles in society, women tend to shift into the labor force, putting less emphasis on child rearing.

Economic factors are tied closely to population growth. Poverty exacerbates population growth, and rapid population growth worsens poverty. This connection is important because 99% of the next billion people to be added to the global population will be born into nations in the developing world (**FIGURE 8.21**). This is unfortunate from a social standpoint, because in some cases these people will be born into nations that are unable to provide for them. It is also unfortunate from an environmental standpoint, because poverty often results in environmental degradation. People who depend on agriculture and live in areas of poor farmland, for instance, may need to farm even if doing so degrades the soil and is not sustainable. This is largely why Africa's Sahel region—like many arid areas of western China—turns to desert during climatically dry periods (**FIGURE 8.22**). Poverty also drives people to cut forests and to deplete biodiversity. For example, impoverished settlers and miners hunt large mammals for "bush meat" in Africa's forests, including the great apes that are now heading toward extinction.

WEIGHING THE ISSUES

ABSTAINING FROM INTERNATIONAL FAMILY PLANNING?
Over the years, the United States has joined 180 other nations in providing millions of dollars to the United Nations Population Fund (UNFPA), which advises governments on family planning, sustainable development, poverty reduction, reproductive health, and AIDS prevention in many nations, including China. Starting in 2001, the George W. Bush administration withheld funds from UNFPA, saying that U.S. law prohibits funding any organization that "supports or participates in the management of a program of coercive abortion or involuntary sterilization," and maintaining that the Chinese government has been implicated in both. Many nations criticized the U.S. decision, and the European Union offered UNFPA additional funding to offset the loss of U.S. contributions. Once President Obama came to office, he reinstated funding to the program. What do you think U.S. policy should be? Should the United States fund family-planning efforts in other nations? What conditions, if any, should it place on the use of such funds?

Expanding wealth can increase the environmental impact per person

Poverty can lead people into environmentally destructive behavior, but wealth can produce even more severe and far-reaching environmental impacts. The affluence of a society such as the United States, Japan, or France is built on levels of resource consumption unprecedented in human history. Much

FIGURE 8.22 In the semi-arid Sahel region of Africa, population may be increasing beyond the land's ability to handle it. Here, drought and dependence on grazing agriculture have led to environmental degradation.

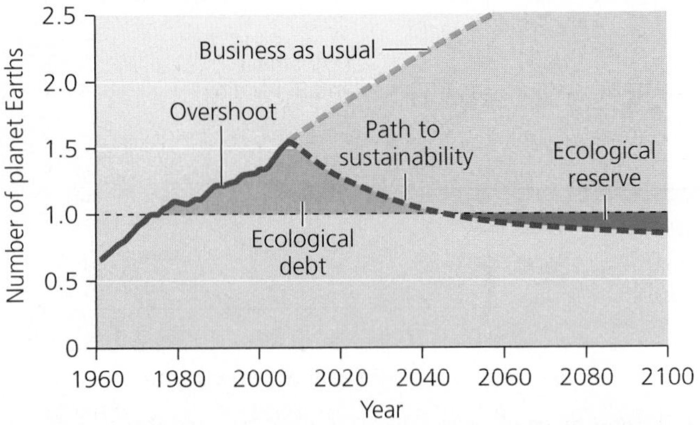

FIGURE 8.23 The global ecological footprint of the human population is estimated to be 50% greater than what Earth can bear. If population and consumption continue to rise (orange dashed line), we will increase our ecological deficit, or degree of overshoot, until systems give out and populations crash. If, instead, we pursue a path to sustainability (red dashed line), we can eventually repay our ecological debt and sustain our civilization. *Adapted from WWF International. 2010.* Living planet report 2010. *Published by WWF-World Wide Fund for Nature. © 2010 WWF (panda.org), Zoological Society of London, and Global Footprint Network.*

of this chapter has dealt with numbers of people rather than with the amount of resources each member of the population consumes or the amount of waste each member produces. The environmental impact of human activities, however, depends not only on the number of people involved but also on the way those people live. Recall the *A* (for affluence) in the IPAT equation. Affluence and consumption are spread unevenly across the world, and wealthy societies generally consume resources from regions far beyond their own.

We have explored the concept of the ecological footprint, the cumulative amount of Earth's surface area required to provide the raw materials a person or population consumes and to dispose of or recycle the waste produced. Individuals from affluent societies leave considerably larger per capita ecological footprints (see Figure 1.12, p. 14). In this sense, the addition of one American to the world has as much environmental impact as the addition of 3.4 Chinese, 8 Indians, or 14 Afghans. This fact reminds us that the "population problem" does not lie solely with the developing world, but is relevant to people everywhere.

Indeed, just as population is rising, so is consumption, and some environmental scientists have calculated that we are already living beyond the planet's means to support us sustainably. One recent analysis concludes that humanity's global ecological footprint surpassed Earth's capacity to support us in 1971 and that our species is now living 50% beyond its means (FIGURE 8.23). This is our overshoot (see Figure 1.4, p. 5). In this analysis, our ecological footprint can be compared to the amount of biologically productive land and sea available to us—an amount termed **biocapacity.** For any given area, if the footprint is greater than the biocapacity, there is an "ecological deficit." If the footprint is less than the biocapacity, there is an "ecological reserve." Because our footprint exceeds our biocapacity by 50% worldwide, we are running a global ecological deficit, gradually draining our planet of its natural capital and its long-term ability to support our civilization.

The richest one-fifth of the world's people possesses over 80 times the income of the poorest one-fifth (FIGURE 8.24). The richest one-fifth also uses 86% of the world's resources. That leaves only 14% of global resources—energy, food, water, and other essentials—for the remaining four-fifths of the world's people to share. It is therefore imperative that we continue and accelerate efforts to promote renewable energy (Chapter 21), "smart" urban design (Chapter 13), and other forms of sustainable development. This way, the rapid industrialization of China, India, and other populous nations can occur with far less environmental impact than that which accompanied the industrialization of developed nations.

HIV/AIDS is exerting major impacts on African populations

The rising material wealth and falling fertility rates of many developed nations today is slowing population growth in accord with the demographic transition model. However, nations where the human immunodeficiency virus (HIV) and acquired immunodeficiency syndrome (AIDS) has taken hold are not following Notestein's script. Instead, in these countries death rates have increased, presenting a scenario more akin to Malthus's fears.

The AIDS epidemic (FIGURE 8.25) is having the greatest impact on human populations of any communicable disease since the Black Death killed roughly one-third of Europe's population in the 14th century and since smallpox and other diseases brought by Europeans to the New World wiped out likely millions of Native Americans.

Africa is being hit hardest. Of the world's 34 million people infected with HIV/AIDS as of 2012, two-thirds live in sub-Saharan Africa. Because HIV is spread by the exchange of bodily fluids during sexual contact, the low rate of contraceptive use that contributes to this region's high TFR also

(a) A family living in the United States

(b) A family living in India

FIGURE 8.24 Material wealth varies widely across countries. A typical U.S. family **(a)** may own a large house with a wealth of material possessions. A typical family in a developing nation such as India **(b)** may live in a small, sparsely furnished dwelling with few material possessions and little money or time for luxuries. Compared with the average resident of India, the average U.S. resident enjoys 14 times more income, uses 11 times more land, and emits 11 times more carbon dioxide emissions. *Data from Population Reference Bureau, 2012.* 2012 World population data sheet *and 2012, World Bank, http://data.worldbank.org/.*

fuels the spread of HIV. One in every 20 adults in sub-Saharan Africa is infected with HIV, and for southern African nations, the figure is 1 in 5. As AIDS takes roughly 3800 lives in sub-Saharan Africa every day, the epidemic unleashes a variety of demographic changes. Infant mortality here has risen to 7 deaths out of 100 live births—14 times the rate in the developed world. Infant mortality and the premature deaths of young adults have caused life expectancy in parts of southern Africa to fall from almost 60 years in the early 1990s back down to 40–50 years, where it stood decades earlier. AIDS also leaves behind millions of orphans.

Africa is not the only region with reason to worry. HIV is well established in the Caribbean and in Southeast Asia, and it is spreading in eastern Europe and central Asia. In China, the government has historically been reluctant to monitor or publicize the status of the disease because of the social stigma that the culture attaches to it. The government has admitted to fewer than 1 million cases, although some international experts fear the number of actual cases is 10 times higher and that vigilance is needed to keep the number from rising, especially with distorted sex ratios in China leaving many men unmarried.

Demographic change has social and economic repercussions

Because it removes young and productive members of society, AIDS undermines the ability of poorer nations to develop. Nations lose billions of dollars in productivity when large numbers of its citizens are battling the disease, and treatment puts a huge burden on health care systems. Children orphaned by AIDS further strain social safety nets, requiring interventions to prevent the cycle of poverty and disease from claiming yet another generation.

These problems are hitting many African countries at a time when their governments are already experiencing **demographic fatigue.** Demographically fatigued governments face overwhelming challenges related to population growth, including educating and employing swelling ranks of young people. With the added stress of HIV/AIDS, these governments face so many demands that they are stretched beyond their capacity to address problems. As a result, the problems grow worse, and citizens lose faith in their governments' abilities to help them. This constellation of challenges is making it difficult for some African nations to advance through the demographic transition.

There is good news, however. Improved public health efforts (including sex education, contraceptives, and intravenous drug abuse policies) across the world have slowed HIV transmission rates, and improved medical treatments are lengthening the lives of people who are infected. Note in Figure 8.25 that the number of AIDS deaths began to decrease in 2006, following a drop in new HIV infections since the late 1990s. We may finally be turning the corner on this challenge, thanks to government policy, international collaboration, medical research, nonprofit aid groups, and the grassroots efforts of patients and their advocates. With continued work, we can venture to hope that the multipronged effort to combat HIV/AIDS might eventually become a success story in humanity's timeless battle against disease.

Population goals support sustainable development

The factors that influence fertility are complex and interacting, so initiatives to slow population growth must be diverse, flexible, and culturally specific. This approach was echoed at the milestone 1994 United Nations conference on population and development in Cairo, Egypt. The conference marked a turn away from older notions of top-down command-and-control

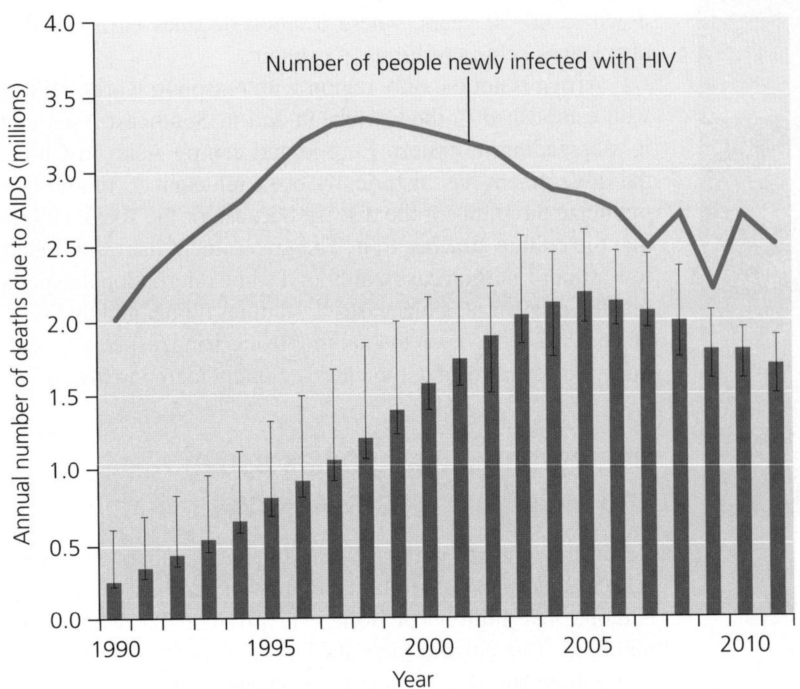

FIGURE 8.25 Nearly 2 million people die from AIDS worldwide each year, and even more are newly infected with HIV each year. However, extensive efforts to combat the disease are beginning to bear fruit; infections have been decreasing since 1998, and deaths began decreasing in 2006. The error bars indicate the range of the number of deaths attributable to AIDS, as not all deaths caused by the disease are reported. *Data from UNAIDS and WHO, 2012.* UNAIDS Report on the global AIDS epidemic. *Geneva, Switzerland.*

FAQ **Why hasn't HIV/AIDS drastically lowered Africa's fertility rates?**

As shown in Table 8.2, Africa's TFR remains the highest, by a wide margin, of all world regions despite high rates of HIV/AIDS infection. This seems contradictory, but it is related to the mode of transmission of HIV/AIDS and the time it takes for the disease to affect its victims' health. Because the disease is typically spread in Africa through sexual behavior, it afflicts individuals who are already in the reproductive age group. Hence, many people already have children by the time they contract the disease and continue having children even after becoming HIV-positive as infected people often live for several years before succumbing to AIDS. As a result, nations with high rates of HIV infection still have extremely high fertility. The southern African nation of Swaziland in 2012, for example, had one of Africa's highest HIV infection rates (21% of females and 30% of males infected) but also one of the region's highest total fertility rates (3.5 children per woman).

(pp. 180–181) population policy geared toward pushing contraception and lowering populations to preset targets. Instead, it urged governments to offer better education and health care and to address social needs that affect population from the bottom up (such as alleviating poverty, disease, and sexism).

The connections we have discussed in this chapter should show that to achieve sustainable development, both population growth and resource consumption levels will need to be addressed. If humanity's overarching goal is to generate a high standard of living and quality of life for all of the world's people, then developing nations must find ways to slow their population growth. However, those of us living in the developed world must also be willing to reduce our consumption. Earth does not hold enough resources to sustain all 7 billion of us at the current North American standard of living, nor can we venture out and bring home extra planets. We must make the best of the one place that supports us all.

Conclusion

Today's human population is larger than at any time in the past. Our growing population and our growing consumption affect the environment and our ability to meet the needs of all the world's people.

However, there are at least two major reasons to be encouraged. First, although global population is still rising, the rate of growth has decreased nearly everywhere, and some countries are even seeing population declines. Most developed nations have passed through the demographic transition, showing that it is possible to lower death rates while stabilizing population and creating more prosperous societies. Second, progress has been made in expanding rights for women worldwide. Although there is still a long way to go, women are obtaining better education, more economic independence, and more ability to control their reproductive decisions. Aside from the clear ethical progress these developments entail, they are helping to slow population growth.

Human population cannot continue to rise forever. The question is how it will stop rising: Will it be through the gentle and benign process of the demographic transition, through restrictive governmental intervention such as China's one-child policy, or through the miserable Malthusian checks of disease and social conflict caused by overcrowding and competition for scarce resources? How we answer this question today will determine not only the quality of the world in which we live, but also the quality of the world we leave to our children and grandchildren.

Reviewing Objectives

You should now be able to:

Perceive the scope of human population growth

- Our global population of nearly 7 billion people adds over 70 million people per year (>2 people every second). (p. 190)
- The global population growth rate peaked at 2.1% in the 1960s and now stands at 1.2%. Growth rates vary among regions. (p. 191)

Assess divergent views on population growth

- Thomas Malthus and Paul Ehrlich warned that overpopulation would deplete resources and harm humanity, whereas Cornucopians see little to fear in population growth (p. 192)
- Rising population can deplete resources, intensify pollution, stress social systems, and degrade ecosystems, such that environmental quality and our quality of life decline. (p. 192)
- Population decline and population aging in some nations have given rise to fears of economic decline. (pp. 192–193)

Evaluate how human population, affluence, and technology affect the environment

- The IPAT model summarizes how environmental impact (*I*) results from interactions among population size (*P*), affluence (*A*), and technology (*T*). (p. 193)
- Rising population and rising affluence may each increase consumption and environmental impact. Technology has frequently worsened environmental degradation, but it can also help mitigate our impacts. (pp. 193–195)

Explain and apply the fundamentals of demography

- Demography applies principles of population ecology to the statistical study of human populations. (p. 196)
- Demographers study size, density, distribution, age structure, and sex ratios of populations, as well as rates of birth, death, immigration, and emigration. (pp. 196–200)
- Total fertility rate (TFR) contributes greatly to change in population size. (pp. 200–201)

Outline and assess the concept of demographic transition

- The demographic transition model explains why population growth has slowed in industrialized nations. Industrialization and urbanization reduce the economic need for children, and education and the empowerment of women decrease unwanted pregnancies. Parents in developed nations choose to invest in quality of life rather than quantity of children. (pp. 201–202)
- The demographic transition may or may not proceed to completion in all of today's developing nations. Whether it does is of immense importance in the quest for sustainability. (p. 202)

Describe how family planning, the status of women, and wealth and poverty affect population growth

- Many birth control methods serve to reduce unwanted pregnancies. (p. 203)
- Family-planning programs and reproductive education have reduced population growth in many nations. (p. 206)
- When women are empowered and achieve equality with men, fertility rates fall, and children tend to be better cared for, healthier, and better educated. (p. 206)
- Poorer societies tend to show faster population growth than do wealthier societies. (p. 207)
- The intensive consumption of affluent societies often makes their ecological impact greater than that of poorer nations with larger populations. (pp. 207–208)

Characterize the dimensions of the HIV/AIDS epidemic

- About 34 million people worldwide are infected with HIV, and 2 million die from AIDS each year. Most live in sub-Saharan Africa. (pp. 208–209)
- Epidemics that claim many young and productive members of society influence population dynamics and can have severe social and economic ramifications. (p. 209)
- We may at last be turning the corner on HIV/AIDS, thanks to a multifaceted effort. (p. 209)

Testing Your Comprehension

1. What is the approximate current human global population? How many people are being added to the population each day?
2. Why has the human population continued to grow despite environmental limitations? Do you think this growth is sustainable? Why or why not?
3. Contrast the views of environmental scientists with those of Cornucopian economists and policymakers regarding whether population growth is a problem. Name several reasons why population growth is commonly viewed as a problem.
4. Explain the IPAT model. How can technology either increase or decrease environmental impact? Provide at least two examples.
5. Describe how demographers use size, density, distribution, age structure, and sex ratio of a population to estimate how it may change. How does each of these factors help determine the impact of human populations on the environment?
6. What is the total fertility rate (TFR)? Why is the replacement fertility for humans approximately 2.1? How is Europe's TFR affecting its rate of natural increase?
7. How does the demographic transition model explain the increase in population growth rates in recent centuries? How does it explain the recent decrease in population growth rates in many countries?
8. Why have fertility rates fallen in many countries?
9. Why are the empowerment of women and the pursuit of gender equality viewed as important to controlling population growth? Describe the aim of family-planning programs.
10. Why do poorer societies have higher population growth rates than wealthier societies? How does poverty affect the environment? How does affluence affect the environment?

Seeking Solutions

1. China's reduction in birth rates is leading to significant change in the nation's age structure. Review Figure 8.12, which shows that the population is growing older, leading to the top-heavy population pyramid for the year 2050. What effects might this ultimately have on Chinese society? What steps could be taken in response?
2. The World Bank estimates that half the world's people survive on less than $2 per day. How do you think this situation affects the political stability of the world? Explain your answer.
3. Apply the IPAT model to the example of China provided in the chapter. How do population, affluence, technology, and ecological sensitivity affect China's environment? Now consider your own country or your own state. How do population, affluence, technology, and ecological sensitivity affect your environment? How can we minimize the environmental impacts of growth in the human population?
4. Do you think that all of today's developing nations will complete the demographic transition and come to enjoy a permanent state of low birth and death rates? Why or why not? What steps might we as a global society take to help ensure that they do? Now think about developed nations such as the United States and Canada. Do you think these nations will continue to lower and stabilize their birth and death rates in a state of prosperity? What factors might affect whether they do so?
5. **THINK IT THROUGH** India's prime minister puts you in charge of that nation's population policy. India has a population growth rate of 1.5% per year, a TFR of 2.5, a 47% rate of contraceptive use, and a population that is 69% rural. What policy steps would you recommend, and why?
6. **THINK IT THROUGH** Now suppose that you have been tapped to design population policy for Germany. Germany is losing population at an annual rate of 0.2%, has a TFR of 1.4, a 66% rate of contraceptive use, and a population that is 73% urban. What policy steps would you recommend, and why?

Calculating Ecological Footprints

A nation's population size and the affluence of its citizens each influence its resource consumption and environmental impact. As of 2012, the world's population passed 7 billion. Average per capita income was $10,030 per year, and the latest estimate for the world's average ecological footprint was 2.7 hectares (ha) per person. The sampling of data in the table will allow you to explore patterns in how population, affluence, and environmental impact are related.

Nation	Population (millions of people)	Affluence (per capita income)[1]	Personal impact (per capita footprint, in ha/person)	Total impact (national footprint, in millions of ha)
Belgium	11.1	$34,760	8.0	89
Brazil	194.3	$10,070	2.9	
China	1350.4	$6,020	2.2	
Ethiopia	87.0	$870	1.1	
India	1259.7	$2,960	0.9	
Japan	127.6	$35,220	4.7	
Mexico	116.1	$14,270	3.0	
Russia	143.2	$15,630	4.4	
United States	313.9	$46,970	8.0	2511

[1]*Measured in GNI PPP (gross national income in purchasing power parity), a measure that standardizes income among nations by converting it to "international" dollars, which indicate the amount of goods and services one could buy in the United States with a given amount of money.*

Data: Population and affluence data are from Population Reference Bureau, 2012. World population data sheet 2012. *Footprint data are for 2007, from WWF International, Zoological Society of London, and Global Footprint Network.* Living planet report 2010.

1. Calculate the total impact (national ecological footprint) for each country.
2. Draw a graph illustrating per capita impact (on the y axis) vs. affluence (on the x axis). What do the results show? Explain why the data look the way they do.
3. Draw a graph illustrating total impact (on the y axis) in relation to population (on the x axis). What do the results suggest to you?
4. Draw a graph illustrating total impact (on the y axis) in relation to affluence (on the x axis). What do the results suggest to you?
5. You have just used three of the four variables in the IPAT equation. Now give one example of how the T (technology) variable could potentially increase the total impact of the United States, and one example of how it could potentially decrease the U.S. impact.

9

Iowa farmers Todd and Arliss Nielsen with the rye grass cover crop that protects their soil between corn harvests

Soil and Agriculture

Upon completing this chapter, you will be able to:

- Explain the importance of soils to agriculture
- Outline major developments in the history of agriculture
- Delineate the fundamentals of soil science, including soil formation and soil properties
- Analyze the causes and impacts of soil erosion and land degradation
- Explain the principles of soil conservation and provide solutions to soil erosion and land degradation
- Describe techniques for watering and fertilizing crops, and offer more sustainable alternatives
- Summarize major policy approaches for pursuing soil conservation and sustainable agriculture

CENTRAL CASE STUDY

Iowa's Farmers Practice No-Till Agriculture

"The nation that destroys its soil destroys itself."

—U.S. President Franklin D. Roosevelt

"There are two spiritual dangers in not owning a farm. One is the danger of supposing that breakfast comes from the grocery, and the other that heat comes from the furnace."

—Conservationist and philosopher Aldo Leopold

Iowa farmers Arliss Nielsen and Todd Nielsen know they need to keep their soil healthy and productive. This father-and-son team cultivates corn and soybeans on 500 ha (1200 acres) in Wright County, Iowa, where rich prairie soils have historically enabled bountiful grain harvests.

Yet the Nielsens also have learned that repeated cycles of plowing and planting have diminished the soil's fertility since farmers first settled the region. Indeed, much of Iowa's topsoil—the valuable surface layer richest in organic matter and nutrients—has been lost to erosion, washed away by water and blown away by wind. Turning the earth by tilling (plowing, disking, harrowing, or chiseling) aerates the soil and works weeds and old crop residue into the soil to nourish it, but tilling also leaves the surface bare, allowing wind and water to erode away precious topsoil.

And so, although Arliss Nielsen had spent half-a-century behind the plow, he and his son Todd abandoned the conventional practice of tilling the soil in 2005 and instead turned to **no-till** farming. Rather than plowing after each harvest, they began leaving crop residues atop their fields, keeping the soil covered with plant material at all times. To plant the next crop, they cut a thin, shallow groove into the soil surface, dropped in seeds, and covered them. By planting seeds of the new crop through the residue of the old, less soil erodes away, organic material accumulates, and the soil soaks up more water—all of which encourages better plant growth.

As a result, the Nielsens are observing the condition of their soil improve. Their tests show that organic matter is increasing in their soil, and earthworms have returned after many years of absence. "We can see the difference in the soil," Arliss Nielsen says.

The Nielsens also find that no-till farming saves time and money. By not tilling, they reduce the number of passes they need to make on the tractor, which saves fuel, time, effort, and wear and tear on equipment.

Along with no-till methods, the Nielsens are using **cover crops,** crops planted to hold the soil in place between times that main food crops are growing. After experimental attempts with clover and vetch, they hired an airplane to scatter rye grass seed. The rye grass prevents erosion when corn and soybeans aren't growing, and it makes planting easier by loosening the soil structure. Moreover, nutrients taken up by the rye make their way to the corn and soybeans once the rye decomposes in the topsoil. Because cover crops such as rye hold nutrients, the Nielsens hope that over time they will be able to reduce the amount of fertilizer they apply.

The Nielsens practice other conservation measures on their land as well. They take soil samples to determine how much fertilizer different areas need, so as not to overapply it. They employ planting methods designed to reduce erosion. They grow grass borders around the farm to keep soil and nutrients from escaping. They plan to install a bioreactor (an underground container full of carbon-rich wood chips) to filter out excess nitrogen so it does not pollute groundwater.

Moreover, the Nielsens avoid farming on lands that are sensitive or have poor soil. Instead, they retire these lands, supported by government programs that pay farmers to take highly erodible land out of cultivation. Arliss Nielsen restored a wetland and set aside 80 acres of woodland and prairie grasses along the White Fox River as wildlife habitat. He planted more than 3200 native trees within this riparian buffer strip. "That 80 acres is marginal farmland at best," he explains. "It's poor soil, some is rocky, and some is wet. This wildlife use is just a better use of the land all the way around." Today it's paying off by providing his family plenty of pheasants to hunt. "I think we can still maintain our profit, with lower input costs using no-till and cover crops, on fewer acres so that we can still have room for wildlife on other parts of our land," Arliss says.

Conservation measures like these are saving the Nielsens—and thousands of other Iowa farmers—money while protecting environmental quality and nurturing soil as an investment for sustainable yields in the future. By enhancing soil conditions and reducing erosion, no-till farming and other conservation measures are benefiting Iowa's people and Iowa's environment, cutting down on pollution in the state's air, waterways, and ecosystems. Similar success is being seen elsewhere in the United States and across the world where these approaches are being applied.

No-till farming is not a magic bullet, and it isn't an ideal solution in every location. But in suitable regions, proponents say it can help make agriculture sustainable. We will need sustainable agriculture if we are to feed the world's human population while protecting the natural environment, including the soils that vitally support our production of food. ■

Soil: The Foundation for Sustainable Agriculture

As the human population has grown, so have the amounts of land and resources we devote to agriculture. We can define **agriculture** as the practice of raising crops and livestock for human use and consumption. We obtain most of our food and fiber from **cropland,** land used to raise plants for human use, and from **rangeland,** or pasture, land used for grazing livestock. Today we commandeer more than one out of every three acres of land on Earth to produce food and fiber for ourselves. Rangeland covers 26% of Earth's land surface, and cropland covers 12%.

Healthy soil is vital for agriculture, as well as for forests (Chapter 12) and for the functioning of Earth's natural systems. **Soil** is not merely lifeless dirt; it is a complex system consisting of disintegrated rock, organic matter, water, gases, nutrients, and microorganisms. Productive soil is a renewable resource. Once depleted, soil may renew itself over time, but renewal generally occurs very slowly. If we abuse soil through careless or uninformed practices, we can greatly reduce its ability to sustain life.

For these reasons, healthy soil is a key component of **sustainable agriculture,** agriculture that we can practice in the same way in the same place far into the future. Sustainable agriculture allows soil to renew its nutrient content and retain its character from one crop to the next. Sustainable agriculture also requires reliable supplies of clean water, minimized use of fossil-fuel-based fertilizers and pesticides, healthy populations of pollinating insects, sustenance of genetic diversity and, arguably, genetic modification (each of which we explore below or in Chapter 10). Because most farming and grazing that people have practiced so far have depleted soils faster than they form, it is imperative for our civilization's future that we develop sustainable methods of working with soil.

Soil supports agriculture

Our agriculture relies on healthy soil in several ways (**FIGURE 9.1**). Crop plants depend on soil that contains organic matter to provide the nutrients they need for growth. Plants also need soil with a structure and texture suitable for roots to penetrate deeply. And plants need soil that retains water and makes water and dissolved

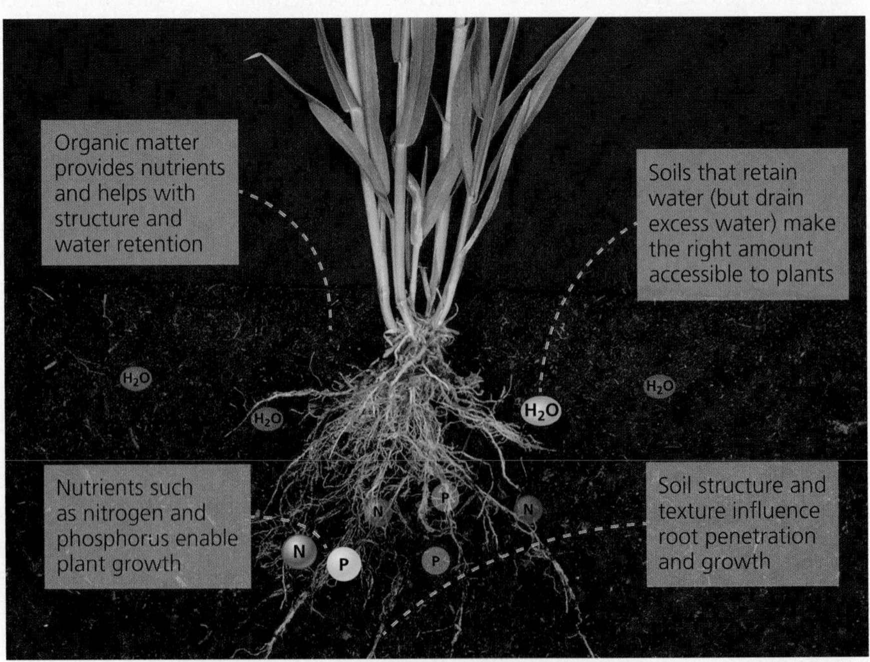

FIGURE 9.1 **Crop plants such as wheat depend on healthy soil for nutrients, organic matter, water retention, and proper root growth.**

nutrients accessible to their roots. Livestock also depend on soil with these characteristics, because livestock eat plants that have grown in the soil. If soil becomes degraded, then agriculture suffers. Because everyone in our society relies directly on agriculture for the meals we eat and the clothing we wear, the quality of our lives is closely tied to the quality of our soil.

Healthy soil has sustained agriculture for thousands of years. When people first began farming, they were able to take advantage of deep, nutrient-rich topsoil that had built up over vast spans of time. Today we face the challenge of producing immense amounts of food from soil that has been farmed many times, while also conserving its fertility for the future. Before we examine soil closely, let's step back and consider how agriculture came about in the first place and how we got to where we are today.

Agriculture arose 10,000 years ago

During most of the human species' 200,000-year existence, we were hunter-gatherers, depending on wild plants and animals for our food and fiber. Then about 10,000 years ago, as the climate warmed and glaciers retreated, people in some cultures began to raise plants from seed and to domesticate animals.

Agriculture most likely began as hunter-gatherers brought wild fruits, grains, and nuts back to their encampments. Some of these foods fell to the ground, were thrown away, or survived passage through the digestive system. The plants that grew from these seeds likely produced fruits larger and tastier than those in the wild, because they sprang from seeds of fruits that people had selected because they were especially large and delicious. As these plants bred with similar ones nearby, they gave rise to subsequent generations of plants with large and flavorful fruits.

Eventually people realized they could guide this process, and our ancestors began intentionally planting seeds from plants whose produce was most desirable. This practice of selective breeding (p. 52) has produced the many hundreds of crops we enjoy today, all of which are artificially selected versions of wild plants. People followed the same process of selective breeding with animals, creating livestock from wild species.

Evidence from archaeology and paleoecology suggests that agriculture was invented independently by different cultures in at least five areas of the world, and possibly 10 or more (**FIGURE 9.2**). The earliest widely accepted evidence for plant and animal domestication is from the "Fertile Crescent" region of the Middle East at least 10,500 years ago. By studying ancient crop remains, scientists have determined that wheat and barley originated here, as did rye, peas, lentils, onions, garlic, carrots, grapes, and other crops. The people of the Fertile Crescent also domesticated goats and sheep. In China, domestication began 9500 years ago, leading to the rice, millet, and pigs we know today. Agriculture in Africa (coffee, yams, sorghum, and more) and the Americas (corn, beans, squash, potatoes, llamas, and more) developed in several regions 4500–7000 years ago, and likely as much as 10,000 years ago.

Once our ancestors learned to cultivate crops and raise animals, they began to settle in more permanent camps and villages, often near water sources. In a self-reinforcing positive feedback cycle (pp. 106–107), the need to harvest crops kept people sedentary, and once they were sedentary, it made sense to plant more crops. As food supplies became more abundant, carrying capacities (pp. 67–68) increased and populations rose. Population increase, in turn, promoted the intensification of agriculture.

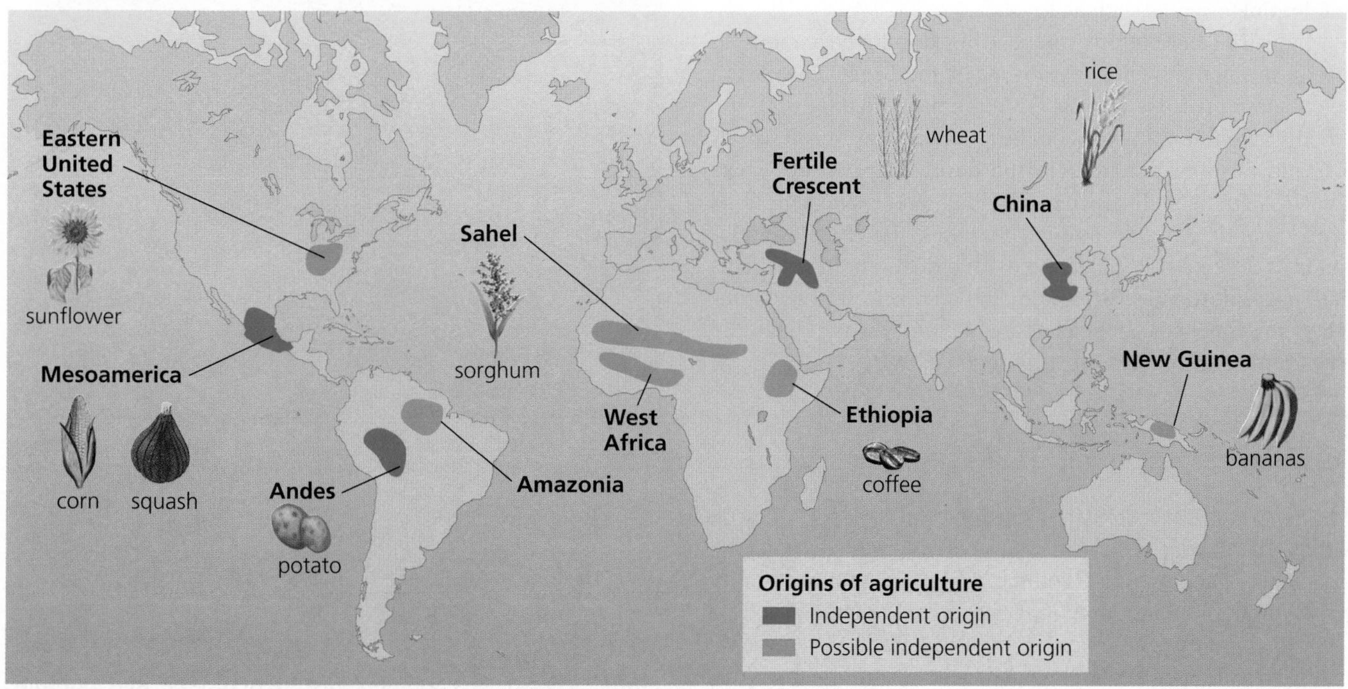

FIGURE 9.2 Agriculture originated independently in multiple regions of the world as different cultures domesticated plants and animals from wild species. *Data from syntheses in Diamond, J., 1997.* Guns, germs, and steel. *New York: W.W. Norton; and Goudie, A., 2000.* The human impact, *5th ed. Cambridge, MA: MIT Press.*

Moreover, the ability to grow excess farm produce enabled some people to live off the food that others produced. This led to the development of professional specialties, commerce, technology, densely populated urban centers, social stratification, and politically powerful elites. For better or worse, the advent of agriculture eventually brought us the civilization we know today.

Industrial agriculture dominates today

For thousands of years, the work of cultivating, harvesting, storing, and distributing crops was performed by human and animal muscle power, along with hand tools and simple machines—an approach known as **traditional agriculture.** In the oldest form of traditional agriculture, known as **subsistence agriculture,** farming families produce only enough food for themselves. As farmers began integrating into market economies and producing excess food to sell, they started using teams of animals for labor and significant quantities of irrigation water and fertilizer.

The industrial revolution (p. 4) introduced large-scale mechanization and fossil fuel combustion to agriculture, just as it did to industry. Farmers replaced horses and oxen with machinery that provided faster and more powerful means of cultivating, harvesting, transporting, and processing crops. Such **industrial agriculture** also boosted yields by intensifying irrigation and by introducing synthetic fertilizers. In addition, the advent of chemical pesticides reduced competition from weeds and herbivory by crop pests. The use of machinery created a need for highly organized approaches to farming, leading us to plant vast areas with single crops in straight orderly rows. Such **monocultures** ("one type") are distinct from the **polycultures** ("many types") typical of traditional agriculture, such as Native American farming systems that mixed maize, beans, squash, and peppers in the same fields. Today, industrial agriculture occupies over 25% of the world's cropland and dominates areas such as Iowa.

Industrial agriculture spread from developed nations to developing nations with the advent of the **Green Revolution** (see Chapter 10; pp. 247–248). Beginning around 1950, the Green Revolution introduced new technology, crop varieties, and farming practices to the developing world. These advances dramatically increased yields and helped millions avoid starvation. Yet despite its successes, the Green Revolution is exacting a price. The intensive cultivation of monocultures using pesticides, irrigation, and chemical fertilizers has many consequences, and can degrade the integrity of soil, the very foundation of our terrestrial food supply.

Soil as a System

We generally overlook the startling complexity of soil. Although it is derived from rock, soil is molded by living organisms (FIGURE 9.3). By volume, soil consists very roughly of 50% mineral matter and up to 5% organic matter. The rest consists of pore space taken up by air or water. The organic matter in soil includes living and dead microorganisms as well as decaying material derived from plants and animals. A single teaspoon of soil can contain millions of bacteria and thousands of fungi, algae, and protists. Soil provides habitat for earthworms, insects, mites, millipedes, centipedes, nematodes, sow bugs, and other invertebrates, as well as for burrowing mammals, amphibians, and reptiles. The composition of a region's soil strongly influences its ecosystems. In fact, because soil is composed of living and nonliving components that interact in complex ways, soil itself meets the definition of an ecosystem (pp. 60, 110).

Soil forms slowly

The formation of soil plays a key role in terrestrial primary succession (p. 85), which begins when the lithosphere's parent material is exposed to the effects of the atmosphere, hydrosphere, and biosphere (pp. 60, 109). **Parent material** is the base geologic material in a particular location. It may be hardened lava or volcanic ash; rock or sediment deposited by glaciers; wind-blown dunes; sediments deposited by rivers, in lakes, or in the ocean; or **bedrock,** the mass of solid rock that makes up

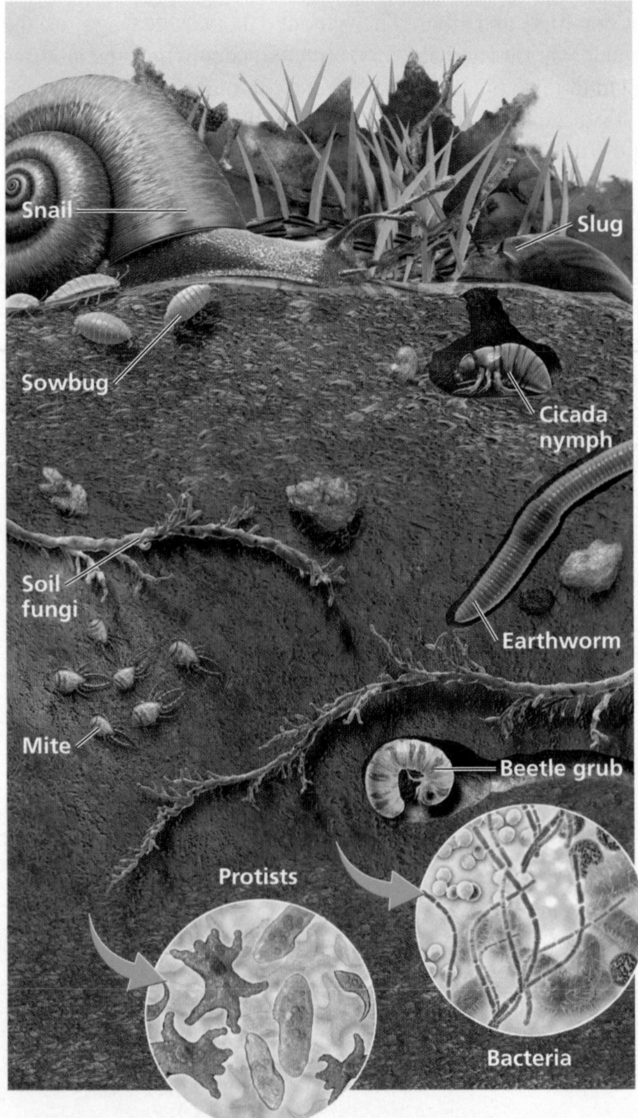

FIGURE 9.3 Soil is a complex mixture of organic and inorganic components and is full of living organisms. In fact, entire ecosystems exist in soil. Most soil organisms decompose organic matter. Some, such as earthworms, also help to aerate the soil.

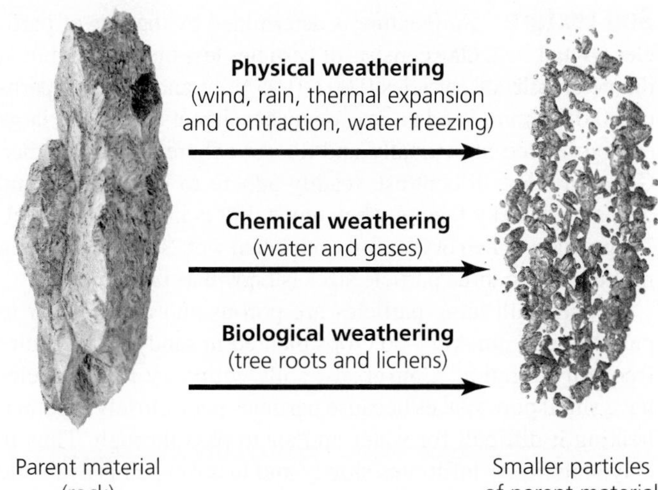

FIGURE 9.4 Weathering breaks down rocks into smaller particles. Physical weathering results from the actions of wind, rain, freezing, and thawing. Chemical weathering occurs as water or gases chemically alter rock. Biological weathering involves living things; for example, lichens (p. 85) produce acid that eats away at rock, and trees' roots rub against rock.

Earth's crust. Parent material is broken down by **weathering,** the physical, chemical, and biological processes that convert large rock particles into smaller particles (FIGURE 9.4).

Once weathering has produced fine particles, biological activity contributes to soil formation through the deposition, decomposition, and accumulation of organic matter. As plants, animals, and microbes die or deposit waste, this material is incorporated amid the weathered rock particles, mixing with minerals. For example, the deciduous trees of temperate forests drop their leaves each fall, and detritivores and decomposers (p. 81) break down this leaf litter and incorporate its nutrients into the soil. In decomposition, complex organic molecules are broken down into simpler ones that plants can take up through their roots. Partial decomposition of organic matter creates **humus,** a dark, spongy, crumbly mass of material made up of complex organic compounds. Soils with high humus content hold moisture well and are productive for plant life.

Weathering and the accumulation and transformation of organic matter are the key processes of soil formation, but these are influenced by five main factors:

- *Climate:* Soil forms faster in warm, wet climates, because heat and moisture speed most physical, chemical, and biological processes.
- *Organisms:* Plants and decomposers add organic matter to soil.
- *Topography:* Hills and valleys affect exposure to sun, wind, and water, and they influence how soil moves.
- *Parent material:* Its attributes influence properties of the soil.
- *Time:* Soil formation can take decades, centuries, or millennia.

Because forming just 1 inch of soil can easily require hundreds or thousands of years, we would be wise to conserve the soil we have. Soil is a renewable resource, but it forms so slowly that for all practical purposes we cannot regain fertile soil once it has been lost.

A soil profile consists of horizons

As wind, water, and organisms move and sort the fine particles that weathering creates, distinct layers eventually develop. Each layer of soil is known as a **horizon,** and the cross-section as a whole, from surface to bedrock, is known as a **soil profile.**

The simplest way to categorize soil horizons is to recognize A, B, and C horizons corresponding respectively to topsoil, subsoil, and parent material. However, soil scientists often recognize at least three additional horizons (FIGURE 9.5). Soils vary by location, and few soil profiles contain all six horizons, but any given soil contains at least some of them.

Generally, the degree of weathering and the concentration of organic matter decrease as one moves downward in a soil profile. Minerals are transported downward as a result of **leaching,** the process whereby solid particles suspended or dissolved in liquid are transported to another location. Soil that undergoes leaching is a bit like coffee grounds in a drip filter. When it rains, water infiltrates the soil, dissolves some of its components, and carries them downward. Minerals commonly

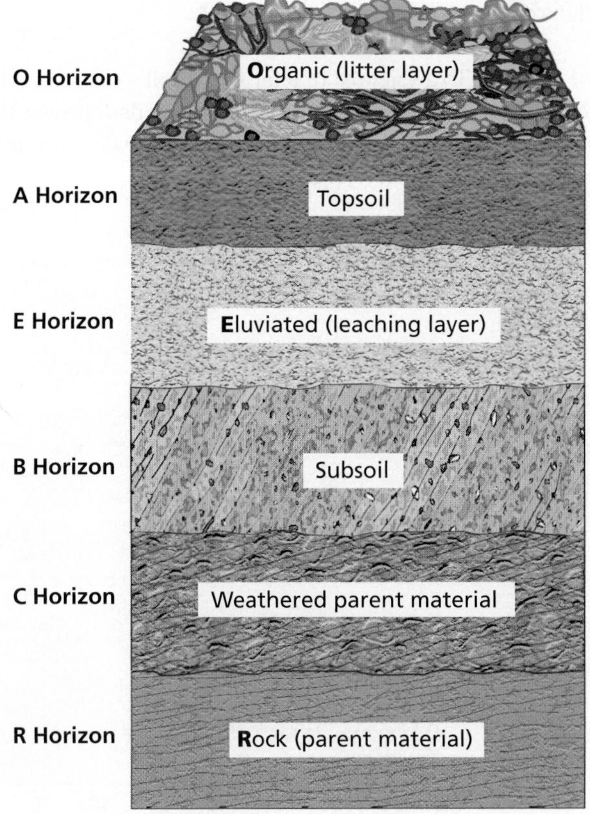

FIGURE 9.5 Mature soil consists of layers, or horizons, that have different compositions and characteristics. Uppermost is the O horizon, or litter layer (O = organic), consisting of organic matter deposited by organisms. Below it lies the A horizon, or topsoil, consisting of some organic material mixed with mineral components. Minerals and organic matter tend to leach out of the E horizon (E = eluviation, or leaching) into the B horizon, or subsoil, where they accumulate. The C horizon of weathered parent material overlies an R horizon (R = rock) of pure parent material.

leached from the E horizon include iron, aluminum, and silicate clay. In some soils, minerals may be leached so rapidly that plants are deprived of nutrients. Minerals that leach from soils may enter groundwater, and some can pose human health risks when the water is extracted for drinking.

A crucial horizon for agriculture and ecosystems is the A horizon, or **topsoil.** Topsoil consists mostly of inorganic mineral components, with organic matter and humus from above mixed in. Topsoil is the portion of the soil that is most nutritive for plants, and it takes its loose texture, dark coloration, and strong water-holding capacity from its humus content. The O and A horizons are home to most of the organisms that give life to soil. Topsoil is vital for agriculture, but agriculture practiced unsustainably over time will deplete organic matter, reducing the soil's fertility and ability to hold water. When a farmer practices no-till farming, he or she essentially creates an O horizon of crop residue to cover the topsoil and then plants seeds of the new crop through this O horizon into the protected topsoil layer.

Soils differ in color, texture, structure, and pH

The six horizons shown in Figure 9.5 depict an idealized soil, but soils display great variety. Scientists classify soils—and farmers judge their quality for farming—based on properties such as color, texture, structure, and pH.

Soil color To a scientist or a farmer, a soil's color can indicate its composition and its fertility. Black or dark brown soils are usually rich in organic matter, whereas a pale color often indicates leaching or low organic content.

Soil texture Soil texture is determined by the size of particles (FIGURE 9.6). **Clay** consists of particles less than 0.002 mm in diameter; **silt,** of particles 0.002–0.05 mm; and **sand,** of particles 0.05–2 mm. Sand grains, as any beachgoer knows, are large enough to see individually and do not adhere to one another. Clay particles, in contrast, readily adhere to one another and give clay a sticky feeling when moist. Silt is intermediate, feeling powdery when dry and smooth when wet. Soil with an even mixture of the three particle sizes is known as **loam.**

Soils with large particles are porous and allow water to pass through quickly—so crops planted in sandy soils require frequent irrigation. Conversely, soils with very fine particles have small pore spaces because particles pack closely together, making it difficult for water and air to pass through. Thus in clay soils water infiltrates slowly and less oxygen is available to soil life. For these reasons, silty soils with medium-sized pores, or loamy soils with a mix of pore sizes, are best for plant growth and agriculture.

Soil structure Soil structure is a measure of the "clumpiness" of soil. An intermediate degree of clumpiness is generally best for plant growth. Repeated tilling can compact soil, reducing its ability to absorb water and inhibiting the penetration of plants' roots.

Soil pH Plants can die in soils that are too acidic or too alkaline, so soils of intermediate pH values (p. 28) are best for most plants. Soil pH influences the availability of nutrients for plants' roots. During leaching, for instance, acids from organic matter may remove some nutrients from the sites of exchange between plant roots and soil particles.

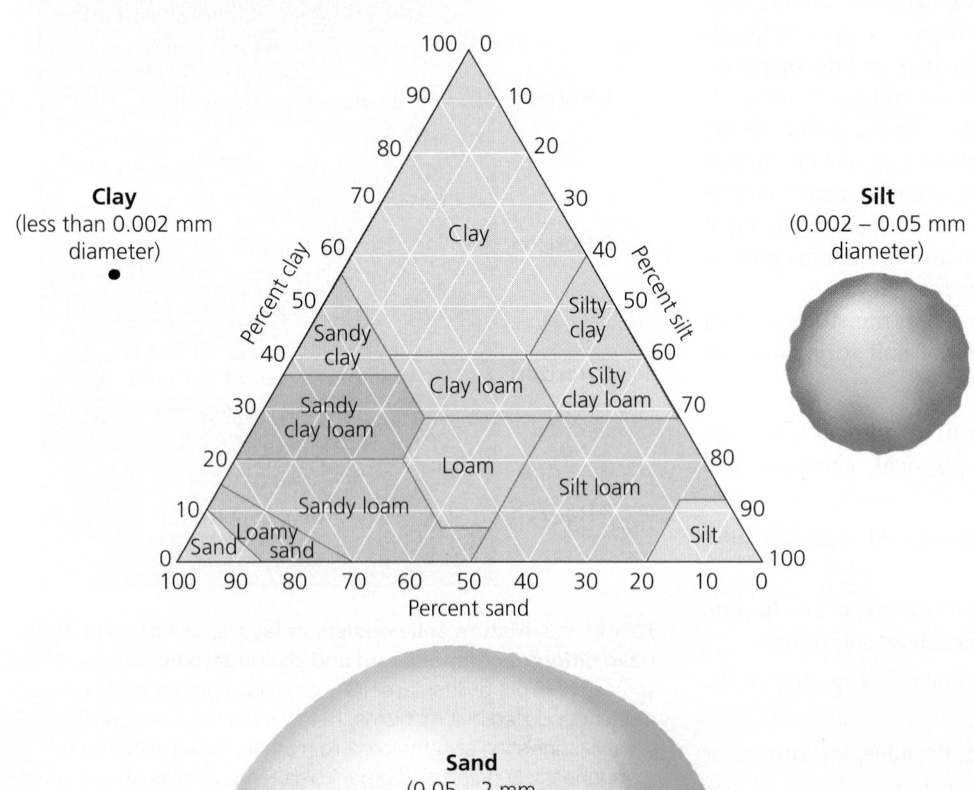

FIGURE 9.6 The texture of soil depends on its mix of particle sizes. Using this diagram, scientists classify soil texture according to the proportions of sand, silt, and clay. After measuring the percentage of each particle size in a soil sample, a scientist can trace the appropriate white lines inward from each side of the triangle to determine texture. Loam is generally best for plant growth, although some plants grow better in other textures.

DATA Q What type of soil contains 20% clay, 60% silt, and 20% sand?

Cation exchange is vital for plant growth

Plants gain many nutrients through a process called **cation exchange.** Soil particle surfaces that are negatively charged hold cations, or positively charged ions (p. 25), such as those of calcium, magnesium, and potassium. In cation exchange, plant roots donate hydrogen ions to the soil in exchange for these nutrient ions, which the soil particles then replenish by exchange with soil water.

Cation exchange capacity expresses a soil's ability to hold cations and prevent them from leaching (thus making them available to plants). This is a useful measure of soil fertility. Soils with fine texture and soils rich in organic matter have high cation exchange capacity. As soil pH becomes lower (more acidic), cation exchange capacity diminishes, nutrients leach away, and soil instead may supply plants with harmful aluminum ions. This is one way in which acid precipitation (pp. 473–475) can damage soils and plant communities.

Regional soil differences affect agriculture

Soil characteristics vary from place to place. For example, it may surprise you to learn that the soil of the Amazon rainforest is much less productive than the soil in Iowa. This is because the enormous amount of rain that falls in the Amazon readily leaches minerals and nutrients out of the topsoil and E horizon and down to the water table, below the reach of plants' roots. At the same time, warm temperatures speed the decomposition of leaf litter and the uptake of nutrients by plants, so the thin topsoil layer contains very little humus.

As a result, when tropical rainforest is cleared for farming, cultivation quickly depletes the soil's fertility. This is why the traditional form of agriculture in tropical forested areas is *swidden* agriculture, in which the farmer cultivates a plot for one to a few years and then moves on to clear another plot, leaving the first to grow back to forest (FIGURE 9.7a). At low population densities this can be sustainable, but with today's dense human populations, soils may not be allowed enough time to regenerate. As a result, agriculture has degraded the soils of many tropical areas.

On the Iowa prairie, in contrast (FIGURE 9.7b), there is less rainfall and less leaching, so nutrients remain within reach of plants' roots. Plants return nutrients to the topsoil as they die, maintaining its fertility. The thick, rich topsoil of temperate grasslands can be farmed repeatedly with minimal loss of fertility if techniques such as no-till farming are used.

FAQ **What is "Slash-and-Burn" Agriculture?**

Soils of tropical rainforests are not well suited for cultivating crops because they contain relatively low levels of plant nutrients. Instead, most nutrients are tied up in the forest's lush vegetation. When farmers cut tropical rainforest for agriculture, they enrich the soil by burning the plants on site. The nutrient-rich ash is tilled into the soil, providing sufficient fertility to grow crops. This practice is called slash-and-burn agriculture. Alas, the nutrients from the ash are usually depleted in one to a few years. At this point, farmers move deeper into the forest and repeat the process, causing further impacts to these productive and biologically diverse ecosystems.

(a) Tropical swidden agriculture on nutrient-poor soil

(b) Industrial agriculture on Iowa's rich topsoil

FIGURE 9.7 **Regional soil differences affect how people farm.** In tropical forested areas such as Indonesia **(a)**, farmers pursue swidden agriculture by the slash-and-burn method because tropical rainforest soils **(inset)** are nutrient-poor and easily depleted. On the Iowa prairie **(b)**, less rainfall means fewer nutrients are leached from the topsoil, while organic matter accumulates, forming a thick, dark topsoil layer **(inset)**.

Conserving Soil

If we are to feed the world's rising human population, we will need to modify our diets or increase agricultural production—and do so sustainably, without degrading our soil and its ability to support agriculture. We cannot simply keep expanding farming and grazing into new areas, because land suitable and available for agriculture is running out. Farming or grazing on unsuitable lands can turn grasslands into deserts; remove ecologically precious forests; diminish biodiversity; encourage invasive species; pollute soil, air, and water with toxic chemicals; and allow fertile soil to be blown and washed away. Instead, we must find ways to improve the efficiency of food production in areas already under cultivation, while pursuing agricultural methods that exert less impact on natural systems.

Damage to soil and land makes conservation vital

Each year, our planet gains 80 million people yet loses 5–7 million ha (12–17 million acres, about the size of West Virginia) of productive cropland. Throughout the world, especially in drier regions, it has become more difficult to raise crops and graze livestock as soils deteriorate in quality and decline in productivity—a process termed **soil degradation** (**FIGURE 9.8a**). Soil degradation results primarily from forest removal, cropland agriculture, and overgrazing of livestock (**FIGURE 9.8b**).

Over the past 50 years, scientists estimate that soil degradation has reduced potential rates of food production by 13% on cropland and 4% on rangeland. By mid-century, there will likely be 2 billion more mouths to feed. For these reasons, it is imperative that we learn to practice agriculture in sustainable ways that maintain the integrity of our soil.

Soil degradation is central to the broader problem known as land degradation. **Land degradation** refers to a general deterioration of land that diminishes its productivity and biodiversity, impairs the functioning of its ecosystems, and reduces the ecosystem services it offers us. Land degradation is caused by the cumulative impacts of unsustainable agriculture, deforestation, and urban development. It is a global phenomenon that affects up to one-third of the world's people. Land degradation is manifested in processes such as soil erosion, nutrient depletion, water scarcity, salinization (p. 234), waterlogging (p. 234), chemical pollution, changes in soil structure and pH, and loss of organic matter from the soil. Scientists, policymakers, farmers, and ranchers are working hard to discover and implement solutions for all of these problems.

Erosion threatens ecosystems and agriculture

Erosion is the removal of material from one place and its transport toward another by the action of wind or water (**FIGURE 9.9**). **Deposition** occurs when eroded material is deposited at a new location. Erosion and deposition are natural processes, and in the long run deposition helps to create soil. Flowing water may deposit freshly eroded sediment rich in nutrients across river valleys and deltas, producing rich and productive soils. This is why floodplains are excellent for farming.

However, erosion is a problem for ecosystems and agriculture because it tends to occur much more quickly than soil

(a) Farmer with degraded soil

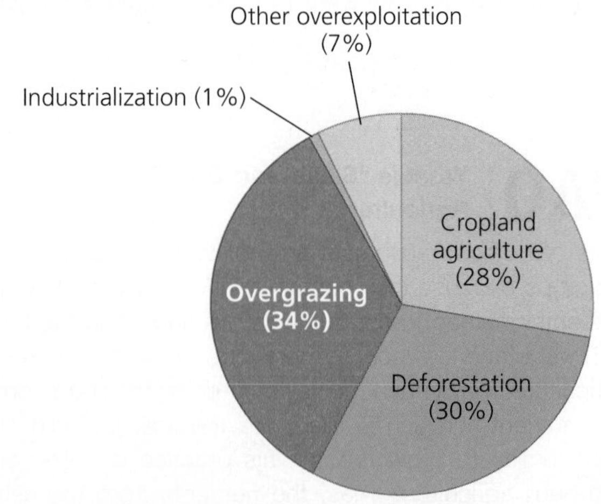

(b) Causes of soil degradation

FIGURE 9.8 We have degraded many soils. A farmer **(a)** shows degraded soil in southern China. Most of the world's soil degradation **(b)** results from cropland agriculture, overgrazing by livestock, and deforestation. *Data (b) from Wali, M.K., et al., 1999. Assessing terrestrial ecosystem sustainability: Usefulness of regional carbon and nitrogen models.* Nature and Resources *35: 21–33.*

FIGURE 9.9 **Water erosion can readily remove soil from areas where soil is exposed, such as farmland.**

is formed. Erosion also tends to remove topsoil, the most valuable soil layer for living things. And when eroded soils are carried out to sea, their nutrients are lost to terrestrial systems. Windy regions with sparse plant cover experience the most wind erosion, whereas areas with steep slopes, high precipitation, and sparse vegetative cover suffer the most water erosion.

People have made land more vulnerable to erosion in three ways:

- Overcultivating fields through poor planning or excessive tilling
- Grazing rangeland with more livestock than the land can support
- Clearing forests on steep slopes or with large clear-cuts (p. 317)

One study determined that at erosion rates typical for the United States, U.S. croplands lose about 2.5 cm (1 in.) of topsoil every 15–30 years, reducing corn yields by 4.7–8.7% and wheat yields by 2.2–9.5%. Erosion can be difficult to detect and measure, even when it is having substantial consequences. For example, losing just a penny's thickness of surface soil may be hard to notice, yet this translates to a loss of fully 12 tons/ha (5 tons/acre) of valuable topsoil.

To minimize erosion, we can erect physical barriers that capture soil. In the long-term and across large areas, however, the growth of vegetation is what prevents soil loss. Vegetation slows wind and water flow while plant roots hold soil in place and take up water. No-till agriculture has been successful because leaving residue on fields after harvest shields topsoil from erosion. Many no-till farmers go a step further and plant cover crops during periods between their main crops. Cover crops serve to cover and anchor the soil during a time when conventional farmers would leave it tilled and bare, susceptible to the ravages of wind and water.

Soil erosion is a global issue

In today's world, people are the primary cause of erosion, and we have accelerated it to unnaturally high rates. In a 2004 study, geologist Bruce Wilkinson analyzed prehistoric erosion rates from the geologic record and compared these with modern rates. He concluded that human activities move over 10 times more soil than all natural processes combined. A 2007 study by soil scientist David Montgomery found an even greater degree of impact (FIGURE 9.10). Montgomery's study also pointed toward a solution, revealing that land farmed under conservation approaches erodes less than land under conventional farming.

More than 19 billion ha (47 billion acres) of the world's croplands suffer from erosion and other forms of soil degradation resulting from human activity. U.S. farmlands lose 5 tons of soil for every ton of grain harvested. In the past decade, China lost an area of arable farmland the size of Indiana. In Kazakhstan, wind eroded tens of millions of hectares after industrial crop agriculture was imposed on land better suited for grazing. In Africa, soil degradation in coming decades could reduce crop yields by half. Couple these declines in soil quality and crop yields with our rapid population growth, and we begin to see why some observers foresee an impending crisis in agriculture.

Desertification reduces productivity of arid lands

Much of the world's population lives and farms in *drylands*, arid and semi-arid environments that cover about

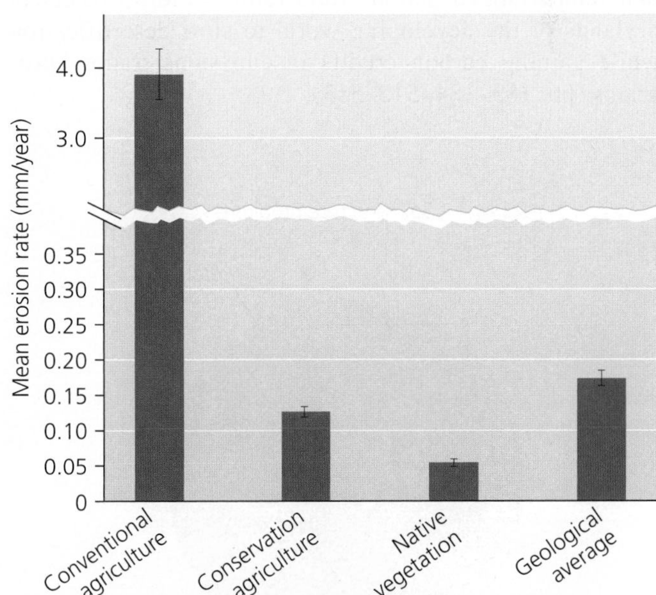

FIGURE 9.10 **Erosion rates from conventional agriculture are high.** They greatly exceed rates in fields farmed under conservation tillage, rates in areas covered by native vegetation, and rates averaged over the geologic record. *Data from Montgomery, D.R., 2007. Soil erosion and agricultural sustainability.* Proc. Natl. Acad. Sci. *104: 13268–13272.*

40% of Earth's land surface. Precipitation in these regions is meager, so drylands are prone to desertification. **Desertification** describes a form of land degradation in which more than 10% of productivity is lost as a result of erosion, soil compaction, forest removal, overgrazing, drought, salinization, climate change, water depletion, and other factors. Most such degradation results from wind and water erosion (FIGURE 9.11). Severe desertification can expand existing desert areas and create new ones (FIGURE 9.12). This process has occurred most dramatically in areas of the Middle East that have been inhabited, farmed, and grazed for long periods—including the Fertile Crescent region, where agriculture first originated (p. 217).

By some estimates, desertification endangers the food supply or well-being of more than 1 billion people in over 100 countries and costs tens of billions of dollars in income each year. China alone loses $6.5 billion annually from desertification. In its western reaches, desert areas are expanding because of overgrazing from over 400 million goats, sheep, and cattle. In the Sistan Basin along the border of Iran and Afghanistan, an oasis that supported a million livestock recently turned barren in just 5 years, and windblown sand buried more than 100 villages. In Kenya, overgrazing and deforestation fueled by rapid population growth has left 80% of its land vulnerable to desertification. Everywhere, soil degradation forces ranchers to crowd onto poorer land and farmers to reduce the fallow periods during which land lies unplanted and can regain nutrients. In a positive feedback cycle (pp. 106–107), both of these actions worsen soil degradation further.

Desertification is expected to grow worse as climate change alters rainfall patterns, making some areas drier. A 2007 United Nations report estimated that 50 million people could soon be displaced. The report also suggested that industrialized nations fund reforestation projects in drylands of the developing world to slow desertification while gaining carbon credits in emissions trading programs (pp. 183–184, 512–513).

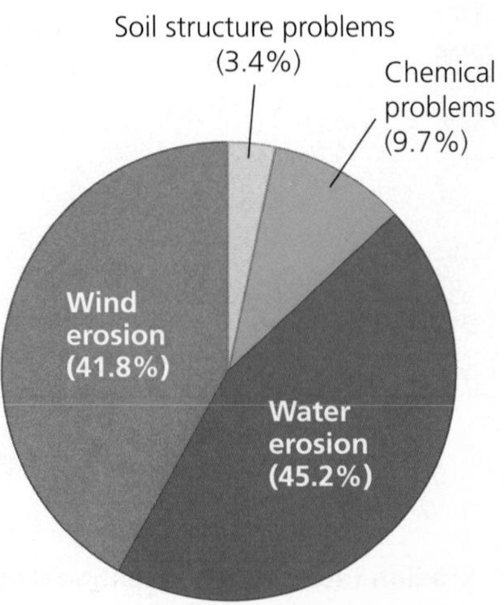

FIGURE 9.11 **Soil degradation on drylands is due primarily to erosion by wind and water.** *Data from U.N. Environment Programme. 2002.* Tackling land degradation and desertification. *Washington and Rome: Global Environment Facility and International Fund for Agricultural Development.*

FIGURE 9.12 **Severe desertification can cause desert areas to expand.** Here, immense sand dunes moving in from the Gobi Desert are burying farm fields in northwestern China.

As a result of desertification, gigantic dust storms from denuded land in China are now blowing across the Pacific Ocean to North America, and dust storms from Africa's Sahara Desert blow across the Atlantic Ocean to the Caribbean Sea (see Figure 17.11c, p. 457). Such massive dust storms occurred in the United States during the early 20th century, when desertification shook American agriculture and society to their very roots.

The Dust Bowl shook the United States

Prior to large-scale cultivation of North America's Great Plains, native prairie grasses of this temperate grassland region held soils in place. In the late 19th and early 20th centuries, many settlers arrived in Oklahoma, Texas, Kansas, New Mexico, and Colorado hoping to make a living there as farmers. Between 1879 and 1929, cultivated area in the region soared from 5 million ha (12 million acres) to 40 million ha (100 million acres). Farmers grew abundant wheat and ranchers grazed many thousands of cattle, sometimes expanding onto unsuitable land and causing erosion by removing native grasses and altering soil structure.

In the early 1930s, a drought worsened the ongoing human impacts, and the region's strong winds began to erode millions of tons of topsoil (FIGURE 9.13). Massive dust storms traveled up to 2000 km (1250 mi), blackening rain and snow as far away as New York and Washington, D.C. Some areas lost as much as 10 cm (4 in.) of topsoil in a few years. The most-affected region in the southern Great Plains became known as the **Dust Bowl,** a term now also used for the historical event itself. The "black blizzards" of the Dust Bowl forced thousands of farmers off their land.

FIGURE 9.13 **Drought and poor agricultural practices devastated millions of U.S. farmers in the 1930s in the Dust Bowl.** The photo **(a)** shows towering clouds of dust approaching homes near Dodge City, Kansas. The map **(b)** shows the Dust Bowl region. Eroded soil from this region blew eastward all the way to the Atlantic Ocean.

(a) Kansas dust storm, 1930s

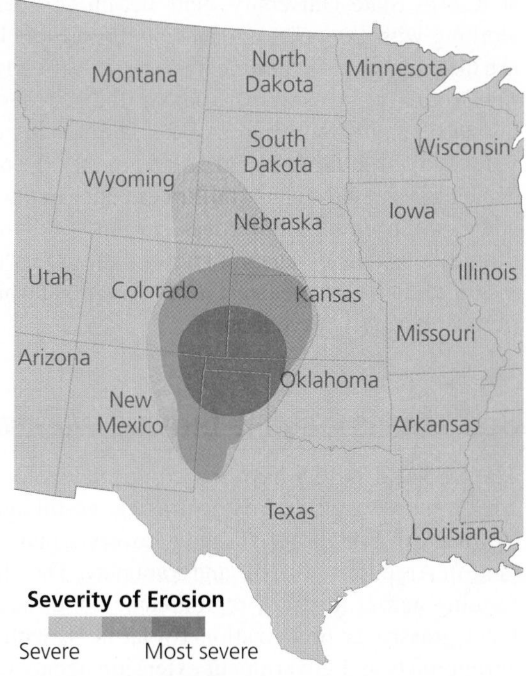

(b) Dust Bowl region

The Soil Conservation Service pioneered measures to address soil degradation

In response to the devastation in the Dust Bowl, the U.S. government, along with state and local governments, increased support for research into soil conservation. The U.S. Congress passed the Soil Conservation Act of 1935, establishing the Soil Conservation Service (SCS). This new agency worked closely with farmers to develop conservation plans for individual farms, using science to assess the land's resources and problems, and collaborating with landowners to ensure that the plans harmonized with landowners' objectives.

The teams formed by the SCS to combat erosion included soil scientists, forestry experts, engineers, economists, and biologists. These teams were among the earliest examples of interdisciplinary environmental problem solving. The first director of the SCS, Hugh Hammond Bennett, was an innovator and evangelist for soil conservation. Under his dynamic leadership, the agency promoted soil conservation practices through county-based **conservation districts.** Organized by the states but operating with federal direction, authorization, and funding, these districts continue today to implement soil conservation programs and empower local residents to plan and set priorities. In 1994 the SCS was renamed the **Natural Resources Conservation Service (NRCS),** and its responsibilities were expanded to include water quality protection and pollution control.

Today most state universities employ *agricultural extension agents*, experts who assist farmers by providing information on new research and by helping them apply this knowledge with new techniques. Extension agents from universities and government agencies help farmers like Todd and Arliss Nielsen determine how best to implement conservation measures on their land (FIGURE 9.14).

In Sioux County, Iowa, Nate and Rachel Ronsiek are among the farmers who are taking advantage of the expertise and resources these experts offer. When Nate took charge of the family farm at age 26 after his father's death, he wanted to put into practice the strategies for conservation that his father had taught him, as well as those he learned as a student

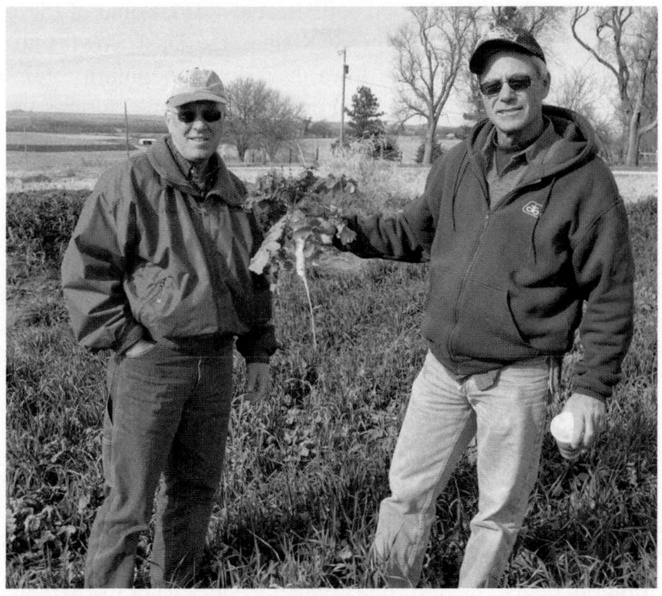

FIGURE 9.14 **Agricultural extension agents assist farmers by providing information on new research and techniques.** In Iowa, NRCS extension agent Greg Mathis (**left**) helps farmer Lowell Forristall implement cover crops around his radishes.

at Kansas State University. Nate became one of 27 farmers working with extension agents from Iowa State University to use his own land to experimentally test the effectiveness of no-till farming versus several types of tilling. After three years, his no-till fields produced as much corn as his conventional fields while requiring less time and money. Ronsiek says he can clearly see how water infiltrates better in the no-till fields and how those fields suffer less erosion. "If there is a way to do it, [the extension agents] know it," he says. "NRCS keeps me up-to-date with the latest in conservation techniques, practices, and ways to save money."

Soil conservation is thriving worldwide

The SCS and NRCS have served as models for efforts elsewhere in the world. In South America, no-till agriculture has exploded in popularity and now covers a majority of farmland in Argentina, Brazil, and Paraguay. The shift to no-till farming across this vast region came about largely through local grassroots organization by farmers, with the help of agronomists and government extension agents who provided them information and resources. Southernmost Brazil alone boasts thousands of "Friends of the Land" clubs in which local farmers collaborate with trained experts.

From Argentina to Iowa, no-till agriculture is one of many approaches to soil conservation. Hugh Hammond Bennett advocated a complex approach, combining techniques such as crop rotation, contour farming, strip cropping, terracing, grazing management, and reforestation, as well as wildlife management. Such measures are now widely applied in many places around the world.

Farmers protect soil in many ways

A number of farming techniques can reduce the impacts of conventional cultivation on soils (**FIGURE 9.15**). Some of these have been promoted by the SCS since the Dust Bowl. Others, like no-till farming in Iowa, have found popularity more recently. Still others have been practiced by some cultures for centuries.

Crop rotation In **crop rotation,** farmers alternate the type of crop grown in a given field from one season or year to the next (**FIGURE 9.15a**). Rotating crops returns nutrients to the soil, minimizes erosion from letting fields lie fallow, and can break cycles of disease associated with continuous cropping. Many U.S. farmers rotate their fields between wheat or corn and soybeans from one year to the next. Soybeans are legumes, plants with specialized bacteria on their roots that fix nitrogen (p. 126), revitalizing soil with nutrients. Crop rotation also reduces insect pests; if an insect is adapted to feed and lay eggs on one crop, planting a different type of crop will leave its offspring with nothing to eat. In a practice similar to crop rotation, many no-till farmers, like the Nielsens in Iowa, plant cover crops such as clover (a nitrogen-replenishing legume) to prevent erosion during times of year when the main crops are not growing.

Contour farming Water running down a hillside with little plant cover can carry soil away, so farmers have developed methods for cultivating slopes. **Contour farming** (**FIGURE 9.15b**) consists of plowing furrows sideways across a hillside, perpendicular to its slope and following the natural contours of the land. In contour farming, the side of each furrow acts as a small dam that slows runoff and captures soil. Contour farming is most effective on gradually sloping land with crops that grow well in rows.

Terracing On very steep terrain, terracing (**FIGURE 9.15c**) is the most effective method for reducing erosion. Terraces are level platforms, sometimes with raised edges, that are cut into steep hillsides to contain water from irrigation or precipitation. **Terracing** transforms slopes into series of steps like a staircase, enabling farmers to cultivate hilly land without losing huge amounts of soil to water erosion. Farmers have used terracing for centuries in mountainous regions, such as the foothills of the Himalayas and the Andes. Terracing is labor-intensive to establish but in the long term is likely the only sustainable way to farm in mountainous terrain.

Intercropping Farmers also minimize erosion by **intercropping,** planting different crops in alternating bands or other spatially mixed arrangements (**FIGURE 9.15d**). Intercropping helps slow erosion by providing more ground cover than does a single crop. Like crop rotation, intercropping reduces vulnerability to insects and disease and, when a nitrogen-fixing legume is used, replenishes the soil. In southern Brazil, some no-till farmers intercrop cover crops with food crops such as maize, soybeans, wheat, onions, cassava, grapes, tomatoes, tobacco, and orchard fruit.

Shelterbelts A widespread technique to reduce erosion from wind is to establish **shelterbelts,** or **windbreaks** (**FIGURE 9.15e**). These are rows of trees or tall shrubs that are planted along the edges of fields to slow the wind. On North America's Great Plains, fast-growing species such as poplars are often used. Shelterbelts can be combined with intercropping by planting mixed crops in rows surrounded by or interspersed with rows of trees that provide fruit, wood, or protection from wind.

Conservation tillage **Conservation tillage** describes an array of approaches that reduce the amount of tilling relative to conventional farming; one common definition is any method of limited tilling that leaves more than 30% of crop residue covering the soil after harvest. No-till farming is the ultimate form of conservation tillage.

No-till farming has many benefits

To plant using the no-till method (**FIGURE 9.15f**), a tractor pulls a "no-till drill" (**FIGURE 9.16**) that cuts furrows through the O horizon of dead weeds and crop residue and the upper levels of the A horizon. The device drops seeds into the furrow and closes the furrow over the seeds, minimizing disturbance to

(a) Crop rotation

(b) Contour farming

(c) Terracing

(d) Intercropping

(e) Shelterbelts

(f) No-till farming

FIGURE 9.15 Farmers have adopted various strategies to conserve soil. Rotating crops **(a)** such as soybeans and corn helps restore soil nutrients and reduce impacts of pests. Contour farming **(b)** reduces erosion on hillsides. Terracing **(c)** minimizes erosion in mountainous areas. Intercropping **(d)** reduces soil loss and maintains soil fertility. Shelterbelts **(e)** protect against wind erosion. In **(f)**, corn grows through the remnants of a cover crop used in no-till agriculture.

the soil. Often a localized dose of fertilizer is added to the soil along with the seed.

By increasing organic matter and soil biota while reducing erosion, no-till farming and conservation tillage can restore soil quality. Based on results in Iowa and elsewhere, proponents of no-till farming credit the practice with a number of benefits (TABLE 9.1). One benefit that looms larger and larger is that of carbon storage (pp. 508, 537–538). To mitigate global climate change (Chapter 18), we must find ways to reduce the atmospheric concentration of carbon dioxide. By

FIGURE 9.16 Farmers practice no-till farming with a no-till drill. The drill 1 cuts a furrow through the soil surface, 2 drops in a seed, and 3 closes the furrow over the seed.

adding organic matter to the soil, no-till farming captures carbon that otherwise would make its way to the atmosphere and instead stores it in the soil. In addition, because no-till farming reduces tractor use, the farmer burns less gasoline. Some farmers have even received money from carbon offsets (p. 513) for using no-till methods. Researchers today are debating just how much impact no-till farming can have on carbon sequestration (see THE SCIENCE BEHIND THE STORY, pp. 230–231).

In the United States today, nearly one-quarter of farmland is under no-till cultivation, and over 40% is farmed using conservation tillage (FIGURE 9.17). Forty percent of soybeans, 21% of corn, and 18% of cotton receive no-till treatment. Iowa ranks second only to Illinois in total area under no-till farming, but its rate of no-till farming (23%) is close to the national average. Tennessee, Virginia, Maryland, and Kentucky use no-till methods on over half their farmland. According to U.S. government figures, erosion rates in the United States declined from 9.1 tons/ha (3.7 tons/acre) in 1982 to 5.9 tons/ha (2.4 tons/acre) in 2003, thanks to conservation tillage and other soil conservation measures.

No-till and conservation tillage methods were pioneered in the United States and the United Kingdom, but are now most widespread in subtropical and temperate South America. In Brazil, Argentina, and Paraguay, over half of all cropland is now under no-till cultivation. In this part of the world, heavy rainfall promotes erosion, causing tilled soils to lose organic matter and nutrients, and hot weather can overheat tilled soil. Thus, the no-till approach is especially helpful here, and results have exceeded those in the United States: Crop yields increased (in some cases they nearly doubled), erosion was reduced, soil quality was enhanced, and pollution declined, all while costs to farmers dropped by roughly 50%.

Critics of no-till farming in the United States note that this approach often requires heavy use of chemical herbicides (because weeds are not physically removed from fields) and synthetic fertilizer (because non-crop plants take up a significant portion of soil nutrients). In many industrialized countries, this has indeed been the case. Proponents, however, point out that in developing regions of South America, farmers have departed from the industrialized model by relying more heavily on *green manures* (dead plants as fertilizer) and by rotating fields with cover crops, including nitrogen-fixing legumes. The manures and legumes nourish the soil, and cover crops reduce weeds by taking up space the weeds might occupy.

Critics counter that green manures are generally not practical for large-scale intensive agriculture. Certainly, conservation tillage methods work better in some areas than in others, and better with some crops than with others. Farmers will do best by educating themselves on the options and choosing what is best for their particular crops on their own land.

TABLE 9.1 Benefits of No-Till Farming
• Reduces erosion
• Helps soil retain moisture
• Enhances soil structure
• Reduces soil compaction
• Stores carbon in soil, lessening its release to the atmosphere
• Improves biological activity in soil
• Helps soil retain nutrients
• Cuts down on water and air pollution
• Requires less time and labor
• Decreases fossil fuel use and costs
• Minimizes wear and tear on equipment
• Tends to increase long-term soil productivity, crop yields, and farming profit

WEIGHING THE ISSUES

HOW WOULD YOU FARM? You are a farmer owning land on both sides of a steep ridge. You want to plant a sun-loving crop on the sunny, but very windy, south slope of the ridge and a crop that needs a great deal of irrigation on the north slope. What farming techniques might best conserve your soil? What factors might you want to know about before you decide to commit to one or more methods?

Plant cover is the key to erosion control

Farming methods to control erosion make use of the general principle that protecting and restoring vegetative cover will protect soils, and this principle is applied widely beyond farming. When grazing livestock, people try to move animals from place to place before the plant cover in any one area is

FIGURE 9.17 No-till farming is practiced on 24% of U.S. farmland. (a) Other forms of conservation tillage are practiced on 18% of land, while "reduced tillage" (15–30% crop residue left on ground after harvest) takes place on 21%, and conventional farming (0–15% residue left on ground) occurs on 37%. Across the United States **(b),** red and orange colors on this county map show where no-till methods are practiced most. *Data are for 2008 (a) and 2004 (b), from Conservation Technology Information Center, National Crop Residue Management Survey.*

DATA Q Can you find your county on this map? What percentage of its farming is no-till? How about your state—what percentages does it show?

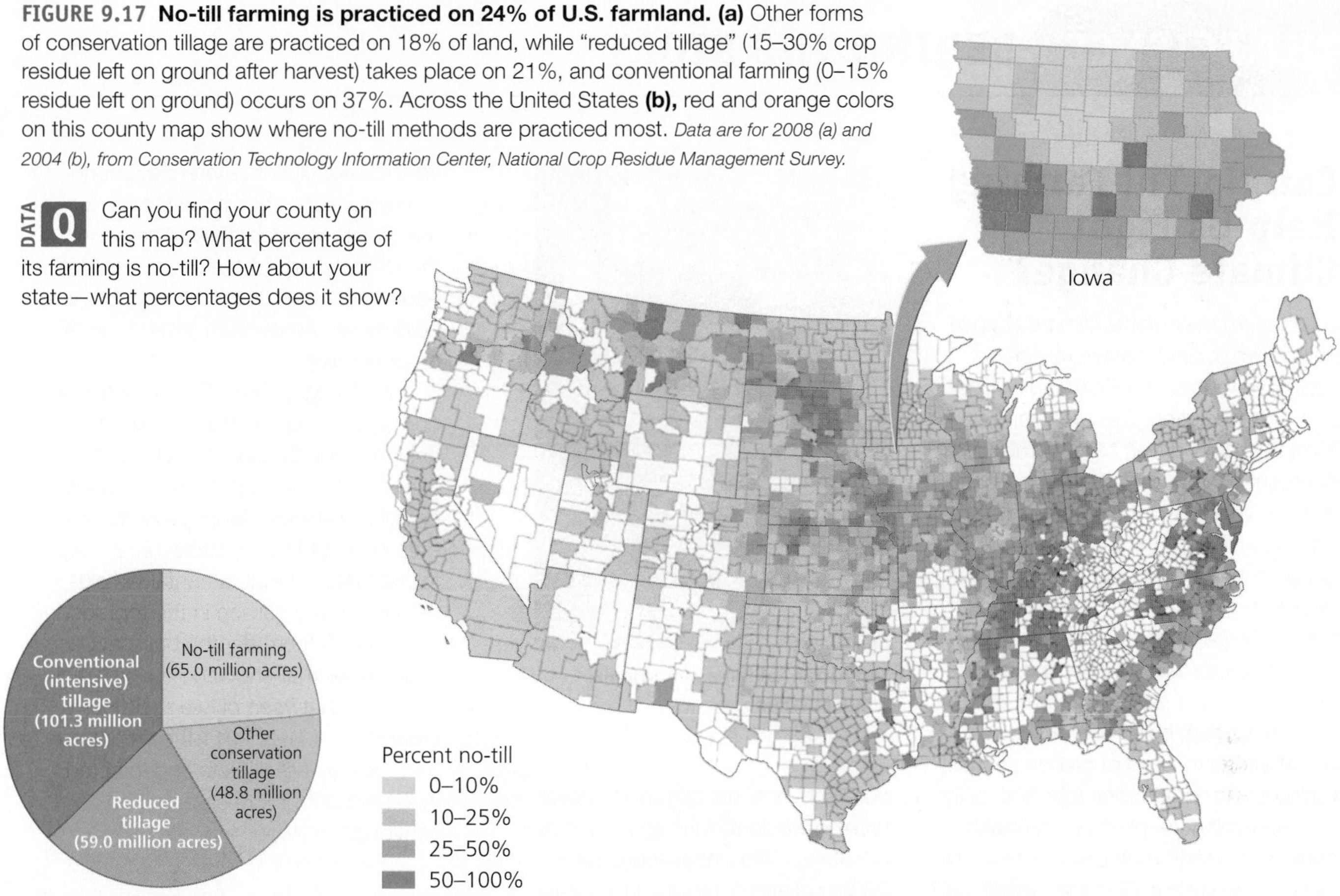

Conventional (intensive) tillage (101.3 million acres)

No-till farming (65.0 million acres)

Other conservation tillage (48.8 million acres)

Reduced tillage (59.0 million acres)

(a) Types of tillage in United States

Iowa

Percent no-till
0–10%
10–25%
25–50%
50–100%

(b) Distribution of no-till farming in United States, and in Iowa

reduced too much. In cultivating trees through the practice of forestry, methods such as clear-cutting—the removal of all trees from an area at once (p. 317)—can lead to severe erosion, but alternative methods that remove fewer trees over longer periods of time help to minimize soil loss. When banks along creeks and roadsides erode, we plant plants to anchor the soil. China has embarked on the world's largest tree-planting program to slow its soil loss (**FIGURE 9.18**). Although such "reforestation" efforts help slow erosion, they do not at the same time produce ecologically functioning forests, because tree species are selected only for their fast growth and are planted in monocultures.

FIGURE 9.18 Vast swaths of countryside in northern and western China have been planted with fast-growing trees such as poplars. These simple tree plantations do not create ecologically functional forests, but they do reduce soil erosion.

Overgrazing can degrade soil

Raising livestock has impacts on soils and ecosystems. When sheep, goats, cattle, or other livestock graze on the open range, they feed primarily on grasses. As long as livestock populations do not exceed the rangeland's carrying capacity (pp. 67–68) and do not consume grass faster than it can regrow, grazing may be sustainable. Moreover, human use of rangeland does not necessarily exclude its use by wildlife or its continued functioning as a grassland ecosystem. However, when too many livestock destroy too much of the plant cover,

Can No-Till Farming Help Us Fight Climate Change?

As concern over global climate change has grown, many agricultural scientists have touted no-till farming as one part of the solution. Climate change (Chapter 18) is driven by our emission of greenhouse gases, such as carbon dioxide, which warm the atmosphere. Because no-till farming reduces soil disturbance, proponents argue that it returns crop residue to the soil faster than soil organisms decompose such organic matter and emit carbon dioxide by respiration (p. 32). In this way, the argument goes, no-till farming could serve as one method of carbon sequestration or carbon storage (pp. 508, 537).

An avalanche of research papers shows that no-till methods increase soil organic carbon (SOC) in the topsoil, and some policymakers were becoming persuaded that polluters should begin paying no-till farmers in carbon-trading markets (pp. 512–513) to offset the polluters' emissions.

Then some researchers began to dig deeper. Literally. A number of soil scientists noticed that very few studies were measuring the organic matter in soil below a depth of 30 cm (1 foot). Yet SOC makes its way downward over time, and many crop plants' roots reach deeper than this. These scientists wondered whether no-till farming was increasing SOC at deeper levels in the soil. If not, then we'd better invest our efforts elsewhere in the urgent search for climate change solutions.

One researcher to raise questions was Rattan Lal of The Ohio State University. Lal had been one of the foremost promoters of the idea that farmers could help fight climate change by using conservation tillage. In a 2004 paper, Lal calculated that converting

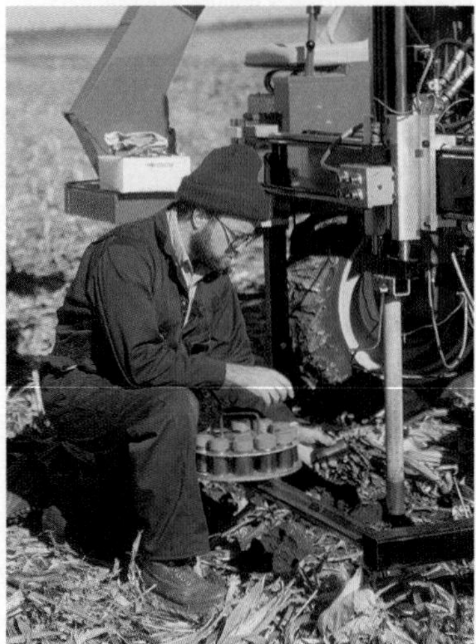

Technician taking a soil sample, Columbus, Ohio.

from conventional plowing to no-till farming would reduce emissions by 30–35 kg C/ha per season. Yet in 2008, Lal and colleague Humberto Blanco-Canqui analyzed soil samples down to 60 cm (2 ft) on farms in Ohio, Pennsylvania, and Kentucky, and they concluded that no-till fields often have *less* carbon at deeper levels than conventionally plowed fields.

Were no-till farming's carbon gains in the topsoil, then, offset by carbon losses deeper in the soil profile? Many researchers began testing this hypothesis. Among them were researchers in Michigan and in Georgia who took advantage of long-term research programs at university-run sites.

At the W.K. Kellogg Biological Station in Michigan, S.P. Syswerda and four colleagues took soil samples down to 1 meter in depth from experimental farm plots established in 1988. They reported in 2011 that plots under no-till farming (as well as plots under organic farming (pp. 257–260) held significantly more carbon at shallow depths than did conventionally plowed plots. At lower depths, there was no significant difference among the plots—and thus no evidence for the offsetting hypothesis. They also found that plots of mature forest that had never been cleared for agriculture held more carbon than either type of cropped plot.

In Georgia, Scott Devine and three colleagues dug 2 meters deep and analyzed soil samples from long-term experimental fields at a site in Athens called Horseshoe Bend. Working with data collected yearly since 1982, they found that no-till plots contained significantly more carbon in the topmost soil layer (0–5 cm depth), but that each layer below that showed no significant difference between plowed and no-till plots (**FIGURE 1**). In 2011 they reported that, overall, no-till plots and forested land held 20% more carbon than conventionally plowed land.

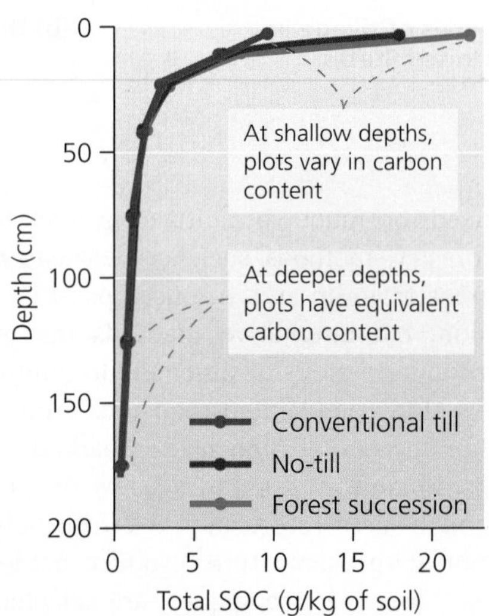

FIGURE 1 In a Georgia study, no-till plots contained more carbon in the topsoil than conventionally plowed plots, while forested plots had the most. At deeper soil levels, there was no significant difference. *Data from Devine, Scott, et al., 2011. Soil carbon change through 2 m during forest succession alongside a 30-year agroecosystem experiment.* Forest Science *57: 36–50.*

Meanwhile, researchers in Brazil—where no-till farming is widespread and has boosted crop yields—were doing similar experiments and finding that no-till was increasing carbon storage at both shallow and deeper depths. Robert Boddey and 12 colleagues from Brazilian universities studied three sites in southern Brazil running long-term (15- to 26-year) farming experiments. They reported in 2010 that sampling soil down to 100 cm reveals no-till methods storing 59% more carbon than is evident when sampling extends to only 30 cm. Boddey maintains that in warm subtropical regions with a lot of rain and well-drained soils, carbon makes its way downward in the soil profile more easily. As a result, whereas shallow sampling in the temperate zone tends to overestimate the carbon-storage benefits of no-till farming, shallow sampling in the tropics may underestimate its benefits.

When many studies show a diversity of results, it can be helpful to perform a *meta-analysis*, a statistical analysis of the results of numerous studies, using each one's results as a single data point. Meta-analyses look for trends and patterns across entire fields of research in order to clarify the big picture.

One meta-analysis in 2010 by Chinese and Australian researchers summarized the relative amounts of carbon at different levels in the soil of no-till, plowed, and unfarmed forest plots, as measured by 69 studies across the world (**FIGURE 2**). Another meta-analysis, by Colorado researchers in 2011, developed maps predicting what changes in carbon storage various areas in the United States would experience if they switched to no-till farming.

Through the hard work of many soil scientists, some things are becoming clear. One is that natural ecosystems sequester more carbon than croplands; forests in eastern North America, for instance, store a great deal of carbon in their vegetation and their humus-rich soil. Another is that no-till farming is not

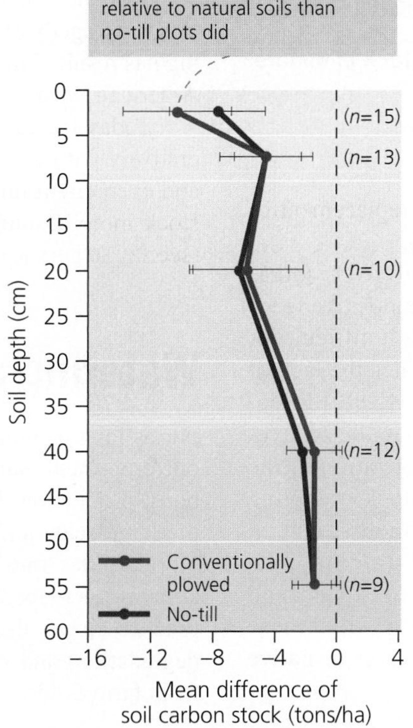

FIGURE 2 In a meta-analysis of 69 studies, more soil carbon was lost in conventionally plowed plots than in no-till plots in the topsoil layer (compared to adjacent natural soils, indicated by a vertical dotted line at the value of zero). At lower depths, there were no significant differences between no-till and conventional plots. *Data from Luo, Zhongkui, et al., 2010. Can no-tillage stimulate carbon sequestration in agricultural soils? A meta-analysis of paired experiments.* Agriculture, Ecosystems, and Environment *139: 224–231.*

DATA Q Which type of plot has more carbon content at a depth of 40 cm?

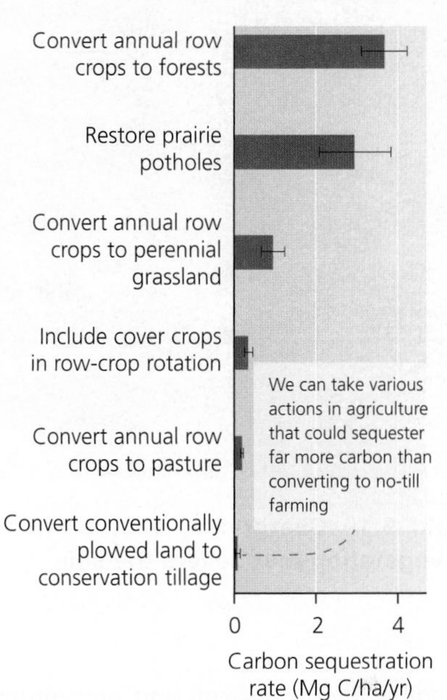

FIGURE 3 Converting to conservation tillage ranks low among actions we can take in agriculture to sequester carbon to fight climate change, based on a review of studies from the Midwest. *Adapted from Fissore, Cinzia, et al., 2010. Limited potential for terrestrial carbon sequestration to offset fossil-fuel emissions in the upper Midwestern U.S.* Frontiers in Ecology and the Environment *8: 409–413.*

a major solution to climate change (**FIGURE 3**); it does help store carbon, but not nearly enough to compensate for the amounts we pump into the atmosphere by burning fossil fuels. A 2010 analysis by Cinzia Fissore and other University of Minnesota researchers calculated that at most, the U.S. Midwest could offset 2% of its region's emissions by switching completely to no-till agriculture.

No-till farming does help lessen fossil fuel combustion by reducing tractor use. Still, scientific research on soil and farming methods is showing that to address global climate change, we had better look elsewhere for major solutions. ■

FIGURE 9.19 **Overgrazing occurs when livestock eliminate the vegetation that covers the soil.**

impeding plant regrowth and preventing the replacement of biomass, the result is **overgrazing** (FIGURE 9.19).

When livestock remove too much plant cover, soil is exposed and made vulnerable to erosion. In a positive feedback cycle (pp. 106–107), soil erosion makes it difficult for vegetation to regrow, a problem that perpetuates the lack of cover and gives rise to more erosion (FIGURE 9.20). Moreover, non-native weedy plants that are unpalatable to livestock may invade and outcompete native vegetation in the new, modified environment. Too many livestock trampling the ground can also compact soils and alter their structure. Soil compaction makes it more difficult for water to infiltrate, for soils to be aerated, for plants' roots to expand, and for roots to conduct cellular respiration (p. 32). All of these effects further decrease the growth and survival of native plants.

Worldwide, overgrazing causes as much soil degradation as cropland agriculture does, and it causes more desertification. Humanity tends more than 3.4 billion cattle, sheep, and goats. Degraded rangeland costs an estimated $23 billion per year in lost productivity. Grazing exceeds the sustainable supply of grass in India by 30% and in parts of China by up to 50%. To relieve pressure on rangelands, both nations are now beginning to feed crop residues to livestock.

Range managers in the United States assess the carrying capacity of rangelands and inform livestock owners of these limits, so that herds are rotated from site to site as needed to conserve grass cover and soil integrity. Managers also can establish and enforce limits on grazing on publicly owned land. Yet U.S. ranchers have traditionally had little incentive to conserve rangelands because most grazing has taken place on public lands leased from the government and because U.S. taxpayers have heavily subsidized grazing. As a result of this classic "tragedy of the commons" situation (p. 165), overgrazing has resulted in extensive environmental impacts across the American West.

Today increasing numbers of ranchers are working cooperatively with government agencies, environmental scientists, and even environmental advocates to find ways to raise livestock more sustainably and safeguard the health of the land (see **THE SCIENCE BEHIND THE STORY**, pp. 234–235).

Watering and Fertilizing Crops

For as long as humanity has been farming, we have provided our crops with supplemental water and nutrients in order to boost production. In the modern age of industrial agriculture, the large-scale and mechanized provision of extra water and nutrients has raised our productivity to unforeseen heights, allowing us to feed a world of 7 billion people. However, these practices have also contributed to soil degradation and land degradation, and can affect ecosystems and people far away from farm fields.

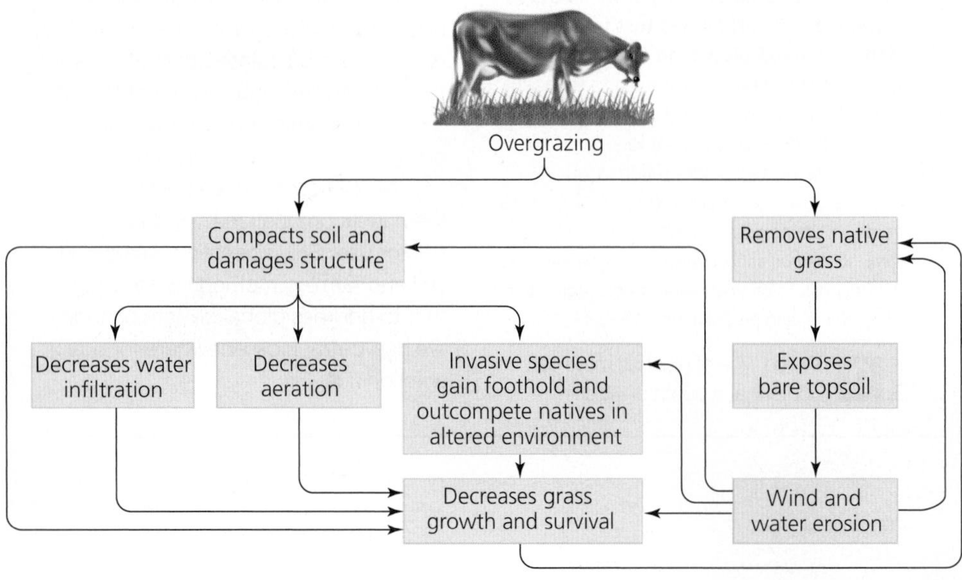

FIGURE 9.20 **Overgrazing has ecological consequences.** When grazing by livestock exceeds the carrying capacity of rangelands and their soil, this can set in motion a series of consequences and positive feedback loops that degrade soils and grassland ecosystems.

DATA Q What is the immediate cause of exposure of bare topsoil? What is the immediate consequence of exposing bare topsoil? How many immediate consequences of wind and water erosion are shown in this diagram?

(a) Flood-and-furrow irrigation of cotton in the southern California desert.

(b) Center-pivot irrigation circles in the Arizona desert.

FIGURE 9.21 **Irrigating water-thirsty crops in arid regions causes us to lose a great deal of water to evaporation.** Examples are **(a)** flood-and-furrow irrigation for cotton in southern California and **(b)** center pivot irrigation (from the air) in Arizona.

Irrigation boosts productivity but can damage soil

The artificial provision of water beyond that which crops receive from rainfall is known as **irrigation.** Some crops, such as rice and cotton, use large amounts of water and generally require irrigation, whereas others, such as beans and wheat, use relatively little. The amount of water required also is influenced by the rate of evaporation and the soil's ability to hold water and make it available to plant roots. If the climate is too dry, or if too much water evaporates or runs off before it can infiltrate the soil, crops may require irrigation. By irrigating crops, people have managed to turn previously dry and unproductive regions into fertile farmland.

Fully 70% of all fresh water that people withdraw is used for irrigation. Irrigated acreage has increased dramatically worldwide, reaching almost 400 million ha (nearly 1 billion acres), twice the area of Mexico. In some cases, withdrawing water for irrigation has depleted aquifers and dried up rivers and lakes. We will examine irrigation further in Chapter 15 (p. 397).

Sustainable approaches to irrigation maximize efficiency

One of the most effective ways to reduce water use in agriculture is to better match crops and climate. Land in many arid regions has been converted into productive farmland through extensive irrigation, often with the support of government subsidies that make irrigation water artificially inexpensive. Some farmers in these areas cultivate crops that require large amounts of water, such as rice and cotton (FIGURE 9.21). This leads to extensive water loss from evaporation in the arid climate. Choosing crops that require far less water could enable these areas to remain agriculturally productive while greatly reducing water use.

Another approach is to embrace technologies that improve efficiency in water use. Currently, irrigation efficiency worldwide is low, as plants end up using only 43% of the water that we apply. The rest evaporates or soaks into the soil away from plant roots (FIGURE 9.22a). Drip irrigation systems that target water directly toward plant roots through hoses can increase efficiencies to over 90% (FIGURE 9.22b). As such systems become more affordable, more farmers are turning to them.

(a) Conventional irrigation

(b) Drip irrigation

FIGURE 9.22 **Irrigation methods vary in their water use.** Conventional methods **(a)** are inefficient. In drip irrigation systems **(b)**, hoses drip water directly onto the plants, so that much less is wasted.

Restoring the Malpai Borderlands

In the high desert of southern Arizona and New Mexico, scientists and cattle ranchers trying to heal the scars left by decades of overgrazing found they had to contend with a creature even more damaging than a hungry steer: Smokey Bear.

Wildfires might seem a natural enemy of grasslands, but researchers in the Malpai Borderlands realized that people's efforts to suppress fire had done far more harm. Before large numbers of settlers and ranchers arrived more than a century ago, this semi-arid Western landscape thrived under an ecological cycle common to many grasslands. Trees such as mesquite grew near creeks. Hardy grasses such as black grama covered the drier plains. Periodic wildfires, usually sparked by lightning, burned back shrubs and trees and kept grasslands open. Deer, rabbits, and bighorn sheep grazed on the grasses but rarely ate enough to deplete the range. Fed by seasonal rains, new grasses sprouted without being overeaten or crowded out by larger plants.

By the 1990s, however, those grasslands were increasingly scarce. Ranchers had brought large cattle herds to the area in the late 1800s. The cows chewed through vast expanses of grass,

Burning brush to re-establish grass cover in the Malpai Borderlands

trampled the soil, and scattered mesquite seed into areas where grasses had dominated. Ranchers fought wildfires to keep their herds safe, and the federal government joined in the firefighting efforts.

The Malpai's ranching families found themselves struggling to make a living. Decades of photos taken by ranchers and by University of Arizona botanist Raymond Turner showed how soil had eroded and how trees and brush had overgrown the grass. The ranchers knew their cattle were part of the problem, but they also suspected that firefighting efforts were to blame.

In 1993, a group of ranchers launched an innovative plan. They formed the Malpai Borderlands Group, designating about 325,000 ha (800,000 acres) of land for protection and study. Ranchers joined government agencies, environmentalists, and scientists to study the region's ecology, bring back grasses, and restore native animal species. They sought to return periodic fire to the landscape by conducting prescribed burns (pp. 320–321) and by allowing natural fires to run their course (**FIGURE 1**).

The group's research efforts have centered on the Gray Ranch, a 121,000-ha (300,000-acre) parcel in the heart of the borderlands. At McKinney Flats within the Gray Ranch, scientists led by Antioch University researcher Charles Curtin divided rangeland into four study areas of about 890 ha (2200 acres) each. Each area is further divided into four "treatments," or areas with varying land management techniques:

- In Treatment 1, land is burned and grazed.
- In Treatment 2, land is burned but not grazed.
- In Treatment 3, land is grazed but not burned.
- In Treatment 4, land is neither grazed nor burned.

Treatments 1 and 3, which allow grazing, also feature small fenced-off areas that prevent animals from eating grass. These "exclosures" allow scientists to make precise side-by-side comparisons of how grazing affects grasses. Scientists measure rainfall in each area and monitor soils for degradation and erosion. Teams of wildlife and vegetation specialists

Salinization and waterlogging are easier to prevent than to correct

If some water is good for plants and soil, it might seem that more must be better. But we can supply too much water. Overirrigated soils saturated with water may experience **waterlogging** when the water table rises to the point that water drowns plant roots, depriving them of access to gases and essentially suffocating them.

A more frequent problem is **salinization,** the buildup of salts in surface soil layers. In dryland areas where precipitation and humidity are low, the evaporation of water from the soil's A horizon may pull water containing dissolved salts up from lower horizons. As the water evaporates at the surface, those salts remain, often turning the soil surface white. Irrigation in arid areas generally hastens salinization, and irrigation water often contains some dissolved salt in the first place, which introduces new salt to the soil. Salinization now inhibits production on one-fifth of all irrigated cropland globally, costing more than $11 billion each year.

Remedying salinization once it has occurred is expensive and difficult, so preventing it in the first place is better. The best way to prevent salinization is to avoid planting

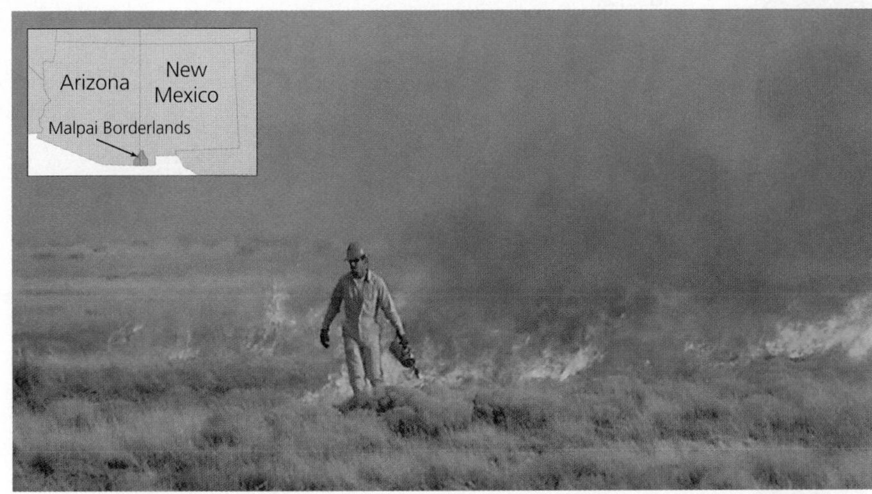

FIGURE 1 To restore native grasslands, the Malpai Borderlands Group reinstated fire as a landscape process, conducting controlled burns. Monitoring indicates that restoring fire has improved ecological conditions in the region.

monitor each treatment for the distribution and abundance of birds, insects, animals, and vegetation (**FIGURE 2**).

By comparing areas where fire is suppressed to those where fire can burn, researchers have documented how the suppression of fire leads to more brush and trees and less grass. Conversely, when fire burns an area, woody plants such as mesquite trees are damaged and their seed production is disrupted, and grass subsequently returns to replace the trees. Such changes follow both natural fires and carefully monitored, deliberately set controlled burns.

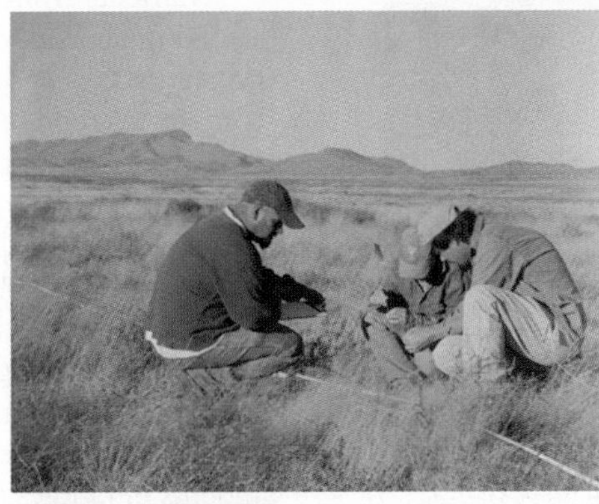

FIGURE 2 Researchers measure vegetation on research plots in the Malpai Borderlands.

All told, the landscape-level (pp. 113–115) McKinney Flats project may be the largest replicated terrestrial ecological experiment in North America. It has helped spur a number of other research initiatives, and the Malpai Borderlands Group now hosts a scientific conference each year featuring research from the region.

The scientists have also helped ranchers develop a cycle of "grassbanking," in which cattle are allowed to graze on shared plots of land while other areas recover. Ranchers must also work with the weather. Scientists have found that controlled burns or grassbanking should track with cycles of rain and drought to bring back grass.

Because of such research, controlled burns are now a regular part of the Malpai landscape. Ranchers have burned more than 100,000 ha (250,000 acres) since 1994, and natural fires are often allowed to run their course with little or no intervention. Damaged areas have been reseeded with native grasses.

Scientists increasingly believe that sustainable ranching, if managed properly, can help restore grasslands damaged by overgrazing. "We cannot assume rangelands will recover on their own," Curtin wrote in a recent study on the Malpai Borderlands. "Conservation of grazed lands requires restoring and sustaining natural processes." ■

crops that require a great deal of water in areas prone to salinization. A second way is to irrigate with water low in salt content. A third way is to irrigate efficiently, supplying no more water than a crop requires. This minimizes the amount of water that evaporates and hence the amount of salt that accumulates in the topsoil.

If salinization has occurred, one potential way to mitigate it is to stop irrigating and wait for rain to flush salts from the soil. However, this solution is unrealistic because in dryland areas where salinization is most often a problem, precipitation is rarely adequate to flush soils. A better option may be to plant salt-tolerant plants, such as barley, that can be used as food or pasture. A third option is to bring in large quantities of less-saline water with which to flush the soil. However, using too much water may cause waterlogging.

Fertilizers boost crop yields but can be overapplied

Plants require nitrogen, phosphorus, and potassium to grow, as well as smaller amounts of over a dozen other nutrients. Plants remove these nutrients from soil as they grow, and leaching likewise removes nutrients. If agricultural soils come to

(a) Inorganic fertilizer

(b) Organic fertilizer

FIGURE 9.23 Two main types of fertilizer exist. Inorganic fertilizer **(a)** consists of synthetically manufactured granules. Organic fertilizer **(b)** includes substances such as compost crawling with earthworms.

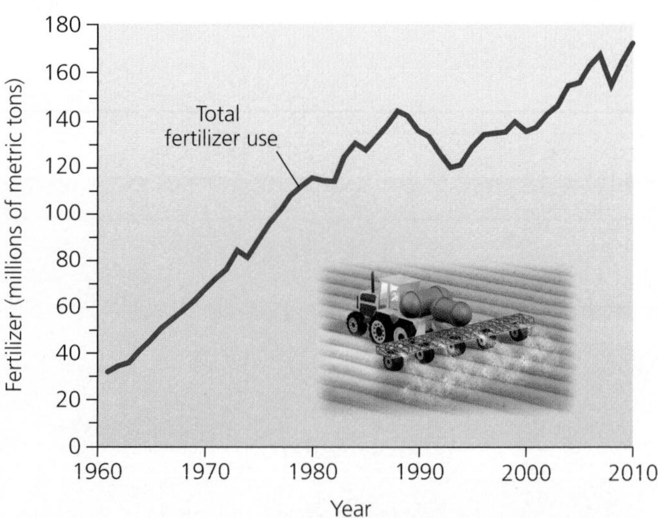

FIGURE 9.24 Use of synthetic, inorganic fertilizers has risen sharply over the past half-century. Today, usage stands at over 170 million metric tons annually. *Data from International Fertilizer Industry Association.*

contain too few nutrients, crop yields decline. For this reason, we go to great lengths to enhance nutrient-limited soils by adding **fertilizer,** substances that contain essential nutrients.

There are two main types of fertilizers (FIGURE 9.23). **Inorganic fertilizers** are mined or synthetically manufactured mineral supplements. **Organic fertilizers** consist of the remains or wastes of organisms and include animal manure; crop residues; fresh vegetation (*green manure*); and **compost,** a mixture produced when decomposers break down organic matter such as food and crop waste in a controlled environment.

Historically people relied on organic fertilizers to replenish soil nutrients, but during the latter half of the 20th century farmers in industrialized nations and nations experiencing the Green Revolution embraced inorganic fertilizers (FIGURE 9.24). The use of inorganic fertilizers has boosted our global food production, but their application has triggered increasingly severe pollution problems. Inorganic fertilizers are generally more susceptible to leaching and runoff than are organic fertilizers, and they more readily contaminate groundwater supplies.

Nutrients from inorganic fertilizers can have impacts far beyond the boundaries of the fields on which they are applied (FIGURE 9.25). For instance, nitrogen and phosphorus runoff from farms and other sources in the Mississippi River basin spurs phytoplankton blooms in the Gulf of Mexico and creates an oxygen-depleted "dead zone" that kills fish and shrimp (see Chapter 15; pp. 410–411). Such eutrophication (pp. 108–109, 412) occurs at countless river mouths, lakes, and ponds throughout the world. Moreover, nitrates readily leach through soil and contaminate groundwater, and components of some nitrogen fertilizers can even volatilize (evaporate) into the air. Through these processes, unnatural amounts of nitrates and phosphates spread through ecosystems and can pose human health risks, including cancer and methemoglobinemia, or blue-baby syndrome, which can asphyxiate and kill infants. Human inputs of nitrogen have modified the nitrogen cycle (pp. 123, 126) and now account for one-half the total nitrogen flux on Earth.

Sustainable fertilizer use involves targeting and monitoring nutrients

Sustainable approaches to fertilizing crops target the delivery of nutrients to plant roots and avoid the overapplication of fertilizer. Farmers using drip irrigation systems can add fertilizer to irrigation water, thereby releasing it only above plant roots. Farmers practicing no-till or conservation tillage often inject fertilizer along with seeds, concentrating it near the developing plant. Farmers can also avoid overapplication by regularly monitoring soil nutrient content and applying fertilizer only when nutrient levels are too low. The Nielsens in Iowa are careful and judicious with their fertilizer use in these ways. They apply fertilizer in spring but not in fall. After two targeted applications, they test for nitrate levels

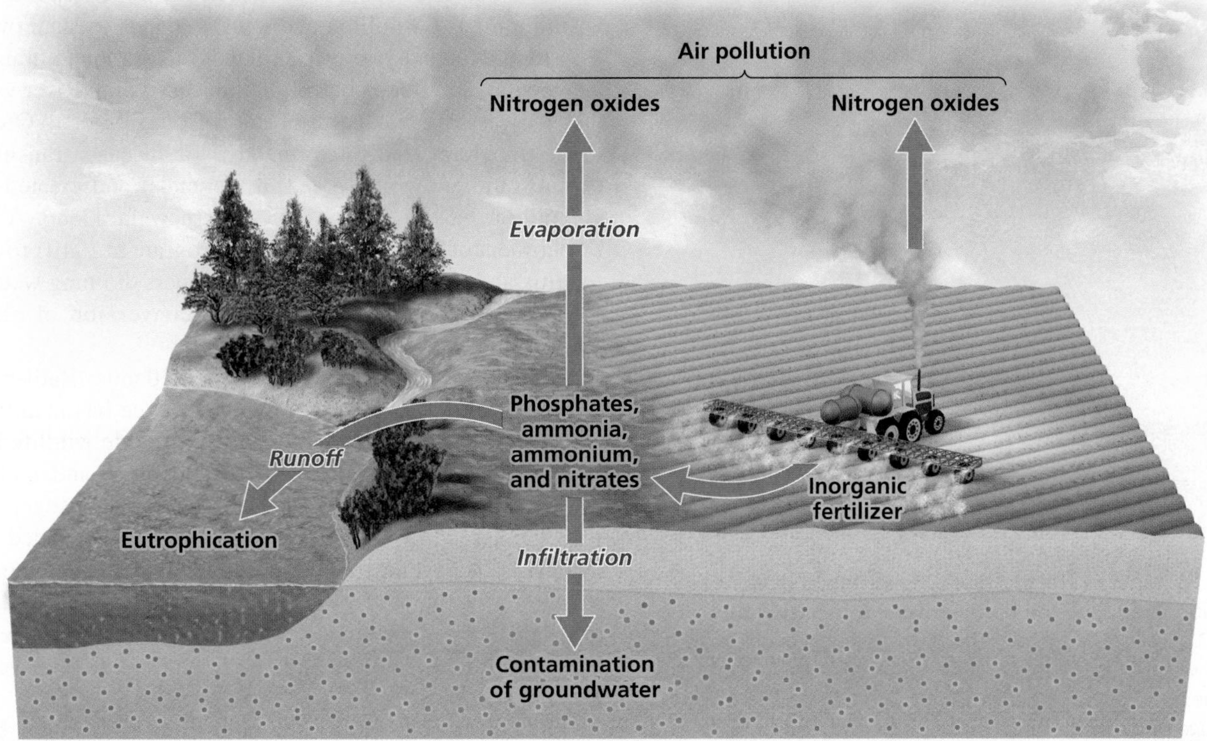

FIGURE 9.25 Overapplication of fertilizers has impacts beyond the farm field, because nutrients not taken up by plants end up elsewhere. Nitrates can leach into aquifers and contaminate drinking water. Runoff of phosphates and nitrogen compounds can alter the ecology of waterways through eutrophication. Compounds such as nitrogen oxides may pollute the air.

once their plants have sprouted to determine how much more to add. "We're putting more nitrogen where the crop can use it and less nitrogen on or in the ground where the crop can't use it," Todd Nielsen says. In addition, buffer strips of vegetation they planted along field edges and along the river on their property help to capture nutrient runoff along with any eroded soils.

Sustainable agriculture embraces the use of organic fertilizers, as they can provide some benefits that inorganic fertilizers cannot. Organic fertilizers provide not only nutrients, but also organic matter that improves soil structure, nutrient retention, and water-retaining capacity. The use of organic fertilizers is not without cost, though. When manure is applied in amounts needed to supply sufficient nitrogen for a crop, it may introduce excess phosphorus, which can run off into waterways. Thus, the Nielsens apply manure, but they are careful not to use too much. Accordingly, sustainable approaches do not rely solely on organic fertilizers, but integrate them with the targeted use of inorganic fertilizer.

Agricultural Policy

Governments have long sought ways to encourage agricultural production. More recently, they have also sought ways to lessen the environmental impacts and external costs (p. 146, 165) of agriculture. In theory, the marketplace should discourage people from farming and grazing using intensive methods that degrade the land they own if such practices are not profitable in the long run. But land degradation often unfolds gradually, whereas farmers and ranchers generally cannot afford to go without profits in the short term, even if they know conservation is in their long-term interests. For this reason, we have increasingly developed public policy to encourage conservation measures.

Some policies worsen land degradation

Many nations spend billions of dollars in government subsidies to promote agricultural production. In many cases these subsidies fund practices that are not economically and environmentally sustainable, such as growing water-thirsty crops in desert regions. Roughly one-fifth of the income of the average U.S. farmer comes from subsidies. Proponents of such subsidies stress that the uncertainties of weather make profits and losses from farming unpredictable from year to year. To persist, these proponents say, farmers need some way of being compensated in bad years. This may be the case, but subsidies can encourage people to cultivate land that would otherwise not be farmed; to produce more food than is needed, driving down prices for other producers; and to practice methods that degrade the land. Thus, opponents of subsidies argue that subsidizing environmentally destructive agricultural practices is unsustainable. They suggest that a better model is for farmers to buy insurance to protect themselves against production shortfalls.

FIGURE 9.26 Most of North America's wetlands have been drained and filled, and the land converted to agriculture. The northern Great Plains are pockmarked with thousands of "prairie potholes," water-filled depressions that provide breeding habitat for most of the continent's waterfowl. Shown are farmlands encroaching on prairie potholes in North Dakota.

Ranchers in the United States also benefit from government subsidies. Most U.S. rangeland is federally owned and managed by the Bureau of Land Management (BLM). The BLM is the nation's largest landowner, with over 100 million ha (248 million acres), mostly in 12 western states (see Figure 12.10, p. 316). Ranchers are allowed to graze livestock on BLM lands for inexpensive fees; a grazing permit in 2013 was just $1.35 per month per "animal unit" (one horse, one cow plus calf, five sheep, or five goats). Such low fees can encourage overgrazing and the degradation of grazing land through the tragedy of the commons scenario.

For this reason, ranchers and environmentalists have traditionally been at loggerheads. In recent years, however, some ranchers and environmentalists have been teaming up to preserve ranchland against what each of them views as a common threat—the encroaching housing developments of suburban sprawl (pp. 339–342). Although developers often pay high prices for ranchland, many ranchers do not want to see the loss of the wide-open spaces and the ranching lifestyle that they cherish.

Wetlands have been drained for farming

Many of our crops grow on the sites of former wetlands (pp. 394–396)—swamps, marshes, bogs, and river floodplains—that people drained and filled in (**FIGURE 9.26**). Today, less than half the original wetlands of the lower 48 U.S. states and southern Canada remain. This loss results from decades of laborious efforts, encouraged by government policy, to drain wetlands for agriculture.

To promote settlement and farming, the United States passed a series of laws known as the Swamp Lands Acts in 1849, 1850, and 1860. The government transferred over 24 million ha (60 million acres) of wetlands to state ownership (and eventually to private hands) to stimulate drainage, conversion, and flood control. In the Mississippi River valley, the Midwest, and a handful of states from Florida to Oregon, these transfers eradicated malaria (a disease transmitted by mosquitoes, which breed in wetlands) and created over 10 million ha (25 million acres) of new farmland. A U.S. Department of Agriculture (USDA) program in 1940 provided funding and technical assistance to farmers draining wetlands on their property, resulting in the conversion of almost 23 million ha (57 million acres).

Today we have a new view of wetlands. Rather than viewing them as worthless swamps, science has made clear that they are valuable ecosystems that provide wildlife habitat, improve water quality, control flooding, and recharge water supplies (Chapter 15). This scientific knowledge, along with the spread of a preservation ethic, has persuaded policymakers to develop regulations intended to safeguard our remaining wetlands. However, because of loopholes, differing state laws, development pressures, and debate over the legal definition of wetlands, many of these vital ecosystems are still being lost.

In current U.S. policy, financial incentives are being used as a means to protect wetlands and influence agricultural land use. Under the *Wetlands Reserve Program*, the U.S. government offers payments to landowners who protect, restore, or enhance wetland areas on their property. Over 2 million acres are currently enrolled in this program nationwide. The Wetlands Reserve Program is one of 15 conservation programs funded by the U.S. Congress in the latest farm bill legislation.

A number of programs promote soil conservation

Every five to six years, the U.S. Congress passes comprehensive legislation that guides agricultural policy. The 2008 "farm bill" funded 15 programs (such as the Wetlands Reserve Program) that encourage the conservation of soil, grasslands, wetlands, wildlife habitat, and other natural resources on agricultural lands. Altogether these conservation programs received nearly 9% of the $288 billion budgeted in the bill. Many of the provisions promoting soil conservation require farmers to adopt soil conservation plans and practices before they can receive government subsidies.

Despite broad agreement in Congress over a 2012 farm bill, and despite the fact that farmers were reeling from severe drought that lowered harvests, the House of Representatives failed to bring the 2012 legislation for a vote. In subsequent negotiations over the so-called "fiscal cliff" in January 2013, Congress approved an extension of the 2008 bill through September 2013. You and your instructor may wish to explore the latest developments regarding this important legislation, which affects so much of America's land, people, and economy.

Todd and Arliss Nielsen's conservation efforts on their Iowa farm are supported by payments from the Environmental Quality Incentives Program and the Conservation Stewardship Program, each of which have subsidized the Nielsens' investments in cover crops. In addition, the Mississippi River Basin Initiative funds the Nielsens for using cover crops and for monitoring and improving water quality. This federal program was enacted to enlist farmers in helping to reduce the nutrient runoff into the Mississippi River that causes a hypoxic dead zone in the Gulf of Mexico each year (see Chapter 15; pp. 410–411).

The Nielsens also are enrolled in the **Conservation Reserve Program,** which pays farmers to stop cultivating damaged and highly erodible cropland and instead make these lands conservation reserves planted with grasses and trees. Lands in this program now cover an area nearly the size of Iowa, and the U.S. Department of Agriculture (USDA) estimates that each dollar invested in the program saves nearly 1 ton of topsoil. Besides reducing erosion, the Conservation Reserve Program generates income for farmers, improves water quality, and provides habitat for wildlife.

Farmers like the Nielsens apply for the program, and the U.S. Farm Service Agency selects the farmers who will be awarded contracts based on an "environmental benefits index." This index includes the predicted benefits to wildlife, water quality, air quality, and the farm through reduced erosion, as well as benefits likely to endure beyond the contract period, with all these balanced against the cost of payments. Contracts generally run for 10–15 years. Of all U.S. states, Iowa is most heavily invested in the Conservation Reserve Program; the state has 102,000 contracts with 52,000 farms worth over $200 million per year. Nationwide, the government pays farmers about $1.8 billion per year for the conservation of lands totaling 11 million ha (27 million acres). The area tends to vary with market prices for crops; for instance, conservation reserves have decreased since 2007 in response to higher food prices as farmers have withdrawn lands from the program and planted them with crops.

Internationally, the United Nations promotes soil conservation and sustainable agriculture through a variety of programs led by the Food and Agriculture Organization (FAO). As just one example, FAO's Farmer-Centered Agricultural Resource Management Program (FARM) supports innovative approaches to resource management and sustainable agriculture in eight Asian nations. This program studies success stories and tries to help farmers elsewhere duplicate successful efforts. Rather than following a top-down, government-controlled approach, the FARM program calls on the creativity of local communities to educate and encourage farmers to conserve soils and secure their food supply. Indeed, we will need improved policies and practices at all levels throughout the world if we are to sustainably provide for our planet's growing human population.

WEIGHING THE ISSUES

SOIL, SUBSIDIES, AND SUSTAINABILITY Do you think that financial incentive programs such as the Conservation Reserve Program and the Wetlands Reserve Program are a good use of taxpayers' money? Are financial incentives more effective than government regulation for promoting certain land use goals? Do you think they can help lead us toward agriculture that is truly sustainable?

Conclusion

Our species has enjoyed a 10,000-year history with agriculture, yet despite all we have learned about land degradation and soil conservation, challenges remain. Many of the policies enacted and the practices developed to combat soil degradation in the United States and worldwide have met with success, particularly in reducing topsoil erosion. It is clear, however, that even the best-conceived soil conservation programs require research, education, funding, and commitment from both farmers and governments if they are to fulfill their potential. In light of continued population growth, we will likely need better technology and wider adoption of soil conservation techniques if we are to achieve sustainable agriculture and feed the 9 billion people expected to crowd our planet by mid-century.

Reviewing Objectives

You should now be able to:

Explain the importance of soils to agriculture

- Successful agriculture requires healthy soil. (p. 216)
- Soil is a renewable resource, but its renewal occurs slowly. (p. 216)
- Our society's future depends on sustainable agriculture, and soil integrity is a key component of this pursuit. (pp. 216–217)

Outline major developments in the history of agriculture

- At least 10,000 years ago, people began breeding crop plants and domesticating animals through the process of selective breeding, or artificial selection. (p. 217)
- Agriculture originated multiple times independently in different cultures across the world. (p. 217)
- Industrial agriculture is replacing traditional agriculture, which largely replaced hunting and gathering. (p. 218)

Delineate the fundamentals of soil science, including soil formation and soil properties

- Soil is a complex system and includes diverse biotic communities that decompose organic matter. (p. 218)
- Weathering and biological activity help form soil, influenced by climate, organisms, topography, parent material, and time. (pp. 218–219)
- Soil profiles consist of distinct horizons (such as the A Horizon, or topsoil) with characteristic properties. (pp. 219–220)
- Soil can be characterized by color, texture, structure, and pH. (p. 220)
- Soil properties differ among regions and affect plant growth and agriculture. (p. 221)

Analyze the causes and impacts of soil erosion and land degradation

- As the human population grows and consumption increases, pressures from agriculture are degrading Earth's soil, and we lose 5–7 million ha (12–17 million acres) of productive cropland annually. (p. 222)
- Soil degradation is the major component of land degradation globally. (p. 222)
- Some agricultural practices have resulted in severe erosion, lowering crop yields. (p. 223)
- Desertification affects soils in the world's arid regions. (pp. 223–224)

Explain the principles of soil conservation and provide solutions to soil erosion and land degradation

- The Dust Bowl in the United States and similar events elsewhere inspired scientists and farmers to develop ways to conserve topsoil. (pp. 224–225)
- Extension agents and agencies such as the National Resources Conservation Service (NRCS) educate and assist farmers. (pp. 225–226)
- Farming techniques such as crop rotation, contour farming, intercropping, terracing, shelterbelts, and conservation tillage enable farmers to reduce soil erosion and boost crop yields. (pp. 226–228)
- As a rule, vegetation helps anchor soil and prevent erosion. (pp. 228–229)
- Overgrazing can cause soil degradation on grasslands and diverse impacts to native ecosystems. (pp. 229, 232)
- Herd rotation, grazing limits, and research cooperation with scientists are some pathways to sustainable grazing. (pp. 232, 234–235)

Describe techniques of watering and fertilizing crops, and offer more sustainable alternatives

- We can conserve water by using efficient irrigation techniques and choosing crops to match climates. (p. 233)
- Overirrigation can cause salinization and waterlogging, which lower crop yields and are difficult to mitigate. (pp. 234–235)
- Overapplying fertilizers can lead to pollution problems, as leaching and runoff transport nutrients that affect ecosystems and human health. (pp. 235–236)
- Fertilizer use can be made more sustainable by targeting fertilizers directly to plants and monitoring when they are needed. (pp. 236–237)

Summarize major policy approaches for pursuing soil conservation and sustainable agriculture

- Some existing policies and subsidies worsen land degradation. (pp. 237–238)
- Past agricultural policy has driven the destruction of wetlands. (p. 238)
- In the United States and across the world, governments are devising agricultural policies and programs for conservation, such as the Conservation Reserve Program and other programs funded in the periodic U.S. farm bills. (pp. 238–239)

Testing Your Comprehension

1. Explain how each of us depends on healthy soil for our day-to-day well-being. What characteristics of soil are important to crop growth and agriculture, and how?
2. Describe the methods used in traditional agriculture. How does industrial agriculture differ from traditional agriculture?
3. What processes are most responsible for the formation of soil? Describe the three types of weathering that contribute to soil formation. Name the five primary factors thought to influence soil formation.
4. How are soil horizons created? How is organic matter distributed in a typical soil profile?
5. Why is erosion generally considered a destructive process? Name three human activities that can promote soil erosion. What factors affect the intensity of water erosion?
6. What was the Dust Bowl? What impacts did it have on the United States? Describe innovations in soil conservation introduced by the Soil Conservation Service (SCS).

7. What farming techniques help reduce the risk of erosion? Explain how no-till farming works. Why does this method reduce soil erosion?
8. Describe the effects of overgrazing on soil. What policies can be linked to the practice of overgrazing? What conditions characterize sustainable grazing practices?
9. How can irrigation practices be made more sustainable? How do fertilizers boost crop growth? How does excess fertilizer added to soil sometimes end up in water supplies and in the atmosphere?
10. Explain the goals and methods of the Conservation Reserve Program.

Seeking Solutions

1. How do you think a farmer can best help to conserve soil? How do you think a scientist can best help to conserve soil? How do you think a national government can best help to conserve soil?
2. How and why might actual soils differ from the idealized six-horizon soil profile presented in the chapter? How might departures from the idealized profile indicate the impact of human activities? Provide at least three examples.
3. Discuss how the methods of no-till farming or conservation tillage, as described in this chapter, can enhance soil quality. What other benefits can these approaches have? What drawbacks or negative effects might no-till or conservation tillage practices have on soil and water quality, and how might these be minimized?
4. Aside from no-till farming, select two other methods or approaches described in this chapter that you feel are promoting the science or practice of soil conservation, and describe how they are accomplishing this.
5. **THINK IT THROUGH** You are a land manager with the U.S. Bureau of Land Management (BLM, p. 238) and have just been put in charge of 500,000 acres of public grasslands that have been degraded by decades of overgrazing. Soil is eroding, creating large gullies. Invasive weeds are replacing native grasses. Shrubs are encroaching on grassland areas because fire was suppressed. Environmentalists want an end to ranching on the land. Ranchers want grazing to continue. What steps would you take to assess the land's condition and begin restoring its soil and vegetation? Would you allow grazing, and if so, would you set limits on it?
6. **THINK IT THROUGH** You are the head of an international granting agency that assists farmers with soil conservation and sustainable agriculture. You have $10 million to disburse. Your agency's staff has decided that the funding should go to (1) farmers in an arid area of Africa prone to salinization, (2) farmers in a fast-growing area of Indonesia where swidden agriculture is practiced, (3) farmers in Argentina practicing no-till agriculture, and (4) farmers in a dryland area of Mongolia undergoing desertification. What types of projects would you recommend funding in each of these areas, how would you apportion your funding among them, and why?

Calculating Ecological Footprints

In the United States, approximately 5 tons of topsoil are lost for every ton of grain harvested. Erosion rates vary greatly with soil type, topography, tillage method, and crop type. For simplicity, let us assume that the 5:1 ratio applies to all plant crops and that a typical diet includes 1 pound of plant material or its derived products (sugar, for example) per day. In the first two columns of the table, calculate the annual topsoil losses associated with growing this food for you and for other groups, assuming the same diet.

	Plant products consumed (lb)	Soil loss at 5:1 ratio (lb)	Soil loss at 3.25:1 ratio (lb)	Reduced soil loss at 3.25:1 relative to 5:1 ratio (lb)
You	365	1825	1186	639
Your class				
Your state				
United States				

1. Improved soil conservation measures reduced erosion in the United States by 35% from 1982 to 2003. If additional measures were again able to reduce the current rate of soil loss by this percentage, the ratio of soil lost to grain harvested would fall from 5:1 to 3.25:1. Calculate the soil losses associated with food production at a 3.25:1 ratio, and record your answers in the third column of the table.
2. Calculate the amount of topsoil hypothetically saved by the additional conservation measures in Question 1, and record your answers in the fourth column of the table.
3. Define a "sustainable" rate of soil loss. Describe how you might determine whether a given farm was practicing sustainable use of soil.

STUDENTS

Go to **MasteringEnvironmentalScience** for assignments, the etext, and the Study Area with practice tests, videos, current events, and activities.

INSTRUCTORS

Go to **MasteringEnvironmentalScience** for automatically graded activities, current events, videos, and reading questions that you can assign to your students, plus Instructor Resources.

A Mexican farmer inspects his maize

10

Agriculture, Biotechnology, and the Future of Food

Upon completing this chapter, you will be able to:

- Explain the challenge of feeding a growing human population
- Identify the goals, methods, and consequences of the Green Revolution
- Assess how we raise animals for food
- Describe approaches for preserving crop diversity
- Discuss the importance of pollination
- Explore strategies for pest management
- Analyze the nature, growth, and potential of organic agriculture
- Describe the science behind genetic engineering
- Evaluate the public debate over genetically modified food
- Summarize pathways to sustainable agriculture

CENTRAL CASE STUDY

Transgenic Maize in Southern Mexico?

"Worrying about starving future generations won't feed them. Food biotechnology will."

—Advertising campaign of the Monsanto Company

"Industrial agriculture has . . . destroyed diverse sources of food, and it has stolen food from other species . . . using huge quantities of fossil fuels and water and toxic chemicals in the process."

—Vandana Shiva, Director of the Research Foundation for Science, Technology, and Natural Resource Policy, India

Corn is a staple of the world's food supply. We can trace its ancestry back 9000 years, when people in the highland valleys of southern Mexico first domesticated that region's wild maize plants. The corn we eat today arose from some of the many varieties that evolved from the selective crop breeding conducted by this region's people.

Today southern Mexico remains a center of biodiversity for maize, with many locally adapted domesticated varieties, called **landraces,** growing in the rich, well-watered soil. Preserving such traditional varieties of crops in their ancestral homelands helps to secure the future of our food supply, scientists say. These varieties serve as reservoirs of genetic diversity that we may need to draw upon to sustain or advance our agriculture.

For this reason, food experts around the world expressed alarm in 2001 when scientists conducting genetic tests of Mexican maize landraces announced that they had turned up DNA (p. 29) that matched genes from genetically modified (GM) corn. GM corn was widely grown in the United States, but Mexico had banned its cultivation in 1998.

Genetically modified foods are foods derived from **genetically modified organisms (GMOs),** organisms that are genetically engineered. **Genetic engineering** is any process whereby scientists directly manipulate an organism's genetic material in the laboratory by adding, deleting, or changing segments of its DNA.

Corn is one of many crops that researchers have genetically engineered to express desirable traits such as large size, fast growth, and resistance to insect pests. To genetically engineer crops, scientists extract genes from the DNA of one organism and transfer them into the DNA of another. The aim is to improve crop performance and feed the world's hungry, but many people worry that **transgenes,** the genes engineered and moved into these **transgenic** plants, may have unintended consequences. One concern is that transgenic crops might crossbreed with local landraces and thereby "contaminate" the genetic makeup of native crops.

Such contamination was exactly what University of California at Berkeley professor Ignacio Chapela and his postdoctoral associate David Quist discovered while testing Mexican maize. They had ventured to the southern Mexican state of Oaxaca (pronounced "wha-HA-ca") and found what they argued were traces of DNA from genetically engineered corn in the genes of native maize plants. They alerted Mexican government scientists, and these scientists independently examined Oaxacan maize and obtained similar results. Quist and Chapela published their research findings in the scientific journal *Nature* in 2001.

Activists opposed to GM food trumpeted the news and urged a ban on imports of transgenic crops into developing nations from producer countries such as the United States. The agrobiotech industry defended the safety of its crops and questioned the validity of the research—as did some of Quist and Chapela's peers. Responding to criticisms from researchers, *Nature* took the unprecedented step of stating that Quist and Chapela's paper should not have been published—a decision that ignited a firestorm of controversy (see **THE SCIENCE BEHIND THE STORY**, pp. 266–267).

Subsequent research by Mexican government scientists reported that in 15 localities, 3–60% of maize contained transgenes and that 37% of maize grains distributed by the government to farmers were transgenic. These results were never published in a peer-reviewed scientific journal, but experts reasoned that corn shipments from the United States contain a mix of GM and non-GM grain, small farmers plant some of this seed, and transgenes entering Mexico can spread by wind pollination and by interbreeding with native landraces of maize.

Meanwhile, teams of Mexican and American scientists were conducting their own surveys of Oaxacan maize. In 2005, one team reported finding no transgenes at all, despite analyzing 154,000 maize seeds. Then in 2009 and 2010, other teams reported evidence of transgenes in native maize at low frequency from scattered locations. Scientists are still debating the issue, but consensus has grown that transgenes have by now spread widely into native landraces.

In recent years Mexico has begun importing more and more of its corn. In 2008 global food prices spiked, and Mexican citizens protested in the streets over the rising cost of tortillas. At this point the Mexican government decided that GM corn might help boost yields, lower prices, and make the country self-sufficient again. The government lifted the ban on testing GM corn and granted permits for 67 small-scale, controlled experimental trials in northern Mexico. Then in 2011, Mexico granted the multinational agrobiotech corporation Monsanto permission to field-test its GM corn on 1 hectare (2.5 acres) of land in northeast Mexico, with no measures to prevent the drift of pollen.

In 2012, Mexico's government stood ready to approve applications from Monsanto, Dupont, and Dow Chemical to plant GM corn on 2.5 million ha (6.1 million acres) of Mexican farmland—an area the size of Vermont or New Hampshire. Nonprofit groups skeptical of GMOs publicized this, thousands of Mexican farmers protested, and the government put off the decision until 2013. "If Mexico allows this crime of historic significance to happen," said Veronica Villa of the nonprofit ETC Group, "GMOs will soon be in the food of the entire Mexican population, and genetic contamination of Mexican peasant varieties will be inevitable."

The potential impact of GM corn on Mexico's native landraces is difficult to assess. A still harder question to answer is how the genetic engineering of crops in general may affect people and the environment (in both positive and negative ways) and whether it constitutes progress toward a goal of sustainable agriculture. In this chapter, we explore the quest for sustainable agriculture as we take a wide-ranging view of the ways we increase our agricultural output and how we might manage the environmental and social consequences of these efforts.

The Race to Feed the World

As the human population continues to grow, we can expect our numbers to swell to 9 billion by the middle of this century. For every four people living today, there will be five in 2050. Feeding 2 billion more mouths while protecting the integrity of soil, water, and ecosystems will require sustainable agriculture. Progressing toward a sustainable model for agriculture may involve a diversity of approaches, ranging from organic farming to genetically modified crops like those that have elicited so much controversy in Mexico.

We are producing more food per person

Over the past half-century, our ability to produce food has grown even faster than global population (**FIGURE 10.1**). We have increased food production by devoting more fossil fuel energy to agriculture; intensifying our use of irrigation, fertilizers, and pesticides; planting and harvesting more frequently; cultivating more land; and developing (through crossbreeding and genetic engineering) more productive crop and livestock varieties.

Improving people's quality of life by producing more food per person is a monumental achievement of which humanity can be proud. However, making our food supply sustainable depends on maintaining healthy soil, water, and biodiversity. Careless agricultural practices can damage environmental systems and diminish the capacity of soils to continue supporting crops and livestock (Chapter 9). Today many of the world's soils are in decline, and most of the planet's arable land has already been claimed. Even though agricultural production has outpaced population growth so far, there is no guarantee that it will continue to do so.

We face undernutrition, overnutrition, and malnutrition

Despite our rising food production, 870 million people worldwide do not have enough to eat. These people suffer from **undernutrition,** receiving fewer calories than the minimum dietary energy requirement. As a result, every 5 seconds, somewhere in the world, a child starves to death. In most cases, the reasons for undernutrition are economic. One-fifth of the world's people live on less than $1 per day, and over half live on less than $2 per day, the World Bank estimates.

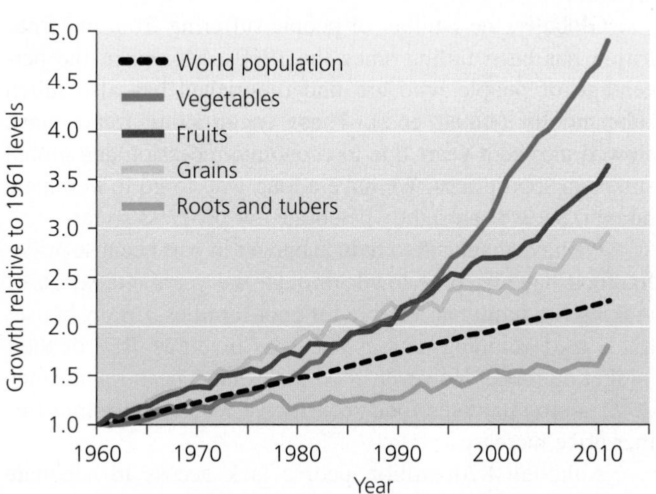

FIGURE 10.1 Global production of most foods has risen more quickly than world population. This means that we have produced more food per person each year. Trend lines show cumulative increases relative to 1961 levels (for example, a value of 2.0 means twice the 1961 amount). Food is measured by weight. *Data from U.N. Food and Agriculture Organization (FAO).*

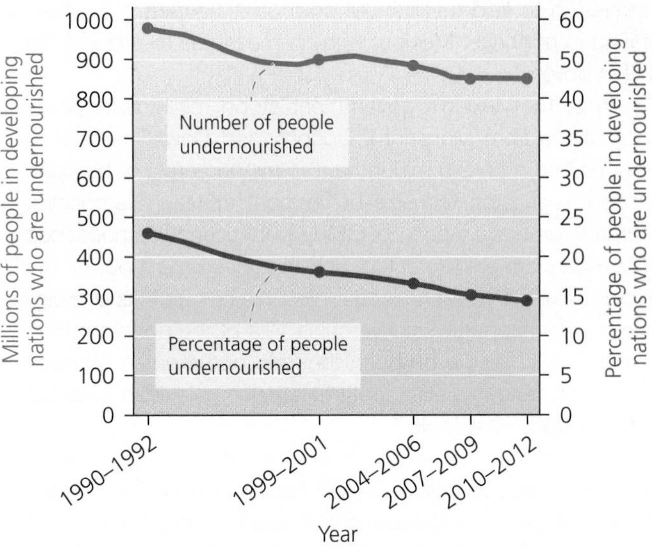

FIGURE 10.2 The number and the percentage of people in the developing world who suffer undernutrition have each been declining. *Data from Food and Agriculture Organization of the United Nations, 2012.* The state of food insecurity in the world, 2012. *FAO, Rome.*

DATA Q Explain how the percentage of undernourished people can have decreased since 2007–2009 while the absolute number of undernourished people stayed the same.

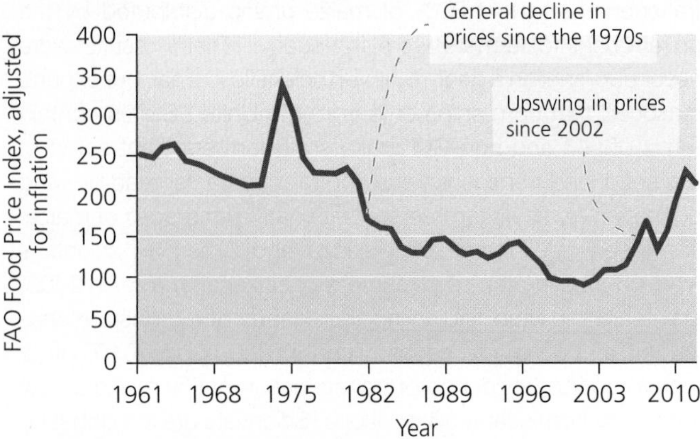

FIGURE 10.3 Food has become less costly over the past half-century, but prices have been rising since 2002. *Data from U.N. Food and Agriculture Organization (FAO).*

Political obstacles, conflict, and inefficiencies in distribution contribute significantly to hunger as well.

Most people who are undernourished live in the developing world. However, hunger is a problem even in the United States, where the U.S. Department of Agriculture (USDA) has classified 50 million Americans as "food insecure," lacking the income required to reliably procure sufficient food. Agricultural scientists and policymakers worldwide pursue a goal of **food security,** the guarantee of an adequate, safe, nutritious, and reliable food supply available to all people at all times.

Globally, the number of people suffering from undernutrition has been falling since the 1960s. Moreover, the percentage of people who are undernourished has also fallen substantially (**FIGURE 10.2**). These encouraging trends have slowed in recent years due to economic recession and global spikes in food prices. We have a long way to go to eliminate hunger, but we can rightly celebrate our progress so far.

We have managed to reduce hunger in part because prices for food have become lower through the years (**FIGURE 10.3**), making food more accessible for poor families. (Many households in developing nations spend as much as 70% of their budget on food.) However, food prices have been rising since 2002, so to guarantee food security we will need to limit further price increases.

Although 870 million people lack access to adequate food, many others consume more than is healthy and suffer from **overnutrition,** receiving too many calories each day. Overnutrition is a problem in developed nations such as the United States, where food is abundant, junk food is cheap, and people tend to lead sedentary lives with little exercise. As a result, more than one in three U.S. adults are obese, according to the Centers for Disease Control and Prevention (CDC). Obesity leads to cardiovascular disease, diabetes, and other health problems, shortening people's lives, impairing their quality of life, and imposing health care costs on society.

Lower-income people in wealthy societies often suffer overnutrition the most, because many rely on inexpensive mass-marketed foods that are calorie-rich but nutrient-poor. However, people of all income levels are eating more processed energy-rich foods that are high in fat and sugar yet low in nutrition, and they are getting less exercise. Worldwide, the World Health Organization estimates that over 1.4 billion adults are overweight and that of these, at least 500 million are obese. Nearly two of every three people live in countries where being overweight causes more deaths each year than being underweight.

Just as the *quantity* of food a person eats is important for health, so is the *quality* of food. **Malnutrition,** a shortage of nutrients the body needs, occurs when a person fails to obtain a complete complement of vitamins and minerals. Malnutrition can easily lead to disease (**FIGURE 10.4**). For example, people who eat a diet that is high in starch but deficient in protein or essential amino acids (p. 29) can develop *kwashiorkor*. Children who have recently stopped breast-feeding are most at risk for developing kwashiorkor, which causes

FIGURE 10.4 Millions of children suffer from forms of malnutrition, such as kwashiorkor and marasmus.

FIGURE 10.5 Norman Borlaug helped launch the Green Revolution. The high-yielding, disease-resistant wheat that he bred helped boost agricultural productivity in many developing countries.

bloating of the abdomen, deterioration and discoloration of hair, mental disability, immune suppression, developmental delays, anemia, and reduced growth. Protein deficiency together with a lack of calories can lead to *marasmus*, which causes wasting or shriveling among millions of children in the developing world.

The Green Revolution boosted agricultural production

The desire for greater quantity and quality of food for our growing population led in the mid- and late 20th century to the **Green Revolution** (first introduced in Chapter 9, p. 218). Realizing that farmers could not go on forever cultivating additional land to increase crop output, agricultural scientists devised methods and technologies to increase crop output per unit area of existing cultivated land. As a result, industrialized nations dramatically increased their per-area yields. The average hectare of U.S. cornfield raised its corn output fivefold during the 20th century, for instance. Many people viewed such growth in production and efficiency as key to ending starvation in developing nations.

The transfer of technology to the developing world that marked the Green Revolution began in the 1940s, when American agricultural scientist **Norman Borlaug** introduced Mexico's farmers to a specially bred type of wheat (FIGURE 10.5). This strain of wheat produced large seed heads, was resistant to diseases, was short in stature to resist wind, and produced high yields. Within two decades of planting this new crop, Mexico tripled its wheat production and began exporting wheat. The stunning success of this program inspired others. Borlaug—who won the Nobel Peace Prize for his work—took his wheat to India and Pakistan and helped transform agriculture there.

Soon many developing countries were doubling, tripling, or quadrupling their yields using selectively bred strains of wheat, rice, corn, and other crops from industrialized nations. When Borlaug died in 2009 at age 95, he was widely celebrated as having "saved more lives than anyone in history"—perhaps as many as a billion.

The Green Revolution brought mixed consequences

Along with the new grains, developing nations imported the methods of industrial agriculture. They began applying large amounts of synthetic fertilizers and chemical pesticides on their fields, irrigating crops generously with water, and using more machinery powered by fossil fuels. From 1900 to 2000, people increased energy inputs into agriculture by 80 times while expanding the world's cultivated area by just 33%.

This high-input agriculture succeeded dramatically in producing more corn, wheat, rice, and soybeans from each hectare of land. Intensified agriculture saved millions in India from starvation in the 1970s and eventually turned that nation into a net exporter of grain (FIGURE 10.6).

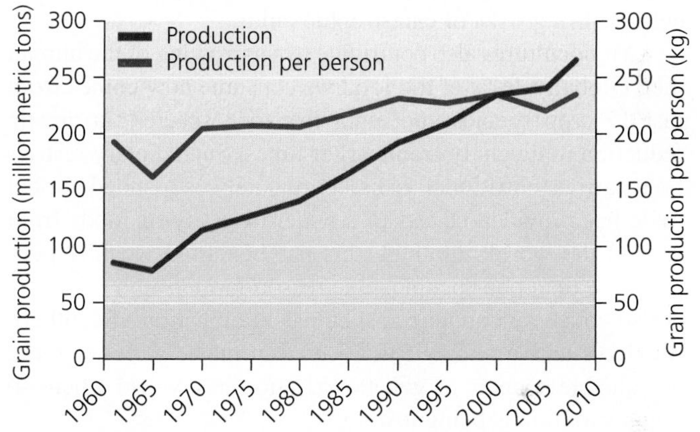

(a) Production and per-capita production rose

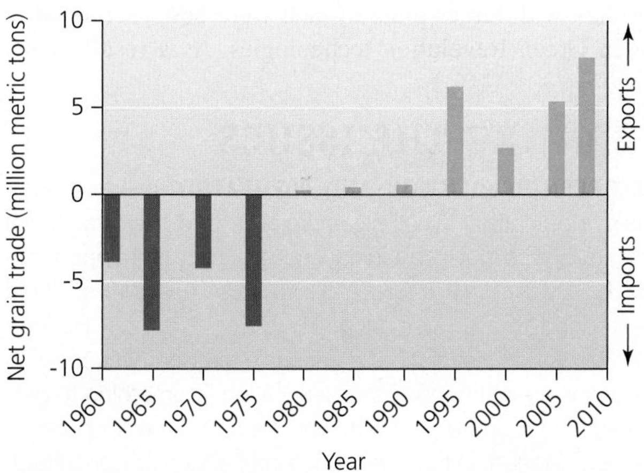

(b) Imports turned to exports

FIGURE 10.6 Green Revolution technology enabled India to boost its grain production. India's production grew faster than its population **(a)**, so that grain per person increased. As a result **(b)**, India was able to stop importing grain and begin exporting it to other nations. *Data from U.N. Food and Agriculture Organization (FAO).*

The effects of these developments on the environment have been mixed. On the positive side, the intensified use of already-cultivated land reduced pressures to convert additional natural lands for new cultivation. Between 1961 and 2010, food production more than tripled and per-person food production rose 48%, while area converted for agriculture increased only 10%. In this way, the Green Revolution prevented some degree of deforestation and habitat conversion and thus helped preserve biodiversity and natural ecosystems. On the negative side, the intensive application of water, fossil fuels, inorganic fertilizers, and synthetic pesticides worsened pollution, topsoil losses, and soil quality (Chapter 9).

FIGURE 10.7 Monocultures improve the efficiency of planting and harvesting but are susceptible to outbreaks of pests. Armyworms (**inset**) may attack this wheat field in Washington.

The planting of crops in **monocultures,** large expanses of single crop types (p. 218; FIGURE 10.7), makes planting and harvesting more efficient and thereby increases output. However, monocultures also reduce biodiversity over large areas, because many fewer wild organisms are able to live in monocultures than in native habitats or in traditional small-scale polycultures. Moreover, when all plants in a field are genetically similar, as in monocultures, all are equally susceptible to viral diseases, fungal pathogens, or insect pests that can spread quickly from plant to plant. For this reason, monocultures bring risks of catastrophic failure.

Monocultures also contribute to a narrowing of the human diet. Globally, 90% of the food we consume now comes from just 15 crop species and eight livestock species—a drastic reduction in diversity from earlier times. Such dietary restriction carries nutritional risks. Fortunately, expanded global trade has provided access to a wider diversity of foods from around the world, although this has benefited wealthy people more than poor people. One reason farmers and scientists are so concerned about transgenic contamination of southern Mexico's native maize is that Mexican maize varieties serve as valuable sources of genetic variation in a world where so much variation is being lost.

Today, yields are declining in some Green Revolution regions, likely due to soil degradation from the heavy use of fertilizers, pesticides, and irrigation. Moreover, wealthier farmers with larger plots of land were best positioned to invest in Green Revolution technologies. As a result, many low-income farmers who could not afford these technologies were driven out of business and moved to cities, adding to the immense migration of poor rural people to urban areas of the developing world (p. 337).

WEIGHING THE ISSUES

THE GREEN REVOLUTION AND POPULATION In the 1960s, India's population was skyrocketing, and its traditional agriculture was not producing enough food to support the growth. By adopting Green Revolution agriculture, India sidestepped mass starvation. However, Norman Borlaug called his Green Revolution methods "a temporary success in man's war against hunger and deprivation," something to give us breathing room to deal with what he called the "Population Monster." Indeed, in the years since intensifying its agriculture, India has added several hundred million more people and continues to suffer widespread poverty and hunger.

Do you think the Green Revolution has solved problems, deferred problems, or created new ones? Which aspects of the Green Revolution do you think help in the quest for sustainability, which do not, and why? Have the Green Revolution's benefits outweighed its costs?

Some biofuels reduce food supplies

Just as the Green Revolution's noble intentions and significant successes gave rise to some problematic side effects, in recent years some well-intentioned efforts to promote renewable energy have had unintended consequences. **Biofuels** (pp. 567, 569–570) are fuels derived from organic materials and used in internal combustion engines as replacements for petroleum. In an effort to shift from fossil fuels to renewable energy sources, policymakers have encouraged the production of biofuels from crops. In the United States, **ethanol** made from corn is the primary biofuel (pp. 569–570). Following expanded subsidies in 2007, U.S. ethanol production nearly doubled as new ethanol facilities opened and farmers began selling their corn for ethanol instead of for food.

This caused a scarcity of corn worldwide, and prices for basic foods (such as tortillas in Mexico) skyrocketed. Prices for other staple grains also rose because farmers shifted fields formerly devoted to other food crops into biofuel production. For low-income people, the steep rise in food prices was frightening. Thousands staged protests, and riots erupted in Mexico and many other nations. The world realized that growing crops for biofuels could compete directly with growing food for people to eat. Scientists today are trying to develop ways of producing biofuels from crop waste and other non-food material (pp. 571–572).

We are moving toward sustainable agriculture

Industrial agriculture has enabled food production to keep pace with our growing population thus far, but it also brings many adverse environmental and social impacts. Industrial agriculture in some form seems necessary to feed our planet's 7 billion people, but most experts feel that to sustain our population in the long run we will need to begin raising animals and crops in ways that are less polluting and less resource-intensive.

Sustainable agriculture is agriculture that does not deplete soils faster than they form (p. 216). It is farming and ranching that does not reduce the amount of healthy soil, clean water, and genetic diversity essential to long-term crop and livestock production. Simply put, sustainable agriculture is agriculture that can be practiced in the same way in the same place far into the future.

No-till farming and other soil conservation methods (Chapter 9) are primary avenues to help make our agriculture more sustainable. Reducing fossil-fuel inputs and the pollution these inputs cause is a key goal of sustainable agriculture. As a result, many approaches essentially move away from the industrial model and toward more traditional models, such as the cultivation of diverse landraces of Mexican maize. Yet plenty of analysts today feel that technology offers our best hope of making agriculture sustainable, and that the genetic engineering of crops and livestock is a vital component of this effort. In this chapter we will survey how our food is produced, the impacts our agriculture exerts, and the many alternative pathways that may help move us toward sustainability in agriculture.

Raising Animals for Food

Food from cropland agriculture makes up the majority of the human diet, but most of us also eat animal products. As our population has grown, consuming animal products has come to have significant environmental, social, agricultural, and economic impacts. How we respond to demand for animal products will have a major effect on the quest for sustainable agriculture and on our society's ecological footprint.

Consumption of animal products is growing

As wealth and global commerce have increased, so has humanity's consumption of meat, milk, eggs, and other animal products (**FIGURE 10.8**). The world population of domesticated animals raised for food rose from 7.3 billion animals to 27.5 billion animals between 1961 and 2011. Most of these animals are chickens. Global meat production has increased fivefold since 1950, and per capita meat consumption has doubled. The United Nations Food and Agriculture Organization (FAO) estimates that as more developing nations go through the demographic transition (pp. 201–202) and become wealthier, total meat consumption will nearly double again by the year 2050.

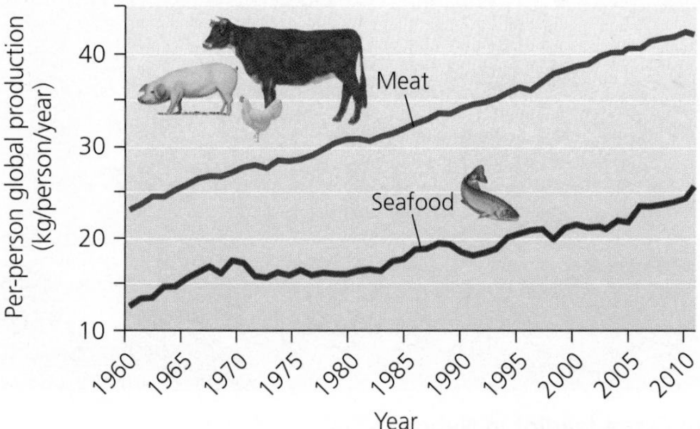

FIGURE 10.8 Per-person production of meat from farmed animals and of seafood has risen steadily worldwide. *Data from U.N. Food and Agriculture Organization (FAO).*

Our food choices are also energy choices

What we choose to eat has ramifications for how we use energy and the land that supports agriculture. Every time that one organism consumes another, only about 10% of the energy moves from one trophic level up to the next, while the great majority of energy is used up in cellular respiration (p. 32). For example, if we feed grain to a cow and then eat beef from the cow, we lose most of the grain's energy to the cow's metabolism. Energy is used up as the cow converts the grain to tissue as it grows, and as the cow conducts cellular respiration on a daily basis to maintain itself. For this reason, eating meat is far less energy-efficient than relying on a vegetarian diet, and it leaves a far greater ecological footprint.

In contrast, if we eat lower on the food chain (a more vegetarian diet), we put a greater proportion of the sun's energy to use as food. The lower on the food chain we eat, the smaller is our ecological footprint, and the more of us Earth can support.

Some animals convert grain feed into milk, eggs, or meat more efficiently than others (**FIGURE 10.9**). Scientists have calculated relative energy-conversion efficiencies for different

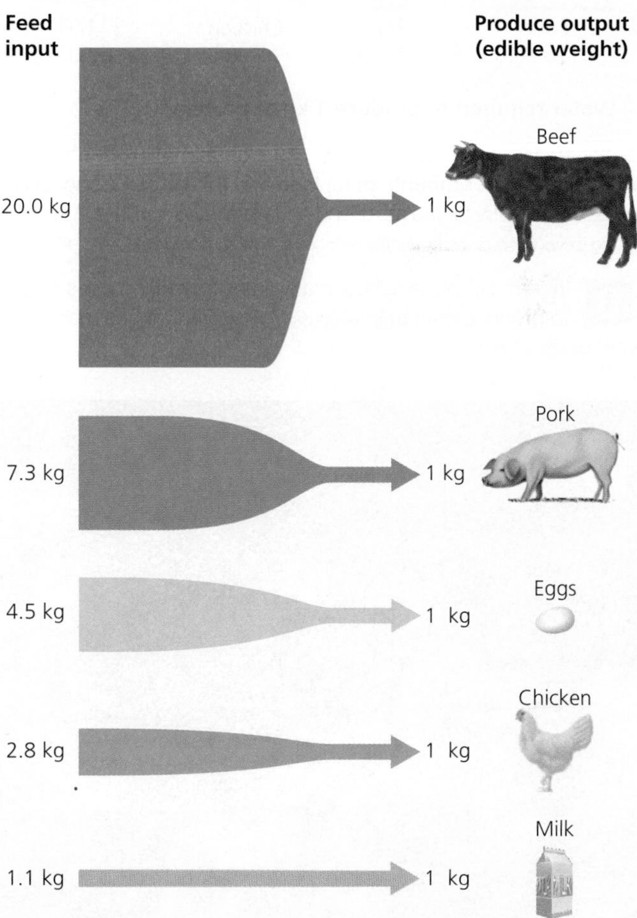

FIGURE 10.9 Producing different animal food products requires different amounts of animal feed. Twenty kilograms of feed must be provided to cattle to produce 1 kg of beef. *Data from Smil, V., 2001.* Feeding the world: A challenge for the twenty-first century. *Cambridge, MA: MIT Press.*

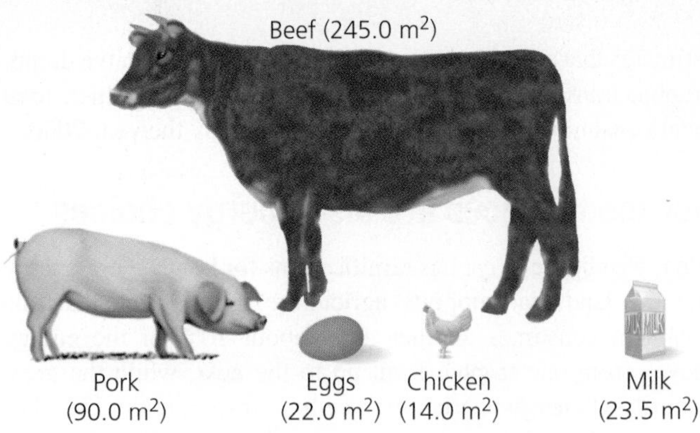

(a) Land required to produce 1 kg of protein

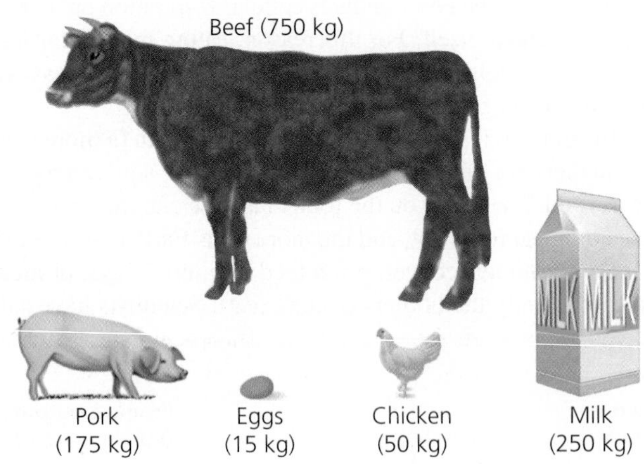

(b) Water required to produce 1 kg of protein

FIGURE 10.10 **Producing different types of animal products requires different amounts of (a) land and (b) water.** Raising cattle for beef requires by far the most land and water. *Data from Smil, V., 2001.* Feeding the world: A challenge for the twenty-first century. *Cambridge, MA: MIT Press.*

DATA Q In terms of protein, how many times more land does it take to produce beef than chicken? How many times more water does it take?

types of animals. Such energy efficiencies have ramifications for land use because land and water are required to raise food for the animals, and some animals require more than others. FIGURE 10.10 shows the area of land and weight of water required to produce 1 kg (2.2 lb) of edible protein for milk, eggs, chicken, pork, and beef. Producing eggs and chicken meat requires the least space and water, whereas producing beef requires the most. Such differences show that when we choose what to eat, we are also indirectly choosing how to make use of resources such as land and water.

Rising demand led to feedlot agriculture

In traditional agriculture, livestock are kept by farming families near their homes or are grazed on open grasslands by nomadic herders or sedentary ranchers. These traditions survive, but the advent of industrial agriculture has brought a new method. **Feedlots,** also known as *factory farms* or *concentrated animal feeding operations (CAFOs),* are essentially huge warehouses or pens designed to deliver energy-rich food to animals living at extremely high densities (FIGURE 10.11). Today nearly half the world's pork and most of its poultry come from feedlots.

Feedlot operations allow for economic efficiency and greater food production, and this makes meat affordable to more people. At the same time, feedlot animals are generally fed grain grown on cropland. One-third of the world's cropland is devoted to growing feed for animals, and 45% of our global grain production goes to livestock and poultry. This elevates the price of staple grains and endangers food security for the very poor.

For environmental quality, feedlots offer one very significant benefit: Taking cattle and other livestock off rangeland and concentrating them in feedlots reduces the grazing impacts they would otherwise exert across large areas of the landscape (pp. 229, 232). Animals that are densely concentrated in feedlots will not contribute to overgrazing. However, intensified animal production through the industrial feedlot model does exert other environmental impacts.

(a) Chicken factory farm in Arkansas

(b) Cattle feedlot in Nebraska

FIGURE 10.11 **Most meat eaten in the United States comes from animals raised in feedlots or factory farms.** These facilities house thousands of **(a)** chickens or **(b)** cattle at high densities. The animals are dosed liberally with antibiotics to control disease.

Livestock agriculture pollutes water and air

Livestock produce prodigious amounts of manure and urine, and their waste can pollute surface water and groundwater. Rich in nitrogen and phosphorus, livestock waste is a common cause of eutrophication (pp. 108–109, 412) in freshwater systems. It can also release a wide array of bacterial and viral pathogens that can sicken people, including *Salmonella*, *E. coli*, *Giardia*, *Microsporidia*, *Pfiesteria*, and pathogens that cause diarrhea, botulism, and parasitic infections.

The crowded conditions under which animals are often kept necessitate heavy use of antibiotics to control disease. Hormones are administered to livestock as well, and feed is spiked with heavy metals that spur growth. Livestock excrete most of these chemicals, which end up in wastewater and may be transferred up the food chain in downstream ecosystems. Some of the chemicals that remain in livestock meat are transferred to us when we eat the meat. In addition, the overuse of antibiotics can cause microbes to evolve resistance to the antibiotics (just as pests evolve resistance to pesticides; p. 255), making these drugs less effective. All in all, the FAO estimates that livestock (including both feedlot and grazed animals) in the United States account for 55% of soil erosion, 37% of pesticide applications, 50% of antibiotics consumed, and one-third of the nitrogen and phosphorus pollution in U.S. waterways.

Feedlot impacts can be minimized when properly managed, and both the EPA and the states regulate U.S. feedlots. Wastewater and manure may be stored in lagoons, where it undergoes a degree of treatment somewhat similar to that of municipal wastewater (pp. 414–415). The resulting sludge may then be applied to farm fields as fertilizer (or injected into the ground where plants need it), reducing the need for chemical fertilizers.

Raising animals for food also results in air pollution. Besides the strong odors that emanate from feedlots and wastewater lagoons, livestock are a major source of greenhouse gases that lead to climate change (pp. 484–486). A comprehensive FAO report in 2006 concluded that livestock agriculture contributes 9% of our carbon dioxide emissions, 37% of our methane emissions, and 65% of our nitrous oxide emissions—altogether, 18% of the emissions driving climate change, a larger share than automobile transportation! Livestock release methane and nitrous oxide in their metabolism and waste. Nitrous oxide is also released from certain feed crops and from fertilizers applied to feed crops. Carbon dioxide is released when forests are cleared for ranching or for growing feed, and when fossil fuels are burned to grow feed, transport animals, and more.

WEIGHING THE ISSUES

FEEDLOTS AND ANIMAL RIGHTS Animal rights activists denounce factory farming because they say it mistreats animals. Chickens, pigs, and cattle are crowded together in small pens their entire lives, fattened up, and slaughtered. Should we concern ourselves with the quality of life of the animals that constitute part of our diet? Do you think animal rights concerns are as important as the environmental issues? Are conditions at feedlots a good reason for being vegetarian?

FIGURE 10.12 People practice many types of aquaculture. Here, fish-farmers tend their animals at a fish farm in China

We raise seafood with aquaculture

Besides growing crops and raising animals on rangeland and in feedlots, we rely on aquatic organisms for food. Wild fish populations are plummeting throughout the world's oceans as increased demand and new technologies lead us to overharvest marine fisheries (pp. 437–442). As a result, raising fish and shellfish on "fish farms" has become necessary to meet our growing demand for these foods (**FIGURE 10.12**).

We call the cultivation of aquatic organisms for food in controlled environments **aquaculture.** Many aquatic species are grown in open water in large, floating net-pens. Others are raised in ponds or holding tanks. Practiced at small scales, community-based aquaculture is a focus of sustainable development efforts in the developing world. At large scales, industrialized aquaculture produces large amounts of food but exerts environmental impacts. People pursue aquaculture with over 220 freshwater and marine species ranging from fish to shrimp to clams to seaweeds (**FIGURE 10.13**).

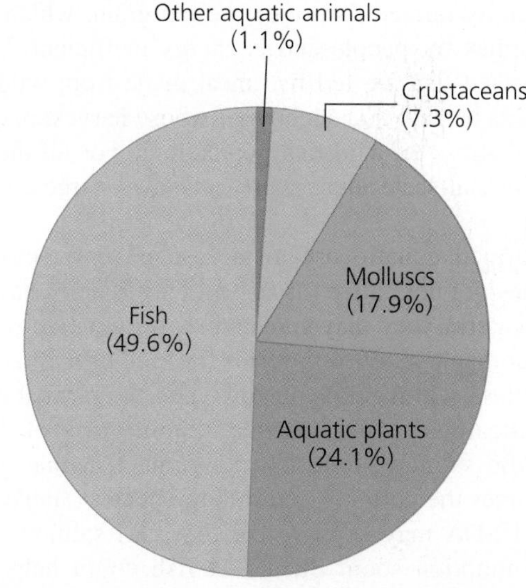

FIGURE 10.13 Aquaculture involves many types of fish, but also a wide diversity of other marine and freshwater organisms. *Data from U.N. Food and Agriculture Organization (FAO).*

Aquaculture is the fastest-growing type of food production; global output has doubled in just the last decade. Most widespread in Asia, aquaculture today produces $125 billion worth of food and provides over three-quarters of the freshwater fish and two-thirds of the shellfish that we eat.

Aquaculture brings benefits

When conducted on a small scale by families or villages, as in China and much of the developing world, aquaculture helps ensure people a reliable protein source. Such small-scale aquaculture can be sustainable, and its compatibility with other activities can make it an excellent path toward sustainable agriculture in general (pp. 269–270). For instance, uneaten fish scraps make excellent fertilizers for crops, and food waste can be fed to fish. At larger scales, aquaculture can help improve a region's or nation's food security by increasing supplies of fish available.

Aquaculture on any scale helps reduce fishing pressure on overharvested and declining wild stocks. Reducing fishing pressure also lessens *bycatch* (p. 439; the unintended catch of nontarget organisms) that results from commercial fishing. Furthermore, aquaculture consumes fewer fossil fuels and provides a safer work environment than does commercial fishing. Fish farming can also be remarkably energy-efficient, producing as much as 10 times more fish per unit area than is harvested from waters of the continental shelf and up to 1000 times more than from the open ocean.

Aquaculture has negative impacts

Along with its benefits, aquaculture has disadvantages. Dense concentrations of farmed animals can increase disease, which reduces food security, necessitates antibiotic treatment, and results in expense. A virus outbreak wiped out half a billion dollars of shrimp in Ecuador in 1999, for instance. Aquaculture can also produce remarkable amounts of waste, both from the farmed organisms and from the feed that goes uneaten and decomposes in the water. Like feedlot livestock, commercially farmed fish often are fed grain, which affects food supplies for people and is energy-inefficient. In other cases, farmed fish are fed fish meal made from wild ocean fish such as herring and anchovies, whose harvest may place additional stress on wild fish populations. For all these reasons, industrial-scale aquaculture can leave a large ecological footprint.

If farmed aquatic organisms escape into ecosystems where they are not native (as several carp species have done in U.S. waters), they may spread disease to native stocks or may outcompete native organisms for food or habitat. These possibilities raise special concern when the farmed animals are genetically modified. Genetic engineering of Atlantic and Pacific salmon produces transgenic fish that grow to several times the normal size for their species (FIGURE 10.14), and the USDA may approve the first GM salmon for sale and consumption soon. Such GM fish could help reduce fishing pressures on wild stocks, but they might also—if they are released or if they escape into the wild—outcompete their wild cousins, interbreed with them, or spread disease to them. Researchers have concluded that in some circumstances, escaped transgenic salmon may increase the extinction risk for native populations of their species, in part because larger male fish (such as those carrying a gene for rapid and excessive growth) have better odds of mating successfully.

FIGURE 10.14 **Transgenic salmon (top) grow considerably larger than wild salmon of the same species.**

Preserving Crop Diversity

Whether one is considering salmon in the Pacific Northwest or maize in Mexico, the prospect of genetically modified organisms crossbreeding with wild relatives and influencing the genetic makeup of the population raises a host of complicated issues. With cropland agriculture, our modern industrial monocultures of genetically similar plants essentially place all our eggs in one basket, such that any single catastrophe might potentially wipe out entire crops. Our wide adoption of industrial agriculture has also reduced the diversity of crops, and today we as a global society rely on a much smaller number of plant types than in decades and centuries past. This is partly why so many people grew concerned about genes from transgenic corn moving into local Mexican landraces of maize.

Crop diversity provides insurance against failure

Preserving the integrity of diverse native crop variants gives us a bulwark against the potential failure of our homogenized commercial crops. The wild relatives of crop plants and their locally adapted landraces contain a diversity of genes that we may someday need to introduce into our commercial crops (through crossbreeding or genetic engineering) to confer resistance to disease or pests or to meet other unforeseen challenges.

Because accidental interbreeding can diminish the diversity of local variants, many scientists argue that we need to protect landraces in areas like southern Mexico that remain important repositories of crop biodiversity. For this reason, the Mexican government helped create the Sierra de Manantlan Biosphere Reserve around an area harboring the world's only population of the plant thought to be the direct

(a) Traditional food plants of the Desert Southwest

(b) Pollination by hand

FIGURE 10.15 Seed banks safeguard the genetic diversity of crop plants. Native Seeds/SEARCH of Tucson, Arizona, preserves seeds of food plants important in traditional diets of Native Americans of the Southwest **(a)**, including chiles, squashes, gourds, maize, lentils, mesquite flour, prickly pear pads, tepary beans, and cholla cactus buds. At the farm where seeds are grown **(b)**, varieties are carefully pollinated by hand to protect their genetic distinctiveness.

ancestor of maize. For this reason, too, it imposed a national moratorium in 1998 on the planting of transgenic corn. However, this ban was lifted in 2009 as multinational agribusiness corporations were allowed to begin experimental plantings in northern Mexico. A Mexican government agency also disburses grain to farmers that includes millions of tons of U.S. corn (one-third of it transgenic), and farmers can also acquire U.S. seed on their own. Given global trade and the increasing use of genetically modified corn worldwide, gene flow between transgenic corn and Mexico's native landraces would seem inevitable at some point.

Worldwide, we have already lost a great deal of genetic diversity in crops in the past century. Only 30% of the maize varieties that grew in Mexico in the 1930s exist today. The number of wheat varieties in China dropped from 10,000 in 1949 to 1000 by the 1970s. In the United States, many fruit and vegetable crops have decreased in diversity by 90% in less than a century. Market forces have discouraged diversity in the appearance of fruits and vegetables: Commercial food processors prefer items to be uniform in size and shape for convenience, and consumers are often wary of unusual-looking food products. However, now that local and organic agriculture are gaining appeal in affluent societies, consumer preferences for diversity (including rare "heirloom varieties" of fruits and vegetables) are increasing.

Seed banks are living museums

Protecting areas and cultures that maintain a wealth of crop diversity is one way to preserve genetic assets for our agriculture. Another is to collect and store seeds from diverse crop varieties. This is the work of **seed banks,** institutions that preserve seed types as a kind of living museum of genetic diversity (**FIGURE 10.15**). These facilities keep seed samples in cold, dry conditions to encourage long-term viability, and they plant and harvest them periodically to renew the stocks.

Major seed banks include the Royal Botanic Garden's Millennium Seed Bank in Britain, the U.S. National Seed Storage Laboratory at Colorado State University, Seed Savers Exchange in Iowa, and the Wheat and Maize Improvement Center in Mexico. In total, 1400 such facilities house 1–2 million distinct types of seeds worldwide.

The most renowned seed bank is the so-called doomsday seed vault established in 2008 on the island of Spitsbergen in Arctic Norway. The internationally funded Svalbard Global Seed Vault (**FIGURE 10.16**) is storing millions of seeds from around the world (spare sets from other seed banks) as a safeguard against global agricultural calamity—"an insurance policy for the world's food supply." This secured, refrigerated facility is built deep into a mountain in an area of permanently frozen ground. The site has no tectonic activity, little natural radiation or humidity, and is high enough above sea level to stay dry even if climate change melts all the planet's ice. The doomsday seed vault is an admirable effort, but we would be well advised not to rely on it to save us. Far better to manage our agriculture wisely and sustainably so that we never need to break into the vault!

FIGURE 10.16 The "doomsday seed vault" in arctic Norway stores seed samples as insurance against global agricultural catastrophe.

Conserving Pollinators, Controlling Pests

One of the most important facets of agriculture is how we handle the many organisms that interact with our crops and livestock. Some organisms are "pests" that pose threats to agriculture, such as insects that feed on crop plants, pathogens that attack livestock, or weeds that compete with crops. Other organisms are beneficial to agriculture. The insects that pollinate crops are among the most vital (yet least appreciated) factors in our food production. Pollinators are the unsung heroes of agriculture. If we are to attain sustainable agriculture, then we will need to find safe and effective ways of limiting losses to pests, and we will need to better conserve the insects that pollinate our crops.

We depend on insects to pollinate crops

Pollination (p. 80) is the process by which male sex cells of a plant (pollen) fertilize female sex cells of a plant (ova, or egg cells); it is the botanical version of sexual intercourse. Plants such as grasses and conifer trees achieve pollination by the wind. Millions of minuscule pollen grains are blown long distances, and by chance a small number land on the female parts of other plants of their species. In contrast, the many kinds of plants that sport showy flowers are typically pollinated by animals, such as hummingbirds, bats, and insects (see Figure 4.8, p. 80). Flowers are, in fact, evolutionary adaptations that function to attract pollinators. The sugary nectar and protein-rich pollen in flowers serve as rewards to lure these sexual intermediaries, and the sweet smells and bright colors of flowers are signals to advertise these rewards.

Our staple grain crops are derived from grasses and are wind-pollinated, but many other crops depend on insects for pollination. The most complete survey to date, by tropical bee biologist Dave Roubik, documented 800 types of cultivated plants that rely on bees and other insects for pollination. An estimated 73% of these types are pollinated by bees, 19% by flies, 5% by wasps, 5% by beetles, and 4% by moths and butterflies. Bats pollinate 6.5%, and birds 4%. Overall, native species of bees in the United States alone are estimated to provide $3 billion of pollination services each year to crop agriculture.

Populations of native pollinators have declined precipitously, however. As one example of many, the U.S. Great Basin states are a world center for the production of alfalfa seed, and alfalfa flowers are pollinated mostly by native alkali bees that live in the soil as larvae. In the 1940s to 1960s, farmers began plowing the land and increasing pesticide use in an effort to boost yields. These measures killed vast numbers of the soil-dwelling bees, and alfalfa seed production plummeted.

Conservation of pollinators is vital

Preserving the biodiversity of native pollinators is especially important today because the domesticated workhorse of pollination, the honeybee (*Apis mellifera*), is also declining. American farmers regularly hire beekeepers to bring colonies of this introduced Old World honeybee to their fields when it is time to pollinate crops (**FIGURE 10.17**). Honeybees pollinate over 100 crops that comprise one-third of the U.S. diet, contributing an estimated $15 billion in services.

FIGURE 10.17 **Beekeepers bring hives of honeybees to farmers' crops when it is time for flowers to be pollinated.**

In recent years, two accidentally introduced parasitic mites have swept through honeybee populations, decimating hives and pushing beekeepers toward financial ruin. On top of this, starting in 2006, entire hives inexplicably began dying off. In each of the last several years, up to one-third of all honeybees in the United States have vanished from what is being called *colony collapse disorder*. Scientists are racing to discover the cause of this mysterious syndrome. Leading hypotheses are insecticide exposure, an unknown new parasite, or a combination of stresses that weaken bees' immune systems and destroy social communication within the hive.

We all can help maintain populations of pollinators by reducing or eliminating the use of chemical pesticides. All insect pollinators are vulnerable to the vast arsenal of insecticides we apply to crops, lawns, and gardens. When people try to control the "bad" bugs that threaten the plants they value, they all too often kill the "good" insects as well. Homeowners can help pollinating insects by planting gardens of flowering plants and by providing nesting sites for bees. Farmers who allow flowering plants (such as clover) to grow around the edges of their fields can maintain a diverse community of insects, some of which will pollinate their crops.

"Pests" and "weeds" hinder agriculture

Although pollinating insects are vital for agriculture, other organisms weaken or destroy our crops or livestock. Throughout the history of agriculture, the insects, fungi, viruses, rats, and weeds that eat or compete with our crops have taken advantage of the ways we cluster food plants into agricultural fields. Pests pose an especially great threat to monocultures, where a pest adapted to specialize on the crop can move easily from plant to plant (see Figure 10.7). From the perspective of an insect that feeds on corn, grapes, or apples, encountering a grain field, vineyard, or orchard is like discovering an endless buffet.

What people term a **pest** is any organism that damages crops that are valuable to us. What we term a **weed** is any plant that competes with our crops. These are subjective categories that we define entirely by our own economic interests. There is nothing inherently malevolent in the behavior of a pest or a weed. These organisms are simply trying to survive and reproduce, but they affect our farm productivity in doing so.

We have developed thousands of chemical pesticides

The highly modified ecosystems of industrial farming limit the ability of natural mechanisms to control pest populations, so farmers need to introduce some type of pest control in order to produce food economically on an industrial scale. In the past half-century, most farmers have turned to chemicals to suppress pests and weeds. In that time we have developed thousands of chemicals to kill insects (*insecticides*), plants (*herbicides*), and fungi (*fungicides*). Such poisons are collectively termed **pesticides.**

All told, roughly 400 million kg (900 million lb) of active ingredients from conventional pesticides are applied in the United States each year. Three-quarters of this total is applied on agricultural land. Since 1960, pesticide use has risen fourfold worldwide. Usage in industrialized nations has leveled off in the past two decades, but it continues to rise in the developing world. Today more than $32 billion is expended annually on pesticides, with one-third of that total spent in the United States. Exposure to synthetic pesticides can have health consequences for people and other organisms under some circumstances (Chapter 14), so their use in food production can have far-reaching effects.

Pests evolve resistance to pesticides

Despite the toxicity of chemical pesticides, their effectiveness tends to decline with time as pests evolve resistance to them. Recall from our discussion of natural selection (pp. 50–53) that organisms within populations vary in their traits. Because most insects, weeds, and microbes can occur in huge numbers, it is likely that a small fraction of individuals may by chance already have genes that enable them to metabolize and detoxify a given pesticide (p. 371). These individuals will survive exposure to the pesticide, while individuals without these genes will not.

Let's say an insecticide application kills 99.99% of the insects in a field. That sounds successful, but it means that 1 in 10,000 survives. If an insect that is genetically resistant to an insecticide survives and mates with other resistant individuals, the genes for insecticide resistance will be passed on to their offspring. As resistant individuals become more prevalent in the pest population, insecticide applications will cease to be effective and the population will grow (**FIGURE 10.18**).

In many cases, industrial chemists are caught up in an evolutionary arms race (p. 79) with the pests they battle, racing to intensify or retarget the toxicity of their chemicals while the armies of pests evolve ever-stronger resistance to their efforts. Because we seem to be stuck in this cyclical process, it has been nicknamed the "pesticide treadmill." As of 2011, among arthropods (insects and their relatives) alone, there were more than 10,000 known cases of resistance by 586 species to over 330 insecticides. Hundreds more weed species and plant diseases have evolved resistance to herbicides and other pesticides. Many species, including insects such as the green peach aphid, Colorado potato beetle, and diamondback moth, have evolved resistance to multiple chemicals.

An additional problem is that pesticides often kill nontarget organisms, including the predators and parasites of the pests. When these valuable natural enemies are eliminated, pest populations become harder to control.

❶ Pests attack crops

❷ Pesticide is applied

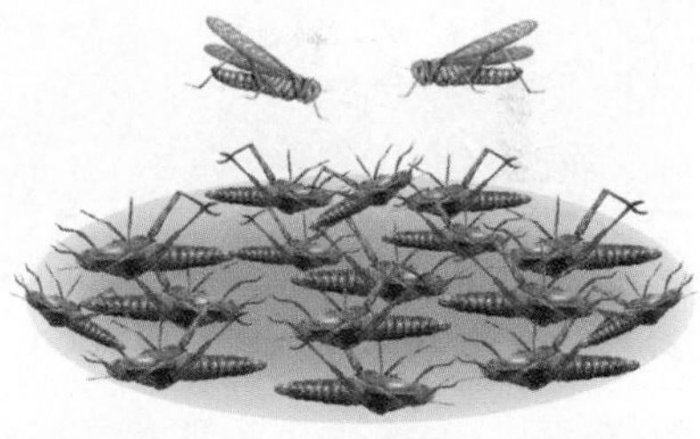
❸ Most pests are killed. A few with innate resistance survive

❹ Survivors breed and produce a pesticide-resistant population

❺ Pesticide is applied again

❻ Pesticide has little effect. New, more toxic, pesticides are developed

FIGURE 10.18 Through the process of natural selection, crop pests often evolve resistance to the poisons we apply to kill them.

Biological control pits one organism against another

Because of pesticide resistance, toxicity to nontarget organisms, and human health risks from some synthetic chemicals, agricultural scientists increasingly battle pests and weeds with organisms that eat or infect them. This strategy, called **biological control** or **biocontrol,** operates on the principle that "the enemy of one's enemy is one's friend." For example, parasitoid wasps (p. 79) are natural enemies of many caterpillars. These wasps lay eggs on a caterpillar, and the larvae that hatch from the eggs feed on the caterpillar, eventually killing it. Parasitoid wasps are frequently used as biocontrol agents and have often succeeded in controlling pests and reducing chemical pesticide use.

One classic case of successful biological control is the introduction of the cactus moth, *Cactoblastis cactorum*, from Argentina to Australia in the 1920s to control invasive prickly pear cactus that was overrunning rangeland. Within just a few years, the moth managed to free millions of hectares of rangeland from the cactus (FIGURE 10.19).

A widespread modern biocontrol effort has been the use of ***Bacillus thuringiensis* (Bt),** a naturally occurring soil bacterium that produces a protein that kills many caterpillars and some fly and beetle larvae. Farmers spray Bt spores on their crops to protect against insect attack. In addition, as we will see (p. 262), scientists have managed to isolate the gene responsible for the bacterium's poison and engineer it into crop plants.

(a) Before cactus moth introduction

(b) After cactus moth introduction

FIGURE 10.19 Photos from the 1920s show an Australian ranch before (a) and after (b) introduction of the cactus moth. Larvae of this moth were used to clear invasive non-native prickly pear cactus from millions of hectares of rangeland.

Biocontrol agents themselves can become pests

When a pest is not native to the region where it is damaging crops, scientists may consider introducing a natural enemy (a predator, parasite, or pathogen) of the pest from its native range, in the expectation that the enemy will attack it. Alternatively, scientists may consider importing a biocontrol agent from abroad that the pest has never encountered, reasoning that the pest has not evolved ways to avoid the biocontrol agent. In either case, this involves introducing an animal or microbe from a foreign ecosystem into a new ecological context. This is risky, because no one can know for certain what effects the biocontrol agent might have. In some cases biocontrol agents have turned invasive and become pests themselves. When this happens, biocontrol organisms are more difficult to manage than chemical controls, because they cannot be "turned off" once they are set loose.

Following the cactus moth's success in Australia, for example, the moth was introduced in other countries to control non-native prickly pear. However, moths introduced to Caribbean islands spread to Florida on their own and are now eating their way through rare native cacti in the southeastern United States. If these moths reach Mexico and the southwestern United States, they could decimate many native and economically important species of prickly pear there.

In Hawaii, wasps and flies have been introduced to control pests at least 122 times over the past century, and biologists Laurie Henneman and Jane Memmott suspected that some of these might be harming native Hawaiian caterpillars that were not pests. They sampled parasitoid wasp larvae from 2000 caterpillars of various species in a remote mountain swamp far from farmland. In this wilderness preserve, they found that fully 83% of the parasitoids were biocontrol agents that had been intended to combat lowland agricultural pests.

Because of concerns about unintended impacts, researchers study biocontrol proposals carefully before putting them into action, and government regulators must approve these efforts. If biological control works as planned, it can be a permanent, effective, and environmentally benign solution. Yet there will never be a sure-fire way of knowing in advance whether a given biocontrol program will work as planned.

Integrated pest management combines biocontrol and chemical methods

As it became clear that both chemical and biocontrol approaches pose risks, agricultural scientists and farmers began developing more sophisticated strategies, trying to combine the best attributes of each approach. **Integrated pest management (IPM)** incorporates numerous techniques, including biocontrol, use of chemicals when needed, close monitoring of populations, habitat alteration, crop rotation, transgenic crops, alternative tillage methods, and mechanical pest removal.

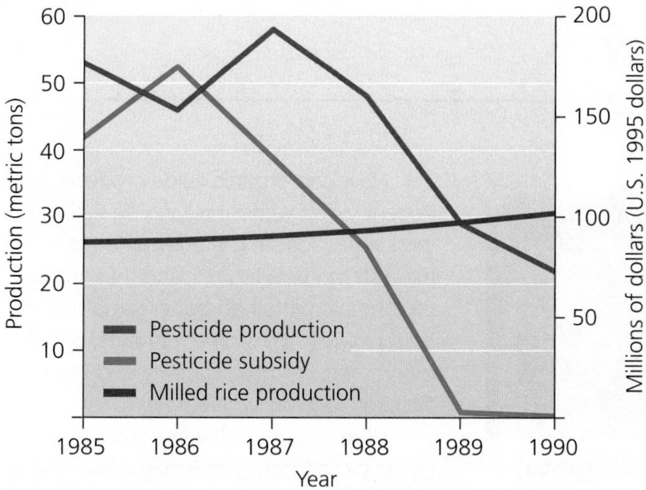

FIGURE 10.20 Once Indonesia threw its weight behind integrated pest management in 1986, pesticide production and imports were reduced, pesticide subsidies were phased out, and yields of rice increased.

IPM has become popular in many parts of the world that are embracing sustainable agriculture. Indonesia stands as an exemplary case (FIGURE 10.20). This nation had subsidized pesticide use heavily for years but came to understand that pesticides were actually making pest problems worse. They were killing the natural enemies of the brown planthopper, which began to devastate rice fields as its populations exploded. Concluding that pesticide subsidies were costing money, causing pollution, and decreasing yields, the Indonesian government in 1986 banned the import of 57 pesticides, slashed pesticide subsidies, and promoted IPM. International experts helped teach Indonesian rice farmers about IPM in "Farmer Field Schools," collaborative groups of farmers who traded information and experimented with new approaches. Within just 4 years, pesticide production fell by half, pesticide imports fell by two-thirds, and subsidies were phased out (saving $179 million annually). Rice yields rose 13%.

Since that time, over 1 million Indonesian farmers have been trained in IPM in Farmer Field Schools, and the approach has spread to dozens of other nations. Across Asia, studies show that farmers in these programs are able to increase crop yields while greatly reducing pesticide use.

Organic Agriculture

Concerns over chemical pesticides and other aspects of industrial agriculture have led many people to support agriculture that involves fewer fossil-fuel inputs and less of the pollution that these inputs cause. *Low-input agriculture* describes farming and ranching that uses lesser amounts of pesticides, fertilizers, growth hormones, antibiotics, water, and fossil fuel energy than in industrial agriculture. This approach seeks to reduce the costs of food production by allowing nature to provide ecosystem services (such as pest control, pollination, and fertilizer) that farmers using industrial methods must pay for themselves. Food-growing practices that use no synthetic fertilizers, insecticides, fungicides, or herbicides—but instead rely on biological approaches such as composting (pp. 616–617) and biocontrol—are termed **organic agriculture.**

FIGURE 10.21 Look for the USDA organic logo to ascertain whether a product is certified organic under the National Organic Program.

Organic approaches reduce inputs and pollution

The bounty of organic agriculture is increasingly available to us (FIGURE 10.21). But what exactly is meant by the term *organic*? In 1990, the U.S. Congress passed the Organic Food Production Act to establish national standards for organic products and facilitate their sale. Under this law, the USDA in 2000 issued criteria by which crops and livestock can be officially certified as organic (TABLE 10.1). These standards went into effect in 2002 as part of the National Organic Program. In crafting a formal definition for the word *organic*, the National Organic Standards Board wrote:

TABLE 10.1 USDA Criteria for Certifying Crops and Livestock as Organic

For crops to be considered organic . . .
• The land must be free of prohibited substances for 3 years.
• Crops must not be genetically engineered.
• Crops must not be irradiated to kill bacteria.
• Sewage sludge cannot be used.
• Organic seeds and planting stock are preferred.
• Farmers must not use synthetic fertilizers. Only crop rotation, cover crops, animal or crop wastes, or approved synthetic materials are allowed.
• Most conventional pesticides are prohibited. Pests, weeds, and diseases should be managed with biocontrol, mechanical practices, or approved synthetic substances.
For livestock to be considered organic . . .
• Mammals must be raised organically from the last third of gestation; poultry, from the second day of life.
• Feed must be 100% organic, although vitamin and mineral supplements are allowed.
• Dairy cows must receive 80% organic feed for 9 months, followed by 3 months of 100% organic feed.
• Hormones and antibiotics are prohibited; vaccines are permitted.
• Animals must have access to the outdoors.

Adapted from the National Organic Program. 2002. Organic production and handling standards. *Washington, DC: U.S. Department of Agriculture.*

THE SCIENCE BEHIND THE STORY

How Productive Is Organic Farming?

Organic farming puts fewer synthetic chemicals into the soil, air, and water than conventional industrial farming does. But can organic farming produce large enough crop yields to feed the human population? The world's two longest-running field experiments on the topic—in Switzerland and in Pennsylvania—suggest that the answer is yes.

Let's first visit Switzerland, where one in every nine hectares of agricultural land is managed organically (the fifth-highest rate in the world). Back in 1977, Swiss researchers established experimental farms at Therwil, near the city of Basel. Long-term studies are rare in agriculture and in ecology. They are highly valuable, because they can reveal slow processes or subtle effects that get swamped out by year-to-year variation in shorter-term studies.

At the Swiss research site, wheat, potatoes, and other crops are grown in plots cultivated in different treatments:

- Conventional farming using chemical pesticides, herbicides, and inorganic fertilizers
- Conventional farming that also uses organic fertilizer (cattle manure)
- Organic farming using only manure, mechanical weeding, and plant extracts to control pests
- Organic farming that also adds natural boosts, such as herbal extracts in compost

Swiss scientist Dr. Paul Mäder (center) visits colleagues at Pennsylvania's Rodale Institute.

Researchers record crop yields at harvest each year. They analyze the soil regularly, measuring nutrient content, pH, structure, and other variables. They also measure the biological diversity and activity of microbes and invertebrates in the soil. Such indicators of soil quality help researchers assess the potential for long-term productivity.

In 2002, Paul Mäder and colleagues from two Swiss research institutes reported in the journal *Science* results from 21 years of data. Over this time, the organic fields yielded 80% of what the conventional fields produced. Organic crops of winter wheat yielded 90% of the conventional yield. Organic potato crops averaged 58–66% of conventional yields because of nutrient deficiency and disease.

Although the organic plots produced 20% less, they did so while receiving 35–50% less fertilizer than the conventional fields and 97% fewer pesticides. Thus, Mäder's team concluded, the organic plots were highly efficient and represent "a realistic alternative to conventional farming systems."

How can organic fields produce decent yields without relying on synthetic chemicals? The answer lies in the soil. Mäder's team found that soil in the organic plots had better structure, better supplies of some nutrients, and much more microbial activity and invertebrate biodiversity (**FIGURE 1**).

Studies are continuing at the Swiss plots today, producing new research results. As one example, Jens Leifeld and two colleagues at a Zurich research institute analyzed soil carbon

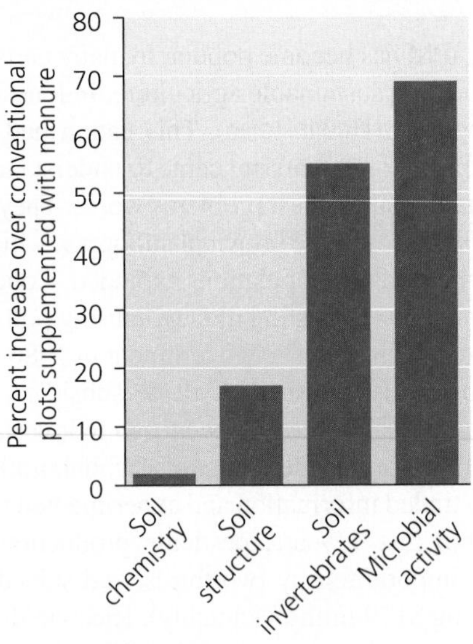

FIGURE 1 Organic fields developed better soil quality than conventional fields supplemented with manure. Values for soil chemistry (6 variables), structure (3 variables), invertebrates (5 variables), and microbial activity (6 variables) were compared. Organic fields outperformed conventional fields without manure (not shown) still more. *Data from Mäder, P., et al., 2002. Soil fertility and biodiversity in organic farming.* Science *296: 1694–1697.*

> Organic agriculture . . . is based on minimal use of off-farm inputs and on management practices that restore, maintain, and enhance ecological harmony [and] minimize pollution from air, soil, and water . . . The primary goal of organic agriculture is to optimize the health and productivity of interdependent communities of soil life, plants, animals, and people.

California, Washington, and Texas established stricter state guidelines for labeling foods organic, and today many U.S. states and over 80 nations have laws spelling out organic standards.

For farmers, organic farming can bring a number of benefits: lower input costs, enhanced income from higher-value produce, and reduced chemical pollution and soil degradation (see **THE SCIENCE BEHIND THE STORY**, above). In many cases more pests attack organic crops because of the lack of chemical pesticides, but biocontrol methods (p. 256) can help keep pests in check. Moreover, the lack of synthetic chemicals maintains

content after 27 years. They found that soil carbon had decreased in all treatments, but that it had declined most in conventional plots. Conventional plots supplemented with manure, however, did just as well as the organic plots in retaining soil carbon.

Organic farming has proved even more successful in Pennsylvania, where the Rodale Institute has compared organic and conventional fields of corn and soybeans in a large-scale experiment running since 1981 on its 330-acre farm. In 2011 it released results from 30 years' worth of data (**FIGURE 2**).

Averaged across the 30 years, yields of organically grown crops equaled yields of conventionally grown crops. Moreover, the organic crops required 30% less energy input. For this reason, raising them released 35% fewer greenhouse gas emissions. Because of lower energy inputs and higher crop prices for organic produce, farming organically created three times the profit for the farmer.

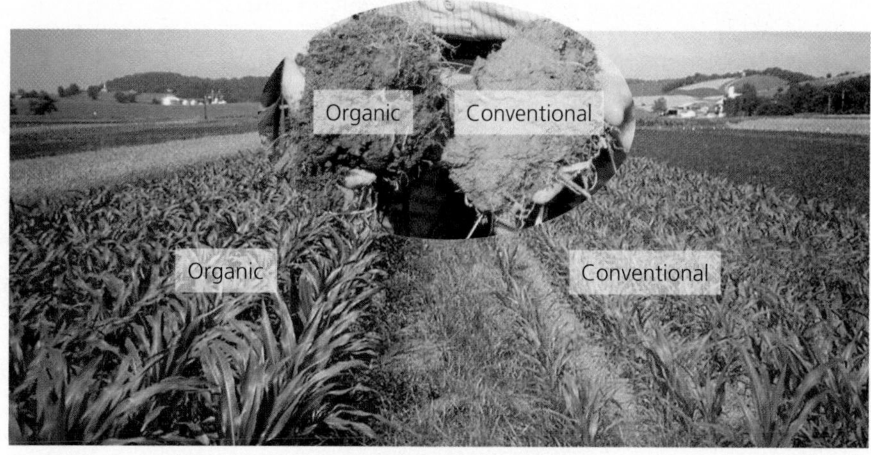

FIGURE 3 Organically grown corn did better than conventionally grown corn during periods of drought in the Rodale Institute experiment. Better-quality soil was part of the reason.

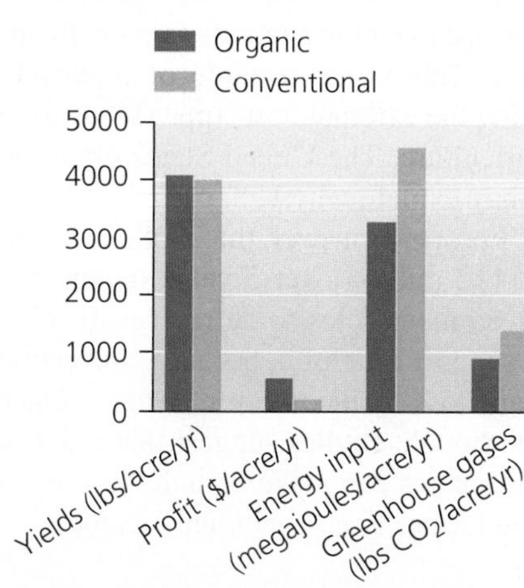

FIGURE 2 Organic crops equaled conventional crops in yields while producing more profit, requiring less energy input, and releasing fewer greenhouse gas emissions. *Data from Rodale Institute, 2011.* The farming systems trial: Celebrating 30 years. *Rodale Institute, Kutztown, Pennsylvania.*

As with the Swiss experiment, the secret lies in the soil. At the Rodale Institute's farm over the years, the soil of the organic fields became visibly darker and better textured than the conventional fields' soil. This helped corn crops in the organic fields to outperform those in the conventional fields during times of drought (**FIGURE 3**).

Shorter-term experiments elsewhere have shown similar results. Researchers comparing organic and conventional farms in North Dakota and Nebraska have found that organic farming produces soils with more microbial life, earthworm activity, water-holding capacity, topsoil depth, and naturally occurring nutrients.

Given such differences in soil quality, researchers expect that organic fields should perform better and better relative to conventional fields as time goes by — in other words, that they are more sustainable. Moreover, the striking yield results of the Rodale experiment suggest that perhaps organic agriculture can feed the world's people every bit as reliably as today's conventional industrial agriculture.

As more long-term data are published and new studies begin, we are learning more and more about the benefits of organic agriculture for soil quality — and about how we might improve conventional methods to maximize crop yields while protecting the long-term sustainability of agriculture. ■

soil quality and encourages pollinating insects. Surveys reveal that farmers who adopt organic techniques do so primarily because they want to practice stewardship toward the land and to safeguard their family's health. Farmers face obstacles to adopting organic methods, however. Foremost among these are the risks and costs of shifting to new methods, particularly during the transition period. For instance, U.S. farmers need to meet standards for three years before their products can be certified.

Many consumers favor organic food out of concern about health risks posed by the pesticides, hormones, and antibiotics used in conventional agriculture. Consumers also buy organic products out of a desire to improve environmental quality. The main obstacle for consumers is price. Organic products tend to be 10–30% more expensive than conventional ones, and some (such as milk) can cost twice as much. However, enough consumers are willing to pay more for organic products that grocers and other businesses are making them more widely available.

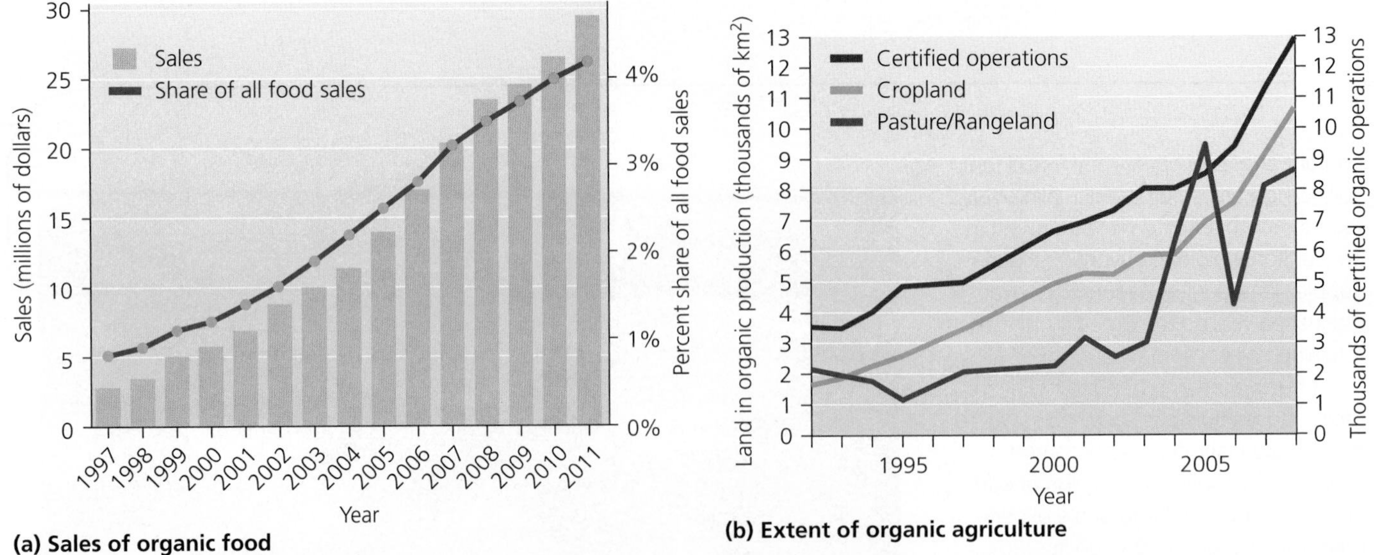

FIGURE 10.22 Organic agriculture is growing. Sales of organic food in the United States **(a)** have increased rapidly, both in total dollar amounts **(bars)** and as a percentage of the overall food market **(line)**. Since the 1990s **(b)**, U.S. acreage devoted to organic crops and livestock has quadrupled, and certified operations have more than tripled. *Sources: (a) Adapted by permission from Willer, Helga (2013)* Organic agriculture worldwide: Current statistics. *FiBL-IFOAM Report. IFOAM, Bonn; FiBL, Frick; ITC, Geneva. Data used with permission from Organic Trade Association OTA: Manufacturer Survey 2012–2013. (b) Data from USDA Economic Research Service.*

Organic agriculture is booming

Just a decade or two ago, few farmers grew organic food, few consumers wanted it, and the only place to buy it was in specialty stores. Today that has changed; three of four Americans buy organic food at least occasionally, more than four of five retail groceries offer it, and Americans are the world's eighth-highest per-person consumers of organic food. In fact, consumer demand for organic food is sometimes so great that farmers cannot keep up with demand, and shortages occur. U.S. consumers spent $29.2 billion on organic food in 2011, amounting to 4.2% of all food sales (FIGURE 10.22a). Worldwide, sales of organic food more than tripled between 2000 and 2010, when sales neared $60 billion.

Production is increasing along with demand (FIGURE 10.22b). Although organic agriculture takes up less than 1% of agricultural land worldwide (37 million ha, or 91 million acres, in 2011), this area is rapidly expanding. Two-thirds of this area is in developed nations, and nearly two-thirds is grazing land. In the United States, nearly 2 million ha (4.8 million acres) are under organic management. Europe boasts still more: 10.6 million ha (26.2 million acres) in 2011, most in nations of the European Union, where 5.4% of the agricultural area is in organic production. Today about 2 million farmers and ranchers in more than 160 nations practice organic agriculture commercially to some extent.

Mexico is a leader in organic agriculture among developing nations. Overall, 1.5% of Mexico's agricultural land is organic, one of the highest percentages for developing nations, and 170,000 Mexican farmers farm organically, the third-highest number in the world. Fully 30% of Mexico's coffee crop is now organic, and Mexico produces more organic coffee than any other nation (FIGURE 10.23). Besides coffee, Mexico grows cocoa and a variety of fruits and vegetables organically. Most of this produce is destined for export to the United States and Europe.

Government initiatives have assisted the growth of organic farming. The European Union supports farmers financially during conversion. This is an example of a subsidy (p. 181) aimed at reducing the external costs (pp. 146, 165) to society of industrial agriculture. The United States offers no such subsidies, which may explain why U.S. organic production lags behind that of Europe. However, the 2008 Farm Bill (p. 238) did set aside $112 million over 5 years for organic agriculture, and the government helps to defray certification expenses. Government support is helpful, because conversion often means a temporary loss in income for farmers. Once conversion is complete, though, studies suggest that reduced inputs and higher market prices can make organic farming at least as profitable for the farmer as conventional methods.

FIGURE 10.23 Mexico is a world leader in the production of organic coffee. These farmers are harvesting coffee beans from an organic plantation of shade-grown coffee in Chiapas, Mexico.

Genetically Modified Food

Organic farming represents one pathway toward sustainable agriculture. **Biotechnology**—the application of biological science to create products derived from organisms—represents another. While organic agriculture seeks to scale down the intensity of industrial agriculture in order to lessen its impacts, biotechnology seeks to scale up the technological aspects of agriculture in order to produce more food at less expense, and to reduce environmental impacts through enhanced efficiency.

The Green Revolution enabled us to feed a greater number and proportion of the world's people, but today relentless population growth is demanding still more innovation. A new set of potential solutions began to arise in the 1980s and 1990s as advances in genetics enabled scientists to directly alter the genes of organisms, including crop plants and livestock. The genetic modification of organisms that provide us food holds promise to enhance nutrition and the efficiency of agriculture while lessening impacts on environmental systems. However, genetic modification may also pose risks that are not yet fully understood, and the increasing role of biotechnology in agriculture has strengthened the influence of multinational corporations over farmers and our food supply. For these reasons, agricultural biotechnology has inspired anxiety and protest by consumer advocates, small farmers, environmental advocates, and critics of big business.

We can genetically modify organisms

The genetic modification of crops and livestock creates genetically modified organisms (GMOs) from which we derive genetically modified (GM) foods. As we learned in our *Central Case Study*, GMOs result from genetic engineering, any process in which scientists directly manipulate an organism's genetic material in the lab by altering segments of its DNA (p. 29). Much of the genetic engineering in agriculture has used **recombinant DNA,** which is DNA that has been patched together from the DNA of multiple organisms. Recombinant DNA technology was developed by scientists studying the bacterium *Escherichia coli*, and **FIGURE 10.24** shows the steps they follow to create genetically modified organisms. The goal is to place genes that produce certain proteins and code for certain desirable traits (such as rapid growth, disease resistance, or high nutritional content) into the genomes of organisms lacking those traits.

For instance, David Quist, Ignacio Chapela, and other researchers examining Mexican maize were looking for a stretch of DNA called the 35S promoter. The 35S promoter was derived from a common natural plant virus, the cauliflower mosaic virus. The 35S promoter directs the expression of many plant genes in nature, and scientists use it in most GM crops that are engineered for resistance to herbicides.

As we learned in our *Central Case Study*, an organism that contains DNA from another species is called a transgenic organism, and the genes that have moved between them are called transgenes. The creation of transgenic organisms is one type of biotechnology. Biotechnology has helped us develop medicines, clean up pollution, understand the causes of cancer and other diseases, dissolve blood clots after heart attacks, and make better beer and cheese. **TABLE 10.2** shows a selection of notable developments in GM foods. The stories behind them illustrate both the promises and pitfalls of food biotechnology.

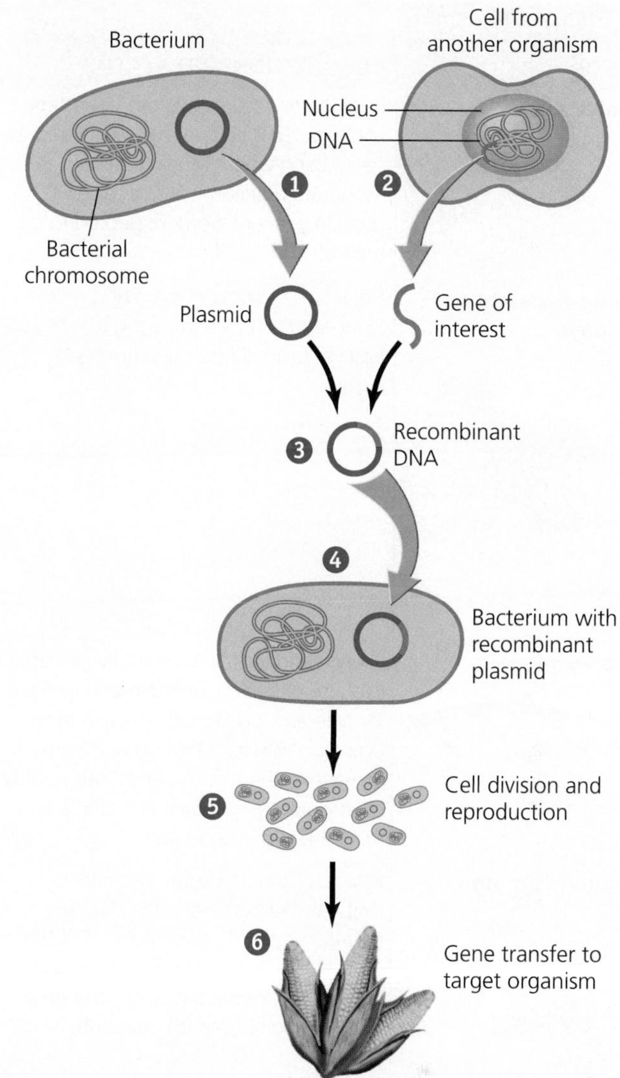

FIGURE 10.24 To create recombinant DNA, scientists follow several steps. First they isolate *plasmids* (1), small circular DNA molecules, from a bacterial culture. DNA containing a gene of interest (2) is then removed from another organism. Scientists insert this gene into the plasmid to form recombinant DNA (3) This recombinant DNA enters new bacteria (4) which reproduce (5) generating many copies of the desired gene. The gene is then transferred to individuals of the target plant or animal (6) and will be expressed in the genetically modified organism as a desirable trait, such as rapid growth or high nutritional content in a food crop.

Genetic engineering is like, and unlike, traditional breeding

The genetic alteration of plants and animals by people is nothing new; through artificial selection (p. 52), we have influenced the genetic makeup of our livestock and crop plants for thousands of years. Our ancestors altered the gene pools of our domesticated plants and animals through selective breeding by preferentially mating individuals with favored traits so

TABLE 10.2 Several Notable Examples of Genetically Modified Food Technologies

CROP	DESCRIPTION AND STATUS	CROP	DESCRIPTION AND STATUS
Golden rice	Engineered to produce beta-carotene to fight vitamin A deficiency in Asia and the developing world. May offer only moderate nutritional enhancement despite years of work. Expected to be marketed in 2013.	**Bt cotton**	Engineered with genes from bacterium *Bacillus thuringiensis* (Bt) that kills insects. Has increased yield, decreased insecticide use, and boosted income for 14 million small farmers in India, China, and other nations.
Virus-resistant papaya	Resistant to ringspot virus and grown in Hawaii. In 2011 became the first biotech crop approved for consumption in Japan.	**Roundup-Ready alfalfa**	One of many crops engineered to tolerate Monsanto's Roundup herbicide (glyphosate). Because the crop can withstand it, the chemical can be applied in great quantities to kill weeds. Unfortunately, many weeds are evolving resistance to glyphosate as a result. Planted in the U.S., 2005–2007, GM alfalfa was then banned because a lawsuit forced the USDA to better assess its environmental impact. Reapproved in 2011.
GM salmon	Engineered for fast growth, large size. Would be the first GM animal approved for sale as food. To prevent fish from breeding with wild salmon and spreading disease to them, company AquaBounty promised to make their fish sterile and raise them in inland pens.	**Roundup-Ready sugar beet**	Tolerant of Monsanto's Roundup herbicide (glyphosate). Swept to dominance (95% of U.S. crop) in just 2 years. As with alfalfa, a lawsuit forced more environmental review, after it had already become widespread. Reapproved in 2012.
Biotech potato	Resistant to late blight, the pathogen that caused the 1845 Irish Potato Famine and that still destoys $7.5 billion of potatoes each year. Being developed by European scientists, but struggling with EU regulations on research.	**Biotech soybean**	The most common GM crop in the world, covering nearly half the cropland devoted to biotech crops. Engineered for herbicide tolerance, insecticidal properties, or both. Like other crops, soybeans may be "stacked" with more than one engineered trait.
Bt corn	Engineered with genes from bacterium *Bacillus thuringiensis* (Bt) that kills insects. One of many Bt crops developed.	**Sunflowers and superweeds**	Research on Bt sunflowers suggests that transgenes might spread to their wild relatives and turn them into vigorous "superweeds" that compete with the crop or invade ecosystems. This is most likely to occur with crops like squash, canola, and sunflowers that can breed with their wild relatives.

that offspring would inherit those traits (p. 217). Early farmers selected plants and animals that grew faster, were more resistant to disease and drought, and produced large amounts of fruit, grain, or meat.

Proponents of GM crops often stress this continuity with our past and say today's GM food is just as safe as selectively bred food. Former USDA head Dan Glickman once remarked:

> Biotechnology's been around almost since the beginning of time. It's cavemen saving seeds of a high-yielding plant. It's Gregor Mendel, the father of genetics, cross-pollinating his garden peas. It's a diabetic's insulin, and the enzymes in your yogurt. . . . Without exception, the biotech products on our shelves have proven safe.

However, as critics are quick to point out, the techniques geneticists use to create GM organisms differ from traditional selective breeding in several ways. For one, selective breeding mixes genes from individuals of the same or similar species, whereas scientists creating recombinant DNA routinely mix genes of organisms as different as viruses and crops, or spiders and goats. For another, selective breeding deals with whole organisms living in the field, whereas genetic engineering works with genetic material in the lab. Third, traditional breeding selects from combinations of genes that come together on their own, whereas genetic engineering creates the novel combinations directly. Thus, traditional breeding changes organisms through the process of selection (pp. 50–53), whereas genetic engineering is more akin to the process of mutation (p. 50).

Biotechnology is transforming the products around us

In just over three decades, GM foods have gone from science fiction to mainstream agribusiness. When recombinant DNA technology was first being developed in the 1970s, scientists debated whether the new methods were safe. They collectively regulated and monitored their own research until most scientists were satisfied that reassembling genes in bacteria did not create dangerous superbacteria. Once the scientific community declared itself confident in the 1980s that the technique was safe, industry leaped at the chance to develop hundreds of applications, from improved medicines (such as hepatitis B vaccine and insulin for diabetes) to designer plants and animals.

Since then, the speed with which GM crops have been adopted and planted across the world has been remarkable (**FIGURE 10.25**). Worldwide, four of every five soybean and cotton plants are now transgenic, as are one of every three corn and canola plants. Globally in 2012, more than 17 million farmers grew GM crops on 170 million ha (420 million acres) of farmland—11% of all cropland in the world.

In the United States today, roughly 90% of corn, soybeans, cotton, and canola consist of genetically modified strains, and close to half of these crops are engineered for more than one trait. It is conservatively estimated that over 70% of processed foods in U.S. stores contain GM ingredients. Thus, it is extremely likely that you consume GM foods on a daily basis.

Most GM crops today are engineered to resist herbicides, so that farmers can apply herbicides to kill weeds without having to worry about killing their crops. Other crops are engineered to resist insect attack. Some are modified for both types of resistance. Resistance to herbicides and pests enables large-scale commercial farmers to grow crops more efficiently, and this is largely why sales of GM seeds to farmers have risen so quickly.

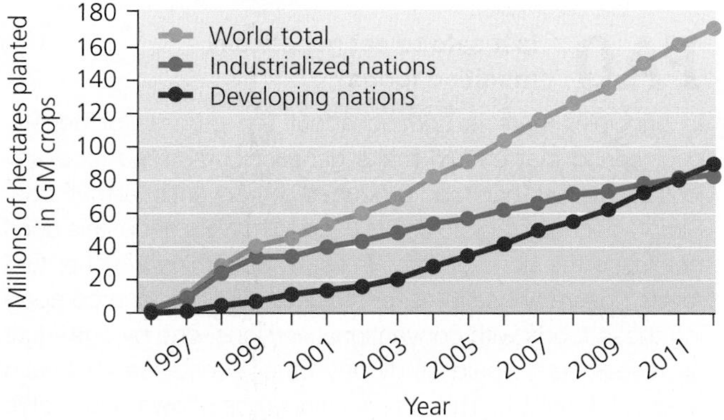

FIGURE 10.25 GM crops have spread with remarkable speed since their commercial introduction in 1996. They now are planted on over 10% of the world's cropland. *Data from the International Service for the Acquisition of Agri-Biotech Applications.*

DATA Q In the last 5 years, have GM crops been growing faster in industrialized nations or in developing nations? If current trends continue, which group of nations will have more GM crops in 2015? Can you estimate how much more they might have?

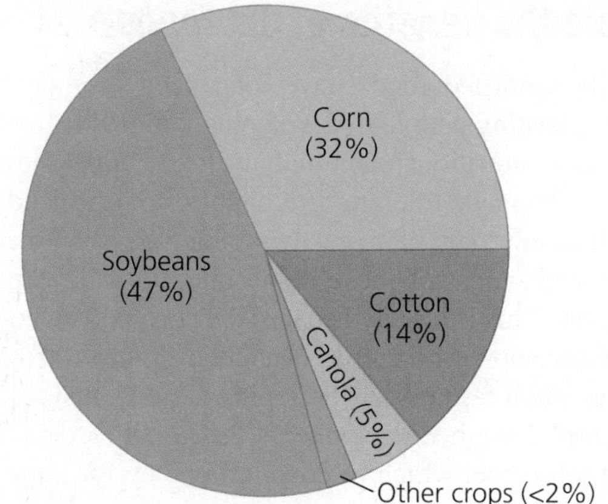

(a) GM crops by type

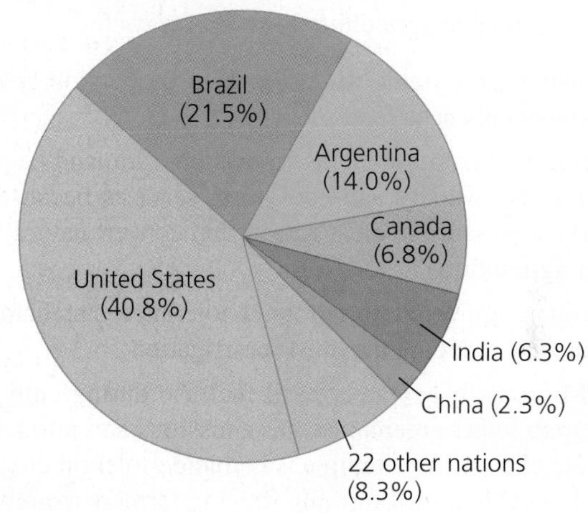

(b) GM crops by nation

FIGURE 10.26 So far, genetic engineering has mainly involved common crops grown in industrialized nations. Of the world's GM crops **(a)**, soybeans are the most common. Of global acreage planted in GM crops **(b)**, the United States devotes the most area. *Data are for 2012, from the International Service for the Acquisition of Agri-Biotech Applications.*

Soybeans are the world's most common GM crop (**FIGURE 10.26a**). Of the 28 nations growing GM crops in 2012, six (the United States, Brazil, Argentina, Canada, India, and China) accounted for over 90% of production, with the United States alone growing 41% of the global total (**FIGURE 10.26b**). However, more than half the world's GM crops are now grown in developing nations.

Many people view biotechnology as a promising avenue toward sustainable agriculture. They feel that GM crops offer economic benefits to poor and rich farmers alike, while reducing environmental impacts. Other citizens, scientists, and policymakers are skeptical and concerned. Some fear the new foods could be dangerous for people to eat. Some are concerned that transgenes may escape and harm nontarget organisms. Others worry that pests will evolve resistance to the supercrops and become "superpests" or that transgenes will be transferred from crops to other plants and turn them into "superweeds." Some, like Quist and Chapela, worry that transgenes might ruin the integrity of native ancestral landraces of crops. What, then, does scientific research tell us about the benefits and the risks of genetically modified organisms?

What are the benefits of GM foods?

Genetically modified foods have long been promoted as a means for assisting people in developing nations. If foods can be made more nutritious, then malnutrition can be alleviated. If crops can be made tolerant of drought or of salinized soils, then small farmers can more easily produce crops on marginal land. However, most of these noble intentions have not yet come to pass. This is largely because the agrobiotech corporations that develop GM varieties cannot easily profit from selling seed to small farmers in developing nations. Instead, most biotech crops have been engineered for insect resistance and herbicide tolerance, which improve efficiency for large-scale industrial farmers.

Regardless, proponents of GM foods maintain that these foods bring environmental and social benefits, assisting efforts toward sustainable agriculture in several ways:

- Raising food yields while lowering production costs for farmers enhances food security for society.
- Crops that increase yields on existing farmland help conserve biodiversity and ecosystem services because they reduce pressure to clear forests and convert natural lands for agriculture.
- Crops engineered for drought tolerance can help save water by reducing the need for irrigation.
- GM crops that reduce fossil fuel use during cultivation help to lower greenhouse gas emissions and mitigate climate change. An example is herbicide-tolerant crops that help enable no-till farming, freeing farmers from having to till to control weeds. No-till farming can reduce erosion, cut down on fossil fuel use, and bring other benefits (pp. 215, 226–228).

In addition, supporters of GM crops maintain that these crops also help to reduce chemical pollution from pesticides. Indeed, adoption of insect-resistant Bt crops appears to reduce the use of chemical insecticides, because farmers use fewer insecticides if their crops do not need them. In India, Bt cotton has enabled 7 million farmers to increase their yields while decreasing their insecticide use.

With herbicide-tolerant crops, however, most studies find that these crops tend to result in *more* herbicide use, because farmers often apply more herbicide if their crops can withstand it. A 2012 study by Washington State University scientist Charles Benbrook using USDA data calculated that as GM crops expanded in the United States between 1996 and 2011, insecticide use declined by 56 million kg (123 million lb), but herbicide use increased by 239 million kg (527 million lb).

This rise in herbicide use is accelerating, according to Benbrook's analysis, and the prime reason is that weeds are evolving resistance to herbicides, causing farmers to apply even more of them. Just as insects can evolve resistance to insecticides (see Figure 10.18, p. 255), weeds can evolve resistance to the chemicals used to kill them. Worldwide, over 200 weed species have evolved resistance to herbicides (**FIGURE 10.27a**).

The most widely used herbicide is Monsanto's Roundup, which contains glyphosate as its active ingredient. Besides producing Roundup, Monsanto engineers seeds for crops that survive spraying with Roundup. For years farmers have loved these Roundup-Ready crops because they reduce labor and boost yields: Simply spray fields with Roundup, and the weeds die while the crops thrive. However, widespread and extended use of Roundup and other glyphosate-based herbicides is resulting in the evolution of resistance (**FIGURE 10.27b**). Scientists have confirmed glyphosate resistance in 24 weed species so far, and farmers feel the number is higher than that. As glyphosate-resistant weeds fill their fields, farmers are dousing their plants with more chemicals, including additional herbicides that are more powerful.

What are the impacts of GM foods?

Although many people worry about health impacts from eating GM foods, persuasive scientific evidence for health impacts has been elusive. Indeed, many millions of Americans eat GM foods every day without obvious signs of harm. For the most part, the public's anxiety over potential health impacts from GM foods has not been corroborated by research. Increased herbicide applications do pose potential health risks, however.

Most scientists feel that ecological impacts of GM foods are a greater issue than health impacts. Many conventional crops can interbreed with their wild relatives (rice can breed with wild rice, for instance), so there seems little reason to believe that transgenic crops would not do the same. In the first confirmed case, GM oilseed rape was found hybridizing with wild mustard. In another case, creeping bentgrass engineered for use on golf courses—a GM plant not yet approved by the USDA—pollinated wild grass up to 21 km (13 mi) away from its test growing site in Oregon. Most scientists think transgenes will inevitably make their way from GM crops into wild plants, but the consequences of this are open to debate.

Because biotechnology is rapidly changing, and because the large-scale introduction of GMOs into our environment is recent, there remains much we don't yet know about the consequences. As a result, it is still too early to dismiss concerns about

FAQ Is it safe to eat genetically modified foods?

In principle, there is nothing about the process of genetic engineering that should make genetically modified food any less safe to eat than food produced by conventional methods. Simply because technology is used to move a gene, this does not make the gene unsafe. Thus, to determine whether GM foods pose any health risks, studies must be done comparing those foods with conventional versions, one by one—just as researchers would study any other substance for health risks (Chapter 14). Thus far, no study has shown undeniable evidence of health impacts from any GM food, but this does not guarantee that all GM foods pose no risks. A great deal of research is controlled by the companies that develop GM foods, and we will never be able to test all foods for all potential effects. However, most scientists agree that the potential health impacts of genetic engineering are likely of less concern than the potential ecological impacts.

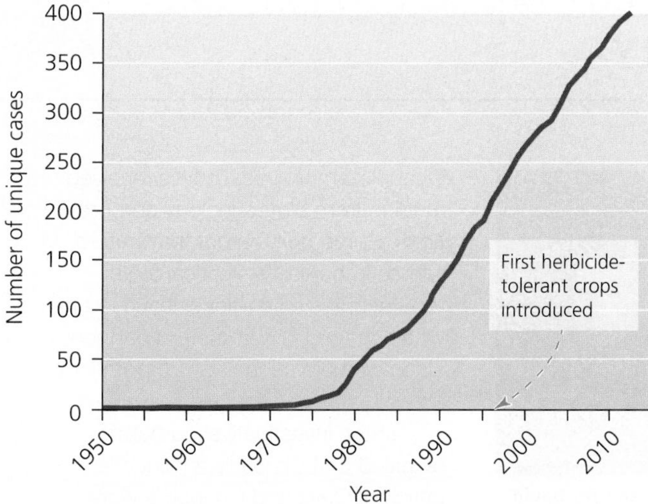

(a) Known cases of herbicide resistance

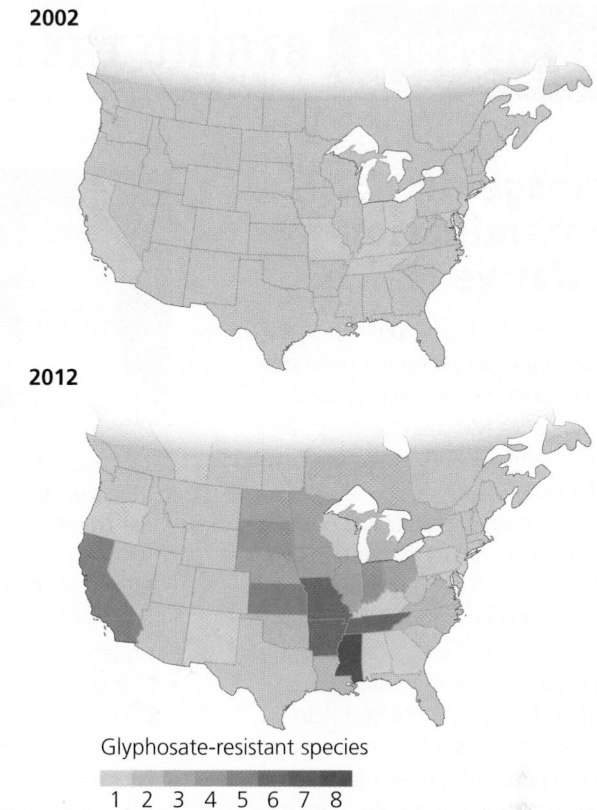

(b) Spread of glyphosate resistance

FIGURE 10.27 Weeds are evolving resistance to herbicides. Documented cases of herbicide resistance have grown **(a)** to 400 biotypes involving over 200 species of plants. In just a decade, weed resistance to glyphosate **(b)** spread across North America. *Data from Heap, I. International Survey of Herbicide Resistant Weeds. June, 2013. www.weedscience.com*

environmental impacts without further scientific research. Many experts feel we should proceed with caution, adopting the **precautionary principle,** the idea that one should not undertake a new action until the ramifications of that action are well understood.

The best efforts so far to scientifically test the ramifications of GM crops were made in Great Britain. The British government, in considering whether to allow the planting of GM crops, commissioned three large-scale studies. The first study found that GM crops could produce long-term economic benefits for Britain, although short-term benefits would be minor. The second study found little to no evidence of harm to human health, but noted that effects on wildlife and ecosystems should be tested before crops are approved. The third study (involving 19 researchers, 200 sites, and $8 million in funding) tested effects on bird and invertebrate populations from four GM crops modified for herbicide resistance. Results showed that fields of GM beets and GM spring oilseed rape supported less biodiversity than fields of their non-GM counterparts. Fields of GM maize supported more, however, and fields of winter oilseed rape showed mixed results. Policymakers had hoped that this study would end all debate, but the science showed that the impacts of GM crops are complex.

Public debate over GM foods continues

Science helps inform us about genetic engineering, but ethical and economic concerns have largely driven the public debate. For many people, the idea of "tinkering" with the food supply seems dangerous or morally wrong. Because every person relies on food for survival and cannot choose *not* to eat, the genetic modification of dietary staples such as corn and rice essentially forces people to consume biotech products or to go to unusual effort to avoid them.

The perceived lack of control over one's own food has driven concern that the global food supply is being dominated by a handful of large corporations that develop GM technologies, among them Monsanto, Syngenta, Bayer CropScience, Dow, DuPont, and BASF. Critics say these multinational corporations threaten the independence and well-being of the family farmer, and that government regulators generally side with big business rather than small farmers. Critics of biotechnology also voice concern that much of the research into the safety of GM organisms is funded, conducted, or influenced by the corporations that stand to benefit if their products are approved.

GM sugar beets provide an example of how the approval process often plays out. After USDA regulators approved Monsanto's Roundup-Ready sugar beets, they were introduced in 2008. The Center for Food Safety, a watchdog organization, brought a lawsuit arguing that the USDA had not performed a full environmental impact statement (EIS; p. 174), and in 2010 a judge agreed and revoked the USDA's approval. However, in the meantime the Roundup-Ready sugar beet had swept the market; 95% of the sugar beets grown in America were now this variety, and half the white sugar on U.S. grocery store shelves came from GM sugar beets. In 2012 the USDA released a full EIS. The EIS admitted that non-GM beets could be contaminated by GM pollen and that weeds were likely to evolve resistance to the herbicide, yet it still argued for full deregulation, allowing GM sugar beets to be planted without any restrictions.

Transgenic Contamination of Native Maize?

David Quist and Ignacio Chapela's *Nature* paper reporting the presence of DNA from genetically engineered corn in native Mexican maize (p. 244) set off controversy the moment it was published.

A number of geneticists questioned their conclusion that transgenes had entered the genomes of native maize landraces in remote areas of Oaxaca. Some of these critics argued that the low levels of transgenic DNA Quist and Chapela detected indicated, at most, first-generation hybrids between local landraces and transgenic varieties. Other critics noted that the technique they relied on, the polymerase chain reaction (PCR), is highly sensitive to contamination and careless laboratory practices.

Quist and Chapela had also suggested that the invading genes had split up and spread throughout the maize genome. They had used a technique called inverse PCR (iPCR) and had reported that the transgenes were surrounded by essentially random sequences of DNA. Several geneticists argued that iPCR was unreliable, that the pair had used insufficient controls, and that their results could have arisen from similarities between the transgenes and stretches of maize DNA.

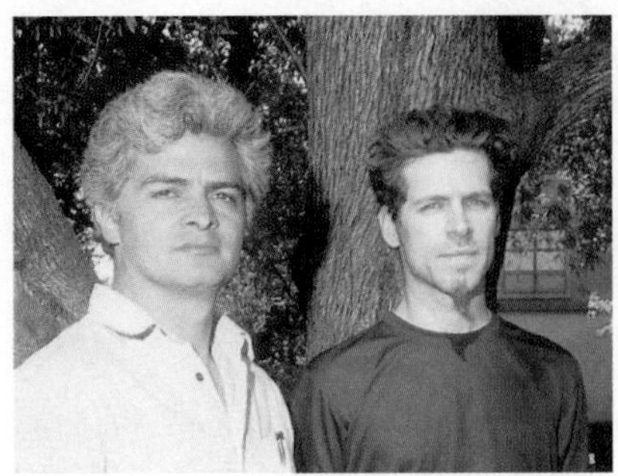

Dr. Ignacio Chapela (left) and Dr. David Quist (right)

In April 2002, *Nature* published two letters from scientists critical of the study and stated that "the evidence available is not sufficient to justify the publication of the original paper."

Quist and Chapela responded by acknowledging that some of their initial findings based on iPCR were likely invalid. But they presented new analyses to support their fundamental claim and pointed to a Mexican government study that also found transgenes in Oaxacan maize.

The correspondence published in *Nature* set off a broader debate on the editorial pages of scientific journals, in newspapers and magazines, on the Internet, and in the halls of academe. GM proponents emphasized that Chapela had spoken out against his university's plan to enter into a $25-million partnership with the biotechnology firm Novartis (now Syngenta), sparking a debate that had divided the UC Berkeley faculty. GM opponents countered that Quist and Chapela's most vocal critics received funding from biotechnology corporations. Some pointed out that *Nature*, whose impartiality is critical to its reputation as a first-tier science journal, was engaged in commercial partnerships with biotechnology corporations as well.

Later, investigative journalists revealed that the earliest influential critics of Quist and Chapela's work were actually biotech PR people hiding behind false identities and posing as scientists. Supporters of Chapela and Quist noted similarities with the way other researchers raising questions about genetic engineering had been personally attacked by industry PR firms and industry-funded peers.

Meanwhile, hoping to resolve whether transgenes had in fact entered Mexican maize, dozens of other researchers began work on the issue. In 2005, a research team led by Sol Ortiz-García of Mexico's Instituto Nacional de Ecologia and Allison Snow of The Ohio State University published results from an extensive survey of maize from across Oaxaca, in the *Proceedings of the National Academy of Sciences of the USA*. They analyzed 154,000 seeds from 870 maize plants in 125 fields at 18 localities in 2003 and 2004, yet they detected no evidence of transgenes. Their statistical analysis indicated with 95% certainty that if transgenes were present at all, they likely occurred in fewer than 1 out of 10,000 seeds.

In 2013, biotech companies appeared to gain even more power when a rider in a budget bill passed by the U.S. Senate stripped the courts of the ability to revoke USDA approval of any GM crop found to be unsafe. Dubbed the "Monsanto Protection Act" by its critics, it inspired a groundswell of opposition from food safety advocates and citizens who viewed it as a violation of the government's system of checks and balances.

So far, GM crops have not lived up to their promise of feeding the world's hungry. Nearly all GM crops are engineered to express either pesticidal properties (e.g., Bt crops) or herbicide tolerance. Crops with traits that might benefit poor farmers of developing countries (such as increased nutrition, drought tolerance, and salinity tolerance) have not been widely commercialized, likely because corporations have less economic incentive to do so. Whereas the Green Revolution was a largely public venture, the "gene revolution" promised by genetic engineering is largely driven by the financial interests of corporations selling proprietary products.

Indeed, corporations patent the transgenes they develop and go to great lengths to protect their investments. Thousands of North American farmers have found themselves at the mercy of the Monsanto Company. This first came to light in Canada, after Monsanto's private investigators took seed samples from canola plants grown by Saskatchewan farmer Percy Schmeiser (**FIGURE 10.28**) and charged him with violating Canada's law that makes it illegal for farmers to reuse patented

To explain how their results could differ so greatly from Quist and Chapela's findings, Ortiz-García's team suggested that transgenes could have become rarer since 2000–2001 either (1) by chance, (2) by hybridization with local landraces, (3) if weeded out by natural selection, or (4) if farmers stopped planting GM seed once extension agents began warning them about it.

Then in 2009, a new research team announced that they had found transgenes in Mexican maize. This team, led by Alma Piñeyro-Nelson and Elena Álvarez-Buylla of the Universidad Nacional Autónoma de México, had collected maize from Oaxaca in 2001, 2002, and 2004, analyzing it with multiple methods. They found evidence of transgenes in three localities. Intensive sampling at two of these localities showed transgenes from 11 of 60 fields, and in roughly 1% of the samples.

In explaining why their team found transgenes whereas Ortiz-García's had not, Piñeyro-Nelson's group mentioned that a company that had run genetic tests for both research teams used methods liable to overlook evidence for transgenes. The company, they maintained, was likely producing "false negatives" by being too conservative.

Their company's reputation on the line, the vice president and the CEO of the firm, Genetic ID, penned a letter to the editor of *Molecular Ecology*, the journal that had published the paper. Nothing was wrong with the firm's techniques, they argued; instead, Piñeyro-Nelson's group had likely contaminated the samples and was claiming "false positives." Piñeyro-Nelson's team responded with a thorough defense and presented new data to support its conclusions. This included tests applying herbicide to plants, revealing that some showed resistance presumably induced by the transgene.

By 2010, two other papers published by more researchers presented additional evidence of transgenes in Mexican maize (**FIGURE 1**). The continuing debate reveals the difficulty researchers sometimes face in achieving certainty in their data and consensus on its interpretation, especially when the stakes are high and powerful commercial interests are involved. However, it increasingly appears that transgenes have in fact moved into native Mexican maize, at low frequencies in scattered locations. Whether this will have substantial effects on agriculture, health, or the environment remains unknown. ■

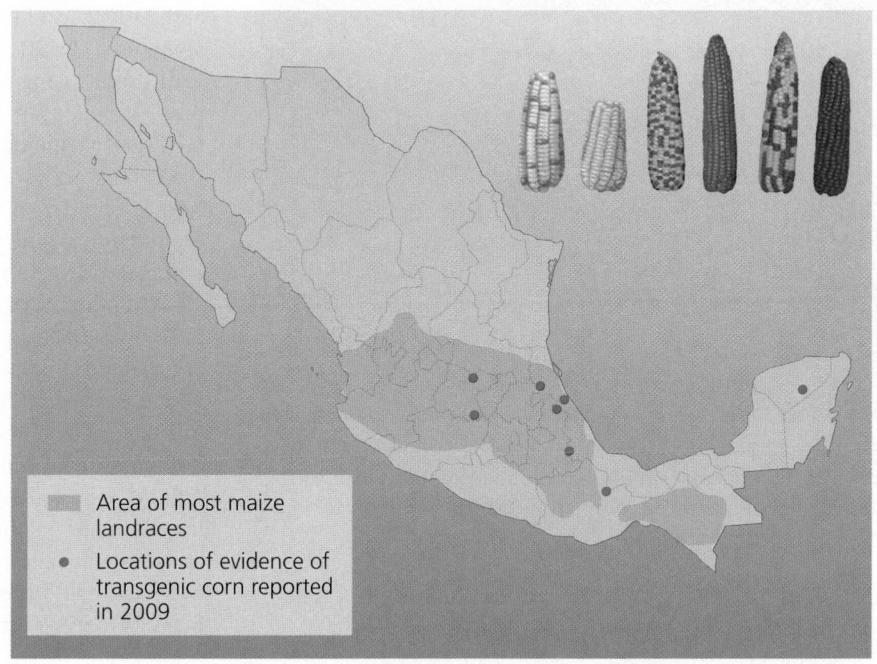

FIGURE 1 Reports of transgenic contamination of Mexican maize have been growing. Mapped are incidences (red dots) reported in a peer-reviewed 2009 research paper. Green areas are regions of greatest density of native maize landraces. *Adapted from Dyer, G.A., et al., 2009. Dispersal of transgenes through maize seed systems in Mexico. Fig 1.* PLoS ONE *4(5) e5734:1–9.*

seed or grow the seed without a contract with the company. Schmeiser maintained that pollen from Monsanto's Roundup-Ready canola used by his neighbors blew onto his land and pollinated his non-GM canola. When Schmeiser harvested his seed and replanted it the next year, many of the plants that grew contained Monsanto's patented herbicide-resistance gene, even though Schmeiser says he did not want this. Monsanto sued Schmeiser, and the court sided with Monsanto, ordering the farmer to pay the corporation $238,000. Schmeiser appealed the case to Canada's Supreme Court, which ruled, 5–4, that Schmeiser had violated Monsanto's patent.

In 2008, Schmeiser and Monsanto reached an out-of-court settlement whereby Monsanto agreed to pay all clean-up costs of the contamination. But the company continued suing other farmers. As of 2010, Monsanto had launched 145 such lawsuits against several hundred farmers and farm companies and had reached out-of-court settlements with several thousand other farmers, resulting in payments of tens of millions of dollars. Most of these farmers were sued for saving seeds from one harvest to another—something farmers have done from time immemorial, yet is now illegal with seeds that contain patented genes. Monsanto says it is merely demanding that farmers heed the patent laws. North Dakota farmer Tom Wiley sees it differently, saying, "Farmers are being sued for having GMOs on their property that they did not buy, do not want, will not use, and cannot sell."

FIGURE 10.28 Saskatchewan farmer Percy Schmeiser became a hero to small farmers and opponents of biotech foods after fighting back against the Monsanto Company in court.

Today there is widespread concern that the burgeoning organic food market will be hindered if organic farms are contaminated with pollen or seed from GM plants. An organic farmer suffering such cross-breeding would not be able to sell his or her produce as organic. For this reason, organic farmers in 2011 launched a class-action lawsuit against Monsanto to try to remove the threat of being sued for unintentional patent infringement. A U.S. Court of Appeals ruled in Monsanto's favor in 2013.

Given such developments, the future of GM foods seems likely to hinge on social, economic, legal, and political factors as well as scientific ones—and these factors vary in different nations. In Europe, consumers have expressed widespread unease about genetic engineering. In contrast, American consumers have largely accepted GMOs, generally without even realizing that the majority of the food they eat contains GM products.

Opposition to GM foods in Europe blocked the import of hundreds of millions of dollars in U.S. agricultural products from 1998 to 2003. The United States complained that the European Union was hindering free trade and brought a successful case before the World Trade Organization (p. 179). In 2013, exports of the $8-billion U.S. wheat crop were threatened after unapproved GM wheat plants were found on an Oregon farm, an apparent vestige of Monsanto field trials. Japan and South Korea suspended purchases of U.S. wheat in response, and Europe geared up to begin testing U.S. wheat. To date, the United States remains one of the few nations not to sign the *Cartagena Protocol on Biosafety*, a treaty that lays out guidelines for open information about exported crops.

Many nations label GM foods

More than 60 nations require that GM foods be labeled so that consumers know what they are buying. In contrast, the United States does not label GM foods and has for years opposed other nations' efforts to do so. As a result, Americans receive less information about the GM content of their food than Europeans or citizens of nations like China, Russia, and Saudi Arabia. This is despite the fact that polls consistently show that a large majority of Americans would like their food to be labeled. A petition requesting labeling sent to the U.S. Food and Drug Administration in 2011 garnered over 1 million signatures. In 2012, a proposition on the statewide ballot in California to mandate labeling of GM food was favored to pass. It ended up failing by a 53–47% vote after food, biotech, and pesticide industries funneled over $45 million into the campaign to defeat the measure, outspending proponents 5 to 1. The defeat in California motivated labeling advocates around the country, however, and soon labeling bills were being considered by legislators in half of U.S. states.

Proponents of labeling argue that consumers have a right to know what's in the food they buy. Opponents of labeling argue that doing so implies that labeled foods are dangerous, whereas research has not shown that to be the case. They also say that consumers wishing to avoid GM foods can do so by buying organic foods and that labeling will entail expense. In nations where labeling has been allowed, stores have ended up eliminating some GM foods from their shelves because of consumer aversion to them. If this were to occur in the United States, it could pressure food producers to avoid using—and farmers to avoid growing—GM foods.

WEIGHING THE ISSUES

DO YOU WANT YOUR FOOD LABELED? How would you have voted in the California ballot initiative to label genetically modified foods? Given that well over 70% of processed food now contains GM ingredients, labeling GM foods would add costs and might force these foods off store shelves. Yet dozens of other nations currently label such foods. Do you want your food to be labeled to indicate whether it is genetically modified? Would you choose among foods based on such labeling? Why or why not?

Sustainable Food Production

Growing enough food for our burgeoning population while maintaining the integrity of the environmental systems that support our agriculture is a tremendous challenge. Fortunately, there are a variety of approaches we can follow.

Employing biotechnology to increase crop yields while reducing environmental impacts is one strategy that can lead toward more sustainable food production. Although the contributions of genetic engineering to sustainable agriculture thus far are debatable, the approach has tremendous potential for the future if we can harness and direct it toward serving people's needs. Biotechnology gives us one way we can make our agriculture more sustainable by modifying aspects within the currently dominant industrial model.

At the other end of the spectrum, organic agriculture promotes sustainability by largely rejecting the industrial model. By eliminating inputs of fossil-fuel-based chemicals and fertilizers, organic agriculture reduces impacts on the land and on our health.

Whether we can feed several billion people in a global urbanized society with organic agriculture is an open question, however.

Approaches toward sustainable agriculture differ greatly, but all aim to address adverse environmental and social impacts. These include soil degradation, overuse of water, fossil fuel combustion, loss of crop diversity, loss of biodiversity, loss of pollinators, overuse of chemical pesticides and fertilizers, and pollution from feedlot and aquaculture operations.

Locally supported agriculture is growing

Many proponents of sustainable agriculture have called attention to the fossil fuel energy we use to transport food long distances. Those who tally "food miles" say the average food product sold in a U.S. supermarket travels at least 1600 km (1000 mi) between the farm and the grocery. Indeed, our extensive transportation infrastructure distributes a wide variety of foods reliably to all areas but, in doing so, burns large amounts of petroleum. Because of the travel time, supermarket produce is often chemically treated to preserve freshness and color.

In response, increasing numbers of farmers and consumers are supporting local small-scale agriculture. Farmers' markets (**FIGURE 10.29**) are springing up throughout North America as people rediscover the joys of fresh, locally grown produce. At **farmers' markets,** consumers buy meats and fresh fruits and vegetables in season from local producers. These markets generally offer a wide choice of organic items and unique local varieties not found in supermarkets.

FIGURE 10.29 Farmers' markets are flourishing as consumers rediscover the benefits of buying fresh, locally grown produce.

Some consumers are partnering with local farmers in **community-supported agriculture (CSA).** In a CSA program, consumers pay farmers in advance for a share of their yield, usually a weekly delivery of produce. Consumers get fresh seasonal produce, while farmers get a guaranteed income stream up front to invest in their crops—a welcome alternative to taking out loans and being at the mercy of the weather.

Farmers markets and community-supported agriculture help strengthen local economies while giving the consumer access to fresher foods. They also often cut down on food miles. However, the number of miles a given item has traveled is not in itself a reliable measure of fossil fuel consumption. In many cases, shipping a food item far away as part of a large shipment may result in *less* fossil fuel use per item than transporting the item a short distance but in small quantities. The heirloom tomato at the farmer's market that was grown 20 miles away may be more delicious than the conventional one at the grocery store that traveled halfway across the country, yet its production, transport, and sale may sometimes involve more fossil fuel use, not less.

To determine which alternative involves less energy use overall, one needs to conduct a **life-cycle analysis,** a quantitative analysis of the inputs and outputs across all stages of an item's production, transport, sale, and use. Performing a full life-cycle analysis (p. 623) is a complex endeavor, and we can expect results to vary from case to case. In general, research so far has shown that fruits and vegetables consume the least fossil fuel energy, that grains use more, that eggs and chicken use still more, and that beef and dairy products use the most.

In the most comprehensive life-cycle analysis of U.S. food production and delivery so far, researchers found that food miles from producer to retailer contributed just 4–5% of total greenhouse gas emissions of the entire process. Fully 83% of emissions resulted from production at the farm or feedlot. As a result, the average consumer can reduce his or her ecological footprint more effectively by shifting dietary choices (such as eating more fruits and vegetables and less meat and dairy) than by eating locally sourced food, these researchers maintain. In **Calculating Ecological Footprints** (pp. 272–273), you will work with some of this data yourself.

Sustainable agriculture mimics natural ecosystems

The best approach for making an agricultural system sustainable is to mimic the way a natural ecosystem functions. Ecosystems operate in cycles and are internally stabilized with negative feedback loops (pp. 106–107). In this way they provide a useful model for agriculture.

One example comes from Japan, where some small-scale rice farmers are reviving ancient traditions and finding them superior to modern industrial methods. Takao Furuno is one such farmer. Starting 20 years ago, he and his wife added a crucial element to their rice paddies: the crossbred aigamo duck. Each spring after they plant rice, the Furunos release hundreds of aigamo ducklings into their paddies (**FIGURE 10.30**). The ducklings eat weeds that compete with the rice, as well as insects and snails that attack the rice. The ducklings also fertilize the rice plants with their waste and oxygenate the water by paddling. Furuno and the scientists and extension agents who have worked with him have found that in rice paddies that have ducks, rice plants grow larger and yield far more rice. Once the rice grains form, the ducks are removed from the paddies (because they would eat the rice grains) and kept in sheds, where they are fed waste grain. They mature, lay eggs, and can be sold at market.

FIGURE 10.30 Ducks help rice crops grow in a sustainable agricultural system developed in Japan, known as the *aigamo* method.

Besides ducks, Furuno raises fish in the paddies, and these provide food and fertilizer as well. He also lets the aquatic fern *Azolla* cover the water surface. This plant fixes nitrogen, feeds the ducks, hides the fish, and provides habitat for insects, plankton, and aquatic invertebrates, which provide further food for the fish and ducks. Because fast-growing *Azolla* can double its biomass in 3 days, surplus plant matter is harvested and used as cattle feed. The end result is a productive ecosystem in which pests and weeds are transformed into resources, and which yields organic rice, eggs, and meat from ducks and fish. From 2 hectares of paddies and 1 hectare of organic vegetables, the Furunos annually produce 7 tons of rice, 300 ducks, 4000 ducklings, and enough vegetables to feed 100 people. At this rate—twice the productivity of the region's conventional farmers—just 2% of Japan's people could supply the nation's food needs.

Takao Furuno wrote a book to popularize the "aigamo method," and today over 10,000 Japanese farmers are using it, increasing their yields 20–50% and regaining huge amounts of time they used to spend manually weeding. The approach is now spreading to other Asian nations. Furuno's approach shows how working with nature rather than against it can produce success in agriculture.

Treating agricultural systems as ecosystems is a key aspect of sustainable agriculture, and this general lesson applies regardless of location, scale, or the crop involved. Southern Mexico's organic coffee farmers promote an ecosystem approach by letting insect pests be eaten by birds, spiders, and other insects, rather than chemically poisoning all of them and their soil as well. Coffee-growers practicing shade-grown techniques, in which coffee bushes are grown under a partial canopy of trees, enhance this sustainable ecosystem model further. Likewise, Mexico's traditional maize farmers who cultivate native landraces are preserving and promoting the genetic diversity that typifies natural systems. Efforts like these, together with other approaches farmers around the world are taking to help make agriculture sustainable, will be crucial for all of us as we progress through the coming century.

Conclusion

Many practices of intensive commercial agriculture exert substantial environmental and social impacts. At the same time, it is important to realize that aspects of industrial agriculture have helped to relieve certain pressures on land or resources. Whether Earth's natural systems would be under more pressure from 7 billion people practicing traditional agriculture or from 7 billion people living under our modern industrialized model is a very complex question.

What is certain is that if our planet is to support 9 billion people by mid-century without further degrading the soil, water, pollinators, and other resources and ecosystem services that support our food production, we must find ways to shift to sustainable agriculture. Approaches such as biological pest control, organic agriculture, pollinator conservation, preservation of native crop diversity, sustainable aquaculture, and likely some degree of careful and responsible genetic modification of food may all be parts of the game plan we will need to set in motion to work together toward a sustainable future.

Reviewing Objectives

You should now be able to:

Explain the challenge of feeding a growing human population

- Our food production has outpaced our population growth, yet 870 million people still go hungry each year. (pp. 245–246)
- Undernutrition, overnutrition, and malnutrition are all challenges to a goal of food security. (pp. 245–246)
- Growing crops for biofuels can result in food shortages and price increases. (p. 248)
- We can work toward sustainable agriculture in a variety of ways. (p. 248–249)

Identify the goals, methods, and consequences of the Green Revolution

- The Green Revolution aimed to enhance agricultural productivity in developing nations. (p. 247)
- Scientists used selective breeding to develop crop strains that grew quickly, were more nutritious, or were resistant to disease or drought. (p. 247)

- The increased efficiency of production fed more people while reducing the amount of natural land converted for farming. (pp. 247–248)
- The expanded use of fossil fuels, chemical fertilizers, and synthetic pesticides enhanced yields but has also increased pollution and soil degradation. (p. 248)

Assess how we raise animals for food

- As wealth has increased, so has consumption of animal products. (p. 249)
- Eating animal products leaves a greater ecological footprint than eating plant products. (pp. 249–250)
- Feedlots create waste and pollution, but they also relieve pressure on lands that could otherwise be overgrazed. (pp. 250–251)
- Aquaculture provides economic benefits and food security, relieves pressures on wild fish stocks, and can be sustainable. (pp. 251–252)
- Aquaculture also gives rise to pollution, habitat loss, and other environmental impacts. (p. 252)

Describe approaches for preserving crop diversity

- Protecting diversity of native crop varieties provides insurance against failure of major crops. (pp. 252–253)
- Seed banks preserve rare and local varieties of seed, acting as storehouses for genetic diversity. (p. 253)

Discuss the importance of pollination

- Insects and other organisms are essential for the reproduction of many crop plants. (p. 254)
- Conservation of pollinating insects is vitally important to our food security. (p. 254)

Explore strategies for pest management

- We kill "pests" and "weeds" with synthetic chemicals that can pollute the environment and pose health hazards. (pp. 254–255)
- Pests can evolve resistance to chemical pesticides, leading us to design more toxic poisons. (p. 255)
- The practice of biological control employs natural enemies of pests against them. (p. 256)
- Integrated pest management combines various techniques and minimizes the use of synthetic chemicals. (pp. 256–257)

Analyze the nature, growth, and potential of organic agriculture

- Organic agriculture, because it reduces chemical and fossil fuel inputs, exerts fewer environmental impacts than industrial agriculture. (pp. 257–259)
- Scientific studies show that organic agriculture is productive and is a realistic alternative to industrial agriculture. (pp. 258–259)
- Organic agriculture comprises a small part of the market but is growing rapidly. (p. 260)

Describe the science behind genetic engineering

- Genetic modification uses recombinant DNA technology to move genes for desirable traits from one type of organism to another. (p. 261)
- Genetic engineering is both like and unlike traditional selective breeding. (pp. 261–262)
- GM foods have spread rapidly and now account for a large portion of our food supply. (p. 263)
- Biotech crops can offer benefits for sustainable agriculture, although these have not reached their potential. (p. 264)
- GM crops can have ecological impacts, including the spread of transgenes and an increase in herbicide pollution. (pp. 264–265)

Evaluate the public debate over genetically modified food

- Many people have ethical qualms about altering food through genetic engineering. (p. 265)
- Development of biotech foods has been controlled by multinational biotechnology corporations, which many critics view as a threat to the independence of small farmers. (pp. 265–268)
- Debate continues over whether to label GM foods in the marketplace. (p. 268)

Summarize pathways to sustainable agriculture

- A variety of approaches—from biotechnology to organic agriculture—give us ways to pursue sustainable agriculture. (p. 265–268)
- Locally supported agriculture, as shown by farmers' markets and community-supported agriculture, is growing in popularity. (p. 269)
- Mimicking natural ecosystems is a key approach to making agriculture sustainable. (pp. 269–270)

Testing Your Comprehension

1. What kinds of techniques have people employed to increase agricultural production? How did Norman Borlaug help inaugurate the Green Revolution?
2. Name several positive and negative environmental consequences of feedlot operations. Why is beef an inefficient food from the perspective of energy consumption?
3. What are some economic benefits of aquaculture? What are some negative environmental impacts?
4. About how many and what types of cultivated plants are known to rely on insects for pollination? Why is it important to preserve the biodiversity of native pollinators?
5. Explain how pesticide resistance occurs.
6. Explain the concept of biocontrol. List several components of a system of integrated pest management (IPM).
7. Define *organic agriculture*, and describe what a farmer must do to get his or her food certified as organic. What factors are driving the growth of organic agriculture?
8. What is recombinant DNA? How is a transgenic organism created? How is genetic engineering different from traditional agricultural breeding? How is it similar?
9. Describe several reasons why many people support the development of genetically modified organisms, and name several uses of such organisms that have been developed so far.
10. Describe the scientific concerns held by opponents of GM crops. Describe some of their other concerns.

Seeking Solutions

1. Assess several ways in which high-input industrial agriculture can be beneficial for environmental quality and several ways in which it can be detrimental. Now suggest several ways in which we might modify industrial agriculture to mitigate its environmental impacts.
2. Describe two fundamental approaches for preserving the genetic diversity of crop variants. Give a specific example of each approach. What benefits and risks do you see in each approach?
3. What factors make for an effective biological control strategy of pest management? What risks are involved in biocontrol? If you had to decide whether to use biocontrol against a particular pest, what questions would you want to have answered before you decide?
4. People who view GM foods as solutions to world hunger and insecticide overuse often want to speed their development and approval. Others adhere to the precautionary principle and want extensive testing for health and environmental safety. How much caution do you think is warranted before a new GM crop is introduced?
5. **THINK IT THROUGH** You are a USDA official and must decide whether to allow the planting of a new genetically modified strain of cabbage that is tolerant to a best-selling herbicide and has twice the vitamin content of regular cabbage. What questions would you ask of scientists before deciding whether to approve the new crop? What scientific data would you want to see? Would you also consult nonscientists or consider ethical, economic, and social factors?
6. **THINK IT THROUGH** You are a farmer in the U.S. Midwest, own 1000 acres, and intend to farm corn for a living. Would you choose to grow genetically modified corn? How would you manage pests and weeds? Would you grow corn for people's food, livestock feed, or biofuels? Think of the various ways corn is grown, purchased, and valued in different places—such as the United States, Europe, and southern Mexico—as you formulate your answer.

Calculating Ecological Footprints

Many people who want to reduce their ecological footprint have focused on how much energy is expended (and how many climate-warming greenhouse gases are emitted) in transporting food from its place of production to its place of sale. The typical grocery store item is shipped by truck, air, and/or sea for many hundreds of miles before reaching the shelves, and this transport consumes oil. This concern over "food-miles" has helped drive the "locavore" movement to buy and eat locally sourced food.

However, food's transport from producer to retailer, as measured by food-miles, is just one source of carbon emissions in the overall process of producing and delivering food.

In 2008, environmental scientists Christopher Weber and H. Scott Mathews conducted a thorough life-cycle analysis of U.S. food production and delivery. By filling in the table below, you will get a better idea of how our dietary choices contribute to climate change.

Food Type	Total emissions[1] across life cycle	Emissions[1] from delivery[2]	Percent emissions from delivery[2]
Fruits & vegetables	0.85	0.10	11.8
Cereals & carbohydrates	0.90	0.07	
Dairy products	1.45	0.03	
Chicken/fish/eggs	0.75	0.03	
Red meat	2.45	0.03	
Beverages	0.50	0.04	

[1] Emissions are measured in metric tons of carbon-dioxide-equivalents, per household per year.

[2] "Delivery" means transport from producer to retailer.

Weber, C.L., and H.S. Mathews, 2008. Food-miles and the relative climate impacts of food choices in the United States. Environmental Science and Technology *42: 3508–3513.*

1. Which type of food is responsible for the most greenhouse gas emissions across its whole life cycle? Which type is responsible for the least emissions?
2. What is the range of values for percent of emissions from transport of food to retailers (delivery)? Weber and Mathews found that 83% of food's total emissions came from its production process on the farm or feedlot. What do these numbers tell you about how you might best reduce your own footprint with regard to food?
3. After measuring mass, energy content, and dollar value for each food type, the researchers calculated emissions per kilogram, calorie, and dollar. In every case, red meat produced the most emissions, followed by dairy products and chicken, fish, and eggs. They then calculated that shifting one's diet from meat and dairy to fruits, vegetables, and grains for just one day per week would reduce emissions as much as eating 100% locally (cutting food-miles to zero) all the time. Knowing all this, how would you choose to reduce your own food footprint? By how much do you think you could reduce it?

STUDENTS

Go to **MasteringEnvironmentalScience** for assignments, the etext, and the Study Area with practice tests, videos, current events, and activities.

INSTRUCTORS

Go to **MasteringEnvironmentalScience** for automatically graded activities, current events, videos, and reading questions that you can assign to your students, plus Instructor Resources.

Wildebeest crossing the vast plains of the Serengeti

11

Biodiversity and Conservation Biology

Upon completing this chapter, you will be able to:

- Characterize the scope of biodiversity on Earth
- Understand today's extinction crisis in geologic context
- Evaluate the primary causes of biodiversity loss
- Specify the benefits that biodiversity brings us
- Assess the science and practice of conservation biology
- Analyze efforts to conserve threatened and endangered species
- Compare and contrast conservation efforts above the species level

CENTRAL CASE STUDY

Will We Slice through the Serengeti?

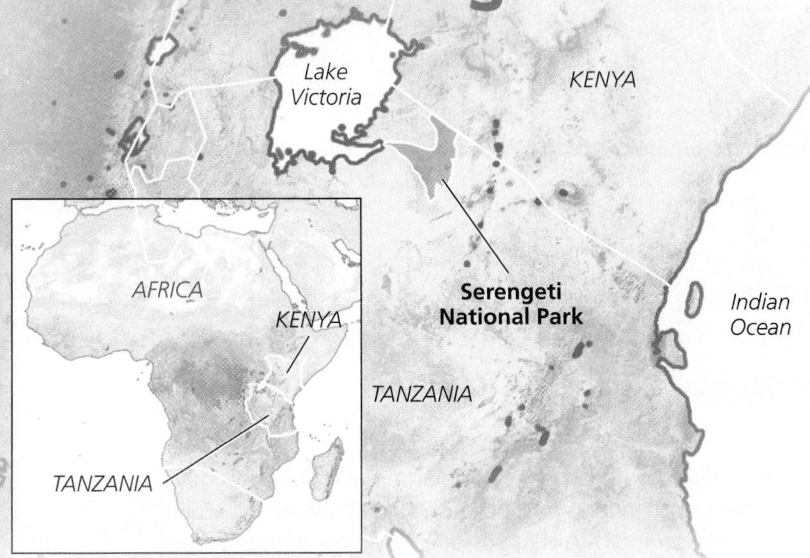

"Construction of the road will be a huge relief for us. We will sell . . . maize and horticultural products to our colleagues in Arusha and they will bring us cows and goats."

—Bizare Mzazi, a small farmer outside Serengeti National Park

"If we construct this road, all our rhinos will disappear. . . . We should strive to conserve our heritage for future generations."

—Sirili Akko, executive officer of the Tanzania Association of Tour Operators

It's been called the greatest wildlife spectacle on Earth. Each year over 2 million wildebeest, zebras, and antelope migrate across the vast plains of the Serengeti in East Africa. After bearing their calves in the wet season, the wildebeest journey north to find fresh grass, in herds that can stretch as far as the eye can see. Packs of lions track the procession and pick off the weak and the unwary. The wildebeest ford rivers where hungry crocodiles wait in ambush. The great herds spend the dry season at the northern end of the Serengeti, and then they turn back southward to complete their cyclical annual journey.

This epic migration, with its dramatic interplay between predators and prey, has inspired countless TV documentaries, and it has cycled on unbroken for millennia. Yet today, many scientists and conservationists are worried that the entire phenomenon is threatened. They are alarmed by a proposal to build a commercial highway across the Serengeti, slicing straight across the animals' migratory route.

Before examining the highway proposal, let's step back for a broad view of the Serengeti. The people native to this region, the Maasai, are semi-nomadic herders who have long raised cattle on the grasslands and savannas. Because the Maasai subsisted on their cattle and lived at low population densities, wildlife thrived here long after it had declined in other parts of Africa.

When East Africa was under colonial rule, the British created game reserves to conserve wildlife. Once Tanzania, Kenya, and other African nations gained independence in the mid-20th century, the British reserves became the basis for today's national protected areas. Serengeti National Park was established in 1951, and the Maasai Mara National Reserve was later established just across the border in Kenya. These two protected areas, together with several adjacent ones, encompass the Serengeti ecosystem. This 30,000-km^2 (11,500-mi^2) region is one of the last places on the planet where an ecosystem remains nearly intact and functional over a vast area.

Today 2 million people visit Tanzania and Kenya each year, most of them ecotourists who visit the parks and protected areas. Serengeti National Park alone receives 800,000 visitors each year. Tourism pumps close to $3 billion into these nations' economies and creates jobs for tens of thousands of local people. As such, the parks have become mainstays of these national economies.

Because the region's people see that functional ecosystems full of wildlife bring foreign dollars into their communities, many are supportive of the parks. Indeed, East Africa has been at the forefront of community-based conservation (p. 302), in which local people steward their own resources, often in collaboration with international conservationists.

However, most people living in northern Tanzania remain desperately poor. Small farmers, villagers, and townspeople along the shores of Lake Victoria feel isolated from the rest of Tanzania by a poor road system. Walled off by Serengeti National Park to their east (which does not allow commercial truck traffic on its few dirt roads), these people have little access to outside markets to buy and sell goods. In response, Tanzania's president Jakaya Kikwete promised them he would build a paved commercial highway across the Serengeti. The highway would connect towns on Lake Victoria with cities to the east and ports on the Indian Ocean. The World Bank and the German government offered to finance the $480-million project, and Chinese contractors stood ready to build it.

Around the world, conservationists reacted with alarm. The proposed highway would slice right through the middle of the wildebeest migration path (**FIGURE 11.1**). Scientists predicted that the road and its vehicles would physically block migration

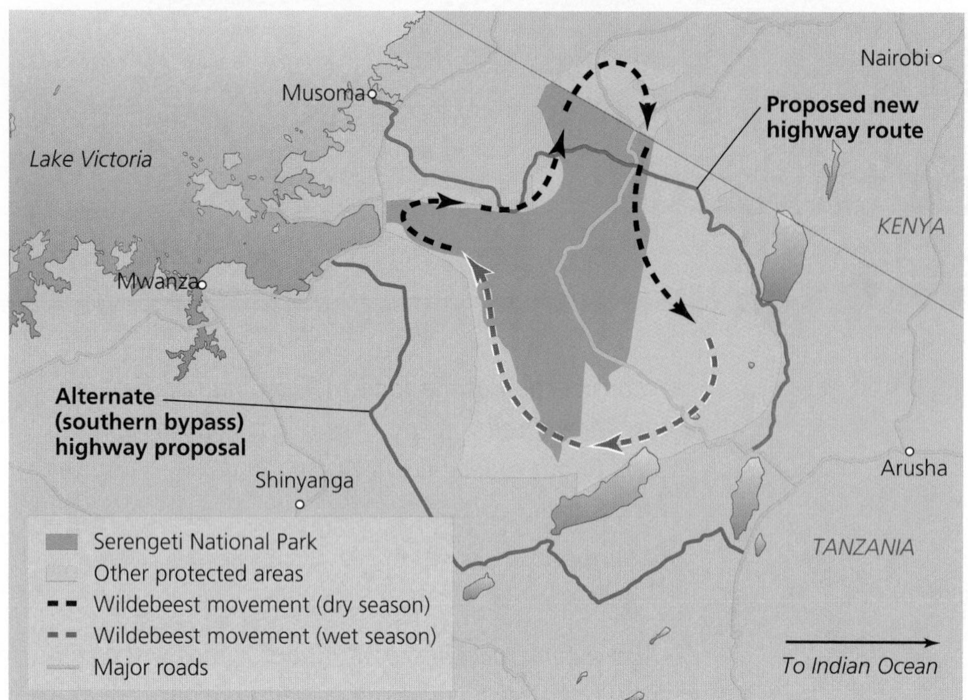

FIGURE 11.1 The proposed highway would run through the northern part of Serengeti National Park. It would connect Tanzanian people on each side of the park but would cut across the migration route for wildebeest and other animals. Highway opponents suggest an alternate route around the park's southern edge.

routes and kill countless animals in collisions. A highway would also provide access for **poaching,** the illegal killing of wildlife for meat or body parts. Likewise, it would allow an entry corridor for exotic plant species that could invade the ecosystem. A highway would encourage human settlement right up to the park boundary, making the park an island of habitat hemmed in by agriculture, housing, and commerce. And by boosting development, the towns on Lake Victoria would grow into large cities, creating demand for still-larger transportation corridors in the future. For all these reasons, experts predicted that the highway would diminish animal populations and possibly destroy the migration spectacle.

Such an outcome could devastate tourism, so the region's tourism operators opposed the highway. So did Kenyans, who feared that the highway would prevent migratory animals from crossing into Kenya's Maasai Mara Reserve. In 2010 a Kenyan nongovernmental organization, the African Network for Animal Welfare, sought to stop the highway with a lawsuit in the East Africa Court of Justice, a body set up to adjudicate international matters in the region. The court in 2011 issued a permanent injunction prohibiting the road project. Tanzania challenged the court's capacity to rule on the matter, but the court held fast to its order.

Meanwhile, international pressure rose on Tanzania to abandon its plans. Highway opponents proposed alternative routes, including a bypass wrapping around the Serengeti's southern end. Although longer, this southern route would pass through more towns, serving five times as many people along the way. The World Bank and the German government offered to finance feasibility studies and help fund this alternative route.

In 2011, the Tanzanian government announced it was abandoning plans to run a paved highway through the park. Instead it would pave the highway outside the park, and leave the 54-km (34-mi) portion running through the park unpaved. Conservationists rejoiced, but in 2012 the region's people called on the government to keep pursuing the paved highway project, and Kikwete announced his support.

The Serengeti is world renowned for its large charismatic animals and for its distinction as one of Earth's last intact and functional large ecosystems. Scientists are finding that Africa's wildlife is declining rapidly (even within its parks), so whatever impacts a Serengeti highway might exert would have global ramifications for how we understand biological diversity on Earth. We would all be impoverished if the Serengeti's biodiversity is lost, so we must hope that Africans can find ways to improve their people's standard of living while conserving the wildlife and natural systems around them. East Africa has helped to pioneer win-win solutions in conservation thus far, so perhaps it will show the way yet again. ■

Our Planet of Life

Rising human population and resource consumption are putting ever-greater pressure on the flora and fauna of our planet. We have been diminishing Earth's diversity of life, the very quality that makes our planet so special. At the same time, many people around the world are working tirelessly to save threatened animals, plants, and ecosystems in efforts to stem the loss of our planet's priceless biological diversity.

Biological diversity, or **biodiversity,** is the variety of life across all levels of biological organization, including the diversity of species, genes, populations, and communities (p. 53). In this chapter we will examine current biodiversity trends and their relevance to our lives. We will then explore science-based solutions for conserving biodiversity.

Biodiversity encompasses multiple levels

Biodiversity is a concept as multifaceted as life itself, and biologists employ different working definitions according to their own aims and philosophies. Yet scientists agree that the concept applies across the major levels in the organization of life (**FIGURE 11.2**). The level that is easiest to visualize and most commonly used is species diversity.

Species diversity A **species** (p. 49) is a distinct type of organism, a set of individuals that uniquely share certain characteristics and can breed with one another and produce fertile offspring. Biologists use differing criteria to distinguish one species from another. Some biologists emphasize characteristics that species share because of common ancestry, whereas others emphasize the ability to interbreed. The spirited debate on this topic could easily fill a textbook in itself! In practice, however, scientists generally agree on species identities.

We can express **species diversity** in terms of the number or variety of species in a particular region. One component of species diversity is *species richness*, the number of species. Another is *evenness* or *relative abundance*, the extent to which species differ in numbers of individuals (greater evenness means they differ less).

Speciation (p. 53) generates new species, whereas extinction (pp. 58–59) diminishes species richness. Immigration, emigration, and local extinction may change species richness locally, but only speciation and extinction change it globally.

Biodiversity exists below the species level in the form of *subspecies*, populations of a species that occur in different geographic areas and differ from one another in some characteristics. Subspecies arise by the same processes that drive speciation (pp. 53–54) but result when divergence stops short of forming separate species. As an example, the black rhinoceros has diversified into four subspecies, each of which live in different parts of Africa. The eastern black rhino, which is native to Kenya and Tanzania, has characteristics slightly different from the western black rhino, the southwestern black rhino, and the southern-central black rhino.

Ecosystem diversity

Species diversity

Genetic diversity

FIGURE 11.2 The concept of biodiversity encompasses multiple levels in the hierarchy of life.

Genetic diversity Scientists designate subspecies when they recognize substantial, genetically based, differences among individuals from different populations of a species. However, all species consist of individuals that vary genetically from one another to some degree, and this genetic diversity is an important component of biodiversity. **Genetic diversity** encompasses the differences in DNA composition (p. 29) among individuals.

Genetic diversity provides the raw material for adaptation to local conditions. In the long term, populations with more genetic diversity may be more likely to persist, because their variation better enables them to cope with environmental change.

Populations with little genetic diversity are vulnerable to environmental change because they may happen to lack genetic variants that would help them adapt to novel conditions. Populations with low genetic diversity may also be more vulnerable to disease and may suffer *inbreeding depression*, which occurs when genetically similar parents mate and produce weak or defective offspring. Scientists have sounded warnings over low genetic diversity in species that have dropped to low population sizes, including American bison, elephant seals, and the cheetahs of the East African plains. The full consequences of reduced diversity in these species remain to be seen. Diminished genetic diversity in our crop plants is a prime concern to humanity (pp. 252–253).

Ecosystem diversity Biodiversity encompasses levels above the species level, as well. **Ecosystem diversity** refers to the number and variety of ecosystems (pp. 60, 110), but biologists may also refer to the diversity of biotic communities (pp. 60, 80) or habitats (p. 61) within some specified area. Scientists may also consider the geographic arrangement of habitats, communities, or ecosystems at the landscape level, including the sizes and shapes of patches and the connections among them (pp. 113–115). Under any of these concepts, a seashore of beaches, forested cliffs, offshore coral reefs, and ocean waters would hold far more biodiversity than the same acreage of a monocultural cornfield. A mountain slope whose vegetation changes with elevation from desert to forest to alpine meadow would hold more biodiversity than a flat area the same size consisting of only desert, forest, or meadow.

The Serengeti's open plains are vast, but the region holds a diversity of habitats, including savanna (p. 97), grassland (p. 95), hilly woodlands, seasonal wetlands, and rock outcroppings called *kopjes*. This habitat diversity contributes to the rich diversity of species in the region.

Some groups hold more species than others

Species are not evenly distributed among taxonomic groups. In number of species, insects show a staggering predominance over all other forms of life (**FIGURE 11.3** and **FIGURE 11.4**). Within insects, about 40% are beetles, and beetles alone outnumber all non-insect animals and all plants. No wonder the British biologist J.B.S. Haldane famously quipped that God must have had "an inordinate fondness for beetles."

In some groups, many species evolved rapidly as populations spread across a variety of environments and adapted to local conditions. Other groups diversified because of a tendency to become subdivided by barriers that promote speciation (pp. 53–54). Still other groups accumulated species through time because of low rates of extinction.

Many species await discovery

Scientists often express biodiversity in terms of species richness because that component is most easily measured and is a good gauge for overall biodiversity. Yet we still are profoundly ignorant of the number of species that exist. So far, scientists have identified and described about 1.8 million species of plants, animals, and microorganisms. However, most of Earth's species remain undiscovered. Estimates for the total number that actually exist range from 3 million to 100 million, with the most widely accepted estimates in the neighborhood of 14 million.

Our knowledge of species numbers is incomplete for several reasons. First, many species are tiny and easily overlooked. These include bacteria, nematodes (roundworms), fungi, protists, and soil-dwelling arthropods. Second, many organisms are so difficult to identify that ones thought to be identical sometimes turn out, once biologists look more closely, to be multiple species. This is frequently the case with

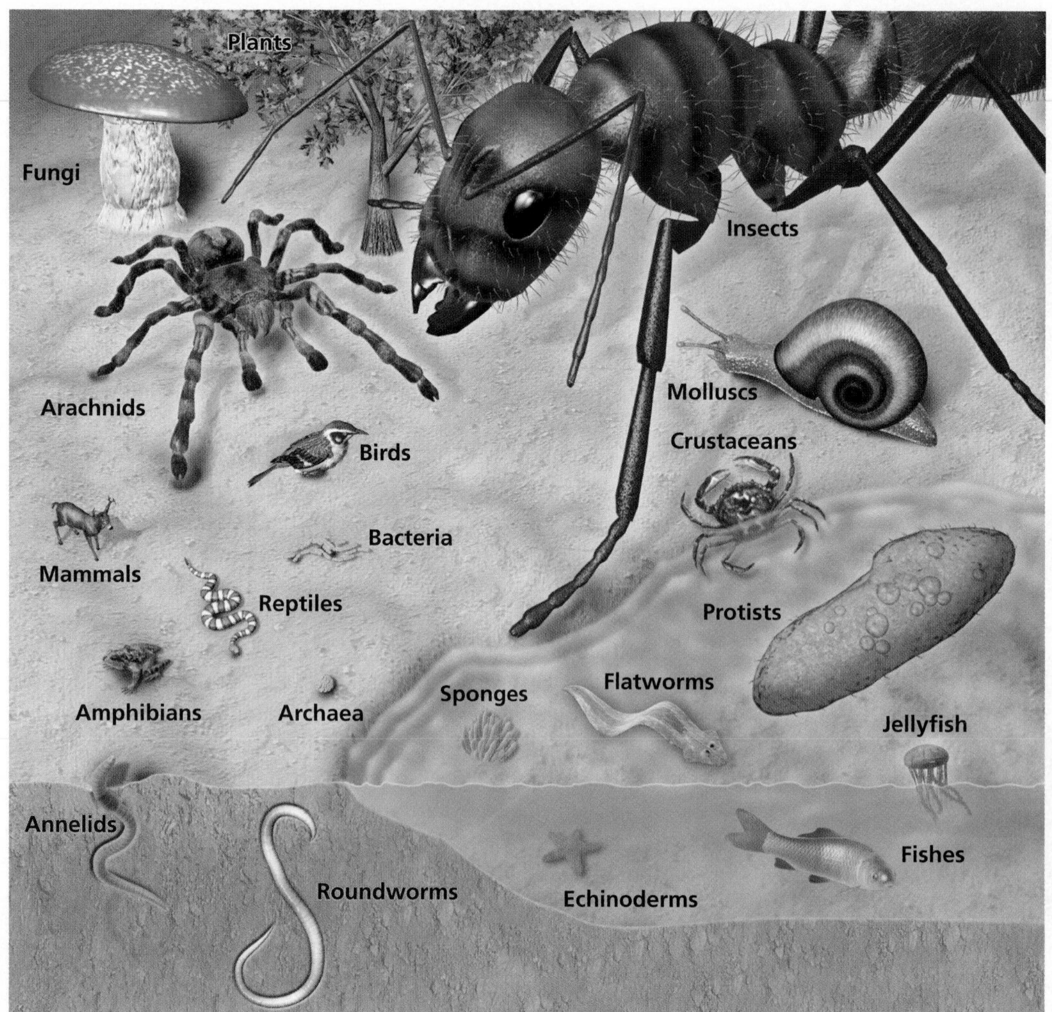

FIGURE 11.3 Some groups contain more species than others. This illustration shows organisms scaled in size to the number of species known from each group, giving a visual sense of their disparity in species richness. Because most species are not yet discovered or described, some groups (such as bacteria, archaea, insects, nematodes, protists, and fungi) may contain far more species than we now know. *Data from Groombridge, B., and M.D. Jenkins, 2002.* Global biodiversity: Earth's living resources in the 21st century. *UNEP-World Conservation Monitoring Centre. Cambridge, U.K.: Hoechst Foundation.*

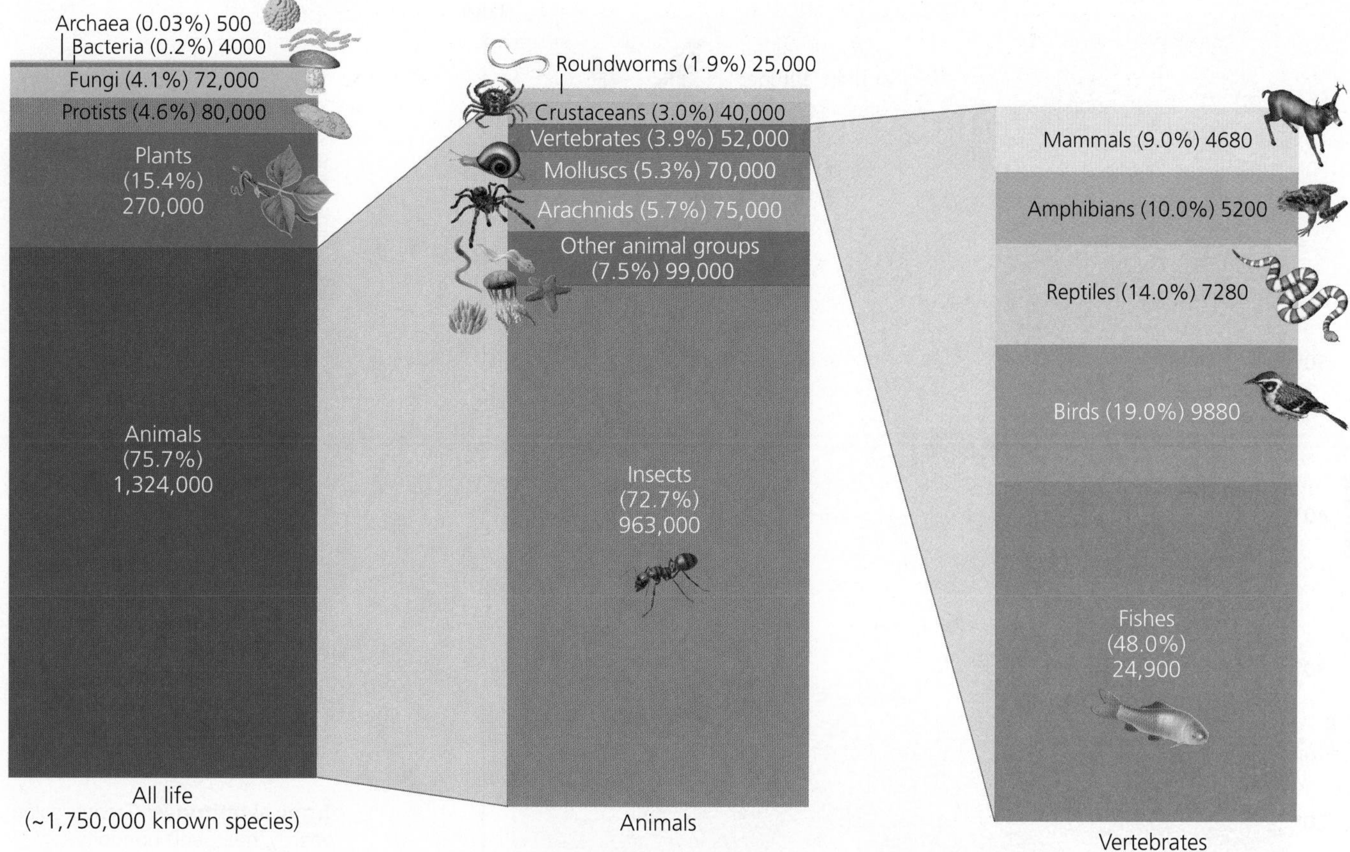

FIGURE 11.4 Insects predominate in number of species. Of all known species, three-quarters are animals. Among animals, nearly three-quarters are insects, whereas vertebrates comprise only 3.9%. Among vertebrates, nearly half are fishes, and mammals comprise only 9%. *Data from Groombridge, B., and M.D. Jenkins, 2002.* Global biodiversity: Earth's living resources in the 21st century. *Cambridge, UK: Hoechst Foundation.*

DATA Q What percentage of the world's total species do mammals comprise?

microbes, fungi, and small insects, but also sometimes with organisms as large as birds, trees, and whales. Third, some areas of Earth remain little explored. We have barely sampled the ocean depths, hydrothermal vents (p. 33), or the tree canopies and soils of tropical forests. As one example, a 2005 expedition to the remote Foja Mountains of New Guinea discovered over 40 new species of vertebrates, plants, and butterflies in less than a month, while research in marine waters nearby turned up another 50 new species.

Smithsonian Institution entomologist Terry Erwin pioneered one method of estimating species numbers. In 1982, Erwin's crews fogged rainforest trees in Central America with clouds of insecticide and then collected insects, spiders, and other arthropods as they fell from the treetops. Erwin concluded that 163 beetle species specialized on the tree species *Luehea seemannii*. If this were typical, he figured, then the world's 50,000 tropical tree species would hold 8,150,000 beetle species and—because beetles represent 40% of all arthropods—20 million arthropod species. If canopies hold two-thirds of all arthropods, then arthropod species in tropical forests alone would number 30 million. Many assumptions were involved in this calculation, and follow-up studies have revised Erwin's estimate downward, but it remains one of the better efforts at estimating species numbers.

Biodiversity is unevenly distributed

Numbers of species tell only part of the story of Earth's biodiversity. Living things are distributed across our planet unevenly. For instance, species richness generally increases as one approaches the equator (**FIGURE 11.5a**). This pattern of variation with latitude, called the *latitudinal gradient*, is one of the most obvious patterns in ecology, yet one of the most difficult for scientists to explain.

Hypotheses abound for the cause of the latitudinal gradient in species richness, but it seems likely that plant productivity and climate stability play key roles (**FIGURE 11.5b**). Greater amounts of solar energy, heat, and humidity at tropical latitudes lead to more plant growth, making areas nearer the equator more productive and able to support larger numbers of animals. The relatively stable climates of equatorial regions—their similar temperatures and rainfall from day to day and season to season—discourage single species from dominating ecosystems and instead allow numerous species to coexist. Whereas variable environmental conditions favor generalists (species that can tolerate a wide range of circumstances), stable conditions favor specialists (species highly adapted to particular circumstances) (p. 61). Additionally, polar and temperate regions may be relatively species-poor because glaciation events repeatedly forced organisms from these regions toward tropical latitudes.

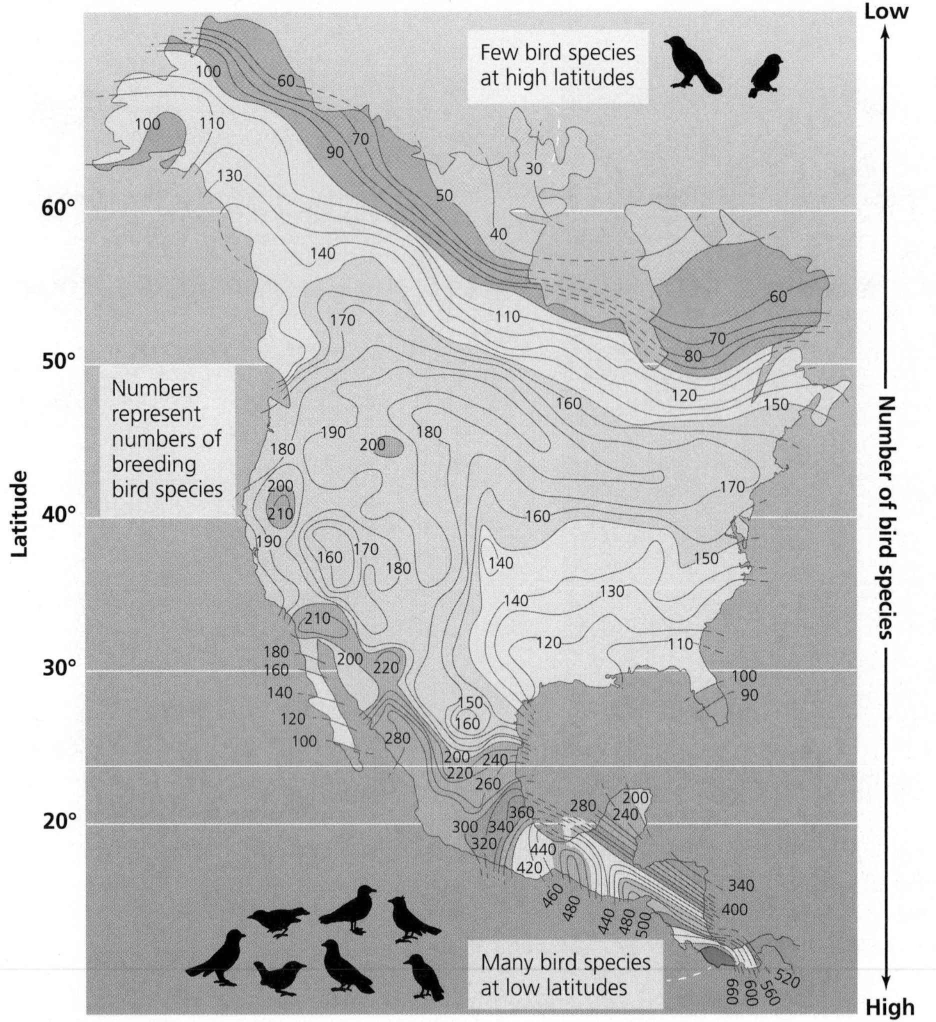

(a) Latitudinal gradient in species richness for birds in North America

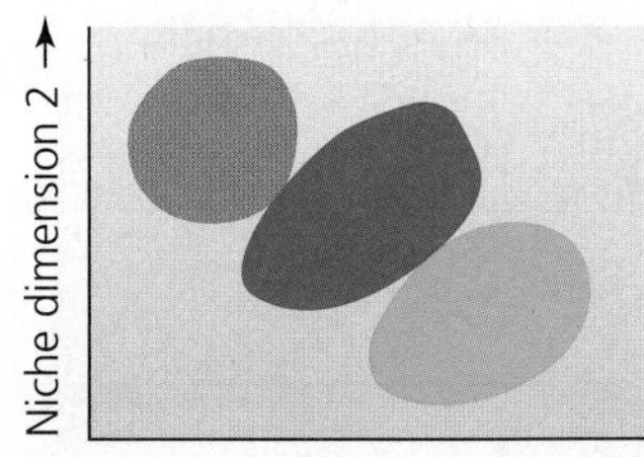

Temperate and polar latitudes: Variable climate favors fewer species, and species that are widespread generalists.

Tropical latitudes: Greater solar energy, heat, and humidity promote more plant growth to support more organisms. Stable climate favors specialist species. Together these encourage greater diversity of species.

(b) One hypothesis to explain the latitudinal gradient

FIGURE 11.5 **Species richness tends to increase toward the equator.** Among birds **(a)**, regions of arctic Canada and Alaska are home to just 30–100 breeding species, whereas areas of Costa Rica and Panama host over 600. A leading hypothesis **(b)** for this pattern explains how climate affects the degree to which organisms specialize. *Adapted (a) from Cook, R.E. et al., 1969. Variation in species density in North American birds.* Systematic Biology *18: 63–84. (Originally published as* Systematic Zoology*). By permission of Oxford University Press.*

The latitudinal gradient influences the species diversity of Earth's biomes (pp. 93–99). Tropical dry forests and rainforests support far more species than tundra and boreal forests, for instance. Tropical biomes typically show more evenness as well, whereas in high-latitude biomes with low species richness, particular species greatly outnumber others. For example, Canada's boreal forest is dominated by immense expanses of black spruce, whereas Panama's tropical forest contains hundreds of tree species, none of which greatly outnumber others.

At smaller scales, diversity varies with habitat type. Structurally diverse habitats tend to allow for more ecological niches (pp. 61, 77) and support greater species richness and evenness. For instance, forests generally support more diversity than grasslands.

For any given area, species diversity tends to increase with diversity of habitats, because each habitat supports a somewhat different set of organisms. Thus, ecotones (transition zones where habitats intermix; p. 113) often support high biodiversity. Human disturbance can sometimes create ecotones or patchwork combinations of habitats. Because this increases habitat diversity locally, species diversity may often rise in disturbed areas. However, this is true only at local scales. At larger scales, human disturbance decreases diversity because it replaces regionally unique habitats with homogenized disturbed habitats, causing specialist species to disappear. Moreover, species that rely on large expanses of habitat disappear when habitats are fragmented by human disturbance.

Understanding patterns of biodiversity is vital for landscape ecology (pp. 113–115), regional planning (p. 343), and forest management (pp. 314–323). These patterns also inform wildlife conservation, as we shall see in this chapter (and in Chapter 12; pp. 323–332), as we explore solutions to the ongoing loss of biodiversity that our planet is experiencing.

Extinction and Biodiversity Loss

Biodiversity at all levels is being lost to human impact, most irretrievably in the extinction of species. Once vanished, a species can never return. **Extinction** (pp. 58–59) occurs when the last member of a species dies and the species ceases to exist. The disappearance of a particular population from a given area, but not the entire species globally, is referred to as **extirpation.** The black rhinoceros has been extirpated from most of its historic range across Africa (FIGURE 11.6), but as a species it is not yet extinct. However, no individuals of the subspecies of black rhino known as the western black rhino have been seen since 2006, and this subspecies is now thought to be extinct. Extirpation is an erosive process that can, over time, lead to extinction.

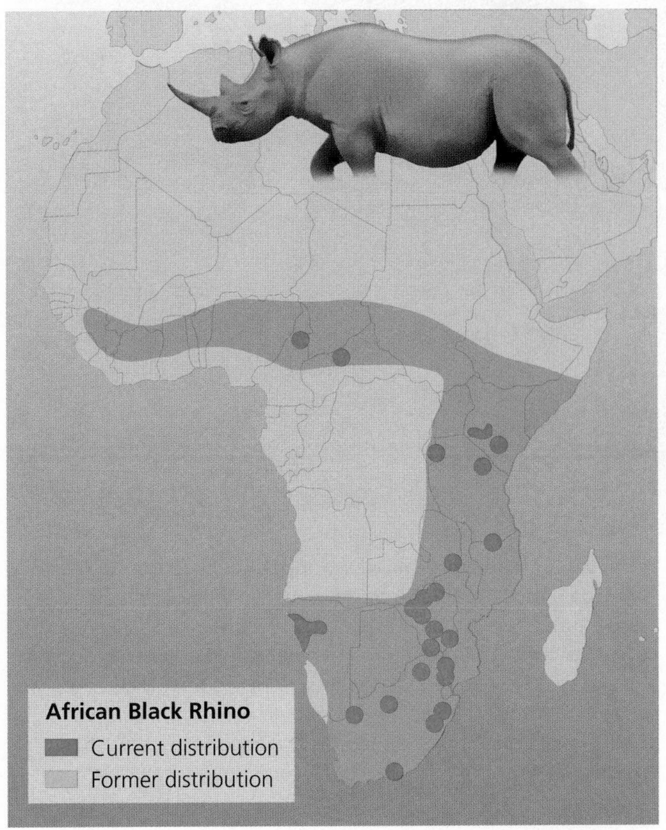

FIGURE 11.6 **The black rhinoceros has disappeared from most of its range across Africa.**

Extinction occurs naturally

Human impact is responsible for most extirpation and extinction today, but these processes also occur naturally, albeit at a much slower rate. If species did not naturally go extinct, we would be up to our ears in dinosaurs, trilobites, ammonites, and the millions of other creatures that vanished from Earth during the immense span of time before humans appeared. Paleontologists estimate that roughly 99% of all species that ever lived are now extinct. Thus, the wealth of species gracing our planet today represents just 1% of the species that have ever existed.

Most extinctions have occurred one by one for independent reasons, at a pace referred to as the **background rate of extinction.** By studying traces of organisms preserved in the fossil record (pp. 55, 58), scientists infer that for mammals and marine animals, each year, on average, 1 species out of every 1–10 million has vanished.

Earth has experienced five mass extinction events

Extinction rates rose far above this background rate at several points in Earth's history. In the past 440 million years, our planet experienced five major **mass extinction events** (p. 59). Each event eliminated more than one-fifth of life's families and at least half its species (TABLE 11.1). The most severe episode occurred at the end of the Permian period (see APPENDIX E). At this time, 248 million years ago, close to 90% of all species went extinct. The best-known episode occurred 65 million years ago at the end of the Cretaceous period, when an asteroid impact (and possibly volcanism) brought an end to the dinosaurs and many other groups. Evidence exists for earlier mass extinctions during and before the Cambrian period, more than half a billion years ago.

TABLE 11.1 Mass Extinctions

EVENT	DATE (MILLIONS OF YEARS AGO [MYA])	CAUSE	TYPES OF LIFE MOST AFFECTED	PERCENTAGE OF LIFE DEPLETED
Ordovician	440 mya	Unknown	Marine organisms; terrestrial record is unknown	>20% of families
Devonian	370 mya	Unknown	Marine organisms; terrestrial record is unknown	>20% of families
Permo-Triassic	250 mya	Possibly volcanism	Marine organisms; terrestrial record is less known	>50% of families; 80–95% of species
End-Triassic	202 mya	Unknown	Marine organisms; terrestrial record is less known	20% of families; 50% of genera
Cretaceous-Tertiary	65 mya	Likely asteroid impact	Marine and terrestrial organisms, including dinosaurs	5% of families; >50% of species
Current	Beginning 0.01 mya	Human impacts	Large animals, specialized organisms, island organisms, organisms harvested by people	Ongoing

FIGURE 11.7 The ivory-billed woodpecker was one of North America's most majestic birds. It lived in old-growth forests throughout the southeastern United States. Forest clearing and timber harvesting eliminated the mature trees it needed for food, shelter, and nesting, and this symbol of the South appeared to go extinct. In recent years, fleeting, controversial observations in Arkansas, Louisiana, and Florida have raised hopes that the species persists, but proof has been elusive.

If current trends continue, the modern era, known as the Quaternary period, may see the extinction of more than half of all species. Although similar in scale to previous mass extinctions, today's ongoing mass extinction is different in two primary respects. First, we are causing it. Second, we will suffer as a result.

We are setting the sixth mass extinction in motion

In the past few centuries alone, we have recorded hundreds of instances of species extinction caused by people. Sailors documented the extinction of the dodo on the Indian Ocean island of Mauritius in the 17th century, for example, and today only a few body parts of this unique bird remain in museums. Among North American birds in the past two centuries, we have driven into extinction the Carolina parakeet, great auk, Labrador duck, passenger pigeon (p. 62), almost certainly the Bachman's warbler and Eskimo curlew, and likely the ivory-billed woodpecker (**FIGURE 11.7**). Several more species, including the whooping crane, Kirtland's warbler, and California condor (pp. 297–298), teeter on the brink of extinction.

However, people may have been hunting species to extinction for thousands of years. Archaeological evidence shows that in case after case, a wave of extinction followed close on the heels of human arrival on islands and continents (**FIGURE 11.8**). After Polynesians reached Hawaii, half its birds

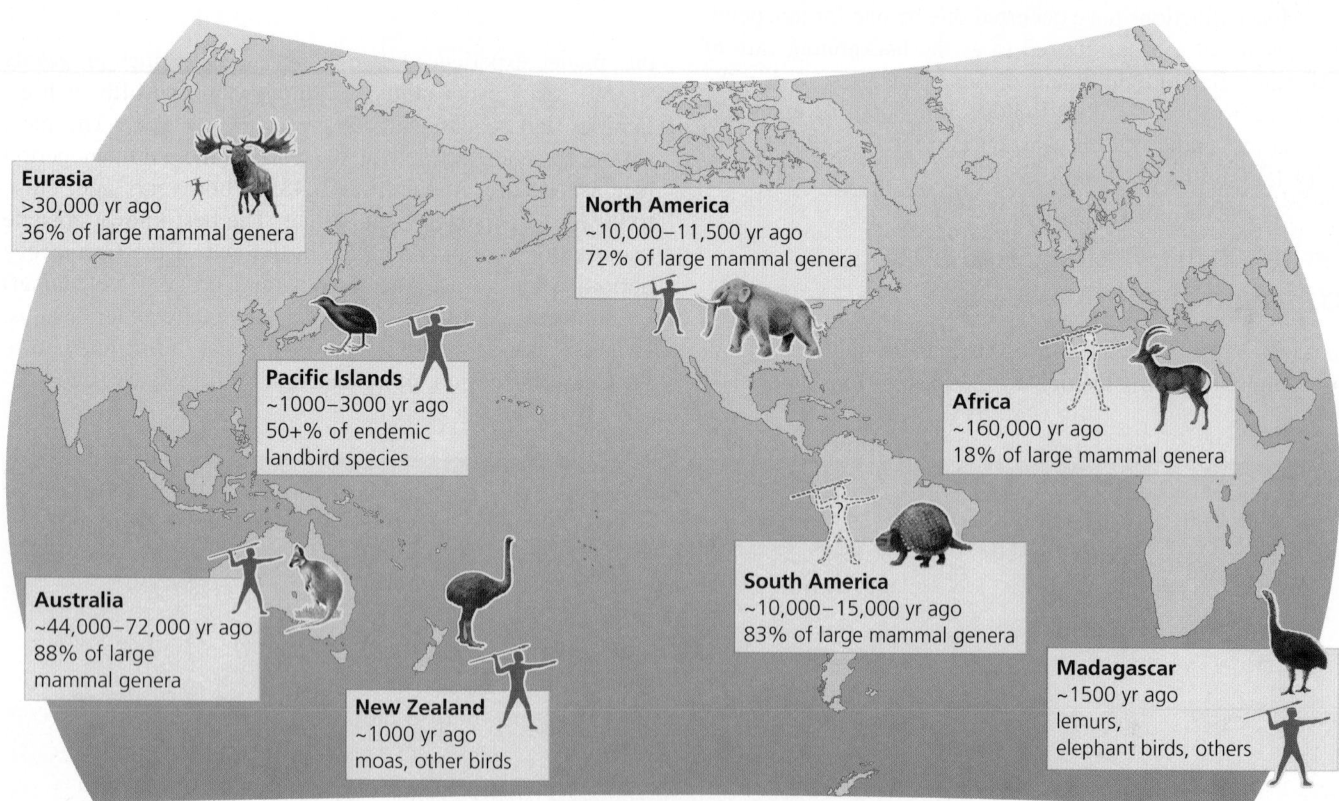

FIGURE 11.8 This map shows when humans arrived in each region, and the extent of extinctions that followed. One extinct animal from each region is illustrated. Larger human hunter icons indicate more evidence and certainty that hunting (as opposed to climate change or other factors) was a primary cause of extinctions. Data for South America and Africa are so far too sparse to be conclusive. *Adapted from Barnosky, A.D., et al., 2004. Assessing the causes of late Pleistocene extinctions on the continents.* Science *306: 70–75; and Wilson, E.O., 1992.* The diversity of life. *Cambridge, MA: Belknap Press.*

went extinct. Birds, mammals, and reptiles vanished following human arrival on many other oceanic islands, including large island masses such as New Zealand and Madagascar. Dozens of species of large vertebrates died off in Australia after people arrived roughly 50,000 years ago, and North America lost 33 genera of large mammals after people arrived more than 10,000 years ago (see Figure 3.9, p. 58).

Current extinction rates are much higher than normal

Today, species loss is accelerating as our population growth and resource consumption put increasing strain on habitats and wildlife. In 2005, scientists with the Millennium Ecosystem Assessment (p. 15) calculated that the current global extinction rate is 100 to 1000 times greater than the background rate. They projected that the rate would increase tenfold or more in future decades.

To monitor threatened and endangered species, the International Union for Conservation of Nature (IUCN) maintains the **Red List,** a continuously updated list of species facing high risks of extinction. In 2012 the Red List reported that 21% (1139) of mammal species, 13% (1313) of bird species, 30% (1933) of amphibian species, and 20% (1867) of fish species were threatened with extinction. For most other groups, scientists do not yet have enough data to make assessments. In the United States alone over the past 500 years, 236 animals and 30 plants are known to have gone extinct. For all these figures, the actual numbers of species are without doubt greater than the known numbers.

FAQ **If a mass extinction is happening, why don't I see species going extinct all around me?**

There are two reasons that most of us don't personally sense the scale of biodiversity loss. First, if you live in a town or city, the plants and animals you see from day to day are generalist species that thrive in disturbed areas. In contrast, the species in trouble are those that rely on less-disturbed habitats, and you may need to go further afield to find them.

Second, a human lifetime is very short! The loss of populations and species over the course of our lifetime may seem a slow process to us, but relative to Earth's timescale it is sudden—almost instantaneous. Because each of us is born into a world that has already lost many species, we don't recognize what's already been lost. Likewise, our grandchildren won't appreciate what we've lost in our lifetimes. Each human generation experiences just a portion of the overall phenomenon, so we have difficulty sensing the big picture. Nonetheless, researchers who study biology and naturalists who spend their time outdoors see biodiversity loss around them all the time—and that's precisely why they feel so passionate about it.

Biodiversity loss involves more than extinction

Extinction is only part of the story of biodiversity loss. The larger part involves declining population sizes. As a species

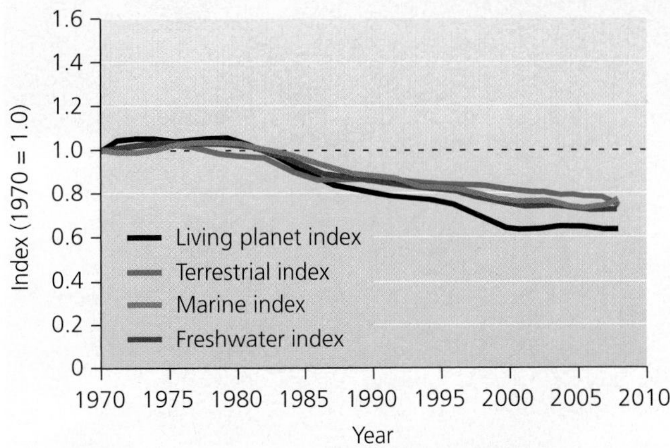

FIGURE 11.9 The Living Planet Index is an indicator of the state of global biodiversity. As of 2012, index values summarized trends for 9014 populations of 2688 vertebrate species. Between 1970 and 2008, the Living Planet Index fell by 28%. The index for terrestrial species fell by 25%; for freshwater species, by 37%; and for marine species, by 22%. *Data from WWF, 2012.* Living planet report 2012. *WWF International, Gland, Switzerland.*

declines in number, its geographic range often shrinks as it is extirpated from parts of its range. Thus, many species today are less numerous and occupy less area than they once did. This is true of nearly all large mammal species in Africa. Indeed, scientific studies have documented significant population declines among large mammals of the Serengeti in recent years (see **THE SCIENCE BEHIND THE STORY**, pp. 286–287). Such declines mean that genetic diversity and ecosystem diversity, as well as species diversity, are being lost.

To measure and quantify such change, scientists at the World Wildlife Fund and the United Nations Environment Programme (UNEP) developed a metric called the *Living Planet Index*. This index expresses how large the average population size of a species is now, relative to its size in the year 1970. The most recent (2012) compilation of information for the Living Planet Index summarized trends in the populations of 1432 terrestrial species, 737 freshwater species, and 675 marine species that are sufficiently monitored to provide reliable data. Between 1970 and 2008, the Living Planet Index fell by 28%—meaning that on average, population sizes are 28% smaller than they were just four decades ago (**FIGURE 11.9**). This decline has been driven primarily by biodiversity losses in tropical regions. In the temperate zones the index rose by 31%, but in tropical regions it decreased by 61%.

Several major causes of biodiversity loss stand out

Scientists have identified four primary causes of population decline and species extinction: habitat loss, invasive species, pollution, and overharvesting. Global climate change (Chapter 18) now is becoming the fifth. Each of these causes is intensified by human population growth and by our increasing per capita consumption of resources.

Habitat loss Habitat loss is the single greatest cause of biodiversity loss today. Species lose their habitats (p. 61) when

FIGURE 11.10 Development of land for housing is one way in which habitat is altered or destroyed.

those habitats are destroyed outright, but habitats are also lost when they are altered through more subtle processes, including fragmentation and other forms of degradation. Because organisms have adapted over thousands or millions of years to the habitats in which they live, any major change in their habitat is likely to render it less suitable for them.

Many human activities alter, degrade, or destroy habitat. Housing development supplants diverse natural ecosystems with simplified human-made ones, driving many species from their homes (**FIGURE 11.10**). Farming replaces diverse natural communities with simplified ones of only a few plant species. Grazing modifies the structure and species composition of grasslands, and it can lead to desertification (p. 223). Clearing forests removes the food, shelter, and other resources that forest-dwelling organisms need to survive. Dams turn rivers into reservoirs upstream and affect water conditions and floodplain communities downstream.

Habitat loss occurs most commonly through gradual, piecemeal degradation, such as **habitat fragmentation** (**FIGURE 11.11**). When farming, logging, road building, or development intrude into an unbroken expanse of forest or grassland, they break up a continuous area of habitat into an array of fragments, or patches. As habitat fragmentation proceeds across a landscape, animals and plants requiring the habitat disappear from one fragment after another. Fragmentation can also prevent animals from moving from place to place, and this is the concern of conservationists fighting the proposal to build a highway through the Serengeti. In response to habitat fragmentation, conservationists have designed landscape-level strategies to optimize the arrangement of areas to be preserved (pp. 327–332).

Habitat loss has affected nearly every biome (**FIGURE 11.12**). Over half of the world's temperate forests, grasslands, and shrublands had been converted by 1950 (mostly for agriculture). Today habitat is being lost most rapidly in tropical rainforests, tropical dry forests, and savannas.

Habitat loss is the primary source of population declines in 83% of threatened mammals and 85% of threatened birds,

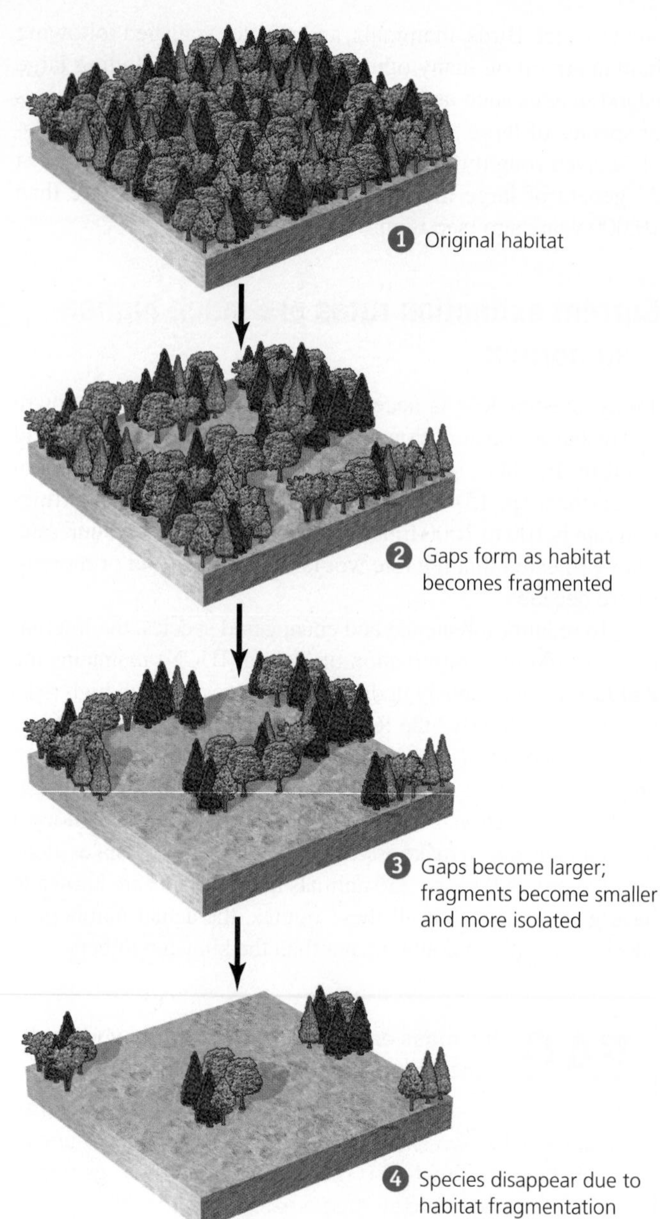

FIGURE 11.11 Habitat fragmentation occurs as human impact creates gaps that expand and eventually come to dominate the landscape, stranding islands of habitat. As habitat becomes fragmented, fewer populations can persist, and numbers of species in the fragments decline.

according to UNEP data. For example, the prairies native to North America's Great Plains are today almost entirely converted to agriculture. Less than 1% of original prairie habitat remains. As a result, grassland bird populations have declined by an estimated 82–99%.

Of course, our habitat alteration benefits some species. Animals such as house sparrows, pigeons, gray squirrels, rats, and cockroaches thrive in cities and towns. However, the species that benefit from our modification of natural habitats are relatively few; for every species that wins, more lose. Furthermore, the species that do well in our midst tend to be weedy generalists that are in little danger of disappearing any time soon.

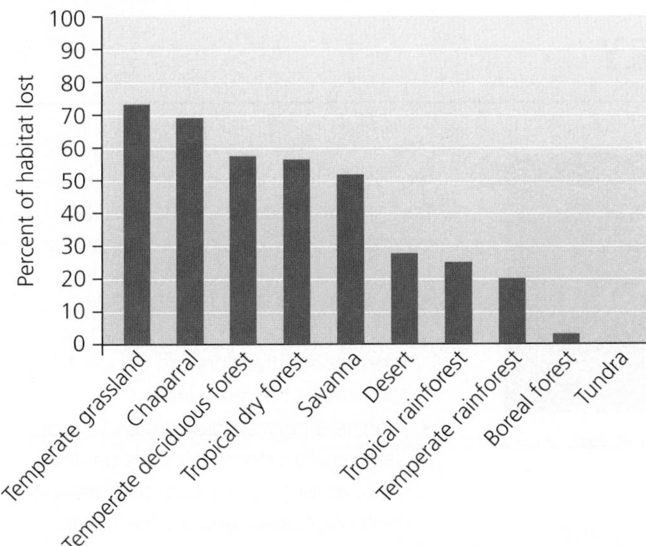

FIGURE 11.12 Human impact has caused habitat loss in all the world's biomes. Shown are percentages of original area that were fully and directly converted for human use through 1990. *Adapted from Millennium Ecosystem Assessment, 2005.* Ecosystems and human well-being: Biodiversity synthesis. *World Resources Institute, Washington, D.C.*

Pollution Pollution can harm organisms in many ways. Air pollution (Chapter 17) degrades forest ecosystems. Water pollution (Chapter 15) impairs fish and amphibians. Agricultural runoff containing fertilizers, pesticides, and sediments (Chapters 5, 9, and 10) harms many terrestrial and aquatic species. Heavy metals, polychlorinated biphenyls (PCBs), endocrine-disrupting compounds, and other toxic chemicals poison people and wildlife (Chapter 14). Plastic garbage in the ocean can strangle, drown, or choke marine creatures (p. 432–436). The effects of oil spills on wildlife (pp. 436–437, 538–541) are dramatic and well known.

Although pollution is a substantial threat, it tends to be less significant than public perception holds it to be. The damage to wildlife and ecosystems caused by pollution can be severe, but this tends to be surpassed by the harm caused by habitat alteration or invasive species.

Overharvesting For most species, hunting or harvesting by people will not in itself pose a threat of extinction, but for some species it can. Large mammals of the African savannas, such as elephants and rhinoceroses, are examples. Large in size, long-lived, and raising few young in their lifetimes—classic K-selected species (p. 69)—elephants and rhinoceroses are just the type of animal to be vulnerable to hunting. People have long killed elephants to extract their tusks for ivory (**FIGURE 11.13**). Because of this, the world's nations enacted a global ban on the commercial trade of ivory. Since imposition of the ban in 1989, elephant numbers have recovered, but thousands still are slaughtered illegally each year, and ivory trade continues on the black market. Similarly, rhinoceroses are killed for their horns, which are sold illegally to Asian nations for traditional medicine and to some Middle Eastern countries for ornamental use as dagger handles.

Over the past century, hunting has led to steep declines in the populations of many K-selected animals. In central Africa, gorillas and other primates are killed for their meat and may face extinction soon. Across Asia, the tiger is threatened by habitat loss and poaching; body parts from one tiger can fetch a poacher $15,000 on the black market, where they are sold as aphrodisiacs in Asian countries. Today half the world's subspecies of tiger are extinct, and most of the remaining animals are crowded onto just 1% of the land they originally occupied. In the oceans, decades of whaling drove the Atlantic gray whale extinct and have left several other whales threatened or endangered. Thousands of sharks are killed each year simply for their fins, which are used in soup. Today the oceans contain only 10% of the large animals they once did (p. 441).

To combat overharvesting, governments have passed laws, signed treaties, and strengthened anti-poaching efforts. Scientists have begun using genetic analyses to expose illegal hunting and wildlife trade. For instance, DNA testing can reveal the geographic origins of elephant ivory and whether whale meat sold in markets is from an animal caught illegally (see **THE SCIENCE BEHIND THE STORY**, pp. 300–301).

FIGURE 11.13 Poachers kill elephants to sell their tusks for ivory. Despite the long-standing ban on ivory trade, 2009–2013 saw the most poaching and ivory confiscation yet. Here, Kenyan officials prepare to set fire to thousands of confiscated tusks at Masai Mara National Park in an effort to dampen the trade.

Wildlife Declines in African Reserves

Tanzania and Kenya have some of the largest and most famous parks and protected areas in the world, with the greatest variety and density of large mammal species to be found anywhere. The parks are generally well managed and well funded. Yet even these places of refuge are not immune to pressures from rising human population, development, and resource use.

For several decades, biologists and park managers have censused wildlife in and around the parks and reserves. In recent years, researchers have begun to analyze these long-term data sets to assess population trends of the large mammals of the East African savanna. These studies are finding that most animals are declining in number—inside the parks and reserves as well as outside.

In Tanzania, several government agencies and nonprofit groups have collaborated to census mammals by airplane. These aerial surveys began in Serengeti National Park in the 1970s and expanded to other parks in the 1980s. In 2006, Chantal Stoner and six colleagues from Tanzania and from the University of California at Davis compiled and analyzed data on 25 species over a 10-year period (roughly 1990–2000) from eight regions, each centered on a major park.

Across the eight regions in wet and dry seasons for all species, this team found that population declines outnumbered increases by more than 10 to 1 (**FIGURE 1**). In the Serengeti region, five species declined while two increased.

Zebras in Serengeti National Park

Population declines were greater outside of parks and in areas that received less protection. However, even within the boundaries of well-protected reserves, many species decreased in number.

Why are animals declining? For decades East Africa has had one of the world's fastest rates of human population growth, and this has intensified pressures on wildlife and ecosystems:

- Settlements have increased as nomadic Maasai herders have become sedentary and as people from elsewhere have moved in.
- Farmers convert grasslands to crops (especially wheat), and this destroys habitat for grazers such as antelope and wildebeest and for the predators that follow them.
- Livestock are competing with wild grazers for food on the grasslands.
- People are killing animals for food. Impoverished residents poach animals for their own subsistence, but "bushmeat" is also sold to wealthy diners in cities as far away as Europe.

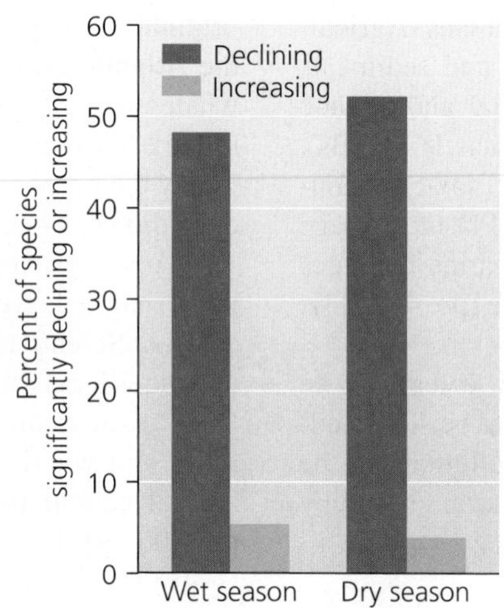

FIGURE 1 Across Tanzania's protected areas, more animals have decreased in number than have increased. *Data are combined from six locations, from Stoner, C., et al., 2006. Changes in large herbivore populations across large areas of Tanzania.* African Journal of Ecology *45: 202–215.*

In Kenya, researchers are seeing similar patterns. As in Tanzania, Kenyan scientists and park managers have been censusing wildlife by airplane, road, and foot for decades, and recently researchers have taken a hard look at the accumulated data.

In 2009, conservation biologist David Western and two colleagues reviewed 30 years of data from aerial surveys conducted from 1977 to 2007 across Kenya's rangelands, which comprise three-quarters of the nation's land area. In most locations, Western's group found downward trends in the populations of most species. Cycles of rainfall and drought affect grazing mammals, but statistical analysis of the data indicated that the long-term declines transcended short-term effects due to rain or drought.

Moreover, the populations surveyed by Western's team were trending

Invasive species Introduction of non-native species to new environments, where some may become invasive (pp. 88–89), can push native species toward extinction (**TABLE 11.2**). Some introductions are accidental. Examples include aquatic organisms, such as zebra mussels, transported in the ballast water of ships (Chapter 4); animals that escape from the pet trade; and weeds whose seeds cling to our socks as we travel from place to place. Other introductions are intentional. In Lake Victoria west of the Serengeti, the Nile perch was introduced as a food fish. Within years it spread throughout the vast lake, preying on and driving extinct dozens of native species of cichlid fish from one of the world's most spectacular adaptive radiations of animals. The Nile perch is providing people food, but at a significant ecological cost. People everywhere have long brought food crops and animals with them as they colonized new places, and today

downward both inside and outside the parks (**FIGURE 2**). Many species, including wildebeest, zebras, and some antelope, migrate into and out of parks, so factors that affect their populations outside the parks will affect numbers counted within park boundaries as well.

Researchers have given particular attention to the Maasai Mara National Reserve, on the Kenyan side of the border adjacent to Serengeti National Park. A decade ago Dutch scientist Wilber Ottichilo analyzed aerial survey data and concluded that nonmigratory mammals had declined by 58% from 1977 to 1997 within the reserve and by a similar amount outside the reserve. Resident wildebeest had fared even worse, showing an 81% decline.

More recently, Joseph Ogutu of the International Livestock Research Institute in Nairobi, Kenya, and his colleagues extended the analysis to 2009 and found that most species had continued declining, both on and off the reserve, and were now at just one-third of their 1977 population sizes (**FIGURE 3**).

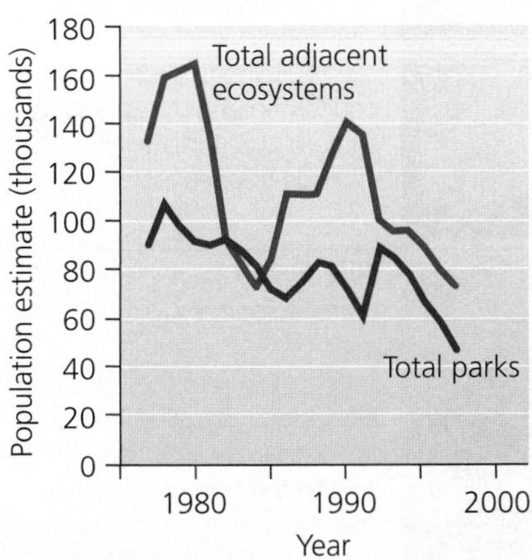

FIGURE 2 Wildlife populations have declined across Kenya, both inside and outside parks. *Data from Western, D., et al., 2009. The status of wildlife in protected areas compared to non-protected areas of Kenya.* PLoS ONE *4(7): e6140.*

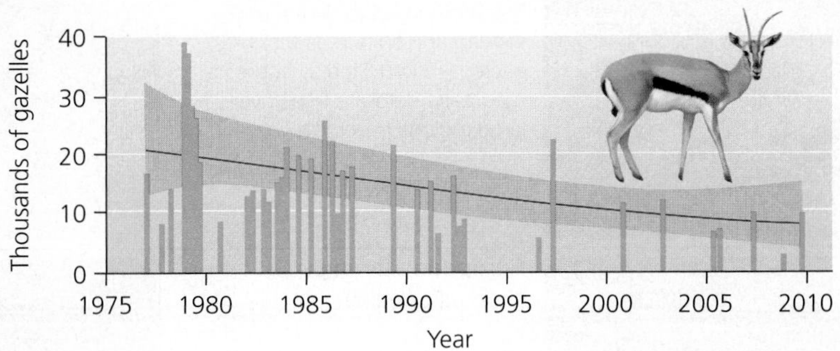

(a) Inside the reserve

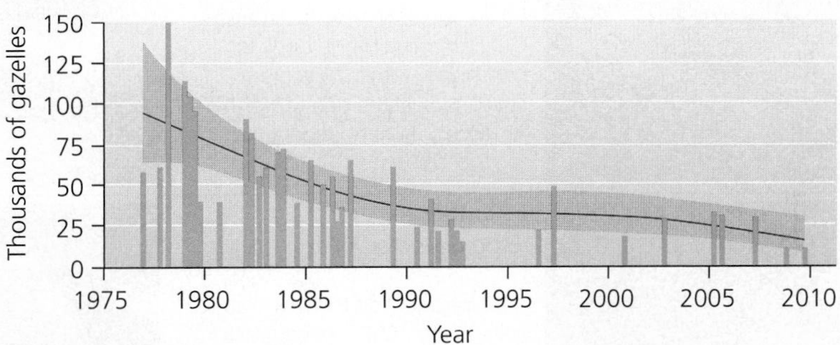

(b) Outside the reserve

FIGURE 3 Thomson's gazelles, the most abundant species at Kenya's Maasai Mara National Reserve, decreased by 59% (a) inside the reserve, and by 77% (b) outside the reserve. *Data from Ogutu, Joseph, et al., 2011. Continuing wildlife population declines and range contraction in the Mara region of Kenya during 1977–2009.* Journal of Zoology *285: 99–109.*

As in Tanzania and elsewhere in Kenya, drought accounted only for short-term fluctuations and not long-term declines. The long-term declines were due to habitat loss to cropland agriculture, intensified human settlement near the parks, poaching, and competition with livestock.

As a result of studies like these, researchers have concluded that merely setting aside parks is not adequate to successfully conserve wildlife and ecosystems. Instead, we need to view the big picture and consider animals' needs across the landscape, both inside and outside of reserves, as well as how human impacts might spill into reserves. Accomplishing conservation goals thus requires working with local people living near protected areas. ■

we continue international trade in exotic pets and ornamental plants, often heedless of the ecological consequences.

Species native to islands are especially vulnerable to disruption from introduced species. Island species have existed in isolation for millennia with relatively few parasites, predators, and competitors; as a result, they have not evolved the defenses necessary to resist invaders that are better adapted to these pressures. For instance, Hawaii's native plants and animals have been under siege from invasive organisms such as rats, pigs, and cats, and this has led to a number of extinctions (Chapter 3).

Most organisms introduced to new areas perish, but the few types that survive may do very well, especially if they are freed from the predators and parasites that attacked them back home or from the competitors that had limited their access to resources. Once released from the limiting factors (p. 67) of predation,

TABLE 11.2 Invasive Species

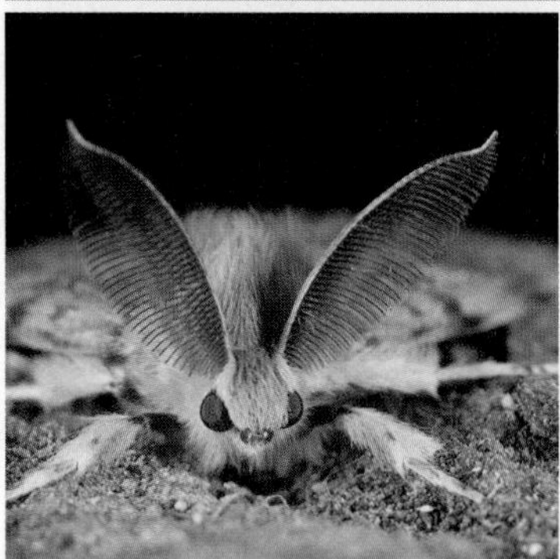

European gypsy moth
(*Lymantria dispar*)

Introduced to Massachusetts in the hope it could produce silk. The moth failed to do so, and instead spread across the eastern United States, where its outbreaks defoliate trees over large regions every few years.

Asian long-horned beetle
(*Anoplophora glabripennis*)

Since the 1990s, has repeatedly arrived in North America in imported lumber. These insects burrow into wood and can kill the majority of trees in an area. Chicago, Seattle, Toronto, New York, and other cities have cleared thousands of trees to eradicate these invaders.

European starling
(*Sturnus vulgaris*)

Introduced to New York City in the 1800s by Shakespeare devotees intent on bringing every bird mentioned in Shakespeare's plays to America. Outcompeting native birds for nest holes, within 75 years starlings became one of North America's most abundant birds.

Emerald ash borer
(*Agrilus planipennis*)

Discovered in Michigan in 2002, this wood-boring insect reached 12 U.S. states and Canada by 2010, killing millions of ash trees in the upper Midwest. Billions of dollars will be spent in trying to control its spread.

Cheatgrass
(*Bromus tectorum*)

After introduction to Washington state in the 1890s, cheatgrass spread across the western United States. It crowds out other plants, uses up the soil's nitrogen, and burns readily. Fire kills many native plants, but not cheatgrass, which grows back stronger without competition.

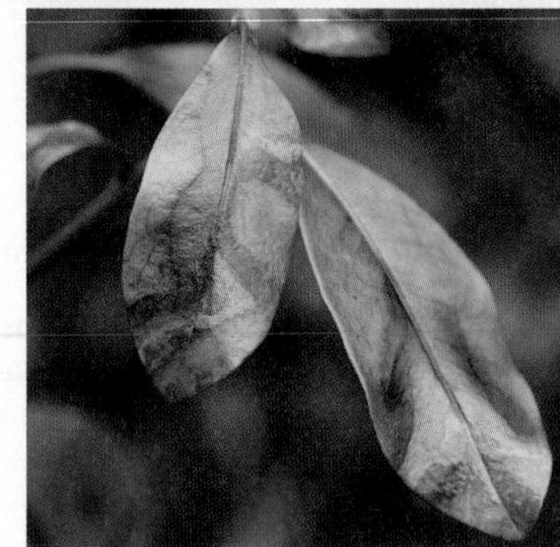

Sudden oak death
(*Phytophthora ramorum*)

This disease has killed over 1 million oak trees in California since the 1990s. The pathogen (a water mold) was likely introduced via infected nursery plants. Scientists are concerned about damage to eastern U.S. forests if it spreads to oaks there.

Brown tree snake
(*Boiga irregularis*)

Nearly every native forest bird on the South Pacific island of Guam has disappeared, eaten by these snakes, which arrived from Asia as stowaways on ships and planes after World War II. Guam's birds had not evolved with snakes, and had no defenses against them.

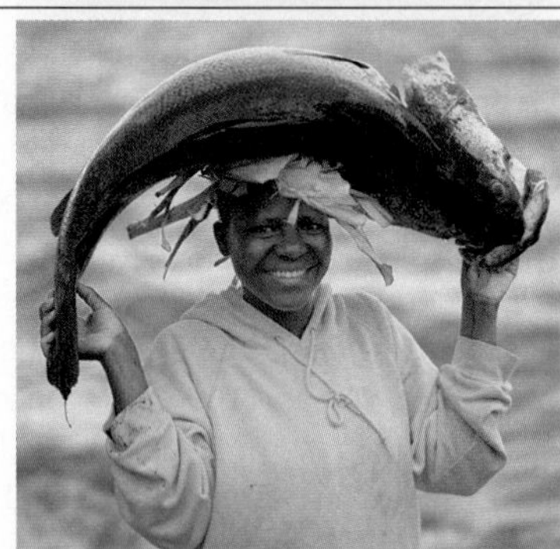

Nile perch
(*Lates niloticus*)

A large fish from the Nile River. Introduced to Lake Victoria in the 1950s, it proceeded to eat its way through hundreds of species of native cichlid fish, driving a number of them extinct (p. 286). People value the perch as food, but it has radically altered the lake's ecology.

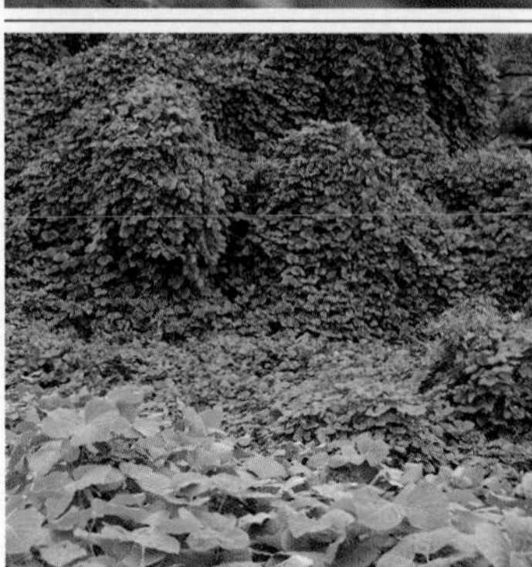

Kudzu
(*Pueraria montana*)

A Japanese vine that can grow 30 m (100 ft) in a single season, the U.S. Soil Conservation Service introduced kudzu in the 1930s to help control erosion. Kudzu took over forests, fields, and roadsides throughout the southeastern United States.

Polynesian rat
(*Rattus exulans*)

One of several rat species that have followed human migrations across the world. Polynesians transported this rat to islands across the Pacific, including Easter Island (pp. 6–7). On each island it caused ecological havoc, and has driven extinct birds, plants, and mammals.

This table shows just a few of the many thousands of invasive species.

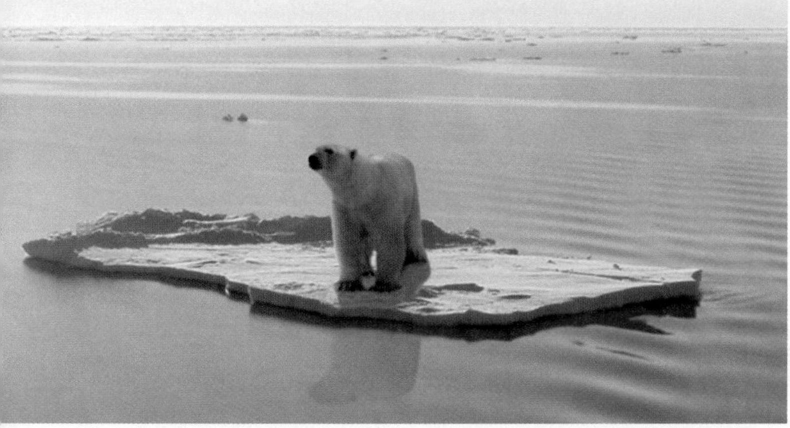

FIGURE 11.14 The polar bear became the first species listed as threatened under the Endangered Species Act as a result of climate change. As Arctic warming melts the sea ice from which they hunt seals, polar bears must swim farther and farther for food.

parasitism, and competition, an introduced species may proliferate and become invasive, displacing native species. Invasive species cause billions of dollars in economic damage each year.

The Serengeti has so far had little problem with invasive species because of its large size, but ecologists fear that a highway built through it would change this by introducing weed seeds from passing vehicles. Park managers there already are concerned about several American plant species, such as datura, parthenium, and prickly poppy, that are toxic to native herbivores and that have spread rapidly in grassland areas of Africa where they have appeared.

Climate change The preceding four types of human impacts affect biodiversity in discrete places and times. In contrast, our manipulation of Earth's climate system (Chapter 18) is having global impacts on biodiversity. As our emissions of greenhouse gases from fossil fuel combustion cause temperatures to warm worldwide, we modify weather patterns and increase the frequency of extreme weather events.

Extreme weather events such as droughts and storms increase stress on populations. In the Arctic, where temperatures have warmed the most, melting sea ice and other impacts (p. 511) are threatening polar bears and people alike (FIGURE 11.14). Across the world, warming temperatures are forcing organisms to shift their geographic ranges toward the poles and higher in altitude. Some species will not be able to adapt. Mountaintop organisms cannot move further upslope to escape warming temperatures, so they may perish. Trees may not be able to move toward the poles fast enough. As ranges shift, animals and plants find themselves among new communities of prey, predators, and parasites to which they are not adapted. All in all, scientists predict that a 1.5–2.5°C (2.7–4.5°F) global temperature rise could put 20–30% of the world's plants and animals at increased risk of extinction.

Amphibians are vanishing

Reasons for the decline of a population or species can be complex and difficult to determine. The worldwide collapse of amphibians provides an example. Today entire populations of frogs, toads, and salamanders are vanishing without a trace. More than 2600 of the 6400 known species of amphibians are in decline, over 1900 are threatened, and roughly 170 species studied just years or decades ago are thought to be extinct (FIGURE 11.15a). As these creatures disappear before our eyes, scientists are racing to discover the reasons. Recent studies implicate a wide array of factors, including habitat destruction, chemical pollution, disease, invasive species, and climate change (FIGURE 11.15b). Biologists suspect that multiple factors may be interacting and multiplying one another's effects.

Many amphibian populations are vanishing in remote and pristine environments when no direct damage is apparent. In many of these cases, the culprit seems to be a fungal disease called chytridiomycosis, caused by the pathogen *Batrachochytrium dendrobatidis*. Researchers do not yet know whether its rapid spread is due to human influence.

As scientists learn more, they are designing responses to amphibian declines. A conservation action plan published by the IUCN recommended protecting and restoring habitat, cracking down on illegal harvesting, enhancing disease monitoring, and establishing captive breeding programs.

All of the main causes of biodiversity loss are intensified by human population growth and rising per capita

(a) Male golden toad from Monteverde, Costa Rica

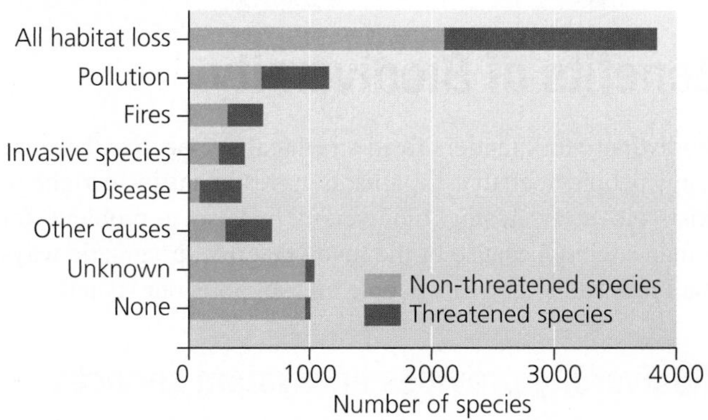

(b) Causes of amphibian declines

FIGURE 11.15 The world's amphibians are declining. The golden toad **(a)** is one of about 170 species of amphibians that have suddenly gone extinct in recent years. This brilliant orange toad of Costa Rican cloud forests disappeared due to drought, climate change, and/or disease. Habitat loss **(b)** is the main reason for amphibian declines, but many declines remain unexplained. *Data from IUCN, 2008.* Global amphibian assessment.

DATA Q What is the second greatest known cause of amphibian declines, after habitat loss? What is the greatest cause for threatened species?

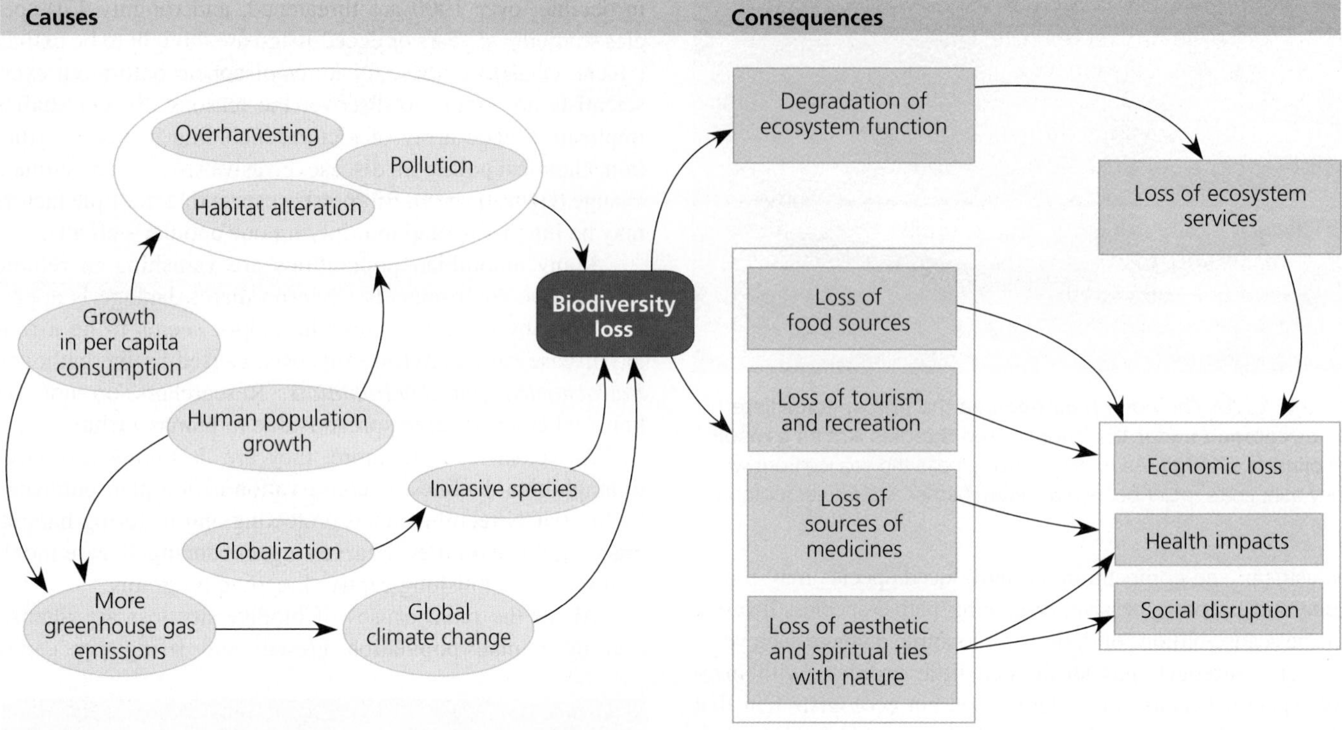

FIGURE 11.16 The loss of biodiversity stems from a variety of causes and results in many consequences for ecological systems and human well-being. Arrows in this concept map lead from causes to consequences. Items grouped within outlines do not necessarily share any special relationship; the outlined shapes are merely to streamline the figure.

Solutions

As you progress through this chapter, try to identify as many solutions to biodiversity loss as you can. What could you personally do to help address this issue? Consider how each action or solution might affect items in the concept map above.

consumption. As researchers gain a solid scientific understanding of the causes of biodiversity loss, we are also grappling with its consequences (**FIGURE 11.16**) as our actions erode the many benefits that biodiversity brings us.

Benefits of Biodiversity

Biodiversity loss matters from an ethical perspective, because many people feel that organisms have an intrinsic right to exist. However, losing biodiversity is also a problem for human society because of the many tangible, pragmatic ways that biodiversity benefits people and supports our society.

Biodiversity provides ecosystem services

Contrary to popular opinion, some things in life can indeed be free—as long as we protect the ecological systems that provide them. Intact forests provide clean air and water, and they buffer hydrologic systems against flooding and drought. Native crop varieties provide insurance against disease and drought. Wildlife can attract tourism and boost economies. Intact ecosystems provide these and other valuable processes, known as ecosystem services (pp. 3, 116–117), for all of us, free of charge.

Maintaining these ecosystem services is one clear benefit of protecting biodiversity. According to UNEP, biodiversity:

- Provides food, fuel, fiber, and shelter
- Purifies air and water
- Detoxifies and decomposes wastes
- Stabilizes Earth's climate
- Moderates floods, droughts, and temperature extremes
- Cycles nutrients and renews soil fertility
- Pollinates plants, including many crops
- Controls pests and diseases
- Maintains genetic resources for crop varieties, livestock breeds, and medicines
- Provides cultural and aesthetic benefits
- Gives us the means to adapt to change

In these ways, organisms and ecosystems support vital processes that people cannot replicate or would need to pay for if nature did not provide them. The annual economic value of just 17 of the world's ecosystem services has been estimated at over $48 trillion per year (p. 152)—more than the gross domestic product of all national economies combined.

Biodiversity helps maintain ecosystem function

Ecological research demonstrates that biodiversity tends to enhance the stability of communities and ecosystems. Research has also found that biodiversity tends to increase the resilience (p. 85) of ecological systems—their ability to weather disturbance, bounce back from stress, or adapt to change. Thus, when we lose biodiversity, this can diminish a natural system's ability to function and to provide services to our society.

Will the loss of a few species really make much difference in an ecosystem's ability to function? Consider a metaphor first offered by Paul and Anne Ehrlich (p. 192): The loss of one rivet from an airplane's wing—or two, or three—may not cause the plane to crash. But at some point as rivets are removed the structure will be compromised, and eventually the loss of just one more rivet will cause it to fail. Keeping this metaphor in mind, we would be wise to preserve as many components of our ecosystems as possible to make sure these systems continue to function.

That said, some species are more vital to an ecosystem than others. Removing a species that can be functionally replaced by others may make little difference in how the system functions. However, as with the keystone that holds together an arch, removing a keystone species (pp. 83–84) can significantly alter an ecological system. If a keystone species is lost, other species may disappear in response.

Top predators are often considered keystone species. Think of lions, leopards, and cheetahs on the Serengeti, or wolves, mountain lions, and grizzly bears at Yellowstone National Park (sometimes nicknamed "America's Serengeti"). A single top predator may prey on many herbivores, each of which may consume many plants. Thus, the removal of a single individual at the top of a food chain can have consequences that multiply as they cascade down the food chain. Moreover, top predators, and large animals in general, are especially vulnerable to human impact. They are frequently hunted and they need large areas of habitat, making them susceptible to habitat loss and fragmentation. Top predators are also vulnerable to the buildup of toxic pollutants in their tissues through the process of biomagnification (p. 374).

"Ecosystem engineers" (p. 84) such as ants and earthworms can be every bit as influential as keystone species, so the loss of an ecosystem engineer from a system can likewise set major changes in motion. A decline in wildebeest on the Serengeti, for instance, would have substantial consequences for grass growth, shrub invasion of grasslands, fire regime, soil quality, and populations of predators and other herbivores. Ecosystems are complex, and it is difficult to predict which particular species may be important. Thus, many people prefer to apply the precautionary principle (p. 265) in the spirit of Aldo Leopold (p. 139), who advised, "To keep every cog and wheel is the first precaution of intelligent tinkering."

Biodiversity enhances food security

Biodiversity provides the food we eat. Throughout our history, human beings have used 7000 plant species and several thousand animal species for food. Today many experts are concerned because industrial agriculture has narrowed our diet. Globally, we now get 90% of our food from just 15 crop species and eight livestock species, and this lack of diversity leaves us vulnerable to failures of particular crops. In a world where nearly 1 billion people go hungry, we can improve food security (the guarantee of an adequate, safe, nutritious, and reliable food supply to all people at all times; p. 246) by finding sustainable ways to capitalize on the nutritional opportunities offered by wild species and rare crop varieties.

Many new or underutilized food sources could be harvested or farmed in sustainable ways. **TABLE 11.3** shows a

TABLE 11.3 Potential New Food Sources

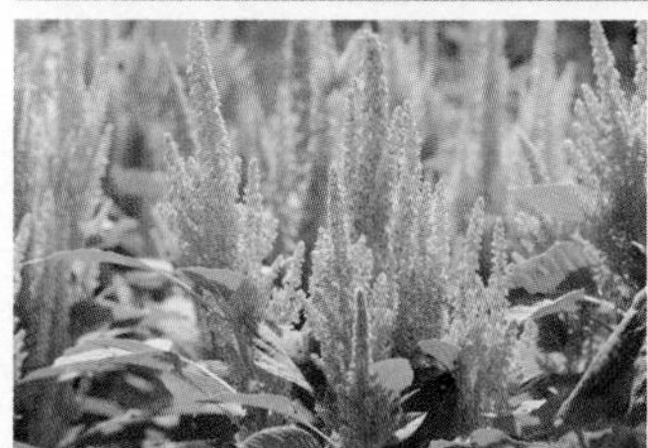

Amaranths
(three species of *Amaranthus*)

Grain and leafy vegetable; livestock feed; rapid growth, drought resistant

Capybara
(*Hydrochoeris hydrochaeris*)

World's largest rodent; meat esteemed; easily ranched in open habitats near water

Buriti palm
(*Mauritia flexuosa*)

"Tree of life" to Amerindians; vitamin-rich fruit; pith as source for bread; palm heart from shoots

Vicuna
(*Lama vicugna*)

Threatened species related to llama; source of meat, fur, and hides; can be profitably ranched

Maca
(*Lepidium meyenii*)

Cold-resistant root vegetable resembling radish, with distinctive flavor; near extinction

Chachalacas
(*Ortalis,* many species)

Tropical birds; adaptable to human habitations; fast-growing

The wild species shown here are just some of the many plants and animals that could supplement our food supply. Adapted from Wilson, E.O., 1992. The diversity of life. *Cambridge, MA: Belknap Press.*

selection of promising food resources from just one region of the world—Central and South America. Plenty more exist elsewhere worldwide.

The babassu palm of the Amazon produces more vegetable oil than any other plant. The serendipity berry generates a sweetener 3000 times sweeter than table sugar. Several species of salt-tolerant grasses and trees are so hardy that farmers can irrigate them with saltwater to produce animal feed, a vegetable oil substitute, and other products.

For our existing crops, having genetic diversity available in crop relatives and wild ancestors is enormously valuable (pp. 252–253). In 1995, Turkey's wheat crops received \$50 billion worth of disease resistance from wild wheat strains. California's barley crops annually receive \$160 million in disease resistance benefits from Ethiopian strains of barley. In the 1970s a researcher discovered a maize species in the mountains of Jalisco, Mexico, known as *Zea diploperennis*. This maize is highly resistant to disease, and it is a perennial, able to grow back year after year without being replanted. Yet we had almost lost this valuable plant; at the time of its discovery, its entire range was limited to a single 10-ha (25-acre) plot of land.

Organisms provide drugs and medicines

People have made medicines from plants for centuries, and many of today's pharmaceuticals are derived from chemical compounds from wild plants (TABLE 11.4). The rosy periwinkle produces compounds that treat Hodgkin's disease and a deadly form of leukemia. Had this plant from Madagascar become extinct, these two fatal diseases would have claimed far more victims. In Australia, a rare species of cork, *Duboisia leichhardtii*, provides hyoscine, a compound that physicians use to treat cancer, stomach disorders, and motion sickness. The Pacific yew of North America's Pacific Northwest produces a compound that forms the basis for the anti-cancer drug taxol. Even aspirin was derived from chemicals found in willows and in meadowsweet. Each year, pharmaceutical products owing their origin to wild species generate up to \$150 billion in sales and save thousands of human lives.

WEIGHING THE ISSUES

BIOPROSPECTING IN COSTA RICA Bioprospectors working for pharmaceutical companies scour biodiversity-rich countries, searching for organisms that can provide new drugs, foods, medicines, or other valuable products. Many have been criticized for "biopiracy"—harvesting indigenous species to create commercial products without compensating the country of origin. To make sure it would not lose the benefits of its own biodiversity, the nation of Costa Rica reached an agreement with the Merck pharmaceutical company in 1991. The nonprofit National Biodiversity Institute of Costa Rica (INBio) allowed Merck to evaluate a number of Costa Rica's species for their commercial potential in return for \$1.1 million, a small royalty rate on any products developed, and training for Costa Rican scientists.

Do you think both sides win in this agreement? What if Merck discovers a compound that it turns into a multi-billion-dollar drug? Does this provide a good model for other countries? For other companies?

TABLE 11.4 Natural Plant Sources of Pharmaceuticals

Image	Plant	Drug	Application
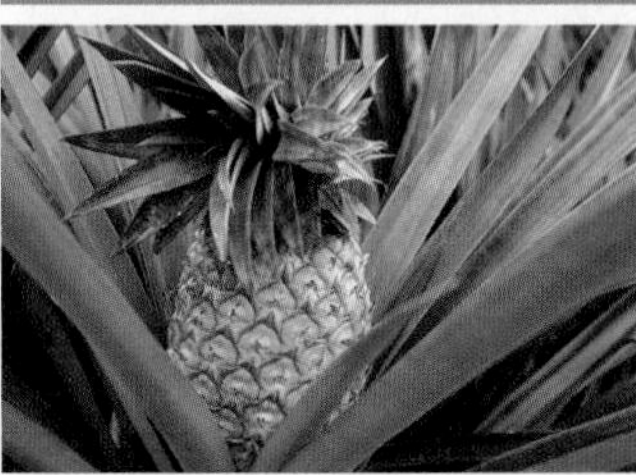	**Pineapple** (*Ananas comosus*)	Bromelain	Controls tissue inflammation
	**Pacific yew** (*Taxus brevifolia*)	Taxol	Anticancer agent (especially ovarian cancer)
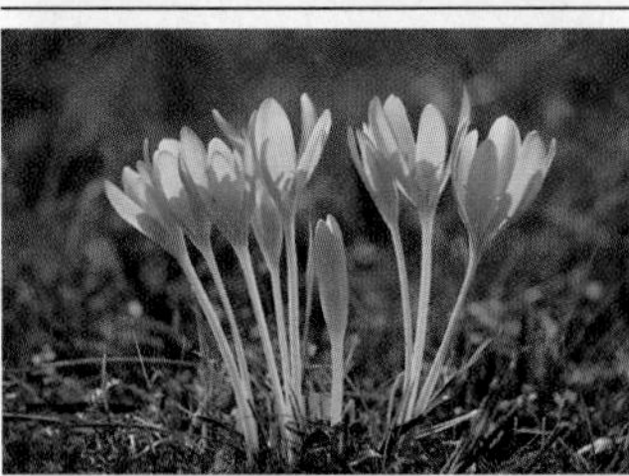	**Autumn crocus** (*Colchicum autumnale*)	Colchicine	Anticancer agent
	**Velvet bean** (*Mucuna deeringiana*)	L-Dopa	Parkinson's disease suppressant
	**Yellow cinchona** (several species of *Cinchona*)	Quinine	Antimalarial agent
	Common foxglove (*Digitalis purpurea*)	Digitoxin	Cardiac stimulant

Shown are just a few of the many plants that provide chemical compounds of medical benefit. Adapted from Wilson, E.O., 1992. The diversity of life. *Cambridge, MA: Belknap Press.*

The world's biodiversity holds a still-greater treasure chest of medicines yet to be discovered. It can truly be said that every species that goes extinct represents one lost opportunity to find a cure for cancer or AIDS. A recent international survey highlighted animals that show particular promise yet may be lost to extinction before we can profit from what they have to offer (**TABLE 11.5**). We lost such an opportunity in the two species of gastric brooding frogs recently discovered in the rainforests of Queensland, Australia (see top photo in Table 11.5). Females of these bizarre frogs raised their young inside their stomachs, where in any other animal, stomach acids would soon destroy them! Apparently the young frogs exuded substances that neutralized their mother's acid production. Any such substance could be of immense use for treating human stomach ulcers, which affect 25 million U.S. citizens. Sadly, both frog species went extinct in the 1980s, taking their medical secrets with them forever.

Biodiversity boosts economies through tourism and recreation

Besides providing for our food and health, biodiversity can generate income through tourism, particularly for developing countries in the tropics that boast impressive species diversity. Many people like to travel to experience protected natural areas, and in so doing they create economic opportunities for residents living near those areas. Visitors spend money at local businesses, hire local people as guides, and support parks that employ local residents. Ecotourism (p. 70) can thereby bring jobs and income to areas that otherwise might suffer poverty.

The parks and wildlife of Kenya and Tanzania are prime examples. Ecotourism brings in fully a quarter of all foreign money entering Tanzania's economy each year. Leaders and citizens in both nations recognize biodiversity's economic benefits, and as a result they have managed their parks and reserves diligently. Ecotourism is a vital source of income for nations such as Costa Rica, with its rainforests; Australia, with its Great Barrier Reef; and Belize, with its reefs, caves, and rainforests. The United States, too, benefits from ecotourism; its national parks draw millions of visitors from around the world.

Ecotourism can serve as a powerful financial incentive for nations, states, and local communities to preserve natural areas and reduce impacts on the landscape and on native species. Yet as ecotourism increases, an overabundance of visitors to natural areas can degrade the outdoor experience and disturb wildlife. Anyone who has been to Yosemite, the Grand Canyon, or the Great Smoky Mountains on a crowded summer weekend can attest to this. As ecotourism continues to grow, so will debate over its costs and benefits for local communities and for biodiversity.

People value connections with nature

Not all of biodiversity's benefits to people can be expressed in the hard numbers of economics or the day-to-day practicalities of food and medicine. Some scientists and philosophers argue that people find a deeper value in biodiversity. Harvard University biologist Edward O. Wilson has popularized the notion of **biophilia,** asserting that human beings

TABLE 11.5 Major Types of Animals at Risk That Offer Potential Medical Uses

SPECIES AT RISK	POTENTIAL MEDICAL USES
Amphibians 30% of all species are threatened with extinction. 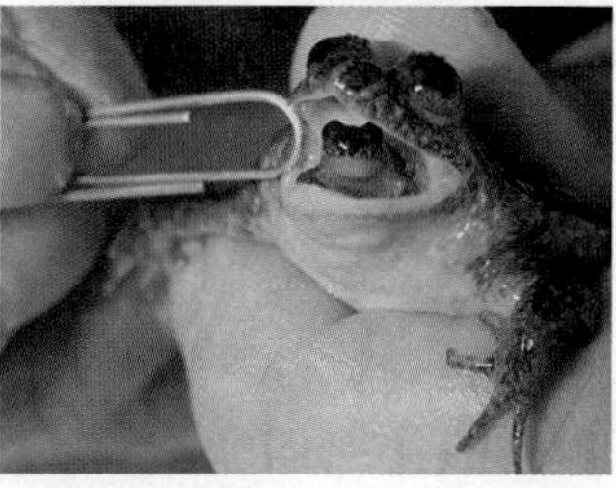	• Antibiotics, alkaloids for painkillers, chemicals for treating heart disease and high blood pressure • Natural adhesives for treating tissue damage • Ability to regenerate organs and tissues could suggest how we might, too. • "Antifreeze" compounds that allow frogs to survive freezing might help us preserve organs for transplants.
Sharks Overfishing has reduced populations of most species. Some risk extinction. 	• Squalamine from sharks' livers could lead to novel antibiotics, appetite-suppressants, drugs to shrink tumors, and drugs to fight vision loss. • Study of salt glands is helping address kidney diseases.
Horseshoe crabs Overfishing is sharply diminishing populations. 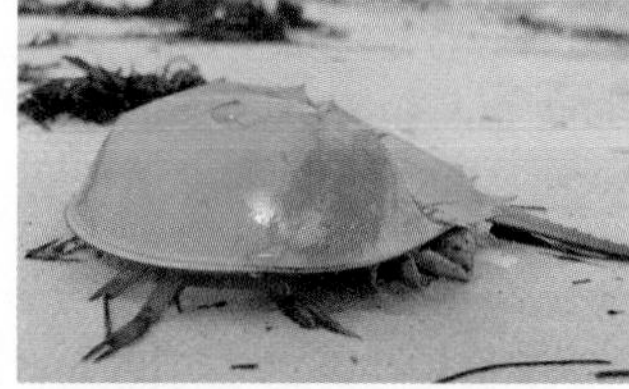	• A number of antibiotics are being developed. • The compound T140 may treat AIDS, arthritis, and several cancers. • Cells from blood can help detect cerebral meningitis in people.
Bears Nine species are at risk of extinction. 	• An acid from bears' gallbladders already treats gallstones and liver disease, and prevents bile buildup during pregnancy. • While hibernating, bears build bone mass. If we learn how, we could treat osteoporosis and hip fractures, which lead to 740,000 deaths per year. • Hibernating bears excrete no waste for months. Learning how could help treat renal disease.
Cone snails Most live in coral reefs, which are threatened ecosystems. 	Compounds from these snails include one that may prevent death of brain cells from head injuries or strokes, and a painkiller 1000 times more potent than morphine. So far just a few hundred of the 70,000–140,000 compounds these snails produce have been studied.

Adapted from Chivian, E., and A. Bernstein, 2008. Sustaining life: How human health depends on biodiversity. *Oxford Univ. Press.*

FIGURE 11.17 **An Indonesian girl peers into a flower of *Rafflesia arnoldii*, the largest flower in the world.** The concept of biophilia holds that human beings have an instinctive love and fascination for nature and a deep-seated desire to affiliate with other living things.

share an instinctive love for nature and feel an emotional bond with other living things (FIGURE 11.17). Wilson and others cite as evidence of biophilia our affinity for parks and wildlife, our love for pets, the high value of real estate with a view of natural landscapes, and our interest in hiking, bird-watching, fishing, hunting, backpacking, and similar outdoor pursuits.

In a 2005 book, writer Richard Louv adds that as today's children are increasingly deprived of outdoor experiences and direct contact with wild organisms, they suffer what he calls "nature-deficit disorder." Louv argues that this alienation from biodiversity and nature damages childhood development and may lie behind many of the emotional and physical problems young people in developed nations face today.

Do we have ethical obligations toward other species?

Aside from all of biodiversity's pragmatic benefits, many people feel that living organisms have an inherent right to exist. In this view, biodiversity conservation is justified on ethical grounds alone.

We human beings are part of nature, and like any other animal we need to use resources and consume other organisms to survive. In that sense, there is nothing immoral about our doing so. However, we also have conscious reasoning ability and are able to control our actions and make deliberate decisions. Our ethical sense has developed from this intelligence and ability to choose. As our society's sphere of ethical consideration has widened over time, and as more of us take up biocentric or ecocentric worldviews (p. 137), more of us have come to feel that other organisms have intrinsic value and an inherent right to exist.

Yet despite our expanding ethical convictions and despite biodiversity's many benefits, the future of biodiversity remains far from secure. The search for solutions to today's biodiversity crisis is urgent, dynamic, and exciting, and scientists are actively developing innovative strategies to maintain Earth's diversity of life.

Conservation Biology: The Search for Solutions

Today, more and more scientists and citizens perceive a need to stop the loss of biodiversity. In his 1994 autobiography, *Naturalist*, E.O. Wilson wrote:

> When the [20th] century began, people still thought of the planet as infinite in its bounty. [Yet] in one lifetime, exploding human populations have reduced wildernesses to threatened nature reserves. Ecosystems and species are vanishing at the fastest rate in 65 million years. Troubled by what we have wrought, we have begun to turn in our role from local conqueror to global steward.

Conservation biology responds to biodiversity loss

The urge to act as responsible stewards of natural systems, and to use science as a tool in this endeavor, sparked the rise of **conservation biology**, a scientific discipline devoted to understanding the factors, forces, and processes that influence the loss, protection, and restoration of biological diversity. It arose as biologists became increasingly alarmed at the degradation of the natural systems they had spent their lives studying. Conservation biologists choose questions and pursue research with the aim of developing solutions to such problems as habitat degradation and species loss (FIGURE 11.18). Conservation biology is thus an applied and goal-oriented science, with implicit values and ethical standards.

Conservation biologists work at multiple levels

Conservation biologists integrate an understanding of evolution and extinction with ecology and the dynamic nature of environmental systems. They use field data, lab data, theory, and experiments to study our impacts on other organisms. They also design, test, and implement ways to alleviate human impact. These researchers address the challenges facing biological diversity at all levels, from genes to species to ecosystems.

At the genetic level, conservation geneticists study genetic attributes of organisms to infer the status of their populations. If two populations of a species are genetically distinct, they may have different ecological needs and may require different types of management. Moreover, as a population dwindles, genetic variation is lost

(a) Sampling insects in Madagascar

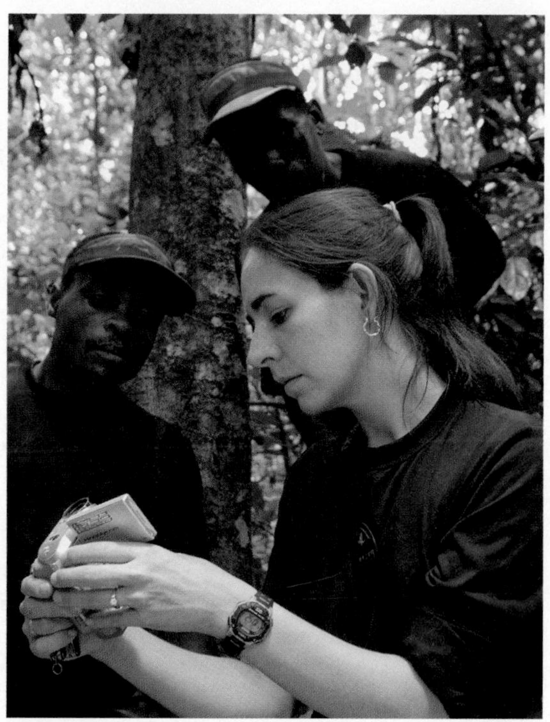

(b) Checking camera traps in Africa

(c) Drawing blood from a Seychelles Magpie Robin

(d) Radiotracking birds in Spain

FIGURE 11.18 Conservation biologists use many approaches to study the loss, protection, and restoration of biodiversity, seeking to develop scientifically sound solutions.

from the gene pool. Conservation geneticists investigate how small a population can become and how much genetic variation it can lose before running into problems such as inbreeding depression (p. 277), whereby genetic similarity causes parents to produce weak or defective offspring. By determining a minimum viable population size, conservation geneticists help wildlife managers decide how vital it may be to increase the population. Problems for populations spell problems for species, because declines and local extirpation can lead to range-wide endangerment and extinction.

Studies of genes, populations, and species inform conservation efforts with habitats, communities, ecosystems, and landscapes. As landscape ecologists know (p. 114), organisms may be distributed across a landscape as a *metapopulation*, or network of subpopulations. Because small and isolated subpopulations are most vulnerable to extirpation, conservation biologists pay special attention to them. By examining how organisms disperse from one habitat patch to another, and how their genes flow among subpopulations, conservation biologists try to learn how likely a population is to persist or succumb in the face of habitat change or other threats.

Endangered species are a focus of conservation efforts

The primary legislation for protecting biodiversity in the United States is the **Endangered Species Act (ESA).** Enacted in 1973, the Endangered Species Act forbids the government and private citizens from taking actions that destroy endangered species or their habitats. The ESA also forbids trade in products made from endangered species. The aim is to prevent extinctions, stabilize declining populations, and enable populations to recover. As of 2013, there were 1118 species in the United States listed as "endangered" and 322 more listed as "threatened," the status considered one notch less severe than endangered. For about 80% of these species, government agencies are running recovery plans to protect them and stabilize or increase their populations.

The ESA has had a number of successes. Following the 1973 ban on the pesticide DDT (p. 369) and years of intensive effort by wildlife managers, the bald eagle, peregrine falcon, brown pelican, and other birds have recovered and are no longer listed as endangered (FIGURE 11.19). Intensive management programs with other species, such as the red-cockaded woodpecker (see Figure 12.15, p. 319), have held populations steady in the face of continued pressure on habitat. Overall, roughly 40% of declining populations have been stabilized.

This success comes despite the fact that the U.S. Fish and Wildlife Service and the National Marine Fisheries Service, the agencies responsible for upholding the ESA, are perennially underfunded for the job. Reauthorization of the ESA faced opposition from the Republican-led Congresses in power from 1994 to 2006. Efforts in 2006 to weaken the ESA by stripping it of its ability to safeguard habitat were narrowly averted after 5700 scientists sent Congress a letter of protest.

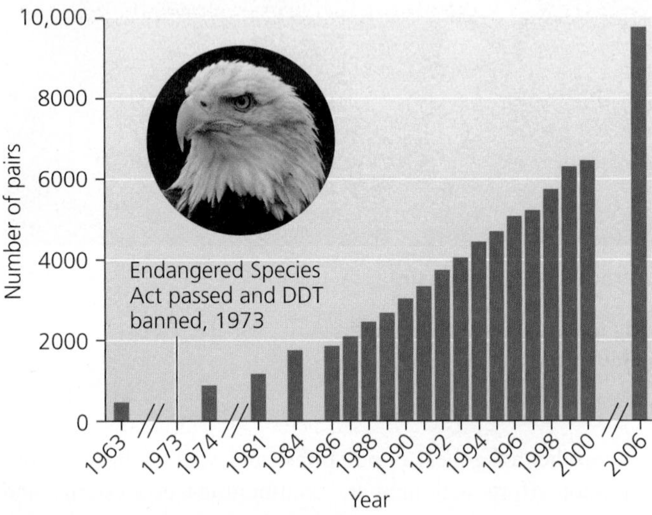

FIGURE 11.19 **The bald eagle's recovery is a success story of the Endangered Species Act.** The U.S. national symbol was close to extinction in the Lower 48 states in the 1960s. Following protection under the ESA and a ban on DDT in 1973, eagles began to rebound. With its Lower-48 population reaching 10,000 pairs in 2007, the bald eagle was declared recovered and was removed from the Endangered Species List. *Data from U.S. Fish and Wildlife Service, based on annual volunteer surveys. The paucity of data after 2000 is because surveys were discontinued once it became clear the eagle was recovering.*

FIGURE 11.20 **The greater sage grouse is one species whose addition to the Endangered Species List is warranted by science yet precluded by a lack of funding.** Sage grouse, like this displaying male, lived in sagebrush habitats throughout the western United States. Their listing could complicate efforts to drill for oil and gas on these lands.

Today, a number of species have been judged by scientists to be in need of ESA protection but have not been added to the endangered species list because the government is not supplying funding to help recover them. Such species are said to be "warranted but precluded"; that is, their listing is warranted by scientific research, but it is precluded by lack of resources (FIGURE 11.20). This has led some environmental advocacy groups to sue the federal government for failing to enforce the law. Dedicated and well-meaning Fish and Wildlife Service staff are frequently caught in a no-win situation, battling lawsuits from the political left while being starved of funds by the political right.

Polls repeatedly show that most Americans support protecting endangered species. Yet some opponents feel that the ESA places more value on the life of an endangered organism than on a person's livelihood. This has been a common perception in the Pacific Northwest, where protection for the northern spotted owl slowed logging in old-growth forests and loggers began to fear for their jobs. In addition, many landowners worry that federal officials will restrict the use of private land on which threatened or endangered species are found. This has led to a practice described as "shoot, shovel, and shut up," among some landowners who want to conceal the presence of such species on their land.

In fact, however, the ESA has stopped few development projects—and a number of its provisions and amendments promote cooperation with landowners. *Habitat conservation plans* and *safe harbor agreements* are arrangements that allow private landowners to harm species in some ways if they voluntarily improve habitat for the species in others.

Today many nations have laws protecting species, although they are not always effective. When Canada enacted its **Species at Risk Act** in 2002, the Canadian government was careful to stress cooperation with landowners and provincial governments, and not to present the law as a decree from the national government. Environmental advocates and scientists protested that the law was weak and failed to protect habitat adequately.

International treaties promote conservation

The United Nations has facilitated several international treaties to protect biodiversity. The 1973 **Convention on International Trade in Endangered Species of Wild Fauna and Flora (CITES)** protects endangered species by banning the international transport of their body parts. When nations enforce it, CITES can protect rhinos, elephants, tigers, and other rare species whose body parts are traded internationally.

In 1992, leaders of many nations agreed to the **Convention on Biological Diversity.** This treaty embodies three goals: to conserve biodiversity, to use biodiversity in a sustainable manner, and to ensure the fair distribution of biodiversity's benefits. The Convention aims to:

- Provide incentives for biodiversity conservation
- Manage access to and use of genetic resources
- Transfer technology, including biotechnology
- Promote scientific cooperation
- Assess the effects of human actions on biodiversity
- Promote biodiversity education and awareness
- Provide funding for critical activities
- Encourage nations to share reports on their conservation efforts

The treaty has helped African nations gain economic benefits from ecotourism with their wildlife preserves. It has also prompted nations worldwide to protect more area in reserves, has enhanced global markets for sustainable crops such as shade-grown coffee, and has moved some rice-growing Asian nations away from pesticide-intensive farming practices. Yet the treaty's overall goal—"to achieve, by 2010, a significant reduction of the current rate of biodiversity loss at the global, regional and national level"—was not met.

Captive breeding, reintroduction, and cloning are being used to save species

In the effort to save threatened and endangered species, zoos and botanical gardens have become centers for **captive breeding,** in which individuals are bred and raised in controlled conditions with the intent of reintroducing them into the wild. The IUCN counts 65 plant and animal species that now exist *only* in captivity or cultivation.

Reintroducing species into areas they used to occur is expensive and resource-intensive, but it often works and can pay big dividends. In 2010 the first of 32 black rhinos were translocated from South Africa to Serengeti National Park to help restore a former population (FIGURE 11.21a). This followed similar reintroduction projects elsewhere in Africa.

In North America, a high-profile example of captive breeding and reintroduction is the program to save the California condor, the continent's largest bird (FIGURE 11.21b). Although they are harmless scavengers of dead animals, condors were shot by people in the early 20th century. They also collided with electrical wires and succumbed to lead poisoning after scavenging carcasses of animals killed with lead shot. By 1982, only 22 condors remained, and biologists made the wrenching decision to take all the birds into captivity, in last-ditch hopes of boosting their numbers and releasing them. Today the ongoing program—a collaboration between the Fish and Wildlife Service and several zoos—is succeeding. As of 2013 there were 170 birds in captivity and 234 birds living in the wild. Condors have been released at sites in

(a) A black rhino is air-lifted into Serengeti National Park

(b) Biologists use hand puppets to nurse condor chicks

FIGURE 11.21 **We can re-establish populations and rescue species by reintroducing them to areas where they have been extirpated.** Black rhinos **(a)** have been helicoptered in to Serengeti National Park from other areas where populations are increasing. To save the California condor **(b)** from extinction, biologists raise chicks in captivity with hand puppets that mimic the heads of adult condors. The chicks are shielded from contact with humans so that when grown, they do not feel an attachment to people.

California, Arizona, and Baja California, and each year they thrill thousands of people lucky enough to sight the huge birds soaring through the skies. Unfortunately, many of these long-lived birds still die from lead poisoning, and wild populations will likely not become sustainable until hunters convert from lead shot to shot made of copper or steel.

Other reintroduction programs have been more controversial. The successful program to reintroduce gray wolves to Yellowstone National Park has proven popular with the American public but continues to meet stiff resistance from ranchers, who fear the wolves will attack their livestock. In Arizona and New Mexico, a wolf reintroduction program has made slow headway, but a number of wolves there have been shot.

One new idea for saving species from extinction is to create individuals by cloning them. In this technique, DNA (p. 29) from an endangered species is inserted into a cultured egg without a nucleus, and the egg is implanted into a female of a closely related species that acts as a surrogate mother. Several mammals have been cloned in this way, with mixed results. Some scientists even talk of recreating extinct species from DNA recovered from preserved body parts. Indeed, in 2009 a subspecies of Pyrenean ibex (a type of mountain goat) was cloned from cells taken from the last surviving individual, which had died in 2000. The cloned baby ibex died shortly after birth, however. Even if cloning can succeed from a technical standpoint, such efforts are not an adequate response to biodiversity loss. Without ample habitat and protection in the wild, having cloned animals in a zoo does little good.

Forensics can help to protect species

To counter poaching and other illegal harvesting, scientists have a new tool at their disposal. **Forensic science,** or **forensics,** involves the scientific analysis of evidence to make an identification or answer a question relating to a crime or an accident. Conservation biologists are now employing forensics to protect species at risk. By analyzing DNA from organisms or their tissues sold at market, researchers can often determine the species or subspecies of organism—and sometimes its geographic origin. This information can help detect illegal activity, enhancing the enforcement of laws protecting wildlife. A prime example is the analysis of whale meat sold in Asian markets (see **The Science behind the Story,** pp. 300–301).

Another example is the effort to track the geographic origin of tusks of African elephants killed for ivory. Trade in ivory has been banned under CITES in an effort to stop the slaughter of elephants. After customs agents seized 6.5 tons of tusks in Singapore in 2002, researchers led by Samuel Wasser of the University of Washington analyzed DNA from the tusks to determine the geographic origin of the elephants that were killed. The researchers sought to find out whether the tusks belonged to savanna elephants killed in Zambia (the origin of the shipment), or whether they came from forest elephants from other locations. The DNA matched known samples from Zambian elephants, indicating that many more elephants were being killed there than Zambia's government had realized. In response, the Zambian government replaced its wildlife director and began imposing harsher sentences on poachers and ivory smugglers.

Some species act as "umbrellas" that protect habitat and communities

Scientists know that protecting species does little good if the larger systems they rely on are not also sustained. Yet no law or treaty exists to protect communities or ecosystems. For these reasons, conservation biologists often use particular species as tools to conserve habitats, communities, and ecosystems. Such species are called *umbrella species* because they serve as a kind of umbrella to protect many other species. Umbrella species often are large animals that roam great distances, as many of the Serengeti's most notable species do. Because such animals require large areas, meeting their habitat needs helps meet those of thousands of less charismatic animals, plants, and fungi that might never elicit as much public interest.

Environmental advocacy organizations have found that using large and charismatic vertebrates as spearheads for biodiversity conservation is an effective strategy. This approach of promoting particular *flagship species* is evident in the longtime symbol of the World Wide Fund for Nature (in North America, the World Wildlife Fund), the panda. A large endangered animal requiring sizeable stands of undisturbed bamboo forest, the panda's lovable appearance has made it a favorite with the public—and an effective vehicle for soliciting support for conservation efforts that protect far more than just the panda.

At the same time, many conservation organizations are moving beyond the single-species approach. The Nature Conservancy focuses on whole communities and landscapes; it purchases lands that preserve important habitats and seeks to connect them with other preserved lands so that ecological processes can function across broad regions.

Parks and protected areas help conserve biodiversity at the ecosystem level

Our practice of setting aside areas of undeveloped land to be preserved in parks and protected areas helps to conserve habitats, communities, ecosystems, and landscapes. Currently we have set aside 13% of the world's land area in national parks, state parks, provincial parks, wilderness areas, biosphere reserves, and other protected areas. Many of these lands are managed for recreation, water quality protection, or other purposes, rather than for biodiversity, and many suffer from illegal logging, poaching, and resource extraction because enforcement is lacking. Yet these areas offer animals and plants a degree of protection from human persecution, and some are large enough to preserve whole natural systems that otherwise would be fragmented, degraded, or destroyed.

Serengeti National Park and the adjacent Maasai Mara National Reserve are two of the world's largest and most famous parks, but Tanzania and Kenya have each set aside a number of other protected areas. Some of the best known include (in Kenya) Amboseli National Park, Tsavo National Park, Mount Kenya National Park, Lake Nakuru National Park, and Kakamega Forest National Reserve; and (in Tanzania) Ngorongoro Conservation Area, Mafia Island Marine Park, Selous Game Reserve, Kilimanjaro National Park, and Gombe

Stream National Park. Altogether roughly 25% of Tanzania's land area and 12% of Kenya's land area is protected.

Alas, protecting land may not be enough to assure effective conservation. Pressures from outside the reserves in Kenya and Tanzania are causing declines in wildlife within the reserves (see **The Science behind the Story,** pp. 286–287). Similar situations occur in North America: Despite the considerable size of Yellowstone National Park, animals such as elk, bears, bison, and wolves roam seasonally in and out of the park. As a result, conservationists have tried to find ways to protect animals and habitats across the Greater Yellowstone Ecosystem, the larger region over which the animals roam.

Moreover, as global climate change (Chapter 18) proceeds, changing conditions are driving many species toward the poles and upward in elevation, often forcing them out of protected areas. Thus, a major challenge today is to link protected areas across the landscape so that species like wildebeest can move in response to climate change as it alters habitats within protected areas. We will explore parks and protected areas and the issues they face more fully in Chapter 12 (pp. 323–332).

Biodiversity hotspots pinpoint regions of high diversity

To prioritize regions for conservation, scientists have mapped biodiversity hotspots (FIGURE 11.22a). A **biodiversity hotspot** is a region that supports an especially great number of species that are endemic (p. 59), found nowhere else in the world (FIGURE 11.22b). To qualify as a hotspot, a region must harbor at least 1500 endemic plant species (0.5% of the world's total plant species). In addition, a hotspot must have already lost 70% of its habitat to human impact and be at risk of losing more.

The ecosystems of the world's biodiversity hotspots together once covered 15.7% of the planet's land surface. Today, because of habitat loss, they cover only 2.3%. This small amount of land is the exclusive home for half the world's plant species and 42% of terrestrial vertebrate species. The hotspot concept motivates us to focus on these areas, where the greatest number of unique species can be protected per unit effort.

We can restore degraded ecosystems

Protecting natural areas before they become degraded is the best way to safeguard biodiversity and ecological systems. However, in some cases we can restore degraded natural systems to some semblance of their former condition, through the practice of **ecological restoration** (pp. 92–93). Ecological restoration aims not simply to bring back populations of animals and plants, but to reestablish the processes—the cycling of matter and the flow of energy—that make an ecosystem function. By restoring complex natural systems such as the Illinois prairies (p. 92), the Florida Everglades (pp. 92–93), or the southeastern longleaf pine forest (p. 319), restoration ecologists aim to recreate functioning systems that filter pollutants, cleanse water and air, build soil, and recharge groundwater, providing habitat for native wildlife and services for people.

In Kenya, efforts are being made to restore the Mau Forests Complex, Kenya's largest remaining forested area and a watershed that provides water for the Maasai Mara Reserve and for the people of the region. Over the years so much forest in this densely populated region has been destroyed by agriculture, settlement, and timber extraction that the water supply for the Serengeti's wildlife and for millions of Kenyan people is now threatened. Kenya's government is working

(a) The world's biodiversity hotspots

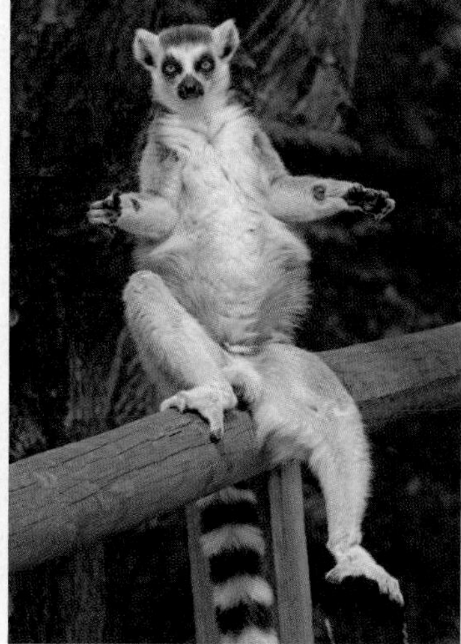

(b) Ring-tailed lemur

FIGURE 11.22 **Biodiversity hotspots are priority areas for habitat preservation because they contain many endemic species.** In red **(a)** are the 34 hotspots mapped by the nongovernmental organization Conservation International. (Only 15% of the areas in red are actually habitat; most area is developed.) These regions are home to species such as the ring-tailed lemur **(b)**, a primate endemic to Madagascar that has lost over 90% of its forest habitat as a result of human population growth and resource extraction. *Data from Conservation International.*

THE SCIENCE BEHIND THE STORY

Using Forensics to Uncover Illegal Whaling

As any television buff knows, forensic science is a crucial tool in solving mysteries and fighting crime. In recent years, conservation biologists have been using forensics to unearth secrets and catch bad guys in the multi-billion-dollar illegal global wildlife trade. One such detective story comes from the Pacific Ocean and Japan.

The meat from whales has long been a delicacy in Japan and other nations. Whaling ships decimated populations of most species of whales in the 20th century through overhunting, and the International Whaling Commission (IWC) outlawed commercial whaling worldwide beginning in 1986. Yet whale meat continues to be sold to wealthy consumers (**FIGURE 1**). This meat comes legally from several sources:

- From scientific hunts. Japan and several other nations negotiated with the IWC to continue to hunt limited numbers of whales for research purposes, and this meat may be sold afterwards.
- From whales killed accidentally when caught in fishing nets meant for other animals (bycatch; p. 439).
- Possibly from stockpiles frozen before the IWC's moratorium.

However, conservation biologists long suspected that much of the whale meat on the market was actually caught illegally for the purpose of selling for food and that fleets from Japan and other nations were killing more whales than international law allowed. Once DNA sequencing technology was developed, scientists could use this tool to find out.

The detectives in this story are conservation geneticists C. Scott Baker, Stephen Palumbi, Frank Cipriano, and their colleagues. For close to two decades they have been traveling to Asia on what have amounted to top-secret grocery shopping trips.

Dr. C. Scott Baker runs genetic analyses in a Japanese hotel room.

It began in 1993, when Baker and Palumbi bought samples of whale meat—all labeled simply as *kujira*, the generic Japanese term for whale meat—from markets in Japan and sequenced DNA from these samples. Law forbids the export of whale meat, so the researchers had to run analyses in their hotel rooms with portable genetic kits. Once back home in the United States, they compared their data with sequences from known whale species.

By analyzing which samples matched which, they concluded that they had sampled meat from nine minke whales, four fin whales, one humpback whale, and two dolphins. Because subspecies of whales from different oceans differ genetically, the researchers were able to analyze the genetic variation in their samples and learn that one fin whale came from the Atlantic, the other three came from the Pacific, and eight of the nine minke whales came from the Southern Hemisphere.

Because several of these species and subspecies were off-limits to hunting, the data suggested that some meat had been hunted, processed, or traded illegally. Baker and Palumbi concluded in a 1994 paper in *Science* that "legal whaling serves as a cover for the sale of illegal whale products." They urged that the international community monitor catches more closely.

Two years later, Baker, Palumbi, and Cipriano presented results from markets in South Korea and Japan. Again their genetic sleuthing revealed a diversity of whale species, and they stated that their data were "difficult to reconcile" with records of legal catches (scientific whaling by Japan and fishing bycatch by South Korea) reported by these nations to the IWC. Among the whales they detected were two specimens of what seemed to be a subspecies or species of whale new to science.

FIGURE 1 Whale meat, much of it illegal, is sold in Japanese and Korean markets.

In 2000, the team analyzed 655 samples labeled as whale meat from Japanese and South Korean markets and found evidence for 12 species or subspecies of whales, along with orcas, porpoises, and dolphins—and even sheep and horses! Seven of the whale species were internationally protected, and together these constituted 10% of the whale meat for sale in Japanese markets.

Genetic analyses of minke whale samples from Japan's markets also indicated that a large percentage came from animals in the Sea of Japan, where Korea and Japan harvested them as fishing bycatch. One-third of the meat on the market was coming from the Sea of Japan, meaning that four times as many whales were being killed there as Japan was reporting (**FIGURE 2**). The research team calculated that Japan and Korea together were taking so many minke whales from the Sea of Japan that they would eventually wipe out the population (**FIGURE 3**).

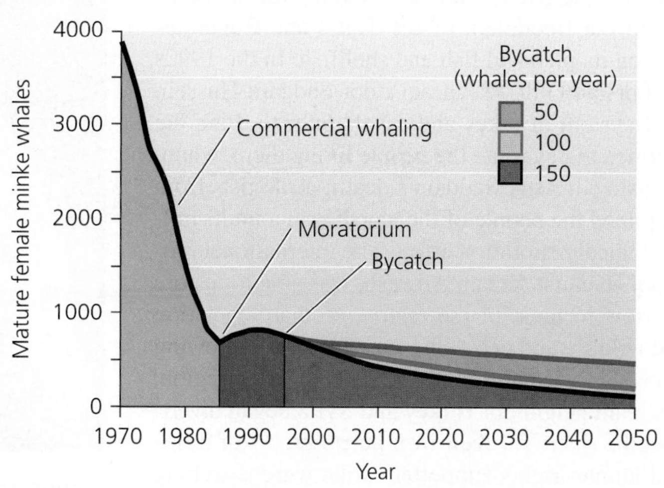

FIGURE 3 Minke whales in the Sea of Japan declined sharply until the 1986 moratorium on their capture. Population models forecast that bycatch of 150, 100, or even 50 minke whales per year would prevent the population's recovery. Data suggest that actual bycatch from the Sea of Japan has been close to 150 per year. *Adapted from Baker, C.S., et al., 2000. Predicted decline of protected whales based on molecular genetic monitoring of Japanese and Korean markets.* Proc. Roy. Soc. Lond. B *267: 1191–1199, Fig 3. By permission of The Royal Society and the author.*

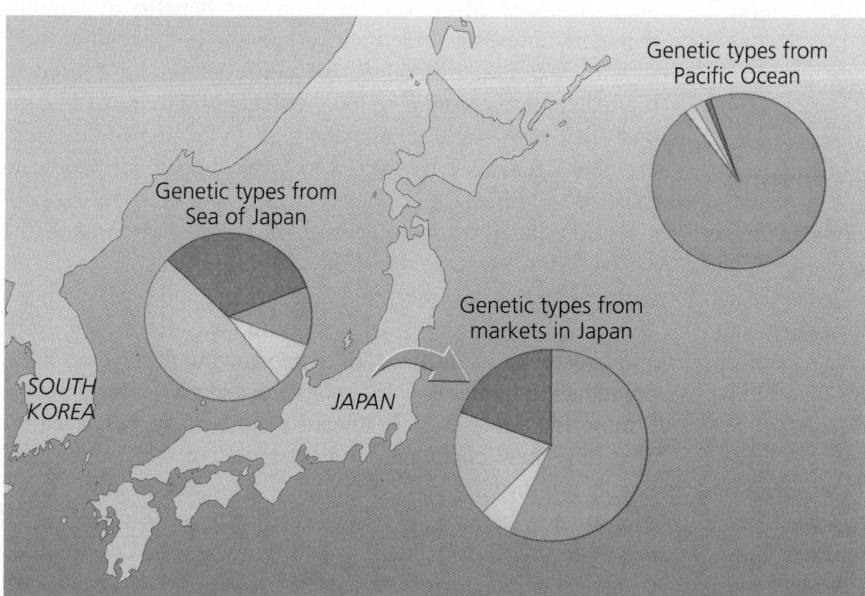

FIGURE 2 Genetic analysis can reveal the origin of animal products. Data on genetic types from minke whale meat in Japanese markets (**center pie chart**) shows evidence of whales from the Sea of Japan (**left chart**) and of whales from the Pacific Ocean (**right chart**). *Adapted from Lukoschek, V., et al., 2009. High proportion of protected minke whales sold on Japanese markets due to illegal, unreported, or unregulated exploitation.* Animal Conservation *12: 385–395, Fig 2. By permission of John Wiley and Sons. www.interscience.wiley.com.*

DATA Q From the data shown, how can we tell that the meat samples from the markets include individuals from both the Sea of Japan and the Pacific Ocean?

In 2007, Baker led a team that combined genetic forensics with ecological methods to estimate numbers of individual whales whose meat was passing through Korean markets. They inferred that meat from 827 minke whales had passed through South Korea's market in five years. The nation had reported catching only 458 minke whales as fishing bycatch, leading the researchers to conclude that the remainder had been taken illegally.

The governments of Japan and South Korea have tried to refute these findings. Yet the technology and approaches that turn scientists into forensic detectives are influencing the debate and negotiation over whaling policy at the international level. ●

with international agencies and with U.S. funding to replant and protect areas of the forest.

The world's highest-profile restoration project today is an effort to restore the vast marshes of southern Iraq. For several thousand years, people lived sustainably among the wetlands of this region in the floodplain of the Tigris and Euphrates rivers, thriving on its plentiful fish and shellfish. In the 1990s following the Persian Gulf War, Iraqi ruler Saddam Hussein ordered a huge system of dikes and canals built to drain the marshes. He aimed to devastate the people living there, whom he viewed as disloyal. After Saddam's death, ecologists from many nations joined the people of the marshes in a multi-million-dollar ecological restoration effort. The international project, led by Iraqi scientist Azzam Allwash, was able to restore natural water flow to most of the region, so that vegetation grew back and wildlife and people began returning. Following this rapid success, however, drought descended on the region, and Iraq's upstream neighbors Turkey and Syria began diverting water from the rivers for their own purposes. As of 2013, ecologists and human rights supporters alike were searching for ways to complete this ambitious restoration project.

WEIGHING THE ISSUES

SINGLE-SPECIES CONSERVATION? What would you say are some advantages of focusing on conserving single species, versus trying to conserve broader communities, ecosystems, or landscapes? What might be some of the disadvantages? Which do you think is the better approach, or should we use both, and why?

Community-based conservation is growing

Helping people, wildlife, and ecosystems all at the same time is the focus of many current efforts in conservation biology. In the past, conservationists from industrialized nations, in their zeal to preserve ecosystems in other parts of the world, often neglected the needs of people in the areas they wanted to protect. Developing nations came to view this as a kind of neocolonialism. Today this has changed, and many conservation biologists actively engage local people in efforts to protect land and wildlife—a cooperative approach called **community-based conservation.** As of 2010, 23% of the world's protected areas were being managed under some kind of community-based conservation.

Community-based conservation has been widely practiced in East Africa. Conservationists and scientists began working with the Maasai and other people of the region years ago, understanding that in order to conserve animals and ecosystems, the local people need to be stewards of the land and feel invested in conservation (**FIGURE 11.23**). This has proved challenging because the parks and reserves were created on land historically used by local people. Sometimes people were forcibly relocated; by some estimates 50,000 Maasai were evicted to create Serengeti National Park. In the view of many local people, the parks were a government land grab, and laws against poaching deprive them of a right to kill wildlife. As human population grew in the region, conflicts between people and wildlife increased. Ranchers worried that wildebeest and buffaloes might spread disease to their cattle. Farmers lost produce when elephants roamed into their fields at night and ate their crops. Moreover, the economic benefits of ecotourism were not being shared with all people in the region.

FIGURE 11.23 Scientists and conservation advocates work cooperatively with local people to conserve wildlife. Biologist Alayne Cotterill of the conservation research group Living with Lions works with Maasai warriors to monitor lion populations near the Serengeti.

In response, proponents of conservation have tried to reallocate tourist dollars to local villages and to transfer some authority over wildlife management to local people. Government agencies have begun to work with local people to manage wildlife jointly, in a type of community-based conservation called *co-management*. In the regions around the Maasai Mara Reserve, the Kenya Wildlife Service and international NGOs have been helping farmers and ranchers build strong electric fences to keep wildlife away from their crops and livestock. Scientific studies are showing that these efforts are reducing human–wildlife conflicts and giving local people a more favorable attitude toward conservation, but that they also raise new challenges. For instance, in some cases elephants are thriving and reproducing on reserves, but because they are now hemmed in by fences, they become overcrowded and overgraze the vegetation, destroying trees and endangered plants the reserves were meant to protect.

Working cooperatively with local people to make conservation beneficial for them requires determination, patience, investment, and trust on all sides. Setting aside land for preservation may deprive local people of access to exploitable resources, but it also helps ensure that those resources can be sustainably managed and will not be used up or sold to foreign corporations. If tourism revenues are adequately spread, people see direct economic benefits of conserving wildlife and may be willing to work alongside government agencies and international conservationists. Community-based conservation has not always been successful, but in a world of rising human population, sustaining biodiversity will require locally based management that sustainably meets people's needs.

Conclusion

Data from scientists worldwide confirm what any naturalist who has watched the habitat change in his or her hometown already knows: From amphibians to zebras, biological diversity is being lost rapidly within our lifetimes. This erosion of biodiversity threatens to result in a mass extinction event equivalent to those of the geologic past. Habitat alteration, invasive species, pollution, overharvesting of biotic resources, and climate change are the primary causes of biodiversity loss. This loss matters, because human society cannot function without biodiversity's pragmatic benefits. Conservation biologists are rising to the challenge of conducting science aimed at saving endangered species, protecting their habitats, recovering populations, and preserving and restoring natural ecosystems. The innovative strategies these scientists are pursuing hold promise to slow, and perhaps reverse, the loss of biodiversity on Earth.

Reviewing Objectives

You should now be able to:

Characterize the scope of biodiversity on Earth

- Biodiversity can be thought of at three levels: species diversity, genetic diversity, and ecosystem diversity. (pp. 276–278)
- Some taxonomic groups (such as insects) hold far more diversity than others. (pp. 278–279)
- Roughly 1.8 million species have been described so far, but scientists agree that the world holds millions more. (pp. 278–279)
- Diversity is unevenly spread across different habitats, biomes, and regions of the world. (pp. 279–280)

Understand today's extinction crisis in geologic context

- Species have gone extinct at a background rate of roughly one species per 1–10 million species each year. Most species that have ever lived are now extinct. (p. 281)
- Earth has experienced five mass extinction events in the past 440 million years. (p. 281)
- Human impact is now initiating a sixth mass extinction. (pp. 282–283)
- Most biodiversity loss consists of a gradual reduction of population sizes and extirpation of populations. (p. 283)

Evaluate the primary causes of biodiversity loss

- Habitat loss (through destruction, alteration, or fragmentation) is the main cause of current biodiversity loss. (pp. 283–285)
- Pollution, overharvesting, and invasive species are also important causes. (pp. 285–289)
- Climate change is becoming a major cause. (p. 289)
- Amphibians are facing a global crisis, probably from a mix of factors. (p. 289)

Specify the benefits that biodiversity brings us

- Biodiversity supports functioning ecosystems and the services they provide us. (pp. 290–291)
- Wild species are sources of food, medicine, and economic development. (pp. 291–293)
- Many people feel that we have a psychological need to connect with the natural world, as well as an ethical duty to preserve nature. (pp. 293–294)

Assess the science and practice of conservation biology

- Conservation biology studies biodiversity loss and seeks ways to protect and restore biodiversity. (p. 294)
- Conservation biologists integrate research at the genetic, population, species, ecosystem, and landscape levels. (pp. 294–295)

Analyze efforts to conserve threatened and endangered species

- The U.S. Endangered Species Act has been largely effective, despite debate over its merits and limitations in funding. (p. 296)
- CITES and the Convention on Biological Diversity are major international treaties to safeguard biodiversity. (p. 297)
- Modern recovery strategies include captive breeding and reintroduction programs. (pp. 297–298)
- Forensics can help us trace products from illegally poached animals. (pp. 298, 300–301)

Compare and contrast conservation efforts above the species level

- Charismatic species are often used in popular appeals to conserve habitats and ecosystems. (p. 298)
- Parks and protected areas conserve biodiversity at the landscape level. (pp. 298–299)
- Biodiversity hotspots help prioritize regions globally for conservation. (p. 299)
- Ecological restoration efforts are restoring degraded ecosystems. (pp. 299, 302)
- Community-based conservation empowers people to invest in conserving their local species and ecosystems. (p. 302)

Testing Your Comprehension

1. What is biodiversity? Describe three levels of biodiversity.
2. What are the five primary causes of biodiversity loss? Give one specific example of each.
3. List three invasive species, and describe their impacts.
4. Define the term ecosystem services. Give three examples of ecosystem services that people would have a hard time replacing if their natural sources were eliminated.
5. What is the relationship between biodiversity and food security? Between biodiversity and pharmaceuticals? Give three examples of potential benefits of biodiversity conservation for food supplies and medicine.
6. List three reasons why people suggest that biodiversity conservation is important.
7. Describe one successful accomplishment of the U.S. Endangered Species Act. Now describe one reason some people have criticized it.
8. Explain how captive breeding can help with endangered species recovery, and give an example. Now explain why cloning could never be, in itself, an effective response to species loss.
9. Name two reasons that a large national park like Serengeti or Yellowstone might not be enough to effectively conserve a population of a threatened species. What solutions exist to address these reasons?
10. Explain the notion of community-based conservation. Why have conservation advocates been turning to this approach? What challenges exist in implementing it?

Seeking Solutions

1. Many arguments have been advanced for the importance of preserving biodiversity. Which argument do you think is most compelling, and why? Which argument do you think is least compelling, and why?
2. Some people declare that we shouldn't worry about endangered species because extinction has always occurred. How would you respond to this view?
3. Conservation advocates from industrialized nations have long pushed to set aside land in biodiversity-rich regions of developing nations. Leaders of developing nations have accused them of neocolonialism. "Your nations attained prosperity and power by overexploiting your environments decades or centuries ago," these leaders ask, "so why should we now sacrifice our development by setting aside our land and resources?" What would you say to these leaders? What would you say to the conservation advocates? Do you see ways that both preservation and development goals might be reached?
4. Compare the approach of setting aside protected areas with the approach of community-based conservation. What are the advantages and disadvantages of each? Can we—and should we—follow both approaches?
5. **THINK IT THROUGH** You are an influential legislator in a nation that has no endangered species act, and you want to introduce legislation to protect your nation's vanishing biodiversity. Consider the U.S. Endangered Species Act and Canada's Species At Risk Act, as well as international efforts such as CITES and the Convention on Biological Diversity. What strategies would you write into your legislation? How would your law be similar to and different from each of these efforts?
6. **THINK IT THROUGH** As a resident of your community and a parent of two young children, you attend a town meeting called to discuss the proposed development of a shopping mall and condominium complex. The development would eliminate a 100-acre stand of forest, the last sizeable forest stand in your town. The developers say the forest loss will not matter because plenty of 1-acre stands still exist scattered throughout the area. Consider the development's possible impacts on the community's biodiversity, children, and quality of life. What will you choose to tell your fellow citizens and the town's decision-makers at this meeting, and why?

Calculating Ecological Footprints

Research shows that much of humanity's footprint on biodiversity comes from our use of grasslands for livestock grazing and forests for timber and other resources. Grasslands and forests contribute different amounts to each nation's biocapacity (pp. 5, 208) depending on how much of these habitats each nation has. Likewise, the per capita biocapacity and per capita ecological footprints of each nation vary further according to their populations. When footprints are equal to or below biocapacity, then resources are being used sustainably. When footprints surpass biocapacity, then resources are being used unsustainably.

In the table, fill in the proportion of each nation's per capita footprint accounted for by use of grazing land and forest land. Then fill in the proportion of each nation's per capita biocapacity provided by grazing land and forest land.

Footprints and biocapacities in hectares per person	Kenya	Tanzania	United States	Canada
Footprint from grazing land	0.27	0.36	0.19	0.42
Footprint from forest land	0.28	0.24	0.86	0.74
Total ecological footprint	0.95	1.19	7.19	6.43
Percent footprint from grazing and forest	58			
Biocapacity of grazing land	0.27	0.39	0.26	0.23
Biocapacity of forest land	0.02	0.13	1.56	8.27
Total biocapacity	0.53	1.02	3.86	14.92
Percent biocapacity from grazing and forest				57

Data from WWF, 2012. Living Planet Report 2012. *WWF International, Gland, Switzerland.*

1. In which nations is grazing land being used sustainably? In which nations is forest land being used sustainably?
2. Do forest use and grazing comprise a larger part of the footprint for temperate-zone industrialized nations such as Canada and the United States, or for tropical-zone developing nations such as Kenya and Tanzania? What do you think accounts for this difference between these two types of nations? What else besides land use of forests and grasslands contributes to an ecological footprint?
3. The Living Planet Index declined 28% since 1970 (p. 283), but its temperate and tropical components were very different. During this period the index for temperate regions increased by 31%, whereas the index for tropical regions declined by 61%. Based on this information, do you predict that biodiversity loss has been steepest in Kenya and Tanzania or in Canada and the United States? Explain your answer.

MasteringENVIRONMENTALSCIENCE™

STUDENTS

Go to **MasteringEnvironmentalScience** for assignments, the etext, and the Study Area with practice tests, videos, current events, and activities.

INSTRUCTORS

Go to **MasteringEnvironmentalScience** for automatically graded activities, current events, videos, and reading questions that you can assign to your students, plus Instructor Resources.

12

A worker examines paper being produced at the mill in Escanaba, Michigan

Forests, Forest Management, and Protected Areas

Upon completing this chapter, you will be able to:

- Summarize the ecological and economic contributions of forests
- Outline the history and current scale of deforestation
- Assess aspects of forest management and describe methods of harvesting timber
- Identify federal land management agencies and the lands they manage
- Recognize types of parks and protected areas and evaluate issues involved in their design

CENTRAL CASE STUDY

Certified Sustainable Paper in Your Textbook

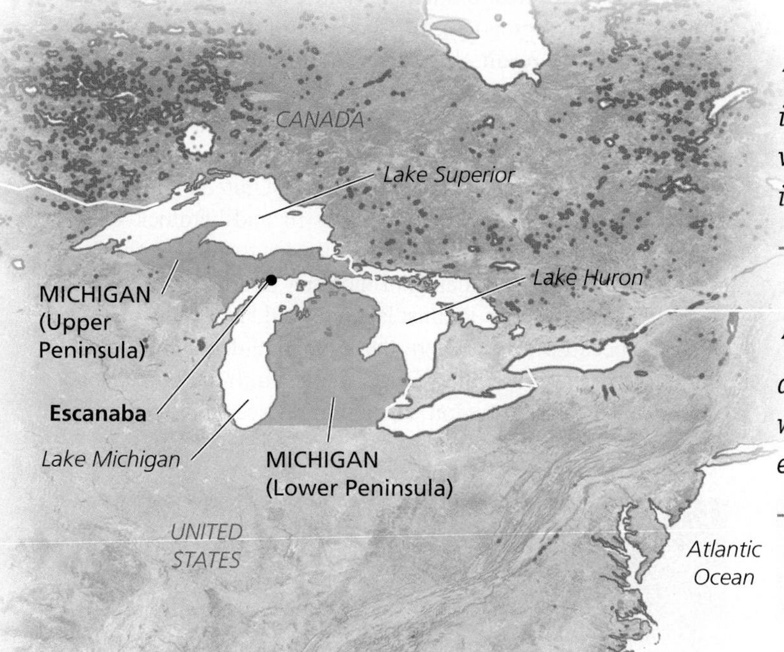

"FSC is the high bar in forest certification, because its standards protect forests of significant conservation value and consistently deliver meaningful improvements in forest management on the ground."

—Kerry Cesareo, deputy director, WWF-US Forests Program

"As a company and as individuals, we must be able to look ourselves in the mirror and know that we are truly doing what is right and good for the environment."

— Rick Willett, president and CEO, NewPage Corporation

As you turn the pages of this textbook, you are handling paper made from trees that were grown, managed, harvested, and processed using certified sustainable practices.

If you were to trace the paper in this book back to its origin, you would find yourself standing in a diverse mixed forest of aspen, birch, beech, maple, spruce, and pine in the Upper Peninsula of Michigan. This sparsely populated region near the Canadian border, flanked by Wisconsin, Lake Michigan, and Lake Superior, remains heavily forested, despite supplying timber to our society for nearly 200 years.

The trees cut to make this book's paper were selected for harvest based on a sustainable management plan designed to avoid depleting the forest of its mature trees or degrading the ecological functions the forest performs. The logs were then transported to a nearby pulp and paper mill at Escanaba, a community of 13,000 people on the shore of Lake Michigan. The mill is Escanaba's largest employer, providing jobs to about 1050 residents.

At the Escanaba mill, the wood is chipped and then fed into a digester, where the chips are cooked with chemicals to break down the wood's molecular bonds. To make white paper, lignin (the polymer that gives wood its sturdiness) is separated out and discarded, leaving only the wood's cellulose fibers. Millworkers bleach, wash, and screen the cellulose fibers and mix them with water and dye. This mixture is poured onto a moving mat, where heavy rollers press it into thin sheets. The sheets of newly formed paper are then dried on heated rollers and made ready to receive any of a variety of coatings. The paper is wound into immense reels, later to be cut into sheets.

In this way, the Escanaba mill produces about 780,000 tons of paper each year. It recycles chemicals and water used in the process, and it combusts discarded lignin to help power the mill.

Each stage in this process has been examined by independent third-party inspectors from the Forest Stewardship Council (FSC) to ensure that practices meet the FSC's strict criteria for sustainable forest management and paper production. The Forest Stewardship Council is an organization that officially certifies forests, companies, and products that meet sustainability standards. FSC-certified timber harvesting operations in Michigan's Upper Peninsula are required to protect rare species and sensitive habitats, safeguard water sources, control erosion, minimize pesticide use, and maintain the diversity of the forest and its ability to regenerate after harvesting. FSC certification is the best way for consumers of forest products to know that they are supporting sustainable practices that protect forests.

The Escanaba mill is the largest of eight mills run by the NewPage Corporation, based in Ohio. NewPage's mills (in Kentucky, Maine, Maryland, Michigan, Minnesota, and Wisconsin) together produce 3.5 million tons of coated paper each year for books, magazines, newspapers, food packaging, and more. Only some of this paper is FSC-certified, but NewPage seeks out suppliers of wood from private land who are certified at least to the less-rigorous standards of the Sustainable Forestry Initiative®. NewPage states that it does not use wood from old-growth forests, rainforests, or forests of exceptional conservation value.

FSC certification for the paper in your textbook was made possible because Pearson Education, the publisher of this book, is striving to follow sustainable practices. Pearson supported FSC paper for this book at the request of your authors and editors because our entire team feels that an environmental science textbook should walk its talk. So, as you flip through this book, you can feel satisfied that you are doing a small part to help safeguard the world's forests by supporting sustainable forestry practices. ■

Forest Ecosystems and Forest Resources

A **forest** is any ecosystem with a high density of trees. Forests provide habitat for countless organisms; help maintain the quality of soil, air, and water; and play key roles in our planet's biogeochemical cycles (pp. 117–128). Forests have also long provided humanity with wood for fuel, construction, paper production, and more.

Many kinds of forests exist

Most of the world's forests occur as boreal forest (p. 99), a biome that stretches across much of Canada, Scandinavia, and Russia; or as tropical rainforest (p. 96), a biome that occurs in Central America, Indonesia, Southeast Asia, northern South America, and equatorial Africa. Temperate deciduous forests (p. 95), temperate rainforests (p. 96), and tropical dry forests (p. 97) also cover large regions of the globe.

Within each forest biome, the nature of the plant community varies from region to region because of differences in soil and climate. As a result, ecologists and forest managers find it useful to classify forests into **forest types**, categories defined by their predominant tree species (**FIGURE 12.1**). The eastern United States contains 10 forest types, ranging from spruce-fir to oak-hickory to longleaf–slash pine. The western United States holds 13 forest types, ranging from Douglas fir and hemlock–sitka spruce forests of the moist Pacific Northwest to ponderosa pine and pinyon-juniper woodlands of the drier interior.

Michigan's Upper Peninsula holds a diverse mix of forest types because it lies where the temperate deciduous forest biome of the eastern United States merges into the boreal forest of Canada. In the region surrounding Escanaba, spruce-fir

(a) Maple-beech-birch forest, Michigan's Upper Peninsula

(b) Oak-hickory forest, West Virginia

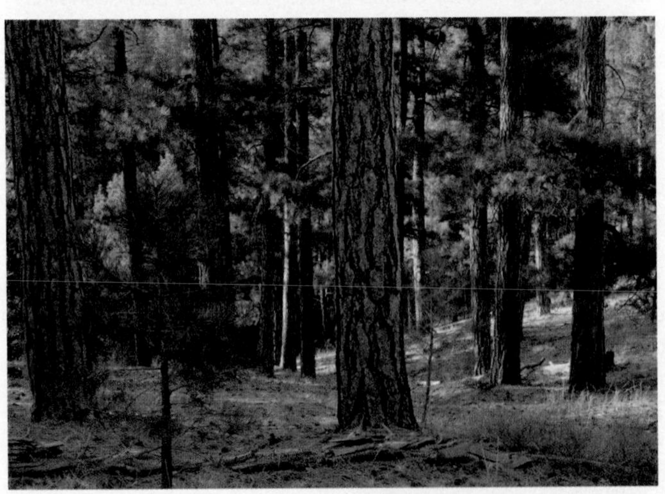

(c) Ponderosa pine forest, northern Arizona

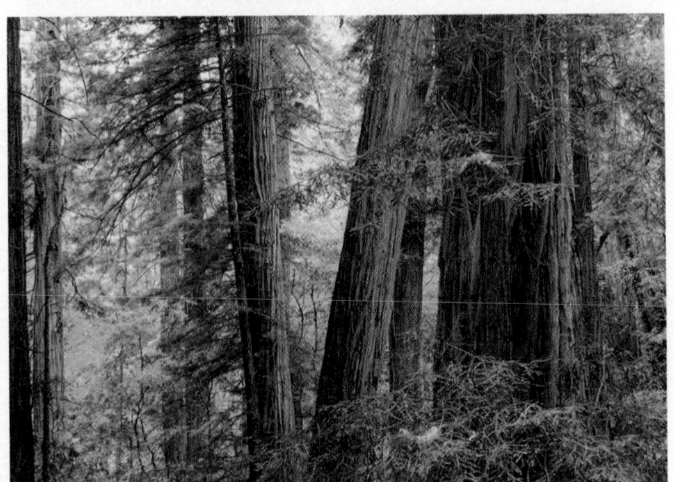

(d) Redwood forest, coastal northern California

FIGURE 12.1 Shown are four of the 23 forest types found in the continental United States.

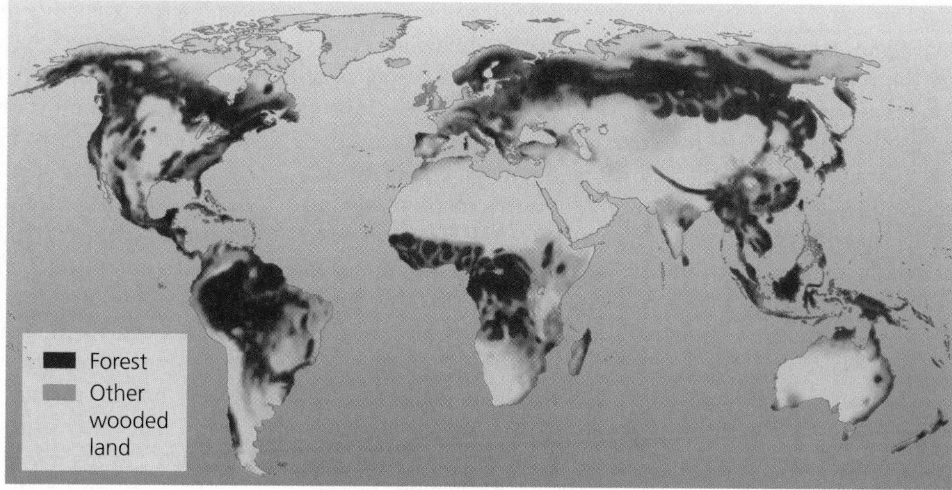

FIGURE 12.2 Forests cover 31% of Earth's land surface Most widespread are boreal forests in the north and tropical forests near the equator. Areas classified as "wooded land" support trees at sparser densities. *Data from Food and Agriculture Organization of the United Nations, 2010.* Global forest resources assessment 2010.

forest intermixes with forests of aspen, birch, maple, and beech, and areas of white, red, and jack pine.

Altogether, forests currently cover 31% of Earth's land surface (**FIGURE 12.2**). Forests occur on all continents except Antarctica.

Forests are ecologically complex

Because of their structural complexity and their capacity to provide many niches for organisms, forests comprise some of the richest ecosystems for biodiversity (**FIGURE 12.3**). The leaves, fruits, and seeds of trees, shrubs, and forest-floor plants furnish food and shelter for an immense diversity of insects, birds, mammals, and other animals. Plants are also colonized by an extensive array of fungi and microbes, in both parasitic and mutualistic relationships (pp. 79–80).

In a forest's **canopy**—the upper level of leaves and branches in the treetops—beetles, caterpillars, and other leaf-eating insects abound, providing food for birds such as warblers and tanagers, while arboreal mammals from squirrels to sloths to monkeys consume fruit and leaves. Animals also live and feed in the **subcanopy** (the middle and lower portions of trees), in shrubs and small trees of the **understory**, and among groundcover plants on the forest floor. Other animals utilize the bark, branches, and trunks of trees. Cavities in trunks and limbs provide nest and shelter sites for a wide variety of animals. Dead and dying trees (called *snags*) are especially valuable; insects that break down the wood provide

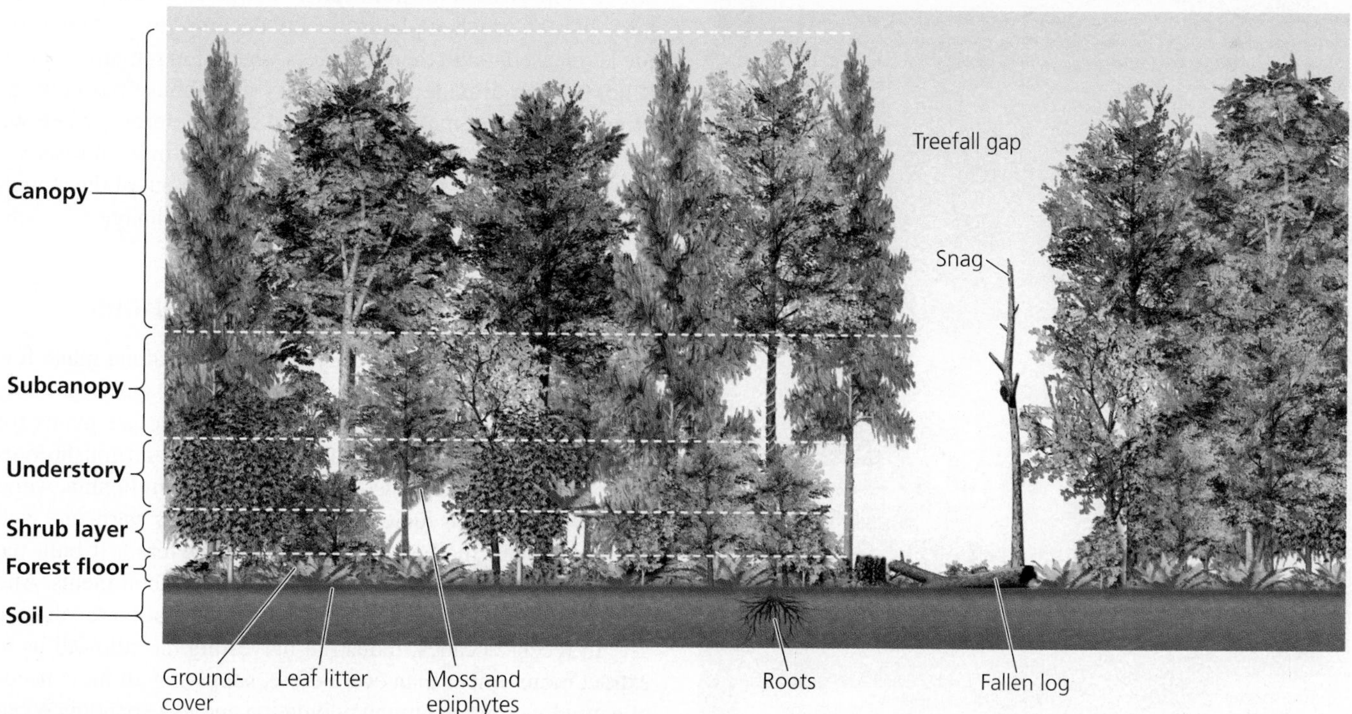

FIGURE 12.3 A cross-sectional diagram shows the complexity of a mature forest. Crowns of tall trees form the canopy, and trees beneath them form the shaded subcanopy and understory. Shrubs and groundcover grow just above the forest floor, and vines, mosses, lichens, and epiphytes cover portions of trees and the forest floor. Snags (dead trees) provide food and nesting sites for woodpeckers and other animals, and logs nourish the soil. Fallen trees create openings called treefall gaps, letting light through and allowing early successional plants to grow.

food for woodpeckers, which create cavities that other animals may later use. Much of a forest's biodiversity resides on the forest floor, where the fallen leaves and branches of the leaf litter nourish the soil. A multitude of soil organisms helps decompose plant material and cycle nutrients (p. 218).

Tropical rainforests feature additional complexity not shown in Figure 12.3. In tropical rainforests, particularly tall individual trees called *emergent trees* protrude here and there above the canopy. *Epiphytes*, plants specialized to grow atop other plants, add to the biomass and species diversity at all levels of a rainforest. Epiphytes include many ferns, mosses, lichens, orchids, and bromeliads.

Forests with a greater diversity of plants tend to host a greater diversity of organisms overall. As forests change over time through the process of succession (pp. 85, 88), their species composition changes along with their structure. In general, old-growth forests host more biodiversity than young forests, because older forests contain more structural diversity and thus more microhabitats and resources for more species. Old-growth forests also are home to more species that today are threatened, endangered, or declining, because old forests have become rare relative to young forests.

Forests provide ecosystem services

Besides hosting biodiversity, forests supply us with many vital ecosystem services (p. 3, 116–117, 152, 290; **FIGURE 12.4**). As plants grow (whether in forest, grassland, or other biomes), their roots stabilize the soil and help to prevent erosion. When rain falls, leaves and leaf litter slow runoff by intercepting water. This prevents flooding, helps water soak into the ground to nourish roots and recharge aquifers, reduces soil erosion, and helps keep streams and rivers clean.

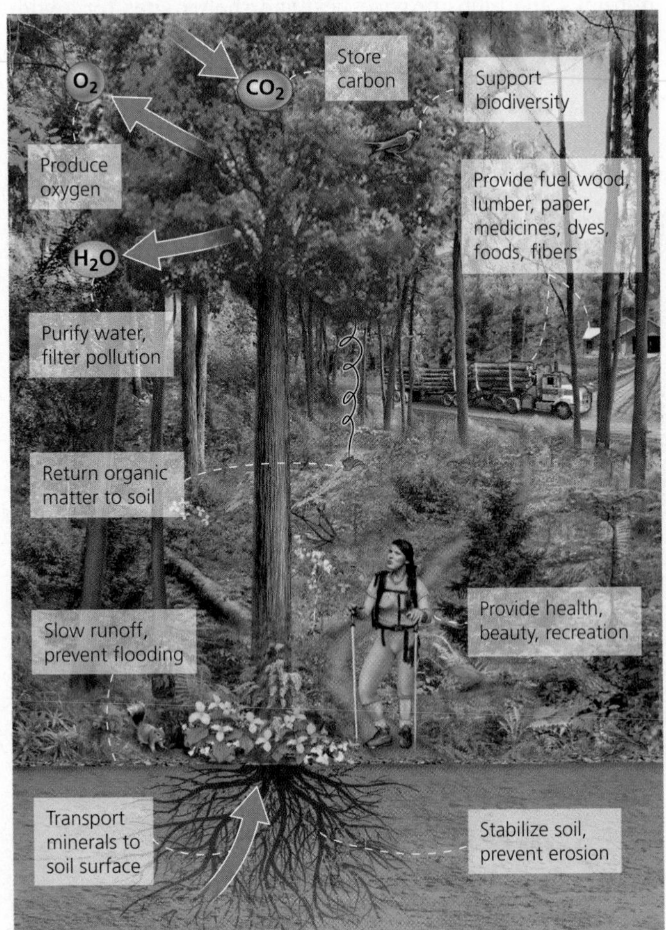

FIGURE 12.4 Forests provide us a diversity of ecosystem services, as well as resources that we can harvest.

Forest plants also filter pollutants and purify water as they take it up from the soil and release it to the atmosphere in transpiration (p. 120). Plants draw carbon dioxide from the air for use in photosynthesis (p. 32), release the oxygen that we breathe, regulate moisture and precipitation, and moderate climate. Trees' roots draw minerals up from deep soil layers and deliver them to surface soil layers where other plants can use them. Plants also return organic material to the topsoil when they die or drop their leaves.

By performing all these ecological functions, forests are indispensable for our survival. Forests also enhance our quality of life by providing us with cultural, aesthetic, health, and recreation values (pp. 150–151). People seek out forests for adventure and for spiritual solace alike—to admire beautiful trees, to observe wildlife, to enjoy clean air, and for many other reasons.

Carbon storage limits climate change

Of all the ecosystem services that forests provide, their storage of carbon has elicited great interest as nations debate how to control global climate change (Chapter 18). Because trees absorb carbon dioxide from the air during photosynthesis and then store carbon in their tissues, forests serve as a major reservoir for carbon. Scientists estimate that the world's forests store over 280 billion metric tons of carbon in living tissue, which is more than the atmosphere contains. When plant matter is burned or when plants die and decompose, carbon dioxide is released—and thereafter less vegetation remains to soak it up. Carbon dioxide is the primary greenhouse gas contributing to global climate change (p. 485). Therefore, when we cut forests, we worsen climate change. The more forests we preserve or restore, the more carbon we keep out of the atmosphere, and the better we can address climate change.

Forests provide us valuable resources

Carbon storage and other ecosystem services alone make forests priceless to our society, but forests also provide many economically valuable resources. Among these are plants for medicines, dyes, and fibers; animals, plants, and mushrooms for food; and, of course, wood from trees. For millennia, wood from forest trees has fueled our fires, keeping us warm and well fed. It has built the houses that keep us sheltered. It built the ships that carried people and cultures between continents. And it gave us paper, the medium of the first information revolution.

In recent decades, industrial harvesting has allowed us to extract more timber than ever before, supplying all these needs of a rapidly growing human population and its expanding economy. The extraction of forest resources has been instrumental in helping our society achieve the standard of living we now enjoy.

Nations maintain and use forests for all these economic and ecological reasons. An international survey in 2010 found that globally, 30% of forests were designated primarily for timber production. Lesser proportions were designated for conservation of biodiversity, protection of soil and water quality, recreation, tourism, education, and conservation of culturally important sites. Most commercial timber extraction today takes place in Canada, Russia, and other nations with large expanses of boreal forest, and in tropical nations with large areas of rainforest, such as Brazil and Indonesia. In the United States, most logging takes place in pine plantations of the South and conifer forests of the West.

Forest Loss

Our demand for wood and paper products and our need for open land for agriculture have led us to clear forested land. When trees are removed more quickly than they can regrow, the result is **deforestation,** the clearing and loss of forests. Deforestation has altered landscapes across much of our planet. In the time it takes you to read this sentence, 2 hectares (5 acres) of tropical forest will have been cleared. As we alter, fragment, and eliminate forests, we lose biodiversity, worsen climate change, and disrupt the ecosystem services that support our societies.

Agriculture and demand for wood put pressure on forests

From the slash-and-burn farmer cutting tropical rainforest to the American suburbanite shopping at a grocery store, we all depend on food and fiber grown on cropland and rangeland—much of which occupies land where forests once grew. All of us also depend in some way on wood, from the subsistence herder in Nepal cutting trees for firewood to the American student consuming reams of paper in the course of getting a degree. To make way for agriculture and to extract wood products, people have been clearing forests for millennia.

Forest clearing has fed our civilization's growth, but unsustainable forest loss has negative consequences, especially as human population grows. Deforestation leads to biodiversity loss, soil degradation, and desertification (pp. 223–224). Moreover, forest loss adds carbon dioxide to the atmosphere, contributing to global climate change.

In 2010, the U.N. Food and Agriculture Organization (FAO) released its latest *Global Forest Resources Assessment.* In this report, researchers combined remote sensing data from satellites, analysis from forest experts, questionnaire responses, and statistical modeling to form a comprehensive picture of the world's forests. The assessment concluded that we are eliminating 13 million hectares (32 million acres) of forest each year. Subtracting annual regrowth from this amount makes for an annual net loss of 5.2 million hectares (12.8 million acres)—an area about half the size of Kentucky or twice the size of Massachusetts. This rate (for the decade 2000–2010) is lower than the deforestation rate for the 1990s, when 8.3 million ha (20.5 million acres) were lost worldwide each year.

We deforested much of North America

Deforestation for timber and farmland propelled the expansion of the United States and Canada westward across the North American continent. The vast deciduous forests of the East were cleared by the mid-19th century, making way for countless small farms. Timber from these forests built the cities of the Atlantic seaboard. Cities such as Chicago, Detroit, and Milwaukee were constructed with timber felled in the vast pine and hardwood forests of Wisconsin and Michigan.

As a farming economy shifted to an industrial one, wood was used to stoke the furnaces of industry. Logging operations moved south to the Ozarks of Missouri and Arkansas, and then to the pine woodlands and bottomland hardwood forests of the South, which were logged and converted to pine plantations. Once mature trees were removed from these areas, timber companies moved west, cutting the continent's biggest trees in the Rocky Mountains, the Sierra Nevada, the Cascade Mountains, and the Pacific Coast ranges.

By the 20th century, very little **primary forest**—natural forest uncut by people—remained in the lower 48 U.S. states, and today even less is left (**FIGURE 12.5**). Nearly all of

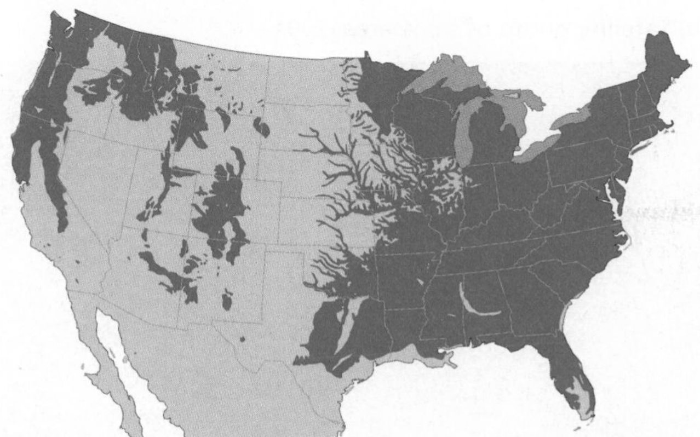

(a) 1620: Areas of primary (uncut) forest

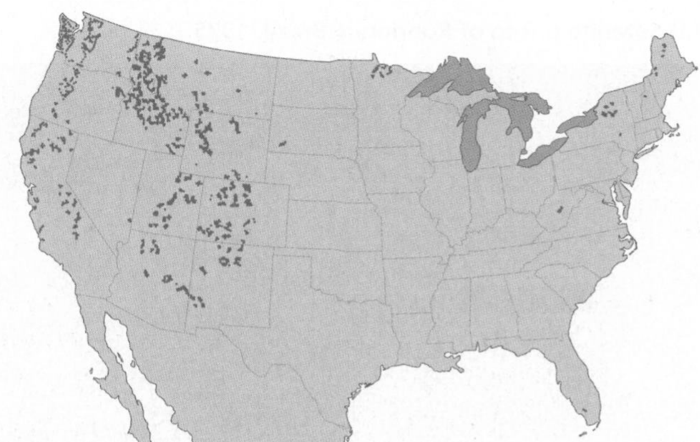

(b) Today: Areas of primary (uncut) forest

FIGURE 12.5 Areas of primary (uncut) forest have been dramatically reduced over the last few hundred years. When Europeans first colonized North America **(a)**, the entire eastern half of the continent and substantial portions of the western half were covered in primary forest (shown in green). Today, nearly all this primary forest is gone **(b)**, having been cut for timber and to make way for agriculture. (Much of the landscape has become reforested with secondary forest.) *Sources: (a) adapted from Greeley, W.B., 1925. The relation of geography to timber supply,* Economic Geography *1:1–11; and (b) map by George Draffan, www.endgame.org.*

the largest oaks and maples found in eastern North America today, and even most redwoods of the California coast, are *second-growth* trees: trees that have sprouted and grown to partial maturity after old-growth trees were cut. Such second-growth trees characterize **secondary forest**. Secondary forest generally contains smaller trees than does primary forest, and the species composition, structure, and nutrient balance of a secondary forest may differ markedly from the primary forest that it replaced.

The fortunes of loggers have risen and fallen with the availability of big trees. As each region was depleted, the timber industry moved on while local loggers lost their jobs. If the remaining ancient trees of North America—most in British Columbia and Alaska—are cut, many loggers will be jobless once again. Their employers may move on to nations of the developing world, as many already have.

Forests are being cleared most rapidly in developing nations

Uncut primary forests still remain in many developing countries, and these nations are in the position the United States and Canada enjoyed a century or two ago: having a resource-rich frontier that they can develop. Today's powerful industrial technologies allow these nations to exploit their resources and push back their frontiers even faster than occurred in North America. As a result, deforestation is rapid in places such as Indonesia, West Africa, Central America, and Brazil (**FIGURE 12.6**).

Indeed, forests are being felled most quickly today in the tropical rainforests and dry forests of Africa and Latin America (**FIGURE 12.7**). Developing nations in these regions are striving to expand settlement for their burgeoning populations and to boost their economies by extracting natural resources. Moreover, many people in these societies harvest fuelwood for their daily cooking and heating needs (p. 567). In contrast, parts of Europe and North America are gaining forested area as they recover from past deforestation. Today's loss of tropical forests in developing nations has global consequences, as these forests are home to far more biodiversity than the temperate forests of North America and Europe.

The South American nation of Brazil, home to most of the vast Amazon rainforest, illustrates success in reducing deforestation but also the continuing pressures upon tropical forests. Not long ago, Brazil was losing forests faster than any other country. Its government was promoting settlement on the forested frontier and was subsidizing an expansion of large-scale cattle ranching and soybean farming to supply demand from wealthy consumers in the United States and Europe. In recent years, Brazil has reduced its deforestation rate significantly. Today 80% of Brazil's forest remains intact, while the nation advances economically and politically as a stable democracy. However, in 2012 Brazil's legislature pushed to weaken the nation's Forest Code, which had helped to slow deforestation, while also pushing for construction of a number of large dams and development projects. Scientists feared these actions would reverse Brazil's progress in slowing forest loss. President Dilma Rousseff vetoed some aspects of the legislation in an effort to balance conservation and development interests.

Developing nations often are desperate enough for economic development and foreign capital that they impose few or no restrictions on logging. Often they allow their timber to be extracted by foreign multinational corporations, which pay fees to the developing nation's government for a **concession**, or right to extract the resource. Once a concession is granted, the corporation has little or no incentive to manage forest resources sustainably. Local people may receive temporary employment from the corporation, but once the timber is gone they no

FIGURE 12.6 Deforestation of Amazonian rainforest has been rapid in recent decades. Satellite images of the state of Rondonia in Brazil show extensive clearing resulting from settlement in the region.

(a) Satellite photo of Rondonia, Brazil, 1975

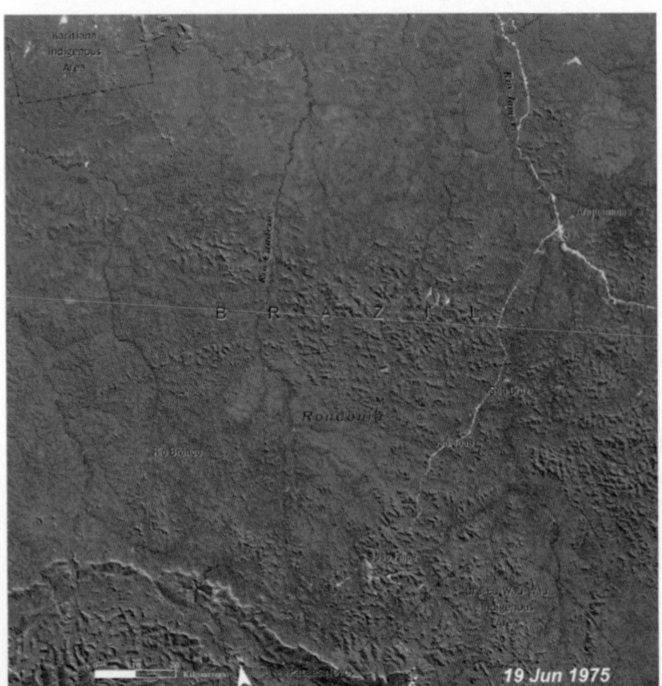

(b) Satellite photo of same area, 2001

(a) Cattle on burned and cleared land

FIGURE 12.7 **Tropical forests are being lost most quickly today.** Many are burned to farm soybeans or to create grazing land for cattle, as shown here **(a)** in Brazil. Most of the soybeans and beef are exported to consumers in wealthier nations. Africa and Latin America are losing forests at rapid rates **(b)**, whereas Europe and North America are slowly gaining secondary forest. In Asia, tree plantations are increasing cover, but natural primary forests are still being lost. *Data from Food and Agriculture Organization of the United States,* Global forest resources assessment 2010. *By permission.*

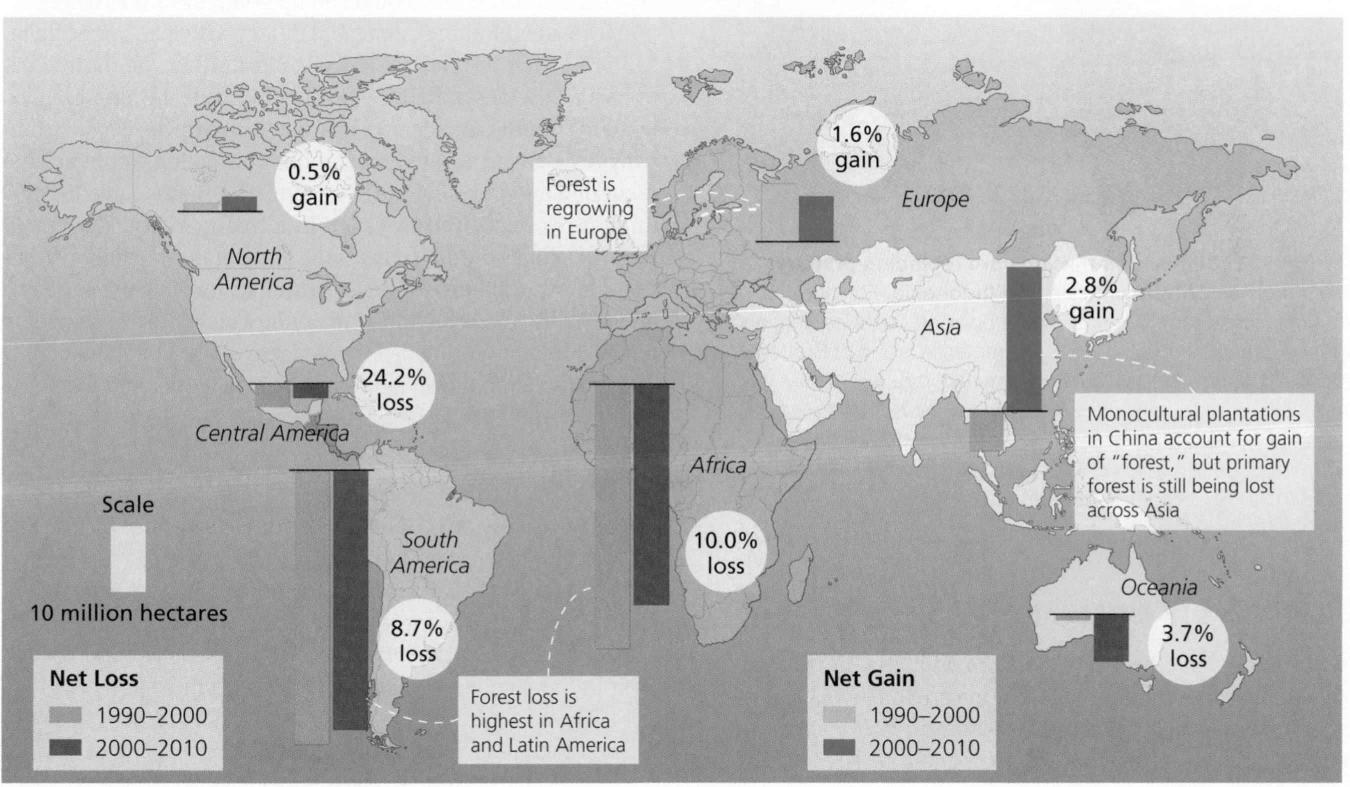

(b) Change in forest area by region

longer have the forest and the ecosystem services it provided. As a result, most economic benefits are short term, reaped not by local residents but by the foreign corporation. Moreover, much of the wood extracted in developing nations is exported to Europe and North America. In this way, our consumption of high-end furniture and other wood products in industrialized nations can drive forest destruction in poorer nations.

Throughout Southeast Asia and Indonesia today, vast swaths of tropical rainforest are being cut to establish plantations of oil palms (FIGURE 12.8). Oil palm fruit produces palm oil, which we use in snack foods, soaps, and cosmetics, and now as a biofuel. In Indonesia, the world's largest palm oil producer, oil palm plantations have displaced over 6 million ha (15 million acres) of rainforest. Clearing for plantations encourages further development and eases access for people to enter the forest and conduct logging illegally.

The palm oil boom represents a conundrum for environmental advocates. Many people eager to fight climate change had urged the development of biofuels (pp. 569–573) to replace fossil fuels. Yet grown at the large scale that our society is demanding, monocultural plantations of biofuel crops such as oil palms are causing severe environmental impacts by displacing natural forests.

WEIGHING THE ISSUES

LOGGING HERE OR THERE Suppose you are an activist protesting the logging of old-growth trees near your hometown. Now let's say you know that if the protest is successful, the company will move to a developing country and cut its primary forest instead. Would you still protest the logging in your hometown? Would you pursue any other approaches?

FIGURE 12.8 Oil palm plantations are replacing primary forest across Southeast Asia and Indonesia. Forest clearing for plantations promotes further development, illegal logging, and forest degradation. Since 1950, the immense island of Borneo (maps at bottom) has lost most of its forest. *Data from Radday, M., WWF-Germany, 2007. Designed by Hugo Ahlenius, UNEP/GRID-Arendal.* Extent of deforestation in Borneo 1950–2001, and projection towards 2020. *http://maps.grida.no/go/graphic/extent-of-deforestation-in-borneo-1950-2005-and-projection-towards-2020.*

Solutions are emerging

New solutions are being proposed to address deforestation in developing nations. Some conservation proponents are pursuing community-based conservation projects (p. 302) that empower local people to act as stewards of their forest resources. In other cases, conservation organizations are buying concessions and using them to preserve forest rather than to cut it down. In such *conservation concessions*, the nation receives money *and* keeps its natural resources intact. The South American nation of Surinam entered into such an agreement with the nongovernmental organization Conservation International and virtually halted logging while gaining $15 million.

In Indonesia, NewPage Corporation, the supplier of this textbook's paper, is funding a project called POTICO (Palm Oil, Timber and Carbon Offsets) that aims to reduce deforestation and illegal logging. In this project, the nonprofit World Resources Institute (WRI) is working with palm oil companies that own concessions to clear primary rainforest and is steering them instead toward land that is already logged and degraded. WRI will then protect the forests that were slated for conversion or else allow FSC-certified sustainable forestry in them. WRI and NewPage hope this will encourage oil palm plantations to become sustainable while preserving primary rainforest. Because primary forest stores more carbon than oil palm plantations, these land swaps can reduce Indonesia's greenhouse gas emissions and qualify for credit via carbon offsets (p. 513).

Carbon offsets are central to emerging international plans to curb deforestation and climate change. Forest loss accounts for at least 12% of the world's greenhouse gas emissions—nearly as much as all the world's vehicles emit—yet the Kyoto Protocol (p. 510) did not address it. Thus, at recent international climate conferences (pp. 510–511), negotiators have outlined a program called *Reducing Emissions from Deforestation and Forest Degradation (REDD),* whereby wealthy industrialized nations pay poorer developing nations to conserve forest. Under this plan, poor nations gain income while rich nations receive carbon credits to offset their emissions in an international cap-and-trade system (pp. 183, 512–513). Although the REDD plan has not been formally agreed to, leaders of rich nations have proposed to transfer $100 billion per year to poor nations by 2020, and much of this could end up going toward REDD.

The small South American nation of Guyana—poor financially but rich in forests—has taken a leading role in international discussions of REDD. Guyana's president Bharrat Jagdeo in 2008 commissioned the consulting firm McKinsey and Company to calculate the amount of money his nation would make if it cut down its forests for agriculture. The figure came to $580 million per year over 25 years. In a free market under purely financial considerations, this is the amount that wealthy nations "should" pay Guyana to forego cutting its forest. Guyana subsequently forged a deal with Norway in which Norway is paying Guyana for conserving its forest—up to $250 million in total by 2015, depending on Guyana's progress in reducing forest loss.

WEIGHING THE ISSUES

PAYING CASH TO CONSERVE FORESTS Some critics say Guyana's suggestion that rich nations pay it to stop cutting forest is a kind of blackmail. But one British tourism operator in Guyana says, "Guyana is a small, impoverished country that's trying to develop itself. If the Western world isn't going to protect the rainforest and start coughing up money to countries like Guyana, then [those countries] are going to have to start using their resources. Just like England did for thousands of years, and just like the States is doing and Canada is doing. You can't be hypocrites about it." What do you think? Is the REDD program a type of blackmail, or is it a fair and just way of curbing climate change? Do you feel such a program would be effective? Why or why not?

Forest Management

Our demand for forest resources and amenities is rising, so we need to take care in managing forests. *Foresters* are professionals who manage forests through the practice of **forestry.** Foresters must balance our society's demand for forest products against the central importance of forests as ecosystems. Today, sustainable forest management practices are spreading as informed consumers demand sustainably produced products. Just as your textbook uses FSC-certified paper from sustainably managed forests, more and more paper, lumber, and other forest products are now made using certified sustainable practices. In this way, consumer choice is influencing the ways forests are managed.

Forest management is one type of resource management

Debates over how to manage forest resources reflect broader questions about how to manage natural resources in general. Resources such as fossil fuels and many minerals are nonrenewable, whereas resources such as the sun's energy are perpetually renewable (pp. 3–4). Between these extremes lie resources that are renewable if they are not exploited too rapidly. These include timber, as well as soils, fresh water, rangeland, wildlife, and fisheries.

Resource management describes our use of strategies to manage and regulate the harvest of renewable resources. Sustainable resource management involves harvesting these resources in ways that do not deplete them. Resource managers are guided by research in the natural sciences, as well as by social, political, and economic factors.

Resource managers help conserve soil resources with farming practices that fight erosion (pp. 226–228); safeguard the supply and quality of surface water and groundwater (pp. 405–416); encourage low-impact grazing to manage rangeland sustainably (p. 232); and protect fisheries and wildlife from overharvesting (p. 285).

Resource managers follow several strategies

A key question in managing resources is whether to focus strictly on the resource of interest or to look more broadly at the environmental system of which it is a part. Taking a broader view often helps avoid degrading the system and thereby helps sustain the resource in the long term.

Maximum sustainable yield

A guiding principle in resource management has been **maximum sustainable yield.** Its aim is to achieve the maximum amount of resource extraction without depleting the resource from one harvest to the next. Recall the logistic growth curve (see Figure 3.17, p. 67), which reflects how limiting factors slow exponential population growth and then cap it at a carrying capacity. The logistic curve indicates that a population grows most quickly when it is at an intermediate size—specifically, at one-half of carrying capacity. A fisheries manager aiming for maximum sustainable yield will therefore prefer to keep fish populations at intermediate levels so that they rebound quickly after each harvest. Doing so should result in the greatest amount of fish harvested over time while sustaining the population indefinitely (**FIGURE 12.9**).

This management approach, however, keeps the fish population at only half its carrying capacity—well below the size it would attain in the absence of fishing. Reducing a population in this way will likely affect other species and alter the food web dynamics of the community. From an ecological point of view, management for maximum sustainable yield may set in motion significant ecological changes.

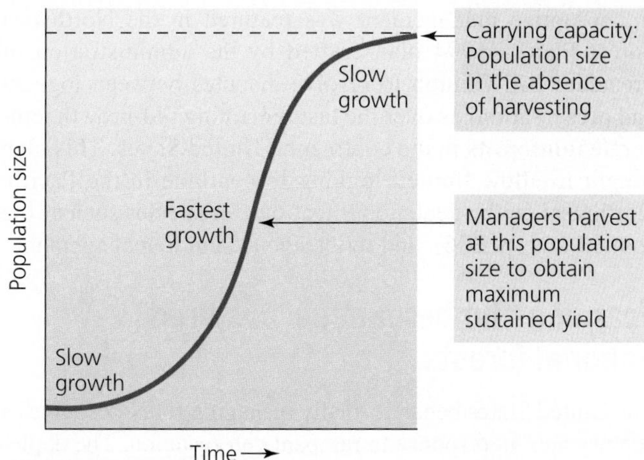

FIGURE 12.9 Maximum sustainable yield maximizes the amount of resource harvested while sustaining the harvest in perpetuity. For a wildlife population or fisheries stock that grows according to logistic growth, managers aim to keep the population at half the carrying capacity, because populations grow fastest at intermediate sizes.

In forestry, maximum sustainable yield argues for cutting trees shortly after they go through their fastest stage of growth. Because trees often grow most quickly at intermediate ages, trees are generally cut long before they have grown as large as they would in the absence of harvesting. This practice maximizes timber production over time, but it also alters forest ecology and eliminates habitat for species that depend on mature trees.

Ecosystem-based management

Because of these dilemmas, more and more managers today espouse **ecosystem-based management,** which aims to minimize impact on the ecosystems and ecological processes that provide the resource. Many certified sustainable forestry plans protect certain forested areas, restore ecologically important habitats, and consider patterns at the landscape level (pp. 113–115), allowing timber harvesting while preserving the functional integrity of the forest ecosystem. This means fostering the area's ecological processes, including succession (pp. 85, 88), in which the forest community naturally changes over time.

It can be challenging, however, to determine how best to implement this type of management. Ecosystems are complex, and our understanding of how they operate is limited. Thus, ecosystem-based management has come to mean different things to different people.

Adaptive management

Some management actions will succeed, and some will fail. A wise manager will try new approaches if old ones are not effective. **Adaptive management** involves systematically testing different approaches and aiming to improve methods through time. For managers, it entails monitoring the results of one's practices and adjusting them as needed, based on what is learned. Adaptive management is intended as a true fusion of science and management, because it explicitly tests hypotheses about how best to manage resources. This process can be time-consuming and complicated but highly effective.

Adaptive management was featured in the Northwest Forest Plan, a 1994 plan crafted by the administration of President Bill Clinton to resolve disputes between loggers and preservationists over the last remaining old-growth temperate rainforests in the continental United States. This plan sought to allow limited logging to continue in the Pacific Northwest, with adequate protections for species such as the spotted owl (p. 296), and to let science guide management.

Fear of a "timber famine" inspired national forests

The United States began formally managing forest resources a century ago, in response to rampant deforestation. The depletion of the eastern U.S. forests had prompted widespread fear of a "timber famine." This led the federal government to form a system of forest reserves: public lands set aside to grow trees, produce timber, and protect water quality. Today the U.S. **national forest** system consists of 77 million ha (191 million acres), managed by the U.S. Forest Service and covering over 8% of the nation's land area (FIGURE 12.10).

The U.S. Forest Service was established in 1905 under the leadership of Gifford Pinchot (p. 138). Pinchot and others developed the concepts of resource management, maximum sustainable yield, and conservation during the Progressive Era, a time of social reform when people began using science to inform public policy. In line with Pinchot's conservation ethic (p. 138–139), the Forest Service aimed to manage the forests for "the greatest good of the greatest number in the long run." Pinchot believed the nation should extract and use resources from its public lands, so timber harvesting was, from the start, a goal of the national forests. But conservation meant planting trees as well as harvesting them, and the Forest Service set out to manage the nation's timber resources wisely.

We extract timber from private and public lands

The vast majority of timber harvesting in the United States today takes place on private land owned by the timber industry or by small landowners. Private timber companies also extract timber from the U.S. national forests and from publicly held state forests. On the national forests, U.S. Forest Service employees manage timber sales and build roads to provide access for logging companies. The Forest Service sells timber below the costs it incurs for marketing and administering the harvest and for building access roads, while the companies go on to sell the timber they harvest for profit. In this way, taxpayers subsidize private timber harvesting on public land (p.182). These subsidies also tend to inflate harvest levels beyond what would occur in a free market.

Timber companies extracted 10.7 million m^3 (378 million ft^3) of live timber from national forests in 2006, the most recent year for which comprehensive data are available. However, this is less than the amount harvested from other public lands, and it is much less than the amount cut on private lands (FIGURE 12.11). At present, in an average year, about 2.1% of U.S. forest acreage is cut for timber. Overall, timber harvesting in the United States and other developed nations has remained stable for the past 45 years, while it has more than doubled in developing countries.

On timber industry land, companies manage their resources in accordance with maximum sustainable yield, so as to obtain maximal profits each year over many years. On public lands, rates of tree removal and growth reflect social and political factors as well as economic ones, and these evolve over time. On the U.S. national forests, private timber extraction increased in the 1950s as the nation experienced a postwar economic boom, paper consumption rose, and the population expanded into newly built suburban homes. Harvests from

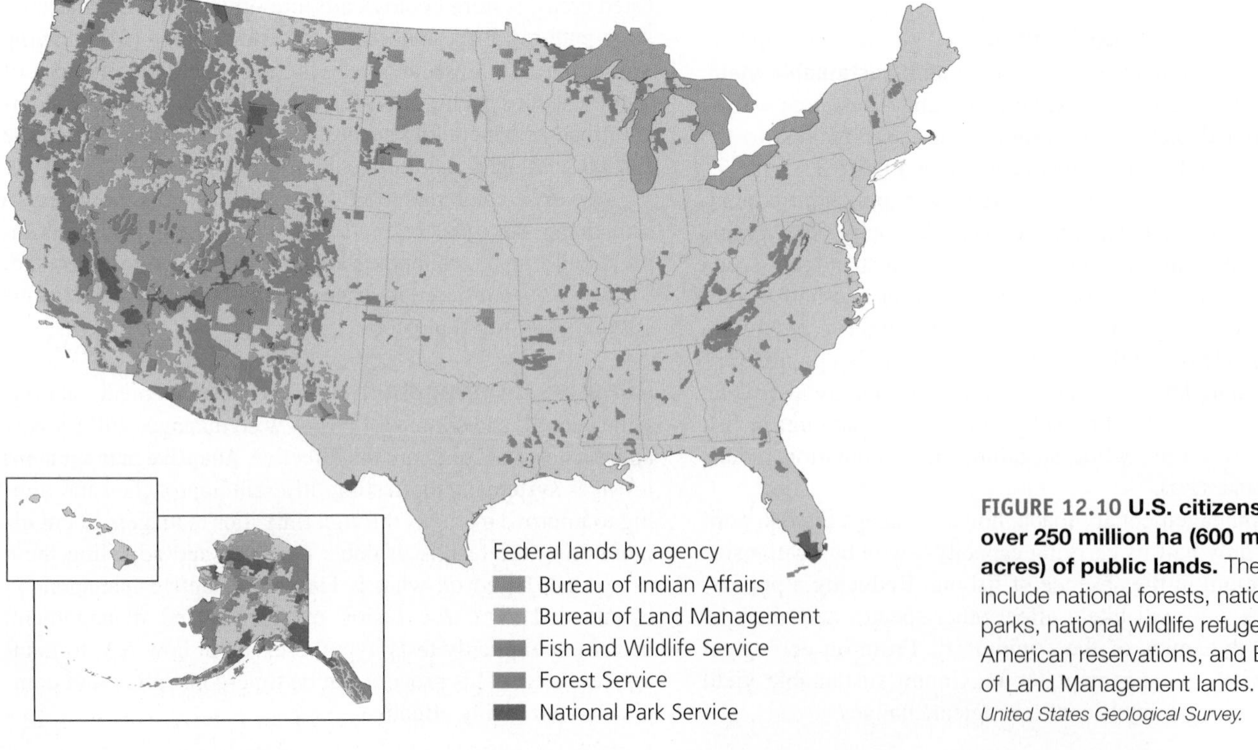

FIGURE 12.10 **U.S. citizens enjoy over 250 million ha (600 million acres) of public lands.** These include national forests, national parks, national wildlife refuges, Native American reservations, and Bureau of Land Management lands. *Data from United States Geological Survey.*

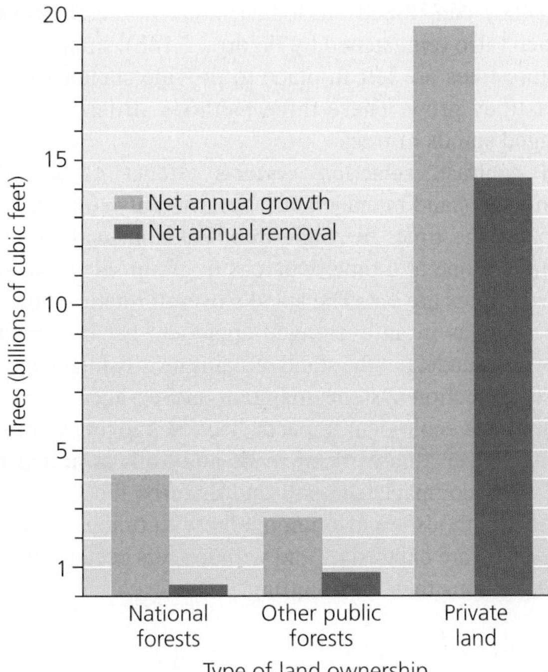

FIGURE 12.11 As the United States recovers from deforestation, trees are growing more quickly than they are being removed. This is particularly true in national forests and on other public lands. However, forests that regrow after logging often differ substantially from the forests that were removed. *"Private land" here combines land owned by the timber industry and by small landholders. Data are for 2006, from USDA Forest Service, 2008.* Forest resources of the United States, 2007. *U.S. Department of Agriculture, Washington, D.C.*

FIGURE 12.12 Even-aged stands differ from uneven-aged stands. Even-aged management, with all trees of equal age, can be seen in the foreground in a plantation regrowing after clear-cutting. Uneven-aged management maintains a mix of tree ages, as seen in the more mature forest.

national forests then began to decrease in the 1980s as economic trends shifted, public concern over clear-cutting grew, and management philosophy evolved. By 2006, regrowth was outpacing removal on these lands by 11 to 1 (see Figure 12.11).

Note, however, that even when regrowth outpaces removal, the character of forests may change. Once primary forest is replaced by younger secondary forest or by single-species plantations, the resulting community may be very different, and generally it is less ecologically valuable.

Plantation forestry has grown

Today the U.S. timber industry focuses on production from plantations of fast-growing tree species planted in single-species monocultures (p. 248). Because all trees in a given stand are planted at the same time, the stands are **even-aged,** with all trees the same age (**FIGURE 12.12**). Stands are cut after a certain number of years (called the *rotation time*), and the land is replanted with seedlings. Such plantation forestry is growing quickly worldwide, and today fully 7% of the world's forests are plantations. One-quarter of these feature non-native tree species.

Ecologists and foresters view plantations as akin to crop agriculture. Because there are few tree species and little variation in tree age, plantations do not offer habitat to many forest organisms. For instance, stands of red pine planted near Escanaba, Michigan, host far less biodiversity than the more-diverse forests of multiple tree species that surround them. Even-aged single-species plantations managed for timber production lack the structural complexity that characterizes a mature natural forest as seen in Figure 12.3 (p. 309). Plantations are also vulnerable to outbreaks of pest species such as bark beetles, as we shall soon see.

For all these reasons, some harvesting methods aim to maintain **uneven-aged** stands, which contain trees of a mix of ages. The greater structural diversity of uneven-aged stands provides superior habitat for most wild species and, if diverse tree species are intermixed, makes these stands more akin to natural, ecologically functional forests.

We harvest timber by several methods

When timber companies harvest trees, they may choose from several methods. In the simplest method, **clear-cutting,** all trees in an area are cut (**FIGURE 12.13**). Clear-cutting is cost-efficient,

FIGURE 12.13 Clear-cutting is cost-efficient for timber companies but has ecological consequences, including soil erosion, water pollution, and altered community composition.

and to some extent it can mimic natural disturbance events such as fires, tornadoes, or windstorms that knock down trees across large areas. However, the ecological impacts of clear-cutting are considerable. An entire ecological community is removed, soil erodes away, and sunlight penetrates to ground level, changing microclimatic conditions such that new types of plants replace those of the original forest. Clear-cutting essentially sets in motion a process of succession (pp. 85, 88) in which the resulting climax community may be quite different from the original climax community. For example, when we clear-cut an eastern U.S. oak–hickory forest, the forest that will regrow may be dominated by maples, beeches, and tulip trees.

Concerns about clear-cutting eventually led foresters and the timber industry to develop alternative harvesting methods (**FIGURE 12.14**). Clear-cutting (**FIGURE 12.14a**) remains the most widely practiced method, but alternative approaches involve cutting some trees while leaving others standing. In the **seed-tree** approach (**FIGURE 12.14b**), small numbers of mature and vigorous seed-producing trees are left standing so that they can reseed the logged area. In the **shelterwood** approach (also represented by Figure 12.14b), small numbers of mature trees are left in place to provide shelter for seedlings as they grow. These three methods all lead to largely even-aged stands of trees.

In contrast, **selection systems** (**FIGURE 12.14c**) allow uneven-aged stand management, because only some trees are cut at any one time. In single-tree selection, widely spaced trees are cut one at a time, whereas in group selection, small patches of trees are cut. The stand's overall rotation time may be the same as in an even-aged approach, because multiple harvests are made, but the stand remains mostly intact between harvests. Selection systems maintain uneven-aged stands, but they still have ecological impacts, because moving trucks and machinery over a network of roads and trails to access individual trees compacts the soil and disturbs the forest floor. Selection methods are also unpopular with timber companies because they are expensive, and with loggers because they are more dangerous than clear-cutting.

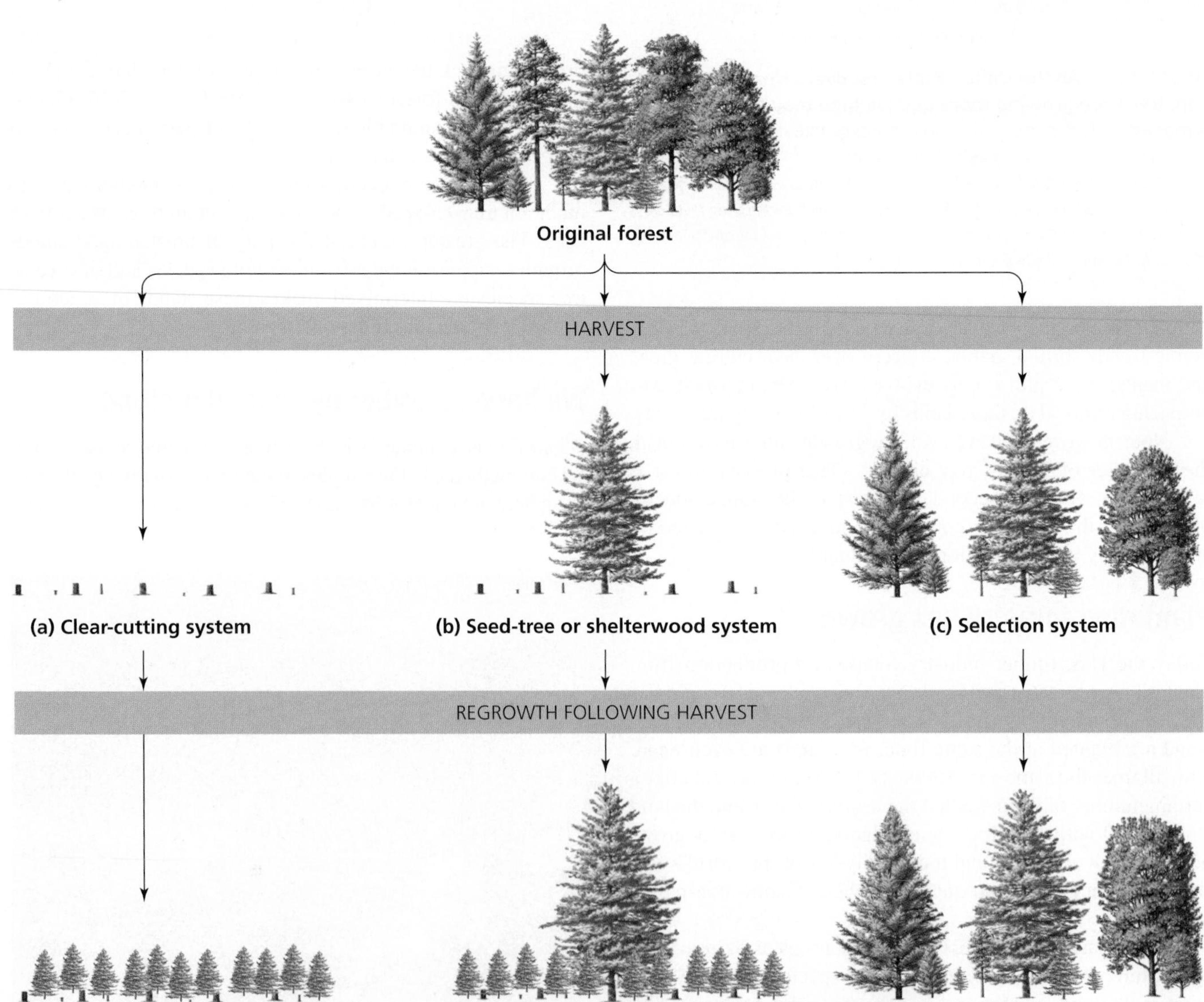

FIGURE 12.14 Foresters have devised several methods to harvest timber. In clear-cutting **(a)**, all trees are cut, extracting a great deal of timber inexpensively but leaving a vastly altered landscape. In seed-tree systems and shelterwood systems **(b)**, a few large trees are left in clear-cuts to reseed the area or provide shelter for seedlings. In selection systems **(c)**, a minority of trees is removed at any one time, while most are left standing.

All timber-harvesting methods disturb soil, alter habitat, and affect plants and animals. All methods modify forest structure and composition. Most methods speed runoff, raise flooding risk, and increase soil erosion, degrading water quality. When steep hillsides are clear-cut, landslides can result. Finding ways to minimize these impacts is important, because timber harvesting is necessary to obtain the wood products that all of us use.

Forest management has evolved over time

For the past half-century, the U.S. Forest Service has nominally been guided by the policy of **multiple use,** meaning that the national forests were to be managed for recreation, wildlife habitat, mineral extraction, and various other uses. In practice, timber production was often the primary use. In recent decades, as people became more aware of the impacts of logging and as development spread across the landscape, citizens began to urge that public forests be managed for recreation, wildlife, and ecosystem integrity, as well as for timber.

In 1976 the U.S. Congress passed the **National Forest Management Act.** This act mandated that every national forest draw up plans for renewable resource management based on the concepts of multiple use and maximum sustainable yield and subject to public input under the National Environmental Policy Act (p. 174). Guidelines specified that these plans:

- Consider environmental factors as well as economic ones, such that profit alone does not determine harvesting decisions.
- Provide for diverse ecological communities and preserve regional diversity of tree species.
- Ensure research and monitoring of management practices.
- Permit increases in harvest levels only if sustainable.
- Ensure that timber is extracted only where impacts have been assessed; cuts are shaped to the terrain; maximum size limits are established; and cuts do not threaten timber regeneration or soil, watershed, fish, wildlife, recreation, or aesthetic resources.

Following passage of the National Forest Management Act, the U.S. Forest Service developed new programs to manage wildlife, non-game animals, and endangered species. It pushed for ecosystem-based management and ran ecological restoration programs to recover plant and animal communities that had been lost or degraded (FIGURE 12.15). Timber harvesting methods were brought more in line with ecosystem-based management goals. A set of approaches dubbed **new forestry** called for timber cuts that mimicked natural disturbances. For instance, "sloppy clear-cuts" that leave a variety of trees standing were intended to mimic the changes a forest might experience if hit by a severe windstorm.

The Hiawatha National Forest provides an example of management under the National Forest Management Act. Headquartered in Escanaba, the Hiawatha National Forest covers 360,000 ha (895,000 acres) in Michigan's Upper Peninsula. Its most recent management plan seeks a balance of uses in the forest (TABLE 12.1).

FIGURE 12.15 **Ecosystem-based management is practiced in this longleaf pine forest in the southeastern United States.** Foresters and biologists restore and nurture mature longleaf pine trees while burning and removing brush from the understory. This habitat is home to a number of specialized species, including the endangered red-cockaded woodpecker.

TABLE 12.1 Goals and Guidelines from the Hiawatha National Forest's Management Plan

Timber harvesting
- Harvest timber for the region's mills at specified rotation ages, then replant seedlings.
- Use clear-cutting, shelterwood, and selection systems, according to location and tree type and in ways that simulate natural disturbances.
- Manage a mix of even-aged and uneven-aged stands of trees for timber and wildlife habitat.
- Protect 52,000 acres of old-growth forest.

Ecosystem management
- Restore 300 acres of wetlands and 9–13 miles of streams.
- Conduct prescribed fire and remove brush on 1000 acres per year.
- Minimize soil damage from logging.
- Do not allow livestock grazing.
- Protect historic and archaeological sites, caves, ponds, rivers, snags, water quality, and wilderness areas.

Fish and wildlife
- Monitor fish and wildlife populations.
- Maintain habitats for threatened and endangered species such as Kirtland's warbler, lakeside daisy, and Canada lynx.
- Control invasive species, especially forest pests.

Recreation and other
- Allow hiking, fishing, boating, snowmobiling, dog mushing, and more, but specify areas for recreation (e.g., no off-road vehicles in some areas).
- Build no new roads; reconstruct 10 miles per year; decommission 5 miles per year.
- Acquire private land (inholdings) within wilderness areas when possible.

Adapted from USDA Forest Service, 2006. Hiawatha National Forest 2006 Forest Plan.

Another national policy milestone that accentuated a shift toward conservation occurred in 2001, when President Bill Clinton issued an executive order that became known as the **roadless rule.** The roadless rule put 23.7 million ha (58.5 million acres)—31% of national forest land and 2% of total U.S. land—off-limits to road construction or maintenance (and thus to logging). The roadless rule received strong popular support, including a record 4.2 million public comments.

The administration of President George W. Bush marked a change in policy direction. In 2004, the Bush administration freed forest managers from many requirements of the National Forest Management Act, granting them more flexibility in managing forests but loosening environmental protections and restricting public oversight. In 2005, the Bush administration repealed the roadless rule, inviting states to decide how national forests within their boundaries should be managed. Some states responded favorably, whereas others sued the administration, asking that the roadless rule be reinstated. Following a series of court rulings, the Obama administration reinstated most of the roadless policy but also negotiated with some states to allow them to develop their own plans. In 2012 the U.S. Supreme Court opted not to hear a challenge to the rule, thereby strengthening its place in federal policy.

Fire can hurt or help forests

Another area of policy debate involves how to handle wildfire. Smokey Bear, the Forest Service's beloved cartoon bear in a ranger's hat, advises us to fight forest fires—and for over a century, the Forest Service and other agencies suppressed fire whenever and wherever it broke out. Yet scientific research now clearly shows that many species and ecological communities depend on fire. Some plants have seeds that germinate only in response to fire, and researchers studying tree rings have documented that North America's grasslands and pine woodlands burned frequently. (Burn marks in a tree's growth rings reveal past fires, giving scientists an accurate history of fire events extending back hundreds or even thousands of years.) Ecosystems dependent on fire are adversely affected when fire is suppressed: Grasslands are invaded by shrubs, and pine woodlands become cluttered with hardwood understory. Invasive plants move in, and animal diversity and abundance decline.

FAQ **Aren't all forest fires bad?**

No. Fire is a natural process that helps to maintain the health of many forests and grasslands. When allowed to occur naturally, low-intensity fires generally burn moderate amounts of material, return nutrients to the soil, and promote lush growth of new vegetation. When we suppress fire, we allow unnaturally large amounts of dead wood, dried grass, and leaf litter to accumulate. This material becomes kindling that eventually can feed a truly damaging fire that grows too big and too hot to control. This is why many land managers today conduct carefully controlled prescribed burns and also allow some natural fires to run their course. By doing so, they aim to help return our fire-dependent ecosystems to a healthier and safer condition.

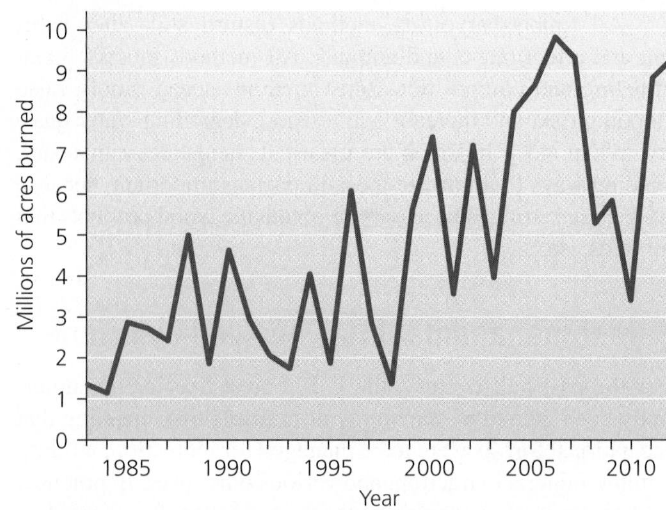

FIGURE 12.16 Wildfires have become larger and more numerous in the United States. Fuel buildup from decades of fire suppression has contributed to this trend. *Data from National Interagency Fire Center.*

In the long term, suppressing frequent low-intensity fires leads to occasional catastrophic fires that damage forests, destroy property, and threaten human lives. This is because fire suppression allows limbs, logs, sticks, and leaf litter to accumulate on the forest floor, producing kindling for a catastrophic fire. Such fuel buildup worsened the 1988 fires in Yellowstone National Park, the 2009 fires in southern California, the 2012 Colorado fires, and thousands of other wildfire episodes. Severe fires have become more numerous in recent years (**FIGURE 12.16**). At the same time, increased residential development alongside forested land—in the **wildland-urban interface**—is placing more homes in fire-prone situations (**FIGURE 12.17**).

To reduce fuel loads, protect property, and improve the condition of forests, land management agencies now burn areas of forest intentionally with low-intensity fires under

FIGURE 12.17 Habitual suppression of fire has led to catastrophic wildfires that damage forests and threaten homes. To avoid these unnaturally severe fires, ecologists suggest allowing natural fires to burn when we can and conducting controlled burns to reduce fuel loads and restore forest ecosystems.

carefully controlled conditions (see Figure 1.8b, p. 10). These **prescribed burns,** or **controlled burns,** clear away fuel loads, nourish the soil with ash, and encourage the vigorous growth of new vegetation. Because they are time-intensive and sometimes are misunderstood by politicians and the public, prescribed burns are conducted on only a small proportion of land (about 2 million acres per year). As a result, vast areas of American forests remain vulnerable to catastrophic fires.

In the wake of major fires in California in 2003, the U.S. Congress passed the Healthy Forests Restoration Act. Although this legislation encouraged some prescribed burning, it primarily promoted the physical removal of small trees, underbrush, and dead trees by timber companies. The removal of dead trees, or snags, following a natural disturbance (such as a fire, windstorm, insect damage, or disease) is called **salvage logging.** From a short-term economic standpoint, salvage logging may seem to make good sense. However, ecologically, snags have immense value; the insects that decay them provide food for wildlife, and many animals depend on holes in snags for nesting and roosting. Removing timber from recently burned land can also cause soil erosion, impede forest regeneration, and promote further wildfire (see THE SCIENCE BEHIND THE STORY, pp. 324–325).

WEIGHING THE ISSUES

HOW TO HANDLE FIRE? A century of fire suppression has left vast areas of North American forests vulnerable to catastrophic wildfires. Prescribed burning helps to alleviate this risk, yet we will never have adequate resources to conduct careful prescribed burning over all these lands. Can you suggest solutions to help protect people's homes in the wildland-urban interface while improving the ecological condition of forests? Do you think people should be allowed to develop homes in fire-prone areas?

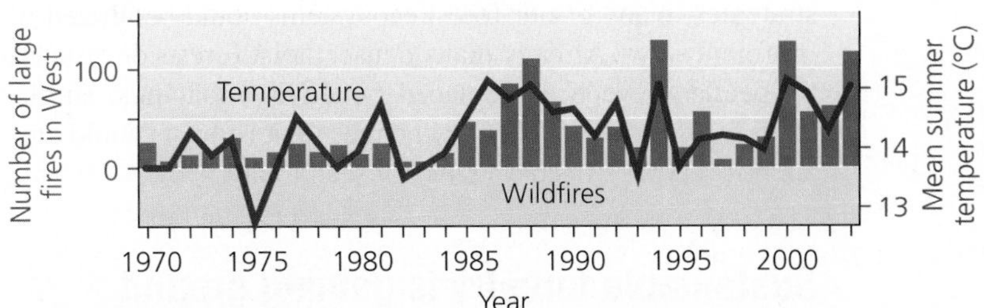

(a) Wildfires correlate with hotter summer temperatures

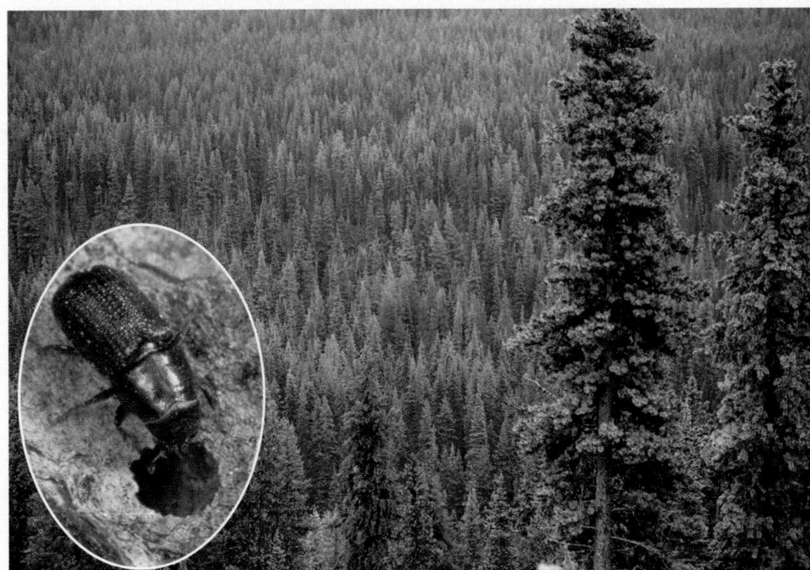

(b) Mountain pine beetles kill more trees in a warmer climate

FIGURE 12.18 Forests face increasing threats as temperatures rise. The number of large wildfires in the western United States **(a)** has risen along with the region's summer temperatures. Warm summers and mild winters have favored bark beetles **(b)**, which are killing vast areas of trees throughout the West. *Source (a): Westerling, A.L., et al., 2006. Warming and earlier spring increase western U.S. forest wildfire activity.* Science *313: 940–943, Fig 1A. Reprinted with permission from AAAS.*

Climate change and pest outbreaks are altering forests

Global climate change (Chapter 18) is now worsening wildfire risk by bringing warmer weather to most of North America and drier weather to much of the American West. Scientific climate models predict further warming and drying (pp. 496, 504). Research suggests that the recent increase in fires in North America has been driven in part by warmer, drier climate (FIGURE 12.18a).

Climate change also is promoting certain pest insects that destroy forest trees. Bark beetles are small beetles that feed within the bark of conifer trees. They attract one another to weakened trees and attack en masse, eating tissue, laying eggs, and bringing with them a small army of fungi, bacteria, and other pathogens (FIGURE 12.18b). Bark beetle infestations can wipe out vast areas of trees with amazing speed, and today's outbreaks are unprecedented. Since the 1990s, beetle outbreaks have devastated more than 11 million ha (27 million acres) of forest in western North America, killing an estimated 30 billion conifer trees.

Scientists studying beetles and their impacts say there are two primary reasons for today's extraordinary outbreaks. One is that past forest management has resulted in even-aged forests across large regions, and many trees in these forests are now at a prime age for beetle infestation. Plantation forests dominated by single species that the beetles prefer are most at risk. The second reason is climate change. Milder winters allow beetles to overwinter further north, and warmer summers speed up their consumption and reproduction. In Alaska, beetles have switched from a two-year life cycle to a one-year cycle. In parts of the Rocky Mountains, they now produce two broods per year instead of one. Meanwhile, droughts like those that have plagued the western and southern United States in recent years have stressed and weakened trees, making them vulnerable to attack. On top of all this, beetle outbreaks create a positive feedback loop (pp. 106–107); by killing trees, they reduce the amount of carbon dioxide pulled from the air, and thereby intensify climate change.

Climate change benefits some species while harming others, and as it interacts with pests, diseases, and management strategies, many of our forest ecosystems could be altered in profound ways. Already many dense, moist forests devastated by beetles have been replaced by drier woodlands, shrublands, or grasslands. Further changes to our forests could create novel types of ecosystems not seen today.

Sustainable forestry is gaining ground

We can help address all of today's challenges to our forests with sustainable forestry practices. Although the world overall is losing forested land, we are making a number of advances toward sustainability in our forestry practices (FIGURE 12.19).

Any company can claim that its timber harvesting practices are sustainable, but how is the purchaser of wood products to know whether they really are? Organizations such as the Forest Stewardship Council (FSC) examine practices and rate them against criteria for sustainability. They grant **sustainable forest certification** to forests, companies, and products produced using methods they judge to be sustainable (FIGURE 12.20).

Several certification organizations exist, but the Forest Stewardship Council is considered to have the strictest standards. So, for example, all of the products produced by NewPage's Escanaba mill attain certification by the Sustainable Forestry Initiative® program, but only some achieve the stricter FSC certification. The paper for this textbook is "chain-of-custody certified" by the FSC, meaning that all steps in the life cycle of the paper's production—from timber harvest to transport to pulping to production—have met strict standards. FSC certification rests on 10 general principles (TABLE 12.2) and 56 more-detailed criteria.

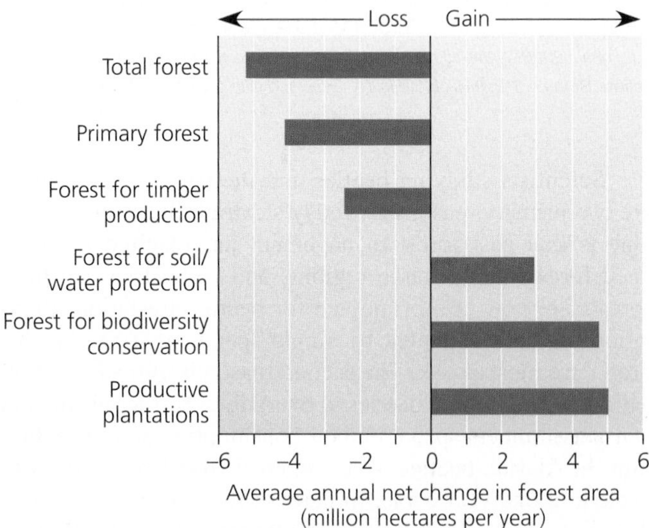

FIGURE 12.19 Progress worldwide toward sustainable forestry is mixed, as shown by six leading indicators. On the one hand, forest uses are shifting toward conservation, and timber production is increasingly coming from plantations. On the other hand, primary forest is still being lost, and global forest area is shrinking.
Data from U.N. Food and Agriculture Organization, 2010. Global forest resources assessment 2010.

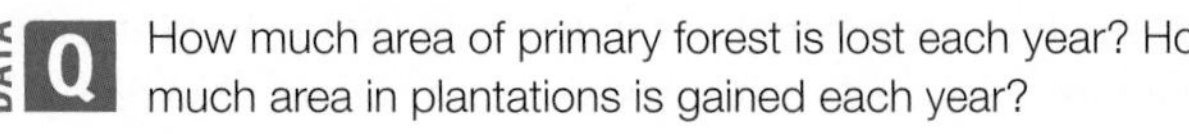

How much area of primary forest is lost each year? How much area in plantations is gained each year?

FIGURE 12.20 Inspectors work with loggers in Michigan's Upper Peninsula to confirm that timber is harvested in accord with FSC criteria for sustainable harvesting.

The number of FSC-certified forests, companies, and products is growing quickly. More than 6% of the world's forests managed for timber production are now FSC certified. This totaled over 179 million ha (440 million acres) in 79 nations as of 2013. More than 1200 operations are certified, and nearly 26,000 companies or projects are chain-of-custody certified. Some of the growth of certification is tied to the increasing construction of green buildings (pp. 348–350).

Pursuing sustainable forestry practices is often more costly for producers in the short term, but producers recoup these costs when consumers are willing to pay more for certified products. In the long term, sustainable practices conserve the resource base, thus holding down costs. When we ask businesses if they carry certified wood or paper,

TABLE 12.2 Ten Principles of Forest Stewardship Council (FSC) Certification

To receive FSC certification, forest product companies must:
• Comply with all laws and treaties.
• Show uncontested, clearly defined, long-term land rights.
• Recognize and respect indigenous peoples' rights.
• Maintain or enhance long-term social and economic well-being of forest workers and local communities, and respect workers' rights.
• Use and share benefits derived from the forest equitably.
• Reduce environmental impacts of logging and maintain the forest's ecological functions.
• Continuously update an appropriate management plan.
• Monitor and assess forest condition, management activities, and social and environmental impacts.
• Maintain forests of high conservation value.
• Promote restoration and conservation of natural forests.

Source: Forest Stewardship Council

we make them aware of our preferences and help drive demand for these products. Consumer demand has led Home Depot and other major retailers to carry sustainable wood, and these retailers' purchasing decisions are influencing timber harvesting practices around the world. You can look for the logos of certifying organizations on forest products where they are sold. If certification standards are kept strong, then we as consumers can exercise choice in the marketplace and thereby help to promote sustainable forestry practices.

Parks and Protected Areas

As our world fills with more people consuming more resources, the conservation and sustainable management of resources from forests and other ecosystems becomes ever more important. So does our need to preserve functional ecosystems by setting aside tracts of undisturbed land to remain forever undeveloped.

Preservation has been part of the American psyche ever since John Muir rallied support for saving scenic lands in the Sierras (pp. 138–139). For ethical reasons as well as pragmatic ecological and economic ones, Americans and people worldwide have chosen to set aside areas of land in perpetuity to be protected from development. Today 12.7% of the world's land area is designated for preservation in various types of parks, reserves, and protected areas.

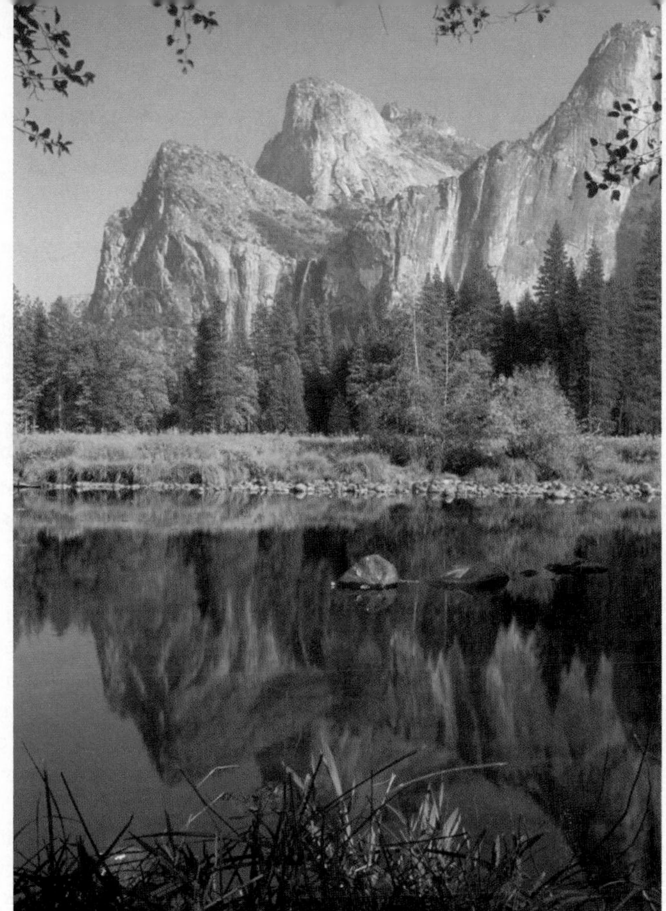

FIGURE 12.21 The awe-inspiring beauty of places such as Yosemite draws millions of people to America's national parks.

Why create parks and reserves?

People establish parks and protected areas for several reasons:

- Enormous or unusual scenic features such as the Grand Canyon, Mount Rainier, or Yosemite Valley inspire people to preserve them (FIGURE 12.21).
- Protected areas offer recreational value for hiking, fishing, hunting, kayaking, bird-watching, and other pursuits.
- Parks generate revenue from ecotourism (pp. 70, 293).
- Undeveloped land offers us peace of mind, health, exploration, wonder, and spiritual solace. Children in particular benefit from healthy exposure to the outdoors (p. 294).
- Protected areas offer utilitarian benefits through ecosystem services. For example, undeveloped watersheds provide cities with clean drinking water and a buffer against floods.
- Reserves protect biodiversity. These islands of habitat help to maintain species, communities, and ecosystems.

Federal parks and reserves began in the United States

The striking scenery of the American West persuaded the U.S. government to create the world's first **national parks,** public lands protected from resource extraction and development but open to nature appreciation and recreation. Yellowstone National Park was established in 1872, followed by Sequoia, General Grant (now Kings Canyon), Yosemite, Mount Rainier, and Crater Lake National Parks. The Antiquities Act of 1906 gave the U.S. president authority to declare selected public lands as *national monuments*, which may later become national parks.

The National Park Service was created in 1916 to administer the growing system of parks and monuments, which today numbers 401 sites totaling 34 million ha (84 million acres) and includes national historic sites, national recreation areas, national wild and scenic rivers, and other areas (see Figure 12.10). The parks receive 280 million reported recreation visits per year—almost one per U.S. resident.

Because America's national parks are open to everyone and showcase the nation's natural beauty in a democratic way, writer Wallace Stegner famously called them "the best idea we ever had." A mill worker in Escanaba can take his or her family for the weekend to Pictured Rocks National Lakeshore, where they can camp along sandstone cliffs on the shore of Lake Superior. They can head across the lake to remote Isle Royale National Park to hike and canoe through lands that are home to moose and wolves. Or they can cross Lake Michigan and enjoy climbing immense sand dunes at Sleeping Bear Dunes National Lakeshore.

Another type of federal protected area in the United States is the **national wildlife refuge.** The system of national wildlife refuges, begun in 1903 by President Theodore Roosevelt, now totals over 560 sites comprising 39 million ha (96 million acres; see Figure 12.10), plus an additional 22 million ha (55 million acres) of ocean and islands within the immense

THE SCIENCE BEHIND THE STORY

Fighting over Fire and Forests

It's not often that a scientific paper throws an entire college into turmoil and lands a graduate student in a Congressional hearing to face hostile questioning from federal lawmakers. But such is the political sensitivity of salvage logging.

When a fire burns a forest, should the killed trees be cut and sold for timber? Proponents of salvage logging say yes: we should not let economically valuable wood go to waste. Opponents of salvage logging counter that the burned wood is more valuable left in place—for erosion control, wildlife habitat (snags provide holes for cavity-dwelling animals and food for insects and birds), and organic material to enhance the soil and nurse future trees.

Proponents of salvage logging argue that forests regenerate best after a fire if they are logged and replanted with seedlings. Moreover, they maintain, salvage logging reduces fire risk by removing woody debris that could fuel the next fire. When the Biscuit Fire consumed 200,000 ha (500,000 acres) in Oregon in 2002, foresters from the College of Forestry at Oregon State University (OSU) made these arguments in support of plans to log portions of the burned area.

Meanwhile, OSU forestry graduate student Daniel Donato and five

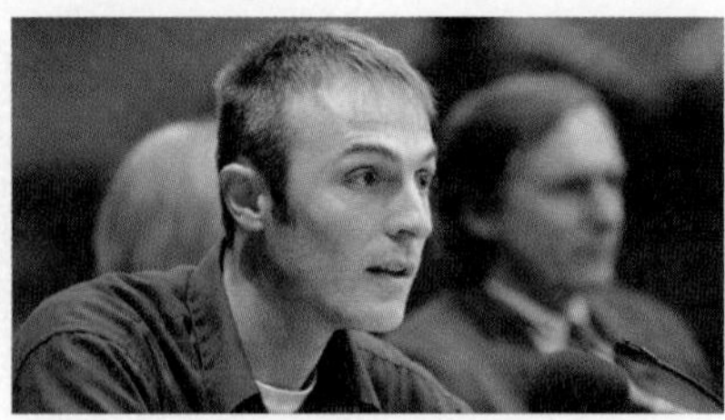

OSU graduate student Daniel Donato testifies in a Congressional hearing

other OSU researchers were setting up research plots in areas burned by the Biscuit Fire to test whether salvage logging really does reduce fire risk and help seedlings regenerate. They measured seedling growth and survival and the amount of woody debris in a number of study plots on burned land before (2004) and after (2005) salvage logging took place and on burned land that was not logged.

The researchers found that conifer seedlings sprouted naturally in the burned areas at densities exceeding what foresters aim for when they replant sites manually. This suggested that manual planting of seedlings may be unnecessary. In contrast, natural seedling densities in logged areas were only 29% as high (**FIGURE 1**). This indicated that salvage logging was hindering seedling survival, presumably because logging disturbs the soil and crushes many seedlings.

Donato's team also found that salvage logging more than tripled the amount of woody debris on the ground relative to unlogged sites (**FIGURE 2**). The research team suggested that the best strategy after fire may be to leave the site alone and leave dead trees standing, so that seedlings regenerate safely.

These conclusions directly contradicted what some OSU forestry professors had argued following the 2002 fire. When they learned that the prestigious journal *Science* had accepted the Donato team's paper for publication, they took the unusual step of asking the journal's editors to reconsider their decision. Claiming that *Science's* peer review process had failed, Professor John Sessions and others tried through back channels to

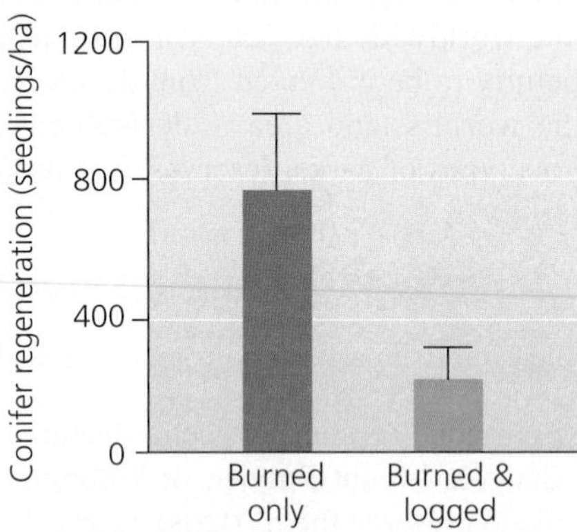

FIGURE 1 Natural growth of conifer seedlings was lower in areas that underwent salvage logging. *Data from Donato, D.C., et al., 2006. Post-wildfire logging hinders regeneration and increases fire risk.* Science *311: 352, Fig 1A. Reprinted with permission from AAAS.*

new Pacific Remote Islands Marine National Monument designated in 2009 that stretches northwest from Hawaii. There are wildlife refuges within an easy drive of nearly every major American city.

The U.S. Fish and Wildlife Service administers the national wildlife refuges, which serve as havens for wildlife but also in many cases encourage hunting, fishing, wildlife observation, photography, environmental education, and other public uses. Some wildlife advocates find it ironic that hunting is allowed at many refuges, but hunters have long been in the forefront of the conservation movement and have traditionally supplied the bulk of funding for land acquisition and habitat management. Many refuges are managed for waterfowl, but managers increasingly consider non-game species, work at the habitat and ecosystem levels, and restore marshes and grasslands.

Wilderness areas are established on federal lands

In response to the public's desire for undeveloped areas of land, in 1964 the U.S. Congress passed the Wilderness Act, which allowed some areas of existing federal lands to be designated as **wilderness areas.** These areas are off-limits to development but are open to hiking, nature study, and other low-impact public recreation (**FIGURE 12.22**).

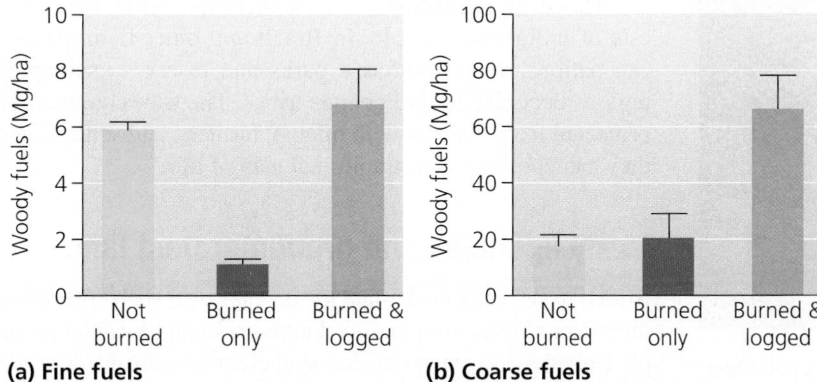

FIGURE 2 Burned sites that were salvage-logged (orange bar) contained more fine (a) and coarse (b) woody debris than unlogged burned sites or sites that did not burn. *Data from Donato, D.C., et al., 2006. Post-wildfire logging hinders regeneration and increases fire risk.* Science *311: 352, Fig 1B. Reprinted with permission from AAAS.*

stop publication of the paper—actions that were widely condemned as an attempt at censorship.

The paper's publication in 2006 unleashed a torrent of bizarre events. U.S. Congressmen Greg Walden of Oregon and Brian Baird of Washington felt the paper threatened legislation they had sponsored to accelerate salvage logging. Walden and Baird called the 29-year-old Donato and others before a hearing of the House of Representatives' Committee on Resources and grilled them before a packed crowd in Medford, Oregon.

The Bureau of Land Management (BLM) then suspended the team's research grant, in what many viewed as a response to political pressure. This highly unusual action drew media attention, and the BLM reinstated the funding.

A heated debate roiled for months in the OSU College of Forestry. The college receives 12% of its funding from taxes on timber sales, leading many to suggest that the college is open to influence from industry. E-mail correspondence was subpoenaed and showed the college's dean, Hal Salwasser, collaborating with timber industry representatives to refute the paper. As publicity built, the college's reputation suffered, and a faculty committee on academic freedom criticized Salwasser for "significant failures of leadership." The dean admitted mistakes, survived a no-confidence vote of the faculty, and pledged to make reforms.

In the pages of *Science* and elsewhere, scientific criticisms of the study's methods were largely rebutted, yet many felt that the study's conclusions stretched beyond what its short-term data could support. Scientists on both sides of the debate agreed that long-term research was needed to fully assess the effects of salvage logging on forest regeneration and fire risk.

Two studies by different OSU forestry scientists soon provided the first such long-term data. Jeffrey Shatford and colleagues documented widespread natural regrowth of conifers across areas of Oregon and northern California that had burned 9–19 years earlier. And Jonathan Thompson and colleagues examined satellite data, aerial photography, and government records for regions within the Biscuit Fire area that had burned in a previous fire 15 years earlier. They found that of the regions burned in that 1987 fire, those that were salvage logged burned more severely in 2002 than regions that were not logged. In a paper in the *Proceedings of the National Academy of Sciences*, Thompson's team concluded that salvage logging increases the risk of severe fires, even when debris is removed and seedlings are manually planted.

As more long-term studies on the impacts of salvage logging are conducted, we should become better able to manage our forests in an age of increasingly frequent wildfire. ■

Congress declared that wilderness areas were needed "to assure that an increasing population, accompanied by expanding settlement and growing mechanization, does not occupy and modify all areas . . . leaving no lands designated for preservation and protection in their natural condition." Despite these words, some preexisting extractive land uses, such as grazing and mining, were allowed to continue within some wilderness areas as a political compromise so the act could be passed.

Wilderness areas are established within national forests, national parks, national wildlife refuges, and land managed by the Bureau of Land Management (BLM). They are overseen by the agencies that administer those areas. In Michigan's Upper Peninsula, several wilderness areas are designated in the Hiawatha and Ottawa National Forests, one in Pictured Rocks National Lakeshore, and one in Seney National Wildlife Refuge. Overall there are more than 750 wilderness areas totaling 44 million ha (110 million acres). These cover 5% of U.S. land area (2.7% if Alaska is excluded).

Not everyone supports land set-asides

The restriction of activities in wilderness areas has generated some opposition to U.S. land protection policies, especially among citizens and policymakers in western states. When these states came into existence, the

FIGURE 12.22 Wilderness areas are preserved and protected from development. Here a child enjoys Pictured Rocks National Lakeshore in Michigan's Upper Peninsula.

federal government retained jurisdiction over much of their acreage. Idaho, Oregon, and Utah control less than half the land within their borders, and in Nevada 80% of the land is federally managed. Some western state governments have sought to obtain land from the federal government and to facilitate resource extraction and development on it. They have been supported by the industries that extract timber, minerals, and fossil fuels, as well as by farmers, ranchers, trappers, and mineral prospectors at the grassroots level. Both wealthy advocates of privatization and people who make their living off the land have supported efforts to secure local private control of public lands, limit government regulation, encourage the extraction of resources, and promote motorized recreation on public lands. These advocates have driven debates over national park policy, such as whether recreational activities that disturb wildlife, such as snowmobiles and jet-skis, should be allowed.

Similar debates take place wherever parks are established. In East Africa, many Maasai people living near Serengeti National Park resent the park because it displaced their families from their traditional land (p. 302). Many people living near the park want a highway to be built through it in order to provide them better mobility and access to trade (pp. 275–276).

Parks and protected areas regularly bring substantial economic benefits to people who live nearby, through ecotourism. However, individuals who do not have jobs related to parks and tourism may feel that a park restricts their economic opportunities. For this reason, proponents of protected areas increasingly try to ensure that the economic benefits of preserving natural land are spread equitably among people in nearby communities.

Groups of indigenous people frequently oppose government actions to set aside land. As with the Maasai in Africa, Native Americans were forced from their land while the U.S. government imposed rules of property ownership that were foreign to most Native American cultures. Some sites used today for recreation in the national parks are sites long sacred to Native cultures. The huge basalt rock formation at Devil's Tower National Monument in Wyoming is sacred to many Plains tribes, yet it is also popular with technical rock climbers. Native Americans view rock climbing as a desecration of the site. As a compromise, the National Park Service asks climbers not to climb the tower in June, when Native Americans travel there for religious ceremonies—an accommodation the U.S. Supreme Court upheld after a lengthy court battle.

However, protected areas sometimes serve the interests of indigenous people. In Brazil and other Latin American nations, some rainforest parks and reserves encompass regions occupied by indigenous tribes. The tribes are thereby protected from conflict with miners, farmers, and settlers, and they can continue their traditional way of life.

Many agencies and groups protect land

Efforts to set aside land—and the debates over such efforts—at the national level are paralleled at regional and local levels. In the United States, each state has agencies that manage resources on public lands, as do many counties and municipalities. When Mackinac Island, the nation's second national park, was transferred to the state of Michigan, it became the first officially designated state park in the nation. Another example is Adirondack State Park in New York. In the 19th century, the state of New York established this park to protect land in a mountainous area where streams converge to form the Hudson River, which thereafter flows south past Albany to New York City. Seeing the need for river water to power industries, keep canals filled, and provide drinking water, the state made a far-sighted decision that has paid dividends through the years. Today these parks are among nearly 7000 state parks across the United States, along with regional parks, county parks, and others.

Private nonprofit groups also preserve land. **Land trusts** are local or regional organizations that purchase land to preserve in its natural condition. The Nature Conservancy, the world's largest land trust, uses science to select areas and ecosystems in greatest need of protection. Smaller land trusts are diverse in their missions and methods. Nearly 1700 local and state land trusts in the United States together own 870,000 ha (2.1 million acres) and have helped preserve an additional 5.6 million ha (13.9 million acres), including scenic areas such as Big Sur on the California coast, Jackson Hole in Wyoming, and Maine's Mount Desert Island. Moreover, thousands of local volunteer groups (often named "Friends of" an area) have organized from the grassroots to help care for protected lands.

Parks and reserves are increasing internationally

Many nations have established systems of protected areas and are benefiting from ecotourism as a result. Tanzania and Kenya are world-renowned for their parks and reserves, including Serengeti National Park and Maasai Mara National Reserve, which protect entire ecosystems and vast populations of large mammals (Chapter 11). Costa Rica, Belize, Ecuador, Thailand, and Australia are also famed for their parks and gain large amounts of foreign capital from tourism.

The total worldwide area in protected parks and reserves has increased nearly sevenfold since 1970, and today the world's 158,000 protected areas cover 12.7% of the planet's land area. However, parks in developing nations do not always receive funding adequate to manage resources, provide for recreation, and protect wildlife from poaching and trees from logging. As a result, many of the world's protected areas are merely *paper parks*—protected on paper but not in reality.

Some types of protected areas fall under national sovereignty but are designated or partly managed by the United Nations. **Biosphere reserves** are tracts of land with exceptional biodiversity that couple preservation with sustainable development to benefit local people. Biosphere reserves are designated by UNESCO (the United Nations Educational, Scientific, and Cultural Organization) following application by local stakeholders. Each biosphere reserve consists of (1) a core area that preserves biodiversity; (2) a buffer zone that allows local activities and limited development; and (3) an outer transition zone where agriculture, human settlement, and other land uses are pursued sustainably (FIGURE 12.23a).

The Maya Biosphere Reserve in Guatemala (FIGURE 12.23b) is an example. Rainforest here is protected in core areas, and in the transition zone timber harvesting takes place in concessions, some of which are FSC certified. A 2008 study by the nonprofit Rainforest Alliance found that FSC certification here gave people economic incentives to conserve the forest. The study found that rates of deforestation and wildfire were much lower in the FSC-certified areas where sustainable logging occurred than in the core area that was supposed to be fully protected.

World heritage sites are another type of international protected area. Nearly 1000 sites in more than 150 countries are listed for their natural or cultural value. One such site is the Serengeti. Another is a mountain gorilla reserve shared by three African countries. This reserve, which integrates national parklands of Rwanda, Uganda, and the Democratic Republic of Congo, is also an example of a *transboundary park*, an area of protected land overlapping national borders. Transboundary parks account for 10% of protected areas worldwide, involving over 100 nations. An example is Waterton–Glacier National Parks on the Canadian–U.S. border.

Some transboundary reserves function as *peace parks*, helping to ease tensions by acting as buffers between nations that quarrel over boundary disputes. This is the case with Peru and Ecuador as well as Costa Rica and Panama, and many people hope that peace parks might also help resolve conflicts between Israel and its neighbors.

Beyond all these efforts on land, the importance of conserving the oceans' natural resources is leading us to establish protected areas and reserves in marine waters (pp. 443–444). Today 1.6% of the world's ocean area and 7.2% of coastal waters fall within designated protected areas.

Economic incentives can help preserve land

Innovative economic strategies can facilitate international efforts to protect natural lands. One strategy is the conservation concession discussed earlier (p. 314). Another is the *debt-for-nature swap*, in which a conservation organization raises money and offers to pay off a portion of a developing nation's international debt in exchange for a promise by the nation to set aside reserves, fund environmental education, and better manage protected areas. The U.S. government committed itself to a program of debt-for-nature swaps through its 1998 Tropical Forest Conservation Act. As of 2011, deals had been struck with 17 developing nations, enabling nearly $400 million in their funds intended for debt payments to go to conservation efforts instead. In the largest such deal yet, the U.S. government forgave Indonesia $30 million of debt while two conservation groups paid Indonesia $2 million. In return, Indonesia will preserve forested areas that are home to the Sumatran tiger and other species.

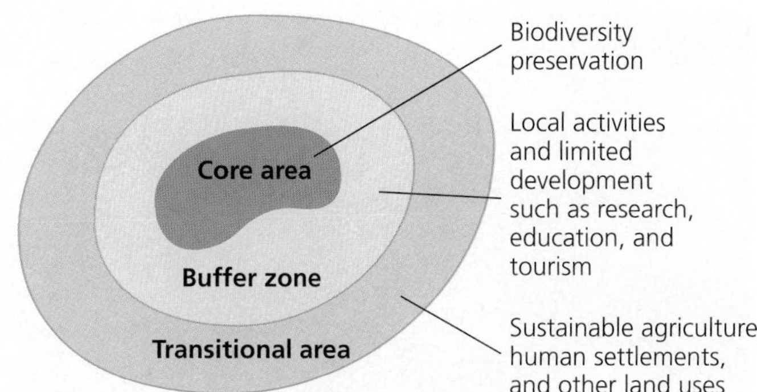

(a) The three zones of a biosphere reserve

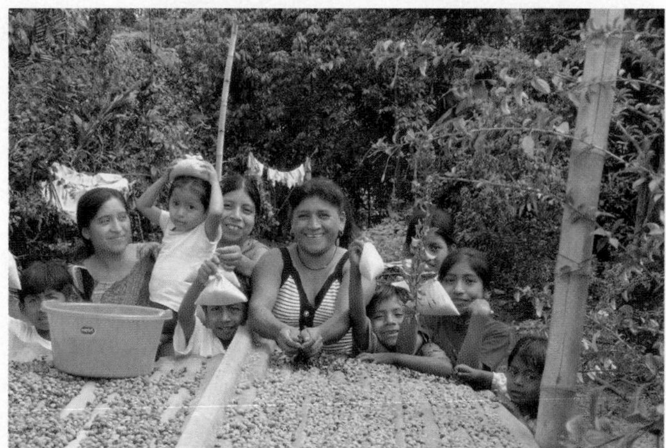

(b) Sustainable harvesting by local people

FIGURE 12.23 **Biosphere reserves couple preservation with sustainable development.** Each biosphere reserve **(a)** includes three zones. At the Maya Biosphere Reserve in Guatemala **(b)**, local women process and sell Maya nuts harvested from rainforest trees. FSC certification in the transition zone here helped prevent illegal deforestation.

Habitat fragmentation makes preserves more vital

Protecting large areas of land has taken on new urgency now that scientists understand the risks posed by habitat fragmentation (p. 284). Often, it is not the outright destruction of forests and other habitats that threatens species and ecosystems, but rather their fragmentation (see THE SCIENCE BEHIND THE STORY, pp. 330–331). Expanding agriculture, spreading cities, highways, logging, and other impacts routinely chop up large contiguous expanses of habitat into small, disconnected ones (see Figure 11.11, p. 284). Forests are being fragmented everywhere these days, as a result of logging (FIGURE 12.24a)

(a) Fragmentation from clear-cuts in Mount Hood National Forest, Oregon

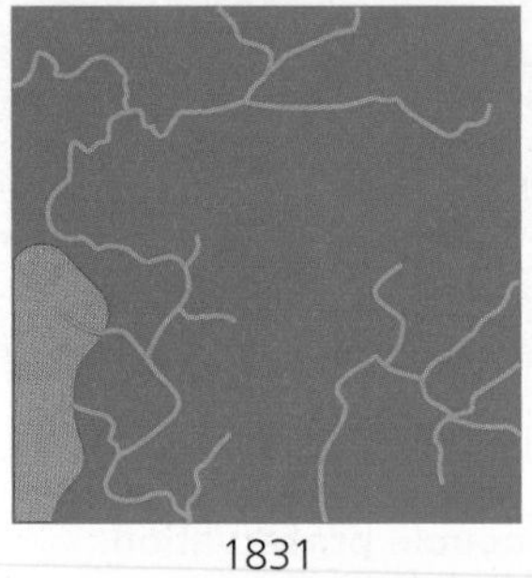

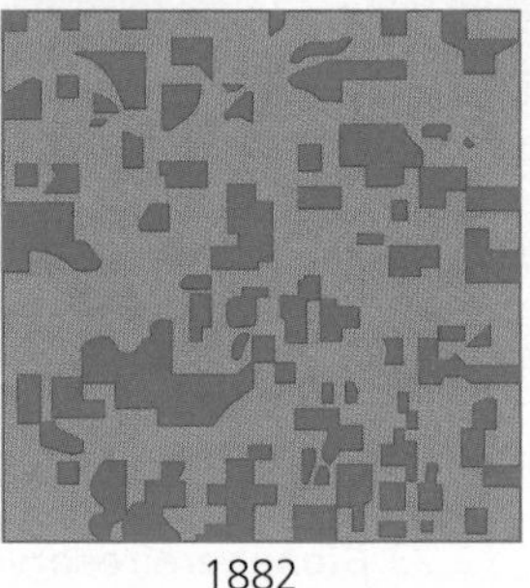

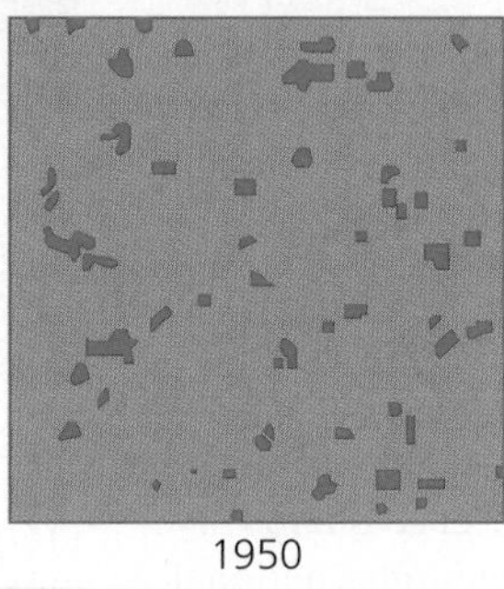

(b) Fragmentation of wooded area (green) in Cadiz Township, Wisconsin

(c) Wood thrush

FIGURE 12.24 **Forests are being fragmented, with ecological consequences.** Fragmentation results from clear-cutting **(a)**, agriculture, and residential development. Shown in **(b)** are historical changes in forested area in a region of Wisconsin. Fragmentation affects forest-dwelling species such as the wood thrush **(c)**, whose nests are parasitized by cowbirds that thrive in surrounding open country. *Source (b): Curtis, J.T., 1956. The modification of mid-latitude grasslands and forests by man. In Thomas, W.L. Jr., Ed.,* Man's role in changing the face of the earth. *©1956. Used by permission of the publisher, University of Chicago Press.*

and other factors, including agriculture and residential development (FIGURE 12.24b). Even in areas where forest cover is increasing (such as the northeastern United States), regrowing forests are becoming fragmented into ever-smaller parcels.

When forests are fragmented, many species suffer. For some, a fragment may simply not contain enough area; bears, mountain lions, and other animals that need large ranges in which to roam may disappear. For other species, the problem may lie with **edge effects,** impacts that result because the conditions along a fragment's edge differ from conditions in the interior. Bird species that thrive in the interior of forests may fail to reproduce when forced near the edge of a fragment (FIGURE 12.24c). Their nests often are attacked by predators and parasites that favor open habitats or travel along habitat edges. Because of edge effects, avian ecologists judge forest fragmentation to be a main reason why populations of many North American songbirds are declining.

Insights from islands warn us of habitat fragmentation

In assessing the impacts of habitat fragmentation on populations, ecologists and conservation biologists have leaned on concepts from **island biogeography theory.** Introduced by E.O. Wilson (p. 293) and ecologist Robert MacArthur in 1963, this theory explains how species come to be distributed among oceanic islands. Since then, researchers have applied it to "habitat islands"—patches of one habitat type isolated within "seas" of others.

Island biogeography theory explains how the number of species on an island results from a balance between the number added by immigration and the number lost through extirpation. It predicts an island's species richness based on the island's size and its distance from the mainland:

- The farther an island lies from a continent, the fewer species tend to find and colonize it. Thus, remote islands host few species because of low immigration rates (FIGURE 12.25a). This is the *distance effect.*
- Larger islands have higher immigration rates because they present fatter targets for dispersing organisms to encounter (FIGURE 12.25b).
- Larger islands have lower extinction rates because more space allows for larger populations, which are less likely to drop to zero by chance (FIGURE 12.25c).

Together, the latter two trends give large islands more species than small islands—a phenomenon called the *area effect.* Large islands also contain more species because they tend to possess more habitats than smaller islands. Very roughly,

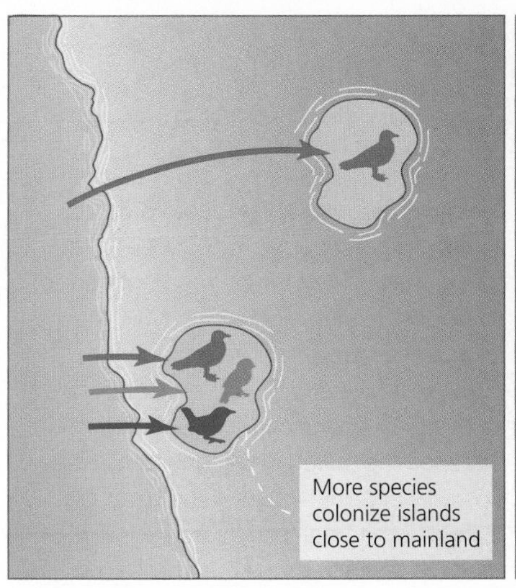

(a) Distance effect

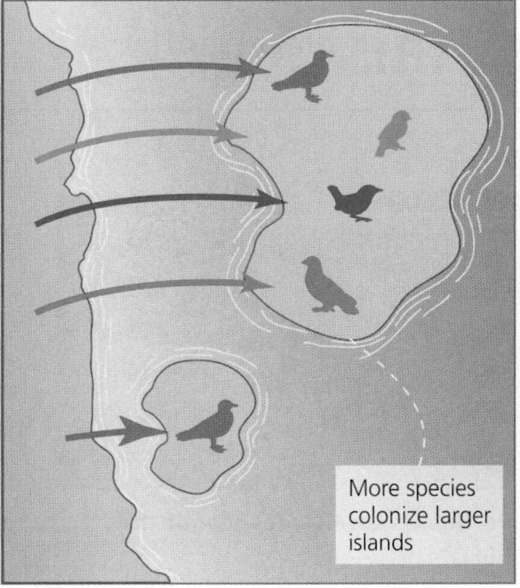

(b) Target size

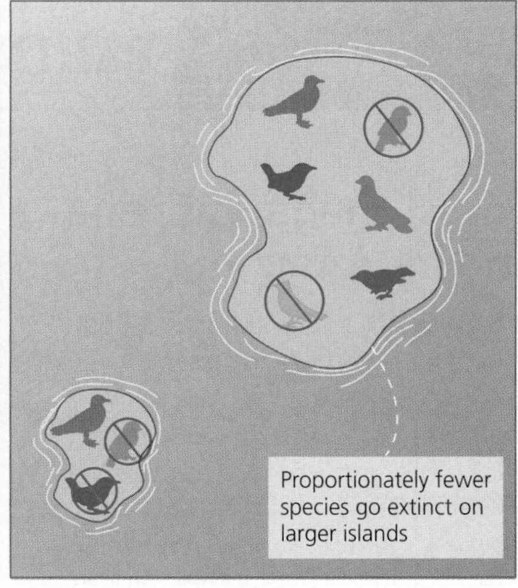

(c) Differential extinction

FIGURE 12.25 Islands that are larger or closer to a mainland support more species. According to island biogeography theory, islands near a continent **(a)** receive more immigrants than distant islands. Larger islands present fatter targets **(b)**, so more species encounter large islands than small islands. Large islands also have lower extinction rates **(c)**, because larger area allows for larger populations.

the number of species on an island is expected to double as island size increases tenfold. This phenomenon is illustrated by **species-area curves** (FIGURE 12.26).

These patterns hold up for terrestrial habitat islands as well, such as forests fragmented by logging and road building. Small "islands" of forest lose diversity fastest, starting with large species that were few in number to begin with. One of the first researchers to show this experimentally was University of Michigan graduate student William Newmark, who in 1983 examined historical records of mammal sightings in North American national parks. The parks, increasingly surrounded by development, were islands of natural habitat isolated by farms, ranches, roads, and cities.

Newmark found that many parks were missing a few species they had held previously. The red fox and river otter had vanished from Sequoia and Kings Canyon National Parks, for example, and the white-tailed jackrabbit and spotted skunk no longer lived in Bryce Canyon National Park. In all, 42 species had disappeared. As island biogeography theory predicted, smaller parks lost more species than larger parks. Species were disappearing because the parks were too small to sustain their populations, Newmark concluded, and because the parks had become too isolated to be recolonized by new arrivals.

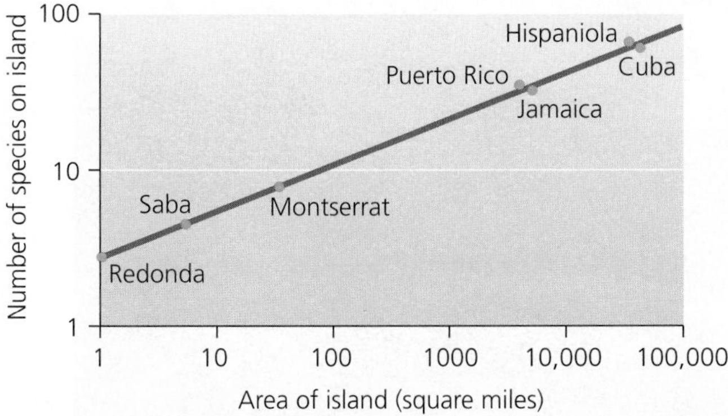

FIGURE 12.26 Larger islands hold more species. This species-area curve shows that the number of amphibian and reptile species on Caribbean islands increases with island area. The increase is not linear, but logarithmic; note the scales of the axes. *Data from MacArthur, R.H, and E.O. Wilson.* The theory of island biogeography. *© 1967 Princeton University Press, 1995 renewed PUP. Reprinted by permission of Princeton University Press.*

Reserve design has consequences for biodiversity

Because habitat fragmentation is such a central issue in biodiversity conservation, and because there are limits on how much land can feasibly be set aside, conservation biologists have argued heatedly about whether it is better to make reserves large in size and few in number, or many in number but small in size. Nicknamed the **SLOSS dilemma,** for "single large or several small," this debate is complex, but it seems clear that large species that roam great distances, such as wildebeest and zebras (Chapter 11), benefit more from the "single large" approach to reserve design. In contrast, creatures such as insects that live as larvae in small areas may do just fine in a number of small isolated reserves, if they can disperse as adults by flying from one reserve to another. The SLOSS debate was one motivation for establishing the Amazonian forest fragmentation project described in our **Science behind the Story** (pp. 330–331).

A related issue is how effectively **corridors** of protected land allow animals to travel between islands of habitat. In theory, connections between fragments provide animals access to more habitat and encourage gene flow to maintain populations in the long term. Many land managers now try to join new reserves to existing reserves for these reasons.

Forest Fragmentation in the Amazon

What happens to animals, plants, and ecosystems when we fragment a forest? A massive experiment smack in the middle of the Amazon rainforest is helping scientists learn the answers.

Stretching across 1000 km^2 (386 mi^2), the Biological Dynamics of Forest Fragments Project (BDFFP) is the world's largest and longest-running experiment on forest fragmentation. For over 30 years, hundreds of researchers have muddied their boots in the rainforest here, publishing over 700 scientific research papers, graduate theses, and books.

The story starts in the 1970s as conservation biologists debated how to apply island biogeography theory to forested landscapes being fragmented by development. Biologist Thomas Lovejoy decided some good hard data were needed. He conceived a huge experimental project to test ideas about forest fragmentation and established it in the heart of the biggest primary rainforest on the planet—South America's Amazon rainforest.

Farmers, ranchers, loggers, and miners were streaming into the Amazon then, and deforestation was rife. If scientists could learn how large fragments had to be to retain their species, it would help them work with policymakers to preserve forests in the face of development pressures.

Lovejoy's team of Brazilians and Americans worked out a deal with Brazil's government: Ranchers could clear some forest within the study area if they left square plots of forest standing as fragments within those clearings. By this process, 11 fragments of three sizes (1 ha [2.5 acres], 10 ha [25 acres], and 100 ha [250 acres]) were left standing, isolated as "islands" of forest, surrounded by "seas" of cattle pasture (**FIGURE 1**). Each fragment was fenced to keep cattle out. Then, 12 study plots (of 1, 10, 100, and 1000 ha) were

Dr. Thomas Lovejoy, founder of the BDFFP

established within the large expanses of continuous forest still surrounding the pastures, to serve as control plots against which the fragments could be compared.

Besides comparing treatments (fragments) and controls (continuous forest), the project also surveyed populations in fragments before and after they were isolated. These data on trees, birds, mammals, amphibians, and invertebrates showed declines in the diversity of most groups (**FIGURE 2**).

As researchers studied the plots over the years, they found that small fragments lost more species, and lost them faster, than large fragments—just as island biogeography theory predicts. To slow down species loss by 10 times, researchers found that a fragment needs to be 1000 times bigger. Even 100-ha fragments were not large enough for some animals, and lost half their species in less than 15 years. Monkeys died out because they need large ranges. So did colonies of army ants and the birds that follow them to eat insects scared up as the ants swarm across the forest floor.

Fragments distant from continuous forest lost more species, but data revealed that even very small openings can stop organisms adapted to deep interior forest from dispersing to recolonize fragments. Many understory birds would not traverse cleared areas of only 30–80 m (100–260 ft). Distances of just 15–100 m (50–330 ft) were insurmountable for some bees, beetles, and tree-dwelling mammals.

Soon, a complication ensued: Ranchers abandoned many of the pastures because the soil was unproductive, and these areas began filling in with secondary forest. As this young forest grew, it made the fragments

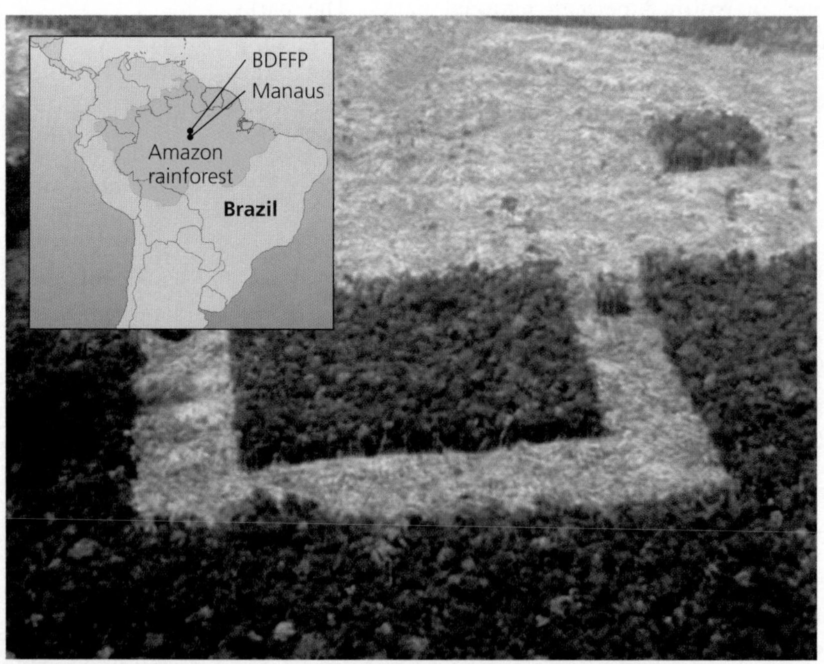

FIGURE 1 **Experimental forest fragments of 1, 10, and 100 ha were created in the BDFFP.**

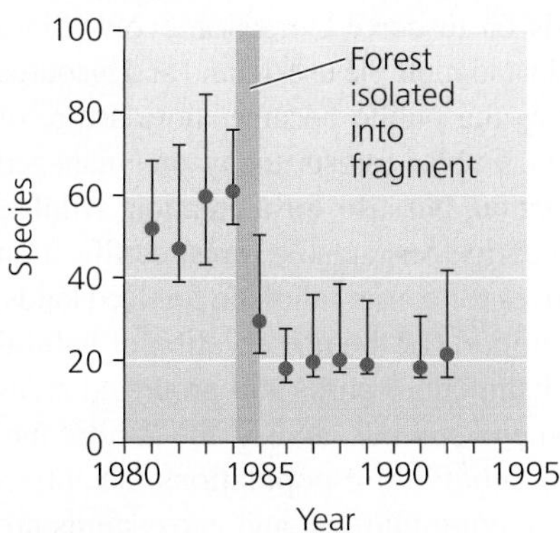

FIGURE 2 Species richness of understory birds declined in this 1-ha forest plot after it was isolated as a fragment in 1984. *Data from Ferraz, G., et al., 2003. Rates of species loss from Amazonian forest fragments.* Proc. Natl. Acad. Sci. *100: 14069–14073. © 2003 National Academy of Sciences. By permission.*

DATA Q On average, about how many bird species were present in this forest plot before fragmentation? How many were present after fragmentation?

less like islands—but it led to new insights. Researchers learned that this habitat can act as a corridor for some species, allowing them to disperse from mature forest and recolonize fragments from which they'd disappeared. By documenting which species did this, scientists learned which might be more resilient to fragmentation. For the Amazon as a whole, it suggested that regrowing forest (as opposed to pasture) can buffer fragments against some species loss.

The secondary forest habitat also introduced new species—generalists adapted to disturbed areas. Frogs, leaf-cutter ants, and small mammals and birds that thrive in second-growth soon became common in adjacent fragments. Open-country butterfly species moved in, displacing interior-forest butterflies.

The invasion of open-country species illustrated one type of edge effect. There were more: Edges receive more sunlight, heat, and wind than interior forest, which can kill trees adapted to the dark, moist interior. Tree death results in the flux of carbon dioxide to the atmosphere, which worsens climate change. Sunlight also promotes growth of vines and shrubs that create a thick, tangled understory along edges. As BDFFP researchers documented these impacts, they found that many edge effects extended far into the forest (**FIGURE 3**). Small fragments essentially became "all edge."

The results on edge effects are relevant for all of Amazonia, because forest clearance and road construction create an immense amount of edge. A satellite image study in the 1990s estimated that for every 2 acres of land deforested, 3 acres were brought within 1 km of a road or pasture edge.

BDFFP scientists emphasize that impacts across the Amazon will be more severe than the impacts revealed by their experiments. This is because most real-life fragments (1) are not protected from hunting, logging, mining, and fires; (2) do not have secondary forest to provide connectivity; (3) are not near large tracts of continuous forest that provide recolonizing species and maintain humidity and rainfall; and (4) are not square in shape, and thus feature more edge. Scientists say the years of data argue for preserving numerous tracts of Amazonian forest that are as large as possible.

The BDFFP has inspired another large-scale, long-term study, this one located in Borneo. The Stability of Altered Forest Ecosystems (SAFE) project is documenting changes that take place in tropical forests as they are logged and converted to oil palm plantations. This project, underway for just a few years, should shortly be providing data and insights that may help us conserve biodiversity amid the rush to oil palm plantations in Southeast Asia.

Ironically, today the BDFFP study site is itself threatened by forest fragmentation. A Brazilian government agency has settled colonists just outside the site and has proposed settling 180 families inside it. Development is proceeding up a new highway from Manaus, a city of 1.7 million people, to Venezuela, and wherever roads are built and people settle, logging, hunting, mining, and burning follow. BDFFP researchers can now hear chainsaws and shotgun blasts from their study plots. "It would be tragic," says leading BDFFP scientist William Laurance, "to see a site that's given us so much information be lost so easily." ■

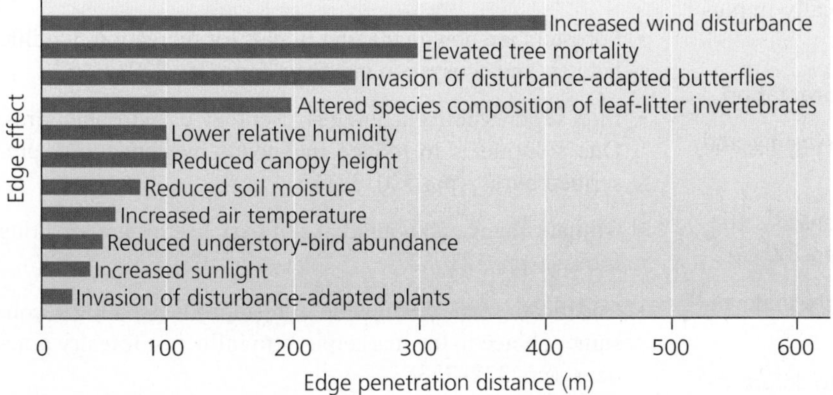

FIGURE 3 Many edge effects documented by the BDFFP extend far into the interior of forest fragments. *Data from studies summarized in Laurance, W.F., et al., 2002. Ecosystem decay of Amazonian forest fragments: A 22-year investigation.* Conservation Biology *16: 605–618, Fig 3, adapted. Reprinted by permission of John Wiley & Sons, Inc.*

DATA Q Would a tree inside a forest fragment, 275 meters in from the edge, be susceptible to any edge effects? If so, which ones?

Climate change threatens protected areas

As if we did not face enough challenges in designing, establishing, and guarding protected areas to preserve species, communities, and ecosystems, global climate change (Chapter 18) now threatens to undo our efforts. As temperatures become warmer, species ranges shift toward cooler climes: toward the poles and upward in elevation (pp. 289, 501).

In a landscape of fragmented habitat, some organisms may be unable to move from one fragment to another. Species we had hoped to protect in parks may, in a warming world, become trapped in them. High-elevation species are most at risk from climate change, because there is nowhere for them to go once a mountaintop becomes too warm or dry. For this reason, corridors to allow movement from place to place become still more important. In response to these challenges, conservation biologists are now looking beyond parks and protected areas as they explore strategies for saving biodiversity.

Conclusion

Forests are ecologically vital and economically valuable, yet we continue to lose them around the world. Forest management in North America reflects trends in land and resource management in general. Early emphasis on resource extraction evolved into policies of sustainable yield and multiple use as land and resource availability declined and as the public became more aware of environmental degradation. Public forests today are managed not only for timber production, but also for recreation, wildlife habitat, and ecosystem integrity. Sustainable forest certification provides economic incentives for conservation on forested lands.

Meanwhile, public support for the preservation of natural lands has led to the establishment of parks and protected areas worldwide. As development spreads across the landscape, fragmenting habitats and subdividing populations, scientists trying to conserve species, communities, and ecosystems are thinking and working at the landscape level.

Reviewing Objectives

You should now be able to:

Summarize the ecological and economic contributions of forests

- Many kinds of forests exist. (p. 308)
- Forests are ecologically complex and support a wealth of biodiversity. (pp. 309–310)
- Forests contribute ecosystem services, including carbon storage. (p. 310)
- Forests provide us timber and other economically important products and resources. (pp. 310–311)

Outline the history and current scale of deforestation

- We have lost forests as a result of timber harvesting and clearance for agriculture. (p. 311)
- Industrialized nations deforested much of their land as settlement, farming, and industrialization proceeded. (pp. 311–312)
- Today deforestation is taking place most rapidly in developing nations. (pp. 312–314)
- Carbon offsets are one new potential solution to deforestation. (p. 314)

Assess aspects of forest management and describe methods of harvesting timber

- Forestry is one type of resource management. (p. 315)
- Resource managers have long managed for maximum sustainable yield and have begun to implement ecosystem-based management and adaptive management. (pp. 315–316)
- The U.S. national forests were established to conserve timber and allow its sustainable extraction. (p. 316)
- Most U.S. timber today comes from private lands. (pp. 316–317)
- Plantation forestry, featuring single-species, even-aged stands, is widespread and growing. (p. 317)
- Harvesting methods include clear-cutting and other even-aged techniques, as well as selection strategies that maintain uneven-aged stands that more closely resemble natural forest. (pp. 317–319)
- Foresters are now managing in part for recreation, wildlife habitat, and ecosystem integrity. (pp. 319–320)
- Fire suppression encourages eventual catastrophic fires. One solution is to reduce fuel loads by conducting prescribed burns. (pp. 320–321)
- Climate change and outbreaks of bark beetles are affecting forests. (pp. 321–322)
- Certification of sustainable forest products allows consumer choice in the marketplace to influence forestry practices. (pp. 322–323)

Identify federal land management agencies and the lands they manage

- The U.S. Forest Service, National Park Service, Fish and Wildlife Service, and Bureau of Land Management manage U.S. national forests, national parks, national wildlife refuges, and BLM land, respectively. (pp. 316, 323–325)

Recognize types of parks and protected areas and evaluate issues involved in their design

- Public demand for preservation and recreation led to the creation of parks, reserves, and wilderness areas. (pp. 323–325)
- Opposition to land preservation policies stems from several sources. (pp. 325–326)
- States, municipalities, and private land trusts all manage protected areas at the state, regional, and local levels. (p. 326)
- Biosphere reserves are one of several types of international protected lands. (pp. 326–327)
- Debt-for-nature swaps provide industrializing nations an economic incentive for land preservation. (p. 327)
- Because habitat fragmentation affects wildlife, conservation biologists are using island biogeography theory to learn how best to design systems of parks and reserves. (pp. 327–331)
- Climate change is posing new threats to protected areas. (p. 332)

Testing Your Comprehension

1. Name at least two reasons why natural primary forests contain more biodiversity than single-species forestry plantations.
2. Describe three ecosystem services that forests provide.
3. Name several major causes of deforestation. Where is deforestation most severe today?
4. Compare and contrast maximum sustainable yield, ecosystem-based management, and adaptive management. How may pursuing maximum sustainable yield sometimes affect populations and communities?
5. Compare and contrast the major methods of timber harvesting. Name an advantage and a disadvantage of each method.
6. Describe several ecological impacts of logging. How has the U.S. Forest Service responded to public concern over these impacts?
7. Are forest fires a bad thing? Explain your answer.
8. Name at least four reasons that people have created parks and reserves. How do national parks differ from national wildlife refuges? What is a wilderness area?
9. What percentage of Earth's land is protected? Describe one type of protected area that has been established outside of North America.
10. Give two examples of how forest fragmentation affects animals. How does island biogeography theory help us design reserves?

Seeking Solutions

1. People in industrialized nations are fond of warning people in industrializing nations to stop destroying rainforest. People of industrializing nations often respond that this is hypocritical, because the industrialized nations became wealthy by deforesting their land and exploiting its resources in the past. What would you say to the president of an industrializing nation, such as Indonesia or Brazil, in which a great deal of forest is being cleared?
2. Do you think maximum sustainable yield represents an appropriate policy for resource managers to follow? Why or why not?
3. What might you tell an opponent of parks and preserves to help him or her understand why a wilderness hiker wants scenic land in Utah federally protected? What might you tell a wilderness hiker to help him or her understand why the park opponent disapproves of the protection? How might you help them find common ground?
4. Given the impacts that climate change may have on species' ranges, if you were trying to preserve an endangered mammal that occurs in a small area and you had generous funding to acquire land to help restore its population, how would you design a protected area for it? Would you use corridors? Would you include a diversity of elevations? Would you design few large reserves or many small ones? Explain your answers.
5. **THINK IT THROUGH** You have just become the supervisor of a national forest. Timber companies are requesting to cut as many trees as you will let them, and environmentalists want no logging at all. Ten percent of your forest is old-growth primary forest, and the remaining 90% is secondary forest. Your forest managers are split among preferring maximum sustainable yield, ecosystem-based management, and adaptive management. What management approach(es) will you take? Will you allow logging of all old-growth trees, some, or none? Will you allow logging of secondary forest? If so, what harvesting strategies will you encourage? What would you ask your scientists before deciding on policies on fire management and salvage logging?
6. **THINK IT THROUGH** You run a major nonprofit environmental advocacy organization and are trying to save an ecologically priceless tract of tropical forest in a poor

industrializing nation. You have worked in this region for years and know and care for the local people, who want to save the forest and its animals but also need to make a living and use the forest's resources. The nation's government plans to sell a concession to a foreign multinational timber corporation to log the entire forest unless your group can work out some other solution. Describe what solution(s) you would you try to arrange. Consider the range of issues and options discussed in this chapter, including government protected areas, private protected areas, biosphere reserves, forest management techniques, carbon offsets, FSC-certified sustainable forestry, and more. Explain reasons for your choice(s).

Calculating Ecological Footprints

We all rely on forest resources. The average North American consumes 225 kg (500 lb) of paper and paperboard each year. Using the estimates of paper and paperboard consumption for each region of the world, calculate the per capita consumption for each region using the population data in the table. Note: 1 metric ton = 2205 pounds.

	Population (millions)	Total paper consumed (millions of metric tons)	Per capita paper consumed (pounds)
Africa	1051	7	15
Asia	4216	189	
Europe	740	95	
Latin America	596	26	
North America	346	78	
Oceania	37	4	
World	6987	400	126

Data are for 2011, from Population Reference Bureau and U.N. Food and Agriculture Organization (FAO).

1. How much paper would be consumed if everyone in the world used as much paper as the average North American?
2. How much paper would North Americans save each year if they consumed paper at the rate of Europeans?
3. North Americans have been reducing their per-person consumption of paper and paperboard by nearly 5% annually in recent years as a result of a shift to online activity, recycling, and reductions in packaging of some products. Name three specific things you personally could do to reduce your own consumption of paper products.
4. Describe three ways in which consuming FSC-certified paper rather than conventional paper can reduce the environmental impacts of paper consumption.

Tom McCall Waterfront Park in Portland, Oregon

13 The Urban Environment: Creating Sustainable Cities

Upon completing this chapter, you will be able to:

- Describe the scale of urbanization
- Assess urban and suburban sprawl
- Outline city and regional planning and land use strategies
- Evaluate transportation options, urban parks, and green buildings
- Analyze environmental impacts and advantages of urban centers
- Assess urban ecology and the pursuit of sustainable cities

CENTRAL CASE STUDY

Managing Growth in Portland, Oregon

"Sagebrush subdivisions, coastal condomania, and the ravenous rampage of suburbia in the Willamette Valley all threaten to mock Oregon's status as the environmental model for the nation."

—Oregon Governor Tom McCall, 1973

"We have planning boards. We have zoning regulations. We have urban growth boundaries and 'smart growth' and sprawl conferences. And we still have sprawl."

—Environmental scientist Donella Meadows, 1999

With the fighting words above, Oregon governor Tom McCall challenged his state's legislature in 1973 to take action against runaway sprawling development, which many Oregon residents feared would ruin the communities and landscapes they loved. McCall was echoing the growing concerns of state residents that farms, forests, and open space were being gobbled up and paved over.

Foreseeing a future of subdivisions, strip malls, and traffic jams engulfing the pastoral Willamette Valley, Oregon acted. The state legislature passed Senate Bill 100, a sweeping land use law that would become the focus of acclaim, criticism, and careful study for years afterward by other states and communities trying to manage their own urban and suburban growth.

Oregon's law required every city and county to draw up a comprehensive land use plan in line with statewide guidelines that had gained popular support from the state's electorate. As part of each land use plan, each metropolitan area had to establish an urban growth boundary (UGB), a line on a map intended to separate areas desired to be urban from areas desired to remain rural. Development for housing, commerce, and industry would be encouraged within these urban growth boundaries but severely restricted beyond them. The intent was to revitalize city centers, prevent suburban sprawl, and protect farmland, forests, and open landscapes around the edges of urbanized areas.

Residents of the area around Portland, the state's largest city, established a new regional planning entity to apportion land in their region. The Metropolitan Service District, or Metro, represents 25 municipalities and three counties. Metro adopted the Portland-area urban growth boundary in 1979 and has tried to focus growth on existing urban centers and to build communities where people can walk, bike, or take mass transit between home, work, and shopping. These policies have largely worked as intended. Portland's downtown and older neighborhoods have thrived, regional urban centers are becoming denser and more community oriented, mass transit has expanded, and development has been limited on land beyond the UGB. Portland began attracting international attention for its "livability."

To many Portlanders today, the UGB remains the key to maintaining quality of life in city and countryside alike. In the view of its critics, however, the "Great Wall of Portland" is an elitist and intrusive government regulatory tool. In 2004, Oregon voters approved a ballot measure that threatened to eviscerate the land use rules that most citizens had backed for three decades. Ballot Measure 37 required the state to compensate certain landowners if government regulation had decreased the value of their land. For example, regulations prevent landowners outside UGBs from subdividing their lots and selling them for housing development. Under Measure 37, the state had to pay these landowners to make up for theoretically lost income or else allow them to ignore the regulations. Because state and local governments did not have enough money to pay such claims, the measure was on track to gut Oregon's zoning, planning, and land use rules.

Landowners filed over 7500 claims for payments or waivers affecting 295,000 ha (730,000 acres). Although the measure had been promoted to voters as a way to protect the rights of small family landowners, most claims were filed by large developers. Neighbors suddenly found themselves confronting the prospect of massive housing subdivisions, gravel mines, strip malls, or industrial facilities being developed next to their homes—and many who had voted for Measure 37 began to have misgivings.

The state legislature, under pressure from opponents and supporters alike, settled on a compromise: to introduce a new ballot measure. Oregon's voters passed Ballot Measure 49 in 2007. It protects the rights of small landowners to gain income from their property by developing small numbers of homes, while restricting large-scale development and development in sensitive natural areas.

In 2010, Metro finalized a historic agreement with representatives and citizens of its region's three counties to determine where urban growth will and will not be allowed over the next 50 years. Metro and the counties apportioned over 121,000 ha (300,000 acres) of undeveloped land into "urban reserves" open for development and "rural reserves" where farmland and forests would be preserved. Boundaries were precisely mapped to give clarity and direction for landowners and governments alike for half a century.

People are confronting similar issues in communities throughout North America, and debates and negotiations like those in Oregon will determine how our cities and landscapes will change in the future. ■

Our Urbanizing World

We have just passed a turning point in human history. Since 2009, for the first time ever, more people are living in urban areas (cities and suburbs) than in rural areas. This shift from the countryside into towns and cities, called **urbanization,** may be the single greatest change our society has undergone since its ancient transition from a nomadic hunter-gatherer lifestyle to a sedentary agricultural one.

As we undergo this shift to urban areas, two pursuits become ever more important. One is to make our urban areas more livable by meeting residents' needs for a safe, clean, healthy urban environment and a high quality of life. The other is to make our urban areas sustainable by creating cities that can prosper in the long term while minimizing our ecological footprint and working with natural systems (rather than against them).

Industrialization has driven urbanization

Since 1950, the world's urban population has multiplied by nearly five times, whereas the rural population has not yet doubled. Urban populations are growing for two reasons: (1) the human population overall is growing (Chapter 8), and (2) more people are moving from farms to cities than are moving from cities to farms.

This shift from country to city began long ago. Agricultural harvests that produced surplus food freed a proportion of citizens from farm life and allowed the rise of specialized manufacturing professions, class structure, political hierarchies, and urban centers (pp. 217–218). The industrial revolution (p. 4) spawned technological innovations that created jobs and opportunities in urban centers for people who were no longer needed on farms. Industrialization and urbanization bred further technological advances that increased production efficiencies, both on the farm and in the city. This process of positive feedback continues today.

The United Nations projects that the urban population will increase by 72% between now and 2050, whereas the rural population will decline by 9%. Trends differ between developed and developing nations, however (**FIGURE 13.1**). In developed nations such as the United States, urbanization has slowed because three of every four people already live in cities, towns, and **suburbs,** the smaller communities that ring cities. Back in 1850, the U.S. Census Bureau classified only 15% of U.S. citizens as urban dwellers. That percentage now stands at 80%. Most U.S. urban dwellers reside in suburbs; fully half the U.S. population today is suburban.

In contrast, today's developing nations, where most people still reside on farms, are urbanizing rapidly. As industrialization diminishes the need for farm labor and increases urban commerce and jobs, rural people are streaming from farms to cities. Sadly, wars, conflict, and ecological degradation are also driving millions of people out of the countryside and into urban centers. For all these reasons, most fast-growing cities today are in the developing world. In cities such as Delhi, India; Lagos, Nigeria; and Karachi, Pakistan, population growth often exceeds economic growth, and the result is overcrowding, pollution, and poverty. United Nations demographers estimate that urban areas of developing nations will absorb nearly all of the world's population growth from now on.

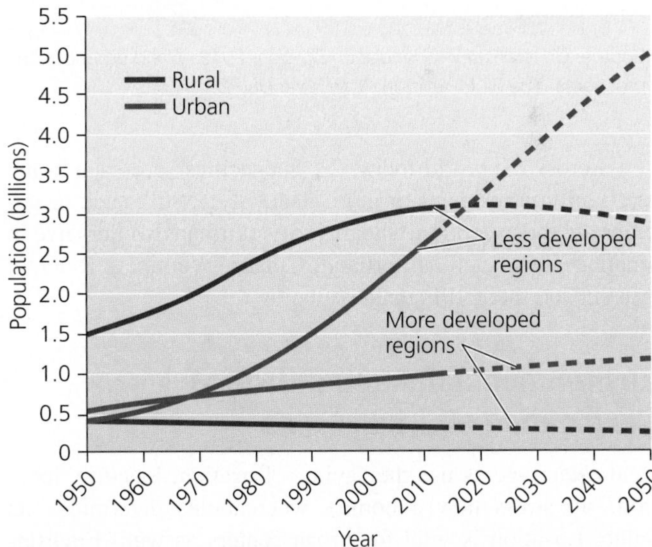

FIGURE 13.1 Population trends differ between poor and wealthy nations. In less-developed regions, urban populations are growing quickly, and rural populations will soon begin declining. More-developed regions are already largely urbanized, so their urban populations are growing slowly, whereas rural populations are falling. Solid lines in the graph indicate past data, and dashed lines indicate future projections. *Data from United Nations Population Division, 2012.* World urbanization prospects: The 2011 revision. *By permission.*

DATA Q Beginning in what decade will the majority of people in less-developed regions be living in urban areas?

TABLE 13.1 Metropolitan Areas with 10 Million Inhabitants or More

City, Country	Millions of people
Tokyo, Japan	37.2
Delhi, India	22.7
Mexico City, Mexico	20.4
New York–Newark, United States	20.4
Shanghai, China	20.2
Sao Paulo, Brazil	19.9
Mumbai (Bombay), India	19.7
Beijing, China	15.6
Dhaka, Bangladesh	15.4
Calcutta, India	14.4
Karachi, Pakistan	13.9
Buenos Aires, Argentina	13.5
Los Angeles–Long Beach–Santa Ana, United States	13.4
Rio de Janeiro, Brazil	12.0
Manila, Philippines	11.9
Moscow, Russia	11.6
Osaka-Kobe, Japan	11.5
Istanbul, Turkey	11.3
Lagos, Nigeria	11.2
Cairo, Egypt	11.2
Guangzhou, China	10.8
Shenzen, China	10.6
Paris, France	10.6

Source: United Nations Population Division, 2012. World urbanization prospects: The 2011 revision. *New York: UNPD.*

Across the world today, 23 "megacities" are each home to 10 million residents or more (TABLE 13.1). Still, such megacities are the exception. The majority of urban dwellers live in smaller cities, such as Portland, Omaha, Winnipeg, Raleigh, Austin, and their still-smaller suburbs.

Environmental factors influence the location of urban areas

Real estate agents use the saying, "Location, location, location," to stress how a home's whereabouts determines its value. Location is vital for urban centers as well. Environmental variables such as climate, topography, and the configuration of waterways influence whether a city will succeed. Think of any major city, and chances are it's situated along a major river, seacoast, railroad, or highway—some corridor for trade that has driven economic growth (FIGURE 13.2).

Many well-located cities have served as linchpins in trading networks, funneling in resources from agricultural regions, processing them, manufacturing products, and shipping those products to other markets. Portland got its start in the mid-19th century as pioneers arriving by the Oregon Trail settled where the Willamette River flowed into the Columbia River. Situated at the juncture of these two major rivers, and just upriver from where the Columbia flows into the Pacific Ocean, Portland had a strategic location for trade. The city grew as it received, processed, and shipped overseas the produce from farms of the river valleys, and as it imported products shipped in from other ports.

We see such geographic patterns again and again with major cities. A prime example is Chicago, which grew with extraordinary speed in the 19th and early 20th centuries as railroads funneled through it the resources from the vast lands of the Midwest and West on their way to consumers and businesses in the populous cities of the East. Chicago became a center for grain processing, livestock slaughtering, meatpacking, and much else.

Today, powerful technologies and cheap transportation enabled by fossil fuels have allowed cities to thrive even in resource-poor regions. The Dallas–Fort Worth area prospers from—and relies on—oil-fueled transportation by interstate highways and a major airport. Southwestern cities such as Los Angeles, Las Vegas, and Phoenix flourish in desert regions by appropriating water from distant sources. Whether such cities can sustain themselves as oil and water become increasingly scarce in the future is an important question.

(a) St. Louis, Missouri

(b) Fort Worth, Texas

FIGURE 13.2 Cities tend to develop along trade corridors. St. Louis **(a)** is situated on the Mississippi River near its confluence with the Missouri River, where river trade drove its growth in the 19th and early 20th centuries. Fort Worth, Texas **(b)**, grew in the late 20th century as a result of the interstate highway system and a major international airport.

In recent years, many cities in the southern and western United States have undergone growth spurts as people (particularly retirees) have moved south and west in search of warmer weather or more space. Between 1990 and 2012, the population of the Dallas–Fort Worth and Houston metropolitan areas each grew by 66%, that of the Atlanta area grew by 84%; that of the Phoenix region grew by 93%; and that of the Las Vegas metropolitan area grew by a whopping 135%.

WEIGHING THE ISSUES

WHAT MADE YOUR CITY? Consider the town or city in which you live, or the major urban center located nearest you. Why do you think it developed where it did? What physical, social, or environmental factors may have aided its growth? Do you think it will prosper in the future? Why or why not?

People have moved to suburbs

American cities grew rapidly in the 19th and early 20th centuries as a result of immigration from abroad and increased trade as the nation expanded westward. The bustling economic activity of downtown districts held people in cities despite growing crowding, poverty, and crime. However, by the mid-20th century, many affluent city dwellers were choosing to move outward to the cleaner, less crowded, and more parklike suburban communities beginning to surround the cities. These people were pursuing more space, better economic opportunities, cheaper real estate, less crime, and better schools for their children.

As affluent people moved outward into the expanding suburbs, jobs followed. This hastened the economic decline of downtown districts, and American cities stagnated. Chicago's population declined to 80% of its peak because so many residents moved to its suburbs. Philadelphia's population fell to 76% of its peak, Washington, D.C.'s to 71%, and Detroit's to just 55%.

Portland followed this trajectory, but also illustrates how some cities have bounced back. Portland's population growth stalled in the 1950s to 1970s as crowding and deteriorating economic conditions drove city dwellers to the suburbs. However, subsequent policies to revitalize the city center helped restart Portland's growth (**FIGURE 13.3**).

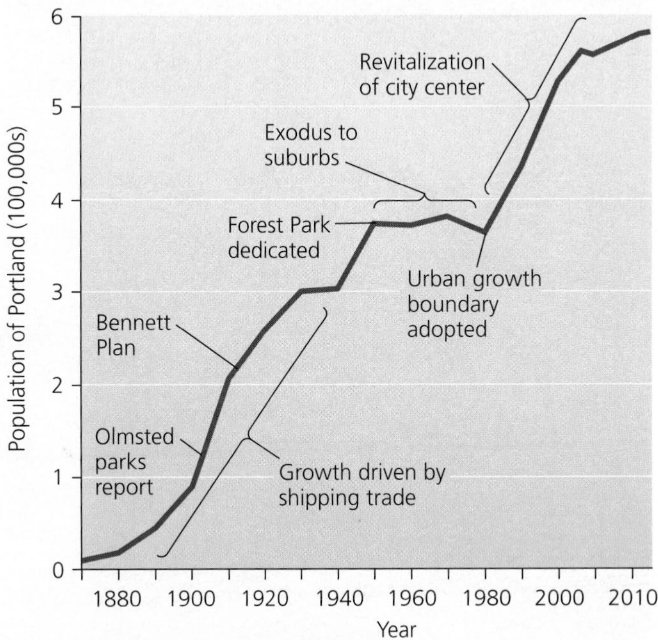

FIGURE 13.3 Portland grew, stabilized, and then grew again. Jobs in the shipping trade helped to boost Portland's economy and population in the 1890s–1920s. City residents began leaving for the suburbs in the 1950s–1970s, but policies to enhance the city center revitalized Portland's growth.

The exodus to the suburbs in 20th-century America was enabled by the rise of the automobile, an expanding road network, and inexpensive and abundant oil. Millions of people could now commute by car to downtown workplaces from new homes in suburban "bedroom communities." By facilitating long-distance transport, fossil fuels and highway networks also made it easier for businesses to import and export resources, goods, and waste. The U.S. government's development of the interstate highway system was pivotal in promoting these trends.

Today, technology continues to reinforce the spread of urban and suburban areas. In our age of the Internet, handheld devices, jet travel, and video-conferencing, being located in a city's downtown on a river or seacoast is no longer so vital to success. As globalization continues to connect distant societies, businesses and individuals can more easily communicate from far-flung locations.

In most ways, suburbs have delivered the qualities people sought in them. The wide spacing of homes, with each one on its own plot of land, gives families room and privacy. However, by allotting more space to each person, suburban growth has spread human impact across the landscape. Natural areas have disappeared as housing developments are constructed. Our extensive road networks ease travel, but suburbanites find themselves needing to climb into a car to get anywhere. People commute longer distances to work and spend more time stuck in traffic. The expanding rings of suburbs surrounding cities have grown larger than the cities themselves, and towns are merging into one another. These aspects of suburban growth inspired a new term: *sprawl*.

Sprawl

The term *sprawl* has become laden with meanings and suggests different things to different people, but we can begin our discussion by giving **sprawl** a simple, nonjudgmental definition: the spread of low-density urban or suburban development outward from an urban center.

Urban areas spread outward

The spatial growth of urban and suburban areas is clear from maps and satellite images of rapidly spreading cities such as Las Vegas (**FIGURE 13.4**). Another example is Chicago, whose metropolitan area now spreads over a region 40 times the size of the city. All in all, houses and roads supplant over 2700 ha (6700 acres) of U.S. land every day.

(a) Las Vegas, Nevada, 1984

(b) Las Vegas, Nevada, 2009

FIGURE 13.4 Satellite images show the rapid urban and suburban expansion that many people have dubbed *sprawl*. Las Vegas, Nevada, is one of the fastest-growing cities in North America. Between 1984 **(a)** and 2009 **(b)**, its population and its developed area each tripled.

Several development approaches can lead to sprawl (FIGURE 13.5). These approaches allot each person more space than in cities. For example, the average resident of Chicago's suburbs takes up 11 times more space than a resident of the city. As a result, the outward spatial growth of suburbs across the landscape generally outpaces growth in numbers of people. In fact, many researchers define *sprawl* as the physical spread of development at a rate that exceeds the rate of population growth. For instance, the population of Phoenix grew 12 times larger between 1950 and 2000, yet its land area grew 27 times larger. Between 1950 and 1990, the population of 58 major U.S. metropolitan areas rose by 80%, but the land area they covered rose by 305%. Even in 11 metro areas where population declined between 1970 and 1990 (for instance, Rust Belt cities such as Detroit, Cleveland, and Pittsburgh), the amount of land covered increased.

Sprawl has several causes

There are two main components of sprawl. One is human population growth—there are simply more of us alive each year (Chapter 8). The other is per capita land consumption—each person is taking up more land than in the past. The amount of sprawl is a function of the number of people added to a region times the amount of land each person occupies.

A study of U.S. metropolitan areas between 1970 and 1990 found that these two factors contribute about equally to sprawl but that cities vary in which is more influential. The Los Angeles metro area increased in population density by 9% between 1970 and 1990, becoming the nation's most densely populated metro area. Increasing density should be a good recipe for preventing sprawl. Yet L.A. grew in size by a whopping 1021 km^2 (394 mi^2) because of an overwhelming influx of new people. In contrast, the Detroit metro area lost 7% of its population between 1970 and 1990, yet it expanded in area by 28%. In this case, sprawl clearly was caused solely by increased per capita land consumption.

Each person is taking up more space these days in part because of factors mentioned earlier: Better highways, inexpensive gasoline, telecommunications, and the Internet have fostered movement away from city centers by freeing businesses from dependence on the centralized infrastructure a major city provides and by giving workers greater flexibility to live where they desire. Given a choice, most people desire space and privacy and prefer living in less congested, more spacious, more affluent communities.

Economists and politicians have encouraged the unbridled spatial expansion of cities and suburbs. The conventional assumption has been that growth is always good and that attracting business, industry, and residents will enhance a community's economic well-being, political power, and cultural influence. Today, this assumption is being challenged as growing numbers of people feel the negative effects of sprawl on their lifestyles.

What is wrong with sprawl?

To some people, the word *sprawl* evokes strip malls, homogenous commercial development, and tracts of cookie-cutter houses encroaching on farmland, ranchland, or forests. It may suggest traffic jams, destruction of wildlife habitat, and loss of open space. For other people, sprawl is simply the collective result of choices made by millions of well-meaning individuals trying to make a better life for their families. In this view, those who criticize sprawl are being elitist and fail to

(a) Uncentered commercial strip development

(b) Low-density single-use development

(c) Scattered, or leapfrog, development

(d) Sparse street network

FIGURE 13.5 **Several standard approaches to suburban development can result in sprawl and necessitate frequent automobile use.** In **(a)**, businesses are arrayed in a long strip along a roadway, with no attempt made to create a centralized community with easy access for consumers. In **(b)**, homes are located on large lots in residential tracts far away from commercial amenities. In **(c)**, developments are created at great distances from a city center and are not integrated. In **(d)**, roads are far enough apart that some areas go undeveloped, but not far enough apart for these areas to function as natural areas or sites for recreation.

appreciate the good things about suburban life. Let's try to leave the emotional debate aside and assess what research can tell us about the impacts of sprawl.

Transportation Most studies show that sprawl constrains transportation options, essentially forcing people to drive cars. In sprawling communities, people need to own a vehicle, to drive it most places, to drive greater distances, and to spend more time in vehicles. Few or no mass transit options exist, and there are more traffic accidents. Across the United States during the 1980s and 1990s the average length of work trips rose by 36%, and total vehicle miles driven rose three times faster than population growth. An automobile-oriented culture encourages congestion, and it also increases dependence on nonrenewable petroleum, with its economic and environmental consequences (pp. 536–545).

Pollution By promoting automobile use, sprawl increases pollution. Carbon dioxide emissions from vehicles contribute to global climate change (Chapter 18) while nitrogen- and sulfur-containing air pollutants lead to tropospheric ozone, urban smog, and acid precipitation (Chapter 17). Runoff of polluted water from paved areas is about 16 times greater than from naturally vegetated areas. Motor oil and road salt from roads and parking lots pollute waterways, posing risks to ecosystems and human health.

Health In addition to the health impacts of pollution, some research suggests that sprawl promotes physical inactivity because driving cars largely takes the place of walking during daily errands. Physical inactivity increases obesity and high blood pressure, which can in turn lead to other ailments. A 2003 study found that people from the most-sprawling U.S. counties weigh 2.7 kg (6 lb) more for their height than people from the least-sprawling U.S. counties and that slightly more people from the most-sprawling counties show high blood pressure.

Land use The spread of low-density development means that more land is developed while less is left as forests, fields, farmland, or ranchland. Of the estimated 1 million ha

(2.5 million acres) of U.S. land converted each year, roughly 60% is agricultural land and 40% is forest. These lands provide vital resources, recreation, aesthetic beauty, wildlife habitat, air and water purification, and other ecosystem services (pp. 3, 116–117, 152, 290). Sprawl generally diminishes all these amenities. Perhaps of most concern, more and more children these days are growing up without the ability to roam through woods, fields, and open space, which used to be a normal part of childhood. Being deprived of access to regular experience with nature as a child, many experts feel, can have psychological and emotional consequences for the individual and for society (p. 294).

Economics Sprawl drains tax dollars from existing communities and funnels money into infrastructure for new development on the fringes of those communities. Funds that could be spent maintaining and improving downtown centers is instead spent on extending the road system, water and sewer system, electricity grid, telephone lines, police and fire service, schools, and libraries. The costs of extending infrastructure are generally paid by taxpayers of the community and are not charged to developers.

One study calculated that sprawling development at Virginia Beach, Virginia, would require 81% more in infrastructure costs and would drain 3.7 times more from the community's general fund each year than compact urban development. Advocates for sprawling development argue that as owners of newly developed homes and businesses pay property taxes, this revenue eventually reimburses the community's investment in extending infrastructure. However, studies have found that in most cases taxpayers continue to subsidize new development unless municipalities specifically require that developers pay new infrastructure costs.

> **WEIGHING THE ISSUES**
>
> **SPRAWL NEAR YOU** Is there sprawl in the area where you live? Does it bother you, or not? Has development in your area had any of the impacts described above? Do you think your city or town should encourage outward growth? Why or why not?

Creating Livable Cities

To respond to the challenges that sprawl presents, architects, planners, developers, and policymakers are trying to restore the vitality of city centers and to plan and manage how urbanizing areas develop. They aim to make cities safer, cleaner, healthier, and more pleasant for their residents.

City and regional planning help to create livable urban areas

How can we design cities so as to maximize their efficiency, functionality, and beauty? These are the questions central to **city planning** (also known as **urban planning**). City planners advise policymakers on development options, transportation needs, public parks, and other matters.

Washington, D.C., is the earliest example of city planning in the United States. President George Washington hired French architect Pierre Charles L'Enfant in 1791 to design a capital city for the new nation on undeveloped land along the Potomac River. L'Enfant laid out a Baroque-style plan of diagonal avenues cutting across a grid of streets, with plenty of space allotted for majestic public monuments, and the city was built according to his plan (**FIGURE 13.6**). A century later, as the city became crowded and dirty, a special commission in 1901 undertook new planning efforts to beautify the city while staying true to the intentions of L'Enfant's original plan. These planners imposed a height restriction on new buildings, which kept the magnificent government edifices and monuments from being crowded and dwarfed by modern skyscrapers. This preserved the spacious, stately feel of the city.

City planning in North America came into its own at the turn of the 20th century, as urban leaders sought to beautify

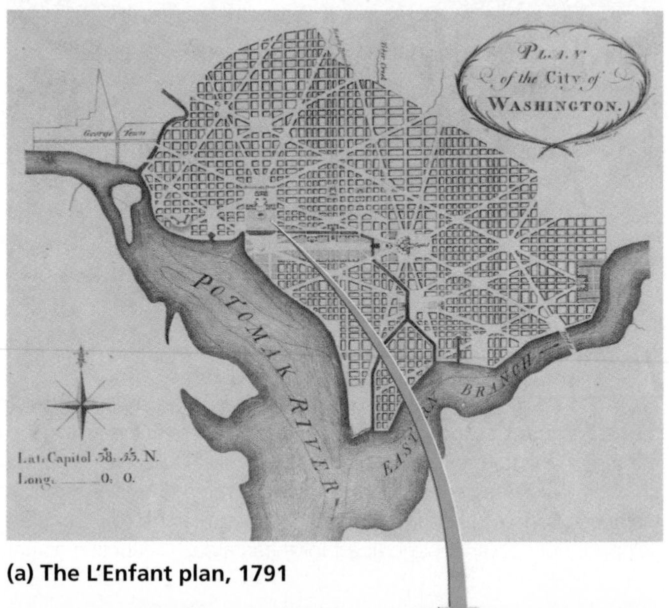

(a) The L'Enfant plan, 1791

(b) Washington, D.C., today

FIGURE 13.6 Washington, D.C., is a prime example of early city planning. The 1791 plan **(a)** for the new U.S. capital laid out a series of splendid diagonal avenues cutting across gridded streets, allowing plenty of space for the magnificent public monuments **(b)** that still grace the city today.

and impose order on fast-growing, unruly cities. Landscape architect Daniel Burnham's 1909 *Plan of Chicago* was perhaps the grandest effort of this time, and it was largely implemented over the following years and decades. Burnham's plan expanded Chicago's city parks and playgrounds, improved neighborhood living conditions, streamlined traffic systems, and cleared industry and railroads from the shore of Lake Michigan to provide public access to the lake.

Portland gained its own comprehensive plan just three years later, in 1912. Edward Bennett's *Greater Portland Plan* recommended rebuilding the harbor; dredging the river channel; constructing new docks, bridges, tunnels, and a waterfront railroad; superimposing wide radial boulevards on the old city street grid; establishing civic centers downtown; and greatly expanding the number of parks. Voters approved the plan by a two-to-one margin, but they defeated a bond issue that would have paid for park development. As the century progressed, several other major planning efforts were conducted, and some ideas, such as establishing a downtown public square, came to fruition.

WEIGHING THE ISSUES

YOUR URBAN AREA Think of your favorite parts of the city you know best. What do you like about them? What do you dislike about your least favorite parts of the city? What could this city do to improve quality of life for its residents?

In today's world of sprawling metropolitan areas, **regional planning** has become at least as important as city planning. Regional planners deal with the same issues as city planners, but they work on broader geographic scales and must coordinate their work with multiple municipal governments. In some places, regional planning has been institutionalized in formal government bodies; the Portland area's Metro is the epitome of such a regional planning entity. When Metro and its region's three counties in 2010 announced their collaborative plan on how to apportion their remaining undeveloped land into "urban reserves" and "rural reserves," it marked a historic accomplishment in regional planning. The agreement, hammered out after two years of complicated negotiations, designates 28,000 acres for urban use and 272,000 acres for rural use. It means that homeowners, farmers, developers, and policymakers alike can feel informed and secure knowing what kinds of land uses lie in store on and near their land over the next half-century.

Zoning is a key tool for planning

One tool that planners use is **zoning,** the practice of classifying areas for different types of development and land use (**FIGURE 13.7**). For instance, to preserve the cleanliness and tranquility of residential neighborhoods, industrial facilities may be kept out of districts zoned for residential use. By specifying zones for different types of development, planners can guide what gets built where. Zoning also gives home buyers and business owners security, because they know in advance what types of development can and cannot be located nearby.

Zoning involves government restriction on the use of private land and represents a top-down constraint on personal property rights. For this reason, some people consider zoning a regulatory taking (p. 170) that violates individual freedoms. Most people defend zoning, however, saying that government has a proper and useful role in setting limits on property rights for the good of the community.

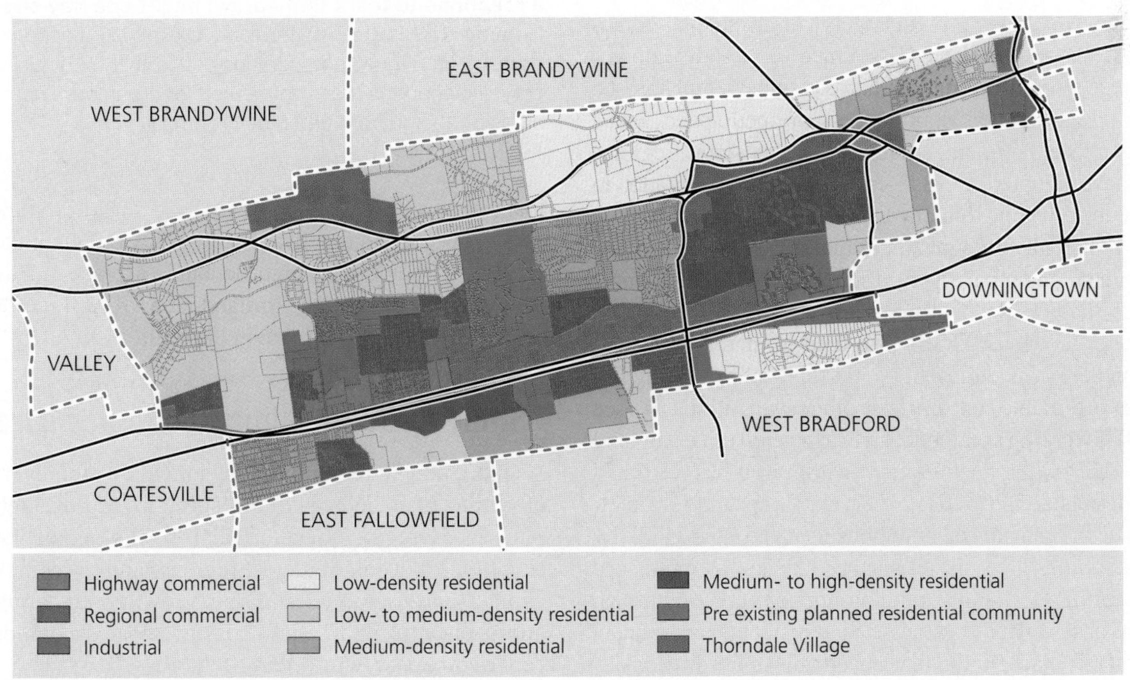

FIGURE 13.7 This zoning map for Caln Township, Pennsylvania, shows several common zoning patterns. Public and institutional uses are clustered together in a downtown area. Industrial uses are clustered away from residential areas. Commercial uses are aligned along major roadways. Residential zones generally are higher in density toward the center of town.

Oregon voters sided with private property rights when in 2004 they passed Ballot Measure 37, which shackled government's ability to enforce zoning regulations with landowners who bought their land before the regulations were enacted. Subsequently, however, many Oregonians began witnessing new development they did not condone, so in 2007 they passed Ballot Measure 49 in order to restore public oversight over development. The passage of Oregon's Measure 37 spawned similar ballot measures in other U.S. states, but voters defeated most of these. For the most part, people have supported zoning over the years because the common good it produces for communities is widely felt to outweigh the restrictions on private use.

WEIGHING THE ISSUES

ZONING AND DEVELOPMENT Imagine you own a 10-acre parcel of land that you want to sell for housing development—but the local zoning board rezones the land so as to prohibit the development. How would you respond?

Now imagine that you live next to someone else's undeveloped 10-acre parcel. You enjoy the peace and privacy it provides—but the local zoning board rezones the land so that it can be developed into a dense housing subdivision. How would you respond?

What factors do you think members of a zoning board should take into consideration when deciding how to zone or rezone land in a community?

Urban growth boundaries are now widely used

Planners intended Oregon's urban growth boundaries (UGBs) to limit sprawl by containing growth largely within existing urbanized areas (**FIGURE 13.8**). The UGBs aimed to revitalize downtowns; protect working farms, orchards, ranches, and forests; and ensure urban dwellers some access to open space. Since Oregon began its experiment, a number of other states, regions, and cities have adopted UGBs—from Boulder, Colorado, to Lancaster, Pennsylvania, to many California communities.

UGBs reduce the amounts that municipalities need to pay for infrastructure, compared with sprawl. The best estimate nationally is that UGBs save taxpayers about 20% on infrastructure costs. However, UGBs also tend to increase housing prices within their boundaries. In the Portland area, housing has become less affordable, but in most other ways its UGB (see Figure 13.8) is working as intended. It has lowered prices for land outside the UGB while raising prices within it. It has restricted development outside the UGB, preserving farms and forests. It has increased the density of new housing inside the UGB by over 50% as homes are built on smaller lots and as multistory apartments fulfill a vision of "building up, not out." Downtown employment has grown as businesses and residents invest anew in the central city. And Portland has been able to absorb considerable immigration while avoiding rampant sprawl.

However, urbanized area still increased by 101 km^2 (39 mi^2) in the decade after Portland's UGB was established, because 146,000 people were added to the population. This fact suggests that relentless population growth may thwart even the best anti-sprawl efforts and that livable cities may fall victim to their own success if they are in high demand as places to live. Indeed, Metro has enlarged the Portland-area UGB three dozen times and plans to expand it more in the future.

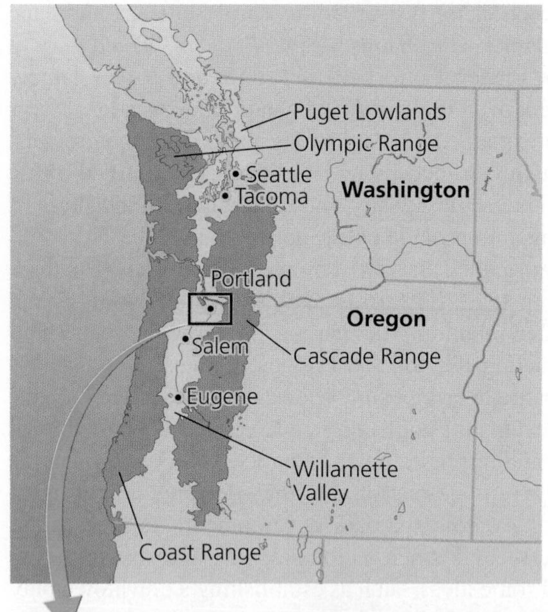

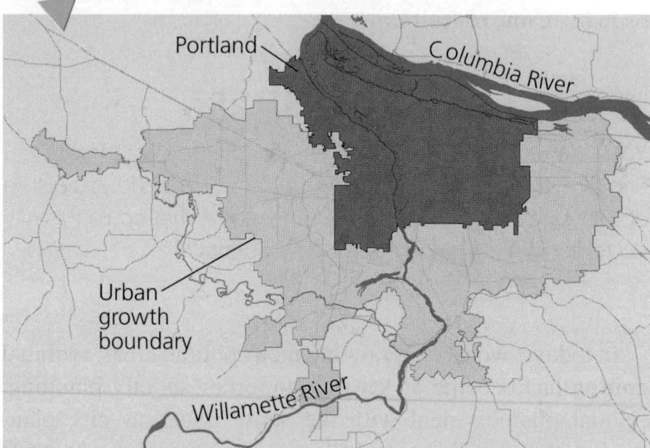

FIGURE 13.8 Oregon's urban growth boundaries (UGBs) were a response to fears that sprawl might one day stretch from Eugene to Seattle. The Portland region's UGB encompasses Portland (dark gray) and portions of 24 other communities (light gray). It separates areas earmarked for urban development from areas protected from urban development.

"Smart growth" aims to counter sprawl

As more people feel impacts of sprawl on their everyday lives, efforts to manage growth are springing up throughout North America. Oregon's Senate Bill 100 was one of the first, and since then dozens of states, regions, and cities have adopted similar land use policies. Urban growth boundaries and other approaches from these policies have coalesced under the concept of **smart growth** (**TABLE 13.2**).

Proponents of smart growth want municipalities to manage the rate, placement, and style of development so as to promote healthy neighborhoods and communities, jobs and economic development, transportation options, and environmental

TABLE 13.2 Ten Principles of "Smart Growth"

- Mix land uses
- Take advantage of compact building design
- Create a range of housing opportunities and choices
- Create walkable neighborhoods
- Foster distinctive, attractive communities with a strong sense of place
- Preserve open space, farmland, natural beauty, and critical environmental areas
- Strengthen existing communities, and direct development toward them
- Provide a variety of transportation choices
- Make development decisions predictable, fair, and cost-effective
- Encourage community and stakeholder collaboration in development decisions

Source: U.S. Environmental Protection Agency.

quality. They aim to rejuvenate the older existing communities that so often are drained and impoverished by sprawl. Smart growth means "building up, not out"—focusing development and economic investment in existing urban centers and favoring multistory shop-houses and high-rises.

"New urbanism" aims to create walkable neighborhoods

A related approach among architects, planners, and developers is **new urbanism.** This approach seeks to design neighborhoods on a walkable scale, with homes, businesses, schools, and other amenities all close together for convenience. The aim is to create functional neighborhoods in which families can meet most of their needs close to home without using a car. Green spaces, trees, a mix of architectural styles, and creative street layouts add to the visual interest of new-urbanist developments (FIGURE 13.9). These developments mimic the traditional urban neighborhoods that existed until the advent of suburbs.

New-urbanist neighborhoods are generally served by public transit systems. In **transit-oriented development,** compact communities in the new-urbanist style are arrayed around stops on a major rail line, enabling people to travel most places they need to go by train and foot alone. Several lines of the Washington, D.C., Metro system are developed in this manner.

Among the nearly 600 communities in the new-urbanist style across North America are Seaside, Florida; Kentlands in Gaithersburg, Maryland; Addison Circle in Addison, Texas; Mashpee Commons in Mashpee, Massachusetts; Harbor Town in Memphis, Tennessee; Celebration in Orlando, Florida; and Orenco Station, west of Portland.

Transit options help cities

Traffic jams on roadways cause air pollution, stress, and countless hours of lost personal time. They cost the U.S. economy an estimated $74 billion each year in fuel and lost productivity. So, a key ingredient in any planner's recipe for improving the quality of urban life is to give residents alternative transportation options.

FIGURE 13.9 **This plaza at Mizner Park in Boca Raton, Florida, is part of a planned community built in the new-urbanist style.** Homes, schools, and businesses are mixed close together in a centered neighborhood so that most amenities are within walking distance.

Bicycle transportation is one key option (FIGURE 13.10). Portland has embraced bicycles like few other American cities, and today 6% of its commuters ride to work by bike (compared to a national average of 0.5%). The city has developed nearly 400 miles of bike lanes and paths, 5000 public bike racks, and special designs at intersections to protect bikers and encourage bike use. Amazingly, all this infrastructure was created for the typical cost of just one mile of urban freeway. Portland also has a bike-sharing program similar to programs in cities such as Montreal, Toronto, Denver, Minneapolis, Miami, San Antonio, Boston, and Washington, D.C.

Other transportation options include **mass transit** systems: public systems of buses, trains, subways, or *light rail* (smaller rail systems powered by electricity) that move large numbers of passengers at once. Mass transit rail systems ease

FIGURE 13.10 **Bicycles provide a healthy transportation alternative to car travel.** Biking to work (here, in San Francisco) and for recreation is increasing throughout North America.

traffic congestion, take up less space than road networks, and emit less pollution than cars. A 2005 study calculated that each year rail systems in U.S. metropolitan areas save taxpayers a total of $67.7 billion in costs for congestion, consumer transportation, parking, road maintenance, and accidents—far more than the $12.5 billion that governments spend to subsidize rail systems each year. As long as an urban center is large enough to support the infrastructure necessary, both train and bus systems are cheaper, more energy-efficient, and cleaner than roadways choked with cars (FIGURE 13.11).

In Portland, the bus system carries 60 million riders per year, and an average bus keeps an estimated 250 cars off the road each day. Portland also boasts downtown streetcars and one of the nation's leading light rail systems (FIGURE 13.12). Metro policy encourages the development of self-sufficient neighborhood communities in the new-urbanist style along the rail lines. In recent years the bus system has faced service cutbacks and fare hikes in the wake of a costly overhaul, but light rail ridership continues to grow as the system expands with new lines to outlying areas.

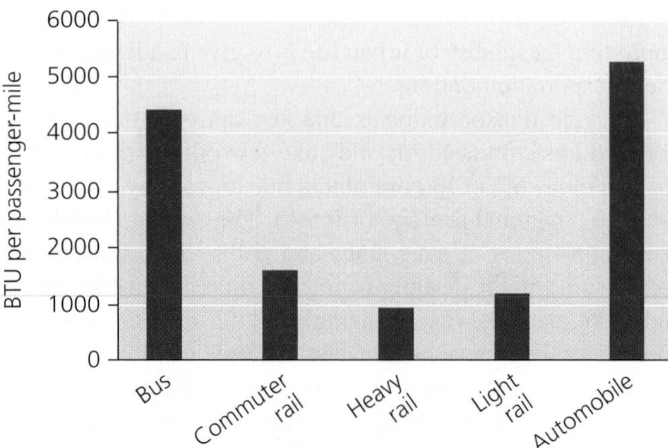

(a) Energy consumption for different modes of transit

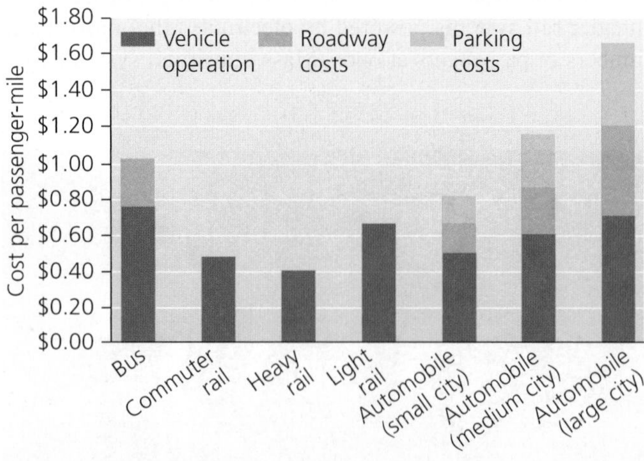

(b) Operating costs for different modes of transit

FIGURE 13.11 Rail transit consumes less energy (a) and costs less (b) per passenger mile than bus or automobile transit.
Data from Litman, T., 2005. Rail transit in America: A comprehensive evaluation of benefits. *© 2005 T. Litman.*

Automobile traffic creates two types of operating costs that are *not* created by rail traffic. What are they?

(a) MAX light rail train

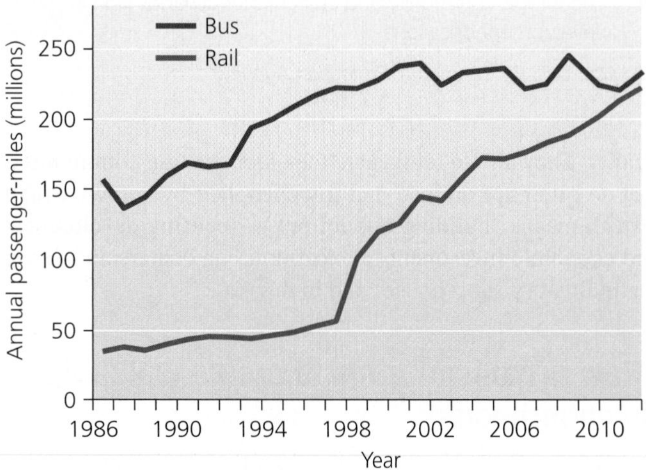

(b) Portland transit ridership trends

FIGURE 13.12 An excellent light rail system is part of an urban planning strategy that has helped make Portland one of America's most livable cities. Here, bus ridership is stable while ridership on the MAX light rail system is growing quickly as the system expands.

The nation's most-used train systems are the extensive heavy rail systems in America's largest cities, such as New York City's subways, Washington, D.C.'s Metro, the T in Boston, and the San Francisco Bay area's BART. Each of these rail systems carries more than one-fourth of its city's daily commuters.

In general, however, the United States lags behind most nations in mass transit. Many countries, rich and poor alike, have extensive and accessible bus systems that ferry citizens within and between towns and cities cheaply and effectively. And whereas Japan and many European nations have developed whole systems of modern high-speed "bullet" trains (FIGURE 13.13), the United States has only one, Amtrak's *Acela Express*. This train connects Boston and Washington, D.C., via New York, Philadelphia, and Baltimore, and it travels more slowly than most bullet trains.

The United States chose instead to invest in road networks for cars and trucks largely because its population density was low and gasoline was cheap. As energy costs and population rise, however, mass transit becomes increasingly appealing, and citizens begin to clamor for train and bus systems in their

FIGURE 13.13 "Bullet trains" of high-speed rail systems in Europe and Asia can travel at 150–220 mph. This Chinese train is speeding through the city of Qingdao. Plans to establish such lines in the United States are just getting underway.

communities. Americans may soon see more high-speed rail; the 2009 stimulus bill passed by Congress set aside $8 billion for developing high-speed rail, and the Obama administration identified 10 potential corridors for development of such trains. Three Republican governors rejected the offer of federal funding for their states, but 32 other states are progressing with rail projects.

Establishing mass transit is not always easy, however. Once a road system is in place, lined with businesses and homes, it can be difficult and expensive to replace or complement it with a mass transit system. Strong and visionary political leadership may be required. Such was the case in Curitiba, Brazil. Faced with an influx of immigrants from outlying farms in the 1970s, city leaders led by Mayor Jaime Lerner decided to pursue an aggressive planning process so that they could direct growth rather than being overwhelmed by it. They established a large fleet of public buses and reconfigured Curitiba's road system to maximize its efficiency. Today this metropolis of 2.5 million people has an outstanding bus system that is used each day by three-quarters of the population. The 340 bus routes, 250 terminals, and 1900 buses accompany measures to encourage bicycles and pedestrians. All of this has resulted in a steep drop in car use, despite the city's rapidly growing population.

To make urban transportation more efficient, policymakers can take a variety of other actions. They can raise fuel taxes, tax inefficient modes of transport, reward carpoolers with carpool lanes, encourage bicycle use and bus ridership, and charge trucks for road damage.

Urban residents need parklands

City dwellers often desire some sense of escape from the noise, commotion, and stress of urban life. Natural lands, public parks, and open space provide greenery, scenic beauty, freedom of movement, and places for recreation. These lands also keep ecological processes functioning by helping to regulate climate, purify air and water, and provide wildlife habitat. The animals and plants of urban and suburban parks and natural lands also serve to satisfy biophilia (pp. 293–294), our natural affinity for contact with other organisms. In the wake of urbanization and sprawl, protecting natural lands and establishing public parks has become more important as many of us come to feel increasingly disconnected from nature.

America's city parks arose in the late 19th century as politicians and citizens yearning to make crowded and dirty cities more livable established public spaces using aesthetic ideals borrowed from European parks, gardens, and royal hunting grounds. The lawns, shaded groves, curved pathways, and pastoral vistas of many American city parks originated with these European ideals, as interpreted by America's leading landscape architect, Frederick Law Olmsted. Olmsted designed Central Park in New York City (**FIGURE 13.14**) and many other urban park systems.

East Coast cities such as New York, Boston, and Philadelphia developed parks early on, but western cities were not far behind. In San Francisco, William Hammond Hall transformed 2500 ha (1000 acres) of the peninsula's natural landscape of dunes into a verdant playground of lawns, trees, gardens, and sports fields. Golden Gate Park remains today one of the world's foremost city parks. Portland's quest for parks began in 1900, when city leaders created a parks commission and hired Olmsted's son, John Olmsted, to design a park system. His 1904 plan recommended acquiring land to ring the city generously with parks, but no action was taken. A full 44 years later, citizens pressured city leaders to create Forest Park along a large forested ridge on the northwest side of the city. At 11 km (7 mi) long, it is today the largest city park in the United States.

Parklands come in various types

Large city parks are vital to a healthy urban environment, but small spaces can make a big difference. Playgrounds give children a place to be active outdoors and interact with their peers. Community gardens allow people to grow vegetables and flowers in a neighborhood setting.

FIGURE 13.14 Central Park in New York City was one of America's first city parks, and it remains one of the largest and finest.

Greenways are strips of land that connect parks or neighborhoods, often run along rivers, streams, or canals, and provide access to networks of walking trails. They can protect water quality, boost property values, and serve as corridors for the movement of wildlife and people alike. The Rails-to-Trails Conservancy has spearheaded the conversion of abandoned railroad rights-of-way into greenways for walking, jogging, and biking. To date, 32,000 km (20,000 mi) of 1600 rail lines have been converted across North America.

A newly developed and instantly popular linear park along an old rail line is The High Line Park on Manhattan's West Side in New York City (**FIGURE 13.15**). An elevated freight line running above the city's streets was going to be demolished, but a group of citizens saw its potential for a park, and they pushed the idea until city leaders came to share their vision. Today thousands of people use the 23-block-long High Line for recreation or on their commute to work.

The concept of the corridor is sometimes implemented on a large scale. **Greenbelts** are long and wide corridors of parklands, often encircling an entire urban area. One example is the system of forest preserves that stretches through Chicago's suburbs like a necklace (**FIGURE 13.16**). In Canada, cities including Toronto, Ottawa, and Vancouver employ greenbelts as urban growth boundaries, containing sprawl and preserving open space for city residents.

Within urban parklands, many cities are working to enhance these areas through ecological restoration (pp. 92–93, 299), the practice of restoring native communities. In Portland's parks, volunteer teams plant native vegetation and remove invasive non-native plants. At some Chicago-area forest preserves, scientists and volunteers use prescribed burns (pp. 320–321) to restore prairie native to the region. In San Francisco's Presidio, areas are being restored to the native dune communities that were displaced by urban development.

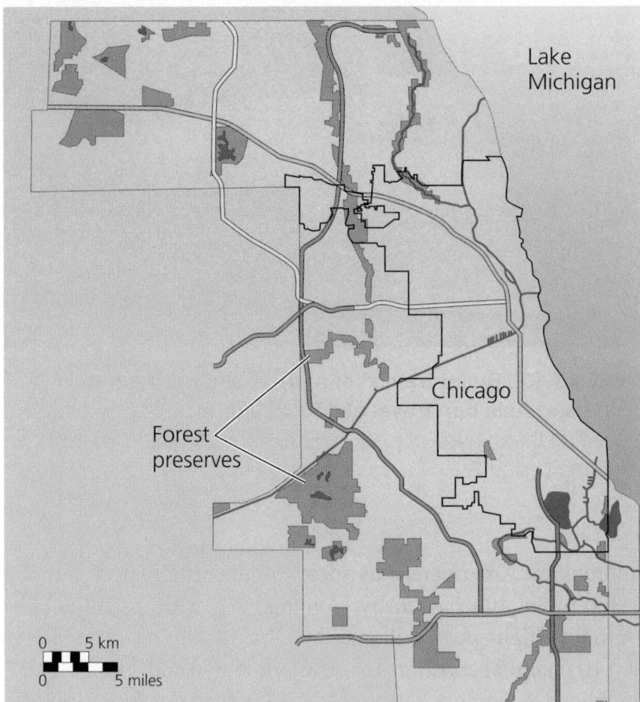

FIGURE 13.16 Forest preserves wind through the suburbs of Chicago. This regional greenbelt system features 40,000 ha (100,000 acres) of woodlands, fields, marshes, and prairies, comprising the largest holding of locally owned public conservation land in the United States.

FIGURE 13.15 Manhattan's new High Line Park was created thanks to a visionary group of citizens. They pushed to make a park out of an abandoned elevated rail line.

Green buildings bring benefits

Although we need parks and open space, we spend most of our time indoors, so our health is affected by the buildings in which we live and work. Moreover, buildings consume 40% of our energy and 70% of our electricity, contributing to the greenhouse gas emissions that drive climate change. As a result, there is a thriving movement in architecture and construction to design and build **green buildings,** structures that use technologies and approaches to minimize the ecological footprint (pp. 4–5) of the building's construction and operation.

Green buildings are built from sustainable materials, limit their use of energy and water, minimize health impacts on their occupants, control pollution, and recycle waste (**FIGURE 13.17**). Constructing or renovating buildings using new efficient technologies is probably the most effective way that cities can reduce energy consumption and greenhouse gas emissions.

The U.S. Green Building Council promotes these efforts by running the **Leadership in Energy and Environmental Design (LEED)** certification program. Buildings (new buildings or renovation projects) apply for certification and, depending on their performance, may be granted silver, gold, or platinum status (**TABLE 13.3**).

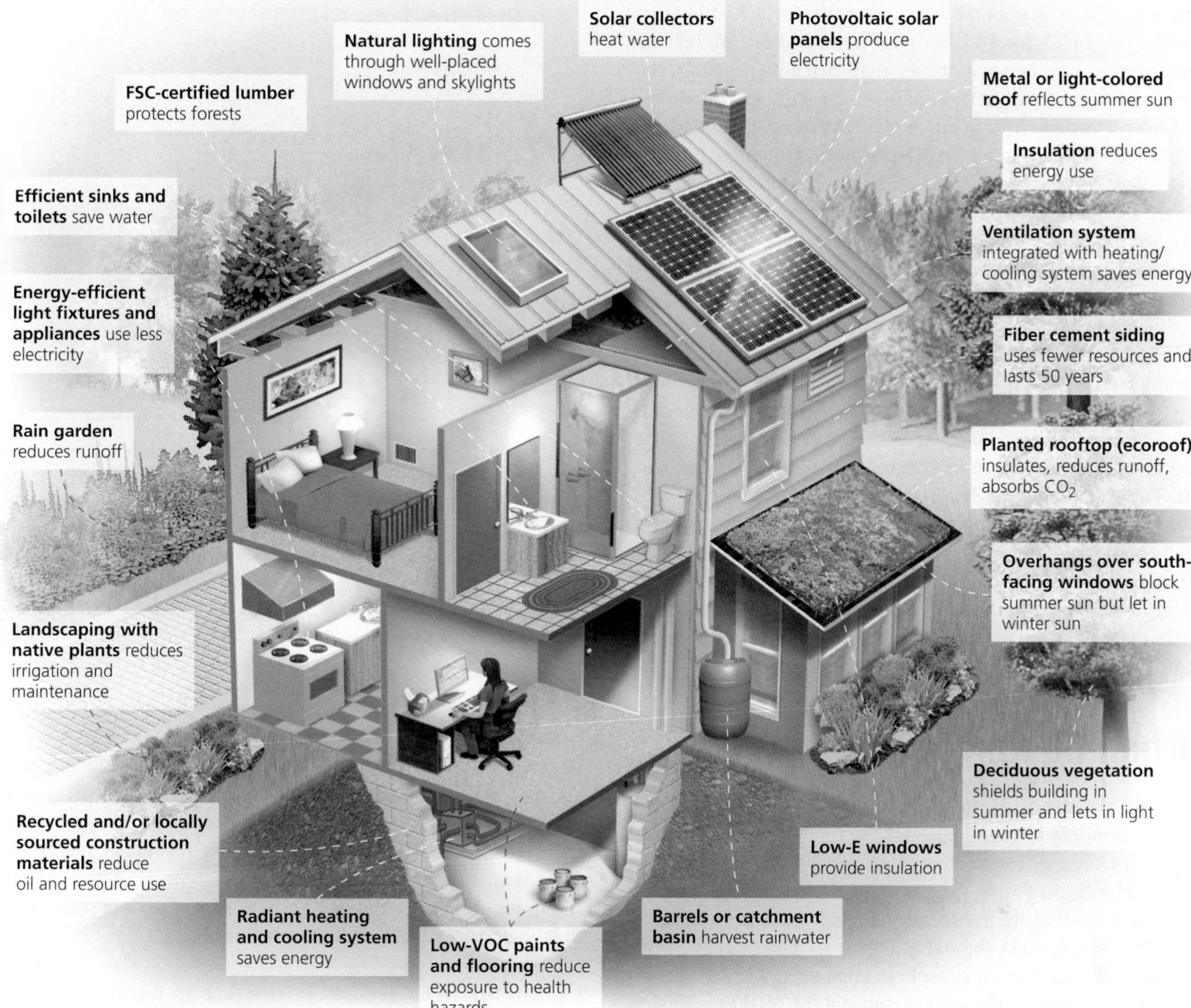

FIGURE 13.17 A green building incorporates design features to minimize its ecological footprint. Made from sustainable materials, green buildings are built to minimize energy use and water use, protect their occupants' health, limit pollution, and recycle waste.

TABLE 13.3 Green Building Approaches for LEED Certification

APPROACHES THAT ARE REWARDED	MAXIMUM POINTS
ENERGY: Monitor energy use; use efficient design, construction, appliances, systems, and lighting; use clean, renewable energy sources	37
THE SITE: Build on previously developed land; minimize erosion, runoff, and water pollution; use regionally appropriate landscaping; integrate with transportation options	21
INDOORS: Improve indoor air quality, provide natural daylight and views; improve acoustics	17
MATERIALS: Use local or sustainably grown, harvested, and produced products; reduce, reuse, and recycle waste	14
WATER USE: Use efficient appliances inside; landscape for water conservation outside	11
POINTS ABOVE MAY TOTAL UP TO 100. THEN, UP TO 10 BONUS POINTS MAY BE AWARDED FOR:	
INNOVATION: New and innovative technologies and strategies to go beyond LEED requirements	6
THE REGION: Addressing environmental concerns most important for one's region	4

Out of 110 possible points, 40 are required for LEED certification, 50 for silver, 60 for gold, and 80 for platinum levels. Point allocation shown is for new construction. Allocation varies slightly for renovations and specific types of buildings.

Green building techniques add expense to construction. The added cost is generally less than 3% for a LEED-silver building and 10% for a LEED-platinum building, but because construction is expensive, these small percentages can represent large sums of money. Nonetheless, LEED certification is booming. Portland now features several dozen LEED-certified buildings, including nine LEED-platinum structures. Portland also boasts the nation's first LEED-gold sports arena (the Rose Garden, where the Trailblazers basketball team plays). The savings on energy, water, and waste at the refurbished Rose Garden paid for the cost of its LEED upgrade after just one year.

Schools, colleges, and universities are leaders in sustainable building. In Portland, the Rosa Parks Elementary School, completed in 2007, is certified LEED-gold. This school building was built with locally sourced and nontoxic materials, uses 24% less energy and water than comparable buildings, and diverted nearly all of its construction waste from the landfill. The schoolchildren learn about renewable energy in their own building by watching a display of the electricity produced by its photovoltaic solar system. Portland State University, the University of Portland, Reed College, and Lewis and Clark College are just a few of the many colleges and universities nationwide that are constructing green buildings as a part of their campus sustainability efforts (Chapter 24).

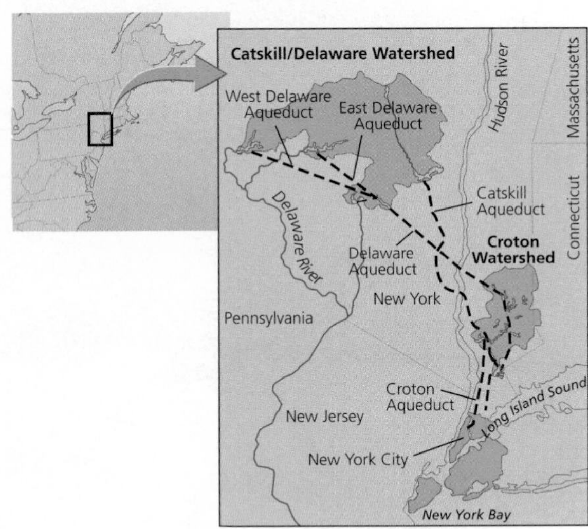

FIGURE 13.18 New York City pipes in its drinking water from upstate reservoirs in the Croton and Catskill/Delaware watersheds. The city acquires, protects, and manages watershed land to minimize pollution of these water sources. When New York City was confronted in 1989 with an order by the Environmental Protection Agency to build a $6-billion filtration plant to protect its citizens against waterborne disease, the city opted instead to purchase and better protect watershed land—for a fraction of the cost.

Urban Sustainability

Most of our efforts to make cities more safe, clean, healthy, and pleasant are also helping to make them more sustainable. A sustainable city is one that can function and prosper over the long term, providing generations of residents a good quality of life far into the future. In part, this entails minimizing the city's impacts on the natural systems and resources that nourish it. Urban centers exert both positive and negative environmental impacts. The extent and nature of these impacts depend on how we utilize resources, produce goods, transport materials, and deal with waste.

Urban resource consumption brings a mix of environmental effects

You might guess that urban living has a greater environmental impact than rural living. However, the picture is not so simple; instead, urbanization brings a complex mix of consequences.

Resource sinks Cities and towns are sinks (p. 117) for resources, having to import from source areas beyond their borders nearly everything they need to feed, clothe, and house their inhabitants and power their commerce. Urban and suburban areas rely on large expanses of land elsewhere to supply food, fiber, water, timber, metal ores, and mined fuels. Urban centers also need areas of natural land to provide ecosystem services, including purification of water and air, nutrient cycling, and waste treatment. Indeed, for their day-to-day survival, major cities such as New York, Boston, San Francisco, and Los Angeles depend on water they pump in from faraway watersheds (**FIGURE 13.18**).

The long-distance transportation of resources and goods from countryside to city requires fossil fuel use and thus has considerable environmental impact. However, imagine that all the world's 3.6 billion urban residents were instead spread evenly across the landscape. What would the transportation requirements be, then, to move all those resources and goods around to all those people? A world without cities would likely require *more* transportation to provide people the same degree of access to resources and goods.

Efficiency Once resources arrive at an urban center, the concentration of people allows goods and services to be delivered efficiently. For instance, providing electricity for densely packed urban homes and apartments is more efficient than providing electricity to far-flung homes in the countryside. The density of cities facilitates the provision of many social services that improve quality of life, including medical services, education, water and sewer systems, waste disposal, and public transportation.

More consumption Because cities draw resources from afar, their ecological footprints are much greater than their actual land areas. For instance, urban scholar Herbert Girardet calculated that the ecological footprint of London, England, extends 125 times larger than the city's actual area. By another estimate, cities take up only 2% of the world's land surface but consume over 75% of its resources.

However, the ecological footprint concept is most meaningful when used on a per capita basis. So, in asking whether urbanization causes increased resource consumption, we must ask whether the average urban or suburban dweller has a larger footprint than the average rural dweller. The answer is yes, but urban and suburban residents also tend to be wealthier than rural residents, and wealth correlates with resource consumption. Thus, although urban and suburban citizens tend to consume more than rural ones, the reason could simply be that they tend to be wealthier.

Urbanization preserves land

Because people pack densely together in cities, more land outside cities is left undeveloped. Indeed, this is the very idea behind urban growth boundaries. If cities did not exist, and if instead all 7 billion of us were evenly spread across the planet's land area, no large blocks of land would be left uninhabited, and we would have much less room for agriculture, wilderness, biodiversity, or privacy. The fact that half the human population is concentrated in discrete locations helps allow room for natural ecosystems to continue functioning and provide the ecosystem services on which all of us, urban and rural, depend.

Urban centers suffer and export pollution

Just as cities import resources, they export wastes, either passively through pollution or actively through trade. In so doing, urban centers transfer the costs of their activities to other regions—and mask the costs from their own residents. Citizens of Indianapolis, Columbus, or Buffalo may not recognize that pollution from nearby coal-fired power plants worsens acid precipitation hundreds of miles to the east. Citizens of New York City may not realize how much garbage their city produces if it is shipped elsewhere for disposal.

However, not all waste and pollution leaves the city. Urban residents are exposed to heavy metals, industrial compounds, and chemicals from manufactured products that accumulate in soil and water. Airborne pollutants cause photochemical smog, industrial smog, and acid precipitation (Chapter 17). Fossil fuel combustion releases greenhouse gases and pollutants that pose health risks.

Urban residents also suffer noise pollution and light pollution. **Noise pollution** consists of undesired ambient sound. Excess noise degrades one's surroundings aesthetically, can cause stress, and at intense levels (such as with prolonged exposure to the sounds of leaf blowers, lawn mowers, and jackhammers) can harm hearing. The glow of **light pollution** from city lights may impair sleep and obscures the night sky, impeding the visibility of stars.

City residents suffer thermal pollution as well, because cities often have ambient temperatures that are several degrees higher than those of surrounding areas. This **urban heat island effect** results from the concentration of heat-generating buildings, vehicles, factories, and people. It also results from the way that buildings and dark paved surfaces absorb heat during the daytime and then release it slowly at night, warming the air and interfering with patterns of convective circulation that would otherwise cool the city (FIGURE 13.19).

These various forms of pollution and the health risks they pose are not evenly shared among urban residents. Those who receive the brunt of the pollution are often those who are too poor to live in cleaner areas. Environmental justice concerns (pp. 140–141) center on the fact that a disproportionate number of people living near, downstream from, or downwind from factories, power plants, and other polluting facilities are people who are poor and, often, people of racial or ethnic minorities.

FAQ **Aren't cities bad for the environment?**

Stand in the middle of a big city and look around. You see concrete, cars, and pollution. Environmentally bad, right? Not necessarily. Cities have a mix of consequences, but the widespread impression that urban living is less environmentally friendly than rural living is largely a misconception. Consider that in a city you can walk to the grocery store instead of driving. You can take the bus or the train. Police, fire, and medical services are close at hand. Water and electricity are easily supplied to your entire neighborhood, and waste is easily collected. In contrast, if you live in the country, resources must be used to transport all these services for long distances, or you need to burn gasoline and time traveling to reach them. By clustering people together, cities allow us to distribute resources efficiently, while also preserving natural lands outside the city. In many ways, each person receives more, with less environmental impact, in a city than in the country.

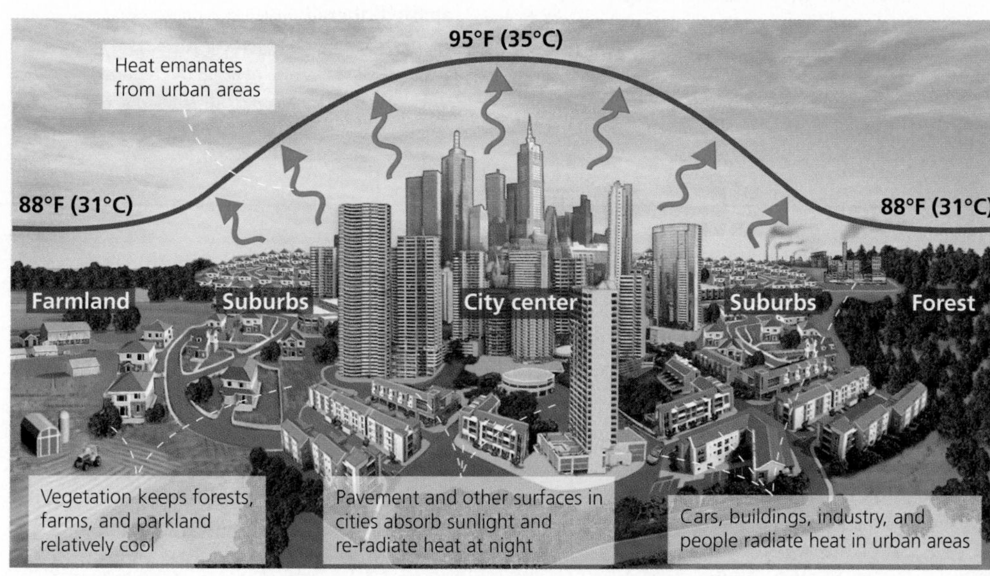

FIGURE 13.19 **Cities produce urban heat islands, creating temperatures warmer than surrounding areas.** People, buildings, vehicles, and factories generate heat, and buildings and dark paved surfaces absorb daytime heat and then release it slowly at night.

Baltimore and Phoenix Showcase Urban Ecology

Researchers in urban ecology examine how ecosystems function in cities and suburbs, how natural systems respond to urbanization, and how people interact with the urban environment. Today, Baltimore and Phoenix are centers for urban ecology.

These two cities are very different: Baltimore is an Atlantic port city on Chesapeake Bay with a long history, whereas Phoenix is a young and fast-growing Southwestern metropolis sprawling across the desert. Each was picked by the U.S. National Science Foundation to serve as a research site in its prestigious Long Term Ecological Research (LTER) program, which funds multi-decade ecological research. Since 1997, hundreds of researchers have studied Baltimore and Phoenix explicitly as ecosystems, examining nutrient cycling, biodiversity, air and water quality, how people react to environmental health threats, and more.

Research teams in both cities are combining old maps, aerial photos, and new remote sensing satellite data to reconstruct the history of landscape change. In Phoenix, one group showed how urban development spread across the desert in a "wave of advance," affecting soils, vegetation, and microclimate as it went. In Baltimore, mapping showed that development fragmented the forest into smaller patches over the past 100 years, even while the overall amount of forest remained the same.

An urban ecologist samples water beneath an overpass in Baltimore.

The study regions designated for each city encompass both heavily urbanized central city areas and rural and natural areas on the urban fringe. To measure the impacts of urbanization, many research projects compare conditions in these two types of areas.

The Baltimore project's scientists can see ecological effects of urbanization just by comparing the urban lower end of their site's watershed with its less-developed upper end. In the lower end, pavement, rooftops, and compacted soil prevent rainfall from infiltrating the soil, so water runs off quickly into streams. The rapid flow cuts streambeds deeply into the earth while leaving the surrounding soil drier. As a result, wetland-adapted trees and shrubs are vanishing, replaced by dry-adapted upland trees and shrubs.

The fast flow of water also worsens pollution. In natural areas, streams and wetlands filter pollution by breaking down nitrogen compounds. But in urban areas, where wetlands dry up and runoff from pavement creates flash floods, streams lose their filtering ability. In Baltimore, the resulting pollution ends up in Chesapeake Bay, which suffers eutrophication and a large hypoxic dead zone (pp. 105, 108–109). Baltimore-project scientists studying nutrient cycling (p. 117) found that urban and suburban watersheds have far more nitrate pollution than natural forests (**FIGURE 1**).

Baltimore research also reveals that applying salt to icy roads in winter has environmental impacts. Road salt makes its way into streams, which become up

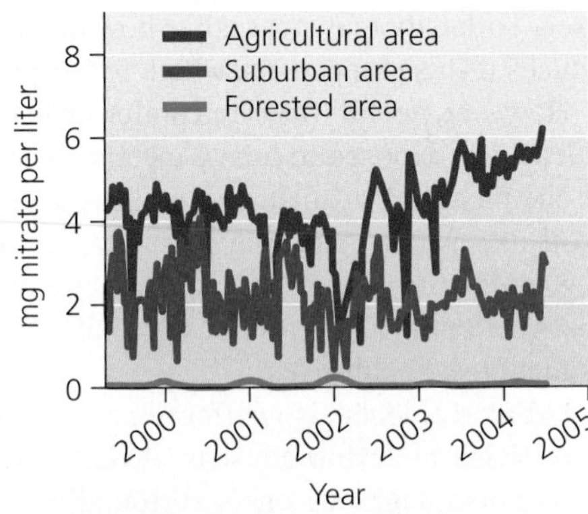

FIGURE 1 Streams in Baltimore's suburbs contain more nitrates than streams in nearby forests, but fewer than those in agricultural areas, where fertilizers are applied liberally. *With kind permission from Springer Science and Business Media and the author, Groffman, P.M., et al., 2004. Nitrogen fluxes and retention in urban watershed ecosystems.* Ecosystems *7: 393–403; Fig 4. Also from Baltimore Ecosystem Study.*

Urban centers foster innovation

Cities promote a flourishing cultural life and, by mixing together diverse people and influences, spark innovation and creativity. The urban environment can promote education and scientific research, and cities have long been viewed as engines of technological and artistic inventiveness. This inventiveness can lead to solutions to societal problems, including ways to reduce environmental impacts.

For instance, research into renewable energy sources is helping us develop ways to replace fossil fuels. Technological advances have helped us reduce pollution. Wealthy and educated urban populations provide markets for low-impact goods, such as organic produce. Recycling programs help reduce the solid waste stream. Environmental education is helping people choose their own ways to live cleaner, healthier, lower-impact lives. All these phenomena arise from the education, innovation, science, and technology that are part of urban culture.

to 100 times saltier, even in summer when salt is not applied. Such high salinity kills organisms (**FIGURE 2**), degrades habitat and water quality, and impairs streams' ability to remove nitrate.

To study contamination of groundwater and drinking water, researchers are using isotopes (p. 24) to trace where salts in the most polluted streams are coming from. Baltimore is now improving water quality substantially with a $900-million upgrade of its sewer system.

Urbanization also affects species and ecological communities. Cities and suburbs facilitate the spread of non-native species, because people introduce exotic ornamental plants and because urbanization's impacts on the soil, climate, and landscape favor weedy generalist species over more specialized native ones. In Baltimore, non-native plant species are most abundant in urban areas. In Phoenix's dry climate, pollen from some non-native plants causes allergy problems for city residents.

Community ecologists studying the wild animals and plants that persist within Phoenix are finding that urbanization alters relationships among them. Compared with natural landscapes, cities offer steady and reliable food resources—think of people's bird feeders, or food scraps from dumpsters.

Growing seasons are extended and seasonal variation is buffered in cities, as well. The urban heat island effect (p. 351) raises nighttime temperatures and makes temperatures more similar year-round. Buildings and

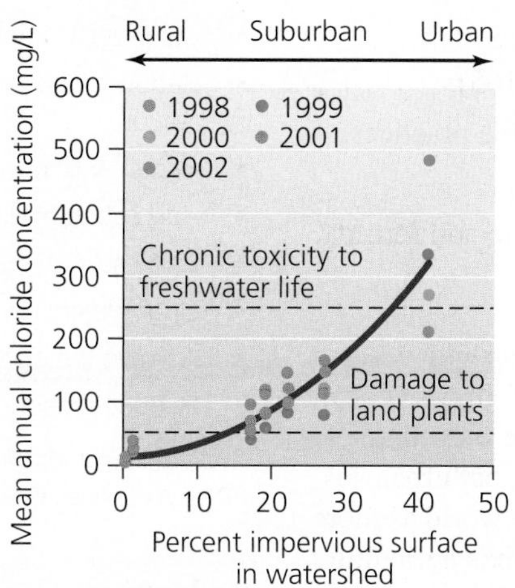

FIGURE 2 Salt concentrations in Baltimore-area streams are high enough to damage plants in the suburbs and to kill aquatic animals in urban areas.
Adapted from Kaushal, S.S., et al., 2005. Increased salinization of fresh water in the northeastern United States. Proc. Natl. Acad. Sci. USA *102: 13517–13520, Fig 2. ©2005 National Academy of Sciences, U.S.A. By permission.*

DATA Q In how many years did mean salt concentrations in urban areas surpass the level of chronic toxicity to freshwater life?

ornamental vegetation shelter animals from extreme conditions, and irrigation in yards and gardens provides water. In a desert city like Phoenix, watering boosts primary productivity and lowers daytime temperatures. Together, all these changes lead to higher population densities of animals but lower species diversity as generalists thrive and displace specialists.

Urban ecologists in Phoenix and Baltimore are studying social and demographic aspects of the urban environment as well. Some research measures how natural amenities affect property values. One study found that proximity to a park increases a home's property values—unless crime is pervasive. If the robbery rate surpasses 6.5 times the national average (as it does in Baltimore) then proximity to a park begins to depress property values.

Other studies focus on environmental justice concerns (p. 140). These have repeatedly found that sources of industrial pollution tend to be located in neighborhoods that are less affluent and that are home to people of racial and ethnic minorities. Phoenix researchers mapped patterns of air pollution and toxic chemical releases and found that minorities and the poor are exposed to a greater share of these hazards. As a result, they suffer from higher rates of childhood asthma.

In Baltimore, researchers found a more complex pattern. Toxic release sites were more likely to be in working-class white neighborhoods than in African American neighborhoods. This, the researchers concluded, was a result of historical inertia. In the past, living close to one's workplace—the factories that release toxic chemicals—was something people preferred, and white workers claimed the privilege of living near their workplaces.

Whether addressing the people, natural communities, or changing ecosystems of the urban environment, studies on urban ecology like those in Phoenix and Baltimore will be vitally informative in our ever-more-urban world. ■

Urban ecology helps cities toward sustainability

Cities that import all their resources and export all their wastes have a linear, one-way metabolism. Linear models of production and consumption tend to destabilize environmental systems and are not sustainable. Proponents of sustainability for cities stress the need to develop circular systems, akin to systems found in nature, which recycle materials and use renewable sources of energy.

Researchers in the field of **urban ecology** hold that cities can be viewed explicitly as ecosystems and that the fundamentals of ecosystem ecology and systems science (Chapter 5) apply to the urban environment. Major urban ecology projects are ongoing in Baltimore and Phoenix, where researchers are studying these cities explicitly as ecological systems (see **THE SCIENCE BEHIND THE STORY**, above).

To help cities reduce their environmental impacts while improving quality of life for their citizens, urban sustainability advocates suggest that cities follow an ecosystem-centered model by striving to:

- Maximize efficient use of resources.
- Recycle as much as possible (pp. 617–619).
- Develop environmentally friendly technologies.
- Account fully for external costs (pp. 146, 165).
- Offer tax incentives to encourage sustainable practices.
- Use locally produced resources.
- Use organic waste and wastewater to restore soil fertility.
- Encourage urban agriculture.

Planners and visionary leaders have come up with designs for entire "eco-cities" built from scratch that follow cyclical, sustainable patterns of resource use and waste recycling. So far, none of these efforts have come to fruition, but urban sustainability is developing step by step across the world as more and more cities adopt sustainable strategies. Urban agriculture is thriving in many places, from Portland to Cuba to Japan. Municipal recycling programs continue to grow. Curitiba, Brazil, shows the kind of success that can result when a city invests in well-planned infrastructure. Besides its highly effective bus transportation network, the city provides recycling, environmental education, job training for the poor, and free health care. Surveys show that its citizens are unusually happy and better off economically than residents of other Brazilian cities.

In 2007, New York City unveiled an ambitious plan that Mayor Michael Bloomberg hoped would make it "the first environmentally sustainable 21st-century city." PlaNYC is a 132-item program to reduce greenhouse gas emissions, improve mass transit, plant trees, clean up polluted land and rivers, and enhance access to parkland. The aim of PlaNYC is to make New York City a better place to live as it accommodates 1 million more people by 2030 (TABLE 13.4). As of 2012, accomplishments included completing 100 energy efficiency projects in city buildings; planting 560,000 trees; opening or renovating 210 school playgrounds; establishing 64 community gardens; upgrading wastewater treatment; acquiring 36,000 acres to protect upstate water supplies; installing bike lanes and bike racks and launching a bike-sharing program; retrofitting ferries to reduce pollution; introducing electric vehicles and converting 30% of the taxi fleet to hybrid vehicles; and reducing greenhouse gas emissions by 13%.

TABLE 13.4 Goals of New York City's PlaNYC Sustainability Program

• Create affordable, sustainable housing for 1 million more people
• Ensure that all New Yorkers live within a 10-minute walk of a park
• Clean up all contaminated land in the city
• Reduce water pollution, and open 90% of waterways for recreation
• Repair and maintain water delivery network
• Add substantial new mass transit capacity
• Bring subways and roads up to full state of good repair
• Upgrade energy infrastructure to provide clean power
• Achieve the cleanest air quality of any big city in America
• Reduce greenhouse gas emissions by 30%

Source: City of New York, Office of the Mayor, PlaNYC Progress Report 2012: A Greener, Greater New York.

Steps toward livability enhance sustainability

The best news of all is that most of the steps being taken to make cities more livable are also helping to make them more sustainable. Planning and zoning are pursuits that specifically entail a long-term vision. By projecting farther into the future than political leaders or businesses generally do, planning and zoning are powerful forces for sustaining urban communities. The principles and practices of smart growth and new urbanism cut down on energy consumption, helping us to address the looming societal challenges of peak oil (pp. 532–533) and climate change (Chapter 18). Moving transportation away from cars and toward mass transit systems reduces gasoline consumption and carbon emissions. Parks offer ecosystem services while promoting residents' health. And green buildings bring a diversity of health and environmental benefits.

Successes from Portland to Curitiba to New York City suggest that cities may one day make urban sustainability a reality. Indeed, because they affect the environment in many positive ways and can encourage efficient resource use, urban centers are a key element in achieving progress toward global sustainability.

Conclusion

As half the human population has shifted from rural to urban lifestyles, the nature of our environmental impact has changed. As urban and suburban dwellers, our impacts are less direct but more far-reaching. Resources must be delivered to us over long distances, requiring the use of still more resources. Limiting the waste of those resources by making urban and suburban areas more sustainable will be vital for our future. Fortunately, the innovative cultural environment that cities foster has helped us develop solutions to alleviate impact and promote sustainability.

Part of seeking urban sustainability lies in making urban areas better places to live. One key component of these efforts involves expanding transportation options to relieve congestion and make cities run more efficiently. Another lies in ensuring access to adequate parklands to keep us from becoming wholly isolated from nature. Proponents of smart growth and new urbanism believe they have solutions to the challenges posed by urban and suburban sprawl. Accomplishments in city and regional planning have improved many American cities, and we should be encouraged about such progress. Continuing experimentation in cities everywhere will help us determine how best to ensure that urban growth improves our quality of life while not degrading the quality of our environment.

Reviewing Objectives

You should now be able to:

Describe the scale of urbanization

- The world's population has become predominantly urban. (p. 337)
- The shift from rural to urban living is driven largely by industrialization and is proceeding fastest in the developing world. (p. 337)
- Nearly all future population growth will be in cities of the developing world. (p. 337)
- Environmental factors influence the location and growth of cities. (p. 338)
- The geography of urban areas is changing as cities decentralize and suburbs grow and expand. (pp. 338–339)

Assess urban and suburban sprawl

- Sprawl covers large areas of land with low-density development. Both population growth and increased per capita land use contribute to sprawl. (pp. 339–340)
- Sprawl results from the home-buying choices of individuals who prefer suburbs to cities, and it has been facilitated by government policy and technological developments. (p. 340)
- Sprawl may lead to negative impacts involving transportation, pollution, health, land use, natural habitat, and economics. (pp. 340–342)

Outline city and regional planning and land use strategies

- City and regional planning and zoning are key tools for improving the quality of urban life. (pp. 342–344)
- Urban growth boundaries, "smart growth," and "new urbanism" attempt to re-create compact and vibrant urban spaces. (pp. 344–345)

Evaluate transportation options, urban parks, and green buildings

- Mass transit systems can enhance the efficiency and sustainability of urban areas. (pp. 345–347)
- The United States lags behind other nations in mass transit, but new efforts are being made. (pp. 346–347)
- Urban parklands provide recreation, soothe the stress of urban life, and keep people in touch with natural areas. (p. 347)
- A variety of types of parks and open space exist, including playgrounds, community gardens, greenways, and greenbelts. (pp. 347–348)
- Green buildings minimize their ecological footprints by using sustainable materials, limiting the use of energy and water, minimizing health impacts on their occupants, controlling pollution, and recycling waste. (pp. 348–349)
- Green building approaches are spreading fast, helped along by the LEED certification program. (pp. 348–350)

Analyze environmental impacts and advantages of urban centers

- Cities are resource sinks with high per capita resource consumption, and they create substantial waste and pollution. (pp. 350–351)
- Cities also maximize efficiency, help preserve natural lands, and foster innovation that can lead to solutions for environmental problems. (pp. 350–352)

Assess urban ecology and the pursuit of sustainable cities

- Linear modes of consumption and production are unsustainable, and more circular modes will be needed to create truly sustainable cities. (pp. 353–354)
- Although a full "eco-city" has yet to be built, many cities worldwide are taking steps to decrease their ecological footprints. (p. 354)
- Most steps taken for urban livability also enhance sustainability. (p. 354)

Testing Your Comprehension

1. What factors lie behind the shift of population from rural areas to urban areas? What types of cities and countries are experiencing the fastest urban growth today, and why?
2. Why have so many city dwellers in the United States, Canada, and other nations moved into suburbs?
3. Give two definitions of *sprawl*. Describe five negative impacts that have been suggested to result from sprawl.
4. What are city planning and regional planning? Contrast planning with zoning. Give examples of some of the suggestions made by early planners such as Daniel Burnham and Edward Bennett.
5. How are some people trying to prevent or slow sprawl? Describe some key elements of "smart growth." What effects, positive and negative, do urban growth boundaries tend to have?

6. Describe several apparent benefits of rail transit systems. What is a potential drawback?
7. How are city parks thought to make urban areas more livable? Give three examples of types of parks or public spaces.
8. What is a green building? Describe several features a LEED-certified building may have.
9. Describe the connection between urban ecology and sustainable cities. List three actions a city can take to enhance its sustainability.
10. Name two positive effects of urban centers on the natural environment.

Seeking Solutions

1. Describe the causes of the spread of suburbs and outline the environmental, social, and economic impacts of sprawl. Overall, do you think the spread of urban and suburban development that is commonly labeled *sprawl* is predominantly a good thing or a bad thing? Do you think it is inevitable? Give reasons for your answers.
2. Would you personally want to live in a neighborhood developed in the new urbanist style? Would you like to live in a city or region with an urban growth boundary? Why or why not?
3. Consider the variety of approaches used to construct or renovate a building using LEED-certified green building techniques. Are there any LEED-certified buildings on your campus? If so, how do they differ from conventional buildings? Think about a building on your campus that you think is unhealthy or environmentally wasteful. Name three specific ways in which green building techniques might be used to improve this building.
4. All things considered, do you feel that cities are a positive thing or a negative thing for environmental quality? How much do you feel we may be able to improve the sustainability of our urban areas?
5. **THINK IT THROUGH** You are the president of your college or university, and students are clamoring for you to help create the world's first fully sustainable campus. Considering how people enhance livability and sustainability in cities, what lessons might you try to apply to your college or university? You are scheduled to give a speech to the campus community about your plans and will need to name five specific actions you plan to take to pursue a sustainable campus. What will they be, and what will you say about each choice to describe its importance?
6. **THINK IT THROUGH** After you earn your college degree, you are offered three equally desirable jobs, in three very different locations. If you accept the first, you will live in the midst of a densely populated city. If you take the second, you will live in a suburb where you have more space but where sprawl may soon surround you for miles. If you select the third, you will live in a rural area with plenty of space but few cultural amenities. You are a person who aims to live in the most ecologically sustainable way you can. Where would you choose to live? Why? What considerations will you factor into your decision?

Calculating Ecological Footprints

One way to reduce your ecological footprint is with alternative transportation. Each gallon of gasoline is converted during combustion to approximately 20 pounds of carbon dioxide (CO_2), which is released into the atmosphere. The table lists typical amounts of CO_2 released per person per mile through various modes of transportation, assuming typical fuel efficiencies.

For an average North American person who travels 12,000 miles per year, calculate and record in the table the CO_2 emitted yearly for each transportation option, and the reduction in CO_2 emissions that one could achieve by relying solely on each option.

Mode of Transport	CO_2 per person per mile	CO_2 per person per year	CO_2 emission reduction	Your estimated mileage per year	Your CO_2 emissions per year
Automobile (driver only)	0.825 lb	9900 lb	0		
Automobile (2 persons)	0.413 lb				
Automobile (4 persons)	0.206 lb				
Vanpool (8 persons)	0.103 lb				
Bus	0.261 lb				
Walking	0.082 lb				
Bicycle	0.049 lb				
				Total = 12,000	

1. Which transportation option provides the most miles traveled per unit of carbon dioxide emitted?
2. Clearly, it is unlikely that any of us will walk or bicycle 12,000 miles per year or travel only in vanpools of eight people. In the last two columns, estimate what proportion of the 12,000 annual miles you think that you actually travel by each method, and then calculate the CO_2 emissions that you are responsible for generating over the course of a year. Which transportation option accounts for the most emissions for you?
3. How could you reduce your CO_2 emissions? How many pounds of emissions do you think you could realistically eliminate over the course of the next year by making changes in your transportation decisions?

MasteringENVIRONMENTALSCIENCE™

STUDENTS

Go to **MasteringEnvironmentalScience** for assignments, the etext, and the Study Area with practice tests, videos, current events, and activities.

INSTRUCTORS

Go to **MasteringEnvironmentalScience** for automatically graded activities, current events, videos, and reading questions that you can assign to your students, plus Instructor Resources.

14

Is this baby ingesting toxic substances?

Environmental Health and Toxicology

Upon completing this chapter, you will be able to:

- Explain the goals of environmental health and identify major environmental health hazards
- Describe the types of toxic substances in the environment, the factors that affect their toxicity, and the defenses that organisms possess against them
- Explain the movements of toxic substances and how they affect organisms and ecosystems
- Discuss the study of hazards and their effects, including wildlife toxicology, case histories, epidemiology, animal testing, and dose-response analysis
- Evaluate risk assessment and risk management
- Compare philosophical approaches to risk and how they relate to regulatory policy

CENTRAL CASE STUDY

Poison in the Bottle: Is Bisphenol A Safe?

Bisphenol A: Worldwide

"Babies in the U.S. are born pre-polluted with BPA. What more evidence do we need to act?"

—Dr. Janet Gray, Director of the Environmental Risks and Breast Cancer Project, Vassar College

"There is no basis for human health concerns from exposure to BPA."

—The American Chemistry Council

How is it that a chemical found to alter reproductive development in animals gets used in baby bottles? How can it be that a substance linked to breast cancer, prostate cancer, and heart disease is routinely used in food and drink containers? The chemical bisphenol A (BPA for short) has been associated with everything from neurological effects to miscarriages. Yet it's in hundreds of products we use every day, and there's a better than 9 in 10 chance that it is coursing through your body right now.

To understand how chemicals that may pose health risks come to be widespread in our society, we need to explore how scientists and policymakers study toxic substances and other environmental health risks—and the vexing challenges these pursuits entail.

Chemists first synthesized bisphenol A, an organic compound (p. 28) with the chemical formula $C_{15}H_{16}O_2$, in 1891. As they began producing plastics in the 1950s, chemists found bisphenol A to be useful in creating epoxy resins used in lacquers and coatings. Epoxy resins containing BPA were soon being used to line the insides of metal food and drink cans and the insides of pipes for our water supply, as well as in enamels, varnishes, adhesives, and even dental sealants for our teeth.

Chemists also found that linking BPA molecules into polymers (p. 29) helped create polycarbonate plastic, a hard, clear type of plastic that soon found use in water bottles, food containers, eating utensils, eyeglass lenses, CDs and DVDs, laptops and other electronics, auto parts, refrigerator shelving, sports equipment, baby bottles, and children's toys. With so many uses, bisphenol A has become one of the world's most-produced chemicals; each year we make about 1 pound of BPA for each person on the planet, and 20 pounds per person in the United States!

Unfortunately, bisphenol A leaches out of its many products and into our food, water, air, and bodies. Fully 93% of Americans carry detectable concentrations in their urine, according to the latest National Health and Nutrition Examination Survey conducted by the Centers for Disease Control and Prevention (CDC). Because most BPA passes through the body within hours, these data suggest that we are receiving almost continuous exposure. Babies and children have higher relative exposure to BPA because they eat more for their body weight and metabolize the chemical less effectively.

What, if anything, is BPA doing to us? To address such questions, scientists run experiments on laboratory animals, administering known doses of the substance and measuring the health impacts that result. Hundreds of studies with rats, mice, and other animals have shown many apparent effects of BPA, including a wide range of reproductive abnormalities. Recent studies suggest humans suffer health impacts from BPA as well (see **THE SCIENCE BEHIND THE STORY**, pp. 362–363).

Many of these effects occur at extremely low doses—much lower than the exposure levels set so far by regulatory agencies for human safety. Scientists say this is because BPA mimics the female sex hormone estrogen; that is, it is structurally similar to estrogen and can induce some of its effects in animals (see Figure 14.9, pp. 370). Hormones such as estrogen function at minute concentrations, so when a synthetic chemical similar to estrogen reaches the body in a similarly low concentration, it can fool the body into responding.

In reaction to the burgeoning research, a growing number of researchers, doctors, and consumer advocates are calling on governments to regulate bisphenol A and for manufacturers to stop using it. The chemical industry insists that BPA is safe, pointing to industry-sponsored research that finds no health impacts. To sort through the debate, several expert panels have convened to assess the fast-growing body of scientific studies. Some panels have found typical BPA exposure to be nothing to worry about, others have expressed concern, and most have struggled with the fact that traditional research methods are not geared to test hormone-mimicking substances that exert effects at low doses. Indeed, dealing with substances like BPA is forcing us toward a challenging paradigm shift in the way we assess environmental health risks.

For instance, the U.S. Food and Drug Administration (FDA) announced in 2008 that it saw no reason to regulate BPA, but its own science advisory committee disagreed, and in 2009 the FDA decided to start a testing program. Soon afterwards, the U.S. Environmental Protection Agency (EPA) announced it would begin its own assessment.

In 2008 Canada banned the use of BPA in baby bottles and in 2010 became the first nation to declare bisphenol A toxic. The use of BPA in products for babies was banned in France and Denmark in 2010, and in 2011 a ban on BPA in baby bottles was adopted across the entire European Union. Turkey has issued a similar ban, and several U.S. states and municipalities have also taken action.

In 2008, the Natural Resources Defense Council petitioned the FDA to ban the use of BPA in food packaging, citing studies showing adverse health impacts on humans. Health advocates were shocked and disappointed when, in 2012, the FDA rejected the proposed ban, stating that there was no compelling scientific evidence to justify new restrictions. Mere days later, the government of Sweden came to a different conclusion from the FDA and announced it would ban the use of BPA in the packaging of food intended for children under the age of 3 (such as the lids of baby food bottles).

In the face of mounting public concern, many companies are choosing to voluntarily remove BPA from their products. The six major U.S. manufacturers of plastic baby bottles promised in 2008 to stop using BPA. Wal-Mart and Toys "R" Us decided to stop carrying children's products with BPA. The manufacturer Sunoco stopped selling BPA to companies that use it in children's products. After these actions, the FDA finally banned the use of BPA in baby bottles and children's cups in 2012.

FIGURE 14.1 Bisphenol A is in many of the foods we eat. Researchers for *Consumer Reports* magazine tested common packaged foods and more in 2009; they found that nearly all of them contained bisphenol A that had leached from the linings of their containers.

As of 2013, concerned parents can now more easily find BPA-free items for their infants and children. The rest of us remain exposed through most food cans, many drink containers, and thousands of other products (**FIGURE 14.1**). But despite the lack of government regulation, food producers are responding to consumer concerns about the safety of BPA. Shortly after the 2011 publication of a study showing consumers of canned soups had urinary BPA concentrations over 1200 times higher than consumers of fresh soups, Campbell's Soup announced that it is transitioning away from the use of BPA in its can liners. Food giants ConAgra, Nestle, and Heinz have also pledged to remove BPA from food packaging. There is precedent for such efforts, as BPA was voluntarily phased out of can liners in Japan starting in the late 1990s.

Bisphenol A is by no means one of our greatest environmental health threats. However, it provides a timely example of how we as a society assess health risks and decide how to manage them. As scientists and government regulators assess BPA's potential risks, their efforts give us a window on how hormone-disrupting chemicals are challenging the way we appraise and control the environmental health risks we face.

Environmental Health

Examining the impacts of human-made chemicals such as bisphenol A is just one aspect of the broad field of environmental health. The study and practice of **environmental health** assesses environmental factors that influence our health and quality of life. These factors include wholly natural aspects of the environment over which we have little or no control, as well as anthropogenic (human-caused) factors. Practitioners of environmental health seek to prevent adverse effects on human health and on the ecological systems that are essential to our well-being.

We face four types of environmental hazards

Many environmental health hazards exist in the world around us. We can categorize them into four main types: physical, chemical, biological, and cultural. For each type of hazard, there is some amount of risk that we cannot avoid—but there is also some amount of risk that we *can* avoid by taking precautions. Much of environmental health consists of taking steps to minimize the risks of encountering hazards and to mitigate the impacts of the hazards we do encounter.

Physical hazards *Physical hazards* arise from processes that occur naturally in our environment and pose risks to human life or health. Some such physical processes are ongoing natural phenomena, such as ultraviolet (UV) radiation from sunlight (**FIGURE 14.2a**). Excessive exposure to UV radiation damages DNA and has been tied to skin cancer, cataracts, and immune suppression. We can reduce our exposure to and risk from UV light by using clothing and sunscreen to shield our skin from intense sunlight.

Other physical hazards include discrete events such as earthquakes, volcanic eruptions, fires, floods, blizzards, landslides, hurricanes, and droughts. We can do little to predict the timing of a natural disaster such as an earthquake, and nothing to prevent one. However, we can minimize risk by preparing ourselves. Scientists can map geologic faults to determine areas at risk of earthquakes, engineers can design

(a) Physical hazard

(b) Chemical hazard

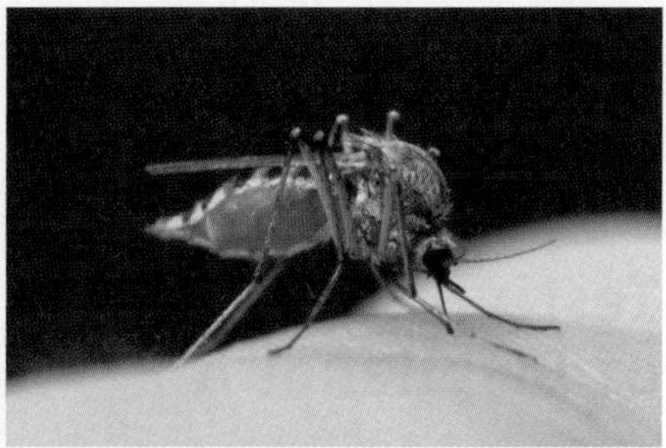

(c) Biological hazard

(d) Cultural hazard

FIGURE 14.2 **Environmental health hazards come in four types.** The sun's ultraviolet radiation is an example of a physical hazard **(a)**. Excessive exposure increases the risk of skin cancer. Chemical hazards **(b)** include both synthetic and natural chemicals. Much of our exposure comes from pesticides and household chemical products. Biological hazards **(c)** include diseases and the organisms that transmit them. Some mosquitoes are vectors for pathogenic microbes, including those that cause malaria. Cultural or lifestyle hazards **(d)** include the behavioral decisions we make, as well as the socioeconomic constraints forced on us. Smoking is a lifestyle choice that raises one's risk of lung cancer and other diseases.

buildings to resist damage, and governments and individuals can create and rehearse emergency plans to prepare for a quake's aftermath.

Some common practices make us vulnerable to certain physical hazards. Clearing forests from slopes makes landslides more likely, for instance, and channelizing rivers promotes flooding in some areas while preventing it in others (p. 400–402). We can reduce risk from such hazards by improving our forestry and flood control practices and by regulating development in areas prone to floods, landslides, fires, or coastal waves.

Chemical hazards *Chemical hazards* include many of the synthetic chemicals that our society manufactures, such as pharmaceuticals, disinfectants, and pesticides (FIGURE 14.2b). Some substances produced naturally by organisms (such as venoms) also can be hazardous, as can many substances that we find in nature and then process for our use (such as hydrocarbons, lead, and asbestos). Following our overview of environmental health, much of this chapter will focus on chemical health hazards and the ways we study and regulate them.

Biological hazards *Biological hazards* result from ecological interactions among organisms (FIGURE 14.2c). When we become sick from a virus, bacterial infection, or other pathogen, we are suffering parasitism (p. 79) by other species that are simply fulfilling their ecological roles. This is what we call **infectious disease**. Some infectious diseases are spread when pathogenic microbes attack us directly. With others, infection occurs through a *vector*, an organism (such as a mosquito) that transfers the pathogen to the host. Infectious diseases such as malaria, cholera, tuberculosis, and influenza (flu) are major environmental health hazards, especially in developing nations with widespread poverty and few resources for health care. As with physical and chemical hazards, it is impossible for us to avoid risk from biological agents completely, but through monitoring, sanitation, and medical treatment we can reduce the likelihood and impacts of infection.

Cultural hazards Hazards that result from our place of residence, our socioeconomic status, our occupation, or our behavioral choices can be thought of as *cultural hazards* or *lifestyle hazards*. We can minimize or prevent some of these cultural or lifestyle hazards, whereas others may be beyond our control. For instance, choosing to smoke cigarettes, or living or working with people who smoke, greatly increases our risk of lung cancer (FIGURE 14.2d). Choosing to smoke is a personal behavioral decision, but exposure to secondhand smoke in the home or workplace may be beyond one's control. The influences of personal choices and "forced" decisions on health can also apply to diet and nutrition, workplace hazards, drug use, criminal behavior, and one's primary mode of transportation. As advocates of environmental justice (pp. 140–141) argue, health factors such as living near toxic waste sites or working with pesticides without proper training and safeguards are often correlated with socioeconomic deprivation. In general, the fewer economic resources or political clout one has, the harder it is to avoid cultural hazards and environmental health risks.

Testing the Safety of Bisphenol A

Of the many studies documenting health impacts of bisphenol A on lab animals, one of the first came about because a lab assistant reached for the wrong soap.

At a laboratory at Case Western Reserve University in Ohio in 1998, geneticist Patricia Hunt (now at Washington State University) was making a routine check of her female lab mice. As she extracted and examined developing eggs from the ovaries, she began to wonder what had gone wrong. About 40% of the eggs showed problems with their chromosomes, and 12% had irregular amounts of genetic material, a dangerous condition called aneuploidy, which can lead to miscarriages or birth defects in mice and people alike.

A bit of sleuthing revealed that a lab assistant had mistakenly washed the lab's plastic mouse cages and water bottles with an especially harsh soap. The soap damaged the cages so badly that parts of them seemed to have melted.

The cages were made from polycarbonate plastic, which contains bisphenol A (BPA). Hunt knew at the time that BPA mimics estrogen and that some studies had linked the chemical to reproductive abnormalities in mice, such as low sperm counts and early sexual development. Other research indicated that BPA leaches out of plastic into water and food when the plastic is treated with heat, acidity, or harsh soap.

Hunt wondered whether the chemical might be adversely affecting the mice in her lab. Deciding to re-create the accidental cage-washing incident in a controlled experiment, Hunt instructed researchers in her lab to wash polycarbonate cages and water bottles using varying levels of the harsh soap. They then compared mice kept in damaged cages with plastic water bottles to mice kept in undamaged cages with glass water bottles.

Dr. Patricia Hunt, Washington State University

The developing eggs of mice exposed to BPA through the deliberately damaged plastic showed significant problems during meiosis, the division of chromosomes during egg formation—just as they had in the original incident (**FIGURE 1**). In contrast, the eggs of mice in the control cages were normal.

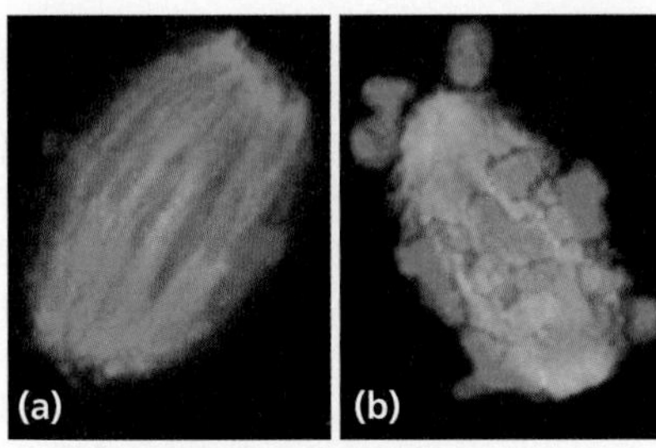

FIGURE 1 Exposure to bisphenol A affects cell division in mice. In normal cell division **(a)**, chromosomes (red) align properly. Exposure to bisphenol A causes abnormal cell division **(b)**, whereby chromosomes scatter and are distributed improperly and unevenly between daughter cells.

In another round of tests, Hunt's team gave sets of female mice daily oral doses of BPA over 3, 5, and 7 days. They observed the same meiotic abnormalities in these mice, although at lower levels (**FIGURE 2**). The mice given BPA for 7 days were most severely affected.

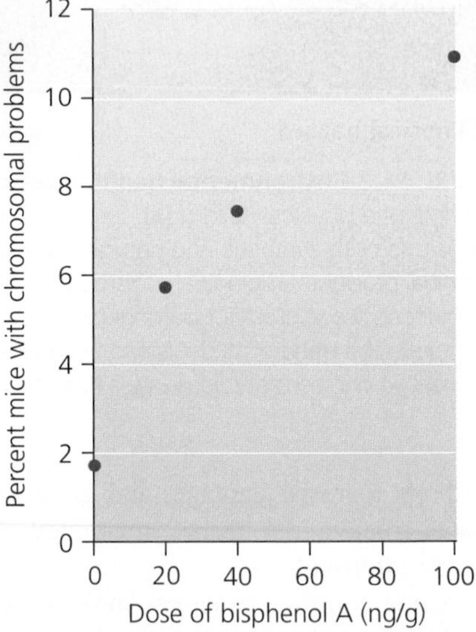

FIGURE 2 In this dose-response experiment, the percentage of mice showing chromosomal problems during cell division rose with increasing dose of bisphenol A. In the United States and Europe, regulators have set safe intake levels for people at a dose of 50 ng/g of body weight per day. *Data from Hunt, P.A., et al., 2003. Bisphenol A exposure causes meiotic aneuploidy in the female mouse.* Current Biology *13: 546–553.*

DATA Q Using the figure, predict the percent of mice in the study that would likely suffer chromosomal problems when exposed to a bisphenol A dosage of 70 ng/g. What would this percentage be?

Disease is a major focus of environmental health

Among the hazards people face, disease stands preeminent. Despite all our technological advances, we still find ourselves battling disease, which causes the vast majority of human deaths worldwide (**FIGURE 14.3a**). Major killers such as cancer, heart disease, and respiratory disorders have some genetic basis, but they are influenced by environmental factors. For instance, whether a person develops asthma depends not only on his or her genes, but also on environmental conditions. Studies have shown that pollutants from fossil fuel combustion worsen asthma, and children raised on farms suffer less asthma than children raised in cities. Malnutrition (p. 246),

Published in 2003 in the journal *Current Biology*, Hunt's findings set off a new wave of concern over the safety of bisphenol A. The findings were disturbing because sex cells of mice and of people divide and function in similar ways. "We have observed meiotic defects in mice at exposure levels close to or even below those considered 'safe' for humans," the research paper stated. "Clearly, the possibility that BPA exposure increases the likelihood of genetically abnormal offspring is too serious to be dismissed without extensive further study."

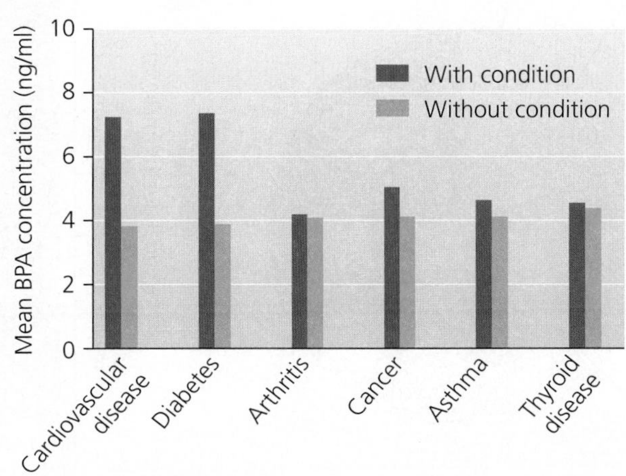

FIGURE 3 In a 2008 epidemiology study, average bisphenol A concentrations were significantly higher for Americans with diabetes and cardiovascular disease. Error bars are 95% confidence intervals. *Adapted from Lang, I.A., et al., 2008. Association of urinary bisphenol A concentration with medical disorders and laboratory abnormalities in adults.* JAMA *300: 1303–1310. Used by permission of the American Medical Association.*

Since that time, dozens of other studies of BPA at low doses have documented harmful effects in lab animals, including reproductive disorders related to estrogen mimicry but also other maladies ranging from thyroid problems to liver damage to elevated anxiety. Scientist Frederick vom Saal, whose research in 1997 had shown the first evidence for BPA's effects, said in 2007, "This chemical is harming snails, insects, lobsters, fish, frogs, reptiles, birds, and rats, and the chemical industry is telling people that because you're human, unless there's human data, you can feel completely safe."

Vom Saal did not have to wait long for the first human study to appear. In 2008, the *Journal of the American Medical Association* published research led by Iain Lang and David Melzer of the Peninsula Medical School, Exeter, U.K. Lang and Melzer's team took an epidemiological approach (pp. 375–376) to assess BPA's possible effects on people. Using data from 1455 participants in the U.S. government's latest National Health and Nutrition Examination Survey, the researchers got a representative sample of adults in the U.S. population. After controlling the data for race/ethnicity, education, income, smoking, body mass, and other variables, they tested for statistical correlations between a series of major health disorders and the concentration of BPA in people's urine.

These researchers' analyses showed that Americans with high BPA concentrations showed high rates of diabetes and cardiovascular disease (**FIGURE 3**), as well as abnormal concentrations of three liver enzymes. The team found no association with a number of other conditions such as cancer, stroke, arthritis, thyroid disease, and respiratory diseases. The researchers also explored correlations with other estrogenic compounds and found that these did not show the associations that BPA showed.

Previous studies of the mechanisms by which BPA acts in cell cultures and in rodents' bodies helped explain how and why BPA might affect liver enzymes and diabetes. However, the reasons for cardiovascular effects remain unclear.

This first direct indication of human health impacts from BPA was a correlative study that does not establish causation. To demonstrate that BPA actually causes the observed effects, researchers would need to track people with low and high BPA levels for years, predict who would most likely get sick, and test these predictions with future data. It will take many years to complete such long-term studies. In the meantime, more and more scientists are urging regulators to restrict BPA based on the evidence already at hand. ■

poverty, and poor hygiene can each foster various illnesses. Lifestyle choices also matter: Smoking can lead to lung cancer, and lack of exercise can lead to heart disease.

Over half the world's deaths result from noninfectious diseases, such as cancer and heart disease, and infectious diseases account for almost 1 of every 4 deaths that occur each year—nearly 15 million people worldwide (**FIGURE 14.3b**). Infectious disease is a greater problem in developing countries, where it accounts for close to half of all deaths. Infectious disease causes many fewer deaths in developed nations because their wealth allows their citizens better public sanitation and hygiene (which reduce the chance of contracting an infectious disease) and access to medicine (which enables people to receive treatment if they contract a disease).

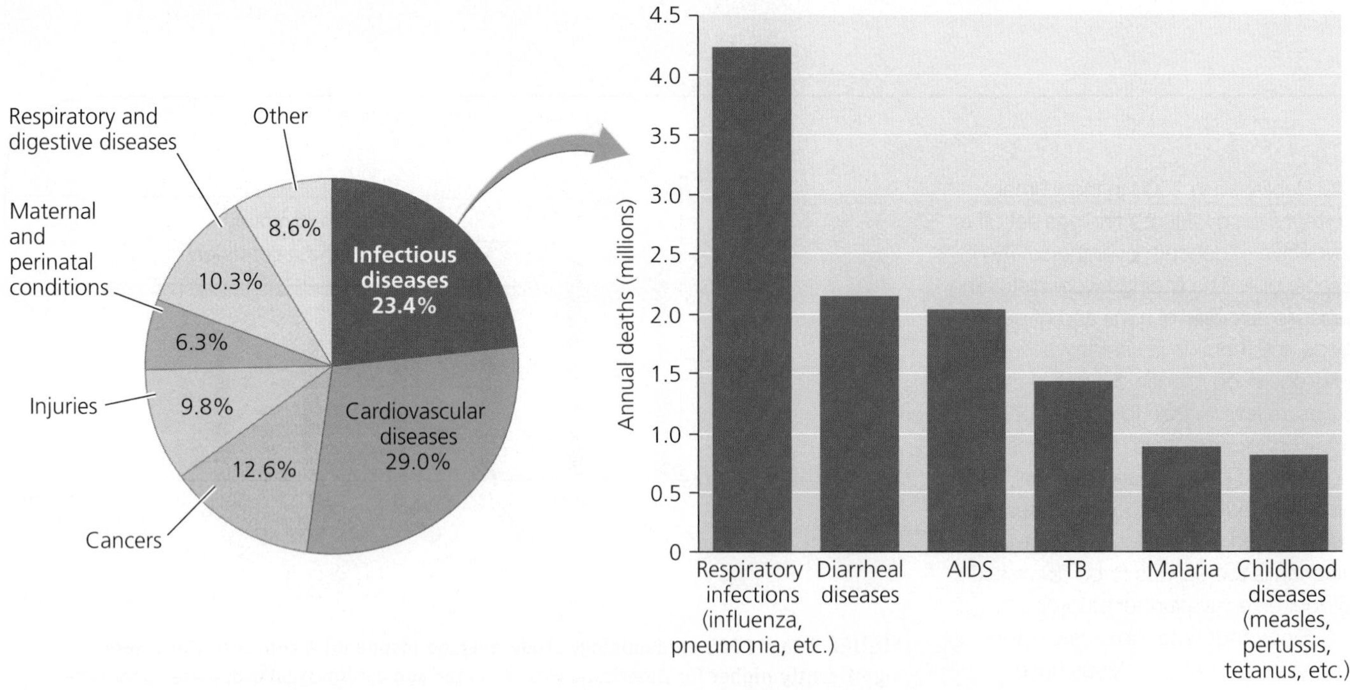

FIGURE 14.3 Infectious diseases are the second-leading cause of death worldwide. Six types of diseases account for 80% of all deaths from infectious disease. *Data from World Health Organization, 2009.* World health statistics 2009. *WHO, Geneva, Switzerland.*

In the United States and other developed nations, lifestyle trends are altering the prevalence of noninfectious disease in ways both good and bad. In the last two decades, the percentage of Americans who smoke cigarettes has dropped by 42%. But as we exercise less and eat fattier diets, the percentage of Americans who suffer obesity has more than doubled (**FIGURE 14.4**). Surprisingly, elevated levels of BPA have been linked to obesity. Studies have found that BPA exposure in mice elevates the production of fat cells in developing embryos and leads to the storage of more fat in organisms, even if they do not eat more. BPA exposure in mice has also been found to cause blood insulin levels to be twice that of unexposed mice. This excess of insulin can, over time, desensitize the body to insulin, leading to weight gain and type 2 diabetes.

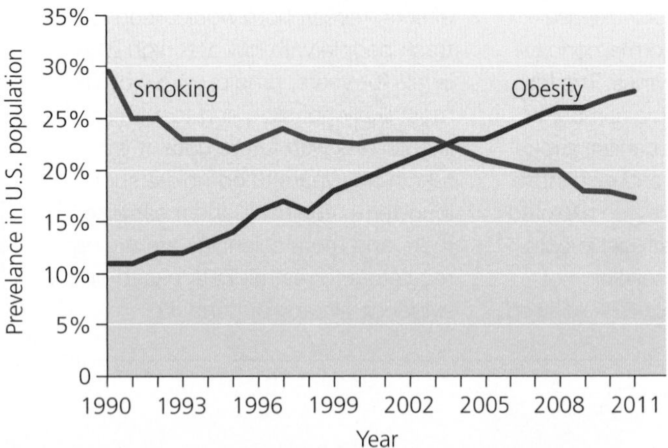

FIGURE 14.4 The prevalence of smoking has decreased in the United States in the past 20 years, but obesity is on the rise. *Data from United Health Foundation, 2011.* America's health rankings, 2011 edition. *Minnetonka, MN.*

DATA Q The population of the United States in mid-2011 was approximately 312 million people. Using the figure, estimate the number of Americans in 2011 who were (a) obese and (b) smokers.

Although infectious disease accounts for fewer deaths than noninfectious disease, infectious disease robs society of more years of human life, because it tends to strike people at all ages, including the very young. The World Health Organization (WHO) estimates that of all the years of life lost to deaths each year, infectious disease takes 51%, whereas noninfectious disease takes 34%. Decades of public health efforts have lessened the impacts of infectious disease and even have eradicated some diseases—yet other diseases are posing new challenges. Some diseases, such as acquired immunodeficiency syndrome (AIDS), continue to spread globally despite concerted efforts to stop them (pp. 208–209). Others, such as tuberculosis and strains of malaria, are evolving resistance to our antibiotics, in the same way that pests evolve resistance to our pesticides (p. 255).

Social and environmental factors can influence the spread of infectious disease

In our world of global mobility and dense human populations, novel diseases (and new strains of old diseases) that emerge in one location are more likely to spread to other locations, even worldwide. A pathogen can now hop continents in a matter of hours by airplane in its human host. As a result, some diseases have moved into new areas.

Recent examples include severe acute respiratory syndrome (SARS) in 2003, the H5N1 avian flu ("bird flu") starting in 2004, and the H1N1 swine flu that spread across the globe in 2009–2010 (**FIGURE 14.5**). Diseases like influenza,

FIGURE 14.5 Disease can spread rapidly in our highly mobile, internationalized world. The H1N1 swine flu outbreak in 2009–2010 showed how a strain of a fast-evolving infectious disease can quickly spread around the globe. As of 2013, over 18,000 people had died of this flu.

whose pathogens evolve rapidly, give rise to a variety of strains, making it more likely that one may turn exceedingly dangerous and threaten a global pandemic.

In addition to natural strains of diseases, there is growing concern over bioterrorism, the intentional genetic manipulation of a pathogenic organism to increase its virulence and/or transmission between humans. This issue received attention in 2012 when, after much debate in the scientific community and security agencies, the journal *Nature* published a study describing how scientists genetically engineered a strain of H5N1 avian flu to make it more transmissible through the air. Critics argued that the journal article provided a "recipe" for creating a bioterror weapon, but supporters contended the information would help governments track and better contain virulent strains of H5N1 by providing information about the genetic factors in the virus that promote airborne transmission.

The changes we cause to our environment can also cause diseases to spread. Human-induced global warming (Chapter 18) is causing tropical diseases such as malaria, dengue, cholera, and yellow fever to begin expanding into the temperate zones. And habitat alteration can affect the abundance, distribution, and movement of certain disease vectors.

To predict and prevent infectious disease, environmental health experts assess the complicated relationships among technology, land use, and ecology. Malaria, an infectious disease that claims an estimated 650,000 lives each year, provides an example. The microscopic protists (four species of *Plasmodium*) that cause malaria depend on mosquitoes as a vector. These protists can sexually reproduce only within a mosquito, and it is the mosquito that injects the protists into a human or other host. Thus, the primary mode to control malaria has been to use insecticides such as dichlorodiphenyl-trichloroethane (DDT) to kill mosquitoes. People using insecticides and draining wetlands in broad-scale eradication projects have eliminated malaria from large areas of the temperate world where it used to occur, such as the southern United States. However, human land disturbance that creates pools of standing water in formerly well-drained areas can boost mosquito populations and allow malaria to reinvade.

We are fighting disease with diverse approaches

We have many ways to fight disease, but one of the most effective is to improve the basic living conditions of the world's poor. Besides providing food security (p. 246), this means ensuring access to safe drinking water and improving sanitation by minimizing exposure to human waste, garbage, and wastewater. In recent years, we have made slow but steady progress in providing adequate drinking water and sanitation to the world's people (**FIGURE 14.6**), which helps reduce the incidence of diseases, such as cholera and dysentery, that are spread through drinking water contaminated with human or animal feces.

Another important pursuit is to expand access to health care. In developing nations, this includes opening clinics, immunizing children against diseases, providing prenatal and postnatal care for mothers and babies, and making generic and inexpensive pharmaceuticals available.

Education campaigns play a vital role in rich and poor nations alike. Public service announcements and government

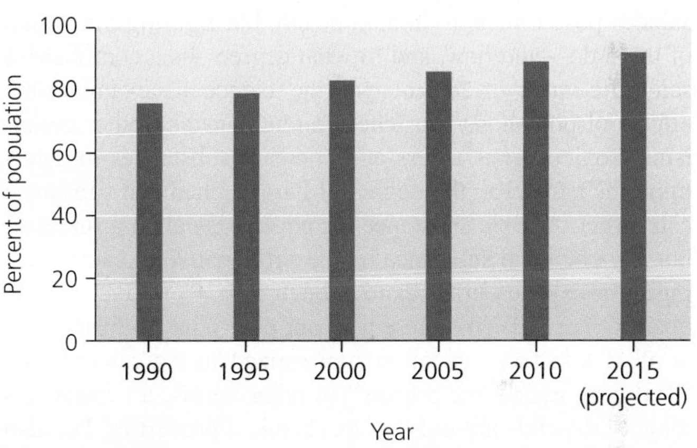

(a) Proportion of global population with access to improved drinking water

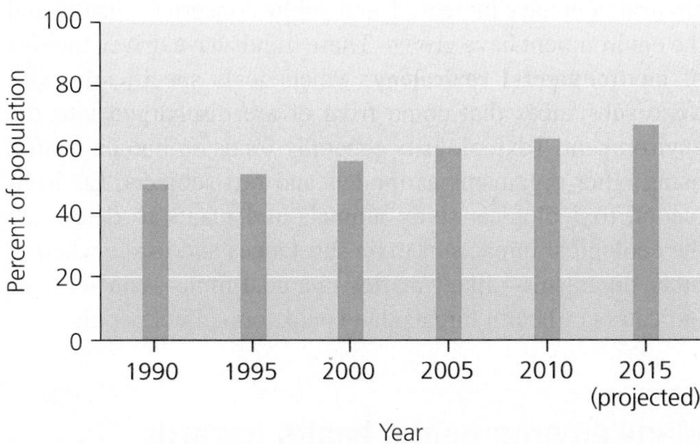

(b) Proportion of global population with access to improved sanitation

FIGURE 14.6 We are gradually providing access to sanitation and clean drinking water for the world's people. The percentage of the global population with access to clean or "improved" drinking water **(a)** and sanitary facilities **(b)** has increased in recent decades. *Data from World Health Organization, 2012.* World health statistics 2012. *WHO, Geneva, Switzerland.*

regulations on packaging and advertising have helped to decrease cigarette use, warn the public of the risks of drugs and alcohol, and advise us on issues of nutrition and exercise. Education on sex and reproductive health is helping women avoid unwanted pregnancies and is slowing population growth (pp. 203, 206), and the promotion of condom use is helping to slow the spread of HIV/AIDS (p. 209).

Such efforts are being spearheaded internationally by the United Nations, the World Health Organization, the U.S. Agency for International Development, and nongovernmental organizations and funding agencies. Private foundations, such as the Bill and Melinda Gates Foundation, have been instrumental in supporting health initiatives around the world. The Gates Foundation alone has donated over $15 billion since 1994 to global health programs promoting disease prevention, prenatal care for pregnant women, and enhanced food security.

Toxicology is the study of chemical hazards

Although most indicators of human health are improving as the world's wealth increases, our modern society is exposing us to more and more synthetic chemicals. Some of these substances pose threats to human health, but figuring out which of them do—and how, and to what degree—is a complicated scientific endeavor. **Toxicology** is the science that examines the effects of poisonous substances on humans and other organisms. Toxicologists assess and compare substances to determine their **toxicity**, the degree of harm a chemical substance can inflict. A toxic substance, or poison, is called a **toxicant**, but any chemical substance may exert negative impacts if we ingest or expose ourselves to enough of it. Conversely, if the quantity is small enough, a toxicant may pose no health risk at all. These facts are often summarized in the catchphrase, "The dose makes the poison." In other words, a substance's toxicity depends not only on its chemical properties, but also on its quantity.

In recent decades, our ability to produce new chemicals has expanded, concentrations of chemical contaminants in the environment have increased, and public concern for health and the environment have grown. These trends have driven the rise of **environmental toxicology**, which deals specifically with toxic substances that come from or are discharged into the environment. Toxicologists generally focus on human health, using other organisms as models and test subjects. Environmental toxicologists study animals and plants to determine the ecological impacts of toxic substances and to see whether other organisms—like canaries in a coal mine—can serve as indicators of health threats that could soon affect people.

Many environmental health hazards exist indoors

Modern Americans spend roughly 90% of their lives indoors. Unfortunately, the spaces inside our homes and workplaces, just like the outdoors, can be rife with environmental hazards (TABLE 14.1; see also pp. 475–478).

Cigarette smoke and radon are leading indoor hazards (p. 476) and are the top two causes of lung cancer in developed nations. *Radon* is a highly toxic radioactive gas that is colorless and undetectable without specialized kits. Radon seeps up from the ground in areas with certain types of bedrock and can accumulate in basements and homes with poor air circulation (FIGURE 14.7). The EPA estimates that slightly less than 1 person in 1000 may contract lung cancer as a result of a lifetime of radon exposure at average levels for U.S. homes, with risks from radon varying according to a region's underlying geology (see Figure. 17.33, p. 476). Another indoor hazard is asbestos. There are several types of *asbestos*, each of which is a mineral that forms long, thin, microscopic fibers. This

TABLE 14.1 Selected Environmental Hazards

Selected Environmental Hazards
Outdoor Air
• Chemicals from automotive exhaust
• Chemicals from industrial pollution
• Photochemical smog (p. 465)
• Pesticide drift
• Dust and particulate matter
Water
• Pesticide and herbicide runoff
• Nitrates and fertilizer runoff
• Mercury, arsenic, and other heavy metals in groundwater and surface water
Food
• Natural toxins
• Pesticide and herbicide residues
Indoors
• Smoking and secondhand smoke
• Radon
• Asbestos
• Lead in paint and pipes
• Toxicants in plastics and consumer products (bisphenol A, PBDEs, phthalates, etc.)
• Toxic compounds produced by mold
• Dust and particulate matter

FIGURE 14.7 People can determine their exposure to radon gas with in-home testing. Air samples are collected in specialized collectors like the one shown, and then mailed to a laboratory for analysis.

structure allows asbestos to trap heat, muffle sound, and resist fire. Because of these qualities, asbestos was used widely as insulation in buildings, as well as in other products. Unfortunately, its fibrous structure also makes asbestos dangerous when inhaled. When asbestos gets lodged in lung tissue, the body produces acid in an attempt to eliminate it. The acid scars the lung tissue but does little to dislodge or dissolve the asbestos. Within a few decades, the scarred lungs may cease to function—a disorder called *asbestosis*. Asbestos can also cause certain types of lung cancer. Because of these risks, asbestos has been removed from many schools and offices. However, the removal process can release some asbestos into the air, increasing people's exposure, so in some cases the best approach is to encase the material in place.

Lead poisoning is another indoor health hazard. When ingested, lead, a heavy metal, can cause damage to the brain, liver, kidney, and stomach; learning problems and behavioral abnormalities; anemia; hearing loss; and even death. Lead poisoning among U.S. children has greatly declined in recent years as a result of education campaigns and government regulations that phased out lead-based paints and leaded gasoline starting in the 1970s. Today lead poisoning can result from drinking water that has passed through the lead pipes common in older homes or from ingesting or inhaling lead-containing dust produced by the slow wearing-away of leaded paint.

We have succeeded in removing significant sources of lead exposure in the United States, but modern Americans are still exposed to elevated levels of lead. For example, millions of children's toys and other items exported from China were found to contain lead-based paint in recent years. The resulting consumer outcry in the United States forced recalls of these items, and China eventually agreed to limit and monitor the use of lead-based paint in its manufacturing.

In 2012, the Centers for Disease Control (CDC) adopted an advisory panel's recommendation and lowered the danger threshold for lead levels in children's blood from 10 micrograms/deciliter to 5 micrograms/deciliter. This decision was based on numerous studies that showed adverse effects of lead on neurological development at blood lead levels below the previous standard.

One recently recognized hazard is a group of chemicals known as *polybrominated diphenyl ethers (PBDEs)*. These compounds provide fire-retardant properties and are used in a diverse array of consumer products, including computers, televisions, plastics, and furniture. They are emitted during production and disposal of products and may also release into the air at very slow rates throughout the lifetime of products. These chemicals persist and accumulate in living tissue, and their abundance in the environment and in people in the United States is doubling every few years.

Like bisphenol A, PBDEs appear to act as hormone disruptors; lab testing with animals shows them to affect thyroid hormones. Animal testing also suggests that PBDEs affect the development of the brain and nervous system and may cause cancer. Concern about PBDEs rose after a study showed that concentrations in the breast milk of Swedish mothers had increased exponentially from 1972 to 1997. U.S. studies also show rising concentrations in breast milk. The European Union decided in 2003 to ban PBDEs, and industries in Europe phased them out. As a result, concentrations in breast milk of European mothers have fallen substantially. In the United States, however, there has so far been little movement to address the issue. The dangers posed by fire retardants such as PBDEs have caused some to question the stringent flammability standards that often make the use of such chemicals necessary. As stated by Linda Birnbaum of the National Institute of Environmental Health Sciences, "I don't question the need for flame retardants in airplanes, but do we need them in nursing pillows and babies' strollers?"

Risks must be balanced against rewards

The job of toxicologists and other scientists who study environmental health hazards is to learn as much as they can about the hazards, but the rest of us need to take this information and weigh it against any benefits we obtain from exposing ourselves to the hazards. With most hazards, there is some tradeoff between risk and reward, and we must judge as best we can how these compare. In regard to bisphenol A, its usefulness for many purposes means that despite its health risks, we may as a society choose to continue using it. The availability of safer and affordable alternatives is important in such decisions. Industry is finding replacements for BPA polymers in baby bottles and water containers, but until a replacement is found for it as an epoxy liner for food cans, it will likely continue to serve this function.

As we review the impacts of toxic substances throughout this chapter, it is important to keep in mind that artificially produced chemicals have played a crucial role in giving us the standard of living we enjoy today. These chemicals have helped create the industrial agriculture that produces our food, the medical advances that protect our health and prolong our lives, and many of the modern materials and conveniences we use every day. It is appropriate to remember these benefits as we examine some of the unfortunate side effects of these advances and as we search for better alternatives.

Toxic Substances and Their Effects on Organisms

Our environment contains countless natural substances that may pose health risks. These include petroleum, oozing naturally from the ground; radon gas, seeping up from bedrock; and **toxins**, toxic chemicals manufactured in the tissues of living organisms. For example, toxins can be chemicals that plants use to ward off herbivores or that insects use to defend themselves from predators. In addition, we are exposed to many synthetic (artificial, or human-made) chemicals, some of which also have toxic properties.

Synthetic chemicals are all around us—and in us

Tens of thousands of synthetic chemicals have been manufactured (TABLE 14.2), and synthetic chemicals surround us in our daily lives. Each year in the United States, we manufacture or import 113 kg (250 lb) of chemical substances for every man, woman, and child. Many of these substances find

TABLE 14.2 Estimated Numbers of Chemicals in Commercial Substances	
TYPE OF CHEMICAL	**ESTIMATED NUMBER**
Chemicals in commerce	100,000
Industrial chemicals	72,000
New chemicals introduced per year	2000
Pesticides (21,000 products)	600
Food additives	8700
Cosmetic ingredients (40,000 products)	7500
Human pharmaceuticals	3300

Data are for the 1990s, from Harrison, P., and F. Pearce, 2000. AAAS atlas of population and environment. *Berkeley, CA: University of California Press.*

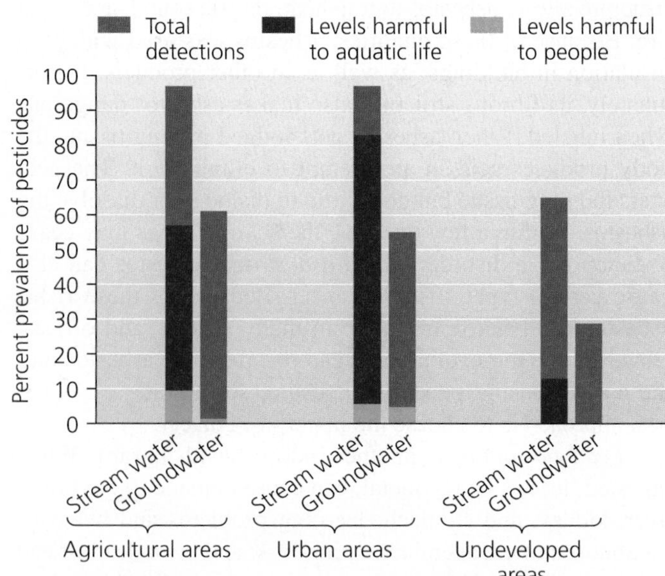

FIGURE 14.8 Nearly all U.S. streams and most aquifers in agricultural and urban areas contain pesticides throughout the year. Fewer than 10% of tested samples violate human health standards, but most violate standards for aquatic life. *Data from Gilliom, Robert J., et al., 2006.* Pesticides in the nation's streams and ground water, 1992–2001. *Circular 1291, National Water-Quality Assessment Program, U.S. Geological Survey.*

their way into soil, air, and water, as revealed by researchers who monitor environmental quality. For instance, scientists at the U.S. Geological Survey's National Water-Quality Assessment Program (NAWQA) have carried out systematic surveys for synthetic chemicals in U.S waterways and aquifers since the 1980s. A 2002 study found that 80% of U.S. streams contain at least trace amounts of 82 wastewater contaminants, including antibiotics, detergents, drugs, steroids, plasticizers, disinfectants, solvents, perfumes, and other substances. A 2006 study of groundwater detected 42 volatile organic compounds (VOCs, p. 459) in 18% of wells and 92% of aquifers tested throughout the nation, although fewer than 2% of samples violated federal health standards for drinking water. (VOCs are emitted from products such as gasoline, paints, and plastics, and they come from many sources, including urban runoff; engine exhaust; industrial emissions; wastewater; and leaky storage tanks, landfills, and septic systems.)

The pesticides we use to kill insects and weeds (p. 255) on farms, lawns, and golf courses are some of the most widespread synthetic chemicals. A 2006 NAWQA study concluded that pesticides are regularly present in streams and groundwater nationwide, finding traces of at least one pesticide in every stream that was tested. The data showed that concentrations were seldom high enough to pose health risks to people, but they were often high enough to affect aquatic life or fish-eating animals (**FIGURE 14.8**). Pesticide contamination is most severe in the farming states of the Midwest and Great Plains.

Synthetic chemicals are in all of our bodies

As a result of all this exposure, every one of us carries traces of hundreds of industrial chemicals in our bodies. The U.S. government's latest National Health and Nutrition Examination Survey (the one that found 93% of Americans showing traces of BPA in their urine; p. 359) gathered data on 148 foreign compounds in Americans' bodies. Among these were several toxic persistent organic pollutants restricted by international treaty (pp. 383–384). Depending on the pollutant, these were detected in 41% to 100% of the people tested. Smaller-scale surveys have found similar results.

In 2009, science writer Arianne Cohen decided to get her own body surveyed to find out what chemicals were present inside her. She worked with researchers, shelled out over $4000 to undergo a battery of tests, and then wrote up the results in the December 2009 issue of *Popular Science* magazine. The verdict: Her body contained BPA, dioxins, and other persistent pollutants, nitrates from food, chemicals from plastics, and plenty more. This wasn't surprising, she pointed out, because we encounter countless chemicals all day, from shampoo in our morning shower to packaging and nonstick pans at meals to pesticides on our lawns in the afternoon to flame retardants on our sheets at bedtime.

Our exposure to synthetic chemicals begins in the womb, as substances our mothers ingested while pregnant were transferred to us. A 2009 study by the nonprofit Environmental Working Group found 232 chemicals in the umbilical cords of 10 newborn babies it tested. Nine of the 10 umbilical cords contained BPA, leading researchers to note that we are born "pre-polluted."

All this should not necessarily be cause for alarm. Not all synthetic chemicals pose health risks, and relatively few are known with certainty to be toxic. However, of the roughly 100,000 synthetic chemicals on the market today, very few have been thoroughly tested. For the vast majority, we simply do not know what effects, if any, they may have.

Silent Spring began the public debate over synthetic chemicals

It was not until the 1960s that people began to seriously consider the risks of exposure to pesticides. One key event in this growing awareness was the publication of Rachel Carson's 1962 book *Silent Spring*. At the time, large amounts of pesticides that were virtually untested for health effects were indiscriminately sprayed over residential neighborhoods and public areas, on an assumption that the chemicals would do no harm to people. Carson brought together a diverse collection of scientific studies, medical case histories, and other data to demonstrate that the insecticide DDT in particular, and artificial pesticides in general, were hazardous to people, wildlife, and ecosystems. Most consumers had no idea that the store-bought chemicals they used in their houses, gardens, and crops might be toxic.

The chemical industry challenged Carson's book vigorously, attempting to discredit the author's science and personal reputation. Carson suffered from cancer as she finished *Silent Spring*, (p. 174) and she lived only briefly after its publication. However, the book was a best-seller and helped generate significant social change in views and actions toward the environment. The use of DDT was banned in the United States in 1973 and is now illegal in a number of nations.

U.S. chemical companies still manufacture and export DDT, because developing countries with tropical climates use DDT to control disease vectors, such as mosquitoes that transmit malaria. In these countries, malaria represents a greater health threat than do the toxic effects of the pesticide. New technologies are promising to reduce the need for pesticides to control malaria, however. In 2012, researchers reported that they genetically modified a type of bacteria found in the digestive tract of mosquitoes so that it produced a protein that impairs the hatching of the malarial parasite from its egg sacs in the mosquito's gut. The bacteria were introduced into mosquitoes by mixing them in a sugar solution for mosquitoes to drink. While 90% of mosquitoes that didn't drink the solution were later found to contain malarial parasites, only 20% of the mosquitoes that drank the solution later contained parasites. The use of genetically modified organisms is not itself without risk (pp. 261–268), but such research may provide non-chemical options for reducing human mortality from malaria and speed the phase-out of DDT around the world.

WEIGHING THE ISSUES

A CIRCLE OF POISON? Although the United States has banned the use of DDT, U.S. companies still manufacture and export the compound to developing nations. Thus, it is possible that pesticide-laden food can be imported back into the United States in what has been called a "circle of poison." How do you feel about this? Is it unethical for one country to sell to others a substance that it has deemed toxic? Or would it be unethical for the United States *not* to sell DDT to African nations if they desire it for controlling malaria?

Not all toxic substances are synthetic, and not all synthetic chemicals are toxic

Although many toxicologists focus on synthetic chemicals, toxic substances also exist naturally in the environment around us and in the foods we eat. Thus, it would be a mistake to assume that all artificial substances are unhealthy and that all natural substances are healthy. In fact, the plants and animals we eat contain many chemicals that can cause us harm. Recall that plants produce toxins to ward off animals that eat them. In domesticating crop plants, we have selected (p. 52) for strains with reduced toxin content, but we have not eliminated these dangers. Furthermore, when we consume animal meat, we ingest toxins the animals obtained from plants or animals they ate. Scientists are actively debating just how much risk natural toxicants pose, and it is clear that more research is required on these questions.

Toxic substances come in different types

Toxicants can be classified based on their particular effects on health. The best-known are **carcinogens**, which are substances or types of radiation that cause cancer. In cancer, malignant cells grow uncontrollably, creating tumors, damaging the body, and often leading to death. Cancer frequently has a genetic component, but a wide variety of environmental factors are thought to raise the risk of cancer. Indeed, in 2010 the President's Cancer Panel concluded that the prevalence of environmentally induced cancer has been "grossly underestimated." In our society today, the greatest number of cancer cases is thought to result from carcinogens contained in cigarette smoke. Polycyclic aromatic hydrocarbons (PAHs; p. 29) comprise some of the carcinogens found in cigarette smoke. PAHs also occur in charred meats and are released from the combustion of coal, oil, and natural gas.

Carcinogens can be difficult to identify because there may be a long lag time between exposure to the agent and the detectable onset of cancer—up to 15–30 years in the case of cigarette smoke. Moreover, as with all risks, only a portion of people exposed to a carcinogen will eventually get cancer. Cancer is a leading cause of death that kills millions and leaves few families untouched. Two of every five Americans are diagnosed with cancer at some time in their lives, and one of every five dies from it. Thus, the study of carcinogens has played a large role in shaping the way that toxicologists pursue their work.

Mutagens are substances that cause genetic mutations in the DNA of organisms (p. 29). Although most mutations have little or no effect, some can lead to severe problems, including cancer and other disorders. If mutations occur in an individual's sperm or egg cells, then the individual's offspring suffer the effects.

Chemicals that cause harm to the unborn are called **teratogens**. Teratogens that affect development of human embryos in the womb can cause birth defects. One example involves the drug thalidomide, developed in the 1950s to aid sleeping and to prevent nausea during pregnancy. Tragically, the drug turned out to be a powerful teratogen, and its use caused birth defects in thousands of babies. Even a single dose during pregnancy could result in limb deformities and organ defects. Thalidomide was banned in the 1960s once scientists recognized its connection with birth defects. Ironically, today the drug shows promise in treating a wide range of diseases, including Alzheimer's disease, AIDS, and various types of cancer.

Other chemical toxicants known as **neurotoxins** assault the nervous system. Neurotoxins include venoms produced by animals such as snakes and stinging insects, heavy metals such as lead and mercury, pesticides, and some chemical weapons developed for use in war. A famous case of neurotoxin poisoning occurred in Japan, where a chemical factory dumped mercury waste into Minamata Bay between the 1930s and 1960s. Thousands of people there ate fish contaminated with the mercury and soon began suffering from slurred speech, loss of muscle control, sudden fits of laughter, and in some cases death. The company and the government eventually paid about $5000 in compensation to each poisoned resident.

The human immune system protects our bodies from disease. Some toxicants weaken the immune system, reducing the body's ability to defend itself against bacteria, viruses, allergy-causing agents, and other attackers. Others, called **allergens**, overactivate the immune system, causing an immune response when one is not necessary. One hypothesis for the increase in asthma in recent years is that allergenic synthetic chemicals are more prevalent in our environment. Allergens are not universally considered toxicants, however, because they affect some people but not others and because one's response does not necessarily correlate with the degree of exposure.

Pathway inhibitors are toxicants that interrupt vital biochemical processes in organisms by blocking one or more steps in important biochemical pathways. Rat poisons, for example, cause internal hemorrhaging in rodents by interfering with the biochemical pathways that create blood clotting proteins. Some herbicides, such as atrazine, kill plants by blocking steps in photosynthesis. Cyanide kills by interrupting chemical pathways that produce energy in mitochondria, thereby depriving cells of life-sustaining energy.

Most recently, scientists have recognized **endocrine disruptors**, toxicants that interfere with the *endocrine system.* The endocrine system consists of chemical messengers (*hormones*) that travel through the bloodstream at extremely low concentrations and have many vital functions. They stimulate growth, development, and sexual maturity, and they regulate brain function, appetite, sex drive, and many other aspects of our physiology and behavior. Some hormone-disrupting toxicants affect an animal's endocrine system by blocking the action of hormones or accelerating their breakdown. Others are so similar to certain hormones in their molecular structure and chemistry that they "mimic" the hormone by interacting with receptor molecules just as the actual hormone would (**FIGURE 14.9**).

Bisphenol A is one of many chemicals that appear to mimic the female sex hormone estrogen and bind to estrogen receptors. Many plastic products also contain another class of hormone-disrupting chemical, called *phthalates*. Used to soften plastics and enhance fragrances, phthalates are used widely in children's toys (**FIGURE 14.10a**), perfumes and cosmetics (**FIGURE 14.10b**), and other items. Health research on phthalates has linked them to birth defects, breast cancer, reduced sperm counts, and other reproductive effects. The European Union and nine other nations have banned phthalates, California and Washington enacted bans for children's toys, and the United States in 2008 banned six types of phthalates in toys. Still, across North America many routes of exposure remain.

It is important to note that any given substance may have multiple effects. For example, bisphenol A is primarily

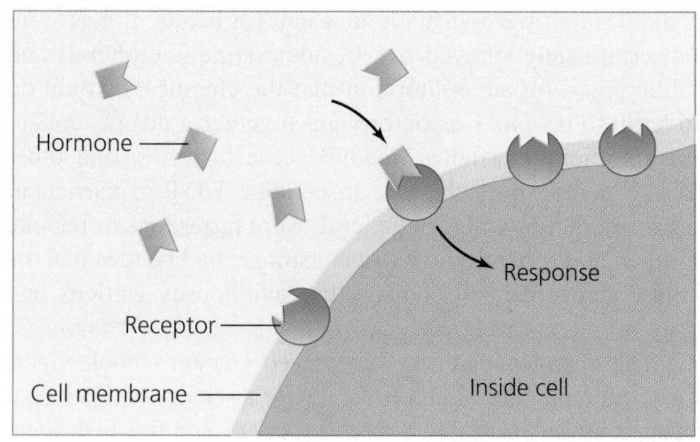

(a) Normal hormone binding

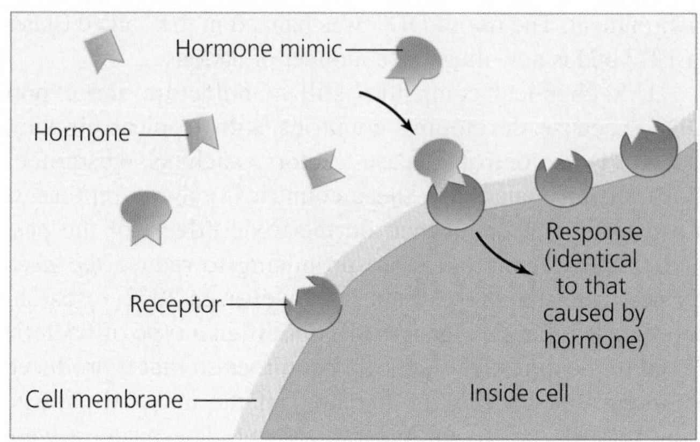

(b) Hormone mimicry

FIGURE 14.9 Many endocrine-disrupting substances mimic the structure of hormone molecules. Like a key similar enough to fit into another key's lock, the hormone mimic binds to a cellular receptor for the hormone, causing the cell to react as though it had encountered the hormone.

known as an endocrine disruptor, but research shows it to have mutagenic and teratogenic effects as well, and because studies link it to breast cancer and prostate cancer in lab animals, it also appears to act as a carcinogen.

Organisms have natural defenses against toxic substances

Although synthetic toxicants are new, organisms have been exposed to natural toxicants for millions of years. Mercury, cadmium, arsenic, and other harmful substances are found naturally in the environment. Some organisms produce biological toxins to avoid predators or capture prey. Examples include venom in poisonous snakes, toxins in sea urchins, and the natural insecticide pyrethrin found in chrysanthemums. These exposures have provided selection pressure (pp. 50–52) for protection from toxins, and over time, organisms able to tolerate these harmful substances have gained an evolutionary advantage.

Barriers such as skin, scales, feathers, and fur are the first line of defense against toxic substances because they help the body to resist uptake from the surrounding environment. However, toxicants can circumvent these barriers and enter the body from

(a) Exposure through toys

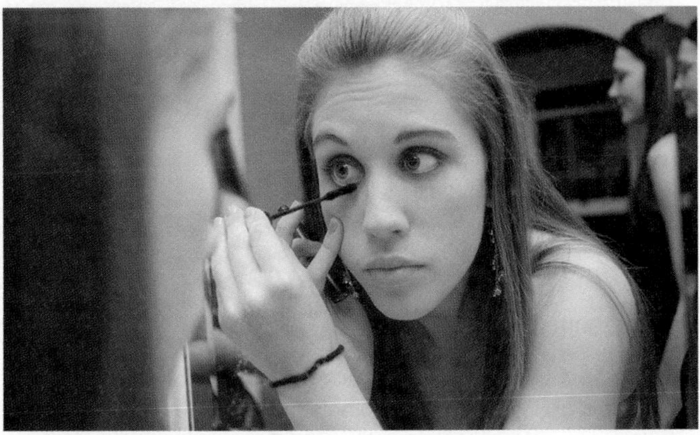

(b) Exposure through cosmetics

FIGURE 14.10 **Many soft plastic children's toys and many cosmetics contain phthalates, a hormone-disrupting chemical.** Banned in Europe, phthalates remain widespread in the United States.

vital activities such as eating, drinking, and breathing. Once in the organism, they are distributed widely by the circulatory and lymph systems in animals, and by the vascular system in plants.

Organisms possess biochemical pathways that use enzymes to detoxify harmful chemicals. Some pathways break down, or metabolize, toxic substances to render them inert. Other pathways make toxic substances water soluble so they are easier to excrete through the urinary system. In humans, many of these pathways are found in the liver, so this organ is disproportionately affected by intake of harmful substances such as excessive alcohol.

Some toxic substances cannot be effectively detoxified or made water soluble by detoxification enzymes. These chemicals are sequestered in fatty tissues and cell membranes to keep them away from vital organs. Heavy metals, dioxins, and some insecticides (including DDT) are stored in body tissue in this manner.

These defenses can protect organisms against low levels of some toxicants but can be overwhelmed if exposure exceeds critical levels. For other toxicants, harm occurs with any exposure if organisms have no defense against the substance. Defense mechanisms for natural toxins have evolved over millions of years. Organisms have not had long-term exposure to the synthetic chemicals that are so prevalent in today's environment, so the impacts of these toxic substances can be severe and unpredictable.

FAQ **Does exposure to a toxic substance cause genetic resistance to the substance?**

When a population of organisms is exposed to a toxicant, such as a pesticide, a few individuals often survive while the vast majority of the population is killed. These individuals survive because they possess genes (which others in the population do not) that code for enzymes that counteract the toxic properties of the toxicant. Because the effects of these genes are expressed only when the pesticide is applied, many people think the toxicant "creates" detoxification genes by mutating the DNA of a small number of individuals. This is not the case. The genes for detoxifying enzymes were present in the DNA of resistant individuals from birth, but their effects were seen only when pesticide exposure caused selective pressure (pp. 50–52) for resistance to the toxic substance.

Individuals vary in their responses to hazards

Some of the defenses described above have a genetic basis. As a result, individuals may respond quite differently to identical exposures to hazards because they happen to have different combinations of genes. Poorer health also makes an individual more sensitive to biological and chemical hazards. Sensitivity also can vary with sex, age, and weight. Because of their smaller size and rapidly developing organ systems, younger organisms (for example, fetuses, infants, and young children) tend to be much more sensitive to toxicants than are adults. Regulatory agencies such as the U.S. Environmental Protection Agency (EPA) typically set human chemical exposure standards for adults and extrapolate downward for infants and children. However, many scientists contend that these linear extrapolations often do not offer adequate protection to fetuses, infants, and children.

The type of exposure can affect the response

The risk posed by a hazard often varies according to whether a person experiences high exposure for short periods of time, known as **acute exposure,** or low exposure over long periods of time, known as **chronic exposure.** Incidences of acute exposure are easier to recognize, because they often stem from discrete events, such as accidental ingestion, an oil spill, a chemical spill, or a nuclear accident. Toxicity tests in laboratories generally reflect acute toxicity effects. However, chronic exposure is more common—and more difficult to detect and diagnose. Chronic exposure often affects organs gradually, as when smoking causes lung cancer, or when alcohol abuse leads to liver or kidney damage. Pesticide residues on food or arsenic in drinking water also pose chronic risk. Because of the long time periods involved, relationships between cause and effect may not be readily apparent.

Toxic Substances and Their Effects on Ecosystems

When toxicants concentrate in environments and harm the health of many individuals, populations (p. 49) of the affected species become smaller. This decline in population can then affect other species. For instance, species that are prey of the organism affected by toxicants could experience population growth because predation levels are lower. Predators of the poisoned species, however, would decline as their food source became less abundant. Cascading impacts can cause changes in the composition of the biological community (p. 60) and threaten ecosystem functioning. There are many ways toxicants can concentrate and persist in ecosystems and affect ecosystem services.

Airborne substances can travel widely

Toxic substances are released around the world from agricultural, industrial, and domestic activities and can sometimes be redistributed by air currents (Chapter 17), exerting impacts on ecosystems far from their emission site.

Because so many substances are carried by the wind, synthetic chemicals are ubiquitous worldwide, even in seemingly pristine areas. Scientists who travel to the most remote alpine lakes in the wilderness of British Columbia find them contaminated with industrial toxicants, such as polychlorinated biphenyls (PCBs), which are by-products of chemicals used in transformers and other electrical equipment. Earth's polar regions are particularly contaminated, because natural patterns of global atmospheric circulation (p. 455) tend to move airborne chemicals toward the poles (FIGURE 14.11). Thus, although we manufacture and apply synthetic substances mainly in temperate and tropical regions, contaminants are strikingly concentrated in the tissues of Arctic polar bears, Antarctic penguins, and people living in Greenland.

Effects can also occur over relatively shorter distances. Pesticides, for example, can be carried by air currents to sites far from agricultural fields in a process called pesticide drift. The Central Valley of California is the world's most productive agricultural region, but because it is naturally arid, food production depends on the intensive use of irrigation, fertilizers, and pesticides. The region's frequent winds often blow airborne pesticide spray—and dust particles containing pesticide residue—for long distances. In the mountains of the Sierra Nevada, research has associated pesticide drift from the Central Valley with population declines in four species of frogs. Families living in towns in the Central Valley suffer health impacts, and advocates for farm workers maintain that hundreds of thousands of the state's residents are at risk.

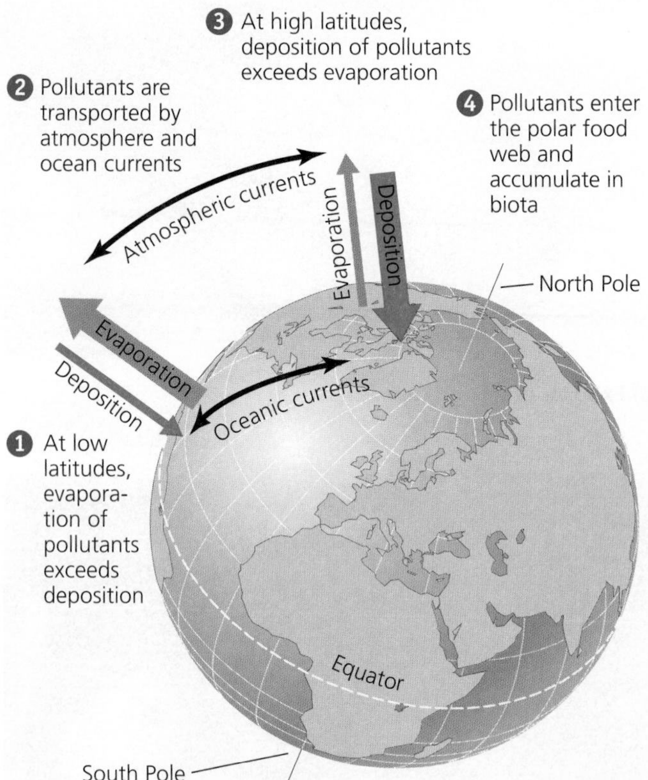

FIGURE 14.11 **Air and water currents direct pollutants to the poles.** In global distillation, pollutants that evaporate and rise high into the atmosphere at lower latitudes 1, or are deposited in the ocean, are carried toward the poles 2 by atmospheric currents of air and oceanic currents of water. This process concentrates pollutants near the poles 3 and causes elevated exposure to toxic substances in polar organisms 4.

Toxic substances may concentrate in water

Toxic substances are not evenly distributed in the environment, and they move about in specific ways (FIGURE 14.12). Water running off from land often transports toxicants from large areas and concentrates them in small volumes of surface water. The NAWQA findings on water quality reflect this concentrating effect. Wastewater treatment plants also add toxins, pharmaceuticals, and detoxification products from humans to waterways. If chemicals persist in soil, they can leach into groundwater and contaminate drinking water supplies.

Many chemicals are soluble in water and enter organisms' tissues through drinking or absorption. For this reason, aquatic animals such as fish, frogs, and stream invertebrates are effective indicators of pollution. When aquatic organisms become sick, we can take it as an early warning that something is amiss. If scientists find low concentrations of pesticides harming frogs, fish, and invertebrates, they view this as a warning that people could be next. The contaminants that wash into streams and rivers also flow and seep into the water we drink and drift through the air we breathe. Once concentrated in waters, toxic substances can move long distances and affect a variety of ecosystems (pp. 391–392).

Some toxicants persist in the environment

Once a toxic substance arrives somewhere, it may degrade quickly and become harmless, or it may remain unaltered and persist for many months, years, or decades. The rate at which a given substance degrades depends on its chemistry and on factors such as temperature, moisture, and sun exposure. The *Bt* toxin (pp. 256, 262) used in biocontrol and genetically modified crops has a very short persistence time, whereas

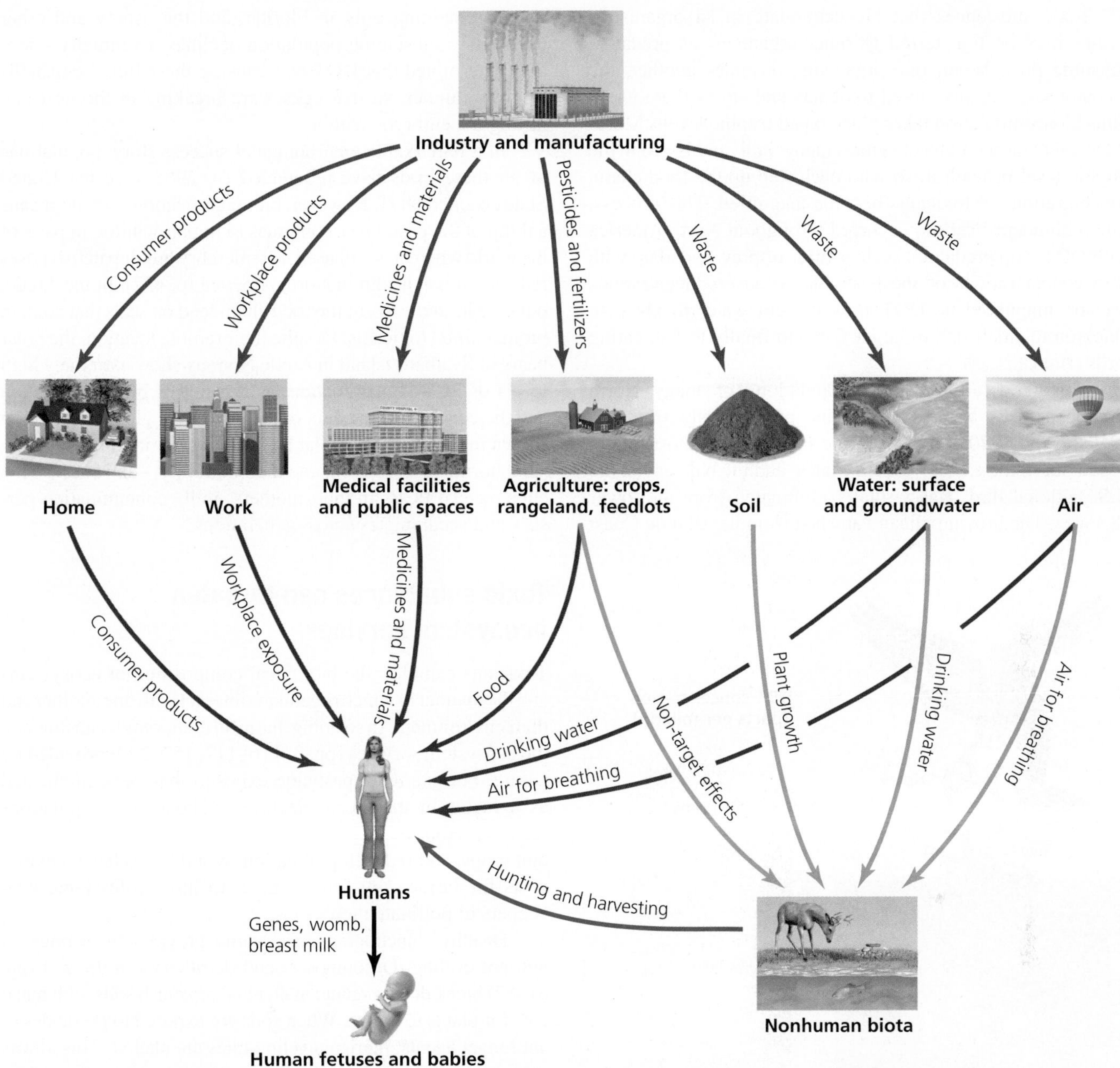

FIGURE 14.12 Synthetic chemicals take many routes in traveling through the environment. People take in only a tiny proportion of these compounds, and many compounds are harmless. However, people receive small amounts of toxicants from many sources, and developing fetuses and babies are particularly sensitive.

chemicals such as DDT and PCBs persist for decades. Atrazine, our most widely used herbicide, is highly variable in its persistence, depending on environmental conditions.

Persistent synthetic chemicals exist in our environment today because we have designed them to persist. The synthetic chemicals used in plastics, for instance, are used precisely because they resist breakdown. Sooner or later, however, most toxicants degrade into simpler compounds called *breakdown products*. Often these are less harmful than the original substance, but sometimes they are just as toxic as the original chemical, or more so. For instance, DDT breaks down into DDE, a highly persistent and toxic compound in its own right. Atrazine produces a large number of breakdown products whose effects have not been fully studied.

Toxicants may accumulate and move up the food chain

Of the toxic substances that organisms absorb, breathe, or consume, some are quickly excreted, and some are degraded into harmless breakdown products. Others persist intact in the body. Substances that are fat-soluble or oil-soluble (including organic compounds such as DDT and DDE) are absorbed and stored in fatty tissues. Substances such as methylmercury (CH_3Hg^+) may be stored in muscle tissue. Such persistent toxicants accumulate in an animal's body in a process termed **bioaccumulation,** such that the animal's tissues have a greater concentration of the substance than exists in the surrounding environment.

Toxic substances that bioaccumulate in an organism's tissues may be transferred to other organisms as predators consume prey. When one organism consumes another, the predator takes in any stored toxicants and stores them itself. Thus bioaccumulation takes place on all trophic levels. Moreover, each individual consumes many individuals from the trophic level beneath it, so with each step up the food chain, concentrations of toxicants become magnified. This process, called **biomagnification,** occurred throughout North America with DDT. Top predators, such as birds of prey, ended up with high concentrations of the pesticide because concentrations became magnified as DDT moved from water to algae to plankton to small fish to larger fish and finally to fish-eating birds (FIGURE 14.13).

Biomagnification caused populations of many North American birds of prey to decline precipitously from the 1950s to the 1970s. The peregrine falcon was almost totally wiped out in the eastern United States, and the bald eagle, the U.S. national bird, was virtually eliminated from the lower 48 states. The brown pelican vanished from its Atlantic Coast range, remaining only in Florida, and the osprey and other hawks saw substantial population declines. Eventually scientists determined that DDT was causing these birds' eggshells to grow thinner, so that eggs were breaking in the nest and killing the embryos within.

In a remarkable environmental success story, populations of all these birds have rebounded (p. 296) since the United States banned DDT. However, biomagnification is by no means a thing of the past. DDT continues to impair wildlife in parts of the world where it is still used, and mercury buildup in fish poses risks to human health in North America (p. 437). In the Arctic, polar bears at the top of the food chain feed on seals that contain biomagnified toxicants. Despite their remote location, the polar bears of Svalbard Island in Arctic Norway show extremely high levels of PCB contamination, as a result of biomagnification and the concentrating effect of the process of global distillation shown in Figure 14.11. Polar bear cubs suffer immune suppression, hormone disruption, and high mortality—and because the cubs receive PCBs in their mothers' milk, contamination persists and accumulates across generations.

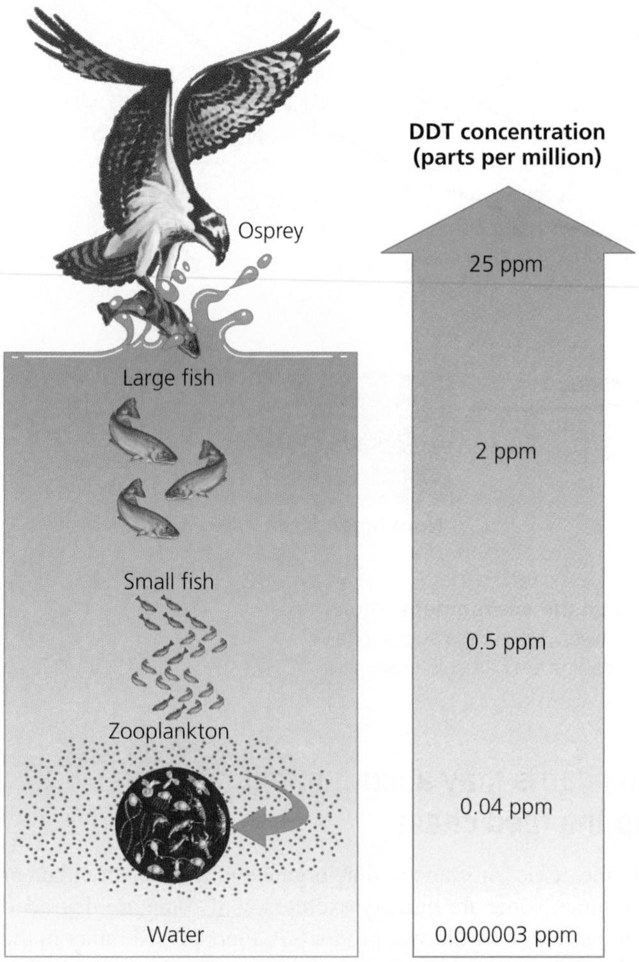

FIGURE 14.13 **In a classic case of biomagnification, DDT moves from zooplankton through various types of fish, becoming highly concentrated in fish-eating birds such as ospreys.** Organisms at the lowest trophic level take in fat-soluble compounds such as DDT from water. As animals at higher trophic levels eat organisms lower on the food chain, each organism passes its load of toxicants up to its consumer, such that organisms on all trophic levels bioaccumulate the substance in their tissues.

Toxic substances can threaten ecosystem services

Toxicants can alter the biological composition of ecosystems and the manner in which organisms interact with one another and their environment. In so doing, harmful compounds can threaten the ecosystem services (pp. 3, 116–117, 152, 290) provided by nature. For example, pesticide exposure has been implicated as a factor in the recent declines in honeybee populations (p. 254). Honeybees pollinate over 100 economically important crops, and reduced pollination by wild bees has increased costs for farmers by forcing them to hire professional beekeepers to pollinate their crops.

Healthy, functioning ecosystems provide the service of nutrient cycling. Decomposers and detritivores in the soil (pp. 81–82) break down organic matter and replenish soils with nutrients for plants to utilize. When soils are exposed to pesticides or antifungal agents, nutrient cycling rates are altered. This affects the quantity of nutrients available to producers, affects their growth, and produces adverse effects throughout the ecosystem.

Studying Effects of Hazards

Determining health effects of particular environmental hazards is a challenging job, especially because any given person or organism has a complex history of exposure to many hazards throughout life. Scientists rely on several different methods with people and with wildlife, ranging from correlative surveys to manipulative experiments (p. 12).

Wildlife studies integrate work in the field and lab

Scientists study the impacts of environmental hazards on wild animals to help conserve animal populations and also to understand potential risks to people. Just as placing the proverbial canary in a coal mine helped miners determine whether the air was safe for them to breathe, studying how wild animals respond to pollution and other hazards can help us detect environmental health threats before they do us too much harm.

Wildlife toxicologists use a variety of approaches in their research. When scientists were zeroing in on the impacts of DDT, one key piece of evidence came from museum collections of wild birds' eggs from the decades before synthetic pesticides were manufactured. Old eggs from museum collections had measurably thicker shells than the eggs scientists were studying in the field from present-day birds, indicating that something was thinning the present-day shells.

Often wildlife toxicologists work in the field with animals to take measurements, document patterns, and generate hypotheses, before heading to the laboratory to run controlled manipulative experiments to test their hypotheses. The work of two of the pioneers in the study of endocrine disruptors illustrates the approaches embraced in wildlife studies.

Biologist Louis Guillette studied alligators in Florida (**FIGURE 14.14a**) and discovered that many showed bizarre reproductive problems. Females had trouble producing viable eggs, young alligators had abnormal gonads, and male hatchlings had too little of the male sex hormone testosterone while female hatchlings had too much of the female sex hormone estrogen. Because certain lakes received agricultural runoff that included insecticides such as DDT and dicofol and herbicides such as atrazine, he hypothesized that chemical contaminants were disrupting the endocrine systems of alligators during their development in the egg. Indeed, when Guillette and his co-workers compared alligators in polluted lakes with those in cleaner lakes, they found the ones in polluted lakes to be suffering far more problems. Moving into the lab, the researchers found that several contaminants detected in alligator eggs and young could bind to receptors for estrogen and reverse the sex of male embryos. Their experiments showed that atrazine appeared to disrupt hormones by inducing production of aromatase, an enzyme that converts testosterone to estrogen.

Following Guillette's work, researcher Tyrone Hayes (**FIGURE 14.14b**) found similar reproductive problems in frogs and attributed them to atrazine. In lab experiments, male frogs raised in water containing very low doses of the herbicide became feminized and hermaphroditic, developing both testes and ovaries. Hayes then moved to the field to look for correlations between herbicide use and reproductive impacts in the wild. His field surveys showed that leopard frogs across North America experienced hormonal problems in areas of heavy atrazine usage. His work indicated that atrazine, which kills plants by blocking biochemical pathways in photosynthesis, can also act as an endocrine disruptor.

Human studies rely on case histories, epidemiology, and animal testing

In studies of human health, we gain much knowledge by studying sickened individuals directly. Medical professionals have long treated victims of poisonings, so the effects of common poisons are well known. Autopsies help us understand what constitutes a lethal dose. This process of observation and analysis of individual patients is known as a **case history** approach. Case histories have advanced our understanding of human illness, but they do not always help us infer the effects of rare hazards, new hazards, or chemicals that exist at low environmental concentrations and exert minor, long-term effects. Case histories also tell us little about probability and risk, such as how many extra deaths we might expect in a population due to a particular cause.

For such questions, which are common in environmental toxicology, we need **epidemiological studies**, large-scale

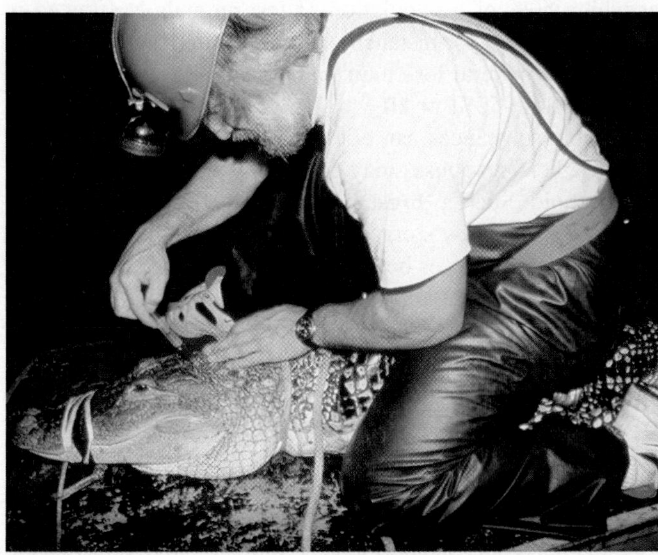

(a) Louis Guillette taking blood sample from alligator

(b) Tyrone Hayes in lab with frog

FIGURE 14.14 Wildlife studies examine the effects of toxic substances in the environment. Researchers Louis Guillette **(a)** and Tyrone Hayes **(b)** found that alligators and frogs, respectively, show reproductive abnormalities that they attribute to endocrine disruption by pesticides.

comparisons among groups of people, usually contrasting a group known to have been exposed to some hazard and a group that has not. Epidemiologists track the fate of all people in the study for a long period of time (often years or decades) and measure the rate at which deaths, cancers, or other health problems occur in each group. The epidemiologist then analyzes the data, looking for observable differences between the groups, and statistically tests hypotheses accounting for differences. When a group exposed to a hazard shows a significantly greater degree of harm, it suggests that the hazard may be responsible. For example, epidemiologists have tracked asbestos miners for evidence of asbestosis, lung cancer, and mesothelioma (cancer of the cells that line the body's internal organs). Survivors of the Chernobyl and Fukushima nuclear disasters have been monitored for thyroid cancer and other illnesses (pp. 562–563). Canadian epidemiologists are now tracking people for impacts of bisphenol A exposure.

The epidemiological process is akin to a natural experiment (p. 12), in which the experimenter studies groups of subjects made possible by some event that has occurred. A similar approach was followed by anthropologist Elizabeth Guillette, who happens to be married to alligator biologist Louis Guillette (see **THE SCIENCE BEHIND THE STORY**, pp. 378–379). The advantages of epidemiological studies are their realism and their ability to yield relatively accurate predictions about risk. Drawbacks include the need to wait a long time for results and an inability to address future effects of new hazards, such as products just coming to market. In addition, participants in epidemiological studies encounter many factors that affect their health besides the one under study. Epidemiological studies measure a statistical association between a health hazard and an effect, but they do not confirm that the hazard *causes* the effect.

To establish causation, manipulative experiments are needed. However, subjecting people to massive doses of toxic substances in a lab experiment would clearly be unethical. This is why researchers have traditionally used nonhuman animals as test subjects. Foremost among these animal models have been laboratory strains of rats, mice, and other mammals (**FIGURE 14.15**). Because of shared evolutionary history, the bodies of all mammals function similarly, so substances that harm mice and rats are reasonably likely to harm us. Some people feel the use of animals for testing is unethical, but animal testing enables scientific and medical advances that would be impossible or far more difficult otherwise. Still, new techniques (with human cell cultures, bacteria, or tissue from chicken eggs) are being devised that may soon replace some live-animal testing.

FIGURE 14.15 Animal testing is used to study toxic substances in the laboratory. Tests with specially bred strains of mice, rats, and other animals allow researchers to study the toxicity of substances, develop safety guidelines, and make medical advances in ways they could not achieve without these animals.

Dose-response analysis is a mainstay of toxicology

The standard method of testing with lab animals in toxicology is **dose-response analysis.** Scientists quantify the toxicity of a substance by measuring the strength of its effects or the number of animals affected at different doses. The **dose** is the amount of substance the test animal receives, and the **response** is the type or magnitude of negative effects the animal exhibits as a result. The response is generally quantified by measuring the proportion of animals exhibiting negative effects. The data are plotted on a graph, with dose on the x axis and response on the y axis (**FIGURE 14.16a**). The resulting curve is called a **dose-response curve.**

Once they have plotted a dose-response curve, toxicologists can calculate a convenient shorthand gauge of a substance's toxicity: the amount of the substance it takes to kill half the population of study animals used. This lethal dose for 50% of individuals is termed the **LD_{50}.** A high LD_{50} indicates low toxicity, and a low LD_{50} indicates high toxicity.

If the experimenter is interested in nonlethal health effects, he or she may want to document the level of toxicant at which 50% of a population of test animals is affected in some other way (for instance, the level of toxicant that causes 50% of lab mice to lose their hair). Such a level is called the effective-dose-50%, or **ED_{50}.**

Some substances can elicit effects at any concentration, but for others, responses may occur only above a certain dose, or threshold. Such a **threshold dose** (**FIGURE 14.16b**) might be expected if the body's organs can fully metabolize or excrete a toxicant at low doses but become overwhelmed at high concentrations. It might also occur if cells can repair damage to their DNA only up to a certain point.

Sometimes a response may *decrease* as a dose increases. Toxicologists are finding that some dose-response curves are U-shaped, J-shaped, or shaped like an inverted U (**FIGURE 14.16c**). Such counterintuitive curves contradict toxicology's traditional assumption that "the dose makes the poison." These unconventional dose-response curves often occur with endocrine disruptors, likely because the hormone system is geared to respond to minute concentrations of substances (normally, hormones in the bloodstream). Because the endocrine system responds to minuscule amounts of chemicals, it may be vulnerable to disruption by contaminants that are dispersed through the environment and

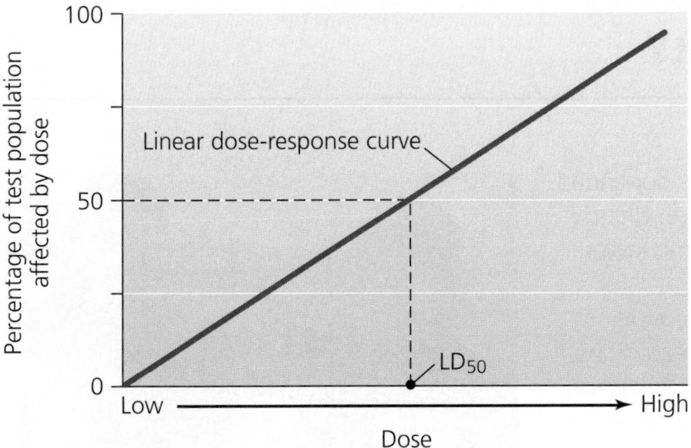

(a) Linear dose-response curve

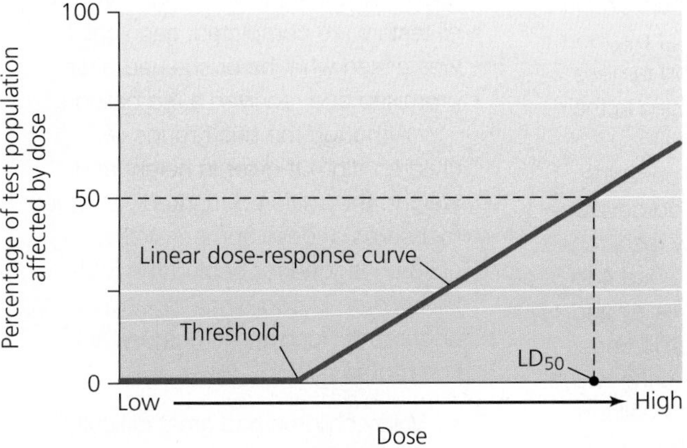

(b) Dose-response curve with threshold

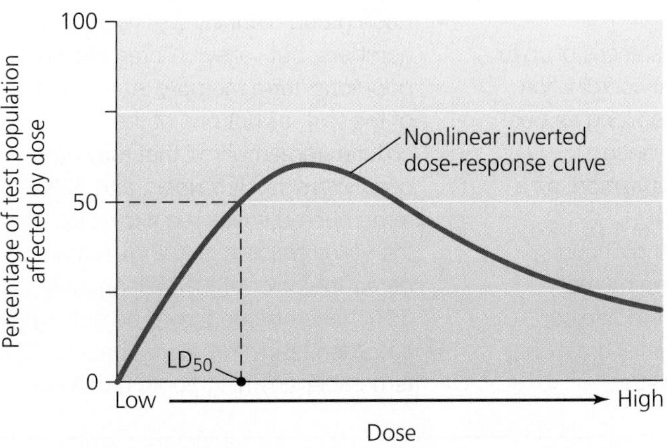

(c) Unconventional dose-response curve

FIGURE 14.16 Dose-response curves show that organisms' responses to toxicants may sometimes be complex. In a classic linear dose-response curve **(a)**, the percentage of animals killed or otherwise affected by a substance rises with the dose. The point at which 50% of the animals are killed is labeled the *lethal-dose-50*, or LD_{50}. For some toxic substances, a threshold dose **(b)** exists, below which doses have no measurable effect. Some substances, in particular endocrine disruptors, show unconventional, nonlinear dose-response curves **(c)** that are U-shaped, J-shaped, or inverted.

that reach our bodies in very low concentrations. In research with bisphenol A, a number of studies with lab animals have found unconventional dose-response curves.

Researchers generally give lab animals much higher doses relative to body mass than people would receive in the environment. This is so that the response is great enough to be measured and so that differences between the effects of small and large doses are evident. Data from a range of doses give shape to the dose-response curve. Once the data from animal tests are plotted, researchers can extrapolate downward to estimate responses to still-lower doses from a hypothetically large population of animals. This way, they can come up with an estimate of, say, what dose causes cancer in 1 mouse in 1 million. A second extrapolation is required to estimate the effect on humans, with our greater body mass. Because these two extrapolations stretch beyond the actual data obtained, they introduce uncertainty into the interpretation of what doses are safe for people.

Mixes may be more than the sum of their parts

It is difficult enough to determine the impact of a single hazard, but the task becomes astronomically more difficult when multiple hazards interact. Chemical substances, when mixed, may act together in ways that cannot be predicted from the effects of each in isolation. Mixed toxicants may sum each other's effects, cancel out each other's effects, or multiply each other's effects. Whole new types of impacts may arise when toxicants are mixed together. Interactive impacts that are greater than the simple sum of their constituent effects are called **synergistic effects**.

With Florida's alligators, lab experiments have indicated that the DDT breakdown product DDE can either help cause sex reversal or inhibit it, depending on the presence of other chemicals. Mice exposed to a mixture of nitrate, atrazine, and aldicarb have been found to show immune, hormone, and nervous system effects that were not evident from exposure to each of these chemicals alone.

Traditionally, environmental health has tackled effects of single hazards one at a time. In toxicology, the complex experimental designs required to test interactions, and the sheer number of chemical combinations, have meant that single-substance tests have received priority. This approach is changing, but the interactive effects of most chemicals are unknown.

Endocrine disruption poses challenges for toxicology

As today's emerging understanding of endocrine disruption leads toxicologists to question their assumptions, unconventional dose-response curves are presenting challenges for scientists studying toxic substances and for policymakers trying to set safety standards for them. Knowing the shape of a dose-response curve is crucial if one is using it to predict responses at doses below those that have been tested. Because so many novel synthetic chemicals exist in very low concentrations over wide areas, many scientists suspect that we may have underestimated the dangers of compounds that exert impacts at low concentrations.

THE SCIENCE BEHIND THE STORY

Pesticides and Child Development in Mexico's Yaqui Valley

With spindly arms and big, round eyes, one set of pictures shows the sorts of stick figures drawn by young children everywhere. Next to them is another group of drawings, mostly disconnected squiggles and lines, resembling nothing. Both sets of pictures are intended to depict people. The main difference identified between the two groups of young artists: long-term pesticide exposure.

Children's drawings are not a typical tool of toxicology, but anthropologist Elizabeth Guillette was interested in the effects of pesticides on children and wanted to try new methods. She devised tests to measure childhood development based on techniques from anthropology and medicine. Searching for a study site, Guillette found the Yaqui valley region of northwestern Mexico.

The Yaqui valley is farming country, worked for generations by the indigenous group that gives the region its name. Synthetic pesticides arrived in the area in the 1940s. Some Yaqui embraced the agricultural innovations, spraying their farms in the valley to increase their yields. Yaqui farmers in the surrounding foothills, however, generally chose to bypass the chemicals and to continue following more traditional farming practices. Although differing in farming techniques, Yaqui in the valley and foothills continued to share the same culture, diet, education system, income levels, and family structure.

At the time of the study, in 1994, valley farmers planted crops twice a year, applying pesticides up to 45 times from planting to harvest. A previous study conducted in the valley in 1990, focusing on areas with the largest farms, had indicated high levels of multiple pesticides in the breast milk of mothers and in the umbilical cord blood of newborn babies. In contrast, foothill families avoided chemical pesticides in their gardens and homes.

To understand how pesticide exposure affects childhood development, Guillette and fellow researchers studied 50 preschoolers aged 4 to 5, of whom 33 were from the valley and 17 from the foothills. Each child underwent a half-hour exam, during which researchers showed a red balloon, promising to give the balloon later as a gift, and using the promise to evaluate long-term memory. Each child was then put through a series of physical and mental tests:

- Catching a ball from distances of up to 3 m (10 ft) away, to test coordination
- Jumping in place for as long as possible, to assess endurance
- Drawing a picture of a person, as a measure of perception
- Repeating a short string of numbers, to test short-term memory
- Dropping raisins into a bottle cap from a height of about 13 cm (5 in.), to gauge fine-motor skills

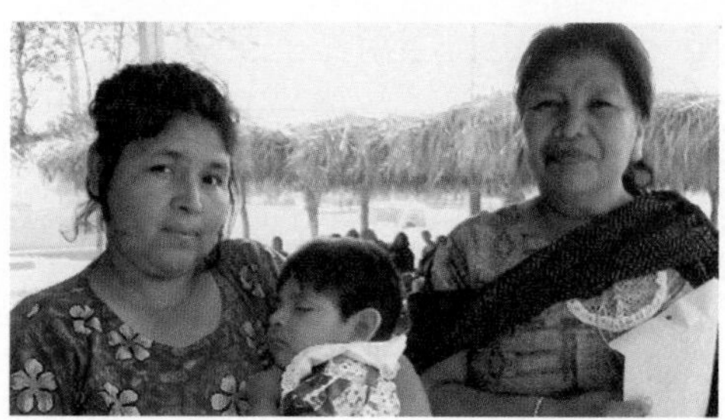

A Yaqui family

The researchers also measured each child's height and weight. When all tests were completed, each child was asked what he or she had been promised and received a red balloon.

Although the two groups of children did not differ in height and weight, they differed markedly in other measures of development. Valley children lagged far behind the foothill children in coordination, physical endurance, long-term memory, and fine motor skills:

- Valley children had great difficulty catching the ball.
- The average valley child could jump for 52 seconds, compared to 88 seconds for foothill children.
- Each group did fairly well repeating numbers, but valley children showed poor long-term memory. At the end of the test, all but one of the foothill children remembered that they had been promised a balloon, and 59% remembered it was red. However, of the valley children only 27% remembered the color of the balloon, only 55% remembered they'd be getting a balloon, and 18% were unable to remember anything about a balloon.

Scientists first noted endocrine-disrupting effects decades ago, but the idea that synthetic chemicals might be altering the hormones of animals was not widely appreciated until the 1996 publication of the book *Our Stolen Future*, by Theo Colburn, Dianne Dumanoski, and J.P. Myers. Like *Silent Spring*, this book integrated scientific work from various fields and presented a unified picture that shocked many readers—and brought criticism from some scientists and from the chemical industry.

Today, thousands of studies have linked hundreds of substances to effects on reproduction, development, immune function, brain and nervous system function, and other hormone-driven processes. Evidence is strongest so far in nonhuman animals, but many studies suggest impacts on people. Some researchers argue that the sharp rise in breast cancer rates (one in eight U.S. women today develops breast cancer) may be due to hormone disruption, because an excess of estrogen appears to feed tumor development in older women. Other studies suggest that endocrine disruptors may account for rising rates of testicular cancer, undescended testicles, and genital birth defects in men. For example, a 2009 study determined

Drawings by children in the foothills

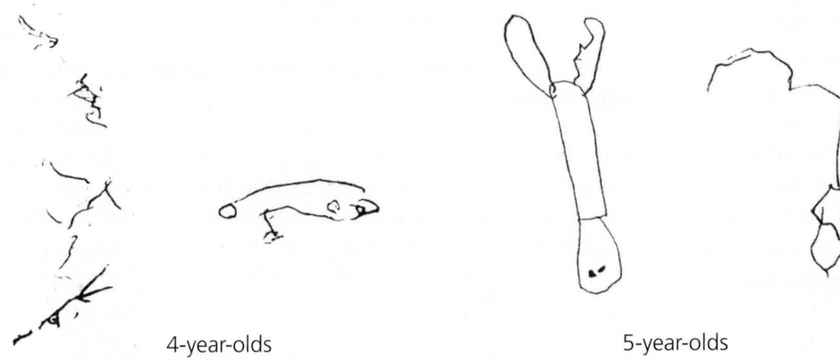

Drawings by children in the valley

FIGURE 1 Data from children in Mexico's Yaqui valley offer a startling example of apparent neurological effects of pesticide poisoning. Young children from foothills areas where pesticides were not commonly used drew recognizable figures of people. Children of the same age from valley areas where pesticides were heavily used drew less-recognizable figures. *Adapted from Guillette, E.A., et al., 1998.* Environmental Health Perspectives *106: 347–353.*

- Most valley children missed the bottle cap when dropping raisins, whereas foothill children dropped them into the caps more often.

The children's drawings exhibited the most dramatic difference between valley and foothill children (**FIGURE 1**). The researchers determined each drawing could earn 5 points, with 1 point each for a recognizable feature: head, body, arms, legs, and facial features. Foothill children drew pictures that looked like people, averaging about 4.5 points per drawing. Valley children, in contrast, averaged 1.6 points per drawing; their scribbles resembled little that looked like a person. By the standards of developmental medicine, the 4- and 5-year-old valley children drew at the level of 2-year-olds.

Some scientists greeted Guillette's study skeptically, pointing out that its sample size was too small to be meaningful. Others said that factors the researchers missed, such as different parenting styles or unknown health problems, could be to blame. Prominent toxicologists argued that because the researchers lacked time and money to take blood or tissue samples to check for pesticides or other toxic substances, the study results couldn't be tied to agricultural chemicals. Regardless of these criticisms, Guillette maintains that her findings show that nontraditional study methods are a valid way to track the effects of environmental toxicants and that pesticides present complex, long-term risks to human growth and health.

In a follow-up study published in 2006, Guillette and her colleagues found that adolescent girls in the valley had developed significantly less mammary tissue than girls from the foothills—a result, they hypothesized, that could be due to higher exposure to endocrine-disrupting chemicals *in utero*. In some cases, the reduction in mammary tissue development was large enough to potentially prevent some girls from being able to breastfeed their children. As this and other studies suggest, the effects of toxic substances can potentially span multiple generations, supporting calls for long-term studies of the effects of toxic chemicals on human health. ■

that workers in Chinese factories that manufactured BPA had four times the rate of erectile dysfunction as workers in factories where BPA wasn't present. A follow-up study in 2010 found that workers with detectable levels of BPA in their urine were two to four times more likely to have reduced sperm counts and poorer sperm quality than workers in which no BPA was detected. While the BPA exposure in these workers was far higher than that experienced by the average American, they represent some of the first studies to link bisphenol A exposure to reproductive abnormalities in humans.

Much of the research into hormone disruption has brought about strident debate. This is partly because a great deal of scientific uncertainty is inherent in any young and developing field. Another reason is that negative findings about chemicals pose an economic threat to the manufacturers of those chemicals. For instance, Tyrone Hayes's work has met with fierce criticism from scientists associated with atrazine's manufacturer, which stands to lose many millions of dollars if its top-selling herbicide were to be banned or restricted in the United States.

Risk Assessment and Risk Management

Policy decisions on whether to ban chemicals or restrict their use generally follow years of rigorous testing for toxicity. Likewise, strategies for combating disease and other health threats are based on extensive scientific research. However, policy and management decisions also incorporate economics and ethics. And all too often, they are influenced by political pressure from powerful interests. The steps between the collection and interpretation of scientific data and the formulation of policy involve assessing and managing risk.

We express risk in terms of probability

Exposure to an environmental health threat does not invariably produce some given effect. Rather, it causes some probability of harm, some statistical chance that damage will result. To understand a health threat, a scientist must know more than just its identity and strength. He or she must also know the chance that one will encounter it, the frequency with which one may encounter it, the amount of substance or degree of threat to which one is exposed, and one's sensitivity to the threat. Such factors help determine the overall risk posed.

Risk can be measured in terms of *probability*, a quantitative description of the likelihood of a certain outcome. The probability that some harmful outcome (for instance, injury, death, environmental damage, or economic loss) will result from a given action, event, or substance expresses the **risk** posed by that phenomenon.

Our perception of risk may not match reality

Every action we take and every decision we make involves some element of risk, some (generally small) probability that things will go wrong. We try in everyday life to behave in ways that minimize risk, but our perceptions of risk do not always match statistical reality (**FIGURE 14.17**). People often worry unduly about negligibly small risks yet happily engage in other activities that pose high risks. For instance, most people perceive flying in an airplane as a riskier activity than driving a car, but according to a 2012 report by the National Safety Council, a person's chance of dying from an automobile accident is 73 times higher than from an airplane crash. Psychologists agree that this difference between perception and reality stems from the fact that we feel more at risk when we are not controlling a situation and more safe when we are "at the wheel"—regardless of the actual risk involved.

This psychology may help account for people's anxiety over nuclear power, toxic waste, and pesticide residues on foods—environmental hazards that are invisible or little understood and whose presence in our lives is largely outside our personal control. In contrast, people are more ready to accept and ignore the risks of smoking cigarettes, overeating, and not exercising—voluntary activities statistically shown to pose far greater risks to health.

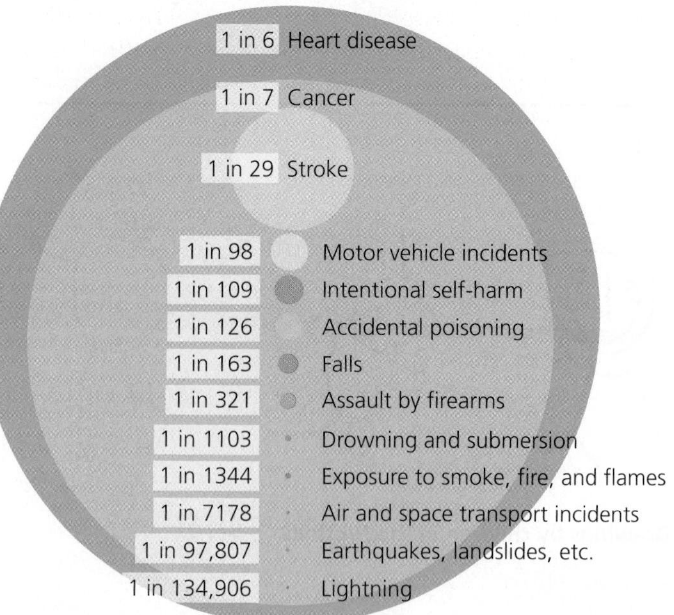

FIGURE 14.17 Our perceptions of risk do not always match the reality of risk. Listed here are several leading causes of death in the United States, along with a measure of the risk each poses. The larger the area of the circle in the figure, the greater the risk of dying from that cause. Note that the risk of death from motor vehicle incidents is over 70 times that from air and space transport incidents. *Data are for 2008 from* Injury Facts, *2012, National Safety Council, Itasca, IL.*

Risk assessment analyzes risk quantitatively

The quantitative measurement of risk and the comparison of risks involved in different activities or substances together are termed **risk assessment.** Risk assessment is a way to identify and outline problems. In environmental health, it helps ascertain which substances and activities pose health threats to people or wildlife and which are largely safe.

Assessing risk for a chemical substance involves several steps. The first steps involve the scientific study of toxicity we examined above—determining whether a substance has toxic effects and, through dose-response analysis, measuring how effects vary with the degree of exposure. Subsequent steps involve assessing the individual's or population's likely extent of exposure to the substance, including the frequency of contact, the concentrations likely encountered, and the length of encounter.

To assess risk from a widely used substance such as bisphenol A, teams of scientific experts may be convened to review hundreds of studies so that regulators and the public can benefit from informed summaries. Between 2001 and 2012, the U.S. government sponsored three such panels on BPA, and the American Plastics Council, an industry group, sponsored two. The 2008 panel, sponsored by the government's National Toxicology Program of the National Institute of Environmental and Health Sciences, assessed BPA as being of intermediate concern for human health, on the third level of their five-level scale. The panel voiced "some concern for effects on brain, behavior, and prostate gland in fetuses, infants, and children at current human exposures," and wrote that "the possibility that bisphenol A may alter human development cannot be dismissed."

The regulatory challenges faced with endocrine disruptors such as BPA were highlighted in this review. Initially, the panel deemed 80 studies appropriate for informing policy on regulating BPA, 70% of which were from academic laboratories (many of which found adverse effects of BPA on organisms) rather than industry laboratories (which typically found no effects of BPA). Shortly thereafter, the panel received a 93-page letter from the American Chemistry Council contending that many of the studies had flaws that made them unsuitable for informing regulatory policy. One common criticism advanced in the letter was that academic studies did not follow Good Laboratory Practice (GLP), which is part of the protocols adopted by regulatory agencies around the world to evaluate potentially toxic substances. After considering these concerns, the panel eliminated many academic studies from consideration, reducing academic studies to a mere 30% of the studies being considered. Given this pool of studies to consider, the panel failed to rate any impacts of BPA as of "Concern for Adverse Impact" or "Serious Concern for Adverse Impact."

Scientists argued that as a federal agency regulating potentially toxic substances, the panel is expected to consider both GLP and non-GLP studies in its deliberations. Also, they contended, very rigid GLP methods were not always appropriate for chemicals such as BPA that show unusual dose-response curves. To aid efforts to standardize research protocols and produce more studies that would qualify as suitable for consideration, the National Institute of Environmental and Health Sciences sponsored a meeting in 2009 of BPA researchers at which common protocols were established. Further, the agency devoted $30 million in 2011–2013 to BPA research to stimulate studies using these protocols, generate additional information, and better inform future panels about potential health impacts of BPA on humans. Because ratings like this heavily influence regulatory decisions, such as the one issued by the FDA in 2012 opting not to restrict BPA use in food packaging (p. 360), initiatives like these can broaden the scientific studies used to evaluate threats to public health.

Risk management combines science and other social factors

Accurate risk assessment is a vital step toward effective **risk management**, which consists of decisions and strategies to minimize risk (FIGURE 14.18). In most nations, risk management is handled largely by federal agencies. In the United States, these include agencies such as the FDA, the EPA, and, the Centers for Disease Control and Prevention (CDC). In risk management, scientific assessments of risk are considered in light of economic, social, and political needs and values. Risk managers assess costs and benefits of addressing risk in various ways with regard to both scientific and nonscientific concerns before making decisions on whether and how to reduce or eliminate risk.

In environmental health and toxicology, comparing costs and benefits (p. 146) can be difficult because the benefits are often economic, whereas the costs often pertain to health. Moreover, economic benefits are generally known,

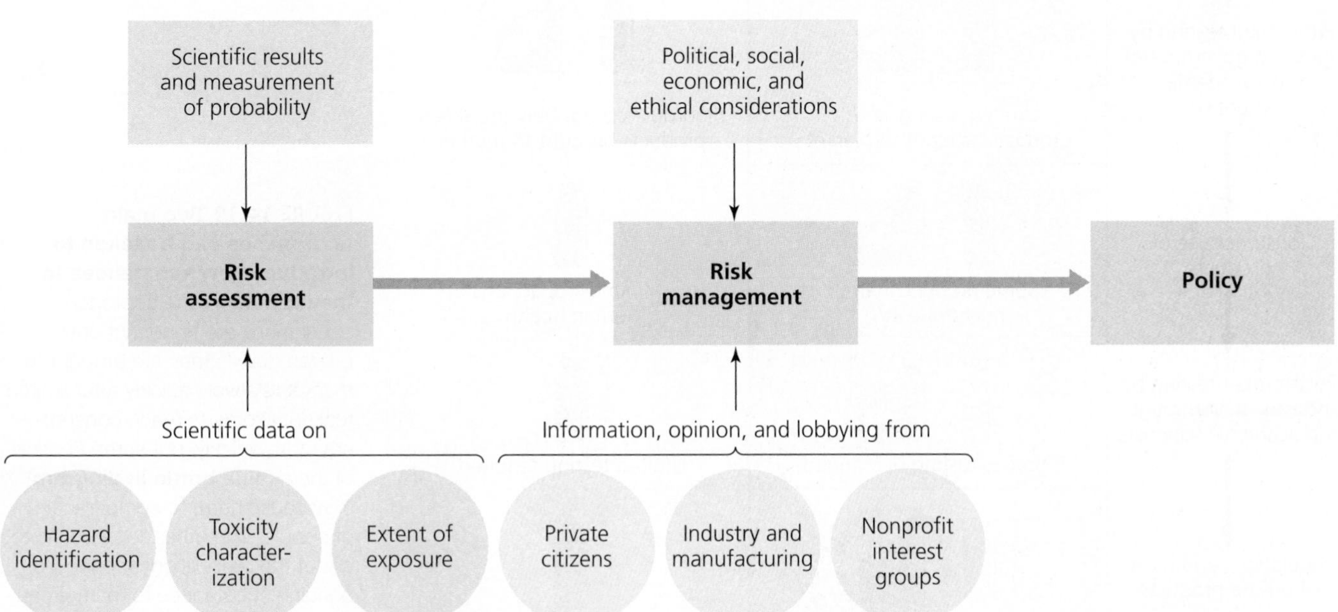

FIGURE 14.18 The first step in addressing the risk of an environmental hazard is risk assessment. Once science identifies and measures risks, then risk management can proceed. In this process, economic, political, social, and ethical issues are considered in light of the scientific data from risk assessment. The consideration of all these types of information is intended to result in policy decisions that minimize the risk of the environmental hazard.

easily quantified, and of a discrete and stable amount, whereas health risks are hard-to-measure probabilities, often involving a small percentage of people likely to suffer greatly and a large majority likely to experience little effect. When a government agency bans a pesticide, it may mean considerable economic loss for the manufacturer and potential economic loss for the farmer, whereas the benefits accrue less predictably over the long term to some percentage of factory workers, farmers, and the general public. Because of the lack of equivalence in the way costs and benefits are measured, risk management frequently tends to stir up debate.

In the case of bisphenol A, eliminating plastic linings in our food and drink cans could do more harm than good, because the linings help prevent metal corrosion and the contamination of food by pathogens. Alternative substances exist for most of BPA's uses, but replacing BPA with alternatives will entail economic costs to industry, and these costs are passed on to consumers in the prices of products. Such complex considerations can make risk management decisions difficult even if the science of risk assessment is fairly clear. This difficulty may help account for the hesitancy of U.S. regulatory agencies to issue restrictions on BPA so far. As of 2013, both the FDA and the EPA were continuing to review options for managing risk from BPA.

Philosophical and Policy Approaches

Because we cannot know a substance's toxicity until we measure and test it, and because there are so many untested chemicals and combinations, science will never eliminate the many uncertainties that accompany risk assessment. In such a world of uncertainty, there are two basic philosophical approaches to categorizing substances as safe or dangerous (**FIGURE 14.19**).

Two approaches exist for determining safety

One approach is to assume that substances are harmless until shown to be harmful. We might nickname this the "innocent-until-proven-guilty" approach. Because thoroughly testing every existing substance (and combination of substances) for its effects is a hopelessly long, complicated, and expensive pursuit, the innocent-until-proven-guilty approach has the virtue of facilitating technological innovation and economic activity. However, it has the disadvantage of putting into wide use some substances that may later turn out to be dangerous, such as happened with the pesticide DDT.

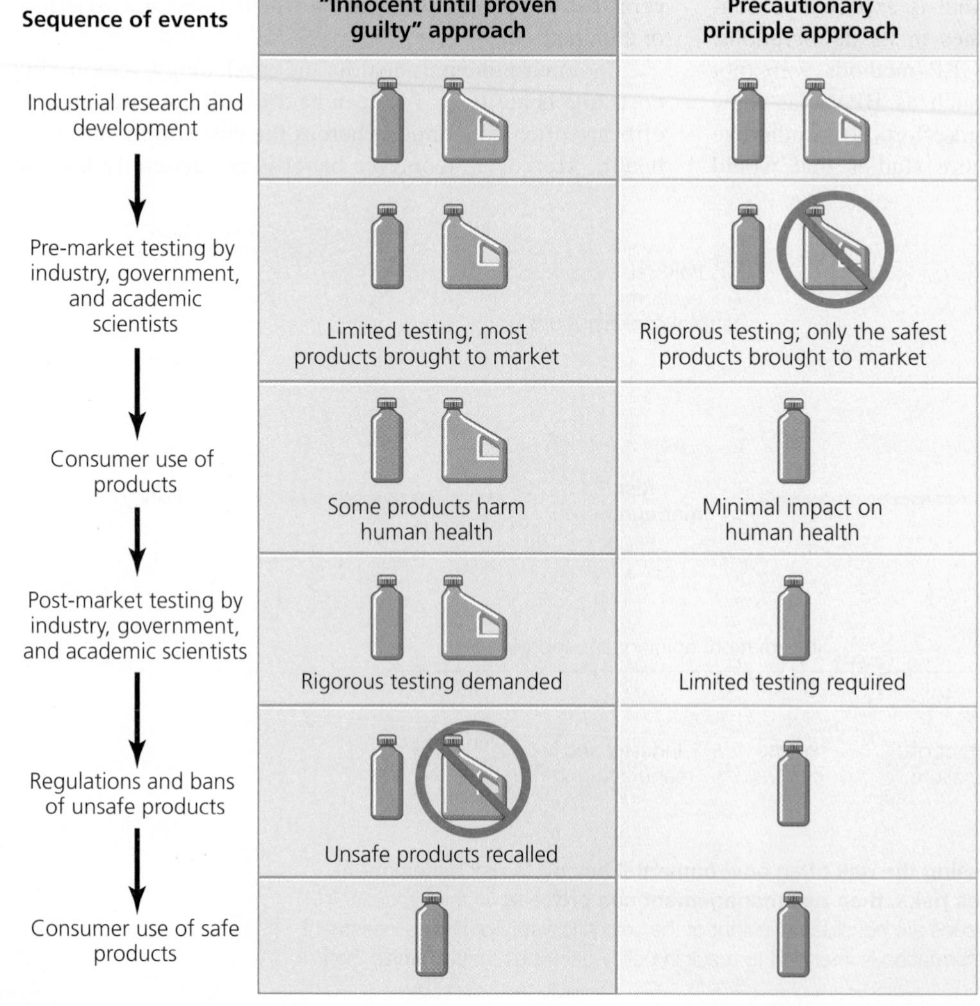

FIGURE 14.19 Two main approaches can be taken to introduce new substances to the market. In one approach, substances are "innocent until proven guilty"; they are brought to market relatively quickly after limited testing. Products reach consumers more quickly, but some fraction of them (**blue bottle in diagram**) may cause harm to some fraction of people. The other approach is to adopt the precautionary principle, bringing substances to market cautiously, only after extensive testing. Products that reach the market should be safe, but many perfectly safe products (**purple bottle in diagram**) will be delayed in reaching consumers.

The other approach is to assume that substances are harmful until shown to be harmless. This approach follows the precautionary principle (p. 265). This more cautious approach should enable us to identify troublesome toxicants before they are released into the environment, but it may also impede the pace of technological and economic advance.

These two approaches are actually two ends of a continuum of possible approaches. The two endpoints differ mainly in where they lay the burden of proof—specifically, whether product manufacturers are required to prove safety or whether government, scientists, or citizens are required to prove danger.

WEIGHING THE ISSUES

THE PRECAUTIONARY PRINCIPLE Industry's critics say chemical manufacturers should bear the burden of proof for the safety of their products before they hit the market. Industry's supporters say that mandating more safety research will hamper the introduction of products that consumers want, increase the price of products as research costs are passed on to consumers, and cause companies to move to nations where standards are more lax. What do you think? Should government follow the precautionary principle and require proof of safety prior to a chemical's introduction into the market?

Philosophical approaches are reflected in policy

This choice of philosophical approach has direct implications for policy, and nations vary in how they blend the two approaches when it comes to regulating synthetic substances. European nations have recently embarked on a policy course that incorporates the precautionary principle, whereas the United States largely follows an innocent-until-proven-guilty approach. For instance, compounds in cosmetics require no FDA review or approval before being sold to the public.

In the United States, several federal agencies apportion responsibility for tracking and regulating synthetic chemicals. The FDA, under the Food, Drug, and Cosmetic Act of 1938 and its subsequent amendments, monitors foods and food additives, cosmetics, drugs, and medical devices. The EPA regulates pesticides under the Federal Insecticide, Fungicide, and Rodenticide Act of 1947 (FIFRA) and its amendments. The Occupational Safety and Health Administration (OSHA) regulates workplace hazards under a 1970 act. Several other agencies regulate other substances. Synthetic chemicals not covered by other laws are regulated by the EPA under the 1976 Toxic Substances Control Act (TSCA).

EPA regulation is only partly effective

The **Toxic Substances Control Act** directs the EPA to monitor thousands of industrial chemicals manufactured in or imported into the United States, ranging from PCBs to lead to asbestos, and including bisphenol A and many other compounds in plastics. The act gives the agency power to regulate these substances and ban them if they are found to pose excessive risk.

However, many public health advocates view TSCA as being far too weak. The 62,000 chemicals existing when TSCA became law were grandfathered in, and the number has since swollen to 83,000, while just five have been restricted. Health advocates note that the screening required of industry is minimal and that to mandate more extensive and meaningful testing, the EPA must show proof of the chemical's toxicity. In other words, the agency is trapped in a Catch-22: To push for studies looking for toxicity, it must have proof of toxicity already. The result is that most synthetic chemicals are not thoroughly tested before being brought to market. Of those that fall under TSCA, only 10% have been thoroughly tested for toxicity; only 2% have been screened for carcinogenicity, mutagenicity, or teratogenicity; fewer than 1% are regulated; and almost none have been tested for endocrine, nervous, or immune system damage, according to the U.S. National Academy of Sciences.

In its regulation of pesticides under FIFRA, the EPA is charged with "registering" each new pesticide that manufacturers propose to bring to market. The registration process involves risk assessment and risk management. The EPA first asks the manufacturer to provide information, including results of safety assessments the company has performed according to EPA guidelines. The EPA examines the company's research and all other relevant scientific research. It examines the product's ingredients and how the product will be used and tries to evaluate whether the chemical poses risks to people, other organisms, or water or air quality. The EPA then approves, denies, or sets limits on the chemical's sale and use. It also must approve language used on the product's label.

Because the registration process takes economic considerations into account, critics say it allows hazardous chemicals to be approved if the economic benefits are judged to outweigh the hazards. Here the challenges of weighing intangible risks involving human health and environmental quality against the tangible and quantitative numbers of economics become apparent.

Toxicants are regulated internationally

The European Union is taking the world's boldest step toward testing and regulating manufactured chemicals. In 2007, the EU's **REACH** program went into effect (*REACH* stands for Registration, Evaluation, Authorization, and Restriction of Chemicals). REACH largely shifts the burden of proof for testing chemical safety from national governments to industry and requires that chemical substances produced or imported in amounts of over 1 metric ton per year be registered with a new European Chemicals Agency. This agency evaluates industry research and decides whether the chemical seems safe and should be approved, whether it is unsafe and should be restricted, or whether more testing is needed. It is expected that REACH will require the registration of about 30,000 substances.

TABLE 14.3 American versus European Approaches to Chemical Regulation

TSCA (UNITED STATES)	REACH (EUROPEAN UNION)
• Government bears burden of proof to show harm	• Industry bears burden of proof to show safety
• Little data on new chemicals are required from industry	• More data on new chemicals are required from industry
• Chemicals in use before 1976 are not regulated	• Chemicals in use before 1981 bear scrutiny like that directed toward newer chemicals
• Prioritizing problems is hampered by lack of data	• Problems are prioritized using data on risk
• Industry is allowed to keep trade secrets from the public	• Database will allow public access to chemical information

Adapted from Schwarzman, M.R., and M.P. Wilson, 2009. New science for chemicals policy. Science *326: 1065–1066.*

The REACH policy also aims to help industry by giving it a single streamlined regulatory system and by exempting it from having to file paperwork on substances under 1 metric ton. By requiring stricter review of major chemicals already in use, exempting chemicals made only in small amounts, and providing financial incentives for innovating new chemicals, the EU hopes to help European industries research and develop safer new chemicals and products while safeguarding human health and the environment. REACH differs markedly from TSCA (**TABLE 14.3**), illustrating how the approach that Europe is pursuing is different from the approach being pursued in the United States.

In an impacts assessment in 2003, EU commissioners estimated that REACH will cost the chemical industry and chemical users 2.8–5.2 billion euros (U.S. \$3.8–7.0 billion) over 11 years but that the health benefits to the public would be roughly 50 billion euros (U.S. \$67 billion) over 30 years. Changes in the program since then have made the predicted cost-to-benefit ratio even better.

The world's nations have also sought to address chemical pollution with international treaties. The *Stockholm Convention on Persistent Organic Pollutants (POPs)* came into force in 2004 and has been ratified by over 150 nations. POPs are toxic chemicals that persist in the environment, bioaccumulate and biomagnify up the food chain, and often can travel long distances. The PCBs and other contaminants found in polar bears are a prime example. Because contaminants often cross international boundaries, an international treaty seemed the best way to deal fairly with such transboundary pollution. The Stockholm Convention aims first to end the use and release of 12 POPs shown to be most dangerous, a group nicknamed the "dirty dozen" (**TABLE 14.4**). It sets guidelines for phasing out these chemicals and encourages transition to safer alternatives.

TABLE 14.4 The "Dirty Dozen" Persistent Organic Pollutants (POPs) Targeted by the Stockholm Convention

TOXICANT	DESCRIPTION
Aldrin	Insecticide to kill termites and crop pests
Chlordane	Insecticide to kill termites and crop pests
DDT	Insecticide to protect against insect-spread disease; still applied in some countries to control malaria
Dieldrin	Insecticide to kill termites, textile pests, crop pests, and disease vectors
Dioxins	By-product of incomplete combustion and chemical manufacturing; released in metal recycling, pulp and paper bleaching, auto exhaust, tobacco smoke, and wood and coal smoke
Endrin	Pesticide to kill rodents and crop insects
Furans	By-product of processes that release dioxins; also present in commercial mixtures of PCBs
Heptachlor	Broad-spectrum insecticide
Hexachlorobenzene	Fungicide for crops; released by chemical manufacture and processes that release dioxins and furans
Mirex	Household insecticide; fire retardant in plastics, rubber, and electronics
PCBs	Industrial chemical used in heat-exchange fluids, electrical transformers and capacitors, paints, sealants, and plastics
Toxaphene	Insecticide to kill crop insects and livestock parasites

Data from United Nations Environment Programme (UNEP), 2001.

Conclusion

International agreements such as REACH and the Stockholm Convention represent a sign that governments may act to protect the world's people, wildlife, and ecosystems from toxic substances and other environmental hazards. At the same time, solutions often come more easily when they do not arise from government regulation alone. Consumer choice exercised through the market can often be an effective way to influence industry's decision making, but this requires consumers have full information from scientific research regarding the risks involved. Once scientific results are in, a society's philosophical approach to risk management will determine what policy decisions are made.

Whether the burden of proof is laid at the door of industry or of government, we will never attain complete scientific knowledge of any risk. Rather, we must make choices based on the information available. Synthetic chemicals have brought us innumerable modern conveniences, a larger food supply, and medical advances that save and extend human lives. Human society would be very different without them. Yet a safer and happier future, one that safeguards the well-being of both people and the environment, depends on knowing the risks that some hazards pose and on having means in place to phase out harmful substances and replace them with safer ones.

Reviewing Objectives

You should now be able to:

Explain the goals of environmental health and identify major environmental health hazards

- Environmental health seeks to assess and mitigate environmental factors that adversely affect human health and ecological systems. (p. 360)
- Environmental health threats include physical, chemical, biological, and cultural hazards. (pp. 360–361)
- Disease is a major focus of environmental health. We have successfully fought some infectious diseases, but others are spreading. (pp. 362–366)
- Sanitation, clean water, food security, education, and access to medical care are strategies to enhance environmental health. (pp. 365–366)
- Toxicology is the study of poisonous substances. (p. 366)
- Several major environmental hazards (cigarette smoke, radon, asbestos, lead, and PBDEs) exist indoors. (pp. 366–367)

Describe the types of toxic substances in the environment, the factors that affect their toxicity, and the defenses that organisms possess against them

- Toxicants may be natural as well as synthetic, and thousands of synthetic substances exist in our environment and in our bodies. (pp. 367–369)
- Toxicants include carcinogens, mutagens, teratogens, allergens, pathway inhibitors, neurotoxins, and endocrine disruptors. (pp. 369–370)
- Organisms possess defenses against some toxins (pp. 370–371)
- Toxicity or strength of response may be influenced by the nature of exposure (acute or chronic) and individual variation. (p. 371)

Explain the movements of toxic substances and how they affect organisms and ecosystems

- Toxic substances may travel long distances through the atmosphere or may concentrate in and move through surface water and groundwater. (p. 372)
- Some chemicals break down very slowly and thus persist in the environment. (pp. 372–373)
- Some toxic substances bioaccumulate and move up the food chain, poisoning consumers at high trophic levels through the process of biomagnification. (pp. 373–374)
- Toxic substances impair ecosystem services by affecting organisms that provide these services, such as honeybee pollinators. (pp. 374)

Discuss the study of hazards and their effects, including wildlife toxicology, case histories, epidemiology, animal testing, and dose-response analysis

- Studies of wildlife can inform initiatives to improve human health. (pp. 374–375)
- In case histories, researchers study health problems in individual people. (p. 375)
- Epidemiology involves gathering data from many people over long-term periods and comparing groups with and without exposure to the factor being assessed. (pp. 375–376)
- In dose-response analysis, scientists measure the response of test animals to various doses of the suspected toxicant. (pp. 376–377)
- Toxicity may be influenced by synergistic interactions among hazards. (p. 377)
- Unconventional dose-response curves from endocrine-disrupting substances are causing toxicologists to reassess some of their assumptions and methods. (pp. 377–379)

Evaluate risk assessment and risk management

- Risk assessment involves quantifying and comparing risks involved in different activities or substances. (pp. 380–381)
- Risk management integrates science with political, social, and economic concerns in order to design strategies to minimize risk. (pp. 381–382)

Compare philosophical approaches to risk and how they relate to regulatory policy

- An innocent-until-proven-guilty approach assumes that a substance is safe unless shown to be harmful. This puts the burden of proof on the public to prove that a substance is harmful. (p. 382)
- A precautionary approach assumes that a substance may be harmful unless proven safe. This puts the burden of proof on the manufacturer to prove that a substance is safe. (p. 383)
- The EPA, CDC, FDA, and OSHA are responsible for regulating environmental health hazards under U.S. policy. (p. 383)
- European nations take a more precautionary approach than does the United States when it comes to testing chemical products, as shown by a comparison of REACH and TSCA. (pp. 383–384)
- The Stockholm Convention aims to ban 12 persistent organic pollutants. (p. 384)

Testing Your Comprehension

1. What four major types of health hazards does research in the field of environmental health encompass?
2. In what way is disease the greatest hazard that people face? What kinds of interrelationships must environmental health experts study to learn about how diseases affect human health?
3. Where does most exposure to lead, asbestos, radon, and PBDEs occur? How has each been addressed?
4. When did concern over the effects of pesticides start to grow in the United States? Describe the argument presented by Rachel Carson in *Silent Spring*. What policy resulted from the book's publication? Where and how is DDT still used?
5. List and describe the general categories of toxic substances described in this chapter.
6. How do toxic substances travel through the environment, and where are they most likely to be found? Describe and contrast the processes of bioaccumulation and biomagnification.
7. What are epidemiological studies, and how are they most often conducted?
8. Why are animals used in laboratory experiments in toxicology? Explain the dose-response curve. Why is a substance with a high LD_{50} considered safer than one with a low LD_{50}?
9. What factors may affect an individual's response to a toxic substance? Why is chronic exposure to toxic agents often more difficult to measure and diagnose than acute exposure? What are synergistic effects, and why are they difficult to measure and diagnose?
10. How do scientists identify and assess risks from substances or activities that may pose health threats?

Seeking Solutions

1. Describe some environmental health hazards that you think you may be living with indoors. How do you think you may have been affected by indoor or outdoor hazards in the past? How could you best deal with these hazards in the future?
2. Do you feel that laboratory animals should be used in experiments in toxicology? Why or why not?
3. Why has research on endocrine disruption spurred so much debate? What steps do you think could be taken to help establish greater consensus among scientists, industry, regulators, policymakers, and the public?
4. Describe differences in the policies of the United States and the European Union toward the study and management of the risks of synthetic chemicals. Which do you believe is better, the policies of the United States or those of the European Union, and why?
5. **THINK IT THROUGH** You are the parent of two young children, and you want to minimize the environmental health risks your kids are exposed to. Name five steps that you could take in your household and in your daily life that would minimize your children's exposure to environmental health hazards.
6. **THINK IT THROUGH** You work for a public health organization and have been asked to educate the public about bisphenol A and to suggest ways to minimize exposure to the chemical. You begin by examining your lifestyle and finding ways to use alternatives to BPA-containing products. Create a list of five ways you are exposed daily to bisphenol A, and then list approaches that would avoid or minimize these exposures. Do these steps require more time and/or money? What are some costs of embracing these changes? What would you tell an interested person about bisphenol A as it relates to human health?

Calculating Ecological Footprints

In 2007, the last year the U.S. EPA reported data on pesticide use, Americans used 1.13 billion pounds of pesticide active ingredients, and world pesticide use totaled 5.21 billion pounds of active ingredients. In that same year, the U.S. population was 302 million, and the world's population totaled 6.63 billion. Pesticides include hundreds of chemicals used as insecticides, fungicides, herbicides, rodenticides, repellants, and disinfectants. They are used by farmers, governments, industries, and individuals. In the table, calculate your share of pesticide use as a U.S. citizen in 2007 and the amount used by (or on behalf of) the average citizen of the world.

	Annual pesticide use (pounds of active ingredient)
You	
Your class	
Your state	
United States	1.13 billion
World (total)	5.21 billion
World (per capita)	

1. What is the ratio of your annual pesticide use to the world's per capita average?

2. In 2007, the average U.S. citizen had an ecological footprint of 8.0 hectares and the average world citizen's footprint was 2.7 hectares (Chapter 1). Compare the ratio of pesticide usage with the ratio of the overall ecological footprints. How do these differ, and how would you account for this?

3. Does the figure for per capita pesticide use for you as a U.S. citizen seem reasonable for you personally? Why or why not? Do you find this figure alarming or of little concern? What else would you like to know to assess the risk associated with this level of pesticide use?

Louisiana's vanishing coastal wetlands support a diversity of wildlife, such as this Great Egret.

Freshwater Systems and Resources

Upon completing this chapter, you will be able to:

- Describe the distribution of fresh water on Earth and the major types of freshwater systems
- Discuss how we use water and alter freshwater systems
- Assess problems of water supply and propose solutions to address depletion of fresh water
- Describe the major classes of water pollution and propose solutions to address water pollution
- Explain how we treat drinking water and wastewater

CENTRAL CASE STUDY

Starving the Louisiana Coast of Sediment

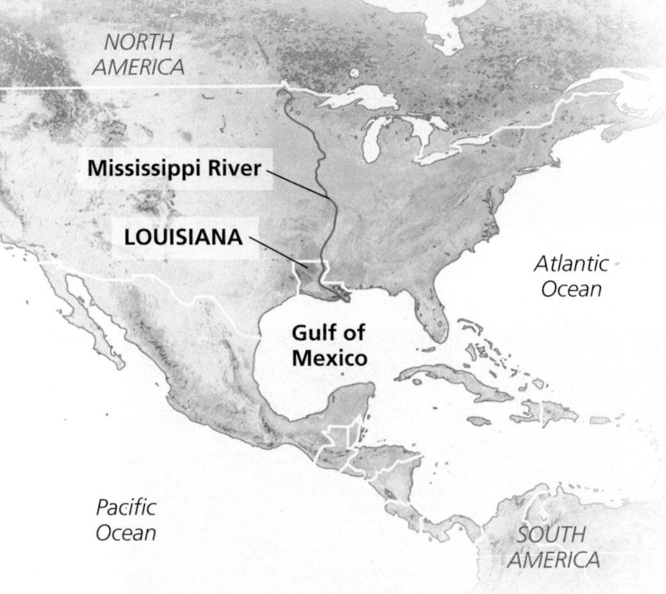

"The Louisiana and Mississippi coastal region is critical to the economic, cultural, and environmental integrity of the nation."

—Nancy Sutley, Chair of the White House Council on Environmental Quality

"What really screwed up the marsh is when they put the levees on the river. They should take the levees out and let the water run; that's what built the land."

—Frank "Blackie" Campo, Resident of Shell Beach, Louisiana

The state of Louisiana is shrinking. Its coastal wetlands straddle the boundary between the land and the ocean, and these wetlands are disappearing beneath the waters of the Gulf of Mexico. Louisiana loses 65 km^2 (25 mi^2) of coastal wetlands each year, and comparisons of wetland area from the mid-1800s to the early 1990s show drastic losses (**FIGURE 15.1a**). Since the 1930s alone, Louisiana has lost nearly 4900 km^2 (1900 mi^2) of coastal wetlands—an area roughly the size of Delaware.

Louisiana's coastal wetlands transition from communities of salt-tolerant grasses at the ocean's edge to freshwater bald cypress swamps further inland. These ecosystems support a diversity of animals, including eagles, pelicans, shrimp, oysters, black bears, alligators, and sea turtles. The state's coastal wetlands also protect cities such as New Orleans and Baton Rouge from damaging storms. Vegetation in these wetlands acts as a windbreak on strong winds and as a water break on waves coming inland from the Gulf.

Louisiana's millions of acres of coastal wetlands were created over the past 7000 years as the Mississippi River fanned out and deposited its sediments at its delta before emptying into the Gulf of Mexico. The Mississippi River accumulates large quantities of sediment from water flowing over land into streams in the river's 3.2-million-km^2 (1.2-million-mi^2) watershed (**FIGURE 15.1b**). Much of this sediment originates from the Missouri River basin that drains America's agricultural heartland.

The salt marshes in the river's delta naturally compact over time. This compaction lowers the level of the marsh bottom and submerges vegetation under increasingly deeper waters. When waters become too deep, the vegetation dies, and soils are then washed away by the ocean. The natural compaction is offset, however, by inputs of sediments from the river and from the deposition of organic matter from marsh grasses. These additions keep soil levels high, water depths relatively stable, and vegetation healthy.

So why are Louisiana's wetlands being swallowed by the sea? It's because people have modified the Mississippi River so extensively that much of its sediments no longer reach the wetlands that need them. The river's basin contains roughly 2000 dams, which slow river flow and allow sediments suspended in the water to settle in reservoirs. This not only prevents sediments from reaching the river's delta, but also slowly fills in each dam's reservoir, decreasing its volume and shortening its life span. Therefore, dams in Minnesota, Montana, and Pennsylvania and other locations throughout the Mississippi basin affect the Louisiana coastline hundreds of miles downriver.

The Mississippi River is also lined with thousands of miles of *levees* (long, raised mounds of earth). These structures prevent small-scale flooding, and the mouth of the Mississippi is lined with levees to provide a deep river channel for shipping into the Gulf of Mexico. These levees prevent the river from fanning out into its delta and turn the lower Mississippi into a "barrel" that shoots sediments off the continental shelf into the deep waters of the Gulf (**FIGURE 15.1c**).

Although Louisiana's economy has benefited from oil and gas extraction, these activities have also promoted wetland losses. The extraction of large quantities of oil, natural gas,

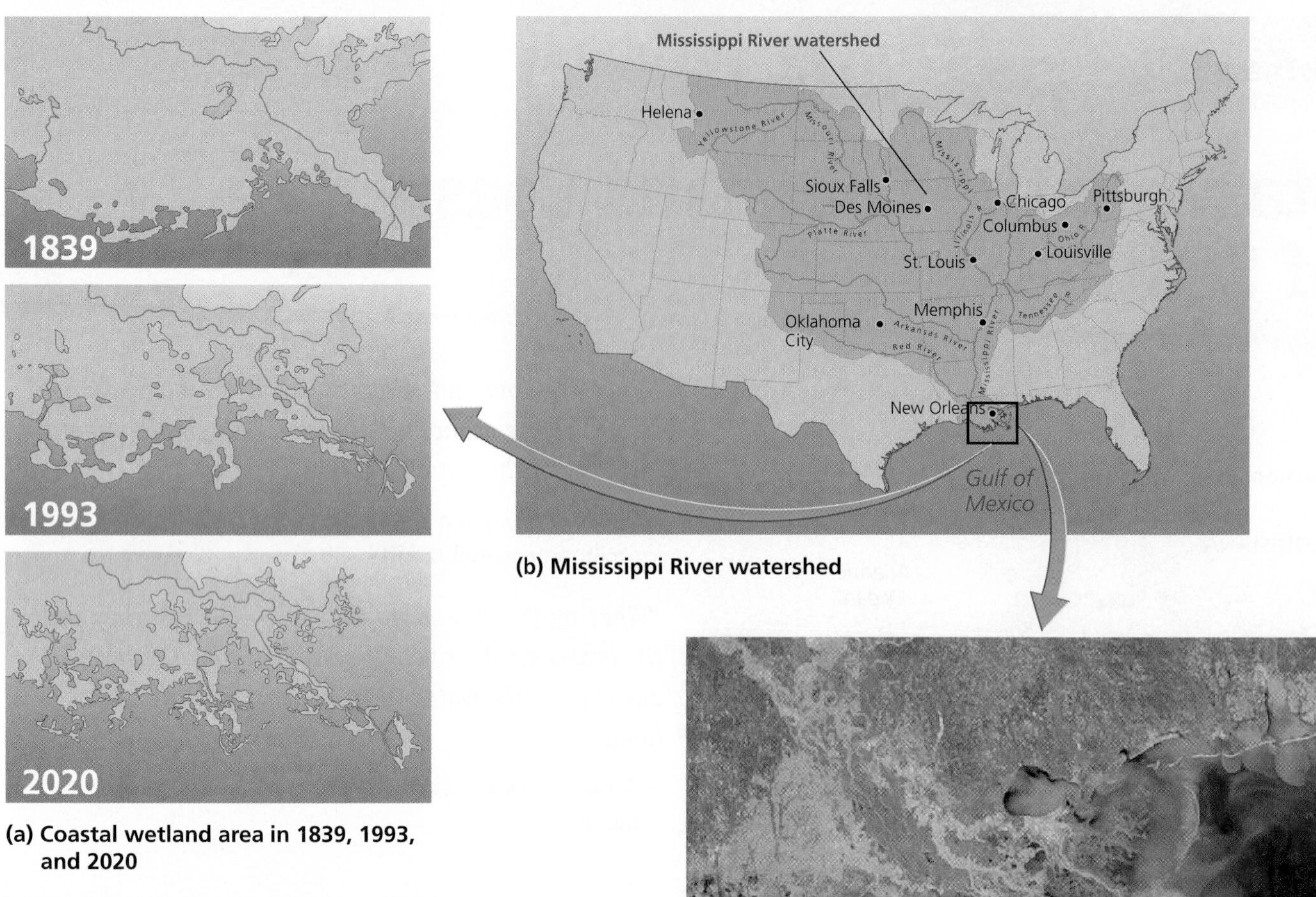

(a) Coastal wetland area in 1839, 1993, and 2020

(b) Mississippi River watershed

(c) Sediment plumes from Mississippi River entering Gulf

FIGURE 15.1 **Human modifications all along the Mississippi River affect coastal wetlands at the river's mouth.** The size of Louisiana's coastal wetlands shrank substantially **(a)** from 1839 to 1993. It is predicted to shrink even more by 2020 because people have modified sediment deposition patterns in the river's delta by constructing extensive levees along the river and blocking sediments upriver behind dams. The Mississippi River system **(b)** is the largest in the United States, draining over 40% of the land area of the lower 48 states. A satellite image of south Louisiana **(c)** shows the brown plumes of sediments being released into the deep waters of the Gulf of Mexico from the Mississippi River (plume on the right) and the Atchafalaya River (plume on the left). *(a) Adapted from Environmental Defense Fund.*

and saline groundwater associated with oil deposits causes the land to compact, lowering soil levels. Additionally, engineers have cut nearly 13,000 km (8000 mi) of canals through coastal wetlands to facilitate shipping and oil and gas exploration. These canals fragment the wetlands and increase erosion rates. They also enable salty ocean water to penetrate inland and damage vegetation and wildlife in freshwater marshes. The 2010 *Deepwater Horizon* oil spill in the Gulf of Mexico (pp. 538–541) also affected Louisiana's marshes by coating marsh grasses with oil and impairing their ability to secure oxygen.

Proposed solutions for coastal erosion center on restoring the system to its natural state by diverting large quantities of water from the Mississippi River into coastal wetlands instead of shotgunning it out into the Gulf in the river's main channel. Proponents of this approach point to the Atchafalaya River, which currently drains one-third of the Mississippi River's volume. The Atchafalaya delta, fed by this water and sediment, is actually gaining coastal land area. Gulf restoration efforts are being bolstered by the 2012 Resources and Ecosystems Sustainability, Tourist Opportunities and Revived Economies of the Gulf Coast States Act (RESTORE Act), legislation passed in the aftermath of the *Deepwater Horizon* spill. The Act creates a comprehensive ecosystem restoration plan for the Gulf Coast, and uses 80% of the fines paid for Clean Water Act (p. 413) violations associated with the spill to support Gulf Coast restoration.

Given the conflicting demands we put on freshwater waterways for water withdrawal, shipping, and flood control, there are no easy solutions to the problems faced in the Mississippi River and southern Louisiana. But how we tackle problems like those in Louisiana's coastal wetlands will help determine the long-term sustainability of our most precious natural resource—water. ■

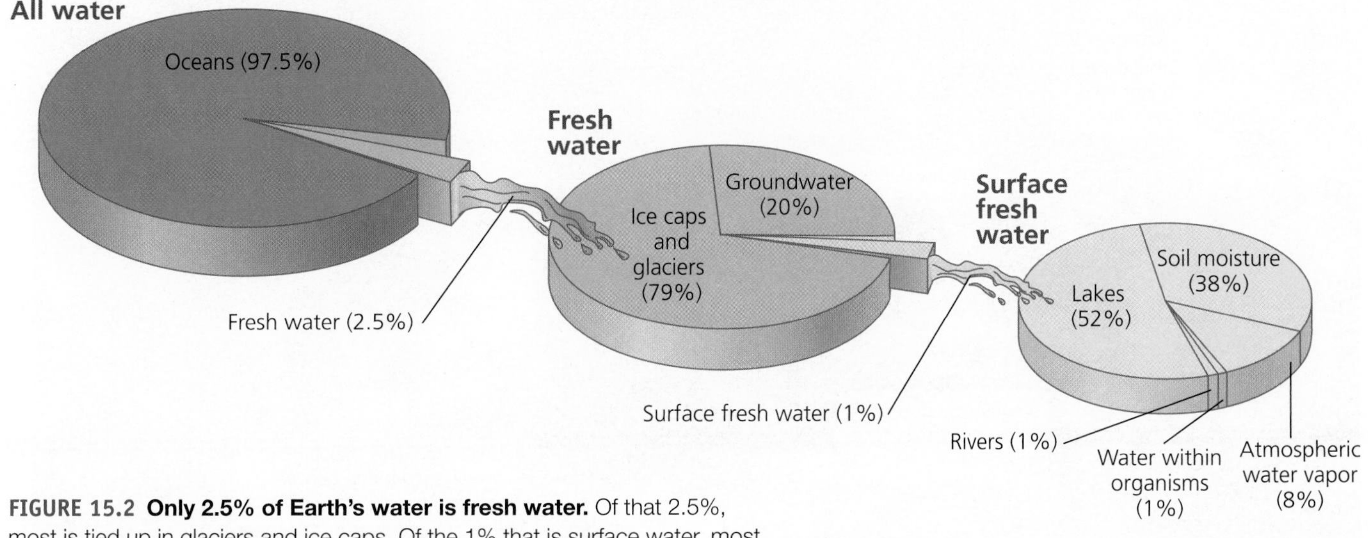

FIGURE 15.2 Only 2.5% of Earth's water is fresh water. Of that 2.5%, most is tied up in glaciers and ice caps. Of the 1% that is surface water, most is in lakes and soil moisture. *Data from United Nations Environment Programme (UNEP) and World Resources Institute.*

What percentage of Earth's water is fresh water in lakes?

Freshwater Systems

"Water, water, everywhere, nor any drop to drink." The well-known line from the poem *The Rime of the Ancient Mariner* describes the situation on our planet well. Water may seem abundant, but water that we can drink is quite rare and limited (**FIGURE 15.2**). About 97.5% of Earth's water resides in the oceans and is too salty to drink or to use to water crops. Only 2.5% is considered **fresh water,** water that is relatively pure, with few dissolved salts. Because most fresh water is tied up in glaciers, icecaps, and underground aquifers, just over 1 part in 10,000 of Earth's water is easily accessible for human use.

Water is renewed and recycled as it moves through the water cycle (pp. 120–121). Precipitation falling from the sky either sinks into the ground or acts as runoff to form rivers, which carry water to the oceans or large inland lakes. As they flow, rivers can interact with ponds, wetlands, and coastal aquatic ecosystems. Underground aquifers exchange water with rivers, ponds, lakes, and the ocean through the sediments on the bottoms of these water bodies. The movement of water in the water cycle creates a web of interconnected freshwater and marine aquatic systems (**FIGURE 15.3**) that exchange water, organisms, sediments, pollutants, and other dissolved substances. What happens in one system therefore affects other systems—even those that are far away. Let's examine the freshwater components of the interconnected system, beginning with groundwater. Marine and coastal components of the system will be examined subsequently (Chapter 16), but note that all these systems interact extensively.

Groundwater plays key roles in the hydrologic cycle

Liquid fresh water occurs either as surface water or groundwater. **Surface water** is water located atop Earth's surface (such as in a river or lake) and **groundwater** is water beneath the surface held within pores in soil or rock. Any precipitation reaching Earth's land surface that does not evaporate, flow into waterways, or get taken up by organisms infiltrates the surface. Groundwater makes up one-fifth of Earth's fresh water supply and plays a key role in meeting human water needs.

Groundwater flows downward and from areas of high pressure to areas of low pressure. However, a typical rate of flow might be only about 1 m (3 ft) per day, so groundwater can remain underground for a long time. When we pump groundwater through wells, we are drawing up ancient water. The average age of groundwater has been estimated at 1400 years, and some is tens of thousands of years old.

Groundwater is contained within **aquifers:** porous, spongelike formations of rock, sand, or gravel that hold water (**FIGURE 15.4**). An aquifer's upper layer, or *zone of aeration*, contains pore spaces partly filled with water. In the lower layer, or *zone of saturation*, the spaces are completely filled with water. The boundary between these two zones is the **water table.** Picture a sponge resting partly submerged in a tray of water; the lower part of the sponge is saturated, whereas the upper portion contains plenty of air in its pores. Any area where water infiltrates Earth's surface and reaches an aquifer below is known as a *recharge zone*.

Like a tray of lasagna, the earth underground consists of layers of materials with different textures and densities. When a porous, water-bearing layer of rock, sand, or gravel is trapped between upper and lower layers of less permeable substrate (often clay), we have a **confined aquifer,** or **artesian aquifer.** In such a situation, the water is under great pressure. In contrast, an **unconfined aquifer** has no impermeable upper layer to confine it, so its water is under less pressure and can be readily recharged by surface water.

The largest known aquifer is the Ogallala Aquifer, which underlies the Great Plains of the United States. Water from this massive aquifer has enabled American farmers to create

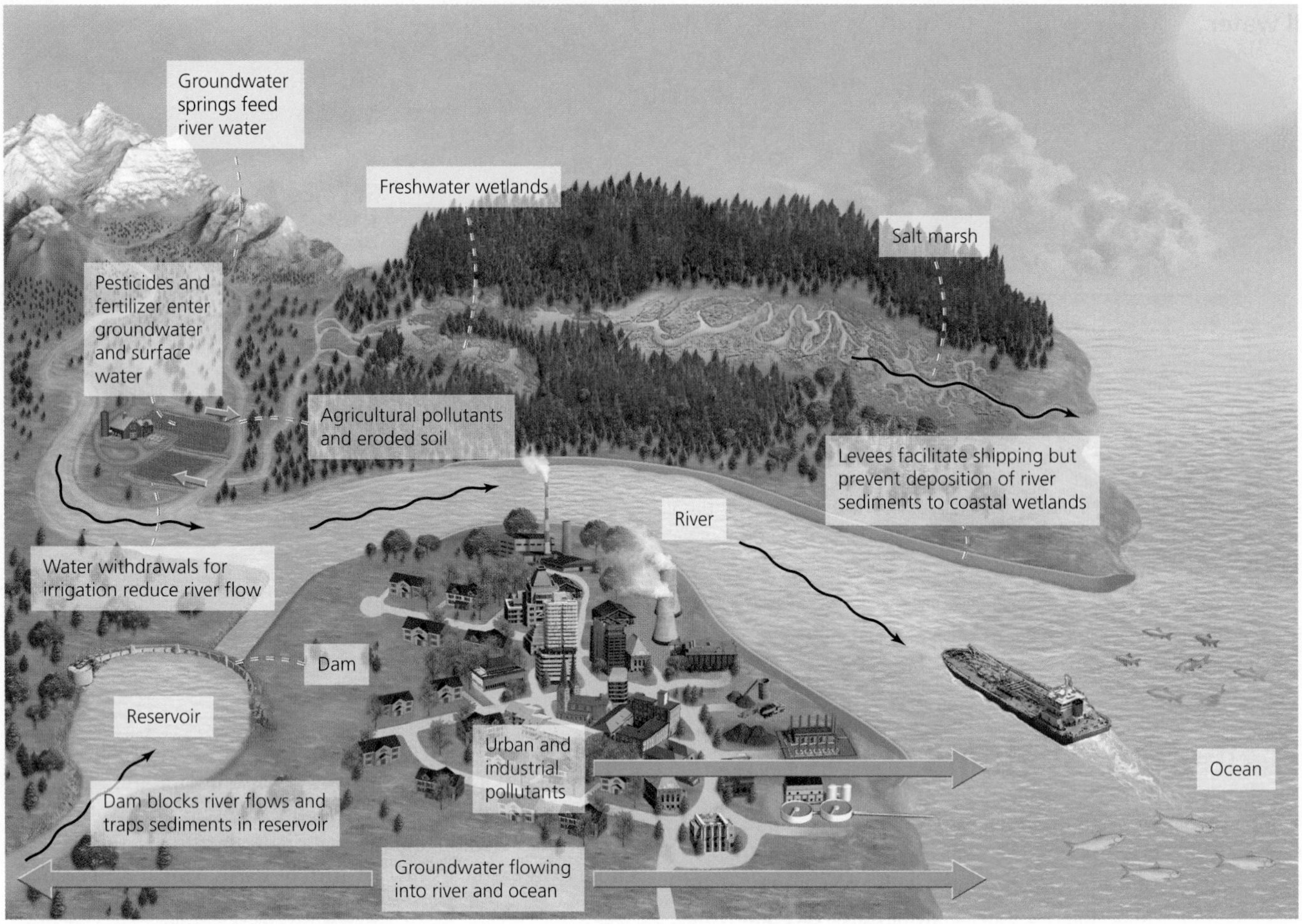

Groundwater springs feed river water
Freshwater wetlands
Salt marsh
Pesticides and fertilizer enter groundwater and surface water
Agricultural pollutants and eroded soil
Levees facilitate shipping but prevent deposition of river sediments to coastal wetlands
River
Water withdrawals for irrigation reduce river flow
Dam
Reservoir
Urban and industrial pollutants
Ocean
Dam blocks river flows and traps sediments in reservoir
Groundwater flowing into river and ocean

FIGURE 15.3 Water flows through freshwater systems and marine and coastal aquatic systems that interact extensively with one another. People affect the components of the system by constructing dams and levees, withdrawing water for human use, and introducing pollutants. Because the systems are closely connected, these impacts can cascade through the system and cause effects far from where they originated.

Aquifer recharge area
Artesian well
Well
Water table
Ground-water
Spring
Unconfined aquifer
Confined aquifer
Upper confining layer (clay)
Lower confining layer (clay)

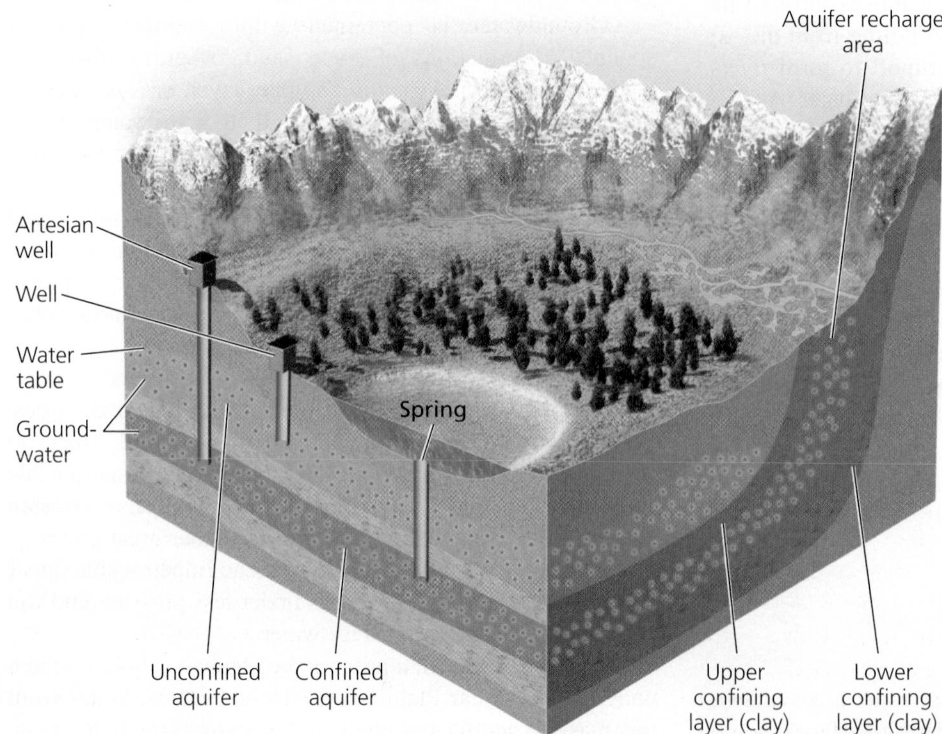

FIGURE 15.4 Groundwater may occur in unconfined aquifers above impermeable layers or in confined aquifers under pressure between impermeable layers. Water may rise naturally to the surface at springs and through the wells we dig. Artesian wells tap into confined aquifers to mine water under pressure.

the most bountiful grain-producing region in the world. However, unsustainable water withdrawals are threatening the long-term use of the aquifer for agriculture (**FIGURE 15.5**).

FAQ **Is groundwater found in huge underground caverns?**

As one of the "out of sight" elements of the water cycle, it's sometimes difficult for people to visualize how water exists underground. Many incorrectly assume that groundwater is found in large, underground caves—essentially lakes beneath Earth's surface. The reality is far less glamorous. If you look at soil under a microscope, you'll see there are small pores between the particles of minerals and organic matter that compose the soil. In the soil near the surface, these pores are filled with air. Beneath the water table, however, the pores are filled with water. Groundwater can even be found within rock. There are many types of rock, such as limestone and sandstone, which have relatively large pores between the particles of minerals that compose the rock. Just as in soil, these pores can be filled with air or, beneath the water table, by water. So when people extract groundwater with wells, we are simply sucking water out of the pores that exist within soil or rock under the surface of the Earth.

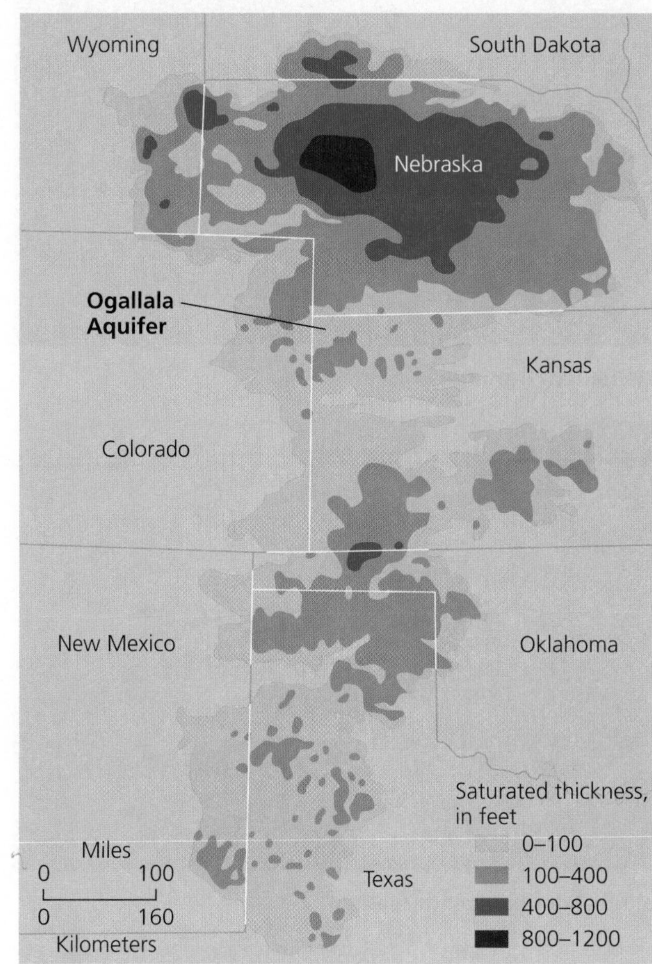

FIGURE 15.5 The Ogallala Aquifer is the world's largest aquifer, and it held 3700 km³ (881 mi³) of water before pumping began. This aquifer underlies 453,000 km² (175,000 mi²) of the Great Plains beneath eight U.S. states. Overpumping for irrigation is currently reducing the volume and extent of this aquifer.

Surface water converges in river and stream ecosystems

Surface water accounts for just 1% of fresh water, but it is vital for our survival and for the planet's ecological systems. Once water falls from the sky as rain, emerges from springs, or melts from snow or glaciers, it may soak into the ground or may flow downhill over land. Water that flows over land is called *runoff* (p. 108).

As it flows downhill, runoff converges where the land dips lowest, forming streams, creeks, or brooks. These small watercourses may merge into rivers, whose water eventually reaches a lake or ocean. A smaller river flowing into a larger one is a *tributary*. The area of land drained by a *river system*—a river and all its tributaries—is that river's **watershed** (p. 105), also called a *drainage basin*. If you could stand at the mouth of the Mississippi River and trace every drop of water in it back to the spot it first fell as precipitation, then you would have delineated the Mississippi River's watershed, the very area shaded in Figure 15.1.

Groundwater and surface water interact extensively. Surface water becomes groundwater by infiltration. Groundwater becomes surface water through springs (and from wells drilled by people), often keeping streams flowing or wetlands moist when surface conditions are otherwise dry. Each day in the United States, 1.9 trillion L (492 billion gal) of groundwater are released into bodies of surface water—nearly as much as the daily flow of water in the entire Mississippi River.

Landscapes determine where rivers flow, but rivers shape the landscapes through which they run. A river that runs through a steeply sloped region and carries a great deal of sediment may flow as an interconnected series of watercourses called a *braided river* (**FIGURE 15.6a**). In flatter regions, most rivers are *meandering rivers* (**FIGURE 15.6b**). In a meandering river, the force of water rounding a bend gradually eats away at the outer shore, eroding soil from the bank. Meanwhile, sediment is deposited along the inside of the bend, where water currents are weaker. Over time, river bends become exaggerated in shape, forming oxbows (**FIGURE 15.6c**). If water erodes a shortcut from one end of the loop to the other, pursuing a direct course, the oxbow is cut off and remains as an isolated, U-shaped water body called an *oxbow lake*.

Over thousands or millions of years, a meandering river may shift from one course to another, back and forth over a large area, carving out a flat valley. Areas nearest a river's course that are flooded periodically are said to be within the river's **floodplain.** Frequent deposition of silt from flooding makes floodplain soils especially fertile. As a result, agriculture thrives in floodplains, and *riparian* (riverside) forests are productive and species-rich. A river's meandering is often driven by large-scale flooding events that scour new channels during periods of high flow. However, extensive damming on the Mississippi and other rivers in its watershed has reduced the rate of river meandering 66–83% from its historic rate. This occurs because floodwaters are trapped by dams and held in reservoirs rather than coursing down the river and creating meanders.

(a) Braided river in Nebraska

(b) Meandering river in Alaska

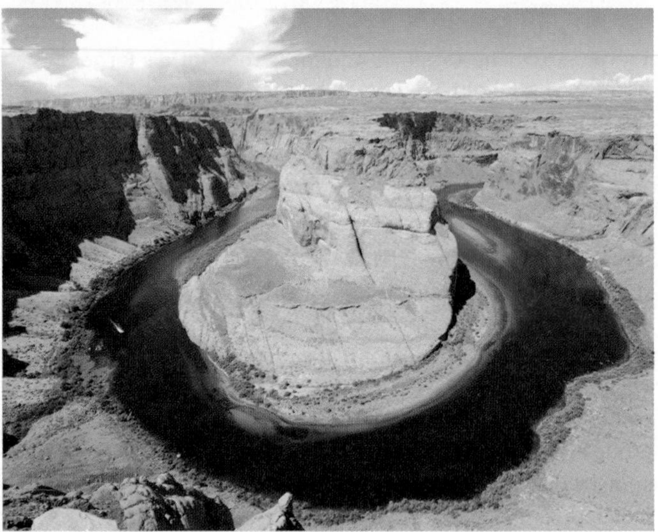

(c) Oxbow river in Arizona

FIGURE 15.6 **Rivers are classified according to their flow across the landscape.** The Platte River in Nebraska **(a)** shows an example of a braided river, whereas the Wood River in Alaska **(b)** meanders across the landscape. The Horseshoe Bend portion of the Colorado River in Arizona **(c)** demonstrates an oxbow.

Rivers and streams host diverse ecological communities. Algae and detritus support many types of invertebrates, from water beetles to crayfish. Insects as diverse as dragonflies, mayflies, and mosquitoes develop as larvae in streams and rivers before maturing into adults that take to the air. Fish and amphibians consume aquatic invertebrates and plants, and birds such as kingfishers, herons, and ospreys dine on fish and amphibians.

Lakes and ponds are ecologically diverse systems

Lakes and ponds are bodies of standing surface water. Their physical conditions and the types of life within them vary with depth and the distance from shore. As a result, scientists have described several zones typical of lakes and ponds (FIGURE 15.7).

Around the nutrient-rich edges of a water body, the water is shallow enough that aquatic plants grow from the mud and reach above the water's surface. This region, named the *littoral zone*, abounds in invertebrates—such as insect larvae, snails, and crayfish—on which fish, birds, turtles, and amphibians feed. The *benthic zone* extends along the bottom of the lake or pond, from shore to the deepest point. Many invertebrates live in the mud, feeding on detritus or on one another. In the open portion of a lake or pond, far from shore, sunlight penetrates shallow waters of the *limnetic zone*. Because light enables photosynthesis (pp. 31–32), the limnetic zone supports phytoplankton (algae, protists, and cyanobacteria; p. 75), which in turn support zooplankton (p. 75), both of which are eaten by fish. Within the limnetic zone, sunlight intensity (and therefore water temperature) decreases with depth. Clear water allows sunlight to penetrate deeply, whereas turbid water (water with suspended solids) does not. Below the limnetic zone lies the *profundal zone*, the volume of open water that sunlight does not reach. This zone lacks photosynthetic life and is lower in dissolved oxygen than upper waters.

Ponds and lakes change over time as streams and runoff bring them sediment and nutrients. *Oligotrophic* lakes and ponds, which are low in nutrients and high in oxygen, may slowly give way to the high nutrient, low oxygen conditions of *eutrophic* water bodies (jump ahead to see Figure 15.23, p. 412). Eventually, water bodies may fill in completely by the process of aquatic succession (p. 88). As lakes or ponds change over time, species of fish, plants, and invertebrates adapted to oligotrophic conditions may give way to those that thrive in eutrophic conditions. These changes occur naturally, but eutrophication can also result from human-caused nutrient pollution (pp. 108–109).

The largest lakes are sometimes known as inland seas. North America's Great Lakes are prime examples. Lake Baikal in Asia is the world's deepest lake, at 1637 m (just over 1 mile) deep. The Caspian Sea is the world's largest body of fresh water, covering nearly as much area as Montana or California.

Freshwater wetlands include marshes, swamps, bogs, and vernal pools

Wetlands are systems in which the soil is saturated with water and which generally feature shallow standing water with ample vegetation. There are many types of wetlands, and most are enormously rich and productive. In *freshwater marshes*, shallow water allows plants such as cattails and bulrushes to grow above the water surface. *Swamps* also consist of shallow water rich in vegetation, but they occur in forested areas.

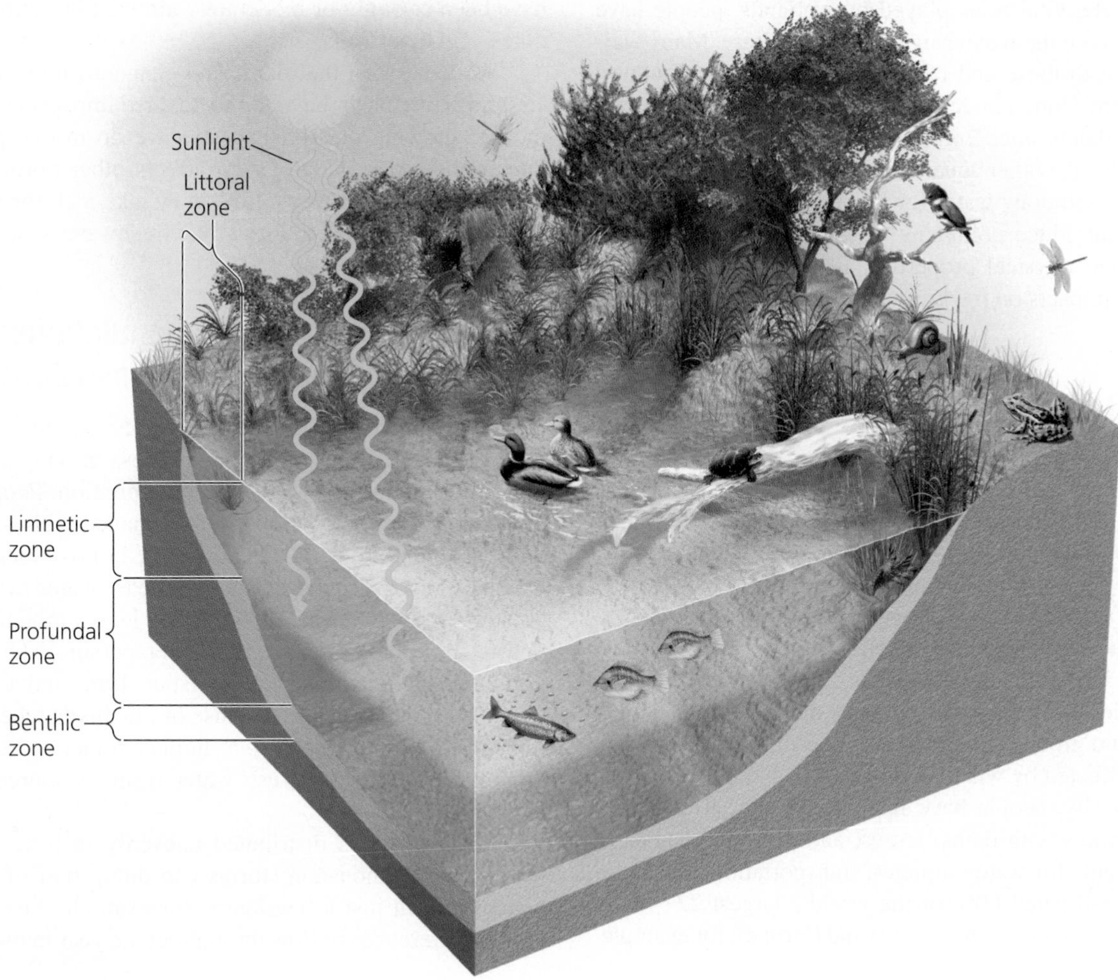

FIGURE 15.7 In lakes and ponds, emergent plants grow along the shoreline in the littoral zone. The limnetic zone is the layer of open, sunlit water, where photosynthesis takes place. Sunlight does not reach the deeper profundal zone. The benthic zone, at the bottom of the water body, often is muddy, rich in detritus and nutrients, and low in oxygen.

In cypress swamps of the southeastern United States, cypress trees grow in standing water (FIGURE 15.8). In northern forests, swamps are created when beavers dam streams with limbs from trees they have cut, flooding wooded areas upstream.

FIGURE 15.8 Freshwater wetlands such as this bald cypress swamp in Louisiana support biologically diverse and productive ecosystems.

Bogs are ponds covered with thick floating mats of vegetation and can represent a stage in aquatic succession.

Some wetlands are seasonal, being wet only at some times of the year. *Vernal pools* are an example. In many parts of North America, these pools form in early spring from rain and snowmelt, and then dry up once weather becomes warmer. Numerous animals have evolved to take advantage of the ephemeral time windows in which they exist each year.

Wetlands are extremely valuable habitat for wildlife. Louisiana's coastal wetlands, for example, provide habitat for approximately 1.8 million migratory waterbirds each year. Wetlands also provide important ecosystem services by slowing runoff, reducing flooding, recharging aquifers, and filtering pollutants. Cypress swamps in coastal Louisiana are home to many rare species that suffer habitat loss when cypress trees are ground up to produce cypress mulch for landscaping. Environmental groups lobbied major home-improvement retailers to cease selling cypress mulch from Louisiana, and scored major victories in the effort to save these forests when in 2007 Walmart, Home Depot, and Lowe's announced they would no longer buy or sell cypress mulch from southern Louisiana.

Despite the vital roles played by wetlands, people have drained and filled them extensively for agriculture. Many wetlands are lost to these and other human activities. Southern Canada and the United States, for example, have lost well over half their wetlands since European colonization.

Wetlands, like other aquatic systems, are affected by people when we withdraw water for human use, build dams and levees, and introduce pollutants that alter water's chemical, biological, and physical properties. Let's now take a closer look at such impacts on freshwater ecosystems.

Human Activities Affect Waterways

Fresh water is one of the world's most precious resources. Not only do we need it to keep our bodies hydrated and healthy, but we also require huge quantities of water for our homes, farms, and factories. Although water is a limited resource, it is also a renewable resource as long as we manage our use sustainably. Unfortunately, people are withdrawing water at unsustainable levels and are depleting many sources of surface water and groundwater. Already, one-third of the world's people are affected by water shortages.

Additionally, people have intensively engineered freshwater waterways with dams, levees, and diversion canals to satisfy demands for water supplies, transportation, and flood control. An estimated 60% of the world's largest 227 rivers (and 77% of those in North America and Europe), for example, have been strongly or moderately affected by artificial dams, dikes, and diversions.

We have seen that dams and channelization in the Mississippi River basin have led to adverse impacts at the river's mouth, showing clearly that what we do in one part of the interconnected aquatic system affects other portions, sometimes in significant ways. Let's now take a closer look at the many ways human activities affect freshwater systems.

Fresh water and human populations are unevenly distributed across Earth

The availability of fresh water varies widely around the world because different regions possess varying amounts of groundwater, surface water, and precipitation. People are not distributed across the globe in accordance with water availability (**FIGURE 15.9**). For example, Asia possesses the most water of any continent but has the least water available per person, whereas Australia, with the least amount of water, boasts the most water available per person. Many densely populated nations, such as Pakistan, Iran, and Egypt, face serious water shortages. Because of the mismatched distribution of water and population, human societies have always struggled to transport fresh water from its source to where people need it.

Fresh water is distributed unevenly in time as well as space. India's monsoon storms can dump half of a region's annual rain in just a few hours, for example. Seasonal rains lead to differences in flow throughout the year in many places.

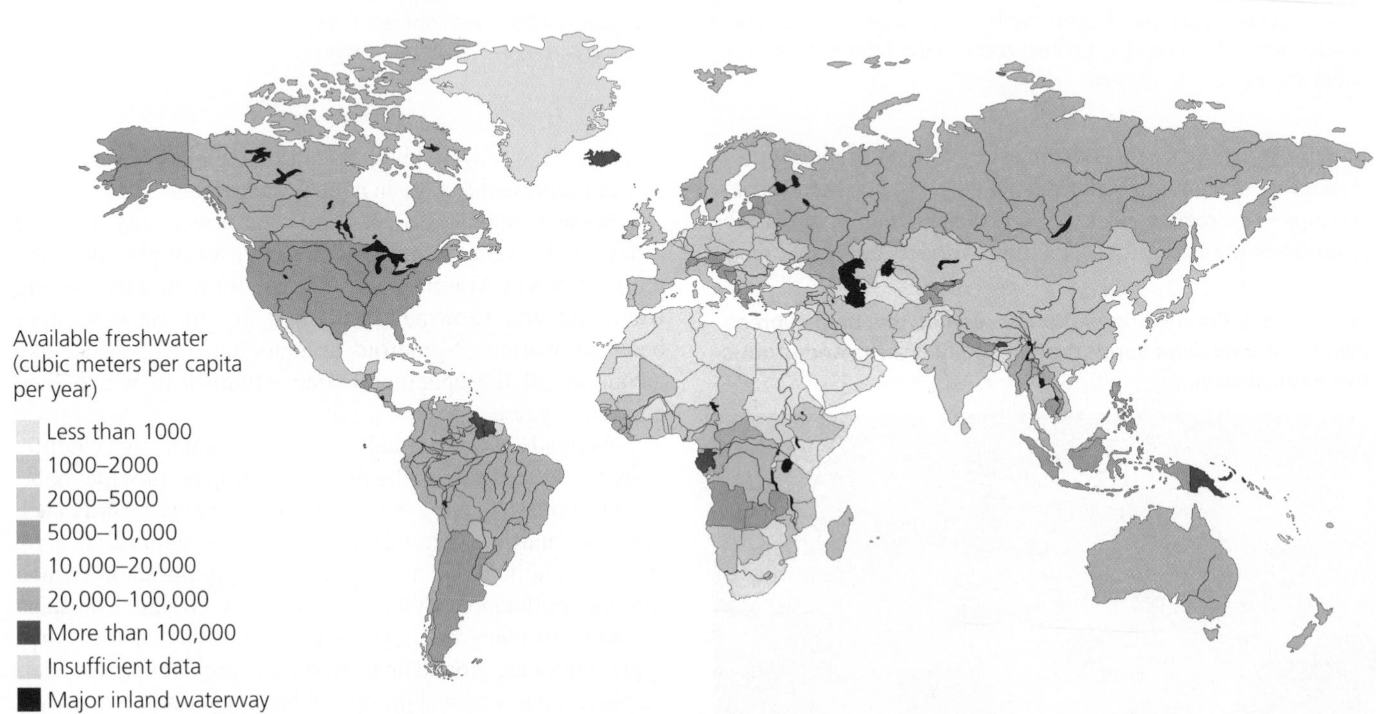

FIGURE 15.9 Nations vary tremendously in the amount of fresh water per capita available to their citizens. For example, Iceland, Papua New Guinea, Gabon, and Guyana (dark blue in this map) each have over 100 times more water per person than do many Middle Eastern and North African countries. *Data from Harrison, P., and F. Pearce, 2000.* AAAS atlas of population and the environment, *edited by the American Association for the Advancement of Science, © 2000 by the American Association for the Advancement of Science.*

For this reason, dams store water from wetter months that can be used in drier times of the year when river flow is reduced.

As if the existing mismatches between water availability and human need were not enough, global climate change (Chapter 18) will worsen conditions in many regions by altering precipitation patterns, melting glaciers, causing early-season runoff, and intensifying droughts and flooding. A 2009 study found that one-third of the world's 925 major rivers experienced reduced flow from 1948 to 2004, with the majority of the reduction attributed to effects of climate change. The reduction in water flow into the Pacific Ocean during this period was equivalent to the output of the mighty Mississippi, demonstrating the scale of changes involved.

Water supplies households, industry, and especially agriculture

We all use water at home for drinking, cooking, and cleaning. Most mining, industrial, and manufacturing processes require water. Farmers and ranchers use water to irrigate crops and water livestock. Globally, we allot about 70% of our annual fresh water use to agriculture. Industry accounts for roughly 20%, and residential and municipal uses for only 10%. Nations with arid climates tend to use a greater share for agriculture, whereas heavily industrialized nations use a greater share for industry.

When we remove water from an aquifer or surface water body and do not return it, this is called **consumptive use.** Our primary consumptive use of water is for irrigation, which is the water applied to crops (p. 233). In contrast, **nonconsumptive use** of water does not remove, or only temporarily removes, water from an aquifer or surface water body. Using water to generate electricity at hydroelectric dams is an example of nonconsumptive use; water is taken in, passed through dam machinery to turn turbines, and released downstream.

Why do we allocate 70% of our water use to agriculture? Our rapid population growth requires us to feed and clothe more people each year, and the intensification of agriculture during the Green Revolution (pp. 218, 247–248) required significant increases in irrigation. As a result, we withdraw 70% more water for irrigation today than we did 50 years ago and have doubled the amount of land under irrigation. So far, expansion of irrigated agriculture has kept pace with population growth; irrigated area per person has remained for four decades at around 460 m^2 (4950 ft^2)—one-tenth of a football field for each of us.

Irrigation can more than double crop yields by allowing farmers to apply water when and where it is needed. The 18% of world farmland that we irrigate yields fully 40% of our produce, including 60% of the global grain crop. Still, most irrigation remains highly inefficient, and crop plants end up using only 40% of the water that we apply. Inefficient "flood and furrow" irrigation, in which fields are liberally flooded with water that may evaporate from standing pools, accounts for 90% of irrigation worldwide. Overirrigation leads to waterlogging and salinization (pp. 234–235), which affect one-fifth of farmland today and reduce farming income worldwide by $11 billion. Such inefficient irrigation is possible because many national governments subsidize irrigation to promote agricultural self-sufficiency, drastically lowering water costs for farmers.

Excessive water withdrawals can drain rivers and lakes

In many places we are withdrawing surface water at unsustainable rates to meet our many demands for fresh water. As a result, many of the world's major rivers regularly run dry before reaching the sea. The Colorado River often fails to reach the Gulf of California after the many withdrawals of its water in the arid western United States and Mexico. This reduction in flow threatens the future of the cities and farms that depend on the river. And it has drastically altered the ecology of the river and its delta, changing plant communities, wiping out populations of fish and invertebrates, and devastating fisheries.

The Colorado's plight is not unique. Several hundred miles to the east, the Rio Grande also frequently runs dry, the victim of overextraction by both Mexican and U.S. farmers in times of drought. China's Yellow River also often fails to reach the sea. Even the river that has nurtured human civilization longer than any other, the Nile in Egypt, now peters out before reaching its mouth.

Nowhere are the effects of surface water depletion so evident as in the Aral Sea. Once the fourth-largest lake on Earth, just larger than Lake Huron, it lost over four-fifths of its volume in just 45 years (**FIGURE 15.10**). This dying inland sea, on the border of present-day Uzbekistan and Kazakhstan, is the victim of irrigation practices. The former Soviet Union instituted industrial cotton farming in this dry region by flooding

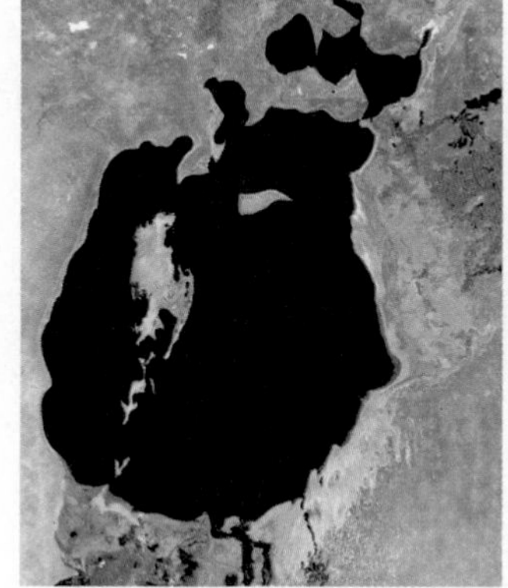

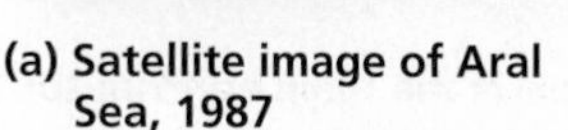

(a) Satellite image of Aral Sea, 1987

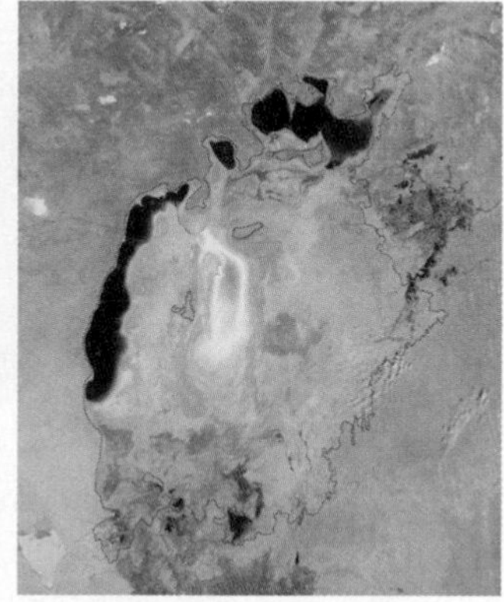

(b) Satellite image of Aral Sea, 2009

FIGURE 15.10 The Aral Sea in central Asia was once the world's fourth largest lake. However, it has been shrinking **(a, b)** because so much water was withdrawn to irrigate cotton crops.

FIGURE 15.11 Ships were stranded along the former shoreline of the Aral Sea because the waters receded so far and so quickly. Today restoration efforts are beginning to reverse the decline in the northern portion of the sea, and waters there are slowly rising.

the land with water from the two rivers that supplied the Aral Sea its water. For a few decades this boosted Soviet cotton production, but it shrunk the Aral Sea, and the irrigated soil became salty and waterlogged. Today 60,000 fishing jobs are gone, winds blow pesticide-laden dust up from the dry lakebed (**FIGURE 15.11**), and little cotton grows on the blighted soil. However, all may not be lost: Scientists, engineers, and local people struggling to save the northern portion of the Aral Sea and its ecosystems may now have begun reversing its decline.

Worldwide, roughly 15–35% of water withdrawals for irrigation are thought to be unsustainable. In areas where agriculture is demanding more fresh water than can be sustainably supplied, *water mining*—withdrawing water faster than it can be replenished—is taking place (**FIGURE 15.12**). In these areas, aquifers are being depleted or surface water is being piped in from other regions.

Groundwater can also be depleted

Groundwater is more easily depleted than surface water because most aquifers recharge very slowly. If we compare an aquifer to a bank account, we are making more withdrawals than deposits, and the balance is shrinking. Today we are mining groundwater, extracting 160 km^3 (5.65 trillion ft^3) more water each year than returns to the ground. This is a problem because one-third of Earth's human population—including 99% of the rural population of the United States—relies on groundwater to meet its needs for water.

As aquifers are mined, water tables drop. Groundwater becomes more difficult and expensive to extract, and eventually it may be depleted. In parts of Mexico, India, China, and many Asian and Middle Eastern nations, water tables are falling 1–3 m (3–10 ft) per year. In the United States, by the late

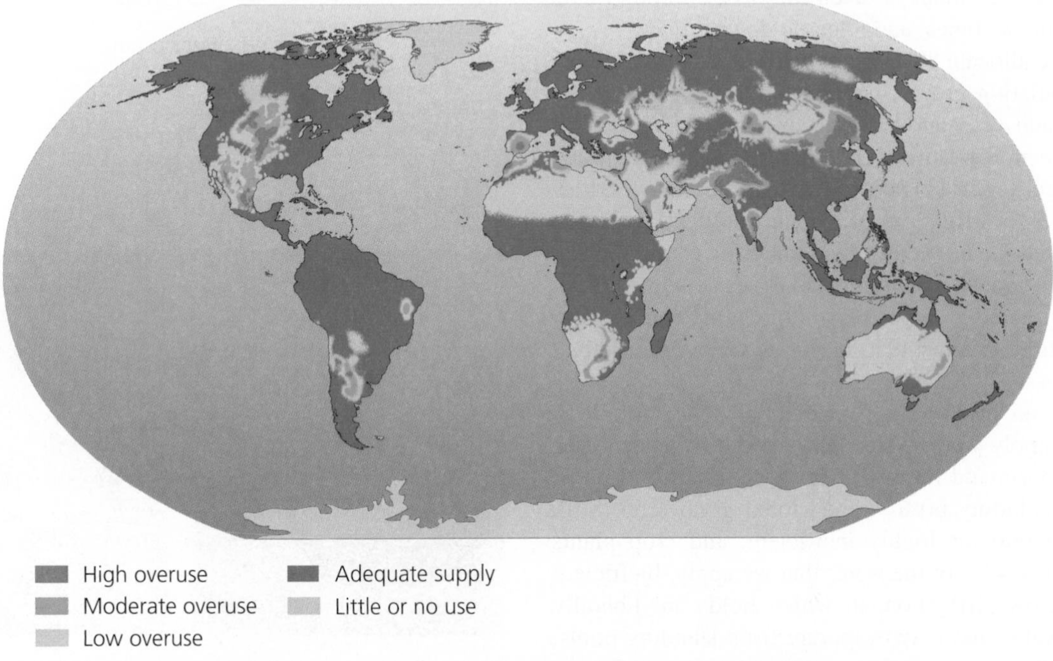

FIGURE 15.12 Irrigation for agriculture is the main contributor to unsustainable water use. Mapped are regions where overall use of fresh water (for agriculture, industry, and domestic use) exceeds the available supply, requiring groundwater depletion or diversion of water from other regions. The map understates the problem, because it does not reflect seasonal shortages. *Data from UNESCO, 2006.* Water: A shared responsibility. *World Water Development Report 2. UNESCO and Berghahn Books.*

(a) 16th-century chapel in Mexico City

(b) Sinkhole in Florida

FIGURE 15.13 When too much groundwater is withdrawn, the land may weaken and subside. This can cause buildings to lean, as seen in Mexico City **(a)**. Large areas of land may sometimes collapse suddenly in sinkholes, as seen here in Florida **(b)**.

1990s overpumping had drawn the Ogallala Aquifer down by 85 trillion gallons, a volume equal to half the annual flow of the Mississippi River.

When groundwater is overpumped in coastal areas, salt water from the ocean can intrude into aquifers, making water undrinkable. This has occurred in Florida, the Middle East, and other locations. Moreover, as aquifers lose water, they can become less capable of supporting overlying strata, and the land surface above may subside. For this reason, cities from Venice to Bangkok to Beijing are slowly sinking. Mexico City's downtown has sunk over 10 m (33 ft) since the time of Spanish arrival; streets are buckled, old buildings lean at angles, and underground pipes break so often that 30% of the system's water is lost to leaks (FIGURE 15.13a).

Sometimes land subsides suddenly, creating **sinkholes,** areas where the ground gives way with little warning, occasionally swallowing homes and businesses (FIGURE 15.13b). Once the ground subsides, soil and rock becomes compacted, losing the porosity that enabled it to hold water. Recharging a depleted aquifer thereafter becomes more difficult. Estimates suggest that compacted aquifers under California's Central Valley have lost storage capacity equal to that of 40% of the state's surface reservoirs.

Falling water tables also do vast ecological harm. Permanent wetlands exist where water tables reach the surface, so when water tables drop, wetland ecosystems dry up. In Jordan, the Azraq Oasis covered 7500 ha (18,500 acres) and enabled hundreds of thousands of migratory birds and other animals to find water in the desert. As a result of excessive groundwater pumping by farmers and the nearby city of Amman, the oasis dried up in the early 1990s, threatening this ecologically valuable site. Today international donors are helping Jordan's government to find alternative sources of water and restore this oasis, and to more efficiently manage groundwater withdrawals, as the water table has fallen some 17 m (56 ft) over the past 20 years.

Groundwater supplies our bottled water

These days, our groundwater is being withdrawn for a new purpose: to be packaged in plastic bottles and sold on supermarket shelves. Bottled water is booming business. The average American drinks 29 gallons of bottled water a year, and in 2012 sales topped $15 billion in the United States and $65 billion worldwide. Americans now drink more bottled water than beer or milk and pay more per gallon for it than for gasoline.

Most people who buy bottled water do so for portability and convenience, or because they believe it will taste better or be safer and healthier than tap water. However, in blind taste tests people think tap water tastes just as good, and chemical analyses show that bottled water is no safer or healthier than tap water (see THE SCIENCE BEHIND THE STORY, pp. 400–401). Much of it is municipal tap water that corporations have simply purified, packaged, and marked up in price to sell. In fact, when you buy a bottle of water, you have no idea where it came from or what its quality really is. The U.S. government strictly regulates the tap water that municipal governments provide us, but it requires little reporting from the corporations that sell us water in bottles at prices up to 1900 times more expensive than water from the tap.

Bottled water also exerts substantial ecological impact because it is heavily packaged and because we transport it long distances using fossil fuels. A 2009 study calculated the energy costs of bottled water to be 1000–2000 times greater than the energy costs of tap water. Most energy was used in manufacturing the bottle and transporting the product. Other studies indicate that producing one liter of bottled water requires the input of a quarter-liter of oil and 3–5 liters of additional water.

After use, at least three out of four bottles in the United States are thrown away, and not recycled. That's 30–40 billion containers per year (5 containers for every human being on Earth), and close to 1.5 million tons of plastic waste.

As a result of the environmental impacts of bottled water, as of 2012 over 90 colleges and universities in the United States and

Is It Better in a Bottle?

Which is safer and healthier for you to drink, tap water or bottled water?

If you said bottled water, you're not alone. Most people think bottled water is safer and healthier, which is why sales have doubled each decade for the past 20 years. But is bottled water really as pure and clean as its marketers want us to think?

It's hard to know the answer, because companies are not required to tell us anything about the quality of the water in their bottles, or even where the water comes from.

Municipalities that provide tap water to their residents need to submit regular reports to the Environmental Protection Agency describing their sources, treatment methods, and contaminants. In contrast, bottled water is regulated much more lightly because it is classifed as a "food" by the Food and Drug Administration (FDA). Bottling companies do not have to inform the public or the government where their water comes from or how it is treated, and they are not required to test samples with certified laboratories or notify the FDA of contamination problems.

In 2009, Congress held a hearing on this discrepancy after the U.S. Government Accountability Office and the nonprofit Environmental Working Group each published reports detailing how consumers cannot get basic information about the bottled water they purchase. The Environmental Working Group (EWG), based in Washington, D.C., surveyed the labels and websites of 188 brands of bottled water. Only two of these brands disclosed information comparable to that required of municipal tap water providers. The FDA told Congress it would step up inspections, but the agency is overworked and understaffed as it is.

Each year, 30–40 billion plastic water bottles are thrown away in the United States.

So, to find out what's in bottled water, scientists have had to do some detective work. In 2008, research scientists at the EWG sent samples of 10 major brands of bottled water to the University of Iowa's Hygienic Laboratory for analysis. The lab's chemists ran a battery of tests and detected 38 chemical pollutants, including traces of heavy metals, radioactive isotopes, caffeine and pharmaceuticals from wastewater pollution, nitrate and ammonia from fertilizer, and various industrial compounds such as solvents and plasticizers (**FIGURE 1**). Each brand contained eight contaminants on average, and two brands had levels of chemicals that

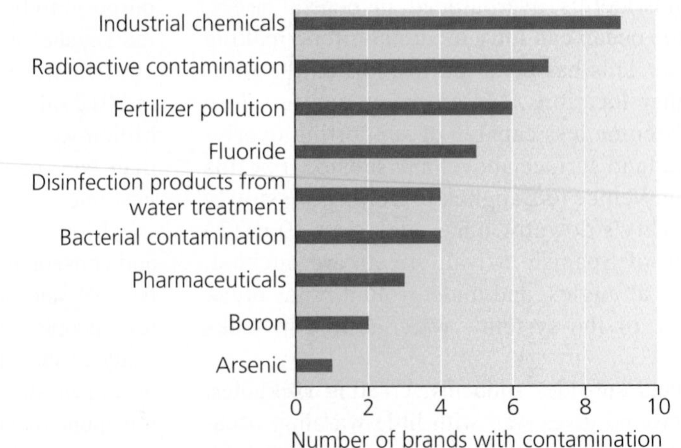

FIGURE 1 Of 10 leading brands of bottled water tested, most contained industrial chemicals, radioactive isotopes, and fertilizer pollution, as well as other contaminants. *Source: Naidenko, O., et al., 2008.* Bottled water quality investigation: 10 major brands, 38 pollutants. *Environmental Working Group, Washington, D.C.*

Canada have banned the sale of bottled water or restricted the use of plastic water bottles on campus. Similarly, major U.S. cities including New York City, San Francisco, and Seattle prohibit using government funds to purchase bottled water, in part due to the cost savings of drinking tap water instead of bottled water.

People build dikes and levees to control floods

Among the reasons we control the movement of fresh water, flood prevention ranks high. People have always been attracted to riverbanks for their water supply and for the flat topography and fertile soil of floodplains. But if one lives in a floodplain, one must be prepared to face flooding. **Flooding** is a normal, natural process that occurs when snowmelt or heavy rain swells the volume of water in a river so that water spills over the river's banks. In the long term, floods are immensely beneficial to both natural systems and human agriculture, because floodwaters build and enrich soil by spreading nutrient-rich sediments over large areas.

In the short term, however, floods can do tremendous damage to the farms, homes, and property of people who

exceeded legal limits in California and industry safety guidelines.

Two brands showed the chemical composition of standard municipal water treatment—including chlorine, fluoride, and other by-products of disinfection—indicating that these were identical to tap water. Indeed, an estimated 25–44% of bottled water *is* tap water. Companies can essentially just turn on the faucet, filter the water, bottle it, and sell it to us at marked-up prices.

EWG also sent water samples to a lab at the University of Missouri, where researchers tested for cancer-causing effects in cell cultures. Their assays measure whether a substance causes an increase in the growth of breast cancer cells. One bottled water brand caused a 78% increase in cancer cell growth relative to the control sample, suggesting that something in the water was carcinogenic.

Other scientists have compared bottled water to tap water in different ways. In 2009, researchers Martin Wagner and Jörg Oehlmann of Johann Wolfgang Goethe University in Frankfurt, Germany, tested 20 brands of bottled water for the presence of hormone-disrupting chemicals that mimic estrogen. Endocrine disruptors such as bisphenol A, phthalates, and other compounds found in plastics can exert a wide array of health impacts, even at very low doses (Chapter 14).

Wagner and Oehlmann compared nine brands packaged in glass bottles, nine brands packaged in plastic bottles (PET, or polyethylene terephthalate, the type of plastic with a #1 symbol on the bottom), and two brands packaged in "Tetra Pak" paperboard boxes with an inner plastic coating. They placed samples (as well as tap-water samples as controls) in a "yeast estrogen screen," a standard test-tube screening procedure that uses yeast cells engineered with genes to change color when exposed to estrogen-mimicking compounds.

The researchers detected estrogenic contamination in 60% of the samples (**FIGURE 2**). Both Tetra Pak brands and seven of nine plastic brands contained hormone-mimicking substances that apparently leached from the packaging. So did three of the glass-bottled brands, presumably from contamination at the bottling plant.

Wagner and Oehlmann then tested whether chemicals in the water would affect a living animal, a type of snail that is known to increase its reproduction when confronted with an estrogenic substance. They raised some snails in PET plastic containers and others in glass containers. After 56 days, the snails in the PET containers had produced 39–122% more embryos than snails in control conditions, whereas snails in glass containers showed no difference. This suggested that estrogenic compounds were leaching from the plastic. Their results were published in the journal *Environmental Science and Pollution Research*.

Research on the quality and safety of bottled water is just getting started, but already these studies and others like them have indicated that bottled water can contain a range of contaminants, some of which may pose health risks. Tap water may also contain plenty of pollutants, but municipalities are required to test and report on their water quality, so it is much easier for consumers to get information.

For maximum safety and health, the EWG recommends drinking filtered tap water (using a filter on your faucet) instead of bottled water. Depending on the quality of your local tap water, it may be as good or better than bottled water, even when unfiltered. You can consult water quality information from your utility to find out what you are drinking. ■

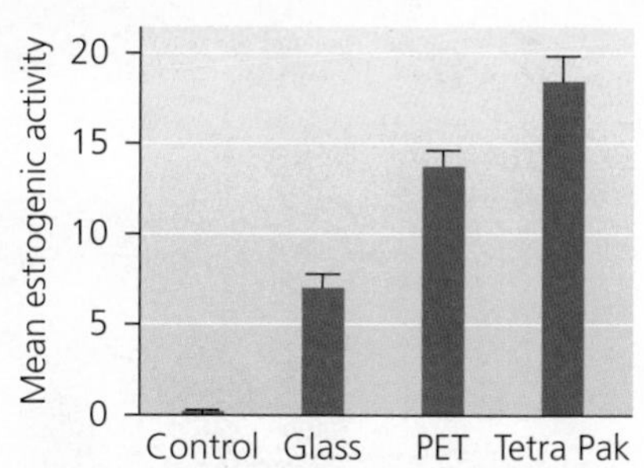

FIGURE 2 Estrogenic activity (as determined by a yeast cell culture test) was highest in bottled water from Tetra Pak containers, followed by PET plastic containers, and then glass containers. A negative control showed no appreciable estrogenic potency. *With kind permission from Springer Science and Business Media and the author, from Wagner, M., and J. Oehlmann, 2009. Endocrine disruptors in bottled mineral water: Total estrogenic burden and migration from plastic bottles*. Environmental Science and Pollution Research *16: 278–286, Fig 3a.*

choose to live in floodplains. To protect against floods, communities and governments have built levees (also called *dikes*) along banks of rivers such as the Mississippi to hold water in main channels. Many dikes are small and locally built, but the U.S. Army Corps of Engineers has constructed thousands of miles of massive levees along major waterways. These structures prevent flooding at most times and places but can sometimes worsen flooding because they force water to stay in channels and accumulate, building up enormous energy and leading to occasional catastrophic overflow events (**FIGURE 15.14**).

When we engineer rivers to stay in their channels, we often end up increasing the frequency of floods in downriver areas. Human development also tends to worsen flooding because pavement and compacted soils speed runoff, sending intense pulses of water into rivers. In contrast, forests and wetlands make flooding less likely because they allow water to spread out while vegetation slows its flow and porous soil soaks it up. Thus, major flood events may become more frequent in developed areas. Indeed, a "100-year flood" is not something that can occur only once every 100 years. Rather, it is a statistical probability: in any given year, there is a 1%

FIGURE 15.14 Unusually high water levels in the Mississippi River in May 2011 caused this levee in East Carroll Parish, Louisiana, to collapse, flooding adjacent farmland. Levees upstream in some less-populated areas were intentionally destroyed and the floodplain inundated in order to lower river volumes and protect more-populated areas downriver.

chance of a 100-year flood. Like spinning a roulette wheel, one occasionally gets the same unlikely result twice in a row. Scientists calculate the likelihood of major floods based on natural historical conditions, but if conditions change (through increased development, habitat loss, engineering and channelization, or global climate change), then the frequency of flooding can change. Many of the levees that currently line the Mississippi River were constructed after one such catastrophic flood struck the lower Mississippi River in 1927, leading to public demand for greater protection from flooding.

We divert surface water to suit our needs

People have long diverted water from rivers and lakes to farm fields, homes, and cities. Water in the Colorado River in the western United States is heavily diverted and utilized as the river flows toward the Pacific Ocean (**FIGURE 15.15**). Early in its course, some Colorado River water is piped through a mountain tunnel and down the Rockies' eastern slope to supply the city of Denver. More is removed for Las Vegas and other cities and for farmland as the water proceeds downriver. When it reaches Parker Dam on the California–Arizona state line, large amounts are diverted into the Colorado River Aqueduct, which brings water to millions of people in the Los Angeles and San Diego areas via a long, open-air canal. From Parker Dam, Arizona also draws water, transporting it in canals of the Central Arizona Project. Further south, water is diverted into the Coachella and All-American Canals, destined for agriculture, mostly in California's Imperial Valley.

The world's largest diversion project is underway in China. There, government leaders are pushing through an ambitious plan to pipe water from the Yangtze River

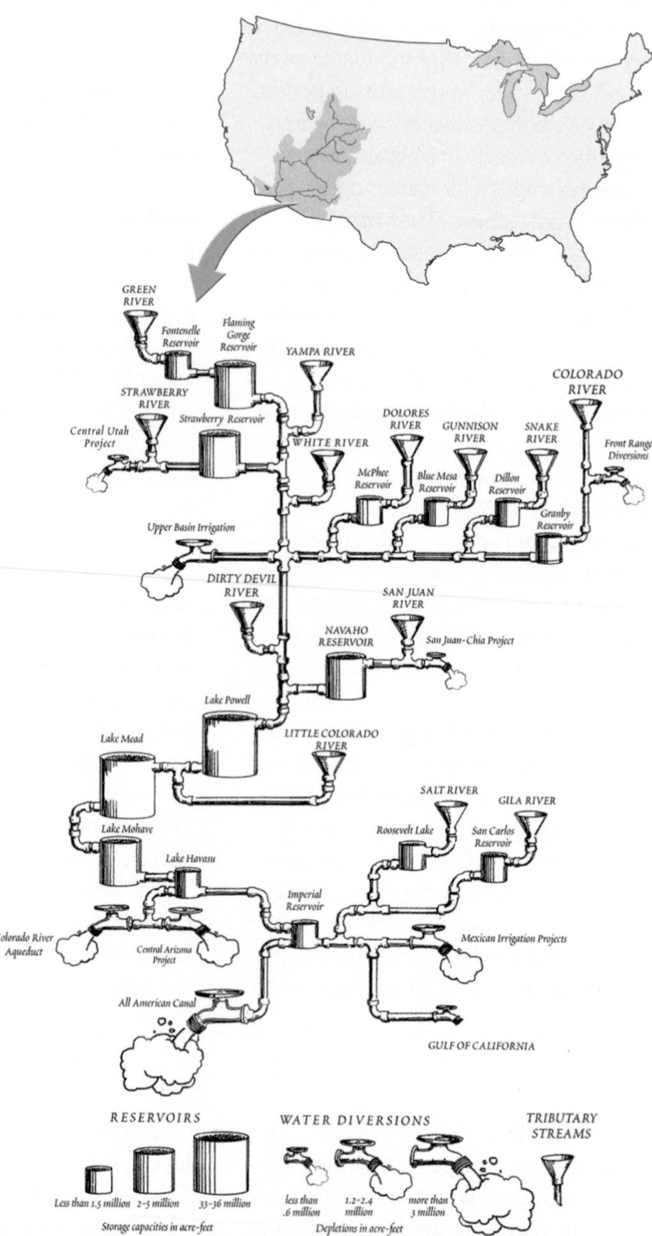

FIGURE 15.15 The degree to which we have engineered the once-wild Colorado River led cartoonist Lester Dore to portray the Colorado and its tributaries as an immense plumbing system. Thirteen major dams pool water in enormous reservoirs along the lengths of the Colorado River and its tributaries. Several major canals divert water from the river, mostly to irrigate crops in desert regions. *From* High Country News, *10 November 1997.*

WEIGHING THE ISSUES

REACHING FOR WATER Los Angeles is not the only controversial diversion of fresh water in the United States. The rapidly growing Las Vegas metropolitan area is exceeding its allotment of water from the Colorado River and has proposed a $3.5 billion project that would divert groundwater from 450 km (280 mi) away to meet the growing demand in Nevada's largest city. Do you think such diversions are ethically justified? If rural communities and wetland ecosystems at the diversion site in eastern Nevada are destroyed by this project, is this an acceptable cost given the economic activity generated in Las Vegas? How else might cities like Las Vegas meet their future water needs?

in southern China (where water is plentiful) to northern China's Yellow River, which routinely dries up at its mouth because the climate is drier and people withdraw much of its water. Three sets of massive aqueducts, totaling 2500 km (1550 mi) in length, are being built to move trillions of gallons of water northward. China's leaders hope the diversions will solve water shortages for northern farms and cities. Many scientists say the $62 billion project won't transfer enough to make a difference, yet would cause extensive environmental impacts and displace hundreds of thousands of people.

In the past, large-scale diversion projects have enabled politically strong yet water-poor regions to forcibly appropriate water from communities too weak to keep it for themselves. The city of Los Angeles grew by commandeering water from the rural Owens Valley 350 km (220 mi) away. In so doing, it turned the environment of this region to desert, creating dustbowls and destroying its economy. Then in 1941, city leaders decided to divert streams feeding into Mono Lake, over 565 km (350 mi) away in northern California. As the lake level fell 14 m (45 ft) over 40 years, salt concentrations doubled, and aquatic communities suffered.

We have erected thousands of dams

A **dam** is any obstruction placed in a river or stream to block its flow. Dams create **reservoirs,** artificial lakes that store water for human use. We build dams to prevent floods, provide drinking water, facilitate irrigation, and generate electricity (pp. 574–575).

Worldwide, we have erected more than 45,000 large dams (greater than 15 m, or 49 ft, high) across rivers in over 140 nations. We have built tens of thousands of smaller dams. Only a few major rivers in the world remain undammed and free-flowing. These run through the tundra and taiga of Canada, Alaska, and Russia and in remote regions of Latin America and Africa.

Dams produce a mix of benefits and costs, as illustrated in **FIGURE 15.16**. As an example of this complex mix, we can consider the world's largest dam project. The Three Gorges Dam on China's Yangtze River, 186 m (610 ft) high and 2.3 km (1.4 mi) wide, was completed in 2008 (**FIGURE 15.17a**). Its reservoir stretches for 616 km (385 mi; as long as Lake Superior). This project provides flood control, enables boats and barges to travel farther upstream, and generates enough hydroelectric power to replace dozens of large coal or nuclear plants.

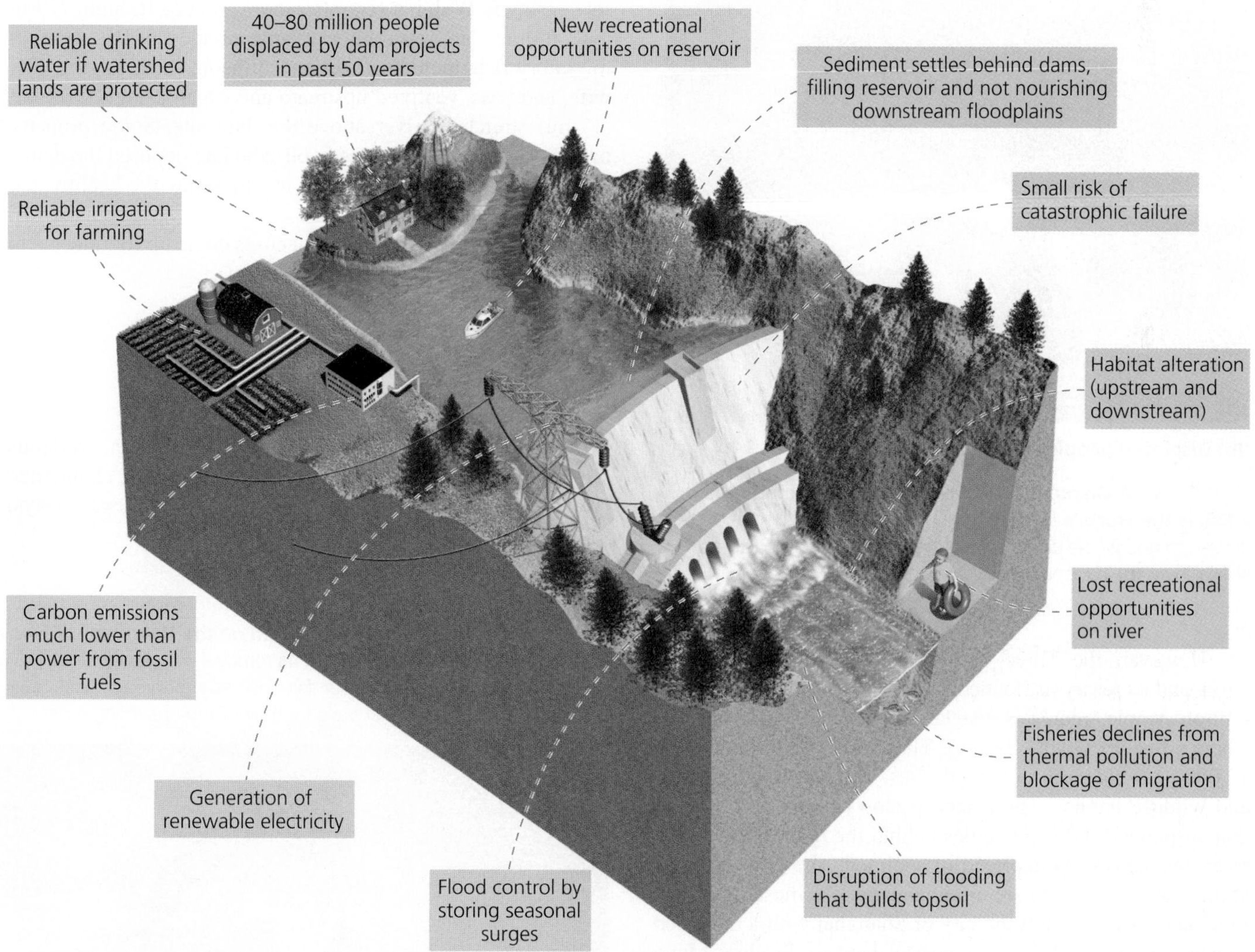

FIGURE 15.16 Damming rivers has diverse consequences for people and the environment. The generation of clean and renewable electricity is one of several major benefits **(green boxes)** of hydroelectric dams. Habitat alteration is one of several negative impacts **(red boxes)**.

(a) The Three Gorges Dam in Yichang, China

(b) Displaced people in Sichuan Province, China

FIGURE 15.17 **China's Three Gorges Dam (a), completed in 2008, is the world's largest dam.** Over 1.2 million people were displaced and whole cities were leveled for its construction, as shown here **(b)** in Sichuan Province.

However, the Three Gorges Dam cost $39 billion to build, and its reservoir flooded 22 cities and the homes of 1.24 million people, requiring the largest resettlement project in China's history (FIGURE 15.17b). The rising water submerged 10,000-year-old archaeological sites, productive farmlands, and wildlife habitat. The reservoir slows the river's flow so that suspended sediment settles behind the dam. Because the river downstream is deprived of sediment, the tidal marshes at the Yangtze's mouth are eroding away, like those in coastal Louisiana. This has left the city of Shanghai with a degraded coastal environment and less coastal land to develop. Many scientists worry that the Yangtze's many pollutants will also be trapped in the reservoir, making the water undrinkable. The Chinese government plans to spend $5 billion building hundreds of sewage treatment and waste disposal facilities. On top of all these worries, earthquakes in southern China in 2008 and again in 2012 raised fears that a future quake could damage the dam, perhaps even leading to a catastrophic collapse.

Some dams are being removed

People who feel that the costs of some dams outweigh their benefits are pushing for such dams to be dismantled (FIGURE 15.18). By removing dams and letting rivers flow free, they say, we can restore riparian ecosystems, reestablish economically valuable fisheries, and revive river recreation such as fly-fishing and rafting. Increasingly, private dam owners and the Federal Energy Regulatory Commission (FERC), the U.S. government agency charged with renewing licenses for dams, have agreed. Many aging dams are in need of costly repairs or have outlived their economic usefulness, and roughly 400 dams have been removed in the United States in the past decade.

The drive to remove dams first gathered steam in 1999 with the dismantling of the Edwards Dam on Maine's Kennebec River. FERC had determined that the environmental benefits of removing the dam outweighed the economic benefits of relicensing it. Within a year after the 7.3-m (24-ft) high, 279-m (917-ft) long dam was removed, large numbers of 10 species of migratory fish, including salmon, sturgeon, shad, herring, alewife, and bass, ventured upstream and began using the 27-km (17-mi) stretch of river above the dam site. Some property owners along the former reservoir who had opposed the dam's removal had a change of heart once they saw the healthy and vibrant river that now ran past their property. More dams will come down as over 500 FERC licenses come up for renewal in the next decade.

Wetlands are affected by human manipulations of waterways

From the Mississippi River Delta to the Aral Sea, wetlands are being lost as we divert and withdraw water, channelize rivers, build dams, and otherwise engineer natural waterways.

FIGURE 15.18 **The Great Works Dam on the Penobscot River in Maine was removed in 2012.** Its removal will provide endangered Atlantic salmon and other fish species improved access to around 1000 miles of waterways for their inland spawning runs.

These actions add to the extensive draining of wetlands for agriculture (p. 238). As wetlands disappear, we lose the many ecosystem services (pp. 3, 116–117, 152, 190) they provide us, such as filtering pollutants, harboring wildlife, controlling floods, and helping to maintain drinking water supplies. A 2010 report estimated the ecosystem services provided by Louisiana's coastal wetlands alone at $12–47 billion a year, showing the economic value in conserving these natural resources.

Fortunately, many people today see the value in wetlands and are trying to protect and restore them. In 1971 an international agreement was reached in Ramsar, Iran, to document and protect wetlands around the world. The *Ramsar Convention on Wetlands of International Importance* seeks the "conservation and wise use of all wetlands" through the "maintenance of their ecological character . . . within the context of sustainable development." Today the Chesapeake Bay estuary (p. 105), Azraq oasis in Jordan, and nearly 1900 other sites covering 185 million ha (714,000 mi^2) across the globe are granted a degree of protection as Ramsar Wetlands—wetlands noted for their ecological, social, and economic importance.

Solutions to Depletion of Fresh Water

Population growth, expansion of irrigated agriculture, and industrial development doubled our annual fresh water use in the last 50 years. We now use an amount equal to 10% of total global runoff. The hydrologic cycle makes fresh water a renewable resource, but if we take more than a lake, river, or aquifer can provide, we must either reduce our use, find another water source, or be prepared to run out of water.

Solutions can address supply or demand

To address shortages of fresh water, we can aim either to increase supply or to reduce demand. We can increase supply temporarily through more intensive extraction, but this is generally not sustainable. Diversions may solve supply problems in one area while causing shortages in others. In contrast, strategies for reducing demand include conservation and efficiency measures. Lowering demand is more difficult politically in the short term but may be necessary in the long term. In the developing world, international aid agencies are increasingly funding demand-based solutions over supply-based solutions, because demand-based solutions offer better economic returns and cause less ecological and social damage.

Desalination "makes" more fresh water

A supply strategy with some potential for sustainability is to generate fresh water by **desalination**, or *desalinization*, the removal of salt from seawater or other water of marginal quality. One method of desalination mimics the hydrologic cycle by evaporating allotments of ocean water with heat and then condensing the vapor—essentially distilling fresh water. Another method forces water through membranes to filter out salts; the most common such process is reverse osmosis. The process converts saline water with up to 35,000 parts per million (ppm) of dissolved salts to fresh water with less than 1000 ppm dissolved salts.

Over 20,000 desalination facilities are operating worldwide. However, desalination is expensive, requires large inputs of fossil fuel energy, kills aquatic life at water intakes, and generates concentrated salty waste. As a result, large-scale desalination is pursued mostly in wealthy oil-rich nations where water is extremely scarce (**FIGURE 15.19**). In Saudi Arabia, desalination produces half the nation's drinking water. The largest facility in the United States is in Tampa, Florida, whose groundwater suffers from saltwater intrusion.

FAQ **Can't we just use desalination to fulfill our demand for water?**

Given the seemingly endless supply of water in Earth's oceans, many people assume that desalination is the answer to our world's water crises. So why aren't we eagerly utilizing this technology everywhere?

Simply put, we lack the abundant, clean energy sources needed to make the widespread use of desalination economically viable and environmentally sustainable. For example, the United States withdraws over 700 billion liters (185 billion gallons) of fresh water every day for use in food production, industry, and public supplies. Diverting the energy necessary to supply even a tiny fraction of this quantity from desalination would cause prices for electricity, gasoline, and other fuels to skyrocket. Using fossil fuels as an energy source for desalination would also drastically increase U.S. emissions of air pollutants and greenhouse gases. Due to these constraints, it is unlikely that desalination will be widely embraced in the United States unless we are able to find abundant, environmentally friendly energy sources.

Agricultural demand can be reduced

Because most water is used for agriculture, it makes sense to look first to agriculture for ways to decrease demand. Farmers can improve efficiency by lining irrigation canals to prevent leaks, leveling fields to minimize runoff, and adopting efficient irrigation methods. Low-pressure spray irrigation squirts water downward toward plants, and drip irrigation systems target individual plants and introduce water directly onto the soil (see Figure 9.22b, p. 233). Both methods reduce water lost to evaporation and runoff. Experts estimate that drip irrigation (in which as little as 10% of water is wasted) could reduce water withdrawals while raising yields by 20–90% and producing $3 billion in extra annual income for farmers of the developing world.

Choosing crops to match the land and climate in which they are farmed can save huge amounts of water. Currently, crops that require a great deal of water, such as cotton, rice, and alfalfa, are often planted in arid areas with government-subsidized irrigation. As a result of the subsidies, the true cost of water is not part of the costs of growing the crop. Eliminating subsidies and growing crops in climates with adequate rainfall could greatly reduce water use.

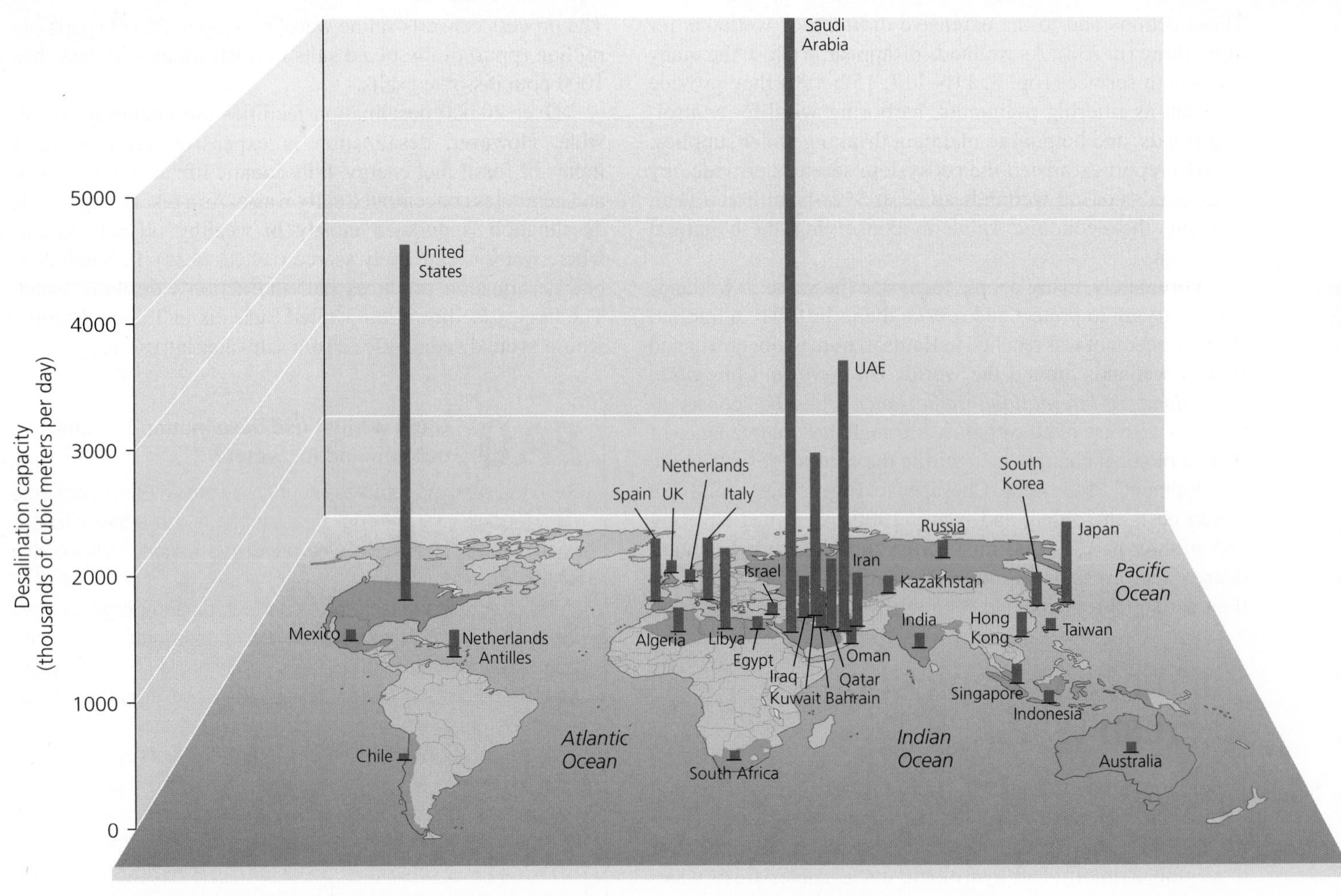

FIGURE 15.19 Much of the world's desalinization occurs in arid nations with few fresh water resources and little access to reliable supplies of energy. In the United States, both seawater and saline groundwater are desalinated to produce fresh water. Only nations with a desalination capacity of 70,000 cubic m per day or more are shown. *Data from Pacific Institute,* The world's water, 2009.

In addition, selective breeding (p. 52) and genetic modification (pp. 261–268) can produce crop varieties that produce high yields with less water.

We can lower residential and industrial water use

In our households, we can reduce water use by installing low-flow faucets, showerheads, washing machines, and toilets. Automatic dishwashers, studies show, use less water than does washing dishes by hand. Catching rain runoff from your roof in a barrel—*rainwater harvesting*—will reduce the amount you need to use from the hose. And if your city allows it, you can use *gray water*—the wastewater from showers and sinks—to water your yard. Better yet, you can replace a water-intensive lawn with native plants adapted to your region's natural precipitation patterns. *Xeriscaping*, landscaping using plants adapted to arid conditions, has become popular in the U.S. Southwest (**FIGURE 15.20**). Each of us can cut our daily water use dozens, hundreds, or even thousands of gallons a day by reexamining aspects of our daily lives.

Industry and municipalities can take water-saving steps as well. Manufacturers are shifting to processes that use less water and in doing so are reducing their costs. Las Vegas is one of many cities that are recycling treated municipal wastewater for irrigation and industrial uses. Governments in Arizona and in England are capturing excess runoff and pumping it into aquifers. Finding and patching leaks in pipes has saved some cities and companies large amounts of water—and money. Boston and its suburbs reduced water demand by 31% over 17 years by patching leaks, retrofitting homes with efficient plumbing, auditing industry, and promoting conservation to the public. This program enabled Massachusetts to avoid an unpopular $500 million river diversion scheme.

Market-based approaches to water conservation are being debated

Economists who want to use market-based strategies to achieve sustainable water use have suggested ending government subsidies of inefficient practices and instead letting water become a commodity whose price reflects the true costs

FIGURE 15.20 Xeriscaping lets homeowners and businesses reduce water consumption by landscaping with attractive, drought-tolerant plants.

of its extraction. Others worry that making water a fully priced commodity would make it less available to the world's poor and increase the gap between rich and poor. Because industrial use of water can be 70 times more profitable than agricultural use, market forces alone might favor uses that would benefit wealthy and industrialized people, companies, and nations at the expense of the rural poor.

Similar concerns surround another potential solution, the privatization of water supplies. During the 1990s, many public water systems were partially or wholly privatized, as governments transferred construction, maintenance, management, or ownership to private companies. This was done to enhance efficiency, but firms have little incentive to allow equitable access to water for rich and poor alike. Already in some developing countries, rural residents without access to public water supplies find themselves forced to buy water from private vendors and pay 12 times more than those connected to public supplies.

Other experiences indicate that decentralization of control over water, from the national level to the local level, can help conserve water. In Mexico, the effectiveness of irrigation systems improved dramatically once they were transferred from public ownership to the control of 386 local water user associations.

Regardless of how demand is addressed, the shift from supply-side to demand-side solutions is paying dividends. In Europe, a new focus on demand (through government mandates and public education) has decreased public water consumption, and industries are becoming more water-efficient. The United States decreased its water consumption by 5% from 1980 to 2005 thanks to conservation measures, even while its population grew 31%.

Nations often cooperate to resolve water disputes

We've seen that nations have unequal access to fresh water supplies, and there are fears that scarcity of this vital resource can lead to conflict. For example, a 2012 report by the Intelligence agencies of the U.S. government concluded that in the subsequent decade, many nations vital to U.S. interests will experience political and economic instability due to water shortages, making freshwater supplies one of the greatest threats to U.S. national security interests. Faced with such instability, one recourse would be to commandeer the water resources of other nations by force, sparking international military conflict. **FIGURE 15.21** shows some waterways for which conflict is a concern in coming decades.

A total of 261 major rivers (whose watersheds cover 45% of the world's land area) cross national borders, and transboundary disagreements are common. Water is already a key element in the hostilities among Israel, the Palestinian people, and neighboring nations. The United States has its share of conflicts over water. The Colorado River's water allocations

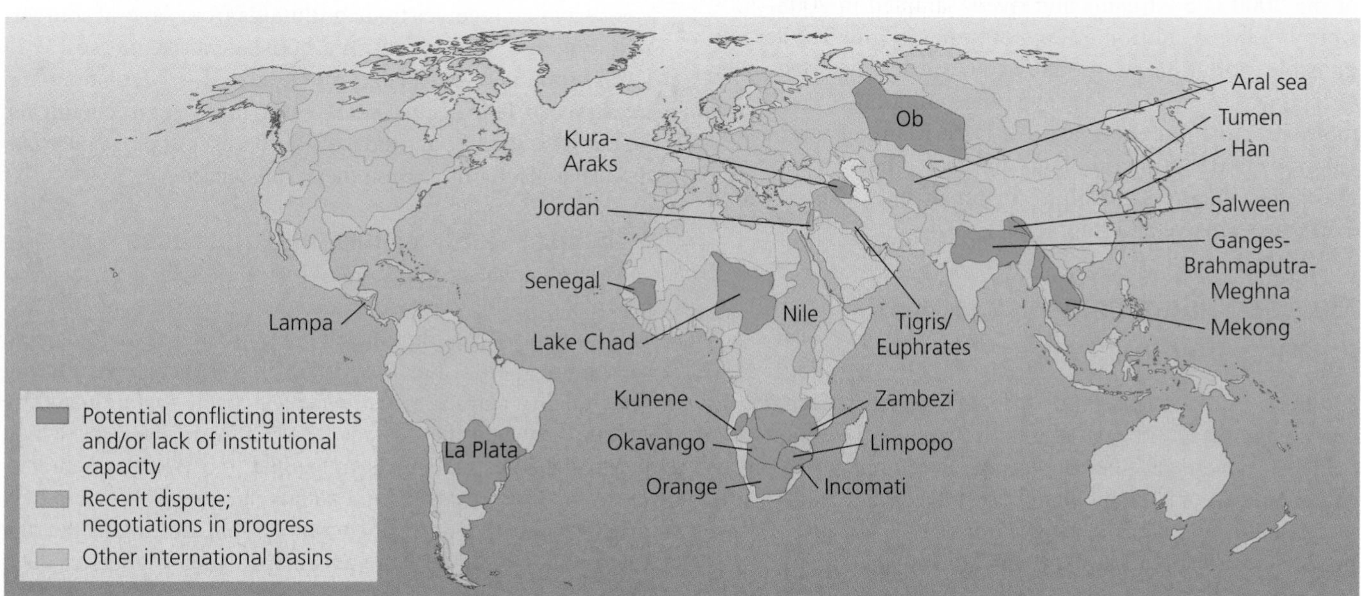

FIGURE 15.21 Water basins that cross national boundaries (yellow) have the potential for conflict if water supplies become scarce. Basins with higher potential for conflict **(red)** are found in regions with growing populations, but negotiations are underway on several international basins to prevent conflict **(orange)**.

have long been a source of conflict between farms and growing cities in the arid West. The states of Georgia, Alabama, and Florida are also currently embroiled in disputes over water withdrawals from shared rivers. Conflicts also visited the Mississippi River during the extreme drought in the Midwest in 2012. Deprived of precipitation, river levels dropped significantly, and this pitted upriver states who wanted to hold river water behind their dams against downriver states who wanted greater releases from those dams to facilitate barge travel and increase water supplies.

Yet on the positive side, so far many nations have cooperated to resolve water disputes. India has struck agreements to co-manage transboundary rivers with Pakistan, Bangladesh, Bhutan, and Nepal. In Europe, nations along the Rhine and Danube rivers have signed water-sharing treaties. Such progress gives reason to hope that "water wars" will be few and far between in coming decades.

Freshwater Pollution and Its Control

We have seen that people affect aquatic systems by withdrawing too much water and by altering the systems' natural processes by engineering waterways with dams, diversions, and levees. However, people also affect aquatic ecosystems and threaten human health when we introduce toxic substances and disease-causing organisms into surface waters and groundwater.

Developed nations have made admirable advances in cleaning up water pollution over the past few decades. Still, the World Commission on Water recently concluded that over half the world's major rivers remain "seriously depleted and polluted, degrading and poisoning the surrounding ecosystems, threatening the health and livelihood of people who depend on them." Levels of impairment are similar in U.S. waterways. In 2013 EPA reported that 55% of the 2000 U.S. streams and rivers sampled in 2008–2009 were in poor condition to support aquatic life. The largely invisible pollution of groundwater, meanwhile, has been termed a "covert crisis." Preventing pollution is easier and more effective than mitigating it later. Many of our current solutions to pollution problems embrace preventative strategies rather than "end-of-pipe" treatment and cleanup, which is often expensive and impractical.

Water pollution comes from point and non-point sources

Pollution is the release into the environment of matter or energy that causes undesirable impacts on the health or well-being of humans or other organisms. Pollution can be physical, chemical, or biological and can affect water, air, or soil. **Water pollution** comes in many forms and can cause diverse impacts on aquatic ecosystems and human health.

Most forms of water pollution are not conspicuous to the human eye, so scientists and technicians measure water's chemical properties (such as pH, nutrient concentrations, and dissolved oxygen concentration), physical characteristics (such as temperature and turbidity—the density of suspended particles in a water sample), and biological properties (such as the presence of harmful microorganisms or the species diversity in aquatic ecosystems).

Some water pollution is emitted from **point sources**—discrete locations, such as a factory or sewer pipe. In contrast, pollution from **non-point sources** is cumulative, arising from multiple inputs over larger areas, such as farms, city streets, and residential neighborhoods (**FIGURE 15.22**). The U.S. Clean Water Act (p. 176) addressed point-source pollution with some success by targeting industrial discharges. As a result, water quality in the United States today suffers most from non-point-source pollution, resulting from countless common activities such as applying fertilizers and pesticides to lawns, applying salt to roads in winter, and changing automobile oil. To minimize non-point-source pollution of drinking water, governments limit development on watershed land surrounding reservoirs.

Water pollution takes many forms

Water pollution comes in many forms that can impair waterways and threaten people and organisms that drink or live in affected waters. Let's survey the major classes of water pollutants affecting waters in the world today.

Toxic chemicals Our waterways have become polluted with toxic organic substances of our own making, including pesticides, petroleum products, and other synthetic chemicals (pp. 367–368). Many of these can poison animals and plants, alter aquatic ecosystems, and cause an array of human health problems, including cancer. In addition, toxic metals such as arsenic, lead, and mercury damage human health and the environment, as do acids from acid precipitation (pp. 473–474) and from acid drainage from mining sites (p. 639). With its massive watershed encompassing rural, urban, and suburban areas, the Mississippi River is one waterway that receives toxic pollutants from several sources, including agriculture, industry, homes, and businesses.

Issuing and enforcing more stringent regulations on industry can help reduce releases of many toxic chemicals. We can also modify our industrial processes and our purchasing decisions to rely less on these substances.

Pathogens and waterborne diseases Disease-causing organisms (pathogenic viruses, protists, and bacteria) can enter drinking water supplies when these are contaminated with human waste from inadequately treated sewage or with animal waste from feedlots (p. 251). Specialists monitoring water quality can tell when water has been contaminated by such waste when they detect fecal coliform bacteria, which live in the intestinal tracts of people and other vertebrates. These bacteria are usually not pathogenic themselves, but they serve as indicators of fecal contamination, alerting us that the water may hold other pathogens that can cause ailments such as giardiasis, typhoid, or hepatitis A.

Biological pollution by pathogens causes more human health problems than any other type of water pollution. In the United States, an estimated 20 million people fall ill

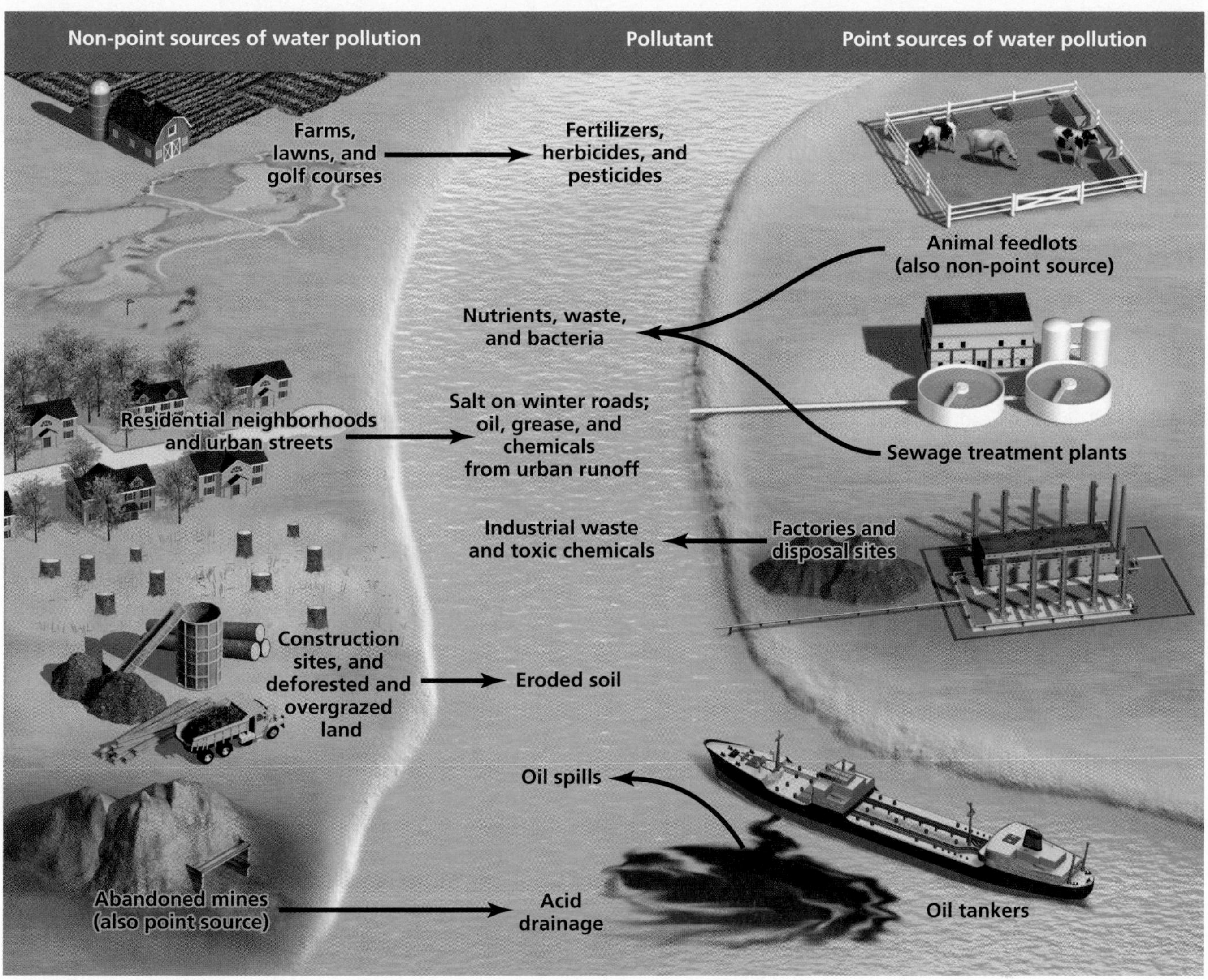

FIGURE 15.22 Point-source pollution (on right) comes from discrete facilities or locations, usually from single outflow pipes. Non-point-source pollution (such as runoff from streets, residential neighborhoods, lawns, and farms; **on left**) originates from numerous sources spread over large areas.

each year from drinking water contaminated with pathogens. Worldwide, the United Nations estimates that 3800 children die every day from diseases associated with unsafe drinking water, such as cholera, dysentery, and typhoid fever. While nearly 800 million people still lack reliable access to safe drinking water and 37% do not have sanitation/sewer facilities, we are making slow but steady progress worldwide in supplying people with these services (p. 365).

We reduce the risks posed by waterborne pathogens by using chemical or other means to disinfect drinking water (p. 414) and by treating wastewater (pp. 414–415). Other measures to lessen health risks include public education to encourage personal hygiene and government enforcement of regulations to ensure the cleanliness of food production, processing, and distribution.

Nutrient pollution The Chesapeake Bay's dead zone shows how nutrient pollution from fertilizers and other sources can lead to eutrophication and hypoxia in surface waters (Chapter 5, pp. 105, 108). When excess nitrogen and/or phosphorus enters a water body, it fertilizes algae and aquatic plants, boosting their growth. Algae then spread and cover the water's surface, depriving underwater plants of sunlight. As algae die off, bacteria consume them. Because this decomposition requires oxygen, the increased bacterial activity drives down levels of dissolved oxygen. These levels can drop too low to support fish and shellfish, leading to dramatic changes in aquatic ecosystems.

A "dead zone"—an area of very low dissolved oxygen levels—appears annually in the northern Gulf of Mexico, fueled by nutrients from Midwest farms carried by the Mississippi and Atchafalaya rivers (see **THE SCIENCE BEHIND THE STORY**, pp. 410–411). The low oxygen conditions have adversely affected marine life and reduced catches of shrimp and fish to half of what they were in the 1980s, impacting people whose livelihoods depended on seafood harvests.

Eutrophication (**FIGURE 15.23**, see p.412) is a natural process, but nutrient input from runoff from farms, golf courses, lawns, and sewage can dramatically increase the rate at which it occurs. We can reduce nutrient pollution by treating wastewater, reducing fertilizer application, using phosphate-free detergents, and planting vegetation and protecting natural areas to increase nutrient uptake.

Hypoxia and the Gulf of Mexico's "Dead Zone"

She was prone to seasickness, but Nancy Rabalais cared too much about the Gulf of Mexico to let that stop her. Leaning over the side of an open boat idling miles from shore, she hauled a water sample aboard—and helped launch efforts to breathe life back into the Gulf's "dead zone."

Since that first expedition in 1985, Rabalais, her colleague and husband Eugene Turner, and fellow scientists at the Louisiana Universities Marine Consortium (LUMCON) and Louisiana State University have made great progress in unraveling the mysteries of the region's hypoxia—and in getting it on the political radar screen.

Rabalais and other researchers began by tracking oxygen levels at nine sites in the Gulf every month and continued those measurements for five years. At dozens of other spots near the shore and in deep water they took less frequent oxygen readings. Sensors, as they are lowered into the water, measure oxygen levels and send continuous readings back to a shipboard computer. Further data come from fixed, submerged oxygen meters that continuously measure dissolved oxygen and store the data.

The team also collected hundreds of water samples, using lab tests to measure levels of nitrogen, salt, bacteria, and phytoplankton. LUMCON scientists logged hundreds of miles in their ships, regularly monitoring more than 70 sites in the Gulf. They also donned scuba gear to view firsthand the condition of shrimp, fish, and other sea life. Such a range of long-term data allowed the researchers to build a "map" of the dead zone, tracking its location and its consequences.

Dr. Nancy Rabalais, LUMCON

In 1991, Rabalais made that map public, earning immediate headlines. That year, her group mapped the size of the zone at more than 10,000 km^2 (about 4000 mi^2). Bottom-dwelling shrimp were stretching out of their burrows, straining for oxygen. Many fish had fled. The bottom waters, infused with sulfur from bacterial decomposition, smelled of rotten eggs.

The group's years of monitoring also enabled them to explain and predict the dead zone's emergence. As rivers rose each spring (and as fertilizers were applied in the Midwestern farm states), oxygen would start to disappear in the northern Gulf. The hypoxia would last through the summer or fall, until seasonal storms mixed oxygen into hypoxic areas.

Over time, monitoring linked the dead zone's size to the volume of river flow and its nutrient load. The 1993 flooding of the Mississippi created a zone much larger than the year before, whereas a drought in 2000 brought low river flows, low nutrient loads, and a small dead zone (**FIGURE 1**). Similar relationships between river flow and dead zone size have been seen ever since. In 2005, the dead zone was predicted to be large, but Hurricanes Katrina and Rita stirred oxygenated surface water into the depths, decreasing the dead zone that year.

The source of the problem, Rabalais said, lay back on land. The Mississippi and Atchafalaya rivers draining into the Gulf were polluted with agricultural runoff, and the nutrient pollution from fertilizers spurred algal blooms whose decomposition by bacteria snuffed out oxygen in wide stretches of ocean water (pp. 108–109). This work had clearly demonstrated the interconnections between freshwater aquatic systems and the Gulf, and how pollutants from farm fields in the upper Midwest could exert effects far away at the mouth of the Mississippi River.

Many Midwestern farming advocates and some scientists, such as Derek Winstanley, chief of the Illinois State Water Survey, challenged the findings. They argued that the Mississippi naturally carries high loads of nitrogen from runoff and that Rabalais's team had not ruled out upwelling in the Gulf as a source of nutrients.

But sediment analyses showed that Mississippi River mud contained many fewer nitrates early in the century, and Rabalais and Turner found that silica residue from phytoplankton blooms increased in Gulf sediments between 1970 and 1989, paralleling rising nitrogen levels. In 2000, a federal integrative assessment team of dozens of scientists laid the blame for the dead zone on nutrients from fertilizers and other sources in the fresh waters emptying into the Gulf.

Then in 2004, while representatives of farmers and fishermen debated political fixes, Environmental Protection Agency water quality scientist Howard Marshall suggested that to alleviate the dead zone we'd be best off reducing phosphorus pollution from industry and sewage treatment. His reasoning: Phytoplankton need both nitrogen and phosphorus, but there is now so much nitrogen in the Gulf that phosphorus has become the limiting factor on phytoplankton growth.

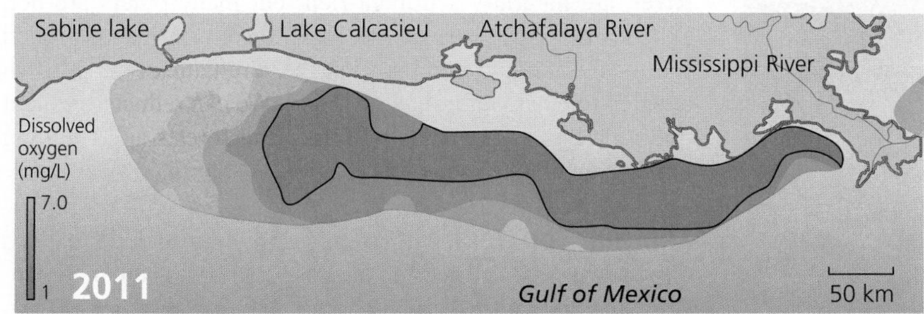

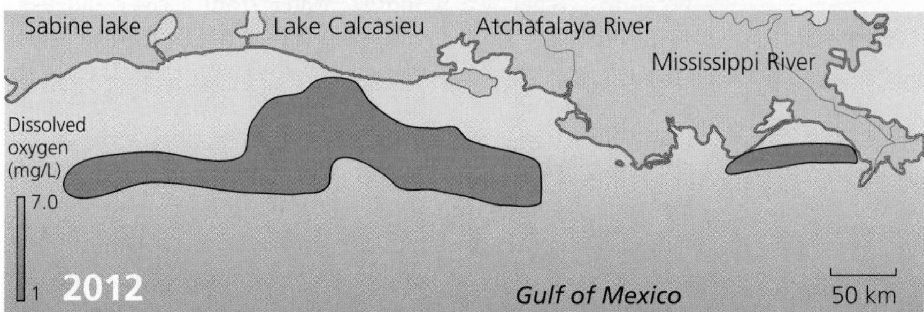

(a) Dissolved oxygen at bottom

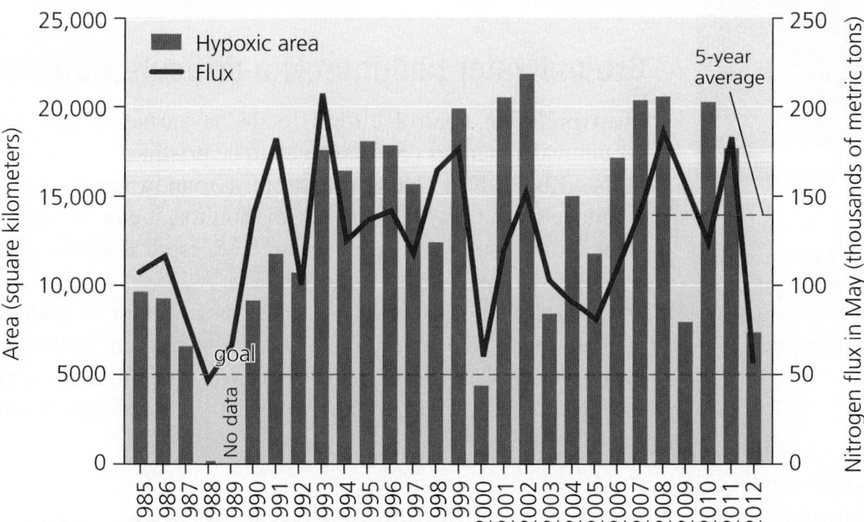

(b) Area of hypoxic zone in the northern Gulf of Mexico

FIGURE 1 The map in (a) shows dissolved oxygen concentrations in bottom waters of the Gulf of Mexico off the Louisiana coast in 2011 and 2012. Areas in red indicate the lowest oxygen levels. Regions considered hypoxic (< 2 mg/L) are encircled with a black line. The dead zone forms to the west of the mouth of the Mississippi River because prevailing currents carry nutrients in that direction. The graph in **(b)** shows that the size of the hypoxic zone (shown by bars) is correlated with the amount of nitrogen pollution entering from the Mississippi River (shown by line). In addition, floods increase the size of the hypoxic zone by bringing additional runoff (as in 2011), whereas droughts decrease its size by reducing nutrient runoff (as in 2012). Tropical storms decrease the size of hypoxic water by mixing oxygen-rich water into the dead zone (as in 2003). Scientists and policymakers aim to reduce the size of the dead zone to 5000 km^2 (1930 mi^2). *Data from Nancy Rabalais, Louisiana Universities Marine Consortium (LUMCON).*

DATA Q In which year was the hypoxic zone largest, and how many square kilometers was it? In which year was nitrogen flux the highest, and what was its approximate value that year?

Since then, research has supported this contention, and scientists now propose that nitrogen and phosphorus should be managed jointly. Moreover, recent research indicates that a federally mandated 30% reduction in nitrogen in the river will not be adequate to eliminate the dead zone. Scientists also maintain that large-scale restoration of wetlands along the river and at the river's delta, like those profiled in the **Central Case Study**, would best filter pollutants before they reach the Gulf.

All this research is guiding a federal plan to reduce farm runoff, clean up the Mississippi, restore coastal wetlands, and shrink the Gulf's dead zone. It has also led to a better understanding of hypoxic zones around the world. ■

(a) Oligotrophic water body

(b) Eutrophic water body

FIGURE 15.23 **Pollution of freshwater bodies by excess nutrients accelerates the process of eutrophication.** An oligotrophic water body **(a)** with clear water and low nutrient content may eventually become a eutrophic water body **(b)** with abundant algae and high nutrient content.

Biodegradable Wastes Introducing large quantities of biodegradable materials into waters decreases dissolved oxygen levels, too. When human wastes, animal manure, paper pulp from paper mills, or yard wastes (grass clippings and leaves) enter waterways, bacterial decomposition escalates as organic material is metabolized. This lowers dissolved oxygen levels in the water, just as in waters receiving elevated inputs of plant nutrients. **Wastewater** is water affected by human activities and can be a source of biodegradable wastes. It includes water from toilets, showers, sinks, dishwashers, and washing machines; water used in manufacturing or industrial cleaning processes; and stormwater runoff. The widespread practice of treating wastewater to remove organic matter has greatly reduced impacts from biodegradable wastes in rivers in developed nations. Oxygen depletion remains a major problem in some developing nations, however, where wastewater treatment is less common.

Sediment As we saw in the **Central Case Study,** eroded soils are carried to rivers by runoff and transported long distances by river currents (FIGURE 15.24). Clear-cutting, mining, clearing land for development, and cultivating farm fields all expose soil to wind and water erosion (pp. 222–223). Some water bodies, such as the Colorado River and China's Yellow River, are naturally sediment-rich, but many others are not. When a clear-water river receives a heavy influx of eroded sediment, aquatic habitat changes dramatically, and fish adapted to clear water may be killed. We can reduce sediment pollution by better managing farms and forests and avoiding large-scale disturbance of vegetation.

Thermal pollution Water's ability to hold dissolved oxygen decreases as temperature rises, so some aquatic organisms may not survive when human activities raise water temperatures. When we withdraw water from a river and use it to cool an industrial facility, we transfer heat from the facility back into the river where the water is returned. People also raise water temperatures by removing streamside vegetation that shades water.

Too little heat can also cause problems. On the Mississippi and many other dammed rivers, water at the bottoms of reservoirs is colder than water at the surface. When dam operators release water from the depths of a reservoir, downstream water temperatures drop suddenly. In some river systems, these pulses of cold water have favored cold-loving invasive fish species over native species adapted to normal river temperatures.

Groundwater pollution is a difficult problem

Most pollution control efforts focus on surface water. Yet groundwater sources once assumed to be pristine are regularly polluted by industry and agriculture. Groundwater pollution is hidden from view and difficult to monitor; it can be out-of-sight, out-of-mind for decades until widespread contamination of drinking supplies is discovered.

Groundwater pollution is also more difficult to address than surface water pollution. Rivers flush their pollutants fairly quickly, but groundwater retains its contaminants until they decompose, which in the case of persistent pollutants can

FIGURE 15.24 **Sediments wash into the Pacific Ocean from a river in Panama.** Farming, construction, and other human activities can cause elevated levels of soil to enter waterways, affecting water quality and aquatic wildlife.

be many years or decades. The long-lived pesticide DDT, for instance, is found widely in U.S. aquifers even though it was banned over 35 years ago. Moreover, chemicals break down much more slowly in aquifers than in surface water or soils. Decomposition is slower in groundwater because it is not exposed to sunlight, contains fewer microbes and minerals, and holds less dissolved oxygen and organic matter. For example, concentrations of the herbicide alachlor decline by half after 20 days in soil, but in groundwater this takes almost four years.

There are many sources of groundwater pollution

Some chemicals that are toxic at high concentrations, including aluminum, fluoride, nitrates, and sulfates, occur naturally in groundwater. After all, groundwater is in contact with rock for thousands of years, and during that time all kinds of compounds, both toxic and benign, may leach into the water.

However, groundwater pollution resulting from human activity is widespread. Industrial, agricultural, and urban wastes—from heavy metals to petroleum products to solvents to pesticides—can leach through soil and seep into aquifers. Pathogens and other pollutants can enter groundwater through improperly designed wells and from the pumping of liquid hazardous waste below ground (p. 627). A recent 17-year study of volatile organic compounds (VOCs; p. 459) detected 42 types of VOCs from manufactured products and industrial processes in nearly all U.S. aquifers and in 18% of wells sampled (p. 368).

Leakage of carcinogenic pollutants (such as chlorinated solvents and gasoline) from underground tanks of oil and industrial chemicals also poses a threat to groundwater. Across the United States, the U.S. Environmental Protection Agency (EPA) has embarked on a nationwide cleanup program to unearth and repair leaky tanks. After more than 15 years of work, by 2013 the EPA had confirmed leaks from 510,000 tanks, and had completed cleanups on over 430,000 of them.

The leaking of radioactive compounds from underground tanks is also as source of groundwater pollution. In 2013, federal and state officials revealed that an underground storage tank at the Hanford Nuclear Reservation in Washington was leaking 568–1135 L (150–300 gal) of radioactive waste into the soil each year (**FIGURE 15.25**). Shortly thereafter, leaks were found in another five tanks. The site is the most radioactively contaminated area in the United States, and stores 60% of the United States' high-level radioactive waste in 177 underground tanks. It has been storing wastes since the 1940s, and billions of dollars have been spent on remediation efforts at the facility. Its cleanup has experienced delays and cost overruns, however, and the radioactive material in aging underground tanks is not scheduled to be completely removed until 2040.

Agriculture contributes to groundwater pollution in several ways. Pesticides were detected in most of the shallow aquifer sites tested in the United States in the 1990s, although levels generally did not violate EPA safety standards for drinking water. Nitrate from fertilizers has leached into aquifers in Canada and in 49 U.S. states. Nitrate in drinking water has been linked to cancers, miscarriages, and "blue-baby" syndrome, which reduces the oxygen-carrying capacity of infants' blood. Agriculture can also contribute pathogens; in 2000, the groundwater supply of Walkerton, Ontario, became contaminated with the bacterium *Escherichia coli*, or *E. coli*. Two thousand people became ill, and seven died.

FIGURE 15.25 Leaky underground storage tanks are a major source of groundwater pollution. Underground tanks housing radioactive waste at the Hanford Nuclear Reservation in Washington were found in 2013 to be leaking, threatening groundwater and the nearby Columbia River.

Legislative and regulatory efforts have helped to reduce pollution

As numerous as our freshwater pollution problems may seem, it is important to remember that many were worse a few decades ago, when the Cuyahoga River repeatedly caught fire (p. 175). Citizen activism and government response during the 1960s and 1970s in the United States resulted in legislation such as the Federal Water Pollution Control Act of 1972 (later amended and renamed the Clean Water Act in 1977). These acts made it illegal to discharge pollution from a point source without a permit, set standards for industrial wastewater, set standards for contaminant levels in surface waters, and funded construction of sewage treatment plants. Thanks to such legislation, point-source pollution in the United States was reduced, and rivers and lakes became notably cleaner.

In the past decade, however, enforcement of water quality laws grew weaker, as underfunded and understaffed state and federal regulatory agencies succumbed to pressure from industry and from politicians who receive money from industry. A comprehensive investigation by *The New York Times* in 2009 revealed that violations of the Clean Water Act have risen and that documented violations now number over 100,000 per year (to say nothing of undocumented instances). The EPA and the states act on only a tiny percentage of these violations, the *Times* found. As a result, 1 in 10 Americans have been exposed to unsafe drinking water—for the most part unknowingly, because many pollutants cannot be detected by smell, taste, or color. In response, EPA Administrator Lisa Jackson promised at the time to strengthen enforcement.

The Great Lakes of Canada and the United States represent an encouraging success story in fighting water pollution. In the 1970s these lakes, which hold 18% of the world's surface fresh water, were badly polluted with wastewater, fertilizers, and toxic chemicals. Algal blooms fouled beaches, and Lake Erie was pronounced "dead." Today, efforts of the Canadian and U.S. governments have paid off. According to Environment Canada, releases of seven toxic chemicals are down by 71%, municipal phosphorus has decreased by 80%, and chlorinated pollutants from paper mills are down by 82%. Levels of PCBs and DDE are down by 78% and 91%, respectively. Bird populations are rebounding, and Lake Erie is now home to the world's largest walleye fishery. The Great Lakes' troubles are by no means over—sediment pollution is still heavy, algal blooms still plague Lake Erie, and fish are not always safe to eat. However, the progress so far shows how conditions can improve when citizens push their governments to take action.

Enforcement of the Clean Water Act, however, has become more difficult due to several recent Supreme Court decisions that exempted certain waters from EPA oversight. These court decisions suggested that waterways that are entirely within a single state, streams that run dry during certain times of year, and lakes that do not connect to larger waterways are not "navigable" (capable of being navigated by ships) and therefore do not fall under the Clean Water Act's directive to limit discharges "into the navigable waters" of the United States. Internal EPA analyses indicate that these decisions could put up to 45% of major polluters outside the jurisdiction of the law and expose up to 117 million Americans to higher levels of pollution in their drinking water. Efforts to change the Act's wording to encompass all waters in the United States was, as of 2013, stalled in the U.S. Congress.

We treat our drinking water

Technological advances as well as government regulation have improved our control of pollution. The treatment of drinking water is a widespread and successful practice in developed nations today. Before being sent to your tap, water from a reservoir or aquifer is treated with chemicals to remove particulate matter; passed through filters of sand, gravel, and charcoal; and/or disinfected with small amounts of an agent such as chlorine. The U.S. EPA sets standards for over 90 drinking water contaminants, which local governments and private water suppliers are obligated to meet.

We treat our wastewater

Wastewater treatment is also now a mainstream practice. Wastewater includes water that carries sewage; water from showers, sinks, washing machines, and dishwashers; water used in manufacturing or industrial cleaning processes; and storm water runoff. Natural systems can process moderate amounts of wastewater, but the large and concentrated amounts that our densely populated areas generate can harm ecosystems and pose health threats. Thus, attempts are now widely made to treat wastewater before it is released into the environment.

In rural areas, **septic systems** are the most popular method of wastewater disposal. In a septic system, wastewater runs from the house to an underground septic tank, inside which solids and oils separate from water. The clarified water proceeds downhill to a drain field of perforated pipes laid horizontally in gravel-filled trenches underground. Microbes decompose pollutants in the wastewater these pipes emit. Periodically, solid waste from the septic tank is pumped out and taken to a landfill.

In more densely populated areas, municipal sewer systems carry wastewater from homes and businesses to centralized treatment locations. There, pollutants are removed by physical, chemical, and biological means (**FIGURE 15.26**). At a treatment facility, **primary treatment,** the physical removal of contaminants in settling tanks or clarifiers, removes about 60% of suspended solids. Wastewater then proceeds to **secondary treatment,** in which water is stirred and aerated so that aerobic bacteria degrade organic pollutants. Roughly 90% of suspended solids may be removed after secondary treatment. Finally, the clarified water is treated with chlorine, and sometimes ultraviolet light, to kill bacteria. Most often, the treated water, called *effluent*, is piped into rivers or the ocean following primary and secondary treatment. However, many municipalities are recycling "reclaimed" water for lawns and golf courses, for irrigation, or for industrial purposes such as cooling water in power plants.

As water is purified throughout the treatment process, the solid material removed is termed *sludge*. Sludge is sent to digesting vats, where microorganisms decompose much of the matter. The result, a wet solution of "biosolids," is then dried and either disposed of in a landfill, incinerated, or used as fertilizer on cropland. Methane-rich gas created by the decomposition process is sometimes burned to generate electricity, helping to offset the cost of treatment. Each year about 6 million dry tons of sludge are generated in the United States.

WEIGHING THE ISSUES

SLUDGE ON THE FARM It is estimated that up to half the biosolids from sewage sludge produced each year is used as fertilizer on farmland. This practice makes productive use of the sludge, increases crop output, and conserves landfill space, but many people have voiced concern over accumulation of toxic metals, proliferation of dangerous pathogens, and odors. Do you feel that this practice represents an efficient use of resources or an unnecessary risk? What further information would you want to know to inform your decision?

Constructed wetlands can aid treatment

Long before people built the first wastewater treatment plants, natural wetlands were filtering and purifying water. Recognizing this, engineers have begun manipulating wetlands and even constructing new wetlands to employ them as tools to cleanse wastewater. Generally in this approach, wastewater that has gone through primary or secondary treatment at a conventional facility is pumped into the wetland, where microbes living amid the algae and aquatic plants decompose the remaining pollutants. Water cleansed in the wetland can then be released into waterways or allowed to percolate underground.

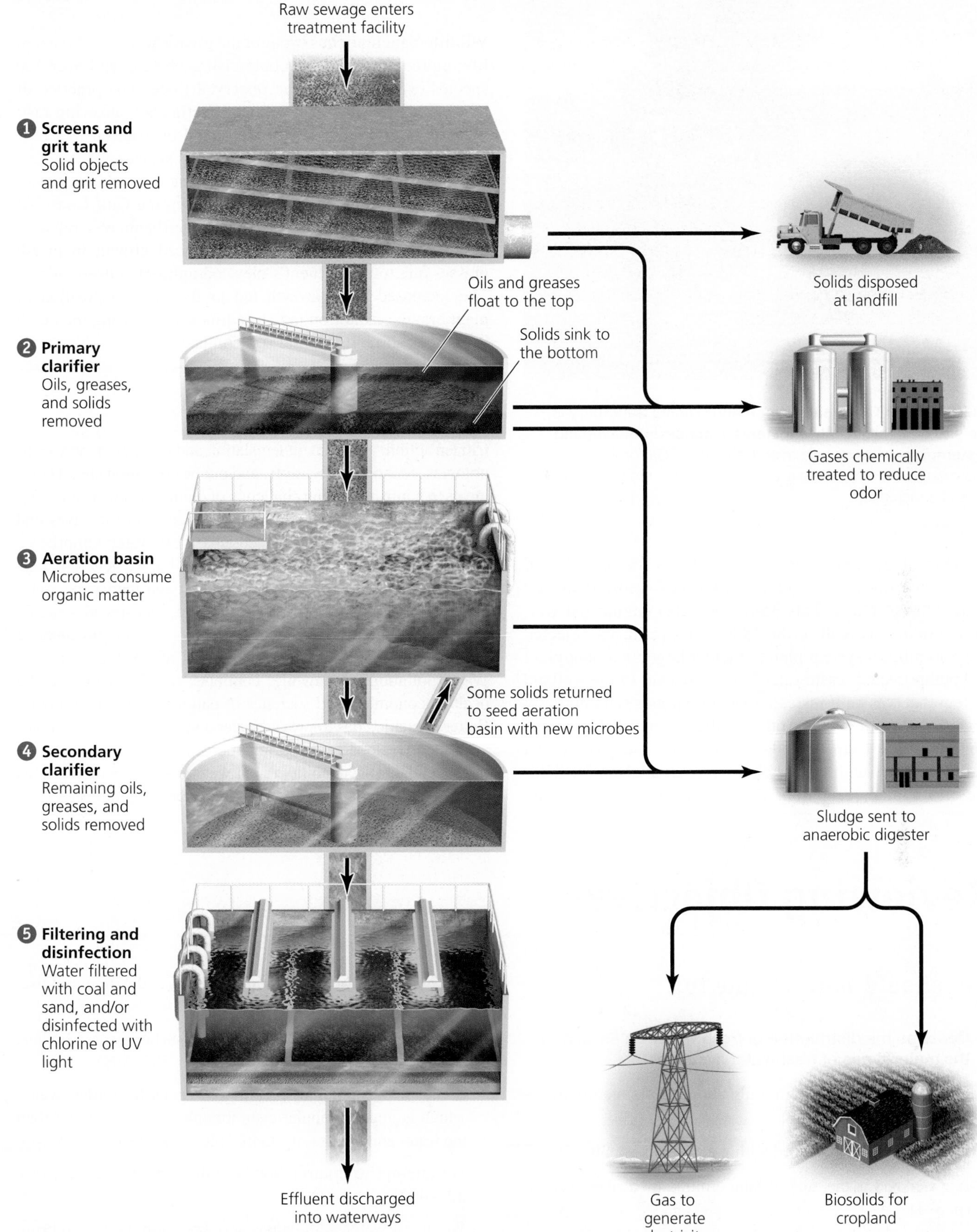

FIGURE 15.26 Shown here is a generalized process from a modern, environmentally sensitive wastewater treatment facility. Wastewater initially passes through screens to remove large debris and into grit tanks to let grit settle ❶. It then enters tanks called primary clarifiers ❷, in which solids settle to the bottom and oils and greases float to the top for removal. Clarified water then proceeds to aeration basins ❸ that oxygenate the water to encourage decomposition by aerobic bacteria. Water then passes into secondary clarifier tanks ❹ for removal of further solids and oils. Next, the water may be purified ❺ by chemical treatment with chlorine, passage through carbon filters, and/or exposure to ultraviolet light. The treated water (called *effluent*) may then be piped into natural water bodies, used for urban irrigation, flowed through a constructed wetland, or used to recharge groundwater. In addition, most treatment facilities use anaerobic bacteria to digest sludge removed from the wastewater. Biosolids from digesters may be sent to farm fields as fertilizer, and gas from digestion may be used to generate electric power.

FIGURE 15.27 **The Arcata Marsh and Wildlife Sanctuary is the site of an artificially engineered, constructed wetland system.** The wetlands help treat this northern California city's wastewater, and upland areas around the marsh are open to the public for recreation.

One of the first constructed wetlands was established in Arcata, a town on northern California's scenic Redwood Coast (FIGURE 15.27). This 35-ha (86-acre) engineered wetland system was built in the 1980s after residents objected to a $50 million system plan to build a large treatment plant and pump treated wastewater into the ocean. In the wetland system that was built instead, oxidation ponds send partially treated wastewater to the wetland, where plants and microbes continue to perform secondary treatment. The project cost just $7 million, and the site also serves as a haven for wildlife and human recreation. In fact, the Arcata Marsh and Wildlife Sanctuary has brought the town's waterfront back to life; more than 100,000 people visit each year, and over 300 species of birds have been observed there. The practice of treating wastewater with artificial wetlands is growing fast; today over 500 artificially constructed or restored wetlands in the United States are performing this service.

The release of wastewater effluent even shows promise for preserving coastal wetlands along the Gulf Coast. At a study site in Louisiana, wastewater effluent was released into coastal wetlands, where it enhanced growth in marsh grasses due to the effluent's elevated nutrient concentrations. The increased plant growth led to increased deposition of plant organic matter on marsh sediments, offsetting the depth increases caused by natural soil compaction.

Conclusion

Citizen action, government legislation and regulation, new technologies, economic incentives, and public education are all helping us to confront a rising challenge of our new century: ensuring adequate quantity and quality of fresh water for ourselves and for the planet's ecosystems. Accessible fresh water comprises a minuscule percentage of the hydrosphere, but we generally take it for granted. Our expanding population and increasing water use are bringing us toward conditions of widespread scarcity. Water depletion has become a serious concern in many areas of the developing world and in arid regions of developed nations. Water pollution, meanwhile, continues to take a toll on the health, economies, and societies of nations both rich and poor. Better regulation has improved water quality in the United States and other developed nations, and there is reason to hope that we may yet attain sustainability in our water use. Potential solutions are numerous, and the issue is too important to ignore.

Reviewing Objectives

You should now be able to:

Describe the distribution of fresh water on Earth and the major types of freshwater systems

- Of all the water on Earth, only about 1% is readily available for our use. (p. 391)
- Groundwater is contained within aquifers. (pp. 391–393)
- A watershed is the area of land drained by a river system. (p. 393)
- The main types of freshwater ecosystems include rivers and streams, lakes and ponds, and wetlands. (pp. 393–396)

Discuss how we use water and alter freshwater systems

- Water is a renewable but limited resource, so we must manage it sustainably. (p. 396)
- We use water for agriculture, industry, and residential use. Globally, 70% is used for agriculture. (p. 397)
- We pump water from aquifers and surface water bodies, sometimes at unsustainable rates. (pp. 397–399)
- Some of our water extraction now goes to bottled water, which is hugely popular even though it is no healthier than tap water and creates substantial plastic waste. (pp. 399–400)
- We attempt to control floods with dikes and levees. (pp. 400–402)
- We divert water with canals and irrigation ditches to bring water where it is desired. (pp. 402–403)
- We have dammed most of the world's rivers. Dams bring a diversity of benefits and costs. (pp. 403–404)
- Some dams are now being removed. (p. 404)
- Many wetlands have been lost, and we are now trying to restore some. (pp. 404–405)

Assess problems of water supply and propose solutions to address depletion of fresh water

- Desalination increases water supply, but it is expensive and energy-intensive. (p. 405)
- Solutions to reduce demand include technology, market-based approaches, and consumer products that increase efficiency in agriculture, industry, and the home. (pp. 405–407)
- Privatization of water supplies is a much-debated issue. (p. 407)
- Political tensions over water may heighten in the near future. (pp. 407–408)

Describe the major classes of water pollution and propose solutions to address water pollution

- Water pollution stems from point sources and non-point sources. (p. 408)
- Water pollutants include toxic chemicals, microbial pathogens, excessive nutrients, biodegradable wastes, sediment, and thermal pollution. (pp. 408–409, 412)
- Groundwater pollution is more persistent and difficult to address than surface water pollution. (pp. 412–413)
- Legislation and regulation have improved water quality in developed nations in recent decades. (p. 413–414)

Explain how we treat drinking water and wastewater

- Municipalities treat drinking water by filtering and disinfection in a multistep process. (p. 414)
- Septic systems help treat wastewater in rural areas. (p. 414)
- Wastewater is treated physically, biologically, and chemically in a series of steps at municipal wastewater treatment facilities. (pp. 414–415)
- Artificial wetlands enhance wastewater treatment while restoring habitat for wildlife. (pp. 414, 416)

Testing Your Comprehension

1. Compare and contrast the main types of freshwater ecosystems. Name and describe the major zones of a typical pond or lake.
2. Why are sources of fresh water unreliable for some people and plentiful for others?
3. Describe three benefits and three costs of damming rivers. What particular environmental, health, and social concerns has China's Three Gorges Dam and its reservoir raised?
4. Why do the Colorado, Rio Grande, Nile, and Yellow rivers now slow to a trickle or run dry before reaching their deltas?
5. Why are water tables dropping around the world? What are some negative impacts of falling water tables?
6. Name three major types of water pollutants, and provide an example of each. Explain which classes of water pollutants you think are most important in your local area.
7. Define *groundwater*. Why do many scientists consider groundwater pollution a greater problem than surface water pollution?
8. What are some anthropogenic (human) sources of groundwater pollution?
9. Describe how drinking water is treated. How does a septic system work?
10. Describe and explain the major steps in the process of wastewater treatment. How can artificial wetlands aid such treatment?

Seeking Solutions

1. How can we lessen agricultural demand for water? Describe some ways we can reduce household water use. How can industrial uses of water be reduced?
2. How might desalination technology help "make" more water? Describe two methods of desalination. Why is this technology mostly being used in places like Saudi Arabia?
3. Describe three ways in which your own actions contribute to water pollution. Now describe three ways in which you could diminish these impacts.
4. Have the provisions of the Clean Water Act been effective? Discuss some of the methods we can adopt, in addition to "end-of-pipe" solutions, to prevent water pollution.
5. **THINK IT THROUGH** Your state's governor has put you in charge of water policy for the state. The aquifer beneath your state has been overpumped, and many wells have run dry. Agricultural production last year decreased for the first time in a generation, and farmers are clamoring for you to do something. Meanwhile, the state's largest

city is growing so fast that more water is needed for its burgeoning urban population. What policies would you consider to restore your state's water supply? Would you try to take steps to increase supply, decrease demand, or both? Explain why you would choose such policies.

6. **THINK IT THROUGH** Having solved the water depletion problem in your state, your next task is to deal with pollution of the groundwater that provides your state's drinking water supply. Recent studies have shown that one-third of the state's groundwater has levels of pollutants that violate EPA standards for human health, and citizens are fearful for their safety. What steps would you consider taking to safeguard the quality of your state's groundwater supply, and why?

Calculating Ecological Footprints

One of the single greatest personal uses of water is for showering. Old-style showerheads that were standard in homes and apartments built before 1992 dispense at least 5 gallons of water per minute, but low-flow showerheads produced after that year dispense just 2.5 gallons per minute. Given an average daily shower time of 8 minutes, calculate the amounts of water used and saved over the course of a year with old standard versus low-flow showerheads, and record your results in the table.

	Annual water use with standard showerheads (gallons)	Annual water use with low-flow showerheads (gallons)	Annual water savings with low-flow showerheads (gallons)
You			
Your class			
Your state			
United States			

1. Beginning in 2010, under its WaterSense program, the EPA promoted showerheads that produce still-lower flows of 2 gallons per minute (gpm). Some cities are already requiring these, and some models today go even lower. How much water would you save per year by using a 2-gpm showerhead instead of a 2.5-gpm showerhead?
2. How much water would you be able to save annually by shortening your average shower time from 8 minutes to 6 minutes? Assume you use a 2.5-gpm showerhead.
3. Compare your answers to questions 1 and 2. Do you save more water by showering 8 minutes with a 2-gpm showerhead or 6 minutes with a 2.5-gpm showerhead?
4. Can you think of any factors that are not being considered in this scenario of water savings? Explain.

Massachusetts cod fishermen haul in a dwindling catch

16

Marine and Coastal Systems and Resources

Upon completing this chapter, you will be able to:

- Identify physical, geographical, chemical, and biological aspects of the marine environment
- Explain how the oceans influence and are influenced by climate
- Describe major types of marine ecosystems
- Assess impacts from marine pollution
- Review the state of ocean fisheries and reasons for their decline
- Evaluate marine protected areas and reserves as innovative solutions

CENTRAL CASE STUDY

Collapse of the Cod Fisheries

"Either we have sustainable fisheries, or we have no fishery."

—Canadian Fisheries Minister David Anderson

"So the human management side of this is critical. We don't want to shoot ourselves in the foot as stocks begin to rebuild. We've got to nurse them along."

—George Rose, Centre for Fisheries Ecosystem Research, Memorial University

No fish has had more impact on human civilization than the Atlantic cod. Europeans exploring the coasts of North America 500 years ago discovered that they could catch these abundant fish merely by dipping baskets over the railings of their ships. The race that ensued to harvest this resource helped lead to the colonization of the New World. Starting in the early 1500s, schooners captured countless millions of cod, and the fish became a dietary staple in cultures on both sides of the Atlantic.

Since then, cod fishing has been the economic engine for hundreds of communities in coastal New England and eastern Canada. Massachusetts honored the fish by naming Cape Cod after it and by erecting a carved wooden cod statue in its statehouse. In many Canadian coastal villages, cod fishing has been a way of life for generations. So it came as a shock when the cod all but disappeared, and governments had to step in and close the fisheries.

The Atlantic cod (*Gadus morhua*) is a type of groundfish, a name given to fish that live or feed on the bottom. People have long coveted groundfish such as halibut, pollack, haddock, and flounder. Adult cod eat smaller fish and invertebrates, commonly grow 60–70 cm long, and can live 20 years. A mature female cod can produce several million eggs. Atlantic cod inhabit cool ocean waters on both sides of the North Atlantic and occur in 24 discrete populations, called stocks. One stock inhabits the Grand Banks off Newfoundland, and another lives on Georges Bank off Massachusetts (**FIGURE 16.1**).

The Grand Banks provided ample fish for centuries. With advancing technology, however, ships became larger and more effective at finding fish. By the 1960s, massive industrial trawlers from Europe were vacuuming up unprecedented numbers of groundfish. In 1977, Canada exercised its legal right to the waters 200 nautical miles from shore, kicked out foreign fleets, and claimed most of the Grand Banks for itself. Canada developed the same industrial technologies and revved up its fishing industry like never before.

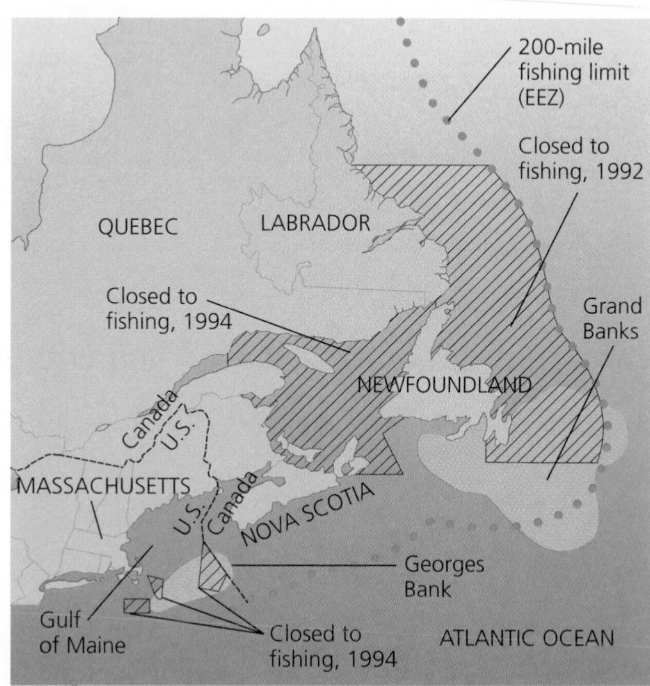

FIGURE 16.1 Populations of Atlantic cod inhabit areas of the northwestern Atlantic Ocean, including the Grand Banks and Georges Bank, regions of shallow water that are especially productive for groundfish. Portions of these and other areas have been closed to fishing in recent years because cod populations have collapsed after being overfished.

Then came the crash. Catches dwindled in the 1980s because too many fish had been harvested and because bottom-trawling (fishing by dragging weighted nets across the seafloor, p. 439) had destroyed huge expanses of the cod's underwater habitat. By 1992 the situation was dire: Scientists reported that mature cod were at just 10% of their long-term abundance. Canadian Fisheries Minister John Crosbie announced a two-year ban on commercial cod fishing off Labrador and Newfoundland, where the $700 million fishery supplied income to 16% of the province's workforce. To compensate fishers, the government offered 10 weekly payments of $225, along with training for new job skills and incentives for early retirement. Over the next two years, 40,000 fishers and processing-plant workers lost their jobs, and some coastal communities faced economic ruin.

Cod stocks did not rebound by 1994, so the government extended the moratorium, enacted bans on all other major cod fisheries, and scrambled to offer more compensation, eventually spending over $4 billion. In 1997–1998, Canada partially reopened some fisheries, but data soon confirmed that the stocks were not recovering. In 2003, the cod fisheries were closed indefinitely, to recreational fishing as well as commercial fishing. It became illegal for Canadians in these areas to catch even one cod for their family's dinner. Fishers challenging the ban were arrested, fined, and jailed.

Then in 2009, a portion of the Grand Banks off the southeastern coast of Newfoundland was reopened to cod fishing after data showed the stock was recovering slightly. Some dreamed of a comeback for the fishery, but others thought the decision ill-advised. Allowable cod harvests were set beyond what a scientific review board had recommended, and many feared another hasty reopening would simply decimate the stock yet again. Fishing continues in this limited area and researchers and resource managers are monitoring populations closely to see how they fare over time.

Across the border in U.S. waters, cod stocks had collapsed in the Gulf of Maine and on Georges Bank. In 1994, the National Marine Fisheries Service (NMFS) closed three prime fishing areas on Georges Bank. Over the next several years, NMFS designed a number of regulations, but these steps were too little, too late. A 2005 report revealed that the cod were not recovering, and further restrictions were enacted. As of 2008, managers announced that the Gulf of Maine stock was 58% of what it needed to be to be sustainable, and the Georges Bank stock was only 12% as large as it needed to be.

There is good news in the cod fishery, however. A 2011 study reported Grand Banks cod populations were at 34% of historical levels, after hovering at around 5% of historical levels for the previous 20 years. When cod were driven to low numbers by overharvesting, populations of forage fish, such as capelin, increased ninefold as a result of reduced cod predation. These forage fish then preyed upon and outcompeted young cod, slowing cod recovery. Populations of forage fish are now in decline, however, because they have outstripped their plankton food supply and are now themselves being harvested by fishermen, giving cod the opportunity to rebound. The good news in the Georges Bank is not limited to cod, though. Seafloor invertebrates have begun to recover in the absence of trawling. Spawning stock of haddock and yellowtail flounder has risen. Sea scallops have increased in biomass 14-fold. Recoveries like these in no-fishing areas are showing scientists, fishers, and policymakers that protecting areas of ocean can help save dwindling marine populations and restore fisheries. ■

The Oceans

It's been said that our planet "Earth" should more properly be named "Ocean." After all, ocean water covers most of our planet's surface. The oceans are an important component of Earth's interconnected aquatic systems (p. 392). The vast majority of rivers empty into oceans (a small number of rivers empty into inland seas), so the oceans receive most of the inputs of water, sediments, pollutants, and organisms carried by freshwater systems.

The oceans touch and are touched by virtually every environmental system and every human endeavor. They shape our planet's climate, teem with biodiversity, provide us resources, and facilitate our transportation and commerce. Even if you live in a landlocked region far from the coast, the oceans affect you. They provide fish for people to eat in Iowa, they supply crude oil for cars in New Mexico, and they influence the weather in Tennessee.

Oceans cover most of Earth's surface

The world's five oceans—Pacific, Atlantic, Indian, Arctic, and Southern—are all connected, comprising a single vast body of water (**FIGURE 16.2**). This one "world ocean" covers 71% of Earth's surface and contains 97.5% of its water. The oceans take up most of the hydrosphere, influence the atmosphere and lithosphere, and encompass much of the biosphere (p. 60). Let's first briefly survey the physical and chemical makeup of the oceans—for although they may look homogenous from a beach, boat, or airplane, marine systems are complex and dynamic.

Seafloor topography can be rugged

Most maps depict oceans as smooth swaths of blue, but when we examine what's beneath the waves we see that the geology of the ocean floor can be intricate. Underwater volcanoes shoot forth enough magma to build islands above sea level, such as the Hawaiian Islands (p. 41). Steep canyons as large as Arizona's Grand Canyon lie just offshore of some continents. The lowest spot in the oceans—the Mariana Trench in the South Pacific—is deeper than Mount Everest is high, by over 2.1 km (1.3 mi). Our planet's longest mountain range is under water: the Mid-Atlantic Ridge (p. 34) runs the length of the Atlantic Ocean.

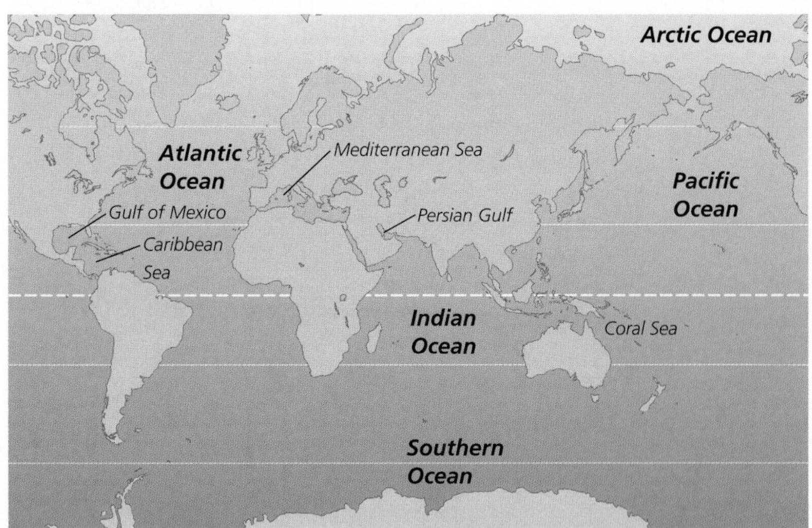

FIGURE 16.2 The world's oceans are connected in a single vast body of water but are given different names. The Pacific Ocean is the largest and, like the Atlantic and Indian Oceans, includes both tropical and temperate waters. The Arctic and Southern Oceans include the waters in the polar regions. Many smaller bodies of water are named as seas or gulfs; a selected few are shown here.

To comprehend underwater geographic features, we can examine a stylized map (FIGURE 16.3) that reflects *bathymetry* (the measurement of ocean depths) and *topography* (physical geography, or the shape and arrangement of landforms). In bathymetric profile, gently sloping **continental shelves** underlie the shallow waters bordering the continents. Continental shelves vary tremendously in width but average 80 km (50 mi) wide, with an average slope of just 1.9 m/km (10 ft/mi). These shelves drop off at the *shelf-slope break*, where the *continental slope* angles more steeply downward to the deep ocean basin below.

Some island chains, such as the Florida Keys, are formed by reefs (pp. 431–432) and lie atop the continental shelf. Others, such as the Aleutian Islands, which curve across the North Pacific from Alaska toward Russia, are volcanic in origin. The Aleutians are also the site of a deep trench that, like the Mariana Trench, formed at a convergent tectonic plate boundary, where one slab of crust dives beneath another in the process of subduction (p. 35).

Wherever reefs, volcanism, or other processes create physical structure underwater, life thrives. Marine animals make use of physical structure as habitat, and topographically complex areas often make for productive fishing grounds. Georges Bank and the Grand Banks are examples; they are essentially huge underwater mounds formed during the ice ages when glaciers dumped debris at their southernmost extent. As climate warmed and the glaciers retreated, sea level rose, and these hilly areas were submerged in the ocean's salty water.

Ocean water contains high concentrations of dissolved salts

Ocean water contains approximately 96.5% H_2O by mass. Most of the remainder consists of ions from dissolved salts

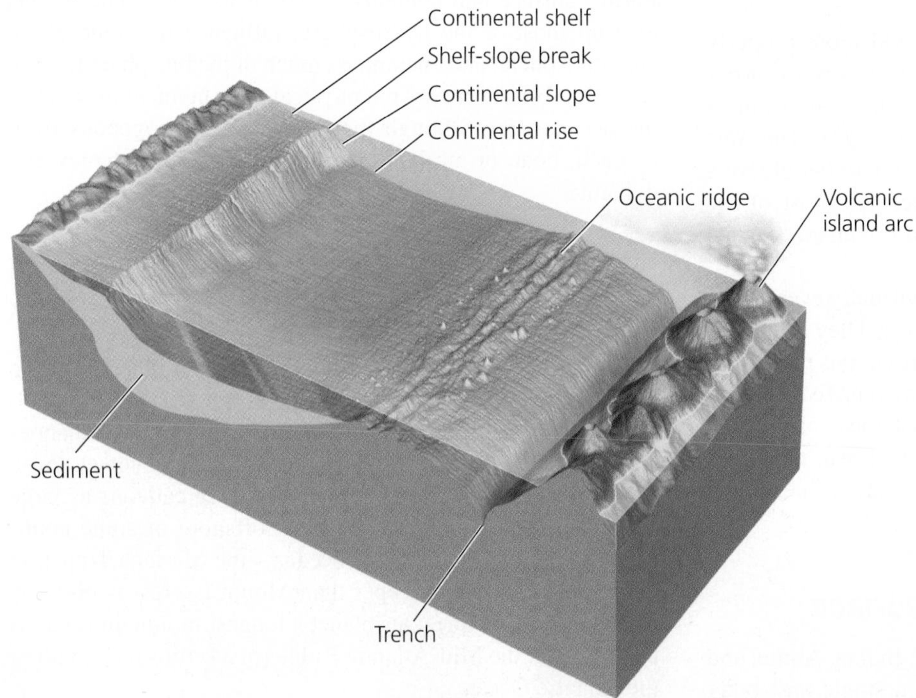

FIGURE 16.3 A stylized bathymetric profile shows key geologic features of the submarine environment. Shallow water exists around the edges of continents over the continental shelf, which drops off at the shelf-slope break. The steep continental slope gives way to the more gradual continental rise, all of which are underlain by sediments from the continents. Vast areas of seafloor are flat abyssal plain. Seafloor spreading occurs at oceanic ridges, and oceanic crust is subducted in trenches (p. 35). Volcanic activity along trenches may give rise to island chains such as the Aleutian Islands. Features on the left side of this diagram are more characteristic of the Atlantic Ocean, and features on the right side of the diagram are more characteristic of the Pacific Ocean. *Thurman, Harold V.; Trujilo, Alan P., 2002. Essentials of Oceanography, 7th Ed. Adapted and Electronically reproduced by permission of Pearson Education, Inc., Upper Saddle River, New Jersey.*

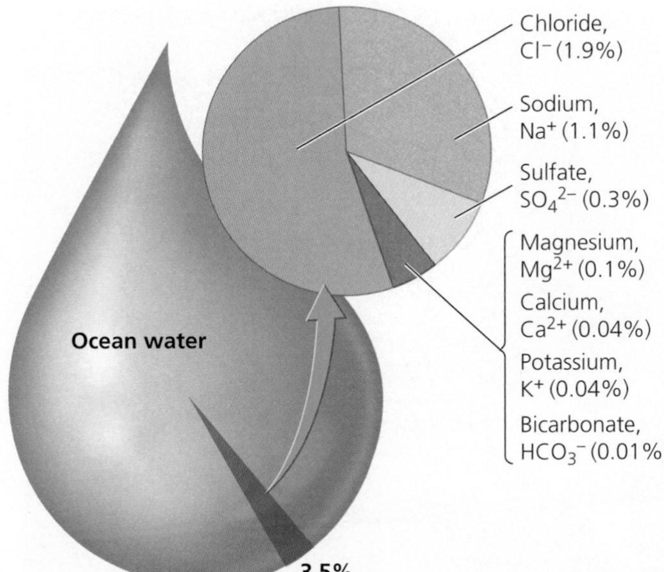

FIGURE 16.4 Ocean water consists of 3.5% salt, by mass. Most of this salt is NaCl in solution, so sodium and chloride ions are abundant. A number of other ions and trace elements are also present.

DATA Q If you had a beaker containing one kilogram (1000 grams) of seawater, how many grams of salt are in the beaker? Of this salt, how many grams are from negatively-charged ions?

(**FIGURE 16.4**). Ocean water is salty primarily because ocean basins are the final repositories for water that runs off the land. Runoff on land collects salts from minerals in weathered rocks and carries them, along with sediments, to the ocean. Wind also blows salts from the land out to sea. Whereas the water in the ocean evaporates, the salts do not, and they accumulate in ocean basins. If we were able to evaporate all the water from the oceans, their basins would be left covered with a layer of dried salt 63 m (207 ft) thick.

The salinity of ocean water generally ranges from 33,000 to 37,000 parts per million, varying from place to place because of differences in evaporation, precipitation, and freshwater runoff (freshwater has less than 500 parts per million salinity) from land and glaciers. Coastal waters are often less saline because of the influx of freshwater runoff. Salinity near the equator is low because this region has a great deal of precipitation, which is relatively salt free. In contrast, surface salinity is high at latitudes roughly 30–35 degrees north and south, where evaporation exceeds precipitation.

Besides dissolved salts, nutrients such as nitrogen and phosphorus occur in seawater in trace amounts (well under 1 part per million) and play essential roles in nutrient cycling (p. 117) in marine ecosystems. Another aspect of ocean chemistry is dissolved gas content. Roughly 36% of the gas dissolved in seawater is oxygen, which is produced by photosynthetic plants, bacteria, and phytoplankton (p. 75) and enters by diffusion from the atmosphere. Oxygen concentrations are highest in the upper layer of the ocean, reaching 13 ml/L of water. Marine animals depend on dissolved oxygen, and if oxygen is depleted, a hypoxic "dead zone" may ensue, killing animals or forcing them to leave (pp. 105, 410–411). Another gas that is soluble in ocean water is carbon dioxide (CO_2). As we pump excess CO_2 into the atmosphere by burning fossil fuels, more CO_2 diffuses into the oceans. We shall soon see (pp. 428–429) how this affects ocean pH (p. 28), turning the water more acidic and posing problems for marine life.

Solar energy structures ocean water from surface to bottom

Sunlight warms the ocean's surface but does not penetrate deeply, so ocean water is warmest at the surface and becomes colder with depth. Surface waters in tropical regions receive more solar radiation and therefore are warmer than surface waters in temperate or polar regions. Warmer water is less dense than cooler water, but water also becomes denser as it gets saltier. This occurs because as salt dissolves in water, it increases solution mass (the mass of water and its dissolved salts) more than it increases solution volume. These relationships give rise to different layers of water: Heavier (colder and saltier) water sinks, whereas lighter (warmer and less salty) water remains nearer the surface. Waters of the surface zone are heated by sunlight and stirred by wind such that they are of similar density down to a depth of about 150 m (490 ft). Below this zone lies the *pycnocline*, a region in which density increases rapidly with depth. The pycnocline contains about 18% of ocean water by volume, compared to the surface zone's 2%. The remaining 80% lies in the deep zone beneath the pycnocline. The dense water in this deep zone is sluggish and unaffected by winds, storms, sunlight, and temperature fluctuations.

Despite the daily heating and cooling of surface waters, ocean temperatures are much more stable than temperatures on land. Midlatitude oceans experience yearly temperature variation of only around 10°C (18°F), and tropical and polar oceans are still more stable. The reason for this stability is that water has a high *heat capacity*, a measure of the heat required to increase temperature by a given amount. It takes more energy to increase the temperature of water than it does to increase the temperature of air. High heat capacity enables ocean water to absorb a tremendous amount of heat from the air. In fact, just the top 2.6 m (8.5 ft) of the oceans holds as much heat as the entire atmosphere! By absorbing heat and releasing it to the atmosphere, the oceans help regulate Earth's climate (Chapter 18). They also influence climate by moving heat from place to place via the ocean's surface circulation, a system of currents that move in the pycnocline and the surface zone.

Surface water flows horizontally in currents

Earth's ocean is composed of vast, riverlike flows driven by density differences, heating and cooling, gravity, and wind. Surface **currents** flow horizontally within the upper 400 m (1300 ft) of water for great distances and in long-lasting patterns across the globe (**FIGURE 16.5**). Warm-water currents carry water heated by the sun from equatorial regions, while cold-water currents carry water cooled in high-latitude regions or from deep below. Some surface currents are very slow. Others, like the Gulf Stream, are rapid and powerful. From the Gulf of Mexico, the Gulf Stream flows up the U.S.

FIGURE 16.5 The upper waters of the oceans flow in surface currents, long-lasting and predictable global patterns of water movement. Warm- and cold-water currents interact with the planet's climate system, and people have used them for centuries to navigate the oceans. *Source: Adapted from Rick Lumpkin (NOAA/AOML).*

DATA Q If you released a special buoy that traveled on the surface ocean currents shown above into the Pacific Ocean from the southeastern coast of Japan, would it likely reach The United States or Australia first? On what currents would it be carried?

Atlantic coast and past the eastern edges of Georges Bank and the Grand Banks at nearly 2 m/sec, or over 4 mph. Averaging 70 km (43 mi) across, the Gulf Stream continues across the North Atlantic, bringing warm water to Europe and moderating that continent's climate (p. 488), which otherwise would be much colder.

Besides influencing climate, ocean currents have aided navigation and shaped human history. Currents helped carry Polynesians to Easter Island (p. 6), Darwin to the Galápagos (p. 50), and Europeans to the New World. Currents transport heat, nutrients, pollution, and the larvae of cod and many other marine species from place to place. Currents in the Pacific Ocean have transported debris from the tsunami that devastated eastern Japan in 2011 all the way to the western coast of the United States (**FIGURE 16.6**).

Vertical movement of water affects marine ecosystems

Surface winds and heating also create vertical currents in seawater. Where horizontal surface currents diverge from one another, cold, deep waters are pulled to the surface in a process called **upwelling.** Upwelled water is rich in nutrients from the bottom, so upwellings are often sites of high primary productivity (p. 111) and lucrative fisheries. Upwellings will occur where strong winds blow away from or parallel to coastlines (**FIGURE 16.7**). An example is the Pacific coast of North America, where north winds and the Coriolis effect (p. 455) move surface waters away from shore, raising nutrient-rich water from below and creating a biologically rich region. The cold water also chills the air along the coast, giving San Francisco its famous fog and cool summers.

FIGURE 16.6 Currents carried a 20-m (65-ft), 188-ton dock from Japan to the Oregon coast. The dock was dislodged by the 2011 tsunami in Japan and washed ashore in Oregon over a year later. The dock algae and invertebrates that had attached to the dock were removed by state wildlife officials to prevent them from becoming invasive species (pp. 88–89) in the western United States.

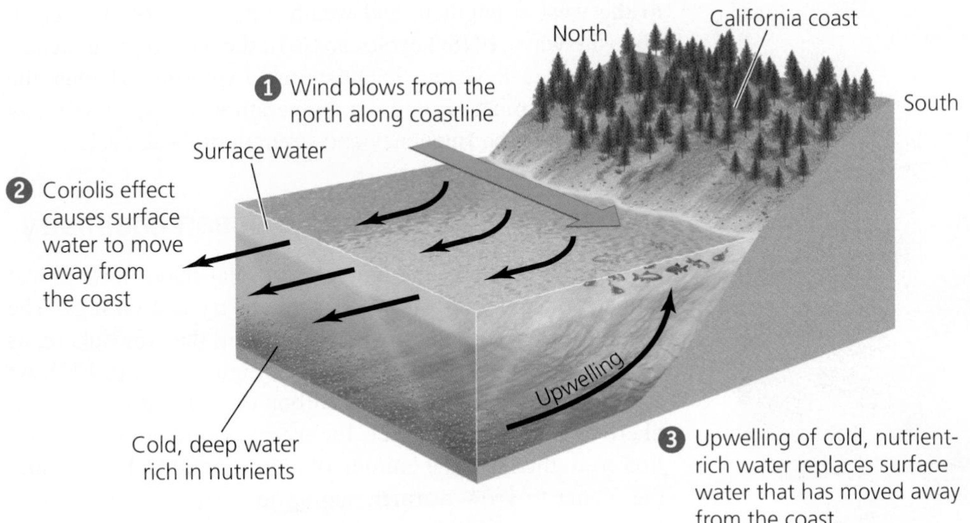

FIGURE 16.7 Upwelling is the movement of bottom waters upward. This often brings nutrients up to the surface, creating productive areas for marine life. For example, north winds blow along the California coastline 1, while the Coriolis effect (p. 455) draws wind and water away from the coast 2. Water is then pulled up from the bottom 3 to replace the water that moves away from shore.

In areas where surface currents converge, or come together, surface water sinks—a process called **downwelling.** Downwelling transports warm surface water rich in dissolved gases to deeper waters, providing an influx of oxygen for deep-water life and "burying" CO_2 in deep ocean waters. Vertical currents also occur in the deep zone, where differences in density can lead to rising and falling convection currents, similar to those in molten rock (p. 34) and in air (p. 455).

Ocean currents affect Earth's climate

The horizontal and vertical movements of ocean water can have far-reaching effects on climate globally and regionally. The **thermohaline circulation** is a worldwide current system in which warmer, fresher water moves along the surface and colder, saltier water (which is denser) moves deep beneath the surface (**FIGURE 16.8**). One segment of this worldwide conveyor-belt system includes the warm surface water in the Gulf Stream that flows across the Atlantic Ocean to Europe. As this water releases heat to the air, keeping Europe warmer than it otherwise would be, the water cools, becomes saltier through evaporation, and thus becomes denser and sinks, creating a region of downwelling known as the *North Atlantic Deep Water (NADW)*.

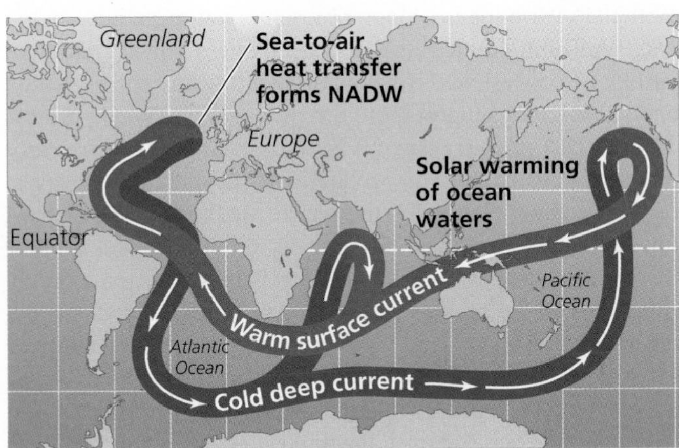

FIGURE 16.8 As part of the oceans' thermohaline circulation, warm surface currents carry heat from equatorial waters northward toward Europe, where they warm the atmosphere. The water then cools and sinks, forming the North Atlantic Deep Water (NADW). Scientists debate whether rapid melting of Greenland's ice sheet could interrupt this heat flow and cause Europe to cool dramatically.

Scientists hypothesize that interrupting the thermohaline circulation could trigger rapid climate change. If global warming (Chapter 18) causes much of Greenland's ice sheet to melt, the resulting freshwater runoff into the North Atlantic would make surface waters less dense (because fresh water is less dense than salt water). This could potentially stop the NADW formation and shut down the northward flow of warm water, causing Europe to cool rapidly. Some data suggest that the thermohaline circulation in this region is already slowing, but other researchers maintain that Greenland will not produce enough runoff to cause a shutdown this century.

Another interaction between ocean currents and the atmosphere that influences climate is the **El Niño–Southern Oscillation (ENSO),** a systematic shift in atmospheric pressure, sea surface temperature, and ocean circulation in the tropical Pacific Ocean. Under normal conditions, prevailing winds blow from east to west along the equator, from a region of high pressure in the eastern Pacific to one of low pressure in the western Pacific, forming a large-scale convective loop in the atmosphere (**FIGURE 16.9a**). The winds push surface waters westward, causing water to "pile up" in the western Pacific. As a result, water near Indonesia can be 50 cm (20 in.) higher and 8°C warmer than water near South America. The westward-moving surface waters allow cold water to rise up from the deep in a nutrient-rich upwelling along the coast of Peru and Ecuador.

El Niño conditions are triggered when air pressure decreases in the eastern Pacific and increases in the western Pacific, weakening the equatorial winds and allowing the warm water to flow eastward (**FIGURE 16.9b**). This suppresses upwelling along the Pacific coast of the Americas, shutting down the delivery of nutrients that support marine life and fisheries. This phenomenon was called El Niño (Spanish for "little boy" or "Christ Child") by Peruvian fishermen because the arrival of warmer waters usually

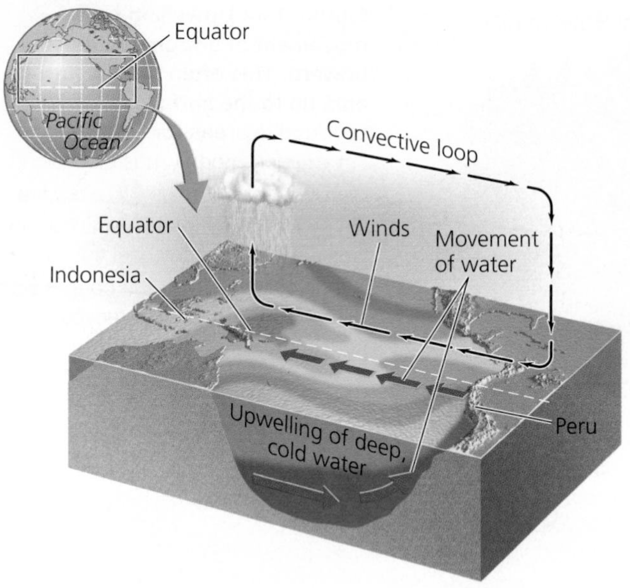

(a) Normal conditions

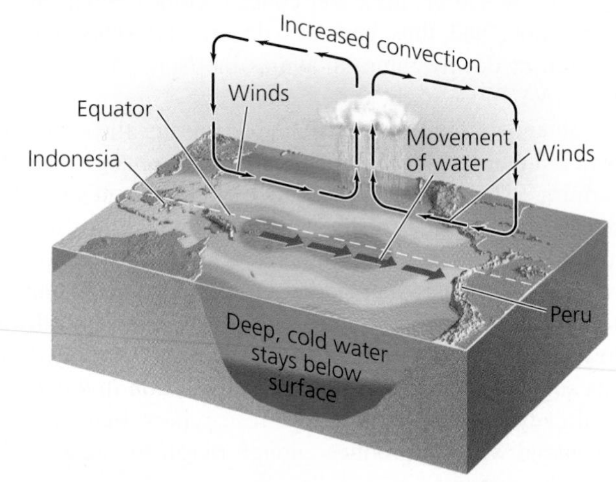

(b) El Niño conditions

FIGURE 16.9 El Niño conditions occur every 2 to 8 years, causing marked changes in weather patterns. In these diagrams, red and orange colors denote warmer water, and blue and green colors denote colder water. Under normal conditions **(a)**, prevailing winds push warm surface waters toward the western Pacific. Under El Niño conditions **(b)**, winds weaken, and the warm water flows back across the Pacific toward South America, like water sloshing in a bathtub. This shuts down upwelling along the American coast and alters precipitation patterns regionally and globally. *Adapted from National Oceanic and Atmospheric Administration, Tropical Atmospheric Ocean Project.*

occurred shortly after Christmas. Coastal industries such as Peru's anchovy fisheries are devastated by El Niño events, and the 1982–1983 El Niño alone caused over $8 billion in economic losses worldwide. El Niño events alter weather patterns around the world, creating rainstorms and floods in areas that are generally dry (such as southern California) and causing drought and fire in regions that are typically moist (such as Indonesia).

La Niña events are the opposite of El Niño events; in a La Niña event, unusually cold waters rise to the surface and extend westward in the equatorial Pacific when winds blowing to the west strengthen, and weather patterns are affected in opposite ways. ENSO cycles are periodic but irregular, occurring every 2–8 years. Scientists are exploring whether the globally warming air and sea temperatures (Chapter 18) may be increasing the frequency and strength of these cycles.

Climate change is altering ocean chemistry

Scientists today are learning a great deal about how global climate change will affect ocean chemistry and biology. The oceans absorb carbon dioxide (CO_2) from the atmosphere, as we first saw in the carbon cycle (see Figure 5.17, p. 122). As our civilization pumps excess carbon dioxide into the atmosphere by burning fossil fuels for energy and removing vegetation from the land, the buildup of atmospheric CO_2 is causing the planet to grow warmer, setting in motion many changes and consequences (Chapter 18).

The oceans have soaked up roughly a third of the excess CO_2 that we've added to the atmosphere, and this has slowed global climate change. However, there are two concerns. One is that the ocean's surface water may soon become saturated with as much CO_2 as it can hold. Once it reaches this limit, then climate change will accelerate as the oceans will no longer remove large amounts of carbon dioxide from the atmosphere.

The second concern is that as ocean water soaks up CO_2, it becomes more acidic. As **ocean acidification** proceeds, many sea creatures have difficulty forming shells because the chemicals they need to create them are less available. This process is harming corals in particular (see **THE SCIENCE BEHIND THE STORY**, pp. 428–429). Corals build reefs, which are hubs for marine biodiversity and provide billions of dollars' worth of ecosystem services. Because of the decline of coral reefs, scientists are warning that ocean acidification is shaping up to be one of most damaging consequences of global climate change.

Marine and Coastal Ecosystems

With their variation in topography, temperature, salinity, nutrients, and sunlight, marine and coastal environments feature a variety of ecosystems. Regions of ocean water differ greatly, and some zones support more life than others. The uppermost 10 m (33 ft) of water absorbs 80% of solar energy, so nearly all of the oceans' primary productivity occurs in the top layer, or **photic zone.** Generally, the warm, shallow waters of continental shelves are most biologically productive and support the greatest species diversity. Habitats and ecosystems occurring between the ocean's surface and floor are termed **pelagic,** whereas those that occur on the ocean floor are called **benthic.** Most marine and coastal ecosystems are powered by solar energy, with sunlight driving photosynthesis by phytoplankton in the photic zone. Yet even the darkest ocean depths host life.

As we survey marine and coastal ecosystems, keep in mind that they are part of a web of interconnected freshwater and marine aquatic systems (see Figure 15.3, p. 392) that exchange water, organisms, sediments, pollutants, and other dissolved substances with one another. Hence, the topics we discuss in this chapter are greatly influenced by those in freshwater ecosystems, and vice versa.

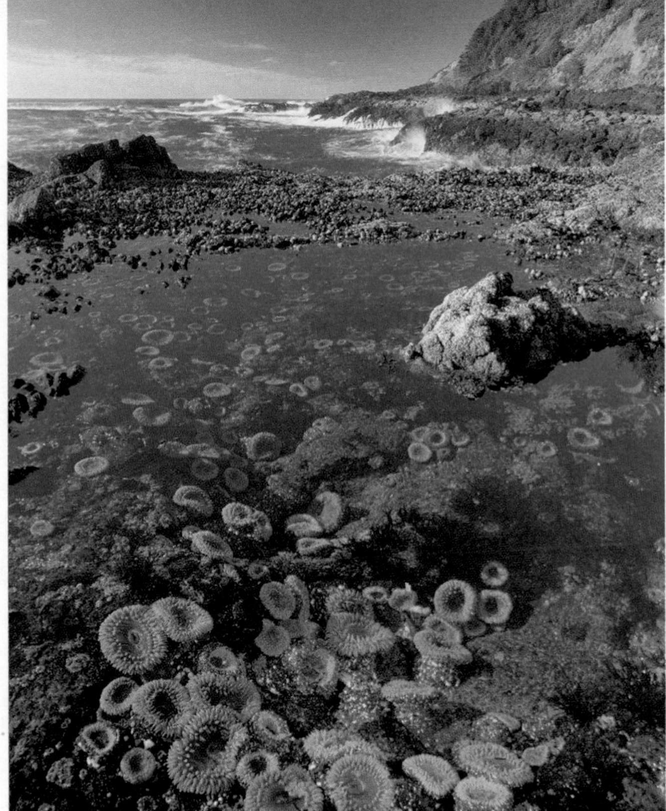

(a) Tide pools at low tide

Intertidal zones undergo constant change

Where the ocean meets the land, **intertidal,** or **littoral,** ecosystems (FIGURE 16.10) spread between the uppermost reach of the high tide and the lowest limit of the low tide. **Tides** are the periodic rising and falling of the ocean's height at a given location, caused by the gravitational pull of the moon and sun (see Figure 21.21, p. 601). High and low tides occur roughly 6 hours apart, so intertidal organisms spend part of each day submerged in water, part of the day exposed to air and sun, and part of the day being lashed by waves. These creatures must also protect themselves from marine predators at high tide and terrestrial predators at low tide.

The intertidal environment is a tough place to make a living, but it is home to a remarkable diversity of organisms. Life abounds in the crevices of rocky shorelines, which provide shelter and pools of water (tide pools) during low tides. Sessile animals such as anemones, mussels, and barnacles live attached to rocks, filter-feeding on plankton in the water that washes over them. Urchins, sea slugs, chitons, and limpets eat intertidal algae or scrape food from the rocks. Sea stars (starfish) creep slowly along, preying on the filter-feeders and herbivores. Crabs clamber around the rocks, scavenging detritus.

The rocky intertidal zone is so diverse because environmental conditions such as temperature, salinity, and moisture change dramatically from the high to the low reaches. This environmental variation gives rise to horizontal bands dominated by different sets of organisms arrayed according to their habitat needs, competitive abilities, and adaptation to exposure. Sandy intertidal areas, such as those of Cape Cod, host less biodiversity, yet plenty of organisms burrow into the sand at low tide to await the return of high tide, when they emerge to feed.

(b) Tidal zones

FIGURE 16.10 The rocky intertidal zone stretches along rocky shorelines between the lowest and highest reaches of the tides. Fish and invertebrates inhabit tidal pools **(a)**, and many types of algae cover the rocks. The intertidal zone **(b)** provides niches for a diversity of organisms, including sea stars (starfish), barnacles, crabs, sea anemones, corals, bryozoans, snails, limpets, chitons, mussels, nudibranchs (sea slugs), and sea urchins. Areas higher on the shoreline are exposed to the air more frequently and for longer periods, so organisms that tolerate exposure best specialize in the upper intertidal zone. The lower intertidal zone is exposed less frequently and for shorter periods, so organisms less tolerant of exposure thrive in this zone.

Will Climate Change Rob Us of Coral Reefs?

The oceans have helped delay the advance of global climate change by soaking up a portion of the excess carbon dioxide that we pump into the atmosphere when we burn fossil fuels and clear forests. But when we alter a flux in a biogeochemical cycle, there are usually consequences. What consequences might there be for this increase in the flux of carbon from the atmosphere to the oceans? Hundreds of scientists in recent years have been working to find out.

Let's start with the chemistry (**FIGURE 1**). Carbon dioxide (CO_2) from the air ❶ can react on the ocean's surface with water (H_2O) to form carbonic acid (H_2CO_3) ❷. This is what happens in soft drinks when carbon dioxide is added to give them fizz:

Tiny corals, the builders of reefs

$$CO_2 + H_2O \rightarrow H_2CO_3$$

This carbonic acid molecule soon dissociates ❸ into a bicarbonate ion (HCO_3^-) and a proton, or hydrogen ion:

$$H_2CO_3 \rightarrow HCO_3^- + H^+$$

The free proton then combines with a carbonate ion (CO_3^{2-}) ❹ to form a second bicarbonate ion:

$$H^+ + CO_3^{2-} \rightarrow HCO_3^-$$

This is bad news for marine organisms such as corals, which depend on carbonate ions in order to form the calcium carbonate ($CaCO_3$) that comprises their hard shells. As the availability of carbonate ions in the water declines, it gets more difficult to build calcium carbonate shells by the process known as calcification. In fact, as carbonic acid and bicarbonate ions become more common in the water, the calcium

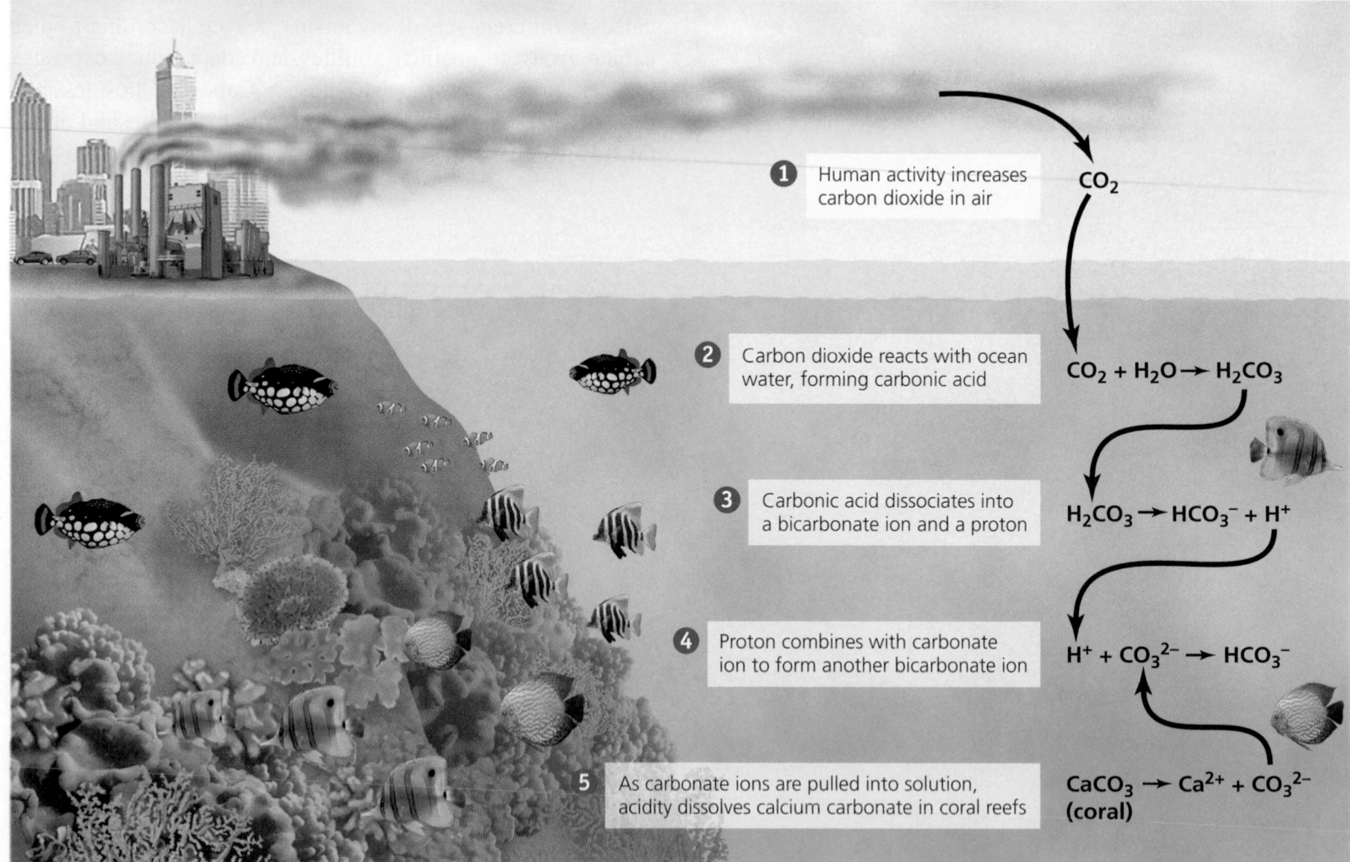

FIGURE 1 Excess carbon dioxide from the atmosphere changes the ocean's chemistry. Carbon dioxide reacts with seawater to create carbonic acid, eventually diminishing the number of carbonate ions available for corals to use to build their shells. *Adapted from Hoegh-Guldberg, O., et al., 2007. Coral reefs under rapid climate change and ocean acidification.* Science *318: 1737–1742. Fig 1a. Reprinted with permission from AAAS.*

carbonate shells of marine creatures like corals even begin dissolving ❺:

$$CaCO_3 \rightarrow Ca^{2+} + CO_3^{2-}$$

Researchers have been testing how increasing acidity affects corals in the lab and in the field. They've also been studying the current global distribution of coral reefs, the massive calcium carbonate structures built by millions of tiny corals that create habitat for so many other marine animals. Through these research avenues they aim to make predictions about how coral reefs will fare in the future as climate change proceeds. In 2007, a team led by Ove Hoegh-Guldberg of the University of Queensland in Australia reviewed existing knowledge of these questions and published its conclusions in the journal *Science*.

Chemistry tests in the lab show that coral shells begin to erode faster than they are built once the carbonate ion concentration falls below 200 micromoles/kg of seawater. Researchers studying coral reefs in the field are finding the same thing: Reefs are growing only in waters with greater than 200 micromoles/kg of carbonate ion availability. A few calculations reveal that this concentration is what we would expect to result from an atmospheric CO_2 concentration of 480 parts per million (ppm). Earth's natural ("preindustrial") level was 280 ppm, and today we have raised it above 380 ppm. Thus we are a little more than halfway to the point at which coral reefs will begin to dissolve in most of the areas they now exist. At the rate CO_2 is accumulating in the atmosphere, we could easily reach 480 ppm around the middle of this century.

Hoegh-Guldberg's team sought some long-term context, so they used data from the Vostok ice core (p. 490), drilled from Antarctic ice, to look for patterns from the past 420,000 years. They found that in all that time, carbonate ion concentrations in the oceans had never been as low as they are today. Indeed, current sea temperatures are 0.7°C warmer and pH is 0.1 unit lower than in the 420,000 years previous. These researchers concluded that our oceans are well on the way to losing coral reefs.

Paleontological studies of the fossil record lend support to this idea, showing that coral reefs and other marine organisms using calcification experienced wide-scale population declines and extinctions in the early Triassic period during an episode of extremely high atmospheric CO_2 levels.

What should we expect in the future? A number of scientists have looked to the *Fourth Assessment Report* of the Intergovernmental Panel on Climate Change (IPCC; p. 492), the 2007 consensus document summarizing scientific knowledge on climate change to that point. Using the IPCC's known and predicted data on atmosphere and ocean temperatures and pH, these researchers modeled how oceanic carbonate ion concentrations should change at various values for atmospheric CO_2.

The results predicted that as atmospheric CO_2 levels rise, coral reefs will shrink in distribution, diversity, and density (**FIGURE 2**). By the time atmospheric CO_2 levels pass 500 ppm, little area of ocean will be left with conditions to support coral reefs. Most of the world's coral reefs, home to so much vibrant biodiversity, "will become rapidly eroding rubble banks," Hoegh-Guldberg's team wrote.

Without living and growing coral reefs, the rich communities of reef-dependent animals will likely collapse. For human society, fisheries will decline, tourism will wither, and coasts will lose protection against storm surges. Developing nations and island nations in the tropics—those already expected to suffer the most from climate change—will be hit hardest.

Scientists are hoping there may be silver linings. Organisms vary in how they react, and not all will suffer from the ocean's chemistry changes. Corals that are more tolerant to temperature changes (such as a variety called *Porites*) may succeed where others fail. Perhaps certain rapidly colonizing species (such as those in the genus *Acropora*) will be able to colonize areas vacated by dying corals.

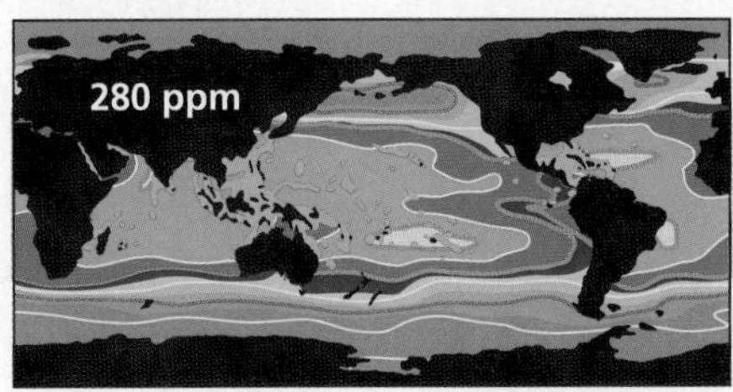

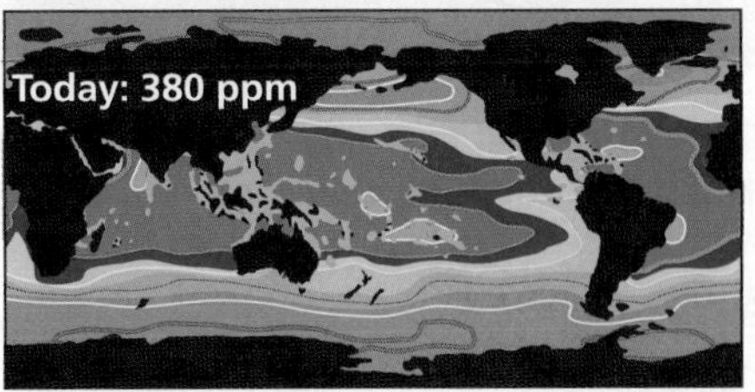

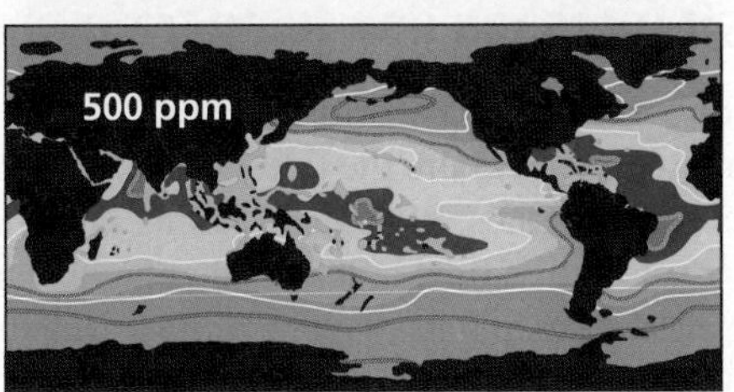

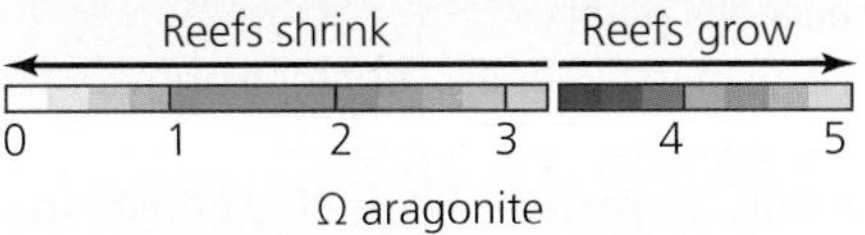

FIGURE 2 Increased atmospheric carbon dioxide levels have decreased the number of ocean areas that support coral reefs. Most of Earth's tropical and subtropical oceans were suitable for the growth of coral reefs (blue colors; **top panel**) before people began emitting carbon dioxide to the atmosphere. Today, at 380 ppm atmospheric CO_2 (**middle panel**), fewer regions are suitable. When the planet reaches 500 ppm (**bottom panel**), very few areas of the ocean will have conditions suitable for coral reefs. *Adapted from Hoegh-Guldberg, O., et al., 2007. Coral reefs under rapid climate change and ocean acidification.* Science *318: 1737–1742. Fig 4. Reprinted with permission from AAAS.*

What can we do about the threats from ocean acidification? Ultimately the only effective solution is to reduce our carbon emissions, and soon. The clock is ticking. In 2013, the Mauna Loa Observatory in Hawaii measured 400 ppm of atmospheric CO_2, the highest such reading for the last nearly 3 million years. ■

FIGURE 16.11 Salt marshes occur in temperate intertidal zones where the substrate is muddy. Tidal waters flow in channels called *tidal creeks* amid flat areas called *benches*, sometimes partially submerging the salt-adapted grasses.

FIGURE 16.12 Mangrove forests line tropical and subtropical coastlines. Mangrove trees, with their unique roots, are adapted for growing in saltwater and provide habitat for many fish, birds, crabs, and other animals.

Salt marshes line temperate shorelines

Along many of the world's coasts at temperate latitudes, **salt marshes** occur where the tides wash over gently sloping sandy or silty substrates. Rising and falling tides flow into and out of channels called tidal creeks and at highest tide spill over onto elevated marsh flats (FIGURE 16.11). Marsh flats grow thick with salt-tolerant grasses, as well as rushes, shrubs, and other herbaceous plants.

Salt marshes boast very high primary productivity and provide critical habitat for shorebirds, waterfowl, and many commercially important fish and shellfish species. Salt marshes also filter pollution and stabilize shorelines against storm surges. However, because people desire to live and do business along coasts, we have altered or destroyed vast expanses of salt marshes to make way for coastal development. When salt marshes are destroyed, we lose the ecosystem services they provide. When Hurricane Katrina struck the Gulf Coast, for instance, the flooding was made worse because vast areas of salt marshes had vanished during the preceding decades as a result of engineering of the Mississippi River, development, and subsidence from oil and gas drilling (Chapter 15).

Mangrove forests line coasts in the tropics and subtropics

In tropical and subtropical latitudes, mangrove forests replace salt marshes along gently sloping sandy and silty coasts. **Mangroves** are some of the few types of trees that are salt tolerant, and they have unique types of roots that curve upward like snorkels to attain oxygen lacking in the mud or curve downward like stilts to support the tree in changing water levels (FIGURE 16.12). Fish, shellfish, crabs, snakes, and other organisms thrive among the root networks, and birds feed and nest in the dense foliage of these coastal forests. Besides serving as nurseries for fish and shellfish that people harvest, mangroves also provide materials that people use for food, medicine, tools, and construction.

From Florida to Mexico to the Philippines, half the world's mangrove forests have been destroyed as people have developed coastal areas. Shrimp farming in particular has driven the conversion of large areas of mangroves. When mangroves are removed, coastal areas lose the ability to slow runoff, filter pollutants, and retain soil. As a result, offshore systems such as coral reefs and eelgrass beds are more readily degraded. Moreover, mangrove forests protect coastal communities against storm surges and tsunamis. The 2004 Indian Ocean tsunami (p. 42) devastated areas where mangroves had been removed but caused less damage where mangroves were intact.

Fresh water meets salt water in estuaries

Many salt marshes and mangrove forests occur in or near **estuaries,** water bodies where rivers flow into the ocean, mixing fresh water with salt water. Estuaries are biologically productive ecosystems that experience fluctuations in salinity as tides and freshwater runoff vary daily and seasonally. Sheltered from crashing surf, the shallow water of estuaries nurtures eelgrass beds and other plant life, producing abundant food and resources. For shorebirds and for many commercially important shellfish species, estuaries provide critical habitat. For fishes such as salmon, which spawn in streams and mature in the ocean, estuaries provide a transitional zone where young fish make the passage from fresh water to salt water.

Estuaries everywhere have been affected by coastal development, water pollution, habitat alteration, and overfishing. Think of any major estuary in the United States, from New York Harbor to Chesapeake Bay (Chapter 5) to Florida Bay to the mouth of the Mississippi River (Chapter 15) to San Francisco Bay. They all have suffered pollution and other impacts from human development. Coastal ecosystems have borne the brunt of human impact because two-thirds of Earth's people choose to live within 160 km (100 mi) of the ocean.

Kelp forests harbor many organisms

Along many temperate coasts, large brown algae, or **kelp,** grow from the floor of continental shelves, reaching up toward

WEIGHING THE ISSUES

COASTAL DEVELOPMENT A developer wants to build a large marina on an estuary in your coastal town. The marina would boost the town's economy but eliminate its salt marshes. What consequences would you expect for property values? For water quality? For wildlife? As a homeowner living adjacent to the marshes, how would you respond? Do you think that developers or town officials should offer homeowners insurance against damage from storm surges when a protective salt marsh is destroyed?

the sunlit surface. Some kelp reaches 60 m (200 ft) in height and can grow 45 cm (18 in.) per day. Dense stands of kelp form underwater "forests" (**FIGURE 16.13**). Kelp forests supply shelter and food for invertebrates and fish, which in turn provide food for predators such as seals and sharks (indeed, kelp forests were the setting for our discussion of keystone species in Chapter 4). Kelp forests also absorb wave energy and protect shorelines from erosion. People in Asian cultures eat some types of kelp, and kelp provides compounds known as alginates, which serve as thickeners in such consumer products as cosmetics, paints, paper, soaps, and ice cream.

Coral reefs are treasure troves of biodiversity

Shallow subtropical and tropical waters are home to coral reefs. A reef is an underwater outcrop of rock, sand, or other material. A **coral reef** is a mass of calcium carbonate composed of the skeletons of tiny marine animals known as *corals*. A coral reef may occur as an extension of a shoreline, along a *barrier island* paralleling a shoreline, or as an *atoll*, a ring around a submerged island.

Corals are tiny invertebrate animals related to sea anemones and jellyfish. They remain attached to rock or existing reef and capture passing food with stinging tentacles (see photo on p. 428). Corals also derive nourishment from symbiotic algae known as *zooxanthellae*, which inhabit their bodies and produce food through photosynthesis. Most corals are colonial, and the colorful surface of a coral reef consists of millions of densely packed individuals. (The colors come from zooxanthellae.) As corals die, their skeletons remain part of the reef while new corals grow atop them, increasing the reef's size.

FIGURE 16.13 "Forests" of tall brown algae known as *kelp* grow from the floor of the continental shelf. Numerous fish and other creatures eat kelp or find refuge among its fronds.

(a) Coral reef community

(b) Bleached coral

FIGURE 16.14 Coral reefs provide food and shelter for a tremendous diversity (a) of fish and other creatures. Today these reefs face multiple stresses from human impacts. Many corals have died as a result of coral bleaching **(b)**, in which corals lose their zooxanthellae. Bleaching is evident in the whitened portion of this coral.

Like kelp forests, coral reefs protect shorelines by absorbing wave energy. They also host tremendous biodiversity (**FIGURE 16.14a**). This is because coral reefs provide complex physical structure (and thus many habitats) in shallow nearshore waters, which are regions of high primary productivity. Besides the staggering diversity of anemones, sponges, hydroids, tubeworms, and other sessile (stationary) invertebrates, innumerable molluscs, flatworms, sea stars, and urchins patrol reefs, while thousands of fish species find food and shelter in reef nooks and crannies.

Coral reefs are experiencing alarming declines worldwide, however. Many have undergone "coral bleaching," a process that occurs when zooxanthellae die or leave the coral, depriving it of nutrition. Corals lacking zooxanthellae lose color and frequently die, leaving behind ghostly white patches in the reef (**FIGURE 16.14b**). Coral bleaching is thought to result from increased sea surface temperatures associated with global climate change, from the influx of pollutants, from unknown natural causes, or from combinations of these factors.

Nutrient pollution in coastal waters also promotes the growth of algae, which are smothering reefs in the Florida Keys and in many other regions. Moreover, coral reefs sustain damage when divers stun fish with cyanide to capture them for food or for the pet trade, a common practice in waters of Indonesia and the Philippines. Finally, as global climate change proceeds, the oceans are becoming more acidic as excess carbon dioxide from the atmosphere reacts with seawater to form carbonic acid. Acidification threatens to deprive corals of the carbonate ions they need to produce their structural parts (see **The Science behind the Story**, pp. 428–429).

A few coral species thrive in waters outside the tropics and build reefs on the ocean floor at depths of 200–500 m (650–1650 ft). These little-known reefs, which occur in cold-water areas off the coasts of Norway, Spain, the British Isles, and elsewhere, are only now beginning to be studied by scientists. Already, however, many have been badly damaged by bottom-trawling (p. 439)—the same practice that has drastically degraded the benthic habitats of groundfish such as the Atlantic cod. Norway and other countries are now beginning to protect some of these deep-water reefs.

Open-ocean ecosystems vary in their biodiversity

Biological diversity in pelagic regions of the open ocean is highly variable in its distribution. Primary production (p. 111) and animal life near the surface are concentrated in regions of nutrient-rich upwelling. Microscopic phytoplankton constitute the base of the marine food chain in the pelagic zone. These photosynthetic algae, protists, and cyanobacteria feed zooplankton (p. 75), which in turn become food for fish, jellyfish, whales, and other free-swimming animals (FIGURE 16.15). Predators at higher trophic levels include larger fish, sea turtles, and sharks. Fish-eating birds such as puffins, petrels, and shearwaters feed at the surface of the open ocean, returning periodically to nesting sites on islands and coastlines.

FIGURE 16.15 The uppermost reaches of ocean water contain billions upon billions of phytoplankton—tiny photosynthetic algae, protists, and bacteria that form the base of the marine food chain. This part of the ocean is also home to zooplankton, small animals and protists that dine on phytoplankton and comprise the next trophic level.

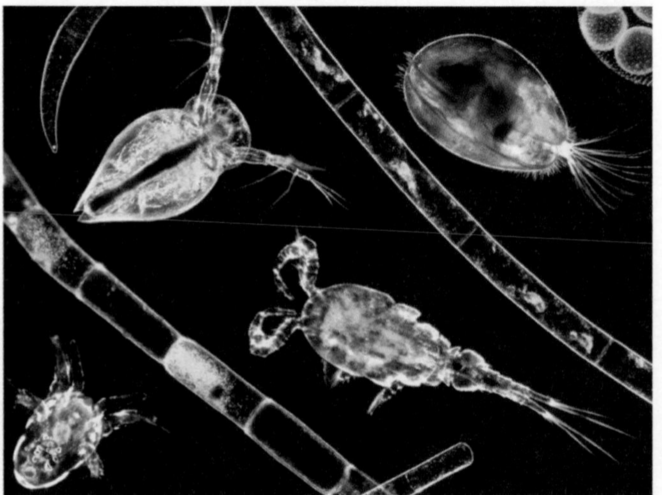

In the little-known deep-water ecosystems, animals have adapted to tolerate extreme water pressures and to live in the dark without food from autotrophs. Many of these often bizarre-looking creatures scavenge carcasses or organic detritus that fall from above. Others are predators, and still others attain food from symbiotic mutualistic (p. 80) bacteria.

Ecosystems form around hydrothermal vents (p. 33) in the ocean, where heated water spurts from the seafloor, carrying minerals that precipitate to form rocky structures. Tubeworms, shrimp, and other creatures in these systems use symbiotic bacteria to derive their energy from chemicals in the heated water rather than from sunlight, a process called *chemosynthesis*. They manage to thrive within amazingly narrow zones between scalding-hot and icy cold water.

Marine Pollution

People have long made the oceans a sink for waste and pollutants. Even into the mid-20th century, it was common for coastal U.S. cities to dump trash and untreated sewage along their shores. The Clean Water Act (p. 176) reduced the discharge of point-source pollution (p. 408) into the oceans. But a great deal of pollution is non-point-source pollution that comes from countless small things we all do and from items we throw away far inland. For the oceans are downstream from everywhere: As long as rivers flow to the sea, much of our pollution on land sooner or later ends up in the oceans.

Oil, plastic, toxic chemicals, and excess nutrients all eventually make their way into the oceans. Sewage and trash from cruise ships and abandoned fishing gear from fishing boats add to the input. The scope of trash in the sea can be gauged by the amount picked up by volunteers who trek beaches in the Ocean Conservancy's annual International Coastal Cleanup. Over 25 years, more than 65 million kg (144 million lb) of waste have been collected from the world's beaches as part of this program.

Plastic debris endangers marine life

Plastic bags and bottles, fishing nets and line, and small particles of plastic and other trash can harm marine creatures. Organisms can become entangled in such debris and drown, or they may die as a result of ingesting material they cannot digest or expel (FIGURE 16.16a). Floating plastic and debris concentrates in *gyres*, areas of the ocean where currents converge (see THE SCIENCE BEHIND THE STORY, pp. 434–435). The North Pacific Gyre contains the **Great Pacific Garbage Patch,** an area larger than Texas, in which tiny pieces of floating plastic outnumber organisms by a 6 to 1 margin.

Because plastic is designed not to break down, it can drift for decades before washing up on beaches. Plastic does degrade slowly in seawater and sunlight, but in doing so, breaks down into smaller and smaller bits that become more and more numerous. This can actually make things worse: The oceans are now filled with uncountable trillions of tiny pellets of plastic floating just under the surface like confetti (FIGURE 16.16b). Some plastics sink, littering the bottom of the ocean where the

(a) Plastics entangling wildlife

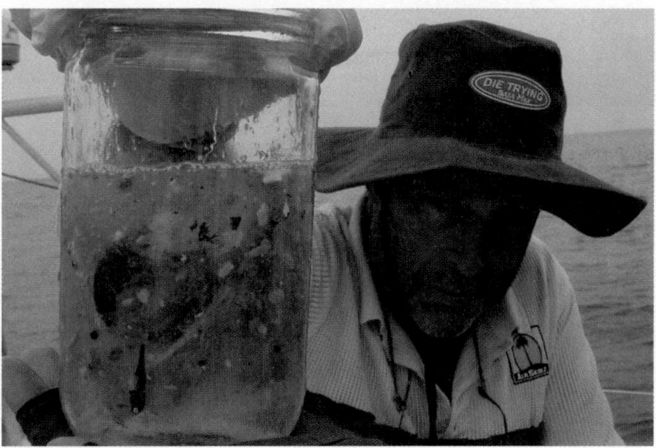

(b) Tiny bits of debris from ocean gyres

(c) Organisms ingesting plastics

(d) Preventing oceanic plastic pollution

FIGURE 16.16 **Plastics affect oceanic life.** Like thousands of marine mammals every year, this grey seal **(a)** became entangled in a discarded fishing net. Small pieces of plastic like this sample from the "Great Pacific Garbage Patch" are also dangerous to organisms **(b)**. Plastics are often mistaken for food items and ingested by wildlife, such as by this dead albatross **(c)**. Prevention is the best approach for addressing oceanic plastic pollution. Here, booms capture trash in the Los Angeles River **(d)** to prevent it from entering the Pacific Ocean.

low-temperature, low-light conditions are not favorable for degradation. In the North Sea, for example, underwater surveys have found over 100 pieces of plastic per km^2 on the ocean floor.

Floating plastics are particularly harmful to marine life. Organisms mistake the small pieces of plastic for plankton or fish eggs, ingest them, and suffer injury or death (FIGURE 16.16c). A 2010 study found that the average plankton-eating fish in the Great Pacific Garbage Patch had 2.1 pieces of plastic in its digestive system. Fish are not the only organisms affected. Seabird chicks are killed when parents unknowingly feed them pieces of plastic when regurgitating food. One study attributed 40% of premature deaths of Albatross chicks in the Pacific islands of Midway to this cause. Some 267 species are affected by marine plastic debris, leading to an estimated 100,000 marine mammals and 1 million seabird deaths each year.

Plastics can have toxic effects on organisms. Plastics contain harmful substances, such as bisphenol A and phthalates (Chapter 14), which leach into ocean water or the digestive tracts of animals (if the plastic is ingested). Pieces of plastic in the ocean can also "grab" and concentrate persistent organic pollutants (POPs; p. 384) such as DDT and PCBs, thereby increasing the toxicity of the plastic to any organism that ingests it.

Floating pieces of plastic also serve as "rafts" and transport species over long distances they could not travel on their own. Seaweeds, sessile (nonswimming) invertebrates, and other creatures have become invasive species when carried to new habitats on floating plastic. For example, two species of bryozoans (a predatory, sessile invertebrate that forms polyps on seaweed and stones) from the Caribbean are reaching the Florida coast on floating plastics and are impacting native species.

Because plastics take 500–1000 years to degrade at sea and there is no viable way to collect the small bits of plastics that litter the oceans, preventing their entry into the oceans is key to remedying oceanic plastic pollution (FIGURE 16.16d). In 2006, the U.S. Congress responded to ocean pollution by passing the Marine Debris Research, Prevention, and Reduction Act, aiding efforts to keep plastics out of marine waters. These and other efforts are ecologically wise but also convey significant economic benefits. In Asia alone, the costs of plastic pollution on fisheries, tourism, and other industries is an estimated $1 billion a year.

Predicting the Oceans' "Garbage Patches"

In 1997, while sailing across a little-traveled portion of the Pacific Ocean as he returned from a recreational yacht race, Captain Charles Moore encountered a huge floating mass of debris he described as a "soupy" collection of items including tires, plastics, chemical drums, coat hangers, fishing nets, and other items. This was the first recorded visual confirmation of what is now called the "Great Pacific Garbage Patch" and brought firmly into the public eye the issue of plastic pollution in the oceans. Knowing where floating debris will accumulate in the oceans is important to marine biologists and oceanographers, but short of stumbling upon these areas in the ocean, as Captain Moore did, is there any way to know where to look for them? Thanks to the work of Nikolai Maximenko and his collaborators, the answer is yes.

Maximenko, a senior researcher of the International Pacific Research Center at the University of Hawaii at Manoa, studies the movements of currents in the oceans. In particular, his work examines *Ekman drift*, wind-driven currents in the water's upper layers first modeled by scientist Walfrid Ekman in the early 20th century. Although the ocean currents that are driven by pressure gradients and the Coriolis force (collectively called *geostrophic currents*) are fairly well described, the movements associated with Ekman currents are not. This lack of information hinders researchers' ability to predict smaller-scale movements of floating material in the oceans—such as plastics and other debris.

Data gathered by the Global Drifter Program of the U.S. National Oceanic and Atmospheric Administration (NOAA) is helping scientists better understand these currents. Since 1979, the program has deployed more than 12,000 buoys in the world's oceans and tracked their movements to determine patterns in ocean currents. Each buoy (**FIGURE 1**) is composed of a 30- to 40-cm (11.8–15.7 in.) floating drifter, which contains sensors for detecting ocean temperature, salinity, and other properties, as well as a transmitter to send its information to satellites overhead. The drifter is tethered to a type of anchor, called a subsurface drogue, which hangs at a depth of 15 m (49.2 ft). The drogue keeps the drifter upright on the surface and provides an underwater "sail" for currents to push. Drifters typically operate for about 400 days before they stop transmitting, and the program aims to maintain an array of 1250 drifters at any given time in the world's oceans.

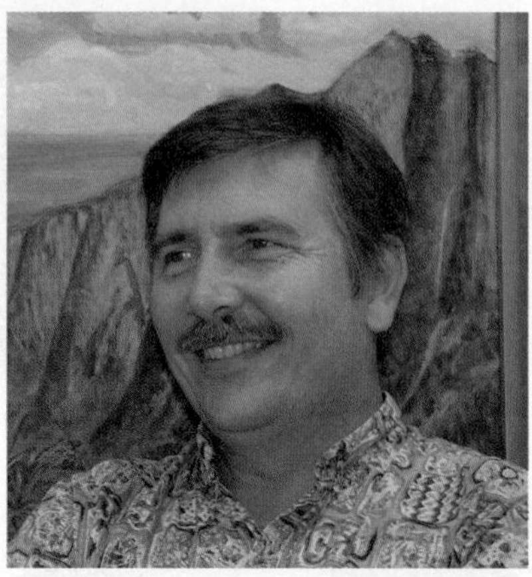

Nikolai Maximenko, International Pacific Research Center at the University of Hawaii at Manoa

In 2008, Maximenko partnered with Peter Niiler of the Scripps Institution of Oceanography to use data from the Global Drifter program to produce a more detailed map of surface ocean currents. The team combined data on drifter movements with satellite altimetry and wind currents, producing a highly detailed map of both geostrophic and Ekman currents. As the issue of oceanic plastic pollution gained attention, Maximenko saw an opportunity to utilize his map to predict the areas of the ocean where floating debris is likely to accumulate. To do this, he partnered with his colleague, Jan Hafner, and created a computer model that predicted the movements of drifters over long time periods in the world's oceans. He then ran a simulation in which he uniformly distributed drifters in the oceans and saw where drifters concentrated over time.

The simulation's results, published in 2012 in *Marine Pollution Bulletin*, revealed that oceanic debris was likely to accumulate in portions of five subtropical gyres (**FIGURE 2**), including the area of the Great Pacific Garbage Patch. The simulation also revealed that 70% of the drifters remained at sea after 10 years, showing the lengthy life span of floating debris in the oceans. One of the model's predictions, which were first released in 2008, was verified when a 2010 study found high concentrations of plastic in

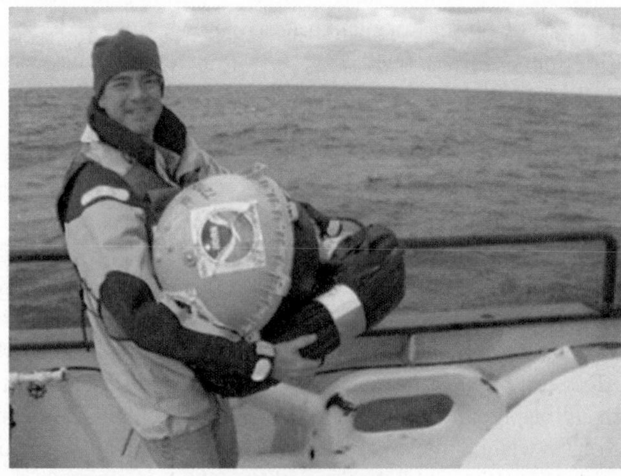

FIGURE 1 A researcher prepares to deploy a buoy for the Global Drifter Program. The cylindrical sea anchor (or drogue) extends vertically in the water column after deployment and is pushed along by shallow ocean currents.

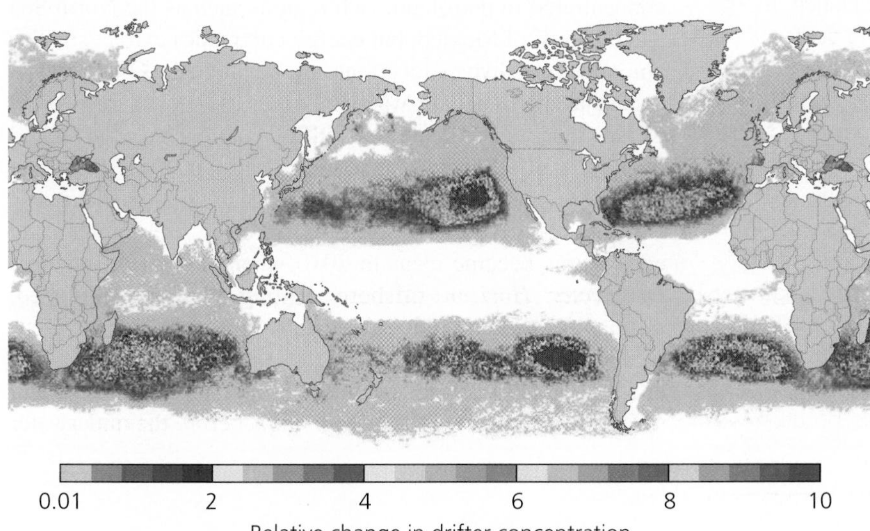

FIGURE 2 Maximenko's computer simulation predicts where plastics and trash accumulate based on oceanic currents. They include the North and South Atlantic Gyres, Indian Ocean Gyre, and the North and South Pacific Gyres. The North Pacific Gyre off the coast of California is home to the "Great Pacific Garbage Patch." *Source: IPRC Climate, 2008.*

the North Atlantic Gyre where Maximenko's model predicted it. Further validation arrived in 2011, when a survey of the South Atlantic Gyre found elevated levels of plastic in locations predicted by the model. The remaining "garbage patches" (South Pacific Ocean and South Indian Ocean) have yet to be verified as of 2012 because these areas, like the other areas identified in the model, lie outside traditional shipping lanes.

Maximenko's model was recently put to another use—predicting the fate of the massive amounts of debris washed into the northern Pacific by the tsunami following the 2011 Tohoku earthquake (pp. 22–23). The model correctly predicted the arrival of significant debris in Hawaii in 2012–2013 and predicts that sizeable amounts of debris will reach the western coast of the United States in 2014 (**FIGURE 3**). The debris will then revisit Hawaii in 2016 in greater quantities and for a longer duration than the first visit.

Thanks to the work of Maximenko and his colleagues, we now better understand the movements of plastics and other debris in the oceans. Armed with this knowledge, we will be better able to properly assess and, we hope, remediate the impacts of plastics on marine life. ■

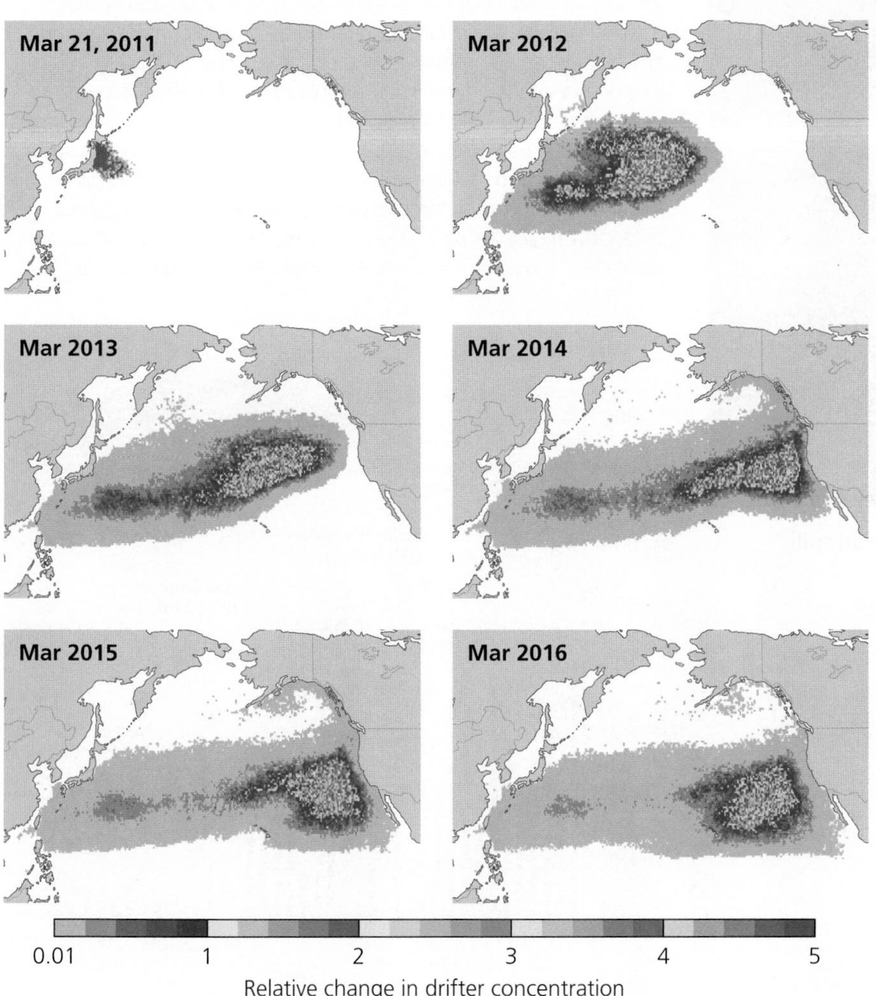

FIGURE 3 Computer models predict the fate of debris from the 2011 tsunami in northern Japan. The bulk of the debris is expected to reach the shores of the western United States in early to mid 2014 and to remain off the coast of California in the North Pacific Gyre. *Source: Nikolai Maximenko, International Pacific Research Center.*

FAQ Is the "Great Pacific Garbage Patch" a huge mass of floating debris?

The vast majority of plastics that accumulate in the ocean are quite small and are dispersed over relatively large areas, so they are difficult to detect with the naked eye. Hence, a person expecting to see dense expanses of floating debris when traveling through the Great Pacific Garbage Patch and other hotspots of oceanic pollution is likely to be disappointed. This can incorrectly lead one to believe the problem of oceanic debris is exaggerated.

But even though the plastic and other debris in these gyres is typically small in size (usually only several millimeters in length), it still poses threats to wildlife from ingestion and toxicity. In many ways, smaller pieces of plastic are more dangerous to marine life than larger pieces, because they resemble food items much more closely than do large items. So while a floating tire or fishing net certainly provides a striking visual impact, it's important to realize that the oceanic plastic pollution we cannot easily see is often of greatest concern.

Oil pollution comes from spills of all sizes

For many people, ocean pollution first brings to mind oil pollution. About 30% of our crude oil and much of our natural gas come from seafloor deposits. Most offshore oil and gas is concentrated in petroleum-rich regions such as the North Sea and the Gulf of Mexico, but energy companies extract smaller amounts from diverse locations, among them the Grand Banks and adjacent Canadian waters. Proposals to drill for oil and gas in Georges Bank and the Gulf of Maine have been stalled by both the U.S. and Canadian governments, in large part because any spilled oil could damage the region's fisheries.

The danger of oil spills to fisheries, economies, and ecosystems became clear in 2010 when British Petroleum's *Deepwater Horizon* offshore drilling platform exploded, killing 11 workers and sinking into the ocean off the Louisiana coast (p. 538). Oil gushed from the underwater well at rates of 1800 gallons per minute, rose to the surface, and spread (**FIGURE 16.17a**) for three months before the underwater

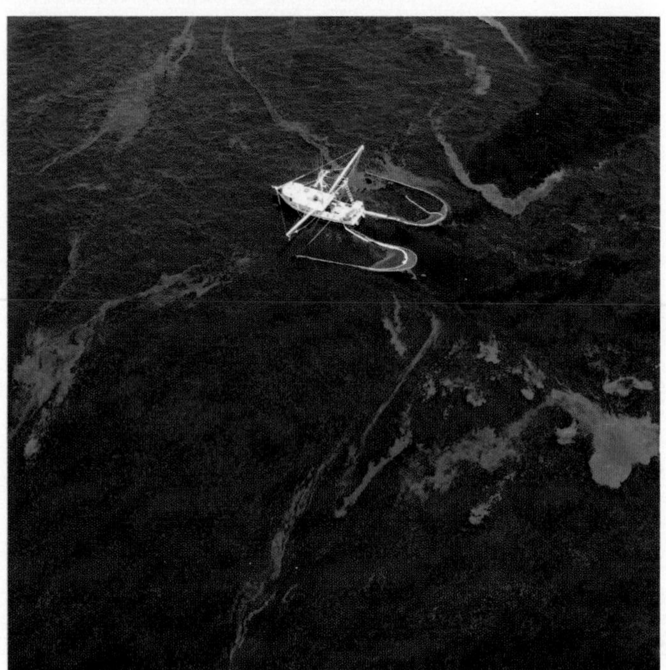
(a) Boat proceeds through *Deepwater Horizon* oil spill

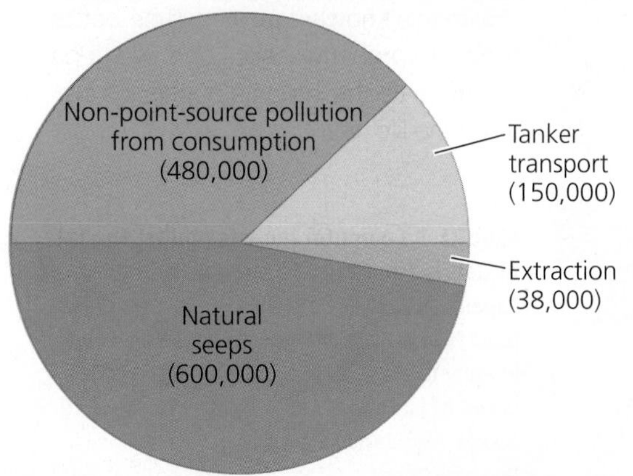

(b) Sources of petroleum input into oceans (metric tons), in 2003

FIGURE 16.17 Oil pollution from both large and small spills affects the oceans. Pollution was severe **(a)** after BP's Deepwater Horizon oil drilling platform exploded and disgorged millions of gallons of crude oil into the Gulf of Mexico in 2010. **(b)** Non-point-source pollution from petroleum consumption by people accounts for 38% of total input into oceans. Sources include numerous diffuse sources, especially runoff from rivers and coastal communities and leakage from two-stroke engines. Less oil is being spilled into ocean waters today in large tanker spills **(c)**, thanks in part to regulations on the oil shipping industry and improved spill response techniques. The bar chart shows cumulative quantities of oil spilled worldwide from nonmilitary spills over 7 metric tons, identifying larger spills by vessel name. *Data from: (b) National Research Council, 2003.* Oil in the sea III. Inputs, fates, and effects. *Washington, DC: National Academies Press; and (c) International Tanker Owners Pollution Federation Ltd.*

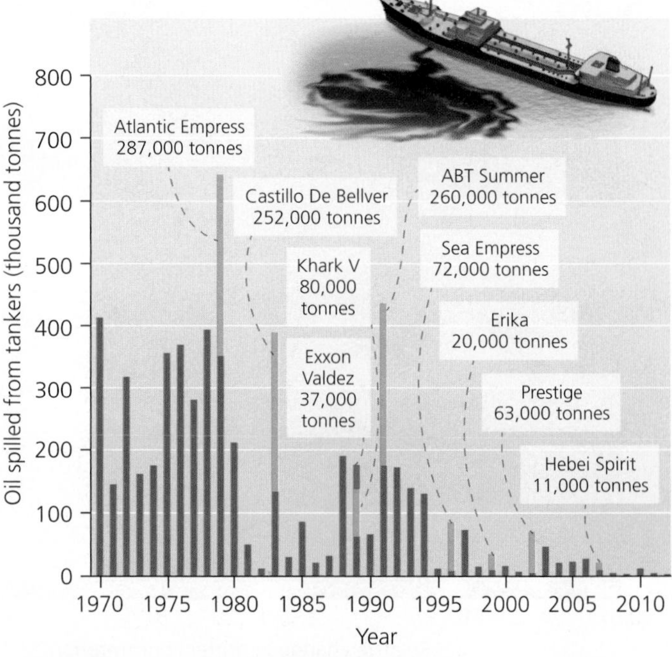

(c) Quantity of petroleum spilled from oil tankers, 1970–2012

well was capped. Eventually the oil polluted hundreds of miles of water and shoreline along the coasts of Louisiana, Mississippi, Alabama, and Florida and affected marshes and many species of wildlife. Even organisms in the deep were affected by this spill. A 2012 study found that corals living 11 km (6.8 mi) from the oil well at a depth of 1370 m (4495 ft) were adversely affected by exposure to an underwater plume of oil emanating from the spill site.

Such major oil spills from platforms and from the tanker ships that transport oil make headlines and cause severe environmental and economic problems. Yet despite the severity of events like the *Deepwater Horizon* oil spill, most oil pollution in the oceans accumulates from innumerable, widely spread, small non-point sources, including leakage from small boats and runoff from human activities on land. Moreover, the amount of petroleum spilled into the oceans in recent years is equaled by the amount that seeps up from naturally occurring seafloor deposits (FIGURE 16.17b).

In response to headline-grabbing oil spills, governments have begun to implement stricter safety standards for tankers, such as requiring industry to pay for tugboat escorts in coastal waters and to develop prevention and response plans for major spills. The U.S. Oil Pollution Act of 1990 created a $1 billion prevention and cleanup fund and required that by 2015 all oil tankers in U.S. waters be equipped with double hulls as a precaution against puncture. Over the past three decades, the amount of oil spilled by tankers worldwide has decreased (FIGURE 16.17c), in part because of an increased emphasis on spill prevention and response. In the wake of the *Deepwater Horizon* spill, the U.S. government is considering tighter regulations on offshore drilling operations.

Toxic pollutants can contaminate seafood

Aside from the harm that pollutants such as petroleum and plastic can do to marine life, toxic pollutants can make some fish and shellfish unsafe for people to eat. One prime concern today is mercury contamination. Mercury is a toxic heavy metal (p. 370) released into the environment from coal combustion (p. 462), mine tailings, and other sources. After settling onto land and water, mercury bioaccumulates in animals' tissues and biomagnifies as it makes its way up the food chain (p. 374). As a result, fish and shellfish at high trophic levels can contain substantial levels of mercury. Eating seafood high in mercury is particularly dangerous for young children and for pregnant or nursing mothers, because the fetus, baby, or child can suffer neurological damage as a result.

Because seafood is an important part of a healthy diet, nutritionists do not advocate avoiding seafood entirely. However, people in at-risk groups should avoid fish high in mercury (such as swordfish, shark, and albacore tuna) while continuing to eat seafood low in mercury (such as catfish, salmon, and canned light tuna). We should also be careful not to eat seafood from local areas where health advisories have been issued.

Excess nutrients cause algal blooms

Pollution from fertilizer runoff or other nutrient inputs can create dead zones in coastal ecosystems, as we saw with the Chesapeake Bay (Chapter 5) and the Gulf of Mexico (pp. 410–411). The release of excess nutrients into surface waters can spur unusually rapid growth of phytoplankton, causing eutrophication (p. 108) in freshwater and saltwater systems.

FIGURE 16.18 **In a harmful algal bloom, certain types of algae multiply to great densities in surface waters and produce toxins that can harm organisms.** Red tides are a type of algal bloom in which the algae produce pigment that turns the water red.

Excessive nutrient concentrations sometimes give rise to population explosions among some species of marine algae. These dinoflagellate algae produce powerful toxins that attack the nervous systems of vertebrates. Blooms of these algae are known as **harmful algal blooms.** Some dinoflagellates produce reddish pigments that discolor surface waters, and blooms of these species are nicknamed **red tides** (FIGURE 16.18). Harmful algal blooms can cause illness and death among zooplankton, birds, fish, marine mammals, and people as their toxins are passed up the food chain. They also cause economic loss for communities that rely on beach tourism and fishing. Reducing nutrient runoff into coastal waters can lessen the frequency of these outbreaks. When they occur, we can minimize their health impacts by monitoring to prevent human consumption of affected organisms.

Emptying the Oceans

As severe as the impacts of marine pollution on marine organisms can be, most scientists concur that the more worrisome dilemma is overharvesting. Sadly, the old cliché that "there are always more fish in the sea" is misleading. The oceans today have been overfished, and as with the groundfish of the Northwest Atlantic, many stocks have been largely depleted.

The oceans and their biological resources have met human needs for thousands of years, but today we are placing unprecedented pressure on marine resources. Over half the world's marine fish populations are fully exploited, meaning that we cannot harvest them more intensively without depleting them, according to the U.N. Food and Agriculture Organization (FAO). An additional 28% of marine fish populations are overexploited and already being driven toward extinction. Only one-fifth of the world's marine fish populations can

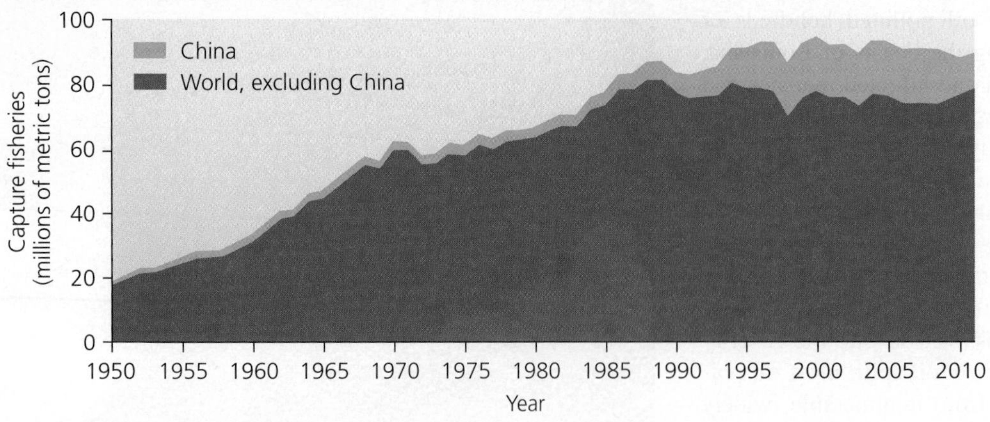

FIGURE 16.19 After rising for decades, the total global fisheries catch has stalled for the past 20 years. Many scientists fear that a global decline is imminent if conservation measures are not taken. The figure shows trends with and without China's data, because research suggests that China's data may be somewhat inflated. *Data from the Food and Agriculture Organization of the United Nations, 2012.* The state of world fisheries and aquaculture 2012. *Fig 3.*

yield more than they are already yielding without being driven into decline.

Total global fisheries catch, after decades of increases, leveled off after about 1988 (FIGURE 16.19), despite increased fishing effort. In 2008, the FAO concluded that "the maximum wild capture fisheries potential from the world's oceans has probably been reached." Fishery collapses such as those off Newfoundland and New England are ecologically devastating and take a severe economic toll on human communities that depend on fishing. A comprehensive 2006 study in the journal *Science* predicted that if current trends continue, populations of *all* ocean species that we fish for today will collapse by the year 2048. If fisheries collapse as predicted, we will lose the ecosystem services they provide. Productivity will be reduced, ecosystems will become more sensitive to disturbance, and the filtering of water by vegetation and organisms (such as oysters) will decline, making harmful algal blooms, dead zones, fish kills, and beach closures more common. Aquaculture (raising fish in tanks or pens) is booming and is helping to relieve pressure on wild stocks, but fish farming comes with its own set of environmental dilemmas (pp. 251–252). All this makes it vital, many scientists and fisheries managers say, that we turn immediately to more sustainable fishing practices.

We have long overfished

People have always harvested fish, shellfish, turtles, seals, and other animals from the oceans. Much of this harvesting was sustainable, but scientists are learning that people began depleting some marine populations centuries or millennia ago. Overfishing then accelerated during the colonial period of European expansion and intensified further in the 20th century. At each stage, improved technologies and increasingly global markets intensified our impact.

Recent syntheses of historical evidence by marine biologists and historians reveal that ancient overharvesting likely affected ecosystems in ways we only partially understand today. Large animals, including the Caribbean monk seal, Steller's sea cow, and Atlantic gray whale, were hunted to extinction prior to the 20th century—before scientists were able to study them and the ecological roles they played. In the Caribbean, green sea turtles ate sea grass and likely kept it cropped low, like a lawn. But with today's turtle population a fraction of what it was, sea grass grows thickly, dies, and rots, giving rise to sea grass wasting disease, which ravaged Florida Bay sea grass in the 1980s. The best-known case of historical overharvesting is the near-extinction of many species of whales. This resulted from commercial whaling that began centuries ago and was curtailed only in 1986. Since then, some species (such as the humpback whale) have been recovering, but others have not.

Groundfish in the Northwest Atlantic historically were so abundant that the people who harvested them never imagined they could be depleted. Yet careful historical analysis of fishing records has revealed that even in the 19th century, fishers repeatedly experienced locally dwindling catches, and each time they needed to introduce some new approach or technology to extend their reach and restore their catch rate.

Fishing has industrialized

Today's industrialized commercial fishing fleets employ fossil fuels, huge vessels, and powerful new technologies to capture fish in great volumes. *Factory fishing* vessels even process and freeze their catches while at sea. The global reach of today's fleets makes our impacts much more rapid and intensive than in the past.

The modern fishing industry uses several methods to capture fish at sea that are highly efficient but also environmentally damaging. Some vessels set out long *driftnets* that span large expanses of water (FIGURE 16.20a). These chains of transparent nylon mesh nets are arrayed to drift with currents so as to capture passing fish, and they are held vertical by floats at the top and weights at the bottom. Driftnetting usually targets species that traverse open water in immense schools, such as herring, sardines, and mackerel. Specialized forms of driftnetting are used for sharks, shrimp, and other animals.

Longline fishing (FIGURE 16.20b) involves setting out extremely long lines (up to 80 km [50 mi] long) with up to several thousand baited hooks spaced along their lengths. Tuna and swordfish are among the species targeted by longline fishing.

(a) Driftnetting

(b) Longlining

(c) Bottom-trawling

FIGURE 16.20 Commercial fishing fleets use several methods of capture. In driftnetting **(a)**, long transparent nylon nets are set out to drift through open water to capture schools of fish. In longlining **(b)**, lines with numerous baited hooks are set out in open water. In bottom-trawling **(c)**, weighted nets are dragged along the floor of the continental shelf. All methods result in large amounts of bycatch, the capture of nontarget animals. The illustrations above are schematic for clarity and do not portray the immense scale that these technologies can attain; for instance, industrial trawling nets can be large enough to engulf multiple Boeing 747 jumbo jets.

Trawling entails dragging immense cone-shaped nets through the water, with weights at the bottom and floats at the top. Trawling in open water captures pelagic fish, whereas *bottom-trawling* (FIGURE 16.20c) involves dragging weighted nets across the floor of the continental shelf to catch groundfish and other benthic organisms, such as scallops.

Fishing practices kill nontarget animals and damage ecosystems

Unfortunately, these fishing practices catch more than just the species they target. *Bycatch*, the accidental capture of animals, accounts for the deaths of millions of fish, sharks, marine mammals, and birds each year. The impact of bycatch can be substantial. A 2011 report from NOAA reported that 17% of all commercially-harvested fish were captured unintentionally.

Driftnetting captures dolphins, seals, and sea turtles, as well as countless nontarget fish. Most of these creatures end up drowning (mammals and turtles need to surface in order to breathe) or dying from air exposure on deck (fish suffocating when kept out of the water). Driftnetting is now banned in international waters because of excessive bycatch, but the practice continues in most national waters.

Similar bycatch problems exist with longline fishing, which kills turtles, sharks, and albatrosses, magnificent seabirds with wingspans up to 3.6 m (12 ft). Several methods are being developed to limit bycatch from longline fishing (such as using flagging to scare birds away from the lines), but an estimated 300,000 seabirds of various species die each year when they become caught on hooks while trying to ingest bait.

The scale of bycatch and solutions to address it are illustrated by the story of dolphins and tuna. Several species of dolphins that associate with yellowfin tuna become caught in purse seines set for the tuna in the tropical Pacific. In purse seining, boats surround schools of tuna with a net and draw the net in, trapping both tuna and dolphins. Hundreds of thousands of dolphins were being needlessly killed each year throughout the 1960s. The U.S. Marine Mammal Protection Act of 1972 forced U.S. fleets to try to free dolphins from seines before they were hauled up and spurred fishing gear to be modified to allow dolphins to escape. Bycatch dropped greatly as a result (FIGURE 16.21a).

However, as other nations' ships began catching tuna, dolphin bycatch rose again. Because U.S. fleets were operating under more restrictions, the U.S. government required that tuna imported from foreign fleets also minimize dolphin bycatch, and it supported ecolabeling efforts (p. 155) to label tuna as "dolphin-safe" if its capture used methods designed to avoid bycatch. These measures helped reduce dolphin deaths from 133,000 in 1986 to less than 2000 per year since 1998. We can celebrate this success story but should also recognize that there is little accountability for the many "dolphin-safe" labels, that sharks and other animals continue to be caught as bycatch, and that dolphin populations have not recovered (FIGURE 16.21b).

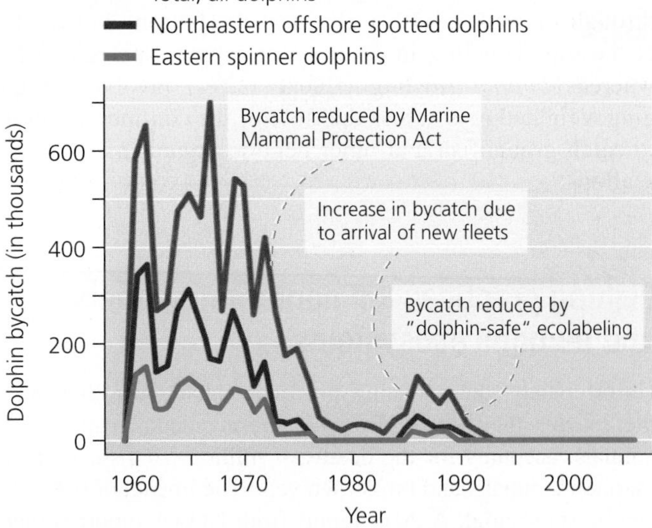

(a) Reduction in dolphin bycatch

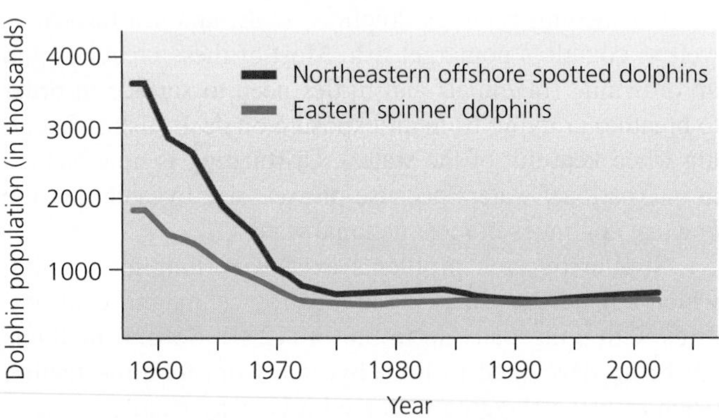

(b) Lack of recovery by dolphin populations

FIGURE 16.21 **Bycatch has severely decreased dolphin populations.** Bycatch of dolphins by tuna fleets **(a)** declined first as a result of regulations following the 1972 U.S. Marine Mammal Protection Act and later, when "dolphin-safe" ecolabeling encouraged fleets to adopt methods that reduced bycatch. However, despite this success, populations of the two dolphin species most affected **(b)** have not rebounded, possibly because there are fewer fish for them to eat. *Data from (a) National Oceanic and Atmospheric Administration, www.noaa.gov, and (b) Wade, P.R., et al., 2007. Depletion of spotted and spinner dolphins in the eastern tropical Pacific: Modeling hypotheses for their lack of recovery.* Mar. Ecol. Prog. Ser. *343: 1–14.*

Bottom-trawling not only results in bycatch but also can destroy entire ecosystems. The weighted nets crush organisms in their path and leave long swaths of damaged sea bottom. Bottom-trawling is especially destructive to structurally complex areas, such as reefs, that provide shelter and habitat for animals. In recent years, underwater photography has begun to reveal the extent of structural and ecological disturbance done by bottom-trawling (FIGURE 16.22). Bottom-trawling is often likened to clear-cutting (pp. 317–318) and strip mining (p. 639). In heavily fished areas, the bottom may be damaged multiple times. At Georges Bank, it is estimated that the average expanse of ocean floor has been trawled three times. Bottom-trawling here is known to destroy young cod as bycatch, and this is thought to be a main reason why the Georges Bank cod stock is not recovering. Reducing bycatch is also an important part of restoring Grand Banks populations; bycatch of cod while fishing for other species in the Grand Banks rose from 600 metric tons in 2006 to 1100 metric tons in 2009, slowing the recovery of cod populations.

(a) Before trawling at Georges Bank

(b) After trawling at Georges Bank

FIGURE 16.22 **Bottom-trawling causes severe structural damage to reefs and benthic habitats, and it can decimate underwater communities and ecosystems.** A photo of an untrawled location **(a)** on the seafloor of Georges Bank shows a vibrant and diverse benthic community. A photo of the same site after trawling **(b)** shows a flattened expanse of sea bottom with only scarce biological diversity and productivity.

Modern fishing fleets deplete marine life rapidly

We can see the effects of large-scale industrialized fishing in the catch records of groundfish from the Northwest Atlantic. Although cod had been harvested since the 1500s on the Grand Banks, catches more than doubled once immense industrial trawlers from Europe, Japan, and the United States appeared in the 1960s (FIGURE 16.23a). These record-high catches lasted only a decade; the industrialized approach removed so many fish that the stock has not recovered. Likewise, on Georges Bank, cod catches rose greatly in the 1960s, remained high for 30 years, and then collapsed (FIGURE 16.23b).

Throughout the world's oceans, today's industrialized fishing fleets are depleting marine populations with astonishing speed. In a 2003 study, Canadian fisheries biologists Ransom Myers and Boris Worm analyzed data from FAO archives and found the same pattern for region after region: In just a decade after the arrival of industrialized fishing, catch rates dropped precipitously, with 90% of large-bodied fish and sharks eliminated. Populations then stabilized at 10% of their former levels. This means, Myers and Worm concluded, that the oceans today contain only one-tenth of the large-bodied animals they once did.

As we have seen (pp. 83–84), when animals at high trophic levels are removed from a food web, the proliferation of their prey can alter the nature of the entire community. Many scientists conclude that most marine communities may have been very different prior to industrial fishing.

Several factors mask declines

Although industrialized fishing has depleted fish stocks in region after region, the overall global catch has remained roughly stable for two decades (see Figure 16.19). This seeming stability can be explained by several factors that mask population declines. Fishing fleets are now traveling longer distances to reach less-fished portions of the ocean. They also are fishing in deeper waters; average depth of catches was 150 m (495 ft) in 1970 and 250 m (820 ft) in 2000. Moreover, fleets are spending more time fishing and are setting out more

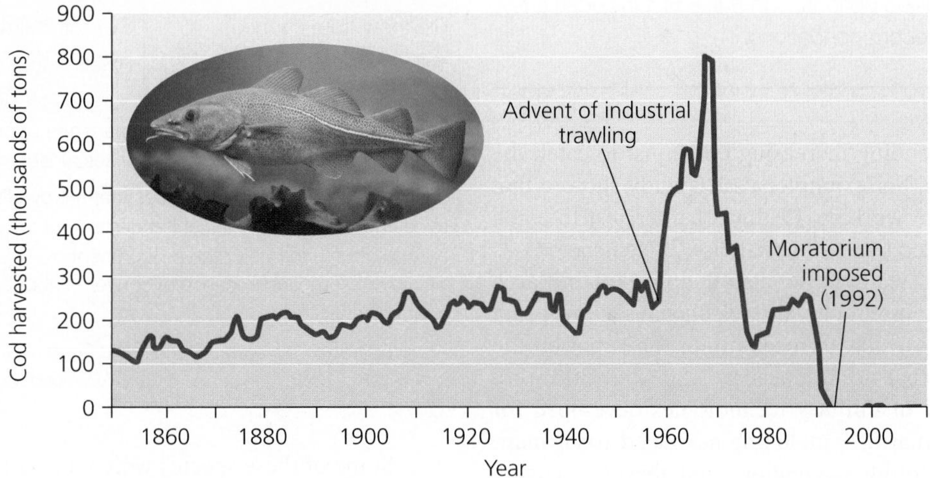

(a) Cod harvested from Grand Banks

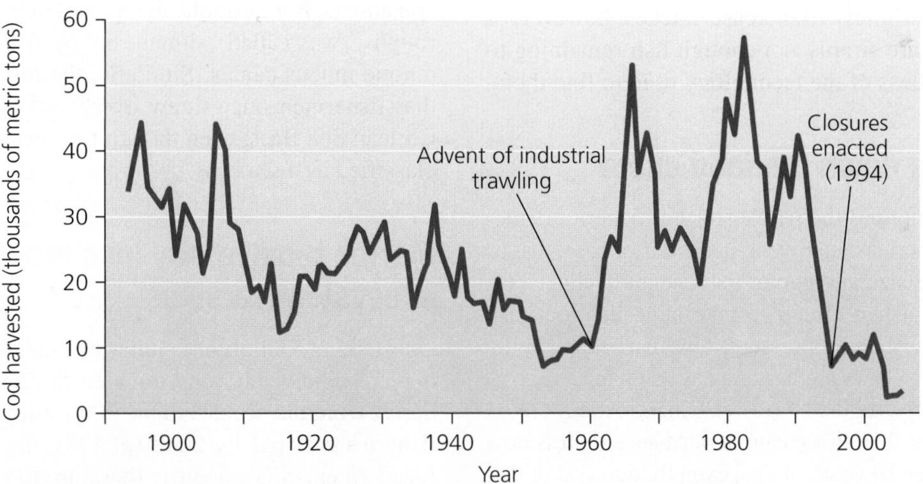

(b) Cod harvested from Georges Bank

FIGURE 16.23 In the North Atlantic off the coast of Newfoundland, commercial catches of Atlantic cod have fluctuated over time. Catches increased with intensified fishing by industrial trawlers in the 1960s and 1970s **(a)**, but the fishery subsequently crashed, and moratoria imposed in 1992 and 2003 have not brought it back. A similar pattern is seen in the cod catches at Georges Bank **(b)**; industrial fishing produced 30 years of high catches, followed by a collapse and the closure of some areas to fishing. Note also that in each case, there is one peak before 1977 and one after 1977. The first peak and decline resulted from foreign fishing fleets, whereas the second peak and decline resulted from Canadian and U.S. fleets, respectively, after they laid claim to their 200-mile exclusive economic zones. *Data (a) from Millennium Ecosystem Assessment, 2005.* Ecosystems and human well-being: synthesis. *World Resources Institute. Washington, DC. Used with permission and (b) O'Brien, L., et al., 2008.* Georges Bank Atlantic cod. An Assessment of 19 Northeast groundfish stocks through 2007. *Northeast Fisheries Science Center, Woods Hole, MA.*

FAQ I love seafood, so how can I make sustainable choices?

To most of us, marine fishing practices may seem a distant phenomenon over which we have no control, especially as more than 80% of seafood sold in the United States is imported. But although we don't have control over the seafood presented to us, we have full control over which items we buy. Finding out how seafood items were caught is difficult because this information is not readily available to consumers in most cases. Thus, several organizations have devised concise guides and smartphone apps to help consumers differentiate fish and shellfish that are overfished or whose capture is ecologically damaging from those that are harvested more sustainably. For instance, the Environmental Defense Fund provides a guide for sustainable seafood options, and **TABLE 16.1** includes a few examples from its recommendations.

TABLE 16.1 Seafood Choices for Consumers

SUSTAINABLE CHOICES[1]	SEAFOOD TO AVOID[2]
Catfish (U.S.-farmed)	
Clams (farmed)	Caviar (sturgeon, imported, wild)
Dungeness crab	Chilean seabass/toothfish*
Pacific halibut	Atlantic halibut, flounders, soles
Mussels (farmed)	Mahimahi/dolphinfish (imported, longline caught)
Oysters (farmed)	Orange roughy*
Salmon (wild-caught, Alaska)	Salmon (farmed or Atlantic)*
Bay scallops (farmed)	Blacktip Shark
Shrimp (U.S.-farmed)	Shrimp (imported)
Tilapia (U.S.-farmed)	Tilapia (farmed in Asia)
Rainbow trout (farmed)	Swordfish (imported)*
Yellowfin Atlantic tuna (troll/pole-caught)	Bluefin tuna*

[1]Fish or shellfish from healthy, well managed populations that are caught or farmed in environmentally sustainable ways.

[2]Fish or shellfish from wild sources that are overfished, have high bycatch, cause extensive habitat damage or are farmed in ways that harm other marine life or the environment.

* Limit consumption because of concerns about mercury or other contaminants.

Adapted from: Environmental Defense Fund, March 2013. Safe seafood and responsible fisheries. http://www.edf.org.

nets and lines—expending increasing effort just to catch the same number of fish. For example, a 2010 study showed that British trawlers were working 17 times harder just to catch the same number of cod and other fish as 120 years ago.

More powerful technology also helps explain large catches despite declining stocks. Today's Japanese, European, Canadian, and U.S. fleets can reach almost any spot on the globe with vessels that attain speeds of 80 kph (50 mph). They boast an array of military technologies developed for locating enemy submarines, including advanced sonar mapping equipment, satellite navigation, and thermal sensing systems. Some fleets rely on airplanes to find schools of commercially valuable fish such as bluefin tuna. Technology cannot continue indefinitely to increase catches, however, as at some point there are simply not enough fish remaining to be harvested, regardless of the technology used to find them.

We are "fishing down the food chain"

Numbers of fish do not tell the whole story of fisheries depletion. Analyses of fisheries data reveal, in case after case, that as fishing increases, the size and age of fish caught decline. This is not only because fishers prefer to take large fish, but also because under intense fishing pressure, few fish escape being caught for very many years, so that few have a chance to grow to large size. Cod caught in the Northwest Atlantic today are on average much smaller than those caught decades ago. It is now rare to find a cod over 10 years of age, even though cod of this age formerly were common. Because large female fish produce far more young than small ones, the intense harvesting of larger fish makes it harder for populations to recover once depleted.

In addition, as particular species become too rare to fish profitably, fleets begin targeting other species that are more abundant. Generally this means shifting from large, desirable species to smaller, less desirable ones. Time and again, fleets have depleted popular food fish (such as cod) and shifted to species of lower value (such as capelin, a smaller fish eaten by cod). Because this often entails catching species at lower trophic levels, this phenomenon has been termed "fishing down the food chain."

Some of these species were undesirable species that fishermen formerly threw back when fishing for more marketable species, but underwent "image makeovers" to aid their sale to consumers. For example, the species of fish now called "orange roughy" was called "slimehead" by fishermen because of its unique mucus canals. Similarly, the toad-colored "toothfish" that fishermen once threw overboard has found new life as Chilean Sea Bass, even though the species is not biologically classified as a sea bass.

Marine biodiversity loss erodes ecosystem services

Overfishing, pollution, habitat change, and other factors that deplete biodiversity can threaten the ecosystem services we derive from the oceans. In the 2006 study that predicted global fisheries collapse by 2048 (p. 438), the study's authors analyzed all existing scientific literature to summarize the effects of biodiversity loss on ecosystem function and ecosystem services. They found that across 32 different controlled experiments conducted by various researchers, systems with reduced species diversity or genetic diversity showed less primary and secondary production and were less able to withstand disturbance.

The team also found that when biodiversity was reduced, so were habitats that serve as nurseries for fish and shellfish. Moreover, biodiversity loss was correlated with reduced filtering and detoxification (as from wetland vegetation and oyster beds). This can lead to harmful algal blooms, dead zones, fish kills, and beach closures.

WEIGHING THE ISSUES

EATING SEAFOOD After reading this chapter, do you plan to alter your decisions about eating seafood in any way? If so, how? If not, why not? Do you think consumer buying choices can exert an influence on fishing practices? On mercury contamination in seafood?

Marine Conservation

Because we bear responsibility and stand to lose a great deal if valuable ecological systems collapse, marine scientists have been working to develop solutions to the problems that threaten the oceans. Many have begun by taking a hard look at the strategies used traditionally in fisheries management.

Fisheries management has been based on maximum sustainable yield

Fisheries managers conduct surveys, study fish population biology, and monitor catches. They then use that knowledge to regulate the timing of harvests, the scale of harvests, and the techniques used to catch fish. The goal is to allow for maximal harvests while keeping fish available for the future—the concept of *maximum sustainable yield* (p. 315). Recall that this means keeping populations at about half the level they would otherwise achieve, because populations growing by logistic growth grow fastest at this size (pp. 67–68). If data indicate that current yields are unsustainable, managers might limit the number or biomass of fish that can be harvested, or they might restrict the type of gear fishers can use.

The science of estimating population sizes of species such as cod is imprecise and can lead to significant changes in maximum sustained yield. For example, the cod population in the Gulf of Maine was sampled in 2008 and its biomass estimated at 34 million kg (74.9 million lb). Allowable harvests by fishermen were set based on this survey, but a subsequent survey in 2011 concluded cod biomass in the Gulf to be only 12 million kg (24.6 million lb). This reduction was not explainable by fish harvests over the three years, so resource managers concluded that the 2008 survey overestimated cod populations based on a few large trawls, many of which have not been indicative of true cod numbers. In light of the new assessment of cod biomass, fishing quotas were lowered, forcing fishermen in the area to quickly adapt to new levels of allowable catch. Similar overestimates occurred in cod populations in the Georges Bank population at the same time, with similar outcomes.

Despite efforts to restrict fish harvests to sustainable levels, a number of fish and shellfish stocks have plummeted. Thus, many scientists and managers feel it is time to shift the focus away from individual species and toward viewing marine resources as elements of larger ecological systems. This perspective means considering the impacts of fishing practices on habitat quality, species interactions, and other factors that may have indirect or long-term effects on populations. One key aspect of such ecosystem-based management (p. 315) is to set aside areas of ocean where systems can function without human interference.

We can protect areas in the ocean

Hundreds of **marine protected areas (MPAs)** have been established, most of them along the coastlines of developed countries. However, despite their name, nearly all MPAs allow fishing or other extractive activities. As one report from an environmental advocacy group put it, MPAs "are dredged, trawled, mowed for kelp, crisscrossed with oil pipelines and fiber-optic cables, and swept through with fishing nets."

Because of the lack of true refuges from fishing pressure, many scientists want to establish areas where fishing is prohibited. Such "no-take" areas have come to be called **marine reserves.** Designed to preserve ecosystems intact, marine reserves are also intended to improve fisheries. Scientists argue that marine reserves can act as production factories for fish for surrounding areas, because fish larvae produced inside reserves will disperse outside and stock other parts of the ocean. By serving both purposes, proponents maintain, marine reserves are a win-win proposition for environmentalists and fishers alike.

Many commercial and recreational fishers dislike the idea of no-take reserves, however, just as most were opposed to the Canadian groundfish moratoria and the Georges Bank closures. Nearly every marine reserve that has been established or proposed has met with opposition from people and businesses who use the area for fishing or recreation. Some protests have turned violent. To protest fishing restrictions, fishermen in the Galápagos Islands destroyed offices at Galápagos National Park and threatened researchers and park managers with death.

Clearly, when marine reserves are established, it pays to be sensitive to the concerns of people of the area. In 2006, President Bush established the Papahanaumokuakea Marine National Monument around the northwestern Hawaiian Islands—at 362,000 km^2 (140,000 mi^2), an area larger than all U.S. national parks combined. In this new monument, native Hawaiians were given fishing rights in areas where fishing was otherwise made off-limits.

In contrast, the United Kingdom in 2010 established the world's largest marine reserve around the Chagos Archipelago, a group of 55 islands it controls in the Indian Ocean. The indigenous people of these islands were forcibly evicted from the main island, Diego Garcia, between 1967 and 1973 so that the United States could build a military base there. The roughly 4000 Chagossian people have been fighting in courts ever since for the right to return to their islands. They now fear that with fishing banned, they will never be able to return because there would be no way to make a living. The U.K. government has

WEIGHING THE ISSUES

PRESERVATION AT SEA Almost 4% of U.S. land area is designated as wilderness, yet before the recent actions in Hawaii and the Chagos Islands, far less than 1% of coastal waters were protected in reserves. Why do you think it is taking so long for the preservation ethic to make the leap to the oceans? What challenges do you see with preserving areas of coastal ocean? Do you think local people should be given fishing rights in marine reserves? What could be done to better protect marine ecosystems, fisheries, and local fishing cultures?

said that if the people return, the fishing bans will be reassessed and that in the meantime, the bans will save the region's fish from being depleted by industrial fleets from other countries.

Reserves can work for both fish and fishers

Over the past two decades, data from marine reserves around the world have been indicating that reserves can work as win-win solutions that benefit ecosystems, fish populations, and fishing economies. A comprehensive review of data from marine reserves in 2001 revealed that just one to two years after their establishment, marine reserves:

- Increased densities of organisms on average by 91%.
- Increased biomass of organisms on average by 192%.
- Increased average size of organisms by 31%.
- Increased species diversity by 23%.

That year, 161 prominent marine scientists signed a "consensus statement" summarizing the effects of marine reserves. Besides boosting fish biomass, total catch, and record-sized fish, the report stated, marine reserves yield several benefits. Within reserve boundaries, they:

- Produce rapid and long-term increases in abundance, diversity, and productivity of marine organisms.
- Decrease mortality and habitat destruction.
- Lessen the likelihood of species extirpation.

Outside reserve boundaries, marine reserves:

- Can create a "spillover effect" as individuals of protected species spread outside reserves.
- Allow larvae of species protected within reserves to "seed the seas" outside reserves.

The consensus statement was backed up by research into reserves worldwide. At Apo Island in the Philippines, biomass of large predators increased eightfold inside a marine reserve, and fishing improved outside the reserve. At two coral reef sites in Kenya, commercially fished species and keystone species were up to 10 times more abundant in the protected area as in the fished area. At Leigh Marine Reserve in New Zealand, snapper increased 40-fold, and spiny lobsters were increasing by 5–11% yearly. Spillover from this reserve improved fishing and ecotourism, and local residents who once opposed the reserve now support it. Since that time, further research has shown that reserves create a fourfold increase in catch per unit effort in fished areas surrounding reserves and that they can greatly increase ecotourism by divers and snorkelers.

On Georges Bank, once commercial trawling was halted in 1994, populations of many organisms began to recover. As benthic invertebrates began to come back, numbers of groundfish such as haddock and yellowtail flounder rose inside the closed areas, and scallops increased by 14 times. Moreover, fish from the closure areas appear to be spilling over into adjacent waters, because fishers have been catching more groundfish from Georges Bank as a whole since the late 1990s. From these and other datasets, increasing numbers of scientists, fishers, and policymakers are advocating the establishment of fully protected marine reserves as a central management tool.

How should reserves be designed?

If marine reserves work in principle, the question becomes how best to design reserves and arrange them into networks. Studies are modeling how to optimize the size and spacing of reserves so that ecosystems are protected, fisheries are sustained, and people are not overly excluded from marine areas (**FIGURE 16.24**). Scientists are asking how large reserves need to be, how many there need to be, and where they need to be placed to take best advantage of ocean currents. Of several dozen studies that have estimated how much area of the ocean should be protected in no-take reserves, estimates range from 10% to 65%, with most falling between 20% and 50%. Most scientists feel that involving fishers directly in the planning process is crucial for coming up with answers to all these questions. If marine reserves can be

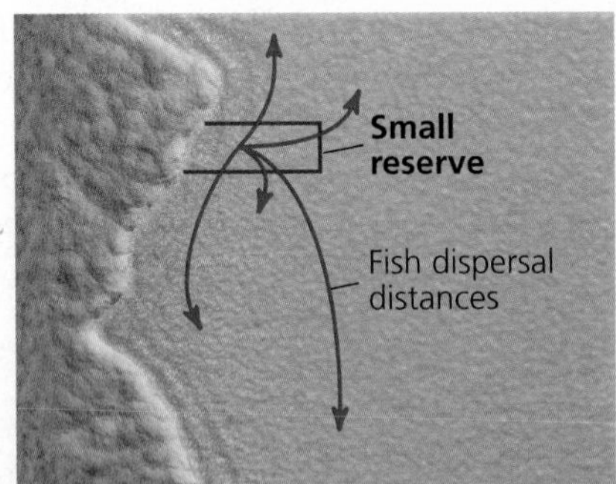

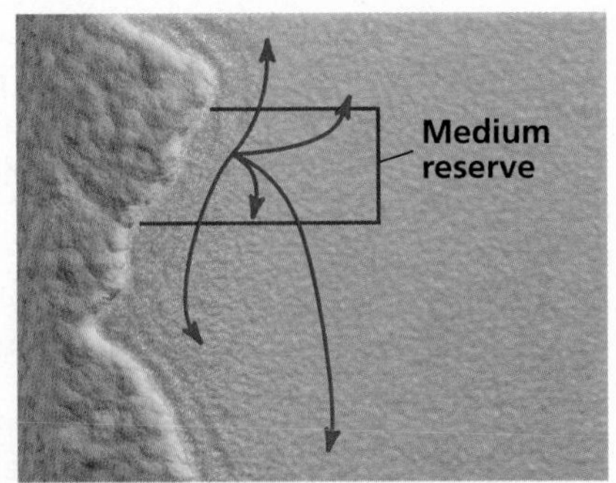

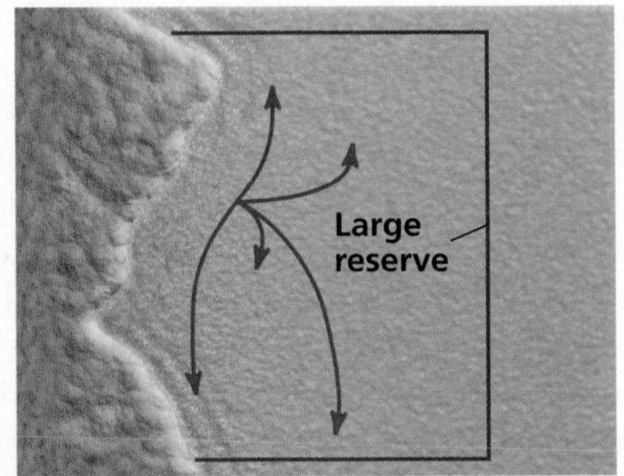

FIGURE 16.24 Marine reserves of different sizes may have varying effects on ecological communities and fisheries. Young and adult fish and shellfish of different species can disperse different distances, as indicated by the red arrows in the figure. A small reserve (**left panel**) may fail to protect animals because too many disperse out of the reserve. A large reserve (**right panel**) may protect fish and shellfish very well but will provide relatively less "spillover" into areas where people can legally fish. Thus medium-sized reserves (**middle panel**) may offer the best hope of preserving species and ecological communities while also providing adequate fish to people. *Source: Adapted from Halpern, B.S., and R.R. Warner, 2003. Matching marine reserve design to reserve objectives.* Proceedings of the Royal Society of London *B 270: 1871–1878, Fig 1. Used by permission of The Royal Society and the author.*

made to work and to be accepted, then they may well seed the seas and help lead us toward solutions to one of our most pressing environmental problems.

Conclusion

Oceans cover most of our planet, and we are still exploring their diverse topography and ecosystems. As we learn more about marine and coastal environments, we continue to intensify our use of their resources and to exert more severe impacts. Ocean acidification, loss of coral reefs, several types of pollution, and the ongoing depletion of many of the world's marine fish stocks are all major challenges that we will need to address. Yet scientists are demonstrating that setting aside protected areas of the ocean can serve to maintain and restore natural systems and also to enhance fisheries. In the meantime, we can all contribute through consumer choices that can help move us toward sustainable fishing practices.

Reviewing Objectives

You should now be able to:

Identify physical, geographical, chemical, and biological aspects of the marine environment

- Oceans cover 71% of Earth's surface and contain over 97% of its water. (p. 421)
- Seafloor topography can be complex. (pp. 421–422)
- Ocean water contains 96.5% H_2O by mass and various dissolved salts. (pp. 422–423)
- Colder, saltier water is denser and sinks. Water temperatures vary with latitude, and temperature variation is greater in surface layers. (p. 423)
- Surface currents move horizontally through the oceans, driven by wind and other factors. (pp. 423–424)
- Vertical water movement includes upwelling and downwelling, which affect the distribution of nutrients and life. (pp. 424–425)

Explain how the oceans influence and are influenced by climate

- The thermohaline circulation shapes regional climate, for instance, keeping Europe warm. Global warming could potentially shut down existing circulation patterns. (p. 425)
- El Niño and La Niña events alter climate and affect fisheries. (pp. 425–426)
- The oceans sequester atmospheric carbon and have slowed global climate change, but they could soon become saturated. (p. 426)
- Absorption of excess carbon dioxide leads to ocean acidification, which hinders corals in forming reefs. (pp. 426, 428–429)

Describe major types of marine ecosystems

- Major types of marine and coastal ecosystems include intertidal zones, salt marshes, mangrove forests, estuaries, kelp forests, coral reefs, and pelagic and deep-water open ocean systems. (pp. 426–427, 430–432)
- Many of these systems are highly productive and rich in biodiversity. Many also suffer heavy impacts from human influence. (pp. 426–427, 430–432)

Assess impacts from marine pollution

- Plastic trash harms marine life and accumulates in ocean regions, where it is trapped by currents. (pp. 432–436)
- Marine oil pollution results from non-point sources on land as well as from spills at sea from tankers and drilling platforms. (pp. 436–437)
- Heavy-metal contaminants in seafood affect human health. (p. 437)
- Nutrient pollution can lead to dead zones and harmful algal blooms. (p. 437)

Review the state of ocean fisheries and reasons for their decline

- Over half the world's marine fish populations are fully exploited, 28% are overexploited, and only 20% can yield more without declining. (pp. 437–438)
- Global fish catches have stopped growing since the late 1980s despite increased fishing effort and improved technologies. (p. 438)
- People began depleting marine resources long ago, but impacts have intensified in recent decades. (p. 438)
- Commercial fishing practices include driftnetting, longline fishing, and trawling, all of which capture nontarget organisms, called bycatch. (pp. 438–439)
- Non-target species are killed when they are captured as bycatch while fishing for commercially-valuable species. (pp. 439–440)
- Today's oceans hold just one-tenth as many large animals as they did before industrialized commercial fishing. (p. 441)
- As fishing intensity increases, fish become smaller and fishermen switch to less-desirable species. (p. 442)

- Consumers can encourage good fishery practices by shopping for sustainable seafood. (p. 442)
- Marine biodiversity loss affects ecosystem services. (p. 442)
- Traditional fisheries management has not stopped declines, so many scientists feel that ecosystem-based management is needed. (p. 443)

Evaluate marine protected areas and reserves as innovative solutions

- We have established fewer protected areas in the oceans than we have on land, and most marine protected areas allow many extractive activities. (pp. 443–444)
- No-take marine reserves can protect ecosystems while also boosting fish populations and making fisheries sustainable. (p. 444)

Testing Your Comprehension

1. What proportion of Earth's surface do oceans cover? What is the average salinity of ocean water? How are density, salinity, and temperature related in each layer of ocean water?
2. What factors drive ocean currents? Give an example of how a surface current affects climate regionally.
3. What is causing ocean acidification? What consequences do scientists expect ocean acidification to bring about?
4. Where in the oceans are productive areas of biological activity likely to be found?
5. Describe three kinds of ecosystems found near coastal areas and the types of life they support.
6. Why are coral reefs biologically valuable? How are they being degraded by human impact? What is causing the disappearance of mangrove forests and salt marshes?
7. Discuss three ways plastics affect marine life.
8. Describe an example of how overfishing can lead to ecological damage and fishery collapse.
9. Name three industrial fishing practices, and explain how they create bycatch and harm marine life.
10. How does a marine reserve differ from a marine protected area? Why do many fishers oppose marine reserves? Explain why many scientists say no-take reserves will be good for fishers.

Seeking Solutions

1. What benefits do you derive from the oceans? How does your behavior affect the oceans? Give specific examples.
2. We have been able to reduce the amount of oil we spill into the oceans, but petroleum-based products such as plastic continue to litter our oceans and shorelines. Discuss some ways that we can reduce this impact on the marine environment.
3. Describe the trends in global fish capture over the past 55 years and over the past 20 years, and explain several factors that account for these trends.
4. Consider what you know about biological productivity in the oceans, about the scientific data on marine reserves, and about the social and political issues surrounding the establishment of marine reserves. What types of ocean regions do you think it would be particularly appropriate to establish as marine reserves? Why?
5. **THINK IT THROUGH** You make your living fishing on the ocean, just as your father and grandfather did, and as most of your neighbors do in your small coastal village. Your region's fishery has just collapsed, however, and everyone is blaming it on overfishing. The government has closed the fishery for three years, and scientists are pushing for a permanent marine reserve to be established on your former fishing grounds. You have no desire to move away from your village, so what steps will you take now? Will you protest the closure? What compensation will you ask of the government if it prevents you from fishing? Will you work with scientists to establish a reserve that improves fishing in the future, or will you oppose their attempts to create a reserve? What data and what assurances will you ask of them?
6. **THINK IT THROUGH** You are mayor of a coastal town where some residents are employed as commercial fishers and others make a living serving ecotourists who come to snorkel and scuba dive at the nearby coral reef. In recent years, several fish stocks have crashed, and ecotourism is dropping off as fish disappear from the increasingly degraded reef. Scientists are urging you to help establish a marine reserve around portions of the reef, but most commercial and recreational fishers are opposed to this idea. What steps would you take to restore your community's economy and environment?

Calculating Ecological Footprints

The relationship between the ecological goods and services that individuals use and the amount of *land* area needed to provide those goods and services is relatively well developed. People also use goods and services from Earth's oceans, where the concept of *area* is less useful. It is clear, however, that our removal of fish from the oceans has an impact, or an ecological footprint.

The table shows data on the mean annual per-person consumption from ocean fisheries for North America, China, and the world as a whole. Using the data provided, calculate the amount of fish each consumer group would consume per year, given the annual per capita consumption rates, for each of these three regions. Record your results in the table.

Annual Consumption

Consumer group	North America (24.1 kg per person)	China (31.9 kg per person)	World (18.4 kg per person)
You			
Your class			
Your state			
United States			
World			

Data from Food and Agriculture Organization of the United Nations, 2012. The state of world fisheries and aquaculture: 2012.

1. Calculate the ratio of North America's per capita fish consumption rate to that of the world. Compare this ratio to the ratio of the per capita ecological footprints for the United States, Canada, and Mexico (see Figure 1.12, p. 14) versus the world average footprint of 2.7 ha/person/year. Can you account for similarities and differences between these ratios?
2. The population of China has grown at an annual rate of 1.1% since 1987, while over the same period fish consumption in China has grown at an annual rate of 8.9%. Speculate on the reasons behind China's rapidly increasing consumption of fish.
3. What ecological concerns do the combined trends of human population growth and increasing per capita fish consumption raise for you? What role might you play in contributing to these concerns or to their solutions?

Mexico City on a smoggy day

Mexico City on a clear day

Atmospheric Science, Air Quality, and Pollution Control

Upon completing this chapter, you will be able to:

- Describe the composition, structure, and function of Earth's atmosphere
- Relate weather and climate to atmospheric conditions
- Identify major pollutants, outline the scope of outdoor air pollution, and assess solutions
- Explain stratospheric ozone depletion and identify steps taken to address it
- Define *acid deposition*, illustrate its consequences, and explain how we are addressing it
- Characterize the scope of indoor air pollution and assess solutions

CENTRAL CASE STUDY

Clearing the Air in L.A. and Mexico City

"I left L.A. in 1970, and one of the reasons I left was the horrible smog. And then they cleaned it up. That was one of the greatest things the government has ever done for me. You have beautiful days now. It's a much, much nicer place to live."

—Actor and comedian Steve Martin, speaking to *Los Angeles Magazine*

"This city can be a model for others."

— Mexico City Mayor Miguel Ángel Mancera

Los Angeles has long symbolized air pollution in Americans' minds. Smog blanketed the city in the 1970s, 1980s, and 1990s, a result of the exhaust from millions of automobiles clogging its freeways. But Los Angeles has improved its air quality, thanks to policy efforts and new technologies. Today L.A. still suffers the nation's worst smog, but its skies are clearer than they have been in decades—and its air is cleaner than in many of its "sister cities" elsewhere in the world.

Devised to foster peace and understanding among nations, the Sister Cities program links hundreds of American cities with foreign counterparts through shared government visits and cultural ties. One of L.A.'s sister cities is Mexico City, the capital of Mexico. Alas, one thing these two cities share is smog.

Not long ago Mexico City suffered the most polluted air in the world. On bad air days throughout the 1990s, residents wore surgical masks on the streets, teachers kept students inside at recess, and outdoor sports events were cancelled. Birds were said to die in flight, and children used brown crayons to color the sky. Each year thousands of deaths and tens of thousands of hospital visits were blamed on pollution. Mexican novelist Carlos Fuentes called his capital "Makesicko City."

As in Los Angeles, traffic generates most of the pollution in Mexico City, where motorists in 6 million cars sputter across miles of urban sprawl. And like L.A., Mexico City lies in a valley surrounded by mountains, so thermal inversions trap pollutants over the city. Moreover, at Mexico City's high altitude—2240 m (7350 ft) above sea level—solar radiation is intense, which worsens the formation of urban smog. Mexico City environmental chemist Armando Retama likens his hometown to "a casserole dish with a lid on top."

Despite all the challenges facing the world's third-largest metropolis, Mexico City's 20 million people have fought back and turned things around. A series of mayors took bold action to clean up the air, and today Mexico City is enjoying a renaissance. As the smog begins to clear, revealing beautiful views of the snow-capped peaks that ring the valley, the city has become a model for other cities seeking to fight pollution.

Efforts began in the 1990s when city leaders shut down an oil refinery and pushed factories and power plants to shift to cleaner-burning natural gas. Lead was phased out from gasoline, the sulfur content of diesel fuel was reduced, and pollution control technologies such as catalytic converters were phased in for new vehicles.

Levels of many pollutants fell, but as the city grew in population and sprawled across the valley, smog from automobile traffic persisted. In response, city officials stepped up vehicle emissions testing and upgraded taxis and city vehicles to cleaner models. To monitor air quality, 34 sampling stations were set up across the city, sending real-time data to city engineers.

In 2007 Mayor Marcelo Ebrard accelerated efforts as part of a 15-year sustainability plan he launched aiming to make Mexico City "the greenest city in the Americas." New lines were added to the subway system, and 800 exhaust-spewing minibuses were replaced with fuel-efficient buses that moved rapidly in designated lanes. Over 450,000 people use these buses each day, and the new system has reduced carbon dioxide emissions by an estimated 80,000 tons per year.

In 2010 Ebrard introduced a bicycle-sharing program in order to free short-distance commuters from dependence on cars. With 1000 bikes at rental stations throughout the city, people can rent a bike cheaply at one location and drop it off at another. In Ebrard's most popular initiative, every Sunday morning the city's main boulevard, the Paseo de la Reforma, is closed to car traffic, creating a safe and pleasant community space for pedestrians, bikers, joggers, and skateboarders.

Mexico City residents were so happy with the progress that in 2012, when Ebrard stepped down because of term limits, voters rewarded his political party by electing his ally Miguel Ángel Mancera by a 44-point landslide. Mancera announced plans to expand the bicycle program to 4000 bikes and to put electric buses and taxis on the roads. He also rolled out a car-sharing program that officials hope can eventually remove 40,000 vehicles from circulation. Private entrepreneurs are now getting in on the act; recent university graduates have launched companies providing car-sharing and carpooling services, some using electric vehicles.

All these changes are paying off with cleaner air. In 1991, the city's air was deemed hazardous to breathe on all but 8 days of the year. Today, most pollutants have been slashed by over 75%, and the air meets health standards on one of every two days.

Plenty of hurdles remain. Rampant development is challenging the city's efforts to plan for sustainable growth. New highway segments under construction may induce more people to drive. The typical driver still spends three hours a day stuck in traffic that averages just 13 mph. Worst of all, smog still contributes to an estimated 4000 deaths each year. But Mexico City is following in the footsteps of Los Angeles, making steady progress toward cleaner air.

Indeed, L.A. and Mexico City today typify cities of developed nations and those of developing nations. Nations that are industrializing as they try to build wealth for their citizens are confronting the same air quality challenges that plagued the United States and other wealthy nations a generation and more ago. Los Angeles has 24 other sister cities, many of which struggle with severe air pollution. Athens, Greece; Jakarta, Indonesia; Mumbai, India; Guangzhou, China; Taipei, Taiwan; San Salvador, El Salvador; and Tehran, Iran all experience poor air quality on a regular basis as new emissions from increased auto exhaust and fossil fuel combustion are added to more traditional pollution sources.

Most of these cities are taking steps to improve their air quality, just as Los Angeles and other American cities have done before them. We will examine the solutions sought in Los Angeles, Mexico City, and elsewhere throughout this chapter as we learn about Earth's atmosphere and how to reduce the pollutants we release into it.

The Atmosphere

Every breath we take reaffirms our connection to the **atmosphere,** the thin layer of gases that envelops our planet. The atmosphere moderates our climate, provides us oxygen, absorbs hazardous solar radiation, burns up incoming meteors, and transports and recycles water and nutrients.

Earth's atmosphere consists of 78% nitrogen (N_2) and 21% oxygen (O_2). The remaining 1% is composed of argon (Ar) and minute concentrations of water vapor and other gases (**FIGURE 17.1**). These include *permanent gases*, which remain at stable concentrations over centuries or millennia, and *variable gases*, which can vary in concentration on shorter time scales or from place to place as a result of natural processes or human activities.

Over our planet's long history, the atmosphere's composition has changed. Long ago our atmosphere was dominated by carbon dioxide (CO_2), nitrogen, carbon monoxide (CO), and hydrogen (H_2), but about 2.7 billion years ago, oxygen began to build up with the emergence of autotrophic microbes that emitted oxygen by photosynthesis (p. 32). Today, human activity is altering the quantities of some atmospheric gases, such as carbon dioxide, methane (CH_4), and ozone (O_3). Before exploring how our pollutants change the air we breathe and how we strive to control pollution, we will begin with an overview of Earth's atmosphere.

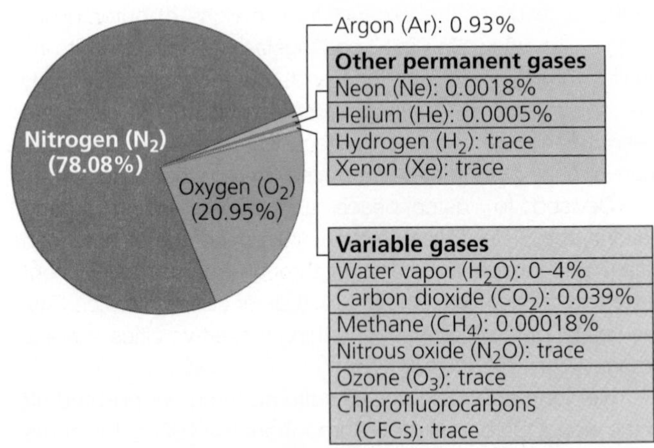

FIGURE 17.1 Earth's atmosphere consists of nitrogen, oxygen, argon, and a mix of gases at dilute concentrations. Permanent gases are fixed in concentration over long periods of time, whereas variable gases vary in concentration. *Data from Ahrens, C.D., 2007.* Meteorology today, *8th ed. Belmont, CA: Brooks/Cole.*

The atmosphere is layered

The atmosphere that stretches so high above us and seems so vast is actually just a thin coating about 1/100 of Earth's

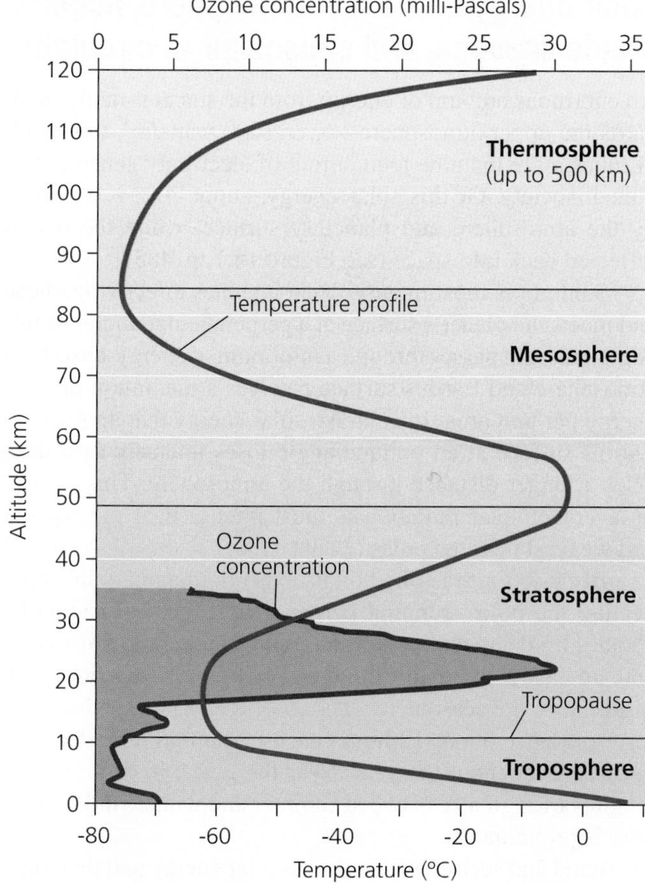

FIGURE 17.2 The atmosphere's layers differ in their properties. Temperature (red line) drops with altitude in the troposphere, rises with altitude in the stratosphere, drops in the mesosphere, and rises in the thermosphere. The tropopause separates the troposphere from the stratosphere. Ozone (blue-shaded area) reaches a peak in the lower stratosphere, giving rise to the term *ozone layer*. *Adapted from Jacobson, M.Z., 2002.* Atmospheric pollution: History, science, and regulation. *Cambridge: Cambridge University Press; and Parson, E.A., 2003.* Protecting the ozone layer: Science and strategy. *Oxford: Oxford University Press.*

diameter, like the fuzzy skin of a peach. It consists of four layers that differ in temperature, density, and composition (**FIGURE 17.2**).

The bottommost layer, the **troposphere,** blankets Earth's surface and provides us the air we breathe. Movement of air within the troposphere also drives the planet's weather. Although it is thin (averaging 11 km or 7 mi high) relative to the atmosphere's other layers, the troposphere contains three-quarters of the atmosphere's mass, because gravity pulls mass downwards, making air denser near Earth's surface. Tropospheric air gets colder with altitude, dropping to roughly –52°C (–62°F) at the top of the troposphere. At this point, temperatures stabilize, marking a boundary called the *tropopause*. The tropopause acts like a cap, limiting mixing between the troposphere and the atmospheric layer above it, the stratosphere.

The **stratosphere** extends 11–50 km (7–31 mi) above sea level. Similar in composition to the troposphere, the stratosphere is 1000 times drier and less dense. Its gases experience little vertical mixing, so once substances (including pollutants) enter it, they tend to remain for a long time. The stratosphere's air becomes warmer with altitude, attaining a maximum temperature of –3°C (27°F) at its top. The reason is that ozone and oxygen absorb and scatter the sun's ultraviolet (UV) radiation (p. 31), so that much of the UV radiation that penetrates the upper stratosphere fails to reach the lower stratosphere. Most of the atmosphere's ozone concentrates in a portion of the stratosphere roughly 17–30 km (10–19 mi) above sea level, a region we have come to call Earth's **ozone layer.** The ozone layer greatly reduces the amount of UV radiation that reaches Earth's surface. Because UV light can damage living tissue and induce mutations in DNA, the ozone layer's protective effects are vital for life on Earth.

Above the stratosphere lies the *mesosphere*, which extends 50–80 km (31–56 mi) above sea level. Air pressure is extremely low here, and temperatures decrease with altitude. The *thermosphere*, our atmosphere's top layer, extends upward to an altitude of 500 km (300 mi).

Temperature, pressure, and humidity vary within the atmosphere

Air moves dynamically within the lower atmosphere as a result of differences in the physical properties of air masses. Among these properties are pressure and density, relative humidity, and temperature.

Gravity pulls gas molecules toward Earth's surface, causing air to be most dense near the surface and less dense as altitude increases. **Atmospheric pressure,** which measures the force per unit area produced by a column of air, also decreases with altitude, because at higher altitudes there are fewer molecules being pulled down by gravity (**FIGURE 17.3**).

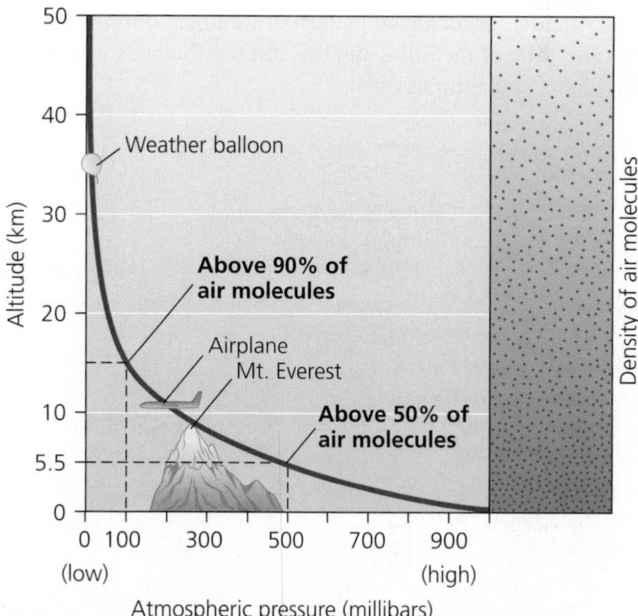

FIGURE 17.3 As one climbs higher through the atmosphere, gas molecules become less densely packed, and atmospheric pressure decreases. One needs to be only 5.5 km (3.4 mi) high to be above half the planet's air molecules. *Adapted from Ahrens, C.D., 2007.* Meteorology today, *8th ed. Fig 1.9. © 2007. Belmont, CA: Brooks/Cole. By permission of Cengage Learning.*

At sea level, atmospheric pressure averages 14.7 $lb/in.^2$ or 1013 millibars (mb). Mountain climbers trekking to Mount Everest, the world's highest mountain, can look up and view their destination from Kala Patthar, a nearby peak, at roughly 5.5 km (18,000 ft) in elevation. At this altitude, pressure is 500 mb, and half the atmosphere's air molecules are above the climber whereas half are below. A climber who reaches Everest's peak at 8.85 km (29,035 ft) in elevation, where the "thin air" is just over 300 mb, stands above two-thirds of the molecules in the atmosphere! When we fly on a commercial jet airliner at a typical cruising altitude of 11 km (36,000 ft), we are above 80% of the atmosphere's molecules.

Another property of air is **relative humidity,** the ratio of water vapor a given volume of air contains to the maximum amount it *could* contain at a given temperature. Average daytime relative humidity in June in the desert at Phoenix, Arizona, is only 31% (meaning that the air contains less than a third of the water vapor possible at its temperature), whereas on the tropical island of Guam, relative humidity rarely drops below 88%. People are sensitive to changes in relative humidity because we perspire to cool our bodies. When humidity is high, the air is already holding nearly as much water vapor as it can, so sweat evaporates slowly and the body cannot cool itself efficiently. This is why high humidity makes it feel hotter than it actually is. Low humidity speeds evaporation and makes it feel cooler.

The temperature of air also varies with location and time. At the global scale, temperature varies over Earth's surface because the sun's rays strike some areas more directly than others. At more local scales, temperature varies because of topography, plant cover, proximity of water to land, and many other factors. Sometimes this local variation is striking; a hillside sheltered from wind or sunlight may have a very different *microclimate*, or localized pattern of weather conditions, than the other side of the hill—and this often influences where certain plants and animals occur.

Solar energy heats the atmosphere, helps create seasons, and causes air to circulate

An enormous amount of energy from the sun constantly bombards the upper atmosphere—over 1000 $watts/m^2$, thousands of times more than the total output of electricity generated by human society. Of this solar energy, about 70% is absorbed by the atmosphere and planetary surface, while the rest is reflected back into space (see Figure 18.1, p. 485).

Sunlight is most intense when it shines directly overhead and meets the planet's surface at a perpendicular angle. At this angle, sunlight passes through a minimum of energy-absorbing atmosphere and Earth's surface receives a maximum of solar energy per unit area. In contrast, solar energy that approaches Earth's surface at an oblique angle loses intensity as it traverses a longer distance through the atmosphere. This is why, on average, solar radiation is most intense near the equator and weakest near the poles (**FIGURE 17.4**).

Because Earth is tilted on its axis (an imaginary line connecting the poles, running perpendicular to the equator) by about 23.5 degrees, the Northern and Southern Hemispheres end up being tilted toward the sun for half of each year, resulting in the seasons (**FIGURE 17.5**). Regions near the equator experience about 12 hours each of sunlight and darkness per day throughout the year. Near the poles, in contrast, day length varies greatly between summer and winter, and seasonality is pronounced.

Land and surface water absorb solar energy and then radiate heat, causing some water to evaporate. Air near Earth's surface therefore tends to be warmer and moister than air at higher altitudes. These differences set into motion a process of **convective circulation** (**FIGURE 17.6**). Warm air, being less dense, rises and creates vertical currents. As air rises into regions of lower atmospheric pressure, it expands and cools. Once the air cools, it descends and becomes denser, replacing warm air that is rising. The descending air picks up heat and moisture near ground level and prepares to rise again, continuing

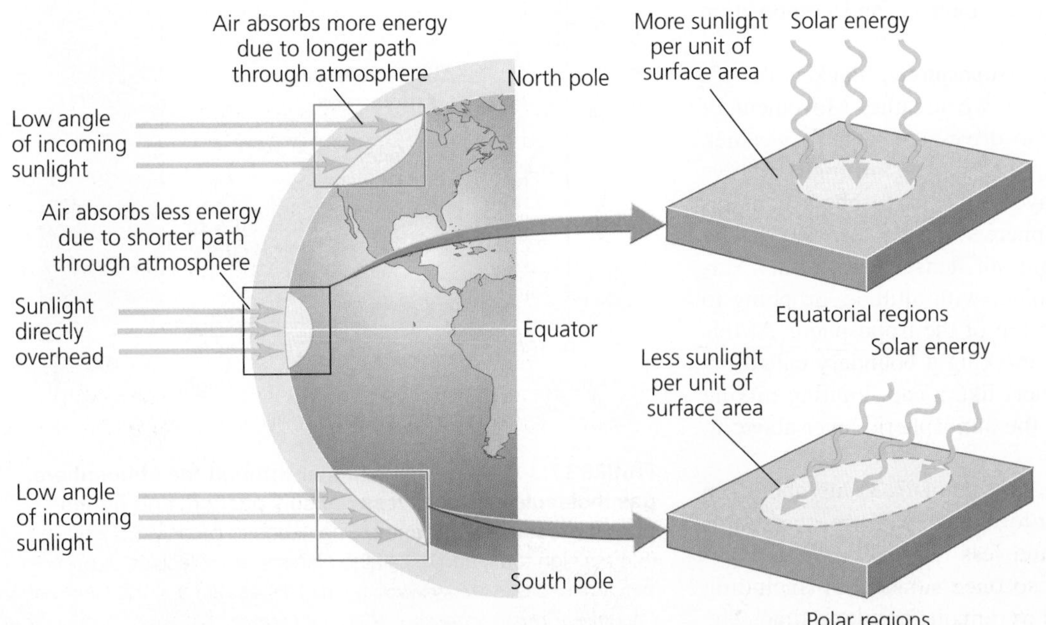

FIGURE 17.4 Because of Earth's curvature, polar regions receive less solar energy than equatorial regions. One reason is that sunlight gets spread over a larger area when striking the surface at an angle. Another reason is that sunlight approaching at a lower angle near the poles must traverse a longer distance through the atmosphere, causing more energy to be absorbed or reflected. These patterns represent year-round averages; the latitude at which radiation approaches the surface perpendicularly varies with the seasons (see Figure 17.5).

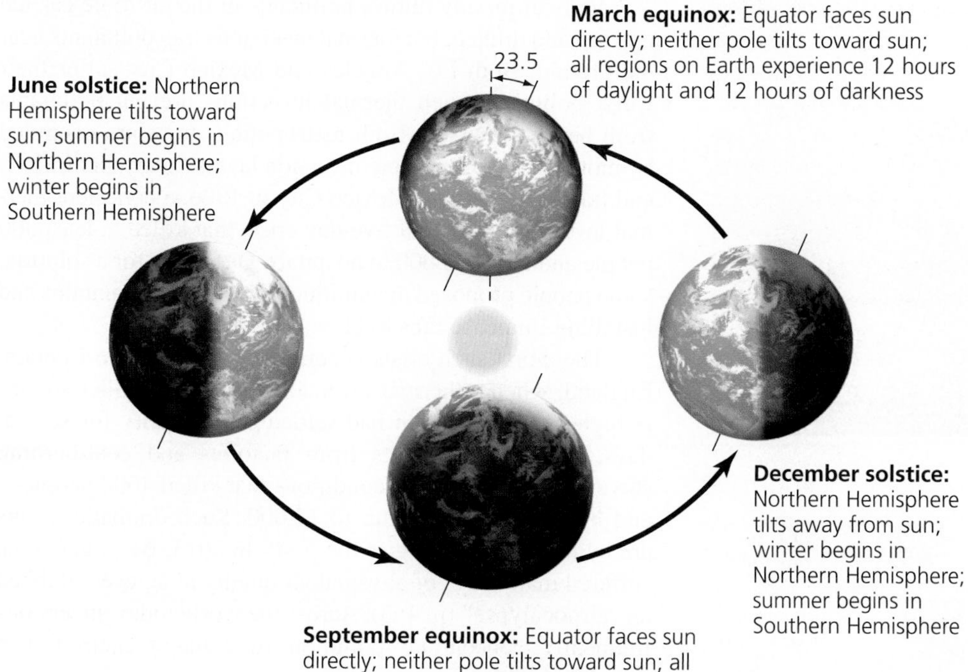

FIGURE 17.5 The seasons occur because Earth is tilted on its axis. As Earth revolves around the sun, the Northern Hemisphere tilts toward the sun for one half of the year, and the Southern Hemisphere tilts toward the sun for the other half of the year. In each hemisphere, summer occurs when the hemisphere receives the most solar energy because of its tilt toward the sun.

the process. Convective circulation patterns occur in ocean waters (pp. 425–426), in magma beneath Earth's surface (p. 35), and even in a simmering pot of soup. Convective circulation influences both weather and climate.

The atmosphere drives weather and climate

Weather and climate each involve the physical properties of the troposphere, such as temperature, pressure, humidity, cloudiness, and wind. **Weather** specifies atmospheric conditions over short time periods, typically hours or days, and within relatively small geographic areas. In contrast, **climate** describes the pattern of atmospheric conditions found across large geographic regions over long periods of time, typically years, decades, or centuries. Mark Twain once noted the distinction by remarking, "Climate is what we expect; weather is what we get." For example, Los Angeles has a "Mediterranean" climate characterized by reliably warm, dry summers and mild, rainy winters, yet on occasional autumn days, dry Santa Ana winds blow in from the desert and bring extremely hot weather.

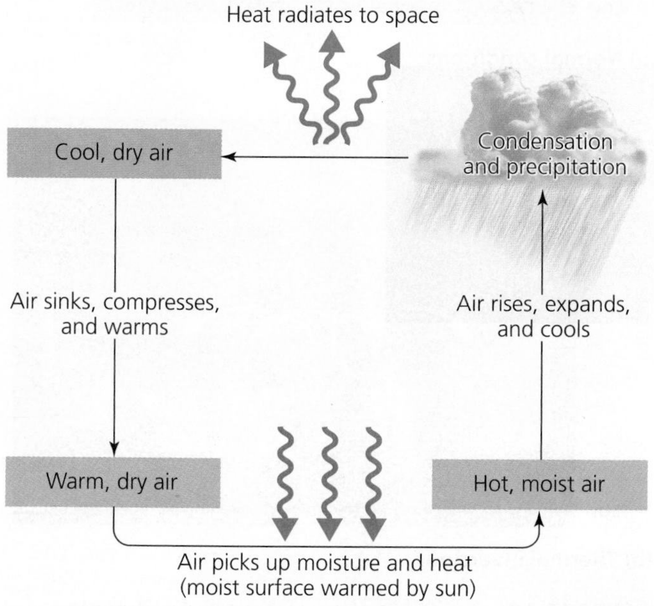

Heat radiates to space

Cool, dry air

Condensation and precipitation

Air sinks, compresses, and warms

Air rises, expands, and cools

Warm, dry air

Hot, moist air

Air picks up moisture and heat (moist surface warmed by sun)

FIGURE 17.6 Convective circulation helps to drive weather. Air heated near Earth's surface picks up moisture and rises. Once aloft, this air cools and moisture condenses, forming clouds and precipitation. Cool, drying air begins to descend, compressing and warming in the process. Warm, dry air near the surface begins the cycle anew.

Air masses interact, producing weather

Weather can change quickly when air masses with different physical properties meet. The boundary between air masses that differ in temperature and moisture (and therefore density) is called a **front.** The boundary along which a mass of warmer, moister air replaces a mass of colder, drier air is termed a **warm front** (**FIGURE 17.7a**). Some of the warm, moist air along the leading edge of a warm front usually rises over the cooler air mass that is blocking its progress. As it rises, the warm air cools and the water vapor within condenses, forming clouds and light rain. A **cold front** (**FIGURE 17.7b**) is the boundary along which a colder, drier air mass displaces a warmer, moister air mass. The colder air, being denser, tends to wedge beneath the warmer air. The warmer air rises, expands, then cools to form clouds and thunderstorms. Once a cold front passes through, the sky usually clears, and temperature and humidity drop.

Adjacent air masses may also differ in atmospheric pressure. A **high-pressure system** contains air that descends because it is cool and then spreads outward as it nears the ground. High-pressure systems typically bring fair weather.

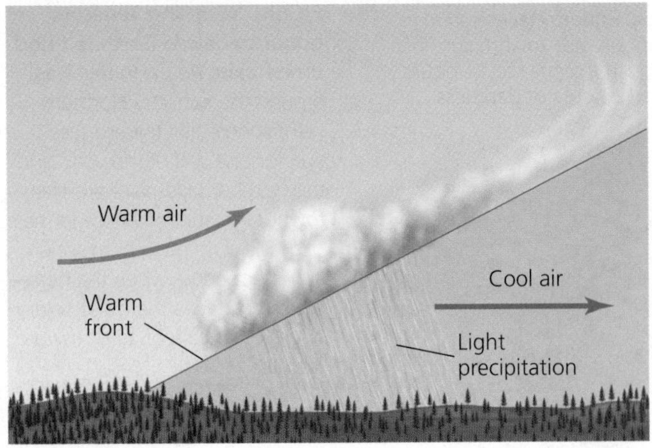

(a) Warm front

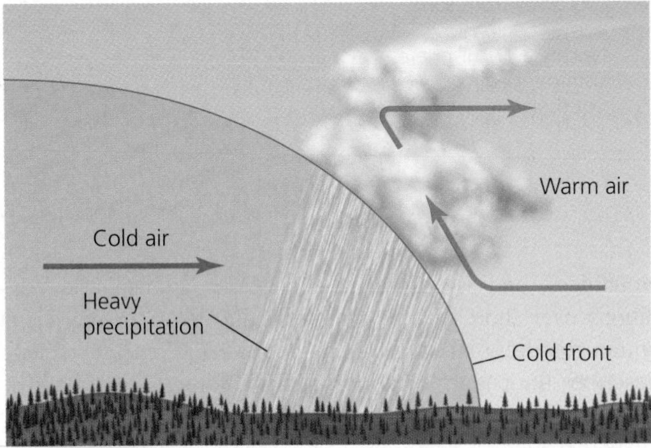

(b) Cold front

FIGURE 17.7 Fronts occur where air masses meet. When a warm front approaches **(a)**, warmer air rises over cooler air, causing light or moderate precipitation as moisture in the warmer air condenses. When a cold front approaches **(b)**, colder air pushes beneath warmer air, and the warmer air rises, resulting in condensation and heavy precipitation.

In a **low-pressure system,** warmer air rises, drawing air inward toward the center of low atmospheric pressure. The rising air expands and cools, and clouds and precipitation often result.

Under most conditions, air in the troposphere becomes cooler as altitude increases. Because warm air rises, vertical mixing results (FIGURE 17.8a). Occasionally, however, a layer of cool air may form beneath a layer of warmer air. This departure from the normal temperature profile is known as a **temperature inversion,** or **thermal inversion** (FIGURE 17.8b). The band of air in which temperature rises with altitude is called an **inversion layer** (because the normal direction of temperature change is inverted). The cooler air at the bottom of the inversion layer is denser than the warmer air above, so it resists vertical mixing and remains stable. Thermal inversions can occur in different ways, sometimes involving cool air at ground level and sometimes producing an inversion layer higher above the ground. One common type of inversion (shown in Figure 17.8b) occurs in mountain valleys where slopes block morning sunlight, keeping ground-level air within the valley shaded and cool.

Vertical mixing allows pollutants in the air to be carried upward and diluted, but thermal inversions trap pollutants near the ground. Both Los Angeles and Mexico City suffer their worst pollution when thermal inversions prevent pollutants from being dispersed. Both metropolitan areas are encircled by mountains that promote inversion layers, interrupt air flow, and trap pollutants. In Mexico City in 1996, a persistent thermal inversion sparked a five-day crisis that killed at least 300 people and sent 400,000 to hospitals. Desperate for a solution, some people proposed dynamiting a hole in the mountains and installing immense fans to blow out the air.

The worst such crisis occurred back in 1952 in London, England, when a thermal inversion spawned a "killer smog." A high-pressure system had settled over the city for several days, trapping pollutants from factories and coal-burning stoves. This created foul conditions that killed 4000 people—and by some estimates up to 12,000. Such dramatic events are by no means a thing of the past: In 2013, Beijing, China, suffered many days of abysmal air quality in an event dubbed an "airpocalypse" (p. 463). Across the world today, inversions frequently concentrate pollution over major metropolitan

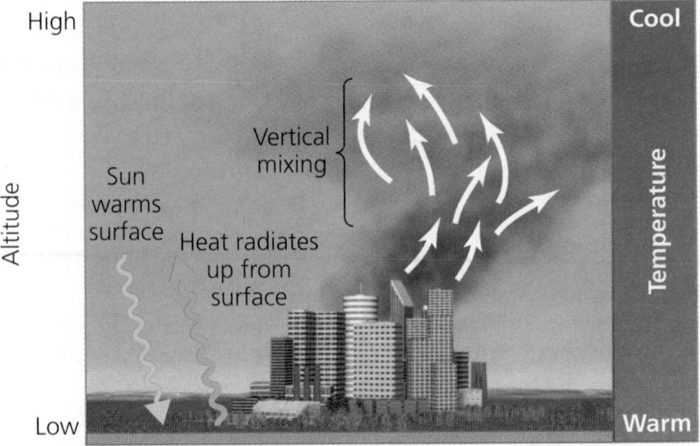

(a) Normal conditions

(b) Thermal inversion

FIGURE 17.8 Thermal inversions trap air and pollutants. Under normal conditions **(a)**, air becomes cooler with altitude and air of different altitudes mixes, dispersing pollutants upward. In a thermal inversion **(b)**, dense cool air remains near the ground, and air warms with altitude within the inversion layer. Little mixing occurs, and pollutants are trapped.

areas in valleys ringed by mountains, from Tehran to Seoul to Río de Janeiro to São Paulo.

Large-scale circulation systems produce global climate patterns

At large geographic scales, convective air currents contribute to broad climate patterns (**FIGURE 17.9a**). Near the equator, solar radiation sets in motion a pair of convective cells known as **Hadley cells.** Here, where sunlight is most intense, surface air warms, rises, and expands. As it does so, it releases moisture, producing the heavy rainfall that gives rise to tropical rainforests near the equator. After releasing much of its moisture, this air diverges and moves in currents heading north and south. The air in these currents cools and descends at about 30 degrees latitude north and south. Because the descending air has low relative humidity, the regions around 30 degrees latitude are quite arid, giving rise to deserts. Two further pairs of convective cells, **Ferrel cells** and **polar cells,** lift air and create precipitation around 60 degrees latitude north and south and cause air to descend at 30 degrees latitude and in the polar regions.

Together these three pairs of convective cells create wet climates near the equator, arid climates near 30° latitude, moist regions near 60° latitude, and dry conditions near the poles. These patterns, combined with temperature variation, help explain why biomes tend to be arrayed in latitudinal bands (see Figure 4.17, p. 93).

The Hadley, Ferrel, and polar cells interact with Earth's rotation to produce global wind patterns (**FIGURE 17.9b**). As Earth rotates on its axis, locations on the equator spin faster than locations near the poles. This means that as air currents of the convective cells flow north or south, some regions of the planet's surface move west to east beneath them more quickly than others. As a result, from the perspective of an Earth-bound observer, these air currents appear to be deflected from a straight path. This deflection is called the **Coriolis effect,** and it results in the curving global wind patterns in Figure 17.9b. Near the equator lies a region with few winds known as the *doldrums*. Between the equator and 30° latitude, *trade winds* blow from east to west. From 30° to 60° latitude, *westerlies* blow from west to east. People made use of these patterns for centuries to facilitate ocean travel by wind-powered sailing ships.

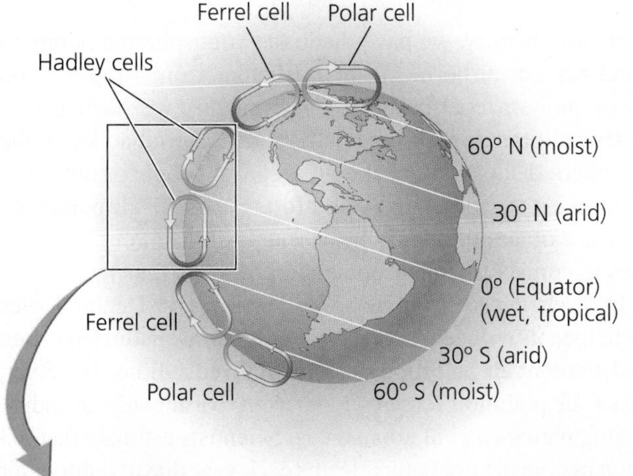

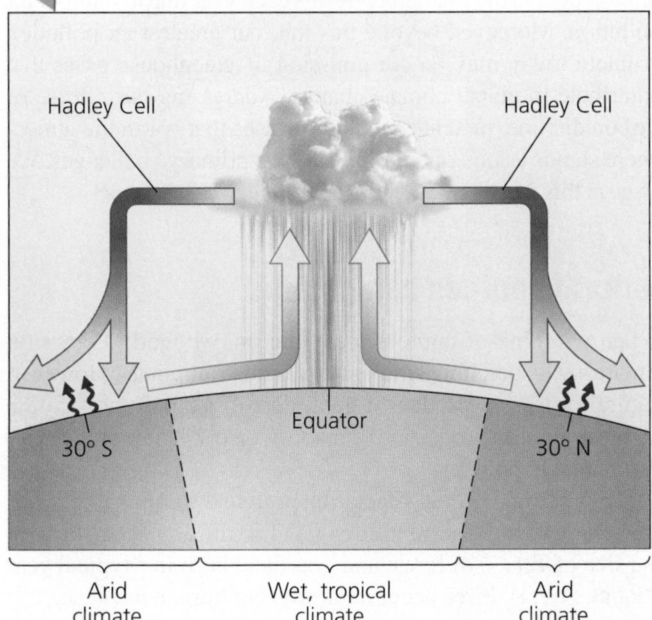

(a) Convection currents

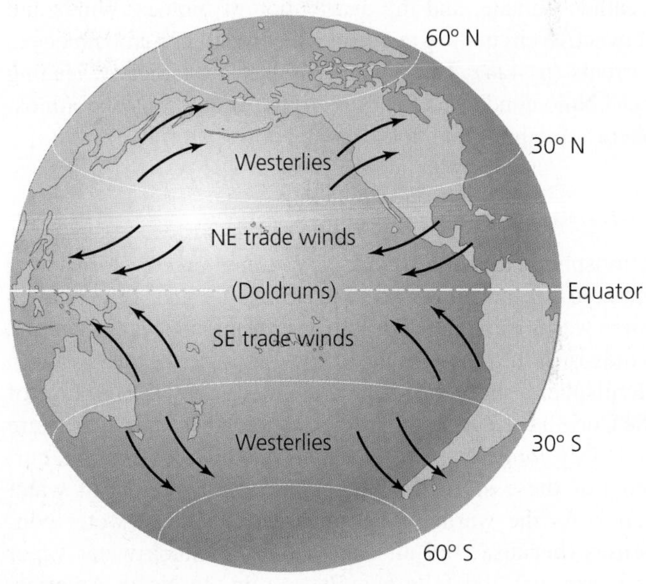

(b) Global wind patterns

FIGURE 17.9 Large-scale convective cells create global patterns in moisture and wind. These cells **(a)** give rise to a wet climate in tropical regions, arid climates around 30° latitude, moist climates around 60°, and dry climates near the poles. Surface air movement of these cells interacts with the Coriolis effect to create **(b)** global wind currents.

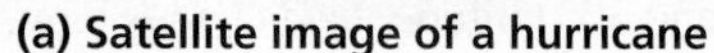

(a) Satellite image of a hurricane

(b) Photograph of a tornado

FIGURE 17.10 Hurricanes and tornadoes are cyclonic storms that pose hazards to our life and property.

The atmosphere interacts with the oceans to affect weather, climate, and the distribution of biomes. Winds and convective circulation in ocean water together maintain ocean currents (p. 424). Trade winds weaken periodically, leading to El Niño conditions (pp. 425–426). And oceans and atmosphere sometimes interact to create violent storms.

Storms pose hazards

Atmospheric conditions can sometimes create storms that threaten life and property. **Hurricanes** (FIGURE 17.10a) form when warm, moisture-laden air over tropical oceans rises and winds rush into these areas of low pressure. In the Northern Hemisphere, these winds turn counterclockwise because of the Coriolis effect. In other regions, such cyclonic storms are called *cyclones* or *typhoons*. The powerful convective currents of these storms draw up immense amounts of water vapor. As the warm moist air rises and cools, water condenses (because cool air cannot hold as much water vapor as warm air) and falls heavily as rain. In North America, the Gulf Coast and Atlantic Coast are most susceptible to hurricanes.

Tornadoes (FIGURE 17.10b) form when a mass of warm air meets a mass of cold air and the warm air rises quickly, setting a powerful convective current in motion. If high-altitude winds are blowing faster and in a different direction from low-altitude winds, the rising column of air may begin to rotate. Eventually the spinning funnel of rising air may lift up soil and objects in its path with winds up to 500 km per hour (310 mph). In North America, tornadoes are most apt to form in the Great Plains and the Southeast, where cold air from Canada and warm air from the Gulf of Mexico frequently meet.

Understanding how the atmosphere functions can help us predict violent storms and warn people of their approach. Such knowledge can also help us comprehend how our pollution of the atmosphere affects climate, ecological systems, economies, and human health.

Outdoor Air Quality

Throughout history, we have made the atmosphere a dumping ground for our airborne wastes. Whether from simple wood fires or modern coal-burning power plants, we have generated **air pollutants,** gases and particulate material added to the atmosphere that can affect climate or harm people or other living things. At the same time, our efforts to control **air pollution,** the release of air pollutants, have brought some of our best successes in confronting environmental problems thus far.

In recent decades, public policy and improved technologies have helped us reduce most types of **outdoor air pollution** (often called **ambient air pollution**) in industrialized nations. However, outdoor air pollution remains a problem, particularly in industrializing nations and in urban areas. Scientists estimate that 3.3 million people die premature deaths each year due to outdoor air pollution. Moreover, beyond this toll, our greatest air pollution problem today may be our emission of greenhouse gases that contribute to global climate change. Addressing our release of carbon dioxide, methane, and other gases that warm the atmosphere stands as one of our civilization's primary challenges. We discuss this issue separately and in depth in Chapter 18.

Natural sources can pollute

When we think of outdoor air pollution, we tend to envision smokestacks belching smoke from industrial plants. However, natural processes produce a great deal of air pollution. Some of these natural impacts are made worse by human activity and land use policies.

Fires from burning vegetation pollute the atmosphere with soot and gases. Over 60 million ha (150 million acres, an area the size of Texas) of forest and grassland burn in a typical year (FIGURE 17.11a). Fires occur naturally, but human influence can make them more severe. In North America, fuel buildup from decades of fire suppression has promoted destructive forest

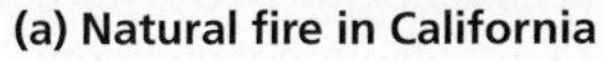

(a) Natural fire in California

(b) Mount Saint Helens eruption, 1980

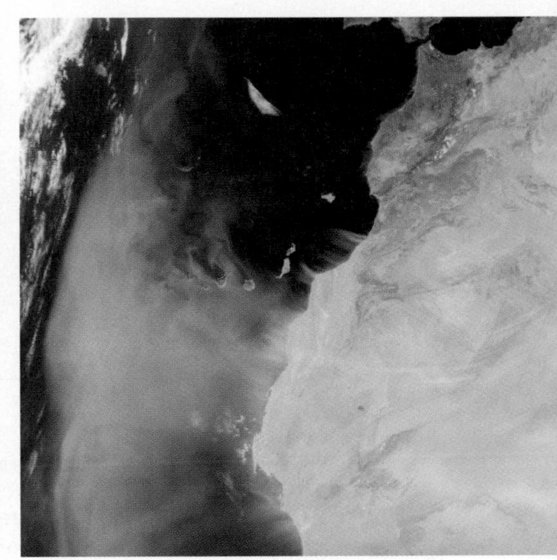

(c) Dust storm blowing dust from Africa to the Americas

FIGURE 17.11 Wildfire, volcanoes, and dust storms are three natural sources of air pollution.

fires in recent years (p. 320). In regions like the Los Angeles basin, where residential development has encroached into chaparral ecosystems (p. 320) that are naturally fire-prone, fires may cause extensive and costly damage. In the tropics, many farmers set fires to clear forest for farming and grazing using a "slash-and-burn" approach (p. 221). In 1997, a severe drought brought on by the 20th century's strongest El Niño event caused unprecedented forest fires in Mexico, Central America, Indonesia, and Africa. Immense swaths of rainforest burned while smoke pollution sickened 20 million Indonesians and caused a plane to crash and ships to collide. Scientists voice concern that global climate change (Chapter 18) is increasing fires as a result of drought in many regions.

Volcanic eruptions (pp. 40–42) release large quantities of particulate matter, as well as sulfur dioxide and other gases, into the troposphere (FIGURE 17.11b). In 2012, residents of Mexico City went on alert as Popocatepetl, a volcano just 70 km (45 mi) from the city, let loose a series of moderate eruptions. Ash from these eruptions added to the region's pollution challenges. Ash from major volcanic eruptions near populated areas can ground airplanes, destroy car engines, and pose respiratory health dangers for millions of people. Major eruptions may also blow matter into the stratosphere, where it can circle the globe for months or years. Sulfur dioxide reacts with water and oxygen and then condenses into fine droplets, called **aerosols,** which reflect sunlight back into space and thereby cool the atmosphere and surface. The 1991 eruption of Mount Pinatubo in the Philippines ejected nearly 20 million tons of ash and aerosols and cooled global temperatures by 0.5°C (0.9°F).

Winds sweeping over arid terrain can send huge amounts of dust aloft. Dust storms occur naturally, but they are made worse by unsustainable farming and grazing practices that strip vegetation from the soil, promote wind erosion, and lead to desertification (pp. 223–224). Continental-scale dust storms took place in the United States in the 1930s, when soil from the drought-plagued Dust Bowl states blew eastward to the Atlantic (pp. 224–225). Today, trade winds blow soil across the Atlantic Ocean from Africa to the Americas (FIGURE 17.11c), carrying fungal and bacterial spores that harm Caribbean coral reefs, but also bringing nutrients to the Amazon rainforest. Strong westerlies sometimes lift soil from deserts in Mongolia and China and blow it all the way across the Pacific Ocean to North America.

We create outdoor air pollution

Human activity introduces many sources of air pollution. As with water pollution, anthropogenic (human-caused) air pollution can emanate from point sources or non-point sources (pp. 408–409). A point source describes a specific location from which large quantities of pollutants are discharged, such as a coal-fired power plant. Non-point sources are more diffuse, consisting of many small, widely spread sources (such as thousands of automobiles).

Primary pollutants, such as soot and carbon monoxide, are pollutants that can cause harm directly or that react chemically to form harmful substances. Harmful substances produced when primary pollutants interact or react with constituents of the atmosphere are called **secondary pollutants.**

Pollutants differ in the amount of time they spend in the atmosphere—called their **residence time**—because substances differ in how readily they react in air and in how quickly they settle to the ground. Pollutants with brief residence times exert localized impacts over short time periods. Most particulate matter and most pollutants from automobile exhaust stay aloft only hours or days, which is why air quality in a city like Mexico City or Los Angeles can change from day to day. In contrast, pollutants with long residence times can exert impacts regionally or globally for long periods, even centuries. The pollutants that drive global climate change and those that deplete Earth's ozone layer (two separate phenomena!—see **FAQ,** p. 472) are each able to cause these global and long-lasting impacts because they persist in

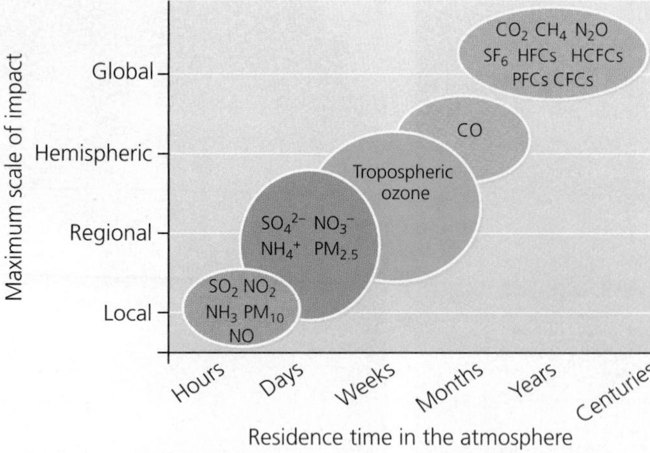

FIGURE 17.12 Substances with short residence times affect air quality locally, whereas those with long residence times affect air quality regionally or globally. *Data from United Nations Environment Programme, 2007.* Global environmental outlook (GEO-4), *Nairobi, Kenya.*

the atmosphere for so long. **FIGURE 17.12** shows this relationship, with examples.

Clean Air Act legislation addresses pollution in the United States

To address air pollution in the United States, Congress has passed a series of laws, beginning with the Air Pollution Control Act of 1955. The Clean Air Act of 1963 funded research into pollution control and encouraged emissions standards for automobiles and for stationary point sources such as industrial plants. Subsequent amendments expanded the legislation's scope and established a nationwide air quality monitoring system.

In 1970, Congress thoroughly revised the law in what came to be known as the **Clean Air Act of 1970.** This legislation set stricter standards for air quality, imposed limits on emissions from new sources, provided new funds for pollution-control research, and enabled citizens to sue parties violating the standards. Some of these goals came to be viewed as too ambitious, so amendments in 1977 loosened some standards and extended some deadlines for compliance.

The **Clean Air Act of 1990** sought to strengthen regulations pertaining to air quality standards, auto emissions, toxic air pollution, acid deposition, and stratospheric ozone depletion. It also introduced an emissions trading program for sulfur dioxide (p. 183). Beginning in 1995, businesses and utilities were allocated permits for emitting this pollutant and could then buy, sell, or trade these allowances. Each year the overall amount of allowed pollution was decreased. This market-based incentive program has helped reduce sulfur dioxide emissions nationally (see Figure 7.16, p. 183). It has also spawned similar cap-and-trade programs for other pollutants, including greenhouse gases (pp. 512–513). The Los Angeles region adopted its own cap-and-trade program in 1994. The RECLAIM (Regional Clean Air Incentives Market) program helped the L.A. basin reduce emissions of sulfur dioxide and nitrogen dioxide by over 70% by 2010.

As a result of Clean Air Act legislation, the U.S. Environmental Protection Agency (EPA) sets nationwide standards for:

- Emissions of several key pollutants
- Concentrations of key pollutants in ambient air

It is largely up to the states to monitor emissions and air quality and to develop, implement, and enforce regulations within their borders. States submit implementation plans to the EPA for approval, and if a state's plans are not adequate, the EPA can take control of enforcement. If a region fails to clean up its air, the EPA can prevent it from receiving federal money for transportation projects.

Agencies monitor emissions

State and local agencies monitor and report to the EPA emissions of six major pollutants—carbon monoxide (CO), sulfur dioxide (SO_2), nitrogen oxides (NO_X), volatile organic compounds (VOCs), particulate matter, and lead (Pb). Across the United States in 2012, human activity polluted the air with 85 million tons of these six monitored pollutants. Carbon monoxide was the most abundant pollutant by mass, followed by VOCs, NO_X, and SO_2 (**FIGURE 17.13**).

Carbon monoxide

Carbon monoxide is a colorless, odorless gas produced primarily by the incomplete combustion of fuel. Vehicles and engines account for most CO emissions in the United States. Other sources include industrial processes, waste combustion, and residential wood burning. Carbon monoxide is hazardous because it can bind irreversibly to hemoglobin in red blood cells, preventing the hemoglobin from binding with oxygen.

Sulfur dioxide

Sulfur dioxide is a colorless gas with a pungent odor. The vast majority of SO_2 pollution results from the combustion of coal for electricity generation and industry. During combustion, elemental sulfur (S) in coal reacts with

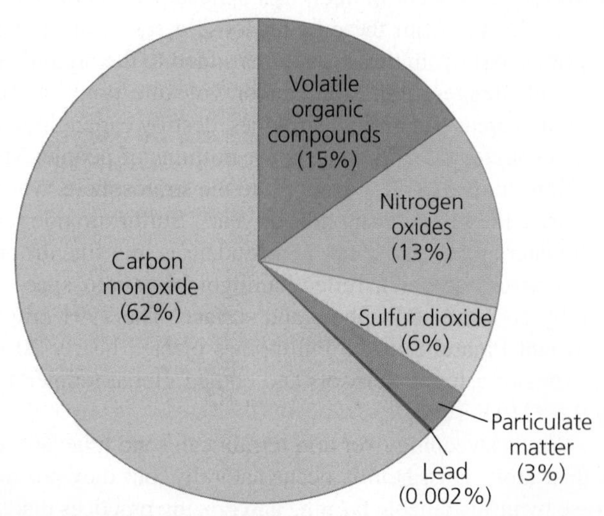

FIGURE 17.13 In 2012, the United States emitted 85 million tons of the six major pollutants whose emissions are monitored by the EPA and state agencies. *These figures omit pollutants from dust and wildfires. Data from U.S. EPA.*

oxygen (O_2) to form SO_2. Once in the atmosphere, SO_2 may react to form sulfur trioxide (SO_3) and sulfuric acid (H_2SO_4), which may then settle back to Earth in acid deposition (p. 473).

Nitrogen oxides **Nitrogen oxides** are a family of compounds that include nitric oxide (NO) and nitrogen dioxide (NO_2). Nitrogen oxides result when atmospheric nitrogen and oxygen react at high temperatures. Most U.S. NO_X emissions result from combustion in vehicle engines. Electrical utility and industrial combustion account for most of the rest. NO_X emissions contribute to smog, acid deposition, and stratospheric ozone depletion.

Volatile organic compounds **Volatile organic compounds (VOCs)** are carbon-containing chemicals used in and emitted by vehicle engines and a wide variety of solvents, industrial processes, household chemicals, and consumer items. One group of VOCs consists of hydrocarbons (p. 28) such as methane (CH_4, the primary component of natural gas), propane (C_3H_8, used as a portable fuel), butane (C_4H_{10}, found in cigarette lighters), and octane (C_8H_{18}, a component of gasoline). Human activities account for about half the VOC emissions in the United States. The remainder comes from natural sources; for example, plants produce isoprene and terpenes, compounds that generate a bluish haze that has given the Blue Ridge Mountains their name. VOCs can react to produce a number of secondary pollutants.

Particulate matter **Particulate matter** is composed of solid or liquid particles small enough to be suspended in the atmosphere and able to damage respiratory tissues when inhaled. Particulate matter includes primary pollutants such as dust and soot, as well as secondary pollutants such as sulfates and nitrates. Scientists classify particulate matter by the size of the particles. PM_{10} pollutants consist of particles less than 10 microns in diameter (one-seventh the width of a human hair), whereas $PM_{2.5}$ pollutants consist of still-finer particles less than 2.5 microns in diameter. Most PM_{10} pollution is from road dust, whereas most $PM_{2.5}$ pollution results from combustion.

Lead **Lead** is a heavy metal that enters the atmosphere as a particulate pollutant. The lead-containing compounds tetraethyl lead and tetramethyl lead, when added to gasoline, improve engine performance. However, exhaust from the combustion of leaded gasoline emits airborne lead, which can be inhaled or can be deposited on land and water. Lead can enter the food chain, accumulate in body tissues, and cause central nervous system malfunction and many other ailments (p. 367). Since the 1980s, leaded gasoline has been phased out in most industrialized nations (p. 8), and as a result lead pollution has plummeted. The United States led the way, Mexico City phased out leaded gasoline in the 1990s, and today most developing nations are following suit, although auto exhaust still creates significant lead pollution in many of them. In developed nations today, the main source of atmospheric lead pollution is industrial metal smelting.

We have reduced pollutant emissions

Since passage of the Clean Air Act of 1970, the United States has reduced emissions of each of the six monitored pollutants substantially (**FIGURE 17.14a**). These dramatic reductions in emissions have occurred despite significant increases in the nation's population, energy consumption, miles traveled by vehicle, and gross domestic product (**FIGURE 17.14b**). Likewise, most other industrialized nations have taken their own steps to reduce emissions and have attained similar results.

We have achieved this success in controlling pollution as a result of policy steps and technological developments, each motivated by grassroots social demand for cleaner air. Cleaner-burning motor vehicle engines and automotive technologies such as catalytic converters (**FIGURE 17.15**) have played a large part, reducing the emissions of carbon monoxide and other pollutants from motor vehicles. In factories, power plants, and refineries, technologies such as baghouse filters, electrostatic precipitators, and **scrubbers** (**FIGURE 17.16**) have been installed to chemically convert or physically remove airborne pollutants before they are emitted from smokestacks. The sulfur-dioxide permit-trading program (p. 183) and clean coal technologies (p. 537) have reduced SO_2 emissions. And phaseouts of leaded gasoline

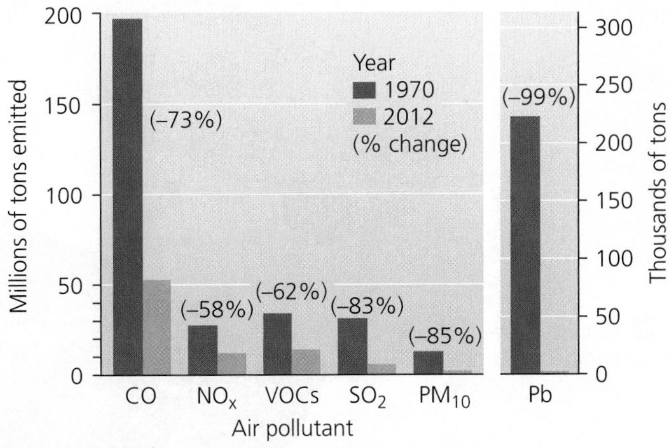

(a) Declines in six major pollutants

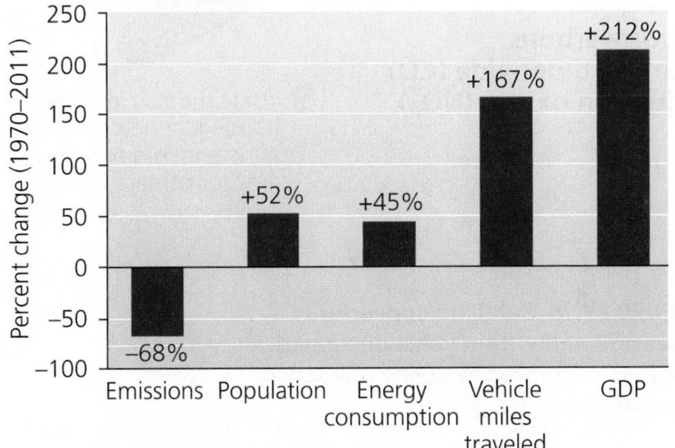

(b) Trends in major indicators

FIGURE 17.14 U.S. emissions have declined sharply since 1970. We have achieved reductions **(a)** in the six major pollutants tracked by the EPA, despite increases **(b)** in U.S. population, energy consumption, vehicle miles traveled, and gross domestic product. *Data from U.S. EPA.*

DATA Q By what percentage has population increased since 1970? By what percentage have emissions decreased? Using these two amounts, calculate the change in emissions per person.

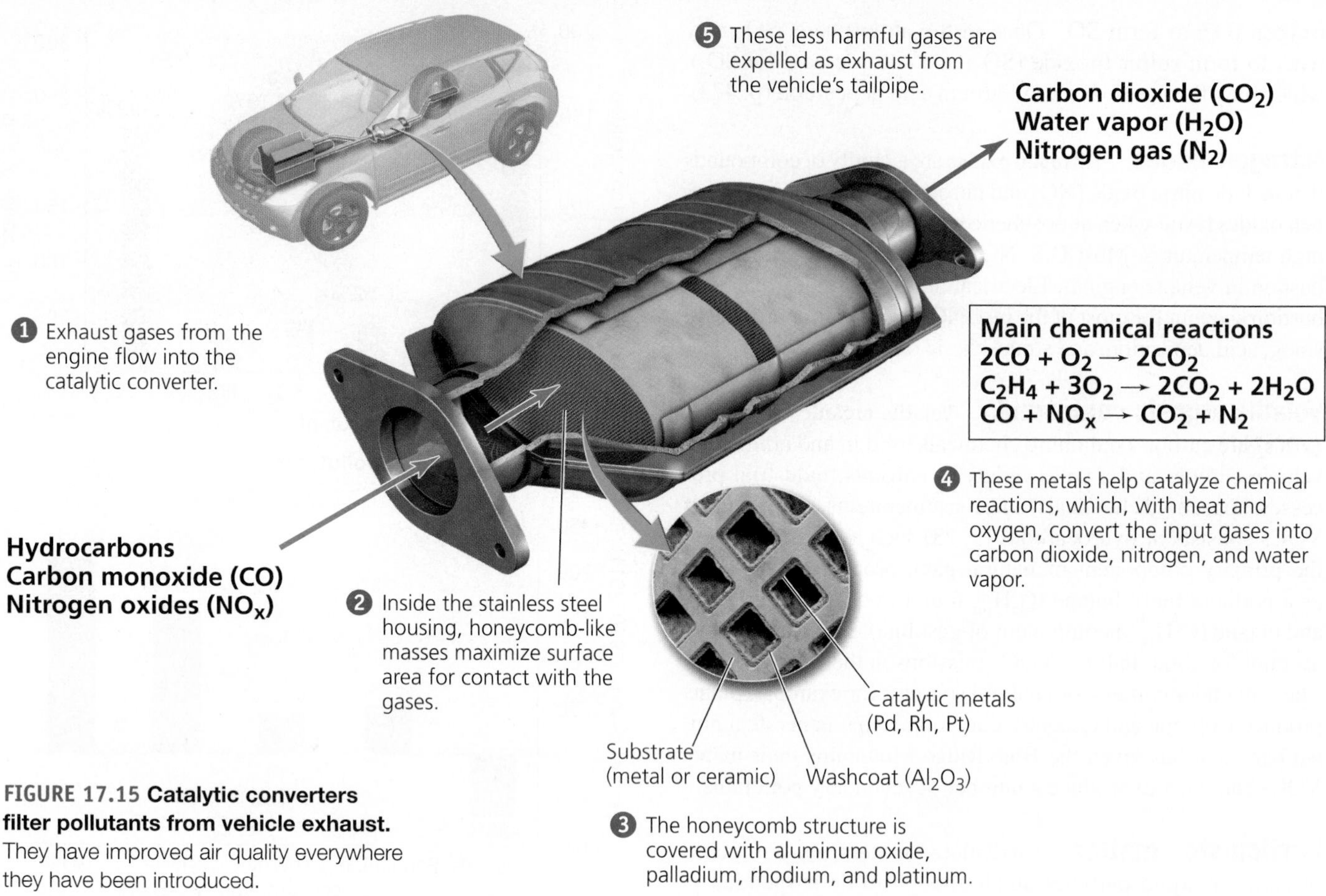

FIGURE 17.15 Catalytic converters filter pollutants from vehicle exhaust. They have improved air quality everywhere they have been introduced.

have caused lead emissions to plummet; in the United States lead emissions fell by 93% in the 1980s alone.

The reduction of emissions in the United States since 1970 has resulted in notably cleaner air throughout the nation. This reduction in outdoor air pollution represents one of the greatest accomplishments of the United States in safeguarding human health and environmental quality. It demonstrates how seemingly intractable problems can be tackled and addressed within a democratic system when government and industry are informed by science and are responsive to the public's demands. The EPA estimates that between 1970 and 1990 alone, clean air regulations and the resulting technological advances in pollution control saved the lives of 200,000 Americans. Similar advances in pollution control have been made by most other wealthy nations in recent years.

Air quality has improved

As a result of emissions reductions, air quality has improved markedly in industrialized nations throughout the world. In the United States, the EPA and the states monitor outdoor air quality by measuring the concentrations of six **criteria pollutants,** pollutants judged to pose substantial risk to human health. For each of these, the EPA has established **national ambient air quality standards (NAAQS),** which are maximum concentrations allowable in ambient outdoor air. The six criteria pollutants include four of the six pollutants whose emissions are monitored—carbon monoxide (CO), sulfur dioxide (SO_2), particulate matter, and lead (Pb)—as well as nitrogen dioxide (NO_2) and tropospheric ozone (O_3).

Nitrogen dioxide is a highly reactive, foul-smelling reddish brown gas that contributes to smog and acid deposition. Along with nitric oxide (NO), NO_2 belongs to the family of compounds called nitrogen oxides (NO_X). Nitric oxide reacts readily in the atmosphere to form NO_2, which is both a primary and secondary pollutant.

Although ozone in the stratosphere shields us from the dangers of UV radiation, ozone from human activity accumulates low in the troposphere. **Tropospheric ozone** (also called *ground-level ozone*) is a secondary pollutant, created by the interaction of sunlight, heat, nitrogen oxides, and volatile carbon-containing chemicals. A major component of smog, this colorless gas poses health risks due to the instability of the O_3 molecule. This triplet of oxygen atoms will readily split into a molecule of oxygen gas (O_2) and a free oxygen atom. The oxygen atom may then participate in reactions that can injure living tissues and cause respiratory problems. Tropospheric ozone is the pollutant that most frequently exceeds its national ambient air quality standard.

Thanks to the actions of scientists, policymakers, industrial leaders, and everyday citizens, air quality today is far better than it was a generation or two ago (**FIGURE 17.17**). However, plenty of room for improvement remains. Concerns over new pollutants are emerging, greenhouse gas emissions are altering the climate, and many Americans live in areas where concentrations of criteria pollutants continue to reach unhealthy levels. Residents of Los Angeles County, for instance, breathe air that violates

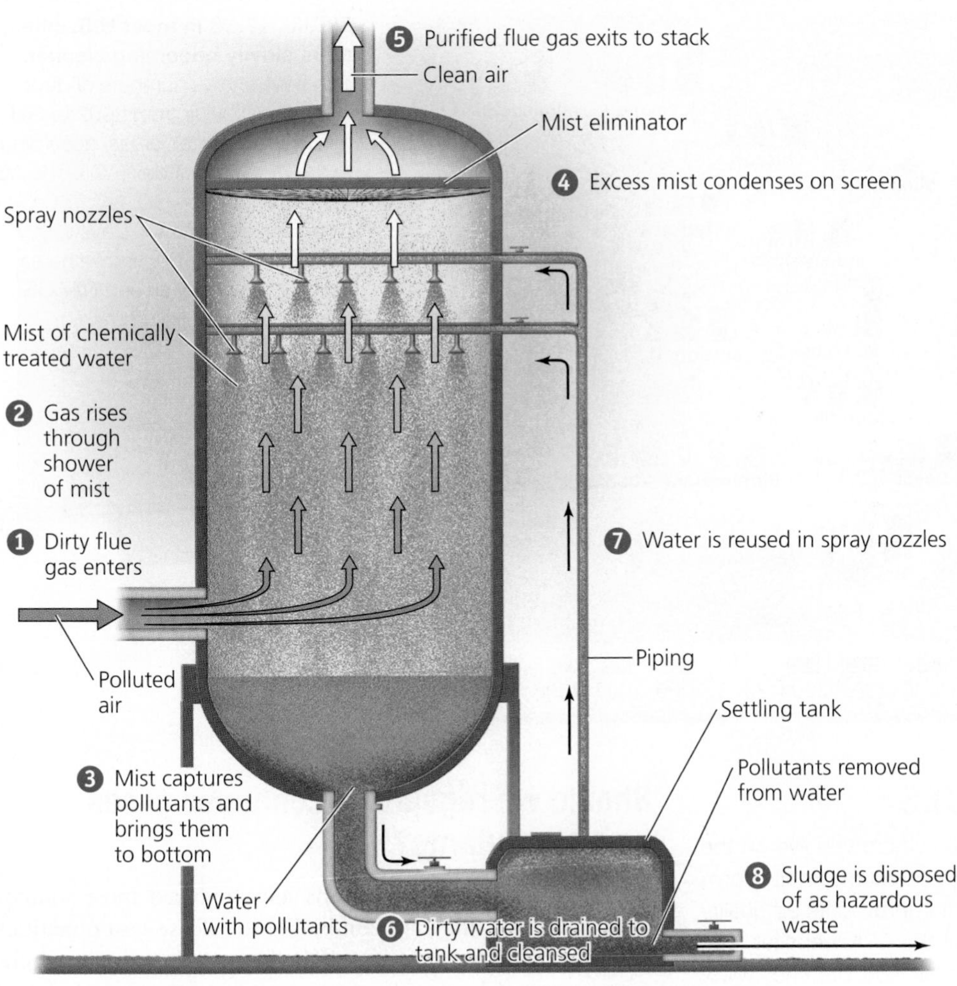

FIGURE 17.16 Scrubbers typically remove at least 90% of particulate matter and gases such as sulfur dioxide. Scrubbers and other pollution control devices come in many designs. In this spray-tower wet scrubber, polluted air rises through a chamber while nozzles spray a mist of water mixed with lime or other active chemicals to capture pollutants and wash them out of the air.

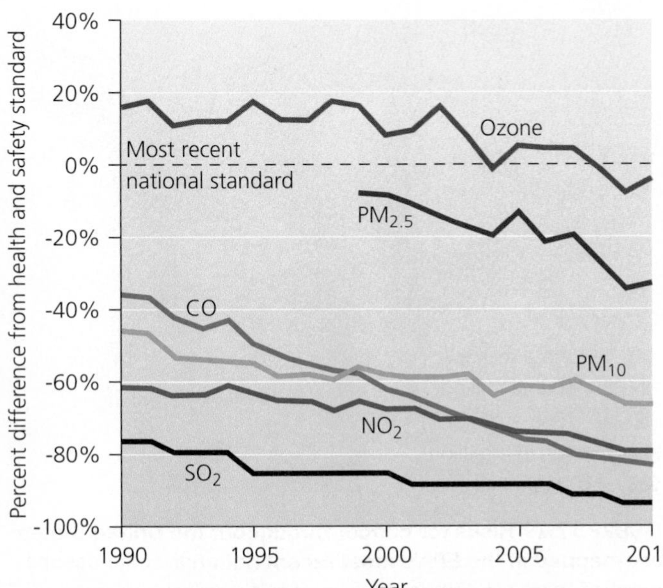

FIGURE 17.17 Concentrations of criteria pollutants in ambient air across the United States have steadily fallen. All except tropospheric ozone now average well below their standards for health and safety set by the EPA. Lead is not shown but has declined from 450% of its standard in 1990. *Data from U.S. EPA.*

national ambient air quality standards for five of the six criteria pollutants, according to 2012 data, while people in four adjacent southern California counties breathe air that violates four of the standards. All together, as of 2010, 124 million Americans lived in counties that violated the national ambient air quality standard for at least one criteria pollutant. Still, even L.A. and other pollution-choked metropolises are making perceptible headway toward cleaner air for their citizens (FIGURE 17.18).

Although we tend to focus on pollution in cities, air quality is a rural issue as well. In rural areas, people suffer from drift of airborne pesticides from farms, as well as industrial pollutants that drift far from cities, factories, and power plants. Air pollution also emanates from feedlots (p. 251) where cattle, hogs, or chickens are raised. The huge numbers of animals densely concentrated at feedlots and the voluminous amounts of waste they produce generate dust as well as methane, hydrogen sulfide, and ammonia. These gases create objectionable odors, and ammonia contributes to nitrogen deposition. Studies show that people working at and living near feedlots have high rates of respiratory illness.

Indeed, some of the worst air quality in the United States occurs in certain rural regions, including California's Central Valley (the nation's agricultural fruit basket) and areas near natural gas extraction sites, where fumes from extraction pollute the air.

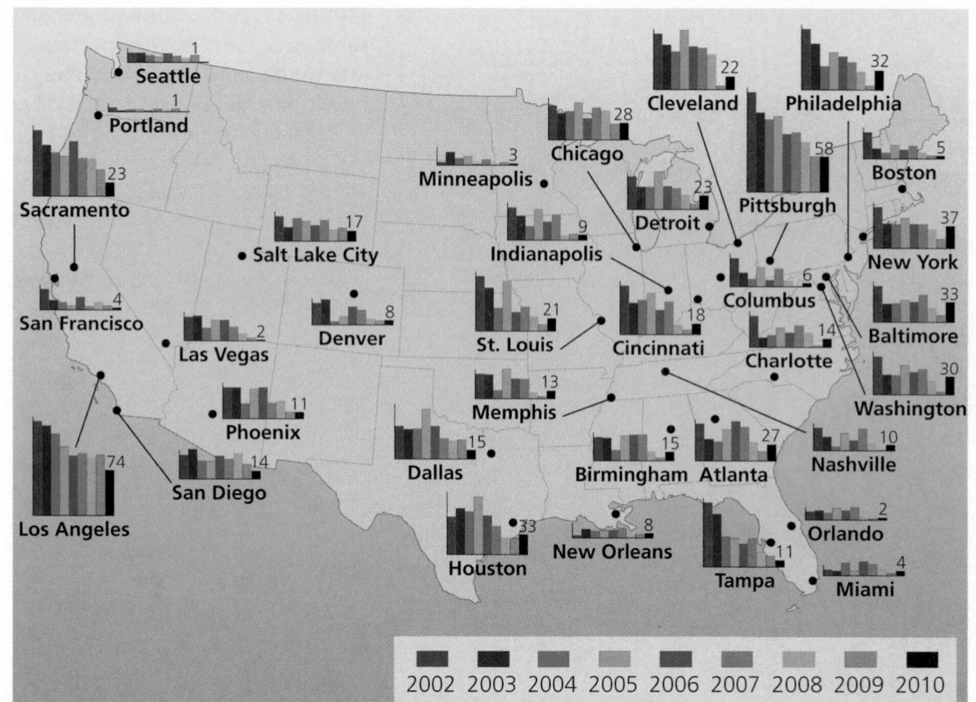

FIGURE 17.18 In most U.S. cities, air is slowly becoming cleaner. This map shows numbers of days with unhealthy air from 2002 to 2010 for 35 metropolitan areas, according to the Air Quality Index (AQI). The AQI combines data on CO, NO_2, SO_2, and particulate matter. Days with AQI values over 100, shown here, indicate unhealthy air and NAAQS violations. *Data from U.S. EPA.*

WEIGHING THE ISSUES

YOUR REGION'S AIR QUALITY Locate where you live on the map in Figure 17.18. How does your nearest major city compare to others in its air quality? Has it improved its air quality in recent years? Now explore one of the EPA websites that lets you browse information on the air you breathe: www.airnow.gov, www.epa.gov/aircompare, or www.epa.gov/air/emissions/where.htm. What factors do you think influence the quality of your region's air? Propose three steps for reducing air pollution in your region.

Toxic pollutants pose health risks

We are also reducing emissions of **toxic air pollutants,** substances known to cause cancer, reproductive defects, or neurological, developmental, immune system, or respiratory problems in people and other organisms. Under the 1990 Clean Air Act, the EPA regulates 187 toxic air pollutants produced by a variety of activities, including metal smelting, sewage treatment, and industrial processes. These pollutants range from the heavy metal mercury (from coal-burning power plant emissions and other sources) to VOCs such as benzene (a component of gasoline) and methylene chloride (found in paint stripper).

Based on monitoring at 300 sites across the United States, scientists estimate that toxic air pollutants cause cancer in 1 out of every 20,000 Americans (50 cancer cases per 1 million people). The latest national assessment mapped how these elevated risks vary geographically for cancer (**FIGURE 17.19**), as well as for non-cancerous respiratory ailments. Areas such as the Los Angeles region experience high health risks, but nationwide the EPA estimates that Clean Air Act regulations on facilities such as chemical plants, waste incinerators, dry cleaners, and coke ovens have helped to reduce emissions of toxic air pollutants since 1990 by more than 42%.

Should we regulate greenhouse gases as air pollutants?

Although industrialized nations have reduced most sources of air pollution, they are continuing to release vast quantities of the greenhouse gases (p. 484) that are driving global climate change (Chapter 18). As a result, our emission of carbon dioxide and other gases that warm the lower atmosphere is arguably today's biggest air pollution problem. Industry and

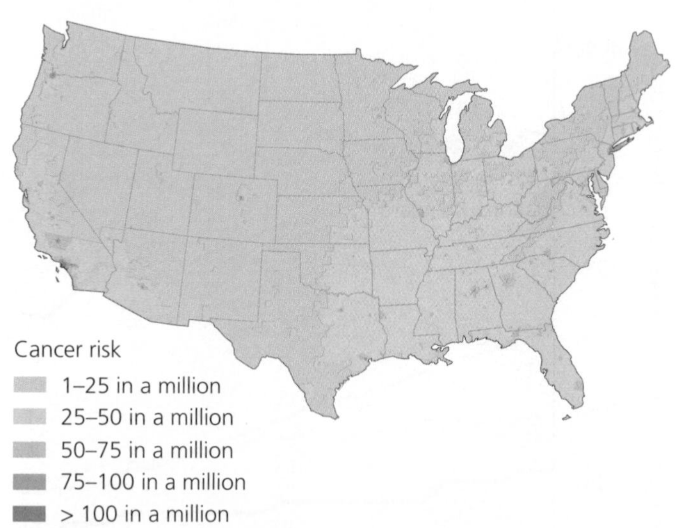

FIGURE 17.19 Risks for cancer throughout the United States are mapped in the EPA's most recent national-scale assessment of toxic air pollutants. Darker colors reflect higher risk. At least 80 toxic air pollutants are carcinogens, including formaldehyde, benzene, naphthalene, and others. Geographic patterns for non-cancerous respiratory ailments are similar. *Data from U.S. EPA, 2011.* 2005 National-Scale Air Toxics Assessment.

DATA Q What is the cancer rate due to toxic pollutants in the location where you live?

utilities generate much of these emissions, but all of us contribute by living carbon-intensive lifestyles. Each year the average U.S. vehicle driver releases close to 6 metric tons of carbon dioxide, 275 kg (605 lb) of methane, and 19 kg (41 lb) of nitrous oxide, all of them greenhouse gases that drive climate change.

In 2007 the U.S. Supreme Court ruled that the EPA has legal authority under the Clean Air Act to regulate carbon dioxide and other greenhouse gases as air pollutants. President Barack Obama voiced his preference that Congress address greenhouse gas emissions through bipartisan legislation. When Congress failed to pass climate change legislation, Obama instructed EPA Administrator Lisa Jackson to begin developing regulations to address greenhouse gas emissions. In 2011, the EPA introduced moderate carbon emission standards for cars and light trucks, and in 2012 it announced that it would limit carbon emissions for new coal-fired power plants and cement factories (but not existing ones). The EPA decided to phase in regulations gradually, beginning with the largest emitters.

The coal-mining and petrochemical industries objected, and these industries and several states sued to stop the regulations. A court of appeals unanimously upheld the EPA regulations in 2012. The automotive industry supported the regulations. U.S. automakers had already begun investing in fuel-efficient vehicles, and were glad to have one set of federal emissions standards, so as not to have to worry about meeting many differing state standards. The public also voiced strong support; 2.1 million Americans sent comments to the EPA in favor of the regulations—a record number of public comments for any federal regulation.

The EPA will no doubt continue to face formidable political opposition from emitting industries and from policymakers who fear that regulations will hamper economic growth. Yet if we were able to reduce emissions of other major pollutants sharply since 1970 while advancing our economy, we can hope to achieve similar results in reducing greenhouse gas emissions.

Indeed, although U.S. carbon dioxide emissions rose by 51% from 1970 to 2007, they decreased by 12% from 2007 to 2012. This decrease in emissions resulted partly from reduced energy use during a time of economic recession, but also from a shift from coal to cleaner-burning natural gas, and from improved fuel-efficiency in automobiles and other technologies.

Industrializing nations are suffering increasing air pollution

Although the United States and other industrialized nations have improved their air quality, outdoor air pollution is growing worse in many industrializing countries. In these societies, proliferating factories and power plants are emitting more pollutants as governments encourage economic growth. Additionally, more citizens own and drive automobiles. At the same time, most people continue to burn traditional sources of fuel, such as wood, charcoal, and coal, for cooking and home heating.

Mexico embodies these trends, and despite the progress in its capital, residents of Mexico City and many other Mexican cities and towns suffer a variety of health impacts from heavily polluted air (see **THE SCIENCE BEHIND THE STORY**, pp. 466–467). Altogether across the world, the World Health Organization (WHO) estimates that outdoor air pollution in cities causes 3.2 million premature deaths each year.

FIGURE 17.20 Air quality is poor in cities of today's developing nations. At Tiananmen Square in Beijing, China, children don face masks during the "airpocalypse" that gripped the city in 2013.

The people of China suffer some of the world's worst air pollution. China has fueled its rapid industrial development with its abundant reserves of coal, the most-polluting fossil fuel. Power plants and factories have sprung up across the nation, often using outdated, inefficient, heavily polluting technology because it is cheaper and quicker to build. Car ownership is skyrocketing; in the capital of Beijing alone 1500 new cars hit the streets each day. As a result, in many Chinese cities, the haze is often too thick for people to see the sun.

In Beijing in January–February 2013, smog became so severe that airplane flights were cancelled and people wore face masks to breathe (**FIGURE 17.20**). Levels of particulate matter were literally off the charts; a pollution monitor atop the U.S. Embassy designed to measure air quality on an index from 0 to 500 detected record-breaking readings up to 755. During this so-called airpocalypse, a fire at a factory went unnoticed for three hours because the smog was so thick that no one could see the smoke from the fire! Countless thousands of people suffered ill health as the pollution soared 30 times past the WHO's safe limits. Conditions were so bad that the government and the official state media were finally forced to admit the problem and begin a public discussion about solutions.

Across China, the health impacts of air pollution day in and day out are enormous. A 2013 international research report blamed outdoor air pollution for 1.2 million premature deaths in China each year. Winds carry some of China's pollution to neighboring nations. Some even travels across the Pacific Ocean to Los Angeles and other western U.S. cities.

China's government is now striving to reduce pollution. It has closed down some heavily polluting factories and mines, phased out some subsidies for polluting industries, installed pollution controls in power plants, and encouraged the development of wind, solar, and nuclear power. It subsidizes people to buy efficient electric heaters for their homes to replace dirty, inefficient coal stoves. It has mandated cleaner formulations for gasoline and diesel and has raised standards for fuel efficiency and emissions for cars above what the United States requires. In Beijing, mass transit is being expanded, many buses run on natural gas, and heavily polluting vehicles are restricted from operating in the central city. China is also aggressively developing cleaner wind, solar, and nuclear power to substitute for power produced by burning coal. The nation likely will need to accelerate such steps, as its 1.35 billion citizens are becoming increasingly fed up with air pollution.

Pollution from autos, industry, agriculture, and wood-burning stoves in China, India, and other industrializing nations of Asia has resulted in a persistent 2-mile-thick layer of pollution that hangs over southern Asia throughout the dry season each December through April. Dubbed the *Asian Brown Cloud*, or *Atmospheric Brown Cloud*, this massive layer of brownish haze is estimated to reduce the sunlight reaching Earth's surface in southern Asia by 10–20%; promote flooding in some areas and drought in others by altering the monsoon; decrease rice productivity by 5–10%; speed the melting of Himalayan glaciers by depositing dark soot that absorbs sunlight; and contribute to many thousands of deaths each year.

Smog poses health risks

Let's now examine one of our most prevalent types of air pollution: smog. *Smog* is an unhealthy mixture of air pollutants that often forms over urban areas as a result of fossil fuel combustion.

Since the onset of the industrial revolution, cities have suffered a type of smog known as **industrial smog.** When coal or oil is burned, some portion is completely combusted, forming CO_2; some is partially combusted, producing CO; and some remains unburned and is released as soot (particles of carbon). Moreover, coal contains contaminants such as mercury and sulfur. Sulfur reacts with oxygen to form sulfur dioxide, which can undergo a series of reactions to form sulfuric acid and ammonium sulfate (**FIGURE 17.21a**). These chemicals and others produced by further reactions, along with soot, are the main components of industrial smog.

FIGURE 17.21 Industrial smog results from fossil fuel combustion. When fossil fuels are burned, sulfur contaminants give rise to sulfur dioxide, which may react with atmospheric gases to produce other sulfur compounds **(a)**. Industrial smog also consists of particulate matter, carbon monoxide, and carbon dioxide. Under certain weather conditions, industrial smog can blanket whole towns or regions, as it did in Donora, Pennsylvania **(b)**, shown in the daytime during its deadly 1948 smog episode.

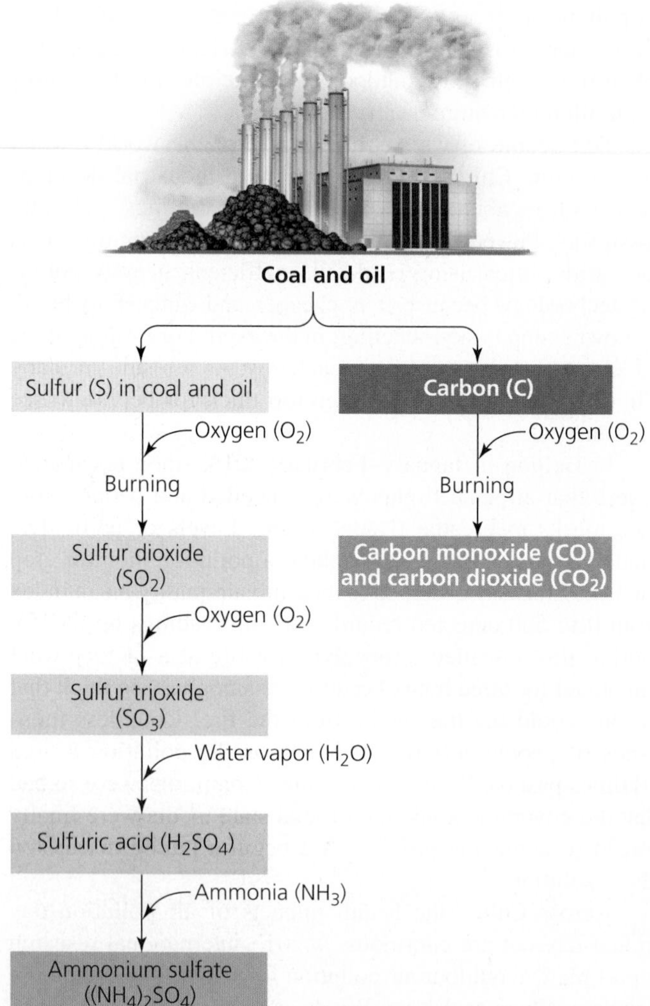

(a) Burning sulfur-rich oil or coal without adequate pollution control technologies

(b) Donora, Pennsylvania, at midday in the 1948 smog event

The thick and blinding "killer smog" of 1952 in London occurred when weather conditions trapped emissions from the coal used to fire the city's industries and heat people's homes. Sulfur dioxide and particulate matter caused most of the 4000–12,000 deaths from that episode. In the wake of this catastrophe and others, the governments of developed nations began regulating industrial emissions and have greatly reduced industrial smog. However, in industrializing regions such as China, India, and eastern Europe, coal burning and lax pollution control result in industrial smog that poses significant health risks.

As we've seen, weather and topography play roles in smog formation. Four years before London's killer smog, a similar event occurred in Pennsylvania in a small town named Donora (FIGURE 17.21b). Donora is located in a mountain valley, and one day after air had cooled during the night, the morning sun did not reach the valley floor to warm and disperse the cold air. The resulting thermal inversion trapped smog containing particulate matter emissions from a steel and wire factory. Twenty-one people died, and over 6000 people—nearly half the town—became ill.

For most urban areas today, however, pollution results largely from automobile exhaust. In Mexico City, vehicles contribute 31% of volatile organic compounds, 50% of sulfur dioxide, and 82% of nitrogen oxides. Cities like Mexico City and Los Angeles also have sunny climates. As a result, such cities suffer from a different type of smog. **Photochemical smog** forms when sunlight drives chemical reactions between primary pollutants and normal atmospheric compounds, producing a mix of over 100 different chemicals, tropospheric ozone often being the most abundant (FIGURE 17.22a). Because it also includes NO_2, photochemical smog generally appears as a brownish haze (FIGURE 17.22b).

Hot, sunny, windless days in urban areas provide perfect conditions for the formation of photochemical smog. On a typical weekday, exhaust from morning traffic releases NO and VOCs into a city's air. Sunlight then promotes the production of ozone and other constituents of photochemical smog. Levels of photochemical pollutants in urban areas typically peak in midafternoon and can irritate people's eyes, noses, and throats.

We can take steps to reduce smog

Los Angeles's struggle with air pollution began in 1943, when the city's first major smog episode cut visibility to three blocks. Since then, L.A. residents have dealt with headaches, eye irritation, asthma, lung damage, and related illnesses. However, Los Angeles confronted its problem and has made great progress in clearing the air since the 1970s.

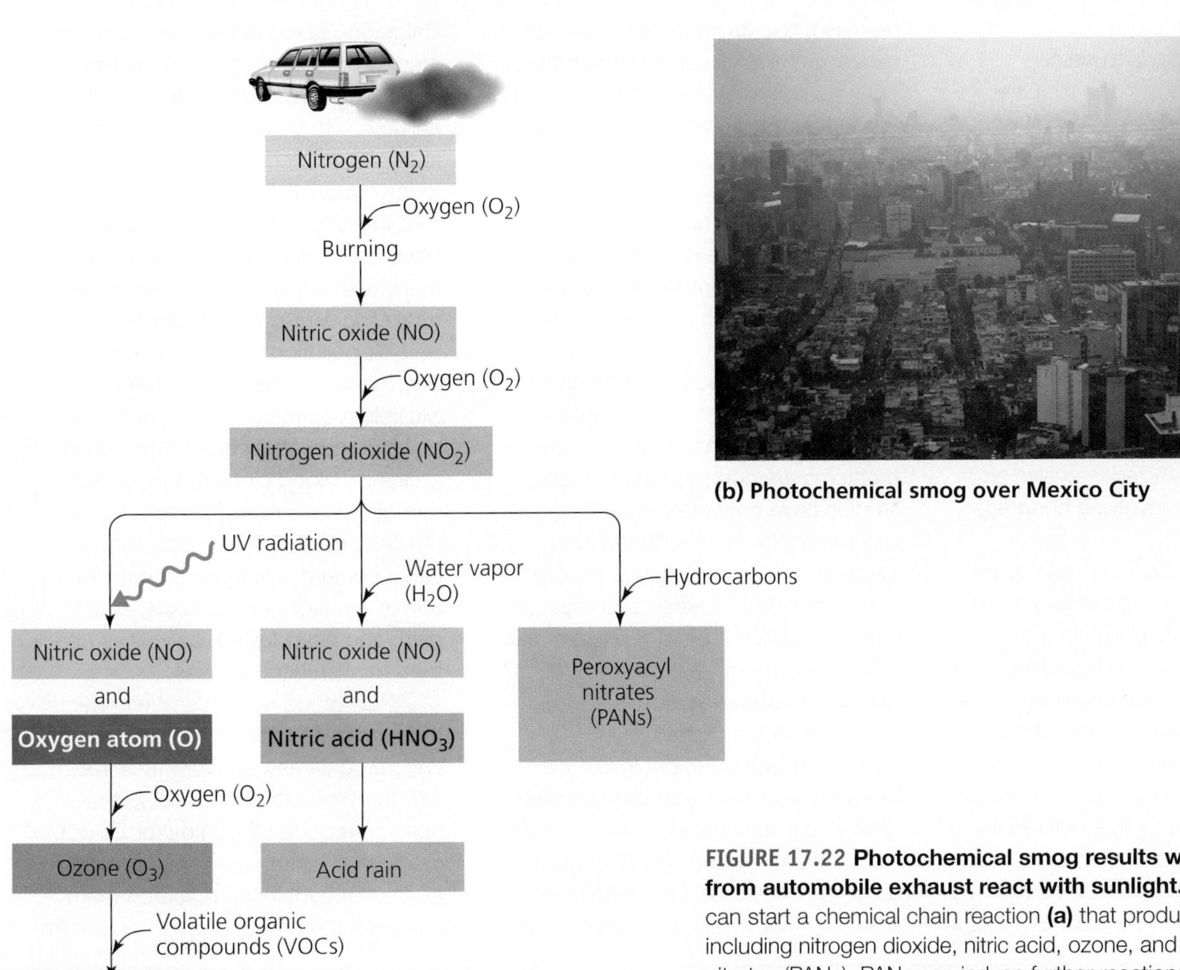

(a) Formation of photochemical smog

(b) Photochemical smog over Mexico City

FIGURE 17.22 **Photochemical smog results when pollutants from automobile exhaust react with sunlight.** Nitric oxide can start a chemical chain reaction **(a)** that produces compounds including nitrogen dioxide, nitric acid, ozone, and peroxyacyl nitrates (PANs). PANs can induce further reactions that damage living tissues. Photochemical smog is common over Mexico City **(b)** and many other urban areas, especially those with hilly topography or frequent inversion layers.

Measuring the Health Impacts of Mexico City's Air Pollution

"I know I'm inhaling poison," a 38-year-old candy vendor named Guadalupe told a reporter amid the fumes of a traffic-choked intersection in Mexico City. "But there is nothing I can do."

For as long as we have polluted our air, people have felt effects on their health. But identifying and quantifying those impacts poses a challenge for scientists. For researchers wanting to understand pollution's health impacts—and design solutions for people like Guadalupe—what better place to go than Mexico City, long home to some of the world's worst air pollution?

A key first step is to determine what's actually in the air. One researcher who has led the way is Mario Molina, the Nobel Prize–winning chemist who helped discover the cause of stratospheric ozone depletion and who appears in this chapter's other **Science behind the Story** feature (pp. 470–471). Molina stepped away from scholarly work at U.S. universities to return to his hometown of Mexico City and help address its pollution issues. Here, in 2003 and 2006, Molina organized intensive air-sampling projects involving hundreds of scientists.

Nearly 200 research publications later, these efforts have clarified many aspects of the city's pollution. For instance, one study used machines that could identify and record individual particles in real time, and found that metal-rich particulates from trash incinerators were peaking in the morning, whereas smoke from fires outside the city blew in during the afternoon. Other researchers discovered that volatile organic compounds control the amount of tropospheric ozone formed in smog (not nitrogen oxides, as was expected). City officials responded by targeting VOC emissions for reduction.

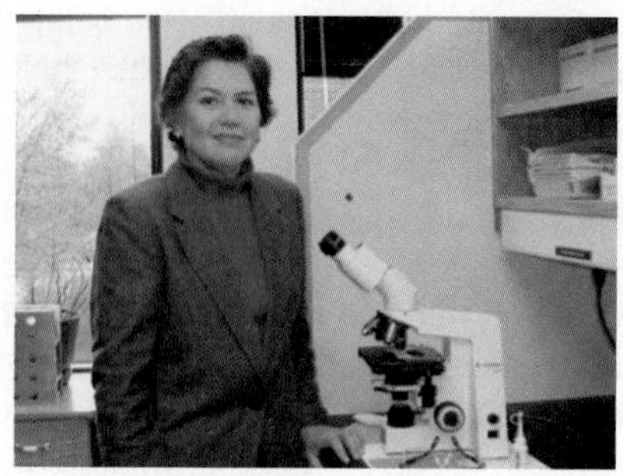

Dr. Lilian Calderón-Garcidueñas

Few people today understand Mexico City's air pollution in more detail than Armando Retama, the city's director of atmospheric monitoring. But he may grasp its impacts best when he leaves town. "I can breathe better. I'm not all dry. My eyes aren't irritated. My skin doesn't crack," he says. "We have chronic symptoms that we aren't aware of."

Most health impacts of urban pollution affect the respiratory system. At high altitudes like Mexico City's, the "thin air" forces people to breathe deeply to obtain enough oxygen. This means they pull more air pollutants into their lungs than people at lower elevations. As result, respiratory problems are commonplace. Many studies have confirmed that Mexico City residents show reduced lung function in comparison with people from less-polluted areas and that respiratory problems become worse and emergency room visits become more numerous when pollution is severe.

Most studies have looked at short-term exposure, but in 2007 a research team led by Isabelle Romieu of Mexico's National Institute of Public Health examined the effects of growing up amid polluted air. Her team measured lung function in 3170 8-year-old children from 39 Mexico City schools across 3 years and correlated this with their exposure to tropospheric ozone, nitrogen dioxide, and particulate matter. The children's ability to inhale and exhale deeply improved as they matured, but children from more polluted areas lagged behind those from cleaner areas, indicating smaller, weaker lungs.

Romieu and her colleagues also showed in a series of studies that the city's pollution worsens asthma in children, particularly those with certain genetic profiles. In 2008 her team analyzed data from 200 asthmatic and healthy children and found that children in areas with more traffic and pollutants coughed, wheezed, and used medication more often.

Another Mexican researcher, Lilian Calderón-Garcidueñas, has led several studies comparing chest X-ray films and medical records of Mexico City children with those of similar children from nonpolluted locations. Her team found hyperinflation and other problems with the lungs of the Mexico City youth. Mexico City children also reported many respiratory problems whereas rural children did not (**FIGURE 1**).

Air pollution harms the heart and the cardiovascular system, too. Multiple recent studies reveal that pollution can affect heart rate, blood pressure, blood clotting, blood vessels, and atherosclerosis. Epidemiological studies (pp. 375–376) show that pollution correlates with emergency room admissions for heart attacks, chest pain, and heart failure, as well as death from heart-related causes.

Part of the reason smog impacts the cardiovascular system is that tiny particulates can work their way into the bloodstream, causing the heart to reduce blood flow or go out of rhythm. Even young people are at risk. One Mexican research team analyzed the hearts of 21 people from

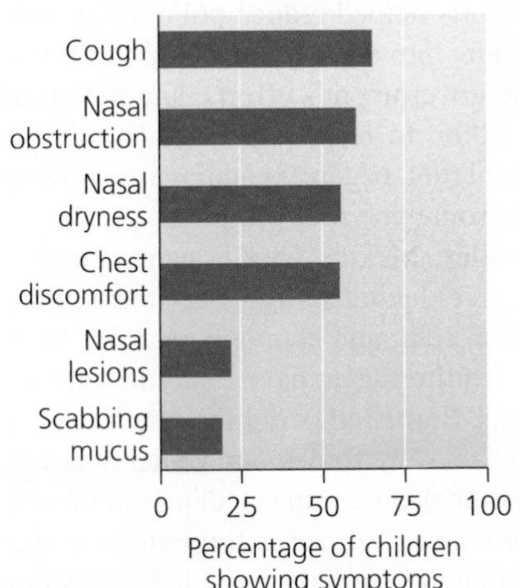

FIGURE 1 Most Mexico City children show respiratory symptoms from air pollution. Similar children from less-polluted areas outside Mexico City show none of these conditions. *Data from Calderón-Garcidueñas, L., et al., 2003. Respiratory damage in children exposed to urban pollution.* Pediatric Pulmonology *36: 148–161.*

Mexico City who had died at an early age. They found that pollution exacts a toll before the age of 18 and that tiny bits of dead bacteria that cling to pollutant particles are part of the problem. The heart mounts an inflammatory response to try to repel the bacteria-laden particles, but because the pollution is persistent, the inflammation becomes chronic and stresses the heart.

Recent research also shows that air pollution affects children's brains, sometimes leaving damage in brain tissue that is similar to that seen in Alzheimer's disease. In one study using brain scans, Calderón-Garcidueñas and her colleagues found that 56% of Mexico City youth had lesions on the brain's prefrontal cortex, whereas fewer than 8% did in a region with clean air. Studies elsewhere, from Italy to Boston, are finding similar results.

In 2011, Calderón-Garcidueñas's team compared 20 children aged 7–8 from Mexico City with 10 similar children from a Mexican city with clean air, measuring their cognitive skills and scanning their brains with magnetic resonance imagery (MRI). Results showed that the Mexico City children performed more poorly on most cognitive tests of reasoning, knowledge, and memory. The differences in cognition were consistent with differences in volume of white matter in key portions of the brain as revealed by the MRIs.

Pollution even damages the sense of smell. In 2006–2008 Robyn Hudson and colleagues ran lab experiments comparing Mexico City residents with residents of a geographically similar rural region with clean air. Hudson's team presented the subjects with various smells at different intensities. Compared with the rural residents, Mexico City residents had trouble smelling distinct and familiar scents such as orange juice and coffee. They also struggled to distinguish between the common Mexican beverages horchata and atole. The city-dwellers even had trouble noticing the stench of rotting food! The researchers attributed this to tissue damage in the noses of city-dwellers from a lifetime of exposure to pollutants. Indeed, other scientists were documenting such tissue damage with electron microscopy.

All these impacts can lead to higher rates of death. Studies by a U.S. and Mexican research team in Mexico City in the late 1990s confirmed this by comparing death certificate records against air pollution measurements. The team found that death rates rose on the day of and the day after severe pollution episodes, especially in response to particulate matter (**FIGURE 2a**). They also found that infant mortality was significantly higher in the days following strong pollution episodes (**FIGURE 2b**).

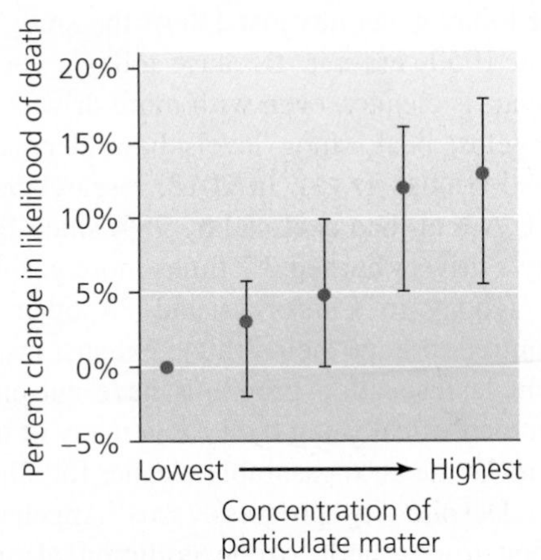

(a) Death rates increase with pollution intensity

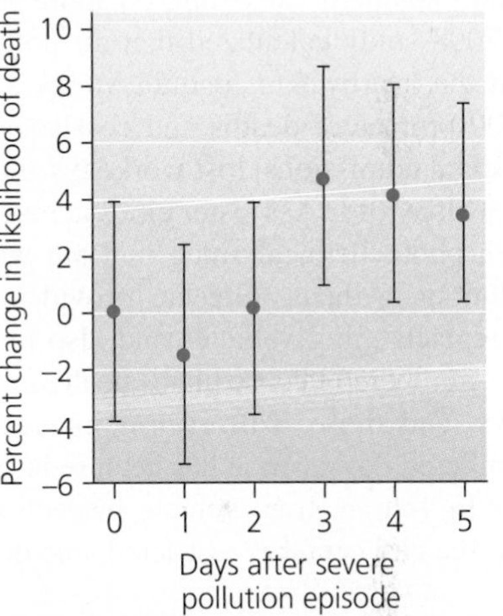

(b) Infant mortality rates are higher after a pollution episode

FIGURE 2 Rates of (a) death and (b) infant mortality each increase in Mexico City with the intensity of air pollution. *Data from (a) Borja-Aburto, V., et al., 1997. Ozone, suspended particulates, and daily mortality in Mexico City.* Am. J. Epidemiology *145: 258–268; and (b) Loomis, D., et al., 1999. Air pollution and infant mortality in Mexico City.* Epidemiology *10: 118–123.*

The extensive research showing a diversity of health impacts from air pollution in Mexico City has caught the attention of city leaders. These scientific findings have helped push them to work hard toward cleaning up their city's air. ■

California took the lead among U.S. states in adopting pollution control technology and setting emissions standards for vehicles. California's demands helped lead the auto industry to develop less-polluting cars. A 2010 study by the nonprofit group Environment California concluded that a new car today generates just 1% of the smog-forming emissions of a 1960s-era car. Because today's cars are 99% cleaner, the air is cleaner, even with more drivers on the road. In Los Angeles, peak smog levels have decreased 60–70% since 1980 (**FIGURE 17.23**). In 2012, researchers added that VOCs in L.A.'s air had declined by 98% since 1960, even while the city's drivers burned 2.7 times more gasoline.

Today in California and 33 other states, drivers are required to have their vehicle exhaust inspected periodically. Vehicle inspection programs have cut emissions leading to photochemical smog by 30% in many of these states, helping to make the air measurably cleaner for all of us.

Despite its progress, Los Angeles still suffers the worst tropospheric ozone pollution of any U.S. metropolitan area, according to 2012 rankings by the American Lung Association. L.A. residents breathe air exceeding California's health standard for ozone on more than 130 days per year. A 2008 study calculated that air pollution in the L.A. basin and the nearby San Joaquin Valley each year caused nearly 3900 premature deaths and cost society $28 billion (due to hospital admissions, lost workdays, etc.).

Many of L.A.'s sister cities across the world also struggle with photochemical smog and are working hard to develop solutions. Athens, Greece, provides its citizens incentives to replace aging vehicles and also mandates that autos with odd-numbered license plates be driven only on odd-numbered days, and those with even-numbered plates only on even-numbered days. Smog has been reduced by 30% as a result.

In Tehran, Iran, vehicle inspections are required, traffic into the city center is restricted, and drivers are paid to turn in old polluting cars for newer cleaner ones. Sulfur was reduced in diesel fuel, lead was removed from gasoline, and buses running on (cleaner-burning) natural gas hit the roads. To raise public awareness, 22 real-time pollution indicator boards were installed around the city, electronically displaying current pollutant levels. All these efforts helped reduce pollution, yet so many people were streaming into the city and buying cars that these trends overtook the government's efforts, and pollution grew worse again after 2006. In response, officials lowered gasoline subsidies, rationed fuel, began expanding the subway system, and strengthened some previous programs.

Of all the world's cities, Mexico City is gaining attention today for its success in reducing smog—once the world's worst—even as population, cars, and economic activity have grown. Regulations now require cars to have catalytic converters and get emissions tests. Some industrial facilities cleaned up their processes, and others were forced out. Under pressure from city leaders, the national oil company Pemex removed lead from gasoline, improved its refineries, imported cleaner gasoline, and removed pollutants from the liquefied petroleum gas that city residents use for cooking and heating. The subway system and a fleet of low-emission buses continue to be expanded today, while residents begin using new bike-sharing and car-sharing programs. As a result, since 1990 smog has been reduced by more than half, and particulate matter is down by 70%, carbon monoxide by 74%, sulfur dioxide by 86%, and lead by 95%.

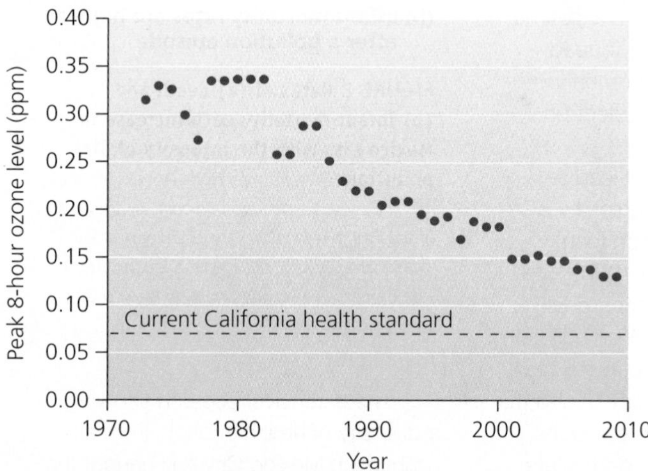

FIGURE 17.23 Peak levels of tropospheric ozone in the Los Angeles region have been reduced since the 1970s, thanks to public policy and improved automotive technology. Ozone pollution still violates the state health standard, however. *Data from Environment California, 2010.* Clean cars in California: Four decades of progress in the unfinished battle to clean up our air.

DATA Q By what percentage has Los Angeles reduced its ozone pollution since the 1970s?

WEIGHING THE ISSUES

SMOG-BUSTING SOLUTIONS Does the city you live in, or the nearest major city to you, suffer from photochemical smog or other air pollution? How is this city responding? What policies do you think it should pursue? What benefits might your city enjoy from such policies? Would they bring any problems?

Ozone Depletion and Recovery

Although ozone in the troposphere is a pollutant in photochemical smog, ozone is a highly beneficial gas in the stratosphere, where it forms the ozone layer (p. 451; see Figure 17.2). In this region of the stratosphere, concentrations of ozone reach only about 12 parts per million. However, ozone molecules are so effective at absorbing incoming ultraviolet (UV) radiation from the sun that even this diffuse concentration helps to protect life on Earth's surface from radiation's damaging effects.

One generation ago, scientists discovered that our planet's stratospheric ozone was being depleted, posing a major threat to human health and the environment. Years of dynamic research by hundreds of scientists (see **THE SCIENCE BEHIND THE STORY**, pp. 470–471) revealed that certain airborne chemicals destroy ozone and that most of these **ozone-depleting substances** are human-made. Our subsequent campaign to halt degradation of the ozone layer stands as one of society's most successful efforts addressing a major environmental problem.

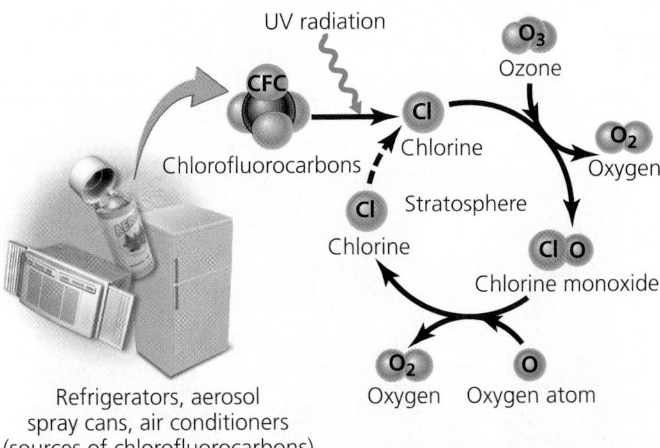

FIGURE 17.24 CFCs destroy ozone in a multistep process, repeated many times. A chlorine atom released from a CFC molecule in the presence of UV radiation reacts with an ozone molecule, forming one molecule of oxygen gas and one chlorine monoxide (ClO) molecule. The oxygen atom of the ClO molecule then binds with a stray oxygen atom to form oxygen gas, leaving the chlorine atom to begin the destructive cycle anew. In this way, a single chlorine atom can destroy up to 100,000 ozone molecules.

Synthetic chemicals deplete stratospheric ozone

Researchers identifying ozone-depleting substances pinpointed primarily **halocarbons**—human-made compounds derived from simple hydrocarbons (p. 28), such as ethane and methane, in which hydrogen atoms are replaced by halogen atoms, such as chlorine, bromine, or fluorine. Industry was mass-producing one type of halocarbon, **chlorofluorocarbons (CFCs)**, at a rate of a million tons per year in the early 1970s, and this rate was growing by 20% a year. CFCs were useful as refrigerants, fire extinguishers, propellants for aerosol spray cans, cleaners for electronics, and for making polystyrene foam. Because CFCs rarely reacted with other chemicals, scientists surmised that they would be harmless to people and the environment.

Alas, the nonreactive qualities that made CFCs ideal for industrial purposes were having disastrous consequences for the ozone layer. Whereas reactive chemicals are broken down quickly in the troposphere, CFCs reach the stratosphere unchanged and can linger there for a century or more. In the stratosphere, intense UV radiation from the sun eventually breaks CFCs into their constituent chlorine and carbon atoms. In a two-step chemical reaction (FIGURE 17.24), each newly freed chlorine atom can split an ozone molecule and then ready itself to split another one. During its long residence time in the stratosphere, each free chlorine atom can catalyze the destruction of as many as 100,000 ozone molecules!

The Antarctic ozone hole appears each spring

In 1985, researchers shocked the world by announcing that stratospheric ozone levels over Antarctica in springtime had declined by nearly half in just the previous decade, leaving a thinned ozone concentration that was soon named the **ozone hole** (FIGURE 17.25). During each Southern Hemisphere spring since then, ozone concentrations over this immense region have dipped to roughly half their historic levels.

Extensive scientific detective work has revealed why seasonal ozone depletion is so severe over Antarctica (and to a lesser extent, the Arctic). During the dark and frigid Antarctic winter, temperatures in the stratosphere dip below –80°C (–112°F), enabling unusual high-altitude *polar stratospheric clouds* to form. Many of these icy clouds contain condensed nitric acid, which splits chlorine atoms off from compounds such as CFCs. The freed chlorine atoms accumulate in the clouds, trapped over Antarctica by wind currents that swirl in a circular *polar vortex* that prevents air from mixing with the rest of Earth's atmosphere.

In the Antarctic spring (starting in September), sunshine returns, and UV radiation dissipates the clouds. This releases the chlorine atoms, which begin destroying ozone. The solar radiation also catalyzes chemical reactions, speeding up ozone depletion as temperatures warm. The ozone hole lingers over Antarctica until December, when warmed air shuts down the polar vortex, allowing ozone-depleted air to diffuse away and ozone-rich air from elsewhere to stream in. The ozone hole vanishes until the following spring.

By the time scientists had worked most of this out, plummeting ozone levels were becoming a serious international concern. Already worried that intensified UV exposure at Earth's surface would lead to more skin cancer, scientists were also predicting damage to crops and to ocean phytoplankton, the base of the marine food chain.

FIGURE 17.25 The "ozone hole" is a vast area of thinned ozone density in the stratosphere over the Antarctic region. It has reappeared seasonally each September in recent decades. This colorized satellite imagery of Earth's Southern Hemisphere from September 24, 2006, shows the ozone hole (**purple/blue**) at its maximal recorded extent to date.

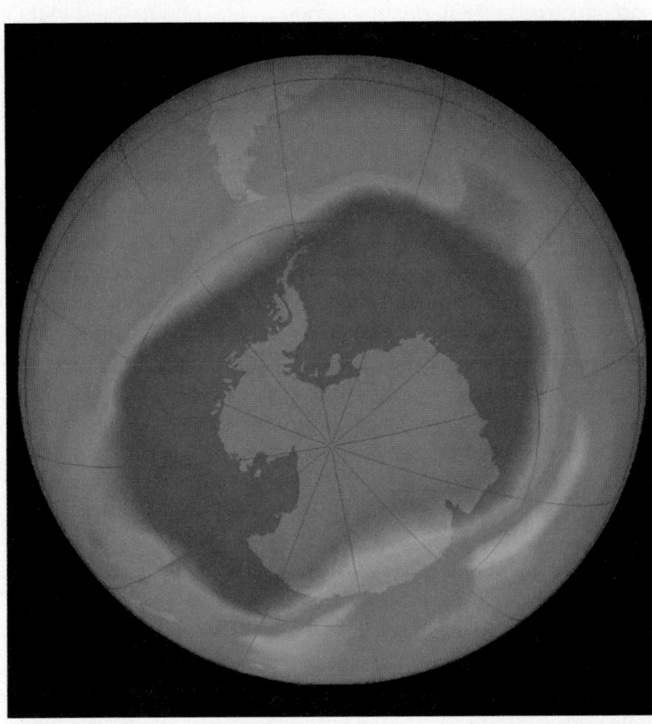

Discovering Ozone Depletion and the Substances Behind It

Mario Molina with F. Sherwood Rowland (L) and Paul Crutzen (R) upon their receipt of the Nobel Prize

In discovering the depletion of stratospheric ozone and coming to understand the roles of halocarbons and other substances, scientists have relied on historical records, field observations, laboratory experiments, computer models, and satellite technology.

The story starts back in 1924, when British scientist G.M.B. Dobson built an instrument that measured atmospheric ozone concentrations by sampling sunlight at ground level and comparing the intensities of wavelengths that ozone does and does not absorb. By the 1970s, the Dobson ozone spectrophotometer was being used by a global network of observation stations.

Meanwhile, atmospheric chemists were learning how stratospheric ozone is created and destroyed. Ozone and oxygen exist in a natural balance, with one occasionally reacting to form the other, and oxygen being far more abundant. Researchers found that certain chemicals naturally present in the atmosphere, such as hydroxyl (OH) and nitric oxide (NO), destroy ozone, keeping the ozone layer thinner than it would otherwise be. And nitrous oxide (N_2O) produced by soil bacteria can make its way to the stratosphere and produce NO, Dutch meteorologist Paul Crutzen reported in 1970. This last observation was important, because some human activities, such as fertilizer application, were increasing emissions of N_2O.

Following Crutzen's report, American scientists Richard Stolarski and Ralph Cicerone showed in 1973 that chlorine atoms can catalyze the destruction of ozone even more effectively than N_2O can. And two years earlier, British scientist James Lovelock had developed an instrument to measure trace amounts of atmospheric gases and found that virtually all the chlorofluorocarbons (CFCs) humanity had produced in the past four decades were still aloft, accumulating in the stratosphere.

This set the stage for the key insight. In 1974, American chemist F. Sherwood Rowland and his Mexican postdoctoral associate Mario Molina took note of all the preceding research and realized that CFCs were rising into the stratosphere, being broken down by UV radiation, and releasing chlorine atoms that ravaged the ozone layer (see Figure 17.24, p. 469). Molina and Rowland's analysis, published in the journal *Nature*, earned them the 1995 Nobel prize in chemistry jointly with Crutzen.

The paper also sparked discussion about setting limits on CFC emissions. Industry leaders attacked the research; DuPont's chairman of the board reportedly called it "a science fiction tale . . . a load of rubbish . . . utter nonsense." But measurements in the lab and in the stratosphere by numerous researchers soon confirmed that CFCs and other halocarbons were indeed depleting ozone. In response, the United States and several other nations banned the use of CFCs in aerosol spray cans in 1978. Other uses continued, however, and by the early 1980s global production of CFCs was again on the rise.

Then, a shocking new finding spurred the international community to take action. In 1985, Joseph Farman and colleagues analyzed data from a British research station in Antarctica that had been recording ozone concentrations since the 1950s. Farman's team reported in *Nature* that springtime Antarctic ozone concentrations had plummeted by 40–60% just since the 1970s (**FIGURE 1a**).

Farman's team had beaten a group of NASA scientists to the punch. The NASA scientists were sitting on reams of data from satellites showing a global drop in ozone levels (**FIGURE 1b**), but they had not yet submitted their analysis for publication.

But why should an "ozone hole" be localized over Antarctica, of all places? And why only in the southern spring? To determine what was causing this odd phenomenon, atmospheric chemists Susan Solomon, James Anderson, Crutzen, and others mounted expeditions in 1986 and 1987 to analyze atmospheric gases using ground stations and high-altitude balloons and aircraft. From their data they figured out how the region's polar stratospheric clouds and circulating winds provide ideal conditions for chlorine from CFCs and other chemicals to set in motion the destruction of massive amounts of ozone.

By 1987, the mass of scientific evidence helped convince the world's nations to agree on the Montreal Protocol, which aimed to cut CFC production in half by 1998. Within two years, further scientific evidence and computer modeling showed that more drastic measures were needed. In 1990, the Montreal Protocol was strengthened to include a complete phaseout of CFCs by 2000, in the first of several follow-up agreements. Today, amounts of ozone-depleting substances in the stratosphere are beginning to level off.

As the ozone layer begins a long-term recovery, scientists continue their research. In 2009, a team led by A.R. Ravishankara of the National Oceanic and Atmospheric Administration determined that nitrous oxide (N_2O) had now become the leading cause of ozone depletion (**FIGURE 2**). Its emissions are not regulated, so its impacts have come to surpass those that the remaining halocarbons currently exert. Ravishankara's team points out that regulating nitrous oxide, which is also a potent greenhouse gas, would help mitigate climate change as well as speed ozone recovery. ■

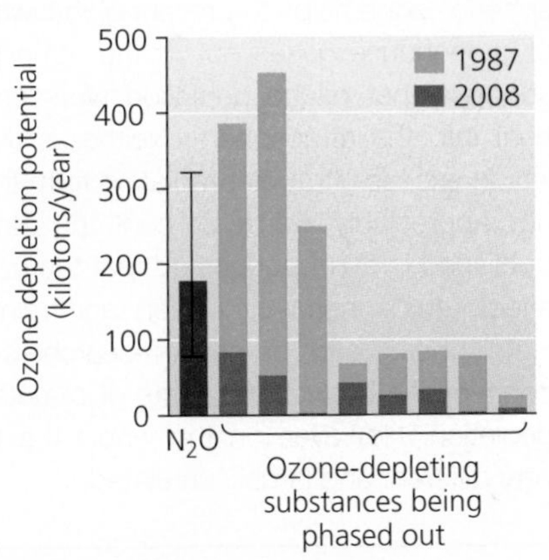

FIGURE 2 Since most ozone-depleting substances were phased out beginning in 1987, nitrous oxide (N_2O; left bar) has become the primary ozone-depleting substance we emit. It has less impact than CFCs and other halocarbons did in 1987 (**full-bar values**), but more impact than any other substance today (**lower portion of bars**). *Adapted from Ravishankara, A.R., et al., 2009. Nitrous oxide (N_2O): The dominant ozone-depleting substance emitted in the 21st century.* Science *326: 123–125, Fig 1. Reprinted by permission of AAAS and the author.*

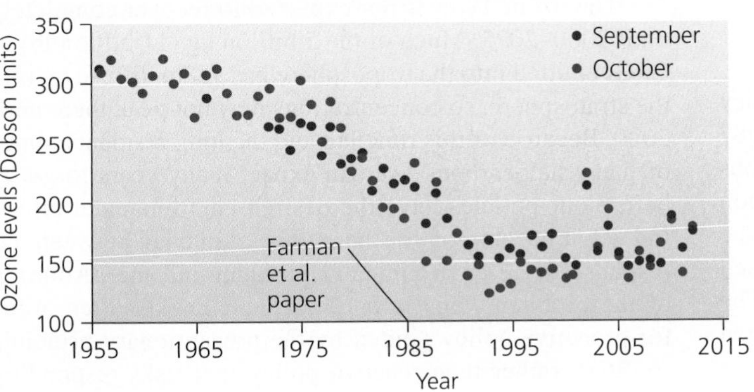

(a) Monthly mean ozone levels at Halley, Antarctica

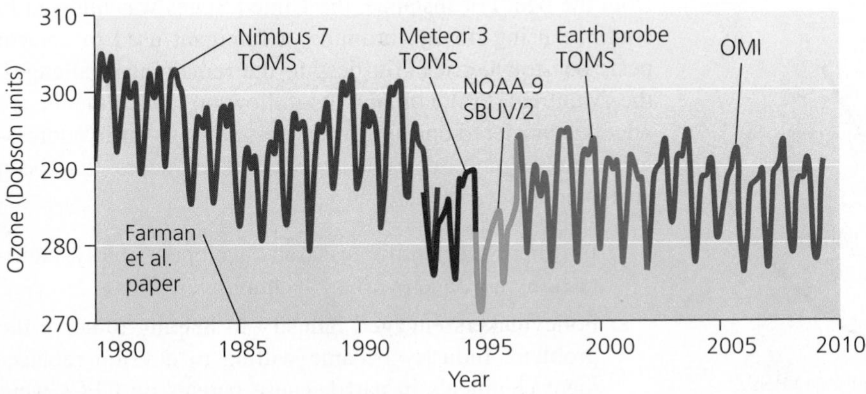

(b) Global ozone readings from 5 satellites

FIGURE 1 Monitoring shows ozone depletion. Data from Halley, Antarctica **(a)**, show a decrease in stratospheric ozone concentrations from the 1960s to 1990. Once ozone-depleting substances began to be phased out, ozone concentrations stopped declining. Ozone decline and stabilization are also evident globally **(b)**, as seen in data from five satellites. *Data from (a) British Antarctic Survey; and (b) NASA.*

FAQ Is the ozone hole related to global warming?

This is a common misconception held by the public. Some people are under the impression that the depletion of stratospheric ozone helps to prevent global warming by letting heat or greenhouse gases out of the atmosphere. Other people suppose that ozone depletion worsens warming by letting heat into the atmosphere. Neither is true. Ozone depletion lets in excess ultraviolet radiation from the sun, but this does not appreciably warm or cool the atmosphere. Research published in 2011 suggested that the ozone hole may in fact affect atmospheric circulation and rainfall in the Southern Hemisphere, and perhaps researchers will discover other connections between aspects of climate change and ozone depletion. However, on the whole the two phenomena are very different and largely unrelated.

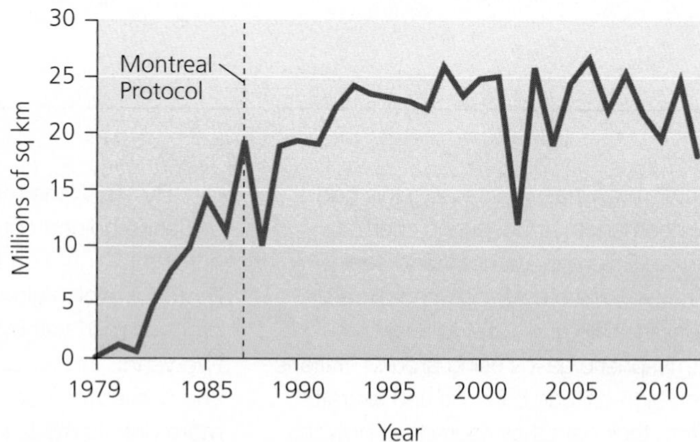

FIGURE 17.27 The Antarctic ozone hole grew quickly after its appearance, but phase-outs of ozone-depleting substances beginning in 1987 have halted its growth. *Data from NASA, reflecting averages from 7 Sept. to 13 Oct. each year.*

We addressed ozone depletion with the Montreal Protocol

In response to the scientific concerns, international policy efforts to restrict production of CFCs bore fruit in 1987 with the **Montreal Protocol.** In this treaty, signatory nations (eventually numbering 196) agreed to cut CFC production in half by 1998. Five follow-up agreements deepened the cuts, advanced timetables for compliance, and addressed additional ozone-depleting substances (FIGURE 17.26). The substances covered by these agreements have now been mostly phased out, and industry has been able to shift to safer alternative chemicals that are inexpensive and efficient. As a result, we have evidently stopped the Antarctic ozone hole from growing worse (FIGURE 17.27)—a success that all humanity can celebrate.

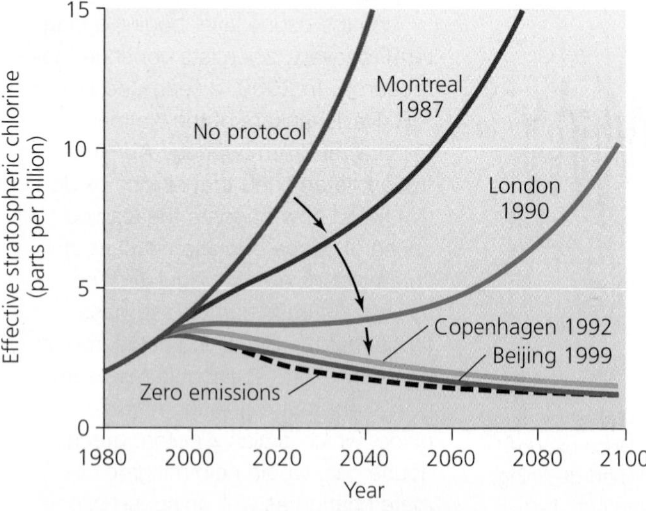

FIGURE 17.26 The Montreal Protocol reduced stratospheric ozone depletion, and follow-up agreements in London, Copenhagen, and Beijing reduced it still more. In this graph, the *y*-axis values give a collective measure of ozone-destructive potential from all substances. *Data from Emmanuelle Bournay, UNEP/GRID-Arendal, http://maps.grida.no/go/graphic/effects-of-the-montreal-protocol-amendment-and-their-phase-out-schedules.*

The ozone layer is not expected to recover completely until 2060–2075. Much of the 5 billion kg (11 billion lb) of CFCs emitted into the troposphere has yet to diffuse up into the stratosphere, so concentrations may not peak there until 2020. Because of this time lag and the long residence times of many halocarbons, we can expect many years to go by before our policies have the desired environmental effect. Indeed, it is common for there to be a time lag between our response to an environmental problem and the resolution of the problem. This is one reason scientists often argue for proactive policy guided by the precautionary principle (p. 265), rather than reactive policy that risks responding too late.

One challenge in restoring the ozone layer is that nations can plead for some ozone-depleting substances to be exempt from the ban. For instance, the United States was allowed to continue using methyl bromide, a fumigant used to control pests on strawberries. Yet despite the remaining challenges, the Montreal Protocol and its follow-up amendments are widely considered our biggest success story so far in addressing a global environmental problem. The success has been attributed to several factors:

1. Informative scientific research developed rapidly, facilitated by new and evolving technologies.
2. Policymakers engaged industry in helping to solve the problem. Industry became willing to develop replacement chemicals in part because patents on CFCs were running out and firms wanted to position themselves to profit from next-generation chemicals.
3. Implementation of the Montreal Protocol after 1987 followed an adaptive management approach (p. 315), adjusting strategies midstream in response to new scientific data, technological advances, or economic figures.

Because of its success in addressing ozone depletion, the Montreal Protocol is widely viewed as a model for international cooperation in addressing other pressing global problems, from biodiversity loss (p. 297) to persistent organic pollutants (p. 384) to climate change (pp. 510–511).

Addressing Acid Deposition

Just as stratospheric ozone depletion crosses political boundaries, so does another atmospheric pollution concern—acid deposition. As with ozone depletion, we are seeing some success in addressing this challenge.

Fossil fuel pollution spreads acidic substances widely

Acid deposition (or *acidic deposition*) refers to the deposition of acidic (pp. 27–28) or acid-forming pollutants from the atmosphere onto Earth's surface. This can take place either by precipitation (commonly referred to as **acid rain,** but also including acid snow, sleet, and hail), by fog, by gases, or by the deposition of dry particles. Acid deposition is one type of **atmospheric deposition,** which refers more broadly to the wet or dry deposition on land of a wide variety of pollutants, including mercury, nitrates, organochlorines, and others.

Acid deposition originates primarily with the emission of sulfur dioxide and nitrogen oxides, largely through fossil fuel combustion by automobiles, electric utilities, and industrial facilities. Once airborne, these pollutants can react with water, oxygen, and oxidants to produce compounds of low pH (p. 28), primarily sulfuric acid and nitric acid. Suspended in the troposphere, droplets of these acids may travel days or weeks for hundreds of kilometers before falling in precipitation (**FIGURE 17.28**). Depending on climate, 20% to 80% of all acidic compounds emitted into the atmosphere may fall in precipitation, with the remainder falling as dry deposition.

Acid deposition has many impacts

Acid deposition has wide-ranging, cumulative detrimental effects on ecosystems and on our infrastructure (**TABLE 17.1**). Acids leach nutrients such as calcium, magnesium, and potassium ions from the topsoil, altering soil chemistry and harming plants and soil organisms. This occurs because hydrogen ions from acid precipitation take the place of calcium, magnesium, and potassium ions in soil compounds, and these valuable nutrients leach into the subsoil, where they become inaccessible to plant roots.

Acid precipitation also "mobilizes" toxic metal ions such as aluminum, zinc, mercury, and copper by chemically converting them from insoluble forms to soluble forms. Elevated soil concentrations of metal ions such as aluminum weaken plants by damaging root tissue, hindering their uptake of water and nutrients. In some areas, acid fog with a pH of 2.3 (equivalent to vinegar, and

TABLE 17.1 Impacts of Acid Deposition

Acid deposition in northeastern forests has . . .
• Accelerated leaching of base cations (ions such as Ca^{2+}, Mg^{2+}, NA^{+}, and K^{+}, which counteract acid deposition) from soil
• Allowed sulfur and nitrogen to accumulate in soil, where excess N can overfertilize native plants and encourage weeds
• Increased dissolved inorganic aluminum in soil, hindering plant uptake of water and nutrients
• Leached calcium from needles of red spruce, causing trees to die from wintertime freezing
• Increased mortality of sugar maples due to leaching of base cations from soil and leaves
• Acidified 41% of Adirondack, New York, lakes and 15% of New England lakes
• Diminished lakes' capacity to neutralize further acids
• Elevated aluminum levels in surface waters
• Reduced species diversity and abundance of aquatic life, affecting entire food webs

Source: Adapted from Driscoll, C.T., et al., 2001. Acid rain revisited. *Hubbard Brook Research Foundation. © 2001 C.T. Driscoll. Used with permission.*

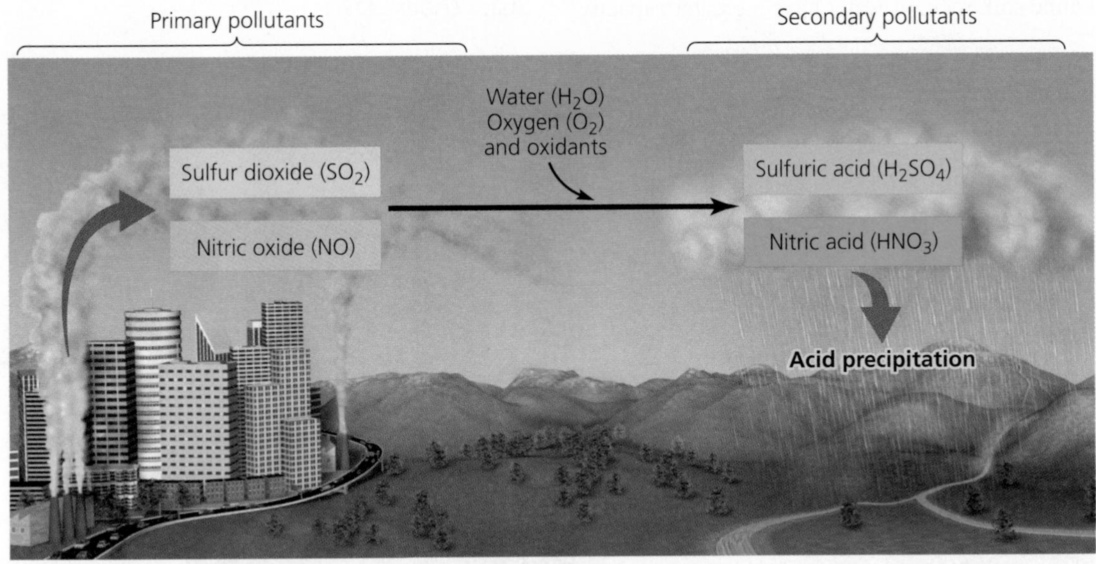

FIGURE 17.28 Acid deposition can have consequences far downwind from its source. Sulfur dioxide and nitric oxide emitted by industries and utilities are transformed into sulfuric acid and nitric acid through chemical reactions in the atmosphere. These acidic compounds descend to Earth's surface in rain, snow, fog, and dry deposition.

to buffer themselves against acid deposition. This also means that once calcium or similar ions are leached from a soil, the soil becomes more sensitive to acidification.

Besides altering natural ecosystems, acid precipitation damages crops, erodes stone buildings, corrodes cars, and erases the writing from tombstones. Ancient cathedrals in Europe, monuments in Washington, D.C., temples in Asia, and stone statues in London are experiencing billions of dollars of damage as their features dissolve away (**FIGURE 17.30**).

Because the pollutants leading to acid deposition can travel long distances, their effects may be felt far from their sources. For instance, much of the pollution from power plants and factories in Pennsylvania, Ohio, and Illinois falls out in states to their east, including New York, Vermont, and New Hampshire, as well as in regions of Canada to the north. As a result, regions of greatest acidification tend to be downwind from heavily industrialized source areas of pollution.

FIGURE 17.29 Acid deposition killed these trees on Mount Mitchell in North Carolina.

over 1000 times more acidic than normal rainwater) has enveloped forests for extended periods, killing trees (**FIGURE 17.29**).

When acidic water runs off from land, it affects streams, rivers, and lakes. Thousands of lakes in Canada, Scandinavia, the United States, and elsewhere have lost their fish because acid precipitation leaches aluminum ions out of soil and rock and into waterways, where they damage the gills of fish and disrupt their salt balance, water balance, breathing, and circulation. Terrestrial animals are affected, too; populations of snails and other invertebrates typically decline, and this reduces the food supply for birds.

The severity of all these effects depends not only on the pH of the deposition, but also on the *acid-neutralizing capacity* of the soil, rock, or water that receives the acidic input. Substrates differ naturally in their chemistry and pH, and regions with more alkaline soil, rock, or water have a greater capacity

We are addressing acid deposition

The emissions trading program for sulfur dioxide established by the Clean Air Act of 1990 has helped us address acid deposition. The economic incentives created by this cap-and-trade program have encouraged polluters to invest in technologies such as scrubbers (p. 461) and to devise other ways to become cleaner and more efficient. As a result, SO_2 emissions across the United States have fallen by 67%, and the program is now preventing the emission of 10 million tons of SO_2 per year (see Figure 7.16, p. 183). The EPA has calculated that the program's economic benefits outweigh its costs by 40 to 1.

As a result of declining SO_2 emissions, average sulfate loads in precipitation across the eastern United States were 51% lower in 2008–2010 than in 1989–1991. Emissions of NO_X also have been lowered significantly, thanks to the emissions-trading program, and wet nitrogen deposition declined between these periods as well. As a result, air and water quality has improved throughout the eastern United States (**FIGURE 17.31**).

(a) Before acid rain damage

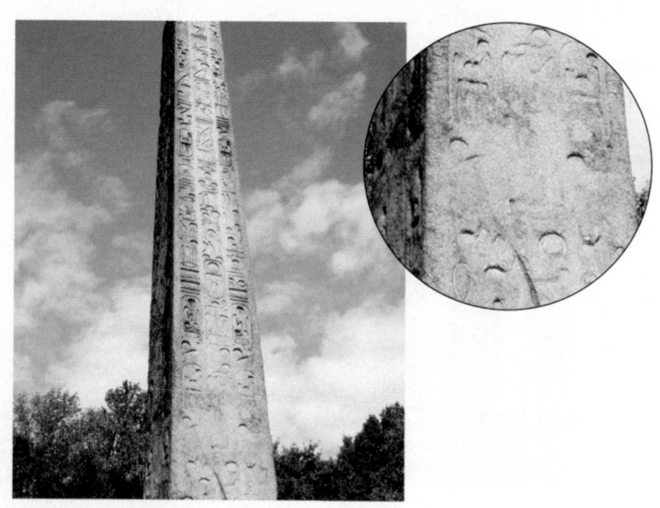

(b) After acid rain damage

FIGURE 17.30 Acid deposition corrodes statues and buildings. Shown is an Egyptian obelisk known as Cleopatra's Needle, in Central Park, New York City, **(a)** before and **(b)** after the onset of significant acid deposition.

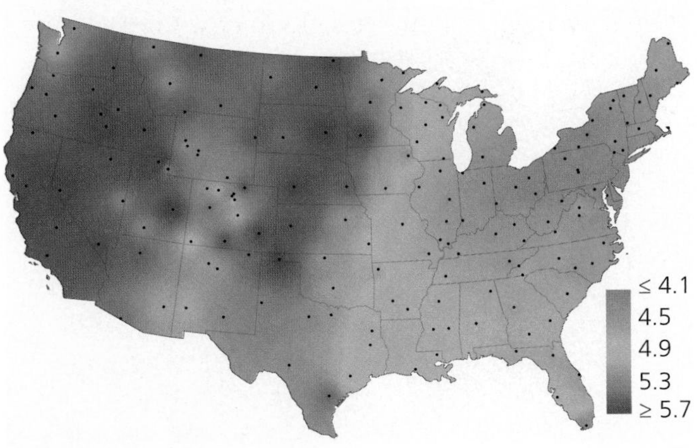

(a) Acid deposition in 1990

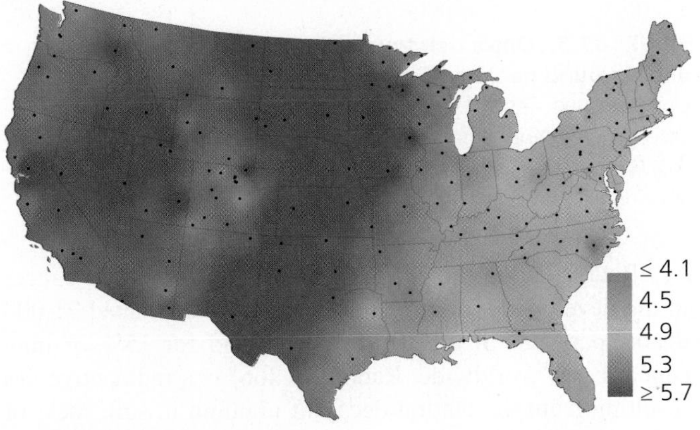

(b) Acid deposition in 2011

FIGURE 17.31 **Precipitation has become less acidic as a result of air quality improvements following the Clean Air Act.** Average pH values for precipitation have risen between **(a)** 1990 and **(b)** 2011. Precipitation remains most acidic in the Northeast and Midwest, near and downwind from (roughly east of) areas of heavy industry. *Data from the National Atmospheric Deposition Program.*

DATA Q In the area where you live, how did the pH of precipitation change between 1990 and 2011? Has precipitation become more acidic or less acidic?

At Hubbard Brook Experimental Forest in New Hampshire, where scientists first studied acid deposition's effects in the United States, researchers jumpstarted a long, slow recovery by using a helicopter to distribute 50 tons of a calcium-containing mineral called wollastonite over one watershed. Within three years of this experimental application, topsoil pH rose from 3.9 to 4.2. Sugar maples (one of the forest's key tree species that had been declining) are now producing healthier foliage, thicker root growth, more seeds, and more surviving seedlings. Over the next 50 years, scientists plan to evaluate the impact of calcium addition on the watershed's soil, water, and life, and compare these results to watersheds where calcium remains depleted.

As with ozone depletion, there is a time lag before the positive consequences of emissions cuts kick in, so it will take time for acidified ecosystems to recover. Research in 2012 indicated that soils across the northeastern United States are showing signs of recovery, but that they remain degraded and vulnerable. Moreover, scientists point out that further pollution reductions are needed if we are to fully restore ecosystems in the Northeast and prevent further damage to property and infrastructure.

While the United States, Canada, and Western Europe are beginning to recover from acid deposition after cutting sulfur emissions, acid deposition is becoming worse in industrializing nations. Today China emits the most sulfur dioxide of any nation and has the world's worst acid rain problem, as a result of extensive coal combustion in power plants and factories that often lack effective pollution control equipment. The government is tackling the issue, but it faces a challenge as the nation's industrial sector continues to expand by leaps and bounds.

Overall, data on acid deposition show that we have made advances in controlling outdoor air pollution, but that more can be done. The same can be said for indoor air pollution, a source of human health threats that is less familiar to most of us, but statistically more dangerous.

Indoor Air Quality

Indoor air generally contains higher concentrations of pollutants than does outdoor air. As a result, the health impacts from **indoor air pollution** in workplaces, schools, and homes outweigh those from outdoor air pollution. The World Health Organization (WHO) attributes nearly 3.5 million premature deaths each year to indoor air pollution (compared with 3.3 million for outdoor air pollution). Indoor air pollution takes nearly 10,000 lives each day.

If this seems surprising, consider that the average U.S. citizen spends at least 90% of his or her time indoors. Then consider the dizzying array of consumer products in our homes and offices that play major roles in our daily lives. Many of these products are made of synthetic materials, and novel synthetic substances are not comprehensively tested for health effects before being brought to market (Chapter 14). Products and materials as diverse as cleaning fluids, insecticides, furniture, carpeting, and the many varieties of plastics all exude volatile chemicals into the air.

Ironically, some attempts to be environmentally prudent during the "energy crises" of the 1970s (p. 544) worsened indoor air quality. To improve energy efficiency by reducing heat loss, building managers sealed off ventilation in buildings, and designers constructed new buildings with limited ventilation and with windows that did not open. These steps saved energy, but they also worsened indoor air quality by trapping stable, unmixed air—and pollutants—inside.

There is good news, however. In both developing and developed nations, we have known and feasible ways to address the primary causes of indoor air pollution.

Burning fuelwood causes indoor pollution in the developing world

Indoor air pollution has by far the greatest impact in the developing world, where poverty forces millions of people

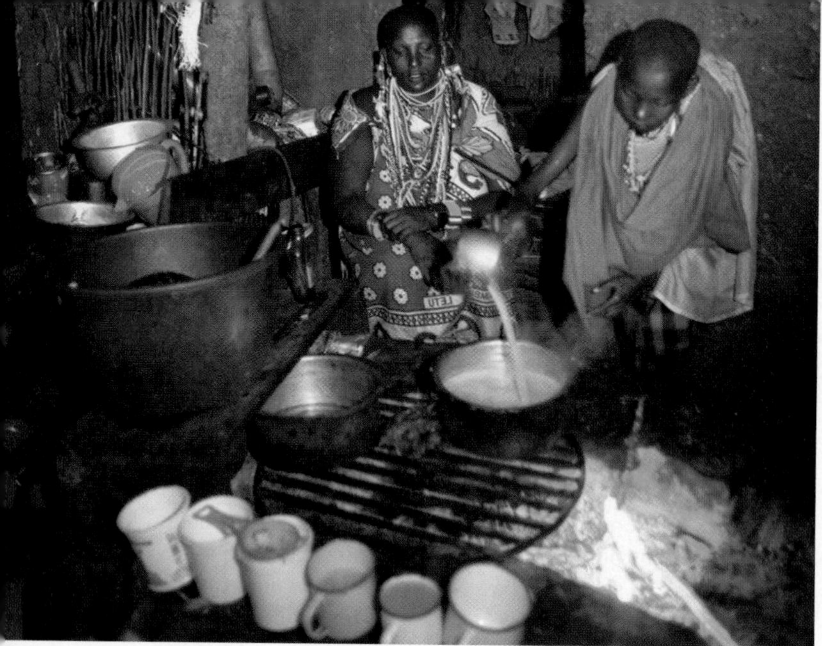

FIGURE 17.32 **In the developing world, many people build fires inside their homes for cooking and heating, as seen here in a Maasai home in Kenya.** Indoor fires expose people to severe pollution from particulate matter and carbon monoxide.

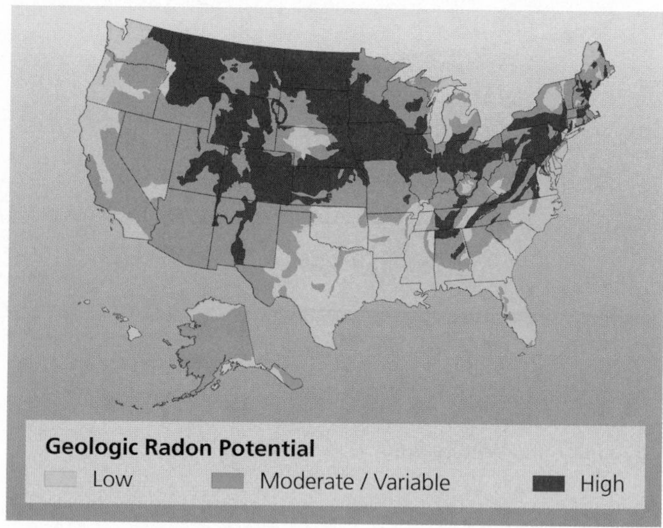

FIGURE 17.33 **One's risk from radon depends largely on underground geology.** This map shows levels of risk across the United States. Testing your home for radon is the surest way to determine whether this colorless, odorless gas could be a problem. *Data from U.S. Geological Survey, 1993.* Generalized geological radon potential of the United States, *1993.*

to burn wood, charcoal, animal dung, or crop waste inside their homes for cooking and heating, with little or no ventilation (FIGURE 17.32). In the air of such homes, the WHO has found that concentrations of particulate matter are commonly 20 times above U.S. EPA standards. As a result, people inhale dangerous amounts of soot, carbon monoxide, and other pollutants, which together increase risks of premature death by pneumonia, bronchitis, and lung cancer, as well as allergies, sinus infections, cataracts, asthma, emphysema, and heart disease. International health researchers estimate that indoor air pollution from burning fuelwood, dung, and coal kills 3.5 million people each year, comprising nearly 7% of all deaths. Many people are not aware of the health risks, and of those who are, many are too poor to have viable alternatives.

Tobacco smoke and radon are the primary indoor pollutants in industrialized nations

In industrialized nations, the primary indoor air health risks are cigarette smoke and radon, a naturally occurring radioactive gas. Smoking cigarettes irritates the eyes, nose, and throat; worsens asthma and other respiratory ailments; and greatly increases the risk of lung cancer. Inhaling secondhand smoke (smoke inhaled by a nonsmoker who is nearby or shares an enclosed airspace with a smoker) causes many of the same problems. This hardly seems surprising when one considers that tobacco smoke is a brew of over 4000 chemical compounds, some of which are known or suspected to be toxic or carcinogenic. Smoking has become less popular in developed nations in recent years as a result of public education campaigns, and many public and private venues now ban smoking. Still, smoking is estimated in the United States alone to cause 160,000 lung cancer deaths per year, and secondhand smoke to cause 3000.

Radon gas is the second-leading cause of lung cancer in the developed world, responsible for an estimated 21,000 deaths per year in the United States and for 15% of lung cancer cases worldwide. Radon (p. 366) is a radioactive gas resulting from the natural decay of uranium in soil, rock, or water. It seeps up from the ground and can infiltrate buildings. Colorless and odorless, radon's presence can be impossible to predict without knowing an area's underlying geology (FIGURE 17.33). The only way to determine whether radon is entering a building is to sample air with a test kit. The EPA estimates that 6% of U.S. homes exceed its safety standard for radon. Since the 1980s, millions of U.S. homes have been tested for radon and close to a million have undergone radon mitigation. New homes are being built with radon-resistant features.

Many VOCs pollute indoor air

In our daily lives at home, we are exposed to many indoor air pollutants (FIGURE 17.34). The most diverse are volatile organic compounds (p. 459). These airborne carbon-containing compounds are released by plastics, oils, perfumes, paints, cleaning fluids, adhesives, and pesticides. VOCs evaporate from furnishings, building materials, color film, carpets, laser printers, fax machines, and sheets of paper. Some products, such as chemically treated furniture, release large amounts of VOCs when new and progressively less as they age. Other items, such as photocopying machines, emit VOCs each time they are used.

Although we are surrounded by products that emit VOCs, they are released in very small amounts. Studies have found total levels of VOCs in buildings to be nearly always less than 1 part per 10 million. This is, however, a much greater concentration than is found outdoors. Moreover, we experience instances of especially high exposure. The "new car smell" that fills the interiors of new automobiles comes from

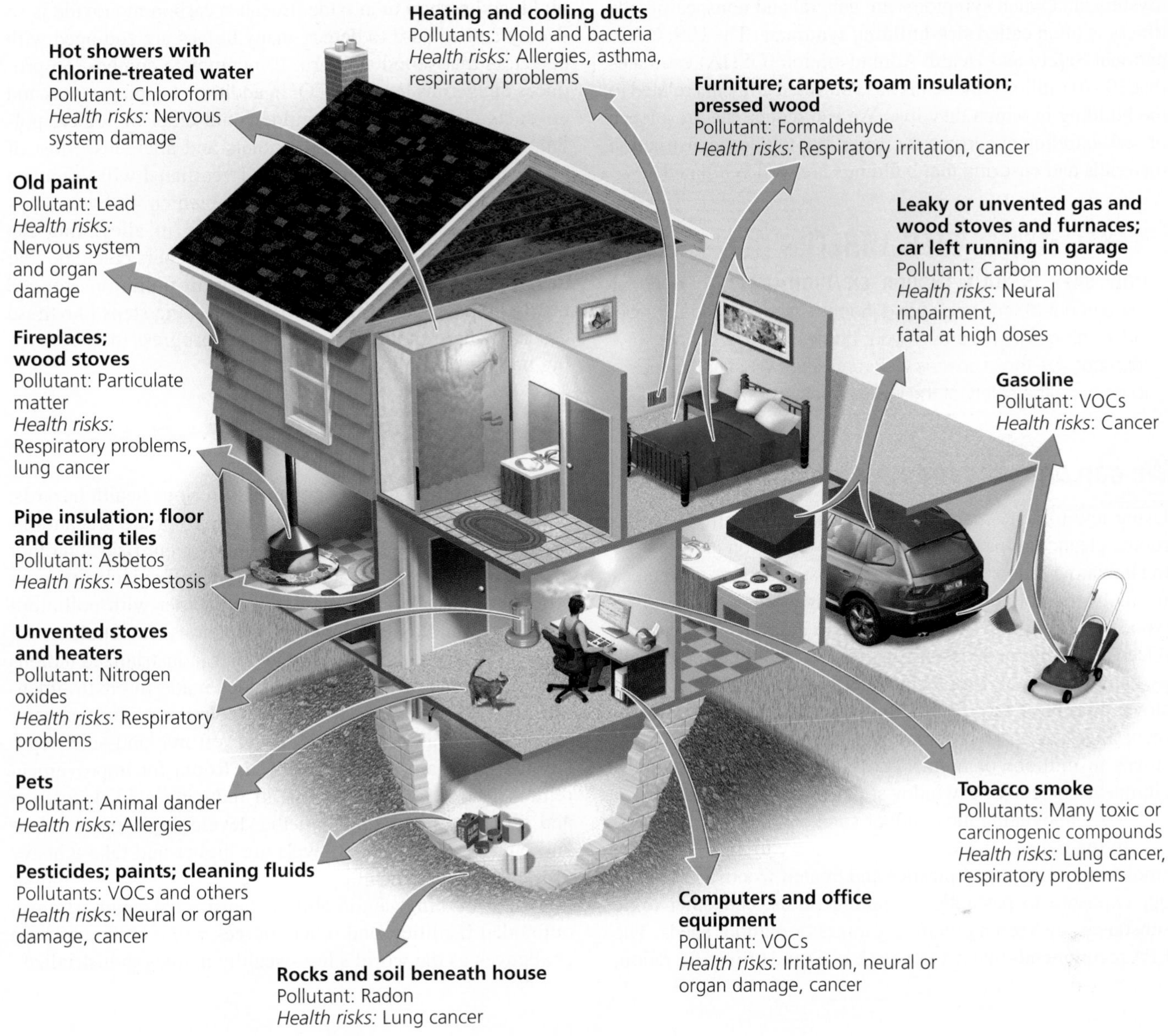

FIGURE 17.34 The typical home contains many sources of indoor air pollution. Shown are common sources, the major pollutants they emit, and some of the health risks they pose.

a complex mix of dozens of VOCs as they outgas from the newly manufactured plastic, metal, and leather components of the car. The smell diminishes with time, but some scientific studies warn of health risks from this brew and recommend that you keep a new car well ventilated.

The implications for human health of chronic exposure to volatile organic compounds are far from clear. Because they generally exist in low concentrations and because individuals regularly are exposed to mixtures of many different types, it is extremely difficult to study the effects of any one pollutant. An exception is *formaldehyde*, a VOC that has clear and known health impacts. Widely used in pressed wood, insulation, and other products, formaldehyde irritates mucous membranes, induces skin allergies, and causes a number of other ailments. The use of plywood has decreased in the last decade because of health concerns over formaldehyde.

Living organisms can pollute

The most widespread source of indoor air pollution in the developed world may be tiny living organisms. Dust mites and animal dander can worsen asthma in children. The airborne spores of some fungi, molds, and mildews can cause allergies, asthma, and other respiratory ailments. Some airborne bacteria can cause infectious disease, including Legionnaires' disease. Of the estimated 10,000–15,000 annual U.S. cases of Legionnaires' disease, 5–15% are fatal. Heating and cooling systems in buildings make ideal breeding grounds for microbes, providing moisture, dust, and foam insulation as substrates, along with air currents to carry the organisms aloft.

Microbes that induce allergic responses are thought to be a major cause of *building-related illness*, a sickness produced by indoor pollution. When the cause of such an illness is a mystery, and when symptoms are general and nonspecific, the illness is often called **sick-building syndrome.** The U.S. Occupational Safety and Health Administration (OSHA) estimates that 30–70 million Americans have suffered ailments related to the building in which they live. We can reduce the prevalence of sick-building syndrome by using low-toxicity construction materials and ensuring that buildings are well ventilated.

WEIGHING THE ISSUES

HOW SAFE IS YOUR INDOOR ENVIRONMENT? Think about the amount of time you spend indoors. Name some potential indoor air quality hazards in your home, work, or school environment. Are these spaces well ventilated? What could you do to improve the safety of the indoor spaces you use?

We can enhance indoor air quality

Using low-toxicity materials, monitoring air quality, keeping rooms clean, and providing adequate ventilation are the keys to alleviating indoor air pollution in most situations. Remedies for fuelwood pollution in the developing world include drying wood before burning (which reduces the amount of smoke produced), cooking outside, shifting to less-polluting fuels (such as natural gas), and replacing inefficient fires with cleaner stoves that burn fuel more efficiently. The Chinese government invested in a program that has placed fuel-efficient stoves in millions of homes across China. Installing hoods, chimneys, or cooking windows can increase ventilation for little cost, alleviating most indoor smoke pollution.

In the industrialized world, we can try to avoid cigarette smoke, limit our use of plastics and treated wood, and restrict our exposure to pesticides, cleaning fluids, and other toxic substances by keeping them in garages or outdoor sheds. The EPA recommends that we test our homes and offices for radon, mold, and carbon monoxide. Because carbon monoxide is so deadly and so hard to detect, many homes are equipped with detectors that sound an alarm if incomplete combustion produces dangerous levels of CO. In addition, keeping rooms and air ducts clean and free of mildew and other biological pollutants will reduce potential irritants and allergens. Most of all, keeping our indoor spaces well ventilated will minimize concentrations of the pollutants among which we live.

Progress is being made worldwide in alleviating the health toll of indoor air pollution. Researchers calculate that rates of premature death from indoor air pollution dropped nearly 40 percent from 1990 to 2010. Taking steps like those described here should bring us further progress in safeguarding people's health.

Conclusion

Indoor air pollution poses potentially serious health hazards, but by keeping informed and taking appropriate precautions on a personal basis, we each can minimize our risks. Outdoor air pollution has been addressed more effectively by government legislation and regulation, together with pollution-control technologies. Indeed, reductions in outdoor air pollution in the United States and other industrialized nations represent some of the greatest strides made in environmental protection to date. The global depletion of stratospheric ozone has been halted thanks to our efforts, and acid deposition is gradually being addressed. Room for improvement remains, however, particularly in reducing acid deposition and photochemical smog. In the developing world, indoor and outdoor air pollutant levels are higher and take a heavy toll on people's health. Reducing pollution from indoor fuelwood burning, automobile exhaust, coal combustion in outmoded facilities, and other sources will continue to pose challenges as the world's less-wealthy nations industrialize.

Reviewing Objectives

You should now be able to:

Describe the composition, structure, and function of Earth's atmosphere

- The atmosphere moderates climate, provides us oxygen, conducts and absorbs solar radiation, and transports and recycles nutrients and waste. (p. 450)
- The atmosphere consists of 78% nitrogen gas, 21% oxygen gas, and a variety of other gases in minute concentrations. (p. 450)
- The atmosphere includes four layers: the troposphere, stratosphere, mesosphere, and thermosphere. Temperature and other characteristics vary across these layers. Ozone is concentrated in the stratosphere. (pp. 450–451)

Relate weather and climate to atmospheric conditions

- The sun's energy heats the atmosphere, drives air circulation, and helps determine weather, climate, and the seasons. (pp. 452–453)
- Weather is a short-term phenomenon, whereas climate is a long-term phenomenon. Fronts, pressure systems, and the interactions among air masses influence weather. (pp. 453–454)
- Global convective cells called Hadley, Ferrel, and polar cells create latitudinal climate zones. (p. 455)
- Hurricanes and tornadoes are types of cyclonic storms that can threaten life and property. (p. 456)

- **Identify major pollutants, outline the scope of outdoor air pollution, and assess solutions**

- Natural sources such as fires, volcanoes, and windblown dust pollute the atmosphere. Human activity can worsen some of these phenomena. (pp. 456–457)
- The pollutants we emit include primary and secondary pollutants from point and non-point sources. (p. 457)
- To safeguard public health under the Clean Air Act, the U.S. EPA and state governments monitor emissions of six major pollutants: carbon monoxide, sulfur dioxide, nitrogen oxides, volatile organic compounds, particulate matter, and lead. (pp. 458–459)
- Agencies also monitor ambient concentrations of the six criteria pollutants: carbon monoxide, sulfur dioxide, nitrogen dioxide, tropospheric ozone, particulate matter, and lead. (p. 460)
- Thanks to public policy and to pollution-control technologies, emissions in the United States have decreased substantially since 1970, and ambient air quality has improved in most respects. (pp. 459–462)
- Emissions of 187 toxic air pollutants are also declining, but they still pose health risks. (p. 462)
- The U.S. EPA is taking early steps toward regulating greenhouse gases as pollutants because they drive climate change. (pp. 462–463)
- Industrializing nations such as China and India are experiencing some of the world's worst air pollution today. (pp. 463–464)
- Industrial smog produced by fossil fuel combustion is still a problem in urban and industrial areas of many developing nations. (pp. 464–465)
- Photochemical smog is created by chemical reactions of pollutants in the presence of sunlight. It impairs visibility and human health in urban areas. (p. 465)
- Cities such as Los Angeles and Mexico City are taking bold steps to address photochemical smog. (pp. 465, 468)

- **Explain stratospheric ozone depletion and identify steps taken to address it**

- CFCs and other persistent human-made compounds destroy stratospheric ozone. Thinning ozone concentrations pose dangers to life because they allow more ultraviolet radiation to reach Earth's surface. (pp. 468–469)
- Ozone depletion is most severe over Antarctica, where an "ozone hole" appears each spring. (p. 469)
- The Montreal Protocol and its follow-up agreements have proven remarkably successful in reducing emissions of ozone-depleting substances. (p. 472)
- The long residence time of CFCs in the atmosphere accounts for a time lag between the protocol and full restoration of stratospheric ozone. (p. 472)

- **Define *acid deposition*, illustrate its consequences, and explain how we are addressing it**

- Acid deposition results when pollutants such as SO_2 and NO react in the atmosphere to produce strong acids that are deposited on Earth's surface. (p. 473)
- Acid deposition may be wet (e.g., "acid rain") or dry, and it may occur a long distance from the source of pollution. (p. 473)
- Acid deposition damages soils, water bodies, plants, animals, ecosystems, and human property and infrastructure. (pp. 473–474)
- Regulation, cap-and-trade programs, and technology are all helping to reduce acid deposition in North America. Industrializing nations will need to tackle the problem as well. (pp. 474–475)

- **Characterize the scope of indoor air pollution and assess solutions**

- Indoor air pollution causes more deaths and health problems worldwide than outdoor air pollution. (p. 475)
- Indoor burning of fuelwood is the developing world's primary indoor air pollution risk. (pp. 475–476)
- Tobacco smoke and radon are the worst indoor pollutants in the developed world. (p. 476)
- Volatile organic compounds and living organisms can pollute indoor air. (pp. 476–477)
- Using low-toxicity materials, keeping spaces clean, monitoring air quality, and maximizing ventilation all help to enhance indoor air quality. (p. 478)

Testing Your Comprehension

1. About how thick is Earth's atmosphere? For each of the four atmospheric layers, name one characteristic.
2. Where is the "ozone layer" located? How and why is stratospheric ozone beneficial for people, whereas tropospheric ozone is harmful?
3. How does solar energy influence weather and climate? Describe how Hadley, Ferrel, and polar cells help to determine long-term climatic patterns and the location of biomes.

4. Describe a thermal inversion. Explain how inversions contribute to severe smog episodes such as the ones in London and in Donora, Pennsylvania.
5. How does a primary pollutant differ from a secondary pollutant? Give an example of each.
6. What has happened with the emissions of major pollutants in the United States in recent decades? What has happened with concentrations of "criteria pollutants" in U.S. ambient air in recent decades? Name one health risk from toxic air pollutants.
7. How does photochemical smog differ from industrial smog? How do the weather and topography influence smog formation?
8. Explain how chlorofluorocarbons (CFCs) deplete stratospheric ozone. Why is this depletion considered a long-term international problem? What was done to address this problem?
9. Why are the effects of acid deposition often felt in areas far from where the primary pollutants are produced? List three impacts of acid deposition.
10. Name three common sources of indoor pollution and their associated health risks. For each pollution source, describe one way to reduce exposure to the source.

Seeking Solutions

1. Consider responses to the photochemical smog pollution that has plagued Los Angeles, Mexico City, and other metropolitan areas. Describe several ways in which major cities have tried to improve their air quality.
2. Name one type of natural air pollution, and discuss how human activity can sometimes worsen it. What potential solutions can you think of to minimize this human impact?
3. Describe how and why emissions of major pollutants have been reduced by well over 50% in the United States since 1970, despite increases in population, energy use, and economic activity.
4. International action through a treaty has helped to halt further stratospheric ozone depletion, but other transboundary pollution issues, including acid deposition, have not yet been addressed as effectively. What types of actions do you feel are appropriate for pollutants that cross political boundaries?
5. **THINK IT THROUGH** You have become the head of your county health department, and the EPA informs you that your county has failed to meet the national ambient air quality standards for ozone, sulfur dioxide, and nitrogen dioxide. Your county is partly rural but is home to a city of 200,000 people and 10 sprawling suburbs. There are several large and aging coal-fired power plants, a number of factories with advanced pollution control technology, and no public transportation system. What steps would you urge the county government to take to meet the air quality standards? Explain how you would prioritize these steps.
6. **THINK IT THROUGH** You have been elected mayor of the largest city in your state. Your city's residents are complaining about photochemical smog and traffic congestion. Traffic engineers and city planners project that population and traffic will grow by 20% in the next decade. Some experts are urging you to restrict traffic into the city, allowing only cars with odd-numbered license plates on odd-numbered days, and those with even-numbered plates on even-numbered days. However, business owners fear losing money should these measures discourage shoppers from visiting. Consider the particulars of your city, and then decide whether you will pursue an odd-day/even-day driving program, and explain why or why not. What other steps would you take to address your city's smog problem?

Calculating Ecological Footprints

"While only some motorists contribute to traffic fatalities, all motorists contribute to air pollution fatalities." So stated a writer for the Earth Policy Institute, pointing out that air pollution kills far more people than vehicle accidents. According to EPA data, emissions of nitrogen oxides in the United States in 2012 totaled 11.3 million tons. Nitrogen oxides come from fuel combustion in motor vehicles, power plants, and other industrial, commercial, and residential sources, but fully 6.4 million tons of the 2012 total came from vehicles. The U.S. Census Bureau estimates the nation's population to have been 313.9 million in 2012 and projects that it will reach 346.7 million in 2025. Considering these data, calculate the missing values in the table below (1 ton = 2000 lb).

	Total NO_x emissions (lb)	NO_x emissions from vehicles (lb)
You		
Your class		
Your state		
United States	22.6 billion	12.8 billion

Data from U.S. EPA.

1. By what percentage is the U.S. population projected to increase between 2012 and 2025? Do you think that NO_X emissions will increase, decrease, or remain the same over that period of time? Why? (You may want to refer to Figure 17.14.)
2. Assume you are an average American driver. Using the 2012 emissions totals, how many pounds of NO_X emissions are you responsible for creating? How many pounds would you prevent if you were to reduce by half the vehicle miles you travel? What percentage of your total NO_X emissions would that be?
3. How might you reduce your vehicle miles traveled by 50%? What other steps could you take to reduce the NO_X emissions for which you are responsible?

MasteringENVIRONMENTALSCIENCE™

STUDENTS

Go to **MasteringEnvironmentalScience** for assignments, the etext, and the Study Area with practice tests, videos, current events, and activities.

INSTRUCTORS

Go to **MasteringEnvironmentalScience** for automatically graded activities, current events, videos, and reading questions that you can assign to your students, plus Instructor Resources.

18

The Maldives' underwater cabinet meeting

Global Climate Change

Upon completing this chapter, you will be able to:

- Describe Earth's climate system and explain the factors influencing global climate
- Characterize human influences on the atmosphere and on climate
- Summarize how researchers study climate
- Outline current and future trends and impacts of global climate change
- Suggest ways we may respond to climate change

CENTRAL CASE STUDY

Rising Seas May Flood the Maldives

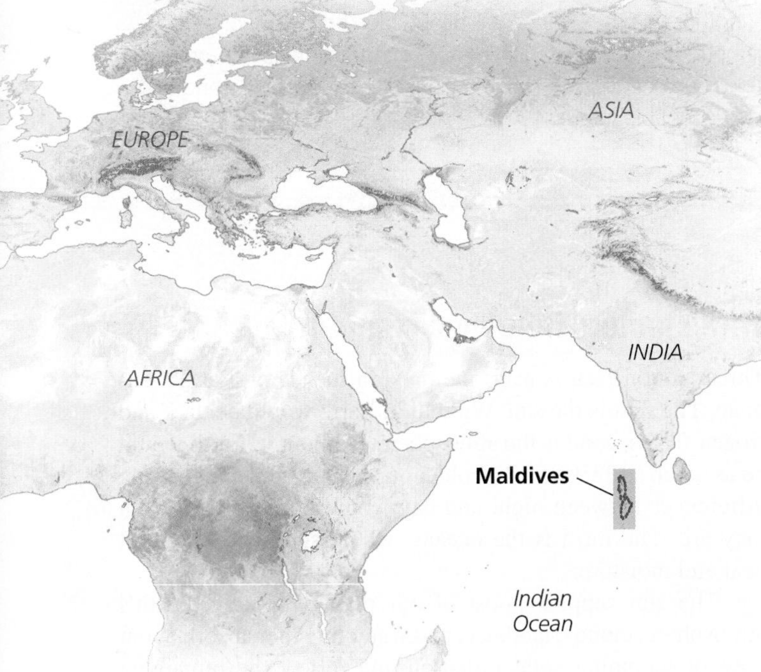

"Climate change threatens the very existence of our country."

—**Mohamed Waheed, president, Maldives**

"If we can't save the Maldives today, we can't save London, New York, or Hong Kong tomorrow."

—**Mohamed Nasheed, former president, Maldives**

With sun-drenched beaches, colorful coral reefs, and a spectacular tropical setting, the Maldives seems a paradise to the many tourists who visit. For its 370,000 residents, this island nation in the Indian Ocean is home. But residents and tourists alike now fear that the Maldives could soon be submerged by the rising seas brought by global climate change.

For a nation of 1200 islands whose highest point is just 2.4 m (8 ft) above sea level, rising seas are a matter of life or death. Four-fifths of the Maldives' land area lies less than 1 m (39 in.) above sea level. The world's oceans rose 10–20 cm (4–8 in.) during the 20th century as warming temperatures expanded ocean water and as melting polar ice discharged water into the ocean. According to current projections, sea level will rise another 18–59 cm (7–23 in.) by the year 2100.

Higher seas are expected to flood large areas of the Maldives and cause salt water to contaminate drinking water supplies. Scientists expect storms intensified by warmer water to erode beaches and damage the coral reefs that are vital to the nation's tourism and fishing industries. "If things go business as usual," former president Mohamed Nasheed has said, "our country will not exist."

Although small island nations like the Maldives are responsible for very few of the carbon emissions that drive global climate change, these nations are the ones bearing the earliest consequences. The Maldives' political leaders have made sure the world knows this. In 2009, President Nasheed donned scuba gear and dove into the blue waters of Girifushi Island lagoon, followed by his entire cabinet. These officials held the world's first underwater cabinet meeting. Sitting at a table beneath the waves, they signed a declaration reading:

> *SOS from the front line:* Climate change is happening and it threatens the rights and security of everyone on Earth. With less than one degree of global warming, the glaciers are melting, the ice sheets collapsing, and low-lying areas are in danger of being swamped. We must unite in a global effort to halt further temperature rises, by slashing carbon dioxide emissions to a safe level of 350 parts per million.

The underwater cabinet meeting was part of a campaign to draw global attention to the impacts of climate change. Nasheed followed this with a high-profile role at international climate talks in Copenhagen, where he pleaded with the United States, China, India, and other major polluting nations to unite in efforts to reduce emissions of gases that warm the atmosphere.

Back home in the Maldives, Nasheed announced a plan to make his nation carbon-neutral by 2020. But already residents had to be evacuated from several of the lowest-lying islands, and Nasheed arranged to begin buying land in mainland nations in case his people one day need to abandon their homeland.

Then in 2012, Nasheed—the nation's first democratically elected president—was forced from power at gunpoint, and the vice-president was installed in his place. Nasheed's supporters called it a coup d'etat. His detractors said he had abused power by illegally imprisoning the nation's chief judge. They put him on trial, which many viewed as simply an attempt to keep him out of the 2013 presidential election.

Despite the political turmoil, the people of the Maldives remain united in their concern over climate change. The new president, Mohamed Waheed, proposed to continue the plan for carbon-neutrality through a voluntary tax on tourists that could fund carbon offsets and investment in renewable energy.

Residents of the Maldives are not alone in their predicament. Other island nations, from the Galápagos to Fiji to the Seychelles, also face a future of encroaching seawater. These island nations have organized themselves to make their concern over climate change known to the world through AOSIS, the Alliance of Small Island States.

Mainland coastal areas across the world will face similar challenges from sea level rise—from the hurricane-battered coasts of Florida, Louisiana, Texas, and the Carolinas to coastal cities such as San Francisco and New York City. Superstorm Sandy in 2012 was a wake-up call for the eastern United States. The lost lives and billions of dollars of damage brought by this massive hurricane and its storm surge in New York, New Jersey, and other states made clear that the costs of rising seas could be enormous.

Storm damage from rising seas is just one of the many imminent consequences of global climate change. In one way or another, climate change will affect each and every one of us for the remainder of our lifetimes. Putting solutions into action stands as a central challenge for our society right now and for the foreseeable future.

Our Dynamic Climate

Climate influences virtually everything around us, from the day's weather to major storms, from crop success to human health, and from national security to the ecosystems that support our economies. If you are a student in your teens or twenties, the accelerating change in our climate today may well be *the* major event of your lifetime and the phenomenon that most shapes your future.

Climate change is also the fastest-developing area of environmental science. New scientific studies that refine our understanding of climate are published every week, and policymakers and businesspeople make decisions and announcements just as quickly. By the time you read this chapter, some of its information will already be out of date. We urge you to explore further, with your instructor and on your own, the most recent information on climate change and the impacts it will have on your future.

What is climate change?

Climate describes an area's long-term atmospheric conditions, including temperature, precipitation, wind, humidity, barometric pressure, solar radiation, and other characteristics. *Climate* differs from *weather* (p. 453) in that weather specifies conditions at localized sites over hours or days, whereas climate describes conditions across broader regions over years, decades, or centuries. **Global climate change** encompasses an array of changes in aspects of Earth's climate, such as temperature, precipitation, and storm frequency and intensity. People often use the term *global warming* synonymously in casual conversation, but **global warming** refers specifically to an increase in Earth's average surface temperature. Global warming is only one aspect of global climate change, but warming does in turn drive other components of climate change.

Over the long term, our planet's climate varies naturally. However, today's climatic changes are unfolding at an exceedingly rapid rate, and they are creating conditions humanity has never experienced. Scientists agree that human activities, notably fossil fuel combustion and deforestation, are largely responsible. Understanding how and why today's climate is changing requires understanding how our planet's climate functions. Thus, we first will examine Earth's climate system—a complex and finely tuned system that has nurtured life for billions of years.

Three factors influence climate

Three natural factors exert the most influence on Earth's climate. The first is the sun. Without it, Earth would be dark and frozen. The second is the atmosphere. Without it, Earth would be as much as 33°C (59°F) colder on average, and temperature differences between night and day would be far greater than they are. The third is the oceans, which store and transport heat and moisture.

The sun supplies most of our planet's energy. Earth's atmosphere, clouds, land, ice, and water together absorb about 70% of incoming solar radiation and reflect the remaining 30% back into space (**FIGURE 18.1**). The 70% that is absorbed powers many of Earth's processes, from winds to waves to evaporation to photosynthesis. We will assess how each major factor influences climate, focusing first on the atmosphere.

Greenhouse gases warm the lower atmosphere

As Earth's surface absorbs solar radiation, the surface increases in temperature and emits infrared radiation (p. 31), radiation with wavelengths longer than those of visible light. Atmospheric gases having three or more atoms in their molecules tend to absorb infrared radiation. These include water vapor, ozone (O_3), carbon dioxide (CO_2), nitrous oxide (N_2O), and methane (CH_4), as well as halocarbons, a diverse group of mostly human-made gases that includes chlorofluorocarbons (CFCs; p. 469). Such gases are known as **greenhouse gases.** After absorbing radiation emitted from the surface, greenhouse gases subsequently re-emit infrared radiation. Some of this re-emitted energy is lost to space, but some travels back downward, warming the lower atmosphere (specifically the troposphere; p. 451) and the surface in a phenomenon known as the **greenhouse effect.**

Greenhouse gases differ in their ability to warm the troposphere and surface. *Global warming potential* refers to the relative ability of one molecule of a given greenhouse gas to contribute to warming. **TABLE 18.1** shows global warming potentials for several greenhouse gases. Values are expressed in relation to carbon dioxide, which is assigned a value of 1. Thus, a molecule of methane is 25 times more potent than a

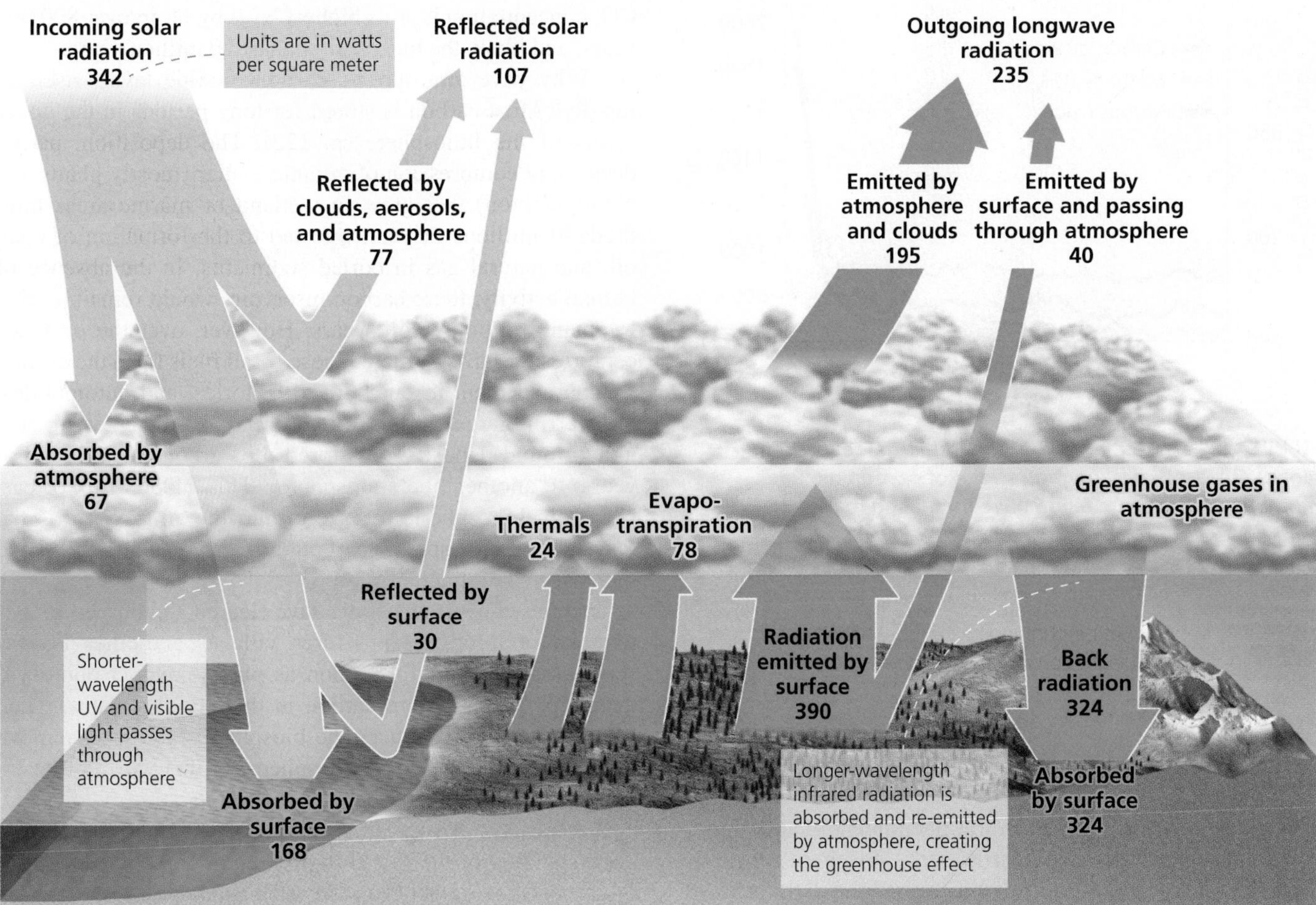

FIGURE 18.1 Our planet receives 342 watts of energy per square meter from the sun, and it naturally reflects and emits this same amount. Earth absorbs nearly 70% of the solar radiation it receives, and reflects the rest back into space (**yellow arrows**). The radiation absorbed is then re-emitted (**orange arrows**) as infrared radiation, which has longer wavelengths. Greenhouse gases in the atmosphere absorb a portion of this long-wavelength radiation and then re-emit it, sending some back downward to warm the atmosphere and the surface by the greenhouse effect. *Data from Kiehl, J.T., and K.E. Trenberth, 1997. Earth's annual global mean energy budget.* Bulletin of the American Meteorological Society *78: 197–208. © American Meteorological Society (AMS). By permission.*

molecule of carbon dioxide, and a molecule of nitrous oxide is 298 times more potent than a CO_2 molecule.

Although carbon dioxide is less potent on a per-molecule basis than methane or nitrous oxide, it is far more abundant in the atmosphere, so it contributes more to the greenhouse effect. Moreover, greenhouse gas emissions from human activity consist mostly of carbon dioxide. And CO_2 has a much longer residence time (pp. 457–458) in the atmosphere than methane, so methane's impact is greater in the short term than in the long term. According to the latest data, CO_2 is causing nearly six times more warming than methane, nitrous oxide, and halocarbons combined.

TABLE 18.1 Global Warming Potentials of Four Greenhouse Gases

GREENHOUSE GAS	RELATIVE HEAT-TRAPPING ABILITY (IN CO_2 EQUIVALENTS)
Carbon dioxide	1
Methane	25
Nitrous oxide	298
Hydrochlorofluorocarbon HFC-23	14,800

Data are for a 100-year time horizon, from IPCC, 2007. Fourth assessment report. Climate change 2007: The physical science basis.

FAQ **The greenhouse effect works just like a greenhouse, right?**

Actually, not quite. A greenhouse helps plants grow because its glass walls trap heat. In contrast, greenhouse gas molecules in our atmosphere absorb particular wavelengths of light reflected up from the surface, then re-emit radiation at different wavelengths. Some of this radiation travels back toward the surface, keeping the surface and lower atmosphere warmer than they would otherwise be. This phenomenon differs from what happens in a greenhouse, but it was called the "greenhouse effect" in the past, and the name has stuck.

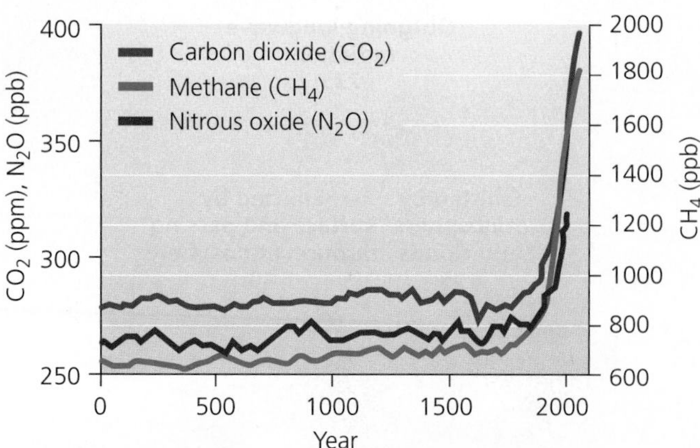

FIGURE 18.2 Since the start of the industrial revolution, global atmospheric concentrations of carbon dioxide, methane, and nitrous oxide have increased markedly. *Data from Intergovernmental Panel on Climate Change (IPCC), 2007.* Fourth assessment report. *FAQ 2.1, Fig 1, in* The physical science basis: Contribution of Working Group I.

DATA Q By about what percentage has atmospheric carbon dioxide concentration increased since 1750?

Greenhouse gas concentrations are rising fast

The greenhouse effect is a natural phenomenon, and greenhouse gases have been present in our atmosphere for all of Earth's history. It's a good thing, too. Without the natural greenhouse effect, our planet would be too cold to support life as we know it. Thus, it is not the natural greenhouse effect that concerns scientists today, but rather the *anthropogenic* (human-generated) intensification of the greenhouse effect. By adding novel greenhouse gases (certain halocarbons) to the atmosphere, and by increasing the concentrations of several natural greenhouse gases over the past 250 years (**FIGURE 18.2**), we are intensifying our planet's greenhouse effect beyond what our species has ever experienced.

We have boosted Earth's atmospheric concentration of carbon dioxide from 280 parts per million (ppm) in the late 1700s to 396 ppm in 2013 (see Figure 18.2). Today's atmospheric CO_2 concentration is at its highest level by far in over 800,000 years, and likely the highest in the last 20 million years.

Why have atmospheric carbon dioxide levels risen so rapidly? Most carbon is stored for long periods in the upper layers of the lithosphere (p. 122). The deposition, partial decay, and compression of organic matter (mostly plants and phytoplankton) that grew in wetland or marine areas hundreds of millions of years ago led to the formation of coal, oil, and natural gas in buried sediments. In the absence of human activity, these carbon reservoirs would remain buried for many millions more years. However, over the past two centuries we have extracted these fossil fuels from the ground and burned them in our homes, factories, and automobiles, transferring large amounts of carbon from one reservoir (the underground deposits that stored the carbon for millions of years) to another (the atmosphere). This sudden flux of carbon from lithospheric reservoirs into the atmosphere is the main reason atmospheric carbon dioxide concentrations have increased so dramatically.

At the same time, people have cleared and burned forests to make room for crops, pastures, villages, and cities. Forests serve as a reservoir for carbon as plants conduct photosynthesis (p. 32) and store carbon in their tissues. Thus, when we clear forests it reduces the biosphere's ability to remove carbon dioxide from the atmosphere. In this way, deforestation (pp. 311–314) contributes to rising atmospheric CO_2 concentrations. **FIGURE 18.3** summarizes scientists' current understanding of the fluxes (both natural and anthropogenic) of carbon dioxide between the atmosphere and reservoirs on Earth's surface.

Methane concentrations are also rising—2.5-fold since 1750 (see Figure 18.2)—and today's atmospheric concentration is the highest by far in over 800,000 years. We release methane by tapping into fossil fuel deposits, raising livestock that emit methane as a metabolic waste product, disposing of organic matter in landfills, and growing crops such as rice.

Human activities have also enhanced atmospheric concentrations of nitrous oxide. This greenhouse gas, a by-product of feedlots, chemical manufacturing plants, auto emissions, and synthetic nitrogen fertilizers, has risen by nearly 20% since 1750 (see Figure 18.2).

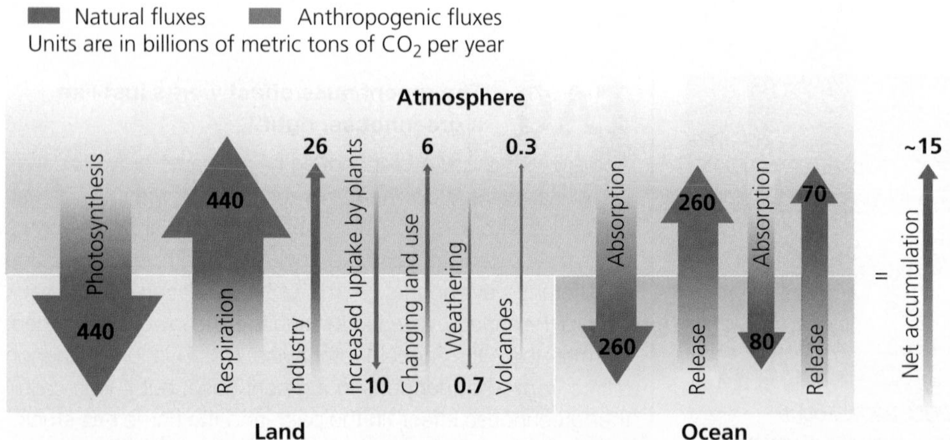

FIGURE 18.3 Human activities are sending more carbon dioxide from Earth's surface to its atmosphere than is moving from the atmosphere to the surface. Shown are all current fluxes of CO_2, with arrows sized according to mass. Green arrows indicate natural fluxes, and red arrows indicate anthropogenic fluxes. *Adapted from IPCC, 2007.* Fourth assessment report.

DATA Q For every metric ton of carbon dioxide we emit due to changing land use (e.g. deforestation), how much do we emit from industry?

Among other greenhouse gases, ozone concentrations in the troposphere have risen roughly 36% since 1750 because of photochemical smog (p. 465). The contribution of halocarbon gases to global warming has begun to slow because of the Montreal Protocol and subsequent controls on their production and use (p. 472).

Water vapor is the most abundant greenhouse gas in our atmosphere and contributes most to the natural greenhouse effect. Its concentrations vary locally, but its global concentration has not changed over recent centuries. Because its concentration has not changed, it is not thought to have driven industrial-age climate change.

Most aerosols exert a cooling effect

Whereas greenhouse gases exert a warming effect on the atmosphere, **aerosols** (p. 457), microscopic droplets and particles, can have either a warming or a cooling effect. Soot particles, or "black carbon aerosols," generally cause warming by absorbing solar energy, but most other tropospheric aerosols cool the atmosphere by reflecting the sun's rays. Sulfate aerosols produced by fossil fuel combustion may slow global warming, at least in the short term. When sulfur dioxide enters the atmosphere, it undergoes various reactions, some of which lead to acid precipitation (pp. 473–475). These reactions can form a sulfur-rich aerosol haze in the upper atmosphere that blocks sunlight. For this reason, aerosols released by major volcanic eruptions can exert cooling effects on Earth's climate for up to several years. This occurred in 1991 with the eruption of Mount Pinatubo in the Philippines (p. 457).

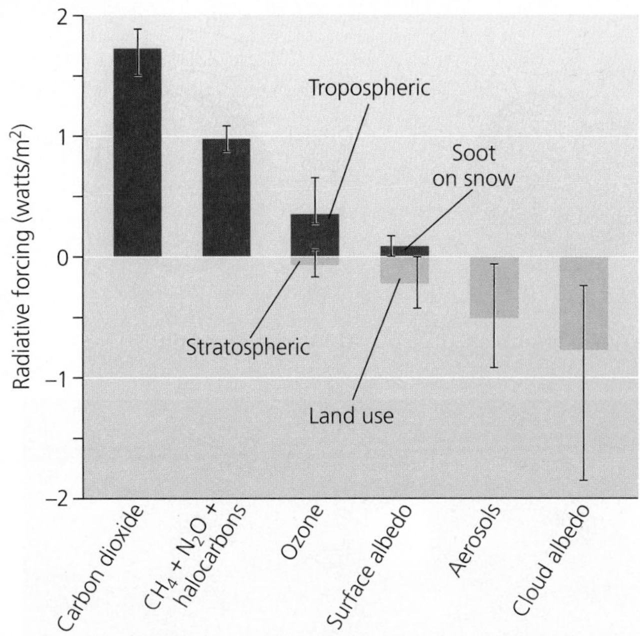

FIGURE 18.4 Radiative forcing measures the degree of influence that aerosols, greenhouse gases, and other factors exert over Earth's energy balance. In this graph, radiative forcing is expressed as the warming or cooling effect that each factor has on temperature today relative to 1750, in watts/m^2. Red bars indicate positive forcing (warming), and blue bars indicate negative forcing (cooling). *Albedo* (p. 498) refers to the reflectivity of a surface. A number of more minor influences are not shown. *Data from IPCC, 2007.* Fourth assessment report.

Radiative forcing expresses change in energy input

To measure the degree of impact that a given factor exerts on Earth's temperature, scientists calculate its **radiative forcing,** the amount of change in thermal energy that the factor causes. Positive forcing warms the surface, whereas negative forcing cools it. FIGURE 18.4 shows researchers' best calculations of the radiative forcing that our planet is experiencing today.

When scientists sum up the effects of all factors, they find that Earth is now experiencing overall radiative forcing of about 1.6 watts/m^2. This means that today's planet is receiving and retaining 1.6 watts/m^2 more thermal energy than it is emitting into space. (By contrast, the pre-industrial Earth of 1750 was in balance, emitting as much radiation as it was receiving.) This extra amount is equivalent to the power converted into heat and light by 140 incandescent lightbulbs (or 650 CFLs) over a football field. For context, look back at Figure 18.1 and note that Earth is estimated naturally to receive and give off 342 watts/m^2 of energy. Although 1.6 may seem like a small proportion of 342, over time it is actually enough to alter climate significantly.

Feedback complicates our predictions

As tropospheric temperatures increase, Earth's water bodies should transfer more water vapor into the atmosphere, but scientists aren't yet sure how this will affect our climate. On one hand, more atmospheric water vapor could lead to more warming, which could lead to more evaporation and water vapor, in a positive feedback loop (pp. 106–107) that would amplify the greenhouse effect. On the other hand, more water vapor could enhance cloudiness, which might, in a negative feedback loop (pp. 106–107), slow global warming by reflecting more solar radiation back into space. In this second scenario, depending on whether low- or high-elevation clouds result, they might either shade and cool Earth (negative feedback) or else contribute to warming and accelerate evaporation and further cloud formation (positive feedback). We simply don't yet know which effect might outweigh the other. Because of feedback loops, minor modifications of components of the atmosphere can potentially lead to major effects on climate. This poses challenges for making accurate predictions of future climate change.

Climate varies naturally for several reasons

Besides atmospheric composition, our climate is influenced by cyclic changes in Earth's rotation and orbit, variation in energy released by the sun, absorption of carbon dioxide by the oceans, and ocean circulation patterns.

Milankovitch cycles In the 1920s, Serbian mathematician Milutin Milankovitch described three types of periodic changes in Earth's rotation and orbit around the sun. Over thousands of years, our planet wobbles on its axis, varies in the tilt of its axis, and experiences change in the shape of its orbit, all in regular long-term cycles of different lengths.

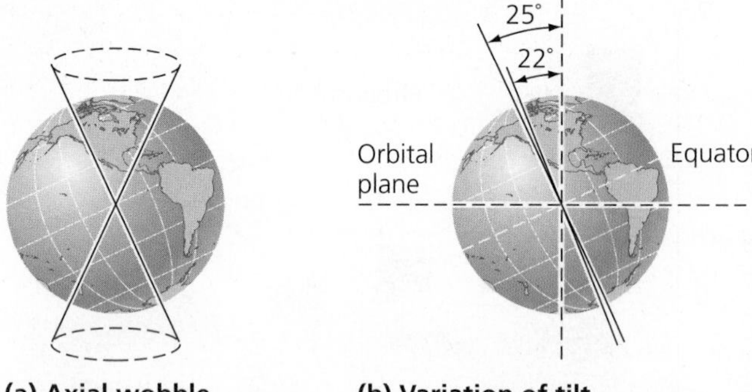

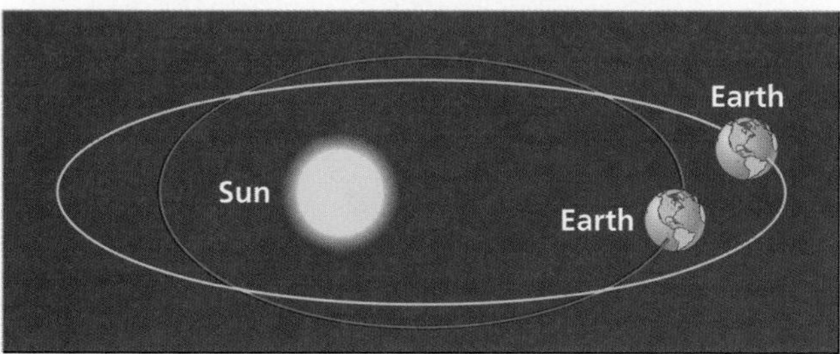

FIGURE 18.5 There are three types of Milankovitch cycles: **(a)** an axial wobble that occurs on a 19,000- to 23,000-year cycle; **(b)** a 3-degree shift in the tilt of Earth's axis that occurs on a 41,000-year cycle; and **(c)** a variation in Earth's orbit from almost circular to more elliptical, which repeats every 100,000 years.

These variations, known as **Milankovitch cycles,** alter the way solar radiation is distributed over Earth's surface (**FIGURE 18.5**). By modifying patterns of atmospheric heating, these cycles trigger long-term climate variation. This includes periodic episodes of *glaciation* during which global surface temperatures drop and ice sheets advance from the poles toward the midlatitudes, as well as intervening warm *interglacial* periods.

Solar output The sun varies in the amount of radiation it emits, over both short and long timescales. However, scientists are concluding that the variation in solar energy reaching our planet in recent centuries has simply not been great enough to drive significant temperature change on Earth's surface. Estimates place the radiative forcing of natural changes in solar output at only about 0.12 watts/m^2—less than any of the anthropogenic causes shown in Figure 18.4. Moreover, solar radiation has been decreasing since the 1970s, not increasing, so it clearly cannot explain Earth's recent warming trend.

Ocean absorption The oceans hold 50 times more carbon than the atmosphere holds. They absorb carbon dioxide from the atmosphere when this gas dissolves directly in water and when marine phytoplankton use it for photosynthesis. However, the oceans are absorbing less CO_2 than we are adding to the atmosphere (see Figure 5.17, p. 122). Thus, carbon absorption by the oceans is slowing global warming but is not preventing it. Moreover, recent evidence indicates that the rate of absorption is decreasing. As ocean water warms, it absorbs less CO_2 because gases are less soluble in warmer water—a positive feedback effect (pp. 106–107) that accelerates warming of the atmosphere.

Ocean circulation Ocean water exchanges heat with the atmosphere, and ocean currents move energy from place to place. In equatorial regions, such as the area around the Maldives, the oceans receive more heat from the sun and atmosphere than they emit. Near the poles, the oceans emit more heat than they receive. Because cooler water is denser than warmer water, the cooler water at the poles tends to sink, and the warmer surface water from the equator moves to take its place. This is one principle underlying global ocean circulation patterns (p. 424).

The oceans' thermohaline circulation system has influential regional effects (p. 425). For example, it moves warm tropical water northward toward Europe, providing the European continent a far milder climate than it would otherwise have. Scientists are studying whether freshwater input from Greenland's melting ice sheet might shut down this warm-water flow (p. 425). Such an occurrence would plunge Europe into much colder conditions.

Multiyear climate variability results from the El Niño–Southern Oscillation (pp. 425–426), which involves systematic shifts in atmospheric pressure, sea surface temperature, and ocean circulation in the tropical Pacific Ocean. These shifts overlie longer-term variability from a phenomenon known as the Pacific Decadal Oscillation. El Niño and La Niña events alter weather patterns from region to region in diverse ways, often leading to rainstorms and floods in dry areas and drought and fire in moist areas. This leads to impacts on wildlife, agriculture, and fisheries.

FAQ **The climate changes naturally, so why worry about climate change?**

Earth's climate does indeed change naturally across very long periods of time. However, no known natural factors can account for the rapid speed of the change we are experiencing today. Moreover, our civilization has never before experienced the sheer amount of change predicted during this century. The quantity by which the world's temperature is forecast to rise is greater than the amount of cooling needed to bring on an ice age. Greenhouse gas concentrations are already higher than they've been in over 800,000 years, and are still rising. Our entire civilization arose only in the last few thousand years during an exceptionally stable period in Earth's climate history. Unless we reduce our emissions, we will soon be challenged by climatic conditions the human species has never lived through before.

Studying Climate Change

To comprehend any phenomenon that is changing, we must study its past, present, and future. Scientists monitor present-day climate, but they also have devised clever means of inferring past change as well as sophisticated methods to predict future conditions.

Proxy indicators tell us about the past

Evidence about **paleoclimate,** climate in the ancient past, is vital for giving us a baseline against which we can measure changes happening in our climate today. To understand paleoclimate, scientists have developed ingenious methods to decipher clues from thousands or millions of years ago by taking advantage of the record-keeping capacity of the natural world. **Proxy indicators** are types of indirect evidence that serve as proxies, or substitutes, for direct measurement and that shed light on past climate.

For example, Earth's ice caps, ice sheets, and glaciers hold clues to climate history. In frigid areas over the poles and atop high mountains, snow falling year after year for millennia compresses into ice. Over the ages, this ice accumulates to great depths, preserving within its layers tiny bubbles of the ancient atmosphere (FIGURE 18.6). Scientists can examine the trapped air bubbles by drilling into the ice and extracting long columns, or cores. The layered ice, accumulating season after season over thousands of years, provides a timescale. By studying the chemistry of the ice and the bubbles in each layer in these ice cores, scientists can determine atmospheric composition, greenhouse gas concentrations, temperature trends, snowfall, solar activity, and even (from trapped soot particles) frequency of forest fires and volcanic eruptions during each time period. By extracting ice cores from Antarctica, scientists have now been able to go back in time 800,000 years, reading Earth's history across eight glacial cycles (see THE SCIENCE BEHIND THE STORY, pp. 490–491.)

Researchers also drill cores into beds of sediment beneath bodies of water. Sediments often preserve pollen grains and other remnants from plants that grew in the past (as we saw with the study of Easter Island; pp. 6–7). Because climate influences the types of plants that grow in an area, knowing what plants were present can tell us a great deal about the climate at that place and time.

Tree rings provide another proxy indicator. The width of each ring of a tree trunk cut in cross-section reveals how much the tree grew in a particular growing season. A wide ring means more growth, generally indicating a wetter year. Long-lived trees such as bristlecone pines can provide records of precipitation and drought going back several thousand years. Tree rings are also used to study fire history, since a charred ring indicates that a fire took place in the region in that year.

In arid regions such as the U.S. Southwest, packrat middens are a valuable source of climate data. Packrats are rodents that carry seeds and plant parts back to their middens, or dens, in caves and rock crevices sheltered from rain. In arid locations, plant parts may be preserved for centuries, allowing researchers to study the past flora of the region.

Researchers gather data on past ocean conditions from coral reefs (pp. 431–432). Living corals take in trace elements and isotope ratios (p. 24) from ocean water as they grow, and they incorporate these chemical clues, layer by layer, into growth bands in the reefs they build.

Proxy indicators often tell us information about local or regional areas. To get a global perspective, scientists need to combine multiple records from various areas. Because the number of available indicators decreases the further back in time we go, estimates of global climate conditions for the recent past tend to be more reliable than those for the distant past.

(a) Ice core

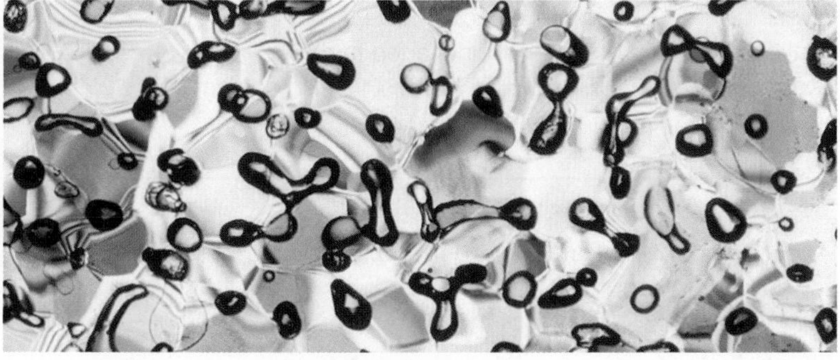

(b) Micrograph of ice core

FIGURE 18.6 **Scientists drill deep into ancient ice sheets and remove cores of ice.** Dr. Gerald Holdsworth of the University of Calgary **(a)** extracts information about past climates from an ice core. Bubbles trapped in the ice **(b)** contain small samples of the ancient atmosphere.

Direct measurements tell us about the present

Today we measure temperature with thermometers, rainfall with rain gauges, wind speed with anemometers, and air pressure with barometers, using computer programs to integrate and analyze this information in real time. With these technologies and more, we document in detail the fluctuations in weather day-by-day and hour-by-hour across the globe. As a result, we have gained an understanding of present-day climate conditions in every region of our planet.

We also measure the chemistry of the atmosphere and the oceans. Direct measurements of carbon dioxide concentrations in the atmosphere reach back to 1958, when scientist Charles Keeling began analyzing hourly air samples from a monitoring station at Hawaii's Mauna Loa Observatory. Here, unpolluted, well-mixed air from over vast stretches of ocean blows across the top of Earth's most massive

THE SCIENCE BEHIND THE STORY

Reading History in the World's Longest Ice Core

An EPICA researcher prepares a Dome C ice core sample for analysis.

In the most frigid reaches of our planet, snow falling year after year for millennia compresses into ice and stacks up into immense sheets that scientists can mine for clues to Earth's climate history. The ice sheets of Antarctica and Greenland trap tiny air bubbles, dust particles, and other proxy indicators (p. 489) of past conditions. By drilling boreholes and extracting ice cores, researchers can tap into these valuable archives.

Recently, researchers drilled and analyzed the deepest core ever. At a remote and pristine site in Antarctica named Dome C, they drilled down 3270 m (10,728 ft) to bedrock and pulled out more than 800,000 years' worth of ice. The longest previous ice core (from Antarctica's Vostok station) had gone back "only" 420,000 years.

Ice near the top of these cores was laid down most recently, and ice at the bottom is oldest, so by analyzing ice at intervals along the core's length, researchers can generate a timeline of environmental change. To date layers of the ice core, researchers first analyze deuterium isotopes (p. 25) to determine the rate of ice accumulation, referencing studies and models of how ice compacts over time. They then calibrate the timeline by matching recent events in the chronology (for example, major volcanic eruptions) with independent data sets from previous cores, tree rings, and other sources.

Dome C, a high summit of the Antarctic ice sheet, is one of the coldest spots on the planet, with an annual mean temperature of –54.5°C (–98.1°F). The Dome C ice core was drilled by the European Project for Ice Coring in Antarctica (EPICA), a consortium of researchers from 10 European nations.

In 2004, this team of 56 researchers published a paper in the journal *Nature*, reporting data across 740,000 years. The researchers obtained data on surface air temperature by measuring the ratio of deuterium isotopes to normal hydrogen in the ice, because this ratio is temperature-dependent.

From 2005 to 2008, five follow-up papers in the journals *Science* and *Nature* reported analyses of greenhouse gas concentrations from the EPICA ice core and extended the gas and temperature data back to cover all 800,000 years. By analyzing air bubbles trapped in the ice, the researchers quantified atmospheric concentrations of carbon dioxide and methane (red line and green line, respectively, in **FIGURE 1**).

These data show that by emitting these greenhouse gases since the industrial revolution, we have brought their atmospheric concentrations well above the highest levels they reached naturally any time in the last 800,000 years. Today's carbon dioxide spike is too recent to show up in the ice core, but its concentration (of 396 ppm in 2013) is far above previous maximum values (of ~300 ppm) shown in the red line of the figure. These data reveal that we as a society have brought ourselves deep into uncharted territory.

The EPICA results also confirm that temperature swings in the past

mountain. These data show that atmospheric CO_2 concentrations have increased from 315 ppm in 1958 to 396 ppm in 2013 (**FIGURE 18.7**).

Direct measurements of climate variables such as temperature and precipitation extend back in time somewhat further. Precise and reliable thermometer measurements cover more than a century. Scientists can also infer past climate conditions from historical records of economic activities affected by climate. Fishers have recorded the timing of sea ice formation, and winemakers have kept meticulous records of precipitation and the length of the growing season. Accurate records of all these types extend back, at most, a few hundred years.

Models help us predict the future

To understand how climate systems function and to predict future climate change, scientists simulate climate processes with sophisticated computer programs. **Climate models** are programs that combine what is known about atmospheric circulation, ocean circulation, atmosphere–ocean interactions, and feedback cycles to simulate climate processes

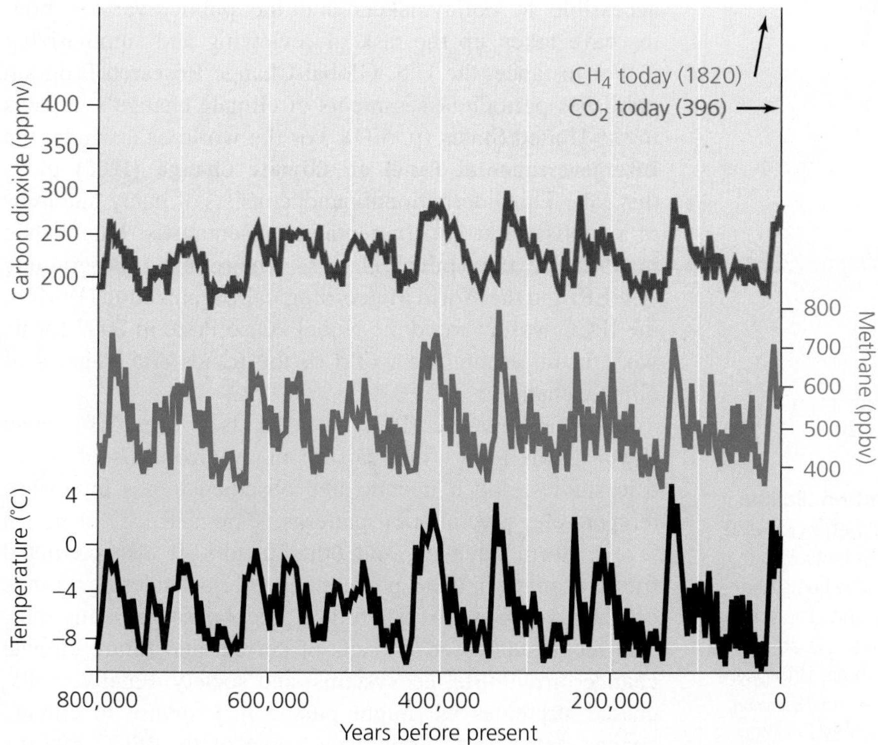

FIGURE 1 Data from the EPICA ice core reveal changes across 800,000 years. Shown are surface temperature (**black line**), atmospheric methane concentration (**green line**), and atmospheric carbon dioxide concentration (**red line**). Concentrations of CO_2 and methane rise and fall in tight correlation with temperature. Today's current values are included at the top right of the graph, for comparison. *Adapted by permission of Macmillan Publishers Ltd: Brook, E. 2008. Paleoclimate: Windows on the greenhouse.* Nature *453: 291-292, Fig 1a. www.nature.com.*

were tightly correlated with concentrations of greenhouse gases (compare the top two datasets in Figure 1 with the temperature dataset at bottom). This finding bolsters the scientific consensus that greenhouse gas emissions are causing our planet to warm today.

Also clear from the data is that temperature has varied with swings in solar radiation due to Milankovitch cycles (pp. 487–488). The complex interplay of the Milankovitch cycles produces periodic temperature fluctuations on Earth resulting in periods of glaciation (when temperate regions of the planet are covered in ice) and in warm interglacial periods. The Dome C ice core spans eight glacial cycles.

Other findings from the ice core are not easily explained. Intriguingly, the early glacial cycles differ from the recent cycles (see the black line in Figure 1). In the recent cycles, glacial periods are long, whereas interglacial periods are brief, with a rapid rise and fall of temperature. Interglacials thus appear on the graph as tall thin spikes. In older glacial cycles, the glacial and interglacial periods are of more equal duration, and the interglacials are not as warm.

This change in the nature of glacial cycles had been noted before by researchers working with oxygen isotope data from the fossils of marine organisms. But why cycles should differ before and after the 450,000-year mark, no one knows.

Today polar scientists are searching for a site that might provide an ice core stretching back more than 1 million years. At that time, data from marine isotopes tell us that glacial cycles switched from a periodicity of roughly 41,000 years (conforming to the influence of planetary tilt) to intervals of about 100,000 years (more similar to orbital changes). An ice core that captures cycles on both sides of the 1-million-year divide might help clarify the influence of Milankovitch cycles or perhaps offer other explanations.

The intriguing patterns revealed by the Dome C ice core show that we still have plenty to learn about our complex climate history. However, the clear relationship between greenhouse gases and temperature evident in the EPICA data suggest that if we want to prevent sudden global warming, we will need to reduce our society's greenhouse emissions. ■

(see **THE SCIENCE BEHIND THE STORY**, pp. 494–495). This requires manipulating vast amounts of data with complex mathematical equations—a task not possible until the advent of modern computers.

Climate modelers essentially provide starting information to the model, set up rules for the simulation, and then let it run. Researchers strive for accuracy by building in as much information as they can from what is understood about how the climate system functions. They then test the efficacy of a model by entering past climate data and running the model toward the present. If a model accurately reconstructs current climate, based on well-established data from the past, then we have reason to believe that it simulates climate mechanisms realistically and that it may accurately predict future climate.

Plenty of challenges remain for climate modelers, because Earth's climate system is so complex and because many uncertainties remain in our understanding of feedback processes. Yet as scientific knowledge of climate improves, as computing power intensifies, and as we glean enhanced data from proxy indicators, climate models are improving in resolution and are predicting climate change region by region across the world.

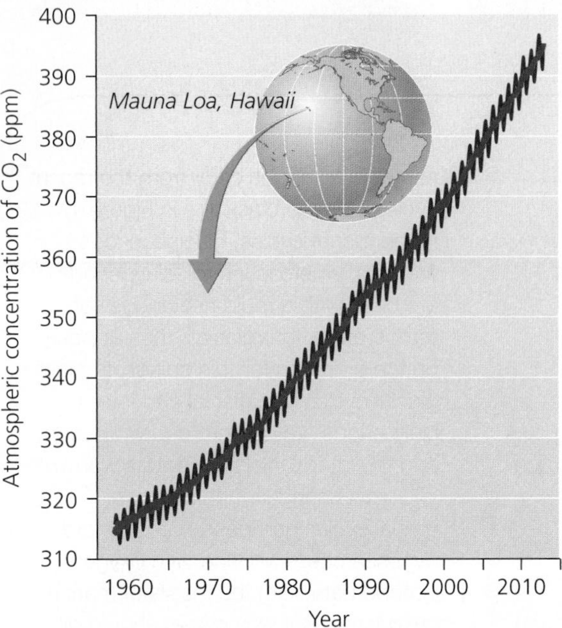

FIGURE 18.7 Atmospheric concentrations of carbon dioxide are rising steeply. Direct long-term measurements began in 1958, when Charles Keeling started collecting these data at Hawaii's Mauna Loa Observatory. The jaggedness of the trend reflects seasonal variation: the Northern Hemisphere has more land area and vegetation than the Southern Hemisphere, so more CO_2 is absorbed during the northern summer, when plants in the north are photosynthetically active. *Data from National Oceanic and Atmospheric Administration, Earth System Research Laboratory, Global Monitoring Division, 2013.*

Current and Future Trends and Impacts

It seems that virtually everyone is noticing changes in the climate these days. Maldives fishermen note the seas encroaching on their home island. Texas ranchers suffer a multiyear drought. Florida homeowners find it difficult to obtain insurance against hurricanes and storm surges. New Yorkers, Bostonians, Chicagoans, and Los Angelenos face one unprecedented weather event after another.

Extreme weather events are indeed part of a real pattern backed by a tremendous volume of scientific evidence. Climate change has already had numerous impacts on the physical properties of our planet, on organisms and ecosystems, and on human well-being. If we continue to emit greenhouse gases into the atmosphere, the consequences of climate change will grow more severe.

Scientific evidence for climate change is extensive

For years, scientists have studied the many aspects of climate change in enormous breadth, depth, and detail. Labs and government agencies have been continuously monitoring climate variables with everything from thermometers to satellites, building up detailed long-term databases for researchers to study. As a result, the scientific literature today is replete with independent published studies, and we have gained a rigorous and reliable understanding of most aspects of climate change.

To make this vast and growing research knowledge accessible to policymakers and the public, various bodies have taken up the task of reviewing and summarizing it. For instance, the U.S. Global Change Research Program publishes periodic assessments of climate change's impacts in the United States (p. 504). For the world as a whole, the **Intergovernmental Panel on Climate Change (IPCC)** plays this role. This international panel consists of many hundreds of scientists and governmental representatives. Established in 1988 by the United Nations Environment Programme (UNEP) and the World Meteorological Organization (WMO), the IPCC was awarded the Nobel Peace Prize in 2007 for its work in informing the world of the trends and impacts of climate change.

In that year the IPCC released its *Fourth Assessment Report*. This report summarized many thousands of scientific studies, and it documented observed trends in surface temperature, precipitation patterns, snow and ice cover, sea levels, storm intensity, and other factors. It also predicted future changes in these phenomena after considering a range of potential scenarios for future greenhouse gas emissions. The report addressed impacts of current and future climate change on wildlife, ecosystems, and society. Finally, it discussed strategies we might pursue in response to climate change. TABLE 18.2 summarizes some of the IPCC report's major observed and predicted trends and impacts.

Like all science, the IPCC's reports deal in uncertainties, so their authors take great care to assign statistical probabilities to conclusions and predictions. Moreover, the IPCC's estimates regarding impacts on society are conservative, because its scientific conclusions need to be approved by representatives of the world's national governments, some of which are reluctant to move away from a fossil-fuel-based economy. Since the publication of the *Fourth Assessment Report*, many aspects of climate change have grown more severe faster than predicted. It appears that the report underestimated many changes, while overestimating very few.

As scientists around the world continue to monitor and model our changing climate, new research is being completed faster than ever. The IPCC's *Fifth Assessment Report* is due out in a series of four documents from late 2013 through late 2014. You and your instructor may wish to download these new reports (they will be publicly accessible online) and explore their coverage of the latest research findings.

Temperatures continue to rise

Average surface temperatures on Earth have risen by about 0.9°C (1.6°F) in the past 100 years (FIGURE 18.8). These include increases both in air temperature over land and in sea surface temperature. Most of this century's increase has occurred recently, just since 1975. The numbers of

TABLE 18.2 Observed and *Predicted* Trends and Impacts of Climate Change, from IPCC *Fourth Assessment Report*
GLOBAL PHYSICAL INDICATORS
Average surface temperature increased 0.74°C (1.33°F) in the past 100 years, *and will rise 1.8–4.0°C (3.2–7.2°F) in the 21st century.*
Oceans absorbed >80% of heat added to the climate system and warmed to depths of at least 3000 m (9800 ft).
Glaciers, snow cover, ice caps, ice sheets, and sea ice are melting *and will continue, contributing to sea level rise.*
Sea level rose by an average of 17 cm (7 in.) in the 20th century *and will rise 18–59 cm (7–23 in.) in the 21st century.*
Ocean water became more acidic by about 0.1 pH unit *and will decrease in pH by 0.14–0.35 units more by century's end.*
Storm surges increased, *and will increase further.*
Carbon uptake by terrestrial ecosystems will peak by mid-century, then weaken or reverse, amplifying climate change.
REGIONAL PHYSICAL INDICATORS
Arctic areas warmed fastest. *Future warming will be greatest in the Arctic and greater over land than over water.*
Summer Arctic sea ice thinned by 7.4% per decade since 1978.
Precipitation will increase at high latitudes and decrease at subtropical latitudes, making wet areas wetter and dry ones drier.
Droughts became longer, more intense, and more widespread since the 1970s, especially in the tropics and subtropics.
Droughts and flooding will increase, leading to agricultural losses.
Hurricanes have intensified in the North Atlantic since 1970[1] *and will continue to intensify.*
SOCIAL INDICATORS
Farmers and foresters have had to adapt to altered growing seasons and disturbance regimes.
Temperate-zone crop yields will rise until 3°C (5.4°F) warming. In dry tropics and subtropics, crop productivity will fall.
Impacts on biodiversity will cause losses of food, water, and other ecosystem goods and services.
Sea level rise will displace people from islands and coasts.
Melting of mountain glaciers will reduce water supplies to millions.
Economic costs will outweigh benefits. Costs could average 1–5% of GDP globally for 4°C (7.2°F) of warming.
Poorer nations and communities are suffering more.
Human health will suffer as increased warm-weather health hazards outweigh decreased cold-weather health hazards.
BIOLOGICAL INDICATORS
Species ranges are shifting toward the poles and upward in elevation, *and they will continue to shift.*
The timing of seasonal phenomena (such as migration and breeding) is shifting *and will continue to shift.*
About 20–30% of species studied so far could face extinction.
Corals will suffer further bleaching and ocean acidification.

Observed phenomena are in plain text. *Predicted future phenomena are in italicized text.* This table expresses mean estimates only. (The IPCC report provides ranges and statistical probabilities as well.) Data from IPCC, 2007. Fourth assessment report.

extremely hot days and heat waves have increased globally, whereas the number of cold days has decreased. According to the WMO, the 18 warmest years on record since global measurements began 150 years ago have all been since 1990. The decade from 2001–2010 was the hottest ever recorded, and since the 1960s each decade has been warmer than the last.

In fact, consider this: If you were born after 1985, you have never in your life lived through a month with average global temperatures lower than the 20th-century average.

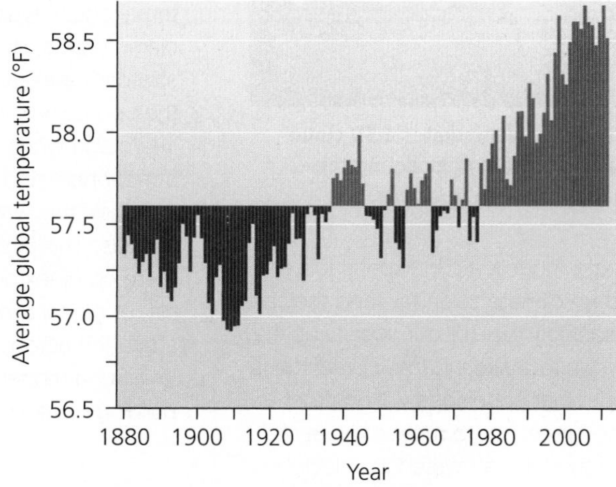

(a) Global temperature measured since 1880

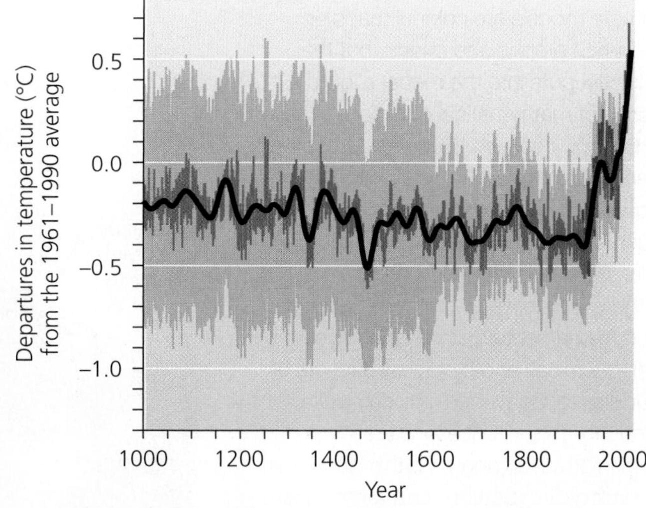

(b) Northern Hemisphere temperature over the past 1000 years

FIGURE 18.8 Global temperatures have risen sharply in the past century. Data from thermometers **(a)** show changes in Earth's average surface temperature since 1880. Since 1976, every single year has been warmer than average. In **(b)**, proxy indicators (**blue line**) and thermometer data (**red line**) together show average temperatures in the Northern Hemisphere over the past 1000 years. The gray-shaded zone represents the 95% confidence range. *Data from (a) NOAA National Climatic Data Center as presented in U.S. Global Change Research Program, 2013.* National climate assessment. *Draft for public review; and (b) IPCC, 2001.* Third assessment report.

How Do Climate Models Work?

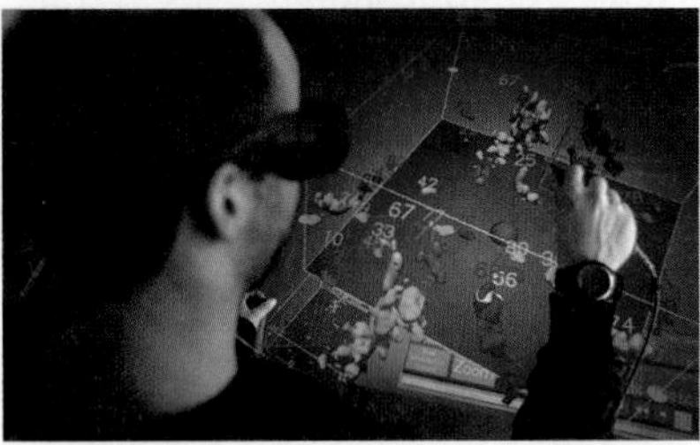

A researcher manipulates a 3D cloud simulator to help in modeling the climatic influence of clouds.

Climate models are indispensable for modern climate science—and they are increasingly vital for our society as they allow us to make informed predictions about what conditions will confront us in the future. Yet to most of us, a climate model is a mysterious black box. So how exactly do scientists go about creating a climate model?

The output we often see from climate models are colorful maps or data-rich graphs and charts, but the scientist puts into the model a long series of mathematical equations. These equations describe how various components of Earth's systems function. Some equations are derived from physical laws such as those on the conservation of mass, energy, and momentum (p. 23). Others are derived from observational and experimental data on the physical, chemical, and biological aspects of our planet. Converted into computing language, these equations are integrated with information about Earth's landforms, hydrology, vegetation, and atmosphere (**FIGURE 1**). All these types of information are the building blocks of a climate model.

The number of these building blocks determines the complexity of the model. Earth's climate system is mind-bogglingly complex, and modelers will never capture all the factors that influence climate. Yet as computers become more powerful and models more sophisticated, they are incorporating more and more of the factors—major and minor, direct and indirect—that affect climate.

To handle the complexity, generally a model is composed of multiple submodels. Essentially, submodels for components of the Earth system—ocean waters, sea ice, glaciers, forests, deserts, troposphere, stratosphere—are each built and are then combined into a global model. Years ago when models first coupled atmosphere and ocean components together, they were called "coupled" models. This is standard in today's far more complex *general circulation models*, or *global climate models* (which share the acronym GCM).

For a model to function, all the building blocks must be given equations to make them behave realistically in space and time. In the real climate system, time is continuous and spatial effects reach down to the level of molecules interacting with one another. But the virtual reality of climate models cannot be so deep and precise—there is simply not enough computer power available. Instead, modelers approximate reality by dividing time up into periods (called time steps) and by dividing the Earth's surface up into cells or boxes according to a grid (called grid boxes) (**FIGURE 2**).

Each grid box contains land, ocean, or atmosphere, much like a digital photograph is comprised of discrete pixels of certain colors. The grid boxes are arrayed in a three-dimensional layer by latitude and longitude, or in equal-sized polygons.

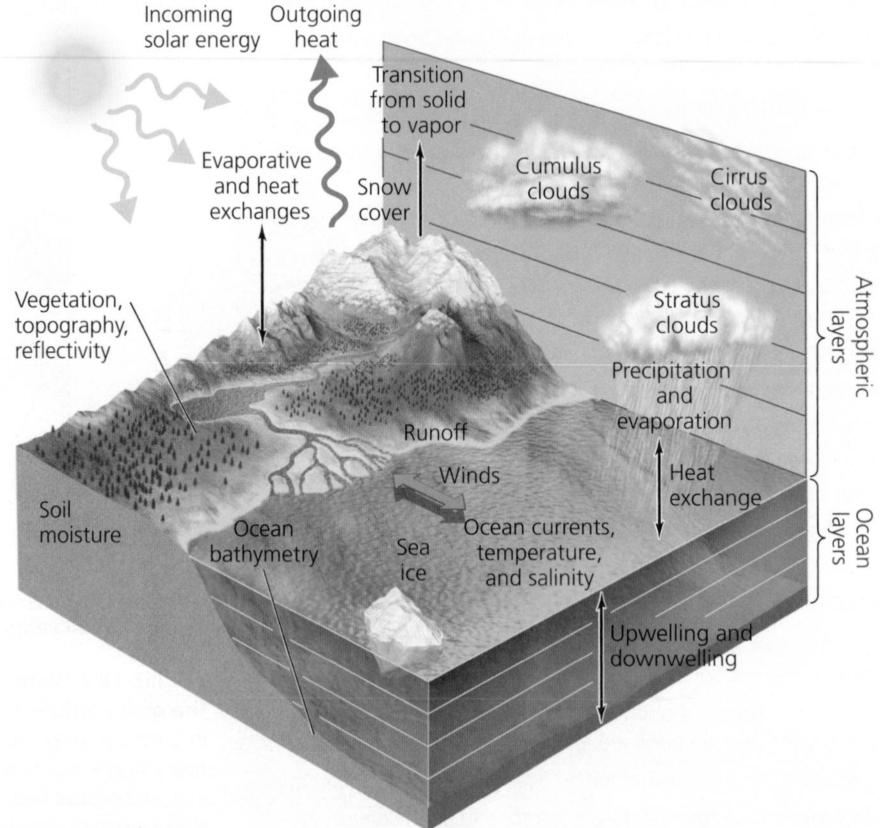

FIGURE 1 Climate models incorporate a diversity of natural factors and processes. Anthropogenic factors can then be added in.

The finer-scale the grid, the greater resolution the model will have, and the better it will be able to predict results region by region. However, more resolution means more computing power is needed, and climate models already strain the most powerful supercomputing networks. The best climate models today feature dozens of grid boxes piled up from the bottom of the ocean to the top of the atmosphere, with each grid box measuring a few dozen miles wide, and time measured in periods of just minutes.

Once the grid is established, the processes that drive climate are assigned to each grid box, with their rates parceled out among the time steps. The model lets the grid boxes interact through time by means of the flux of materials and energy into and out of each grid box.

Once modelers have input all this information, learned from our study of Earth and the climate system, they let the model run through time and simulate climate, from the past through the present and into the future. If the computer simulation accurately reconstructs past and present climate, then that gives us confidence that it may predict future climate accurately as well.

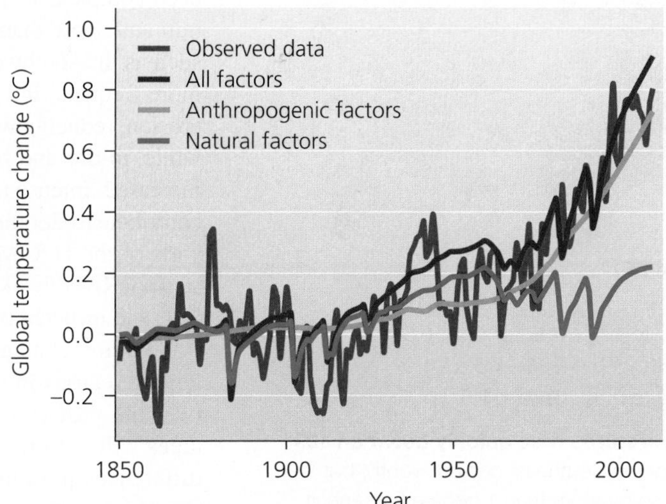

FIGURE 3 Models that incorporate both natural and anthropogenic factors predict observed climate trends best. *Adapted from U.S. Global Change Research Program, 2013.* National climate assessment. *Draft for public review.*

A number of studies have compared model runs that include only natural processes, model runs that include only human-generated processes, and model runs that combine both. Repeatedly these studies have found that the model runs that incorporate both human and natural processes are the ones that fit the real-world climate observations the best (**FIGURE 3**). This supports the idea that human activities are influencing our climate.

The major human influence on climate is our emission of greenhouse gases, and modelers need to select values to enter for future emissions if they want to predict future climate. Generally they will run their simulations multiple times, each time with a different emission amount according to a specified scenario. Differences between the results from such scenarios tell us what influence these different actions would have. You can see results from such a comparison in Figure 18.24 (p. 507).

Throughout this chapter, you will see figures that show results of various models. In crafting its assessment reports, the IPCC consults nearly two dozen major models, considers the strengths and weaknesses of each, and presents the best summary its authors can muster.

Researchers are constantly testing and evaluating their climate models. They continually improve them by incorporating what is learned from new research and by taking advantage of what increasingly powerful computing technologies will allow. As their work proceeds, we can expect better and better predictions about what climate conditions we will encounter in the future. ■

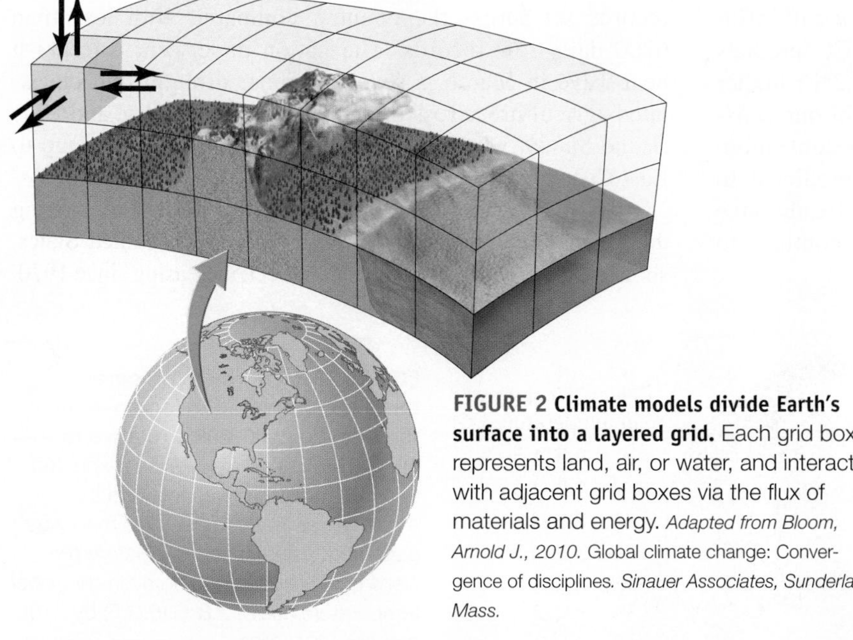

FIGURE 2 Climate models divide Earth's surface into a layered grid. Each grid box represents land, air, or water, and interacts with adjacent grid boxes via the flux of materials and energy. *Adapted from Bloom, Arnold J., 2010.* Global climate change: Convergence of disciplines. *Sinauer Associates, Sunderland, Mass.*

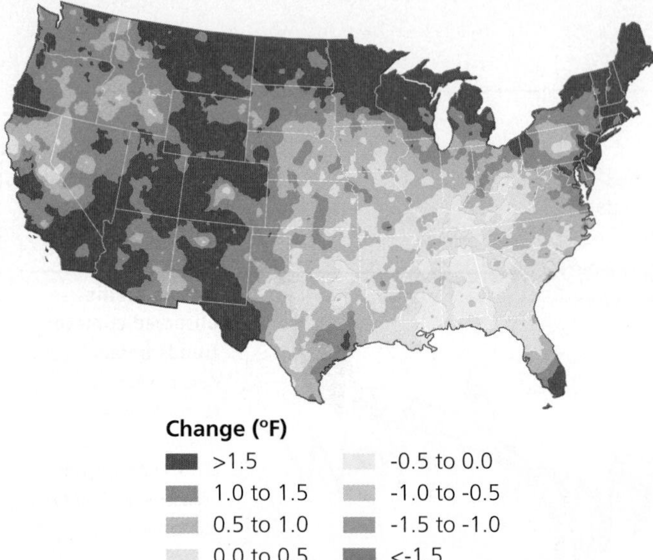

FIGURE 18.9 U.S. temperatures rose quickly between 1991 and 2011. Some areas of the Southeast cooled slightly, but most of the nation warmed by more than 1 degree Fahrenheit (0.6°C).*Data from NOAA National Climatic Data Center as presented in U.S. Global Change Research Program, 2013.* National climate assessment. *Draft for public review.*

DATA Q By how much did the average temperature rise or fall where you live? How does this compare with changes in other parts of the country?

From just 1991 to 2011, most areas of the United States experienced temperature increases of more than 1 degree Fahrenheit (FIGURE 18.9).

In the next two decades, we can expect average global surface temperatures to rise another 0.4°C (0.7°F), according to IPCC analysis. Even if we were to halt all our greenhouse gas emissions today, temperatures would still rise 0.1°C (0.2°F) per decade because of a time lag: Some gases already in the atmosphere have yet to exert their full influence. At the end of the 21st century, the IPCC predicts global temperatures will be 1.8–4.0°C (3.2–7.2°F) higher than today's, depending on how well we control our emissions. Unusually hot days and heat waves will become more frequent. Future changes in temperature are predicted to vary from region to region in ways that they already have (FIGURE 18.10). For example, polar regions will continue to experience the most intense warming.

Precipitation is changing, too

A warmer atmosphere speeds evaporation and holds more water vapor, and precipitation has increased worldwide by 2% over the past century. However, changes in precipitation patterns have been complex, with some regions of the world receiving more rain and snow than usual and others receiving less. In regions such as the southwestern United States, droughts have become more frequent and severe, harming agriculture, worsening soil erosion, reducing water supplies, and triggering wildfire. Meanwhile, in dry and humid regions alike, heavy rain events have increased. Intense rainstorms, combined with land use changes, contribute to flooding, such as the 2008 floods in Iowa and other parts of the U.S. Midwest and the 2011 floods along the Mississippi River that killed dozens of people, left thousands homeless, and inflicted billions of dollars in damage.

Future changes in precipitation are predicted to vary among regions in ways that parallel regional differences seen over the past century (FIGURE 18.11). In many regions, rainy areas will get rainier and dry areas will get drier, magnifying differences in rainfall that already exist and worsening water shortages in many developing countries of the arid subtropics. In many areas, heavy precipitation events will become more frequent, increasing the risk of flooding. Some of these patterns may relate to a possible expansion or intensification of Earth's Hadley cells (p. 455); this is currently an active area of study.

Extreme weather is becoming "the new normal"

The sheer number of extreme weather events—droughts, floods, tornadoes, hurricanes, snowstorms, heat waves—in recent years has caught everyone's attention, and weather records are being broken left and right. In the United States, 2012 was the hottest year ever recorded—1 full degree Fahrenheit hotter than average, with over 34,000 daily high records set across the country, compared with less than 6700 daily low records. The nation underwent a freakish heat wave in March, a severe summer drought that devastated agriculture across three-fifths of the country, and Hurricane Sandy, which caused over $60 billion in damage to East Coast states.

The U.S. Climate Extremes Index, an index summarizing the frequency of extreme weather events in the United States, shows that these types of events have been increasing since 1970.

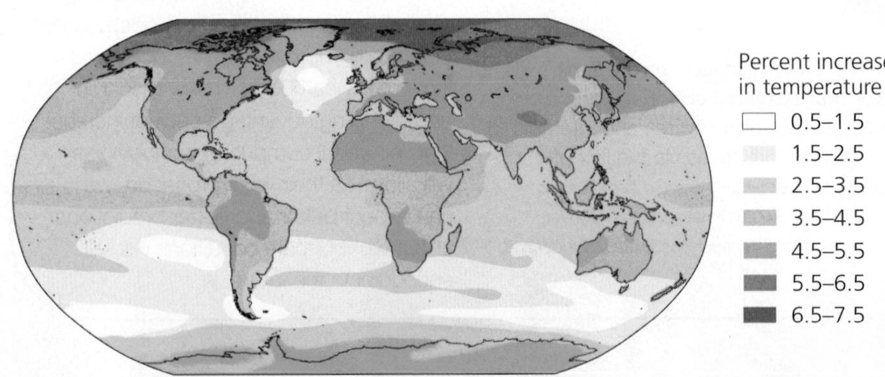

FIGURE 18.10 Surface temperatures are projected to increase for the decade 2090–2099, relative to 1980–1999. Land masses are expected to warm more than oceans, and the Arctic will warm the most. This map was generated using an intermediate emissions scenario involving an average global temperature rise of 2.8°C (5.0°F) by 2100. *Data from IPCC, 2007.* Fourth assessment report. Climate Change 2007: Synthesis Report, *Fig SPM.6.*

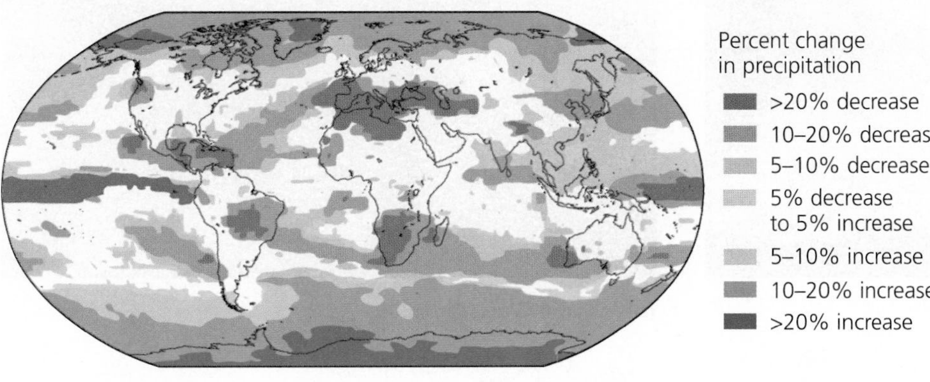

FIGURE 18.11 Precipitation (June–Aug.) is projected to change for the decade 2090–2099, relative to 1980–1999. Browner shades indicate less precipitation, and bluer shades indicate more. White indicates areas where models did not agree. This map was generated using an intermediate emissions scenario involving an average global temperature rise of 2.8°C (5.0°F) by 2100. *Data from IPCC, 2007.* Fourth assessment report. Climate Change 2007: Synthesis Report, *Fig 3.3.*

Moreover, a 2011 study by climate scientist James Hansen and others revealed that summer temperatures since the 1950s have become not only warmer, but also more variable. As a result, extreme summers (some of them unusually cool, more of them unusually warm) have occurred more and more frequently.

Scientists are not the only ones to notice the increase in extreme weather events. The insurance industry is finely attuned to such patterns, since insurers are the ones paying out money each time a major storm, drought, or flood hits. The major German insurer Munich Re calculated that from 1980 to 2011, extreme weather events causing losses increased fully 5 times in North America. They rose by 4 times in Asia, 2.5 in Africa, 2 in Europe, and 1.5 in South America. And this was not even counting the events of 2012.

Researchers have long conservatively stated that although climate trends influence the probability of what the weather may be like on any given day, no single particular weather event can reliably be attributed to climate change. However, as we gain a better understanding of what causes extreme weather events, it is starting to become possible to link certain weather events to climate change.

In 2012, a research paper by Jennifer Francis of Rutgers University and Stephen Vavrus of the University of Wisconsin outlined a mechanism that may explain how global warming may commonly lead to more extreme weather. Their analysis indicated that because warming has been greater in the Arctic than at lower latitudes, this has weakened the intensity of the Northern Hemisphere's polar jet stream. This *jet stream* is a high-altitude air current that blows west-to-east and meanders north and south, influencing much of the weather from day to day across North America and Eurasia.

As the jet stream slows down, its meandering loops become longer. These long lazy loops move west to east more slowly, and may get stuck in a north–south orientation for long periods of time. This is what meteorologists call an *atmospheric blocking pattern*, because it can block the eastward movement of weather systems and hold them in place (FIGURE 18.12). When this happens, a rainy system that would normally move past a city in a day or two might instead be held in place for several days, resulting in flooding. Or dry conditions over a farming region might last two weeks instead of two days, causing a drought. Cold spells last longer, and hot spells last longer, too.

Indeed, the record-breaking heat wave that roasted the eastern United States in March 2012 resulted after the jet stream became stuck in place (see Figure 18.12b). Atmospheric blocking patterns also were associated with the 2011 drought in Texas, the 2012 wildfires in Colorado, both floods and heat waves in Europe, and various other extreme weather events. Additional independent research has supported Francis and Vavrus's jet stream hypothesis, and if it continues to be strengthened by further data, it would provide a valuable explanatory mechanism for how climate change brings extreme weather to North America and Europe.

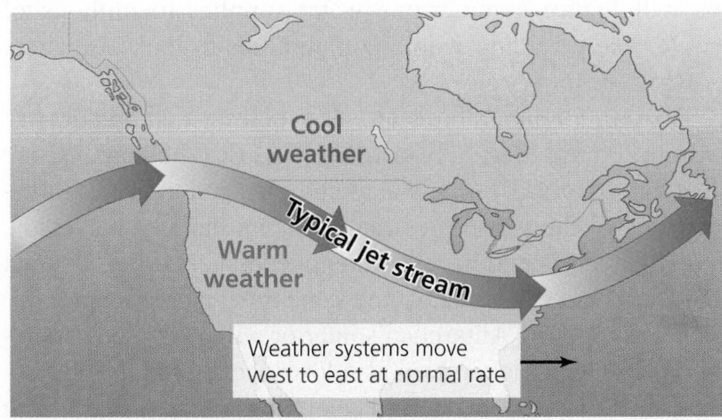

(a) Normal jet stream

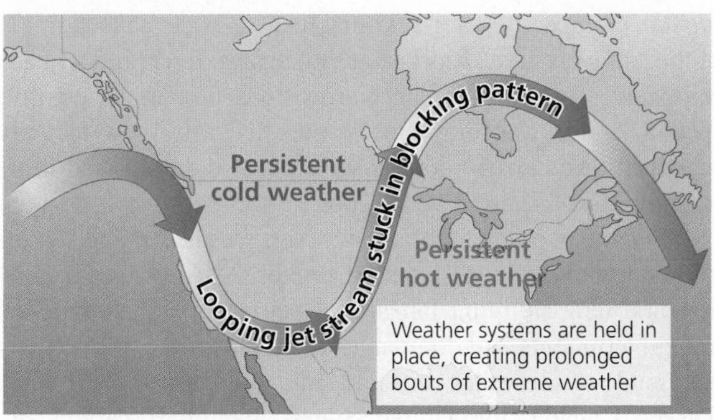

(b) Jet stream in March 2012

FIGURE 18.12 Changes in the jet stream can cause extreme weather events. When Arctic warming slows the jet stream, it departs from its normal configuration **(a)** and goes into a blocking pattern **(b)** that stalls weather systems in place, leading to extreme weather events. The blocking pattern shown here brought record-breaking heat to the eastern United States in March 2012.

(a) Jackson Glacier in 1911

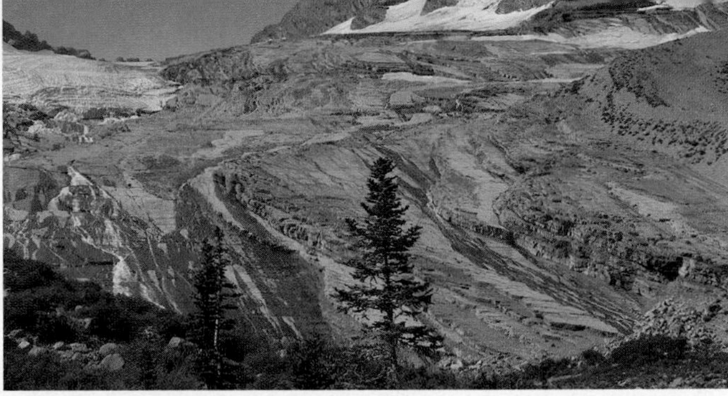

(b) Jackson Glacier in 2009

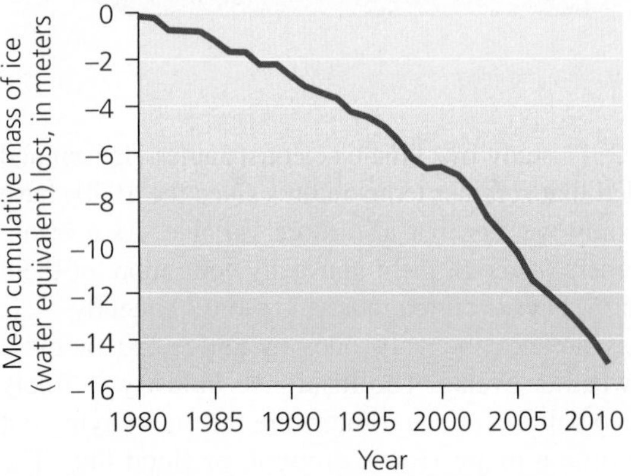

(c) The world's major glaciers are losing mass

FIGURE 18.13 Glaciers are melting rapidly as global warming proceeds. The Jackson Glacier in Glacier National Park, Montana, retreated substantially between **(a)** 1911 and **(b)** 2009. The graph **(c)** shows average declines in mass in 37 of the world's major glaciers monitored since 1980. *Data from World Glacier Monitoring Service.*

Melting ice has far-reaching effects

As the world warms, mountaintop glaciers are disappearing (**FIGURE 18.13**). Between 1980 and 2011, the World Glacier Monitoring Service estimates that the world's glaciers on average have each lost mass equivalent to 15 m (49 ft) vertical thickness of water. Many glaciers on tropical mountaintops have disappeared already. In Glacier National Park in Montana, only 25 of 150 glaciers present at the park's inception remain, and scientists estimate that by 2030 even these will be gone.

Mountains accumulate snow in the winter and release meltwater gradually during the summer. Over one-sixth of the world's people live in regions that depend on mountain meltwater. As warming temperatures diminish mountain glaciers, this will reduce summertime water supplies to millions of people, likely forcing whole communities to look elsewhere for water, or to move.

Warming temperatures are also melting vast amounts of ice in the Arctic. Recent research reveals that the immense ice sheet that covers Greenland is melting faster and faster. At the other end of the world, in Antarctica, coastal ice shelves the size of Rhode Island have disintegrated as a result of contact with warmer ocean water. At the same time, increased precipitation is supplying Antarctica's interior with extra snow, making its ice sheet thicker even as it loses ice around its edges.

One reason warming is accelerating in the Arctic is that as snow and ice melt, darker, less reflective surfaces (such as bare ground and pools of meltwater) are exposed, and Earth's *albedo*, or capacity to reflect light, decreases. As a result, more of the sun's rays are absorbed at the surface, fewer reflect back into space, and the surface warms. In a process of positive feedback, this warming causes more ice and snow to melt, which in turn causes more absorption of radiation and more warming (see Figure 5.1b, p. 106).

Scientists predict that snow cover, ice sheets, and sea ice will continue to diminish near the poles. As Arctic sea ice disappears, new shipping lanes are opening up for commerce, and governments and companies are rushing to exploit newly accessible underwater oil and mineral reserves. Already, Russia, Canada, the United States, and other nations are jockeying for position, using new survey data to try to lay claim to regions of the Arctic as the ice melts.

Warmer temperatures in the Arctic are also causing *permafrost* (permanently frozen ground) to thaw. As ice crystals within permafrost melt, the thawing soil settles, destabilizing buildings, pipelines, and other infrastructure. When permafrost thaws, it also can release methane that has been stored for thousands of years. Because methane is a potent greenhouse gas, its release acts as a positive feedback mechanism that intensifies climate change. No one knows how much methane lies beneath Arctic permafrost, but it is a very large amount, and some researchers worry that its release could drive climate change beyond our control.

Rising sea levels may affect hundreds of millions of people

As glaciers and ice sheets melt, increased runoff into the oceans causes sea levels to rise. Sea levels also are rising because ocean water is warming, and water expands in volume as its temperature increases. In fact, recent sea level rise has resulted primarily from the thermal expansion of seawater. Worldwide, average sea levels have risen 21 cm (8.3 in.) in the past 130 years (**FIGURE 18.14**). Studies indicate that seas rose by 1.8 mm/year from 1961 to 2003 and 2.9–3.4 mm/year from 1993 to 2010. These numbers represent vertical rises in water level, and on most coastlines a vertical rise of a few inches means many feet of horizontal incursion inland.

Higher sea levels lead to beach erosion, coastal flooding, intrusion of salt water into aquifers, and greater impacts from storm surges. A *storm surge* is a temporary and localized rise in sea level brought on by the high tides and winds associated with storms. The higher that sea level is to begin with, the further inland a storm surge can reach. In 1987, unusually high

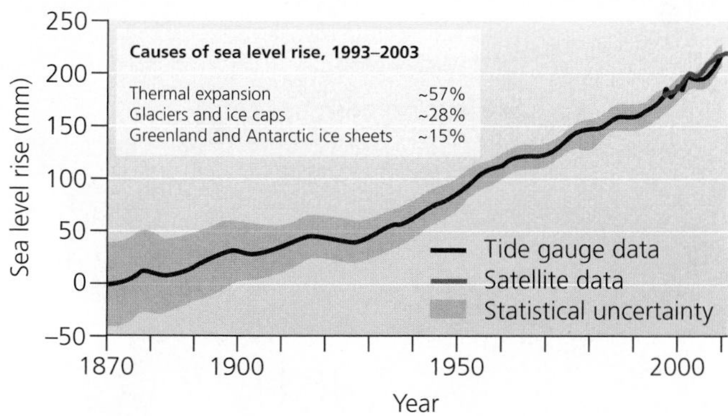

FIGURE 18.14 Global average sea level has risen over 200 mm (7.9 in.) since 1870. Data from tide gauges and satellite observations each confirm the same trend. Thermal expansion of water accounts for most sea level rise. *Data from Intergovernmental Panel on Climate Change, 2007.* Fourth assessment report; *and CSIRO.*

waves struck the Maldives and triggered a campaign to build a large seawall around Malé, the nation's capital. "The Great Wall of Malé" is intended to protect buildings and roads by dissipating the energy of incoming waves during storm surges.

On December 26, 2004, the Maldives was hit by a massive tsunami (pp. 42–43) that devastated coastal areas throughout the Indian Ocean. The tsunami was triggered by an earthquake, not by climate change—yet as sea level rises, the damage that such natural events can inflict increases considerably. The tsunami killed 100 Maldives residents and left 20,000 homeless. Schools, boats, tourist resorts, hospitals, and transportation and communication infrastructure were damaged or destroyed. The World Bank estimated that direct damage in the Maldives totaled $470 million, an astounding 62% of the nation's gross domestic product (GDP). Indirect damage from soil erosion, saltwater contamination of aquifers, and other impacts continues to cause further economic losses today.

The Maldives has actually fared better against sea level rise than many other island nations. It has seen sea levels rise about 3 mm per year since 1990, but most Pacific islands are experiencing greater rises in sea level, some up to 9 mm/year. Regions experience differing amounts of sea level change because land may be rising or subsiding naturally, depending on local geological conditions.

In the United States, Hurricane Sandy demonstrated the impact that storm surges can have even on highly developed metropolitan areas (**FIGURE 18.15**). This massive hurricane battered the eastern part of the nation in October 2012, causing over $60 billion in damage and leaving over 130 people dead and thousands homeless. New York City and the New Jersey coast bore the brunt of the storm. In New Jersey, thousands of beach houses were destroyed, iconic boardwalks were washed away, and whole coastal communities were inundated with salt water and tons of sand thrown up by the storm. In Manhattan, economic activity ground to a halt as tunnels and

FIGURE 18.15 Climate change contributes to the power and reach of storms like Hurricane Sandy. Damage was extensive in the superstorm's aftermath. The map shows areas in New York City flooded by the storm. The graph shows sea level rise in New York City in the past century. *Map data from* The New York Times *as adapted from federal agencies; graph data from New York City Panel on Climate Change 2010.*

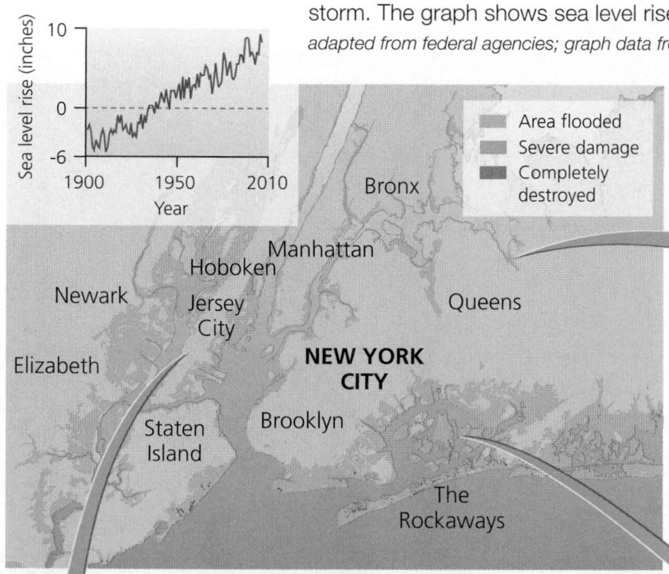

subway stations flooded and countless vehicles and buildings suffered damage. A fire broke out amid flooded homes in Queens and destroyed an entire neighborhood.

Hurricane Sandy was not directly and solely caused by global warming, but in a statistical sense it was certainly facilitated and strengthened by it. The warmer ocean water that has resulted from climate change increases the chances of large and powerful hurricanes. The warmer atmosphere retains more moisture that a hurricane can dump onto land. A blocking pattern in the jet stream contributed to Sandy's energy. And higher sea levels magnify the damage caused by storm surges.

In Sandy's aftermath, an apt metaphor spread across the Internet: When a baseball player takes steroids and starts hitting more home runs, you can't attribute any one particular home run to the steroids, but you *can* conclude that steroids were responsible for the increase in home runs. Our greenhouse gas emissions are like steroids that are supercharging our climate and making extreme weather events more likely.

Seven years before Sandy, the United States was hit by an even more costly storm. Hurricane Katrina (followed by Hurricane Rita) slammed into New Orleans and the Gulf Coast in 2005, killing more than 1800 people and causing over $80 billion in damage. Outside New Orleans today, marshes of the Mississippi River delta are being lost rapidly as rising seas eat away at coastal vegetation (pp. 389–390). These coastal wetlands are also being lost because dams upriver hold back silt that once maintained the delta, because petroleum extraction has caused land to subside, and because salt water is encroaching up channelized waterways, killing freshwater plants. All told, more than 2.5 million ha (1 million acres) of Louisiana's coastal wetlands have vanished since 1940. Continued wetland loss will deprive New Orleans (much of which is below sea level, safeguarded only by levees) of protection against future storm surges.

Across the United States, 53% of the population lives in coastal counties, so a great many people are vulnerable to impacts from storm surges. Different stretches of coast are experiencing different degrees of sea level rise (**FIGURE 18.16**).

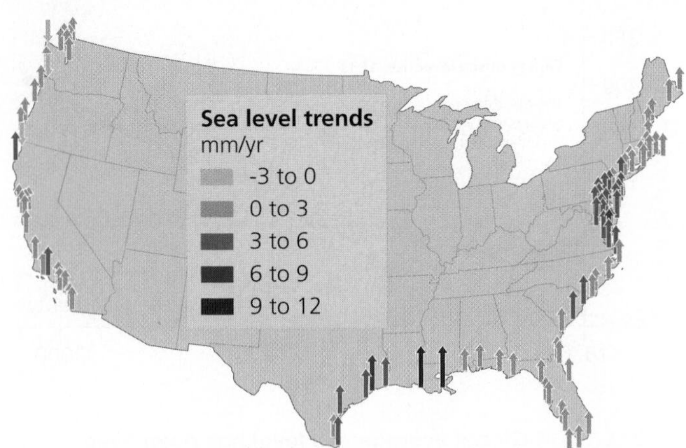

FIGURE 18.16 Sea level is rising at different rates along stretches of the U.S. coast. Rates are highest where land is subsiding. Most vulnerable to future damage are the Gulf Coast and the central Atlantic Seaboard. *Data from National Oceanic and Atmospheric Administration.*

In its 2007 assessment report, the IPCC predicted that mean sea level would rise 18–59 cm (7–23 in.) higher by the end of the 21st century, depending on our level of emissions. However, new research is finding that Greenland's ice is melting at an accelerating rate. If polar melting continues to accelerate, then sea levels will rise more quickly.

Indeed, in 2009 a European research team used a different approach to estimate future sea level rise by carefully examining the rates at which it has risen and fallen in the past. This team forecast that sea level would rise 0.9–1.3 meters by the year 2100. Some researchers are now using a figure of a 1-meter rise to assess risk. Jeremy Weiss of the University of Arizona and his colleagues have pointed out that 3.9 million Americans live within 1 vertical meter of the high tide line, and they estimate that a 1-m rise threatens 180 U.S. cities with losing an average of 9% of their land area. Miami, Tampa, New Orleans, and Virginia Beach were judged most at risk (**FIGURE 18.17**).

Whether sea levels this century rise 18 cm or 1 m, hundreds of millions of people will be displaced or will need to invest in costly efforts to protect against high tides and storm surges. Densely populated regions on low-lying river deltas, such as Bangladesh, would be most affected. So would storm-prone regions such as Florida, coastal cities such as Houston and Charleston, and areas where land is subsiding, such as the U.S. Gulf Coast. Many Pacific islands would need to be evacuated. Already some nations such as Tuvalu and the

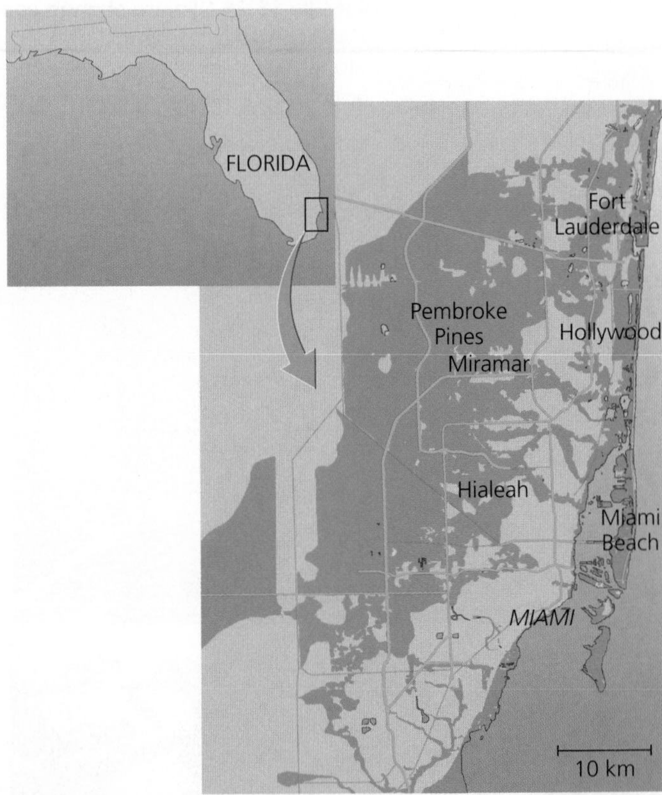

FIGURE 18.17 Miami, Florida, is one of many U.S. cities vulnerable to sea level rise. Shown are areas of the Miami region that would be flooded by a 1-meter rise in sea level. *Data from Jeremy Weiss, Environmental Studies Laboratory, Department of Geosciences, University of Arizona.*

Maldives fear for their very existence. In the meantime, island nations such as the Maldives are likely to suffer shortages of fresh water as rising seas bring salt water into aquifers. The contamination of groundwater and soils by seawater also threatens coastal areas such as Tampa, Florida, which depend on small lenses of fresh water that float atop saline groundwater.

WEIGHING THE ISSUES

ENVIRONMENTAL REFUGEES The Pacific island nation of Tuvalu has been losing 9 cm (3.5 in.) of elevation per decade to rising seas. Appeals from Tuvalu's 11,000 citizens were heard by New Zealand, which began accepting "environmental refugees" from Tuvalu in 2003. Do you think the rest of the world should grant such environmental refugees international status and assume some responsibility for taking care of them? Do you think a national culture can survive if its entire population is relocated? Think of the tens of thousands of refugees from Hurricane Katrina. How did their lives and culture fare in the wake of that tragedy?

Climate change threatens coral reefs

Maldives residents also worry about damage to their coral reefs (pp. 431–432), marine ecosystems that are critical for their economy. Coral reefs provide habitat for important fish that are consumed locally and exported. They offer snorkeling and scuba diving sites for tourism. Reefs also reduce wave intensity, protecting coastlines from erosion. Around the world, rising seas are eating away at the coral reefs, mangrove forests, and salt marshes that serve as barriers protecting our coasts (pp. 430–431).

Climate change poses two additional threats to coral reefs. First, warmer waters contribute to coral bleaching (p. 431), which kills corals. Second, enhanced CO_2 concentrations in the atmosphere alter ocean chemistry, leading to **ocean acidification** (pp. 426, 428–429). As ocean water absorbs atmospheric CO_2, it becomes more acidic, and this impairs the ability of corals and other organisms to build exoskeletons of calcium carbonate. Indeed, ocean acidification and the potential loss of coral reefs worldwide (explained in **The Science behind the Story**, Chapter 16, pp. 428–429) threaten to become one of the most serious and far-reaching impacts of global climate change. The oceans have already decreased by 0.1 pH unit, and they are predicted to decline in pH by 0.14–0.35 more units over the next 100 years. This could easily be enough to destroy most of our planet's living coral reefs. Such destruction could be catastrophic for marine biodiversity and fisheries, because so many organisms depend on living coral reefs for food and shelter.

Climate change affects organisms and ecosystems

As the coral reef crisis shows, changes in Earth's physical systems have consequences for living things. Organisms are adapted to their environments, so they are affected when we alter those environments. As global warming proceeds, it is modifying all manner of biological phenomena that are regulated by temperature. In the spring, plants are now leafing out earlier, insects are hatching earlier, birds are migrating earlier, and animals are breeding earlier. These shifts can create mismatches in seasonal timing with phenomena that are regulated by other factors. For example, European birds known as great tits had evolved to time their breeding so as to raise their young when caterpillars peak in abundance. Now caterpillars are peaking earlier, but the birds have been unable to adjust, and fewer young birds are surviving.

Biologists are also recording spatial shifts in the ranges of organisms, as plants and animals move toward the poles or upward in elevation (i.e., toward cooler regions) as temperatures warm (**FIGURE 18.18**). As these trends continue, some organisms will not be able to cope, and the IPCC estimates that as many as 20–30% of all plant and animal

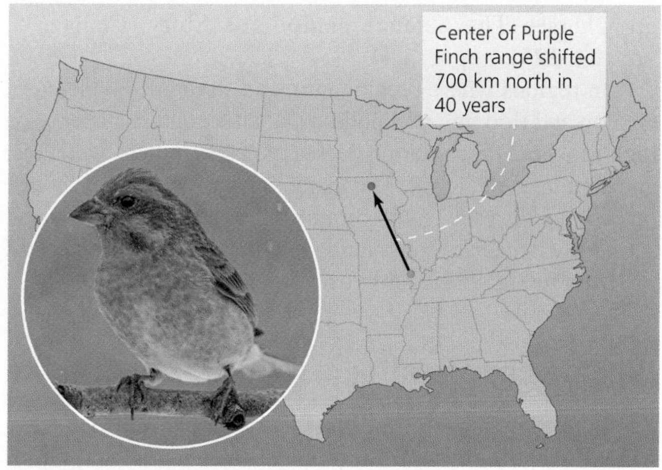

(a) Birds are moving north

(b) Pikas are being forced upslope

FIGURE 18.18 Animal populations are shifting toward the poles and upward in elevation. Fully 177 out of 305 North American bird species have shifted their winter ranges significantly northward in the past 40 years, according to a 2009 analysis of Christmas Bird Count data by National Audubon Society researchers. The purple finch **(a)** has shown the greatest shift; its center of abundance moved 697 km (433 mi) north. Montane animals such as the pika **(b)**, a unique mammal that lives at high elevations in western North America, are being forced upslope (into more limited habitat) as temperatures warm. Many pika populations in the Great Basin have disappeared from mountains already.

species could be threatened with extinction. Trees may not be able to shift their distributions fast enough. Rare species may be forced out of preserves and into developed areas where they cannot survive. Animals and plants adapted to mountainous environments may be forced uphill until there is nowhere left to go.

Effects on plant communities comprise an important component of climate change, because by drawing in CO_2 for photosynthesis, plants act as reservoirs for carbon. If higher CO_2 concentrations enhance plant growth, then more CO_2 might be removed from the air, helping to mitigate carbon emissions, in a process of negative feedback. However, if climate change decreases plant growth (through drought, fire, or disease, for instance), then carbon flux to the atmosphere could increase, in a process of positive feedback. Free-Air CO_2 Enrichment (FACE) experiments are revealing complex answers, showing that extra carbon dioxide can bring both positive and negative results for plant growth (see **The Science behind the Story**, Chapter 5, pp. 124–125).

In regions where precipitation and stream flow increase, erosion and flooding will pollute and alter aquatic systems. In regions where precipitation decreases, lakes, ponds, wetlands, and streams will shrink. The many impacts of climate change on ecological systems will diminish the ecosystem goods and services we receive from nature and that our societies depend on, from food to clean air to drinking water.

Climate change affects society

Drought, flooding, storm surges, and sea level rise have already taken a toll on the lives and livelihoods of millions of people. However, climate change will have still more consequences. These include impacts on agriculture, forestry, health, and economics.

Agriculture For some crops in the temperate zones, moderate warming may slightly increase production because growing seasons become longer. The availability of additional carbon dioxide to plants for photosynthesis may also increase yields (although as mentioned above, elevated CO_2 can have mixed results). However, some research shows that crops become less nutritious when supplied with more carbon dioxide. And if rainfall shifts in space and time, intensified droughts and floods will likely cut into agricultural productivity (**FIGURE 18.19**).

Considering all factors together, the IPCC predicts global crop yields to increase somewhat, but beyond a rise of 3°C (5.4°F), it expects crop yields to decline. In seasonally dry tropical and subtropical regions, growing seasons may be shortened and harvests may be more susceptible to drought. Thus, scientists predict that crop production will fall in these regions even with minor warming. This would worsen hunger in many of the world's developing nations.

Forestry In the forests that provide our timber and paper products, enriched atmospheric CO_2 may spur greater growth in the near term, but other climatic effects such as drought, fire, and disease may eliminate these gains. For example,

FIGURE 18.19 Drought induced by climate change will likely decrease crop yields. Withered corn fields like this one in Illinois were a common sight in 2012, when the U.S. government declared 1000 counties across 26 states to be disaster areas due to drought.

droughts brought about by a strong El Niño in 1997–1998 allowed immense forest fires to destroy vast areas of rainforest in Indonesia, Brazil, and Mexico. In North America, forest managers increasingly find themselves battling catastrophic fires, invasive species, and insect and disease outbreaks. Catastrophic fires are caused in part by decades of fire suppression (p. 321) but are also promoted by longer, warmer, drier fire seasons (see Figure 12.18a, p. 321). Milder winters and hotter, drier summers are promoting outbreaks of bark beetles that are destroying millions of acres of trees (see Figure 12.18b, p. 321).

Health As climate change proceeds, we will face more heat waves—and heat stress can cause death, especially among older adults. A 1995 heat wave in Chicago killed at least 485 people, and a 2003 heat wave in Europe killed 35,000 people. A warmer climate also exposes us to other health problems:

- Respiratory ailments from air pollution, as hotter temperatures promote formation of photochemical smog (p. 465)
- Expansion of tropical diseases, such as dengue fever, into temperate regions as vectors of infectious disease (such as mosquitoes) move toward the poles
- Disease and sanitation problems when floods overcome sewage treatment systems
- Injuries and drowning if storms become more frequent or intense

Health hazards from cold weather will decrease, but most researchers feel that the increase in warm-weather hazards will more than offset these gains.

Economics People will experience a variety of economic costs and benefits from the many impacts of climate change, but on the whole researchers predict that costs will outweigh benefits, especially as climate change grows more severe. Climate change is also expected to widen the gap between rich and poor, both within and among nations. Poorer people have less wealth and technology with which to adapt to climate change, and they rely more on resources

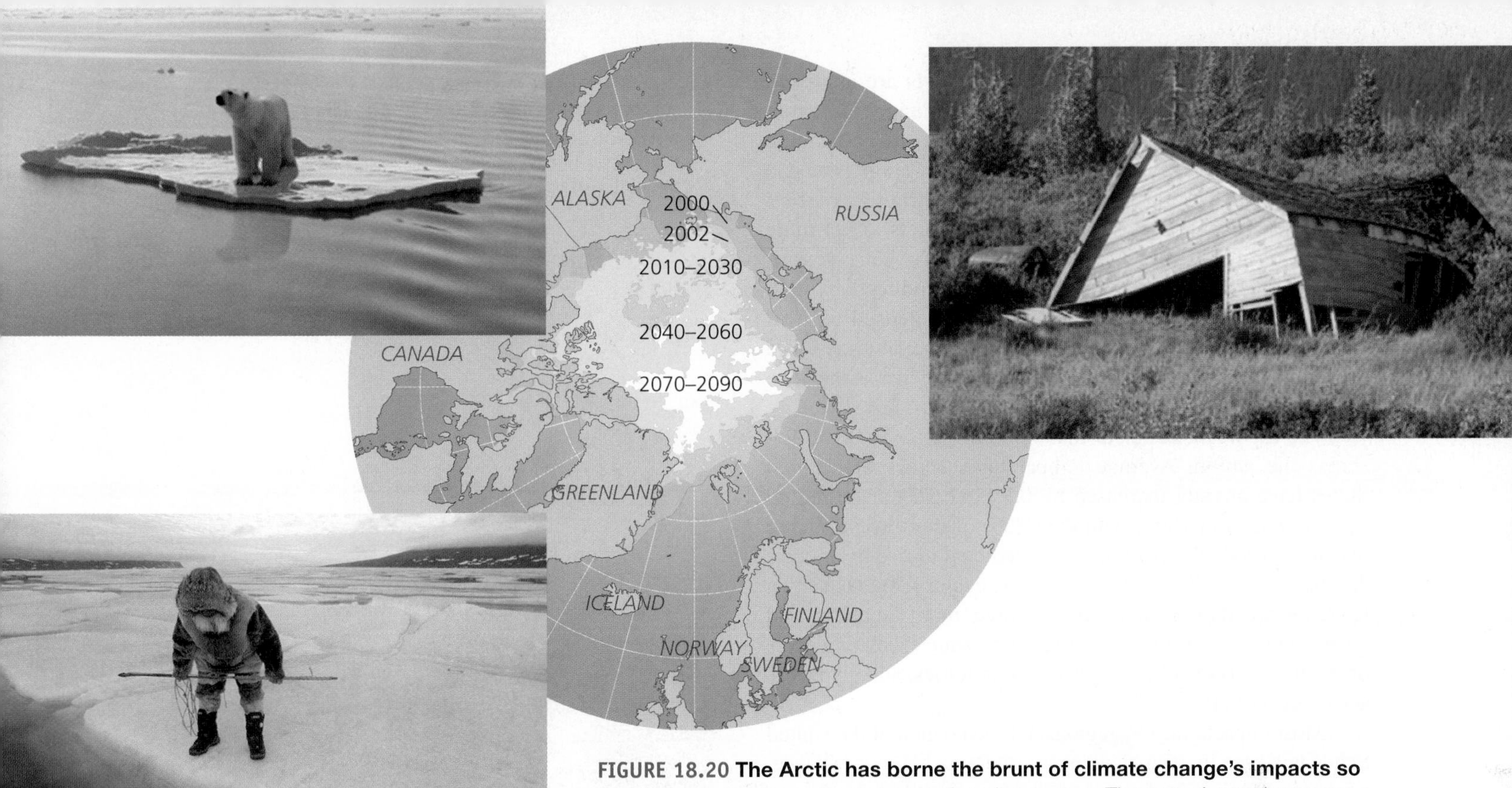

FIGURE 18.20 The Arctic has borne the brunt of climate change's impacts so far. As Arctic sea ice melts, it recedes from large areas. The map shows the mean minimum summertime extent of sea ice for the recent past, present, and future. Inuit people find it difficult to hunt and travel in their traditional ways, and polar bears starve because they are less able to hunt seals. Structures are damaged as permafrost thaws beneath them: Buildings can lean, buckle, crack, and fall. *Map data from National Center for Atmospheric Research and National Snow and Ice Data Center.*

(such as local food and water) that are sensitive to climatic conditions.

From a variety of economic studies, the IPCC estimated that climate change will cost 1–5% of GDP on average globally, with poor nations losing proportionally more than rich nations. Economists trying to quantify damages from climate change by measuring its external costs (pp. 146, 165) have proposed costs of anywhere from \$10 to \$350 per ton of carbon. The highest-profile economic study to date has been the *Stern Review* commissioned by the British government (see **The Science behind the Story**, Chapter 6, pp. 148–149). This exhaustive review concluded that climate change could cost us roughly 5–20% of GDP by the year 2200, but that investing just 1% of GDP starting now could enable us to avoid these future costs. Regardless of the precise numbers, many economists and policymakers are concluding that spending money now to mitigate climate change will save us a great deal more money in the future.

Impacts vary by region

The impacts of climate change are subject to regional variation, so the way each of us experiences these impacts will depend on where we live. Temperature changes have been greatest in the Arctic (**FIGURE 18.20**). Here, ice sheets are melting, sea ice is thinning, storms are increasing, and altered conditions are posing challenges for people and wildlife. As sea ice melts earlier, freezes later, and recedes from shore, it becomes harder for Inuit people and for polar bears alike to hunt the seals they each rely on for food. Thin sea ice is dangerous for people to travel and hunt upon, and in recent years, polar bears have been dying of exhaustion and starvation as they try to swim long distances between ice floes. Permafrost is thawing in the Arctic, destabilizing countless buildings. The strong Arctic warming is melting ice caps and ice sheets, contributing to sea level rise.

WEIGHING THE ISSUES

CLIMATE CHANGE AND HUMAN RIGHTS In 2005, a group representing North America's Inuit people sent a legal petition to the Inter-American Commission on Human Rights, demanding that the United States restrict its greenhouse gas emissions, which the Inuit maintained were destroying their way of life in the Arctic. The Commission dismissed the petition. Do you think Arctic-living people deserve some sort of compensation from industrialized nations whose emissions have caused climate change that is disproportionately affecting the Arctic? What ethical or human rights issues, if any, do you think climate change presents? How could these best be resolved?

For the United States, potential impacts are analyzed and summarized by the U.S. Global Change Research Program, which Congress created in 1990 to coordinate federal climate research. In 2013, this program issued a comprehensive 1200-page report summarizing current research, present trends, and future impacts of climate change on the United States (TABLE 18.3). This report, the *National Climate Assessment*, was produced by 240 scientists overseen by a 60-person staff. Released in draft form online for public comment, it will be finalized and presented to Congress, the president, and the American people in 2014.

The report makes clear that some impacts are being felt across the nation. Average temperatures across the United States have already increased by 0.8°C (1.5°F) since record keeping began in 1895, with over 80% of this rise occurring just since 1980. Temperatures are predicted to rise by another 2.1–6.2°C (4–11°F) by the end of this century (FIGURE 18.21). Extreme weather events have become more frequent and costly, and will continue to worsen, imposing escalating costs on farmers, city-dwellers, coastal communities, and taxpayers across the country.

Most impacts vary by region, and each region of the United States will face its own challenges (FIGURE 18.22). For instance, winter and spring precipitation is projected to decrease across the South but increase across the North. Drought may strike in some regions and flooding in others. Sea level rise will likely affect the Atlantic and Gulf Coasts more than the West Coast. Agriculture, forests, wildlife, and human health may experience a wide array of impacts that will vary from one region to another. You may learn more about the scientific predictions for your own region by consulting this publicly accessible report online.

Reduced Emissions Scenario
Projected Temperature Change (°F)
End-of-Century (2071–2099 average)

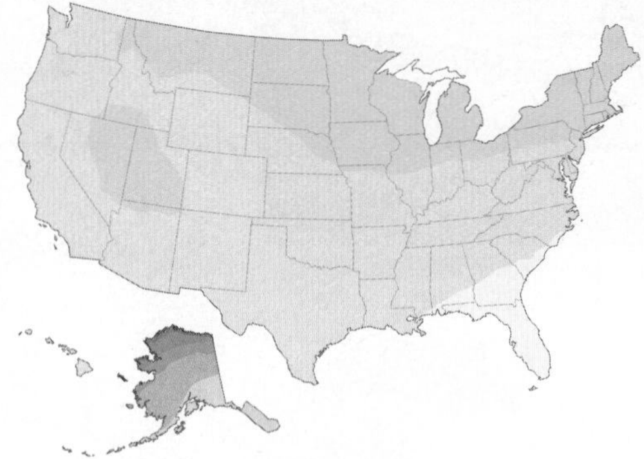

Continued Emissions Scenario
Projected Temperature Change (°F)
End-of-Century (2071–2099 average)

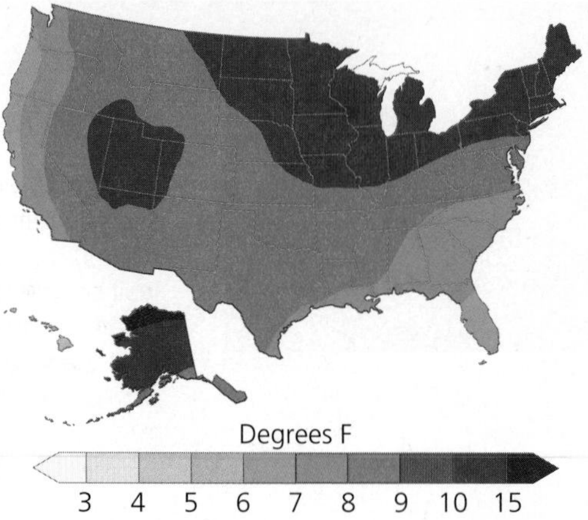

FIGURE 18.21 Average temperatures across the United States are predicted to rise by the end of this century. Even under a scenario of sharply reduced emissions (**top**), temperatures are predicted to rise by 3–4°F. Under a scenario of business-as-usual emissions (**bottom**), temperatures are predicted to rise by 7–11°F. *Data from U.S. Global Change Research Program, 2013.* National climate assessment. *Draft for public review.*

TABLE 18.3 Some Predicted Impacts of Climate Change in the United States

- Average temperatures will rise 2.1–6.2°C (4–11°F) further by the year 2100.
- Droughts, flooding, and wildfire will worsen; dry areas will get drier and wet areas wetter.
- Extreme weather events will become more frequent. The costs they impose on society will grow.
- Sea level will rise an additional 0.3–1.2 m (1–4 ft) by 2100.
- Storm surges will continue to erode beaches and coastal wetlands, destroy real estate, and damage infrastructure.
- Health problems due to heat stress, disease, and pollution will rise. Some tropical diseases will spread north.
- Drought, fire, and pest outbreaks will continue to alter forests.
- Marine ecosystems and fisheries will be affected by ocean acidification.
- Although enhanced CO_2 and longer growing seasons favor crops, increased drought, heat stress, pests, and diseases will decrease most yields.
- Snowpack will decrease in the West; water shortages will worsen in many areas.
- Alpine ecosystems and barrier islands will begin to vanish.
- Northeast forests will lose sugar maples; Southwest ecosystems will turn more desertlike.
- Melting permafrost will undermine Alaskan buildings and roads.

Adapted from U.S. Global Change Research Program, 2013. National climate assessment. *Draft for public review. http://ncadac.globalchange.gov.*

All these impacts of climate change are projected consequences of the warming effect of our greenhouse gas emissions (FIGURE 18.23, see p. 506). We are bound to experience further consequences, but by addressing the root causes of anthropogenic climate change now, we may still be able to prevent the most severe future impacts.

Are we responsible for climate change?

Scientists agree that most or all of today's global warming is due to the well-documented recent increase in greenhouse gas concentrations in our atmosphere. They also agree that this rise in greenhouse gases results primarily

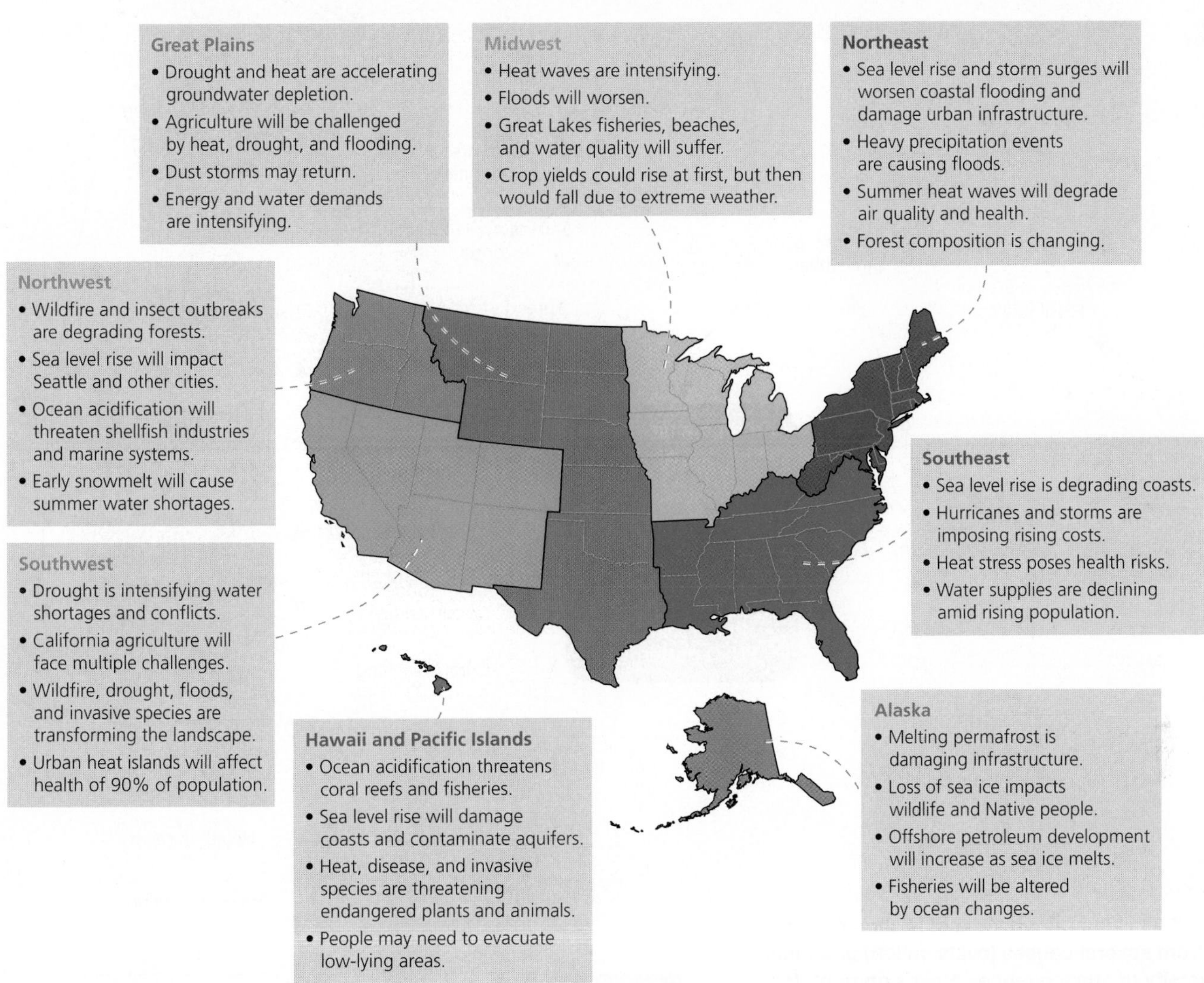

FIGURE 18.22 Impacts of climate change will vary by region. Shown are a few of the most important impacts scientists expect for each region by the end of the century. *Data from U.S. Global Change Research Program, 2013.* National climate assessment. *Draft for public review.*

from our combustion of fossil fuels for energy and secondarily from the loss of carbon-absorbing vegetation due to deforestation.

A decade ago, many scientists had already become concerned enough about the consequences of climate change to put themselves on record urging governments to address the issue. In 2005, the national academies of science from 11 nations (Brazil, Canada, China, France, Germany, India, Italy, Japan, Russia, the United Kingdom, and the United States) issued a joint statement urging political leaders to take action. The statement read, in part:

> The scientific understanding of climate change is now sufficiently clear to justify nations taking prompt action [to reduce] global greenhouse gas emissions. . . . A lack of full scientific certainty about some aspects of climate change is not a reason for delaying an immediate response that will, at a reasonable cost, prevent dangerous anthropogenic interference with the climate system.

Such a clear consensus statement from the world's scientists was virtually unprecedented, on any issue—and scientists' concerns have only grown since that time.

Yet despite the overwhelming evidence for climate change and its impacts, many people, especially in the United States, have long tried to deny that it is happening. Many of these naysayers now admit that the climate is changing but doubt that we are the cause. Indeed, while most of the world's nations moved forward to confront climate change through international dialogue, in the United States public discussion of climate change remained mired in outdated debates over whether the phenomenon was real and whether humans were to blame. These debates have been fanned by spokespeople from think tanks and a handful of scientists, many funded by fossil fuel industries. These individuals, together with corporate interests, have aimed to cast doubt on the scientific consensus, and their views are amplified by the American news media, which seeks to present two sides to every issue, even when the sides' arguments are not equally supported by evidence.

For many Americans, former Vice President Al Gore's 2006 film and book, *An Inconvenient Truth*, presented an eye-opening summary of the science of climate change and a compelling call to action. Awareness of climate change grew as the 2007 IPCC report was made publicly available on the Internet and was widely covered in the media, and later as Gore and

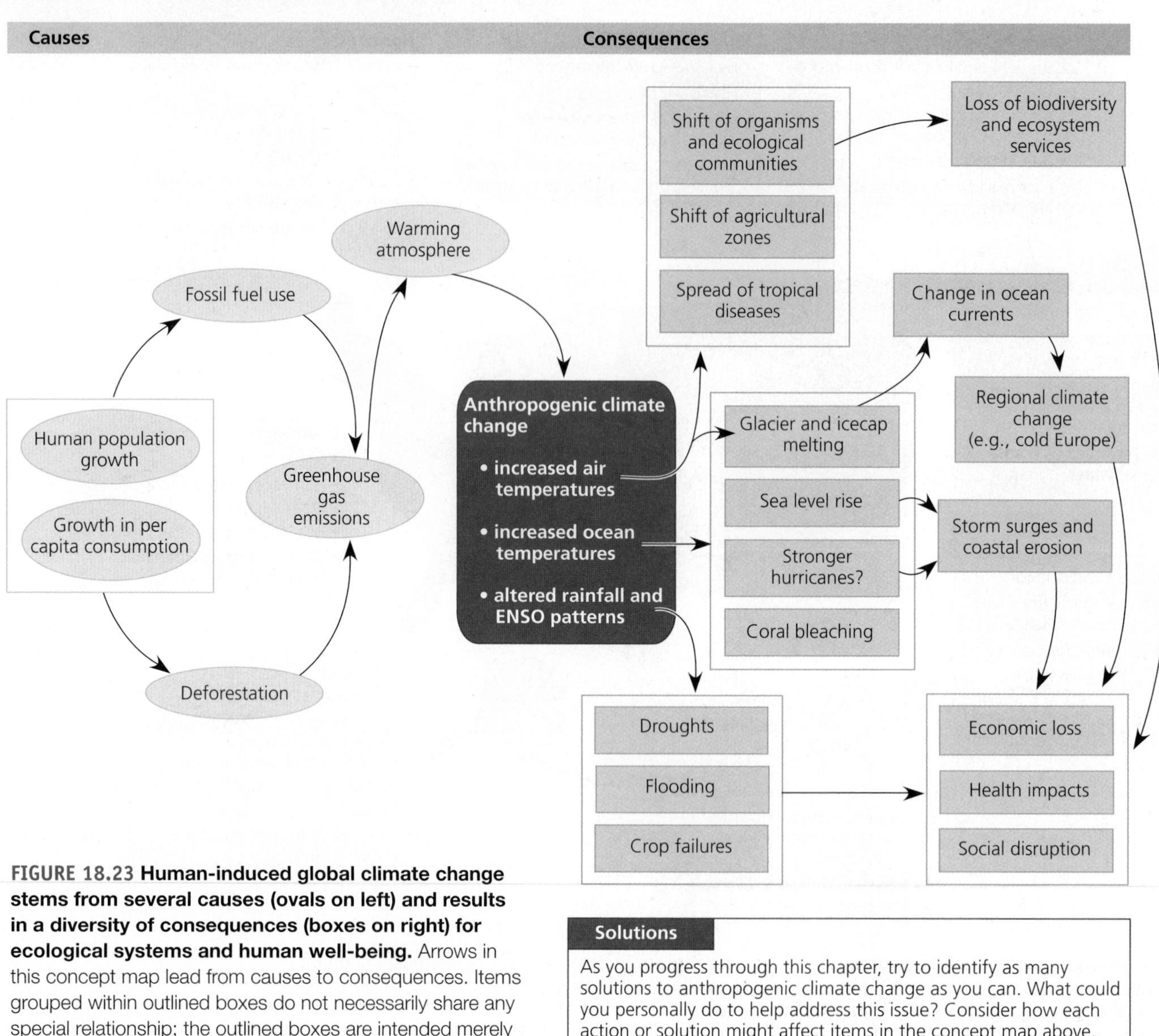

FIGURE 18.23 Human-induced global climate change stems from several causes (ovals on left) and results in a diversity of consequences (boxes on right) for ecological systems and human well-being. Arrows in this concept map lead from causes to consequences. Items grouped within outlined boxes do not necessarily share any special relationship; the outlined boxes are intended merely to streamline the figure.

Solutions

As you progress through this chapter, try to identify as many solutions to anthropogenic climate change as you can. What could you personally do to help address this issue? Consider how each action or solution might affect items in the concept map above.

the IPCC were jointly awarded the Nobel Peace Prize. At the same time, however, many Americans who disliked Gore's politics also came to reject his message on climate change.

In 2009, a hacker illegally broke into computers at the University of East Anglia, U.K., and made public several thousand documents, including over 1000 private emails among a handful of climate scientists. A few of these messages appeared to show questionable behavior in the use of data and the treatment of other researchers. Climate-change deniers named the incident "Climategate" and used it to accuse the entire scientific establishment of wrongdoing and conspiracy. The news media disseminated the story widely.

However, subsequent investigations into the affair by six different independent panels all cleared the climate scientists, concluding that there was no evidence of wrongdoing. Each panel concluded that some individuals may have exercised poor taste or judgment and that some practices could be improved, but they also found that many media accounts trumpeting the news had misrepresented the content of the emails. The panels agreed that the hacked messages among a few individuals in no way called into question the vast array of research results compiled by thousands of hard-working independent climate scientists over several decades.

Questions were raised the same year about some of the IPCC report's conclusions, and subsequent inquiry revealed that several statements were inadequately backed by evidence or otherwise misstated or misleading. It is hardly surprising in such a vast collaborative effort that a few statements out of thousands would be in error. Yet scientists wanted to ensure that the IPCC's reputation for reliability not be tarnished, and so reforms were set in motion to strengthen the IPCC's process during the preparation of its Fifth Assessment.

Responding to Climate Change

Today most of the world's people recognize that our fossil fuel consumption is altering the planet that our children will inherit. From this point onward, our society will be focusing on how best to respond to the challenges of climate change. The good news is that everyone—not just leaders in government and business, but everyday people, and especially today's youth—can play a part in this all-important search for solutions.

Shall we pursue mitigation or adaptation?

We can respond to climate change in two fundamental ways. One is to pursue actions that reduce greenhouse gas emissions, so as to lessen the severity of climate change. This strategy is called **mitigation** because the aim is to mitigate, or alleviate, the problem. Examples include improving energy efficiency, switching to clean and renewable energy sources, preventing deforestation, recovering landfill gas, and encouraging farm practices that protect soil quality.

The second type of response is to pursue strategies to cushion ourselves from the impacts of climate change. This strategy is called **adaptation** because the goal is to adapt to change. Erecting a seawall, as Maldives residents did with the Great Wall of Malé, is an example of adaptation. Some people of Tuvalu also adapted, by leaving their island to make a new life in New Zealand (see **Weighing the Issues**, p. 501). Other examples of adaptation include restricting coastal development; adjusting farming practices to cope with drought; and modifying water management practices to deal with reduced river flows, glacial outburst floods, or salt contamination of groundwater.

Both adaptation and mitigation are necessary. Adaptation is needed because even if we could halt all our emissions right now, the greenhouse gas pollution already in the atmosphere would continue driving global warming until the planet's systems reach a new equilibrium, with temperature rising an estimated 0.6°C (1.0°F) more by the end of the century. Because we will face this change no matter what we do, it is wise to develop ways to minimize its impacts.

We also need to pursue mitigation, because if we do nothing to diminish climate change, it will eventually overwhelm any efforts we might make to adapt. To leave a sustainable future for our civilization and to safeguard the living planet that we know, we will need to pursue mitigation. The sooner we begin reducing our emissions, the lower the level at which they will peak, and the less we will alter climate (FIGURE 18.24). We will spend the remainder of our chapter examining approaches for the mitigation of climate change.

We are developing solutions in electricity generation

From cooking to heating to lighting, much of what we do each day depends on electricity. The generation of electricity produces the largest portion (40%) of U.S. carbon dioxide emissions. Fossil fuel combustion generates 70% of U.S. electricity, and coal accounts for most of the resulting emissions. There are two ways to reduce the amount of fossil fuels

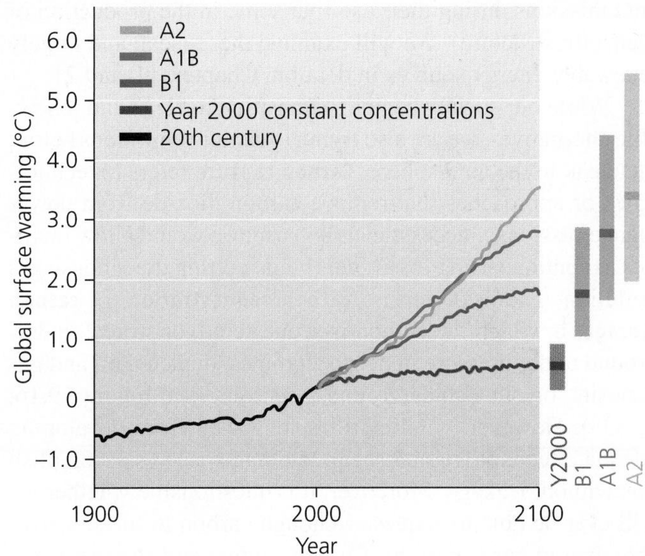

FIGURE 18.24 The sooner we stabilize our emissions, the less climate change we will cause. The red line shows temperature change we could expect if we were to limit our yearly carbon emissions to the level they were in the year 2000. The blue, green, and yellow lines show the change expected under scenarios of rapid, medium, and slow control of emissions, respectively. The vertical bars show means and ranges of year-2100 temperatures for each scenario. Predictions are based on a large number of climate models. *Data from IPCC, 2007.* Fourth assessment report.

we burn to generate electricity: (1) encouraging conservation and efficiency (pp. 546–548) and (2) switching to cleaner and renewable energy sources (Chapters 20 and 21).

Conservation and efficiency As individuals we all can make lifestyle choices to reduce electricity consumption. New energy-efficient technologies make it easier to conserve. Replacing standard light bulbs with compact fluorescent lights reduces energy use for lighting by 40%. The U.S. Environmental Protection Agency's Energy Star Program rates household appliances, lights, windows, fans, office equipment, and heating and cooling systems by their energy efficiency. Replacing an old washing machine with an Energy Star washing machine can cut your CO_2 emissions by 200 kg (440 lb) annually. Energy Star homes use highly efficient windows, ducts, insulation, and heating and cooling systems to reduce energy use and emissions by 30% or more. Such technological solutions also save consumers money by reducing utility bills.

Manufacturers can make the same types of choices as consumers in their purchases, and can manufacture energy-efficient products. Power producers can use approaches such as cogeneration (p. 546) to produce fewer emissions per unit of energy generated.

Sources of electricity We can also reduce greenhouse gas emissions by switching to cleaner energy sources. Natural gas generates the same amount of energy as coal, with half the emissions. Cleaner still are alternatives to fossil fuels, including nuclear power (pp. 555–566), bioenergy, hydroelectric power, geothermal power, solar photovoltaic cells, wind power, and ocean sources. These energy sources give off no net emissions during their use (but some in the production of their infrastructure). We will examine these clean and largely renewable energy sources in detail in Chapters 20 and 21.

While our society begins to transition to clean and renewable alternatives, we are also trying to capture emissions before they leak to the atmosphere. **Carbon capture** refers to technologies or approaches that remove carbon dioxide from power plant emissions. Successful carbon capture would allow facilities to continue using fossil fuels while cutting greenhouse gas pollution. The next step is **carbon sequestration,** or **carbon storage,** in which the carbon is sequestered, or stored, underground under pressure in deep salt mines, depleted oil and gas deposits, or other underground reservoirs (see Figure 19.16, p. 538). However, we are still a long way from developing adequate technology and secure storage space to accomplish this without leakage. Moreover, it is questionable whether we will ever be able to sequester enough carbon to make a sizeable dent in our emissions. Carbon capture and storage is discussed in more detail in Chapter 19 (pp. 537–538).

Transportation solutions are at hand

Can you imagine life without a car? Most Americans probably can't—a reason why transportation is the second-largest source of U.S. greenhouse gas emissions. The average American family makes 10 trips by car each day, and U.S. taxpayers spend over $200 million per day on road construction and repairs for the nation's 250 million registered automobiles.

Automotive technology The typical automobile is highly inefficient. Over 85% of the fuel you pump into your gas tank does something other than move your car down the road (FIGURE 18.25). The technology exists to reduce these losses and make our vehicles far more fuel-efficient. Indeed, the vehicles of many nations are more fuel-efficient than those of the United States. More aerodynamic designs, increased engine efficiency, and improved tire design all can help. Recent government mandates are encouraging greater fuel efficiency in American-made vehicles (pp. 547–548), and as gasoline prices rise, consumer demand for more fuel-efficient automobiles will intensify.

Advancing technology is also bringing us alternatives to the traditional combustion-engine automobile. These include hybrid vehicles that combine electric motors and gasoline-powered engines for greater efficiency (p. 547). They also include fully electric vehicles, alternative fuels such as compressed natural gas and biodiesel (pp. 570–572), and hydrogen fuel cells that use oxygen and hydrogen and produce only water as a waste product (pp. 603–604).

Transportation choices We can make lifestyle choices that reduce our reliance on cars. Some people are choosing to live nearer to their workplaces. Others use mass transit such as buses, subway trains, and light rail. Still others bike or walk to work or on errands (FIGURE 18.26). Public transportation in the United States currently serves 3–4% of passenger trips, reducing gasoline use by 4.2 billion gallons each year and saving 37 million metric tons of CO_2 emissions, the American Public Transportation Association estimates. If U.S. residents were to increase their use of mass transit to the levels of Canadians (7% of daily travel needs) or Europeans (10% of daily travel needs), the United States could cut its air pollution, its dependence on imported oil, and its contribution to climate change.

Unfortunately, reliable and convenient public transit is not yet available in many U.S. communities. Making automobile-based cities and suburbs more friendly to pedestrian and bicycle traffic and improving people's access to public transportation stand as central challenges for city and regional planners (pp. 345–347).

We will need to follow multiple strategies

Advances in agriculture, forestry, and waste management can help us mitigate climate change. In agriculture, sustainable management of cropland and rangeland enables soil to store more carbon. New techniques reduce the emission of methane from rice cultivation and from cattle and their manure, and reduce nitrous oxide emissions from fertilizer. We can also grow renewable biofuel crops, although whether these decrease or increase emissions is an active area of research (pp. 569–573).

In forest management, preserving existing forests, reforesting cleared areas, and pursuing sustainable forestry practices (p. 322) all help to absorb carbon from the air. Waste managers are doing their part to cut emissions by treating wastewater (pp. 414–416), generating energy from waste in incinerators (p. 616), and recovering methane seeping from landfills (p. 616). Individuals, communities, and waste haulers also help

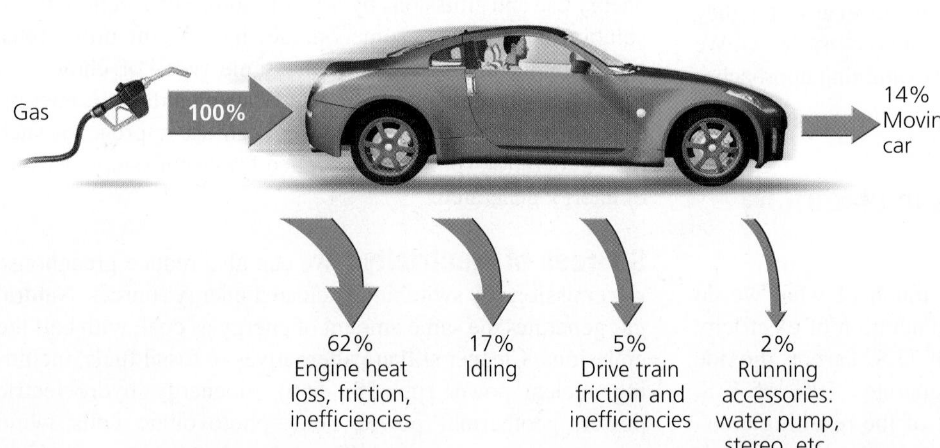

FIGURE 18.25 Conventional automobiles are fuel-inefficient. Only about 13–14% of the energy from a tank of gas actually moves the typical car down the road. Nearly 85% of useful energy is lost, primarily as heat. *Data from U.S. Department of Energy.*

FIGURE 18.26 **Commuting by bike or on foot greatly reduces one's transportation-related greenhouse gas emissions.**

reduce emissions when they encourage recycling, composting, and the reuse of materials and products (pp. 616–619).

We should not expect to find a single "magic bullet" for mitigating climate change. Reducing emissions will require many steps by many people and institutions across many sectors of our economy. The good news is that most reductions can be achieved using current technology and that we can begin implementing these changes right away. Environmental scientists Stephen Pacala and Robert Socolow advise that we follow some age-old wisdom: When the job is big, break it into small parts. Pacala and Socolow have proposed that we adopt a portfolio of strategies that together can stabilize our CO_2 emissions at current levels (FIGURE 18.27). They identify 15 strategies that could each eliminate 1 billion tons of carbon per year by 2050 if deployed at a large scale. Achieving just 7 of these 15 aims would stabilize our emissions. If we achieve more, then we reduce emissions.

What role should government play?

Even if people agree on strategies and technologies to reduce emissions, they may disagree on what role government should play in encouraging those strategies and technologies: Should it mandate change through laws and regulations? Should it impose no policies at all and hope that private enterprise will develop solutions on its own? Should it take the middle ground and design programs that give private entities financial incentives to reduce emissions? This debate has been vigorous in the United States and Canada, where many business leaders and politicians have opposed all government action to address climate change, fearing that emissions reductions will impose economic costs on industry and consumers.

In 2007, the U.S. Supreme Court ruled that carbon dioxide was a pollutant that the Environmental Protection Agency (EPA) could regulate under the Clean Air Act (p. 458). When Barack Obama became president, he instead urged that Congress craft legislation to address emissions. In 2009, the Democratic-led House of Representatives passed legislation to create a **cap-and-trade** system (p. 183) in which industries and utilities would compete to reduce emissions for financial gain, and under which emissions were mandated to decrease 17% by 2020. However, similar legislation did not pass in the Senate.

As a result, responsibility for addressing emissions passed to the EPA, which began developing regulations in 2011. The EPA aimed to phase in emissions limits on industry and utilities gradually over many years, hoping to spur energy efficiency retrofits and renewable energy use at a pace that will minimize economic impacts and political opposition.

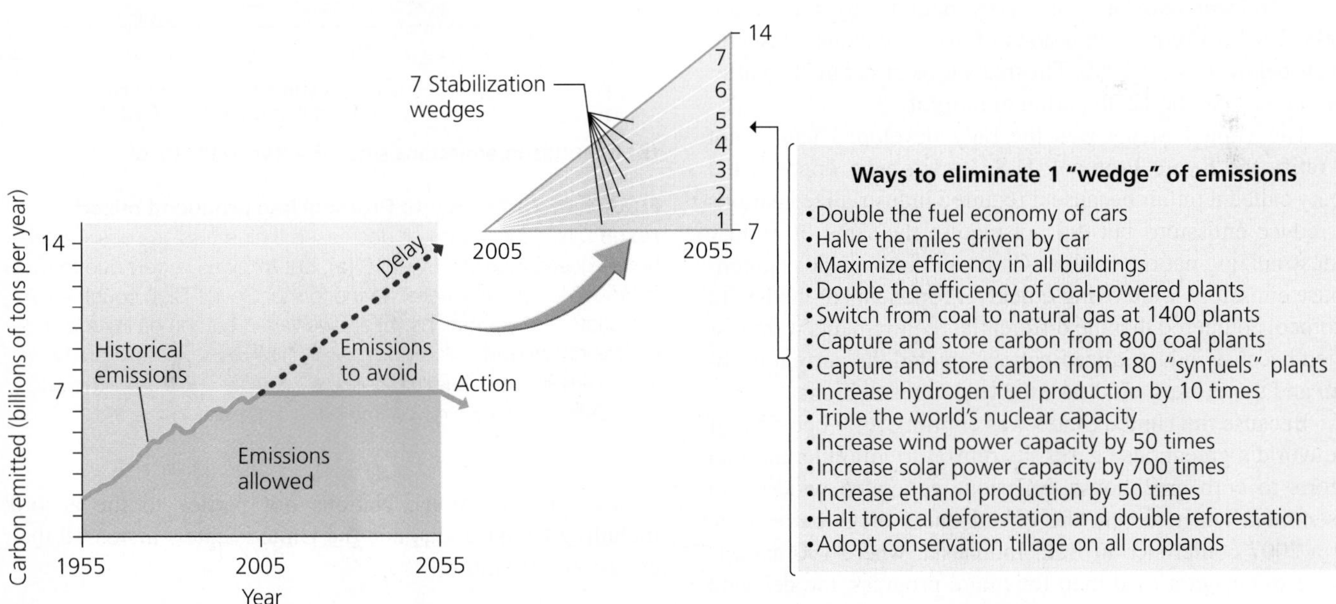

FIGURE 18.27 **We can accomplish the large job of stabilizing emissions by breaking it into smaller steps.** Environmental scientists Stephen Pacala and Robert Socolow began with a standard graph showing the doubling of CO_2 emissions that scientists expect to occur from 2005 to 2055. They added a flat line to represent the trend if emissions were held constant and then separated the graph into emissions allowed (below the line) and emissions to be avoided (the triangular area above the line). They then divided this "stabilization triangle" into seven equal-sized portions. Each of these "stabilization wedges" represents 1 billion tons of CO_2 emissions in 2055 to be avoided. Finally, they identified a series of strategies, each of which could take care of one wedge. If we accomplish just 7 of these strategies, we could halt our growth in emissions for the next half-century. *Adapted from Pacala, S., and R. Socolow, 2004. Stabilization wedges: Solving the climate problem for the next 50 years with current technologies.* Science *305: 968–972. Reprinted by permission of AAAS and the author.*

In June 2013 President Obama gave a speech at Georgetown University announcing that, because of legislative gridlock, he would take steps to address climate change using his executive authority. His "climate action plan" urged the EPA to speed its regulation of new power plants and to begin regulating existing power plants. It also aimed to jumpstart renewable energy development, modernize the electric grid, finance clean coal and carbon storage efforts, improve automotive fuel economy, protect and restore forests, and encourage energy efficiency. At the same time, the president's plan sought to prepare the nation to adapt to the impacts of climate change, and to better engage with other countries to address greenhouse gas emissions.

The Kyoto Protocol sought to limit emissions

Climate change is a global problem, so global cooperation is needed to forge effective solutions. This is why the world's policymakers have tried to tackle climate change with international treaties. In 1992 at the U.N. Conference on Environment and Development Earth Summit in Río de Janeiro, Brazil, most of the world's nations signed the **United Nations Framework Convention on Climate Change (FCCC).** This agreement outlined a plan for reducing greenhouse gas emissions to 1990 levels by the year 2000 through a voluntary, nation-by-nation approach.

Emissions kept rising, however, so nations came together to forge a binding treaty to *require* emissions reductions. An outgrowth of the FCCC drafted in 1997 in Kyoto, Japan, the **Kyoto Protocol** mandated signatory nations, by the period 2008–2012, to reduce emissions of six greenhouse gases to levels below those of 1990. The treaty took effect in 2005 after Russia became the 127th nation to ratify it.

The United States was the only developed nation not to ratify the Kyoto Protocol. U.S. leaders who opposed the treaty called it unfair because it required industrialized nations to reduce emissions but did not require the same of rapidly industrializing nations such as China and India, whose greenhouse emissions were rising quickly. Proponents of the Kyoto Protocol countered that the differential requirements were justified because industrialized nations created the current problem and therefore should take the lead in resolving it.

Because the United States was emitting fully one-fifth of the world's greenhouse gases, its refusal to join international efforts to curb greenhouse emissions generated widespread resentment and undercut the effectiveness of global efforts. At a 2007 conference in Bali, Indonesia, where 190 nations strove to design a road map for future progress, the delegate from Papua New Guinea drew thunderous applause when he requested of the U.S. delegation, "If for some reason you are not willing to lead . . . please get out of the way."

As of 2010 (the most recent year with full international data), nations that signed the Kyoto Protocol had decreased their emissions by 8.9% from 1990 levels (**FIGURE 18.28**). However, much of this reduction was due to economic contraction in Russia and nations of the former Soviet Bloc following the breakup of the Soviet Union. When these nations are factored out, the remaining signatories showed a 4.9% *increase* in emissions. Nations not parties to the accord, including China, India, and the United States, increased their emissions still more.

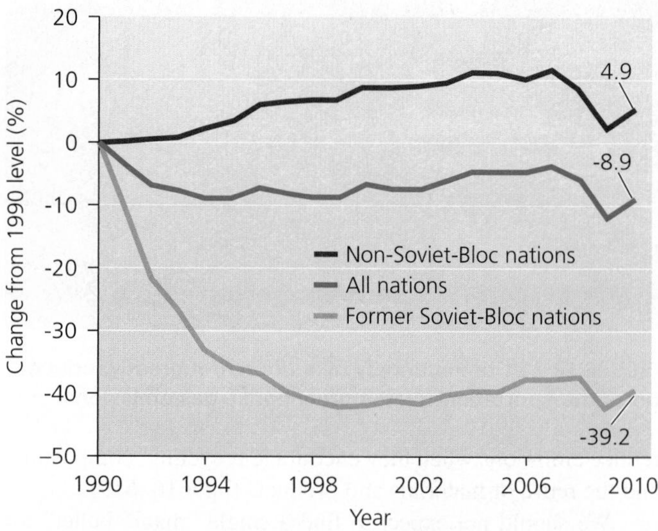

(a) Emissions through time since the Kyoto Protocol

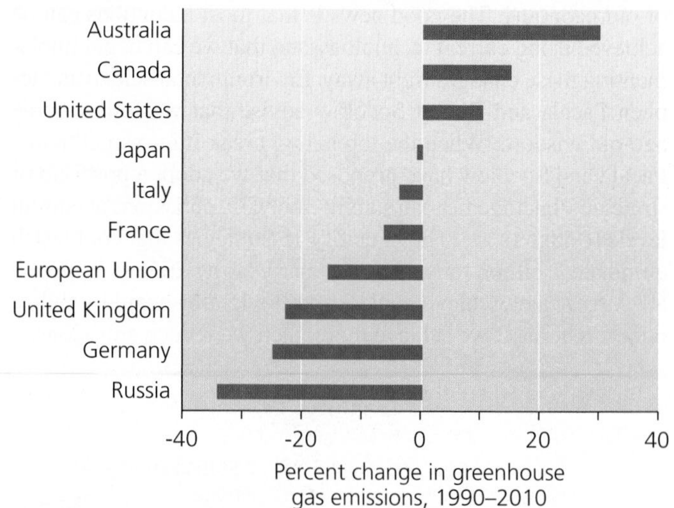

(b) Changes in emissions since the Kyoto Protocol

FIGURE 18.28 The Kyoto Protocol has produced mixed results. Nations ratifying it decreased their emissions of six greenhouse gases by 8.9% by 2010 **(a)**, but this was largely due to unrelated economic contraction in the former Soviet-Bloc countries. A selection of major nations **(b)** shows varied outcomes in reducing emissions. *The United States did not ratify the Protocol, Australia joined it late, and Canada left early. Values do not include influences of land use and forest cover. Data from U.N. Framework Convention on Climate Change, 2013.*

International climate negotiations seek a way forward

In recent years, representatives of the world's nations have met at a series of international conferences, trying to design a treaty to succeed the Kyoto Protocol. At their 2009 meeting in Copenhagen, Denmark, these climate negotiators failed to reach consensus (**FIGURE 18.29**). China promised steep emissions cuts but would not allow international monitoring to confirm them. U.S. President Obama favored emissions cuts but would not promise the international community more than the U.S. Congress would agree to. (The Senate had rejected climate legislation, and Senate approval is needed for U.S. ratification of any treaty.) On the final day of the conference, leaders of major nations put together an agreement, the Copenhagen Accord. However, it failed to win a unanimous vote, and the conference ended without specific targets or solid commitments. In a dramatic final session, several nations denounced the accord, so that the conference could not even formally adopt it by consensus and had to merely "take note" of it. Tuvalu and the Maldives took opposite sides in the debate. Tuvalu insisted on protesting the accord on principle, whereas the Maldives concluded that a weak accord was better than no accord.

The process got back on track in Cancun, Mexico, in 2010, where nations fleshed out plans based on the Copenhagen Accord. Developed nations promised to pay developing nations to help with their mitigation and adaptation efforts—up to $100 billion per year by 2020—through a fund overseen by the World Bank. Nations broadly agreed upon a plan, nicknamed *REDD* (p. 314), to help tropical nations reduce forest loss. Developed nations agreed to transfer clean energy technology to developing nations. And rapidly industrializing nations such as China and India agreed in principle to emission targets and international monitoring. Moreover, nations shared how they seek to reduce greenhouse gas pollution; China would accelerate its renewable energy efforts, for example, and Brazil would limit deforestation and encourage no-till agriculture (pp. 226–228).

In Durban, South Africa, in 2011, negotiators failed to design a new treaty but did succeed in getting all nations—including the top emitters China, the United States, and India—to agree to a "road map" toward a legally binding international deal in 2015, which would come into force only after 2020. This plan was reaffirmed at the 2012 conference in Doha, Qatar, where negotiators also extended the Kyoto Protocol until 2020. A number of nations backed out of the Kyoto extension, however, and this treaty now applies to only about 15% of the world's emissions. Neither the $100 billion "Green Climate Fund" nor the REDD program made progress in Durban or Doha, because nations could not agree on how to finance them.

Although the Durban and Doha conferences produced a plan that might bear fruit years down the road, most scientists reacted with disappointment and alarm, because waiting until 2020 for a meaningful agreement creates a "lost decade" during which climate change could grow far worse. The United Nations estimates there is now a gap of more than 6 billion metric tons of emissions between what the world has pledged to cut and the degree of cuts that science tells us is needed to limit climate change to 2°C of warming.

Reaching consensus among 200 nations through the treaty process is a daunting challenge. As a result, experts now predict that most success in mitigating climate change will come from technological advances, economic incentives, and national, regional, and local initiatives. Business and industry are beginning to accelerate renewable energy and energy-efficiency efforts, and policymakers are looking to create environments in which private-sector efforts can generate productive solutions.

FIGURE 18.29 Activists have kept pressure on climate negotiators at each international conference. At Copenhagen, activists showed support for island nations such as Tuvalu and the Maldives and for bringing the atmospheric carbon dioxide level down to 350 parts per million.

Will emissions cuts hurt the economy?

Many U.S. policymakers have opposed emissions reductions out of fear that they will dampen the national economy. China, India, and other large industrializing nations have so far resisted emissions cuts under the same assumption. This is understandable, given that our current economies depend so heavily on fossil fuels. Yet other nations have demonstrated that economic vitality does not require ever-growing emissions.

For example, Germany has the third most technologically advanced economy in the world and is a leading producer of iron, steel, coal, chemicals, automobiles, machine tools, electronics, textiles, and other goods—yet it managed between 1990 and 2010 to reduce its greenhouse gas emissions by 25%. In the same period, the United Kingdom cut its emissions by 23%, and the European Union as a whole cut its emissions by 15% (see Figure 18.28). Today the citizens of many nations enjoy standards of living comparable to that of U.S. citizens, yet with far fewer emissions. In fact, on a per-person basis, wealthy nations from France to Denmark to New Zealand to Hong Kong to Switzerland to Sweden emit greenhouse gases at less than half of the U.S. rate.

Because resource use and per capita emissions are high in the United States, policymakers and industries often assume the United States has more to lose economically from restrictions on emissions than developing nations do. However, industrialized nations are also the ones most likely to *gain* economically from major energy transitions, because they are best positioned to invent, develop, and market new technologies to power the world in a post-fossil-fuel era.

Germany, Japan, and China have realized this and are now leading the world in production, deployment, and sales of solar energy technology (**FIGURE 18.30**). China recently surpassed the United States to become the world's biggest greenhouse gas emitter (although it still emits far less per person than the United States). Yet China has also embarked on a number of initiatives to develop and sell renewable energy technologies on a scale beyond what any other nation has attempted. If the

FIGURE 18.30 China is racing to develop renewable energy technology. It is on track to surpass the United States as a leader in green energy. Here, workers at a Chinese factory produce photovoltaic solar panels.

United States does not act quickly to develop energy technologies for the future, then the future could belong to nations like China, Germany, and Japan.

States and cities are advancing climate change policy

In the absence of legislative action by the U.S. federal government to address climate change, state and local governments across the country are advancing policies to limit emissions. Mayors from over 1000 cities from all 50 U.S. states have signed on to the U.S. Mayors Climate Protection Agreement, initiated by former Seattle mayor Greg Nickels. Under this agreement, mayors commit their cities to pursue policies to "meet or beat" Kyoto Protocol guidelines.

A number of U.S. states have enacted targets or mandates for renewable energy production, seeking to boost cleaner alternatives to fossil fuels. Many more states and cities have adopted plans for adapting to climate change's impacts. New York City Mayor Michael Bloomberg launched the New York City Panel on Climate Change in 2008 as part of the PlaNYC sustainability plan (p. 354). Its deliberations and its 2010 report gave New York a head start in tackling the challenges posed by Superstorm Sandy and those that await it in the new era of sea level rise.

At the state level, the boldest action so far has come in California. In 2006 that state's legislature worked with Governor Arnold Schwarzenegger to pass the Global Warming Solutions Act, which aims to cut California's greenhouse gas emissions 25% by the year 2020. This law was the first state legislation with penalties for noncompliance and followed earlier efforts in California to mandate higher fuel efficiency for automobiles. Among other approaches it has established a permit trading program for carbon emissions.

Action is also being taken by nine northeastern states collaborating in the Regional Greenhouse Gas Initiative. In this effort, Connecticut, Delaware, Maine, Maryland, Massachusetts, New Hampshire, New York, Rhode Island, and Vermont run a joint cap-and-trade program for carbon emissions from power plants. A similar effort, the Western Climate Initiative, involves British Columbia, California, Manitoba, Ontario, and Quebec. These emissions trading programs (pp. 183–184) show how government can engage the market economy to pursue public policy goals.

Market mechanisms are being used to address climate change

Permit trading programs aim to harness the economic efficiency of market capitalism to achieve public policy goals while allowing business, industry, or utilities flexibility in how they meet those goals (p. 183). Supporters of permit trading programs argue that they provide the fairest, least expensive, and most-effective method of reducing emissions. Polluters choose how to cut their emissions and are given financial incentives for reducing emissions below the legally required amount (**FIGURE 18.31**).

As an example of how a cap-and-trade emissions trading program for carbon emissions works, consider the approach of the Regional Greenhouse Gas Initiative:

1. Each state decided what polluting sources it would require to participate.
2. Each state set a cap on the total CO_2 emissions it would allow, equal to its 2009 levels.
3. Each state distributed to each emissions source one permit for each ton it emits, up to the amount of the cap.
4. Each state will lower its cap progressively—10% by 2018.
5. Sources with too few permits to cover their pollution must reduce their emissions, buy permits from other sources, or pay for credits through a carbon offset project (p. 513). Sources with excess permits may sell them.
6. Any source emitting more than its permitted amount faces penalties.

Once up and running, it is hoped that the system will be self-sustaining. The price of a permit is meant to fluctuate freely in the market, creating the same kinds of financial incentives as any other commodity that is bought and sold in our capitalist system.

The world's largest cap-and-trade program is the European Union Emission Trading Scheme. This market got off to a successful start in 2005—until investors discovered that national governments had allocated too many permits to their industries. The overallocation gave companies little incentive to reduce emissions, so permits lost their value and prices in the market took a nosedive. Europeans partly addressed these problems by making emitters pay for permits and setting emissions caps across the entire European Union while expanding the program to include more greenhouse gases, more emissions sources, and additional members.

Similar difficulties have befallen the Regional Greenhouse Gas Initiative, as well as the world's first emissions trading program for greenhouse gas reduction, the Chicago Climate Exchange, which operated from 2003 to 2010 and involved several hundred corporations, institutions, and municipalities. All these early experiments in carbon markets are providing

FIGURE 18.31 A cap-and-trade emissions trading system harnesses the power of market capitalism to achieve the goal of reducing emissions. In such a system, ① government first sets an overall cap on emissions. As polluting facilities respond, some will have better success reducing emissions than others. In this figure, ② Plant A succeeds in cutting its emissions well below the cap, whereas ③ Plant B fails to cut its emissions at all. As a result, ④ Plant B pays money to Plant A to purchase allowances that Plant A is no longer using. Plant A profits from this sale, and the government cap is met, reducing pollution overall. Over time, the cap can be progressively lowered to achieve further emissions cuts.

lessons for how to set up effective and sustainable programs in the future. One lesson is that in the long run, permits will be valuable and the market will work only if government policies are in place to limit emissions.

Carbon taxes are another option

As the world's carbon trading markets show mixed results early in their growth, some economists, scientists, and policymakers are saying that cap-and-trade systems are not effective enough, don't work quickly enough, or leave too much to chance. Many of these critics would prefer that governments enact a **carbon tax** instead. In this approach, governments charge polluters a fee for each unit of greenhouse gases they emit. This gives polluters a financial incentive to reduce emissions.

Such a tax can be implemented in various ways; it can be charged to energy producers, utilities, or motor vehicle users, and it can be scaled according to energy efficiency. Carbon taxes of various types have so far been introduced in roughly 20 nations. In the United States, Boulder, Colorado, implemented a tax on electricity consumption, and Montgomery County, Maryland, is taxing polluting power plants.

The downside of a carbon tax is that most polluters simply pass the cost along to consumers by charging higher prices for the products or services they sell. Proponents of carbon taxes have responded by proposing an approach called **fee-and-dividend**. In this approach, funds from the carbon tax, or "fee," paid to government by polluters are transferred as a tax refund, or "dividend," to taxpayers. This way, if polluters pass their costs along to consumers, those consumers will be reimbursed for those costs by the tax refund they receive. In theory, the system should provide polluters a financial incentive to reduce emissions while imposing no financial burden on taxpayers.

WEIGHING THE ISSUES

CAP-AND-TRADE OR A CARBON TAX? What advantages and disadvantages do you see in using a cap-and-trade system to reduce greenhouse gas emissions? What pros and cons do you see in using carbon taxes to achieve this goal? What do you think of the idea of a "fee-and-dividend" program? If you were a U.S. senator, what type of policy would you support in order to address emissions in the United States, and why?

Carbon offsets are popular

Emissions trading programs generally allow participants to buy **carbon offsets,** voluntary payments intended to enable another entity to help reduce the emissions that one is unable to reduce. The payment thus offsets one's own emissions. For example, a coal-burning power plant could pay a reforestation project to plant trees that will soak up as much carbon dioxide as the coal plant emits. Or a university could fund the development of clean and renewable energy projects to make up for fossil fuel energy the university uses. Carbon offsets have fast become popular among utilities, businesses, universities, governments, and individuals trying to achieve **carbon-neutrality,** a condition in which no net carbon is emitted.

In principle, carbon offsets seem a great idea, but rigorous oversight is needed to make sure that the offset money actually accomplishes what it is intended for. Offsets are effective only if they fund emissions reductions that would not occur otherwise. And because trees can soak up only so much carbon dioxide, at some point our ability to reduce emissions by funding reforestation could reach its limit. Efforts to create a transparent and enforceable system for verifying the effectiveness of offsets are ongoing. If these efforts succeed, then carbon offsets could become an important means of mitigating climate change.

Corporations are going carbon-neutral

Carbon offsets are a major route toward carbon-neutrality among businesses and corporations seeking to make their practices more sustainable (pp. 155–156), but corporations also can find ways to reduce their carbon footprints directly. An excellent example is Pearson Education, the publisher of your textbook!

In 2009 Pearson achieved carbon-neutrality after a concerted two-year effort (p. 155). Pearson reduced its energy consumption and carbon footprint directly by 12% by upgrading buildings for energy efficiency, designing more efficient computer servers, reducing the number of vehicles in its fleets, increasing the proportion of hybrid vehicles, and cutting back on employee business travel while enhancing the use of video conferencing. Pearson eliminated a further 47% of its emissions by purchasing clean renewable energy instead of fossil fuel energy and by installing large solar panel arrays at two of its sites in New Jersey and a wind turbine at a Minnesota site. To offset the remaining 41% of its emissions, the company is funding a number of programs to preserve forest and replant trees in various areas of the world, from England to Costa Rica.

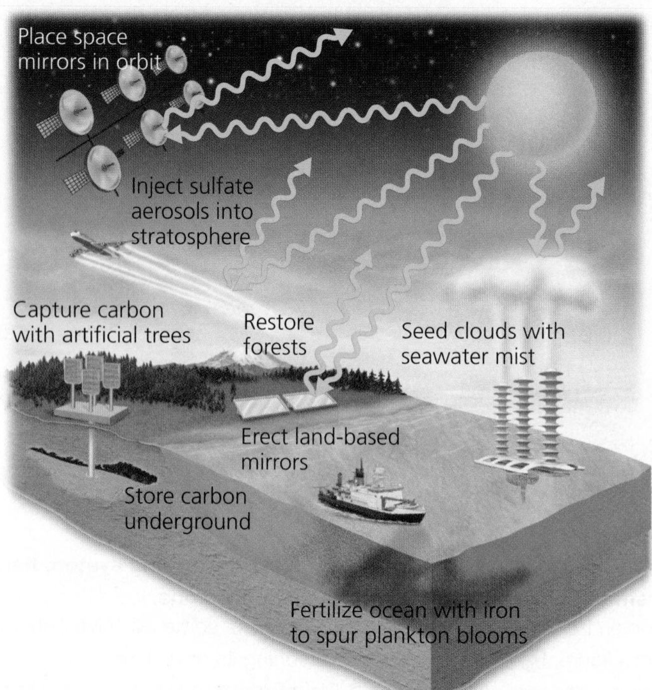

FIGURE 18.32 Geoengineering proposals seek to cool the climate by removing carbon dioxide from the air or reflecting sunlight away from Earth. However, most geoengineering ideas are untested, would take years to develop, may not work well, or might cause undesirable side effects. Thus, they are not a substitute for reducing emissions.

Should we engineer the climate?

What if all our efforts to reduce emissions are not adequate to rein in climate change? As severe climate change begins looking more and more likely, some scientists and engineers are reluctantly considering drastic, assertive steps to alter Earth's climate in a last-ditch attempt to reverse global warming—an approach called **geoengineering** (FIGURE 18.32).

One geoengineering approach would be to suck carbon dioxide out of the air. For example, we might enhance photosynthesis in natural systems by planting trees or by fertilizing ocean phytoplankton with nutrients like iron. A more high-tech method would be to design "artificial trees," structures that chemically filter CO_2 from the air.

A second geoengineering approach would be to block sunlight before it reaches Earth, thus cooling the planet. We might deflect sunlight by injecting sulfates or other fine dust particles into the stratosphere, by seeding clouds with seawater, or by deploying fleets of reflecting mirrors on land, at sea, or in orbit in space.

Scientists have long been reluctant even to discuss the notion of geoengineering. The potential methods are technically daunting and would take years or decades to develop, and some could pose unforeseen environmental risks. Moreover, blocking sunlight does not reduce greenhouse gas concentrations, so ocean acidification would continue. In addition, many experts are wary of promulgating hope for easy technological fixes, lest politicians lose incentive to develop policy to reduce emissions.

However, as climate change intensifies, more scientists are becoming willing to contemplate geoengineering as a back-up plan. Respected researchers and scientific institutions are beginning to assess the risks and benefits of geoengineering options, so that we can be ready to take well-informed action if climate change becomes severe enough to justify it.

You can address climate change

National policies, international treaties, emissions trading programs, corporate actions, and technological innovations—and perhaps even geoengineering—will all play roles in mitigating climate change. But in the end, the most influential factor may be the collective decisions of millions of regular people. Just as we each have an ecological footprint (pp. 4–5), we each have a **carbon footprint** that expresses the amount of carbon we are responsible for emitting. To help reduce emissions, each of us can take steps in our everyday lives—from turning off lights and choosing energy-efficient appliances, to eating a less meat-oriented diet, to deciding where to live and how to get to work.

College students are vital to driving the personal and societal changes needed to reduce carbon footprints and address climate change—both through everyday lifestyle choices and through lobbying and activism. Today a groundswell of interest is sweeping across campuses, and many students are pressing their administrations to seek carbon-neutrality or to divest from fossil fuel investments and promote renewable energy (pp. 16–18, 655–665). Campus action on climate change first made news on January 31, 2008, when over 1900 schools participated in the Focus the Nation teach-in on global warming. Young people have played a large part in subsequent grassroots events and organizations, including 350.org's International Day of Climate Action on October 24, 2009. This event—kicked off by the Maldives' underwater cabinet meeting—featured 5200 events in 181 nations and was called "the most widespread day of political action in the planet's history."

Global climate change may be the biggest challenge we face, but halting it would be our biggest victory. With concerted action, there is still time to avert the most severe impacts and sustain a livable climate. Through outreach, education, innovation, and lifestyle choices, today's youth have the power to turn the tables on climate change and help bring about a bright future for humanity and our planet.

Conclusion

Many factors influence Earth's climate, and human activities have come to play a major role. Climate change is well underway, and further greenhouse gas emissions will intensify global warming and cause increasingly severe and diverse impacts. Sea level rise and other consequences of global climate change will affect locations worldwide from the Maldives to Bangladesh to Alaska to New York to Florida. As scientists and policymakers come to better understand anthropogenic climate change and its environmental, economic, and social consequences, more and more of them are urging immediate action. Reducing greenhouse gas emissions and taking other steps to mitigate and adapt to climate change represent the foremost challenges for our society in the coming years.

Reviewing Objectives

You should now be able to:

Describe Earth's climate system and explain the factors influencing global climate

- Earth's climate changes naturally over time, but it is now changing rapidly because of human influence. (p. 484)
- The sun provides most of Earth's energy. Earth absorbs 70% of incoming solar radiation and reflects 30% back into space. (pp. 484–485)
- Greenhouse gases such as carbon dioxide, methane, water vapor, nitrous oxide, ozone, and halocarbons warm the atmosphere by absorbing and re-emitting infrared radiation. (pp. 484–485)
- Earth is experiencing radiative forcing of 1.6 watts/m^2 of thermal energy above what it was experiencing 250 years ago. (p. 487)
- Milankovitch cycles, solar radiation, ocean absorption, and ocean circulation all influence climate. (pp. 487–488)

Characterize human influences on the atmosphere and on climate

- By burning fossil fuels and clearing forests, people are increasing atmospheric concentrations of greenhouse gases. (p. 486)
- Increased greenhouse gas emissions enhance the greenhouse effect. (p. 486)
- Input of aerosols into the atmosphere exerts a variable but slight cooling effect. (p. 487)

Summarize how researchers study climate

- Proxy indicators—such as data from ice cores, sediment cores, tree rings, packrat middens, and coral reefs—reveal information about past climate. (pp. 488–491)
- Direct measurements of temperature, precipitation, and other conditions tell us about current climate. (pp. 489–490, 492)
- Climate models serve to predict future changes in climate. (pp. 490–491, 494–495)

Outline current and future trends and impacts of global climate change

- Climate science is a huge body of research. The Intergovernmental Panel on Climate Change (IPCC) synthesizes current research, and its periodic reports represent the consensus of the scientific community. (pp. 492–493)
- Temperatures on Earth have warmed by an average of 0.74°C (1.33°F) over the past century and are predicted to rise 1.8–4.0°C (3.2–7.2°F) over the next century. (pp. 492–493, 496)
- Changes in precipitation have varied by region. (pp. 496–497)
- Extreme weather events are becoming more frequent, likely in part as a result of modification of the jet stream. (pp. 496–497)
- Melting glaciers will diminish water supplies, and melting ice sheets are adding to sea level rise. (p. 498)
- Sea level has risen an average of 21 cm (8.3 in.) over the past 130 years, and will rise by more in the coming century. (pp. 498–500)
- Storm surges worsened by higher sea levels threaten the mainland United States, and not just oceanic islands. (pp. 498–501)
- Ocean acidification may be one of the most far-reaching impacts of our greenhouse gas emissions. (p. 501)
- Climate change exerts impacts on organisms and ecosystems, as well as on agriculture, forestry, health, and economics. (pp. 501–503)
- Climate change and its impacts vary regionally. (pp. 503–505)

- Despite some remaining uncertainties, the scientific community feels that evidence for humans' role in influencing climate is strong enough to justify taking action to reduce emissions. (pp. 504–505)

Suggest ways we may respond to climate change

- Both adaptation and mitigation are necessary. (p. 507)
- Conservation, energy efficiency, and new clean and renewable energy sources will help reduce greenhouse gas emissions. (pp. 507–508)
- New automotive technologies and investment in public transportation will help reduce emissions. (pp. 508–509)
- Addressing climate change will require multiple strategies. (p. 509)
- The Kyoto Protocol provided a first step for nations to begin addressing climate change. (p. 510)
- International efforts to design a treaty to follow the Kyoto Protocol have made some progress, but have fallen far short of what is needed to limit climate change. (pp. 510–511)
- Developing renewable energy technologies presents economic opportunities. (pp. 511–512)
- Some U.S. states and cities are acting to address emissions. (p. 512)
- Emissions trading programs provide a way to harness the free market and engage industry in reducing emissions. (pp. 512–513)
- A carbon tax, specifically a fee-and-dividend approach, is another option. (p. 513)
- Individuals and corporations are increasingly exploring carbon offsets and other means of reducing personal carbon footprints. (pp. 513–515)
- Some scientists are so anxious about our lack of response to climate change that they are now studying potential geoengineering options. (p. 514)

Testing Your Comprehension

1. What happens to solar radiation after it reaches Earth? How do greenhouse gases warm the lower atmosphere?
2. Why is carbon dioxide considered the main greenhouse gas? Why are carbon dioxide concentrations increasing in the atmosphere?
3. How do scientists study the ancient atmosphere? Describe what a proxy indicator is, and give two examples.
4. Has simulating climate change with computer programs been effective in helping us predict climate? Briefly describe how these programs work.
5. List three major trends in climate that scientists have documented so far. Now list three future trends that they are predicting, along with their potential impacts.
6. Describe how rising sea levels, caused by global warming, can create problems for people. How is climate change affecting marine ecosystems?
7. How might a warmer climate affect agriculture? How is it affecting distributions of plants and animals? How might it affect human health?
8. What are the largest two sources of greenhouse gas emissions in the United States? How can we reduce these emissions?
9. What roles have international treaties played in addressing climate change? Give two specific examples.
10. Describe one market-based approach for reducing greenhouse gas emissions. Explain one reason it may work well and one reason it may not work well.

Seeking Solutions

1. Some people argue that we need "more proof" or "better science" before we commit to substantial changes in our energy economy. How much certainty do you think we need before we should take action regarding climate change? How much certainty do you need in your own life before you make a major decision? Should nations and elected officials follow a different standard? Do you believe that the precautionary principle (pp. 265, 383) is an appropriate standard in the case of global climate change? Why or why not?
2. Describe several ways in which we can reduce greenhouse gas emissions from transportation. Which approach do you think is most realistic, which approach do you think is least realistic, and why?
3. Suppose that you would like to make your own lifestyle carbon-neutral and that you aim to begin by reducing the emissions you are responsible for by 25%. What three actions would you take first to achieve this reduction?

4. Think about the many ways in which your campus might reduce its greenhouse gas emissions. Come up with three concrete proposals for ways to reduce emissions on your campus that you feel would be effective and feasible. How would you present these proposals to your campus administration to get its support?
5. **THINK IT THROUGH** You have been appointed as the United States representative to an international conference to negotiate terms of the emissions reduction treaty to replace the Kyoto Protocol in 2020. The U.S. government has instructed you to take a leading role in designing the new treaty and to engage constructively with other nations' representatives while protecting your nation's economic and political interests. What type of agreement will you try to shape? Describe at least three components that you would propose or agree to, and at least one that you would oppose.
6. **THINK IT THROUGH** You have just been elected governor of a medium-sized U.S. state. Polls show that the public wants you to take bold action to reduce greenhouse gas emissions. However, polls also show that the public does not want prices of gasoline or electricity to rise. Carbon-emitting industries in your state are wary of emissions reductions being required of them but are willing to explore ideas with you. Your state legislature will support you in your efforts as long as you remain popular with voters. The state to your west has just passed ambitious legislation mandating steep greenhouse gas emissions reductions. The state to your east has just joined a new regional emissions trading consortium. What actions will you take in your first year as governor?

Calculating Ecological Footprints

Global climate change is something to which we all contribute, because fossil fuel combustion plays such a large role in supporting the lifestyles we lead. Conversely, as individuals, each one of us can help to mitigate climate change through personal decisions and actions in how we live our lives.

Several online calculators enable you to calculate your own personal carbon footprint, the amount of carbon emissions for which you are responsible. Go to http://www.nature.org/greenliving/carboncalculator/, take the quiz, and enter the relevant data in the table.

	Carbon footprint (tons per person per year)
World average	
U.S. average	
Your footprint	
Your footprint with three changes (see Question 3)	

1. How does your personal carbon footprint compare to that of the average U.S. resident? How does it compare to that of the average person in the world? Why do you think your footprint differs from these in the ways it does?
2. As you took the quiz and noted the impacts of various choices and activities, which one surprised you the most?
3. Think of three changes you could make in your lifestyle that would lower your carbon footprint. Now take the footprint quiz again, incorporating these three changes. Enter your resulting footprint in the table. By how much did you reduce your yearly emissions?
4. What do you think would be an admirable yet realistic goal for you to set as a target value for your own footprint? Would you choose to purchase carbon offsets to help reduce your impact? Why or why not?

Extraction of oil sands at a mine in Alberta

19 Fossil Fuels, Their Impacts, and Energy Conservation

Upon completing this chapter, you will be able to:

- Identify the energy sources we use
- Describe the origin and nature of major types of fossil fuels
- Explain how we extract and use fossil fuels
- Evaluate peak oil and the challenges it may pose
- Examine how we are reaching further for fossil fuels
- Outline and assess environmental impacts of fossil fuel use, and explore solutions
- Evaluate political, social, and economic aspects of fossil fuel use
- Specify strategies for conserving energy and enhancing efficiency

CENTRAL CASE STUDY

Alberta's Oil Sands and the Keystone XL Pipeline

"It's good for our country, and it's good for our economy, and it's good for the American people, especially those who are looking for work."

—House Speaker John Boehner (R-Ohio)

"It will be game over for the climate."

—Climate scientist James Hansen

Everything about Canada's oil sands is huge. These fossil fuel deposits cover a region the size of Illinois, within boreal forests that span the width of the continent. The open pit mines dug to extract the fuel are miles wide; the vehicles moving inside them like ants are million-pound haul trucks with 14-foot tires and shovels that are five stories high. The economic value of the extracted oil is astounding. Last but not least, burning all this fuel will alter the very climate of our planet.

Oil sands, also called **tar sands,** are layers of sand or clay saturated with a viscous, tarry type of petroleum called *bitumen*. Huge areas of these wet blackish deposits underlie a thinly populated region of northern Alberta, and the implications of mining them for oil are momentous. To some people the oil sands represent wealth and security, a key to maintaining our fossil-fuel-based lifestyle far into the future. To others they are a source of appalling pollution and threaten to radically alter Earth's climate.

To extract oil from oil sands, companies clear the boreal forest and then strip-mine the land, peeling back layers of peat and creating open pits 215 m (400 ft) deep. The gooey deposits are mixed with hot water and chemicals to separate the bitumen from the sand, and the bitumen is removed and processed, while wastewater is dumped into toxic tailings lakes that are even larger than the mines. In locations where oil sands are more deeply buried, hot water is injected down deep shafts to liquefy, separate, and extract the bitumen *in situ*.

Mining for oil sands began in Alberta in 1967, but for many years it was hard to make money extracting these low-quality deposits. Rising oil prices in recent years have now turned it into a profitable venture, and today dozens of companies are mining here. Canadian oil sands are producing 1.7 million barrels of oil per day, more than half of Canada's petroleum production. Thanks to the oil sands, Canada boasts the world's third-largest proven reserves of oil, after Saudi Arabia and Venezuela. Each truckload of oil sands that leaves a mine carries oil worth close to $20,000 at 2013 prices.

Canada looked for buyers south of its border first, seeking to capitalize on the United States' insatiable appetite for oil. TransCanada Corporation built the Keystone Pipeline to ship diluted bitumen into the United States. This pipeline system began operating in 2010, bringing oil from Alberta nearly 3500 km (2200 mi) to Illinois and Oklahoma. At the Oklahoma terminus in the town of Cushing, a bottleneck created a glut of oil that was unable to reach refineries on the Texas coast fast enough to meet demand. TransCanada proposed the Keystone XL extension, a two-part project consisting of (1) a southern leg to connect Cushing to the Texas refineries and (2) a northern leg that would cut across the Great Plains to shave off distance and add capacity to the existing line.

The Keystone XL pipeline proposal soon met opposition from people living along the proposed route who were concerned about health, environmental protection, and property rights. It also faced nationwide opposition from advocates of action to address global climate change.

Pipeline proponents feel the Keystone XL project will create jobs for workers in the U.S. heartland and will guarantee a dependable oil supply for decades to come. They stress that buying oil from Canada—a stable, friendly, democratic neighbor—could help end U.S. reliance on oil-producing nations such as Saudi Arabia and Venezuela that have had authoritarian governments and poor human rights records.

Opponents of the pipeline extension express dismay at the destruction of boreal forest and anxiety about transporting oil over the continent's largest aquifer, where spills could contaminate drinking water for millions of people and irrigation water for America's breadbasket. They also seek to avoid extracting a vast new source of fossil fuels whose combustion would release immense amounts of greenhouse gases that will intensify climate change. By buying a source of oil that is energy-intensive to extract and that burns less cleanly than conventional fuels, they maintain, the United States would be prolonging fossil fuel dependence and worsening climate change when it should instead be transitioning to clean renewable energy.

Under pressure from all sides, the administration of U.S. President Barack Obama walked a fine line. Because the northern leg of the Keystone XL extension crosses an international border, it requires a presidential permit from the U.S. Department of State—which TransCanada applied for in 2008. After three years of review, the State Department hesitated to approve the project because of concerns about damage to the ecologically sensitive Sandhills area of Nebraska and potential contamination of the Ogallala Aquifer. Facing street protests at the White House (**FIGURE 19.1**), Obama in November 2011 postponed the permit decision.

Republicans in Congress reacted by demanding a decision in 60 days and attaching this mandate to legislation for a payroll tax cut. Obama responded by announcing that the application would be denied because of insufficient time to review the pipeline's impact. However, Obama encouraged TransCanada to renew its application with a revised route avoiding the areas of concern in Nebraska and to proceed with the southern leg of the pipeline, which requires no permit. TransCanada did both. Meanwhile, Canadian officials grew irritated and began considering building a pipeline west to British Columbia and selling the oil to China instead.

In March 2013 the State Department released a draft environmental impact statement for the new route. The draft EIS gave little indication that it would stand in the way of pipeline development. The State Department planned to issue a final EIS after receiving public comment, after which Obama would decide whether to approve the pipeline.

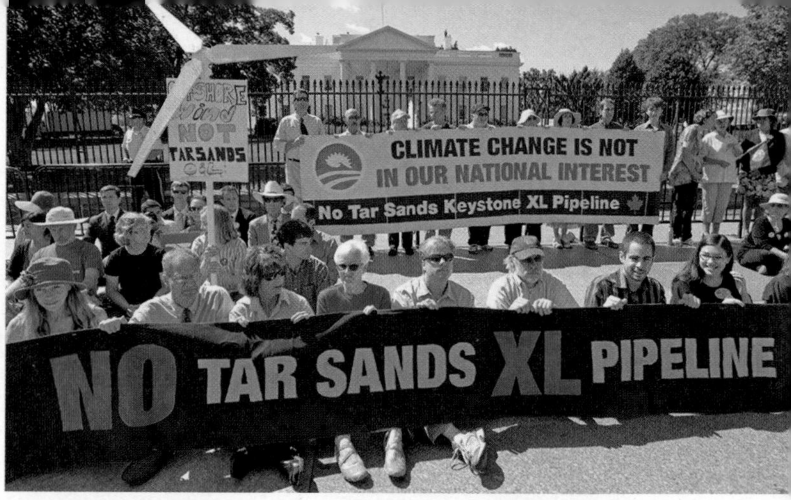

FIGURE 19.1 Many Americans have opposed the Keystone XL pipeline extension. Tens of thousands of activists protested in front of the U.S. White House in increasingly large rallies in 2011, 2012, and 2013.

Throughout 2013 the debate intensified. In February, tens of thousands of Americans protested against the pipeline in front of the White House. These protestors viewed the decision on Keystone XL as a test of Obama's vow to deal with climate change, made in his inauguration speech a month earlier. Pipeline proponents countered that Canada would find a way to extract and sell its oil in any case, so the United States might as well take advantage of the trade benefits of buying it and reselling it on the world market.

Later that year, Obama announced that he would approve the pipeline only if it "does not significantly exacerbate the problem of carbon pollution." What this signaled was unclear, however: The draft EIS suggested the pipeline would not do so, but the Environmental Protection Agency had judged the EIS to be inadequate.

As this book went to press, a decision to approve or deny the Keystone XL pipeline had not yet been made. We will leave it to you and your instructor to flesh out the rest of this story!

The divergent views on Canada's oil sands reflect our confounding relationship with fossil fuels. These energy sources power our civilization and have enabled our modern standard of living—yet as climate change worsens, we face the need to wean ourselves from them and shift to clean renewable energy sources. The way in which we handle this complex transition will determine a great deal about the quality of our lives and the future of our society and our planet.

Sources of Energy

Humanity has devised many ways to harness the renewable and nonrenewable forms of energy available on our planet (**TABLE 19.1**). We use these energy sources to heat and light our homes; power our machinery; fuel our vehicles; produce plastics, pharmaceuticals, and synthetic fibers; and provide the comforts and conveniences to which we've grown accustomed in the industrial age.

Nature offers us a variety of energy sources

Most of Earth's energy comes from the sun. We can harness energy from the sun's radiation directly by using solar power technologies. Solar radiation also helps drive wind and the water cycle, enabling us to harness wind power and hydroelectric power. And of course, sunlight drives photosynthesis (p. 32) and the growth of plants, from which we take wood and other biomass as a fuel source. When plants and other organisms die and are buried in sediments under particular conditions, their stored chemical energy may eventually be transferred to **fossil fuels,** highly combustible substances formed from the remains of organisms from past geologic ages. Today we rely on three main fossil fuels, in the form of a solid (coal), liquid (oil), and gas (natural gas).

In addition, a great deal of energy emanates from Earth's core, making geothermal power available for our use. Energy also results from the gravitational pull of the moon and sun, and we are just beginning to harness power from the ocean tides that these forces generate. Finally, an immense amount of energy resides within the bonds among protons and neutrons in atoms, and this energy provides us with nuclear power. We explore all these energy sources as alternatives to fossil fuels in Chapters 20 and 21.

Energy sources such as sunlight, geothermal energy, and tidal energy are considered perpetually renewable because they are readily replenished, so we can keep using them without depleting them (pp. 3–4). In contrast, energy sources such as coal, oil, and natural gas are considered nonrenewable. These nonrenewable fuels result from ongoing natural processes, but it takes so long for fossil fuels to form that, once depleted, they cannot be replaced within any time span useful to our civilization. It takes a thousand years for the biosphere to generate the amount of organic matter that must be buried to produce a single day's worth of fossil fuels for our society. To replenish the fossil fuels we have depleted so far would take many millions of years.

TABLE 19.1 Energy Sources We Use

ENERGY SOURCE	DESCRIPTION	TYPE OF ENERGY
Crude oil	Fossil fuel extracted from ground (liquid)	Nonrenewable
Natural gas	Fossil fuel extracted from ground (gas)	Nonrenewable
Coal	Fossil fuel extracted from ground (solid)	Nonrenewable
Nuclear energy	Energy from atomic nuclei of uranium	Nonrenewable
Biomass energy	Energy stored in plant matter from photosynthesis	Renewable
Hydropower	Energy from running water	Renewable
Solar energy	Energy from sunlight directly	Renewable
Wind energy	Energy from wind	Renewable
Geothermal energy	Earth's internal heat rising from core	Renewable
Tidal and wave energy	Energy from tides and ocean waves	Renewable

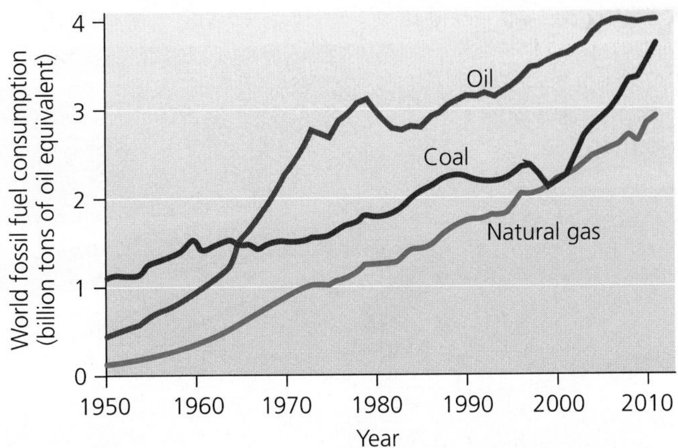

FIGURE 19.2 Annual global consumption of fossil fuels has risen greatly over the past half-century. Oil remains our leading energy source. *Data from U.S. Energy Information Administration; International Energy Agency; and BP p.l.c., 2012,* Statistical review of world energy 2012.

DATA Q By roughly what percentage has the annual consumption of oil risen since the year you were born?

At our accelerating rate of consumption, we will use up Earth's easily accessible store of conventional fossil fuels in just decades. For this reason, and because fossil fuels exert severe environmental impacts, renewable energy sources increasingly are being developed as alternatives to fossil fuels (Chapters 20 and 21).

Fossil fuels dominate our energy use

Since the industrial revolution, fossil fuels have replaced biomass as our society's dominant source of energy. Global consumption of coal, oil, and natural gas has risen for years and is now at its highest level ever (**FIGURE 19.2**). The high energy content of fossil fuels makes them efficient to burn, ship, and store. We use these fuels for transportation, manufacturing, heating, and cooking and also to generate **electricity,** a secondary form of energy that is convenient to transfer over long distances and apply to a variety of uses. Each type of fuel has its own mix of uses, and each contributes in different ways to our economies and our daily needs. For instance, oil is used mostly for transportation, whereas coal is used mostly to generate electricity.

Societies differ in how they use energy. Industrialized nations apportion roughly one-third of their energy to transportation, one-third to industry, and one-third to all other uses. In contrast, industrializing nations devote a greater proportion of energy to subsistence activities such as agriculture, food preparation, and home heating, and much less to transportation. Moreover, people in developing countries often rely on manual or animal energy sources instead of automated ones. For instance, most rice farmers in Southeast Asia plant rice by hand, but industrial rice growers in California use airplanes. Because industrialized nations rely more on equipment and technology, they use more fossil fuels. In the United States, oil, coal, and natural gas together supply 82% of energy demand (**FIGURE 19.3**).

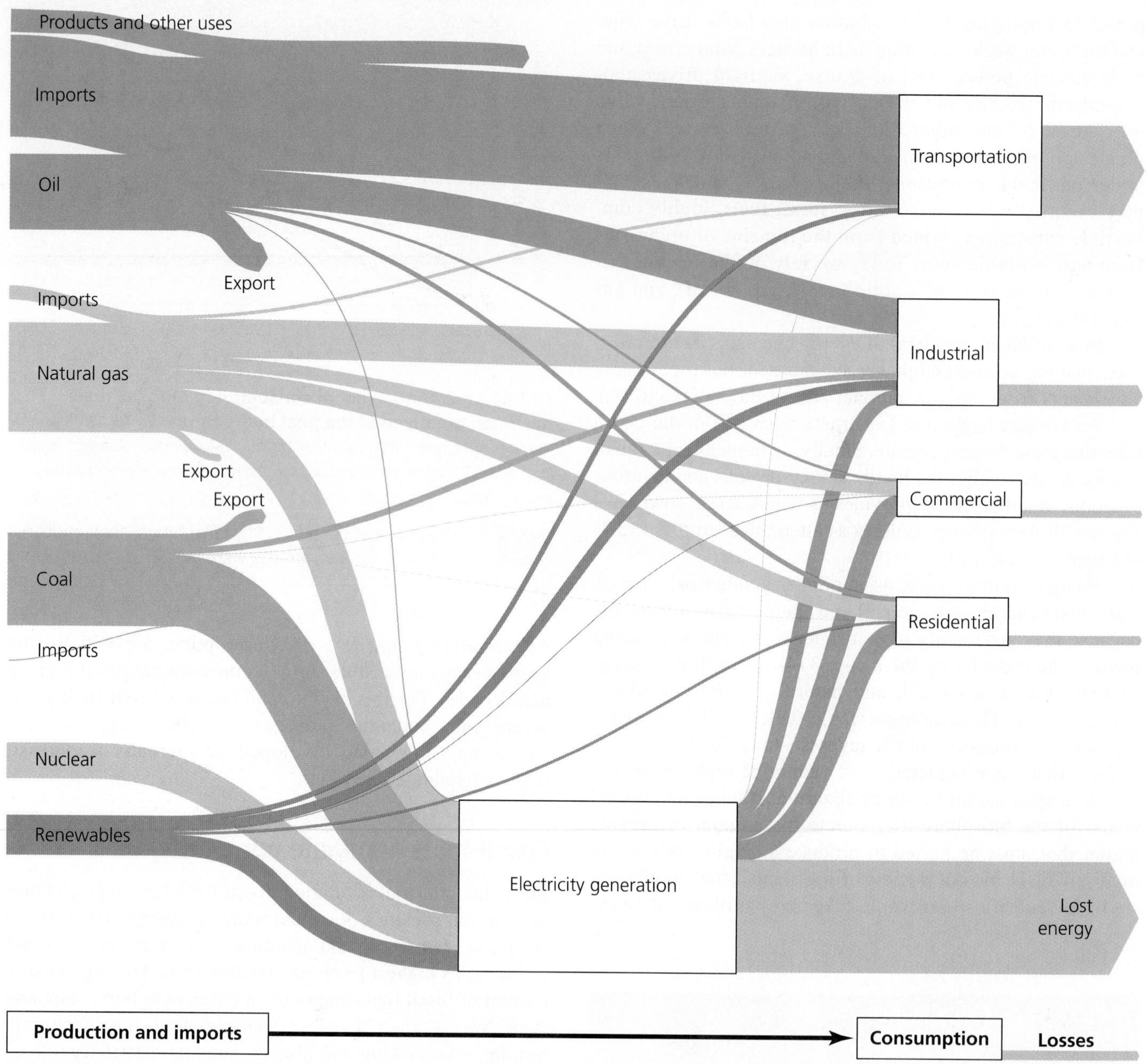

FIGURE 19.3 Total energy flow of the United States. Amounts are represented by the thickness of each bar. Domestic production and imports are shown on the left, and destinations of the energy are shown on the right. Portions of each energy source are used directly in the residential, commercial, industrial, and transportation sectors. Other portions are used to generate electricity, which in turn powers these sectors. The large amounts of energy lost as waste heat are shown on the right. *Data are for 2012, from U.S. Energy Information Administration and Lawrence Livermore National Laboratory.*

Energy sources and consumption are unevenly distributed

A world map of per capita consumption rates shows that citizens of developed regions generally consume far more energy than do those of developing regions (**FIGURE 19.4**). Per person, the most industrialized nations use up to 100 times more energy than do the least industrialized nations. The United States has only 4.4% of the world's population, but it consumes nearly 19% of the world's energy.

The origins of most energy sources also are unevenly distributed over Earth's surface. Some regions have substantial reserves of oil, coal, or natural gas, whereas others have very few. Half the world's proven reserves of crude oil lie in the Middle East. The Middle East is also rich in natural gas, but Russia holds more natural gas than any other country. Russia is also rich in coal, as is China, but the United States possesses the most coal of any nation (**TABLE 19.2**).

It takes energy to make energy

We do not simply get energy for free. To harness, extract, process, and deliver the energy we use, we need to invest

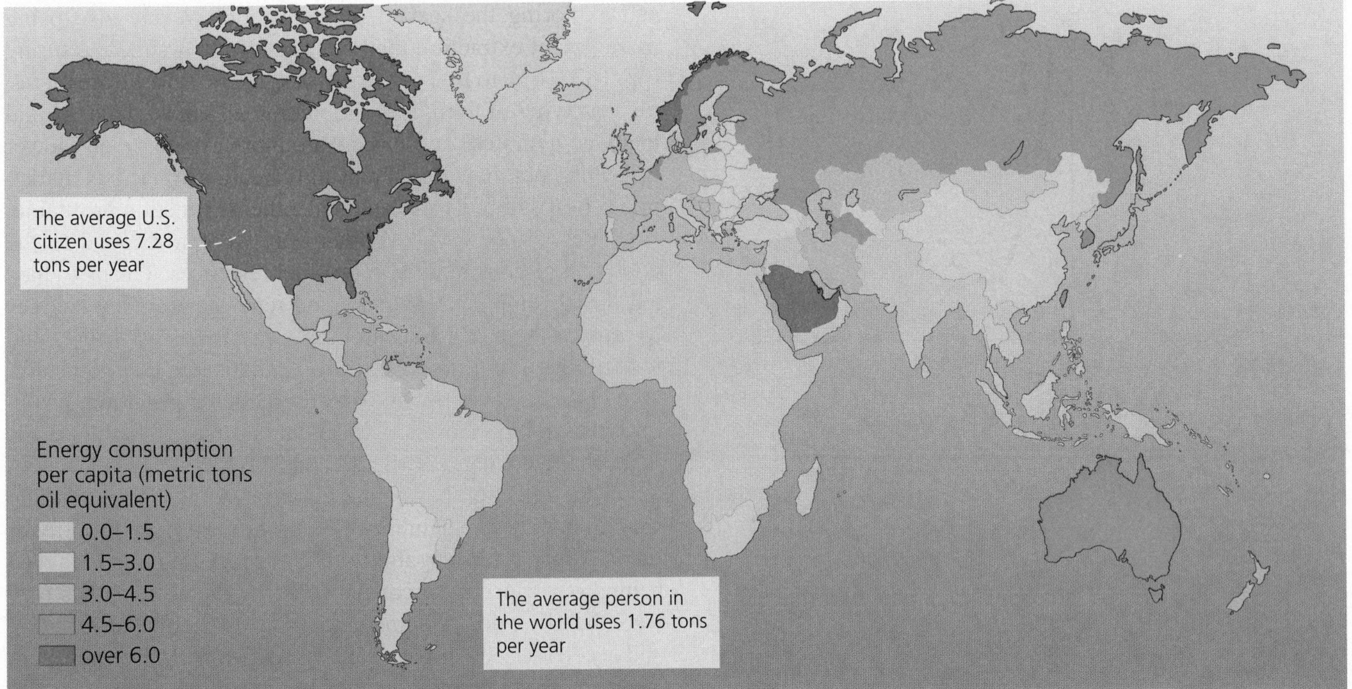

FIGURE 19.4 People in wealthy industrialized nations tend to consume the most energy per person. This map combines all types of energy, standardized to metric tons of "oil equivalent," that is, the amount of fuel needed to produce the energy gained from combusting one metric ton of crude oil. *Data from BP p.l.c., 2012.* Statistical review of world energy 2012.

DATA Q How many times more energy does the average U.S. citizen use than the average person in the world?

substantial inputs of energy. For instance, mining oil sands in Alberta requires extensive use of powerful vehicles and heavy machinery, as well as construction of an immense infrastructure of roads, pipelines, waste ponds, storage tanks, water intakes, processing facilities, housing for workers, and more—all requiring the use of energy. Natural gas must be burned to heat the water that is used to separate the bitumen from the sand. Processing and piping the oil away from the extraction site, and then refining it into products we can use, requires further energy inputs. Thus, when evaluating an energy source, it is important to take energy inputs into consideration by subtracting costs in energy invested from the benefits in energy received. **Net energy** expresses the difference between energy returned and energy invested:

$$\text{Net energy} = \text{Energy returned} - \text{Energy invested}$$

When assessing energy sources, it is useful to use a ratio often denoted as **EROI—energy returned on investment.** EROI ratios are calculated as follows:

$$\text{EROI} = \text{Energy returned/ Energy invested}$$

Higher EROI ratios mean that we receive more energy from each unit of energy that we invest. Fossil fuels are widely used because their EROI ratios have historically been high. However, EROI ratios can change over time. Ratios rise as the technology to extract and process fossil fuels improves, and they fall as resources are depleted and remaining resources become harder to extract.

EROI ratios for producing conventional oil and natural gas in the United States declined from roughly 30:1 in the 1950s to about 20:1 in the 1970s, and today they hover around 11:1

TABLE 19.2 Nations with the Largest Proven Reserves of Fossil Fuels

OIL (% world reserves)		NATURAL GAS (% world reserves)		COAL (% world reserves)	
Venezuela*	17.9	Russia	21.4	United States	27.6
Saudi Arabia	16.1	Iran	15.9	Russia	18.2
Canada*	10.6	Qatar	12.0	China	13.3
Iran	9.1	Turkmenistan	11.7	Australia	8.9
Iraq	8.7	United States	4.1	India	7.0

*Most reserves in Venezuela and Canada consist of oil sands, which are included in these figures.
Data from BP p.l.c., 2012. Statistical review of world energy 2012.

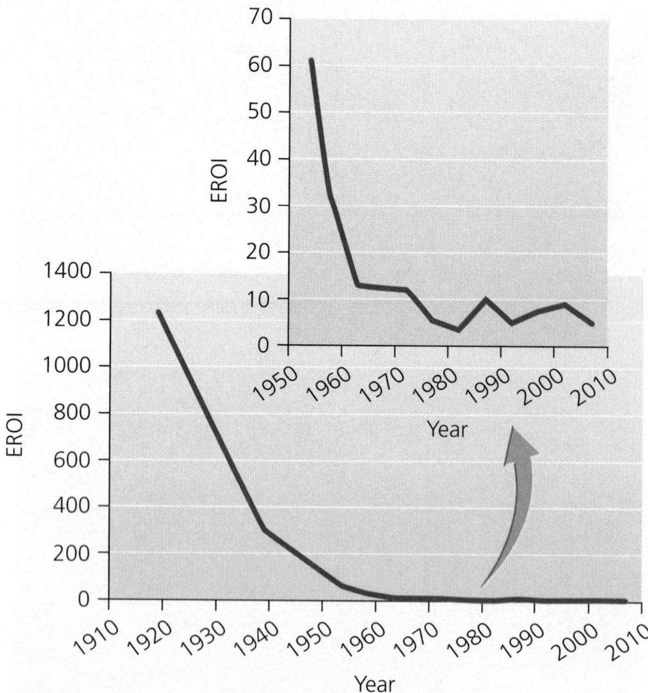

FIGURE 19.5 EROI values for discovering oil and gas in the United States have declined over the past century. *Data from Guilford, M., et al., 2011. A new long term assessment of energy return on investment (EROI) for U.S. oil and gas discovery and production. Pp. 115–136 in* Sustainability, Special Issue, 2011, *eds. C. Hall and D. Hansen,* New studies in EROI (Energy return on investment).

(**FIGURE 19.5**). This means that we used to be able to gain 30 units of energy for every unit of energy expended, but now we can gain only 11. EROI ratios for oil and gas declined because we extracted the easiest deposits first and now must work harder and harder to extract the remaining amounts. For the Alberta oil sands, EROI ratios are still lower, because oil sands are a low-quality fuel that requires a great deal of energy to extract and process. EROI estimates for oil sands from studies so far range from 1.5:1 to 9:1, with most estimates around 3:1 to 5:1.

Where will we turn in the future for energy?

Throughout the 20th century, abundant and inexpensive coal, oil, and natural gas powered the astonishing advances of our civilization. These extraordinarily rich sources of energy helped to bring us a standard of living our ancestors could scarcely have imagined.

We began by extracting the fossil fuel deposits that were readily located and accessed, and we took advantage of boundless energy at cheap prices. Yet because fossil fuel deposits are finite and nonrenewable, we gradually began depleting them. As easily accessible supplies of the three main fossil fuels became depleted, EROI ratios rose, and fuels became more expensive.

In response, we have developed technology to reach deeper and farther, expending more money and energy in order to continue obtaining fossil fuel energy. We have returned to sites that were already extracted, bringing powerful new machinery and approaches to squeeze more fuels from known locations. We are now reaching into formerly inaccessible places by drilling deeper, moving further offshore, and exploring the seabed of the Arctic. We are also using more potent extraction methods, such as hydraulic fracturing (pp. 162–163), to free gas from rock layers. And we are pursuing new types of fossil fuels, including oil sands, shale oil, and methane hydrates. These fuels are more expensive and lower in quality, but they are increasingly being extracted as market prices of fossil fuels rise and make their extraction profitable.

There is, however, another way we can respond to the depletion of conventional fossil fuel resources. This is to hasten the development of clean and renewable energy sources to replace them. By transitioning away from fossil fuels and toward renewable sources, we can gain energy that is sustainable in the long term while greatly reducing pollution, health impacts, and the emission of greenhouse gases that drive climate change (Chapters 18, 20, and 21).

This transition has begun to occur, but it is apparent that we will continue to gain much of our future energy from fossil fuels. Alas, so far our ability to control pollution has lagged behind our capacity to consume energy. Many scientists now warn that if we do not immediately step up energy conservation and accelerate our shift to renewables, we will drive our planet's climate into unprecedented territory, threatening impacts on our economy, our quality of life, and our society's future.

Fossil Fuels and Their Extraction

The three conventional fossil fuels on which our modern industrial society was built, and on which we rely today, are coal, natural gas, and oil. Additional fossil fuels we are beginning to extract or considering for the future include oil sands, shale oil, and methane hydrates. We will first consider how each of these fossil fuels is formed, how we locate deposits, how we extract these resources, and how our society puts them to use. We will then examine some of their environmental and social impacts.

Fossil fuels are formed from ancient organic matter

The fossil fuels we burn today in our vehicles, homes, industries, and power plants were formed from the tissues of organisms that lived 100–500 million years ago. The energy these fuels contain came originally from the sun and was converted to chemical-bond energy by photosynthesis. The chemical energy in these organisms' tissues then became concentrated as these tissues decomposed and their hydrocarbon compounds were altered and compressed (**FIGURE 19.6**).

Most organisms, after death, do not end up as part of a coal, gas, or oil deposit. A tree that falls and decays as a rotting log on the forest floor undergoes mostly **aerobic** decomposition; in the presence of air, bacteria and other organisms that use oxygen break down plant and animal remains into simpler carbon molecules that are recycled through the ecosystem. Fossil fuels are produced only when organic material is broken down in an **anaerobic** environment, one that has little or no oxygen. Such environments include the bottoms of lakes, swamps, and shallow seas. Over millions of years, organic matter that accumulates at the bottoms of such water bodies may be converted into crude oil, natural gas, or coal, depending on (1) the

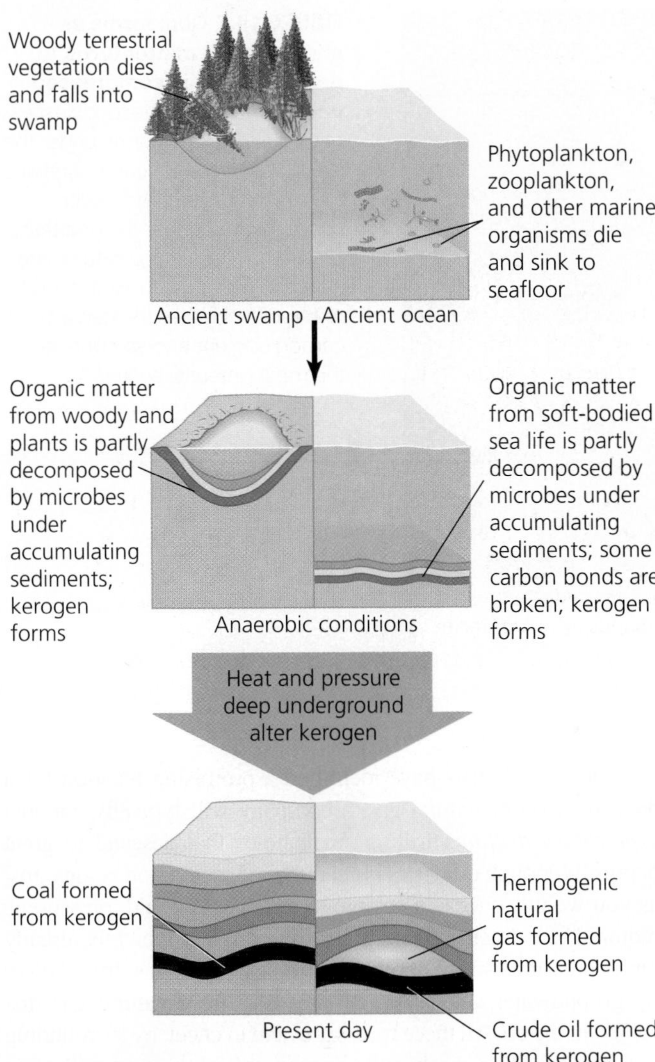

FIGURE 19.6 Fossil fuels begin to form when organisms die and end up in oxygen-poor conditions. This can occur when trees fall into lakes and are buried by sediment, or when phytoplankton and zooplankton drift to the seafloor and are buried (**top diagram**). Organic matter that undergoes slow anaerobic decomposition deep under sediments forms kerogen (**middle diagram**). Coal results when plant matter is compacted so tightly that there is little decomposition (**bottom left diagram**). The action of geothermal heating on kerogen may create crude oil and natural gas (**bottom right diagram**), which come to reside in porous rock layers beneath dense, impervious layers.

chemical composition of the material, (2) the temperatures and pressures to which it is subjected, (3) the presence or absence of anaerobic decomposers, and (4) the passage of time.

Coal The world's most abundant fossil fuel is **coal,** a hard blackish substance formed from organic matter (generally woody plant material) that was compressed under very high pressure, creating dense, solid carbon structures (FIGURE 19.7). Coal typically results when little decomposition takes place because the material cannot be digested or appropriate decomposers are not present. The proliferation 300–400 million years ago of swampy environments where organic material was buried has created coal deposits throughout the world.

Coal varies from deposit to deposit in the amount of water, carbon, and potential energy it contains. Organic material that is broken down anaerobically but remains wet, near the surface, and not well compressed is called *peat*. As peat decomposes further, as it is buried under sediments, as heat and pressure increase, and as time passes, water is squeezed out and carbon compounds are packed more tightly together, forming coal. Scientists classify coal into four types (see Figure 19.7). The more coal is compressed, the greater is its carbon content and the greater is the energy content per unit volume.

Oil and natural gas The sludgelike liquid we know as **oil,** or **crude oil,** contains a mixture of hundreds of different types of hydrocarbon molecules (p. 28). **Natural gas** is a gas consisting primarily of methane (CH_4) and including varying amounts of other volatile hydrocarbons. Oil is also known as **petroleum,** although this term is commonly used to refer to oil and natural gas collectively.

Both oil and natural gas are formed from organic material (especially dead plankton) that drifted down through coastal marine waters millions of years ago and was buried in sediments on the ocean floor. As organic matter is buried more deeply, the pressure exerted by overlying sediments grows, and temperatures increase. Carbon bonds in the organic matter begin breaking, and the organic matter turns to a substance called *kerogen*, which acts as a source material for both natural gas and crude oil. Further heat and pressure act on the kerogen to degrade complex organic molecules into simpler hydrocarbon molecules. Oil tends to form under temperature and pressure conditions often found 1.5–3 km (1–2 mi) below the surface. At depths below 3 km (1.9 mi), the high temperatures and pressures tend to form natural gas.

Natural gas that forms in this way from compression and heat deep underground is called *thermogenic* gas. Thermogenic gas may be formed directly, or from coal or oil that is altered by heating. Most gas extracted commercially is thermogenic and is found above deposits of crude oil or seams of coal, so its extraction often accompanies the extraction of those fossil fuels.

Natural gas is also formed by a second process; *biogenic* gas is created at shallow depths by the anaerobic decomposition of organic matter by bacteria. An example is the "swamp gas" you may smell when stepping into the muck of a swamp. One source of biogenic natural gas is the decay process in landfills, and many landfill operators are now capturing this gas to sell as fuel (p. 616). Biogenic gas is nearly pure methane, whereas thermogenic gas contains small amounts of other gases as well as methane.

Oil sands As we've seen, oil sands (also called tar sands) consist of moist sand and clay containing 1–20% bitumen, a thick and heavy form of petroleum that is rich in carbon and poor in hydrogen. Oil sands represent crude oil deposits that have been degraded and chemically altered by water erosion and bacterial decomposition. The leading scientific hypothesis to explain Alberta's oil sands is that geological changes tens of millions of years ago as the Rocky Mountains were uplifted allowed crude oil deposits to migrate northeastward and upward until they saturated rock and soil in what is now

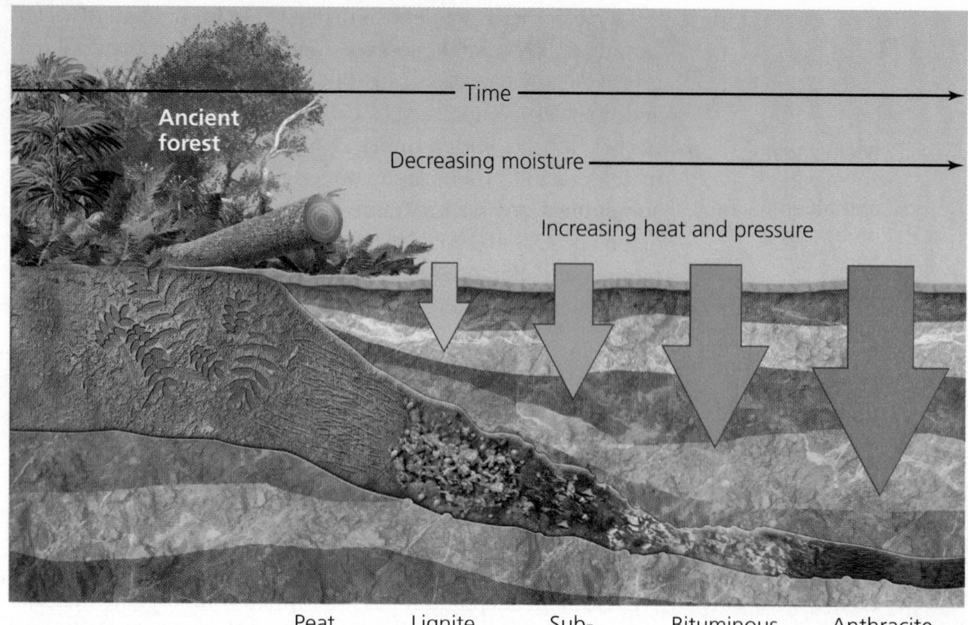

FIGURE 19.7 Coal forms as ancient plant matter turns to peat and then is compressed underground. Of the four classes of coal, lignite coal forms under the least pressure and heat and retains the most moisture. Anthracite coal is formed under the greatest pressure, where temperatures are high and moisture content is low. Anthracite coal has the densest carbon content and so contains the most potential energy.

northeast Alberta. Microorganisms (which are abundant near the surface but absent at depth) began to consume the oil, particularly the lighter components, leaving degraded heavy bitumen.

Oil shale **Oil shale** is sedimentary rock filled with kerogen that can be processed to produce a liquid form of petroleum called **shale oil.** Oil shale is formed by the same processes that form crude oil but occurs when kerogen was not buried deeply enough or subjected to enough heat and pressure to form oil.

Methane hydrate **Methane hydrate** occurs in sediments in the Arctic and on the ocean floor. Also called *methane clathrate* or *methane ice*, this fossil fuel is an ice-like solid consisting of molecules of methane embedded in a crystal lattice of water molecules. Methane hydrate is stable at temperature and pressure conditions found in many sediments in the Arctic and on the seafloor. Most methane in these gas hydrates formed from bacterial decomposition in anaerobic environments, but some resulted from thermogenic formation deeper below the surface.

We mine and drill for fossil fuels

Because fossil fuels of each type form only under certain conditions, they occur in isolated deposits. For instance, oil and natural gas tend to rise upward through cracks and fissures in porous rock until meeting a dense impermeable rock layer that traps them. As a result, geologists search for fossil fuels by drilling cores and conducting ground, air, and seismic surveys to map underground rock formations. With knowledge of underground geology, they can predict where fossil fuel deposits might lie (see **THE SCIENCE BEHIND THE STORY**, pp. 530–531).

Once geologists have identified a promising location for a deposit of oil or natural gas, a company will typically conduct *exploratory drilling*, drilling small holes that descend to great depths. If enough oil or gas is encountered, extraction begins. Just as you would squeeze a sponge to remove its liquid, pressure is required to extract oil from porous rock. Oil is typically already under pressure—from above by rock or trapped gas, from below by groundwater, and at times internally from natural gas dissolved in the oil. All these forces are held in check by surrounding rock until drilling reaches the deposit, whereupon oil will often rise to the surface of its own accord. Once pressure is relieved and some oil or gas has risen to the surface, however, the remainder becomes more difficult to extract and needs to be pumped out.

Coal is a solid, and so we mine it rather than drilling for it. For coal deposits near the surface, we use *strip mining*, in which heavy machinery scrapes away huge amounts of earth to expose the coal. For deposits deep underground, we use *subsurface mining*, digging vertical shafts and blasting out networks of horizontal tunnels to follow seams, or layers, of coal. (Strip mining and subsurface mining are illustrated in Figure 23.6, p. 639.) We are also now mining coal on immense scales in the Appalachian Mountains, essentially scraping off entire mountaintops in a process called *mountaintop removal mining* (pp. 641–645).

Oil from oil sands is extracted by two main methods. For deposits near the surface (**FIGURE 19.8a**), a process akin to strip mining for coal or open-pit mining for minerals (p. 640) is used. Shovel-trucks peel back layers of peat and soil and then dig out vast quantities of bitumen-soaked sand or clay. This is mixed with hot water and piped to an extraction facility, where sand sinks to the bottom of tanks while bitumen floats to the top. The bitumen is skimmed off, solvent is added, and the mixture is spun in a centrifuge to further purify the bitumen, which is then processed into crude oil. Three barrels of water are required to extract each barrel of oil, and the resulting toxic wastewater is discharged into vast tailings lakes.

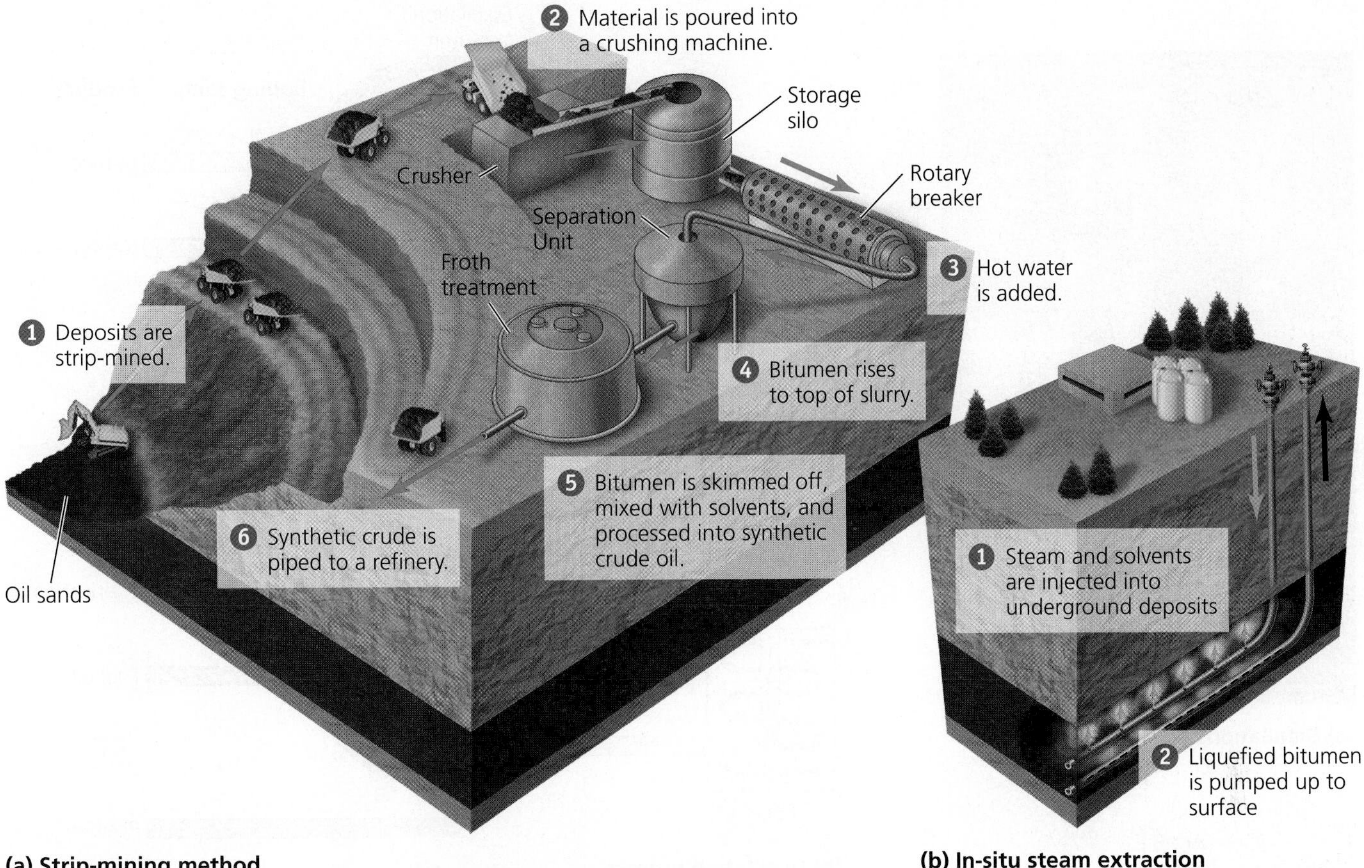

FIGURE 19.8 Oil sands are extracted by two processes. Near-surface deposits of oil sands **(a)** are strip-mined. The deposits are first dug out ❶ with gigantic shovels and trucks and then poured into a crushing machine ❷. The material is then mixed with hot water ❸, and the slurry is piped to a facility where the bitumen floats in a froth atop water in a tank ❹, while sand and clay settle out. The bitumen froth is skimmed off, mixed with chemical solvents, and processed into synthetic crude oil ❺; it is then sent in a pipeline ❻ to a refinery. Deeper deposits of oil sands **(b)** are extracted through well shafts. Pressurized steam is injected down the well ❶ into the oil sand formation, liquefying the bitumen and allowing it to be pumped ❷ to the surface.

Oil sands that are deeper underground (FIGURE 19.8b) are extracted by drilling shafts down to the deposit and injecting steam and solvents down to liquefy and isolate the bitumen in place, then pump it out. After extraction, the bitumen must be refined using chemical reactions that add hydrogen or remove carbon, thus upgrading it into more valuable synthetic crude oil (called *syncrude*).

We mine oil shale using strip mines or subsurface mines. Once mined, oil shale can be burned directly like coal, or it can be baked in the presence of hydrogen and in the absence of air to extract liquid petroleum (a process called *pyrolysis*).

Economics determines how much will be extracted

As we develop more powerful technologies for locating and extracting fossil fuels, the amounts of these fuels that are physically accessible to us—the "technically recoverable" amounts—tend to increase. However, whereas technology determines how much of a fossil fuel *can* be extracted, economics determines how much *will* be extracted. This is because extraction becomes increasingly costly as a resource is removed, so companies will not find it profitable to extract the entire amount. Instead, a company will consider the costs of extraction (and other expenses), and balance these against the income it expects from sale of the fuel. Because market prices of fuel fluctuate, the portion of fuel from a given deposit that is "economically recoverable" fluctuates as well. As market prices rise, economically recoverable amounts approach technically recoverable amounts.

The amount of a fossil fuel that is technologically and economically feasible to remove under current conditions is termed its **proven recoverable reserve.** Proven reserves increase as extraction technology improves or as market prices of the fuel rise. Proven reserves decrease as fuel deposits are depleted by extraction or as market prices fall (making extraction uneconomical).

Refining produces a diversity of fuels

Once we extract oil or gas, it must be processed and refined before we can use it. Let's examine what happens when crude

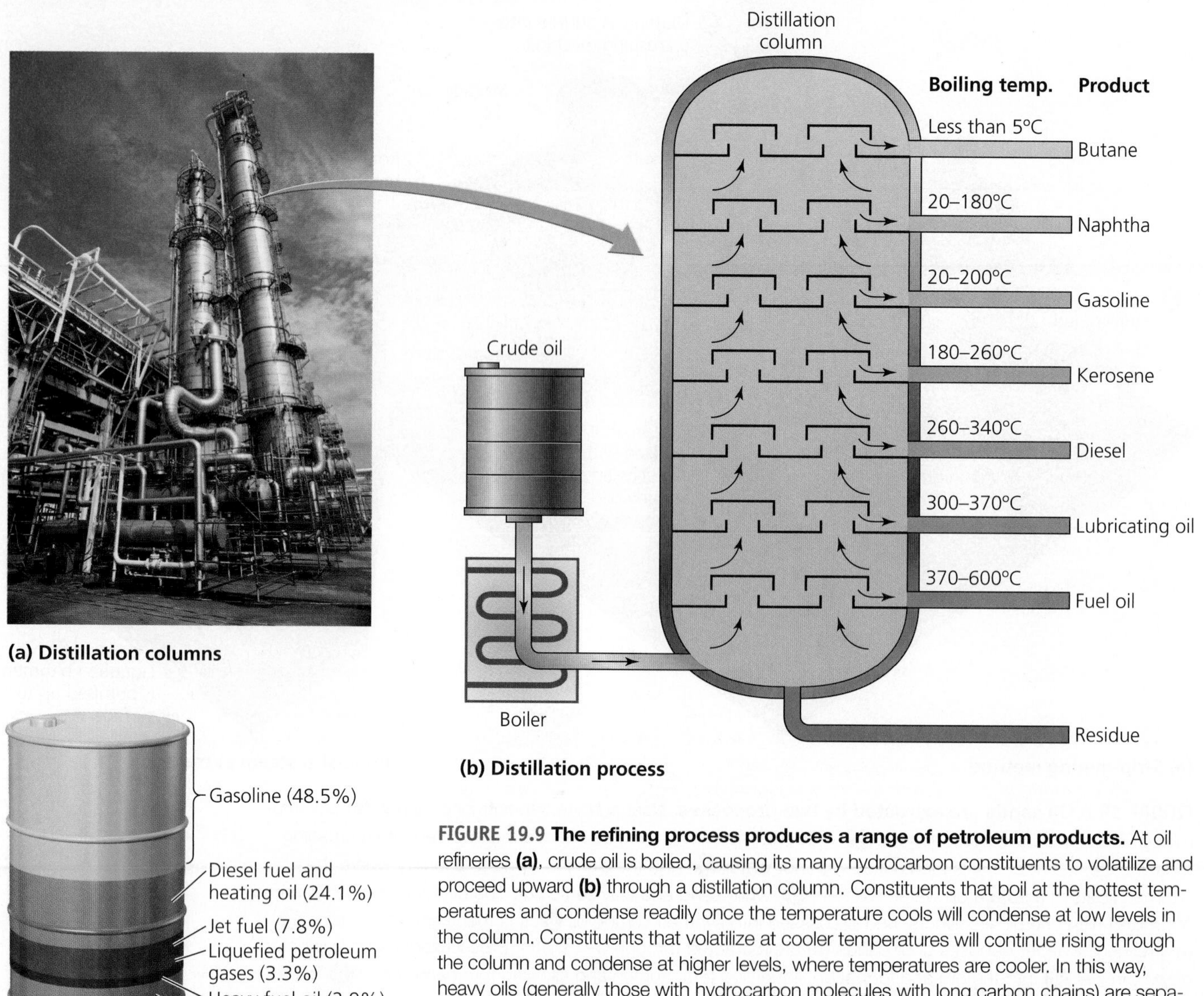

(a) Distillation columns

(b) Distillation process

(c) Typical composition of refined oil

FIGURE 19.9 The refining process produces a range of petroleum products. At oil refineries **(a)**, crude oil is boiled, causing its many hydrocarbon constituents to volatilize and proceed upward **(b)** through a distillation column. Constituents that boil at the hottest temperatures and condense readily once the temperature cools will condense at low levels in the column. Constituents that volatilize at cooler temperatures will continue rising through the column and condense at higher levels, where temperatures are cooler. In this way, heavy oils (generally those with hydrocarbon molecules with long carbon chains) are separated from lighter oils (generally those with short-chain hydrocarbon molecules). Shown in **(c)** are percentages of each major category of product typically generated from a barrel of crude oil. *Data (c) from U.S. Energy Information Administration, 2012.* Annual energy review 2011.

oil is shipped to a refinery (FIGURE 19.9). Because crude oil is a complex mix of hydrocarbons, we can create many types of petroleum products by separating its various components. The many types of hydrocarbon molecules in crude oil have carbon chains of different lengths (p. 28). A chain's length affects its chemical properties, and this has consequences for our use, such as whether a given fuel burns cleanly in a car engine. Through the process of **refining,** hydrocarbon molecules are separated into different size classes and are chemically transformed to create specialized fuels for heating, cooking, and transportation, and to create lubricating oils, asphalts, and the precursors of plastics and other petrochemical products.

Fossil fuels have many uses

As we saw in Figure 19.3, each major type of fossil fuel has its own mix of uses. Let's survey how coal, oil, and natural gas each are used in our society today.

Coal People have burned coal to cook food, heat homes, and fire pottery for thousands of years and in many cultures, from ancient China to the Roman Empire to the Hopi Nation. Coal-fired steam engines helped drive the industrial revolution by powering factories, agriculture, trains, and ships, and by fueling the furnaces of the steel industry. Today we burn coal largely to generate electricity. In coal-fired power plants, coal combustion converts water to steam, which turns a turbine to create electricity (FIGURE 19.10). Coal provides 40% of the electrical generating capacity of the United States, and it powers China's surging economy. China is now the world's primary producer and consumer of coal (TABLE 19.3).

Natural gas Versatile and clean-burning, natural gas emits just half as much carbon dioxide per unit of energy produced as coal and two-thirds as much as oil. We use natural gas to generate electricity in power plants, to heat and cook in our homes, and for much else. Converted to a liquid

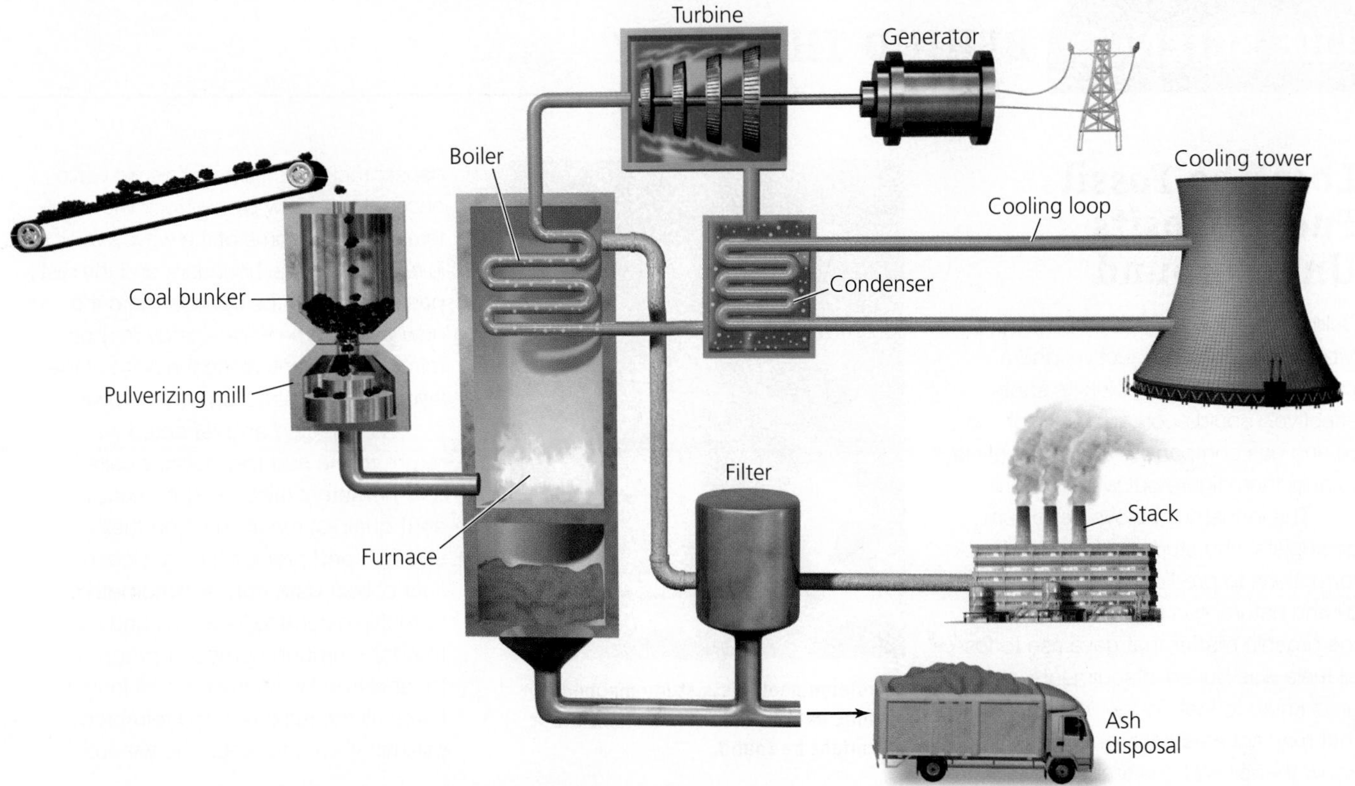

FIGURE 19.10 At a coal-fired power plant, coal is pulverized and blown into a high-temperature furnace. Heat from the combustion boils water, and the resulting steam turns a turbine, generating electricity by passing magnets past copper coils. The steam is then cooled and condensed in a cooling loop and returned to the furnace. "Clean coal" technologies (p. 537) help filter pollutants from the combustion process, and toxic ash residue is taken to hazardous waste disposal sites.

TABLE 19.3 Top Producers and Consumers of Fossil Fuels

PRODUCTION (% world production)		CONSUMPTION (% world consumption)	
COAL			
China	49.5	China	49.4
United States	14.1	United States	13.5
Australia	5.8	India	7.9
India	5.6	Japan	3.2
Indonesia	5.1	South Africa	2.5
OIL			
Saudi Arabia	13.2	United States	20.5
Russia	12.8	China	11.4
United States	8.8	Japan	5.0
Iran	5.2	India	4.0
China	5.1	Russia	3.4
NATURAL GAS			
United States	20.0	United States	21.5
Russia	18.5	Russia	13.2
Canada	4.9	Iran	4.7
Iran	4.6	China	4.0
Qatar	4.5	Japan	3.3

Data from BP p.l.c., 2012. Statistical review of world energy 2012.

at low temperatures (*liquefied natural gas*, or *LNG*), it can be shipped long distances in refrigerated tankers. When we replace coal with natural gas to generate electricity, this cuts carbon emissions in half. For this reason, many energy experts view natural gas as a "bridge fuel"—a bridge leading from today's polluting fossil-fuel economy toward a clean renewable energy economy for the future. The United States and Russia lead the world in gas production and gas consumption (see Table 19.3).

Oil The modern use of oil for energy began after 1859, when the world's first oil well was drilled in Titusville, Pennsylvania. Over the next 40 years, Pennsylvania's oil fields produced half the world's oil supply and helped establish a fossil-fuel-based economy that would hold sway for decades to come.

Today our global society produces and consumes nearly 750 L (200 gal) of oil each year for every man, woman, and child. The majority is used as fuel for vehicles, including gasoline for cars, diesel for trucks, and jet fuel for airplanes. Fewer homes burn oil for heating these days, but industry and manufacturing still account for a great deal of oil use.

Over the past several decades, refining techniques and chemical manufacturing have greatly expanded our uses of petroleum to include a wide array of products and applications, from plastics to lubricants to fabrics to pharmaceuticals. In today's world, petroleum-based products are all around us

Locating Fossil Fuel Deposits Underground

Drilling for oil or gas is risky business: Most wells are unproductive, and a company that doesn't pick its spots effectively could soon go bankrupt. So oil and gas companies turn to scientists to help them figure out where to drill.

The industry employs petroleum geologists who study underground rock formations to predict where deposits of oil and natural gas might lie. Because the organic matter that gave rise to fossil fuels was buried in sediments, geologists know to look for sedimentary rock that may act as a source. They also know that oil and gas tends to seep upward through porous rock until being trapped by impermeable layers.

To map subsurface rock layers, petroleum geologists first survey the landscape on the ground and from airplanes, studying rocks on the surface. Because rock layers often become tilted over geologic time, these strata may protrude at the surface, giving geologists an informative "side-on" view.

But to really understand what's deep beneath the surface, scientists need to conduct seismic surveys. In seismic surveying, a base station creates powerful vibrations at the surface by exploding dynamite, thumping the ground with a large weight, or using an electric vibrating machine (**FIGURE 1**). This sends seismic waves down and outward in all directions through the ground, just as ripples spread when a pebble is dropped into a pond.

As they travel, the waves encounter layers of different types of rock. The waves travel more quickly through denser rock. Each time a seismic wave encounters a new type of rock with a different density, some of the wave's energy is reflected off the boundary, and the rest passes through the boundary into the new layer. Some wave energy may be refracted, or bent, along the edge of the layer, sending refraction waves upward.

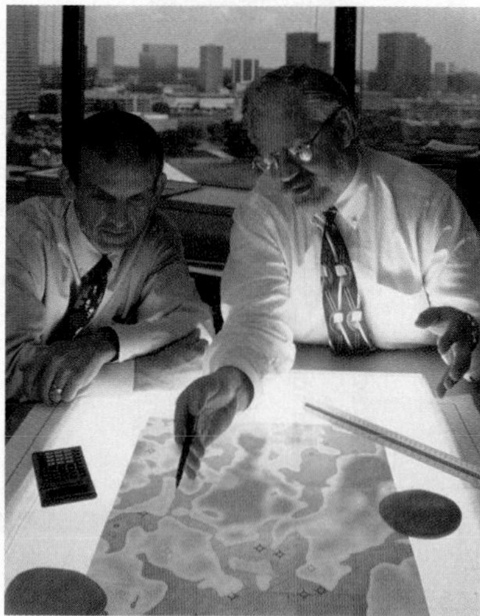

Petroleum geologists study mapped seismic data to determine where oil or gas might be found.

As reflected and refracted waves return to the surface, devices called seismometers (also used to measure earthquakes) record data on their strength and precise timing. Scientists collect data from seismometers at multiple surface locations and run the data through computer programs for analysis. By analyzing how long it takes all the reflected and refracted seismic waves to reach the various

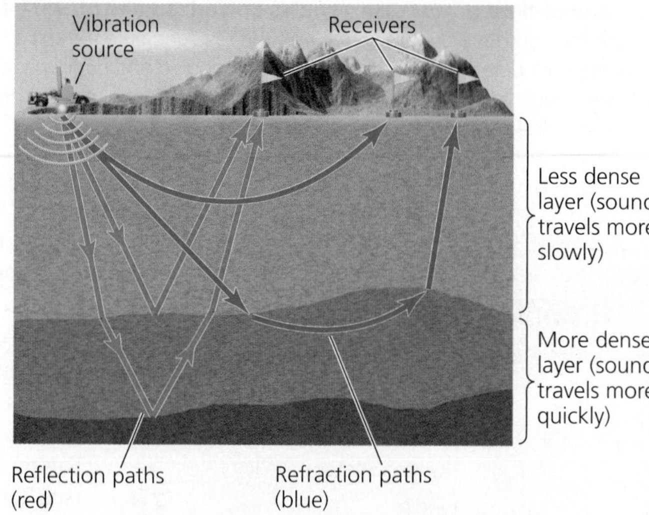

FIGURE 1 Seismic surveying provides clues to the location and size of fossil fuel deposits. Powerful vibrations are created, and receivers measure how long it takes seismic waves to reach other locations. Waves travel more quickly through denser layers, and density differences cause waves to reflect or refract. Scientists interpret the patterns of wave reception to infer the densities, thicknesses, and locations of underlying rock layers.

in our everyday lives (**FIGURE 19.11**). The fact that petroleum is used to create so many items and materials we have come to rely on makes it vital that we take care to conserve our remaining oil reserves.

The United States consumes one-fifth of the world's oil, but rapidly industrializing populous nations such as China and India are increasingly driving world demand (see Table 19.3).

We are gradually depleting fossil fuel reserves

Because fossil fuels are nonrenewable, the total amount available on Earth declines as we use them. Many scientists and oil industry analysts calculate that we have already extracted half the world's conventional oil reserves. So far we have used up about 1.1 trillion barrels of oil, and most estimates hold that

receiving stations, and how strong the waves are at each site, researchers can triangulate and infer the densities, thicknesses, and locations of underlying geologic layers.

Seismic surveying is similar to how we use sonar in water or how bats use echolocation as they fly. It is also used for finding coal deposits, salt and mineral deposits, and geothermal energy hotspots, as well as for studying faults, aquifers, and engineering sites.

Still, even with good survey data, oil companies may have only a 10% success rate in finding oil or gas. They need to conduct exploratory drilling to confirm whether oil or gas actually exists in any given location.

Using data from such techniques, geologists with the U.S. Geological Survey (USGS) in 1998 assessed the subsurface geology of the Arctic National Wildlife Refuge on Alaska's North Slope to predict how much oil it may hold (**FIGURE 2**). Over three years, dozens of scientists conducted fieldwork and combined their results with a reanalysis of 2300 km (1400 mi) of seismic survey data that industry had collected in the 1980s.

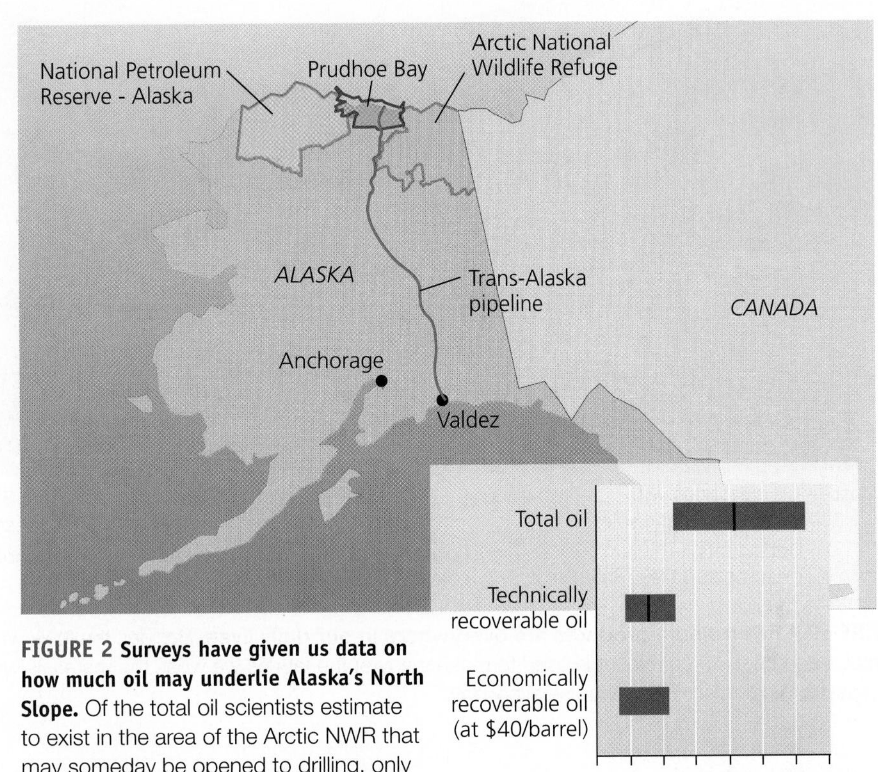

FIGURE 2 Surveys have given us data on how much oil may underlie Alaska's North Slope. Of the total oil scientists estimate to exist in the area of the Arctic NWR that may someday be opened to drilling, only some is technically recoverable, and less is economically recoverable. *Data from U.S. Geological Survey.*

After studying their resulting subsurface maps, USGS scientists concluded, with 95% certainty, that between 11.6 and 31.5 billion barrels of oil lay underneath the region of the refuge that Congress has debated opening for drilling. The mean estimate of 20.7 billion barrels is enough to supply the United States for 3 years at its current rate of consumption.

However, some portion of oil from any deposit is impossible to extract using current technology, so geologists estimate "technically recoverable" amounts of fuels. In its estimate for the Arctic Refuge, the USGS calculated technically recoverable oil to total 4.3–11.8 billion barrels, with a mean estimate of 7.7 billion barrels (just over 1 year of U.S. consumption).

The portion of this oil that is "economically recoverable" depends on the costs of extracting it and the price of oil on the world market. USGS scientists calculated that at a price of $40 per barrel, 3.4–10.8 billion barrels would be economically worthwhile to recover. At today's much higher prices, the economically recoverable amount would be closer to the technically recoverable amount.

In 2002, the USGS conducted similar analyses for the National Petroleum Reserve–Alaska, a vast parcel of tundra to the west of the Arctic Refuge and Prudhoe Bay that the U.S. government set aside 90 years ago as an emergency reserve of petroleum. USGS scientists estimated that this region contained 9.3 billion barrels of technically recoverable oil.

Across the world, petroleum geologists are using similar methods to try to determine how much oil remains to extract. Their work is of vital importance as the world's nations struggle to pursue well-informed energy policies. ■

slightly more than 1 trillion barrels of proven reserves remain. Adding proven reserves of oil sands from Canada and Venezuela brings the total remaining to just over 1.6 trillion barrels.

To estimate how long this remaining oil will last, analysts calculate the **reserves-to-production ratio,** or **R/P ratio,** by dividing the amount of total remaining reserves by the annual rate of production (i.e., extraction and processing). At current levels of production (30 billion barrels globally per year), 1.6 trillion barrels would last about 54 more years. Applying the R/P ratio to natural gas, we find that the world's proven reserves of this resource would last 64 more years. For coal, the latest R/P ratio estimate is 112 years.

The true number of years remaining for these fuels may be less than these figures suggest, however, because our demand and production have been increasing, not constant. Yet at the same time, the true number of years may be *more*

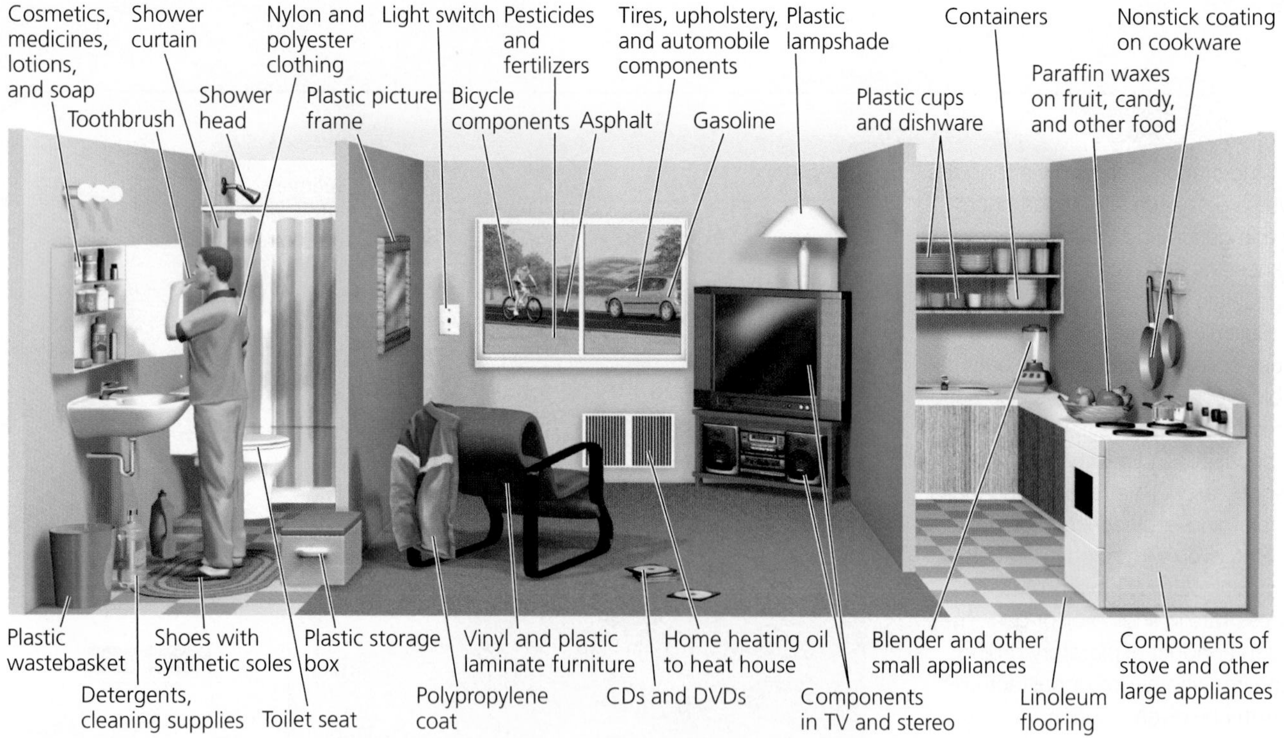

FIGURE 19.11 Petroleum products are everywhere in our daily lives. Besides the fuels we use for transportation and heating, petroleum is used to make many of the fabrics we wear, the materials we consume, and the plastics in countless items we use every day.

than these figures suggest, because proven reserves tend to increase as technology becomes more powerful and as market prices rise. Indeed, mining for oil sands and hydraulic fracturing for natural gas in recent years have increased the proven reserves of these fuels in North America substantially. In fact, advances in oil and gas extraction in the United States in the past several years have been so great that the International Energy Agency in 2012 predicted a resurgence would make the United States the world's biggest oil producer into the 2020s.

Regardless of how many years' worth of a resource we might calculate to be left, a society dependent on that resource will face a crisis not when the last bit of it is extracted from the ground, but rather once the rate of its production comes to a peak and begins to decline. In general, production of a resource tends to decline once reserves are depleted halfway. Past this production peak, if demand for the resource holds steady or continues to increase while production declines, a shortage will result. Many scientists and economists have been anxious about this phenomenon with oil, and the scenario has come to be nicknamed **peak oil.** Because we have already used roughly half of Earth's conventional oil reserves, many experts calculate that a peak oil crisis may well begin in the very near future.

Peak oil will pose challenges

To understand concerns about peak oil, we need to turn back the clock to 1956. In that year, Shell Oil geologist M. King Hubbert calculated that U.S. oil production would peak around 1970. His prediction was ridiculed at the time, but it proved to be accurate; U.S. production peaked in that very year and has fallen since then (**FIGURE 19.12a**). The peak in production came to be known as **Hubbert's peak.**

In 1974, Hubbert analyzed data on technology, economics, and geology, and predicted that global oil production would peak in 1995. It grew past 1995, but many scientists using newer data today predict that at some point in the coming decade, production will begin to decline (**FIGURE 19.12b**). Discoveries of new oil fields peaked 30 years ago, and since then we have been extracting and consuming more oil than we have been discovering.

FAQ **Why should I worry about "peak oil" if there are still years of oil left?**

Bear in mind that the term *peak oil* doesn't refer to running out of oil. It refers to the point at which our production of oil comes to a peak. Once we pass this peak and production begins to decline, the economics of supply and demand take over. Supply will fall, with some estimates putting the decline at 5% per year. Demand, meanwhile, is forecast to continue rising, especially as China, India, and other industrializing nations put millions of new vehicles on the road. The resulting divergence of supply and demand would drive up oil prices, causing substantial economic ripple effects. Although high oil prices will provide financial incentive to develop alternative energy sources and better conservation measures, we may be challenged in a depressed economy to find adequate time and resources to develop new renewable sources.

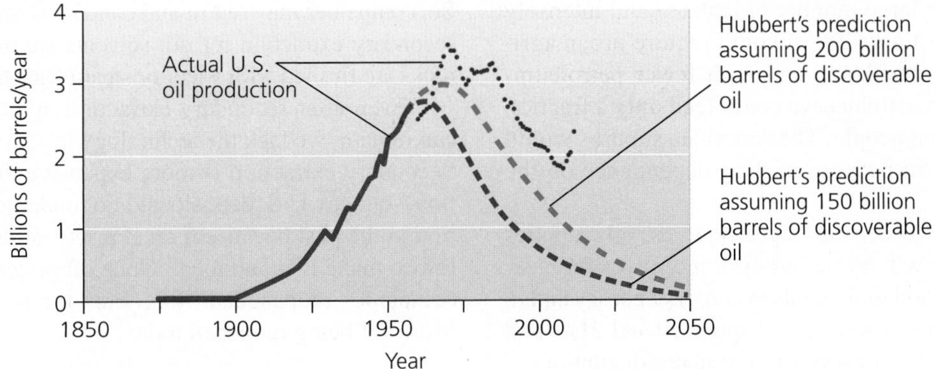

(a) Hubbert's prediction of peak in U.S. oil production, with actual data

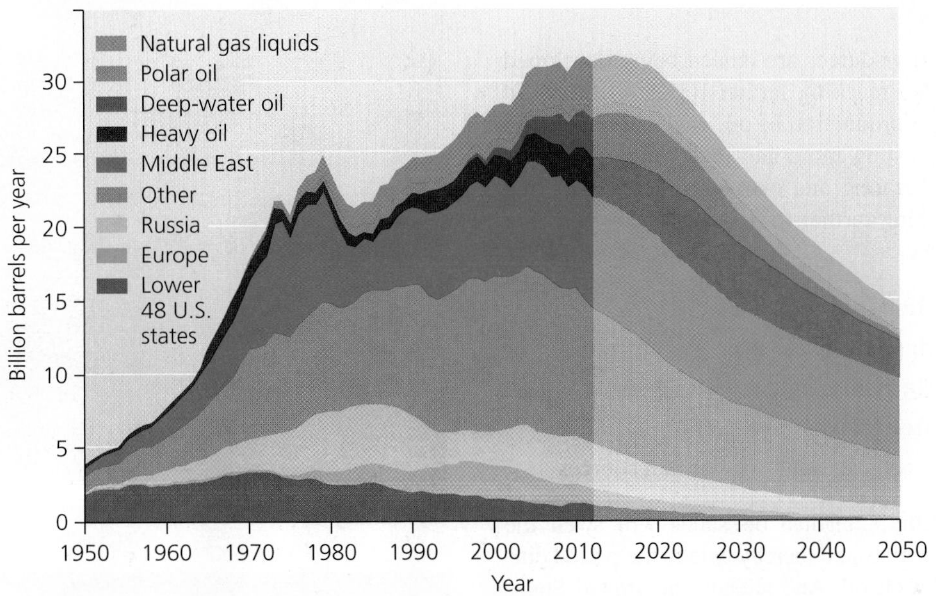

(b) Modern prediction of peak in global oil production

FIGURE 19.12 Peak oil describes a peak in production. U.S. oil production peaked in 1970 **(a)**, just as geologist M. King Hubbert had predicted. Since then, success in Alaska, in the Gulf of Mexico, and with secondary extraction has enhanced production during the decline. (Note that curves shift to the right and peaks increase in height as more oil is discovered.) Today U.S. oil production is spiking upward as deep offshore drilling and hydraulic fracturing make new deposits accessible. Still, some analysts believe global oil production will soon peak. Shown in **(b)** is one recent projection, from a 2011 analysis by scientists at the Association for the Study of Peak Oil. *Data from (a) Hubbert, M.K., 1956.* Nuclear energy and the fossil fuels. *Shell Development Co. Publ. No. 95, Houston, TX; and U.S. Energy Information Administration; and (b) Campbell, C.J., and Association for the Study of Peak Oil. By permission of Dr. Colin Campbell.*

Discovery or development of wholly new sources of oil can delay the peak by boosting our overall proven reserves. This is what is happening as we exploit Canada's oil sands. People counting on a series of new sources in the future tend to believe we can continue relying on oil for a long time, and this may indeed turn out to be the case. However, at some point we will reach peak oil—and peak gas and peak coal. The question is when—and whether we will be prepared to deal with the resulting challenges.

Predicting an exact date for peak oil is difficult. Many companies and governments do not reveal their true data on oil reserves, and estimates differ as to how much oil we can extract secondarily from existing deposits. Indeed, a recent U.S. Geological Survey report estimated 2 trillion barrels of conventional oil remaining in the world, rather than 1 trillion, and some estimates predict still greater amounts. A 2007 report by the U.S. General Accounting Office reviewed 21 studies and found that most estimates for the timing of the oil production peak ranged from now through 2040.

Whenever it occurs, the divergence of supply and demand could have momentous economic, social, and political consequences that will profoundly affect our lives. One prophet of peak oil, writer James Howard Kunstler, has sketched a frightening scenario of our post-peak world during what he calls "the long emergency": Lacking cheap oil with which to transport goods long distances, today's globalized economy would collapse into a large number of isolated and intensely localized economies. Large cities would require urban agriculture to feed their residents, and with fewer petroleum-based fertilizers and pesticides we could feed only a fraction of the world's 7 billion people. The American suburbs would be hit particularly hard because of their dependence on the automobile.

More optimistic observers argue that as oil supplies dwindle, rising prices will create powerful incentives for businesses, governments, and individuals to conserve energy and to develop alternative energy sources (Chapters 20 and 21)—and that these developments will save us from major disruptions.

Or might we end up with "too much" fossil fuel energy?

Today renewable energy sources are indeed being developed faster—but we are also reaching farther for fossil fuels. To stave off the day when production of oil, gas, and coal begin to decline, we are investing more and more money, energy, and technology into locating and extracting new fossil fuel deposits. We are reaching further for fossil fuels by pursuing several main approaches:

- Secondary extraction from existing wells
- Hydraulic fracturing for oil and shale gas
- Offshore drilling in increasingly deep waters
- Moving into ice-free waters of the Arctic
- Exploiting new "unconventional" fossil fuel sources

Development of the Canadian oil sands will swell the amount of oil available to us and thereby extend the period during which we can rely on oil. And already the United States is seeing increased production from deep offshore drilling, hydraulic fracturing, and secondary extraction—so much so that energy experts now project that within a decade the United States may be exporting more oil than Saudi Arabia! As we extend our reach into less-accessible places to obtain fuel that is harder to extract, we expand our proven reserves and postpone the threat of peak production. However, we also reduce the EROI ratios of our fuels, drive up fuel prices for consumers, and intensify pollution and climate change. Indeed, the threat of climate change is serious enough that some scientists are beginning to wonder if we should plan to intentionally leave most oil, gas, and coal in the ground.

In the long term, to achieve a sustainable society we will need to switch to renewable energy sources. Investments in energy efficiency and conservation (pp. 546–548) are vital because they extend the time we have to make this transition.

Secondary extraction produces more fuel

One way we are reaching further for fossil fuels is by returning to sites where we have already removed easily accessible oil or gas and applying new technology or approaches to extract the remaining amounts. At a typical oil or gas well, as much as two-thirds of a deposit may remain in the ground after **primary extraction,** the initial drilling and pumping of oil or gas (**FIGURE 19.13a**). So, companies may return and conduct **secondary extraction.** In secondary extraction for oil, solvents are used or underground rocks are flushed with water or steam (**FIGURE 19.13b**).

Even after secondary extraction, quite a bit of oil or gas can remain; we lack the technology to remove every last drop. Secondary extraction is more expensive than primary extraction, so many U.S. deposits did not undergo secondary extraction in the past because market prices of oil and gas were too low to make it economical. Once oil prices rose in the 1970s, companies reopened drilling sites for secondary extraction. More are being reopened today.

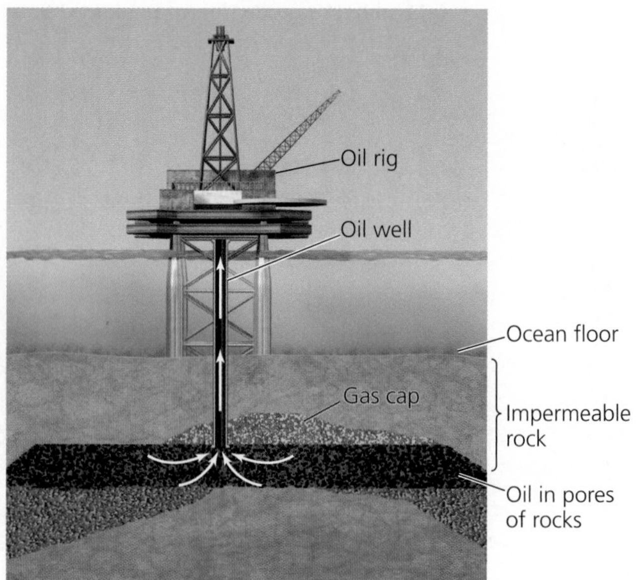

(a) Primary extraction of oil

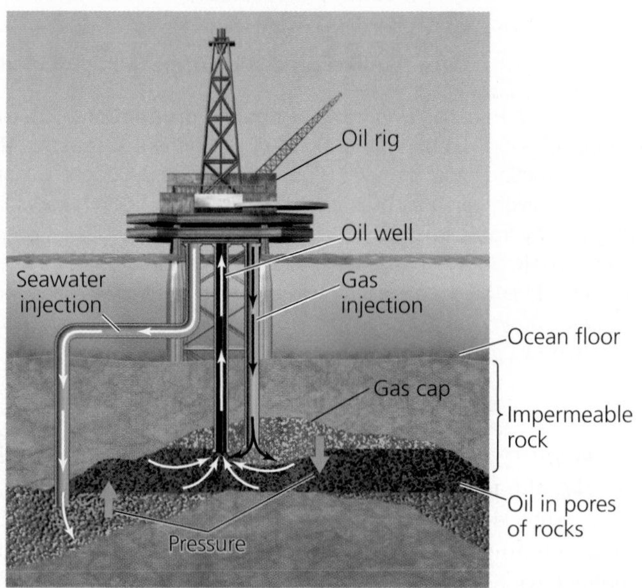

(b) Secondary extraction of oil

FIGURE 19.13 Secondary extraction removes oil not removed by primary extraction. In primary extraction **(a)**, oil is drawn up through a well by keeping pressure at the top lower than pressure at the bottom. Once pressure in the deposit drops, however, material must be injected to increase the pressure. Secondary extraction **(b)** involves injecting seawater beneath the oil and/or injecting gases just above the oil to force more oil up and out of the deposit.

Hydraulic fracturing expands our access to oil and gas

For oil and for natural gas trapped tightly in impermeable shale deposits, we are now using **hydraulic fracturing** (see Figure 7.1, p. 163) to break into rock formations and pump the oil or gas to the surface. Hydraulic fracturing (also called *hydrofracking*, or *fracking*) is being used for secondary extraction and also to tap into new deposits. This technique involves pumping chemically treated water under high pressure into deep layers of shale to crack them. Sand or small glass beads are inserted to hold the cracks open as the water is withdrawn. Gas or oil then travels upward, with pressure and pumping, through the newly created system of fractures.

Hydraulic fracturing allows us to extract gas and oil that is so dispersed through shale formations that it cannot be pumped out by standard drilling. By making formerly inaccessible deposits accessible, hydrofracking has raised proven reserves and has ignited a boom in natural gas extraction in the United States. Natural gas prices have fallen, and gas usage in the United States has risen.

Fracking has engendered debate among people living in each area where it has occurred (**FIGURE 19.14**). For example, hydrofracking of the massive Marcellus Shale deposit is affecting the landscapes, economies, politics, and everyday lives of people in Pennsylvania, New York, and neighboring states (see Chapter 7). The choices people face between financial gain and impacts to their health, drinking water, and environment have been dramatized in popular films such as *Promised Land* and *Gasland*. As with Alberta's oil sands, and like all energy booms before it, today's natural gas rush brings jobs and money to small towns but can also spark social upheaval and leave communities with a legacy of pollution.

FIGURE 19.14 Hydraulic fracturing is expanding U.S. production of oil and natural gas, but it is sparking debates within communities where it is taking place. This drill rig is hydrofracking a shale formation on private land among homes in the rural Hopewell Township of Pennsylvania. Here, some residents support drilling and hope for financial benefits whereas others oppose drilling and fear contamination of their drinking water and damage to their quality of life.

We are drilling farther and farther offshore

Today we drill for oil and natural gas not only on land but also below the seafloor on the continental shelves. Offshore drilling platforms must withstand wind, waves, and ocean currents. Some are fixed, standing platforms built with unusual strength. Others are resilient floating platforms anchored in place above the drilling site. Roughly 35% of the oil and 10% of the natural gas extracted in the United States today comes from offshore sites, primarily in the Gulf of Mexico and secondarily off southern California. The Gulf today is home to 90 drilling rigs and 3500 production platforms. Geologists estimate that most U.S. gas and oil remaining to be extracted occurs offshore and that deepwater sites in the Gulf of Mexico alone may hold 59 billion barrels of oil.

We have been drilling in shallow water for several decades, but as oil and gas are depleted at shallow-water sites and as drilling technology improves, the industry is moving into deeper and deeper water. This poses risks; the *Deepwater Horizon* oil spill of 2010 (pp. 436–437, 538–541) occurred at a deepwater site. In that event, faulty equipment allowed natural gas accompanying the oil deposit to shoot up the well shaft. It ignited atop the platform, killing 11 workers and leading to the largest accidental oil spill in history. British Petroleum's Macondo well, where the accident took place, lay beneath 1500 m (5000 ft) of water. The deepest wells in the Gulf of Mexico are now twice that depth.

Globally, recent discoveries off the coasts of Brazil, Angola, Nigeria, and other nations suggest that a great deal of oil and gas could lie well offshore, and companies are racing one another to get there. Unfortunately, our ability to drill in deep water has outpaced our capacity to deal with accidents there. The fact that it took 86 days for BP to plug the leak at its Macondo well demonstrates the challenge of addressing an emergency situation a mile or more beneath the surface of the sea.

Today all eyes are on the Arctic. As global climate change melts the sea ice that covers the Arctic Ocean (pp. 498, 503), new shipping lanes are opening and nations and companies are scrambling to lay claim to patches of ocean that could hold fossil fuels and other resources. The oil and gas industry plans to drill offshore in deep water—something that has environmental advocates very worried. The Arctic's frigid temperatures, ice floes, winds, waves, and brutal storms make conditions harsh and challenging and make accidents more likely.

In 2008, responding to rising gasoline prices and a desire to lessen dependence on foreign oil, the U.S. Congress lifted a long-standing moratorium on offshore drilling along much of the nation's coastline. The Obama administration in 2010 followed through by designating vast areas open for drilling that had formerly been closed. These included most waters along the Atlantic coast from Delaware south to central Florida, a region of the eastern Gulf of Mexico, and most waters off Alaska's North Slope. However, just weeks after this announcement, the *Deepwater Horizon* spill occurred. Public reaction forced the Obama administration to backtrack, canceling offshore drilling projects it had approved and putting a hold on further approvals until new safety measures could be devised.

In 2011, after weighing economic and environmental concerns, the administration issued a five-year plan that opened access to 75% of technically recoverable offshore oil and gas reserves while banning drilling offshore from states that did not want it. Drilling leases were expanded off Alaska and in the Gulf of Mexico, but areas along the East and West Coasts were not opened to drilling.

The risks from expansion into Arctic waters were highlighted in 2012–2013, when Royal Dutch Shell's *Kulluk* drilling rig ran aground while being towed south from Alaska during a winter storm. Damage to the rig and examination by public officials and the media raised fresh questions as to whether Arctic Ocean drilling can be conducted safely.

We are exploiting new fossil fuel sources

As sources of conventional fossil fuels decline and as prices rise, we will turn increasingly to newer alternatives. These include at least three further sources of fossil fuels that exist in large amounts: oil sands, oil shale, and methane hydrate. The oil sands of Alberta—and those in Venezuela, which likely hold even more oil—are already significantly increasing the amount of oil available to our society. Oil from shale is not yet economical to extract. Gas from methane hydrate is only now becoming technically feasible to extract.

The world's known deposits of oil shale may contain as much as 3 trillion barrels of oil (more than all the conventional crude oil in the world), but most of this will not easily be extracted. About 40% of global oil shale reserves are in the United States, mostly on federally owned land in Colorado, Wyoming, and Utah. Shale oil is costly to extract, and its EROI is very low, with the best estimates ranging from just 1.1:1 to 4:1. Historically, low prices for crude oil have kept investors away from shale oil, but every time crude oil prices rise, shale oil again attracts attention.

As for methane hydrate, scientists believe there are immense amounts of this substance on Earth, holding perhaps twice as much carbon as all known deposits of oil, coal, and natural gas combined. Japan recently showed that it could extract methane hydrate from the seafloor by sending down a pipe and lowering pressure within it so that the methane turned to gas and rose to the surface. However, we do not yet know whether such extraction is safe and reliable. Destabilizing a methane hydrate deposit on the seafloor during extraction could lead to a catastrophic release of gas. This could cause a massive landslide and tsunami and would release huge amounts of methane, a potent greenhouse gas, into the atmosphere, worsening global climate change.

Oil sands, oil shale, and methane hydrate are abundant, but they are no panacea for our energy challenges. For one thing, their net energy values are low, because they are expensive to extract and process. Thus, their EROI ratios are low. Moreover, these fuels exert severe environmental impacts. We will now turn to some of the impacts of our fossil fuel use—environmental, economic, social, and political—and examine potential solutions to these impacts.

Addressing Impacts of Fossil Fuel Use

Our society's love affair with fossil fuels and the many petrochemical products we develop from them has helped to ease constraints on travel, lengthen our life spans, and boost our material standard of living beyond what our ancestors could have dreamed. However, it also causes harm to the environment and human health, and can lead to political and economic instability. Concern over these impacts is a prime reason many scientists, environmental advocates, businesspeople, and policymakers are increasingly looking toward clean and renewable sources of energy.

Fossil fuel emissions pollute air and drive climate change

When we burn fossil fuels, we alter fluxes in Earth's carbon cycle (pp. 121–123). We essentially take carbon that has been retired into a long-term reservoir underground and release it into the air. This occurs as carbon from the hydrocarbon molecules of fossil fuels unites with oxygen from the atmosphere during combustion, producing carbon dioxide (CO_2). Carbon dioxide is a greenhouse gas (p. 484), and CO_2 released from fossil fuel combustion warms our planet and drives changes in global climate (Chapter 18).

Because global climate change is beginning to have diverse, severe, and widespread ecological and socioeconomic impacts, carbon dioxide pollution (**FIGURE 19.15**) is becoming recognized as the greatest environmental impact of fossil fuel use. Moreover, methane is a potent greenhouse gas that drives climate warming. Switching to new fossil fuel sources may

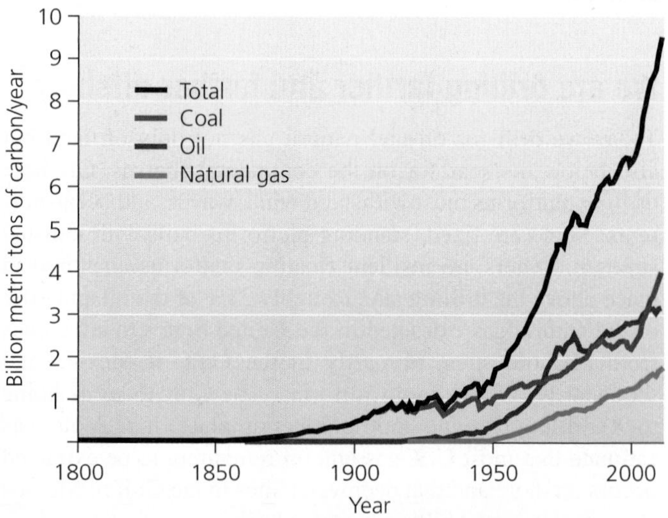

FIGURE 19.15 Emissions from fossil fuel combustion have risen dramatically as nations have industrialized and as population and consumption have grown. Here, global emissions of carbon from carbon dioxide are subdivided by source (oil, coal, or natural gas). *Data from Carbon Dioxide Information Analysis Center, Oak Ridge National Laboratory, U.S. Department of Energy, Oak Ridge, TN.*

DATA Q By what percentage have carbon emissions risen since the year your mother or father was born?

only make emissions worse; oil sands are estimated to generate 14–20% more greenhouse gas emissions than conventional oil, and shale oil is still more polluting.

Besides modifying our climate, fossil fuel emissions affect human health. Combusting coal high in mercury content emits mercury that can bioaccumulate in organisms' tissues, poisoning animals as it moves up food chains (pp. 373–374) and presenting health risks to people. Gasoline combustion in automobiles releases pollutants that irritate the nose, throat, and lungs. Some hydrocarbons, such as benzene and toluene, cause cancer in laboratory animals and likely in people. Gases such as hydrogen sulfide can evaporate from crude oil, irritate the eyes and throat, and cause asphyxiation. Crude oil also often contains trace amounts of known poisons such as lead and arsenic. As a result, workers at drilling operations, refineries, and in other jobs that entail frequent exposure to oil or its products can develop serious health problems, including cancer.

The combustion of oil in our vehicles and coal in our power plants releases sulfur dioxide and nitrogen oxides, which contribute to industrial and photochemical smog (pp. 464–465) and to acid deposition (pp. 473–475). Fossil fuel pollution is intensifying in developing nations that are industrializing rapidly as they grow in population. In contrast, air pollution from fossil fuel combustion has been reduced in developed nations in recent decades as a result of laws such as the U.S. Clean Air Act and government regulations to protect public health (Chapter 17). In these nations, public policy has encouraged industry to develop and install technologies that reduce pollution, such as catalytic converters that cleanse vehicle exhaust (see Figure 17.15, p. 460). Wider adoption of such technologies in the developing world would reduce pollution there considerably.

Clean coal technologies aim to reduce air pollution from coal

Burning coal emits a variety of pollutants unless effective pollution control measures are in place. The composition of emissions from coal combustion depends on chemical impurities in the coal, and coal deposits vary in the impurities they contain, including sulfur, mercury, arsenic, and other trace metals. Coal from the eastern United States tends to be high in sulfur because it was formed in marine sediments, and sulfur is present in seawater. Most coal in China is even more sulfur-rich.

At coal-fired power plants, scientists and engineers are seeking ways to cleanse coal of sulfur, mercury, and other impurities. **Clean coal technologies** refer to an array of techniques, equipment, and approaches that aim to remove chemical contaminants during the process of generating electricity from coal. Among these technologies are various types of scrubbers, devices that chemically convert or physically remove pollutants (see Figure 17.16, p. 461). Some scrubbers use minerals such as magnesium, sodium, or calcium in reactions to remove sulfur dioxide (SO_2) from smokestack emissions. Others use chemical reactions to strip away nitrogen oxides (NO_X), breaking them down into elemental nitrogen and water. Multilayered filtering devices are used to capture tiny ash particles.

Another clean coal approach is to dry coal that has high water content in order to make it cleaner-burning. We can also gain more power from coal with less pollution through a process called *gasification*, in which coal is converted into a cleaner synthesis gas, or *syngas*, by reacting it with oxygen and steam at a high temperature. In an "integrated gasification combined cycle" process, syngas from coal is used to turn a gas turbine and also to heat water to turn a steam turbine.

The U.S. government and the coal industry have each invested billions of dollars in clean coal technologies for new power plants, and these have helped to reduce air pollution from sulfates, nitrogen oxides, mercury, and particulate matter (pp. 458–459). If these technologies were applied to the many older plants that still pollute our air, they could help even more. At the same time, the coal industry spends a great deal of money fighting regulations and mandates on its practices. As a result, many power plants are built with few clean coal technologies, and these plants will continue polluting our air for decades. Moreover, many energy analysts emphasize that these technologies may make for "cleaner" coal but will never result in energy production that is completely clean. Some argue that coal is an inherently dirty way of generating power and should be replaced outright with cleaner energy sources.

Can we capture and store carbon?

Even if clean coal technologies were able to remove every last chemical contaminant from power plant emissions, coal combustion would still pump huge amounts of carbon dioxide (CO_2) into the air, intensifying the greenhouse effect and worsening global climate change. This is why many current efforts focus on **carbon capture** followed by **carbon storage** or **carbon sequestration** (p. 508). This approach consists of capturing carbon dioxide emissions, converting the gas to a liquid form, and then sequestering (storing) it in the ocean or underground in a geologically stable rock formation (**FIGURE 19.16**).

Carbon capture and storage (abbreviated as CCS) is being attempted at a variety of new and retrofitted facilities. The world's first coal-fired power plant to approach zero emissions opened in 2008 in Germany. This Swedish-built plant removes its sulfate pollutants and captures its carbon dioxide, then compresses the CO_2 into liquid form, trucks it 160 km (100 mi) away, and injects it 900 m (3000 ft) underground into a depleted natural gas field.

In North Dakota, the Great Plains Synfuels Plant gasifies its coal, then sends half the CO_2 through a pipeline into Canada, where a Canadian oil company buys the gas to inject into an oilfield to help it pump out the remaining oil. The North Dakota plant also captures, isolates, and sells seven other types of gases for various purposes.

Currently the U.S. Department of Energy is teaming up with seven energy companies to build a prototype of a near-zero-emissions coal-fired power plant. The $1.3-billion *FutureGen* project aims to design, construct, and operate a power plant that burns coal, produces electricity, captures 90% of its carbon dioxide emissions, and sequesters the CO_2 underground. The project, located in Meredosia, Illinois, will pump CO_2 more than 1200 m (three-quarters of a mile) underground beneath layers of impermeable rock. If this showcase project succeeds, it could be a model for a new generation of power plants across the world.

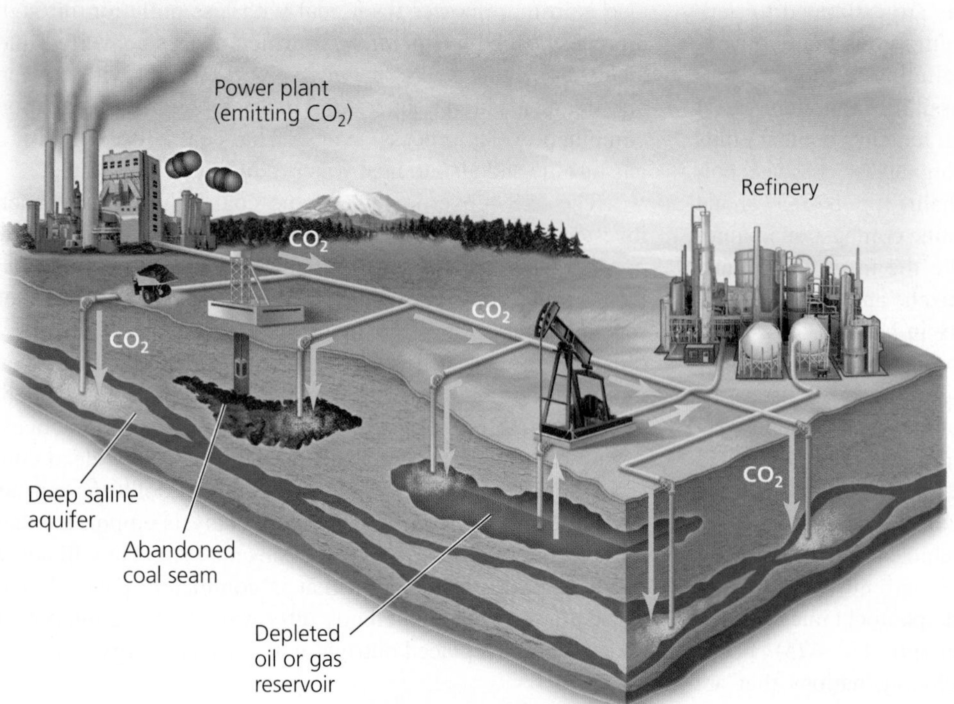

FIGURE 19.16 Carbon capture and storage schemes propose to inject liquefied carbon dioxide emissions underground. The CO_2 may be injected into depleted fossil fuel deposits, deep saline aquifers, or oil or gas deposits undergoing secondary extraction.

At this point, however, carbon capture and storage is too unproven to be the central focus of a clean energy strategy. We do not know whether we can ensure that carbon dioxide will stay underground once injected there or whether these attempts might trigger earthquakes. Injection might in some cases contaminate groundwater supplies, and injecting carbon dioxide into the ocean would further acidify its waters (pp. 426, 428–429, 501). Moreover, CCS is energy-intensive and decreases the EROI of coal, adding to its cost and the amount we consume. Finally, many renewable energy advocates fear that the CCS approach takes the burden off emitters and prolongs our dependence on fossil fuels rather than facilitating a shift to renewables.

WEIGHING THE ISSUES

CLEAN COAL AND CARBON CAPTURE Do you think we should be spending billions of dollars to try to find ways to burn coal cleanly and to sequester carbon emissions from fossil fuels? Or is our money better spent on developing new clean and renewable energy sources that don't yet have enough infrastructure to produce power at the scale that coal can? What pros and cons do you see in each approach?

Oil spills pollute oceans and coasts

Of the many ways that our fossil fuel use pollutes water, what comes to mind most readily for people is the pollution that occurs when oil from tanker ships or drilling platforms fouls coastal waters and beaches. In 2010, BP's *Deepwater Horizon* offshore drilling platform exploded and sank off the coast of Louisiana (**FIGURE 19.17**). Eleven workers were killed, and oil gushed out of a broken pipe on the ocean floor a mile beneath the surface at a rate of 62,000 barrels per day. Emergency shut-off systems had failed, and BP engineers tried one solution after another to stop the flow of oil and gas, which continued for three months, spilling roughly 4.9 million barrels (206 million gallons) of oil.

The crisis proved difficult to control because we had never had to deal with a spill so deep underwater. It revealed that offshore drilling presents serious risks of environmental impact that may be difficult to address, even with our best engineering.

As the oil spread through the Gulf of Mexico and washed ashore, the region suffered a wide array of impacts (**FIGURE 19.18**). Of the countless animals killed, most conspicuous were birds, which cannot regulate their body temperature once their feathers become coated with oil. However, the underwater nature of the BP spill meant that unknown numbers of fish, shrimp, corals,

FIGURE 19.17 The explosion on BP's *Deepwater Horizon* drilling platform in 2010 unleashed the world's biggest-ever accidental oil spill. Here vessels try to put out the blaze.

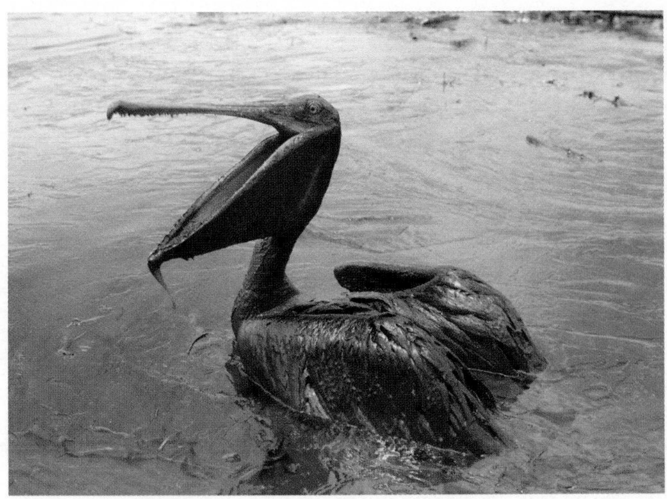

(a) Brown pelican coated in oil

(b) Beach cleanup

FIGURE 19.18 Impacts of the *Deepwater Horizon* spill were numerous and severe. This brown pelican, coated in oil **(a)**, was one of countless animals killed. For months, volunteers and workers paid by BP labored **(b)** to clean oil from the Gulf's beaches.

and other marine animals were also killed, affecting coastal and ocean ecosystems in complex ways. Plants in coastal marshes died, and the resulting erosion of marshes put New Orleans and other coastal cities at greater risk from storm surges and flooding. Gulf Coast fisheries, which supply much of the nation's seafood, were hit hard by the spill, with thousands of fishermen and shrimpers put out of work. Beach tourism suffered, and indirect economic and social impacts were expected to last for years. Throughout this process, scientists have been studying aspects of the spill and its impact on the region's people and natural systems (see THE SCIENCE BEHIND THE STORY, pp. 540–541).

The *Deepwater Horizon* spill was the largest accidental oil spill in world history, far eclipsing the spill from the *Exxon Valdez* tanker in 1989. In that event, oil from Alaska's North Slope, piped to the port of Valdez through the trans-Alaska pipeline, caused long-term damage to ecosystems and economies in Alaska's Prince William Sound when the tanker ran aground. Two-and-a-half decades later, a layer of oil remains just inches beneath the sand of the region's beaches.

Today as climate change melts sea ice in the Arctic, opening new shipping lanes, nations are jockeying for position, hoping to stake claim to oil and gas deposits that lie beneath the seafloor. Offshore drilling in Arctic waters, however, poses severe pollution risks, because if a spill were to occur, icebergs, pack ice, storms, cold temperatures, and wintertime darkness would hamper response efforts, while frigid water temperatures would slow the natural breakdown of oil. In U.S. Arctic waters that are now open to oil and gas leasing, some sites are 1000 miles away from the nearest Coast Guard station. The Obama administration approved these leases despite admitting that infrastructure to respond to a major spill there does not exist. Natural Resources Defense Council president Frances Beinecke, a member of the commission Obama set up to draw lessons from the BP spill, called such Arctic leases a "reckless gamble" and lamented that neither the administration, Congress, nor industry had improved safety measures in any meaningful way since the BP spill.

Fortunately, pollution from large spills has decreased greatly in recent decades (pp. 436–437), thanks to government regulations (such as requirements for double-hulled ships) and improved spill response efforts. And although large catastrophic oil spills have significant impacts on the marine environment, it is important to recognize that most water pollution from oil results from countless small non-point sources (pp. 408–409) to which all of our actions contribute (see Figure 16.17b, p. 436). Oil from automobiles, homes, industries, gas stations, and businesses runs off roadways and enters rivers and wastewater facilities, being discharged eventually into the ocean. Oil can also contaminate groundwater supplies when pipelines rupture or when underground storage tanks (p. 413) containing petroleum products leak. In addition, atmospheric deposition of pollutants from the combustion of fossil fuels exerts many impacts on freshwater ecosystems. Water pollution from industrial point sources has been greatly reduced in the United States following the Clean Water Act (Chapter 15), and many solutions exist to address non-point-source pollution.

Hydrofracking poses new concerns

Extracting oil or natural gas by hydraulic fracturing (in which chemicals are mixed with the pressurized water and sand that is injected deep underground) presents risks of water pollution that are not yet completely understood. One risk is that the chemicals (often called fracking fluids) may leak out of the drilling shafts and into aquifers that people use for drinking water. Another concern is that methane may contaminate groundwater used for drinking if it travels up the fractures or leaks through the shaft (see **The Science behind the Story**, Chapter 7, pp. 166–167).

Fracking sites also create air pollution as methane and volatile toxic components of fracking fluids seep up from drilling locations. In fact, some of the unhealthiest air pollution in

Discovering Impacts of the Gulf Oil Spill

President Barack Obama echoed the perceptions of many Americans when he called the *Deepwater Horizon* oil spill "the worst environmental disaster America has ever faced." But what has scientific research told us about the actual impacts of the Gulf oil spill?

We don't yet have all the answers, because the deep waters affected by the spill have been difficult for scientists to study. A great deal will remain unknown. Yet the intense and focused scientific response to the spill demonstrates the dynamic way in which science can assist society.

As the spill took place, government agencies called on scientists to help determine how much oil was leaking. Researchers eventually determined the rate reached 62,000 barrels per day. Using underwater imaging, aerial surveys, and shipboard water samples, researchers tracked the movement of oil up through the water column and across the Gulf. These data helped predict when and where oil might reach shore, thereby helping to direct prevention and cleanup efforts.

Meanwhile, as engineers struggled to seal off the well using remotely operated submersibles, researchers helped government agencies assess the fate of the oil (**FIGURE 1**). This data would help inform studies of the oil's impacts on marine life and human communities.

Tracking movement of the oil underwater was challenging. University of Georgia biochemist Mandy Joye, who had studied natural seeps in the Gulf for years, documented that the leaking wellhead was creating a plume of oil the size of Manhattan. She also found evidence of low oxygen concentrations, or hypoxia (pp. 105, 410), because some bacteria consume oil and gas, depleting oxygen from the water and making it uninhabitable for fish and other creatures.

Joye and other researchers feared that the thinly dispersed oil might prove devastating to plankton (the base of the marine food chain) and to the tiny larvae of shrimp, fish, and oysters (the pillars of the fishing industry). Scientists taking water samples documented sharp drops in plankton during the spill, but it will take several years to learn whether the impact on larvae will diminish populations of adult fish and shellfish. Studies on the condition of living fish in the region are now being published, some of which show gill damage, tail rot, lesions, and reproductive problems at much higher levels than is typical.

A scientist rescues an oiled Kemp's ridley sea turtle.

What was happening to life on the seafloor was a mystery, because there are only a handful of submersible vehicles in the world able to travel to the crushing pressures of the deep sea. Luckily, a team of researchers led by Charles Fisher of Penn State University was scheduled to embark on a regular survey of deepwater coral across the Gulf of Mexico in late 2010. Using the three-person submersible *Alvin* and the robotic vehicles *Jason* and *Sentry*, the team found healthy coral communities at sites far away from the Macondo well but found dying corals and brittlestars covered in a brown material at a site 11 km from the Macondo well.

Eager to determine whether this community was contaminated by the BP oil spill, the research team added chemist Helen White of Haverford College and returned a month later, thanks to a National Science Foundation program that funds rapid response research. On this trip, chemical analysis of the brown material showed it to match oil from the BP spill, rather than from any other known source.

Other questions revolve around impacts of the chemical dispersant that BP used to break up the oil, a compound called Corexit 9500. Work by biologist Philippe Bodin following

FIGURE 1 Scientists helped track oil from the *Deepwater Horizon* spill. The map **(a)** shows areas polluted by oil. The pie chart **(b)** gives a breakdown of the oil's fate. *Source (a): National Geographic and NOAA; (b) NOAA.*

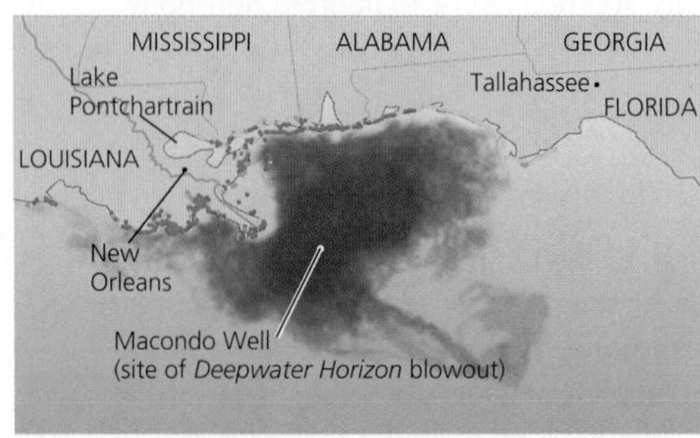

Oil on shoreline
- Very light
- Light
- Medium
- Heavy

Oil on water surface
- 1–10 days
- 10–30 days
- More than 30 days

(a) Extent of the spill

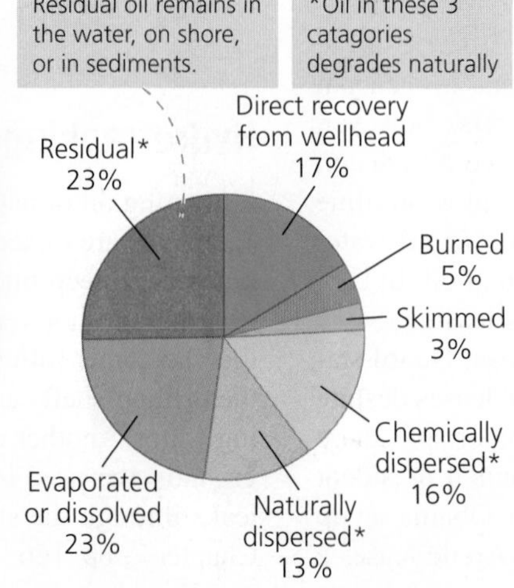

(b) Fate of the oil

the *Amoco Cadiz* oil spill in France in 1978 had found that Corexit 9500 appeared more toxic to marine life than the oil itself. BP threw an unprecedented amount of the chemical at the *Deepwater Horizon* spill, injecting a great deal directly into the path of the oil at the wellhead. This caused the oil to dissociate into trillions of tiny droplets that dispersed across large regions. Many scientists worried that this expanded the oil's reach, affecting more plankton, larvae, and fish.

Impacts of the oil on birds, sea turtles, and marine mammals were easier to assess. Officially confirmed deaths numbered 6104 birds, 605 turtles, and 97 mammals—and hundreds more animals were cleaned and saved by wildlife rescue teams—but a much larger, unknown, number succumbed to the oil. What impacts this mortality may have on populations in coming years is unclear. (After the *Exxon Valdez* spill in Alaska in 1989, populations of some species rebounded, but populations of others have never come back.) Researchers are following the movements of marine animals in the Gulf with radio transmitters to try to learn what effects the oil may have had.

As images of oil-coated marshes saturated the media, researchers worried that widespread death of marsh grass would leave the shoreline vulnerable to severe erosion by waves. Louisiana has already lost many coastal wetlands to subsidence, dredging, sea level rise, and silt capture by dams on the Mississippi River (pp. 389–390). Fortunately, researchers found that oil did not penetrate to the roots of most plants and that oiled grasses were sending up new growth. Indeed, Louisiana State University researcher Eugene Turner said that loss of marshland from the oil "pales in comparison" with marshland lost each year due to other factors.

The ecological impacts of the spill had measurable impacts on people. The region's mighty fisheries were shut down, forcing thousands of fishermen out of work. The government tested fish and shellfish for contamination and reopened fishing once they were found to be safe, but consumers balked at buying Gulf seafood. Beach tourism remained low all summer as visitors avoided the region. Together, losses in fishing and tourism totaled billions of dollars.

Scientists expect some impacts from the Gulf spill to be long-lasting. Oil from the similar *Ixtoc* blowout off Mexico's coast in 1979 still lies in sediments near dead coral reefs, and fishermen there say it took 15–20 years for catches to return to normal. After the *Amoco Cadiz* spill, it took seven years for oysters and other marine species to recover. In Alaska, oil from the *Exxon Valdez* spill remains embedded in beach sand today.

However, many researchers are hopeful about the Gulf of Mexico's recovery from the *Deepwater Horizon* spill. The Gulf's warm waters and sunny climate speed the natural breakdown of oil. In hot sunlight, volatile components of oil evaporate from the surface and degrade in the water, so that fewer toxic compounds such as benzene, naphthalene, and toluene reach marine life.

In addition, bacteria that consume hydrocarbons thrive in the Gulf because some oil has always seeped naturally from the seafloor and because leakage from platforms, tankers, and pipelines is common. These microbes give the region a natural self-cleaning capacity.

Researchers continue to conduct a wide range of scientific studies (**FIGURE 2**). A consortium of federal and state agencies is coordinating research and restoration efforts in the largest ever Natural Resource Damage Assessment, a process mandated under the Oil Pollution Act of 1990. Answers to questions will come in gradually as long-term impacts become clear. ■

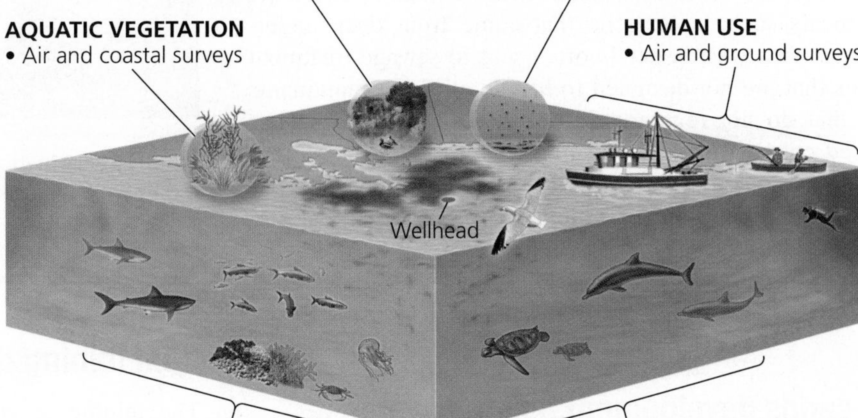

FIGURE 2 Thousands of researchers continue to help assess damage to natural resources from the *Deepwater Horizon* oil spill. They are surveying habitats, collecting samples and testing them in the lab, tracking wildlife, monitoring populations, and more.

the United States was found to be far away from the nearest city, in a little-populated region of Wyoming home to extensive fracking operations.

Many residents of areas near hydraulic fracturing sites have experienced polluted air and fouled drinking water (pp. 162–164), but more research is needed to assess the extent of such pollution and to quantify the health risks.

Hydrofracking also produces immense volumes of wastewater. Injected water often returns to the surface laced with salts, radioactive elements such as radium, and toxic chemicals such as benzene that come from deep underground. This wastewater is often sent to sewage treatment plants that are not designed to handle all the contaminants and that do not regularly test for radioactivity. This has caused concern in Pennsylvania, where a boom in natural gas extraction from the vast Marcellus Shale deposit continues to send millions of gallons of drilling waste to treatment plants, which then release their water into rivers that supply drinking water for people in Pittsburgh, Harrisburg, and other cities.

Oil sands development pollutes water

Similar concerns are being voiced about the extraction and transport of oil from oil sands. People living along the route of the Keystone XL pipeline worry that if oil were to spill from a leak in the pipeline, it would sink into the area's porous ground and quickly reach the region's shallow water table, contaminating the Ogallala Aquifer. This aquifer (p. 393) provides 2 million Americans with drinking water and irrigates a large portion of U.S. agriculture. The pipeline was originally slated to cross the Sandhills region of Nebraska, an ecologically valuable area that hosts most of the world's Sandhill cranes as well as other migratory birds. At the behest of government regulators, TransCanada agreed to move the proposed route eastward to skirt around the edge of the Sandhills region and the Ogallala Aquifer.

Pipeline leaks are a legitimate concern, as oil from oil sands is more corrosive than conventional crude oil. Recent leaks along the Kalamazoo River in Michigan, in a residential neighborhood of Mayflower, Arkansas, and in other locations have caused severe contamination.

In Alberta where the oil sands are mined, the process uses immense amounts of water, and the polluted wastewater that results is left to sit in gigantic reservoirs. The Syncrude company's massive tailings pond near Fort McMurray, Alberta, is so large that it is held back by the world's second-largest dam. Migratory waterfowl land on water bodies like this and are killed as the oily water gums up their feathers and impairs their ability to insulate themselves. These water pollution impacts come on top of the deforestation required to mine the fuels in the first place.

Industry representatives counter that the area deforested so far amounts to just 0.1% of Canada's vast boreal forest. They also point out they are mandated to attempt restoration afterwards. However, effective reclamation has not yet been demonstrated, and regions denuded by the very first oil sand mine in Alberta 30 years ago have still not recovered.

FIGURE 19.19 **In mountaintop removal mining for coal, entire mountain peaks are leveled and fill is dumped into adjacent valleys,** as shown here over many square miles in West Virginia. This can cause erosion and acid drainage into waterways that flow into surrounding valleys, affecting ecosystems and people over large areas.

Coal mining devastates natural systems

The mining of coal exerts substantial impacts on natural systems and human well-being (pp. 639–640). Strip mining destroys large swaths of habitat and causes extensive soil erosion. It also can cause chemical runoff into waterways through the process of **acid drainage** (pp. 639–640). This occurs when sulfide minerals in newly exposed rock surfaces react with oxygen and rainwater to produce sulfuric acid. As the sulfuric acid runs off, it leaches metals from the rocks, many of which are toxic. Acid drainage is a natural phenomenon, but mining greatly accelerates the rate at which it occurs by exposing many new rock surfaces at once. Regulations in the United States require mining companies to restore strip-mined land following mining, but complete restoration is impossible, and ecological modifications are severe and long-lasting (pp. 643, 646). Most other nations exercise less oversight.

Mountaintop removal mining (FIGURE 19.19 and pp. 641–645) has impacts that exceed even conventional strip mining. When countless tons of rock and soil are removed from the top of a mountain, material slides downhill, where immense areas of habitat can be degraded or destroyed and creek beds can be clogged and polluted. Loosening of U.S. government regulations in 2002 enabled mining companies to legally dump mountaintop rock and soil into valleys and rivers below, regardless of the consequences for ecosystems, wildlife, and local residents.

Oil and gas extraction modify the environment

To drill for conventional oil or gas on land, road networks must be constructed and many sites may be explored in the course of prospecting. The extensive infrastructure needed to support a full-scale drilling operation typically includes housing for workers, access roads, transport pipelines, and waste piles for removed soil. Ponds may be constructed to collect the toxic sludge that remains after the useful components of oil have been removed. These activities can pollute the soil, air, and water, fragment habitats, and disturb wildlife. All these impacts have been documented on the tundra of Alaska's North Slope, where policymakers continue to debate whether to open the Arctic National Wildlife Refuge to drilling.

Fortunately, drilling technology is more environmentally sensitive than in the past. **Directional drilling** allows drillers to bore down vertically and then curve to drill horizontally. This enables extraction companies to follow the course of horizontal layered deposits to extract the most they can from them. It allows drilling to reach a large underground area (up to several thousand meters in radius) around a drill pad. As a result, fewer drill pads are needed, and the surface footprint of drilling is smaller.

We all pay external costs

The costs of alleviating the many health and environmental impacts of fossil fuel extraction and use are generally not internalized in the market prices of fossil fuels. Instead, we all pay these external costs (pp. 146, 165) through medical expenses, costs of environmental cleanup, and impacts on our quality of life. Moreover, the prices we pay at the gas pump or on our monthly utility bill do not even cover the financial costs of fossil fuel production. Rather, fossil fuel prices have been kept inexpensive as a result of government subsidies to extraction companies. The profitable and well-established fossil fuel industries still receive far more financial support from taxpayers than do the young and struggling renewable energy sources (Figure 7.14, p. 182; Figure 21.6, p. 585). Thus, we all pay extra for our fossil fuel energy through our taxes, generally without even realizing it.

Fossil fuel extraction has mixed consequences for local people

Across the world today, people living in fossil-fuel-bearing regions must weigh the environmental, health, and social drawbacks of fossil fuel development against the financial benefits that they and their families may gain. Communities where fossil fuel extraction is taking place generally experience a flush of high-paying jobs and economic activity, and for many people these economic benefits far outweigh other concerns. Perceptions may change with time, however. Economic booms often prove temporary, and residents may be left with the legacy of a polluted environment for generations to come.

Fort McMurray is the hub of Alberta's oil sands boom. Fort McMurray's population has skyrocketed from 2000 in the 1960s to over 100,000 today as people have flocked here looking for jobs. Most residents are men, averaging 32 years of age, and the city boasts the highest birthrate in Canada. Salaries are high, but so are rents and home prices. A well-disciplined worker can become wealthy, but others fall behind, victims of drug abuse, alcoholism, or gambling. Like all boomtowns, Fort McMurray is outgrowing its infrastructure, and it is likely to experience a bust when the price of its principal resource falls.

Along the route the oil would take out of Alberta, the Keystone XL pipeline extension would create 20,000 "job-years," TransCanada estimates—6500 construction jobs for two years plus 7000 one-year jobs for manufacturers of supplies. For landowners, the pipeline project has mixed consequences. TransCanada has had to negotiate with thousands of landowners along the Keystone XL route, offering them money for the right to install the pipeline across their land. Many were happy to accept payments, but landowners who declined TransCanada's offers found their land rights taken away by **eminent domain**—the policy by which courts set aside private property rights to make way for projects judged to be for the public good. Following a 2005 U.S. Supreme Court ruling, even private companies can usurp land rights. The landowner is paid an amount determined by a court to be fair and cannot appeal the decision. As John Harter, a South Dakota rancher along the pipeline route, put it, "I found out that they have more rights to my property than I do. It makes me very angry when I paid for it . . . and take care of it."

In Alaska, to gain support for oil drilling among state residents, the oil industry pays the Alaskan government a portion of its revenues. Since the 1970s, the state of Alaska has received over $65 billion in oil revenues. Alaska's state constitution requires that one-quarter of state oil revenues be placed in the Permanent Fund, which pays yearly dividends to all residents. Since 1982, each Alaska resident has received annual payouts ranging from $331 to $2069.

Such distribution of revenue among citizens is unusual; in most parts of the world where fossil fuels are extracted, local residents suffer pollution without compensation. Even when multinational gas or oil corporations pay developing nations for access to extract oil or gas, the money generally does not trickle down from the government to the people who live where the extraction takes place. Moreover, oil-rich developing nations such as Ecuador, Venezuela, and Nigeria tend to have few environmental regulations, and governments may not enforce regulations if there is risk of losing the large sums of money associated with oil development.

In Nigeria, the Shell Oil Company extracted $30 billion of oil from land of the native Ogoni people, yet the Ogoni still live in poverty, with no running water or electricity. Profits from the oil extraction went to Shell and to the military dictatorships of Nigeria. The development resulted in oil spills, noise, and constantly burning gas flares—all of which caused illness among people living nearby. Starting in 1962, Ogoni activist and leader Ken Saro-Wiwa worked for fair compensation to the Ogoni. After years of persecution by the Nigerian government, Saro-Wiwa was arrested in 1994, given a trial universally regarded as a sham, and put to death by military tribunal.

Wherever in the world fossil fuel extraction comes to communities, people seem to find themselves divided over whether the short-term economic benefits are worth the long-term health and environmental impacts. Today this debate is occurring in North Dakota and parts of the West in response to oil and gas drilling, and in Pennsylvania, New York, and other states above the Marcellus Shale where the petroleum industry is hydrofracking for gas (Chapter 7). The debate has gone on for years in Appalachia over mountaintop removal mining. There are no easy answers, but impacts would be lessened if extraction industries were to put more health and environmental safeguards in place for workers and residents.

Dependence on foreign energy affects the economies of nations

Putting all your eggs in one basket is always a risky strategy. Because virtually all our modern technologies and

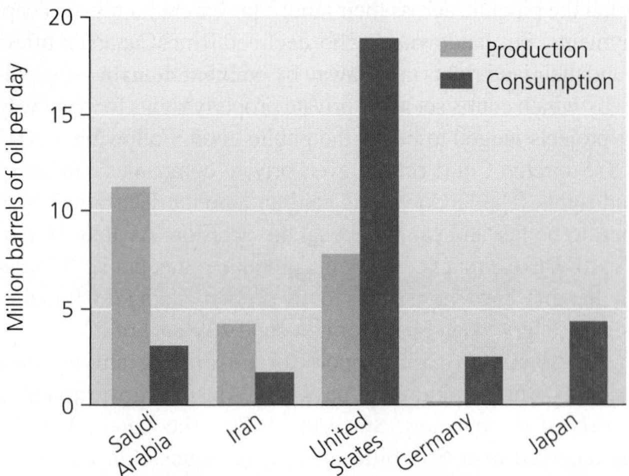

FIGURE 19.20 Japan, Germany, and the United States are among nations that consume far more oil than they produce. Iran and Saudi Arabia produce more oil than they consume and are able to export oil to high-consumption countries. *Data from BP p.l.c., 2012.* Statistical review of world energy 2012.

DATA **Q** For every barrel of oil produced in the United States, how many barrels are consumed in the United States?

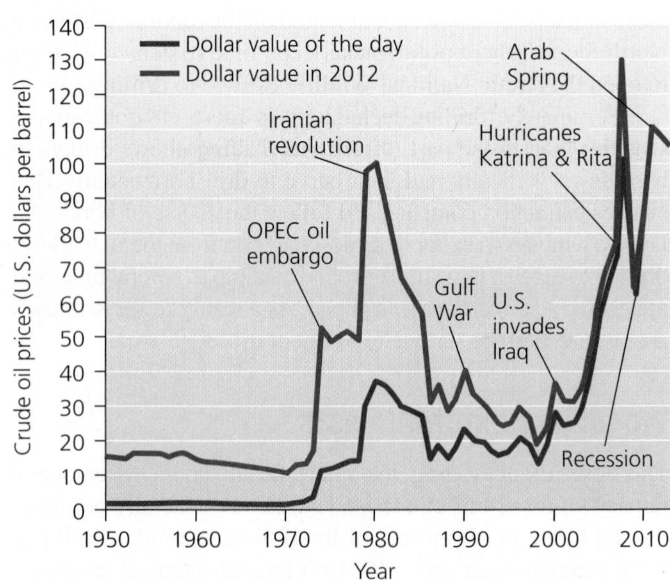

FIGURE 19.21 World oil prices have gyrated over the decades. Often this has resulted from political and economic events in oil-producing countries, particularly in the Middle East. *Data from U.S. Energy Information Administration and BP p.l.c., 2012,* Statistical review of world energy 2012.

services depend in some way on fossil fuels, we are vulnerable to supplies becoming costly or unavailable. Nations that lack adequate fossil fuel reserves of their own are especially vulnerable. For instance, Germany, France, South Korea, and Japan consume far more energy than they produce and thus rely almost entirely on imports (**FIGURE 19.20**). Since its 1970 oil production peak, the United States has relied more on foreign energy, and today imports nearly half of its oil.

Such reliance means that seller nations can control energy prices, forcing buyer nations to pay more as supplies dwindle. This became clear in 1973, when the *Organization of Petroleum Exporting Countries (OPEC)* resolved to stop selling oil to the United States. The predominantly Arab nations of OPEC opposed U.S. support of Israel in the Arab–Israeli Yom Kippur War and sought to raise prices by restricting supply. The embargo created panic in the West and caused oil prices to skyrocket (**FIGURE 19.21**), spurring inflation. Fear of oil shortages drove American consumers to wait in long lines at gas pumps. A similar supply shock followed in 1979 in response to the Iranian revolution.

With the majority of world oil reserves located in the politically volatile Middle East, crises in this region are a constant concern for U.S. policymakers. The democratic street uprisings of the "Arab Spring" that began in 2011 in Tunisia and Egypt and spread elsewhere in the region put leaders of the United States and other Western nations in an awkward position, because they had long supported many of the region's autocratic rulers. These rulers had facilitated Western access to oil, even as they suppressed democracy in their own societies. The Arab Spring uprisings were only the most recent in a long history of events that have affected oil prices and global access to oil, stretching back through the U.S.–led wars in Iraq and the Iran–Iraq war of the 1980s to the 1973 OPEC embargo.

From this perspective, turning to Canada's oil sands as a primary source of oil represents a perfect solution for the United States. Canada is a stable, friendly, democratic neighboring country that is already the United States' biggest trading partner. Trading with Canada for petroleum from the oil sands would lessen U.S. reliance on Middle Eastern oil, and this is a major reason that many American policymakers and citizens favor building the Keystone XL pipeline. Indeed, in recent years the United States has already diversified its sources of imported petroleum considerably and now receives most from non–Middle Eastern nations, including Canada, Mexico, Venezuela, and Nigeria (**FIGURE 19.22**).

Diversifying sources of foreign oil was one way in which the U.S. government responded to the 1973 embargo. The United States also enacted conservation measures, capped the price that domestic producers could charge for oil, funded research into renewable energy sources, and urged oil companies to pursue secondary extraction at old wells. It established an emergency

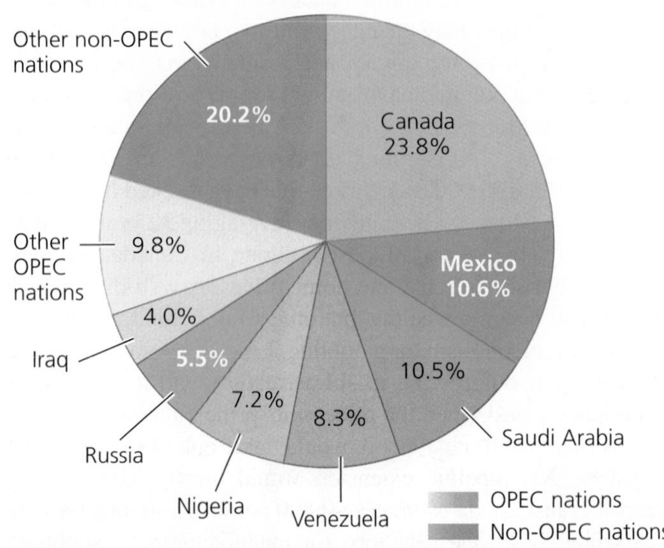

FIGURE 19.22 The United States now imports most oil from non-OPEC nations and from non-Middle-Eastern nations. *Data from U.S. Energy Information Administration, 2012.* Annual energy review 2011.

stockpile (which today stores one month of oil) deep underground in salt caverns in Louisiana, called the Strategic Petroleum Reserve. And it called for developing additional domestic sources, including offshore oil from the Gulf of Mexico.

Since then, the desire to reduce reliance on foreign oil by boosting domestic production has driven the expansion of offshore drilling into deeper and deeper water. It has repeatedly driven a proposal to open the Arctic National Wildlife Refuge on Alaska's North Slope to oil extraction, despite critics' charges that drilling there would spoil America's last true wilderness while adding little to the nation's oil supply. Today it is driving the push to drill for oil offshore in the waters of the Arctic, despite the environmental risks. As domestic U.S. production of oil and gas increase due to enhanced drilling efforts, the United States becomes freer to make geopolitical decisions without being hamstrung by dependence on foreign energy imports. As pressure for increased drilling intensifies in coming years, our society will be debating the complex mix of social, political, economic, and environmental costs and benefits.

WEIGHING THE ISSUES

DRILL, BABY, DRILL? Do you think the United States should open more of its offshore waters to oil and gas extraction? Would the benefits exceed the potential costs? What factors should the government consider when deciding which areas to lease for drilling? How strongly should government regulate oil and gas extraction once drilling begins? Give reasons for your answers.

How will we convert to renewable energy?

Fossil fuels are limited in supply, and their use has health, environmental, political, and socioeconomic consequences (**FIGURE 19.23**). For these reasons, many scientists, environmental advocates, businesspeople, and policymakers have concluded that fossil fuels are not a sustainable long-term solution to our energy needs. They further conclude that we need to shift to clean and renewable sources of energy that exert less impact on natural systems and human health. Many nations are moving far faster than the United States. France relies on nuclear power for its energy needs, Germany is investing in solar power (pp. 581–582), and China is forging ahead and developing multiple renewable energy technologies.

As we make the transition to renewable energy sources, it will benefit us to extend the availability of fossil fuels. We can prolong our access to fossil fuels by instituting measures to

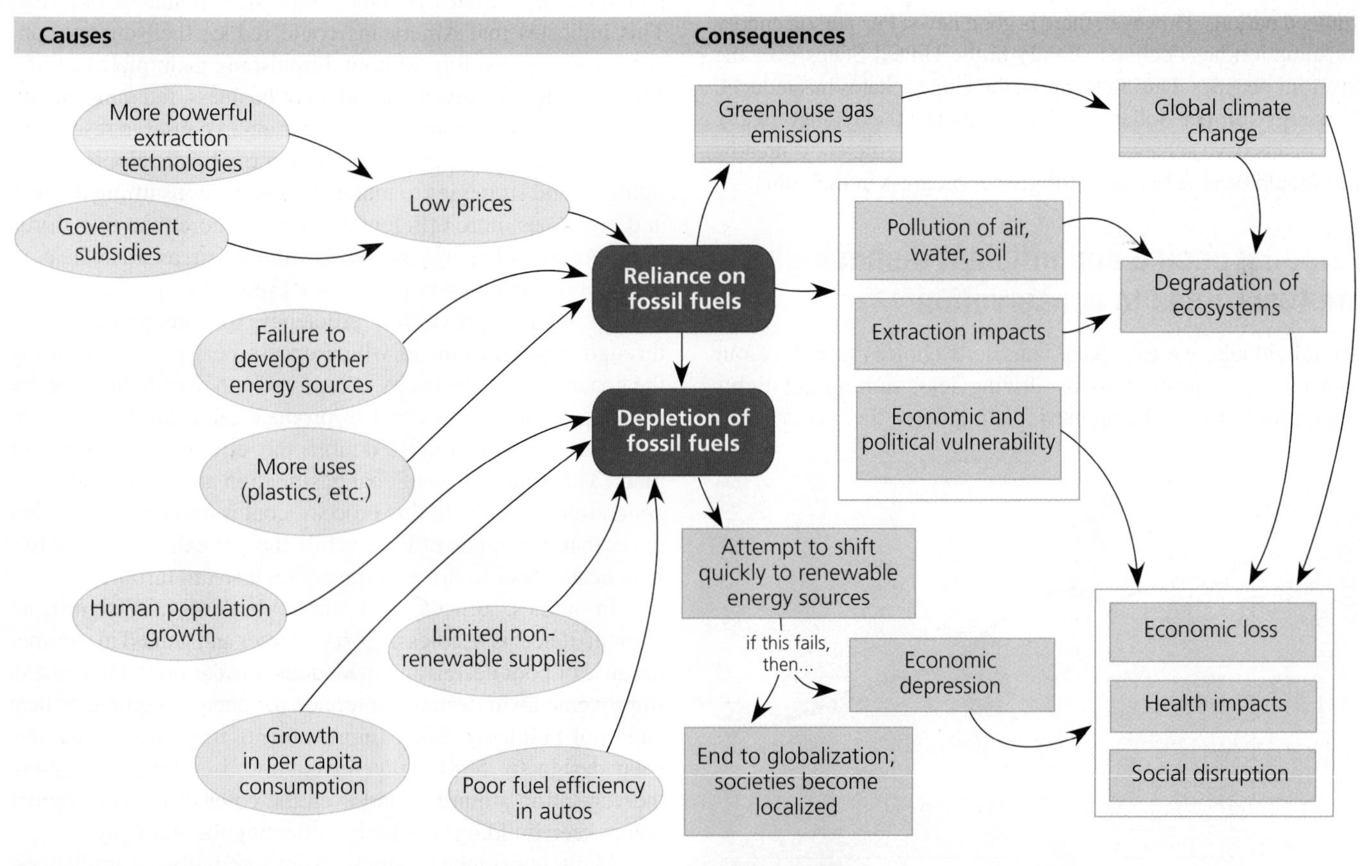

FIGURE 19.23 Our reliance on and depletion of fossil fuels have many causes (ovals on left) and consequences (boxes on right). Arrows in this concept map lead from causes to consequences. Note that items grouped within outlined boxes do not necessarily share any special relationship; the outlined boxes are merely intended to streamline the figure.

Solutions

As you progress through this chapter, try to identify as many solutions to our reliance on and depletion of fossil fuels as you can. What could you personally do to help address this issue? Consider how each action or solution might affect items in the concept map above.

conserve energy through lifestyle changes that reduce energy use, and through technological advances that improve efficiency.

Energy Efficiency and Conservation

Until our society makes the transition to renewable energy sources, we will need to find ways to minimize and extend the use of our precious fossil fuel resources. **Energy efficiency** describes the ability to obtain a given result or amount of output while using less energy input. **Energy conservation** describes the practice of reducing wasteful or unnecessary energy use. In general, efficiency results from technological improvements, whereas conservation stems from behavioral choices. Because greater efficiency allows us to reduce energy use, efficiency is one primary means of conservation.

Efficiency and conservation allow us to be less wasteful and to reduce our environmental impact. Moreover, by enabling us to extend the lifetimes of our nonrenewable energy supplies, efficiency and conservation help to alleviate many of the difficult individual choices and divisive societal debates related to fossil fuels, from oil sands development and the Keystone XL pipeline to Arctic drilling to hydraulic fracturing.

The United States burns through twice as much energy per dollar of Gross Domestic Product (GDP) as do most other industrialized nations. However, there is good news: Per-person energy consumption has declined slightly in the United States over the past four decades. During this time the United States has reduced its energy use per dollar of GDP by about 50% (**FIGURE 19.24**). Americans have achieved tremendous gains in efficiency already, and should be able to make still-greater progress in the future.

Personal choice and efficient technologies are two routes to conservation

As individuals, we can make conscious choices to reduce our own energy consumption by driving less, turning off lights when rooms are not being used, dialing down thermostats, and cutting back on the use of energy-intensive machines and appliances. Many European nations use far less energy per capita than the United States yet enjoy equivalent standards of living. This indicates that Americans could reduce their energy consumption considerably without diminishing their quality of life. Moreover, for any given individual or business, reducing energy consumption saves money while helping to conserve resources.

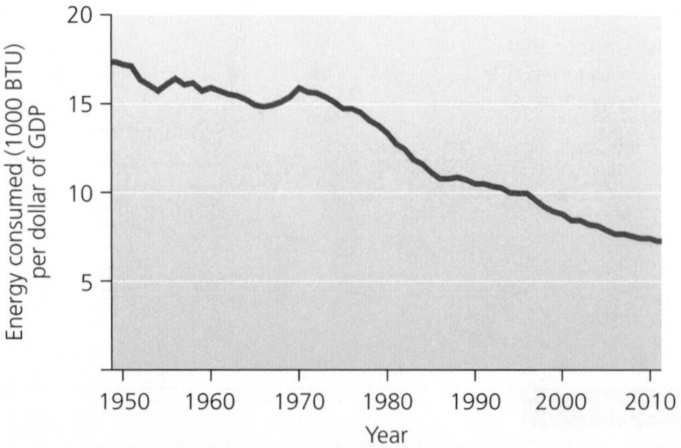

FIGURE 19.24 The United States has been producing more economic output per energy input. This graph shows how energy consumption per inflation-adjusted dollar of GDP has fallen. *Data from U.S. Energy Information Administration, 2012.* Annual energy review 2011.

FIGURE 19.25 A thermogram reveals heat loss from buildings by recording energy in the infrared portion of the electromagnetic spectrum (p. 31). In this image, one house is uninsulated; its red color signifies warm temperatures where heat is escaping, whereas green shades signify cool temperatures where heat is being conserved. Also note that in all houses, more heat is escaping from windows than from walls.

As a society, we can conserve energy by developing technologies and strategies to make our energy-consuming devices and processes more efficient. Currently, more than two-thirds of the fossil fuel energy we use is simply lost, as waste heat, in automobiles and power plants (see Figure 19.3).

We can improve the efficiency of our power plants through **cogeneration,** in which excess heat produced during the generation of electricity is captured and used to heat nearby workplaces and homes and to produce other kinds of power. Cogeneration can almost double the efficiency of a power plant. The same is true of coal gasification and combined cycle generation (p. 537). In this process, coal is treated to create hot gases that turn a gas turbine, while the hot exhaust of this turbine heats water to drive a conventional steam turbine.

In homes, offices, and public buildings, a significant amount of heat is needlessly lost in winter and gained in summer because of poor design and inadequate insulation (**FIGURE 19.25**). Improvements in design can reduce the energy required to heat and cool buildings. Such improvements may involve passive solar design (p. 588), better insulation, a building's location, the vegetation around it, and even the color of its roof (lighter colors keep buildings cooler by reflecting the sun's rays).

Many consumer products, from lightbulbs to appliances, have been reengineered through the years to enhance efficiency. Energy-efficient lighting, for example, can reduce energy use by 80%. Compact fluorescent bulbs are much more efficient than incandescent light bulbs, and many governments are phasing out incandescent bulbs for this reason; the U.S. phaseout is scheduled to be complete in 2014.

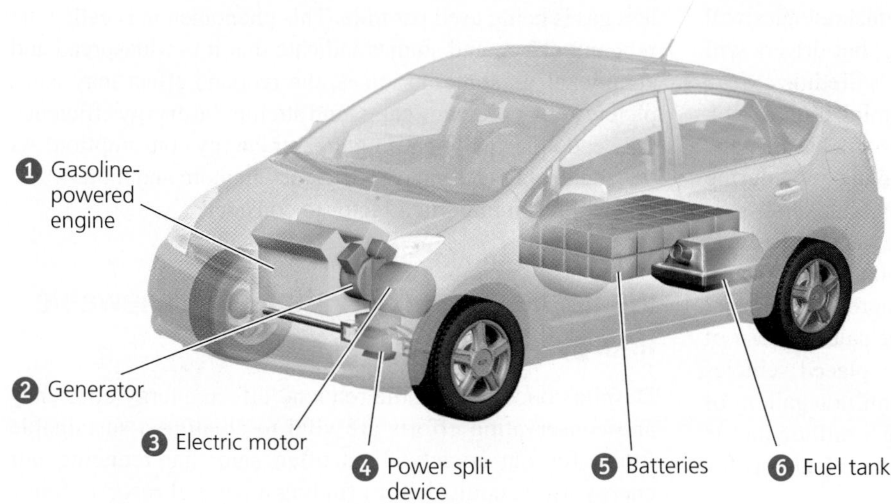

FIGURE 19.26 Hybrid cars have high fuel efficiencies. The Toyota Prius diagrammed here uses a small, clean, and efficient gasoline-powered engine ❶ to produce power that the generator ❷ can convert to electricity to drive the electric motor ❸. The power split device ❹ integrates the engine, generator, and motor, serving as a continuously variable transmission. The car automatically switches among all-electrical power, all-gas power, and a mix of the two, depending on the demands being placed on the engine. Typically, the motor provides power for low-speed city driving and adds extra power on hills. The motor and generator charge a pack of nickel-metal-hydride batteries ❺, which can in turn supply power to the motor. Energy for the engine comes from gasoline carried in a typical fuel tank ❻.

Federal standards for energy-efficient appliances have already reduced per-person home electricity use below what it was in the 1970s. The Energy Star program (p. 507) labels refrigerators, dishwashers, and other appliances for their energy efficiency, enabling consumers to take energy use into account when shopping for these items. For the consumer, studies show that the slightly higher cost of buying energy-efficient washing machines is rapidly offset by savings on water and electricity bills. The U.S. Environmental Protection Agency (EPA) estimates that if all U.S. households purchased energy-efficient appliances, the national annual energy expenditure would be reduced by $200 billion.

Automotive technology represents perhaps our best opportunity to conserve large amounts of fossil fuels fairly easily. We can accomplish this with alternative-technology vehicles such as electric cars, electric/gasoline hybrids (**FIGURE 19.26**), or vehicles that use hydrogen fuel cells (pp. 602–604). Among electric/gasoline hybrids, current U.S. models of the Toyota Prius and the Chevrolet Volt average fuel-economy ratings of 50 miles per gallon (mpg) and 60 mpg, respectively—two to three times better than the average American car. Even without alternative vehicles, we already possess the means to enhance fuel efficiency for gasoline-powered vehicles by using lightweight materials, continuously variable transmissions, and more efficient gasoline engines.

Automobile fuel efficiency is a key to conservation

Among the measures enacted by the U.S. government in response to the OPEC embargo of 1973–1974 were a mandated increase in the mile-per-gallon (mpg) fuel efficiency of automobiles and a reduction in the national speed limit to 55 miles per hour. These measures notably reduced U.S. dependence on Middle Eastern oil and held down greenhouse gas emissions. Over the next three decades, however, many of the conservation initiatives of this time were abandoned. Without high market prices and an immediate threat of shortages, people lacked economic motivation to conserve. Government funding for research into alternative energy sources dwindled, speed limits rose, and U.S. policymakers repeatedly failed to raise the *corporate average fuel efficiency (CAFE) standards*, which set benchmarks for auto manufacturers to meet. The average fuel efficiency of new vehicles fell from 22.0 mpg in 1987 to 19.3 mpg in 2004, as sales of sport-utility vehicles increased relative to sales of cars.

Since then, fuel economy climbed to 22.8 mpg in 2011 (**FIGURE 19.27**) after Congress passed legislation in 2007 mandating that automakers raise average fuel efficiency to 35 mpg by the year 2020. This was a substantial advance, yet even after this boost, American automobiles would still have lagged well behind the efficiency of the vehicles of most other developed nations. As a result, when automakers required a government bailout during the recent recession, President Obama negotiated with them, forcing a series of agreements that ended with automakers agreeing to boost average fuel economies to

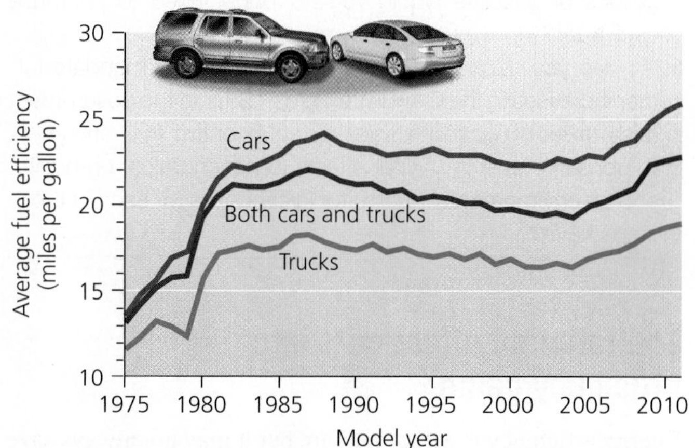

FIGURE 19.27 Automotive fuel efficiencies have responded to public policy. Fuel efficiency for automobiles in the United States rose dramatically in the late 1970s as a result of legislative mandates but then stagnated once no further laws were enacted to improve fuel economy. Recent legislation is now improving it again. *Data from U.S. Environmental Protection Agency, 2012.* Light-duty automotive technology, carbon dioxide emissions, and fuel economy trends: 1975 through 2011.

54.5 mpg by 2025. If this strong improvement comes to pass, it will enable a huge reduction in oil use. New technologies will add over $2000 to the average price of a car, but drivers will save perhaps $6000 in fuel costs over the car's lifetime.

In 2009, Congress and the Obama administration took another major step to improve automobile fuel efficiency while stimulating economic activity and saving jobs during a severe recession. The popular "Cash for Clunkers" program—formally named the Consumer Assistance to Recycle and Save (CARS) Act—paid Americans $3500 or $4500 each to turn in old vehicles and purchase newer, more fuel-efficient ones. The $3-billion program subsidized the sale or lease of 678,000 vehicles averaging 24.9 mpg that replaced vehicles averaging 15.8 mpg. It is estimated that 824 million gallons of gasoline will be saved as a result, preventing 9 million metric tons of greenhouse gas emissions and creating social benefits worth $278 million.

U.S. policymakers could do still more to encourage oil conservation. So far the United States has kept its taxes on gasoline extremely low, relative to most other nations. Americans pay two to three times *less* per gallon of gas than drivers in many European countries. In fact, gasoline in the United States is sold more cheaply than bottled water! As a result, U.S. gasoline prices do not account for the substantial external costs (pp. 146, 165) that oil production and consumption impose on society. Some experts have estimated that if all costs to society were taken into account, the price of gasoline would exceed $13/gallon. Instead, our artificially low gas prices diminish our economic incentives to conserve.

WEIGHING THE ISSUES

MORE MILES, LESS GAS If you drive an automobile, what gas mileage does it get? How does it compare to the vehicle averages in Figure 19.27? If your vehicle's fuel efficiency were 10 mpg greater, and if you drove the same amount, how many gallons of gasoline would you no longer need to purchase each year? How much money would you save?

Do you think the U.S. government should mandate further increases in the CAFE standards? Should the government raise taxes on gasoline sales as an incentive for consumers to conserve energy? What effects (on economics, on health, and on environmental quality, for instance) might each of these steps have?

The rebound effect cuts into efficiency gains

Energy efficiency is a vital pursuit, but it may not always save as much energy as we expect. This is because gains in efficiency from better technology can be partly offset if people engage in more energy-consuming behavior as a result. For instance, a person who buys a fuel-efficient car may choose to drive more because he or she feels it's okay to do so now that less gas is being used per mile. This phenomenon is called the **rebound effect,** and studies indicate that it is widespread and significant. In some instances, the rebound effect may completely erase efficiency gains, and attempts at energy efficiency may end up actually causing greater energy consumption! As our society pursues energy efficiency in more and more ways, this will be an important factor to consider.

We need both conservation and renewable energy

Despite concerns over the rebound effect, energy efficiency and conservation efforts are vital to creating a sustainable future for our society. It is often said that reducing our energy use is equivalent to finding a new oil reserve. Some estimates hold that effective energy conservation and efficiency in the United States could save 6 million barrels of oil a day—nearly the amount gained from all offshore drilling, and considerably more than would be gained from Canada's oil sands. In fact, conserving energy is *better* than finding a new reserve because it alleviates health and environmental impacts while at the same time extending our future access to fossil fuels.

Yet regardless of how effectively we conserve, we will still need energy to power our civilization, and it will need to come from somewhere. Most energy experts feel that the only sustainable way of guaranteeing ourselves a reliable long-term supply of energy is to ensure sufficiently rapid development of renewable energy sources (Chapters 20 and 21).

Conclusion

Over the past two centuries, fossil fuels have helped us build the complex industrialized societies we enjoy today. Yet sometime soon our production of conventional fossil fuels will begin to decline. We can respond to this challenge by seeking out new sources of fossil fuels and continuing our way of life but paying ever-higher economic, health, and environmental costs. Or, we can encourage conservation and efficiency while aggressively developing alternative clean and renewable energy sources. The path we choose will have far-reaching consequences for human health and well-being, for Earth's climate, and for the stability and progress of our civilization.

The debate over the Canadian oil sands and the Keystone XL pipeline is a microcosm of this debate over our energy future. Fortunately, we are not caught in a simple trade-off between fossil fuels' economic benefits and their impacts on the environment, climate, and health. Instead, as renewable energy sources become increasingly feasible and economical, it becomes easier to envision freeing ourselves from a reliance on fossil fuels and charting a bright future for humanity and the planet with renewable energy.

Reviewing Objectives

You should now be able to:

Identify the energy sources we use

- Many renewable and nonrenewable energy sources are available to us. (pp. 520–521)
- Since the industrial revolution, nonrenewable fossil fuels—including coal, natural gas, and oil—have become our primary sources of energy. (p. 521)
- Energy sources and energy consumption are each unevenly distributed across the world. (pp. 522–523)
- The concepts of net energy and EROI allow us to compare the amount of energy obtained from a source with the amount invested in its extraction and production. (pp. 522–524)
- We face a choice in whether to pursue new low-quality fossil fuel sources such as oil sands or whether to develop alternative sources of clean renewable energy. (p. 524)

Describe the origin and nature of major types of fossil fuels

- Fossil fuels are formed very slowly as buried organic matter is chemically transformed by heat, pressure, and/or anaerobic decomposition. (pp. 524–525)
- Coal, our most abundant fossil fuel, results from organic matter that undergoes compression but little decomposition. (pp. 525–526)
- Crude oil is a thick, liquid mixture of hydrocarbons that is formed underground under high temperature and pressure. (p. 525)
- Natural gas consists mostly of methane and can be formed in two ways. (p. 525)
- Oil sands contain bitumen, a tarry substance formed from oil that was degraded by bacteria. This can be processed into synthetic crude oil. (p. 525)
- Shale oil and methane hydrate are fossil fuel sources with potential for future use (p. 526)

Explain how we extract and use fossil fuels

- Scientists locate fossil fuel deposits by analyzing subterranean geology. We then estimate the technically and economically recoverable portions of those reserves. (pp. 526–527, 530–531)
- Coal is mined underground and strip-mined from the land surface, whereas we drill wells to pump out oil and gas. Oil sands may be strip-mined or dissolved underground and extracted through well shafts. (pp. 526–527)
- Components of crude oil are separated in refineries to produce a wide variety of fuel types. (pp. 527–528)
- Coal is used today principally to generate electricity. (pp. 528–529)
- Natural gas is cleaner-burning than coal or oil. (pp. 528–529)
- Oil powers transportation and also is used to create a diversity of petroleum-based products that are everywhere in our daily lives. (pp. 529, 532)

Evaluate peak oil and the challenges it may pose

- R/P ratios help indicate how long a resource may last, but they tell only part of the story. (pp. 531–532)
- Any nonrenewable resource can be depleted, and we have depleted nearly half the world's conventional oil. (pp. 530–533)
- Once we pass the peak of oil production, the gap between rising demand and falling supply may pose immense economic and social challenges for our society. (pp. 532–534)

Examine how we are reaching further for fossil fuels

- Primary extraction may be followed by secondary extraction, in which gas or liquid is injected into the ground to help force up additional oil or gas. (p. 534)
- Hydraulic fracturing is producing natural gas from shale deposits. (p. 535)
- We are drilling for oil and gas further offshore, in deeper water, and are moving into the Arctic. (pp. 535–536)
- New types of fossil fuels we may exploit include oil sands, shale oil, and methane hydrate (p. 536)

Outline and assess environmental impacts of fossil fuel use, and explore solutions

- Emissions from fossil fuel combustion pollute air, pose human health risks, and drive global climate change. (pp. 536–537)
- Public policy and advances in pollution control technology have reduced many of these emissions, but much more remains to be done. (p. 537)
- Clean coal technologies aim to reduce pollution from coal combustion. (p. 537)
- If we could safely and effectively capture carbon dioxide and sequester it underground, this would mitigate a primary drawback of fossil fuels. Carbon capture and storage remain unproven so far, however. (pp. 537–538)
- Oil is a major contributor to water pollution. (pp. 538–539)
- Hydrofracking poses pollution concerns. (pp. 539, 542)
- Oil sands mining and transport cause deforestation, water pollution, and other impacts. (p. 542)
- Coal mining can devastate ecosystems and pollute waterways. (p. 542)

- Oil and gas extraction exert various impacts, but directional drilling has eased them. (pp. 542–543)

Evaluate political, social, and economic aspects of fossil fuel use

- Fossil fuels impose external costs. (p. 543)
- Fossil fuel extraction creates jobs but leaves pollution. People living in areas of fossil fuel extraction experience a range of consequences. (p. 543)
- Today's societies are so reliant on fossil fuel energy that sudden restrictions in oil supplies can have major economic consequences. (pp. 543–545)
- Nations that consume far more fossil fuels than they produce are especially vulnerable to supply restrictions. (p. 544)

Specify strategies for conserving energy and enhancing efficiency

- Energy conservation involves both personal choices and efficient technologies. (p. 546)
- Efficiency in power plant combustion, lighting, and consumer appliances, as well as changes in public policy, can help us conserve. (pp. 546–547)
- Automotive fuel efficiency plays a key role in conserving energy. (pp. 547–548)
- The rebound effect can partly negate our conservation efforts. (p. 548)
- Conservation lengthens our access to fossil fuels and reduces environmental impact, but to build a sustainable society we will also need to shift to renewable energy sources. (p. 548)

Testing Your Comprehension

1. Why are fossil fuels our most prevalent source of energy today? Why are they considered nonrenewable sources of energy?
2. Describe how net energy differs from energy returned on investment (EROI). Why are these concepts important in the evaluation of energy sources?
3. How are fossil fuels formed? How do environmental conditions determine which type of fossil fuel is formed in a given location? Why are fossil fuels often concentrated in localized deposits?
4. How do geologists find oil and gas deposits and estimate amounts of oil or gas available? How do "technically recoverable" and "economically recoverable" amounts of a resource differ?
5. Describe how coal is used to generate electricity.
6. How do we create petroleum products? Provide examples of several of these products.
7. What is meant by *peak oil*? Why do many experts think we are about to pass the global production peak for conventional oil? What consequences could there be for our society?
8. Describe three environmental impacts of fossil fuel production and consumption. Compare contrasting views regarding the impacts of extracting oil from Alberta's oil sands and shipping it by pipeline to the U.S. Gulf Coast.
9. Give an example of a clean coal technology. Now describe how carbon capture and storage is intended to work.
10. Describe one specific example of how technological advances can improve energy efficiency. Now describe one specific action you could take to conserve energy.

Seeking Solutions

1. Summarize the main arguments for and against the Keystone XL pipeline extension. What problems might it help solve? What problems might it create? Do you personally think Canada should continue to develop Alberta's oil sands? Should the United States have approved construction of the Keystone XL pipeline extension? Give reasons for your answers.
2. What impacts might you expect on your lifestyle once our society arrives at peak oil? What lessons do you think we can take from the conservation methods adopted by the United States in response to the "energy crisis" of 1973–1974? What steps do you think we should take to avoid energy shortages in a post-peak-oil future?
3. Compare the health and environmental effects of coal extraction and consumption with those of oil and gas extraction and consumption. What steps could governments, industries, and individuals take to alleviate some of these impacts?
4. Contrast the experiences of the Ogoni people of Nigeria with those of the citizens of Alaska. How have they been similar and different? Do you think businesses or governments should take steps to ensure that local people benefit from oil drilling operations? How could they do so?
5. **THINK IT THROUGH** You have been elected governor of the state of Florida as the federal government is debating opening new waters to offshore drilling for oil and

natural gas. Drilling in Florida waters would create jobs for Florida citizens as well as revenue for the state in the form of royalty payments from oil and gas companies. However, there is always the risk of a catastrophic oil spill, with its ecological, social, and economic impacts. Would you support or oppose offshore drilling off the Florida coastline? Why? What, if any, regulations would you insist be imposed on such development? What questions would you ask of scientists before making your decision? What factors would you consider in making your decision?

6. **THINK IT THROUGH** You are the mayor of a rural Nebraska town along the route of the proposed Keystone XL pipeline extension. Some of your town's residents are eager to have jobs they believe the pipeline will bring. Others are fearful that oil leaks from the pipeline could contaminate the water supply. Some of your town's landowners are looking forward to receiving payments from TransCanada for use of their land, whereas others dread the prospect of noise, pollution, and trees being cut on their property. If the company receives too much local opposition it says it may move the pipeline route away from your town. What information would you seek from TransCanada, from your state regulators, and from scientists and engineers before deciding whether support for the pipeline is in the best interest of your town? How would you make your decision? How might you try to address the diverse preferences of your town's residents?

Calculating Ecological Footprints

Scientists at the Global Footprint Network calculate the energy component of our ecological footprint by estimating the amount of ecologically productive land and sea required to absorb the carbon released from fossil fuel combustion. This translates into 4.9 ha of the average American's 7.2-ha ecological footprint. Another way to think about our footprint, however, is to estimate how much land would be needed to grow biomass with an energy content equal to that of the fossil fuel we burn.

Assume that you are an average American who burns about 6.3 metric tons of oil-equivalent in fossil fuels each year and that average terrestrial net primary productivity (p. 111) can be expressed as 0.0037 metric tons/ha/year. Calculate how many hectares of land it would take to supply our fuel use by present-day photosynthetic production.

	Hectares of land for fuel production
You	1703
Your class	
Your state	
United States	

1. Compare the energy component of your ecological footprint calculated in this way with the 4.9 ha calculated using the method of the Global Footprint Network. Explain why results from the two methods may differ.

2. Earth's total land area is approximately 15 billion hectares. Compare this to the hectares of land for fuel production from the table.

3. In the absence of stored energy from fossil fuels, how large a human population could Earth support at the level of consumption of the average American, if all of Earth's area were devoted to fuel production? Do you consider this realistic? Provide two reasons why or why not.

20

Barsebäck nuclear power plant, Sweden

Conventional Energy Alternatives

Upon completing this chapter, you will be able to:

- Discuss the reasons for seeking energy alternatives to fossil fuels
- Summarize the contributions to world energy supplies of conventional alternatives to fossil fuels
- Describe nuclear energy and how it is harnessed
- Outline the societal debate over nuclear power
- Describe the major sources, scale, and impacts of bioenergy
- Describe the scale, methods, and impacts of hydroelectric power

CENTRAL CASE STUDY

Sweden's Search for Alternative Energy

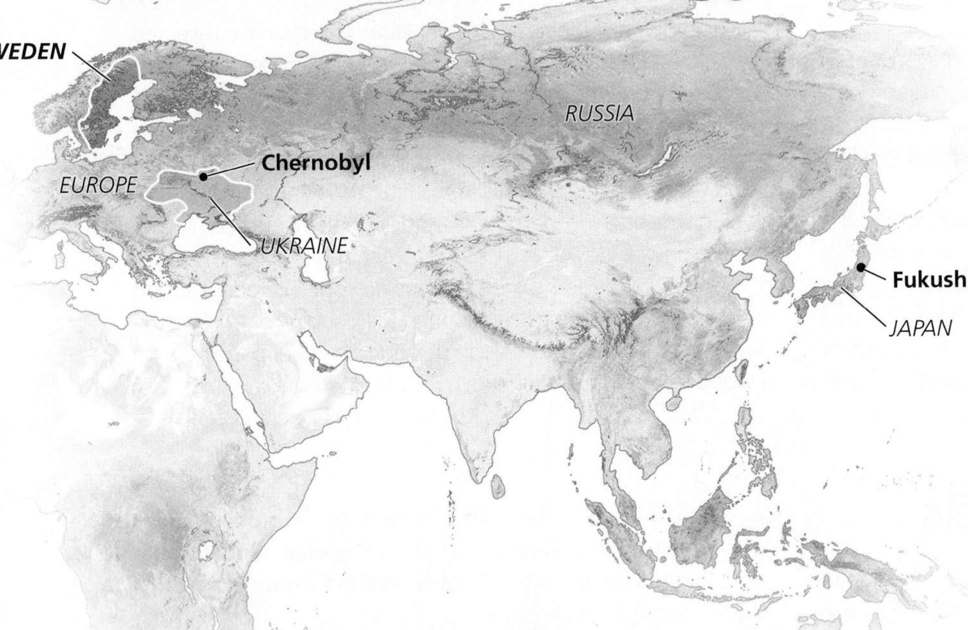

"Nowhere has the public debate over nuclear power plants been more severely contested than Sweden."

—Writer and analyst Michael Valenti

"If [Sweden] phases out nuclear power, then it will be virtually impossible for the country to keep its climate-change commitments."

—Yale University economist William Nordhaus

Flashback to 1986: On the morning of April 28, workers at a nuclear power plant in Sweden detected unusually high radiation levels. The increased radioactivity was not coming from within the plant, but had spread through the atmosphere from the direction of the Soviet Union. They had discovered the outside world's first evidence of the disaster at Chernobyl, more than 1200 km (750 mi) away in what is now the nation of Ukraine. Chernobyl's nuclear reactor had exploded two days earlier, but the Soviet government had not yet admitted it.

Fast-forward to 2011: On March 11, seismometers in Sweden and throughout the world recorded vibrations from the shaking of the massive Tohoku Earthquake off the coast of Japan. Soon thereafter, a tsunami inundated the Japanese coast, destroying entire towns, killing 20,000 people, and disabling nuclear reactors at the Fukushima Daiichi power plant. For the next several weeks, the world would stand on edge as Japanese authorities made frantic efforts to control the leakage of deadly radiation and avert further catastrophe.

The events at both Chernobyl and Fukushima had broad and long-lasting repercussions. Each altered the course of the world's use of nuclear energy, one of our main alternatives to fossil fuels. People in Ukraine and Japan will continue to struggle with the legacies of these events for years to come, but the global debate over nuclear power affects all of us.

In Sweden in the days after Chernobyl, low levels of radioactive fallout rained down on the countryside, contaminating crops and cows' milk. For many Swedes, this confirmed the decision they had made collectively six years earlier, in a 1980 referendum, to phase out their country's nuclear power program, shutting down all nuclear plants by the year 2010.

But trying to phase out nuclear power proved difficult. Aware of the environmental impacts of fossil fuels, Sweden's government and citizens did not want to increase their use of coal, oil, and gas. Sweden has been one of the few nations to decrease use of fossil fuels since the 1970s. But it did so largely by replacing them with nuclear power. As 2010 drew nearer, Sweden was still relying on 10 nuclear reactors for over 30% of its energy supply and over 40% of its electricity. If nuclear plants were to be shut down as planned, then the nation would need to boost its investment in alternative energy sources dramatically.

Sweden, therefore, promoted research and development of renewable energy sources. Hydroelectric power was already supplying most of the rest of the nation's electricity and could not be expanded much more. Instead, the government hoped that energy from biomass sources and wind power could fill the gap. Sweden boosted wind and bioenergy by applying a carbon tax (p. 513) to fossil fuels and by subsidizing renewable energy through a certificate program in which producers and users of fossil-fuel electricity were mandated to buy certificates from producers and users of renewable electricity. In these ways the government used market forces to make fossil fuels more expensive and renewable energy more affordable.

Sweden soon made itself an international leader in renewable energy alternatives, and today gets nearly half its energy from renewable sources. Still, renewables were taking longer to develop than hoped, so policymakers repeatedly postponed the nuclear phaseout. Then in 2009, the government announced that it was reversing its policy and would not phase out nuclear power after all. Instead, new power plants would be built as existing ones needed replacing.

In announcing this decision, government leaders said it would be fiscally and socially irresponsible to dismantle the nation's nuclear program without a ready replacement. Without an abundance of clean renewable power at hand, a nuclear phaseout would mean a return to fossil fuels, or else would require cutting down immense areas of forest to combust biomass. Policymakers also cited Sweden's international obligations to hold down its carbon emissions under the Kyoto Protocol (p. 510). Nuclear power is free of atmospheric pollution and is an effective way to minimize greenhouse gas emissions.

Then Fukushima occurred. As the drama played out in Japan, anti-nuclear protestors all over the world staged demonstrations. Many national governments reassessed their commitments to nuclear power, ran safety checks of existing plants, and halted plans for new plants.

Germany responded most strongly. Just years before, the government of German Chancellor Angela Merkel—herself a physicist and a proponent of nuclear power—had postponed its own planned phaseout of nuclear power, much like Sweden had. But in the wake of Fukushima, hundreds of thousands of demonstrators took to the streets before state elections, and Merkel's government made a stunning U-turn. It reinstated the plan to phase out all nuclear plants by 2022 and immediately shut down its seven oldest plants.

Sweden's environment minister criticized Germany's reaction, saying it would impede Germany's ability to move away from fossil fuels. Meanwhile, the Swedish state–owned firm Vattenfall, a major contractor building German plants, sued the German government on the grounds that its phaseout, together with new taxes applied to the plants, breached its contract.

The Swedish government's shifting policies have reflected the tension in public sentiment between the benefits of fighting climate change by developing clean energy sources and the risk of supporting nuclear power while these new energy sources are being developed. Public opinion polls from 2003 onward showed substantial majorities of Swedish citizens in favor of maintaining or increasing their nation's production of nuclear power. However, in 2012, a year after Fukushima, 44% of Swedes wanted to phase out nuclear power. Today, strong majorities of Swedish citizens continue to support boosting the development of renewable energy sources to keep their nation a world leader in the shift away from fossil fuels. ■

Alternatives to Fossil Fuels

Fossil fuels helped to drive the industrial revolution and to create the unprecedented material prosperity we enjoy today. Our global economy is largely powered by fossil fuels; over 80% of our energy comes from oil, coal, and natural gas (**FIGURE 20.1a**). These three fuels also generate two-thirds of the world's electricity (**FIGURE 20.1b**). However, these nonrenewable energy sources will not last forever. Easily extractable supplies of oil and natural gas are in decline, and we are expending more and more energy and money to extract them (pp. 533–536). Moreover, the use of coal, oil, and natural gas drives global climate change and entails many other health and environmental impacts (Chapters 17, 18, and 19).

For these reasons, most energy experts accept that we will need to shift from fossil fuels to energy sources that are less easily depleted and gentler on our health and environment. Developing alternatives to fossil fuels has the added benefit of helping to diversify an economy's mix of energy, thus lessening price volatility and dependence on foreign fuel imports.

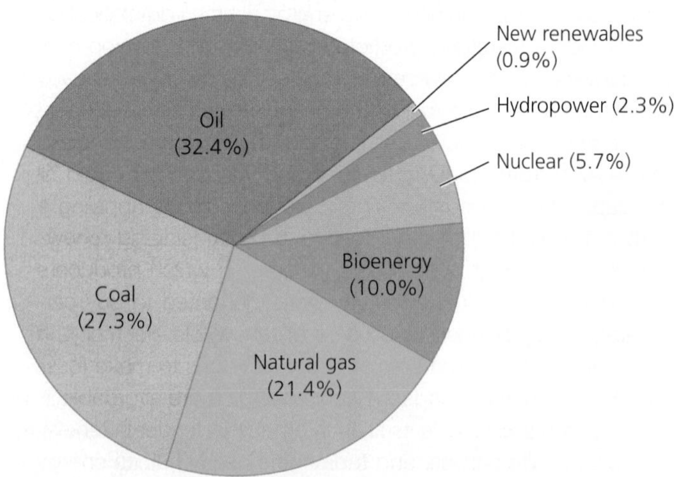

(a) World energy production, by source

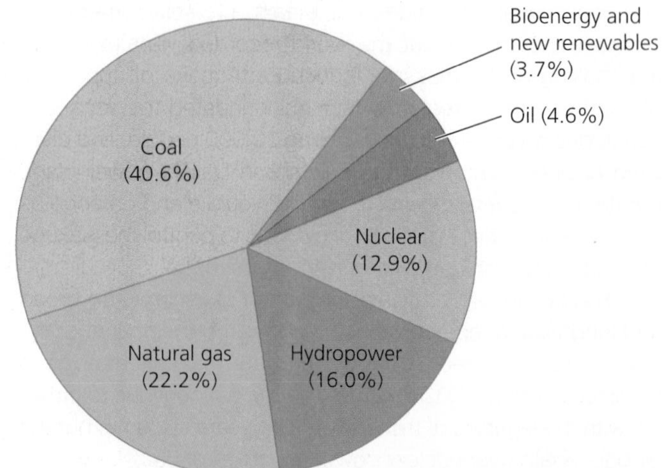

(b) World electricity generation, by source

FIGURE 20.1 Fossil fuels dominate the global energy supply.
Together, oil, coal, and natural gas account for 81% **(a)** of the world's energy production. Nuclear power and hydroelectric power contribute substantially to global electricity generation **(b)**, but fossil fuels still power two-thirds of our electricity. *Data from International Energy Agency, 2012.* Key world energy statistics 2012. *Paris: IEA.*

We have developed a range of alternatives to fossil fuels (see Table 19.1, p. 521). Most of these energy sources are renewable, and most have less impact on health and the environment than oil, coal, or natural gas. At this time most remain more expensive than fossil fuels, at least in the short term and when external costs (pp. 146, 165) are not included in market prices. As technologies develop and as we invest in infrastructure to better transmit power from renewable sources, prices will come down further and help us transition toward these new energy sources.

Nuclear power, bioenergy, and hydropower are conventional alternatives

Three alternative energy sources are currently the most developed and most widely used: nuclear power, hydroelectric power, and energy from biomass. Each of these well-established energy sources plays a substantial role in our energy and electricity budgets today. We can therefore call nuclear power, hydropower, and biomass energy (bioenergy) "conventional alternatives" to fossil fuels.

Each of these three conventional energy alternatives are generally considered to exert less environmental impact than fossil fuels, but more impact than the "new renewable" alternatives (Chapter 21). Yet as we will see, they each involve a unique and complex mix of benefits and drawbacks. Nuclear power is commonly considered a nonrenewable energy source, and hydropower and bioenergy are generally described as renewable, but the reality is more complicated. Each of these energy sources is perhaps best viewed as an intermediate along a continuum of renewability.

Conventional alternatives provide much of our electricity

Fuelwood and other bioenergy sources provide 10% of the world's energy, nuclear power provides about 6%, and hydropower provides about 2%. The less established renewable energy sources together account for less than 1% (see Figure 20.1a). Although their global contributions to our overall energy supply are minor, alternatives to fossil fuels do contribute greatly to our generation of electricity. Nuclear energy and hydropower together account for nearly 30% of the world's electricity generation (see Figure 20.1b).

Energy consumption patterns in the United States (**FIGURE 20.2a**) are similar to those globally, except that the United States relies less on fuelwood and slightly more on fossil fuels and nuclear power than most other countries. A graph showing trends in energy consumption in the United States over the past 60 years (**FIGURE 20.2b**) reveals two things. First, conventional alternatives play minor yet substantial roles in overall energy use. Second, use of conventional alternatives has been growing more slowly than use of fossil fuels.

Sweden, however, has shown that it is possible for a wealthy and advanced economy to replace fossil fuels gradually with alternative sources while continuing to raise living standards for its citizens. Since 1970, Sweden has decreased its fossil fuel use from 81% to 38% of its national energy budget. Today nuclear power, bioenergy, and hydropower together provide Sweden with over 60% of its energy and virtually all of its electricity.

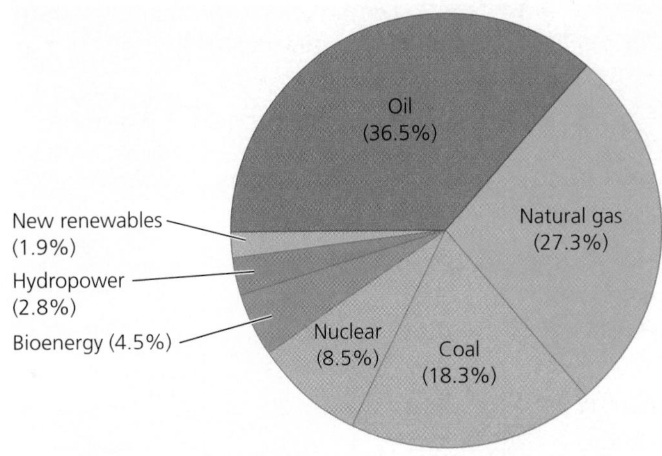

(a) U.S. energy consumption, by source

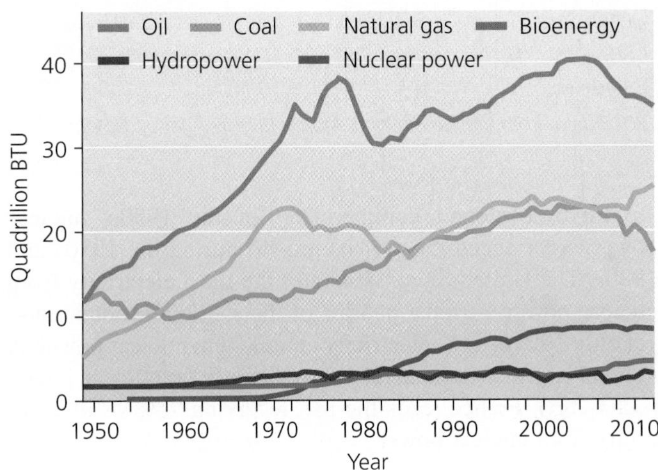

(b) U.S. energy consumption, 1949–2012

FIGURE 20.2 Fossil fuels predominate in the United States. Together, oil, natural gas, and coal account for 82% **(a)** of U.S. energy consumption. Over the past 60 years **(b)**, U.S. consumption of fossil fuels has grown faster than that of bioenergy or hydropower. Nuclear power grew considerably between 1970 and 2000. *Data from U.S. Energy Information Administration, 2013.*

Nuclear Power

Nuclear power occupies an odd and conflicted position in our modern debate over energy. It is free of the air pollution produced by fossil fuel combustion, so it has long been put forth as an environmentally friendly alternative to fossil fuels, and it remains one of our most influential solutions to climate change. Yet nuclear power's great promise has been clouded by nuclear weaponry, the dilemma of radioactive waste disposal, and the long shadow of Chernobyl and now Fukushima. As such, public safety concerns and the costs of addressing them have constrained nuclear power's spread.

NATION	NUCLEAR POWER CAPACITY (GIGA-WATTS)	NUMBER OF REACTORS	PERCENTAGE ELECTRICITY FROM NUCLEAR POWER
United States	102.1	104	19.0
France	63.1	58	74.8
Japan	44.2	50	2.1
Russia	23.6	33	17.8
South Korea	20.7	23	30.4
Canada	14.1	20	15.3
Ukraine	13.1	15	46.2
China	12.9	17	2.0
Germany	12.1	9	16.1
United Kingdom	9.9	18	18.1
Sweden	9.4	10	38.1

TABLE 20.1 Top Producers of Nuclear Power

2012 data, from the International Atomic Energy Agency, 2013.

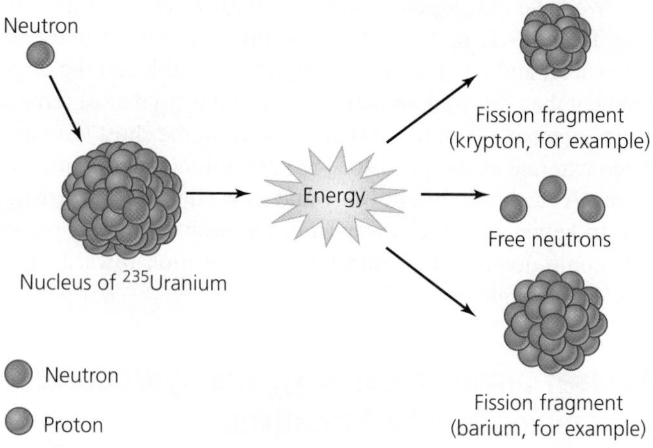

FIGURE 20.3 Nuclear fission drives modern nuclear power. In nuclear fission, the nucleus of an atom of uranium-235 is bombarded with a neutron. The collision splits the uranium atom into smaller atoms and releases two or three neutrons, along with energy in the form of heat, light, and radiation. The neutrons can continue to split uranium atoms and set in motion a runaway chain reaction, so engineers at nuclear plants must absorb excess neutrons with control rods to regulate the rate of the reaction.

First developed commercially in the 1950s, nuclear power experienced most of its growth during the 1970s and 1980s. The United States generates the most electricity from nuclear power—over a quarter of the world's production—yet only 19% of U.S. electricity comes from nuclear power. A number of other nations rely more heavily on nuclear power (TABLE 20.1). France leads the list, receiving 75% of its electricity from nuclear power.

Fission releases nuclear energy in reactors to generate electricity

Strictly defined, **nuclear energy** is the energy that holds together protons and neutrons (p. 24) within the nucleus of an atom. We harness this energy by converting it to thermal energy inside **nuclear reactors,** facilities contained within nuclear power plants. This thermal energy is then used to generate electricity.

The reaction that drives the release of nuclear energy inside nuclear reactors is **nuclear fission,** the splitting apart of atomic nuclei (FIGURE 20.3). In fission, the nuclei of large, heavy atoms, such as uranium or plutonium, are bombarded with neutrons. Ordinarily neutrons move too quickly to split nuclei when they collide with them, but if neutrons are slowed down they can break apart nuclei. Each split nucleus emits energy in the form of heat, light, and radiation, and also releases multiple neutrons. These neutrons (two to three in the case of uranium-235) can in turn bombard other nearby uranium-235 (^{235}U) atoms, resulting in a self-sustaining chain reaction.

If not controlled, this chain reaction becomes a runaway process of positive feedback (pp. 106–107)—the process that creates the explosive power of a nuclear bomb. Inside a nuclear power plant, however, fission is controlled so that, on average, only one of the two or three neutrons emitted with each fission event goes on to induce another fission event. In this way, the chain reaction maintains a constant output of energy at a controlled rate.

For fission to begin in a nuclear reactor, the neutrons bombarding uranium are slowed down with a substance called a *moderator*, most often water or graphite. As fission proceeds, it becomes necessary to soak up the excess neutrons produced when uranium nuclei divide, so that on average only a single neutron from each nucleus goes on to split another nucleus. For this purpose, *control rods*, made of a metallic alloy that absorbs neutrons, are placed into the reactor among the water-bathed fuel rods. Engineers move these control rods into and out of the water to maintain the fission reaction at the desired rate.

All this takes place within the reactor core and is the first step in the electricity-generating process of a nuclear power plant (FIGURE 20.4). The reactor core is housed within a reactor vessel, and the vessel, steam generator, and associated plumbing are often protected within a containment building. Containment buildings, with their meter-thick concrete and steel walls, are constructed to prevent leaks of radioactivity due to accidents or natural catastrophes such as earthquakes. Not all nations require containment buildings, which points out the key role that government regulation plays in protecting public safety.

Nuclear energy comes from processed and enriched uranium

We use the element uranium for nuclear power because it is radioactive. Radioactive isotopes, or *radioisotopes* (p. 24), emit subatomic particles and high-energy radiation as they decay into lighter radioisotopes until they ultimately become stable isotopes. The isotope uranium-235 decays into a series of daughter isotopes, eventually forming lead-207. Each radioisotope decays at a rate determined by that isotope's *half-life* (p. 24), the time it takes for half of the atoms to give off radiation and decay. The half-life of ^{235}U is about 700 million years.

We obtain uranium from various minerals in naturally occurring ore (*ore* is rock that contains minerals of economic

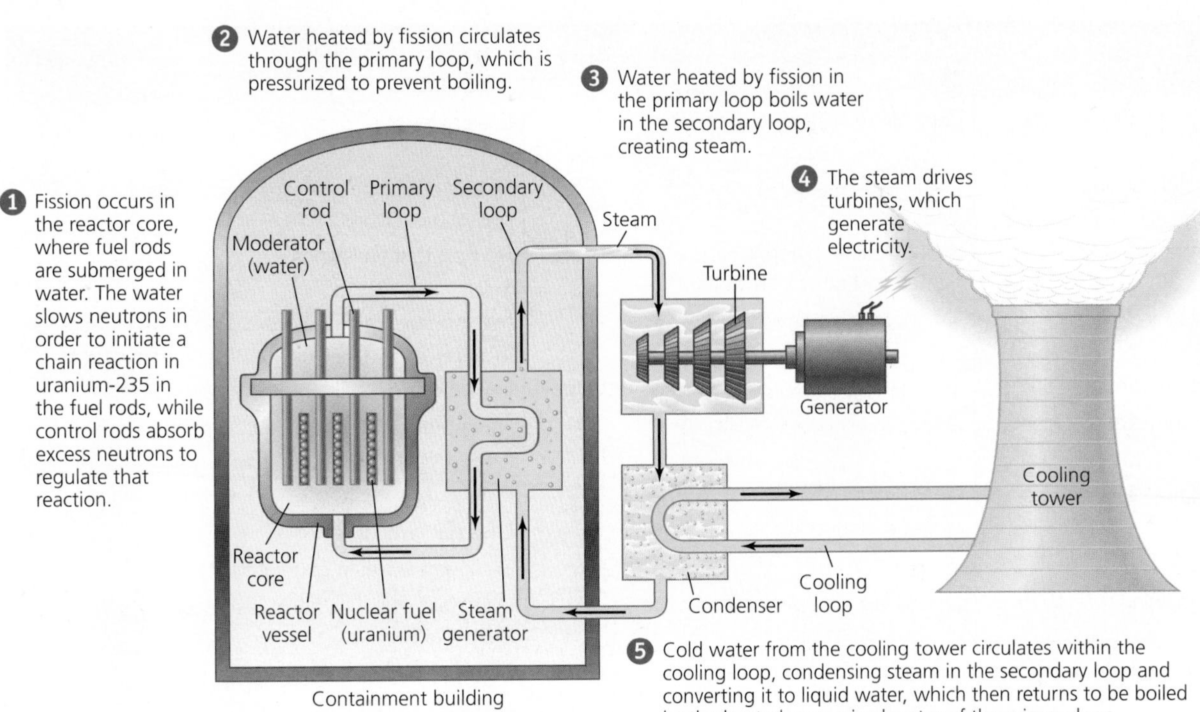

FIGURE 20.4 Nuclear reactors produce electricity. In a pressurized light water reactor (the most common type of nuclear reactor), uranium fuel rods are placed in water, which slows neutrons so that fission can occur 1. Control rods are moved into and out of the reactor core, absorbing excess neutrons to regulate the chain reaction. Water heated by fission circulates through the primary loop 2 and warms water in the secondary loop, which turns to steam 3. Steam drives turbines, which generate electricity 4. The steam is then cooled in the cooling tower by water from an adjacent river or lake and returns to the containment building 5, to be heated again by heat from the primary loop.

interest [p. 637]). Uranium-containing minerals are uncommon, and uranium ore is in finite supply, so nuclear power is generally considered a nonrenewable energy source.

Over 99% of the uranium in nature occurs as the isotope uranium-238. Uranium-235 (with three fewer neutrons) makes up less than 1% of the total. Because ^{238}U does not emit enough neutrons to maintain a chain reaction when fissioned, we use ^{235}U for commercial nuclear power. Therefore, we must process the ore we mine to enrich the concentration of ^{235}U to at least 3%. The enriched uranium is formed into pellets of uranium dioxide (UO_2), which are incorporated into metallic tubes called *fuel rods* (**FIGURE 20.5**) that are used in nuclear reactors.

After several years in a reactor, enough uranium has decayed so that the fuel no longer generates adequate energy, and it must be replaced with new fuel. In some countries, the spent fuel is reprocessed to recover the remaining usable energy. However, this process is costly relative to the low prices of uranium on the world market in recent years, so most spent fuel is disposed of as radioactive waste (pp. 564–566).

Nuclear power delivers energy more cleanly than fossil fuels

Using fission, nuclear power plants generate electricity without creating air pollution from stack emissions. In contrast, combusting fossil fuels can emit sulfur dioxide, which contributes to acid deposition; particulate matter, which threatens human health; and carbon dioxide and other greenhouse gases, which drive global climate change. Even considering all the steps involved in building plants and generating power, researchers from the International Atomic Energy Agency (IAEA) have calculated that nuclear power releases 4–150 times fewer emissions than fossil fuel combustion. Scientists estimate that

FIGURE 20.5 Enriched uranium fuel is packaged into fuel rods. These rods are encased in metal and used to power fission inside the cores of nuclear reactors. In this photo, fuel rods are being loaded into a circular arrangement beneath a pool of water inside a nuclear reactor in Brazil.

TABLE 20.2 Risks and Impacts of Coal-fired versus Nuclear Power Plants

TYPE OF IMPACT	COAL	NUCLEAR
Land and ecosystem disturbance from mining	Extensive, on surface or underground	Less extensive
Greenhouse gas emissions	Considerable emissions	None from plant operation; much less than coal over the entire life cycle
Other air pollutants	Sulfur dioxide, nitrogen oxides, particulate matter, and other pollutants	No pollutant emissions
Radioactive emissions	No appreciable emissions	No appreciable emissions during normal operation; possibility of emissions during severe accident
Occupational health among workers	More known health problems and fatalities	Fewer known health problems and fatalities
Health impacts on nearby residents	Air pollution impairs health	No appreciable known health impacts under normal operation
Effects of accident or sabotage	No widespread effects	Potentially catastrophic widespread effects
Solid waste	More generated	Less generated
Radioactive waste	None	Radioactive waste generated
Fuel supplies remaining	Should last several hundred more years	Uncertain; supplies could last longer or shorter than coal supplies

For each type of impact, the more severe impact is indicated in red.

nuclear power helps the United States avoid emitting 600 million metric tons of carbon dioxide each year, equivalent to the CO_2 emissions of almost all passenger cars in the nation and 11% of total U.S. CO_2 emissions. Worldwide, nuclear power avoids emissions of 2.5 billion metric tons of carbon dioxide per year, about 7% of global CO_2 emissions.

Nuclear power has additional advantages over fossil fuels—coal in particular. For residents living downwind from power plants, scientists calculate that nuclear power poses far fewer chronic health risks from pollution than does fossil fuel combustion. For instance, nuclear power prevents the emission of half a million tons of nitrogen oxide and 1.4 million tons of sulfur dioxide each year that might otherwise be generated by coal plants. And because uranium generates far more power than coal by weight or volume, less of it needs to be mined, so uranium mining causes less damage to landscapes and generates less solid waste than coal mining. Moreover, in the course of normal operation, nuclear power plants are safer for workers than are coal-fired plants.

Nuclear power also has drawbacks. One is that the waste it produces is radioactive, and arranging for safe disposal of this waste is challenging. The second main drawback is that if an accident occurs at a power plant, or if a plant is sabotaged, the consequences can potentially be catastrophic.

Given this mix of advantages and disadvantages (TABLE 20.2), most governments (although not necessarily most citizens) have judged the good to outweigh the bad, and today the world has 435 operating nuclear plants in 31 nations.

WEIGHING THE ISSUES

CHOOSE YOUR RISK Examine Table 20.2. Given the choice of living next to a nuclear power plant or living next to a coal-fired power plant, which would you choose? What would concern you most about each option?

Fusion remains a dream

For as long as scientists and engineers have generated power from nuclear fission, they have tried to figure out how to harness nuclear fusion instead. **Nuclear fusion**—the process that drives our sun's vast output of energy, and the force behind hydrogen bombs (thermonuclear bombs)—involves forcing together the small nuclei of lightweight elements under extremely high temperature and pressure. The hydrogen isotopes deuterium and tritium can be fused together to create helium, releasing a neutron and a tremendous amount of energy (FIGURE 20.6).

Overcoming the mutually repulsive forces of protons in a controlled manner is difficult, and fusion requires temperatures of many millions of degrees Celsius. Thus, researchers have not yet developed this process for commercial power generation. Despite billions of dollars of funding and decades of research, fusion experiments in the lab still require scientists to input more energy than they produce from the process. That is, they experience a loss in *net energy* (p. 523) and a ratio of *energy returned on investment* (*EROI*; pp. 523–524) lower than 1.

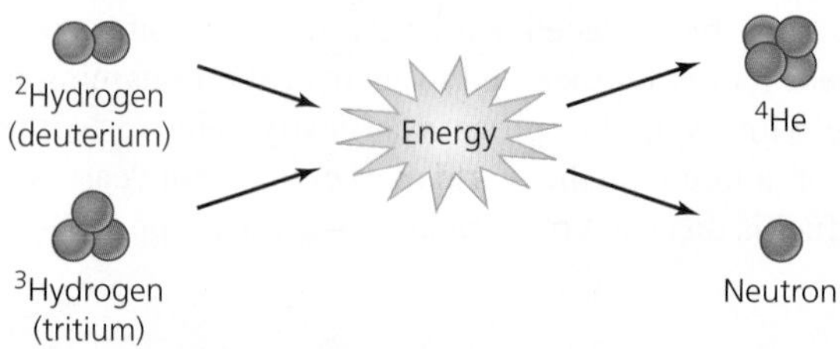

FIGURE 20.6 Can we harness nuclear fusion? In nuclear fusion, two small atoms, such as the hydrogen isotopes deuterium and tritium, are fused together, releasing energy along with a helium nucleus and a free neutron. So far, scientists have not been able to fuse atoms without supplying far more energy than the reaction produces, so this process is not used commercially.

FIGURE 20.7 **The Three Mile Island nuclear power plant near Harrisburg, Pennsylvania, suffered a partial meltdown in 1979.** This emergency was a "near-miss"—radiation was released but was mostly contained, and no health impacts were confirmed. The incident put the world on notice, however, that a major accident could potentially occur.

If we could find a way to control fusion in a reactor, the potential payoffs would be immense: We could produce vast amounts of energy using water as a fuel, and the process would create only low-level radioactive wastes, without pollutant emissions or the risk of dangerous accidents, sabotage, or weapons proliferation. A consortium of industrialized nations is collaborating to build a prototype fusion reactor called the International Thermonuclear Experimental Reactor (ITER) in southern France. It aims to achieve an EROI of 10:1. Even if this multi-billion-dollar effort succeeds, however, power from fusion seems likely to remain many years in the future.

Nuclear power poses small risks of large accidents

Although nuclear power delivers energy more cleanly than fossil fuels, the possibility of catastrophic accidents has spawned a great deal of public anxiety. Three events have been most influential in shaping public opinion about nuclear energy: Three Mile Island in the United States was a near-miss; Chernobyl, the world's most severe accident, shocked the world; and most recently, Fukushima Daiichi followed the 2011 Japanese earthquake and tsunami.

Three Mile Island At the **Three Mile Island** plant in Pennsylvania in 1979 (FIGURE 20.7), a combination of mechanical failure and human error caused coolant water to begin draining from the reactor vessel, temperatures to rise inside the reactor core, and metal surrounding the uranium fuel rods to start melting, releasing radiation. This process is termed a **meltdown,** and it proceeded through half of one reactor core at Three Mile Island. Residents of Harrisburg and nearby towns stood ready to be evacuated as the nation held its breath, but fortunately most radiation remained trapped inside the containment building.

The accident was brought under control within days, the damaged reactor was shut down, and multi-billion-dollar cleanup efforts stretched on for years. Three Mile Island is best regarded as a near-miss; the emergency could have been far worse had the meltdown proceeded through the entire stock of uranium fuel or had the containment building not contained the radiation. Although residents have shown no significant health impacts in the years since, the event raised safety concerns in the United States and abroad. It was Three Mile Island that inspired Sweden's referendum in 1980 that resulted in its national vote for a phaseout of nuclear power.

Chernobyl In 1986 an explosion at the **Chernobyl** plant in Ukraine (part of the Soviet Union at the time) caused the most severe nuclear power plant accident the world has yet seen (FIGURE 20.8A). Engineers had turned off safety systems to

FIGURE 20.8 **The world's worst nuclear accident unfolded in 1986 at Chernobyl.** The destroyed reactor **(a)** was later encased in a massive concrete sarcophagus to contain further radiation leakage. Technicians scoured the landscape **(b)**, measuring radiation levels, removing soil, and scrubbing roads and buildings.

(a) The destroyed reactor at Chernobyl

(b) Technicians measuring radiation

conduct tests, and human error, combined with unsafe reactor design, led to explosions that destroyed the reactor and sent clouds of radioactive debris billowing into the atmosphere. For 10 days radiation escaped from the plant while emergency crews risked their lives putting out fires (some later died from radiation exposure). Most residents of the surrounding countryside remained at home for these 10 days, exposed to radiation, before the Soviet government belatedly began evacuating more than 100,000 people.

In the months and years afterwards, workers erected a gigantic concrete sarcophagus around the demolished reactor, scrubbed buildings and roads, and removed irradiated materials FIGURE 20.8b). However, the landscape for at least 30 km (19 mi) around the plant remains contaminated, the demolished reactor is still full of dangerous fuel and debris, and radioactivity leaks from the hastily built and quickly deteriorating sarcophagus. Today an international team is trying to build a larger sarcophagus around the original one to prevent a catastrophic re-release of radiation.

The accident killed 31 people directly and sickened or caused cancer in thousands more. Exact numbers are uncertain because of inadequate data and the difficulty of determining long-term radiation effects (see THE SCIENCE BEHIND THE STORY, pp. 562–563). Health authorities estimate that most of the 6000-plus cases of thyroid cancer diagnosed in people who were children at the time resulted from radioactive iodine spread by the accident. Estimates for the total number of cancer cases attributable to Chernobyl, past and future, vary widely, but an international consensus effort 20 years after the event estimated that radiation may have raised the cancer rate among exposed people by as much as a few percentage points, possibly resulting in several thousand fatal cancer cases.

Atmospheric currents carried radioactive fallout from Chernobyl across much of the Northern Hemisphere, particularly Ukraine, Belarus, and parts of Russia and Europe (FIGURE 20.9). Fallout was greatest where rainstorms brought radioisotopes down from the radioactive cloud. Parts of Sweden received high amounts of fallout, and the accident reinforced the Swedish public's fears about nuclear power. A poll taken after the event showed that nearly half of Swedish citizens now regretted their own nation's investment in nuclear power.

Fukushima Daiichi On March 11, 2011, a magnitude 9.0 earthquake struck eastern Japan and sent an immense tsunami roaring onshore (pp. 22–23). Over 20,000 people were killed, and many thousands of buildings were destroyed. This natural disaster affected the operation of several of Japan's nuclear plants, most notably the **Fukushima Daiichi** nuclear power plant. Here, the earthquake shut down power and the tsunami flooded the plant's emergency power generators. The plant was protected by a 5.7-m (19-ft) seawall, but the tsunami reached 14 m (46 ft) high, and the generators were located in the basement of the plant (FIGURE 22.10a). Without electricity, workers could not use moderators and control rods to cool the uranium fuel, and the fuel began to overheat as fission proceeded, uncontrolled.

Amid the damage and chaos across the region, help was slow to arrive, and workers had to begin flooding the reactors with seawater in a desperate effort to prevent meltdowns. Several explosions and fires occurred over the next few days, and eventually three reactors experienced full meltdowns, while the plant's other three reactors were seriously damaged. Parts of the plant remained inaccessible for months because of radioactive water, and it will likely require decades to fully clean up the site.

Radioactivity was released during and after these events at levels about one-tenth of those from Chernobyl. Much of the radioactivity spread by air or water into the Pacific Ocean, and trace amounts were detected around the world (FIGURE 20.10b). Thousands of residents of areas near the plant were evacuated and screened for radiation effects (FIGURE 22.10c), and restrictions were placed on food and water from the region. Over two years after the event, minor releases of radioactivity continued, and TEPCO (Tokyo Electric Power Company)—the

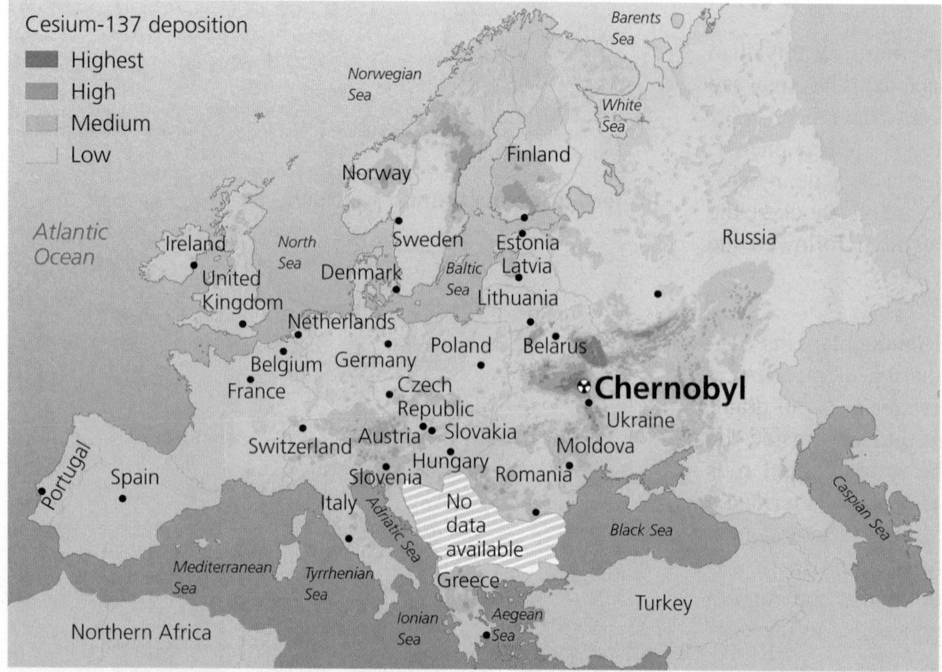

FIGURE 20.9 Radioactive fallout from Chernobyl was deposited across Europe in complex patterns. Patterns of cesium-137 deposition resulted from atmospheric currents and rainstorms in the days following the accident. Although Chernobyl produced 100 times more fallout than the U.S. bombs dropped on Hiroshima and Nagasaki in World War II, it was distributed over a much wider area. Thus, levels of contamination in any given place outside of Ukraine, Belarus, and western Russia were relatively low; during these several days, the average European received less than the amount of radiation a person receives naturally in a year. *Data from chernobyl.info, Swiss Agency for Development and Cooperation, Bern, 2005.*

(a) The tsunami barrels toward the Fukushima reactors

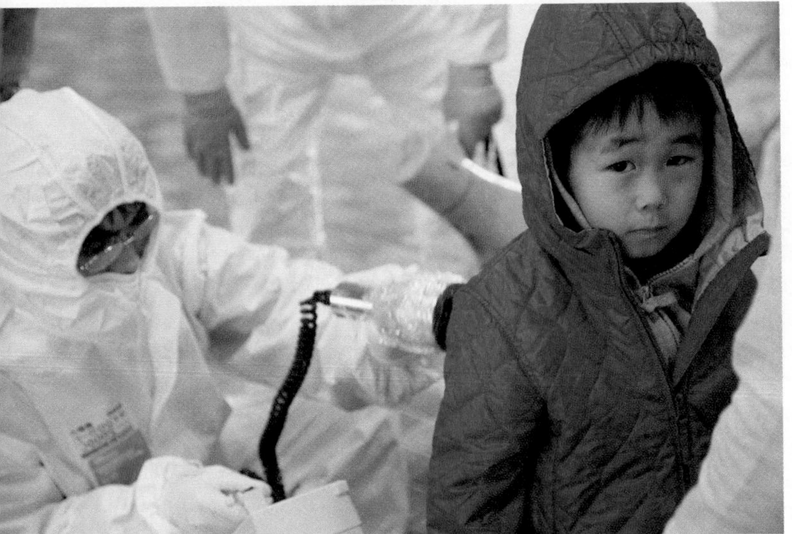

(c) A Japanese child is screened for radiation

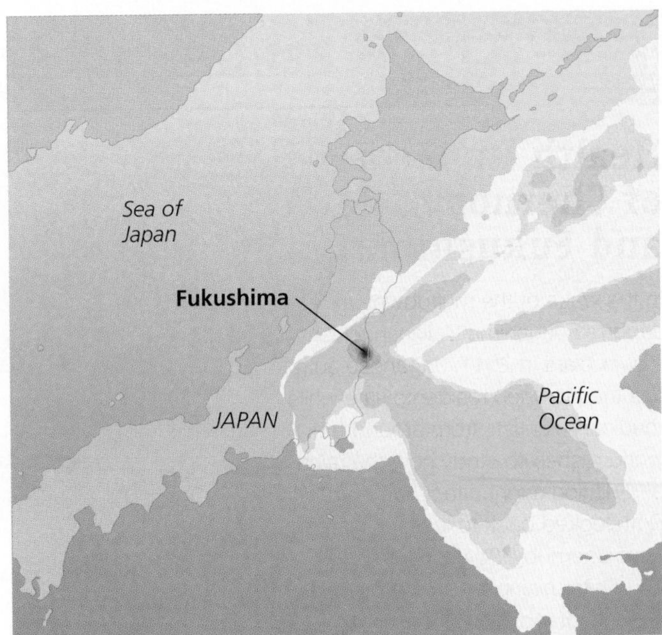

Spread of radiation fallout outward from the damaged plant site

0.9–1 | 0.4–0.6 | 0.01–0.1
0.7–0.9 | 0.3–0.4 | 0.005–0.01
0.6–0.7 | 0.1–0.3 | 0.001–0.005

(b) Most radiation drifted eastward over the ocean

FIGURE 20.10 The Fukushima Daiichi crisis was unleashed after an earthquake generated a massive tsunami. The tsunami tore through a seawall **(a)** and inundated the plant's nuclear reactors. About 80% of the radiation that escaped from the plant drifted over the ocean, as shown in this map **(b)** of cesium-137 isotopes in the 9 days following the accident. Children evacuated from the region **(c)** were screened for radiation exposure. *Data in (b) from Yasunari, T.J., et al. 2011. Cesium-137 deposition and contamination of Japanese soils due to the Fukushima nuclear accident.* Proc. Natl. Acad. Sci. *108: 19530–19534.*

plant's owner—continued to ascertain what precisely went wrong. Long-term health effects on the region's people remain uncertain and debated (see **The Science behind the Story,** pp. 562–563).

Amid the extraordinary challenges of responding to the earthquake, tsunami, and nuclear crisis, the Japanese government and TEPCO were criticized for not sharing full and accurate information with the public. This, combined with the fact that TEPCO had failed to respond to earlier warnings that the plant was vulnerable to a large tsunami, shook the Japanese people's confidence in their leaders and in nuclear power. In the aftermath of the disaster, the government idled all 50 of the nation's nuclear reactors and embarked on safety inspections. In 2012, the government restarted a number of plants, arguing that electricity from them was necessary to avoid summertime blackouts. In response, large crowds of people demonstrated in protest, and some urged a national referendum on phasing out nuclear power. Across the world, many nations reassessed their nuclear programs, and Germany, Belgium, and Spain proposed to phase out nuclear power.

We are managing risks from nuclear power and nuclear weapons with some success

It is fortunate that we have not experienced more accidents on the scale of Fukushima or Chernobyl. Yet smaller-scale incidents have occurred. A 1999 accident at a plant in Tokaimura, Japan, killed two workers and exposed over 400 others to leaked radiation. And Sweden experienced a near-miss in 2006, when the Forsmark plant north of Stockholm narrowly avoided a meltdown after only two of four generators started up following a power outage. In 2011 and 2012, U.S. reactors had nine emergency shutdowns in response to tornadoes, hurricanes, earthquakes, and flooding.

Thankfully, the designs of most modern reactors are safer than Chernobyl's, and designs for future plants promise more safety features. And in most emergencies at reactors around the world, safety systems have functioned well. For instance, Japan's Onagawa power plant was closer to the epicenter of the 2011 quake, yet its safety systems protected it from serious damage.

However, as plants around the world age, they require more maintenance and become less safe. There is also the

Health Impacts of Chernobyl and Fukushima

In the wake of the meltdowns at Japan's Fukushima Daiichi nuclear power plant in 2011, Japanese authorities tried to keep residents safe while medical scientists from around the world rushed to study how the release of radiation might affect human health. Both looked back to lessons learned from Chernobyl 25 years earlier.

Determining long-term health impacts of radiation exposure is enormously difficult, and initial attempts to predict Fukushima's health impacts have stirred vigorous debate. Most scientists expect that the reactor failures at Fukushima will have less severe health consequences than those at Chernobyl, because less radiation was emitted at Fukushima and because most of it drifted over the ocean away from populated areas (see Figure 20.10b and pp. 26–27). However, judging Fukushima's impacts will be challenging and will require long-term study.

Let's turn first to Chernobyl. The hundreds of researchers who have tried to pin down Chernobyl's health impacts have sometimes differed in their conclusions. In an effort to reach consensus, the World Health Organization (WHO) engaged 100 experts to review all studies through 2006 and issue a report summarizing what scientists had learned in the 20 years since the accident. In 2008, a United Nations committee issued a comprehensive report as well.

The most severe effects were documented in emergency workers who battled to contain the incident in its initial days. Medical staff treated and recorded the progress of 134 workers hospitalized with acute radiation sickness (ARS). Radiation destroys cells in the body, and if the destruction outpaces the body's abilities to repair the damage, the person will soon die. Symptoms of ARS include vomiting, fever, diarrhea, thermal burns, mucous membrane damage, and weakening of the immune system. In total, 28 people died from ARS soon after the accident. Those who died had the greatest estimated exposure to radiation.

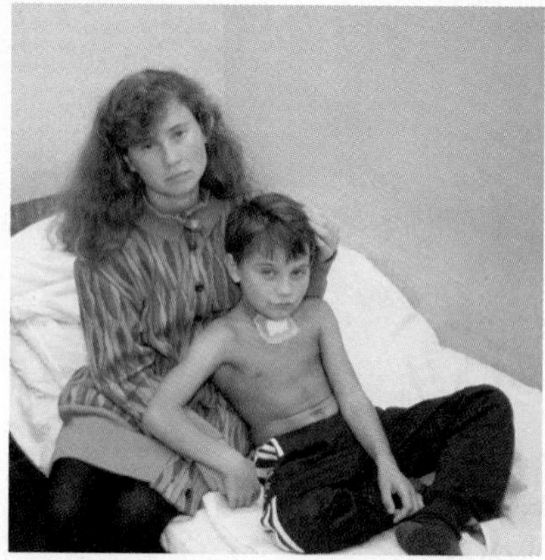

Chernobyl-area boy and his mother after his surgery for thyroid cancer.

The major health consequence of Chernobyl's radiation, however, has been thyroid cancer in children. The thyroid gland is where our bodies concentrate iodine, and one of the most common radioactive isotopes released early in the disaster was iodine-131 (^{131}I). Children have large and active thyroid glands, so they are especially vulnerable to thyroid cancer induced by radioisotopes.

Predicting that thyroid cancer might be a problem, medical workers measured iodine activity in the thyroid glands of several hundred thousand people in Russia, Ukraine, and Belarus following the accident. They also measured food contamination and surveyed people on their food consumption. These data showed that drinking milk from cows that had grazed on contaminated grass was the main route of exposure to ^{131}I, although fresh vegetables also contributed.

As doctors had feared, rates of thyroid cancer rose among children in regions of highest exposure (**FIGURE 1**). Multiple studies found linear dose-response relationships (p. 376) in data from Ukraine and Belarus. By 2006, medical professionals estimated the number of cases at 6000 and rising. Fortunately, treatment of thyroid cancer has a high success rate, so as of that time, only 15 children had died from it.

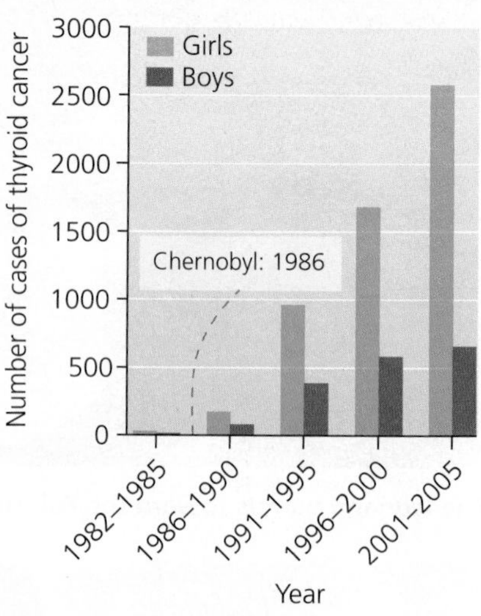

FIGURE 1 The incidence of thyroid cancer in girls and boys below age 18 began rising in regions of Ukraine, Belarus, and Russia that experienced the heaviest fallout of radioactive iodine from Chernobyl. *Data from U.N. Scientific Committee on the Effects of Atomic Radiation, 2008.* Sources and effects of ionizing radiation, Vol. 2. *New York, 2011.*

Studies addressing other health aspects of the accident have found limited impact. Some research has shown an increase in cataracts due to radiation, especially among emergency workers. But neither the WHO nor U.N. assessments found evidence that rates of leukemia or any other cancer (aside from thyroid cancer) had risen among people exposed to Chernobyl's radiation. Still, some cancers may appear decades after exposure, so it is possible that many illnesses have yet to arise. Moreover, conducting epidemiological studies (pp. 375–376) with enough statistical power to detect minor increases in rare events requires observing huge numbers of people over many years.

When the Fukushima crisis struck, Japanese authorities applied lessons from Chernobyl. They provided iodine tablets to young people evacuated from the region and they restricted agriculture near the plant, stopping contaminated food and milk from entering the market. These measures to prevent people's uptake of iodine-131 apparently were effective. In late March 2011, 1000 evacuated children were tested for ^{131}I exposure, and none showed evidence of a high dose affecting the thyroid.

As the crisis was gradually brought under control, some researchers began predicting what long-term health impacts might arise. Using data on cancer rates from studies on survivors of the atomic bombs dropped on Hiroshima and Nagasaki, researchers extrapolated to predict cancer rates at the lower radiation doses from Fukushima. Frank Von Hippel in a 2011 study calculated that about 1 million people were exposed to more than 1 curie per km^2 of radiation from cesium-137, which should result in a 0.1% increase in cancer risk among them, or 1000 cancer cases.

In 2012, John Ten Hoeve and Mark Jacobson of Stanford University used a global atmospheric model together with data on radiation, population dispersion, and weather in the days following the event to estimate radiation levels people received (**FIGURE 2**). In the journal *Energy and Environmental Science*, they predicted that Fukushima's radiation would eventually produce 125 cancer-related deaths and 178 non-fatal cancer cases. Large uncertainties attended both estimates; the projected number of total cases ranged from 39 to 2900.

Ten Hoeve and Jacobson then used their methods to predict the consequences of a hypothetical Fukushima-scale event taking place in the United States. They simulated Fukushima's radiation releases as though they had occurred at the Diablo Canyon Power Plant in California, which lies on a major earthquake fault. Their computer simulations indicated that because of different weather patterns, such an accident might cause 25% more

(a) After 36 hours

(b) After 180 hours

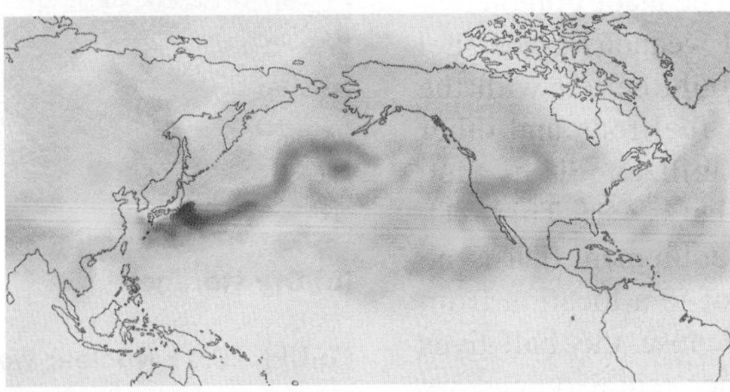

(c) After 324 hours

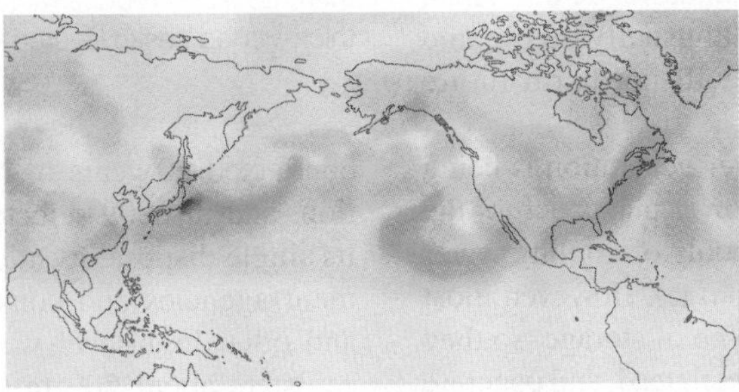

(d) After 468 hours

FIGURE 2 Radiation from cesium-137 spread around the world from Fukushima following the accident. Atmospheric modeling shows how this radioisotope spread for three weeks after the accident. *Data from Ten Hoeve, J., and M. Jacobson, 2012. Worldwide health effects of the Fukushima Daiichi nuclear accident.* Energy & Environmental Science, *DOI 10.1039/c2ee22019a.*

deaths than in Japan, despite a much lower population density.

To generate their estimates, Ten Hoeve and Jacobson used a dose-response curve (p. 377) that extrapolates effects from high to low doses in a linear manner. However, some researchers believe there could be a threshold below which radiation has no health effects. These researchers felt that Ten Hoeve and Jacobson's cancer estimates were likely too high.

Meanwhile, their Stanford colleague Burton Richter wrote to the journal that—far from suggesting that nuclear power is dangerous—the number of fatalities from Fukushima is far less than Japan would have experienced from air pollution had its electricity been produced by fossil fuels instead!

Time will tell what health impacts the Fukushima crisis causes. It will prove difficult to measure statistically rare instances of cancer in a large population exposed to very low doses, but researchers will try. They will be assisted by Japan's government, which budgeted $1.2 billion for coordinated research into long-term health effects of radiation.

Questionnaires were sent to all 2 million residents of Fukushima Prefecture, asking where they were in the days after the accident, and what they ate and drank. Thyroid exams are being given to all 360,000 children and teens from the region, and 20,000 pregnant women and their babies will be closely monitored. All 200,000 people evacuated from the area will get exams, and mental health support will be offered. If the research program can extend for 30 years, as the government intends, it could provide valuable information on the health effects of low-dose radiation. ■

concern that radioactive material could be stolen from plants and used in terrorist attacks. This possibility has been especially worrisome in the cash-strapped nations of the former Soviet Union, where hundreds of former nuclear sites have gone without adequate security for years. Finally, there is the ever-present concern that more nations may develop nuclear weapons.

To address concerns about stolen fuel and to reduce the world's nuclear weapons stockpiles, the United States and Russia embarked on a remarkably successful program called *Megatons to Megawatts*. In this cooperative international agreement, the United States has been buying up weapons-grade uranium and plutonium from Russia, letting Russia process it into lower-enriched fuel, and diverting it to peaceful use in power generation. In recent years, up to 10% of America's electricity has been generated from fuel recycled from Russian warheads that used to be atop missiles pointed at American cities! In 2013 it is expected that the last of 500 metric tons of highly enriched uranium will be processed and transferred, after which Russia and the United States may negotiate some sort of continuation of the program.

Waste disposal remains a challenge

Even if nuclear power generation could be made completely safe, and even if we could recycle all weapons-grade fuel into fuel for power plants, we still would be left with the conundrum of what to do with spent fuel rods and other radioactive waste. Recall that fission utilizes ^{235}U as fuel, leaving as waste the 97% of uranium that is ^{238}U. This ^{238}U, as well as all irradiated material and equipment that is no longer being used, must be disposed of in a location from which radiation will not escape. Because the half-lives of uranium, plutonium, and many other radioisotopes are far longer than multiple human lifetimes, this waste will continue emitting radiation for thousands of years. Thus, radioactive waste must be placed in unusually stable and secure locations where radioactivity will not harm future generations.

Currently, nuclear waste from power generation is being held in temporary storage at nuclear power plants across the world. Spent fuel rods are sunken in pools of cooling water to minimize radiation leakage (**FIGURE 20.11a**). However, most U.S. plants have no room left for this type of storage, so they are now storing waste in thick casks of steel, lead, and concrete (**FIGURE 20.11b**). In total, U.S. power plants are storing nearly 70,000 metric tons of high-level radioactive waste—enough to fill a football field to the depth of 7 m (21 ft)—as well as much more low-level radioactive waste. This waste is held at more than 120 sites spread across 39 states (**FIGURE 20.12**). A 2005 report from the National Academy of Sciences judged that most of these sites were vulnerable to terrorist attacks. Over 161 million U.S. citizens live within 125 km (75 mi) of temporarily stored waste.

Because storing waste at many dispersed sites creates a large number of potential hazards, nuclear waste managers would prefer to send all waste to a central repository that can be heavily guarded. In Sweden, that nation's nuclear industry conducted 20 years of research looking for a suitable location, and in 2009 selected the Forsmark power plant site as its single disposal location. If the site is approved by government agencies and constructed as planned, spent fuel rods and other high-level waste will be systematically buried in canisters about 500 m (1650 ft) underground within stable bedrock.

(a) Wet storage

(b) Dry storage

FIGURE 20.11 Nuclear waste is stored at nuclear power plants, because no central repository yet exists. Spent fuel rods are kept in "wet storage" in pools of water **(a)**, which keep them cool and reduce radiation release, or in "dry storage" **(b)** in thick-walled casks layered with lead, concrete, and steel.

In the United States, the multiyear search homed in on Yucca Mountain, a remote site in the desert of southern Nevada, 160 km (100 mi) from Las Vegas (**FIGURE 20.13a**). Choice of this site followed extensive study by government scientists (**FIGURE 20.13b**), and $13 billion was spent on its development, although most Nevadans were not happy about the choice. In 2010, as the site was awaiting approval from the Nuclear Regulatory Commission, President Barack Obama's administration ended support for the project. Ironically this came just days after Obama had urged expanding nuclear power in his State of the Union address. Most political observers agree that the opposition of Senate

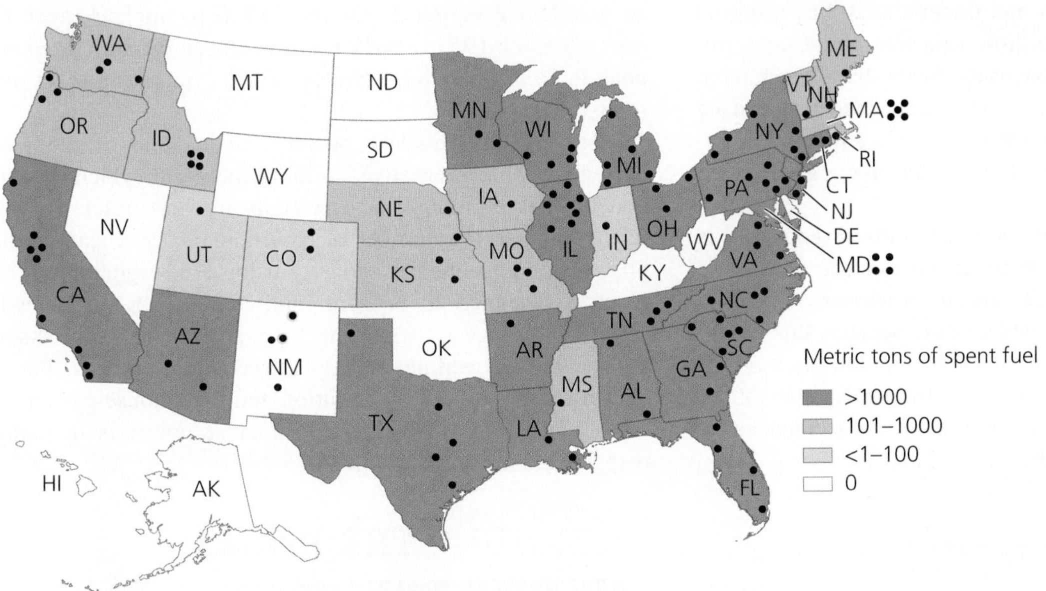

FIGURE 20.12 High-level radioactive waste from civilian reactors is currently stored at over 120 sites in 39 states across the United States. In this map, dots indicate storage sites and colors indicate the amount of waste stored in each state. *Data from Nuclear Energy Institute.*

Majority Leader Harry Reid, who represents Nevada, was a key reason for the change in policy. As of 2013, the issue remains unresolved, as a number of lawsuits are in progress challenging this decision. Without Yucca Mountain, the United States has no place designated to dispose of its radioactive waste from commercial nuclear power plants, so this waste will remain at its numerous current locations across the country.

At Yucca Mountain, waste would be stored in a network of tunnels 300 m (1000 ft) underground, yet 300 m (1000 ft) above the water table (**FIGURE 20.13c**). Scientists and policymakers chose the Yucca Mountain site because they

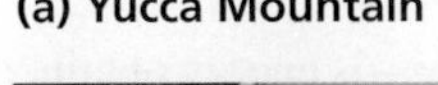

(a) Yucca Mountain

(b) Scientific testing

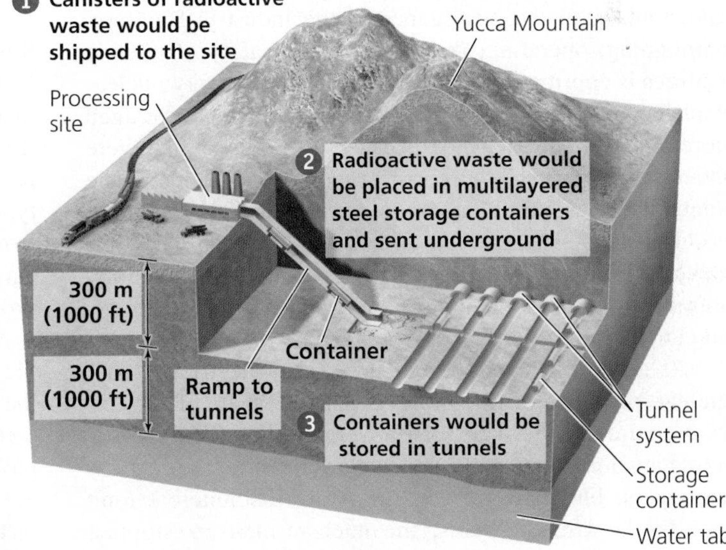

(c) Proposed design

FIGURE 20.13 Yucca Mountain has not been approved. This site **(a)**, in a remote part of Nevada, was being developed as the central repository for all commercial nuclear waste in the United States until support was withdrawn in 2010. Here **(b)**, technicians test the effects of extreme heat from radioactive decay on the stability of rock. Waste was to be buried **(c)** in a network of tunnels deep underground, yet still high above the water table.

determined that it is remote and unpopulated, has minimal risk of earthquakes, receives little rain that could contaminate groundwater with radioactivity, has a deep water table atop an isolated aquifer, and is on federal land that can be protected from sabotage. However, some scientists, antinuclear activists, and concerned Nevadans have challenged these conclusions.

Another concern with any other centralized repository is that nuclear waste will need to be transported there from the 120-some current storage areas and from current and future nuclear plants and military installations. Because this would involve many thousands of shipments by rail and truck across hundreds of public highways through almost every state of the union, many people worry that the risk of an accident or of sabotage is unacceptably high.

WEIGHING THE ISSUES

HOW TO STORE WASTE? Which do you think is a better option—to transport nuclear waste cross-country to a single repository or to store it permanently at numerous power plants and military bases scattered across the nation? Would your opinion be affected if you lived near the repository site? Near a power plant? On a highway route along which waste is transported?

Multiple dilemmas have slowed nuclear power's growth

Dogged by concerns over waste disposal, safety, and expensive cost overruns, nuclear power's growth has slowed. Since the late 1980s, nuclear power has grown by 2.5% per year worldwide, about the same rate as electricity generation overall. Public anxiety in the wake of Chernobyl made utilities less willing to invest in new plants, and reaction to Fukushima stalled a resurgence in the industry. Building, maintaining, operating, and ensuring the safety of nuclear facilities is enormously expensive, and almost every nuclear plant has overrun its budget. In addition, plants have aged more quickly than expected because of problems that were underestimated, such as corrosion in coolant pipes. The plants that have been shut down—well over 100 around the world to date—have served on average less than half their expected lifetimes. Moreover, shutting down, or decommissioning, a plant is sometimes more expensive than the original construction.

As a result of these financial issues, electricity from nuclear power remains more expensive than electricity from coal and other sources. Governments are still subsidizing nuclear power to keep electricity prices down for ratepayers, but many private investors lost interest long ago. In the United States, the nuclear industry stopped building plants following Three Mile Island, and public opposition scuttled many that were under construction. The $5.5-billion Shoreham Nuclear Power Plant on New York's Long Island was shut down just two months after being licensed because officials determined that evacuation would be impossible in this densely populated area should an accident ever occur. Of the 259 U.S. nuclear reactors ordered since 1957, nearly half have been cancelled. At its peak in 1990, the United States had 112 operating reactors; today it has 104.

Nonetheless, nuclear power remains one of the few currently viable alternatives to fossil fuels with which we can generate large amounts of electricity in short order. This is why an increasing number of environmental advocates propose expanding nuclear capacity using a new generation of reactors designed to be safer and less expensive. Indeed, nuclear power was beginning to experience a renaissance before the Fukushima tragedy raised new concerns. For a nation wishing to cut its pollution and greenhouse gas emissions quickly and substantially, nuclear power is in many respects the leading option.

WEIGHING THE ISSUES

MORE NUCLEAR POWER? Sweden and many other European nations have reduced or slowed their carbon emissions by expanding their use of nuclear power to replace fossil fuels. Do you think the United States should expand its nuclear power program? Why or why not?

With slow growth expected for nuclear power, fossil fuels in limited supply, and climate change worsening, where will we turn for clean and sustainable energy? Increasingly, people are turning to renewable sources of energy: energy sources that cannot be depleted by our use. Many promising renewable sources are still early in their development (Chapter 21), but two of them—bioenergy and hydroelectric power—are already well developed and widely used.

Bioenergy

Bioenergy—also known as **biomass energy**—is energy obtained from biomass resources. **Biomass** (p. 82) consists of organic material derived from living or recently living organisms, and it contains chemical energy (p. 30) that originated with sunlight and photosynthesis. We harness bioenergy from many types of plant matter, including wood from trees, charcoal from wood charred in the absence of oxygen, and matter from agricultural crops, as well as from combustible animal waste products such as cattle manure.

The great attraction of bioenergy is that—in principle—it is renewable and releases no net carbon dioxide into the atmosphere. Although burning biomass emits plenty of carbon, this is balanced by the fact that photosynthesis had pulled this amount of carbon from the atmosphere to create the biomass just years, months, or weeks before. Therefore, in theory, when we replace fossil fuels with bioenergy, we reduce net carbon flux to the atmosphere, helping to alleviate global climate change (Chapter 18). However, in practice it is not so simple, and judging the sustainability of any given bioenergy strategy requires careful consideration of which type of biomass source we are using and how we gain energy from it.

We gain bioenergy from many sources

To a poor farmer in Africa, bioenergy entails cutting wood from trees or collecting livestock manure by hand and burning it to heat and cook for her family. To an industrialized farmer in Iowa, bioenergy means shipping his grain to a hi-tech refinery that converts it to liquid fuel to run automobiles. The diversity of sources and approaches involved in bioenergy (TABLE 20.3) gives us many ways to address our energy challenges.

Over 1 billion people use wood from trees as their principal energy source. In developing nations, especially in rural areas, families gather fuelwood to burn in their homes for heating, cooking, and lighting (FIGURE 20.14; also see Figure 17.32, p. 476). Although fossil fuels are replacing traditional energy sources as developing nations industrialize, fuelwood, charcoal, and manure still account for one-third of energy use in these nations—and up to 90% in the poorest nations.

Fuelwood and other traditional biomass sources constitute three-quarters of all renewable energy used worldwide. However, biomass is renewable only if it is not overharvested. Harvesting fuelwood at unsustainably rapid rates can lead to deforestation, soil erosion, and desertification (pp. 311, 222, 223), thereby damaging landscapes, diminishing biodiversity, and impoverishing human societies. Heavily populated arid regions that support meager woodlands are most vulnerable to overharvesting, and these include many regions of Africa and Asia. Another drawback of burning fuelwood and other biomass for cooking and heating is that it leads to health hazards from indoor air pollution (pp. 475–476).

Although much of the world still relies on fuelwood, charcoal, and manure, new bioenergy approaches are being developed using a variety of materials to provide innovative types of energy (see Table 20.3). Some of these biomass sources can be burned in power plants to produce **biopower,** generating heat and electricity. Other sources can be converted into **biofuels,** liquid fuels used primarily to power automobiles. Because many of these novel biofuels and biopower strategies depend on technologies resulting from extensive research and development, they are being developed primarily in wealthier industrialized nations, such as Sweden and the United States.

FIGURE 20.14 **Well over a billion people in developing countries rely on wood from trees for heating and cooking.** In theory, biomass is renewable, but in practice it may not be if forests are overharvested.

TABLE 20.3
Direct combustion for heating
• Wood cut from trees (fuelwood)
• Charcoal
• Manure from farm animals
Biofuels for powering vehicles
• Corn grown for ethanol
• Bagasse (sugarcane residue) grown for ethanol
• Soybeans, rapeseed, and other crops grown for biodiesel
• Used cooking oil for biodiesel
• Plant matter treated with enzymes to produce cellulosic ethanol
• Algae grown for biofuels
Biopower for generating electricity
• Crop residues (such as cornstalks) burned at power plants
• Forestry residues (such as wood waste from logging) burned at power plants
• Processing wastes (such as solid or liquid waste from sawmills, pulp mills, and paper mills) burned at power plants
• "Landfill gas" burned at power plants
• Livestock waste from feedlots for gas from anaerobic digesters
• Organic components of municipal solid waste from landfills

Biopower generates electricity from biomass

We harness biopower by combusting biomass to generate electricity in the same way that we burn coal for power (see Figure 19.10, p. 529). This can be done using a variety of sources and techniques.

Waste products Some waste products can be used as sources for biopower. The forest products industry generates large amounts of woody debris in logging operations and at sawmills, pulp mills, and paper mills (FIGURE 20.15). Sweden's efforts to promote bioenergy have focused largely on using forestry residues. Because so much of the nation is forested and the timber industry is a major part of the Swedish economy, plenty of forestry waste is available.

Other waste sources used for biopower include residue from agricultural crops (such as cornstalks and corn husks), animal waste from feedlots, and organic waste from municipal landfills. The anaerobic bacterial breakdown of waste in landfills produces methane and other components, and this "landfill gas" is being captured and sold as fuel (p. 616). Methane and other gases can also be produced in a more controlled way in anaerobic digestion facilities. This *biogas* can then be burned in a power plant's boiler to generate electricity.

FIGURE 20.15 **Forestry residues are a major source of material for biopower in some regions.** Here a Swedish logging operation gathers woody residue.

Bioenergy crops We are beginning to grow certain types of plants as crops to generate biopower. These include fast-growing grasses such as bamboo, fescue, and switchgrass, as well as trees such as specially bred willows and poplars (FIGURE 20.16). Many of these plants are also being grown to produce liquid biofuels.

Combustion strategies Power plants built to combust biomass operate like those fired by fossil fuels; combustion heats water, creating steam to turn turbines and generators, thereby generating electricity. Much of the biopower produced so far comes from power plants that use cogeneration (p. 546) to generate both electricity and heating. These plants are often located where they can take advantage of forestry waste.

In some coal-fired power plants, wood chips, wood pellets, or other biomass is introduced with coal into a high-efficiency boiler in a process called *co-firing*. We can substitute biomass for up to 15% of the coal with only minor equipment modification and no appreciable loss of efficiency. Co-firing is a relatively easy and inexpensive way for utilities to expand their use of renewable energy.

We also harness biopower through *gasification* (p. 537), in which biomass is vaporized at extremely high temperatures in the absence of oxygen, creating a gaseous mixture including hydrogen, carbon monoxide, carbon dioxide, and methane. This mixture can generate electricity when used to turn a gas turbine to propel a generator in a power plant. We can also treat gas from gasification in various ways to produce methanol, synthesize a type of diesel fuel, or isolate hydrogen for use in hydrogen fuel cells (pp. 603–604). An alternative method of heating biomass in the absence of oxygen results in *pyrolysis* (p. 527), which produces a mix of solids, gases, and liquids. This includes a liquid fuel called pyrolysis oil, which can be burned to generate electricity.

Scales of production At small scales, farmers, ranchers, or villages can operate modular biopower systems that use livestock manure to generate electricity. Small household biodigesters provide portable and decentralized energy production for remote rural areas. At large scales, industries such as the forest products industry are using their waste to generate power, and industrialized farmers are growing bioenergy crops. In Sweden, over one-fifth of the nation's energy supply now comes from biomass, and biomass provides more fuel for electricity generation than coal, oil, or natural gas. Pulp mill liquors are the main source, but solid wood waste, municipal solid waste, and biogas from digestion are all used. In the United States, several dozen biomass-fueled power plants are now operating, and several dozen coal-fired plants are experimenting with co-firing.

Benefits and drawbacks By enhancing energy efficiency and recycling waste products, biopower helps move our utilities and industries in a sustainable direction. The U.S. forest products industry now obtains over half its energy by combusting the waste it recycles, including woody waste and liquor from pulp mill processing. Biopower also helps mitigate climate change by reducing carbon dioxide emissions, and capturing landfill gas reduces emissions of methane, a potent greenhouse gas. Relative to fossil fuels, biopower also benefits human health. By replacing coal in co-firing and direct combustion, biopower reduces emissions of sulfur dioxide because plant matter, unlike coal, contains no appreciable sulfur content. In addition, biomass resources tend to be geographically widespread and well dispersed, so using them

FIGURE 20.16 **These hybrid poplars, specially bred for fast growth in dense plantations, are harvested for biopower.** This is renewable energy, but it has environmental impacts: These monocultural plantations may replace natural systems over large areas and do not function ecologically as forests.

can help support rural economies and reduce many nations' dependence on imported fuels.

A disadvantage of biomass power is that when we burn crops or plant matter for power, we deprive the soil of the nutrients it would have gained from the plant matter's decomposition. We essentially draw fertility from the soil and never return it, so that the soil becomes progressively depleted. This also is the case when we burn crops or plant matter as a fuel, as discussed next. The depletion of soil fertility is a major long-term problem for bioenergy and is one reason that relying solely on bioenergy is not a sustainable option.

Ethanol can power automobiles

Liquid fuels from biomass sources are powering millions of vehicles on today's roads. The two primary biofuels developed so far are ethanol (for gasoline engines) and biodiesel (for diesel engines).

Ethanol is the alcohol in beer, wine, and liquor. It is produced as a biofuel by fermenting biomass, generally from carbohydrate-rich crops, in a process similar to brewing beer. In fermentation, carbohydrates are converted to sugars and then to ethanol. Spurred by the 1990 Clean Air Act amendments and generous government subsidies, ethanol is widely added to gasoline in the United States to reduce automotive emissions. In 2012 in the United States, over 50 billion L (13.3 billion gal) of ethanol were produced, mostly from corn (FIGURE 20.17). This amount has grown rapidly, and more than 200 U.S. ethanol production facilities are now operating. Further growth is assured, as the Energy Independence and Security Act passed by Congress in 2007 mandates production and use of 136 billion L (36 billion gal) per year of ethanol by 2022.

Any vehicle with a gasoline engine runs well on gasoline blended with up to 10% ethanol, but more and more vehicles are being produced that can run primarily on ethanol. Sweden has many such public buses, and the U.S. "big three" automakers are now producing *flexible-fuel vehicles* that run on E-85, a mix of up to 85% ethanol and 15% gasoline. Over 9 million such cars are on U.S. roads today. Most gas stations do not yet offer E-85 (FIGURE 20.18), so drivers often fill these cars with conventional gasoline, but this situation is changing as infrastructure for ethanol increases.

In Brazil, sugarcane residue is crushed to make *bagasse*, a material that is then used to make ethanol. Half of all new Brazilian cars are flexible-fuel vehicles, and ethanol from sugarcane accounts for 40% of all automotive fuel that Brazil's drivers use.

(a) Corn grown for ethanol

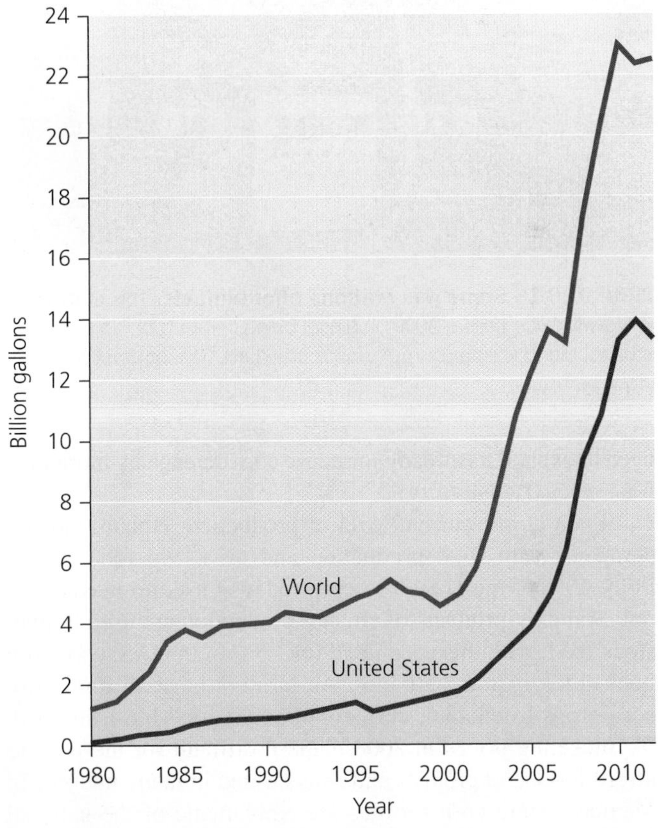

(b) Ethanol production, 1980–2012

FIGURE 20.17 Ethanol is booming. About 40% of the U.S. corn crop **(a)** is used to produce ethanol to add to gasoline (28% after by-products are put to use). Brazil produces most of the rest of the world's ethanol, from bagasse (sugarcane residue). Ethanol production **(b)** has grown rapidly in recent years. *Data (b) from Renewable Fuels Association.*

DATA Q Roughly what percentage of the world's ethanol is produced by the United States?

Ethanol is not our most sustainable energy choice

The enthusiasm for corn-based ethanol shown by U.S policymakers is not widely shared by environmental scientists. Growing corn to produce ethanol exerts considerable impacts on ecosystems, including pesticide use, fertilizer use, fresh water depletion, and other consequences of monocultural industrialized agriculture (pp. 218, 247–248). Corn ethanol crops take up precious land that might otherwise be left in its natural condition or developed for other purposes. If we were to try to produce all the automotive fuel currently used in the United States with ethanol from U.S. corn, the nation would

FIGURE 20.18 **Some gas stations offer biofuels.** This station in New Mexico sells a 20% biodiesel blend (left pump), an 85% ethanol blend (center pump), and a standard 10% ethanol blend (right pump).

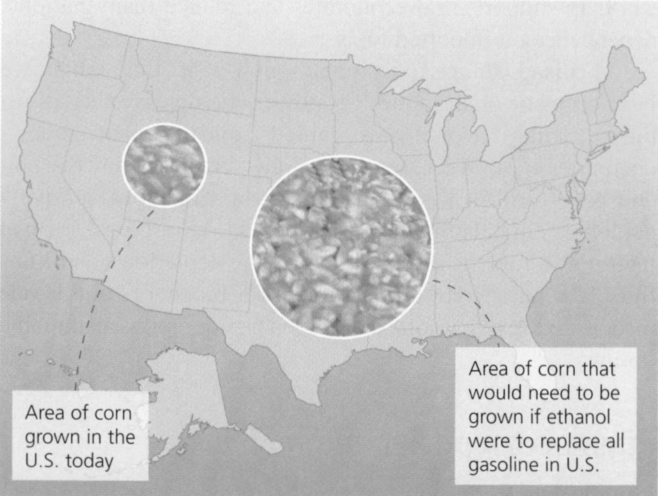

FIGURE 20.19 **Growing corn to produce ethanol uses up a great deal of land.** To produce enough ethanol to replace all gasoline used by U.S. drivers, the United States would need to more than quadruple its area planted in corn.

need to expand its already immense corn acreage by more than four times (FIGURE 20.19).

Even at our current level of production, ethanol already competes with food production and drives up food prices. Fully 40% of the U.S. corn crop today is used to make ethanol. Some by-products of ethanol production are used in livestock feed; with this accounted for, 28% of the U.S. corn crop goes solely toward ethanol. As farmers have shifted more corn crops to ethanol, corn supplies for food have dropped. Skyrocketing prices in 2008 made it difficult for the poor to afford food, and protests and riots ensued in many nations. In Mexico, where corn tortillas are emblematic of the national cuisine, working-class citizens found themselves struggling to buy their staple food, and protests erupted across the country. In 2012, as drought cut into U.S. corn production, international leaders urged the United States to cut back its ethanol production so as to free up more corn for export to nations in need of food.

Growing corn for ethanol also requires substantial inputs of fossil fuel energy (for operating farm equipment, making petroleum-based pesticides and fertilizers, transporting corn to processing plants, and heating water in refineries to distill ethanol). Thus, simply shifting from gasoline to corn ethanol for our transportation needs would not eliminate our reliance on fossil fuels.

Moreover, corn ethanol yields only a modest amount of energy relative to the energy that needs to be input. The EROI (energy returned on investment) ratio (p. 523) for corn-based ethanol is variable and controversial, but recent estimates place it around 1.3:1 (see THE SCIENCE BEHIND THE STORY, pp. 572–573). This means that to gain 1.3 units of energy from ethanol, we need to expend 1 unit of energy. The EROI of Brazilian bagasse ethanol is considerably higher, but the extremely low ratio for corn ethanol makes this fuel inefficient. For this reason, many critics do not view corn ethanol as an effective path to sustainable energy use.

FAQ **If we substitute ethanol for gasoline, won't that solve most of our problems with oil dependence?**

In the United States, government subsidies for corn-based ethanol have been politically popular, and many people believe that the more ethanol we produce and substitute for gasoline, the better off we'll be. Increasing the proportion of ethanol in gasoline does indeed help to conserve oil and reduce reliance on foreign imports. However, obtaining the amount of corn ethanol needed to replace gasoline entirely would require that impractically large amounts of land be converted to corn production. Moreover, so much corn would likely be diverted from food to fuel that food prices would rise sharply. This is why researchers are studying other plants as more efficient sources of ethanol and are trying to develop ways of producing cellulosic ethanol from crop and forestry wastes.

Biodiesel powers diesel engines

Drivers of diesel-fueled vehicles can use **biodiesel**, a fuel produced from vegetable oil, used cooking grease, or animal fat. The oil or fat is mixed with small amounts of ethanol or methanol (wood alcohol) in the presence of a chemical catalyst. In Europe, where most biodiesel is used, rapeseed oil is the oil of choice, whereas U.S. biodiesel producers use mostly soybean oil. Vehicles with diesel engines can run on 100%

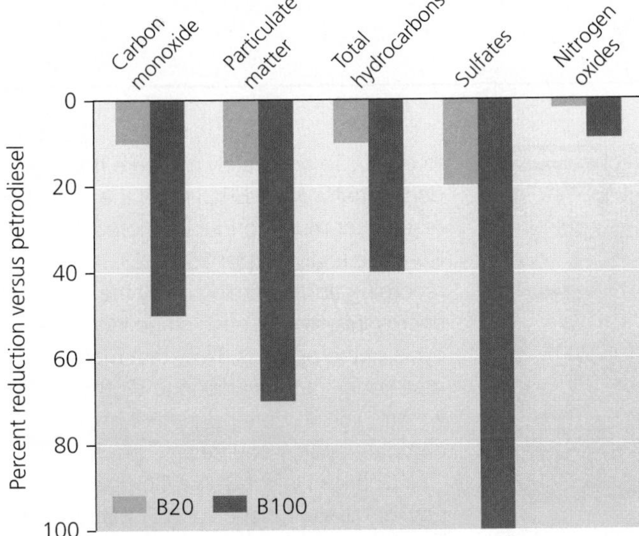

FIGURE 20.20 Burning biodiesel in a diesel engine emits less pollution than burning conventional petroleum-based diesel. Shown are percentage reductions in several major pollutants that one can attain by using B20 (a mix of 20% biodiesel and 80% petrodiesel) and B100 (pure biodiesel). *Data from U.S. Environmental Protection Agency.*

DATA Q Which type of pollutant does use of biodiesel help reduce most effectively, relative to petrodiesel?

biodiesel, or biodiesel can be mixed with conventional petrodiesel; a 20% biodiesel mix (called B20) is common today.

Biodiesel cuts down on emissions compared with petrodiesel (FIGURE 20.20). Its fuel economy is almost as good, it costs just slightly more, and it is nontoxic and biodegradable. Increasing numbers of people are fueling their cars with biodiesel from waste oils (FIGURE 20.21) Some buses and recycling trucks now run on biodiesel, and many state and federal fleets use biodiesel blends.

Some enthusiasts have taken biofuel use further. Eliminating the processing step that biodiesel requires, they use straight vegetable oil in their diesel engines (which requires modifying the engine).

Using waste oil as a biofuel is sustainable, but most biodiesel today, like most ethanol, comes from crops grown specifically for the purpose—and these crops have environmental impacts. Growing soybeans in Brazil and oil palms in Southeast Asia hastens the loss of tropical rainforest (pp. 312–314). Growing soybeans in the United States and rapeseed in Europe takes up large areas of land as well. Because the major crops grown for biodiesel and for ethanol exert heavy impacts on the land, farmers and agricultural scientists are experimenting with a variety of other crops, from wheat, sorghum, cassava, and sugar beets to less known plants such as hemp, jatropha, and the grass miscanthus.

Novel biofuels are being developed

One promising next-generation biofuel crop is algae (better known to most of us as green pond scum)! Several species of these photosynthetic organisms produce large amounts

FIGURE 20.21 At Loyola University Chicago, students and staff produce biodiesel from waste vegetable oil from the dining halls and use it to fuel this van. They transport this mini-diesel reactor to local high schools to teach students about alternative fuels.

of lipids that can be converted to biodiesel. Alternatively, carbohydrates in algae can be fermented to create ethanol. In fact, a variety of fuels, including jet fuel, can be produced from algae.

Algae can be grown in open ponds, including "raceway" ponds that circulate algae around a racetrack. More controlled and intensive production can be achieved in closed tanks or in closed transparent tubes called photobioreactors (FIGURE 20.22). Algae grow much faster than terrestrial crops, can be harvested every few days, and produce much more oil than other biofuel crops. Algae farms can use wastewater, ocean water, or saline water. Because algae need nutrients, wastewater from sewage treatment plants can actually be a good source of water. And because piping in carbon dioxide speeds their growth, algae can make use of smokestack emissions. Indeed, placing algae

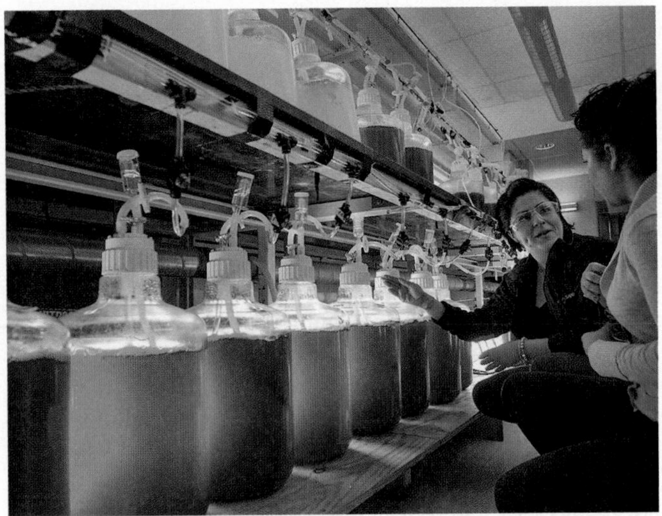

FIGURE 20.22 Algae are a leading candidate for next-generation biofuels. Here algae are seen growing at a demonstration facility. Algae grow quickly and productively, can be farmed in many places (including locations other crops cannot be grown), and can produce biodiesel, ethanol, or other fuels.

THE SCIENCE BEHIND THE STORY

Assessing EROI Values of Energy Sources

As our society develops energy sources as alternatives to fossil fuels, we need to be able to compare them so that we can make informed decisions about which energy sources to pursue and prioritize. One important aspect to compare is the amount of greenhouse gas emissions and other pollution each source produces. In this respect, nuclear power, bioenergy, and hydroelectric power compare very favorably against coal, oil, and natural gas. Market prices of energy sources reflect differences in various other aspects, including their availability, reliability, energy content, and readiness for use with existing pipeline and transmission line infrastructure.

One crucial aspect that has been overlooked to a surprising extent over the years is the measure called *EROI*, or *energy returned on investment* (pp. 523–524). As you'll recall, EROI is a ratio that indicates how much energy is produced for each unit of energy that is invested in the energy-producing activity.

Scientists began calculating estimates of EROI widely in the 1970s, when the Western world was confronted with its first oil shortages (p. 544). Yet as those

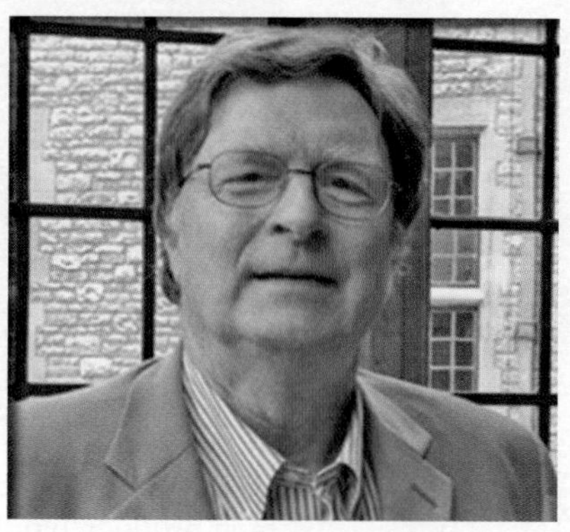

Dr. Charles Hall, a pioneer of EROI research

temporary shortages passed and prices fell, researchers lost interest in studies of EROI values. Today, however, we find ourselves expending more and more effort searching for and extracting fossil fuels. As we start to develop a diverse array of renewable energy options, knowing how much energy it takes to obtain energy has come to seem important again, and EROI studies are blossoming.

It's fairly straightforward to measure how much energy is produced when we combust a fuel or when we let water or wind turn a turbine. The difficult part is calculating how much energy is invested in all the diverse aspects of finding, extracting, transporting, and using a resource, and in manufacturing the equipment needed to do so. To accurately measure the denominator in an EROI ratio, the researcher must conduct a thorough life-cycle analysis (pp. 269, 623), recording and summing up all the energy invested at each stage in the life cycle of the process. Now that more researchers are beginning to do this, we are gaining better data that helps us compare energy sources.

Charles Hall of SUNY College of Environmental Science and Forestry in Syracuse, New York, was one of the pioneers in calculating EROI values, and in 2010 he and David Murphy of Northern Illinois University surveyed recent research and compiled EROI estimates for major energy sources (**FIGURE 1**). Their summary revealed a wide variation in the efficiency of our major energy sources.

Hydroelectric power had the best EROI value, at 100:1, meaning that for each unit of energy invested, we get 100 units of energy from hydropower. At the other end of the spectrum were biodiesel and corn ethanol, with an EROI of only 1.3:1. For these biofuels, we receive just 1.3 units of energy for each unit that we invest. This is because large amounts of equipment, fertilizer, pesticides, and irrigation—all powered by or made from fossil fuels—are needed to grow monocultural fields

---

farms next to power plants as a source of carbon capture (pp. 508, 537–538) is a promising combination. So far, producing biofuels from algae is expensive, but recent investment from the private sector may bring costs down soon.

Because relying on any monocultural crop for energy may not be a sustainable strategy, researchers are refining techniques to use enzymes to produce **cellulosic ethanol** by using enzymes to produce ethanol from the cellulose that gives structure to all plant material. This would be a substantial advance because ethanol as currently made from corn or sugarcane uses starch, which is valuable to us as food. Cellulose, in contrast, is of no food value to people yet is abundant in all plants. If we can produce cellulosic ethanol in commercially feasible ways, then ethanol could be made from low-value crop waste (residues such as corn stalks and husks), rather than from high-value crops.

Certain plants hold promise as crops specifically for cellulosic ethanol. Switchgrass (**FIGURE 20.23**) could potentially be a sustainable choice for the United States because it is a native grass of the North American prairies. Thus it could provide wildlife habitat while serving as a crop, especially if planted in a polyculture mixed with other species. Cellulosic ethanol from switchgrass has an EROI ratio of 5:1 (much better than corn ethanol's), and growing this perennial grass could prevent the soil erosion and depletion of soil carbon that would result from removing too much crop waste for ethanol.

Is bioenergy carbon-neutral?

In principle, energy from biomass is carbon-neutral, releasing no net carbon into the atmosphere. This is because

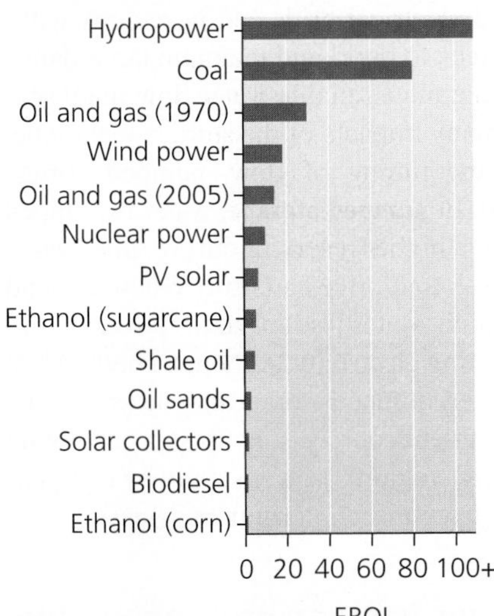

FIGURE 1 EROI values vary from over 100:1 for hydropower down through 1.3:1 for biodiesel and corn ethanol. *Adapted from Murphy, D., and C. Hall, 2010. Year in review—EROI or energy return on (energy) invested.* Annals of the New York Academy of Sciences *1185: 102–118.*

of corn or soybeans and then transport the yields to refineries.

Indeed, ethanol has received more research attention than any other fuel in regard to EROI, and results have varied, but researchers invariably find the EROI of corn ethanol to be remarkably low. Low EROI values are a large part of the reason many environmental scientists oppose U.S. government subsidies for corn ethanol. Ethanol from sugarcane in Brazil is somewhat more efficient, showing a value of 5:1 in Murphy and Hall's review.

EROI values for fossil fuels were historically very high but have been dropping through the years (see Figure 19.5, p. 524) because we have extracted the easy deposits already and now need to expend more and more energy to reach the less-accessible deposits that remain. New fossil fuel resources such as oil sands and shale oil have low EROI values. This means that if we switch to these sources as conventional oil begins to run out, we will deplete them rapidly, because some of their energy will need to be devoted to extracting and processing them.

Solar power technologies also tend to have low EROI values, but as these young technologies are further developed and become more efficient, we can expect EROI values of solar power to rise.

Nuclear power places in the middle of the pack in Murphy and Hall's study, with an EROI of 10:1. Nuclear power produces a large amount of electricity, but it also requires substantial investment in uranium mining and in the facilities, equipment, and maintenance of power plants.

In 2011, Hall and Doug Hansen of Hansen Financial Management in San Diego teamed up to edit a collection of studies on the EROI of various energy sources. Published in a special issue of the journal *Sustainability*, 33 scientists addressed cutting-edge issues in 20 articles.

In one paper from this collection, Hall, Bruce Dale, and David Pimentel sought to clarify why different studies had come up with different EROI estimates for corn ethanol. By carefully comparing the methods of each study, they determined that the differences were due to how the researchers measured energy investment in the many steps involved in producing ethanol. They recommended that future researchers be more transparent and explicit about their methods as they try to work out the EROI values of newer biofuels, such as those from switchgrass and cellulosic ethanol.

As studies of EROI become more numerous and of higher quality, they will help scientists, engineers, and policymakers compare the relative advantages of different energy sources. This will help us to prioritize and make better decisions as we chart our energy future and try to shift from fossil fuels to alternative and renewable energy sources. ■

burning biomass releases carbon dioxide that plants had recently pulled from the atmosphere during photosynthesis. However, burning biomass for energy is not carbon-neutral if forests are destroyed in order to plant bioenergy crops. Forests sequester more carbon (in vegetation and in soil) than do croplands (pp. 230-231), so cutting forests to plant crops will increase carbon flux to the atmosphere. Bioenergy also fails to be carbon-neutral if we need to use fossil fuel energy to produce the biomass (for instance, by driving tractors, using fertilizers, and applying pesticides to grow biofuel crops).

International climate change policy so far has failed to encourage sustainable bioenergy approaches. The Kyoto Protocol (p. 510) required nations to submit data on emissions from energy use and from changes in land use, but only the emissions from energy use were "counted" toward judging how well nations were controlling their emissions under the treaty. Negotiators trying to design a follow-up treaty to Kyoto have been trying to address this (pp. 510–511). In the meantime, researchers are trying to develop means of using bioenergy that are truly renewable and carbon-neutral. With continued research and careful decision making, our many

WEIGHING THE ISSUES

BIOFUELS Do you think producing and using ethanol from corn is a good idea? Do the benefits outweigh the drawbacks? Do you think we should invest billions of dollars into developing next-generation biofuels such as algae and cellulosic ethanol? Can you suggest ways of using biofuels that would minimize environmental impacts?

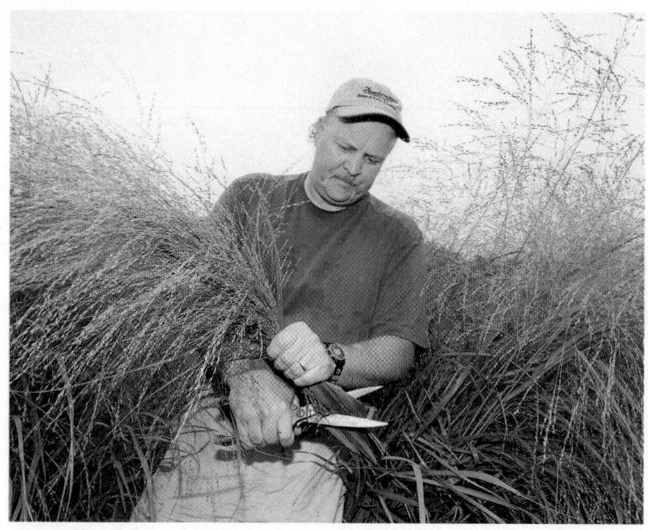

FIGURE 20.23 Switchgrass provides fuel for biopower and is being studied as a crop to provide cellulosic ethanol.

bioenergy options may provide promising avenues for sustainably replacing fossil fuels.

Hydroelectric Power

Next to biomass, we draw more renewable energy from the motion of water than from any other resource. In **hydroelectric power,** or **hydropower,** we use the kinetic energy of moving water to turn turbines and generate electricity. We examined hydropower and its environmental impacts in our discussion of fresh water resources (pp. 403–404). Now we will take a closer look at hydropower as an energy source.

Modern hydropower uses three approaches

Most of our hydroelectric power today comes from impounding water in reservoirs behind concrete dams that block the flow of river water and then letting that water pass through the dam. Because immense amounts of water are stored behind dams, this is called the **storage** technique.

As reservoir water passes through a dam, it turns the blades of turbines, which cause a generator to generate electricity (FIGURE 20.24). Electricity generated in the powerhouse of a dam is transmitted to the electric grid by transmission lines, while the water flows into the riverbed below the dam and continues downriver. The amount of power generated depends on the distance the water falls and the volume of water released. By storing water in reservoirs, dam operators can ensure a steady and predictable supply of electricity at all times, even during periods of naturally low river flow.

An alternative approach is the **run-of-river** technique, which generates electricity without greatly disrupting the flow of river water. Several methods can be used; one is to divert a portion of a river's flow through a pipe or channel, passing it through a powerhouse and then returning it to the river (FIGURE 20.25). This can be done with or without a small reservoir that pools water temporarily, and the pipe or channel can be run along the surface or underground. Another method is to flow river water over a dam small enough not to impede fish passage, siphoning off water to turn turbines, and then returning the water to the river. Run-of-river systems are useful in areas remote from established electrical grids and in regions without the economic resources to build and maintain large dams. This approach cannot guarantee reliable water flow in all seasons, but it minimizes many impacts of the storage technique.

To better control the timing of flow, pumped-storage hydropower can be used. In **pumped storage,** water is pumped from a lower reservoir to a higher reservoir during times when demand for power is weak and prices are low. When demand is strong and prices are high, water is allowed to flow downhill through a turbine, generating electricity. Although energy must be input to pump the water, pumped storage can be profitable. When paired with intermittent sources such as solar and wind power, it can help balance a region's power supply by compensating for dips in power from these intermittent sources.

Hydroelectric power is clean and renewable, but also has impacts

For producing electricity, hydropower has three clear advantages over fossil fuels. First, it is renewable; as long as precipitation falls from the sky and fills rivers and reservoirs, we can use water to turn turbines.

Second, hydropower is efficient. It is thought to have an EROI (energy returned on investment) ratio of 100:1 or more, higher than any other modern-day energy source.

Third, no carbon compounds are burned in the production of hydropower, so no carbon dioxide or other pollutants are emitted into the atmosphere, and this helps safeguard air quality, climate, and human health. Fossil fuels *are* used in constructing and maintaining dams—and recent evidence indicates that large reservoirs release the greenhouse gas methane as a result of anaerobic decay in deep water. But overall, hydropower accounts for only a small fraction of the greenhouse gas emissions typical of fossil fuel combustion.

Although it is renewable, efficient, and produces little air pollution, hydropower does exert negative impacts. Damming rivers (p. 403) destroys habitat for wildlife as riparian areas above dam sites are submerged and those below dam sites often are starved of water. Because water discharge is regulated to optimize electricity generation, the natural flooding cycles of rivers are disrupted. Suppressing flooding prevents river floodplains from receiving fresh nutrient-laden sediments. Instead, sediments become trapped behind dams, where they begin filling the reservoir.

Dams also cause thermal pollution (p. 412), because water downstream may become unusually warm if water levels are kept unnaturally shallow. Moreover, periodic flushes of cold water occur from the release of reservoir water. Such thermal shocks, together with habitat alteration, have diminished or eliminated native fish populations in many dammed waterways. In addition, dams generally block the passage of fish and other aquatic creatures, effectively fragmenting the river and reducing biodiversity in each stretch.

(a) Ice Harbor Dam, Snake River, Washington

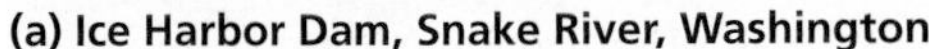

(b) Turbine generator inside McNary Dam, Columbia River

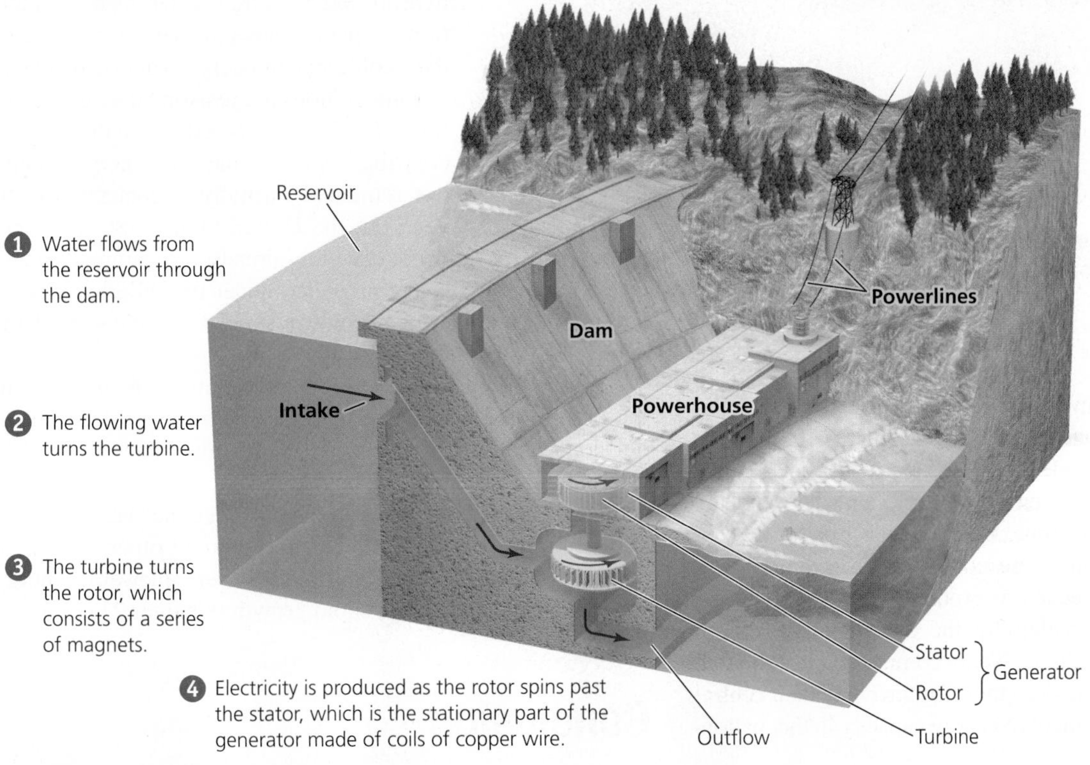

(c) Hydroelectric power

FIGURE 20.24 Large dams generate substantial amounts of hydroelectric power. Inside these dams **(a)**, flowing water is used to turn turbines **(b)** and generate electricity. Water is funneled from the reservoir through a portion of the dam **(c)** to rotate turbines, which turn rotors containing magnets. The spinning rotors generate electricity as their magnets pass coils of copper wire. Electrical current is transmitted through power lines, and the river's water flows out through the base of the dam.

All these ecological impacts generally translate into negative social and economic impacts on local communities. We discussed the benefits, drawbacks, and impacts of dams more fully in Chapter 15 (pp. 403–404).

Hydropower is widely used

Hydropower accounts for 16% of the world's electricity production (see Figure 20.1b). For nations with large amounts of river water and the economic resources to build dams, hydroelectric power has been a keystone of their development and wealth. Canada, Brazil, Norway, Austria, Switzerland, Venezuela, and many other nations today obtain large amounts of their energy from hydropower (TABLE 20.4).

Sweden receives 11% of its total energy and nearly half its electricity from hydropower. In the wake of the nation's 1980 referendum to phase out nuclear power, many people had hoped hydropower could play a still larger role and compensate for the electrical capacity that would be lost. However, Sweden has already dammed so many of its rivers that it cannot gain much additional hydropower by erecting more dams. Moreover, Swedish citizens have made it clear that they want some rivers to remain undammed, preserved in their natural state.

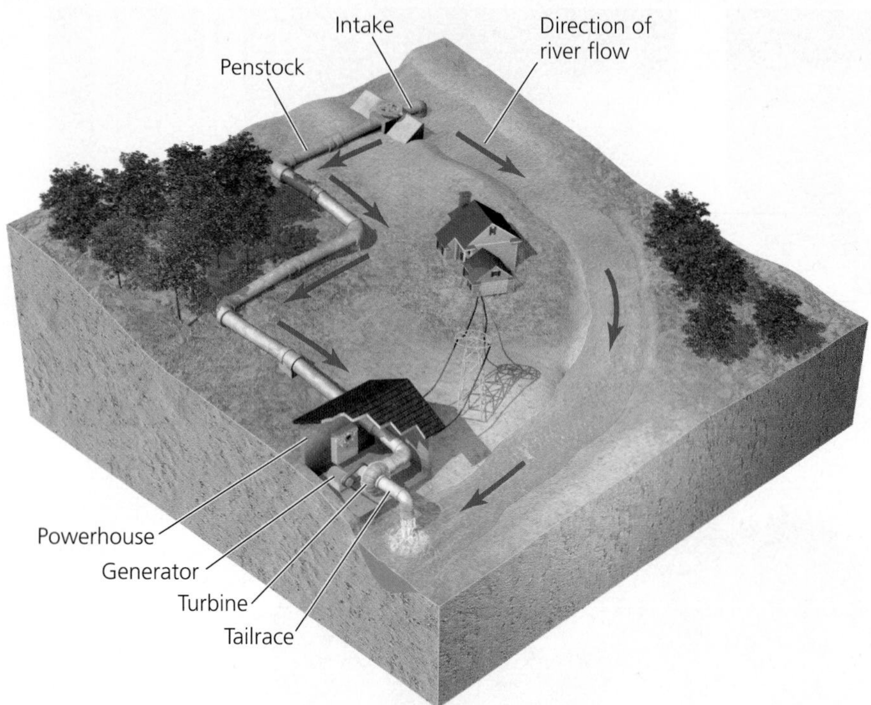

FIGURE 20.25 Run-of-river systems divert a portion of a river's water. Water may be piped downhill through a powerhouse and released downriver, as shown. Other designs use water as it flows over shallow dams.

The great age of dam building for hydroelectric power (as well as for flood control and irrigation) began in the 1930s in the United States, when the federal government constructed dams as public projects, partly to employ people and help end the economic depression of the time. U.S. dam construction peaked in 1960, when 3123 dams were completed in a single year. American engineers subsequently exported their dam-building technologies, notably to nations of the developing world. In India, Prime Minister Jawaharlal Nehru commented in 1963 that "dams are the temples of modern India," referring to their central importance in the development of energy capacity in that nation.

TABLE 20.4 Top Producers of Hydropower

NATION	HYDROPOWER PRODUCED (Percentage of world total)	PERCENTAGE OF NATION'S ELECTRICITY GENERATION FROM HYDROPOWER
China	20.5	17.2
Brazil	11.5	78.2
Canada	10.0	57.8
United States	8.1	6.5
Russia	4.8	16.2
Norway	3.4	94.7
India	3.3	11.9
Japan	2.6	8.1
Venezuela	2.2	64.9
France	1.9	11.7
Rest of world	31.7	15.4

Data from the International Energy Agency.

Hydropower's expansion is limited

Today the world is witnessing some gargantuan hydroelectric projects. China's recently completed Three Gorges Dam (pp. 403–404) is the world's largest. Its reservoir displaced 1.3 million people, and the dam is now generating as much electricity as dozens of coal-fired or nuclear plants.

However, hydropower has less potential for expansion than some other renewable energy sources. The main reason is that, as in Sweden, most of the world's large rivers that offer excellent opportunities for hydropower are already dammed. In addition, as people have grown more aware of the ecological impacts of dams, residents in some regions are resisting dam construction. Indeed, in Sweden, hydropower's contribution to the national energy budget has remained virtually unchanged for 40 years. In the United States, 98% of rivers appropriate for dam construction already are dammed, many of the remaining 2% are protected under the Wild and Scenic Rivers Act, and many people now want to dismantle some dams and restore river habitats (p. 404).

Overall, hydropower will likely continue to increase in developing nations that have yet to dam their rivers, but in developed nations hydropower growth may slow down. The International Energy Agency forecasts that hydropower's share of electricity generation will decline very slightly between now and 2035, whereas the share of other renewable energy sources will nearly triple. Newer renewable energy sources have potential for rapid growth (Chapter 21).

Conclusion

Given limited supplies of fossil fuels and their considerable environmental impacts, many nations have sought to diversify their energy portfolios with alternative energy sources. The three most developed and widely used alternatives so far are nuclear power, bioenergy, and hydropower.

Nuclear power showed promise at the outset to be a pollution-free and highly efficient form of energy. But escalating costs and public fears over safety in the wake of accidents at Three Mile Island, Chernobyl, and Fukushima stalled its growth, and a few nations are attempting to phase it out completely. Bioenergy sources are diverse and range from traditional fuelwood to newer biofuels and various means of generating biopower. These sources can be carbon-neutral but are not all strictly renewable. Hydropower is a renewable, pollution-free alternative, but it can expand only so much further and also involves substantial ecological impacts. Some nations, such as Sweden, already rely heavily on these three conventional alternatives, but it appears that our global society will need further renewable sources of energy as well.

Reviewing Objectives

You should now be able to:

Discuss the reasons for seeking energy alternatives to fossil fuels

- Fossil fuels are nonrenewable resources, and we are gradually depleting them. (p. 554)
- Fossil fuel combustion causes air pollution that results in many environmental and health impacts and contributes to global climate change. (p. 554)

Summarize the contributions to world energy supplies of conventional alternatives to fossil fuels

- Conventional energy alternatives are the alternatives to fossil fuels that are most widely used. They include nuclear power, bioenergy, and hydroelectric power. In their renewability and environmental impact, these sources fall between fossil fuels and the less widely used "new renewable" sources. (p. 555)
- Biomass provides 10% of global primary energy use, nuclear power provides 6%, and hydropower provides 2%. (pp. 554–555)
- Nuclear power generates 13% of the world's electricity, and hydropower generates 16%. (pp. 554–555)

Describe nuclear energy and how it is harnessed

- We gain nuclear power by converting the energy of subatomic bonds into thermal energy, using uranium isotopes. (pp. 556–557)
- By controlling the reaction rate of nuclear fission, nuclear power plant engineers produce heat that powers electricity generation. (pp. 556–557)
- Uranium is mined, enriched, processed into pellets and fuel rods, and used in nuclear reactors. (pp. 556–557)

Outline the societal debate over nuclear power

- Many advocates of clean energy support nuclear power because it does not emit the pollutants that fossil fuels do. (pp. 557–558)
- For many people, the risk of a major power plant accident, like those at Chernobyl and Fukushima, outweighs the benefits of clean energy. (pp. 559–564)
- The disposal of nuclear waste remains a major dilemma. Temporary storage and single-repository approaches each involve health, security, and environmental risks. (pp. 564–566)
- Economic factors and cost overruns have slowed the nuclear industry's growth. (p. 566)

Describe the major sources, scale, and impacts of bioenergy

- Bioenergy theoretically is renewable and adds no net carbon to the atmosphere. But overharvesting of wood can lead to deforestation, and growing crops solely for fuel production can be inefficient and ecologically damaging. (pp. 566–567, 572–574)
- Fuelwood remains the major source of bioenergy today, especially in developing nations. (p. 567)
- We use biomass to generate electrical power (biopower) in several ways. Sources include special crops as well as waste products from farming and forestry. (pp. 567–569)
- Biofuels, including ethanol and biodiesel, are used to power automobiles. Some crops are grown specifically for this purpose, and waste oils are also used. (pp. 569–572)
- Ethanol figures prominently in scientists' debates over EROI, the measure of how much energy is returned on investment. (pp. 570, 572–573)
- Next-generation biofuels such as algae and cellulosic ethanol need further research and development, but hold promise for more sustainable biofuel production. (p. 571–572)

Describe the scale, methods, and impacts of hydroelectric power

- Hydroelectric power is generated when water from a river runs through a powerhouse and turns turbines. (pp. 574–575)
- Three major hydropower approaches (storage, run-of-river, and pumped storage) each offer advantages in certain situations. (pp. 574–576)
- Hydropower produces little air pollution, but dams and reservoirs can greatly alter riverine ecology and local economies. (pp. 574–575)
- Hydropower's long-term growth is more limited than that of other renewable sources. (p. 576)

Testing Your Comprehension

1. How much of our global energy supply do nuclear power, bioenergy, and hydroelectric power contribute? How much of our global electricity do these three conventional energy alternatives generate?
2. Describe how nuclear fission works. How do nuclear plant engineers control fission and prevent a runaway chain reaction?
3. In terms of greenhouse gas emissions, how does nuclear power compare to coal, oil, and natural gas? How do hydropower and bioenergy compare?
4. In what ways did the events at Three Mile Island, Chernobyl, and Fukushima Daiichi differ? What consequences resulted from each incident?
5. List several concerns about the disposal of radioactive waste. What has been done so far about its disposal?
6. List five sources of bioenergy. What is the world's most used source of bioenergy? How does bioenergy use differ between developed and developing nations?
7. Describe two biofuels, where each comes from, and how each is used.
8. Evaluate two potential benefits and two potential drawbacks of bioenergy.
9. Compare and contrast the three major approaches to generating hydroelectric power.
10. Assess two benefits and two negative environmental impacts of hydroelectric power.

Seeking Solutions

1. Given what you have learned about fossil fuels (Chapter 19) and about some of the conventional alternatives (this chapter), do you think it is important for us to minimize our use of fossil fuels and maximize our use of alternatives? If such a shift were to require massive public subsidies, how much should we subsidize this? What challenges or obstacles would we need to overcome in order to shift to such alternatives?
2. Nuclear power has by now been widely used for half a century, and the world has experienced only two major accidents (Chernobyl and Fukushima) responsible for any significant number of injuries or deaths. Would you call this a good safety record? Should we maintain, decrease, or increase our reliance on nuclear power? Why might safety at nuclear power plants be better in the future? Why might it be worse?
3. How serious a problem do you think the disposal of radioactive waste represents? How would you like to see this issue addressed?
4. There are many different sources of biomass and many ways of harnessing energy from biomass. Discuss one that seems particularly beneficial to you, and one with which you see problems. What bioenergy sources and strategies do you think our society should focus on investing in?
5. **THINK IT THROUGH** You are the head of the national department of energy in a nation that has just experienced a minor accident at one of its nuclear plants. A partial meltdown released radiation, but the radiation was fully contained inside the containment building, and there were no health impacts on area residents. However, citizens are terrified, and the media is highlighting the dangers of nuclear power. Your nation relies on its five nuclear plants for 25% of its energy and 50% of its electricity needs. It has no fossil fuel deposits and recently began a promising but still-young program to develop renewable energy options. What will you tell the public at your next press conference, and what policy steps will you recommend taking to ensure a safe and reliable national energy supply?
6. **THINK IT THROUGH** You are an investor and would like to invest in alternative energy. You are considering buying stock in companies that (1) construct nuclear reactors, (2) build turbines for hydroelectric dams, (3) build corn ethanol refineries, (4) are ready to supply farm and forestry waste for cellulosic ethanol, and (5) are developing algae farms. For each of these companies, what questions would you research before deciding how to invest your money? How do you expect you might apportion your investments, and why?

Calculating Ecological Footprints

Our three conventional energy alternatives vary tremendously in their energy returned on investment (EROI). Examine the data on EROI for each of the energy sources as provided in the figure from **The Science behind the Story** on p. 573, and enter the data in the table below.

ENERGY SOURCE	EROI
Coal	80
Oil and gas (2005)	
Oil sands	
Nuclear power	
Hydroelectric power	
Ethanol from corn	
Biodiesel	
Wind power	

1. How many units of energy would you generate by investing 1 unit of energy into producing hydroelectric power? How many units of energy would you need to invest into producing nuclear power if you wanted to generate that same amount of energy? How many units of energy would you need to invest into producing corn ethanol if you wanted to generate that same amount of energy?
2. Based on EROI values, is it more efficient to get petroleum from conventional oil or from oil sands? Which source would you guess has a larger ecological footprint, based on EROI values?
3. Let's say you wanted to generate 130 units of energy from biodiesel. How many units of energy would you need to invest?
4. Based on EROI values alone, which energy sources would you advocate that your nation further develop? Which would you urge your nation to avoid? What other issues, besides EROI, are worth considering when comparing energy sources?

21

Homes in Freiburg, Germany, which produce excess solar power and sell it to the grid

New Renewable Energy Alternatives

Upon completing this chapter, you will be able to:

- Outline the major sources of renewable energy, summarize their benefits, and assess their potential for growth
- Describe solar energy and the ways it is harnessed, and evaluate its advantages and disadvantages
- Describe wind power and how we harness it, and evaluate its benefits and drawbacks
- Describe geothermal energy and the ways we make use of it, and assess its advantages and disadvantages
- Describe ocean energy sources and how we can harness them
- Explain hydrogen fuel cells and weigh options for energy storage and transportation

CENTRAL CASE STUDY

Germany Goes Solar

"Someday we will harness the rise and fall of the tides and imprison the rays of the sun."

—**Thomas A. Edison, 1921**

"[Renewable energy] will provide millions of new jobs. It will halt global warming. It will create a more fair and just world. It will clean our environment and make our lives healthier."

—**Hermann Scheer, energy expert and member of the German parliament, 2009**

When we think of solar energy, most of us envision a warm sunny place like Arizona or southern California. Yet the country that produces the most solar power in the world is Germany, a northern European nation that receives less sun than Alaska! Germany is the world's top generator of photovoltaic (PV) solar power, a technology that produces electricity from sunshine. In recent years Germany has installed nearly half the world's total of this technology. Germany now obtains more of its energy from solar power than any other nation, and the amount is growing each year.

How is this happening in such a cool and cloudy country? A bold federal policy is using economic incentives to promote solar power and other forms of renewable energy. Germany has a **feed-in tariff** system whereby utilities are required to buy power from anyone who can generate power from renewable energy sources and feed it into the electric grid. Under this system, utilities must pay guaranteed premium prices for this power under long-term contract. As a result, German homeowners and businesses have rushed to install PV panels and are selling their excess solar power to the utilities at a profit.

The feed-in tariffs apply to all forms of renewable energy. As a result of these incentives, Germany established itself as the world leader in wind power in the 1990s, until being overtaken by the United States and China. It ranks third in the world in using power from biomass, and second in solar water heating. Overall, Germany ranks third in the world in electrical power capacity from renewable sources, trailing only China and the United States, which have far more people and businesses. In renewable energy generated per person, Germany is number one in the world.

Germany's push for renewable energy goes back to 1990. In the wake of the Chernobyl nuclear disaster (pp. 559–560), Germany decided to phase out its nuclear power plants. However, by shutting these down, the nation would lose virtually all its clean energy and would become utterly dependent on oil, gas, and coal imported from Russia and the Middle East.

Enter Hermann Scheer, a German parliament member and an expert on renewable energy. While everyone else assumed that technologies for harnessing solar, wind, and geothermal energy were costly, risky, and underdeveloped, Scheer saw them as a great economic opportunity—and as the only long-term answer. In 1990, Scheer helped push through a landmark law establishing feed-in tariffs. Ten years later, the law was revised and strengthened: The Renewable Energy Sources Act of 2000 aimed to promote renewable energy production and use, enhance the security of the energy supply, reduce carbon emissions, and lessen the many external costs (pp. 146, 165) of fossil fuel use.

Under this landmark law, each renewable energy source is assigned its own payment rate according to market considerations, and most rates are reduced year by year in order to encourage increasingly efficient means of producing power. In 2004 and in 2009, the government adjusted the amounts utilities were required to pay homeowners and businesses for their energy production.

In recent years, the feed-in-tariffs were adding about 2 cents per kilowatt-hour onto the average German citizen's electric bill, amounting to an annual cost of about $60 per person. To reduce the cost of these subsidies, the German government in 2010 slashed PV solar tariff rates by 16%. The new lower rates eased the burden on ratepayers, while homeowners still had economic incentive to invest in solar panels because the feed-in-tariffs had already helped to cut PV market prices in half.

In response to the prospect of lower tariffs, sales of PV modules skyrocketed as Germans rushed to lock in the old rates. Since then, a series of reductions in the payment rates have each spurred rushes to install more PV systems. In 2010, 2011, and 2012, Germans installed over 7 gigawatts of PV solar capacity each year—an amount equal to the total cumulative capacity of the United States. In 2012 the government proposed to end subsidies once the nation reaches 52 gigawatts of cumulative capacity, roughly twice its total as of 2011.

By reducing the subsidies gradually, the government aims to create a stronger solar industry that can sustain growth over the long term and can compete with foreign companies for international business. Indeed, boosted by domestic demand, German industries have become global leaders in "green tech," designing and selling renewable energy technologies around the world. Germany is second in PV production behind China, leads the world in production of biodiesel, and has developed several cellulosic ethanol (p. 572) facilities. Renewable energy industries in Germany today employ nearly 400,000 citizens.

By 2020, Germany aims to obtain 35% of its electricity from renewable sources, and plans to increase this percentage to 50% by 2030, 65% by 2040, and 80% by 2050. To make this happen, the government has been allotting more public money to renewable energy than any other nation—over $25 billion annually in recent years. Following the Fukushima nuclear disaster in Japan in 2011 (pp. 22–23, 560–561), the German government responded to anti-nuclear demonstrations by shutting down 7 of its 15 nuclear power plants and promising to phase out the rest by 2022. This dramatic reduction in electricity supply from nuclear power is causing rates to rise and puts added pressure on Germany to develop alternative energy sources quickly.

By developing renewable energy to replace some of its fossil fuel use, Germany has decreased its emissions of carbon dioxide by 140 million tons per year—equal to taking 24 million cars off the road. Since 1990, carbon dioxide emissions from German energy sources have fallen by 24%, and emissions of seven other major pollutants (CH_4, N_2O, SO_2, NO_X, CO, VOCs, and dust) have been reduced by 12–95%. At least half of these reductions are thought to be due to renewable energy paid for under the feed-in tariff system.

Germany's success is serving as a model for other countries in Europe and around the world. As of 2013, more than 90 nations had implemented some sort of feed-in tariff. Spain and Italy ignited their wind and solar development as a result. In North America, Vermont and Ontario established feed-in tariff systems similar to Germany's, while California, Hawaii, Oregon, Washington, and New York conduct more-limited programs. In 2010, Gainesville, Florida, became the first U.S. city to establish feed-in tariffs, and solar power is growing quickly there as a result. Moreover, utilities in 46 U.S. states now offer net metering, in which utilities credit customers who produce renewable power and feed it into the grid. As more nations, states, and cities encourage renewable energy, we may soon experience a historic transition in the way we meet our energy demands.

"New" Renewable Energy Sources

Germany's bold federal policy is just one facet of a global shift toward renewable energy. Across the world, nations are searching for ways to move away from fossil fuels while ensuring a reliable and affordable supply of energy for their economies. This is because the economic and social costs, security risks, and health and environmental impacts of fossil fuel dependence (Chapter 19) are all continuing to intensify.

The two renewable energy sources that are most developed and widely used are bioenergy, the energy from combustion of biomass (wood and other plant matter), and hydropower, the energy from running water (Chapter 20). These conventional alternatives to fossil fuels are renewable, but they can be depleted with overuse and exert some undesirable environmental impacts.

In this chapter we explore a group of alternative energy sources that are often called "new renewables." These diverse sources include energy from the sun, from wind, from Earth's geothermal heat, and from ocean water. These energy sources are not truly new. In fact, they are as old as our planet, and people have used them for millennia. We commonly refer to them as "new" because (1) they are just beginning to be used on a wide scale in our modern industrial society, (2) they are harnessed using technologies still in a rapid phase of development, and (3) they will likely play much larger roles in the future.

New renewable sources are growing fast

As with other energy sources, the new renewables provide energy for three types of applications: (1) to generate electricity, (2) to heat air or water, and (3) to fuel vehicles. Their potential is enormous, yet so far their contribution to our society's overall energy budget remains small. Today we obtain just 1% of our global energy from the new renewable energy sources. Fossil fuels provide 81% of the world's energy and nuclear power provides 6%, while biomass and hydropower supply nearly all of the 13% that renewable energy sources provide (see Figure 20.1a, p. 554).

The new renewables make a similarly small contribution to our global generation of electricity thus far (see Figure 20.1b, p. 554). Only 20% of our electricity comes from renewable energy, and hydropower accounts for four-fifths of this amount.

Nations and regions vary in the renewable sources they use. In the United States, most renewable energy comes from biomass and hydropower. As of 2012, wind power accounted for 15%, solar energy for 3%, and geothermal energy for 3% (**FIGURE 21.1a**). Of electricity generated in the United States from renewables, hydropower accounts for most, while as of 2012 wind power accounted for 29%, geothermal for 3%, and solar for just 1% (**FIGURE 21.1b**).

Although they comprise a small proportion of our energy budget, the new renewable energy sources are growing quickly. Over the past four decades, solar, wind, and geothermal energy sources have grown far faster than has

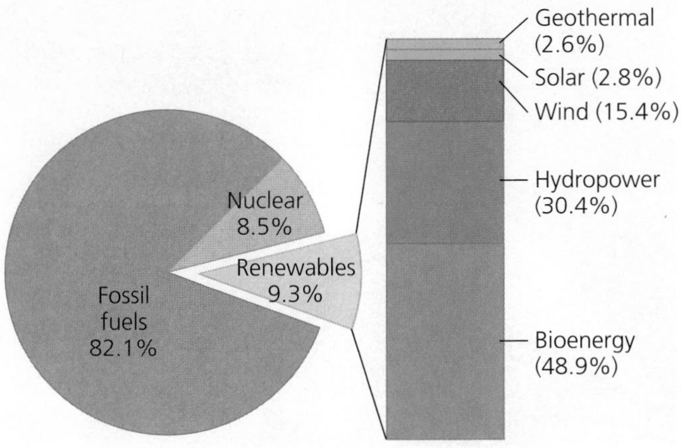

(a) U.S. consumption of renewable energy, by source

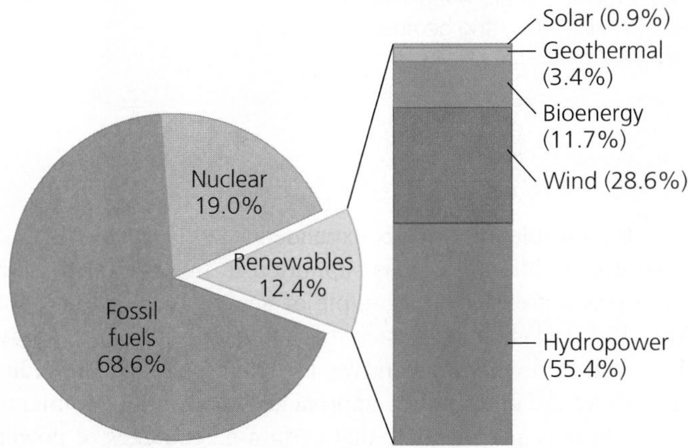

(b) U.S. electricity generation from renewable sources

FIGURE 21.1 **Nine percent of the energy consumed in the United States each year comes from renewable sources.** Of this amount **(a)**, most derives from bioenergy and hydropower. Wind power, solar energy, and geothermal energy together account for just 21% of this amount. Similarly, just 12% of electricity generated in the United States **(b)** comes from renewable energy sources, predominantly hydropower. *Data are for 2012, from Energy Information Administration, U.S. Department of Energy, 2013.*

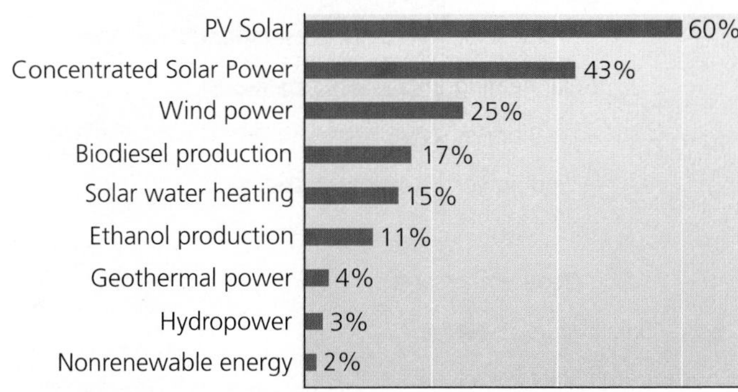

FIGURE 21.2 **The "new renewable" energy sources are growing far faster than conventional energy sources.** Shown are average rates of growth each year between 2007 and 2012. *Data from REN21, 2013.* Renewables 2013: Global status report. *REN21, UNEP, Paris.*

DATA Q At the yearly growth rate shown, if you began with 10 units of PV solar capacity, how much would there be after 5 years?

the overall energy supply. The long-term leader in growth is wind power, which has expanded by nearly 50% *each year* since the 1970s. In recent years, solar power has grown faster than wind (FIGURE 21.2). Because these sources started from such low levels of use, however, it will take them some time to catch up to conventional sources. The absolute amount of energy added by a 50% increase in wind power today is less than the amount added by just a 1% increase in fossil fuels.

The new renewables offer advantages

Use of new renewables has been expanding because of growing concerns over diminishing fossil fuel supplies and because of the many environmental and health impacts of fossil fuel combustion (Chapter 19). Advances in technology are also making it easier and less expensive to harness renewable energy sources.

The new renewables promise several benefits over fossil fuels. They help alleviate many types of air pollution (Chapter 17). In particular, they help reduce the greenhouse gas emissions that drive global climate change (Chapter 18) (FIGURE 21.3). Unlike fossil fuels, renewable sources are inexhaustible on time scales relevant to our society. Renewables also can diversify an economy's energy mix, helping to reduce price volatility and buffer us against restrictions in supply of imported fuels. Moreover, novel energy sources can generate income and property tax for communities, especially in rural areas passed over by other economic development.

Shifting to renewable energy also creates new employment opportunities. The design, installation, maintenance, and

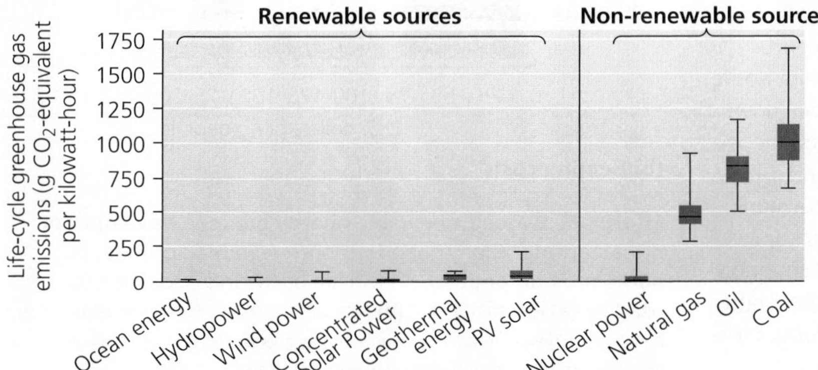

FIGURE 21.3 **Renewable energy sources release far fewer greenhouse gas emissions than fossil fuels.** Shown are ranges of estimates from scientific studies of each source when used to generate electricity. *Data from Intergovernmental Panel on Climate Change, 2012.* Renewable energy sources and climate change mitigation. *Special report. Cambridge Univ. Press, New York.*

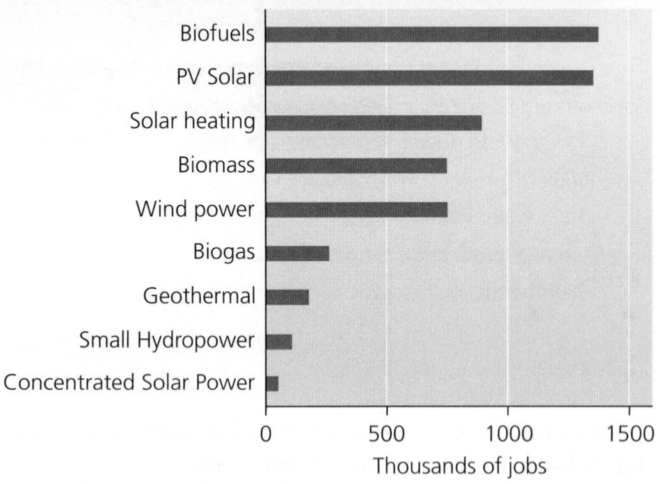

(a) Global jobs in each sector

(b) A wind power technician in Texas

FIGURE 21.4 **Renewable energy creates new green-collar jobs.** Nearly 6 million people worldwide were employed in renewable energy jobs as of 2012, with more than half of these in solar, wind, and geothermal energy. *Data from REN21, 2013.* Renewables 2013: Global status report. *REN21, UNEP, Paris.*

management required to develop technologies and rebuild and operate our society's energy infrastructure promise to be major sources of employment for young people today, through **green-collar jobs.** Nearly 6 million people work in renewable energy jobs around the world already, and the number is rising (FIGURE 21.4).

As renewable energy technologies evolve and as novel energy sources are harnessed and developed, we can better assess the particular benefits and drawbacks of each approach. As we learn more about the contributions and impacts of each source and technique, we can better judge how to prioritize our energy choices (see THE SCIENCE BEHIND THE STORY, pp. 586–587).

Policy and investment can accelerate our transition

Rapid growth in renewable energy seems likely to continue as global population and consumption rise, easily accessible fossil fuel supplies decline, and people demand cleaner environments. Yet we cannot wholly convert to renewable energy sources overnight, because we lack the infrastructure needed to transfer huge volumes of power from renewable sources inexpensively on a continent-wide scale. Advances in recent years, though, suggest that we can overcome these technological and economic barriers if renewables are lent political support.

Renewable energy efforts received support when some national governments responded to the global financial downturn of 2008–2009 by enacting stimulus packages and boosting spending on green energy programs to help create jobs. As more governments, utilities, corporations, and consumers promote and use renewable energy, market prices of renewables should continue to fall, further hastening their adoption. At this point, most renewable energy is priced more highly than fossil fuel energy, but some sources have become cost-competitive (FIGURE 21.5).

Renewable energy has expanded quickly whenever and wherever public policy has supported it. Feed-in-tariffs like Germany's are a prime example of an economic policy tool (pp. 180–184) that can hasten the spread of renewable energy by creating financial incentives for businesses and individuals. There are other policy approaches as well. Governments can set goals and mandate that certain percentages of power come from renewable sources by a certain date. As of 2013,

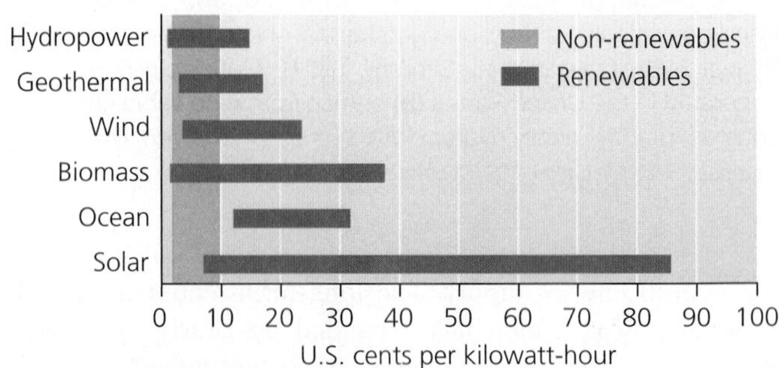

(a) Electricity costs

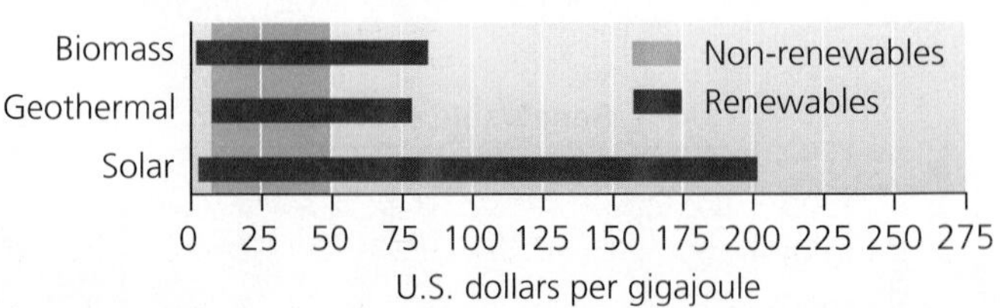

(b) Heating costs

FIGURE 21.5 **Most renewable energy sources have market prices greater than those for nonrenewable sources, but some are competitive.** Shown are price ranges for each source used for **(a)** electricity and **(b)** heating. *Data from Intergovernmental Panel on Climate Change, 2012.* Renewable energy sources and climate change mitigation. *Special report. Cambridge Univ. Press, New York.*

120 nations and 36 U.S. states had set official targets for renewable energy use. Governments also invest in research and development of technologies, lend money to businesses as they start up, and offer tax credits and tax rebates to companies and individuals who produce or buy renewable energy.

When a government boosts an industry with such policies, the private sector often responds by investing in the industry as private investors recognize an enhanced chance of success and profit. As a result, throughout the world in recent years, advances in renewable energy have been tightly linked to national, state, or local policies that encourage them and to the investment funding that flows more freely as a result. Global investment (public plus private) in renewable energy rose to $244 billion in 2012—nearly six times the amount just seven years earlier. In a feedback process, investment breeds success, and success breeds further investment.

Yet the economics of renewable energy has been erratic. Technologies evolve quickly, and policies vary from place to place and can change unpredictably. As a result, the industry has been volatile, experiencing steep price fluctuations, bankruptcies of promising companies, and bursts of rapid growth. Moreover, we are still learning how to apply policies wisely. In 2007 Spain adopted feed-in tariffs that were so generous (58 cents per kilowatt-hour) that thousands of people rushed to set up solar facilities. Within a year, Spain increased its solar capacity fivefold, accounting for half the solar power installed worldwide that year. However, the generous payments attracted a flood of unqualified contractors and speculators; as a result, many solar plants were hastily built, poorly designed, or badly located. The government realized that many of these new plants would never be economically self-sufficient, so it abruptly slashed the tariff rates.

Critics of government subsidies for renewable energy complain that funneling taxpayer money to particular energy sources is inefficient and skews the market. Instead, they propose, we should let energy sources compete in a free market. Proponents of renewable energy point out that governments have long subsidized fossil fuels and nuclear power far more than they are now subsidizing renewable energy. As a result, there has never been a level playing field, nor a truly free market.

A 2011 study by venture capitalist Nancy Pfund and Yale University graduate student Ben Healey dug deep into data on the U.S. government's many subsidies and tax breaks (pp. 181–182) for all energy sources, from the early days of oil and gas in 1918 through to the present. Their report revealed three main findings:

1. In total, oil and gas have received 75 times more subsidies than new renewable energy sources (**FIGURE 21.6a**). Nuclear power has received 31 times more subsidies than new renewable energy. This alone is not terribly surprising, because oil, gas, and nuclear power have been major contributors to the U.S. energy supply for a longer time period than solar, wind, and geothermal sources.
2. Per year, oil and gas have received 13 times more subsidies than new renewable energy sources, and nuclear power has received over 9 times more support than new renewable energy (**FIGURE 21.6b**). This is notable, because this per-year comparison controls for the amount of time each source has been used and thereby shows that oil, gas, and nuclear power have received far more support than solar, wind, and geothermal power.
3. Even in the earliest years of each energy source (when subsidies are most useful), oil, gas, and nuclear power received far more in subsidy support than the new renewables have received. In fact, only in one single year have solar, wind, and geothermal combined *ever* received as much support as the *lowest* amount ever offered to oil, gas, or nuclear power.

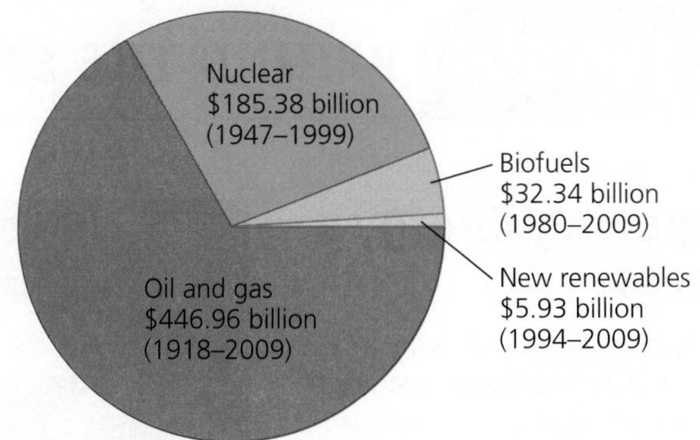

(a) Total subsidies

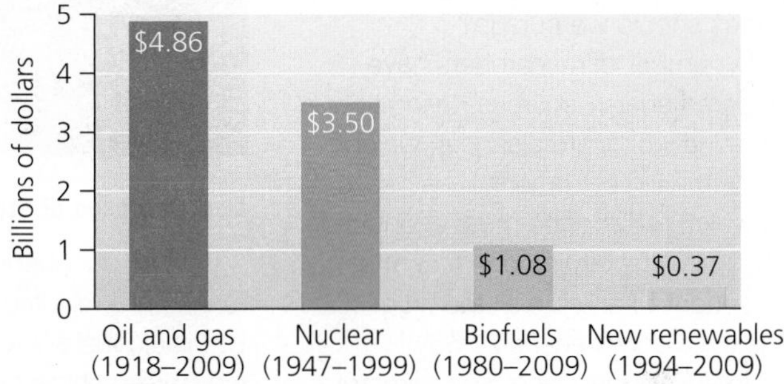

(b) Per-year subsidies

FIGURE 21.6 Fossil fuels and nuclear power have received far more in U.S. government subsidies than have renewable energy sources. This is true both for **(a)** total amounts over the past century and for **(b)** average amounts per year. *Data are in 2010 dollars, from Pfund, N., and B. Healey, 2011.* What would Jefferson do? The historical role of federal subsidies in shaping America's energy future. *DBL Investors.*

These findings corroborate the notion that renewable energy sources have not received nearly the amount of support in subsidies and tax breaks that our long-established nonrenewable energy sources continue to enjoy.

Many feel that the subsidies showered on fossil fuels have helped to enhance America's economy, national security, and international influence by establishing thriving global energy industries dominated by U.S. firms. By this logic, if America wants to be a global leader in the transition to clean and renewable energy, then political and financial support will likely need to be redirected toward these new energy sources. In his 2010 State of the Union speech, U.S. President Barack Obama declared: "The nation that leads the clean energy economy will be the nation that leads the global economy." Whether the United States will be that nation remains to be seen.

Comparing Energy Sources

In a world facing climate change, air pollution, and risks to energy security, the choices we make as a society about energy are far-reaching. With the stakes so high, and with a variety of energy sources to choose from, which sources should we pursue?

A number of researchers have compared energy sources, trying to determine which are cleanest, which release the fewest greenhouse gases, and which provide the most economic stability. Let's examine one recent study that sought to analyze all the biggest issues in a comprehensive way.

In 2009 Mark Jacobson, director of the Atmosphere/Energy Program at Stanford University, undertook a broad review of renewable energy sources, digging deep into the published literature to calculate impacts of different energy sources across their entire life cycles (including their materials, construction, operation, and so on). Reviewing over 100 existing studies, his life-cycle analysis (p. 269, 623) was published in the journal *Energy and Environmental Science*.

It is difficult to compare energy sources because some are used primarily to generate electricity whereas others are used mostly for heating and cooling, while still others provide fuel for transportation. To make an apples-to-apples comparison, Jacobson calculated impacts for each source as if it alone were being used to power all the vehicles in the United States. For example, for electricity-producing sources such as wind or PV solar, he calculated how much was needed to power electric vehicles for the entire nation, and he compared the impacts to those from powering the same number of combustion-engine vehicles with liquid fuels.

Mark Jacobson of Stanford University

In all, Jacobson evaluated and ranked nine electric power sources and two liquid fuel sources, in conjunction with three vehicle types: battery-electric vehicles (p. 547), hydrogen-fuel-cell vehicles (p. 604), and E85 flex-fuel vehicles (p. 569).

He first examined carbon dioxide emissions. His analysis showed that wind power was lowest in CO_2 emissions but that nearly all renewable options eliminated the vast majority of emissions from vehicles, thus reducing total U.S. emissions by close to 30%. The exceptions were corn-based ethanol and cellulosic ethanol (pp. 569–570, 572), both of which, he found, would be likely to *increase* emissions.

Next, Jacobson analyzed air pollution from seven major pollutants. Using data on U.S. deaths attributed to fossil fuel pollution each year, he calculated the number of deaths expected to result from using each energy source to fuel vehicles. The number of deaths was close to zero for all renewable sources except for biofuels. Deaths from ethanol emissions were equivalent in number to those currently attributed to gasoline. Nuclear power was equivalent to most renewables in having a low death rate, but Jacobson also calculated impacts from potential nuclear proliferation and weapons use by a government or by terrorists. When these factors were included, nuclear power emerged as the least safe option.

Jacobson next looked at the footprint in land area that each energy source takes up. He calculated area taken up by energy infrastructure directly, as well as area influenced by the spacing of infrastructure. All electricity-generating renewables were very modest in their land demand, with hydropower and wind power taking up greater areas than the rest. Again, biofuels were high-impact; both corn ethanol and cellulosic ethanol required land areas greater than the size of California.

Jacobson continued by assessing water consumption, effects on wildlife, thermal pollution, chemical pollution of water, radioactive waste, operating reliability, and energy supply disruption.

With regard to supply disruption, he concluded that although many renewables are intermittent in their supply, they are least prone to major disruptions because they are less centralized than coal-fired plants, nuclear plants, or oil and ethanol refineries. Moreover, he argued that intermittency could be overcome by integrating sources over large geographic areas and by combining them strategically through the day, taking advantage of peaks in power from solar and wind and balancing the troughs with geothermal, pumped-storage hydropower, batteries, hydrogen fuel, and other means (**FIGURE 1**).

When Jacobson at last went to assess overall results and rank the energy sources by a weighted average of all the above issues, he found that wind power was most desirable, followed by concentrated solar power, geothermal power, tidal power, and

PV solar (**FIGURE 2**). Worst were corn ethanol and cellulosic ethanol.

Is it realistic to think we could power all our vehicles with wind power? Jacobson says yes. He calculated that to power all U.S. vehicles would require between 73,000 and 144,000 5-megawatt wind turbines. This sounds like a lot, yet it is fewer than the 300,000 airplanes that the United States built during World War II, and fewer than the number of (mostly smaller) turbines already erected worldwide. One reason so few turbines are needed is that battery-electric vehicles make use of power much more efficiently than gasoline-powered vehicles, so that far less energy is required.

As his research paper was being published, Jacobson also co-authored an article in the popular magazine *Scientific American* with Mark Delucchi of the University of California–Davis, analyzing how the world could meet all its energy needs from renewable energy. The pair concluded that our needs could be met with a combination of:

- 490,000 tidal turbines
- 5350 geothermal plants
- 900 hydroelectric plants
- 3.8 million wind turbines
- 720,000 wave converters
- 1.7 billion rooftop PV systems
- 49,000 concentrated solar power plants
- 40,000 PV power plants

These researchers judged that our main potential barrier was not the scale of development but rather the availability of a handful of rare-earth metals in limited supply (e.g., silver, platinum, lithium, indium, tellurium, and neodymium).

Critics of Jacobson's work have pointed to such articles in popular magazines, and to public speaking events, to question Jacobson's objectivity and paint him as an advocate. Critics who objected to his inclusion of impacts from a nuclear weapons disaster in his comparison of energy sources pointed out that he has spoken out against nuclear power in public forums.

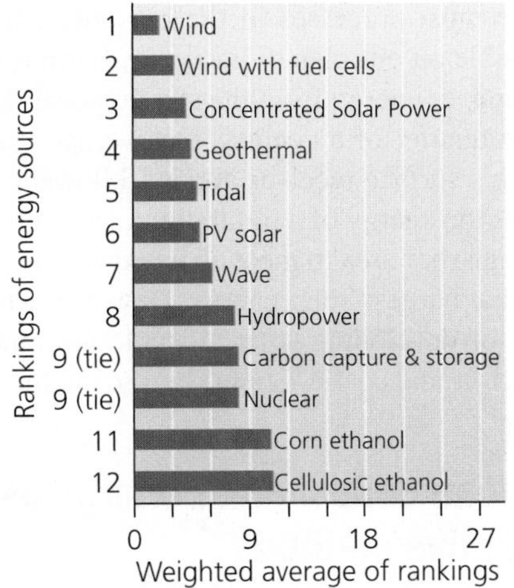

FIGURE 2 Jacobson found wind power to be the most desirable renewable energy source overall, and ethanol to be least desirable. *Data from Jacobson, M., 2009. Review of solutions to global warming, air pollution, and energy security.* Energy & Environmental Science *2009 (2): 148–173.*

Any scientist who steps out of the lab, field, or office and advocates publicly for particular solutions is liable to be criticized by those who disagree with his or her positions. Whether such criticisms are fair—and whether they are in society's best interest—are complicated questions. On one hand, remaining objective and free of advocacy gives science its strength and credibility. On the other hand, don't we want our advocates to be the people who have studied the issues themselves and are the best informed about them?

Regardless of how one views the recent research by Jacobson and his colleagues, the number of such energy comparisons is increasing, and policymakers today have a better and better array of information to draw on to help chart our energy future. ●

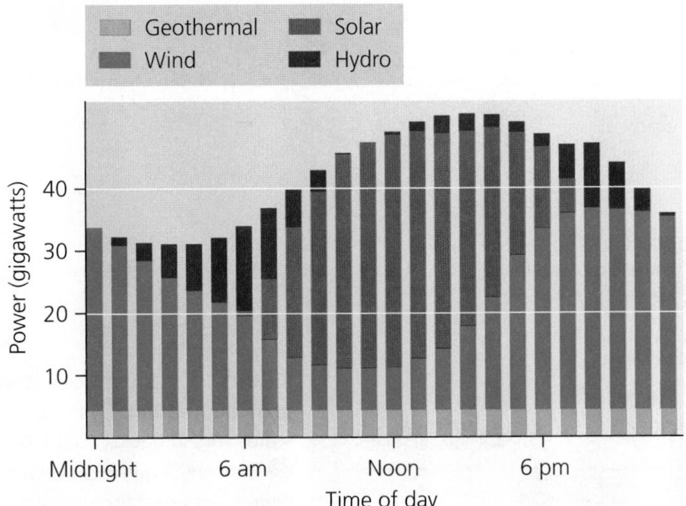

FIGURE 1 Intermittent wind and solar power can be balanced with stable and flexible geothermal and hydropower. This model for California shows how such a combination might reliably supply our daily energy needs. *Data from Jacobson, M., and M. Delucchi, 2009. A path to sustainable energy by 2030.* Scientific American *Nov. 2009 pp. 58–65.*

Solar Energy

The sun releases astounding amounts of energy by converting hydrogen to helium through nuclear fusion (p. 558). The tiny proportion of this energy that reaches Earth is enough to drive most processes in the biosphere, helping to make life possible on our planet. Each day, Earth receives enough **solar energy,** or energy from the sun, to power human consumption for a quarter of a century. On average, each square meter of Earth's surface receives about 1 kilowatt of solar energy—17 times the energy of a lightbulb. As a result, a typical home has enough roof area to meet all its power needs with rooftop panels that harness solar energy. However, we are still in the process of developing solar technologies and learning the most effective and cost-efficient ways to put the sun's energy to use.

We can collect solar energy using passive or active methods

The simplest way to harness solar energy is through **passive solar energy collection.** In this approach, buildings are designed to maximize absorption of sunlight in winter and to keep the interior cool in the heat of summer. Installing south-facing windows maximizes the capture of winter sunlight. Overhangs shade windows in summer, when the sun is high in the sky and when cooling, not heating, is desired.

Passive solar techniques also use materials that absorb heat, store it, and release it later. Such *thermal mass* (of straw, brick, concrete, or other materials) often makes up floors, roofs, and walls, or can be used in portable blocks. Planting vegetation around a building to buffer the structure from temperature swings is another passive solar approach. Passive solar approaches are an important component of sustainable buildings (pp. 348–350). By heating buildings in cold weather and cooling them in warm weather, passive solar methods conserve energy and reduce energy costs.

Active solar energy collection makes use of devices to focus, move, or store solar energy. We can use such technologies to heat water and air in our homes and businesses. One common method involves installing *flat plate solar collectors* on rooftops. These panels generally consist of dark-colored, heat-absorbing metal plates mounted in flat glass-covered boxes. Water, air, or antifreeze runs through tubes that pass through the collectors, transferring heat from the collectors to the building or its water tank (**FIGURE 21.7**). Heated water can be stored for later use and passed through pipes designed to release the heat into the building.

Over 200 million households worldwide heat water with solar collectors. Only 1.5 million of these are in the United States, and most of these are used for swimming pools. Active solar heating is used more widely in China and in Europe, where Germans motivated by feed-in tariffs installed 200,000 new systems in 2008 alone. Because solar heating systems do not need to be connected to an electrical grid, they are useful in isolated locations. In Gaviotas, a remote town on the high plains of Colombia in South America, residents use active solar technology for heating, cooling, and water purification. This shows that solar energy need not be confined to wealthy communities or to regions that are always sunny.

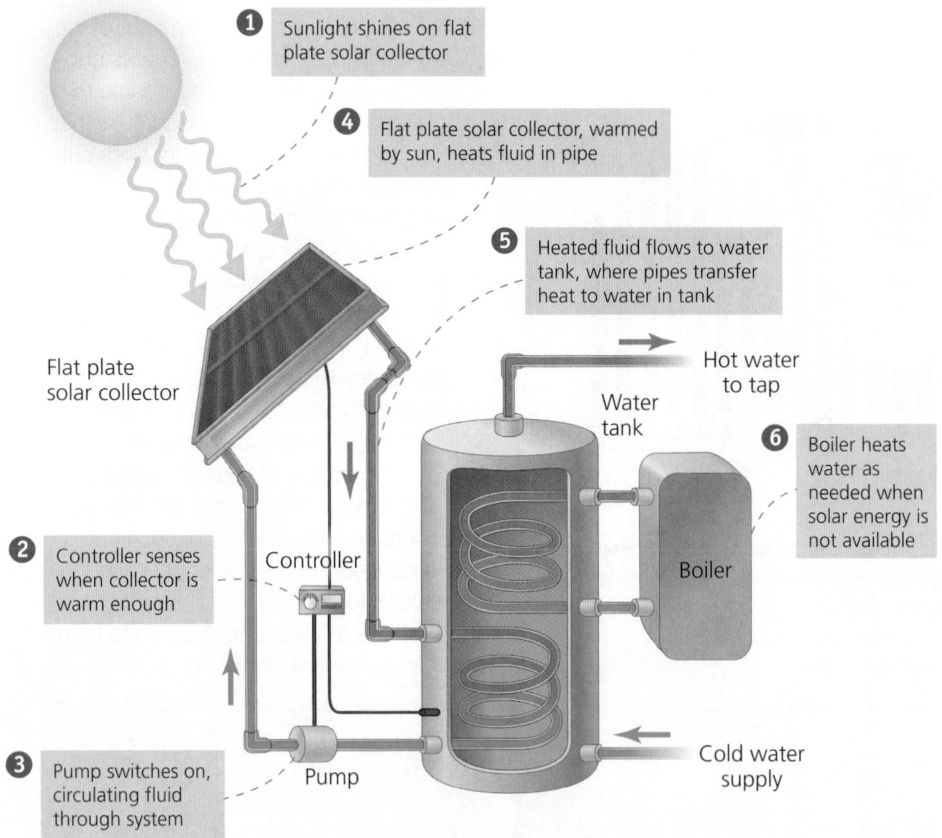

FIGURE 21.7 Solar collectors can provide heating. Systems for heating water vary in design, but typically 1 sunlight is gathered on a flat plate solar collector, until a controller 2 switches on a pump 3 to circulate fluid through pipes to the collector. The sunlit collector heats the fluid 4, which flows through pipes 5 to a water tank. The hot fluid in the pipes transfers heat to the water in the tank, and this heated water is available for the taps of the home or business. Generally an external boiler 6 kicks in to heat water when solar energy is not available.

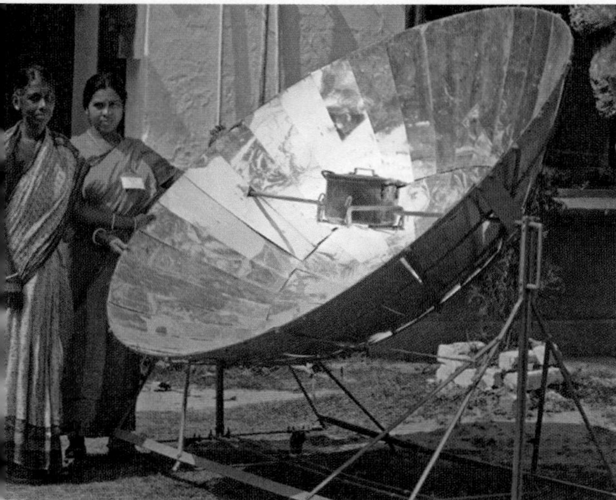

(a) Solar cooker in India

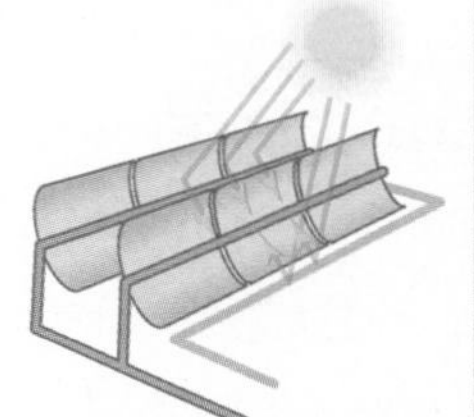

Curved reflectors heat liquid in horizontal tubes

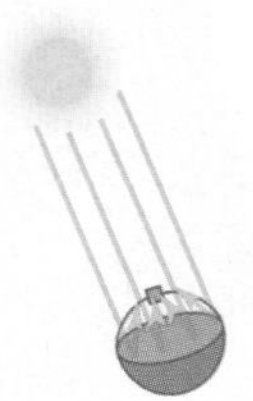

Each curved reflector focuses light on its own small receiver

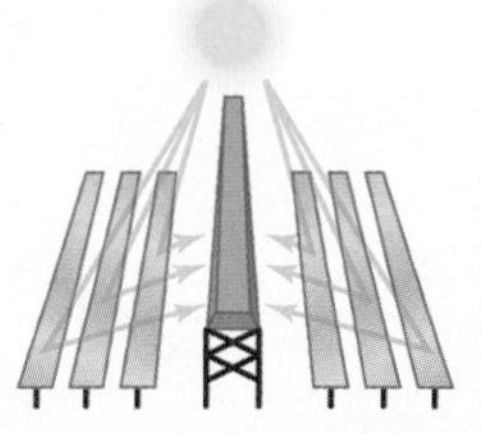

Curved mirrors reflect light onto absorber tube

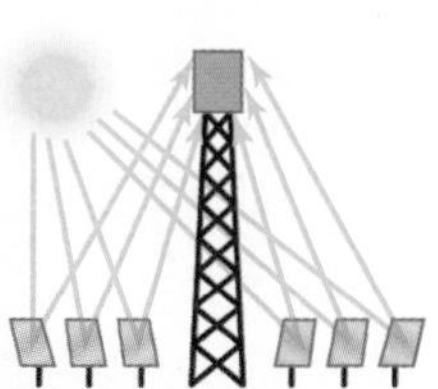

Field of heliostats focus light on central power tower

(b) Four methods of concentrating solar power

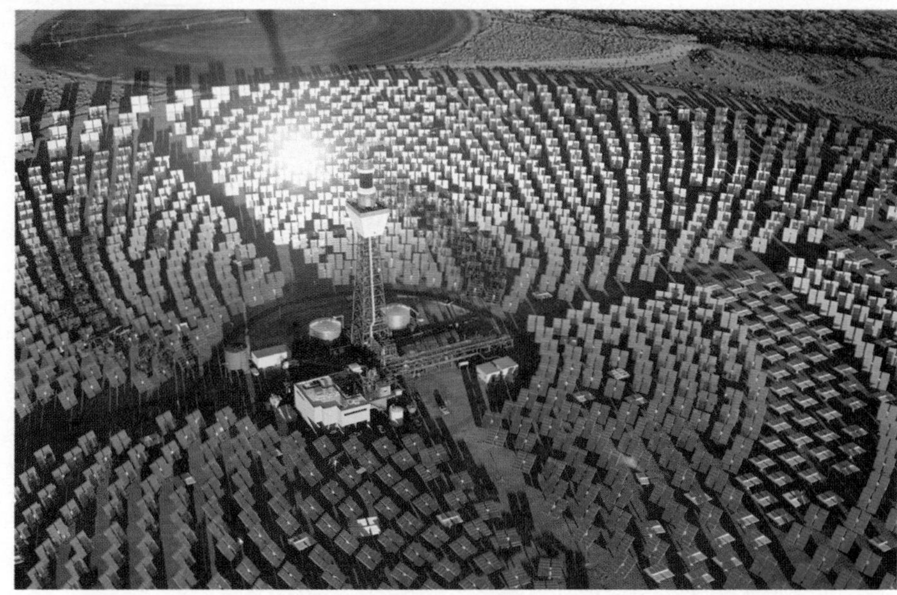

(c) The Solar Two power tower facility in California

FIGURE 21.8 By concentrating solar energy, we can provide heat and electricity. Solar cookers **(a)** focus solar radiation to cook food. Utilities concentrate solar power with several approaches **(b)** to generate electricity at large scales. At the Solar Two facility in the southern California desert **(c)**, hundreds of mirrors reflect sunlight onto a central receiver atop a power tower, producing electricity for 10,000 households.

Concentrating solar rays magnifies energy

As any mischievous young boy who has killed ants with a magnifying glass knows, we can intensify solar energy by gathering sunlight from a wide area and focusing it on a single point. This is the principle behind *solar cookers*, simple portable ovens that use reflectors to focus sunlight onto food and cook it (FIGURE 21.8a). Such cookers are proving extremely useful in the developing world.

At much larger scales, utilities are using the principle behind solar cookers to generate electricity. **Concentrated solar power (CSP)** is being harnessed by several methods (FIGURE 21.8b) in sunny regions in Spain, the U.S. Southwest, and elsewhere. The dominant technology so far is the parabolic trough approach (leftmost diagram in Figure 21.8b), in which curved mirrors focus sunlight onto synthetic oil in pipes. The superheated oil is piped to an adjacent facility where it creates steam that drives turbines to generate electricity. In another approach, numerous mirrors concentrate sunlight onto a receiver atop a tall "power tower" (FIGURE 21.8c). From this central receiver, heat is transported by air or fluids (often molten salts) and piped to a steam-driven generator to create electricity. CSP facilities can harness light from lenses or mirrors spread across large areas of land, and the lenses or mirrors may move to track the sun's movement.

CSP facilities need to be located in sunny areas, but they have great potential. The International Energy Agency estimates that just 260 km^2 (100 mi^2) of Nevada desert could generate enough electricity to power the entire U.S. economy. Another study estimated that CSP could fulfill one-quarter of global electricity demand by 2050 if we step up investment. German industrialists and investors have spearheaded an ambitious effort to create an immense CSP facility in Africa's Sahara Desert. In this planned $775-billion project, called Desertec, thousands of mirrors spread across vast areas of desert in Morocco would harness the Sahara's sunlight and transmit electricity to Europe, the Middle East, and North Africa. Critics of the project—including Hermann Scheer—say it would be vulnerable to sandstorms and political disputes and would be less reliable and more expensive than the decentralized production from rooftop panels that feed-in tariffs are promoting in Europe. Moreover, many people are growing wary of the environmental impacts that such large-scale developments may pose (see THE SCIENCE BEHIND THE STORY, pp. 592–593).

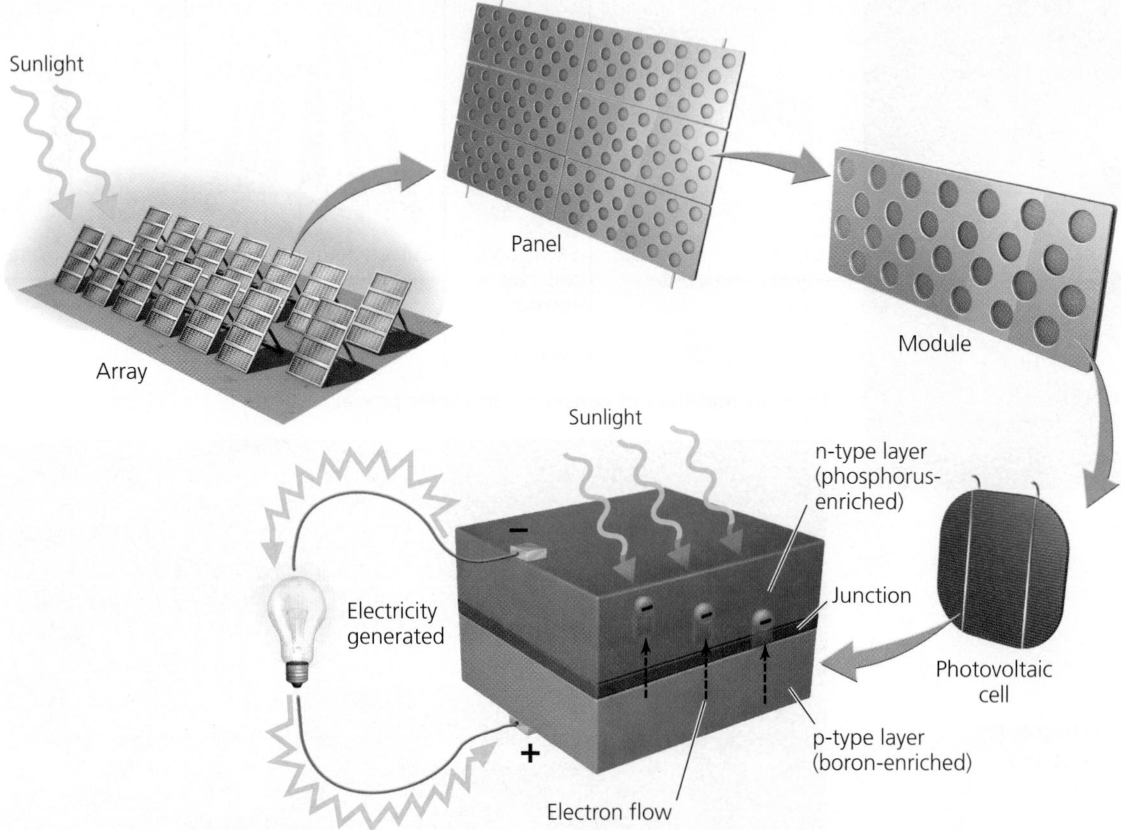

FIGURE 21.9 A photovoltaic (PV) cell converts sunlight to electrical energy. When sunlight hits the silicon layers of the cell, electrons are knocked loose from some of the silicon atoms and tend to move from the boron-enriched "p-type" layer toward the phosphorus-enriched "n-type" layer. Connecting the two layers with wiring remedies this imbalance as electrical current flows from the n-type layer back to the p-type layer. This direct current (DC) is converted to alternating current (AC) to produce usable electricity. PV cells are grouped in modules, comprising panels, which can be erected in arrays.

Photovoltaic cells generate electricity directly

The most direct way to produce electricity from sunlight involves photovoltaic (PV) systems. **Photovoltaic (PV) cells** convert sunlight to electrical energy by making use of the *photovoltaic effect*, or *photoelectric effect*. This effect occurs when light reaches the PV cell and strikes one of a pair of plates made primarily of silicon, a semiconductor that conducts electricity. The light causes one plate to release electrons, which are attracted by electrostatic forces to the opposing plate. Connecting the two plates with wires enables the electrons to flow back to the original plate, creating an electrical current (direct current, DC), which can be converted into alternating current (AC) and used for residential and commercial electrical power (**FIGURE 21.9**). Small PV cells may already power your watch or your calculator. Atop the roofs of homes and other buildings, PV cells are arranged in modules, which comprise panels, which can be gathered together in arrays. Arrays of PV panels can be seen on the roofs of the German houses in the photo that opens this chapter (p. 580).

Researchers are experimenting with variations on PV technology, and manufacturers today are developing **thin-film solar cells,** photovoltaic materials compressed into ultra-thin sheets. Thin-film solar cells are lightweight and far less bulky than the standard crystalline silicon cells shown in Figure 21.9. Although less efficient at converting sunlight to electricity, they are cheaper to produce. Thin-film technologies can be incorporated into roofing shingles and potentially many other types of surfaces, even highways! For these reasons, many people view thin-film solar technologies as a promising direction for the future.

Photovoltaic cells of all types can be connected to batteries that store the accumulated charge until it is needed. Alternatively, producers of PV electricity can sell their power to their local utility if they are connected to the regional electric grid. In parts of 46 U.S. states, homeowners can sell power to their utility in a process called **net metering,** in which the value of the power the consumer provides is subtracted from the consumer's monthly utility bill. Feed-in tariff systems like Germany's go a step further by paying producers more than the market price of the power. Feed-in tariffs thereby generally offer power producers the hope of turning a profit.

Solar energy is expanding

Although active solar technology dates from the 18th century, it was pushed to the sidelines as fossil fuels came to dominate our energy economy. Funding for research and development of solar technology has been erratic. After the 1973 oil embargo (p. 544), the U.S. Department of Energy funded the installation and testing of over 3000 PV systems, providing a boost to fledgling companies in the solar industry. But later, as oil prices declined, so did government support for solar power.

Largely because of the lack of investment, solar energy contributes just 0.22%—22 parts in 10,000—of the U.S. energy supply, and just 0.1% of U.S. electricity generation. Even in Germany, which gets more of its energy from solar than any other nation, the percentage is only 5%. However, solar energy use has grown by 30% each year worldwide in the past four decades, a growth rate second only to that of wind power. Solar energy is proving especially attractive in developing countries, many of which are rich in sun but poor in infrastructure, and where hundreds of millions of people still live without electricity.

PV technology is the fastest-growing power generation technology today, having recently doubled every two years (**FIGURE 21.10**). China leads the world in yearly production of PV cells, followed by Germany and Japan. Germany leads the world in installation of PV technology, and German rooftops host over one-third of all PV cells in the world. Germany's investment began in 1998 when Hermann Scheer spearheaded a "100,000 Rooftops" program to install PV panels atop 100,000 German roofs. The popular program ended up easily surpassing this goal, and today over *1.2 million* German rooftops have PV systems.

The United States ranks fifth in production of PV cells, and U.S. firms account for only 4% of the industry. Recent federal tax credits and state-level initiatives may help the United States recover the leadership it lost to other nations in this technology, but China is moving faster and is dominating the market. In fact, the Chinese government's support of its solar industry has led to so much production that supply has outstripped global demand in recent years. Highly subsidized Chinese firms are selling solar products abroad at low prices (often at a loss), driving American and European solar manufacturers out of business. In response, the United States and European nations have slapped tariffs on Chinese imports, while both sides have filed complaints with the World Trade Organization in an escalating global trade dispute.

Despite volatility in the industry as firms try to deal with swings in national policies, global production of PV cells continues to rise sharply, while prices fall (see Figure 21.10). At the same time, efficiencies are increasing, making each unit more powerful. Use of solar technology should continue to expand as prices fall, technologies improve, and governments enact economic incentives to spur investment.

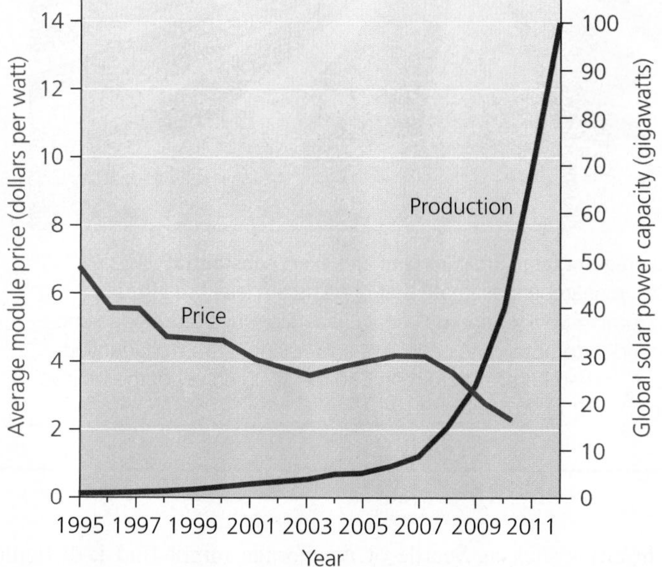

FIGURE 21.10 Global production of PV cells has been growing exponentially, and prices have fallen rapidly. *Data from REN21, 2013.* Renewables 2013: Global status report. *REN21, UNEP, Paris; and U.S. Department of Energy, EERE, 2011.* 2010 Solar technologies market report.

Solar energy offers many benefits

The fact that the sun will continue burning for another 4–5 billion years makes it practically inexhaustible as an energy source for human civilization. Moreover, the amount of solar energy reaching Earth should be enough to power our civilization once we develop technology adequate to harness it. These advantages of solar energy are clear, but the technologies themselves also provide benefits. PV cells and other solar technologies use no fuel, are quiet and safe, contain no moving parts, require little maintenance, and do not require a turbine or generator to create electricity. An average unit can produce energy for 20–30 years.

Solar systems also allow for local, decentralized control over power. Homes, businesses, and isolated communities can use solar power to produce electricity without being near a power plant or connected to a grid. This is especially helpful in developing nations, where solar cookers (see Figure 21.8a) enable families to cook food without gathering fuelwood. This lessens people's daily workload and helps reduce deforestation. In refugee camps, solar cookers are helping to relieve social and environmental stress. The low cost of solar cookers—many can be built locally for \$2–10 each—has made them accessible in many impoverished areas. In developed nations, most PV systems are connected to the regional electric grid. As a result, homeowners with PV systems can sell their excess solar energy to their local utility through feed-in tariffs or net metering.

The development and deployment of solar systems are producing many new green-collar jobs. Currently, among major energy sources, PV technology employs the most people per unit energy output, resulting in over 800,000 jobs worldwide (see Figure 21.4).

A major advantage of solar power over fossil fuels is its reduction of greenhouse gas emissions and other air pollutants (see Figure 21.3). The manufacture of photovoltaic cells *does* currently require fossil fuel use, but once up and running, a PV system produces no emissions. Consumers can access online calculators offered by the U.S. Department of Energy and the U.S. Environmental Protection Agency to estimate the economic and environmental effects of installing a PV system. At the time of this writing, these calculators estimated that installing a 5-kilowatt PV system in a home in Fort Worth, Texas, to provide over half of its annual power needs would save the homeowners \$681 per year on energy bills and would prevent over 5 tons of carbon dioxide emissions per year—as much CO_2 as results from burning 570 gallons of gasoline. Even in overcast Seattle, Washington, a 5-kilowatt system producing half a home's energy needs can save \$310 per year and prevent over 3.5 tons of CO_2 emissions (equal to 390 gallons of gas).

THE SCIENCE BEHIND THE STORY

What Are the Impacts of Solar and Wind Development?

Renewable energy sources and technologies alleviate many of the negative impacts of fossil fuel combustion, and may one day sustainably fulfill our energy needs. However, this does not mean that renewable energy is a panacea free of costs. As our society decides how to pursue energy sources such as solar and wind power, we will need to consider their impacts as well as their benefits.

This has become clear in recent years as the Cape Wind project in Massachusetts and a number of large solar projects in California have brought to the fore a host of issues that some clean-energy proponents had not considered. Scientific study of these impacts is just getting underway, and will be important as energy development proceeds.

The several dozen solar power installations currently under review for the Mojave Desert and other arid regions of California would, if constructed, cover many thousands of acres of land (**FIGURE 1**).

Desert environments are particularly sensitive, so researchers say we should expect substantial impacts. Besides altering the pristine appearance of an undeveloped landscape, arrays of thousands of mirrors or panels affect communities of plants and animals by casting shade and altering microclimate. Altered conditions tend to hurt native desert-adapted species while helping invasive weeds. At existing solar facilities, the sites are graded and sprayed with herbicide, eliminating plants and damaging fragile soils. Human presence increases as workers maintain

Local residents demonstrate for and against the proposed Cape Wind farm.

the facilities. Solar power plants also require water for cooling and cleaning, and water is scarce in the arid regions hosting most of these facilities. All these impacts will have consequences for plants, animals, and ecosystems.

Large-scale projects need government approval and are subject to the environmental impact statement process (pp. 174–175). As a result, teams of researchers study conditions at each site to determine what impacts energy development may have. If impacts are judged to be severe enough, then government agencies can insist that plans be amended.

For instance, the California Energy Commission asked for limits on the proposed Calico Solar Project in southern California in 2010, after biologists concluded that the project would damage habitat of the desert tortoise and bighorn sheep. The company agreed to reduce the size of its footprint by nearly half, reducing estimated impacts to wildlife by 80%, and the Commission approved the project.

In central California, a solar project underwent 18 months of environmental analysis and was approved after the Solargen company agreed to purchase and set aside 23,000 acres of preserved land as "mitigation" for the 3200 acres it was developing. A third California solar project, the Topaz Solar Farm, was scaled back in size after researchers found that scaling back was needed

FIGURE 1 Solar power farms require large areas of land and exert substantial environmental impacts. Still, researchers estimate that impacts are less than those demanded by fossil fuels; burning coal for energy uses at least as much land, once one includes the strip mining needed to obtain the coal. This solar plant in Kramer Junction, California, is one of nine that spread across more than 650 ha (1600 acres) of the Mojave Desert, providing power for over 230,000 homes.

Location, timing, and cost can be drawbacks

Solar energy currently has three major disadvantages. One is that not all regions are equally sunny (**FIGURE 21.11**). People in cities such as Seattle or Anchorage might find it difficult to harness enough sunlight most of the year to rely on solar power. (However, observe in Figure 21.11 that Germany receives even less sun than Alaska, and yet it is the world leader in solar power!)

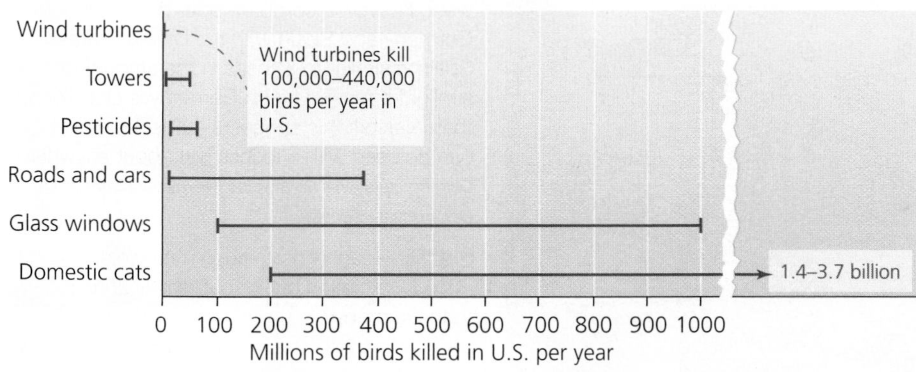

FIGURE 2 Wind turbines kill birds that fly into them. Yet far more birds are killed by other human causes. Shown are ranges of recent estimates of yearly bird mortality in the United States from several main causes. Habitat alteration is responsible for still more than any of the causes shown. *Data from American Bird Conservancy, from sources dating through 2010; and Loss, S.R. et al., 2012. The impact of free-ranging domestic cats on wildlife of the United States.* Nature Communications *4: Article 1396.*

to protect farmland; minimize aesthetic impacts; and lessen disturbance to tule elk, kit foxes, pronghorn antelope, burrowing owls, and seasonal freshwater pools.

On the U.S. Great Plains, researchers from the National Renewable Energy Laboratory are currently studying impacts of solar installations on prairie ecosystems by comparing a developed site and an undeveloped control site.

Given the impacts of large-scale solar facilities, researchers have determined that installing photovoltaic panels on rooftops of buildings is a low-impact alternative. Simply adding PV panels or roofing tiles to a rooftop has no effect on the landscape. One study, led by five Dutch, German, and American researchers, compared impacts of various ground-based and rooftop PV systems in Germany and in Arizona. The researchers assessed impacts over the systems' entire life cycles (from production to installation to operation). They found that besides avoiding land use impacts, the rooftop systems also emitted significantly fewer greenhouse gases.

A different study in 2008 measured the amount of energy required by PV cells throughout their life cycles and found that replacing fossil fuel energy with PV solar power would prevent 89–98% of greenhouse gas emissions.

Scientists have also studied health impacts of PV panels. PV cells can pose risks to workers manufacturing them, because they contain toxic chemicals such as cadmium and arsenic and release fine silicon dust that can damage lung tissue. Researchers have concluded that proper attention to worker safety can greatly reduce these risks and that the exposure risks are not much different from other exposure risks we all face in our modern industrialized society.

The overall messages from studies so far are that (1) solar power, even with its impacts, is still cleaner and more sustainable than fossil fuel power; and that (2) we can minimize the impacts of solar power by using rooftop panels and developing better technologies.

Similar messages are emerging from the science on wind power. One major concern is that birds and bats are killed when they fly into the spinning blades of turbines. At California's Altamont Pass wind farm, turbines killed dozens of golden eagles and other raptors in the 1990s. Studies since then at other sites suggest that bird deaths may be a less severe problem than was initially feared, but uncertainty remains. For instance, one European study indicated that migrating seabirds fly past offshore turbines without problem, but other data show that resident seabird densities have declined near turbines.

On land, the wind industry estimates that about two birds are killed per 1-megawatt-turbine per year. This is far fewer than the hundreds of millions of birds being killed each year by television, radio, and cell phone towers; pesticides; automobiles; glass windows; and domestic cats (**FIGURE 2**). If you own a cat and let it outside, you may be killing more wildlife than most wind farms.

At this point, bat mortality appears to be a more severe problem at wind turbines, but further research is needed. We can reduce wildlife impacts by avoiding development at sites on known migratory flyways or amid prime habitat for bat and bird species that are likely to fly into the blades.

Continued studies on the impacts of wind and solar development should help us find ways to harness renewable energy and attain a sustainable energy future while minimizing the environmental and social impacts of this development. ■

A second drawback is that solar energy is an intermittent resource. Daily or seasonal variation in sunlight can limit stand-alone solar systems if storage capacity in batteries or fuel cells is not adequate or if backup power is not available from a municipal electric grid. Pumped-storage hydropower (p. 574) can sometimes help by compensating for periods of low solar power production.

The third disadvantage of current solar technology is the up-front cost of the equipment. Because of the investment cost, solar power remains the most expensive way to produce

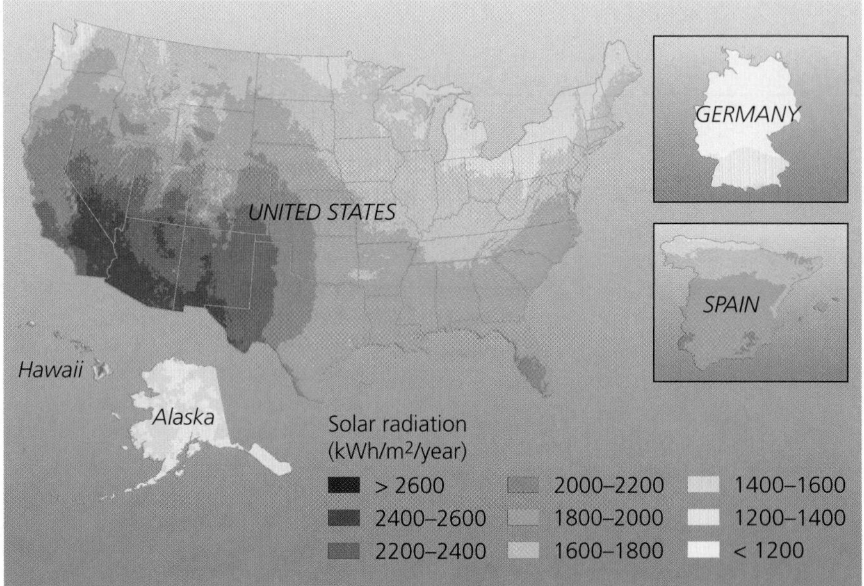

FIGURE 21.11 Solar radiation varies from place to place. Harnessing solar energy is more profitable in sunny regions such as the southwestern United States than in cloudier regions such as Alaska and the Pacific Northwest. However, compare solar power leaders Germany and Spain with the United States. Spain is similar to Kansas in the amount of sunlight it receives, and Germany is cloudier than Alaska! This suggests that solar power can be used with success just about anywhere. *Data from National Renewable Energy Laboratory, U.S. Department of Energy.*

DATA Q Roughly how much more solar radiation does southern Arizona receive than Germany?

electricity (see Figure 21.5). It may take 20 years or more for most homeowners to break even on an investment in PV arrays or solar collectors. The high costs are due to the fact that the technologies are young and developing. Moreover, they are competing against energy sources (fossil fuels and nuclear power) that have remained relatively cheap as a result of decades of taxpayer support and whose external costs (pp. 146, 165) are not included in market prices. As a result, market prices have given governments, businesses, and consumers little economic incentive to switch to solar energy thus far.

However, declines in price and improvements in efficiency of solar technologies so far are encouraging, even in the absence of significant funding from government and industry. At their advent in the 1950s, solar technologies had efficiencies of around 6% while costing $600 per watt. Today, PV cells are showing up to 20% efficiency commercially and 40% efficiency in lab research, suggesting that future solar cells could be more efficient than any energy technologies we have today. Solar systems are becoming less expensive and now can sometimes pay for themselves in less than 10–20 years. After that time, they provide energy virtually for free as long as the equipment lasts.

Wind Power

As the sun heats air in the atmosphere, the movement of differentially heated air masses produces wind. We can harness **wind power** from air's movement by using **wind turbines,** mechanical assemblies that convert wind's kinetic energy (p. 30), or energy of motion, into electrical energy.

Wind turbines convert kinetic energy to electrical energy

Today's wind turbines have their roots in Europe, where wooden windmills were used for 800 years to grind grain and pump water. The first wind turbine built to generate electricity was constructed in the late 1800s in Cleveland, Ohio. However, it was not until after the 1973 oil embargo that governments and industry in North America and Europe began funding research and development for wind power.

In a modern wind turbine, wind turns the blades of the rotor, which rotate machinery inside a compartment called a *nacelle*, which sits atop a tower (**FIGURE 21.12**). Inside the nacelle

FIGURE 21.12 A wind turbine converts wind's energy of motion into electrical energy. Wind causes a turbine's blades to spin, turning a shaft that extends into the nacelle. Inside the nacelle, a gearbox converts the rotational speed of the blades, which can be up to 20 revolutions per minute (rpm) or more, into much higher speeds (over 1500 rpm). This provides adequate motion for the generator to produce electricity.

are a gearbox, a generator, and equipment to monitor and control the turbine's activity. Today's towers average 80 m (260 ft) in height, and the largest are taller than a football field is long. Higher is generally better, to minimize turbulence (and potential damage) while maximizing wind speed. Turbines are often erected in groups; such a development is called a **wind farm.** The world's largest wind farms contain hundreds of turbines spread across the landscape.

Engineers design turbines to yaw, or rotate back and forth in response to changes in wind direction, ensuring that the motor faces into the wind at all times. Some turbines are designed to generate low levels of electricity by turning in light breezes. Others are programmed to rotate only in strong winds, generating large amounts of electricity in short time periods. Slight differences in wind speed yield substantial differences in power output, for two reasons. First, the energy content of wind increases as the square of its velocity; thus if wind velocity doubles, energy quadruples. Second, an increase in wind speed causes more air molecules to pass through the wind turbine per unit time, making power output equal to wind velocity cubed. Thus a doubled wind velocity results in an eightfold increase in power output.

Wind power is growing fast

Like solar energy, wind provides just a small proportion of the world's power needs, but wind power is growing fast—doubling every three years (FIGURE 21.13). Five nations account for three-quarters of the world's wind power output (FIGURE 21.14a), but dozens of nations now produce wind power. Germany had long produced the most, but the United States overtook it in 2008, and China surpassed the United States two years later.

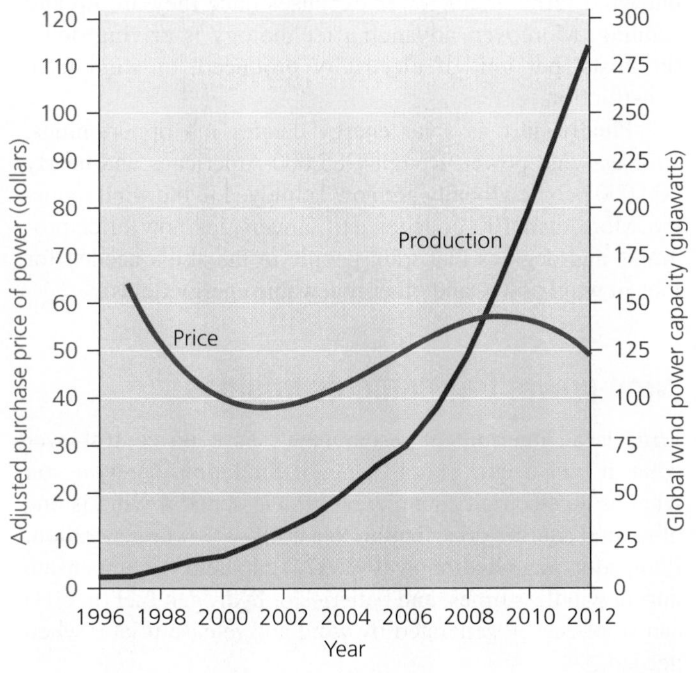

FIGURE 21.13 Global production of wind power has been doubling every three years in recent years, and prices have fallen slightly. *Data from Global Wind Energy Council; and U.S. Department of Energy, EERE, 2012.* 2011 Wind technologies market report.

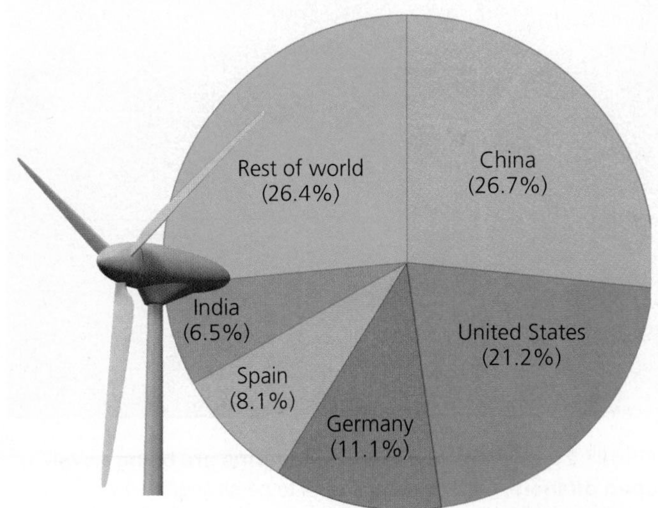

(a) Percentage of global wind power in each nation

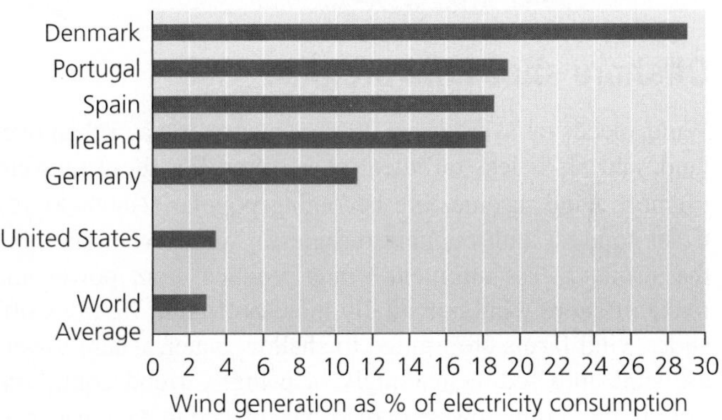

(b) Leading nations in proportion of electricity from wind power

FIGURE 21.14 Several nations are leaders in wind power. Most of the world's wind power capacity **(a)** is concentrated in a handful of nations led by China, the United States, and Germany. Yet tiny Denmark **(b)** obtains the highest percentage of its electricity needs from wind. *Data from (a) Global Wind Energy Council, 2013.* Global wind report: Annual market update 2012. *GWEC, Brussels, Belgium; and (b) U.S. Department of Energy, EERE, 2012.* 2011 Wind technologies market report.

Denmark leads the world in obtaining the greatest percentage of its energy from wind power. In this small European nation, wind farms supply nearly 30% of Danish electricity needs (FIGURE 21.14b). Germany is fifth in this respect, and the United States is 13th. Texas generates the most wind power of all U.S. states, while Iowa and South Dakota each obtain nearly 25% of their electricity from wind.

Wind power's growth in the United States has been haphazard because Congress has not committed to a long-term federal tax credit for wind development, but instead has passed a series of short-term renewals, leaving the industry uncertain about how much to invest. However, experts agree that wind power's growth will continue, because only a small portion of this resource is currently being tapped and because wind power at favorable locations already generates electricity at prices nearly as low as fossil fuels (see Figure 21.5). A 2008 report by a consortium of experts outlined how the United States could meet fully one-fifth of its electrical demands with wind power by 2030.

FIGURE 21.15 **More and more wind farms are being developed offshore.** Offshore winds tend to be stronger yet less turbulent.

Offshore sites hold promise

Wind speeds on average are 20% greater over water than over land, and air is less turbulent over water. For these reasons, offshore wind turbines are becoming popular (FIGURE 21.15). Costs to erect and maintain turbines in water are higher, but the stronger, less turbulent winds produce more power and make offshore wind potentially more profitable. Today's offshore wind farms are limited to shallow water, where towers are sunk into sediments singly or using a tripod configuration. In the future, towers may also be placed in deep water on floating pads anchored to the seafloor.

Denmark erected the first offshore wind farm in 1991, and soon more came into operation across northern Europe, where the North Sea and Baltic Sea offer strong winds. Once Germany raised its feed-in tariff rate for offshore wind from 9 cents to 15 cents per kilowatt-hour in 2009, many projects began construction and today several are operating. By 2013, over 1800 wind turbines were operating in 65 wind farms in the waters of 10 European nations.

In the United States, no offshore wind farms have yet been constructed, but development of the first was approved in 2010 after nine years of debate. The Cape Wind offshore wind farm, if constructed, will feature 130 turbines rising from Nantucket Sound 8 km (5 mi) off the coast of Cape Cod in Massachusetts. In announcing the government's approval, U.S. Interior Secretary Ken Salazar predicted that it would be "the first of many projects up and down the Atlantic coast." Indeed, as of 2013, eight offshore wind developments were in the planning stages off the Northeast coast, one off the Texas coast, and one in Lake Erie.

Wind power has many benefits

Like solar power, wind power produces no emissions once the equipment is manufactured and installed. As a replacement for fossil fuel combustion in the average U.S. power plant, running a 1-megawatt wind turbine for 1 year prevents the release of more than 1500 tons of carbon dioxide, 6.5 tons of sulfur dioxide, 3.2 tons of nitrogen oxides, and 60 lb of mercury, according to the U.S. Environmental Protection Agency. The amount of carbon pollution that all U.S. wind turbines together prevent from entering the atmosphere is equal to the emissions from nearly 10 million cars, or from combusting the cargo of a 750-car freight train of coal each and every day.

Under optimal conditions, wind power appears efficient in its energy returned on investment (EROI; pp. 523–524, 572–573). Most studies find that wind turbines produce roughly 20 times more energy than they consume. This EROI value is better than that from most other energy sources. Wind farms also use less water than do conventional power plants.

Wind turbine technology can be used on many scales, from a single tower for local use to farms of hundreds that supply large regions. Small-scale turbine development can help make local areas more self-sufficient, just as solar energy can. For instance, the Rosebud Sioux Tribe of Native Americans set up a single turbine on its reservation in South Dakota. The turbine has been producing electricity for 200 homes and brings the tribe $15,000 per year in revenue. The tribe has now developed a 30-megawatt wind farm nearby, and 20 more turbines are slated to be added soon.

Another benefit of wind power is that farmers and ranchers can lease their land for wind development. A single large turbine can bring in $2000 to $4500 in annual royalties while occupying just a quarter-acre of land. Most of the land can still be used for agriculture. Royalties from the wind power company provide the farmer or rancher revenue while also increasing property tax income for their rural community.

Wind power involves up-front expenses to erect turbines and to expand infrastructure to allow electricity distribution, but over the lifetime of a project it requires only maintenance costs. Unlike fossil-fuel power plants, wind turbines incur no ongoing fuel costs. Startup costs of wind farms generally are higher than those of fossil-fuel plants, but wind farms incur fewer expenses once they are up and running. Moreover, advancing technology is driving down the costs, per unit of electricity produced, of wind farm construction.

Finally, just as solar energy creates job opportunities, so does wind power. Roughly 85,000 Americans and nearly 700,000 people globally are now employed in the wind industry. More than 100 colleges and universities now offer programs and degrees that train people in the skills needed for jobs in wind power and other renewable energy fields.

Wind power has some downsides

Wind is an intermittent resource; we have no control over when it will occur. This is a major limitation in relying on wind as an electricity source, but it is lessened if wind is one of several sources contributing to a utility's power generation. Pumped-storage hydropower (p. 574) can help to compensate during windless times, and batteries or hydrogen fuel (p. 603) can store energy generated by wind and release it later when needed.

Just as wind varies from time to time, it varies from place to place; some areas are windier than others. Global wind patterns combine with local topography—mountains, hills, water bodies, forests, cities—to create local wind patterns.

Resource planners and wind power companies study these patterns closely before planning a wind farm. Meteorological research has given us data with which to judge prime areas for locating wind farms. A map of average wind speeds across the United States (**FIGURE 21.16a**) reveals that mountainous regions are best, along with areas of the Great Plains. Based on such information, the wind power industry has located much of its generating capacity in states with high wind speeds (**FIGURE 21.16b**). Provided that wind farms are strategically erected in optimal locations, an estimated 15% of U.S. energy demand could be met using only 43,000 km^2 (16,600 mi^2) of land (with less than 5% of this land area actually occupied by turbines, equipment, and access roads).

However, most of North America's people live near the coasts, far from the Great Plains and mountain regions that have the best wind resources. Thus, continent-wide transmission networks would need to be enhanced to send wind power's electricity to these population centers.

When wind farms *are* proposed near population centers, local residents often oppose them. Turbines are generally located in exposed, conspicuous sites, and some people object to wind farms for aesthetic reasons, feeling that the structures clutter the landscape. Although polls show wide public approval of existing wind projects and of the concept of wind power, newly proposed wind projects often elicit the **not-in-my-backyard (NIMBY)** syndrome among people living

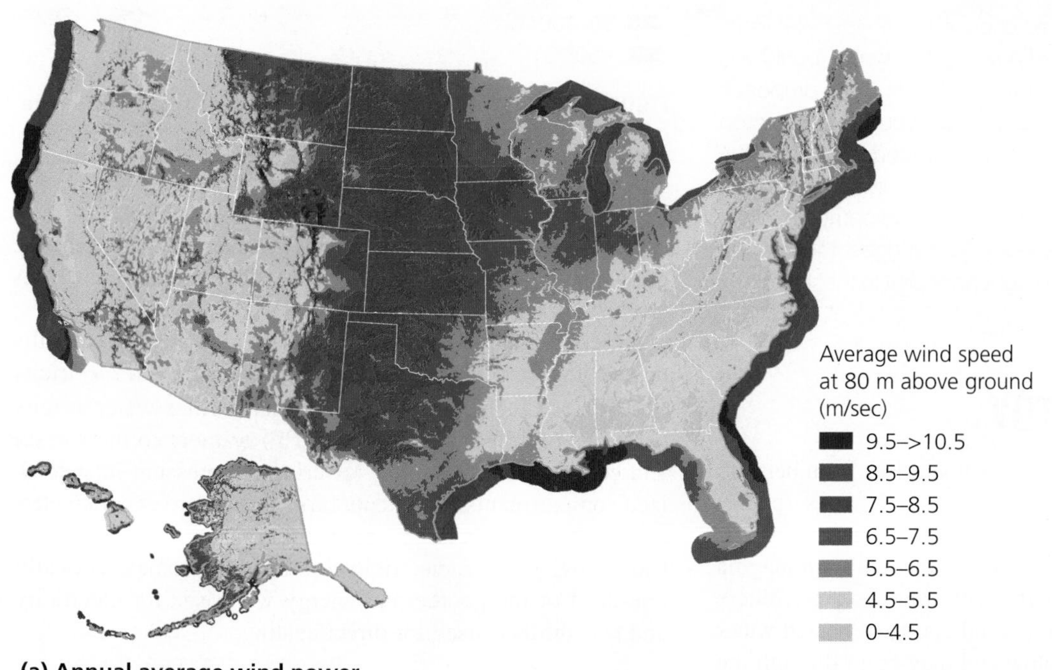

(a) Annual average wind power

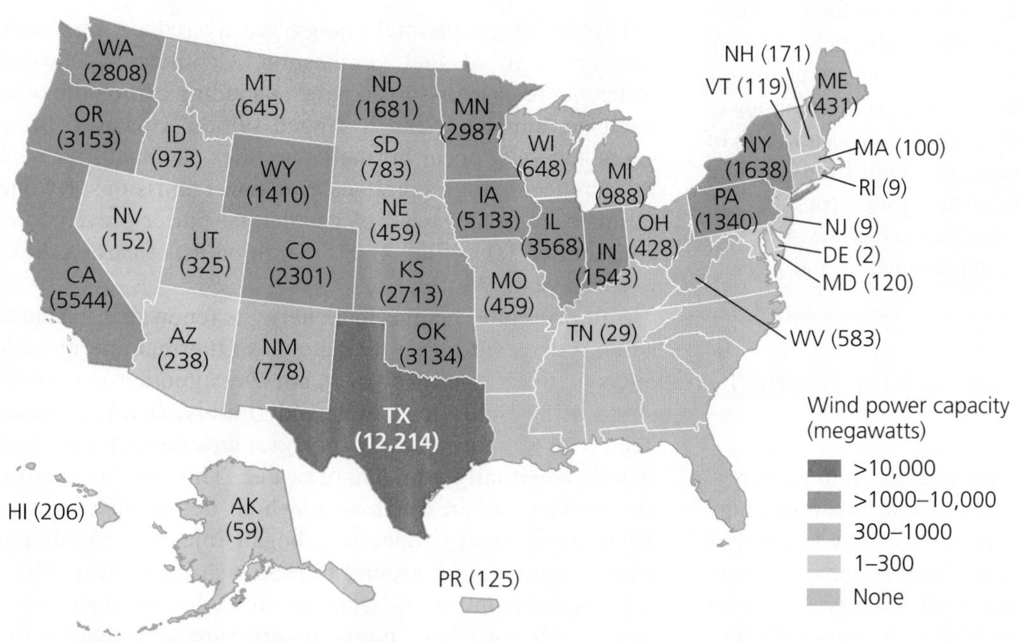

(b) Wind generating capacity, 2013

FIGURE 21.16 Wind speed varies from place to place. Maps of average wind speeds **(a)** help guide the placement of wind farms. Another map **(b)** shows megawatts of wind-power-generating capacity developed in each U.S. state through early 2013. *Sources: (a) U.S. National Renewable Energy Laboratory; (b) American Wind Energy Association, 2013.* 1st Quarter 2013 Market Report.

DATA Q Compare parts (a) and (b). Which states or regions have high wind speeds but are not yet heavily developed with commercial wind power?

nearby. For instance, the Cape Wind project has faced years of opposition from wealthy residents of Cape Cod, Nantucket, and Martha's Vineyard, even though many of these residents consider themselves progressive environmentalists.

Wind turbines also pose a threat to birds and bats, which are killed when they fly into the rotating blades. More research on wildlife impacts is urgently needed (see **The Science behind the Story**, pp. 592–593). One strategy for protecting birds and bats may be selecting sites that are not on migratory flyways or in the midst of prime habitat for species that are likely to fly into the blades.

WEIGHING THE ISSUES

WIND AND NIMBY If you could choose to get your electricity from a wind farm or a coal-fired power plant, which would you choose? How would you react if the electric utility proposed to build the wind farm that would generate your electricity atop a ridge running in back of your neighborhood, such that the turbines would be clearly visible from your living room window? Would you support or oppose the development? Why? If you would oppose it, where would you suggest the farm be located? Do you think anyone might oppose it in that location?

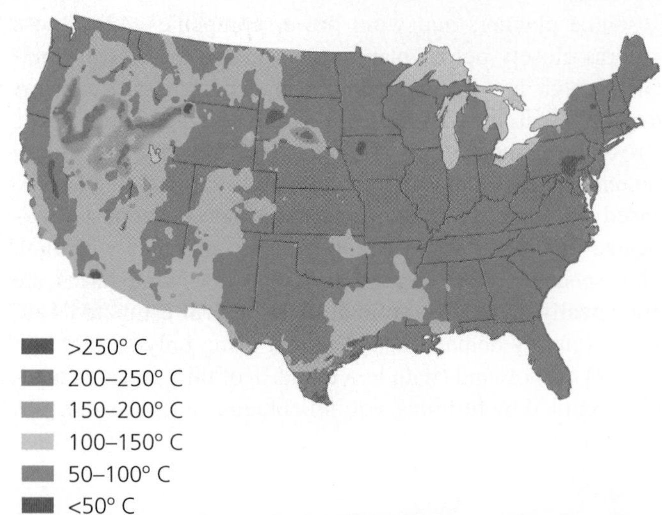

FIGURE 21.17 Geothermal resources in the United States are greatest in the western states. This map shows water temperatures 3 km (1.9 mi) belowground. *Data from Idaho National Laboratory.*

Geothermal Energy

Geothermal energy is thermal energy that arises from beneath Earth's surface. The radioactive decay of elements (p. 24) amid the extremely high pressures deep in the interior of our planet generates heat that rises to the surface through magma (molten rock, p. 34) and through cracks and fissures. Where this energy heats groundwater, natural spurts of heated water and steam are sent up from below and may erupt through the surface as terrestrial geysers (p. 33) or submarine hydrothermal vents (p. 33).

Geothermal energy manifests itself at the surface in these ways only in certain areas, and regions vary in their geothermal resources (FIGURE 21.17). One geothermally rich area is in California near Napa Valley's wine country. There, engineers have for years operated the world's largest geothermal power plants, The Geysers. The nation of Iceland also has a wealth of geothermal energy resources because it is built from lava that extruded and cooled at the Mid-Atlantic Ridge (pp. 34–35), along the spreading boundary of two tectonic plates. Because of the geothermal heat in this region, volcanoes and geysers are numerous in Iceland.

We harness geothermal energy for heating and electricity

Geothermal energy can be harnessed directly from geysers at the surface, but most often wells must be drilled down hundreds or thousands of meters toward heated groundwater. Hot groundwater can be used directly for heating homes, offices, and greenhouses; for driving industrial processes; and for drying crops. Iceland heats nearly 90% of its homes through direct heating with piped hot water. Direct use of naturally heated water is efficient and inexpensive, but it is feasible only where geothermal energy is readily available and does not need to be transported far.

Geothermal power plants harness the energy of naturally heated underground water and steam to generate electricity (FIGURE 21.18). Generally, a power plant brings water at temperatures of 150–370°C (300–700°F) or more to the surface and converts it to steam by lowering the pressure in specialized compartments. The steam turns turbines to generate electricity. The world's largest geothermal plants, The Geysers in California, provide electricity for 725,000 homes. Globally, one-third of the geothermal energy we use is for electricity, and two-thirds is used for direct heating.

Geothermal power has benefits and limitations

All forms of geothermal energy greatly reduce emissions relative to fossil fuel combustion. Geothermally heated water can release dissolved gases, including carbon dioxide, methane, ammonia, and hydrogen sulfide. However, these gases generally occur in small quantities, and facilities with filtering technologies produce even fewer emissions. By one estimate, each megawatt of geothermal power prevents the emission of 7.0 million kg (15.5 million lb) of carbon dioxide each year.

In principle, geothermal energy is renewable, because using it does not affect the amount of thermal energy produced underground. However, not every geothermal power plant will be able to operate indefinitely. If a plant uses heated water more quickly than groundwater is recharged, it will eventually run out of water. This was occurring at The Geysers in California, which began operation in 1960. In response, operators began injecting municipal wastewater into the ground to replenish the supply. More geothermal plants worldwide are now injecting water back into aquifers to help maintain pressure and sustain the resource.

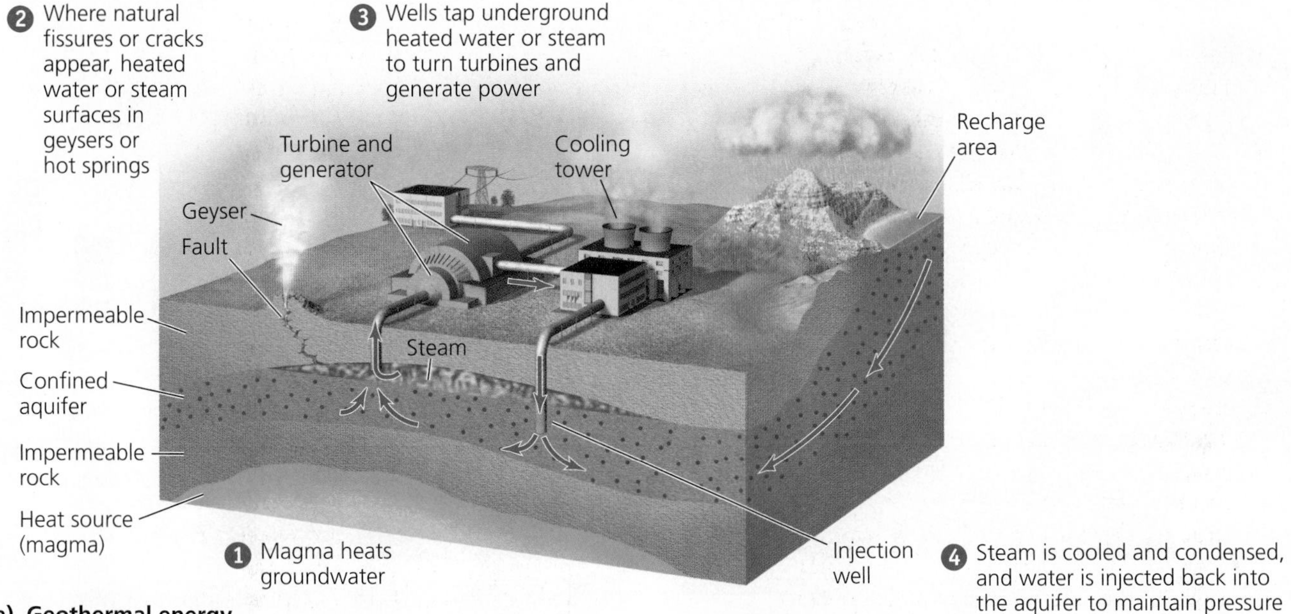

(a) Geothermal energy

(b) Nesjavellir geothermal power station, Iceland

FIGURE 21.18 Geothermal power plants generate electricity using naturally heated water from underground. With geothermal energy **(a)**, magma heats groundwater ①, some of which escapes through surface vents such as geysers ②. A power plant may tap into heated water and channel steam through turbines to generate electricity ③. Once used, steam may be condensed into water and pumped back into the aquifer to maintain pressure ④. At Iceland's Nesjavellir power station **(b)**, steam piped in from wells heats water from a lake. The heated water is sent through an insulated pipeline to the capital city, where residents use it for washing and space heating.

A second reason geothermal energy is not always renewable is that patterns of geothermal activity in Earth's crust shift naturally over time. As a result, an area that produces hot groundwater now may not always do so. In addition, the water of many hot springs is laced with salts and minerals that corrode equipment and pollute the air. These factors may shorten the lifetime of plants, increase maintenance costs, and add to pollution.

The greatest limitation of geothermal power is its restriction to regions where we can tap energy from naturally heated groundwater. Places like Iceland, northern California, and Yellowstone National Park are rich in naturally heated groundwater, but most areas of the world are not.

Enhanced geothermal systems might widen our reach

To overcome the limitation of geothermal energy to areas where naturally heated groundwater occurs, engineers are currently developing **enhanced geothermal systems (EGS).** In the EGS approach, engineers drill deeply into dry rock, fracture the rock, and pump in cold water. The water becomes heated deep underground and is then drawn back up and used to generate power. EGS thereby uses natural thermal energy underground but supplies the water, which can be reused repeatedly. In theory we could use EGS widely in many locations. Germany, for instance, has little heated groundwater, but feed-in-tariffs have enabled an EGS facility to operate profitably there.

EGS technology shows significant promise, and a 2006 report estimated that heat resources below the United States alone are enough to power the world's energy demands for several millennia. However, EGS also appears to trigger minor earthquakes. Unless we can develop ways to use EGS safely and reliably, our use of geothermal power will remain more localized than our use of solar, wind, bioenergy, or hydropower.

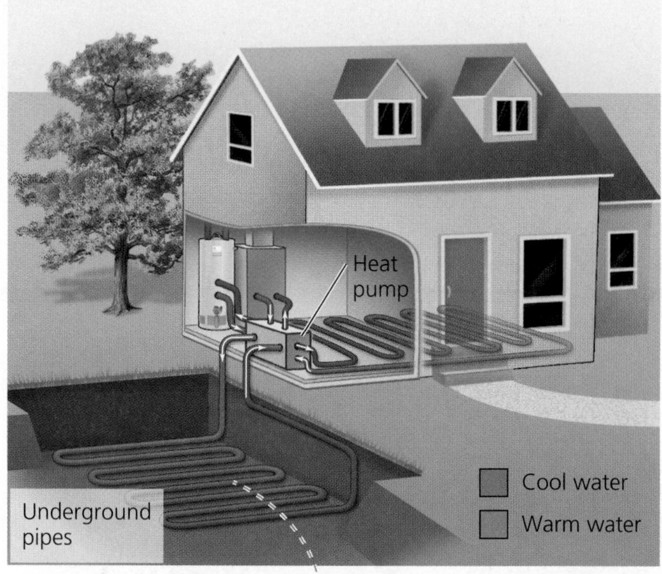

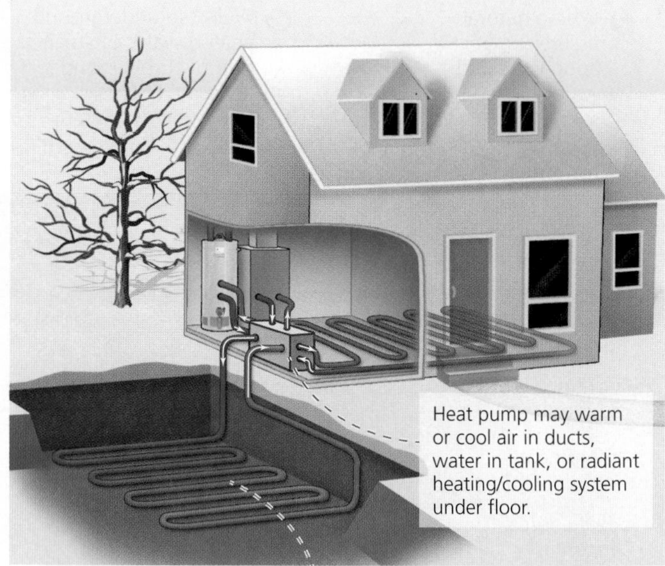

FIGURE 21.19 **Ground-source heat pumps provide an efficient way to heat and cool air and water in one's home.** A network of pipes filled with water and antifreeze extend underground. Soil is cooler than air in the summer (**left**), and warmer than air in the winter (**right**), so by running fluid between the house and the ground, these systems adjust temperatures inside.

Heat pumps make use of temperature differences above and below ground

Although heated groundwater is available only in certain areas, we can take advantage of the temperature differences that exist naturally between the soil and the air just about anywhere. Soil varies in temperature from season to season less than air does, because it absorbs and releases heat more slowly and because warmth and cold do not penetrate deeply belowground. Just several inches below the surface, temperatures are nearly constant year-round. Geothermal heat pumps, or **ground-source heat pumps (GSHPs)**, make use of this phenomenon.

Ground-source heat pumps provide heating in the winter by transferring heat from the ground into buildings, and they provide cooling in the summer by transferring heat from buildings into the ground. This heat transfer is accomplished with a network of underground plastic pipes that circulate water and antifreeze (FIGURE 21.19). Because heat is simply moved from place to place rather than being produced using outside energy inputs, heat pumps can be highly energy-efficient.

More than 600,000 U.S. homes use GSHPs. Compared to conventional electric heating and cooling systems, GSHPs heat spaces 50–70% more efficiently, cool them 20–40% more efficiently, can reduce electricity use by 25–60%, and can reduce emissions by up to 70%.

Ocean Energy Sources

The oceans are home to several underexploited energy sources resulting from continuous natural processes. Of the four approaches being developed, three involve motion, and one involves temperature.

We can harness energy from tides, waves, and currents

Just as dams on rivers use flowing fresh water to generate hydroelectric power, we can use kinetic energy from the natural motion of ocean water to generate electrical power.

Scientists and engineers are working to harness the motion of ocean waves and convert this mechanical energy into electricity. Many designs for machinery to harness **wave energy** have been invented, but few have been adequately tested. Some designs for offshore facilities involve floating devices that move up and down with the waves. A prime example is the snake-like wave energy converter shown on the front cover of this book. Built by the Scottish company Pelamis (a Latin word denoting a genus of sea snake), variations on this jointed, columnar design have been deployed in several parts of the world. Machinery and hydraulic fluids inside the floating columns use wave motion to generate electricity, which is transmitted to shore via undersea cables.

Wave energy is greatest at deep-ocean sites, but transmitting electricity to shore is expensive. Some designs for coastal onshore facilities funnel waves from large areas into narrow channels and elevated reservoirs, from which water then flows out, generating electricity as hydroelectric dams do. Other coastal designs use rising and falling waves to push air into and out of chambers, turning turbines (FIGURE 21.20). The first commercially operating wave energy facility began operating in 2011 in Spain, using technology tested for a decade in

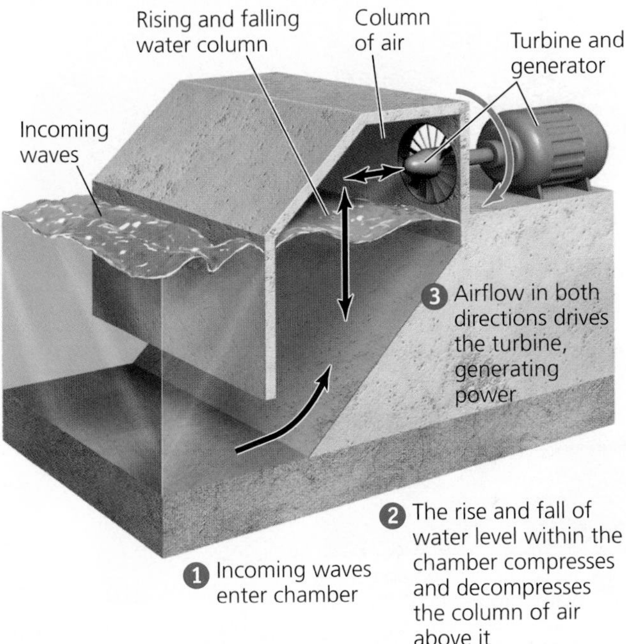

FIGURE 21.20 Coastal facilities harness energy from ocean waves. In one design, as waves enter and exit a chamber 1, the air inside is alternately compressed and decompressed 2, creating airflow that rotates turbines 3 to generate electricity.

Scotland. Demonstration projects exist in Europe, Japan, and Oregon.

We are also developing ways of harnessing energy from tides. The rise and fall of ocean tides (p. 427) twice each day moves large amounts of water past any given point on the world's coastlines. Differences in height between low and high tides are especially great in long, narrow bays such as Alaska's Cook Inlet or the Bay of Fundy between New Brunswick and Nova Scotia (**FIGURE 21.21**). Such locations are best for harnessing **tidal energy,** which is accomplished by erecting dams across the outlets of tidal basins. The incoming tide flows through sluice gates and is trapped behind them. Then, as the outgoing tide passes through the gates, it turns turbines to generate electricity (**FIGURE 21.22**). Some designs generate electricity from water moving in both directions.

The world's largest tidal generating station is South Korea's Sihwa Lake facility (see Figure 21.22, inset photo). This power station opened in 2011 and is just larger than the La Rance tidal facility in France, which has operated for nearly 50 years. Smaller facilities operate in Canada, China, Russia, and the United Kingdom. The first U.S. tidal station began operating in 2012 in Maine, and one is scheduled to be built in New York City's East River starting in 2013. Tidal stations release few or no pollutant emissions, but they can affect the ecology of estuaries and tidal basins. Five more tidal stations are planned in South Korea, but some have been delayed in the wake of concerns over environmental impacts.

A third way to harness marine kinetic energy is to use the motion of ocean currents (p. 424), such as the Gulf Stream. Devices like underwater wind turbines have been erected in European waters to test this approach.

The ocean stores thermal energy

Each day the tropical oceans absorb solar radiation with the heat content of 250 billion barrels of oil—enough to provide 20,000 times the electricity used daily in the United States. The ocean's sun-warmed surface is warmer than its deep water, and **ocean thermal energy conversion (OTEC)** is based on this gradient in temperature.

In the *closed cycle* approach, warm surface water is piped into a facility to evaporate chemicals, such as ammonia, that boil at low temperatures. These evaporated gases spin turbines to generate electricity. Cold water piped up from ocean depths then condenses the gases so they can be reused. In the *open cycle* approach, warm surface water is evaporated in a

FIGURE 21.21 Ocean tides change roughly every six hours. Tides are extreme at Canada's Bay of Fundy, where boats docked at high tide **(a)** become stranded on the mud at low tide **(b)** as the water recedes.

(a) High tide

(b) Low tide

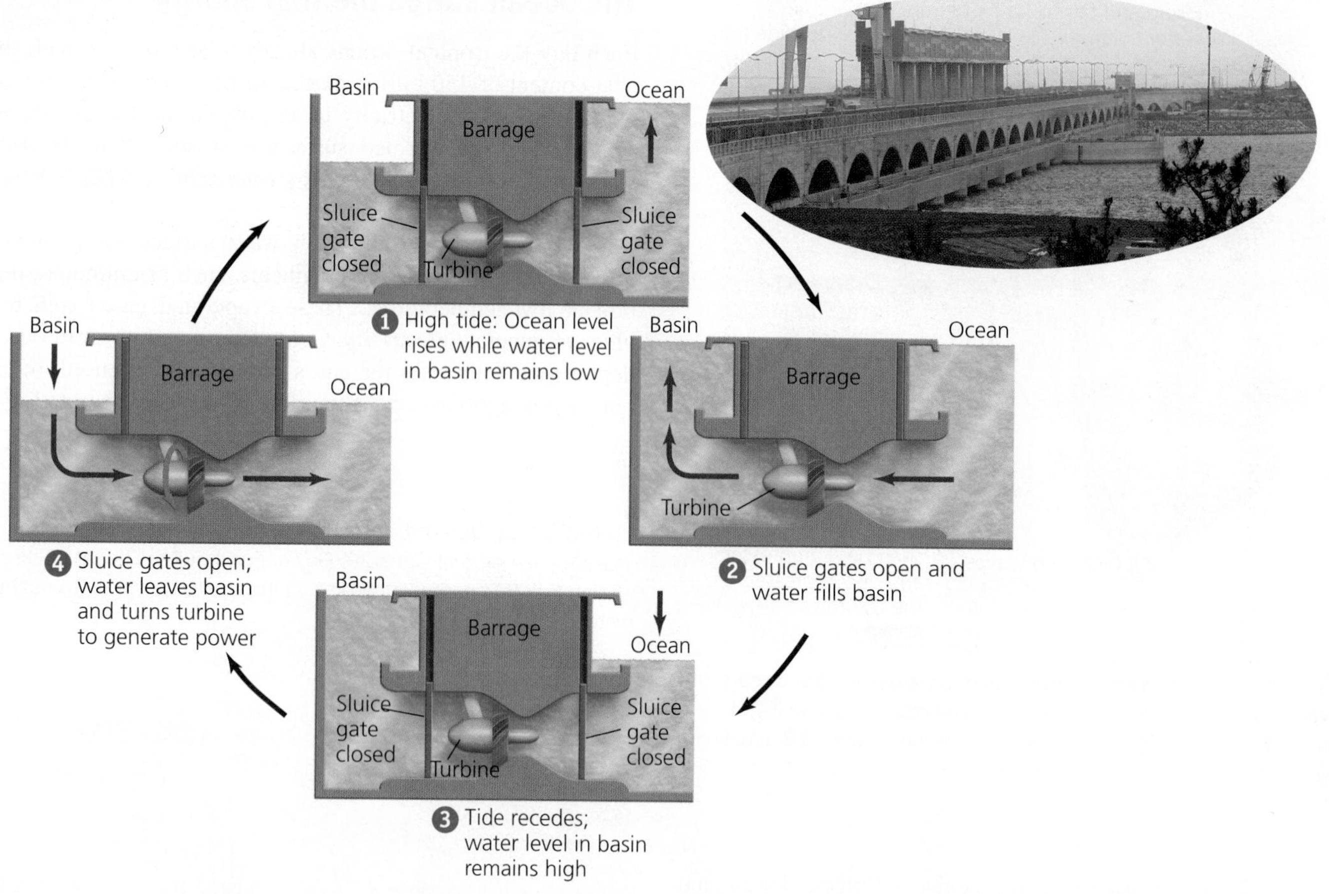

FIGURE 21.22 We can harness tidal energy by using bulb turbines in concert with the outgoing tide. At high tide 1, ocean water is let through the sluice gates, filling an interior basin 2. At low tide 3, the basin water is let out into the ocean, spinning turbines to generate electricity 4. This technology is used at the new Sihwa Lake tidal generating station (**photo**) in South Korea.

vacuum, and its steam turns turbines and then is condensed by cold water. Because ocean water loses salts as it evaporates, the water can be recovered, condensed, and sold as desalinized fresh water for drinking or agriculture. OTEC research has been conducted in Hawaii and elsewhere, but costs remain high, and so far no facility operates commercially.

WEIGHING THE ISSUES

YOUR NATION'S ENERGY? You are the president of a nation the size of Germany, and your nation's congress is calling on you to propose a national energy policy. Your country is located along a tropical coastline. Your geologists do not yet know whether there are fossil fuel deposits or geothermal resources under your land, but your country gets a lot of sunlight and a fair amount of wind, and broad, shallow shelf regions line its coasts. Your nation's population is moderately wealthy but is growing fast, and importing fossil fuels from other nations is becoming expensive.

What approaches would you propose in your energy policy? Name some specific steps you would urge your congress to fund. Are there trade relationships you would seek to establish with other countries? What questions would you fund your nation's scientists to research?

Hydrogen

Each of the renewable energy sources we have discussed can be used to generate electricity more cleanly than can fossil fuels. However, electricity cannot be stored easily in large quantities for use when and where it is needed. This is why most vehicles rely on gasoline from oil for power. The development of fuel cells and of fuel consisting of hydrogen—the universe's simplest and most abundant element—shows promise as a way to store considerable quantities of energy conveniently, cleanly, and efficiently. Like electricity and like batteries, hydrogen is an energy carrier, not a primary energy source. It carries energy that can be converted for use at later times and in different places.

Some yearn for a "hydrogen economy"

Some energy experts envision that hydrogen fuel, together with electricity, could serve as the basis for a clean, safe, and efficient energy system. In such a system, electricity generated from intermittent renewable sources, such as wind or solar energy, could be used to produce hydrogen. Fuel cells—essentially, hydrogen batteries (**FIGURE 21.23**)—could then use hydrogen to produce electrical energy to power vehicles,

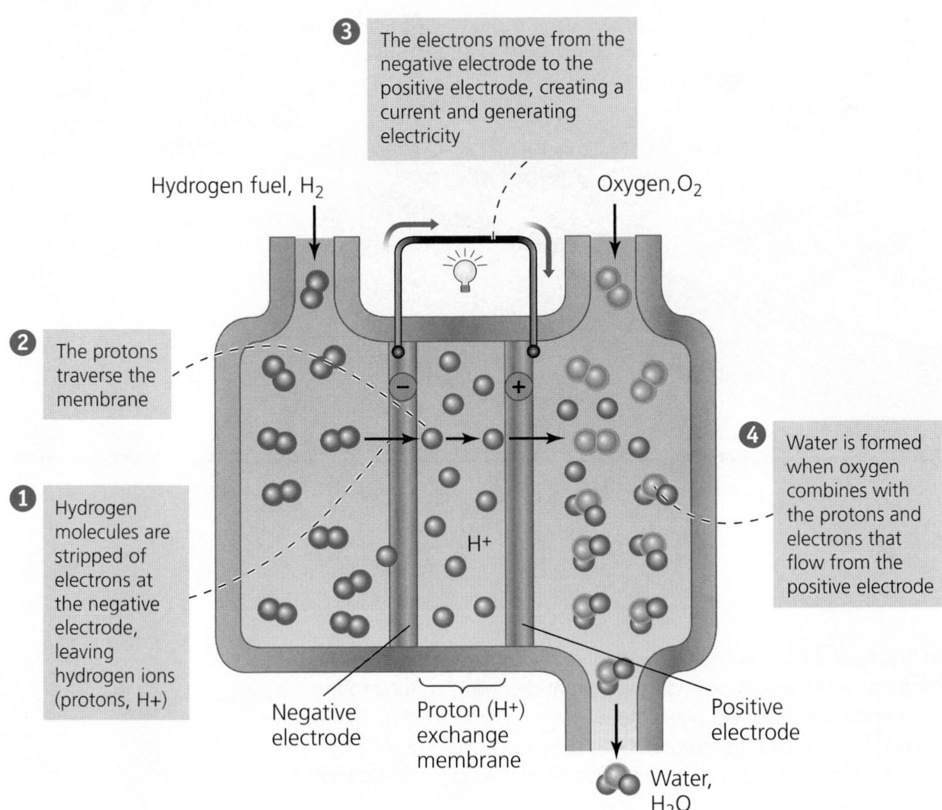

FIGURE 21.23 Hydrogen fuel drives electricity generation in a fuel cell, creating water as a waste product. Atoms of hydrogen are split 1 into protons and electrons. The protons, or hydrogen ions 2, pass through a proton exchange membrane. The electrons, meanwhile, move from a negative electrode to a positive one via an external circuit 3, creating a current and generating electricity. The protons and electrons then combine with oxygen 4 to form water molecules.

computers, cell phones, home heating, and other applications. NASA's space programs have used fuel-cell technology since the 1960s.

Basing an energy system on hydrogen could alleviate dependence on foreign fuels and help fight climate change. For these reasons, governments are funding research into hydrogen and fuel-cell technology, and automobile companies are investing in research and development to produce vehicles that run on hydrogen. A decade ago the island nation of Iceland decided to move toward a "hydrogen economy" and to set an example for the rest of the world to follow. Iceland achieved several early steps in its 30- to 50-year plan to phase out fossil fuels, such as converting buses in the capital city of Reykjavik to run on hydrogen fuel. But the global economic downturn in 2008–2009 and delays in manufacturing hydrogen cars have delayed its progress. Meanwhile, Germany is one of several other nations with hydrogen-fueled city buses (**FIGURE 21.24**), and it plans by 2015 to launch a network of hydrogen filling stations for hydrogen cars that are being designed.

Hydrogen fuel may be produced from water or from other matter

Hydrogen gas (H_2) does not tend to exist freely on Earth. Instead, hydrogen atoms bind to other molecules, becoming incorporated in everything from water to organic molecules. To obtain hydrogen gas for fuel, we must force these substances to release their hydrogen atoms, and this requires an input of energy. Scientists are studying several potential ways of producing hydrogen. In the process of **electrolysis,** electricity is input to split hydrogen atoms from the oxygen atoms of water molecules:

$$2H_2O \rightarrow 2H_2 + O_2$$

Electrolysis produces pure hydrogen, and it does so without emitting the carbon- or nitrogen-based pollutants of fossil fuel combustion. However, whether this strategy for producing hydrogen will cause pollution over its entire life cycle depends on the source of the electricity used for the electrolysis. If coal is burned to generate the electricity, then the process will not reduce emissions compared with reliance on fossil fuels. However, if the electricity is produced by some less-polluting renewable source, then hydrogen production by electrolysis would create much less pollution and greenhouse warming than reliance on fossil fuels. The "cleanliness" of a future hydrogen economy would, therefore, depend largely on the source of electricity used in electrolysis.

The environmental impact of hydrogen production will also depend on the source material for the hydrogen. Besides water, hydrogen can be obtained from biomass and from fossil fuels. Obtaining hydrogen from these sources generally requires less energy input but results in emissions of carbon-based pollutants. For instance, extracting hydrogen from the methane (CH_4) in natural gas entails producing one molecule of the greenhouse gas carbon dioxide for every four molecules of hydrogen gas:

$$CH_4 + 2H_2O \rightarrow 4H_2 + CO_2$$

Thus, whether a hydrogen-based energy system is cleaner than a fossil fuel system will depend on how the hydrogen is extracted.

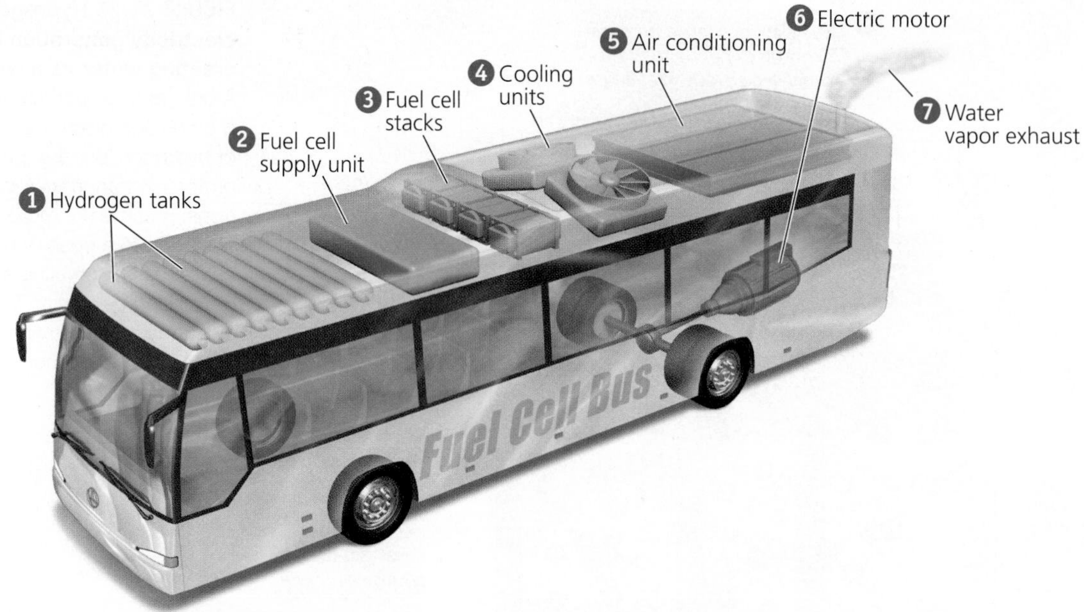

FIGURE 21.24 In one type of hydrogen-fueled bus operating in some German cities, hydrogen is stored in nine fuel tanks 1. The fuel cell supply unit 2 controls the flow of hydrogen, air, and cooling water into the fuel cell stacks 3. Cooling units 4 and the air conditioning unit 5 dissipate waste heat produced by the fuel cells. Electricity generated by the fuel cells is changed from direct current (DC) to alternating current (AC) by an inverter, and it is transmitted to the electric motor 6, which powers the operation of the bus. The vehicle's exhaust 7 consists simply of water vapor.

Fuel cells produce electricity by joining hydrogen and oxygen

Once isolated, hydrogen gas can be used as a fuel to produce electricity within fuel cells. The chemical reaction involved in a fuel cell is simply the reverse of that for electrolysis. An oxygen molecule and two hydrogen molecules each split, and their atoms bind to form two water molecules:

$$2H_2 + O_2 \rightarrow 2H_2O$$

Figure 21.23 shows how this occurs within one common type of fuel cell. As shown in the diagram, hydrogen gas (usually compressed and stored in an attached fuel tank) is allowed into one side of the cell, and the movement of the hydrogen's electrons from one electrode to the other creates the output of electricity.

Hydrogen and fuel cells have costs and benefits

One major drawback of hydrogen at this point is a lack of infrastructure. To convert a nation such as Germany or the United States to hydrogen would require massive and costly development of facilities to transport, store, and provide the fuel.

Another concern is that some research suggests that leakage of hydrogen from its production, transport, and use could potentially deplete stratospheric ozone (pp. 468–472) and lengthen the atmospheric lifetime of the greenhouse gas methane. Research into these questions is ongoing, because scientists do not want society to switch from fossil fuels to hydrogen without first knowing the risks.

Hydrogen's benefits include the fact that we will never run out of it, because it is the most abundant element in the universe. Hydrogen can be clean and nontoxic to use, and—depending on its source and the source of electricity for its extraction—it may produce few greenhouse gases and other pollutants. Water and heat are the only waste products from a hydrogen fuel cell, along with negligible traces of other compounds. In terms of safety for transport and storage, hydrogen can catch fire and explode, but if kept under pressure, it may be no less safe than gasoline in tanks.

Hydrogen fuel cells are energy-efficient. Depending on the type of fuel cell, 35–70% of the energy released in the reaction can be used. If the system is designed to capture heat as well as electricity, then the energy efficiency of fuel cells can rise to 90%. These rates are comparable or superior to most nonrenewable alternatives.

Fuel cells are also silent and nonpolluting. Unlike batteries (which also produce electricity through chemical reactions), fuel cells will generate electricity whenever hydrogen fuel is supplied, without ever needing recharging. For all these reasons, hydrogen fuel cells may soon be used to power cars, much as they are already powering buses operating on the streets of some German cities.

Conclusion

Rising concern over air pollution, global climate change, health impacts, and security risks resulting from our dependence on fossil fuels—as well as concerns over supplies of oil and natural gas—have convinced many people that we need to shift to renewable energy sources that pollute far less and that will not run out. Renewable sources with promise for sustaining our civilization far into the future without greatly degrading our environment include solar energy, wind power, geothermal energy, and ocean energy sources. Moreover, by using electricity from renewable sources to produce hydrogen fuel, we may be able to use fuel cells to produce electricity when and where it is needed, helping to create a nonpolluting and renewable transportation sector.

Renewable energy sources have been held back by limited funding for research and development and by competition with established nonrenewable fuels whose market prices do not cover external costs. Despite these obstacles, renewable technologies have progressed far enough to offer hope that we can shift from fossil fuels to renewable energy. To what degree we can also limit environmental impact will depend on how soon, how quickly, and how carefully we make the transition—and to what extent we put efficiency and conservation measures into place.

Reviewing Objectives

You should now be able to:

Outline the major sources of renewable energy, summarize their benefits, and assess their potential for growth

- The "new renewable" energy sources include solar, wind, geothermal, and ocean energy sources. They are not truly "new," but rather are in a stage of rapid development. (p. 582)
- The new renewables currently provide far less energy and electricity than we obtain from fossil fuels or other conventional energy sources. (pp. 582–583)
- Use of new renewables is growing quickly, and this growth is expected to continue as people seek to move away from fossil fuels. (pp. 582–583)
- Relative to fossil fuels, the new renewables alleviate air pollution, reduce greenhouse gas emissions, and can self-renew. They can diversify a society's energy mix and bring income and jobs to communities. (pp. 583–584)
- Government subsidies have long favored nonrenewable energy, but investment and public policies such as feed-in tariffs can speed our transition to renewable sources. (pp. 584–585)

Describe solar energy and the ways it is harnessed, and evaluate its advantages and disadvantages

- Energy from the sun's radiation can be harnessed using passive methods or by active methods involving powered technology. (p. 588)
- Solar technologies include flat plate collectors for heating water and air, mirrors to concentrate solar rays, and photovoltaic (PV) cells to generate electricity. (pp. 588–590)
- PV solar is today's fastest-growing energy source. (p. 591)
- Solar energy is perpetually renewable, creates no emissions, and enables decentralized power. (p. 591)
- Solar radiation varies in intensity from place to place and time to time, and harnessing solar energy remains expensive. (pp. 592–594)

Describe wind power and how we harness it, and evaluate its benefits and drawbacks

- Energy from wind is harnessed using wind turbines mounted on towers. (pp. 594–595)
- Turbines are often erected in arrays at wind farms located on land or offshore, in locations with optimal wind conditions. (pp. 595–597)
- Wind power is one of today's fastest-growing energy sources. (p. 595)
- Wind energy is renewable, turbine operation creates no emissions, wind farms can generate economic benefits, and the cost of wind power is competitive with that of electricity from fossil fuels. (p. 596)
- Wind is an intermittent resource and is adequate only in some locations. Turbines kill some birds and bats, and wind farms often face opposition from local residents. (pp. 593, 596–598)

Describe geothermal energy and the ways we make use of it, and assess its advantages and disadvantages

- Thermal energy from radioactive decay in Earth's core rises toward the surface and heats groundwater. This energy may be harnessed by geothermal power plants and used to directly heat water and air and to generate electricity. (pp. 598–599)
- Geothermal energy can be efficient, clean, and renewable. However, naturally heated water occurs near the surface

only in certain areas, and this water may be exhausted if it is overpumped. (pp. 598–599)

- Enhanced geothermal systems allow us to gain geothermal energy from more regions, but the approach can trigger earthquakes. (p. 599)
- Ground-source heat pumps heat and cool homes and businesses by circulating water whose temperature is moderated underground. (p. 600)

Describe ocean energy sources and how we can harness them

- Ocean energy sources include the motion of tides, waves, and currents, and the thermal heat of ocean water. (pp. 600–602)
- Ocean energy is perpetually renewable and holds much promise, but technologies have seen limited development thus far. (pp. 600–602)

Explain hydrogen fuel cells and weigh options for energy storage and transportation

- Hydrogen can serve as a fuel to store and transport energy, so that electricity generated by renewable sources can be made portable and used to power vehicles. (pp. 602–603)
- Hydrogen can be produced cleanly through electrolysis, but also by use of fossil fuels—in which case its environmental benefits are reduced. (p. 603)
- Fuel cells create electricity by controlling an interaction between hydrogen and oxygen, and they produce only water as a waste product. (pp. 602–604)
- Hydrogen infrastructure requires much more development, but hydrogen can be clean, safe, and efficient. Fuel cells are silent and nonpolluting, and they do not need recharging. (p. 604)

Testing Your Comprehension

1. What proportion of our energy now comes from renewable sources? What is the most prevalent form of renewable energy we use? What form of renewable energy is most used to generate electricity?
2. What factors and concerns are causing renewable energy use to expand? Which two renewable sources are experiencing the most rapid growth?
3. Contrast passive and active solar heating. Describe how each works, and give examples of each.
4. Define *photoelectric effect*. Explain how photovoltaic (PV) cells function and are used.
5. Describe several environmental and economic advantages of solar power. What are some disadvantages?
6. How do modern wind turbines generate electricity? How does wind speed affect the process? What factors affect where we place wind turbines?
7. Describe several environmental and economic benefits of wind power. What are some drawbacks?
8. Define *geothermal energy*, and explain three main ways in which it is obtained and used. Describe one sense in which it is renewable and one sense in which it is not renewable.
9. List and describe four approaches for obtaining energy from ocean water.
10. How is hydrogen fuel produced? Is this a clean process? What factors determine the amount of pollutants hydrogen production will emit?

Seeking Solutions

1. Explain how Germany accelerated its development of renewable energy by establishing a system of feed-in tariffs. Do you think the United States should adopt a similar system? Why or why not?
2. For each source of renewable energy discussed in this chapter, what factors stand in the way of an expedient transition from fossil fuel use? What could be done in each case to ease a shift to these renewable sources?
3. Do you think we can develop renewable energy resources to replace fossil fuels without significant social, economic, and environmental disruption? What steps would we need to take? Will market forces alone suffice to bring about this transition, or will we also need government? Do you think such a shift will be good for our economy? Why or why not?
4. Explain the circumstances under which using hydrogen fuel could be helpful in moving toward a low-emission future. If you could advise policymakers of Iceland, Germany, or the United States on hydrogen, what would you tell them?
5. **THINK IT THROUGH** You have just graduated from college, gotten married, landed a good job, and purchased your first home. You and your spouse plan to stay in this home for the foreseeable future and are considering installing flat plate solar collectors and/or PV panels on your roof. What factors will you consider, and what

questions will you ask before deciding whether to make the investment in solar energy?

6. **THINK IT THROUGH** You are the CEO of a company that develops wind farms. Your staff has presented you with three options, listed below, for sites for your next development. Describe at least one likely advantage and at least one likely disadvantage you would expect to encounter with each option. What further information would you like to know before deciding which to pursue?
   - Option A: A remote rural site in North Dakota
   - Option B: A ridge-top site in the suburbs of Philadelphia
   - Option C: An offshore site off the Florida coast

Calculating Ecological Footprints

Assume that average per capita residential consumption of electricity is 12 kilowatt-hours per day, that photovoltaic cells have an electrical output of 15% incident solar radiation, and that PV panels cost $1000 per square meter. Now refer to Figure 21.11 on p. 594 and estimate the area and cost of the PV panels needed to provide all of the residential electricity used by each group in the table.

	Area of photovoltaic cells	Cost of photovoltaic cells
You		
A resident of Arizona		
A resident of Alaska		
Total for all U.S. residents		

1. What additional information would you need to increase the accuracy of your estimates for the areas in the table above?
2. Considering the distribution of solar radiation in the United States, where do you think it will be most feasible to greatly increase the percentage of electricity generated from photovoltaic solar cells?
3. The purchase price of a photovoltaic system is considerable. What other costs and benefits should you consider, in addition to the purchase price, when contemplating "going solar"?

22

Managing Our Waste

Upon completing this chapter, you will be able to:

- Summarize and compare the types of waste we generate
- List the major approaches to managing waste
- Delineate the scale of the waste dilemma
- Describe conventional waste disposal methods: landfills and incineration
- Evaluate approaches for reducing waste: source reduction, reuse, composting, and recycling
- Discuss management of industrial solid waste and principles of industrial ecology
- Assess issues in managing hazardous waste

CENTRAL CASE STUDY

Transforming New York's Fresh Kills Landfill

"An extraterrestrial observer might conclude that conversion of raw materials to wastes is the real purpose of human economic activity."

—Gary Gardner and Payal Sampat, Worldwatch Institute

"Recycling is one of the best environmental success stories of the late 20th century."

—U.S. Environmental Protection Agency

The closure of a landfill is not the kind of event that normally draws politicians and the press, but the Fresh Kills Landfill was no ordinary dump. Said to be the largest human-made structure on Earth, Fresh Kills was the primary repository of New York City's garbage for half a century. On March 22, 2001, New York City Mayor Rudolph Giuliani and New York Governor George Pataki were on hand to celebrate as a barge arrived on the shore of New York City's Staten Island and dumped the final load of trash at Fresh Kills.

The landfill's closure was a blessing for Staten Island's 450,000 residents, who had long viewed Fresh Kills as a foul-smelling eyesore, health threat, and civic blemish. The 890-ha (2200-acre) landfill featured six gigantic mounds of trash and soil. The highest, at 69 m (225 ft), was higher than the nearby Statue of Liberty.

New York City had grandiose plans for the site. It planned to transform the old landfill into a world-class public park—a verdant landscape of ball fields, playgrounds, jogging trails, rolling hills, and wetlands teeming with wildlife. Almost three times bigger than Manhattan's Central Park, the site hosted the region's largest remaining complex of salt marshes and freshwater creeks, while the mounds offered panoramic views of the Manhattan skyline. The city sponsored an international competition to select landscape architects to design plans for the new park.

Meanwhile, with its only landfill closed, New York City began exporting its waste. The city began plans to develop an efficient network of stations to package and transfer the waste and ship it outward by barge and railroad. However, these plans soon fell apart amid neighborhood opposition, economic misjudgments, and accusations of political favoritism and mob influence. New York City instead found itself paying contractors exorbitant prices to haul its garbage away one truckload at a time. In the years following the Fresh Kills closure, trucks full of trash rumbled through neighborhood streets, carrying 12,000 tons of waste each day bound for 26 different landfills and incinerators in New York, New Jersey, Virginia, Pennsylvania, and Ohio. The city sanitation department's budget nearly doubled, and budget woes caused the city to scale back its recycling program. Some New Yorkers suggested reopening Fresh Kills.

The landfill was reopened, but not for a reason anyone could have foreseen. After the September 11, 2001, terrorist attacks, 1.8 million tons of rubble from the collapsed World Trade Center towers, including unrecoverable human remains, was taken by barge to Fresh Kills. A monument will be erected as part of the new park.

Today, park development is progressing. Roads, ball fields, sculptures, and in-line skating rinks are being designed. Wetlands are being restored. People will be able to bicycle on trails alongside the region's largest estuary and reach stunning vistas atop the hills. Recreation areas named Owl Hollow Fields and Schmul Park opened in 2012, and work is underway on further parcels, including North Park, which overlooks an adjacent wildlife refuge.

The full conversion of Fresh Kills Landfill into Fresh Kills Park—one of the largest public works projects in the world—will take 30 years. Its progress has been hampered by bureaucratic, regulatory, and financial challenges. But in the end, this longtime symbol of waste will be transformed into a world-class center for recreation and urban ecological restoration.

In the meantime, the site has already compensated Staten Islanders for some of its stench through the years; in 2012 its mounds and wetlands helped to absorb the force of Hurricane Sandy's storm surge (p. 499), thereby shielding many Staten Island homes from the brunt of the storm.

Meanwhile, though, New York City continues to truck its trash to out-of-state landfills. It has built a transfer station at Fresh Kills to compact the waste and ship it outward by barge and railroad at less expense, but it is still paying a premium to rid itself of its garbage. Even if you close your landfill, there is no true "away" for the materials that we do not reuse or recycle. For New Yorkers—and for the rest of us—the waste we discard needs to go somewhere. Fortunately, we have sustainable alternatives to just throwing things away. Reducing waste at its source, reusing products, and recycling and composting materials all help us to manage and minimize our waste.

Approaches to Waste Management

As the world's population rises, and as we produce and consume more material goods, we generate more waste. **Waste** refers to any unwanted material or substance that results from a human activity or process.

For management purposes, we divide waste into several main categories. **Municipal solid waste** is nonliquid waste that comes from homes, institutions, and small businesses. **Industrial solid waste** includes waste from production of consumer goods, mining, agriculture, and petroleum extraction and refining. **Hazardous waste** refers to solid or liquid waste that is toxic, chemically reactive, flammable, or corrosive. It can include everything from paint and household cleaners to medical waste to industrial solvents. Another type of waste is *wastewater*, water we use in our households, businesses, industries, or public facilities and drain or flush down our pipes, as well as the polluted runoff from our streets and storm drains (pp. 412, 414–415).

We have several aims in managing waste

Waste can degrade water quality, soil quality, and air quality, thereby impairing human health and the environment. Waste is also unpleasant aesthetically. Moreover, waste is a measure of inefficiency—so reducing waste can save money and resources. For all these reasons, waste management has become a vital pursuit.

There are three main components of **waste management:**

1. Minimizing the amount of waste we generate
2. Recovering discarded materials and finding ways to recycle them
3. Disposing of waste safely and effectively

Minimizing waste at its source—called **source reduction**—is the preferred approach. There are several ways to reduce the amount of material that enters the **waste stream,** the flow of waste as it moves from its sources toward disposal destinations (**FIGURE 22.1**). Manufacturers can use materials more efficiently. Consumers can buy fewer goods, buy goods with less

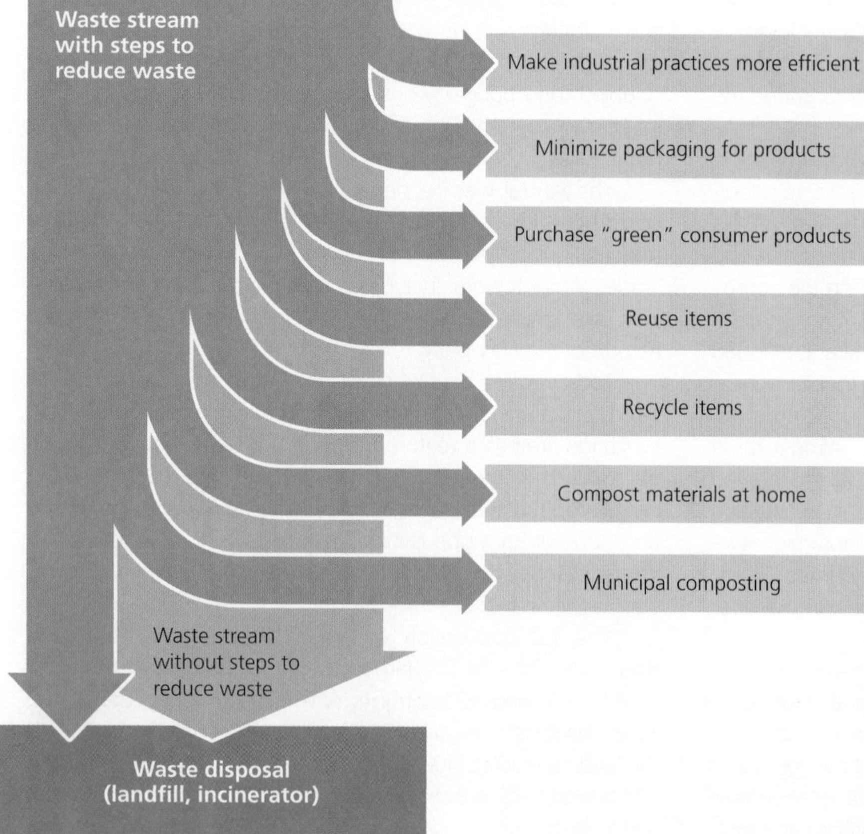

FIGURE 22.1 The most effective way to manage waste is to minimize the amount of material that enters the waste stream. To do this, manufacturers can increase efficiency, and consumers can buy "green" products that have minimal packaging or are produced in ways that minimize waste.

packaging, and use those goods longer. Reusing goods you already own, purchasing used items, and donating your used items for others all help reduce the amount of material entering the waste stream.

The next-best strategy in waste management is **recovery,** which consists of recovering, or removing, waste from the waste stream. Recovery includes both recycling and composting. **Recycling** is the process of collecting used goods and sending them to facilities that extract and reprocess raw materials that can then be used to manufacture new goods. **Composting** is the practice of recovering organic waste (such as food and yard waste) by converting it to mulch or humus (p. 219) through natural biological processes of decomposition. People did not invent recycling and composting; rather, they are fundamental features of the way natural systems function. All materials in nature are broken down at some point, and matter cycles through ecosystems (Chapter 5). People have taken these concepts from nature and applied them to help cut down on waste and conserve resources.

Regardless of how well we reduce our waste stream through source reduction and recovery, there will likely always be some waste left to dispose of. Disposal methods include burying waste in landfills and burning waste in incinerators. Waste managers attempt to find disposal methods that minimize impact on human health and environmental quality. In this chapter we first examine how the three major approaches are used to manage municipal solid waste, and then we address industrial solid waste and hazardous waste.

Municipal Solid Waste

Municipal solid waste—the waste we generate in our homes, in public facilities, and in small businesses—is what we commonly refer to as "trash" or "garbage." In the United States, paper, food scraps, yard trimmings, and plastics are the principal components of municipal solid waste, together accounting for 68% of what enters the waste stream (**FIGURE 22.2a**). Paper is recycled at a high rate and yard trimmings are composted at a high rate, so after recycling and composting reduce the waste stream, food scraps and plastics are left as the largest components of U.S. municipal solid waste (**FIGURE 22.2b**). In developing nations, food scraps are often the primary component, and paper makes up a smaller proportion.

Most municipal solid waste comes from packaging and nondurable goods (products meant to be discarded after a short period of use). In addition, consumers throw away old durable goods and outdated equipment as they purchase new products. Plastics, which came into wide consumer use only after 1970, have accounted for the greatest relative increase in the waste stream during the last several decades.

Consumption leads to waste

As we acquire more goods, we generate more waste. In the United States since 1960, waste generation (before recovery) has increased (**FIGURE 22.3**) by 2.8 times, and per capita waste generation has risen by 65%. In 2010, U.S. citizens produced 250 million tons of municipal solid waste (before recovery)—close to 1 ton per person. The average American generates 2.0 kg (4.4 lb) of trash per day—considerably more than people in most other industrialized nations. The relative wastefulness of the U.S. lifestyle, with its excess packaging and reliance on nondurable goods, has caused critics to label the United States "the throwaway society."

However, Americans are beginning to turn this around. Thanks to source reduction and reuse (especially by businesses looking to cut costs), total waste generation decreased slightly between 2005 and 2010. Americans now generate slightly less waste per capita than they did in 1990.

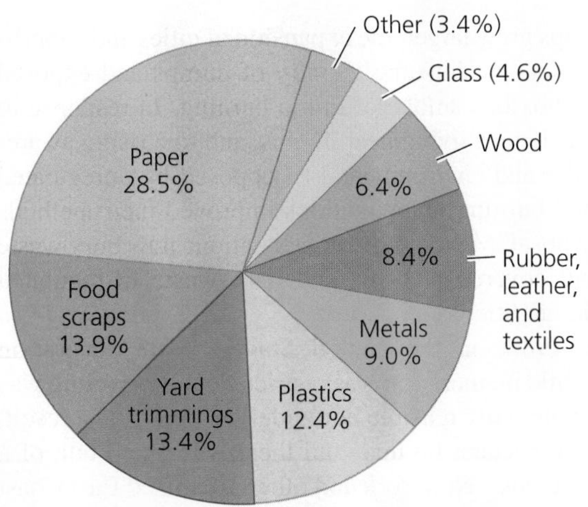

(a) Before recycling and composting

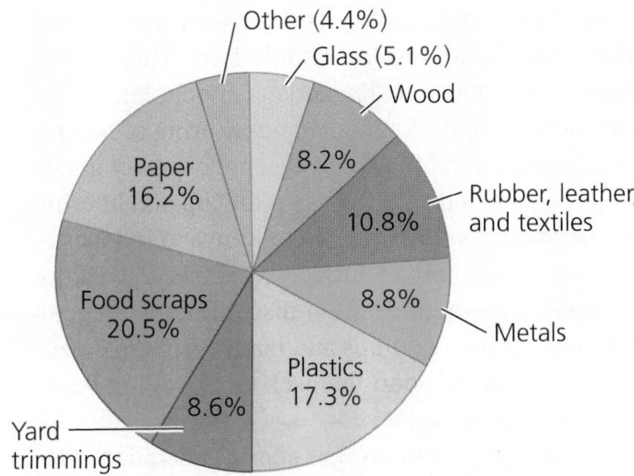

(b) After recycling and composting

FIGURE 22.2 Components of the municipal solid waste stream in the United States. Paper products comprise the greatest portion by weight **(a)**, but after recycling and composting removes many items **(b)**, the waste stream becomes one-third smaller. Food scraps are now the largest contributor, because so much paper is recycled and yard waste is composted. *Data from U.S. Environmental Protection Agency, 2011.* Municipal solid waste generation, recycling, and disposal in the United States: Facts and figures for 2010. *EPA, Washington, D.C.*

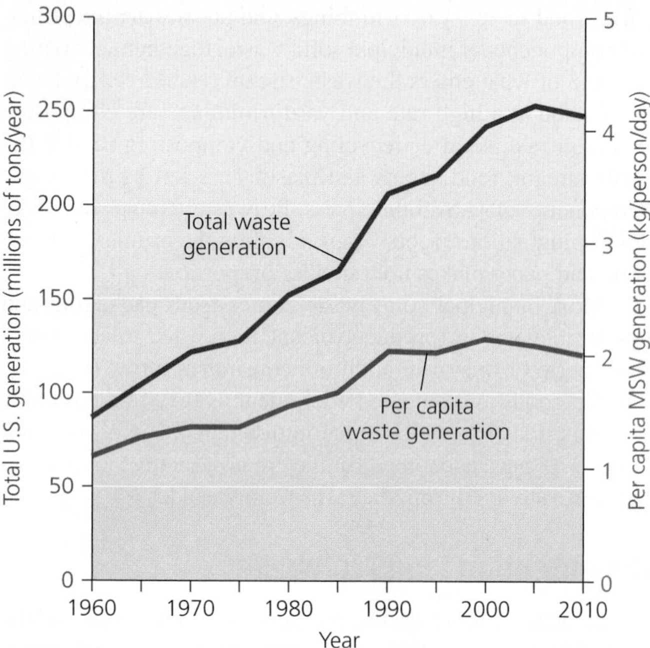

FIGURE 22.3 Rising U.S. waste generation has now leveled off. Total U.S. waste generation before recycling (**blue line**) nearly tripled after 1960, and U.S. per capita waste generation before recycling (**red line**) rose by 65%. In recent years, total and per-person waste generation have each leveled off, largely thanks to source-reduction efforts. *Data from U.S. Environmental Protection Agency, 2011.* Municipal solid waste generation, recycling, and disposal in the United States: Facts and figures for 2010. *EPA, Washington, D.C.*

DATA Q From these data, what would you infer has happened to the population size of the United States between 1990 and 2010? Explain how you know this.

FIGURE 22.4 Affluent consumers discard so much usable material that some people in developing nations support themselves by scavenging items from dumps. Tens of thousands of people used to scavenge each day from this dump outside Manila in the Philippines, finding items for themselves and selling material to junk dealers for 100–200 pesos (U.S. $2–$4) per day. The dump was closed in 2000 after an avalanche of trash killed hundreds of people.

In developing nations, people consume fewer resources and goods, and as a result, generate less waste. However, the intensive consumption that has long characterized wealthy nations is spreading in developing nations as they become more affluent, and these nations are now generating increasing amounts of waste. Over the past three decades, per capita waste generation rates have more than doubled in Latin American nations and have risen more than fivefold in the Middle East. This growth in waste reflects rising material standards of living, but it also results from an increase in packaging, manufacturing of nondurable goods, and production of inexpensive, poor-quality goods that wear out quickly. As a result, trash is piling up and littering the landscapes of countries from Mexico to Kenya to Indonesia.

Like U.S. consumers in the "throwaway society," wealthy consumers in developing nations often discard items that can still be used. In fact, at many dumps and landfills in the developing world, poor people support themselves by selling items that they scavenge (**FIGURE 22.4**).

In many industrialized nations, per capita generation rates have begun to decline in recent years. Wealthier nations also can afford to invest more in waste collection and disposal, so they are often better able to manage their waste and minimize impacts on human health and the environment. Moreover, enhanced recycling and composting efforts have reduced the waste stream. We will examine reduction, reuse, recycling, and composting shortly, but let's first assess how we dispose of waste.

Open dumping has led to improved disposal methods

Historically, people dumped their trash wherever it suited them. Until the mid-19th century, New York City's official method of garbage disposal was to dump it into the East River. As population densities increased, municipalities took on the task of consolidating trash into open dumps at specified locations in order to keep other areas clean. To decrease the volume of trash, these dumps were burned from time to time. Open dumping and burning still occur throughout much of the world.

As dumps grew larger, the expansion of cities and suburbs forced more people into the vicinity of dumps and exposed them to the noxious smoke of dump burning. In response to outcry from citizens living near dumps, and to a rising awareness of health and environmental risks posed by unregulated dumping and burning, many nations improved their methods of waste disposal. Most industrialized nations now bury waste in lined and covered landfills and burn waste in regulated incineration facilities.

In the 1980s in the United States, waste generation increased while incineration was restricted, and recycling was neither economically feasible nor widely popular. As a result, landfill space became limited, and there was much talk of a "solid waste crisis." New York and other urbanized East Coast areas felt this situation most acutely; Fresh Kills Landfill accepted its all-time annual peak of trash, 29,000 tons, in 1986–1987. Since the late 1980s, however, recovery of materials for recycling has expanded, decreasing the pressure on landfills (**FIGURE 22.5**). As of 2010, U.S. waste managers were landfilling 54% of municipal solid waste, incinerating 12%, and recovering 34% for composting and recycling.

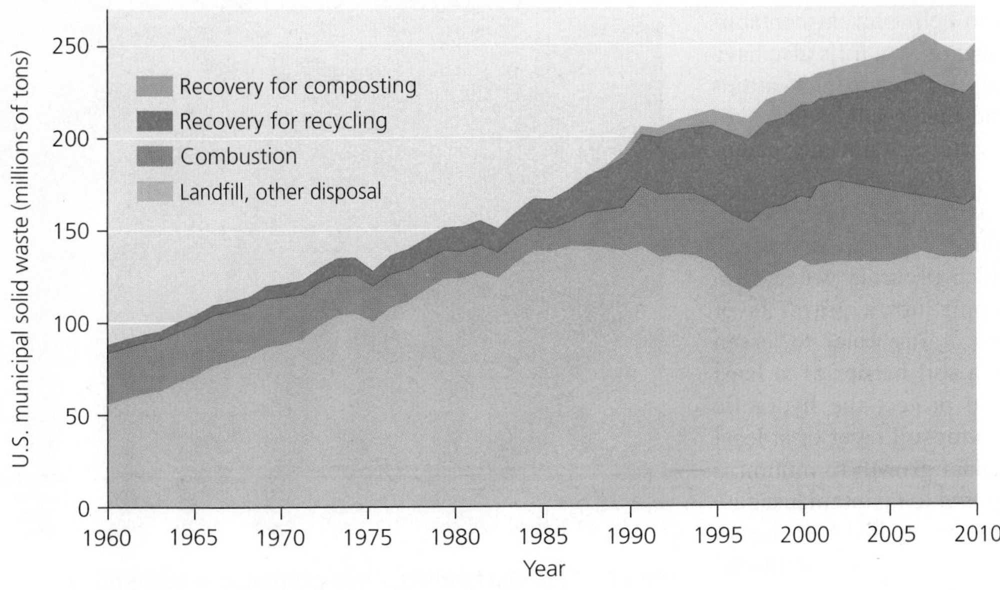

FIGURE 22.5 Since the 1980s, recycling and composting have grown in the United States; as a result, the proportion of waste going to landfills has declined. As of 2010, 54% of U.S. municipal solid waste went to landfills and 12% to incinerators, whereas 34% was recovered for composting and recycling. *Data from U.S. Environmental Protection Agency, 2011.* Municipal solid waste generation, recycling, and disposal in the United States: Facts and figures for 2010. *EPA, Washington, D.C.*

DATA Q Has the amount of solid waste that is combusted (incinerated) increased or decreased since 1960?

Sanitary landfills are regulated with health and environmental guidelines

In modern **sanitary landfills,** waste is buried in the ground or piled up in large mounds engineered to prevent waste from contaminating the environment and threatening public health (FIGURE 22.6). Most municipal landfills in the United States are regulated locally or by the states, but they must meet national standards set by the U.S. Environmental Protection Agency (EPA) under the **Resource Conservation and Recovery Act** (p. 176), a major federal law enacted in 1976 and amended in 1984.

In a sanitary landfill, waste is partially decomposed by bacteria and compresses under its own weight to take up less space. Soil is layered along with the waste to speed decomposition, reduce odor, and lessen infestation by pests. Some infiltration of rainwater into the landfill is good, because it encourages biodegradation by bacteria—yet too much is not good, because contaminants can escape if water carries them out.

To protect against environmental contamination, U.S. regulations require that landfills be located away from wetlands and earthquake-prone faults and be at least 6 m (20 ft) above the water table. The bottoms and sides of sanitary landfills must be lined with heavy-duty plastic and 60–120

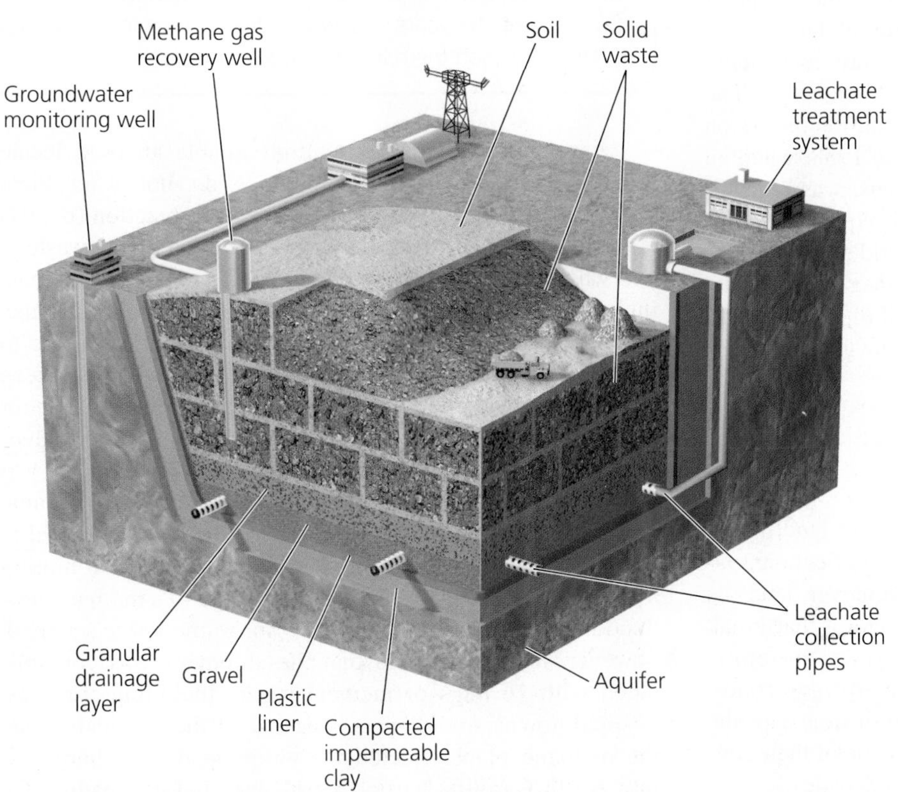

FIGURE 22.6 Sanitary landfills are engineered to prevent waste from contaminating soil and groundwater. Waste is laid in a large depression lined with plastic and impervious clay designed to prevent liquids from leaching out. Pipes draw out these liquids from the bottom of the landfill. Waste is layered with soil, filling the depression, and then is built into a mound until the landfill is capped. Landfill gas produced by anaerobic bacteria may be recovered, and waste managers monitor groundwater for contamination.

cm (2–4 ft) of impermeable clay to help prevent contaminants from seeping into aquifers. Sanitary landfills also have systems of pipes, collection ponds, and treatment facilities to collect and treat **leachate,** liquid that results when substances from the trash dissolve in water as rainwater percolates downward.

Once a landfill is closed, it is capped with an engineered cover that must be maintained. This cap consists of a hydraulic barrier of plastic, which prevents water from seeping down and gas from seeping up; a gravel layer above the hydraulic barrier, which drains water to lessen pressure on the hydraulic barrier; a soil barrier of at least 60 cm (24 in.) to store water and protect the hydraulic layer from weather extremes; and a topsoil layer of at least 15 cm (6 in.), which encourages plant growth to minimize erosion. Landfill managers are required to maintain leachate collection systems for 30 years after a landfill has closed. Regulations also require that groundwater be monitored regularly for contamination.

The Fresh Kills Landfill was considered a model for advanced landfill technology at the time of its construction, but it was built before most of the EPA guidelines. As a result, it caused some environmental contamination. However, engineers have retrofitted the landfill with clay liners and a sophisticated leachate collection system. Three of the six mounds have been capped with a "final cover," and the remaining mounds will soon be capped. Indeed, the city's investment of several hundred million dollars in an existing landfill to bring environmental protection measures up to modern standards is unprecedented. Because these safeguards need to be maintained for 30 years after closure, designs for a public park at Fresh Kills have had to work around them.

In 1988 the United States had nearly 8000 landfills, but today it has fewer than 2000. Waste managers have consolidated the waste stream into fewer landfills of larger size. In many cities, landfills that were closed are now being converted into public parks or other uses (FIGURE 22.7). The Fresh Kills endeavor is the world's largest landfill conversion project, but such efforts date back at least to 1938, when an ash landfill at Flushing Meadows, in Queens, was redeveloped for the 1939 World's Fair. The site subsequently hosted the United Nations and the 1964–1965 World's Fair. Designated a park in 1967, today the site hosts Shea Stadium, the Queens Museum of Art, the New York Hall of Science, and the Queens Botanical Garden, as well as playgrounds, wetlands, and festival events.

Landfills have drawbacks

Despite improvements in liner technology and landfill siting, liners can be punctured, and leachate collection systems eventually cease to be maintained. Moreover, landfills are kept dry to reduce leachate, but the bacteria that break down material thrive in wet conditions. Dryness, therefore, slows waste decomposition. In fact, the low-oxygen conditions of most landfills turns trash into a sort of time capsule. Researchers examining landfills often find some of their contents perfectly preserved, even after years or decades.

FIGURE 22.7 **Old landfills, once capped, can serve other purposes.** Visitors to Fresh Kills Park will enjoy this panoramic view of the Manhattan skyline from atop what used to be an immense mound of trash.

FAQ **How much does garbage decompose in a landfill?**

You might assume that a banana peel you throw in the trash will soon decay away to nothing in a landfill. However, it just might survive longer than you do! This is because surprisingly little decomposition occurs in landfills. Researcher William Rathje, a recently retired archaeologist known as "the Indiana Jones of Solid Waste," made a career out of burrowing into landfills and examining their contents to learn about what we consume and what we throw away. His research teams would routinely come across whole hot dogs, intact pastries that were decades old, and grass clippings that were still green. Newspapers 40 years old were often still legible, and the researchers used them to date layers of trash.

A second challenge is to find suitable areas to locate landfills, because most communities do not want them nearby. This not-in-my-backyard (NIMBY) reaction (p. 597) is largely why New York City decided to export its waste—and why residents of states receiving that waste are increasingly unhappy about it. As a result of the NIMBY syndrome, landfills are rarely sited in neighborhoods that are home to wealthy and educated people with the political clout to keep them out. Instead, they end up disproportionately sited in poor and minority communities, as environmental justice advocates (pp. 140–141) have frequently pointed out.

The reluctance of communities to accept waste became apparent with the famed case of the "garbage barge." In Islip, New York, in 1987, the town's landfills were full, prompting town administrators to ship waste by barge to a methane production plant in North Carolina. Prior to the barge's arrival, however, it was revealed that the shipment was contaminated with 16 bags of medical waste, including syringes, hospital gowns, and diapers. Because of the medical waste, the methane plant rejected the entire load. The barge sat in a North Carolina harbor for 11 days before heading for

Louisiana. However, Louisiana would not permit the barge to dock. The barge traveled toward Mexico, but the Mexican navy prevented it from entering that nation's waters. In the end, the barge traveled 9700 km (6000 mi) before eventually returning to New York, where, after several court battles, the waste was incinerated at a facility in Queens.

Incinerating trash reduces pressure on landfills

Just as sanitary landfills are an improvement over open dumping, incineration in specially constructed facilities is an improvement over open-air burning of trash. **Incineration,** or combustion, is a controlled process in which garbage is burned at very high temperatures (FIGURE 22.8). At incineration facilities, waste is generally sorted and metals removed. Metal-free waste is chopped into small pieces to aid combustion and then is burned in a furnace. Incinerating waste reduces its weight by up to 75% and its volume by up to 90%.

The ash remaining after trash is incinerated contains toxic components and therefore must be disposed of in hazardous waste landfills (p. 627). Moreover, when trash is burned, hazardous chemicals—including dioxins, heavy metals, and polychlorinated biphenyls (PCBs) (Chapter 14)—can be created and released into the atmosphere. Such emissions caused a backlash against incineration from citizens concerned about health hazards.

Most developed nations now regulate incinerator emissions, and some have banned incineration outright. Engineers have also developed technologies to mitigate emissions. Scrubbers (see Figure 17.16, p. 461) chemically treat the gases produced in combustion to remove hazardous components and neutralize acidic gases, such as sulfur dioxide and hydrochloric acid, turning them into water and salt. Scrubbers generally do this either by spraying liquids formulated to neutralize the gases or by passing the gases through dry lime.

Particulate matter, called *fly ash*, often contains some of the worst dioxin and heavy metal pollutants in incinerator emissions. To physically remove these tiny particles, facilities may use a huge system of filters known as a *baghouse*. In addition, burning garbage at especially high temperatures can destroy certain pollutants, such as PCBs. Even all these measures, however, do not fully eliminate toxic emissions.

WEIGHING THE ISSUES

ENVIRONMENTAL JUSTICE? Do you know where your trash goes? Where is your landfill or incinerator located? Are the people who live closest to the facility wealthy, poor, or middle class? What race or ethnicity are they? Do you know whether the people of this neighborhood protested against the introduction of the landfill or incinerator?

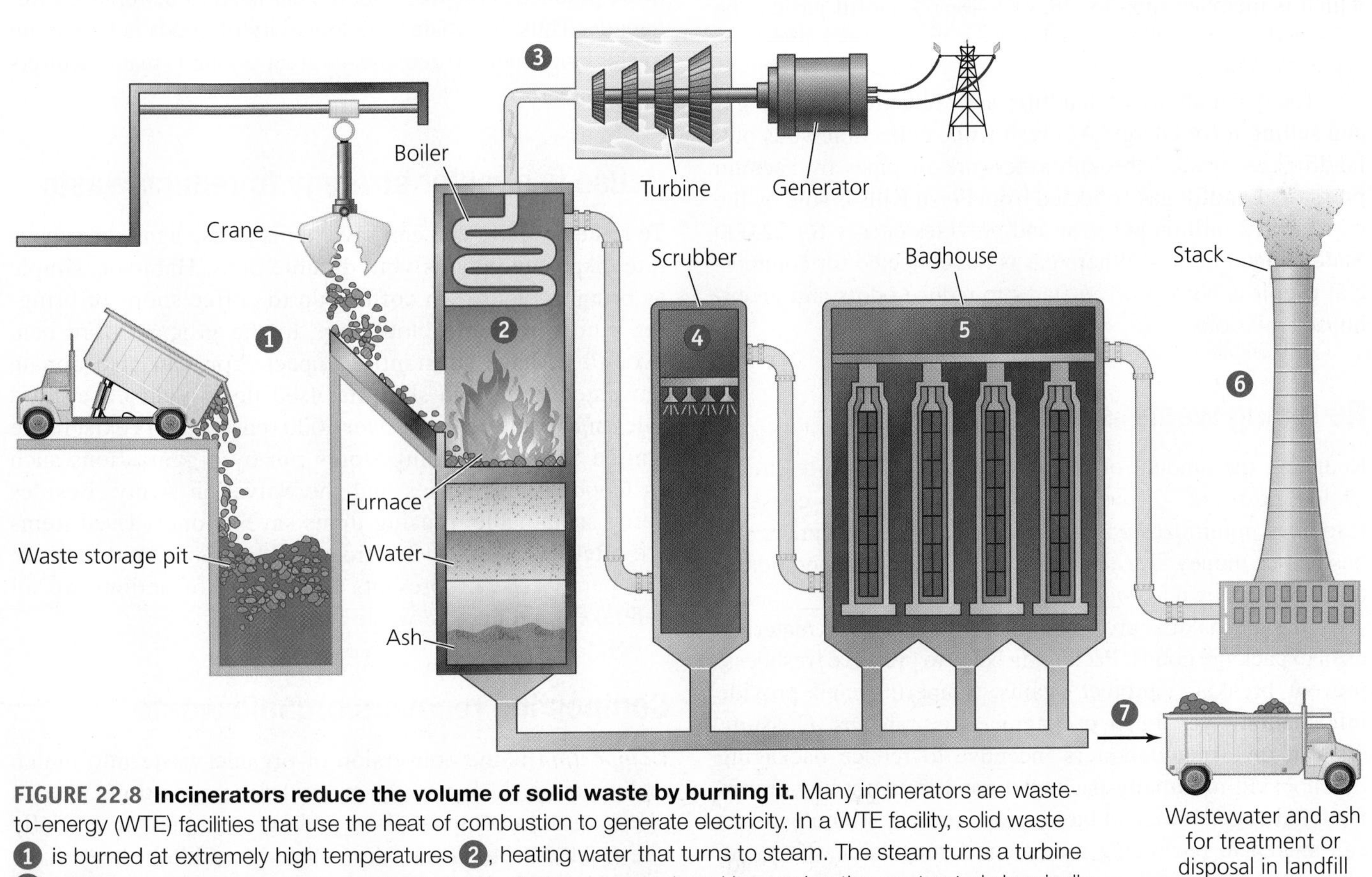

FIGURE 22.8 Incinerators reduce the volume of solid waste by burning it. Many incinerators are waste-to-energy (WTE) facilities that use the heat of combustion to generate electricity. In a WTE facility, solid waste ❶ is burned at extremely high temperatures ❷, heating water that turns to steam. The steam turns a turbine ❸, which powers a generator to create electricity. Toxic gases produced by combustion are treated chemically by a scrubber ❹, and airborne particulate matter is filtered physically in a baghouse ❺ before air is emitted from the stack ❻. Residual ash is disposed of ❼ in a landfill.

We can gain energy from incineration

Incineration was initially practiced simply to reduce the volume of waste, but today it often serves to generate electricity as well. Most incinerators now are **waste-to-energy (WTE) facilities** that use the heat produced by waste combustion to boil water, creating steam that drives electricity generation or that fuels heating systems. When burned, waste generates approximately 35% of the energy generated by burning coal. About 115 WTE facilities are operating across the United States (mostly in the Northeast and South), with a total capacity to process nearly 100,000 tons of waste per day.

Revenues from power generation, however, are usually not enough to offset the considerable financial cost of building and running incinerators. Because it can take many years for a WTE facility to become profitable, many companies that build and operate these facilities require communities contracting with them to guarantee the facility a minimum amount of garbage. On occasion, such long-term commitments have interfered with communities' subsequent efforts to reduce their waste through recycling and source reduction.

Landfills can produce gas for energy

Combustion in WTE plants is not the only way to gain energy from waste. Deep inside landfills, bacteria decompose waste in an oxygen-deficient environment. This anaerobic decomposition produces **landfill gas,** a mix of gases, roughly half of which is methane (pp. 26, 28, 484–486). Landfill gas can be collected, processed, and used in the same way as natural gas (pp. 525, 528).

Today hundreds of landfills are collecting landfill gas and selling it for energy. At Fresh Kills, collection wells pull landfill gas upward through a network of pipes by vacuum pressure. Landfill gas collected from Fresh Kills is sold by the city for $12 million per year and provides energy for 22,000 Staten Island homes. Where gas is not collected for commercial use, it is burned off in flares to reduce odors and greenhouse emissions.

Reducing waste is our best option

Reducing the amount of material entering the waste stream avoids costs of disposal and recycling, helps conserve resources, minimizes pollution, and can save consumers and businesses money. Recall that preventing waste generation in this way is known as *source reduction.*

One means of source reduction is to lessen the materials used to package goods. Packaging helps to preserve freshness, prevent breakage, protect against tampering, and provide information—yet much packaging is extraneous. Consumers can give manufacturers incentive to reduce packaging by choosing minimally packaged goods, buying unwrapped fruit and vegetables, and buying food in bulk. Manufacturers can switch to packaging that is more recyclable. They can also reduce the size or weight of goods and materials, as they already have with many items, from aluminum cans to plastic soft drink bottles to personal computers.

WEIGHING THE ISSUES

REDUCING PACKAGING: IS IT A WRAP? Reducing packaging cuts down on the waste stream, but packaging can serve worthwhile purposes, such as safeguarding consumer health and safety. Can you think of a product for which you would not want to see less packaging? Why? Can you name a product for which packaging could easily be reduced without ill effect to the consumer? Would you be any more or less likely to buy these products if they had less packaging?

Some governments have recently taken aim at a major source of waste and litter—plastic grocery bags. These lightweight polyethylene bags can persist for centuries in the environment, choking and entangling wildlife and littering the landscape—yet Americans discard 100 billion of them each year. A number of major cities and over 20 nations have now enacted bans or limits on their use. In 2007, San Francisco became the first U.S. city to ban nonbiodegradable plastic bags. The city's action saves 14,000 bags every day. Financial incentives are also effective. When Ireland began taxing these bags, their use dropped 90%. The Ikea company began charging for them and saw similar drops in usage. Increasing numbers of stores now give discounts if you bring your own reusable bags.

Increasing the longevity of goods also helps reduce waste. Consumers generally choose goods that last longer, all else being equal. To maximize sales, however, companies often produce short-lived goods that need to be replaced frequently. Thus, increasing the longevity of goods is largely up to the consumer. If demand is great enough, manufacturers will respond.

Reuse is a major strategy to reduce waste

To reduce waste, you can save items to use again or substitute disposable goods with durable ones. Habits as simple as bringing your own coffee cup to coffee shops or bringing sturdy reusable cloth bags to the grocery store can, over time, have substantial impact. You can also donate unwanted items and shop for used items yourself at yard sales and resale centers. Over 6000 reuse centers exist in the United States, including stores run by organizations such as Goodwill Industries and the Salvation Army. Besides being sustainable, reusing items saves money. Used items are often every bit as functional as new ones, and they are cheaper. TABLE 22.1 presents a sampling of actions we all can take to reduce the waste we generate.

Composting recovers organic waste

Composting is the conversion of organic waste into mulch or humus (p. 219) through natural decomposition. We can place waste in compost piles, underground pits, or specially constructed containers. As waste is added, heat from microbial action builds in the interior and decomposition proceeds. Banana peels, coffee grounds, grass clippings, autumn leaves, and other organic items can be converted

TABLE 22.1 Some Everyday Things You Can Do to Reduce and Reuse
• Donate used items to charity
• Reuse boxes, paper, plastic wrap, plastic containers, aluminum foil, bags, wrapping paper, fabric, packing material, etc.
• Rent or borrow items instead of buying them, when possible . . . and lend your items to friends
• Buy groceries in bulk
• Bring reusable cloth bags shopping
• Make double-sided photocopies
• Keep electronic documents rather than printing items out
• Bring your own coffee cup to coffee shops
• Pay a bit extra for durable, long-lasting, reusable goods rather than disposable ones
• Buy rechargeable batteries
• Select goods with less packaging
• Compost kitchen and yard wastes
• Buy clothing and other items at resale stores and garage sales
• Use cloth napkins and rags rather than paper napkins and towels
• Tell businesses what you think about their packaging and products
• When solid waste policy is being debated, let your government representatives know your thoughts
• Support organizations that promote waste reduction

Adapted from U.S. Environmental Protection Agency.

into rich, high-quality compost through the actions of earthworms, bacteria, soil mites, sow bugs, and other detritivores and decomposers (pp. 81, 218). The compost is then used to enrich soil. Home composting is a prime example of how we can live more sustainably by mimicking natural cycles and incorporating them into our daily lives.

Municipal composting programs—3000 across the United States at last count—divert yard debris from the waste stream and send it to central composting facilities, where it decomposes into mulch that community residents can use for gardens and landscaping. Nearly half of U.S. states now ban yard waste from the municipal waste stream, helping to accelerate the drive toward composting. Some now accept food scraps for composting, along with yard debris. All in all, about one-fifth of the U.S. waste stream is made up of materials that can easily be composted. Composting reduces landfill waste, enriches soil, enhances soil biodiversity, helps soil to resist erosion, makes for healthier plants and more pleasing gardens, and reduces the need for chemical fertilizers.

Recycling consists of three steps

Recycling, too, offers many benefits. It involves collecting used items and breaking them down so that their materials can be reprocessed to manufacture new items. Recycling diverted 65 million tons of materials away from incinerators and landfills in the United States in 2010.

The recycling loop consists of three basic steps (**FIGURE 22.9**). The first step is to collect and process used goods and materials. Some communities designate locations where residents can drop off recyclables or receive money for them. Others offer the more convenient option of curbside recycling, in which trucks pick up recyclable items in front of homes, usually in conjunction with municipal trash collection.

Items collected are taken to **materials recovery facilities (MRFs)**, where workers and machines sort items using automated processes including magnetic pulleys, optical sensors, water currents, and air classifiers that separate items by weight and size. The facilities clean the materials, shred them, and prepare them for reprocessing.

Once readied, these materials are used to manufacture new goods. Newspapers and many other paper products use recycled paper, many glass and metal containers are now made from recycled materials, and some plastic containers are of recycled origin. Benches, bridges, and walkways in city parks may now be made from recycled plastics, and glass can be mixed with asphalt (creating "glassphalt") to pave roads and paths. The paper used in this textbook, besides being FSC-certified (Forest Stewardship Council; pp. 307–308, 322), also contains 10% recycled post-consumer waste.

If the recycling loop is to function, consumers and businesses must complete the third step in the cycle by purchasing ecolabeled products (p. 155) made from recycled materials. By buying recycled goods, consumers provide economic incentive for industries to recycle materials and for recycling facilities to expand. As markets for products made with recycled materials expand, prices continue to fall.

Recycling has grown rapidly

Today 9000 curbside recycling programs across all 50 U.S. states serve nearly half of all Americans. These programs, and the 500 MRFs in operation today, have sprung up only in the last 25 years. Recycling in the United States has risen

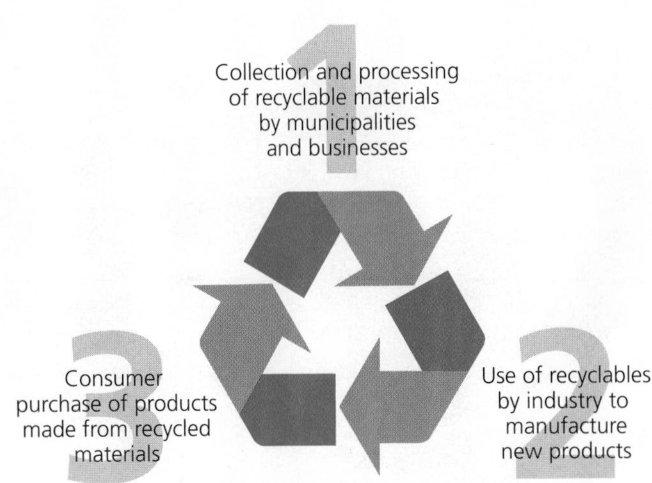

FIGURE 22.9 The familiar recycling symbol represents the three components of a sustainable recycling strategy. After recyclable materials are collected and processed, they are used to make new products, which are then purchased by consumers.

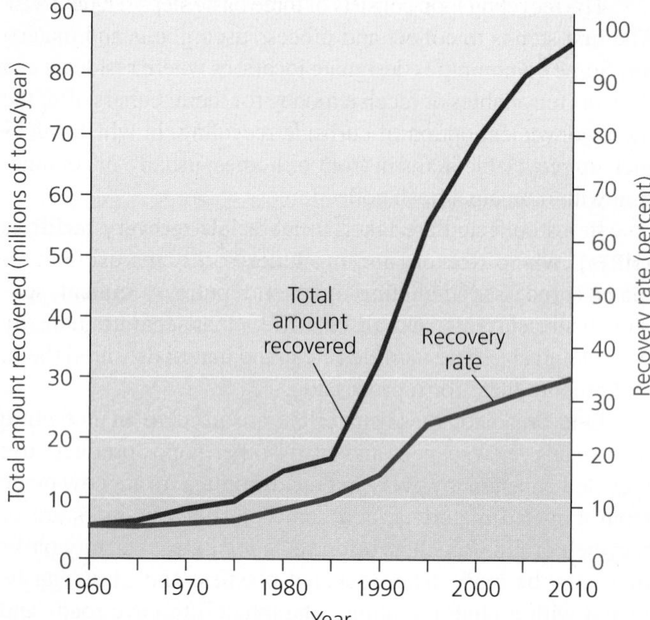

FIGURE 22.10 Recovery has risen sharply in the United States. Today over 85 million tons of material are recovered (65 million tons by recycling and 20 million tons by municipal composting), comprising one-third of the waste stream. *Data from U.S. Environmental Protection Agency.*

TABLE 22.2 Recovery Rates for Various Materials in the United States

MATERIAL	PERCENTAGE THAT IS RECYCLED OR COMPOSTED
Lead-acid batteries	96
Newspapers	72
Paper and paperboard	63
Yard trimmings	58
Aluminum cans	50
Glass containers	33
Total plastics	8

Data are for 2010, from U.S. Environmental Protection Agency.

from 6.4% of the waste stream in 1960 to 26.0% in 2010 (and 34.1% if composting is included) (**FIGURE 22.10**). The EPA calls the growth of recycling "one of the best environmental success stories of the late 20th century."

Recycling rates vary greatly from one product or material type to another, ranging from nearly zero to almost 100% (**TABLE 22.2**). Recycling rates among U.S. states also vary greatly (**FIGURE 22.11**). Highly variable rates from city to city depend largely on how heavily a city invests in making recycling convenient for its citizens. Despite New York City's credentials overall as a sustainable city, its recycling program leaves much to be desired, and it can be hard to find a recycling bin on a New York street. In contrast, cities from Austin to Durham to Fresno to Pittsburgh to Portland run well-funded programs with high participation rates.

Recycling's growth has been propelled in part by economic forces as businesses see prospects to save money and as entrepreneurs see opportunities to start new businesses. It has also been driven by the desire of municipal leaders to reduce waste and by the satisfaction people take in recycling. These latter two forces have driven the rise of recycling even when it has not been financially profitable. In fact, many of our popular municipal recycling programs are run at an economic loss. The expense required to collect, sort, and process recycled goods is often more than recyclables are worth in the marketplace. Additionally, the more people recycle, the more glass, paper, and plastic is available to manufacturers for purchase, which drives down prices. And transporting items to recycling facilities can sometimes involve surprisingly long distances (see **THE SCIENCE BEHIND THE STORY**, pp. 620–621).

Recycling advocates, however, point out that market prices do not take into account external costs (pp. 146, 165)—in particular, the environmental and health impacts of *not* recycling. For instance, it has been estimated that globally, recycling saves enough energy to power more than 6 million households per year. Each year in the United States, recycling

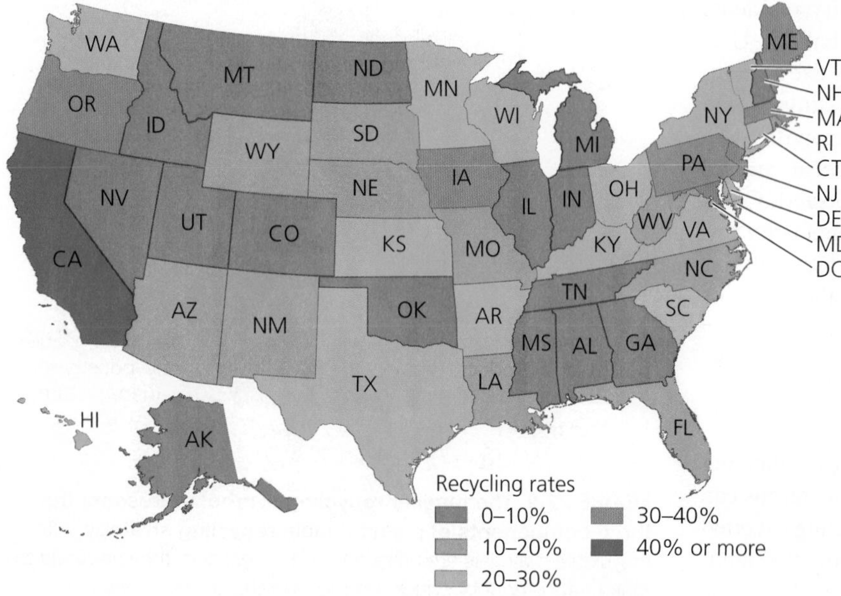

FIGURE 22.11 U.S. states vary greatly in the rates at which their citizens recycle. *Data are for 2008 (with earlier data for several states), from van Haaren, R., et al., 2010. The state of garbage in America.* BioCycle *Oct. 2010: 16–23.*

and composting together save energy equal to that of 230 million barrels of oil, and they prevent carbon dioxide emissions equal to those of 36 million cars. Recycling aluminum cans saves 95% of the energy required to make the same amount of aluminum from mined virgin bauxite, its source material.

As more manufacturers use recycled products and as more technologies and methods are developed to use recycled materials in new ways, markets should continue to expand, and new business opportunities may arise. We are just beginning to shift from an economy that moves linearly, from raw materials to products to waste, to a more sustainable economy that moves circularly, using waste products as raw materials for new manufacturing. The steps we have taken in recycling so far are central to this transition, which many analysts view as key to building a sustainable economy.

WEIGHING THE ISSUES

COSTS OF RECYCLING AND NOT RECYCLING Should recycling programs be subsidized by governments even if they are run at an economic loss? What types of external costs—costs not reflected in market prices—do you think would be involved in not recycling, say, aluminum cans? Do you feel these costs justify sponsoring recycling programs even when they are not financially self-supporting? Why or why not?

We can recycle material from landfills

With improved technology for sorting rubbish and recyclables, businesses and entrepreneurs are weighing the economic benefits and costs of rummaging through landfills to salvage materials of value that can be recycled. Steel, aluminum, copper, and other metals are abundant enough in some landfills to make salvage operations profitable when market prices for the metals are high enough. For instance, Americans throw out so many aluminum cans that at 2013 prices for aluminum, the nation buries $1.7 billion of this metal in landfills each year. If we could retrieve all the aluminum from U.S. landfills, it would exceed the amount the world produces from a year's worth of mining ore.

Besides metals, landfills also offer organic waste that can be mined and sold as premium compost. Old landfill waste can also be incinerated in newer, cleaner-burning WTE facilities to produce energy. Some companies are even looking into gaining carbon credits (p. 513) by harvesting methane leaking from open dumps in developing nations.

Such approaches have been tried in places from New York to Israel to Sweden to Singapore, and they can be profitable when market prices are high enough. However, the costs of mining landfills and meeting regulatory requirements while commodity prices change unpredictably have meant that investing in landfill mining has been risky. This could change in the future, though, if prices rise and technologies improve.

Financial incentives help address waste

Waste managers offer consumers economic incentives to reduce the waste stream. In "pay-as-you-throw" garbage collection programs, municipalities charge residents for home trash pickup according to the amount of trash they put out. The less waste one generates, the less one has to pay. Over 7000 such programs operate in the United States, serving more than one of every four communities.

Bottle bills represent another approach that hinges on financial incentive. In the 10 U.S. states and 23 nations that have these laws, consumers pay a deposit on bottles or cans upon purchase—often 5 cents per container—and then receive a refund when they return them to stores after use. The first bottle bills were passed in the 1970s to cut down on litter, but they have also served to decrease the waste stream. Where these laws have been enacted, they have proved effective and popular (**FIGURE 22.12**). U.S. states with bottle bills report that their beverage container litter has decreased by 69–84%, their total litter has decreased by 30–65%, and their per capita container recycling rates have risen 2.6-fold. Beverage container recycling rates for states with bottle bills are 2.5 times higher than for states without them.

As of 2013, bottle bill advocates in 11 states were seeking to establish or expand programs. It is a testament to the lobbying influence of the beverage industries and grocery retailers, which have traditionally opposed passage of bottle bills, that more states do not have such legislation.

States with bottle bills now face two challenges. One is to amend these laws to include new kinds of containers. In New York State, where polls showed that 70% of the public favored expanding their state's law to include more types of containers, the legislature in 2009 added bottled water containers to the law's coverage. It also mandated that most of the $140 million in unclaimed deposit money (for unreturned containers) go to the state rather than to the bottling industry.

The second challenge for bottle-bill states is to adjust refunds for inflation. In the 42 years since Oregon passed the

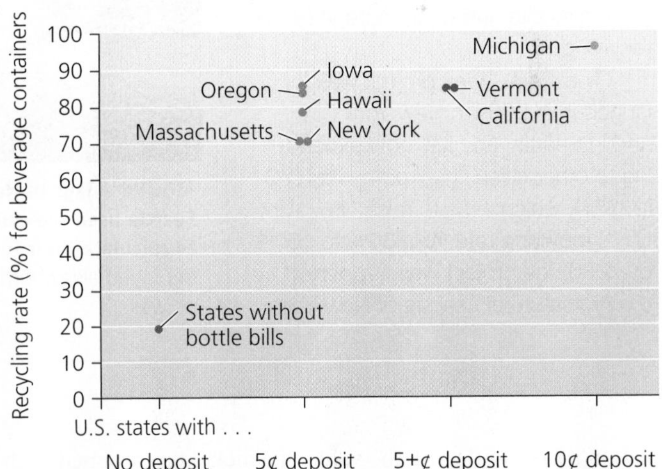

FIGURE 22.12 Bottle bills increase recycling rates, and higher redemption amounts boost these rates further, data suggest. States with bottle bills have much higher recycling rates for beverage containers than states without bottle bills. Michigan offers the highest redemption rate and has the highest recycling rate. *Data from Container Recycling Institute, Arlington, VA, 2010, using most recent data from the states, mostly from 2009. Maine and Connecticut also have bottle bills but have not kept detailed data on recycling rates.*

DATA Q How many times greater is Michigan's bottle recycling rate than that of states without bottle bills?

THE SCIENCE BEHIND THE STORY

Tracking Trash

Where will it go?

Where does your trash go once you throw it away? What happens to your recycling? How far might it travel, and how much energy might it take to get rid of it?

With the help of the latest tracking technology, we can find out. Researchers from the SENSEable City Lab at the Massachusetts Institute of Technology (MIT) are affixing tiny sensors to everyday items in our trash and monitoring them to reveal their hidden travels. By documenting what actually happens to trash and to recyclables, they hope to help make the trash removal process more effective and to encourage better recycling.

The Trash Track project was launched in 2009 in New York City and in Seattle, and it is now expanding to other cities. Inspired by PlaNYC (p. 354), which aims to raise New York City's recycling rate from 30% to 100% by 2030, the project was supported by the Architectural League of New York. Early results were unveiled at public exhibitions in New York and Seattle.

Here's how trash-tracking works: Project director Carlo Ratti and associate director Assaf Biderman, both architects at MIT, organize research teams and local volunteers in the target city to affix tiny electronic tags (**FIGURE 1**) to hundreds of items being thrown away. As each item makes its final journey through the waste stream, its tag calculates its location every few hours and relays the information via the cell phone network to a central server at MIT. A computer plots the movements atop satellite maps, helping the researchers to visualize and interpret the migration of trash.

Each trash tag calculates its position by measuring the signal strength from nearby cell phone towers and comparing this to a map of tower locations. Tags used in New York City and Seattle are not as accurate as a global positioning system (GPS), but their signals carry better through such barriers as building roofs, walls of garbage trucks, and deep piles of trash. Second-generation tags the project is now using are more accurate, combining GPS technology with better cell network triangulation.

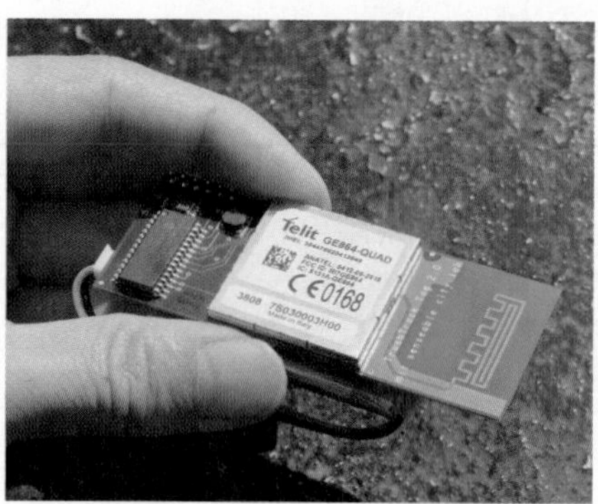

FIGURE 1 Tags in New York City and Seattle used the cell phone network to calculate their location. Tags being deployed elsewhere use GPS technology as well.

As an example, a plastic container of liquid soap was tagged on September 5, 2009, and placed in the trash at 457 Madison Avenue in Manhattan (**FIGURE 2**). Mapping reveals that the truck that picked it up looped through the city's streets a few times on its route, crossed the Hudson River via the Lincoln Tunnel, and then headed to Rutherford, New Jersey. Here the truck turned south and was in transit along the Bellevue Turnpike in Kearny, New Jersey, three days later, when the tag's battery gave out.

As of 2013, the project had posted mapped data of all its Seattle items online for the public to view and was preparing to do the same for its New York items.

The results from Seattle reveal some expected patterns but also some odd surprises. Of the 760 items tagged, about 200 ended up at the city's Allied Waste Recycling Center and Transfer Station after brief journeys. Smaller numbers were transported to other city or regional landfills and recycling centers. But some items followed surprisingly circuitous routes, being transferred from one waste center to another, or back and forth between cities. Some

---

nation's first bottle bill, the value of a nickel has dropped such that today, the refund would need to be 28 cents to reflect the refund's original intended value. Proponents argue that increasing refund amounts will raise return rates, and available data support this view (see Figure 22.12).

One Canadian city showcases reduction and recycling

Edmonton, Alberta, has created one of the world's top waste management programs. Just 40% of the city's waste goes to its sanitary landfill, whereas 15% is recycled and an impressive 45% is composted.

When Edmonton's residents put their trash out at the curb, city trucks take it to their co-composting plant—at the size of eight football fields, the largest in North America. The waste is dumped on the floor of the facility, and large items, such as furniture, are removed and landfilled. The bulk of the waste is mixed with dried sewage sludge for 1–2 days in five large rotating drums, each the length of six buses. The resulting mix travels on a conveyor to a screen that removes nonbiodegradable items. It is aerated for several weeks in the largest stainless steel building in North

ended up in seemingly random places, perhaps having fallen off a garbage truck along a roadside.

A few items made very long journeys. Two printer cartridges were driven down Interstate 5 to California's border with Mexico, perhaps to be disassembled at a factory. Two cell phones were transported halfway across the country to Dallas (one flown directly and one making apparent stops at three other cities). A compact fluorescent bulb went to St. Louis after traveling to Portland, Oregon, and back to Seattle. Chicago received a coffee cup that apparently made its way east on the nation's interstates. Shipments of batteries were flown 1500 miles to Minneapolis, 2500 miles to Pittsburgh, and 2600 miles to Atlanta.

The longest-traveling piece of trash was a cell phone that was transported over 3000 miles from Seattle to the other corner of the United States, ending up near Ocala, Florida.

In general, hazardous waste items and electronic waste items tended to travel farthest because they often needed to reach special facilities for handling. This raises the question of whether this special handling is worthwhile, given the impacts of the extra gasoline use and greenhouse gas emissions it entails. Posing this question in a 2012 research paper, the Trash Track team suggested that we find ways to make our processing systems for hazardous and electronic waste as efficient as those for curbside waste collection and recycling.

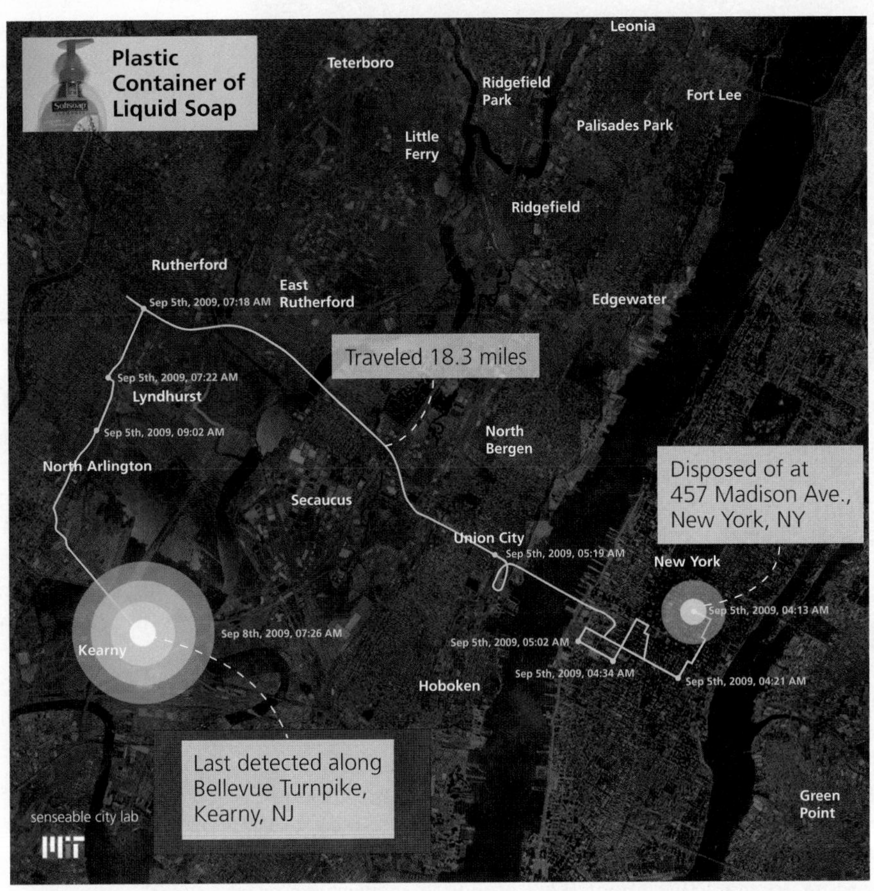

FIGURE 2 Follow that trash truck! A plastic soap container put out with the trash on Madison Avenue in New York City looped through midtown Manhattan, crossed the Hudson River, and was last detected traveling down the Bellevue Turnpike in New Jersey.

Why did each of the tracked items migrate so far? How much gasoline was used to transport them? When we move items great distances to dispose of them or recycle their parts, does that reflect efficiency from an economy of scale—or does it indicate an excessive waste of resources? Do we need more recycling and disassembly facilities nearer to each major city? Researchers and waste managers hope to use data from the Trash Track project to address such questions and improve the way we handle waste. ■

America (FIGURE 22.13). The mix is then passed through a finer screen and left outside for 4–6 months. The resulting compost—80,000 tons annually—is made available to farmers and gardeners. The facility even filters the air it emits to eliminate odors.

Edmonton's program includes a state-of-the-art MRF, a leachate treatment plant, a research center, public education programs, and a wetland and landfill revegetation program. In addition, 100 pipes collect enough landfill gas to power 4600 homes, bringing thousands of dollars to the city and helping to power the new waste management center. Five area businesses reprocess the city's recycled items. Newsprint and magazines are turned into new newsprint and cellulose insulation, and cardboard and paper are converted into building paper and shingles. Household metal is made into rebar and blades for tractors and graders, and recycled glass is used for reflective paint and signs.

Edmonton has just built a new integrated processing and transfer facility to handle compostable and recyclable waste from homes and businesses, and the city will soon complete a biofuels facility to create ethanol (p. 569) from waste that cannot be recycled or composted. With these new facilities, Edmonton hopes to achieve 90% recovery in 2013.

FIGURE 22.13 Edmonton, Alberta, boasts one of North America's best waste management programs. Inside Edmonton's aeration building, which is the size of 14 professional hockey rinks, a mix of solid waste and sewage sludge is exposed to oxygen and composted for 14–21 days.

Industrial Solid Waste

In the United States, *industrial solid waste* is defined as solid waste that is considered neither municipal solid waste nor hazardous waste under the Resource Conservation and Recovery Act. This includes waste from factories, mining activities, agriculture, petroleum extraction, and more. Each year, U.S. industrial facilities generate about 7.6 billion tons of waste, according to the EPA, about 97% of which is wastewater. Thus, very roughly, 230 million or so tons of solid waste are generated by 60,000 facilities each year—an amount similar to that of municipal solid waste. Waste is generated at various points along the process from raw materials extraction to manufacturing to sale and distribution (FIGURE 22.14).

Regulation and economics each influence industrial waste generation

Most methods and strategies of waste disposal, reduction, and recycling by industry are similar to those for municipal solid waste. Businesses that dispose of their own waste on site must design and manage their landfills in ways that meet state, local, or tribal guidelines. Other businesses pay to have their waste disposed of at municipal disposal sites. Whereas the federal government regulates municipal solid waste, state or local governments regulate industrial solid waste (with federal guidance). Regulation varies greatly from place to place, but in most cases, state and local regulation of industrial solid waste is less strict than federal regulation of municipal solid waste. In many areas, industries are not required to have permits, install landfill liners or leachate collection systems, or monitor groundwater for contamination.

The amount of waste generated by a manufacturing process is a good measure of its efficiency; the less waste produced per unit or volume of product, the more efficient that process is, from a physical standpoint. However, physical efficiency is not always reflected in economic efficiency. Often it is cheaper for industry to manufacture its products or perform its services quickly but messily. That is, it can be cheaper to generate waste than to avoid generating waste. In such cases, economic efficiency is maximized, but physical efficiency is not. Because our market system awards only economic efficiency, all too often industry has no financial incentive to achieve physical efficiency. The frequent mismatch between these two types of efficiency is a major reason why the output of industrial waste is so great.

Rising costs of waste disposal enhance the financial incentive to decrease waste. Once either government or the market makes the physically efficient use of raw materials also economically efficient, businesses gain financial incentives to reduce their waste.

Industrial ecology seeks to make industry more sustainable

To reduce waste, growing numbers of industries today are experimenting with industrial ecology. A holistic approach that integrates principles from engineering, chemistry, ecology, and economics, **industrial ecology** seeks to redesign industrial systems to reduce resource inputs and to maximize both physical

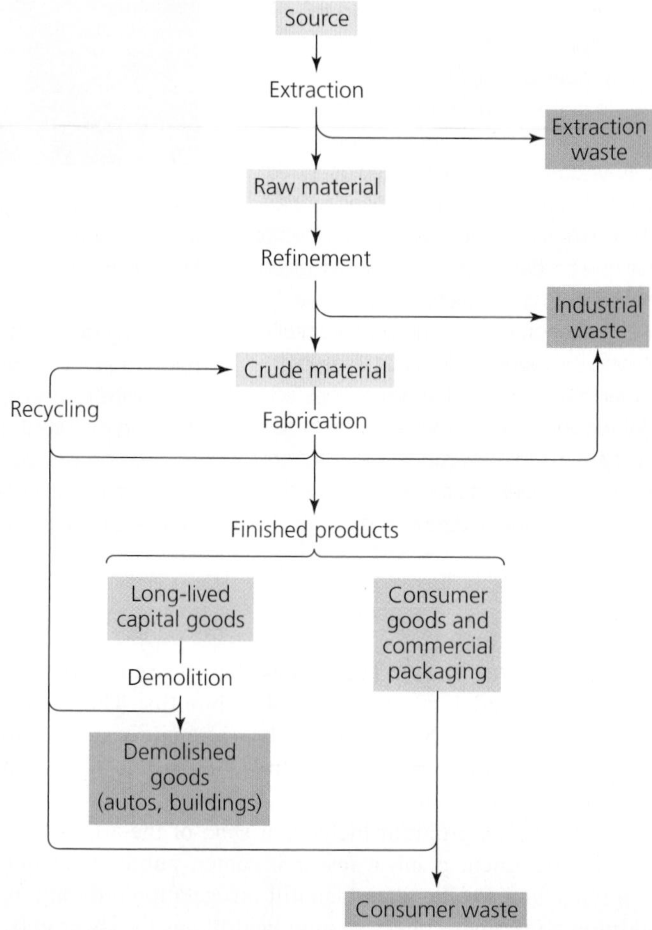

FIGURE 22.14 Waste is generated at multiple stages in the life cycles of products. Each stage presents opportunities for efficiency improvements, waste reduction, or recycling.

and economic efficiency. Industrial ecologists would reshape industry so that nearly everything produced in a manufacturing process is used, either within that process or in a different one.

The larger idea behind industrial ecology is that industrial systems should function more like ecological systems, in which organisms use almost everything that is produced. This principle brings industry closer to the ideal of ecological economists, in which economies function in a circular fashion rather than a linear one (p. 150).

Industrial ecologists pursue their goals in several ways:

- They examine the entire life cycle of a given product—from its origins in raw materials, through its manufacturing, to its use, and finally its disposal—and look for ways to make the process more efficient. This strategy is called **life-cycle analysis** (p. 269).
- They try to identify how waste products from one manufacturing process might be used as raw materials for another. For instance, used plastic beverage containers can be shredded and reprocessed to make other plastic items, such as benches, tables, and decks.
- They examine industrial processes with an eye toward eliminating environmentally harmful products and materials.
- They study the flow of materials through industrial systems to look for ways to create products that are more durable, recyclable, or reusable. For instance, they seek to design computers, automobiles, and appliances to be easily disassembled so more of their component parts can be reused or recycled.

Businesses are adopting industrial ecology

Attentive businesses are taking advantage of the insights of industrial ecology to save money while reducing waste. For example, American Airlines switched from hazardous to nonhazardous materials in its Chicago facility, decreasing its need to secure permits from the EPA. The company used over 50,000 reusable plastic containers to ship goods, reducing packaging waste by 90%. Its Dallas–Fort Worth headquarters recycled enough aluminum cans and white paper in five years to save $205,000 and recycled 3000 broken baggage containers into lawn furniture. A program to gather suggestions from employees brought over 700 ideas to reduce waste—and 15 of these ideas saved the company over $8 million.

The Swiss Zero Emissions Research and Initiatives (ZERI) Foundation sponsors dozens of innovative projects worldwide that attempt to create goods and services without generating waste. One example involves breweries in Canada, Sweden, Japan, and Namibia (**FIGURE 22.15**). Brewers in these projects take waste from the beer-brewing process and use it to fuel other processes. As a result, the brewer can make money from bread, mushrooms, pigs, gas, and fish, as well as beer, all while producing little waste. Although most ZERI projects are not fully closed-loop systems, they attempt to approach this ideal. In so doing, they cut down on waste while increasing output and income, often generating new jobs as well.

Few businesses have taken industrial ecology to heart as much as the carpet tile company Interface, which founder Ray Anderson set on the road to sustainability over a decade ago (p. 155). Interface asks customers to return used tiles for recycling and for reuse as backing for new carpet. It modified its tile design and production methods to reduce waste. It adapted its boilers to use landfill gas for energy. Through such steps, Anderson's company cut its waste generation by 80%, its fossil fuel use by 45%, and its water use by 70%—all while saving $30 million per year, holding prices steady for its customers, and raising profits by 49%.

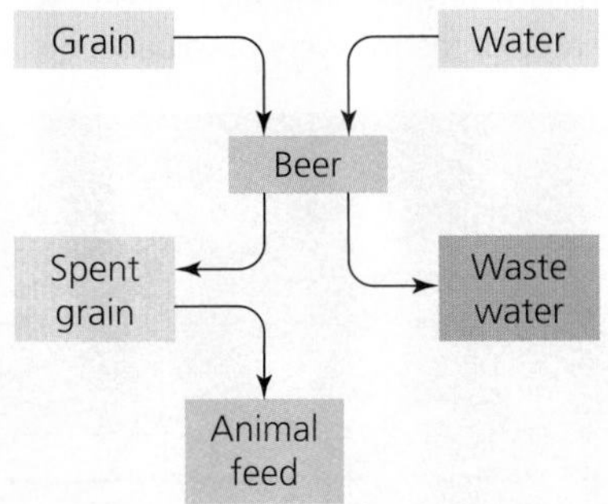

(a) Traditional brewery process

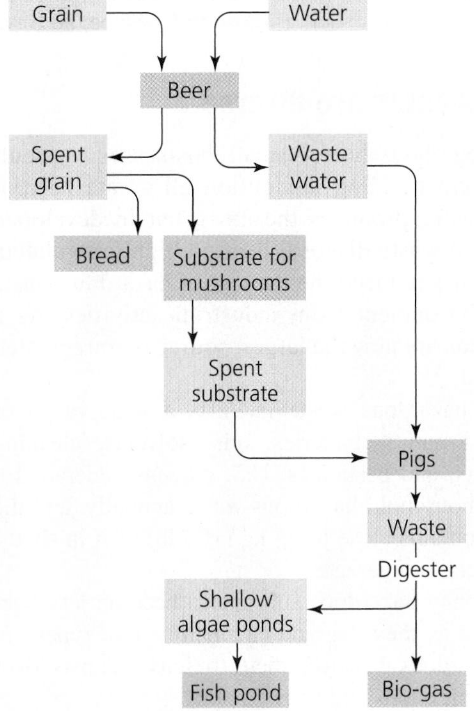

(b) ZERI brewery process

FIGURE 22.15 Creative use of waste products can help us approach zero-waste systems. Traditional breweries **(a)** produce only beer while generating much waste, some of which goes toward animal feed. Breweries sponsored by the Zero Emissions Research and Initiatives (ZERI) Foundation **(b)** use their waste grain to make bread and to farm mushrooms. Waste from the mushroom farming, along with brewery wastewater, goes to feed pigs. The pigs' waste is digested in containers that capture natural gas and collect nutrients used to nourish algae for growing fish in fish farms. The brewer derives income from bread, mushrooms, pigs, gas, and fish, as well as beer.

For businesses, governments, and individuals alike, there are plenty of ways to reduce waste and mitigate the impacts of our waste generation—and quite often, doing so brings economic benefits. This is true both for solid waste and for hazardous waste.

Hazardous Waste

Hazardous wastes are diverse in their chemical composition and may be liquid, solid, or gaseous. By EPA definition, *hazardous waste* is waste that is one of the following:

- *Ignitable.* Likely to catch fire (for example, gasoline or alcohol).
- *Corrosive.* Apt to corrode metals in storage tanks or equipment (for example, strong acids or bases).
- *Reactive.* Chemically unstable and readily able to react with other compounds, often explosively or by producing noxious fumes (for example, ammonia reacting with chlorine bleach).
- *Toxic.* Harmful to human health when inhaled, ingested, or touched (for example, pesticides or heavy metals).

Substances with these characteristics can harm human health and environmental quality. Flammable and explosive materials can cause ecological damage and atmospheric pollution. For instance, fires at large tire dumps in California's Central Valley have caused air pollution and highway closures. Toxic wastes in lakes and rivers have caused fish die-offs and closed important fisheries, such as those in Chesapeake Bay.

Hazardous wastes are diverse

Industry, mining, households, small businesses, agriculture, utilities, and building demolition all create hazardous waste. Industry produces the most, but in developed nations industrial waste disposal is often highly regulated. This regulation has reduced the amount of hazardous waste entering the environment from industrial activities. As a result, households are now the largest source of unregulated hazardous waste.

Household hazardous waste includes a wide range of items, including paints, batteries, oils, solvents, cleaning agents, lubricants, and pesticides. U.S. citizens generate 1.6 million tons of household hazardous waste annually, and the average home contains close to 45 kg (100 lb) of it in sheds, basements, closets, and garages.

Although many hazardous substances become less hazardous over time as they degrade chemically, two types are particularly hazardous because their toxicity persists over time: organic compounds and heavy metals.

Organic compounds and heavy metals pose hazards

In our daily lives, we rely on synthetic organic compounds and petroleum-derived compounds to resist bacterial, fungal, and insect activity. Plastic containers, rubber tires, pesticides, solvents, and wood preservatives are useful to us precisely because they resist decomposition. We use these substances to protect buildings from decay, kill pests that attack crops, and keep stored goods intact. However, these compounds' capacity to resist decay is a double-edged sword, for it also makes them persistent pollutants. Many synthetic organic compounds are toxic because they are readily absorbed through the skin and can act as mutagens, carcinogens, teratogens, and endocrine disruptors (p. 370).

Heavy metals such as lead, chromium, mercury, arsenic, cadmium, tin, and copper are used widely in industry for wiring, electronics, metal plating, metal fabrication, pigments, and dyes. Heavy metals enter the environment when paints, electronic devices, batteries, and other materials are disposed of improperly. Lead from fishing weights and from hunting ammunition accumulates in rivers, lakes, and forests. In older homes, lead from pipes contaminates drinking water, and lead paint remains a problem, especially for infants. Heavy metals that are fat-soluble and break down slowly are prone to bioaccumulate and biomagnify (pp. 373–374). In California's Coast Range, for instance, mercury washed downstream from abandoned mercury mines enters lakes and rivers, is consumed by bacteria and invertebrates, and accumulates in increasingly large quantities up the food chain, poisoning organisms at higher trophic levels and making fish unsafe to eat.

E-waste is growing

Today's proliferation of computers, printers, smartphones, TVs, DVD players, MP3 players, and other electronic technology has created a substantial new source of waste (**FIGURE 22.16**). These products have short life spans before people judge them obsolete, and most are discarded after just a few years. The amount of this **electronic waste**—often called **e-waste**—has grown rapidly, and now comprises 2% of the U.S. solid waste stream (**FIGURE 22.17a**). Over 7 billion electronic devices have been sold in the United States since

FIGURE 22.16 Each day, Americans throw away half a million cell phones. Phones that enter waste stream can leach toxic heavy metals into the environment. Alternatively, we can recycle them for reuse and for the recovery of valuable metals.

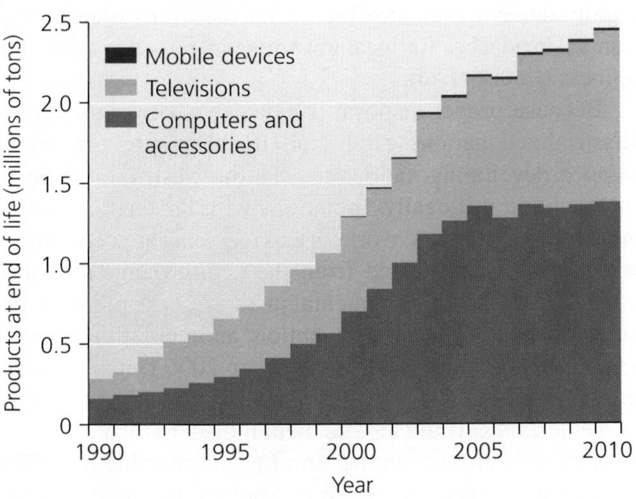

(a) Products at end of life each year

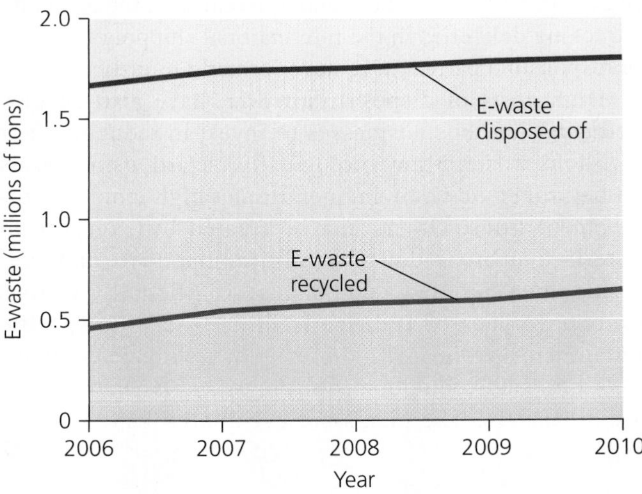

(b) E-waste disposed of and recycled each year

FIGURE 22.17 **Electronic waste is increasing, but so is its recycling.** The amount of electronic products at the end of their lives each year **(a)** in the United States has skyrocketed. The amounts disposed of and recycled each year **(b)** are both growing. *Data from U.S. Environmental Protection Agency, 2011.* Electronic waste management in the United States through 2009. *EPA, Washington, D.C.*

1980, and U.S. households discard more than 300 million per year—two-thirds of them still in working order. Fortunately, e-waste recycling is expanding along with e-waste disposal, and Americans now recycle one-fourth of their e-waste, by weight (FIGURE 22.17b).

Of the electronic items we discard, most end up in conventional sanitary landfills and incinerators. However, electronic products contain heavy metals and toxic flame-retardants, and research suggests that e-waste should instead be treated as hazardous waste (see THE SCIENCE BEHIND THE STORY, pp. 628–629). The EPA and a number of states are now taking steps to keep e-waste out of conventional sanitary landfills and incinerators and instead treat it as hazardous waste.

Increasingly, used electronics are collected by businesses, nonprofit organizations, or municipal services and are processed for reuse or recycling. When e-waste is recycled, the devices are taken apart, and parts and materials are refurbished and reused in new products. There are serious concerns, however, about health risks that recycling may pose to workers doing the disassembly. Wealthy nations ship much of their e-waste to developing countries, where low-income workers disassemble the devices and handle toxic materials with minimal safety regulations.

Another challenge is that the recent conversion of television and computer monitor technology from cathode ray tubes to LCD and plasma screens has meant that there is no longer much demand for recycled cathode ray tubes. As a result, old cathode ray tubes (rich in toxic lead) are piling up in recyclers' warehouses and are at risk of never being recycled.

Besides keeping toxic substances out of our waste stream, e-waste recycling helps us recover trace metals used in electronics that are rare and lucrative. A typical cell phone contains up to \$2.50 worth of precious metals (p. 649). By one estimate, 1 ton of computer scrap contains more gold than 16 tons of mined ore from a gold mine. Every ounce of metal we can recycle from a manufactured item is an ounce of metal we don't need to mine from the ground. Thus, "mining" e-waste for metals helps reduce the environmental impacts of mining the earth.

In one of the more intriguing efforts to promote sustainability through such recycling, the 2010 Winter Olympic Games in Vancouver produced its stylish gold, silver, and bronze medals (FIGURE 22.18) from metals recovered from recycled and processed e-waste!

WEIGHING THE ISSUES

TOXIC ELECTRONICS? The cathode ray tubes in older televisions and desktop monitors held up to 5 kg (8 lb) of heavy metals, such as lead and cadmium. These represent the second-largest source of lead in U.S. landfills today, behind auto batteries. The new flat-panel LCD and plasma screen technologies that are replacing them alleviate these risks, but they contain mercury, are made using the potent greenhouse gas nitrogen trifluoride, and are (so far) less recyclable. Smartphones and other handheld devices require fewer materials, yet more devices are being purchased worldwide all the time. Considering the rapid turnover of these products, what future waste problems and environmental health issues might you expect? What steps do you think we should take to best handle the reuse, recycling, and disposal of these products?

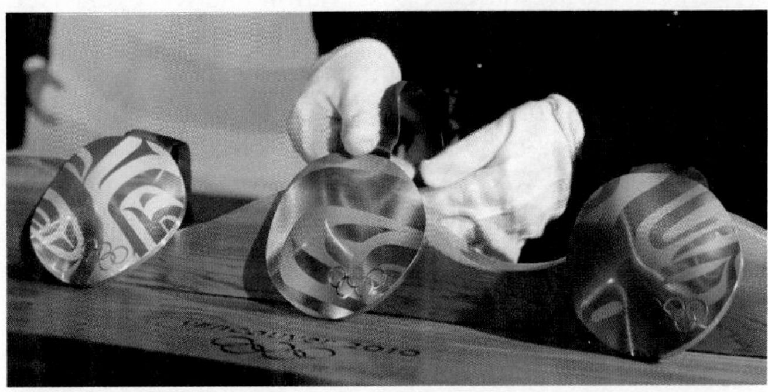

FIGURE 22.18 **The medals awarded to athletes at the 2010 Winter Olympic Games in Vancouver were manufactured partly from precious metals recycled from discarded e-waste.**

FIGURE 22.19 **Many communities designate collection sites or collection days for household hazardous waste.** Here, workers handle waste from a collection event near Los Angeles.

Several steps precede the disposal of hazardous waste

For many years we discarded hazardous waste without special treatment. In many cases, people did not know that certain substances were harmful to human health. In other cases, it was assumed that the substances would disappear or be sufficiently diluted in the environment. The resurfacing of toxic chemicals in the 1970s in a residential area at Love Canal (p. 627) in upstate New York, years after their burial, convinced the public that hazardous waste deserves special attention and treatment.

Many communities now designate sites or special collection days to gather household hazardous waste, or designate facilities for the exchange and reuse of substances (FIGURE 22.19). Once consolidated, the waste is transported for treatment and disposal.

Under the Resource Conservation and Recovery Act, the EPA sets standards by which states manage hazardous waste. The Act also requires large generators of hazardous waste to obtain permits. Finally, it mandates that hazardous materials be tracked "from cradle to grave." As hazardous waste is generated, transported, and disposed of, the producer, carrier, and disposal facility must each report to the EPA the type and amount of material generated; its location, origin, and destination; and the way it is handled. This process is intended to prevent illegal dumping and to encourage the use of reputable waste carriers and disposal facilities.

Because current U.S. law makes disposing of hazardous waste quite costly, irresponsible companies sometimes illegally dump waste, creating health risks for residents and financial headaches for local governments forced to deal with the mess (FIGURE 22.20).

Because proper disposal is expensive, companies from industrialized nations often find it cheaper to pay cash-strapped developing nations to take the waste—or cheaper still, to dump it illegally. In nations with lax environmental and health regulations, workers and residents are often uninformed of or unprotected from the health dangers of this waste. This global environmental justice issue (pp. 140–141) occurs despite the Basel Convention, an international treaty to limit such practices. For instance, in 2006 Dutch authorities informed the owners of a ship carrying toxic waste that the Netherlands would charge them money to dispose of the waste in Amsterdam. So the ship instead traveled to Africa and secretly dumped its waste in Abidjan, the capital of the Ivory Coast. The waste caused several deaths and thousands of illnesses in Abidjan, and street protests forced the government to resign over the scandal. Because of the difficulty of tracking deliveries in the international shipping industry, the responsible parties were never brought to justice.

High costs of disposal, however, have also encouraged conscientious businesses to invest in reducing their hazardous waste. Many biologically hazardous materials can be broken down by incineration at high temperatures in cement kilns. Others can be treated by exposure to bacteria that break down harmful components and synthesize them into new compounds. Additionally, various plants have been bred or engineered to take up specific contaminants from soil and break down organic contaminants into safer compounds or concentrate heavy metals in their tissues. The plants are eventually harvested and disposed of.

FIGURE 22.20 **Unscrupulous individuals or businesses sometimes dump hazardous waste illegally to avoid disposal costs.**

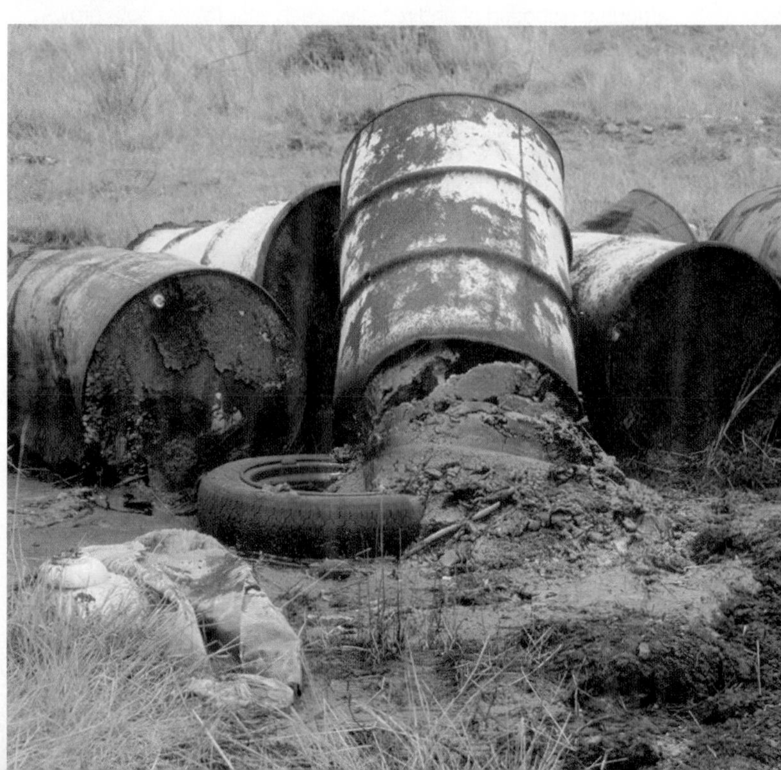

We use three disposal methods for hazardous waste

We have developed three primary means of hazardous waste disposal: landfills, surface impoundments, and injection wells. These do nothing to lessen the hazards of the substances, but they help keep the waste isolated from people, wildlife, and ecosystems. Design and construction standards for landfills that receive hazardous waste are stricter than those for ordinary sanitary landfills. Hazardous waste landfills must have several impervious liners and leachate removal systems and must be located far from aquifers. Dumping of hazardous waste in ordinary landfills has long been a problem. In New York City, Fresh Kills largely managed to keep hazardous waste out, but most of the city's older landfills were declared to be hazardous sites because of past toxic waste dumping.

Liquid hazardous waste, or waste in dissolved form, may be stored in **surface impoundments,** shallow depressions lined with plastic and an impervious material, such as clay. The liquid or slurry is placed in the pond and water is allowed to evaporate, leaving a residue of solid hazardous waste on the bottom. This process is repeated and eventually the dry residue is removed and transported elsewhere for permanent disposal. Impoundments are not ideal. The underlying layer can crack and leak waste. Some material may evaporate or blow into surrounding areas. Rainstorms may cause waste to overflow and contaminate nearby areas. For these reasons, surface impoundments are used only for temporary storage.

The third method is intended for long-term disposal. In **deep-well injection,** a well is drilled deep beneath the water table into porous rock, and wastes are injected into it (FIGURE 22.21). The waste is meant to remain deep underground, isolated from groundwater and human contact. However, wells can corrode and can leak wastes into soil, contaminating aquifers. Roughly 34 billion L (9 billion gal) of hazardous waste are placed in U.S. injection wells each year.

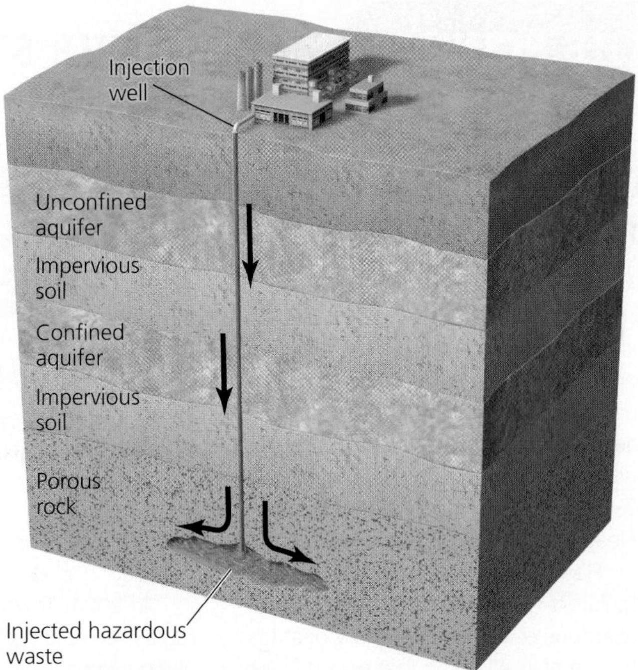

FIGURE 22.21 **Liquid hazardous waste is pumped deep underground by deep-well injection.** The well must be drilled below any aquifers, into porous rock isolated by impervious clay. The technique is expensive, and waste may sometimes leak from the well shaft into groundwater.

Radioactive waste is especially hazardous

Radioactive waste is particularly dangerous to human health and is persistent in the environment. The dilemma of disposal has dogged the nuclear energy industry and the U.S. military for decades. The United States has no designated single site to dispose of its commercial nuclear waste if Yucca Mountain in Nevada is removed from consideration (pp. 564–566). Instead, waste will continue to accumulate at the many nuclear power plants spread through the nation (see Figure 20.12, p. 565).

Currently, a site in the Chihuahuan Desert in New Mexico serves as a permanent disposal location for radioactive waste. The Waste Isolation Pilot Plant is the world's first underground repository for transuranic waste from nuclear weapons development. The mined caverns holding the waste are located 655 m (2150 ft) belowground in a huge salt formation thought to be geologically stable. This site became operational in 1999 and receives thousands of shipments of waste from 23 other locations.

Contaminated sites are being cleaned up, slowly

Many thousands of former military and industrial sites remain contaminated with hazardous waste in the United States and virtually every other nation on Earth. For most nations, dealing with these messes is simply too difficult, time-consuming, and expensive. In 1980, however, the U.S. Congress passed the Comprehensive Environmental Response Compensation and Liability Act (CERCLA). This law established a federal program to clean up U.S. sites polluted with hazardous waste. The EPA administers this cleanup program, called the **Superfund.** Under EPA auspices, experts identify sites polluted with hazardous chemicals, take action to protect groundwater, and clean up the pollution. Later laws also charged the EPA with cleaning up **brownfields,** lands whose reuse or development is complicated by the presence of hazardous materials.

Two well-publicized events spurred creation of the Superfund legislation. In *Love Canal*, a residential neighborhood in Niagara Falls, New York, families were evacuated in 1978–1980 after toxic chemicals buried by a company and the city in past decades rose to the surface, contaminating homes and an elementary school. In Missouri, the entire town of *Times Beach* was evacuated and its buildings demolished after being contaminated in the 1970s by dioxin (p. 384) from waste oil sprayed on its roads.

Once a Superfund site is identified, EPA scientists evaluate how near the site is to homes, whether wastes are confined

Testing the Toxicity of "E-Waste"

Most electronic waste, or "e-waste," is disposed of in conventional sanitary landfills. However, most electronics contain heavy metals, flame retardants, and other materials with the potential to cause environmental contamination and public health risks. For instance, over 6% of a typical computer is composed of lead.

Researchers, engineers, and regulators have long debated how hazardous e-waste is and how best to dispose of it. All agree that encouraging reuse and recycling is vital, but what should we do with electronics and their components (**FIGURE 1**) once they've truly reached their "end of life"?

One early set of studies was conducted when the U.S. EPA funded Dr. Timothy Townsend's lab at the University of Florida at Gainesville to determine whether e-waste is toxic enough to be classified as hazardous waste under the Resource Conservation and Recovery Act.

With students and colleagues, Townsend determined in 1999–2000 that cathode ray tubes (CRTs) from computer monitors and televisions leach an average of 18.5 mg/L of lead, far above the regulatory threshold of 5 mg/L. Following this research, the EPA proposed classifying CRTs as hazardous waste, and several U.S. states banned these items from conventional landfills.

Dr. Brajesh Dubey (L) and Dr. Timothy Townsend (R) preparing the TCLP test

In 2004, Townsend's lab group ran experiments on 12 other types of electronic devices. To measure toxicity, they used the EPA's standard test, the Toxicity Characteristic Leaching Procedure (TCLP), designed to mimic the process by which chemicals leach out of solid waste in landfills. In the TCLP, waste is ground up into fine pieces, and 100 g (3.5 oz) of it is put in a container with 2 L (0.53 gal) of an acidic fluid. The container is rotated for 18 hours, after which the leachate is chemically analyzed. Researchers look for eight heavy metals—arsenic, barium, cadmium, chromium, lead, mercury, selenium, and silver—and determine for each whether their concentration in the leachate exceeds that allowed by EPA regulations.

To conduct the standard TCLP, Townsend's team ground up the central processing units (CPUs) of personal computers, creating a jumbled mix made up by weight of 16% circuit board, 8% plastic, 68% ferrous metal, 5% nonferrous metal, and 3% wire and cable. However, grinding up a computer into small bits is no easy task, and it is hard to obtain a sample that accurately represents all components and materials. So the researchers also designed a modified TCLP test in which they placed whole CPUs—with the parts disassembled but not ground up—in a rotating 55-gallon drum full of acidic liquid. They tested their 12 types of devices using both standard and modified TCLP methods.

The team found lead to be the only heavy metal that exceeded the EPA's

FIGURE 1 Discarded electronic waste can leach heavy metals.

or likely to spread, and whether the pollution threatens drinking water supplies. Sites judged to be harmful are placed on the National Priorities List, ranked according to the risk to human health that they pose. Cleanup proceeds as funds are available. Throughout the process, the EPA is required to hold public hearings to inform area residents of its findings and to receive feedback.

The objective of CERCLA was to charge the polluting parties for the cleanup of their sites, according to the *polluter-pays principle* (p. 168). For many sites, however, the responsible parties cannot be found or held liable, and in such cases—roughly 30% so far—cleanups have been covered by taxpayers and from a trust fund established by a federal tax on industries producing petroleum and chemical raw materials. However, Congress let the tax expire and the trust fund went bankrupt in 2004, so taxpayers are now shouldering the entire burden. As the remaining cleanup jobs become more expensive, fewer are being completed.

As of 2013, 1320 Superfund sites remained on the National Priorities List, and only 365 have been cleaned up or otherwise deleted from the list. The average cleanup has cost over $25 million and has taken nearly 15 years. Many sites are contaminated with hazardous chemicals we have no effective way to deal with. In such cases, cleanups simply aim to isolate waste from human contact, either by building trenches and barriers around a site or by excavating contaminated material

regulatory threshold, but this threshold (5 mg/L) was surpassed in the majority of trials. Desktop computer monitors leached the most lead (48 mg/L on average), because older-style monitors include cathode ray tubes. However, laptops, TVs, smoke detectors, cell phones, and computer mice also leached high levels of lead. Next came remote controls, VCRs, keyboards, and printers, all of which leached more lead on average than the EPA threshold—and in 50% or more of the trials.

The researchers found that items containing more ferrous metals (such as iron) tended to leach less lead. For instance, CPUs contain 68% ferrous metals (compared to just 7% in laptops), and laptops leached seven times as much lead as CPUs. Further experiments confirmed that ferrous metals chemically react with lead and stop it from leaching.

Since that time, researchers have conducted a number of other studies. Taken together, they have shown similar results: that most electronics in most cases leach lead beyond the EPA threshold (**FIGURE 2**).

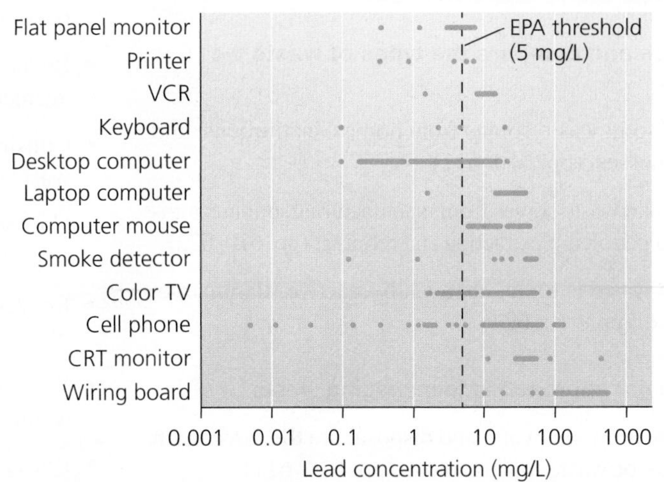

FIGURE 2 Most devices tested have exceeded the EPA safety standard for lead leachate. Data points for each type of device represent data from TCLP tests in different scientific studies between 2000 and 2008. *Adapted from Townsend, Timothy G., 2011. Environmental issues and management strategies for waste electronic and electrical equipment.* J. Air & Waste Mgmt. Assoc. *61: 587–610.*

Townsend and others say this research suggests that many electronic devices should be classified as hazardous waste. However, lab tests may or may not accurately reflect what actually happens in landfills. For this reason, Townsend's team has filled large tubular columns with e-waste and buried them in Florida landfills. The researchers are testing leachate from these tubes, and are reporting results for various materials as time passes. For example, a 2011 paper showed that the leaching of several heavy metals was influenced by whether the landfill was experiencing aerobic or anaerobic decomposition.

Scientists are also testing new materials that electronics manufacturers are using as they attempt to create safer alternatives. For instance, in 2008 Townsend and his colleagues found alternative solders to leach less than traditional tin/lead solders used on printed wire boards.

Scientific results from such projects are helping regulators and policymakers decide how best to deal with e-waste. Currently federal regulations require special handling and disposal of the most hazardous types of e-waste, but exemptions are made to encourage reuse and recycling. Some states consider most e-waste to be hazardous and exercise strict controls, whereas others do not.

In our new age of smartphones and so much more, the number and variety of electronic devices continue to expand, and further research will be needed as our technologies evolve. ■

and shipping it to a hazardous waste disposal facility. For all these reasons, the current emphasis in the United States and elsewhere is on preventing hazardous waste contamination in the first place.

Conclusion

We have made great strides in addressing our waste problems. Modern methods of waste management are far safer for people and gentler on the environment than past practices of open dumping and open burning. Recycling and composting efforts are advancing steadily, and Americans now divert one-third of all solid waste away from disposal. The continuing growth of recycling, driven by market forces, public policy, and consumer behavior, shows potential to further alleviate our waste problems.

Despite these advances, our prodigious consumption habits have created more waste than ever before. Our waste management efforts are marked by a number of difficult challenges, including the cleanup of Superfund sites, safe disposal of hazardous and radioactive waste, and frequent local opposition to disposal sites. These dilemmas make clear that the best solution is to reduce our generation of waste. Finding ways to reduce, reuse, and efficiently recycle the materials and goods that we use stands as a key ongoing challenge for our society.

Reviewing Objectives

You should now be able to:

Summarize and compare the types of waste we generate

- Municipal solid waste comes from homes, institutions, and small businesses. (pp. 610, 611)
- Industrial solid waste comes from manufacturing, mining, agriculture, and petroleum extraction and refining. (pp. 610, 622)
- Hazardous waste is toxic, chemically reactive, flammable, or corrosive. (pp. 610, 624)

List the major approaches to managing waste

- Source reduction, recovery, and disposal are the three main components of waste management. (pp. 610–611)

Delineate the scale of the waste dilemma

- Developed nations generate far more waste than developing nations, but they are beginning to decrease their waste. (pp. 611–613)
- Waste in developing nations is increasing as population and consumption grow. (p. 612)
- Open dumping and burning continue in developing nations. (p. 612)

Describe conventional waste disposal methods: landfills and incineration

- Sanitary landfills guard against contamination of groundwater, air, and soil. Nonetheless, such contamination can occur. (pp. 613–614)
- Incinerators reduce waste volume by burning it. Pollution control technology removes most pollutants from emissions, but some escape, and toxic ash needs to be disposed of in landfills. (pp. 615–616)
- We are harnessing energy from landfill gas and generating electricity from incineration. (p. 616)

Evaluate approaches for reducing waste: source reduction, reuse, composting, and recycling

- Reducing waste before it is generated is the best waste management approach. (p. 616)
- Consumers can take an array of simple steps to reduce their waste output. (pp. 616–617)
- Composting reduces waste while creating organic matter for gardening and farming. (p. 617)
- Recycling has grown in recent years and now removes 26% of the U.S. waste stream. (pp. 617–619)
- We can recycle materials from landfills, given the right technology and high-enough market prices. (p. 619)
- Bottle bills are one way in which financial incentives can motivate people to reduce waste. (pp. 619–620)

Discuss management of industrial solid waste and principles of industrial ecology

- Regulations differ, but industrial waste management is similar to that for municipal solid waste. (p. 622)
- Industrial ecology studies how industrial systems can mimic ecological systems and provides ways for industry to enhance efficiency. (pp. 622–624)

Assess issues in managing hazardous waste

- Hazardous wastes are diverse, and households are a major source of it. (p. 624)
- Organic compounds, heavy metals, and radioactive waste are three common types of hazardous waste. (pp. 624, 627)
- Electronic waste may be considered hazardous and constitutes a growing waste source. (pp. 624–625, 628–629)
- Hazardous waste is regulated and monitored, yet illegal dumping remains a problem. (p. 626)
- No fully satisfactory method of disposing of hazardous waste has yet been devised. (p. 627)
- The Superfund program cleans up hazardous waste sites, but cleanup is a long and expensive process. (pp. 627–629)

Testing Your Comprehension

1. Describe five major methods of managing waste. Why do we practice waste management?
2. Why have some people labeled the United States "the throwaway society"? How much solid waste do Americans generate, and how does this amount compare to that of people from other countries?
3. Name several guidelines by which sanitary landfills are regulated. Describe three problems with landfills.
4. Describe the process of incineration or combustion. What happens to the resulting ash? What is one drawback of incineration?

5. What is composting, and how does it help reduce the waste stream?
6. What are the three elements of a sustainable process of recycling?
7. In your own words, describe the goals of industrial ecology.
8. What four criteria are used to define hazardous waste? What makes heavy metals and synthetic organic compounds particularly hazardous?
9. What are the largest sources of hazardous waste? Describe three ways to dispose of hazardous waste.
10. What is the Superfund program? How does it work?

Seeking Solutions

1. How much waste do you generate? Look into your waste bin at the end of the day and categorize and measure the waste there. List all other waste you may have generated in other places throughout the day. How much of this waste could you have avoided generating? How much could have been reused or recycled?
2. Some people have criticized current waste management practices as merely moving waste from one medium to another. How might this criticism apply to the methods now in practice? What are some potential solutions?
3. Of the various waste management approaches covered in this chapter, which ones are your community or campus pursuing, and which are they not pursuing? Would you suggest that your community or campus start pursuing any new approaches? If so, which ones, and why?
4. Can manufacturers and businesses benefit from source reduction if consumers were to buy fewer products as a result? How? Given what you know about industrial ecology, what do you think the future of sustainable manufacturing may look like?
5. **THINK IT THROUGH** You are the CEO of a major corporation that produces containers for soft drinks and a wide variety of other consumer products. Your company's shareholders are asking that you improve the company's image—while not cutting into profits—by taking steps to reduce waste. What steps would you consider taking?
6. **THINK IT THROUGH** You are the president of your college or university. Your trustees want you to engage with local businesses and industries in ways that benefit both the school and the community. Your faculty and students want you to make the school a leader in waste reduction and industrial ecology. Consider the industries and businesses in your community and the ways they interact with facilities on your campus. Bearing in mind the principles of industrial ecology, can you think of any novel ways in which your school and local businesses might mutually benefit from one another's services, products, or waste materials? Are there waste products from one business, industry, or campus facility that another might put to good use? What steps would you propose to take as president?

Calculating Ecological Footprints

The 17th biennial "State of Garbage in America" survey documents the ability of U.S. residents to generate prodigious amounts of municipal solid waste (MSW). According to the survey, on a per capita basis, Missouri residents generate the least MSW (4.5 lb/day), and Hawaii residents (and Hawaii's many tourists) generate the most (15.8 lb/day). The average for the entire country is 7.0 lb MSW per person per day. Calculate the amount of MSW generated in 1 day and in 1 year by each of the groups indicated, at each of the rates shown in the table.

Groups generating municipal solid waste	Per capita MSW generation rates					
	U.S. average (7.0 lb/day)		Missouri (4.5 lb/day)		Hawaii (15.8 lb/day)	
	Day	Year	Day	Year	Day	Year
You	7.0	2555				
Your class						
Your state						
United States						
World						

Data from van Haaren, R., et al., 2010. The state of garbage in America. BioCycle *51: 16–23.*

1. Suppose your town of 50,000 people has just approved construction of a landfill nearby. Estimates are that it will accommodate 1 million tons of MSW. Assuming the landfill is serving only your town, and that your town's residents generate waste at the U.S. average rate, for how many years will it accept waste before filling up? How much longer would a landfill of the same capacity serve a town of the same size in Missouri?
2. One recent study estimated that the average world citizen generates 1.47 pounds of trash per day. How many times more does the average U.S. citizen generate?
3. The same study showed that the average resident of a low-income nation generates 1.17 pounds of waste per day and that the average resident of a high-income nation generates 2.64 pounds per day. Why do you think U.S. residents generate so much more MSW than people in other "high-income" countries, when standards of living in those countries are comparable?

STUDENTS

Go to **MasteringEnvironmentalScience** for assignments, the etext, and the Study Area with practice tests, videos, current events, and activities.

INSTRUCTORS

Go to **MasteringEnvironmentalScience** for automatically graded activities, current events, videos, and reading questions that you can assign to your students, plus Instructor Resources.

Coltan miners in eastern Congo

23

Minerals and Mining

Upon completing this chapter, you will be able to:

- Outline types of mineral resources and how they contribute to our products and society
- Describe the major methods of mining
- Characterize the environmental and social impacts of mining
- Assess reclamation efforts and mining policy
- Evaluate ways to encourage sustainable use of mineral resources

CENTRAL CASE STUDY

Mining for . . . Cell Phones?

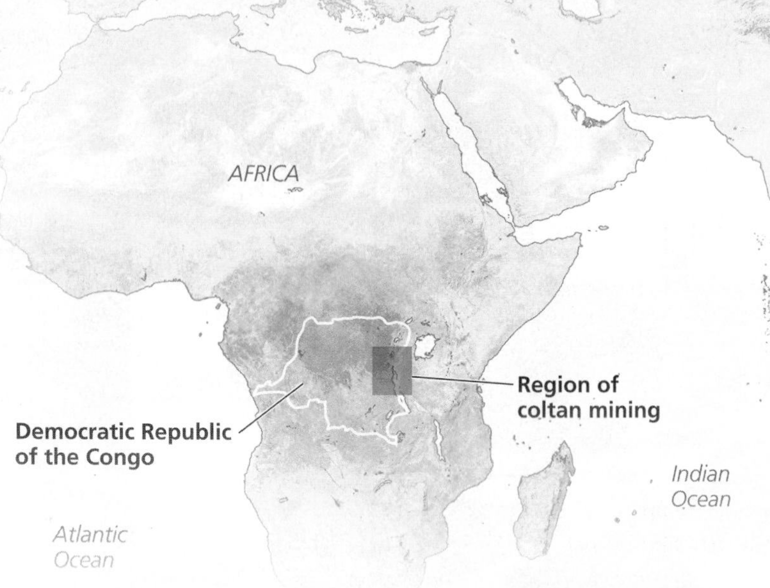

"The conflict in the Democratic Republic of the Congo has become mainly about access, control, and trade of five key mineral resources: coltan, diamonds, copper, cobalt, and gold."

—Report to the United Nations Security Council, April 2001

"Coltan . . . is not helping the local people. In fact, it is the curse of the Congo."

—African journalist Kofi Akosah-Sarpong

Pulling a cell phone from her pocket, a student on a college campus in the United States dials a friend. Inside her phone is a little-known metal called tantalum—just a tiny amount, but no cell phone could operate without it.

Half a world away, a miner in the heart of Africa toils all day in a jungle streambed, sifting sediment for nuggets of coltan ore, which contain tantalum. At nightfall, rebel soldiers take most of his ore, leaving him to sell what little remains to buy food for his family at the squalid mining camp where they live.

In bedeviling ways, tantalum links our glossy global high-tech economy with one of the most abused regions on Earth. The Democratic Republic of the Congo has long been embroiled in a sprawling conflict that has involved six nations and various rebel militias. Over 5.4 million people have lost their lives in this war since 1998. For its population size, this is as though Congo were being hit by three September 11 terrorist attacks every day for a decade. It is the latest chapter in the sad history of a nation rich in natural resources—copper, cobalt, gold, diamonds, uranium, and timber—whose impoverished people keep losing control of those resources to others.

At the center of the recent conflict is tantalum (Ta), element number 73 on the periodic table (**APPENDIX D**). We rely on this metal for our cell phones, computer chips, DVD players, game consoles, and digital cameras. Tantalum powder is ideal for capacitors (the components that store energy and regulate current in miniature circuit boards) because it is highly heat resistant and readily conducts electricity.

Tantalum comes from a dull blackish mineral called tantalite, which often occurs with a mineral called columbite—so the ore is referred to as columbite–tantalite, or *coltan* for short. In eastern Congo, men dig craters in rainforest streambeds, panning for coltan much as early California miners panned for gold.

As information technology boomed in the late 1990s, global demand for tantalum rose, and market prices for the metal shot up to \$500/kg (\$230/lb) in 2001. High prices led some Congolese men to mine coltan by choice, but many more were forced to work as miners. As the war began in 1998, local militias, supported by forces from neighboring Rwanda and Uganda, overran eastern Congo. Farmers were chased off their land, villages were burned, and civilians were raped, tortured, and killed. Soldiers from each army seized control of mining operations. They forced farmers, refugees, prisoners, and children to work, and the soldiers skimmed profits from the coltan the people mined. Children and teachers abandoned school and worked in the mines, while prostitution spread AIDS and sexually transmitted disease through the mining camps. The turmoil also caused ecological havoc as miners and soldiers streamed into national parks, clearing rainforests and killing wildlife for food, including forest elephants, hippopotamuses, endangered gorillas, and the okapi, a rare relative of the giraffe.

Most miners ended up with little, while rebels, soldiers, and bandits enriched themselves selling coltan to traders, who sold it to processing companies in Europe and the United States. These companies refine and sell tantalum powder to capacitor manufacturers, which in turn sell capacitors to Nokia, Motorola, Sony, Intel, Compaq, Dell, and other high-tech corporations.

In 2001, an expert panel commissioned by the United Nations Security Council concluded that coltan riches were fueling, financing, and prolonging the war. The panel urged a U.N. embargo on coltan and other minerals smuggled from Congo and exported by neighboring nations. A grass-roots activist movement advanced the slogan "No blood on my cell phone!"

Sony, Nokia, Ericsson, and other corporations rushed to assure consumers that they were not using tantalum from eastern Congo—and the region was in fact producing less than 10% of the world's supply. Meanwhile, some observers felt an embargo could hurt the long-suffering Congolese people, rather than help them. The mining life may be miserable, they said, but it pays better than most alternatives in a land where the average income is 20 cents a day.

Soon, however, the high-tech boom went bust, and global demand for tantalum diminished. This occurred just as Australia and other countries were ramping up industrial-scale tantalite mining. As supply outpaced demand, the market price of tantalum fell, and several major producers quit mining tantalum. But nations began to work through their stockpiles, and by 2010 demand had grown, driving prices up once again. The demand for tantalum is predicted to grow by 6–7% annually, ensuring a thriving market for the mineral for the foreseeable future.

Today, the war is declared over, and foreign troops are out of Congo. Even so, internal factions continue to fight, and thousands of people continue to die or to flee their homes. Western electronics companies avoid knowingly purchasing tantalum from Congo, but trade in minerals has many middlemen, so it is sometimes difficult for companies to be certain of the origin of the tantalum they use.

Progress is being made on addressing these issues, however. In 2010 the U.S. Congress included in the Dodd-Frank financial reform bill an amendment requiring all electronics companies to report the origin of the tantalum in the products they sell. Additionally, the Public-Private Alliance for Responsible Minerals Trade was formed in 2011 to help private companies, national governments, and nongovernmental organizations create a system to trace and certify minerals such as tantalum, gold, and tin as "conflict-free." Participating companies include Nokia, Sprint, Intel, Motorola, and Verizon. This alliance provides an opportunity to significantly reduce trade in conflict minerals while promoting trade of minerals sourced from legitimate mines in poor nations such as Congo.

Unfortunately, Congo is not the only source of conflict minerals in the world today. A thriving black market in coltan is emerging in remote portions of Brazil, Colombia, and Venezuela in the northern Amazon jungle. Armed gangs and narcotics smugglers in the region are accused of using women, children, and indigenous people to mine and smuggle coltan ore. The recent discovery of vast mineral reserves in Afghanistan (pp. 647), coupled with that nation's political unrest, suggests that it too could become a significant source of conflict minerals in the near future. So while the trade in conflict minerals from these and other nations continues, it is hoped that ongoing regulatory efforts to certify minerals will provide a framework that satisfies our intense demand for mineral resources while ensuring those resources were mined in a responsible manner. ■

Earth's Mineral Resources

Coltan provides just one example of how we extract raw materials from beneath our planet's surface and turn them into products we use in our everyday lives. In the lithosphere (p. 34), the region that includes the uppermost layers of rock near Earth's surface, the rock cycle (p. 36) creates new rock and alters existing rock. Plate tectonics (pp. 34–36) builds mountains; shapes the geography of oceans, islands, and continents; and gives rise to earthquakes and volcanoes. The coltan mining areas of eastern Congo are situated along the western edge of Africa's Great Rift Valley system, a region where the African tectonic plate is slowly pulling itself apart. Some of the world's largest lakes have formed in the immense valley floors, far below towering volcanoes such as Mount Kilimanjaro. Here and throughout the world, geological processes are fundamental to shaping the world around us.

We will now take a closer look at how rock and the resources of the lithosphere contribute directly to our economies and our lives. We will first examine the mineral resources we mine and the products they provide us. Next we will study the various ways we extract minerals from the earth. We will then examine the many social and environmental impacts that our mining efforts exert and consider steps we can take to mitigate these impacts. Finally, because mineral resources are nonrenewable on human timescales, we need to be attentive to finite and decreasing supplies of economically important minerals. Thus we will examine solutions we can pursue to make our mineral use more sustainable.

Rocks provide the minerals we use

A **rock** is a solid aggregation of minerals, and a **mineral** is a naturally occurring solid chemical element or inorganic compound with a crystal structure, a specific chemical composition, and distinct physical properties (p. 36). (See the periodic table in **APPENDIX D** for chemical elements.) For instance, the mineral tantalite consists of the elements tantalum, oxygen, iron, and manganese. Tantalite occurs most commonly in pegmatite, a type of igneous rock (p. 37) similar to granite. In addition to tantalite, pegmatite generally contains the minerals feldspar, quartz, and mica, and occasionally it even includes gemstones and other rare minerals. Geologic processes influence the distribution of rocks and minerals in the lithosphere and their availability to us.

We depend on a wide array of mineral resources as raw materials for our products, so we mine and process these resources. Just consider a typical scene from a student lounge at a college or university (**FIGURE 23.1**) and note how many items are made with elements from the minerals we take from the earth. Without the resources from beneath the ground that we use to make building materials, wiring, clothing, appliances, fertilizers for crops, and so much more, civilization as we know it could not exist.

We obtain minerals by mining

We obtain the minerals we use in all these ways through the process of mining. The term *mining* in the broad sense

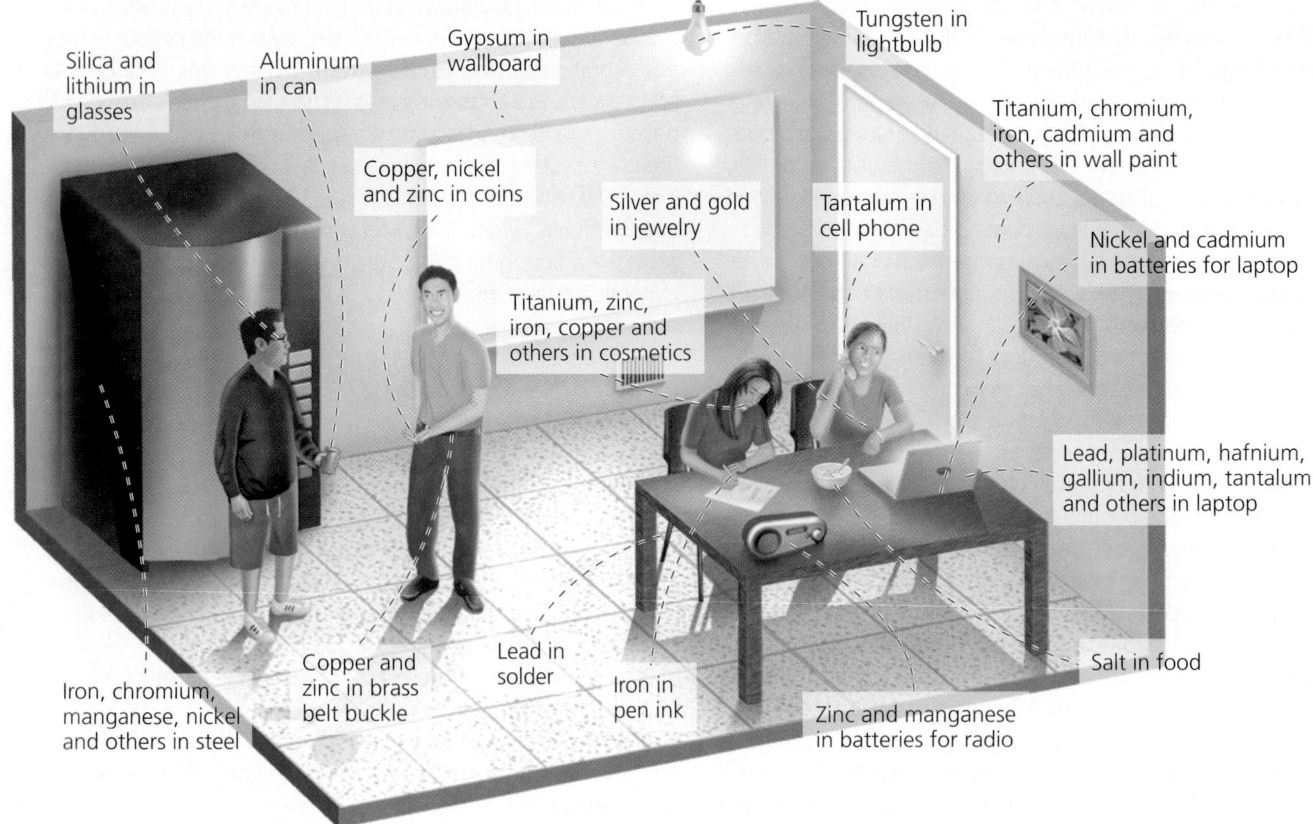

FIGURE 23.1 **Elements from minerals that we mine are everywhere in the products we use in our everyday lives.** This scene from a typical college student lounge points out just a few of the many elements from minerals that surround us.

describes the extraction of any resource that is nonrenewable on the timescale of our society. In this sense, we mine fossil fuels and groundwater, as well as minerals. When used specifically in relation to minerals, **mining** refers to the systematic removal of rock, soil, or other material for the purpose of extracting minerals of economic interest. Because most minerals of interest are widely spread but in low concentrations, miners and mining geologists first try to locate concentrated sources of minerals before mining begins.

FAQ **How do geologists "see" large mineral deposits below the ground?**

Finding large reserves of underground minerals, called *prospecting*, can be done in a number of ways. The earliest prospectors searched promising areas on foot, looking for exposed seams of mineral-containing rocks or minerals (such as gold) carried into streams by runoff. As our technological sophistication grew, along with our demand for minerals, prospecting became more complex. Today, geologists direct vibrations into underground rock strata and capture them with sensors as they reflect back. This enables scientists to visualize the underlying rock layers and identify potential reserves, just as they do for fossil fuel deposits (pp. 530–531). Geologists measure the magnetic fields in rock layers to look for metal ores and conduct chemical analyses of stream water to detect specific minerals. If promising sites are located, cores are often drilled deep into the ground and inspected for the desired mineral. Despite all these advances, however, prospecting remains a difficult endeavor because we are unable to clearly "see" valuable minerals under the ground.

We use mined materials extensively

We often don't notice how many mined resources we use every day. Using data from the U.S. government, the Minerals Education Coalition estimated that in 2013 the average American consumed more than 17,200 kg (38,000 lb) of new minerals and fuels every year. At current rates of use, a child born in 2012 will use over 1.3 million kg (2.9 million lb) during his or her lifetime (FIGURE 23.2).

More than half of the annual mineral and fuel use is from the coal, oil, and natural gas used to supply our intensive demands for energy. Much of the remaining mineral use is attributable to the sand, gravel, and stone used in constructing our buildings, roads, bridges, and parking lots. Metal use is dwarfed by these other two categories, but the average American will still use more than 3 tons of aluminum over his or her lifetime. This level of consumption clearly shows the potential of recycling and reuse (such as recycling stone and gravel from old highways into new construction) to make our modern, mineral-intensive lifestyle more sustainable.

Metals are extracted from ores

Some minerals can be mined for metals. A **metal** is a type of chemical element, or a mass of such an element, that typically

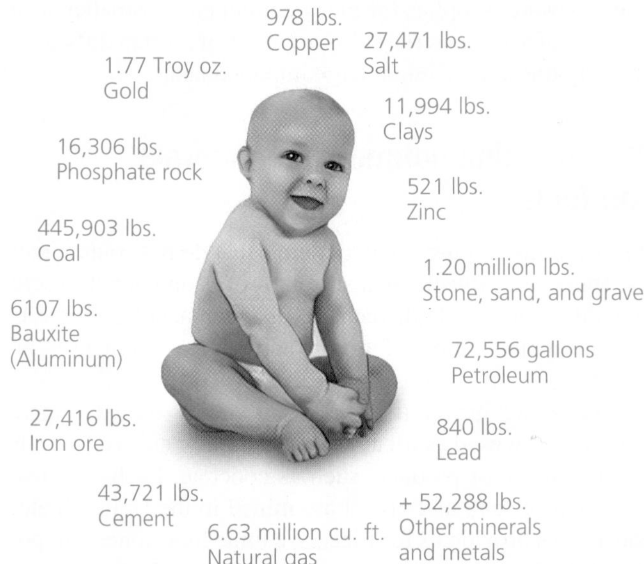

FIGURE 23.2 At current rates of use, a baby born in 2012 is predicted to use over 1.3 million kg (3 million lb) of minerals over his or her lifetime. *Data from Minerals Education Coalition, 2013.*

is lustrous, opaque, malleable, and can conduct heat and electricity. Most metals are not found in a pure state in Earth's crust, but instead are present within **ore,** a mineral or grouping of minerals from which we extract metals.

Copper, iron, lead, gold, and aluminum are among the many economically valuable metals we extract from mined ore. These metals and others serve so many purposes that our modern lives would be impossible without them. The tantalum used in the electronic components of computers, cell phones, video game consoles, and other devices, as you may recall from the case study, is a metal that comes from the mineral tantalite (FIGURE 23.3). In nature, tantalite is often found with the mineral columbite within the ore called coltan. Columbite also contains the metal niobium, formerly known as columbium, which is utilized in ways similar to tantalum.

We process metals after mining ore

Extracting minerals from the ground is the first step in putting them to use. However, most minerals need to be processed in some way to become useful for our products. For example, after ores are mined, the rock is crushed and pulverized, and the desired metals are isolated by chemical or physical means. The material is then processed to purify the metals we desire. With coltan, processing facilities use acid solvents to separate tantalite from columbite. Other chemicals are then used to produce metallic tantalum powder. This powder can be consolidated by various melting techniques and can be shaped into wire, sheets, or other forms.

Sometimes we mix, melt, and fuse a metal with another metal or a nonmetal substance to form an *alloy*. For example, steel is an alloy of the metal iron that has been fused with a small quantity of carbon. The strength and malleability of this particular alloy make steel ideal for its many applications in buildings, vehicles, appliances, and more. To make steel, we first mine iron ore, which consists of iron-containing compounds such as iron oxide. Steelmakers then heat the ore and chemically extract the iron with carbon in a process known as **smelting** (heating ore beyond its melting point and combining it with other metals or chemicals). They then melt and reprocess the mixture, removing precise amounts of carbon and shaping the product into rods, sheets, or wires. During this melting process, certain other metals may be added to modify the strength, malleability, or other characteristics of the steel, as desired.

Processing minerals exerts environmental impacts. Most methods are water-intensive and energy-intensive. Moreover, many chemical reactions and heating processes used for extracting metals from ore emit air pollution, and smelting plants in particular have long been hotspots of toxic air pollution (Chapter 17). In addition, soil and water commonly become polluted by **tailings,** portions of ore left over after metals have been extracted. Tailings may leach heavy metals present in the ore waste as well as chemicals applied in the extraction process. For instance, we use cyanide to extract gold from ore, and we use sulfuric acid to extract copper. Mining operations often pump a toxic slurry of tailings into

FIGURE 23.3 Tantalum is used to manufacture electronics. Coltan ore **(a)** is mined from the ground and then processed to extract the pure metal tantalum. This metal is used in capacitors **(b)** and other electronic components in computer chips, cell phones, and many other devices.

(a) Coltan ore

(b) Capacitor containing tantalum

FIGURE 23.4 **This surface impoundment at the Upper Big Branch mine in West Virginia holds coal tailings from a surface mining operation.**

large reservoirs called **surface impoundments** (FIGURE 23.4). Impoundment walls are designed to prevent leaks and collapse, but accidents can occur if the structural integrity of the impoundment is compromised. In 2000, a breach of a coal tailings impoundment near Inez, Kentucky, released over 1 billion liters (250–300 million gal) of coal slurry, blackening 120 km (75 mi) of streams, killing aquatic wildlife, and affecting drinking water supplies for many communities. Smaller-scale leaching of toxic materials also occurs, as it is often difficult to properly line and maintain large impoundments.

We also mine nonmetallic minerals and fuels

We also mine and use many minerals that do not contain metals. FIGURE 23.5 illustrates a selection of economically useful mineral resources, both metallic and nonmetallic. For each one, its major nation of origin and several main uses are shown.

Sand and gravel (the most commonly mined mineral resources) are used as fill and as construction materials for the manufacturing of products such as concrete. Each year over $7 billion of sand and gravel are mined in the United States. Phosphates provide us fertilizer. We mine limestone, salt, potash, and other minerals for a number of diverse purposes.

Gemstones are treasured for their rarity and beauty. For instance, diamonds have long been prized—and like coltan, they have fueled resource wars. Besides the conflict in eastern Congo, the diamond trade has acted to fund, prolong, and intensify wars in Angola, Sierra Leone, Liberia, and elsewhere, as armies exploit local people for mine labor, then sell the diamonds for profit. This is why you may hear the phrase "blood diamonds," just as coltan has been called a "conflict mineral." The more that we in developed nations consume, the more our economic demand affects people elsewhere. Throughout

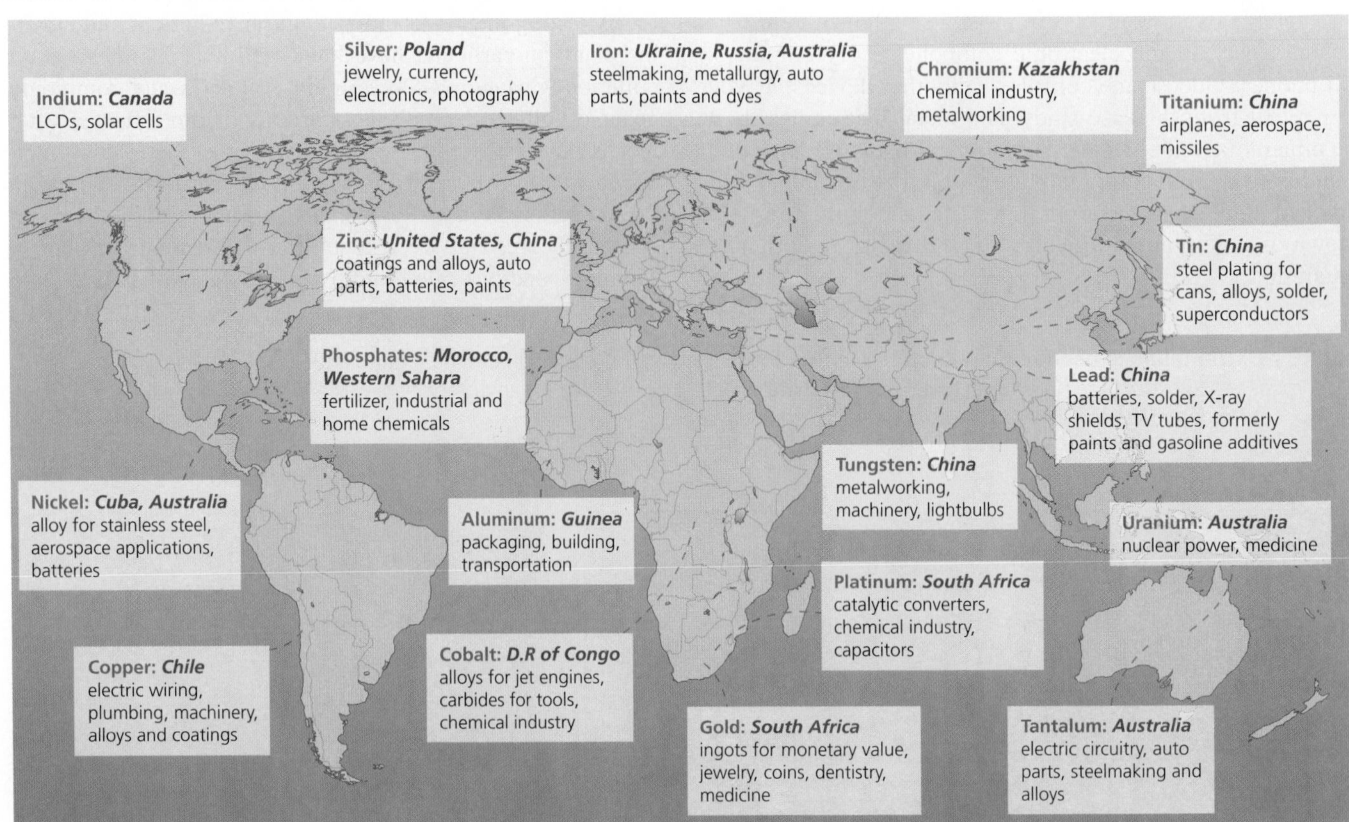

FIGURE 23.5 **The minerals we use come from all over the world.** Shown is a selection of economically important minerals (mostly metals, with several nonmetals), together with their major uses and their main nation of origin. Only a minority of minerals, uses, and origins is shown.

the world, people in developing regions from Africa to Asia to the Amazon often suffer unintended consequences of the developed world's appetite for mineral resources.

We also mine substances we use for fuel (Chapter 19). Uranium ore is a mineral from which we extract the metal uranium, which we use in nuclear power (pp. 555–557). One of the most common fuels we mine is coal. Coal (p. 524) is the modified remains of ancient swamp plants and is comprised of the mineral carbon. Other fossil fuels—petroleum, natural gas, and alternative fossil fuels such as oil sands, oil shale, and methane hydrates—are also organic and are extracted from the earth (Chapter 19).

Mining Methods and Their Impacts

Mining for minerals is an important industry that provides jobs for people and revenue for communities in many regions. Mining supplies us raw materials for countless products we use daily, so it is necessary for the lives we lead. In 2012, raw materials from mining contributed $76 billion to the U.S. economy, and after processing, mineral materials contributed $704 billion. About 175,000 Americans were employed directly in mining for coal and metals in 2012, and the mining industry, together with processors and manufacturers of products from mined materials, employed over 1.2 million people.

At the same time, mining also exerts a price in environmental and social impacts. Because minerals of interest often make up only a small portion of the rock in a given area, typically very large amounts of material must be removed in order to obtain the desired minerals. This frequently means that mining disturbs large areas of land, thereby exerting severe impacts on the environment and on people living nearby.

Depending on the nature of the mineral deposit, any of several mining methods may be employed to extract the resource from the ground. Mining companies select which method to use based largely on its economic efficiency. We will examine seven major mining approaches commonly used throughout the world, and will also take note of the impacts of each approach as we proceed.

(a) Strip mining

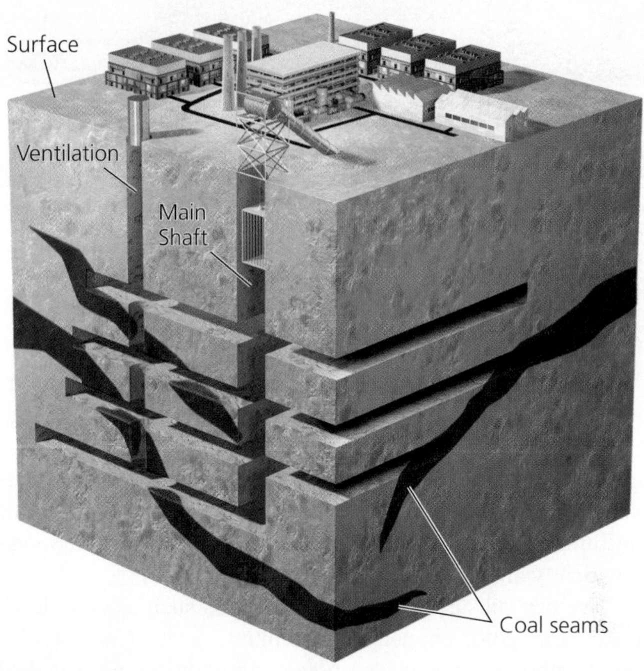

(b) Subsurface mining

FIGURE 23.6 Coal mining illustrates two types of mining approaches. In strip mining **(a)**, soil is removed from the surface in strips, exposing seams from which coal is mined. In subsurface mining **(b)**, miners work belowground in shafts and tunnels blasted through the rock. These passageways provide access to underground seams of coal or minerals.

Strip mining removes surface layers of soil and rock

When a resource occurs in shallow horizontal deposits near the surface, the most effective mining method is often **strip mining,** whereby layers of surface soil and rock are removed from large areas to expose the resource. Heavy machinery removes the overlying soil and rock (termed *overburden*) from a strip of land, and the resource is extracted. This strip is then refilled with the overburden that had been removed, and miners proceed to an adjacent strip of land and repeat the process. Strip mining is commonly used for coal (p. 525; **FIGURE 23.6a**) and oil sands (pp. 525–526), and sometimes for sand and gravel.

Strip mining for coal and oil sands can be economically efficient, but it causes severe environmental impacts. By completely removing vegetative cover and nutrient-rich topsoil, strip mining obliterates natural communities over large areas. Soil from refilled areas can easily erode away. Strip mining also pollutes waterways through the process of **acid drainage,** which occurs when sulfide minerals in newly exposed rock surfaces react with oxygen and rainwater to produce sulfuric acid. As the sulfuric acid runs off, it leaches metals from the rocks, many of which are toxic to organisms (**FIGURE 23.7**). This toxic liquid is called *leachate*, a term used to describe the toxic liquids that form in landfills (pp. 613–614). Whereas leachates from landfills are caused by anaerobic reactions in organic substances, leachates from mining sites result from aerobic reactions in inorganic substances. Acid drainage is a natural phenomenon, but mining greatly accelerates this process by exposing many new rock surfaces at once.

FIGURE 23.7 **Acidic drainage from an underground coal mine streams down a slope in West Virginia.** The yellow-orange color is due to iron from the drainage settling out on the soil surface and forming rust.

In subsurface mining, miners work underground

When a resource occurs in concentrated pockets or seams deep underground, and the earth allows for safe tunneling, then mining companies pursue **subsurface mining.** In this approach, shafts are excavated deep into the ground, and networks of tunnels are dug or blasted out to follow deposits of the mineral (FIGURE 23.6b). Miners remove the resource systematically and ship it to the surface.

We use subsurface mining for metals such as zinc, lead, nickel, tin, gold, copper, and uranium, as well as for diamonds, phosphate, salt, and potash. In addition, a great deal of coal is mined using the subsurface technique. The scale of subsurface mining can be mind-boggling; the world's deepest mines (certain gold mines in South Africa) extend nearly 4 km (2.5 mi) underground.

Subsurface mining is the most dangerous form of mining and indeed one of society's most dangerous occupations. Fatal accidents are not unusual. In China, coal-mining conditions are so dangerous that in 2012 alone nearly 1400 miners lost their lives. Besides risking injury or death from dynamite blasts, natural gas explosions, and collapsing shafts and tunnels, miners inhale toxic fumes and coal dust, which can lead to respiratory diseases, including fatal black lung disease.

Occasionally subsurface mines can affect people years after they are closed. The collapse of tunnels at the Retsof Salt Mine in Genessee Valley, New York, after a minor earthquake in 1994 created sinkholes at the surface that damaged roads, bridges, and homes and sucked groundwater from neighborhood wells. Coal veins in abandoned mines underneath Centralia, Pennsylvania, caught fire in 1961 and are still burning today. This once-thriving city is now a ghost town, as nearly all its residents accepted buyouts in the 1980s and relocated when it became clear the smoldering fires beneath them could not be contained.

FAQ **Why would anyone work in a mine when it's such dangerous work?**

It seems that mining accidents appear regularly in the headlines. In late 2010, for example, 33 miners in Chile were rescued after being trapped 600 m (2000 ft) underground for 69 days in a gold and copper mine. In 2013, more than 60 lives were lost after the collapse of a gold mine in Sudan. Gas explosions claimed the lives of 29 miners in a coal mine in West Virginia in 2010 and 17 miners in Russia in 2013. Given these dangers, and the chronic health impacts of working in an underground mine, it's reasonable to wonder why anyone would accept such a job.

Many of the people who work in mines do so because they have few other options. Underground mining often occurs in economically depressed areas, such as Appalachia in the United States, where mining is one of the few well-paying jobs. And for most mining jobs, people can begin working right out of high school. So although the work is dangerous, many miners are willing to accept those risks to provide for themselves and their families because few other career opportunities are available.

In terms of environmental impact, subsurface mining creates acid drainage just as surface mining does, and toxic leachate can make its way from mining sites to nearby groundwater. Abandoned mine sites can continue polluting groundwater long after mining has ceased.

Open pit mining creates immense holes in the ground

When a mineral is spread widely and evenly throughout a rock formation, or when the earth is unsuitable for tunneling, the method of choice is **open pit mining.** This essentially involves digging a gigantic hole and removing the desired ore, along with waste rock that surrounds the ore. Some open pit mines are inconceivably enormous. The world's largest, the Bingham Canyon Mine near Salt Lake City, Utah, is 4 km (2.5 mi) across and 1.2 km (0.75 mi) deep (FIGURE 23.8).

FIGURE 23.8 **The Bingham Canyon open pit mine outside Salt Lake City, Utah, is the world's largest human-made hole in the ground.** This immense mine produces mostly copper.

Conveyor systems and immense trucks with tires taller than a person carry out nearly half a million tons of ore and waste rock each day.

Open pit mines are terraced so that men and machinery can move about, and waste rock is left in massive heaps outside the pit. The pit is expanded until the resource runs out or becomes unprofitable to mine. Open pit mining is used for copper, iron, gold, diamonds, and coal, among other resources. We also use this technique to extract clay, gravel, sand, and stone such as limestone, granite, marble, and slate, but we generally call these pits *quarries*.

Open pit mines are so large because huge volumes of waste rock need to be removed in order to extract relatively small amounts of ore, which in turn contain still smaller traces of valuable minerals. The sheer size of these mines means that the degree of habitat loss and aesthetic degradation is considerable. Another impact is chemical contamination from acid drainage as water runs off the waste heaps or collects in the pit. Once mining is complete, abandoned pits generally fill up with groundwater, which soon becomes toxic as water and oxygen react with sulfides from the ore and produce sulfuric acid. Acidic water from the pit can harm wildlife and can percolate into aquifers and spread through the region. Regulations in developed nations require that waste heaps be capped with clay, then soil, and then planted with vegetation once mines are closed. However, many dumps will likely leach acid for hundreds or thousands of years.

The Berkeley Pit, a former copper mine near Butte, Montana, is today one of the largest Superfund toxic waste cleanup sites (pp. 627–629) in the United States. After its closure in 1982, it filled with groundwater and became so acidic (pH of 2.2) and concentrated with toxic metals that microbiologists discovered new species of microbes in the water—the harsh conditions were so rare in nature that scientists had never before encountered microbes adapted to them!

Placer mining uses running water to isolate minerals

Some metals and gems accumulate in riverbed deposits, having been displaced from elsewhere and carried along by flowing water. To search for these metals and gems, miners sift through material in modern or ancient riverbed deposits, generally using running water to separate lightweight mud and gravel from heavier minerals of value (FIGURE 23.9). This technique is called **placer mining** (pronounced "plasser").

Placer mining is the method used by Congo's coltan miners, who wade through streambeds, sifting through large amounts of debris by hand with a pan or simple tools, searching for high-density tantalite that settles to the bottom while low-density material washes away. Today's African miners practice small-scale placer mining similar to the method used by American miners long ago who ventured to California in the Gold Rush of 1849, and later to Alaska in the Klondike Gold Rush of 1896–1899. Indeed, placer mining for gold is still practiced in areas of Alaska and Canada, although today it uses large dredges and heavy machinery.

Besides the many social impacts of placer mining in places such as Congo, placer mining is environmentally destructive because most methods wash large amounts of debris into

FIGURE 23.9 **Miners in eastern Congo find coltan by placer mining.** Sediment is placed in plastic tubs, and water is run through them. A mixing motion allows the sediment to be poured off while the heavy coltan settles to the bottom.

streams, making them uninhabitable for fish and other life for many miles downstream. Gold mining in northern California's rivers in the decades following the Gold Rush washed so much debris all the way to San Francisco Bay that a U.S. district court ruling in 1884 finally halted this mining practice. Placer mining also disturbs stream banks, causing erosion and harming ecologically important riparian plant communities.

Mountaintop mining reshapes ridges and can fill valleys

When a resource occurs in underground seams near the tops of ridges or mountains, the mining method of choice may be **mountaintop removal mining,** in which several hundred vertical feet of mountaintop are removed to allow recovery of entire seams of the resource (FIGURE 23.10a). This method of mining is used primarily for coal in the Appalachian Mountains of the eastern United States. In mountaintop removal mining, a mountain's forests are clear-cut and the timber is sold, topsoil is removed, and then rock is repeatedly blasted away to expose the coal for extraction. Overburden is placed back onto the mountaintop, but this waste rock is unstable and typically takes up more volume than the original rock, so generally a great deal of waste rock is dumped into adjacent valleys (a practice called "valley filling"). So far, mountaintop removal has blasted away an area the size of Delaware and has buried nearly 3200 km (2000 mi) of streams.

Mountaintop removal mining has recently expanded because it is an economically efficient way for companies to extract coal. In addition, 1990 Clean Air Act amendments that encouraged clean-burning low-sulfur coal led to more mining in Appalachia, where low-sulfur coal predominates. Moreover, the most

(a) Mountaintop mining in eastern Kentucky

(b) Train hauls coal past homes in West Virginia

(c) Flood damage below a West Virginia mine site

(d) People protesting mountaintop mining

FIGURE 23.10 Mountaintop mining has social and environmental impacts. In the Appalachians, mountaintop mining for coal **(a)** takes place on massive scales. Communities near these sites **(b)** experience environmental impacts such as flash floods below mining sites **(c)**. Mountaintop mining provides employment opportunities, but some residents, such as these in Frankfort, Kentucky **(d)**, oppose the process because of its environmental costs.

easily accessible coal, in level areas near the surface, has already been mined. Thus, the mining industry says mountaintop removal is necessary if we are to obtain coal from the Appalachian region in order to provide the electricity we all use. The practice grew in the 1990s when the Clinton administration opposed strengthening regulations on valley filling, and the G.W. Bush administration then loosened regulations to allow coal companies to dump waste rock directly into valleys and streams.

Scientists are finding that dumping tons of debris into valleys degrades or destroys immense areas of habitat, clogs streams and rivers, and pollutes waterways with acid drainage. With slopes deforested and valleys filled with debris, erosion intensifies, mudslides become frequent, and flash floods ravage the lower valleys (see THE SCIENCE BEHIND THE STORY, pp. 644–645).

People living in communities near mining sites experience social and health impacts (FIGURE 23.10b). Blasts from mines crack house foundations and wells, loose rock tumbles down into yards and homes, overloaded coal trucks speed down once-peaceful rural roads, and floods tear through properties (FIGURE 23.10c). Coal dust causes respiratory ailments, and contaminated water unleashes a variety of health problems. Some landowners are pressured to sell their land to coal companies, and people lose the forests and landscapes they have lived with since childhood. Opposition by some local residents (FIGURE 23.10d) to mining operations is often the result.

Although the people of Appalachia have long relied on the coal industry for employment, mountaintop removal is highly mechanized, so fewer workers are needed for mining. As a result, employment has declined in recent years even as coal extraction has risen. In all these ways, the people of Appalachia—already among the poorest in the United States—suffer substantial external costs (pp. 146, 165) from mining. Meanwhile, the rest of us benefit from the electricity that we produce with their coal.

Critics of mountaintop removal mining argue that valley filling violates the Clean Water Act because runoff flowing through waste rocks in valleys often contains high levels of salts and toxic metals that degrade water quality and impact aquatic organisms. While hundreds of permits for mountaintop mining were issued during the Clinton and G.W. Bush administrations, in 2010 the U.S. Environmental Protection Agency (EPA) announced new guidelines that prohibit valley fills unless strict measures of water quality can be attained. Critics of the policy argue that these new guidelines will essentially end the practice of mountaintop mining. In 2011, the EPA revoked the permit of an existing mountaintop mining operation in West Virginia, citing these new guidelines, and was reexamining other permits across Appalachia.

Solution mining dissolves and extracts resources in place

When a deposit is especially deep and the resource can be dissolved in a liquid, miners may use a technique called *solution mining* or *in-situ recovery*. In this technique, a narrow borehole is drilled deep into the ground to reach the deposit, and water, acid, or another liquid is injected down the borehole to leach the resource from the surrounding rock and dissolve it in the liquid. The resulting solution is then sucked out, and the desired resource is isolated. Salts can be mined in this way; water is pumped into deep salt caverns, the salt dissolves in the water, and the salty solution is extracted. Besides sodium chloride (table salt), such salts include lithium, boron, bromine, magnesium, and potash. In-situ recovery also is sometimes used for copper (dissolved with acids) and uranium (dissolved with acids or carbonates).

Solution mining generally exerts less environmental impact than other mining techniques, because less area at the surface is disturbed. The main potential impacts involve accidental leakage of acids into groundwater surrounding the borehole and the contamination of aquifers with acids, heavy metals, or uranium leached from the rock.

Some mining occurs in the ocean

The oceans hold many minerals useful to our society. We extract some minerals from seawater, such as magnesium from salts held in solution. We extract other minerals from the ocean floor. Using large vacuum-cleaner-like hydraulic dredges, miners collect sand and gravel from beneath the sea. They extract sulfur from salt deposits in the Gulf of Mexico and phosphate from offshore areas near the California coast and elsewhere. Other valuable minerals found on or beneath the seafloor include calcium carbonate (used in making cement) and silica (used as fire-resistant insulation and in manufacturing glass), as well as copper, zinc, silver, and gold ore. Many minerals are concentrated in manganese nodules, small ball-shaped accretions that are scattered across parts of the ocean floor. Over 1.5 trillion tons of manganese nodules may exist in the Pacific Ocean alone, and their reserves of metal may exceed all terrestrial reserves. The logistical difficulty of mining them, however, has kept their extraction uneconomical so far.

As land resources become scarcer and as undersea mining technology develops, mining companies may turn increasingly to the seas. Some companies already are exploring hydrothermal vents (p. 33) as potentially concentrated sources of metals such as gold, silver, and zinc, because these vents emit dissolved metals resulting from underground volcanic activity.

Impacts of undersea mining are largely unknown, but increases in such mining would undoubtedly destroy marine habitats and organisms that have not yet been studied. It would also likely cause some metals to diffuse into the water column at toxic concentrations and enter the food chain.

Restoration helps to reclaim mine sites

Because of the environmental impacts of mining, governments of the United States and other developed nations now require that mining companies restore, or reclaim, surface-mined sites following mining. The aim of such restoration, or **reclamation,** is to restore the site to a condition similar to its condition before mining. To restore a site, companies are required to remove buildings and other structures used for mining, replace overburden, fill in shafts, and replant the area with vegetation (**FIGURE 23.11**). In the United States, the 1977 *Surface Mining Control and Reclamation Act* mandates restoration efforts, requiring companies to post bonds to cover reclamation costs before mining can be approved. This ensures that if the company fails to restore the land for any reason, the government will have the money to do so. Most other nations exercise less oversight, and in nations such as Congo, there is no regulation at all.

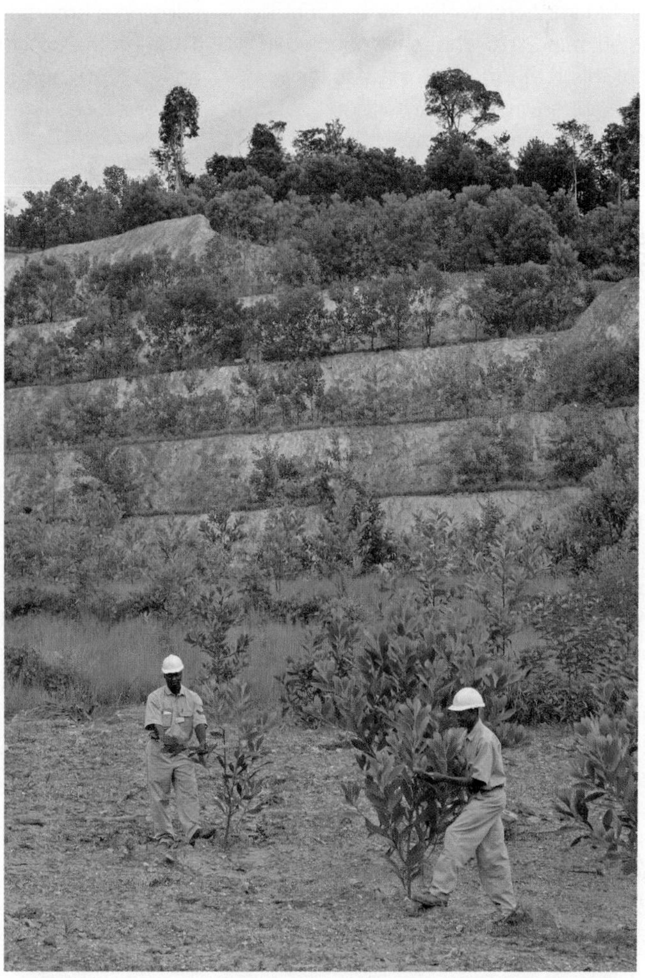

FIGURE 23.11 **More mine sites are being restored today, but restoration rarely is able to recreate the natural community present before mining.** Here, reclamation workers in Ghana, West Africa, plant trees on the benches and floor of an abandoned gold-mining pit.

Mountaintop Removal Mining: Assessing the Environmental Impacts

Mountaintop removal mining has attracted a great deal of criticism from environmental activists. But what do scientific studies tell us about the impacts of this mining method? Is the outcry justified?

Stream ecologist Margaret Palmer of the University of Maryland and a team of 11 other scientists aimed to find out. They reviewed all the scientific literature on mountaintop mining impacts. Writing in the journal *Science* in 2010, the team concluded that this method of mining causes "serious environmental impacts that mitigation practices cannot successfully address." The team also concluded that published studies indicate "a high potential for human health impacts."

The Appalachian forests that are cleared in mountaintop mining are some of the richest forests for biodiversity in the nation. Yet forest clearance—and the loss of biodiversity and ecosystem services—is merely the most obvious impact. Palmer's team stressed that many more consequences result when waste material from mountaintop removal is dumped into adjacent valleys, burying streams and trees.

Once buried, headwater streams are lost as ecosystems. Gone with them are certain rare endemic species, as well as the ecosystem's ability to cycle nutrients and produce organic matter for downstream systems. Flash floods also result: Because mining removes vegetation

Dr. Margaret Palmer, University of Maryland

and topsoil, alters topography, and compacts soil, less rain infiltrates the ground, and instead rain runs off quickly, causing flooding downstream. Moreover, when a valley is filled with mined material, water that runs through the fill emerges at the bottom carrying a brew of toxic substances. Through the process of acid drainage (p. 639), the acidic water carries dissolved heavy metals leached from the rock.

Researchers with the United States Geological Survey (USGS) carried out extensive water quality surveys in Appalachian streams in the late 1990s. For example, separate teams led by Katherine Paybins and by James Sams took hundreds of samples from diverse areas, chemically analyzed the samples, and then mapped their data, looking for geographic patterns. Their data clearly showed that concentrations of pollutants such as sulfates were strongly linked to upstream mining activities. They also showed that this pollution and its impacts (such as biodiversity loss) are long-lasting. Indeed, many studies show biodiversity declines in streams disturbed by mining, and no study has yet documented a recovery of stream life in the years after mining pollution.

Palmer's team tested for water quality impacts by tapping into a large database of measurements made by staff of the West Virginia Department of Environmental Protection. These workers had measured concentrations of pollutants in the waters of 1058 streams in West Virginia and had recorded numbers and types of aquatic insects in the streams.

Palmer's team first confirmed that sulfate concentrations were tightly linked with upstream mining activity. They then looked for statistical correlations between sulfate concentrations and other water quality indicators, such as concentrations of selenium, iron, aluminum, and manganese. They found that concentrations of all these minerals rose with concentrations of sulfates (**FIGURE 1, TOP**).

The researchers then graphed their insect data against sulfate concentrations (**FIGURE 1, BOTTOM**). All types of insects decreased as sulfate concentrations rose, suggesting that the changes in water chemistry brought about by mining were diminishing insect diversity in the streams.

Water pollution from mining can affect organisms higher on the food chain as well. Selenium is known to bioaccumulate (p. 373) in organisms and to cause birth defects (**FIGURE 2**). EPA scientists surveying 78 streams near mines in 2002 found that 73 had selenium concentrations above the level at which studies have documented risks to insects, fish, and the birds that eat them.

Scientific research also documents health impacts on people that may be due to mountaintop mining. Health impacts come from inhaling air pollution and dust, drinking contaminated groundwater, and eating fish contaminated with selenium and other toxic substances. A 2009 study documented that people in mountaintop mining areas show elevated

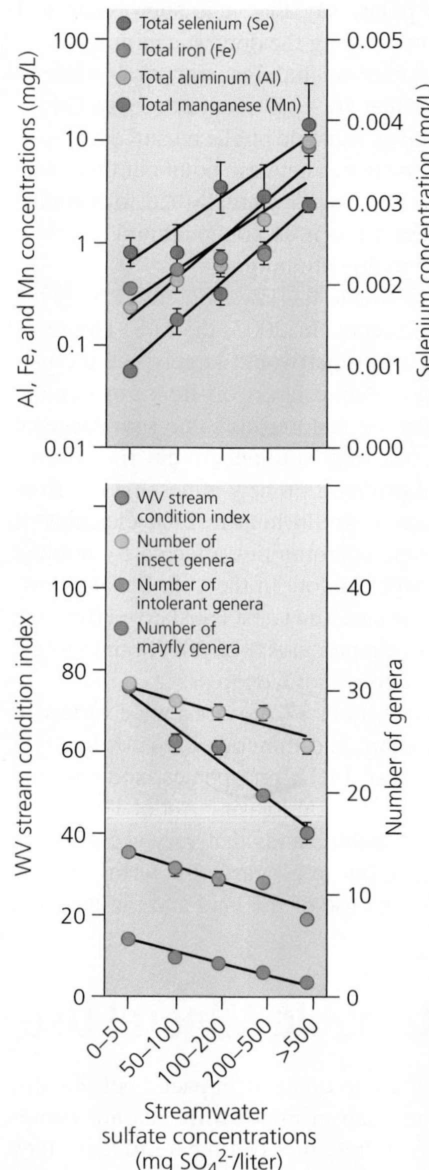

FIGURE 1 Mining increases pollutant concentrations in streams and affects aquatic organisms. Data from over 1000 West Virginia streams show that concentrations of four pollutants (selenium, iron, aluminum, and manganese) rise along with the increased sulfate concentrations that result from mining **(top)**. Data also show that insect diversity decreases with rising sulfate concentrations **(bottom)**. *Source: Palmer, M.A., et al., 2010. Mountaintop mining consequences.* Science *327: 148–149. Reprinted with permission of AAAS.*

levels of lung cancer, heart disease, kidney disease, pulmonary disorders, hypertension, and mortality. The public health researchers who carried out this study, Michael Hendryx and Melissa Ahern, found that these health problems occurred in women as well as men, so they could not be the result of direct occupational exposure among coal miners.

How much damage is too much, when it comes to a watershed? Various studies over the years have found that when over 5–10% of a watershed's area is disturbed by some sort of human land alteration, water quality and biodiversity in a stream decline. Mountaintop mining typically disturbs a greater percentage than this. For instance, of eight mountaintop mining permits issued in West Virginia in 2008, mining was allowed to cover 17–51% of each watershed.

Can mined mountaintops, filled valleys, and human health be restored to their original condition after mining? The science so far says no. A 2006 USGS study showed that even after reclamation efforts, groundwater from people's wells near mining sites still contains higher levels of mine-derived chemicals than water from wells in unmined areas.

Reclamation efforts have traditionally focused on planting grasses and herbs—a far less diverse plant assemblage than was destroyed. Moreover, the degraded, compacted soils do not hold water, nutrients, or organic matter well, so trees have a hard time establishing. A 2008 study found little or no regrowth of trees and shrubs a full 15 years after reclamation was completed. Another study projected that even 60 years later, the soil would hold only 77% as much carbon as it did originally.

FIGURE 2 High selenium concentrations in streamwater can lead to birth defects. This embryo of a fish has a severely curved spine as a result of selenium exposure downstream from a mine. *Source: Lemly, A.D., 2008. Aquatic hazard of selenium pollution from coal mining, pp. 167–183, in G.B. Fosdyke, ed.,* Coal mining: Research, technology, and safety. *Nova Science Publishers, Inc.*

Palmer's team concluded that the U.S. government is failing to enforce the Clean Water Act and the Surface Mining Control and Reclamation Act, and that these laws may not be strong enough to prevent severe impacts from mountaintop mining. The researchers urged the U.S. government to strengthen regulation of mountaintop mining practices. "Regulators," they wrote, "should no longer ignore rigorous science."

Just months after publication of this paper, the U.S. Environmental Protection Agency announced that it was strengthening its regulations on mountaintop mining. After reviewing EPA-sponsored and independent scientific studies, the agency adopted rules aimed at protecting 95% of aquatic life in Appalachian streams, and began revisiting permits granted in Appalachia for mountaintop mining operations. ■

The mining industry has made great strides in reclaiming mined land and employs many hard-working ecologists and engineers to conduct these efforts. However, even on sites that are restored, impacts from mining (such as soil and water damage from acid drainage) can be severe and long-lasting. Moreover, reclaimed sites do not generally regain the same biotic communities that were naturally present before mining. One reason is that fast-growing grasses are generally used to initiate and anchor restoration efforts. This helps control erosion quickly from the outset, but it can hinder the longer-term establishment of forests, wetlands, or other complex natural communities. Instead, grasses may outcompete slower-growing native plants in the acidic, compacted, nutrient-poor soils that usually result from mining. Moreover, many inconspicuous but vital symbiotic relationships (p. 80) that maintain ecosystems—such as specialized relationships between plants and fungi and plants and insects—are eliminated by mining and are very difficult to restore.

Water polluted by mining and acid drainage can also be reclaimed, if pH can be moderated and if toxic heavy metals can be removed. Like the reclamation of land, this is a challenging and imperfect process, but researchers and the mining industry are making progress in improving techniques. The need for treatment can be long-lasting. Mines in Spain from the era of the Roman Empire still leach acid drainage into waterways today.

WEIGHING THE ISSUES

RESTORING MINED AREAS Mining has severe environmental impacts, but restoring mined sites to their pre-mining condition is costly and difficult. How much do you think we should require mining companies to restore after a mine is shut down, and what criteria should we use to guide restoration? Should we require complete restoration? No restoration? What should our priorities be—to minimize water pollution, health impacts, biodiversity loss, soil damage, or other factors? Should the amount of restoration we require depend on how much money the company made from the mine? Explain your recommendations.

An 1872 law still guides U.S. mining policy

Government policy plays a role in the ways that mining companies stake claims and use land. In the United States, this has been controversial because policy is still guided by a law that is well over a century old. The **General Mining Act of 1872** encourages people and companies to prospect for minerals on federally owned land by allowing any U.S. citizen or any company with permission to do business in the United States to stake a claim on any plot of public land open to mining. The person or company owning the claim gains the sole right to take minerals from the area. The claim-holder can also patent the claim (i.e., buy the land) for only about $5 per acre. Regardless of the profits they might make on minerals they extract, the law requires no payments of any kind to the public, and until recently no restoration of the land after mining was required.

The General Mining Act of 1872 was enacted partly in response to the chaos of the California Gold Rush and other episodes, and it was designed to bring some order to mining activities. It also aimed to promote mining at a time when the government was trying to hasten settlement of the West in an orderly way. The law may have made good sense in 1872, but the United States has changed a great deal since then, and many question the law's suitability for today's nation.

Supporters of the policy say that it is appropriate and desirable to continue encouraging the domestic mining industry, which must undertake substantial financial risk and investment to locate resources that are vital to our economy. Critics counter that the policy gives valuable public resources away to private interests nearly for free. They also point out that many claims made under this law have eventually led to lucrative land development schemes (such as condominium development) that have nothing to do with mining.

Critics have tried to amend the law many times over the years, mostly without success. In 2007, the U.S. House of Representatives passed a bill that would largely end the patenting process, put some public lands off-limits to mining, mandate that mined sites be restored to some semblance of their former condition, and require miners to pay the government royalties of 4% of profits from new mines and 8% from existing mines. The money would help to fund cleanups of mined sites and reimburse communities affected by mining. This bill was never brought to vote in the U.S. Senate, however, and failed to become law. The latest legislative effort, the Hardrock Mining and Reclamation Act of 2009, failed to get out of committee in both houses of Congress.

The General Mining Act of 1872 covers a wide variety of metals, gemstones, uranium, and minerals used for building materials. In contrast, fossil fuels, phosphates, sodium, and sulfur are governed by the Mineral Leasing Act of 1920. This law sets terms for leasing public lands that vary according to the resource being mined, but in all cases the terms include the payment of rents for the use of the land and the payment of royalties on profits.

Toward Sustainable Mineral Use

Mining exerts plenty of environmental impacts, but we also have another concern to keep in mind: Minerals are nonrenewable resources (p. 3) in finite supply. Like fossil fuels, they form far more slowly than we use them, and if we continue to mine them, they will eventually be depleted. As a result, it will benefit us to find ways to conserve the supplies we have left and to make them last. Reducing waste and developing means of recovering and recycling used mineral resources are ways we can pursue the use of mineral resources more sustainably. We will likely never achieve 100% recovery, but we can do much better than we are doing today.

Minerals are nonrenewable resources in limited supply

Unlike sunlight or water or forests, minerals do not regenerate fast enough to provide us a new supply once we have mined all known reserves. They are therefore considered nonrenewable resources. Some minerals we use are abundant in their supply and will likely never run out, but others are rare enough that they could soon become unavailable.

For instance, geologists in 2013 calculated that the world's known reserves of tantalum will last about 200 more years at today's rate of consumption. If demand for tantalum increases, it could run out faster. And if everyone in the world began consuming tantalum at the rate of U.S. citizens, then it would last for only 15 years! Most pressing may be dwindling supplies of indium. This obscure metal, which is used for LCD screens, might last only another 30 years. Because of these supply concerns and price volatility, industries now are working hard to develop ways of substituting other materials for indium. A lack of indium and gallium would threaten the production of high-efficiency cells for solar power. Platinum is dwindling too, and if it became unavailable, it would be harder to develop fuel cells for vehicles. However, platinum's high market price encourages recycling, which may keep it available, albeit as an expensive metal.

FIGURE 23.12 shows estimated years remaining for several selected minerals at today's consumption rates. Note that each bar in the figure consists of two parts. The left (red) portion of the bar shows the number of years researchers calculate that we will have the mineral available under today's economic conditions. As minerals become scarcer, demand for them increases and price rises. Higher market prices make it more profitable for companies to mine the resource, so they become willing to spend more to reach further deposits that were not economically worthwhile originally. Thus the entire length of each bar in Figure 23.12 shows the number of years researchers calculate that we will have the mineral available in total, as prices rise. These two categories used by the U.S. Geological Survey (USGS) are similar to the way the USGS classifies fossil fuel deposits as being either "economically recoverable" or "technically recoverable" (pp. 527, 531).

Several factors affect how long mineral deposits may last

Calculating how long a given mineral resource will be available to us, as shown by the data in Figure 23.12, is beset by a great deal of uncertainty. There are several major reasons why such estimates may increase or decrease over time.

Discovery of new reserves As we discover new deposits of a mineral, the known reserves—and thus the years this mineral is available to us—increase. For this reason, some previously predicted shortages have not come to pass, and we may have access to these minerals for longer than currently estimated. As one recent example, in 2010 geologists associated with the U.S. military discovered that Afghanistan holds immense mineral riches that were previously unknown. The newly discovered reserves of iron, copper, niobium, lithium, and many other metals are estimated to be worth over $1 trillion—enough to realign the entire Afghan economy around mining. (Note, however, that such riches are not guaranteed to make Afghanistan a wealthy nation; history teaches us that regions rich in nonrenewable resources, such as Congo and Appalachia, have often been unable to prosper from them.)

New extraction technologies Just as rising prices of scarce minerals encourage companies to expend more effort

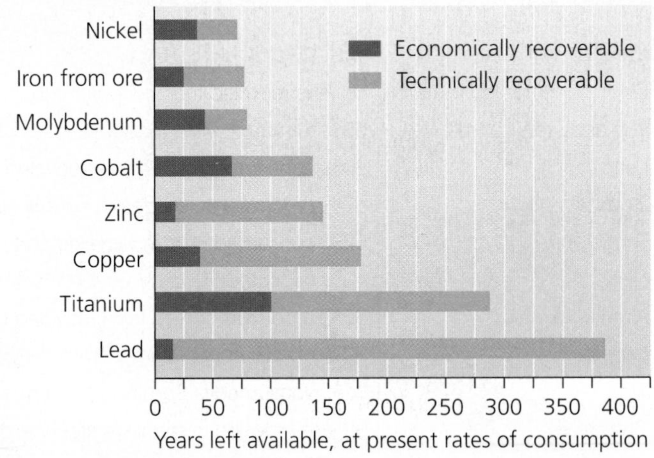

FIGURE 23.12 Minerals are nonrenewable resources, so supplies of metals are limited. Shown in red are the numbers of remaining years that certain metals are estimated to be economically recoverable at current prices. The entire lengths of the bars (red plus orange) show how long certain metals are estimated to be available using current technology on all known deposits, whether economically recoverable or not. *Data are for 2012, from U.S. Geological Survey, 2013.* Mineral commodity summaries 2013. *USGS, Washington, D.C.*

DATA Q Which metal has the highest proportion of its technically recoverable reserves that are currently economically recoverable? Approximately what percentage is economically recoverable? Which metal has the smallest proportion of its technically recoverable reserves that are currently economically recoverable and what is this value?

to reach difficult deposits, rising prices also may favor the development of enhanced mining technologies that can reach more minerals at less expense. If more powerful technologies are developed, then these may increase the amounts of minerals that are technically feasible for us to mine.

Changing social and technological dynamics New societal developments and new technologies in the marketplace can modify demand for minerals in unpredictable ways. Just as cell phones and computer chips boosted demand for tantalum, fiber-optic cables decreased demand for copper as they replaced copper wiring in communications applications. Today lithium–ion batteries are replacing nickel–cadmium batteries in many devices. Synthetically produced diamonds are driving down prices of natural diamonds and extending their availability. Additionally, health concerns sometimes motivate change: We have replaced toxic substances such as lead and mercury with safer materials in many applications, for example.

Changing consumption patterns Changes in the rates and patterns of consumption also alter the speed with which we exploit mineral resources. For instance, economic recession depressed demand and led to a decrease in production and consumption of most minerals from 2007 to 2009 after rising for long stretches. However, over the long term, demand has been rising. This is especially true today as China, India, and other major industrializing nations rapidly increase their consumption.

TABLE 23.1 Recycled minerals in the United States

MINERAL	U.S. RECYCLING RATE
Gold	More is recycled than is consumed
Iron and steel scrap	95% for autos, 90% for appliances, 70–98% for construction materials, 71% for cans
Lead	80% consumed comes from recycled post-consumer items
Zinc	57% produced is recovered, mostly from recycled materials used in processing
Tungsten	52% consumed is from recycled scrap
Nickel	43% consumed is from recycled nickel
Aluminum	35% produced comes from recycled post-consumer items
Copper	33% of U.S. supply comes from various recycled sources
Tin	31% consumed is from recycled tin
Chromium	30% is recycled in stainless steel production
Germanium	30% consumed worldwide is recycled. Optical device manufacturing recycles over 60%
Molybdenum	About 30% gets recycled as part of steel scrap that is recycled
Silver	25% consumed is from recycled silver. U.S. recovers as much as it produces
Cobalt	25% consumed comes from recycled scrap
Niobium (columbium)	Perhaps 20% gets recycled as part of steel scrap that is recycled
Platinum-group metals	About 11% consumed is recycled
Bismuth	All scrap metal containing bismuth is recycled, providing 10% of consumption
Diamond (industrial)	7% of production is from recycled diamond dust, grit, and stone

Data are for 2012, from U.S. Geological Survey, 2013. Mineral commodity summaries 2013. *USGS, Washington, D.C.*

Recycling Advances in recycling technologies and the extent of recycling have been helping us to extend the lifetimes of many mineral resources. Further progress in recycling will likely continue to do so.

Despite these sources of uncertainty, we would be wise to be concerned about Earth's finite supplies of mineral resources and to try to use them more sustainably. Sustainable use will benefit future generations by conserving resources for them to use. It will also benefit us today, because conserving mineral resources through reuse and recycling can prevent price hikes that result from reduced supply. And as we saw with fossil fuels (Chapter 19), domestic conservation of resources helps make a national economy less vulnerable in instances when other nations decide to withhold resources. Currently, the United States is 100% dependent on imports from other nations for 17 of the 63 major minerals on which the USGS reports annually. For 24 more of these minerals, the United States relies on imports for 50% or more of its supply.

We can make our mineral use more sustainable

We can address both major challenges facing us regarding mineral resources—finite supply and environmental damage—by encouraging recycling of these resources. Metal-processing industries regularly save resources and money by reusing some of the waste products produced during their refining processes. In addition, municipal recycling programs help provide metals by handling used items that we as consumers place in recycling bins and divert from the waste stream. In 2010, fully 35% of metals in the U.S. municipal solid waste stream were diverted for recycling. For example, 80% of the lead we consume today comes from recycled materials, in particular recycled car batteries. Similarly, 33% of our copper comes from recycled copper sources such as pipes and wires. We recycle steel, iron, platinum, and other metals from auto parts. Altogether, we have found ways to recycle much of our gold, lead, iron and steel scrap, chromium, zinc, aluminum, and nickel. TABLE 23.1 shows minerals that currently boast high recycling rates in the United States.

In many cases, recycling can decrease energy use substantially. For instance, making steel by remelting recycled iron and steel scrap requires much less energy than producing steel from virgin iron ore. Because this practice saves money, the steel industry today is designed to make efficient use of iron and steel scrap. Over half its scrap comes from discarded consumer items such as cars, cans, and appliances; scrap produced within their plants and scrap produced by other types of plants in the industry each account for nearly one-fourth of the total. Similarly, over 40% of the aluminum in the United States today is recycled (FIGURE 23.13). This is a good thing because it takes over 20 times more energy to extract virgin aluminum from ore (bauxite) than it does to obtain it from recycled sources. Every ton of aluminum cans your community recycles saves the energy equivalent of over 1600 gallons of gasoline.

Saving energy means cutting down on pollution from fossil fuel combustion. All in all, the 7 million tons of metals that U.S. consumers recycle each year reduce greenhouse gas emissions by 25 million metric tons of carbon dioxide, the Environmental Protection Agency estimates. This reduction is like taking more than the yearly emissions of 4.5 million cars out of the atmosphere.

Tantalum is recycled from scrap by-products generated during the manufacture of electronic components and also from scrap from tantalum-containing alloys and manufactured materials. Currently the industry estimates that recycling accounts for 20–25% of the tantalum available for use in products. This percentage has been growing quickly, but its future growth will depend on how quickly we expand recycling efforts for used cell phones and other electronic waste (pp. 624–625) and on how well we enable recycling facilities to recover metals from these products.

FIGURE 23.13 When you recycle aluminum cans, you contribute to valuable efforts to save mineral resources, money, and energy.

We can recycle metals from e-waste

Electronic waste, or e-waste, from discarded computers, printers, cell phones, handheld devices, and other electronic products is rising fast—and that e-waste contains hazardous substances (p. 624). Recycling old electronic devices helps keep them out of landfills and also helps us conserve valuable minerals such as tantalum. As Dr. Timothy Townsend (featured in Chapter 22's **Science behind the Story**, pp. 628–629) has put it, "If we know these metals are, overall, bad for us, it doesn't make sense to keep digging them up from the earth's crust and bringing them into the biosphere while—at the same time—we're taking the ones we've already got and burying them."

In fact, each of the 1.7 billion cell phones sold each year contains about 200 chemical compounds and close to a dollar's worth of precious metals (FIGURE 23.14). Upgrades and improvements render over 130 million cell phones obsolete each year in the United States alone, and an estimated 500 million old cell phones are currently lying inactive in people's homes and offices.

When you turn in your old phone to a recycling and reuse center rather than discarding it, the phone may be refurbished and resold in a developing country. People in African nations in particular readily buy used cell phones, because they are inexpensive and because land-line phone service does not always exist in poor and rural areas. Alternatively, the phone may be dismantled in a developing country and the various parts refurbished and reused, or recycled for their metals. Either way, you are helping to extend the availability of resources through reuse and recycling and to decrease waste of valuable minerals.

Today only about 10% percent of old cell phones are recycled. That leaves a long way to go! As more of us recycle our phones, computers, and other electronic items, more tantalum and other metals may be recovered and reused. By recycling more, we can reduce demand for virgin ore and decrease pressure on African people and ecosystems where coltan is mined. Throughout the world, recycling to make better use of the mineral resources we have already mined will help minimize the impacts of mining and assure us access to resources farther into the future.

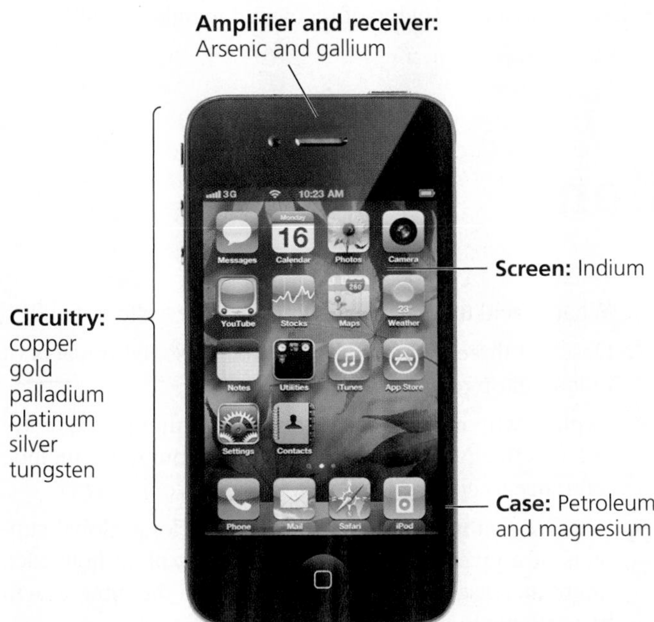

FIGURE 23.14 Your cell phone contains a diversity of mined materials from around the world. Figure 23.5 lists the nations that are major producers of these minerals.

Conclusion

We depend on a diversity of minerals and metals to help manufacture products widely used in our society. We mine these nonrenewable resources by various methods, according to how the minerals are distributed. Economically efficient mining methods have greatly contributed to our material wealth, but they have also resulted in extensive environmental impacts, ranging from habitat loss to acid drainage. Restoration efforts and enhanced regulation help to minimize the environmental and social impacts of mining, although to some extent these impacts will always exist. We can lengthen our access to mineral resources and make our mineral use more sustainable by maximizing the recovery and recycling of key minerals.

Reviewing Objectives

You should now be able to:

Outline types of mineral resources and how they contribute to our products and society

- Minerals we mine from the earth provide raw materials for most of the products we use every day. (pp. 635–636)
- We mine ore for metals, as well as nonmetallic minerals and fuels. (pp. 636–639)
- Processing and refining metals through methods such as smelting are important steps between mining ore and manufacturing products. (pp. 637–638)

Describe the major methods of mining

- Strip mining removes surface layers of soil and rock to expose resources. (p. 639)
- In subsurface mining, miners tunnel underground. (p. 640)
- Open pit mining involves digging gigantic holes. (pp. 640–641)
- Placer mining uses running water to isolate minerals. (p. 641)
- Mountaintop removal mining removes immense amounts of rock from mountaintops and dumps it into valleys below. (pp. 641–643)
- Solution mining uses water, acid, or other liquids to dissolve minerals in place and extract them. (p. 643)
- A great deal of mineral wealth exists in the oceans but is mostly uneconomical (so far) to reach. (p. 643)

Characterize the environmental and social impacts of mining

- Many methods of mining completely remove vegetation, soil, and habitat. (pp. 639–643)
- Acid drainage occurs when water leaches compounds from freshly exposed waste rock. It is often toxic to aquatic organisms. (p. 639)
- Mining may have health impacts on miners and diverse social impacts on people living near mines. (pp. 640–642)
- Mountaintop removal for coal destroys forests, mountaintops, and adjacent valleys and streams. (pp. 641–643)

Assess reclamation efforts and mining policy

- Reclamation efforts generally fall short of effective ecological restoration. (pp. 643, 646)
- U.S. mining policy remains dominated by a law passed in 1872. (p. 646)

Evaluate ways to encourage sustainable use of mineral resources

- Minerals are nonrenewable resources, and some are limited in supply. (pp. 646–647)
- Several factors affect how long a given mineral resource will last. (pp. 647–648)
- Reuse and recycling by industry and consumers are the keys to more sustainable practices of mineral use. (pp. 648–649)
- E-waste can be a source of recovered metals. (p. 649)

Testing Your Comprehension

1. Define each of the following and contrast them with one another: (1) mineral, (2) metal, (3) ore, (4) alloy.
2. A mining geologist locates a horizontal seam of coal very near the surface of the land. What type of mining method will the mining company use to extract it? What is one common environmental impact of this type of mining?
3. How does strip mining differ from subsurface mining? How does each of these approaches differ from open pit mining?
4. What type of mining is used for both coltan and gold? What does a miner do to conduct this type of mining?
5. Describe and contrast how water is used in placer mining and in solution mining.
6. What is acid drainage, and why can it be toxic to fish?
7. Describe three major environmental or social impacts of mountaintop removal mining.
8. Explain why reclamation efforts after mining frequently fail to effectively restore natural communities. Include reference to both soil and vegetation in your answer.
9. List five factors that can influence how long global supplies of a given mineral will last, and explain how each might increase or decrease the time span the mineral will be available to us.
10. Name three types of metal that we currently recycle, and identify the products or materials that are recycled to recover these metals.

Seeking Solutions

1. List three impacts of mining on the natural environment, and describe how particular mining practices can lead to each of these impacts. How are these impacts being addressed? Can you think of additional solutions to prevent, reduce, or mitigate these impacts?
2. List three impacts of mining on people's health, lifestyles, or well-being, and describe how particular mining practices can lead to each of these impacts. How are these impacts being addressed? Can you think of additional solutions to prevent, reduce, or mitigate these impacts?
3. You have won a grant from the EPA to work with a mining company to develop a more effective way of restoring a mine site that is about to be abandoned. Describe a few preliminary ideas for carrying out restoration better than it is typically being done. Now describe a field experiment you would like to run to test one of your ideas.
4. Present one common argument in favor of retaining the General Mining Act of 1872. Now present one common argument in favor of repealing or reforming the law. Describe how you think the United States today should balance our need for minerals with our need to protect public lands and public interests. Would you retain, repeal, or reform the 1872 law?
5. **THINK IT THROUGH** The story of coltan in the Congo is just one example of how an abundance of exploitable resources can often worsen or prolong military conflicts in nations that are too poor or ineffectively governed to protect these resources. In such "resource wars," civilians often suffer the most as civil society breaks down. Suppose you are the head of an international aid agency that has earmarked $10 million to help address conflicts related to mining in the Democratic Republic of the Congo. You have access to government and rebel leaders in Congo and neighboring countries, to ambassadors of the world's nations in the United Nations, and to representatives of international mining corporations. Based on what you know from this chapter, what steps would you consider taking to help improve the situation in the Congo?
6. **THINK IT THROUGH** As you finish your college degree, you learn that the mountains behind your childhood home in the hills of Kentucky are slated to be mined for coal using the mountaintop removal method. Your parents, who still live there, are worried for their health and safety and do not want to lose the beautiful forested creek and ravine behind their property. However, your brother is out of work and could use a mining job. What would you attempt to do in this situation?

Calculating Ecological Footprints

As we saw in Figure 23.12, the supplies of some metals are limited enough that, at today's prices, these metals could be available to us for only a few more decades. After that, prices will rise as they become scarcer. The number of years of total availability (at all prices) depends on a number of factors: On the one hand, metals will be available for longer if new deposits are discovered, new mining technologies are developed, or recycling efforts are improved. On the other hand, if our consumption of metals increases, the number of years we have left to use them will decrease.

Currently the United States consumes metals at a much higher per-person rate than the world does as a whole. If one goal of humanity is to lift the rest of the world up to U.S. living standards, then this will sharply increase pressures on mineral supplies.

The chart shows currently known economically recoverable global reserves for several metals, together with the amount used per year (each figure in thousands of metric tons). For each metal, calculate and enter in the fourth column the years of supply left at current prices by dividing the reserves by the amount used annually.

The fifth column shows the amount that the world would use if everyone in the world consumed the metal at the rate that Americans do. Now calculate the years of supply left at current prices for each metal if the world were to consume the metals at the U.S. rate, and enter these values in the sixth column.

Metal	Known economic reserves	Amount used per year	Years of economic supply left	Amount used per year if everyone consumed at U.S. rate	Years of economic supply left if everyone consumed at U.S. rate
Titanium	700,000	7000		31,703	
Copper	680,000	17,000		39,798	
Nickel	75,000	2100		5014	
Tin	4900	230		951	
Tungsten	3200	73		369	
Antimony	1800	180		519	
Silver	540	24		133	
Gold	52	2.7		3.37	

Data are for 2012, from U.S. Geological Survey, 2013. Mineral commodity summaries 2013. *USGS, Washington, D.C. All numbers are in thousands of metric tons. World consumption data are assumed equal to world production data. "Known economic reserves" include extractable amounts under current economic conditions. Additional reserves exist that could be mined at greater cost.*

1. Which of these eight metals will last the longest under current economic conditions and at current rates of global consumption? For which of these metals will economic reserves be depleted fastest?
2. If the average citizen of the world consumed metals at the rate that the average U.S. citizen does, the economic reserves of which of these eight metals would last the longest? Which would be depleted fastest?
3. In this chart, our calculations of years of supply left do not factor in population growth. All else being equal, how do you think population growth will affect these numbers?
4. Describe two general ways that we could increase the years of supply left for these metals. What do you think it will take to accomplish this?

STUDENTS

Go to **MasteringEnvironmentalScience** for assignments, the etext, and the Study Area with practice tests, videos, current events, and activities.

INSTRUCTORS

Go to **MasteringEnvironmentalScience** for automatically graded activities, current events, videos, and reading questions that you can assign to your students, plus Instructor Resources.

24

De Anza College students in front of their LEED-certified Science Center

Sustainable Solutions

Upon completing this chapter, you will be able to:

- Describe approaches being taken on college and university campuses to promote sustainability
- Explain the concept of sustainable development
- Discuss how environmental protection can enhance economic well-being
- Assess key approaches to designing sustainable solutions
- Explain how time is limited yet human potential to solve problems is tremendous

CENTRAL CASE STUDY

De Anza College Strives for a Sustainable Campus

"Sustainability considers our impact not just on air, land, and water, but includes our impact on community vibrancy, environmental stewardship, social equity, and financial responsibility."

—De Anza College's Sustainability Management Plan

"Sustainability isn't a distant, unattainable concept to be discussed in the abstract. It is a very real way of making decisions that can be integrated into every single person's daily actions."

—Yale University student Jacquelyn Maitram Truong

California's Silicon Valley has long been a hub for innovation. So perhaps it's no surprise that a college in the heart of this region is a leader in the burgeoning movement for campus sustainability.

De Anza College, located in Cupertino, California (home of Apple Computer, Inc.), serves 24,000 students, making it one of North America's largest community colleges. Today De Anza has become one of the "greenest" community college campuses as well, thanks to the ongoing commitment of its students, faculty, staff, and administrators. Today more and more colleges and universities are serving as models for our larger society by taking steps to address their resource consumption, pollution, and ecological footprints.

At De Anza College, sustainability efforts reach back to 1990, when faculty member Julie Phillips began incorporating concepts of sustainability into the curriculum and energized students with the idea of a green building project. Soon the College Environmental Advisory Group (CEAG) was created, bringing together students, faculty, staff, administrators, and community members to encourage green building and other sustainable practices on campus. Eventually, those early students brought their classroom experiences, sustainable practices, and green building vision to reality by urging the De Anza Associated Student Body to allocate $180,000 for the conceptual design of what would become the Kirsch Center for Environmental Studies.

In 2005, the Kirsch Center opened for instruction, as the nation's first LEED-Platinum sustainable building (pp. 348–350, 657) at a community college. The Kirsch Center shows students and the public how to merge energy efficiency, pollution prevention, and biodiversity protection. Showcasing hi-tech and innovative labs and classrooms that promote visual and hands-on learning, the Kirsch Center aims to be "a building that teaches about energy, resources, and stewardship."

Indeed, the building does teach. Displays inside the entrance show real-time data on the energy generated by the 36.5-kilowatt photovoltaic energy system on the rooftop. Other monitors show current temperature readings that guide the advanced radiant heating and cooling system, while red and green lights advise when windows should be opened and closed. Students use labs, classrooms, and open study stations that are bathed in natural lighting from outdoors, which saves on electricity while creating a bright and pleasant environment that helps students learn better. Everything from carpeting to furniture to structural steel to toilet seats is made with recycled materials, and features to conserve water and energy abound.

Located adjacent to the Kirsch Center is a 1.5-acre arboretum called the Cheeseman Environmental Study Area. Here, 12 California native plant communities are represented, with over 400 species of native plants. Five other LEED-certified green buildings grace De Anza's campus as well.

In 2006, De Anza's administrators signed a sustainability policy, committing to green building renovation and construction and to the selection of vendors experienced with sustainability.

The school took another major step in 2007 when it adopted the Sustainability Management Plan that CEAG had spearheaded. This plan helps to identify environmental and health risks and to prioritize opportunities for addressing them. It specifies six focus areas:

- Reducing solid and hazardous waste
- Conserving energy and reducing carbon emissions
- Conserving water
- Making sustainable purchasing decisions
- Pursuing ecologically responsible landscaping and maintenance
- Undertaking green building practices in construction and renovation

This plan guides De Anza's greening efforts, while CEAG monitors progress and makes policy recommendations.

Challenges abound, and the main ones today are financial as California and other states suffer severe budget cutbacks in higher education. In tough financial times, more and more young people are enrolling in colleges, yet as state budgets and private endowments shrink, schools receive fewer resources. Sustainability initiatives often cost money, and their proponents generally have to argue their case to penny-conscious administrators time after time. But De Anza's administration, like increasing numbers of others, has recognized that many of the short-term costs associated with sustainability efforts are actually investments that can save substantial amounts of money in the long term.

Despite these challenges, De Anza continues to make strides. Today, 60% of the campus's waste is diverted from landfills. Ninety percent of the chemicals used by the custodial staff are environmentally friendly. The dining services are using biodegradable containers and local and organic food. Students receive transit passes from the local bus service and can rent bicycles for use on and off campus. De Anza's sustainability proponents have produced a policy handbook on energy efficiency and are leading workshops for other community colleges.

FIGURE 24.1 **Sustainability-oriented curricula extend far beyond the classroom.** In California's Coyote Valley, De Anza College instructor Ryan Phillips trains students in tracking and surveying wildlife as they help perform research into the importance of habitat corridors.

Moreover, thousands of students enroll in De Anza's environmental studies and environmental sciences courses each year. The learning extends far beyond the classroom. In one new certificate program, students are trained in wildlife tracking and help conduct research on wildlife populations and habitat corridors in the Coyote Valley and other areas in the region. The data they collect informs the work of regional planners and brings pragmatic benefits to their region, as the students themselves gain hands-on experience in the science of conservation biology (FIGURE 24.1). All in all, De Anza is providing an inspiring model for other community colleges in California and for colleges and universities nationwide as they strive for campus sustainability.

Sustainability on Campus

Whether on campus or around the world, sustainability means living in a way that can be lived far into the future. Sustainability involves conserving resources to prevent their depletion, reducing waste and pollution, and safeguarding ecological processes and ecosystem services, so as to ensure that our society's practices can continue and our civilization can endure. Truly sustainable solutions will satisfy all three pillars of sustainability: environmental quality, economic well-being, and social justice (pp. 156–157).

If we are to attain a sustainable civilization, we will need to make efforts at every level, from the individual to the household to the community to the nation to the world. Governments, corporations, and organizations must all encourage and pursue sustainable practices. Among the institutions that can contribute to sustainability efforts are colleges and universities (pp. 16–17). In today's quest for sustainable solutions, students at colleges and universities are playing a crucial role. They are creating models for the wider world by leading sustainability initiatives on their campuses (FIGURE 24.2).

An audit is a useful way to begin

Campus sustainability efforts often begin with a quantitative assessment of the institution's operations. An audit provides baseline information on what an institution is doing or how much it is consuming. Audits also help set priorities and goals. Students can conduct an audit themselves, as when University of Vermont graduate student Erika Swahn measured heating, transportation, electricity, waste, food, and water use to calculate her school's ecological footprint (it turned out to be 4.5 acres for each student, instructor, and staff member). If you set out to do an audit yourself, one useful tool is a handheld device called a "Kill-A-Watt" meter (FIGURE 24.3), which can help measure energy use plug by plug and room by room.

Sometimes an audit can be done as part of a class, as when students at Stetson University in Florida conducted a greenhouse gas inventory of their campus. In contrast, Harford Community College in Maryland hired specialists to audit its energy use and pollutant emissions, which then served as the basis for setting reduction goals. Students at Georgian Court University in New Jersey worked with consultants from the Environmental Protection Agency to assess their campus's carbon footprint, then remeasured it after instituting changes.

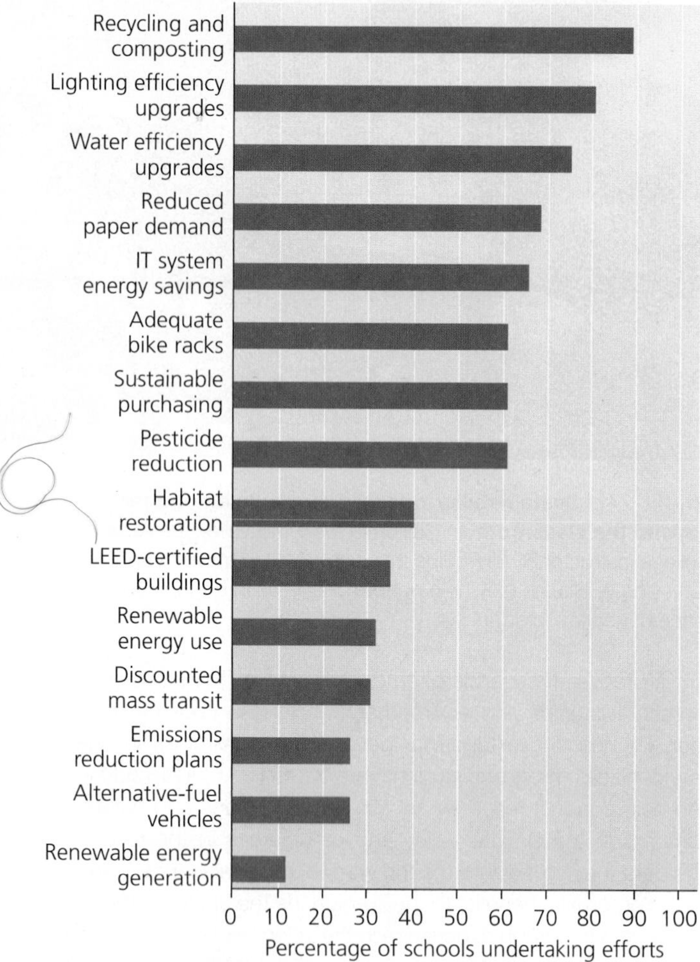

FIGURE 24.2 Campus sustainability efforts are diverse. Shown are the frequency of different pursuits at schools across North America that responded to a recent comprehensive survey.
Source: *McIntosh, M., et al., 2008.* Campus environment 2008: A national report card on sustainability in higher education. *National Wildlife Federation Campus Ecology; survey conducted by Princeton Survey Research Associates International.*

Their suite of actions prevented the emission of nearly 5500 million tons of CO_2, as much as is emitted by 1100 cars in a year. At De Anza College, students helped conduct a comprehensive audit of various buildings on campus, describing the buildings' inefficiencies and identifying opportunities for improvement. The information from this audit was then included in the college's Sustainability Management Plan.

It is most useful in an audit to target items that can lead directly to specific recommendations. For instance, an audit should quantify the performance of individual appliances so that decision makers can identify particular ones to replace. Once changes are implemented, the institution can monitor progress by comparing future measurements to the audit's baseline data.

FIGURE 24.3 A valuable tool for auditing energy use on campus is a Kill-a-Watt meter. This device measures the electrical current drawn by appliances and fixtures.

Recycling and waste reduction are common campus efforts

Campus sustainability efforts frequently involve waste reduction, recycling, or composting. According to the most recent comprehensive survey of campus sustainability efforts (see Figure 24.2), most schools recycle or compost at least some waste, and the average recycling rate reported was 29%. Waste management initiatives are relatively easy to conduct because they offer many opportunities for small-scale improvements and because people generally enjoy recycling and reducing waste.

The best-known collegiate waste management event is RecycleMania, a 10-week competition among schools to see which can recycle the most. This annual competition was started by students, and grew to involve 523 schools by its 13th year in 2013. University of Missouri–Kansas City won the overall award that year, recycling 86% of its waste. Students at California State University–San Marcos won the title for per capita recycling, with 24 kg (53 lb) of material recycled per student. Other top-performing schools in 2013 included Baldwin-Wallace College, Baylor College of Medicine, Franklin and Marshall College, Johnson and Wales University–Denver, Kendall College of Art and Design, Rutgers University, and Valencia College.

"Trash audits" or "landfill on the lawn" events involve tipping dumpsters onto a campus open space and sorting out recyclable items (**FIGURE 24.4**). When students at Ashland University in Ohio audited their waste, they found that 70% was recyclable, and they used this data to press their administration to support recycling programs. Louisiana State

FIGURE 24.4 In "landfill on the lawn" events, students sort through rubbish and separate out recyclables. Events like this one at the University of North Carolina at Greensboro demonstrate to passersby just how many recyclable items are needlessly thrown away.

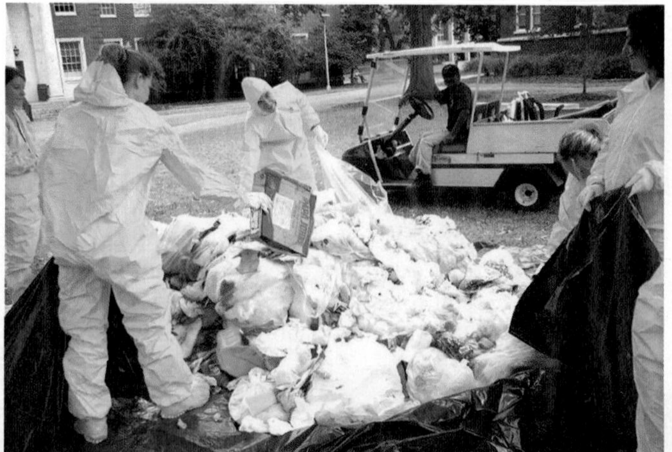

University students initiated recycling efforts at home football games, and over three seasons they recycled 68 tons of refuse that otherwise would have gone to the landfill.

Composting (pp. 616–617) is becoming popular as well. Ball State University in Indiana composts bulky wood waste, shredding surplus furniture and wood pallets and making them into mulch to nourish campus plantings. Appalachian State University in North Carolina updated its composting facilities as part of a drive to become a "zero waste" campus by 2022. A composting program at Texas State University designed by a graduate student combines food scraps, paper waste, and waste from a nearby chicken feedlot—and then adds water hyacinth, an invasive aquatic plant unwanted in the area. The program aims to determine whether the heat generated by composting can kill seeds of the water hyacinth while producing top-quality compost. At Ithaca College in New York, 44% of the food waste is composted, saving the college $11,500 each year in landfill disposal fees. The compost is used on campus plantings, and experiments showed that the plantings grew better with the compost mix than with chemical soil amendments.

Reuse is more sustainable than recycling, so students at some campuses systematically collect unwanted items and donate them to charity or resell them to returning students in the fall. Students at the University of Texas at Austin run a "Trash to Treasure" program. Each May, they collect 40–50 tons of items that students discard as they leave and then resell them at low prices in August to arriving students. This keeps waste out of the landfill, provides arriving students with items they need at low cost, and raises $10,000–20,000 per year that gets plowed back into campus sustainability efforts. Hamilton College in New York runs a similar program, called "Cram & Scram." It reduces Hamilton's landfill waste by 28% (about 90 tons) each May.

FIGURE 24.5 The Kirsch Center for Environmental Studies at De Anza College is a climate-responsive and energy-efficient building. It conserves water, generates renewable energy, uses recycled and nontoxic materials, and features outdoor learning spaces and labs. The spacious interior gathers natural daylight, providing a welcoming learning environment.

Green building design is a key to sustainable campuses

Buildings are responsible for 70–90% of a campus's greenhouse gas emissions, so making them more efficient can make a big difference. Many campuses now boast *green buildings* (pp. 348–350) that are constructed from sustainable building materials and whose design and technologies reduce pollution, use renewable energy, and encourage efficiency in energy and water use. The Leadership in Energy and Environmental Design (LEED) standards (pp. 348–349), developed and maintained by the nonprofit U.S. Green Building Council, guide the design and certification of new construction and the renovation of existing structures.

De Anza College's Kirsch Center for Environmental Studies (FIGURE 24.5) has achieved a "platinum" LEED ranking (the highest possible). Besides the features described earlier (p. 654), the building contains recycled steel in its beams, cabinets made with FSC-certified (Forest Stewardship Council) wood, tiles made from car windshields, toilet seat lids made from recycled water bottles, concrete floors containing fly ash from coal-fired power plants, and carpeting made from recycled materials with no PVCs, vinyl, or toxic adhesives. The building's layout and orientation utilize passive solar design concepts (p. 588), natural daylighting, and energy-efficiency strategies. In the winter, south-facing windows let in sunlight. In the summer, the high solar angle, exterior sunshades, and deciduous trees effectively block the sun.

Students helped launch the Kirsch Center when the De Anza Associated Student Body contributed $180,000 for the conceptual design of the building. The Steve and Michele Kirsch Foundation contributed $2 million toward construction, and local bond funds provided $8 million. Its many energy- and water-saving features help the college save money in the long run, relative to a conventional building. Arup, the architectural firm responsible for the energy and mechanical design of the building, has calculated that the Kirsch Center's energy efficiency features save $65,000 per year (the total regulated energy cost is reduced by 88%) when compared to a typical building.

(a) Lewis Center at Oberlin

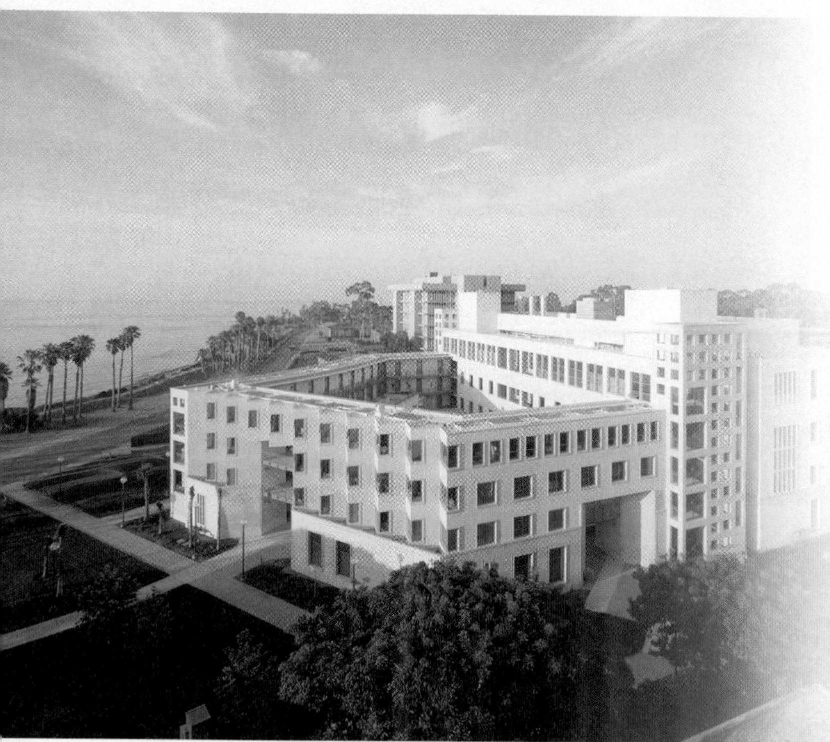

(b) Bren Hall at UC Santa Barbara

FIGURE 24.6 Green buildings are becoming numerous on American campuses. Two of the earliest and best known are **(a)** Oberlin College's Adam Joseph Lewis Center for Environmental Studies and **(b)** Bren Hall at the University of Santa Barbara, California.

One of the first green buildings on a college campus was the Adam Joseph Lewis Center for Environmental Studies at Oberlin College in Ohio (FIGURE 24.6a). This building was constructed using materials that were recycled or reused, took little energy to produce, or were locally harvested, produced, or distributed. Carpeting materials are leased and then returned to the company for recycling when they wear out. The Lewis Center contains energy-efficient lighting, heating, and appliances, and it maximizes indoor air quality with a state-of-the-art ventilation system, as well as paints, adhesives, and carpeting that emit few volatile organic compounds. The structure is powered largely by solar energy from photovoltaic (PV) panels on the roof, active solar heating, and passive solar heating from south-facing walls of glass and a tiled slate floor that acts as a thermal mass (p. 588). Over 150 sensors throughout the building monitor conditions such as temperature and air quality.

Bren Hall at the University of Santa Barbara in California (FIGURE 24.6b) uses an innovative and efficient heating and cooling system integrated with solar panels and white roofing material to reflect sunlight. The structural steel and other materials in the building are composed of 40% recycled content, and over 90% of construction waste was recycled. The building contains no formaldehyde, asbestos, or CFCs, and other toxic substances are kept to a minimum. Bren Hall conserves water with low-flow fixtures, waterless urinals, reclaimed water for toilets, and automatic sensors. Sustainable materials and approaches added only 2% to the $26 million construction cost, and this added cost is easily being recovered through energy savings.

The demand for "green buildings" is growing fast. More than one-third of schools responding to a 2008 survey had at least one LEED-certified building or retrofit, and over half planned to pursue them in the future. Butler University in Indiana even has a LEED-certified fraternity house. The University of Florida has built or started construction on 18 green buildings since 2003. Pacific Union College in California is seeking to build an entire "ecovillage" of dorms designed for sustainability. At Catawba College in North Carolina, students convinced their board of trustees to commit to construct only sustainable buildings.

Sustainable architecture doesn't stop at a building's walls. The landscaping around the Kirsch Center and around Bren Hall each features drought-tolerant native plants that require little watering. Oberlin's Lewis Center is set among orchards, gardens, and a restored wetland that helps filter wastewater, and with urban agriculture and lawns of grass specially bred to require less chemical care. Careful design of campus landscaping can create livable spaces that promote social interaction and where plantings supply shade, prevent soil erosion, create attractive settings, and provide wildlife habitat. It has been said, in fact, that groundskeepers are more vital to colleges' recruiting efforts than are vice presidents!

Water conservation is important

Conserving water is a key element of sustainable campuses—especially in arid regions, as students at the University of Arizona in Tucson know. Despite Tucson's dry desert climate, rain falls in torrents during the late-summer monsoon season, when water pours off paved surfaces and surges down riverbeds, causing erosion, carrying pollution, and flowing too swiftly to sink into the ground and recharge aquifers. So UA students sought to redirect these floodwaters and put them to use. With the help of a grant written by Dr. James Riley and student Chet Phillips, the group created an independent study course to design and implement a rainwater harvesting project

FIGURE 24.7 **University of Arizona students reengineered the landscape on parts of their campus to prevent flooding and harvest rainwater.**

on campus (FIGURE 24.7). Students surveyed sites, researched their hydrology, and worked with staff to design and engineer channels, dams, berms, and basins to slow the water down and direct it into swales, where it can nourish plants and sink in to recharge the aquifer.

Many other campuses are pursuing outdoor water conservation projects. George Mason University built a rain garden. Eastern Mennonite University constructed a stormwater management system. Emory University is building a "bluehouse," a facility like a greenhouse that treats wastewater biologically and makes the cleansed water available for irrigation, boilers, and other nonpotable uses.

Water conservation is just as important indoors. A student organization called Greeks Going Green pursues sustainable solutions for students in fraternities and sororities. In its water conservation program at the University of Washington, students installed several hundred 5-minute shower timers, low-flow showerheads, and low-flow aerators for sink faucets. Water-saving technologies such as waterless urinals and "living machines" to treat wastewater are being installed on many campuses. The University of British Columbia in Vancouver, Canada, sank $35 million into retrofitting 300 campus buildings with water- and energy-efficient upgrades that now save the school $2.6 million annually. The upgrades reduce water use by 30% each year—enough water for 12,000 homes.

Students at Reading Area Community College in Pennsylvania installed water-bottle fillers at drinking fountains in college buildings and mounted a campaign informing their peers of the environmental impacts of bottled water (pp. 399-401) and urging them to fill used bottles with tap water instead. Loyola University Chicago went one step further: After installing water refill stations (FIGURE 24.8) and mounting an information campaign on the impacts of bottled water, the Loyola student body voted in 2012 to end the sale of bottled water on their campuses—a phase-out to be completed in 2013.

At Denison University in Ohio, students staged a month-long "water wars" competition among dorms. The dorm that reduced its water use the most received a financial reward, which was donated to a community charity voted on by residents of the dorm. As more schools began such competitions, these efforts coalesced into a national program similar to RecycleMania, called the Campus Conservation Nationals (CCN). In 2013, CCN's second full year, 185 schools participated, with residence halls pitted against residence halls. At the end of the three-week period, these schools cumulatively had saved 1.68 million gallons of water—equal to 11,200 hours in the shower or 10.8 million 20-ounce water bottles. Top schools averaged up to 10% water savings.

Energy efficiency is easy to improve

In the Campus Conservation Nationals, students compete to conserve energy as well as water. Over the three weeks in 2013, the 185 participating schools saved a total of 2.1 million kilowatts of electricity and more than 1200 tons of carbon dioxide emissions—the equivalent of taking 187 U.S. homes off the grid for a year.

The CCN competition is a natural outgrowth of the many similar programs that were being independently run on campuses across North America. Students at Williams College in Massachusetts were some of the first to transform energy conservation into a kind of intramural sport. Their "Do It in the Dark" competition pitted residence halls against one another and soon produced a 13% cut in energy use as students got caught up in the fun. Connecticut College students ran a similar competition and then took 25% of the money saved by energy conservation and used it to fund a concert for the whole student body. At Oberlin College, students developed monitoring systems that display in real time the energy use in a dorm compared to energy use in other dorms. The students who invented the technology went on to found a company that markets similar devices for various uses.

Students engaged in these efforts know that there are many ways to conserve energy and that, often, simple steps can pay big dividends. Students at SUNY-Purchase in New York saved their school $86,000 per year simply by turning down hot water temperatures by 5°F (2.8°C). Students at California State University, Chico, voted to urge the university to adjust building thermostats by 3°F (1.7°C). This simple adjustment

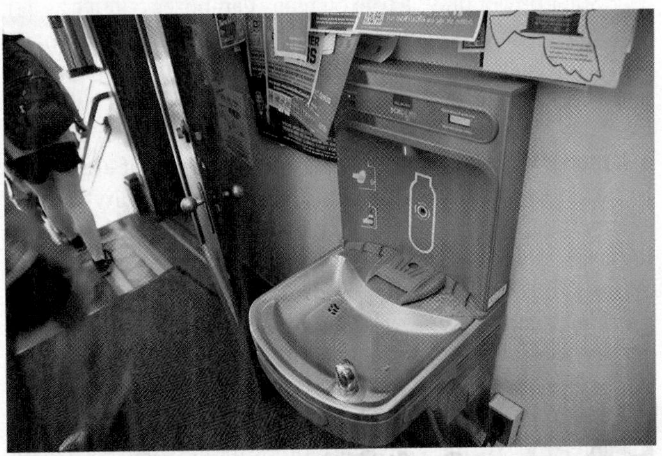

FIGURE 24.8 **Water refill stations enabled Loyola University Chicago to phase out bottled water sales on campus.**

saves the institution $151,000 each year and prevents the emission of 1100 tons of carbon dioxide. The College of Saint Benedict saved $50,000 per year and 425 tons of CO_2 emissions simply by replacing 35-watt bulbs with 25-watt bulbs. At Colorado College, students launched a four-month public awareness campaign, using "eco-reps" in dorms to give advice on saving energy, as well as e-mails, murals, sculptures, and energy-saving equipment. In those four months, the college saved $100,000 in utility costs and reduced its greenhouse gas emissions by 10%. At a number of schools, students have distributed thousands of energy-efficient compact fluorescent bulbs to their peers and to community members, in exchange for standard incandescent lightbulbs.

Campuses can harness large energy savings simply by not powering unused buildings. Central Florida Community College booked its summer classes into a minimum number of efficient buildings so that other buildings could be shut down. Central Florida students also surveyed staff and then removed lightbulbs the staff said they did not need. The college also installed motion detectors to turn off lights when people are not in rooms. At Augusta State University in Georgia, students studied lighting in the science building and concluded that installing automated sensors would save $1200 per year in electricity bills and save 6.4 tons of CO_2 emissions. At this rate, the cost of the sensors would be recouped in just four years.

Students are promoting renewable energy

Campuses can reduce fossil fuel consumption and emissions by altering the type of energy they use. Ball State University replaced its coal-fired heating plant with cleaner and more efficient fluidized-bed combustion technology. Middlebury College in Vermont switched its central plant from fuel oil to carbon-neutral wood chips, reducing emissions by 40%. Furman University in South Carolina installed ground-source heat pumps (p. 600) that should cut energy costs by $55,000 each year. Eastern Illinois University replaced its coal-fired power plant with a Renewable Energy Center at which gasification of biomass has cut carbon emissions by 80% and is saving $7 million in energy and operating costs each year.

Solar power plays a role on many campuses (**FIGURE 24.9a**). Schools in sunny climates such as De Anza College are a natural fit for solar power, including solar thermal and photovoltaic (PV) systems. These applications provide a portion of De Anza's energy, and the Foothill-De Anza District recently installed PV solar panels producing 1 megawatt of electricity. Butte College in California gets 28% of its electricity from its solar array. California State University, Long Beach, is installing a PV solar system that will decrease emissions equal to taking 50,000 cars off the road. At the University of South Florida, solar systems were built using funds from a green energy fee that 70% of the student body had voted to impose on themselves.

Solar power also works in cooler and cloudier climes (Chapter 21). At the University of Vermont, solar PV panels provide enough electricity for nine desktop computers or 95 energy-efficient lightbulbs for 10 hours each day. Appalachian State University students designed and installed a combined PV and solar thermal system. Hopkinsville Community College in Kentucky is integrating its recently installed solar array into its science curriculum and is training students to maintain the system. William Paterson University in New Jersey has begun installing what could become the largest single-campus PV solar installation in the nation, which could provide 14% of campus electricity needs and save $4.4 million over 15 years.

(a) Installing solar panels at Arizona State

(b) Wind turbine at University of Maine at Presque Isle

FIGURE 24.9 Campuses are generating renewable energy. PV solar panels **(a)** at Arizona State University generate electricity from sunlight. A wind turbine **(b)** at the University of Maine at Presque Isle saves $100,000 per year in electricity costs while reducing CO_2 emissions and serving as an educational resource.

Middlebury College and Minnesota's Macalester College were pioneers in installing wind turbines to help meet their energy needs, and today more colleges are doing so. The University of Maine at Presque Isle installed a wind turbine that will save $100,000 each year in electricity expenses with a corresponding reduction in carbon emissions (FIGURE 24.9b). St. Olaf College in Minnesota gets one-third of its electricity from its wind turbine. Massachusetts Maritime Academy is erecting a wind turbine expected to generate 25% of the school's electricity and save $300,000 per year.

Campuses that do not generate renewable energy themselves can still invest in renewable energy by purchasing "green tags," or carbon offsets (p. 513), which subsidize renewable energy sources. Western Washington University, College of the Atlantic in Maine, Georgian Court University, and other schools now offset 100% of their greenhouse gas emissions by buying green tags for renewable energy. The California State University system buys 20% of its power from renewable sources, and students in the University of California system were integral in convincing the UC regents to pass a system-wide policy back in 2003 to encourage renewable energy. Renewable power generally costs more than power from fossil fuels, but on dozens of campuses so far, students have voted to increase student fees—to tax themselves—to help fund the purchase of renewable energy. At the University of Tennessee, Knoxville, green fees brought in $1.4 million to buy green power and make energy-efficiency upgrades.

Some students even design renewable energy technology! In 2011, teams of students from 20 universities competed in the fifth Solar Decathlon. In this remarkable biennial event, teams of students travel to the National Mall in Washington, D.C., bringing material for solar-powered homes they have spent months designing. The teams erect their homes on the Mall, where they are open to the public for 10 days (FIGURE 24.10). The homes are judged on various criteria, and prizes are awarded to winners in each category. A team from the University of Maryland won the 2011 competition, where the houses were inspected by 350,000 visitors. Since 2002, a total of 112 collegiate teams have competed. The event moves to Irvine, California, in 2013, and it has inspired similar events in China and Europe.

FIGURE 24.10 **In October 2011, college and university teams converged on Washington, D.C., for the fifth Solar Decathlon.** Each team designed and erected an entire house fully powered by solar energy.

Dining services and campus farms can promote more sustainable food

For those of us who can't hit a nail with a hammer, let alone build a house, we can still make a difference three times a day—each time we eat. One way is by cutting down on waste, estimated at 25% of food that students take. Composting food scraps is an effective method of recycling waste once it is created, but trayless dining can reduce waste at its source. Coe College in Iowa is one of many schools where dining halls are eliminating trays and asking students to carry plates individually. Not having to wash trays saves water, detergent, and energy, and Coe's dining manager calculated that the trayless system prevents 200–300 pounds of food waste per week.

Campus food services also can buy organic produce, purchase food in bulk or with less packaging, and buy locally grown or produced food. At Sterling College in Vermont, many foods are organic, grown by local farmers, or produced by Vermont-based companies. Some foods are grown and breads are baked on the Sterling campus, and food shipped in is purchased in bulk. Dish soap is biodegradable, and kitchen scraps are composted along with unbleached paper products.

Some college campuses even have gardens or farms where students help to grow food that is eaten on campus (FIGURE 24.11). Students help run sizeable farms now at many dozens of campuses—all of which provide food for the dining halls and often for local communities. San Diego City College converted an expanse of lawn into a vibrant organic farm. University of Montana students tend a 10-acre farm that produces 20,000 pounds of organic produce that is given to a local food bank. At Dickinson College in Pennsylvania, a small garden originating with a class project grew to three-fourths of an acre and began supplying a local food bank. It then moved off-campus to a 20-acre site, where it now produces $28,000 worth of food for community-supported agriculture (CSA) members (p. 269), as well as for the campus food service.

George Mason University runs a CSA for its students, staff, and faculty, and it also established an apiary for honeybees. Loyola University Chicago is raising bees to produce honey on its campus as well, and Loyola students and staff are going into the community and tending community gardens to feed the homeless. They also are planting fruit- and nut-bearing trees in urban neighborhoods in Chicago to help establish homegrown food sources for working-class residents.

At Kennesaw State University in Georgia, a new "Farm-to-Campus" program is providing 20% of the produce served in the KSU Commons dining hall and plans to boost that figure to 50% by 2014. Newly acquired lands are producing organic, heirloom, and non-GM produce, and include an apiary, apple orchard, and greenhouses—all of which KSU faculty and students use as an outdoor learning lab.

(a) Winona State University

(b) Yale University

FIGURE 24.11 In campus gardens, students can grow organic produce for their dining halls. At Winona State University in Minnesota **(a)**, Dr. Bruno Borsari shows students herbs grown on campus. At Yale University **(b)**, student Claire Bucholz harvests beets from a campus farm.

Purchasing decisions wield influence

Purchasing decisions like those made in dining halls favoring local food, organic food, and biodegradable products can be applied across the entire spectrum of a campus's needs. When campus purchasing departments buy recycled paper, certified sustainable wood, energy-efficient appliances, goods with less packaging, and other ecolabeled products, they send signals to manufacturers and increase the demand for such items.

At Chatham College in Pennsylvania, students chose to honor their school's best-known alumnus, Rachel Carson (pp. 174, 369), by seeking to eliminate toxic chemicals on campus. Administrators agreed, provided that alternative products to replace the toxic ones worked just as well and were not more expensive. Students brought in the CEO of a company that produces nontoxic cleaning products, who demonstrated to the janitorial staff that his company's products were superior. The university switched to the nontoxic products, which were also cheaper, and proceeded to save $10,000 per year. Chatham students then found a company offering paint without volatile organic compounds and negotiated with it for a free paint job and discounted prices on later purchases. Students also worked with grounds staff to eliminate herbicides and fertilizers used on campus lawns and to find alternative treatments.

Transportation alternatives are many

Many campuses struggle with traffic congestion, parking shortages, commuting delays, and pollution from vehicle exhaust. Indeed, commuting to and from campuses in vehicles accounts for over half of the carbon emissions of the average college or university (FIGURE 24.12). Some institutions are addressing these issues by establishing or expanding bus and shuttle systems; introducing alternative vehicles to university fleets; and encouraging walking, carpooling, and bicycling (FIGURE 24.13).

Juniata College in Pennsylvania runs a bike-sharing program in which students can borrow bikes from a fleet. University of Texas at Austin students refurbish donated bicycles and then provide them free to new students. Passing bikes from one generation to the next reduces traffic congestion, pollution, and parking costs while promoting a healthy mode of transportation.

Humboldt State University in California grants its students free unlimited bus service. As a result, enough students shifted from cars to buses that a planned parking garage did not need to be built. At the University of Montana, students actually run a mass transit system of several buses that provide several hundred thousand free rides per year. In 1999, students took over a small van service, and it has steadily grown, fed by student demand and funded by self-imposed student fees.

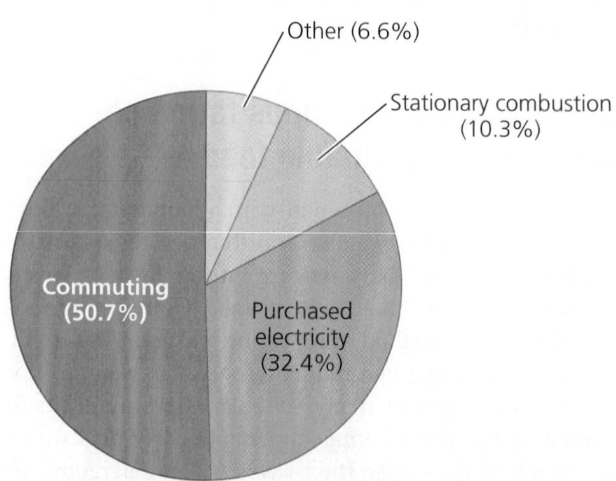

FIGURE 24.12 Commuting to and from campuses in automobiles accounts for half the greenhouse gas emissions of the average college or university. *Data, measured in CO_2-equivalents, from American College and University Presidents' Climate Commitment.*

FIGURE 24.13 Bike-sharing programs like this one at University of Rhode Island provide healthy, convenient transportation.

Alternative vehicles and alternative fuels are playing larger and larger roles. Butte College in California operates three buses that run on natural gas and 10 that run on biodiesel. These buses keep 1100 cars off campus each day. These buses were funded through a student-approved fee. At University of California–Irvine, students pushed for the bus system to be converted to biodiesel, and the 20 converted buses today save 480 tons of carbon emissions per year. The SUNY College of Environmental Science and Forestry in Syracuse acquired electric vehicles, a gas-electric hybrid car, and a delivery van that runs on compressed natural gas while also converting its buses to biodiesel.

Middlebury College students began Project Bio Bus, which has crisscrossed North America each summer in a biodiesel bus spreading the gospel of this alternative fuel. Dartmouth College students took their own Big Green Bus on the road in a similar effort. Students at Rice University, MIT, Loyola University Chicago, the University of Central Florida, and elsewhere are producing biodiesel (**FIGURE 24.14**). These students are collecting waste cooking oil from dining halls and restaurants, and brewing biodiesel for campus bus fleets.

FIGURE 24.14 Biodiesel can be produced from waste vegetable oil from campus dining halls. Here, it's used to fuel a vehicle at Dickinson College.

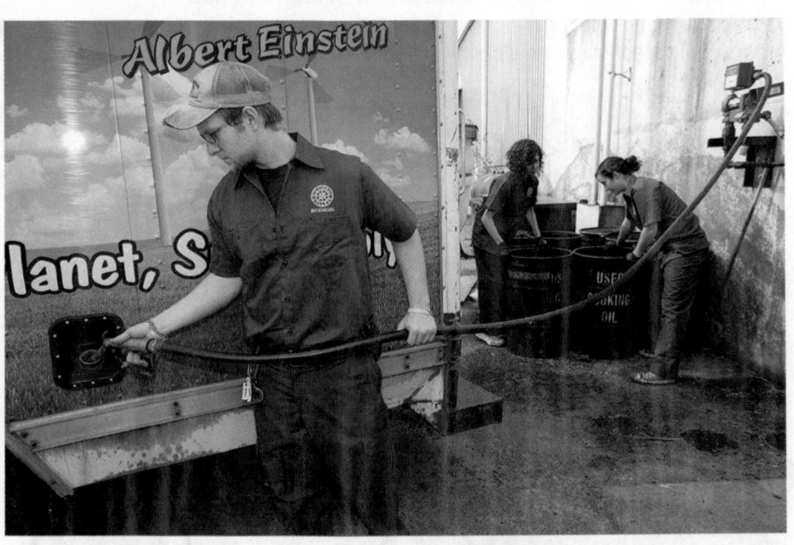

The University of Washington is a leader in transportation efforts. It provides unlimited mass transit access, discounts on bicycle equipment, rides home in emergencies, merchant discounts, subsidies for carpooling, rentals of hybrid vehicles, and more. Each day, three-quarters of the population commutes by some means other than driving alone in a car. The program has kept peak traffic below 1990 levels, despite 23% growth in the campus population.

Campuses are restoring native plants, habitats, and landscapes

No campus sustainability program would be complete without some effort to enhance the campus's natural environment. Such efforts remove invasive species, restore native plants and communities, improve habitat for wildlife, enhance soil and water quality, and create healthier, more attractive surroundings.

These efforts are diverse in their scale and methods. Seattle University in Washington landscapes its grounds with native plants and has not used pesticides since 1986. Its campus includes an ethno-botanical garden and areas for wildlife. At Warren Wilson College in North Carolina, students and the landscaping supervisor built a greenhouse, expanded an arboretum, and are propagating local grasses and wildflowers. Ohio State University and the New College of California in San Francisco have rooftop gardens. Students at Loyola University Chicago recommended modifications to glass-windowed buildings on campus after quantifying the number of migratory birds that were killed by flying into them.

At De Anza College, the Cheeseman Environmental Study Area maintains small-scale examples of entire plant communities native to California. Docents lead schoolchildren on field trips here, thousands of environmental studies students gain knowledge of California native plants, and the demonstration garden receives thousands of visitors interested in learning about native landscaping.

Some schools have embarked on ambitious projects of ecological restoration (pp. 92–93, 299, 302). The University of Central Florida manages 500 of its 1400 acres for biodiversity conservation. It is using prescribed fire (pp. 320–321) and has so far restored 100 acres for fire-dependent species while protecting the campus against out-of-control wildfires (**FIGURE 24.15**). In Washington, Cascadia Community College and the University of Washington–Bothell together restored a 58-acre wetland, dismantling levees and ditches to restore natural water flow and reintroducing native vegetation. At Northland College in Wisconsin, sustainability advocates relandscaped half their campus, replacing invasive plants with native ones and planting meadows that capture stormwater runoff and filter pollution. Students at North Hennepin Community College in Minnesota planted over 35,000 seedlings and are transforming 2.8 ha (7 acres) of lawn and channelized ponds into marsh, prairie/savanna, and native forest. Besides providing wildlife habitat, these restored areas reduce erosion and maintenance costs and provide educational opportunities.

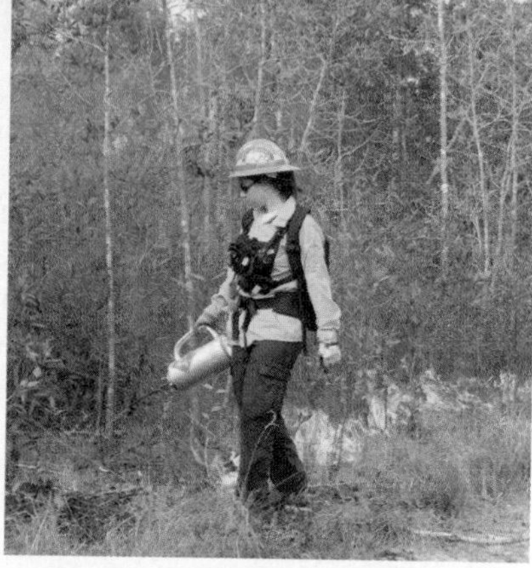

FIGURE 24.15 Some schools run ecological restoration projects to beautify their campuses, provide wildlife habitat, restore native plants, and filter water runoff. Here, staff and students at the University of Central Florida conduct prescribed burns to restore healthy pine woodland.

Carbon-neutrality is a major goal

Now that global climate change has vaulted to the forefront of society's concerns, reducing greenhouse gas emissions from fossil fuel combustion has become a top priority for campus sustainability proponents. Today many campuses are aiming to become carbon-neutral.

Students at Lewis and Clark College in Oregon began the trend in 2002. After conducting a campus audit, student leaders reduced greenhouse gas emissions by the percentage that would be required under the Kyoto Protocol (p. 510), largely by purchasing carbon offsets from a nonprofit that funds energy efficiency and revegetation projects. While U.S. leaders were citing economic expense in refusing to cut emissions at the national level, Lewis and Clark students found that becoming Kyoto-compliant on their campus cost only $10 per student per year. Similarly, Western Washington University's carbon offset program costs just $10.50 per student per year—an amount that students voted overwhelmingly to pay.

The University of Minnesota, Morris, aims to be carbon-neutral without relying on offsets. It already gets half its electricity from a wind turbine on campus and is constructing a second turbine to cover the other half. A newly built biomass gasification plant should offset nearly all campus fossil fuel use with local biomass feedstocks.

Today, student pressure and petitions at many campuses are nudging administrators and trustees to set targets for reducing greenhouse gas emissions. As of 2013, nearly 700 university presidents had signed onto the American College and University Presidents' Climate Commitment. Presidents taking this pledge commit to inventory emissions, set target dates and milestones for becoming carbon-neutral, and take immediate steps to lower emissions with short-term actions, while also integrating sustainability into the curriculum.

Education and advocacy on climate change is spilling out from the campus into the general society. On January 31, 2008, members of the public joined an estimated 240,000 students on 1900 American campuses as part of Focus the Nation, a national "teach-in" on solutions to global climate change. The next year, Focus the Nation sponsored a Nationwide Town Hall on America's Energy Future that aimed to educate the public and influence lawmakers. In 2009, 12,000 students converged on Washington, D.C., for the second Power Shift conference, networking with one another and urging legislators to promote clean renewable energy and green-collar jobs. Students were also integral in 350.org's International Day of Climate Action on October 24, 2009 (p. 515), a massive effort to urge action on climate change.

Students are demanding divestment from fossil fuel corporations

Most recently, students are leading a movement urging universities to divest from stock holdings in fossil fuel corporations. Endowment money in U.S. higher education is estimated at over $400 billion, and colleges and universities invest much of this money in the stock market, including in stocks of corporations in the coal, oil, and gas industries. Because fossil fuel combustion is the main driver of climate change, and because fossil fuel industries have consistently and effectively lobbied politicians to avoid action to reduce fossil fuel use, students are aiming to convince their campus administrations, boards of trustees, and fund managers to sell off stocks in fossil fuel companies (FIGURE 24.16).

The fossil fuel divestment movement is modeled on efforts to divest from companies doing business in apartheid-era South Africa—a movement that surged across American campuses in the 1980s. Activists at that time succeeded in focusing international attention on South Africa's system of racial discrimination, helping to overturn that system and set in motion an inspiring transition to a fuller and more just democracy.

The fossil fuel divestment movement blossomed with remarkable speed in late 2012, and by 2013, student campaigns were active at over 200 colleges and universities. In response, Unity College and Hampshire College became the first colleges to formally divest. Harvard University set up a socially responsible investment fund (p. 155) for alumni donations after 72% of the student body voted for divestment in America's first campuswide referendum on the topic. Even in its infancy, the movement has served as a model for the wider society; within

FIGURE 24.16 The fossil fuel divestment movement has grown quickly. Here, students at Tufts University urge their administration to divest stock holdings from fossil fuel corporations.

weeks, policymakers for some cities and states were talking about fossil fuel divestment in their portfolios as well.

A respected voice from South Africa hailed the new movement. Nobel Peace Prize winner Archbishop Desmond Tutu noted that the anti-apartheid movement on American campuses in the 1980s "played a key role in helping liberate South Africa. The corporations understood the logic of money even when they weren't swayed by the dictates of morality. Climate change is a deeply moral issue, too, of course. . . . Once again, we can join together as a world and put pressure where it counts."

WEIGHING THE ISSUES

SUSTAINABILITY ON YOUR CAMPUS What are students doing on your campus to promote sustainability? What changes have come about so far as a result of these efforts? Which of the many types of initiatives described above do you think could be successful on your campus? How could you become involved?

Organizations assist campus efforts

Many campus sustainability initiatives are supported by organizations such as the Association for the Advancement of Sustainability in Higher Education and the Campus Ecology program of the National Wildlife Federation. These organizations act as information clearinghouses for campus sustainability efforts. Each year the Campus Ecology program awards the most successful campus sustainability initiatives and posts hundreds of case studies on its website. In addition, national and international conferences are growing, such as the biennial Greening of the Campus conferences at Ball State University. With the assistance of these organizations and events, it is easier than ever to start sustainability efforts on your own campus. You can find links to resources for campus sustainability efforts in the Selected Sources and References section at the back of this book.

Strategies for Sustainability

The pursuit of sustainable solutions on college and university campuses parallels efforts in the world at large. As more people come to appreciate Earth's limited capacity to accommodate our rising population and consumption, they are voicing concern that we will need to modify our behaviors, institutions, and technologies if we wish to sustain our civilization and the natural environment on which it depends. In the quest for sustainability, the strategies pursued on campuses reflect those pursued in the wider society, and they also can serve as models.

Sustainable development aims to achieve a triple bottom line

As we explored sustainable development (pp.156–157), we considered the definition put forth by the United Nations: "development that meets the needs of the present without compromising the ability of future generations to meet their own needs." Sustaining human institutions in a healthy and functional condition requires sustaining ecological systems in a healthy and functional condition. This is because the contributions of biodiversity (pp. 290–294) and ecosystem goods and services (pp. 3, 116–117, 152, 290) are fundamental to human welfare. Yet sustainability means promoting social justice and economic well-being as well as environmental protection. Meeting this triple bottom line (p. 156) is the goal of modern sustainable development. It is our primary challenge for this century and likely for the rest of our species' time on Earth.

Environmental protection can enhance economic opportunity

Our society has long labored under the misconception that economic well-being and environmental protection are in conflict. In reality, the opposite is true: Our well-being depends on a healthy environment, and protecting environmental quality can improve our economic bottom line.

For individuals, businesses, and institutions, reducing resource consumption and waste often saves money, as many colleges and universities discover when they embark on sustainability initiatives. Sometimes savings accrue immediately, and other times an up-front investment brings long-term savings. For society as a whole, promoting environmental quality can enhance economic opportunity by providing new types of employment. As we transition to a more sustainable economy, some industries will decline while others spring up to take their place. As jobs in logging, mining, and manufacturing have disappeared in developed nations in recent decades, jobs have proliferated in service occupations and high-technology sectors. As we decrease our dependence on fossil fuels, green-collar jobs (p. 584) and investment opportunities are opening up in renewable energy sectors (**FIGURE 24.17**).

Moreover, people desire to live in areas that have clean air, clean water, intact forests, and parks and open space. Environmental protection increases a region's appeal, drawing residents, increasing property values, and boosting the tax

FIGURE 24.17 Green-collar jobs, such as employment as a wind power technician, increase as we progress toward a sustainable economy.

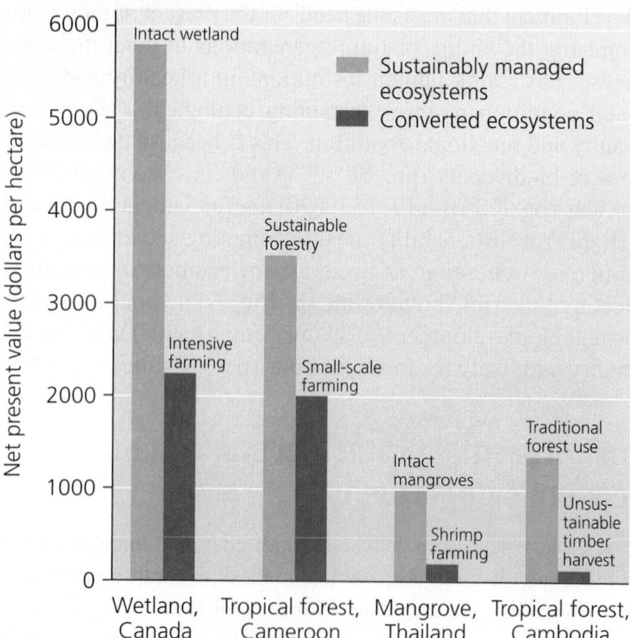

FIGURE 24.18 The economic value of sustainably managed ecosystems generally exceeds that of ecosystems converted for intensive private resource harvesting, once external costs and benefits are factored in. Shown are land values calculated by researchers in four such comparisons from sites around the world. *Adapted from Millennium Ecosystem Assessment, 2005.* Ecosystems and human well-being: biodiversity synthesis. *World Resources Institute, Washington, DC. By permission.*

DATA Q How does the value of 1 hectare of intact wetland in Canada compare to the value of 1 hectare of intensively farmed land there?

revenues that fund social services. As a result, regions that act to protect their environments are generally the ones that retain and enhance their wealth, health, and quality of life.

A recent U.S. government review concluded that the economic benefits of environmental regulations greatly exceed their economic costs (see *Calculating Ecological Footprints*, p. 186). Indeed, the U.S. and global economies have each expanded rapidly in the past 45 years, the very period during which environmental protection measures have proliferated. What's more, if we look beyond conventional economic accounting (which measures only private economic gain and loss) and include external costs and benefits (pp. 146, 165) that affect people at large, then environmental protection is recognized as still more valuable. Take several studies reviewed by the Millennium Ecosystem Assessment (p. 15): They each show how overall economic value is maximized by conserving natural resources rather than exploiting them for short-term private gain (FIGURE 24.18).

We are part of our environment

On a day-to-day basis, it is easy to feel disconnected from our natural environment, particularly in industrialized nations and large cities. We live inside houses, work in shuttered buildings, travel in enclosed vehicles, and generally know little about the plants and animals around us. Millions of urban citizens have never set foot in an undeveloped area. Just a few centuries or even decades ago, most of the world's people were able to name the plants and animals that lived nearby and describe their habits. They knew exactly where their food, water, and clothing came from. Today it seems that water comes from the faucet, clothing from the mall, and food from the grocery store. It's little wonder we have lost track of the connections that tie us to our natural environment.

However, this doesn't make those connections any less real. Consider a thoroughly un-"natural" (yet delicious!) invention of the human species: the banana split (FIGURE 24.19). Even in this triumph of human creation, seemingly concocted *de novo* at an ice cream shop, each and every element has ties to the resources of the natural environment, and our extraction or harvesting of each exerts environmental impacts.

Once we learn to consider where the things we use and value each day actually come from, it becomes easier to see how we are part of our environment. And once we reestablish this connection, it becomes readily apparent that our own interests are best served by preservation or responsible stewardship of the natural systems around us. Because what is good for the environment can also be good for people, win-win solutions are very much within reach, if we learn from what science can teach us, think creatively, and act on our ideas.

We can follow a number of strategies toward sustainable solutions

Sustainable solutions to environmental problems are numerous, and we have seen specific examples throughout this book. We can now summarize several broad strategies or approaches that may help move us toward truly sustainable solutions (TABLE 24.1).

Political engagement Sustainable solutions often require policymakers to usher them through, and policymakers respond to whoever exerts influence. Corporations and interest groups employ lobbyists to influence politicians all the time. Citizens in a democratic republic have the same power, if they choose to exercise it. You can exercise your power at the ballot box, by attending public hearings, by donating to advocacy groups that promote positions you favor, and by writing letters and making phone calls to officeholders.

TABLE 24.1 Some Major Approaches to Sustainability
• Be politically active
• Vote with our wallets
• Pursue quality of life, not just economic growth
• Limit population growth
• Encourage green technologies
• Mimic natural systems by promoting closed-loop industrial processes
• Enhance local self-sufficiency, yet embrace some aspects of globalization
• Pursue systemic solutions
• Think in the long term
• Promote research and education

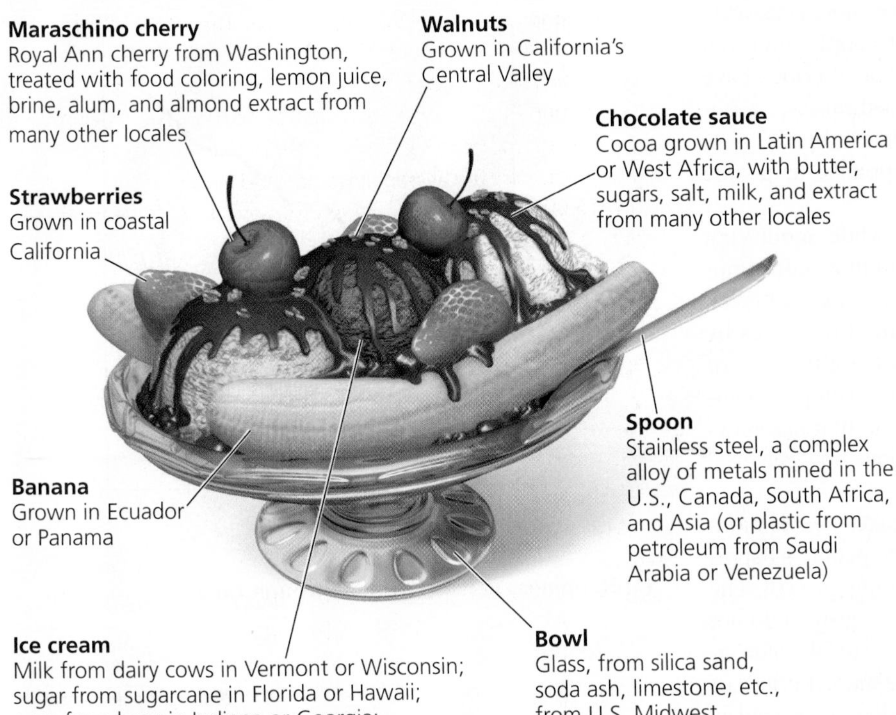

FIGURE 24.19 **A banana split eaten at an ice cream shop in Tulsa, Denver, or Des Moines consists of ingredients from around the world, whose production has impacts on the environments of many locations.** Ice cream requires milk from dairy cows that graze pastures or are raised in feedlots on grain grown in industrial monocultures. Ice cream is sweetened with sugar from sugar beet farms or sugarcane plantations. The banana was shipped thousands of miles by oil-fueled transport from a tropical country, where it grew on a plantation that displaced rainforest and where it was liberally treated with fertilizers and fungicides. Fruits and nuts grown in California's Central Valley were irrigated generously with water piped in from the Sierras. The spoon originated with metal ores mined along with thousands of tons of soil and processed into stainless steel using energy from fossil fuels.

Today's environmental and consumer protection laws came about because citizens pressured their representatives to act. The raft of legislation enacted in the 1960s and 1970s in the United States and other nations might never have come about had ordinary citizens not stepped up and demanded action. We owe it to future generations to be engaged and to act responsibly now so that they have a better world in which to live. The words of anthropologist Margaret Mead are worth repeating: "Never doubt that a small group of thoughtful, committed people can change the world. Indeed, it's the only thing that ever has."

Consumer power Each of us also wields influence through the choices we make as consumers. When products produced sustainably are ecolabeled (p. 155), consumers can "vote with their wallets" by purchasing these products. Consumer choice has helped drive sales of everything from recycled paper to organic produce to sustainable seafood.

Individuals can multiply their own influence by promoting sustainable purchasing habits at their school or workplace. We saw how purchasing power at colleges and universities has spurred sales of certified sustainable wood, organic food, energy-efficient appliances, and more. Employees in businesses and government agencies can often promote change within those institutions by voicing their preferences in purchasing decisions.

Quality of life It is conventional among economists and policymakers to speak of economic growth as an ultimate goal. Many politicians view nurturing economic expansion as their prime responsibility while in office. Yet economic growth is merely a tool with which we try to attain the real goal of maximizing human happiness.

Economic growth can result from gains in efficiency, but it has largely been driven by rising consumption of material goods and services (and thus the use of resources involved in their manufacture) (FIGURE 24.20). Our tendency to believe that more, bigger, and faster are always better is reinforced by advertisers seeking to sell us more goods more quickly. Consumption has grown tremendously, with the wealthiest nations leading the way. The United States, with less than 5% of the world's population, consumes 20% of world energy resources. U.S. homes are larger than ever, sport-utility vehicles remain popular, and many citizens have more material belongings than they know what to do with.

FIGURE 24.20 **Citizens of the United States consume more than the people of any other nation.** Unless we find ways to increase the sustainability of our manufacturing processes, our rising rate of consumption cannot be sustained in the long run.

Yet the accumulation of possessions does not necessarily bring contentment. Observing how affluent people often fail to find happiness in their material wealth, social critics have given this phenomenon a name like a dreaded disease: affluenza (p. 147). Indeed, scientific research backs up the contention that money cannot buy as much happiness as people typically believe (**FIGURE 24.21**).

We can enhance our quality of life while modifying our behavior, attitudes, and lifestyles to minimize consumption and thereby squeeze more from less. On a societal level, industry can produce goods using fewer natural resources by improving the technology of materials and the efficiency of manufacturing processes. Industry can also develop sustainable manufacturing systems—ones that are circular and recycle, in which the waste from a process becomes raw material for input into that process or others (p. 623).

Economists and policymakers can also take steps to modify our economic approaches so as to promote people's happiness and quality of life, and not simply economic growth. The data that economists use to calculate economic growth do not incorporate external costs (pp. 146, 165), those social, environmental, and personal costs not included in the market prices of goods and services. Currently, goods and services are priced as though pollution and resource extraction involved no costs to society. If, instead, we can introduce green taxes (p. 181), eliminate harmful subsidies (pp. 181–182), and make our accounting practices reflect indirect consequences to the public, then we will provide a clearer view of the full costs and benefits of a given action, decision, or product. The market could then become a truly free market and a powerful tool for improving environmental quality, our economy, and our quality of life.

Population stability Just as continued growth in consumption is not sustainable, neither is growth in the human population. We have seen (pp. 66–68) that populations may grow exponentially for a time but eventually encounter limiting factors and level off or decline. We have used technology to increase Earth's carrying capacity for our species, but our population cannot continue growing forever; sooner or later, human population growth will cease. The question is how: through war, plagues, and famine, or through voluntary means as a result of wealth and education?

The demographic transition (pp. 201–202) is already far along in many developed nations thanks to urbanization, wealth, education, and the empowerment of women. If today's developing nations also pass through the demographic transition, then there is hope that humanity may halt its population growth while creating a more prosperous and equitable society.

Green technologies It is largely technology—developed with the agricultural revolution, the industrial revolution, and advances in medicine and health—that has spurred our population increase. Technology has magnified our impacts on Earth's environmental systems, yet it can also give us ways to reduce our impact. The *IPAT* equation (p. 193), summarizes human environmental impact (*I*) as the interaction of population (*P*), consumption or affluence (*A*), and technology (*T*). Technology can exert either a positive or a negative value in this equation. The shortsighted use of technology often gets us into a mess, but wiser use of environmentally friendly, or "green," technologies can help get us out.

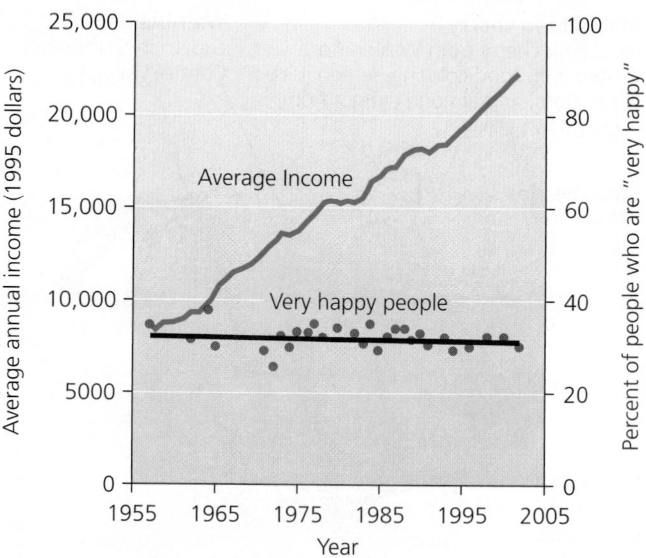

(a) Happiness versus income, through time

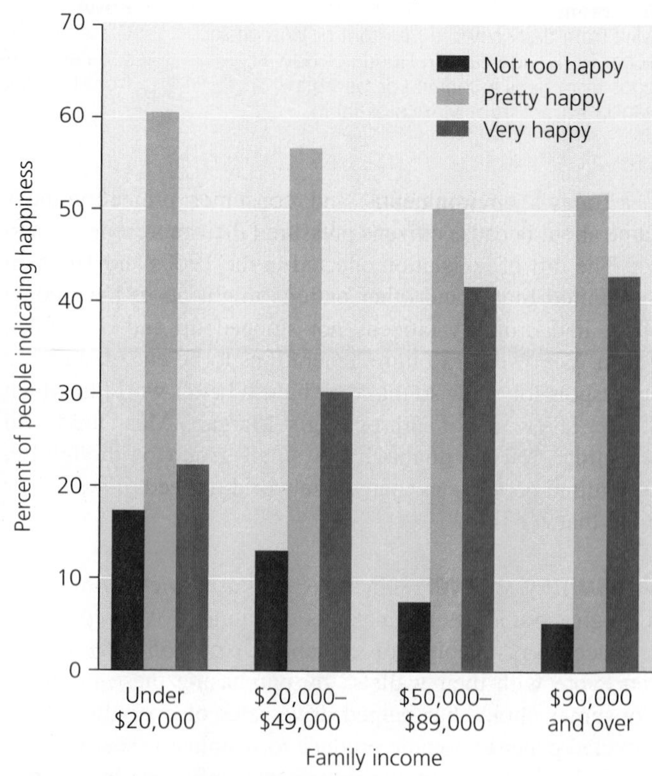

(b) Happiness versus family income

FIGURE 24.21 Money is not the only key to happiness. Although average U.S. income rose steadily in the past half-century **(a)**, the percentage of people reporting themselves as being "very happy" remained stable or declined slightly. A different study **(b)** found that Americans in higher income brackets were more likely to report themselves as "very happy" and less likely to report themselves as "not too happy." However, the difference was far less than people generally expect. For instance, when asked how much a fivefold increase in income improves a person's mood day to day, respondents guessed on average 32%, but the actual improvement, from respondents' self-reporting, was only 12%. *Data in (a) from Gardner, G., and E. Assadourian, 2004. "Rethinking the good life." Worldwatch Institute,* State of the World 2004. www.worldwatch.org. *By permission. Data in (b) from Kahneman, D., et al., 2006. Would you be happier if you were richer? A focusing illusion.* Science *312: 1908–1910.*

In recent years, we have intensified environmental impacts in developing countries by exporting industrial technologies from the developed world to poorer nations eager to industrialize. In developed nations, meanwhile, we have begun using green technologies to mitigate our impacts. Catalytic converters on cars have reduced emissions (see Figure 17.15, p. 460), as have scrubbers on smokestacks (see Figure 17.16, p. 461). Recycling technology and wastewater treatment are reducing our waste output. Solar, wind, and geothermal energy technologies are producing cleaner, renewable energy. Technological advances such as these help explain why people of the United States and western Europe today enjoy cleaner environments—although they consume far more—than people of eastern Europe or rapidly industrializing nations such as China.

Mimicking nature As industries seek to develop green technologies and sustainable practices, they have an excellent model: nature itself. Environmental systems tend to operate in cycles featuring feedback loops and the circular flow of materials. In natural systems, output is recycled into input. In contrast, human manufacturing processes have run on a linear model in which raw materials are input and processed to create products while by-products and waste are generated and discarded. Some forward-thinking industrialists are making their processes more sustainable by transforming linear pathways into circular ones, in which waste is recycled and reused (pp. 622–623). For instance, several companies now produce carpets that can be retrieved from the consumer when they wear out, and these materials are then recycled to create new carpeting (p. 623). Some automobile manufacturers are planning cars that can be disassembled and recycled into new cars. Proponents of this industrial model see little reason why virtually all products cannot be recycled, given the right technology. Their ultimate vision is to create truly closed-loop industrial processes, generating no waste.

The local and the global As our societies become more globally interconnected, we experience a diversity of impacts, positive and negative. Encouraging local self-sufficiency is one important element of building sustainable societies. When people feel closely tied to the area in which they live, they tend to value the area and seek to sustain its environment and its human community. Moreover, relying on locally made products cuts down on fossil fuel use from long-distance transportation. This argument is frequently made in relation to the cultivation and distribution of food, in encouraging local organic or sustainable agriculture (p. 269).

Many advocates of local self-sufficiency criticize globalization. They are troubled by the homogenization of the world's societies, in which a few cultures and worldviews displace many others. For instance, many of the world's languages are going extinct as large and powerful industrialized societies displace smaller traditional societies. Traditional ways of life in many areas are being abandoned as more people take up the material and cultural trappings of a few dominant cultures, particularly those of the United States.

The growing power of large multinational corporations is a key driver of these trends. In today's globalizing world, multinational corporations have attained greater and greater power over global trade while governments retain less and less. Critics of globalization consider corporations more likely than governments to promote a high-consumption lifestyle and less likely than governments to support environmental protection, so they feel that globalization will hinder progress toward sustainability.

In recent years, many people have reacted against the power of multinational corporations and the homogenizing effects of globalization. This movement coalesced in Seattle in 1999, when thousands of protestors picketed a meeting of the World Trade Organization (**FIGURE 24.22**). Since that year's "Battle in Seattle," protestors have picketed every WTO meeting.

However, globalization is a complicated phenomenon with diverse consequences. One very positive aspect is the way that people of the world's diverse cultures are increasingly communicating and learning about one another. Air travel, books, television, and the Internet have made us more aware of one another's cultures and more likely to respect and celebrate, rather than fear, differences among cultures.

Moreover, globalization may foster sustainability because Western democracy, as imperfect as it is, serves as a model and a beacon for people living under repressive governments. Open societies allow for entrepreneurship and the

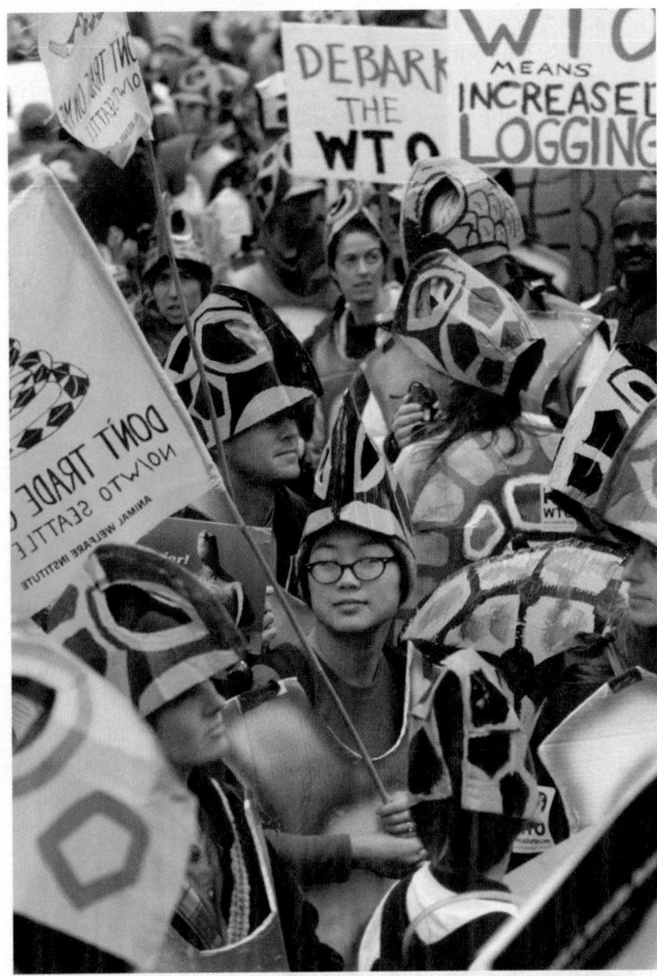

FIGURE 24.22 Globalization has inspired concern. Thousands of protesters picketed the WTO meeting in Seattle in 1999, criticizing the homogenizing effects of globalization, as well as relaxations in labor and environmental protections brought about by free trade.

flowering of creativity in business, art, science, and education. Millions of free minds thinking about issues are more likely to come up with sustainable solutions than the minds of a few holding authoritarian power.

WEIGHING THE ISSUES

GLOBALIZATION From your own experience, what advantages and disadvantages do you see in globalization? Have you personally benefited or been hurt by it in any way? In what ways might promoting local self-sufficiency be helpful for the pursuit of global sustainability? In what ways might it not?

Systemic solutions There are two general ways to respond to a problem. One is to address the symptoms of the problem, and the other is to address the root cause of the problem. All too often our society takes a symptomatic approach (addressing the symptoms as we view things one part at a time) rather than a systemic approach (viewing the sum of the parts as a whole system and addressing the root cause). Addressing symptoms one by one as they appear is often easier in the short-term, but is generally not effective in resolving the problem.

For instance, when pests evolve resistance to a pesticide (p. 255), we generally call on chemists to develop more potent pesticides. This addresses a symptom but does not resolve the overall problem, because pests will likely proceed to evolve resistance to the new chemical as well. A systemic solution would be to develop agricultural approaches that rely less heavily on chemical pesticides. Similarly, as we deplete easily accessible fossil fuel deposits, we are choosing to reach farther and deeper for new fossil fuel sources, even as environmental and health repercussions mount. A systemic solution to our energy demands would involve developing clean renewable energy sources instead. For a great many issues that face our society, it will prove worthwhile to pursue systemic solutions.

Long-term perspective Whatever solutions we pursue, we must base our decisions on long-term thinking, because to be sustainable, a solution must work in the long term. Often the best long-term solution is not the best short-term solution, which explains why much of what we currently do is not sustainable. Policymakers in democracies often act for short-term good because they aim to produce quick results that will help them be reelected. Yet many environmental dilemmas are cumulative, worsen gradually, and can be resolved only over long periods. Often the costs of addressing an environmental problem are short term, whereas the benefits are long term, giving politicians little incentive to tackle the problem. In such a situation, citizen pressure on policymakers is especially vital.

Research and education Finally, we each can magnify our influence by educating others and by serving as role models through our actions. The campus sustainability efforts at De Anza College and so many other colleges and universities accomplish both approaches. The discipline of environmental science plays a key role in providing information that people can use to make wise decisions about a wide diversity of issues. By promoting scientific research and by educating the public about environmental science, we can all assist in the pursuit of sustainable solutions.

Precious Time

By shifting our conventional patterns of thinking and following the approaches outlined above, we can bring sustainable solutions within reach. However, the natural systems we depend on are changing quickly. Human impacts continue to intensify, including deforestation, overfishing, wetland loss, resource extraction, and climate change. Our window of opportunity for turning these trends around is getting short. Even if we can visualize sustainable solutions to our many problems, how can we find the time to implement them before we do irreparable damage to our environment and our future?

We need to reach again for the moon

In 1961, U.S. President John F. Kennedy announced that within a decade the United States would be "landing a man on the moon and returning him safely to the Earth" (**FIGURE 24.23**). It was a bold and astonishing statement; the technology to achieve this unprecedented, almost unimaginable, feat did not yet exist. Kennedy's directive had powerful motivation behind it, however. The United States was dueling with the Soviet Union in the Cold War. In this competition for global hegemony, the two nations, held mutually at bay with nuclear weapons, tried to prove their mettle by other means, and the race for dominance in space became the centerpiece of the rivalry. The prospect of "losing" the space race prompted

FIGURE 24.23 Human ingenuity can be powerful when we want to meet a challenge. After all, astronauts reached the moon a mere eight years after U.S. President John F. Kennedy called on Congress to fund that space program. This provides hope that we will be able to muster the vision, resolve, and commitment needed to meet the larger challenge of living sustainably on Earth.

Kennedy's administration to set a national goal on an ambitious timeline. Congress supplied funding, NASA performed the science and engineering, and in 1969 astronauts walked on the moon.

The United States accomplished this milestone in human history by building public support for a goal and giving its scientists and engineers the wherewithal to develop technology and strategies to meet the goal. Similarly great challenges were met when the United States confronted the Great Depression, when it threw itself into World War II, and when it conducted the Marshall Plan after that war to help rebuild Western Europe.

Today humanity faces a challenge more important than any previous one—achieving sustainability. Attaining sustainability is a larger and more complex process than traveling to the moon. However, it is one to which every person on Earth can contribute; in which government, industry, and citizens can all cooperate; and toward which all nations can work together. If America was able to reach the moon in a mere eight years, then certainly humanity can begin down the road to sustainability with comparable speed. Human ingenuity is capable of it; we merely need to rally public resolve and engage our governments, institutions, and entrepreneurs in the race.

We must think of Earth as an island

We began this book with the vision of Earth as an island, and indeed that is what it is (**FIGURE 24.24**). Islands can be paradise, as Easter Island (pp. 6–7) likely was when early Polynesians first reached it. But when Europeans arrived there centuries later, they witnessed the aftermath of a civilization that had collapsed once its island's resources were depleted and its environment degraded.

It would be tragic folly to let such a fate befall our planet as a whole. By recognizing this, by seeking ways to modify our individual behaviors and our cultural institutions to encourage sustainable practices, and by employing science to help us achieve these ends, we may yet be able to live happily and sustainably on our wondrous island, Earth.

FIGURE 24.24 **This photo of Earth, taken by astronauts orbiting the moon, shows our planet as it truly is—an island in space.** Everything we know, need, love, and value comes from and resides on this small sphere, so we had best treat it well.

Conclusion

In any society facing dwindling resources and environmental degradation, there will be those who raise alarms and those who ignore them. Fortunately, in our global society today we have many thousands of scientists who study Earth's processes and resources. For this reason, we are amassing a detailed knowledge and an ever-developing understanding of our dynamic planet, what it offers us, and what impacts it can bear. The challenge for our global society today, our one-world island of humanity, is to support scientific endeavor so that we may be able to identify concerns and to distinguish false alarms from real problems. Environmental science, this study of Earth and of ourselves, offers us hope for our future.

Reviewing Objectives

You should now be able to:

Describe approaches being taken on college and university campuses to promote sustainability

- Audits produce baseline data on how much a campus consumes and pollutes. (pp. 655–656)
- Recycling and waste reduction are common campus sustainability efforts. (pp. 656–657)
- Green buildings are being constructed on a growing number of campuses. (pp. 657–658)
- There are many ways to reduce water use, and these efforts often save money. (pp. 658–659)
- Students have many feasible ways to conserve energy and to promote renewable energy sources. (pp. 659–661)
- Dining services and campus farms and gardens can help provide local food and reduce waste. (pp. 661–662)

- Colleges and universities can favor sustainable products in institutional purchasing. (p. 662)
- Campuses can use alternative fuels and vehicles and encourage bicycling, walking, and public transportation. (pp. 662–663)
- Restoration of plants, habitat, and landscapes are popular campus sustainability activities. (pp. 663–664)
- To address climate change, a current drive is to make campuses "carbon-neutral." (p. 664)
- A movement urging divestment from fossil fuel corporations is gaining momentum quickly (pp. 664–665)

Explain the concept of sustainable development

- Sustainable development entails environmental protection, economic development, and social justice. (p. 665)
- Proponents of sustainable development feel that economic development and environmental quality can enhance one another. (pp. 665–666)

Discuss how environmental protection can enhance economic well-being

- Environmental protection and green technologies and industries can create rich sources of new jobs. (pp. 665–666)
- Safeguarding environmental quality enhances a community's desirability and economic well-being. (p. 666)

Assess key approaches to designing sustainable solutions

- Ten general approaches can inspire specific sustainable solutions. (pp. 666–670)

Explain how time is limited yet human potential to solve problems is tremendous

- Time for turning around our environmental impacts is running short. (p. 670)
- The United States and other nations have met tremendous challenges before, so we have reason to hope that we will be able to attain a sustainable society. (pp. 670–671)

Testing Your Comprehension

1. In what ways are campus sustainability efforts relevant to sustainability efforts in our broader society?
2. Describe one way in which campus sustainability proponents have addressed each of the following issues: (1) recycling and waste reduction, (2) "green" building, and (3) water conservation.
3. Describe one way in which campus sustainability proponents have addressed each of the following issues: (1) energy efficiency, (2) renewable energy, and (3) global climate change.
4. Describe one way in which campus sustainability proponents have addressed each of the following issues: (1) dining services, (2) institutional purchasing, (3) trans-portation, and (4) habitat restoration.
5. What do environmental scientists mean by sustainable development?
6. Describe three ways in which environmental protection can enhance economic well-being.
7. Why are many people now living at the highest level of material prosperity in history? Is this level of consumption sustainable? How can we consume less while improving our quality of life?
8. In what ways can technology help us achieve sustainability? How do natural processes provide good models of sustainability for manufacturing? Provide examples.
9. Why do many people feel that local self-sufficiency is important? What consequences of globalization may threaten sustainability? How can open democratic societies help to promote sustainability?
10. How can thinking of Earth as an island help us in the quest for a sustainable future?

Seeking Solutions

1. What sustainability initiatives would you like to see attempted on your campus? If you were to take the lead in promoting such initiatives, how would you go about it? What obstacles would you expect to face, and how would you deal with them?
2. Choose one item or product that you enjoy, and consider how it came to be. Think of as many components of the item or product as you can, and determine how each of them was obtained or created. Now refer to Figure 24.19 as a guide. What steps were involved in creating your item's components, and where did the raw materials come from? How was your item manufactured? How was it delivered to you?

3. Do you think that we can increase our quality of life through development while also protecting the integrity of the environment? Discuss examples from your course or from other chapters of this book that illustrate possible win-win solutions. Are you familiar with any cases in your community or at your college that bear on this issue? Describe such a case, and state what lessons you would draw from it.
4. If we accept that all people depend on environmental systems for sustenance, what resources and strategies do we have to ensure that the actions of a few do not determine the outcome for all and that sustainable solutions are a common global goal?
5. **THINK IT THROUGH** You have been elected president of your college class, and your school's administrators promise to be responsive to student concerns. Many of your fellow students are asking you to promote sustainability initiatives on your campus. Consider the many approaches and activities pursued by the colleges and universities mentioned in this chapter, and now think about your own school. Which of these approaches and activities are most needed at your school? Which might be most effective? What ideas would you prioritize and promote during your term as president of your class?
6. **THINK IT THROUGH** In our final "Think It Through" question, you are . . . you! In this chapter and throughout this book, you have encountered a diversity of ideas for sustainable solutions to environmental problems. Many of these are approaches you can pursue in your own life. Name at least three ways in which you think you can make a difference—and would most like to make a difference—in helping to attain a more sustainable society. For each approach, describe one specific thing you could do today or tomorrow or next week to begin.

Calculating Ecological Footprints

Individuals can contribute to sustainable solutions for our society and our planet in many ways. Some of these involve advocating for change at high levels of government or business or academia. But plenty of others involve the countless small choices we make in how we live our lives day to day. Where we live, what we buy, how we travel—these types of choices we each make as citizens, consumers, and human beings determine how we affect the environment and the people around us. As you know, such personal choices are summarized (crudely, but usefully) in an ecological footprint.

Turn back to the *Calculating Ecological Footprints* exercise in Chapter 1 (p. 20), and recover the numerical value of your own personal ecological footprint that you calculated at the beginning of your course. Enter it in the table. Now return to the same online ecological footprint calculator that you used at that time. For many of you, this will have been www.myfootprint.org or http://www.footprintnetwork.org/en/index.php/GFN/page/personal_footprint. Take the footprint quiz again, and calculate your current footprint.

	Footprint value (hectares per person)
World average	2.7
U.S. average	9.4
Your footprint from Chapter 1	
Your footprint now	
Your footprint with three more changes	

1. Enter your current footprint, as determined by the online calculator, in the table. If you calculated your footprint at the beginning of the course, how does this value compare? By what percentage did your footprint decrease or increase? If it changed, why do you think it changed? What changes have you made in your lifestyle since beginning this course that influence your environmental impact?
2. How does your personal footprint compare to the average footprint of a U.S. resident? How does it compare to that of the average person in the world? What do you think would be an admirable yet realistic goal for you to set as a target value for your own footprint?
3. Now think of three changes in your lifestyle that would lower your footprint. These should be changes that you would like to make and that you believe you could reasonably make. Take the footprint quiz again, incorporating these three changes. Enter the resulting footprint in the table.
4. Now set as a goal reducing your footprint by 25% and experiment by changing various answers in your footprint quiz. What changes would allow you to attain a 25% reduction in your footprint? What changes would be needed to reduce your footprint to the hypothetical target value you set in Question 2?

STUDENTS

Go to **MasteringEnvironmentalScience** for assignments, the etext, and the Study Area with practice tests, videos, current events, and activities.

INSTRUCTORS

Go to **MasteringEnvironmentalScience** for automatically graded activities, current events, videos, and reading questions that you can assign to your students, plus Instructor Resources.

Answers to Data Analysis Questions

APPENDIX A

Chapter 1

Fig. 1.4 The global ecological footprint today is roughly 1.5 planet Earths. The global ecological footprint half a century ago (1961) was roughly 0.75 planet Earths. This makes for a difference of 0.75 planet Earths, and it means that today's footprint is twice the size of the footprint half a century ago.

Fig. 1.12 Of the nations shown in the figure, the United States has the largest per capita footprint, and Afghanistan has the smallest per capita footprint. The U.S. footprint (7.2 ha) is 14.4 times larger than Afghanistan's footprint (0.5 ha), because 7.2 divided by 0.5 equals 14.4.

Fig. 1.13 Answers will vary. Red and orange areas experience more-than-average impact, green areas experience less-than-average impact, and yellow areas experience average impact.

Chapter 2

Fig. 1 (SBS1) Drawing a straight line through the clusters of points from 1960 to 2010 produces a downward-sloping line that shows a decreasing trend in ocean radioactivity from cesium-137 over time. (This is because input of radioactive cesium into the ocean was reduced once nations stopped conducting nuclear weapons tests in the atmosphere.) Inputs of cesium-137 to the oceans from the Chernobyl accident did not significantly alter this downward trend, as radioactivity levels continued to decline following this "pulse" of increased radioactivity. Similar dilution of radioactivity over time of the radioactive material released from the Fukushima Daiichi accident would therefore be expected, leading to continued declines in radioactivity from cesium-137 in coming decades.

Fig. 2.20 Comparison of the two figures reveals that this belt of intense earthquake and volcanic activity corresponds closely to the subduction zones at the boundaries of the tectonic plates that surround the Pacific Ocean. As shown in Figure 2.16, convergent plate boundaries dominate the length of the ring of fire. Note that other locations that experience earthquakes and volcanic activity—such as Indonesia, Iran, Turkey, and southern Italy—are similarly located along convergent plate boundaries.

Chapter 3

Fig. 3.15 A human being has the highest rate of survival at a young age. As shown by the type I survivorship curve, the vast majority of individuals survive during youth, and then mortality rates rise during older age.

Fig. 3.18 Because exponential growth cannot last forever, we would expect that growth of the western U.S. population of Eurasian collared doves will eventually slow down and that the population will reach carrying capacity. Thus the population growth graph for the western United States would come to have a shape more like the current Florida graph, showing logistic growth.

Chapter 4

Fig. 4.10 In the generalized example shown, there are 10 rodents for every 100 grasshoppers; therefore, we can say there are 1/10 (or 1:10) as many rodents as grasshoppers. Thus, if there were 3000 grasshoppers, we would expect 300 rodents (300/3000, a 1:10 ratio).

Fig. 2 (SBS2) The mudflow substrate gained the highest species richness, as seen by the fact that its curve is highest on the graph in part (a). The maximum number of species it had in any one year was 22 or 23, about 17–18 years after the eruption. The pumice substrate showed the slowest increase in percent plant cover, as seen by the fact that its curve is the lowest in the graph in part (b).

Fig. 4.20 Note that the temperature curve is above the precipitation curve between July and September. High temperatures lead to increased evaporation. Thus, even though precipitation is roughly average at this time, the warm temperatures cause increased evaporation and thus dry conditions.

Chapter 5

Fig. 5.6 The width of the arrows in each figure represents magnitude, so the wider the arrow, the larger the value. For chemical energy, the widest arrow goes from producers to detritus. This means the largest directional flux of chemical energy in the ecosystem is from producers to detritus. For nutrients, a comparison of arrow widths shows the flux from detritus to producers to be, by far, the largest.

Fig. 5.21 Reducing nitrogen through enhanced nutrient management programs costs $21.90 per pound, whereas a 1-pound reduction from forest buffers costs only $3.10. Dividing $21.90 per pound by $3.10 per pound shows us that we could keep about 7 pounds of nitrogen out of waterways with forested buffers for the same price as a 1-pound reduction from using nutrient management programs.

Chapter 6

Fig. 6.13 If resources were being used more intensively, then they would be depleted more quickly, so the resource (green) line would drop more steeply. Faster resource consumption would likely lead to faster generation of food, industrial output, and population, so these three lines (orange, red, and dark blue) would rise more steeply at first, but then fall sooner and more steeply as resources became depleted. The line for pollution (purple) would be expected to rise sooner and more steeply.

Fig. 6.18 The average U.S. citizen has just slightly fewer happy life years, yet a far greater ecological footprint, than the average Costa Rican citizen. In fact, the U.S. ecological footprint is so high that it drags down its Happy Planet Index score sharply, despite the general happiness of U.S. citizens. As a result, the United States ranks 105th out of 151 nations in terms of this indicator.

Chapter 7

Fig. 2 (SBS1) Based on the data in the graph, we can predict that water in the well 5000 m from a drilling site would be very unlikely to contain methane, because all the existing data points in the vicinity of 5000 m are near zero. For the well just 250 m from the drilling site, existing values at that distance range from zero to 65 mg/L, with about half below the action level for hazard mitigation and half above the action level. Thus, your best prediction would be a roughly 50% likelihood that the new well would contain water above the action level for hazard mitigation.

Fig. 7.14 Once direct spending and tax breaks are added together to get total subsidies, fossil fuels received about $72 billion, while renewable energy received about $13 billion in the United States during the period covered by the graph. If we divide 72 by 13, we get 5.54. Thus, about $5.50 is spent on fossil fuel subsidies for every $1.00 that goes to renewable energy.

Chapter 8

Fig. 8.4 By examining the key that links colors on the map to growth rates, we see that red indicates the highest growth rates, and that Africa has the highest overall growth rate of any region. Within Africa, nations in southern Africa have the slowest growth, followed by nations in northern Africa, where rates of increase are slightly higher. Annual population growth rates are highest in central Africa and typically are 2.25% or higher.

Fig. 8.19 The best approach to answer a question such as this one is to draw a "best-fit" line through the points on the figure that minimizes the distance between each point and the line you draw. Doing this yields a line that slopes downward from left to right, suggesting a negative relationship between total fertility rate and rate of enrollment of girls in secondary school. This relationship makes sense— one would expect that as more girls pursue education, they delay childbirth and reduce the nation's TFR.

Chapter 9

Fig. 9.6 Following the white lines in the diagram inward from these three values to where they intersect brings us to a point within the light turquoise area in the lower right portion of the triangle, within the region for "silt loam." The soil in question is therefore silt loam.

Fig. 9.17b Answers will vary. Consultation of the color key will reveal percentage of no-till farming for each region.

Fig. 2 (SBS1) Conventionally plowed plots show slightly more carbon content than no-till plots at 40 cm of depth. We know this because the data point for conventionally plowed plots is further to the right, closer to the vertical dashed line representing the carbon content of natural (uncropped) soils. Thus the conventionally plowed soils at this depth have retained more carbon than the no-till soils—a result opposite to that seen in the shallowest soil depths. In reality, however, the values are so close together, and the error bars show such substantial variation, that these two values may not actually differ with statistical significance.

Fig. 9.20 In the diagram, arrows point from causes to consequences. Thus, according to the diagram, the immediate cause of exposure of bare topsoil is the removal of native grass. Likewise, the immediate consequence of exposing bare topsoil is wind and water erosion. Four arrows lead away from wind and water erosion to other items, so therefore wind and water erosion has four immediate consequences.

Chapter 10

Fig. 10.2 Overall growth of the human population provides the answer. Since the period 2007–2009, our global population has increased by several hundred million people, with the vast majority of this growth occurring in developing nations. Thus, although the absolute number of undernourished people in the developing world has remained the same (852 million), many people have been added to the total population, so the percentage of people who are undernourished has fallen.

Fig. 10.10 Beef requires 17.5 times more land to produce than chicken (245.0 m^2/14.0 m^2 = 17.5). Beef also requires 15 times more water to produce than chicken (750 kg/50 kg = 15).

Fig. 10.25 In the last 5 years, GM crops have been growing faster in developing nations, as indicated by the fact that the developing nations' curve rises upward more steeply than the curve for industrialized nations. In 2012, developing nations for the first time produced more GM crops than industrialized nations. If current trends continue, developing nations should be growing more GM crops than industrialized nations in 2015. To estimate how much more GM crops they might be growing in 2015, we can extend the two trend lines forward 3 years, keeping their slopes the same as in the previous several years. Doing this leads *approximately* to values of 110 million hectares for developing nations and 90 million ha for industrialized nations. The spread between these two values might be even greater if, as the data suggest, developing nations are speeding up their rate of adoption and industrialized nations are slowing down theirs.

Chapter 11

Fig. 11.4 The right portion of the figure shows that there are 4680 species of mammals. The left portion of the figure shows that there are about 1,750,000 known and described species of organisms in total. Because 4680 is 0.27% of 1,750,000, this means that mammals comprise just 0.27% of all organisms (or about 1 out of 400). In reality, the percentage is actually much lower than this, because virtually all mammal species have already been discovered, yet most species of other types of organisms have not yet been discovered.

Fig. 11.15b The bar for pollution stretches to a value of nearly 1200 species, second only to habitat loss; this indicates that pollution is the second greatest cause of amphibian declines overall. For threatened species alone, we need to look at the red portions of the bars. Comparing red portions of the bars, we can see that habitat loss is the primary cause of declines for threatened species of amphibians.

Fig. 2 (SBS2) In the pie chart for the markets, note that the proportions of the genetic types are intermediate between those of each of the two geographic areas. The amounts of red, orange, pink, and yellow coloration in the market pie chart are midway between those of the other two pie charts. This indicates that the market pie chart likely consists of a combination of individuals from the other two areas.

Chapter 12

Fig. 12.19 According to the data in the graph, each year a little more than 4 million hectares of primary forest is lost, while about 5 million hectares in plantations is gained.

Fig. 2 (SBS2) According to the data in the graph, the forest plot held about 55 bird species in the 4 years it was censused before its fragmentation in 1984. After the plot became a fragment, the average number of bird species dropped to about 20 species.

Fig. 3 (SBS2) According to the data in the graph, a tree 275 meters in from the edge of a forest fragment would be susceptible to elevated tree mortality (an edge effect that extends 300 m in) and increased wind disturbance (which extends 400 m in).

Chapter 13

Fig. 13.1 The dashed red line (which projects the urban population) in less-developed regions surpasses the dashed blue line (which projects the rural population) in less-developed regions between the year 2010 and the year 2020.

Fig. 13.11b Roadway costs and parking costs are created by automobile traffic but not by rail traffic. Note that the bars for automobile traffic in part (b) of the figure contain yellow and orange portions, as well as the red portions shared by rail traffic. These roadway costs and parking costs make automobile traffic more costly overall than rail traffic.

Fig. 2 (SBS) Among the four data points for years for areas labeled "urban" (with 40% impervious surface), three of them lie above the level at which there is chronic toxicity to freshwater life.

Chapter 14

Fig. 2 (SBS1) Begin by connecting the data points in the figure to form a dose-response curve like that shown in Figure 14.16c. Mark the spot on the curve directly above 70 ng/g on the X-axis. Then reference the Y-axis at the site of your mark, which indicates that an estimated 9% of the mice would likely suffer chromosomal effects at that dose.

Fig. 14.4 Consulting the figure, note that about 28% of Americans were obese and 17% were smokers in 2011. For the calculations in part (a), multiply 0.28 (28%) by 312 million people and find that roughly 87,360,000 Americans were obese in 2011. Similarly, for part (b), multiply 0.17 (17%) by 312 million people, and find that about 53,040,000 Americans in 2011 were smokers.

Chapter 15

Fig. 15.2 Consulting the figure, note that 2.5% of the water on Earth is fresh water and that 1% of all fresh water is surface water. Within this surface water, 52% is found in lakes. To determine the percentage of Earth's water found in freshwater lakes, multiply 2.5% (.025) by 1% (0.01) and by 52% (0.52). Multiplying these three values (and then multiplying the answer by 100 to convert it to a percentage) reveals that although freshwater lakes (such as the Great Lakes) seem massive, all of the world's freshwater lakes *combined* contain only 0.013% of Earth's water.

Fig. 1b (SBS 2) To interpret figures with two *y*-axis values, such as this one, carefully note which axis corresponds with which value on the graph. In this figure, the left *y*-axis represents the area of the hypoxic zone (the bars on the figure), and the right axis represents nitrogen flux (the line on the figure). To determine the year with the largest hypoxic zone, look for the highest bar (2002), and then consult the left *y*-axis at its height. You'll see that the hypoxic zone that year was around 22,000 km^2. To determine the year with the largest nitrogen flux, find the highest point on the line on the figure (1993), and follow its value to the right *y*-axis. You'll see that roughly 210,000 of metric tons of nitrogen entered the northern Gulf that year.

Chapter 16

Fig. 16.4 Multiplying 1000 g by 3.5% (0.035) reveals that there are 35 g of salts in the 1000 g sample of seawater in the beaker. To determine the grams of negatively-charged ions in the sample, sum the values for chloride (1.9%), sulfate (0.3%), and bicarbonate (0.01%) and find that 2.21% of the sample (or 22.1 g) is from such ions.

Fig. 16.6 Released off the southeast coast of Japan, the buoy would be carried northeast by the Kuroshio Current and then eastward across the ocean on the North Pacific Current. Upon reaching North America, it could turn southward on the California Current and float by the western coast of the United States, passing Washington, Oregon, and California. Alternatively, the buoy might follow the Alaska Current northward upon reaching North America and pass Alaska, then return to Japan. So although Japan is closer to Australia, ocean currents would carry the buoy to the United States first.

Chapter 17

Fig. 17.14b Population has increased by 52% since 1970. Emissions have decreased by 68%. Thus, emissions per person have decreased by nearly five times. (Imagine a population rising from 100 to 152, and emissions dropping from 100 to 32. 32/152 = 0.21, or one-fifth of the original 1-unit-per-person rate.)

Fig.17.19 Answers will vary. But for example, a person living in Los Angeles would face a risk of 50–75 in 1 million of developing cancer due to toxic air pollution.

Fig.17.23 According to the data in the graph, in the 1970s ozone levels averaged a bit above 0.30 ppm. In the most recent years, ozone levels have been about 0.14 ppm. This represents about a 55% reduction.

Fig. 17.31 Answers will vary, but in virtually all locations, precipitation has become less acidic. For example, in many parts of the northeastern United States, pH has increased from about 4.3 to about 4.9.

Chapter 18

Fig. 18.2 Since 1750 the atmospheric carbon dioxide concentration has increased from about 280 ppm to about 396 ppm—a 41% increase.

Fig. 18.3 Changing land use accounts for 6 metric tons of carbon dioxide emissions per year, and industry emits 26 metric tons of carbon dioxide annually. Therefore, 26/6 = 4.33. This means that for every 1 metric ton released by changing land use, 4.33 tons are released by industry.

Fig. 18.9 Answers will vary. In most regions temperature rose. In some areas of the Southeast it was stable or fell slightly.

Chapter 19

Fig. 19.2 Answers will vary. One should take the value at the far right end of the data line for oil (4.06 billion tons in 2011) and divide it by the value of that line in the year one was born. For a person born in 1995, when oil consumption was about 3.25 billion tons per year, the percent change by 2011 would be 4.06 / 3.25 = 1.25, or roughly a 25% increase.

Fig. 19.4 According to the data in the text boxes, the average U.S. citizen uses 7.28 tons per year, whereas the average person in the world uses 1.76 tons per year. Dividing 7.28 by 1.76, we see that the average U.S. citizen uses 4.14 times more energy than the average person in the world.

Fig. 19.15 Answers will vary. One should take the value at the far right end of the black ("Total") data line and divide it by the value of that line in the year one's mother or father was born. For example, if one's parent was born in 1965, when emissions were about 3.0 billion tons per year, then the percent change by 2011 would be about 9.47 / 3.0 = 3.16, or roughly a 216% increase.

Fig. 19.20 The United States produces 7.8 million barrels of oil per day and consumes 18.8 million barrels per day, as indicated by the graph's orange and red bars, respectively. Therefore, for every barrel that is produced, 18.8 / 7.8 = 2.4 barrels are consumed.

Chapter 20

Fig. 20.17 In 2012, 22.48 billion gallons of ethanol were produced across the world (red line in graph), of which the United States produced 13.30 billion gallons (blue line in graph). Thus, U.S. production is 13.30/22.48 = 0.59, making up 59% of the world total.

Fig. 20.21 Biodiesel reduces sulfates most effectively relative to petrodiesel. For both B20 and B100, the percent reduction for sulfates is greater than for any other pollutant shown.

Chapter 21

Fig 21.2 The yearly growth rate of PV solar is 60%, so you would multiply 10 units by 1.60, then multiply that number by 1.60, and so on, for 5 years. The result is 104.9 units.

Fig 21.11 Match the colors in the key with the colors in the maps for these two regions. On average, southern Arizona receives roughly 2400–2600 kilowatt-hours of solar radiation per square meter per year, whereas most of Germany receives fewer than 1200 kilowatt-hours /m^2/yr. Thus, southern Arizona receives more than twice as much sunlight as does Germany.

Fig 21.16 Answers will vary, but regions that appear underutilized for wind power include (on land) South Dakota, Nebraska, Montana, and New Mexico; and (offshore) the entire Atlantic and Gulf coasts.

Chapter 22

Fig. 22.3 Between 1990 and 2010, the population size of the United States has grown at a rate that roughly matches the rate of growth in total waste generation. We can infer this from the data in the graph alone because the per capita waste generation rate has remained fairly steady, even though total generation has risen.

Fig. 22.5 The amount of solid waste that is combusted (incinerated) is shown in green. To determine increase or decrease, note whether the green band widens or narrows over time. (This is separate from the overall height of the graph's data, which reflects cumulative, summed, totals of the four categories each year.) Examining the green band alone, we see that the amount of solid waste that is combusted (incinerated) decreased from 1960 until about 1985, and then increased until about 2002. It then decreased slightly, and as of 2010, the amount was roughly equal to the amount back in 1960.

Fig. 22.12 Michigan's recycling rate is nearly 100%, whereas states without bottle bills have a 20% rate. Thus, Michigan's rate is nearly 5 times higher.

Chapter 23

Fig. 23.12 At present rates of consumption, molybdenum has technically recoverable reserves that would last around 77 years, of which about 44 years of reserves are economically recoverable. Dividing 44 years of economically recoverable reserves by 77 years of technically recoverable reserves yields the value 0.57, indicating that 57% (multiply 0.57 by 100 to present the value as a percentage) of molybdenum reserves are economically recoverable. Nickel (51%) and cobalt (49%) come in a close second and third when similar calculations are performed.

The metal with the lowest percentage of economically recoverable reserves is lead. Lead's 17 years of economically recoverable reserves divided by its 385 years of technically recoverable reserves finds that only 4% of the world's known lead reserves are currently economically recoverable.

Chapter 24

Fig. 24.18 A hectare of intact wetland in Canada is worth about $5800 (once external costs and benefits are considered), whereas a hectare of intensive farmland is worth about $2250. Thus, a hectare of wetland is worth about $3550 more in absolute terms, or about 58% more in percentage terms.

How to Interpret Graphs

APPENDIX B

Presenting data in ways that help make trends and patterns visually apparent is a vital element of science. For scientists, businesspeople, policymakers, and others, the primary tool for expressing patterns in data is the graph. Thus, the ability to interpret graphs is a skill that you will want to cultivate. This appendix guides you in how to read graphs, introduces a few vital conceptual points, and surveys the most common types of graphs, giving rationales for their use.

Navigating a Graph

A graph is a diagram that shows relationships among *variables*, which are factors that can change in value. The most common types of graphs relate values of a *dependent variable* to those of an *independent variable*. As explained in Chapter 1 (p. 11), a dependent variable is so named because its values "depend on" the values of an independent variable. In other words, as the values of an independent variable change, the values of the dependent variable change in response. In a manipulative experiment (p. 12), changes that a researcher specifies in the value of the independent variable *cause* changes in the value of the dependent variable. In observational studies, there may be no causal relationship, and scientists may plot a correlation (p. 12). In a positive correlation, values of one variable go up or down along with values of another. In a negative correlation, values of one variable go up when values of the other go down. Whether we are graphing a correlation or a causal relationship, the values of the independent variable are known or specified by the researcher, whereas the values of the dependent variable are unknown until the research has taken place. The values of the dependent variable are what we are interested in observing or measuring.

By convention, independent variables are generally represented on the horizontal axis, or *x-axis*, of a graph, while dependent variables are represented on the vertical axis, or *y-axis*. Numerical values of variables generally become larger as one proceeds rightward on the *x-axis* or upward on the *y-axis*. Note that the tick marks along the axes must be uniformly spaced so that when the data are plotted, the graph gives an accurate visual representation of the scale of quantitative change in the data.

In many cases, independent variables are not numbers, but categories. For example, in a graph that presents population sizes of several nations, the nations would comprise a categorical independent variable, whereas population size would be a numerical dependent variable.

As a simple example, **FIGURE B.1** shows data from a classic early lab experiment that measured population growth among yeast cells. The *x*-axis shows values of the independent variable, which in this case was time, expressed in units of hours. The researcher was interested in how many yeast cells would propagate over time, so the dependent variable, presented on the *y*-axis, is the number of yeast cells present. For each hour at which data were measured during the experiment, a data point on the graph is plotted to show the number of yeast cells present. In this particular graph, a line (red curve) was then drawn through the actual data points (orange dots), showing how closely the empirical data matched the logistic growth curve (p. 67), a theoretical phenomenon of importance in ecology.

Now that you're familiar with the basic building blocks of a graph, let's survey the most common types of graphs you'll see, and examine a few vital concepts in graphing.

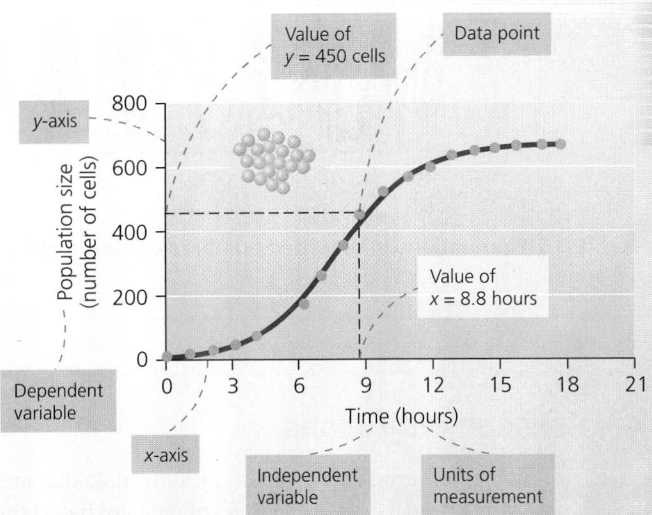

FIGURE B.1 Logistic population growth, demonstrated by the growth of yeast cells over time. (Figure 3.19a, p. 69)

Mastering ENVIRONMENTALSCIENCE™

Once you've explored this appendix, take advantage of the graphing resources at **www.masteringenvironmentalscience.com**. The GRAPHit! tutorials allow you to plot your own data. The **Interpreting Graphs and Data** exercises guide you through critical-thinking questions on graphed data from recent research in environmental science. The **Data Analysis Questions** help you hone your skills in reading graphed data. All these features will help you expand your comprehension and use of graphs.

Graph Type: Line Graph

A line graph is used when a data set involves a sequence of some kind, such as a series of values that occur one by one and change through time or across distance. In a line graph, a line runs from one data point to the next. Line graphs are most appropriate when the *y*-axis expresses a continuous numerical variable, and the *x*-axis expresses either continuous numerical data or discrete sequential categories (such as years). **FIGURE B.2** shows values for the size of the ozone hole over Antarctica in recent years. Note how the data show that the size of the hole increases until 1987, when the Montreal Protocol (p. 472) came into force, and then begins to stabilize afterwards.

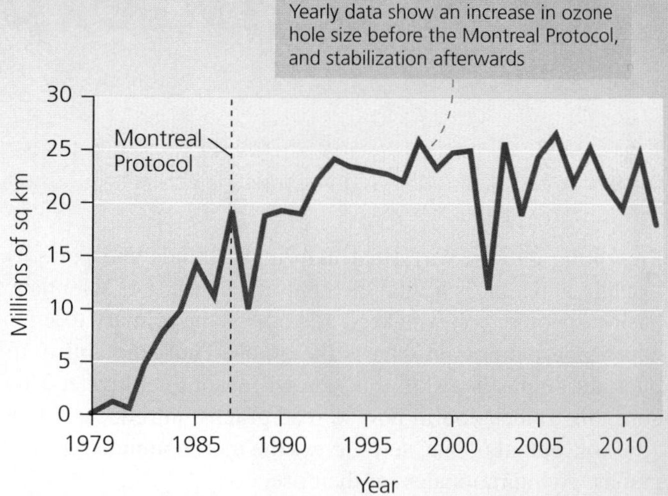

FIGURE B.2 Size of the Antarctic ozone hole before and after a treaty that was designed to address it. (Figure 17.27, p. 472)

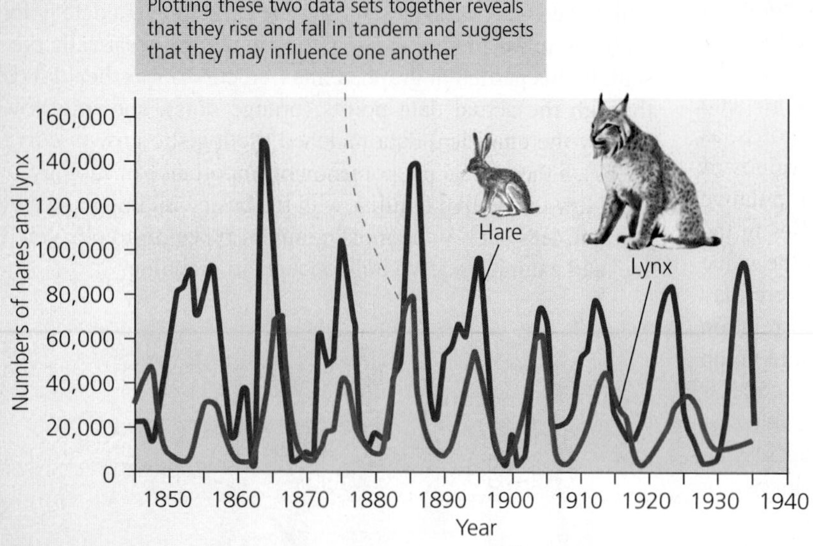

FIGURE B.3 Fluctuations in recorded numbers of hare and lynx in Canada. (Figure 4.4, p. 78)

One useful technique is to plot two or more data sets together on the same graph. This allows us to compare trends in the data sets to see whether they may be related, and if so, the nature of that relationship. In **FIGURE B.3**, recorded numbers of a predator species rise and fall immediately following those of its prey, suggesting a possible connection.

Key Concept: Projections

Besides showing observed data, graphs can show data that are predicted for the future. Such *projections* of data are based on models, simulations, or extrapolations from past data, but they are only as good as the information that goes into them—and future trends may not hold if conditions change in unforeseen ways. Thus, in this textbook, projected future data are shown with dashed lines, as in **FIGURE B.4**, to indicate that they are less certain than data that have already been observed. Be careful when interpreting graphs in the popular media and on the Internet, however; often newspapers, magazines, web sites, and advertisements will show projected future data in the same way as known past data!

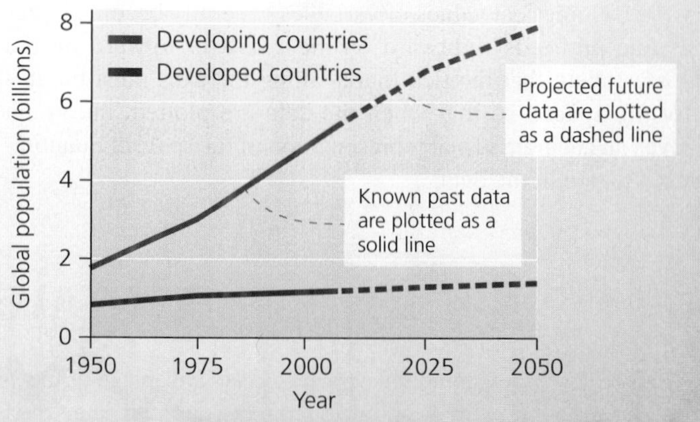

FIGURE B.4 Past and projected population growth for developing and developed countries. (Figure 8.21, p. 207)

Graph Type: Bar Chart

A bar chart is most often used when one variable is a category and the other is a number. In such a chart, the height (or length) of each bar represents the numerical value of a given category. Higher or longer bars mean larger values. In **FIGURE B.5**, the bar for the category "Automobile" is higher than that for "Light rail," indicating that automobiles use more energy per passenger-mile (the numerical variable on the *y*-axis) than light rail systems do.

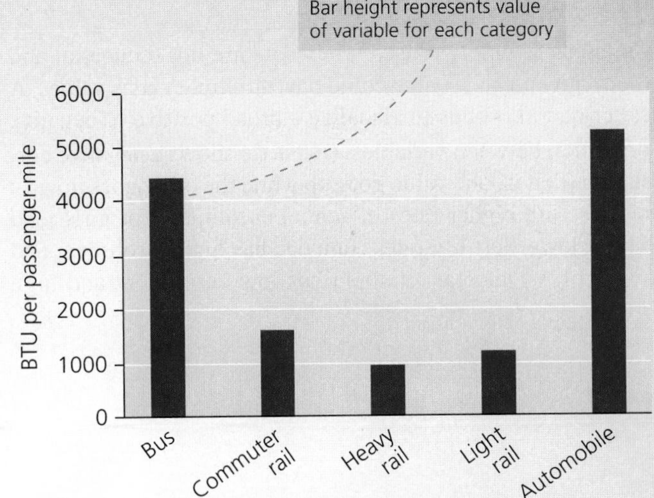

FIGURE B.5 Energy consumption for different modes of transit. (Figure 13.11a, p. 346)

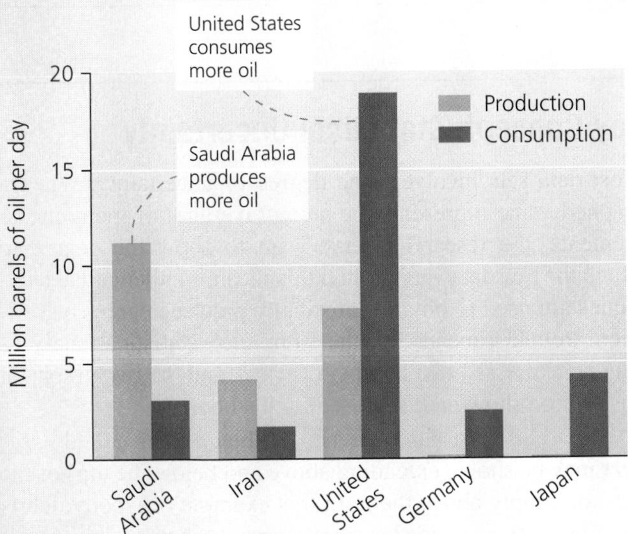

FIGURE B.6 Oil production and consumption by selected nations. (Figure 19.20, p. 544)

As we saw with line graphs, it is often instructive to graph two or more data sets together to reveal patterns and relationships. A bar chart such as **FIGURE B.6** lets us compare two data sets (oil production and oil consumption) both within and among nations. A graph that does double duty in this way allows for higher-level analysis (in this case, suggesting which nations depend on others for petroleum imports). Most bar charts in this book illustrate multiple types of information at once in this manner.

Graph Type: Pie Chart

A pie chart is used when we wish to compare the numerical proportions of some whole that are taken up by each of several categories. Each category is represented visually like a slice from a pie, with the size of the slice reflecting the percentage of the whole that is taken up by that category. For example, **FIGURE B.7** shows the percentages of genetically modified crops worldwide that are soybeans, corn, cotton, and canola.

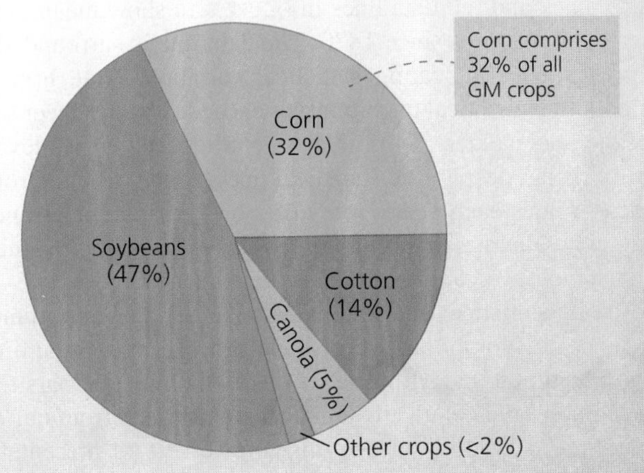

FIGURE B.7 Genetically modified crops grown worldwide, by type. (Figure 10.26a, p. 263)

Graph Type: Scatter Plot

A scatter plot is often used when data are not sequential and when a given x-axis value could have multiple y-axis values. A scatter plot allows us to visualize a broad positive or negative correlation between variables. **FIGURE B.8** shows a negative correlation (that is, one value goes up while the other goes down): Nations with higher rates of school enrollment for girls tend to have lower fertility rates. Jamaica has high enrollment and low fertility, whereas Ethiopia has low enrollment and high fertility.

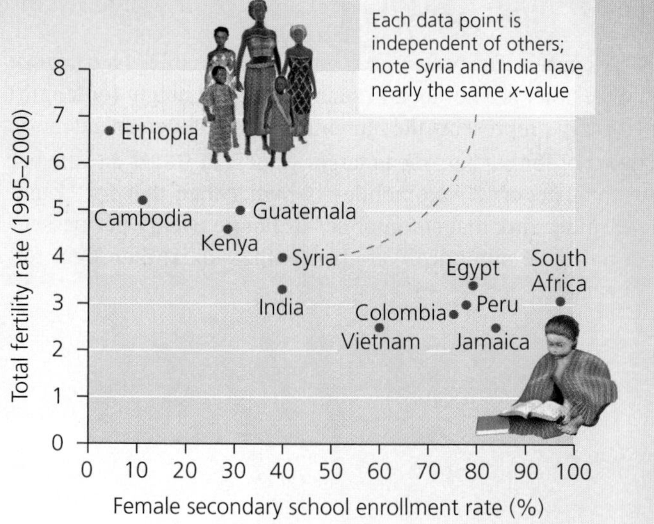

FIGURE B.8 Fertility rate and female education. (Figure 8.19, p. 206)

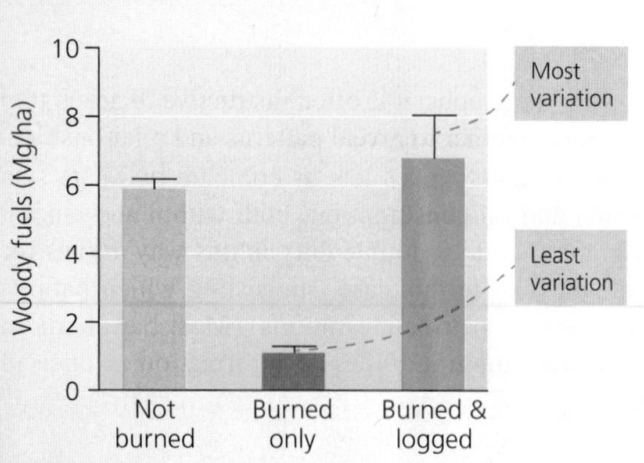

FIGURE B.9 Fine-scale woody debris left after treatments in a salvage logging study. (Figure 2a, p. 325)

Key Concept: Statistical Uncertainty

Most data sets involve some degree of uncertainty. When a graphed value represents the *mean* (average) of many measurements, the researcher may want to show the degree to which the raw data vary around this mean. Mathematical techniques are used to obtain statistically precise degrees of variation around a mean. Results from such statistical analyses may be expressed in a number of ways, and the two graphs in this section show methods used in this book.

In a bar chart (**FIGURE B.9**), thin black lines called *error bars* may be shown extending above and below the tops of the bars, or simply above them. In this example of woody debris remaining after salvage logging, error bars show the most variation in measurements occurred at "Burned and logged" sites.

Sometimes shading is used to express variation around a mean. The black and red data lines in **FIGURE B.10** show mean global sea level readings since 1870. The data line is surrounded by gray shading indicating statistical variation. Note how the amount of statistical uncertainty is exceeded by the sheer scale of the sea level rise. This gives us confidence that sea level is truly rising, despite the statistical uncertainty we find around mean values each year. Note also how the amount of uncertainty has decreased through time. This reflects improvements in technology enabling more accurate measurements.

The statistical analysis of data is critically important in science. In this book, we provide a broad and streamlined introduction to many topics, so we often omit error bars from our graphs and details of statistical significance from our discussions. Bear in mind that this is for clarity of presentation only; the research we discuss analyzes its data in far more depth than any textbook could possibly cover.

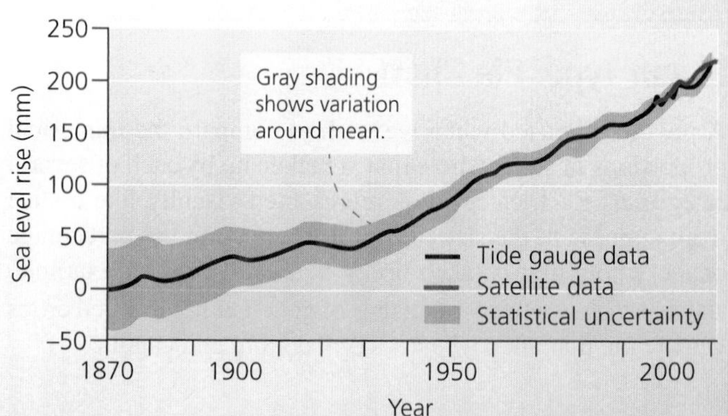

FIGURE B.10 Change in global sea level, measured since 1870. (Figure 18.14, p. 499)

Metric System

APPENDIX C

Measurement	Unit and Abbreviation	Metric Equivalent	Metric to English Conversion Factor	English to Metric Conversion Factor
Length	1 kilometer (km)	= 1,000 (10^3) meters	1 km = 0.62 mile	1 mile = 1.61 km
	1 meter (m)	= 100 (10^2) centimeters	1 m = 1.09 yards	1 yard = 0.914 m
		= 1,000 millimeters	1 m = 3.28 feet	1 foot = 0.305 m
			1 m = 39.37 inches	
	1 centimeter (cm)	= 0.01 (10^{-2}) meter	1 cm = 0.394 inch	1 foot = 30.5 cm
				1 inch = 2.54 cm
	1 millimeter (mm)	= 0.001 (10^{23}) meter	1 mm = 0.039 inch	
Area	1 square meter (m^2)	= 10,000 square centimeters	1 m^2 = 1.1960 square yards	1 square yard = 0.8361 m^2
			1 m^2 = 10.764 square feet	1 square foot = 0.0929 m^2
			1 cm^2 = 0.155 square inch	1 square inch = 6.4516 cm^2
	1 square centimeter (cm^2)	= 100 square millimeters		
Mass	1 metric ton (t)	= 1,000 kilograms	1 t = 1.103 tons	1 ton = 0.907 t
	1 kilogram (kg)	= 1,000 grams	1 kg = 2.205 pounds	1 pound = 0.4536 kg
	1 gram (g)	= 1,000 milligrams	1 g = 0.0353 ounce	1 ounce = 28.35 g
	1 milligram (mg)	= 0.001 gram		
Volume (solids)	1 cubic meter (m^3)	= 1,000,000 cubic centimeters	1 m^3 = 1.3080 cubic yards	1 cubic yard = 0.7646 m^3
			1 m^3 = 35.315 cubic feet	1 cubic foot = 0.0283 m^3
	1 cubic centimeter (cm^3 or cc)	= 0.000001 cubic meter	1 cm^3 = 0.0610 cubic inch	1 cubic inch = 16.387 cm^3
		= 1 milliliter		
	1 cubic millimeter (mm^3)	= 0.000000001 cubic meter		
Volume (liquids and gases)	1 kiloliter (kl or kL)	= 1,000 liters	1 kL = 264.17 gallons	1 gallon = 3.785 L
	1 liter (l or L)	= 1,000 milliliters	1 L = 0.264 gallons	1 quart = 0.946 L
			1 L = 1.057 quarts	
	1 milliliter (ml or mL)	= 0.001 liter	1 ml = 0.034 fluid ounce	1 quart = 946 ml
		= 1 cubic centimeter	1 ml = approximately $\frac{1}{4}$ teaspoon	1 pint = 473 ml
				1 fluid ounce = 29.57 ml
				1 teaspoon = approx. 5 ml
Time	1 millisecond (ms)	= 0.001 second		
Temperature	Degrees Celsius (°C)		°C = $(\frac{5}{9}$F − 32)	°F = $\frac{9}{5}$ °C + 32
Energy and Power	1 kilowatt-hour	= 34,113 BTUs = 860,421 calories		
	1 watt	= 3.413 BTUs/hr		
		= 14.34 calories/min		
	1 calorie	= the amount of heat necessary to raise the temperature of 1 gram (1 cm^3) of water 1 degree Celsius		
	1 horsepower	= 7.457×102 watts		
	1 joule	= 9.481×10^{-4} BTUs		
		= 0.239 cal		
		= 2.778×10^{-7} kilowatt-hours		
Pressure	1 pound per square inch (psi)	= 6894.757 pascals (Pa)		
		= 0.068045961 atmosphere (atm)		
		= 51.71493 millimeters of mercury (mm hg = Torr)		
		= 68.94757 millibars (mbar)		
	1 atmosphere (atm)	= 6.894757 kilopascals (kPa)		
		= 101.325 kilopascals (kPa)		

APPENDIX D

Periodic Table of the Elements

Representative (main group) elements: IA, IIA; Transition metals: IIIB–IIB; Representative (main group) elements: IIIA–VIIIA

Period	IA	IIA	IIIB	IVB	VB	VIB	VIIB	VIIIB	VIIIB	VIIIB	IB	IIB	IIIA	IVA	VA	VIA	VIIA	VIIIA
1	1 **H** 1.0079 Hydrogen																	2 **He** 4.003 Helium
2	3 **Li** 6.941 Lithium	4 **Be** 9.012 Beryllium											5 **B** 10.811 Boron	6 **C** 12.011 Carbon	7 **N** 14.007 Nitrogen	8 **O** 15.999 Oxygen	9 **F** 18.998 Fluorine	10 **Ne** 20.180 Neon
3	11 **Na** 22.990 Sodium	12 **Mg** 24.305 Magnesium											13 **Al** 26.982 Aluminum	14 **Si** 28.086 Silicon	15 **P** 30.974 Phosphorus	16 **S** 32.066 Sulfur	17 **Cl** 35.453 Chlorine	18 **Ar** 39.948 Argon
4	19 **K** 39.098 Potassium	20 **Ca** 40.078 Calcium	21 **Sc** 44.956 Scandium	22 **Ti** 47.88 Titanium	23 **V** 50.942 Vanadium	24 **Cr** 51.996 Chromium	25 **Mn** 54.938 Manganese	26 **Fe** 55.845 Iron	27 **Co** 58.933 Cobalt	28 **Ni** 58.69 Nickel	29 **Cu** 63.546 Copper	30 **Zn** 65.39 Zinc	31 **Ga** 69.723 Gallium	32 **Ge** 72.61 Germanium	33 **As** 74.922 Arsenic	34 **Se** 78.96 Selenium	35 **Br** 79.904 Bromine	36 **Kr** 83.8 Krypton
5	37 **Rb** 85.468 Rubidium	38 **Sr** 87.62 Strontium	39 **Y** 88.906 Yttrium	40 **Zr** 91.224 Zirconium	41 **Nb** 92.906 Niobium	42 **Mo** 95.94 Molybdenum	43 **Tc** 98 Technetium	44 **Ru** 101.07 Ruthenium	45 **Rh** 102.906 Rhodium	46 **Pd** 106.42 Palladium	47 **Ag** 107.868 Silver	48 **Cd** 112.411 Cadmium	49 **In** 114.82 Indium	50 **Sn** 118.71 Tin	51 **Sb** 121.76 Antimony	52 **Te** 127.60 Tellurium	53 **I** 126.905 Iodine	54 **Xe** 131.29 Xenon
6	55 **Cs** 132.905 Cesium	56 **Ba** 137.327 Barium	57 **La** 138.906 Lanthanum	72 **Hf** 178.49 Hafnium	73 **Ta** 180.948 Tantalum	74 **W** 183.84 Tungsten	75 **Re** 186.207 Rhenium	76 **Os** 190.23 Osmium	77 **Ir** 192.22 Iridium	78 **Pt** 195.08 Platinum	79 **Au** 196.967 Gold	80 **Hg** 200.59 Mercury	81 **Tl** 204.383 Thallium	82 **Pb** 207.2 Lead	83 **Bi** 208.980 Bismuth	84 **Po** 209 Polonium	85 **At** 210 Astatine	86 **Rn** 222 Radon
7	87 **Fr** 223 Francium	88 **Ra** 226.025 Radium	89 **Ac** 227.028 Actinium	104 **Rf** 267 Rutherfordium	105 **Db** 268 Dubnium	106 **Sg** 269 Seaborgium	107 **Bh** 270 Bohrium	108 **Hs** 269 Hassium	109 **Mt** 278 Meitnerium	110 **Ds** 281 Darmstadtium	111 **Rg** 281 Roentgenium	112 **Cn** 285 Copernicium		114 **Fl** 289 Flerovium		116 **Lv** 293 Livermorium		

Rare earth elements

Lanthanides	58 **Ce** 140.115 Cerium	59 **Pr** 140.908 Praseodymium	60 **Nd** 144.24 Neodymium	61 **Pm** 145 Promethium	62 **Sm** 150.36 Samarium	63 **Eu** 151.964 Europium	64 **Gd** 157.25 Gadolinium	65 **Tb** 158.925 Terbium	66 **Dy** 162.5 Dysprosium	67 **Ho** 164.93 Holmium	68 **Er** 167.26 Erbium	69 **Tm** 168.934 Thulium	70 **Yb** 173.04 Ytterbium	71 **Lu** 174.967 Lutetium
Actinides	90 **Th** 232.038 Thorium	91 **Pa** 231.036 Protactinium	92 **U** 238.029 Uranium	93 **Np** 237.048 Neptunium	94 **Pu** 244 Plutonium	95 **Am** 243 Americium	96 **Cm** 247 Curium	97 **Bk** 247 Berkelium	98 **Cf** 251 Californium	99 **Es** 252 Einsteinium	100 **Fm** 257 Fermium	101 **Md** 258 Mendelevium	102 **No** 259 Nobelium	103 **Lr** 262 Lawrencium

The periodic table arranges elements by atomic number and atomic weight into horizontal rows called periods and vertical columns called groups.

Elements of each group in Class A have similar chemical and physical properties. This reflects the fact that members of a particular group have the same number of valence shell electrons, which is indicated by the group's number. For example, group IA elements have one valence shell electron, group IIA elements have two, and group VA elements have five. In contrast, as you progress across a period from left to right, properties of the elements change, varying from the very metallic properties of groups IA and IIA to the nonmetallic properties of group VIIA to the inert elements (noble gases) in group VIIIA. This reflects changes in the number of valence shell electrons.

Class B elements, or transition elements, are metals, and generally have one or two valence shell electrons. In these elements, some electrons occupy more distant electron shells before the deeper shells are filled.

In this periodic table, elements with symbols printed in black exist as solids under standard conditions (25 °C and 1 atmosphere of pressure), whereas elements in red exist as gases, and those in dark blue as liquids. Elements with symbols in green do not exist in nature and must be created by some type of nuclear reaction.

Geologic Time Scale

APPENDIX E

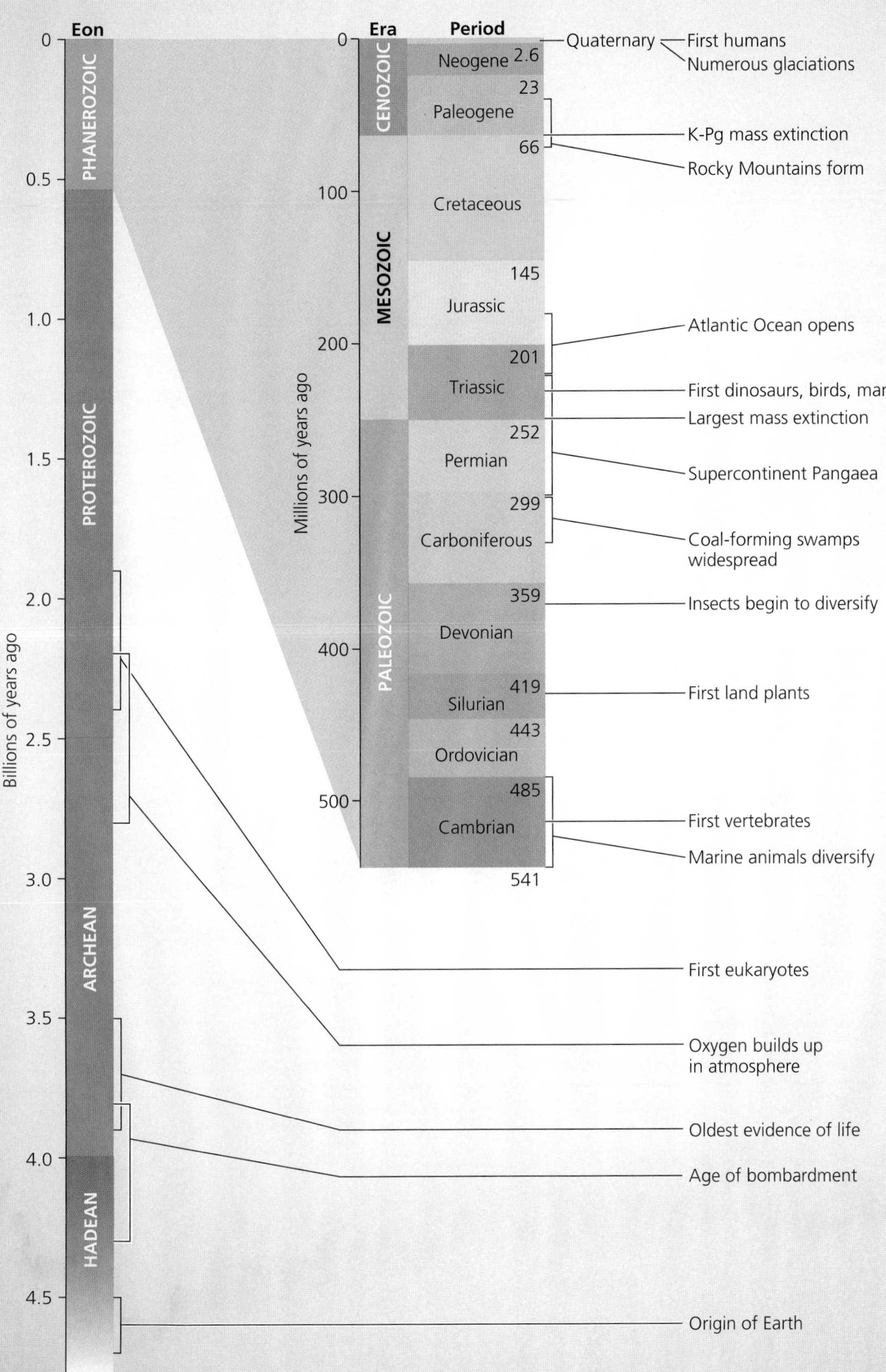

Glossary

acid deposition The settling of acidic or acid-forming pollutants from the *atmosphere* onto Earth's surface. This may take place by precipitation, fog, gases, or the deposition of dry particles. Compare *acid rain*.

acid drainage A process in which sulfide minerals in newly exposed rock surfaces react with *oxygen* and rainwater to produce sulfuric acid, which causes chemical *runoff* as it *leaches* metals from the rocks. Acid drainage is a natural phenomenon, but mining greatly accelerates it by exposing many new surfaces.

acidic The property of a *solution* in which the concentration of *hydrogen* (H^+) *ions* is greater than the concentration of hydroxide (OH^-) ions. Compare *basic*.

acid-neutralizing capacity The capacity of soil, rock, or water to resist *pH* change from *acid deposition*, due to its alkaline chemistry.

acid rain *Acid deposition* that takes place through rain.

active solar energy collection An approach in which technological devices are used to focus, move, or store *solar energy*. Compare *passive solar energy collection*.

acute exposure Exposure to a *toxicant* occurring in high amounts for short periods of time. Compare *chronic exposure*.

adaptation (re: *climate change*) The pursuit of strategies to protect ourselves from the impacts of climate change. Compare *mitigation*.

adaptation (re: *evolution*) (1) The process by which *traits* that lead to increased reproductive success in a given environment evolve in a *population* through *natural selection*. (2) See *adaptive trait*.

adaptive management The systematic testing of different management approaches to improve methods over time.

adaptive trait A trait that confers greater likelihood that an individual will reproduce. An *adaptive trait* is also called an *adaptation*.

aerobic Occurring in an *environment* where *oxygen* is present. For example, the decay of a rotting log proceeds by aerobic decomposition. Compare *anaerobic*.

aerosols Very fine liquid droplets or solid particles aloft in the atmosphere.

affluenza Term coined by social critics to describe the failure of material goods to bring happiness to people who have the financial means to afford them.

age distribution The relative numbers of organisms of each age within a *population*. Age distributions can have a strong effect on rates of population growth or decline and are often expressed as a ratio of age classes, consisting of organisms (1) not yet mature enough to reproduce, (2) capable of reproduction, and (3) beyond their reproductive years.

age structure See *age distribution*.

agricultural revolution The shift around 10,000 years ago from a hunter-gatherer lifestyle to an agricultural way of life in which people began to grow crops and raise domestic animals. Compare *industrial revolution*.

agriculture The practice of cultivating *soil*, producing crops, and raising livestock for human use and consumption.

air pollutants Gases and particulate material added to the atmosphere that can affect *climate* or harm people or other organisms.

air pollution The act of polluting the air, or the condition of being polluted by *air pollutants*.

airshed The geographic area that produces air pollutants likely to end up in a waterway.

albedo The capacity of a surface to reflect light. Higher albedo values refer to greater reflectivity.

allergen A *toxicant* that overactivates the immune system, causing an immune response when one is not necessary.

allopatric speciation Species formation due to the physical separation of populations over some geographic distance. Compare *sympatric speciation*.

alloy A substance created by fusing a *metal* with other metals or nonmetals. Bronze is an alloy of the metals copper and tin, and steel is an alloy of iron and the nonmetal *carbon*.

ambient air pollution See *outdoor air pollution*.

anaerobic Occurring in an *environment* that has little or no *oxygen*. The conversion of organic matter to *fossil fuels* at the bottom of a deep lake, swamp, or shallow sea is an example of anaerobic decomposition. Compare *aerobic*.

anthropocentrism A human-centered view of our relationship with the *environment*. Compare *biocentrism* and *ecocentrism*.

anthropogenic Caused by human beings. For example, anthropogenic climate change, as opposed to natural climate change.

aquaculture The cultivation of aquatic organisms for food in controlled *environments*.

aquifer An underground water reservoir.

archipelago A group of islands, most often in a linear arrangement.

area effect In *island biogeography theory*, the pattern that large islands host more species than smaller islands, because larger islands provide larger targets for immigration and because extinction rates are reduced.

artesian aquifer See *confined aquifer*.

artificial selection *Natural selection* conducted under human direction. Examples include the *selective breeding* of crop plants, pets, and livestock.

asbestos Any of several types of *mineral* that form long, thin microscopic fibers—a structure that allows asbestos to insulate buildings for heat, muffle sound, and resist fire. When inhaled and lodged in lung tissue, asbestos scars the tissue and may eventually lead to lung cancer or *asbestosis*.

asbestosis A disorder resulting from lung tissue scarred by acid following prolonged inhalation of *asbestos*.

Asian Brown Cloud A persistent 2-mile-thick layer of *air pollution* from southern Asia that hangs over the Indian subcontinent throughout the dry season, each December through April. Also called *Atmospheric Brown Cloud*.

asthenosphere A layer of the upper *mantle*, just below the *lithosphere*, consisting of especially soft rock.

atmosphere The thin layer of gases surrounding planet Earth. Compare *biosphere; hydrosphere; lithosphere*.

atmospheric blocking pattern A condition in which the atmosphere's *jet stream* slows and meanders widely into a north–south orientation that blocks the eastward movement of weather systems across the midlatitudes.

Atmospheric Brown Cloud See *Asian Brown Cloud*.

atmospheric deposition The wet or dry deposition on land of a wide variety of pollutants, including mercury, nitrates, organochlorines, and others. *Acid deposition* is one type of atmospheric deposition.

atmospheric pressure The weight per unit area produced by a column of air.

atoll A ring-shaped island (generally of *coral reef*) surrounding an older submerged island area.

atom The smallest component of an *element* that maintains the chemical properties of that element.

autotroph (primary producer) An organism that can use the energy from sunlight to produce its own food. Includes green plants, algae, and cyanobacteria.

***Bacillus thuringiensis* (Bt)** A naturally occurring *soil* bacterium that produces a protein that kills many pests, including caterpillars and the larvae of some flies and beetles.

background rate of extinction The average rate of *extinction* that occurred before the appearance of humans. For example, the *fossil record* indicates that for both birds and mammals, one *species* in the world typically became extinct every 500–1,000 years. Compare *mass extinction event*.

baghouse A system of large filters that physically removes *particulate matter* from *incinerator emissions*.

barrier island A long thin island that parallels a shoreline. Generally of sand or *coral reef*, barrier islands protect coasts from storms.

basic The property of a *solution* in which the concentration of *hydroxide* (OH^-) *ions* is greater than the concentration of hydrogen (H^+) ions. Compare *acidic*.

bathymetry The study of ocean depths.

bedrock The continuous mass of solid rock that makes up Earth's *crust*.

benthic Of, relating to, or living on the bottom of a water body. Compare *pelagic*.

benthic zone The bottom layer of a water body. Compare *littoral zone; limnetic zone; profundal zone*.

bioaccumulation The buildup of *toxicants* in the tissues of an animal.

biocapacity A term in *ecological footprint* accounting meaning the amount of biologically productive land and sea available to us.

biocentrism A philosophy that ascribes relative values to actions, entities, or properties on the basis of their effects on all living things or on the integrity of the *biotic* realm in general. The biocentrist evaluates an action in terms of its overall impact on living things, including—but not exclusively focusing on—human beings. Compare *anthropocentrism* and *ecocentrism*.

biodiesel Diesel fuel produced by mixing vegetable oil, used cooking grease, or animal fat with small amounts of *ethanol* or methanol (wood alcohol) in the presence of a chemical catalyst.

biodiversity (biological diversity) The variety of life across all levels of biological organization, including the diversity of *species*, their *genes*, their *populations*, and their *communities*.

biodiversity hotspot An area that supports an especially great diversity of *species*, particularly species that are *endemic* to the area.

bioenergy (biomass energy) *Energy* harnessed from plant and animal matter, including wood from trees, charcoal from burned wood, and combustible animal waste products, such as cattle manure. *Fossil fuels* are not considered biomass energy sources because their organic matter has not been part of living organisms for millions of years and has undergone considerable chemical alteration since that time.

biofuel Fuel produced from *biomass energy* sources and used primarily to power automobiles. Examples include *ethanol* and *biodiesel*.

biogas Methane-rich gas produced by bacterial action in *anaerobic* digestion facilities. Can be burned in a power plant to generate *electricity*.

biogenic Type of *natural gas* created at shallow depths by the anaerobic decomposition of organic matter by bacteria. Consists of nearly pure *methane*. Compare *thermogenic*.

biogeochemical cycle See *nutrient cycle*.

biological control (biocontrol) Control of pests and weeds with organisms that prey on or parasitize them, rather than with *pesticides*.

biological hazard Human health hazards that result from ecological interactions among organisms. These include *parasitism* by viruses, bacteria, or other *pathogens*. Compare *infectious disease; chemical hazard; cultural hazard; physical hazard*.

biological diversity See *biodiversity*.

biomagnification The magnification of the concentration of *toxicants* in an organism caused by its consumption of other organisms in which toxicants have *bioaccumulated*.

biomass (1) In ecology, organic material that makes up living organisms; the collective mass of living matter in a given place and time. (2) In energy, organic material derived from living or recently living organisms, containing chemical *energy* that originated with *photosynthesis*.

biomass energy See *bioenergy*.

biome A major regional complex of similar plant *communities;* a large *ecological* unit defined by its dominant plant type and vegetation structure.

biophilia An inherent love for and fascination with nature and an instinctive desire people have to affiliate with other living things. Defined by biologist E.O. Wilson as "the connections that human beings subconsciously seek with the rest of life."

biopower Power attained by combusting *bioenergy* sources to generate *electricity*.

biosphere The sum total of all the planet's living organisms and the *abiotic* portions of the *environment* with which they interact.

biosphere reserve A tract of land with exceptional *biodiversity* that couples preservation with *sustainable development* to benefit local people. Biosphere reserves are designated by UNESCO (the *United Nations* Educational, Scientific, and Cultural Organization) following application by local stakeholders.

biotechnology The material application of biological *science* to create products derived from organisms. The creation of *transgenic* organisms is one type of biotechnology.

biotic potential An organism's capacity to produce offspring.

birth control The effort to control the number of children one bears, particularly by reducing the frequency of pregnancy. Compare *contraception, family planning*.

bitumen A thick and heavy form of *petroleum* rich in *carbon* and poor in *hydrogen*. The fossil-fuel component of *oil sands*.

bog A type of *wetland* in which a pond is thoroughly covered with a thick floating mat of vegetation. Compare *freshwater marsh; swamp*.

boreal forest A *biome* of northern coniferous forest that stretches in a broad band across much of Canada, Alaska, Russia, and Scandinavia. Also known as *taiga*, boreal forest consists of a limited number of *species* of evergreen trees, such as black spruce, that dominate large regions of forests interspersed with occasional bogs and lakes.

Borlaug, Norman (1914–2009) American agricultural scientist who introduced specially bred crops to developing nations in the 20th century, helping to spur the *Green Revolution*.

bottleneck A step in a process that limits the progress of the overall process.

bottom-trawling Fishing practice that involves dragging weighted nets across the seafloor to catch *benthic* organisms. Trawling crushes many organisms in its path and leaves long swaths of damaged sea bottom.

braided river A river that flows as an interconnected series of watercourses because it runs through a steeply sloped region or carries a great deal of sediment. Compare *meandering river*.

breakdown product *A compound* that results from the degradation of a toxicant.

brownfield An area of land whose redevelopment or reuse is complicated by the presence or potential presence of hazardous material.

building-related illness Any sickness caused by indoor pollution.

bycatch (1) The accidental capture of nontarget organisms while fishing for target species. (2) That portion of a commercial fishing catch consisting of animals caught unintentionally. Bycatch kills many thousands of fish, sharks, marine mammals, and birds each year.

cap-and-trade A *permit trading* system in which government determines an acceptable level of *pollution* and then issues polluting parties permits to pollute. A company receives credit for amounts it does not emit and can then sell this credit to other companies. A type of *emissions trading system*.

capitalist market economy An *economy* in which buyers and sellers interact to determine which *goods* and *services* to produce, how much of them to produce, and how to distribute them. Compare *centrally planned economy*.

captive breeding The practice of capturing members of threatened and endangered *species* so that their young can be bred and raised in controlled *environments* and subsequently reintroduced into the wild.

canopy The upper level of tree leaves and branches in a *forest*.

carbohydrate An *organic compound* consisting of *atoms* of *carbon*, *hydrogen*, and *oxygen*.

carbon The chemical *element* with six protons and six neutrons. A key element in *organic compounds*.

carbon capture Technologies or approaches that remove *carbon dioxide* from power plant or other emissions, in an effort to mitigate *global climate change*.

carbon cycle A major *nutrient cycle* consisting of the routes that *carbon atoms* take through the nested networks of environmental *systems*.

carbon dioxide (CO_2) A colorless gas used by plants for *photosynthesis*, given off by *respiration*, and released by burning *fossil fuels*. A primary *greenhouse gas* whose buildup contributes to *global climate change*.

carbon footprint The cumulative amount of carbon, or *carbon dioxide*, that a person or institution emits, and is indirectly responsible for emitting, into the *atmosphere*, contributing to *global climate change*. Compare *ecological footprint*.

carbon monoxide (CO) A colorless, odorless gas produced primarily by the incomplete combustion of fuel. An EPA *criteria pollutant*.

carbon-neutrality The state in which an individual, business, or institution emits no net carbon to the atmosphere. This may be achieved by reducing carbon emissions and/or employing *carbon offsets* to offset emissions.

carbon offset A voluntary payment to another entity intended to enable that entity to reduce the *greenhouse gas* emissions that one is unable or unwilling to reduce oneself. The payment thus offsets one's own emissions.

carbon sequestration Technologies or approaches to sequester, or store, *carbon dioxide* from industrial emissions (e.g., underground under pressure in locations where it will not seep out) in an effort to mitigate *global climate change*. We are still a long way from developing adequate technology and secure storage space to accomplish this.

carbon storage See *carbon sequestration*.

carbon tax A fee charged to entities that pollute by emitting *carbon dioxide*. A carbon tax gives polluters a financial incentive to reduce pollution and is thus foreseen as a way to address *global climate change*. Compare *fee-and-dividend*.

carcinogen A chemical or type of radiation that causes cancer.

carnivore An organism that consumes animals. Compare *herbivore*; *omnivore*.

carrying capacity The maximum *population size* that a given *environment* can sustain.

case history Medical approach involving the observation and analysis of individual patients.

case law A body of law made up of cumulative decisions rendered by courts.

Cassandra A worldview (or a person holding the worldview) that predicts doom and disaster as a result of our environmental impacts. In Greek mythology, Cassandra was the princess of Troy with the gift of prophecy, whose dire predictions were not believed. Compare *Cornucopian*.

categorical imperative An *ethical standard* described by Immanuel Kant, which roughly approximates Christianity's "golden rule": to treat others as you would prefer to be treated yourself.

cation exchange Process by which plants' roots donate *hydrogen ions* to the soil in exchange for cations (positively charged ions) such as those of calcium, magnesium, and potassium, which plants use as *nutrients*. The soil particles then replenish these cations by exchange with soil water.

cation exchange capacity A soil's ability to hold cations, preventing them from *leaching*, and thus making them available to plants. A useful measure of soil fertility.

cellular respiration The process by which a *cell* uses the chemical reactivity of *oxygen* to split glucose into its constituent parts, water and *carbon dioxide*, and thereby release chemical energy that can be used to form chemical bonds or to perform other tasks within the cell. Compare *photosynthesis*.

cellulosic ethanol *Ethanol* produced from the cellulose in plant tissues by treating it with enzymes. Techniques for producing cellulosic ethanol are under development because of the desire to make ethanol from low-value crop waste (residues such as corn stalks and husks), rather than from the sugars of high-value crops.

centrally planned economy An *economy* in which a nation's government determines how to allocate resources in a top-down manner. Also called a "state socialist economy." Compare *capitalist market economy*.

chaparral A *biome* consisting mostly of densely thicketed evergreen shrubs occurring in limited small patches. Its "Mediterranean" *climate* of mild, wet winters and warm, dry summers is induced by oceanic influences. In addition to ringing the Mediterranean Sea, chaparral occurs along the coasts of California, Chile, and southern Australia.

character displacement A phenomenon resulting from *competition* among *species* in which competing species evolve characteristics that better adapt them to specialize on the portion of the resource they use. The species essentially become more different from one another, reducing their competition.

chemical hazard Chemicals that pose human health hazards. These include *toxins* produced naturally, as well as many of the disinfectants, *pesticides*, and other synthetic chemicals that our society produces. Compare *biological hazard; cultural hazard; physical hazard.*

chemistry The study of the different types of matter and how they interact.

chemosynthesis The process by which bacteria in *hydrothermal vents* use the chemical energy of hydrogen sulfide (H_2S) to transform inorganic *carbon* into *organic compounds*. Compare *photosynthesis*.

Chernobyl Site of a nuclear power plant in Ukraine (then part of the Soviet Union), where in 1986 an explosion caused the most severe *nuclear reactor* accident the world has yet seen. As with *Three Mile Island* and *Fukushima*, people often use the term to denote the accident itself. Compare *Fukushima Daiichi, Three Mile Island.*

chlorofluorocarbon (CFC) A type of *halocarbon* consisting of only chlorine, fluorine, carbon, and hydrogen. CFCs were used as refrigerants, fire extinguishers, propellants for aerosol spray cans, cleaners for electronics, and for making polystyrene foam. They were phased out under the *Montreal Protocol* because they are *ozone-depleting substances* that destroy stratospheric *ozone*.

chronic exposure Exposure for long periods of time to a *toxicant* occurring in low amounts. Compare *acute exposure*.

city planning The professional pursuit that attempts to design cities in such a way as to maximize their efficiency, functionality, and beauty. Also known as *urban planning*.

classical economics Founded by *Adam Smith*, the study of the behavior of buyers and sellers in a *capitalist market economy*. Holds that individuals acting in their own self-interest may benefit society, provided that their behavior is constrained by the rule of law and by private property rights and operates within competitive markets. See also *neoclassical economics*.

clay *Sediment* consisting of particles less than 0.002 mm in diameter. Compare *sand; silt*.

Clean Air Act of 1970 Revision of prior U.S. *legislation* to control *air pollution* that set stricter standards for air quality, imposed limits on emissions from new stationary and mobile sources, provided new funds for *pollution* control research, and enabled citizens to sue parties violating the standards.

Clean Air Act of 1990 U.S. *legislation* that strengthened *regulations* pertaining to air quality standards, auto emissions, toxic *air pollution, acid deposition*, and depletion of the *ozone layer*, while also introducing market-based incentives to reduce *pollution*.

clean coal technologies A wide array of techniques, equipment, and approaches that seek to remove chemical contaminants (such as sulfur) during the process of generating *electricity* from *coal*.

clear-cutting The harvesting of timber by cutting all the trees in an area. Although it is the most cost-efficient method, clear-cutting is also the most ecologically damaging.

climate The pattern of atmospheric conditions found across large geographic regions over long periods of time. Compare *weather*.

climate change See *global climate change*.

climate diagram A visual representation of a region's average monthly temperature and *precipitation*. Also know as a *climatograph*.

climate model A computer program that combines what is known about weather patterns, atmospheric circulation, atmosphere–ocean interactions, and feedback mechanisms, in order to simulate *climate* processes.

climatographs See *climate diagram*.

climax community In the traditional view of ecological *succession*, a *community* that remains in place with little modification until *disturbance* restarts the successional process. Today, ecologists recognize that community change is more variable and less predictable than originally thought and that assemblages of *species* may instead form complex mosaics in space and time.

closed cycle An approach in *ocean thermal energy conversion* in which warm surface water is used to evaporate chemicals that boil at low temperatures. These evaporated gases spin turbines to generate *electricity*. Cold water piped in from ocean depths then condenses the gases so they can be reused.

clumped distribution Distribution pattern in which organisms arrange themselves in patches, generally according to the availability of the resources they need.

coal Our most abundant *fossil fuel*. A hard blackish substance formed from organic matter (generally woody plant material) that was compressed under very high pressure and with little decomposition, creating dense, solid carbon structures.

coevolution Process by which two or more species evolve in response to one another. Parasites and hosts may coevolve, as may flowering plants and their pollinators.

co-firing A process in which *biomass* is combined with *coal* in coal-fired power plants. Can be a relatively easy and inexpensive way for *fossil-fuel*-based utilities to expand their use of *renewable energy*.

cogeneration A practice in which the extra heat generated in the production of *electricity* is captured and put to use heating workplaces and homes, as well as producing other kinds of power.

cold front The boundary where a mass of cold air displaces a mass of warmer air. Compare *warm front*.

co-management A type of *community-based conservation* in which government agencies work with local people to jointly and cooperatively manage a protected area and its resources.

command-and-control A top-down approach to policy, in which a legislative body or a regulating agency sets rules, standards, or limits and threatens punishment for violations of those limits.

community In *ecology*, an assemblage of *populations* of organisms that live in the same place at the same time.

community-based conservation The practice of engaging local people to protect land and wildlife in their own region.

community ecology The scientific study of patterns of species diversity and interactions among *species*, from one-to-one interactions to complex interrelationships involving entire *communities*.

community-supported agriculture (CSA) A system in which consumers pay farmers in advance for a share of their yield, usually in the form of weekly deliveries of produce.

competition A relationship in which multiple organisms seek the same limited resource.

competitive exclusion An outcome of *interspecific competition* in which one *species* excludes another species from resource use entirely.

compost A mixture produced when decomposers break down organic matter, such as food and crop waste, in a controlled environment.

composting The conversion of organic *waste* into mulch or *humus* by encouraging, in a controlled manner, the natural biological processes of decomposition.

compound A *molecule* whose *atoms* are composed of two or more *elements*.

concentrated solar power (CSP) A means of generating *electricity* at a large scale by focusing sunlight from a large area onto a smaller area. Several approaches are used.

concession The right to extract a resource, granted by a government to a corporation. Compare *conservation concession*.

confined (artesian) aquifer A water-bearing, porous layer of rock, *sand*, or gravel that is trapped between an upper and lower layer of less permeable substrate, such as *clay*. The water in a confined aquifer is under pressure because it is trapped between two impermeable layers. Compare *unconfined aquifer*.

conservation biology A scientific discipline devoted to understanding the factors, forces, and processes that influence the loss, protection, and restoration of *biodiversity* within and among *ecosystems*.

conservation concession A type of *concession* in which a conservation organization purchases the right to prevent resource extraction in an area of land, generally to preserve habitat in developing nations.

conservation district A county-based entity created by the Soil Conservation Service (now the Natural Resources Conservation Service) to promote practices to conserve *soil*.

conservation ethic An *ethic* holding that people should put *natural resources* to use but also have a responsibility to manage them wisely. Compare *preservation ethic*.

Conservation Reserve Program U.S. policy in farm bills since 1985 that pays farmers to stop cultivating highly erodible cropland and instead place it in conservation reserves planted with grasses and trees.

conservation tillage *Agriculture* that limits the amount of tilling (plowing, disking, harrowing, or chiseling) of *soil*. Compare *no-till*.

consumptive use Use of *fresh water* in which water is removed from a particular *aquifer* or surface water body and is not returned to it. *Irrigation* for *agriculture* is an example of consumptive use. Compare *nonconsumptive use*.

continental collision The meeting of two tectonic plates of continental *lithosphere* at a *convergent plate boundary*, wherein the continental *crust* on both sides resists *subduction* and instead crushes together, bending, buckling, and deforming layers of rock and forcing portions of the buckled crust upward, often creating mountain ranges.

continental shelf The gently sloping underwater edge of a continent, varying in width from 100 m (330 ft) to 1300 km (800 mi), with an average slope of 1.9 m/km (10 ft/mi).

continental slope The portion of the ocean floor that angles somewhat steeply downward, connecting the *continental shelf* to the deep ocean basin below.

contingent valuation A technique that uses surveys to determine how much people would be willing to pay to protect a resource or to restore it after damage has been done.

contour farming The practice of plowing furrows sideways across a hillside, perpendicular to its slope, to help prevent the formation of rills and gullies. The technique is so named because the furrows follow the natural contours of the land.

contraception The deliberate attempt to prevent pregnancy despite sexual intercourse. Compare *birth control*.

control The portion of an *experiment* in which a *variable* has been left unmanipulated, to serve as a point of comparison with the *treatment*.

controlled burn See *prescribed burn*.

controlled experiment An *experiment* in which a *treatment* is compared against a *control* in order to test the effect of a *variable*.

control rods Rods made of a metallic alloy that absorbs *neutrons*, which are placed in a *nuclear reactor* among the water-bathed *fuel rods* of uranium. Engineers move these control rods into and out of the water to maintain the *fission* reaction at the desired rate.

convective circulation A circular *current* (of air, water, magma, etc.) driven by temperature differences. In the atmosphere, warm air rises into regions of lower *atmospheric pressure*, where it expands and cools and then descends and becomes denser, replacing warm air that is rising. The air picks up heat and moisture near ground level and prepares to rise again, continuing the process.

convention A *treaty* or binding agreement among national governments.

conventional law International law that arises from *conventions*, or treaties, that nations agree to enter into. Compare *customary law*.

Convention on Biological Diversity An international treaty that aims to conserve *biodiversity*, use biodiversity in a *sustainable* manner, and ensure the fair distribution of biodiversity's benefits.

Convention on International Trade in Endangered Species of Wild Fauna and Flora (CITES) A 1973 treaty facilitated by the *United Nations* that protects endangered *species* by banning the international transport of their body parts.

convergent evolution The evolutionary process by which very unrelated species acquire similar traits as they adapt to selective pressures from similar environments.

convergent plate boundary The area where tectonic plates converge or come together. Can result in *subduction* or *continental collision*. Compare *divergent plate boundary* and *transform plate boundary*.

coral Tiny marine animals that build *coral reefs*. Corals attach to rock or existing reef and capture passing food with stinging tentacles. They also derive nourishment from photosynthetic symbiotic algae known as *zooxanthellae*.

coral reef A mass of calcium carbonate composed of the skeletons of tiny colonial marine organisms called *corals*.

core The innermost part of Earth, made up mostly of iron, that lies beneath the *crust* and *mantle*.

Coriolis effect The apparent deflection of north–south air *currents* to a partly east–west direction, caused by the faster spin of regions near the equator than of regions near the poles as a result of Earth's rotation.

Cornucopian A worldview (or a person holding the worldview) that we will find ways to make Earth's natural resources meet all of our needs indefinitely and that human ingenuity will see us through any difficulty. In Greek mythology, cornucopia—literally "horn of plenty"—is the name for a magical goat's horn that overflowed with grain, fruit, and flowers. Compare *Cassandra*.

corporate average fuel efficiency (CAFE) standards Miles-per-gallon fuel efficiency standards set by the U.S. Congress for auto manufacturers to meet, by a sales-weighted average of all models of the manufacturer's fleet.

correlation A statistical association among *variables*.

corridor A passageway of protected land established to allow animals to travel between islands of protected *habitat*.

cost-benefit analysis A method commonly used by *neoclassical economists*, in which estimated costs for a proposed action are totaled and then compared to the sum of benefits estimated to result from the action.

covalent bond A type of chemical bonding where atoms share electrons in chemical bonds. An example is a water molecule, which forms when an oxygen atom shares electons with two hydrogen atoms.

cover crop A crop that covers and anchors the *soil* during times between main crops, intended to reduce *erosion*.

criteria pollutants Six *air pollutants—carbon monoxide, sulfur dioxide, nitrogen dioxide, tropospheric ozone, particulate matter*, and *lead*—for which the *Environmental Protection Agency* has established maximum allowable concentrations in ambient outdoor air because of the threats they pose to human health.

cropland Land that people use to raise plants for food and fiber.

crop rotation The practice of alternating the kind of crop grown in a particular field from one season or year to the next.

crude birth rate The number of births per 1,000 individuals for a given time period.

crude death rate The number of deaths per 1,000 individuals for a given time period.

crude oil (petroleum) A *fossil fuel* produced by the conversion of *organic compounds* by heat and pressure. Crude oil is a mixture of hundreds of different types of *hydrocarbon* molecules characterized by *carbon* chains of different lengths.

crust The lightweight outer layer of the Earth, consisting of rock that floats atop the malleable *mantle*, which in turn surrounds a mostly iron *core*.

cultural hazard Human health hazards that result from the place we live, our socioeconomic status, our occupation, or our behavioral choices. These include choosing to smoke cigarettes, or living or working with people who do. Also known as *lifestyle hazard*. Compare *biological hazard; chemical hazard; physical hazard*.

culture The overall ensemble of knowledge, beliefs, values, and learned ways of life shared by a group of people.

current The flow of a liquid or gas in a certain direction.

customary law International law that arises from long-standing practices, or customs, held in common by most *cultures*. Compare *conventional law*.

cyclone A cyclonic storm that forms over the ocean but can do damage upon its arrival on land.

Daly, Herman Contemporary American ecological economist and well-known proponent of a *steady-state economy*.

dam Any obstruction placed in a river or stream to block the flow of water so that water can be stored in a *reservoir*. Dams are built to prevent floods, provide drinking water, facilitate *irrigation*, and generate electricity.

Darwin, Charles (1809–1882) English naturalist who proposed the concept of *natural selection* as a mechanism for *evolution* and as a way to explain the great variety of living things. Compare *Wallace, Alfred Russel*.

data Information, generally quantitative information.

debt-for-nature swap A transaction in which a conservation organization pays off a portion of a developing nation's international debt in exchange for the nation's promise to set aside reserves, fund environmental education, and better manage protected areas.

deciduous Describes a plant that loses its leaves each fall and goes dormant during the winter.

decomposer An organism, such as a fungus or bacterium, that breaks down leaf litter and other nonliving matter into simple constituents that can be taken up and used by plants. Compare *detritivore*.

Deepwater Horizon The British Petroleum offshore drilling platform that sank in 2010, causing the largest oil spill in U.S. history.

deep-well injection A *hazardous waste* disposal method in which a well is drilled deep beneath an area's *water table* into porous rock below an impervious *soil* layer. Wastes are then injected into the well, so that they will be absorbed into the porous rock and remain deep underground, isolated from *groundwater* and human contact. Compare *surface impoundment*.

deforestation The clearing and loss of *forests*.

demand The amount of a product people will buy at a given price if free to do so. Compare *supply*.

demographer A social scientist who studies the population size, density, distribution, age structure, sex ratio, and rates of birth, death, immigration, and emigration of human populations. See *demography*.

demographic fatigue An inability on the part of governments to address overwhelming challenges related to population growth.

demographic transition A theoretical *model* of economic and cultural change that explains the declining death rates and birth rates that occurred in Western nations as they became industrialized. The model holds that industrialization caused these rates to fall naturally by decreasing mortality and by lessening the need for large families. Parents would thereafter choose to invest in quality of life rather than quantity of children.

demography A *social science* that applies the principles of *population ecology* to the study of statistical change in human *populations*.

denitrifying bacteria Bacteria that convert the nitrates in *soil* or water to gaseous *nitrogen* and release it back into the *atmosphere*.

density-dependent factor A *limiting factor* whose effects on a *population* increase or decrease depending on the *population density*. Compare *density-independent factor*.

density-independent factor A *limiting factor* whose effects on a *population* are constant regardless of *population density*. Compare *density-dependent factor*.

deoxyribonucleic acid (DNA) A double-stranded *nucleic acid* composed of four nucleotides, each of which contains a sugar (deoxyribose), a phosphate group, and a nitrogenous base. DNA carries the hereditary information for living organisms and is responsible for passing traits from parents to offspring. Compare *RNA*.

dependent variable The *variable* that is affected by manipulation of the *independent variable* in an *experiment*.

deposition The arrival of eroded *soil* at a new location. Compare *erosion*.

desalination (desalinization) The removal of salt from seawater.

descriptive science Research in which scientists gather basic information about organisms, materials, systems, or processes that are not yet well known. Compare *hypothesis-driven science*.

desert The driest *biome* on Earth, with annual *precipitation* of less than 25 cm. Because deserts have relatively little vegetation to insulate them from temperature extremes, sunlight readily heats them in the daytime, but daytime heat is quickly lost at night, so temperatures vary widely from day to night and in different seasons.

desertification A form of *land degradation* in which more than 10% of a land's productivity is lost due to *erosion*, soil compaction, forest removal, *overgrazing*, drought, *salinization*, *climate* change, water depletion, or other factors. Severe desertification can result in the expansion of desert areas or creation of new ones. Compare *land degradation*; *soil degradation*.

detritivore An organism, such as a millipede or soil insect, that scavenges the waste products or dead bodies of other community members. Compare *decomposer*.

development The use of natural resources for economic advancement (as opposed to simple subsistence, or survival).

dike A long raised mound of earth erected along a river bank to protect against floods by holding rising water in the main channel.

directional drilling A drilling technique (e.g., for *oil* or *natural gas*) in which a drill bores down vertically and then bends horizontally in order to follow layered deposits for long distances from the drilling site. This enables us to extract more fossil fuels with less environmental impact at the surface.

discounting A practice in *neoclassical economics* by which short-term costs and benefits are granted more importance than long-term costs and benefits. Future effects are thereby "discounted," under the notion that an impact far in the future should count much less than one in the present.

distance effect In *island biogeography theory*, the pattern that islands far from a mainland host fewer *species* because fewer species tend to find and colonize it.

disturbance An event that affects environmental conditions rapidly and drastically, resulting in changes to the *community* and *ecosystem*. Disturbance can be natural or can be caused by people.

divergent plate boundary The area where tectonic plates push apart from one another as *magma* rises upward to the surface, creating new *lithosphere* as it cools and spreads. A prime example is the Mid-Atlantic Ridge. Compare *convergent plate boundary* and *transform plate boundary*.

DNA See *deoxyribonucleic acid*.

doldrums A region near the equator with little wind activity.

dose The amount of *toxicant* a test animal receives in a dose-response test. Compare *response*.

dose-response analysis A set of experiments that measure the *response* of test animals to different *doses* of a *toxicant*. The response is generally quantified by measuring the proportion of animals exhibiting negative effects.

dose-response curve A curve that plots the *response* of test animals to different *doses* of a *toxicant*, as a result of *dose-response analysis*.

downwelling In the ocean, the flow of warm surface water toward the ocean floor. Downwelling occurs where surface *currents* converge. Compare *upwelling*.

drainage basin See *watershed*.

driftnet Fishing net that spans large expanses of water, arrayed strategically to drift with currents so as to capture passing fish, and held vertical by floats at the top and weights at the bottom. Driftnetting results in substantial *bycatch* of dolphins, seals, sea turtles, and nontarget fish.

drylands Arid and semi-arid environments that are prone to *desertification* and that cover about 40% of Earth's land surface.

Dust Bowl An area that loses huge amounts of *topsoil* to wind *erosion* as a result of drought and/or human impact. First used to name the region in the North American Great Plains severely affected by drought and topsoil loss in the 1930s. The term is now also used to describe that historical event and others like it.

dynamic equilibrium The state reached when processes within a *system* are moving in opposing directions at equivalent rates so that their effects balance out.

earthquake A release of energy that occurs as Earth relieves accumulated pressure between masses of lithosphere and that results in shaking at the surface.

ecocentrism A philosophy that considers actions in terms of their damage or benefit to the integrity of whole ecological systems, including both living and nonliving elements. For an ecocentrist, the well-being of an individual is less important than the long-term well-being of a larger integrated ecological system. Compare *anthropocentrism* and *biocentrism*.

ecolabeling The practice of designating on a product's label how the product was grown, harvested, or manufactured, so that consumers are aware of the processes involved and can judge which brands use more sustainable processes.

ecological economics A developing school of *economics* that applies the principles of *ecology* and *systems* thinking to the description and analysis of *economies*. Compare *environmental economics; neoclassical economics*.

ecological footprint The cumulative area of biologically productive land and water required to provide the raw materials a person or *population* consumes and to dispose of or *recycle* the *waste* that is produced.

ecological modeling The practice of constructing and testing *models* that aim to explain and predict how ecological systems function.

ecological restoration Efforts to reverse the effects of human disruption of ecological systems and to restore *communities* to their condition before the disruption. The practice that applies principles of *restoration ecology*.

ecology The *science* that deals with the distribution and abundance of organisms, the interactions among them, and the interactions between organisms and their *abiotic environments*.

economic development Improvement in the efficiency of production due to better technologies and approaches that allow us to produce more goods with fewer inputs.

economic growth An increase in an economy's activity—that is, an increase in the production and consumption of goods and services.

economics The study of how we decide to use scarce resources to satisfy demand for *goods* and *services*.

economy A social *system* that converts resources into *goods* and *services*.

ecosystem All organisms and nonliving entities that occur and interact in a particular area at the same time.

ecosystem-based management The attempt to manage the harvesting of resources in ways that minimize impact on the *ecosystems* and ecological processes that provide the resources.

ecosystem diversity The number and variety of ecosystems in a particular area. One way to express *biodiversity*. Related concepts consider the geographic arrangement of *habitats*, *communities*, or *ecosystems* at the landscape level, including the sizes, shapes, and interconnectedness of patches of these entities.

ecosystem ecology The study of how the living and nonliving components of *ecosystems* interact.

ecosystem service An essential service an *ecosystem* provides that supports life and makes *economic* activity possible. For example, ecosystems naturally purify air and water, cycle *nutrients*, provide for plants to be *pollinated* by animals, and receive and recycle the *waste* we generate.

ecotone A transitional zone where *ecosystems* meet.

ecotourism Visitation of natural areas for tourism and recreation. Most often involves tourism by more-affluent people, which may generate *economic* benefits for less-affluent communities near natural areas and thus provide economic incentives for conservation of natural areas.

ED_{50} (effective dose–50%) The amount of a *toxicant* it takes to affect 50% of a *population* of test animals. Compare *threshold dose; LD_{50}*.

edge effect An impact on organisms, populations, or communities that results because conditions along the edge of a habitat fragment differ from conditions in the interior.

electricity A secondary form of energy that can be transferred over long distances and applied for a variety of uses.

electrolysis A process in which electrical current is passed through a *compound* to release *ions*. Electrolysis offers one way to produce *hydrogen* for use as fuel: Electrical current is passed through water, splitting the water *molecules* into hydrogen and *oxygen atoms*.

electronic waste (e-waste) Discarded electronic products such as computers, monitors, printers, televisions, DVD players, cell phones, and other devices. *Heavy metals* in these products mean that this waste may be judged hazardous.

Ekman drift Wind-driven currents in the ocean's upper layers. These currents are named for scientist Walfrid Ekman, who first modeled them in the early 20th century. Compare *geostrophic currents*.

electron A negatively charged particle that moves around the nucleus of an *atom*.

element A fundamental type of matter; a chemical substance with a given set of properties, which cannot be broken down into substances with other properties. Chemists currently recognize 92 elements that occur in nature, as well as more than 20 others that have been artificially created.

El Niño The exceptionally strong warming of the eastern Pacific Ocean that occurs every 2 to 7 years and depresses local fish and bird *populations* by altering the marine *food web* in the area. Originally, the name that Spanish-speaking fishermen gave to an unusually warm surface *current* that sometimes arrived near the Pacific coast of South America around Christmas time. Compare *La Niña*.

El Niño–Southern Oscillation (ENSO) A systematic shift in atmospheric pressure, sea surface temperature, and ocean circulation in the tropical Pacific Ocean. ENSO cycles give rise to *El Niño* and *La Niña* conditions.

emergent property A characteristic that is not evident in a *system*'s components.

emergent tree An especially tall tree that protrudes above the *canopy* of a *tropical rainforest*.

Emerson, Ralph Waldo (1803–1882) American author, poet, and philosopher who espoused *transcendentalism*.

emigration The departure of individuals from a *population*.

eminent domain A policy in which a government pays landowners for their land at market rates and the landowners have no recourse to refuse. In eminent domain, courts set aside private property rights to make way for projects judged to be for the public good.

emissions trading system A *permit trading* system for emissions in which a government issues marketable emissions permits to conduct environmentally harmful activities. Under a *cap-and-trade* system, the government determines an acceptable level of *pollution* and then issues permits to pollute. A company receives credit for amounts it does not emit and can then sell this credit to other companies. Compare *cap-and-trade*.

Endangered Species Act (ESA) The primary *legislation*, enacted in 1973, for protecting *biodiversity* in the United States. It forbids the government and private citizens from taking actions (such as developing land) that would destroy endangered *species* or their *habitats*, and it prohibits trade in products made from endangered species.

endemic Native or restricted to a particular geographic region. An endemic species occurs in one area and nowhere else on Earth.

endocrine disruptor A *toxicant* that interferes with the *endocrine (hormone) system*.

endocrine system The body's *hormone* system.

energy The capacity to change the position, physical composition, or temperature of matter; a force that can accomplish work.

energy conservation The practice of reducing *energy* use as a way of extending the lifetime of our *fossil fuel* supplies, of being less wasteful, and of reducing our impact on the *environment*. Conservation can result from behavioral decisions or from technologies that demonstrate *energy efficiency*.

energy conversion efficiency The ratio of the useful output of *energy* to the amount that needs to be input. See also *EROI* and *net energy*.

energy efficiency The ability to obtain a given result or amount of output while using less energy input. Technologies permitting greater energy efficiency are one main route to *energy conservation*.

energy returned on investment See *EROI*.

enhanced geothermal systems (EGS) A new approach whereby engineers drill deeply into rock, fracture it, pump in water, and then pump it out once it is heated below ground. This approach would enable us to obtain *geothermal energy* in many locations.

environment The sum total of our surroundings, including all of the living things and nonliving things with which we interact.

environmental economics A developing school of *economics* that modifies the principles of *neoclassical economics* to address environmental challenges. Most environmental economists believe that we can attain *sustainability* within our current economic systems. Whereas ecological economists call for revolution, environmental economists call for reform. Compare *ecological economics; neoclassical economics*.

environmental ethics The application of *ethical standards* to environmental questions.

environmental health The study of environmental factors that influence human health and quality of life and the health of *ecological* systems essential to environmental quality and long-term human well-being.

environmental history The study of the history of environmental change and the history of human interactions with the *environment*. Includes the study of how people influence their environments over time and how nature influences human *culture* and society.

environmental impact statement (EIS) A report of results from detailed studies that assess the potential effects on the *environment* that would likely result from development projects or other actions undertaken by the government.

environmental justice The fair and equitable treatment of all people with respect to environmental policy and practice, regardless of their income, race, or ethnicity. Responds to the perception that minorities and the poor suffer more pollution than whites and the rich.

environmental policy *Public policy* that pertains to human interactions with the *environment*. It generally aims to regulate resource use or reduce *pollution* to promote human welfare and/or protect natural systems.

Environmental Protection Agency (EPA) An administrative agency charged with conducting and evaluating research, monitoring environmental quality, setting standards, enforcing those standards, assisting the states in meeting standards and goals for environmental protection, and educating the public.

environmental science The scientific study of how the natural world functions, how our *environment* affects us, and how we affect our environment.

environmental studies An academic *environmental science* program that emphasizes the social sciences as well as the natural sciences.

environmental toxicology The study of *toxicants* that come from or are discharged into the *environment*, including the study of health effects on humans, other animals, and *ecosystems*.

environmentalism A social movement dedicated to protecting the natural world, and by extension, people.

epidemiological study A study that involves large-scale comparisons among groups of people, usually contrasting a group known to have been exposed to some *toxicant* and a group that has not.

epiphyte A plant that grows atop another plant, rather than from soil. Many ferns, mosses, lichens, orchids, and bromeliads in tropical rainforests and cloudforests are epiphytes.

EROI (energy returned on investment) The ratio determined by dividing the quantity of *energy* returned from a process by the quantity of energy invested in the process. Higher EROI ratios mean that more energy is produced from each unit of energy invested. See also *net energy*.

erosion The removal of material from one place and its transport to another by the action of wind or water.

estuary An area where a river flows into the ocean, mixing *fresh water* with salt water.

ethanol The alcohol in beer, wine, and liquor, produced as a *biofuel* by fermenting biomass, generally from *carbohydrate*-rich crops such as corn.

ethical standard A criterion that helps differentiate right from wrong.

ethics The academic study of good and bad, right and wrong. The term can also refer to a person's or group's set of moral principles or values.

European Union (EU) Political and economic organization formed after World War II to promote Europe's economic and social progress. As of 2013, the EU consisted of 27 member nations.

eutrophic Term describing a water body that has high nutrient and low oxygen conditions. Compare *oligotrophic*.

eutrophication The process of *nutrient* enrichment, increased production of organic matter, and subsequent *ecosystem* degradation in a water body.

evaporation The conversion of a substance from a liquid to a gaseous form.

even-aged Condition of timber plantations—generally *monocultures* of a single *species*—in which all trees are of the same age. Most *ecologists* view plantations of even-aged stands more as crop *agriculture* than as ecologically functional *forests*. Compare *uneven-aged*.

evenness See *relative abundance*.

evolution Genetically based change in the appearance, functioning, and/or behavior of organisms across generations, often by the process of *natural selection*.

executive order A specific legal instruction for government agencies ordered by the U.S. president.

experiment An activity designed to test the validity of a *hypothesis* by manipulating *variables*. See *controlled experiment*, *manipulative experiment*, and *natural experiment*.

exploratory drilling Drilling that takes place after a *fossil fuel* deposit has been identified, in order to gauge how much of the fuel exists and whether extraction will prove worthwhile. Involves drilling small holes that descend to great depths.

exponential growth The increase of a *population* (or of anything) by a fixed percentage each year.

external cost A cost borne by someone not involved in an economic transaction. Examples include harm to citizens from *water pollution* or *air pollution* discharged by nearby factories.

extinction The disappearance of an entire *species* from Earth. Compare *extirpation*.

extirpation The disappearance of a particular *population* from a given area, but not the entire *species* globally. Compare *extinction*.

factory fishing A highly industrialized approach to commercial fishing, employing fossil fuels, huge vessels, and powerful new technologies to capture fish in immense volumes. Factory fishing vessels even process and freeze their catches while at sea.

family planning The effort to plan the number and spacing of one's children so as to offer children and parents the best quality of life possible.

farmers' market A market at which local farmers and food producers sell fresh locally grown items.

fee-and-dividend A program of *carbon taxes* in which proceeds from the taxes are paid to consumers as a tax refund or "dividend." This strategy seeks to prevent consumers from losing money if polluters pass their costs along to them.

feedback loop A circular process in which a *system*'s output serves as input to that same system. See *negative feedback loop; positive feedback loop*.

feed-in tariff A program of public policy intended to promote renewable energy investment, whereby utilities are mandated to purchase electricity from homeowners or businesses that generate power from renewable energy sources and feed it into the electric grid. Under such a system, utilities must pay guaranteed premium prices for this power under long-term contract. Compare *net metering*.

feedlot A huge barn or outdoor pen designed to deliver *energy*-rich food to animals living at extremely high densities. Also called a *factory farm* or *concentrated animal feeding operation (CAFO)*.

Ferrel cell One of a pair of cells of *convective circulation* between 30° and 60° north and south latitude that influence global *climate* patterns. Compare *Hadley cell; polar cell*.

fertilizer A substance that promotes plant growth by supplying essential *nutrients* such as *nitrogen* or *phosphorus*.

Fifth Assessment Report A report from the *Intergovernmental Panel on Climate Change* summarizing the current state of climate change research, due out in 2013–2014. This report represents the consensus of scientific climate research from around the world, and supercedes the IPCC's *Fourth Assessment Report*.

first law of thermodynamics The physical law stating that *energy* can change from one form to another, but cannot be created or lost. The total energy in the universe remains constant and is said to be conserved.

flagship species A *species* that has wide appeal with the public and that can be used to promote conservation efforts that also benefit other, less charismatic, species.

flat plate solar collectors Panels for collecting and transmitting *solar energy* for the purpose of heating or cooling. Generally these are rooftop panels made of dark heat-absorbing metal plates mounted in flat glass-covered boxes.

flooding The spillage of water over a river's banks due to heavy rain or snowmelt.

floodplain The region of land over which a river has historically wandered and periodically floods.

flux The movement of nutrients among *pools* or *reservoirs* in a *nutrient cycle*.

fly ash *Particulate matter* from incinerator emissions, often containing pollutants.

food chain A linear series of feeding relationships. As organisms feed on one another, energy is transferred from lower to higher *trophic levels*. Compare *food web*.

food security An adequate, reliable, and available food supply to all people at all times.

food web A visual representation of feeding interactions within an *ecological community* that shows an array of relationships between organisms at different *trophic levels*. Compare *food chain*.

forensics See *forensic science*.

forensic science (forensics) The scientific analysis of evidence to make an identification or answer a question relating to a crime or an accident.

forest Any ecosystem characterized by a high density of trees.

forester A professional who manages *forests* through the practice of *forestry*.

forestry The professional management of *forests*.

forest type A category of *forest* defined by its predominant tree *species*.

formaldehyde A common synthetically made *volatile organic compound* that causes a variety of health impacts.

fossil The remains, impression, or trace of an animal or plant of past geological ages that has been preserved in rock or *sediments*.

fossil fuel A *nonrenewable natural resource*, such as *crude oil, natural gas*, or *coal*, produced by the decomposition and compression of organic matter from ancient life.

fossil record The cumulative body of *fossils* worldwide, which paleontologists study to infer the history of past life on Earth.

Fourth Assessment Report A 2007 report from the *Intergovernmental Panel on Climate Change* that summarized thousands of scientific studies, documenting observed trends in surface temperature, precipitation patterns, snow and ice cover, sea levels, storm intensity, and other factors. It also predicted future changes in these phenomena under a range of emission scenarios; addressed impacts of *climate change* on wildlife, ecosystems, and human societies; and discussed strategies we might pursue in response. The Fourth Assessment Report represented the consensus of scientific climate research from around the world. It will be superceded by the IPCC's *Fifth Assessment Report*, due out in 2013–2014.

fracking See *hydraulic fracturing*.

free rider A party that fails to invest in controlling *pollution* or carrying out other *environmentally* responsible activities and instead relies on the efforts of other parties to do so. For example, a factory that fails to control its emissions gets a "free ride" on the efforts of other factories that do make the sacrifices necessary to reduce emissions.

fresh water Water that is relatively pure, holding very few dissolved salts.

freshwater marsh A type of *wetland* in which shallow water allows plants such as cattails to grow above the water surface. Compare *swamp; bog*.

front The boundary between air masses that differ in temperature and moisture (and therefore density). See *warm front; cold front*.

fuel rods Rods of uranium that supply the fuel for nuclear *fission* and are kept bathed in a *moderator* in a *nuclear reactor*.

Fukushima Daiichi Japanese nuclear power plant severely damaged by the tsunami associated with the March 2011 Tohoku earthquake that rocked Japan. Most radiation drifted over the ocean away from population centers, but the event was history's second most serious nuclear accident. Compare *Chernobyl, Three Mile Island*.

full cost accounting An accounting approach that attempts to summarize all costs and benefits by assigning monetary values to entities without market prices and then generally subtracting costs from benefits. Examples include the *Genuine Progress Indicator*, the Happy Planet Index, and others. Also called *true cost accounting*.

fundamental niche The full *niche* of a *species*. Compare *realized niche*.

fungicide A type of chemical *pesticide* that kills fungi.

gasification A process in which *biomass* is vaporized at extremely high temperatures in the absence of *oxygen*, creating a gaseous mixture including *hydrogen*, *carbon monoxide*, and *methane*, in order to produce *biopower* or *biofuels*. In coal gasification, *coal* is converted to a *syngas* by reacting it with oxygen and steam at a high temperature.

gene A stretch of *DNA* that represents a unit of hereditary information.

genera Plural of *genus*.

general circulation model See *climate model*.

General Mining Act of 1872 U.S. law that legalized and promoted *mining* by private individuals on public lands for just $5 per acre, subject to local customs, with no government oversight.

generalist A *species* that can survive in a wide array of *habitats* or use a wide array of resources. Compare *specialist*.

genetically modified food Food derived from a *genetically modified organism*.

genetically modified organism (GMO) An organism that has been *genetically engineered* using *recombinant DNA* technology.

genetic diversity A measurement of the differences in *DNA* composition among individuals within a given *species*.

genetic engineering Any process scientists use to manipulate an organism's genetic material in the lab by adding, deleting, or changing segments of its *DNA*.

Genuine Progress Indicator (GPI) An *economic* indicator that attempts to differentiate between desirable and undesirable economic activity. The GPI accounts for benefits such as volunteerism and for costs such as environmental degradation and social upheaval. Compare *Gross Domestic Product (GDP)*.

genus A taxonomic level in the Linnaean classification system that is above *species* and below family. A genus is made up of one or more closely related species.

geoengineering Any of a suite of proposed efforts to cool Earth's climate by removing carbon dioxide from the atmosphere or reflecting sunlight away from Earth's surface. Such ideas are controversial and are not nearly ready to implement.

geographic information system (GIS) Computer software that takes multiple types of data (for instance, on geology, hydrology, vegetation, animal species, and human development) and overlays them on a common set of geographic coordinates. The idea is to create a complete picture of a landscape and to analyze how elements of the different datasets are arrayed spatially and how they may be correlated. A common tool of geographers, landscape ecologists, resource managers, and conservation biologists.

geology The scientific study of Earth's physical features, processes, and history.

geostrophic currents Oceanic currents driven by differences in atmospheric pressure and the Coriolis force. Compare *Ekman drift*.

geothermal energy Energy that arises from beneath Earth's surface, ultimately from the radioactive decay of elements amid high pressures deep underground. Can be used to generate electrical power in power plants, for direct heating via piped water, or in *ground-source heat pumps*.

glaciation The spread of ice sheets from the polar regions far into Earth's temperate zones during cold periods of Earth's history.

global climate change Systematic change in aspects of Earth's *climate*, such as temperature, *precipitation*, and storm intensity. Generally refers today to the current warming trend in global temperatures and the many associated climatic changes. Compare *global warming*.

global climate model See *climate model*.

global warming An increase in Earth's average surface temperature. The term is most frequently used in reference to the pronounced warming trend of recent years and decades. Global warming is one aspect of *global climate change* and in turn drives other components of climate change.

global warming potential A quantity that specifies the ability of one *molecule* of a given *greenhouse gas* to contribute to atmospheric warming, relative to *carbon dioxide*.

good A material commodity manufactured for and bought by individuals and businesses.

gray water Wastewater from showers and sinks.

Great Pacific Garbage Patch A portion of the North Pacific *gyre* where currents concentrate plastics and other floating debris that pose danger to marine organisms.

greenbelt A long and wide corridor of parkland, often encircling an entire urban area.

green building (1) A structure that minimizes the ecological footprint of its construction and operation by using sustainable materials, using minimal energy and water, reducing health impacts, limiting pollution, and recycling waste. (2) The pursuit of constructing or renovating such buildings.

green-collar jobs Jobs resulting from new employment opportunities in a more sustainably oriented *economy*, such as jobs in *renewable energy*.

greenhouse effect The warming of Earth's surface and *atmosphere* (especially the *troposphere*) caused by the *energy* emitted by *greenhouse gases*.

greenhouse gas A gas that absorbs infrared radiation released by Earth's surface and then warms the surface and *troposphere* by emitting *energy*, thus giving rise to the *greenhouse effect*. Greenhouse gases include *carbon dioxide* (CO_2), water vapor, *ozone* (O_3), nitrous oxide (N_2O), halocarbon gases, and *methane* (CH_4).

green manure *Organic fertilizer* composed of freshly dead plant material.

Green Revolution An intensification of the industrialization of *agriculture* in the developing world in the latter half of the 20th century that has dramatically increased crop yields produced per unit area of farmland. Practices include devoting large areas to *monocultures* of crops specially bred for high yields and rapid growth; heavy use of *fertilizers, pesticides*, and *irrigation* water; and sowing and harvesting on the same piece of land more than once per year or per season.

green tax A levy on *environmentally* harmful activities and products aimed at providing a market-based incentive to correct for *market failure*. Compare *subsidy*.

greenwashing A public-relations effort by a corporation or institution to mislead customers or the public into thinking it is acting more sustainably than it actually is.

greenway A strip of park land that connects parks or neighborhoods; often located along rivers, streams, or canals.

Gross Domestic Product (GDP) The total monetary value of final *goods* and *services* produced in a country each year. GDP sums all *economic* activity, whether good or bad, and does not account for benefits such as volunteerism or for *external costs* such as environmental degradation and social upheaval. Compare *Genuine Progress Indicator (GPI)*.

gross primary production The *energy* that results when *autotrophs* convert solar energy (sunlight) to energy of chemical bonds in sugars through *photosynthesis*. Autotrophs use a portion of this production to power their own metabolism, which entails oxidizing *organic compounds* by *cellular respiration*. Compare *net primary production*.

ground-level ozone See *tropospheric ozone*.

ground-source heat pump A pump that harnesses *geothermal energy* from near-surface sources of earth and water, in order to heat and cool buildings. Operates on the principle that temperatures below ground are more stable than temperatures above ground.

groundwater Water held in aquifers underground. Compare *surface water*.

gyre An area of the ocean where currents converge and floating debris accumulates.

Haber-Bosch process A process to synthesize ammonia on an industrial scale. Developed by German chemists Fritz Haber and Carl Bosch, the process has enabled humans to double the natural rate of *nitrogen fixation* on Earth and thereby increase *agricultural* productivity, but it has also dramatically altered the *nitrogen cycle*.

habitat The specific *environment* in which an organism lives, including both *biotic* and *abiotic factors*.

habitat conservation plan An arrangement under the Endangered Species Act that allows a landholder or agency to harm an endangered species if it also mitigates the harm by improving habitat for the species.

habitat fragmentation The process by which an expanse of natural *habitat* becomes broken up into discontinuous fragments, often as a result of farming, logging, road building, and other types of human development and land use.

habitat selection The process by which organisms select *habitats* from among the range of options they encounter.

habitat use The process by which organisms use *habitats* from among the range of options they encounter.

Hadley cell One of a pair of cells of *convective circulation* between the equator and 30° north and south latitude that influence global *climate* patterns. Compare *Ferrel cell; polar cell*.

half-life The amount of time it takes for one-half the atoms of a *radioisotope* to emit radiation and decay. Different radioisotopes have different half-lives, ranging from fractions of a second to billions of years.

halocarbon A class of human-made chemical *compounds* derived from simple *hydrocarbons* in which *hydrogen* atoms are replaced by halogen atoms such as bromine, fluorine, or chlorine. Many halocarbons are *ozone-depleting substances* and/or *greenhouse gases*.

harmful algal bloom A *population* explosion of toxic algae caused by excessive *nutrient* concentrations.

hazardous waste Liquid or solid *waste* that is toxic, chemically reactive, flammable, or corrosive. Compare *industrial solid waste; municipal solid waste*.

heat capacity A measure of the heat energy required to increase the temperature of a given substance by a given amount.

herbicide A type of chemical *pesticide* that kills plants.

herbivore An organism that consumes plants. Compare *carnivore*; *omnivore*.

herbivory The consumption of plants by animals.

heterotroph (consumer) An organism that consumes other organisms. Includes most animals, as well as fungi and microbes that decompose organic matter.

high-pressure system An air mass with elevated *atmospheric pressure*, containing air that descends, typically bringing fair *weather*. Compare *low-pressure system*.

homeostasis The tendency of a *system* to maintain constant or stable internal conditions.

horizon A distinct layer of *soil*.

hormone A chemical messenger that travels though the bloodstream to stimulate growth, development, and sexual maturity and to regulate brain function, appetite, sexual drive, and many other aspects of physiology and behavior.

host The organism in a parasitic relationship that suffers harm while providing the *parasite* nourishment or some other benefit.

Hubbert's peak The peak in production of *crude oil* in the United States, which occurred in 1970 just as Shell Oil geologist M. King Hubbert had predicted in 1956.

humus A dark, spongy, crumbly mass of material made up of complex organic compounds, resulting from the partial decomposition of organic matter.

hurricane A cyclonic storm that forms over the ocean but can do damage upon its arrival on land. A type of *cyclone* or *typhoon* that usually forms over the Atlantic Ocean.

hydraulic fracturing A process to extract *shale gas*, in which a drill is sent deep underground and angled horizontally into a shale formation; water, sand, and chemicals are pumped in under great pressure, fracturing the rock; and gas migrates up through the drilling pipe as sand holds the fractures open. Also called *hydrofracking* or simply *fracking*.

hydrocarbon An *organic compound* consisting solely of *hydrogen* and *carbon atoms*.

hydroelectric power (hydropower) The generation of *electricity* using the *kinetic energy* of moving water.

hydrofracking See *hydraulic fracturing*.

hydrogen The chemical *element* with one proton. The most abundant element in the universe. Also a possible fuel for our future economy.

hydrologic cycle The flow of water—in liquid, gaseous, and solid forms—through our *biotic* and *abiotic environment*. Also called the *water cycle*.

hydropower See *hydroelectric power*.

hydrosphere All water—salt or fresh, liquid, ice, or vapor—in surface bodies, underground, and in the *atmosphere*. Compare *biosphere; lithosphere*.

hydrothermal vent A location in the deep ocean where heated water spurts from the seafloor, carrying *minerals* that precipitate to form rocky structures. Unique and recently discovered *ecosystems* cluster around these vents; tubeworms, shrimp, and other creatures here use symbiotic bacteria to derive their energy from chemicals in the heated water rather than from sunlight.

hypothesis A statement that attempts to explain a phenomenon or answer a *scientific* question. Compare *theory*.

hypothesis-driven science Research in which scientists pose questions that seek to explain how and why things are the way they are. Generally proceeds in a somewhat structured manner, using *experiments* to test *hypotheses*. Compare *descriptive science*.

hypoxia The condition of extremely low dissolved *oxygen* concentrations in a body of water.

igneous rock One of the three main categories of rock. Formed from cooling *magma*. Granite and basalt are examples of igneous rock. Compare *metamorphic rock* and *sedimentary rock*.

immigration The arrival of individuals from outside a *population*.

in-situ recovery See *solution mining*.

inbreeding depression A state that occurs in a *population* when genetically similar parents mate and produce weak or defective offspring as a result.

incineration A controlled process of burning solid *waste* for disposal in which mixed garbage is combusted at very high temperatures. Compare *sanitary landfill*.

independent variable The *variable* that the scientist manipulates in an *experiment*.

indoor air pollution *Air pollution* that occurs indoors.

industrial agriculture A form of *agriculture* that uses large-scale mechanization and *fossil fuel* combustion, enabling farmers to replace horses and oxen with faster and more powerful means of cultivating, harvesting, transporting, and processing crops. Other aspects include *irrigation* and the use of *inorganic fertilizers*. Use of chemical herbicides and *pesticides* reduces *competition* from weeds and *herbivory* by insects. Compare *traditional agriculture*.

industrial ecology A holistic approach to industry that integrates principles from engineering, chemistry, *ecology, economics*, and other disciplines and seeks to redesign industrial *systems* in order to reduce resource inputs and minimize inefficiency.

industrial revolution The shift beginning in the mid-1700s from rural life, animal-powered agriculture, and manufacturing by craftsmen to an urban society powered by *fossil fuels*. Compare *agricultural revolution*.

industrial smog "Gray-air" smog caused by the incomplete combustion of *coal* or oil when burned. Compare *photochemical smog*.

industrial solid waste Nonliquid *waste* that is not especially hazardous and that comes from production of consumer goods, mining, *petroleum* extraction and *refining*, and *agriculture*. Compare *hazardous waste; municipal solid waste*.

industrial stage The third stage of the *demographic transition* model, characterized by falling birth rates that close the gap with falling death rates and reduce the rate of *population* growth. Compare *pre-industrial stage, post-industrial stage, transitional stage*.

infant mortality rate The number of deaths of infants under one year of age per 1000 live births in a population.

infectious disease A disease in which a pathogen attacks a host.

inherent value See *intrinsic value*.

inorganic fertilizer A *fertilizer* that consists of mined or synthetically manufactured mineral supplements. Inorganic fertilizers are generally more susceptible than *organic fertilizers* to *leaching* and *runoff* and may be more likely to cause unintended off-site impacts.

insecticide A type of chemical *pesticide* that kills insects.

instrumental value Value ascribed to something for the pragmatic benefits it brings us if we put it to use. Synonymous with *utilitarian value*.

integrated pest management (IPM) The use of multiple techniques in combination to achieve long-term suppression of pests, including *biocontrol*, use of *pesticides*, close monitoring of *populations*, *habitat* alteration, *crop rotation, transgenic* crops, alternative tillage methods, and mechanical pest removal.

intercropping Planting different types of crops in alternating bands or other spatially mixed arrangements.

interdisciplinary field A field that borrows techniques from multiple traditional fields of study and brings together research results from these fields into a broad synthesis.

interest group A small group of people seeking private gain that may work against the larger public interest.

interglacial Term describing relatively warm periods in Earth history that occur between periods of *glaciation*.

Intergovernmental Panel on Climate Change (IPCC) An international panel of *climate* scientists and government officials established in 1988 by the *United Nations Environment Programme* and the World Meteorological Organization. The IPCC's mission is to assess and synthesize scientific research on *global climate change* and to offer guidance to the world's policymakers, primarily through periodic published reports. See *Fourth Assessment Report; Fifth Assessment Report*.

interspecific competition *Competition* that takes place among members of two or more different *species*. Compare *intraspecific competition*.

intertidal Of, relating to, or living along shorelines between the highest reach of the highest *tide* and the lowest reach of the lowest tide.

intraspecific competition *Competition* that takes place among members of the same *species*. Compare *interspecific competition*.

intrinsic value Value ascribed to something for its intrinsic worth; the notion that the thing has a right to exist and is valuable for its own sake. Synonymous with *inherent value*.

introduced species Species introduced by human beings from one place to another (whether intentionally or by accident). A minority of introduced species may become *invasive species*.

invasive species A *species* that spreads widely and rapidly becomes dominant in a *community*, interfering with the community's normal functioning.

inversion layer In a *temperature inversion*, the band of air in which temperature rises with altitude (instead of falling with altitude, as temperature does normally).

ion An electrically charged *atom* or combination of atoms.

ionic bond A type of chemical bonding where electrons are transferred between atoms, creating oppositely charged ions that bond due to their differing electrical charges. Table salt, sodium chloride, is formed by the bonding of positively charged sodium ions with negatively charged chloride ions.

ionizing radiation A high-energy form of radiation that can damage the cells of living things. Sources of this radiation include the sun and radioactive particles from *nuclear energy* and natural sources.

IPAT model A formula that represents how humans' total impact (I) on the *environment* results from the interaction among three factors: *population* (P), affluence (A), and technology (T).

irrigation The artificial provision of water to support *agriculture*.

island biogeography theory *Theory* initially applied to oceanic islands to explain how *species* come to be distributed among them. Researchers have increasingly applied the theory to islands of *habitat* (patches of one type of habitat isolated within "seas" of others). Aspects of the theory include *immigration* and *extinction* rates, the effect of island size (*area effect*), and the effect of distance from the mainland (*distance effect*). Full name is the equilibrium theory of island biogeography.

isotope One of several forms of an *element* having differing numbers of *neutrons* in the nucleus of its *atoms*. Chemically, isotopes of an element behave almost identically, but they have different physical properties because they differ in mass.

jet stream A high-altitude air current that blows west to east, meandering north and south, and which influences weather across much of North America and Eurasia.

kelp Large brown algae or seaweed that can form underwater "forests," providing habitat for marine organisms.

kerogen A substance derived from deeply buried organic matter that acts as a source material for both *natural gas* and *crude oil*.

keystone species A *species* that has an especially far-reaching effect on a *community*.

kinetic energy *Energy* of motion. Compare *potential energy*.

K–selected Term denoting a *species* with low *biotic potential* whose members produce a small number of offspring and take a long time to gestate and raise each of their young, but invest heavily in promoting the survival and growth of these few offspring. *Populations* of K–selected species are generally regulated by *density-dependent factors*. Compare *r–selected*.

kwashiorkor A form of *malnutrition* that results from a high-*starch* diet with inadequate *protein* or *amino acids*. In children, causes bloating of the abdomen, deterioration and discoloration of hair, mental disability, immune suppression, developmental delays, anemia, and reduced growth.

Kyoto Protocol An international agreement drafted in 1997 that called for reducing, by 2012, emissions of six *greenhouse gases* to levels lower than their levels in 1990. It has been extended to 2020 until a replacement treaty can be reached. An outgrowth of the *United Nations Framework Convention on Climate Change*.

land degradation A general deterioration of land that diminishes its productivity and biodiversity, impairs the functioning of its ecosystems, and reduces the ecosystem services the land can offer us. Compare *desertification*; *soil degradation*.

landfill gas A mix of gases that consists of roughly half *methane* produced by anaerobic decomposition deep inside *landfills*.

landrace A locally adapted domesticated variety of an agricultural crop native to a particular area.

landscape ecology The study of how landscape structure affects the abundance, distribution, and interaction of organisms. This approach to the study of organisms and their *environments* at the landscape scale focuses on broad geographical areas that include multiple *ecosystems*.

landslide The collapse and downhill flow of large amounts of rock or soil. A severe and sudden form of *mass wasting*.

land trust A local or regional organization that preserves lands valued by its members. In most cases, land trusts purchase land outright with the aim of preserving it in its natural condition.

La Niña An exceptionally strong cooling of surface water in the equatorial Pacific Ocean that occurs every 2 to 7 years and has widespread climatic consequences. Compare *El Niño*.

latitudinal gradient The increase in *species richness* as one approaches the equator. This pattern of variation with latitude has been one of the most obvious patterns in *ecology*, but one of the most difficult ones for scientists to explain.

lava *Magma* that is released from the *lithosphere* and flows or spatters across Earth's surface.

law of conservation of matter The physical law stating that *matter* may be transformed from one type of substance into others, but that it cannot be created or destroyed.

LD_{50} (lethal dose–50%) The amount of a *toxicant* it takes to kill 50% of a *population* of test animals. Compare *ED_{50}*; *threshold dose*.

leachate Liquids that seep through liners of a *sanitary landfill* and leach into the *soil* underneath.

leaching The process by which solid materials such as *minerals* are dissolved in a liquid (usually water) and transported to another location.

lead A heavy metal that may be ingested through water or paint, or that may enter the *atmosphere* as a particulate pollutant through combustion of leaded gasoline or other processes. Atmospheric lead deposited on land and water can enter the *food chain*, accumulate within body tissues, and cause *lead poisoning* in animals and people. An EPA *criteria pollutant*.

Leadership in Energy and Environmental Design (LEED) The leading set of standards for sustainable building. Compare *green building*.

lead poisoning Poisoning by ingestion or inhalation of the heavy metal *lead*, causing an array of maladies including damage to the brain, liver, kidney, and stomach; learning problems and behavioral abnormalities; anemia; hearing loss; and even death. Lead poisoning can result from drinking water that passes through old lead pipes or ingesting dust or chips of old lead-based paint.

legislation Statutory law.

Leopold, Aldo (1887–1949) American scientist, scholar, philosopher, and author. His book *The Land Ethic* argued that humans should view themselves and the land itself as members of the same *community* and that humans are obligated to treat the land *ethically*.

levee See *dike*.

life-cycle analysis A quantitative analysis of inputs and outputs across the entire life cycle of a product—from its origins, through its production, transport, sale, and use, and finally its disposal—in an attempt to judge the sustainability of the process and make it more ecologically efficient.

life expectancy The average number of years that individuals in particular age groups are likely to continue to live.

lifestyle hazard See *cultural hazard*.

light pollution Pollution from urban or suburban lights that obscures the night sky, impairing people's visibility of stars.

light rail A *mass transit* rail system of trains powered by electricity, often at a moderate scale.

limiting factor A physical, chemical, or biological characteristic of the *environment* that restrains *population* growth.

limnetic zone In a water body, the layer of open water through which sunlight penetrates. Compare *littoral zone; benthic zone; profundal zone*.

lipids A class of chemical compounds that do not dissolve in water and are used in organisms for energy storage, for structural support, and as key components of cellular membranes.

liquefied natural gas (LNG) *Natural gas* that has been converted to a liquid at low temperatures and that can be shipped long distances in refrigerated tankers.

lithosols *Soils* with high mineral content and low organic matter content. Common in *deserts*.

lithosphere The outer layer of Earth, consisting of *crust* and uppermost *mantle* and located just above the *asthenosphere*. More generally, the solid part of Earth, including the rocks, *sediment*, and *soil* at the surface and extending down many miles underground. Compare *atmosphere*, *biosphere*, and *hydrosphere*.

littoral Near the shore of a water body. See *intertidal*.

littoral zone The region ringing the edge of a water body. Compare *benthic zone; limnetic zone; profundal zone*.

loam *Soil* with a relatively even mixture of *clay-*, *silt-*, and *sand*-sized particles.

lobbying The expenditure of time or money in an attempt to influence an elected official.

logistic growth curve A plot that shows how the initial *exponential growth* of a *population* is slowed and finally brought to a standstill by *limiting factors*.

longline fishing Fishing practice that involves setting out extremely long lines with up to several thousand baited hooks spaced along their lengths. Kills turtles, sharks, and an estimated 300,000 seabirds each year as *bycatch*.

low-input agriculture *Agriculture* that uses smaller amounts of *pesticides*, *fertilizers*, growth hormones, water, and *fossil fuel* energy than are used in *industrial agriculture*. Compare *sustainable agriculture*; *organic agriculture*.

low-pressure system An air mass in which air moves toward the low *atmospheric pressure* at the center of the system and spirals upward, typically bringing clouds and *precipitation*. Compare *high-pressure system*.

macromolecule A very large molecule, such as a *protein, nucleic acid, carbohydrate*, or lipid.

macronutrients Elements and compounds required in relatively large amounts by organisms. Examples include nitrogen, carbon, and phosphorus.

magma Molten, liquid rock.

malnutrition The condition of lacking *nutrients* the body needs, including a complete complement of vitamins and minerals.

Malthus, Thomas (1766–1834) British economist who maintained that increasing human *population* would eventually deplete the available food supply until starvation, war, or disease arose and reduced the population.

mangrove A tree with a unique type of roots that curve upward to obtain *oxygen*, which is lacking in the mud in which they grow, and that serve as stilts to support the tree in changing water levels. Mangrove forests grow on the coastlines of the tropics and subtropics.

manipulative experiment An *experiment* in which the researcher actively chooses and manipulates the *independent variable*. Compare *natural experiment*.

mantle The malleable layer of rock that lies beneath Earth's *crust* and surrounds a mostly iron *core*.

marasmus A form of *malnutrition* that results from *protein* deficiency together with a lack of calories, causing wasting or shriveling among millions of children in the developing world.

marine protected area (MPA) An area of the ocean set aside to protect marine life from fishing pressures. An MPA may be protected from some human activities but be open to others. Compare *marine reserve*.

marine reserve An area of the ocean designated as a "no-fishing" zone, allowing no extractive activities. Compare *marine protected area (MPA)*.

market failure The failure of markets to take into account the *environment*'s positive effects on *economies* (for example, *ecosystem services*) or to reflect the negative effects of economic activity on the environment and thereby on people (*external costs*).

mass extinction event The *extinction* of a large proportion of the world's *species* in a very short time period due to some extreme and rapid change or catastrophic event. Earth has seen five mass extinction events in the past half-billion years.

mass transit A public transportation system for a metropolitan area that moves large numbers of people at once. Buses, trains, subways, streetcars, trolleys, and *light rail* are types of mass transit.

mass wasting The downslope movement of soil and rock due to gravity. Compare *landslide*.

materials recovery facility (MRF) A *recycling* facility where items are sorted, cleaned, shredded, and prepared for reprocessing into new items.

matter All material in the universe that has mass and occupies space. See *law of conservation of matter*.

maximum sustainable yield The maximal harvest of a particular *renewable natural resource* that can be accomplished while still keeping the resource available for the future.

meandering river A river that winds its way across its floodplain, depositing sediment on the riverbank on the insides of curves and eroding the riverbanks on the outside of curves. Compare *braided river*.

meltdown The accidental melting of the uranium *fuel rods* inside the core of a *nuclear reactor*, causing the release of radiation.

mesosphere The atmospheric layer above the *stratosphere*, extending 50–80 km (31–56 mi) above sea level.

metal A type of chemical *element*, or a mass of such an element, that typically is lustrous, opaque, and malleable and that can conduct heat and electricity.

metamorphic rock One of the three main categories of rock. Formed by great heat and/or pressure that reshapes crystals within the rock and changes its appearance and physical properties. Common metamorphic rocks include marble and slate. Compare *igneous rock* and *sedimentary rock*.

metapopulation A network of subpopulations, most of whose members stay within their respective landscape *patches*, but some of whom move among patches or mate with members of other patches.

methane (CH_4) A colorless gas produced primarily by *anaerobic* decomposition. The major constituent of *natural gas* and a *greenhouse gas* that is molecule-for-molecule more potent than *carbon dioxide*.

methane hydrate An ice-like solid consisting of molecules of *methane* embedded in a crystal lattice of water molecules. Most is found in sediments on the continental shelves and in the Arctic. Methane hydrate is a potential alternative *fossil fuel*.

microclimate A small-scale localized pattern of weather conditions.

micronutrients Elements and compounds required in relatively small amounts by organisms. Examples include zinc, copper, and iron.

Milankovitch cycle One of three types of variations in Earth's rotation and orbit around the sun that result in slight changes in the relative amount of solar radiation reaching Earth's surface at different latitudes. As the cycles proceed, they change the way solar radiation is distributed over Earth's surface and contribute to changes in *atmospheric* heating and circulation that have triggered *glaciations* and other *climate* changes.

Millennium Development Goals A program of targets for *sustainable development* set by the international community through the *United Nations* at the turn of this century.

mineral A naturally occurring solid *element* or inorganic *compound* with a crystal structure, a specific chemical composition, and distinct physical properties. Compare *ore* and *rock*.

mining (1) In the broad sense, the extraction of any resource that is nonrenewable on the timescale of our society (such as *fossil fuels* or *groundwater*). (2) In relation to *mineral* resources, the systematic removal of *rock*, *soil*, or other material for the purpose of extracting minerals of economic interest.

mitigation The pursuit of strategies to lessen the severity of climate change, notably by reducing emissions of greenhouse gases. Compare *adaptation*.

mixed economy An economy that combines elements of a *capitalist market economy* and a *centrally planned economy*.

model A simplified representation of a complex natural process, designed by scientists to help understand how the process occurs and to make predictions.

moderator Within a *nuclear reactor*, a substance, most often water or graphite, that slows the *neutrons* bombarding uranium so that *fission* can begin.

molecule A combination of two or more *atoms*.

monoculture The uniform planting of a single crop over a large area. Characterizes *industrial agriculture*. Compare *polyculture*.

Montreal Protocol International treaty ratified in 1987 in which 180 (now 196) signatory nations agreed to restrict production of *chlorofluorocarbons (CFCs)* in order to halt stratospheric *ozone* depletion. This was a protocol of the Vienna Convention for the Protection of the Ozone Layer. The Montreal Protocol is widely considered the most successful effort to date in addressing a global *environmental* problem.

mortality Rate of death within a *population*.

mosaic In *landscape ecology*, a spatial configuration of *patches* arrayed across a landscape.

mountaintop removal mining A large-scale form of *coal* mining in which entire mountaintops are leveled. The technique exerts extreme environmental impact on surrounding *ecosystems* and human residents.

Muir, John (1838–1914) Scottish immigrant to the United States who eventually settled in California and made the Yosemite Valley his wilderness home. Today, he is most strongly associated with the *preservation ethic*. He argued that nature deserved protection for its own inherent values (an *ecocentrist* argument) but also claimed that nature facilitated human happiness and fulfillment (an *anthropocentrist* argument).

multiple use A principle guiding management policy for *national forests* specifying that forests be managed for recreation, wildlife habitat, mineral extraction, water quality, and other uses, as well as for timber extraction.

municipal solid waste Nonliquid *waste* that is not especially hazardous and that comes from homes, institutions, and small businesses. Compare *hazardous waste; industrial solid waste*.

mutagen A *toxicant* that causes *mutations* in the *DNA* of organisms.

mutation An accidental change in *DNA* that may range in magnitude from the deletion, substitution, or addition of a single nucleotide to a change affecting entire sets of chromosomes. Mutations provide the raw material for evolutionary change.

mutualism A relationship in which all participating organisms benefit from their interaction. Compare *parasitism*.

nacelle Compartment in a *wind turbine* containing machinery for generating power.

natality Rate of birth within a *population*.

national ambient air quality standards (NAAQS) Maximum allowable concentrations of *criteria pollutants* in ambient outdoor air, set by the U.S. *EPA*.

National Environmental Policy Act (NEPA) A U.S. law enacted on January 1, 1970, that created an agency called the Council on Environmental Quality and required that an *environmental impact statement* be prepared for any major federal action.

national forest An area of forested public land managed by the U.S. Forest Service. The system consists of 191 million acres (more than 8% of the nation's land area) in many tracts spread across all but a few states.

National Forest Management Act *Legislation* passed by the U.S. Congress in 1976, mandating that plans for renewable resource management be drawn up for every *national forest*. These plans were to be explicitly based on the concepts of *multiple use* and *maximum sustainable yield* and be open to broad public participation.

national monument An area of designated public land that may later become a *national park*.

national park A scenic area set aside for recreation and enjoyment by the public and managed by the National Park Service. The U.S. national park system today numbers 397 sites totaling 84 million acres and includes national historic sites, national recreation areas, national wild and scenic rivers, and other areas.

national wildlife refuge An area of public land set aside to serve as a haven for wildlife and also sometimes to encourage hunting, fishing, wildlife observation, photography, environmental education, and other uses. The system of 550 sites is managed by the U.S. Fish and Wildlife Service.

natural capital Earth's accumulated wealth of resources.

natural experiment An *experiment* in which the researcher cannot directly manipulate the *variables* and therefore must observe nature, comparing conditions in which variables differ, and interpret the results. Compare *manipulative experiment*.

natural gas A *fossil fuel* consisting primarily of *methane* (CH_4) and including varying amounts of other volatile hydrocarbons.

natural rate of population change See *rate of natural increase*.

natural resource Any of the various substances and *energy* sources that we take from our environment and that we need in order to survive.

Natural Resources Conservation Service (NRCS) U.S. agency that promotes soil conservation, as well as water quality protection and pollution control. Prior to 1994, known as the Soil Conservation Service.

natural sciences Academic disciplines that study the natural world. Compare *social sciences*.

natural selection The process by which traits that enhance survival and reproduction are passed on more frequently to future generations of organisms than traits that do not, thus altering the genetic makeup of populations through time. Natural selection acts on genetic variation and is a primary driver of *evolution*.

negative feedback loop A *feedback loop* in which output of one type acts as input that moves the *system* in the opposite direction. The input and output essentially neutralize each other's effects, stabilizing the system. Compare *positive feedback loop*.

neoclassical economics A mainstream economic school of thought that explains market prices in terms of consumer preferences for units of particular commodities and that uses *cost-benefit analysis*. Compare *ecological economics; environmental economics*.

net energy The quantitative difference between *energy* returned from a process and *energy* invested in the process. Positive net energy values mean that a process produces more energy than is invested. See also *energy conversion efficiency; EROI*.

net metering Process by which homeowners or business with *photovoltaic* systems or *wind turbines* can sell their excess *solar energy* or *wind power* to their local utility. Whereas *feed-in tariffs* award producers with prices above market rates, net metering offers market-rate prices.

net primary production The *energy* or biomass that remains in an ecosystem after *autotrophs* have metabolized enough for their own maintenance through *cellular respiration*. Net primary production is the energy or biomass available for consumption by *heterotrophs*. Compare *gross primary production; secondary production*.

net primary productivity The rate at which *net primary production* is produced. See *productivity; gross primary production; net primary production; secondary production*.

neurotoxin A *toxicant* that assaults the nervous system. Neurotoxins include heavy metals, *pesticides*, and some chemical weapons developed for use in war.

neutron An electrically neutral (uncharged) particle in the nucleus of an *atom*.

new forestry A set of *ecosystem-based management* approaches for harvesting timber that explicitly mimic natural disturbances. For instance, "sloppy clear-cuts" that leave a variety of trees standing mimic the changes a *forest* might experience if hit by a severe windstorm.

new urbanism An approach among architects, planners, and developers that seeks to design neighborhoods in which homes, businesses, schools, and other amenities are within walking distance of one another. Proponents of new urbanism aim to combat *sprawl* by creating functional neighborhoods in which families can meet most of their needs close to home without the use of a car.

niche The functional role of a *species* in a *community*.

nitrification The conversion by bacteria of ammonium ions (NH_4^+) first into nitrite ions (NO_2^-) and then into nitrate ions (NO_3^-).

nitrogen The chemical *element* with seven *protons* and seven *neutrons*. The most abundant element in the *atmosphere*, a key element in *macromolecules*, and a crucial plant *nutrient*.

nitrogen cycle A major *nutrient cycle* consisting of the routes that *nitrogen atoms* take through the nested networks of environmental *systems*.

nitrogen dioxide (NO_2) A foul-smelling reddish brown gas that contributes to *smog* and *acid deposition*. It results when atmospheric *nitrogen* and *oxygen* react at the high temperatures created by combustion engines. An EPA *criteria pollutant*.

nitrogen fixation The process by which inert *nitrogen* gas combines with *hydrogen* to form ammonium ions (NH_4^+), which are chemically and biologically active and can be taken up by plants.

nitrogen-fixing bacteria Bacteria that live independently in the soil or water, or those that form *mutualistic* relationships with many types of plants and provide *nutrients* to the plants by converting gaseous *nitrogen* to a usable form.

nitrogen oxide (NO_x) One of a family of compounds that includes nitric oxide (NO) and nitrogen dioxide (NO_2).

no-analog community (novel community) An ecological *community* composed of a novel mixture of organisms, with no current analog or historical precedent.

noise pollution Undesired ambient sound.

nonconsumptive use *Fresh water* use in which the water from a particular *aquifer* or surface water body either is not removed or is removed only temporarily and then returned. The use of water to generate electricity in hydroelectric *dams* is an example. Compare *consumptive use*.

nongovernmental organization (NGO) An organization not affiliated with any national government, and frequently international in scope, that pursues a particular mission or advocates for a particular cause.

nonmarket value A value that is not usually included in the price of a *good* or *service*.

non-point source A diffuse source of *pollutants*, often consisting of many small sources. Compare *point source*.

nonrenewable natural resources *Natural resources* that are in limited supply and are formed much more slowly than we use them. Compare *renewable natural resources*.

North American Free Trade Agreement (NAFTA) A 1994 treaty among Canada, Mexico, and the United States that reduced or eliminated barriers to trade (such as tariffs) among these nations. Side agreements were negotiated to minimize the degree to which protections for workers and the environment were undermined.

North Atlantic Deep Water (NADW) The deep portion of the *thermohaline circulation* in the northern Atlantic Ocean.

no-till *Agriculture* that does not involve tilling (plowing, disking, harrowing, or chiseling) the *soil*. The most intensive form of *conservation tillage*.

not-in-my-backyard (NIMBY) Syndrome in which people do not want something (e.g., a polluting facility) near where they live, even if they may want or need the thing to exist somewhere else.

novel community See *no-analog community*.

nuclear energy The *energy* that holds together *protons* and *neutrons* within the nucleus of an *atom*. Several processes, each of which involves transforming *isotopes* of one *element* into isotopes of other elements, can convert nuclear energy into thermal energy, which is then used to generate *electricity*. See also *nuclear fission; nuclear reactor*.

nuclear fission The conversion of the *energy* within an *atom*'s nucleus to usable thermal energy by splitting apart atomic nuclei. Compare *nuclear fusion*.

nuclear fusion The conversion of the *energy* within an *atom*'s nucleus to usable thermal energy by forcing together the small nuclei of lightweight *elements* under extremely high temperature and pressure. Developing a commercially viable method of nuclear fusion remains an elusive goal.

nuclear reactor A facility within a nuclear power plant that initiates and controls the process of *nuclear fission* in order to generate electricity.

nucleic acid A *macromolecule* that directs the production of *proteins*. Includes *DNA* and *RNA*.

nutrient An *element* or *compound* that organisms consume and require for survival.

nutrient cycle The comprehensive set of cyclical pathways by which a given *nutrient* moves through the *environment*.

observational science See *descriptive science*.

ocean acidification The process by which today's oceans are becoming more *acidic* (attaining lower *pH*) as a result of increased *carbon dioxide* concentrations in the atmosphere. Ocean acidification occurs as ocean water absorbs CO_2 from the air and forms carbonic acid. This impairs the ability of corals and other organisms to build exoskeletons of calcium carbonate, imperiling coral reefs and the many organisms that depend on them.

ocean thermal energy conversion (OTEC) A potential *energy* source that involves harnessing the solar radiation absorbed by tropical ocean water. See *closed cycle*; *open cycle*.

oil See *petroleum*.

oil sands (tar sands) Deposits that can be mined from the ground, consisting of moist sand and clay containing 1–20% *bitumen*. Oil sands represent crude oil deposits that have been degraded and chemically altered by water *erosion* and bacterial decomposition. Widely envisioned as a replacement for *crude oil* as this resource is depleted.

oil shale *Sedimentary rock* filled with *kerogen* that can be processed to produce liquid *petroleum*. Oil shale is formed by the same processes that form *crude oil* but occurs when kerogen was not buried deeply enough or subjected to enough heat and pressure to form oil.

oligotrophic Term describing a water body that has low nutrient and high oxygen conditions. Compare *eutrophic*.

omnivore An organism that consumes both plants and animals. Compare *carnivore*; *herbivore*.

open cycle An approach in *ocean thermal energy conversion* whereby warm surface water is evaporated in a vacuum, its steam turns turbines, and it is then condensed by cold water.

open pit mining A *mining* technique that involves digging a gigantic hole and removing the desired *ore*, along with waste *rock* that surrounds the ore.

ore A *mineral* or grouping of minerals from which we extract *metals*.

organic agriculture *Agriculture* that uses no synthetic *fertilizers* or *pesticides* but instead relies on biological approaches such as *composting* and *biocontrol*. Compare *low-input agriculture*; *sustainable agriculture*.

organic compound A *compound* made up of *carbon atoms* (and, generally, *hydrogen* atoms) joined by covalent bonds and sometimes including other *elements*, such as *nitrogen*, *oxygen*, sulfur, or *phosphorus*. The unusual ability of carbon to build elaborate *molecules* has resulted in millions of different organic compounds showing various degrees of complexity.

organic fertilizer A *fertilizer* made up of natural materials (largely the remains or wastes of organisms), including animal manure; crop residues, fresh vegetation, and compost. Compare *inorganic fertilizer*.

organismal ecology The scientific study of an organism and its relationships to its environment.

Organization of Petroleum Exporting Countries (OPEC) Cartel of predominantly Arab nations that in 1973 embargoed oil shipments to the United States and other nations supporting Israel, setting off the nation's first oil shortage.

outdoor air pollution *Air pollution* that occurs outdoors. Also called *ambient air pollution*.

overburden The *rock* and *soil* that are removed from a site to mine the *minerals* underneath. See *strip mining*.

overgrazing The consumption by too many animals of plant cover, impeding plant regrowth and the replacement of *biomass*. Overgrazing can exacerbate damage to *soils*, natural *communities*, and the land's productivity for further grazing.

overnutrition A condition of excessive food intake in which people receive more than their daily caloric needs.

overshoot The amount by which humanity's resource use, as measured by its *ecological footprint*, has surpassed Earth's long-term capacity to support us.

oxbow lake A U-shaped water body that becomes isolated from a river when river water erodes a shortcut from one end of an oxbow to the other, so that the river pursues a direct course.

oxygen The chemical *element* with eight *protons* and eight *neutrons*. A key element in the *atmosphere* that is produced by *photosynthesis*.

ozone (O_3) A *molecule* consisting of three atoms of *oxygen*. Absorbs ultraviolet radiation in the *stratosphere*. Compare *ozone layer* and *tropospheric ozone*.

ozone-depleting substances Airborne chemicals, such as *halocarbons*, that destroy *ozone* molecules and thin the *ozone layer* in the *stratosphere*.

ozone hole Term popularly used to describe the thinning of the stratospheric *ozone layer* that occurs over Antarctica each year, as a result of *chlorofluorocarbons (CFCs)* and other *ozone-depleting substances*.

ozone layer A portion of the *stratosphere*, roughly 17–30 km (10–19 mi) above sea level, that contains most of the *ozone* in the *atmosphere*.

paleoclimate Climate in the geologic past.

paper park A term referring to parks that are protected "on paper" but not in reality.

paradigm A dominant philosophical and theoretical framework within a scientific discipline.

parasite The organism in a parasitic relationship that extracts nourishment or some other benefit from the *host*.

parasitism A relationship in which one organism, the *parasite*, depends on another, the *host*, for nourishment or some other benefit while simultaneously doing the host harm. Compare *mutualism*.

parasitoid An insect that parasitizes other insects, generally causing eventual death of the *host*. Compare *parasite*.

parent material The base geologic material in a particular location.

particulate matter Solid or liquid particles small enough to be suspended in the *atmosphere* and able to damage respiratory tissues when inhaled. Includes *primary pollutants* such as dust and soot as well as *secondary pollutants* such as sulfates and nitrates. An EPA *criteria pollutant*.

passive solar energy collection An approach in which buildings are designed and building materials are chosen to maximize their direct absorption of sunlight in winter and to keep the interior cool in the summer. Compare *active solar energy collection*.

patch In *landscape ecology*, spatial areas within a landscape. Depending on a researcher's perspective, patches may consist of habitat for a particular organism, or communities, or ecosystems. An array of patches forms a *mosaic*.

pathogen A *parasite* that causes disease in its host.

pathway inhibitor A *toxicant* that interrupts vital biochemical processes in organisms by blocking one or more steps in important biochemical pathways. Compounds in the herbicide atrazine kill plants by blocking key steps in the process of *photosynthesis*.

peace park A *transboundary park* intended to help ease tensions by acting as a buffer between nations that have quarreled over boundary disputes.

peak oil Term used to describe the point of maximum production of *petroleum* in the world (or for a given nation), after which oil production declines. This is also expected to be roughly the midway point of extraction of the world's oil supplies. Compare *Hubbert's peak*.

peat A kind of precursor stage to *coal*, produced when organic material that is broken down by *anaerobic* decomposition remains wet, near the surface, and poorly compressed.

peer review The process by which a manuscript submitted for publication in an academic journal is examined by specialists in the field, who provide comments and criticism (generally anonymously) and judge whether the work merits publication in the journal.

pelagic Of, relating to, or living between the surface and floor of the ocean. Compare *benthic*.

permafrost In *tundra*, underground soil that remains more or less permanently frozen.

permanent gas A gas that remains at a stable concentration in the *atmosphere*.

permit trading The practice of buying and selling government-issued marketable emissions permits to conduct environmentally harmful activities. Under a *cap-and-trade* system, the government determines an acceptable level of *pollution* and then issues permits to pollute. A company receives credit for amounts it does not emit and can then sell this credit to other companies. Compare *emissions trading system*.

pest A pejorative term for any organism that damages crops that are valuable to us. The term is subjective and defined by our own economic interests and is not biologically meaningful. Compare *weed*.

pesticide An artificial chemical used to kill insects (*insecticide*), plants (*herbicide*), or fungi (*fungicide*).

petroleum See *crude oil*. However, the term is also used to refer to both *oil* and *natural gas* together.

pH A measure of the concentration of *hydrogen ions* in a *solution*. The pH scale ranges from 0 to 14: A solution with a pH of 7 is neutral; solutions with a pH below 7 are *acidic*, and those with a pH higher than 7 are *basic*. Because the pH scale is logarithmic, each step on the scale represents a 10-fold difference in hydrogen ion concentration.

phase shift A fundamental shift in the overall character of an ecological community, generally occurring after some extreme disturbance, and after which the community may not return to its original state. Also known as a *regime shift*.

phosphorus cycle A major *nutrient cycle* consisting of the routes that *phosphorus atoms* take through the nested networks of environmental *systems*.

photic zone In the ocean or a freshwater body, the well-lit top layer of water where *photosynthesis* occurs.

photochemical smog "Brown-air" smog caused by light-driven reactions of *primary pollutants* with normal atmospheric *compounds* that produce a mix of over 100 different chemicals, ground-level *ozone* often being the most abundant among them. Compare *industrial smog*.

photoelectric effect Effect that occurs when light strikes one of a pair of metal plates in a *photovoltaic cell*, causing the release of *electrons*, which are attracted by electrostatic forces to the opposing plate. The flow of electrons from one plate to the other creates an electrical current.

photosynthesis The process by which *autotrophs* produce their own food. Sunlight powers a series of chemical reactions that convert *carbon dioxide* and water into sugar (glucose), thus transforming low-quality *energy* from the sun into high-quality energy the organism can use. Compare *cellular respiration*.

photovoltaic (PV) cell A device designed to collect sunlight and directly convert it to electrical *energy* by making use of the *photoelectric effect*.

photovoltaic effect See *photoelectric effect*.

phylogenetic tree A treelike diagram that represents the history of divergence of *species* or other taxonomic groups of organisms.

physical hazard Physical processes that occur naturally in our environment and pose human health hazards. These include discrete events such as *earthquakes*, *volcanic eruptions*, fires, floods, blizzards, *landslides*, hurricanes, and droughts, as well as ongoing natural phenomena such as ultraviolet radiation from sunlight. Compare *biological hazard; chemical hazard; cultural hazard*.

phytoplankton Microscopic photosynthetic algae, protists, and cyanobacteria that drift near the surface of water bodies and generally form the first *trophic level* in an aquatic *food chain*. Compare *zooplankton*.

Pinchot, Gifford (1865–1946) The first professionally trained American *forester*, Pinchot helped establish the U.S. Forest Service. Today, he is the person most closely associated with the *conservation ethic*.

pioneer species A *species* that arrives earliest, beginning the ecological process of *succession* in a terrestrial or aquatic *community*.

placer mining A *mining* technique that involves sifting through material in modern or ancient riverbed deposits, generally using running water to separate lightweight mud and gravel from heavier *minerals* of value.

plastics Synthetic (human-made) *polymers* used in numerous manufactured products.

plate tectonics The process by which Earth's surface is shaped by the extremely slow movement of tectonic plates, or sections of *crust*. Earth's surface includes about 15 major tectonic plates. Their interaction gives rise to processes that build mountains, cause *earthquakes*, and otherwise influence the landscape.

poaching The illegal killing of wildlife, usually for meat or body parts.

point source A specific spot—such as a factory—where large quantities of *air pollutants* or *water pollutants* are discharged. Compare *non-point source*.

polar cell One of a pair of cells of *convective circulation* between the poles and 60° north and south latitude that influence global *climate* patterns. Compare *Ferrel cell; Hadley cell*.

polar stratospheric clouds High-altitude icy clouds containing condensed nitric acid, which enhance the destruction of stratospheric *ozone* in the spring.

polar vortex Circular wind currents that trap air over Antarctica during winter, worsening *ozone* depletion over the continent.

policy A rule or guideline that directs individual, organizational, or societal behavior.

pollination A plant-animal interaction in which one organism (for example, a bee or a hummingbird) transfers pollen (containing male sex cells) from flower to flower, fertilizing ovaries (containing female sex cells) that grow into fruits with seeds.

polluter-pays principle Principle specifying that the party responsible for producing *pollution* should pay the costs of cleaning up the pollution or mitigating its impacts.

pollution The release of matter or *energy* into the *environment* that causes undesirable impacts on the health and well-being of humans or other organisms. Pollution can be physical, chemical, or biological, and it can affect water, air, or soil.

polybrominated diphenyl ethers (PBDEs) Synthetic compounds that provide fire-retardant properties and are used in a diverse array of consumer products, including computers, televisions, plastics, and furniture. Released during production, disposal, and use of products, these chemicals persist and accumulate in living tissue and appear to be *endocrine disruptors*.

polyculture The planting of multiple crops in a mixed arrangement or in close proximity. An example is some traditional Native American farming that mixed maize, beans, squash, and peppers. Compare *monoculture*.

polymer A chemical *compound* or mixture of compounds consisting of long chains of repeated *molecules*. Important biological molecules, such as *DNA* and *proteins*, are examples of polymers.

pool A location in which nutrients in a *biogeochemical cycle* remain for a period of time before moving to another pool. Can be living or nonliving entities. Synonymous with *reservoir*. Compare *flux; residence time*.

population A group of organisms of the same *species* that live in the same area. Species are often composed of multiple populations.

population density The number of individuals within a *population* per unit area. Compare *population size*.

population distribution The spatial arrangement of organisms within a particular area.

population ecology The study of the quantitative dynamics of population change and the factors that affect the distribution and abundance of members of a population.

population growth rate The rate of change in a *population*'s size per unit time (generally expressed in percent per year), taking into accounts births, deaths, immigration, and emigration. Compare *rate of natural increase*.

population size The number of individual organisms present at a given time in a *population*.

positive feedback loop A *feedback loop* in which output of one type acts as input that moves the *system* in the same direction. The input and output drive the system further toward one extreme or another. Compare *negative feedback loop*.

post-industrial stage The fourth and final stage of the *demographic transition* model, in which both birth and death rates have fallen to a low level and remain stable there, and *populations* may even decline slightly. Compare *industrial stage, pre-industrial stage, transition stage*.

potential energy *Energy* of position. Compare *kinetic energy*.

prairie See *temperate grassland*.

precautionary principle The idea that one should not undertake a new action until the ramifications of that action are well understood.

precedent A legal ruling that serves as a guide for later cases, steering judicial decisions through time.

precipitation Water that condenses out of the *atmosphere* and falls to Earth in droplets or crystals.

pre-industrial stage The first stage of the *demographic transition* model, characterized by conditions that defined most of human history. In pre-industrial societies, both death rates and birth rates are high. Compare *industrial stage, post-industrial stage, transitional stage*.

predation The process in which one *species* (the *predator*) hunts, tracks, captures, and ultimately kills its *prey*.

predator An organism that hunts, captures, kills, and consumes individuals of another species, the *prey*.

prediction A specific statement, generally arising from a *hypothesis*, that can be tested directly and unequivocally.

prescribed (controlled) burns The practice of burning areas of *forest* or grassland under carefully controlled conditions to improve the health of *ecosystems*, return them to a more natural state, reduce fuel loads, and help prevent uncontrolled catastrophic fires.

preservation ethic An ethic holding that we should protect the natural *environment* in a pristine, unaltered state. Compare *conservation ethic*.

prey An organism that is killed and consumed by a *predator*.

primary consumer An organism that consumes *producers* and feeds at the second *trophic level*.

primary extraction The initial drilling and pumping of the most easily accessible *crude oil*. Compare *secondary extraction*.

primary forest *Forest* uncut by people. Compare *secondary forest*.

primary pollutant A hazardous substance, such as soot or *carbon monoxide*, that is emitted into the *troposphere* in a form that is directly harmful. Compare *secondary pollutant*.

primary producer See *autotroph*.

primary production The conversion of solar energy to the energy of chemical bonds in sugars during *photosynthesis*, performed by *autotrophs*. Compare *secondary production*.

primary succession A stereotypical series of changes as an *ecological community* develops over time, beginning with a lifeless substrate. In terrestrial *systems*, primary succession begins when a bare expanse of rock, *sand*, or *sediment* becomes newly exposed to the atmosphere and *pioneer species* arrive. Compare *secondary succession*.

primary treatment A stage of *wastewater* treatment in which contaminants are physically removed. Wastewater flows into tanks in which sewage solids, grit, and particulate matter settle to the bottom. Greases and oils float to the surface and can be skimmed off. Compare *secondary treatment*.

principle of utility An *ethical standard*, elaborated by British philosophers Jeremy Bentham and John Stuart Mill, holding that something is right when it produces the greatest practical benefits for the most people.

probability A quantitative description of the likelihood of a certain outcome.

producer (autotroph) An organism that uses energy from sunlight to produce its own food. Includes green plants, algae, and cyanobacteria. See *autotroph*.

productivity The rate at which plants convert solar *energy* (sunlight) to *biomass*. *Ecosystems* whose plants convert solar energy to biomass rapidly are said to have high productivity. See *net primary productivity; gross primary production; net primary production*.

profundal zone In a water body, the volume of open water that sunlight does not reach. Compare *littoral zone; benthic zone; limnetic zone*.

protein A *macromolecule* made up of long chains of amino acids.

proton A positively charged particle in the nucleus of an *atom*.

proven recoverable reserve The amount of a given *fossil fuel* in a deposit that is technologically and economically feasible to remove under current conditions.

proxy indicator A source of indirect evidence that serves as a proxy, or substitute, for direct measurement and that sheds light on past climate. Examples include data from ice cores, sediment cores, tree rings, packrat middens, and coral reefs.

public policy *Policy* made by governments, including those at the local, state, federal, and international levels; it consists of *legislation*, *regulations*, orders, incentives, and practices intended to advance societal welfare. See also *environmental policy*.

pumped storage A technique used to generate *hydroelectric power*, in which water is pumped from a lower reservoir to a higher reservoir when power demand is weak and prices are low. When demand is strong and prices are high, water is allowed to flow downhill through a turbine, generating *electricity*. Compare *run-of-river*, *storage*.

pycnocline A zone of the ocean beneath the surface in which density increases rapidly with depth.

pyrolysis The chemical breakdown of organic matter (*biomass*, *oil shale*, and so on) by heating in the absence of *oxygen*, which often produces materials that can be more easily converted to usable energy. A number of variations on this process exist.

quarry A pit used to extract *mineral* resources such as clay, gravel, sand, or stone (for example, limestone, granite, marble, or slate).

radiative forcing The amount of change in thermal energy that a factor (such as a *greenhouse gas* or an *aerosol*) causes in influencing Earth's temperature. Positive forcing warms Earth's surface, whereas negative forcing cools it.

radioactive The quality by which some *isotopes* "decay," changing their chemical identity as they shed atomic particles and emit high-energy radiation.

radioisotope A radioactive *isotope* that emits subatomic particles and high-*energy* radiation as it "decays" into progressively lighter isotopes until becoming a stable isotope.

radon A highly *toxic*, radioactive, colorless gas that seeps up from the ground in areas with certain types of bedrock and that can build up inside basements and homes with poor air circulation.

rainshadow A region on one side of a mountain or mountain range that experiences arid climate. This occurs because moisture-laden air rising over the terrain from the opposite direction releases precipitation on the windward slope as it cools, leaving the air's humidity low as it descends over the peak and into the rainshadow region.

rainwater harvesting The act of capturing rainwater *runoff* from a rooftop in a barrel, in order to conserve water.

Ramsar Convention on Wetlands of International Importance A 1971 international treaty to inventory and protect important wetlands around the world.

random distribution Distribution pattern in which individuals are located haphazardly in space in no particular pattern (often when needed resources are spread throughout an area and other organisms do not strongly influence where individuals settle).

rangeland Land used for grazing livestock.

rate of natural increase The rate of change in a *population's* size resulting from birth and death rates alone, excluding migration. Also called *natural rate of population change*. Compare *population growth rate*.

REACH Program of the European Union that shifts the burden of proof for testing chemical safety from national governments to industry and requires that chemical substances produced or imported in amounts of over 1 metric ton per year be registered with a new European Chemicals Agency. *REACH*, which stands for Registration, Evaluation, Authorisation and Restriction of Chemicals, went into effect in 2007.

realized niche The portion of the *fundamental niche* that is fully realized (used) by a *species*.

rebound effect The phenomenon by which gains in efficiency from better technology are partly offset when people engage in more energy-consuming behavior as a result. This common psychological effect can sometimes reduce conservation and efficiency efforts substantially.

recharge zone An area where water infiltrates Earth's surface and reaches an *aquifer* below.

reclamation The act of restoring a *mining* site to an approximation of its pre-mining condition. To reclaim a site, companies are required to remove buildings and other structures used for mining, replace *overburden*, fill in shafts, and replant the area with vegetation.

recombinant DNA *DNA* that has been patched together from the DNA of multiple organisms in an attempt to produce desirable traits (such as rapid growth, disease and pest resistance, or higher nutritional content) in organisms lacking those traits.

recombination The process by which the genetic material of male and female organisms becomes mixed and recombined in sexual reproduction, generating novel combinations of genes in the offspring.

recovery Waste management strategy composed of *recycling* and *composting*.

recycling The collection of materials that can be broken down and reprocessed to manufacture new items.

REDD Acronym for *Reducing Emissions from Deforestation and Forest Degradation*. A proposed international program to help address climate change in which wealthy industrialized nations would pay poorer developing nations to conserve forest. Under this plan, poor nations would gain income while rich nations receive carbon credits to offset their emissions in an international cap-and-trade system.

Red List An updated list of *species* facing unusually high risks of *extinction*. The list is maintained by the World Conservation Union.

red tide A *harmful algal bloom* consisting of algae that produce reddish pigments that discolor surface waters.

refining Process of separating the *molecules* of the various *hydrocarbons* in *crude oil* into different-sized classes and transforming them into various fuels and other petrochemical products.

regime shift See *phase shift*.

regional planning Planning similar to *city planning* but conducted across broader geographic scales, generally involving multiple municipal governments.

regulation A specific rule issued by an administrative agency, based on the more broadly written statutory law passed by Congress and enacted by the president.

regulatory taking The deprivation of a property's owner, by means of a law or *regulation*, of most or all economic uses of that property.

relative abundance The extent to which numbers of individuals of different species are equal or skewed. One way to express species diversity. See *evenness*; compare *species richness*.

relative humidity The ratio of the water vapor contained in a given volume of air to the maximum amount the air could contain, for a given temperature.

relativist An ethicist who maintains that *ethics* do and should vary with social context. Compare *universalist*.

renewable natural resources *Natural resources* that are virtually unlimited or that are replenished by the *environment* over relatively short periods (hours to weeks to years). Compare *nonrenewable natural resources*.

replacement fertility The *total fertility rate (TFR)* that maintains a stable *population* size.

reproductive window The portion of a woman's life between sexual maturity and menopause during which she may become pregnant.

reserves-to-production ratio (R/P ratio) The total remaining reserves of a *fossil fuel* divided by the annual rate of production (extraction and processing).

reservoir (1) An artificial water body behind a dam that stores water for human use. (2) See *pool*.

residence time (1) In a biogeochemical cycle, the amount of time a nutrient remains in a given *pool* or *reservoir* before moving to another. Compare *flux*. (2) In the *atmosphere*, the amount of time a gas molecule or a *pollutant* remains aloft.

resilience The ability of an ecological *community* to change in response to disturbance but later return to its original state. Compare *resistance*.

resistance The ability of an ecological *community* to remain stable in the presence of a disturbance. Compare *resilience*.

Resource Conservation and Recovery Act (RCRA) Congressional *legislation* (enacted in 1976 and amended in 1984) that specifies, among other things, how to manage *sanitary landfills* to protect against environmental contamination.

resource management Strategic decision making about how to extract resources, so that resources are used wisely and conserved for the future.

resource partitioning The process by which *species* adapt to *competition* by evolving to use slightly different resources, or to use their shared resources in different ways, thus minimizing interference with one another.

response The type or magnitude of negative effects an animal exhibits in response to a *dose* of *toxicant* in a *dose-response analysis*. Compare *dose*.

restoration ecology The study of the historical conditions of *ecological communities* as they existed before humans altered them. Principles of restoration ecology are applied in the practice of *ecological restoration*.

revolving door The movement of powerful officials between the private sector and government agencies.

ribonucleic acid (RNA) A usually single-stranded *nucleic acid* composed of four nucleotides, each of which contains a sugar (ribose), a phosphate group, and a nitrogenous base. RNA carries the hereditary information for living organisms and is responsible for passing traits from parents to offspring. Compare *DNA*.

riparian Relating to a river or the area along a river.

risk The mathematical probability that some harmful outcome (for instance, injury, death, *environmental* damage, or *economic* loss) will result from a given action, event, or substance.

risk assessment The quantitative measurement of *risk*, together with the comparison of risks involved in different activities or substances.

risk management The process of considering information from scientific *risk assessment* in light of economic, social, and political needs and values, in order to make decisions and design strategies to minimize *risk*.

river system A river and all its tributaries. River systems drain *watersheds*.

RNA See *ribonucleic acid*.

roadless rule A 2001 Clinton Administration executive order that put 31% of national forest land off-limits to road construction or maintenance.

rock A solid aggregation of *minerals*.

rock cycle The very slow process in which *rocks* and the *minerals* that make them up are heated, melted, cooled, broken, and reassembled, forming *igneous*, *sedimentary*, and *metamorphic* rocks.

rotation time The number of years that pass between the time a *forest* stand is cut for timber and the next time it is cut.

r–selected Term denoting a *species* with high *biotic potential* whose members produce a large number of offspring in a relatively short time but do not care for their young after birth. *Populations* of r–selected species are generally regulated by *density-independent factors*. Compare *K–selected*.

runoff The water from *precipitation* that flows into streams, rivers, lakes, and ponds, and (in many cases) eventually to the ocean.

run-of-river Any of several methods used to generate *hydroelectric power* without greatly disrupting the flow of river water. Run-of-river approaches eliminate much of the *environmental* impact of large *dams*. Compare *pumped storage*, *storage*.

safe harbor agreement A cooperative agreement that allows landowners to harm threatened or endangered species in some ways if they voluntarily improve habitat for the species in others.

salinization The buildup of salts in surface *soil* layers.

salt marsh Flat land that is intermittently flooded by the ocean where the *tide* reaches inland. Salt marshes occur along temperate coastlines and are thickly vegetated with grasses, rushes, shrubs, and other herbaceous plants.

salvage logging The removal of dead trees following a natural disturbance. Although it may be economically beneficial, salvage logging can be ecologically destructive, because *snags* provide food and shelter for wildlife and because removing timber from recently burned land can cause *erosion* and damage to *soil*.

sand *Sediment* consisting of particles 0.005–2.0 mm in diameter. Compare *clay; silt*.

sanitary landfill A site at which solid waste is buried in the ground or piled up in large mounds for disposal, designed to prevent the waste from contaminating the *environment*. Compare *incineration*.

savanna A *biome* characterized by grassland interspersed with clusters of acacias and other trees. Savanna is found across parts of Africa (where it was the ancestral home of our *species*), South America, Australia, India, and other dry tropical regions.

science (1) A systematic process for learning about the world and testing our understanding of it. (2) The accumulated body of knowledge that arises from this dynamic process.

scientific method A formalized method for testing ideas with observations that involves a more-or-less consistent series of interrelated steps.

scrubber Technology to chemically treat gases produced in combustion in order to reduce smokestack emissions. These devices typically remove hazardous components and neutralize acidic gases, such as *sulfur dioxide* and hydrochloric acid, turning them into water and salt.

secondary consumer An organism that consumes *primary consumers* and feeds at the third *trophic level*.

secondary extraction The extraction of *crude oil* remaining after *primary extraction* by using solvents or by flushing underground rocks with water or steam. Compare *primary extraction*.

secondary forest *Forest* that has grown back after *primary forest* has been cut. Consists of *second-growth* trees.

secondary pollutant A hazardous substance produced through the reaction of substances added to the *atmosphere* with chemicals normally found in the atmosphere. Compare *primary pollutant*.

secondary production The total *biomass* that *heterotrophs* generate by consuming *autotrophs*. Compare *primary production*.

secondary succession A stereotypical series of changes as an *ecological community* develops over time, beginning when some event disrupts or dramatically alters an existing community. Compare *primary succession*.

secondary treatment A stage of *wastewater* treatment in which biological means are used to remove contaminants remaining after *primary treatment*. Wastewater is stirred up in the presence of *aerobic* bacteria, which degrade organic pollutants in the water. The wastewater then passes to another settling tank, where remaining solids drift to the bottom. Compare *primary treatment*.

second-growth Term describing trees that have sprouted and grown to partial maturity after virgin timber has been cut.

second law of thermodynamics The physical law stating that the nature of *energy* tends to change from a more-ordered state to a less-ordered state; that is, *entropy* increases.

sediment The eroded remains of rocks.

sedimentary rock One of the three main categories of rock. Formed when dissolved *minerals* seep through *sediment* layers and act as a kind of glue, crystallizing and binding sediment particles together. Sandstone and shale are examples of sedimentary rock. Compare *igneous rock* and *metamorphic rock*.

seed bank A storehouse for samples of the world's crop diversity.

seed-tree Timber harvesting approach that leaves small numbers of mature and vigorous seed-producing trees standing so that they can reseed a logged area.

selection system Method of timber harvesting whereby single trees or groups of trees are selectively cut while others are left, creating an *uneven-aged* stand.

selective breeding See *artificial selection*.

septic system *A wastewater* disposal method, common in rural areas, consisting of an underground tank and series of drainpipes. Wastewater runs from the house to the tank, where solids precipitate out. The water proceeds downhill to a drain field of perforated pipes laid horizontally in gravel-filled trenches, where microbes decompose the remaining waste.

service Work done for others as a form of business.

sex ratio The proportion of males to females in a *population*.

shale gas *Natural gas* trapped deep underground in tiny bubbles dispersed throughout formations of shale, a type of *sedimentary rock*. Shale gas is often extracted by *hydraulic fracturing*.

shale oil A liquid form of petroleum extracted from deposits of *oil shale*.

shelf-slope break The portion of the ocean floor where the *continental shelf* drops off with relative suddenness.

shelterbelt A row of trees or other tall perennial plants that are planted along the edges of farm fields to break the wind and thereby minimize wind *erosion*.

shelterwood Timber harvesting approach that leaves small numbers of mature trees in place to provide shelter for seedlings as they grow.

sick-building syndrome A *building-related illness* produced by indoor *pollution* in which the specific cause is not identifiable.

silicon The chemical *element* with 14 *protons* and 14 *neutrons*. An abundant element in rocks in Earth's crust.

silt *Sediment* consisting of particles 0.002–0.005 mm in diameter. Compare *clay; sand*.

sink In a *nutrient cycle*, a *pool* that accepts more *nutrients* than it releases.

sinkhole An area where the ground has given way with little warning as a result of subsidence caused by depletion of water from an *aquifer*.

SLOSS (Single Large or Several Small) dilemma The debate over whether it is better to make reserves large in size and few in number or many in number but small in size.

smart growth A *city planning* concept in which a community's growth is managed in ways intended to limit *sprawl* and maintain or improve residents' quality of life.

smelting A process in which *ore* is heated beyond its melting point and combined with other *metals* or chemicals, in order to form metal with desired characteristics. Steel is created by smelting iron ore with carbon, for example.

Smith, Adam (1723–1790) Scottish philosopher known today as the father of *classical economics*. He believed that when people are free to pursue their own economic self-interest in a competitive marketplace, the marketplace will behave as if guided by "an invisible hand" that ensures that their actions will benefit society as a whole.

smog Term popularly used to describe unhealthy mixtures of air *pollutants* that often form over urban areas. See *industrial smog*; *photochemical smog*.

snag A dead tree that is still standing. Snags are valuable for wildlife.

social sciences Academic disciplines that study human interactions and institutions. Compare *natural sciences*.

socially responsible investing Investing in companies that have met criteria for environmental or social sustainability.

soil A complex plant-supporting *system* consisting of disintegrated rock, organic matter, air, water, *nutrients*, and microorganisms.

soil degradation A deterioration of soil quality and decline in soil productivity, resulting primarily from forest removal, cropland agriculture, and *overgrazing* of livestock. Compare *desertification*; *land degradation*.

soil profile The cross-section of a *soil* as a whole, from the surface to the *bedrock*.

solar cooker A simple portable oven that uses reflectors to focus sunlight onto food and cook it.

solar energy Energy from the sun. It is perpetually renewable and may be harnessed in several ways.

solution mining A *mining* technique in which a narrow borehole is drilled deep into the ground to reach a *mineral* deposit, and water, acid, or another liquid is injected down the borehole to *leach* the resource from the surrounding rock and dissolve it in the liquid. The resulting solution is then sucked out, and the desired resource is isolated. Also called *in-situ recovery*.

source In a *nutrient cycle*, a *pool* that releases more *nutrients* than it accepts.

source reduction The reduction of the amount of material that enters the *waste stream* to avoid the costs of disposal and *recycling*, help conserve resources, minimize *pollution*, and save consumers and businesses money.

specialist A *species* that can survive only in a narrow range of *habitats* that contain very specific resources. Compare *generalist*.

speciation The process by which new *species* are generated.

species A *population* or group of populations of a particular type of organism whose members share certain characteristics and can breed freely with one another and produce fertile offspring. Different biologists may have different approaches to diagnosing species boundaries.

species-area curve A graph showing how number of *species* varies with the geographic area of a landmass or water body. *Species richness* commonly doubles as area increases tenfold.

Species at Risk Act Canada's endangered species protection law, enacted in 2002.

species coexistence An outcome of *interspecific competition* in which no competing *species* fully excludes others and the species continue to live side by side.

species diversity The number and variety of *species* in the world or in a particular region.

species richness The number of species in a particular region. One way to express species diversity. Compare *evenness*; *relative abundance*.

sprawl The unrestrained spread of urban or suburban development outward from a city center and across the landscape. Often specified as growth in which the area of development outpaces *population* growth.

steady-state economy An *economy* that does not grow or shrink but remains stable.

steppe See *temperate grassland*.

Stockholm Convention on Persistent Organic Pollutants (POPs) A 2004 international *treaty* that aims to end the use of 12 persistent organic *pollutants* nicknamed the "dirty dozen."

storage Technique used to generate *hydroelectric power*, in which large amounts of water are impounded in a reservoir behind a concrete *dam* and then passed through the dam to turn *turbines* that generate *electricity*. Compare *pumped storage*, *run-of-river*.

storm surge A temporary and localized rise in sea level brought on by the high tides and winds associated with storms.

stratosphere The layer of the *atmosphere* above the *troposphere* and below the *mesosphere*; it extends from 11 km (7 mi) to 50 km (31 mi) above sea level.

strip mining The use of heavy machinery to remove huge amounts of earth to expose *coal* or *minerals*, which are mined out directly. Compare *subsurface mining*.

strong sustainability A school of thought that argues that human-made capital cannot always substitute for *natural capital* and that we must not allow *natural capital* to diminish. Compare *weak sustainability*.

subcanopy The middle and lower levels of trees in a *forest*, beneath the *canopy*.

subduction The *plate tectonic* process by which denser *crust* slides beneath lighter crust at a *convergent plate boundary*. Often results in *volcanism*.

subsidy A government grant of money or resources to a private entity, intended to support and promote an industry or activity.

subsistence agriculture The oldest form of *traditional agriculture*, in which farming families produce only enough food for themselves.

subsistence economy A survival *economy*, one in which people meet most or all of their daily needs directly from nature and do not purchase or trade for most of life's necessities.

subspecies *Populations* of a *species* that occur in different geographic areas and vary from one another in some characteristics. Subspecies are formed by the same processes that drive *speciation* but result when divergence does not proceed far enough to create separate species.

subsurface mining Method of *mining* underground deposits of *coal*, *minerals*, or fuels, in which shafts are dug deeply into the ground and networks of tunnels are dug or blasted out to follow coal seams. Compare *strip mining*.

suburb A smaller community located at the outskirts of a city.

succession A stereotypical series of changes in the composition and structure of an *ecological community* through time. See *primary succession; secondary succession*.

sulfur dioxide (SO_2) A colorless gas that can result from the combustion of *coal*. In the *atmosphere*, it may react to form sulfur trioxide and sulfuric acid, which may return to Earth in *acid deposition*. An EPA *criteria pollutant*.

Superfund A program administered by the *Environmental Protection Agency* in which experts identify sites polluted with hazardous chemicals, protect *groundwater* near these sites, and clean up the *pollution*. Established by the Comprehensive Environmental Response Compensation and Liability Act (CERCLA) in 1980.

supply The amount of a product offered for sale at a given price. Compare *demand*.

surface impoundment (1) A disposal method for *hazardous waste* or mining waste in which waste in liquid or slurry form is placed into a shallow depression lined with impervious material such as clay and allowed to evaporate, leaving a solid residue on the bottom. (2) The site of such disposal. Compare *deep-well injection*.

surface water Water located atop Earth's surface. Compare *groundwater*.

survivorship curve A graph that shows how the likelihood of death for members of a *population* varies with age.

sustainability A guiding principle of *environmental science*, entailing conserving resources, maintaining functional ecological systems, and developing long-term solutions, such that Earth can sustain our civilization and all life for the future, allowing our descendants to live at least as well as we have lived.

sustainable agriculture *Agriculture* that can be practiced in the same way and in the same place far into the future. Sustainable agriculture does not deplete *soils* faster than they form, nor reduce the clean water and genetic diversity essential to long-term crop and livestock production.

sustainable development Development that satisfies our current needs without compromising the future availability of *natural capital* or our future quality of life.

sustainable forest certification A form of *ecolabeling* that identifies timber products that have been produced using *sustainable* methods. The Forest Stewardship Council and several other organizations issue such certification.

swamp A type of *wetland* consisting of shallow water rich with vegetation, occurring in a forested area. Compare *bog; freshwater marsh*.

swidden The traditional form of *agriculture* in tropical forested areas, in which the farmer cultivates a plot for one year to a few years and then moves on to clear another plot, leaving the first to grow back to *forest*. When the forest is burned, this may be called "slash-and-burn" agriculture.

symbiosis A relationship between different *species* of organisms that live in close physical proximity. People most often use the term "symbiosis" when referring to a mutualism, but symbiotic relationships can be either *parasitic* or *mutualistic*.

sympatric speciation Speciation that takes place without the geographic separation of populations. Compare *allopatric speciation*.

syncrude Synthetic *crude oil*.

synergistic effect An interactive effect (as of *toxicants*) that is more than or different from the simple sum of their constituent effects.

syngas Synthesis gas created from *gasification* of *coal*.

system A network of relationships among a group of parts, elements, or components that interact with and influence one another through the exchange of *energy*, matter, and/or information.

taiga See *boreal forest*.

tailings Portions of *ore* left over after *metals* have been extracted in *mining*.

tar sands See *oil sands*.

taxonomist A scientist who classifies species using an organism's physical appearance and/or genetic makeup, and who groups species by their similarity into a hierarchy of categories meant to reflect evolutionary relationships.

temperate deciduous forest A *biome* consisting of midlatitude *forests* characterized by broad-leafed trees that lose their leaves each fall and remain dormant during winter. These forests occur in areas where *precipitation* is spread relatively evenly throughout the year: much of Europe, eastern China, and eastern North America.

temperate grassland A *biome* whose vegetation is dominated by grasses and features more extreme temperature differences between winter and summer and less *precipitation* than *temperate deciduous forests*. Also known as *steppe, prairie*.

temperate rainforest A *biome* consisting of tall coniferous trees, cooler and less *species*-rich than *tropical rainforest* and milder and wetter than *temperate deciduous forest*.

temperature inversion A departure from the normal temperature distribution in the *atmosphere*, in which a pocket of relatively cold air occurs near the ground, with warmer air above it. The cold air, denser than the air above it, traps *pollutants* near the ground and can thereby cause a buildup of *smog*. Also called *thermal inversion*.

teratogen A *toxicant* that causes harm to the unborn, resulting in birth defects.

terracing The cutting of level platforms, sometimes with raised edges, into steep hillsides to contain water from *irrigation* and *precipitation*. Terracing transforms slopes into series of steps like a staircase, enabling farmers to cultivate hilly land while minimizing their loss of *soil* to water *erosion*.

tertiary consumer An organism that consumes *secondary consumers* and feeds at the fourth *trophic level*.

theory A widely accepted, well-tested explanation of one or more cause-and-effect relationships that has been extensively validated by a great amount of research. Compare *hypothesis*.

thermal inversion See *temperature inversion*.

thermal mass Construction materials that absorb heat, store it, and release it later, for use in *passive solar energy* approaches.

thermogenic Type of *natural gas* created by compression and heat deep underground. Contains *methane* and small amounts of other *hydrocarbon* gases. Compare *biogenic*.

thermohaline circulation A worldwide system of ocean currents in which warmer, fresher water moves along the surface and colder, saltier water (which is denser) moves deep beneath the surface.

thermosphere The *atmosphere*'s top layer, extending upward to an altitude of 500 km (300 mi).

thin-film solar cells Photovoltaic materials compressed into ultra-thin lightweight sheets that may be incorporated into various surfaces to produce photovoltaic solar power.

Thoreau, Henry David (1817–1862) American *transcendentalist* author, poet, and philosopher. His book *Walden*, recording his observations and thoughts while he lived at Walden Pond away from the bustle of urban Massachusetts, remains a classic of American literature.

Three Mile Island Nuclear power plant in Pennsylvania that in 1979 experienced a partial *meltdown*. The term is often used to denote the accident itself, the most serious *nuclear reactor* malfunction that the United States has thus far experienced. Compare *Chernobyl, Fukushima Daiichi.*

threshold dose The amount of a *toxicant* at which it begins to affect a *population* of test animals. Compare ED_{50}; LD_{50}.

tidal energy *Energy* harnessed by erecting a *dam* across the outlet of a tidal basin. Water flowing with the incoming or outgoing *tide* through sluices in the dam turns turbines to generate *electricity*.

tide The periodic rise and fall of the ocean's height at a given location, caused by the gravitational pull of the moon and sun.

topography The study of the shape and arrangement of landforms. Compare *bathymetry*.

topsoil That portion of the *soil* that is most nutritive for plants and is thus of the most direct importance to *ecosystems* and to *agriculture*. Also known as the A horizon.

tornado A type of cyclonic storm in which funnel clouds pick up soil and objects, threatening life and great damage to property.

tort law A system of law addressing harm caused by one entity to another, which operates primarily through lawsuits.

total fertility rate (TFR) The average number of children born per female member of a *population* during her lifetime.

toxic air pollutant *Air pollutant* that is known to cause cancer, reproductive defects, or neurological, developmental, immune system, or respiratory problems in humans, and/or to cause substantial *ecological* harm by affecting the health of nonhuman animals and plants. The *Clean Air Act of 1990* identifies 188 toxic air pollutants, ranging from the heavy metal mercury to *volatile organic compounds (VOCs)* such as benzene and methylene chloride.

toxicant A substance that acts as a poison to humans or wildlife.

toxicity The degree of harm a chemical substance can inflict.

toxicology The scientific field that examines the effects of poisonous chemicals and other agents on humans and other organisms.

Toxic Substances Control Act A 1976 U.S. law that directs the *Environmental Protection Agency* to monitor thousands of industrial chemicals and gives the EPA authority to regulate and ban substances found to pose excessive risk.

toxin A *toxic* chemical stored or manufactured in the tissues of living organisms. For example, a chemical that plants use to ward off *herbivores* or that insects use to deter *predators*.

trade winds Prevailing winds between the equator and 30° latitude that blow from east to west.

traditional agriculture Biologically powered *agriculture*, in which human and animal muscle power, along with hand tools and simple machines, perform the work of cultivating, harvesting, storing, and distributing crops. Compare *industrial agriculture*.

tragedy of the commons The process by which publicly accessible resources open to unregulated use tend to become damaged and depleted through overuse. Coined by Garrett Hardin and widely applicable to resource issues.

transboundary park A reserve of protected land that overlaps national borders.

transcendentalism A philosophical movement that flourished in the United States in the 1840s. Transcendentalist writers such as *Henry David Thoreau*, *Ralph Waldo Emerson*, and *Walt Whitman* viewed nature as a manifestation of the divine, championed a spiritual approach to life, and critiqued society's focus on material goods.

transform plate boundary The area where two tectonic plates meet and slip and grind alongside one another, creating *earthquakes*. For example, the Pacific Plate and the North American Plate rub against each other along California's San Andreas Fault. Compare *convergent plate boundary* and *divergent plate boundary*.

transgene A *gene* that has been extracted from the *DNA* of one organism and transferred into the DNA of an organism of another *species*.

transgenic Term describing an organism that contains *DNA* from another *species*.

transitional stage The second stage of the *demographic transition* model, which occurs during the transition from the *pre-industrial stage* to the *industrial stage*. It is characterized by declining death rates but continued high birth rates. See also *post-industrial stage*. Compare *industrial stage, post-industrial stage, pre-industrial stage*.

transit-oriented development A development approach in which compact communities in the *new urbanism* style are arrayed around stops on a major rail transit line.

transpiration The release of water vapor by plants through their leaves.

trawling Fishing method that entails dragging immense cone-shaped nets through the water, with weights at the bottom and floats at the top to keep the nets open. Compare *bottom-trawling*.

treatment The portion of an *experiment* in which a *variable* has been manipulated in order to test its effect. Compare *control*.

treaty See *convention*.

tributary A smaller river that flows into a larger one.

triple bottom line An approach to sustainability that attempts to meet environmental, economic, and social goals simultaneously.

trophic cascade A series of changes in the *population* sizes of organisms at different *trophic levels* in a *food chain*, occurring when *predators* at high trophic levels indirectly promote populations of organisms at low trophic levels by keeping species at intermediate trophic levels in check. Trophic cascades may become apparent when a top predator is eliminated from a system.

trophic level Rank in the feeding hierarchy of a *food chain*. Organisms at higher trophic levels consume those at lower trophic levels.

tropical deciduous forest See *tropical dry forest*.

tropical dry forest A *biome* that consists of deciduous trees and occurs at tropical and subtropical latitudes where wet and dry seasons each span about half the year. Widespread in India, Africa, South America, and northern Australia. Also known as *tropical deciduous forest*.

tropical rainforest A *biome* characterized by year-round rain and uniformly warm temperatures. Found in Central America, South America, Southeast Asia, west Africa, and other tropical regions. Tropical rainforests have dark, damp interiors; lush vegetation; and highly diverse *biotic communities*.

tropopause The boundary between the *troposphere* and the *stratosphere*. Acts like a cap, limiting mixing between these atmospheric layers.

troposphere The bottommost layer of the *atmosphere;* it extends to 11 km (7 mi) above sea level.

tropospheric ozone *Ozone* that occurs in the *troposphere*, where it is a *secondary pollutant* created by the interaction of sunlight, heat, *nitrogen oxides*, and volatile *carbon*-containing chemicals. A major component of *smog*, it can injure living tissues and cause respiratory problems. An EPA *criteria pollutant*.

true cost accounting See *full cost accounting*.

tsunami An immense swell, or wave, of ocean water triggered by an *earthquake*, *volcano*, or *landslide* that can travel long distances across oceans and inundate coasts.

tundra A *biome* that is nearly as dry as *desert* but is located at very high latitudes along the northern edges of Russia, Canada, and Scandinavia. Extremely cold winters with little daylight and moderately cool summers with lengthy days characterize this landscape of lichens and low, scrubby vegetation.

typhoon See *cyclone*.

umbrella species A *species* for which meeting its *habitat* needs automatically helps meet those of many other species. Umbrella species generally are species that require large areas of habitat.

unconfined aquifer A water-bearing, porous layer of rock, *sand*, or gravel that lies atop a less-permeable substrate. The water in an unconfined aquifer is not under pressure because there is no impermeable upper layer to confine it. Compare *confined aquifer*.

undernutrition A condition of insufficient *nutrition* in which people receive less than 90% of their daily caloric needs.

understory The layer of a *forest* consisting of small shrubs and trees above the forest floor and below the *subcanopy*.

uneven-aged Term describing stands consisting of trees of different ages. Uneven-aged stands more closely approximate a natural *forest* than do *even-aged* stands.

uniform distribution Distribution pattern in which individuals are evenly spaced (as when individuals hold territories or otherwise compete for space).

United Nations (U.N.) Organization founded in 1945 to promote international peace and to cooperate in solving international economic, social, cultural, and humanitarian problems.

United Nations Framework Convention on Climate Change An international treaty signed in 1992 outlining a plan to reduce emissions of *greenhouse gases*. Gave rise to the *Kyoto Protocol*.

universalist An ethicist who maintains that there exist objective notions of right and wrong that hold across cultures and situations. Compare *relativist*.

upwelling In the ocean, the flow of cold, deep water toward the surface. Upwelling occurs in areas where surface *currents* diverge. Compare *downwelling*.

uranium The chemical *element* with 92 *protons* and 92 *neutrons*. Uranium is used as a fuel source to produce electricity with *nuclear energy*.

urban ecology A scientific field of study that views cities explicitly as *ecosystems*. Researchers in this field apply the fundamentals of *ecosystem ecology* and *systems* science to urban areas.

urban heat island effect The phenomenon whereby a city becomes warmer than outlying areas because of the concentration of heat-generating buildings, vehicles, and people, and because buildings and dark paved surfaces absorb heat and releae it at night.

urbanization A population's shift from rural living to city and suburban living.

urban planning See *city planning*.

utilitarian value See *instrumental value*.

variable In an *experiment*, a condition that can change. See *dependent variable* and *independent variable*.

vector An organism that transfers a *pathogen* to its *host*. An example is a mosquito that transfers the malaria pathogen to humans.

vernal pool A type of seasonal wetland that forms in spring from rain and snowmelt and then dries up later in the year.

vested interest A direct interest (from an individual or *interest group*) in some condition or policy change due to the prospect for personal or financial benefit, even if this counteracts the common good.

volatile organic compound (VOC) One of a large group of potentially harmful organic chemicals used in industrial processes. One of six major pollutants whose emissions are monitored by the *EPA* and state agencies.

volcano A site where molten rock, hot gas, or ash erupts through Earth's surface, often creating a mountain over time as cooled *lava* accumulates.

Wallace, Alfred Russel (1823–1913) English naturalist who proposed, independently of *Charles Darwin*, the concept of *natural selection* as a mechanism for *evolution* and as a way to explain the great variety of living things.

warm front The boundary where a mass of warm air displaces a mass of colder air. Compare *cold front*.

waste Any unwanted product that results from a human activity or process.

waste management Strategic decision making to minimize the amount of *waste* generated and to dispose of waste safely and effectively.

waste stream The flow of *waste* as it moves from its sources toward disposal destinations.

waste-to-energy (WTE) facility An incinerator that uses heat from its furnace to boil water to create steam that drives *electricity* generation or that fuels heating systems.

wastewater Any water that is used in households, businesses, industries, or public facilities and is drained or flushed down pipes, as well as the polluted *runoff* from streets and storm drains.

water A *compound* composed of two hydrogen atoms bonded to one oxygen atom, denoted by the chemical formula H_2O.

water cycle See *hydrologic cycle*.

waterlogging The saturation of *soil* by water, in which the *water table* is raised to the point that water bathes plant roots. Waterlogging deprives roots of access to gases, essentially suffocating them and eventually damaging or killing the plants.

water mining The withdrawal of water at a rate faster than it can be replenished.

water pollution The act of polluting water, or the condition of being polluted by water pollutants.

watershed The entire area of land from which water drains into a given river.

water table The upper limit of *groundwater* held in an *aquifer*.

wave energy *Energy* harnessed from the motion of ocean waves. Many designs for machinery to harness wave energy have been invented, but few have been adequately tested.

weak sustainability A school of thought that argues that we can allow *natural capital* to decline over time as long as human-made capital increases to compensate for it. Compare *strong sustainability*.

weather The local physical properties of the *troposphere*, such as temperature, pressure, humidity, cloudiness, and wind, over relatively short time periods. Compare *climate*.

weathering The process by which rocks and *minerals* are broken down, turning large particles into smaller particles. Weathering may proceed by physical, chemical, or biological means.

weed A pejorative term for any plant that competes with our crops. The term is subjective and defined by our own economic interests, and is not biologically meaningful. Compare *pest*.

westerlies Prevailing winds from 30° to 60° latitude that blow from west to east.

wetland A system in which the soil is saturated with water and that generally features shallow standing water with ample vegetation. These biologically productive systems include *freshwater marshes*, *swamps*, *bogs*, and seasonal wetlands such as *vernal pools*.

Whitman, Walt (1819–1892) American poet who espoused *transcendentalism*.

wilderness area Federal land that is designated off-limits to development of any kind but is open to public recreation, such as hiking, nature study, and other activities that have minimal impact on the land.

wildland-urban interface A region where urban or suburban development meets forested or undeveloped lands.

windbreak See *shelterbelt*.

wind farm A development involving a group of *wind turbines*.

wind power A source of *renewable energy*, in which *kinetic energy* from the passage of wind through *wind turbines* is used to generate *electricity*.

wind turbine A mechanical assembly that converts the wind's *kinetic energy*, or energy of motion, into electrical energy.

World Bank Institution founded in 1944 that serves as one of the globe's largest sources of funding for *economic* development, including such major projects as *dams*, *irrigation* infrastructure, and other undertakings.

world heritage site A location internationally designated by the *United Nations* for its cultural or natural value. There are over 900 such sites worldwide.

World Trade Organization (WTO) Organization based in Geneva, Switzerland, that represents multinational corporations and promotes free trade by reducing obstacles to international commerce and enforcing fairness among nations in trading practices.

worldview A way of looking at the world that reflects a person's (or a group's) beliefs about the meaning, purpose, operation, and essence of the world.

xeriscaping Landscaping using plants that are adapted to arid conditions.

zone of aeration The upper soil layers above the *water table*, containing pore spaces filled with both air and water. Compare *zone of saturation*.

zone of saturation The lower soil layers of an *aquifer*, beneath the *water table*, containing pore spaces completely filled with water. Compare *zone of aeration*.

zoning The practice of classifying areas for different types of development and land use.

zooplankton Tiny aquatic animals that feed on *phytoplankton* and generally comprise the second *trophic level* in an aquatic *food chain*. Compare *phytoplankton*.

zooxanthellae Symbiotic algae that inhabit the bodies of *corals* and produce food through *photosynthesis*.

Selected Sources and References for Further Reading

Chapter 1

Bahn, Paul, and John Flenley. 1992. *Easter Island, Earth island*. Thames and Hudson, London.

Bowler, Peter J. 1993. *The Norton history of the environmental sciences*. W.W. Norton, New York.

Brown, Lester R. 2009. *Plan B 4.0: Mobilizing to save civilization*. Earth Policy Institute and W.W. Norton, New York.

Brown, Lester R. 2011. *World on the edge: How to prevent environmental and economic collapse*. Earth Policy Institute and W.W. Norton, New York.

Campus Ecology. National Wildlife Federation. www.nwf.org/campusecology.

Diamond, Jared. 2005. *Collapse: How societies choose to fail or succeed*. Viking, New York.

Flenley, John, and Paul Bahn. 2003. *The enigmas of Easter Island*. Oxford Univ. Press.

Global Footprint Network, 2011. *What happens when an infinite-growth economy runs into a finite planet?: Global Footprint Network 2011 annual report*. Oakland, Calif.

Global Footprint Network, 2010. *The ecological wealth of nations*. Oakland, Calif.

Goudie, Andrew. 2013. *The human impact on the natural environment: Past, present, and future*. 7th ed. Blackwell Publishing, London.

Harrison, Paul, and Fred Pearce, eds. 2000. *AAAS atlas of population & environment*. Univ. of California Press, Berkeley.

Hunt, Terry L., and Carl P. Lipo. 2006. Late colonization of Easter Island. *Science* 311: 1603–1606.

Hunt, Terry L., and Carl P. Lipo. 2011. *The statues that walked: Unraveling the mystery of Easter Island*. Simon & Schuster.

Kuhn, Thomas S. 1962. *The structure of scientific revolutions*, 2nd ed., 1970. Univ. of Chicago Press, Chicago.

Lomborg, Bjorn. 2001. *The skeptical environmentalist: Measuring the real state of the world*. Cambridge Univ. Press.

Millennium Ecosystem Assessment. 2005. *Ecosystems and human well-being: General synthesis*. Millennium Ecosystem Assessment and World Resources Institute.

Musser, George. 2005. The climax of humanity. *Scientific American* 293(3): 44–47.

Ponting, Clive. 1991. *A green history of the world: The environment and the collapse of great civilizations*. Penguin Books, New York.

Popper, Karl R. 1959. *The logic of scientific discovery*. Hutchinson, London.

Redman, Charles R. 1999. *Human impact on ancient environments*. Univ. of Arizona Press, Tucson.

Sagan, Carl. 1997. *The demon-haunted world: Science as a candle in the dark*. Ballantine Books, New York.

Siever, Raymond. 1968. Science: Observational, experimental, historical. *American Scientist* 56: 70–77.

U.N. Environment Programme. 2011. *Keeping track of our changing environment: From Rio to Rio+20 (1992–2012)*. UNEP, Nairobi.

U.N. Environment Programme. 2012. *Global environment outlook 5 (GEO-5)*. UNEP, Nairobi.

Valiela, Ivan. 2001. *Doing science: Design, analysis, and communication of scientific research*. Oxford Univ. Press.

Van Tilburg, Jo Anne. 1994. *Easter Island: Archaeology, ecology, and culture*. Smithsonian Institution Press, Washington, D.C.

Wackernagel, Mathis, and William Rees. 1996. *Our ecological footprint: Reducing human impact on the Earth*. New Society Publishers, Gabriola Island, British Columbia, Canada.

Wackernagel, Mathis, et al. 2002. Tracking the ecological overshoot of the human economy. *PNAS* 99: 9266–9271.

Worldwatch Institute. 2012. *State of the world 2012: Moving toward sustainable prosperity*. Worldwatch Institute, Washington, D.C.

Worldwatch Institute. 2012. *Vital signs 2012*. Worldwatch Institute, Washington, D.C. http://vitalsigns.worldwatch.org.

Worldwatch Institute. 2013. *State of the world 2013: Is sustainability still possible?* Worldwatch Institute, Washington, D.C.

WWF–World Wide Fund for Nature. 2012. *Living planet report 2012*. WWF, Gland, Switzerland.

Chapter 2

Berardelli, Phil. 2008. Human-driven planet: Time to make it official? *ScienceNOW*. 24 Jan. 2008. http://sciencenow.sciencemag.org/cgi/content/full/2008/124/1.

Berry, R. Stephen. 1991. *Understanding energy: Energy, entropy and thermodynamics for every man*. World Scientific Publishing Co.

Buesseler, Ken O., et al. 2012. Fishing for answers off Fukushima. *Science* 26: 480–482.

Buesseler, Ken O., et al. 2012. Fukushima-derived radionuclides in the ocean and biota off Japan. *PNAS* 109: 5984–5988.

Christopherson, Robert W. 2011. *Geosystems: An introduction to physical geography*, 8th ed. Prentice Hall, Upper Saddle River, N.J.

Craig, James R., David J. Vaughan, and Brian J. Skinner. 2010. *Earth resources and the environment*, 4th ed. Benjamin Cummings, San Francisco.

Keller, Edward A. 2011. *Introduction to environmental geology*, 8th ed. Prentice Hall, Upper Saddle River, N.J.

Keller, Edward A., and Robert H. Blodgett. 2011. *Natural hazards: Earth's processes as hazards, disasters, and catastrophes*, 3rd ed. Prentice Hall, Upper Saddle River, N.J.

Keller, Edward A., and Nicholas Pinter. 2001. *Active tectonics: Earthquakes, uplift, and landscape*, 2nd ed. Prentice Hall, Upper Saddle River, N.J.

Lancaster, Mike, et al. 2010. *Green chemistry: An introductory text*. Royal Society of Chemistry, London.

Manahan, Stanley E. 2009. *Environmental chemistry*, 9th ed. Lewis Publishers, CRC Press, Boca Raton, Fla.

Massachusetts Institute of Technology. 2006. *The future of geothermal energy: Impact of enhanced geothermal systems (EGS) on the United States in the 21st century.* Idaho National Laboratory, Idaho Falls.

McMurry, John E. 2011. *Organic chemistry*, 8th ed. Brooks/Cole, San Francisco.

Montgomery, Carla. 2013. *Environmental geology*, 10th ed. McGraw-Hill, New York.

Skinner, Brian J., and Stephen C. Porter. 2003. *The dynamic earth: An introduction to physical geology*, 5th ed. Wiley and Sons, Hoboken, N.J.

Tarbuck, Edward J., Frederick K. Lutgens, and Dennis Tasa. 2011. *Earth science*, 13th ed. Prentice Hall, Upper Saddle River, N.J.

Timberlake, Karen C. 2011. *Chemistry: An introduction to general, organic, and biological chemistry*, 11th ed. Pearson Benjamin Cummings, San Francisco.

Van Dover, Cindy Lee. 2000. *The ecology of deep-sea hydrothermal vents*. Princeton Univ. Press.

Van Ness, H.C. 1983. *Understanding thermodynamics*. Dover Publications, Mineola, New York.

Zalasiewicz, Jan, et al. 2008. Are we now living in the Anthropocene? *GSA Today* 18(2) (Feb. 2008): 4–8.

Chapter 3

Alvarez, Luis W., et al. 1980. Extraterrestrial cause for the Cretaceous-Tertiary extinction. *Science* 208: 1095–1108.

Atkinson, Carter T., and Dennis A. LaPointe. 2009. Introduced avian diseases, climate change, and the future of Hawaiian honeycreepers. *J. Avian Med. Surgery* 23: 53–63.

Baldwin, Bruce, and Michael Sanderson. 1998. Age and rate of diversification of the Hawaiian silversword alliance (Compositae). *Proc. Natl. Acad. Sci. USA*. 95: 9402–9406.

Begon, Michael, Colin R. Townsend, and John L. Harper. 2006. *Ecology: From individuals to ecosystems*, 4th ed. Wiley-Blackwell Publishing, U.K.

Camp, Richard J., et al. 2009. *Passerine bird trends at Hakalau Forest National Wildlife Refuge, Hawaii*. Hawaii Cooperative Studies Unit Tech. Rep. HCSU-011. University of Hawaii at Hilo.

Camp, Richard J., et al. 2010. Population trends of forest birds at Hakalau Forest National Wildlife Refuge, Hawaii. *Condor* 112: 196–212.

Campbell, Neil A., and Jane B. Reece. 2010. *Biology*, 9th ed. Benjamin Cummings, San Francisco.

Darwin, Charles. 1859. *The origin of species by means of natural selection*. John Murray, London.

Endler, John A. 1986. *Natural selection in the wild*. Monographs in Population Biology 21, Princeton Univ. Press.

Herron, Jon C., and Scott Freeman. 2013. *Evolutionary analysis*. 5th ed. Pearson Prentice Hall, Upper Saddle River, N.J.

Freed, Leonard A. and Rebecca L. Cann. 2009. Negative effects of an introduced bird species on growth and survival in a native bird community. *Current Biology* 19: 1736–1740.

Freed, Leonard A. and Rebecca L. Cann. 2010. Misleading trend analysis and decline of Hawaiian forest birds. *Condor* 112: 213–221.

Futuyma, Douglas J. 2013. *Evolution*, 3rd ed. Sinauer Associates, Sunderland, Mass.

Krebs, Charles J. 2009. *Ecology: The experimental analysis of distribution and abundance*, 6th ed. Benjamin Cummings, San Francisco.

Lerner, Heather R.L., et al. 2011. Multilocus resolution of phylogeny and timescale in the extant adaptive radiation of Hawaiian honeycreepers. *Current Biology* 21: 1838–1844.

MacLeod, Norman. 2013. *The great extinctions: What causes them and how they shape life*. Firefly Books, Richmond Hill, Ontario.

Molles, Manuel C. 2010. *Ecology: Concepts and applications*, 5th ed. McGraw-Hill, Boston.

Olson, Storrs L., and Helen F. James. 1982. Fossil birds from the Hawaiian islands: Evidence for wholesale extinction by man before Western contact. *Science* 217: 633–635.

Powell, James L. 1998. *Night comes to the Cretaceous: Dinosaur extinction and the transformation of modern geology*. W.H. Freeman, New York.

Price, Jonathan P., and David A. Clague. 2002. How old is the Hawaiian biota? Geology and phylogeny suggest recent divergence. *Proc. Roy. Soc. B* 269: 2429–2435.

Raup, David M. 1991. *Extinction: Bad genes or bad luck?* W.W. Norton, New York.

Ricklefs, Robert E., and Gary L. Miller. 2000. *Ecology*, 4th ed. W.H. Freeman, New York.

Ricklefs, Robert E., and Dolph Schluter, eds. 1993. *Species diversity in ecological communities*. Univ. of Chicago Press, Chicago.

Scott, J. Michael, et al. 1988. Conservation of Hawaii's vanishing avifauna. *BioScience* 38: 238–252.

Smith, Thomas M., and Robert L. Smith. 2009. *Elements of ecology*, 8th ed. Benjamin Cummings, San Francisco.

Steadman, David W. 1995. Prehistoric extinctions of Pacific island birds: Biodiversity meets zooarchaeology. *Science* 267: 1123–1131.

U.S. Fish and Wildlife Service. 2010. *Hakalau Forest National Wildlife Refuge Comprehensive Conservation Plan*. USFWS, Hilo, Hawaii.

Wagner, Warren L., and Vicki A. Funk. 1995. *Hawaiian biogeography: Evolution on a hot spot archipelago*. Smithsonian Institution Press, Washington, D.C.

Wilson, Edward O. 1992. *The diversity of life*. Harvard Univ. Press.

Chapter 4

Baskin, Yvonne. 2002. *A plague of rats and rubbervines: The growing threat of species invasions*. Island Press, Washington, D.C.

Breckle, Siegmar-Walter. 2002. *Walter's vegetation of the Earth: The ecological systems of the geo-biosphere*, 4th ed. Springer-Verlag, Berlin.

Bright, Chris. 1998. *Life out of bounds: Bioinvasion in a borderless world*. Worldwatch Institute and W.W. Norton, Washington, D.C., and New York.

Bronstein, Judith L. 1994. Our current understanding of mutualism. *Quarterly Journal of Biology* 69: 31–51.

Clewell, Andre F., and James Aronson. 2013. *Ecological restoration: Principles, values, and structure of an emerging profession*. 2nd ed. Island Press.

Connell, Joseph H., and Ralph O. Slatyer, 1977. Mechanisms of succession in natural communities. *American Naturalist* 111: 1119–1144.

Dale, Virginia H., Frederick J. Swanson, and Charles M. Crisafulli, eds. 2005. *Ecological responses to the 1980 eruption of Mount St. Helens*. Springer, New York.

Drake, John M., and Jonathan M. Bossenbroek. 2004. The potential distribution of zebra mussels in the United States. *BioScience* 54: 931–941.

Estes, James A., et al. 2011. Trophic downgrading of planet Earth. *Science* 333: 301–306.

Falk, Donald A., et al., eds. 2006. *Foundations of restoration ecology*. Island Press. Washington, D.C.

Hobbs, Richard J., Eric S. Higgs, and Carol M. Hall. 2013. *Novel ecosystems: Intervening in the new ecological world order*. Wiley-Blackwell, West Sussex, U.K.

Menge, Bruce A., et al. 1994. The keystone species concept: Variation in interaction strength in a rocky intertidal habitat. *Ecological Monographs* 64: 249–286.

Molles, Manuel C. 2010. *Ecology: Concepts and applications*, 5th ed. McGraw-Hill, Boston.

Nijhuis, Michelle. 2007. Wish you weren't here. *High Country News*, 5 Mar. 2007.

Pace, M.L., et al. 2010. Recovery of native zooplankton associated with increased mortality of an invasive mussel. *Ecosphere* 1(1) Article 3.

Pimentel, David, et al. 2005. Update on the environmental and economic costs associated with alien-invasive species in the United States. *Ecological Economics* 52: 273–288.

Power, Mary E., et al. 1996. Challenges in the quest for keystones. *BioScience* 46: 609–620.

Ricklefs, Robert E. 2010. *The economy of nature*, 6th ed. W.H. Freeman and Co., New York.

Sax, Dov F., John J. Stachowicz, and Steven D. Gaines, eds. 2012. *Species invasions: Insights into ecology, evolution, and biogeography*. Sinauer, Sunderland, Mass.

Shea, Katriona, and Peter Chesson. 2002. Community ecology theory as a framework for biological invasions. *Trends in Ecology and Evolutionary Biology* 17: 170–176.

Smith, Robert L., and Thomas M. Smith. 2001. *Ecology and field biology*, 6th ed. Benjamin Cummings, San Francisco.

Stokstad, Erik. 2007. Feared quagga mussel turns up in western United States. *Science* 315: 453.

Strayer, David L. 2009. Twenty years of zebra mussels: Lessons from the mollusk that made headlines. *Frontiers in Ecology and the Environment* 7: 135–141.

Strayer, David L. 2010. Alien species in fresh waters: Ecological effects, interactions with other stressors, and prospects for the future. *Freshwater Biology* 55(Suppl 1): 152–174.

Strayer, David L., et al. 1999. Transformation of freshwater ecosystems by bivalves: A case study of zebra mussels in the Hudson River. *BioScience* 49: 19–27.

Strayer, David L., et al. 2004. Effects of an invasive bivalve (*Dreissena polymorpha*) on fish in the Hudson River estuary. *Canadian Journal of Fisheries and Aquatic Sciences* 61: 924–941.

Thompson, John N. 1999. The evolution of species interactions. *Science* 284: 2116–2118.

U.S. Geological Survey. Zebra and quagga mussel information resource page. http://nas.er.usgs.gov/taxgroup/mollusks/zebramussel.

Van Andel, Jelte, and James Aronson. 2012. *Restoration ecology: The new frontier*. 2nd ed. Wiley-Blackwell, West Sussex, U.K.

Weigel, Marlene, ed. 1999. *Encyclopedia of biomes*. UXL, Farmington Hills, Michigan.

Whittaker, Robert H., and William A. Niering. 1965. Vegetation of the Santa Catalina Mountains, Arizona: A gradient analysis of the south slope. *Ecology* 46: 429–452.

Woodward, Susan L. 2003. *Biomes of Earth: Terrestrial, aquatic, and human-dominated*. Greenwood Publishing, Westport, Conn.

Chapter 5

Alexander, Richard B., et al. 2008. Differences in phosphorus and nitrogen delivery to the Gulf of Mexico from the Mississippi River Basin. *Env. Sci. Technol*. 42: 822–830.

Carpenter, Edward J., and Douglas G. Capone, eds. 1983. *Nitrogen in the marine environment*. Academic Press, New York.

Chesapeake Bay Foundation. 2012. *2012 State of the Bay report*. Chesapeake Bay Foundation, Annapolis, MD.

Chesapeake Bay Program. 2011. *Bay barometer: A health and restoration assessment of the Chesapeake Bay and Watershed in 2010*. Chesapeake Bay Program, Annapolis, MD.

Committee on Environment and Natural Resources. 2000. *An integrated assessment: Hypoxia in the northern Gulf of Mexico*. CENR, National Science and Technology Council, Washington, D.C.

Committee on the Mississippi River and the Clean Water Act. 2009. *Nutrient control actions for improving water quality in the Mississippi River basin and northern Gulf of Mexico*. National Academies Press, Washington, D.C.

Diaz, Robert J., and Rutger Rosenberg. 2008. Spreading dead zones and consequences for marine ecosystems. *Science* 321: 926–929.

Field, Christopher B., et al. 1998. Primary production of the biosphere: Integrating terrestrial and oceanic components. *Science* 281: 237–240.

Gruber, Nicolas, and James N. Galloway. 2008. An Earth-system perspective of the global nitrogen cycle. *Nature* 451: 293–296.

Isebrands, J.G., et al. 2001. Growth responses of *Populus tremuloides* clones to interacting elevated carbon dioxide and tropospheric ozone. *Environmental Pollution* 115: 359–371.

Jacobson, Michael, et al. 2000. *Earth system science from biogeochemical cycles to global changes*. Academic Press.

Mississippi River/Gulf of Mexico Watershed Nutrient Task Force. 2008. *Gulf hypoxia action plan 2008*. U.S. EPA, Washington, D.C.

Mississippi River/Gulf of Mexico Watershed Nutrient Task Force. 2011. *Moving forward on Gulf hypoxia: Annual report 2011*. U.S. EPA, Washington, D.C.

Mitsch, William J., et al. 2001. Reducing nitrogen loading to the Gulf of Mexico from the Mississippi River Basin: Strategies to counter a persistent ecological problem. *BioScience* 51: 373–388.

National Science and Technology Council, Committee on Environment and Natural Resources. 2003. *An assessment of coastal hypoxia and eutrophication in U.S. waters*. National Science and Technology Council, Washington, D.C.

Raloff, Janet. 2004. Dead waters: Massive oxygen-starved zones are developing along the world's coasts. *Science News* 165: 360–362.

Raloff, Janet. 2004. Limiting dead zones: How to curb river pollution and save the Gulf of Mexico. *Science News* 165: 378–380.

Ricklefs, Robert E. 2010. *The economy of nature*, 6th ed. W.H. Freeman and Co., New York.

Schlesinger, William H. 2013. *Biogeochemistry: An analysis of global change*, 3rd ed. Academic Press, London.

Schulte, D.M., et al. 2009. Unprecedented restoration of a native oyster metapopulation. *Science* 325: 1124–1128.

Smith, Thomas M., and Robert L. Smith. 2012. *Elements of ecology*, 8th ed. Pearson Benjamin Cummings, San Francisco.

Takahashi, Taro. 2004. The fate of industrial carbon dioxide. *Science* 305: 352–353.

Turner, Monica, et al. 2003. *Landscape ecology in theory and practice: Pattern and process*. Springer.

U.S. Department of Energy. 2002. *An evaluation of the Department of Energy's free-air carbon dioxide enrichment (FACE) experiments as scientific user facilities*. U.S. DOE, Washington, D.C.

Vitousek, Peter M., et al. 1997. Human alteration of the global nitrogen cycle: Sources and consequences. *Ecological Applications* 7: 737–750.

Whittaker, Robert H. 1975. *Communities and ecosystems*, 2nd ed. Macmillan, New York.

Wu, Jianguo, and Richard J. Hobbs, eds. 2007. *Key topics in landscape ecology*. Cambridge Univ. Press.

Chapter 6

Balmford, Andrew, et al. 2002. Economic reasons for conserving wild nature. *Science* 297: 950–953.

Barbour, Ian G. 1992. *Ethics in an age of technology*. Harper Collins, San Francisco.

Beddoe, Rachael, et al. 2009. Overcoming systemic roadblocks to sustainability: The evolutionary redesign of worldviews, institutions, and technologies. *PNAS* 106: 2483–2489.

Blewitt, John. 2008. *Understanding sustainable development*. Earthscan, London.

Brown, Lester. 2001. *Eco-economy: Building an economy for the Earth*. Earth Policy Institute and W.W. Norton, New York.

Carson, Richard T., et al. 1994. Valuing the preservation of Australia's Kakadu Conservation Zone. *Oxford Economic Papers* 46: 727–749.

Cole, Luke W., and Sheila R. Foster. 2001. *From the ground up: Environmental racism and the rise of the environmental justice movement*. New York Univ. Press.

Costanza, Robert, et al. 2013. *An introduction to ecological economics*. 2nd ed. CRC Press, Boca Raton, Fla.

Costanza, Robert, et al. 1997. The value of the world's ecosystem services and natural capital. *Nature* 387: 253–260.

Daily, Gretchen, ed. 1997. *Nature's services: Societal dependence on natural ecosystems*. Island Press, Washington, D.C.

Daly, Herman E. 1996. *Beyond growth: The economics of sustainable development*. Beacon Press, Boston.

Daly, Herman E. 2005. Economics in a full world. *Scientific American* 293(3): 100–107.

Daniels, Amy E., et al. 2010. Understanding the impacts of Costa Rica's PES: Are we asking the right questions? *Ecological Economics* 69: 2116–2126.

De Graaf, John, et al. 2002. *Affluenza: The all-consuming epidemic*. Berrett-Koehler Publishers, San Francisco.

Esty, Daniel C., and Andrew S. Winston. 2006. *Green to gold: How smart companies use environmental strategy to innovate, create value, and build competitive advantage*. Yale Univ. Press, New Haven, Conn.

Fletcher, Robert and Jan Breitling. 2012. Market mechanism or subsidy in disguise? Governing payment for environmental services in Costa Rica. *Geoforum* 43: 402–411.

Fox, Stephen. 1985. *The American conservation movement: John Muir and his legacy*. Univ. of Wisconsin Press, Madison.

Goodstein, Eban. 1999. *The tradeoff myth: Fact and fiction about jobs and the environment*. Island Press, Washington, D.C.

Goodstein, Eban. 2010. *Economics and the environment*, 6th ed. Wiley & Sons, Hoboken, N.J.

Greiber, Thomas, and Simone Schiele, eds. 2011. *Governance of ecosystem services: Lessons learned from Cameroon, China, Costa Rica, and Ecuador*. IUCN, Gland, Switzerland.

Hawken, Paul, Amory Lovins, and L. Hunter Lovins. 1999. *Natural capitalism*. Little, Brown, and Co., Boston.

HM Treasury. 2006. *Stern review on the economics of climate change*. HM Treasury and Cambridge Univ. Press.

Kinzig, Ann P., et al. 2011. Paying for ecosystem services—promise and peril. *Science* 334: 603–604.

Kolstad, Charles D. 2010. *Environmental economics*, 2nd ed. Oxford Univ. Press.

Leopold, Aldo. 1949. *A Sand County almanac, and sketches here and there*. Oxford Univ. Press.

McCauley, Douglas J. 2006. Selling out on nature. *Nature* 443: 27–28.

Meadows, Donella, Jørgen Randers, and Dennis Meadows. 2004. *Limits to growth: The 30-year update*. Chelsea Green Publishing Co., White River Junction, Vermont.

Millennium Ecosystem Assessment. 2005. *Ecosystems and human well-being: Opportunities and challenges for business and industry*. Millennium Ecosystem Assessment and World Resources Institute.

Nash, Roderick F. 1989. *The rights of nature*. Univ. of Wisconsin Press, Madison.

Nash, Roderick F. 1990. *American environmentalism: Readings in conservation history*, 3rd ed. McGraw-Hill, New York.

National Research Council, Board on Sustainable Development. 1999. *Our common journey: A transition toward sustainability*. National Academies Press, Washington, D.C.

Nordhaus, William. 2007. Critical assumptions in the Stern Review on Climate Change. *Science* 317: 201–202.

O'Neill, John O., et al., eds. 2002. *Environmental ethics and philosophy*. Edward Elgar, Cheltenham, U.K.

Overseas Development Institute. 2011. *Costa Rica's sustainable resource management: Successfully tackling tropical deforestation*. ODI Publications, London.

Pagiola, Stefano. 2007. Payments for environmental services in Costa Rica. *Ecological Economics* 65: 712–724.

Renner, Michael. 2008. *Green jobs: Working for people and the environment*. Worldwatch Report 177. Worldwatch Institute, Washington, D.C.

Ricketts, Taylor, et al. 2004. Economic value of tropical forest to coffee production. *PNAS* 101: 12579–12582.

Singer, Peter, ed. 1993. *A companion to ethics*. Blackwell Publishers, Oxford.

Smith, Adam. 1776. *An inquiry into the nature and causes of the wealth of nations*. 1993 ed., Oxford Univ. Press.

Stone, Christopher D. 1972. Should trees have standing? Towards legal rights for natural objects. *Southern California Law Review* 1972: 450–501.

Talberth, John, et al. 2007. *The genuine progress indicator 2006: A tool for sustainable development*. Redefining Progress, Oakland, Calif.

TEEB. 2010. *The economics of ecosystems and biodiversity: Mainstreaming the economics of nature: A synthesis of the approach, conclusions, and recommendations of TEEB*. The Economics of Ecosystems and Biodiversity.

TEEB. 2012. *Nature and its role in the transition to a green economy*. The Economics of Ecosystems and Biodiversity.

Tietenberg, Tom, and Lynne Lewis. 2010. *Environmental economics and policy*, 6th ed. Prentice Hall, Upper Saddle River, N.J.

United Nations. 2012. *Report of the United Nations Conference on Sustainable Development*. Rio de Janeiro, Brazil, 20–22 June 2012. http://www.uncsd2012.org/content/documents/814UNCSD%20REPORT%20final%20revs.pdf

United Nations. 2013. *The Millennium Development Goals Report 2013*. U.N., New York.

Walker, Gordon. 2012. *Environmental justice: Concepts, evidence, and politics*. Routledge, New York.

Wenz, Peter S. 2001. *Environmental ethics today*. Oxford Univ. Press.

White, Lynn. 1967. The historic roots of our ecologic crisis. *Science* 155: 1203–1207.

Worldwatch Institute. 2008. *State of the world 2008: Innovations for a sustainable economy*. Worldwatch Institute, Washington, D.C.

Worldwatch Institute. 2010. *Transforming cultures*. Worldwatch Institute, Washington, D.C.

Worldwatch Institute. 2012. *State of the world 2012: Moving toward sustainable prosperity*. Worldwatch Institute, Washington, D.C.

Wunder, Sven, et al. 2008. A comparative analysis of payments for environmental services programs in developed and developing countries. *Ecological Economics* 65: 834–852.

Wunscher, Tobias, et al. 2008. Spatial targeting of payments for environmental services: A tool for boosting conservation benefits. *Ecological Economics* 65: 822–833.

Chapter 7

Bogojevic, Sanja. 2013. *Emissions trading schemes: Markets, states, and laws*. Hart Publishing, Oxford, U.K.

Boyer, Elizabeth W., et al. 2012. *The impact of Marcellus gas drilling on rural drinking water supplies*. The Center for Rural Pennsylvania, a legislative agency of the Pennsylvania General Assembly, Harrisburg, Penn.

Clark, Ray, and Larry Canter. 1997. *Environmental policy and NEPA: Past, present, and future*. St. Lucie Press, Boca Raton, Fla.

Dietz, Thomas, et al. 2003. The struggle to govern the global commons. *Science* 302: 1907–1912.

Energy Information Administration, U.S. Department of Energy. 2013. *Technically recoverable shale oil and shale gas resources: An assessment of 137 shale formations in 41 countries outside the United States*. Washington, D.C.

Environmental Law Institute. 2009. *Estimating U.S. government subsidies to energy sources: 2002–2008*. ELI, Washington, D.C.

European Parliament, 2011. *Impacts of shale gas and shale oil extraction on the environment and on human health*. Directorate-General for Internal Policies.

Fogleman, Valerie M. 1990. *Guide to the National Environmental Policy Act*. Quorum Books, New York.

Green Scissors. 2012. *Green Scissors 2012: Cutting wasteful and environmentally harmful spending*. Friends of the Earth, Taxpayers for Common Sense, R Street Institute.

Hansjurgens, Bernd, ed. 2005. *Emissions trading for climate policy: US and European perspectives*. Cambridge Univ. Press, Cambridge, U.K.

Hardin, Garrett. 1968. The tragedy of the commons. *Science* 162: 1243–1248.

Houck, Oliver. 2003. Tales from a troubled marriage: Science and law in environmental policy. *Science* 302: 1926–1928.

Jackson, Robert B., et al. 2011. *Research and policy recommendations for hydraulic fracturing and shale-gas extraction*. Center on Global Change, Nicholas Schools of the Environment, Duke University.

Jacoby, Henry D., et al. 2012. The influence of shale gas on U.S. energy and environmental policy. *Economics of Energy and Environmental Policy* 1: 37–51.

Kubasek, Nancy K., and Gary S. Silverman. 2013. *Environmental law*, 8th ed. Pearson Prentice Hall, Upper Saddle River, N.J.

Myers, Norman, and Jennifer Kent. 2001. *Perverse subsidies: How misused tax dollars harm the environment and the economy*. Island Press, Washington, D.C.

The National Environmental Policy Act of 1969, as amended. http://ceq.hss.doe.gov/nepa/regs/nepa/nepaeqia.htm.

Nordhaus, Ted, and Michael Shellenberger. 2007. *Break Through: From the death of environmentalism to the politics of possibility*. Houghton Mifflin, Boston.

OECD. 2012. Inventory of estimated budgetary support and tax expenditures for fossil fuels 2013. OECD Publishing.

Osborn, Stephen G., et al. 2011. Methane contamination of drinking water accompanying gas-well drilling and hydraulic fracturing. *Proc. Natl. Acad. Sci. USA* 108: 8172–8176.

Percival, Robert V., et al. 2013. *Environmental regulation: Law, science, and policy*, 7th ed. Aspen Publishers, New York.

Phillips, Susan. 2012. Dimock: A town divided. StateImpact Pennsylvania. 28 March 2012. http://stateimpact.npr.org/pennsylvania/2012/03/28/dimock-a-town-divided/.

Rosenbaum, Walter. 2013. *Environmental politics and policy*, 9th ed. CQ Press, Congressional Quarterly, Inc., Washington, D.C.

Schmidt, Charles W. 2011. Blind rush? Shale gas boom proceeds amid human health questions. *Environ. Health Persp.* 119: A348–A353.

Shellenberger, Michael, and Ted Nordhaus. 2004. *The death of environmentalism: Global warming politics in a post-environmental world*. Presented at the Environmental Grantmakers Association meeting, Oct. 2004.

Tietenberg, Tom H. 2006. *Emissions trading: Principles and practice*. 2nd ed. Resources for the Future, Washington, D.C.

Tietenberg, Tom, and Lynne Lewis. 2010. *Environmental economics and policy*, 6th ed. Prentice Hall, Upper Saddle River, N.J.

U.N. General Assembly. 2012. *The future we want*. Outcome document from Rio+20 Conference. Resolution 66/288. http://www.un.org/ga/search/view_doc.asp?symbol=A/RES/66/288&Lang=E

U.S. Environmental Protection Agency. Summary of the National Environmental Policy Act. http://www2.epa.gov/laws-regulations/summary-national-environmental-policy-act.

U.S. Environmental Protection Agency. 2011. *Clean Air Interstate Rule, Acid Rain Program, and Former NO_x Budget Trading Program: 2010 progress report: Emission, compliance, and market analyses*. EPA, Washington, D.C.

U.S. Environmental Protection Agency. 2011. *Plan to study the potential impacts of hydraulic fracturing on drinking water resources*. EPA Office of research and Development, Washington, D.C.

U.S. Office of Management and Budget, 2011. *2011 report to Congress on the benefits and costs of regulations and unfunded mandates on state, local, and tribal entities*. Executive Office of the President of the United States, Washington, D.C.

Urbina, Ian. 2011. Regulation lax as gas wells' tainted water hits rivers. *New York Times*, 26 Feb. 2011.

Vig, Norman J., and Michael E. Kraft, eds. 2009. *Environmental policy: New directions for the twenty-first century*, 7th ed. CQ Press, Congressional Quarterly, Inc., Washington, D.C.

Wilber, Tom. Shale Gas Review. http://tomwilber.blogspot.com/.

Wilber, Tom. 2012. *Under the surface: Fracking, fortunes, and the fate of the Marcellus Shale*. Cornell Univ. Press, Ithaca, N.Y.

World Bank. 2010. *World development indicators 2013*. World Bank, Washington, D.C. http://data.worldbank.org/products/wdi.

Chapter 8

Ausubel, Jesse. H. 1996. Can technology spare the earth? *American Scientist* 84: 166–178.

Ball, Philip. 2008. Where have all the flowers gone? [China's one-child legacy.] *Nature* 454: 374–375.

Balter, Michael. 2006. The baby deficit. *Science* 312: 1894–1897.

Cohen, Joel E. 1995. *How many people can the Earth support?* W.W. Norton, New York.

Cohen, Joel E. 2005. Human population grows up. *Scientific American* 293(3): 48–55.

De Souza, Roger-Mark, et al. 2003. Critical links: Population, health, and the environment. *Population Bulletin* 58(3). Population Reference Bureau, Washington, D.C.

Eberstadt, Nicholas. 2000. China's population prospects: Problems ahead. *Problems of Post-Communism* 47: 28.

Ehrlich, Paul. 1968. *The population bomb*. 1997 reprint, Buccaneer Books, Cutchogue, New York.

Ehrlich, Paul R., and Anne H. Ehrlich. 1990. *The population explosion*. Touchstone, New York.

Ehrlich, Paul R., and John P. Holdren. 1971. Impact of population growth: Complacency concerning this component of man's predicament is unjustified and counterproductive. *Science* 171: 1212–1217.

Engelman, Robert. 2008. *More: Population, nature, and what women want*. Island Press, Washington D.C.

Engelman, Robert. 2010. *Population, climate change, and women's lives*. Worldwatch Report #183. Worldwatch Institute, Washington, D.C.

Greenhalgh, Susan. 2001. Fresh winds in Beijing: Chinese feminists speak out on the one-child policy and women's lives. *Signs: Journal of Women in Culture & Society* 26: 847–887.

Gribble, James N. and J. Bremner. 2012. Acheiving a demographic dividend. *Population Bulletin* 67 (2). Population Reference Bureau, Washington, D.C.

Haberl, Helmut, et al. 2007. Quantifying and mapping the human appropriation of net primary production in earth's terrestrial ecosystems. *PNAS* 104: 12942–12947.

Harrison, Paul, and Fred Pearce, eds. 2000. *AAAS atlas of population & environment*. Univ. of California Press, Berkeley.

Haub, Carl, and James N. Gribble. 2011. The world at 7 billion. *Population Bulletin* 66 (2). Population Reference Bureau, Washington, D.C.

Haub, Carl, and O.P. Sharma. 2006. India's population reality: Reconciling change and tradition. *Population Bulletin* 61(3). Population Reference Bureau, Washington, D.C.

Holdren, John P., and Paul R. Ehrlich. 1974. Human population and the global environment. *American Scientist* 62: 282–292.

Imhoff, Marc L., et al. 2004. Global patterns in human consumption of net primary production. *Nature* 429: 870–873.

Kane, Penny. 1987. *The second billion: Population and family planning in China*. Penguin Books, Australia, Ringwood, Victoria.

Kane, Penny, and Ching Y. Choi. 1999. China's one child family policy. *British Medical Journal* 319: 992.

Kent, Mary M., and Carl Haub. 2005. Global demographic divide. *Population Bulletin* 60(4). Population Reference Bureau, Washington, D.C.

La Ferrara, E., et al., 2012. Soap operas and fertility: Evidence from Brazil. *American Economic Journal: Applied Economics* 4: 1–31.

Lamptey, Peter R., et al. 2006. The global challenge of HIV and AIDS. *Population Bulletin* 61(1). Population Reference Bureau, Washington, D.C.

Malthus, Thomas R. *An essay on the principle of population*. 1983 ed. Penguin USA, New York.

Notestein, Frank. 1953. Economic problems of population change. Pp. 13–31 in *Proceedings of the Eighth International Conference of Agricultural Economists*. Oxford Univ. Press.

Population Reference Bureau. 2012. *2012 World population data sheet*. Population Reference Bureau, Washington, D.C.

Riley, Nancy E. 2004. *China's population: New trends and challenges*. *Population Bulletin* 59(2). Population Reference Bureau, Washington, D.C.

UNAIDS and World Health Organization. 2012. *UNAIDS report on the global AIDS epidemic*. UNAIDS and WHO, Geneva, Switzerland.

U.N. Population Division. 2009. *World population ageing 2009*. UNPD, New York.

U.N. Population Division. 2011. *World population prospects: The 2010 revision*. UNPD, New York.

U.N. Population Fund. 2010. *The millennium development goals report 2010*. UNFPA.

U.S. Census Bureau. www.census.gov.

Wackernagel, Mathis, and William Rees. 1996. *Our ecological footprint: Reducing human impact on the earth*. New Society Publishers, Gabriola Island, British Columbia, Canada.

Chapter 9

Ashman, Mark R., and Geeta Puri. 2002. *Essential soil science: A clear and concise introduction to soil science*. Wiley-Blackwell.

Curtin, Charles G. 2002. Integration of science and community-based conservation in the Mexico/U.S. borderlands. *Conservation Biology* 16: 880–886.

Diamond, Jared. 1999. *Guns, germs, and steel: The fates of human societies*. W.W. Norton, New York.

Diamond, Jared, and Peter Bellwood. 2003. Farmers and their languages: The first expansions. *Science* 300: 597–603.

Egan, Timothy. 2006. *The worst hard time: The untold story of those who survived the great American dust bowl*. Mariner Books, Houghton-Mifflin, Boston and New York.

Garcia-Tejero, Iva, et al. 2011. *Water and sustainable agriculture*. Springer, New York.

Glanz, James. 1995. *Saving our soil: Solutions for sustaining Earth's vital resource*. Johnson Books, Boulder, Colorado.

Huggins, David R., and John P. Reganold. 2008. No-till: How farmers are saving the soil by parking their plows. *Scientific American*, 30 June 2008.

Imeson, Anton. 2012. *Desertification, land degradation, and sustainability: Paradigms, processes, principles, and policies*. Wiley-Blackwell, Sussex, U.K.

Iowa State University Extension. 2009. *Considerations in selecting no-till*. Iowa State University Extension and NRCS.

Jenny, Hans. 1941. *Factors of soil formation: A system of quantitative pedology*. McGraw-Hill, New York.

Kaiser, Jocelyn. 2004. Wounding Earth's fragile skin. *Science* 304: 1616–1618.

Lal, Rattan, ed. 1998. *Soil quality and agricultural sustainability*. Ann Arbor Press, Chelsea, Mich.

Luo, Zhongkui, et al. 2010. Can no-tillage stimulate carbon sequestration in agricultural soils? A meta-analysis of paired experiments. *Agriculture, Ecosystems and Environment* 139: 224–231.

Malpai Borderlands Group. Malpai Borderlands Group. www.malpaiborderlandsgroup.org.

Millennium Ecosystem Assessment. 2005. *Ecosystems and human well-being: Desertification synthesis*. Millennium Ecosystem Assessment and World Resources Institute.

Morgan, R.P.C. 2010. *Soil erosion and conservation*. 3rd ed. Blackwell Science, Malden, Mass.

Montgomery, David R. 2007. *Dirt: The erosion of civilizations*. University of California Press, Berkeley and Los Angeles.

Montgomery, David R. 2007. Soil erosion and agricultural sustainability. *PNAS* 104: 13268–13272.

Morgan, R.P.C. 2005. *Soil erosion and conservation*, 3rd ed. Blackwell, London.

Natural Resources Conservation Service. 2010. *2007 national resources inventory: Soil erosion on cropland*. NRCS, USDA, Washington, D.C.

Natural Resources Conservation Service. Soils. NRCS, USDA. www.soils.usda.gov.

Ogle, Stephen M., et al. 2012. No-till management impacts on crop productivity, carbon input, and soil carbon sequestration. *Agriculture, Ecosystems and Environment* 149: 37–49.

Pieri, Christian, et al. 2002. *No-till farming for sustainable rural development*. Agriculture & Rural Development Working Paper. International Bank for Reconstruction and Development, Washington, D.C.

Pierzynski, Gary M., et al. 2005. *Soils and environmental quality*, 3rd ed. CRC Press, Boca Raton, Fla.

Soil Science Society of America. About soils. https://www.soils.org/about-soils.

Trimble, Stanley W., and Pierre Crosson. 2000. U.S. soil erosion rates—myth and reality. *Science* 289: 248–250.

Troeh, Frederick R., and Louis M. Thompson. 2005. *Soil and soil fertility*, 6th ed. Blackwell Publishing, London.

U.N. Environment Programme. 2012. "Land." Chapter 3 in *Global environment outlook 5 (GEO-5)*. UNEP, Nairobi.

Uri, Noel D. 2001. The environmental implications of soil erosion in the United States. *Env. Monitoring and Assessment* 66: 293–312.

Wilkinson, Bruce H. 2005. Humans as geologic agents: A deep-time perspective. *Geology* 33: 161–164.

Chapter 10

Benbrook, Charles M. 2012. Impacts of genetically engineered crops on pesticide use in the U.S.—the first sixteen years. *Environmental Sciences Europe* 24: 24, 13 pp.

Brar, Satinder Kaur. 2012. *Biocontrol: Management, processes, and challenges*. Nova Science Publishers, Hauppauge, New York.

Brown, Lester R. 2004. *Outgrowing the Earth: The food security challenge in an age of falling water tables and rising temperatures*. Earth Policy Institute, Washington, D.C.

Brown, Lester R. 2012. *Full planet, empty plates: The new geopolitics of food scarcity*. Earth Policy Institute, Washington, D.C.

Buchmann, Stephen L., and Gary Paul Nabhan. 1996. *The forgotten pollinators*. Island Press/Shearwater Books, Washington, D.C./Covelo, California.

Center for Food Safety. 2007. *Monsanto vs. U.S. farmers: November 2007 update*. Center for Food Safety, Washington, D.C.

Cerdeira, Antonio L., and Stephen O. Duke. 2006. The current status and environmental impacts of glyphosate-resistant crops: A review. *J. Environ. Quality* 35: 1633–1658.

Commission for Environmental Cooperation. 2004. *Maize and biodiversity: The effects of transgenic maize in Mexico*. CEC Secretariat.

[Correspondence to *Nature*, various authors]. 2002. *Nature* 416: 600–602, and 417: 897–898.

Dimitri, Carolyn, and Lydia Oberholtzer. 2009. *Marketing U.S. organic foods*. Economic Research Service, USDA, Washington, D.C.

Dyer, George A., et al. 2009. Dispersal of transgenes through maize seed systems in Mexico. *PLoS One* 4: e5734, pp. 1–9.

The Farm Scale Evaluations of spring-sown genetically modified crops. 2003. A themed issue from *Philosophical Transactions of the Royal Society of London B: Biological Sciences* 358 (1439), 29 Nov. 2003.

Fedoroff, Nina, and Nancy Marie Brown, 2004. *Mendel in the kitchen: A scientist's view of genetically modified foods*. National Academies Press, Washington, D.C.

Food and Agriculture Organization. 2006. *Livestock's long shadow: Environmental issues and options*. FAO, Rome.

Food and Agriculture Organization. 2013. *The state of food insecurity in the world 2012*. FAO, Rome.

Food and Agriculture Organization. 2013. *The state of world fisheries and aquaculture 2012*. FAO Fisheries and Aquaculture Department, Rome.

Gardner, Gary, and Brian Halweil. 2000. *Underfed and overfed: The global epidemic of malnutrition*. Worldwatch Paper #150. Worldwatch Institute, Washington, D.C.

Gurian-Sherman, Doug. 2009. *Failure to yield: Evaluating the performance of genetically engineered crops*. Union of Concerned Scientists, Cambridge, Mass.

Halweil, Brian. 2004. *Eat here: Reclaiming homegrown pleasures in a global supermarket*. Worldwatch Institute, Washington, D.C.

Halweil, Brian. 2008. *Farming fish for the future*. Worldwatch Report 176. Worldwatch Institute, Washington, D.C.

James, Clive. 2013. *Global status of commercialized biotech/GM crops: 2012*. International Service for the Acquisition of Agri-biotech Applications.

Kristiansen, P., et al., eds. 2006. *Organic agriculture: A global perspective*. CABI Publishing, Oxfordshire, U.K.

Liebig, Mark A., and John W. Doran. 1999. Impact of organic production practices on soil quality indicators. *J. Env. Qual.* 28: 1601–1609.

Maeder, Paul, et al. 2002. Soil fertility and biodiversity in organic farming. *Science* 296: 1694–1697.

Mann, Charles C. 2002. Transgene data deemed unconvincing. *Science* 296: 236–237.

Manning, Richard. 2000. *Food's frontier: The next green revolution*. North Point Press, New York.

Miller, Henry I., and Gregory Conko. 2004. *The frankenfood myth: How protest and politics threaten the biotech revolution*. Praeger Publishers, Westport, Connecticut.

Nierenberg, Danielle. 2005. *Happier meals: Rethinking the global meat industry*. Worldwatch Paper #171. Worldwatch Institute, Washington, D.C.

Nierenberg, Danielle, and Brian Halweil. 2005. Cultivating food security. Pp. 62–79 in *State of the world 2005*. Worldwatch Institute, Washington, D.C.

Nierenberg, Danielle, and Laura Reynolds. 2012. *Innovations in sustainable agriculture: Supporting climate-friendly food production*. Worldwatch Report #188. Worldwatch Institute, Washington, D.C.

Norris, Robert F., et al. 2003. *Concepts in integrated pest management*. Prentice Hall, Upper Saddle River, N.J.

Ortiz-García, Sol, et al. 2005. Absence of detectable transgenes in local landraces of maize in Oaxaca, Mexico (2003–2004). *PNAS* 102: 12338–12343.

Paoletti, Maurizio G., and David Pimentel. 1996. Genetic engineering in agriculture and the environment: Assessing risks and benefits. *BioScience* 46: 665–673.

Pearce, Fred. 2002. The great Mexican maize scandal. *New Scientist* 174: 14.

Pedigo, Larry P., and Marlin Rice. 2009. *Entomology and pest management*, 6th ed. Prentice Hall, Upper Saddle River, N.J.

Piñeyro-Nelson, A., et al. 2009. Transgenes in Mexican maize: Molecular evidence and methodological considerations for GMO detection in landrace populations. *Molecular Ecology* 18: 750–761.

Polak, Paul. 2005. The big potential of small farms. *Scientific American* 293(3): 84–91.

Pollan, Michael. 2009. *In defense of food: An eater's manifesto*. Penguin Press, New York.

Pretty, Jules. 2007. Agricultural sustainability: Concepts, principles, and evidence. *Phil. Trans. Roy. Soc. B* 363: 447–465.

Pretty, Jules. 2007. *Sustainable agriculture and food*. Earthscan, London, U.K.

Pringle, Peter. 2003. *Food, Inc.: Mendel to Monsanto—The promises and perils of the biotech harvest*. Simon and Schuster, New York.

Quist, David, and Ignacio H. Chapela. 2001. Transgenic DNA introgressed into traditional maize landraces in Oaxaca, Mexico. *Nature* 414: 541–543.

Roberts, Paul. 2008. *The end of food*. Houghton Mifflin, Boston.

Rodale Institute. 2011. *The farming systems trial: Celebrating 30 years*. Rodale Institute, Kutztown, Penn.

Ruse, Michael, and David Castle, eds. 2002. *Genetically modified foods: Debating technology*. Prometheus Books, Amherst, N.Y.

Shiva, Vandana. 2000. *Stolen harvest: The hijacking of the global food supply*. South End Press, Cambridge, Mass.

Smil, Vaclav. 2001. *Feeding the world: A challenge for the twenty-first century*. MIT Press, Cambridge, Mass.

Soleri, Daniela, et al. 2006. Transgenic crops and crop varietal diversity: The case of maize in Mexico. *BioScience* 56.

Stewart, C. Neal. 2004. *Genetically modified planet: Environmental impacts of genetically engineered plants*. Oxford Univ. Press, New York.

Sustainable Agriculture Network. 2010. *The new American farmer: Profiles of agricultural innovation*, 2nd ed. Sustainable Agriculture Network, Beltsville, MD.

Sustainable Agriculture Research and Education (SARE). 2010. *Exploring sustainability in agriculture*. SARE.

U.S. Department of Agriculture. 2009. *2007 Census of agriculture*. USDA, Washington, D.C.

Xie, Jian, et al. 2011. Ecological mechanisms underlying the sustainability of the agricultural heritage rice-fish coculture system. *Proc. Natl. Acad. Sci. USA* 108: E1381-E1387.

Weasel, Lisa. 2008. *Food fray: Inside the controversy over genetically modified food*. AMACOM Books.

Weber, Christopher L., and H. Scott Mathews, 2008. Food-miles and the relative climate impacts of food choices in the United States. *Environmental Science and Technology* 42: 3508–3513.

Willer, Helga. 2012. *Organic agriculture worldwide: Current statistics*. FiBL and IFOAM, Bonn and Frick.

Willer, Helga, 2012. *Organic agriculture worldwide: Key results from the survey on organic agriculture worldwide 2012*. FiBL and IFOAM, Bonn and Frick.

Wolfenbarger, L. La Reesa. 2000. The ecological risks and benefits of genetically engineered plants. *Science* 290: 2088.

Worldwatch Institute. 2011. *State of the world 2011: Innovations that nourish the planet*. Worldwatch Institute, Washington, D.C.

Chapter 11

Baker, C. Scott, et al. 2000. Scientific whaling: Source of illegal products for market? *Science* 290: 1695.

Baker, C. Scott, Frank Cipriano, and Stephen R. Palumbi. 1996. Molecular genetic identification of whale and dolphin products from commerical markets in Korea and Japan. *Molecular Ecology* 5: 671–685.

Balmford, Andrew, et al. 2002. Economic reasons for conserving wild nature. *Science* 297: 950–953.

Barnosky, Anthony D., et al. 2004. Assessing the causes of late Pleistocene extinctions on the continents. *Science* 306: 70–75.

Baskin, Yvonne. 1997. *The work of nature: How the diversity of life sustains us*. Island Press, Washington, D.C.

Brodie, Jedediah F., et al., eds. 2013. *Wildlife conservation in a changing climate*. University of Chicago Press, Chicago.

Chivian, Eric and Aaron Berstein, eds., 2008. *Sustaining life: How human health depends on biodiversity.* Oxford Univ. Press, New York.

CITES Secretariat. Convention on International Trade in Endangered Species of Wild Fauna and Flora. www.cites.org.

Collins, James P. and Martha L. Crump. 2009. *Extinction in our times: Global amphibian decline*. Oxford University Press, U.K.

Convention on Biological Diversity. www.biodiv.org.

Cooper, John E. and Margaret E. Cooper. 2013. Wildlife forensic investigation: Principles and practice. CRC Press, Boca Raton, Fla.

Daily, Gretchen C., ed. 1997. *Nature's services: Societal dependence on natural ecosystems*. Island Press, Washington, D.C.

Ehrenfeld, David W. 1970. *Biological conservation*. Holt, Rinehart, and Winston, New York.

Estes, R.D., et al. 2006. Downward trends in Ngorongoro Crater ungulate populations 1986–2005: Conservation concerns and the need for ecological research. *Biological Conservation* 131: 106-120.

Galatowitsch, Susan. 2012. *Ecological restoration*. Sinauer Associates, Sunderland, Mass.

Gascon, Claude, et al., eds. 2007. Amphibian conservation action plan. IUCN, Gland, Switzerland.

Gaston, Kevin J., and John I. Spicer. 2004. *Biodiversity: An introduction*, 2nd ed. Blackwell, London.

Gettleman, Jeffrey. 2012. To save wildlife, and tourism, Kenyans take up arms. *New York Times,* 29 Dec. 2012.

Groom, Martha J., et al. 2005. *Principles of conservation biology*, 3rd ed. Sinauer Associates, Sunderland, Mass.

Groombridge, Brian, and Martin D. Jenkins. 2002. *Global biodiversity: Earth's living resources in the 21st century*. UNEP, World Conservation Monitoring Centre, and Aventis Foundation; World Conservation Press, Cambridge, U.K.

Groombridge, Brian, and Martin D. Jenkins. 2002. *World atlas of biodiversity: Earth's living resources in the 21st century*. Univ. of California Press, Berkeley.

Hambler, Clive and Susan M. Canney. 2013. *Conservation*. 2nd ed. Cambridge University Press., U.K.

Hanken, James. 1999. Why are there so many new amphibian species when amphibians are declining? *Trends in Ecology and Evolution* 14: 7–8.

Jenkins, Martin. 2003. Prospects for biodiversity. *Science* 302: 1175–1177.

Kaufman, Leslie. 2012. Zoos' bitter choice: To save some species, letting others die. *New York Times*. 27 May 2012.

Louv, Richard. 2005. *Last child in the woods: Saving our children from nature-deficit disorder*. Algonquin Books, Chapel Hill, North Carolina.

Lovejoy, Thomas E., and Lee Hannah, eds. 2006. *Climate change and biodiversity*. Yale Univ. Press, New Haven, Conn.

MacArthur, Robert H., and Edward O. Wilson. 1967. *The theory of island biogeography*. Princeton Univ. Press.

Mburu, John, and Regina Birner. 2007. Emergence, adoption, and implementation of collaborative wildlife management or wildlife partnerships in Kenya: A look at conditions for success. *Society and Natural Resources* 20: 379–395.

Millennium Ecosystem Assessment. 2005. *Ecosystems and human well-being: Biodiversity synthesis*. Millennium Ecosystem Assessment and World Resources Institute.

Mooney, Harold A., and Richard J. Hobbs, eds. 2000. *Invasive species in a changing world*. Island Press, Washington, D.C.

Mora, Camilo, et al. 2011. How many species are there on Earth and in the ocean? *PLoS Biology* 9(8): e1001127.

Ogutu, Joseph O., et al. 2011. Continuing wildlife population declines and range contraction in the Mara region of kenya during 1977–2009. *Journal of Zoology* 285: 99–109.

Pisupati, Balakrishna, and Renata Rubian. 2008. *MDG on reducing biodiversity loss and the CBD's 2010 target*. UNU-IAS report, Yokohama, Japan.

Primack, Richard B. 2010. *Essentials of conservation biology*, 5th ed. Sinauer Associates, Sunderland, Mass.

Quammen, David. 1996. *The song of the dodo: Island biogeography in an age of extinction*. Touchstone, New York.

Rohr, Jason R., et al. 2008. Evaluating the links between climate, disease spread, and amphibian declines. *PNAS* 105:17436–17441.

Rosenzweig, Michael L. 1995. *Species diversity in space and time*. Cambridge Univ. Press.

Sepkoski, John J. 1984. A kinetic model of Phanerozoic taxonomic diversity. *Paleobiology* 10: 246–267.

Simberloff, Daniel. 1998. Flagships, umbrellas, and keystones: Is single-species management passé in the landscape era? *Biological Conservation* 83: 247–257.

Soulé, Michael E. 1986. *Conservation biology: The science of scarcity and diversity*. Sinauer Associates, Sunderland, Mass.

Stoner, Chantal, et al. 2007. Assessment of effectiveness of protection strategies in Tanzania based on a decade of survey data for large herbivores. *Conservation Biology* 21: 635–646.

Takacs, David. 1996. *The idea of biodiversity: Philosophies of paradise*. Johns Hopkins Univ. Press, Baltimore.

TEEB. 2010. *The economics of ecosystems and biodiversity: Mainstreaming the economics of nature: A synthesis of the approach, conclusions, and recommendations of TEEB*. The Economics of Ecosystems and Biodiversity.

U.N. Environment Programme. 2012. "Biodiversity." Chapter 5 in *Global environment outlook 5 (GEO-5)*. UNEP, Nairobi.

U.S. Environmental Protection Agency. Summary of the Endangered Species Act. http://www2.epa.gov/laws-regulations/summary-endangered-species-act.

U.S. Fish and Wildlife Service. Endangered species program. www.fws.gov/endangered.

Ward, Peter D. 2007. *Under a green sky: Global warming, the mass extinctions of the past, and what they can tell us about our future*. Smithsonian Books and Collins, Washington D.C. and New York.

Western, David. 2003. Conservation science in Africa and the role of international collaboration. *Conservation Biology* 17: 11–19.

Western, David, et al. 2009. The status of wildlife in protected areas compared to non-protected areas of Kenya. *PLoS One* 4(7): e6140.

Wilson, Edward O. 1984. *Biophilia*. Harvard Univ. Press.

Wilson, Edward O. 1992. *The diversity of life*. Harvard Univ. Press.

Wilson, Edward O. 2002. *The future of life*. Alfred A. Knopf, New York.

World Conservation Union. IUCN Red List. www.iucnredlist.org.

WWF–World Wide Fund for Nature. 2012. *Living planet report 2012*. WWF, Gland, Switzerland.

Chapter 12

Agronne, Dianna M. 2013. *Deforestation and climate change*. Nova Science Publishers, Hauppauge, New York.

Aubertin, Catherine and Estienne Rodary, eds. 2011. *Protected areas, sustainable land?* Ashgate Publishing, Surrey, U.K. and Burlington, Vermont.

Bettinger, Pete, et al. 2008. *Forest management and planning*. Academic Press, New York.

Bratkovich, Steve, et al. 2012. *Forests of the United States: Understanding trends and challenges*. Dovetail Partners Inc., Minneapolis, Minn.

British Columbia Ministry of Forests. Introduction to Silvicultural Systems. www.for.gov.bc.ca/hfd/pubs/SSIntroworkbook/index.htm. B.C. Ministry of Forests, Victoria, B.C.

Clary, David. 1986. *Timber and the Forest Service*. Univ. Press of Kansas, Lawrence.

Donato, Daniel C., et al. 2006. Post-wildfire logging hinders regeneration and increases fire risk. *Science* 311: 352.

Duncan, Dayton, and Ken Burns. 2009. *The national parks: America's best idea*. Alfred A. Knopf, New York.

Ferraz, Goncalo N., et al. 2007. A large-scale deforestation experiment: Effects of patch area and isolation on Amazon birds. *Science* 315: 238–241.

Food and Agriculture Organization. 2012. *State of the world's forests 2012*. FAO Forestry Department, Rome.

Food and Agriculture Organization. 2010. *Global forest resources assessment*. FAO Forestry Department, Rome.

Forest Stewardship Council. www.fsc.org.

Gorte, Ross W. and Pervaze A. Sheikh. 2013. *Deforestation and climate change*. Congressional Research Service, Washington, D.C.

Harris, Larry D. 1984. *The fragmented forest: Island biogeography theory and the preservation of biotic diversity*. Univ. of Chicago Press, Chicago.

Jensen, Sara E. and Guy R. McPherson. 2008. *Living with fire: Fire ecology and policy for the twenty-first century*. University of California Press, Berkeley and Los Angeles.

Land Trust Alliance. 2011. *2010 National land trust census report: A look at voluntary land conservation in America*. Land Trust Alliance, Washington, D.C.

Laurance, William F., et al. 2011. The fate of Amazonian forest fragments: A 32-year investigation. *Biological Conservation* 144(1): 56-67.

Lockwood, Michael, et al. 2006. *Managing protected areas: A global guide*. IUCN and Earthscan Publishing, London, U.K.

MacArthur, Robert H., and Edward O. Wilson. 1967. *The theory of island biogeography*. Princeton Univ. Press.

McKenzie, Donald, Carol Miller, and Donald A. Falk, eds. 2011. The landscape ecology of fire. *Ecological Studies* 213. Springer, New York.

National Forest Management Act of 1976. http://www.fs.fed.us/emc/nfma/includes/NFMA1976.pdf.

National Interagency Fire Center. http://www.nifc.gov/.

Newmark, William D. 1987. A land-bridge perspective on mammal extinctions in western North American parks. *Nature* 325: 430.

Pimm, Stuart L. 1998. The forest fragment classic. *Nature* 393: 23–24.

Pyne, Stephen. 2001. *Fire: A brief history*. University of Washington Press, Seattle.

Pye, Oliver, and Jayati Bhattacharya, eds. 2012. *The palm oil controversy in Southeast Asia: A transnational perspective*. ISEAS Publishing, Singapore.

Runte, Alfred. 1979. *National parks and the American experience*. Univ. of Nebraska Press, Lincoln.

Sedjo, Robert A. 2000. *A vision for the U.S. Forest Service*. Resources for the Future, Washington, D.C.

Shatford, Jeffrey, et al. 2007. Conifer regeneration after forest fire in the Klamath-Siskiyous: How much, how soon? *J. Forestry* 105: 139–146.

Smith, David M., et al. 1996. *The practice of silviculture: Applied forest ecology*, 9th ed. Wiley, New York.

Smithsonian Tropical Research Institute. Biological Dynamics of Forest Fragments Project. http://www.stri.si.edu/english/research/programs/programs_information/biological_dynamics_forest_fragments.php.

Soulé, Michael E., and John Terborgh, eds. 1999. *Continental conservation*. Island Press, Washington, D.C.

Stegner, Wallace. 1954. *Beyond the hundredth meridian: John Wesley Powell and the second opening of the West*. Houghton Mifflin, Boston.

Thompson, Jonathan R., et al. 2007. Re-burn severity in managed and unmanaged vegetation in a large wildfire. *PNAS* 104: 10743–10748.

U.N. Environment Programme. 2012. "Land." Chapter 3 in *Global environment outlook 5 (GEO-5)*. UNEP, Nairobi.

U.N. Environment Programme and World Conservation Monitoring Centre. 2012. *Protected planet report 2012: Tracking progress towards global targets for protected areas*. Cambridge, U.K.

USDA Forest Service. 2009. *Forest resources of the United States, 2007*. Gen. Tech. Rep. WO-78. U.S. Department of Agriculture, Washington, D.C.

USDA Forest Service. 2011. *National report on sustainable forests – 2010*. FS-979. U.S. Department of Agriculture, Washington, D.C.

Westerling, A.L., et al. 2006. Warming and earlier spring increase western U.S. forest wildfire activity. *Science* 313: 940–943.

Williams, Michael. 2006. *Deforesting the Earth: From prehistory to global crisis: An abridgment*. University of Chicago Press, Chicago.

Young, Raymond A. and Ronald L. Giese. 2002. *Introduction to forest ecosystem science and management*. 3rd ed. Wiley and Sons, New York.

Chapter 13

Abbott, Carl. 2001. *Greater Portland: Urban life and landscape in the Pacific Northwest*. Univ. of Pennsylvania Press.

Abbott, Carl. 2002. Planning a sustainable city. Pp. 207–235 in Squires, Gregory D., ed. *Urban sprawl: Causes, consequences, and policy responses*. Urban Institute Press, Washington, D.C.

Breuste, Jurgen, et al. 1998. *Urban ecology*. Springer-Verlag, Berlin.

Brockerhoff, Martin P. 2000. An urbanizing world. *Population Bulletin* 55(3). Population Reference Bureau, Washington, D.C.

Calthorpe, Peter. 2010. *Urbanism in the age of climate change*. Island Press, Washington D.C.

Cronon, William. 1991. *Nature's metropolis: Chicago and the great West*. W.W. Norton, New York.

Duany, Andres, et al. 2001. *Suburban nation: The rise of sprawl and the decline of the American dream*. North Point Press, New York.

Duany, Andres, et al. 2009. *The smart growth manual*. McGraw Hill Professional, New York.

Ellin, Nan. 2012. *Good urbanism*. Island Press, Washington, D.C.

Ewing, Reid, et al. 2002. *Measuring sprawl and its impact*. Smart Growth America.

Girardet, Herbert. 2004. *Cities people planet: Livable cities for a sustainable world*. Academy Press.

Hall, Kenneth B., and Gerald A. Porterfield. 2001. *Community by design: New urbanism for suburbs and small communities*. McGraw-Hill, New York.

Harnik, Peter. 2010. *Urban green: Innovative parks for resurgent cities*. The Trust for Public Land and Island Press, Washington, D.C.

Jacobs, Jane. 1992. *The death and life of great American cities*. Vintage.

Kalnay, Eugenia, and Ming Cai. 2003. Impact of urbanization and land-use change on climate. *Nature* 423: 528–531.

Kriken, John Lund. 2010. *City building: Nine planning principles for the twenty-first century*. Princeton Architectural Press, New York.

Litman, Todd. 2004. *Rail transit in America: A comprehensive evaluation of benefits*. Victoria Transport Policy Institute and American Public Transportation Association.

Long Term Ecological Research Network: Baltimore Ecosystem Study. http://www.lternet.edu/sites/bes.

Long Term Ecological Research Network: Central Arizona – Phoeniz LTER. http://www.lternet.edu/sites/cap.

Mayor's Office of Long-term Planning and Sustainability, New York. 2013. *PlaNYC: Progress report 2012: A greener, greater New York*. The City of New York.

Metro. www.metro-region.org.

New Urbanism. www.newurbanism.org.

Northwest Environment Watch. 2004. *The Portland exception: A comparison of sprawl, smart growth, and rural land loss in 15 U.S. cities*. Northwest Environment Watch, Seattle.

Oppenheimer, Laura. 2006. Measure 37 changed state, but how much? *Oregonian*, 3 Dec. 2006: 1 and A15.

Pearce, Fred. 2005. Cities lead the way to a greener world. *New Scientist*, 4 June 2005: 8–9.

Pugh, Cedric, ed. 1996. *Sustainability, the environment, and urbanization*. Earthscan Publications, London.

Renner, Michael, and Gary Gardner. 2010. *Global competitiveness in the rail and transit industry*. Worldwatch Institute, Washington, D.C.

Sheehan, Molly O'Meara. 2001. *City limits: Putting the brakes on sprawl*. Worldwatch Paper #156. Worldwatch Institute, Washington, D.C.

Speck, Jeff. 2012. *Walkable city: How downtown can save America, one step at a time*. Farrar, Straus, and Giroux., New York.

TEEB. 2011. *TEEB manual for cities: Ecosystem services in urban management*. The Economics of Ecosystems and Biodiversity.

U.N. Population Division. 2012. *World urbanization prospects: The 2011 revision*. UNPD, New York.

U.S. Green Building Council. www.usgbc.org.

Worldwatch Institute. 2007. *State of the world 2007: Our urban future*. Worldwatch Institute, Washington, D.C.

Chapter 14

Ames, Bruce N., et al. 1990. Nature's chemicals and synthetic chemicals: Comparative toxicology. *PNAS* 87: 7782–7786.

Bloom, Barry. 2005. Public health in transition. *Scientific American* 293(3): 92–99.

Carson, Rachel. 1962. *Silent spring*. Houghton Mifflin, Boston.

Carwhile, Jenny R., et al. 2011. Canned soup consumption and urinary bisphenol A: A randomized crossover trial. *Journal of the American Medical Association* 306: 2218–2220.

Colburn, Theo, Dianne Dumanoski, and John P. Myers. 1996. *Our stolen future*. Penguin USA, New York.

Consumer Reports. 2009. Concern over canned foods: Our tests find wide range of bisphenol A in soups, juice, and more. *Consumer Reports* Dec. 2009.

Curtis, Kathleen, and Bobbi Chase Wilding. 2010. *Is it in us? Chemical contamination in our bodies*. Body Burden Working Group and Commonweal Biomonitoring Resource Center.

Environmental Defence. 2008. *Toxic baby bottles in Canada.* Environmental Defence, Toronto, Ontario.

European Commission, Environment Directorate General. 2007. *REACH in brief.* European Commission.

Gilliom, Robert J., et al. 2006. *Pesticides in the nation's streams and ground water, 1992–2001—A summary.* USGS National Water-Quality Assessment Program Circular 1291.

Gross, Liza. 2007. The toxic origins of disease. *PLoS Biology* 5(7): e193. doi:10.1371/journal.pbio.0050193.

Guillette, Elizabeth A., et al. 1998. An anthropological approach to the evaluation of preschool children exposed to pesticides in Mexico. *Env. Health Perspectives* 106: 347–353.

Guillette, Louis J. Jr., et al. 2000. Alligators and endocrine disrupting contaminants: A current perspective. *American Zoologist* 40: 438–452.

Hayes, Tyrone, et al. 2003. Atrazine-induced hermaphroditism at 0.1 PPB in American leopard frogs (*Rana pipiens*): Laboratory and field evidence. *Env. Health Perspectives* 111: 568–575.

Hunt, Patricia A., et al. 2003. Bisphenol A exposure causes meiotic aneuploidy in the female mouse. *Current Biology* 13: 546–553.

Kent, Mary M., and Sandra Yin, 2006. Controlling infectious diseases. *Population Bulletin* 61(2), 24 pp. Population Reference Bureau, Washington, D.C.

Kolpin, Dana W., et al. 2002. Pharmaceuticals, hormones, and other organic wastewater contaminants in U.S. streams, 1999–2000: A national reconnaissance. *Env. Sci. Technol.* 36: 1202–1211.

Landis, Wayne G., et al. 2010. *Introduction to environmental toxicology*, 4th ed. CRC Press, Boca Raton, Fla.

Lang, Iain A., et al. 2008. Association of urinary bisphenol A concentration with medical disorders and laboratory abnormalities in adults. *JAMA* 300: 1303–1310.

Loewenberg, Samuel. 2003. E.U. starts a chemical reaction. *Science* 300: 405.

Manahan, Stanley E. 2009. *Environmental chemistry*, 9th ed. Lewis Publishers, CRC Press, Boca Raton, Fla.

McGinn, Anne Platt. 2000. *Why poison ourselves? A precautionary approach to synthetic chemicals.* Worldwatch Paper #153. Worldwatch Institute, Washington, D.C.

Millennium Ecosystem Assessment. 2005. *Ecosystems and human well-being: Health synthesis.* World Health Organization.

Moeller, Dade. 2011. *Environmental health*, 4th ed. Harvard Univ. Press.

Myers, Samuel S. 2009. *Global environmental change: The threat to human health.* Worldwatch Report #181. Worldwatch Institute, Washington, D.C.

National Center for Environmental Health; U.S. Centers for Disease Control and Prevention. 2005. *Third national report on human exposure to environmental chemicals.* NCEH Pub. No. 05-0570, Atlanta.

National Center for Health Statistics. 2009. *Health, United States, 2009, with special feature on medical technology.* Hyattsville, Maryland.

Our Stolen Future. http://www.ourstolenfuture.org/.

Pirages, Dennis. 2005. Containing infectious disease. Pp. 42–61 in *State of the world 2005.* Worldwatch Institute, Washington, D.C.

President's Cancer Panel. 2010. *Reducing environmental cancer risk.* President's Cancer Panel 2008–2009 annual report. U.S. Department of Health and Human Services, Washington, D.C.

Renner, Rebecca. 2002. Conflict brewing over herbicide's link to frog deformities. *Science* 298: 938–939.

Rodricks, Joseph V. 1994. *Calculated risks: Understanding the toxicity of chemicals in our environment.* Cambridge Univ. Press.

Stancel, George, et al. 2001. Report of the bisphenol A sub-panel. Chapter 1 in *National Toxicology Program's report of the endocrine disruptors low-dose peer review.* U.S. EPA and NIEHS, NIH.

Stockholm Convention on Persistent Organic Pollutants. www.pops.int.

United Health Foundation. 2013. *America's health rankings: 2012 annual report.* United Health Foundation, Minnetonka, Minn.

U.N. Environment Programme. 2012. "Chemicals and Waste." Chapter 6 in *Global environment outlook 5 (GEO-5).* UNEP, Nairobi.

U.S. Environmental Protection Agency. Summary of the Toxic Substances Control Act. http://www2.epa.gov/laws-regulations/summary-resource-conservation-and-recovery-act.

Vogel, Sarah. 2009. The politics of plastic: The making and unmaking of bisphenol A "safety". *Framing Health Matters, Am. J. Public Health* Suppl 3, vol. 99 #S3, S559–S566.

vom Saal, Frederick S., et al. 2012. The estrogenic endocrine disrupting chemical bisphenol A (BPA) and obesity. *Molecular and Cellular Endocrinology* 354: 74–84.

World Health Organization. 2008. *The global burden of disease: 2004 update.* WHO, Geneva, Switzerland.

World Health Organization. 2009. *Global health risks: Mortality and burden of disease attributable to selected major risks.* WHO, Geneva, Switzerland.

World Health Organization. 2013. *World health statistics 2013.* WHO, Geneva, Switzerland.

Zogorski, John S., et al. 2006. *Volatile organic compounds in the nation's ground water and drinking-water supply wells.* USGS National Water-Quality Assessment Program Circular 1292.

Chapter 15

American Rivers. 2002. *The ecology of dam removal: A summary of benefits and impacts.* American Rivers, Washington, D.C.

Coastal Protection and Restoration Authority. 2012. *2012 coastal master plan.* Coastal Protection and Restoration Authority of Louisiana. Baton Rouge, LA.

De Villiers, Marq. 2000. *Water: The fate of our most precious resource.* Mariner Books.

Gleick, P.H., and H.S. Cooley. 2009. Energy implications of bottled water. *Environ. Res. Lett.* 4: 014009 (6 pp).

Gleick, Peter H., et al. 2009. *The world's water 2008–2009: The biennial report on freshwater resources.* Island Press, Washington, D.C.

Gleick, Peter H. 2010. *Bottled and sold: The story behind our obsession with bottled water.* Island Press, Washington, D.C.

Gulf Coast Ecosystem Restoration Council. 2013. Draft initial comprehensive plan: *Restoring the Gulf Coast's ecosystem and economy.* http://www.restorethegulf.gov.

Jenkins, Matt. 2006. Running on empty in Sin City. *High Country News* 38(17), 18 Sept. 2006.

Millennium Ecosystem Assessment. 2005. *Ecosystems and human well-being: Wetlands and water synthesis.* Millennium Ecosystem Assessment and World Resources Institute.

Naidenko, Olga, et al. 2008. *Bottled water quality investigation: 10 major brands, 38 pollutants.* Environmental Working Group. www.ewg.org.

New York Times. 2009–2010. *Toxic Waters: A series about the worsening pollution in American waters and regulators' response. New York Times*, http://projects.nytimes.com/toxic-waters.

Nickson, Ross, et al. 1998. Arsenic poisoning of Bangladesh groundwater. *Nature* 395: 338.

Pala, Christopher. 2006. Once a terminal case, the North Aral Sea shows new signs of life. *Science* 312: 183.

Postel, Sandra. 1999. *Pillar of sand: Can the irrigation miracle last?* W.W. Norton, New York.

Postel, Sandra. 2003. *Rivers for life: Managing water for people and nature.* Island Press, Washington, D.C.

Postel, Sandra. 2005. *Liquid assets: The critical need to safeguard freshwater ecosystems.* Worldwatch Paper #170. Worldwatch Institute, Washington, D.C.

Rabalais, Nancy N., et al. 2002. Beyond science into policy: Gulf of Mexico hypoxia and the Mississippi River. *BioScience* 52: 129–142.

Rabalais, Nancy N., et al. 2002. Hypoxia in the Gulf of Mexico, a.k.a. "The dead zone." *Annual Review of Ecology and Systematics* 33: 235–263.

Reisner, Marc. 1986. *Cadillac desert: The American West and its disappearing water.* Viking Penguin, New York.

Stone, Richard. 1999. Coming to grips with the Aral Sea's grim legacy. *Science* 284: 30–33.

TEEB. 2013. *The economics of ecosystems and biodiversity (TEEB) for water and wetlands.* The Economics of Ecosystems and Biodiversity. IEEP, London and Brussels, Ramsar Secretariat, Gland.

U.N. Environment Programme. 2012. "Water." Chapter 4 in *Global environment outlook 5 (GEO-5).* UNEP, Nairobi.

U.N. Environment Programme. 2008. *Water quality for ecosystem and human health*, 2nd ed. UNEP Global Environment Monitoring System (GEMS)/Water Programme, Burlington, Ontario.

U.N. World Water Assessment Programme. 2012. *U.N. world water development report: Managing water under uncertainty and risk.* Paris, New York, and Oxford, UNESCO and Berghahn Books.

U.S. Environmental Protection Agency. 2004. *Primer for municipal wastewater treatment systems*. EPA 832-R-04-001. Office of Wastewater Management, Washington D.C.

U.S. Environmental Protection Agency. Summary of the Clean Water Act. http://www2.epa.gov/laws-regulations/summary-clean-water-act.

U.S. Environmental Protection Agency. 2009. *Water on tap: What you need to know*. EPA 816-K-09-002. Office of Water, Washington, D.C.

U.S. Government Accountability Office (GAO). 2009. *Bottled water: FDA safety and consumer protections are often less stringent than comparable EPA protections for tap water.* GAO Report to Congressional Requesters, GAO-09-610.

Wagner, Martin, and Jörg Oehlmann. 2009. Endocrine disruptors in bottled mineral water: Total estrogenic burden and migration from plastic bottles. *Env. Sci. Pollut. Res.* 16: 278–286.

Worldwatch Institute. 2011. *State of the world 2011: Innovations that nourish the planet*. Worldwatch Institute, Washington, D.C.

Chapter 16

Allsopp, Michelle, et al. 2007. *Oceans in peril: Protecting marine biodiversity*. Worldwatch Report 174. Worldwatch Institute, Washington, D.C.

Bellwood, David R., et al. 2004. Confronting the coral reef crisis. *Nature* 429: 827–833.

[Correspondence to *Science*, various authors]. 2001. *Science* 295: 1233–1235.

Food and Agriculture Organization. 2012. *The state of world fisheries and aquaculture 2012*. FAO Fisheries and Aquaculture Department, Rome.

Frank, Kenneth T., et al. 2005. Trophic cascades in a formerly cod-dominated ecosystem. *Science* 308: 1621–1623.

Garrison, Tom. 2012. *Oceanography: An invitation to marine science*, 8th ed. Brooks/Cole, San Francisco.

Gell, Fiona R., and Callum M. Roberts. 2003. Benefits beyond boundaries: The fishery effects of marine reserves. *Trends in Ecology and Evolution* 18: 448–455.

Halweil, Brian. 2006. *Catch of the day: Choosing seafood for healthier oceans*. Worldwatch Paper #172. Worldwatch Institute, Washington, D.C.

Henderson, Caspar. 2006. Ocean acidification: The *other* CO_2 problem. *New Scientist* 5 Aug. 2006.

Hoegh-Guldberg, O., et al. 2007. Coral reefs under rapid climate change and ocean acidification. *Science* 318: 1737–1742.

International Pacific Research Center. 2008. Tracking ocean debris. *IPRC Climate* 8: 14–16.

Jackson, Jeremy B.C., et al. 2001. Historical overfishing and the recent collapse of coastal ecosystems. *Science* 293: 629–638.

Kurlansky, Mark. 1998. *Cod: A biography of the fish that changed the world.* Penguin Books, New York.

Lotze, Heike L., et al. 2006. Depletion, degradation, and recovery potential of estuaries and coastal seas. *Science* 312: 1806–1809.

Maximenko, N.A., et al. 2012: Pathways of marine debris from trajectories of Lagrangian drifters. *Marine Pollution Bulletin*, 65: 51–62.

Mayo, Ralph, and Loretta O'Brien. 2006. Status of fishery resources off the northeastern U.S. NEFSC. http://www.nefsc.noaa.gov/sos/spsyn/pg/cod/.

Moore, Charles James. 2008. Synthetic polymers in the marine environment: A rapidly increasing, long-term threat. *Environmental Research* 108: 131–139.

Morrissey, John F. and James L. Sumich. 2010. *Introduction to the biology of marine life*, 10th ed. Jones & Bartlett, Boston.

Myers, Ransom A., and Boris Worm. 2003. Rapid worldwide depletion of predatory fish communities. *Nature* 423: 280–283.

National Center for Ecological Analysis and Synthesis (NCEAS) and Communication Partnership for Science and the Sea (COMPASS), sponsors. 2001. *Scientific consensus statement on marine reserves and marine protected areas*. www.nceas.ucsb.edu/consensus.

National Research Council. 2003. *Oil in the sea III: Inputs, fates, and effects*. National Academies Press, Washington, D.C.

Norse, Elliott, and Larry B. Crowder, eds. 2005. *Marine conservation biology: The science of maintaining the sea's biodiversity*. Island Press, Washington, D.C.

Nybakken, James W., and Mark D. Bertness. 2004. *Marine biology: An ecological approach*, 6th ed. Benjamin Cummings, San Francisco.

Orr, James C. 2005. Anthropogenic ocean acidification over the twenty-first century and its impact on calcifying organisms. *Nature* 437: 681–686.

Pauly, Daniel, et al. 2002. Towards sustainability in world fisheries. *Nature* 418: 689–695.

Pauly, Daniel, et al. 2003. The future for fisheries. *Science* 302: 1359–1361.

Pew Oceans Commission. 2003. *America's living oceans: Charting a course for sea change*. A report to the nation. Pew Oceans Commission, Arlington, Va.

Roberts, Callum M., et al. 2001. Effects of marine reserves on adjacent fisheries. *Science* 294: 1920–1923.

Rosenberg, A., et al. 2006. Rebuilding U.S. fisheries: Progress and problems. *Frontiers in Ecology and the Environment* 4(6).

TEEB. 2012. *Why value the oceans—A discussion paper.* The Economics of Ecosystems and Biodiversity.

Trujillo, Alan P., and Harold V. Thurman. 2013. *Essentials of oceanography*, 11th ed. Prentice Hall, Upper Saddle River, N.J.

U.S. Commission on Ocean Policy. 2004. *An ocean blueprint for the 21st century*. Final report. Washington, D.C.

U.S. Department of Commerce and U.S. Department of the Interior. Marine protected areas of the United States. www.mpa.gov.

Weber, Michael L. 2001. *From abundance to scarcity: A history of U.S. marine fisheries policy*. Island Press, Washington, D.C.

Weiss, Kenneth R., and Usha Lee McFarling. 2006. Altered oceans (a special five-part series). *Los Angeles Times*, 30 July–3 August, 2006. Available at: http://www.latimes.com/news/local/la-oceans-series,0,7783938.special.

Worm, Boris, et al. 2006. Impacts of biodiversity loss on ocean ecosystem services. *Science* 314: 787–790.

Chapter 17

Ahrens, C. Donald. 2008. *Meteorology today*, 9th ed. Brooks/Cole, San Francisco.

Akimoto, Hajime. 2003. Global air quality and pollution. *Science* 302: 1716–1719.

Bernard, Susan M., et al. 2001. The potential impacts of climate variability and change on air pollution-related health effects in the United States. *Env. Health Perspectives* 109(Suppl 2): 199–209.

Bruce, Nigel, Rogelio Perez-Padilla, and Rachel Albalak. 2000. Indoor air pollution in developing countries: A major environmental and public health challenge. *Bull. World Health Organization* 78: 1078–1092.

Calderón-Garcidueñas, Lilian, et al. 2003. Respiratory damage in children exposed to urban pollution. *Pediatric Pulmonology* 36: 148–161.

Cooper, C. David, and F. C. Alley. 2010. *Air pollution control: A design approach*, 4th ed. Waveland Press.

Davis, Devra. 2002. *When smoke ran like water: Tales of environmental deception and the battle against pollution*. Basic Books, New York.

Davis, Devra L., et al. 2002. A look back at the London smog of 1952 and the half century since. *Env. Health Perspectives* 110: A734.

Driscoll, Charles T., et al. 2001. *Acid rain revisited: Advances in scientific understanding since the passage of the 1970 and 1990 Clean Air Act Amendments*. Hubbard Brook Research Foundation.

Godish, Thad. 2004. *Air quality*, 4th ed. CRC Press, Boca Raton, Fla.

Hall, Jane V., et al. 2008. *The benefits of meeting federal clean air standards in the South Coast and San Joaquin Valley air basins*. William and Flora Hewlett Foundation.

Hoffman, Matthew J. 2005. *Ozone depletion and climate change: Constructing a global response*. SUNY Press, New York.

Jacobson, Mark Z. 2002. *Atmospheric pollution: History, science, and regulation*. Cambridge Univ. Press.

Kaiman, Jonathan. 2013. Chinese struggle through 'airpocalypse' smog. *The Observer*. 16 Feb. 2013.

Likens, Gene E. 2004. Some perspectives on long-term biogeochemical research from the Hubbard Brook ecosystem study. *Ecology* 85: 2355–2362.

Loomis, Dana, et al. 1999. Air pollution and infant mortality in Mexico City. *Epidemiology* 10: 118–123.

Lutgens, Frederick K., Edward J. Tarbuck, and Dennis Tasa. 2013. *The atmosphere: An introduction to meteorology*, 12th ed. Pearson Education.

Molina, Mario J., and F. Sherwood Rowland. 1974. Stratospheric sink for chlorofluoromethanes: Chlorine atom catalyzed destruction of ozone. *Nature* 249: 810–812.

Parson, Edward A. 2003. *Protecting the ozone layer: Science and strategy*. Oxford Univ. Press.

U.N. Environment Programme. 2012. "Atmosphere." Chapter 2 in *Global environment outlook 5 (GEO-5)*. UNEP, Nairobi.

U.N. Environment Programme, Ozone Secretariat. Montreal Protocol. http://ozone.unep.org/new_site/en/index.php.

U.S. Environmental Protection Agency, Office of Air and Radiation. www.epa.gov/air.

U.S. Environmental Protection Agency. Summary of the Clean Air Act. http://www2.epa.gov/laws-regulations/summary-clean-air-act.

U.S. Environmental Protection Agency. 2011. *Clean Air Interstate Rule, Acid Rain Program, and Former NO_x Budget Trading Program: 2010 progress report: Emission, compliance, and market analyses*. EPA, Washington, D.C.

U.S. Environmental Protection Agency. 2012. *Clean Air Interstate Rule, Acid Rain Program, and Former NO_x Budget Trading Program: 2010 progress report: Environmental and health results*. EPA, Washington, D.C.

U.S. Environmental Protection Agency. 2012. *National air quality: Status and trends through 2010*. EPA, Washington, D.C.

World Health Organization. Indoor air pollution. www.who.int/indoorair/en/index.html.

Chapter 18

Alley, Richard B. 2000. *The two-mile time machine: Ice cores, abrupt climate change, and our future*. Princeton Univ. Press.

Bianco, Nicholas M., and Franz T. Litz. 2010. *Reducing greenhouse gas emissions in the United States using existing federal authorities and state action*. WRI Report, World Resources Institute, Washington, D.C.

Bloom, Arnold J. 2009. *Global climate change: Convergence of disciplines*. Sinauer Associates, Sunderland, Mass.

Bogojevic, Sanja. 2013. *Emissions trading schemes: Markets, states, and laws*. Hart Publishing, Oxford, U.K.

Borja-Aburto, Victor H., et al. 1997. Ozone, suspended particulates, and daily mortality in Mexico City. *Am. J. Epidemiology* 145: 258–268.

Burney, Nelson E., ed. 2010. *Carbon tax and cap-and-trade tools: Market-based approaches for controlling greenhouse gases*. Nova Science Publishers, Hauppauge, New York.

Burroughs, William James. 2007. *Climate change: A multidisciplinary approach*, 2nd ed. Cambridge Univ. Press.

Caldeira, Kenneth, and Michael E. Wickett. 2003. Anthropogenic carbon and ocean pH. *Nature* 425: 365.

Center for Climate and Energy Solutions. http://www.c2es.org/.

Edenhofer, Ottmar, et al., eds. 2012. *Renewable energy sources and climate change mitigation*. Special Report of the Intergovernmental Panel on Climate Change. IPCC and Cambridge Univ. Press.

EPICA community members. 2004. Eight glacial cycles from an Antarctic ice core. *Nature* 429: 623–628.

Field, Christopher, et al., eds. 2012. *Managing the risks of extreme events and disasters to advance climate change adaptation*. Special Report of the Intergovernmental Panel on Climate Change. IPCC and Cambridge Univ. Press.

Flannery, Tim. 2005. *The weather makers: The history and future impact of climate change*. Text Publishing, Melbourne, Australia.

Francis, Jennifer A., and Stephen J. Vavrus. 2012. Evidence linking Arctic amplification to extreme weather in mid-latitudes. *Geophysical Research Letters* 39: L06801, 6 pp.

Gelbspan, Ross. 1997. *The heat is on: The climate crisis, the cover-up, the prescription*. Perseus Books, New York.

Gelbspan, Ross. 2004. *Boiling point: How politicians, big oil and coal, journalists, and activists are fueling the climate crisis—and what we can do to avert disaster*. Basic Books, New York.

Goodstein, Eban, 2007. *Fighting for love in the century of extinction: How passion and politics can stop global warming*. Univ. Press of New England, Lebanon, N.H.

Gore, Al. 2006. *An inconvenient truth: The planetary emergency of global warming and what we can do about it*. Rodale Press and Melcher Media, New York.

Hansen, James, Makiko Sato, and Reto Ruedy. 2012. Perception of climate change. *Proc. Natl. Acad. Sci. USA* E2415-E2423, 6 Aug. 2012.

Hansjurgens, Bernd, ed. 2005. *Emissions trading for climate policy: US and European perspectives*. Cambridge Univ. Press, Cambridge, U.K.

InsideClimate News. http://insideclimatenews.org/.

Intergovernmental Panel on Climate Change. 2001. *IPCC third assessment report—Climate change 2001: Synthesis report*. World Meteorological Organization and U.N. Environment Programme, Geneva, Switzerland.

Intergovernmental Panel on Climate Change. 2007. *IPCC fourth assessment report—Climate change 2007: The AR4 synthesis report*. World Meteorological Organization and U.N. Environment Programme, Geneva, Switzerland.

Intergovernmental Panel on Climate Change. 2007. *Climate change 2007: The physical science basis*. Contribution of Working Group I to the fourth assessment report of the IPCC. WMO and UNEP.

Intergovernmental Panel on Climate Change. 2007. *Climate change 2007: Impacts, adaptation, and vulnerability*. Contribution of Working Group II to the fourth assessment report of the IPCC. WMO and UNEP.

Intergovernmental Panel on Climate Change. 2007. *Climate change 2007: Mitigation of climate change*. Contribution of Working Group III to the fourth assessment report of the IPCC. WMO and UNEP.

Intergovernmental Panel on Climate Change. www.ipcc.ch.

International Energy Agency. 2012. *CO2 emissions from fuel combustion: Highlights: 2012* edition. IEA, Paris.

Jonzén, Niclas, et al. 2006. Rapid advance of spring arrival dates in long-distance migratory birds. *Science* 312: 1959–1961.

Karl, Thomas R., and Kevin E. Trenberth. 2003. Modern global climate change. *Science* 302: 1719–1723.

Kerr, Richard A. 2006. A tempestuous birth for hurricane climatology. *Science* 312: 676–678.

Kraska, James, ed. 2013. *Arctic security in an age of climate change*. Cambridge Univ. Press, Cambridge, U.K.

Mann, Michael, and Lee R. Kump. 2008. *Dire Predictions: Understanding global warming*. DK Publishing and Pearson Education, New York.

Mayewski, Paul A., and Frank White. 2002. *The ice chronicles: The quest to understand global climate change*. Univ. Press of New England, Hanover, N.H.

McKibben, Bill. 2012. Global warming's terrifying new math. *Rolling Stone*. 2 Aug. 2012.

National Geographic. 2008. Changing climate. *National Geographic special report*, 22 June 2008.

NOAA Climate.gov. http://www.climate.gov/.

Pacala, Stephen, and Robert Socolow. 2004. Stabilization wedges: Solving the climate problem for the next 50 years with current technologies. *Science* 305: 968–972.

Parmesan, Camille, and Gary Yohe. 2003. A globally coherent fingerprint of climate change impacts across natural systems. *Nature* 421: 37–42.

Real Climate. www.realclimate.org.

Root, Terry L., et al. 2003. Fingerprints of global warming on wild animals and plants. *Nature* 421: 57–60.

The Royal Society. 2005. *Ocean acidification due to increasing atmospheric carbon dioxide*. The Royal Society, London, U.K., June 2005.

Scherr, Sara J., and Sajal Sthapit. 2009. *Mitigating climate change through food and land use*. Worldwatch Report 179. Worldwatch Institute, Washington, D.C.

Schiermeier, Quirin. 2006. Climate credits. *Nature* 444: 976–977.

Schneider, Stephen H., and Terry L. Root, eds. 2002. *Wildlife responses to climate change: North American case studies*. Island Press, Washington, D.C.

Service, Robert. 2012. Rising acidity brings an ocean of trouble. *Science* 337: 146–148.

Taylor, David. 2003. Small islands threatened by sea level rise. Pp. 84–85 in *Vital signs 2003*. Worldwatch Institute, Washington D.C.

Tietenberg, Tom H. 2006. *Emissions trading: Principles and practice*. 2nd ed. Resources for the Future, Washington, D.C.

U.N. Environment Programme. 2011. *Bridging the emissions gap*. A UNEP Synthesis Report.

U.S. Climate Change Science Program. 2008. *Scientific assessment of the effects of global change on the United States*. Committee on Environment and Natural Resources, National Science and Technology Council, Washington, D.C.

U.S. Climate Change Science Program. 2008. *Climate models: An assessment of strengths and limitations*. Synthesis and Assessment Product 3.1. Washington, D.C.

U.S. Climate Change Science Program. 2009. *Coastal sensitivity to sea-level rise: A focus on the mid-Atlantic region*. Synthesis and Assessment Product 4.1, Washington, D.C.

U.S. Global Change Research Program (Karl, Thomas R., Jerry M. Melillo, and Thomas C. Peterson, eds.). 2009. *Global climate change impacts in the United States*. U.S. Global Change Research Program and Cambridge Univ. Press.

U.S. Global Change Research Program. 2013. *National climate assessment*. Draft for Public Review, 11 Jan. 2013.

United Nations. U.N. Framework Convention on Climate Change. http://unfccc.int/2860.php.

United Nations. Kyoto Protocol. http://unfccc.int/kyoto_protocol/items/2830.php.

Victor, David G., et al. 2005. A Madisonian approach to climate policy. *Science* 309: 1820–1821.

Walsh, Bryan. 2008. How to win the war on global warming. *Time*, 28 Apr. 2008.

Wilcox, Jennifer. 2012. *Carbon capture*. Springer, New York.

World Bank. 2012. *Turn down the heat: Why a 4° warmer world must be avoided.* A report for the World Bank by the Potsdam Institute for Climate Impact Research and Climate Analytics. World Bank, Washington D.C.

World Meteorological Organization. 2013. *WMO statement on the status of the global climate in 2012.* WMO No. 1108. Geneva, Switzerland.

Worldwatch Institute. 2009. *State of the world 2009: Into a warming world.* Worldwatch Institute, Washington, D.C.

Chapter 19

Al-Fattah, Saud M., et al. 2011. *Carbon capture and storage: Technologies, policies, economics, and implementation strategies*. CRC Press, Boca Raton, Fla.

American Petroleum Institute. 2009. *Offshore access to oil and natural gas reserves.* American Petroleum Institute, Washington, D.C.

Association for the Study of Peak Oil and Gas. www.peakoil.net.

Avery, Samuel. 2013. *The pipeline and the paradigm: Keystone XL, tar sands, and the battle to defuse the carbon bomb.* Ruka Press, Washington, D.C.

British Petroleum. 2013. *BP statistical review of world energy 2013*. BP, London.

Campbell, Colin J. 1997. *The coming oil crisis*. Multi-Science Publishing Co., Essex, U.K.

Deffeyes, Kenneth S. 2005. *Beyond oil: The view from Hubbert's peak*. Farrar, Straus, and Giroux, New York.

Energy Information Administration, U.S. Department of Energy. www.eia.doe.gov.

Energy Information Administration, U.S. Department of Energy. 1999. *Petroleum: An energy profile, 1999.* DOE/EIA-0545(99).

Energy Information Administration, U.S. Department of Energy. 2013. *Annual energy outlook 2013*. Washington, D.C.

Energy Information Administration, U.S. Department of Energy. 2013. *Annual energy review 2012*. DOE/EIA, Washington, D.C.

Energy Information Administration, U.S. Department of Energy. 2013. *Technically recoverable shale oil and shale gas resources: An assessment of 137 shale formations in 41 countries outside the United States.* Washington, D.C.

Findley, David. 2010. *Do-it-yourself home energy audits: 140 simple solutions to lower energy costs, increase your home's efficiency, and save the environment*. McGraw Hill, New York.

Hall, Charles A.S. and Doug Hansen. 2011. New Studies in EROI (Energy Return on Investment). *Sustainability*, Special Issue. MDPI AG, Basel, Switzerland, 2011.

Haerens, Margaret. 2010. *Offshore drilling (opposing viewpoints)*. Greenhaven Press, Farmington Hills, Michigan.

Herring, Horace, and Steve Sorrell, eds. 2009. *Energy efficiency and sustainable consumption: The rebound effect*. Palgrave Macmillan, London.

Hubbert, M. King. 1956. *Nuclear energy and the fossil fuels.* Publication No. 95, Shell Development Company, Houston, Texas.

International Energy Agency. 2012. *Key world energy statistics 2012*. IEA, Paris.

International Energy Agency. 2012. *World energy outlook 2012*. IEA, Paris.

Kitasei, Saya. 2010. *Powering the low-carbon economy: The once and future roles of renewable energy and natural gas*. Worldwatch Report #184. Worldwatch Institute, Washington, D.C.

Kunstler, James H. 2005. *The long emergency*. Atlantic Monthly Press, New York.

Kunzig, Robert. 2009. Scraping bottom. *National Geographic*, March 2009.

Leffler, William L., et al., 2011. *Deepwater petroleum exploration and production: A nontechnical guide*. 2nd ed. Pennwell Corp. Publishing, Tulsa, Okla.

Levant, Ezra. 2011. *Ethical oil: The case for Canada's oil sands*. McClelland & Stewart, Toronto, Ontario.

Lovins, Amory B. 2005. More profit with less carbon. *Scientific American* 293(3): 74–83.

Lovins, Amory B., et al. 2004. *Winning the oil endgame: Innovation for profits, jobs, and security*. Rocky Mountain Institute, Snowmass, Colorado.

MacKay, David J.C. 2009. *Sustainable energy—without the hot air*. UIT Cambridge.

Miller, Bruce G. 2010. *Clean coal engineering technology*. Butterworth-Heinemann, Elsevier.

National Commission on the BP Deepwater Horizon Oil Spill and Offshore Drilling. 2011. *Deep water: The Gulf oil disaster and the future of offshore drilling*. Report to the President. Oil Spill Commission.

National Oceanic and Atmospheric Administration. 2012. *Natural resource damage assessment for the Deepwater Horizon oil spill: April 2012 status update*. NOAA, Washington, D.C.

OECD. 2012. Inventory of estimated budgetary support and tax expenditures for fossil fuels 2013. OECD Publishing.

Parfomak, Paul, et al., 2012. *Keystone XL pipeline project: Key issues*. Congressional Research Service CRS Report for Congress 7-5700 R41668.

Rao, Vikram. 2012. *Shale gas: The promise and the peril*. RTI International, RTI Press, Research Triangle Institute, N.C.

Ristinen, Robert A., and Jack J. Kraushaar, 2005. *Energy and the environment*, 2nd ed. Wiley and Sons, New York.

Renne, John L., and Billy Fields. 2013. *Transport beyond oil: Policy choices for a multimodal future*. Island Press, Washington, D.C.

Roberts, Paul. 2004. *The end of oil: On the edge of a perilous new world*. Houghton Mifflin, Boston.

Spellman, Frank R. 2013. *Environmental impacts of hydraulic fracturing*. CRC Press, Boca Raton, Fla.

Sumper, Andreas, and Angelo Baggini. 2012. *Electrical energy efficiency: Technologies and applications*. John Wiley & Sons, Chichester, U.K.

U.S. Department of State. 2013. *Draft supplemental environmental impact statement for the Keystone XL project*. 1 March 2013.

U.S. Environmental Protection Agency. 2012. *Light-duty automotive technology, carbon dioxide emissions, and fuel economy trends: 1975 through 2011*. EPA420-R-12-001a. EPA Office of Transportation and Air Quality, Washington, D.C.

U.S. Fish and Wildlife Service. 2001. Potential impacts of proposed oil and gas development on the Arctic Refuge's coastal plain: Historical overview and issues of concern. http://arctic.fws.gov/issues1.htm.

U.S. Geological Survey. 2001. *Arctic National Wildlife Refuge, 1002 Area, petroleum assessment, 1998, including economic analysis*. USGS Fact Sheet FS-028-01.

U.S. Geological Survey. 2002. *Petroleum resource assessment of the National Petroleum Reserve Alaska (NPRA)*. USGS, Washington, D.C.

U.S. Government Accountability Office (GAO). 2007. *Crude oil: Uncertainty about future oil supply makes it important to develop a strategy for addressing a peak and decline in oil production*. Report to Congressional Requesters.

White, Helen K., et al. 2012. Impact of the *Deepwater Horizon* oil spill on a deep-water coral community in the Gulf of Mexico. *Proc. Natl. Acad. Sci. USA* 109: 20303-20308.

Wilcox, Jennifer. 2012. *Carbon capture*. Springer, New York.

Chapter 20

Ayres, Robert, et al., eds. 2004. *Encyclopedia of Energy*. Elsevier.

Biomass Research and Development Board. 2008. *National biofuels action plan*. Biomass Research and Development Initiative, U.S. DOE and USDA.

British Petroleum. 2013. *BP statistical review of world energy 2013*. BP, London.

The Chernobyl Forum. 2006. *Chernobyl's legacy: Health, environmental and socio-economic impacts* and *recommendations to the governments of Belarus, the Russian Federation and Ukraine. The Chernobyl Forum: 2003–2005*. Second revised version. World Health Organization and International Atomic Energy Agency, Vienna.

Earley, Jane, and Alice McKeown. 2009. *Red, white, and green: Transforming U.S. biofuels*. Worldwatch Report #180. Worldwatch Institute, Washington, D.C.

Energy Efficiency and Renewable Energy, U.S. Department of Energy. www.eere.energy.gov.

Energy Information Administration, U.S. Department of Energy. www.eia.doe.gov.

Energy Information Administration, U.S. Department of Energy. 2013. *Annual energy outlook 2013*. Washington, D.C.

Energy Information Administration, U.S. Department of Energy. 2013. *Annual energy review 2012*. DOE/EIA, Washington, D.C.

European Commission/International Atomic Energy Agency/World Health Organization. 1996. One decade after Chernobyl: Summing up the consequences of the accident. Summary of the conference results. Vienna, Austria, 8–12 April 1996. EC/IAEA/WHO.

Flavin, Christopher. 2008. *Low-carbon energy: A roadmap*. Worldwatch Report 178. Worldwatch Institute, Washington, D.C.

Hall, Charles A.S. and Doug Hansen. 2011. New Studies in EROI (Energy Return on Investment). *Sustainability*, Special Issue. MDPI AG, Basel, Switzerland, 2011.

International Atomic Energy Agency. *Nuclear power and sustainable development*. IAEA Information Series 02-01574/FS Series 3/01/E/Rev.1. Vienna, Austria.

International Atomic Energy Agency. 2006. *Environmental consequences of the Chernobyl accident and their remediation: Twenty years of experience*. Report of the U.N. Chernobyl Forum Expert Group "Environment." IAEA, Vienna.

International Energy Agency. 2007. *Biomass for power generation and CHP*. IEA, Paris.

International Energy Agency. 2010. *Technology roadmap: Nuclear power*. IEA, Paris.

International Energy Agency. 2012. *Key world energy statistics 2012*. IEA, Paris.

International Energy Agency. 2012. *World energy outlook 2012*. IEA, Paris.

Murphy, David J. and Charles A.S. Hall. 2010. Year in review—EROI or energy return on (energy) invested. *Annals of the New York Academy of Sciences* 1185: 102–118.

National Renewable Energy Lab, U.S. Department of Energy. www.nrel.gov.

Nature. 2006. Special report: Chernobyl and the future. *Nature* 440: 982–989.

Normile, Dennis. 2011. Fukushima revives the low-dose debate. *Science* 332: 908–910.

Nuclear Energy Agency. 2002. *Chernobyl: Assessment of radiological and health impacts*. 2002 update of *Chernobyl: Ten years on*. OECD, Paris.

Pearce, Fred. 2006. Fuels gold: Are biofuels really the greenhouse-busting answer to our energy woes? *New Scientist*, 23 Sept. 2006: 36–41.

REN21 Renewable Energy Policy Network for the 21st Century. 2013. *Renewables 2013 global status report*. REN21 Secretariat, Paris.

REN21 Renewable Energy Policy Network for the 21st Century. 2013. *Renewables global futures report*. REN21 Secretariat, Paris.

Renewable Fuels Association. 2010. *Climate of opportunity: 2010 ethanol industry outlook*. RFA, Washington, D.C.

Sawin, Janet L., and William R. Moomaw. 2009. *Renewable revolution: Low-carbon energy by 2030*. Worldwatch Report 182. Worldwatch Institute, Washington, D.C.

Schneider, Mycle, et al. 2011. *The world nuclear industry status report 2010–2011: Nuclear power in a post-Fukushima world*. Worldwatch Institute, Washington, D.C.

Science. 2005. News Focus: Rethinking nuclear power. *Science* 309: 1168–1179.

Spadaro, Joseph V., et al. 2000. Greenhouse gas emissions of electricity generation chains: Assessing the difference. *IAEA Bulletin* 42(2).

Swedish Bioenergy Association (SVEBIO). 2003. *Focus: Bioenergy*. Nos. 1–10. SVEBIO, Stockholm.

Swedish Energy Agency. 2013. *Energy in Sweden 2012*. Swedish Energy Agency, Eskilstuna, Sweden.

Swedish Energy Agency. 2012. *Sustainable biofuels 2011*. Swedish Energy Agency, Eskilstuna, Sweden.

Ten Hoeve, John E., and Mark Z. Jacobson. 2012. Worldwide health effects of the Fukushima Daiichi nuclear accident. *Energy & Environmental Science* 5: 8743-8757.

U.N. Environment Programme. 2009. *Assessing biofuels*. UNEP, Nairobi, Kenya.

U.N. Scientific Committee on the Effects of Atomic Radiation. 2011. *Sources and effects of ionizing radiation*. UNSCEAR 2008 Report to the General Assembly with Scientific Annexes, Vol. II. United Nations, New York.

World Health Organization. 2006. *Health effects of the Chernobyl accident and special health care programmes*. Report of the U.N. Chernobyl Forum Expert Group "Health." WHO, Geneva.

World Nuclear Association. 2013. *Nuclear power in Sweden*. http://www.world-nuclear.org/info/Country-Profiles/Countries-O-S/Sweden/#.UdLWiuCbIbA.

Worldwatch Institute. 2007. *Biofuels for transport: Global potential and implications for sustainable agriculture and energy in the 21st century*. Worldwatch Institute, Washington, D.C.

Worldwatch Institute and Center for American Progress. 2006. *American energy: The renewable path to energy security*. Washington, D.C.

Yasunari, Teppei J. 2011. Cesium-137 deposition and contamination of Japanese soils due to the Fukushima nuclear accident. *Proc Natl. Acad. Sci. USA* 108: 19530–19534.

Chapter 21

American Wind Energy Association. 2012. *U.S. wind industry annual market report 2012*. AWEA, Washington, D.C.

Ananthaswamy, Anil. 2003. Reality bites for the dream of a hydrogen economy. *New Scientist*, 15 Nov. 2003: 6–7.

Ayres, Robert, et al., eds. 2004. *Encyclopedia of Energy*. Elsevier.

Barbose, Galen, et al. 2011. *Tracking the sun: An historical summary of the installed costs of photovoltaics in the United States from 1998 to 2010*. Lawrence Berkeley National Laboratory, Berkeley, Calif.

Boyle, Godfrey. 2012. *Renewable energy: Power for a sustainable future*. 3rd ed. Oxford Univ. Press USA, New York.

Chow, Jeffrey, et al. 2003. Energy resources and global development. *Science* 302: 1528–1531.

Davidson, Osha Gray. 2012. *Clean break: The story of Germany's energy transformation and what Americans can learn from it*. Inside Climate News.

Dunn, Seth. 2000. The hydrogen experiment. *World Watch* 13: 14–25.

Economist. 2012. Germany's energy transformation: Energiewende. *The Economist*, 28 July 2012.

Energy Efficiency and Renewable Energy, U.S. Department of Energy. www.eere.energy.gov.

Energy Efficiency and Renewable Energy, U.S. Department of Energy. 2011. *2010 solar technologies market report*. EERE, Washington, D.C.

Energy Efficiency and Renewable Energy, U.S. Department of Energy. 2012. *2011 wind technologies market report*. EERE, Washington, D.C.

Energy Information Administration, U.S. Department of Energy. www.eia.doe.gov.

Energy Information Administration, U.S. Department of Energy. 2013. *Annual energy outlook 2013*. Washington, D.C.

Energy Information Administration, U.S. Department of Energy. 2013. *Annual energy review 2012*. DOE/EIA, Washington, D.C.

Environmental Law Institute. 2009. *Estimating U.S. government subsidies to energy sources: 2002–2008*. ELI, Washington, D.C.

Federal Ministry for the Environment, Nature Conservation, and Nuclear Safety [Germany]. http://www.erneuerbare-energien.de/en/.

Federal Ministry for the Environment, Nature Conservation, and Nuclear Safety [Germany]. 2009. *Electricity from renewable energy sources: What does it cost?* Berlin.

Federal Ministry for the Environment, Nature Conservation, and Nuclear Safety [Germany]. 2011. *Renewable energy sources in figures: National and international development*. Berlin.

Federal Ministry for the Environment, Nature Conservation, and Nuclear Safety [Germany]. 2012. *Innovation through research: 2011 annual report on research funding in the renewable energies sector*. Berlin.

Federal Ministry of Economics and Technology [Germany]. http://www.renewables-made-in-germany.com/index.php?id=50&L=1.

Flavin, Christopher. 2008. *Low-carbon energy: A roadmap*. Worldwatch Report 178. Worldwatch Institute, Washington, D.C.

Global Wind Energy Council. 2013. *Global wind report 2012*. GWEC, Brussels, Belgium.

Grant, Paul M., et al. 2006. A power grid for the hydrogen economy. *Scientific American*, July 2006: 77–83.

Hall, Charles A.S. and Doug Hansen. 2011. New Studies in EROI (Energy Return on Investment). *Sustainability*, Special Issue. MDPI AG, Basel, Switzerland, 2011.

International Energy Agency. 2007. *Renewables in global energy supply: An IEA fact sheet*. IEA, Paris.

International Energy Agency. 2012. *Key world energy statistics 2012*. IEA, Paris.

International Energy Agency. 2009. *Technology roadmap: Wind energy*. IEA, Paris.

International Energy Agency. 2012. *World energy outlook 2012*. IEA, Paris.

International Energy Agency. 2012. *Renewables information 2012*. IEA, Paris.

Jacobson, Mark Z. 2009. Review of solutions to global warming, air pollution, and energy security. *Energy & Environmental Science* 2: 148–173.

Jacobson, Mark Z. and Mark A. Delucchi. 2009. A path to sustainable energy by 2030. *Scientific American* Nov. 2009: 58–65.

Jacobson, Mark Z., et al. 2005. Cleaning the air and improving health with hydrogen fuel-cell vehicles. *Science* 308: 1901–1905.

Kitasei, Saya. 2010. *Powering the low-carbon economy: The once and future roles of renewable energy and natural gas*. Worldwatch Report #184. Worldwatch Institute, Washington, D.C.

Knott, Michelle. 2003. Power from the waves. *New Scientist*, 20 Sept. 2003: 33–35.

McNamee, Gregory. 2008. *Careers in renewable energy: Get a green energy job*. Pixy Jack Press, Masonville, Colo.

National Renewable Energy Lab, U.S. Department of Energy. www.nrel.gov.

Ochs, Alexander, and Shakuntala Makhijan. 2012. *Sustainable energy roadmaps: Guiding the global shift to domestic renewables*. Worldwatch Report #187. Worldwatch Institute, Washington, D.C.

OECD. 2012. Inventory of estimated budgetary support and tax expenditures for fossil fuels 2013. OECD Publishing.

Pan, Jihua, et al. 2011. *Green economy and green jobs in China: Current status and potentials for 2020*. Worldwatch Report #185. Worldwatch Institute, Washington, D.C.

Pfund, Nancy, and Ben Healey. 2011. *What would Jefferson do? The historical role of federal subsidies in shaping America's energy future*. DBL Investors, San Francisco.

REN21 Renewable Energy Policy Network for the 21st Century. 2013. *Renewables 2013 global status report*. REN21 Secretariat, Paris.

REN21 Renewable Energy Policy Network for the 21st Century. 2013. *Renewables global futures report*. REN21 Secretariat, Paris.

Quashning, Volker. 2010. *Renewable energy and climate change*. Wiley-IEEE Press.

Ristinen, Robert A., and Jack J. Kraushaar, 2005. *Energy and the environment*, 2nd ed. Wiley and Sons, New York.

Sawin, Janet. 2004. *Mainstreaming renewable energy in the 21st century*. Worldwatch Paper 169. Worldwatch Institute, Washington, D.C.

Sawin, Janet L., and William R. Moomaw. 2009. *Renewable revolution: Low-carbon energy by 2030*. Worldwatch Report 182. Worldwatch Institute, Washington, D.C.

Weisman, Alan. 1998. *Gaviotas: A village to reinvent the world*. Chelsea Green Publishing Co., White River Junction, Vermont.

World Alliance for Decentralized Energy. http://www.localpower.org.

World Future Council. 2010. *Feed-in tariffs—Boosting energy for our future*. Earthscan Publications.

Worldwatch Institute and Center for American Progress. 2006. *American energy: The renewable path to energy security*. Washington, D.C.

Chapter 22

Ayres, Robert U., and Leslie W. Ayres. 1996. *Industrial ecology: Towards closing the materials cycle*. Edward Elgar Press, Cheltenham, U.K.

Beede, David N., and David E. Bloom. 1995. The economics of municipal solid waste. *World Bank Research Observer* 10: 113–150.

Campbell, Stu. 1998. *Let it rot! The gardener's guide to composting*. 3rd ed. Storey Publishing.

Container Recycling Institute. http://www.container-recycling.org.

Douglas, Ed. 2009. There's gold in them there landfills. *New Scientist*, 1 Oct. 2008.

Edmonton, City of. Edmonton Waste Management Centre. http://www.edmonton.ca/for_residents/garbage_recycling/edmonton-waste-management-centre.aspx.

Graedel, Thomas E., and Braden R. Allenby. 2009. *Industrial ecology and sustainable engineering*. Prentice Hall, Upper Saddle River, N.J.

Leonard, Annie. 2010. *The story of stuff*. Free Press, New York.

Lilienfeld, Robert, and William Rathje. 1998. *Use less stuff: Environmental solutions for who we really are*. Ballantine, New York.

Manahan, Stanley E. 1999. *Industrial ecology: Environmental chemistry and hazardous waste*. Lewis Publishers, CRC Press, Boca Raton, Fla.

McDonough, William, and Michael Braungart. 2002. *Cradle to cradle: Remaking the way we make things*. North Point Press, New York.

Navarro, Mireya. 2011. Lunch, landfills, and what I tossed. *New York Times*, 21 Oct. 2011.

New York City Department of Parks and Recreation. Fresh Kills Park. www.nycgovparks.org/sub_your_park/fresh_kills_park/html/fresh_kills_park.html.

New York City Department of Planning. Fresh Kills Park Project. www.nyc.gov/html/dcp/html/fkl/fkl3.shtml.

Rathje, William, and Colleen Murphy. 2001. *Rubbish! The archeology of garbage*. Univ. of Arizona Press.

Scott, Nicky. 2007. *Reduce, reuse, recycle: An easy household guide*. Chelsea Green Publishing Co., White River Junction, Vermont.

Spalvins, E., B. Dubey, and T. Townsend, T. 2008. Impact of electronic waste disposal on lead concentrations in landfill leachate. *Environmental Science and Technology* 42: 7452–7458.

Townsend, Timothy. 2011. Environmental issues and management strategies for waste electronic and electrical equipment. *J. Air & Waste Manage. Assoc.* 61: 587–561.

Trash Track. http://senseable.mit.edu/trashtrack/.

U.N. Environment Programme. 2012. "Chemicals and Waste." Chapter 6 in *Global environment outlook 5 (GEO-5)*. UNEP, Nairobi.

U.S. Environmental Protection Agency. 2011. *Electronics waste management in the United States through 2009*. EPA530-R-11-002. EPA Office of Resource Conservation and Recovery, Washington, D.C.

U.S. Environmental Protection Agency. 2013. *Municipal solid waste generation, recycling, and disposal in the United States: Facts and figures for 2011*. EPA530-F-13-001. EPA Office of Solid Waste and Emergency Response, Washington, D.C.

U.S. Environmental Protection Agency. Summary of the Comprehensive Environmental Response, Compensation, and Liability Act (Superfund). http://www2.epa.gov/laws-regulations/summary-comprehensive-environmental-response-compensation-and-liability-act.

U.S. Environmental Protection Agency. Summary of the Resource Conservation and Recovery Act. http://www2.epa.gov/laws-regulations/summary-resource-conservation-and-recovery-act.

van Haaren, R., et al. 2010. The state of garbage in America. *BioCycle* 51: 16–23.

Chapter 23

Brugge, Doug, and Rob Goble. 2002. The history of uranium mining and the Navajo people. *Am. J. Public Health* 92(9): 1410–1419.

Christopherson, Robert W. 2011. *Geosystems: An introduction to physical geography*, 8th ed. Prentice Hall, Upper Saddle River, N.J.

Cohen, David. 2007. Earth audit. *New Scientist*, 26 May 2007: pp. 34–41.

Craig, James R., David J. Vaughan, and Brian J. Skinner. 2010. *Earth resources and the environment*, 4th ed. Benjamin Cummings, San Francisco.

Essick, Kristi. 2001. Guns, money, and cell phones. *The Industry Standard Magazine*, 11 Jun. 2001.

Gordon, R.B., et al. 2006. Metal stocks and sustainability. *PNAS* 103(5): 1209–1214.

Hendryx, Michael, and Melissa M. Ahern. 2009. Mortality in Appalachian coal mining regions: The value of statistical life lost. *Public Health Reports* 124: 541–550.

Keller, Edward A. 2011. *Introduction to environmental geology*, 5th ed. Prentice Hall, Upper Saddle River, N.J.

Lovgren, Stefan. 2006. Can cell-phone recycling help African gorillas? *National Geographic News*, 20 Jan. 2006.

Mineral Education Coalition. Mine Reclamation [extensive examples of mine reclamation efforts with photographs]. http://www.mii.org/recl.html.

Mooallem, Jon. 2008. The afterlife of cellphones. *New York Times Magazine* 13 Jan. 2008.

Mountain Justice. What is mountaintop removal mining? http://www.mountainjusticesummer.org/facts/steps.php.

Palmer, Margaret, et al. 2010. Mountaintop mining consequences. *Science* 327: 148–149.

Perkins, Dexter. 2010. *Mineralogy*, 3rd ed. Prentice Hall, Upper Saddle River, N.J.

Rogich, D.G., and G.R. Matos. 2008. *The global flows of metals and minerals*. U.S. Geological Survey Open-File Report 2008–1355, 11 pp.

Sibley, Scott F., ed. 2004. *Flow studies for recycling metal commodities in the United States*. USGS Circular 1196-A-M. U.S. Geological Survey, Reston, Va.

Skinner, Brian J., and Stephen C. Porter. 2003. *The dynamic earth: An introduction to physical geology*, 5th ed. Wiley and Sons, Hoboken, N.J.

Sullivan, Daniel E. 2006. *Recycled cell phones—A treasure trove of valuable metals.* USGS Fact Sheet 2006-3097. U.S. Geological Survey, Denver, Colo.

Tarbuck, Edward J., Frederick K. Lutgens, and Dennis Tasa. 2011. *Earth science*, 13th ed. Prentice Hall, Upper Saddle River, N.J.

U.N. Security Council. 2001. *Report of the panel of experts on the illegal exploitation of natural resources and other forms of wealth of the Democratic Republic of the Congo.* U.N. Security Council, 12 Apr. 2001.

U.N. Security Council. 2007. *Interim report of the group of experts on the Democratic Republic of the Congo, pursuant to Security Council resolution 1698 (2006).* U.N. Security Council, 25 Jan. 2007.

U.S. Department of the Interior. 2003. *Surface coal mining reclamation: 25 years of progress, 1977–2002.* Office of Surface Mining, Washington D.C.

U.S. Geological Survey, 2013. *Mineral commodity summaries.* USGS, Washington, D.C.

U.S. Geological Survey. Minerals information. http://minerals.usgs.gov/minerals.

Chapter 24

Bartlett, Peggy, and Geoffrey W. Chase, eds. 2004. *Sustainability on campus: Stories and strategies for change.* MIT Press, Cambridge, Mass.

Blewitt, John. 2008. *Understanding sustainable development.* Earthscan, London.

Brower, Michael, and Warren Leon. 1999. *The consumer's guide to effective environmental choices: Practical advice from the Union of Concerned Scientists.* Three Rivers Press, New York.

Brown, Lester R. 2009. *Plan B 4.0: Mobilizing to save civilization.* Earth Policy Institute and W.W. Norton, New York.

Brown, Lester R. 2011. *World on the edge: How to prevent environmental and economic collapse.* Earth Policy Institute and W.W. Norton, New York.

Campus Ecology. National Wildlife Federation. www.nwf.org/campusecology.

Campus Ecology. 2008. *Campus environment 2008: A national report card on sustainability in higher education.* Campus Ecology program of the National Wildlife Federation.

Creighton, Sarah Hammond. 1998. *Greening the ivory tower: Improving the environmental track record of universities, colleges, and other institutions.* MIT Press, Cambridge, Mass.

Daly, Herman E. 1996. *Beyond growth: The economics of sustainable development.* Beacon Press, Boston.

Dasgupta, Partha, et al. 2000. Economic pathways to ecological sustainability. *BioScience* 50: 339–345.

De Anza College. Sustainability. http://www.deanza.edu/sustainability.

Durning, Alan. 1992. *How much is enough? The consumer society and the future of the Earth.* Worldwatch Institute, Washington, D.C.

French, Hilary. 2004. Linking globalization, consumption, and governance. Pp. 144–163 in *State of the world 2004.* Worldwatch Institute, Washington, D.C.

Gardner, Gary. 2011. *Creating sustainable prosperity in the United States: The need for innovation and leadership.* Worldwatch Report #186. Worldwatch Institute, Washington, D.C.

Hawken, Paul. 1994. *The ecology of commerce: A declaration of sustainability.* Harper Business, New York.

Kahneman, Daniel, et al. 2006. Would you be happier if you were richer? A focusing illusion. *Science* 312: 1908–1910.

Keniry, Julian. 1995. *Ecodemia: Campus environmental stewardship at the turn of the 21st century.* National Wildlife Federation, Washington, D.C.

McMichael, A.J., et al. 2003. New visions for addressing sustainability. *Science* 302: 1919–1921.

Meadows, Donella, Jørgen Randers, and Dennis Meadows. 2004. *Limits to growth: The 30-year update.* Chelsea Green Publishing Co., White River Junction, Vermont.

Millennium Ecosystem Assessment. 2005. *Ecosystems and human well-being: General synthesis.* Millennium Ecosystem Assessment and World Resources Institute.

National Research Council, Board on Sustainable Development. 1999. *Our common journey: A transition toward sustainability.* National Academies Press, Washington, D.C.

Renner, Michael. 2008. *Green jobs: Working for people and the environment.* Worldwatch Report 177. Worldwatch Institute, Washington, D.C.

Sanderson, Eric W., et al. 2002. The human footprint and the last of the wild. *BioScience* 52: 891–904.

Schor, Juliet B., and Betsy Taylor, eds. 2002. *Sustainable planet: Solutions for the twenty-first century.* The Center for a New American Dream. Beacon Press, Boston.

United Nations. 2002. *Report of the World Summit on Sustainable Development, Johannesburg, South Africa, 26 August–4 September 2002.* U.N., New York.

United Nations. 2012. *Report of the United Nations Conference on Sustainable Development.* Rio de Janeiro, Brazil, 20–22 June 2012. http://www.uncsd2012.org/content/documents/814UNCSD%20REPORT%20final%20revs.pdf.

United Nations. 2013. *The Millennium Development Goals Report 2013.* U.N., New York.

U.N. Development Programme. 2013. *Human development report 2013.* Oxford Univ. Press.

U.N. Environment Programme. 2012. *Global environment outlook 5 (GEO-5).* UNEP, Nairobi.

U.N. General Assembly. 2012. *The future we want.* Outcome document from Rio+20 Conference. Resolution 66/288. http://www.un.org/ga/search/view_doc.asp?symbol=A/RES/66/288&Lang=E

World Bank. 2010. *World development indicators 2013.* World Bank, Washington, D.C. http://data.worldbank.org/products/wdi.

World Commission on Environment and Development. 1987. *Our common future.* Oxford Univ. Press.

Worldwatch Institute. 2008. *State of the world 2008: Innovations for a sustainable economy.* Worldwatch Institute, Washington, D.C.

Worldwatch Institute. 2010. *State of the world 2010: Transforming cultures.* Worldwatch Institute, Washington, D.C.

Worldwatch Institute. 2012. *State of the world 2012: Moving toward sustainable prosperity.* Worldwatch Institute, Washington, D.C.

Worldwatch Institute. 2012. *Vital signs 2012.* Worldwatch Institute, Washington, D.C. http://vitalsigns.worldwatch.org.

Worldwatch Institute. 2013. *State of the world 2013: Is sustainability still possible?* Worldwatch Institute, Washington, D.C.

Campus Sustainability Resources

350.org. *Fossil Free: A campus guide to fossil fuel divestment.* https://s3.amazonaws.com/s3.350.org/images/350_FossilFreeBooklet_LO4.pdf. 350.org.

American College and University Presidents' Climate Commitment. http://www.presidentsclimatecommitment.org/.

Association for the Advancement of Sustainability in Higher Education. http://www.aashe.org.

Association for the Advancement of Sustainability in Higher Education. 2010. *Creating a campus sustainability revolving loan fund: A guide for students.* AASHE, Denver, Colorado. http://www.aashe.org/files/A_Call_to_Action_final%282%29.pdf.

Association for the Advancement of Sustainability in Higher Education. 2011 *Higher education sustainability review.* AASHE, Lexington, Kentucky.

Ball State University: Greening of the Campus [conference series.] http://cms.bsu.edu/academics/centersandinstitutes/goc.

Barth, Matthias. 2013. *Implementing sustainability in higher education: Learning in an age of transformation.* Routledge, New York.

Bartlett, Peggy, and Geoffrey W. Chase, eds. 2004. *Sustainability on campus: Stories and strategies for change.* MIT Press, Cambridge, Mass.

Bartlett, Peggy, and Geoffrey W. Chase, eds. 2013. *Sustainability in higher education: Stories and strategies for transformation.* MIT Press, Cambridge, Mass.

Campus Conservation Nationals. http://www.competetoreduce.org/.

Campus Ecology. National Wildlife Federation. www.nwf.org/campusecology.

Campus Ecology. 2008. *Campus environment 2008: A national report card on sustainability in higher education.* Campus Ecology program of the National Wildlife Federation.

Campus Transport Management: Trip reduction programs on college, university, and research campuses. Online TDM Encyclopedia. Victoria Transport Policy Institute. http://www.vtpi.org/tdm/tdm5.htm.

Carlson, Scott. 2006. In search of the sustainable campus: With eyes on the future, universities try to clean up their acts. *Chronicle of Higher Education* 53: A10.

Clean Air Cool Planet: Sustainable Campuses. http://cleanair-coolplanet.org/sustainable-campuses/.

Clean Air Cool Planet: Campus Carbon Calculator. http://cleanair-coolplanet.org/campus-carbon-calculator/.

Creighton, Sarah Hammond. 1998. *Greening the ivory tower: Improving the environmental track record of universities, colleges, and other institutions*. MIT Press, Cambridge, Mass.

Diebolt, Asa, and Timothy Den Herder-Thomas. 2007. *Sustainability curriculum in higher education: A call to action*. AASHE, Lexington, Kentucky. http://www.aashe.org/documents/resources/pdf/CERF.pdf.

Erickson, Christina, and David J. Eagan. 2009. *Generation E: Students leading for a sustainable, clean energy future*. Campus Ecology program of the National Wildlife Federation.

Fossil Free. http://gofossilfree.org/. 350.org.

Higher Education Associations Sustainability Consortium. http://heasc.aashe.org/.

Indvik, Joe, et al. 2013. *Green revolving funds: An introductory guide to implementation and management*. Sustainable Endowments Institute and Association for the Advancement of Sustainability in Higher Education. Cambridge, Mass.

Keniry, Julian. 1995. *Ecodemia: Campus environmental stewardship at the turn of the 21st century*. National Wildlife Federation, Washington, D.C.

Koester, Robert J., James Eflin, and John Vann. 2006. Greening of the campus: A whole-systems approach. *Journal of Cleaner Production* 14: 769–779.

LEED Campus Program. http://www.usgbc.org/leed/certification/programs/campus. U.S. Green Building Council.

Newman, Julie. 2009. *Reaching beyond compliance: The challenges of achieving campus sustainability*. VDM Verlag.

Recyclemania. http://recyclemaniacs.org/.

Sierra. Ten Coolest Schools. http://www.sierraclub.org/sierra/201209/coolschools/. *Sierra*.

Simpson, Walter, ed. 2008. *The green campus: Meeting the challenge of environmental sustainability*. Appa Association of Higher Education.

Simpson, Walter. 2009. *Cool campus! How-to guide for college and university climate action planning*. AASHE, Lexington, Kentucky. http://www.aashe.org/files/resources/cool-campus-climate-planning-guide.pdf.

Sustainable Endowments Institute: College Sustainability Report Card. http://www.greenreportcard.org/.

Sustainable Endowments Institute. 2012. *Greening the bottom line*. Sustainable Endowments Institute, Cambridge, Mass.

U.S. Department of Energy: Solar Decathlon. http://www.solardecathlon.gov/.

Toor, Will, and Spenser W. Havlick. 2004. *Transportation and sustainable campus communities: Issues, examples, solutions*. Island Press, Washington, D.C.

University Leaders for a Sustainable Future. www.ulsf.org.

Photo Credits

Cover © Mark Ferguson/Alamy

Chapter 1 Part One Opening Photo Venture Media Group/ Aurora Open/Alamy **Opening Photo** NASA/Johnson Space Center **The Science Behind the Story** Andrzej Gibasiewicz/ Shutterstock **The Science Behind the Story** vasen/Shutterstuck **1.6** George Konig/Hulton Archive/Getty Images **1.7** Jay Mallin/ZUMAPRESS/Newscom **1.8A** Ethan Miller/Getty Images **1.8B** Jay Withgott **1.10** Brian J. Skerry/National Geographic Stock **1.14** JayKay57/iStockphoto **1.15A** Alex Leff-Hampshire College **1.15B** Samuel Masinter, Amherst College **1.15C** Davidson College **1.15D** © Bob Daemmrich/ Alamy

Chapter 2 Opening Photo Paula Bronstein/Getty Images **The Science Behind the Story (left)** Photo by Tom Kleindinst, Woods Hole Oceanographic Institution **(right)** Photo by Ken Kostel, Woods Hole Oceanographic Institution **2.5** JupiterImages **2.9A** Mcpics/Dreamstime LLC-Royalty Free **2.9B** Jupiterimages/Thinkstock **2.13** Jack Dykinga/Getty Images Inc. – Stone Allstock **2.14** NOAA/Science Source **2.19A** Gregritchie/Dreamstime LLC-Royalty Free **2.19B** Martin Gabriel/Nature Picture Library **2.19C** Jack Dykinga /Nature Picture Library **2.19D** © Tom Till/Alamy **The Science Behind the Story** Dr. Sarah C. Sherlock **2.21** Julie Jacobson/AP Wide World Photos **2.22** © RGB Ventures LLC dba SuperStock/Alamy **2.23** zumalive/Newscom **2.24** Bay Ismoyo/Getty Images, AFP

Chapter 3 Opening Photo (inset) Jack Jeffrey Photography **3.1A** Jack Jeffrey Photography **3.1B** John Elk/Lonely Planet Images/Getty **3.1C** David Sanger/Getty **3.1D** Darlyne Murawski/National Geographic Stock **3.3A** H. Douglas Pratt **3.3B** © desertsolitaire/Fotolia and A. Jagel/AGE Fotostock America, Inc. **The Science Behind the Story** Heather Lerner **3.8** REUTERS/Fayaz Kabli **3.11A** NASA/Johnson Space Center **3.11B** © Danita Delimont/Alamy **3.11D** © David Fleetham/Alamy **3.11** (turtle swimming) Monica and Michael Sweet/Getty Images **3.12** Jack Jeffrey Photography **3.13A** G.I. Bernard/Science Source **3.14A** Darren Green /Fotolia **3.14B** Art Wolfe/ Getty Images **3.14C** Georgette Douwma/Getty Images, Jack Jeffrey Photography **3.16** Mandar Khandilkar/Alamy **The Science Behind the Story** Jack Jeffrey Photography **The Science Behind the Story** © Douglas Peebles Photography/Alamy **3.20** © Bluegreen Pictures /Alamy

Chapter 4 Opening Photo Peter Yates/Time & Life Pictures/ Gety Images **4.1A** Peter Yates/Science Source **4.1B** Randy Westbrooks, Biological Resources Division, U.S. Geological Survey **4.5A** © hakoar/Fotolia **4.5C** Photoshot Holdings, Ltd/Alamy **4.6** James L. Amos/National Geographic **4.7** Andrew Darrington/Alamy **4.8** Chesapeake Images /Shutterstock **4.13** Miroslav Hladik/Shutterstock **The Science Behind the Story** David Strayer **The Science Behind the Story** U.S. Geological Survey **The Science Behind the Story** tusharkoley/Shutterstock **4.16** George Thompson /MCT/Newscom **4.19** JayKay57/iStockphoto **4.20** Gerrit Vyn/Nature Picture Library **4.21** © Charles Mauzy/CORBIS **4.22** © David Samuel Robbins/Corbis **4.23** Geoffrey Gallice **4.24** Cheryl-Samantha Owen/Nature Picture Library **4.25** Getty Images/Digital Vision **4.26** Radius Image/Alamy **4.27** © Bill Brooks/Alamy **4.28** Earl Scott/Photo Researchers /Getty Images

Chapter 5 5.9 David W. Schindler **5.11A** ©FLPA/Bob Gibbons/AGE Fotostock **5.11B** © Nikolay Dimitrov/AGE Fotostock **5.11C** © ION/AmanaimagesRF/AGE Fotostock **5.11D** © George H.H. Huey/AGE Fotostock **5.11E** murata-photo com/Shutterstock **5.13** © Ashley Cooper/Corbis **The Science Behind the Story** USDA/NRCS/Natural Resources Conservation Service **The Science Behind the Story** USDA/NRCS/ Natural Conservation Service **5.19 (left)** Gregory G. Dimijian, M.D./Science Source **5.19 (right)** Milos Kalab/Custom Medical Stock Photo, Inc.

Chapter 6 Opening Photo Alberto Font/Tico Times **6.2A** Yadid Levy/Robert Harding **6.2B** © Maxime bessieres/ ethnoscape.net**6.5** Corbis **6.6** Corbis **6.7** Bettmann/Corbis **6.8** Mario Tama/Getty **6.11** REUTERS/Sigit Pamungkas **6.12** Laurin Rinder/Shutterstock **The Science Behind the Story** Associated Press **The Science Behind the Story** © Scott Griessel/Fotolia **6.14A** thinair28/iStockphoto **6.14B** © Belizar/Dreamstime.com**6.14C** BigWest1/iStockphoto **6.14D** Corbis **6.14F** © Image Source/Corbis **6.14G** Pgiam/iStockphoto **6.16** © Accent Alaska.com/Alamy

Chapter 7 Opening Photo REUTERS/Tim Shaffer **7.3** AP Photo/Scranton Times & Tribune, Michael J. Mullen **The Science Behind the Story** Jim Lo Scalzo/EPA/Newscom **7.6** AP Photo/Scranton Times & Tribune, Michael J. Mullen **7.7A** Bettmann/Corbis **7.7B** CORBIS **7.7C** University of Washington Libraries **7.8** Erich Hartmann/Magnum Photos, Inc. **7.9** Corbis **7.12B** Jennifer Szymaszek/Reuters **7.15** Joel Sartore/NGS Image Collection

Chapter 8 Part Two Opening Photo Dave King © Dorling Kindersley **Chapter Opening** © Bob Krist/Corbis **8.1** Adrian Bradshaw/EPA/Newscom **8.5A** © GL Archive/Getty Images **8.5B** John C. Ogen **8.13C** Guang Niu/Getty Images **8.13D**

AFP/AFP/Getty Images **8.14** Getty Images **The Science Behind the Story** AP Photo/Felipe Dana **8.22** Photononstop /SuperStock **8.24A** Peter Menzel Photography **8.24B** Peter Ginter

Chapter 9 Opening Photo Lynn Betts **9.1** Nigel Cattlin/Science Source **9.7A** © Environmental Images/AGE Fotostock **9.7B** Scott Sinklier/Agstockusa/AGE Fotostock America Inc. **9.8** © Xinhua/Photoshot **9.12** Carl Purcell/ Science Source **9.14** Jason Johnson/Iowa NRCS **9.15A** Scott Sinklier/AGE Fotostock America, Inc. **9.15B** Kevin Horan/Stone/Getty Images **9.15C** Keren Su/Stone/Getty Images **9.15D** © Inga Spence/Alamy **9.15E** © David Wall/ Alamy **9.15F** Associated Press **9.18** China Photos/Getty **The Science Behind the Story** Doug Martin/Science Source **9.19** © Photoshot Holdings Ltd./Alamy **9.21A** Tony Hertz/Alamy Images **9.21B** © Jim Wark/AGE Fotostock **9.22A** Photodisc/Getty Images **9.22B** James L. Stanfield/ National Georgraphic/Getty Images **9.23A** © Paul Debois/ AGE Fotostock **The Science Behind the Story** Malpai Borderlands Group **9.23B** © Rachel Husband/Alamy **9.26** NHPA/SuperStock

Chapter 10 Opening Photo © Chuck Place/Alamy **10.4** AP Photo/Jean-March Bouju **10.5** Art Rickerby/Time Life Pictures/Getty **10.7** Jack Dykinga/Getty **10.7(inset)** Nigel Cattlin/Alamy **10.11A** IndexStock/Photo Library/Getty **10.11B** Mark Newman/AGE Fotostock **10.12** Imaginechina via AP Images **10.15** Native Seeds/SEARCH **10.16** MORVAN/SIPA/Newscom **10.17** © FLPA/Richard Becker/AGE Fotostock **10.19** Australia Lands Department **The Science Behind the Photo** (c) 2013 Rodale Institute **Table 10.2A** Pigdevil Photo/Shutterstock **Table 10.2B** Peter Zijlstra/ Shutterstock **Table 10.2C** Krasowit/Shutterstock **Table 10.2D** oksix/Shutterstock **Table 10.2E** Maks Narodenko/ Shutterstock **Table 10.2F** mexrix/Shutterstock **Table 10.2G** dabjola/Shutterstock **Table 10.2H** Luis Carlos Jimensz del rio/Shutterstock **Table 10.2I** Jiang Hongyan/Shutterstock **Table 10.2J** Sergio33/Shutterstock **The Science Behind the Story** Peg Skorpinski Photography **10.29** Associated Press

Chapter 11 Opening Photo Karl Ammann/Getty Images **11.7** William Leaman/Alamy **11.10** Photobank/Shutterstock **11.13A** John Warburton-Lee/Getty Images **11.13B** J&A Scott/NHPA/Photoshot/Newscom **The Science Behind the Story** © Martin Harvey/Alamy **Table 11.2A** © Les Gibbon/ Alamy **Table 11.2B** Mircea Bezergheanu/Shutterstock **Table 11.2C** © Nigel Cattlin/Alamy **Table 11.2D** Jurgen Freund/ Nature Picture Library **Table 11.2E** © Greg Wright/Alamy **Table 11.2F** © John T. Fowler/Alamy **Table 11.2G** Science Source **Table 11.2** David Cappaert, Michigan State University, Bugwood.org**Table 11.2H** © Dave Bevan/Alamy **Table 11.2I** © John Warburton-Lee Photography/Alamy **Table 11.2J** Brian Enting/Science Source **11.14** Jan Martin Will/ Shutterstock **11.15** Handout/MCT/Newscom **Table 11.3A** © Bildagentur Walkhaeus/AGE Fotostock **Table 11.3B** © Kuttig – Travel 2/Alamy **Table 11.3C** © Hemis/Alamy **Table 13.3D** © ARCO/C. Wermter **Table 11.3E** © ARCO/ C. Wermter **Table 11.3F** © imageroker/Alamy **Table 11.3G** © Richard Mittleman/Gon2Foto/Alamy **Table 11.4B** © Bob Gibbons/Alamy **Table 11.4D** © blickwinkel/Alamy **Table 11.4E** © Dinodia/AGE Fotostock **Table 11.4F** © Plantography/Alamy **Table 11.5A** Ant/NHPH/Photoshot **Table 11.5B** © Stephen Frink Collection/Alamy **Table 11.5C** © Nick Greaves/Alamy **Table 11.5D** © Gary Dublanko/ Alamy **Table 11.5E** © Brandon Cole Marine Photography/ Alamy **11.17** Jeremy Holden **11.18A** Phillippa Psaila/ Photo Researchers, Inc. **11.18B** Ian Nichols/National Geographic Creative **11.18D** Photoshot/NHPA Limited **11.19** Drnickburton/Dreamstime LLC **11.20** Tom Reichner/Shutterstock **11.21A** Green Renaissance **The Science Behind the Story** Scott Baker **11.22B** Alamy Images **The Science Behind the Story** AP Photo/Itsuo Inouye **11.23** Pitamitz/ SIPA/Newscom

Chapter 12 Opening Photo Newpage Corporation **12.1D** Tetra Images/Alamy **12.6** NASA Earth Observing System **12.7A** Michael Nichols/National Geographic Stock **12.8** National Geographic/Getty Images **12.12** Weyerhaeuser Company **12.15** Scott J. Ferrell/Alamy **12.15(inset)** Brian E. Small © Dorling Kindersley **12.17** Karl Mondon/MCT/Landov Media **12.18** Tracy Ferrero/Alamy Images **12.18(inset)** British Columbia Province of B.C.-Ministry of Citizens' Services/Dion Manastryrski **12.20** © Jim West/Alamy **The Science Behind the Story** Ross Hamilton/The Oregonian **12.22** Aaron Peterson/Alamy Images **12.24C** Photo Researchers/ Getty Images **The Science Behind the Story** Courtesy of Thomas Lovejoy **The Science Behind the Story** Courtesy of R.O. Bierregaard Jr.

Chapter 13 Opening Photo © Joe McLaughlin/Getty Images **13.2A** Chad Palmer/Shutterstock **13.2B** JupiterImages/ Thickstock/Alamy Images **13.4** NASA Earth Observing System **13.5A** Harrison Shull/Getty Images **13.5C** haveseen/ Shutterstock **13.5D** Aldo Torelli/Stone/Getty Images **13.6A** Courtesy of the Library of Congress **13.9** Cooper Carry and Association **13.10** Justin Sullivan/Getty Images **13.12A** Steve Semler/Alamy Images **13.13** Wu Hong/U.S. Environmental Protection Agency Headquarters **13.14** Peter Scholey/ Getty Images **13.15** Joy Scheller/Photoshot/Newscom **The Science Behind the Story** Steward Pickett/Cary Institute of Ecosystems Studies

Chapter 14 14.1 Gary Porter/Newscom **14.2A** Ingram Publishing/Alamy Images **14.2B** Spencer Grant/PhotoEdit Inc. **14.2C** Knorre/Dreamstime LLC **14.2D**thinkstock/ iStockphoto.com**The Science Behind the Story** Patricia Hunt **The Science Behind the Story** Reproduced by permission from Hunt, PA KE Koehler, M Susiarjo, CA Hodges, A Illagan, RC Voight, S Thomas, BF Thomas and TJ Hassold. Bispenol: A exposure causes meiotic aneuploidy in the female mouse. Current Biology 13, 546-553. Copyright © 2003 by Elsevier Science Ltd. **14.5** AP Photo/Khin Maung Win **14.7** AP Photo/Danielle Peterson-Statesman Journal **14.10A** Stockdisc/Getty/Photolibrary.com **14.10B** Rosanne Olson/

Getty/Stone Allstock **14.14A** Howard K. Suzuki **14.14B** Peg Skorpinski Photography **14.15** Masterfile **The Science Behind the Story** Jeff Conant **The Science Behind the Story** Elizabeth A. Guillette

Chapter 15 15.6A NEBRASKAland Magazine **15.6C** Alexey Stiop/Shutterstock **15.10** NASA Earth Observing System **15.11C** Vyacheslav Oseledko/AFP/Getty Images/Newscom **15.13A** gianmarco maggiolini/AGE Fotostock America, Inc. **15.17A** AP Wide World Photos **15.17B** Ian Berry/Magnum Photos Inc. **15.18** Craig Dilger/The York New Times/Redux Pictures **15.20** constantgardener/Getty Images **The Science Behind the Story** © David Pearson/Alamy **The Science Behind the Story** Louisiana Universities Marine Consortium **15.24** © Aerial Archives/Alamy **15.25** The U.S. Department of Energy **15.27** Gary Crabbe/Alamy Images

Chapter 16 16.6 © Robin Loznak/ZUMA Press/Corbis **The Science Behind the Story** Jamie Collier/Alamy Images **16.11** David Shale/Nature Picture Library **16.13** Joe Dovala/Getty Images **16.14A** Getty Images Inc. – Image Bank **16.14B** Melvin Lee/Shutterstock **16.15** Laguna Design/Science Photo Library/Science Source **16.16A** Dave Bevan/Alamy Images **16.16B** Chris Jordan/Algalita Marine Research Foundation **16.16C** Matt Cramer/Algalita Marine Research Foundation **16.16D** Bill Macdonald/Algalita Marine Research Foundation **The Science Behind the Story** IPRC/SOFST, University of Hawaii **The Science Behind the Story** NOAA **The Science Behind the Story** GDP **16.17A** Eric Gay/AP Wide World Photos **16.18** © Don Paulson/AGE Fotostock **16.22** Newscom **16.23** The Science Behind the Story IPRC/SOFST, University of Hawaii Wildfire GmbH/Alamy Images

Chapter 17 Opening Photo AP Photo/Dario Lopez Mills (Top) Joe Cavaretta (Bottom) **17.10A** NASA/Goddard Space Flight Center **17.11A** © Design Pics Inc./Alamy **17.11B** Biological Resources Division, U.S. Geological Survey **17.11C** NASA Earth Observing System **17.20** © Lou Linwei/Alamy **17.21** Pittsburgh Post-Gazette **17.22** Keith Dannemiller/ZUMAPRESS/Newscom **The Science Behind the Story** Courtesy of Lilian Calderón-Garcidueñas **17.25** NASA **The Science Behind the Story** AP Wide World Photos **17.30A** © Bettmann/CORBIS **17.30B** Cordelia Molloy Science Source **17.32** © Collin McPherson/CORBIS

Chapter 18 Opening Photo wennoddpix/Newscom **18.6B** Centers for Disease Control & Prevention **The Science Behind the Story** Pasquale Sorrentino/Science Source **The Science Behind the Story** REUTERS/Jerry Lampen **18.13A** Glacier National Park, Montana, near the foot of Blackfoot Glacier, Morton J. Elrod/Archives & Special Collections, Mansfield Library, The University of Montans **18.13B** AP Photo/U.S. Geological Service, Lisa McKeon **18.15A** Stan Honda/AFP/Getty Images/Newscom **18.15B** Michael Bocchieri/Getty Images **18.15C** © US Coast Guard Photo/Alamy **18.19** Daniel Acker/Bloomberg via Getty Images **18.20A** Jan Martin Will/Shutterstock **18.20B** Gordon Wiltsie/National Geographic Stock **18.20C** Ashley Cooper/Photoshot Archive Creative **18.29** AP Wide World Photos

Chapter 19 19.1 AP Photo/J. Scott Applewhite **19.9A** Corbis **The Science Behind the Story** AGE Fotostock America, Inc. **19.14** Scott Goldsmith/National Geographic Stock **19.17** Newscom **19.18A** REUTERS/Sean Gardner **19.18B** AP Photo/David Martin **The Science Behind the Story** NOAA/Rex Features/Associated Press **The Science Behind the Story** NOAA **19.19** Jim West/AGE Fotostock America, Inc. **19.25** pixac/123RF

Chapter 20 Opening Photo Kentaroo Tryman/Getty Images **20.5** © BrazilPhotos.com/Alamy**20.7** John S. Zeedick/Getty Images, Inc. **20.8A** AP Wide World Photos **20.8B** Reuters/Corbis **20.10A** Ho New/Reuters **20.10B** © Aflo Foto Agency/Alamy **20.12B** Sean Gallup/Getty Images **20.13A** Nuclear Energy Institute **20.13B** U.S. Department of Energy **20.14** Chris Steele-Perkinds/Magnum Photos Inc. **20.15** Lars-Erik Larsson/Svebio **20.16** GreenWood Resource, Inc./Jake Eaton **20.17** Curt Maas/AGE Fotostock America, Inc. **20.18** Charles Bensinger/Renewable Energy Partners **20.21** Brent Baker **20.22** AP Wide World Photos **The Science Behind the Story** Charles Hall **20.23** USDA/ARS/Agricultural Research Service **20.24A** Earl Roberge/Science Source

Chapter 21 Opening Photo Alamy Images **21.4** Robert Nickelsberg/Getty Images **The Science Behind the Story** Mark Jacobson **21.8A** Joerg Reuther/Alamy Images **The Science Behind the Story** AP Photo/Julia Cumes **21.15** Joern Pollex/Getty Images **21.18B** Simon Fraser/Science Photo Library/Science Source **21.21** Laszlo Podor/Alamy Images **21.22** Yonhap News/YNA/Newscom

Chapter 22 Opening Photo Irvin Silverstain/Landov Media **22.4** Romeo GACAD Agence France Presse/Newscom **22.7** Timothy A. Clary/AFP/Getty Images/Newscom **22.13B** City of Edmonton **22.16** © Jim West/Alamy **22.18** AP Photo/The Canadian Press/Jonathan Hayward **22.19** Joe Sohm/Alamy Images **22.20** Robert Brook/Science Source **The Science Behind the Story** SENSEable City Laboratory **The Science Behind the Story** Timothy Townsend **The Science Behind the Story** Walter Bieri, Keystone/AP Wide World Photos

Chapter 23 23.3A DEA/R APPIANI/agefotostock **23.3B** Renn Sminkey, Pearson Education **23.4** Vivian Stockman/OHVEC **23.7** Picture Desk, Inc. **23.8** David R. Frazier/The Image Works **23.10C** Appalachian Voices **23.10D** Ricky Carioti/The Washington Post/Getty Images **23.11** Greenshoots Communications/Alamy Images **The Science Behind the Story** Dr. Margaret Palmer **The Science Behind the Story** USDA Forest Service **23.14** © Oleksiy Maksymenko/AGE Fotostock

Chapter 24 Opening Photo Gino De Grandis/De Anza College **24.1** Ryan Phillips/De Anza College **24.3** Michael

Lometti **24.4** University of North Carolina at Greensboro **24.5** Kent Clemenco/De Anza College **24.6A** Oberlin College Archives **24.6B** Courtesy of the Bren School of Environmental Science & Mangement, University of California, Santa Barbara **24.7** Prabjit Virdee **24.8** Mark Beane/Loyola University Chicago **24.9B** University of Maine at Presque Isle **24.10** U.S. Department of Energy **24.11A** John McKeith/Winona State University **24.11B** Bob Child/AP Wide World Photos **24.13** Victoria Arocho/AP Wide World Photos **24.15** Alaina Bernard/University of Central Florida **24.20** Costco Wholesale **24.22** AP Wide World Photos **24.23** CORBIS-NY **24.24** NASA/Langley Research Center

Index

Note to the reader: Page numbers in **bold** refer to definitions of key terms; page numbers followed by *f* refer to figures; page numbers followed by *t* refer to tables.

B

D

E

G

H

N

O